2008 International Conference on Electronic Packaging Technology & High Density Packaging

Shanghai, China
28-31 July 2008

Pages 1-550

IEEE Catalog Number: CFP08553-PRT
ISBN 13: 978-1-4244-2739-0

Copyright © 2008 by The Institute of Electrical and Electronics Engineers, Inc.
All Rights Reserved

Copyright and Reprint Permissions: Abstracting is permitted with credit to the source. Libraries are permitted to photocopy beyond the limit of U.S. copyright law for private use of patrons those articles in this volume that carry a code at the bottom of the first page, provided the per-copy fee indicated in the code is paid through Copyright Clearance Center, 222 Rosewood Drive, Danvers, MA 01923.

For other copying, reprint or republications permission, write to IEEE Copyrights Manager, IEEE Operations Center, 445 Hoes Lane, Piscataway, New Jersey USA 08854. All rights reserved.

IEEE Catalog Number:	CFP08553-PRT
ISBN 13:	978-1-4244-2739-0
LOC:	2008906231

Additional Copies of This Publication Are Available from:

IEEE Service Center
445 Hoes Lane
Piscataway, NJ 08854

Phone:	(800) 678-IEEE
	(732) 981-1393
Fax:	(732) 981-9667
E-mail:	customer-service@ieee.org

2008 International Conference on Electronic Packaging Technology & High Density Packaging

Shanghai, China
28-31 July 2008

IEEE Catalog Number: CFP08553-POD
ISBN: 978-1-42442-739-0

Directed by
China Institute of Electronics, China
Science & Technology Department, Ministry of Education, China
Foreign Affairs Office of Shanghai Municipality, China
Department of High and New Technology, Development & Industrialization, Ministry of Science and Technology, China
Ministry of Industry and Information Technology, China
China International Culture Exchange Center

Sponsored by
China Electronics Packaging Society, of China Institute of Electronics, China (CEPS)
Fudan University, China (Fudan University)
The Component, Packaging, and Manufacturing Technology Society of IEEE (IEEE-CPMT)
The International Microelectronics and Packaging Society (IMAPS)

Organized by
China Electronics Packaging Society of China Institute of Electronics (CEPS)
Shanghai Integrated Circuit Industry Association (SCIA), China
Fudan University, China (Fudan University)
Beijing Faith Consulting Co., Ltd.

Co-organized by
Hong Kong Applied Science & Technology Research Institute (ASTRI), China
Shanghai Zhangjiang Group Co., Ltd, China
Shanghai Zhangjiang Hi-Tech Park Management Committee
Shanghai Pudong New Area Science & Technology Association, China

Editors
Keyun Bi
China Semiconductor Industrial Association

Fei Xiao
Department of Materials Science, Fudan University, Shanghai, 200433, China

Table of Contents

Qualification for Product Development 1
Weiqiang Wang, Michael H. Azarian and Michael Pecht

High Density 3D Integration 13
Roy Yu

A Study of Thermal Performance for the Panel Base Package (PBPTM) Technology 23
Ming-Chih Yew, Chun-Fai Yu, Mars Tsai, Dyi-Chung Hu, Wen-Kung Yang and Kuo-Ning Chiang

FMEA of System-in-Package (SiP) -based Tire Pressure Monitoring System 28
Man-Lung Sham, Tung-Chin Lui, Ziyang Gao, Tom Chung

Ultrasonic Bonding of Polymer Microfluidic Chips 34
Zong-bo Zhang, Yi Luo, Xiao-Dong Wang,, Zhen-Qiang Zhang, Li-Ding Wang

Mold Array Package for POP Applications 39
Aching Lee, Louie Huang, Mike Hung

New Technologies for advanced high density 3D packaging by using TSV process 43
Paul Kettner, Bioh Kim, Stefan Pargfrieder, Swen Zhu

Recent Progress of Ohmic Contact on ZnO 46
Yao Lv, Lixi Wan

Low-cost High-efficiency 4 Channel Pluggable Parallel Optical Transceiver Using Optoelectronic MCM Packaging Technologies 50
Baoxia Li, Lixi Wan, Chengyue Yang, Wei Gao, Yao Lv, Zhihua Li, and Xu Zhang

SiP/SoP Technology and Its Implementation 54
Lixi Wan

Packaging of Polymer Based Microfluidic Systems Using Low Frequency Induction Heating (LFIH) 57
Benedikt J. Knauf, D. Patrick Webb, Changqing Liu, Paul P. Conway

Analysis of the Reliability of Package-on-Package Devices Manufactured Using Various Underfill Methods 63
Vicky Wang and Dan Maslyk

Meeting Thermal Performance and Reliability Challenges for a Thermally Enhanced Ball Grid Array Package (TEBGA 66
Quan Qi

Numerical Simulation of the Micro-channel Heat Sink on Non-uniform Heat Source 72
Zhang Minliang, Wang Xiaojing, Liu Hongjun, Wang Guoliang

Heat Transfer Simulation of Nanofluids in Micro Channel Cooler 75
Liu Hongjun, Wang Xiaojing, Zhang Minliang, Zhang Wen, Liu Johan

Flip-chip on Board packaging of a Thermal Wind Sensor 80
Guang-ping Shen, Ming Qin, Qing-An Huang, Hua Zhang, Jian Wu

System-in-Package Solutions with Embedded Active and Passive Components 84
Rolf Aschenbrenner

Design Advisor for Package-on-Package (PoP) Manufacturing 86
Bin Xie, Peng Sun and Daniel Shi

Warpage Reduction of Package-on-Package (PoP) Module by Material Selection & Process Optimization 93
Peng Sun, Vincent Chi-Kuen Leung, Bin Xie, Vivian Wei Ma, Daniel Xun-Qing Shi

A New Process to Fabricate Cavities in Pyrex7740 Glass for High Density Packaging of Micro- System 99
Junwen Liu, Qing-an Huang, Jintang Shang, Jing Song, and Jieying Tang

DPA Tests on SiP Device 103
Jin Ling, Hu Jun, Han Li

Table of Contents

Ultrasonic Features in Wire Bonding and Thermosonic Flip Chip..106
Junhui Li, Lei Han, Jue Zhong

Development of Three-dimensional Multichip Module Based on Embedded Substrate with Multiple
Interconnections...111
Gaowei Xu, Yanhong Wu, Fei Geng, Qiuping Huang, Jian Zhou, Le Luo

Novel Low-Temperature Micro-insert Bonding Technology for 3D Package....................................117
Po Xu, Anming Hu, Zhuo Chen, Ming Li, Dali Mao

High-Q On-Chip Inductors Embedded in Wafer-Level Package for RFIC Applications.....................121
Tao Feng, Jian Cai, Henri HK Kwon, Qian Wang, Xinyu Dou

Thermo and Mechanical Study on Integrated High-density Packaging..125
Qian Chen, Yi-ping Wu, Feng-shun Wu, Zhi-jun Zhu, Yuan Tao

Research on the Contact Resistance and Reliability of Flexible RFID Tag Inlays Packaged by Anisotropic
Conductive Paste..129
Xiong-hui Cai, Bing An, Yi-ping Wu, Feng-shun Wu, Xiao-wei Lai

Ultra-Fine Via Pitch on Flexible Substrate for High Density Interconnect (HDI)............................134
Kelvin Pun, C.Q. Cui, T.F. Chung

Introduction of Microelectronics Manufacturing Engineering into Professional Education: a Joint Effort
among Industry, Government and Universities 140
Kailin Pan, Daoguo Yang, Qiao Kai, Kouchi Zhang, Lianfa Yang

FEA Based Reliability Prediction for Different Sn-Based Solders Subjected to Fast Shear and Fatigue
Loadings 145
Rainer Dudek, Eberhard Kaulfersch, Sven Rzepka, Mike Röllig, Bernd Michel

Thermo-Mechanical Analysis of a Wafer Level Packaging by Induction Heating 152
Wenming Liu, Mingxiang Chen, Yanyan Xi, Changyong Lin, Sheng Liu,

Synthesizing SPICE-Compatible Models of Power Delivery Networks with Resonance Effect by Time-
domain Waveforms..157
Chen-Chao Wang, Shu-Qiang Zhang, Hung-Hsiang Cheng, Tzu-Chih Lin, Chi-Tsung Chiu, and Chih-Pin Hung

Optimization of CAVLC Algorithm and Its FPGA Implementation 161
Xu Meihua, Li Ke, Xuan Xiangguang, Fan Yule

Thermal Management and testing of MCM with embedded chip in Silicon Substrate......................165
Fei Geng, Jia-jie Tang, Le Luo

Strain rate effect and Johnson-Cook models of lead-free solder alloys 171
Qin Fei, An Tong, Chen Na

A Scale Reduced Computation Scheme for Peeling Stress of Solder Joints under Drop Impact.......178
An Tong, Qin Fei

Dynamic Bending Tests and Numerical Simulation of Board Level Electronic Package....................183
Qin Fei, Wang Yngve, Liu Bin, An Tong, Jin Ling

A New Method for The Investigation of Strip Warpage of MAP-QFN 187
Guohua Gao, Honghui Wang, Guoji Yang, Haiqing Zhu

Numerical Investigation on the Effect of Filler Distribution on Effective Thermal Conductivity of Thermal
Interface Material...192
Cong Yue, Yan Zhang, Johan Liu,, Zhaonian Cheng, Jing-yu Fan

Development of Moisture Automation Analysis System for Microelectronic Packaging Structures.....197
Yangjian Xia, Yuanxiang Zhang, Lihua Liang

High Speed Package Design and Electrical Performance Analysis...204
Shu-Qiang Zhang, Hung-Hsiang Cheng, Yin-Guang Zheng, Chang-Lin Yeh

Table of Contents

Methodology for Modeling Simultaneous Switching Noise in BGA Packaging 208
Song Li, Xue-tao Weng

A Lowpass Filter with An Embedded Capacitor for Wideband Noise Suppression in Multi-GHz PCBs 211
Wei Gao, Lixi Wan, Jun Li

A Simplified Thermal Resistance Network Model for High Power LED Street Lamp 216
Xiaobing Luo, Wei Xiong, Sheng Liu

Improvement of Power Integrity with novel Segmented Power Bus Structures in RF/Digital SOP 223
Jun Li, Lixi Wan, Wei Gao, Cheng Liao

Study of RF Front-End Filters with Embedded Capacitor Technology 227
Yunfeng Wang, Lixi Wan, Lei Li

The Design of the Cache Crossbar based on OpenSPRAC Architecture 231
Xi-chuan Wang, Bin-Feng Qian

Interpolating Algorithm Optimization and FPGA Implementation in Image Scaling Engine 235
Feng Ran, Jing Liu, Meihua Xu

The Research of the Inclusive Cache used in Multi-Core Processor 239
Bin-feng Qian, Li-Min Yan

Effects of Phosphor's Location on LED Packaging Performance 243
Zongyuan Liu, Sheng Liu, Kai Wang, Xiaobing Luo

Computer Simulation of Crack Propagation in Power Electronics Module Solder Joints 250
Hua Lu, Steve Ridout, Chris Bailey, Wei Sun Loh, Agyakwa Pearl, and Mark Johnson

Process Simulation of DRIE and its Application in Tapered TSV Fabrication 256
Min Miao, Hongguang Liao, Xin Wan, Liwei Zhao, Yunxia Guo, Yufeng Jin

Simulation Study on the Warpage Behavior and Board-level Temperature Cycling Reliability of PoP Potentially for High-speed Memory Packaging 260
Wei Sun, W.H. Zhu, Kriangsak Sae Le and H.B. Tan

Miniaturization Design of Backside-Via Structures Underneath Collector-Up HBTs Using A 3-D Finite-Element Model 268
H. C. Tseng, P. H. Lee, and J. H. Chou

Modeling Ion Transport through Molding Compounds and its Relation to Product Reliability 271
M. van Soestbergen, R.T.H. Rongen, L.J. Ernst and G.Q. Zhang,

Reliability Analysis of Copper Interconnections of System-in-Packaging Structure using Finite Element Method 279
Shih-Ying Chiang, Shin-Yueh Yang, Chan-Yen Chou, Ming-Chih Yew, and Kuo-Ning Chiang

A Multi-scale Interfacial Delamination Model of Cu-SAM-Epoxy Systems 285
H. B. Fan, Cell K.Y. Wong, Matthew M.F. Yuen

Three Dimensional Corner Delamination Analysis for Fan-Out Chip Scale Package 290
Yu-Ren Chen, G. S. Shen

Frequency Dielectric Constant and Loss Tangent Extracting of Organic Material Using Multi-length Microstrip 295
Sung-Mao Wu, Chi-Chang Lai, Hung-Hsiang Cheng, Yu-Che Tai, Chen-Chao Wang

Thermo-mechanical Design of Large Die Fine Pitch Copper/Low-k FCBGA and Lead-free Interconnections 299
Kalyan Biswas, Shiguo Liu, Xiaowu Zhang, TC Chai

Study on Non-Uniformity of Through-Mask Electroplated Ni Thin-Film 304
Jun Tang, Hong Wang, Rui Liu, Shengping Mao, Xiaolin Zhao, Guifu Ding

Table of Contents

Simulation and Analysis for Backward Compatibility of Solder Joints under Thermal Cycle 308
Ning Ye-xiang, Pan kai-lin, Li Ni

First-Principles Study on the Elastic Anisotropy of Au-Sn Intermetallic Compounds 314
Rong An, Chunqing Wang, Yanhong Tian

Numerical Analysis of Interfacial Delamination in Thin Array Plastic Package 319
Ke Xue, Jingshen Wu, Haibin Chen, Yongqiao Sun, Kenneth Kwan ,John Yuen, Angus Lam

Design and Implementation of LED Daylight Lamp Lighting System 324
Rongfeng Guan, Dalei Tian, Xing Wang

Effect of Bonding on the Packaged RF MEMS Switch 327
Le Yang, Xiao-Ping Liao, Jing Song

Genetic Algorithm Optimization in TFT-LCD Drive System 331
Zhi-Jie Tang Ran Feng Mei-Hua Xu

The design of the Ku band Dielectric Resonator Oscillator 334
Guoguang Yan

Simulation of Multi-bit Digital Delta-Sigma Modulator 337
Wen-Rong Yang, Yuan-Yuan Cheng, Jiong-ming Wang

The Research of Sub-space Partition Strategy and Bit Scanning Control Method Based on Human's Vision Nonlinearity Rule 340
Li-min Yan, Shen-nan Qiu

Layout Optimization and Modeling of an ESD-protection n-MOSFET in 0.13um Silicide CMOS Technology 344
Jiang Yuxi, Li Jiao, Ran Feng and Dian Yang

A Verification Method Based On A Mixed-Signal System For MV06 350
HuYue-li, ZhangYi-chi, XuanXiang-guang

Effects of Phosphor's Thickness and Concentration on Performance of White LEDs 354
Zongyuan Liu, Sheng Liu, Kai Wang, Xiaobing Luo

Digital Dimmable Controller in CCFL Module based on Variable Frequency Technique 360
Feng Ran, Tiezhu Li, Meihua Xu, Jian Wu

Design of a Low Voltage Band-gap Reference Circuit for OLED-On-Silicon 364
Meihua Xu, Jian Wu, Feng Ran, Tiezhu Li

Effects of Cu on the electromigration behavior of Al interconnect by using First-Principles method 368
Chun Yu, Hao Lu

A Mixed- Signal Physical Design and Its Verification 371
Hu Yue-li, Yan Ke

Hardware /Software Co-design for Viterbi Decoder 375
Ming Li, Tao Wen

Attenuators Using Thin Film Resistors for RF Application 379
Yiqin Sun,Lei Li, Han Lin, Zhiyuan Yu, Mian Huang, Lixi Wan

Task scheduling and management in single-chip multi-processor system 382
HuYue-li, WangYao-ming, XuanXiang-guang

Analysis of Factors Affecting Color Distribution of White LEDs 386
Zongyuan Liu, Sheng Liu, Kai Wang, Xiaobing Luo

Development of OLED demonstration system based on SD card 394
Chen Zhangjin, Jin Chen, Wang Hao

Table of Contents

Combinational Test Generation for Transition Faults in Acyclic Sequential Circuits......................398
Shi Hui, Ran Feng, Zhang Jinyi

Design of a CMOS Charge Pump for high-performance phase-locked loop...............................403
Xiangguang Xuan, Feng Ran, Meihua Xu

Performance Simulation for EVPD with Equivalent Circuit Models 407
Zhihua Li, Lixi Wan

A March-CL Test for Interconnection Faults of SOC...410
Zhang Jinyi, Yang Xiaodong, Yang Yi, Zhang Dong, Dong hui

Multi-clock SOC Test schedule based on TWC&S..415
Zhang Jinyi, Jiang Yanhui, Lin Feng, Wang Jia, Sun yan

Research on the Characteristics Theory of Reverse SoC TAM Design Based on Dual-Balanced Strategy..................419
Zhang Jinyi, Wang Jia, Lin Feng, Jiang Yanhui, Zhou yi kai

Optimization of hierarchical SOC test time based on genetic algorithm....................................424
Li Jiao, Zhang Jinyi, Shi Hui, Luo xiao wei

Effective Dielectric Constant Method for Trace Impedance Control 428
Te-Chun Wang and Yin-Guang Zheng

Study on No-fillet SMT Solder Joint Reliability Based on Solder Joint Shape CAD 431
Wu Zhao, Hua Zhou, DeJian Huang, ChunYue

Simulation and Experiment Study of Dispensing Patterns Influence on Underfill Filling Process...............438
Jinghua Xie, Guiling Deng, Fei Geng, Junquan Wang

Study on Thermal Simulation Technology for SMA in Lead-free Reflow Soldering 443
Wu Zhaohua Zhou Dejian

A Novel High Effective Envelope-tracking Amplifier for OFDM Systems..................................449
Yingliang Li; Jide-Zhao

A Dual-band MEMS PA Study for Mobile Communication Systems..452
Ji-de Zhao, Ying-liang Li, De-fang Wei

Optical Analysis of A 3W Light-Emitting Diode (LED) MR16 Lamp 456
Kai Wang, Sheng Liu, Xiaobing Luo, Zhongyuan Liu, and Fei Chen

Research on the Cascaded Inverters Based on Simplex DC Power Source..................................461
Yibo Xin, wenqing Chen, huajing Fang

Simulation on Thermal Characteristics of LED Chips for Design Optimization 464
Ting Cheng, Xiaobing Luo,, Suyi Huang, Zhiyin Gan, Sheng Liu

Application of BP Neural Network in FBG Sensing System Performance Improvement...............468
Zhang Jiana, Zhao Hongb ,Rong Xian-weia

Numerical Simulation of Solder Spreading and Solidification during Solder Jet Bumping Process 472
Dewen Tian, Chunqing Wang, Yanhong Tian

Research of Design-for-Testability of CMOS Image Sensor...479
Zhaohui Ou, Feng Lin

A 10-bit 40MSPS Pipeline Analog-to-Digital Converter 483
Cai Jun, Ran Feng, Xu Meihua

Analytical Analysis on the Effect of Time Duration of Acceleration Pulse to a JEDEC Board in Drop Test 487
Jiang Zhou

Thermal Management of A Multi-core Master Processing Unit (MPU) for An Ultrascalable Computing Platform.................493
Ting Cheng, Wei Xiong, Xiaobing Luo, Suyi Huang, Zhiyin Gan, Sheng Liu

Table of Contents

Investigation of Thermal Performance of Various Power-Device Packages 496
Xuejun Fan,

Wafer Level LED Packaging with Integrated DRIE Trenches for Encapsulation 503
Rong Zhang and S. W. Ricky Lee

Impact of Assembly Process Technologies on Electronic Packaging Materials 509
T. Tilford, C. Bailey, A.K. Parrott, J. Rizvi, C. Yin, K.I. Sinclair and M.P.Y. Desmulliez

Board Level Assembly of Inertial MEMS Devices Based on Surface Mounting Technology 515
Kejia Li, Young Sun, Hongguang Liao, Yufeng Jin

Encapsulation of Organic Light-emitting Devices for the Application of Display 519
C Y Li, B Wei and J H Zhang,

A Study of Fluid Coolant with Carbon Nanotube Suspension for Microchannel Coolers 523
Yi Fan, Yifeng Fu, Yan Zhang, Teng Wang, Xiaojing Wang, Zhaonian Cheng and Johan Liu,

Novel Pore-sealing Technology in the Preparation of Low-k Underfill Materials for RF Applications 528
Kuo-Yuan Hsu and Jihperng Leu

A Reliability Investigation of MEMS Transducers with Comb Structures 534
Ping An, Yandong He, Yufeng Jin, Yilong Hao

Low Temperature Sinterable (Zn,Mg)TiO3 Microwave Dielectrics 540
Lih-Shan Chen, Ming-Liang Hsieh, Hsiang-Chen Hsu and Shen-Li Fu

Vibration and Buckling of a Carbon Nanotube Inserted with a Carbon Chain 544
Zhili Hu, X.M.Guo, Johan Liu

Analysis of Electromagnetic Wave Propagation Characteristics in Rotating Environments 551
Yi Leng, Yingfeng Pan, Qingxia Li, Sheng Liu

Electrodeposition of Palladium Films on Ni-Co Coatings 558
Yanping He, Dongyan Ding, Xiang Gao, Zhi Chen, Ming Li, Dali Mao

Titania Nanostructures Fabricated Through Anodization of Ti6Al4V Alloy 563
Yan Li, Dongyan Ding, Shuo Bai, Ming Li, Dali Mao

The Influence of Low level Doping of Ni on the Microstructure and Reliability of SAC Solder Joint 567
Zhenqing Zhao, Lei Wang, Xiaoqiang Xie, Qian Wang, Jaisung Lee

BGA Assembly Process Development for 45nm ELK CUP Devices 572
Andy Tseng, Bryan Lin, Louie Huang, Mike Hung

C4NP for Pb-Free Solder Wafer Bumping and 3D Fine-Pitch Applications 577
D.-Y. Shih, B. Dang, P. Gruber, M. Lu, S. Kang, S. Buchwalter, J. Knickerbocker, E. Perfecto J. Garant,
S. Knickerbocker, K. Semkow, B. Sundlof, J. Busby, R. Weisman K. Ruhmer and E. Hughlett

Manufacture of Hourglass-shaped Solder Joint by Induction Heating Reflow 584
Hongbo Xu, Mingyu Li, Liqing Zhang, Jongmyung Kim and Hongbae Kim

The Effects of Ni Nanoparticles Addition on Shear Behavior and Microstructure of Sn-Ag Lead-free Solder 589
Fangjuan Qi, Li Sun, Zhezhe Hou, JianqiangWang, Cha Qin

Nanoscale Analysis of Ultrasonic Wedge Bond Interface by Using High-Resolution Transmission Electron Microscopy 592
Hongjun Ji, Mingyu Li, Chunqing Wang, Jongmyung Kim and Daewon Kim

Limited ß-Sn grain number of miniaturized Sn-Ag-Cu solder joints 597
Shihua Yang, Chunqing Wang, Yanhong Tian, Pengrong Lin, Le Liang

Table of Contents

Manufacture, Microstructure and Microhardness Analysis of Sn-Bi Lead-Free Solder Reinforced with Sn-Ag-Cu Nano-particles 600
Lili Zhang, Wenkai Tao, Johan Liu, Yan Zhang, Zhaonian Cheng, Cristina Andersson, Yulai Gao and Qijie Zhai

A comparison study of two different methods to synthesize magnetic slurry for the fabrication of magnetic films 605
Xu Zheng, Rong Sun, Shuhui Yu, Lei Li, Mian Huang, Guangfu Yin, Ruxu Du, Lixi Wan

Microstructure and Properties of Barium Strontium Titanate Thin Films Prepared on Copper Foils via Addition of PEG to the Sol Precursor 609
Yanhua Fan, Shuhui Yu, Rong Sun, Lei Li, Mian Huang, Yansheng Yin, Lixi Wan

Dynamic Behavior Tests of Lead-free Solders at High Strain Rates by the SHPB Technique 613
Qin Fei, An Tong, Chen Na

Formation of Double-layer Cu3Sn in Solid-State Aging Process at the Interface of Eutectic SnBi Solder and (100) Single Crystal Cu 617
Pan-Ju Shang, Zhi-Quan Liu, Dou-Xing Li, and Jian-Ku Shang

Synthesis of High Purity O-cresol Novolac Epoxy Resins 621
Xiao-Wei Tian, Zhen-Guo Yang

Defect Analysis of Copper Ball Bonding 626
Huabin Chu, Jun Hu, Ling Jin, Yingliang Jie

Fundamental Influence of Segregated Bi on the Mechanical Properties of Interconnect of Bismuth-containing Solder and Copper 629
Xue-Yong Pang, Zhi-Quan Liu, Shao-Qing Wang, and Jian-Ku Shang

Effects of Bi and Ni Addition on Wettability and Melting Point of Sn-0.3Ag-0.7Cu Low-Ag Pb-free Solder 632
Y. Liu, F. L Sun, T. L.Yan, W. G. Hu

Moisture Diffusion Model Verification of Packaging Materials 636
Xiaosong Ma, K.M.B. Jansen, L.J. Ernst, W.D van Driel, O. van der Sluis, G.Q.Zhang, Charles Regard, Christian Gautier, Hélène Frémont

Recent Progress of Carbon Nanotubes as Cooling Fins in Electronic Packaging 641
Johan Liu, Yifeng Fu, and Teng Wang

Environmentally Friendly Electronics for High Reliability 648
J. B. McElroy, R. C. Pfahl

Morphology, Evolution and Performance of IMC in SAC105 Solder/UBM (Ni (P)-Au) 651
F. L. Sun, P Hochstenbach, W. D. Van Driel , G. Q. Zhang

Development of High Temperature Stable Isotropic Conductive Adhesives 655
Zhikun Zhang, Sijia Jiang, Johan Liu, and Masahiro Inoue

Microstructural and physical characteristics of Sn-Ag-Cu-Mg Lead-free Solders 660
Sheng Lu, Fei Luo, Jing Chen, Baohua Wang

Microstructural Investigation on the Interfacial Evolution of SnBi/Cu Interconnect during Reflow and Solid-State Aging 664
Zhi-Quan Liu, Pan-Ju Shang, Xue-Yong Pang, and Jian-Ku Shang

Dynamic Mechanical Properties of the Transparent Silicone Resin for High Power LED Packaging 667
Qin Zhang, Xiu Mu, Kai Wang, Zhiyin Gan, Xiaobing Luo and Sheng Liua

Study of Five Substrate Pad Finishes for the Co-design of Solder Joint Reliability under Board-level Drop and Temperature Cycling Test Conditions 671
Wei Sun, W.H. Zhu, Edith S. W. Poh, H.B. Tan and Richard Te Gan

The Influence of Heat Treatment on the Adhesion between Molding Compound and Lead Frame 679
Wei Tan, Xinyu Du, Guangchao Xie, Suqiong Qin, Hujie Mei, XingMing Cheng

Table of Contents

Shear Fracture Behavior of Sn3.0Ag0.5Cu Solder joints on Cu Pads with Different Solder Volumes........................684
Yanhong Tian, Chunqing Wang, Shihua Yang, Penrong Lin, Le Liang

Effect of Moisture and Temperature on Al-Cu Interfacial Strength 688
Hui Teng, Huiliang Zhang, Hongbo Yang, Ming Zhou, Anthony C. Tsui

Thiol based chemical treatment as adhesion promoter for Cu-epoxy interface..........................692
Cell K Y Wong and Matthew M F Yuen

Design of Testing Chip for Measuring Mechanical Properties of Thin Films..........................699
Rui Liu, Hong Wang, Xueping Li, Jun Tang, Shengping Mao, Guifu Ding

The Design and Fabrication of RF Band Pass Filter by LTCC Technology 703
Yuanxun Li, Yingli Liu, Huaiwu Zhang, Dafu Lu, Lifei Bian, Zongbao Yang

Cu Out-Diffusion Kinetics in Pre-plated Cu-alloy Leadframes Investigated by a Developed EDX-based Oxidation Test..707
Lilin Liu, Ran Fu, Deming Liu, Tong-Yi Zhang

Corrosion Performance of Pb-free Sn-Zn Solders in Salt Spray...713
Huang Dan, Zhou Jian, Li Pei Pei

The Synthesis and Curing Kinetics of the Silicon-containing Epoxy Resin with Environmentally Friendship Flame Retardance 717
Ming-shan, Yang, Jie He, Lin-kai Li, Zhen Liu

Nonlinear Optical Fluorinated Polyimide/Inorganic Composites for Optical Interconnections and Devices..............720
Li Guoyuan, Tang Chuan and Ren Li

Nano-Thermal Interface Material with CNT Nano-Particles For Heat Dissipation Application........................725
Li Xu, Cong Yue, Johan Liu, Yan Zhang, Xiu Zhen Lu and Zhaonian Cheng

The Humidity and Thermal Characteristics of Die-Attach (DA) and its Impact on the Package Reliability 729
Chen Ning, Ma Xiao-song, Jiang Haihua

Influence of Crystal Orientation on the Oxidation Failure of Copper for IC Package 735
Jie Gao, Anmin Hu, Ming Li, Dali Mao

The Influence of Pre-heat Treatment on Peeling Resistance of Oxide Film of Copper Alloy Lead Frames..................740
Jiwang Mao, Xi, Chen, Anmin Hu, Ming Li, Dali Mao

Interfacial Reactions and Reliability of Sn-Zn-Bi-XCr Solder Joints with Cu Pads 744
Yidong Shen, Anming Hu, Xi Chen, Ming Li, Dali Mao

The Thermal and Electric Conduction Properties of High Dense TiB2P/Cu Composites Fabricated by Squeeze Casting Technology..748
Guoqin Chen, Ziyang Xiu, Songhe Meng, Gaohui Wu, Su Chen

Microstructure and Properties of Environmental-friendly Sip/1199Al Composites Used for Electronic Packaging 751
Ziyang Xiu, Guoqin Chen, Zongquan Deng, Gaohui Wu

Preparation of Polysulfoneamide Electrospinning Nanofibers...754
Li Liu, Ying Shi, Qinghua Jiao, Yanan Wu, Zhikun Zhang, Johan Liu

Liquid-State Interfacial Reaction of Sn-10Sb-5Cu High Temperature Lead-free Solder and Cu Substrate..............758
Qiulian Zeng, Jianjun Guo, Xiaolong Gu, Xinbing Zhao

Application of Embedded Components in Package Substrate..763
Lingwen Kong, Zhiqin Yang, Jianhui Liu, Minfei Lu

Roadmap of Reliability: Methodology and Application in Guangdong's Lead-Free Technology........................765
Xiaoyan Li, Ling Wang, Yonggao Fu, Lu Zeng

Table of Contents

Effect of Isothermal Aging on Interfacial IMC Growth and Fracture Behavior of SnAgCu/Cu Soldered Joints 769
Xiaoyan Li, Xiaohua Yang, Fenghui Li

Electrochemical Corrosion Behaviors of ITO Films at Anodic and Cathodic Polarization in Sodium Hydroxide Solution 774
Wang Hao, Zhong Cheng, Li Jin, Jiang Yiming

Effects of Bonding Pressure on Nonlinear Dynamic Characteristic of the Ultrasonic Wire Bonding System 778
Lei Lv, Lei Han

Effects of Ni addition on Microstructure and the Shear Strength of Sn-3.0Ag-0.5Cu/Cu Solders Joints 783
Lifeng Wang, Xuwei Shen, Fenglian Sun, Yang Liu

An Investigation of Capillary Vibration during Wire Bonding Process 788
Jian Gao, Robert Kelly, Zhijun Yang, Xin Chen

The Optimization of Hierarchical SOC Test Architecture to Reduce Test Time 794
Xu Chuan-pei, Dai Kui

The Influence of Heating Temperature on Alignment Precision in Thermosonic Flip-chip Bonding 798
Li-na Zhang, Lei Han

A High Speed Image Preprocessing Method for IC Wafer Inspection 804
Wei Chen, Li-Ming Wu, Shan-Ling Cui

The Influence Discipline of Temperature of High Viscosity Fluid Jetting 807
Hu Hao, Deng Gui-ling

The Influence of Structural Parameters of Electromagnetic Fluid Jetting Dispenser 813
Wang junquan, Deng Guiling

Dynamic Phase-frequency Characteristic of Thermosonic Wire Bonder Transducer 819
Fuliang Wang, Changhui Zou, Jiaping Qiao

Study of Prepress Force on Piezoelectric Transducer of Wire Bonding 823
Li zhanhui, Wu yunxin, Long zhili

Simulation and Experimental Research on Water-jet guided Laser Cutting Silicon Wafer 826
Wang Yang, Li Ling, Yang Lijun, Liu Bei, Wang Zhe

Research on Solder Joint Intelligent Optical Inspection Analysis 832
Li Ni, Pan Kai-lin, Li Peng

Productivity Improvement of Stack Package Line through Die Bonding Process & Scheme Optimization 836
Xing Jin, Ming Li

Developing the Stencil Printing Process for 01005 Lead-Free Assemblies 841
Yong-Won Lee, Keun-Soo Kim, Katsuaki Suganuma and Jong-Hoon Kim

Shear of Sn-3.8Ag-0.7Cu Solder Balls on Electrodeposited FeNi Layer 848
Q. S. Zhu, J. J. Guo, Z. G. Wang, Z. F. Zhang and J. K. Shang

IMC Formation between Electroless Ni/Pd/Au Surface Finish and SnAgCu Solder 852
Jiwang Mao, Bin Liu, Ming Li, Yu Wang, Dali Mao

Optimization of the Fatigue Life of Epoxy Molding Compounds based on BP Neural Network Prediction Model 856
Miao Cai, Dao-guo Yang, Quan-yong Li, Li-jun Zhong

Electromigration Behavior of the Ni/SnZn/Cu Solder Interconnect 863
X.F.Zhang, J.D.Guo, J.K.Shang,

Analysis on Cracking Blind Vias of PCB for Mobile Phones 867
Li-Na Ji, and Zhen-Guo Yang

Table of Contents

Interfacial Reaction and Failure Mechanism for SnAgCu Solder Bump with Ni(V)/Cu Under Bump Metallization During Aging 873
Kai Jheng Wang and Jenq Gong Duh

Effect of displacement rate on lap shear test of SAC solder ball joints 878
X.J. Wang, Z.G. Wang, J.K. Shang

Capability Study on the Destructive Pull Test of 1 Mil Gold Wire Bond and Its Asymmetric Distribution 882
Jin Peng

Mechanical Reliability Estimation for μBGA Solder Joints Based on Heating Factor Q. 885
Tao Bo, Yin Zhouping, Ding Han and Wu Yiping

Electromigration Simulation with Consideration of the Atomic Concentration Gradient 889
Xuefan Chen, Lihua Liang

Finite Element Analysis of Reliability on Compliant Wafer Level Packaging With Compliant Layer 896
Peng Li, Kai-lin Pan, Ye-xiang Ning

On the Study of the In-Use Stability of a DCA Assembled MEMS Device 900
Liyuan Xu, Jing Song, Jieying Tang

IC Chip Crack Issues due to Mounting Process For Ultra-thin IC Smart Card Module 905
Pingyue Fan, Jiaji Wang

Optimization of Interface Strength for SCSP Based on Uniform Experimental Design 909
Gongke Li, D. G. Yang, Liancheng Qin, Fuxi Yi

Evaluate Anti-Shock Property of Solder Bumps by Impact Test 914
Hongjia Xi, Minyi Lou, Bing An, Fengshun Wu, Yiping Wu

Cavitation instability in Valanis-Landel hyperelastic IC packaging material 920
Li Zhigang, Shu Xuefeng

Moisture Absorption and Void Growing Effects on Failure of Electronic Packaging 923
Zhao Zhendong, Li Zhigang, Zhang Yu, Shu Xuefeng

Mechanical Test after Temperature Cycling on Lead-free Sn-3Ag-0.5Cu Solder Joint 927
Chung-Nan Peng, Jenq-Gong Duh

Reliability Challenges and Design Considerations for Wafer-Level Packages 931
Xuejun Fan, Qiang Han

Mixed Mode Interface Characterization Considering Thermal Residual Stress 937
A. Xiao, G. Schlottig, H. Pape, B. Wunderle, K M. B. Jansen, L. J. Ernst

The Effect of Strain Rate and Strain Range on Bending Fatigue Test 944
Minyi Lou, Long Wen, Zhengrong Chen, Jianwei Zhou, Qian Wang, Jaisung Lee

Dynamic Properties Testing of Solders and Modeling of Electronic Packages Subjected to Drop Impact 949
Long Wen, Xingming Fu, Jianwei Zhou, Qian Wang, Jaisung Lee

Electromigration in Pb-free Solders 956
Minhua Lu, Da-Yuan Shih, Paul Lauro

Effect of Stand-off Height on the Microstructure and Fracture Mode of Cu/Sn-9Zn/Cu Solder Joint under Tensile Test 964
Bo Wang, Fengshun Wu, Bin Du, Bing An, Yiping Wu

Failure Mode Analysis of Lead-free Solder Joints under Differential Reflow Profiles by High Speed Impact Testing 968
C. Y. Lin, Y. R. Chen, and G.S. Shen D. S. Liu, C. Y. Kuo, And C. L. Hsu

Influence of Underfill Methods on the Solder Joint Fatigue of Wafer Level Packaging 974
Charles Regard, Christian Gautier, Hélène Fremont, Patrick Poirier

xiii

Table of Contents

The Role of the Molecular Simulation Approach for IC-backend Developments 980
C. Yuan, O. van der Sluis, W. D. van Driel, G. Q. Zhang

Strain-rate and Impact Velocity Effects on Joint Adhesion Strength 984
Chang-Lin Yeh, Yi-Shao Lai

Parametric Study on Board-level Electronic Test Device Subjected to JEDEC Vibration Loads 988
Chang-Lin Yeh,, Yi-Shao Lai, Ching-Chun Wang

Modeling Techniques for Board Level Drop Test for a Wafer-Level Package 994
Harpreet S. Dhiman, Xuejun Fan, Tiao Zhou

Effect of Shear Rate on Lead Free Solder Joint Strength 1003
Zheming Zhang, Jingshen Wu, Adam R. Zbrzezny, Neil Mclellan

Analysis and Comparison of Thermal Stress and Hygrothermal Stress of SiP Device By QFN Packaging 1008
Jiang Haihua, Ma Xiao-song, Chen Ning

**Crack Growth Analysis of Ball Grid Array Resistor's Solder Joint Subjected to Thermal Cycling and 4
Point Cycling Bending** 1013
Xiangzhao, Ye-Yuming, Sun-Fujiang, Tu-Yunhua, Liusang

A Dual-Output Voltage Reference for High-Accuracy Pipelined ADC 1017
Dongfang Cheng, Xiaohui Li, Jue Zhang, Jiongming Wang

Study of Interface Reliability in QFN Device under Hygro-Thermal Environment 1021
Ting-biao Jiang, Hong-mi Nong, Chao Du

In-situ Observation on Electrochemical Migration of Lead-free Solder Joints under Water Drop Test 1026
Y.H. Xia, W. Jillek, E. Schmitt

**The Reliability Study of Sub 100 Microns SnAg Flip Chip Solder Bump on FR4 Substrate under Thermal
Cycling** 1031
Xiaoqin Lin, Le Luo

Analysis and Solving of the EMI effect on LC-VCO in mixed-signal ICs 1036
Wenrong Yang, Jiongming Wang, Jue zhang, Xiaohui Li

**Thermal Behavior Analysis of Lead-free Flip-Chip Ball Grid Array Packages with Different Underfill
Material Properties** 1040
Hsin-yuan Chena, Kuo-yuan Hsua, Tsung-shu Linb and Jihperng Leua

**Enrichment and Removal of Heavy Metals Contained in PCB Boards by Multiwalled Carbon Nanotubes
for WEEE Directive** 1047
L. Hua, H. N. Hou

A Study on Application of N&K Analyzer in OLED Failure Analysis 1052
Qiang Fang, Yafang Peng, HK Yu

Failure Analysis of the First Wire's Bond 1055
Ren Chunling, Huang Qiang, Ding Rongzheng, Jiang Changshun

Investigation of Electromigration in Copper Interconnection of ULSI 1059
Dechun Lu, Shengxiang Bao, Lili Ma, Zhibo Du

**Study on MCM Interconnect Test Generation using Ant Algorithm and Particle Swarm Optimization
Algorithm** 1063
Chen Lei

MCM Interconnect Test Scheme based on Adaptive Genetic Algorithm 1067
Chen Lei

Investigation on Fatigue-Creep Interaction Damage Model for Solder 1070
Na Liu, Xiaoyan Li, Yongchang Yan

xiv

Table of Contents

Study of Plasticity Damage Mechanics Constitutive Model for SnAgCu Solder Joint.................................1073
Xiao-yan Li, Yong-chang Yan,Na Liu

A Design for Increasing the Immunity to RFI of Protection IC of Lithium-ion Battery...........................1077
Dongfang Cheng Jue Zhang Xiaohui Li Jiongming Wang

The Effect of the Different Teflon Films on Anisotropic Conductive Adhesive Film (ACF) Bonding........................1082
Jun Zhang, Y.C. LIN, Liugang Huang

Reliability Study of Flexible Display Module by Experiments.......................................1086
Quayle Chen, Leon Xu, Antti Salo

Thermal Fatigue Life Analysis and Forecast of PBGA Solder Joints On the Flexible PCB Based on Finite Element Analysis..........................1092
Huang Chunyue

xvi

Qualification for Product Development

Weiqiang Wang, Michael H. Azarian and Michael Pecht
Center for Advanced Life Cycle Engineering (CALCE)
University of Maryland
College Park, MD 20742, USA
Email: pecht@calce.umd.edu;
Telephone number: (01)301-405-5323;
Fax number: (01)301-314-9269

Abstract

The aim of qualification is to verify whether a product meets or exceeds the reliability and quality requirements of its intended application. Qualification plays an important role in the process of product development. It can be classified by its specific purpose at different stages of the product development process. In this paper, a new methodology of product qualification is proposed based on physics-of-failure. This methodology consists of: product configuration and material information collection; application requirement information collection; strength limits and margins; failure modes, mechanisms and effects analysis; definition of qualification requirements; qualification test planning; testing; failure analysis and verification; and quality and reliability assessment. This approach to qualification ensures that it successfully addresses the failure mechanisms applicable to the product's specific design, manufacture, and application conditions.

1 Introduction

Qualification is the process of demonstrating that an entity or process is capable of meeting or exceeding the specified requirements [1]. Qualification includes activities which ensure that the nominal design and manufacturing process will meet or exceed the specified targets. The purpose of qualification is to obtain the acceptable range of variability for all critical product parameters affected by design and manufacturing, such as geometric dimensions and material properties. Attributes that fall outside the acceptable range are termed as "defects" because they have the potential to make the product fail to meet the specified requirements [2].

Qualification can be used in process development and product development. Process qualification involves a set of procedures which validate that a process used to manufacture a product meets specified performance requirements [3]. It is used to provide assurance that a particular process is under control and known to produce qualified products [4]. Product qualification aims to evaluate performance of products under specified operating and environmental conditions within a specified period of time, which will be the focus of this paper.

The performance of a product consists of quality (which includes function) and reliability, whose requirements are set during product design. The qualification process is intended to examine whether the products' performance can meet the design requirements. The product qualification includes the verification of functions, the assessment of reliability in application conditions, and the validation in the system application if the product is a component of a system.

Product qualification can be used to baseline the design, materials and processes. It determines the product performance degradation under normal application conditions. It can also be used to compare different designs to help make design decisions. Product qualification is used to meet the requirements of customers with consideration of the intended application and application conditions.

2 Qualification in product development process

Qualification occurs in different stages of the product development process, as shown in **Figure 1**. Qualification activities in different stages have different purposes. Virtual qualification is to evaluate the functional and reliability performance of the product design without any physical testing on the product. Virtual qualification involves using computer-assisted modeling and simulation based on physics-of-failure (PoF) [5]. Product qualification is to evaluate the product based on the physical testing on the manufactured prototype. The purpose is to verify whether the product has met or exceeded its intended quality and reliability requirements. After virtual and product qualification, the products are mass produced. During and after the manufacturing process, the products can be inspected and tested to evaluate their quality and defected parts can be screened out. This process can be considered as a third stage in the overall qualification process, and it is more commonly referred to as quality assurance testing.

Virtual and product qualification efforts are part of a larger process of product design and development. At various intersections of the process, maturity levels can be assigned to indicate progress and specific readiness for the next phase. The design and product qualification process may include feedback iterations shown in **Figure 2**. If the product design is found to be unqualified during the virtual qualification process, it is modified and then virtually re-qualified before proceeding to the next phase. Similarly, when a design has successfully passed through the virtual qualification process, but does not meet the qualification requirements during product qualification stage, feedback iterations may be necessary. In this case, the virtual qualification process and specifically, the physics-of-failure based models may have to be re-evaluated and modified. After design completion, the product is manufactured in high volume and subjected to quality assurance testing during and after the process.

978-1-4244-2739-0/08/$25.00 ©2008 IEEE

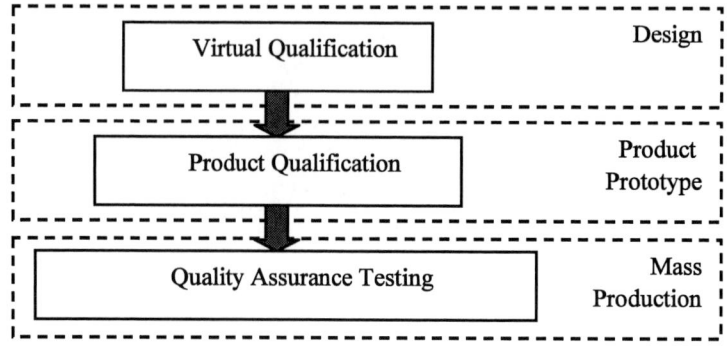

Figure 1: Qualification activities in different stages of a product development process

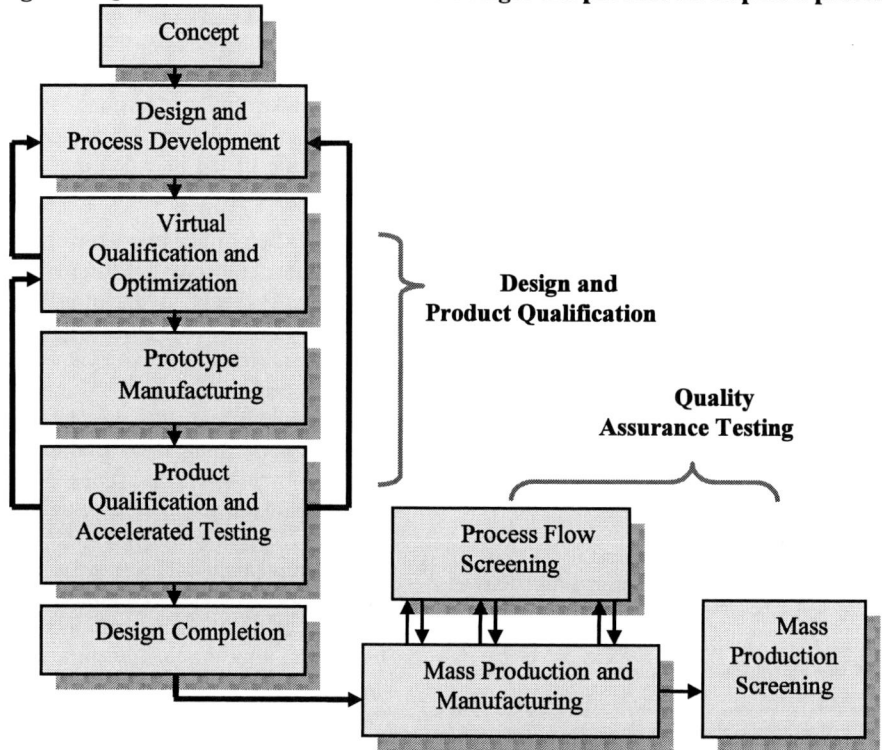

Figure 2: Qualification and quality assurance testing within the product design and manufacturing process flow including iterative feedback process

The objectives of qualification testing are to (a) evaluate the quality of a product to see if it meets the design requirements, (b) develop information on the integrity of a product and its structure, (c) estimate the expected service life and reliability and (d) evaluate the effectiveness of materials, processes, and designs,. Qualification tests estimate expected life and design integrity of a product. Most tests are not conducted under the normal application conditions, but at accelerated levels of stresses to accelerate potential failure mechanisms at associated sites in a product.

Successful qualification of a sampling of a product does not assure that all products made by the same manufacturer to the same specifications will also meet the qualification requirements. Qualification should be conducted by the manufacturer, although the customer may do so for special applications. Data from all possible sources should be used in qualification. These sources include material and component suppliers' test data,

qualification data from similar items, and accelerated test data from materials, components, and subassemblies.

2.1 Virtual qualification

Virtual qualification is a methodology for assessing and improving the reliability of products through the use of validated failure models and simulation tools [6] [7]. It is also an important step in developing effective physical tests to verify product reliability. The application of virtual qualification has led to significant cost savings for commercial and military organizations.

Virtual qualification is the first stage of the overall qualification process. It is the application of PoF based reliability assessment to determine if a proposed product can survive its anticipated life cycle [8]. Virtual qualification (also called simulation-assisted reliability assessment) assesses whether a part or system can meet its reliability goals under anticipated life cycle profiles based on its materials, geometry, and operating characteristics. The technique involves the application of

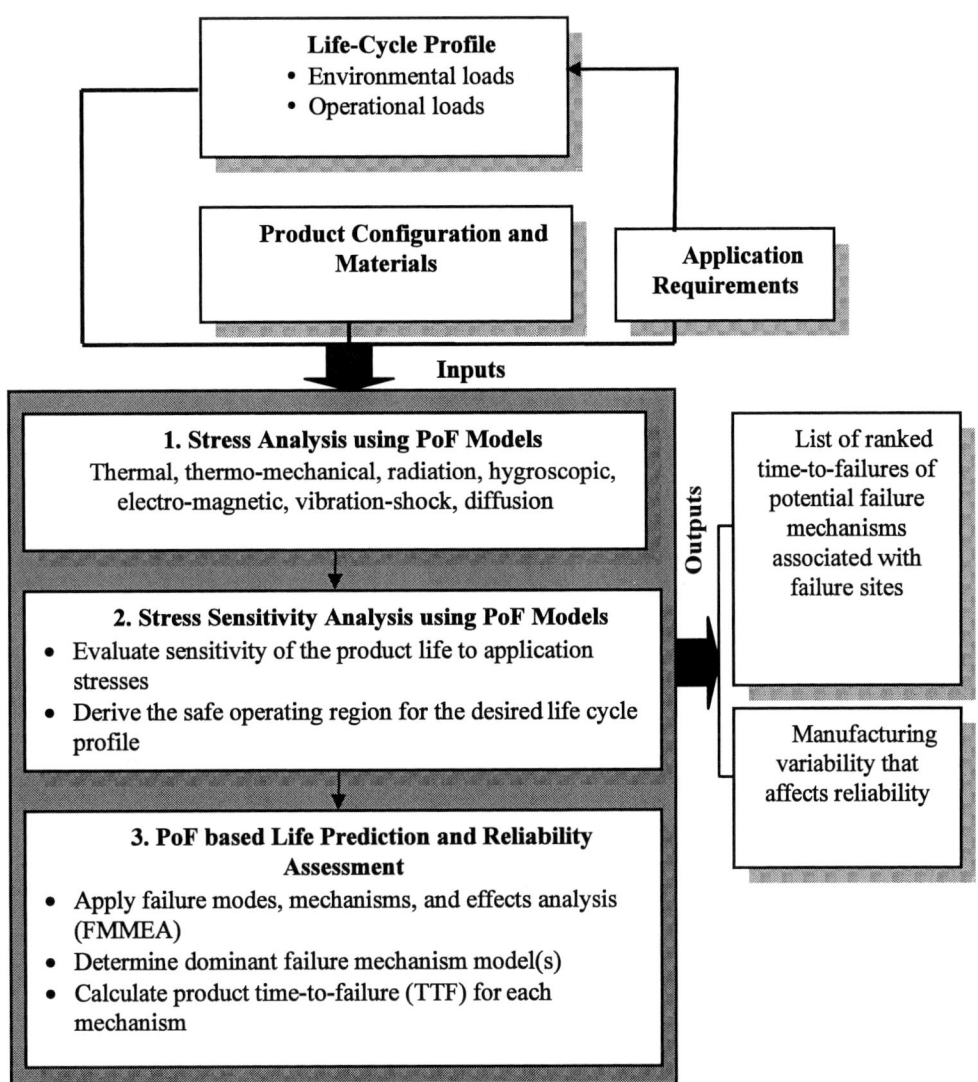

Figure 3: Flowchart of virtual qualification

simulation software to model physical hardware to determine the probability of the system's meeting desired life goals [1] [9] [10] [11].

Engineers can realize significant time savings by developing a flow-through process of life-cycle characterization, product modeling, load transformation, and failure assessment to qualify products. Virtual qualification can be applied at design stages and hence it helps move reliability assessment process into the design phase [6] [7]. It allows the design team to consider qualification at the initial stages of design, technology and functional definition, and supplier selection. A flowchart of virtual qualification is shown in Figure 3. This system takes advantage of advances in computer-aided engineering software tools that permit components and systems to be qualified based on an analysis of the susceptibility of their designs to failure due to a number of fundamental physical and chemical mechanisms [12] [13]. The reliability assessment tool assesses the candidate and existing product designs for reliability in many different environments using a database of fully validated PoF models. It calculates time-to-failure of fundamental mechanisms that cause failures and evaluates the effects of

different manufacturing processes on reliability by calculating the time-to-failure as a function of typical manufacturing tolerances and defects [14] [15]. It facilitates the selection of cost-effective test parameters for validating reliability assessment and design and also aids the selection of high-volume commercial off-the-shelf components by permitting their virtual qualification [16].

The inputs consist of life cycle profile and product characteristics. The life cycle profile can be further categorized as environmental and operational stresses as shown. The inputs are fed into physics-of-failure (PoF) model and simulation software where stress analysis, reliability assessment and stresses sensitivity analysis are performed. The outputs of virtual qualification are predicted time-to-failures (TTF) based on the most dominant failure mechanisms, stress margin conditions, and screening and accelerated testing conditions.

In addition to time-to-failure prediction and reliability assessment, virtual qualification combined with advanced optimization techniques can be used to optimize the design criteria including cost, electrical performance, thermal

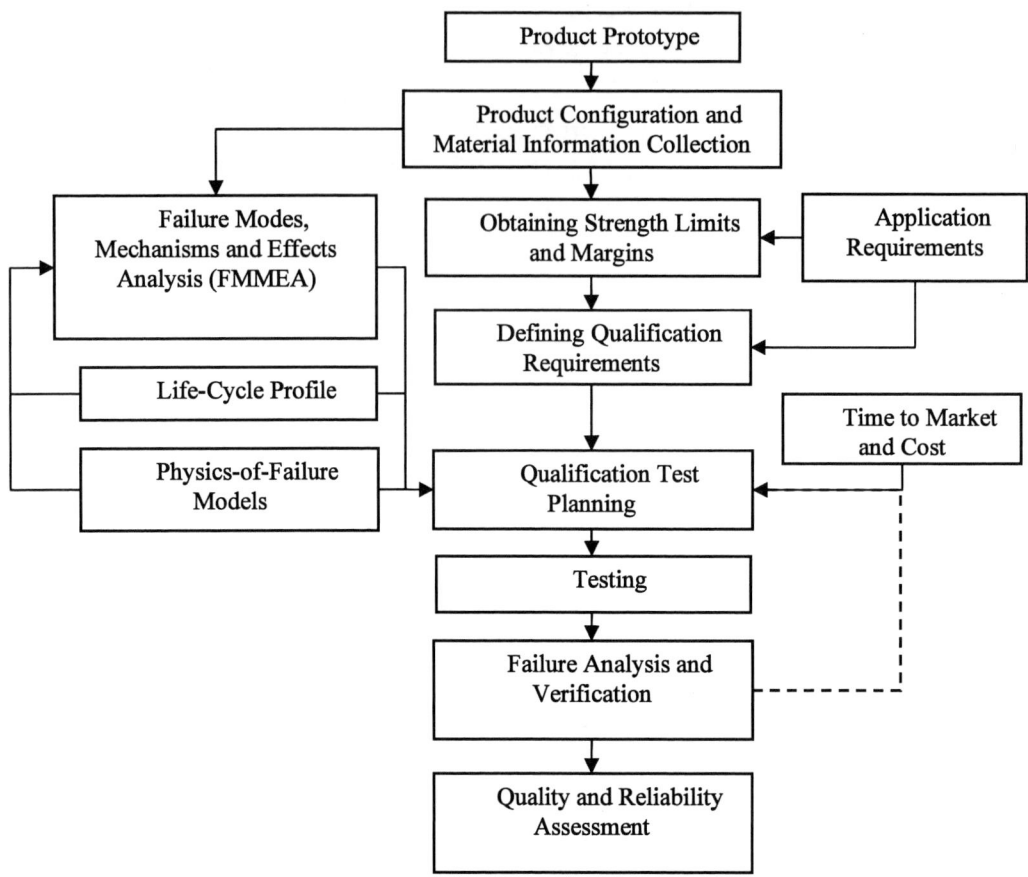

Figure 4: Flowchart of methodology of product qualification

management, physical attributes, and reliability. By examining potential trade-offs between the aforementioned criteria, ideal values can be achieved for specific applications.

In the virtual qualification process, it is imperative to use the most accurate inputs including material properties, design configuration, dimensions, and operational and environmental conditions. Furthermore, the failure mechanism models used in time-to-failure (TTF) prediction and reliability assessment must be valid. If the data or models on which the virtual qualification is performed is inaccurate or unreliable, any qualification results based on the data or models are suspicious.

2.2 Product qualification

Product qualification is the evaluation of products after prototype manufacturing. It is intended to qualify a product before its mass production. The qualification includes the verification of their function and performance, the validation in the system application (if applicable) and the qualification for processability and reliability. After the qualification, products' specifications that do not meet the design and customer requirements should be reported to the design team for correction action. Parameters in qualification tests, failure modes and failure mechanisms of products during qualification tests will be provided to the design team as feedback on how to improve the design or the manufacturing process. Product qualification will be the main focus of this paper. It will be detailed in section 3.

2.3 Quality assurance testing during mass production

The qualification tests that take place during mass production and before the products are shipped to the customer are more properly considered to be quality assurance testing that ensure that the products are manufactured according to the design within allowable tolerances. Furthermore, accelerated stresses are applied to the products to accelerate early failures that are caused by manufacturing defects. These qualification tests ensure the quality of products that will be used in field applications. Manufacturing defects are screened out of the product shipping list.

3 Methodology of product qualification

Upon completion of virtual qualification, the product prototype is manufactured, and then product qualification process begins. In the product qualification process, physical tests are applied to the manufactured prototype to verify whether it meets its functionality and reliability requirements. If the design and manufacturing processes that were initially considered during the virtual qualification process has not been modified, then product qualification process essentially begins with strength limit testing or HALT. Conversely, any changes made to the product characteristics outside the design and manufacturing tolerance ranges requires virtual re-qualification or a product qualification process that includes the re-definition of product characteristics and the repeat of FMMEA process. The center for advanced life cycling engineering (CALCE) developed a methodology for product qualification to make the process of qualification clearer to engineers. This methodology includes collecting product configuration and material information, collecting product life cycle profile, strength limit and margins, failure mode, mechanisms and effects analysis (FMMEA), defining qualification requirements, qualification test planning, testing, failure analysis and verification and quality and reliability assessment. A flowchart is shown in Figure 4.

3.1 Product configuration and materials

One of the most fundamental steps in the product qualification is to characterize the product in terms of its configuration and materials. Information about the configuration and materials of the product provides engineers with the basic knowledge for the qualification. The information includes the architecture of product, the materials used to manufacture the product and the process that the materials have experienced during the manufacturing process.

A product consists of a number of components and subassemblies working together to deliver the overall function of the product. Each subassembly may consist of lower level assemblies that are also interconnected. The architecture of a product describes the physical and functional relations between the subassemblies. The hardware configuration of the product describes the design of the components and subassemblies and the product architecture. It may also include the effects of the manufacturing processes on the final product in the form of tolerances on the dimensions and material properties.

The hardware of electronic equipments includes electronic parts, printed circuit boards, connectors, and enclosures. An electronic part may be a semiconductor chip and the package that provides power and ground inputs, signal communication paths to the outside and protection from the environment. An electronic part can also be a passive component such as a resistor or capacitor. The part geometry and structure, the sub-component geometry, and the connection methods, such as wirebonds or solder balls, will also be characterized. A printed circuit board description include the materials, layer stacks, the connections between layers, the additions to the layers, such as heat spreaders, and elements like stiffeners.

Materials used to construct a product influence the level of stress on the product due to external and internal loads and the process of damage accumulation [17] [18]. To the extent that materials influence stress and damage, their physical properties should be characterized [19] [20]. For example, a failure in a solder joint may be driven by stress arising from repeated temperature excursions. In this situation, the coefficient of thermal expansion of a material is needed to determine the cyclic stress state. In another situation, a failure may occur due to a reduction in the contact force between connector elements. This situation may require the elastic modulus of the connector elements, loading elements and their housings to determine the contact force and its degradation pattern. Properties for common materials used in electronic products can be found in references [19] [20] [21].

Products are not normally produced by a single manufacturing process. They often require a sequence of different processes to achieve all the required attributes of the final product. The manufacturing process applies stresses on materials, may produce residual stress, and may even modify some of material properties. For example, a lead-free reflow profile can change the thermo-physical properties of a printed circuit board. The variations in geometry and material properties caused by different manufacturing processes need to be characterized.

3.2 Life-cycle profile

The second step in the product qualification is to understand the life-cycle profile (LCP) of products. The LCP is the base for selecting product qualification test conditions, including types and severity levels. The major task in understanding the LCP is to characterize the loads applied onto the product during its life cycle. The environmental loading to a component should be considered to be from its surrounding environment as well as from within, but not from the system level environment. For example, when a silicon chip is working, the temperature and humidity of its environment will affect its function and reliability, as does the heat generation within the chip.

A LCP is a time history of events and conditions associated with a product from its release from manufacturing to its removal from service. The life cycle includes various phases that an item will encounter in its life, such as: handling, shipping, and storage prior to use; mission profiles while in use[1]; phases between missions, such as stand-by or storage, transfer to and from repair sites and alternate locations; geographical locations of expected deployment and maintenance; and maintenance and repair procedures for the system and the component.

Loads applied to the product during its life cycle drive the processes that lead to product degradation and failure. The life cycle of a product includes manufacturing and assembling, testing, reworking, storing, transporting and handling, operating (e.g., modes of operation, on-off cycles), and repairing. The life cycle loads include assembly/installation related loads, environmental loads and operational loads. These loads can be thermal [22], mechanical, chemical, physical, and/or operational loading conditions. Various combinations and levels of these loads can influence the reliability of the product. The extent and rate of product degradation depend upon the nature, magnitude and duration of exposure to such loads.

Since a product may experience numerous loads, it is necessary to identify the most critical ones to its function and reliability. Some of the loads will play major roles in activating and accelerating the failure of the product, while others can be ignored. For example, low levels of radiation can often be ignored for ground-based electronic products, since this rarely causes dysfunction or damage to products. Whether the loads can or cannot be ignored depends on the critical failure mechanisms that are identified in the analysis, which consider the life cycle conditions.

---

[1]In some cases, the environmental factors experienced by constituents of the product begin before manufacturing -e.g., storage of parts (material) far in advance of their use in manufacturing.

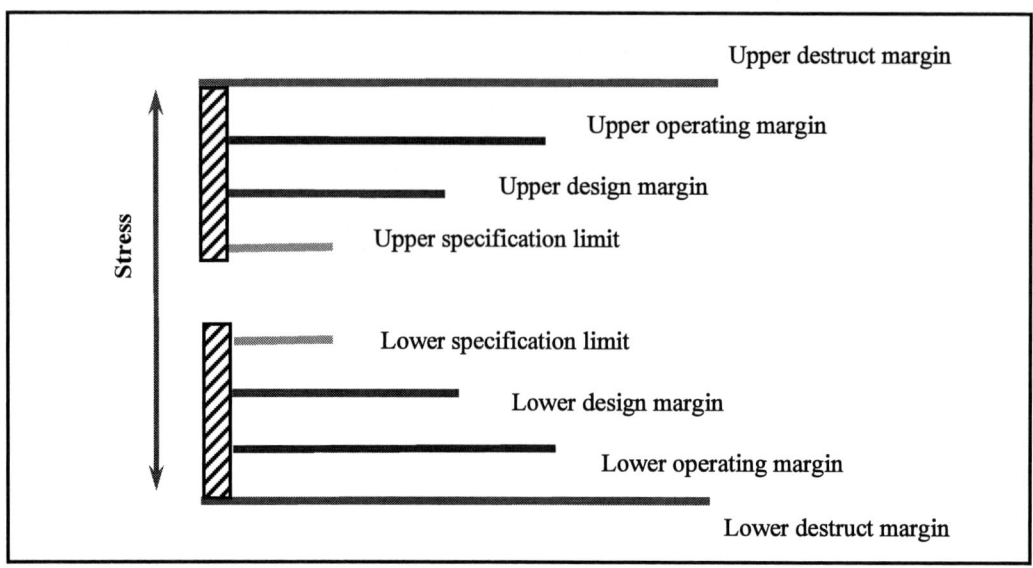

Figure 5: Stress limits and margins diagram

3.3 Strength limits and margins

Strength limits are obtained by following the methodologies of the highly accelerated life test (HALT). The purpose of HALT is to expose design weakness by iteratively subjecting the product to increasingly higher levels of stress and then learn what should be improved in

the upper operating limit and the upper destruct limit. The specification limits are provided by the manufacturer to limit the use conditions by the customer. The design limits are the stress conditions at which the product is designed to survive. The operational limits of the product are reached when the product can no longer function at the accelerated conditions due to a recoverable failure. The stress value at which the product fails permanently and catastrophically is identified as the destruct limit. Generally, large margins are desired between the operational and destruct limits, and between the actual performance stresses and the specification limits of the product, ensuring higher inherent reliability.

Accurate mean strength limits and margins can be identified only if sufficient numbers of samples are tested to reveal complete distribution characteristics. The strength limits obtained from HALT can be used in planning the accelerated test and screening conditions. The destruct limits can be used as the baseline for highly stress screening tests (HASS) during production level qualification. If the product demonstrates survivability well beyond its operational limits or the limits of screening equipment, then the search for destruct limits can be terminated.

For designs, the primary goal of the HALT is to place as much margin between the products specified or guaranteed operating limits and the observed operating limits during the HALT. Studies have consistently shown that products with generous performance margins between the specification and actual performance are inherently more reliable. The determination of the destruct limits are used to ensure that a sufficient margin exists between the operating and destruct limits to provide insight into how a product design/process can be improved and to establish a

the specific product. HALT is the first physical testing performed during the product qualification stage.

In product qualification, HALT can be used to identify the operational and destruct limits and margins, known as the 'strength limits' as shown in **Figure 5**. The limits include the upper and lower specification limit, the upper design margin, baseline for a production level highly accelerated stress screen (HASS). For some products, the search for the destruct limits may be aborted when the product exhibits survivability well beyond the previously determined operating limits or survivability at the limits of the screening equipment.

3.4 Failure modes, mechanisms, and effects analysis (FMMEA)

It is important to identify the critical failure mechanisms of a product induced by life-cycle loads. This will provide options for the selection of qualification tests. Only tests targeting the identified critical failure mechanisms should be selected as qualification tests. A cross-functional team (design, manufacturing, reliability, etc) will allow better identification of issues and criticality.

FMMEA is a methodology used to identify critical failure mechanisms. FMMEA utilizes the basic steps in developing a traditional FMEA in combination with knowledge of the physics of failure [23]. It then uses a life cycle profile to identify active stresses and to select the potential failure mechanisms. Knowledge of load type, level, and frequency combined with the failure site are used to prioritize failure mechanisms according to their severity and likelihood of occurrence. Figure 6 is a schematic diagram of FMMEA. FMMEA is based on understanding the relationships between product requirements and the physical characteristics of the product (and their variations in the production process), the interactions of product materials with loads (stresses at application conditions), and their influence on the product's susceptibility to failure. Potential failure mechanisms are determined based on appropriate available

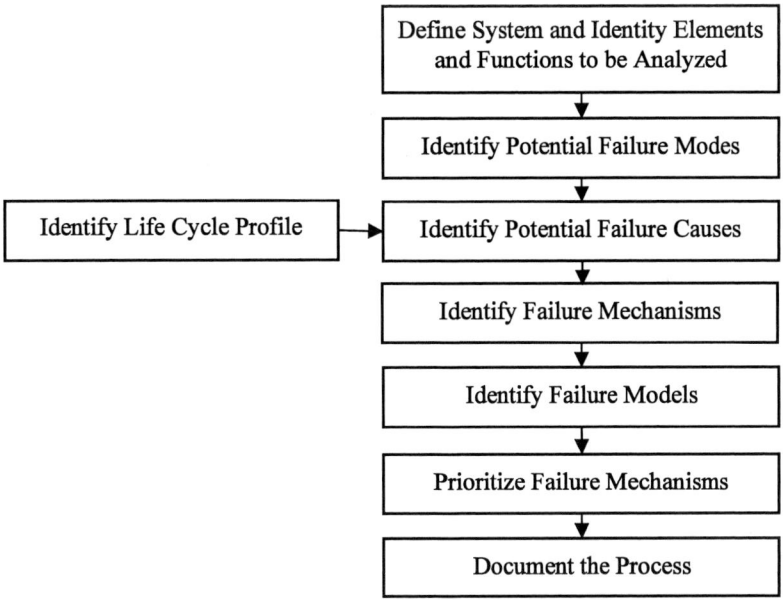

Figure 6: FMMEA Methodology

mechanisms corresponding to the material system, stresses, failure modes, and causes. FMMEA prioritizes the failure mechanisms based on their occurrence and severity to provide guidelines for determining the major operational stresses and environmental and operational parameters that must be either accounted for in the design or controlled. The high-priority failure mechanisms identified through the combination of occurrence and severity are the critical mechanisms. Critical failure mechanisms are the priority mechanisms considered in qualification tests. The failure sites, modes and causes associated with the critical failure mechanisms will provide information used to select the qualification test conditions.

The basic categories of failures are overstress (i.e., based on stress strength interference) and wear-out (i.e., based on damage accumulation); they are often identified through a mode that goes beyond performance tolerance (e.g., excessive propagation delays). Overstress and wear-out failures generally result from irreversible material damage; however, some overstress failures can be caused by reversible material damage (e.g., elastic deformation).

Failure models are used as tools to assess failure propensity. In PoF models, the stresses and the various stress parameters and their relationships to materials, geometry, and product life are considered. Each potential failure mechanism is represented by one or more of the prevalent models. A model should provide repeatable results, reflect the variables and interactions that are causing failures, and predict the behavior of the product over the entire domain of its operational environment. This type of a model allows development of accelerated testing and may help to reduce the number of test runs. Many PoF models, such as the Arrhenius model, the Coffin-Mason model and the Steinberg model, exist for predicting the behavior of components and products. Different models have different associated assumptions, which limit their applications to specific ranges of conditions.

3.5 Qualification requirements

Qualification requirements are the quality and reliability properties of the product suited to demonstrate compliance to the application requirements [1]. Qualification requirements must define the objectives and contents of the qualification activities. Qualification requirements are derived from the application requirements of the customer or the application segments [24]. They are based on the application requirements specified by the customer including functional performance, application conditions and time (use condition profile), processing conditions, robustness against random external stresses and expected statistical reliability properties such as tolerable infant mortality failures. Different products have different application requirements, varying from benign environments and short term use to harsh environmental conditions and long term use. Qualification requirements should reflect the requirements of the application and should ensure that the qualified product can survive and perform its function reliably under the application conditions. Qualification requirements have also to be defined based on the life cycle load profile of the product. These loads include what the product experiences during its life cycle including manufacturing, assembly, storage, transportation and operation. Depending on the individual experience and biases of the person performing the evaluation, qualification can be classified by the following four levels, as shown in **Figure 7**.

Similarity: Similarity is the lowest form of reliability qualification. Processes, products and packages may be qualified by being similar to something that has previously been qualified to a higher level. This level is usually accomplished with an engineering argument based upon logic. For example, if a certain package style is capable of completing a series of environmental tests, then it is likely to pass similar tests, regardless of the design of the die. Therefore, many different integrated circuits (ICs) using the same type of package could be "qualified by similarity." This level is accomplished with the lowest amount of resources, but also carries the highest risk of omission since tests are not actually conducted for each qualification.

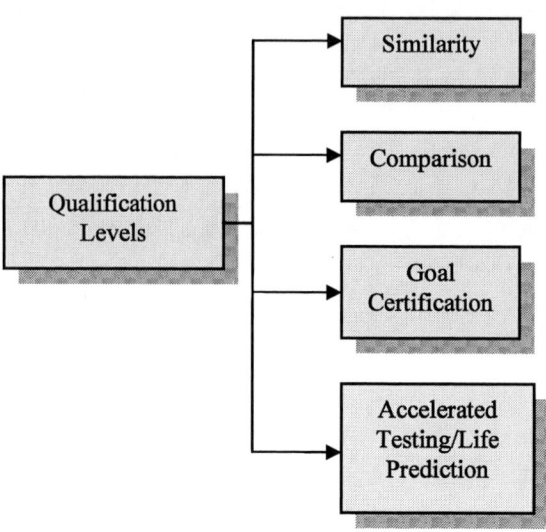

Figure 7: Qualification levels

Comparison: Comparison is the next level of qualification. It involves specific testing that compares results, but it does not necessarily meet goals of reliability for definable time periods. These tests generally collect attributes, not variables. For example, many package-related qualifications involve "standard" testing that does not have accepted acceleration factors that can be used to compare with normal life or expected use conditions, such as lead fatigue or temperature cycling. This is not to say that such testing doesn't have value, but favorable results usually can't be easily translated into specific statements about expected lifetimes. For example, what lifetime corresponds to 100 temperature cycles? We all know that it is good for a device to pass 100 temperature cycles without failure, but there is not necessarily an acceleration factor that can be used to translate the result to a failure rate ormedian life. Another form of comparison testing involves information about thermal performance, absolute maximum ratings, or other parameters that generally measure quality attributes that don't have direct correlation to reliability. In fact, results of comparison testing are often measured by attributes, not variables. In its most liberal form, qualification by comparison is achieved by matching results of previously performed tests or baselines

Goal certification: The goal certification level of qualification involves meeting specific goals involving life. For example, life tests confirming reliability performance during warranty periods or expected use periods may confirm the goals. This level is discerned from the comparison level since the results can be directly translated into longevity. This level is different from the next higher level since actual end-of-life need not be measured. For example, a one-year-long life test may be conducted without acceleration to cover an expected warranty period of 1 year. Knowledge of degradation modes, failure mechanisms, lifetimes, or acceleration factors will not likely be discovered. However, the data could be convincing in terms of knowing that products can survive reliably during the warranty period. Various goals can be certified using this level of qualification and the pre-determined acceleration factors. This is the most common form of qualification, particularly in military specifications.

Accelerated testing/life prediction: The fourth and highest level of qualification involves accelerated testing and the time-to-failure measurement. Tests are conducted on the electronic package to simulate the life cycle environmental stresses under accelerated conditions for time compression. In this level of qualification, failure modes and mechanisms associated with failure sites are likely to be identified, and the median lifetime and reliability can be determined from the tests results and acceleration factors.

3.6 Qualification Tests

After the qualification requirements are defined, qualification tests will be conducted to examine whether the product can meet the requirements. However, before the tests are conducted, it is important to do test planning, including test condition selection and stress level selection.

3.6.1 Qualification test planning

Before qualification test planning is conducted, the qualification requirements have to be understood and converted to target values and tolerances that the product or the elements of the product have to meet. Relevant parameters and target values and tolerances are determined in order to meet or exceed the requirements. For example, a useful life requirement of a product has to be met by all its elements. By analysis, a weak element which has potential to fail earlier than others can be determined. Then a useful life requirement can be assigned to this element to make it meet a target useful life value and tolerance.

The qualification test conditions and stress levels are determined from the LCP of the product, FMMEA, experience from previous similar products and the qualification requirements. Target values and tolerances have to be met by the product or product elements under the test conditions selected in respect to the LCP [25]. Critical failure mechanisms, identified by FMMEA, associated with failure sites and failure modes will be considered in the qualification tests. When these failure mechanisms are combined with the reliability requirements, the targeted failure mechanisms in the qualification tests can be determined. If a failure mechanism identified to be critical in the FMMEA has no significant impact on the product's life within the reliability requirements, then this failure mechanism will not be considered in the qualification test. For example, if in the FMMEA, corrosion on printed circuit boards alone will cause failure to the product in 10 years, but the reliability requirement is that the product operates without failure for 5 years, then the corrosion won't be a concern in the qualification test.

The selection of the qualification test stress level should ensure that the stress will induce the same failure mechanism as it would under operating conditions without introducing any new failure mechanism. And the transformation between the qualification test results and the actual life under application conditions should be able to be realized based on acceleration factor.

The sample size selection is a critical issue in reliability qualification test planning. Sample sizes should be adequate for the characterization of the failure distribution. If the testing properties are systematically common to all products of a similar type: the sample size can be small. If the testing is for a small fraction of defective products, the sample size should be large corresponding to the small fraction to be determined.

Time to market and cost are two important constraints that have to be applied during the qualification test planning. Time to market is an important factor for a product to conquer more market share, especially for some new technologies. So it is better to compress the qualification testing time to be as short as possible in term of time to market. However, the qualification testing time depends on the testing conditions, stress level and sample size that are implemented during the testing. The testing conditions, stress levels and sample size have to be selected within allowable range based on the rules discussed in previous paragraphs. Cost is another factor that constrains the selection of stress types and stress levels. Higher stress, longer testing time or larger sample size may increase the cost of the qualification testing. So tradeoffs have to be made between time to market and cost without violating the underlying scientific rules that are governing the testing.

3.6.2 Accelerated testing

Many qualification tests are accelerated tests, since the higher level of stress can cause failures within a short period of time over the life cycle of the product. Accelerated testing allows for reduced test times by providing test conditions that "speed up" the evolution of failures, thus saving the time-to-market of a product. Accelerated testing involves measuring the performance of the test product at loads or stresses that are more severe than would normally be encountered, in order to enhance the damage accumulation rate within a reduced time period. The goal of such testing is to accelerate the time-dependent failure mechanisms and the damage accumulation rate to reduce the time to failure. The failure mechanisms and modes in the accelerated environment must be the same as (or quantitatively correlated with) those observed under actual usage conditions, and it must be possible to quantitatively extrapolate from the accelerated environment to the usage environment with some reasonable degree of assurance.

There are two types of accelerated testing: qualitative accelerated testing and quantitative accelerated testing. In quantitative accelerated testing, the engineer is mostly interested in identifying failures and failure modes without attempting to make any predictions as to the product's life under normal use conditions. In quantitative accelerated life testing, the engineer is interested in predicting the life of the product at normal use conditions from data obtained in an accelerated life test.

Accelerated testing should be targeted at the critical failure mechanisms that have been determined. The load parameter that directly causes the time-dependent failure is selected as the acceleration parameter and is commonly called the accelerated load [26]. Common accelerated loads include thermal loads, such as temperature; chemical loads, such as corrosion; electrical loads, such as voltage; and mechanical loads, such as vibration. The accelerated test conditions may include a combination of these loads. Interpretation of the results for combined loads requires a quantitative understanding of their relative interactions and the contribution of each load to the overall damage. Failure due to a particular mechanism can be induced by several acceleration parameters, for example, corrosion can be accelerated by both temperature and humidity; and creep can be accelerated by both mechanical stress and temperature. Furthermore, a single acceleration stress can induce failure by several wearout mechanisms, simultaneously. For example, temperature can accelerate wearout damage accumulation not only by electromigration, but also by corrosion, creep and so on. Failure mechanisms that dominate under usual operating conditions may lose their dominance as the stress is elevated. Conversely, failure mechanisms that are dormant under normal use conditions may contribute to device failure under accelerated conditions. Thus, accelerated tests require careful planning in order to represent the actual usage environments and operating conditions without introducing extraneous failure mechanisms or non-representative physical or material behavior. The degree of stress acceleration is usually controlled by an acceleration factor, defined as the ratio of the life under normal use conditions to that under the accelerated condition. The acceleration factor should be tailored to the hardware in question, and should be estimated from a functional relationship between the accelerated stress and reduced life, in terms of all the hardware parameters.

Once the dominant failure mechanisms are identified, it is necessary to select the appropriate acceleration load; to determine the test procedures and the stress levels; to determine the test method, such as constant stress acceleration or step-stress acceleration; to perform the tests; and to interpret the test data, which includes extrapolating the accelerated test results to normal operating conditions [26] [27]. The test results provide qualitative failure information for improving the hardware through design and/or process changes. Accelerated testing thus includes accelerated test planning and development, test vehicle characterization, accelerated life testing, and life assessment. Accelerated test planning and development is used to develop a test program that focuses on the potential failure mechanisms and modes that were identified in this phase, including design of the test matrix and test loads, analysis, design and preparation of the test vehicle, setting up the test facilities (e.g., test platforms, stress monitoring schemes, failure collection and post-processing schemes), fixture design, effective sensor placement, data collection and post-processing schemes. Test vehicle characterization is used to identify the contribution of the environment on the test vehicle in accelerated life tests. Accelerated life testing is used to evaluate the vulnerability of the product to applied life cycle durability only if it is preceded by the steps discussed above. Without these steps, accelerated life testing can only provide comparison between alternate designs if the same failure mechanism is precipitated. Lastly, reliability assessment is used to provide a scientific and rational method to understand and extrapolate accelerated life testing failure data to estimate the life of the product in the field environment.

3.7 Failure analysis and verification

Detailed failure analysis of failed samples is a crucial step in the qualification and validation program. Without such analyses and feedback to the design team for corrective action, the purpose of the qualification program is defeated. In other words, it is not adequate simply to collect failure data. The key is to use the test results to provide insights into, and consequent control over, relevant failure mechanisms and to prevent them, cost effectively.

The purpose of doing failure analysis on samples which failed during the qualification testing is to verify that the failures were caused by failure mechanisms expected before the testing. No other failure mechanisms were induced during the testing. It insures the validity of the qualification testing that was intended to focus on specified failure mechanisms that would occur under application conditions.

If a product with new technologies was developed, the failure mechanisms before the testing may not be clear. The failure analysis is a chance to reveal the actual failure mechanisms that caused the failures. Thus the right PoF models can be selected to calculate the reliability of the product based on testing results in the following step.

3.8 Quality and reliability assessment

The reliability assessment is performed based on the accelerated test data and the PoF models. The reliability of the product is determined in terms of time-to-failure at the identified failure sites for a specific failure mechanism due to specific load condition. With the failure sites, stress inputs, and failure models, the reliability of a product is estimated and reported in terms of time to failure of the identified failure sites. Most failure models define time to failure under a specific loading condition. In the qualification test, the reliability of products is defined in order to meet the specified reliability requirement under qualification test conditions.

For most products, the life cycle profile consists of multiple loading conditions. As a result, methods for evaluating time to failure over multiple loading conditions must be derived. One approach is to cast the time to failure for a specific failure mechanisms in terms of the ratio of exposure time to the stress condition over time to failure for the stress condition. This ratio is often referred to as the damage ratio. If the exposure time is equivalent to the time to failure, then the ratio would equal to one. If one assumes that damage accumulates in a linear fashion, the damage ratios for the same failure site and mechanism can be added over multiple defined stress conditions. It is then assumed that once the accumulated damage ratio equals

one, failure at the site would occur. For the same site and the same failure mechanism and for fixed duration load events, a specific damage ratio can be determined. For example, drop of a hand held device from a certain height may result in a loss of ten percent of the life of a solder interconnect. In this case, each drop will result in an increment of 0.1 damage ratio for the solder interconnect. For repetitive events, a damage rate may be established by the uses the appropriate failure models to estimate the number of events required to produce failure. The damage rate is then defined as one over the estimated number of survivable events. For example, if a failure model estimates that a solder interconnect can survive 2000 temperature cycles, then the damage rate per cycle is 0.0005.

In general, time to failure data is obtained as a distribution for each failure site and failure mechanism. This distribution on time to failure is achieved by considering the inputs parameters to the failure models as distributions. In reality all dimensional and material properties are distributed about a nominal value as a result of variations in manufacturing. The same is true for the environmental loads. The physics of failure based reliability assessment allows for utilization of these natural variations in the reliability assessment. With the time to failure distribution on each site known, reliability can be evaluated in different metrics such as hazard rate, warranty return rate, or mean time to failure.

In addition to evaluating time to failure, the use of failure models allows for the examination of time to failure sensitivity to material, geometry, and life cycle profile. By considering the impact of the identified material and product geometries and loading conditions, the most influential parameters can be identified. This information can be used to improve design through closer attention to critical design parameters.

Conclusions

Qualification is an application-specific process involving the evaluation of the product with respect to its quality and reliability. The aim of the qualification process is to verify whether the product meets or exceeds reliability and quality requirements of the intended application.

Qualification plays an important role in the product development process. It occurs in different stages of the product development process with different purpose: design qualification, product qualification, quality assurance testing. Virtual qualification is based on PoF model predictions on the life of the design without any physical testing. Virtual qualification is relatively less expensive and less time-consuming than product qualification.

Product qualification involves physical tests on the manufacturer prototype, including highly accelerated life testing (HALT) to determine the strength limits of the product, and accelerated testing for reliability assessment. In product qualification, product configurations and materials and the life-cycle profile are captured as inputs to identify critical potential failure mechanisms associated with failure sites of the product under application conditions. Application requirements from customers are converted to determine the qualification requirements. In consideration of the life-cycle profile and qualification requirements, qualification test conditions, stress levels and sample size are selected to meet the function and reliability requirements. Accelerated tests are used in the qualification process due to the advantage of time compression. They accelerate the failure mechanisms in time due to the higher levels of stress than what the product would experience in the application environment. However, it should be ensured that only the failure mechanism under evaluation is being accelerated, while no other failure mechanisms are introduced. Quality and reliability of the product are assessed based the qualification test result. The assessment provides feedback to the design team to improve the product with tradeoffs.

References

[1] JEDEC, JEP 148, "Reliability Qualification of Semiconductor Devices Based on Physics of Failure Risk and Opportunity Assessment", April 2004.

[2] Pecht, M., Dasgupta, A., Evans, J., and Evans, J., Quality Conformance and Qualification of Microelectronic Packages and Interconnects, John Wiley & Sons (New York, NY, 1994).

[3] IPC, IPC-9701, "Performance Test Methods and Qualification Requirements for Surface Mount Solder Attachments", January, 2002.

[4] Stark, B., and Kayali, S., "Qualification Testing Protocols for MEMS", NASA Document, http://parts.jpl.nasa.gov/docs/JPL%20PUB%2099-1J.pdf.

[5] Osterman, M., and Stadterman, T., "Failure Assessment Software for Circuit Card Assemblies", *Proceedings of the IEEE Annual Reliability and Maintainability Symposium*, (1999), pp. 269-276.

[6] Caruso, H., and Dasgupta, A., "A Fundamental Overview of Analytical Accelerated Testing Models", *Journal of the Institute of Environmental Sciences*, Vol. 41, No.1, January/February (1998), pp. 16-30.

[7] Hu, J., Barker, D., Dasgupta, A., and Arora, A., "The Role of Failure Mechanism Identification in Accelerated Testing", *Journal of the Institute of Environmental Sciences*, Vol. 36, No. 4, July (1993), pp. 39-45.

[8] Pecht, M., Radojcic, R., and Rao, G., Guidebook for Managing Silicon Chip Reliability, CRC Press (Boca Raton, FL, 1999).

[9] Cunningham, J., Valentin, R., Hillman, C., Dasgupta, A., and Osterman, M., "A Demonstration of Virtual Qualification for the Design of Electronic Hardware", *Proceedings of the Institute of Environmental Sciences and Technology Meeting*, Phoenix, AZ, April 2001.

[10] Cushing, M., Mortin, D., Stadterman, T., and Malhotra, A., "Comparison of Electronics-Reliability Assessment Approaches", *IEEE Transactions on Reliability*, Vol. 42, No. 4, December (1993), pp. 542-546.

[11] Larson, T., and Newell, J., "Test Philosophies for the New Millennium", *Journal of the Institute of Environmental Sciences*, Vol. 40, No. 3, May/June (1997), pp 22-27.

[12] GRCI Inc., "Reliability Assessment Process Improvement Demonstration (RAPID)", Contract No: F33615-96-D-5302, Delivery Order 041, Subtask: 3.3, prepared for ESC/DIT, 1998.

[13] Cunningham, J., Valentin, R., Hillman, C., Dasgupta, A., and Osterman, M., "A Demonstration of Virtual Qualification for the Design of Electronic Hardware," *Proceedings of the Institute of Environmental Sciences and Technology Meeting*, April 24 (2001).

[14] Pecht, M., Dasgupta, A., & Barker, D., "The Reliability Physics Approach to Failure Prediction Modeling," *Quality and Reliability Engineering International*, Vol. 6, Iss. 4 (1990), pp. 276-273.

[15] Pecht, M., and Dasgupta, A., "Physics-of-Failure: An Approach to Reliable Product Development," *Journal of the Institute of Environmental Sciences*, Vol. 38, No. 5 (1995), pp. 30-34.

[16] McCluskey, P., Pecht, M., and Azarm, S., "Reducing Time-to-Market Using Virtual Qualification", *Proceedings of the Institute of Environmental Sciences Conference*, 1997, pp. 148-152.

[17] Dasgupta, A., and Pecht, M., "Material Failure Mechanisms and Damage Models," *IEEE Transactions on Reliability*, Vol. 40 (5), Dec. (1991), pp. 531-536.

[18] Pecht, M., Handbook of Electronic Package Design, Marcell Dekker Inc. (New York, NY, 1991).

[19] Pecht, M., Agarwal, R., McCluskey, P., Dishongh, T., Javadpour, S., and Mahajan, R., "Electronic Packaging Materials and their Properties", CRC Press, (Boca Raton, FL, 1999).

[20] Ganesan, S., and Pecht, M., Lead-free Electronics, Second Edition, John Wiley & Sons, Inc. (New York, NY, 2006).

[21] Pecht, M., Nguyen, L., and Hakim, E., "Plastic Encapsulated Microelectronics: Materials, Processes, Quality, Reliability, and Applications", John Wiley Publishing Co., (New York, NY, 1995).

[22] Lall, P., Pecht, M., and Hakim, E., "Influence of Temperature on Microelectronics and System Reliability", CRC Press, (New York, 1997).

[23] Ganesan, S., Eveloy, V., Das, D., and Pecht, M., "Identification and Utilization of Failure Mechanisms to Enhance FMEA and FMECA," *Proceedings of the IEEE Workshop on Accelerated Stress Testing and Reliability (ASTR)*, Austin, Texas, October 2-5 (2005).

[24] Cluff, D. K., and Osterman, M., "Defining Accelerated Test Requirements for PWBs: A Physics-Based Approach," IPC Printed Circuits Expo, (Long Beach Convention Center, Long Beach, CA, March 24-28, 2002).

[25] Snook, I., Marshall, J., and Newman, R., "Physics of Failure as an Integrated Part of Design for Reliability", *Proceedings of the IEEE Annual Reliability and Maintainability Symposium*, (2003), pp. 46-54.

[26] Lall, P., Pecht, M., and Cushing, M., "A Physics-of-Failure (PoF) Approach to Addressing Device Reliability in Accelerated Testing," *5th European Symposium on Reliability of Electron Devices*,

Failure Physics and Analysis, Glasgow, Scotland, October 4-7, (1994).

[27] Upadhyayula, K., and Dasgupta, A., "Physics-of-Failure Guidelines for Accelerated Qualification of Electronic Systems," *Quality and Reliability Engineering International*, Vol. 14, Iss. 6 (1998), pp. 433-447.

High Density 3D Integration

Roy Yu

IBM T.J. Watson Research center, P.O. Box 218, Yorktown Heights, NY 10598, USA
yur@us.ibm.com

Abstract

This paper discusses the current and future needs in continued CMOS scaling, reviews the status of the transfer and joining (TJ) technology for MCM-D and wafer level 3DI integration, and explores the opportunities of the TJ technology in the realm of the "More than Moore" era.

I. Introduction

The continued CMOS device scaling in the past decades has enabled over a million fold increase in device content per chip and has exponentially increased device performance and reduced the device cost per bit by as much in the trend known as the "Moore's law" [1]. The recent inclusion and revision of "More Moore" and "More than Moore" directions in the ITRS road map for semiconductor industries [2] is a clear recognition that more content with better integration on the system level can be achieved through continued device scaling as well as through hybrid system integration for at least another decade. While transistor speed is gaining ground by switching to high–K gate dielectric and metal-gate coupled with SiGe gate stressor [3], the memory latency and wiring RC delay continue to challenge the 2D configuration in chip architecture. The current drive in device architecture toward the "More Moore" concept is largely the result of the shift from the frequency scaling toward the parallel scaling in order to alleviate power consumption and memory bottleneck while enabling the continued device size reduction [4]. The recent rise in device multi-cores approach has allowed better system performance architecture and a more balanced power management [5] while maintaining the trend in device density increase. The performance gain by multi-cores is well established by the "Amdahl's law" in the mainframe high performance computing [6]. It is recently that the multi-cores in device enabled the chip performance enhancement to continue while reducing the costly power consumption, and to some extent the memory bottleneck [7]. The eight cores Cell chip [8], IBM PowerPC architecture [9], and the Intel tera-scale 80-core processor [10], are some of the multi-cores examples. Multi-cores architecture helps to reduce some of the wiring delays while maintaining the device performance by parallelism [11]. Each generation device dimension scaling increases the proportion of RC delay between the CPU and the memory cache and reduces the efficiency of CPU access to the memory [12]. To place the memory closer to the CPU with a higher bandwidth access bus is critical to increase the chip overall performance. 3D architecture, placing the memory directly over the CPU using the through-Si-via (TSV) architecture offers the nearly perfect memory access (shortest distance and widest bus) for the CPU and the best chip overall performance [13]. In a 3D configuration, with a mere re-arrangement of the position of the cache to the logic from 2D to 3D, the device performance can be increased by as much as 30%, which is equivalent to complete generation of 2D device scaling (18 months) with everything else being equal [13].

Therefore, 3D integration will be a critical part of the device integration for CMOS scaling to continue in the future. With the recent rapid development in hybrid 3D integration technologies, the device performance at system level by innovative system schemes is expected to advance concurrently with or over the 2D scaling. Such cross-over transitions between the integrated (serial, homogeneous, device-centric) and the hybridized (parallel, heterogeneous, component-centric) systems had repeated many times since the very beginning of the IC revolution: Vacuum tubes (hybrids), transistors (integrated), discrete circuits (hybrid), integrated circuits (integrated), multi-chip modules (hybrid), system on chip (integrated), and now 3D system integration (hybrid). Technology maturity and the cost often dictate the prevalent technology options. The integrated and hybridized architectures both enable a much higher system density, and therefore system performance but in their unique ways: The integrated architecture allows a streamlined process, a high volume production, a better system scaling, and a diverse components specialization. The hybridized architecture offers new functionality, better system contents, specialized applications, and new market opportunities. We are in the middle of another such cross-over transition.

It must be noted that the underlying driving force to adopt of any new technology is its ability to concurrently enhance performance and reduce cost. CMOS 2D scaling in the past decades has been able to increase the chip content while reducing the cost per bit by a million times. Any successful 3D technology will also have to meet such a challenge in order to be a viable contender to maintain the continuation of the CMOS scaling as well as to enable and lead the "More than Moore" advancement.

This paper will discuss the 3D integration needs in both system hybrids and system stacks. The current status of chip-to-chip, chip-to-wafer, and wafer-to-wafer architectures will be reviewed. In particular, transfer and joining, a fine pitch system level 3D integration method will be presented. This transfer and joining method for 3D integration was initially developed for IBM high end system thin film multi-chip module (MCM-D). The technology was then further enhanced into a wafer level bonding, thinning, and fine pitch transfer joining connection technology with and/or without the aide of a glass carrier. We will demonstrate the transfer and joining technology in wafer level applications in fine pitch 2D CMOS integration, in CMOS wafer thinning, in CMOS to MEMS application, and in a proposed CMOS to opto-electronics application. We will review the results in 2D chip-to-wafer

978-1-4244-2739-0/08/$25.00 ©2008 IEEE

(C2W) applications for chip-to-wafer placement, bonding, polishing, and wiring over the gaps between chips. We will also review the results of 3D fine pitch transfer-join contacts yield, reliability with varying contact via sizes. Finally, we will discuss whether fine pitch 3DI as applicable to packaging, as interconnect, or wafer level integration.

II. 3DI Overview

3DI introduces a new integration dimension in devices and various methods have been proposed to address the associated challenges [14]. The approaches can be grouped into package-centric (chip-chip, or C2C) and BEOL-centric (wafer-wafer, or W2W). In the package-centric approach the emphasis is on known good die (KGD), and on chip-chip or passive-chip integration where the wiring density improvement is secondary to functionality. Multi-stacking and lower cost are possible but with relatively higher difficulty [15]. In the BEOL-centric approach the focus is on wafer level connections and potentially with a high TSV density and lower cost [16]. The following is a brief overview of the commonly used 3DI approaches:

3DI with lateral connections (chip stacking):

Chip stacking is one of the earliest 3D integration concepts [17] in which memory chips are stacked together and the signal leads are taken out from the chip edges and are wire bonded to a logic chip. Such stacks are widely used in mobile devices today for low power and lower IO count devices to make the package with a higher density for information storage. The advantages of the chip stack methodology are its compactness and its requirement of limited power and IO density. Recently, through-Si-vias are incorporated in the chip stacks and allow memory-logic with either face-to-face (F2F) through C4 flip chip connection and form the basic building block in C2C and in W2W 3DI scheme [18]. More recently the system-in-a-cube (e-cube) concept is targeting high volumetric density for mobile and medical application [19].

3DI with through-connections:

For desktop and server application where IO band width and fast switching speed are required the 3DI technology favors through-Si-via (TSV) with a wider bandwidth. In such cases wafer-wafer (W2W), chip-chip (C2C) and chip-wafer (C2W) direct bonding and IO connections offer integration alternatives with potential for higher performance and cost reduction.

3DI covers several key components outlined below: 1) through-Si-via (TSV), 2) wafer bonding and thinning, and 3) interface-via-connection (IVC). Each 3D integration scheme addresses these aspects in its own unique ways:

1) Through-Si-Via (TSV):

TSV provides a direct, short and wide bus path for device-device communication in a 3DI stack in either C2C or W2W format [20]. Since it is more suitably formed at wafer level, TSV has brought the wafer level integration to the forefront. TSV process is strongly coupled with wafer bonding/thinning capability. For wafers with thickness 20-100 µm tungsten TSV is preferred. Tungsten has a closer thermal coefficient of expansion (TCE) to Si [15] than other metals, a high aspect ratio via fill by the CVD process, and is capable of a tight TSV pitch. However, tungsten's high resistivity, high stress and high temperature deposition requirements limit its wide use. 50:1 aspect ratio W via-fill has been demonstrated in Si as thick as 100 µm [15]. The W TSV pitch is normally limited by the wafer thickness and resistance requirements. Currently, W TSV is mostly considered applicable to Si thickness above 20 - 50 µm with the functional pitch about in about the same range.

When wafers can be thinned below 20 µm other TSV fill materials like Ni and Cu become alternatives to W. This is because the TSV size can be reduced and Ni and Cu TCE become less of an issue. Cu plating fill aspect ratio at 5:1 or less also restricts Cu to such a low thickness. Cu or Ni TSV filling is desirable since they conform to the back-of-the-line (BEOL) process and allow a greater TSV compatibility with the BEOL flow.

2) Wafer and chip bonding:

A reliable 3DI wafer bonding is vital to enable post-bonding Si thinning, TSV choice, interface via connection (IVC), post-bonding BEOL, and multi-layer device stacks. The primary options for wafer bonding are: A) metal bonding, B) dielectric bonding, and C) metal/dielectric hybrid. The type of bonding interface also determines the TSV build before the bonding (via-first) or after (via-last). There is also the stack direction. Wafers can be connected face to face (F2F) or face to back (F2B). This stacking direction is dictated by how the stacks are carried in the process, the registration and functionality required. Memory stacks tend to use F2B stacking while memory-logic prefers F2F stacking. F2F allows a better managed registration and performance while F2B enables a better managed design. In this aspect W2W and C2C share the requirements. The commonly used bonding methods and their salient features are summarized below:

A1) **Metal bonding with C4 or micro-C4:** C4 bonding uses solder for metal connection. It is widely used in chip stacking and TSV C2C with relatively thick 3DI chips with a more relaxed connection pitch. To reduce the TSV thickness (for smaller pitch) an underfill to secure the thin chip is normally required. For low connection density (> 50 µm pitch), this approach can be effective, especially if the wafer level yield is low and the known-good-die assembly KGD is necessary. This approach normally requires a via-first option. It is possible but can be quite complex for multiple chip stacks in this way [15].

A2) **Metal bonding with Cu-Cu thermal compression:** This type connection extends the connectivity ability of flip chip solder connection and can be used for both C2C and W2W. The metal contacts form interface connection as well as mechanical bonding between wafers. The metal interface is simple to form and has a relatively high strength. Normally this structure has a good vertical stress relief on the contacting surfaces during thermal cycles. They can have a tighter pitch down to a few microns and typically work well for thick wafers [21]. For tighter pitch 3DI with wafer thinning to less than 20 µm, the wafers tend to bend and crack along the bonding edges due to the lateral stress build-up. Since metal is not as compliant as much as adhesive, it is relatively difficult to achieve a wafer level bonding with metal contacts alone.

14

This approach is relatively successful for chip level bonding with a thick 3D chip. This approach also prefers a via-first option.

B1) Dielectric interface with low temperature oxide-oxide bonding: Oxide-oxide bonding stems from bulk SOI bonding [22]. In SOI the bonding oxide goes through over 1000C and forms a bonding interface. Low temperature oxide bonding takes advantage of surface activation to achieve the bonding strength [23]. Due to the nature of the bonds oxide-oxide bonding has a relatively low surface adhesion strength of several J/m^2, which is sufficient for wafer grinding and polish but might be marginal for continued BEOL and ability to sustain chip-package interaction (CPI). With a high content of through-via connections using via-last process the oxide-oxide bonding can be enhanced for continued BEOL and CPI. Oxide-oxide bonding normally starts from the wafer center and progresses toward the edge and provision is required for air to escape. Care must be given for alignment drift during such bonding.

B2) Dielectric interface with adhesive bonding: This alternative to oxide-oxide bonding, adhesive bonding originates from packaging technology. A layer of compliant (or B-staged) dielectric material or a dry adhesive is used to bond the wafers together. This approach frequently requires a vacuum bonder to ensure the interfaces are free of air pockets. Similar to oxide-oxide, adhesive bonding requires a via-last process for power and signal IO with slightly higher TSV aspect ratio due to the adhesive thickness of several μm. BCB and polyimide adhesives are commonly used as high temperature adhesives [24]. More recently, BEOL low-k dielectrics are being explored for enhanced thermal budgets [25]. After the bonded wafers are thinned the connections are added through the thinned top wafer in a BEOL-like process. For a fine pitch connection, this approach requires the top wafer to be thinned to less than 10 μm. Compared to oxide-oxide bonding, dielectric adhesive bonding has better topography conformity but has lesser registration accuracy, through-vias connection pitch, and BEOL thermal budget.

C1) Hybrid with Metal/oxide: In this case the metal contacts and the oxide dielectric are finely polished and are bonded together in one step. Due to planarity requirement this approach is suitable for C2C or C2W bonding. It is relatively difficult to achieve a wafer-wafer bonding with this method. Sometimes the oxide is replaced with BEOL inter-level dielectric (ILD) based on the bonding methods.

C2) Hybrid with planar Metal/adhesive: This method is similar to metal/oxide hybrid but replaces the oxide with an adhesive. The surface planarity in this approach is less stringent than that of oxide/metal hybrid due to the compliant nature of the adhesive [26].

C3) Hybrid with interlocking metal/adhesive: Another way to make the hybrid joining surface is to use an interlocking structure. Compared to the planar hybrid structure, the interlocking hybrid structure provides several additional features for wafer bonding. The first is the enabling of the direct metal-metal contact to improve connection yield. The second is the maintaining of registration without slippage during lamination. The third is the facilitation of air evacuation during lamination prior to pressurization. The fourth is the securing of the entire interface for the thinned wafer. In this paper we will focus in more detail on the hybrid option with interlocking adhesive/metal structure bonding. This metal/adhesive interlocking hybrid approach was developed during the earlier work for MCM-D structure made by thin film transfer and joining (TFTJ). In the case of wafer-wafer bonding, areas without bonding will have different stress than areas with bonding (either with metal or dielectric). This is particularly the issue for a 3D structure with a thinned top wafer for high density through-Si-connection. In the case of flip chip C4 connection the stress is maintained on the metal contacts due to the thickness of the chip and the stress concentration can be alleviated by increased C4 heights and the use of underfill. In the case of 3DI, the metal contacts alone can have only a maximum contact area of 25% which concentrates thermal stress and would lead to Si early fatigue fracture. With adhesive between the metal contacts this stress concentration can be distributed to the entire surface. The dielectric between the contacts has another (5th) function. It is to reduce the metal corrosion or migration due to the environment. The adhesive can prevent such issues and potentially increase reliability.

There are several ways to form this hybrid wafer level interconnection. This paper will review one of the methods we have practiced, transfer-join (TJ) method. We will review from its early application in MCM-D to recent results in wafer level 3D integration.

3) Interface vias connection (IVC) (Imbedded and through-vias): With oxide and adhesive bonding the via-last process is necessary to form the inter-wafer connections. This type via connection is also called through-via connections. Since the ability to form the TSV also depending on the top Si thickness, via-last is normally a good option for thin top Si (< 20 μm) and for power and ground connections to the bottom wafer with a low resistance and inductance. Theoretically a via-last through-via offers a better wiring pitch as the vias are lithographically defined. In reality, a high density through-via with via-last approach is density limited by the blocked wiring channels in the inter-level-dielectric (ILD) they pass-through and by the via fill aspect ratio. In addition, the wafer-wafer overlay accuracy in wafer bonding limits the pitch of the through-vias. Unless these practical issues are resolved, via-last offers less than the ideally projected through-via density advantages over the via-first approach. For via-first approach, the through-vias are formed as part of the BEOL layer build and are patterned with base wafer lithography and can allow better wiring channels. Since the wafer to wafer connections are made at the interface layer, via-last process is also called imbedded vias. Therefore, via-first is better suited for high density signals for the inter-wafer connections.

4) 3DI as BEOL (W2W), as packaging (C2C), or as a separate discipline:

Current C2C 3D favors adoption of packaging (C2C) approach while W2W 3DI tends to use BEOL-like flow. Both approaches have advantages and limitations. In the case of C2C 3DI, the materials and cost limit their use in board level final assembly, therefore it is difficult to add additional structure beyond 2 or 3 levels. However, C2C has the known-good-die (KGD) advantage which is critical for early

15

production. Wafer level 3DI in theory can lower the cost and build more stacks. However, the cumulative yield can be a major concern. In addition, to re-introduce the bonded wafers back to BEOL processing limits the material selection for an optimum wafer bonding. Alternatively, using C2W with wider choice of packaging materials in a dedicated line might be a better option for 3DI. This 3DI line would also enable "More than Moore" integration with functions beyond CMOS which will be discussed later.

III. Transfer and join (TJ) for MCM-D and for wafer level 3DI:

1) MCM/SCM Thin film Transfer and join (TFTJ) with face-up and face-down build:

In this section the recent results in transfer and join (TJ) technology from MCM-D and SCM to wafer level 3DI will be reviewed. The wafer level transfer and join (TJ) technology was based on the thin film transfer and joining (TFTJ) technology developed for an IBM high-end thin film multi-chip-module-deposited dielectric (MCM-D) [27]. In order to enhance the MCM-ceramic (MCM-C) wiring density, a low-K polyimide thin film high density and fine pitch wiring plane pair was added to the package [28]. This thin

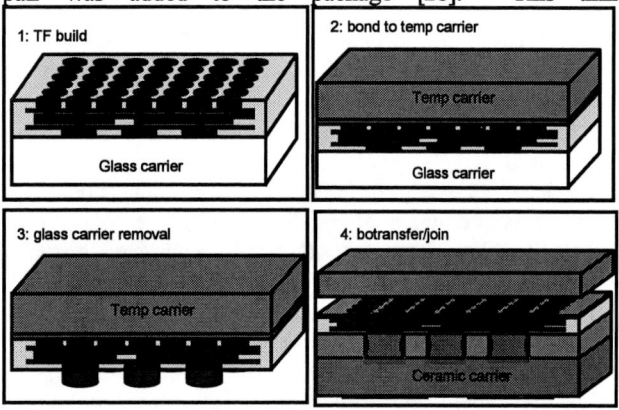

Figure 1: Top-surface up build [35]

film package was originally added to the MCM-C base through a serial build process. In order to simplify this process a parallel build for the thin film (TF) layers with glass plate as carrier was developed [29]. The TF was built on the glass carrier with either a face-up [30] or a face-down [31] build approach. The face-up approach is for complex systems with full level test and repair capability from both top and bottom surfaces of the TF modules. To better manage the film distortion during the build a two-step transfer method was developed. First the face-up TF (top surface up) was built, tested and flip transferred to a temporary glass carrier using a Teflon-PFA adhesion layer. The connecting IOs (bottom surface) are then prepared after the flip and tested for the bonding interface and the TF is then transferred from the temporary glass carrier to the MCM-C base through the bottom surface IO. The Teflon adhesion layer on the top surface also serves as a compliant layer during the lamination to the base MCM-C carrier. This "flip" approach ensures the TF always attaches to a rigid carrier and is the first successful film transfer method on a module level (Figure 1 from [35]). For the simpler single chip modules (SCM) the TF is

preferred to build face-down (bottom surface up) and the TF is transferred directly from the glass carrier to the MCM-C base in one transfer step (Figure 2, also from [35]).

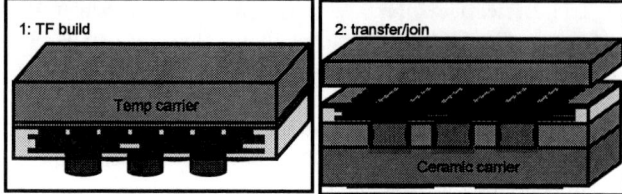

Figure 2: Bottom surface up build [35]

2) MCM/SCM Discrete and integrated interlocking bonding interface:

In the final transfer lamination of the TF to its MCM-C base, two bonding structures were developed. The first one is the discrete spacer structure in which a Kapton-EKJ 200 composite adhesion film (from DuPont) with punched through-vias was used as the bonding structure. Figure 3 shows the schematic of the discrete spacer structure. Figure 4

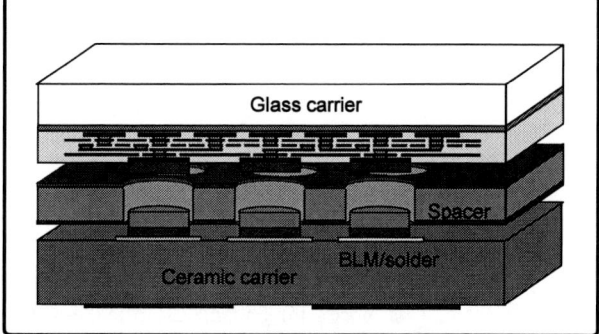

Figure 3: Discrete spacer (from [35])

Figure 4: Modules built with discrete spacer [35]

shows some of the MCM-D modules built with discrete spacer and Figure 5 is a typical SEM cross-section of the spacer structure. The Kapton-EKJ 200 film has a 37 μm kapton core and 6 μm EKJ adhesive on both sides of the core. The IO metal uses a normal C4 Ni ball limiting metal (BLM) and the solder is applied using 400 mesh grade 90Pb/10Sn paste with a glass molder transfer method [32]. The TF solder balls are inserted into the punched through-vias in the spacer for both alignment and IO connection to the MCM-C module

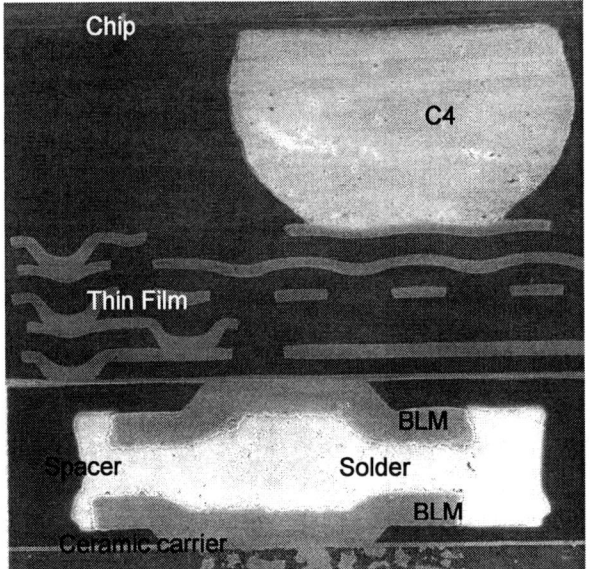

Figure 5: Discrete spacer x-section [35]

by a lamination. The spacer bonds TF to the base MCM-C and functions similar to C4 underfill structure. During the lamination the IO C4 BLM thickness is found to have a profound impact on the TF drift (or slippage) during the bonding lamination. When the BLM is thinner than the adhesive thickness (6 μm) the position of the TF IO BLM would slip out of the through-vias and the IO registration could not be maintained within the spacer through-vias. When the BLM thickness is increased to 12 μm (2x the adhesive thickness) the IO BLM and solder position is always contained within the spacer through-vias. The kapton core of

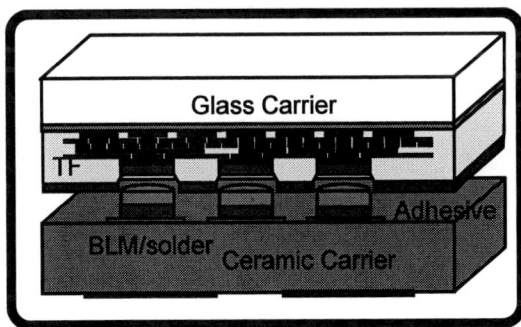

Figure 6: Integrated spacer [35]

the EKJ film and the BLM Ni metal form an interlocking structure when BLM is thicker than 6 μm and it is this interlock that provides C4 IO positioning stability during lamination. To control such a lamination slippage a lock and key interface structure was therefore always provided to confine the metal contacts in their allowed capture IO via area during lamination.

The original interlocking structure in discrete spacer was extended to an integrated spacer structure. Figure 6 show the schematic of the integrated spacer. Figure 7 shows some of the SCM modules built with integrated spacer structure and Figure 8 is a typical SEM cross-section for the integrated

structure. In the integrated spacer a metal protrusion is

Figure 7: SCM modules built with integrated spacer [35]

provided on one interface. The other interface is a double

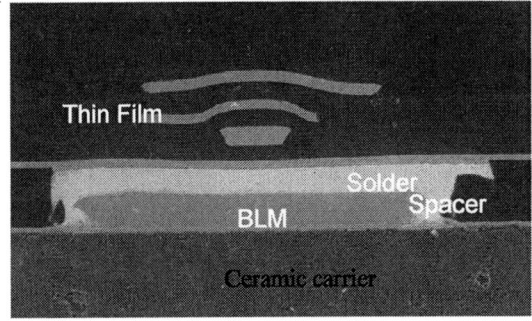

Figure 8: Integrated spacer x-section [35]

layer recess: a fully cured polyimide (such as PI5878G from HD-microsystems) and a thermoplastic polyimide adhesive (such as PI3003X1, also from HD-microsystems). The metal

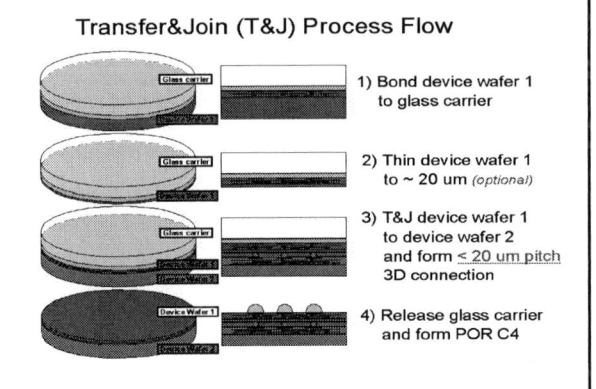

Figure 9: Wafer level bonding and thinning

prior to pressurization for bonding. This integrated approach was used for single chip module (SCM) for both ceramic and laminate carriers.

3) 3DI Wafer-level bonding and thinning

The bonding and transferring of polyimide thin film module using glass carrier formed the basis for wafer level 3DI bonding, thinning, transferring, and joining. Figure 9 shows the schematic of extending TFTJ film handling technology into wafer level bonding, thinning and transfer.

17

Figure 10: A 5 μm stand-alone wafer membrane [35].

The process can be accomplished with or without a glass carrier. Figure 10 shows an 8" device wafer initially bonded to a carrier with the adhesive, was then thinned to 5 μm, and is released from the carrier as a stand-alone wafer membrane.

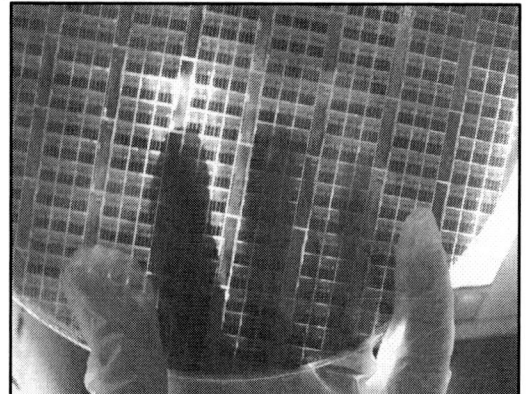

Figure 11: A bonded and thinned device wafer [22]

Figure 11 shows that the bonded wafer can be polished to remove all the bulk Si to its buried oxide (BOX) surface with devices only. Both Teflon-PFA and HD-3003X1 are evaluated for use in lamination at and beyond 300C [33, 35, 37a]. Kapton-KJ adhesive had shown to hold the structure through 370C chip reflow [35]. The works are under way to extend the wafer level bonding and thinning capability to 300 mm wafers.

4) Through-Si-via (TSV)

With the wafer bonding and thinning to sub-20 μm capability, TSV process with high density connection becomes possible. Several ways of deep Si via fill have been evaluated. Initial work showed that a 15 μm CD 45 μm pitch at 20 um deep TSV with electroless Ni is possible [35, 37]. Other TSV metal fills are also possible in this thickness regime [37c]. The through-Si via typically is formed before the wafer bonding (via-first). They can also be formed after the wafer thinning (via-last). Via-first normally allows for a wider material and processing choice as there is less constraint on the process temperature. Via-last, on the other hand, allows the TSV and interface via connection (IVC) form in one step, normally with a better production yield albeit at a lower via density. In most of our work, via-first is used with an imbedded IVC for inter-wafer connection.

5) W2W and C2W interface via connection (IVC):

One of the key 3DI aspects is the ability to interconnect stacked wafers. Thinning wafers to below 20 μm has enabled TSV via-last build which is typically better for power/ground through-via connections. For wiring critical applications the TSV via-first build is preferred for better wiring channel density. The integrated spacer concept developed for the MCM/SCM transfer bonding has been advanced to a finer pitch and was extended to wafer level transfer and joining (TJ) technology [33]. One of the initial via/stud interlocking interfaces demonstrating the alignment and connection

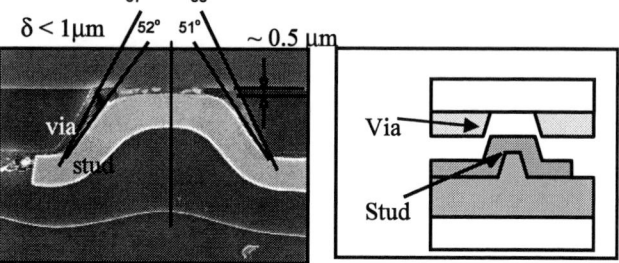

Figure 12: Via/stud (lock/key) alignment and accuracy [33]

capability is shown in Figure 12. One side of the interface is built with Cu protrusions (stud) and the other interface with recesses (via). The overlay accuracy by using the stud/via lock/key structure on a wafer level has been demonstrated to be within 1 μm as shown in the figure. The via-recess is normally a double layer structure with the lower layer as a distortion restrainer and the top layer an adhesive. The via/stud size has be demonstrated ranging from 1 μm to 20 μm with a typical 5 μm CD and a height of 3 to 5 μm in most cases. The distortion restrainer (lower layer of the double layer) can be any material with sufficient rigidity at processing temperature. We typically use a cross-linked

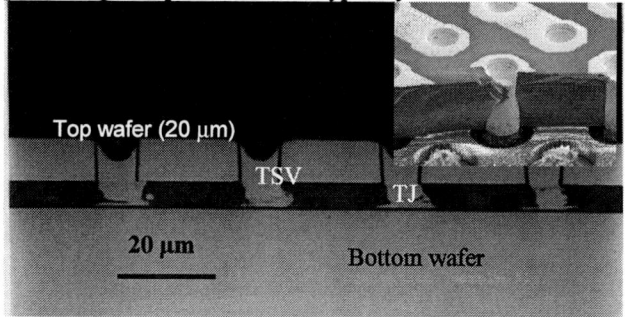

Figure 14: 20 μm thru-Si via-chain [35, 37a]

polyimide or a CVD oxide. The adhesive layer is a modified polyimide film with a Tg and Tm to allow flow during lamination without thermal decomposition [34]. The TJ assembly prior to full lamination has several features that address the major wafer bonding issues. The direct contact of the metal stud/via (lock/key) allows a metal bonding with a locally higher than applied pressure. The double layer recess controls the distortion and alignment during the heat ramp. A slight gap between the two mating surfaces in bonding enables outgas of the moisture trapped on the surfaces and in the bulk. A less obvious feature of the TJ assembly is that it releases the stress between the interfaces. The small portion of the metal contact allows the two surfaces to remain stress free until bonding takes place. The adhesive absorbs any thermal and mechanical stress and protects the thinned wafer structure. This is very similar to C4 under fill used to spread the stress across the entire surface. Figure 14 shows a complete 20 μm CD 45 μm pitch Ni-TSV via-chain using TJ connects to a base wafer [33, 35, 37].

The bonding strength of the polyimide adhesive is very high. It typically exceeds the Si 4-point bend fracture strength. Figure 15 [from 21, 23, 34c]] compares the adhesion strength for several commonly used bonding surfaces in 3DI. Oxide-oxide is normally in the range of 1-3 J/m² while Cu-Cu can be as high as 20-30 J/m². However, polyimide adhesive strength is typically beyond 4-point bend measurement

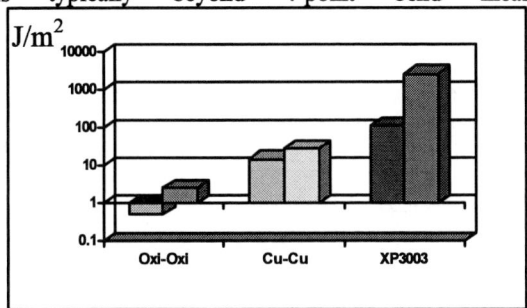

Figure 15: Bonding strength for various materials [21, 23, and 34c]

capability and Si beam break normally results. The adhesion strength is indirectly obtained by using peel test instead, which translates to over 100 J/m².

Because of the interfaces are interlocked TJ bonding preferably done in an iso-static vacuum assisted lamination process. Figure 16 shows a schematic of the lamination fixture. A conforming layer of air-tight materials is used as vacuum and pressure seal. Due to the self alignment nature of the lock and key structure, TJ structure has a high alignment accuracy and via connection yield, as mentioned earlier.

The TJ structure with a temporary bonding has been

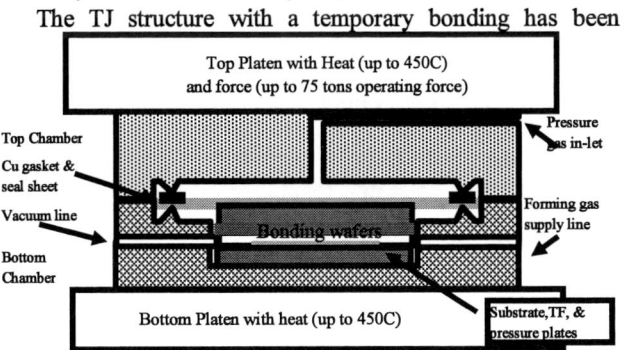

Figure 16: TJ lamination fixture schematic [35]

evaluated for chip-wafer placement for known-good-die (KGD) 3DI bonding at wafer level [35, 37b]. In this exercise we took two wafers with TJ mating structures. One of the wafers is diced into chips and the chips are re-aligned to the un-diced wafer using the TJ via/stud as alignment. The chips are tacked down to the surface wiring of the base wafer temporarily using a micro-tipped tack-down adhesive. After the chips are placed they are laminated to the base wafer and bonded to the base wafer using via/stud both as registration and as IVC via connection. The surface adhesive between the

chips and the base wafer form the final permanent bond to secure the chips to the base wafer. In order to enable the polishing of the chip-wafer assembly after bonding, the edge chips, normally discarded after dicing, are put into the assembly as well to reconstruct the uniform surface required for lamination and for wafer thinning. After normal backside grinding step there is no damage to the chip corners if the chips are well bonded to the base wafer. There is no need for dicing channel back-fill and no concern about polish slurry and chemical getting to the joined surface as TJ joining seals the entire contact surfaces.

6) TJ Data and discussion:

Dozens of IBM MCM-D modules have been built by the TJ process and evaluated for construction and reliability, as well as manufacturability. A fully functional MCM module with 30 chips was assembled and that module was put in user condition for an extended period for field performance study. The reliability of the joining as well as the potential electrical impact in system level was evaluated. No performance disparity, due to the joint structure, was noted. The spacer thickness was found to have significant impact on the joints reliability. The thickness reduction from a 75 um spacer (EKJ320) to a 50 um spacer (EKJ200) significantly enhanced the thermal cycle life. TFTJ was also extended for use in SCM. This is a thin film patch concept which applies thin film only to a localized area of a chip carrier or a card where high density wiring is required. Based on the data from MCM modules SCM on alumina ceramic base carriers the spacer thickness below 20 μm was studied and no stress failure was reported in thermal cycles within the module expected lifetime. This local patch arrangement by TFTJ combines the advantages of low cost base carrier with nominally low complexity and wiring density provided with high wiring density of thin film wiring only in a small local area. This enables optimized chip carrier performance at system level while reducing over all module cost. This is possible because with TFTJ, the base carrier can remain at a relatively coarse ground rule with a better electrical resistivity, high production yield, and low cost. With the small area thin film patch, more patches can be made on each temporary panel reducing cost per patch especially when compared with serially additive processing of the same local pattern on the base carrier one at a time. Most significantly, TFTJ allows thin film to be processed on a large panel line such as flat panel display line which would further reduce processing cost. One interesting application space is to add a TFTJ patch directly to a low density micro card. In this way any low density card can be transformed into higher density by a local patch.

For wafer level 3DI application the TJ joining was refined based on TFTJ and was developed to enable a much finer pitch and registration accuracy. The typical applications are for join pitches below 50 μm. This tight pitch is designed for high density, high bandwidth 3DI applications. Figure 17 shows such a transfer of interconnects 10 μm lines on 20 μm pitch. There are two sets of wiring in the stitching pattern. The x-wirings are formed on the base wafer. The ends of the

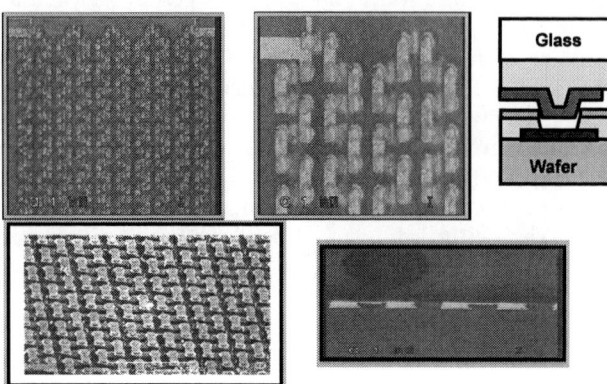

Figure 17: 10 on 20 µm TJ transferred via-chains [33,35]

x-wirings are protruding studs of 7 µm in height. The y-lines are formed on a transferring wafer or glass with the y-wiring

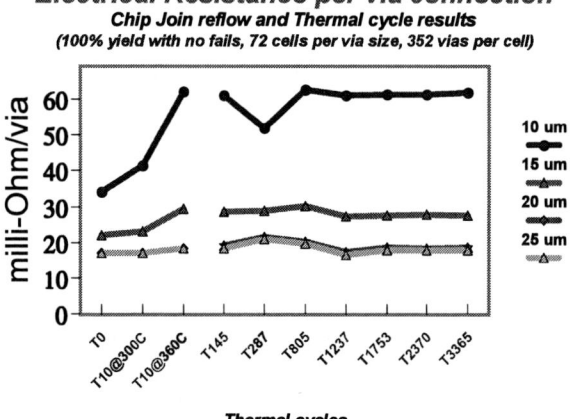

Figure 18: Chip join reflow and Thermal cycles results of the transferred TJ via-chains [33, 35]

ends with a recess of 5 µm. The x and y lines from both wafers are brought together with a lamination. The carrying wafer for the y-wirings is then removed. The y-wirings are then transferred to the base wafer and connect to the x-wirings and form via-chains. The transferred structures have the stitch sizes of 10, 15, 20, and 25 µm. After the transfer the joining yield was assessed to be 100% in 72 such transfers in each via size with 352 stitches in each chain. The transferred structures were then subjected to 13 times chip reflow cycles at 300C, then 350C to simulate module rework. The via-chain resistance is measured after the cycles are shown in Figure 18. The 10 µm stitch was found to have a resistance per via increase from 30 milli-ohms to 60 milli-ohms during the 350C reflow. For other via sizes the resistance is stable. The stitches were then put through 3650 cycles of 25 C to 150C thermal cycles for thermal fatigue test. The via-resistance values are plotted as a function of thermal stress cycles time. There is no appreciable change in the via-chain resistance during thermal cycles.

For chip-wafer 2D applications, the wiring resistance yield across the space between the chips dicing channels are measured before and after the lamination. No noticeable

resistance change and yield loss was found for line width down to 5 µm.

The general approach described can in principle be extended to larger wafer sizes by using appropriate tooling option.

7) "More than Moore" applications

The ability of TJ technology to re-assemble chips into a pseudo-wafer with high registration accuracy enables the technology to be applied to much wider application spaces beyond CMOS scaling. ITRS roadmap anticipates that future CMOS advancement to follow "More Moore" and "More than Moore" parallel paths. TJ has been demonstrated a viable technology for "More Moore" CMOS 3DI applications. It is equally suitable for "More than Moore" integration. We will show some of the demonstrated applications with TJ:

A) **MEMS:** Using the same principle outlined in the previously discussions the micro electro-mechanical systems (MEMS) can be assembled accurately with CMOS devices.

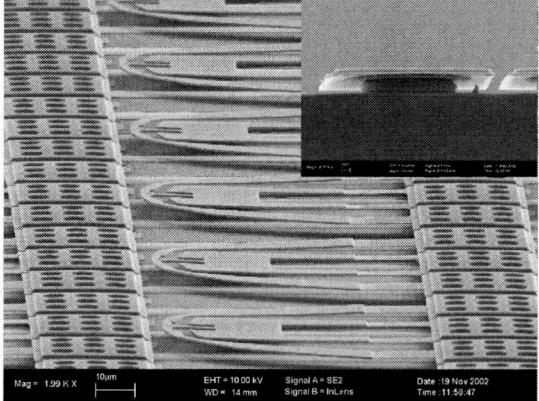

Figure 19: TJ for MEMS transfer [37a, 37b]

Millipede is an IBM project using nano-size dots as a recording device for high density low power mobile storage device. The central focus of the integration is to use a thermal-electrical logic device to drive micron scale mechanical levers for surface data read and write. TJ technology is used here to transfer the high density mechanical levers from one wafer to the surface of a CMOS wafer. Figure 19 shows the wafer level transfer of the levers with the levers sitting on the TJ studs. A high transfer and joining levers yield (>95%) and high planarity uniformity (20 nm / 100 µm) was demonstrated and the device successfully enters data system test [37a and 37b].

B) **Opto-electronics:** Optical interconnect plays a major role for box-box communication. With the advancement of integration it is conceivable that they might replace some of the on-chip wiring to increase the data bandwidth. One of the issues to incorporate optical channels into the electrical system is the placement accuracy and the fine size of the fiber channels with respect to the light emitters. TJ technology was demonstrated to be able to resolve both issues. Figure 20 shows an array of fine pitch optical channels formed by using polyimide conduits with metal reflector to form total reflection [38]. The channels are about 10 µm in size and 25 µm in pitch. Similar channels using oxide on Si wafers can be made similar to the Millipede MEMS device. Figure 20 also

shows a schematics of how a

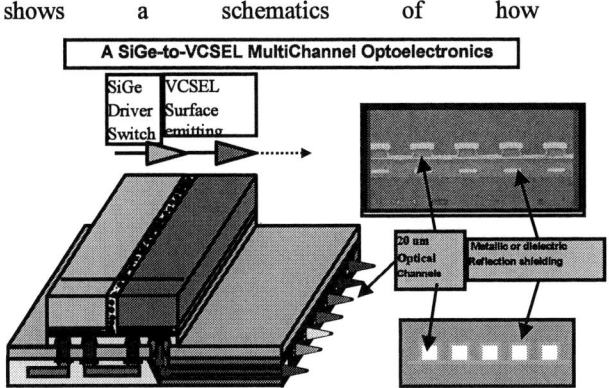

Figure 20: TJ for opto-electronics [33,35]

CMOS (SiGe) high speed logic and the laser drive (VCSEL) and the optical channels can be positioned accurately to send the laser signals into the multi-channels optical bus. The

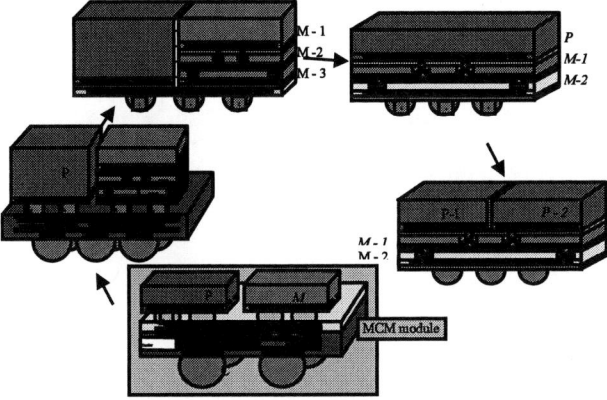

Figure 21: TJ enablement of hybrid systems [35]

receiving end of the optical channels will be a light detector device with the output to drive a logic CMOS, SiGe, or a III-V device. With the readily available TSV technology the optical channels can be used with vertical laser from wafer-wafer through 3DI connections.

C) **Hybridized device:** With the ability to integrating device both in 2D and 3D, it is possible now to construct a high density

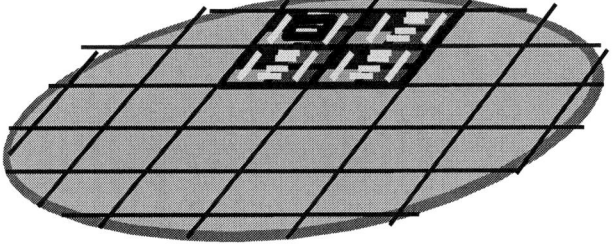

Figure 22: Wafer level Chip hybridization [33, 37b]

device in a wafer level with either chip-wafer or wafer-wafer to integrate SiGe, GaAs, III-V, logic/Memory in variety of ways. Figure 21 schematically shows some of the potential combinations of the 2D/3D to achieve the best circuit speed and or special functionality. Due to the material and substrate

size compatibility issues the different materials normally can not be processed sequentially on the same wafer. With the T&J method the wafers can be separately processed and diced and then recombined on to a base carrier with proper connection as in discussed and illustrated schematically in Figure 22 (see [33, 37b, 39-42]).

IV. Summary:

The MCM-D thin film transfer and join (TFTJ) technology and the related wafer level transfer and joining (TJ) technology are reviewed. These integration technologies can enable high density fine pitch 2D and 3D integrations at wafer-wafer and chip-wafer level for "More Moore" scaling. The advances are also applicable for "more than Moore" applications in MEMS, opto-electrinics, hybrid devices. These advances can ultimately bring SOC and SIP together, achieving SOC performance with SIP versatility for the future CMOS.

Acknowledgments

The Author wishes to thank Dr. S. Purushothaman for his critical review and comments of this manuscript.

V. References

[1] G. Moore, Electronics, V38, pp114, 1965

[2] 2007 ITRS, http://www.ITRS.org/

[3] S. Joshi, et al, DRC2007 pp.53 – 54

[4] T.C. Chen, ICSICT2006, pp4-7.

[5] G.G. Shahidi, IEEE-CICC 2007, pp413-416.

[6] G. Amdahl, AFIPS Conference Proceedings, (30), pp.483-485, 1967.

[7] J.E. Fritts and E.D. Chamberlain, proc. 35th Simulations Symposium, 2002. Page(s):352 - 362

[8] S. Kapoor; HLVDT2007. Page(s):48 - 52

[9] B. stolt et al, V43No.1, Jan. 2008 pp21 – 28, JSSC.2007.

[10] S.R Vangal et al, V43No.1, 2008, pp29 – 41, JSSC.2007.

[11] L. Chai, et al, CCGRID2007. pp471-478

[12] R. Bergamaschi et al, ASPDAC2008. pp.708-713

[13] P. Jacob et al, IEEE Design&test, V22,No.6, 2005, pp540-547

[14] P. Leduc et al, Proc. IEEE (IITC), pp. 210-212, 2007.

[15] P.S. Andry, et al, Proc. of the 56th ECTC, pp. 831-837, 2006.

[16] P. R. Morrow et al, *IEEE Elect. Dev. Lett.* **27**, 335-337 (2006).

[17a] Val, C.; Lemoine, T.; IEEE Trans. on Components, Packaging, and Manufacturing Technology, Part A, B, C] V13, 14, Dec. 1990 Page(s):814 - 821

[17b] M. Karnezos; IEMT.2004. pp64-67

[18] K. Takahashi, et al; ECTC Vol.1,2004 Pp.601 - 609

[19] E. Beyne; VTSA.2006.pp1-9

[20] S. C. Jonson, *Semiconductor International*, pp. 40-45, June, 2007.

[21] K.-N. Chen, et al, *IEDM Tech. Digest*, 20-22 (2006).

[22] K. W. Guarini, et al, *IEDM Tech. Digest*, 943-946 (2003).

[23] J.A. Burns, et al, IEEE tran. Elec. Devices, V53, No.10, pp2507-2515, Oct. 2006.

[24] M. Wiemer et al, ESTC2006 pp.1401-1405

[25] D. Temple, et al, IEDM2006, pp:1 - 4

[26] R.S. Patti, Proc. Of IEEE V. 94, I6, June 2006 Pp1214 – 1224

[27] G.A. Katopis, et al, IBM J. Res.develop. Vol 43, No. 5/6, (1999), p621.

[28] K. Prasad and E. Perfecto, IEEE trans. Comp. Pkg. Mfg. Techno. Part B, Adv. Pkg., Vol 17, No.1 (1994) p38.

[29] C. Narayan, et al., IEEE trans. Comp. Pkg. Mfg. Tech., Part B, Vol 18, Issue 1, 1995, p42

[30]United States Patent 6,281,452, Prasad , et al. August 28, 2001

[31] E. Perfecto, et al, IEEE trans.Adv.Pkg., Vol 23, No.3 (2000), p1.

[32] United States Patent 6,099,935, Brearley , et al. August 8, 2000.

[33] H. B. Pogge et al., Proc. AMC'2001, pp.129-136,2001;

[34a] K. Kanakarajan and John A. Kreuz, US patent 5298331 (1994).

[34b] R. K. Kanakarajan, IPC national conf. On Flex Circuits, 1995, Minneapolis, MN

[34c] K. Patel et al, Proc. 10th Symposium on Polymers for Microelectronics, May 2002, Wilmington, DE, Session III-3.

[35] R. Yu, proc. VMIC2007, p223-230.

[36] F. Liu, et al, to be published.

[37a] M. Despont, et al; TRANSDUCERS, Solid-State Sensors, Actuators and Microsystems, 12th International Conference on, 2003,Volume 2, 8-12 June 2003 Page(s):1907 - 1910 vol.2

[37b] R. Guerre, et al, JMEMS, 2007, V17, No1, p157-165.

[37c] M.J.Wolf et al; ECTC2008.Page(s):563 – 570 via fill

[38] United States Patent 6,640,021, Pogge , et al. October 28, 2003

[39] United States Patent 6,444,560, Pogge, et al. September 3, 2002

[40] United States Patent 6,090,633, Yu , et al. July 18, 2000.

[41] United States Patent 6,835,589, Pogge , et al.

[42] United States Patent 6,856,025, Pogge , et al. February 15, 2005

[43]D. Gerty, (THERMINIC 2007), Budapest, Hungary, Sep. 17-19, 2007.

A Study of Thermal Performance for the Panel Base Package (PBP[TM]) Technology

Ming-Chih Yew[1], Chun-Fai Yu[2], Mars Tsai[2], Dyi-Chung Hu[2], Wen-Kung Yang[2] and Kuo-Ning Chiang[1,*]

[1]Advanced Microsystem Packaging and Nano-Mechanics Research Lab.,
Department of Power Mechanical Engineering
Advanced Packaging Research Center, National Tsing Hua University, HsinChu, Taiwan, China 30013
[2]Advanced Chip Engineering Technology Inc.
No. 65, Kuang-Fu North Rd., Hsin-Chu Industrial Park, Hu-Kou, HsinChu, Taiwan, China
*Phone: 886-3-5742925   *Fax: 886-3-5745377   *Email: knchiang@pme.nthu.edu.tw

Abstract

A new panel base package (PBP) technology that was developed based on the concepts of the wafer level package (WLP) has been proposed in order to obtain the signal fan-out capability for the fine-pitched integrated circuit (IC). In the PBP, the chip is attached to a selected chip carrier, and the volume of IC devices is extended for the redistribution of the original die pads. In this study, the thermal performance of the PBP technology was investigated and discussed through three-dimensional finite element (FE) analysis. In order to compare the thermal performance between conventional WLP and the proposed PBP, the junction temperature of WLP was also recorded through the modified FE model. The results showed that due to the larger packaging size of the PBP structure, the added solder bumps can be used as thermal balls. Moreover, they can effectively reduce the packaging thermal resistance (from 55°C/W to 41°C/W). It is expected that thermal performance could be further improved by applying solder paste between the chip and chip carrier. The study likewise discussed the condition of forced convection and developed the PBP technology for high-density IC devices. In light of the results obtained from this study, we believe in our new PBB technology's great potential for future applications.

Introduction

The important factors for packaging technology include the following: IC package costs, the impact of the package on the circuit and system performance, the reliability of the package, and packaging thermal performance. Nowadays, thermal management for semiconductor devices is consistently becoming critical because the density of the transistor together with the metallization in each die increases progressively. A substantial number of studies are available on the subject. In one of these studies, Park et al. [1] studied the thermal performance of WLP by means of a thermal model utilizing the FE method and computational fluid dynamics (CFD). The thermal resistance of the tested WLP is between 25°C/W to 29°C/W under natural convection. In 2003, Chen et al. [2] proposed an effective methodology that integrates infrared (IR) thermography measurement and FE model for the thermal characterization of the packages. Chen et al.'s study used the thin quad flat package (TQFP) as the test vehicle, and the proposed methodology was benchmarked by a thermal test die measurement. On the other hand, Chen et al. [3] and Chang et al. [4] utilized the FE method to predict

the thermal resistance of the flip chip-plastic ball grid array (FC-PBGA) and the quad flat non-leaded (QFN) package, respectively. The computer-aided engineering (CAE) methodology is effective in the conduct of research and development of new products as well as in the improvement of existing packages. As a matter of fact, it has been extensively applied in recent years.

As the IC manufacturing process improves, the difference in pad pitch between the die side and the substrate side becomes more and more obvious due to the limitation of fabrication technology and the cost of the substrate. Moreover, the thermal management for semiconductor devices has also become critical. In this study, a new packaging technology which retains the advantages of WLP, the PBP technology, is proposed to develop the packaging process capability of signals fan-out for the fine-pitched IC. Furthermore, it also seeks to improve thermal dissipation performance through the designed back side chip carrier. In the PBP, the filler material is selected to fill the trench around the chip and to provide a smooth surface for the redistribution lines. Fig. 1 shows that the solder bumps are located on both the filler and the chip surface, and the pitch of the chip side is fanned out.

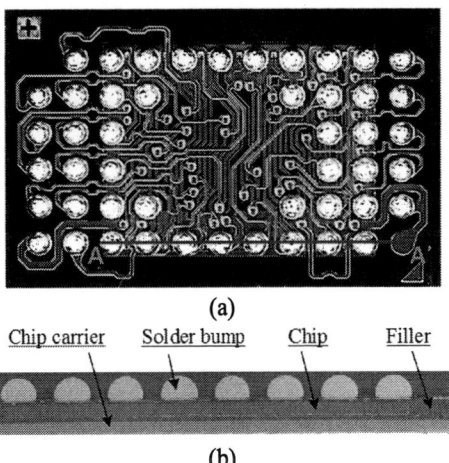

(a)

Chip carrier   Solder bump   Chip   Filler

(b)

Fig. 1 The Panel Base Package technology. (a) top view; (b) AA' cross-sectional view

Our previous research [5-7] shows that the thermo-mechanical behavior of the PBP is different from that of conventional WLP due to the designed packaging structure. The applied soft filler and lamination material can provide a

978-1-4244-2739-0/08/$25.00 ©2008 IEEE

stress buffer layer under solder bumps. Therefore, the solder joint reliability of the PBP is outstanding. On the other hand, the accumulated stress/strain from the coefficient of thermal expansion (CTE) mismatch at the metal lines can be effectively released through a suitable layout technique. This study focuses on the thermal performance of the PBP technology. The route of heat dissipation is observed through three-dimensional FE analysis. Likewise, the effects of the main design parameters in the PBP technology are extracted. Based on the direct comparison of the thermal performance between conventional WLP and the proposed PBP, the thermal characteristic of the fan-out type package is discussed.

Design Specifications of the PBP

The proposed PBP technology applies the filler material to broaden the area at the chip side, thus providing the additional base material when proceeding with the interconnect redistribution. Moreover, the extended volume near the original chip may assist in the heat dissipation for high-power IC. The quasi-wafer-level type of manufacturing process is performed on the panel which is used as the supporting layer. Fig. 2 illustrates the fabrication of the proposed PBP with fan-out capability. In this study, IC devices with a chip size of 2mm × 3.5mm × 0.11mm were selected to conduct the developed PBP technology. The exterior dimension after packaging is 4mm × 6mm × 0.325mm, and the pitch of the contact solder pads is 0.5mm. The solder pads on the chip are designed to be solder mask defined with a diameter of 0.29mm. Besides, the 95.5Sn/3.8Ag/0.7Cu lead-free solders with a diameter of 0.3mm are applied. It should be noted that the original pitch size of the chip is about 0.1mm and has been successfully extended to 0.5mm by the batch process.

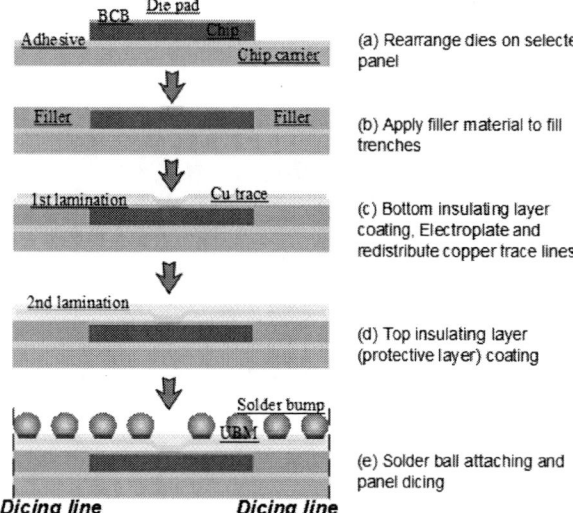

(a) Rearrange dies on selected panel

(b) Apply filler material to fill trenches

(c) Bottom insulating layer coating, Electroplate and redistribute copper trace lines

(d) Top insulating layer (protective layer) coating

(e) Solder ball attaching and panel dicing

Fig. 2 Schematic illustration of the manufacturing process for the PBP technology

FE modeling and Thermal Properties Determination

Fig. 3 shows the established FE model based on the real test sample. Due to the symmetry of the package, the quarter package containing 12 solders and one-half solder is modeled. The IC device is embedded by the PBP technology, and the package is mounted on the standard thermal test board [8]. In the FE analysis, the boundary conditions consisting of heat transfer coefficient are applied to the top and bottom of the package, the top and bottom of the test board, and the side wall of the solder joints. The heat transfer coefficient is a non-linear function of external surface temperature in this model that applied to natural and forced convection regime and includes the contribution of radiation. The total heat transfer coefficient, h_T, is calculated from the following equations [2, 9].

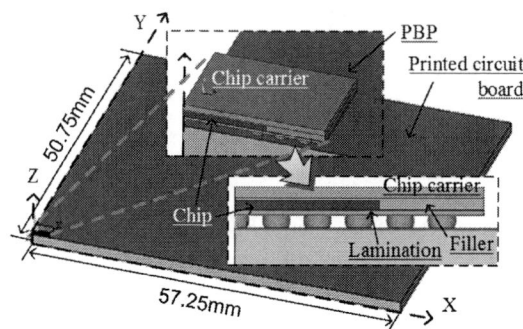

Fig. 3 The established three-dimensional finite element model for thermal characterization of the PBP technology

$$h_T = \left(h_{NC}^3 + h_{FC}^3 \right)^{1/3} + h_{rad} \tag{1}$$

$$h_{NC} = a \left(\frac{T_s - T_{amb}}{L_{ch}} \right)^n \tag{2}$$

$$h_{FC} = 3.79 \sqrt{V / L_{FC}} \tag{3}$$

$$h_{RAD} = \varepsilon \sigma \left[\left(T_s + T_{amb} \right) \left(T_s^2 + T_{amb}^2 \right) \right] \tag{4}$$

where T_s and T_{amb} are the external surface temperature of the package and the ambient temperature, respectively, ε is emissivity, σ is the Stefan-Boltzmann constant, and L_{ch} is the characteristic length. The emissivity of the packaging device is set at 0.9. For horizontal plates, $L_{ch}=0.5WL/(W+L)$, where W and L are the width and length of the plate, respectively. For vertical plates, $L_{ch}=H$, where H is the vertical height of the plate. Moreover, the constant a and n are given as a=0.83 and n=0.33 for a horizontal plate facing upward, a=0.415 and n=0.33 for a horizontal plate facing downward, and a=1.09 and n=0.35 for a vertical plate. In the forced convection term (Eq. 3), V represents the free-stream air velocity. L_{FC} is the characteristic length (set as equal to the package length).

According to the JEDEC standard [8], the test board is a four-layer printed circuit board (PCB) in which copper-patterns are printed in both of the surfaces. The effective thermal conductivity of the test board can be calculated through the following equations [1]:

$$k_{in-plane} = \frac{\sum_{i=0}^{N} k_i \times t_i}{\sum_{i=0}^{N} t_i} , \ k_{cross-plane} = \frac{\sum_{i=0}^{N} t_i}{\sum_{i=0}^{N} \frac{t_i}{k_i}} \quad (5, 6)$$

where t_i and k_i are the thickness and thermal conductivity of each layer, respectively.

Table 1. Material properties applied in the FE model

Material	Thermal Conductivity, K (W/m°C)
Chip carrier	17
Filler	0.2
Copper trace, pad	380
Chip	150
Lamination	0.2
Solder mask	0.2
Adhesive	0.2
SAC387 solder	57
PCB (x, y / z)	(22.8 / 0.34)

The applied thermal material properties in the FE analysis are listed in Table 1. Before the characterization of the PBP is conducted, the aforementioned analytic methodology was validated through a conventional WLP structure and then compared with its experimental data [1, 10]. The calculated thermal resistance consists with the standard thermal testing results under both natural and forced convection condition.

FE Analytic Results and Discussion

By applying the established FE model of PBP (Fig. 3), the numerical analysis is initially performed as the power dissipation is 0.5W. The ambient temperature is set at 25°C, and the predicted stable temperature distribution is shown in Fig. 4 (a). The results show that the junction temperature under natural convection is 45.8°C. In addition, the temperature difference of PBP is 4.8°C. According to the temperature pattern on the test board, the solder joints in the chip region provide a better route of heat dispersion in the PBP, as shown in Fig. 4 (b).

Traditionally, the thermal performance of a given package is represented by the junction-to-ambient thermal resistance, θ_{ja}, defined below:

$$\theta_{ja} = \frac{T_j - T_{amb}}{P} \quad (7)$$

where T_j and T_{amb} are the junction and ambient temperature, respectively, and P is the dissipation power at the given chip. In this study, the calculated thermal resistance of the PBP structure with aforementioned chip/package ratio is 41°C/W.

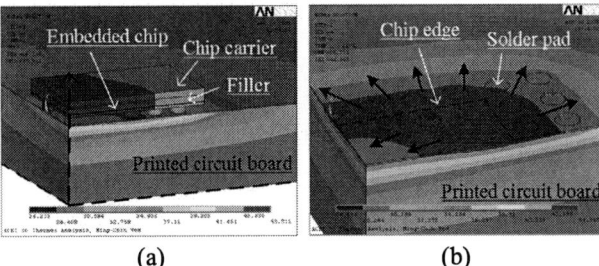

(a)　　　　　　　　(b)

Fig. 4 Thermal performance analysis of the PBP technology. (a) predicted temperature distribution; (b) temperature distribution on the printed circuit board (nature convection, power dissipation = 0.5W)

In the PBP, the filler material is applied to fill the trench around the chip. In addition, the lamination provides electrical insulation for the redistribution lines. Both of them have relatively poor thermal conductivity (K), and their effects are studied through FE analysis. Fig. 5 (a) shows the predicted packaging thermal resistance as the thermal conductivity of the filler is modified. The results show a minor filler conductivity effect of a 4.2% improvement as the K increases from 0.2 to 100 W/m°C. On the other hand, the effective thermal conductivity of the lamination layer consisting of dielectric and metal lines is calculated by Eq. (5, 6). As shown in Fig. 5 (b), the ratio of metal/lamination directly affects the thermal conductivity of the lamination layer. The heat dissipated through solders is easier as the lamination has larger thermal conductivity, i.e. 11% improvement of thermal performance as the metal ratio increases from 10% to 70%.

Referring to the design concepts of the PBP technology, packages with the same chip carrier may contain chips with different sizes. As the size of the embedded chip decreases, a larger portion of the filler material needs to be applied. In order to study the effect of the chip/package ratio in the PBP technology, different sizes of chips with the same power dissipation, i.e. 1W, are analyzed. Fig. 6 shows that the thermal resistance increases by 33% as the ratio of the chip/package size changes from 80% to 8%. Basically, an IC device with a smaller size has a larger power dissipation density. In the PBP, materials are attached around the chip through the batch process, and this could provide effective thermal channels from the package to the PCB.

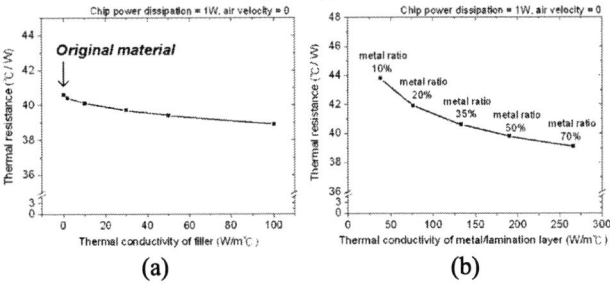

(a)　　　　　　　　(b)

Fig. 5 Thermal conductivity effects of materials in the PBP. (a) filler material; (b) metal/lamination layer

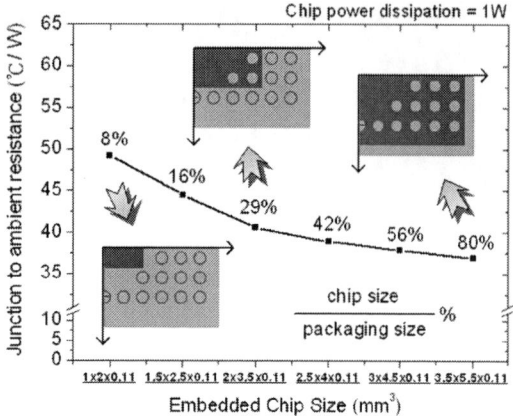

Fig. 6 Thermal resistance value under different (chip size/package size) ratios

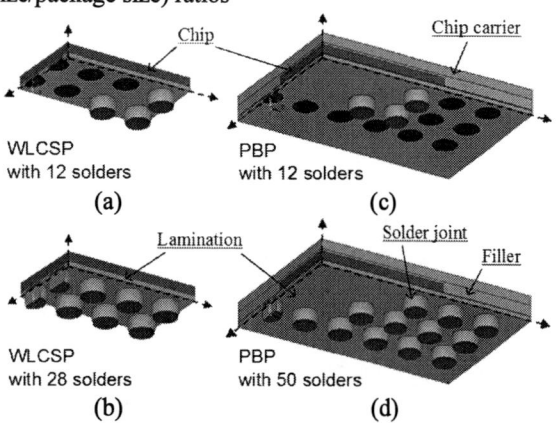

Fig. 7 Established models of WLCSP and PBP in FE analysis (chip size = 2 × 3.5 × 0.11 mm³, packaging size of PBP = 4 × 6 × 0.33 mm³)

The IC manufacturing process improves progressively and thus shrinks the final size of the semiconductor device. Generally, chips with high power density need better thermal management. It should be noted that the PBP technology is one of the possible solutions which can be completed in the batch process. In this study, the thermal performance between the conventional wafer level chip level package (WLCSP) and the proposed PBP is investigated. Based on the established PBP model (Fig. 3), the added materials around the chip are removed, and the numbers of solder bump between the package and PCB are varied. Fig. 7 shows four modified packaging structures and their corresponding solder bump layouts. The results show that the thermal resistance of WLCSP with 12 solders is 63.8°C/W, as shown in Fig. 8. By adding 16 solders as the thermal balls, the additional thermal channels could improve its thermal performance. , i.e. the thermal resistance decreases to 55°C/W.

The application of a thermal ball is cumbersome for ICs with high power density due to chip size restriction. A comparison of the results from the models of Fig. 7(a) and Fig. 7(c) shows that as the bump number is fixed, an increase in the packaging size through the PBP technology could also enhance the capability of heat dissipation. Furthermore, the thermal resistance can successfully decrease to 40.6°C/W as the bump number increases to 50.

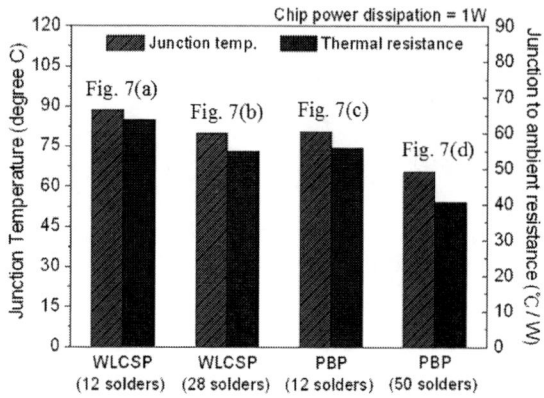

Fig. 8 Thermal performance analysis among different packaging structures

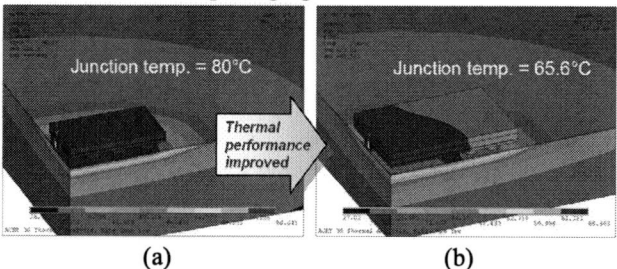

Fig. 9 Predicted temperature distribution in FE analysis. (a) WLCSP with 28 solders; (b) PBP with 50 solders

Fig. 9 shows the predicted temperature distribution in the models of Fig. 7(b) and Fig. 7(d). The WLCSP has its limitation of thermal performance due to small packaging size. If the applied material around the original chip is achievable, the thermal balls can be added effectively. Furthermore, the generated heat can be dissipated through the PCB much easier. Fig. 10(a) shows the difference in thermal resistance between conventional WLCSP (28 solders) and proposed PBP technology (50 solders) under different power dissipation. Although the PBP technology has a larger package size, i.e. $4 \times 6 \times 0.25$ mm³ $> 2 \times 3.5 \times 0.15$ mm³, the chip carrier as well as the added solder bumps can effectively reduce the packaging thermal resistance. On the other hand, thermal performance under forced convection is also compared in Fig. 10(b). Due to the relatively uniform temperature gradient, the PBP structure has a larger air flow effect. The improvement of thermal performance by applying the PBP technology to conventional IC may progressively increase to more than 30%.

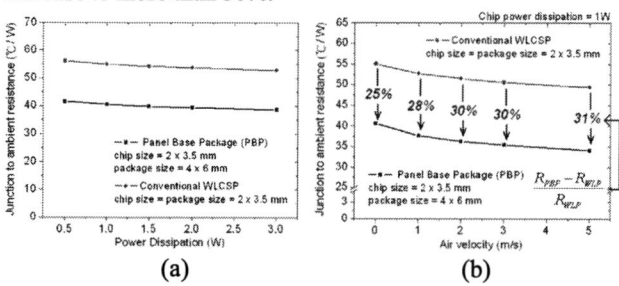

Fig. 10 Thermal performance analysis between conventional WLCSP (28 solders) and proposed PBP technology (50 solders). (a) effect of power dissipation; (b) effect of air velocity

Conclusions

The PBP technology accomplished by batch process has been developed for high-density IC devices. In the PBP, the I/Os with small pitch can be expanded by using the redistribution layers on the applied filler material. On the other hand, the extended materials around the chip may improve the performance of packaging heat dissipation. Based on the analytic results in this study, the effect of filler with small thermal conductivity is minor. In addition, the thermal resistance of the PBP varies as it has different chip/package ratios. Generally, PCB provides the most important route for the heat dispersion of IC devices. By applying the PBP technology, a larger chip side area together with the designed thermal balls can effectively conduct the accumulated heat to PCB. Therefore, the thermal performance of the PBP is outstanding, and it is especially suitable for applications on IC devices with high power density.

Acknowledgments

The authors would like to thank the National Science Council (Project NSC96-2628-E-007-033-MY3) for supporting the research on advanced electronic packages, and the members of the Advanced Chip Engineering Technology (ACET) for supplying the data on the PBP technology. Note: The PBP is the tread mark of Advanced Chip Engineering Technology.

References

1. S. W. Park, J. M. Kim, H. G. Baik, S. H. Kim, J. K. Hong, and H. S. Chun, "Thermal and Electrical Performance for Wafer Level Package," *Proc 50th Electronic Components and Technology Conf*, Las Vegas, USA, May 2000, pp. 301-310.

2. W. H. Chen, H. C. Cheng and H. A. Shen, "An Effective Methodology for Thermal Characterization of Electronic Packaging," *IEEE Trans on Components and Packaging Technologies*, Vol. 26, No. 1 (2003), pp. 222-232.

3. K. M. Chen, K. H. Houng and K. N. Chiang, "Thermal resistance analysis and validation of flip chip PBGA packages," *Microelectronics Reliability*, Vol. 46 (2006), pp. 440-448.

4. C. L. Chang and Y. Y. Hsieh, "Thermal Analysis of QFN Packages Using Finite Element Method," *Proc 5th EuroSimE*, Brussels, Belgium, May 2004, pp. 499-503.

5. M. C. Yew, C. Yuan, C. N. Han, C. S. Huang, W. K. Yang and K. N. Chiang, "Factorial Analysis of Chip-on-Metal WLCSP Technology with Fan-Out Capability," *Proc 13th International Symposium on the Physical and Failure Analysis of Integrated Circuits*, Singapore, July 2006, pp. 223-228.

6. M. C. Yew, H. P. Wei, C. S. Huang, D. C. Hu, W. K. Yang and K. N. Chiang, "A Study of Failure Mechanism and Reliability Assessment for the Panel Level Package (PLP) Technology," *Proc 8th EuroSimE*, London, England, April 2007, pp. 475-482.

7. M. C. Yew, C. Y. Chou and K. N. Chiang, "Reliability Assessment for Solders with a Stress Buffer Layer using Ball Shear Strength Test and Board-level Finite Element Analysis," *Microelectronics Reliability*, Vol. 47 (2007), pp. 1658-1662.

8. EIA/JEDEC Standard, "Test Boards for Area Array Surface Mount Package Thermal Measurements," Tech. Rep., EIA/JESD51-9, Arlington, VA, 2000.

9. G. N. Ellison, Thermal Computations for Electronic Equipment, R. E. Krieger Publishing Company (Malabar, FL, 1989)

10. C. Y. Chou, C. J. Wu, H. P. Wei, M. C. Yew, C. C. Chiu and K. N. Chiang, "Thermal management on hot spot elimination / junction temperature reduction for high power density system in package structure," *Proc International Electronic Packaging Technical Conference and Exhibition*, Vancouver, Canada, July 2007.

FMEA of System-in-Package (SiP) –based Tire Pressure Monitoring System

Man-Lung SHAM[1], Tung-Chin LUI[2], Ziyang GAO & Tom CHUNG
Advanced Packaging Technologies, Material and Packaging Technologies
Hong Kong Applied Science & Technology Research Institute (ASTRI), Hong Kong Science Park, Shatin, Hong Kong, China
[1]ivansham@astri.org [2]hollylui@astri.org

Abstract

For transferring R&D efforts into real product manufacturing, proper product reliability qualification is one of the most critical considerations during product development in addition to assembly yield prediction. It is particularly important for automotive electronics because the operating conditions are extremely harsh (e.g. -20°C ~ 105°C) and a number of applications are even related to human safety.

Failure mode and effects analysis (FMEA) of SiP-based Tire Pressure Monitoring System (TPMS) is selected in this paper as an illustration of the process for transferring R&D efforts into real product. FMEA is proven as a useful tool in the early design stage to identify any potential design and/or process –related failure modes, corresponding effects, root causes followed by corrective actions. Better quality and reliability, shorter system development time and cost, as well as early identification and elimination of potential failure modes can therefore be achieved. In addition, numerical analysis was performed during the course of FMEA in order to address the potential risks and therefore to provide proper recommendations.

1. Introduction

With the ever-growing demands and complexity of integrated chip (IC), the requirements of nowadays electronic packaging design are extremely stringent, including reasonable performance, small form-factor, efficient thermal dissipation, good reliability, and certainly, low-cost. It goes without saying that if the design is proven to be robust in the early stage by all possible means, the whole development and commercialization cycle can be smoother and cost-effective, where Failure Mode and Effects Analysis (FMEA) is considered to be an useful approach to improve the robustness of the design [1-5]. The key purposes of FMEA are to identify all the potential failure locations and modes, respective risk levels, possible effects due to the failures, etc. during the production and operation.

In principle, failures can be ranked according to their seriousness to the performance and reliability of the final product, occurrence frequency, detectability, etc. Once the FMEA is defined in the package design stage, engineers can always refer to the recommendations from the FMEA to review their design and perform necessary corrections before prototype implementation. Of course, the contents of the FMEA can be continuously updated along with the characteristics of the evolving design and the manufacturing procedures. Undoubtedly, FMEA is to eliminate or at least reduce the risks in encountering failures, and FMEA should also recommend the corresponding remedial actions in case failure is encountered.

Tire pressure monitoring system (TPMS) is regarded as one of the key safety-related automotive electronics applications going to be regulated all over the world, firstly regulated in US from September 2007. The main function of TPMS is to warn the drivers when any tire has been out of a normal tire pressure range such that unnecessary damages to the tires can be reduced and more importantly traffic accidents due to inappropriate tire pressure can be avoided. It is reported that the TPMS market is going to be enormous in the coming years, reaching 169 million units and over US$2 billion by 2011 [6].

Typically a TPMS contains sensors (pressure sensor, temperature sensor, accelerometer, etc.), microcontroller, transceiver/RF chip, antenna and battery, as illustrated in Fig. 1. In fact, state-of-the-art TPMS-related IC package is actually based on System-in-Package (SiP) approach to integrate the microcontroller chip, RF chip, sensor chips into a single IC package such that the dimensions and costs of the final TPMS end-product can be substantially reduced. Detailed discussions about the packaging technologies of TPMS have been described elsewhere [7].

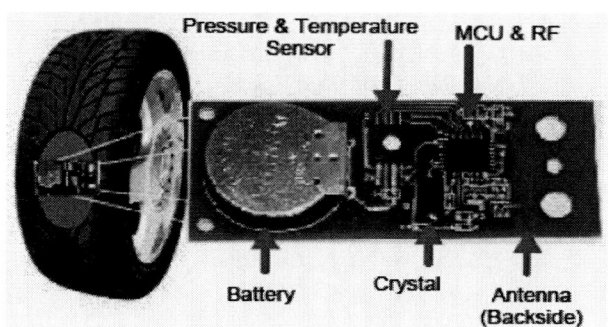

Fig. 1 Typical components in a TPMS [6].

The objective of this paper is to perform a FMEA for a SiP-based electronic package for TPMS application, and parametric studies are performed by using numerical analysis in order to identify the weakest links of the package and their subsequent effects on the overall product reliability. The possible remedial solutions are also included.

2. Experimental Procedures

A simplified Small-Outline-Package with 28 leads (SOP-28) was considered in this study, as shown in Fig. 2. Leadframe-based package indeed is still the major packaging format for automotive electronics -related applications due to its proven reliability. Commercial off-the-shelf components were integrated in this SOP-28 design, including packaged pressure sensor, MCU chip, RF chip and passive component. Moreover, two different SOP-28 package designs (Design A and Design B) were compared. Provided that both designs

978-1-4244-2739-0/08/$25.00 ©2008 IEEE   28

were compatible to conventional surface mount technology (SMT) pick & place process (i.e. having a flat top surface sufficient for common vacuum nozzle to land), Design B can significantly save over 40% of molding compound material per unit when compared to Design A.

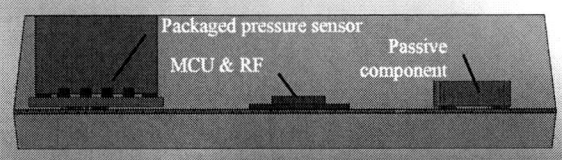

Design A

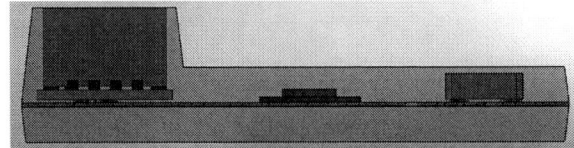

Design B

Fig. 2 SOP-28 design for TPMS application

Except the solder interconnections, all the other components are modeled as isotropic and linear elastic as summarized in Table 1 whereas the Young's modulus of the solder interconnections were assumed thermo-elastic-plastic following Equation 1 [8-11], and temperature dependent bilinear isotropic plastic property was employed to describe the elastic-plastic behavior of the solder as plotted in Fig. 3.

$$E(T) = 74.84 - 0.08T \ (MPa) \qquad (1)$$

where T is temperature in K.

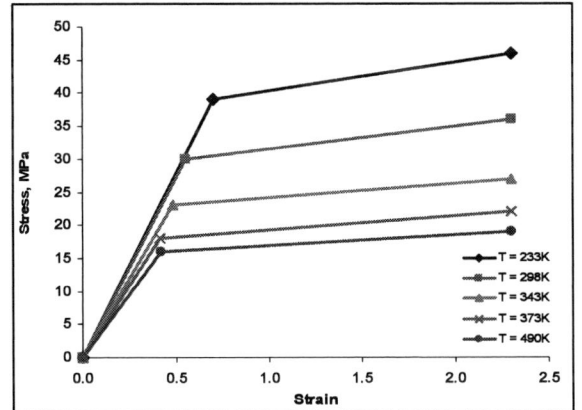

Fig. 3 Bilinear isotropic hardening properties of solder joint under different temperature.

Steady creep behavior of Sn-Ag-Cu -based solder was modeled by Garofalo constitutive equation (sinh law model), as described in Equation 2.

$$\frac{\partial \varepsilon}{\partial t} = C_1 \left[\sinh(C_2\sigma) \right]^{C_3} \exp\left(-\frac{C_4}{T} \right) \qquad (2)$$

where t and T are time and temperature, and C_1=4.41 x 10^5 (1/s); C_2=5 x 10^{-9}(Pa); C_3=4.2 and C_4=5,412(K) [12].

2-D numerical analyses were performed by ANSYS 11.0, and 8-nodes PLANE 82 elements were used for all the package components.

Table 1. Materials properties of the package components.

	Elastic Modulus (GPa)	Poisson's Ratio	Density (kg/m³)	CTE* (ppm/°C)
Leadframe (Copper)	110	0.343	8960	16.4 at 293K 18.5 at 523K
Mold compound (EME7730)	28.22 at 233K 23.52 at 298K 1.764 at 513K	0.25	1900	9 at 233K 10 at 403K 17 at 418K 22 at 423K 27 at 428K 34 at 443K 35 at 473K
Silicon die	112.4	0.28	2329	2.49 at 293K 3.61 at 523K
Die Attach (ABLEBOND 84-MVB)	12.18 at 233K 6.77 at 298K 0.207 at 473K	0.35	720	44 at 233K 45 at 353K 79 at 363k 89 at 368K 99 at 373K 133 at 383K 134 at 473K
Mold compound of pressure sensor	24	0.37	1900	8; T ≤398K 34; T ≥398K
PCB FR4 (Component of pressure	22	0.28	1800	18.5
Sn-3.9Ag-0.6Cu (Component of pressure sensor)	74.84 – 0.08T	0.40	7500	16.66 +0.017T
Underfill (Component of pressure sensor)	11	0.3	1190	21
Nickel (Component of	207	0.31	8880	13.1
Ceramic (Component of	0.177	0.284	5606	2.9

* CTE – Coefficient of thermal expansion.

3. FMEA strategy

The flow of the whole assembly process is summarized in Fig. 4.

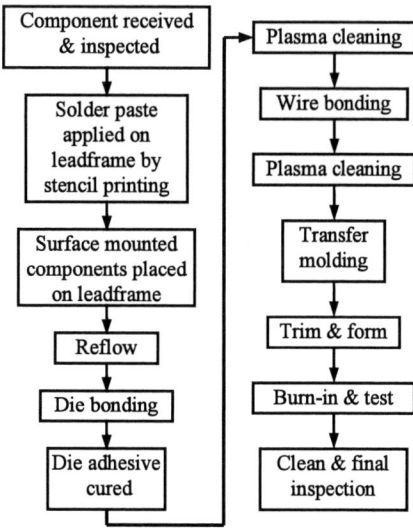

Fig. 4 Process flow of the SiP.

Based on the above process flow, examples of some of the individual process FMEAs are summarized in Table 2.

In the following sections we will demonstrate the approaches in adopting numerical analysis to provide more information to package design engineers in optimizing the package.

4. Results and Discussions

4.1 Effects of package design on overall stress distribution

As abovementioned, Design B can significantly save over 40% of the molding compound, and it is important in understanding how the package design affects the overall stress distribution. Fig. 5 shows the SiP with various geometries, i.e. different draft angles as defined in the figure. To expedite the computation in this section, thermal excursions as shown in Fig. 6 were applied without considering the solder-creep properties and the stress distributions at 233K during the last thermal cycle were compared for different designs, as illustrated in Fig. 5 as well. Essentially, stress concentration regions for all cases were identified in the region underneath the bottom die and the leadframe as shown in Fig. 5. Considering that the thickness of die adhesive in-between the die and the leadframe (<20μm) was comparatively smaller than the thickness of the silicon die (~150μm) and the leadframe (127μm), the local region can be approximated as bi-material strip structure and the elevated stress (σ_x) due to the temperature change (ΔT) can be estimated by Equation 3 [13]:

$$\left[\Delta \sigma_x\right]_i = \frac{E_i \Delta T}{(1 - \upsilon_i)}\left(\alpha_j - \alpha_i\right) \qquad (3)$$

where E is the modulus, α is the CTE, υ is the Poisson's ratio and i refers to the material under investigation while j refers to the material adhering on material i. Apparently, apart from the CTE mismatch ($\alpha_j - \alpha_i$) and the temperature change, the modulus and Poisson's ratio will also contribute to the resultant stress. It is noticed that the maximum stress near the

bottom die region were insensitive to the draft angle, and the maximum stress along the bottom die interface with the die adhesive was 150MPa.

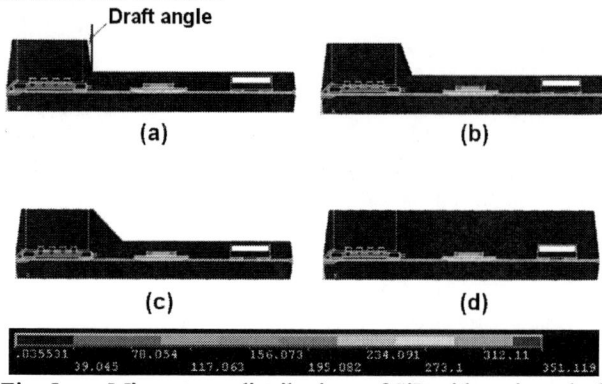

Fig. 5 von Mises stress distributions of SiP with various draft angle: (a) 7°; (b) 20°; (c) 45°; and (d) without any draft angle.

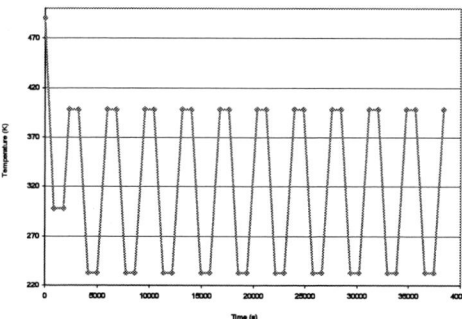

Fig. 6 Thermal excursions between 233K (-40°C) and 398K (125°C).

4.2 Effects of package design on solder joint reliability

It is of interest to investigate if the solder joint fatigue life was affected by package design, and the packages with 7° and without any draft angle were studied under the same thermal excursions as previously shown in Fig. 6. Solder creep property was considered in the modela. Comparing the von Mises stresses inside the solder joints of the packages, it is noticed that the stress levels underneath the packaged pressure sensor were higher than the stresses under the passive component on the other side of the package, and the stress distribution near the pressure sensor of the two designs are shown in Fig. 7.

Although it appears that the stress distributions particularly inside the solder joints were very similar, the maximum shear stress levels and the creep strain levels were indeed very different as shown in Fig. 8. The maximum shear stress and creep shear strain of the 7° draft angle package were approximately 2 times and 6 times higher than the package without any draft angle at the solder joint under the pressure sensor respectively. Accordingly, the lifetime of the solder joint would be dramatically affected due to the package design.

4.3 Effects of delamination in solder joint reliability

Since the top surface of the packaged pressure sensor has to be exposed in order to determine the environment pressure, the top surface of the packaged pressure sensor is therefore at

30

the same level as the top surface of the SiP as illustrated previously. Yet, the exposed interfaces between the packaged pressure sensor and the molding compound of the SiP would induce stress singularities which in fact are the potential sites for initiating delamination. Thence, it is important to investigate how the package design affects the solder joint reliability of the pressure sensor in this particular SiP.

Elements with air properties were introduced along both sides of the pressure sensor with the length equal to half of the pressure sensor height. Other materials' models and thermal excursions were the same as in previous sections.

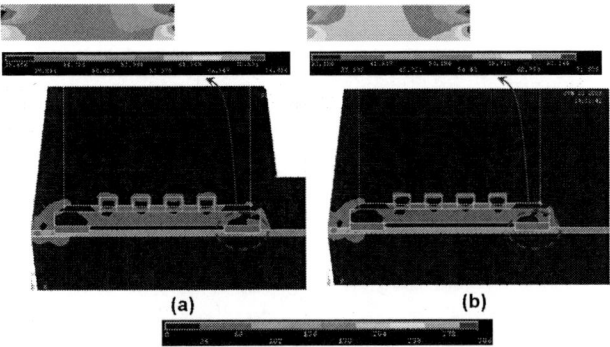

Fig. 7. von Mises stress distribution near the pressure sensor region: (a) 7° draft angle package; and (b) without any draft angle.

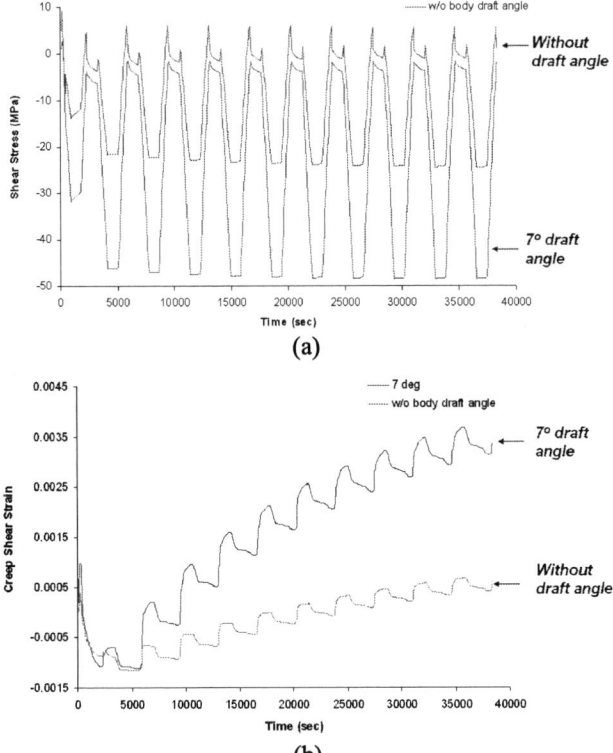

(a)

(b)

Fig. 8. Maximum stressing condition inside the solder joint of packaged pressure sensor inside the 7° draft angle package with respect to the thermal loading history: (a) shear stress; and (b) creep shear strain.

Figure 9 compares the hysteresis maximum shear stress-creep shear strain loop of the 7° draft angle package with and without delamination along the interfaces between packaged pressure sensor-molding compound of the SiP. It is noticed that although the stress levels of the delaminated condition were slightly lower than the perfect-adhesion condition, presumably because of the overall stress re-distribution, the amounts of creep shear strain of the delaminated condition indeed were continuously higher than the perfect-adhesion condition with respect to the thermal excursions. This implicates that the lifetime of the solder joints would be reduced if there exist delamination along the packaged pressure sensor-molding compound interfaces. Special attention has to be paid to enhance the interfacial adhesion, for example, plasma treatment before the mold transfer process.

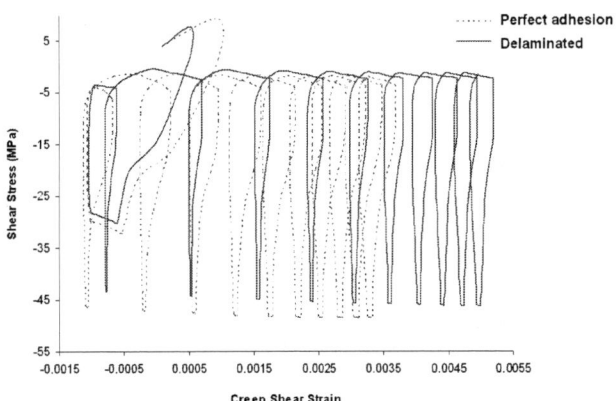

Fig. 9. Hysteresis shear stress-creep shear strain loop of 7° draft angle package with thermal excursions.

5. Concluding Remarks

FMEA of a SiP design for TPMS applications had been conducted, and numerical analyses were also performed to determine the potential risks of failure of the package Based on the above discussions following remarks are concluded:

By performing a comprehensive FMEA, the package design can be effectively reviewed and more robust before prototype manufacturing, and the time-to-market cycle can be therefore shortened effectively.

The draft angle of package had insignificant effects on the maximum stress of the SiP, but it affected the reliability of solder joints underneath the packaged pressure sensor.

Once delamination was encountered between the package pressure sensor and molding compound of the SiP, the maximum creep shear strain inside the solder joints of the packaged pressure sensor showed significant increment. Proper interfacial bonding enhancement technique should be considered during volume-manufacturing.

Acknowledgements

The authors would like to express the gratitude to Mr. Vincent Leung, Lourdito Olleres, Mr. Gomer Egnisaban, Mr. Antonio Mangente and Dr. Lydia Leung of ASTRI for their advices on package design development throughout this study.

31

References

[1] K. Pickard, T. Leopold, P. Müller, B. Bertsche, Electronic Failures and Monitoring Strategies in Automotive Control Units, IEEE 2007, pp. 17-21.

[2] M. Kennedy, Failure Modes & Effects Analysis (FMEA) of Flip Chip Devices Attached to Printed Wiring Boards (PWB), Int'l Electronics Manufacturing Technology, 1998, pp. 232-239.

[3] C. J. Price, N. S. Taylor, FMEA for Multiple Failures, Proceedings of IEEE Annual Reliability and Maintainability, 1998, pp. 43-47.

[4] S. Prasad, Improving Manufacturing Reliability in IC Package Assembly Using the FMEA Technique, IEEE Transactions on Components, Hybrids and Manufacturing Technology, vol. 14, no. 3. Sep. 1991.

[5] I. D. Wolf, Reliability of MEMS, 7th Int. Conf. on Thermal Mechanical and Multiphysics Simulation and Experiments in Micro-Electronics and Micro-System, EuroSimE 2006, pp. 1-6.

[6] "2005-2006年中国TPMS产业研究报告," Pday Research, 2007/08.

[7] M.L. Sham, Z. Gao, Lydia L. Leung, Y.C. Chen & T. Chung, Advanced Packaging Technologies for Automotive Electronics, ICEPT 2007, pp. 1-5.

[8] Y. Qi, H. R. Ghorbani, J. K. Spelt, Thermal Fatigue of SnPb and SAC Resistor Joints: Analysis of Stress-Strain as a Function of Cycle Parameters", IEEE Transactions on Advanced Packaging, vol. 29, no. 4, Nov. 2006

[9] I. Kim, S. Lee, Fatigue Life Evaluation of Lead-free Solder under Thermal and Mechanical Loads, IEEE Electronics Components and Technology Conference, 2007, pp. 95-102.

[10] Y. Qi et al, Accelerated Thermal Fatigue of Lead-Free Solder Joints as a Function of Reflow Cooling Rate, Electronic Materials, vol. 33, no. 12, Dec. 2004.

[11] B. A. Zahn, Impact of Ball via Configurations on Solder Joint Reliability in Tape Based Chip-Scale Packages, ChipPAC Inc. Arizona.

[12] J. Lau, W. Dauksher, J. Smetana, HDPUG's Design for Lead-Free Solder Joint Reliability of High-Density Packages, IPC SMEMA Council APEX, 2003, pp. S42-2-1 – S42-2-12.

[13] M.L. Sham, J.K. Kim, Evolution of residual stresses in modified epoxy resins for electronic packaging applications, Composites A, vol. 35, 2004, pp. 537-546

[14] R.J. Hannemann, A.D. Kraus, M. Pecht, Semiconductor Packaging: A Multidisciplinary Approach, John Wiley & Sons Inc. Publ., Canada, 1994.

[15] M.L. Sham, J.K. Kim, Improved underfill adhesion in flip-chip packages by means of ultraviolet light-ozone treatment, IEEE IEEE Transactions on Advanced Packaging, vol. 27, no. 1, 2004, pp.179-187.

Table 2. FMEA of SiP.

Item	Failure Mode	Failure Effect(s)	Failure Causes	Preventive Steps
Wirebond [14]	• Gap in the wire. • Rupture of the wire at the heel of the bond. • Life-off of the deformed wire.	Broken down the electrical connections.	• Excessive current overheating and melting the wire. • Thermo-mechanical stress due to the CTE mismatches between the components • Excessive thermal stress lifting the weakened bond or fracturing of the silicon under the bond.	• Increase the wire diameter for high current applications • Optimized the material selections to minimize the CTE mismatches.
Die-attach (DA)	Die fracture after thermal excursions.	• Crack arises from the die bottom to the die edge and corners. • Vertical and horizontal die cracks.	• Vertical cracks from the die backside are primarily due to the presence of voids in the DA. • Horizontal cracks are due to the elevated interfacial stresses near the die edges [14].	• Avoid the presence of voids in DA by dispensing proper amounts of DA and appropriate dispense pattern. • Select proper DA for minimum residual stress.
SMT components	<u>Class I</u> • Solderability • Electrical Failure <u>Class II</u> Pre-mature solder joint failure due to voids inside.	<u>Class I</u> • Excess paste • insufficient paste • misalignment <u>Class II</u> • Improper reflow profile • Exaggerated stress concentration at certain locations.	<u>Class I</u> • Printer setup • Damaged screen • Incorrect process setting, e.g. print speed. • Incorrect stencil design, e.g. aspect ratio of the opening • Viscosity of paste <u>Class II</u> • Insufficient dwell time for flux activation and evaporation • Too fast ramp up rate from dwell to max. reflow temperature • Un-optimized package design.	<u>Class I</u> Proper process parameters have to be determined. At the beginning of each shift a regular setup proceeding should be performed, including the inspection of tooling, equipment and screen. <u>Class II</u> • The solder reflow profile should match the reflow characteristics • For every new design reflow oven calibration is recommended • Retailed material analysis should be performed to predict the straining condition of the package and the reliability tests.
Die	Corrosion (Die fracture will be discussed in die adhesive section)	• Moisture diffusion into the molding compound • Uncontrolled assembly environment • Poor adhesive between the die and molding compound	• Improper molding compound • Diffusion through the passivation layer • Transport of ionic contaminants to the potential corrosion site • Electrochemical reaction between water and the various ionic constituents • Low filler loading • Trim and form process damages the leads and introduces microcracks at the epoxy-leadframe interface.	• Select proper molding compound(s) with sufficient resistant against the moisture level as specified in the corresponding standard's requirements (e.g. AES Q100) • Based on the characteristics of the passivation layer of the die, select proper molding compound(s) for superior adhesion (Detailed experimental conditions have been reported elsewhere [15]

Ultrasonic Bonding of Polymer Microfluidic Chips

Zong-bo ZHANG[1*], Yi LUO[1,2], Xiao-dong WANG[1,2*], Zhen-qiang ZHANG[1], Li-ding WANG[1,2]

[1] Key Laboratory for Dalian University of Technology Precision & Non-traditional Machining of Ministry of Education, Dalian, China

[2] Key Laboratory for Micro/Nano Technology and System of Liao Ning Province, Dalian, China

xdwang@dlut.edu.cn, zzb001_0@163.com

Abstract:

Bonding is an essential step to enclose microchannels or microchambers in lab-on-a-chip. Ultrasonic bonding was studied as a deformation-free technique to realize high efficiency bonding of microfluidic chips. Based on viscoelastic dissipation theory, the main influential factors of heat generation rate during ultrasonic bonding was theoretically analyzed and numerically calculated using finite element method. According to the results, micro energy directors were designed to concentrate the ultrasonic power and to control the location of the joint. To demonstrate the performance of this bonding method, specially designed PMMA substrates of microfluidic chips were fabricated by means of hot embossing. With the ultrasonic bonding technique, the chips were reliably and hermetically bonded within less than a second.

Keywords: Ultrasonic bonding; microfluidic chips; finite element method; hot embossing

1. Introduction

Microfluidic chips have attracted a great deal of attention in the field of pharmaceutics, biotechnology as well as life sciences owning to its advantages such as: fast separate speed and low consumption of analyte[1,2]. And thermoplastic polymers have been extensively investigated over the past decade as substrates for the fabrication of microfluidic chips because of its low-cost, disposability as well as ease of fabrication. In a typical process, open microchannels and other micro structures are formed on a thermoplastic substrate using one of several techniques such as hot embossing[3], micro-injection molding[4] or laser ablation[5]. Thereafter, a cover plate is bonded to the substrate to enclose the microstructures. A key and challenging technique in almost all chip fabrication methods is the hermetical enclosure without deforming and clogging of microchannels or microchambers by bonding.

Several techniques for bonding of polymer substrates have been studied in the past decade, including thermal bonding[6,7,8], adhesive bonding[9], laser/microwave bonding[10,11], solvent bonding[12] and so on. However, bonding is still a bottleneck problem in the process of chip manufacturing due to the limitations in efficiency, quality, chemical compatibility and biological fitness of the bonding techniques mentioned above[13]. In this regard, some superior bonding methods are urgently needed to be found. The ultrasonic bonding which was firstly published to seal microchannels in 2006 was a promising solution to this bottleneck problem[14], owning to its advantages such as: foreign substances free, little damage to devices, short cycle time and well defined local heating of only the joint area and so on[15]. However, the minimum channel bonded by ultrasonic bonding till now was $500 \mu m \times 500 \mu m$[13,15],

which was too large for micro-devices. Moreover, as a novel technique, ultrasonic bonding of polymer microfluidic chips still lacks systematic research, especially on the design and fabrication of auxiliary bonding structures, the control of the bonding process as well as the improving of the accuracy. In this paper, a new type of microfluidic chip whose characteristic dimension was less than $100 \mu m$ with auxiliary bonding structures were fabricated. Thereafter, with a standard plastic ultrasonic welding machine, this study presented the efficient and high performance of the ultrasonic bonding method for the bonding of PMMA-based microfluidic chips without deformation of microstructures.

2. Heat generation during ultrasonic bonding

A viscoelastic material that is subjected to a sinusoidal strain dissipates some energy into viscoelastic heat through intermolecular friction. This is the most dominant heat source for ultrasonic bonding of thermoplastics[16]. In ultrasonic bonding process, the heat transfer can be ignored, because polymer is a kind of typical thermal insulation material and the duration of the process is relatively short.

2.1. Theoretical analysis

Consider a linear viscoelastic material bar subjected to a sinusoidal strain:

$$\varepsilon(t) = \mathrm{Re}(\varepsilon_0 e^{i\omega t}) = \varepsilon_0 \cos \omega t \qquad (1)$$

In which ε_0 is the amplitude of strain, ω is the angular frequency of the driving force and t is time. Then, the strain in the material is related to the stress through the complex modulus E^*:

$$\sigma(t) = \mathrm{Re}(E^*(\omega)\varepsilon(t)) = \varepsilon_0(E' \cos \omega t - E'' \sin \omega t) \qquad (2)$$

Where $E' = E \cos \delta$ is the storage modulus, which is an in-phase modulus to measure the ability of storing energy, and $E'' = E \sin \delta$ is the loss modulus, which is an out-of-phase modulus to measure the ability of dissipating energy. E and δ are the modulus and loss angle of the material respectively.

Therefore, the total strain energy generated within time t is:

$$
\begin{aligned}
W &= \int_0^t \sigma(t) d\varepsilon(t) \\
&= \int^{\cos \omega t} \varepsilon_0^2 (E' \cos \omega t - E'' \sin \omega t) d\cos \omega t \\
&= W_1 + W_2
\end{aligned} \qquad (3)
$$

Where $W_1 = E'\varepsilon_0^2 (\cos 2\omega t - 1)/4$,

$W_2 = E''\varepsilon_0^2 (2\omega t - \sin 2\omega t)$.

W_1 is a pure sinusoidal function, that is to say, the elastic in-phase components produce no net work in a cycle. While the out-of-phase components expressed in W_2 dissipates mechanical energy into heat. According to the analysis above, the heat generation rate Q in per volume is:

$$Q = \frac{1}{4}E''\varepsilon_0^2(2\omega - 2\omega\cos 2\omega t) = \frac{1}{4}E\sin\delta\varepsilon_0^2(2\omega - 2\omega\cos 2\omega t) \quad (4)$$

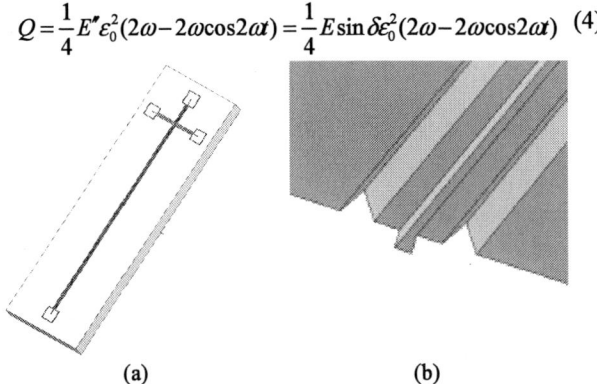

(a)   (b)

Fig. 1 The designed chip (a) integrated graph; (b) cross-section graph

Thus, the heat generation rate is proportional to loss modulus E'' and the square of strain amplitude ε_0. Assuming the material is isotropical, it is predictable that the temperature of the area where the strain concentrates will rise fastest. Therefore, it is advisable to design some special structures near the expected joint to concentrate the strain during bonding. In this study, trilateral shaped microstructures named 'energy directors' were designed, as shown in Fig. 1.

2.2. Numerical calculation

In order to demonstrate the temperature rising process during bonding, a PMMA chip with rectangle-shaped energy directors was calculated by Finite Element Method (FEM), in the current work. The assumptions for this calculation are as follows:

(1) The material is isotropic;
(2) Convective and conductive heat exchange is ignored due to very short boning time and low thermoconductivity of PMMA;
(3) PMMA is linear viscoelastic material.

The FEM calculation was performed by commercial software Ansys. As a typical type of viscoelastic material, the mechanical property of PMMA differs very much in different temperature range. Moreover, this is a nonlinear calculation process. So the material models in the software can not simulate the temperature-dependent properties of PMMA properly. In order to characterize the temperature-dependent properties of PMMA, the software was re-developed by UPFs tool.

The initial temperature of this calculation was 100 ℃, which was slightly below T_g of PMMA (T_g=105℃). Under a bonding amplitude of 30 μm and a bond pressure of 0.4MPa, the results are shown in Fig. 2.

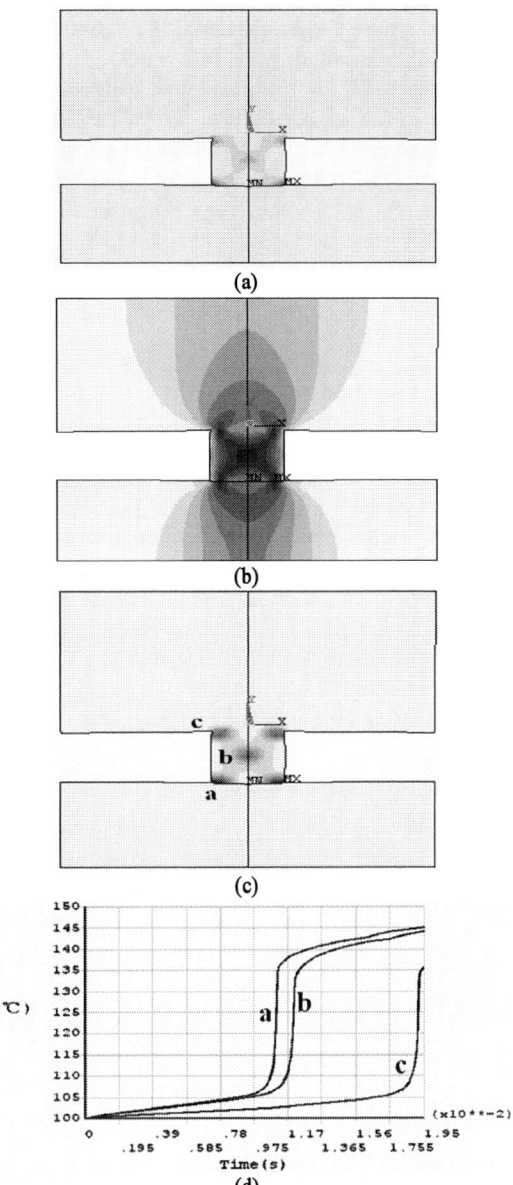

Fig. 2 Calculation results (a) contour of strain; (b) contour of stress; (c) contour of risen temperature caused by viscoelastic heating; (d) curve of rising temperature of different locations

The temperature contour coincides well with the strain and stress contours, seen in Fig. 2(a,b,c). In other words, acoustic energy dissipation is highest in the location where the stresses and strains concentrate. So it is reasonable to design some focal structures near the joint zone to direct and focus the acoustic energy. Fig. 2 (c) also illustrates that the obvious risen temperature areas are in the four corners as well as the center part of the energy director. And the risen temperature of the four corners rises earlier than that of the center part, as shown in Fig. 2 (d). Moreover, when the temperature reach Tg, the rising rate of temperature dramatically increases and then slows down when the temperature reach a certain threshold. This can be explained from Eq. (4), which shows that the dissipated energy depends much on the modulus E and loss

35

angle δ. In a purely elastic material the loss angle is $0°$, so that there will be no dissipation. In a purely viscous material the loss angle is $90°$, but the modulus vanishes, as a result there will be no dissipation, neither. When the temperature is below Tg, the loss angle of the material is so small that there is little energy dissipation. On heating, the plastic softens and E drops near Tg, at the same time, δ increases drastically, resulting in dramatic increasing of temperature. Hereafter, the plastic will become almost a pure viscous material with very low E and approximately $90°$ loss angle, so the rising rate of temperature slows down.

3. Experiments and results

3.1. Preparation of designed substrates

In this paper, the significant micro-hot-embossing method was employed to fabricate the micro channels and micro energy directors on PMMA substrates. Because there were both convex energy directors and concave channels on the chip, its mold was fabricated by multi-step lithography technology and two times of wet-etching processes on a silicon wafer<100>. The size of the mold was $27mm \times 54mm \times 2mm$. The channel was $40\,\mu m$ in depth and $70\,\mu m$ in width. The cross-section of energy director was isosceles triangular with $150\,\mu m$ base long and $106\,\mu m$ high. Fig. 3 shows the overall appearance and the details of the mold.

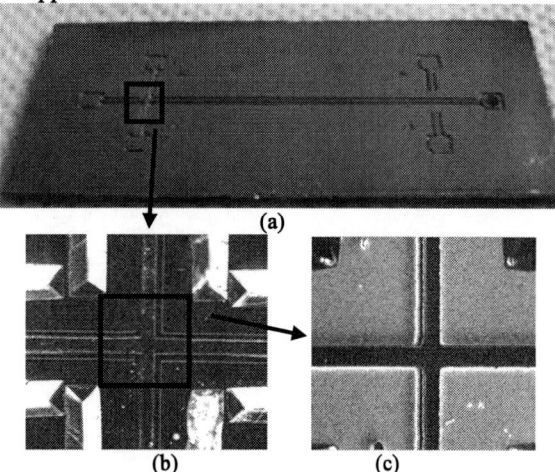

Fig. 3 Photographs of silicon mold (a) integrated photo of the mold; (b) local micrograph of the mold; (c) micrograph of channels

The PMMA (polymethyl methacrylate) used for hot embossing, provided by Asahi Kasei Corporation, is a widely used material in microfluidic devices, with a glass transition temperature of approximately $105°C$. Each substrate was cut into a size of $50mm \times 25mm$ with a thickness of $2mm$. A self-developed micro hot embossing apparatus was used to emboss the designed microstructures on PMMA substrates. In this process, embossing temperature, embossing pressure and embossing time were significant parameters for high replication accuracy. The pressure should not be too high because of the brittleness of silicon mold. Moreover, the concave energy directors on the mold made the embossing temperature at least $10°C$ higher than traditional processing. The profiles of the embossed chips are shown in Fig. 4.

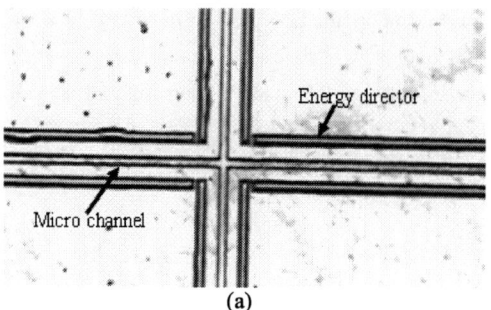

(a)

(b)

Fig. 4 Micrographs of embossed substrate (a) microchannel and micro energy directors on the embossed substrate; (b) the cross-section view of embossed substrate

3.2. Ultrasonic bonding of microfluidic chips

The bonding process was performed on a standard ultrasonic welding system 2000 f/aef from Branson company of USA, with a generator frequency of $20KHZ$ and a power of $1500W$, as shown in Fig. 5. The embossed substrate and a cover plate of same material were clamped on a leveling anvil. Thereafter, a static pressure and the ultrasonic range oscillating compress were applied to the chip by sonotrode. The ultrasonic compress then propagates into the specimen in the form of structure borne sound. The sound energy was concentrated in the energy directors as a consequence of the quasi line shaped auxiliary bonding structures. Intermolecular and interfacial friction results in the heating up and melting of the energy directors.

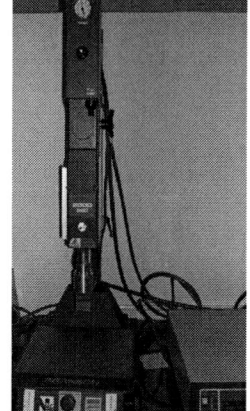

Fig. 5 The using ultrasonic welding system

Bonding force is an important parameter in this process. And the collapse distance of energy directors is a vital representation of bonding quality. Fig. 6 shows the bonding force, collapse distance profile and the corresponding status of energy director during ultrasonic bonding. In this process, the bonding and holding time were $0.3s$ and $0.5s$. The trigger force, the bonding force as well as the holding force were $40N$, $300N$ and $180N$, respectively.

The detailed processing conditions were as follows:
(1) The substrates were clamped on the anvil. Hereafter, the Sonotrode went down with a speed of about $50mm/s$. The stress in the energy director started to

36

increase, which caused mechanical collapse as status (b) shows in Fig. 6.

(2) When the stress reached the trigger force, an ultrasonic range vibration with a amplitude of 30 μm was applied to the chip by sonotrode as the force was linearly increasing till came up to the bonding force. In this stage, the temperature of energy directors increased rapidly to glass transition temperature or even melting temperature. Consequently, energy directors became so soft and adhesive that the energy directors collapsed and fused rapidly as shown in (c) and (d) of Fig. 6.

(3) Then the ultrasound was shutdown and the force declined to the holding force, the squeezing flow was cooled off by heat diffusion. As a result, the two parts of the chip were bonded together.

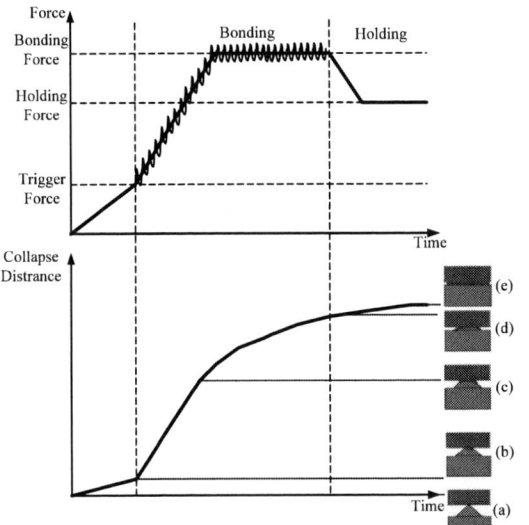

Fig. 6 Force-collapse distance profile for ultrasonic bonding

Thermal bonding technique is regarded one of the most popular bonding techniques for polymer microfluidic chips, in which the bonding strength increases with the bonding temperature and pressure. However, the high temperature and pressure applied may result in undesirable deformation or even collapse of structures, as shown in Fig. 7[7]. While in ultrasonic bonding technique, most part of the chip remain cool except for energy directors on which the vibration energy focuses. So there is no deformation of microstructures during this cycle.

In order to examine the cross-section of the micro channels, the chips were cut and observed under electron microscope, as shown in Fig. 8. Cross-section micrographs of microchannels before and after bonding indicated that the microtopography of the channels was well kept.

Then the bonding strength is tested by tensile strength experiment. The material near the bonding joint fractured before the joint's fragmentation. This phenomenon showed that the joint was even stronger than the body material. Then, to confirm the opening and the hermeticity of the microchannels, fluorescence ink was injected into the chip. Bonded regions showed a transparent color since the ink could not be injected into those regions, whereas opened regions show an orange color due to the injected ink as show

in Fig. 9. It also showed that by means of ultrasonic bonding the microchannels can be hermetically sealed.

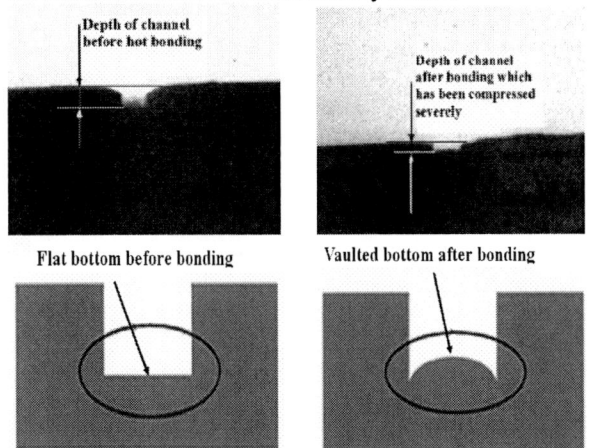

Fig. 7 The probable deformations of microchannel in hot bonding process [7]

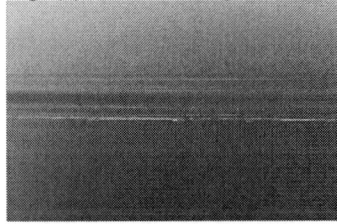
(a)   (b)

Fig. 8 Micrographs of cross-section view of microchannels (a) Before bonding; (b) After bonding

Fig. 9 Leak testing image

It might be finally mentioned that the whole bonding process by ultrasonic had last for only less than a second, which could dramatically improve the bonding efficiency of microfluidic chips.

4. Conclusions

In this paper, a functional-microstructures deformation free bonding technique to realize high efficiency bonding of microfluidic chips is presented. With theoretical analysis the auxiliary micro structures for ultrasonic bonding were designed. And a quick and simple method was established to fabricate the mold of the designed chip. Then the PMMA substrates of the chips were manufactured by hot embossing. Because of the special structures and the properties of the mold the embossing temperature at least 10℃ higher than traditional processing. Thereafter, these substrates and cover plates of same material were successfully boned with properly selected parameters by standard ultrasonic welding system.

The bonding strength is even higher than initial strength of the material, which are strong enough for most microfluidic applications. Further, the deformation of functional structures deforms very little because of location heat in the process.

37

Moreover, opening and hermeticity test of the microchannels demonstrates that all channels on the chip are successfully sealed. Finally, the whole process takes only less than a second, which is much shorter than traditional thermal bonding technique. Overall, the current ultrasonic bonding technique offers a number of attractive features, including excellent bonding strength, little damage to the functional structures, high hermeticity and good compatibility. Most importantly, the manufacturing efficiency of this method is much higher than that of traditional techniques.

Acknowledgements

The research was supported by Program for New Century Excellent Talents in University (NCET-06-0279) from Education Ministry of the People's Republic of China and granted by National Natural Science Foundation of China (No.50775024).

Thanks are also given to Peking Branch of Branson Ultrasonic Corporation for providing experimental ultrasonic welding equipment.

References

[1] M. Heckele, W. Bacher and K. D. Muller. Hot embossing the molding technique for plastic microstructures. Microsyst. Technol. 1998,4 (3):122–124.

[2] H. Becker and U. Heim. Hot embossing as a method for the fabrication of polymer high aspect ratio structures. Sensors Actuators A. 2000,83 (5):130–135.

[3] L. Martynova, L. E. Locascio, M. Gaitan, et. Fabrication of plastic microfluid channels by imprinting methods. Anal. Chem. 1997, 69(23):4783–4789.

[4] R. M. McCormick, R. J. Nelson, M. G. Alonso-Amigo, et. Microchannel electrophoretic separations of DNA in injection-molded plastic substrates. Anal. Chem. 1997, 69(14): 2626–2630.

[5] M. A. Roberts, J. S. Rossier, P. Bercier and H. Girault, UV Laser Machined Polymer Substrates for the Development of Microdiagnostic Systems. Anal. Chem. 1997, 69(18): 2035–2042.

[6] Xuelin Zhu, Gang Liu, Yuhua Guo, Yangchao Tian. Study of PMMA thermal bonding. Microsyst Technol. 2007, 24(13): 403–407.

[7] J M Li, C Liu, H C Qiao, L Y Zhu, G Chen and X D Dai.Hot embossing/bonding of a poly (ethylene terephthalate) (PET).microfluidic chip. J. Micromech. Microeng. 2007,28(11):1-10.

[8] C. W. Tsao, L. Hromada, J. Liu, P. Kumar and D. L. DeVoe. Low temperature bonding of PMMA and COC microfluidic substrates using UV/ozone surface treatment. Lab on a Chip. 2007, 7:499–505.

[9] A. Gerlach, H. Lambach, D. Seidel. Propagation of adhesives in joints during capillary adhesive bonding of micro- components. Microsyst. Technol. 1999,6:19–22.

[10] H. Klein, E. Haberstroh. Laser beam welding of plastic micro parts. Annual Technical Conference of the Society of Plastics Engineers. ANTEC99, SPE. 1999,1(1): 1406–1410.

[11] A. A. Yussuf, I. Sbarski, J .P. Hayes, M. Solomon, N. Tran. Microwave welding of polymeric-microfluidic devices. J. Micromech. Microeng. 2005,15:1692–1699.

[12] Che-Hsin Lin, Chien-Hsiang Chao, Che-Wei Lan. Low azeotropic solvent for bonding of PMMA microfluidic devices. Sensors and Actuators B. 2007, 121:698–705.

[13] R. Truckenmueller, R.Ahrens. An ultrasonic welding based process for building up a new class of inert fluidic micro sensors and actuators from polymers. Sensors and Actuators. 2006,132:385-392.

[14] Wei He, Wang Xiaodong. Bonding simulation of ultrasonic welding method for plastic microfluidic chip. China Mechanical Engineering. 2005,16(7):82－85.

[15] R. Truckenmueller, Y. Cheng, R. A. hrens, H. Bahrs, G. Fischer, J. Lehmann. Micro ultrasonic welding: joining of chemically inert polymer microparts for single material fluidic components and systems. Microsys Technology. 2006,12:1027-1029.

[16] A. Benatar, R.V. Eswaran, S.K.Nayar. Ultrasonic welding of thermoplastics in the near-field. Ploymer Engineering and Science. 1989,29(23):1689-1698.

Mold Array Package for POP Applications

Aching Lee, Louie Huang, Mike Hung

Aching_Lee@aseglobal.com,
louie_huang@aseglobal.com, mike_hung@aseglobal.com,

(TEL) +886-7-361-7131. (FAX) +886-7-361-3094.
ASE (Advanced Semiconductor Engineering Inc), Taiwan
26. Chin 3rd RD., Nantze Export Processing Zon, Kaohsiung, Taiwan 811, China
Advanced Semiconductor Engineering Group, Kaohsiung, Taiwan, China

ABSTRACT

The Package on Package (POP) stacking is getting more and more popular for system in package (SIP) applications. But during the assembly process, the POP had encountered the challenge of packages stacking yield loss, especially when top package and bottom package stacking. The key factors are the mount height of top package, the mold cap of bottom package, and the metallized ball land on the top surface of bottom package. JEDEC JC-11 has defined the rules of two packages stacking. However, the fine pitch package stacking application will meet the process capability limitation, including thinner mold cap, wafer thinning and the lowest wire bond loop height challenges. The POP used the top gate mold chase for the bottom package to expose the metallized ball land on the top side of package which is a dedicated molding tooling. Also, some process are used for solving yield loss issues such as a POP with interposer between top package and bottom package, or a bottom package with pre-mounted the solder ball on chip side ready for top package to attach. Those are customized tooling and not a prevailing tooling that increases the developing cost and timing.

To resolve the stacking process yield loss issue, a MAPPOP solution had been revealed for eliminating the limitation between the top and bottom package stacking. The assembly process of mold array package (MAP) for fine pitch BGA has been implemented for MAPPOP applications. In the paper, the package design rules, and assembly process of exposed metallized ball land on the top surface of bottom package had been discussed. Finally the warpage performance and the packaging level reliability had also been discussed and analyzed.

1 Introduction

MAPPOP bottom package's molding tool is same as FBGA (side gate mold - DOFU mold type), but different from TRDPOP (Top gate mold). Table1.1 is the bottom package process comparison between MAPPOP and TRDPOP(Top gate mold POP is traditional POP = TRDPOP). The main difference between MAPPOP and TRDPOP is the interconnection implant on the top surface of bottom package , and the half cut process to expose the metallized land. Figure 1.1 is the MAPPOP bottom package configuration after half cut process.

Table 1.1 Process Comparison

Process Station	POP	MAP POP
Interconnection Implant	X	V
Die Attach	V	V
Wire Bond	V	V
Molding	V TOP Gate Mold	V DOFU Mold
Laser Marking	V	V
Ball Mount	V	V
Half Cut	X	V
Singulation	V	V

Table1.2 MapPOP v.s. TRDPOP Advantage Comparison

Feature	TRD PoP (FBGA)	MAP PoP (FBGA)
Mold System	Top Gate System	Side Gate System
Mold Chase Investment	More (Mold Area & Cap)	Less (Mold Thickness)
EMC Consumption	More	Less
PKG Height	Same	
Defect Mode	Mold Flash/ Incomplete Fill	Same as FBGA
Warpage	Larger	Smaller
Form Factor	Larger	Smaller
Interconnect Pitch	Larger (0.65mm/ 0.80mm)	Smaller (0.50mm)
Pre-Stack Yield	Lower	Higher
Test / Handling	Same	
Unit Cost	Same	

Fig 1.1 MAPPOP bottom package configuration

Table 1.2 compare the advantages between MAPPOP and TRDPOP, and MAPPOP is more flexible in many features, for instance, Figure 1.3 shows package stacking comparison, a top package with 0.5mm ball pitch, 0.35mm ball size and the stand off is 0.27mm +/- 0.05mm. Bottom package mold cap is 0.35mm, MAPPOP can be stacked with top package, but TRDPOP has met the solder joint open problem.

978-1-4244-2739-0/08/$25.00 ©2008 IEEE

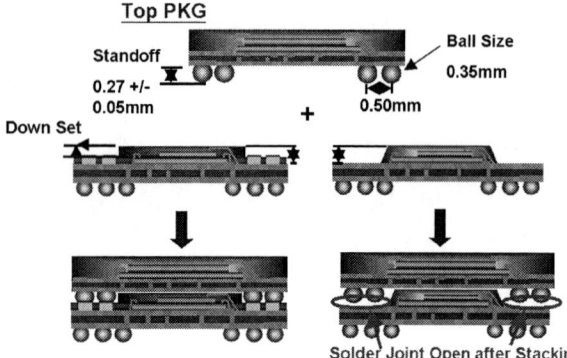

Fig 1.3 Stacking comparison

2 MAPPOP DESIGN FOR STACKING

2.1 Key Dimension

The key dimensions of MAPPOP are : 1. Top Ball Height(TBH), 2. Bottom Cap(BC), 3. Remain Compound(RC), 4. Solder Pad Opening(SPO), 5. Interconnection layout, 6. Die Thickness, and 7. Wire Loop Height. The basic design requirement is TBH > BC for good stacking yield (Fig 2.1).

Fig 2.1 MAPPOP key dimension

2.2 Bottom Cap Design

The stacked configuration Design FMEA is described in Table2.1, the top package solder ball diameter is 0.35mm on a 0.3mm ball pad opening, and TBH is 0.22mm minimum. To use a 0.25mm mold cap for MAPPOP application. The first step is to design remain compound (RC) height. Due to Minimum RC = Maximum Bottom Cap, if BC > TBH, will encounter interconnection open issue during package stacking. Base on table 2.1, the RC design provide a reasonable stacking configuration design, the interconnection open risk is the lowest.

Table 2.1 MAPPOP configuration design for 0.5mm interconnection

Ball Size(nom) of top package	RC(Nominal)- Bottom Package (+/-40um)	Risk(Level 1-10) Level 1:Low; Level 10: High
0.35mm	C	Solder Joint Open(9)
	B	Solder Joint Open(5)
	A	Solder Joint Open(1)

A test vehicle of 14x14 MAPPOP with 2 rows 0.5mm interconnection pitch, package outline design as follows, 0.25mm mold cap, and RC2 is implemented.

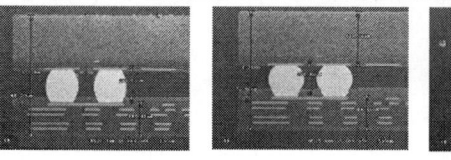

Cell	RC	Interconnection	Solder pad Opening
1	RC1 min.	Good	SPO1
2	RC2 Nominal	Good	SPO2
3	RC3 Max.	Good	SPO3

Fig 2.2 SEM photo of all the interconnections

The SEM photo in fig 2.2 reveals all the interconnection are good even though the minimum and maximum RC.

MAPPOP mold cap design rule is defined in Table 2.2. TRDPOP mold cap design limits the package stacking which must use a thinner mold cap, will face the challenge of thinner wafer, and lower wire loop. On the other hand, MAPPOP provides a more flexible design in mold cap option.

Base on above DFMEA, RC design rule is described in table 2.3. Regarding 0.4mm pitch interconnection design the mold cap is 0.2mm and RC is 100um at least.

Table 2.2 MAPPOP Mold Cap design rule

Ball Pitch of top package	A5(max) – bottom package	TRD POP Mold Cap - bottom package	MAP POP ASE Map- Mold cap Bottom package
0.65 mm	0.33 mm (JEDEC)	0.3 mm	0.3 /0.35 /0.4
0.50 mm	0.22 mm (JEDEC)	0.2 mm	0.2 / 0.25 / 0.3 mm
	N/A	N/A	0.2 mm / 0.25 mm
0.40 mm	N/A	N/A	0.2

Table 2.3 MAPPOP Remain Compound design

PITCH (mm)	MAPPOP RC
0.5	A

2.3 Interconnection Design

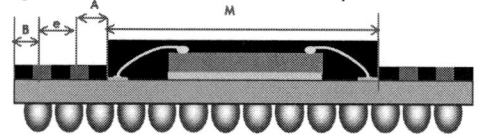

Figure 2.3 MAPPOP interconnection and bottom cap area size

Table 2.4 2 rows Interconnection design for a 14x14 pakcage

	Map-PoP	
e	500µm	650µm
A	350µm	400µm
B	500µm	500µm
M max	11300µm	10900µm
Interconnection count	200	152

40

Table 2.5 MAPPOP Interconnection Ball Court vs Mold cap area

Package Size	0.5mm pitch Inter connection ball count			Max Mold Cap Area (mm)		
	1 Row	2 Rows	3 Rows	1 Row	2 Rows	3 Rows
10x10	72	136	192	8.3x8.3	7.3x7.3	6.3x6.3
11x11	80	152	216	9.3x9.3	8.3x8.3	7.3x7.3
12x12	88	168	240	10.3x10.3	9.3x9.3	8.3x8.3
13x13	96	184	264	11.3x11.3	10.3x10.3	9.3x9.3
14x14	104	200	288	12.3x12.3	11.3x11.3	10.3x10.3

Remark : e.g. for 0.5mm inter connect pitch
1. Package size range : 8x8~21x21
2. Interconnection row : 1 row~3 rows
3. Max. mold cap area depends on A(350um) /B(500um) /C(500um)

The next step is to design the interconnection counts, it can be calculated by package size and pitch size, if the B (distance from edge to ball center) and A bottom cap to ball center) can be defined, then the M (bottom cap area) can be defined, too. The M value is related to the die size design. Table 2.4 is a 2 rows interconnection design for a 14x14 package size. The maximum interconnection is 200 terminals for a 0.5mm ball pitch top package with 2 rows ball peripheral design. A 3 rows is up to 288 ball counts.(Table 2.5)

3 Exposed the metallized ball land process development

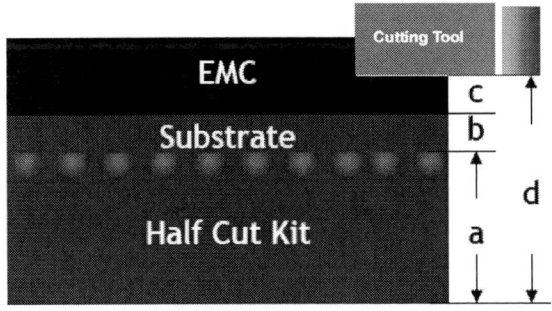

Fig 3.1 Half cut setting

The methodology to exposed the metalized ball pad here is that mechanical cutting process. There are 3 key components : 1. Half cut kit height (a), 2.substrate thickness (b), 3. Remain compound (c). (Fig 3.1)

The cutting height setting d = a + b + c base on the design value of package outline dimension. RC tolerance is related to substrate thickness tolerance, because of a, and c both are the fix values. Fig 3.2 is the optical photo of MAPPOP after half cut implement.

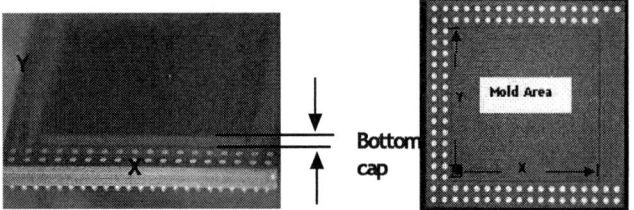

Fig 3.2　Metalized ball land exposed after Half cut setting

4　MAPPOP Warpage performance

To design a MAPPOP, the first step is to design a TBH > BC package outline dimension. The next step is to realize the performance of package warpage, this will facilitate the interconnection joint together or open. Generally, a package is always warpage.

A package is the composition of many materials, including substrate, EMC, silicon chip, die glue…etc. These materials with different properties will conduct the package warpage, due to mismatch of thermal expansion. To realize the influence of these factors, a L8 design of experience with 2 levels, 5 factors is implemented. The output is analyzed by JUMP. The best collocation is C: die size / D: EMC shrinkage rate / A: mold cap thickness / E: substrate thickness / B: die thickness.

Table 4.1 Warpage DOE

	Level 1	Level 2
A: Mold cap thickness	270	350
B: Die thickness	75	100
C: Die size	6.5x6.5	8.8x8.6
D: EMC shrinkage rate	SR1	SR2
E: Substrate thickness	260	300

Cell No.	Die size	Die thickness	EMC type	Mold cap	Substrate thickness
1	6.5x6.5	75	SR1	270	300
2	6.5x6.5	75	SR2	350	260
3	6.5x6.5	100	SR1	350	260
4	6.5x6.5	100	SR2	270	300
5	8.8x8.6	75	SR1	350	300
6	8.8x8.6	75	SR2	270	260
7	8.8x8.6	100	SR1	270	260
8	8.8x8.6	100	SR2	350	300

Table 4.2 DOE result response chart

Effects	A	B	C	D	E
L1	-155.151288	-153.523166	-147.541668	-156.600184	-155.106086
L2	-152.351022	-153.979144	-159.960642	-150.902126	-152.396224
ABS(L1-L2)	2.800266	0.455978	12.418974	5.698058	2.709862
Rank	3	5	1	2	4

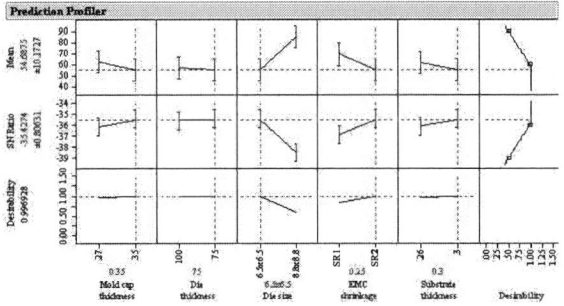

MAPPOP is superior to TRDPOP in coplanarity had been revealed in previous study[1]. The following study focuses on the remaining compound versus coplanarity. This study use a 0.3mm mold cap MAPPOP, and remain compound setting value are 50 / 100 /150 um, output is coplanarity. Figure 4.2 shows that low correlation between RC and coplanarity.

Base on above study, what's the main factor to improve the MAPPOP package warpage. It is the EMC property after fixed the configuration of die size, thickness, mold cap thickness, and substrate design. The study result uses different EMC for a 14x14 MAPPOP warpage improvement is that high CTE EMC (CTE A<B) with better warpage performance during high temperature as shown in Fig 4.2. For the high stacking yield, both (top and bottom) packages' warpage performance must to be considered during the solder ball molten.

Reliability

A and B compounds are built in green BOM for package level reliability test. Both of them pass MSL3, TCT1000 cycles, THT 1000hrs, HTST 1000hrs with 77 samples size. For board level reliability test, the Sn/Ag/ Cu/ Ni solder ball with OSP substrate surface finish is built for drop test, and it pass the 30 drops. It passes TCT 1000cycles, too.

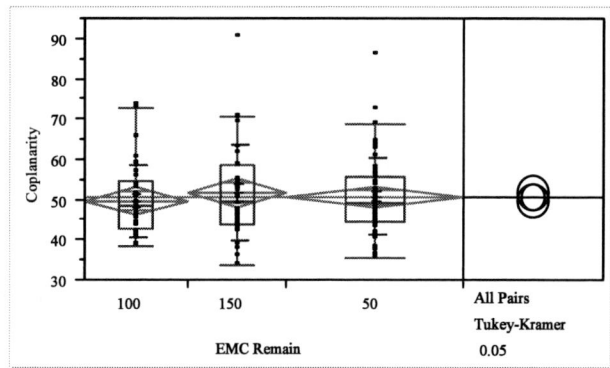

Fig 4.1 Remain compound versus coplanarity

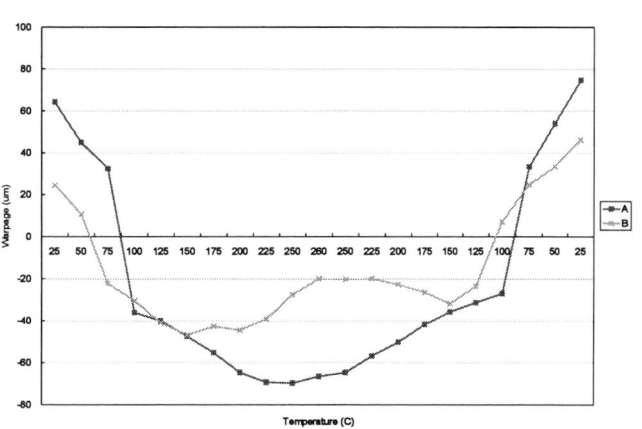

Fig 4.2 Shadow moir'e profile of EMC A and B

Conclusion

Mold array package for POP application is more flexible than top gate mold package. The MAPPOP key configuration

is the half cut depth that can be implemented by the half cut process. The configuration design must consider about top package stacking availability. There are two key points, the first one is Top package ball height > Bottom cap, and the second one is warpage performance during reflow. MAPPOP's half cut configuration is not related to the package warpage. All materials composed in the package are the major factor of warpage. Both top and bottom package warpage performance must be considered for the high stacking yield. Smaller the gap between top and bottom package during solder ball molten will have a higher stacking yield, and the package warpage performance can be control by a optimum material composition.

Acknowledgements

The author would like to appreciate the MAPPOP team members from ASE Kaohsiung Central Engineering and ASE group. The ASE corp. R&D lab. for conducting package level reliability test, and ASEK QA lab. for conducting package reliability test, too.

References

1. Wei Chung Wang, " Platform of 3D Package Integration "2007.ECTC conference.

New Technologies for advanced high density 3D packaging by using TSV process

Paul Kettner, Bioh Kim, Stefan Pargfrieder, Swen Zhu
EV Group
E. Thallner Str. 1, A-4782 St. Florian/Inn, Austria
p.kettner@EVGroup.com, +43 7712 5311 0

Abstract

There is no question that 3D integration will be the next generation of packaging. This requires new technologies from ultra thin wafer handling to wafer to wafer bonding with 3D inter substrate connections.

TSV is a process in which wafers are thinned, stacked and interconnected to significantly improve electrical performance such as signal transmission, interconnect density, reduced power consumption, form factor and manufacturing costs.

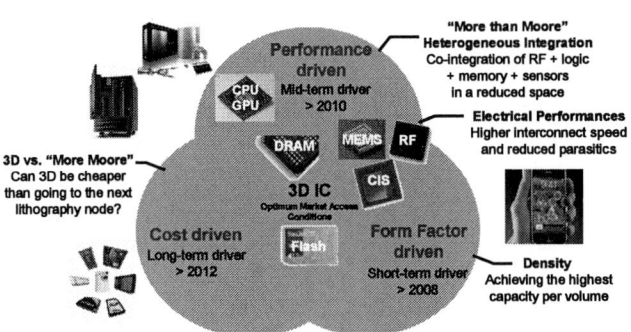

Graph 1: Market drivers for 3D application
source *Yole* Development

TSV – the next generation of packaging

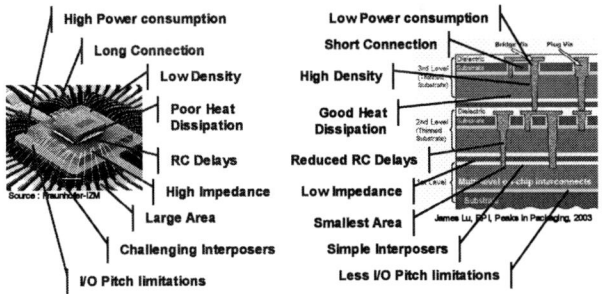

Graph 2: comparison wire bonding to 3D stacking

The TSV process requires ultra thin wafers with less than 50μm to reduce the stack thickness and improve the performance of the IC Device. EVG has developed a solution for temporary bonding of wafers to carriers to be able to run in standard IC Fab's all processes for back grinding, etching, metallization, wafer bonding and solder ball formation.

Ultra thin wafer handling solutions

Depending on the process requirements 2 different solutions are on the market:
- lamination of tapes on carriers, Si or glass wafers; disadvantage of tapes are the limitation of process temperature of usually max 170°C as well as the poor edge protection
- spin on adhesive materials on carriers with the advantage of high process temperature up to 250°C with excellent edge protection.

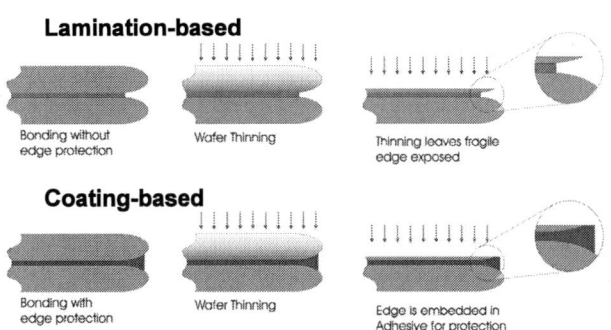

Graph 3: edge protection by tape versa spin on material

EVG has developed systems for a fully automated lamination and bond process as well as de-bonding systems for high volume manufacturing.

When high process temperature e.g. for metallization is required we prefer spin on materials from our partner Brewer Science. The materials from Brewer Science HT250 and HT10.10 have high chemical resistance as well as high thermal resistance up to 250°C.

Chemical	Bath Temperature	Time	Results
1% HF	room temp	30 min	excellent
49% HF	room temp	15 min	3 mm of etch at edge excellent bond
NMP	85°C	1 hr	1 mm of etch at edge
Acetone	room temp	30 min	excellent
IPA	room temp	30 min	excellent
RCA1	75°C	30 min	excellent
RCA2	75°C	30 min	excellent
15% H_2O_2	60°C	30 min	excellent
25% TMAH	60°C	30 min	excellent
10% NH_4OH	60°C	30 min	excellent
*excellent = no measurable dissolution of the material			

Graph 4: Brewer Science HT250 material
www.Brewerscience.com

Packaging houses request sometime to perform back side processes on fully processed IC wafer including bumps on the front side. In this case we are able to spray coat the BSI material over the high structure and protect the bumps in the interface to the carrier

978-1-4244-2739-0/08/$25.00 ©2008 IEEE

Graph 5: back side process for IC wafers with bumps

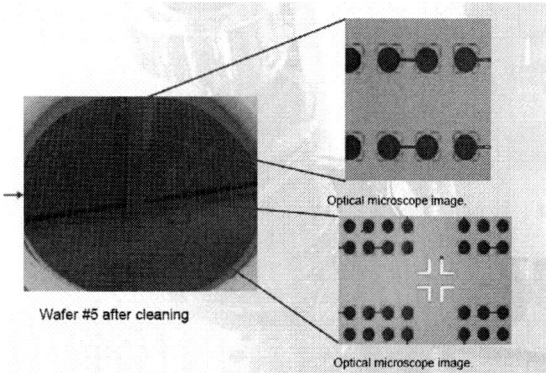

Picture 1: de-bonded wafer front side after back side process

By using a special "slid-off" process at elevated temperature we are able to de-bond the thin IC wafer without damaging the bumps and release to a cleaning station where by solvent rinse the remaining HT material can be cleaned off the wafer surface, see Picture 1.

Chip to Wafer or Wafer to Wafer bonding

3D Technology started with Tezzaron in Singapore 8 years ago on wafer to wafer bonding using Cu-Cu direct bonds. W2W bonds have the advantage of high alignment accuracy which enables to reduce the design of the vias to 2µm. In the last years more bonding technologies with or without an intermediate layer emerged. Due to the thermal budget of CMOS devices, the bonding processes are limited to such as solder-based bonding, low-temperature fusion bonding by plasma activated Si surface, polymer bonding or hybrid bonding. (see picture 2)

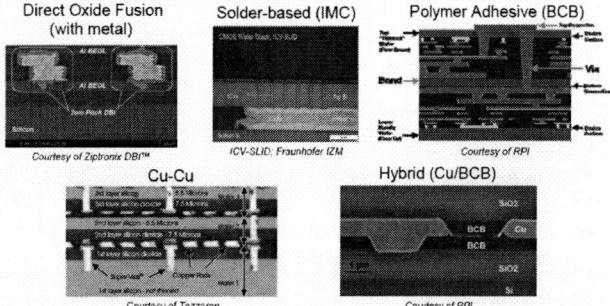

Picture 2: wafer to wafer bonding examples for 3D

Direct Oxide Fusion bonds require very strict surface conditions getting sufficient bond strength at room temperature following with a batch annealing process. Cu-Cu direct bonds need the highest temperature and high contact force close to the limit of IC wafers. Advantage of both bonding procedures is the high post bond alignment accuracy.

Polymer as well as hybrid bonds are easier from the surface conditions but more complicated by additional lithography steps which increase the total costs.

Solder based (IMC) bonds require larger visa size and pad size but can be easily integrated in existing manufacturing lines. Beside Cu-Sn-Cu and Au-Sn connects which form new metal structures by diffusion at lower temperature with high melting point so that multiple wafer bonds can be performed.

	Wafer-to-Wafer	Chip-to-Wafer
Wafer size	Common size Wafer and Die	Dissimilar wafer size Dissimilar die size ✓
Economy / Throughput	Wafer scale throughput ✓	Die scale throughput
System compatibility	Wafer handling ✓	Wafer and die handling
Yields	Lower than lowest yield wafer	Select known good die (KGD) ✓
Alignment	<2µm global alignment ✓	~10µm >1000dph < 2µm <100dph
Fabrication site	Fab or packaging ✓	Packaging

Graph 6: EVG comparison W2W versa C2W

Depending on the required process and available production facility both technologies will be in the future on the market. E.g. Wafer to wafer approach for image wafer integration by TSV technology is the firth HVM application.

EMC -3D Consortium

To bring the total process from R&D to HVM EVG has started a consortium for 3D integration where design and process technology, materials as well as system suppliers working in a platform. In this consortium a total process will be evaluated with the final costs target down to 200 USD/wafer based on 200mm wafers.

www.emc3D.org

Conclusions

Three primary factors are driving the adoption of 3-D ICs and associated packaging technologies: increased functionality, size shrinks and a lower cost. For through-silicon via (TSV) technology to be adopted in 3-D ICs, the cost has to be right — and there's a lot of research underway right now to try to drive it down below the $200 mark.

CMOS image sensors using TSV technology are already in production, but memory is not far behind. Once the cost target is met, the use of TSV technology in 3-D ICs is expected to eventually gain widespread adoption from manufacturers, because it significantly reduces the interconnect distance and removes any real limitation on the number of die that can be stacked. This makes smaller form factor devices possible while also improving speed and functionality.

References

1. Yole Development, EMC-3D road show February 2008, "Market drivers for 3D application?".
2. Brewer Science Inc. HT Material for temporary bonding www. Brewerscience.com

Recent Progress of Ohmic Contact on ZnO

Yao Lv, Lixi Wan

Institute of Microelectronics, Chinese Academy of Sciences
Beijing, 100029, China
tomato8241984@163.com, 010-82995601

Abstract

ZnO as an excellent candidate for UV light emitters, varistors, transparent high-power electronics, surface acoustic wave devices, piezoelectric transducers, and chemical and gas sensors could be integrated in a SiP (System-in-Package). The SiP could be a critical part in sensor nodes in a sensor network. Normally, the ZnO device in SiP is fabricated with nanoscale films which can be compatible with other materials and processing in a package. However, despite the great potential for electron and photonic applications, ZnO device fabrication is difficult to obtain good ohmic contact. The low resistance and thermal stable ohmic contacts is critical to realize high-performance ZnO-based devices. In this paper, the recent advances of ohmic contacts on ZnO are analyzed and reviewed. The mechanism of the energy band bending at the interface of the semiconductor and the metal is discussed. The factors of forming good quality ZnO films such as the choice of the substrate and the method to deposit ZnO film, the effect of the contact resistance and thermal stability of ohmic contacts are summarized.

1 Introduction

ZnO is a versatile II–VI and wide-bandgap (3.37eV at RT) compound semiconductor with wurtzite structure [1]. It has gained substantial interest because of its large exciton binding energy (60meV) which could lead to lasing action based on exciton recombination even above room temperature [2]. In addition, ZnO has excellent thermal and chemical stability. Moreover, there is a much simple crystal-growth technology leading to potentially lower cost for ZnO-based devices. With these advantages above, ZnO is more and more widely used in UV light emitters, varistors, transparent high-power electronics, surface acoustic wave devices, piezoelectric transducers and gas sensors, integrated in a SiP[3] (System-in-Package). Normally, the ZnO device in SiP is fabricated in nanoscale films which can be compatible with other materials and processes in the package. However, despite the great potential of the applications in electron and photonic, there are difficulties in fabrication of the ZnO devices. One of them is metal-semiconductor contact. There are two main contacts, one is schottky contact, and the other is ohmic contact [4]. A schottky contact has the similar property with p-n junction diode, which means it exhibits unilateral conductivity but has extra advantages: lower-power and super-high speed. With these predominations, schottky contact is often used in high-frequency, low-voltage, and heavy-current rectifier diode. An Ohmic contact can be

_____

This work was supported by the National High-Technology Research Development Program (863). Contract number: 2006AA01Z236, 2007AA01Z200. National Science Foundation of China, contract number: 90607006.

defined as having a linear and symmetric current-voltage relationship for both positive and negative voltages and is so important for carrying electrical current into and out of the semiconductor, ideally with no parasitic resistance[1]. In order to obtain high-performance devices, it is essential to achieve ohmic contact that have both low resistance and are thermally stable and reliable [5].-[7]. As the band gap of ZnO is three times than Si, besides, the unique material structure and the electrical performance, all of these distinguish ZnO from the traditional possessing. Take the doping for example: in order to gain p-type Si, B or Al is often used, but these doping will change ZnO into n-type. In addition, the etching processing of Al and Cu cannot be used in ZnO devices. So the primary task is to develop ZnO's own possessing. High-quality ohmic contact is critical in possessing. In the paper, the authors pay more attention on the n-type ZnO's ohmic contact.

2 The physics model of metal-semiconductor for ohmic contact

2.1 The formation mechanism of the ohmic contact

The mechanism of the ohmic contact can be explained by the electronic transport properties[8]. Take the n-type semiconductor-metal contact for an example: there are four kinds of transport properties. We can see clearly from the figure 1: (1) Thermionic emission over the barrier is one of the dominant transport mechanisms when the barrier thickness is large. (2) When the barrier width is reduced, tunneling through the barrier can become the important transport mechanism. (3) In the space-charge region, the

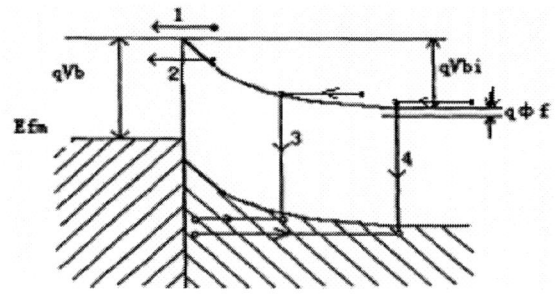

Fig.1 The sketch of the electronic transport

the hole recombine. (4) In neutral region, the electron and the hole recombine. Among these four kinds of transport properties, the electron has to cross over the schottky barrier which is so called Thermionic emission. The I-V curve will exhibit rectify behavior. Concerning the second mechanism, we can divide it into to section: the electrons are tunneling the barrier with the energy near the Fermi level, or the electrons are stimulated to a higher energy level and the chance of tunneling increase because of the reduced barrier. They are named filed emission and Thermionic filed emission respectively. For ohmic contact, the first emission is much more important. The third and the forth mechanism can be neglected when forming the ohmic contact. An ideal ohmic contact is small compared with the parasitic resistance of the device. It is difficult to measure the exact value of the contact resistance. So the contact resistivity is introduced to describe the value, presented as $\Omega cm2$. The definitions are as follows:

$$R_c = \left(\frac{\partial J}{\partial V}\right)^{-1}_{V \to 0} = \frac{k}{qA^*T}\exp\left(\frac{q\phi_{Bn}}{kT}\right)(low\ doping)$$

$$\approx \exp\left[\frac{4\pi\sqrt{\xi_s m^*}}{h}\left(\frac{\phi_{Bn}}{\sqrt{N_D}}\right)\right](high\ doping)$$

In general, Thermionic emission will be dominant for semiconductors with low doping concentration, and tunneling will be dominant for high doping concentration. In order to obtain low resistant ohmic contact, there are two common methods: reduce the barrier height or increase of the effective carrier concentration of the surface.

2.2 ZnO-metal ohmic contact

ZnO normally forms in the hexagonal (wurtzite) crystal structure with a=3.25 Å and c=5.12 Å. The Zn atoms are tetrahedrally coordinated to four O atoms, where the Zn d-electrons hybridize with the O p-electrons. Layers occupied by zinc atoms alternate with layers occupied by oxygen atoms [1]. Electron doping in nominally undoped ZnO has been attributed to Zn interstitials, oxygen vacancies, or hydrogen. The intrinsic defect levels that lead to n-type doping lie approximately 0.01–0.05eV below the conduction band. That's why it is easier to form ohmic contact than schottky contact. For intrinsic ZnO, the electron affinity$\chi$is 4.35eV. As the work function can be express as:

$$W_s = \chi + E_n, \quad (3)$$

Where the E_n is the value of E_c-$(E_F)_s$. From the equation (3), we can calculate that for n-type doping ZnO, the work function rang from 4.35eV to 5eV. In order to take shape good ohmic contact for n-type ZnO, the work function of metal (W_m) should be lower than W_s; that is to say, the W_m should be less than 4.35eV. From the table of the work function of the metal, we can find the priority is In, secondly, Ti and Al, etc. Base on this theory, researchers devoted to trying many kinds of material or ways hoping to find out an excellent ohmic contact with ZnO, such as Pt-Ga[5], Ti/Au[9], nonalloyed Al[10], Ti/Al[6], Al / Pt[11] and Ta/Au[7]、In/Ag、In/Al, and so on.

3 The processing of ZnO ohmic contact
3.1 The choice for ZnO substrate and the growth method

There are quite a few methods for ZnO growth. Based on the different application, there are different request for the ZnO film, such as Crystal Orientation, surface roughness, conductivity, Piezoelectricity, optical performance, all of which are conducted by the technological parameter. At present, there are quite a few methods to grow ZnO film: RF Magnetron Sputtering [9], pulse laser deposition [12], Molecular-beam epitaxy [13], Metal Organic Chemical Vapor Deposition [7], spray, Sol-Gel, Atomic Layer epitaxy, etc. Table 1 shows the methods and the substrate of fabrication of

ZnO in the labs. From this table, it is clearly that in order to obtain high oriented, smooth ZnO film, the researchers prefer to take sapphire for ZnO's substrate. Despite a relatively large lattice mismatch, C-plane sapphire is the most commonly used substrate for ZnO epitaxial growth because ZnO can grow along the (0001) direction. The R-plane sapphire was chosen as a substrate instead of the commonly used C-plane sapphire as it offers two advantages: (1) the c-axis of ZnO lies in the surface plane, with the lattice mismatch parallel to the c-axis of ZnO being lower than the mismatch perpendicular to the c-axis (1.53% versus 18.3%), resulting in overall less mismatch;

Table.1. The substrate and the growth method that the researchers used in the labs

Material of substrate	Growth method	Ref
C-plane sapphire	RF magnetron sputtering	[9]
C-plane sapphire	RF magnetron sputtering	[14]
C-plane sapphire	RF magnetron sputtering	[11]
C-plane sapphire	PLD	[12]
C-plane sapphire	PLD	[15]
C-plane sapphire	MOCVD	[6]
R-plane sapphire	MOCVD	[10]
R-plane sapphire	MOCVD	[16]
R-plane sapphire	MOCVD	[7]
GaN/C-plane sapphire	MBE	[13]
Si(111)	PLD	[17]

and (2) in-plane anisotropy in the ZnO/R-Al2O3 system can be used to make novel optical, electrical, and piezoelectric devices. GaN has the similar lattice constant with the ZnO (GaN: 0.319nm; ZnO: 0.32496nm). Although Si's lattice is quite different from ZnO, Si is still the most common material for ZnO growth, because Si is accommodating with the mature integrated circuit craft. Besides substrate, the growth methods are also the important factors for the growth of ZnO film. Researchers are always looking for balance point between the economic condition and the quality of ZnO film. From table 1, we can see the RF magnetron sputtering and MOCVD are the most popular technique. The advantage of the RF magnetron sputtering is the controllable and repeatable film thickness, easy to get high oriented (0002) and high surface roughness. The merit of MOCVD is that the experimenters can easily and accurately dominate the thickness, constituent and the interface of the film. In addition, it can produce large area of uniform film, suitable for large quantities of industrial production. The PLD and MBE are also the candidates for depositing ZnO. However, PLD has a shortcoming in doping control and smoothing film surface, meanwhile MBE is too expensive to fabricate in large scale.

In general, the RF magnetron sputtering and MOCVD are the most favorite ways for researchers.

3.2 The preprocessing and post processing for ZnO ohmic contact

Ohmic-contact metallization should be one of the main goals in improving device performance which plays an important role in device. Although a low resistance Ohmic contact on wide-band-gap semiconductors can be obtained by thermal annealing, surface roughness and structural degradation of the interface can be induced during the thermal annealing process, resulting in poor device performance and reliability. Therefore, it is necessary to search for novel approaches for ZnO ohmic contact's pre-treatment and post-treatment. Table 2 display labs use

Table.2. The different preprocessing and post-processing taken in labs

Deposition Method	preprocessing	post-processing	Specific contact resistance (Ωcm^2)	Ref
hydrothermal	KrF excimer laser irradiation	-------------	7×10^{-1}	[18]
Rf magnetron sputtering	-------------	N_2 300˙ RTA 1min	2×10^{-4}	[9]
Rf magnetron sputtering	-------------	N_2 300˙ RTA 1min	2×10^{-4}	[19]
Rf magnetron sputtering	BOE solution treatment	-------------	10^{-1}-10^{-5}	[11]
Rf magnetron sputtering	ICP hydrogen and argon plasma		4.3×10^{-5} (H_2) 5×10^{-6} (Ar)	[14]
Rf magnetron sputtering	-------------	Pt capping layer	1.2×10^{-5}	[20]
Rf magnetron sputtering	BOE solution treatment	Pt capping layer N_2 300˙ RTA	2×10^{-6}	[21]
Sputtering	-------------	N_2 500˙ RTA	5×10^{-3}	[22]
PLD	Ga FIB	-------------	3×10^{-4}	[5]
PLD	-------------	N_2 200˙ RTA 1min	8×10^{-7}	[12]
MOCVD	-------------	300˙ RTA	9×10^{-7}	[6]
MOCVD	-------------	Pt capping layer	2.5×10^{-5}	[10]
MOCVD	-------------	N_2 300˙ RTA 30s	5.4×10^{-6}	[7]
Bulk ZnO	-------------	-------------	3×10^{-4}	[23]

different means to improve the ohmic contact. As can be seen from table 2, there are quite a few ways to improve ZnO ohmic contact varying from specific contact resistance. Generally speaking, it is a prerequisite using organic solution for Ultrasonic cleaning. In this step, acetone, ethanol or methanol are used to remove surface contamination. Besides ultrasonic cleaning, other means can be divided into two parts: one is plasma treatment, called dry processing. For example, Lee[14] introduces Inductively Coupled Plasma (ICP) to clean the ZnO surface. The Ti/Au ohmic contact resistance has reduced one order of magnitude after the nitrogen plasma treated. It is well known that energetic ion bombardment can induce preferential sputtering of the relatively light atom from the surfaces of II–VI semiconductors as a result of physical momentum transfer between ions in the plasma and atoms on the material surface. The nonstoichiometric surface, induced by ion bombardment, can increase the surface carrier concentration, thus decreasing the specific contact resistivity

due to the increase in the concentration of shallow donors which are localized at the metal/semiconductor interface. For hydrogen

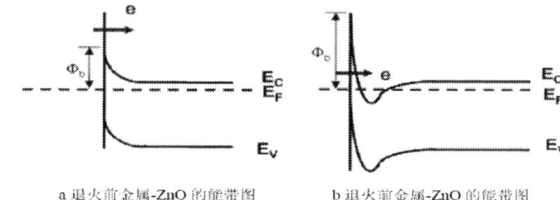

a 退火前金属-ZnO 的能带图   b 退火前金属-ZnO 的能带图

Fig.2. The band bending was observed after the annealing and the quantum well was formed.

plasma treatment, the mechanism is different from nitrogen because of small momentum of hydrogen. O cannot be bombarded out of the ZnO by hydrogen ion. So there must be some other reason that the resistance can be reduced by two orders magnitude after the hydrogen ion plasma treated. Scientists use photoluminescence spectrum, detecting that the deep level of ZnO was decreased after the hydrogen bombarded ZnO surface. Even some researchers have proposed that the hydrogen ion itself act as carrier in ZnO, increasing the carrier concentration, reducing the contact resistance [24]. Another typical cleaning technique is represented by BOE solution, called wet processing. By this method, any layers of native contamination can be removed. In Kim's [11] work, for example, after treat the surface ZnO with BOE, the contact resistance drop to 10^{-5}. After the metal deposited on ZnO, rapid annealing is adopted for some experiment. Based on the different condition, the temperature for annealing range from 200˙ to 500˙ in different labs. However, something in common is that after annealing, diffusion is detected and the new phases are developed. This process brings about the increasing of the ratio of Zn/O. The chemical reactions take place between O in ZnO and contact metal, specially the active metals, such as Ti, Al. They can form new phases. The enthalpy of TiO is much lower than ZnO, so Ti can easily react with O in ZnO, even before the annealing. With the help of this reaction, the vacancies of O increase which act as donor in ZnO, so the carrier concentration augment. At the same time, the work function raise also because of the formation of TiO and the schottky barrier height increase too. However, the increasing carrier concentration is dominant. As we can see from the figure 2, the filed emission mechanism is the main factor for electric transportation. Although the new phase elevates the barrier, the carrier concentrations in the well increased that will make the electric tunnel the barrier much easier and lower the contact resistance. However, something must be noticed that the temperature for annealing have to be limited within a range. Ti/Au annealing temperature over 300 Celsius will lead to performance degradation, because the high temperature will make atoms of metal diffuse faster. This will lead to roughness in the interface and decrease the contact area. In order to prevent diffusion as far as possible, Au or Pt are often used above the active metal protecting them from being oxidized. This measure is proved to be an efficient way to lower the contact resistance.

4 Conclusions

In this paper, the ohmic contact for n-ZnO is discussed. Generally, there are two main ways for ZnO to improve its ohmic contact: One is surface cleaning. Through this method, the impurities are removed and the barrier is reduced. Electrics can pass through the barrier by filed emission, another is doping. By this way, the carrier concentrations near the surface increase which will make the probability of tunneling increase too. Although the active metals have low work function, ZnO as a semiconductor contain O element by itself. O can easily react with active metals, producing oxides. Even at the room temperature, the diffusion is observed. That's why ZnO is unique from other semiconductor material. By controlling the reaction appropriately, there will be a thin oxide film on the surface. The formation of O vacancy plays as a donor in ZnO and increase the carrier concentration. This is the major step for researchers to lower the contact resistance at present. However, thermal treatment also has a negative impact. A novel means needs to be found in the future.

References

1. Ü.Özgür, Ya. I. Alivov, C. Liu, et al, "A comprehensive review of ZnO materials and devices," *Journal of Applied Physics*, Vol.95, No.041301 (2005), pp.1-103.
2. K.Ip, G.T. Thaler, Hyucksoo Yang, et al. "Contacts to ZnO," *Journal of Crystal Growth*, Vol.287, (2006), pp.149-156.
3. Yang Xiaotian, Liu Boyang, Ma Yan, et al. A study on the manufacture of ZnO-based ultra-violet detector. Chinese Journal of Luminescence. Vol.25, No.2 (2004), pp.156-158.
4. Liu Enke, Zhu Bingsheng, Luo Jinsheng, et al. Semiconductor Physics, National Defence Industry Press. (1994), pp.178-194.
5. A. Inumpudi, A.A. Iliadis, S. Krishnamoorthy, et al, "Pt-Ga Ohmic contacts to n-ZnO using focused ion beams," *Solid-State Electronics*, Vol.46 (2002),pp.1665-1668.
6. Soo Young Kim, Ho Won Jang, Jong Kyu Kim, et al, "Low-Resistance Ti/Al Ohmic Contact on Undoped ZnO", *Journal of Electronic Materials*, Vol.31,No.8 (2002),pp.868-871.
7. H. Sheng, N.W. Emanetoglu, S.Muthukumar, et al," Ta/Au Ohmic Contacts to n-Type ZnO," *Journal of Electronic Materials*, Vol.32, No.9 (2003) ,pp.935-938.
8. Wu Dingfen, Yan Benda. The principle, measurement and process of Ohm contact at the interface between metal and semiconductor. Shanghai Jiaotong University Press. (1989), pp. 4-5.
9. Han-Ki Kim, Sang-Heon Han, and Tae-Yeon Seong. "Low-resistance Ti/Au ohmic contacts to Al-doped ZnO layers，" *Applied Physics Letters*, Vol.77, No.11 (2000), pp.1647-1649.
10. H. Sheng, N.W. Emanetoglu, S.Muthukumar, et al. "Nonalloyed Al Ohmic Contacts to MgxZn1-xO", Journal of Electronic Materials, Vol.31, No.7 (2002),pp.811-814.
11. Han-Ki Kim, Kyoung-Kook Kim, Seong-Ju Park, et al. "Formation of low resistance nonalloyed Al/Pt ohmic contacts on n-type ZnO epitaxial layer," *Journal Of Applied Physics*, Vol.94,No.6 (2003),pp.4225-4227.
12. K.Ip, Y. W. Heo, K. H. Baik, et al. "Carrier concentration dependence of Ti/Al/Pt/Au contact resistance on n-type ZnO," *Applied Physics Letters*, Vol.84, No.4 (2004), pp.544-546.
13. Jau-Jiun Chen, Soohwan Jang, F. Ren et, al. "Thermal stability of Ti/Al/Pt/Au and Ti/Au Ohmic contacts on n-type ZnCdO," *Applied Surface Science*, Vol.253 (2006),pp.746-752.
14. Ji-Myon Lee, Kyoung-Kook Kim, Seong-Ju Park et al . "Low-resistance and nonalloyed ohmic contacts to plasma treated ZnO," *Applied Physics Letters*, Vol. 78, No. 24(2001), pp. 3842-3844.
15. A.A. Iliadis, R.D. Vispute, T. Venkatesan et al. "Ohmic metallization technology for wide band-gap semiconductors," *Thin Solid Films*, Vol. 420 (2002), pp. 478-486.
16. Y. LIU, C.R. GORLA, S. LIANG et al. "Ultraviolet Detectors Based on Epitaxial ZnO Films Grown by MOCVD,"*Journal of Electronic Materials*, Vol. 29, No. 1 (2000), pp.69-74.
17. Ye Zhizhen, Zhang Yinzhu, Chen Hanhong, et al. A study on the preparation and characteristics of ZnO photoconduction ultra-violet detector. Chinese Journal of Electronics. Vol. 31, No. 11 (2003), pp.691605-1607.
18. Toshimitsu Akane, Koji Sugioka, Katsumi Midorikawa. "Nonalloy Ohmic contact fabrication in a hydrothermally grown n-ZnO (0001) substrate by KrF excimer laser irradiation," *J. Vac. Sci. Technol. B*, Vol. 18, No. 3 (2000), pp.1406-1408.
19. Han-Ki Kim,Sang-Heon Han, Tae-Yeon Seong et al. "Electrical and Structural Properties of Ti/Au Ohmic Contacts to n-ZnO,"*Journal of the Electrochemical Society*, Vol. 148, No. 3 (2001), pp.G114-G117.
20. Han-Ki Kim, Ji-Myon Lee. "Low resistance nonalloyed Al-based ohmic contacts on n-ZnO: Al," *Superlattices and Microstructures*, Vol. 42(2007), pp.255-258.
21. Han-Ki Kim, I. Adesida, K.K. Kim et al. "Study of the Electrical and Structural Characteristics of Al/Pt Ohmic Contacts on n-Type ZnO Epitaxial Layer," *Journal of The Electrochemical Society*, Vol. 151, No. 4(2004), pp.G223-G226.
22. Sun Tengda, Xie Jiachun, Liang jin, et al. The preparation of ZnO Ohm electrode and the ultra-violet photoelectricity characteristics of ZnO/p-Si heterogeneity joint. Journal of University Of Science and Technology of China, Vol. 36, No. 43(2006), pp.328-332.
23. Hyuck Soo Yang, D. P. Norton, S. J. Pearton et al. "Ti/Au n-type Ohmic contacts to bulk ZnO substrates," *Applied Physics Letters*, Vol. 87, (2005), pp.212106.
24. C. G. Van de Walle, "Hydrogen as a Cause of Doping in Zinc Oxide," *Phys. Rev. Lett*, Vol. 85(2000), pp.1012.

Low-cost High-efficiency 4 Channel Pluggable Parallel Optical Transceiver Using Optoelectronic MCM Packaging Technologies

Baoxia Li, Lixi Wan, Chengyue Yang, Wei Gao, Yao Lv, Zhihua Li , and Xu Zhang
Institute of Microelectronics, Chinese Academy of Sciences
3#, BEITUCHENG West Road, CHAOYANG District, Beijing, 100029, China
libaoxia@ime.ac.cn, 86-010-82995593

Abstract

A compact 4×2-channel parallel optical MCM transceiver with data rates up to 3.125 Gb/s per channel was studied for Very Short Reach (VSR) interconnection. The transceiver was based on 1×4 VCSEL and PD arrays of 850nm wavelength, and a 12-fiber-ribbon as the transmission medium. Greatly relaxed alignment tolerance and high coupling efficiency between optoelectronic (OE) device arrays and fiber arrays were achieved. The eye-diagram at 2.5Gb/s was measured under 2^{31}-1 pseudorandom bit stream (PRBS).

Introduction

Optical communication technology based on fibers has gotten a howling success in long-distance telecom network, and has changed people's life completely all over the word. With the development of higher bandwidth demands of peripheral equipments and high-capacity data center, fibers have gradually been making its way into shorter-distance and replacing metal wires to carry data between racks, boxes and boards. Photons can carry higher data densities to longer distances with less power consumption. Furthermore, photons are electromagnetic interference free and no cross-talking. For ultra-short-reach, such as chip-to-chip on-board, the copper based electrical interconnect technology still remains the cheapest solution up to 10 Gb/s rate right now. Considering the ever-growing clock frequency in processors and data exchanges between processor and memory or between multiprocessors, copper wires will be facing their physical limitation sooner or later, optical interconnect technology is known as the most promising approach to meet the future bandwidth requirements.

Vertical cavity surface emitting lasers (VCSELs) and are conventional optical source in parallel optical interconnects, and they have many advantages such as low threshold current, good beam quality and low-cost 1-D or 2-D monolithic arrays. For edge-emitter lasers only 1-D array are possible. Planar PIN photodetectors (PDs) are widely used light receiver. In 2007, IBM reported its 16-channel optical solution at a data rate of 10Gb/s per channel for chip-to-chip on-board application [1]. High-speed CMOS IC with flip-chip attached 4×4 VCSEL and PD arrays was mounted on an organic carrier. The operating wavelength was 985nm, and the optical coupling structure was based on 2 lens arrays and 45°reflection mirrors. This year, IBM improved its optical solution to 24-channel with 850nm wavelength, and the

The work has been supported by National Natural Science Foundation of China, No. 90607006 and Hi-tech Research and Development Program of China (863 Program) No. 2006AA01Z236 and No.2007AA01Z2a6.

optical system was similar to the former [2]. Comparing with optical multimode polymer waveguide in/on PCB board limited only for ultra short straight line, multimode fibers (MMFs) provide a minimal loss in signal propagation and considerable distance-bandwidth product. Therefore, optical interconnect module based on MMF still is an efficient approach. Many researches were focus on methods to make the OE-to-fiber coupling assembly more cost-effective, particularly manufactureable[3-8].

In this paper, we report a parallel optical transceiver module with 4 transmitter and 4 receiver channels specifically for VSR links at the board- and backplane level, and each channel has a potential to operate at a data rate of 3.125 Gb/s. A Silicon Optical Bench (SOB) was fabricated to provide a precision alignment aid and a basic platform for four 4-channel array components: a VCSELs chip, a PDs chip, a high-speed laser diode driver and a high-speed transimpedance amplifier plus limiting amplifier (TIA/LA). A direct OE-to-fiber coupling structure was applied to eliminate discrete optical components such as microlens and turning mirrors. A MPO optical connector and a high-speed 133-pin pluggable electrical connector were used in the transceiver for optical and electrical plugable interfaces respectively.

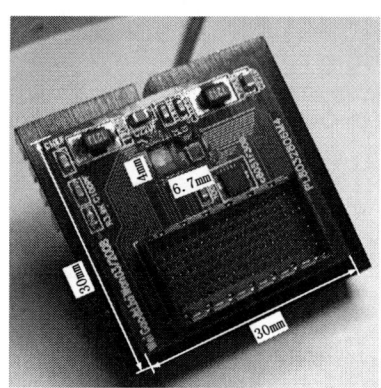

Fig.1. Photograph of an assembled optical interconnect transceiver

Structure

As shown in Fig. 1, the optical transceiver module consists of a SOB of 4 mm in width and 6.7 mm in length, which is the most critical part of the transceiver module and provides mechanical support and electrical connections for chips attached on it. Silicon is a unique subcarrier material due to its excellent thermal conductivity, high mechanical strength, and suitable coefficient of thermal expansion (CTE). Silicon wafers are extremely flat, smooth and dimensionally scalable.

978-1-4244-2739-0/08/$25.00 ©2008 IEEE

Perhaps the most advantage of all is its mature process technology, which offers great flexibility of assembly structure. The SOB in this module adopted thin film MCM-D processing using PI coating as dielectric layer. High-resistivity p-type (111) Si substrates with oxide thickness about 1μm were employed. The VCSEL and PD arrays were flip-chip bonded on the top surface of SOB using microsolder bumps with 90μm diameter and 25μm height, which were fabricated by electroplating. Reflow process was conducted to self-align two OE arrays to right position, which benefited from controlled microsolder arrays with precise vertical and horizontal geometry. The two high-speed ICs were adhesive onto the SOB using heat conduction glue. Wire-bonding technology was used for electric connects between high-speed ICs and SOB.

A 4-layer organic circuit board is 3cm×3cm and an opening cut on it to house the SOB. A Microcontroller chip, a 133-pin electrical connector and passive components for power supply filtering were mounted on the surface of organic circuit board using the conventional SMT package process. 100-Ω differential transmission lines were applied on board for routing high-speed signals from inputs/outputs of high-speed ICs to the bond pads of electrical connector. A 12-channel 62.5μm multimode-fiber (MMF) ribbon was align and fixed on the back side of SOB.

Optical Coupling

A major problem of the optical interconnects for ultra-short-reach is cost. The alignment of optic-to-fiber or optic-to-waveguide takes a large part of the cost. The alignment contains 6-degree of freedom motion, namely, lateral (x-axis and y-axis), longitude (z-axis) and angular (rotation around x/y/z-axis). Optical coupling alignment is a much more fragile matter and less manufacturability. Simplifying optical coupling processing is one of the most efficient approaches to cut cost of an optical module. We developed a simple passive optical coupling approach in the MCM structure without any additional optical devices, such as 45°reflection mirrors or microlenses.

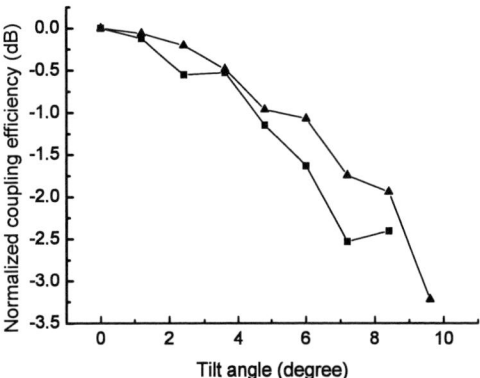

Fig.3. Effects of angular misalignment on coupling

Since assembly tolerance is crucial to low cost mass production in the optical module, it is essential to identify and evaluate the possible misalignment effects on the optical system. We studied lateral, longitude and angular alignment tolerance in our VCSEL-to-fiber coupling structure. The dependence of the coupling efficiency on misalignment was determined by varying the relative position of a fiber end to a VCSEL chip. VCSEL chips were mounted on a silicon carrier with fixed position, and MMFs with 62.5μm-core were placed on 5-degree Newport Precision Stage by which fiber ends were adjusted slightly. The output light power was detected at another end of fiber, and the operating current of VCSEL remained constant at a value of 6mA during the test. Fig.2 shows the detail profile of the normalized coupling

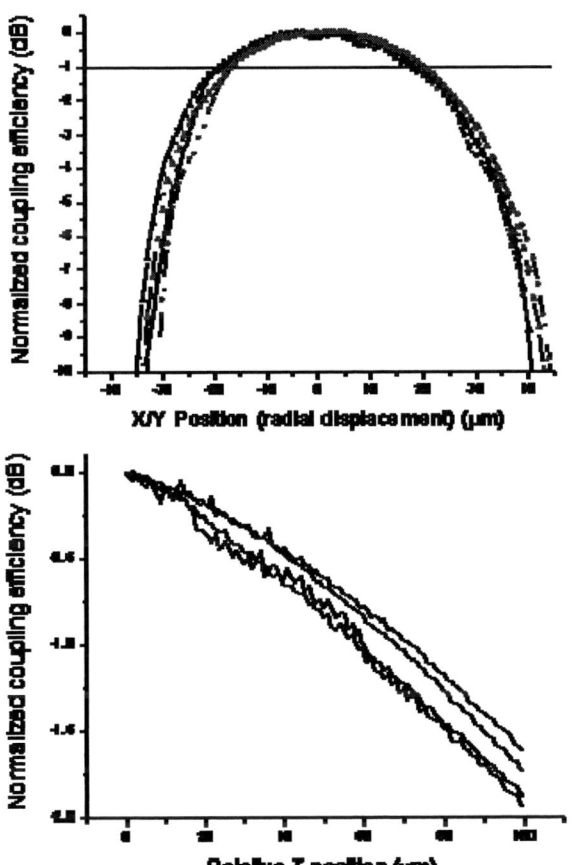

Fig.2. Measured alignment tolerances in the lateral (top) and longitude (bottom) directions

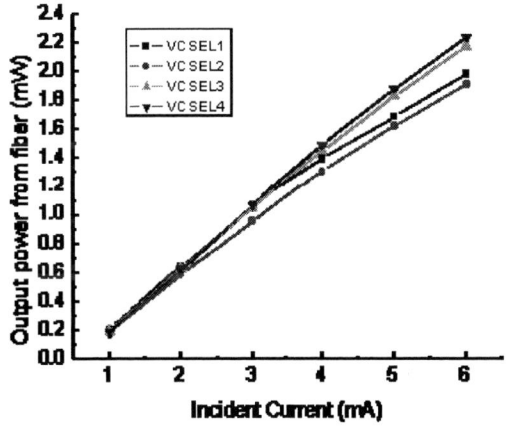

Fig.4. MMF output characteristics for 1×4 VCSEL array

efficiency dependence in the lateral and axial directions. The measured lateral alignment tolerances of horizontal x-axis (solid line) direction are consistent with that of vertical y-axis (dot line) direction, and the alignment displacement shows about ±14µm at -0.5dB and ±18µm at -1dB coupling efficiency. The dependence of the light coupling efficiency in the axial direction is not critical providing a much more loose alignment offset of nearly 70µm at -1dB coupling efficiency. Fig.3 shows the influences of the angle between VCSEL emitting direction and fiber axis. The angular alignment tolerance for 0.5dB excess loss is more than 2°.

In this optical transceiver module, a 1×4 VCSEL array and a 1×4 PD arrays was placed in line at SOB surface with 500µm interval, which coupled with the 8 among 12 fibers of a 12-fiber-ribbon. Light coupling were verified, and the measurement was carried out using DC probes at room temperature. Fig.4 shows the dependence of output powers of

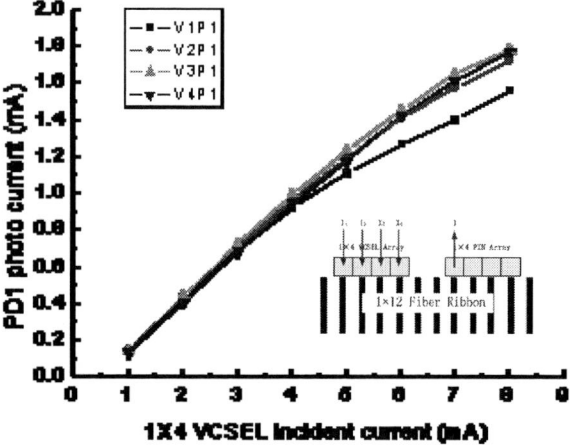

Fig.5. Photocurrents of PD1 at different incident currents of VCSEL1-4

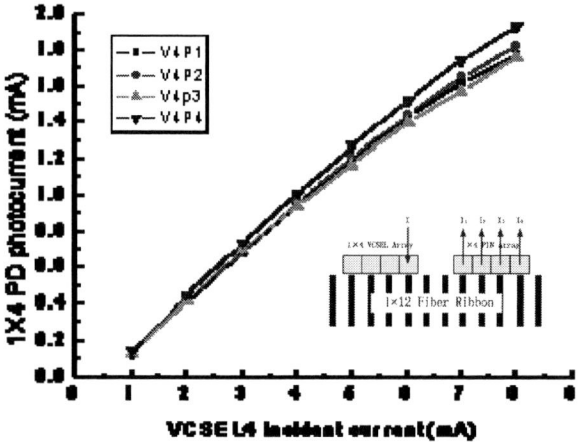

Fig.6. Photocurrents of PD1-4 at different incident currents of VCSEL4

from MMF connecters on VCSELs input currents. Comparing with the datasheet of VCSEL chips, in which the typical optical output power of VCSEL chips is about 2.2mW at 6mA operating current, we can estimate the lowest coupling efficiency is more than 85%. The photocurrents of PD array as a function of input light at fiber end were measured as well.

The results showed greater uniform channel-by-channel and the slope of current-light (I-L) curves was about 0.63, and the coupling loss between the 62.5-um MMF and the PD was less than 10%.

To test the total DC loss in one transmitter-to-receiver link between electric signals of VCSEL input and that of PD output, four VCSEL diodes were emitted channel by channel, and optical fiber output of the operating VCSEL was connected to same PD diode. Photocurrents of PD were observed at different VCSEL source and different VCSEL input currents, and the data was shown in Fig. 5. Similarly, in the next measurement, the light from certain VCSEL was guided to four PD diodes in turn, and Fig. 6 presents the relationship of photocurrents of different PD diodes on bias currents of VCSEL4. From Fig. 5 and Fig. 6, we can see the uniformity of coupling efficiency among four PD diodes is better than that among four VCSEL lasers. Photocurrents of PD increased linearly with the input currents of VCSEL up to 6mA, and the slop is about 0.25. So the total DC loss $\alpha_{\text{VCSEL-PD}}$ from VCSEL input to PD output within a transceiver link is about 6dB, which is associated with VCSEL slope efficiency η_{VCSEL}, optical coupling efficiency between VCSEL and MMF $\eta_{\text{VCSEL-to-MMF}}$, insert loss of optical connector $\alpha_{\text{connector}}$, optical coupling efficiency from MMF to PD diode $\eta_{\text{MMF-to-VCSEL}}$, and PD responsivity η_{PD}. namely,

$$\alpha_{\text{VCSEL-PD}}$$
$$= 1 - \eta_{\text{VCSEL}} \bullet \eta_{\text{VCSEL-to-MMF}}$$
$$\bullet (1 - \alpha_{\text{Connector}}) \bullet \eta_{\text{MMF-to-VCSEL}} \bullet \eta_{\text{PD}}$$

Fiber loss is very little and is ignored in equation. Maximum value of our VCSEL slope efficiency η_{VCSEL} is 0.5 and typical value is 0.4. For PD, only minimum of η_{PD} can be got from our PD datasheet, which is 0.5. According to the data given by Landmark Optoelectronics Corporation, a supplier of GaAs PD Epiwafer in Taiwan, the typical η_{PD} of similar PIN structure and planar aperture is 0.65 at 850nm. Insert loss of optical connector $\alpha_{\text{connector}}$ is 0.15dB typically. The typical value of $\eta_{\text{VCSEL}} \bullet (1 - \alpha_{\text{Connector}}) \bullet \eta_{\text{PD}}$ is -6dB. Therefore, the total coupling loss in one transceiver link in our module is very little.

High-speed performance

The back-to-back eye-diagram was measured as shown in Fig.7. In the case, a $2^{31}-1$ pseudorandom bit stream (PRBS) generator drove the VCSELs to produce optical signals. Through the optical coupling structure and the multimode fiber and optical coupling structure again, optical signals was converted back to electrical signals at PDs within the same transceiver module. Limited by the speed of generator output, the data rate was 2.5Gb/s in the measurement.

Conclusions

In this paper, we present a transceiver module for parallel optical interconnects application. Not only 4 receiver channels but also 4 transmitter channels exhibit high-efficient and good uniform light-coupling performance with MMF. The ±18μm lateral and 70μm longitude alignment tolerances for 1dB coupling loss in VCSEL-to-MMF optical system provide a reasonable margin for assembly offset and enable automatic alignment in fiber-to-optic coupling process. Eye-diagram at a data rate of 2.5Gb/s for 2^{31}-1 NRZ-PRBS of one transmitter-to-receiver link is shown.

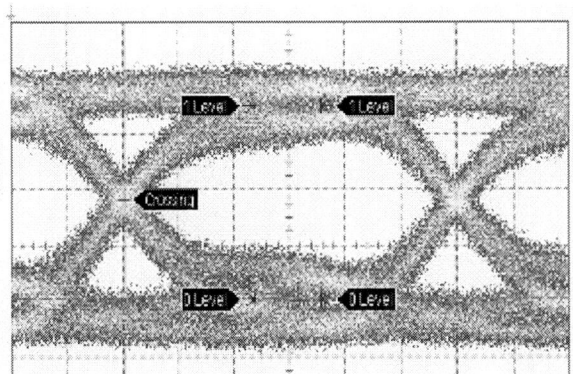

Fig.7. 2.5Gb/s eye-diagram of the links between 1 transmitter channel and 1 receiver channel in a single transceiver

Acknowledgments

The authors would like to thank Zhongchao Fan of the Institute of Semiconductors, Chinese Academy of Sciences, for his help in SOB process.

References

1. Fuad E. D., Clint L. S. et a , "160-Gb/s Bidirectional Parallel Optical Transceiver Module for Board-Level Interconnects Using a Single-Chip CMOS IC", Electric Components and Technology Conference, 2007, pp. 1256-1261.

2. Schow C. L., Doany F. E. et al, "300-Gb/s, 24-Channel Full-Duplex, 850-nm, CMOS-Based Optical Transceivers", OFC/NFOEC, 2008, OMK5.

3. Takaaki I., Atsushi S. et al, "High-density and Low-cost 10Gpbs ×12 ch Optical Modules for High-end Optical Interconnect Application", OFC/NFOEC, 2008, OMK6.

4. Heikkinen V., Alajoki T. *et al*, "Fiber-Optic Transceiver Module for High-Speed Intrasatellite Networks", *J. Lightwave Technol.*, Vol.25, No.5 (2007), pp.1213-1223.

5. Hwang S. H., An J. Y. *et al*, "VCSEL Array Module Using (111) Facet Mirrors of a V-Grooved Silicon Optical Bench and Angled Fibers", *IEEE Photon Technol. Lett.*, Vol.17,No.2 (2005), pp.477-479.

6. Takahara H., Tanaka N. *et al*, "Passively Aligned LD/PD Array Submodules by using Micro-Caoillaries", *IEEE Transactions on advanced packaging*, Vol.23, No.2 (2000), pp. 323-327.

7. Rosinski B., Chi J. W. D. *et al*, "Multichannel Transmission of a Multicore Fiber Coupled with Vertical-Cavity Surface-Emitting Lasers", *J. Lightwave Technol.*, Vol.17, No.5 (1999), pp.807-810.

8. Ouchi T., Imada A. *et al*, "Direct coupling of VCSELs to Plactic Optical Fibers Using Guide Holes Patterned in a Think Photoresist", *IEEE Photon Technol. Lett.*, Vol.14, No.3 (2002), pp263-265.

SiP/SoP Technology and Its Implementation

Lixi Wan

Institute of Microelectronics, Chinese Academy of Sciences
Beijing, China, 100029
E-mail: lixiwan@ime.ac.cn

Abstract

"System-in-Package"(SiP) and "System-on-Package" (SoP) are different but similar in concepts. SiP and SoP definition were found in many open sources. SoP promises much more technologies and functions over SiP, leads to too many and more complicated research areas, and long time to develop, which could lost patience and interest from industry. Module-in-Package(MiP) was proposed as a replacement of SoP for real implementations.

Key word: System-in-Package, System-on-Package, Module-in-Package

Introduction

System-in-Package (SiP) or System-on-Package (SoP) concept has been proposed for over ten years. ITRS defined SiP as "a combination of multiple active electronic components of different functionality, assembled into a single unit, that provides multiple functions associated with a system or sub-system. A SiP may optionally contain passives, MEMS, optical components and other packages and devices."[1] While SoP was defined by Rao Tummala, director of the Package Research Center (PRC), Georgia Institute of Technology, as "the Second Law of Electronics for System Integration. SOP is a highly miniaturized system technology combining computing, communication, consumer, and bio-electronic functions in a single package or module. It accomplishes this miniaturization by package integration of system-level components at microscale in the short term and nanoscale in the long term." [2] Unfortunately, there have not been any SoP products on market even a prototype in the laboratory of the center after more than15 years the director and his colleagues has put all of their efforts. It is unbelievable that the SoP concept stands on system, but it was proposed by a not system expert. In the SiP, some products claimed using SiP technology have been on market for years. Actually, all of the products are either based on 3D stacked same dies such as memory or Multi-Chip-Module (MCM) – similar to the SoC. They are not real "system in package" but a kind of MCMs.

Author, as a former researcher of PRC, tried to analyze the reason why there have not been products on the market through summarization of SoP research areas and challenges since over ten years to develop in an effort in both industry and academia. There are four parts in the paper: comparison between SiP and SoP, research areas of SoP, module-in-

This work was supported by the High-Technology Research and Development Program of China (863), and the National Natural Science Foundation of China. Contract number: 2006AA01Z236, 2007AA01Z200 and 90607006

package, and followed by a conclusion.

Comparison between SiP and SoP

SoP could be distinguished from SiP by the concept

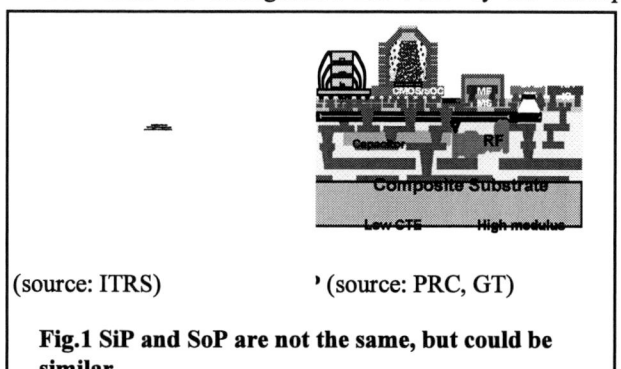

(source: ITRS)   ' (source: PRC, GT)

Fig.1 SiP and SoP are not the same, but could be similar.

driven: SiP concept was driven by extending mono-function on a single chip to multi-function in a package through multi-chips combination, similar to the concept of "System-on-Chip" (SoC); while SoP was driven by an entire system function. It places emphasis on not ICs only in a package, but also passives, especially on film passive component integration in the package, as shown in the Fig. 1. There must be a lot of passives in a system, normally 60% or more over ICs in number and 70% over ICs in room taken, to implement a system function. Therefore, to implement SoP will face more challenges than that in SiP, since SiP is "OPTIONALLY contain passives, …". And SiP did not try to use the embedded technology neither actives nor passives. If a substrate or interposer of a SiP was fabricated with film embedded components, the SiP would be exactly as same as SoP from the concept point.

Research areas of SoP

To implement SoP, there are a lot challenges in six areas: Design, Material, Processing, Assembly, Availability, and Test, as shown in Fig.2.

Design In addition of traditional package design, SOP utilizes a lot a film technologies, such as film capacitor, resistor, inductor, filters and balus or antenna, etc., as embedded devices, to reduce the number of SMD, and save room for more ICs for adding more functions in the estate, or to keep the same function but shrinking a system area. Due to many film devices present in a package with a lot of complicated structures like vias, traces, pads, power/ground networks, etc. the electromagnetic field coupling between them become strong and complicated. Signal integrity and power integrity have to be paid attention carefully, especially in a mixed-signal or high speed/frequency SoP. Similarly to SoC, Design For Test (DFT) and Design For Manufacture (DFM) are needed. Another challenge is co-design of IC-package-PCB. Placement of high density actives in a SoP challenges a conventional thermal management.

Material Many new materials for embedded passives are needed in SoP technology. The capacitance density of the films for embedded capacitors on the market is not high enough for most decoupling applications. While the embedded inductors need high permeability film, but there is no such material available on the market as well. Both the high capacitance density and permeability materials should be

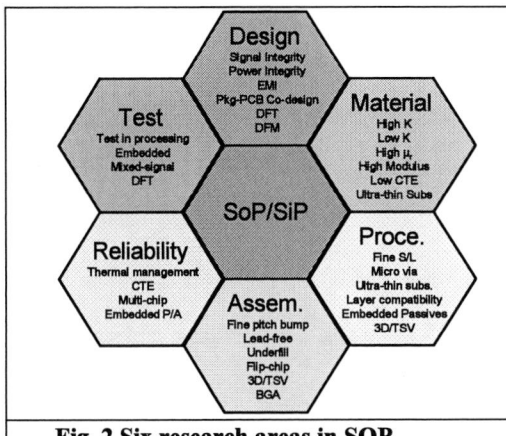

Fig. 2 Six research areas in SOP

low loss. Our research showed the loss of the thin films is imperative even for the embedded decoupling capacitors. If antenna would be expected to be integrated in a SoP by embedded technology in a small area a material with equal value in both permittivity and permeability, with very low loss, would be expected. Due to ultra-thin substrate – say, less than 50 um in total substrate thickness - can provide ultra-low parasitic inductance and resistance in the signal via which is critical for the high speed and frequency systems. But ultra-thin substrate will face problems like warpage, low modulus, CTE mismatching, etc. Many new materials are required also for the emerging package requirements: Thermal conductivity for dielectrics and materials interfaces, for instance.

Molding compounds compatible with copper and other new materials; Resistance to electromigration as temperature and current density continue to rise; Dielectrics with improved fracture toughness and interfacial adhesion; Green materials that meet requlatory, cost and reliability, etc.

Processing Substrate with various embedded passives must be a challenge in processing. A typical example is the high permittivity material for embedded capacitors. For different application, the materials could be either organic or non-organic material, and processing could be either high temperature or low temperature. In addition, it could be very thin, as thin as down to a few micrometers or even less. Fine line/space is always required in substrate industry, but paid more attention in SOP technology, just because of higher density integration than conventional substrate. If active ICs are expected to be embedded into a substrate for some particular reasons, the substrate processing could be changed a lot in procedure as substrate processing-assembly-substrate

processing. Big improvements are required in smaller blind via size and through hole, ultra-thin substrate without core for interposer for 3-D packages, precise impedance controlled transmission lines, precise feature for some components like balus, antenna, filter, etc.

Assembly We suppose to be able to handle a fine pitch BGA die with tens thousands I/Os in near future SOP assembly process. Meanwhile, stacked die, or 3-D assembly have been facing challenges in alignment, interconnect, interface compatibility, different size, known-good-die (KGD), though-silicon-via (TSV), etc. MEMS, bio-chips, digital/RF/optical chips etc. may be placed on a substrate. Wire-bound and Flip-chip could be utilized in the same SOP. Underfiller must work for the fine pitch of solder balls and high density BGA. The list could be longer than above, but people have solved a little but less than new problems emerging leading to even longer list.

Reliability It is even a severer issue in the SoP in which it contains high density structures with multi-material, multi-interface, multi-chip and possible 3-D stacked chips. The small features/structures, multi-interface and multi-materials may cause mismatching problems by CTEs between materials. Heat distribution with multi-chips present in such a small room could be complicated, and some spots may be too hot to give rise to a fail in interconnects. In a 3-D structure, if the heat could not move out of the structure the heat could make the system crash before it damaged from overheat. But it is difficult to design an efficient heat sinker and disperser in such small room. In addition to the heat, humidity, vibration, and chemical and bio environment may be failure factors to the SoP. They may be more harmful to the SoP if it has some sensors integrated.

Test We know the test in ICs can utilize JTAG, but there is not the same approaches or tools in SoP. Test strategy in PCBs is hard to be used in substrates because of so density and small room. Meanwhile, many embedded components in substrate processing may needed to be confirmed and possible to fix up to right value, so that a fast and precise test is necessary. Test methodology for SoP must be created, and new testers may need to be developed.

We just listed the challenges which we thought SoP would face up. They are definitely not all of them. Since SoP concept promised too much in functions and integration, there are too many barriers and difficulties in the development. In other words, the "system" defined in the SoP is too big and complicated to realize. We doubt PRC's system of SoP will happen in the future. So that, we preferred to refer SoP as a dream instead of a real product.

Module-in-Package

Although SoP is too far away from implementation in near future, it still works for us in our recent projects of sub-system integration by embedded passives technology which is really benefit to improve performance through reducing parasitic parameters, and saving room and SMD components. We replaced the "system" of SoP with "module", i.e. Module-in-Package (MiP). It is easier to implement than PRC's SoP, and make sense to our projects. We take the embedded passive, multi-chip and high density integration concepts from SoP. The MiP concept makes sense at: **Reality** It usually contains some of passives, not too many to be embedded; **Low cost in development** Since the technologies limit, we can utilize the technologies available, at least we have not to develop any technology for MiP; **Short time to market** There are two factors for the point: short develop time and origin module may has existed on the market which may save marketing time. **Eliminate confusion on term of "system" in SoP** Term "System" in SoP always rises a problem in discussions of the technology. The "system" definition could change the technology from reality to impossibility. In MiP, it can implement at most cases. As an example, we developed an optical interconnect module which has a low-pass filter for power supply. It contains 8 SMD components as shown in Fig.3. Utilized an embedded capacitor which is a planar version and took two layers in the substrate, the filter network become very simple – one embedded capacitor and a 100 uF SMD, shown in Fig.4. The feature of the new filter network was better than that of the origin one in high frequency band up to multi-GHz, and as good as that in low frequency band, as shown in Fig. 4. Therefore, an embedded filter was able to replace the SMD one in our module. MiP is better in practice than SoP, able to guide us to work more efficiently, and maybe lead to SoP finally. But if we stake in SoP, industry, and finally costumers would lose their interest and patient, and result in failure in technology improvement in miniaturization.

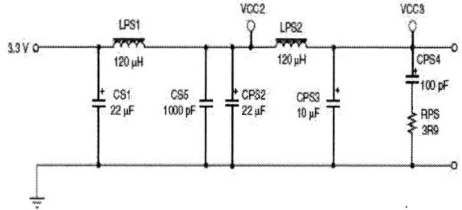

Fig.3 A low-pass filter for a power supply in an optical interconnect module

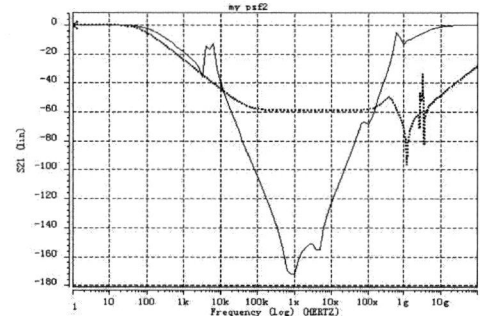

Fig.4 A simulation result. Solid curve: origin low-pass filter, and dash curve: new filter.

Conclusion

SiP concept is just as SoC, it is to extend the chip function or adding more functions in package level. A typical SiP is 3-D stacked dies with either TSV or interposer technology. SoP promises too much functions and too complicated technologies. Six main research areas were summarized in the paper. To integrate them in SoP as PRC definition seems impossible to realize in the future. So, we prefer to refer it as a dream. With author's practice in China, SoP concept does not work for any case. In author's opinion, people may not make a real "system on package". But the negative comment on SoP does not block us to practice SoP concept. SiP or SoP, if it means 3D stacked dies or sub-system, could be implemented in module level. "Module-in-Package" (MiP) could replace SoP for reality. An optical MiP, as an example, was presented in the paper.

Reference

[1] ITRS Executive Summary, 2007 Edition

[2] Home page of the Package Research Center, Georgia Tech. www.prc.gatech.edu

Packaging of Polymer Based Microfluidic Systems Using Low Frequency Induction Heating (LFIH)

Benedikt J. Knauf, D. Patrick Webb, Changqing Liu & Paul P. Conway
Wolfson School of Mechanical & Manufacturing Engineering
Loughborough University, Loughborough, LE11 3TU, UK
B.J.Knauf@lboro.ac.uk

Abstract

Microfluidic systems are being used in more and more areas and the demand for such systems is growing every day. Hence, a cheap and rapid method for sealing these microfluidic platforms which can be used for mass manufacture is needed.

In this paper low frequency induction heating (LFIH) is presented as technique for the packaging of polymer based microfluidic systems. Thin metal layers serving as susceptors are introduced between a stack of polymer slides and heated inductively. The generated heat melts the surrounding polymer and creates a bond. Preliminary work reported here has demonstrated such bonds are able to withstand a pressure of up to 590 kPa, that both ferro- and paramagnetic susceptors are suitable for the bonding process, and that even small metal features can be rapidly heated to a temperature of 200 °C.

Introduction

As micro systems are getting smaller and smaller while demand for their use in harsh environments is growing, the requirements on the packaging are becoming ever more stringent. The packaging must protect the system against dirt, humidity, stresses, etc and, depending on the application, provide electrical, fluidic or optical interconnection and thermal management.

A special task is the packaging of microfluidic systems. Microfluidic systems are networks of channels with width and depth in the micron scale, designed to do continuous flow chemistry with small volumes of fluid. Microfluidic devices are also referred to as lab-on-a-chip (LOC) or, if they are more complex, micro total analysis systems (μTAS). In this application packaging must not only withstand external influences, but also internal pressures. The demand on the world-to-chip interface, the interconnection scaling the fluid delivery network from macro down to micro dimensions and coupling them into the microfluidic system, are also different to those of "normal" MEMS. These interfaces have to be strong and flexible, must provide good sealing, and must connect reservoirs of millilitre or litre volumes to systems with a capacity of micro or even nanolitres.

Most of the existing sealing and interconnecting techniques are cost-intensive and slow compared to the other manufacturing steps of the microfluidic system, so that they form a bottleneck in mass production. Hence, rapid and cheap techniques to seal and interconnect microfluidic chips are needed. A single technique being able to do both sealing and interconnecting, either sequentially or even simultaneously, would be the optimum solution.

In this paper low frequency induction heating (LFIH) is introduced as technique for the sealing and packaging of polymer microfluidic systems. Induction heating is well established in the steel industry for hardening, melting, soldering, welding, and annealing, and increasingly is finding application in other areas like heating fillings in dental medicine, and bottle cap sealing. The advantages of this technique are manifold. It is a very rapid heating process with a small scaling loss, the start-up is very fast, it is energy efficient, and the technique is capable of high production rates.

In this work the results of feasibility trials to prove that LFIH is capable of heating thin susceptors suitable for use in microfluidic modules are reported. Nickel and aluminium susceptors a few microns thick, sputtered or evaporated onto a PMMA substrate and functioning as an intermediate coating between two PMMA layers representative of a microfluidic system, were heated inductively. It was found that the surrounding polymer material melted and formed an interlayer bond.

Different solutions based on induction heating for sealing and interconnecting microfluidic systems which will be realised in subsequent work are described in this paper, as well as methods to test the strength of the bonds so created. A good bond strength should be achieved while avoiding distortion in the microfluidic platform. To be able to handle the process it is important to manage the heat dissipation. Both issues are discussed in this paper. A work programme to investigate how the process can be controlled is underway.

Induction Heating

The discovery of electromagnetic induction by Michael Faraday in 1831 led to the development of electric motors, generators, transformers and wireless communications devices. Heat loss has been always a major factor reducing the efficiency of these systems and researchers have generally sought to minimize it. In the early 20th century the heat loss was utilized for the first time. This utilization was referred to as induction heating and nowadays it is used for many applications, such as melting, hardening, and welding of metals, and many more. The main benefits of induction heating over other heating methods are the selectivity of the heated area, the fast response time, and a good efficiency.

An induction heating unit consists of a power generator, a tank circuit and coil, and a water cooling system. The tank circuit, which is connected to the power generator and the cooling system, provides power and cooling connections for the coil and is equipped with four or more capacitors. These capacitors are connected to the coil to create a resonant circuit. During calibration the power generator checks the resonance frequency of this circuit and generates an alternating current of the same frequency. During operation the work piece is immersed into the magnetic field generated by the coil and absorbs energy from it, leading to rapid heating.

978-1-4244-2739-0/08/$25.00 ©2008 IEEE

The fundamental theory of induction heating is similar to that for transformers. The work coil used in induction heating is equivalent to the first coil in a transformer, while the load functions as a short-wired second coil.

The premise of induction is that a change in magnetic flux induces a current in a circuit or conductor. The change in magnetic flux can be achieved by either altering the magnetic field, or moving the conductor in the magnetic field. The principle is expressed by Faraday's law:

$$E = -N \frac{d\phi}{dt} \qquad (1)$$

Where
E: Induced voltage [V]
N: Number of windings of the coil
Φ: Magnetic flux through a single winding [Vs]
t: Time [s]

The minus sign means that the induced voltage E will cause a current to flow that generates a magnetic field counteracting the change in the inducing field (Lenz's Law). These eddy currents heat the load according to the Joule effect.

Every conductor offers resistance to a flow of a current which causes loss of power. The loss of power is converted to heat energy and is described in Joule's law:

$$P = R \cdot I^2 \qquad (2)$$

Where
P: Power dissipated in the conductor [W]
R: Resistance of the conductor [Ω]
I: Current induced in the conductor [A]

This effect is also referred to as the Joule effect. In most induction heating applications there is a non-uniform distribution of current induced in the conductor. Equation 2, however, gives us an idea of which parameters affect the heating rate. As resistance is determined by the resistivity ρ, materials with high ρ can be heated more efficiently with induction heating.

For ferromagnetic materials in an alternating magnetic field a second heating effect occurs. The magnetic orientation of the domains of the susceptor align with and attempt to follow the rapidly varying field The friction of this movement in the crystal plane heats up the metal and is referred to as hysteresis heating. If a ferromagnetic material is heated to its Curie temperature it becomes paramagnetic and hysteresis heating ceases

Alternating currents tend to flow preferentially on the outside of a conductor. This "skin-effect" is characterized by its penetration depth δ, defined as the thickness of the layer, measured from the outside, in which 87% of the power is developed. [1]

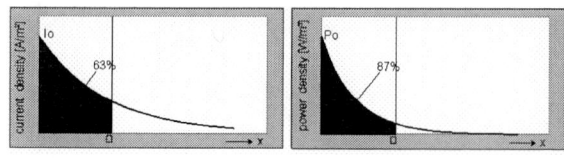

Fig. 1 Penetration depth [1]

For an alternating current of frequency f the penetration depth is given by

$$\delta = \sqrt{\frac{2 \cdot \rho}{\mu \cdot \omega}} \qquad (3), \qquad \text{with } \omega = 2\pi f$$

$$\rightarrow \quad \delta = \sqrt{\frac{\rho}{\pi \cdot \mu \cdot f}} \qquad (4)$$

Where
δ: Penetration depth [m]
ρ: Resistivity [Ω·m]
μ: Permeability [H/m]
f: Frequency [Hz]

Permeability μ is the product of the magnetic field constant μ_0 and the relative permeability μ_r of the conductor

$$\mu = \mu_0 \cdot \mu_r \qquad (5), \qquad \text{with}$$

$$\mu_r = \frac{B}{B_0} \qquad (6)$$

Where
B: Magnetic flux density in the conductor [Vs/m]
B_0: Magnetic flux density in vacuum [Vs/m]

μ_r is less than 1 for diamagnetic materials, and slightly greater than 1 and many times greater than 1 for paramagnetic and ferromagnetic materials, respectively.

The following conditions can be identified for high heating rates in micro work pieces:
- High magnetic field strength should be used.
- Distance between windings of the inductor coil should be as small as possible
- Distance between inductor and load should be as small as possible
- Load should have high resistivity.
- Load should have high permeability.
- High frequency of the applied alternating current is useful.

The following materials were identified as being the most likely to heat easily as thin films by induction:
- Nickel alloys (high permeability, high resistivity)
- Steels (high permeability, high resistivity)
- Titanium alloys (high resistivity)

Considered to be more difficult but still workable materials were:

- Aluminium
- Copper alloys

Microfluidic Systems

In principle microfluidic systems are platforms containing microfluidic channel networks with a lid sealing those channels, as shown in Fig. 2a. Lid and platform have a thickness of a few mm while the channels in the platform can have a width and depth smaller than 100 μm. To be able to access the microfluidic channels holes with a diameter of about 500 μm are drilled into the lid working as ports for micro tubing as shown in Fig. 2b.

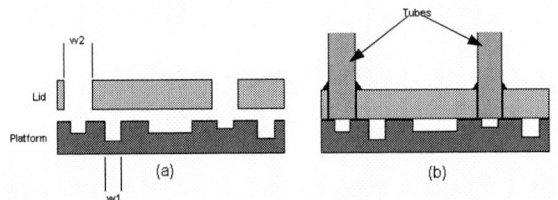

Fig. 2 Principle of the assembly of microfluidic devices with w1 = channel width and w2 = port width. (a) lidding, (b) tube connection

The materials used for microfluidic devices depend on the demands of the applications. Glass is chemically inert but it is expensive and hard to process. Silicon can be used to make active parts like micro valves or pumps but it is expensive. Polymers are capable only of passive channel networks, but due to being cheap in bulk and easy to process, they are suited to mass manufacture. In this work the focus is on polymer based microfluidic systems.

Nowadays microfluidic systems can be found everywhere. Ink-jet printers rely on the technology as well as micro fuel cells, and many applications in biotechnology, such as chips for environmental monitoring [2], for analysing DNA, proteins, cells [3], and so on. In the automotive industry microfluidics find their application in fuel injection, oil quality monitoring, and exhaust gas analysis, and they can also be used for local cooling of electronics. Microfluidics are used in analytical chemistry for chemical monitoring [4], micro reactors and mixers. They are used for drug delivery and diagnostic systems (e.g. blood sugar or alcohol tests) in medical applications and even agriculture benefits from the technology with micro lubrication devices for instance. High pressure systems are used in micro engine and micro rocket devices [5] and for high precision spacecraft thrusters.

As micro system technology becomes more and more sophisticated, more functions can be integrated in microfluidic systems. They are no longer designed for single tasks like mixing liquids, but contain mixers, micro pumps, valves, reaction chambers, optoelectrical analysis systems [6], micro dispensers, heaters, filters, and are able to do a complete analysis or processing (micro separation, mixing, etc.) of a liquid.

Experimental Details

Polymethyl methacrylate (PMMA) is commonly used for manufacturing microfluidic devices. It is benign to most of the fluids used in medical applications, it is soluble only in a few solvents (e.g. acetone), and PMMA is highly transparent so it allows reactions and fluid flow to be monitored optically. During this work Repsol Glass® PMMA slides with a thickness of 2 mm and an edge length of 36 mm were used.

Nickel was chosen as intermediate layer as it is a good susceptor and can be evaporated onto plastic substrates. In the first trial a nickel foil with an area of 25 mm x 25 mm and a thickness of 7.5 μm was placed between two PMMA substrates.

In the second trial a bonded nickel layer was created on PMMA substrates by plating 5 μm thick nickel layers. This configuration is closer to the that of the expected practical implementation of inductive joining of microfluidic components.

A third experiment was prepared to gain an indication of the minimum feasible susceptor thickness. A 50-100 nm thick nickel layer was coated onto a PMMA substrate and was heated using different coils and frequencies.

To test how the area of the susceptor effects the heat dissipation, a small piece of the 7.5 μm thick nickel foil was placed between to PMMA substrates.

Two samples were prepared to be able to carry out some leak tests after bonding. The samples were made of a nickel foil with a punched hole which was aligned with a hole drilled through on of the bonding pair of substrates. After induction bonding a tube was inserted into the hole and sealed to the sample with an epoxy. The sample was then air pressurised via the tube and leakage looked for by immersion of the assembly in water.

Substrates with 5 μm thick aluminium layers were prepared to investigate the relative magnitude of hysteresis and eddy currents in heating the susceptor. Aluminium is diamagnetic so there is no hysteresis heating.

To apply pressure between the two pieces of PMMA to be joined an ordinary plastic G-clamp was used, as shown in Fig. 3.

Fig. 3 Setup during bonding

Apart from one experiment with a Hüttinger IG 2 MHz unit all test were carried out with a Hüttinger AXIO 10-450 with a maximum power of 10 kW and frequencies of up to 800 kHz. A pyrometer was used to measure the temperature on top of the upper substrate (i.e. not at the susceptor directly). The pyrometer output was used in a feedback loop with the induction coil power supply to control the heating profile of the samples. Where not otherwise stated this was constant power until the temperature on top of the upper substrate reached a setpoint.

Results and Discussion

The 7.5 μm nickel foil sample after heating at a frequency of about 250 kHz and 80% output is shown in Figure 4. The nickel foil melted the surrounding polymer in a few seconds and created a bond resistant to manually applied shear force.

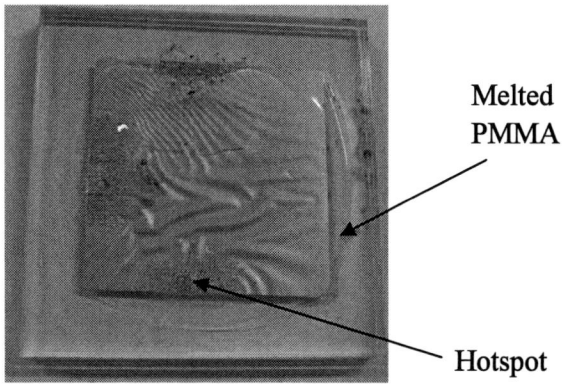

Fig. 4 Bonded PMMA slides with 7.5 μm Nickel foil

The hotspots indicated in the figure occurred due to the circular inductor coil interacting with a square shaped susceptor. A cross section of the bond between the polymer slides is shown in Fig. 5:

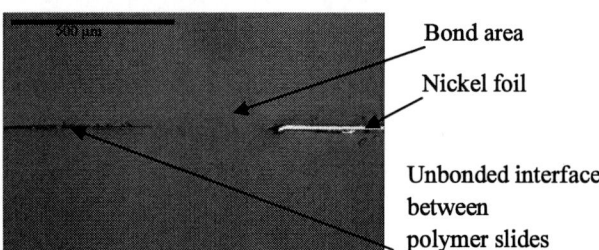

Fig. 5 Bond Area

For the 5 μm nickel plated sample, due to the reduced thickness of the susceptor layer the output power had to be increased to 100% to create bonds between PMMA and Nickel. A bond was created but at the hotspots the nickel was observed to glow and the PMMA was burned (Fig. 6). In this experiment the temperature of the top layer reached 103 °C.

Further evidence of the high temperature achieved throughout the thickness of the sample is that the substrates were bent by the pressure of the clamp, showing that the forming temperature of PMMA was reached (~ 105 °C).

Another 5 μm nickel plated sample test with a similar setup was processed but instead of using the previous, constant power control method a slow heating profile was implemented. After slowly heating up the complete system to about 50 °C, small pulses of full power were applied to make the polymer facing the susceptor melt without heating up the whole substrate. Using this method bond was created in a

shorter time without deforming the substrates. This indicates that the polymer can be heated and melted at the bonding area using high power pulses with the same or even better results in terms of bond strength, and less distortion in the microfluidic platform, compared to the application of slow heating ramps.

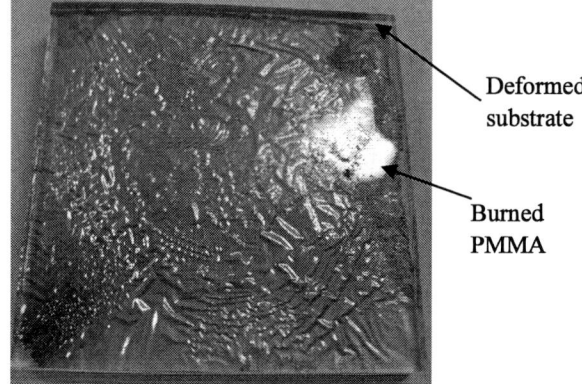

Fig. 6 PMMA bonded to Nickel

The substrate with the 50-100 nm thick nickel layer was heated up with different coils at frequencies between 150 and 700 kHz (AXIO) and at 2.2 MHz (IG). Because some of the coils were cage types that obscure line of sight contact with the substrate, a rod mounted thermocouple was used to measure temperature. To avoid physical contact with and damage to the thin nickel layer, the temperature was measured on the backside of the substrate after switching off the power generator. The temperature on the back of the substrate reached between 50 and 60 °C every time, independent of the coil type or frequencies used. This indicates that frequency and penetration depth didn't affect the heating rate which is unsurprising as the penetration depths were many times bigger than the susceptor thickness. For nickel the penetration depths are 14 μm and 4 μm, for frequencies of 150 kHz and 2.2 MHz respectively.

To test the response of a limited area receptor to the AXIO system a small piece of nickel foil was put between two substrates. The longest dimension of the piece was about 6mm (Fig. 7).

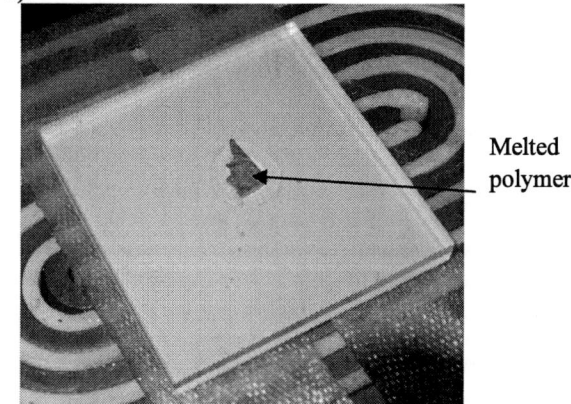

Fig. 7 Bond created by small area nickel susceptor

The experiment was stopped when the temperature on the back of the lower substrate reached more than 220 °C as the

PMMA was melting over a large area around the piece of foil. This test proves that it is possible to heat up small susceptor structures efficiently, and melt the surrounding material creating a bond. In Fig. 7 the melted area can clearly be seen.

Fig. 8 Sample a for pressure test

For bonding the pressure test samples two different parameters were used during bonding. First 80% of the output power was used for 6.6 s (Sample a) then 60% of the power was applied for 27.75 s (Sample b). Both profiles produced an apparent seal by visual inspection. Sample a is shown in Figure 8 after bonding.

Sample b withstood an air pressure of 70 kPa but showed leakage at 140 kPa, while Sample a withstood a pressure of up to 590 kPa, at which point the adhesive joint to the connector tube failed. (Fig.9).

Fig.9 Sample b sample showing failure of the tube to sample joint at 590kPa applied pressure.

This result may also be taken as an indication of the better performance of short pulses of high energy compared to application of low power for a longer time. However the G-clamp used does not apply pressure homogenously over the whole sample so further investigation is required with an improved design of clamping mechanism.

At the end of the series of experiments the aluminium coated substrates were tested. A good heating rate in the samples was observed for a variety of coil types. For instance using minimum coupling distance (sample lying on coil), a

pancake shaped coil, a frequency of 250 kHz and 50% maximum output power the aluminium layer was heated to an apparent temperature of 175 °C in 5 s. The temperature measured is not directly comparable with the earlier experiments because of the emissivity of aluminium, i.e. aluminium is bright while the pyrometer is optimised for dark surfaces. Overall, while the efficiency achieved was not as good as with a nickel susceptor, the experiment clearly demonstrated that even thin layers of diamagnetic metals can be heated using LFIH.

Design Consideration for Microfluidic Interconnects

The design of the microfluidic device and susceptor is likely to be critical to realise good lid sealing. We have identified two main classes of design as shown in Fig.10.

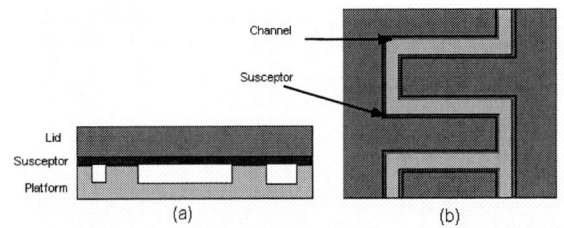

Fig.10 (a) Cross section view of design 1 and (b) Top view of design 2

Design 1 uses a lid with a continuous susceptor layer which is bonded to the microfluidic platform. Alternatively the susceptor layer could be coated onto the platform before structuring to create the channels if this is a subtractive process, in order to maintain visibility of the fluid flow and avoid liquid contact with the susceptor material. In this design the bond is made between polymer and metal. In design 2 the susceptor is only a thin line following the microfluidic channels on both sides. This could be realised using lithography or an ink jet printer with conductive ink. In addition to using less susceptor material, not obstructing line of sight to the channels and avoiding wetting of the susceptor, this design has the advantage of creating a bond between platform and lid directly as the polymer melts around the susceptor. The disadvantage is that more complex processing and alignment are needed compared to design 1.

In addition to design of the susceptor and coil the choice of the susceptor material is also important aspect. While nickel has performed well in the tests described above other materials like copper, which is very easy to deposit, and aluminium, which can be sputtered onto polymers due to its low melting temperature, might work almost as well. Using alternative to nickel would make the technique more flexible and even more suitable for mass manufacture due to lower material costs.

For world-to-chip interconnection of microfluidic systems using induction heating joining again two basic design ideas have been identified and will be realised, tested, and optimised in the course of the work programme currently underway.

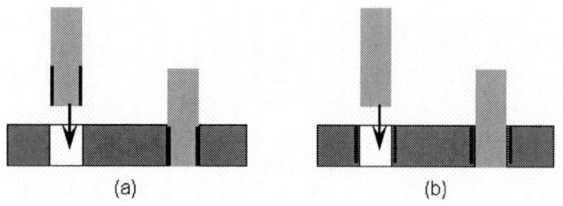

Fig.11 Interconnection with (a) Susceptor on tubes and (b) Integrated susceptor

The first design uses normal holes drilled into the lid of the microfluidic device. The tip of the micro-tube or capillary to be connected is coated with a susceptor and put into the hole as shown in Fig.11a. The susceptor is heated inductively and creates a bond between tubing and lid. To reinforce the bond the susceptor could be structured (e.g. with corrugations). In the second design shown in Fig.11b the susceptor is integrated with the microfluidic ports. The main advantage of the second design is that ordinary tubing could be used without pre-treatment, but the manufacturing of the lid is more challenging.

The spatial selectivity of induction heating offers the prospect that in a mature process lid sealing and connectorisation could be done simultaneously. In addition a stack of polymer layers could be bonded simultaneously to create a multi-layer microfluidic device. The concept is illustrated in Fig.12.

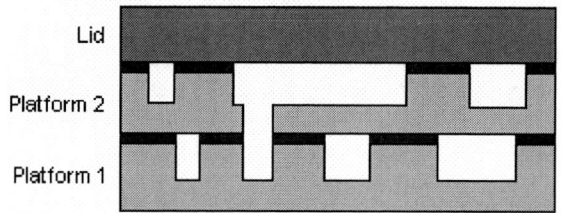

Fig.12 Multi layer sealing

Conclusions and Future Work

From our initial trials on utilising the induction heating method for microfluidic bonding, the following conclusions can be drawn:

The method is a potential basis of a cost-effective, rapid volume production method for polymer microfluidic device lidding.

The initial trials have demonstrated the possibility of joining polymers with a thin metal layer – both ferromagnetic and paramagnetic – using low frequency induction heating.

It is possible to bond polymers using small area metal features as susceptor.

Air pressurisation tests on bonded PMMA plates have shown the joint withstand a pressure of up to 590 kPa without failure.

Pulses of high power were found to perform better than the application of constant power in terms of bond strength and distortion.

The major challenges involved in creating a robust process using induction heating for microfluidics are in the areas of materials handling and process optimisation. In order to meet the challenges detailed process characterisation experiments are underway to enable optimal heat delivery and dissipation. The effect of both geometry and the material of the susceptor on heating will be determined, and joints created by induction heating will be assessed in terms of bond strength, uniformity, and microstructure. In addition exemplars of the designs described in this paper for sealing and interconnecting microfluidic devices will be realised.

Acknowledgments

This work is part of the 3D Mintegration Grand Challenge, funded by the UK EPSRC. The bonding experiments were carried out at Hüttinger Elektronik in Freiburg, Germany.

References

1. J. Callebaut, "Induction Heating," *Power Quality & Utilisation Guide,* 02/07. 2007.
2. J. Wang, "Lab-on-valve mesofluidic analytical system and its perspectives as a "world-to-chip" front-end," *Analytical and Bioanalytical Chemistry,* vol. 381, pp. 809-811, 02/01. 2005.
3. P. Abgrall and A. M. Gué, "Lab-on-chip technologies: making a microfluidic network and coupling it into a complete microsystem - a review," *J Micromech Microengineering,* vol. 17, pp. R15-R49, 2007.
4. D. R. Reyes, D. Iossifidis, P. A. Auroux and A. Manz, "Micro total analysis systems. 1. Introduction, theory, and technology," *Anal. Chem.,* vol. 74, pp. 2623-2636, 06/02. 2002.
5. Y. Peles, V. T. Srikar, T. S. Harrison, C. Protz, A. Mracek and S. M. Spearing, "Fluidic packaging of microengine and microrocket devices for high-pressure and high-temperature operation," *Microelectromechanical Systems, Journal of,* vol. 13, pp. 31-40, 2004.
6. C. H. Lin, G. B. Lee, L. M. Fu and S. H. Chen, "Integrated optical-fiber capillary electrophoresis microchips with novel spin-on-glass surface modification," *Biosensors and Bioelectronics,* vol. 20, pp. 83-90, 07/04. 2004.

Analysis of the Reliability of Package-on-Package Devices Manufactured Using Various Underfill Methods

Vicky Wang and Dan Maslyk*
Henkel Loctite (China) Co., Ltd, Yantai, Shandong, China 264006
*Henkel Corporation, Irvine, CA, USA
Vicky.Wang@cn.henkel.com

Abstract

As next-generation electronic packages continue to dictate smaller devices and more functionality, package-on-package (POP) configurations have started to gain popularity in the SMT industry. These stacked package devices enable board space savings, simplified system design, enhanced performance, and lower pin count. Although POPs are experiencing rapid growth for certain applications such as mobile handsets, digital cameras, PDAs, and MP3 players, concerns over POP drop test and thermal cycling performance reliability issues have been raised. Recently, the electronics industry has gathered a great deal of POP reliability data to help optimize the POP manufacturing and application process. A number of studies and tests have been conducted to investigate the board-level reliability of POPs in relation to drop test and thermal cycling performance. The test conditions have examined packages manufactured with and without underfill and have also analyzed the impact of different underfill dispensing patterns (i.e. full underfill, cornerbond and edgebond) However, few papers discuss the effects of the underfilling strategy -- such as undefilling the bottom component only or underfilling both top and bottom components, or the effects of solder alloy choice on the reliability of POP packaging.

In this paper, the effects of underfill dispensing type and POP ball alloy type on the reliability of POP devices during drop testing and thermal cycle testing were evaluated. It was found that both underfill dispensing type and alloy type have a profound effect on POP reliability. The study results revealed that underfilling only the bottom component seems to have no significant contribution to POP drop test reliability. Underfilling both the top and bottom components yields better drop test performance than underfilling only the bottom component. In addition, the SAC105 (98.5%Sn + 1.0%Ag + 0.5%Cu) bump alloy shows better drop test performance than the SAC305 (96.5%Sn + 3.0%Ag + 0.5%Cu) alloy.

Introduction

Today's consumers continue to push the electronics industry to deliver mobile multimedia products with smaller and thinner designs, more functionality, better performance, and lower costs [1].

Because of the relentless push to put more and more functions into highly miniaturized products, POP device configurations have emerged as the leading 3D packaging platform to address these challenging requirements.

Currently, POP packages are used predominately to integrate a high density digital logic device in the bottom (base) package with high capacity or combination memory devices (i.e. DRAM and flash) in the top (stacked) package as shown in Fig.1 [1].

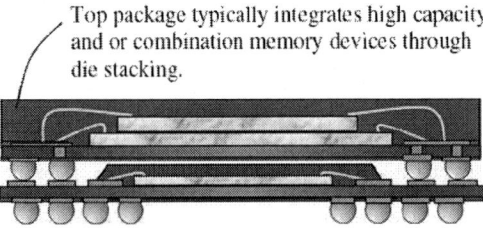

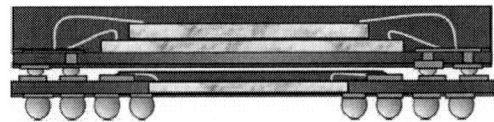

Fig. 1 Schematic of cross-section images of a typical package on package (POP) device, which includes a high- density PSvfBGA base structure (top) and lower profile cavity type PSetCSP structure (bottom) [1].

Although POP technology is experiencing fast growth in portable electronic products such as mobile phones, digital cameras, PDAs, portable players, gaming, and other mobile applications [2][3], its reliability issues are concerning many researchers and electronic manufacturers.

With the lower profile cavity type PSetCSP stacking, board-level reliability was reported by Yoshida and Ishibashi [4][5]. Yoshida et al. reported a detailed study on stacking yield versus the warpage impact for a 14 mm x 14 mm POP configuration [6].

Dreiza et al. compared different Pb-free ball alloys and BGA substrate pad finishes to determine which solder joint and BGA pad finish structures offered the best board-level reliability, cost, and performance balance for the BGA interfaces [2]. Lee et al. reported the effect of underfill materials and dispensing patterns to the POP reliability performance under thermal cycling and drop test [7]. Toleno and Maslyk analyzed three different attachment methods for the POP assembly process [3].

In this study, the effects of underfill dispensing types and POP ball alloys on the package reliability were studied.

Experimental & Results

Test Vehicles

For this analysis, Amkor 0.5-pitch 12x12mm PSvfBGA305 components were used for the bottom package and 0.65-pitch 12x12mm FBGA128 components were used for the top package.

The test printed circuit board (PCB) used in the study was designed according to the JEDEC standard [8]. Components were then mounted on the 15 non-via-in-pad component locations shown in Fig. 2. The board build-up structure was a 1-6-1 FR4 substrate material with a Cu-OSP surface finish.

Fig.2. Drop Test PCB as per JESD22-B104

Lead-free alloys evaluated in this experiment include SAC125 (98.3%Sn + 1.2%Ag + 0.5%Cu) for the bottom component, SAC105 (98.5%Sn + 1.0%Ag + 0.5%Cu) for the top component, and SAC305 (96.5%Sn + 3.0%Ag + 0.5%Cu) for both the top and bottom components. Three underfilling types -- no underfill, bottom only underfill, and both top and bottom underfill were analyzed. The detailed experimental design is shown in Table 1.

Table 1. Design of POP experiment

Board#	Bottom/ Alloy	Top / Alloy	UF Type	Remarks
1	SAC125	SAC105	Bottom only	for drop test
2	SAC125	SAC105	Bottom only	for drop test
3	SAC125	SAC105	Bottom only	for TC
4	SAC305	SAC305	Bottom only	for drop test
5	SAC305	SAC305	Bottom only	for drop test
6	SAC305	SAC305	Bottom only	for TC
7	SAC125	SAC105	Top and Bottom	for drop test
8	SAC125	SAC105	Top and Bottom	for drop test
9	SAC125	SAC105	Top and Bottom	for TC
10	SAC305	SAC305	Top and Bottom	for drop test
11	SAC305	SAC305	Top and Bottom	for drop test
12	SAC305	SAC305	Top and Bottom	for TC
13	SAC125	SAC105	No Underfill	For Drop Test
14	SAC125	SAC105	No Underfill	For Drop Test
15	SAC125	SAC105	No Underfill	For TC
16	SAC305	SAC305	No Underfill	For Drop Test
17	SAC305	SAC305	No Underfill	For Drop Test
18	SAC305	SAC305	No Underfill	For TC

The PCB and POP components were assembled in the SMT assembly line at Henkel's advanced SMT laboratory in Irvine, California, USA. Equipment in the line includes a DEK Viking Screen Printer, Universal Advantix Pick and Place machine and a Heller 1700 reflow oven.

After the POP packages were assembled, underfill was applied using a dispensing machine. The underfill used was a reworkable CSP and BGA underfill with low viscosity, fast flow rate at room temperature, and fast curing at low temperature to minimize thermal stress to other components. Detailed underfill properties are shown in Table 2. For boards

numbered 1 to 6, only the bottom components were underfilled. For boards 7 through 12, both the top and bottom components were underfilled. Underfill dispensing was conducted using the parameters listed in Table 3. Underfill dispensing type is schematically shown in Fig. 3. Following dispensing, the underfills were cured using a reflow process. Drop testing was conducted according to the JEDEC JESD22-B111 specification using a Landsmont model 15-D shock tester. Test boards were placed on the fixture with the component side down and then dropped, as shown in Fig.4. Electrical resistance was measured after each drop, and a change of resistance to infinite was defined to be a failure in this test.

Table 2. Basic Properties of Underfill

	Typical Value
Viscosity @ 25°C, mPas, Brookfield, CP52/20	1,848
Glass Transition, (Tg), °C, by TMA	53
Coefficient of Thermal Expansion cm/cm/°C,	
α_1 (<Tg)	63×10^{-6}
α_2 (>Tg)	178×10^{-6}
Storage Modulus, GPa	3.5

Table 3. POP packages dispensing parameters

Underfill	Henkel Loctite
Dispensing pump	Auger
Dispensing needle	23G
Substrate preheat temperature	70° C
Dispensing type	Single Line type, 80% component length
Bottom only dispensing	2 passes (15, 15mg), with a 5 sec wait time for each pass
Top and bottom dispensing	4 passes ((15, 18, 18 mg) with a 5 sec wait time for each pass

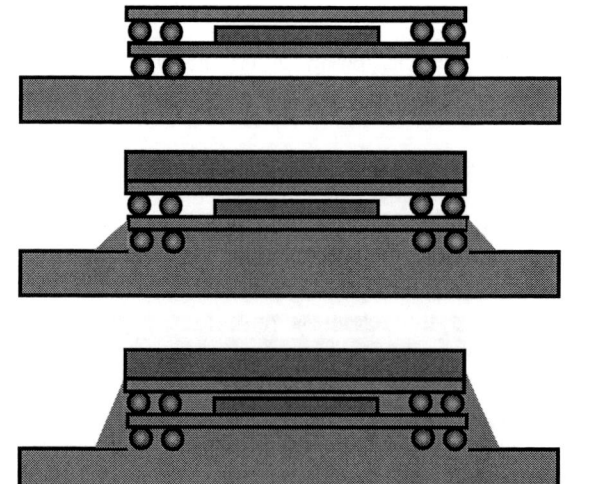

Fig.3. Underfill dispensing types: no underfill (top), bottom only (middle), and top and bottom (bottom)

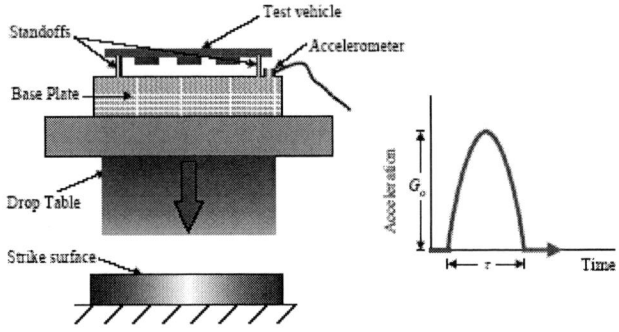

Fig.4 Drop test set up and shock pulse

POP Drop Test

The drop test was performed according to the JEDEC JESD22-B111 standard, and the test conditions include shock pulse 1500G peak acceleration and 0.5ms pulse duration, as shown in Fig. 5.

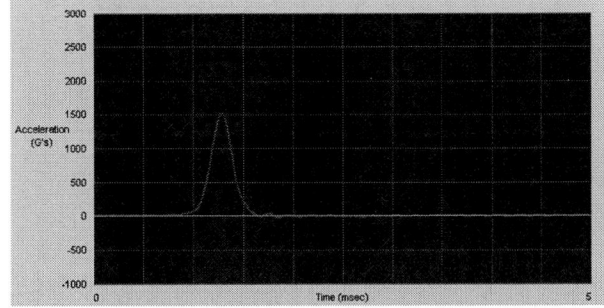

Fig.5. Drop response during the drop test (1500G peak acceleration of 0.5ms pulse duration)

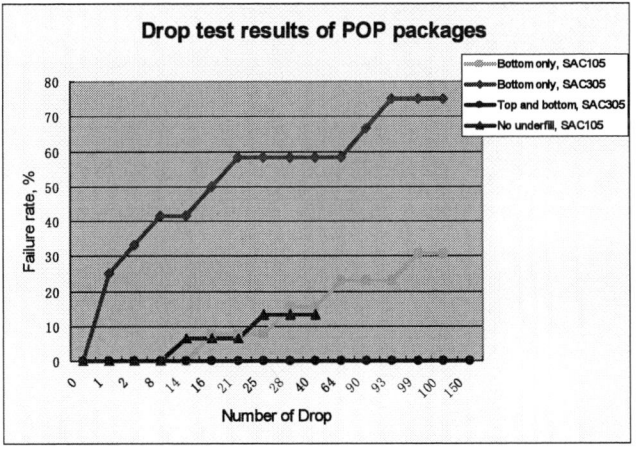

Fig.6. POP drop test results

Electrical resistance measurements showed that no failure occurred in the POP package with both top and bottom underfilling. All failures occurred in the top package of the POP devices that had bottom-only underfill.

For the SAC105 alloy, drop test results after 40 drops for the package without underfill were similar to the package with bottom-only underfill, indicating that underfilling the bottom component only does not improve POP reliability.

After 150 drops, components that were underfilled on both the top and bottom showed no failure as compared to the failures noted with bottom-only underfill. This suggests that underfilling both the top and bottom devices of the POP offers better drop test performance than underfilling the bottom

device only. Ball alloy SAC105 had lower failure rates during the drop test than the SAC305 alloy. This result is consistent with other POP drop test work [2].

Conclusions

The results of this study show:
1. No failures occurred in the POP packages where both the top and bottom components were underfilled. All the failures occurred in the POP top package when bottom-only underfill was used.
2. Top and bottom package underfilling delivers better drop performance than underfilling the bottom package only.
3. For the POPs where only the bottom package was underfilled, the SAC105 sphere alloy offered better drop performance than the SAC305 alloy.

Acknowledgments

The authors would like to thank the application engineers Beejal Mistry, Jeremy Alonte, Julie Bradbury, and Brian Toleno in Henkel's Irvine, California advanced research lab for providing the POP test vehicles, design input, and thermal cycle testing.

References

1. Dreiza, M., Yoshida, A., Micksch, J. and Smith, L., "Stacked Package-on-Package Design Guidelines," International Wafler Level Packaging Congress, San Jose, CA November 2005
2. Dreiza, M., Smith, L., *et al.* " Package on Package Stacking and Board Level Reliability, Results of Joint Industry Study," www.Amkor.com/products/notes_papers/POP_Stacking_IMAPS_0306.pdf
3. Toleno, B., Maslyk D. " Process and Assembly Methods for Increased Yield of Package on Package Devices," SMTA Pan Pacific Symposium, Kauai, Hawaii, January 22-24, 2008
4. Yoshida, Akito and Ishibashi, Kazuo, "An Extremely Thin, BGA Format Chip-Scale Package and Its Board Level Reliabilty" pp. 1335–1340 IEEE, Electronic Components and Technology Conference (ECTC), 2002
5. Yoshida, Akito and Ishibashi, Kazuo, "Design and Stacking of An Extremely Thin Chip-Scale Package" pp. 1095–1100 IEEE, Electronic Components and Technology Conference (ECTC), May 2003
6. Yoshida, Akito et. al. "A Study on Package Stacking Process for Package-on-Package (PoP)", Electronic Components and Technology Conference (ECTC), coming in May 2006, San Diego, CA
7. Lee, J.Y., Hwang T.K., *et al.* " Study on the Board Level Reliability Test of Package-on-Package (POP) with 2nd level underfill," 2007 IEEE, Electronic Components and Technology Conference (ECTC), 2007 Proceedings
8. JEDEC Standard No. 22-B111, JEDEC Solid State Technology Association, July 2003

Meeting Thermal Performance and Reliability Challenges for a Thermally Enhanced Ball Grid Array Package (TEBGA)

By Quan Qi, Ph.D.*

quan.qi1@yahoo.com, 170 West Tasman Drive

San Jose, CA 95134-1706 USA

*Currently with CISCO System, Inc., work published here was performed while the author was at Maxtek.

Abstract

For devices with challenging power management requirement, thermally enhanced ball grid array package (TEBGA) offers a good solution, where the device is attached to a heat spreader, usually made of copper, with a thermally conductive epoxy to ensure a good conductive path for heat to escape from the die. The top die surface and bonding wires are covered with an overmolding compound for environmental protection such that heat dissipation is typically limited in that direction.

However, TEBGA is not without its unique challenges. In this paper, we present a study on the challenges of meeting the thermal performance and reliability requirements for a ASIC packaged with TEBGA. A localized deformation or "dimple" of the TEBGA package is discovered during the package assembly process, where the heat-spreader is noted to have deformed under the die shadow, which results in a circular shaped indentation. This raises concerns about the impact on the thermal performance of the subsequent package to heat sink interface when it is integrated into the system. Solution to this potential problem rests on balancing thermal performance, reducing package stress level & understanding potential long term package reliability. Deformation of the package with each process step will be first described and particular attention will be given to the change of package profile after the die attach process; then a finite element analysis of the stress and deformation of the die attach process is discussed and important parameters affecting the deformation and stress are shown; moreover, a thermal resistance model assessing the thermal budget for this package in a system environment is reviewed and confirmation with numerical analysis & validation by experimental analysis are highlighted; furthermore, an interactive analysis is subsequently performed based on the FEA model for package stress/deformation and thermal resistance model to optimize the packaging solution; finally, balanced solution through this interactive optimization process is summarized and demonstrated in the manufacturing process.

1. Introduction

Thermally enhanced ball-grid-array (TEBGA) package is a wire-bonded BGA package that provides improved thermal performance over the conventional wire-bonded BGA packages. Such an enhancement is achieved with the application of a heat-spreader, usually made with Cu or Al, onto which the active device is attached with thermally conducting adhesives. Wire bonds are formed on the active side of the device and the package is then covered with overmolding materials to protect bonding wires. The overmolding material is typically epoxy based and has poor thermal conductivities. Consequently, the primary thermal path for the device is through the back side of the device via thermal conducting attach layer, heat spreader and then a heat sink, all defined as part of a system-level thermal solution. A typical TEBGA design with multi-tier wire bond pads is shown in Figure 1.

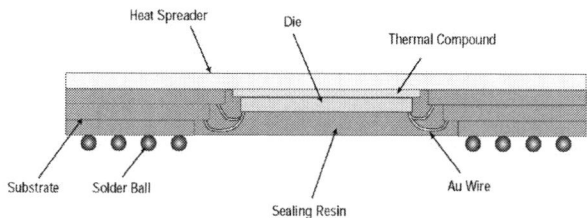

Fig. 1 A characteristic, thermally enhanced ball-grid array (TEBGA) package construction with multi-tier wire bonding pads. Sealing resin is synonymous with overmolding compound.

2. Package failures and deformation

The package of particular interest to us is of medium sized (25 X 25 mm^2) with only a single row of wire bonding pads. The device packaged with this package is 7.5 X 7.5 mm^2. The layout outline of the package is shown in Figure 2.

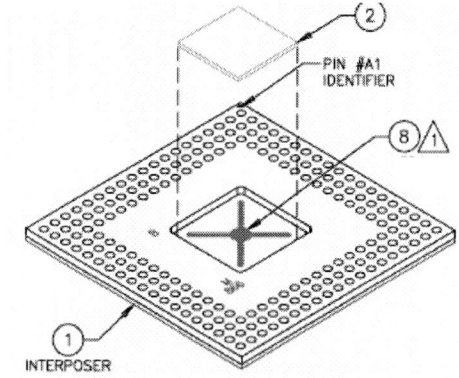

Fig. 2 Layer-out of the particular package under investigation.

Electrical tests indicate failure of some these packages and a series of investigations were launched to understand the root cause of the failure. CSAM, X-sections, TDR, warpage measurement, FEA analysis, etc were all applied in the study. A typical CSAM and x-sectioning pictures are shown in Figure 3.

978-1-4244-2739-0/08/$25.00 ©2008 IEEE

It is of great interest to note that device in the picture appears to have cracked in the middle and this has, for some time, been the focus of the investigation. Attempt had been made to link its location to the height of thermally conductive adhesive used to attach the device to the metal spreader and careful examination of the cross-sections of other similar devices reveal that it is more tied with the way the cross-section technique used, particularly the potting process used instead of device failure. In addition to investigating suspected failures experienced by the packaged device, package deformation was also carefully measured against the manufacturing process and the goal was to identify what processing step has the largest impact on the package deformation. The TEBGA package manufacturing process steps are shown in Figure 4a and the associated package deformation, as measured from the smooth, heat spreader surface is shown in Figure 4b.

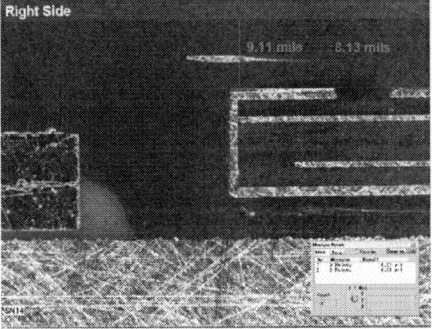

Fig. 3 X-sectional view of the TEBGA package on the right: Si device is shown on the left middle of the picture, cavity carrier is shown on the right, and heat spreader is shown at the bottom. Thermally conductive epoxy, partial Au bonding wire embedded in the overmolding materials is also seen.

It is of great interest to note that device in the picture appears to have cracked in the middle and this has, for some time, been the focus of the investigation. Attempt had been made to link its location to the height of thermally conductive adhesive used to attach the device to the metal spreader and careful examination of the cross-sections of other similar devices reveal that it is more tied with the way the cross-section technique used, particularly the potting process used instead of device failure. In addition to investigating suspected failures experienced by the packaged device, package deformation was also carefully measured against the manufacturing process and the goal was to identify what processing step has the largest impact on the

package deformation. The TEBGA package manufacturing process steps are shown in Figure 4a and the associated package deformation, as measured from the smooth, heat spreader surface is shown in Figure 4b.

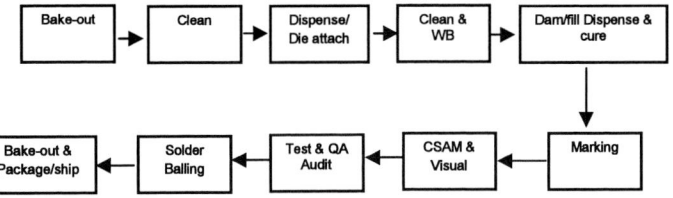

Fig. 4a TEBGA package assembly processing steps – die attach process is the primary factor.

It may be inferred from these measurements that die attach and subsequent curing of adhesive, wire bonding process, and overmolding and subsequent curing all contribute to the warpage of the TEBGA package. TEBGA package deformation and its impact on package thermal performance and reliability will be the focus of the subsequent investigations.

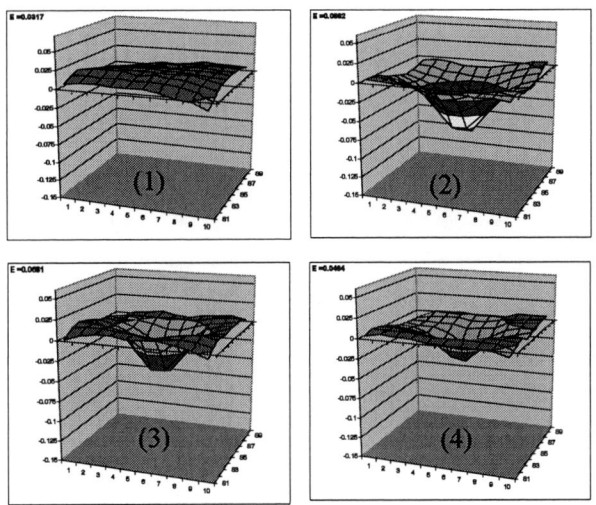

Fig. 4b Measured package deformation or warpage after each processing step. The numbers are also indicated in Figure 4a to show the correlations.

3. Interaction of package warpage with thermal management and package reliability

3.1 Deformation analysis

On closer examination, it becomes clear that die attach process contributes most significantly to the package warpage (step (2) in Figure 4). A finite element analysis is consequently undertaken to quantify the package deformation and the associated stress level induced during the die attached process. ANSYS is used for this purpose with solid 3D mesh by assuming orthotropic, linear materials properties and considering glass transition effect for both thermal coefficient of expansion (CTE) and Young's modulus (E). Due to symmetry, only a quarterly symmetrical model is needed and the meshed model is shown in Figure 5.

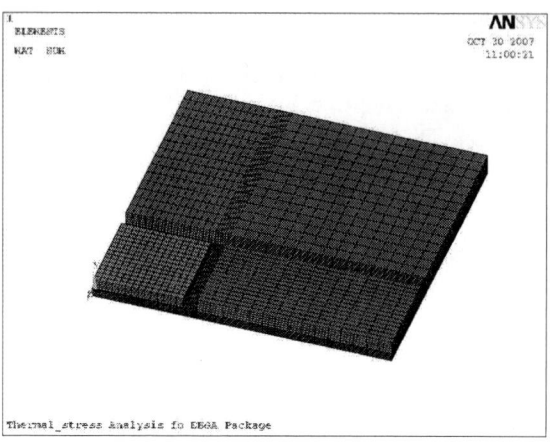

Fig. 5 Quarterly symmetry FEA model of the TEBGA package.

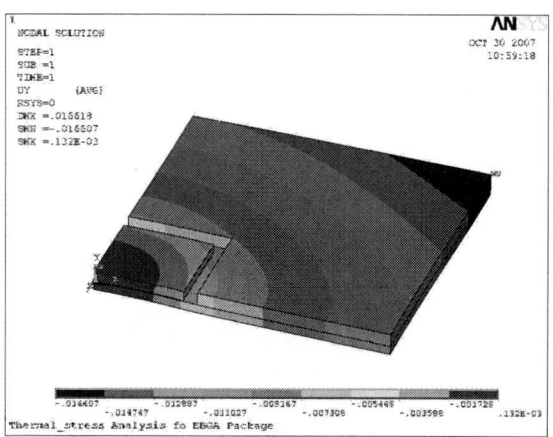

Fig. 6 Deformation of the assembly under the temperature excursion.

that impact warpage most significantly. To simulate the die attached process, all components are assumed to be at the neutral state without any deformation at 150°C (die attach curing temperature) and the whole assembly, including silicon device, heat spreader, substrate & substrate attach, is then cooled down to the room temperature of 25°C. The resulting deformation is shown in Figure 6.

Overall, due to the difference of CTE's of Cu substrate, organic substrate materials and silicon most significant displacements occur at the corners of the package away from the center of the device under the temperature excursion, with the primary driver being the CTE mismatch between silicon device and the Cu heat spreader. Cu heat spreader contracts more than silicon with the reducing temperature and causes the overall package to warp. Locally under the device shadow, deformation is more interesting, as shown in Figure 7, where it is observed that Cu heat spreader displays a circular deformation.

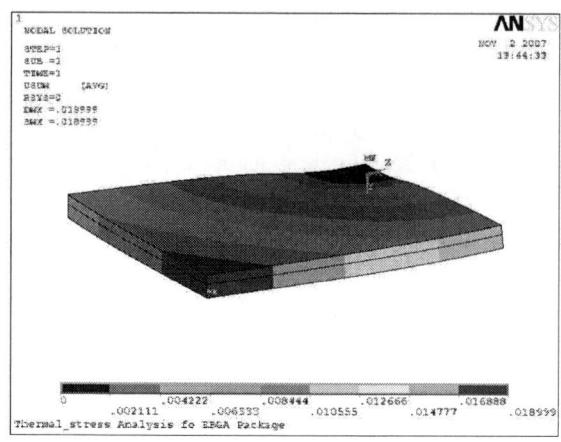

Fig. 7 Local deformations under the die shadow—note the circular shape or "dimple."

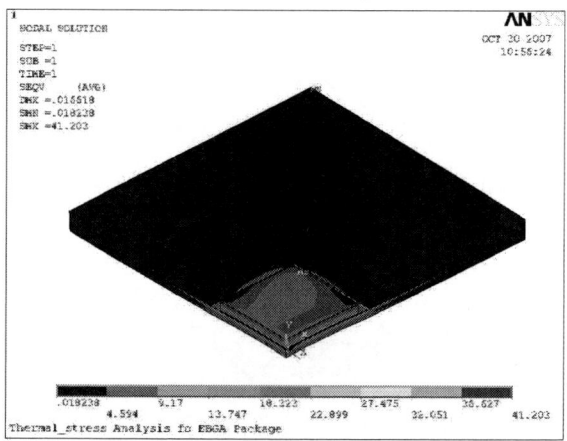

Fig. 8 Von Mises stress distribution in the package—largest stress is at the corner of the die attach layer, consistent with the minor delamination shown in the CSAM picture earlier (left image in Figure 3).

3.2. Thermal Analysis

In order to assess the impact of "dimple" deformation on the thermal performance of the package, independent thermal analyses are required. First, a simple resistor network model was established and it is graphically shown in Figure 9.

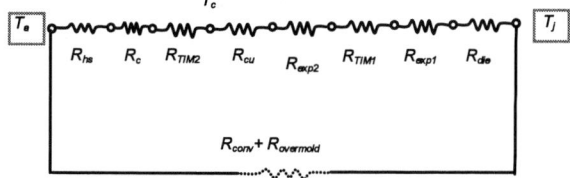

Fig. 9 Resistor network model for thermal analysis. R_{TIM2} represents the interface between the heat spreader and heat sink and its value depends on fractional void volume, which in turn is related to the dimple deformation.

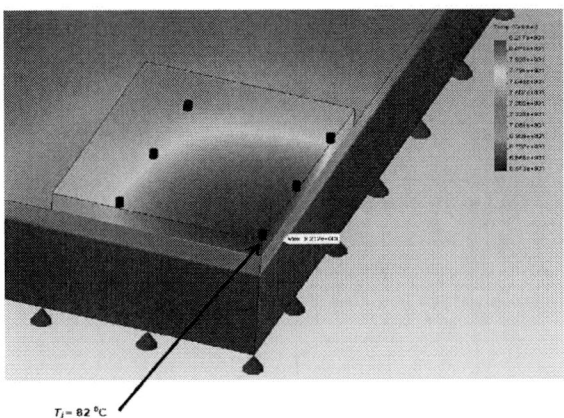

$T_j = 82\ ^{\circ}C$

Fig.10 COSMOS model for the thermal analysis and validation of resistor network model.

Similarly several parameters are adjustable in resistor-network based thermal analysis as well, such as contact resistance at different interfaces, expansion angles and fractional void volume. An independent validation of the model is consequently required. For this purpose, a finite element thermal model was established with COSMOS package and it is shown in Figure 10. For this particular calculation, the results from the resistor network model gives:

Die Data:		
Die L=	0.295	[IN]
Die W=	0.295	[IN]
Nominal Power =	8	[W]
Power tolerance =	-------->	
Actual power area =	0.043513	[in^2]
Actual power density =	184	[W/in^2]
Ambient Conditions:		
Ambient =	50	[C]
Forced convection	400	LFMP
Alpha Novatech heat sink	UB25-20B	
Heat sink resistance	1.85	C/W
Temperatures:		
Die Tj =	82	[C]
Tolerance =	8	[C]
Upper Tj limit =	90	[C]
Lower Tj limit =	74	[C]

Fig. 11 Predicted junction temperatures with resistor network model.

Comparison with the numerical results suggests acceptable accuracy of the resistor network model. It can be shown that the fractional void volume between heat spreader and heat sink contained in R_{TIM2} may be used as a convenient vehicle to study the impact of different level of "dimple" deformation on the thermal performance of the package.

3.3 Dependence on "dimple" deformation and process variations

The resistor network model has been shown to produce the acceptable results. Increased dimple deformation at the heat

spreader and heat sink interface may be accounted for conveniently by applying the concept of fractional void volume based on the fact that more deformation of the heat spreader leads to more trapped air at the interface and consequently reduced capability to conduct heat across that interface (hence a larger R_{TIM2} in the model). A rather straightforward formulation can take into account of the reduced effective area due to presence of void or air gaps. In Figure 12, thermal resistance variation with the fractional void volume is plotted. It clearly seen that thermal resistance has a nonlinear dependence on the fractional void volume (FVV) and resistance can increase almost 30 folds when FVV varies from 20% to 70%!

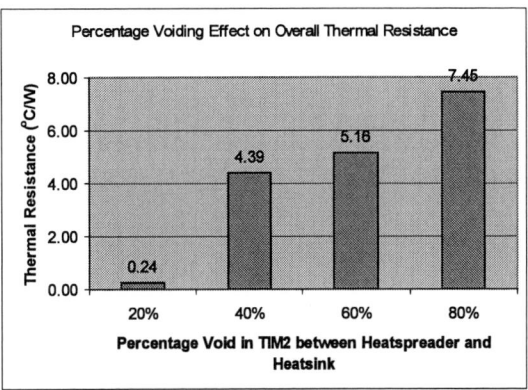

Fig.12 Package thermal resistance variations with the fractional void volume.

Another important parameter is the die-attach thickness, which affects "dimple" deformation, package stress distribution, and the package thermal performance. It is one of the process related parameter that can be optimized. For these reasons, its impact on the package thermal is calculated (Figure 13).

Additionally, process variations are assessed as well in conjunction with these two parameters. Both upper and lower junction temperature bonds are shown in Figure 14 under pre-assumed ambient conditions. These results are based on the assumed, accumulative process capability for each of the processing steps at a specific assembly site.

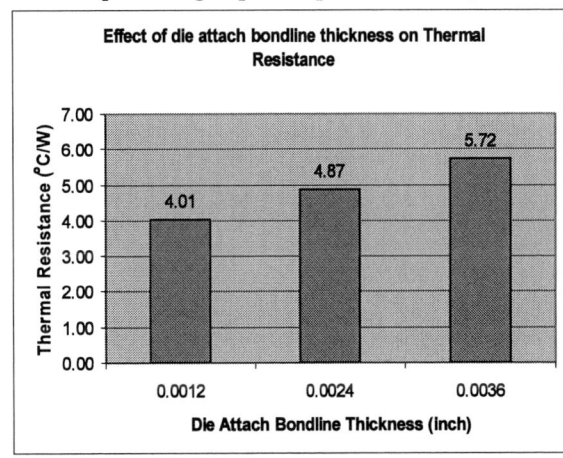

Fig.13 Effect of bond-line thickness on the package thermal resistance.

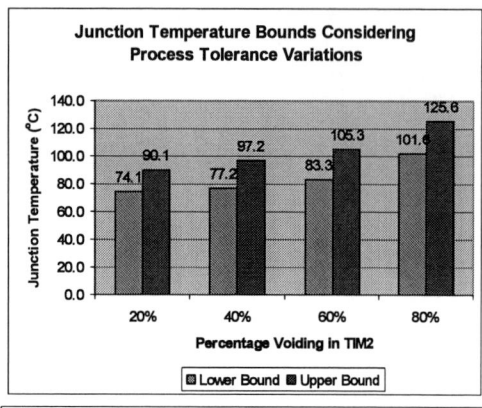

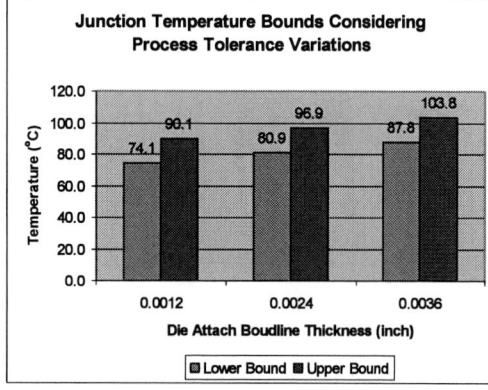

Fig.14 Process variations and junction temperature bonds.

3.4 Iterative trade-off analysis

Now that both modeling approached are (partially) validated and basic trends understood, we can proceed to conduct an iterative analysis to search for a balanced solution which results in acceptable package thermal performance as well as die attach layer stresses. The iterative process goes as follows:

1. Deformation and stress analyses are performed for a given set of die attach materials and different thickness – maximum stress level in die attach and maximum dimple deformation are recorded (Figure 15)
2. Conducting the thermal analysis by correlating the degree of "dimple" deformation with the fractional void volume to assess the corresponding thermal resistance of the package (Figures 13-14)
3. Evaluate the junction temperature with the obtained thermal resistance, taking into account the process variations to ensure none of the existing thermal constraints is violated (Figure 15)
4. If the thermal performance is acceptable, assess the maximum Von Mises stress distribution to ensure that it meets the reliability requirement (determined through another set of independent test); if not, repeat the process

Stress and deformation analyses conducted indicate that soft and thicker die attach materials (TIM1) tend to result in low maximum stress in the die attach layer but they also tend to have somewhat poorer heat conducting properties. Similarly, thicker TIM2 material tends to fill the gap (thus minimizing the dimple deformation impact) better while offering increased thermal resistance compared with thinner

ones. These factors are the very reason that thermal performance and mechanical deformations and consequently stress distributions in the package are intimately coupled.

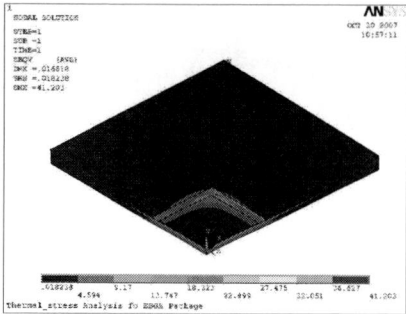

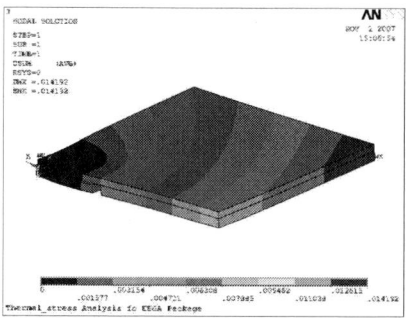

Fig. 15 Maximum Von Mises stress in die attach layer and dimple deformation

It is through this iterative process that we were able to achieve a balanced packaging solution that included proper selections of die attach materials (TIM1) and appropriate thickness as well as mechanical properties, thermal interface materials with heat sink (TIM2) and appropriate thickness as well as mechanical properties. The solution has been implemented in the production setting and the design has to date survived component level reliability testing. Since many proprietary design parameters are involved, we'll not get into the details here.

3.5 Comments

Following evaluation via experiment of different die attach epoxies and varying epoxy cure schedules (time and temperature) – no change was made to the original package configuration. By using a lower temperature cure adhesive, we were able to reduce the magnitude of the "dimple." The magnitude of reduction was approximately 0.001".

The "dimple" phenomenon was determined not to pose a significant issue for this particular product, although the long term product reliability impact is yet to be fully understood – it would have been of interest to perform a more detailed system level testing to validate performance and this is left for future investigations.

With respect to the CSAM image showing epoxy delamination under the device (Figure 3) – it was determined that insufficient epoxy coverage under the IC was responsible although stress concentration is usually anticipated at the die corner. Later experiments demonstrated that 100% epoxy coverage under the IC can help minimize delamination.

Finally, with respect to the IC cracking shown on the right of figure 3, it was determined that cracking was indeed a byproduct of the x-sectioning process and not the result of stress placed on the IC resulting from the "dimple" phenomenon.

It should be mentioned in passing that in order to achieve a meaningful thermal solution; a system-level perspective is clearly needed. For example, the thermal analysis presented earlier depended on assumed ambient conditions, heat sink configuration, etc. Without the information, thermal analysis would loose its relevance for a particular application. In this sense, the optimized solution is only meaningful in the context of this particular design. Nevertheless, the methodology proposed here is applicable in applications involving trade-offs of the thermal and mechanical designs of an electronics package.

4. Summary and Conclusions

A combined thermal and mechanical analysis of a TEBGA package is presented in this paper. While the primary goal was to present a consistent methodology for the trade-off analysis, an actual package and its assembly processing steps are used as the vehicle to present the idea. With the input of system level thermal solution information, an iterative solution procedure is demonstrated with an ANSYS model for stress analysis and a validated resistor network model for thermal analysis.

Key parameters investigated include the two thermal interfaces, TIM1 and TIM2. TIM1 is also the die attach materials layer and it provides both a thermal interface and a mechanical interface to mitigate the stress induced by the (significant) CTE mismatch of silicon device and Cu or Al heat spreader and temperature excursion. Process related measurement also show that the largest concentric deformation of the package, or dimple, is associated with the die attaches and subsequent curing step and this deformation has implication for the system thermal performance of the device. Different TIM1 materials and thickness impact the "dimple" deformation differently. On the other hand, TIM2 material serves as the thermal interface of the package with the system. Thinner TIM2 layer provide better thermal performance but less flexibility to make up for the dimple-deformation related void. It is this trade-off of thermal and mechanical performance that is the focus of this presentation.

The procedure is suggested that shows a way to achieve a balanced solution that lead to rational materials selections and process optimizations while considering impacts to thermal performance, stress distribution and potentially long term reliabilities of the TEBGA packages.

Numerical Simulation of the Micro-channel Heat Sink on Non-uniform Heat Source

Zhang Minliang, Wang Xiaojing, Liu Hongjun, Wang Guoliang
Shanghai University
Shanghai University, 224mail box, 149 Yan Chang RD., Shanghai, 20072, China
bright13792@hotmail.com 086-13795480400

Abstract

In this paper, the different heat source distributions are studied in the micro-channel heat sink (MCHS) cooler. The simulation model is established to analyze the heat distributing and temperature rise of the MCHS with different channel width-dimensions. Uniform and non-uniform heat source are used to offer a constant heat flux, and the total heat flux amount is 270W. Water is chosen as the coolant for its superior hot properties and the velocity range is from 0.01m/s to 10m/s. The results show that tradition type MCHS will be under a low efficiency for cooling a non-uniform heat source. And the non-equal displacement of fins can effective decrease the temperature rise about 10% under the same cooling conditions cooling a non-uniform heat source. We also find that it is impossible to enhance the cooling effect through adding the flow rate if it has surpassed it's limitation.

1 Introduction

Since Gordon Moore proposed that the density of transistors on a chip will double every eighteen months, increased demands for dissipation high heat fluxes from electronic, power, and laser devices creates the need for new cooling technologies as well as improvements in existing technologies. Micro-cooler has become one of the basic devices in micro-mechanics and micro-engineer. The analysis of heat dissipation is one of the most important essential in the design of MEMS system.

Many efforts have been devoted to this area of research. Tuckerman et al[1] first demonstrated the configuration of micro cooler with many micro channels inside can remove 790W/cm² with forced convective water at a substrate-to-coolant temperature difference of 71K. Fig.1 shows the schematic configuration of the MCHS. These heat sink consist of an extruded or machined base, which is flat on the module facing side, and grooved on the fin side. The fins are then exposed into the grooves. These interface, reduces the overall thermal resistance of the system.

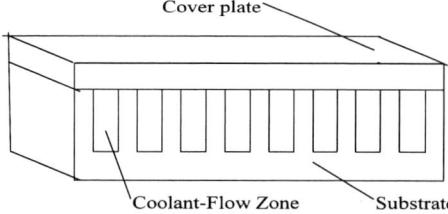

Fig.1. Schematic configuration of micro-channel heat sink system

Heat transfer in micro-channels is complicated by the geometry of the micro-channels, requiring three-dimensional analysis of heat transfer in both solid and liquid phases. Computational Fluid Dynamics (CFD) models were implemented in order to study and optimize the thermal and hydraulic performance of micro-channel heat sink[2-5].

2 Theoretical analysis

2.1 Governing Equations

Based on the computational domain showed in fig.2, the conductions and convection in the micro-pool are simulated in the same time to capture the heat transfer. Computational fluid dynamics (CFD) is used to tackle this problem. The fluid is assumed to be incompressible and the flow field is continuous (Knudsen number is small enough in the present study). Steady state continuity, momentum and energy equations are solved.

Continuity equation:

$$\frac{\partial U}{\partial x} + \frac{\partial V}{\partial y} + \frac{\partial W}{\partial z} = 0 \tag{1}$$

Momentum equation:

$$\begin{cases} U\frac{\partial U}{\partial X} + V\frac{\partial U}{\partial Y} + W\frac{\partial U}{\partial Z} = \frac{1}{\text{Re}}(\frac{\partial^2 U}{\partial X^2} + \frac{\partial^2 U}{\partial Y^2} + \frac{\partial^2 U}{\partial Z^2}) - \frac{\partial P}{\partial X} \\ U\frac{\partial V}{\partial X} + V\frac{\partial V}{\partial Y} + W\frac{\partial V}{\partial Z} = \frac{1}{\text{Re}}(\frac{\partial^2 V}{\partial X^2} + \frac{\partial^2 V}{\partial Y^2} + \frac{\partial^2 V}{\partial Z^2}) - \frac{\partial P}{\partial Y} \\ U\frac{\partial W}{\partial X} + V\frac{\partial W}{\partial Y} + W\frac{\partial W}{\partial Z} = \frac{1}{\text{Re}}(\frac{\partial^2 W}{\partial X^2} + \frac{\partial^2 W}{\partial Y^2} + \frac{\partial^2 W}{\partial Z^2}) - \frac{\partial P}{\partial Z} \end{cases} \tag{2}$$

Energy equation:

$$\frac{\partial}{\partial t}(\rho E) + \nabla \cdot (\vec{v}(\rho E + p)) = \nabla \cdot \left(k_{\mathit{eff}} \nabla T - \sum_j h_j \vec{J}_j + \left(\overline{\overline{\tau}}_{\mathit{eff}} \cdot \vec{v} \right) \right) + S_h \tag{3}$$

Where k_{eff} is the effective conductivity ($k_{eff} = k + k_t$, where k_t is the turbulent thermal conductivity, defined according to the turbulence model being used), and J_j is the diffusion flux of species j. The first three terms on the right-hand side of Equation (3) represent energy transfer due to conduction, species diffusion, and viscous dissipation, respectively. S_h includes the heat of chemical reaction, and any other volumetric heat sources user have defined.

2.2 Boundary conditions

The uniform velocity and temperature are applied in the inlet of the micro-channel. The total flow rate can be determined by the total number of micro-channels used in a micro-channel heat sink module and the area and velocity of individual micro-channel. The constant pressure of 1atm is applied in the exit of the micro-channel. The heat flux of 270 W/cm² is applied at the bottom of the heat sink. Heat is assumed to be removed by working fluid only. For the uniform heat source simulation the heat flux is applied at the bottom (3cm×3cm)of the heat sink average, and for the non-uniform one the heat flux concentrate on the center zone(1cm ×1cm) of the bottom.

978-1-4244-2739-0/08/$25.00 ©2008 IEEE

3 Simulation Results

3.1 Comparison between with uniform and non-uniform heat source

The discription of the example MCHS in our simulation are as follows, the substrate copper plate of 30mm × 30mm × 2mm, eight fins with 26mm length and 2mm width and 5mm height distribute on the plate uniformly.

Table1 Temperature-rise under different inlet velocity

Inlet velocity (m/s)	Temperature-rise with uniform heat source (k)	Temperature-rise with non-uniform heat source(k)
0.005	117	N/A
0.01	77	104
0.02	52	82
0.03	44	74
0.04	40	69
0.05	37	67
0.1	31	61
0.5	26	56
1	25	56
5	24	55
10	24	55

The effect of inlet velocity with different heat source is list in table 1. It can be seen that the temperature with concentrate heat flux is much higher than the uniform one. With the velocity increasing, the temperature drops quickly. Then with the huge inlet velocity, the temperature varies small. Fig.2 is the temperature distribution with different heat source. Fig.2 (a) shows the temperature variation of a MCHS with uniform heat flux input in which the inlet velocity is 0.5m/s and the heat flux input is 270W. And Fig.2 (b) shows the temperature distribution of the same MCHS with non-uniform heat flux input in which the inlet velocity and the heat flux input are the same as fig.2(a). The total heat flux and mass transfer of the two MCHSs are identical. Through comparison we can find that the coolant flow in the margin is not so effect. The low temperature rise of the fins in the margin weakens the capability of the heat dissipation. Fig.3 shows the contrast when MCHS faces the two different kinds of heat source under the same flow rate from 0.01m/s to 10m/s.

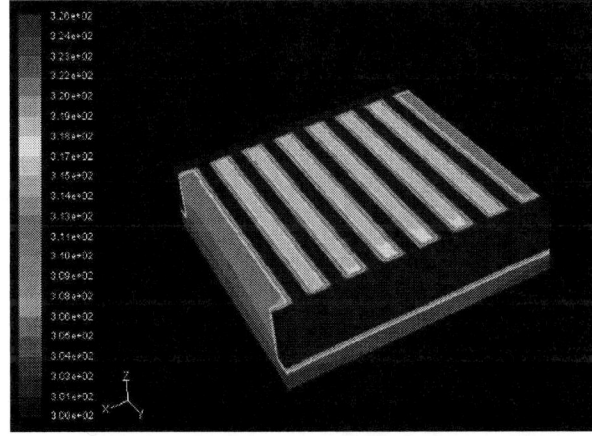

(a) Temperature distribution of a MCHS with average heat flux input

(b) Temperature distribution of a MCHS with concentrate heat flux input

Fig.2 Temperature distribution of a MCHS with different heat flux input

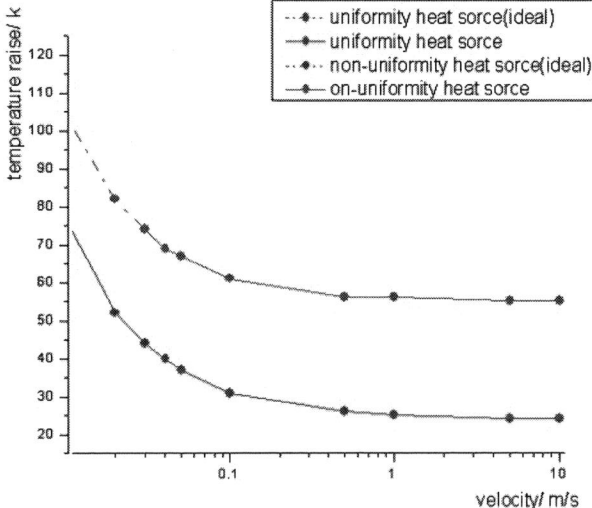

Fig.3 Comparison of temperature-rise between the MCHSs on uniform heat source and non-uniform heat source

3.2 Optimization of the MCHS with concentrate heat flux

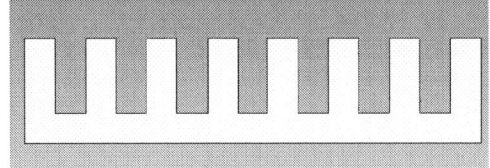

（a）seven channels with 2mm×5mm cross-section (type1)

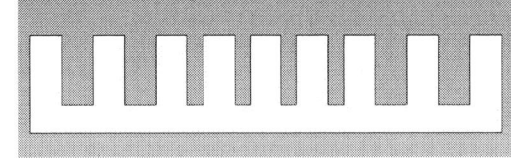

（b）four channels with 2mm×5mm cross-section in the center and four channels with 2mm×5mm cross-section adjust(type 2)

Fig.4 the optimation geometry

To weaken the extra temperature rise caused by non-uniform heat source, a new type of MCHS is designed. Fig.4 shows the cross section of the two types of MCHS. Type2 MCHS is similar with the type1, and the difference of them is the width of the four channels in the center. The width is only half(1mm) of those in type1.

Table2 Temperature-rise comparison between type1 and type2 MCHSs on non-uniform heat source under different inlet velocity

Inlet velocity (m/s)	Temperature-rise of Type1 (k)	Temperature-rise of Type2 (k)
0.005	N/A	151
0.01	104	103
0.02	82	77
0.03	74	68
0.04	69	63
0.05	67	61
0.1	61	56
0.5	56	52
1	56	52
5	55	50
10	55	N/A

We can also learn from the Table2 that temperature rise of type2 MCHS is lower than type1 about 5° under the same flow rate and heat input condition. It improves that the distributing of channels should be designed with pertinence to transfer the heat from the hot points.

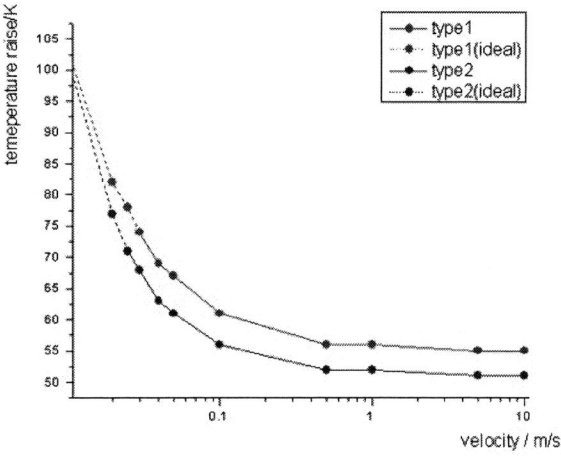

Fig.5 Variation of the temperature rise with the coolant inlet velocity of the type1 and type2 MCHS: virtual lines indicate the ideal state when the temperature is higher than the boiling point of the coolant

In fig.5, the temperature curve has two parts: Virtual-line part and real-line part. Virtual lines indicate the ideal state when the temperature is higher than the boiling point of the coolant. Many researchers have declared that the heat dissipation will be enhanced with the increasing of flow rate[6-7]. The result of simulation has the same trend with the existing research. When the velocity increased from 0.01m/s to 0.5m/s, the temperature rise of type1 MCHS decreased from 75K to 60K, and type2 decreased from 70K to 55K. From the figure it can be learned that the decrease of temperature rise has it's limitation through add the flow rate.

4 Conclusions

The work focuses on the simulation of three types of the micro-channel heat sink. Three models are established and computed through the Fluent software. By analyzing the results, key findings from the study are as follows:

Even if the pressure drop is ignored in these simulations, the problem of heat transfer still can't be solved through using high power pump which provides high flow rate. Higher velocity rate can't bring out more heat flux when surpass the limitation;

A reasonable varying of the array dimension of channel, such as narrow the channel beside the hot point, is effective to increase the heat dissipation about 10% when the heat source is non-uniform without add the flow rate;

The performance increase through using different type MCHS is limited, so new micro-cooler should be designed to meet the need of higher heat dissipation demanding.

Acknowledgments

The authors acknowledge the financial support from 863 program (No.2008AA04Z301) and Shanghai Municipal Education Commission (No.08YZ15).

References

1. D.B.Tuckerman, R.F.W.Peaser, "High performance heat sinking for VLSI, " *IEEE Electron Device Letters* Vol.Edl-2, No.5 (1981) , pp.126－129.
2. W. Qu, I. Mudawar, "Analysis of three-dimensional heat transfer in microchannel heat sink," *International Journal of Heat and Mass Transfer* Vol 45 (2002)
3. Amit Shah, Bahgat G. Sammakia, Hari Srihari, and Koneru Ramakrishna, "A numerical study of the thermal performance of anImpingement heat sink—fin shape optimization," *IEEE Transations on components and packaging technologies,,* Vol.27, No.4(2004), pp.710-717
4. Y. Mishan, A. Mosyak, E. Pogrebnyak, G. Hetsroni, "Effect of developing flow and thermal regime on momentum and heat transfer in micro-scale heat sink," *International Journal of Heat and Mass Transfer,* Vol.50, Issues15-16.(2007) pp. 3100－3114
5. Y.J. Cheng, "Numerical simulation of stacked micro-channel heat sink with mixing-enhanced passive structure," *International Communications in Heat and Mass Transfer ,* Vol.34, (2007) pp.295–303
6. M.B.Bowers, I.Mudawar, "High flux boiling in low flow rate, low pressure drop mini-channel and micro-channel heat sink" *International Journal of Heat and Mass Transfer,* Vol.37, No.2(1994), pp. 321－332.
7. D. Klein, G. Hetsroni, A. Mosyak, "Heat transfer characteristics of water and APG surfactant solution in a micro-channel heat sink," *International Journal of Multiphase Flow* Vol.31 (2005), pp.393－415

Heat Transfer Simulation of Nanofluids in Micro Channel Cooler

Liu Hongjun, Wang Xiaojing, Zhang Minliang, Zhang Wen, Liu Johan
Shanghai University
Shanghai University, 224mail box, 149 Yan Chang RD.,Shanghai, 20072, China
arjmy@shu.edu.cn , 13611948764

Abstract

Since the pioneering work by Tuckerman & Pease, lots of publications about heat sink has been researched in the last decade. Many enhancements are proposed in order to increase the thermal conductivity of micro-channels including nanofluids, special shapes, and two phase flows. The nanofluid is a solid-liquid mixture which is composed of nanoparticles and a basic liquid. In this paper, the nanofluids with suspended multiwalled carbon nanotube and other metallic or nonmetallic particles are compared to enhance the heat transfer performance. The thermal resistance of the heat sink with nanofluids is simulated.

1. Introduction

As the technical development of very-large-scale integrated (VLSI) circuits, the advanced electronic devices require better cooling systems because of the high level of heat dispassion. The work is inspired by Tuckerman and Pease [1] who supposed a high-performance micro-channel heat sink (MCHS) for advanced electronic devices. They found out the very compact water-cooled heat sink can make a minimum thermal resistance of $0.09\ ^0C/W$ over 1-cm^2 area. Since then, many optimizations to micro-channels are reported by previous investigations (Goldberg [2], Phillips [3], Knight [4]). These researchers focused mainly on the geometry optimizations based on theoretical analysis and numerical modeling. One of the methods for enhancing the heat transfer is the application of nanofluids as the working liquid. First reported by Choi [5] in 1995, nanofluids in which metallic or nonmetallic nanoparticles are suspended exhibit higher thermal conductivity values [6,7]. After that, numerous experiments are done to predict the heat transfer thermal conductivities of nanofluids. Xuan and Roetzel [8] pointed out two approaches treating the nanofluid as a single-phase fluid or a two-phase mixture respectively. Xuan and Li [9] discussed some factors such as the volume fraction, dimensions and shapes of the nanoparticles. Multiwalled carbon nanotube (MWCNT) nanofluids can greatly enhance the heat transfer performance for their extremely high thermal conductivities which can reach 3000W/(m-k) at room temperature[10]. Recently, few studies focus on the performance of MCHS using MWCNT nanofluids, so in this paper, performance of MCHS using MWCNT nanofluids treated as a single-phase flow is simulated compared with other liquid based on FLUENT.

2. Micro-channel heat sink with MCNT nanofluids

2.1 Geometry

Figure 1 shows the schematic of a high-performance heat sink. In our simulation three different kinds of liquid as the working coolants are compared flowing through numerous rectangular channels. The width, height and length of the typical heat sink are 1cm, 400um and 1cm respectively [1]. The top face which is made of glass is insulated while the base of the heat sink is heated with a constant heat flux q". The fluid flows through the channel with a varied inlet velocity and constant temperature 300K. Figure 2 describes a representative micro-channel and the three dimensional mesh. The width of the channel and the fin are both 50um, the height of the channel is set to be 300um.

2.2 Characteristics of the heat sink

The hydraulic diameter, D_h, defined as

$$D_h = \frac{4A_c}{p} \qquad (1)$$

In this case, the hydraulic diameter is 300um. The performance of a heat sink is measured by its thermal resistance R_{th}. Knight [4] expresses it as

$$R_{th} = \frac{T_{w,\max} - T_{f,in}}{q''} \qquad (2)$$

where $T_{w,\max}$ and $T_{f,in}$ stand for the maximum wall temperature and the temperature of inlet flowing liquid. Usually, the thermal resistance is calculated under certain power, P_p, defined as

$$P_p = \Delta p \cdot \dot{V} \qquad (3)$$

where Δp and $\dot{V}$ are pressure drop and volume flow rate respectively.

2.3 Governing equations

In order to simulate the heat transfer performance of the heat sink with nanofluids, the energy equation for both the coolant and the fin is solved employing FLUENT. In our model, the flow is assumed to be laminar, incompressible and both hydrodynamically and thermally fully developed so the governing equations are as follows:

Continuity equation:

$$\nabla \cdot \vec{u} = 0 \qquad (4)$$

Momentum equation:

$$(\vec{u} \cdot \nabla)\vec{u} = -\frac{1}{\rho}\nabla p + \frac{u_{eff}}{\rho}\nabla^2 \vec{u} \qquad (5)$$

978-1-4244-2739-0/08/$25.00 ©2008 IEEE

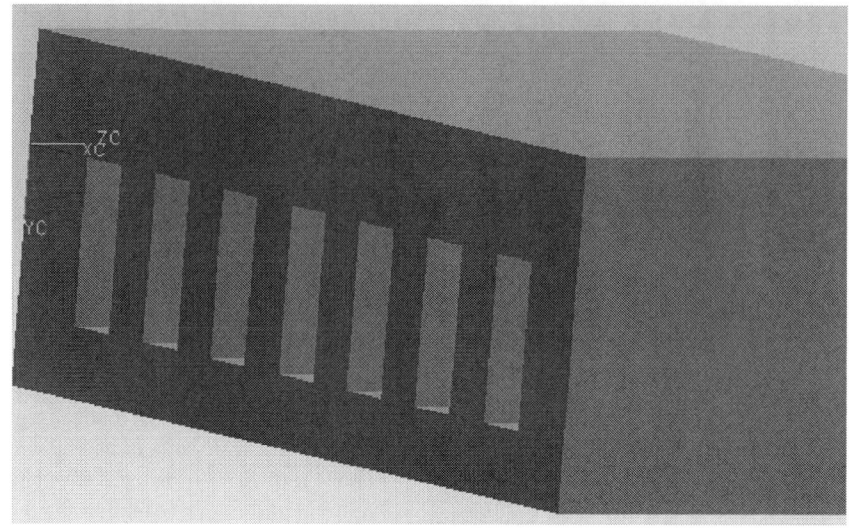

Fig.1 Geometry of the heat sink

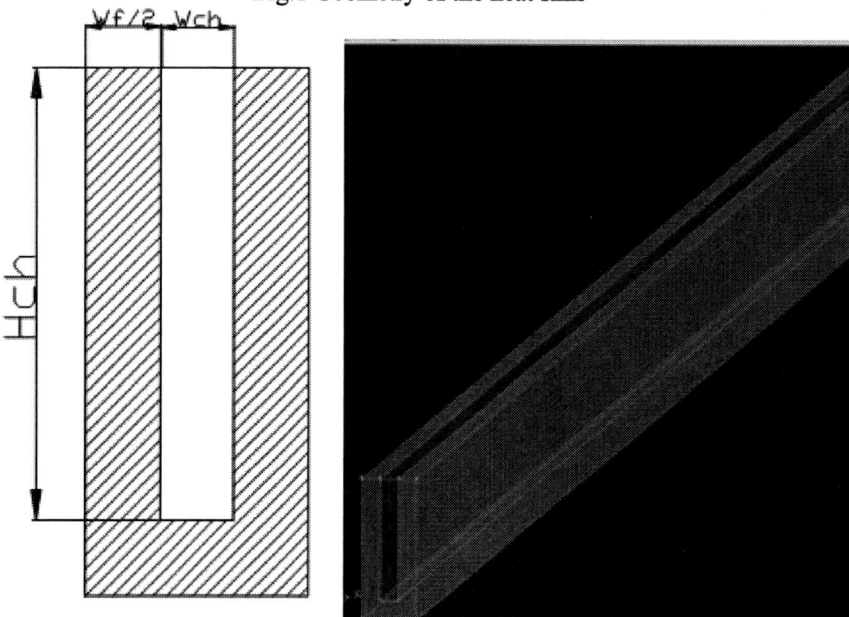

Fig.2 Representative one micro-channel and the three-dimensional mesh

Energy equation for solid:

$$(\vec{u} \cdot \nabla)T = \frac{k_{eff}}{\rho C_p} \nabla^2 T + \frac{u_{eff}}{\rho C_p} \Phi \tag{6}$$

Where,

$$\Phi = (\frac{\partial u_i}{\partial x_j} + \frac{\partial u_j}{\partial x_i}) \frac{\partial u_i}{\partial x_j} \tag{7}$$

Energy equation for fluid:

$$\nabla^2 T = 0 \tag{8}$$

2.4 Theoretical model for the thermal properties of nanofluids

In this study, the effective thermal conductivity model based on kinetics, Kapitza resistance, and convection developed by Jang and Choi [11] is considered.

$$k_{nanofluids} = k_{BF}(1-f) + \beta k_{particle} f + 3C_1 \frac{d_{BF}}{d_{nano}} k_{BF} \text{Re}_{nano}^2 P_r \tag{9}$$

According to the Einstein model [12],

$$\mu_{eff} = \mu_f(1+1.25f) \quad \text{for} \quad f \leq 0.05 \tag{10}$$

In addition, the effective specific heat and density of nanofluids are written as follows by mixing theory [13, 14]

$$C_{p,eff} = C_{p,f}(1-f) + C_{p,particle} f \tag{11}$$

$$\rho_{eff} = \rho_f(1-f) + \rho_{particle} f \tag{12}$$

where u_f, u_{eff}, $C_{p,f}$, $C_{p,particle}$, $C_{p,eff}$, ρ_f, $\rho_{particle}$ and ρ_{eff} are viscosity of base fluid, effective viscosity of nanofluids, specific heats of base fluid, nanoparticles, effective specific heat of nanofluids, density of base fluid, nanoparticles, and the effective density of nanofluids, respectively.

Table 1 Geometry dimensions and boundary conditions

H_{ch}	W_{ch}	W_f	$T_{f,in}$	u_{in}
300um	50um	50um	300K	6.8m/s

Table 2 Comparison between the thermal resistance of Tuckerman [1] and that of simulated results

	Experimental tests by Tuckerman and Pease[1]	Simulated results
Volume flow rate cm^3/s	8.6	8.6
H_{ch} um	302	300
Heat flux W/cm^2	790	790
Thermal resistance $^0C/W$	0.09	0.10

3. Results and discussion

3.1 Simulation results compared with experimental test [1]

Table 1 shows the simulated data such as the inlet temperature $T_{f,in}$, inlet velocity u_{in} and etc. From table 2, the difference between experimental tests [1] and simulated results is $0.01\,^0C/W$, cause in this study, only one typical channel is calculated and also some other factors. From the comparison, our simulated results have a good agreement with experimental results.

3.2 Nanofluids enhancement

Fig.3 shows that as the volume flow rate increases, the temperature changes slowly, and b) displays the temperature contour.

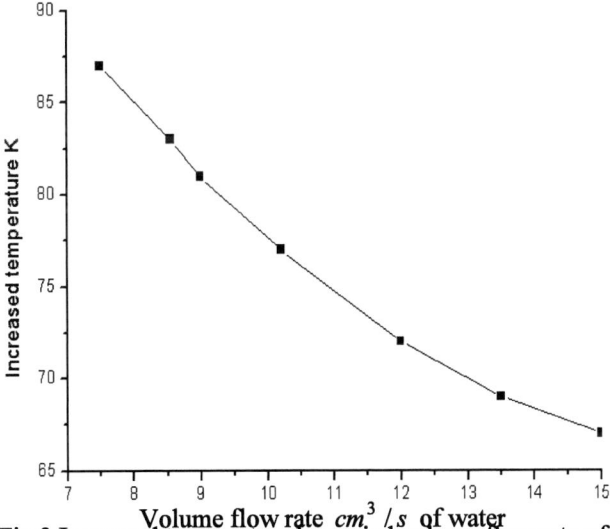

Fig.3 Increased temperature by varied volume flow rate of water

The calculated thermal resistance of the heat sink for the CuO-water mixture at different volume fractions are given in Fig.4.Nanofluids can improve the thermal performance of the MCHS: the larger the volume fraction of nanoparticles, the higher is the thermal performance. A volume fraction of 1% and 4% CuO-water nanofluid shows an average 2.813% and 12.6% enhancement of thermal performance respectively.

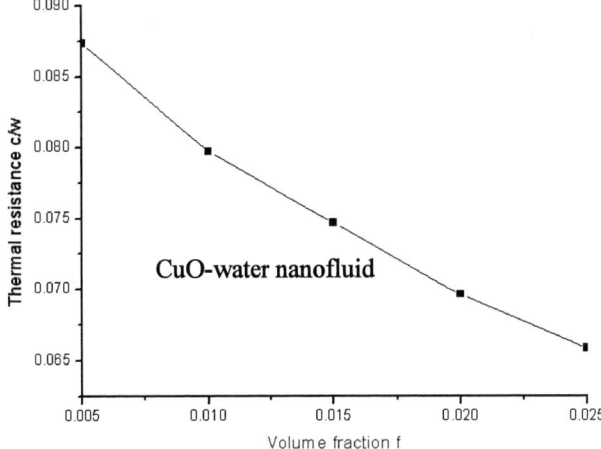

Fig.4 CuO-water mixture enhancement

3.3 Micro-channel heat sink with MWCNT nanofluids

Fig.5 shows the temperature contour of one channel heat sink with MWCNT nanofluid and Fig.6 gives us the outlet temperature displacement. The thermal conductivity enhancement of water-based MWCNT nanofluid is increased up to 11.3% at a volume fraction of 0.01[15]. The heat performance of the heat sink with different nanofluids is compared in Fig.7 under the condition that only the thermal conductivities of these liquid are different from each other for the difficulty to predict the heat capacity of MWCNT nanofluids. The water-based MWCTNT nanofluids enhance by 5.626% than pure water. CuO and SiO$_2$ nanofluids almost have the same enhancement by 2.813% compared to pure water.

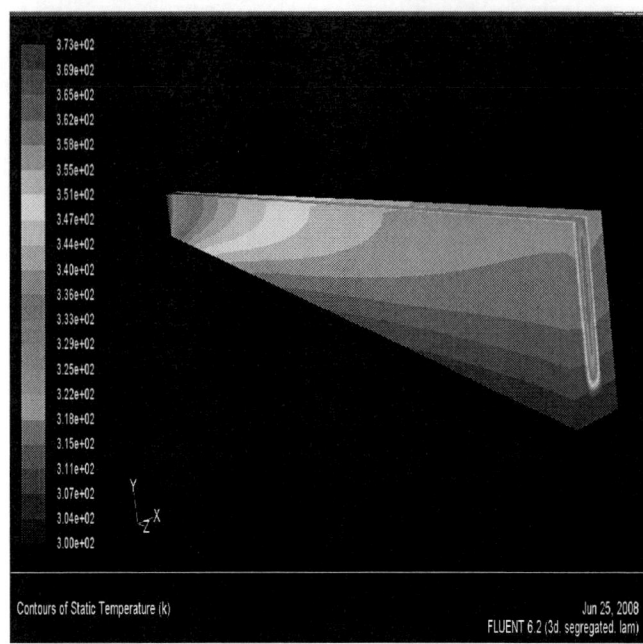

Fig.5 Temperature contour of one channel heat sink with MWCNT nanofluid

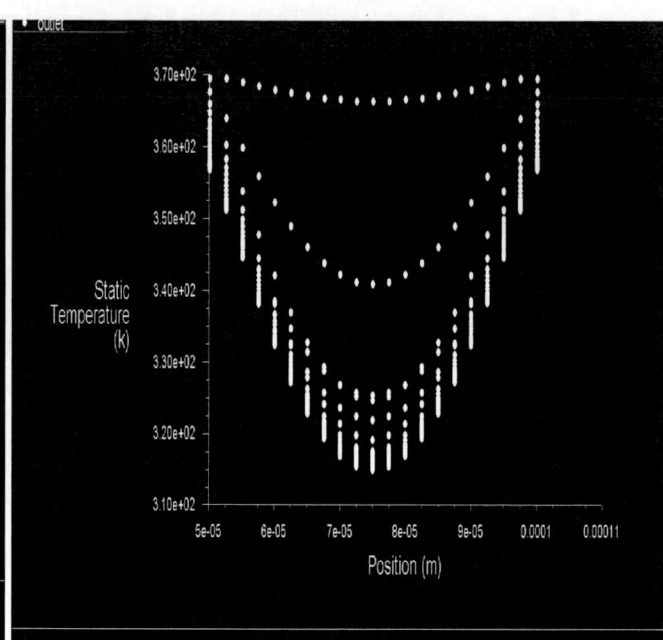

Fig.6 The outlet temperature displacement

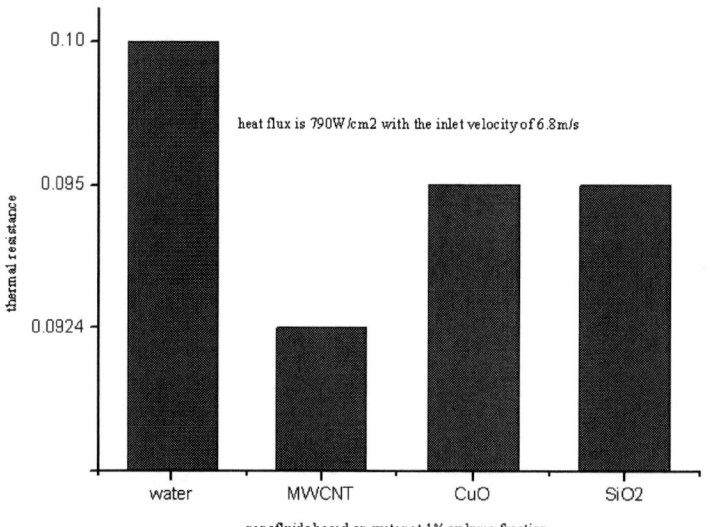

Fig.7 Comparison of different coolants

4. Conclusions

In this paper, the heat performance of heat sink with nanofluids is simulated. The results show that nanofluids do enhance the thermal performance and the micro-channel heat sink with MWCNT nanofluids have the best thermal performance than other nanofluids such as CuO and SiO_2 nanofluids for the high thermal conductivity of MWCNT.

Acknowledgments

The authors acknowledge the financial support from 863 program (No.2008AA04Z301) and Shanghai Municipal Education Commission (No.08YZ15).

References

1. Tuckerman D.B. et al, "High-performance heat sinking for VLSI", IEEE Electronic Devices Letters, EDL 2 (1981) 126–129.

2. Goldberg N. "Narrow channel forced air heat sink" IEEE Trans. Comp. Hybrids Manuf. Technol., vol. CHMT-7, pp. 154-159, Mar. 1984.

3. Phillips. R.J. "Microchannel heat sinks" in: A. Bar-Cohen, A.D. Krous (Eds.), Advances in Thermal Modeling of Electronic Components and Systems, vol. 2, ASME, New York, 1990.

4. Knight R.W. et al, "Optimal thermal design of forced convection heat sinks-Analytical", ASME J. Electron. Packag. 113 (1991) 313–321.

5. Choi S.U.S., "Enhancing thermal conductivity of fluids with nanoparticles". ASME FED 231, 99±103.

6. Koo J., Kleinstreuer C., "A new thermal conductivity model for nanofluids", J. Nanoparticle Res. 6 (2004), in press.

7. Xuan Y., Li Q., "Heat transfer enhancement of nanofluids", Int. J. Heat Fluid Flow 21 (2000) 58–64.

8. Xuan Y., Roetzel W., "Conceptions for heat transfer correlation of nanofluids", Int. J. Heat Mass Transfer 43 (2000)3701–3707.

9. Xuan Y., Li Q., "Investigation on convective heat transfer and flow features of nanofluids", ASME J. Heat Transfer125 (2003) 151–155.

10. Xie H. et al, "Nanofluids containing multiwalled carbon nanotubes and their enhanced thermal conductivities", J. Appl. Phys. 94 (8) (2003) 4967– 4971.

11. Jang S.P., Choi S.U.S., "The role of Brownian motion in the enhanced thermal conductivity of nanofluids", Appl. Phys. Lett. 84 (2004) 4316– 4318.

12. Einstein A., "Investigation on the Theory of Brownian Movement", Dover, New York, 1956.

13. Smith J.M., Van Ness H.C., "Introduction to Chemical Engineering Thermodynamics", McGraw-Hill, New York, 1987.

14. Jang S.P., Choi S.U.S., "Free convection in rectangular cavity (Benard Convection) with nanofluids", in: Proc. IMECE, Anaheim, USA, 2004.

15. Hwang Y.J. et al, "Investigation on characteristics of thermal conductivity enhancement of nanofluids", J. Appl. Phys. Lett. 6 (2006) 1068– 1071.

Flip-chip on Board packaging of a Thermal Wind Sensor

Guang-ping Shen, Ming Qin, Qing-An Huang, Hua Zhang, Jian Wu
Key Laboratory of MEMS of Ministry of Education, Southeast University
210096 Nanjing China
sgpapple@hotmail.com

Abstract

A two dimensional wind sensor was designed, fabricated and packaged on ceramic substrate instead of silicon substrate. The Ti/Pt heater and thermistors were fabricated using single lift-off process. The gold bumps were then sputtered and patterned on the chip using lift-off process again. Correspondingly, the Pb/Sn bumps were fabricated on the FR4 substrate using stencil printing method after metallization. The sensor chip was flip-chip packaged on the FR4 substrate, and the gap was filled with epoxy-based underfill to improve the structure strength and thermal isolation. The wind velocity and direction offsets of the sensor were analyzed and compensated using software and hardware calibration. The packaged sensor was tested in wind tunnel in constant power mode. Both the simulation and test results show that the thermal wind sensor can measure wind speeds up to 10m/s with an accuracy of 0.5m/s, and wind direction in a full range of 360° with a resolution within 5°.

Introduction

The thermal flow sensors have been accepted as a promising technology for industrial process, weather forecast, and biomedical applications in the past 10 years. For the thermal wind sensor packaging, the chip has to be in contact with the surrounding air while being protected against mechanical contact and undesired media. On the one hand, the sensitive parts must be partially transparent to the air flow; on the other hand, both the processing circuit and sensor need good environment isolation. The traditional packaging method is gluing the senor chip to the backside of a thin ceramic carrier, and wire bonding is also applied on the backside to eliminate the problem of disturbance of the flow by the bonding wires [1]. The flip-chip packaging for anemometers was first proposed by H. Battles [2], the chip is flip-chip mounted on a ceramic substrate and an opening is aligned with the sensor that enables free convection. The flip-chip packaging was improved by JB Sun [3] using copper pillar bump technology.

In order to simplify the fabrication and packaging processes, a two dimensional wind sensor fabricated on ceramic substrate using lift-off process is presented in this paper. The sensor is composed of Ti/Pt heaters and thermistors, and it measures the wind velocity and direction with the typical calorimetric principle. Gold and Pb/Sn bumps were fabricated on sensor chip and FR4 substrate, respectively. The sensor chip was flip-chip packaged on the FR4 substrate. After the software and hardware calibration of the packaged wind senor, the wind tunnel test was carried out in constant power (CP) mode. The thermal wind sensor can measure wind speeds up to 10m/s with an accuracy of 0.5m/s, and wind direction in a full range of 360° with a resolution within 5°.

Sensor structure and fabrication process

As shown in Fig.1, a two dimensional wind sensor was fabricated on ceramic substrate, which consists of two heaters in the center and four symmetrically located thermistors. The chip size is 6mm×6mm, the heater size is 1mm2, the gap between heaters and thermistors is 300μm, and the pattern width of the resistors is 25μm. Further, as shown in Fig.3, the sensor was flip-chip packaged on the FR4 substrate. The wind sensor senses the air flow with its backside, while the heaters and thermistors are on the front side. The ceramic chip does not only provide the smooth surface for the flow, but also protects the sensing parts. For the moderate thermal conduction of ceramic substrate, the principle of calorimetric flow sensor still works.

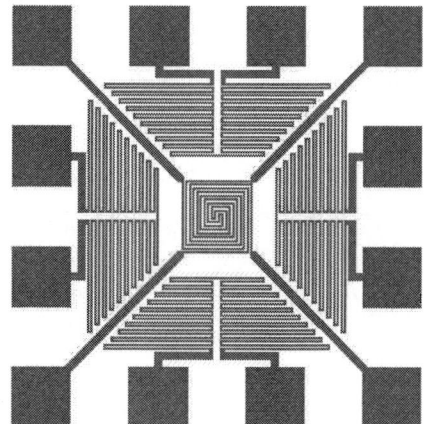

Fig.1 Structure of the thermal wind sensor

The designed sensor was fabricated using microfabrication technology [4]. After the pre-cleaning process, 2μm thick photoresist was spin-coated and patterned on the ceramic substrate. Once the Ti 500Å and Pt 2000Å were sputtered, unwanted Ti/Pt layers were removed by a lift-off process, leaving the heaters and thermistors resistors on the ceramic substrate. The Ti layer was used as an interlayer in order to improve the adhesion of Pt resistors. The resistance of the fabricated heaters and sensing thermistor is 500ohm and 1.5kohm at room temperature, respectively. The temperature coefficient of Ti/Pt resistance is 1050ppm in the range of 25°Cto 60°C.

In order to achieve low-cost package of the thermal wind sensor, a flip-chip on board (FCOB) packaging method of the thermal wind sensor is proposed as shown in Fig.7. As a typical chip size package (CSP), the unique interconnect structure of FCOB does not only offer enhanced performance and functionality, but also greatly minimizes the circuitry, weight and cost [5-6]. The interconnect metallurgy of the FCOB technology comprises gold bumps on the ceramic chip pads and the stencil printed 63Pb/37Sn bumps on the FR4

978-1-4244-2739-0/08/$25.00 ©2008 IEEE

substrate. As Ti/Pt layers have already been deposited on the sensor chip, 3000 Å gold bump was sputtered and patterned on the chip pads using lift-off process. Meanwhile, the 1μm 63Pb/37Sn eutectic bump was fabricated using stencil printing process after the metallization of the PCB circuit layout. These eutectic bumps must exactly match to gold bumps on the sensor chip for better alignment during the flip chip package process. The size of the bumps is 500μm, and the bump pitch is 500μm.

As shown in Fig.8, the flip chip packaging process was carried out on the EVGTM SEC 850 Flip-chip machine. The FR4 board and ceramic chip were picked up and assembled using a vacuum chuck sequentially. The alignment of the bumps was accomplished by a beam splitter prism. The solder reflows were performed with a hotplate. The peak temperature is 240°C, and the hotplate temperature remained above solidus temperature of 184°C for 20 seconds. To improve the mechanical strength of the flip-chip construction and thermal isolation from sensor chip to FR4 substrate, epoxy-based underfill was applied [7]. After dispensing the underfill material along one or two edges of the chip, the gap between sensor chip and FR4 was filled by surface tension capillary flow.

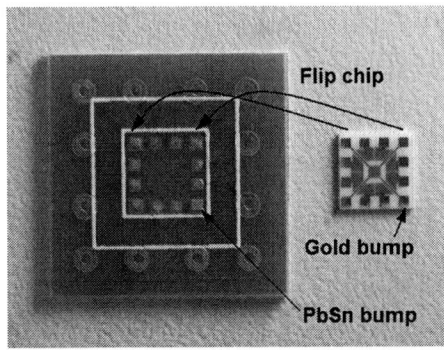

Fig.2 Flip chip packaging of the wind sensor

FEM simulation of the packaged sensor

Considering the low thermal conduction of the FR4 substrate and envelope resin, only the sensor chip and air flow are included in FEM simulation. Fig.3 is the simplified ANSYS model of the packaged thermal wind sensor, where the thickness of the ceramic chip and air flow is 0.25mm and 0.5mm, respectively.

In order to achieve good mesh quality, the solid model is meshed using block mesh method in ICEM CFD, where air flow, heater R_h, thermistors R_{1-4} and ceramic chip were defined as blocks and meshed individually. The air flow and the chip were defined as fluid zones and solid zones in ANSYS CFX, as well as the Fluid-Solid Interface (FSI). Once the inflow and outflow boundary conditions and heat generation were set, the temperature distribution of the sensor and airflow in CP mode can be given by steady-state simulation.

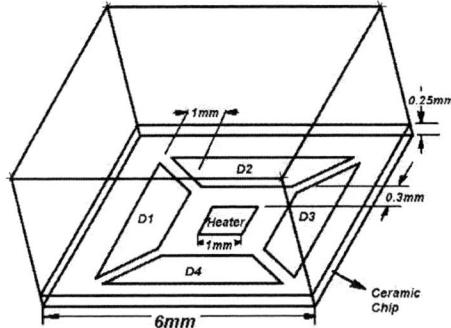

Fig.3 The FEM model of the packaged sensor

In order to investigate the thermal behavior of the sensor, the temperature distribution of the sensor under different power, wind speed and direction were simulated systematically. When the wind flow through the sensor, the air flow will take the heat to the fluid downstream, making the temperature gradients of the chip is no longer symmetrical. Fig. 4 shows that when wind speed increases, the temperature difference between upstream and downstream increases until saturated. As shown in Fig. 5, the temperature difference in 2m/s is cosine (sine) function of wind direction Besides, the simulation results show that the increase of heating power can significantly improve sensor sensitivity, and there must be a compromise between sensitivity and power consumption for handheld wind sensors.

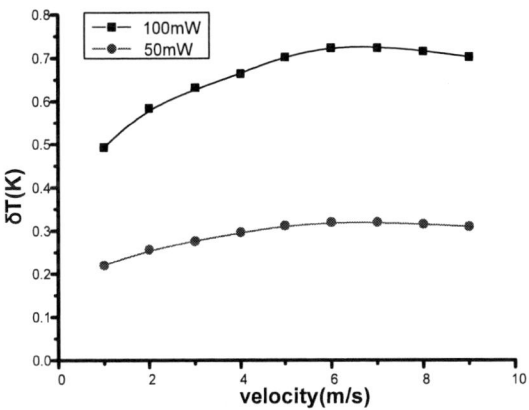

Fig.4 Simulated temperature difference of the sensor surface versus gas flow speed.

For an ideal sensor with symmetrical structure, a hotspot occurs in the middle of the sensor in absence of flow. Because of the manufacturing and dicing tolerances, the sensor chip becomes structurally and thermally asymmetrical and the hotspot shifts from its central position.

In order to investigate the influence of the chip location on the ceramic, the FEM model in Fig.3 was revised to predict the offset due to 1-D structure asymmetry. As shown in Fig.6, when the wind direction changes, T_{EW} remains a sine curve with an offset of , and T_{NS} remains the same for an offset of m along the X direction. The systematical simulation results show that the offset can be approximated to be a linear relationship, as shown in Fig.7.

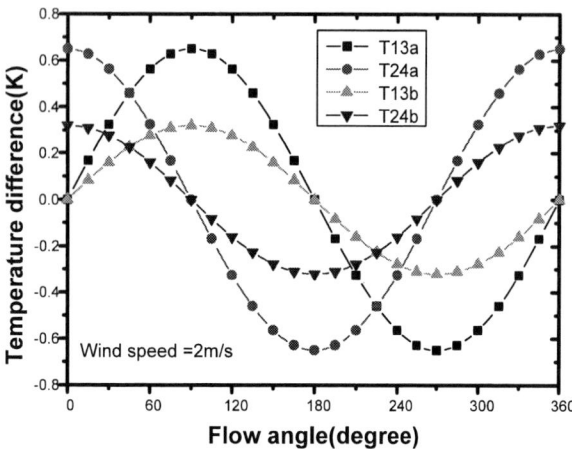

Fig.5 Simulated temperature difference of the sensor surface versus direction .

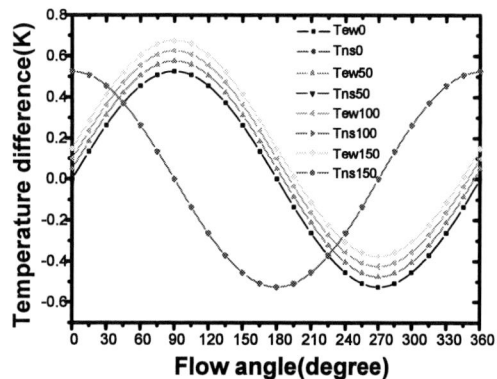

Fig.6 Temperature difference in the sensor versus air flow angle with displacement

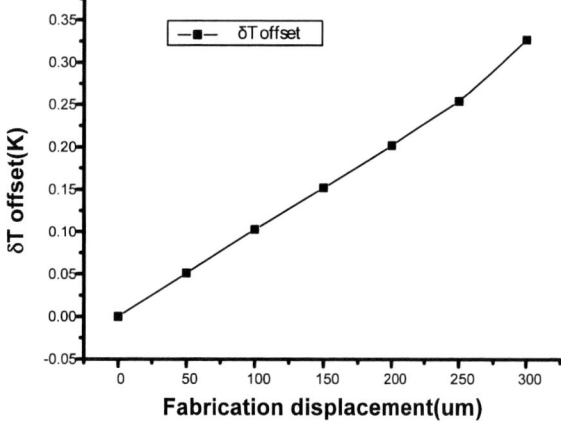

Fig.7 Temperature difference in the sensor versus fabrication offset

In order to verify the simulation results, a wind sensor with a dicing error of 200μm was measured. As shown in Fig.13, the output of the unheated chip is 1.812 V in the absence of wind flow. The middle point of the bridge output increases to 1.961 V when the heating power is 100 mW at the wind speed of 5m/s. The voltage offset is about 0.15 V, which corresponding to a temperature difference of 0.24 K.

The test results agree well with the simulation results shown in Fig. 7.

Experimental results and discussions

As shown in Fig.8, the block diagram of the wind sensor is composed of conditioning circuits and microprocessors. In real applications, the CP mode is approximated with constant voltage (CV) mode. The microvolt-level bridge outputs are boosted by low-offset instrument amplifier AD623 with 500 times amplification. The analog signals were digitalized by the 10 bit ADC of the micro-processor C8051F330.

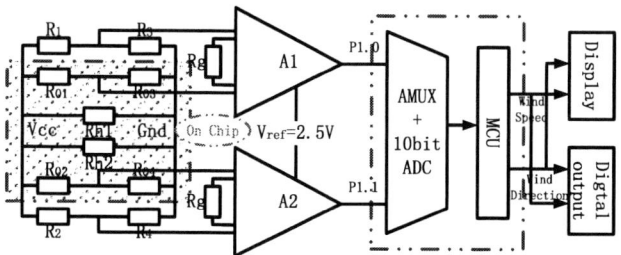

Fig.8 The hardware block of the wind sensor system

After the software calibration, the wind speed was approximated by linear interpolation of $V_{NS}^2 + V_{EW}^2$, and the wind direction was computed by $actan(V_{EW}/V_{NS})$. It is worth noting that there will be a quadrant error in direction description. Finally, the wind speed and direction values were transferred to digital output and LCD display. Fig. 9 shows the wind sensor microsystems including the conditioning circuits and microprocessor.

Fig. 9 Photograph of the wind sensor micorsystem

The wind sensor system was tested in the wind tunnel, including both the wind speed and direction tests. As shown in Fig.10, when the wind speed increases from 0 to 8m/s, the bridge output will increase until saturated. Furthermore, when the heating power increases, higher sensitivity can be easily achieved. Fig.11 shows that the bridge output vs. flow angle at 5m/s with an interval of 15°. The two outputs were sine (cosine) curve of the flow angle, with a_{NS} =0.05V, a_{EW} =0.01Vand θ_0=3°. After software calibration, as shown in Fig.18, the wind direction resolution is with 5°.

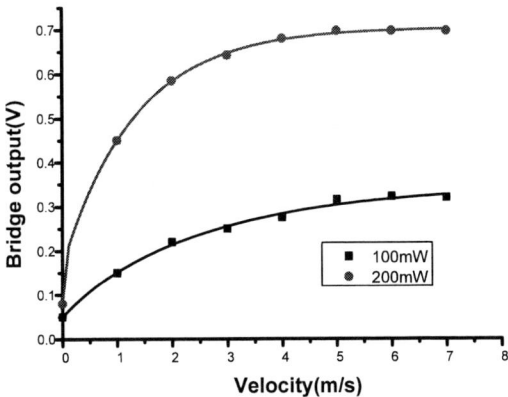

Fig. 10 Bridge output versus air flow speed

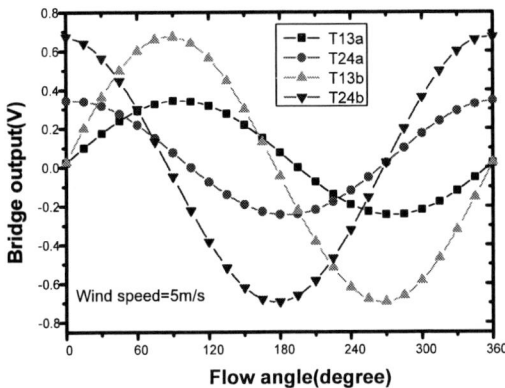

Fig. 11 Bridge output versus air flow direction

Conclusion

This paper presents the design, fabrication, packaging and calibration of a 2-D wind sensor. In order to improve the sensitivity, the wind sensor was fabricated on ceramic substrate instead of silicon substrate. It also presents a flip-chip on board packaging technology for the 2-D wind sensor. With this package, the backside of sensor chip is directly exposed to the airflow, and the sensing parts are protected by the ceramic chip. Both the simulation and experimental results show that it still keeps the calorimetric thermal flow sensor principle. Besides, three main offsets of the wind sensor were analyzed and calibrated to improve the resolution. Finally, the wind sensor was tested in the wind tunnel. The measurement of wind speeds up to 10m/s with an accuracy of 0.5m/s, and wind direction in a full range of 360° with a resolution within 5° is demonstrated.

Acknowledgment

Project supported by Natural Science Foundation of China under Contract No.90607002 and 60476019.

References

1. Oudheusden BW 1988 Silicon flow sensors. *Control Theory and Applications, IEE Proceedings* 373-380
2. Mayer, F., Paul, O., and Baltes, H.: 'Flip-chip packaging for thermal CMOS anemometers', Micro Electro Mechanical Systems, 1997. *MEMS'97, Proceedings, IEEE., Tenth Annual International Workshop on*, 1997, pp. 203-208
3. Sun JB Qin M Huang QA 2007 Flip-Chip packaging for a two-dimensional thermal flow sensor using a copper pillar bump technology *IEEE Sensors Journal* **7** 990-995
4. Shen, G.P., Wu, J., Zhang, H., Qin, M., and Huang, Q.A: 'Design of a 2 D Thermal Wind Sensor Based on MEMS Process', Bandaoti Xuebao(Chinese Journal of Semiconductors), 2007, 28, (11), pp. 1830-1835
5. Lau J 2000 *Low Cost Flip Chip Technologies for DCA, WLCSP, and PBGA Assemblies* (McGraw-Hill Professional) 183-210
6. Yegnasubramanian S. Deshmukh R *etal* 1997 Flip-chip-on-board (FCOB) assembly and reliability *Electronics Manufacturing Technology Symposium Twenty-First IEEE/CPMT International* 32-36
7. Gamota DR Melton CM 1997 The development of reflowable materials systems to integrate the reflow and underfill dispensing processes for DCA/FCOB assembly *Components, Packaging, and Manufacturing Technology* **20** 183-187

System-in-Package Solutions with Embedded Active and Passive Components

Rolf Aschenbrenner
Fraunhofer Institute for Reliability and Microintegration
Gustav-Meyer-Allee 25
13355 Berlin, Germany
e-mail: rolf.aschenbrenner@izm.fraunhofer.de, Tel. +49 30 464 03-164

Abstract

Future generations of electronic products require further developments of integration and packaging technologies. The reasons for this are higher signal frequencies and the increasing functional density at acceptable costs. With the existing technologies organic substrates with high-density built-up layers with microvias can be produced. On both sides of the substrates, passive and active components can be assembled. The surface demand at the side of a printed circuit board for active components can be reduced to a minimum by the application of CSPs (Chip Size Packages) or flip chips. However a further miniaturization requires a three-dimensional integration of the components. Advanced packages contain stacked chips, which are connected by bond wires with an interposer or a lead frame. Apart from the miniaturization the new applications require signal frequencies of several GHz, which can only be recalled with difficult due to the long bond wires and the extensive connection paths on the printed circuit board. Signal integrity requires connections that are much shorter and impedance-matched. This can be reached by embedded components. Embedding signifies that the conductor is not only located under the embedded components, but also on top of them. This enables to continue three-dimensional packaging on top of the embedded component as well. The component is electrically connected with the upper or lower conducting layer or with both of them, e. g. as it is the case in power ICs with contacts on both sides.

The Fraunhofer IZM and the Technical University Berlin jointly develop advanced technologies for embedding of active chips for system-in-package (SiP) applications. The first development was the so-called chip-in-polymer technology, which enables the realization of SiPs as well as printed circuit boards with integrated components. It is based on the embedding of thin chips into built-up layers under the consequent use of printed circuit board technologies (PCB technologies). Electrical contacts to the chips are realized by laser-drilled and metallized microvias.

Chip in Polymer

A new concept for the integration of active components is the so called chip in polymer technology [1]. It is based on the embedding of ultra-thin chips into build-up layers of printed circuit boards (PCBs). The interconnect structure, which is neither a flip chip nor a wire bond, is shown in figure 1.

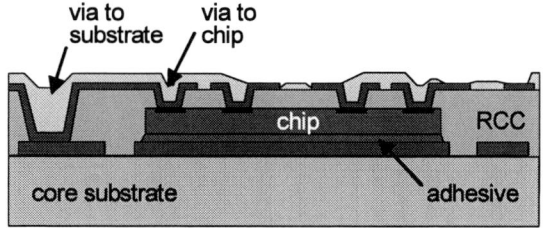

Figure 1: Principle of interconnection of an embedded chip into a PCB built-up layer

The basic idea of Chip in Polymer is that a thin semiconductor chip fits into a standard PCB construction and the technology can be used for fabricating 3D-stacks of multiple dies. In order to achieve a surface which is compatible to a PCB metallization process, the Al contact pads are covered by Cu bumps. Then wafers are thinned down to 50 µm in order to make the bare dies suitable for embedding into build-up layers. The next step is to die bond the chips using an adhesive. Precise thickness control of the bond line is essential for maintaining uniform thicknesses of the build-up dielectric on top of the chip. Therefore different thin chip handling and assembly solutions using die attach film or adhesive paste printing have been explored. RCC (resin coated copper) layers with thin Cu are used for the lamination. Process parameters had to be tuned in order to avoid damage to the chips during lamination. Contacts to the chip are made by laser drilled microvias followed by a PCB-compatible Cu plating. All the process steps in this technology are set up for large scale manufacturing, using panel sizes of 18" x 24" in a combination with high accuracy positioning methods using local fiducials for die placement, laser drilling and laser direct imaging. Figure 2 shows a cross-section of a Cu interconnect to an embedded chip.

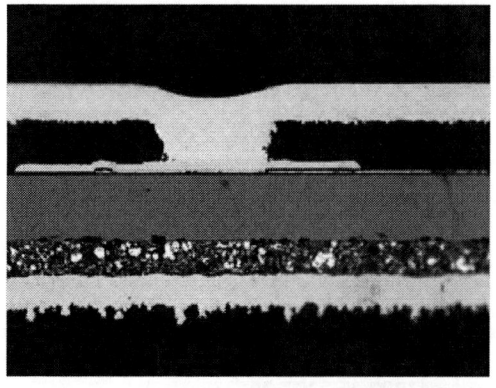

Figure 2: Cu interconnect to embedded chip.

978-1-4244-2739-0/08/$25.00 ©2008 IEEE

Reliability

In order to realize reliability tests, printed circuit boards for testing were prepared with chips of the size of 2,5 mm. The FR4 core substrates had a thickness of 0,5 mm. For chip bonding an Ag-filled adhesive was used. After the chip bonding the 50μm test chip was embedded into an RCC-layer with a dielectric thickness of 80 μm. The in- and outputs of the chip were connected with a daisy chain. After the test, the daisy chains were electrically tested and finally cross sections were created. The following tests were realized:

- Temperature storage at 125 °C for 1000 hours
- Thermal shock condition air-to-air -55 / +125 °C. 2000 cycles
- Humidity storage 85 °C / 85 % rel. humidity for 2000 hours

All tests were passed without contact interruption or a damage of the daisy chain resistors. The cross sections did not show any damage like delamination. Another investigation was a test with regard to susceptibility of moisture during reflow soldering. The samples were tested according to JEDEC level 3, e. g. 168 hours at 30 °C and 60% relative humidity followed by three reflows at 260 °C peak temperature (lead-free condition). No damages could be detected either. Finally during a test according to JEDEC Level 1 (168 hours at 85 °C and assuming 85% relative air humidity) a delamination between chip and chip bond adhesive could be observed. In spite of that these chips did not suffer an interruption of the electrical circuit. Apart from the delamination the daisy chains were intact.

In addition to the experimental evaluation of the reliability the 3D FEM modelling and the simulation of the thermo-mechanical behavior of embedded chips were also realized and published [2]. To sum up the investigations it can be said that these simulations did not show any critical points in the chip in polymer packages.

The main application of the chip in polymer technology is expected in the production of small packages such as stackable chips, SiPs or small modules with only a few chips. The maximum number of embedded chips is determined by the revenue, i. e. of the final costs.

It has to be emphasized that only known good dies should be used.

The first functional device which has been realized in the frame of the project HIDING DIES is a chip card module. It contains a Philips chip with a size of 3,2 x 2,9 mm² and ten connected contacts. The 50 μm chips were bonded onto a 100 μm FR4 core and afterwards the RCC was laminated onto both sides, so that the module has four Cu-layers. The entire thickness of the module is 300 μm. The functionality of the control chip was successfully tested after the realization of the module.

References

[1] R. Aschenbrenner, Andreas Ostmann, Alexander Neumann, Herbert Reichl, "Process Flow and Manufacturing Concept for Embedding Active Devices", December 8-10, 2004, EPTC 2004, Singapore

[2] J.-P. Sommer, B. Michel, A. Ostmann, „Electronic Assemblies with Hidden Dies – Design Support by Means of FE Analysis", Proceedings ECTC September 5-7, 2006, Dresden, Germany

Design Advisor for Package-on-Package (PoP) Manufacturing

Bin Xie, Peng Sun and Daniel Shi
Hong Kong Applied Science and Technology Research Institute (ASTRI)
1st Floor, 2 Science Park East Avenue, Hong Kong Science Park, Shatin, NT, Hong Kong, China
Tel: (852) 3406-2773   Fax: (852) 3406-2805   Email: bxie@astri.org

Abstract

The needs to integrate devices into portable products with smaller form factor and more functionality have fueled enormous growth of 3D packaging technology. The package-on-package (PoP) is one of the 3D packaging solutions, by which the packaging and assembly houses can achieve a lower cost, faster turn benefits and testing prior to assembly. PoP is a complicated system with multi-layered structure, which induces more manufacturability issues, such as stand-off height issue and top & bottom packages having different types of warpage. In order to reduce R&D cost, achieve fast time-to-market and address most of the manufacturability issues during the development of a new PoP, a design advisor for PoP manufacturing has been developed based on the design for manufacturability (DFM) methodology. The key components of this design advisor are the validated numerical models, comprehensive materials library, design guidelines of PoP packaging and novel finite element analysis (FEA) techniques. With the developed novel FEA techniques for curing process simulation and seamless packaging process simulation, complete numerical models for PoP manufacturing were developed and validated, which can simulate the whole PoP manufacturing processes. The design advisor is easy to use by selecting package geometries, material properties and process parameters. By running the envelope-based design advisor for normal package design or the FEA-based design advisor for special package design, the detailed analysis reports can be generated, including simulation results, design evaluations and recommendations to ensure the first-time success of package design. Therefore, the design advisor can help improve the yield of complex PoP manufacturing processes leading to higher quality and confidence of manufacturing processes, faster time-to-market and lower overall manufacturing cost.

Introduction

3D packaging technology grows rapidly because of the needs to integrate devices into portable products with smaller form factor, more functionality, better performance and lower cost [1]. The package-on-package (PoP) is one of the 3D packaging solutions, which typically integrates a high-density digital logic processor in the bottom package with high-capacity memory dies in the top package [1-2]. By adopting PoP packaging technology, the packaging and assembly houses can achieve a lower cost and faster turn benefits, when these two components are sourced from different IC suppliers and stacked on the printed circuit board (PCB) without releasing confidential IPs. It also allows for burn-in and testing prior to assembly in order to mitigate the known-good-die (KGD) related issue. Thus, PoP has been rapidly used in portable products due to its flexibility and testability [3-4].

Since PoP is a complicated system with multi-layered structure, the packaging industry encounters more and more manufacturability issues during the development of PoP packages [5]. In order to stack two packages sourced by different device manufacturers and reflow simultaneously on the PCB, the package-stacking process needs to be carefully developed. The process should enable package-to-package connection with a higher yield even though both the top and bottom packages have different types of warpage, i.e., convex vs. concave [6-8].

In order to reduce the R&D cost, achieve higher confidence of manufacturing processes and shorten the time-to-market, finite element analysis (FEA) was adopted to help package design in previous studies. Tzeng *et al.* studied the effects of package geometries and materials on package warpage under thermal cycling [9]. Luan investigated the solder joint performance of PoP under drop testing using FEA [10]. Amagai *et al.* and Suzuki *et al.* studied the effects of package geometries and materials on package warpage in reflow process [11-12]. However, most of these works adopted FEA to help address the manufacturability issues which have occurred in one or two manufacturing processes. Few FEA works were involved in the initial stage of package design to study the potential manufacturing issues. In addition, single process simulation ignores the impacts of deformation and stress caused by earlier manufacturing processes. Few works adopted FEA techniques to simulate and study the whole PoP manufacturing processes, including die stacking (DS), wire bonding (WB), molding, reflow, underfill, etc. The simulation for entire PoP manufacturing process is important because it not only reveals the stress evolution and deformation history of a PoP package, but also discovers potential manufacturability issues in the whole PoP manufacturing process prior to a real package fabrication. In addition, PoP manufacturing processes are standard and mature. Therefore, a design advisor for PoP manufacturing has been developed based on the design for manufacturability (DFM) methodology to ensure the first-time success of package design.

Concept and Structure of Design Advisor

The real packaging manufacturing processes, such as DS, WB, molding and reflow, are sequential and continuous. As shown in Fig. 1, the applied thermal and mechanical loadings on PoP are various but continuous in different manufacturing processes. Therefore, the deformation level of packages will be accumulated during the whole manufacturing processes (as shown in Fig. 2), inducing the stress accumulation as well. Based on the facts that applied loadings, deformation variation and stress evolution are sequential and continuous, this work developed a design advisor, which integrates the modeling of all the key PoP manufacturing processes into one

978-1-4244-2739-0/08/$25.00 ©2008 IEEE

indicator. In addition, the standardization and maturity of PoP manufacturing processes ensure the feasibility of the design advisor. The design advisor includes a large envelope of package geometry, material property and process parameter. Within the envelope, the envelope-based design advisor can generate the evaluations of package design immediately. The user can also define the special package design and obtain the evaluations from FEA-based design advisor if the package design is beyond the envelope. In summary, the design advisor integrates the methodology of design for manufacturability (DFM), helps address most manufacturability issues in the initial stage of package design and ensures the first-time success of package design.

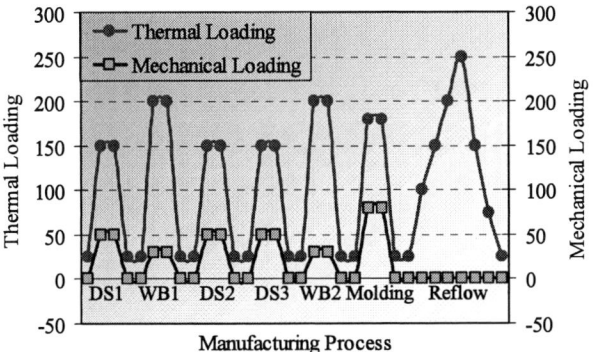

Fig. 1: Schematic diagram of applied thermal and mechanical loadings on PoP in real manufacturing processes.

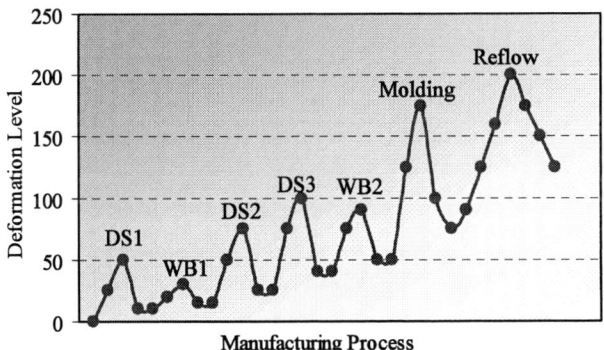

Fig. 2: Schematic diagram of deformation level of packages in real manufacturing processes.

The interface of the design advisor for PoP manufacturing is shown in Fig. 3. In the section of Model Generation, user can define the geometries of key materials of PoP module, i.e., substrate, die-attach film, die, molding compound (MC) and solder ball. The section of Material Selection is established with a comprehensive material library. By selecting the material properties provided in the library, material properties can be assigned to each material. Typically, PoP assembly processes compose of DS, WB, molding, reflow and underfill. After defining the process parameters given in the section of Process Setting, the design advisor can analyze and evaluate the package design. By clicking the button of "Reports & Recommendations", a

report will be generated immediately to list the details of package evaluations. If the current design of PoP can not meet the design guidelines or may create potential manufacturability issues, the report can provide the recommendations for package redesign, material re-selection and process re-setting. For most cases, the design advisor can address most manufacturability issues and provide final package design prior to the real PoP manufacturing.

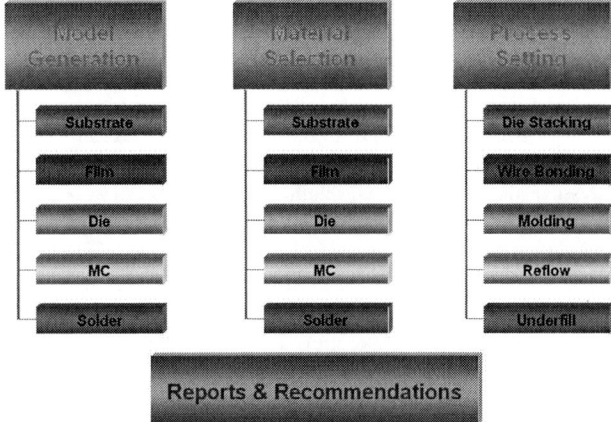

Fig. 3: Interface of the design advisor for PoP manufacturing.

The key components of this design advisor are the validated numerical models, the comprehensive materials library, the design guidelines of PoP packaging and the novel FEA techniques. The numerical models, which were validated by moiré technique, can simulate all the key PoP manufacturing processes, i.e., DS, WB, molding, reflow, underfill, etc. The material properties used in the design advisor came from the materials library, which contains the properties of all materials in PoP module characterized by both vendors and packaging houses, such as elastic, plastic, thermal and moisture related properties. The design guidelines of PoP packaging were developed based on the specifications of leading companies. They are the key criteria for the evaluation of the package design, material selection and process parameter setting. In the development of design advisor, two novel FEA techniques were introduced to enable the PoP manufacturing process simulation in a more realistic manner than ever before. This paper mainly focuses on the development and implementation of novel FEA techniques using ABAQUS version 6.7-3.

Novel Finite Element Analysis Techniques

The numerical models simulating the whole PoP manufacturing processes are the keys of the design advisor. To implement the whole process simulation, two difficulties should be overcome. One difficulty exists in the simulation of MC curing process, because MC curing is a chemical reaction and existing commercial FEA software can not handle the simulation of chemical reaction. The other difficulty is to link all the manufacturing processes seamlessly to ensure the deformation variance and stress evolution continuous. Therefore, this work developed two novel FEA techniques to address two difficulties respectively. The following topics give the development of novel FEA techniques and implementation of the techniques to a test vehicle, which was validated by moiré technique.

1. Curing Process Simulation

MC curing process is an important process, which determines the warpage pattern and value of a package after it cools down from molding temperature to room temperature. At the beginning of MC curing process, MC is extremely soft with negligible modulus. After MC fully curing, its modulus increases significantly at the GPa level even for the same high temperature. Currently, few works were reported to simulate the curing process because it is a chemical reaction process and the existing commercial FEA software can not simulate the chemical reaction. To simulate the package performance during curing process at high molding temperature, this work transfers the chemical manner of curing process to the physical manner. The methodology of the simulation is to set the modulus of MC as negligible value (e.g., at the MPa level) in the beginning of curing, while set it as a much larger value (e.g., at the GPa level) after fully curing under the same molding temperature (e.g., 175°C). The different MC modulus can not be set as temperature-dependent material property because the negligible modulus and larger modulus are assigned to MC under the same temperature. Therefore, the different MC moduli under the same temperature are set as the field-dependent material properties.

To implement the FEA technique, a typical PoP numerical model (1/4 model) is shown in Fig. 4. The whole model includes top package, bottom package, PCB, solder balls between top & bottom packages, solder balls between bottom package and PCB, and underfill filled in the gaps among top & bottom packages and PCB.

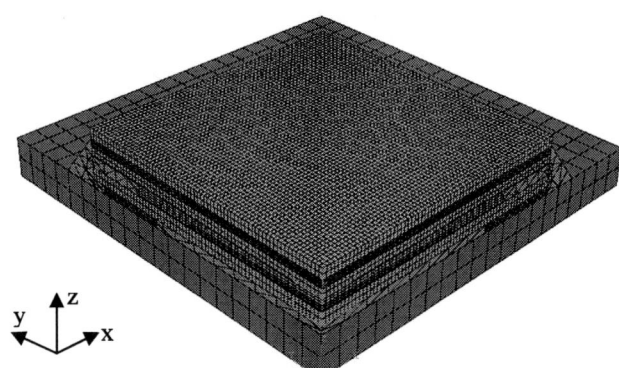

Fig. 4: PoP numerical model (1/4 model) on board level.

The top package and the bottom package are shown in Figs. 5 and 6 respectively. The top package is a fine-pitch ball grid array (FBGA) assembly and the bottom package is a plastic ball grid array (PBGA) assembly. The top and bottom package models (1/4 model) consist of 36 and 92 solder balls, respectively. The detailed geometries of top & bottom packages are listed in Table 1. Temperature-dependent material properties used in the simulations of molding process are listed in Table 2. Please note that the modulus of MC

increases from 1MPa to 2.3GPa for the same molding temperature 175°C.

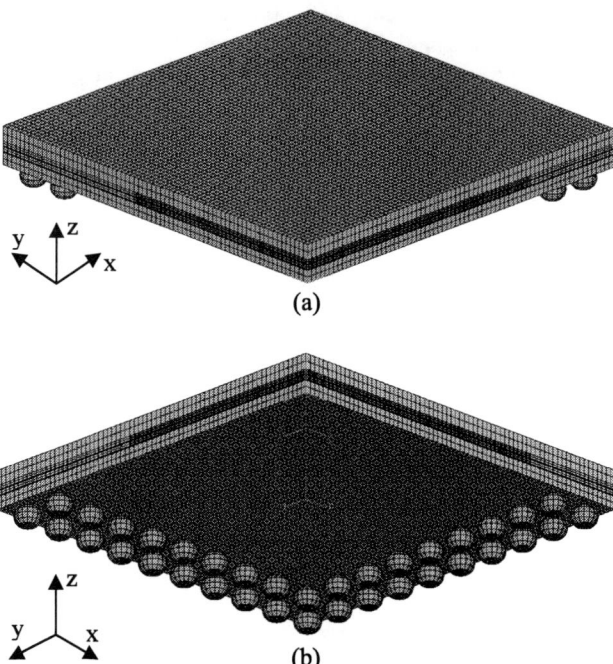

Fig. 5: Numerical model of top package (a) top side and (b) bottom side.

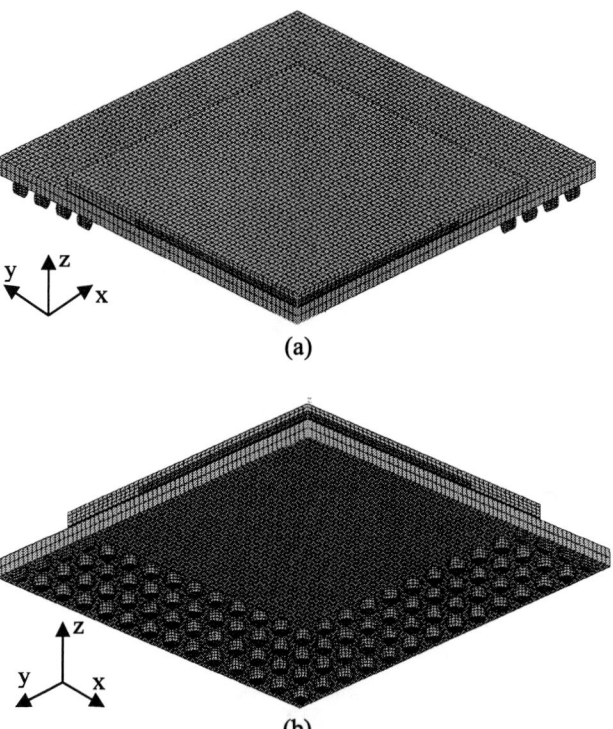

Fig. 6: Numerical model of bottom package (a) top side and (b) bottom side.

Table 1: Geometries of all materials of PoP[1]

	Materials	Length (mm)	Width (mm)	Thickness (mm)
Top Package	Substrate	14	14	0.21
	1st Die-attach Film	10.2	8	0.02
	1st Die	10.2	8	0.075
	2nd Die-attach Film	10.2	8	0.02
	2nd Die	10.2	8	0.075
	Mold Cap	14	14	0.45
Bottom Package	Substrate	14	14	0.3
	Die-attach Epoxy	7.6	7.6	0.03
	Die	7.6	7.6	0.12
	Mold Cap	10.8	10.8	0.27

Note 1: geometries in Table 1 were selected from the section of Model Generation of the design advisor.

Table 2: Temperature-dependent material properties used for the modeling of molding process[2]

	Elastic Modulus (GPa)	Poisson's Ratio	CTE (ppm/^{o}C)
Substrate	22.6 @ 25°C 19 @ 125°C	0.25	20 (x,y), 80 (z)
Die-attach Film	4.8 @ 25°C 0.008 @ 100°C 0.0045 @ 200°C	0.4	245 @ 25°C 300 @ 100°C 300 @ 125°C
Die	130 @ 25°C 129 @ 125°C	0.3	2.81 @ 25°C 3.3 @ 152°C
Molding Compound	28.5 @ 25°C 3.3 @ 125°C 0.001$\rightarrow$2.3 @ 175°C	0.25	13.58 @ 25°C 24.74 @ 125°C 35 @ 144°C 35 @ 175°C

Note 2: material properties in Table 2 were selected from the section of Material Selection of the design advisor.

Table 3: Temperature-dependent material properties used for the modeling of reflow process[3]

	Elastic Modulus (GPa)	Poisson's Ratio	CTE (ppm/^{o}C)
Substrate	22.6 @ 25°C 19 @ 125°C	0.25	20 (x,y), 80 (z)
Die-attach Film	4.8 @ 25°C 0.008 @ 100°C 0.0045 @ 200°C	0.4	245 @ 25°C 300 @ 100°C 300 @ 125°C
Die	130 @ 25°C 129 @ 125°C	0.3	2.81 @ 25°C 3.3 @ 152°C
Molding Compound	28.5 @ 25°C 3.3 @ 125°C 2.3 @ 175°C 0.5 @ 260°C	0.25	13.58 @ 25°C 24.74 @ 125°C 35 @ 144°C 35 @ 175°C 50 @ 260°C

Note 3: material properties in Table 3 were selected from the section of Material Selection of the design advisor.

To validate the numerical models, moiré technique was adopted to measure the package warpage after cooling down to room temperature of 25°C. The warpage pattern of top package on substrate side at 25°C obtained from numerical simulation is shown in Fig. 7(a). Experimental measurement was performed for four times and one of the measurement results is shown in Fig. 7(b). As shown, the warpage pattern obtained from the simulation was similar with that obtained from the measurement. The warpage pattern shows the corners of substrate surface are convex while the center of substrate surface is concave, i.e., convex warpage (or "crying warpage") [13]. The warpage values obtained from the measurement and simulation are summarized in Table 4. As shown, the warpage value from simulation was similar with the average value obtained from the measurement as well. The error between the measurement and simulation is around 6%, indicating the novel FEA technique of curing process simulation was implemented to top package successfully.

The procedure of modeling validation for bottom package is similar with that for top package. The error between the measurement and simulation is around 1%.

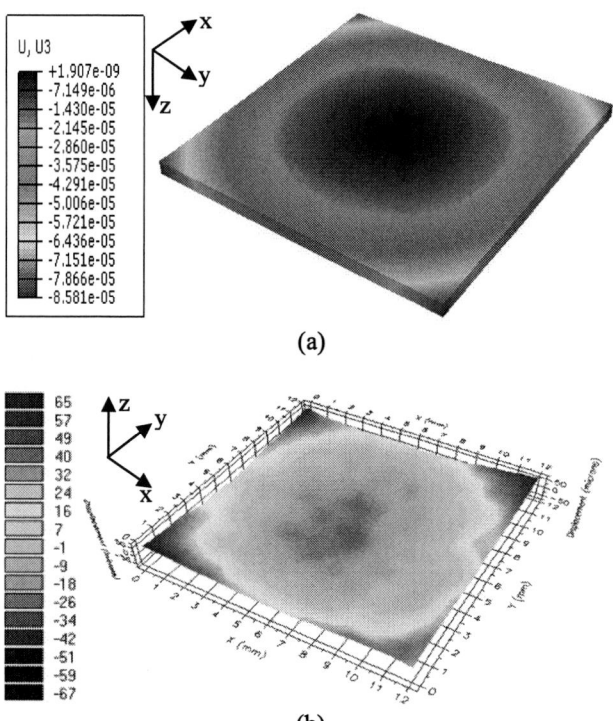

(a)

(b)

Fig. 7: Warpage patterns (at 25°C) of top package on substrate side obtained from (a) the numerical simulation and (b) the experimental measurement.

Table 4: Warpage of top package on substrate side obtained the from measurement and simulation (at 25°C)

Measurement (um)					Simulation (um)	Error
M1	M2	M3	M4	Ave.	86	6%
96	71	71	86	**81**		

2. Seamless Packaging Process Simulation

After MC fully curing at 175°C, the packages cool down to room temperature of 25°C, then encounter the solder reflow process with the temperature increasing to high reflow temperature, e.g., 260°C. Before the reflow process, the packages have been deformed with the stress inside after molding process. Therefore, the simulations of molding process and reflow process are sequential and should be connected. However, the sequential processes are difficult to be set as sequential steps in one simulation job, because the material properties are not only field-dependent (e.g., modulus of MC in curing process), but also temperature-dependent (e.g., materials encountering temperature drop and rising as 175°C→25°C→260°C). The easiest way to connect molding process and reflow process is to set the sequential packaging manufacturing processes as various sequential simulation jobs. The main advantage of this way is that different field-dependent and temperature-dependent material properties can be assigned to according simulation jobs to avoid the confliction of material assignment.

To connect all the sequential simulation jobs, a novel FEA technique for seamless packaging process simulation was developed in this work by linking the initial state of current simulation job with the final state of previous simulation job. The FEA technique of seamless packaging process simulation is essential for whole PoP manufacturing process simulation, because material properties highly depend on the manufacturing processes. For certain processes, some material properties are field-dependent (e.g., modulus of MC for MC curing process and modulus of underfill for underfill curing process); while for the most processes, most material properties are temperature-dependent due to the fluctuations of thermal loadings as shown in Fig. 1. It is noted that the technique of seamless packaging process simulation was implemented to the whole PoP manufacturing process simulations. To demonstrate the technique implementation, this paper adopted two typical sequential manufacturing processes—molding process followed by reflow process.

Using the technique of seamless packaging process simulation, the reflow simulation was performed after the molding simulation. The same test vehicle as shown in Fig. 4 was adopted. The material properties in the reflow simulation are listed in Table 3. By applying a reflow loading selected from the section of Process Setting onto the top package (as shown in Fig. 8), the warpage pattern (at 260°C) of the top package on the substrate side obtained from the numerical simulation is shown in Fig. 9(a). Experimental measurement was performed twice and one of the measurement results is shown in Fig. 9(b). As shown, the warpage pattern from the simulation was similar with that from the measurement. The warpage pattern shows the corners of substrate surface are concave while the center of substrate surface is convex, i.e., concave warpage (or "smiling warpage") [13]. The warpage values from the measurement and simulation are summarized in Table 5. As shown, the warpage value from the simulation was similar with the average value from the measurement as well. The error between the measurement and simulation is around 6%. The matches of warpage patterns and values between the measurement and the simulation indicate the modeling of reflow process at 260°C for top package was validated. Fig. 10 shows the warpage values at various reflow temperatures from the measurement and simulation. The simulation results match well with the measurement results, indicating the technique of seamless packaging process simulation was implemented to top package successfully.

Similarly, the measurement and the simulation were performed to bottom package. The warpage values at various reflow temperatures from the measurement and simulation for the bottom package are shown in Fig. 11. Also, the simulation results match well with the measurement results.

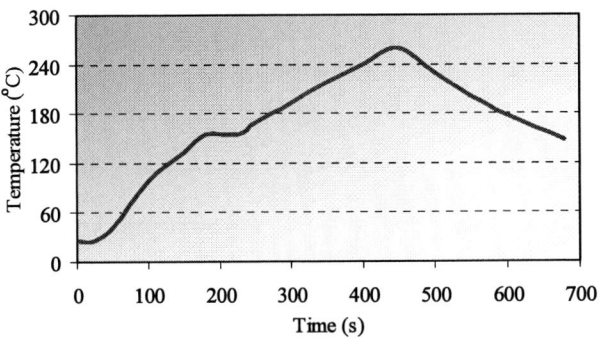

Fig. 8: Reflow loading profile for both numerical simulation and experimental measurement.

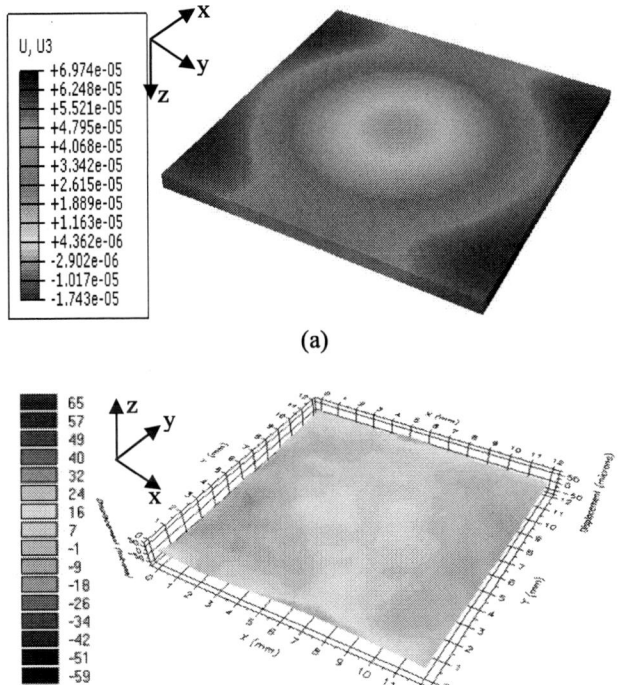

Fig. 9: Warpage patterns (at 260°C) of top package on substrate side obtained from (a) the numerical simulation and (b) the experimental measurement.

Table 5: Warpage of the top package on the substrate side from the measurement and simulation (at 260°C)

Measurement (um)			Simulation (um)	Error
M1	M2	Ave.	66	6%
51	73	62		

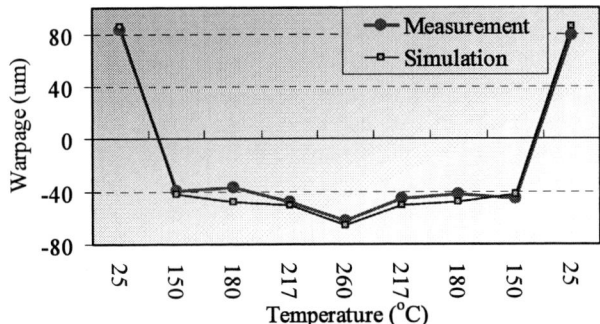

Fig. 10: Warpage values of the top package on the substrate side from the numerical simulation and the experimental measurement during whole reflow process.

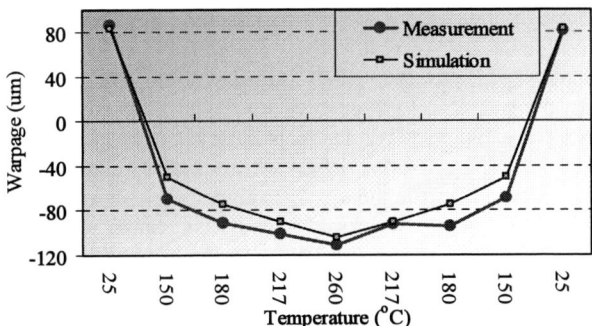

Fig. 11: Warpage values of the bottom package on the substrate side from the numerical simulation and the experimental measurement during whole reflow process.

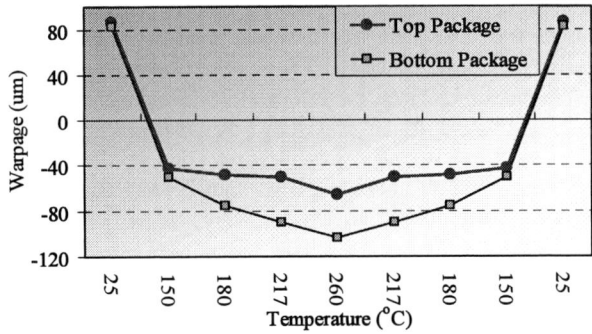

Fig. 12: Comparison of the warpage values of the top & bottom packages during whole reflow process.

Typically, a PoP is assembled on a PCB by one-reflow process. The main issue for PoP assembly is the disconnection between the solder balls of the top package and bottom package, and/or the disconnection between the solder balls of the bottom package and the PCB at a high reflow temperature. For the former disconnection, it is mainly caused by the warpage of the top & bottom packages in the opposite direction, or large warpage difference between top & bottom packages in the same direction. For the later disconnection, it is mainly caused by the large warpage of bottom package if assuming the PCB with low warpage.

To evaluate the risk of disconnection in one-reflow process, the warpage values of the top package and bottom package during whole reflow process are summarized in Fig. 12. As shown, the warpage of top package is in the same direction with that of bottom package during whole reflow process. The largest warpage difference between the top package and bottom package occurred at 260°C with the value of 38um. Disconnection may occur between the top package and bottom package but with low risk. High risk of disconnection exists between the bottom package and PCB, because the warpage difference is more than 100um if assuming the PCB is flat.

Conclusions

Because of the maturity of PoP manufacturing processes, a design advisor for PoP manufacturing has been established and demonstrated based on the DFM methodology to help improve the yield of PoP manufacturing, achieve higher confidence of manufacturing processes, shorten time-to-market and reduce overall manufacturing cost. The key components of this design advisor are the validated numerical models, comprehensive materials library, design guidelines of PoP packaging and novel FEA techniques. By implementing the novel FEA techniques of curing process simulation and seamless packaging process simulation to a whole PoP manufacturing process simulation, complete numerical models of PoP manufacturing processes were established. This paper adopted two typical manufacturing processes, e.g., molding process and reflow process, to demonstrate the implementations of two techniques. By selecting package geometries, material properties and process parameters, the evaluations of PoP package design can be generated by the envelope-based design advisor to ensure the first-time success of package design. If the package design is beyond of the envelope of design advisor, the evaluations of special package design can be generated by FEA-based design advisor.

In the future study, the envelope of design advisor will be extended to ensure the evaluations of most package designs. The design advisor with extended envelope can improve the efficiency of PoP package design significantly and save the design cycle-time dramatically.

References

1. International Technology Roadmap for Semiconductors (ITRS) – Assembly and Packaging, 2007 Edition.
2. Advanced IC Packaging, Electronic Trend Publications, Inc., 2007.
3. W. C. Wang, F. Lee, G. L. Weng, *et al.*, "Platform of 3D Package Integration", *Electronic Components and Technology Conference*, 2007, pp. 743-747.
4. J. Y. Lee, T. K. Hwang, J. Y. Kim, *et al.*, "Study on the Board Level Reliability Test of Package on Package (PoP) with 2nd Level Underfill", *Electronic Components and Technology Conference*, 2007, pp. 1905-1910.
5. F. Carson and S. M. Lee, "Controlling Top Package Warpage for POP Applications", *Electronic Components and Technology Conference*, 2007, pp. 737-742.
6. A. Yoshida, J. Taniguchi, K. Murata, *et al.*, "A Study on Package Stacking Process for Package-on-Package (PoP)", *Electronic Components and Technology Conference*, 2006, pp. 825-830.
7. M. Dreiza, A. Yoshida, K. Ishibashi, *et al.*, "High Density PoP (Package-on-Package) and Package Stacking Development", *Electronic Components and Technology Conference*, 2007, pp. 1397-1402.
8. K. Ishibashi, "PoP (Package-on-Package) Stacking Yield Loss Study", *Electronic Components and Technology Conference*, 2007, pp. 1403-1408.
9. Y. L. Tzeng, N. Kao, E. Chen, *et al.*, "Warpage and Stress Characteristic Analyses on Package-on-Package (PoP) Structure", *Electronics Packaging Technology Conference*, 2007, pp. 482-487.
10. J. E. Luan, "Design for Improvement of Drop Impact Performance of Package-on-Package", *Electronics Packaging Technology Conference*, 2007, pp. 937-942.
11. M. Amagai, Y. Suzuki, K. Abe, *et al.*, "Package-on-Package Mechanical Reliability Characterization", *Electronics Packaging Technology Conference*, 2007, pp. 557-570.
12. Y. Suzuki, Y. Kayashima, T. Maeda, *et al.*, "Development of Thin Flip-Chip BGA for Package on Package", *Electronic Components and Technology Conference*, 2007, pp. 8-14.
13. JESD22B112, "High Temperature Package Warpage Measurement Methodology", JEDEC Solid State Technology Association, 2005.

Warpage Reduction of Package-on-Package (PoP) Module by Material Selection & Process Optimization

Peng SUN, Vincent Chi-Kuen LEUNG, Bin XIE, Vivian Wei MA, Daniel Xun-Qing SHI
Hong Kong Applied Science & Technology Research Institute, ASTRI
1st Floor, 2 Science Park East Avenue, Hong Kong Science Park, Shatin, New Territories, Hong Kong, China
Tel: (852)-34062534; Email: psun@astri.org

Abstract

The package technology has matured significantly over the past several years, shifting from conventional components and direct board level assembly to chip or package level system integration. Two major commonly used approaches are System-on-Chip (SoC) and System-in-Package (SiP). Package-on-Package (PoP) that integrates logic die in the bottom package and memory die in the top package into a single 3D package is one of the promising SiP solutions.

The major advantage of PoP packaging is that the top and bottom packages, which are usually designed with FBGA and PBGA package formats, can be tested individually before they are assembled. The yield loss of the whole PoP module can be reduced significantly. However, due to the Coefficient of Thermal Expansion (CTE) mismatch and the stiffness mismatch exist among EMC, substrate and silicon chip, warpages on both top and bottom packages are often observed. Large warpage could cause solder joint open failure and substrate delamination, leading to the electrical connection failure of the assembled module.

Theoretically, three approaches can be used to solve the warpage issue of two BGA packages contained in a PoP module: package design, material selection and process optimization. Developing a new package or changing the existing design usually involves many efforts and needs long cycle time, which can not meet the needs of competitive microelectronics industry. The material selection and process optimization are often adopted by industry to achieve the goal of shortening time to market.

In this paper, Finite Element (FE) simulation is performed firstly. The CTE of epoxy molding compound (EMC) is found to make an important contribution to the warpage of PoP. The guideline for materials selection is proposed. Based on this guideline, one type of "Green" EMC is selected. Material properties of EMC including filler content, curing degree, CTE and Tg are characterized with thermo-gravimetric analysis (TGA), differential scanning calorimetry (DSC) and thermo-mechanical analysis (TMA) respectively. The effect of the material properties and the post mold curing (PMC) process on the warpage behavior of FBGA package is investigated. Shadow moiré system is employed to characterize the warpage of the molded block and signal units.

1. Introduction

Package-on-Package (PoP) is one of the major 3D packaging approaches. It consists of one top package and one bottom package which are joined together by solder joints. The top package contains staked memory dies; typically it is a standard Fine-pitch Ball Grid Array (FBGA) package. The bottom package can contain a logic processor; and normally it is a Plastic Ball Grid Array (PBGA) package. The FBGA provides big solder balls on the back side of the substrate and the PBGA has the bonding pads on the top side of substrate, when the FBGA and PBGA packages are joined together by surface mount technology (SMT), the solder balls provide clearance room for the mold cap of the bottom PBGA package. All PoP packages, FBGA and PBGA can be tested individually as normal IC packages [1]. Moreover, PoP package is a low cost packaging solution for integrating the logic and memory devices together. Recently, the PoP technologies are attracting more interests, especially for portable electronics product applications [2,3].

Definitely, package warpage is still a great challenge to package/assembly engineers and material suppliers [4]. Package warpage is mainly generated during the curing and cooling process of EMC of the package due to the thermo-mechanical stress induced by the coefficient of thermal expansion (CTE) mismatch among the respective materials [4-6]. It's difficult to control the warpage, especially for the large and thin packages [7.8]. Warpage of package after encapsulation is also a potential reliability concern because of the risk of passivation cracking, aluminum metallization deformation on the die, bond wire breakage, die cracking and EMC cracking [6,7,9]. Furthermore, excess warpage may affect the downstream IC assembly processes and production yield due to non-coplanarity of package which could result in the solder joint opening during the board level assembly process [6,10-12]. Hence, it's important to control the PoP warpage during the assembly and stacking processes.

Since the package warpage depends on the package geometry, material property and assembly process, three methods could be employed to reduce the warpage, namely, package design, material selection and process optimization [7,9,11,13]. Compared with package geometry modification, material selection and process optimization provide large flexibility and low cost manufacturable solution.

For material selection, reducing the mold shrinkage of the EMC that consists of chemical shrinkage and thermal shrinkage or adjusting the CTE of EMC to match with the substrate could reduce the warpage of BGA packages [14]. Chemical shrinkage of EMCs is mostly dominated by polymerization methods and curing kinetics, while the CTE of EMCs is largely controlled by filler contents [4]. A typical EMC consists of epoxy resin, hardeners, catalyst, filler, stress releasing agent, flame retardant, coupling agent and coloring additive [13-15]. For the typical EMC, filler content is always over 80% in weight percentage. It is also known that increasing the filler content can reduce the CTE of EMC.

In this study, finite element (FE) simulation is performed firstly. It is understood that the CTE property of EMC is the primary factor for warpage performance of FBGA and PBGA packages used in this work. Base on this, one type of "Green" EMC is chosen. Material properties of EMC, such as filler content, curing degree, Tg and CTE, are characterized by thermo-gravimetric analysis (TGA), differential scanning calorimetry (DSC) and thermo-mechanical analysis (TMA) respectively. A series of measurement on molded block warpage following the components' assembly flow is fulfilled and PoP warpage after stacking reflow process is also evaluated based on the traditional flux dipping method. This paper focuses on the effect of post mold curing (PMC) process on the block warpage of PoP top package FBGA components.

2. Numerical Modeling and Experimental Details

2.1 Finite element (FE) simulation

In this work, to study the effects of material properties (e.g., CTE and modulus) of EMC and substrate on the warpage performance of top package (FBGA) and bottom package (PBGA) after transfer molding process, the 3D FE simulation is performed for the FBGA and PBGA packages respectively. The ¼ numerical model of FBGA package is shown in Figure 1(a). The FBGA package composes of two dies (10mm x 10mm x 0.075mm and 10mm x 6mm x 0.075mm), one spacer, die attach film, molding compound and substrate (12mm x 12mm x 0.2mm). The numerical model of PBGA package is shown in Figure 1(b). The PBGA package composes of one die (6mm x 6mm x 0.075mm), die-attach film, molding compound and substrate (12mm x 12mm x 0.2mm).

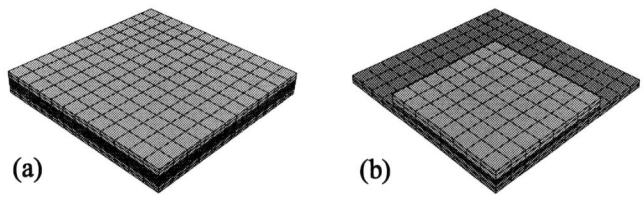

Figure 1. Numerical models of (a) top package and (b) bottom package.

This paper focuses on the warpage values after transfer molding and cooling down to the room temperature. The FE simulation plan is listed in Table 1. When one material property is varied, the others are fixed to study the effects of various properties on the warpage after transfer molding. The material properties of molding compound and substrate for top and bottom packages are listed in Table 1.

Based on the FE simulation plan, the initial warpage of the FBGA components after transfer molding process under each condition is evaluated and the results are shown in Figure 2. As seen, CTE has more impact on the warpage variety than elastic modulus. The CTE of epoxy molding compound has greater effect than that of substrate in terms of the warpage variation.

Table 1. FE simulation plan and material properties

No.	Package	Molding compound		Substrate	
		Modulus	CTE α_1	Modulus	CTE α_2
1	FBGA	Variable	8.5ppm/°C	27.5GPa	14ppm/°C
2	FBGA	21.5GPa	Variable	27.5GPa	14ppm/°C
3	FBGA	21.5GPa	8.5ppm/°C	Variable	14ppm/°C
4	FBGA	21.5GPa	8.5ppm/°C	27.5GPa	Variable
5	PBGA	Variable	8.5ppm/°C	27.5GPa	14ppm/°C
6	PBGA	21.5GPa	Variable	27.5GPa	14ppm/°C
7	PBGA	21.5GPa	8.5ppm/°C	Variable	14ppm/°C
8	PBGA	21.5GPa	8.5ppm/°C	27.5GPa	Variable

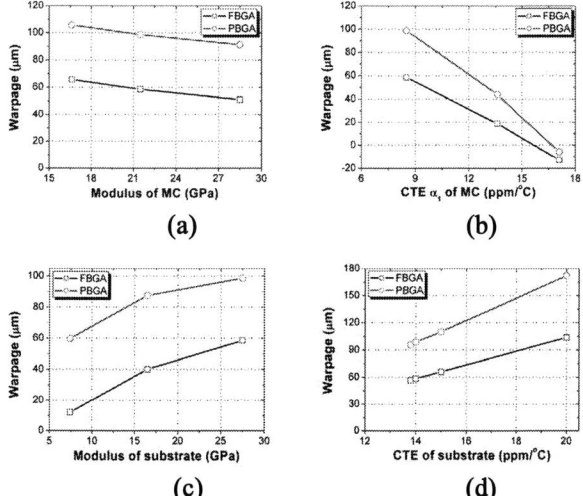

Figure 2. (a) Warpage evolution in cases 1 and 5 when E of EMC is varied, (b) warpage evolution in case 2 and 6 when CTE of EMC is varied, (c) warpage evolution in case 3 and 7 when E of substrate is varied and (d) warpage evolution in case 4 and 8 when CTE of substrate is varied.

2.2 Test vehicle description

Following the FE simulation work, the prototypes of FBGA and PBGA components are manufactured using the same epoxy molding compound and assembly processes.

The top PoP package is one over-molded FBGA with two stacked dies inside and its body size is 14mm x 14mm x 0.7mm (Length x Width x Height). The FBGA substrate consists of two Cu layers and one layer of BT/prepreg core material, Mitsubishi HL-0832. Each block is made up by 3 x 3 package matrix. The mold window size is 46mm x 46mm. After singulation process, the dimension of nine singulated FBAG components is 14mm x 14mm. The internal testing FBGA at several procedures for this study is shown in Figure 3.

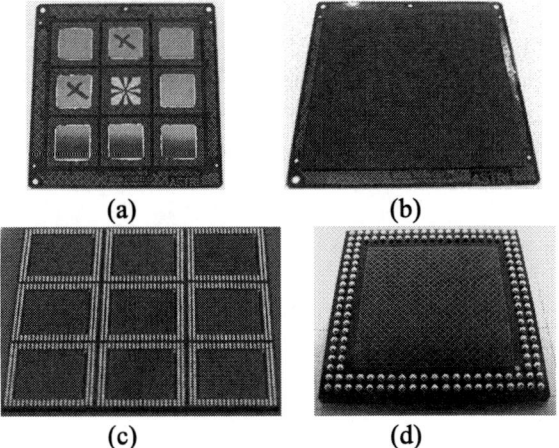

Figure 3. FBGA package (a) top view after die attach, (b) top view after transfer molding, (c) bottom view after ball attach and (d) bottom view of single FBGA unit.

The bottom PoP package is one PBGA with one die inside and the body size is also 14mm x 14mm. The PBGA substrate consists of four Cu layers, one layer of BT core material, Mitsubishi HL0832 and two layers of glass fibers prepreg material. The geometry information of FBGA and PBGA package is summarized in Table 2.

Table 2. Package geometry of FBGA and PBGA components

	FBGA	PBGA
Die size (mm)	9 x 9 x 0.1	5 x 5 x 0.1
Number of Die	2	1
Mold body size (mm)	14 x 14 x 0.5	10.5 x 10.5 x 0.3
BT core substrate size (mm)	14 x 14 x 0.2	14 x 14 x 0.3
Die attach material	Sliver epoxy and attach film	Sliver epoxy
Ball pitch (mm)	0.65	0.50
Ball diameter (mm)	0.45	0.35
Alloy of ball material	SAC305	SAC405
Number of ball	152	352

Several material properties of this "Green" EMC are characterized individually. The filler content of EMC is determined by (TGA) and result is defined as the residue left after the carbon based materials completely decomposed. The curing degree is determined by the enthalpy change measured by DSC. The glass transition temperature (Tg) and the coefficients of thermal expansions (CTEs) are characterized by TMA. The testing results on these material properties are listed in Table 3.

Table 3. Molding compound material properties

Property	Filler	CTE α_1	CTE α_2	Tg by TMA	Modulus at RT
	85.9%	10ppm/°C	40ppm/°C	140°C	21 GPa

2.3 Process Flow

The assembly processes and warpage measurement used in this study for both FBGA and PBGA packages are provided in Figure 4.

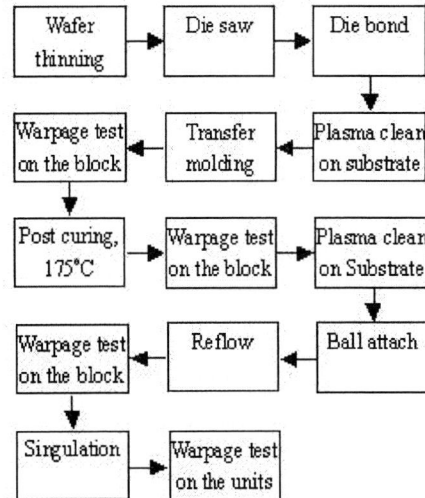

Figure 4. PoP package/assembly process flow.

As mentioned above, two dies are stacked in the FBGA component. For the first die attach, the silver epoxy is used; for the second die attach, the die attach film is adopted. The cross section image of die attach silver epoxy and die attach film are presented in Figure 5.

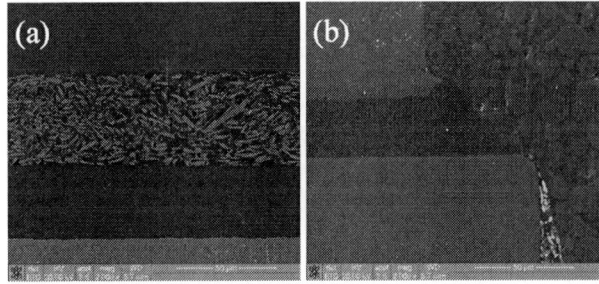

Figure 5. Cross-sectional view of die attach (a) silver epoxy between the first die and substrate, (b) film between the second and first die.

Before the transfer molding process, BT core substrate should be cleaned by Ar plasma in order to enhance the adhesion of the epoxy molding compound/substrate interface [14,16]. After the transfer molding process, the warpage measurement is performed immediately. Following the post mold curing, ball attach and singulation process, the warpage is also measured on the molded block and/or signal unit respectively, as shown in Figure 4.

2.4 Warpage measurement

In this paper, all the warpage is measured at room temperature by 'AkroMetrix' LineMoiré PS-16M and accuracy of measurement is ±2.5μm. The warpage direction definition is shown in Figure 6. The concave shape is defined as smiling warpage and the convex shape is defined as crying warpage. The data of concave shape warpage is negative value and data of convex shape warpage is positive one.

(a)   (b)

Figure 6. Definition of package warpage (a) concave and (b) convex.

3. Results and Discussion

3.1 Molded block warpage

The typical warpage results of molded block after transfer molding, post mold curing and reflow process are shown in Figure 7 (a) to (c). The initial molded block warpage after the transfer molding process is a convex shape and results vary in the range from 300μm to 600μm, while the final molded block warpage after ball attach reflow process before singulation is stable at about 200μm. A typical warpage measurement result on signal FBGA unit is presented in Figure 7 (d) and the average value of the unit warpage is about 35μm. It meets the internal specification that is 100μm or 4mils for 14mm x 14mm FBGA package under room temperature.

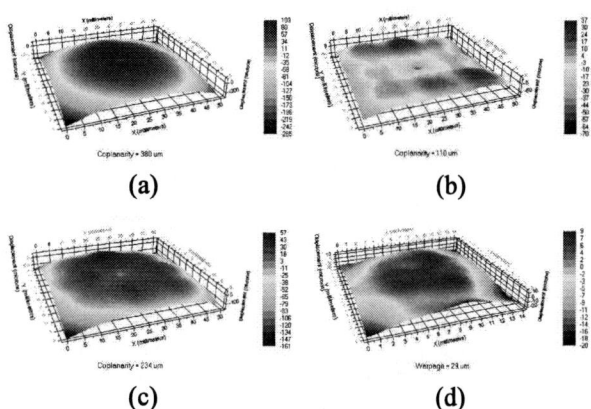

(a)   (b)

(c)   (d)

Figure 7. Warpage pattern of the FBGA package (a) block after transfer molding, (b) block after PMC, (c) block after ball attach reflow and (d) single unit after singulation.

3.2 Effect of PMC treatment on PoP warpage

It is well known that the epoxy mold compound could not reach 100% chemical conversion in the transfer molding process and the reaction rate will be slow at the later stages of conversion, so the PMC treatment is usually adopted in the industry to fulfill the curing and achieve the stable chemical & physical properties when 'complete' conversion is reached [10,11]. Curing rate of each PMC condition for FBGA is measured and results as a function of PMC time are plotted in Figure 8.

Figure 9 presents the warpage results as a function of PMC time from the initial condition without PMC process to the final status with 6 hours PMC treatment. It is shown that 4 hours PMC treatment is sufficient to reduce the warpage by 20%. According to the results of Figures 8 and 9, it is

understood that the finial warpage and full curing degree can be achieved within 4 hours PMC treatment. The warpage results of 4 hours and 6 hours PMC do not show any obvious difference. It indicates that 6 hours PMC adopted in this study may not be necessary in terms of warpage reduction.

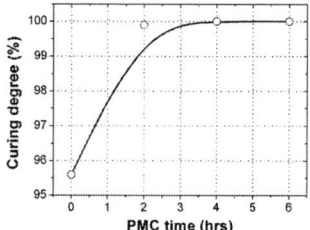

Figure 8. Degree of curing after transfer molding with various post mold curing time by using DSC measurement.

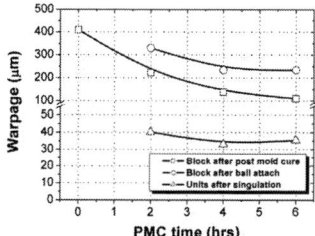

Figure 9. Block or unit warpage under different post mold curing times.

As reported in literatures, it is well known that the EMC shrinkage will introduce warpage. The total shrinkage of EMC is composed by chemical shrinkage and thermal shrinkage when it cools down from transfer molding temperature to room temperature [7,15]. Generally, the chemical shrinkage depends on the resin system and filler content in EMC [11]. EMC with higher filler content typically has lower chemical shrinkage [7]. The thermal shrinkage during the processes will be treated as a key parameter. In this case, the Tg of EMC is about 140°C and low than the curing temperature (175°C), then the thermal shrinkage can be calculated by equation 1 when it cools down from the transfer molding temperature to room temperature [8].

$$s = \int \alpha(T)dt \approx \alpha_1(T_g - 25^oC) + \alpha_2(175^oC - T_g) \quad (1)$$

for $25^oC < Tg < 175^oC$

From equation 1, it can be seen that lowering the coefficient of thermal expansion (CTE) α_1 and α_2 or by enhancing the glass transition temperature (Tg) could be the useful method to reduce the thermal shrinkage of EMC. It has been reported that the final Tg could be raised gradually during the PMC process. The longer PMC time and higher PMC temperature could result in the higher Tg of EMC [10]. Figure 10 demonstrates that CTE and Tg of EMC evolve as a function of PMC time in this work. It is shown that the CTE values do not have obvious change during the PMC process. The values of CTE α_1 and α_2 are stabilized at 10ppm/°C and 40ppm/°C respectively. While Tg temperature shows little rising during the PMC process. After 4 hours PMC treatment, the Tg is 2°C higher than that of initial sample without PMC. This phenomenon is consistent with the experimental observation and theoretical analysis mentioned above that the warpage of molded block or single unit after PMC is lower than that without PMC due to Tg increasing. After 4 hours PMC, the average value of FBGA unit warpage is about 35μm. The result indicates that EMC with low Tg (e.g., 140°C) could meet the warpage requirement when the high Tg (e.g., 170°C) BT core substrate is used.

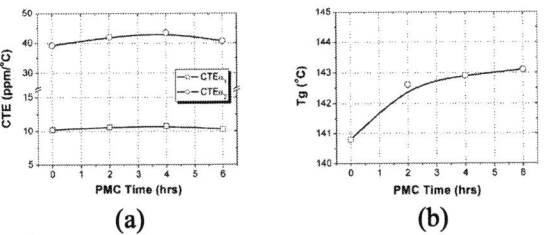

(a)   (b)

Figure 10. (a) CTE and (b) Tg after transfer molding with different post mold curing times by using TMA measurement.

3.3 FBGA warpage in PoP stacking

Following the FBGA & PBGA assembly flow, the PoP warpage after stacking process is also evaluated basing on the traditional flux dipping method [1,16] as shown in Figure 11. Firstly, solder paste is screen printed on the testing PCB with a steel stencil and the bottom package PBGA is mounted on the board. Then the top package FBGA is dipped into a tacky flux tray. After flux dipping process, the top package FBGA is placed on the top of the bottom package PBGA. When the mounting process is fulfilled, the top and bottom packages on the testing board are reflowed at the same time. The PoP stacking process is performed in the same reflow oven as the ball attach process.

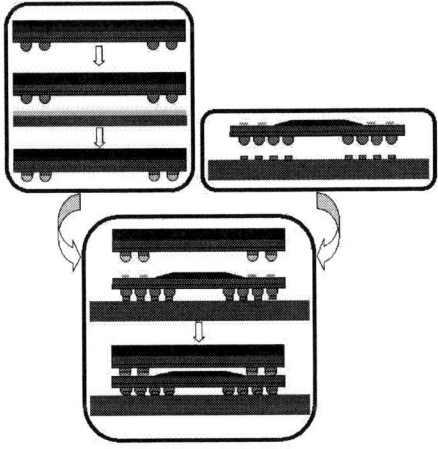

Figure 11. Typical PoP stacking process

After PoP stacking reflow, the FBGA warpage could be observed by visual inspection. While for the bottom PBGA package, it's not easy to estimate the deformation. Following the visual observation, cross-section, SEM and X-ray inspection are adopted. Some typical failure modes including FBGA solder joints' opening (Figure 12(a)), FBGA elongated solder joints (Figure 12(b) and (c)), voids in PBGA solder joints, PBGA solder ball missing and bridging (Figure (f)) have been observed. In these failure modes, the primary one is FBGA solder joints' opening, which is induced by FBGA package warpage. For the bottom PBGA package, notable warpage is not observed and the shape of solder joints located at either the left or right side is acceptable (Figure (a), (b) and (d)). It seems that the coplanarity of bottom package meets the requirement for achieving the yield performance of solder ball interconnection on the testing board. Based on above results, it is understood that FBGA warpage is the big challenge in PoP interconnection due to the primary failure mode it caused. To improve the interconnection quality between FBGA and PBGA with FBGA warpage, process optimization (e.g., reflow soldering profile) and material selection (e.g., adopting dip solder paste instead of tacky flux) have been conducted. Form experimental results, it's seem that these methods could improve the interconnection performance between two packages. This part work will be presented in another paper.

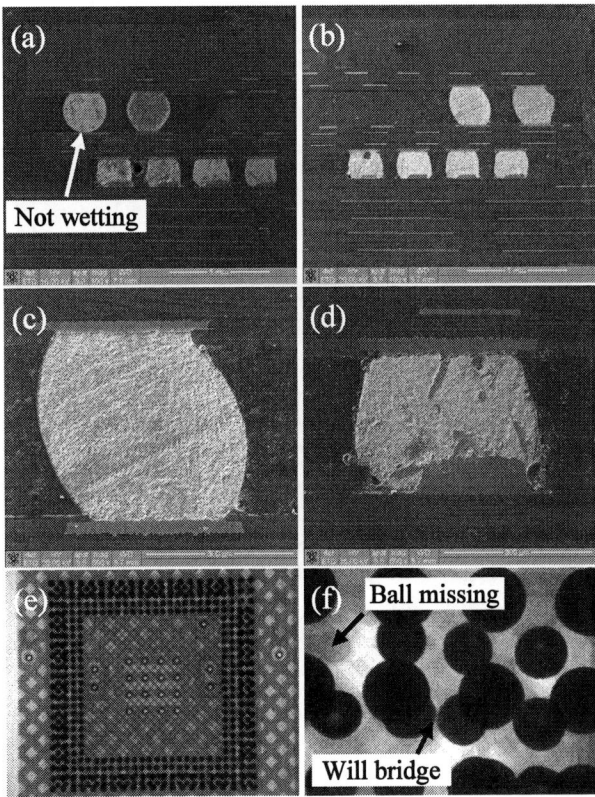

Figure 12. Cross-sectional views of PoP assembly and interconnections: (a) left edge of stacking PoP, (b) right edge of stacking PoP, (c) FBGA solder joint, (d) PBGA solder joint, (e) X-ray image of PoP and (f) X-ray image of PBGA solder joints.

Conclusions

Finite Element (FE) simulation analysis on given FBGA and PBGA packaging structures is performed firstly for material selection in this study. It is shown that CTE properties of epoxy molding compound (EMC) and substrate have higher impacts on the warpage than other key material properties, e.g., elastic modulus. According to the warpage values, it is also understood that CTE of EMC is the first factor which has more impact on the warpage behavior in such configurations.

Following FE simulation, experimental work is conducted to study the effects of process parameters and material properties of EMC on the warpage of FBGA components. The results indicate that adopting PMC treatment could reduce the block warpage of FBGA. In this work, 4 hours PMC treatment is sufficient to reduce the warpage by 20% and fulfill the curing process. It is demonstrated that long time PMC process commonly used may not be necessary for meeting the warpage requirement. It is also noted that EMC with low Tg (e.g., 140°C) could meet the FBGA warpage specification when the high Tg (e.g., 170°C) BT core substrate is used. Following the package assembly process, PoP stacking has been attempted by dipping flux process. Traditional lead free soldering equipment and process could be adopted in the current PoP stacking. It is understood that FBGA warpage is the big challenge in PoP interconnection because of the yield loss caused by FBGA solder joint opening failure. To improve the interconnection performance between FBGA and PBGA, optimizing the reflow profile and employing the dipping solder paste instead tacky flux have been investigated. The results showed that the solder joint opening failure could be addressed. The work will be presented in another paper.

References

1. David Geiger, Dongkai Shangguan, Samuel Tam, Dan Rooney, "Package Stacking in SMT for 3D PCB Assembly", *28th IEEE/CPMT/SEMI International Symposium on Electronics Manufacturing Technology*, 2003, pp. 261-264

2. Moody Dreiza, Akito Yoshida, Jonathan Micksch, Lee Smith, "Stacked Package-on-Package Design Guidelines", *www.amkor.com/products/notes_papers/Package-on-Package_PoP_Article_0705.pdf*

3. Joanna Kristine Wildhart, Moody Dreiza, "Challenges for high density PoP (package on package) utilizing SoP (solder on pad)", *Global SMT & Packaging*, April 2008, pp. 42-51

4. Hao Tang, Jonathan Nguyen, Jack Zhang, Irving Chien, " Warpage Study of a Package on Package Configuration", *International Symposium on High Density Packaging and Microsystem Integration*, HDP'07, 2007, pp.1-5

5. G. Kelly, C.Lyden, W. Lawton, J. Barrett, "Accurate prediction of PQFP warpage", *Proceeding of 44th Electronic Components and Technology Conference*, ECTC 1994, 1994, pp.102-106

6. G. Kelly, C.Lyden W., Lawton J. Barrett, " The Importance of Molding Compound Chemical Shrinkage in the Stress and Warpage Analysis of PQFPs", *Proceeding of 45th Electronic Components and Technology Conference*, ECTC 1995, 1995, pp.977-981

7. Laurene Yip, Ahmad Hamzehdoost, "Package Warpage Evaluation for High Performance PQFP", *Proceeding of 45th Electronic Components and Technology Conference* , ECTC 1995, 1995, pp. 229-233

8. Janet W.Y.Kong, Jang-Kyo Kim, Matthew M.F.Yuen, "Warpage in Plastic Packages: Effects of Process Conditions, Geometry and Materials", *IEEE Transitions on Electronics Packaging Manufacturing*, Vol.26, No.3(2003), pp.245-252

9. D.G.Yang, L.J.Ernst, K.M.B.Jansen, C.van't Hof, C.Q.Zhang, W.van Driel, H.J.L.Bressers, "Fully Cure-Dependent Polymer Molding and Application to QFN-Package Warpage", *Proceeding of 6th Electronic Packaging Technology Conference*, EPTC 2004, 2004, pp. 87-91

10. Bill Kiang, Janice Wittmershaus, Rudra Kar, Neil Sugai, "Package Warpage Evaluation for Multi-Layer Molded PQFP", *11th IEEE/CPMT International Symposium Electronics on Manufacturing Technology*, 1991, pp.89-93

11. Minjin Ko, Dongsuk Shin, Myungsun Moon, "The Effect of Mold Compounds on Warpage in LOC Package", *Proceeding of 49th Electronic Components and Technology Conference*, ECTC 1999, 1999, pp.1196-1200

12. Rattana Ingkanisorn, Anocha Sriyarunya, "RoHS-Compliant Molding Compound Evaluation and Manufacturability For FBGA Packages", *Proceeding of 6th Electronic Packaging Technology Conference*, EPTC 2004, 2004, pp. 479-482

13. K Irving Y. Chien, Jack Zhang, Lou Rector, Michael Todd, " Low Warpage Molding Compound Development for Array Packages", *Proceeding of 1st Electronics Systemintegration Technology Conference*, ESTC 2006, 2006, pp.1001-1006

14. Wen-Li Yang, Daniel M.S.Yin, "The Effects of Epoxy Molding Compound Composition on the Warpage and PoPcorn Resistance of PBGA", *Proceeding of 49th Electronic Components and Technology Conference*, ECTC 1999, 1999, pp.721-726

15. Fenny Liu, C.T.Yao, Don Son Jiang, Yu Po Wang, C.S.Hsiao, "Halogen-Free Mold Compound Development for Ultra-Thin Packages", *Proceeding of 57th Electronic Components and Technology Conference*, ECTC 2007, 2007, pp.1051-1055

16. Joon-Yeob Lee, Tae-Kyung Hwang, Jin-Yong Kim, Min Yoo, Eun-Sook Sohn, Ji-Young Chung, Moody Dreiza, "Study on the Board Level Reliability Test of Package on Package (PoP) with 2nd Level Underfill", *Proceeding of 57th Electronic Components and Technology Conference*, ECTC 2007, 2007, pp.1905-1910

A New Process to Fabricate Cavities in Pyrex7740 Glass for High Density Packaging of Micro-System

Junwen Liu, Qing-an Huang, Jintang Shang, Jing Song, and Jieying Tang

Key Laboratory of MEMS of Ministry of Education, Southeast University

Nanjing, Jiangsu Province, China

stonejee@sohu.com

Abstract

In the domain of manufacturing and packaging of micro-system, the Pyrex7740 glass is a widely-used material since its coefficient of thermal expansion is similar to that of silicon, and its good optical performance for biosensors and optical sensors. But the use of Pyrex7740 glass is limited for its isotropic etching characteristic of tradition micro-machining. In this paper, we present a new process to fabricate deep grooves in Pyrex7740 glass. The process is based on the anodic bonding, and it uses the Si substrate as the mold for forming the shape of the cavity. Finally the cavities were formed by the atmospheric pressure after the special heat treatment. The Pyrex7740 glass with cavities could be used for high density packaging of micro-system by anodic bonding or adhesive bonding.

The approach of fabricating deep cavities in Pyrex7740 glass is a key technology, which has seldom studied before. We have experimentally verified the feasibility of this new process. First of all, we fabricated the array of desired shape of cavities on silicon substrate by wet etching or dry etching. It is much easier to get the precise shape on the silicon substrate by micro machining than in Pyrex7740 glass. In our experiment, we had chosen several different side length of the square as a pattern. Then we bonded the Pyrex7740 glass and the silicon substrate together under the vacuum environment by anodic bonding. After that twice heat treatments were taken to the bonding wafer. One was to form the Pyrex7740 glass into desired shape by the silicon mold with the temperature up to the softening point. Another was to release thermal stress of the anodic bonding and the first heat treatment. The placement of the wafer during the heat treatment must be taken attention to. Finally, the bonding wafer with cavities was finished for the high density packaging of micro-system.

Key words: Pyrex7740 glass, cavity, high density packaging

1. Introduction

MEMS (Micro-Electronics-Mechanical-System) meets the requirement of modern electronic device, such as small in size, light weight, low energy consumption and intelligence integration. But successful commercial applications of MEMS are still limited. It has still encountered great technical challenges while being commercialized. Unlike the ICs, the micro-structures usually contain freestanding moving parts resulting from the removal of the sacrificial layer [1]. They interact with the environments to sense and actuate. One of the challenges is how to protect these delicate micro-structures well with accepted cost and provide ways for non-electric signal input/outputs at the same time. The packaging technology has become the most key point for MEMS applications. Due to the motion characteristics of micro-devices, the hermetic package mode has become to be the most important package scheme for some MEMS devices. In a vacuum environment, the moving parts exhibit a high Q value of vibration and show a convincing stability for their analog function. But as the MEMS structures' particularity of packaging process, the packaging cost accounts for the overall production cost of 50%~80% [1]-[3]. Regarding the movable structure, it usually needs to make the a cavity environment to come to carry on the packaging process, the RF MEMS switch component, the DMD optical device, the accelerometer and the MEMS gyroscope's resonator, the pressure sensor and so on need fabricate the cavity structure on Si substrate or other materials, guaranteed that the moving parts obtain the enough activity space, and provides the machinery protection and the environment maintenance. These parts also need to move in the vacuum environment, the vacuum environment may enable the least air damping of the structures, to act a long-term reliable performance [4]-[7].

The Pyrex7740 glass is a widely-used material since its coefficient of thermal expansion is similar to that of silicon, and its good optical performance for biosensors and optical sensors. But the use of Pyrex7740 glass is limited for its isotropic etching characteristic of tradition micro-machining. In this paper, we present a new process to fabricate deep grooves in Pyrex7740 glass which could be used for high density packaging of micro-system.

2. Principle

How to fabricate a cavity structure for packaging is an important goal for MEMS industry. At present, the MEMS hermetic packaging process has two mature methods to fabricate the micro cavity structure as shown in Fig.1.

The first way, we can use the low pressure chemical vapor deposition (LPCVD) to deposit the sacrificial layer for covering the MEMS structures to sustain the shape of micro cavity, then the processing makes the outer seal shell of the micro cavity. The key process of the way is releasing process and it removes the PSG sacrificial layer through the release hole by wet or dry etching process. Finally, we seal the release hole in vacuum environment or in the specifically atmosphere. The second way, we use the bulk micromachining process directly to the materials such as silicon, Pryex7740 glass or polymer, that the micro cavity structure is fabricated for the next surface bonding process. Before bonding, we must carry on the releasing process to the structures, and then the surface bonding process is taking place in vacuum environment or in the specifically atmosphere, at the end the packaging process is completing. Because of adopting the surface bonding process, there leaves a less influence to the moving MEMS structures and provides moreover accuracy control of shape of the micro cavity. It also avoids getting failure through the twice seal processes which is related to the release hole. As the relatively simple reliably, such way provides the fundamental process of MEMS packaging.

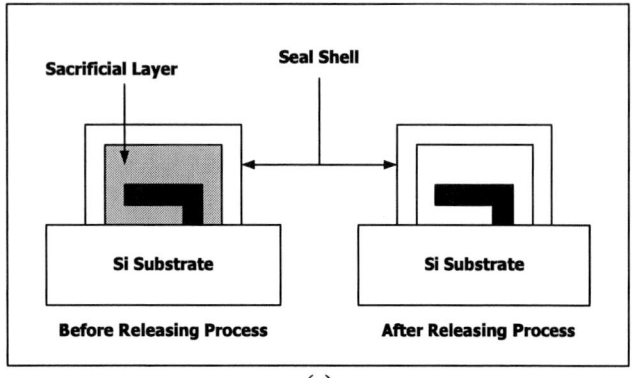

(a)

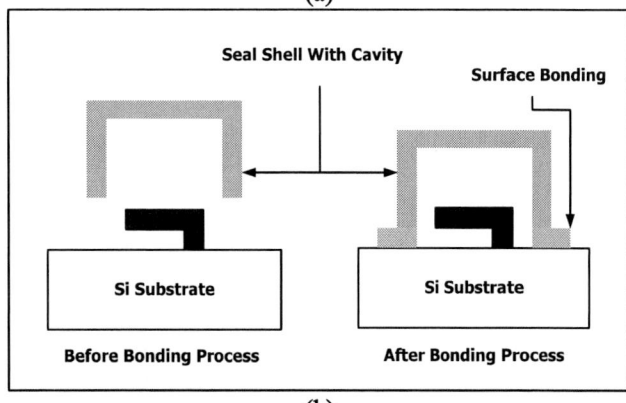

(b)

Fig.1 Two methods of hermetic packaging process

The Pyrex7740 glass is an important material of MEMS micro-machining since it has the similar thermal-expansion coefficient with the silicon and the good optical character, but it is actually very difficult to machining anyhow. If we use a wet etching process, it presents the isotropic etching characteristic, and forms the graph boundary is fuzzy. It is unable to provide an accuracy control size, and etching for more than 10μm is difficulty and so on. By using dry etching process, because of the lithographic mask's question, it causes the process to be too complex, the cost is too high, and it also could not get the deep cavity as the wet etching process.

This paper proposed a new heat expansion Pryex7740 glass micro cavity manufacture process, not only may form the precise wanted measurement, but also may moreover form the enough deep cavities at all. It is a new MEMS process with Pyrex7740 which is never mentioned before.

The Pyrex7740 glass widely applies in the anode bonding technology, the comparison between the anode bonding and the other surface bonding processes is shown in Tab.1. The anode bonding process is mature, it has a medium process temperature, and the bonding strength is high even more than the crack strength of Si itself. It also has a low coefficient of thermal expansion (CTE) and it can provide a good hermetic characteristic, so that it becomes a main MEMS manufacture and packaging process.

Tab.1 The comparison of surface bonding processes

Bonding	Temperature	Bonding Strength	CTE	Hermetic
Adhesive Bonding	Low	Low	Bad	Bad
Eutectic Bonding	High	Medium	Bad	Good
Anodic Bonding	Medium	High	Good	Very Good
Si-Si Bonding	High	High	Good	Very Good

Because of the low coefficient of thermal expansion (CTE) and high hermetic characteristic for packaging process, the anodic bonding process can realize our new process idea. We use the Si substrate as the mold layer, and carries on the anode bonding process with the Pyrex7740 glass under the vacuum environment, then puts the bonding wafer into the heat-treatment furnace to take a heat treatment under the standard atmospheric pressure environment to nearby the soft point temperature of Pyrex7740 glass. The process uses pressure differences between inside and outside of the furnace, forms the wanted shape of the Pyrex7740 glass by in the temperature to soften the glass naturally. Then we carry on an annealing process again to eliminate the residual stress of bonding wafer as far as possible. Finally there form Pyrex7740 glass micro cavities, and the schematic diagram of process principle is shown by Fig.2.

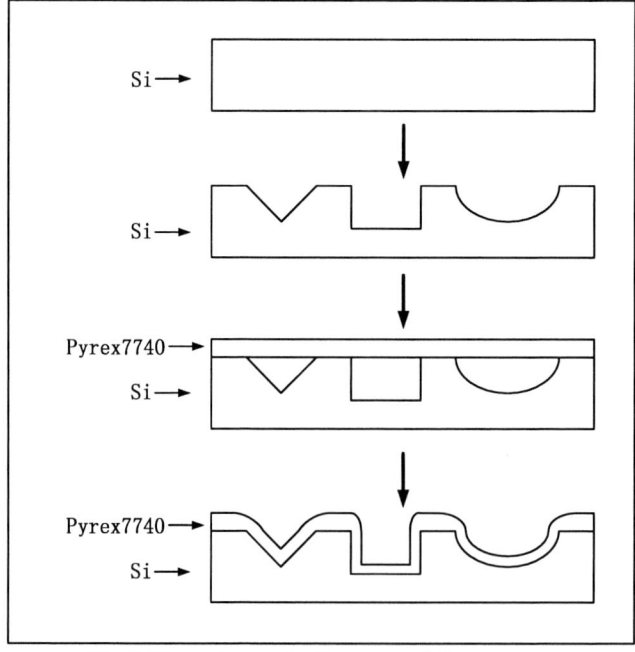

Fig.2 The schematic diagram of the process

3. Process and result

The main experimental material of process is base on the single-crystal Si substrate and Pyrex7740 glass. The single-crystal Si is P+, 4 inches, single-side polishing, and the thickness is 525μm, crystal orientation is (1,0,0). The Pyrex7740 glass is 4 inches, double-side polishing, and the thickness is 500μm.

3.1. Wet etching process

This experiment uses the wet etching process and it takes on the anisotropy characteristic in the surface of the Si wafer. The Si wafer layer is acting as a mold lay for the Pryex7740 glass in this process. The SiO_2 layer as the lithographic mask is used for the Si wet etching, and it is created by the wet oxidation process. The concrete step is as follows:

(1) Oxidation process: wet oxidation process, SiO_2 layer thickness is 600nm.

(2) Lithography process: UV light expose, BHF solution (NF:NH_4:H_2O = 31.25ml:62.5g:100ml) in 40℃,150s, etch the wanted regions of SiO2 mask layer.

(3) Etching process: 12.5%TMAH solution, 90℃,7 hour, forms the wanted deep cavities in Si substrate..

(4) Replace SiO_2 layer process: the same BHF solution to replace the whole SiO_2 layer. Again, and the depth of the mold cavity of the Si substrate is about 280μm.

3.2 Anodic bonding process

Pyrex4470 glass and Si anodic bonding the laboratory procedure is as follows:

(1) Precleaning process: common Precleaning process to Si mold layer and the Pyrex7740 glass. Boil the bonding wafer the 1# acid solution, after the 1# acid solution ebullition, boil again for 10 minutes, then uses the deionized water to flush 40 minutes to the wafer.

(2) Anodic bonding: The Pyrex7740 glass and the Si substrate align at graphite pad and put into the EVG501 anode bonding machine (shown by Fig.3). The bonding condition supposes for temperature 400℃, direct current 800V, pressure 5*10⁻³Pa, bonding time is 10min. After all, it turns to the cooling process. Then the Si mold layer with deep cavities is bonding to the Pyrex7740 glass under a vacuum environment.

Fig.3 EVG501 anode bonding machine

3.3 Heat treatment

The bonding wafer with a array of cavities before the heat treatment process is shown by Fig.4, and the Fig.5 shows the top view of one cavity before heat treatment.

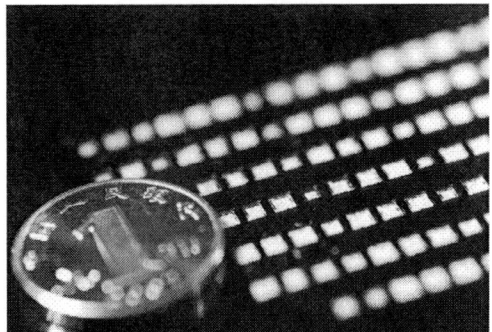

Fig.4 An array of cavities before the heat treatment process

Fig.5 Top view of one cavity before heat treatment process

We send the bonding wafer into the heat treatment furnace under the standard atmospheric pressure. The heating rate is 35℃/min and the final heat treatment is 850℃ which is above the Pyrex7740 glass softening point at 810℃. The heat treatment process maintains 5 minutes. As the pressure in the micro cavity being close to vacuum, the differential pressure between inside and outside causes the distortion to the soften glass, and finally there form the glass micro cavities. After heat treatment process we carry on the cooling process. One micro cavity of bonding wafer after cooling process is shown by Fig.6.

Fig.6 Top view of one micro cavity after cooling process

After completing of the formation heat treatment process, we send the bonding wafer into the heat treatment furnace to carry on once more annealing process, the conditions for 350 ℃and 10 minutes, cooling down in a slow rate. This process can eliminate the thermal stress of the bonding wafer.

3.4 Result

In order to facilitate the use of the next packaging process, we use the wet etching process to remove the Si layer by placing the bonding wafer in the 10% KOH solution under the temperature 80℃. After the Si layer is removed completely, we carry on the cleaning process to the Pyrex7740 glass wafer. It could be seen that the Pyrex7740 glass surface has the array of designed and wanted pattern micro cavities, our final process goal achieves. The top view of one micro cavity after removal of Si mold layer is shown by Fig.7, and the cross section SEM photo of a 2mm diameter glass micro cavity is shown by Fig.8.

Fig.7 Top view of one micro cavity after removal of Si mold layer

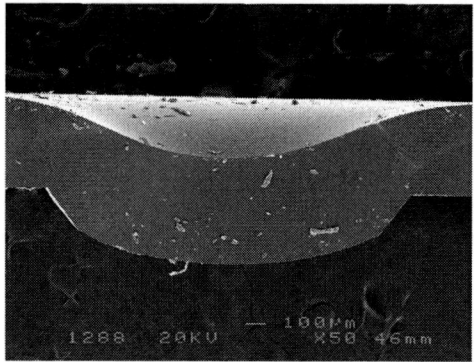

Fig.8 SEM photo of a 2mm diameter Pyrex7740 glass micro cavity

4. Conclusion

As to the difficulty in the micro-machining of glass, we present a new process to fabricate deep cavities in Pyrex7740 glass. The process is based on the anodic bonding, and it uses the Si substrate as the mold for forming the shape of the cavity by the atmospheric pressure after the special heat treatment. The Pyrex7740 glass with cavities could be used for high density packaging of micro-system by anodic bonding or adhesive bonding, and may also benefit the application in the micro flow channel system. In brief, this is a kind of new MEMS micro-machining process.

Project supported by:

(1) National Natural Science Foundation of China (Grant No.50775038).

(2) National Hi-Tech Research and Development Program of China (Grant No.2007AA04Z320).

References

1. Najafi K 2003 Micro packaging technologies for integrated microsystems: applications to MEMS and MOEMS Proc. SPIE 4982 9–27.

2. N. Maluf，An Introduction to MEMS Engineering. Boston，MA: Artech House，2000.

3. A. R. Mirza, "Wafer-level packaging technology for MEMS," in The7th IEEE Intersoc. Conf. Thermal and Thermomechanical Phenomena in Electron. Syst., vol. 2，2000，pp. 113–119.

4. H. A. C. Tilmans，D. J. van de Peer，and E. Beyne, "The indent reflow sealing (IRS) technique-a method for the fabrication of sealed cavities for mems devices," IEEE/ASME J. Microelectromech. Syst., vol. 9，no. 2，pp. 206–217，Jun. 2000.

5. A. V. Chavan and K. D. Wise, "A monolithic fully-integrated vacuumsealed CMOS pressure sensor," IEEE Trans. Electron Devices，vol. 49，no. 1，pp. 164–169，Jan. 2002.

6. Y. T. Cheng，L. Lin，and K. Najafi, "Localized silicon fusion and eutectic bonding for MEMS fabrication and packaging," in Proc. IEEE/ASME J.Microelectromech. Syst.，vol. 9，2000，pp. 3–8.

7. Y. Awatani，Y. Matsumoto，and K. Kato, "Damage free dicing method for mems devices," in Proc. IEEE/LEOS Int. Conf. Opt. MEMs，Aug. 2002，pp. 137–138.

DPA Tests on SiP Device

Jin Ling, Hu Jun, Han Li
Guangdong Yuejing High-Tech Co, Ltd., Guangzhou 510663,China

Abstract

The reliability of multi-dice in package was studied in this paper; DPA tests were operated on qualified devices to distribute early failure from overstress failure. Then factorial experiments manipulated on early failure samples for failure analysis. Comparing the C-SAM images before and after a series reliability tests, the reliability of SiP devices was confirmed.

I. Introduction

The miniaturization and integration have brought to the hot of SiP. SiP not only reduced size, but also solved the inherent defects of SoC. SoC technologies can't integrate different functions on the same based materials because the compound semiconductor product has very different materials, processes, metal finishes and passivations and it also take more design period time than SiP [1][2]. Meanwhile, Sip also behaves high reliability. All these advantages lead to the popularity of SiP today. Recently, DPA tests are widely operated on devices which used on certain condition or for specific use. It filtrate early failure components according to bath tub curve. This paper focuses on the reliability of SiP devices which were produced on SOP24 production line [2]. DPA tests were carried on qualified devices to prove reliability, optimize structure design and techniques after analyzing the failure devices. The aim of this research is to provide pre-work data for ongoing project of automobile safety electronics which faces ultra strict reliability standard. SiP devices always are constituted of passive parts and semiconductor components, the complex internal connect, encapsulation condition and SMT assembly technique application all these facts will affect reliability. Therefore, DPA tests are necessary for SiP devices especially used on automobile. This experimentation was done with qualified devices.

II. Sample description

YJ6221 was produced on SOP24 production line[2]. Inside the package, there were HT6221 die, 8050 transistor, 47 ohm resistor and 1K ohm resistor. The samples were drawn from products store after 100 days of reservation under normal store condition. Of course, the chosen samples all passed function test.

III. Experiment

1. C-SAM images pre-tests

The samples were scanned by Sonoscan® D6000, from upper and back two sides to inspect interfacial delamination. The test follows MIL-STD-1580B.

The red region indicates delamination formation. Compare with the picture before molding as below, the delamination is intended to happen at 8050 transistor and resistor position.

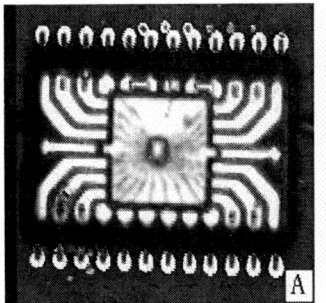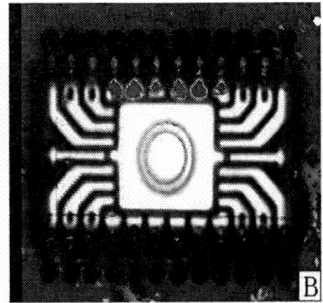

Fig. 1 C-SAM images of samples. IC active layer and molding compound (a), components adhesive with lead frame layer (b)

The C-SAM images show no delamination happen on the IC's active layer and the adhesive layer for every functional sample. But delaminations were found between the resistor-molding compound and transistor-molding compound.

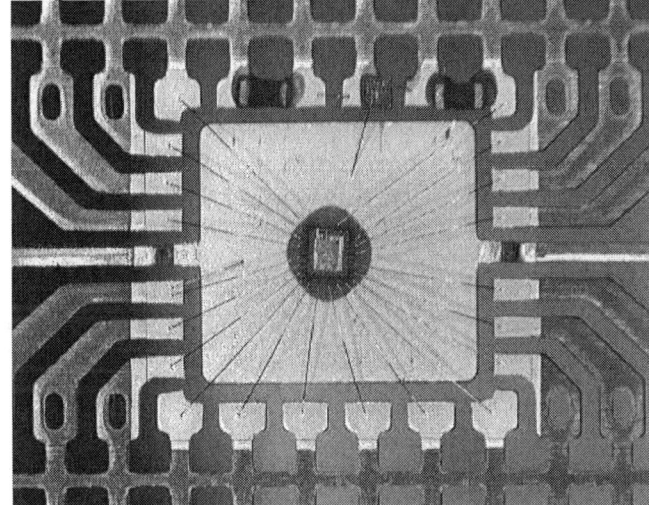

Fig. 2 Sample picture before molding

2. Reliability Tests and Result
Tab. 1 Test items and test condition

	Items	Test condition
Group1	Preconditioning	260℃ lead-free reflow, 3 times
Group2	HTSL (High Temperature Storage Life)	150℃ 1000 hours
Group3	TS (thermal shock)	-55℃ to +150℃, 6 minutes per cycle, 720times
Group4	THB (Temperature Humidity-Bias)	85℃/85 % RH for 168 hours

According to American Standard for Testing and Materials, ASTM, MIL-STD-883 and Joint Electronic Devices Engineering Council, JEDEC (Electronic Industries Associations, EIA) JESD-22 &JESD-26, the testing conditions were settled in Tab.1

The samples were divided into four groups to taken these tests, 22 for each. The results are listed in Tab.2

Tab. 2   Tests results

	Group1	Group2	Group3	Group4
Function failure sample	1	0	0	0

The failure sample comes from the preconditioning test.

IV.   Failure analysis
1. X-Ray inspection

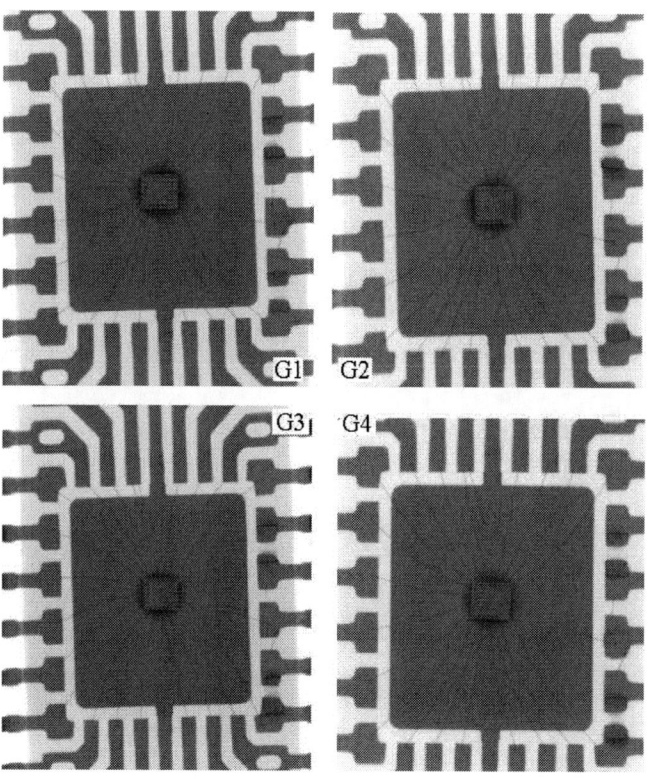

Fig. 3 X-Ray images after DPA tests (G1: group 1; G2: group 2; G3: group 3; G4: group 4)

X-Ray pictures which take by Shmadzu ®SMX 1000 show no wire was brushed off after destructive tests. So wirebond production was qualified, no dummy wirebond pad.

2. C-SAM images

Compare the C-SAM images with pre-tests. Fatal delamintion was found in Group 1.

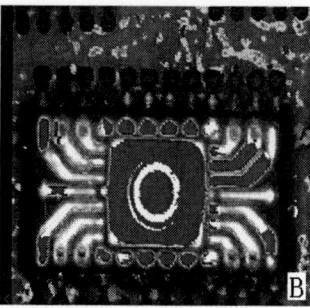

Fig. 4 C-SAM images after 2 times reflow (A: front, B: back)

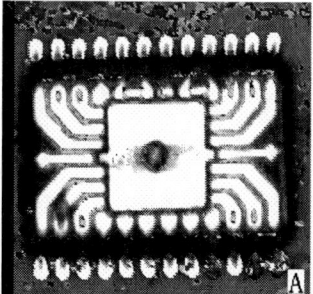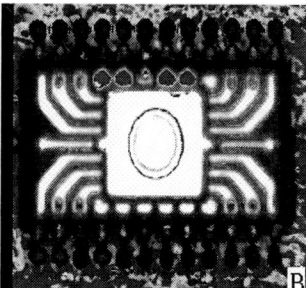

Fig. 5 C-SAM images after High Temperature Storage Life (A: front, B: back)

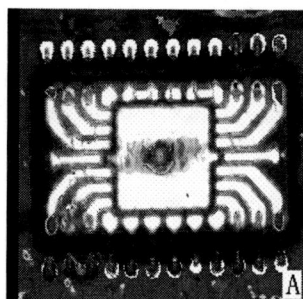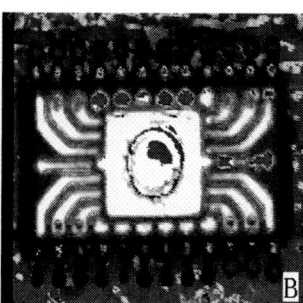

Fig. 6 C-SAM images after thermal shock (A: front, B: back)

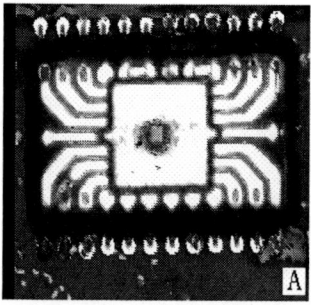

Fig. 7 C-SAM images after Temperature Humidity-Bias (A: front, B: back)

3. Sample decapsulation

The failure sample was decapped by Nisene® decapsulator. And the images were pictured by Keyence® 3D digital microscope as below:

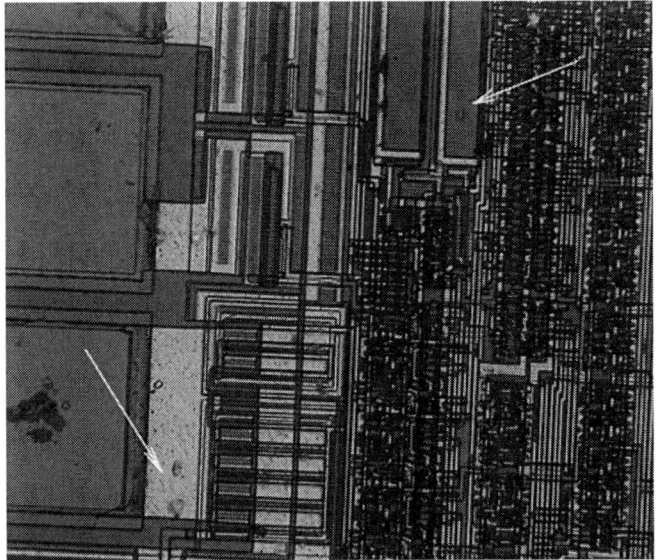

(a) Electromigration

(b) Open circuit caused by over heat

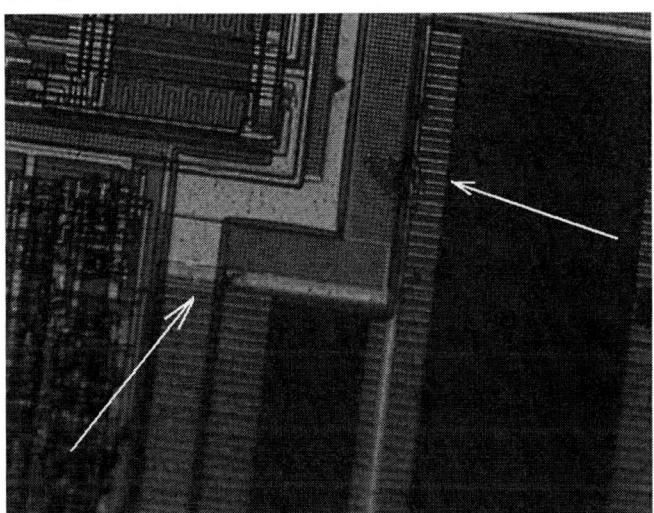

(c) Left arrow: pn junction transverse breakdown; right arrow: over heat damage

Fig. 8   Failure samples' IC surface images after decap

4. Failure analysis

From the C-SAM images after reflow, popcorn phenomena large area delamination between IC chip and leadframe result in the failure evidently. Delamination under transistor and resistor are not paramount in causing failure.

On microscope pictures after decapsulation, over heat, short circuit and thermal migration phenomena were found as showed in Figure 8. These types failure are all derived from delamination[3].

The way to solve these drawbacks is choosing appropriate molding compound which behave higher MSL and the CTE should match the leadframe material.

V.   Conclusion

Compare of the test results, SiP performances reliability is better than single die packaged device. These tests unveil that reflow influence most on reliability, the next is thermal shock.

Failure analysis announces the major factor influences the reliability of this SiP device is material matching between molding compound and leadframe which was similar to single die package.

The conclusion is that SiP won't reduce the reliability of packaging if we choose suitable materials, structure design and producing process, namely, the influences can be deduced to very low level. However on global solution level, SiP will surely be more reliable than PCB solution, because it has predigested process to a large extent.

References:

1. Wu J h, Anderson M J,COLLER D, et al, RF SIP technology innovation through integration[A].5th International conference on Electronic Packaging Technology[C]. New Jersey, USA, 2003, 484-491
2. ZHU Junshan, JIN Ling, Wu Shengping, SiP based on SOP production technology, Electronics and packaging, 2008, Vol.8, No. 4, 20-22
3. ZHU Junshan, JIN Ling, Package Failure Mechanism study for Interfacial Delamination, Electronics and packaging, 2008,Vol.8, No.3, 4-7

Ultrasonic Features in Wire Bonding and Thermosonic Flip Chip

Junhui Li[a, b, c, *], Lei Han[a, b], Jue Zhong[a, b]

[a]School of Mechanical and Electronical Engineering, Central South University,
ChangSha, 410083, China
[b]Key Laboratory of Modern Complex Equipment Design and Extreme Manufacturing, Ministry of Education, ChangSha,
410083, China
[c]State Key Lab of Digital Manufacturing Equipment & Technology, Wuhan, 430074, China

Aastract: Driving voltage and current signals of Piezoceramic Transducer (PZT) were measured directly by designing circuits from ultrasonic generator and using a data acquisition software system. Input impedance and power of PZT were investigated by using Root Mean Square (RMS) calculation. Vibration driven by high frequency was tested by laser Doppler vibrometer (PSV-400-M2). Thermosonic bonding features were observed by scanning electron microscope (JSM-6360LV). Results show that the input power of bonding is lower than that of no load. The input impedance of bonding is greater than that of no load. Nonlinear phase, plastic flow and expansion period, and strengthening process were shown in impedance and power curves. The ultrasonic power is proportion to vibration displacement driven by the power, and greater displacements driven by high-power result in welding failure phenomena, such as crack, break, and peel off in wedge bonding,. For thermosonic flip chip bonding, the high power decreases position precision of bonding or results in slippage and rotation phenomena of bumps. To improve reliability and precision of thermosonic bonding, the low ultrasonic power should be chosen.

Key words: input impedance and power; wedge bonding; thermosonic flip chip; failure

1 INTRODUCTION

It has a long history that ultrasonic bonding technology has been applied to microelectronic packaging industry. At present, wire bonding is still used commonly bonding interconnection technique in the first level microelectronic package[1,2]. As ultrasonic bonding processes have unique advantages, in recent years, thermosonic flip chip technology is increasingly used in low pin counts applications, such as smart card, light-emitting diode (LED) and Surface Acoustic Wave (SAW) filter in telecommunication applications[3]. This package technology is promising since it is clean, lead-free, adhesive-free and solder-less for area array interconnection. Thermosonic gold bonding provides strong metallurgical joining, which is thought to be more reliable than conductive adhesive and comparable to solder interconnection[4,5].

The three machine variables of load, power, and time are often used to characterize bond formation during ultrasonic bonding. Final bond integrity is dependent on all three variables. A proper load allows the wedge movement to be transmitted to the wire and subsequently to the bond interface. The amount of wire deformation during ultrasonic bonding is more sensitive to changes in power than to changes in machine load. The observed linear increase in groove spacing with power correlates with the wedge motion that also occurs with increased power. The tip-to-tip movement of the wedge

is also dependent upon the fixturing of the wedge to the transducer horn. If the wedge is not suitably tightened, an unstable wave form will be transmitted to the wire. Fixturing is so critical that replacement of a wedge to the same length with the same screw and the same tightness does not necessarily result in the same tip-to-tip displacement, with a variation of 20% being quite possible[6-8].

Ultrasonic energy driven by high frequency is a very important parameter. Frequency is one of important power parameters. Vibration and welding characteristics of complex vibration ultrasonic welding systems of 27 and 40 kHz were studied[9].The vibration characteristics of longitudinal–complex transverse vibration systems with multiple resonance frequencies of 350–980 kHz for ultrasonic wire bonding of IC, LSI or electronic devices were studied. The complex vibration systems can be applied for direct welding of semiconductor tips (face-down bonding, flip-chip bonding) and packaging of electronic devices. The vibration distributions along ceramic and stainless steel welding tips were measured at up to 980 kHz. A high-frequency vibration system with a height of 20.7 mm and a weight of less than 15 g was obtained[10]. For welding Systems with high frequencies of 55 kHz, 95 kHz, 190 kHz, 330 kHz, and 600 kHz[11-14], lots of bonding frequency characteristics, such as welding temperature rise, the vibration locus of the welding tip, and bonding strength, have been investigated. Current work has provided an important insight into the high frequency thermosonic bonding process.

The frequency of ultrasonic vibration is high, and the changing of pressure is fast. During thermosonic bonding process, the bonding between gold ball and substrate must experience many complex processes, such as contact, softening, slippage, and penetration[15]. Those processes must reflect to the power of bonding process and the structure features. In this paper, according to the relationship between the driving electric signal and power, driving electric signals were measured by using a data collecting system, vibration features driven by high frequency were tested by using PSV-400-M2 laser Doppler vibrometer, and features of bonding were observed by using scanning electron microscope (JSM-6360LV). Based on obtained images, the power and bonding characteristics were discussed. These are significant to enhance reliability of ultrasonic bonding.

2 EXPERIMENTAL

2.1 Power Data Collecting System

Nickel plated on a copper plate was selected as a bond surface for ultrasonic aluminum wedge bonds, and the major portion of this bonding program was performed using 500μm diameter-aluminum wire and LW500 wedge capillary on

U3000 ultrasonic wedge bonder (made in Shenzhen) with 60KHz frequency.

According to ultrasonic bonding, PZT is driven by electric signal from ultrasonic generator. In the experiment, the driving voltage signal was measured directly by designing circuits from ultrasonic generator and using data acquisition software-labview shown in Fig.1, and the current signal can be tested by using a resistance (R=9Ω). The electric signals acquired by labview may be stored in text file, and further process, such as analysis or display, can be performed with a Matlab program.

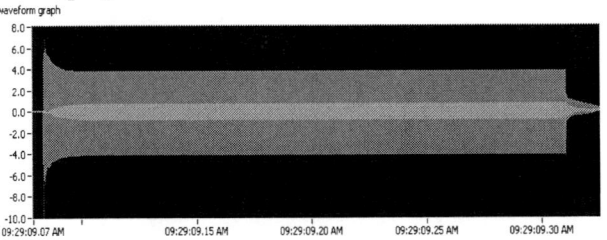

Fig.1 Measurement of driving signals of PZT

Characteristics of vibration parameters were tested by using PSV-400-M2 high frequency (1.5 MHz) laser Doppler vibrometer in Fig.2. Relationship between power and vibration velocity driven by different power are tested.

2.2 Flip Chip Bed

To research power features of thermosonic flip chip, flip chip bonding bed was integrated by using a T/S-2100 ultrasonic wire bonder and an U3000 ultrasonic wedge bonder. The following process parameters ranges of the assembly bed are: 60 KHz for frequency, 0-15 W for ultrasonic power, 20-500 ms for bonding time, 0.30-12 N for bonding force, and room temperature -400 ℃ for heating.

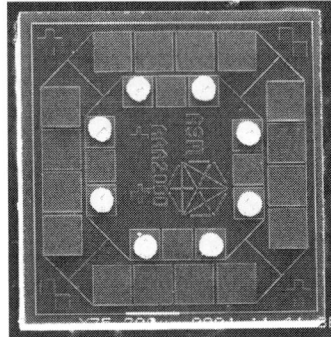

Fig.2 Chip with 8 gold bumps

Testing chips (come from ASM) have 8-I/O connection gold bumps with 80 micron in diameter that was formed by using a ball bonder (See Fig.2).

Characteristics of Thermosonic bonding were investigated by using scanning electron microscope (JSM-6360LV), after flip chip bonding points were lifted off by using Dage 4000 bonding tester.

3 RESULTS AND DISCUSSION

3.1 Input impedance and power

During ultrasonic bonding process, the input impedance of PZT transducer is dynamic and nonlinear, unless transducer assembly is operating at its resonance. However,

we can use the concept of generalized impedance and power to indicate the changing of the input impedance and power of PZT. The input impedance of PZT is equation 1,

$$Z = U_0 / I_0 = \sqrt{\sum_{k=1}^{k=N} u(k)^2 / N} \Big/ \sqrt{\sum_{k=1}^{k=N} i(k)^2 / N} \quad (1)$$

The input power of PZT is equation 2,

$$P = U_0 \bullet I_0 = \sqrt{\sum_{k=1}^{k=N} u(k)^2 \Big/ N} \bullet \sqrt{\sum_{k=1}^{k=N} i(k)^2 \Big/ N} \quad (2)$$

U_0 and I_0 are the root mean square (RMS) calculation, where $u_{(k)}$ and $i_{(k)}$ are the sampling data in certain time, N is the sampling numbers, and N must be more than the data numbers of electric signal in a period. In the experiment, for sampling frequency is 500 kSa/s, the data numbers of electric signal in a period is 18.

So, we can use RMS calculation to analyze the input impedance and power of PZT. For the given bonding machine variables (time is 100ms, power is 5W, and force is 3.6N), curves of voltage and current signals are shown in screen of Fig.1. The RMS calculation results by using Matlab on no load and bonding are shown in Fig.4 and Fig.5, where N=18.

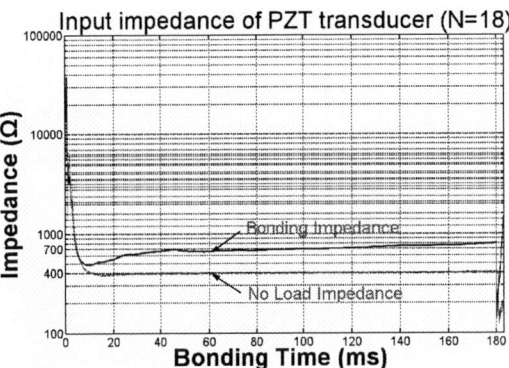

Fig.3 The input impedance curves of PZT in no load and bonding process

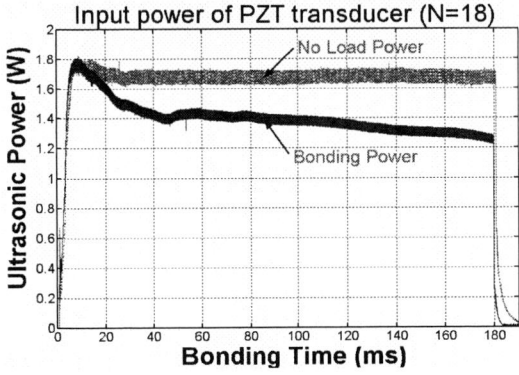

Fig.4 The power curves of PZT in no load and bonding process

Fig.3 shows that the input impedance of bonding is greater than that of no load. This is because the damping effects increase impedance of transmitting energy system when the wire contacts the substrate during bonding process.

Fig.4 shows that the input power of bonding is lower than that of no load. On the other hand, the starting of curves on the input impedance and power is nonlinear phase, then decreases and remains smooth finally. Falling period of the power may be indicated material plastic flow and bonding expansion. The penetration of bonding interface may be strengthened in later stable time. Features of input impedance and power gotten by voltage and current signals show the process during thermosonic bonding.

3.2 Vibration and power

When power was 3.5 W, utltrasonic vibration velocity at tool tip of flip chip is shown in Fig. 5. Being similar to power, the start of vibration driven by the high frequency is nonlinear. When vibration remain stable, the peak value of vibration velocity is A=1.3 m/s. Fig. 6 is the curve of vibration velocity from 11.3 ms to 11.5 ms, and shows that ultrasonic vibration is sinusoid,

$$v_{(t)} = A\omega\sin(\omega t + \varphi + \frac{\pi}{2}) \qquad (3)$$

Where, ω is resonance angle frequency, φ is phase. Fig.7 is result of Fast Fourier Transform (FFT) from Fig. 5, and shows that resonance frequency is f=62.73 KHz. Thus,

$$\omega = 2\pi f = 2 \times 3.14 \times 62.73 \times 10^3 \approx 3.94 \times 10^5 \quad (4)$$

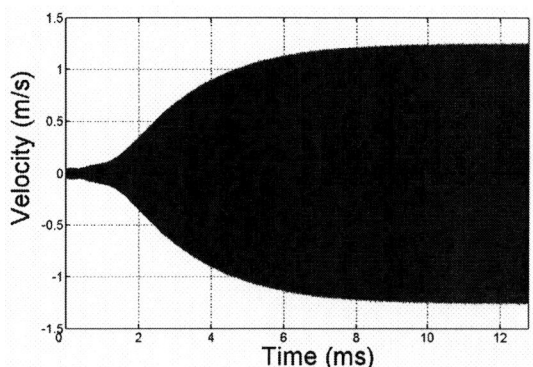

Fig.5 Result of vibration velocity at tool tip

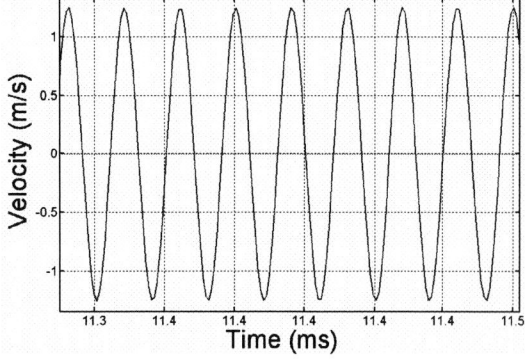

Fig.6 Graph of vibration velocity from 11.3 ms to 11.5ms

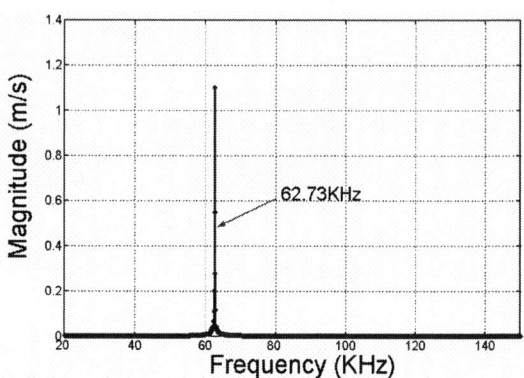

Fig.7 FFT result of vibration velocity at tool tip

Then the ultrasonic vibration displacement is,

$$y_{(t)} = A\sin(\omega t + \varphi) \qquad (5)$$

The peak value of the vibration displacement is proportion to that of vibration velocity.

For the different power, the testing peak value of vibration velocity filled in Fig.8 where shows that the power is directly proportion to vibration velocity. So, the power is proportion to vibration displacement driven by the power.

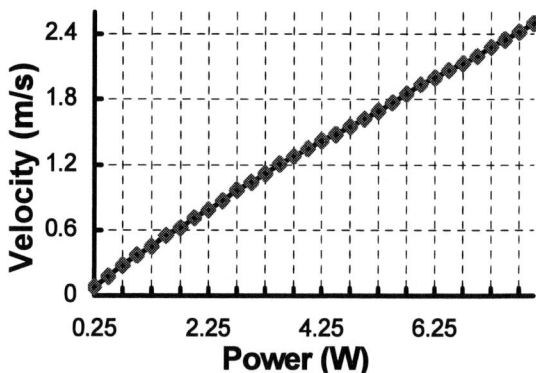

Fig.8   Relationship between vibration velocity and power

3.3 Failure of greater power

When power is 8 W, the peak value of vibration velocity is 2.50 m/s (See Fig.8), the peak value of vibration displacement is,

$$
\begin{aligned}
A &= \frac{2.50\text{m/s}}{\omega} = \frac{2.50\text{m/s}}{2\pi f} \\
&= \frac{2.50 \times 10^6 \mu\text{m/s}}{2 \times 3.14 \times 62.73 \times 10^3 /\text{s}} \approx 6.35\mu\text{m}
\end{aligned}
\qquad (6)
$$

It indicates that the greater displacements driven by high-power decrease characteristics of thermosonic bonding. Fig.9 shows that failures of over high power (8 W) happen, such as crack, break, and peel off in wedge bonding. However, when the power is 4 W, Fig.10 shows that ultrasonic bonding is very good.

108

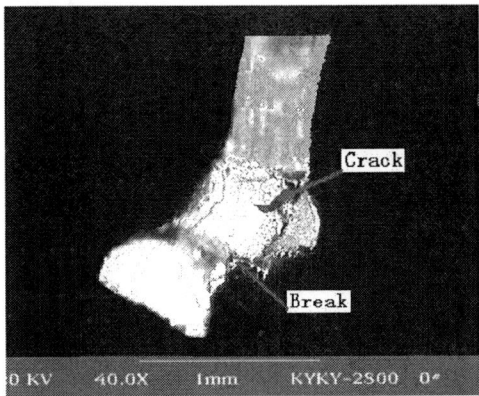

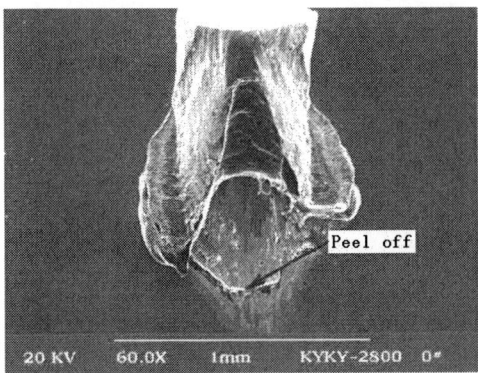

Fig.9 Failure pattern of high power

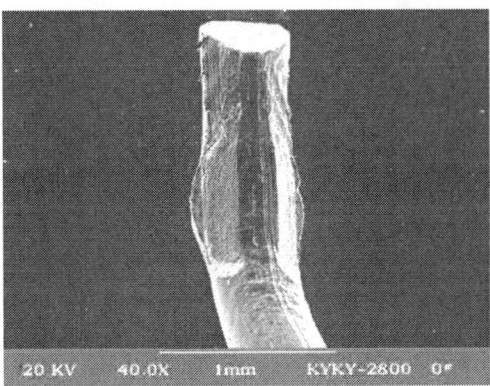

Fig.10 Successful wedge bonding

Similary, Fig.11 shows that thermosonic flip chip bonding is better for power 1.5 W during thermosonic flip chip bonding. However, Fig.12 indicates that the greater displacements driven by high-power (6 W) decrease bump positioning precision of flip chip bonding or result in slippage and rotation phenomena of bumps.

To improve reliability and precision of ultrasonic bonding, the lower power should be chosen.

4 CONCLUSION

1) The input impedance and power curves of PZT analyzed by use RMS (Root Mean Square) calculation show the nonlinear phase, plastic flow and expansion period, and strengthening process during ultrasonic bonding. The input power of bonding is lower than that of no load. The input impedance of bonding is greater than that of no load.

2) Ultrasonic vibration driven by high frequency sine signals is sinusoid, the power is proportion to vibration

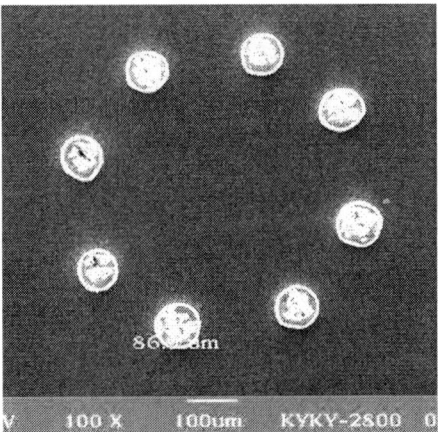

Fig.11 Lift off characteristics for optimal power parameter in flip chip

Fig.12 Lift off characteristics of high power in flip chip

displacement driven by the power. The greater displacements driven by high power lead to the welding failure phenomena, such as crack, break, and peel off during wedge bonding. Similary, the high power decreases precision of bumps or results in slippage and rotation phenomena of bumps during thermosonic flip chip bonding. To obtain high-reliability and high-precision of bonding, the low ultrasonic power should be selected during thermosonic flip chip.

ACKNOWLEDGMENTS

This work was supported by National Natural Science Foundation of China (50675227, 50575229, 50705098), Hunan Natural Science Foundation of China (07JJ3091), State Key Lab of Digital Manufacturing Equipment & Technology (2007001), China High Technology R&D Program 973(2003CB716202) , and Program for Changjiang Scholars and Innovative Research Team in University (Grant No. IRT0549).

REFERENCES

1. LI J, HAN L, DUAN J, ZHONG J. Interface mechanism of ultrasonic flip chip bonding, Applied Physics Letters[J], 2007, 90: 242902.
2. LI Jun-hui, HAN Lei, ZHONG Jue. The characteristics of ultrasonic vibration transmission and coupling in bonding technology, Proceedings of the Sixth IEEE

CPMT Conference on High Density Microsystem Design and Packaging and Component Failure Analysis, HDP'04[J], Shanghai, 2004: 311-315.

3. Elger G, Hutter M, Oppermann H. Development of an assembly process and reliability investigations for flip-chip LEDs using the AuSn soldering[J], Microsystem Technology, 2002, 7(5): 239-243.

4. Taizo T. Thermosonic flip chip bonding for low cost packaging, 2002 International Symposium on Microelectronics[J], Denver, CO, United States, Sep, 2002: 360-365.

5. LI J, HAN L, ZHONG J. Interface structure of ultrasonic wedge bonding joints of Ni/Al, Transactions of Nonferrous Metals Society of China[J] 2005,15 (4): 846-850

6. Winchell V H, Berg H M. Enhancing ultrasonic bond development, IEEE Trans. Hybrids, and Manuf. Technol.[J] Chmt-1 1978, 3: 211-220

7. LI Jun-hui., HAN Lei, ZHONG Jue. Microstructural characteristics of Au/Al bonded interface, Materials Characterization[J]. 2007, 58(2): 103-107

8. LI Jun-hui., HAN Lei, ZHONG Jue. Characteristics of ultrasonic vibration transmission in bonding process, Journal of Central South University of Technology[J], 2005, 12 (5): 567-571

9. Tsujino J, Sano T, Ogata H, et al. Complex vibration ultrasonic welding systems with large area welding tips, Ultrasonics[J], 2002, 40(3): 361–364

10. Tsujino J, Yoshihara H, Sano T, et al. High-frequency ultrasonic wire bonding systems, Ultrasonics[J], 2000, 38(3): 77–80

11. LI Jun-hui., HAN Lei, ZHONG Jue. Features of machine variables in thermosonic flip chip, Key Engineering Materials[J], 2007, 339: 257-262

12. Tsujino J, Hongoh M, Tanaka R, et al. Ultrasonic plastic welding using fundamental and higher resonance frequencies, Ultrasonics[J], 2000, 40(2): 375-378

13. Tsujino J, Koichi H. Ultrasonic wire bonding using high frequency 330, 600 kHz and complex vibration 190 kHz welding systems, Ultrasonics[J], 1996,34 (4): 223-228.

14. Tsujino J, Yoshihara H, Kamimoto K, et al. Welding characteristics and temperature rise of high frequency and complex vibration ultrasonic wire bonding, Ultrasonics[J],1998, 36 (2): 59-65.

15. Kang S F, Williams P M, Mclaren T S. Studies of thermosonic bonding for flip-chip assembly[J]. Materials Chemistry and Physics[J], 1995, 42(3): 31-37

Development of Three-dimensional Multichip Module Based on Embedded Substrate with Multiple Interconnections

Gaowei Xu, Yanhong Wu, Fei Geng, Qiuping Huang, Jian Zhou, Le Luo
Shanghai Institute of Microsystem and Information Technology, Chinese Academy of Sciences,
Shanghai 200050, China
xugw@sina.com

Abstract

A new type of 3D multichip module (3D-MCM) for wireless sensor network was developed based on a kind of embedded FR-4 substrate for the wireless sensor network, in which FCOB (flip-chip on board), COB(chip on board), BGA(ball grid array) technologies, wirebonding and flip-chip interconnection technologies were combined together. The PBGA device and bare die were hybrid-integrated on the embedded multi-layer FR-4 substrate. By solder ball placement and reflow the BGA was formed at the bottom of 3D-MCM, and solder balls with different melting points were used for initial and final vertical interconnections for the sake of compatibility of all levels interconnections of BGA by reflow soldering. The application of embedded substrate solved the problem that the top surface of the encapsulated chip overtops the solder balls in the condition that the chip was assembled in the same side of substrate with BGA. The thermal management was conducted and the thermal related reliability of 3D-MCM were simulated and evaluated respectively. This kind of packaging structure satisfies the electrical performance and thermal requirement, and meets the challenge of minimization, high reliability and low cost of the package design for the wireless network.

Key words: 3D-MCM, embedded substrate, combination of multiple interconnections, compatibility of solder melting, thermo-mechanical reliability.

1 Introduction

To meet the demand of lightweight, minimization and system integration for electronic products, all sorts of new packaging structures are constantly developed [1]. As one of advanced package type, the three-dimensional (3D) package becomes noticeable. It has many advantages such as small volume, lightweight, and can improve assembly efficiency, fasten signal speed. In addition, it also has multi-function, high reliability and low cost etc. In despite of those advantages its structure is rather complicated and challenge to thermal design, electric characteristics, thermo-mechanical reliability control, assembly process etc.

Most study for 3D-MCM gives attention to single-type-chip package and accordingly one interconnection method. Laminated structure and embedded structure for 3D-MCM are generally adopted. However the embedded components are generally limited to passive film ones. Typically, glass ceramic substrate is used and 1000°C sintering temperature is also needed, therefore the active device e.g. IC cannot be embedded. Therefore embedding active device (IC) into organic substrate will be challengeable.

The authors once developed a 3D-MCM with BGA (ball grid array) based on general organic substrate. The problem is that the height of encapsulant (glob top) on chip may overtop the height of solder ball of BGA, which absolutely affects the BGA interconnection between 3D-MCM and motherboard. Therefore embedded organic substrate will be taken into account. In fact, for 3D-MCM with different IC types, multiple interconnections and embedded substrate, the assembly process is further difficult and complicated.

For the moment, there are many studies about 3D-MCM. The supermulti-I/O and super-fine-pitch MCM structure was developed in literature [2]. The interconnection technology for the high-density three- dimension-integrated microwave module based on LTCC technology was discussed in literature [3]; Zhang performed the thermo-mechanical analysis using finite element (FE) method on laminated substrate [4]; Xueren Zhang et al studied the warpage of the stacked BGA structure using FE method[5]; Literature [6] simulated the thermal characteristics of stacked SRAM 3D-MCM by FE method. However, above studies for 3D-MCM were limited to one kind of chips with the same I/O pin type, and to one kind of interconnection technology, e.g. the chip connection of 3D-MCM was generally performed by means of single technique such as flip chip [7] or wirebonding [8] on ceramic or organic substrate.

Stacked packaging structure is generally adopted for 3D-MCM[9]. Embedded structure is also used, however, the embedded devices are limited to passive film devices [10]. The typical process is to print passive devise i.e. C, R and L on glass-ceramic raw slices, perform the stacking, prepressing and once-firing process with about 1000°C sintering temperature. Therefore the fabricable devices are restricted and the active device such as IC cannot be embedded. It was also reported that both passive devices i.e. L and C were embedded into BT resinous substrate so as to decrease ghost effect. Tee et al once studied the stacked package with "FC+WB" mixed interconnection and predicted the fatigue life of the stacked soldering ball[11]. For 3D-MCM with multiple interconnections, multiform packaging chip are integrated, multiform interconnections are adopted, the chips may be embedded into high-density multilayer substrate, and the multilevel BGA process must be performed using multiple reflowing technology, so many process factors are involved and the fabrication is difficult, therefore, technique and reliability of 3D-MCM are more challengeable.

Study for 3D-MCM with multiple interconnection technologies was reported in literature [12]. In this work, a packaged PBGA chip and a ASIC chip were assembled in both sides of the multilayer organic substrate by wirebonding and flip-chip interconnection methods respectively, and BGA was adopted for I/O pin mode. However, the disadvantage of

978-1-4244-2739-0/08/$25.00 ©2008 IEEE

this kind of package structure is visible: the top surface of glob-top of chip is easy to overtop the height of soldering balls on the bottom of substrate, which may affect the interconnection reliability between 3D-MCM and motherboard. In order to overcome this limitation, in this paper the embedded organic substrate was developed to embed the wirebonding chip. This kind of 3D-MCM, with multiform packaging chips and multiple interconnections and based on embedded substrate, had not been reported as yet.

2 Design and Results

2.1 Design of 3D-MCM

The Terminal of wireless sensor network includes a few functional modules, such as baseband ASIC, general-purpose DSP chip and other support circuits, and generally, ASIC and DSP are mounted on upside of PCB, and form a 2D-MCM, as shown in Fig.1. Baseband ASIC is a bare chip, which performs modulation & demodulation function, and DSP is a PBGA, which performs protocol & control function, as shown in Fig.2. Table 1 shows the main parameter of ASIC and DSP. The problems and disadvantages of the existing 2D-MCM are as following:

- Multiple independently packaged devices are mounted on surface of PCB, with longer wiring length;
- The surface area and volume of 2D-MCM are larger;
- Density of integration is lower.

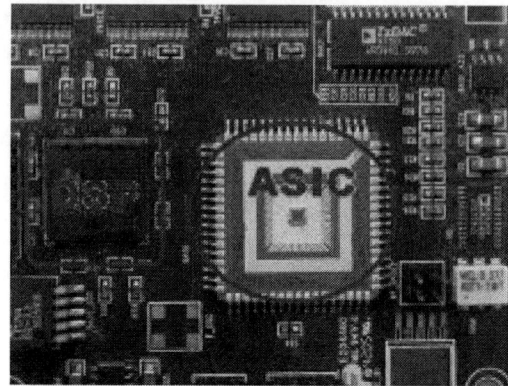

Fig.1 2D layout of the circuit (2D-PCB)

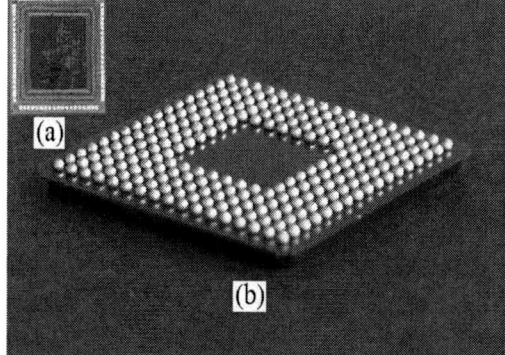

Fig.2 (a) Microscope photo of ASIC (b) Photo of DSP (peripheral BGA: 17x17)

Table 1 Important parameters of ASIC and DSP

Item	Package	Dimension (mm)	Descriptions
ASIC	Silicon die	2.22×2.07×0.30	63Pads, Pad width 0.076mm, pitch 0.094mm(min.), heat dissipation 30mW
DSP	PBGA	15×15×1.4	240-Ball, peripheral BGA 17×17, 63Sn37Pb, polyimide, Heat dissipation 39mW

In order to enlarge the applied range of Terminal of network, it is necessary to develop new packaging type so as to decrease dimension and volume of circuit board. In this paper, the SiP (system in package) conception was inducted, and the "3D-MCM" structure was developed by combining FCOB (flip-chip on board), COB (chip on board), BGA (ball grid array) technologies, wirebonding and flip-chip interconnection technologies together. The substrate is made of modified FR-4 material and is designed as embedded configuration and multilayer so as to embed ASIC and to perform vertical interconnection respectively. The PBGA device and bare chip are hybrid-integrated on/in the substrate. The DSP device is mounted (by flip chip interconnection) on the upside of substrate and the BGA of DSP is underfilled. The ASIC is mounted in the embedded cavity of the bottom of the substrate by wirebonding and glob-top technologies. By solder ball placement the BGA was formed at the bottom side of 3D-MCM. The dimension of substrate is 18mm×18mm. The schematic of 3D-MCM is shown in Fig.3.

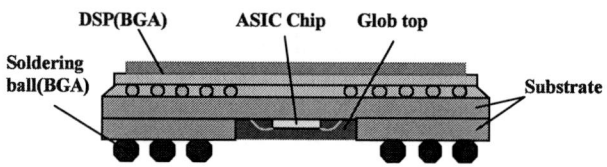

Fig.3 Schematic of 3D-MCM

2.2 Thermal Simulation

Considering the thermal reliability the thermal simulation was conducted so as to optimize the thermal management. ANSYS software was adopted to simulate and evaluate the original package solution. Heat transmission was assumed as natural convection and radiation. Ambient temperature and coefficient of heat convection were assumed as 35℃ and 10W/m^2·℃ respectively. 3D-MCM was assumed mounted on the motherboard (PCB) so as to get the real temperature field of 3D-MCM. The detailed temperature distributions of 3D-MCM are shown in Fig.4. From the simulation result it can be seen that the maximum temperature of the whole 3D-MCM is 44.0°C and the maximum temperature of upside of 3D-MCM is 42.7°C. From Fig.4 it also can be seen that the maximum-

112

temperature point lies in the center of ASIC and the minimum one on the outer corner of the PCB.

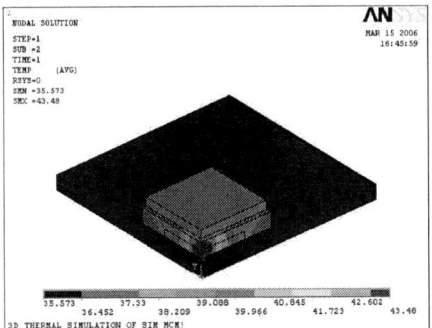

Fig.4(a) Contour of 3D-MCM temperature (1/4 model)

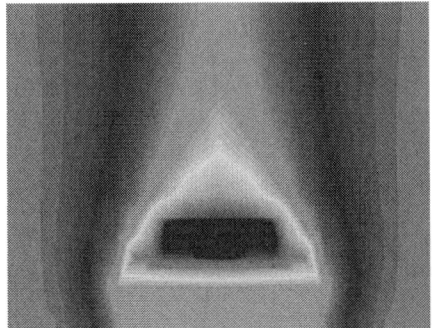

Fig.4(b) Sectional view of 3D-MCM temperature (Overall model)

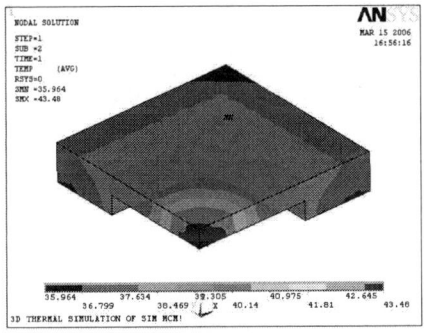

Fig.4(c) Temperature distribution of 3D-MCM substrate (1/4 model)

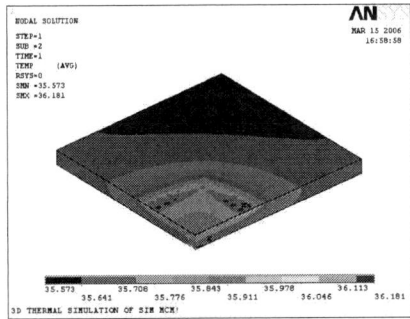

Fig.4(d) Temperature distribution of PCB(1/4 model)

Table 2 shows the percentage of heat dissipation of 3D-MCM. The amount of heat dissipation by means of convection and radiation was equivalent to the heat power of the thermal sources by error 1.8%. From Table 2 it can be seen that the quantity of heat convected from PCB accounts for 39.25%, the largest percentage, which can be attributed to the good conduction performance of soldering balls under the substrate and to the largest interface area between PCB and atmosphere. The quantity of heat dissipated from DSP accounts for 32.81%, which can be attributed to the larger interface area between DSP and atmosphere and the higher temperature of DSP. The quantity of heat dissipated from substrate accounts for 4.23%, which is the smallest. It also can be seen that the convection account for 63.19%, which is the leading heat dissipation mode, and for radiation 36.81%.

Table 2 Percentage of heat dissipation of 3D-MCM

Item	DSP	Substrate	PCB
Convection	21.31%	2.63%	39.25%
Radiation	11.50%	1.60%	23.71%

The simulation result demonstrates that under ambient temperature the temperature of 3D-MCM is below 85°C. It shows that the design of 3D-MCM satisfies the requirement for thermal reliability.

2.3 Package

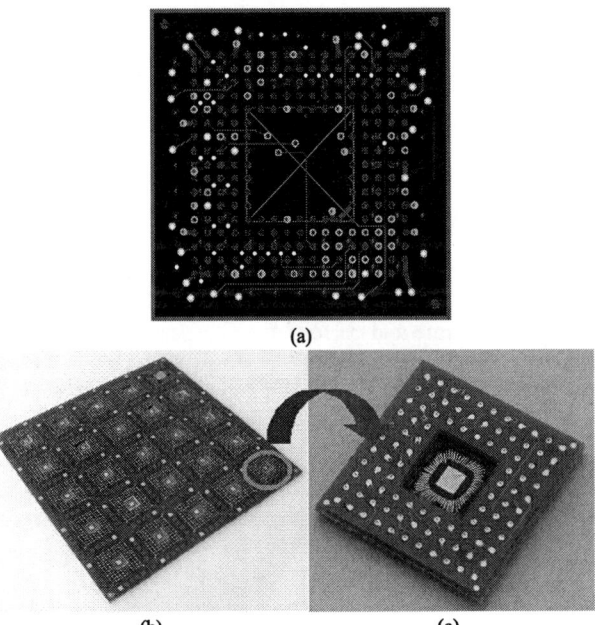

Fig.5 Embedded substrate (a) View of substrate design (b) photo of substrate sheet; (c) single substrate

Fabrication of substrate is a pivotal step. The dimension of DSP is 15mm×15mm, however, the expected dimension of 3D-MCM is 18mm×18mm, which means that the all I/O and the communication between DSP and ASIC must be realized through the via holes of substrate; Two kind of device with different interconnection will be mounted on the two side of substrate, and their direct current voltages are various, e.g. 3.3V, 2.6V, 1.5V; On the bottom side of substrate there are not only 63 pins of ASIC to be wirebonded, but also 96 solder balls (leading-out ends) of 3D-MCM to be arranged; Besides, the issue of electro magnetic compatibility (EMC) must be taken into account. All above facts and factors demand the high level of substrate fabrication.

Considering above factors, a 10-layer FR-4 substrate with high-density interconnection (HDI) was designed. Wire width and wire interval of substrate are small down to $100 \mu m$. Via diameter is small down to $150 \mu m$, which reaches at the utmost of via-machining ability. Therefore, a series of interconnection design and manufacture technologies for high-density embedded substrate were introduced. Besides through-hole technology (THT) various inner via holes (IVH) such as blind via hole and buried via hole are adopted. In order to decrease pad density the POH (pad on hole) technology was also used. Fig.5 shows the layout and photo of embedded substrate.

Generally, from packaging process to being mounted to motherboard, 3D-MCM is exposed to three times of reflow soldering: first, flip chip of DSP, second, planting of soldering balls (63Sn37Pb), and third, mounting 3D-MCM on motherboard. Owing to its eutectic component, the soldering balls of DSP melt during the second and the third reflow. Furthermore, during the ball planting and the second reflow the mounted DSP and substrate must be upside down, so the second reflow may deform the soldering ball of DSP and damage its interconnection performance. Experience shows that multiple reflow may introduce the deformation of the solder balls of DSP, even if underfill, and result in the high probability of short-circuit or open-circuit of interconnection for 3D-MCM. Fig.6 shows the shape comparison of solder balls of DSP after first reflow and multiple reflow respectively.

In order to solve above problem, the lead-free soldering ball (Sn96.5Ag3.5), whose melting point is higher than that of eutectic one, was introduced. Accordingly, above traditional multiple reflow technology was improved, and a new solution i.e. combinatory application of solder balls with multilevel melting points: firstly plant the lead-free soldering ball on the bottom of substrate and perform the first reflow process, then mount DSP on substrate and perform the second reflow process. It can be seen that this solution decrease the reflow times of eutectic ball of DSP, therefore the quality of eutectic balls is ensured.

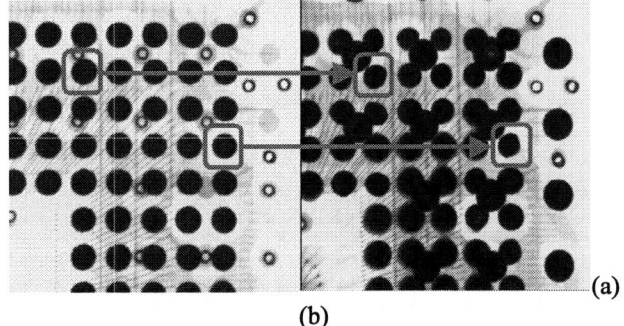

(b)

(a)

Fig.6 Comparison of 63Sn37Pb ball shapes (under X-ray) after (a) one-time reflow and (b) multiple reflow

Owing to the problem of multiple reflow soldering, traditional SMT (surface mounting technology) are not suitable for 3D-MCM. It is necessary to consider the structure of 3D-MCM and the temperature behaviors of various materials especially the soldering balls, and to optimize the process flow, including design of HDI substrate, characteristics of solder paste, printing process of solder paste, mounting process, suitable temperature curve of reflow soldering [13] and application of underfill [14] etc. The comparison between two kinds of process-flows based on Sn63Pb37 and Sn96.5Ag3.5 respectively indicated that if combinatory application of solder balls with multilevel melting points is adopted, accordingly, the maximum reflow temperatures of the multiple reflows decrease (ideally) in succession like sidestep could improve the percentage of pass of 3D-MCM assembling largely.

3. Experiments and Results

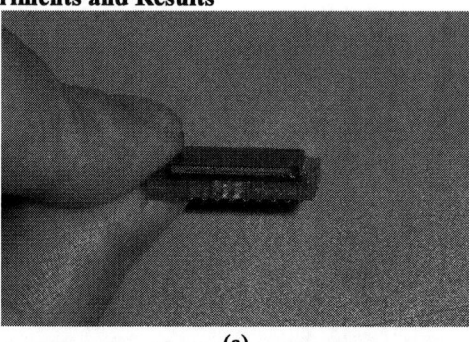

(a)

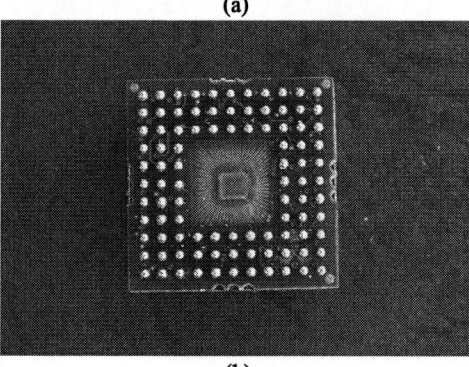

(b)

Fig.7 Photo of (a) Whole view and (b) Bottom view of 3D-MCM

The developed 3D-MCM is shown in Fig.7. The specifications of 3D-MCM is as followings:
- Substrate: Multilayer (10layers), Embedded; FR-4
- Leading-out terminal: Peripheral BGA (11×11)
- Dimension: 15mm×15mm, which is equivalent to about 23% of that of 2D-PCB (52mm×28mm) and is down to about 140% of that of DSP device (15mm×15mm)
- Assemble efficiency: Up to 70%
- Volume is greatly decreased than 2D-MCM

3.1 Function Test

In order to test its electrical function, the specimen 3D-MCM was mounted on a special test board (motherboard), the schematic circuit diagram of which is the same with that of 2D-PCB. The schematic diagram of electric function test for 3D-MCM is shown in Fig.8. Two test boards with 3D-MCM were regarded as transmitter and receiver respectively. The transmitted and received data were observed and compared by means of oscilloscope. The comparison showed that the received data conform to the transmitted data, as shown in Fig.9. It can be concluded that the electrical function of 3D-MCM conforms to that of 2D-PCB.

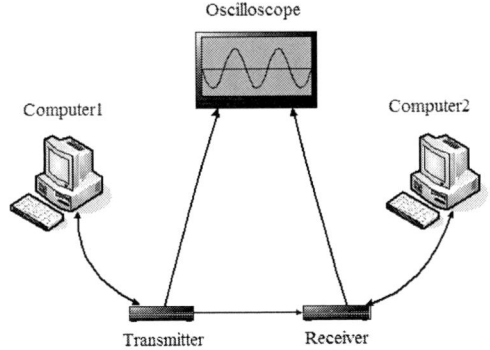

Fig.8 Schematic diagram of function test of 3D-MCM

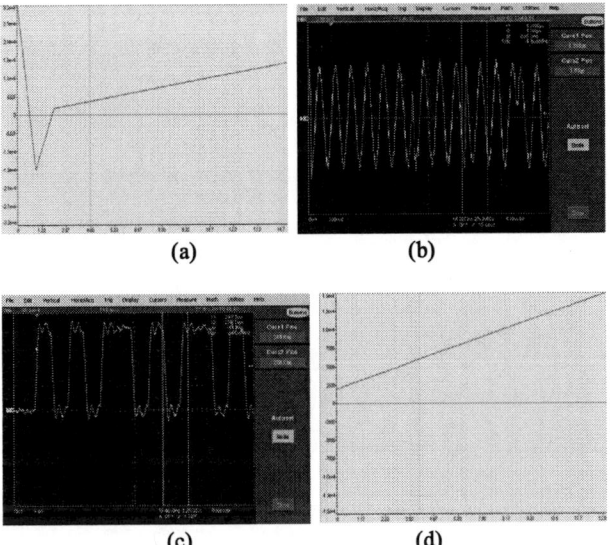

(a)   (b)

(c)   (d)

Fig.9 Transmitting and receiving test for 3D-MCM. (a) Transmitted data wave from computer 1#. (b) Output wave of transmitter DA. (c) Output wave of receiver AD. (d) Received data wave from computer 2#.

3.2 Temperature measurement

The steady temperature distribution of 3D-MCM was measured. 3D-MCM was mounted test board, and the temperatures of a few of positions of the module surface in normal wording state were measured by thermocouple. The ambient temperature was 35℃. The results were shown in Table 3. The temperature on the center of upside of DSP was measured as 42.0℃ and that simulated was 42.7℃. It can be seen that the error between measured and simulated results was not more than 2%. Owing to the precise modeling, the exact material attributes and the conformance of boundary conditions between simulation and experiment the simulated results conform well to measured ones.

Table 3 Comparison between simulation and experiment results of the temperature distribution of 3D-MCM

Position			Simulation (℃)	Measurement (℃)
Upside of DSP	Center		42.7	42.0
	Corner	1	42.4	41.9
		2		42.0
		3		41.9
		4		41.8
Upside of substrate	Corner	1	42.5	41.9
		2		42.1
		3		42.0
		4		41.7
Upside of test board	Corner of Sub.	1	40.8	40.1
		2		40.4
		3		41.1
		4		40.6
	Corner of test board	1	36.6	36.0
		2		36.2
		3		36.4
		4		36.8
Downside of test board (center of MCM)			38.6	37.9

4 Conclusions

According to the multiformity of existing chips/modules for the Terminal of wireless sensor network, a solution of three-dimensional mixed integration with multiple interconnections was put forward, and a kind of high density embedded organic substrate was design and fabricated, therefore a customer-oriented SiP with 3D-MCM package was developed. The compatibility problem of various interconnection processes was solved by optimized multiple reflow technology. The assemble efficiency reaches up to 70%. The application of embedded substrate solved the problem that the top surface of the encapsulated chip overtops the solder balls in the condition that the chip was assembled in the same side of substrate with BGA. The combinatory application of various solder-balls with multilevel melting point and the multiple-reflow technology (i.e. the maximum reflow temperatures of the multiple reflows decrease in succession like sidestep) can improve the percentage of pass of 3D-MCM assembling largely. The thermal management was conducted and the thermal related reliability of 3D-MCM were simulated and evaluated respectively. This kind of packaging structure satisfies the electrical performance and thermal requirement, and meets the challenge of minimization, high reliability and low cost of the package design for the wireless network.

References

[1] Tian Min Bo, Electronic Packaging Technology. Beijing: Tsinghua University Press, 2003.662-684.

[2] Cheng Yingjun, Xu Gaowei, Zhu Dapent, et al, Thermo-mechanical Reliability Study of High I/Os Flip Chip On Laminated Substrate Based on FEA, RSM and Interfacial Fracture Mechanics, Sixth International Conference on Electronics Packaging Technology, Shenzhen, China, 2005: 459-465.

[3] Yan Wei, Yu Shenglin, Fang Xunlei, Three Dimensional Integrated Microwave Modules Based on LTCC Technology. ACTA ELECTRONICA SINICA, 2005, 33(11): 2009-2012.

[4] Zhang X. W., Thermo-mechanical Finite Element Analysis in a Multichip Build up Substrate Based Package Design, Microelectronics Reliability, 2004(44): 611-619.

[5] Zhang Xueren, Tee Tong Yan, and Zhou Jiang, Novel Process Warpage Modeling of Matrix Stacked-Die BGA, IEEE Transactions on Advanced Packaging, 2006, VOL. 29, NO. 2, pp. 232-239.

[6] Cao Yusheng, Yu Haiping, Shi Fazhong. The Study for Heat Analysis Technology of Stacked Three-dimension Multi-chip Module, Control & Automation, 2006, 22(4-2), pp. 191-194.

[7] Chong Desmond Y. R., Lim B K, Rebibis Kenneth J., et al, Development of a New Improved High Performance Flip Chip BGA Package, IEEE 2004 Electronic Components and Technology Conference, 2004: 1174-1180.

[8] George G. Harman, Wire Bonding in Microelectronics, Second Edition, McGraw-Hill, New York, 1997: 266-269.

[9] Gu Jing, Wang Jun, Lu Zhen, et al, Failure Analysis and Thermal Stress simulation in a Stacked Die Package, Chinese Journal of Semiconductors. 2005, 26(6): 1273-1277.

[10] Tian Minbo, Lin Jindu, Zhu Datong, Substrates for High Density Package. Beijing: Tsinghua University Press. 2003:761-766.

[11] Tee Tong Yan, Mayhuan Lim, Ng hun Shen et al, Dsign Analysis of Solder Joint Reliability for Stacked Die Mixed Flip-Chip and Wirebond BGA, 2002 Electronics Packaging Technology Conference, 2002: 391-397.

[12] Xu Gaowei, Wu Yanhong, Zhu Minghua, et al. Development of 3D Multichip Module for Wireless Sensor Net, Proceedings of 2006 International Forum of Electronic Interconnecting Technology and Materials, Shanghai, 2006: 35-39.

[13] Tang Qinghua, Pan Xiaoguang, Chen Y C. The Effect of Process Condition on Electronical Property of Solder Joint in BGA Assemble[J], Journal of Huazhong University of Science & Technology, 1998, 26(9): 78-80.

[14] Wang Tie, Chew T. H., Lum Colin, Chew Y. X., Miao P. and Foo L., Assessment of Flip Chip Assembly and Reliability via Reflowable Underfill. 2001 Electronic Components and Technology Conference. 2001. 803-809.

Novel Low-Temperature Micro-insert Bonding Technology for 3D Package

Po Xu，Anming Hu，Zhuo Chen，Ming Li，Dali Mao

Lab of Microelectronic Materials & Technology, School of Materials Science and Engineering, Shanghai Jiao Tong
University, Shanghai 200240, China
Email: huanmin@sjtu.edu.cn, Tel: +86-21-34202542

Abstract

In this paper, a novel low-temperature micro-insert bonding technology for 3D package has been reported. Nickel microcone arrays (MCA) fabricated on the bonding pad was used as the under bump metallization (UBM). The bonding temperature is below the melting point of the solder. At certain temperature and pressure, the MCA inserted into the lead-free Sn-Ag-Cu solder bumps to achieve a good adhesion. The bonding of the joint is realized by the mechanical interlocking and the diffusion between the MCA and the solder. The nickel microcone arrays were prepared by directional electrodeposition (DEP) method on the Cu substrates in the solution with inorganic additives. And then hundreds of bumps were bonded on the substrates at different temperatures (150℃-210℃) and different bonding pressure (450, 560, 750gf/p). Subsequently, ball shear testings were performed to evaluate the mechanical reliability and failure mode of the solder joints. After the shear testings, the microstructures of the fracture interfaces were investigated by SEM.

1. Introduction

In recent years, three-dimensional (3D) integration technology has attracted much more attention since it offers the possibility that some electronic interconnection problem would be solved. 3D integration technology provides great advantage such as high package density, parallel processing, short wire length and so on. Nowadays, the fabrication technology of 3D packaging include chip to chip (C2C), chip to wafer (C2W), wafer to wafer (W2W), and micro-bumps etc. Currently, there are a variety of ways to compactly connect multiple chips using peripheral wiring technologies, but the packaging density of IC is limited by the wire-bonding technology. The Through-Silicon-Via (TSV) technology is one of the critical and practical technologies for 3D integration. The benefits of 3D integration with TSV technology include reduced interconnect delay due to shorter chip to chip interconnection length, smaller die size, and ability to use distinct, even heterogeneous technologies on separate vertically interconnected layers to build complex systems. So the face to face lead-free interconnection is a key technology for the 3D integration to assemble homogeneous or heterogeneous systems, fulfill high speed and reliability of signal transmission. Traditional approach uses lead-free interconnections, which have either a binary or ternary eutectic compositions, result in good bonding yield and later form an over-melting temperature to bond the joint. But there are some disadvantages for the traditional approach when facing the high performance and ultrahigh density interconnections that require high temperature process. The high temperature process will cause difficulties in the bump

alignment and lead to extra residual thermal stresses. Thus, a low temperature bonding method has to be developed.

T. Suga et al. [1] developed a room temperature bonding method using surface activated bonding (SAB) method. SAB is a process for bonding metals, semiconductors or insulators at surfaces that are cleaned and activated by ion beam bombardment or plasma irradiation prior to bonding. The concept of the method is based on the reactivity of atomically clean surface of solids and formation of chemical bonds on contact between such clean and activated surfaces. The bonding procedure consists of cleaning process and contact in ultra-high vacuum or in a certain ambient atmosphere. The highly activated surfaces enable to bond them together at a temperature lower than that of conventional bonding processes.

W. Satoru et al [2] reported a low-temperature chip on chip interconnection through using low melting point metals (e.g., In and InSn) for micro-solder bumps.

This work reports a novel method to bond the joint by using nickel microcone arrays (on the bonding pad) as the Under Bump Metallization at low temperature. Due to modulus difference between the solder and microcone arrays, the bonding process was realized by the microcone arrays' being inserted into the solder bumps.

2. Experimental

2.1 Preparation of the Ni microcone arrays as Under Bump Metallization

The Ni microcone arrays were fabricated by electrodeposition onto pure Cu sustrates. The electrodeposition solution was composed of analytical pure $NiCl_2 \cdot 6H_2O$, H_3BO_3, crystallization conditioning agent with concentration of 200g/L, 35g/L, 200g/L, respectively. A pure Ni plate (99.9%) was used as an anode, the solution temperature was kept at 60℃ and pH value of the solution was 4.0. The current density for deposition was 1.25A/dm^2 and the deposition time was 18min. And then a thin layer of Au (as the antioxidant layer) was deposited onto the surface [3].

2.2 Bump bonding

Sn-Ag-Cu bumps (760μm) were bonded on the Cu substrates which had the Ni microcone arrays on the surfaces. Different bonding temperature (150℃-210℃) and pressure (450, 560, 750gf/p) were used. Fig. 1 illustrates the bonding process through hot-pressing. Table 1 presents the parameters of the hot-pressing process.

2.3 Ball shear testings

To evaluate the integrity of the solder balls bonded on the Cu substrate, the ball shear test was conducted by using a

978-1-4244-2739-0/08/$25.00 ©2008 IEEE

RHESCA PTR-5000 Axial-Servohydraulic Dynamic Tester. The applied shear rate was 30mm/min. The measuring method is shown in Fig. 2.

The analysis was performed by means of SEM, to inspect the interface between the solder and Ni microcone arrays and the fracture morphology of the sheared solder.

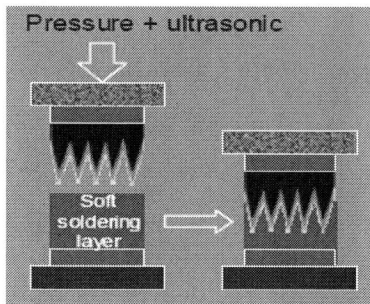

Figure 1　Demonstration of the bond process.

Table 1 Parameters of the hot-pressing process.

Temperature(℃)	Bonding times of different pressure		
	750gf/p	560gf/p	450gf/p
150-170	3min	3min	—
180-185	1mm	1min	—
190-200	1min	1min	1min
205-210	1min	1min	1min

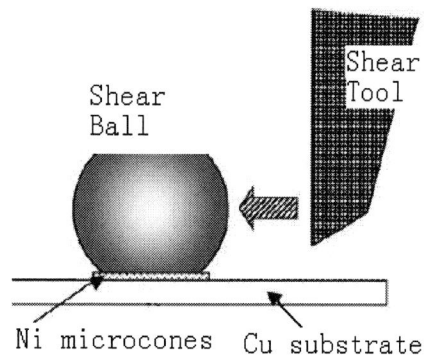

Figure 2　Method for measuring the solder ball shear strength.

3 Results and discussion

3.1 Morphology of Ni microcone arrays

Fig. 3 shows the low and high-magnification SEM images of Ni microcone arrays electrodeposited for 18 min on the Cu surface. From Fig. 3 it can be seen that the roots of the microcone arrays have an average diameter of about 150 nm, and that the tips' average diameter ranges from several to tens of nanometers. The height of the microcone arrays is about 500 nm, while the average apex angle is about 30°. It can be also seen that most of the array has the same growth direction.

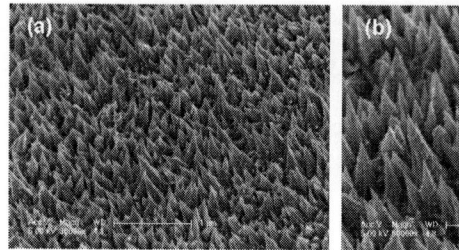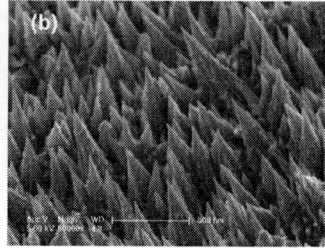

Figure 3　SEM images of the Ni microcone arrays. (a) Low magnification, (b) high magnification.

3.2 Morphology of the interface

As shown in Fig. 4, there is a crack at the interface suggesting that the ball was spreaded during the hot-pressing. Figure 5 shows the morphology of the interface between the bump and the layer of microcone arrays. It could be found that the microcone arrays have inserted into the solder bump. When the ball contacted the layer of microcone arrays, the bottom of the ball touched first, and then spreaded to the surrounding areas. By comparison of Figs. 5a, b and Figs. 5 c, d, it can be seen that the microcone arrays incompletely insert into the bump at the edge of the bump. The insertion at the center of the bump was better than that at the the edge. There are not obvious voids at the interface.

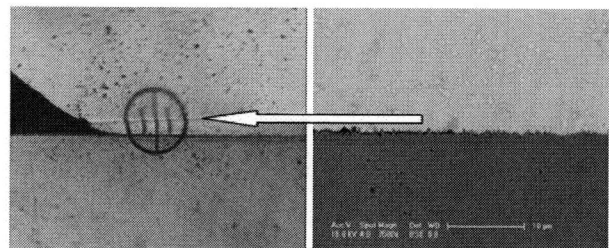

Figure 4　Crack at the interface.

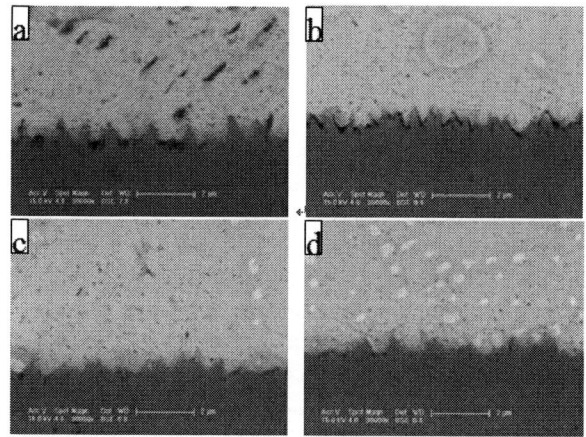

Figure 5 SEM image of the interface. (a, b) edge of the bump, (c, d) center of the bump.

3.3 Diffusion at the interface

The cross-section of the bump after aging at 150℃ for 24h was shown in Fig. 6. From the figure it can be seen that there is an intermetallic compound layer between the solder and Ni microcones. And the tips of the microcone arrays disappeared in the solder.

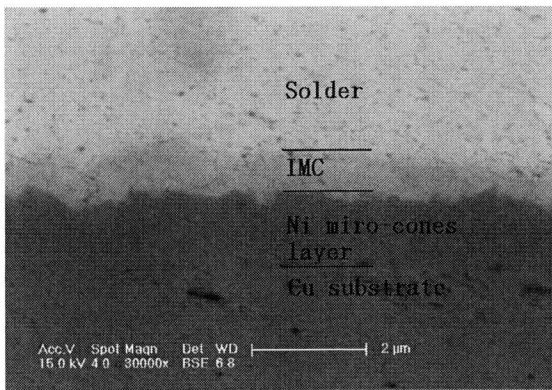

Figure 6 SEM image of the cross-section.

3.4 Shear test

As can be seen in Table 2, the typical spreading area with the pressure 560gf/p increased with increase of the temperature. From the corresponding typical fracture morphology, it can be found that ductile shear failure in the bulk solder was the main failure mode when the bonding temperature was above 180℃. Brittle shear failure in the pad interface was the main failure mode when the bonding temperature was below 180℃. As shown in Fig. 7, the shear strength increased with increase of the temperature.

Table 2 Typical spreading areas and fracture morphology of solder balls (bonding pressure is 560gf/p).

160℃	170℃	180℃	190℃	200℃
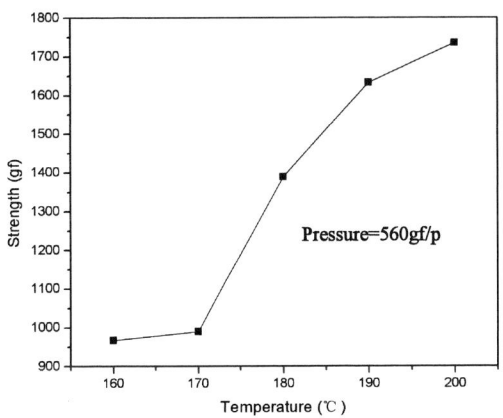

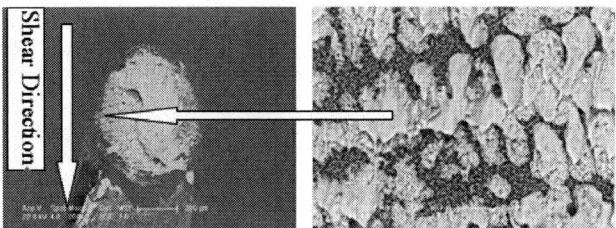

Figure 8 SEM morphology of edge fracture.

Fig. 9 shows the shear strength of the bumps varying with the bonding temperature and pressure. The shear strength increased with increase of the bonding temperature. When the temperature was below 180℃, the strength with the bonding pressure of 750gf/p improved more than 50% percent compared to the bonding pressure of 560gf/p. The shear strength mostly depended on the mechanical occlusion. When the temperature was over 180℃, the shear strength of the two kinds of pressure was closer. So there was not obvious effect of the bonding pressure to the shear strength when the bonding temperature was over 180℃. The shear strength mostly depended on the diffusion under high temperature. The strength with bonding pressure of 450gf/p was much lower than the others. The bonding pressure also has effected the diffusion. Anyway, the formation of the layer of the IMC was the key to the low-temperature bonding reliability.

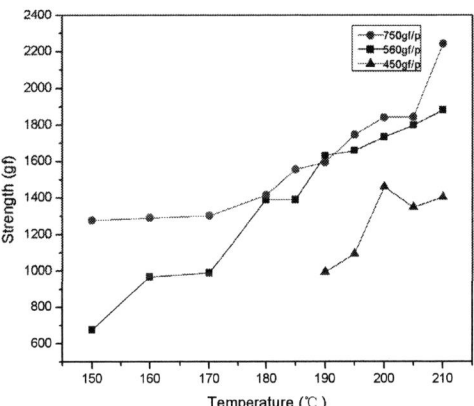

Figure 9 Shear strength of the bumps.

4. Conclusions

We developed a novel low-temperature micro-insert bonding technology featuring a preparation the MCA as the UBM for 3D package. The temperature in the bonding process could be controlled below the melting point of the solder. In conclusion, we get the optimized conditions of the bonding process including the bonding temperature and pressure to obtain sufficient bonding ability. The further study of the application of this technology is progressing in the lab.

Acknowledgments

We thank the support from China International Science and Technology Cooperation (Contact No. 20073774).

Figure 7 Shear strength of the bumps.

Fig. 8 shows the magnified edge fracture morphology at 180℃ and 560gf/p. It can be found that the solder discontinuously adhered to the microcones. It also testifies to the inserting process.

This work is sponsored by International Science and Technology Cooperation Program of China (ISCP) (Contact No. 2008DFA51680).

References

1. Suga, T. et al, Feasibility of surface activated bonding for ultrafine pitch interconnection—a new concept of bumpless direct bonding for system level packaging. In: *Proceeding of the 50th Electronic Components and Technology Conference, ECTC2000*, Las Vegas, NV, May 2000. pp. 702-5.

2. Satoru, W. et al, "Novel Low-Temperature CoC Interconnection Technology for Multichip LSI(MCL)," *Proceeding of Elctronic Components and Technology Conf 2007*, pp. 610-615.

3. T Hang. et al, "Characterization of nikel nanocones routed by electrodeposition without any templat," *Nanotechnology*, Vol. 19, No. 3 (2008), pp. 035201-035205.

High-Q On-Chip Inductors Embedded in Wafer-Level Package for RFIC Applications

Tao Feng[1], Jian Cai[1], Henri HK Kwon[2], Qian Wang[2], Xinyu Dou[1]

[1]Tsinghua University, China [2]Samsung Electronics Co., Korea

Ft1983@gmail.com

Abstract

Wafer level packaging (WLP) technology has been used to integrate high-Q inductor on Si substrate. These inductors consist of a thick Cu electroplated rerouting to reduce series resistance and a thick dielectric layer to separate the inductors from Si substrate. The measured results show that the peak Q-factor is 30 at 4GHz for a 0.77nH inductor, which is good agreement with the simulated performance by HFSS. Therefore, this technology realizes embedded high-Q inductors in WLP and can improve the performance of RF system.

Introduction

On-Chip inductors are one of the most important and promising passive components in RFIC Applications [1]. The performances of inductors, such as quality factor (Q), frequency at peak-Q and self-resonance frequency (f_{res}), still need to be improved to meet the demanding requirements of various low loss RF circuits, such as low noise amplifier (LNA) and voltage-controlled oscillator (VCO) [2-3]. However, the Q-factor is often comparatively low for on-chip inductors mainly because of the large DC resistance of the thin metal layer (usually Al) and energy losses in the silicon substrate due to magnetic field induced currents.

Many efforts [4-6] have been made to overcome these problems, such as utilizing ground pattern, optimum routing pattern, and multilayer inductors as well as MEMS technology, but all these technologies have own drawbacks. A more reasonable solution is to realize the inductors above dielectric layer using wafer-level packaging (WLP) techniques. This WLP structure consists of dual thick Cu metal layers, dual dielectric layers and an encapsulation layer. The embedding inductors are located by using the first and second metal layer via the second dielectric layer. Fig. 1 shows the cross section of WLP structure. Therefore, the attractive advantages of this technology are obvious [7]:

(1) The induced eddy currents in Si substrate can be reduced by the thick dielectric layer separating the conductive layer from Si substrate.

(2) Low resistance of the inductors can be obtained by electroplating thick Cu metal layer.

(3) Additional inductance from wire bonding can be eliminated by the lead-free solder for mounting and interconnection.

In addition, WLP technology has already achieved thermal stabilization, high reliability and cost efficiency [7].

Therefore, there are many researches realizing on-chip inductor using WLP technology have been published [3], [7-10]. In this paper, the details of fabrication process of WLP inductors, and experimental results analysis as well as the simulated values have been presented.

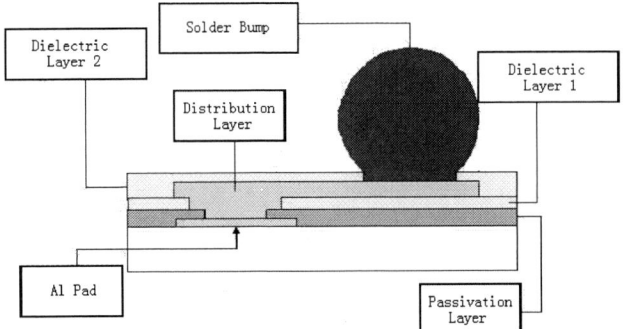

Fig. 1: The cross section of WLP structure.

Model Analysis and Calculation

Physical model and lumped-element equivalent circuit of WLP spiral inductor are shown in Fig. 2(a) and Fig. 2(b), respectively. L_S is self-inductance of the spiral inductor, R_S is series-resistance of inductor itself, C_S is coupling capacitance between spiral and underpass pattern, C_{OX} is dielectric capacitance between spiral metal and Si substrate, Csi and Rsi are Si substrate capacitance and resistance respectively.

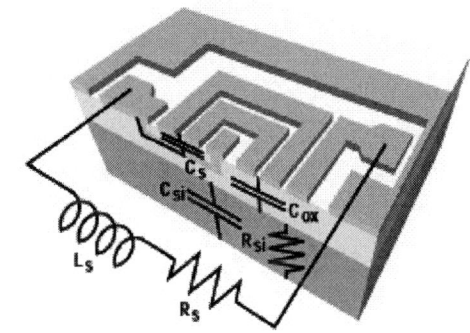

Fig. 2(a): Physical model of WLP spiral inductor.

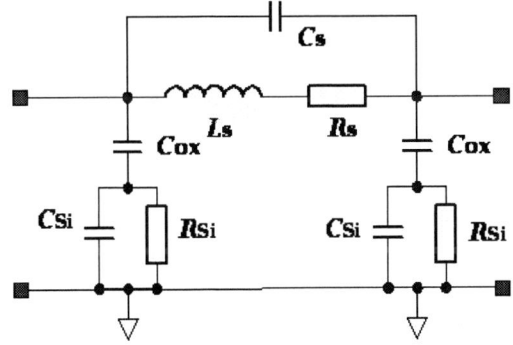

Fig. 2(b): Lumped-element equivalent circuit of WLP spiral inductor.

The lumped-element equivalent circuit is changed into two-port network to calculate Y parameters from which the L and Q of the inductors can be drawn [11].

978-1-4244-2739-0/08/$25.00 ©2008 IEEE

$$Q = \frac{\text{Im}(1/Y_{11})}{\text{Re}(1/Y_{11})} \qquad (1)$$

$$L = \frac{\text{Im}(1/Y_{11})}{\omega} \qquad (2)$$

For the performance of the spiral inductors, L is mainly determined by the geometrical spiral pattern, while Q and f_{res} are chiefly affected by the cross section structure of the inductors, such as the thickness and electrical properties of the substrate and dielectric layer. To improve the performance of the inductors, therefore, the dielectric layer with low dielectric constant should be employed. In this work, photosensitive PI (Polyimide, Dielectric Constant = 3.2, Dissipation Constant = 0.001@1GHz) has been used as dielectric layer and encapsulation layer.

Fabrication Process

Two key processes of the fabrication of WLP inductors are PI curing and Cu electroplating. PI curing includes coating, soft bake, exposure, development, and cure, while Cu electroplating area and thickness should be taken into consideration before electroplating. Fig. 3 represents the schematic diagram of the fabrication process of the WLP inductors. The detailed process is described as follows:

(1). Coating a 7μm thickness PI on Si substrate with SiO$_2$ layer as the 1st dielectric layer.

(2). Sputtering Ti/Cu as a seed-layer and coating a 14μm thickness AZ-4620 on this seed-layer as photoresist. Subsequently, this photoresist is developed to form an underpass pattern.

(3). Electroplating a 5μm-thickness Cu as the underpass conductive layer.

(4). Removing the photoresist and etching the seed-layer.

(5). Coating 10μm-thickness PI as the 2nd dielectric layer and then curing the 2nd PI to form contact via having an area of $30 \times 30\mu m^2$.

(6). Repeating (2), while this photoresist is developed to form spiral pattern.

(7). Electroplating a 5μm-thickness Cu as the spiral conductive layer.

(8). Removing the photoresist and etching the seed-layer.

(9). Coating 10μm-thickness PI as the encapsulation layer and then curing the 3rd PI to form contact via for testing.

A variety of geometries are designed and simulated by HFSS to obtain 0.8nH inductor at 2GHz in a 1.5 turn rectangular spiral. Table 1 lists the geometrical parameters of the fabricated inductors.

Table 1: Inductor Layout parameters

Inner Radius (μm)	40-80
Width (μm)	25
Space (μm)	20
Cu Thickness (μm)	5/5 (1st/2nd)
PI Thickness (μm)	7/10/10 (1st/2nd/3rd)

Results and Discussion

Fig. 4 shows the optical photo of the fabricated 1.5 turn square spiral inductor. Firstly, DC resistance is evaluated to predict the performance of the fabricated inductors. The typical DC resistance of a 1.5 turn spiral inductor in this work is less than 0.5Ω. Then the L and Q are measured by using Cascade microprobe station and an HP8722ES network analyzer. The measurements are performed by the network analyzer to obtain S-parameter of the inductors and open pattern from 100MHz to 10GHz with SOLT calibration. The open pattern is applied to de-embed the effect of G-S-G pads. After the S-parameters of the inductors are de-embedded from the open pattern and then converted to Y-parameters, the L and Q can be drawn from formula (1) and (2).

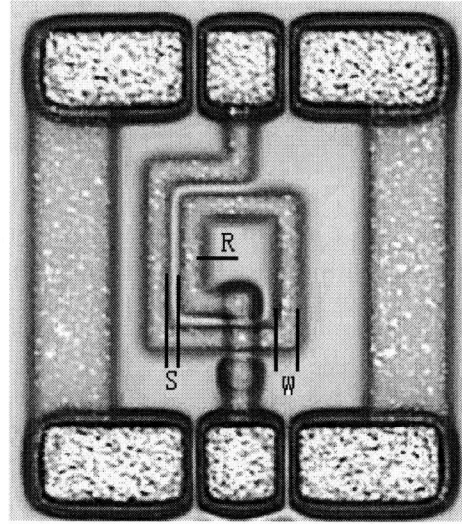

Fig. 4: The optical photo of the fabricated inductor.

Fig. 5 shows measured L as a function of frequency. Table 2 lists the inductance of all the inductors with radius varying from 40 to 80μm. It is obvious that the inductance improve 0.1nH with the augmentation of 10μm for the inner radius. Also, it is important that the inductor with the radius of 60μm is 0.77nH, meeting the design object of this work.

Fig. 6 shows the comparison of L between measurement and simulation by HFSS for 0.77nH inductor. It is obvious that the simulated value corresponded well with measured result with an accuracy of 98%.

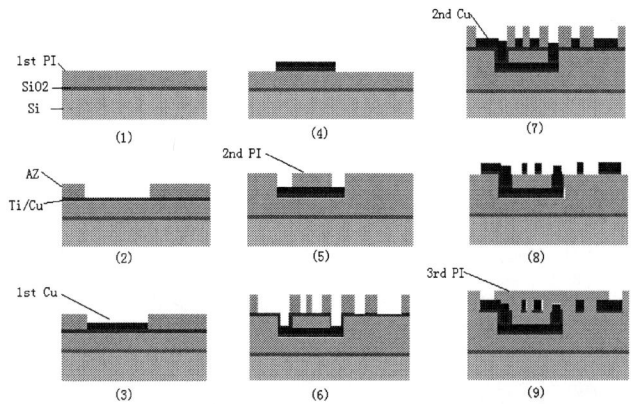

Fig. 3: The schematic diagram of fabrication process

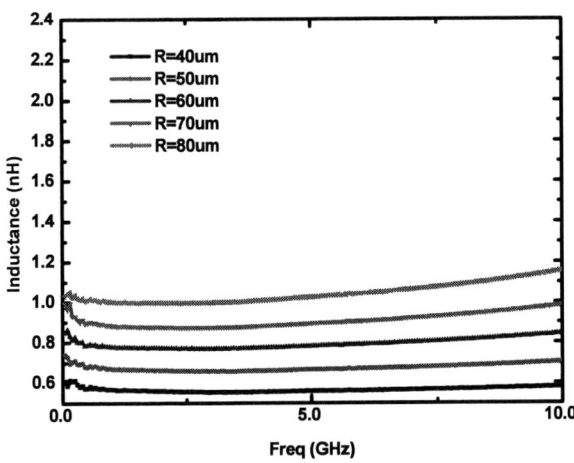

Fig. 5: Measured Inductance of all fabricated inductors

Table 2: List of measured inductance

Inner Radius (μm)	Inductance (nH)
40	0.56
50	0.66
60	0.77
70	0.87
80	0.99

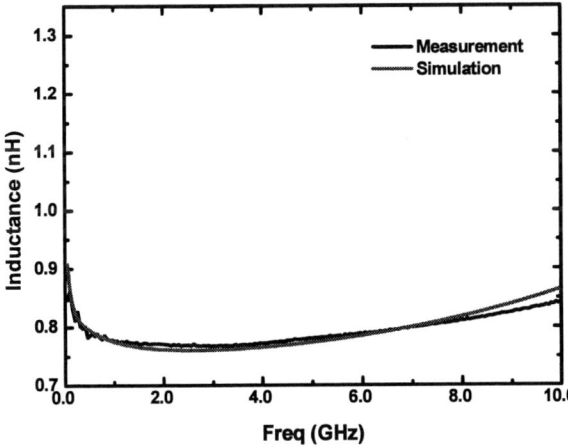

Fig. 6 Comparison of L between measurement and simulation for 0.77nH inductor.

Fig. 7 shows the comparison of Q between measurement and simulation for 0.77nH inductor. The measured Q-factor is 24.5 at 2GHz and peak-Q is 30 at 4GHz as well as f_{res} is more than 10GHz. Also, it is above 20 from 1 to 10GHz, which guarantees a good performance for RF Applications in a wide frequency range. The difference in Q between simulation and measurement at 2GHz is 3%.

In short, it is well demonstrated that HFSS is an effective tool to design the WLP inductor using thick dielectric layer and Cu conductor.

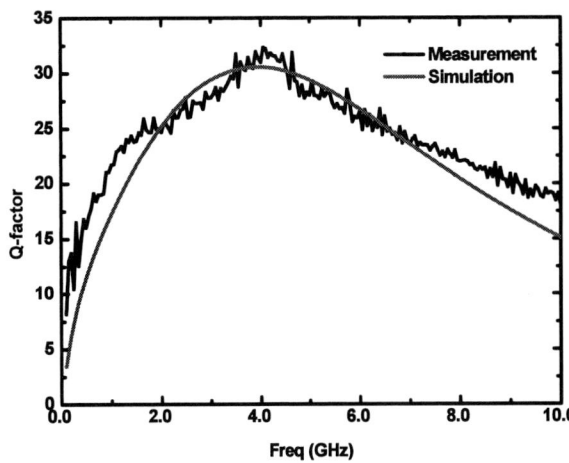

Fig. 7 Comparison of Q between measurement and simulation for 0.77nH inductor.

According to the lumped-element equivalent circuit of WLP spiral inductor, these lumped parameters can be extracted from Y-matrix as following, shown in Table 3:

$$L_S = \frac{1}{2\pi f} \text{Im}\left\{\frac{1}{Y_{21}}\right\} \qquad (3)$$

$$R_S = \text{Re}\left\{-\frac{1}{Y_{21}}\right\} \qquad (4)$$

$$C_{Si} = \frac{1}{2\pi f} \text{Im}\{Y_{11} + Y_{12}\} \qquad (5)$$

$$R_{Si} = \frac{1}{\text{Re}\{Y_{11} + Y_{12}\}} \qquad (6)$$

Table 3: The extracted parameters of 0.77nH inductor at 2GHz

L_s (nH)	0.76
R_s (Ω)	0.38
C_{si} (fF)	71.5
R_{si} (kΩ)	8.62

In table 3, it is noted that R_{si} is 8.62 kΩ, nearly 13 times higher than that of a conventional on-chip inductor on Si substrate [12], considerably reducing the eddy loss. Therefore, it is obviously proved that the significant features of the inductors embedded using WLP technology are for high performance RF applications.

Conclusion

It has been shown that WLP technology enables the realization of high-Q inductor as an alternative solution for high performance inductor integrated RF CMOS circuit. This WLP structure consisted of a dual Cu electroplated layer as a low-loss conductor and dual thick dielectric layers to reduce the substrate loss. The inductance of 0.77nH with Q-factor of 24.5 is obtained for a 1.5 turn rectangular spiral inductor at

2GHz. In addition, the fabricated results well accord with the simulated values, indicating that the electromagnetic simulator may be an effective tool to design WLP inductor. This fabrication process can replace wire-bonding inductor and reduce series resistance of inductors itself by wafer-level package technology as well as achieve a chip-scale package.

Acknowledgments

The authors would like to acknowledge the support from Samsung Electronics Co. Ltd and JCAP Co. Ltd.

References

1. Robert J., et al., "Integration of Silicon with Passive Devices Yields Advantages in Wireless Design," *High Frequency Electronics*, 2003, pp. 56-61.

2. K.S. Yeo, "RFIC Designs for WLAN Applications using CMOS Technologies" *Proc. of Microwave workshop* (2004), pp.1-4.

3. Sang-Woong Yoon, et al., "A 0.35μm CMOS 2-GHz VCO in Wafer-Level Package," *IEEE Microwave and Wireless Components letters*, Vol. 15, No. 4 (2005), pp. 229-231.

4. C. Patrck Yue, et al., "On-Chip Spiral Indcutors with Patterned Ground Shields for Si-Based RF IC's," *IEEE JSSC*, Vol. 33, No. 5 (1998), pp. 743-752.

5. Joachim N. Burghartz, et al., "Multilevel-Spiral Indcutors Using VLSI Interconnect Technology," *IEEE Electron Device Letters*, Vol. 17, No.9 (1996), pp. 428-430.

6. J.B.Yoon, et al., "3-D Construction of Monolithic Passive Components for RF Micromachining Technology," *IEEE Trans. Microwave Theory Tech.*, Vol. 51, No. 1 (2003), pp. 289-296.

7. Kazuhisa Itoi, et al., "Comparison of Compact On-Chip Inductors Embedded in Wafer-Level Package," *Electronic Components and Technology Conference* 2005, pp. 1578-1583.

8. Geert J. Carchon, et al., "Wafer-Level Packaging Technology for High-Q On-Chip Inductors and Transmission Lines," *IEEE Transcations on Microwave Theory and Techniques*, Vol.52, No.4 (2004), pp. 1244-1251.

9. Tae-Je Cho, et al., "Design of CMOS Voltage Controlled Oscillators Using Package Inductor," *Electronic Components and Technology Conference* 2005, pp. 1682-1686.

10. P.M.Mendes, et al., "Wafer-Level Integration of On-Chip Antennas and RF Passives Using High-Resistivity Polysilicon Substrate Technology," *Electronic Components and Technology Conference* 2004, pp. 1879-1884.

11. A. M.Niknejad, et al., "Analysis, Design and Optimization of Spiral Inductors and Transformers for Si RFIC's," *IEEE JSSC*, Vol.33, No.10(1998), pp. 1470-1481.

12. Tung-Sheng Chen, et al., "Improved Performance of Si-Based Spiral Inductors," *IEEE Microwave and Wirelee Compenents letters*, Vol. 14, No.10(2004), pp. 466-468.

Thermo and Mechanical Study on Integrated High-density Packaging

Qian Chen[a,b], Yi-ping Wu[a,b], Feng-shun Wu[a,b], Zhi-jun Zhu[a,b], Yuan Tao[a,b]

a National Laboratory for Optoelectronics, Wuhan, Hubei, China. 430074
b State Key Laboratory of Material Processing and Die & Mould Technology,
Huazhong University of Science and Technology, Wuhan, Hubei, China. 430074
Corresponding author: Yi-ping Wu. E-mail: ypwu@mail.hust.edu.cn Tel.: +86-27-87792402

Abstract

With the demanding of market, the electronic portable products can be characterized by increasing signal frequencies and higher density of functions. The electronic products are expected to be produced smaller and smaller. There is one way to meet the requirement, which is a three-dimensional integration of components. A new concept of a packaging structure is proposed based on an embedded chip structure. Chips and other components can be embedded directly into epoxy and then interconnections are realized by laser drilling and metallization.

In the process, components are placed on a removable tacky film on a temporary or permanent base. The film and base temporarily immobilize them until the structure is encapsulated. The entire array of tested and burned-in components thereby becomes a monolithic assembly, with each component now permanently immobilized by every part of it. The bottoms of these terminations can be exposed by removing the temporary base and film or by making holes in a permanent one by such means as mechanical abrasion, water-jet material removal, or laser ablation. The assembly is now ready to be metalized with copper using standard printed circuit additive (build-up) processing methods, with circuit patterns created to make the required interconnections among leads of all of the components. In most cases, more than one layer will be required, so an insulation layer is placed over it and the process is repeated until all required interconnections are made.

Thermal and thermo-mechanical reliability are taken into account in this structure due to the encapsulation on the chips. For this purpose, numerical study by means of finite element analysis is available to check the desired properties. Thermal analysis aims to give the first insight to characterize the thermal performance for such structure under steady and transient condition. Epoxy around the chip also plays an important role in the heat dissipation. The thickness of copper has large influence on the thermal resistance. Besides, the copper in the top metallization layer of the package is also taken into account. Thermal stress between the copper and the epoxy is also considered in this paper.

Introduction

The coming generations of portable products will require significant improvements of integration and packaging technologies, mainly due to the increasing demand for higher density of functions 1. The current technology named "Chip in polymer" provides organic substrates with high-density build-up layers and micro-vias, equipped on both sides with surface mount passive and active components. A further micromation however requires 3D integration of components. In order to maintain signal integrity, much shorter interconnects between chips and passive components are required. The new concept of packaging technology was introduced by Fraunhofer IZM and TU Berlin. It is based on the embedding of ordinary chips into build-up layers of printed circuit boards 2-4. The structure is shown in figure 1.

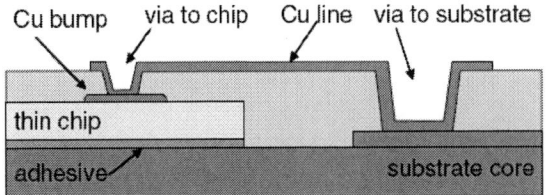

Fig.1: Interconnect principle of an embedded chip in a PCB build-up layer

In the thermal management perspective, this embedded chip structure is expected to have poor thermal performance because of the chips is surrounded by epoxy. It is necessary to understand the thermal behavior of the structure since both the reliability and life expectancy of electronic equipment are inversely related to the component temperature. The aim of the present work is to create 3D Finite Element Modeling (FEM) to investigate the thermal behavior of the structure which includes both passive and active components.

Chip in Polymer

The technology developed at TU Berlin works without a cavity layer in the wire board 5. Here very thin chips are directly placed on top of the core substrate. Chips thinned down to 50 μm are placed and attached onto the board with high precision using adhesive or die attach film. Then the chip is embedded in a polymer layer by vacuum lamination. Vias to the chip contact pads and to the Cu-routing on the board are laser drilled and metalized. Finally the top Cu-layer is structured. The structure now allows to position conventional components directly over the embedded chip 6.

Embedded components

As described earlier, the packaging evolution toward SOPs shows a clear convergence toward a full integration of components and substrates into a monolithic package. Significant efforts were made to integrate resistors and capacitors in PCBs using various techniques 7. It is expected that integrated active will be used in applications with either extreme miniaturization or lowest cost requirements, while a broad replacements of discrete will still take several years due to the limitations in reliability and range of achievable values. A number of approaches to integrate active components into a substrate have also been published. Embedding of active components means to apply the connective metal lines on top of them, which is originally called the "chip first" approach, e.g., by GE. Even for high-performance microprocessors a concept for chip embedding

978-1-4244-2739-0/08/$25.00 ©2008 IEEE 125

has been presented recently. The main advantages of integrating active chips into a substrate are the increased electrical performance by shorter interconnections and therefore lower inductances, increased reliability by lower stresses compared to flip chip assembly and a higher interconnect density. A big challenge however is the required changes in the manufacturing flow compared to a conventional assembly process 8-9.

Concept

The High-density packaging technology is based on the embedding of both positive and active devices into the build-up layers of epoxy. Chips are first bonded on substrate. They are subsequently embedded into a build-up dielectric and vias are drilled to the chip contacts as well as to the substrate Cu lines, as show in Fig. 2-3. Finally, the interconnecting Cu is applied by electroless deposition and electroplating, as shown in Fig.4, thereby connecting electrically the chips to the substrate. On this build-up layer, further layers can be applied, shown in Fig.5 or SMDs can be mounted. In contrast to flip chip assembly, the mechanical attachment of the chips is separated from the electrical contracting, which allows better reliability without the use of processes like soldering. A further benefit includes the very short electrical contacts which have been calculated to have even lower impedance than a flip chip or any other packaging structure, allowing an increased electrical bandwidth. Finally, the embedding of chips into the substrate without increasing its thickness opens the way to much higher functionality by 3-D integration. Furthermore, it is possible to embed both active and passive circuitry for multiple RF and optical functions. The process flow is shown in fig2-4.

Fig.2 The chips are embedded in organic materials

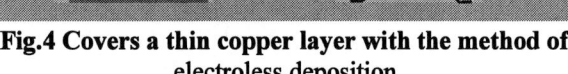

Fig.3 Laser drilling to the chip contacts as well as to the substrate Cu lines

Fig.4 Covers a thin copper layer with the method of electroless deposition

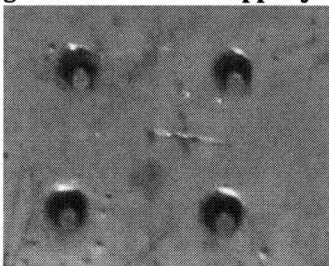

Fig.5 Form multi-layer circuit lines

After laser drilling the micro-vias are cleaned. This is followed by Pd activation and electroless Cu deposition. The thickness of Cu layer is 1- 2 μm and the Cu layer acts as a seed layer for the subsequent galvanic plating. A minimum Cu thickness of 10 μm is required in the micro-vias. Some experiments have been done with electroplating in order to fill the micro-vias completely with Cu, shown in Fig.5-6.

Fig.5 cross section of copper joint

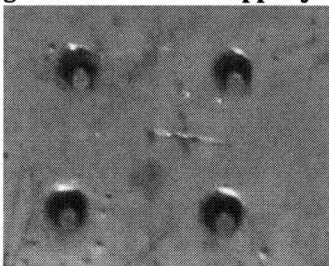

Fig.6 Micro-vias copper deposition with both electroless and electroplating method

Finite Element Design

The mode is created to study the thermal behavior of the embedded chip structure. For simplify, only one chip and one copper layer is introduced. In view of symmetry, only one quarter of the structure is introduced, as shown in figure 2. The chip is embedded in epoxy. The chip is 10×10mm square and 1mm thick; The copper thread is 300μm wide. The connections between the chip and copper track are rectangular pads 100μm square.

Details of materials properties are provided in Table 1. A content value is chosen for each material, no temperature dependent relationship is considered in present work.

Tab.1: Thermal and structure properties of materials used in this simulation

Material	copper	epoxy	chip
Thermal conductivity (W/mK)	383	1	150
Density(kg/m3)	8900	1100	2330
Specific heat(J/kgK)	390	1960	703
CTE(K-1)	1.75E-5	4.62E-5	2.6E-6
Elastic(pa)	1.19E11	3.6E9	2E10
Poisson's Ratio	0.326	0.37	8.36
Yield stress(pa)	9E8	8.36E7	1.69E11
Tang Mods(pa)	1.03E10	2.99E9	8E9

The mode is created in ANSYS workbench, shown in figure 2. The chip is embedded in epoxy. Vias are formed by laser drilling or any other methods, then the vias are metalized by both electroless plating and electroplate.

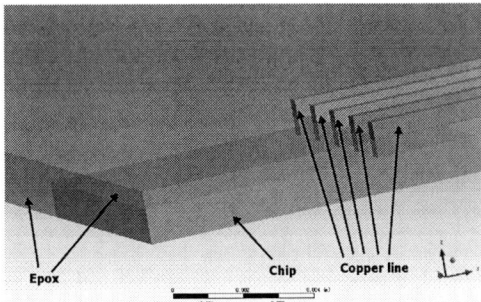

Fig. 2 3D mode of embedded chip structure

Analysis with active component

Assuming that the chip is a positive component, 2-watt heat dissipation is uniformly distributed over the volume of the chip. Considering that the chip is $10 \times 10 \times 1$mm, so the quarter heat generate is 5×106 W/m2 . Natural convection with a heat transfer coefficient 15 W/m2K is applied to all the surfaces exposed to the ambient air. The initial ambient temperature is $25°C$.

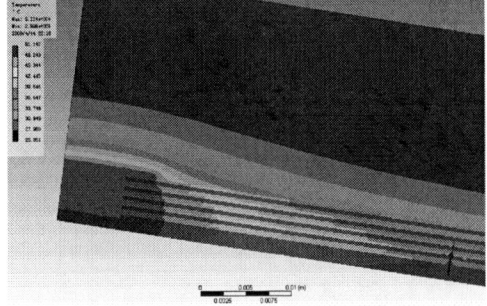

Fig. 3 Temperature distribution of steady-state analysis

Temperature distribution of the steady-state is shown in Fig. 3. The highest temperature is $51.142°C$. The structure has heat stress on the interface of different materials due to the differences in coefficient of thermal expansion, as shown in Fig. 4. The place where has the greatest stress is the corner of the copper line contacting with the epoxy. The maximum value is 1.87×107Pa, which is less than the yield stress of the three materials.

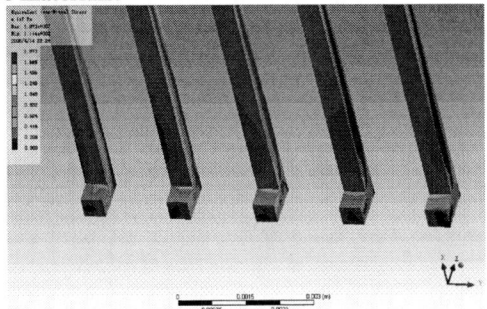

Fig. 4 Von-Mises Stress of the copper with positive component

Heat stress analysis with uniform temperature

Heat stress generated by uniform temperature is also taken into account. Three different temperature$100°C$, $150°C$, $200°C$ are considered in this analysis. The von-Mises stress is shown in figure 5 to 7. The chip is embedded at the temperature $25°C$.

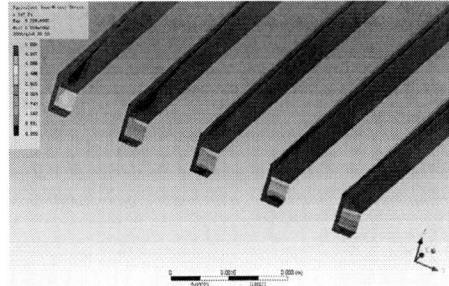

Fig. 5 Von-Mises stress in temperature 100°C

The maximum von-Mises stress in uniform temperature $100°C$ is 5.228×107Pa, which is nearly the yield stress of epoxy. Stress concentration happenes at the corner of copper line contacting with the epoxy.

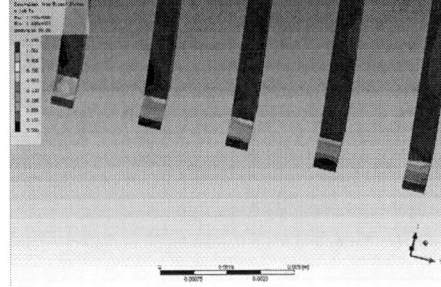

Fig. 6 Von-Mises stress in temperature 150°C

The maximum von-Mises stress in uniform temperature $150°C$ is 8.58×107Pa, which is larger than the yield stress of epoxy. Desquamation may happene at the interface.

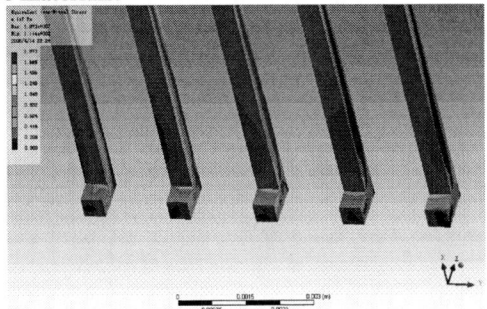

Fig. 7 Von-Mises stress in temperature 200°C

The maximum von-Mises stress in uniform temperature $200°C$ is 1.193×108Pa, which is much larger than the yield stress of epoxy. Serious desquamate may happened at the interface.

The thermal performance is expected to be very poor at a higher temperature since the difference in the coefficient of thermal expansion and bad thermal conductivity of epoxy. Epoxy is very important in fixing the chip position, and to give a stress relaxation to protect the chip, the influence of epoxy is need further study. Epoxy plays an important role in removing the heat. Fatigue 8 shows the influence of the different thermal conductivity of epoxy.

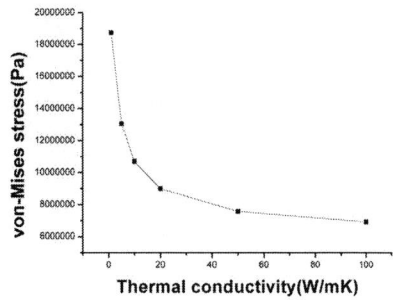

Fig. 8 Influence of epoxy thermal conductivity

It is obviously that with the thermal conductivity decrease the von-Mises stress becomes smaller and smaller, since the heat could dissipate quickly as a result the overall temperature is less than the previous ones, so the stress is decrease obviously.

Conclusions

The finite element analysis was confirmed to be a qualified numerical tool to forecast and analyze the thermal and the thermo-mechanical behavior of complex assemblies including embedded chips.

The numerical results directly influenced the further design of the structure and the chosen of materials and will incorporate into the package design for the project demonstrators. In present work the CTE mismatch between the copper line and the epoxy is the main reason of the failure in embedded chip structure which is concluded in FEM simulation. Most of the insights into the materials, whose compatibility was desired, came from the comparison of the results for different variants, which is to say from relative evaluations. For reliability predictions, ultimate stress or strain values must be available, which have to be measured carefully for all materials and interfaces of interest for future work.

Acknowledgements

The authors acknowledge the financial support by National Natural Science Foundation of China (No. 60776033), National High Technology Research and Development Program of China (863 Program) (No. 2006AA04A110).

References

1. Ostmann, A. Neumann, S. Weser, E. Jung, L. Böttcher and H. Reichl: Realization of a Stackable Package Using Chip in Polymer Technology, Polytronic Conference, June 23.-26. 2002, Zalaegerszeg, Hungary

2. Tummala, R., and Madisetti, V. K., 1999, "System on Chip or System on Package?", IEEE Design & Test of Computers, April-June 1999, pp. 48-56.

3. Ostmann, A., De Baets, J., Kriechbaum, A., Kostner, H. Neumann, A.: Technology for Embedding Active Dies, European Microelectronics Conference 2005, Brugge, Belgium, June 12 - 15, 200.

4. Davidson, E., 2001, "SoC or SoP? A Balanced Approach!" Proc. 51' Electronic Components and Technology Conference, pp. 529-534.

5. Aschenbrenner R., Ostmann A., Neumann A., and Reichl H., Process Flow and Concept for Embedding Active Devices. EPTC 2004, Singapore, 2004.

6. A. Ostmann, A. Neumann, J. Auersperg, C. Ghahremani, G. Sommer, Aschenbrenner R., and Reichl H., Integration of Passive and Active Components into Build-Up Layers, EPTC 2002, Singapore, 2002.

7. Tuominen R. Kivilathi J.K., A novel IMB Technology for Integrating active and passive Components, 4th International Conference on Adhesive Joining & Coating Technololgy in Electronic Manufacturing, 18 – 21 June, p 269, Helsinki, 2000.

8. Löher T., Pahl B., Huang M., Ostmann A., Aschenbrenner R., and Reichl H., An Approach in Microbonding for ultra fine Pitch Applications: Technology and Metallurgy, The 7th VLSI Packaging Workshop of Japan, Kyoto, 2004.

9. Tuominen, R., and Kivilahti, J. K., 2000, ' A Novel IMB Technology for Integrating Active and Passive Components', Adhesive Joining and Coating Technology in Electronic Manufacturing, Proceedings of 4th International Conference onAdhesives in Electronics, pp. 269-273.

Research on the Contact Resistance and Reliability of Flexible RFID Tag Inlays Packaged by Anisotropic Conductive Paste

Xiong-hui Cai, Bing An, Yi-ping Wu*, Feng-shun Wu, Xiao-wei Lai
Wuhan National Laboratory for Optoelectronics,
State Key Laboratory of Material Processing and Die & Mould Technology,
Huazhong University of Science and Technology, Wuhan, 430074, China
ypwu@mail.hust.edu.cn , 86-27-87792402

Abstract

In this work, ACA was prepared by mixing micro-sized spherical Ag particles and latent curing agent into thermo-set epoxy resin, and RFID flip chips were assembled on the Al/PET, printed Ag/PET and printed Ag/Paper antennae through hot-press bonding process. The contact resistance and shear bonding strength before and after the reliability tests (hot humidity test, 85 °C, RH 85%), and degradation mechanism of ACA interconnection for flip-chip-on-flex (FCOF) assembly were studied using modified RFIC and the three kinds of antennae mentioned above. It was found that the contact resistance changed after the reliability test, it was caused by the total results of oxidation of Al/PET antennae and conductive particles, mismatch of coefficient of thermal expansion (CTE) between the ACA adhesive, antennae and flip-chips and post curing of resin. And the bonding strength also affect by the further curing of paste, strain release accumulated in resin and the microstructure change caused by moisture absorption during the reliability test. It was concluded that it was benefit to improve the reliability of FCOF assembly packaged by ACA by introducing the post curing process. And it was suggested that selecting the anti-oxidation conductor and the anti-heat substrate of antennae could decrease the shift of contact resistance, which was especially favored for ultra-high frequency RFID tag. Therefore, the printed Ag/Paper antenna was preferred to large scale, cheap and rapid manufacturing RFID tags.
Keywords: anisotropic conductive adhesive, radio frequency identification, reliability

1. Introduction

Radio frequency identification (RFID) is a small tag containing an integrated circuit chip and an antenna, and has the ability to respond to radio waves transmitted from the RFID reader. The principal advantages of RFID system are the non-contact, non-line-of-sight characteristics of the technology. Tags can be read through a variety of visually and environmentally challenging conditions such as snow, ice, fog, paint, grime, inside containers and vehicles and while in storage. With a response time of less than 100 ms, an RFID reader can read many (several hundreds) tags virtually instantaneously. For these advantages, it is used for a wide variety of applications ranging from the familiar building access control proximity cards to supply chain tracking, toll collections, vehicle parking access control, retail stock managements, ski lift access, tracking library books, theft prevention, vehicle immobilizer systems and railway rolling stock identification and movement tracking. [1] But in order to large scale, cheap and rapid manufacturing RFID tags, the substrate of antennae in RFID system is often flexible and

heat-sensitive polymer. It is inconvenient or costly if adopting traditional bonding material and technology.

In the last few years, anisotropic conductive adhesives (ACAs) have gained increased popularity in flip chip packaging applications.[2-6] And compared with traditional bond materials, ACAs has numerous advantages, such as fewer processing steps, reducing processing cost, lower processing temperature which have made the heat-sensitive and low cost components and substrates possible.[7]–[10] And for this, ACAs meet the requirement of low cost and large-scale manufacturing of RFID tag inlays. So assembly the RFID tag inlays using anisotropic conductive adhesives (ACAs) through flip-chip technology is a preponderant technique to accomplish this manufacturing process.

There are mainly two types of ACAs, anisotropic conductive pastes (ACPs) and anisotropic conductive films (ACFs). However, for the nature of polymer adhesive and flexible RFID antennae, the reliability of flexible RFID tag inlays packaged by ACAs is still a critical problem. And although the cost of ACFs is much higher than ACPs, the reliability of tags is lower. Some researcher had used ACF flip chip technologies to assemble RFID tags and explored the reliabilities. [6, 11]

In this paper, an ACP was prepared by uniformly mixing the uniform micro-sized spherical conductive particles, latent curing agent and other additives in the thermo-set epoxy resin. Modified RFID flip chips were assembled on the Al/PET, printed Ag/PET and printed Ag/paper antennae through hot-press process. Change of contact resistance and shear strength of these RFID tag inlays were detected during the aging test (hot-humidity test, 85°C, RH85%). And the reliability and degradation mechanism of ACA interconnection for flip-chip-on-flex (FCOF) assembly were discussed.

2. Experiment

2.1 Materials

The RFID tag inlays packages were made up pf three different materials: ACP, the silicon chip and flexible substrate.

An ACP was prepared by uniformly mixing the uniform micro-sized spherical conductive particles (the mean diameter was 3μm), latent curing agent and other additives into thermo-set epoxy.

The test chips were modified normal radio frequency integrated circuit (RFIC). The bumps on the chip were not connected each other. They are $1.1 \times 0.8 mm^2$ with four $60 \times 60 \mu m^2$ square-shaped bumps on the periphery. The schematic of the bump pattern was shown in Fig. 1a. Bump metallization of chip was electroless nickel and Au, and the height was 20 μm. The pitch widths were 650μm and 960μm. The chips were sputtered a gold film with 1μm thickness on the side of bumps by plasma spraying.

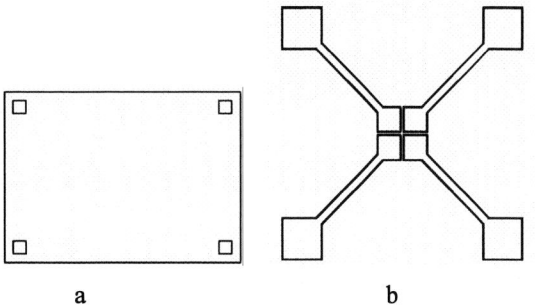

Fig.1 Schematic of the bump pattern (a) and substrate (b)

The flexible substrates used for FCOF assembly were Al/PET, printed Ag/PET and Ag/paper, and the schematic of the substrate was shown in Figure.1b. They were usual substrate used in RFID system. Al/PET is a PET film with etched aluminum line. Printed Ag/PET and Ag/paper antennae contained a PET film and a paper as a base material respectively, printed Ag trace as the conductor.

2.2 Bonding process

In short, FCOF assemblies were prepared by hot-pressing the RFIC on the antenna pads with the ACA. [12] In short, there were generally three process steps for the ACA flip chip assembly bonding. First, the gold bumps on the chip and the pads of the test substrates were aligned. Next, ACA was dispensed onto the substrate using a manual dispensing machine. Finally, hot-press process was performed of 2.4 MPa at a temperature of 170 °C for 15 sec to bond the chips on the antennae.

2.3 Reliability test

In this study, three different assemblies using Al/PET, printed Ag/PET and Ag/Paper antennae respectively were tested according to standard hot humidity test for 192h. The number of one kind of assembly was 5.Each assembly was taken out to measure the contact resistance every 48h. And aging test, the samples were tested according to the shear strength test.

When studying the reliability of conductive adhesive joints, the most important characteristics are contact resistance of single joints and the shear strength of chips. In this method, there was no need to design a special chip as mentioned in some reports. [13,14] The schematic circuitry to measure the contact resistance using the four-point probe method was shown Fig.2.When 1mA constant current was input in the pad 1 and out from pad 2, the voltage between pad 1and 3 was read from a multimeter. The total resistance between pad 1 and 3 was obtained by Ohm's law, R=V/I. It included the resistance of antenna conductor between pad 1and 3, it was constant (it was tested and not shown in the paper.), contact resistance of joints and resistance of bumps and gold film sputtered on chips under this special electric field. The resistance of bumps and gold film sputtered on chips were constant during the whole aging test (it was tested and not shown in the paper.). And when the probes of the multimeter were placed around the corner of chips, the thickness of antenna was so enough that the resistance of

antenna conductor between pad 1and 3 could be neglected (It was tested and not shown in the paper too.). So the contact resistance of bonding joints could be calculated.

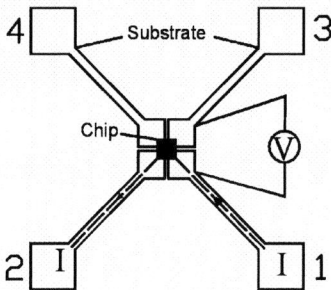

Fig.2 Schematic circuitry to measure the contact resistance using the four-point probe method

Shear strength is an important parameter to evaluate the reliability of bonding joints, and it always detected by shear test. In this paper, three different assemblies using Al/PET, printed Ag/PET and Ag/Paper antennae respectively were test on six axes equipment. And the relationship between force and displacement was recorded. The shear strength was obtained by dividing the area of chip with maximum force. The shear rate was 1mm/s.

After the hot-humidity test, the morphology of the ACP joints was studied. The test samples were molded and polished to prepare for microstructure study of the interface. Scanning electron microscopy (SEM, Hitachi S3000-N, Hitachi High Technologies America, Inc) was used to characterize the interface before and after the aging test. And Fourier Transform Infrared Spectrometry (FT-IR, VERTEX 70, Bruker Optics GmbH) was used to characterize the curing degree change of resin during the hot humidity test.

3. Results and discussion

The change of contact resistance of FCOF assemblies in hot-humidity test was shown in Fig.3. When the antenna was Al/PET, the contact resistance increased rapidly with the increase of aging time. But when the printed Ag-Paper was used as the antenna, the contact resistance of assemblies decreased firstly and then increased during the aging test. And if the printed Ag/PET was selected as the assemblies' antenna, the contact resistance increased smoothly with the increase of test time.

It was clear that the shear strength of three different assemblies each increased after hot humidity test. The shear strength of the assemblies using Al-PET as the antenna increased more than those of others assemblies.

The chips were assembled on the antennae with ACPs through fast hot-press process, and the assemblies were tested in hot humidity test. During the aging test, there were three main causes to affect the change of contact resistance: the mismatch of coefficient of thermal expansion (CTE) of chips, antennae and paste, the oxidation of joints interface between the Ag particles, bumps and antennae, both of them led the increase of contact resistance, and the change of microstructure of polymer matrix caused by the post curing and moisture absorption. Usually, post curing is benefit for polymer matrix to form uniform and stable structure. From this point of view, post curing could improve the packaging reliability. Moisture absorption would lead resin become soft, which decrease the reliability of assemblies. The main three

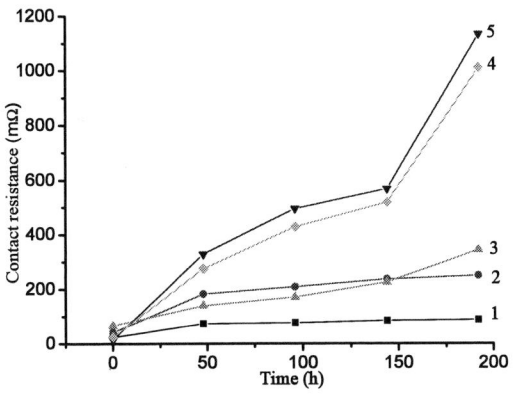

a

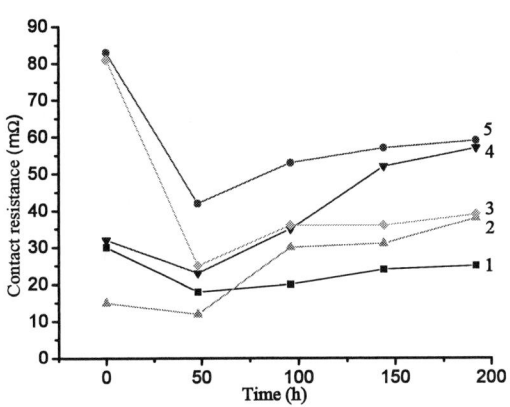

b

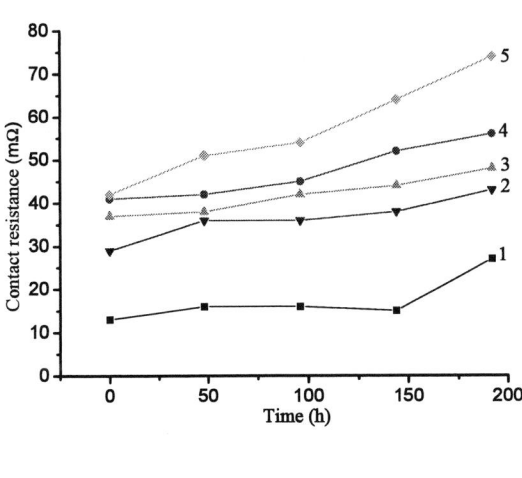

c

Fig.3 The evolution of contact resistance vs time in hot-humidity test (The numbers on the figures represented the series number of samples)
(.a. Al/PET antenna , b. Printed Ag/Paper antenna, c. Printed Ag/PET antenna)

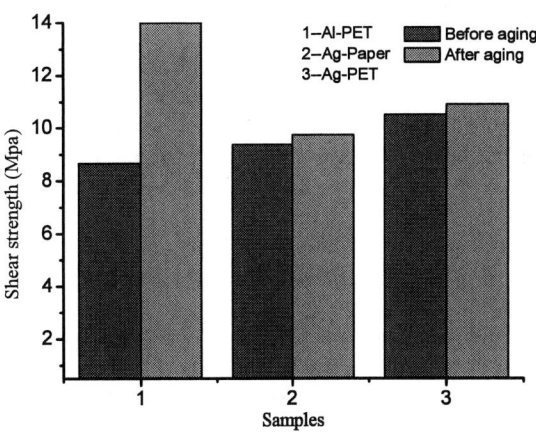

Fig.4 The results of shear strength of three different assemblies

causes competed among them. And if one of them was dominant, the contact resistance was predominated by it. From the results shown in Fig.3a and Fig.3c, it could conclude that the rapid increase of contact resistance of assemblies using the Al/PET as the antenna mainly caused by the oxidation of bonding joints in the hot humidity test. It was because that Al is easy to be oxidated to Al_2O_3 (electrical nonconductive material), and Ag is harder to be oxidated to Ag_2O than Al, even though this was happen, Ag_2O is a electrical conductive material. Then oxidation reaction occurred at the interface between the Ag particles and Al pads, and mainly occurred at the side of al pads. The mismatch of CTE was another cause to lead the increase of contact resistance. It had much less effect than oxidation of joints interface. It was verified by the results of Fig.3c. The curves of contact resistance of assemblies using printed Ag/PET as the antenna vs aging time were smooth. Compared with Fig.3a, the antennas had the same substrate (PET), but the contact resistance increased rapidly in Fig.3a and smoothly in Fig.3c. From the Fig.3c and Fig.3b, it could conclude that the CTE of printed Ag/PET substrate was bigger than printed Ag/Paper substrate. It was because the antennae trace conductor and the aging condition were the same, but the change of contact resistance was different, which was only caused by mismatch of material CTE. In other words, the paper was more anti-heat than PET. Although the effect of mismatch of CTE was less than oxidation of joints interface, it was bigger than that of post curing. In the Fig.3b, the contact resistance firstly decreased and then increased. After chips were bonded on the printed Ag/paper with ACP, and when assemblies were test according to standard hot-humidity, the post curing happened during the aging test, which made better contact between particles and printed Ag pad. In this condition, because CTE of paper was less than that of PET .The post curing process was same. Then it led the first decrease of contact resistance as shown in Fig.3b. It was concluded that the effect of mismatch of CTE of materials was less than that of post curing on shift of contact resistance. So it was benefit to improve the product reliability by introducing the post curing process.

The results of shear strength of three different assemblies were shown in Fig.4. It was clear that the shear strength of three different assemblies each increased after hot humidity test. It accorded with the post curing process mentioned above. To attain high bonding strength, it was necessary to paste to have enough time to wet the material surface. In hot press process, the time of chip bonding was so less that the paste had not enough time to wet the substrate surface. Then the bonding strength between the polymer matrix and the surface was poor. But during aging test, the paste cure further and had enough time to wet the substrate surfaces, which led the increase of shear strength. The shear strength of assemblies using the Al/PET as the antenna increased more than those of others assemblies. It was because that the Al_2O_3 was formed by the oxidation of Al. The hydrogenous bond could maybe be formed between the Al_2O_3 and the hydroxides of cured epoxy.

The change of microstructure of polymer matrix was characterized by FT-IR and SEM, the results was shown in Fig.5 and Fig.6. The wavenumber of 3300cm^{-1} is the characteristic absorption peak of hydroxyl. It increased with the curing degree of epoxy resin. From the intensity of characteristic absorption peak, it could conclude the curing degree of paste. It was clear that the polymer matrix cured further during 48h aging test, and basically cured completely after 48h aging test. It was concluded that after rapid hot-press process, the polymer matrix was far not uniform and stable. During this period, many change happened, including physical and chemical change, which was certified by FT-IR as shown in Fig.5. And the reliability of assembly before curing completely was doubtable. Therefore, a post curing process was suggested to improve it.

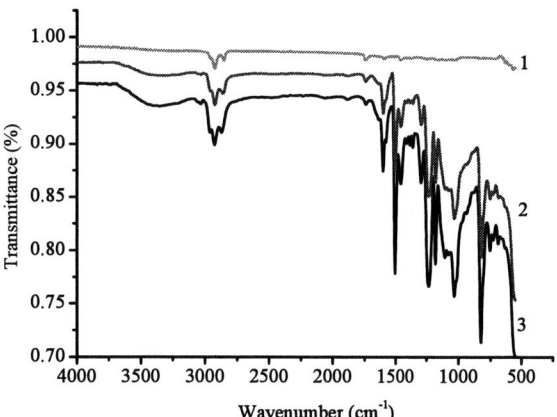

Fig.5 The evolution of FT-IR of ACP in hot humidity test (1. The sample of aging tested for 0 h, 2. The sample of aging tested for 48 h, 3. The sample of aging tested for 96 h)

To explore the effect of hot-humidity test on the microstructure of polymer matrix, the assemblies using Ag/PET as antenna were molded and polished to prepare for microstructure study of the polymer matrix under SEM. The results was shown in Fig.6.It was clear that there was not apparent cracks in the cured resin and cracking and fracture of the conductive particles did not happen. It was different with the ACAs filled the metal-coating-polymer conductive

particles.[15] And because the epoxy did not cure completely, the cure polymer did not form the uniform and stable matrix (Fig.6a). After aging test, the resin cured further and formed the uniform and stable microstructure (Fig.6b).

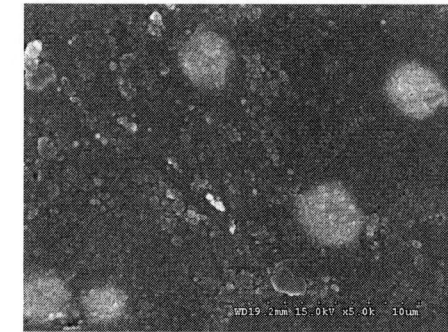

Fig.6 The typical SEM pictures of cured ACP (a.The sample before aging test, b. The sample after aging test.)

Conclusions

An anisotropic conductive paste was prepared. Flexible RFID tag inlays were assembled with the ACP through flip-chip technology. A simple and cheap contact resistance measure method was present. The reliability of three different assemblies using Al/PET, printed Ag/PET and Ag/Paper as antennae respectively was studied. A mechanism of affect the reliability was suggested. It was decided by the total results of the mismatch of CTE of materials, the oxidation of joints interface and the change of microstructure of polymer matrix. And it was found that it was benefit to improve the reliability of RFID tags to introduce the post curing steps. And the printed Ag/Paper antenna was preferred to large scale, cheap and rapid manufacturing RFID tags.

Acknowledgements

The authors acknowledge the financial support by National High Technology Research and Development Program of China (863 Program) (No. 2006AA04A110), and National Natural Science Foundation of China (No. 60776033).

References

1. C.M. Roberts, "Radio frequency identification (RFID)," *Computers & security*, vol. 25(2006), pp. 18-26.
2. A.M.Lynos, C.P.Wong, Conductive adhesives for electronics packaging, Electrochemical Publication, (Edited by Johan Liu, London,:1999), chapter 8, pp. 183-211.

3. Petri Savolainen, llkka Saarinen and Outi Rusanen, "High-density Interconnections in Mobile Phones Using ACF," *Polytronics 2004 IEEE International Conference on Polymers&Adhesives*, Portland, USA, September, 2004, pp.99-104.

4. Gang Zou, Hans Grödnqvist, Zonghe Lai, Ulf Södervall, Johan Liu, "High Frequency Flip Chip Interconnection on Liquid Crystal Polymer Substrate Using Anisotropic Conductive Adhesive," *Polytronics 2004 IEEE International Conference on Polymers&Adhesives*, Portland, USA, September ,2004, pp.137-140.

5. Johan Liu, Yu Wang, James Morris and Helge Kristiansen, "Development of ontology for the anisotropic conductive adhesive interconnect technology in electronics applications," *10th International Symposium and Exhibition on Advanced Packaging Materials*, Irvine, CA, USA, March, 2005 ,pp. 8.1.

6. Myung Jin Yim, In Ho Jeong, Hyung-Kyu Choi, Jin-Sang Hwang, Jin-Yong Ahn, Woonseong Kwon, and Kyung-Wook Paik, "Flip Chip Interconnection using Anisotropic Conductive Adhesives for RF and High Frequency Applications," *IEEE Transactions on Components. Packaging and Technology.* vol. 28 (2005), pp.789-794.

7. Y. Li, K. Moon, and C. P. Wong, Electronics without lead, *Science*, vol. 308(2005), pp. 1419-1420.

8. J. S. Hwang, Ed., Environment-Friendly Electronics: Lead-free Technology, Electrochemical Publications, Ltd , (Port Erin, 2001), chapter. 1, pp. 4-10.

9. J. Lau, C. P. Wong, N. C. Lee, and S. W. R. Lee, Electronics Manufacturing: With Lead-Free, Halogen-Free, and Conductive-Adhesive Materials, McGraw Hill, (New York, 2002), pp. 18.1-18.18

10. Y. Li and C. P. Wong, "Recent advances of conductive adhesives as a lead-free alternative in electronic packaging: materials, processing, reliability, and applications," *Material. Science Engineering Report*, vol. 51(2006) ,pp. 1-35.

11. Jad S. Rasul, Chip on paper technology utilizing anisotropically conductive adhesive for smart label applications, *Microelectronics Reliability*, vol.44 (2004), pp. 135–140.

12. Xiong-hui Cai, Bing An, Xiao-wei Lai, Yi-ping Wu, Feng-shun Wu, "Reliability Evaluation on Flexible RFID Tag Inlay Packaged by Anisotropic Conductive Adhesive," *8th International Conference on Electronics Packaging Technology*, Shanghai, China, August, 2007, pp.716-719.

13. J.H.ZHANG and Y.C.Chan.Research on the contact resistance, reliability, and degradation mechanisms of anisotropically conductive film interconnection for flip-chip on flex applications. *Journal of electronics materials*, vol. 32 No. 4(2003), pp.228-234.

14. LiQiang Cao, Shiming Li, Zonghe Lai, and Johan Liu, Formulation and characterization of anistropic conductive adhesive paste for microelectronics packaging applications, *Journal of electronic materials*, vol. 34, No.11(2005), pp.1420-1427.

15. H. Kristiansen, Z. L. Zhang and J. Liu, "Characterization of Mechanical Properties of Metalcoated Polymer Spheres for Anisotropic Conductive," *10th International Symposium on Advanced Packaging Materials: Processes, Properties and Interfaces* , Irvine, CA, March, 2005, pp.209- 213.

Ultra-Fine Via Pitch on Flexible Substrate for High Density Interconnect (HDI)

Kelvin Pun, C.Q. Cui, T.F. Chung
Compass Technology Co., Ltd.
Suite 10, 5/F, Chiaphua Centre
12 Siu Lek Yuen Road
Shatin, N.T., Hong Kong, China
(852) 2688 8910; Fax: (852) 2636 5626 , kelvin_pun@cgth.com

Abstract

In the trend of miniaturization, low cost, and the performance of electronics, high density interconnect has been required for interfacing with very fine pitch BGA, CSP and SIP. This raises a great challenge to the substrate technology and related interconnect technology in electronic packaging for high density, small feature size and high performance. Miniaturization in electronics means finer lines and smaller vias in substrate technology. Very fine lines on the substrate are difficult to produce by thicker copper layer in conventional CO_2 laser blind via. In this paper, ultra-fine blind via with solid Cu filled at an entry diameter of 20 μm, over the current blind via size of 50-200 μm by CO_2 laser drilling, is demonstrated on polyimide (PI) based flexible substrate. A via pitch at 30 μm for the blind via has been developed for next generation of stack die packaging accompanying with dimpless design, which ameliorates the void entrapment failure caused by soldering and direct flip-chip (FC) bonding, and strengthens interfacial bond strength. In the meantime, thinner Cu conductor at top and bottom side could be achieved for high circuit density. The reliability of the ultra-fine blind vias has been assessed in daisy chain modules at substrate level, subjected to JEDEC air-to-air thermal cycle and thermal shock, and low/high temperature storage tests. Applications in direct FC bonding and their virtues including high electrical and thermal performances, and feasible of various metals surface finishing, will be discussed. In the end, the ultra-fine Cu filled blind via technology has introduced to the production in Compass for SIP, stack die CSP, 2-metal layer chip-on-flex (COF) and multi-layer buildup flex, etc

Key Words: Blind via, High-density, flexible substrate, IC packaging.

Introduction

Low cost, lightweight, handheld, multi-functional and high performance electronic products continue to challenge and drive the electronic packaging and assembly technology. As micro-systems continue to move toward for higher speed and micro-miniaturization, the demand for interconnection density both on the IC, package, and the substrate levels increase tremendously. Flexible HDI substrate is an emerging candidate that utterly fulfilled as-mentioned requirements, compared with conventional printed circuit board (PCB). Therefore, the demand of HDI Flex in high-tech electronic product is expected to grow at a tremendous rate [1,2]. Versatile and denser integrated circuit (IC) chip, and high-density interconnect (HDI) substrate technology are striving.

Because of the obstacle of producing fine pitch interconnects, the substrate portion of the IC packaging industry continues fall behind.

The International Technology Roadmap for Semiconductors (ITRS) has predicted that by 2016 the number of input-output (I/Os) is expected to increase to more than 10,000 by 2016 [3]. The routing of future ICs with 10,000 I/Os requires ultra fine feature sizes of at least 10 um lines/space widths and 35 μm pad diameters [4-6]. This has further aggravated the technology gap between IC chip and substrate technology. In order to minimize the technology gap between IC industry and packaging, researchers and engineers should develop technologies to achieve fine pitch interconnects by shrinking trace width, micro-via size and pitch, and implementing innovative design with landless structure. In this paper, we demonstrate that the Cu fully filled ultra-fine via technology [7] can be developed into ultra-fine via pitch of ultimately 30 μm at 25 μm polyimide (PI). Padless structure with pitch of 30 – 75 μm is achieved by fabricating copper traces hung over the via opening. This capability is completely satisfied to the next generation of low profile flip-chip (FC) CSP that requires pitch scaling down to 20-60 μm with via diameter of 20-50 μm [8]. Their reliability and relevant applications for FC packaging will also be discussed.

Experimental

Schematic flow of producing ultra-fine via pitch substrate was illustrated in figure 1. Adhesiveless PI base films at a sandwich structure with 12 μm Cu metal layers on both sides UBE were used in the evaluation. The thickness of PI dielectric was 25μm. A solid state Nd:YAG laser from ESI 5330 was used for the experiment with the specification (beam size: 25μm, pulse rate 30-70 kHz, peak power >5.7 W at 30 kHz, alignment accuracy +/- 20 μm.) Metallization and direct Cu plating were applied after chemical cleaning process. Vias were filled and both sides of Cu were built up to ~25 μm. In order to make fine trace pattern, both sides of Cu layer were attenuated prior to photolithography process. All of fabrication processes for the substrate with the ultra-fine blind vias were done at reel to reel production line. Scanning Electron Microscope (SEM) was used to examine on via interconnects, surface morphology of traces, and any failure on via and substrate after reliability tests. 3D Laser scanning microscope (LSM) was further used to analyze on dimple depth profile.

978-1-4244-2739-0/08/$25.00 ©2008 IEEE   134

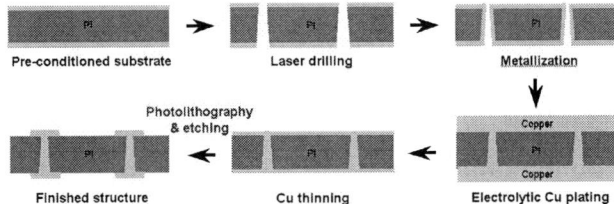

Fig. 1 Schematic process flow of ultra-fine via pitch substrate

The reliability of ultra-fine via was examined in daisy chain module and sample sizes were chosen with JEDEC specification accordingly. To assess the ultra-fine via with fine via pitch pitch reliability, 2 layers structure daisy chain module containing ultra via at 20 μm entry via size was fabricated, as shown in figure 2. In the daisy chain module, the top and bottom conductors were connected by the Cu solid filled ultra-fine blind vias. Both side Cu thickness and PI are controlled to be 25 μm and 50 μm, respectively. One daisy chain consisted of 2000 ultra-fine blind vias in the daisy chain design. Electrical test on the daisy chain was conducted to examine the reliability of the ultra-fine blind vias, before and after subjected to thermal cycling (TC), high (HTS) and low temperature storage (LTS) test and air-to-air thermal shock (TS) testing.

Fig. 2 Daisy chain module made with 2000 ultra fine blind vias

Result and Discussion
Characteristic of ultra-fine blind via

A. Ultra-fine via size and profile

To date, a high routing density substrate can be achieved by advanced metallization and micro-via technology. Several kinds of micro-via technology and conventional drilling are summarized in table 1, and they are implemented in the production of flex substrate depending on the requirement for the compatibility of design structure and the clearance of the capture pad.

Tab. 1 Micro-via technology in Compass.

Description	NC drill	CO₂ Laser	UV YAG Laser
Via Diameter	200 – 350 μm	50 – 240 μm	10 – 50 μm
Via Pitch	>200 μm	>75 μm	<=30 μm
Core Dielectric	12.5, 25, 38, 50, 75, 100 μm	12.5, 25, 38, 50, 75, 100 μm	12.5, 25, 38, 50, 75, 100 μm
Features		High efficiency	Landless, Utterly fitted via

High routing density substrate used in advanced electronic consumer product is typically hard to be satisfied by NC drilled via and CO2 via, due to the limitation in via size and relatively smaller in aspect ratio. In order to meet with the requirement of modern ICs and flip-chip technology [3], UV YAG laser drilling is attained to create via size as small as several ten micrometers, has already implemented for ultra-fine via production in Compass technology.

The minimum via sizes done by mechanical and CO2 laser are normally at 250 μm and 75μm, respectively, while of ultra fine via produced with solid state UV YAG at 355nm could go to the entry via size as small as 20μm. The aspect ratio of mechanical drilling and CO2 laser via are relatively small and usually not large than 1 [7], where the ultra fine via produced with through holes geometry can be reached as high as 1~10. As shown in figure 3, the entry via size of 20 μm Cu fully filled via is demonstrated on the flexible PI substrate at 25μm, 50μm and 75μm PI thickness respectively. The Cu thickness on both side of 2ML interconnection is directly proportional to the half of via size for a solid blind via. With the solid filled Cu plating, the vias are reliable and also suitable to be served as a thermal via with its excellent thermal conductivity of plated copper, 390 W/m °K, effectively through substrate thermal conductivity for heat dissipation.

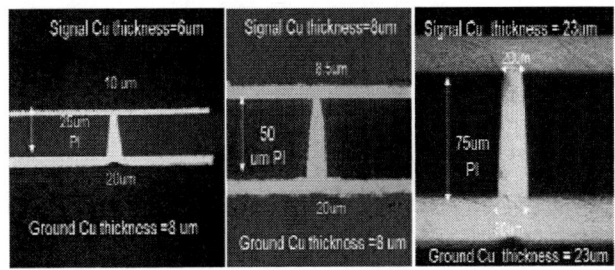

Fig. 3 20 μm ultra fine via with 25 μm, 50 μm and 75 μm PI base film

Ultra fine via with finer via pitch allows better electrical performance and more interconnects in a tiny area with the reduction in the resistance and inductance due to shorter pathway and the increased in parasitic capacitance due to smaller via length. As shown in figure 4, ultra fine via pitch at 30-40μm is demonstrated on the flexible PI substrate at 25 μm PI thickness. The size and pitch of blind via are significant reduced to 20 μm and 30 - 40, respectively. Also, a sharp via profile with nearly same entry (20 μm) and exit (19 μm) dimension is attained using the current YAG laser technology, which maintains a dielectric barrier of 10um space. As shown in figure 4, YAG laser via at a wavelength of 355nm conversely provides a clean cut profile with minimal effect on the surrounding Cu and PI for the benefits in fine via pitch production. In addition, the through-hole geometry of ultra fine via structure facilities the chemistry flow and agitation to assure good z-axis plating without parting line.

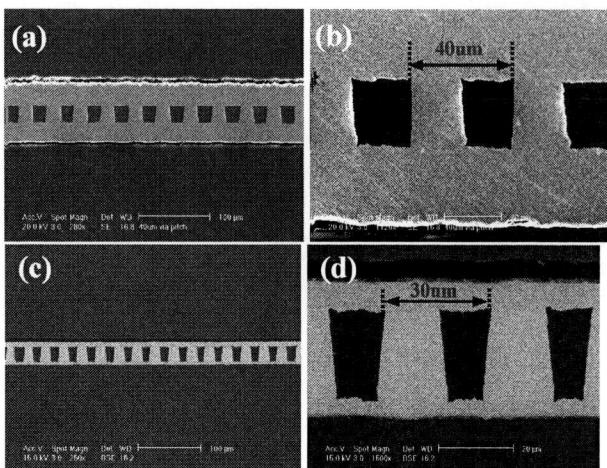

Fig. 4 Cross-sectional SEM images of (a) & (b) 40 μm via pitch, and (c) & (d) 30 μm via pitch with 25 μm PI base film.

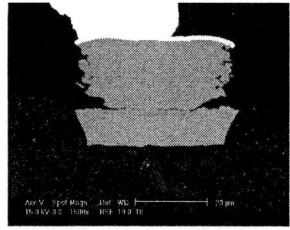

Fig..5 Cross-sectional SEM photo of CO2 via profile at 70μm via pitch.

High magnification cross-sectional images (b) and (d) prove that the quality of via is excellent as there is no plating void and parting line found inside the blind via for 30um via pitch production. In the case of CO2 laser via, misshapen via profile and the presence of PI residues of redeposit resin on capture pad will be resulted from attempted to reduce the via pitch. As shown in figure 5, the drilling profile produced by CO2 laser is at 70 μm via pitch is not sharp. The PI residues are potential to be embedded in the metallization as a parting line, which would be a cause for via reliability failure. The results reveal that a robust via with via pitch down to 30 μm is demonstrated by our ultra fine via technology. This capability enables the development of next generation of HDI flexible circuitry with the requirement of fine via pitch. The design rule of fine via pitch is shown in table 2. The manufacturing of fine via pitch of 70 μm has already commercialized last year.

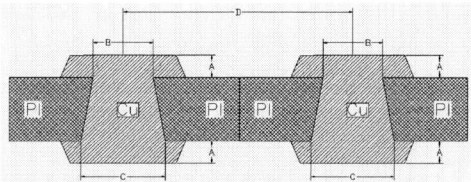

Tab. 2 Comparison between fine via pitch design rules. All units are in μm scale

Blind Via Pitch	30	40	50	75
TOP Via Opening (B)	15	15	15	15
BTM Via Opening (C)	20	20	20	20
Pad width/space (D) @ Cu(A)	20/10@8	20/20@12	20/30@12	35/40@18

B. Dimpless structure

It is of great importance to mention that dimples on plated via are minimized by further plating both sides of Cu metal as shown in figure 6. Dimple is an inherent feature found on the top of plated via. Dimple or incompletely filled of via cavity could lead to fatal failure problem because void entrapment induced failure often happens in conventional CO2 laser via for via in pad design [7], during soldering and direct bonding applications. Therefore, it is crucial to eliminate dimple on the electrical trace surface to provide better interfacial bond strength. Figure 6 depicts the surface morphology of both sides Cu traces of ultra-fine via pitch sample. In a perspective view, figures 6(a) and 6(b), slight dimples could be found on the surface of both sides. As shown in high magnification figures 6(c) and 6(d), there is no via sidewall or gap in the ultra fine via structure, both metal surfaces over via area are plated to form a flat or close to flat panel, which can served as landing pad in the interconnect to board level, as well as bonding pad for chip to package level interconnect or SMD mounting pad for module level application.

In order to effectively examine the dimple feature, 3D LSM was used, as it allows better surface analysis. Surface morphology of the dimple is shown in figure 7. Negligible cone shape dimple with height of less than 2 μm and base diameter of ~35 μm is obvious identified. The depth of dimple is superficial compared to the total thickness of substrate (> 50 μm). Also, the photos illustrate that the neighbor Cu surface is absolutely even. Both SEM and LSM results prove that the dimple which attributes to the laser drilling process, could be nearly minimized by our filled via technology. As shown in figure 8, the via-on pad structure with the ultra-fine blind via has demonstrated to be not only solderable, but also wire bondable, and this conductive bond interface would be beneficial to handle various interconnect configurations including flip chip, stacked die, package on package (PoP) interface and passive component integration.

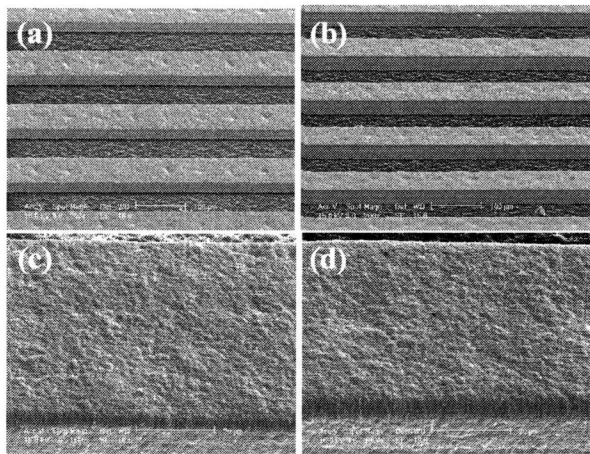

Fig. 6 SEM images of both side plated Cu of ultra-fine via pitch sample. (a) Entry via side, (b) Exit via side, (c) and (d) Surface morphology of vias shown in (a) and (b), respectively.

136

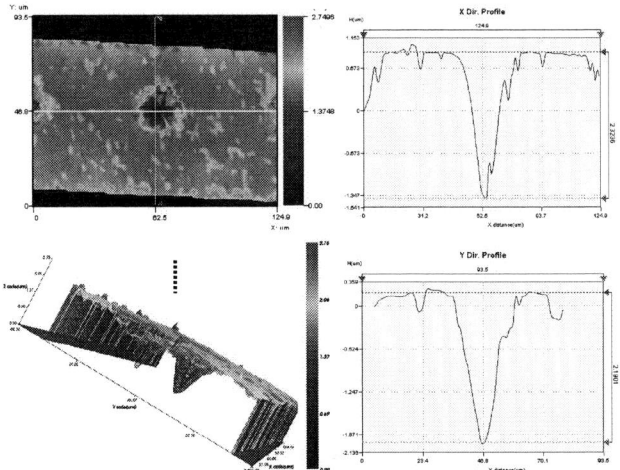

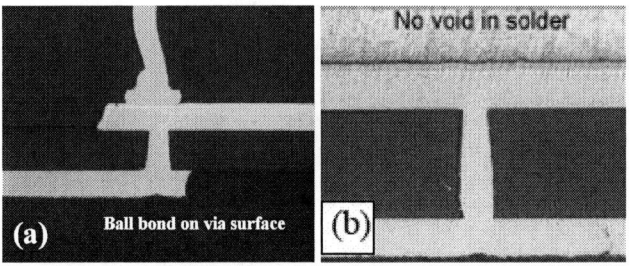

Fig.7 The surface morphology of dimple and its profile. The dimple is located on the top of via with 23 μm diameter and the PI thickness of 25 μm.

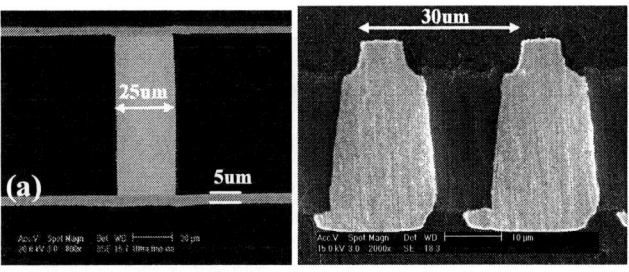

Fig.8 Demonstration of (a) wire bonding and (b) soldering on ultra fine via.

C. Landless structure

Except fine lines, spaces and tiny via features, landless structure is another promising interconnection feature for the next generation substrate that requires high routing density.

Landless structure provides ability to route many more channels on a given plane by eliminating the via capture pad, through V-OUT (via – over/under trace) concepts. The Cu fully filled ultra-fine blind via is examined on the feasibility of landless structure for interconnect. As shown in figure 9, closely packed and dedicated landless structure can be achieved with the trace etched over the Cu fully filled blind via at ~20 μm entry via size. The pitch of the via-in-pad structure is merely 75 μm and the trace width is ~40 μm. Prototype Padless landing with fine pitch of ultimately 30 μm has been achieved with 15um line and 15um space and with 20um vias to underlying layers. As shown in figure 10, thinner Cu conductor down to 5 μm with trace pitch as small as 30 μm is achievable over the blind via, serves as an immediate solution for the applications of fine pitch 2-ML COF. With filled via technology, it is important to mention that it is feasible to fabricate narrower trace over via opening without deteriorating the robustness of via. As shown in figure 10(b), although a little mis-alignment and over-etch occurs for the Cu pattern etched. With the same etch rate of plated copper in the ultra-fine blind via built up structure, both side Cu traces are etched at the stop point without impacting the integrity of blind via. Furthermore, fine via size also leads

to a better registration tolerance, which significant enhances the manufacturability of landless via to meet the future demand of high routing density.

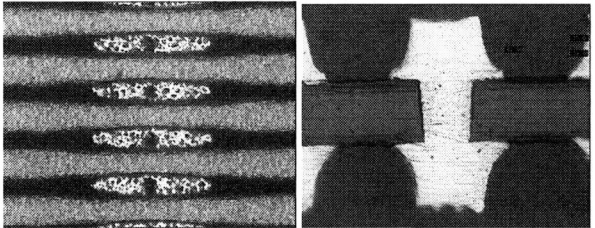

Fig.9 Via-on-pad structure with fine feature of 75 μm pitch

Fig.10 Cross-sectional SEM photos of fine via (a) with diameter of 25 μm fabricated at 75 μm PI film, (b) with via pitch & trace pitch of 30 μm fabricated at 25 μm PI base film

D. Reliability performance

Ultra-fine via possesses more reliable interconnects than CO_2 blind via because of the advantage of through-hole geometry. Through-hole design provides better via cleanliness, uniform metallization and coverage in via that result in a good reliability performance. Daisy chain module shown in figure 2 was applied to investigate the ultra fine via interconnect reliability under circumstances acceptable to the IC packaging industry. Several reliability tests that based on JEDEC's recommendation were conducted and the results are summarized in table 3.

Tab. 3 Summary of the reliability test results

Test	Test Condition	Judge Criteria	Result
TC	JESD22-A104, Condition M -40°C, 15min/ +150°C, 15min, 1000 cycles	Resistance change within +/- 10%	Pass
TS	JESD22-A106, Condition D -55°C, 15min./ +125°C, 15 min, 1000 cycles	Resistance change within +/- 10%	Pass
LTS	JESD22-A103-A -40°C, 1000hrs	Resistance change within +/- 10%	Pass
HTS	JESD22-A103-A 150°C, 1000hrs	Resistance change within +/- 10%	Pass

The resistance transition of the daisy chain in hot segment and cold segment measurements were checked and found to be stable within +/-10% after 1000 cycles of air-to-air TC condition M and TS condition D, and 1000 hrs of HTS and LTS, as shown in figure 11. Additionally, none of electrical failure case was observed in all tests (5 daisy chain units per test). Furthermore, interconnection of ultra fine via was confirmed by cross sectioning illustrated in figure 11(d). There is no visible crack on vias and without delamination in substrate after air-to-air TS for 1000 cycles. Ultra fine via structure, therefore, was proved to be robust and passed the standard reliability requirements.

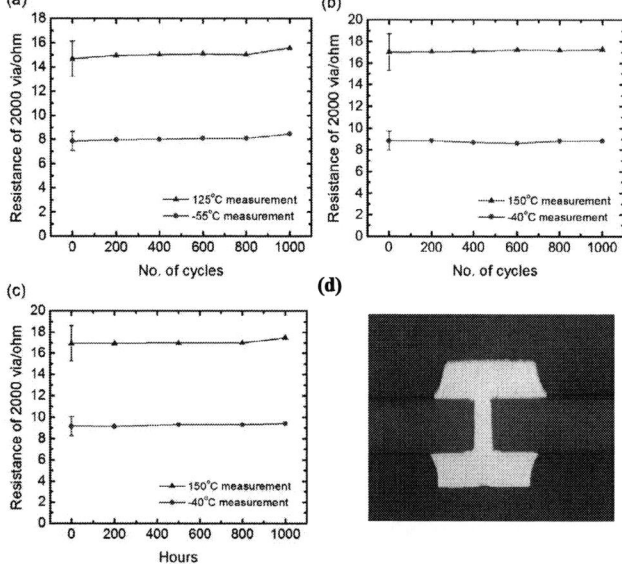

Fig. 11 Resistance measurement result of daisy chain module after (a) TS condition D, (b) TC condition M, and (c) HTS and LTS. (d) Via cross-section after 1000 cycles air-to-air TS test.

E. Direct flip-chip bonding architecture

Market trends require the migration to flip chip interconnects for higher pin count and electrical performance demanded in next-generation of processors. Future applications are demanding a PoP base package with increased interconnect density, reduced pitch, reduced package size and low profile. Flip chip bonding, which is a general technique applying in CSP, SIP and PoP packages, has many merits compared to the most prevalent wire bonding technique. Joint-in-Via architecture provides a solution for the substrate to be used in flip-chip packaging [9]. The approach consolidates the landing pads, micro-vias, and flip-chip joint into one common element, and circumvents the routing of Cu trace and pad landing.

Our fully Cu filled ultra-fine via technology possesses fine via size and ultra fine via pitch, and fine trace capabilities for the next generation flip-chip requirements. The via size and bond pitch as small as 20 μm and 30 μm, respectively, significantly enhance the chip-to-chip spacing, inter-chip wiring, and I/O densities. Schematic diagram of ultra-fine via pitch substrate for flip-chip bonding is illustrated in figure 12. IC chip is firmly attached to flex substrate via Au stud bump using today available flip-chip bonder. And a redistributed trace, for instance, power signal is connected to outside package that ensures sufficient power supply. It is also worth to note that the dimpless feature of our ultra-fine via substrate significantly reduces the chance of void formation for

enhancing the interfacial bond strength for direct bonding. Besides, various types of bumping are apropos, including in Au, Ag and Sn plating finish for FC connection, depicted in figure 12(b). Diversity of surface finishing provides a great flexibility and utilization to different applications. Furthermore, fully Cu filled via architecture guarantees good electrical performance and high thermal dissipation that is a critical issue in modern high-density electronic packaging. Not only the as-mentioned advantages, fine via size and pitch characteristics also enhance registration tolerance and trace routing capacity, that result in smaller package size, small form factor and cost-efficient. The surface of the 2ML with ultra-fine via inter-connection, which are bondable and solderable, provides the chip to be faced top and or down for multi-die packaging and stacked-die chip-scale packages (CSPs) that requiring increased I/O density for a small form factor.

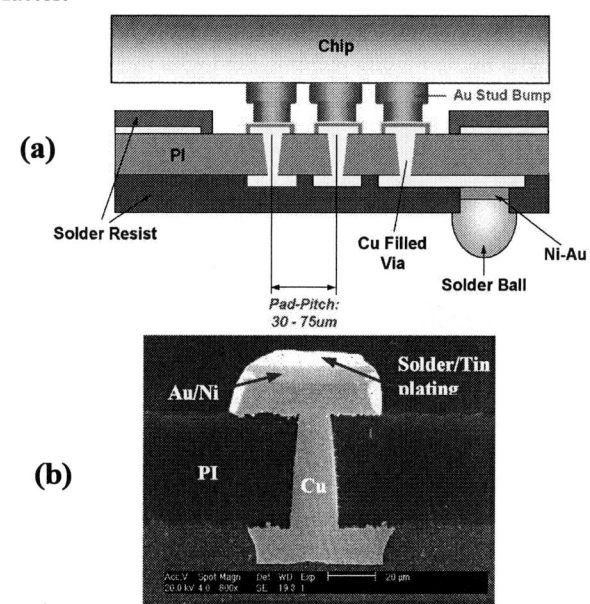

Fig. 12 (a) Schematic diagram of a chip bonded with the ultra fine via pitch substrate via Au-Au bumping. (b) SEM image of a fine via capped by Sn-Ag-Cu solder.

Conclusions

Finer via pitch down to 30um is ready by using YAG laser drilling and fully Cu filled technology, for higher interconnect density. It is not only fulfilled contemporary HDI electronic packaging industry, but also satisfied with the demand in future technology development. Using the ultra-fine blind via pitch technique, a padless structure with fine pitch of 30 – 75 μm was developed in Compass technology. More importantly, the fully Cu filled blind via is proved to be robust and free of any failure in all of the reliability tests including air to air thermal stock, thermal cycling, and high temperature and low temperature storages based on JEDEC's specifications.

The advantages of this ultra-fine blind via with pitch of 30 – 75 μm are given below.

The dimpless and good flatness interconnect provide void-free soldering and high efficient direct bonding.

Via pitch down to 30um is achieved for flip chip bonding.

- Fully filled via with features fine via size is dedicated for landless via structure for high routing density.
- High electrical and thermal performances attributed to fully Cu filled via channel.
- Diversity of metal surface finishing such as Au and Sn is suitable for different applications serves as bond interface.

The intriguing advantages, manufacturing capability and reliable of ultra-fine blind via will be explored for more applications in several areas such as CSP, SIP, COF, Flip Chip and PoP, etc.

Acknowledgments

The authors would like to give the thanks to Electronic Packaging & Analysis (EPA) Center, City University of Hong Kong, for their kind supports in the analysis and reliability instruments.

References

1. BPA Consulting (2006), "Worldwide High-Speed Electronics Technology and Market Trends for the Years 2006-2016," Executive Market and Technology Forum.
2. BCC Research Group, Flexible Circuits: Materials and Applications, Electronics.ca, January, 2006.
3. International Technology Roadmap for Semiconductors (http://public.itrs.net)
4. Tummala, R. R., Swaminathan, M., Tentzeris, M., Laskar, J., Chung, G. K., Sitaraman, S., Keezer, D., Guidotti, D., Huang, R., Lim, K., Wan, L., Bhattacharya, S., Sundaram, V., Liu, F., Wan, L., Bhattacharya, S., Sundaram, V., "SOP for miniaturized mixed-signal computing, communication and consumer systems of the next decade," IEEE Trans-CPMT-A, Vol. 17, No. 3 (1995), pp. 346-351. [A reference to a journal article ...]
5. Guinn, K. V., and Frye, R. C., "Flip Chip and Chip Scale I/O density Requirements and Printed Wiring Board Capabilities," Proc 47th Electronic Components and Technology Conf, San Jose, CA, USA, May. 1997, pp. 649-655.
6. Sundaram, V., Tummala, R. R., Liu, F., Kohl, P. A., Li, J, Allen, S. A. B., and Fukoka, Y., "Next-generation microvia and global wiring technologies for SOP," IEEE Trans. Adv. Pkg., Vol. 27, No. 2 (May, 2004), pp. 315-325.
7. Cui, C. Q., Pun, Kelvin, "Cu Fully Filled Ultra-Fine Blind Via on Flexible Substrate for High Density Interconnect," Proc 31st Inter. Electronic Manufacturing and Technology Conf, Petaling, Jaya, Nov. 2007, pp. 13-19.
8. Shu, William K., "Next Generation Flip Chip and Substrate Technology – Program Development Workshop," Oct, 2007.
9. Lee, T. K., Zhang, Sam, Wong, C. C., Tan, A. C., "A Novel Joint-in-Via, Flip-Chip Chip-Scale Package," IEEE Trans. Adv. Pkg. 54th Electronic Components and Technology Conf, Vol. 29, No.1 (2006), pp. 186-194.

Introduction of Microelectronics Manufacturing Engineering into Professional Education: a Joint Effort among Industry, Government and Universities

Kailin Pan[1], Daoguo Yang[1], Qiao Kai[2], Kouchi Zhang[1,3], Lianfa Yang[1]

1. School of Mechanical Engineering
Guilin University of Electronic Technology, Guilin, 541004, China
Tele: ++86-773-5601311; Fax: ++86-773-5605683
E-mail: d.g.yang@guet.edu.cn pankl@guet.edu.cn
2. Intel Semiconductors (Shanghai) kai.qiao@intel.com
3. Delft University of Technology, the Netherlands g.q.zhang@nxp.com

Abstract

In order to promote the professional education and meet the requirements for the well-qualified technicians of the microelectronics industry , the higher education division of the education ministry and Intel (China) jointly initiated and setup a new mode of higher professional education faculty training:"2008 Microelectronics Manufacturing Engineering Faculty Training Camp", of which the "Intel-GUET Microelectronics Packaging & Assembly Technology Faculty Training Camp" was organized and implemented by Guilin University of Electronic Technology (GUET). The program includes three parts: basic theories and experiments, one week's industry visit & investigation, and one week's education reform discussion & forum. In this paper, the initiatives, curriculum structure, and hands-on training, etc. were presented. The activities and the achievements of this joint faculty training are summarized.

Introduction

China's electronics industry has seen a significant development in the last decades. According to the bulletin from the information ministry, the electronics manufacturing industry has become the country's largest industry. Many well-qualified technicians with the basic knowledge of electronics manufacturing and skills in the major technology and processes are needed to meet the demands of the microelectronics industry and research.

In order to promote the professional education and meet the requirements for the well-qualified technicians of the microelectronics industry , the higher education division of the education ministry and Intel (China) jointly initiated and setup a new mode of higher professional education faculty training:"2008 Microelectronics Manufacturing Engineering Faculty Training Camp". The aim is to educate well-qualified lecturers with not only basic knowledge system of advanced microelectronic manufacturing engineering but also practical skills. This program is divided into two parts:"Wafer manufacturing Technology Faculty Training Camp"and "Microelectronics Packaging & Assembly Technology Faculty Training Camp" which are undertaken correspondingly by Peking University and Guilin University of Electronic Technology (GUET). This paper will focus on the last program

which was held by the School of Mechanical Engineering of GUET

"Intel-GUET Microelectronics Packaging & Assembly Technology Faculty Training Camp" was hold in March and April 2008. This program includes three parts: basic theories and experiments, one week's industry visit & investigation, and one week's education reform discussion & forum. For basic theories and experiments, it is intensive training of advanced microelectronic manufacturing engineering, and the core courses content Introduction to Microelectronics Manufacturing Engineering, Microelectronics Packaging Technology, Microelectronics Assembly Processes, Electronics Equipment and Applications, Reliability Engineering for Microelectronics, and Lead-free Electronics Manufacturing Technology.

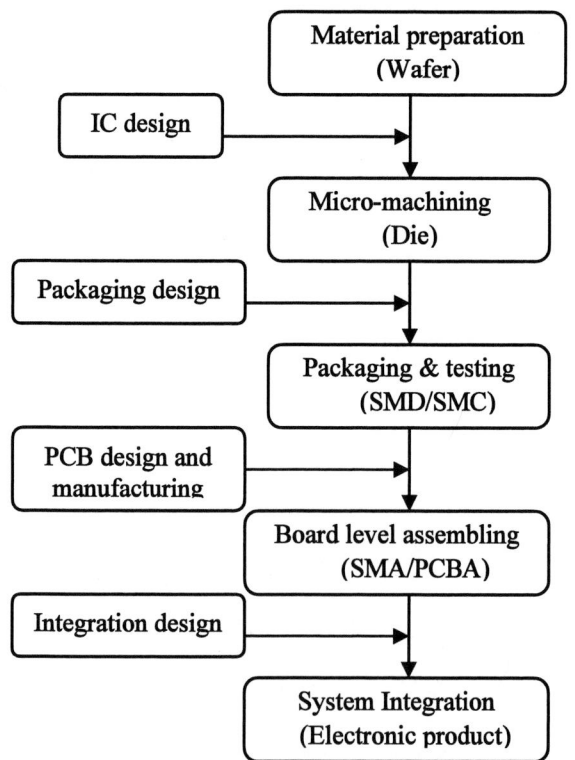

Fig.1 the flow of electronic manufacturing engineering

Concept & system of Advanced Electronic Manufacturing Engineering

Since ultimate target of the program is to educate well-qualified technicians with practical skills for electronic industry. Let's see how one electronic product is manufactured. Form the viewpoint of advanced manufacturing technology, one electronic product is finished by a series of manufacturing process from the wafer manufacturing to system integration as shown in Fig.1.

According to the above electronic industry chains, the Advanced Electronic Manufacturing Engineering includes wafer preparation, wafer manufacturing (semiconductors manufacturing), electronic packaging, Printed Circuit Board (PCB) design & manufacturing, board level assembly, and system integration. The three key processes are wafer manufacturing, electronic packaging and board level assembly. Now, China becomes the biggest electronic manufacturing country in the world. In China, the above three industry share the biggest markets and play a very important role in the Chinese information industry.

Training program

According to the above analysis, in order to meet well the real need of the electronic manufacturing industry and higher professional education, basic theories and experiments, one week's industry visit & investigation, and one week's education reform discussion & forum are systematically planned and successfully implemented. In addition, a series of senior lectures are provided during every week, which cover different process technology, equipment principle & application, reliability engineering, electronic process material, product line management & operation. The lecturers also covered professors, senior engineers from company, senior manager, and specialist from provider of electronic material & equipment. In the following sections, we will discuss the major contents in details.

Part 1: Basic theories and experiments

For basic theories and experiments, we focus on the basic process technology and related equipments operation. The electronic materials performance & application are also included. The arranged courses & related contents are shown Tab.1

Tab.1 Training program part 1: basic theories and experiments

Course	Main contents
Introduction to Microelectronics Manufacturing Engineering	Systematic introduction on the complete process flow of electronics manufacturing, an overview of the major technologies involved (including IC manufacturing, CMOS integration process, etc.), materials for electronics engineering, etc.
Microelectronics Packaging Technology	First level packaging: wire-bonding, TAB, and FC; 2nd level packaging: structure, materials, process; advanced packaging technologies
Microelectronics Assembly Processes	Focus on SMT key processes, including solder printing, SMD/SMC placement, soldering, cleaning, AOI, qualification standard for board level assembly and testing, etc.
Electronics Equipment and Applications	Including lithographer, Ion plant equipment, Wire-bonding machine, solder printing, wire-bonding, pick and placement, solder reflowing equipment. Their structure, principles, control, applications and maintenance.
Reliability Engineering for Microelectronics	Basic knowledge for reliability, major failure modes and failure mechanism, failure analysis, reliability test and data analysis, design for reliability (DFR)
Lead-free Electronics Manufacturing Technology	International laws for green electronics manufacturing, lead-free solder, lead-free packaging lead-free processing, defects in lead-free products, reliability for lead-free assembly, hybrid assembly and reliability, etc.

In order to well understand the nonobjective process and equipment structure, the general visit and corresponding experiments are planned. First, the general visit of assembly line is arranged, and some basic terms and process flow are established. Then, basic process technology and equipments are educated with related experiments. For the Electronics Equipment and Applications, we arrange one-day's field teaching. Every member can see the basic structure and operation process step by step. At the aid of some demonstration videos such as IPC training videos, the running principle, basic operation, maintenance, and basic trouble shooting are shown.

Part 2: Industry visit & investigation

For this part, there are two targets. The first one is, as a professional teacher, to know how the practical electronic manufacturing industry runs, which type of technicians it really needs and what skills it really needs. The second is to further understand the basic theories of related design, process, equipments, materials, product line management and running.

In order to attain above targets, industry visit & investigation are planned and implemented as shown in Tab.2. As you can see, we extend the range of industry visit and investigation, which cover college/professional institute, research institute, packaging fab, OEM, SMT fab, and equipment manufacturing company etc. In different company, we have different focuses. For examples, in Shenzhen Polytechnic, we focus on the SMT lab construction, how the lab runs, and related key courses construction. All of them are representative and famous in one field of the electronic industry or research institute. Combined with senior lectures about process development, DOE (design of experiment), reliability analysis, product line plan & construction, equipment management and maintenances, the industry visit & investigation are more comprehensive and in-depth. Then, the members upgrade their understanding of electronic manufacturing engineering.

Tab.2 Training plan part 2: Industry visit and investigation

Date	Destination	Focus
Day 1	China Electronic Product Reliability and Environmental Testing Research Institute, MII (China CEPREI Laboratories)	R & D equipments Reliability analysis
Day 2	Guang Dong Yuejing High Technology Co., Ltd.	Discrete device packaging
	Shenzhen Polytechnic	School lab plan Courses construction
Day 3	Huawei Technologies Co., Ltd.	communication product Process development
Day 4	ZTE Co., Ltd.; KONKA Co., Ltd	Assembly line; mobile phone assembly
	Shenzhen STS Microelectronics Co., Ltd.	IC packaging
Day 5	Folungwin Automatic Equipment Co., Ltd.	SMT equipments
Day 6	Hui Zhou Daya Bay Guanghong Electronics Co., Ltd.	Flexible PCB assembly

Part 3: Education reform discussion & forum

After part1 and part2, members master the basic knowledge of electronic manufacturing engineering. They analyze and summarize the knowledge system and structure of electronic manufacturing engineering, and three key courses are listed. They are Introduction to Microelectronics Manufacturing Engineering, Electronic Package Technology and Surface Mount Technology. Further, for the last two courses, the teaching layout and syllabus are carefully planned. In addition, all members brain storm and implement the whole presentations. Once they came back their institutes and can use those directly to set up those two courses.

Secondly, based on the skill and knowledge requirements from the industry visit and investigation, how to introduce the electronic manufacturing engineering into professional education are discussed in details. Three different levels introduction are suggested. The primary level is to set up only one course called Introduction to Microelectronics Manufacturing Engineering. The purpose of this one course is to develop the basic idea of how the electronic products produce and how their major knowledge are used in practical electronic manufacturing process, which make the students easier to adapt themselves to career life. This level is suitable for all information technology related major and easiest to carry out. The middle level is to set up major direction. There are two options recommended, i.e. electronic package and surface mount technology (SMT). The corresponding major plan and teaching plan are developed, and from three to five courses are planed for each major direction. This is suitable for some old major such as mechanical engineering. It doesn't mean change the major but apply for new field. In addition, some investments for basic instrument are needed, such as SMT assembly line or wire bonder. The last one is to set up completely new major, i.e. Microelectronics Manufacturing Engineering. For the beginner of microelectronics manufacturing engineering education, it is not recommended. Under the references to the current teaching plan which are successfully implemented in the university, a completely new teaching plan & syllabus are developed.

Finally, the higher education division of the education ministry organized one public forum about information technology related majors education reform. More than 200 staff attended the conference held in Beijing. During the conference, the delegates of the training members summarized their harvest and reported above thinking of major construction. Not only the training members but also attendees high praised this advanced program.

By the way, all material including the presentations, major plan, courses syllabus are free to download from the official website of the higher education division of the education ministry.

Conclusions

Under the great supports of Intel (shanghai) and guidance of the higher education division of the education ministry, Microelectronics Packaging & Assembly Technology Faculty Training Camp was systematically planned and implemented in GUET. This comprehensive training program covers basic theories & experiments, industry visit & investigation, and education reform discussion and forum of introduction to of microelectronics manufacturing engineering into professional education. Three different levels of introduction microelectronics manufacturing engineering into professional education are suggested. There is no doubt that this creative joint program among the industry, government and university is an instructive trial and will promote the Chinese professional education of electronic manufacturing engineering. For one common target: Chinese electronic manufacturing industry's development, we strongly call for more interaction and cooperation among the industry, government and universities.

Acknowledgments

Thanks to Intel (Shanghai) for their fund & great lecturers supports.

Thanks to the higher education division of the education ministry for their great cooperation & kind guidance.

Special thanks to China CEPREI Laboratories, Guang Dong Yuejing High Technology Co., Ltd., Shenzhen Polytechnic, Huawei Technologies Co., Ltd., ZTE Co., Ltd., KONKA Co., Ltd, Shenzhen STS Microelectronics Co., Ltd., Folungwin Automatic Equipment Co., Ltd., Hui Zhou Daya Bay Guanghong Electronics Co., Ltd. for their great supports of visit & investigation.

References

1. http://www.miit.gov.cn/art/2008/02/19/art_4366_40547.html

2. Simon W. and El-abd, H., "Why Electronics in China—A Business Perspective", IEEE Trans. Comp. and Pack. Technol., vol. 26, no.1, 2000, pp. 276-280.

3. Bi Keyun, "Rapid Development of Electronic Packaging Industry in China", Proceedings of the Fifth International Conference on Electronic Packaging Technology (ICEPT2003'), Shanghai, China.

4. "A report on the first joint conference on electronic packaging education", IEEE Trans. Comp., Hyb. and Manufe. Technol., Vol. 16, No.3 , 1993, pp. 242-246.

5. Wesling, P., "Electronics packaging education: NSF and IEEE initiatives and modules", Proceedings of the 51st Electronic Components & Technology Conference, 2001.

6. Rao Tummala, Leyla Conrad and Gary May, "The Needs, evolution, status and challenges of microelectronics packaging education in the U.S.", Proceedings of the 49th Electronic Components & Technology Conference, 1999.

7. Rao Tummala and Leyla Conrad, "Undergraduate microsystems packaging education: needs, status and challenges", Proceedings of the 51st Electronic Components & Technology Conference, 2001.

8. Chan H.A., et al., "Requirements of advanced packaging curriculum", Proceedings of EuroSIME2003, France, 2003.

9. Chen Guan-fang and Yang Dao-guo, "On the education and training of Surface Mount Technology (SMT) engineers", Proceedings of the Fourth National Symposium on SMT/SMD, 1997, pp. 327-329.

10. Yang, D.G., N. Sun and M.X. Song, "Multi-level Education Curriculum of Electronic Manufacturing Engineering in GUET", Proceedings of 54th Electronic Components & Technology Conference, pp: 1716-1719, June 2004, USA.

11. Yang, D.G., "A New Educational Curriculum for Microelectronic Manufacturing Engineering Program", Proceedings of 55th Electronic Components & Technology Conference, June 2005, USA.

12. Pan kai-lin, "Course System Construction for Microelectronic Manufacturing Engineering", 2005. Journal of North China Institute of Astronautic Engineering

13. D.G. Yang , K.L. Pan, X. S. Ma, L. P. Wei. University Education Program on Microelectronics Packaging and Assembly: Facing the Challenging of Fast Developing Electronics Industry, 2006 ICEPT.

14. Pan kai-lin, Advanced Electronic Manufacturing Platform Construction of Practice Teaching. Electronic Process Technology, 2007(6).

FEA Based Reliability Prediction for Different Sn-Based Solders Subjected to Fast Shear and Fatigue Loadings

Rainer Dudek*, Eberhard Kaulfersch*, Sven Rzepka**,
Mike Röllig***, Bernd Michel*
* Micro Materials Center Berlin and Chemnitz
at Fraunhofer- IZM and ENAS, Germany
e-mail: rainer.dudek@che.izm.fraunhofer.de
** Qimonda Dresden GmbH & Co. OHG, Department: QD BET CMI
*** Dresden University of Technology, Electronic Packaging Laboratory, Dresden

Abstract

Recent studies revealed that there is no simple "drop in" solution for the lead-free replacement of SnPb joints, instead different Sn-based solders are advantageous for different use conditions, which can be dominated either by drop loading or by thermal cyclic loading in harsh use conditions. By way of high-speed shear testing reliability assessments of components during drop and shock events can be studied in a simplified manner. Dynamic 3-D finite element simulations have been performed applying explicit FEA to replicate the shear tests virtually. It was shown in this way that SAC 1305 solder outperformed SAC 387 solder. The low cycle fatigue behavior of different SAC alloys is additionally of interest. Fatigue life predictions require both the constitutive description of the lead-free solders and a fatigue hypothesis linked to the material selected. Based on recently measured creep properties the solder joint creep strain and creep dissipation responses were analyzed for several components and thermal cycling conditions. The results based upon non-linear finite element calculations indicate different trends for creep strain and energy dissipation: while the first is clearly increasing with lowered alloying Ag-content, the latter is almost stable and does only slightly vary. Furthermore, these trends are different for different test- and field cycling conditions as well as the different components studied.

Keywords: Lead-free SACxx solders, finite element analysis, fast shear test modeling, creep modeling, lifetime prediction, solder fatigue

Introduction

The lead-free soldering technology raises a couple of questions concerning the joint reliability in different loading situations, for example when joints are subjected to fast mechanical loadings like drop or vibration, or to slow thermal cyclic loadings, or combinations of these loading types. Generally, the questions are further related to the field performance in relation to accelerated testing, because due to the short field use, in particular in harsh environments, little experience on use performance has been gathered during the last decade.

As the growing use of virtual design methodologies in packaging applications is another important point in current developments, theoretical life prediction of solder joints is an important subject of world-wide research, see e.g. [1]-[4]. Almost all of them are based upon non-linear finite element calculation results. Due to increase of computational power

rather complex three-dimensional situations can be handled and very common simplifications like plane or slice models are rarely needed today. However, many hypotheses are still used in modeling of solder interconnects. Neither the complex microstructure and their evolution during loading nor the full number of loading cycles can be accounted for in the simulations with the mostly limited computer resources available. Additionally, realistic constitutive modeling is a limiting factor due to dependencies of the solders response on initial alloying contents and alloying reaction with the surfaces, interfacial intermetallics, solidification and evolving microstructure during loading [5],[6].

Studies on the performance of different Sn based solder alloys applied for large to small sized solder interconnects were undertaken. From the theoretical and experimental investigations on creep, fatigue and brittle fracture behaviors the paper focuses on modeling of the joint behavior under fast shear loading and on the low cycle fatigue performance in test and field thermal environments.

SAC fast shear performance evaluated by FEA and testing

By way of high-speed shear testing reliability assessments of components during drop and shock events can be studied in a simplified manner. Dynamic 3-D finite element simulations were performed applying explicit FEA to replicate the shear tests virtually.

Fast shear experiments of single joints were made by use of a DAGE® tester to BGA solder balls being attached to Cu+Ni/Au pads with recording the peak shear forces and fracture surface plots. With reflowed solder balls, which were 450 μm in diameter after second level packaging, characteristic threshold magnitudes at which the failure mode is changing from fracture in the solder bulk to fracture in the intermetallics were to be obtained. The pad on the substrate side consists of copper with a nickel top layer and a gold flash, covered partially by the solder mask layer. The joint interface is formed by a NiSn intermetallics compound (IMC) of about 1,5 μm thickness, whereas the Ni, IMC and solder layers are usually detached from solder mask. The ball is formed either of SAC105 or of SAC3575 solder.

In Fig. 1 the actual shape of the ball after reflow and the face area of the tool are visualized. Dimensions of the ball shear tool as shown in Fig. 1 below are a shear tool depth of >2.0 mm and a bottom width of 0.75 mm. It is made of ceramic composite material and can be assumed to be a rigid body in comparison to solder bulk material.

978-1-4244-2739-0/08/$25.00 ©2008 IEEE

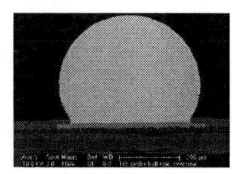

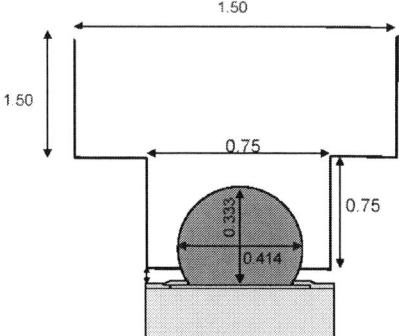

Fig. 1: Single ball and shear tool dimensions.

During shear experiments, peak shear forces and the fracture surface plots were recorded. The tool speed ranged from 0.35 to 2,000 mm/s and included the following values: 0.35, 10, 500, 1000 and 2000 mm/s. For SAC 105, Fig. 2 shows the observed transition from bulk solder shear via partial solder fracture to so called grey plot fracture with all the damage occurring within the IMC. In Fig. 3 a comparison of the corresponding shear strengths is given for both SAC 105 and SAV 3575 solders, obviously SAC 105 reaches higher shear strength at high shear velocities. The reason for the higher strength for SAC 105 is late a transition from bulk fracture to IMC damage, which takes place at a shear speed range from 1000 to 2000 mm/s, shown in Fig. 4.

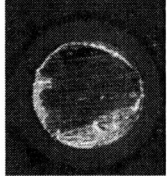 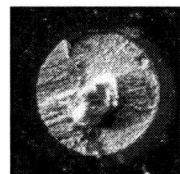 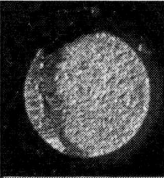

Fig. 2: Fracture plots at 10 mm/s (left, bulk fracture), 1000 mm/s (middle, partial bulk fracture) and 2000 mm/s (right, IMC fracture), SAC 105.

Transient 3D-FEM techniques were applied to simulate a fast ball shear test under different shear speeds/conditions. The simulations involved a rate dependent model for the solder material as well as a cohesive zone approach for the fracture in the IMC layer [7]. The cohesive zone interface was introduced within the interfacial intermetallic layer as can be observed from Fig. 5.

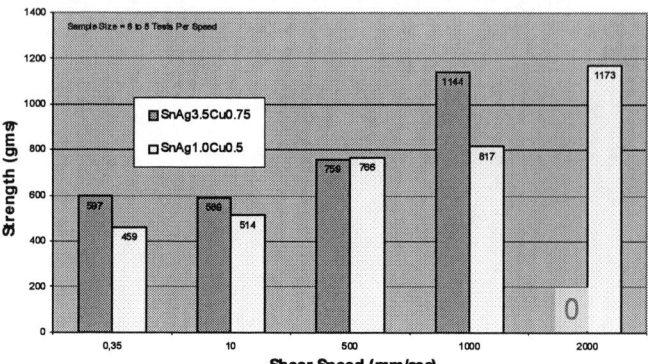

Fig. 3: Solder ball shear strength at different shear speeds for SAC 105 vs. SAC 3575

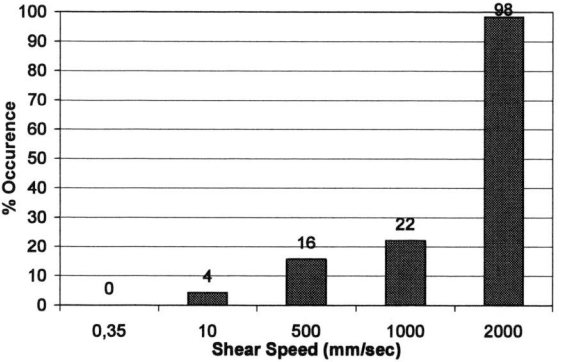

Fig. 4: % brittle fractures after solder ball shear test at different shear speeds for SAC 105

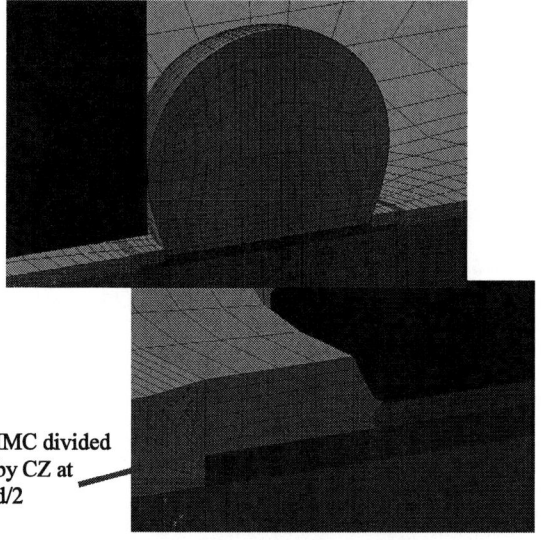

IMC divided by CZ at d/2

Fig. 5: Single ball shear symmetric finite element model with zero thickness cohesive zone dividing IMC layer

The aim of the study was twofold: as the rate-dependent plastic behavior of SAC 105 at a real joint, i.e. including the alloying reaction, was only approximately known, the fit of the measured force deflection response allowed the verification of the true ball solder behavior. Once known, the model is appropriate to simulate the brittle joint fracture behavior in similar loading cases, e.g. in drop tests.

The material model developed had to be capable of identifying changes in failure mode locations (bulk solder, IMC fracture, etc.) depending on input loading conditions from the experimental investigations. It is known that the cohesive zone approach has been successfully employed in such applications.

The material data fit procedure performed here included the following IMC damage model: The cohesive zone was based on a cohesive traction separation approach with uncoupled behavior for the normal and shear forces, with a linear traction-separation response (Fig. 6). The characteristics of the cohesive elements are determined by two parameters, cohesive strength and separation energy. The criterion for damage initiation was chosen to be a maximum nominal stress criterion. An assumption for the corresponding shear stress maximum was a factor of 2 times the normal stress. Damage evolution used linear stiffness degrading down to zero stiffness.

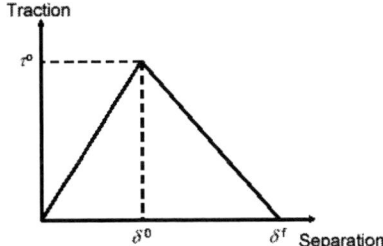

Fig. 6: Linear traction-separation response

Failure mode and mechanical behavior of solder alloys are highly dependent on strain rate. In Fig. 7 the computed force-displacement curves are compared to the measured peak force values at varying shear speed.

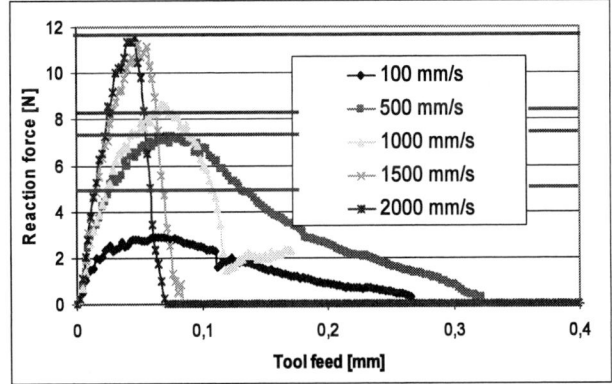

Fig. 7: Calculated force-displacement curves for various shear speeds compared to the measured peak forces (see Fig. 3)

For solder bulk damage a shear criterion with strain rate dependence of critical plastic strain but with neglected dependence on shear stress ratio has been used: Critical plastic strains for damage initiation were 15% at 0.1 s^{-1}, 5% at 100 s^{-1} and 1% at 1000 s^{-1}

The peak force values ranging from the ones in the transition region from bulk damage to IMC fracture towards those at higher velocities are determined solely by the IMC strength, the values below mainly by the solder ductility and bulk damage criterion. The corresponding fracture behavior can be seen from Fig. 8, which depicts the bulk remainder on the pad at a shear speed of 1000 mm/s and a complete fraction of the IMC with no remaining solder at 2000 mm/s. From a

transition region between shear speeds of 1000 and 2000 mm/s, respectively, a high normal stress maximum of 350 MPa could be derived for the IMC strength.

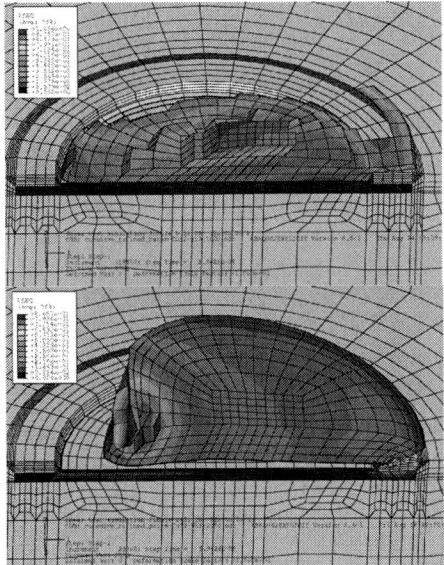

Fig. 8: Bulk remainder plots at 1000 mm/s (top) and 2000 mm/s (bottom)

From the material data fit procedure a constitutive law of the following form has been found whereas the actual numbers of the coefficients remain concealed for preservation of company propriety information:

$$\sigma_y = A \, (d\varepsilon/dt)^n + B \, (d\varepsilon/dt)^m$$

with an initial yield strength of 25 MPa. Results seem to be relatively independent on the initial yield strength in the tested interval from 25 to 40 MPa. The IMC strength was stated to be 350 MPa for tension separation.

In this way the fracture mode transition from the solder bulk to the IMC could be reproduced numerically. The interface turned out to be of high strength. Applying the constitutive relationship, the simulated and measured peak forces are coinciding well. These validated models now allow highly dynamic events of BGA modules to be assessed realistically by means of FEM.

Analyses on thermal fatigue

SAC solder creep in dependence on the alloying content As in the case of brittle fracture the correct description of solder joint constitutive behavior is the basic prerequisite of thermal fatigue modeling, i.e. the temperature dependent creep behavior has to be known in advance. The dilemma of all kinds of fatigue models, also the sophisticated ones which try to include damage propagation by continuum damage techniques, is the strong microstructure dependence of the solder joint constitutive behavior. Already the secondary creep rates tend to differ by up to two orders of magnitude for the same (eutectic) alloying composition [8].

To consider the real joint behavior rather than the bulk behavior, a creep measurement technique using miniaturized ball type specimens had been developed [9]. It allows creep data measurements at specimens with 4 micro solder balls, where a standoff height of 200 µm was used. To avoid

147

influences of the compliance of the substrates on the measuring results, which are the more likely the higher the creep resistance of the alloy under investigation is, ceramic substrates were chosen for the test specimens.

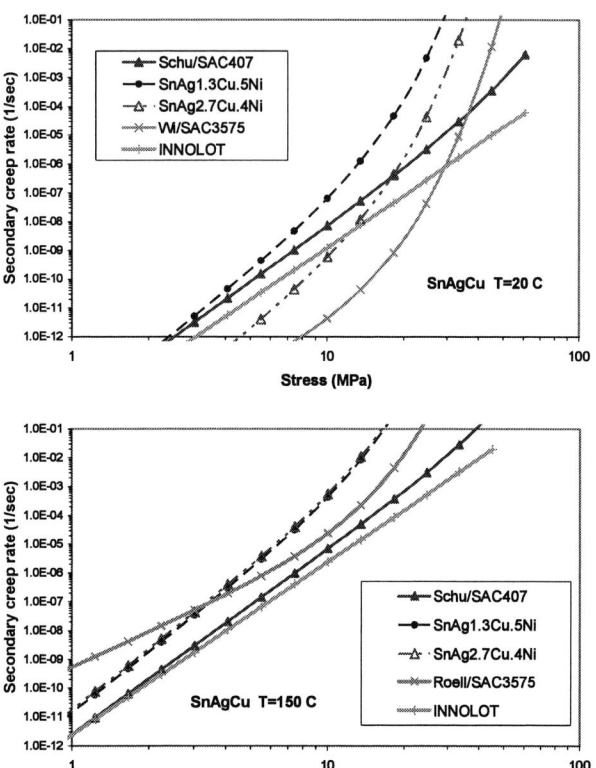

Fig. 9: Secondary creep strain rate versus tensile stress for different SACxx alloys at room temperature and at 150 °C

The secondary creep rates for different alloys investigated, with Ag content 1.3%, 2.7 %, and 3.8% , as well as the special purpose newly developed Innolot alloy (SnAg3.8Cu0.7Bi3.0Sb1.4Ni0.2) are depicted in Fig. 9 for two different temperatures. The reduction in creep rate with increasing alloying Ag content can clearly be seen, particularly for lower temperatures. However, Innolot shows the highest creep resistance within the whole temperature range. It is noted that the small differences in Cu content for the SAC solders become meaningless, since the Cu content always reaches its saturated value of approx. 1.1 % by diffusional flow from the pads if no barrier layer is present.

For FEA two creep laws used are described in more detail. Both of them, the one for SAC3575 as well as the one for SAC135, were newly measured by Röllig et al. [9] at miniaturized ball type joints soldered on Cu pads with Sn finish.

The secondary creep rates are described by the expression,

$$\dot{\varepsilon}_{ss} = C_1 \cdot \sinh(C_2 \cdot \sigma_M)^{C_3} \cdot \exp\left(\frac{-C_4}{R \cdot T[K]}\right) \quad (1),$$

with the two sets of constants given in Table I.

Table I: Coefficients for the secondary creep law (1)

solder	C1	C2	C3	C4
SnAg1.3Cu0.5 joint data	3.0E4s^{-1}	0.111 MPa^{-1}	7	67000
SnAg3.5Cu0.75 joint data	1.5e+9 s^{-1}	0.1538 MPa^{-1}	4	123000

Effects of solder joint composition on cyclic joint stresses for different components

It is well-known that both the inelastic strain (creep strain) and dissipated strain energy density represent suitable indicators to evaluate cyclic damage.

Based on creep properties the solder joint creep strain and creep dissipation responses were analyzed for several components and thermal cycling conditions. Phenomenological models were applied to study the component and cyclic regime dependent creep straining and creep dissipation in several joints to assess solder failure.

In the following reference is made to a ceramic resistor 1206 and a LFBGA 345, the former being relative rigid and the latter relatively compliant with respect to the solder joint loading. The symmetric FE models of both components are depicted in Fig. 10. An air to air cycle of 125 °C to –40 °C with cycle time of approx. 1 hour was chosen for the comparison of different solder alloys.

For a stabilized cycle at the solder joints of a ceramic resistor 1206, the characteristic distribution patterns with maximum straining within the solder gap (standoff) were obtained for each of the joints, irrespective of the kind of the alloy. For purposes of comparison, values of creep strain were averaged along the damage path within the solder gap. Fig. 11 depicts the averaged values of cyclic equiv. creep strain for the CR 1206. The drop of creep strain with increasing Ag content is obvious and for Innolot a significant lower amount is calculated. For comparison, the generic creep law of Schubert et al. [10], which includes additionally also primary creep, was added to the comparison.

In case of the relatively rigid assembly it matches the results for the new, joint-based creep law after Roellig relatively well. The tendencies for the creep energy density, dissipated per cycle, are different from those of the creep strains. As can be observed from Fig. 12, all different SAC solders show approximately the same value. Even for the generic creep law of Schubert et al. including primary creep contributions, the same amount of averaged creep dissipation density is calculated. Only for the Innolot solder, a slightly lower value is obtained.

The analogous results for the LFBGA are depicted in Fig. 13 and Fig. 14. Both the averaged value along the damage path at the interface ball-interposer and a maximum value, averaged only in a small region of expected crack initiation, are given. For this package maximum straining was always calculated at the thermal edge ball, which is subjected to high shearing loads from the mismatched die because of its location in the "die shadow".

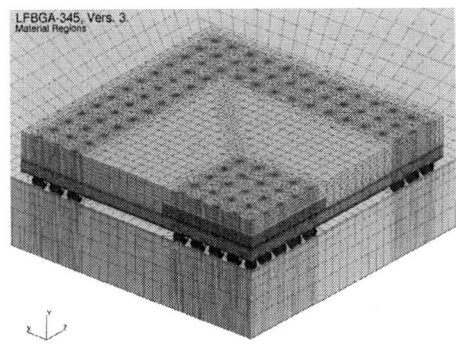

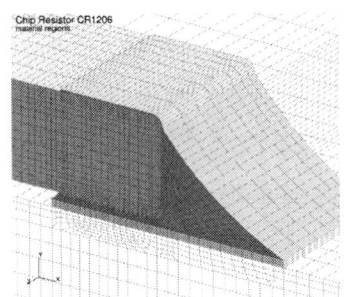

Fig. 10: LFBGA and chip resistor 1206 symmetric finite element models

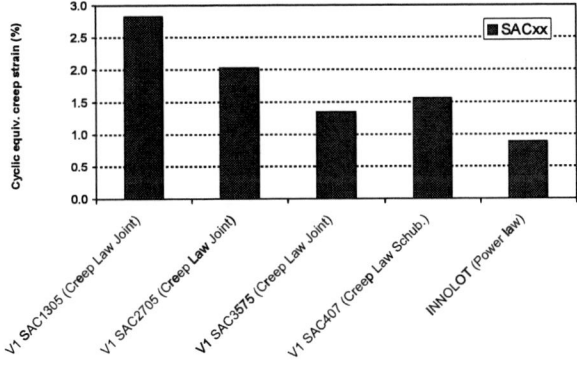

Fig. 11: Cyclic equiv. creep strains, averaged along the damage path in the solder gap, for the CR1206 for different SAC solders.

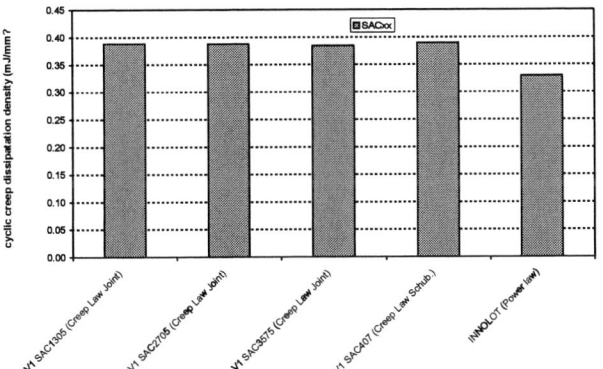

Fig. 12: Cyclic creep dissipation densities, averaged along the damage path in the solder gap, for the CR1206 for different SAC solders.

For both ceramic and CSP components a tendency of increased creep strain accumulation with lowered Ag content was calculated, while the creep dissipation remained relatively constant. In case of the resistors, for "Innolot" the cyclic creep dissipation decreased less than for CSP joints, while the creep strain was approx. three times lower for the resistor but five to six times lower for SAC 1305.

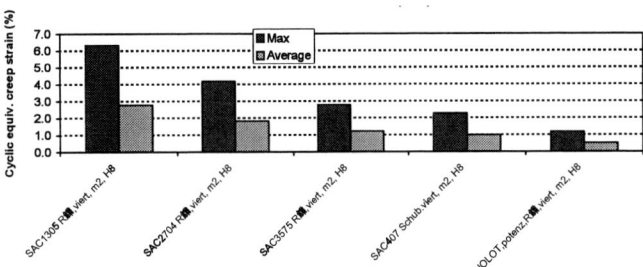

Fig. 13: Cyclic equiv. creep strains for the LFBGA 345 at the maximum strained ball for different SAC solders.

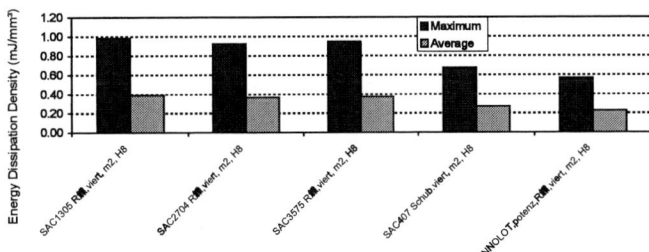

Fig. 14: Cyclic creep dissipation densities for the LFBGA 345 for different SAC solders.

It is obvious from the comparisons that the effects of different solder alloys on creep straining or creep dissipation, respectively, are not the same if different components are considered. In case more rigid components are regarded, like the ceramic chip resistors, joint straining is more displacement controlled and depends less on the creep properties of the solder material. In the examples shown in Figs. 11 to 14 obviously the cyclic creep strain and creep dissipation drops more for the relative flexible component LFBGA in comparison to the rigid resistors. Analogously, the choice of another creep law for the same alloy, SAC 378, affects the calculation results more for the LFBGA joint.

It can be concluded that care has to be taken when a ranking of the different alloys is given concerning their "fatigue resistance": data from one component type must not automatically match the data to be expected for another one.

Effects of different thermal cycle conditions

A question similar to the component dependence on effects of the choice of solder alloying compositions is that on effects of different thermal cyclic conditions, i.e. does the cycle condition affect the ranking of solders concerning their solder fatigue behavior.

Both the temperature range and the cycle duration were varied. Only air to air cycling was applied. The cyclic regimes included -40 °C to 150 °C, with either fast ramps of the environmental temperatures (TS) and 1 h cycle time or slow ramps (TW) and 104 min cycle time, analogous variants for the cycle -40 °C to 125 °C, a cycle -20 °C to 90 °C, 1 h, and a field cycle 0°C to 80 °C with 7 h cycle time (1 h ramp up and 6 h ramp down duration). The calculated cyclic creep strains for the different cyclic regimes are depicted in Fig. 15 for the solder alloys with the lowest and with the highest creep resistance of the alloys tested, i.e. SAC135 and Innolot. The analogous calculation results concerning the cyclic creep dissipation densities are given in Fig. 16.

Obviously, a similar trend of much lower creep straining is seen for Innolot when compared to SAC135 for all thermal cycles. However, the tendency of lower creep straining of Innolot reduces gradually with longer cycle time, the highest difference of approx. 1:3 is seen for the TS 150 cycle which reduces to approx. 1:2 for the slow filed cycle 0 to 80 °C. For the energy dissipation the highest difference is only 1.1 : 1 for the TS 150 and TS 125 cycle, while for the TS 90 and the field cycle the difference between the two solders vanishes.

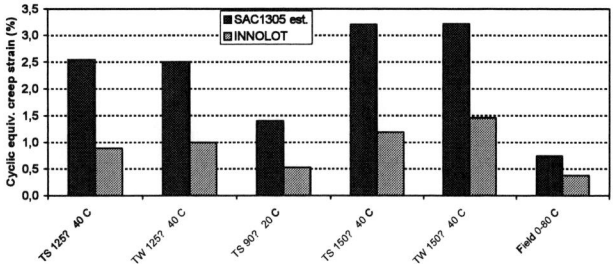

Fig. 15: Cyclic equiv. creep strains for the CR1206 for different thermal cycles.

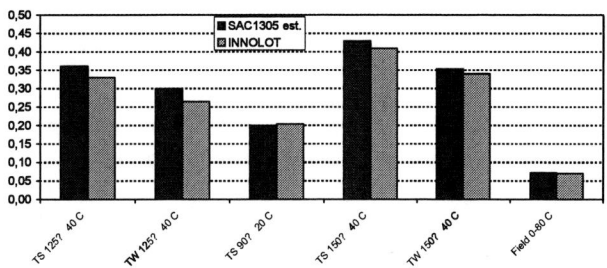

Fig. 16: Cyclic creep dissipation densities for a CR1206 for different thermal cycles.

It can be concluded from these calculation results that the tendencies observed for the test cycles do not necessarily be valid for service conditions.

Summary

By way of high-speed shear testing reliability assessments of components during drop and shock events were studied in a simplified manner. Transient 3-D finite element simulations were performed applying explicit FEA to replicate the shear tests virtually and to adjust the solder joint materials data to a real joint situation. From testing it became obvious that that low Ag content solders like SAC 105 show a superior brittle failure resistance.

Considering thermal fatigue, dependencies of creep properties on alloying content were analyzed. Based on these properties, the solder joint creep strain and creep dissipation responses were calculated for chip resistors and LFBGA

component joints with different solder compositions and different thermal cycling conditions.

The FEA results indicate different general trends for creep strain and energy dissipation: while the first is clearly increasing with lowered Ag alloying content, the latter is more stable and varies only slightly. Differences are seen between the two component types, the creep strain and creep dissipation varies less for more rigid ceramic components. Analogously, the choice of different cycle conditions result in different creep strain or dissipation responses, the lower creep strain accumulation of alloys with high Ag content gradually disappears for cycles approaching field use. Finally, a special purpose SAC alloy called Innolot (SnAg3.8Cu0.7Bi3.0Sb1.4Ni0.2) shows a very high creep resistance and the lowest joint creep straining of all candidates.

Acknowledgement

Funding of fatigue related results taken from the project LIVE by the Federal Ministry of Education and Research of the Federal Republic of Germany (Project No. NKIT03114602) was gratefully appreciated.

References

[1] R. Dudek,, Walter, H., Döring, R., Michel, B., „Thermal Fatigue Modelling for SnAgCu and SnPb Solder Joints", Proceedings EuroSimE 2004, Brussels, Belgium, May 2004, pp. 557-564

[2] Lau, J. H., Dauksher, W., Vianco, P., „Acceleration models, constitutive equations, and reliability of lead-free solders and joints", Proceedings, 54th Electronic Components & Technology Conference, 2003, pp. 229-236.

[3] Syed, A. R., „Accumulated Creep Strain and Energy Density Based Thermal Fatigue Life Prediction Models for SnAgCu Solder Joints," 54st ECTC, Las Vegas, June 2004, pp. 737-746.

[4] D. Bhate, G. Subbarayan," A NONLINEAR FRACTURE MECHANICS PERSPECTIVE ON SOLDER JOINT FAILURE: GOING BEYOND THE COFFIN-MANSON EQUATION", Proc. ITherm, San Diego, 2006

[5] S. Wiese, M. Roellig, M. Mueller, S. Bennemann, M. Petzold, K.-J. Wolter, "The Size Effect on the Creep Properties of SnAgCu-Solder Alloys," Proceedings, 57th Electronic Components & Technology Conference, 2007

[6] S. Wiese, M. Roellig, M. Mueller, K.-J. Wolter; „The Effect of Downscaling the Dimensions of Solder Interconnects on their Creep Properties", Proceedings EuroSimE 2007, London, GB, April 2007

[7] E. Kaulfersch, S. Rzepka, V. Ganeshan, A. Mueller, B. Michel, „Dynamic Mechanical Behavior of SnAgCu BGA Solder Joints Determined by Fast Shear Tests and FEM Simulations", Proceedings of EuroSIME 2007, April 2007, London/Great Britain, pp. 172-176

[8] R. Dudek, W. Faust, S. Wiese, M. Röllig, B. Michel, "Low-cycle Fatigue of Ag-Based Solders Dependent on Alloying Composition and Thermal Cycle Conditions," EPTC 2007, Signapore, Dec. 2007, pp. 14 – 20

[9] Roellig, M.; Wiese, S.; Meier, K.; Wolter, K.-J.: Creep Measurements of 200 μm – 400 μm Solder Joints.

Proceedings of EuroSimE 2007, London 15. -18. April, 2007

[10] Schubert, A., Dudek, R., Auerswald, E., Gollhardt, A., Michel, B., Reichl, B., „Fatigue Life Models for SnAgCu and SnPb Solder Joints Evaluated by Experiments and Simulation", Proceedings, 53rd Electronic Components & Technology Conference, 2003, pp. 603-610

Thermo-Mechanical Analysis of a Wafer Level Packaging by Induction Heating

Wenming Liu[1,2], Mingxiang Chen[1,2], Yanyan Xi [1], Changyong Lin[1], Sheng Liu[1,2]*

1. Wuhan National Lab for Optoelectronics, Huazhong University of Science & Technology, Wuhan, China, 430074
2. Institute of Microsystems, Huazhong University of Science & Technology, Wuhan, China, 430074

* corresponding author

shengliu63@yahoo.com, +86-27-8754260

Abstract

In this paper, a non-linear and one-directional coupled finite element framework has been implemented to simulate induction heating process of wafer-level packaging. Based on numerical results of induction heating, thermally-caused warpages and stresses of the single-sided ceramic wafer have been evaluated. Some primary experiments have also been conducted to verify the numerical method. Using three-dimensional models, the temperature distribution, thermally-caused warpages and stress in the single-sided ceramic wafer subjected to induction heating can be clearly defined. In addition, the temperature-dependent material properties are considered in the modeling. From the finite element analysis, it is found that the induction heating is selective, that is, the temperature in the wafer is lower than that of Cu-loops during the induction heating process; the temperature variation on the Cu-loops, as well as the difference of the temperature between the Cu-loops and the wafer is related with the wafer material properties; the maximum thermal-stresses caused by the induced Joule heating occur on the middle-edge areas of the single-sided ceramic wafer. On the other hand, in order to prove the soundness of the framework established in this paper, the test results obtained by infrared radiometer are compared to that achieved from the proposed numerical analysis method. It is shown that the temperature variation and locations of initial cracks caused by thermal-stresses during the induction heating are in a good agreement with those obtained from the test.

Introduction

Wafer-level packaging can offer a significant cost reduction for zero-level MEMS packaging [1-7], and it has increasingly become a key technology for materials integration in various areas of MEMS, microelectronics and optoelectronics. It is noted that most MEMS devices are packaged after the release of their dynamic elements or the completion of the integrated circuitry which both are temperature-sensitive. Hence, temperatures to which wafers are exposed during the bonding process and the exposure duration should be carefully controlled. Some low-temperature-bonding techniques, such as low-temperature solder bonding, adhesive bonding and surface activation bonding [8-9], are developed to avoid thermal effect problems and damages caused by high bonding temperatures. Unfortunately, most low-temperature-bonding processes do not offer high quality bonds because bonding strength is also a function of temperature.

Localized heating and bonding technology, therefore, was presented [10-11]. That is, only the bonding area was heated and the other features were still kept at a low temperature during the bonding process. Induction heating, owing to its reliable, selective, accurate and energy-efficient heating in minimal amount of time, provides many advantages over other localized heating methods [4] and is commonly used for bonding applications [12-13]. Nevertheless, much large temperature gradient in the localized induction heating exists in the bonding interface between the wafers and bonding areas. In addition, there is large difference in coefficient of thermal expansion (CTE) between the wafer and bonding materials. As a result, significant thermal stresses and deformation in the interfaces and wafers are induced and may lead to various failure modes, such as wafer cracking, bonding voiding and solder cracking [14-16]. Therefore, modeling and experimental evaluation of the temperature distribution and bonding stresses and warpages in wafers are of critical importance for the determination of the package reliability in selective induction heating packaging.

In this paper, the temperature distribution, thermal-stresses and warpages of the single-side ceramic wafer in induction heating are investigated in detail. On one hand, a non-linear and indirect finite element framework is implemented to simulate the wafer-level packaging. On the other hand, some single-sided ceramic wafer bonding experiments are also conducted to verify the finite element framework.

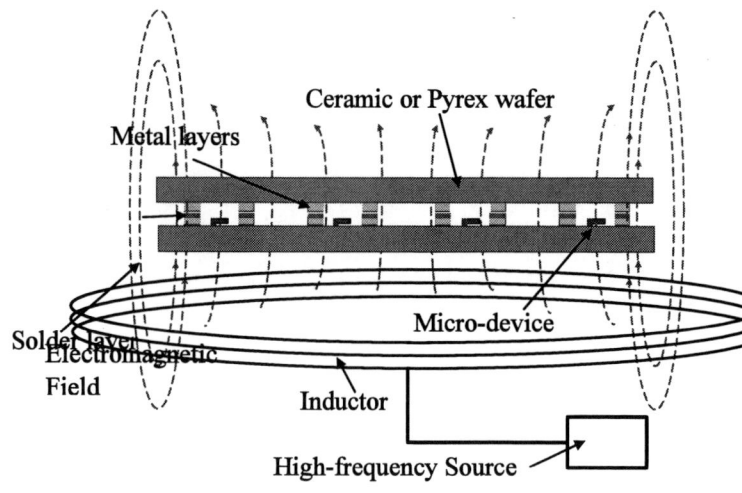

Figure 1. Induction heating for Wafer-level bonding.

978-1-4244-2739-0/08/$25.00 ©2008 IEEE

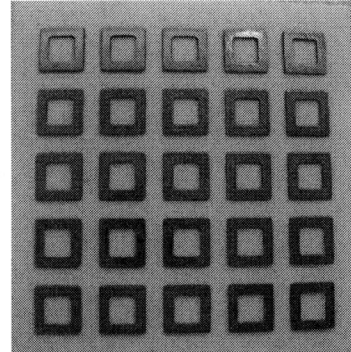

Figure 2(a). Single-sided ceramic wafer. This kind of wafer with Cu-loops is manufactured by DCB (Direct Bond Copper) process, in which the Cu rings were directly sintered on the ceramic wafer.

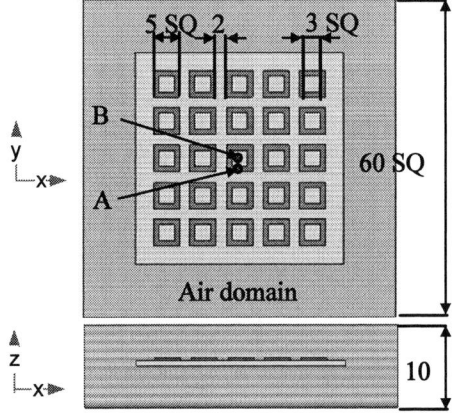

Figure 2(b). Boundary conditions, dimension and geometry of the single-sided ceramic wafer (unit: mm). SQ means the shape signed is square-size. In the harmonic electromagnetic field analysis, far-away air domain is considered; in sequent thermal analysis, the air domain is deactivated and adiabatic boundaries are defined to the single-side ceramic wafer. The point A and point B are located on the Cu-loop and ceramic wafer, respectively.

Models and materials

Modeling

The wafer-level-packaging induction heating process is a coupling of an electromagnetic field analysis and a thermal analysis, schematized in Figure 1. Due to the symmetry along the plane where the solder layer is located, only one half of the model is investigated. Therefore, a single-sided ceramic wafer deposited with the Cu-loop array as shown in Figure 2(a) is taken into consideration. The boundary conditions, dimensions and geometry of the single-sided ceramic wafer model, two parts which consist of both the single-sided ceramic wafer and an air domain, are illustrated in Figure 2(b) (not drawn to scale). In order to simplify the model and reduce the quantities of finite elements, the heating inductor is replaced by loading magnetic vector potentials directly. According the Stokes formulation, the magnetic flux density can be given by the following equation:

$$B = rotA$$
$$= (\frac{\partial A_z}{\partial y} - \frac{\partial A_y}{\partial z})i + (\frac{\partial A_x}{\partial z} - \frac{\partial A_z}{\partial x})j + (\frac{\partial A_y}{\partial x} - \frac{\partial A_x}{\partial y})k$$

where B is the magnetic flux density, A the magnetic vector potential; Ax, Ay, and Az are the magnetic vector potential components. The heating inductor was designed to induce a z-direction magnetic field, and no components along x and y. Therefore, the magnetic flux density component,

$$B_z = (\frac{\partial A_y}{\partial x} - \frac{\partial A_x}{\partial y})k$$

Magnetic field intensity,

$$H = \frac{B}{\mu_r \mu_0}$$

The magnetic field intensity H induced by the inductor can be measured by a magnetic field intensity gauges. Hence, the loads in the numerical analysis can be related with the power supply in the test. The problem is idealized by a finite element mesh, as shown in Figure 3(a). A close-up view of fine mesh is plotted in Figure 3(b). Because of the skin effect, the elements located near the out-surface of each Cu-loop are refined, and their size is set as 0.5*δ (the skin depth). The δ is defined as,

$$\delta = (\rho/\pi\mu f)^{1/2}.$$

where ρ and μ are the resistivity and permeability of the cu-loop, respectively; f the frequency of the high-frequency source.

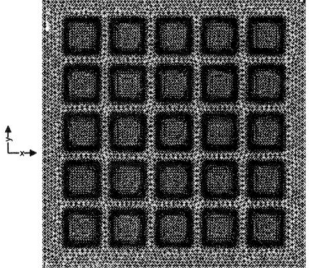

Figure 3(a). Finite element mesh for single-side ceramic wafer.

Figure 3(b). Local general finite element fine mesh for single-side ceramic wafer.

The material behaviors of the single-sided ceramic wafer are assumed to be temperature-dependent. The temperature dependence of material characteristics such as the electrical conductivity, the thermal conductivity and so on, is considered in the induction heating analysis. The measured data of material characteristics concerned are shown in Figures. 4(a), 4(b), 4(c), 4(d), 4(e), 4(f) and 4(g), respectively.

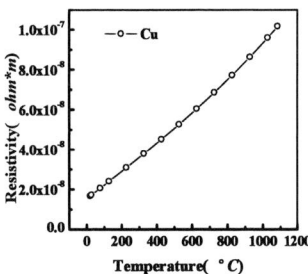

Figure 4(a). Resistivity of Cu with temperature.

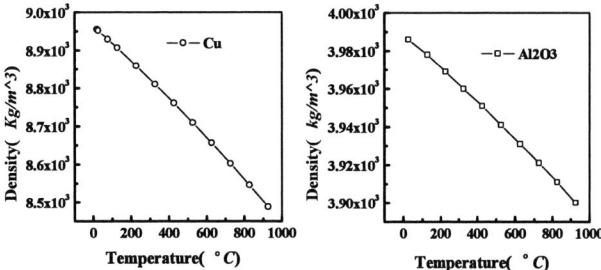

Figure 4(b). Density with temperature.

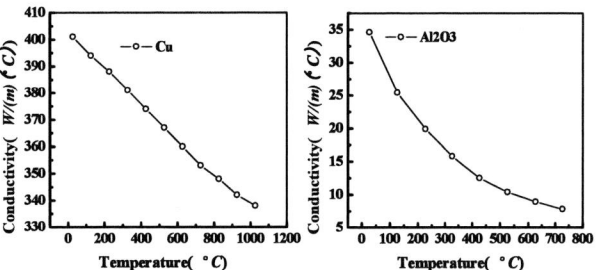

Figure 4(c). Conductivity with temperature.

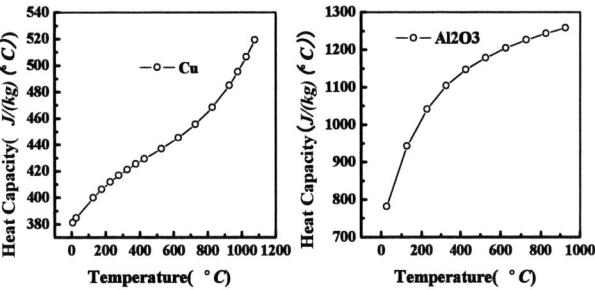

Figure 4(d). Heat Capacity with temperature.

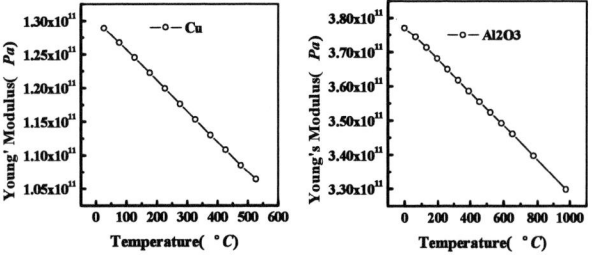

Figure 4(e). Young's Modulus with temperature.

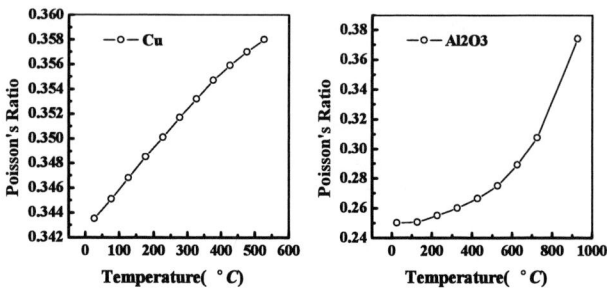

Figure 4(f). Poisson's Ratio with temperature.

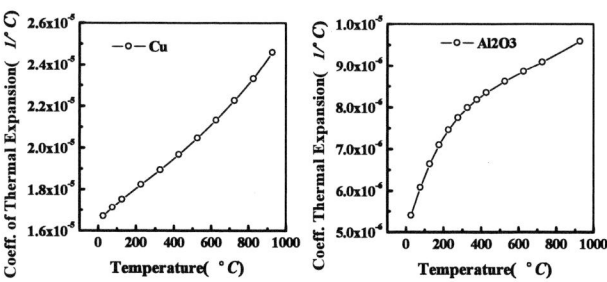

Figure 4(g). CET with temperature.

Method of solution

One of the major features of induction heating computation deals with the fact that both the electromagnetic and thermal phenomena are tightly coupled because of the interrelated nature of the material properties. Obviously, these variations of physical properties make the induction heating problem nonlinear. This interrelated feature dictates the necessity of developing a special computational framework that is able to deal with these coupled effects.

An indirect coupling framework, in this paper, is implemented to solve induction heating problems. This method calls for an iteration process shown in Figure 5(a). An iteration process consists of an electromagnetic computation and then recalculation of heat sources in order to provide a heat transfer simulation. This approach assumes that temperature variations are not significant during certain stages of heating cycle. That is, the electromagnetic properties remain approximately the same, and during certain times the heat transfer process continues to solve without correcting the heat sources. The temperature distribution in the single-sided ceramic wafer obtained from the time-stepped heat transfer computation is used to update the values of material properties at each time step. As soon as the time step is suitable and the heat source variations do not become significant, the convergence condition will be satisfied and the induction heating computation will continue.

After completion of induction heating, the temperature distribution in the single-side ceramic wafer is figured out, and then the induction heating results are loaded as initial conditions of the thermal-stress analysis of the single-side ceramic wafer model. As the last global convergence condition is satisfied, the thermal-stress convergence condition will be satisfied, as shown in Figure 5(b).

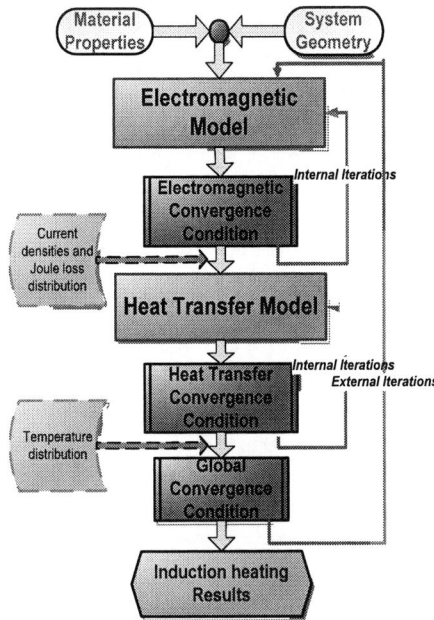

Figure 5(a). Indirect coupling process of induction heating.

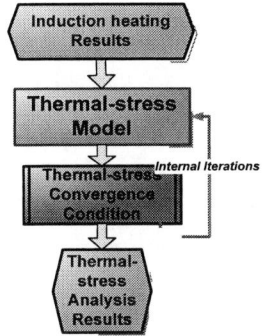

Figure 5(b). Indirect thermal-structure process.

Figure 6. Induction heating system.

Experiment validation

In order to prove the soundness of the framework established in this paper, the results achieved from the proposed numerical analysis vehicle are validated by comparison to the results obtained by using the induction heating experiment. By selection of the same dimension single-sided ceramic wafer specimen, an experimental work is conducted under an Nitrogen atmosphere condition, preventing the oxidization of the specimen. The induction heating system consists of a high-frequency power supply operating at 350 kHz with an output power from 0 to 1 kW, an induction heating conductor made from copper tube and cooled with water, as shown in Figure 6. A uniform harmonic magnetic field through the wafer vertically is induced this specially-designed conductor. In addition, an infrared radiometer is used to monitor the temperature on the Cu-loops.

When the power supply sends an alternative current of 350 kHz to the conductor, a uniform alternative magnetic field with the same frequency is created by the conductor traverses perpendicular to the single-sided ceramic wafer, so the Cu-loops on the ceramic wafer are heated very quickly by induction heating. Because the magnetic field in the conductor is proportional to the input power, when changing the power supply, we monitor the magnetic field by magnetic field intensity gauges.

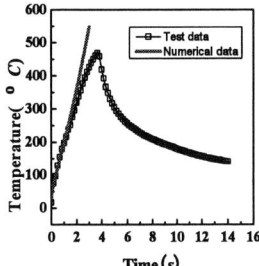

Figure 7. Comparison of temperature variation between test and numerical results.

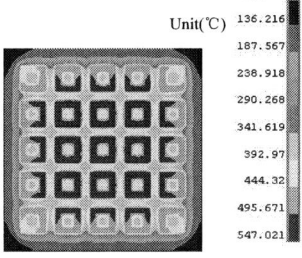

Figure 8. Temperature contours in the single-side ceramic wafer.

Results and discussion

In order to compare the results between the numerical work predicted by the finite element analysis with the test conducted by the induction heating system in the current study, the heating process of a single-sided ceramic wafer is recorded by the infrared radiometer during the test. Figure 7 shows that the temperature variation at the defined point A, on the specimen obtained from the finite element analysis and the test. As can be seen, the temperature at the Cu-loops increases quickly to about 500 ℃ at the end of the induction heating. Figure 8 shows a great temperature difference between the Cu-loops and their central area on ceramic wafer, displaying the selective heating effect of induction heating. It is also found, in the test, that the measured infrared radiometric data shows an approximate 2-second delay compared to the numerical prediction, similar to the conclusion reported by other researchers [17]. The possible reason is that both the induction heating system and the infrared radiometer exist a response delay compared to the timing-meter. In addition, it is found that the temperature curve obtained from the infrared radiometer shows much of the same pattern compared with that modeled by the finite element method. It is proven that the results obtained from the finite element framework established in the paper are reliable and could thereby be used to accurately predict the temperature distribution and variation during the wafer-level-packaging induction heating.

As to the thermal-stress and thermal-deformation in the single-sided ceramic wafer, as can be seen from Fig. 9, large thermal-stresses exist on middle areas of each wafer edge and inner corners of loops. These findings accord with the locations where the cracks caused by the overload of the thermal-stress occur during the induction heating test, as shown on Fig. 10 and Fig. 11. The deformation contours simulated by the finite element method are shown in Fig. 12. Obviously, the warpages of the wafer will result in the different gap between wafers in the wafer-level bonding process, which may lead to a bonding voiding.

Conclusions

Modeling and preliminary experiment of wafer level packaging in selective induction heating has been successfully demonstrated in this study. Numerical results from the single-side ceramic wafer were successfully compared to experimentally measured data, and the finite element framework presented in this paper was proved to be reliable and reasonable. The finite element analysis showed significant difference in temperature variation in the metal loops and center areas on the wafer during heating. Thermal stresses and warpages could lead to wafer cracking and bonding voiding. A further experiment should be conducted with the wafer-level bonding in order to generalize the effects of thermal-stresses and warpages.

Acknowledgments

The authors gratefully acknowledge the support of the National High Technology Research and Development Program of China(No. 2006AA4Z328).

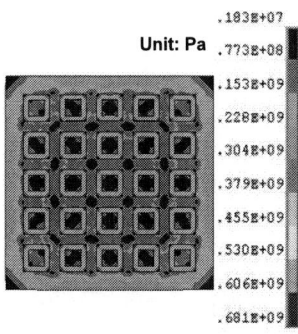

Figure 9. Thermal stresses in single-sided ceramic wafer at 3.5 s.

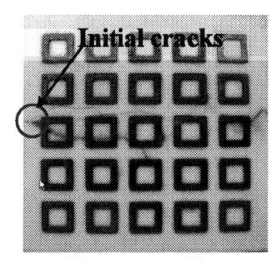

Figure 10. Cracks caused by thermal-stresses in single-sideed ceramic wafer with cu-loops.

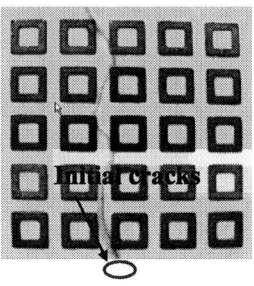

Figure 11. Cracks caused by thermal-stresses in single-side ceramic wafer. 100-μ m Pb63Sn37 is electroplated on Cu-loops.

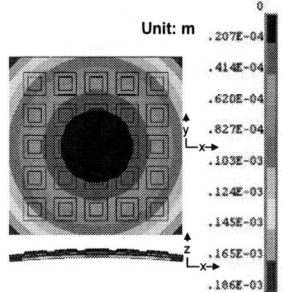

Figure 12. warpages of single-side ceramic wafer at 3.5 s induction heating. Four corners have maximal displacement.

References

1. Schmidt, Martin A., "Wafer-to-wafer bonding for microstructure formation Proceedings", IEEE Vol. 86 (1998), pp. 1575-85.

2. Wen, H. Ko., "Review: Packaging of microfabricated devices and systems Materials chemistry and physics", Vol. 42 (2000), pp. 169-75.

3. Tong, Q Y and Gosele, U., "Semiconductor wafer bonding:Science and technology (New York: John Wiley & Sons) 1999.

4. Chen, M.X., Liu, S. and Gan, Z.Y., "Selective induction heating for microsystem packaging", IEEE International Conference on Electronics Packaging Technology, 2006.

5. Chen M. X., "Research on MEMS packaging by induction heating and its appliction", Ph. D Dissertation.

6. Chen, M. X., et al., "Selective indution heating for sensor packaging", Proc. of ICEPT2005.

7. Yang, H.A., Wu, M.C. and Fang, W.L. "Localized induction heating solder bonding for wafer level MEMS packaging", J. Micromech. Microeng., Vol. 15 (2005), pp. 394-9.

8. Chen, M.X., Yuan L.L. and Liu S., "Research on low-temperature anodic bonding using induction heating", Sensors and Actuators A. Vol. 133 (2007), pp. 266-269.

9. Zhang, X.X. and Raskin, J.P., "Low-Temperature wafer Bonding: A study of void formation and influence on bonding strength", J. MEMS, Vol. 14 (2005), pp.368-382.

10. Mabesa, J. R., Scott1, A. J. and Wu, X., "Localized Heating/Bonding Techniques in MEMS Packaging", Proceedings of SPIE 5804 (2004), pp.700-5.

11. Lin, L.W., "MEMS Post-Packaging by Localized Heating and Bonding", IEEE Transaction on adv. Packaging, Vol. 23 (2000), pp.608-16.

12. Thompson, K., "Direct silicon-silicon bonding by electromagnetic induction heating", J. Microelectromech. Syst, Vol. 11 (2002), pp.285-292.

13. Yang, H.A., Wu, M.C. and Fang. W.L., "Localized induction heating solder bonding for wafer level MEMS packaging", J. Micromech. Microeng, Vol. 15 (2005), pp.394-99.

14. Liu, S. et al., "Investigation of crack propagation in ceramic/adhesive/glass system", IEEE Trans. Comp., Packag.,Manufact. Technol., Vol. 18 (1995), pp.627-633.

15. Liu S., et al., "Bimaterial interfacial crack growth as a function of modemixity", IEEE Trans. Comp., Packag., Manufact. Technol., Vol. 18 (1995), pp.618-626.

16. Liu, S. and Y. Mei, "Behaviors of delaminated plastic IC packages subjected to encapsulation cooling, moisture absorption and wave soldering", IEEE Trans. Comp., Packag., Manufact. Technol., Vol. 18 (1995), pp. 634-645.

17. Clendenin, J., et al., "Microwave bonding of silicon dies with thin metal films for MEMS applications", IEEE, 2003 ECTC.

Synthesizing SPICE-Compatible Models of Power Delivery Networks with Resonance Effect by Time-domain Waveforms

[1]Chen-Chao Wang , [2]Shu-Qiang Zhang, [1]Hung-Hsiang Cheng, [1]Tzu-Chih Lin, [1]Chi-Tsung Chiu, and [1]Chih-Pin Hung
[1]Electrical Laboratory, Corporation Design Division, Corporate R&D, Advanced Semiconductor Engineering Inc.,
Kaohsiung 811, Taiwan, China
[2]Advanced Semiconductor Engineering Group, ShangHai, China
26. Chin 3rd RD., Nantze Export Processing Zon, Kaohsiung, Taiwan 811, China
alexcc_wang@aseglobal.com

Abstract

A novel time-domain approach is proposed to synthesize the broadband equivalent circuit model of the power delivery network based on time-domain reflected (TDR)/transmitted (TDT) waveforms either through time-domain reflectometry measurement or finite-difference time-domain (FDTD) simulation. The step responses of the power delivery network are represented in terms of rational functions by the generalized Pencil-of-Matrix (GPOM) method. According to the step responses, the macro-π model with each element represented by the optimum pole-residue forms is derived to model the power delivery networks. The equivalent circuits of the macro-π model are synthesized by a systematic lumped-model extraction technique. The accuracy of this approach is demonstrated both in frequency- and time-domain.

I. INTRODUCTION

In high-speed digital circuits with operating frequency increasing and working voltage decreasing, a sudden current change incurred by active devices on printed circuit board (PCB) will result in fluctuation of the high frequency voltage. The fluctuating voltage, commonly known as the ground bounce noise or simultaneous switching noise, will propagate outward between the power and ground planes. In order to predict this simultaneous switching noise effectively, a simple and accurate two-port power/ground-planes model is warranted. Many numerical methods have been developed to analyze the power/ground planes. In the early development, the power delivery network consisting of the power/ground planes was modeled as a lumped inductor or inductive network [1]. At high frequencies where the wave phenomenon is more significant, more accurate modeling of the power delivery network is necessary. Various numerical methods based on the cavity models [2] or partial element equivalent circuit method [3], among others, were proposed to model the power/ground planes. Full-wave numerical methods such as finite-element (FEM), finite-difference time-domain (FDTD) and moment methods (MoM) were also adapted to analyze the complicated resonance effect of the simultaneous switching noise [4]-[6]. The full-wave methods provide detailed information of the power/ground planes but consume a large amount of computer resources to obtain the results. Therefore, obtaining accurate broadband models in terms of lumped circuit elements of the power delivery networks advantageous for efficient PCB designs.

In this paper, a novel systematic time-domain approach is proposed for synthesizing the SPICE-compatible circuits of

the power/ground planes based on a macro-π model. Time-domain waveforms from both TDR and TDT simulations or measurements are utilized to construct a complex images representation of the step response through the generalized pencil of matrix method [7]. The macro-π model in terms of the rational function pairs is obtained through a two-port Y matrix transformation. The equivalent circuits of the macro-π model are finally synthesized by a systematic lumped-model extraction technique. The proposed method provides a fast and efficient way to extract the two-port SPICE-compatible models of the power/ground planes without the use of FFT and convolution. Accuracy of the extracted models is confirmed from comparisons with other approaches.

II. THEORY OF MACRO-Π MODEL

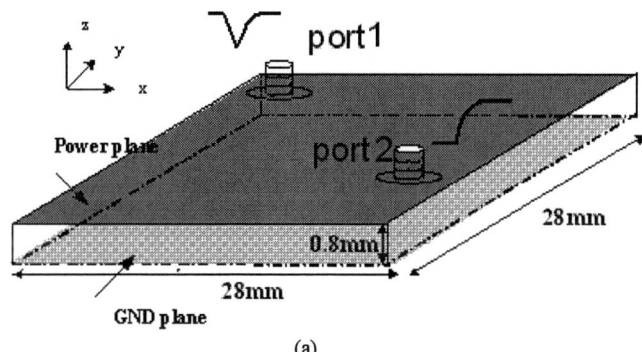

(a)

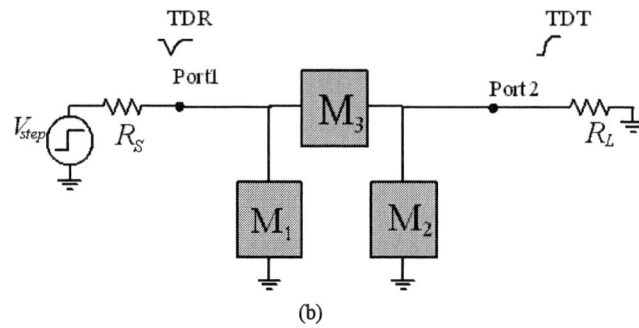

(b)

Figure. 1 (a) The investigated structure of power/ground planes in FR4 board (er = 4.3) with dimensions 28mm 28mm 0.8mm. (b) The proposed equivalent macro- model module of the power/ground planes.

(A) Simulation or measurement setup

Time-domain waveforms are employed to obtain the step response of the power/ground planes. The time domain waveforms are made available either through FDTD simulation or a TDR/TDT measurement. Fig. 1 shows an example of the power/ground planes for a two-layer board (er

978-1-4244-2739-0/08/$25.00 ©2008 IEEE

= 4.3). For the simulation setup, a resistive voltage source injects a step waveform into the power/ground planes at port1, and the port2 is for the time-domain waveform receiving. This setup is used for the power/ground planes model extraction both from the FDTD simulation or the TDR/TDT measurement.

(B) Synthesis of macro-π model

As shown in Fig. 1(a), a step source with a rise time (t_r) is injected into port1 and the reflected and transmitted voltage waveforms are recorded at port1 and port2, respectively. The step response of the power/ground planes at these two ports can be obtained by normalizing the reflected and transmitted waveforms with the incident wave amplitude directly, and are denoted as $y_1(t)$ and $y_2(t)$, respectively. By employing the GPOM, the step responses can be synthesized by the summations of complex exponential terms as

$$y_n(t) = \sum_{i=1}^{L_n} r_{i,n} \exp(-p_{i,n} t) \qquad (1)$$

where $n = 1$ or 2, $r_{i,n}$ and $p_{i,n}$ are the residues and poles in the Z-plane, and L_n is the number of exponential terms synthesized by the GPOM. The corresponding impulse response of the power/ground planes in frequency domain is obtained through the Laplace transformation of (1) as the rational functions

$$\psi_n(s) = s \sum_{i=1}^{L_n} \frac{r_{i,n}}{s + p_{i,n}} \qquad (2)$$

As shown in Fig. 1(b), the power delivery networks is synthesized as a macro-π model with three macro-elements M_1, M_2, and M_3. Through the two-port transformation and the definition of the impulse response in (2), the three macro-elements can be derived as

$$M_1(s) = \frac{1}{Z_0} \frac{(1-\psi_{11})(1+\psi_{22})+\psi_{21}^2-2\psi_{21}}{(1+\psi_{11})(1+\psi_{22})-\psi_{21}^2} \qquad (3a)$$

$$M_2(s) = \frac{1}{Z_0} \frac{(1+\psi_{11})(1-\psi_{22})+\psi_{21}^2-2\psi_{21}}{(1+\psi_{11})(1+\psi_{22})-\psi_{21}^2} \qquad (3b)$$

$$M_3(s) = \frac{1}{Z_0} \frac{2\psi_{21}}{(1+\psi_{11})(1+\psi_{22})-\psi_{21}^2} \qquad (3c)$$

, where Z_0 is the characteristic impedance of the transmission line. Finally, the macro-elements $M_i(s)$'s are used to synthesize the equivalent lumped circuits by a systematic lumped-model extraction technique. In this technique, $M_1(s)$ is first re-arranged in terms of admittance as

$$M_j(s) = s \sum_{\substack{i=1 \\ q_i>0}}^{K_0} \frac{q_i}{s+h_i} + s \sum_{\substack{i=1 \\ v_i>0}}^{K_1} \frac{r_i s + v_i}{s^2 + u_i s + m_i} + Q(s) \qquad (4)$$

, where q_i, h_i, r_i, v_i, u_i, m_i are real numbers. The corresponding circuit models for the first-order and second-order terms in (4)

are shown in Fig. 2(a) and 2(b), respectively. The corresponding values can be derived as $R_{Ci}=1/q_i$ and $C_i=1/h_i R_{Ci}$ for the first-order terms and $C_i=v_i/m_i$, $R_{Ci}=(u_i-r_i/C_i)/v_i$, $R_{Li}=1/r_i-R_{Ci}$, and $L_i=R_{Li}*r_i/(C_i*m_i)$ for the second-order terms. The remaining term $Q(s)$ in (4) is a summation of the first-order and second-order terms with negative values qi and vi, respectively. By employing the idea of voltage-control-voltage-source (VCVS), they can be synthesized as the models shown in Fig. 2(c) and 2(d), respectively, for the first-order and second-order terms. The corresponding values can be derived as $R_{Ci}=1/q_i$ and $C_i=-1/h_i R_{Ci}$ for the first-order terms and $C_i=-v_i/m_i$, $R_{Ci}=(u_i+r_i/C_i)/v_i$, $R_{Li}=1/r_i-R_{Ci}$, and $L_i=R_{Li}*r_i/(C_i*m_i)$. $V_C(s)$ and $V_L(s)$ are the voltages across the corresponding capacitor and inductor, respectively. The $M_2(s)$, and $M_3(s)$ are re-arranged in terms of admittance in (4) directly, the corresponding equivalent lumped circuits are synthesized in a similar way.

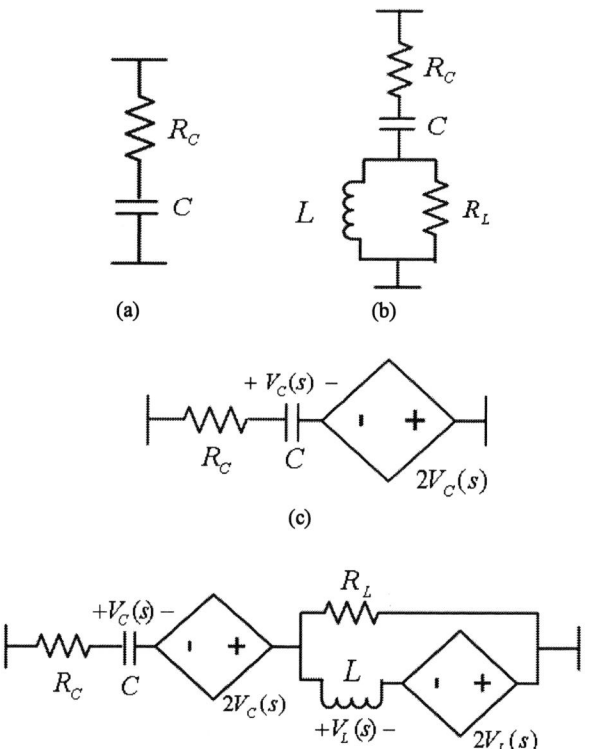

Fig. 2 Four types of equivalent model extracted from the pole-residual representations of the input impedance and admittance.

III. RESULTS

An FDTD simulation with a 0.4V step source of a 35ps rise-time was launched to port1 to obtain the time domain response at port1 and port2. Based on the proposed approach, the macro-π model of the power/ground planes can be synthesized with $M1(s)$, $M2(s)$, and $M3(s)$. The accuracy of the equivalent lumped model is checked both in time- and frequency-domain. Using the proposed approach, the equivalent model of the power/ground planes can be extracted successful. Fig. 3 shows the reconstructed time-domain waveforms using the extracted model in Agilent ADS and the original waveform computed by the FDTD method.

Difference between the two results is almost indistinguishable. For comparing the results in the frequency-domain, S11 and S21 calculated by the extracted model in ADS and direct simulation from Ansoft HFSS is shown in Fig. 4. Both the magnitude and phase agree reasonably well. It is noted that the agreement both in time and frequency domain is also good for the other two ports response. They are not shown here due to the space limit.

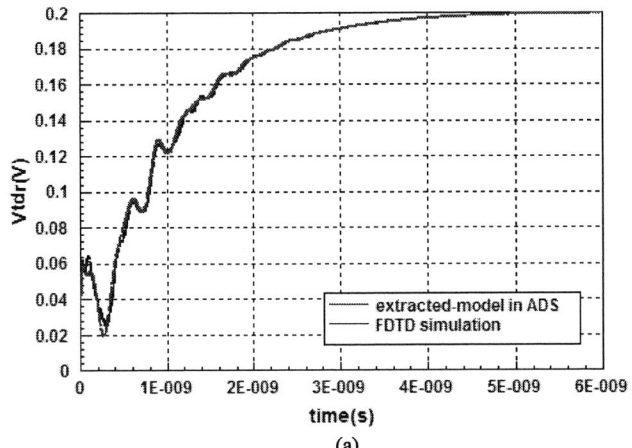

(a)

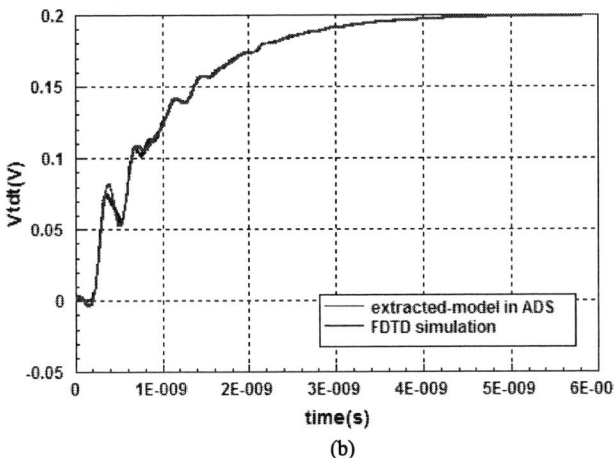

(b)

Fig. 3 Comparison of time-domain response between the extracted lumped model in ADS and the direct FDTD simulated results at (a) port1 (b) port2.

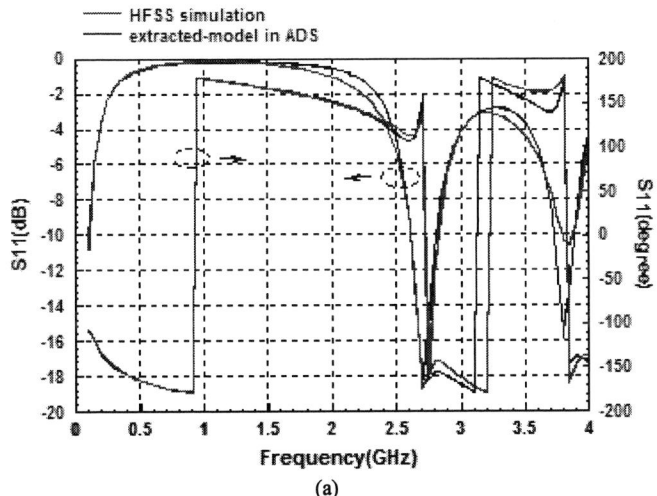

(a)

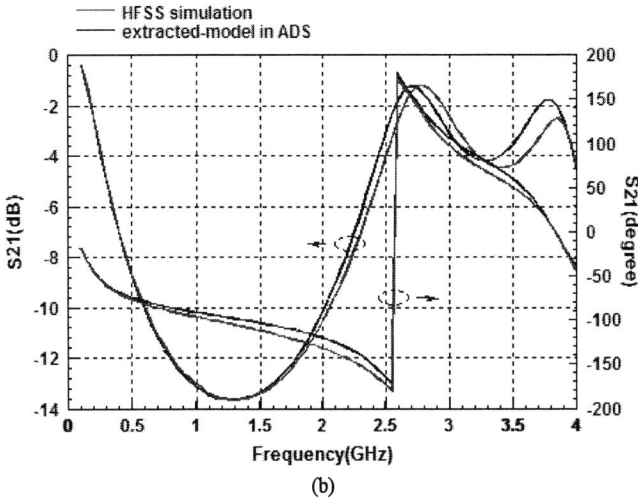

(b)

Fig. 4 Comparison of S-parameters between the extracted lumped model in ADS and the Ansoft HFSS simulated results. (a) S_{11} and (b) S_{21}.

Conclusions

A novel systematic time-domain approach is proposed for synthesizing the SPICE-compatible circuits of the power/ground planes based on a macro-π model. According to either the measured or simulated time-domain waveforms, an efficient and systematic approach is proposed to extract the SPICE-compatible models of the power delivery networks. The extracted models accurately account for the resonance behavior of the power/ground planes over a wide frequency band. Accuracy of the method was verified by comparison with commercial tool. The extracted models can be efficiently incorporated into the HSPICE simulator for the ground bounce noise in high-speed circuits.

References

1 A. E. Ruehli and H. Heeb, "Circuit models for three-dimensional geometries including dielectrics", IEEE Trans. Microw. Theory Tech., vol. 40, pp. 1507-1516, July 1992.

2 N. Na, J. Choi, S. Chun, M. Swaminathan, and J. Srinivasan, "Modeling and transient analysis of planes in electronic packages," IEEE Trans. Adv. Packag., vol. 23, no. 3, pp. 340–352, Aug. 2000.

3 W. Pinello, A. C. Cangellaris, and A. Ruehli, "Hybrid electromagnetic modeling of noise interactions in packages electronics based on the partial-element equivalent-circuit formulation," IEEE Trans. Microwave Theory Tech., vol. 45, pp. 1889–1896, Oct. 1997.

4 S. Van der Berghe, F. Olyslager, D. De Zutter, J. De Moerloose, and W. Temmerman, "Study of the ground bounce caused by power plane resonances," IEEE Trans. Electromagn. Compat., vol. 40, no. 2, pp. 111–119, May 1998.

5 J. Yun, and T. H. Hubing, "On the modeling of a gapped power-bus structure using a hybrid FEM/MoM approach," IEEE Trans. Electromagn. Compat., vol. 44, pp. 566-569, Nov. 2002.

6 W. Shi and J. Fang, "New efficient method of modeling electronics packages with layered power/ground planes," IEEE Trans. Adv. Packag., vol. 25, no. 3, pp. 417–423, Aug. 2002.

7 T. K. Sarkar, and O. Pereira, "Using the matrix pencil to estimate the parameters of a sum of complex exponentials," IEEE Antenna and Propag. Mag., vol. 37, pp. 48-55, 1995

Optimization of CAVLC Algorithm and Its FPGA Implementation

Xu Meihua1, Li Ke 1, Xuan Xiangguang2, Fan Yule1

1. School of Mechatronical Engineering and Automation, Shanghai University
2. Technology Center of Shanghai FeiLo Limited Company
Campus P.O.B. 110, 149 Yanchang Rd., Shanghai 200072, China
mhxu@staff.shu.edu.cn

Abstract

As a new generation of video frequency coding standard, H.264/AVC is excellent in compression performance, while its complexity is much higher than common encoder. Based on the detailed analysis of CAVLC algorithm, this paper first points out the "bottleneck" of CAVLC encoder implementation, then presents the optimization scheme for the major modules of CAVLC encoder, which includes VLC table prediction with multiple reference blocks, fast look-up table matching, and arithmetic eliminating method etc. It is successfully synthesized and simulated with EDA tools and implemented in FPGA of Cyclone II EP2C20F484, and the speed of the coding module is up to 165MHz. The experimental results show that the improved design scheme will be helpful to achieve the real-time processing purpose by saving the hardware resource together with the increasing coding rate.

1. Introduction

With the development of network technique in recent years, the bandwidth of communication channel has been improved. But this change still cannot meet the channel requirements from different kinds of video frequency transmission. Thus, the video frequency compression technique[1][2] becomes an attractive research problem.

As a new generation of video frequency coding standard devised by ISO/IEC and ITU-T, H.264/AVC[3][4] aims at designing and developing an understandable video frequency compression scheme. It can adapt to video frequency transmission in different networks together with raising the compression ratio. The basic system of H.264/AVC is on-limits, which can adapt to the use of IP and wireless network commendably. Compared with the previous video frequency coding standard, H.264/AVC introduced plenty of advanced kernel techniques, whose compression performance was greatly increased. It has an extensive application future. There are two kinds of entropy coding method adopted by remain data in H.264/AVC. One is CAVLC (Context-based Adaptive Variable Length Coding), and the other is CABAC (Context-based Adaptive Binary Arithmetic Coding).

The design of CAVLC encoder includes NC module, Statistic Buffer and Encoding Phase. The Statistic Buffer consists of various counters and buffers, while the coding section includes 5 sub-coding modules. Multiple reference blocks in VLC table prediction is employed for NC module design in this paper, which promotes the forecasting of Coeff_token united character code table. The fast LUT matching (FLM) and arithmetic table elimination (ATE) are adopted to design Total_token and Total_zeros modules instead of LUT matching. It not only saves the hardware resource, but also increases the coding rate. In the design of Level and Run_before coding modules, two characters can be coded simultaneously to increase coding rate with Forward-based Parallel Coding method. The speed of function module exceeds 150MHz, which can greatly improve the real time performance for CAVLC encoder, and its design scheme and experimental result will contribute to further research and development of relevant products.

2. Architecture of CAVLC Encoder

CAVLC[5][6][7] is a residual data coding for brightness and chroma in H.264/AVC. The research of CAVLC algorithm indicates that the coding process of CAVLC is just the one of 5 syntax elements which are Coeff_token, Trailingone_sign_flag, Level, Total_zeros and Run_before. Coeff_token syntax element coding, among syntax elements, is the united coding of Total_coeffs and TrailingOnes. The symbol of TrailingOnes is coded by Trailingone_sign_flag syntax element coding. Level syntax element coding encodes the range of Total_coeffs except TrailingOnes. Total_zeros syntax element coding is the one that codes the number of zeros before the last Total_coeffs. Run_before syntax element will code the number of zeros before every Total_coeffs. Five syntax elements coding can be executed concurrently, so that this paper presents a parallel architecture of CAVLC encoder whose framework is shown in Fig. 1.

For the character coding of CAVLC is operated conversely, a reverse scan method is used to avoid unnecessary resource wastes such as FILO buffer. Zig-Zag Scan module takes charge of converse zig scan for every 4×4 module. NC module is utilized to calculate the value of the current block. TC Counter can calculate the number of Total_coeffs contained by a 4×4 block. T1 Counter calculates and inspects the number of Tail_one coefficients and Level coefficients included by 4×4 data block. Then the detected Tail_one symbol is written into Trailones Reg and Level symbol is written into SIPO Level buffer, which makes preparations for the implementation of T1 Encoder and Level Encoder parallel coding subsequently. The number of every zero coefficients after the first Total_coeffs in a 4×4 data block scan-order is calculated by TZ Counter. Run Counter calculates the number of zero after every Total_coeffs in a 4×4 data block, and then sends the outcome to the SIPO Runbef buffer, which makes preparations for Run-before character coding afterwards. Finally, the code stream is exported through MUX module.

978-1-4244-2739-0/08/$25.00 ©2008 IEEE

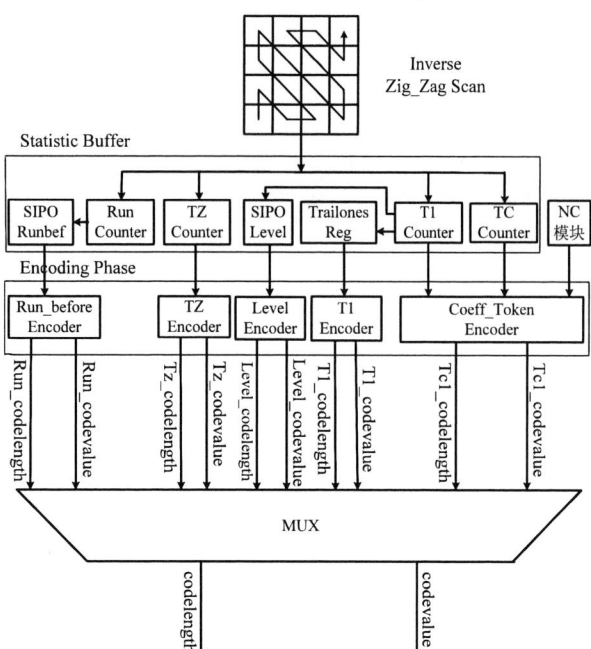

Fig.1 Framework of CAVLC encoder

3. Optimizing Design for NC module

The disadvantage of CAVLC algorithm in H.264/AVC standard is that the probability of choosing VLC code table correctly is too small. For the sake of enhancement of efficiency for VLC code table choosing, NC module is optimized in this paper. In some cases, for example, when one reference block (left block or up block) is available, or the current block is at the edge of a mobile. If two reference blocks (the left block and right block of current block) are used to forecast the number of Total_coeffs of the current block, there is little probability of correctly forecasting the code table. Thus, this paper proposes multiple reference blocks in VLC table prediction, namely, NP in previous frame, NL and NU in current frame are utilized to predict the number of Total_coeffs coefficients. The relationship between NP and three reference blocks is shown in Fig. 2 and Tab. 1.

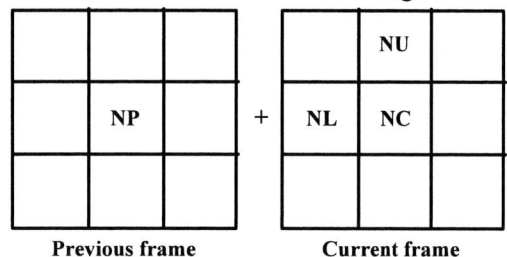

Fig.2 Predicted block of NC

Tab.1 Relationship between NC and its predicted block

NP	NL	NU	NC
X	X	X	(NP+NL+NU)/3
X			NP
	X		NL
		X	NU
			0

4. Coeff_token Coding Module with High Efficiency

In CAVLC algorithm of H.264/AVC standard, two syntax elements need to be encoded by LUT, which are Coeff_token and Total_zeros. However, the coding method based on LUT needs plenty of memory resources. Memory accessing goes against real-time processing, because it not only needs big power drain, but also influences the coding speed. So the coding method which can reduce the number of memory accessing must be adopted. Based on the analysis and summarization of Coeff_token and Total zeros coding table, the method of FLM and ATE combination instead of LUT used to achieve optimal design for Coeff_token and Total_zeros coding modules is presented in the paper.

The idea of the fast LUT matching is to divide the original LUT into the small LUTs, then parallel distribute the address for every small LUTs, only one of which is available. The application of FLM can reduce the LUT area, and increase the speed of LUT at the same time.

Arithmetic table elimination is a coding method which employs arithmetic operation instead of LUT. With the adoption of arithmetic table elimination, the hardware consumption can be greatly decreased and the coding speed can be raised as well.

The syntax element Coeff_token can be coded by ATE when $8 \leqslant NC$. Its arithmetic expression is shown as follows:

$$Codevalue = \begin{cases} 3 & , \ if TC = 0 \\ \{TC-1, T1\} & , \ if TC \ != 0 \end{cases}$$

$$Codelength = 6$$

In the expression, TC represents Total_coeffs, and Tl represents Trailingones.

The structure figure, simulation figure and sequence figure of Coeff_token coding module are shown in Fig. 3, 4 and 5.

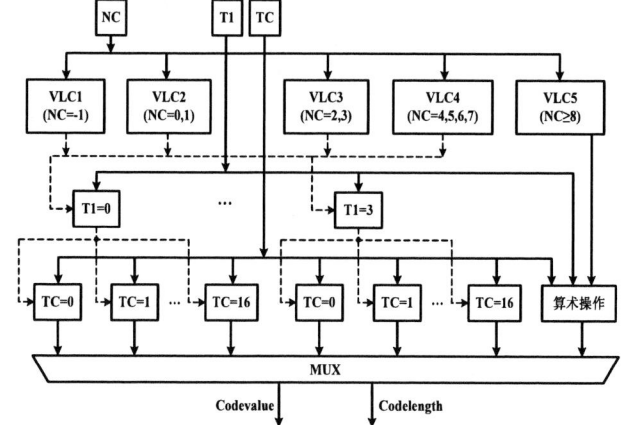

Fig.3 Coeff_token coding module structure

The FLM and ATE methods are applicable for syntax element Total_zeros as well.

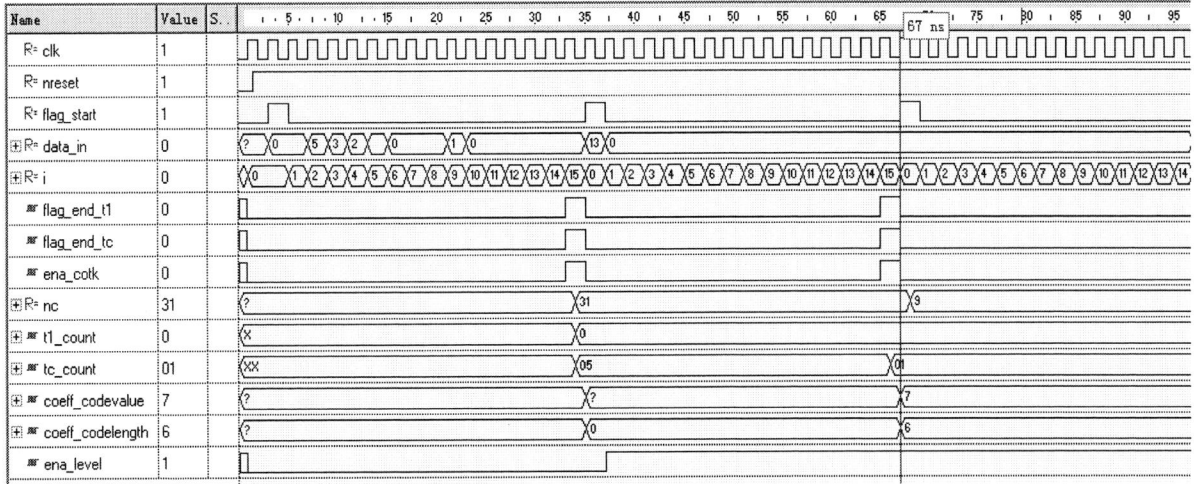

Fig.4 Coeff_token coding module simulation

Starting Clock	Requested Frequency	Estimated Frequency	Requested Period	Estimated Period	Slack	Clock Type	Clock Group
clk	120.0 MHz	86.7 MHz	7.143	11.528	2.193	declared	default_clkgroup_0

Fig.5 Coeff_token coding module sequence

5. Fast Level Coding Module

Since the clock period of Level syntax element coding is determined by the number of Level elements in a 4×4 block, the time spent on Level syntax element in different 4×4 blocks is diverse. For the reason that before the coding of a 4×4 block, it must be Zig-Zag scanned, and the number of Total_coeffs in a 4×4 block is variable, the clock period of a 4×4 block coding may be more than its scanning clock period, which brings instable throughput rate. Moreover, each Level element character coding has context-based adaptive, namely, the coding of next character is related to the current coded character, thus, the characteristic among every characters of Level element restricts the parallel coding among them. For the solution of this problem, Forward-based Parallel Coding is presented to operate the parallel coding for two Level characters. The precondition of FPC is that Writing Serial-Reading Parallel Buffer must be adopted, because SIPO can export two Level characters at one time, which can carry out parallel coding.

The proposed FPC method and SIPO method also can be applied for Run_before syntax element. The example of SIPO is shown in Fig. 6.

From Fig. 6 we can see that the scanned character is written into SIPO serially. Then, the coding unit read two characters from SIPO concurrently to achieve parallel coding together with the improvement of CAVLC coding speed.

The structure figure, simulation figure and sequence figure of Level coding module are shown in Fig. 7, 8and 9.

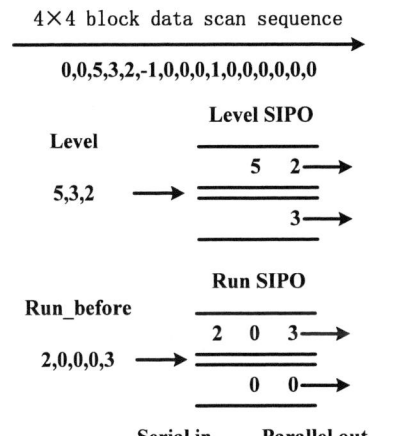

Fig.6 SIPO Example

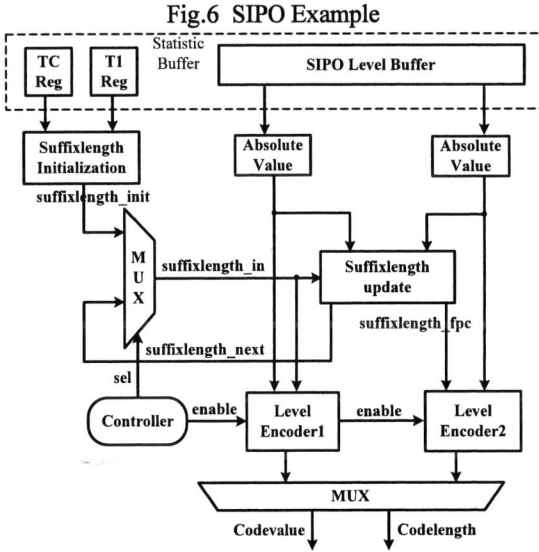

Fig.7 Level coding module structure

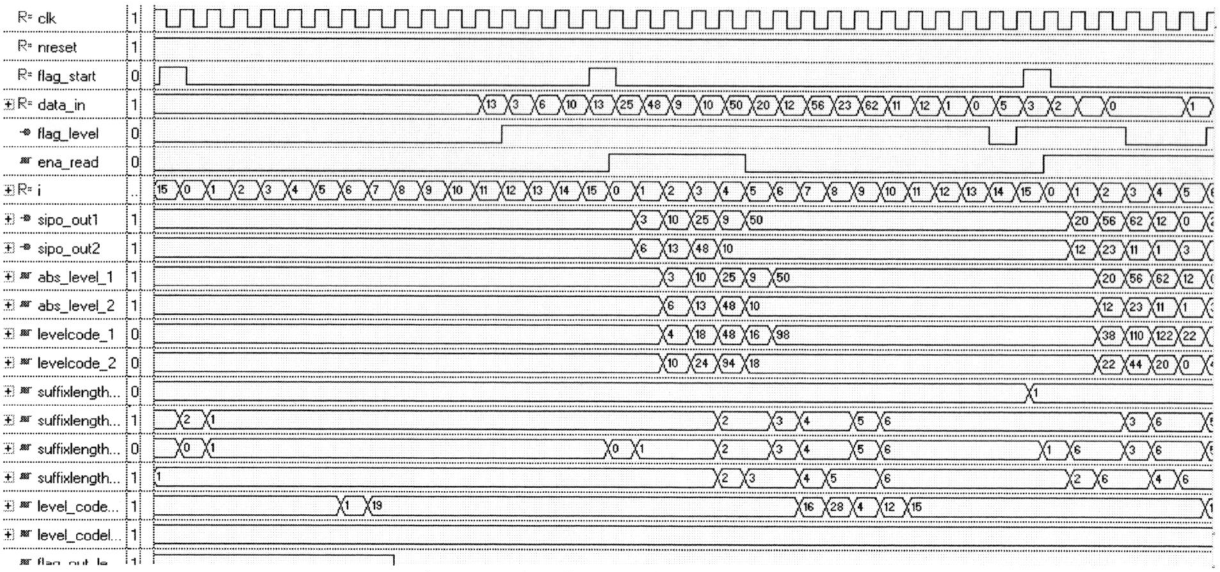

Fig.8 Level coding module simulation

```
00539  Worst slack in design: 0.064
00540
00541                 Requested     Estimated     Requested     Estimated
00542  Starting Clock  Frequency     Frequency      Period        Period        Slack
00543  ----------------------------------------------------------------------------------
00544  clk            150.0 MHz     153.0 MHz       6.667         6.538         0.064
00545  ==================================================================================
```

Fig.9 Level coding module sequence

Conclusions

As a new generation of video frequency coding standard, H.264/AVC is excellent in compression performance, but its complexity is higher than common encoder. Based on the research of H.264/AVC, the paper analyzed CAVLC algorithm in detail, and pointed out the "bottleneck" of CAVLC encoder implementation. Then, the optimization designs for each sub-module of CAVLC encoder are presented. Finally, the main coding units of CAVLC has achieved in FPGA.

Every coding unit in the design is only verified on FPGA. The realized highest clock frequency is up to 165 MHz approximately. If the same hardware structure with better technique library and design restriction is transplanted on ASIC, higher clock frequency and better performance must be obtained. Thus, the research can be used by the design of H.264 encoder chip for reference.

Acknowledgments

The authors would like to acknowledge the National Natural Science Foundation of China for providing financial support for this work under Grant No. 60773081 and Grant No. 60777018, and also to acknowledge the financial support by the Shanghai Municipal Committee under Grant No.AZ028

References

1. T.Wiegand, "Study of Final Committee Draft of Joint Video Specification Draft2," Doc. JVT-F100d2, *Joint Video Team(JVT) of ISO/IEC MPEG&ITU-T VCEG*, Dec. 2002.

2. T.Stockhammer, Hannuksela, T.Wiegand, "H.264/AVC in wireless Environments," *IEEE Transactions on Circuits and Systems For Video Technology*,Vol. 13, No. 7, (July 2003), pp. 657-673.

3. Wiegand, T., Sullivan, G., Bjontegaard, G., and Luthra, "An Overview of the H.264/AVC Video Coding Standard," *.IEEE Transactions on Circuits and Systems For Video Technology*, Vol. 13, N0. 7, (July 2003), pp. 560-576.

4. O. Peng ,J. Jing, "H.264 Codec system-on-chip design and verification," *5th IEEE International Conference on ASIC*, Vol. 2, Oct. 2003, pp. 922-925.

5. G. Bjoietegaard, K. Lillcvold, "Context-adaptive VLC (CAVLC) coding of coefficients," Doc.JVT-028, *JVT of ISO MPEG&ITU VCEG, 3rd Meeting*, Rairfax, Virginia, USA, May. 2002.

6. Zhu-Dong Dong, Dai-Qiong Hai. "Improvement of CAVLC code LUT algorithm in H.264 encoder.," *Television Technique*, Vol. 1(2004), pp. 26-27.

7. Zhang Ding, Zhang Ming, Zhang Jin, Zheng Wei. "A new kind of Adaptive Variable Length Coding algorithm," *Zhe Jiang Unierisity Transaction*, Vol. 40, No. 5, May.2006, pp. 783-786.

Thermal Management and testing of MCM with embedded chip in Silicon Substrate

Fei Geng, Jia-jie Tang, Le Luo

ShangHai Institute of Microsystem and Information Technology, Chinese Academy of Sciences, Shanghai, China,
Changning Road 865, Email: gengfei@mail.sim.ac.cn, 021-62511070-5464

Abstract

In this paper the thermal management and testing of MCM with embedded chip in Si substrate was performed. Regarding the interconnection structure of the module, a benzocyclobutene (BCB) film covers the multichip modules as dielectric layer and Au soldered balls are utilized for multi-layer vertical interconnection. The thermal resistance and junction temperature are simulated and tested respectively. The results show that the thermal resistance is between 0.7℃/W and 1℃/W dependent on the use of a additional heat sink and the optimization of the parameters

1. Introduction

In order to meet high power and packaging density, microelectronic packaging technology has developed from dual-in-line technology, surface mount technology, ball grid array to 3D Multichip Modules (MCM). The 3D packaging technology leads to high concentration of heat source elements, which results in high thermal flux at the junction of packaging structure. Thus, thermal design and management play an important role in 3D multi-layer packaging technology. There has been ways to test the thermal resistance of traditional packaging structures, such as BGA, CSP and Flip-chip, whose thermal resistance value normally higher than 20℃/W[1]. In order to meet the demand of higher thermal dissipation and conduction, several novel cooling methods, such as forced air cooling and compact liquid cooling solution were applied in packaging system.

In this paper, a new 3D silicon substrate multi-layer packaging technology has been presented, which uses bumps and spun-on BCB films to eliminate the need for laser drilling and via plating. A silicon substrate with micro-machined cavities is used. Monolithic microwave IC's (MMIC's) are embedded in recessed cavities. A BCB film covers the multichip modules as dielectric layer and Au solder balls are utilized for multi-layer vertical interconnection. At present, international researches mainly focus on two-layer BCB packaging structure, and BCB usually is used for flatting[3-5]. For multi-layer interconnection (above three), BCB films is normally etched through a photoresist mask by plasma etching or reactive ion etching (RIE)[6]. And then metal deposition and electric plating technology are applied[7]. 3D multi-layer packaging technology is more complicated, intractable, and needs pay more attention to its thermal management[8-9].

Packaging configuration, dielectric material, substrate material and monolithic microwave IC's (MMIC's) relative position make a full impact on the thermal design and management of this packaging structure[10-11]. These factors are optimized to get an ideal structure with lowest thermal resistance. In this paper, the commercial simulation software FLOTHERM 4.2 is used for thermal simulation of this 3D packaging structure. In order to track thermal transmission in different materials the VOF model is applied. According to the datum of temperature changes in the interface between BCB, silicon substrate and the power module, the thermal resistance of this 3D multi-layer packaging structure will be calculated. In this paper, two types of thermal testing chip are fabricated for thermal resistance testing. One testing chip is embedded in silicon substrate used as power module. The other is mounted on the BCB film and exterior surface of the packaging structure. The testing chips are used as temperature-sensing elements which can measure the temperature changing in different places accurately.

Both in simulation and experiment process, chip relative position, power, size of heat sink and packaging materials are all key factors to thermal design of this packaging structure. In order to optimize the experiment results, these factors are changed in the simulation process. The simulation and experiment results show that the thermal resistance can be controlled about 1℃/W with optimized parameters. And the temperature of the power module can be controlled below 80℃.The thermal resistance becomes lower with increasing of substrate thickness and heat sink size. The testing chips fabricated in the experiment process can be sensitive to temperature changes in different interface.

2. Concept of thermal resistance

2.1 Definition of thermal resistance

Thermal resistance is used to evaluate electronic modules' thermal performance. Accurate measurement of MCM modules' thermal resistance is significant for microelectronic packaging technology. The concept of thermal resistance can be expressed by the following equation [12]:

$$\theta_{JX} = \frac{T_J - T_X}{P} \qquad (1)$$

Where:

θ_{JX} is the thermal resistance from junction to specific position;

T_J is the junction temperature;

T_X is the temperature of appointed position;

P is the power dissipation of the packaging structure

2.2 Principle of thermal resistance testing

Because of 3D embedded packaging structure, it is hard to measure the junction temperature and thermal resistance. We design and fabricate a kind of testing chip in which integrates diode and film resistance. The characteristic of diode that forward voltage drops with temperature increment can help to test junction temperature of this packaging structure.

The Electrical Test Method, described herein, makes use of a temperature-sensitive parameter (TSP) to sense the change in temperature of the junction operating area due to the application of electrical power to the device-under-test (DUT). In equation terms (2),

$$\Delta T_J = K \cdot \Delta TSP \quad (2)$$

Where

TSP = change in temperature-sensitive parameter value [mV]

K = constant defining relationship between changes in T_J and TSP [C/mV]

3. Geometrical model and numerical simulation

3.1 Geometrical model of the packaging structure

The commercial simulation software FLOTHERM 4.2 is used for thermal simulation of this 3D packaging structure. Geometrical model is shown in Fig.2

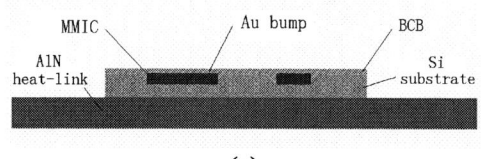

(a)

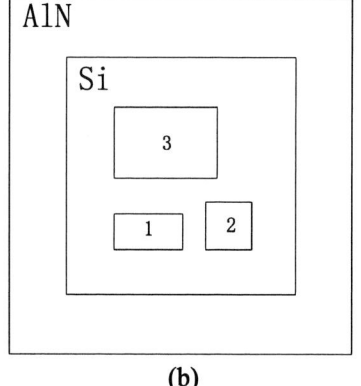

(b)

Fig.2 Geometrical model of MCM
Modules: (a) lateral view; (b) top view

Three MMIC are embedded in silicon substrate. Chip 1 and 3 are power amplifier and the external dimension is $3 \times 1.5 \times 0.1$mm and $4.5 \times 3 \times 0.1$mm respectively. Chip 2 is broadband mixer and the external dimension is $2 \times 2 \times 0.1$mm. The distance between chip 1 and 2 is 1mm. The distance between chip 2 and 3 is also 1mm. The ambient temperature is 35 ℃.The thermal conductivity of specific material is shown in Tab.1

Tab.1

Materials	AsGa	BCB	Si	AlN
Thermal conductivity (W/mk)	48.30	0.29	117.50	170.00

The institution of the boundary condition as following: thermal sources are chip 1 and 3 whose power dissipation is 0.1W and 1W respectively. The thermal conduction among interior materials follows Fourier law. The thermal is transferred to ambient through convection and radiation. And the thermal conductivity is 5W/ (m².K) in the condition of natural convection.

3.2. Discussion of simulation results

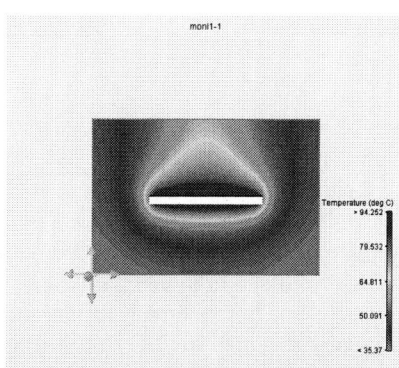

Fig 3 Temperature contour of this packaging structure

Tab.2 temperature in different interface

Cuboid Name	Minimum (deg C)	Maximum (deg C)	Mean (deg C)	Stand.Dev. (deg C)	Volume (m^3)
GaAs	94.1547	94.2524	94.2087	0.0364647	4.49999e-010
bcb2	94.1612	94.2481	94.2096	0.0361173	9.00007e-011
si	93.7657	94.2206	93.9287	0.112096	1.7325e-008

The results shown in Fig.3 and Tab.2 have reflected: the maximum temperature is 180.59℃; the mean temperature of BCB film's top surface (T_B) is 178.36 ℃ ; the mean temperature of AsGa chip's two sides is 178.36℃(T_{G1}) and 178.78 ℃(T_{G2}); the mean temperature of silicon substrate's lower surface(T_S) is 178.84℃. The thermal resistance of this packaging structure θ is composed of two parts θ$_{GS}$ and θ$_{GB}$. θ$_{GS}$ is the thermal resistance from junction to silicon substrate's exterior surface. θ$_{GB}$ is the thermal resistance from junction to BCB film's exterior surface. Accord to equation (1):

$$\theta_{GB} = \frac{T_{G1} - T_B}{P} = \frac{178.36 - 178.26}{0.14418} = 0.6936°C/W$$

$$\theta_{GS} = \frac{T_{G2} - T_S}{P} = \frac{178.78 - 178.74}{0.95007} = 0.0421W°C/W$$

$$\theta = \theta_{GB} + \theta_{GS} = 0.4546 + 0.0944 = 0.7357°C/W$$

The calculation results show that thermal resistance of this packaging structure is less than 1℃/W. But the junction temperature is close to 180℃ which is higher than safe temperature. Because of BCB's low thermal conductivity capability, the thermal resistance mainly concentrates on BCB films. And the thermal resistance offered by silicon substrate is only 5.7% of the total thermal resistance.

In order to reduce the junction temperature, AlN substrate is applied as heat sink. Its external dimensions is $30 \times 30 \times 0.2$mm. The results after adding AlN substrate are shown as following:

$$\theta_{GB} = \frac{T_{G1} - T_B}{P} = \frac{103.71 - 103.69}{0.029838} = 0.6703°C/W$$

$$\theta_{GS} = \frac{T_{G2} - T_S}{P} = \frac{103.78 - 103.74}{1.06907} = 0.0374°C/W$$

$$\theta = \theta_{GB} + \theta_{GS} = 0.6703 + 0.0374 = 0.7077 = 0.7077$$

Because of adding AlN substrate, the mean junction temperature reduce to 103 ℃ which meet operating requirement. In order to further studies on the changes of thermal resistance, the distance between chip 1 and 3 is increased to 4.5mm. And the calculation results are shown as following:

$$\theta_{GB} = \frac{T_G - T_B}{P} = \frac{178.3 - 178.2}{0.14397} = 0.6946°C/W$$

$$\theta_{GS} = \frac{T_{G2} - T_S}{P} = \frac{178.69 - 178.66}{0.9503} = 0.0316°C/W$$

$$\theta = \theta_{GB} + \theta_{GS} = 0.6946 + 0.0316 = 0.7262°C/W = 0.$$

The results show that the change of power chips' distance in small range can only change the thermal resistance and junction temperature a little.

4. Discussion of experimental method and results

4.1 testing method

The thermal resistance testing structure is shown in Fig.5. Testing chip 2 is mounted on silicon substrate with silver epoxy, which is used for power dissipation and temperature testing. And Test chip 1 is mounted on BCB film, which is only used for accurate temperature testing. The thermal resistance of the packaging structure can be calculated base on equation (1).

As shown in Fig.4, pin 1 and 3 are input in constant voltage source; pin 5 and 6 are input in constant current source; pin 2 and 4 are output port.

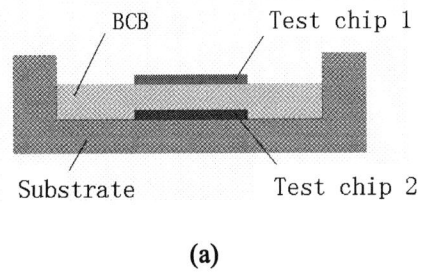

(a)

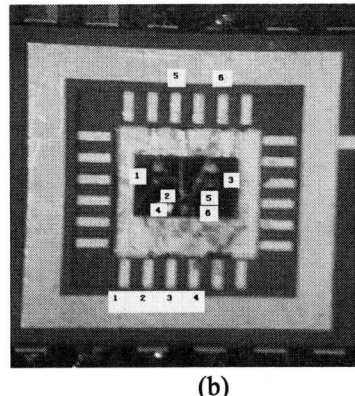

(b)

Fig.4 thermal resistance testing structure

The testing chip fabrication process is shown as follows: silicon oxidization; phosphorus diffusion: boron diffusion; fabrication of interconnection holes; annealing and alloying.

The diagrammatic sketch of fabrication process is shown in Fig.5. And the finished testing chips are shown in Fig.6.

Fig.5 Diagrammatic sketch of fabrication process

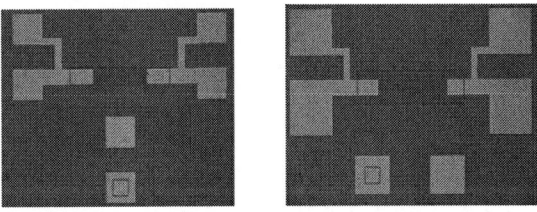

Fig.6 Thermal resistance testing chips

4.2 Testing system

Testing system mainly contains electrical convection oven, ATTEN constant voltage source, Agilent 34970A and Keithley MTS 4500.The maximum temperature of the electrical convection oven can reach to 250℃. The maximum voltage of the constant voltage source can reach to 10V. Agilent 34970A is used for testing film resistance and forward voltage drop. And Keithley MTS 4500 is applied as a constant current source.

4.3 Testing circuit

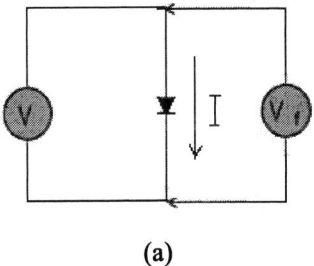

(a)

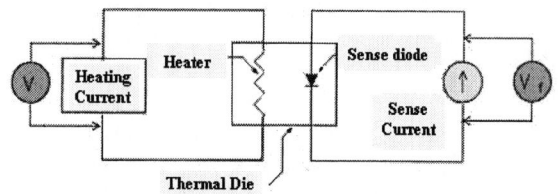

(b)

Fig. 7 Schematic circuit diagram:
(a)Temperature characteristic testing circuit;
(b)Thermal resistance testing circuit

4.4 Testing results

4.4.1 Temperature characteristic test of diode

Testing I-V characteristics curve of at different temperature on 4-probe station. The testing datum show that the forward voltage of diode drops with temperature increment when current below 15mA. When current above 25mA, diode shows positive temperature characteristic just like diffused resistor. The experimental findings show that diode presents good negative temperature characteristic when current is 1mA. So the forward current is fixed on 1mA.

Testing modules are put in a oven at 30℃,45℃,60℃,80℃,100℃ respectively. When temperature becomes stabile, the temperature characteristics can be fitted from testing datum. Testing result is shown in Tab.3

Tab.3 temperature characteristic of diode

T(℃)	34.019	47.982	62.651	82.354	102.308
V_1(mV)	804.423	772.679	738.679	693.447	650.236
V_2(mV)	828.267	796.770	763.275	718.152	675.145

Fitting curve as shown in Fig.8

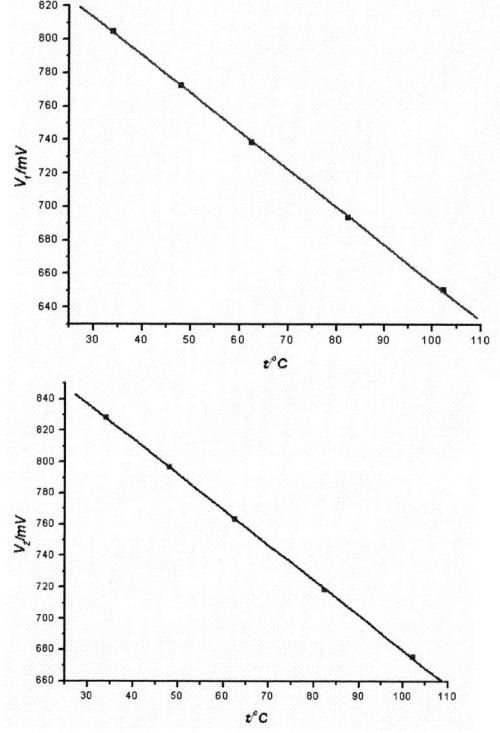

Fig.8 Fitting curves of temperature and voltage

T-V variable formula as following:

V_1=-2.26564t+881.11 V_2=-2.24995t+904.51

K_1=2.26564 K_2=2.24995

K_1 is the calibration factor of testing chip 1 and K_2 is the calibration factor of testing chip 2.

4.4.2 Thermal resistance test

When constant voltage V=7.5V, power resistor R=50.3 Ω, namely P=1.12W,thermal resistance testing result as shown in Tab.4

Tab.4

	Oven temperature(℃)	V_1(mV)	V_2(mV)
Beginning	31.195	811.618	835.010
Steady state	34.093	737.314	758.494

According to T-V Fitting curve shown in Fig.8 Junction temperature:

Tj= (904.51-758.494)/ K2=64.897℃

$\Delta V_1 = 74.304\text{mV} \Rightarrow \Delta t_1 = 32.796°C$

$\Delta V_2 = 76.516\text{mV} \Rightarrow \Delta t_2 = 34.008°C$

Thermal resistance

$$\theta_{ja} = \frac{64.897 - 34.093}{1.12} = 27.5041°C / W$$

$$\theta_{jc} = \frac{34.008 - 32.796}{1.12} = 1.082°C / W$$

θ_{ja} is the thermal resistance from chip junction to ambient air; θ_{jc} is the thermal resistance from chip junction to packaging surface.

4.4.3 Influences of heat link

In order to get the impact of heat link on thermal resistance, the capacity and thickness of AlN substrate will be change.

1) testing thermal resistance without heatlink

When constant voltage V=8.5V, I=0.17A, namely P=1.445W, change of voltage as shown in Tab.6

Tab.6

	V_1(mV)	V_2(mV)
Beginning	848.47	862.48
Steady state	728.64	746.45

$\Delta V_1 = 119.83\text{mV}, \Delta t_1 = 53.041°C$

$\Delta V_2 = 116.03\text{mV}, \Delta t_2 = 51.676°C$

Junction temperature：Tj=80.778 ℃

Steady Ambient temperature：T=27.5055 ℃

$$\theta_{ja} = \frac{80.778 - 27.5055}{1.445} = 36.867\,°C/W$$

$$\theta_{jc} = \frac{53.041 - 51.676}{1.445} = 0.945\,°C/W$$

2) Testing thermal resistance with 2.5cm × 1.3cm × 2mm AlN substrate

When constant voltage V=8.6V, I=0.2A, namely P=1.720W, change of voltage as shown in Tab.7

Tab.7

	V_1(mV)	V_2(mV)
Beginning	843.74	856.86
Steady state	731.38	748.57

$\Delta V_1 = 112.36\text{mV}$, $\Delta t_1 = 49.735°C$

$\Delta V_2 = 108.29\text{mV}$, $\Delta t_2 = 48.509°C$

Junction temperature: Tj =79.564℃

Steady Ambient temperature: T=27.3938℃

$$\theta_{ja} = \frac{79.564 - 27.3938}{1.72} = 30.332\,°C/W$$

$$\theta_{jc} = \frac{49.735 - 48.509}{1.72} = 0.713\,°C/W$$

In same way, the thermal resistance of packaging structure with different heat link dimension can be calculated as following:

3) 5cm × 1.3cm × 2mm AlN substrate:

V=8.5V, I=0.2A, P=1.7w

$$\theta_{ja} = \frac{79.538 - 27.053}{1.7} = 30.87\,°C/W$$

$$\theta_{jc} = \frac{48.287 - 45.973}{1.7} = 1.361\,°C/W$$

4) 5cm × 1.3cm × 4mm AlN substrate:

V=8.5V, I=0.18A, P=1.53w

$$\theta_{ja} = \frac{75.453 - 32.5006}{1.53} = 29.073\,°C/W$$

$$\theta_{jc} = \frac{53.041 - 51.976}{1.53} = 0.448\,°C/W$$

Experiment results:

The thermal resistance of packaging structure without heat link can be controlled between 1 and 2 ℃/W. But the junction temperature is too high. After adding AlN heat link, the junction temperature reduces to safe range and thermal resistance can be controlled below 0.75 ℃/W. And the temperature of the power module can be controlled below 80℃.The thermal resistance becomes lower following the increase in base plate thickness and heat sink size. The testing chips prepared in the experiment process can be sensitive to temperature changes in different interface.

5. Conclusion

1. The three-dimensional simulation model has been established The simulation way used in this paper can get the temperature in different interface of this packaging structure .Thermal resistance can be controlled between 0.7℃/W and 1℃/W with optimized parameters. And the temperature of the power module can be controlled below 80℃ with heat link;

2. The thermal resistance testing system has been set up. The testing result shows that thermal resistance can be controlled below 1 ℃/W, which is compatible with the simulation result. The thermal resistance becomes lower with increasing of substrate thickness and heat sink size. The testing chips fabricated in the experiment process can be sensitive to temperature changes in different interface

References

1. Teoh King Long, Goh Mei Li, K.N. Seetharamu, Ahmad Yusoff Hassan, "A fresh look at thermal resistance in electronic packages". Electronics Packaging Technology Conference[C]. 2000, 124-130

2. Mark. A. Kuhlman, Hasan Sehitoglu, "A thermal evaluation of multichip module (MCM) materials and designs". Eighth IEEE SEMI-THERM Symposium, 110-118

3. Rodrigo Carrillo-Ramirez and RobertW. Jackson, "a Technique for Interconnecting Millimeter Wave Integrated Circuits Using BCB and Bump Bonds", IEEE Microwave and Wireless Components Letters, 2003, 13(6):196-198

4. Ilse Christiaens, Student Member, "Thin-Film Devices Fabricated With Benzocyclobutene Adhesive Wafer Bonding", Journal of Lightwave Technology, 2005, 23(2): 517-523

5. J. J. McMahon, J.-Q. Lu, and R. J. Gutmann, "Wafer Bonding of Damascene-Patterned Metal/Adhesive Redistribution Layers for Via-First Three-Dimensional (3D) Interconnect", 2005 Electronic Components and Technology Conference,2005, 331-336

6. Percy B. Chinoy, Member, IEEE, and James Tajadod, "Processing and Microwave Characterization of Multilevel Interconnects Using Benzocyclobutene Dielectric", IEEE Transactions on Components, Hybrids, and Manufacturing Technology, 1993, 16(7): 714-719

7. Steven A. Vitale, Heeyeop Chae, and Herbert H. Sawin, "Etching chemistry of benzocyclobutene (BCB) low-k dielectric films in F2+O2 and Cl2+O2 high density plasmas", American Vacuum Society, 2000, 18(6):2770-2778

8. Vikram B. Krishnamurthy, Member, IEEE, H. S. Cole, and T. Sitnik-Nieters, "Use of BCB in high frequency MCM interconnects", IEEE Transactions on Components, Packaging and Manufacturing Technology, 1996, 19(1):42-47

9. Rosten, Harvey, Lasance C J M. "The Development of Libraries of Thermal Models of Electronics Components for an Integrated Design Environment". Proceedings of the IEPS Conference [C]. Atlanta, Georgia.

10. Philip E. Garrou Lwona Turlik, "Multichip Module Technology Handbook". Publishing house of Electronics industry, 2006 [M]

11. Joiner B, Adams V. "Measurement and Simulation of Junction to Board Thermal Resistance and Its Application in Thermal Modeling". Proc of SEMITHERM [C]. 1999, San Diego.

12. EIA/JEDEC STANDARD Integrated Circuits Thermal Measurement Method – Electrical Test Method (Single Semiconductor Device) EIA/JESD51-1 DECEMBER 1995 Electronic Industries Association Engineering Department.

Strain rate effect and Johnson-Cook models of lead-free solder alloys

Qin Fei, An Tong, Chen Na
College of Mechanical Engineering and Applied Electronics Technology,
Beijing University of Technology, Beijing 100124, China
Email:qfei@bjut.edu.cn, Phone: +86-10-67392173

Abstract

Drop/impact causes high strain rate deformation in solder joints of microelectronics package. It is important to understand mechanical behavior of solder joints under high strain rate for reliability design of products. In this paper mechanical behaviors of two lead-free solder alloys, Sn3.5Ag and Sn3.0Ag0.5Cu, were investigated by quasi-static tests and the split Hopkinson tension/pressure bar testing technique under high strain rate ($600\sim2200s^{-1}$). The experimental results show that the two materials are sensitive to strain rate and their dynamic flow stresses are much greater than their static flow stresses. The higher the strain rate, the greater their tensile strength but less their fracture strains. Based on the experimental data, constitutive models in the Johnson-Cook form for the two lead-free alloys were derived. The models were then used to simulate the testing process by incorporating it into ABAQUS. The good agreement between the numerical simulations and the experiments indicates that the presented Johnson-Cook models are suitable and reliable to describe dynamic behavior of the two solder alloys under high strain rates. Finally, the proposed constitutive models were used to compute peeling stress and plastic strain of solder joints under drop/impact loadings, and the results were compared with that by linear elastic model and tri-linear elastic-plastic model to show the necessity of using a strain rate dependent material model in simulation of drop/impact of solder joints.

1 Introduction

As the widely application of portable electronic products nowadays, there has increasing concern about the reliability of those products when they are subjected to accidental drop/impact. Under drop/impact loadings, solder joints in board level packages of the products, which are key parts functioned as electric connections and mechanical supports[1], experience deformation in strain rate of 10^3 s^{-1}[2]. However, there is not enough understanding to behavior of the solder joints under so high strain rates currently.

There are a few research reports on high strain rate behavior of solders. Dynamic behavior of Pb solder was investigated by Lee and Dai [3] using the split Hopkinson torsing bar technique. Wang and Yi [4] obtained the stress-strain curves of 63Sn37Pb solder at high strain rates by using the split Hopkinson pressure bar test (SHPB). More recently, the SHPB technique was used by Siviour et al.[5] to investigate stress-strain behavior of Pb and Pb-free solders at high strain rates and in various temperature conditions. However, big diversity exists in the test data from those researches. And, up to now, there is no material models being proposed which can be used to describe the high strain rate behavior of solder joints. As a result, for drop/impact simulations, most of the research works apply linear elastic or elastic-plastic material models to solder joints[6,7,8]. This models neglect the strain rate effect completely and obviously lead to inaccurate estimate of stresses and strains in solder joints.

In this paper, the stress-strain behaviors of two kinds of lead-free solder alloys, Sn3.5Ag and Sn3.0Ag0.5Cu, at strain rate range from $600s^{-1}$ to $2200s^{-1}$, were investigated by using the split Hopkinson pressure/tension bar technique, and the Johnson-Cook models for the two solders were obtained based on the test data. The Johnson-Cook model was applied to simulate behavior of the solder joints in a board level drop impact test. The strain rate, the peeling stress and the equivalent plastic strain of solder joints during the drop impact were analyzed and discussed.

2 Experimental Results and Discussion

Commercial ingots of Sn37Pb (the SnPb), Sn3.5Ag (the SnAg) and Sn3.0Ag0.5Cu (the SnAgCu) were melt and molded into casting rods. The rods were machined into specimens in specific shape and dimensions, then quasi-static (the strain rate is $0.001s^{-1}$) and dynamic (strain rate is $600\sim2200s^{-1}$) tension and compression tests were carried out respectively. The quasi-static tension/compression tests were conducted with a MTS-809 material tester. The split Hopkinson pressure/tension bar setup was designed to be capable to test dynamic behavior of materials over a range of strain rates from 10^2s^{-1} to10^4s^{-1}.

Specimens for quasi-static tension and pressure tests were in size of $\Phi8\times56$ mm and $\Phi12\times12$ mm respectively. For the split Hopkinson bar tests, the specimens have size of $\Phi12\times6$ mm for pressure and $\Phi6\times8$ mm for tension.

In the quasi-static tests, three samples for each kind of materials were tested, and then the applicable data were averaged to acquire the final data. For the split Hopkinson tension/pressure bar tests, three different level strain rates for each solder alloy and five samples for each strain rate were tested.

The quasi-static mechanical properties of three solder alloys are listed in Table 1, in which σ_s, σ_b are the yield stress and ultimate stress respectively, ε_f is the fracture strain. It can be seen that, in the quasi-static tests, there is no apparent difference of the ultimate stresses between the SnAg and the SnPb, while the yield stress of the SnAg is slightly less than that of SnPb, and the ultimate stress of the SnAgCu is obviously greater than that of the others. Dynamic mechanical properties of the two lead-free solder alloys are presented in Table 2. It indicates that the dynamic yield stress of the SnAg is quite greater than its static, and is almost as same as that of SnAgCu. This means that SnAg is much more sensitive to strain rate than the others. Figs1~4 show the true stress-strain relations of the two lead-free solders obtained from the Hopkinson tension/pressure bar tests at various strain

978-1-4244-2739-0/08/$25.00 ©2008 IEEE   171

rates. Strain softening can be observed for the two lead-free solders when the higher strain rate was applied. This can be interpreted as material thermal softening due to temperature rise during the adiabatic plastic deforming process or damage process. The elevated strain rate contributes to increasing of the tensile strength but decreasing of the fracture strain. This implies that the alloys tend to brittle failure at high strain rate.

Table 1 Quasi-static Properties (0.001s⁻¹)

Materials	Young's Modulus, E (GPa)	Yield Stress, σ_s (MPa)	Ultimate Stress, σ_b (MPa)	Fracture Strain, ε_f
SnPb	30	35	43	0.16
SnAg	45	29	42	0.20
SnAgCu	54	38	51	0.16

Table 2 Dynamic Mechanical Properties

Strain rate (s^{-1})	SnAg			SnAgCu		
	σ_s (MPa)	σ_b (MPa)	ε_f	σ_s (MPa)	σ_b (MPa)	ε_f
600	66	146	-	73	154	-
1200	72	158	0.116	83	162	0.100
1800	77	173	0.092	87	172	0.092

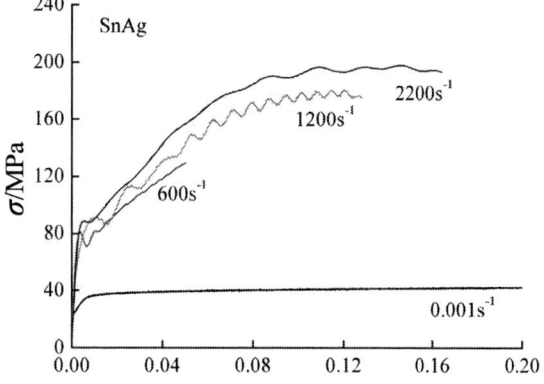

Fig.1 SnAg stress-strain curves in compression

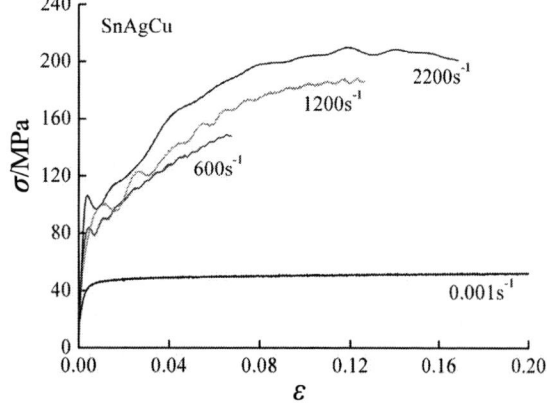

Fig.2 SnAgCu stress-strain curves in compression

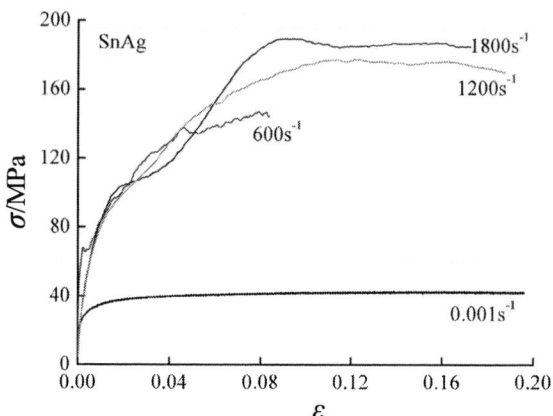

Fig.3 SnAg stress-strain curves in tension

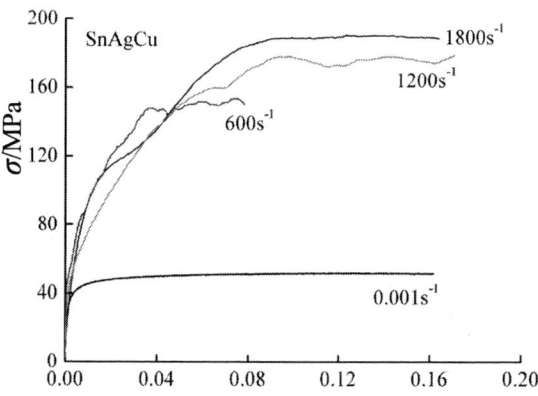

Fig.4 SnAgCu stress-strain curves in tension

3 Johnson-Cook Model for Lead-free Solders

In 1983, G. R. Johnson and W. H. Cook proposed Johnson-Cook (J-C) constitutive model[9] for metals subjected to large strains, high strain rates and high temperatures. Although empirical in nature, J-C constitutive model has been widely used to simulate large strain and high strain rate deformation process[10,11,12] because of its simplicity and fairly agreement with experimental results. A general form of J-C model can be expressed as

$$\sigma = \left[A + B\left(\varepsilon^p\right)^n \right] \left(1 + C \ln \dot{\varepsilon}^*\right) \left(1 - T^{*^m}\right) \quad (1)$$

where σ is the von Mises flow stress, ε^p is the equivalent plastic strain, $\dot{\varepsilon}^* = \dot{\varepsilon}^p / \dot{\varepsilon}_0$ is the dimensionless strain rate and $\dot{\varepsilon}_0 = 0.001$ is reference strain rate in this research. $T^* = (T - T_r)/(T_m - T_r)$ is the homologous temperature, in which T_r and T_m are the reference temperature and melting temperature of the material respectively. The reference temperature was assigned to be room temperature here. The five material constants in Eq.(1) are A, B, C, m and n. Here A is the yield stress defined by the quasi-static strain-stress data, B and n represent the effects of strain hardening, C is used to describe the strain rate effect, and m describes the effect of thermal softening.

The J-C model constants can be determined by considering the three sets of brackets of Eq.(1) separately. The first step is to set out the constants in the first set of brackets. For the quasi-static test conducted at room temperature, $\dot{\varepsilon}^* = 1$ and $T = 0$, thus Eq.(1) becomes

$$\sigma = A + B\left(\varepsilon^{\mathrm{p}}\right)^n \tag{2}$$

The logarithm conversion of the Eq. (2) is

$$\ln(\sigma - A) = \ln B + n \ln \varepsilon^{\mathrm{p}} \tag{3}$$

where A is the initial yield stress, which can be derived from the stress-strain curve obtained from the quasi-static test directly. Here, for the SnAg and SnAgCu, their A constants are 29 MPa and 38 MPa, respectively, as listed in Table 1. Then pick up the points located between the yield point and the ultimate point on the stress-strain curves, and let $x = \ln \varepsilon^{\mathrm{p}}$ and $y = \ln(\sigma - A)$, plot those points in x-y coordinate plane and fit those points by a straight line $y = b_1 + k_1 x$. The slope of the line, k_1, is identified as the material constant n. The intercept, b_1, is identified as $\ln B$.

The next step is to determine the constant C. As for $\varepsilon^{\mathrm{p}} = 0$, and $T = 0$, Eq. (1) becomes

$$\sigma/A = 1 + C \ln \dot{\varepsilon}^* \tag{4}$$

where σ is the dynamic yield stress under current strain rate, which is derived from three dynamic compression tests. The dynamic yield stresses for the SnAg and SnAgCu under different strain rates are listed in Table 2. Let $x = \ln \dot{\varepsilon}^*$ and $y = \sigma/A$, for the three strain rates 600s^{-1}, 1200s^{-1} and 2200s^{-1}, three points can be plotted in x-y coordinate plane. Fitting them by a straight line $y = b_2 + k_2 x$, then we can identified the material constant C as k_2.

By the procedure described above and the experimental data of the SnAg and SnAgCu, the J-C constitutive model constants for the two lead-free solders are derived and presented in Table 3. The material constant m given in Table 3 was not strictly based on the experimental data and was picked up according to values used for most metal materials.

Table 3 J-C model Constants

	A (MPa)	B (MPa)	C	n	m
SnAg	29	243	0.0956	0.703	0.8
SnAgCu	38	275	0.0713	0.710	0.7

4 Validation of the J-C models and Discussion

In order to validate the J-C constitutive models given in this paper, the material models defined by Table 3 were incorporated into ABAQUS finite-element software to simulate the SHPB tests. The materials and dimensions of input, output bars and specimen were exactly the same as that of the experiment. The input and output steel bars are 1000 mm in length and 16 mm in diameter. They were modeled by element type CAX4R, which is a continuum axial symmetric two-dimensional four-node quadrilateral element in ABAQUS. In the pressure test, the contact between the bar and specimen was defined as rigid contact without friction. In the tensile simulation, there was a tie constrain defined between the bar and specimen. Fig.5 shows the finite element

models used for the pressure/tension test simulations. The stress impulses applied to the simulation were derived from the real test data. Two examples of those stress impulses are plotted in Fig.6 to show main features of the stress waves.

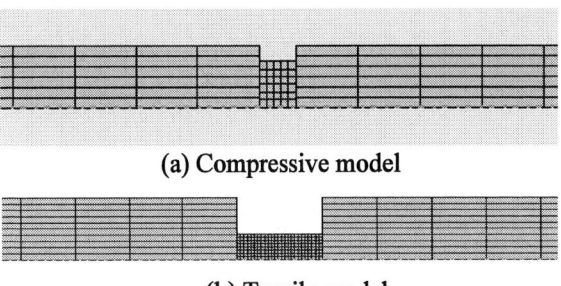

(a) Compressive model

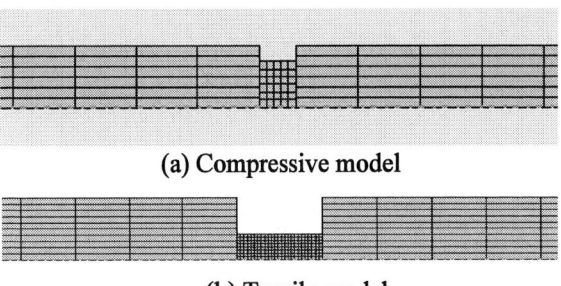

(b) Tensile model

Fig.5 Part mesh of the finite element model

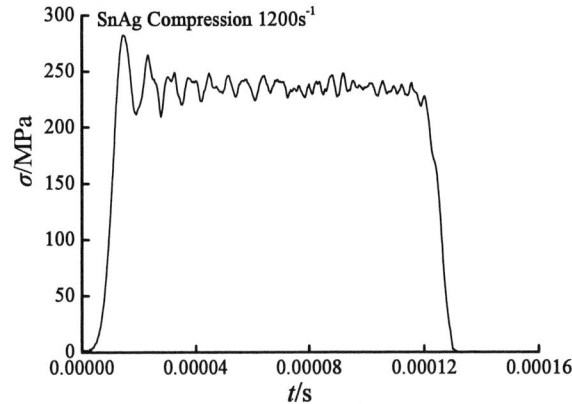

(a) SnAg compressive

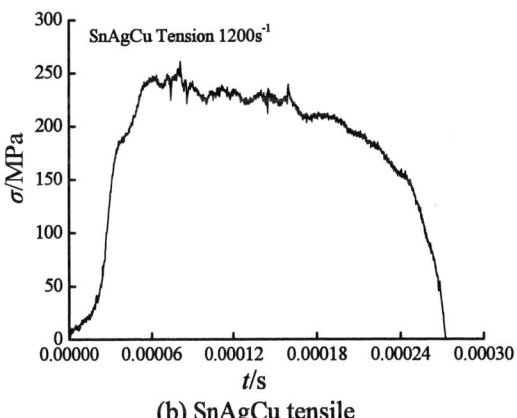

(b) SnAgCu tensile

Fig.6 Input stress pulse (strain rate 1200 s^{-1})

The simulation results indicate that the stress wave arrives at specimen at time $t = 2.25 \times 10^{-4}$ s, and the specimen reached an uniform deformation state at time $t = 2.75 \times 10^{-4}$ s. The normal components of stress and strain in the axial direction of the specimen at were picked up to plot the true stress and strain curve. To avoid boundary effects, the stress and strain at the integral point of the elements located in the middle of the specimen were used. The stress-strain curves from the numerical simulations were compared with the test data in Figs 7~10.

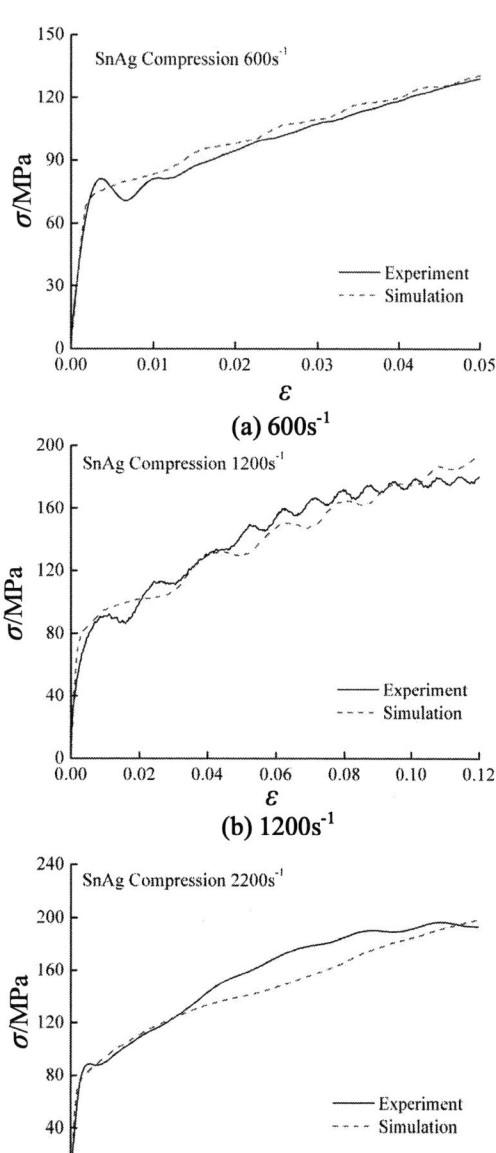

(a) 600s⁻¹

(b) 1200s⁻¹

(c) 2200s⁻¹

Fig.7 True stress-true strain relations of
the SnAg in compression

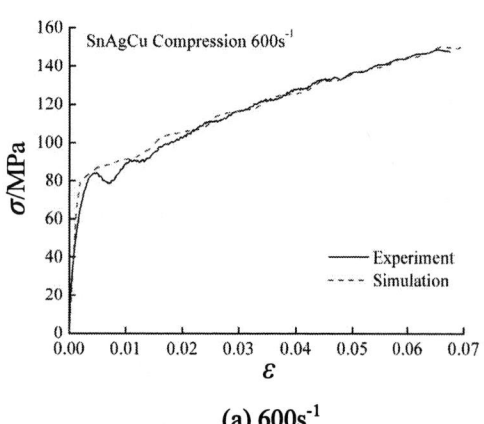

(a) 600s⁻¹

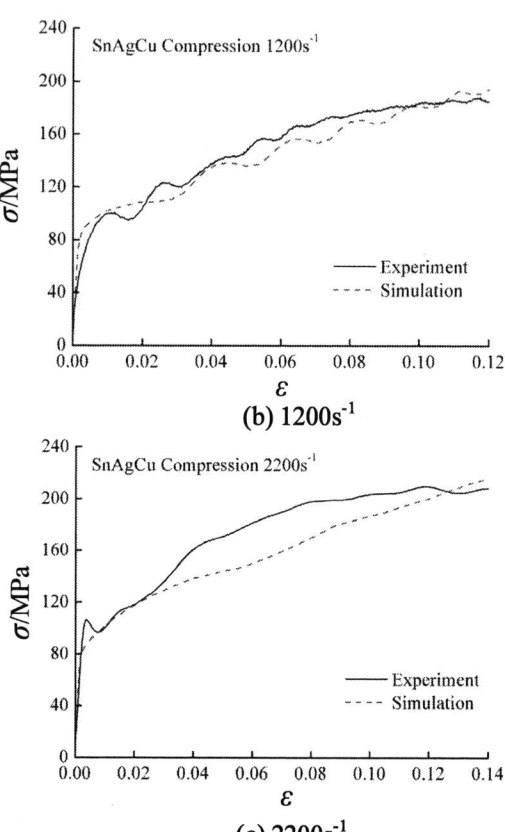

(b) 1200s⁻¹

(c) 2200s⁻¹

Fig.8 True stress-true strain relations of
the SnAgCu in compression

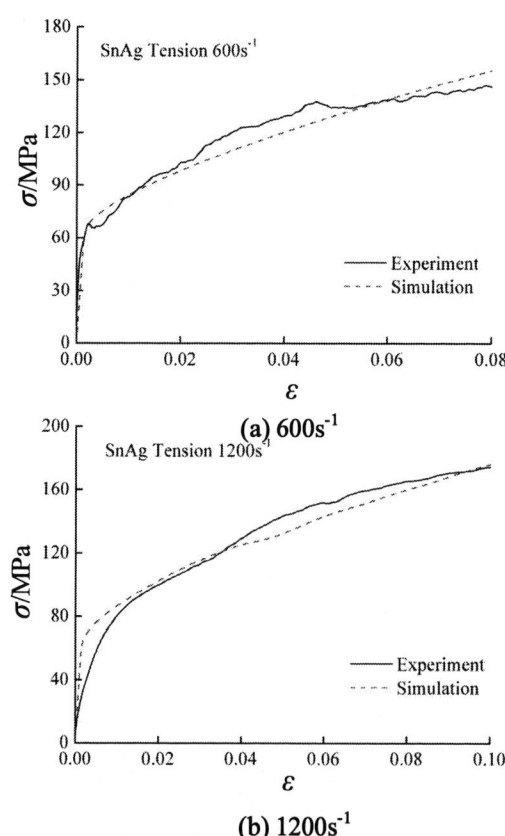

(a) 600s⁻¹

(b) 1200s⁻¹

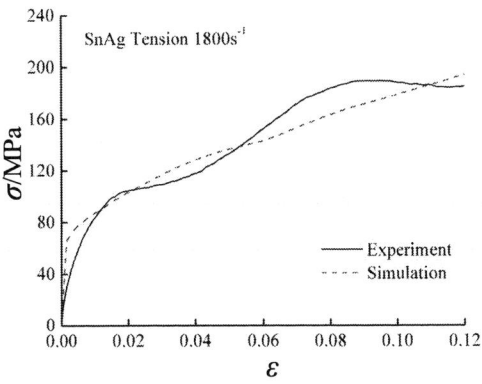

(c) 1800s⁻¹

Fig.9 True stress-true strain relations of the SnAg in tension

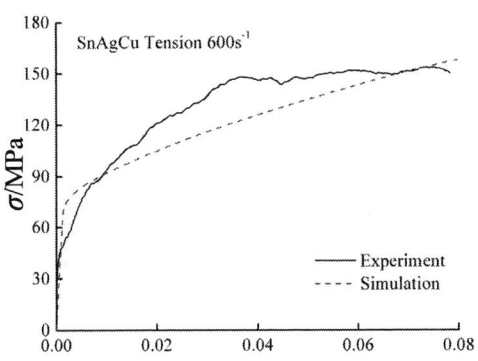

(a) 600s⁻¹

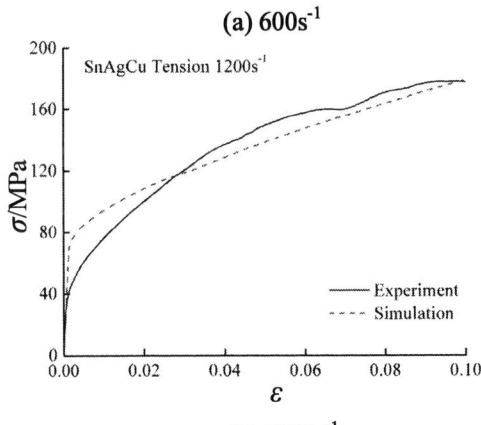

(b) 1200s⁻¹

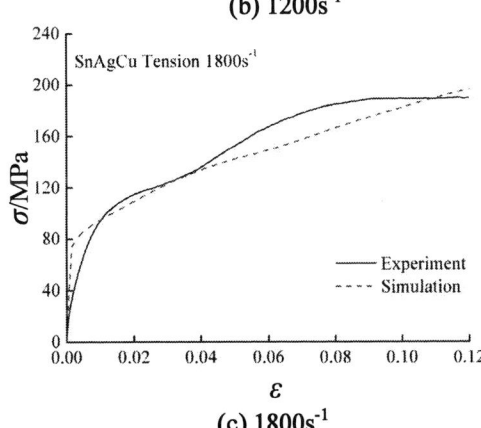

(c) 1800s⁻¹

Fig.10 True stress-true strain relations of the SnAgCu in tension

Figs 7 and 8 compare the simulation results with the test data of the SnAg and SnAgCu under the dynamic compressive loadings, and fairly good agreements can be observed, especially for the cases of strain rate lower than $1000s^{-1}$. Figs 9 and 10 show that under tension loading conditions, the agreements are acceptable but not so good as that under the compression, especially for the cases of strain rate greater than $1000s^{-1}$. Those disagreements are possibly due to errors from both the experiment and the simulation. Firstly, softening behavior was observed in most specimens. However, this softening was not been modeled in the numerical simulations, and the J-C model is intrinsically not able to describe behavior of work softening, more detailed discussion of this can be found in the critical review by Liang and Khan [13]. The disagreements found in Figs 7(c), 8(c), 9(b~c) and Figs10(a~c) can be explained by this point. Actually, damage might occur in the specimens in the final stage of deformation and this effect should be taken into account in numerical models. Secondly, unlike in the compressive conditions, in the tension experiments the specimens were connected into the steel bars by screw threads at two end of the specimen. Because of the specimen materials are much "soft" than the steel bar, as a result, significant plastic deformation may occur in the screws. This leads to energy lose and affects the signal quality in the output bar.

5 Drop/Impact Simulation of Board Level Package

Drop impact of a board level package was modeled as two cantilever beam system as shown in Fig.11. Since the PCB (the top beam) has much larger warpage in the length direction than in the width direction[14] and symmetry of the geometry and the loading condition in drop impact test, this model is feasible. The Input G method[15] was used and the input acceleration $G(t)$ was applied at the far end of the PCB, as depicted in Fig.11.This model can also be regarded as a sliced 3D model. The mesh used is shown in Fig.12.

Fig.11 Mechanical model for the board level packaging

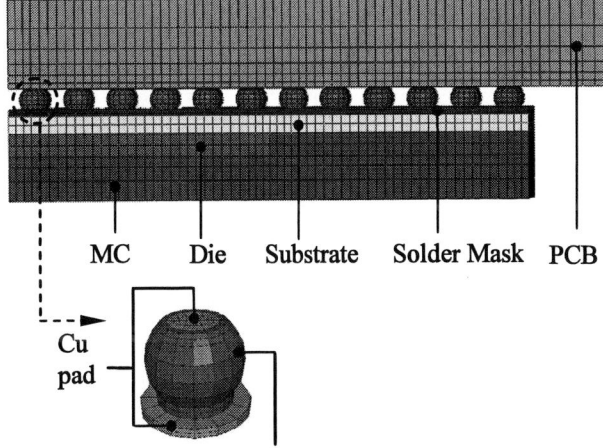

Fig.12 The finite element mesh of the board level package

175

In this analysis, a $6\times0.5\times1.02mm^3$ package interconnects to a $50\times0.5\times1$ mm^3 PCB through Sn3.0Ag0.5Cu (the SnAgCu) solder joints. The package contains a bare $3\times0.5\times0.26$ mm^3 silicon die, the solder mask, substrate and Cu pads are taken into account also. Diameter and standoff of a solder joint are 0.35 and 0.28 mm, respectively. The pitch between adjacent solder joints is 0.5 mm. The pad design is SMD on component side and NSMD on PCB side.

The element type is C3D8R defined by ABAQUS finite element software, and there are 22052 elements and 26258 nodes totally. The Johnson-Cook constitutive model of the SnAgCu solder alloy in Table 3 was used. Other parameters used in the model are presented in Table 4. The impact acceleration is the test condition H defined by the JEDEC standard[16], which is featured as an half-sine impact acceleration pulse with a peak of 2900 G and duration of 0.3 ms.

In order to show the influence of material models on the numerical simulation results, elastic, elastic-plastic and strain rate dependent Johnson-cook models were implemented in this analysis. The elastic-plastic model is a tri-linear form, which is based on the quasi-static tensile test results of the SnAgCu. It is defined by three points: $\varepsilon1=0.065\%$, $\sigma1=45$ MPa, $\varepsilon2=4.3\%$, $\sigma2=50.4$ MPa and $\varepsilon3=40\%$, $\sigma3=55$MPa. The model is shown in Fig.13, and is compared with the test data.

Table 4 Parameters for FE Model

Materials	Young's Modulus (MPa)	Poisson's Ratio	Density $(10^3 g/mm^3)$
Die	131000	0.23	2.33
MC	28000	0.35	1.97
Cu pad	117000	0.34	8.94
PCB	20000	0.11	1.91
Solder mask	5000	0.3	1.15
Substrate	26000	0.11	2

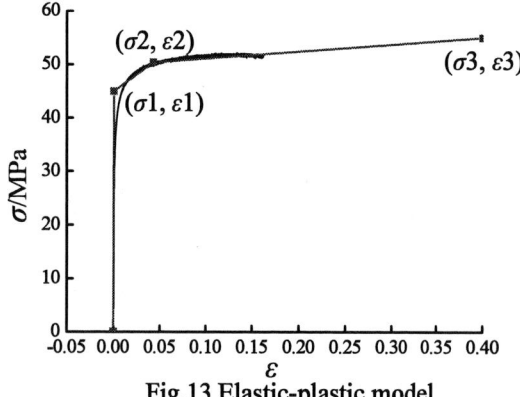

Fig.13 Elastic-plastic model

Numerical results indicate that a strain rate level of 1220 s^{-1} is reached in solder joints during the drop/impact. As discussed in Section 2, the lead-free solders are sensitive to strain rate, therefore, strain rate dependent material model

such as the Johnson-Cook model is critical important to predict more realistic stress and strain in solder joints under drop impact loading.

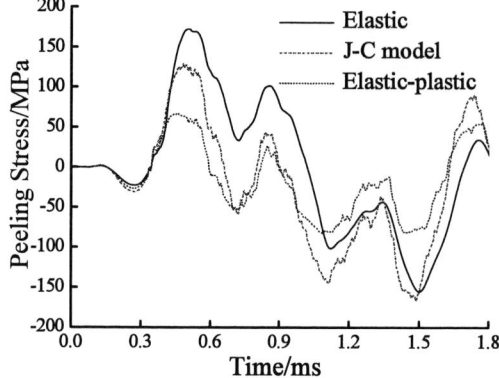

Fig.14 Stress historis for different solder material models

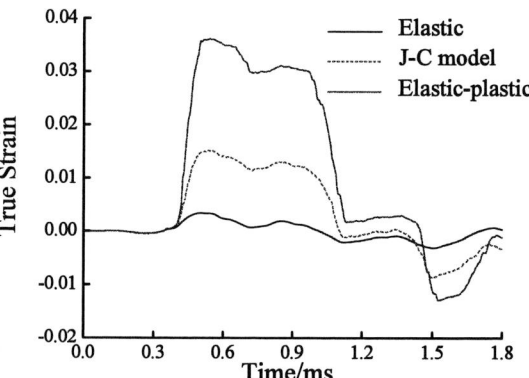

Fig.15 Strain histories for different solder material models

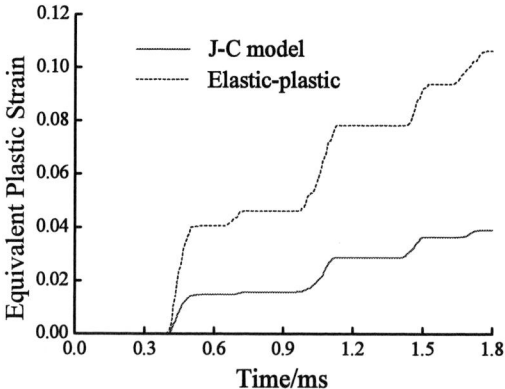

Fig.16 Equivalent plastic strain histories for different solder material models

The peeling stress is regarded as dominant stress component leading to the failure of solder joints[16]. The maximum peeling stress occurs at the solder/PCB interface in the most right solder joint in the model. The histories of the peeling stress, true strains and equivalent plastic strains computed by different material models at the critical point in the critical solder joint are plotted in Figs14~16. It indicates that among the three material models, the elastic model predicts the greatest peeling stress and the least strain, while the elastic-plastic model predicts the least stress but the greatest strain. Besides, the elastic-plastic model predicts over estimated equivalent plastic strain than the J-C model.

6 Conclusions

In this paper stress-strain behaviors under various strain rates of two lead-free solder alloys, Sn3.5Ag and Sn3.0Ag0.5Cu, were investigated by the split Hopkinson tension/pressure bar test, and Johnson-Cook models for the two solders were derived and were applied to board level drop impact simulation. Some conclusions can be drawn as followings:

1) The two lead-free solders are more sensitive to strain rate compared with the SnPb solder. Their yield stresses and ultimate stresses increase as strain rates increase but their fracture strains decrease, which means that they tend to brittle failure at high strain rate.

2) The Johnson-Cook constitutive models proposed in this paper are validated and they can be used to model mechanical behavior of the two lead-free solders in the case of high strain rate deformation.

3) During drop impact of a board level package, the solder joints experience a strain rate of $1220s^{-1}$. This suggests that strain rate dependent material model is necessary in drop impact simulation of electronic packages.

Acknowledgments

The authors would like to thank the financial supports from the National Natural Science Foundation of China (NSFC) under the Grant No.10572010 and the Science & Technology Development Project of Beijing Education Committee under Contract No.KM200610005013.

References

1. Rao R. Tummala, et al. <u>Microelectronics Packaging Handbook</u> (Second edition). 1997, Chapman & Hall, New York

2. Zhou Chunyang, Yu Tongxi, Lee Shiwei Ricky. Review of Shock Response of Portable Electronic Products under Drop Impact—Tests, Simulations and Theories. *Advances in Mechanics*, 2006,36(2): 239-246 (in Chinese)

3. Lee S W R, Dai L H. Characterization of strain rate-dependent behavior of 63Sn-37Pb solder using Split Hopkinson Torsional Bars(SHTB). *Proc 13th Symposium on Mechanics of SMT & Photonic Structures*, New York, USA, Nov., 2001:1-6

4. Wang B, Yi S. Dynamic plastic behavior of 63wt%Sn37wt%Pb eutectic solder under high strain rate. *Material Science Letters*, 2002, 21: 697-698

5. Siviour C R, Walley S M, Proud W G, J E Field. Mechanical properties of SnPb and lead-free solders at high rates of strain. *Physics D: Applied Physics*, 2005, 38: 4131-4139

6. Tee Tongyan, Ng H S, Lim C T, et al. Impact life prediction modeling of TFBGA packages under board level drop test. *Microelectronics Reliability*, 2004, 44(7): 1131~1142

7. Desmond Y. R. Chong, F.X. Che, et al. Drop impact reliability testing for lead-free and lead-based solder IC packages. *Microelectronics Reliability*, 2006, 46: 1160 ~1171

8. Tsung-Yueh Tsai, Chang-Lin Yeh, Yi-Shao Lai, Rong-Sheng Chen. Transient Submodeling Analysis for Board-Level Drop Tests of Electronic Packages. *IEEE Transactions on Electronics Packaging Manufacturing*, 2007, 30(1): 54-62

9. Gordon R. Johnson, William H. Cook. A constitutive model and data for metals subjected to large strains,high strain rates and high temperatures. *Proc 7th Int. Symp. on Ballistics*, The Hague, Netherlands, 1983: 541-547

10. S. Yadav, K. T. Ramesh. The mechanical properties of tungsten-based composites at very high strain rates. *Material Science and Engineering*, 1995:140-153

11. D. Umbrello, R. M'Saoubi, J. C. Outeiro. The influence of Johnson–Cook material constants on finite element simulation of machining of AISI 316L steel. *International Journal of Machine Tools and Manufacture*, 2007, 47: 462-470

12. S. Dey, T. Borvik, O. S. Hopperstad, M. Langseth. On the influence of constitutive relation in projectile impact of steel plate. *International Journal of Impact Engineering*, 2007, 34: 464-486

13. R Liang , A S Khan. A critical review of experimental results and constitutive models for BBC and FCC metals over a wide range of strain rates and temperatures. *Inter. J. Plast.*, 1999,15:963-980

14. Tee Tongyan, Luan Jing-en, Pek E, Lim C T, Zhong Zhaowei. Novel Numerical and Experimental Analysis of Dynamic Responses under Board Level Drop Test. *Proc. 5th Int. Conf. Therm. Mech. Simul. Exp. Microelectron. Microsyst.* New York, 2004: 133~140

15. Tee Tongyan, Luan J.E., Pek E., Lim C.T., and Zhong Z.W. Novel Numerical and Experimental Analysis of Dynamic Responses under Board Level Drop Test. *Proc EuroSime Conference*, 2004: 133-140

16. JEDEC Standard, JESD22-B111, Board Level Drop Test Method of Components for Handheld Electronic Products, 2003

A Scale Reduced Computation Scheme for Peeling Stress of Solder Joints under Drop Impact

An Tong, Qin Fei
College of Mechanical Engineering and Applied Electronics Technology,
Beijing University of Technology, Beijing 100124, China
Email:qfei@bjut.edu.cn, Phone: +86-106739217

Abstract

A beam model of board level electronic package was used to investigate effects of the moment, axial force and shear force induced during drop/impact on the peeling stress of the soldered joints. The peeling stresses in soldered joints were evaluated under static and dynamic bending of the PCB. It shows that the peeling stress is dominated by the bending stress and the maximum occurs at the PCB end. In the soldered joint array, only a few soldered joints closed to the far end of the packaging are stressed and the most joints inside the array are almost stress free. Based on this observation, an approach was proposed to reduce the computation scale. By the approach, only 3 or 4 soldered joints are necessary to be included in the computational model.

1 Introduction

Solder joints functioned as mechanical, thermal and electrical interconnections between electronic packages and the printed circuit board (PCB), their failure can lead to critical malfunction of electronic products directly. In recent years, with the increasing demand and popularity of portable telecommunication devices such as mobile phones and PDAs, the mechanical behavior of solder joints under drop/impact conditions has drawn more and more attentions of researchers and engineers.

Behavior of solder joints must be investigated in the assembly of PCB, solder joints and components. The analysis models applied to the assembly include beam model[1,2], shell-plate model[3,4], and 3-D finite-element model[5,6]. Wong et al.[3] modeled the PCB as a beam or a plate, and obtained analytical solutions of the dynamic response of the PCB. However, the solder joints were not taken into account in the research. In Suhir's work[7], the PCB and the component assembly was treated as a system of plates and springs, and the axial stress solutions of solder joints under static bending moment were given. By Suhir's model, the effect of bending moment and shearing force on stress of solder joints, which is regarded as the primary stress to the failure of solder joints, cannot be evaluated.

In board level packages, components are mounted to PCB through hundreds of solder joints. The diameter of solder ball is 0.35 mm and its height is 0.3 mm in a BGA, while the length of a PCB is usually about 100 mm. This induces about 10^4 order dimension difference in the same assembly. This leads to large amount element needed and dramatic increasing in computation costs, especially for a dynamic response analysis and for 3-D model, which usually is necessary in order to obtain detail stress distribution in the critical joints. Some efforts were done to attack this difficulty. Ren et al.[3,4] modeled the PCB and solder joints as shell and beam elements to reduce the computational cost, but the detailed stress distributions across the solder joints are not available in this model. Zhu[5] has analyzed the solder stresses of a BGA package subjected to static loading by using the submodeling or the global-local method. However, submodeling analysis is not efficient for path-dependent dynamic problems[8]. More efforts are needed to balance the computational costs and result accuracy in numerical simulation.

In this paper, a beam model of board level electronic package was established and used to investigate the effect of PCB deflection induced by drop/impact loading on the stress of the soldered joit array. Based on the analysis, an approach was proposed to reduce the computational scale of the problems, and its feasibility was discussed finally.

2 Modeling of Board Level Assembly under Drop/Impact

A typical board level drop impact test setup is depicted in Fig.1, which is recommended by JEDEC[9]. The PCB and component assembly is mounted onto a metal base via screw bolts. In the test, the entire assembly is subjected to free fall along guide rods from a prescribed height, and the metal base impacts onto a rigid foundation and an impact loading is produced. A prescribed half-sine acceleration impulse can be achived by manipulating the drop height and the dimension or material of cushion pad. Since the stiffness of metal base is exceedingly greater than that of the PCB, the half-sine acceleration impulse resulting from impact predominantly transmits to the PCB via the metal base and screw bolts with little distortion. Therefore the analysis of a board level drop impact test can be simplified by a model in which the PCB is alone subjected to the half-sine acceleration impulse at its points of mounting to the screw bolts, and the model is shown in Fig.2(a), in which $G(t)$ is the acceleration impulse. This approach was proposed by Tee et al.[6], is called the Input-G method, which is very convenient to simulate the drop/impact process.

The PCB and component bend under the act of $G(t)$. However, the difference of the bending stiffness between the PCB and component make the solder joints deforming under tensile/compressive stresses during PCB bends upward/downward. It has been indicated that the peeling stress, i.e., the normal stress vertical to the PCB, is the dominant stress component leading to failure of solder joints[10]. The bending deformation of the PCB and the component has great influence on the magnitude of the peeling stress.

Since the PCB has much larger warpage in its length direction than in the width direction[11], the PCB and component can be modeled as two strips of plate, as shown in Fig.2(b), solder balls are modeled as cylindrical beams. Furthermore, due to the symmetry only one half of the model is analyzed, as shown in Fig.2(c). In this paper, the PCB strip and the component strip are modeled by Timoshenko beams and a bending moment $M(t)$ is applied to the right end of the PCB.

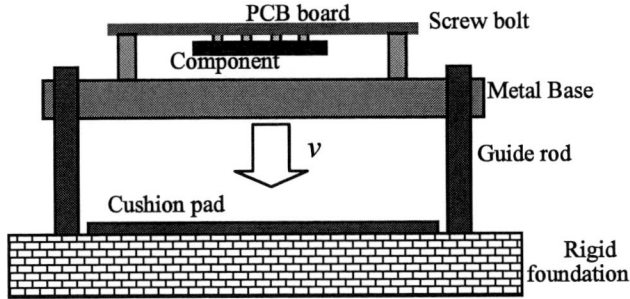

Fig. 1 Test setup of board level drop/impact

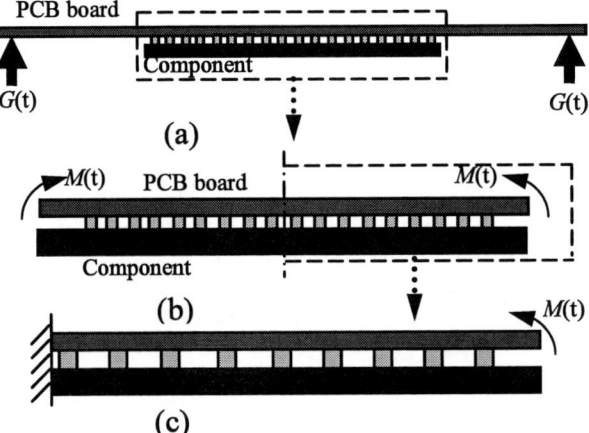

Fig. 2 Mechanical model for the board level packaging

3 Stress Analysis of Solder Joints

A finite element model was built according to Fig.2(c), wherein the PCB, package and solder joints are modeled with Timoshenko beam elements, and the model was analyzed by ABAQUS. The model parameters such as dimensions and materials, are presented in Table 1, in which b and h are the width and thickness of the PCB and the component, d is the diameter of solder column, E and v are the Young's modulus and Poisson's ratio of materials, and ρ is the material density. In order to compare, both the FE static and dynamic analyses have been carried out. In the static analysis, the load was a constant bending moment of 15 N·mm, while in the dynamic analysis, a triangular impulse of bending moment with a peak of 15 N·mm and a duration of 0.3 ms was applied.

Table 1 Model Parameters (mm)

	L	b,d	h	E (GPa)	v	ρ (g/cm³)
PCB board	10	1	1	24	0.28	2
Component	10	1	1	100	0.39	2.5
Solder joint		0.5	0.5	25	0.36	10

The model was firstly verified by comparing the axial stresses in solder joints from this simulation with Suhir's equation[7]. Suhir' model is similar to Fig.2(b), except that it treats the solder joints as spring links with stiffness K, which only axial force $p(x)$ can be sustained

$$p(x) = \frac{kML^2}{2D_1 u^2}\left(\frac{u^2\chi(u)}{3}V_0(\alpha x) - \phi(u)V_2(\alpha x)\right) \quad (1)$$

Where $\alpha = \sqrt[4]{k\dfrac{D}{4D_1 D_2}}$, $u = \alpha L$, $D = D_1 + D_2$, D_1 and D_2 are the bending stiffness of the PCB and package respectively:

$$D_1 = \frac{E_1 h_1^3}{12(1-v_1^2)}, \quad D_2 = \frac{E_2 h_2^3}{12(1-v_2^2)} \quad (2)$$

The other terms in Eq.(1) are

$$V_0(\alpha x) = \cosh\alpha x \cos\alpha x , \quad V_2(\alpha x) = \sinh\alpha x \sin\alpha x \quad (3a)$$

$$\chi(u) = \frac{6}{u^2}\frac{\cosh u \sin u - \sinh u \cos u}{\sinh 2u + \sin 2u} \quad (3b)$$

$$\phi(u) = \frac{\cosh u \sin u + \sinh u \cos u}{\sinh 2u + \sin 2u} \quad (3c)$$

The axial stresses in solder joints by Eq. (1), the static and the dynamic FEA are shown in Fig.3. It shows that the FE results agree quite well with Suhir's, both in the distribution and the maximum value of the stresses. The maximum by Eq.(1) is 46 MPa, while that by the FE static is 41.5 MPa.

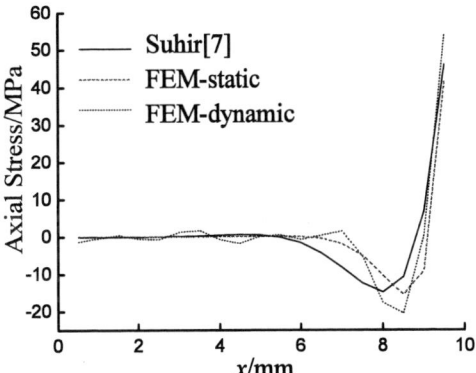

Fig. 3 Axial stress in solder joints

The solder joints are subjected to bending moment, axial force, and shear force when the PCB subjected to bending moment. The static and dynamic stresses induced by the moment, axial force and shear force are presented in Figs 4 and 5. Fig.4 shows that the maximums of bending stress, axial stress and shear stress at the PCB end are 130.5 MPa, 41.5 MPa and 23.6 MPa respectively. This suggests that the bending moment produces much greater stress than the axial or shear force. All the maximum stresses occur at the PCB end of the most outer solder joint. At the component (package) end, the axial stress is the greatest in the all, but it is in a lower level comparing with the stresses induced by bending moment at the PCB end. The bending stress plays a dominant role in solder joints.

There is no significant difference in stress distribution pattern between the FE static and dynamic analysis at the two ends of solder joints. The bending stress by the dynamic simulation is 90 MPa, which is 31% less than that from static, while the stresses induced by axial and shear forces are 55.2 MPa and 29.5 MPa respectively, which are about 33% and 25% greater than that by the static analysis.

As the peeling stress dominates the failure of solder joints, the distribution of the peeling stresses in each solder joint are presented in Fig.6. It shows that the majority of solder joints are under low stress level, there are only a few solder joints at the outer position enduring large peeling stress both in the static and dynamic simulation. The maximum peeling stress from static analysis occurs at the PCB end, and it is 172.1 MPa or 16% greater than that of 144.1 MPa from dynamic analysis. Thus in the following section, only static analyses are conducted to calculate the peeling stress.

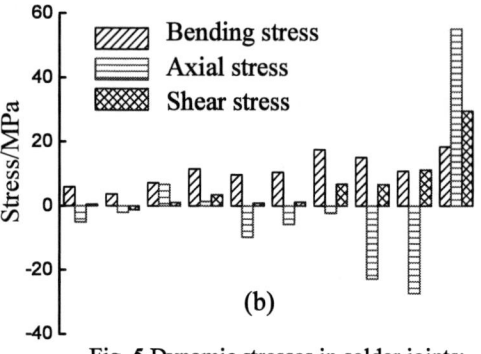

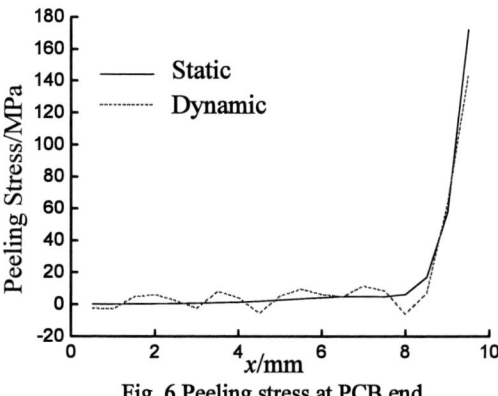

Fig. 5 Dynamic stresses in solder joints:
(a) PCB end; (b) component end

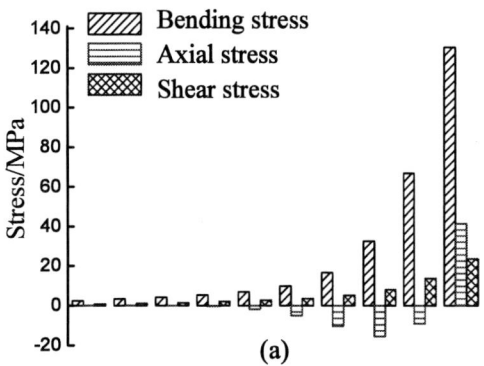

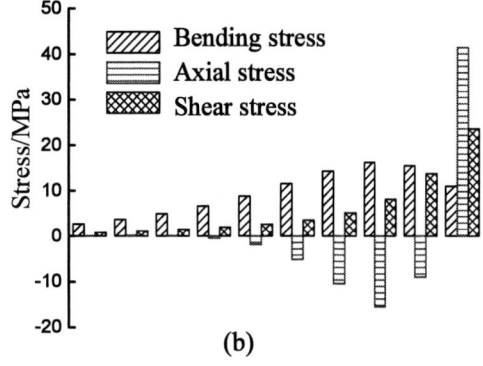

Fig. 4 Static stresses in solder joints:
(a) PCB end; (b) component end

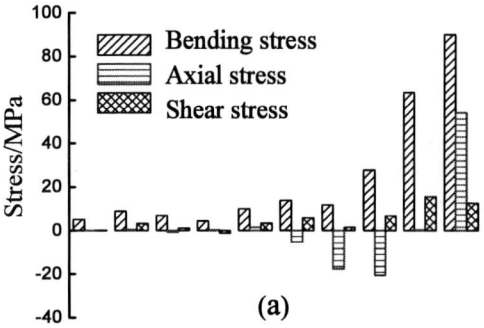

Fig. 6 Peeling stress at PCB end

4 Reduced Model for Stress Computation of Solder Joints

As discussed in Section 3, only a few solder joints close to the edge of the component experience significant peeling stresses while the other majority are at quite low stress level that can be neglected. Based on this observation, one approach was proposed in this paper to reduce the computational model. By the approach, only a few outer solder joints are modeled and the other inner solder joints can be neglected from the model.

In order to validate the approach, models containing 10, 5, 4 and 3 solder joints were computed, respectively. The peeling stress and the PCB deflection of those models were compared carefully with a full model containing 19 solder joints. The comparison of stress distribution in each solder joint was presented in Fig.7, and there is no significant difference between these models. The difference of the peeling stress in percentage between the reduced and the full model is shown in Fig.8. The peeling stress obtained from the model with 10 solder joints is 0.06% greater than that from the full model, and models containing 5 and 4 solder joints predict 1.5% and 3.7% greater peeling stresses than that from the full model. Obviously, only 4 or 5 solder joints being included in the model is enough to predict satisfied peeling stress. The deflection curves of the PCB derived from various models are shown in Fig.9, which also suggests that it is no significant effect on the deflection of the PCB to neglect most of the inner solder joints.

It indicates that most of inner solder joints can be neglected from the model when the maximum peeling stress or the PCB deflection are computed. The reduced model can not only guarantee the computational accuracy but also greatly reduce computational cost, especially when a 3D model is used.

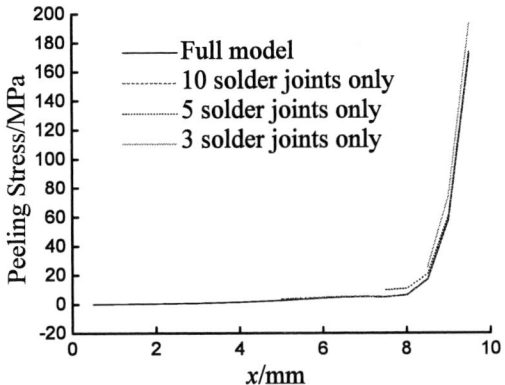

Fig. 7 Peeling stress computed by different models

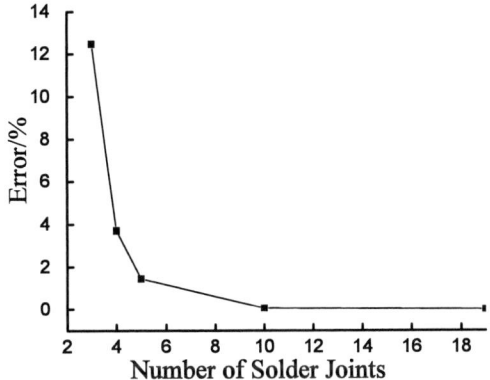

Fig. 8 Relative errors of peeling stress by different models

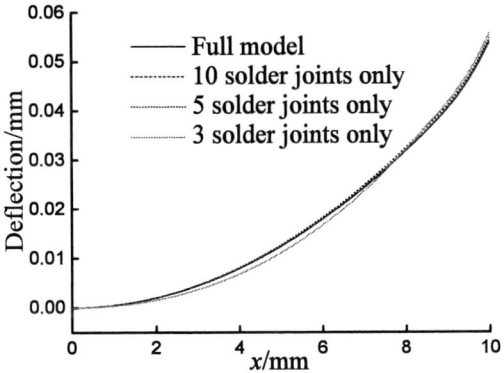

Fig. 9 Deflection of PCB by different models

5 Conclusions

The stress distribution in solder joint array was analyzed by the finite element method and the proposed beam model of a board level assembly under drop/impact loading conditions. The maximum peeling stress occurs at the PCB end of the outer solder joint, and the stress induced by bending moment is dominant. The numerical results show that only a few solder joints close to the edge of the component experience greater peeling stresses while the other majority is under quite low stress level.

An approach to reduce the computational cost was proposed and by the approach, only 4 outer solder joints of total 19 are necessary to be modeled and the other majority solder joints can be neglected from the model.

Acknowledgments

The authors would like to thank financial support from the National Natural Science Foundation of China (NSFC) under the Grant No.10572010 and the Science & Technology Development Project of Beijing Education Committee under Contract No.KM200610005013.

References

1. Wong E H. Dynamics of Board Level Drop Impact, *Trans. ASME, J. Electron. Packag.*, 2005, 127: 200–207
2. Wong E H, Mai Y W, Seah S K W. Board level drop impact-fundamental and parametric analysis. *Trans. ASME, J. Electron. Packag.*, 2005, 129: 496~502
3. Ren Wei, Wang Jianjun, Shell-Based Simplified Electronic Package Model Development and its Application for Reliability Analysis. *Proc.5th Electronics Packaging Technology Conference*, Piscataway, NJ, 2003: 217~222
4. Ren Wei, Wang Jianjun, Reinikainen T. Application of ABAQUS/Explicit Submodeling Technique in Drop Simulation of System Assembly. *Proc. 6th Electronics Packaging Technology Conference*, United States, 2004: 541~546
5. Zhu Liping. Submodeling Technique for BGA Reliability Analysis of CSP Packaging Subjected to an Impact Loading. *Proc. Advances in Electronic Packaging*, New York, 2001: 1401~1409
6. Tee Tongyan, Luan Jing-en, Pek E, Lim C T, Zhong Zhaowei. Novel Numerical and Experimental Analysis of Dynamic Responses under Board Level Drop Test. *Proc. 5th Int. Conf. Therm. Mech. Simul. Exp. Microelectron. Microsyst.* New York, 2004:133~140
7. Suhir E. On a Paradoxical Phenomenon Related to Beams on Elastic Foundation: Could External Compliant Leads Reduce the Strength of a Surface-Mounted Device?, *Trans. ASME, J. Appl. Mech.*, 1988, 55(4): 818-821
8. Lai Yishao, Wang Tonghong. Verification of submodeling technique in thermomechanical reliability assessment of flip-chip package assembly. *Microelectronics Reliability*, 2005, 45(3-4): 575~82
9. JEDEC Standard, JESD22-B111, Board Level Drop Test Method of Components for Handheld Electronic Products, 2003
10. Tee Tongyan, Luan Jing-en, Pek E, Lim C T, Zhong Zhaowei. Advanced experimental and simulation techniques for analysis of dynamic responses during drop impact. *Proc. 54th Electronic Components and Technology Conference*. Piscataway, NJ, 2004: 1088 ~ 1094

11. Tee Tongyan, Ng H S, Lim C T, Pek E, Zhong Zhaowei. Board level drop test and simulation of TFBGA packages for telecommunication applications. *Proc. 53rd Electronic Components and Technology Conference*, New Orleans, LA, 2003: 121~129

Dynamic Bending Tests and Numerical Simulation of Board Level Electronic Package

Qin Fei[1], Wang Yngve[2], Liu Bin[2], An Tong[1], Jin Ling[1]

1 College of Mechanical Engineering and Applied Electronic Technology,
Beijing University of Technology, Beijing 100124, China;
2 Intel Technology Development (Shanghai) Ltd, Shanghai 200131, China
Email:qfei@bjut.edu.cn, Phone: +86-10-67392173

Abstract

4-point dynamic bending tests of board level electronic packages were carried out in order to investigate the reliability of solder joints. A high speed camera and the digital image correlation method were used to measure the deflection of the PCB board. A finite element model to simulate the test was built up and was validated by the test data. A parameter study was subsequently implemented. The results show that at certain value of PCB stiffness the peeling stress reaches its peak. The package installation angel has significant effect on the peeling stress of the solder joints.

1 Introduction

Solder joints, which serve as mechanical, thermal and electrical interconnections between the electronic packages and the printed circuit board (PCB), are the most vulnerable part in a portable product when the product is subjected to accidental drop impact. As a result, a comprehensive study of mechanical behavior of solder joints is indispensable for understanding the reliability of portable electronic products.

There have been two kinds of typical drop tests, product level and board level drop test. Since the repeatability of the product level drop test is poor, the board level drop test is more efficient for research. JEDEC standard have recommended the procedure and conditions[1, 2] for board level drop test of portable electronic products. However, it seems that it is expensive and time-consuming. Thus, manufacturers and researchers are trying to develop simpler and cheaper test methods. Those efforts include the 4-point board level dynamic bending (4PDB) tests [3] which were reported to be used by Motorola corporation, the ball impact test (BIT) proposed by Yeh Chang-Lin[4], and the cold bump pull (CBP) performed by NXP Semiconductors[5]. Compared with JEDEC procedure and the methods mentioned above, the 4-point dynamic bending test needs not so much expensive equipments and is simple. Furthermore, the mechanical behavior of solder joints by the 4PDB is reported to be consistent with that in the drop test prescribed by JEDEC[3].

For reliability design of portable electronic products, it is critical important to perform drop impact tests. However, those tests are costive and time consuming. Contrastively, numerical simulation, which is based on test validated models, can provide more detailed data for reliability design fast and cheaply. Many simulation works[6,7,8] following JEDEC drop impact procedure were reported. However, there are few reports on simulation work related to the 4-point dynamic bending tests.

In this paper, a 4-point dynamic bending test of board level electronic packages was carried out firstly, and a high speed camera and the digital image correlation method was

used to measure the deflection of the PCB, then a very detailed finite element model of the 4-point dynamic bending test was established, and the accuracy of the model was verified by carefully comparing the results of experiment and simulation. The parameters, including rigid ball drop height, PCB stiffness and package installation angel were examined for their effects on the peeling stress in solder joints.

2 Experiment Setup and Measurement of PCB Deflection

Typical setup of 4-point dynamic bending tester is shown in Fig.1. It consists of a PCB with packages installed facing downwards. The PCB is placed on two rolling supporters, and a rigid trestle table is attached to the PCB, which allows the dynamic load induced by a special designed instrument being transmitted to the PCB. There is a cushion pad on the rigid trestle table. Sizes of the PCB and the package are intentionally omitted here. The package is mounted at the center of the PCB, and has an installation angel α, as shown in Fig.2.

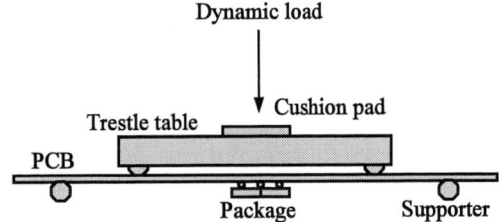

Fig.1 Setup of 4-point dynamic bending test

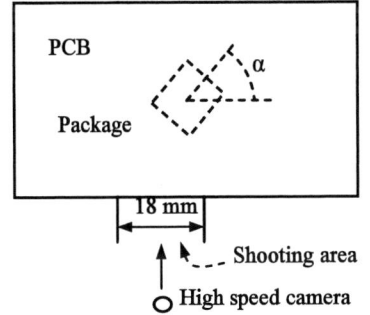

Fig.2 Layout of package and shooting area

During the test, two strain gauges were mounted at the center of the PCB to measure the dynamic strains in the length and width directions of the PCB. In order to measure deflection of the PCB, speckles were sprayed onto the side surface of it, and a high speed camera was used to capture the deformation images of the speckle area. As shown in Fig.2, the shooting area was placed near the center of PCB so that the max deflection can be captured. The speed of the camera

978-1-4244-2739-0/08/$25.00 ©2008 IEEE

is 8000 frame/s, that means, the interval Δt between two images is 0.125 ms.

Fig.3 presents a typical speckle image of the deformed PCB. From the image, obvious large deflection occurred to the PCB. In order to obtain better results, the deflection of the PCB at moment t, $w(t)$, can be calculated by the following procedure: (1) Start from the first image at time t_0, two contiguous images at time t_{i-1} and t_i were calculated by the digital image correlation software DICA1.0 developed by us, and the deflection between the two moments, Δw_i, can then be obtained. (2) Continue this process until the image at time t was calculated by the software. (3) As $t = n\Delta t$, totally $(n+1)$ images were calculated and number of n Δw_i were obtained at the end of Step (2). Sum up all the Δw_i to get the deflection of the PCB at time, that is

$$w(t) = \sum_{i=1}^{n} \Delta w_i \qquad (1)$$

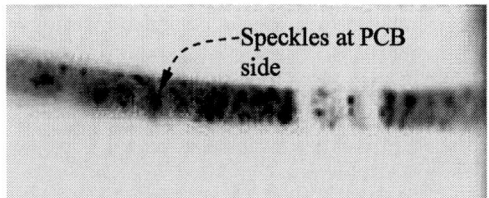

Fig.3 PCB side image by high speed camera

3 Numerical Simulation for the 4-Point Dynamic Bending

3.1 Load and Boundary Conditions

The dynamic load is from a free fall steel ball at height h. At the moment the ball contacts the cushion pad, its velocity is

$$\upsilon = \sqrt{2gh} \qquad (2)$$

In order to simplify the simulation, the free falling of the steel ball is neglected and instead of that, a near-contact velocity is applied as an initial condition. Since the stiffness of the rolling supporters, the trestle table and the steel ball is much greater than the PCB, they are assigned as rigid bodies. Friction free contact elements are placed to the contact surfaces between the PCB and the trestle table, the PCB and the rolling supporters. The cushion pad is treated as incompressible rubber-like materials, which are modeled by Mooney-Rivlin model[9], with mass density ρ =5x10^{-5} g/mm^3, Poisson's ratio v =0.499 and two model constants identified as C_{10}=0.5516 MPa and C_{01}=0.1379 MPa.

3.2 Finite Element Model

Generally, there are hundreds of solder joints used to connect packages and the PCB. The diameter and height of solder balls are 0.35 mm and 0.3 mm, respectively. It is very costive to model all the solder joints by 3D elements. In this paper, the solder joints were modeled as a uniform solder layer. The research work done by Tsai et al.[10] have shown that the solder layer model can predict accurate PCB deflection.

Fig.4 shows the finite element mesh of the board level package, in which the PCB, Cu pad, substrate, die and mold

compound were taken into account. The element type is SOLID164 in ANSYS, and there are 21182 elements and 19068 nodes totally. The PCB was modeled as orthotropic elastic material[11]. The solder layer was modeled as tri-linearly elastic-plastic material with specific strain-stress relation as σ_1 =55.3 MPa, σ_2 =76 MPa, σ_3 =2.2 GPa, ε_1 =0.14%, ε_2 =0.4%, ε_3 =100%[10]. Elastic properties used for other components are presented in Table 1.

3.3 Validation of the Finite Element Model

In order to validate the finite element model, deflection and strain at center location of the PCB obtained both from the experiment and the simulation were compared carefully. Fig.5 shows the deflection at different points along the PCB length direction, and the time evolution of the deflection curve. It indicates that the numerical results agreed well with the experimental, and the PCB deflection reaches its maximum value at 3.5 ms. Fig.6 gives the strain history at center of the PCB when the drop height is 80 mm, and Fig.7 gives the maximum strain at PCB center with various drop heights. These indicate that the numerical model used is feasible and reliable.

4 Parameter Analysis

Due to lack of proper methods to measure stress or strain in solder joints directly, the numerical model was used to compute those quantities. The effects of design parameters, such as drop height, PCB stiffness and package installation angel α, were investigated for their effects on the solder joint stress.

4.1 Drop Height

Fig.8 shows the maximum PCB deflection and the peeling stress in solder joints for 4 different drop heights, h=80 mm, 110 mm, 163 mm and 304 mm. The maximum deflection and the peeling stress increase monotonically with increasing drop height, while after the drop height reaches 168 mm, the increasing of stress tends to be slow down, this implies the start of plastic deformation in solder joints. Fig.9 shows that the PCB defection and the peeling stress achieve their maximum at the same time moment, says 3.5 ms. This indicates that the peeling stress in solder joints is mainly caused by the deflection of the PCB during drop/impact, and this conclusion is consistent with the simulation by following the JEDEC standard [12].

4.2 PCB Stiffness

Bend stiffness of PCB, D, has a relation with its thickness T as

$$D = \frac{ET^3}{12(1-v^2)} \qquad (3)$$

Eq.(3) says that the stiffness increases with increasing thickness. Fig.10 shows the maximum PCB deflection and the peeling stress in solder joints for 4 different PCB thickness

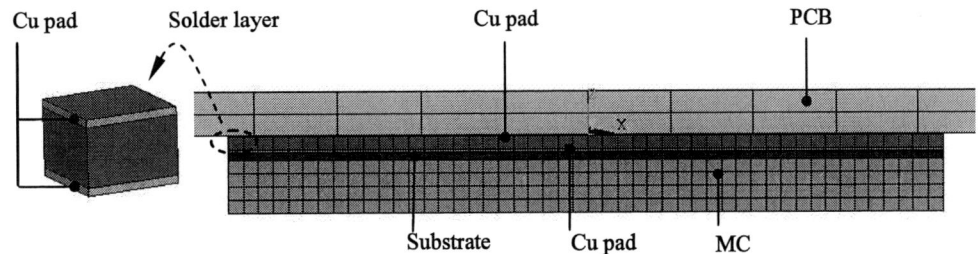

Fig. 4 Finite element mesh of PCB and package

Table 1 Material properties

Materials	Young's Modulus (MPa)	Poisson's Ratio	Density (10^{-3}g/mm^3)
Die	131000	0.3	2.33
Mold Comp.	25506	0.3	1.97
Cu Pad	117000	0.3	8.94
PCB	E_x, E_y:25400 E_z: 11000	v_{xz}, v_{yz}:0.39 v_{xy}: 0.11	2.00
Solder Alloy	39500	0.3	7.44
Substrate	3499	0.3	3.97

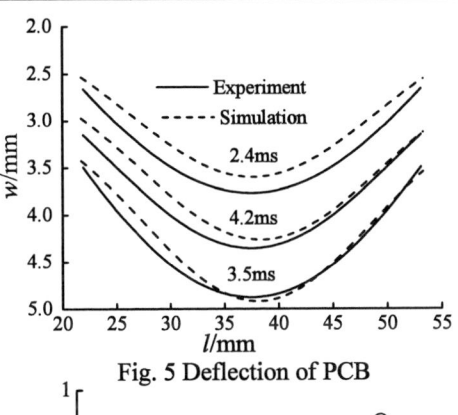

Fig. 5 Deflection of PCB

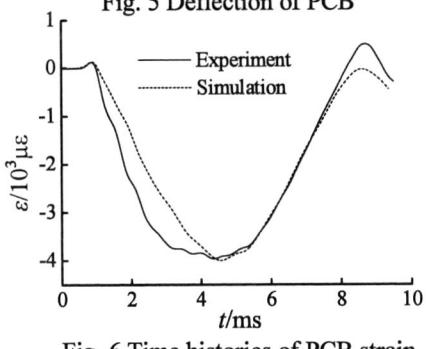

Fig. 6 Time histories of PCB strain

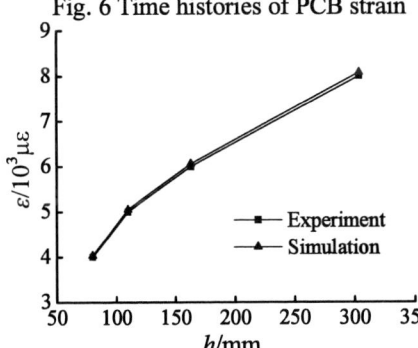

Fig. 7 Effect of dropping height on PCB strain

T=0.6 mm, 0.8 mm, 1.0 mm and 1.2 mm. It shows that the PCB deflection decreases steadily with increasing thickness. For the maximum peeling stress, when the thickness is less than 0.8 mm, it increases with the increasing of the thickness, however, as the thickness is greater than 0.8 mm, it begins to decrease rapidly with the thickness increasing. It is believed that the stiffness of PCB and the mounted package have significant effects on peeling stress. A detailed discussion on this can be found in Ref.[13].

4.3 Package Installation Angel

The solder balls at the outer most corner usually have max. peeling stress in the ball array. For different installation angles, the distance between the outermost ball and the PCB center is different. Therefore, the package installation angel α has obviously influence on the maximum peeling stress in solder joints. Fig.11 shows the maximum peeling stress for $\alpha = 0°$, $30°$, $45°$ and $60°$. It shows that the peeling stress reaches its maximum and minimum as $\alpha = 45°$ and $\alpha = 0°$, respectively, and the greatest is almost double the least.

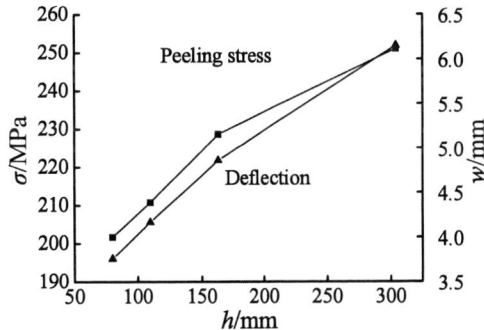

Fig. 8 Effect of drop height on PCB deflection and solder stress

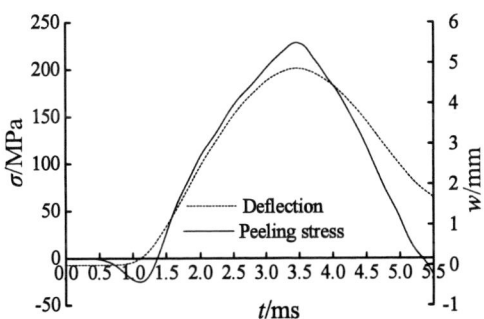

Fig. 9 Time histories of solder stress and PCB deflection

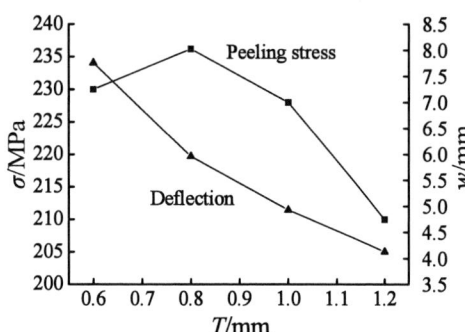

Fig. 10 Effect of PCB thickness on solder stress and PCB deflection

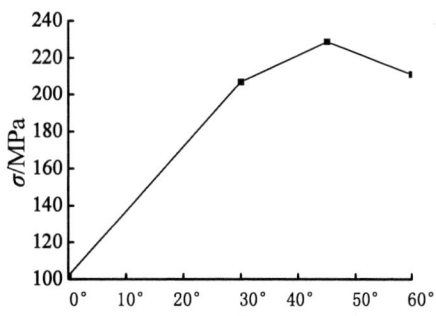

Fig. 11 Effect of package set-angel on solder stress

5 Conclusions

4-point dynamic bending tests were carried out to investigate reliability of board level package under drop/impact. A high speed camera and the digital image correlation method were used to measure the deflection of the PCB. The finite element model to simulate the 4-point dynamic bending test was built up and was validated by the experiment results. A parameter study was performed. The following conclusions can be drawn:

(1) The methodology by using high speed camera and the digital image correlation method to measure the deflection of the PCB during drop/impact are feasible.

(2) The solder joint stress and the PCB deflection increase with the increasing of drop height, however, after the drop height reaches 168 mm, the stress increasing inclines to slow down.

(3) The PCB stiffness significantly affects the solder joint stress, and as the stiffness equals to a specific value, the stress reaches its peak, after that, the stress decreases rapidly with the increasing of PCB stiffness.

(4) The package installation angel should be as small as possible so as to reduce the peeling stress under drop/impact.

Acknowledgments

The authors would like to thank the financial support from the National Natural Science Foundation of China (NSFC) under the Grant No.10572010, the Science & Technology Development Project of Beijing Education Committee under Contract No.KM200610005013 and the Intel High Education Program.

References

1. JEDEC Standard, JESD22-B111, Board Level Drop Test Method of Components for Handheld Electronic Products, 2003
2. JEDEC Standard JESD22-B110, Subassembly Mechanical Shock, 2001
3. Dave Reiff and Edwin Bradley.A Novel Mechanical Shock Test Method to Evaluate Lead-Free BGA Solder Joint Reliability. *Proc. 55th Electronic Components and Technology Conference*, 2005:1519-1525
4. Chang-Lin Yeh,Yi-Shao Lai. Design Guideline for Ball Impact Test Apparatus. *Tran. ASME, Journal of Electronic Packaging*, 2007,129:98-104
5. Zaal J J M, Van Driel W D, et al.Testing Solder Layer Connect Reliabilies Under Drop Impact Loading Conditions. *Proc. 7th ICEPT*, Shanghai, China, August, 2007: 234-239
6. Wu J., Song G., Yeh C., and Wyatt K. Drop/Impact Simulation and Test Validation of Telecommunication Products. *Proc. of InterSociety Conference on Thermal Phenomena*, 1998:330-336
7. Tee, T Y, Luan, J E,Pek,E, Lim,C T, and Zhong, Z W. Advanced Experimental and Simulation Techniques for Analysis of Dynamic Responses during Drop Impact. *Proc. 54th ECTC Conference*, 2004: 1088-1094
8. Tee, T Y , Luan, J E, Pek, E, Lim, C T, and Zhong, Z W. Novel Numerical and Experimental Analysis of Dynamic Responses under Board Level Drop Test, *Proc EuroSime Conference*, 2004:133-140
9. John Swanson. Release 10.0 Documentation for ANSYS. Pennsylvanian: ANSYS company, 2006
10. Tsung-Yueh Tsai, Chang-Lin Yeh, Yi-Shao Lai, and Rong-Sheng Chen. Transient Submodeling Analysis for Board-Level Drop Tests of Electronic Packages. *Transactions on Electronics Packaging Manufacturing*, 2007,30(1):54-62
11. Tee,T Y, Hun Shen Ng, and Zhong, Z W.Design for Enhanced Solder Joint Reliability of Integrated Passives Device under Board Level Drop Test and Thermal Cycling Test. *Proc. 5th EPTC Conference*, Singapore, 2 003:210-216
12. Qin Fei, Bai Jie, An Tong. Drop/Impact Stress Analysis of Solder Joints in Board Level Electronics Package. *Journal of Beijing University of Technology*, 2007,33(10):1038-43 (in Chinese)
13. Qin Fei, An Tong. Effect of PCB Bending Stiffness on Peeling Stress of Solder Joints in Electronics Packaging. *Journal of Beijing University of Technology*, 2008, 34(Sup.) (in Chinese)

A New Method for The Investigation of Strip Warpage of MAP-QFN

Guohua Gao, Honghui Wang, Guoji Yang, Haiqing Zhu

Nantong Fujitsu Microelectronics Co., Ltd

No. 288, Chongchuan Road, Chongchuan Economic Development Area, Nantong, Jiangsu, P. R. China

Email: gao.gh@fujitsu-nt.com

Abstract

Electronic package plays an important part in the development of IC industry. As we all known, strip warpage is a critical issue for the MAP-QFN manufacturing, results from the package structure, the thermal mismatch of materials and the manufacturing process. In this paper, a new finite element model was used to predict the warpage of one MAP-QFN block. And in this model, the temperature-dependent parameters of materials were characterized by DMA and DSC measurements, and the boundary condition was set up to be close to the real deforming situation. Furthermore, by the geometric triangle principle, the calculated warpage was also verified with the experiment measurement data. Finally we accomplished the optimization of related package parameters, and decreased the warpage of MAP-QFN by appling the dead weight on the strip block in the post mould curing process.

Introduction

QFN moulding array package (MAP-QFN) is a package type built-up with an IC chip attached to the exposed pad, then encapsulated with moulding compound. It has several potential advantages over others for a thin, cost reduction, thermal and high frequency package solution. In this paper, the QFN is a copper leadframe package with 48 leads, 12 leads along each edge, and 1024 QFNs are moulded in 4 so-called MAP-QFN blocks. This kind of package uses more molding compound, which will result in the strip warpage after molding.

Taking into account the singulation and solderability, the warpage is one of the major concerns in MAP-QFN manufacturing process. Recent improvements in the warpage control of MAP-QFN have intensified interest in the package geometry, materials properties and manufacturing process. Most researches about package warpage favor in developing the finite element analysis (FEA) method, which has been employed for years to investigate the warpage behavior in plastic encapsulating packages [1~2]. Oota and Shigeno [3] indicated that EMC with low CTE, high glass transition temperature, and low elastic modulus could reduce the package warpage. With simulation viewpoint, Kelly et al. [4] included the chemical shrinkage effect into the analysis due to the solidation behavior of EMC, and more reasonable warpage were predicted. D.G. Yang [5] presented a cure-dependent viscoelastic constitutive model to describe the evolution of material properties during cure.

In this paper, a new finite element model with nonlinear large deformation simulation was used to predict MAP-QFNs' warpage more reasonably. In order to reproduce the warpage, and with the target to eventually diminish this warpage to some extent, stress-free initial condition and chemical shrinkage of molding compound are implemented in this work, and a dead weight forcing the MAP-QFN block remaining flat [6] in the post mould curing process was included. Comparing with the measurements, it is found this novel FEM model can effectively validate the strip warpage behavior of MAP-QFN in the curing and cooling down process.

FEM MAP-QFN model and Materials characterization

Characterized by DMA and DSC measurements, glass transition temperature and temperature-dependent parameters of molding compound and die attach are determined. The mechanical properties of all materials are taken to be homogenous and listed in Table 1. From Table 1, we can find the package material properties for the die and leadframe are treated as temperature-independent linear elastic materials besides molding compound and die attach. Besides of thermal shrinkage, the chemical shrinkage is used in the simulation, its value is 0.15%, offered by the supplier. Moreover, the thickness of MAP-QFN's several layers are shown in the same table, and the total package thickness is fixed to 750 μm.

Table 1 Thickness and materials properties of MAP-QFN

Materials	E (GPa)		CTE (10^{-6}/°C)	Poisson ratio
Molding Compound	550 (μm)	T=25°C: E=29 T=260°C: E=0.9	T<135°C: CTE =7 T>135°C: CTE =34	0.3
Die	200 (μm)	136	2.15	0.278
Die Attach	25 (μm)	T=25°C: E=4.4 T=100°C: E=0.41 T=175°C: E=0.2	T<104°C: CTE =31 T>104°C: CTE =81	0.3
Leadframe	200 (μm)	121	16.8(16.3)	0.3

Because of the time-consuming calculation about complex copper leadframe, there is no necessary to build the leadframe model exactly as it should be. Instead, it was considered as a homogenous mixture of copper leadframe and molding compound. The effective CTE of leadframe was computed using Equation (1), where α_i and E_i are the CTE and Young's modulus values of the various materials and V_i is the volume fraction of the respective materials [7]. By this method, the model will be much simpler without losing accuracy.

$$CTE = \sum (\alpha_i V_i E_i) / \sum (V_i E_i) \quad (1)$$

In the factory, it is found that the strip warpage is generally crossbow after molding, as shown in Fig.1. Thus, the simulation will be focused on one block of MAP-QFN, also on the one-quarter of conventional model and one-quarter of detail model with die and die adhesive due to the symmetry. These three different 3D parametric FEM models are shown in Fig.2 respectively. In this research, we will pay more attention to model 1 and bring up a new idea to calculate the warpage of MAP-QFN. And the typical meshes for three models are also plotted in Fig. 2.

Fig.1 Photograph of MAP-QFN strip warpage

(a)

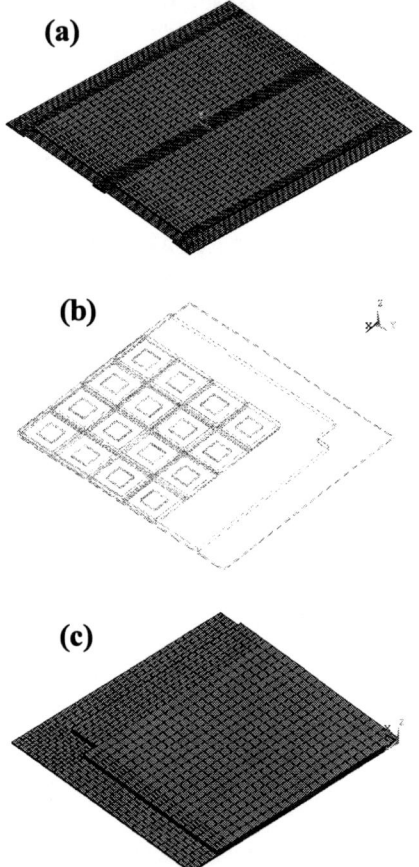

(b)

(c)

Fig.2 3D FEA models and meshes for MAP-QFN: (a) one block model; (b) one-quarter of detail model; (c) one-quarter of bimaterial model

Because warpage of the molding package block is caused by the chemical and thermal shrinkage of the package constituents during curing and cooling down. In this research, the simulation starts at the end of the molding process, assuming that MAP-QFN block is perfectly flat after molding

at 175°C, then it will go through curing and cooling down process in the air within 2 minutes. Lastly, in the post mould curing process, 3 kg dead weight was put on the blocks to try decreasing the strip warpage.

MAP-QFN package warpage measurements

In this paper, we put one smiling face MAP-QFN block on the flatbase, measure the height of four upward corners, then calculate their maximum displacement in the vertical direction. This displacement is defined as the "warpage", as shown in Table 2. The measurements were taken after the packages were cooled from the mould curing temperature to ambient room temperature. The most important, this MAP-QFN for measurement is packaged without die and die attach. So we just take into account the bimaterial model in the numerical simulation.

Table 2 MAP-QFN warpage measurements

	displacement (mm)
1st near corner of the mould inlet	0.98
2nd near corner of the mould inlet	0.93
1st far corner of the mould inlet	1.05
2nd far corner of the mould inlet	1.01

From data presented in Table 2, It could be seen that the two corners' height are 0.98 and 0.93 mm when the block corner is near the inlet of mould injection, but both far corners' height are larger. At last the average warpage of the MAP-QFN block is 0.962 mm. The reason about the difference four corners' height can be attributed to the mold flowing condition and the fixation points position in the process of cooling down from curing temperature. Furthermore, the measured data would also be served as a form of validation for the finite element simulation results.

Discussion 1 warpage simulation with one-quarter of model

Note that all the initial and boundary conditions were kept constant for all the models. Due to the different CTE properties of the materials (die, molding compound, die attach and leadframe), MAP-QFN strip will be principally subjected to thermally induced stresses and strains when the package is under curing and cooling down process. Stresses would be experienced at the interfaces between materials and result in the strip warpage and potential failure.

Using the conventional strip warpage FEA method, only one-quarter of the MAP-QFN block is modeled, based on the structure symmetry, as seen in Fig.2 (b) and (c). The model (b) is simulated like the real MAP-QFN except bonding wire for its tiny influence on the strip warpage; and model (c) is just a bimaterial model. About both models, the same boundary conditions are applied: corresponding symmetric boundary conditions are used for the two symmetric planes, the node at the center axes bottom is fixed in all directions; the map mould are kept flat during the curing process, but free during cooling down.

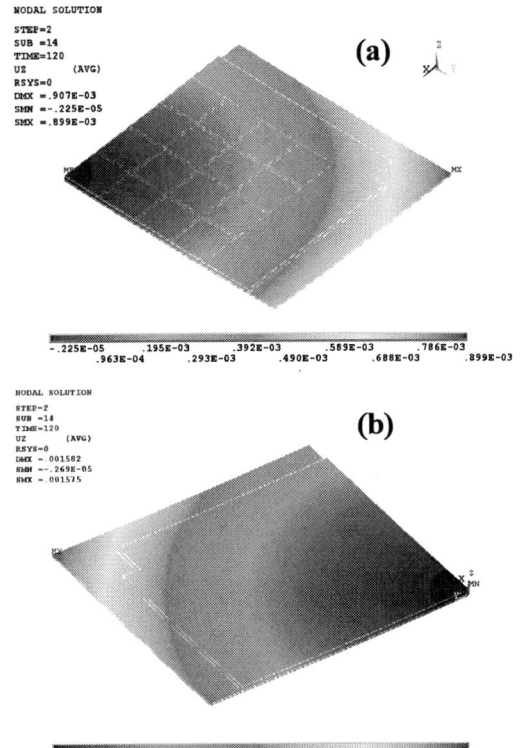

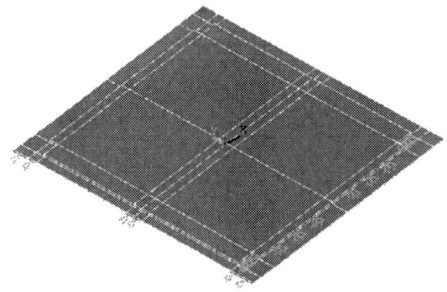

Fig.4 The novel boundary condition set-up of model (a)

The distributions of z displacement for the whole block model are shown in Fig.5. Also the z displacement data of Fig.5 measured along y axes are plotted in Fig.6, and zero point coordinate is the central point of this block. As seen in Fig.5, the block bends upward, just the same as the result of one-quarter of models simulation. But it shows clear crossbow warpage and slight coilset warpage, which agrees well with the actual situation. Here comes another problem: in this study, we just know the nodes' maximum z displacement is 3.377 mm, but how to get the accurate warpage of MAP-QFN?

Fig.3 Simulated warpage patterns of MAP-QFN: (a) one-quarter of detail model; (b) one-quarter of bimaterial model

Fig.3(a) and (b) show the warpage patterns for the one-quarter of block model after been cooling down to the room temperature. It can be seen that both models bend upwards by curing and thermal stress as well. The main reason is that the CTE of the encapsulated epoxy is much larger than the coefficient of the base copper leadframe. Here we take the maximum z displacement as indication of the strip warpage, and the warpages are 0.90mm and 1.13mm.

Comparing the warpages of two models, the bimaterial model (c) performs larger than detail model (b). But we know that simulated calculation of the warpage with detail model will be more complicated, and need high-level computer and time-consuming. So in order to gain more reasonable warpages by simpler model, we present a novel simple-bimaterial model to do simulation of MAP-QFN block.

Discussion 2 warpage simulation with a new model

In this new method, the simple-bimaterial model is still used, but this time the whole MAP-QFN block model is built, including half of slot between two blocks. So in the simulation, the slot can behave its destressing function, and this function must be helpful to decrease the MAP-QFNs' warpage. About this whole block model, the novel boundary conditions are set up as shown in Fig.4: firstly, the nodes at the cross-section of slot are fixed in x direction; secondly, the nodes at the inlet of mould injection are fixed in y and z direction, because actually this inlet is fixed in the curing and cooling down process. This boundary condition set-up will be more close to the real deforming situation in the factory manufacturing, and can predict the strip warpage of MAP-QFN more reasonablely.

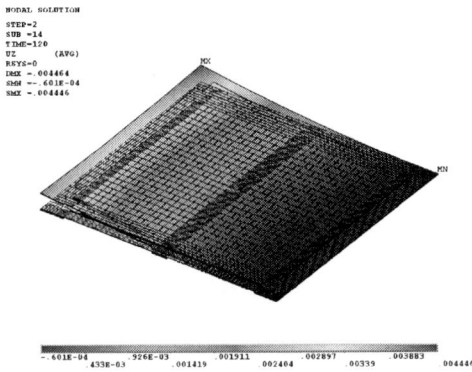

Fig.5 Simulated warpage pattern of MAP-QFN

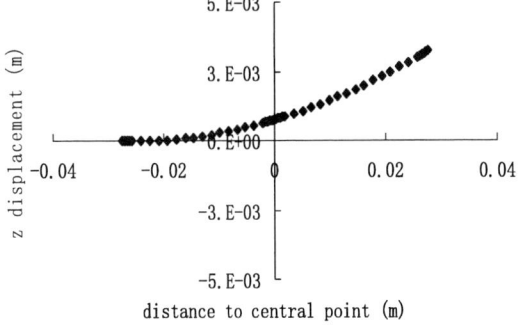

Fig.6 FEM nodal z diplacement of MAP-QFN

So we present a novel method: using the geometric triangle principle to calculate the warpage. The schematic diagram is shown in Fig.7, and the corresponding mathmatical expression is Equation (2).

189

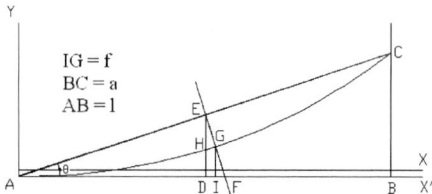

Fig.7 Schematic diagram of Equation (2)

$$warpage = (\frac{a}{2} - f)(\cos arctg \frac{a}{l})^{-1} \quad (2)$$

According to the traces of MAP-QFN blocks' warpage, we can find the diagonal line is a grate circula arc after bending. The circula arc is like the arc AC in the Fig.7, so the G and I point is the central zero point of the block. Line IG and line BC are defined as f and a in the z displacement from numerical calculation result. The other known value is line AB. In the Equation (2), AB is l, which is the length of the diagonal line of one block map. By substituting the numerical value for the variable in the Equation (2), we can calculate the warpages of the two diagonal cross lines, then get the average warpage of the MAP-QFN.

Based on Equation (2), that the average warpage of the MAP-QFN is 0.988mm can be concluded. The calculation result shows great agreement between the experimental measurements and FEM predictions, and this whole block model behaves better and more reasonable than one-quarter of model in the warpage simulation.

In order to decrease the strip warpage of MAP-QFN, many methods have been considered, and the most effective method is to put the dead weight on the MAP-QFN strips in the PMC process [6]. This process was also simulated with one block model to check its validity. In the factory, 3 kg dead weight was put on the whole strip in the PMC process, and in the simulation, 3 kg equivalent compressing stress was applied to the block for the simplicity consideration. The Simulated warpage pattern of MAP-QFN under 3 kg dead weight can be seen in Fig.8, comparing it with the Fig.5, both in the same model and the same mesh, but this simulated pattern shows flater. This 3 kg dead weight has effectively corrected the warpage problem to some extent. As shown in Fig.9, there is less warpage under the dead weight compared with the data plotted in Fig.6.

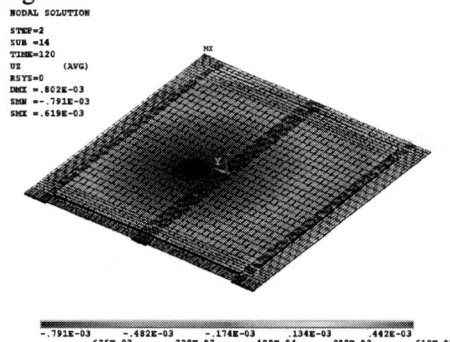

Fig.8 Simulated warpage pattern of MAP-QFN under 3 kg dead weight

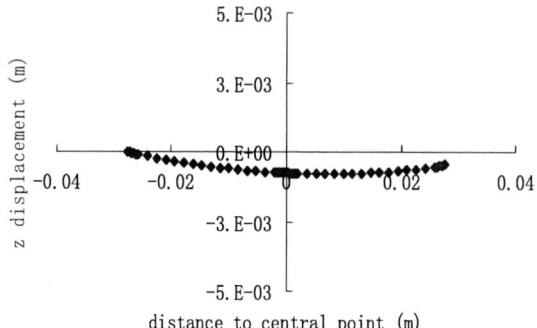

Fig.9 FEM nodal z diplacement of MAP-QFN with 3 kg dead weight

In terms of the standpoint of the dead weight, cooling down process will always occur in a flat strip situation during PMC. In this process, the stresses caused by the curing and cooling down are forced to relax to a relatively flat stress free situation when the temperature is level down [6]. Therefore, a less effect of the shrinkage on the warpage may be achieved when using a dead weight. But without dead weight, the chemical and thermal shrinkage during the period of cooling down can induce some warpage through the coefficient of the expansion and rubber modulus, especially in the MAP-QFN without die and die attach.

Conclusions

In this research, we employed a finite element analysis with a novel model of MAP-QFN to forecast the warpage effect in the curing and cooling down process. Using the geometric triangle principle, the average warpage by FEA simulation was calculated. The warpage result of new model was shown in reasonable agreement with the experiment measured values. In addition, we took account of all the package materials and the specific package structure in one-quarter of detail model, also present 3 kg dead weight compressing on the MAP-QFN block to effectively decrease the strip warpage. So, with FEA method, the numerical simulating analysis provids an adequate guideline for the actual manufacturing optimization and efficiently control the warpage in various molding array packages.

Acknowledgments

The authors would like to acknowledge my colleagues in R&D center (Shen Haijun, Shi Jiangen) for the experiment support.

References

1. G. Q. Zhang, "The challenges of virtual prototyping and qualification for future microelectronics",Microelectronics Reliability, 43, 2003, pp. 1777-1785.
2. Irving Y. Chien, *et al*, "Low-warpage molding compound development for array packages", Global SMT & Packaging, Jan. 2007, pp. 296-300.
3. Oota, K. and Shigeno, K., "Development of Molding Compounds for BGA", *45th Electronic Components Technical Conference*, 1995, pp. 78-85.
4. Kelly G., *et al*, "Investigation of Molding Compound Chemical Shrinkage in the Stress and Warpage Analysis of PQFPs", *IEEE Transactions on Components,*

Packaging, and Manufacturing Technology, Part B, Vol. 19, No. 2 (1996), pp. 296-300.

5. D.G. Yang, *et al*, "Prediction of Process-Induced Warpage of IC Packages Encapsulated with hermosetting Polymers", *Proceedings of 54th Electronic Components & Technology Conference*, 2004, pp. 98-105.

6. Beijer1, J. G. J. *et al*, "Warpage minimization of the HVQFN map mould", *6th Int. Conf. on Thermal, Mechanical and Multiphysics Simulation and Experiments in Micro-Electronics and Micro-Systems, EuroSimE*, 2005, pp. 1-7.

7. Yuan Li, "Accurate predictions of flip chip BGA warpage", *53rd Electronic Components and Technology Conference*, May. 2003, pp. 549-553.

Numerical Investigation on the Effect of Filler Distribution on Effective Thermal Conductivity of Thermal Interface Material

Cong Yue[1], Yan Zhang[1], Johan Liu[1,2*], Zhaonian Cheng[2], Jing-yu Fan[3]

[1]Key Laboratory of Advanced Display and System Applications, Ministry of Education & SMIT Center, School of Mechatronics Engineering and Automation, Shanghai University, P.O.Box 282, Shanghai 200072, China
[2]SMIT Center & Bionano Systems Laboratory, Department of Microtechnology and Nanoscience, Chalmers University of Technology, SE-412 96 Gothenburg, Sweden
[3]Shanghai Institute of Applied Mathematics and Mechanics, Shanghai University, P.O.Box 189, Shanghai 200072, China
*Email: jliu@chalmers.se; Tel: +86-21-56331599; Fax: +86-21-56332054

Abstract

Thermal interface materials have been widely adopted in the thermal management for electronics system. Most of the thermal interface materials are made of polymers with thermally conductive particles distributed inside to enhance the thermal conductivity. Thus it is essential to figure out the effective thermal conductivity of this composite material. In the present paper, a parameterized cubic cell model had been developed and implemented by the finite element method. The numerical simulations were carried out to investigate the effect of the filler distributions on the effective thermal conductivity of the thermal interface materials. The volume percentage loadings of the particles ranging from 13% to 74% had been considered, and different particle distribution patterns had also been analyzed. The simulation results were compared with the experimental data as well as other models. A fairly good agreement was obtained for the particle volume percentage loading under consideration, which verified the developed cubic cell model.

Introduction

With the increasing microprocessor powers, the role of thermal management for electronics system becomes more crucial to the overall system performance as the packaging density becomes much higher while IC power increases at the same time. Various heat dissipation methods and devices have been developed in order to reduce the thermal resistance across the interface, either to increase the actual contact area (e.g. by increasing contact pressure or grinding and polishing of the surfaces to improve flatness and roughness values) or to fill the remaining air gaps with suitable thermal interface materials [1].

Thermal interface materials (TIMs) have been widely adopted to minimize the thermal interface resistance between the rough surfaces of heat generating components and the heat dissipation devices. Most of TIMs are made of polymers with thermally conductive particles distributed inside to enhance the thermal conductivity. And there are various kinds of approaches to calculate the effective thermal conductivity of the two-phase composite systems. The most well-known method to measure the TIM thermal conductivity was based on the ASTM standard [2], and the test device had also been set up. The common mathematic methods for calculating thermal conductivity of TIM include Maxwell's equation [3], Bruggeman's equation [4] and other modified equations [5]. And there are also other models to obtain the thermal conductivity of the TIMs, such as Maxwell-Eudken equation

(EMT) [6], Nelsen Model [7], effective unit cell model (EUCM) [8] and percolation model [9].

The EUCM model had been used for two-dimensional finite element simulation to predict the thermal conductivity of filler composite material, while the percolation model introduced the percolation theory and the matrix method to calculate the values of the effective conductivity. All the approaches were developed to obtain the thermal conductivity of the TIMs in addition to the relatively complex and time-consuming experimental investigations. But the results obtained from those equations/methods did not agree well enough with the experimental data, especially when the filler volume fraction was high. For example, the EMT and Nelsen Model's results were not accurate when the filler volume fraction was higher than 50%. The EUCM models and Percolation Model were able to obtain the accurate results. However, the EUCM model was based on a two-dimensional simulation and the particles were assumed to be cylindrical. The percolation model presumed the particles to be cubic, and consequently a correction factor had to be introduced for the geometric difference.

In fact, most of the particles filled in TIMs are spheres. Then it is nature to consider the thermal interface material as matrix with spherical fillers inside during model developing. In the present paper, a cubic cell model (CCM) had been developed in order to predict the effective thermal conductivity of the considered TIM with randomly distributed particles. The model was then implemented by finite element method (FEM). Various particle volume fractions had been considered to obtain the corresponding effective conductivity. Furthermore, different filler distribution patterns had also been simulated to investigate the influences of the filler distribution on the effective thermal conductivity of thermal interface material.

Numerical Method

In the present paper, we developed a cubic cell model to obtain the thermal conductivity of the thermal interface material. As shown in Fig.1, the considered thermal interface material was trapped between two components of different materials. A unit cubic cell was chosen from the TIM. This was also the reason why the model was named the cubic cell model.

The heat flux through the cubic cell from bottom to top was assumed to be constant and without dissipation from the side surfaces. Two classical expressions were employed in the cubic cell model as:

978-1-4244-2739-0/08/$25.00 ©2008 IEEE

$$R = \frac{\Delta T}{Q} \qquad (1)$$

$$R = \frac{H}{K \cdot A} \qquad (2)$$

where R is the thermal resistance, ΔT is the temperature difference between the bottom and top surfaces of the cubic cell, Q is the heat flux through the composite, H is the overall length of the composite, K is the thermal conductivity and A is the cross section area.

From Eq. (1) and (2) it could be obtained that:

$$K = \frac{H \cdot Q}{\Delta T \cdot A} \qquad (3)$$

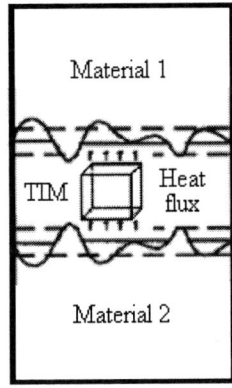

Fig.1 Sketch of the cubic cell in TIM

There were thermal conductive particles distributed in the cubic cell material matrix. In the present work, the particles inside were supposed to be spheres with the same diameter. The material parameters for the considered filler and matrix were listed in Table I.

Table I Material parameters of TIM

	Thermal conductivity (W/m·K)	Specific heat (J/kg·K)	Density (g/cm³)
Filler	12.17	705	2.2
Matrix	0.36	2100	0.92

The way of arranging particles in the matrix and the filler fraction were two important factors for the thermal interface material. And these were also key points during the model development.

In the case of spherical particles in the matrix, there appeared various structures about how to fill the spheres. The most commonly adopted methods included the body-centered cubic crystal structure (BCC), the face-centered cubic crystal structure (FCC) and the hexagonal close-packed crystal structure (HCP). In order to set up a method with a wide range of the filler volume fractions, the cubic cell model was built with the structure of FCC, which possessed the maximum packing fraction of 0.7405 [9].

The model had been implemented by the finite element method in ANSYS environment. Fig.2 (a) showed the particle fillers arranged in the FCC structure, and Fig.2 (b) is the cubic

cell with FCC particle distributed inside. The fillers in the matrix were chosen to be spheres with 10 μm in diameter.

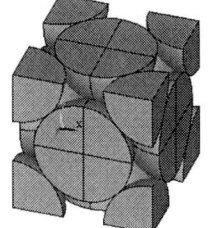

Fig.2 (a) FCC structure

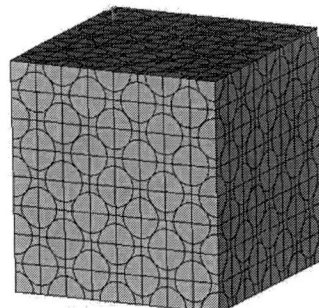

Fig.2 (b) FCC unit

Fig.2 The utilized Face-Centered Cubic structure

In the simulation, the heat flux loaded on the top and bottom surfaces were set to be $100 W/cm^2$, which was a common value in the present apparatus. So there was a constant heat flux through the cell from its bottom to top. During the computation, a constant temperature T_0 was defined on the bottom surface of the cubic cell.

Various filler fractions had been studied in the computation, as listed in Table II. The volume percentage loading of particles ranged from 13% to 74%. Also, different particle distributions at each volume fraction had been investigated in the case study in order to understand the influence of the particle distribution on the effective thermal conductivity of TIM.

Table II Filler fractions in computation

Filler	No.1	No.2	No.3	No.4	No.5	No.6	No.7	No.8
Fraction (%)	74.0	65.3	56.6	47.98	39.3	30.6	21.9	13.3
Number	256	226	196	166	136	106	76	46

In the present finite element analysis, a parameterization method was utilized in the model establishment. This made the simulation process high efficient and the computational cost quite acceptable.

Simulation Results and Analyses

The thermal conductivity values of the filler composite material with the prescribed filler volume fraction had been calculated.

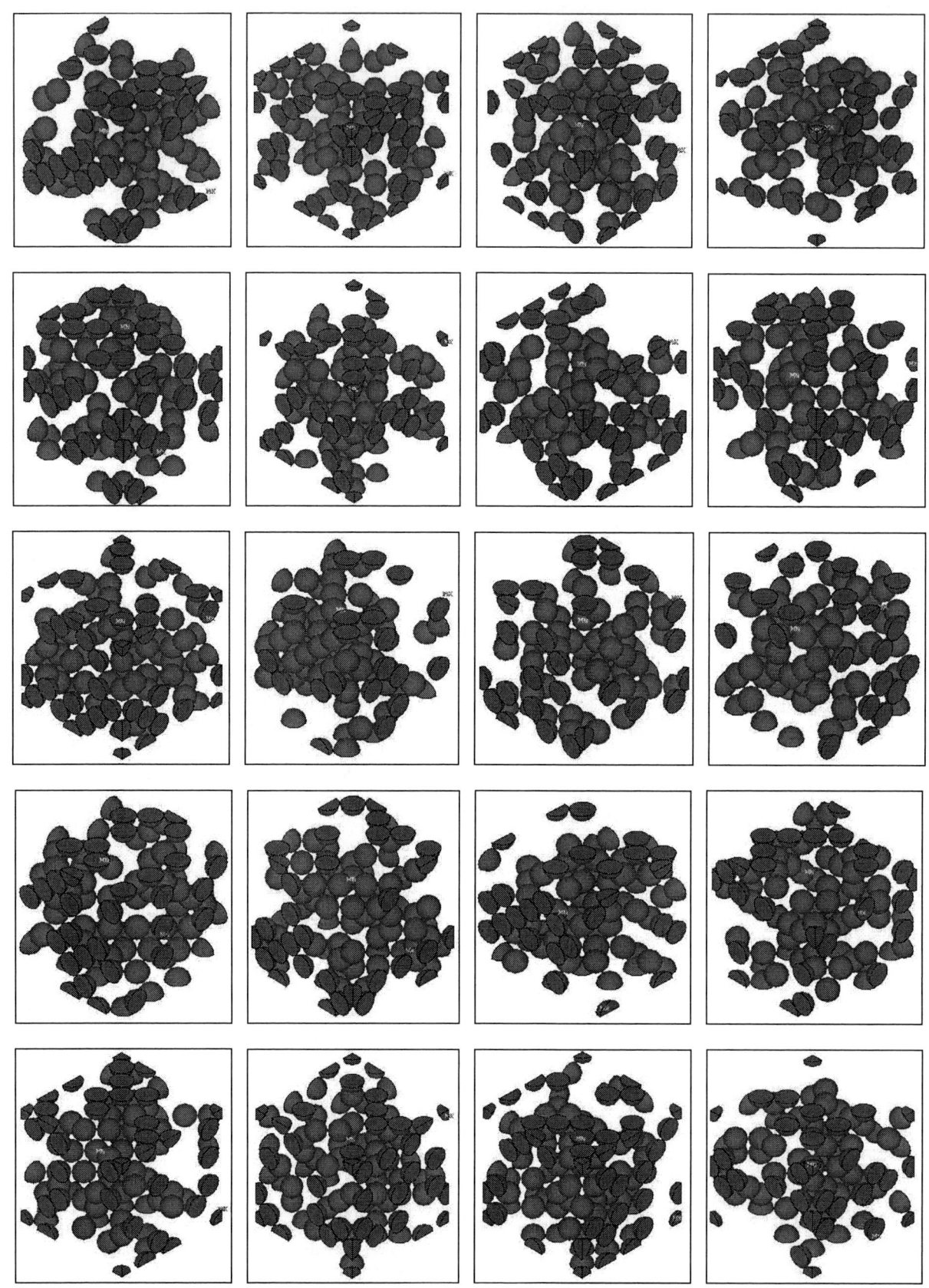

Fig.3 Randomly distributed particles (with 21.9% volume fraction)

In order to study the stability of the simulations with different particle distributions, each filler fraction case had been computed for 20 randomly distributed patterns. Namely the computation process had been conducted for twenty different filler distributions in each filler fraction case. Fig.3 shows the twenty kinds of particle distributions at the filler fraction 21.9%. For each filler fraction case, the calculated value of the effective thermal conductivity is the average of 20 values of the 20 different particle distribution patterns. The model is developed through a parameterization method so that the computation is of high efficiency, and consequently the time expense is still quite acceptable.

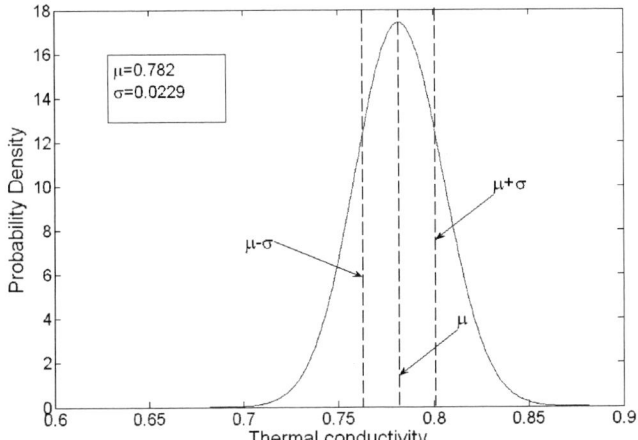

Fig. 4 The normal distribution of effective thermal conductivity of TIM with 21.9% filler fraction

Fig.4 shows the normal distribution of the effective thermal conductivity values of various particle distributions in the case of 21.9% filler volume fraction. Here μ and σ are the mean value and the standard deviation, respectively. It can be inferred from the small σ value that the simulation results are of good stability. This indicates that, for the considered material parameters, the particle distribution patterns only have a limited influence on the bulk thermal conductivity. When it comes to cases with other fillers, such as Cu or Ag with much higher thermal conductivity values than that of the matrix, the particle distribution effect may be more remarkable and need to be considered carefully.

During the computation, the particles in the cubic cell are assumed to contact perfectly with the matrix. While in the practical situation, the contact between the fillers and the matrix of different materials is not so ideal. So a coefficient of contact loss α has been introduced into the model as:

$$K_e = \alpha \cdot K \qquad (4)$$

where K_e is the bulk effective thermal conductivity of the thermal interface material. The value of the coefficient α depends on the filler and the matrix materials. For the present case, $\alpha = 0.7$.

The simulation results obtained from the established cubic cell model have been compared with the results calculated from the percolation model [9] as well as with the experimental measurement [10]. As shown in Fig.5, a fairly good agreement has been obtained, verifying the developed cubic cell model.

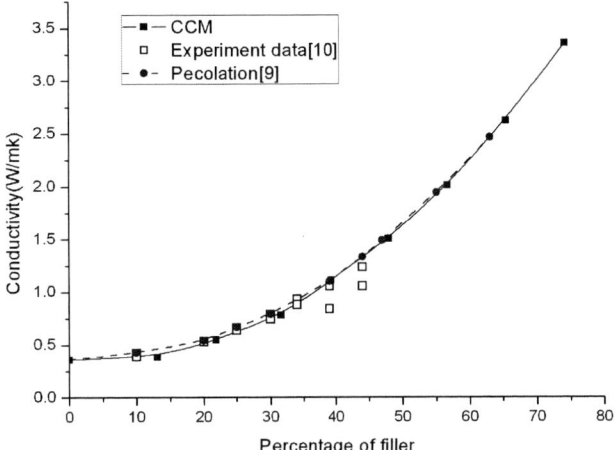

Fig. 5 Comparison of CCM with the percolation model and the experimental data

Conclusions

In the present paper, a parameterized cubic cell model had been developed to obtain the effective thermal conductivity of thermal interface material. Numerical simulations based on the developed model had been carried out to calculate the effective thermal conductivity of matrixes with various filler contents. Furthermore, the influence of the filler distributions on the effective thermal conductivity had also been investigated. Comparison of the results obtained by the developed model with those by the percolation model and experimental data showed good agreement, which verified the developed cubic cell model.

Acknowledgments

This work is supported by the National Natural Science Foundation of China (grant no. 10702037). Supports from the Innovative Foundation of Shanghai University and Shanghai Leading Academic Discipline Project (grant no. Y0103) are also appreciated. The work at Chalmers University of Technology is supported by the Seventh Framework Program of the European Union (Project name: Nano Packaging Technology for Interconnection and Heat Dissipation under the contract No: 216 176) and partly by the EU EC-GEPRO project.

References

1. Wataru Nakayama, E. Bergles, "Thermal interfacing techniques for electronic equipment- a perspective," Journal of Electronic Packaging, Vol. 125, No. 2 (2003), pp. 192–199.
2. J. R. Culham, P. Teertstra, I. Savija, and M.M. Yovanovich, "Design, assembly and commissioning of a test apparatus for characterizing thermal interface materials," Inter Society Conference On Thermal Phenomena, 2002, pp. 128-135.
3. Maxwell, J. C., A Treatise on Electricity and Magnetism, Dover (New York, 1954) Vol. 1 pp. 435-441.

4. Bruggeman DAG., "The prediction of the thermal conductivity of heterogeneous mixtures," Ann Phys, Vol. 24 (1935), pp. 636-664.

5. Wang Jiajun, Yi Xiao-Su, "Effects of interfacial thermal barrier resistance and particle shape and size on the thermal conductivity of AlN/PI composites" Composites Science and Technology, Vol. 64 (2004), pp. 1623–1628.

6. Nielsen, L.E, "The Thermal and electrical conductivity of Two-phase systems," Ind. Eng. Chem. Fundam., Vol. 13, No.1 (1974), pp. 17-20.

7. Davis, L. C., Artz, B.E., "Thermal Conductivity of Metal-Matrix Composites," Journal of Applied Physics, Vol. 77, No. 10 (1995), pp. 4954-4860.

8. Ptrick E.Phelan. Ravi Prasher, "An Effective Unit Cell Approach to Compute the Thermal Conductivity of Composites With Cylindrical Particles," Journal of Heat Transfer, Vol. 127 (2005), pp. 553-559.

9. Amit Devpura, Ptrick E. Phelan, "Percolation Theory Applied to the Analysis of Thermal Interface Materials in Flip-Chip Technology", Intel Society Conference on Thermal phenomena, 2000, pp. 21-28.

10. Agari, Y., Ueda A., Tanaka, M. and Nagai, S., "Thermal Conductivity of a Polymer Filled with Particles in the Wide Range from Low to Super-High Volume Content," Journal of Applied Polymer Science, Vol. 40 (1990), pp. 929-941.

Development of Moisture Automation Analysis System for Microelectronic Packaging Structures

Yangjian Xia, Yuanxiang Zhang, Lihua Liang

Fairchild-ZJUT Microelectronic Packaging Joint Lab, Zhejiang University of Technology, Hangzhou, China

Yong Liu, Scott Irving, Timwah Luk

Fairchild Semiconductor Corp., S.Portland, Maine, USA

0086-571-88320294, lianglihua@zjut.edu.cn

Abstract

Moisture sensitivity of packages is an area of great concern for the electronics industry. The differential swelling of materials in a non-hermetic package during manufacture, handling, storage, assembly, and then also during its lifetime can cause stresses large enough to damage the package. A moisture automation analysis system is developed based on ANSYS Workbench and Excel platform in this paper. The goal of this paper is to develop an analysis system for moisture diffusion, hygro-mechanical stress and vapor pressure analysis automatically for different package family The application of moisture automation analysis system for moisture diffusion, vapor pressure diffusion, thermal-mechanical stress; hygro-mechanical stress; vapor equivalent thermal mismatch stress are performed. The comparisons of some results based on moisture automation analysis system with those from ANSYS are given. The results from this paper agree with those from pure ANSYS.

1. Introduction

Electronic packages are known to absorb moisture when exposed to humid ambient conditions. The presence of moisture in the packages induces hygroscopic stress through differential swelling, induces vapor pressure that is responsible for the eventual popcorn cracking in reflow. Moisture induces the interfacial stresses generated between the die attach and die, die and mold compound, as well as leadframe with mold compound. This may finally lead to die delaminating and package cracking. Therefore, moisture analysis plays an important role in the integrity and reliability of electronic packaging. More and more researches are focusing on the study of moisture and its related reliability of electronic package. [1]- [2]

Obtaining the moisture simulation results (such as hygro-mechanical stress, vapor pressure) is very important to improve the IC package design and reliability. However, most IC package development engineers and reliability engineers may not have the abilities to perform such analysis, because they are familiar with the product design and process but poor in moisture theory and FEA skills. Therefore, it is necessary to develop a fully automation interface that has the ability to complete the package moisture analysis automatically.

ANSYS Workbench is a new-generation platform used for developing and managing FEA simulations. It not only offers highly integrated engineering simulation platform and bi-directional parametric integration with most available CAD systems, but also provides multi-tiered customization tools to support a variety of development efforts. ANSYS Workbench Software Development Kit (SDK) is an open architecture platform that allows customers to develop and integrate application architecture on Workbench environment. In nowadays, Microsoft Excel spreadsheet has been widely used in many fields because its widespread availability and easy-of-use. Especially, Visual Basic for Application (VBA) expands the capability of Excel to realize many automated and complex tasks. Zhang and his co-workers had developed a highly efficiency simulation system (AutoSim) to carry out the thermal analysis automatically in Excel spreadsheet cooperating with Workbench in the past year [3]. A user is only required to input the basic parameters in Excel interface. And then the JScript application, which is stored background, will drive and integrate Workbench components automatically to perform the thermal analysis on package according to existing procedure.

The goal of this paper is to expand the capability of moisture analysis in Excel spreadsheet, include moisture diffusion analysis, hygro-mechanical stress and vapor pressure analysis. The user should be required to input the title of simulation and the user name, select the geometry models from CAD Library, select the moisture analysis type and so on. The Executable Excel Spreadsheet will collect all the information that the user inputs to perform the whole FEA simulation and automatically calculate the moisture analysis results and save all results. This customized system has been applied for moisture diffusion analysis of a Fairchild MLP package family and has obtained accurate results as compared to pure FEA.

2. Moisture Analysis Theory

2.1 Moisture diffusion and hygroswelling

According to the moisture theory, the transient moisture diffusion equation is analogous to the transient heat conduction equation and it can be described by Fick's Law as:

$$\frac{\partial C}{\partial t} = D\left(\frac{\partial^2 C}{\partial x^2} + \frac{\partial^2 C}{\partial y^2} + \frac{\partial^2 C}{\partial z^2}\right) \qquad (1)$$

where C is the local moisture concentration, x, y and z are the spatial coordinates, D is the diffusivity which measures the rate of diffusion. However, unlike temperature, the moisture concentration is discontinuous along the bi-material interface. Moisture concentration discontinuity across bi-material interfaces can be overcome with the use of continuous field variables such as "wetness" [1]. w, as the field variable, which is also continuous across multilaterals interface [2]. It is defined as:

978-1-4244-2739-0/08/$25.00 ©2008 IEEE

$$w = \frac{C}{C_{sat}}, \quad 1 \geq w \geq 0 \qquad (2)$$

where C_{sat} is the maximum moisture concentration that can be absorbed by the material, the lower limit $w = 0$ means it is dry, and the upper limit $w = 1$ means it is fully saturated with moisture. The Eq. (1) can also be rewritten as:

$$\frac{\partial w}{\partial t} = D \left(\frac{\partial^2 w}{\partial x^2} + \frac{\partial^2 w}{\partial y^2} + \frac{\partial^2 w}{\partial z^2} \right) \qquad (3)$$

After that it can be solved as a typical heat transfer problem using commercial FEA softwares. For moisture absorption modeling, the initial condition is $w = 0$ for the whole package, and the boundary condition is $w = 1$ at the external surfaces which are exposed to the ambient moisture. $w = 1$, implies saturated wetness, and $w = 0$ implies complete dryness [4].

Due to CME (Coefficient of Moisture Expansion) mismatch among various materials, the hygro-swelling stress is induced. The concept is analogous to CTE mismatch and thermo-mechanical stress which we are more familiar with. The hygro-mechanical problem can be solved using the same procedure as thermo-mechanical solution. It can be described as equation (4). The moisture loading is from 0% to 100% moisture concentration.

$$\varepsilon_h = \beta \cdot C \qquad (4)$$

where ε_h is the hygrostrain, β is the CME, and C is the moisture concentration. The values of CME for EMC and Die attach materials are shown in Table 1. Die, leadframe, and pad materials absorb no moisture, have no hygroswelling. And the CME value of them should set to be zero, but if they were that, the process of ANSYS Workbench running will not continue. So they were assumed to have very small value compared to EMC.

2.2 Vapor pressure introduction

The moisture exists everywhere in polymer materials in an electronic package after preconditioning. During solder reflow in surface mounting, the temperature of the package body is raised up to 220-260℃. The moisture absorbed in the plastic package becomes vaporized and exerts a pressure on the internal package body. The induced vapor pressure coupled with the thermal stress, hygroswelling stress could result in a failure mechanism often referred to as "popcorn" phenomenon. It is particularly important to analysis the vapor pressure distributions and variations with temperature and moisture concentration.

There are three distinct cases for the vapor pressure evolution from the preconditioning temperature T_0 to the current reflow temperature T [5].

In case 1, the moisture in the void is in the single vapor phase at T_0, thus the vapor pressure at T follows the ideal gas law.

$$\text{When} \quad \frac{C_0}{f_0} \leq \rho_g(T_0), \quad P(T) = \frac{C_0 p_g(T_0) T}{\rho_g(T_0) f T_0} \qquad (5)$$

where, P is the pressure and p_g is the saturated vapor pressure.

In case 2, the moisture is not fully vaporized even at reflow temperature T. The moisture in the void is in the mixed liquid-vapor phase at the temperature from T_0 to T. Thus the vapor pressure maintains the saturated vapor pressure during the course of the temperature rise.

$$\text{When} \frac{C_0}{f} \geq \rho_g(T), \quad P(T) = p_g(T) \qquad (6)$$

In case 3, it is an intermediate case between case 1 and case 2, where the moisture is in the mixed liquid-vapor phase at preconditioning temperature T_0, but in the single vapor phase at T. The moisture is fully vaporized at a temperature between preconditioning temperature T_0, and the peak reflow temperature T.

$$\text{When} \begin{cases} \dfrac{C_0}{f_0} > \rho_g(T_0) \\ \dfrac{C_0}{f} < \rho_g(T) \end{cases}, \quad P(T) = p_g(T_1) \frac{T}{T_1} \frac{f(T_1)}{f} \qquad (7)$$

where T_1 is phase transition temperature at which the moisture can be fully vaporized.

2.3 Equivalent thermal stress theory

The solution of thermal model is analogous to moisture diffusion. However, the thermal diffusivity is much faster than the moisture diffusivity. When the external surface is heated to a reflow temperature, the internal package reaches this uniform temperature within a few seconds. Therefore, in the subsequent thermo-mechanical and vapor pressure models, temperature distribution during reflow can be assumed to be uniform throughout the package body. The temperature loading applied is from the stress free reference temperature (usually it selects a curing temperature or a glass transition temperature of the mold compound) to the reflow temperature.

Linear-elastic thermo-mechanical stress model is applied in this paper. We know the basic thermal expansion mismatch strain can be expressed as

$$\varepsilon_T = \Delta T \cdot \alpha \qquad (8)$$

where ε_T is the thermal strain, α is the CTE, and ΔT is the temperature changes.

The linear vapor pressure induced strain may be equivalently estimated by [6]

$$\varepsilon_p = \frac{1-2v}{E}P \qquad (9)$$

where v is the Poisson's ratio, E is the modulus, and P is the average vapor pressure. The modulus of mold compound drops a few orders at the reflow temperature, thus the vapor pressure strain may become as important as thermal or hygrostrain in reflow.

3. Development of Moisture Automation Analysis System

3.1 Structure of moisture automation simulation system

The moisture automation simulation (AutoSim) system developed here includes three modules: a Package Model Information Library, an Executable Wizard System, and a Package Material Library. User is only required to input basic data in a Wizard interface (an Excel Spreadsheet) and it will link to Workbench and automate the whole steps of moisture and vapor simulation. At last, the analysis results will also export into Wizard interface. Figure 1 shows the modules of AutoSim.

The Wizard System is composed of three parts, which is used for moisture analysis, vapor analysis and combination analysis respectively. The Wizard System for moisture analysis performs the simulation of moisture diffusions and hygroswelling analysis. Other Wizard systems are used to perform the vapor pressure analysis and the combination analysis which includes the stress analysis of thermal-mechanical; hygro-mechanical; vapor induced equivalent thermal mismatch and integrated.

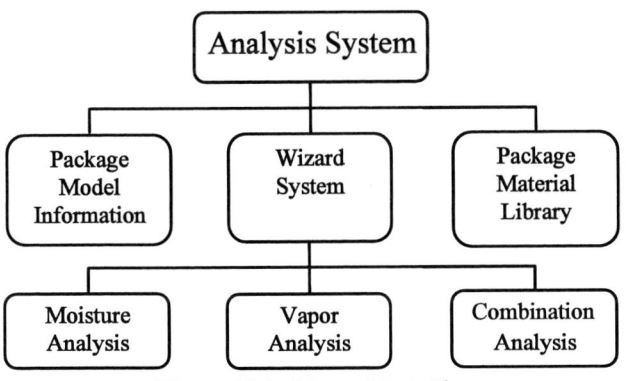

Figure 1 Modules of AutoSim

3.2 Module introduction of moisture automation analysis system

The Package Model Information Library plays a very important role in the whole process of the analysis automatically. It is the core element of moisture automation analysis system and provides an interface for user to store CAD models easily, include the model's name and material. It is shown in figure 3 and figure 4. When the user clicks the "export" button, it will automatically export all components'

names which are defined in CAD software. Then, user may select corresponding materials for every component from Package Material Library. All the information is saved in a database which will be transferred by Wizard System.

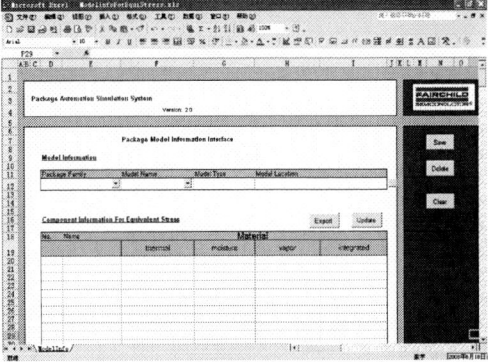

Figure 2 Package Model Information Interface for equivalent thermal stress

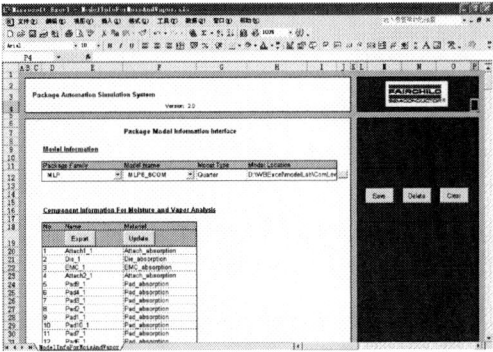

Figure 3 Package Model Information Interface for moisture and vapor distribution analysis

Wizard System is divided into three parts; there are Wizard System for Moisture Diffusion and Hygroswelling Analysis (Figure 4); Wizard System for Vapor Pressure Analysis (Figure 5); Wizard System for Equivalent Thermal Stress Analysis (Figure 6). The Wizard System uses Excel Spreadsheet as the user interface. All the Wizard System require user to input the basic data. An example is to input user name, title of simulation, job name, working directory and so on. User is also required to select a model to be simulated, and then the model information stored in Package Model Information Library will be exported into Wizard interface. The Wizard System will collect all the information that user inputs and perform the whole corresponding simulation automatically according to the predefined procedure in background.

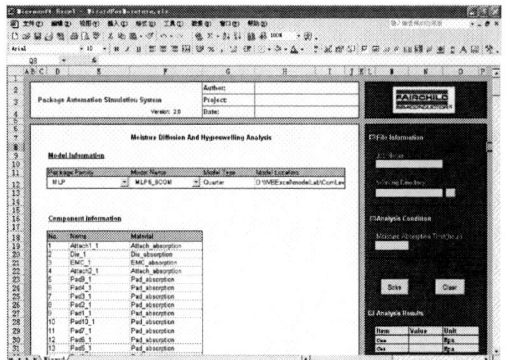

Figure 4 Wizard interface for moisture diffusion and hygroswelling analysis

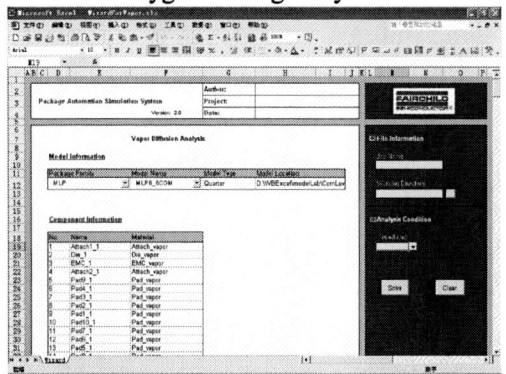

Figure 5 Wizard Interface for vapor pressure distribution analysis

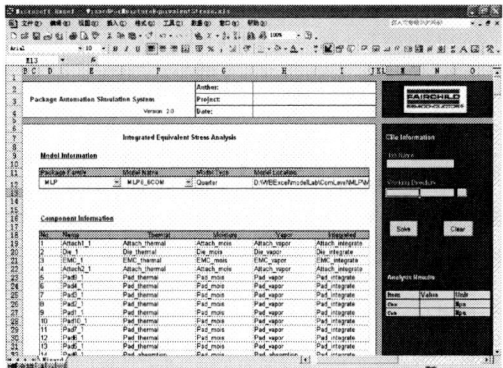

Figure 6 Wizard Interface for equivalent thermal stress analysis

4. Application of Moisture Automation Analysis System

To test the development work, the AutoSim has been applied to moisture and vapor analysis for MLP family packages from Fairchild Semiconductor. In this paper, MLP 6×6 (quarter model) is chosen for analysis, as shown in figure 7.

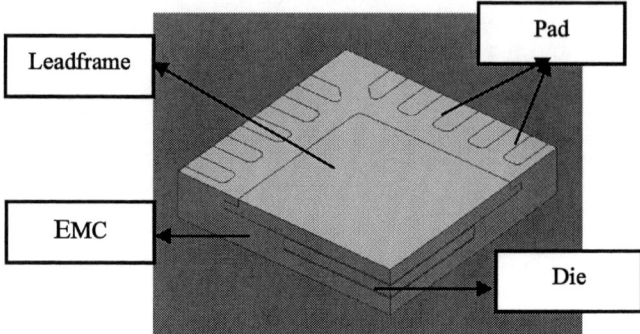

Figure 7: MLP6x6 (quarter) model in Inventor

4.1 Application for moisture analysis

The moisture properties, i.e., diffusivity and C_{sat}, characterized under 85℃/85%RH, are listed in Table 1.

Table 1 Moisture absorption and hygroswelling properties of MLP6x6

Material	D (mm²/s)	C_{sat} (mg/mm³)	CME (mm³/mg)
EMC	4.73e-7	7.06e-3	0.222
Die attach	1.25e-5	6.20e-3	0.520

Figure 7 shows the comparisons of the transient moisture wetness distribution between pure ANSYS and moisture automation analysis system. From Figure 7, it can be seen that the results agree well in different time in both moisture automation system and ANSYS.

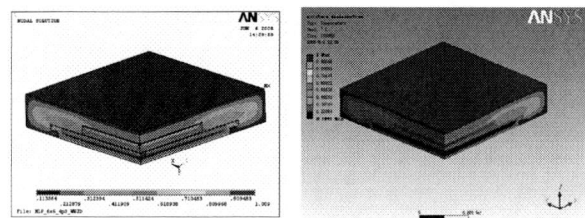

Moisture absorption for 48 hours

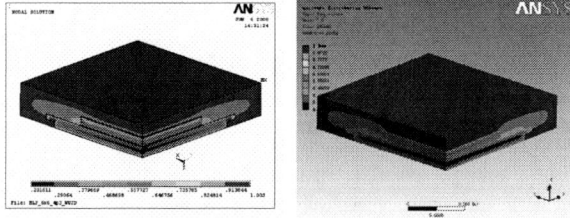

Moisture absorption for 96 hours

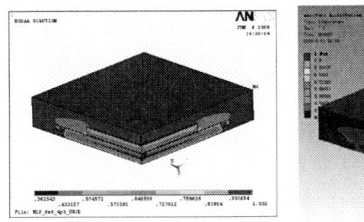 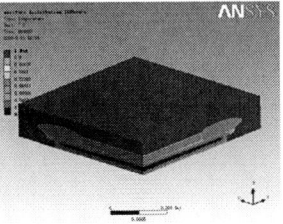

Moisture absorption for 168 hours

Result in classic ANSYS　　Result in AutoSim

Figure 7 Transient moisture wetness distribution.

Figure 8 compares the hygroswelling deformation and Von Mises stress between ANSYS and moisture automation analysis system. Table 2 lists the maximum Von Mises stress and total deformation values of ANSYS and moisture automation analysis system. Form Figure 8 and Tab. 2, the maximum value and corresponding location gotten by moisture automation analysis system are same as those from ANSYS.

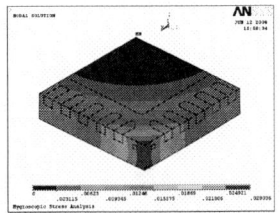 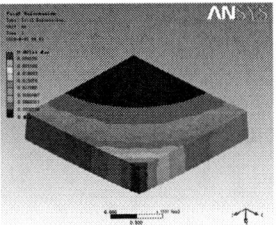

Hygroswelling deformation

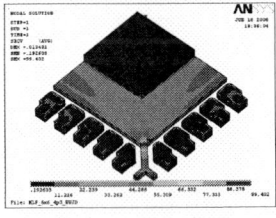 

Hygroswelling Von Mises stress

Result in classic ANSYS　　Result in AutoSim

Figure 8 Hygroswelling deformation and Von Mises stress

Table 2: Hygroswelling results in ANSYS and AutoSim

	Maximum Von Mises stress (MPa)	Maximum Total deformation (mm)
ANSYS	99.402	0.02836
AutoSim	104.65	0.02714

4.2 Application of AutoSim for vapor pressure

Figure 9 shows the comparison of the vapor pressure distribution at reflow between ANSYS and moisture automation analysis system. From Figure 9, we can see that the results were agreed well to each other between in Workbench and ANSYS. At reflow temperature 220C, the saturated vapor pressure is 2.32 MPa. If the moisture is not fully vaporized at reflow, the vapor pressure will maintain the saturated value no matter how much moisture is absorbed.

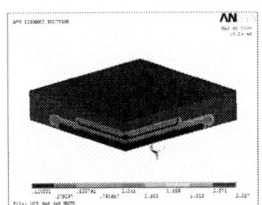 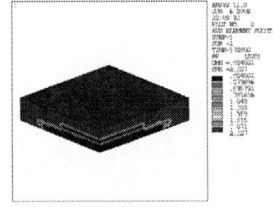

Result in pure ANSYS　　Result in AutoSim

Figure 9 Vapor pressure distributions at reflow temperature

4.3 Application of AutoSim equivalent thermal stress

Linear-elastic thermo-mechanical stress model is applied. The temperature applied is from 175 ℃ to the reflow temperature of 220℃. The thermo-mechanical material properties used in the modeling are shown in Table 3. Poisson's ratio is assumed to be 0.3 for all materials

Table 3　Thermo-mechanical material properties (220℃)

	E (MP)	CTE (ppm/℃)	V
EMC	1100	34	0.3
Die	131000	2.8	0.3
Die Attach	43	170	0.3
Lead frame	127400	17.4	0.3
Pad	127400	17.4	0.3

Figure 10 compares the hygro-mechanical contours of total deformation and Von Mises stress between ANSYS and moisture automation analysis system. Table 4 lists the maximum hygro-mechanical Von Mises stress and total deformation values of ANSYS and moisture automation analysis system. Form Figure 10 and Tab. 4, the maximum value and corresponding location gotten by moisture automation analysis system agree with those from ANSYS.

	Maximum Von Mises stress (MPa)	Maximum total deformation (mm)
ANSYS	99.402	0.013431
AutoSim	104.65	0.013470

The total strains induced by thermo-mechanical, hygro-mechanical, and vapor pressure loadings in mold compound and die attach are listed in Table 7. For hygro-mechanical and vapor pressure induced strain, they are converted into equivalent CTE with temperature from 175C to 220 ℃, so that all three models can be integrated in a thermo-mechanical model with equivalent strain. The total equivalent CTE is much larger than only considering individual model. So the thermal-, hygro- and vapor pressure induced stresses need to be considered to give the total stress in a package, and to allow realistic stress analysis for prediction of damage and failure.

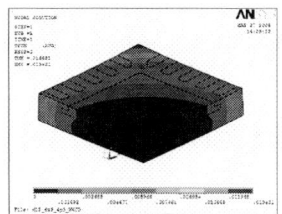
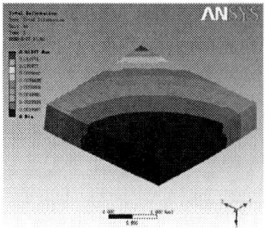

Hygro -mechanical total deformation

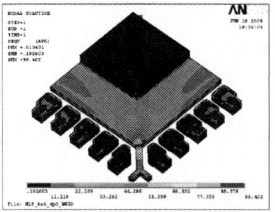
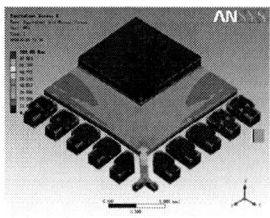

Hygro -mechanical Von Mises stress
Result in classic ANSYS   Result in AutoSim
Figure 10 Hygro-mechanical deformation and stress

Table 4 Hygro-mechanical results in ANSYS and AutoSim
Figure 11 compares the deformation and stress induced by vapor pressure only. Table 5 compares the corresponding maximum values based on the moisture automation analysis system and ANSYS.

Figure 13 compares the integrated contours of total deformation and Von Mises stress between normal ANSYS and AutoSim. Table 6 compares the integrated numeric results between normal ANSYS and AutoSim.

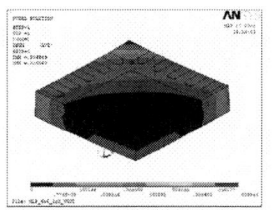
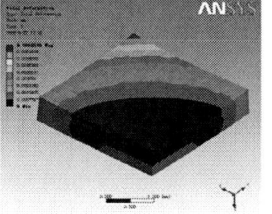

Vapor pressure induced deformation

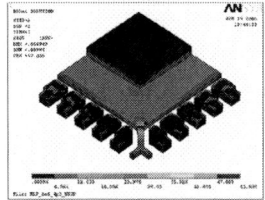
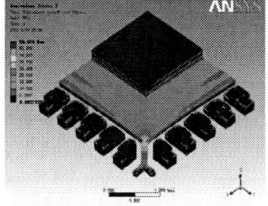

Vapor pressure induced Von Mises stress
Result in classic ANSYS   Result in AutoSim
Figure 11 Vapor pressure induced deformation and stress

Table 5 Vapor pressure induced equivalent stress results in ANSYS and AutoSim

	Maximum Von Mises stress (MPa)	Maximum total deformation (mm)
ANSYS	53.839	0.006949
AutoSim	56.676	0.006954

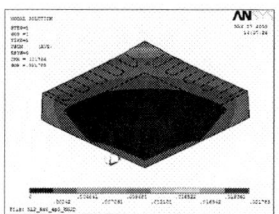
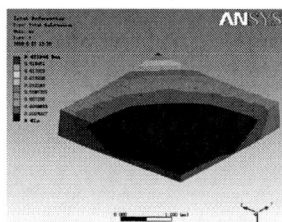

Integrated total deformation

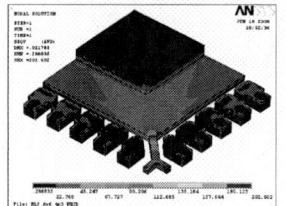

Integrated Von Mises stress
Result in classic ANSYS   Result in Workbench
Figure 13 Integrated deformation and stress

Table 6 Integrated results in ANSYS and AutoSim

	Maximum Von Mises stress (MPa)	Maximum total deformation (mm)
ANSYS	202.602	0.021783
AutoSim	213.13	0.021894

Table 7 Integrated equivalent CTEs in ANSYS and AutoSim

	EMC		Die Attach	
	Total strain	Equivalent mean CTE (ppm/℃)	Total strain	Equivalent mean CTE (ppm/℃)
Thermo-mechanical	1.53E-03	34	7.65E-03	170
Hygro-mechanical	1.57E-03	34.8	3.22E-03	71.6
Vapor-mechanical	8.44E-04	18.7	2.16E-02	480
Integrated (total)	3.94E-03	87.5	3.25E-02	721.6

4. Conclusions

The developed AutoSim focuses on the analysis of simulation automation for plastic package moisture diffusion, vapor pressure distribution, thermal-mechanical stress; hygromechanical stress; vapor induced equivalent thermal mismatch stress and integrated stress. Engineers who do not know FEA and moisture theories can run the simulation easily by using this system.

It should be noted that the AutoSim hasn't automatically generated mapped mesh for a complicated package model. For the stress simulation, there were some differences between ANSYS and the AutoSim based Workbench due to current limitation of Workbench itself.

Acknowledgments

The authors would like to express their gratitude to Fairchild Semiconductor Corporation. The work was supported by the Zhejiang Science Foundation (No. Y107365).

References

1. Wong, E. H., Teo, Y. C. and Lim, T. B., "Moisture diffusion and vapor pressure modeling of IC packaging," *Proc 48th Electronic Components and Technology Conf*, 1998, pp. 1372-1378.

2. Wong, E. H., Chan, K. C., et al, "Comprehensive treatment of moisture induced failure in IC packaging," *Proc 3th IEMT/IMC*, Tokyo, Japan, 1999. pp. 176-181.

3. Zhang, Y. X., et al, "Modeling automation system for electronic package thermal analysis using Excel spreadsheet," *Proc 8th International Conference on Electronic Packaging Technology*, Shanghai, August. 2007, pp. 147-150.

4. Tee, T. Y., Zhong, Z. W., "Integrated vapor pressure, hygroswelling, and thermo-mechanical stress modeling of QFN package during reflow with interfacial fracture mechanics analysis," *Microelectronics Reliability*, Vol. 44, No. 1 (2004) , pp. 105-114.

5. Fan, X. J., Zhou, J., "A micromechanics-based vapor pressure model in electronic packages," *Journal of Electronic Packaging*, Vol. 127 (2005). pp. 262-267.

6. Fan, X. J., "Moisture Related Reliability in Electronic Package," ECTC short course, 2007.

High Speed Package Design and Electrical Performance Analysis

Shu-Qiang Zhang[1]*, Hung-Hsiang Cheng[2], Yin-Guang Zheng[1], Chang-Lin Yeh[1]

[1] Labs of Shanghai Engineering, ASE Assembly & Test (Shanghai) Limited.
No.669, Guoshoujing Rd., Zhangjiang Hi-tech Park, Pudong Area, Shanghai 201203, P.R.C.
*Email: hogain_chang@aseglobal.com
[2] Corporate Design Dep. Advanced Semiconductor Engineering, Inc.

Abstract

More and more high-speed data transmission formats such as Rapid IO, Hyper transport, Gigabit Ethernet, Serial ATA etc. are becoming prevalent. As substrate interconnection density and channel data rate are getting increasingly higher, various 3D effects, crosstalk, and discontinuity – induced ISI are playing a much more important role, for both signal channels and power distribution networks . The substrate interconnection structures are becoming the major bandwidth constraint for most of the package designer.

As data rates continue to increase, transitioning to solder bump or Au stud bump flip-chip interconnects or low-loss substrate materials results in excessive cost. It is therefore increasingly important to provide a high performance and low-cost packaging solution. The aim of this paper is to proposal a design solution of normal Plastic Ball Grid Array (PBGA) package for high speed devices. The electrical simulation method of the effects of variation of package design parameters such as signal path structure, wire bonding, through hole via, ball placement tactic and plating stub are highlighted. The conclusion of this paper is recommendation for high-speed package electrical design.

Introduction

As the CMOS technology development, today's internal circuit can run at tens of gigabits per second (Gbps), but the bandwidth of the channel – the physical medium through which the signal propagates from transmitter output to receiver input- limits link performance. Among channel components, the package is becoming a major bandwidth constraint.[1] However, on the cost saving aspect, using the carefully design of normal ball grid array (BGA) packages to substitute for flip-chip or low loss substrate materials for high speed solution is needed.

Focusing on characteristics of BGA packages using organic substrate, maximum applicable frequency range can be changeable due to the layout variation and package structure e.g. flip-chip. As a result of using organic laminate substrate, design of trace, via, plane and finger/ bump/ ball locations can be dominant on high frequency electrical performance. [2]

In this paper, we take a PBGA package (4 layer substrate) design as an example to illustrate how to design a high speed serial links based on electrical simulation results. The electrical simulation method of the effects of variation of package design parameters such as signal path structure, wire bonding, through hole via, ball placement tactic and plating stub are highlighted. Previous works have studied the use of PBGA packages up to 10 Gbps or even 40 Gbps using frequency domain electrical performance analysis method.[1][3] But it is difficult to find which part is the main

contributed part to the bandwidth degradation , thus, many extra works are needed to modify the drawing and perform the electrical analysis. We propose another electrical simulation method to solve this problem. The time domain TDR waveform is simulated to show the need modification department of the high speed signal serial channel. This method can significantly reduce the design work and cycle time. The proposed design compared with the conventional design electrical simulation results will be demonstrated in frequency domain.

Discontinuity consideration

Fig. 1 shows a typical channel in a PBGA package, which including bonding wires, traces, through hole vias, plating stubs and balls.

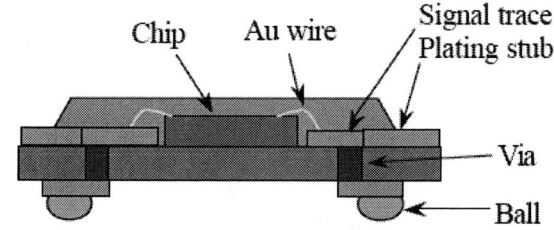

Fig. 1: Typical BGA cross-section

The package traces form normal uniform transmission lines whose characteristic impedance can be easily controlled through careful design. So the package traces' bandwidth only depend on the substrate materials property.

An inductance or capacitance produces reflection when it impairs a transmission line. The main discontinuity departments of the signal channels are bonding wires, through hole vias and plating stubs. The discontinuity departments of the signal channel are either caused by excess inductance or excess capacitance. Thus, reflections occur at these discontinuities. Reflections in the channel degrade the bandwidth. We categorized the bonding wires as the inductive region, because the bonding wire's impedance is lager than the carefully designed transmission line (generally 50 Ohm for single end signals and 100Ohm for differential pair signals). For the through hole vias, solder balls and plating stubs, we categorize them as capacitive region. The inductive reflections are mainly contributed by bonding wires, while the capacitive reflections are mainly contributed by through hole vias, plating stubs and solder balls. [4] To avoid reflection, a straightforward solution is to find which part of the signal channel is the main factor for the bandwidth degradation. Thus, the feasible design solution can be proposed to balance

978-1-4244-2739-0/08/$25.00 ©2008 IEEE

the capacitance with inductance, thereby cancelling any excess reactance.

Package Design methodology

The actual layout substrate cross-section is shown as Fig. 2. The 3D models for EM software analysis are built up according to the actual substrate cross-section.

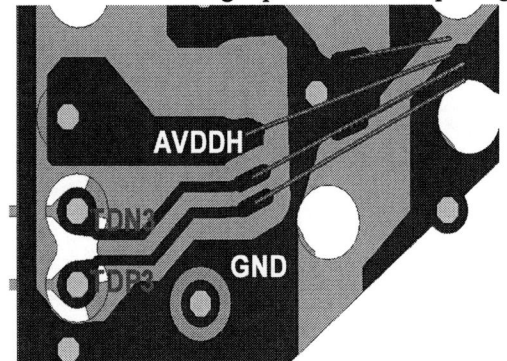

Fig. 2: The actual layout substrate cross-section

Fig.3 shows the conventional design and the simulated TDR waveform of the high speed differential pair signal.

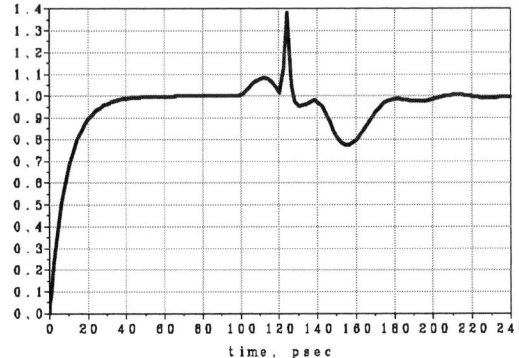

Fig. 3: Conventional high speed signal design and simulated TDR waveform.

From the TDR waveform, there is huge excess inductive reflection. Thus, from previous analysis, the bonding wires should be firstly concerned. We propose short bonding wires for lower inductance. The signal structure of the roughly design is Ground-Signal-Signal (GSS). To decrease the inductance, the proposed signal structure will be GSSG. For this signal structure the induced cross-talk noise from other signals can also be reduced. Fig.4 is the proposed design and the simulated TDR waveform compared with the conventional design. The excess inductive reflection is remarkably reduced for the proposed design.

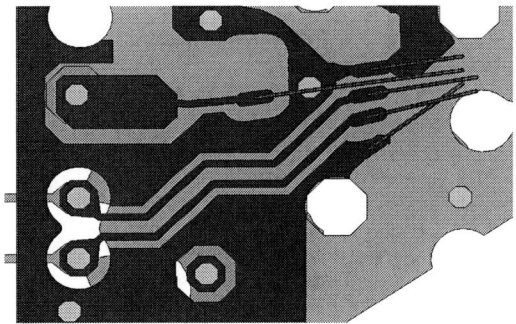

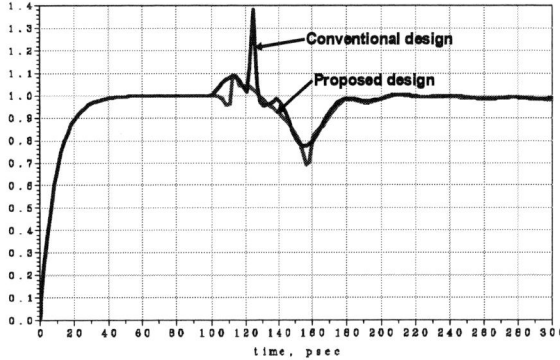

Fig. 4: The proposed signal structure and bonding wires for high speed signal design and simulated TDR waveform.

The excess capacitive reflection now is becoming predominant. Previous studies have showed that enlarge the through hole vias' anti-pad size and void the solder ball pad's above metal layer can reduce the parasitic capacitance, thus, the capacitive reflections can be reduced correspondingly.

Sine plating stub is existing due substrate fabrication need (for gold plating), extra effect is added to the signal transmission path. Research for plating stub effect can be conducted as an open-end micro-strip for discontinuities analysis. [5] Here we perform the time domain simulation for the plating stubs to predict how to design the plating stubs. Four kinds of plating stub situation are simulated. The first situation is none-plating stub, the second is the wide width and long plating stub, the third one is the thin width and long plating stub and the fourth situation is the thin width and short plating stub. The second and third situations with the same plating stub length, but different width. The third and fourth situations with the same plating stub width, but different length. Fig.5 shows the simulation circuits and the simulation results.

The none-plating stub situation has the best electrical performance. However, the cost also increases for the none-plating stub design. On the cost saving aspect, the thin width and short plating stub would be the proposed design choice.

From above analysis, the proposed design to decrease the excess capacitive reflection including: enlarging anti-pad size, voiding the solder ball pad's above the metal layer, using the short and thin plating stub. Fig.6 shows the proposed design and simulated TDR waveform compared with the last proposed design.

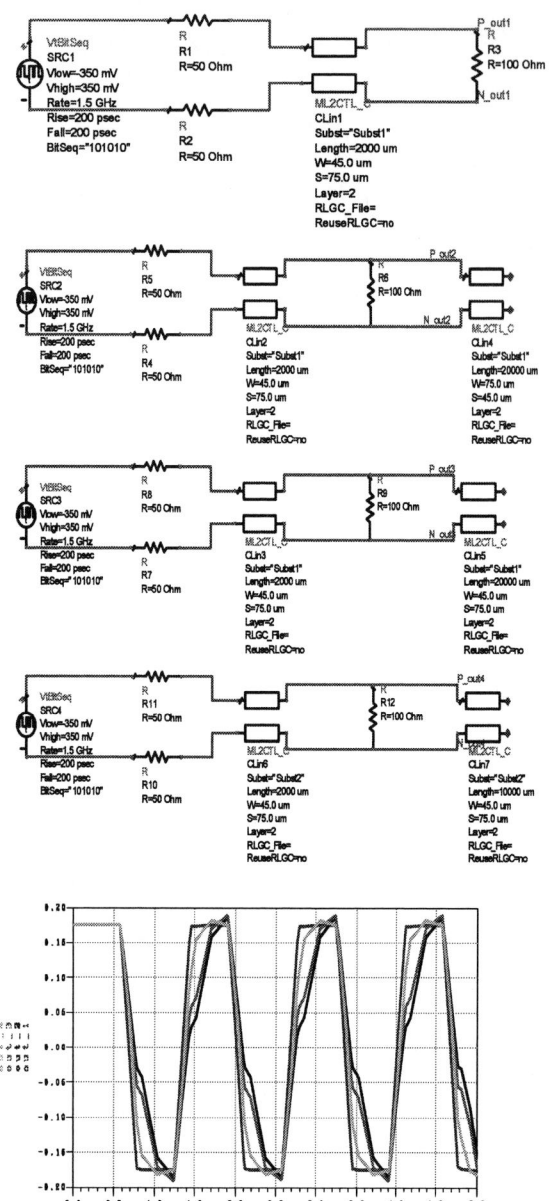

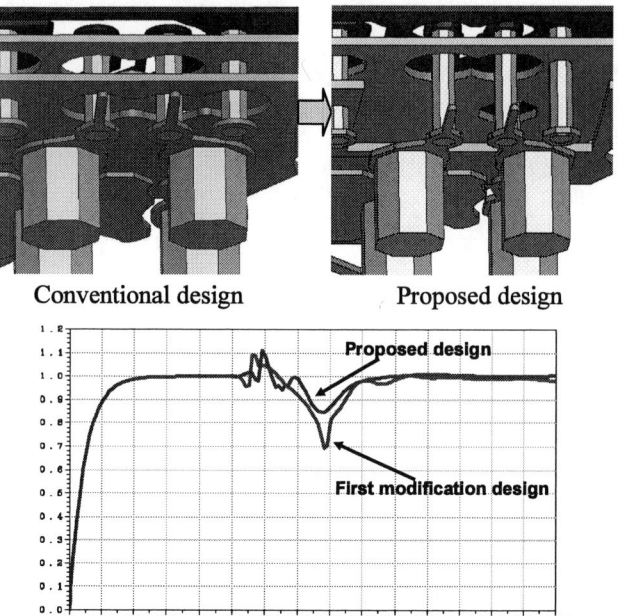

Conventional design   Proposed design

Fig. 6: Proposed via anti-pad, ball pad's above metal layer void, thin plating stub design and simulated TDR waveform compared with last design.

For every signal current, there is an equal and opposite return current. Thus, a current loop is formed as the current return to the source through the return path. For high speed devices, how to design the minimum and clean current return path is as important as signal current for signal integrity. There are a lot of previous studies on power/ ground ball ratio, power/ ground plane stack-up structure.[1][6] Here we just follow the previous studies for ball placement tactics and power/ ground plane stack-up structure.

Electrical performance analysis

After identified each discontinuities' effect, the conventional and the proposed design frequency domain S-parameter simulation are performed to show the entire package channel's electrical performance. Fig. 7 shows the electrical simulation results comparison for the two version design. We performed the simulation up to 40GHz. The proposed design's insertion loss is below 4dB at up to 40GHz. The 3-dB frequency is up to 36GHz. The proposed design shows significantly enhanced performance than the conventional design.

Fig. 5: Plating stub simulation circuit and simulation results.

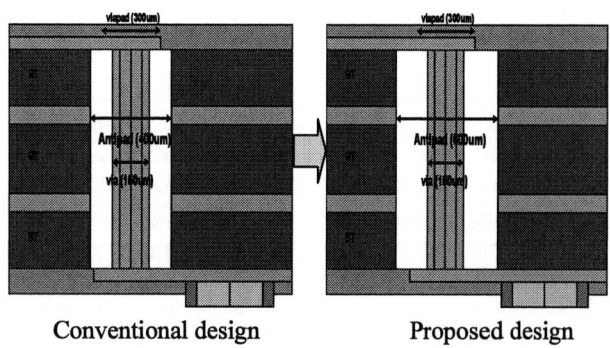

Conventional design   Proposed design

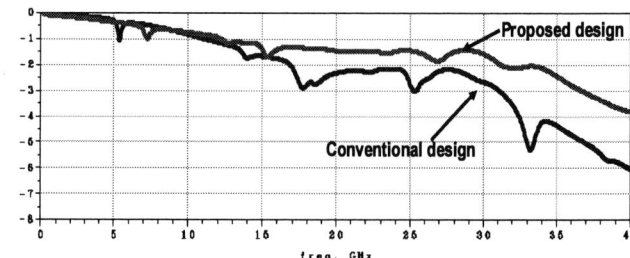

Fig. 7: Conventional, proposed design 3D models and simulation results

Conclusions

With the normal PBGA package structure, the high speed serial links design and electrical analysis methodologies are highlighted. Without using low-loss substrate materials, flip-chip BGA packaging technology   and none-plating stub process, our PBGA packages can provide low-cost packaging solution for future high speed serial links.

The design methodology and electrical analysis method are recommended for high speed package design.

References

1. Dong Gun Kam. et al.,"Packaging a 40-Gbps Serial Link Using a Wire-Bonded Plastic Ball Grid Array", *IEEE Design&Test of Computers* 2006,PP.212-219.

2. C.P.Hung, et al., "20GHz Electrical Characterization and Limit of Organic BGA Packages With and Without Plating Stub" *6th VLSI-PKG-WS*,2002.

3. R. Emigh, "Electrical Design for High Data Rate Signals in Conventional, BT-Based PBGA Substrates Using Wire Bonded Interconnection," Proc. *IEEE Electronics Packaging, Technology Conf. (EPTC 03)*, *IEEE* Press, 2003, pp517-522.

4. S. Hall et al., High-Speed Digital System Design: A Handbook of Interconnect Theory and Design Practices, John Wiley & Sons, 2000.

5. Alexopoulos, N.G. and Wu, S.C., "Frequency-Independent Equivalent Circuit Model for Microstrip Open-End and Gap Discontinuities," *IEEE Trans-MTT*, Vol. 42, No. 7 (1994), pp. 1268-1272.

6. Chia-yu Jin, Chia-hsing Chou. et.al, "Improving signal integrity by optimal design of power/ground plane stack-up structure," Proc. *IEEE Electronics Packaging, Technology Conf. (EPTC 03)*, *IEEE Press, 2006*, pp. 853-859.

Methodology for Modeling Simultaneous Switching Noise in BGA Packaging

Song Li[a], Xue-tao Weng[b]
[a] Wuhan University of Technology, Hubei, Wuhan, China,
[b]Naval University of Engineering, Hubei, Wuhan, 430033, China
E-mail:songli_008@163.com, Tel: 0-13476180491

Abstract

Advances in BGA packaging technologies have led to a dramatic increase in the performance of integrated circuits. Noise source such as supply bounce, signal coupling, and reflections results in reduced performance. The work presents techniques to model and improve performance the performance of BGA designs without moving toward advanced packaging. A single, unified mathematical framework is presented that predicts the performance of a given package depending on the package parasitics. Using 3-Bit Bus example about the package, a methodology is presented to analyse mutual inductive signal coupling and mutual capacitive signal coupling and return current. The performance model illustrate that the per-pin performance is significantly reduced, where the number of switching signals is increasing.

1. Introduction

As package complexity increases, interconnect becomes the dominant factor over transistor sizes in determining the overall performance. Package interconnect has historically been designed to meet mechanical, thermal, and cost objectives. Advances in BGA packaging technologies have led to a dramatic increase in the performance of integrated circuits. The transistor delay in an integrated circuit is no longer the bottleneck to system performance. Unfortunately, the electrical parasitics is found to be a major obstacle for the performance of the packaging interconnect. [1-3] Noise source such as supply bounce, signal coupling, and reflections all results in reduced performance. Due to the parasitic inductance and capacitance of the packaging interconnect, some new problem arise. While advanced packaging can aid in reducing the parasitics, the design of a new package is often not suited for the majority of BGA designs. This makes the task of altering the interconnect structures within the package (to increase electrical performance) very difficult. [4-5]

This paper provides a methodology for modeling simultaneous switching noise. We are primarily interested in modeling electrical parasitics, which is used in BGA packaging. An efficient parasitics algorithm model based on BGA packaging is proposed.

2. Package Construction and Electrical Modeling

Modern Packages

The most popular package in use today is Ball Grid Array with Wire Bonding (BGA-WB).The next step in the evolution is Ball Grid Array with Flip-Chip bumping (BGA-FC). The latter package addresses the electrical parasitics of the wire bond by implementing flip-chip technology as its level 1 connection. When combined with the electrical improvements gained by using a level 2 BGA interconnect, the BGA-FC promises to meet the needs of Very Large Scale Integrated (VLSL) systems into the next decade. Figure 1 shows the cross-section of a BGA-FC package that is mounted to a system PCB. [6]

The BGA-FC package can achieve very high signal counts due to its array style pad pattern. This package is addressed by implementing an array style interconnect between the IC and the package substrate. This allows higher signal counts and redundant power and ground connections which reduces the package noise and increase system performance. Packages with up to 2000 contacts have been successfully implemented using ball grid array with flip-chip bumping technology. [7]

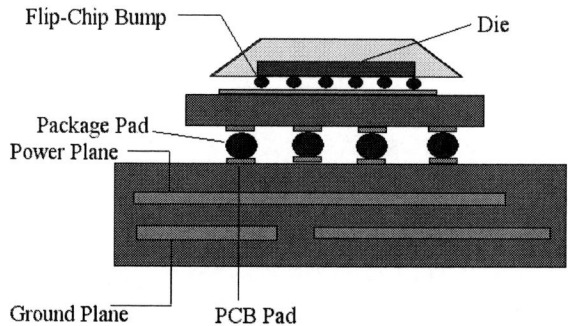

Fig. 1 Cross-Section of System with BGA Flip-Chip Package

2.2 Signal Coupling

Signals transmits data from one circuit to another by bus. When designing a bus that traverses a package, the number of physical interconnect paths is very important. The total number of signal pins are needed, and the number of power and ground pins associated with the bus are also considered.

2.2.1 Mutual Inductive Signal Coupling

The mutual inductive coupling between signals represents the inductive coupling magnitude between an arbitrary pin to its adjacent neighboring pins. The inductive coupling can span segment boundaries. Mutual inductive coupling exists between any two current carrying interconnects. The term P_L represents the number of neighbors on either side of pin V_i^j for which inductive coupling is considered. By considering only P_L neighbors on either side of the signal of interest, was simplified the analysis. [2, 7] The magnitude of the mutual inductive coupling voltage between inductive interconnects is governed by the Equation 1:

$$V_M^j = M_{1K}\left(\frac{di_K}{dt}\right) \qquad (1)$$

M_{1K} is the mutual inductance between conductor j and conductor k. Figure 2 illustrates this idea. In this Figure, P_L =2, which means that 2 neighbor's mutual inductive coupling coefficients (to the left and to the right) is calculated, and the rest is ignored.

2.2.2 Mutual Capacitive signal Coupling

The mutual capacitive coupling between signals C_{ik} represents the capacitive coupling magnitude between an arbitrary pin V_i^j to its adjacent neighboring pins. Mutual capacitive coupling can span segment boundaries. When capacitive coupling is considered, the term P_C is used to describe how many pins away from the pin of interest. [2, 6] Equation 2 describes the voltage magnitude that will be coupled onto a victim line (V_{vic}) from a voltage change on an aggressor line (V_{aggr}).

$$V_{vic} = \left(\frac{C_{1K}}{C_{1K} + C_0} \right) \left(\Delta V_{aggr}^k \right) \qquad (2)$$

Figure 2 illustrates this idea. In this figure, P_C =2, which means that 2 neighbors mutual capacitive coupling (to the left and right) is calculated, and the rest is ignored.

2.2.3 Return Current

When a signal line is charged or discharged, return current will follow through the V_{SS} or V_{DD} pins in the package. When inductance is present in the V_{SS} or V_{DD} paths, supply bounce will result. In BGA packaging the return current for multiple signal pins will share a single supply pin. The return current will always seek the path of least resistance when traversing the package. This means that the return current will seek the closest supply or ground pin. [2, 6]

Figure 2 shows how the return current for a signal can span segment boundaries.

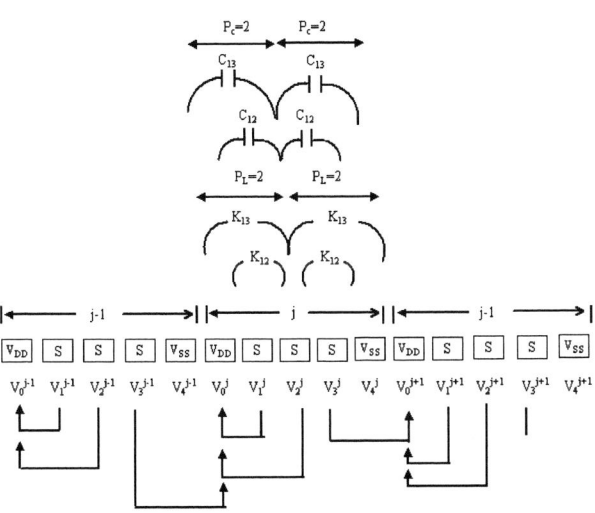

Fig. 2 3-Bit Bus Example

3. Simulation

Equation 3 expresses the relationship of minimum Unit Interval (UI_{min}) and the maximum Datarate (DR) of an individual pin.

$$DR_{\max} = \frac{1}{UI_{\min}} \qquad (3)$$

According to multiple signal pins, the maximum Throughput (TP) of the bus segment can be found using Equation (4).

$$TP_{\max} = DR_{\max} \cdot W_{bus} \qquad (4)$$

UI_{min} can be expressed in terms of risetime(Equation 5).

$$UI_{\min} = 1.5 t_{rise} \qquad (5)$$

Equation 6 expresses slewrate in terms of the rate of change in current or voltage.

$$slewrate = \frac{dv}{dt} = \frac{di}{dt} \cdot Z_0 \qquad (6)$$

Risetime can be expressed in terms of slewrate by Equation 7.

$$t_{rise} = \frac{0.8 V_{DD}}{slewrate} \qquad (7)$$

The Signal-to-Power-Ground (SPG) metric is defined as $W_{bus} : N_p : N_g$. The three SPG configurations used are SPG=7:1:1, SPG=5:1:1, SPG=3:1:1. In this work, the number of switching signal pins within a segment are defined from W_{bus} =1 to W_{bus} =16.

Figure 3 shows the per-pin datarate for the BGA package as a function of the number of signal pins within the segment. This result illustrate that the per-pin datarate is reduced as channels are added to the bus segment. This plot also shows the performance of the segment is influenced by the amount of ground and power pins. In addition, $\frac{di}{dt}$ through each supply pin is reduced, therefore a lower SPG obtains a higher datarate. Figure 4 shows the bus throughput for the BGA package as a function of the number of signals pins. The throughput experiences a less than linear increase as channels are added; instead.

The experimental results illustrate how the electrical parasitics of the IC package limit the overall system performance. The effect of increasing the number of simultaneously switching signals is that the per-pin performance is significantly reduced.

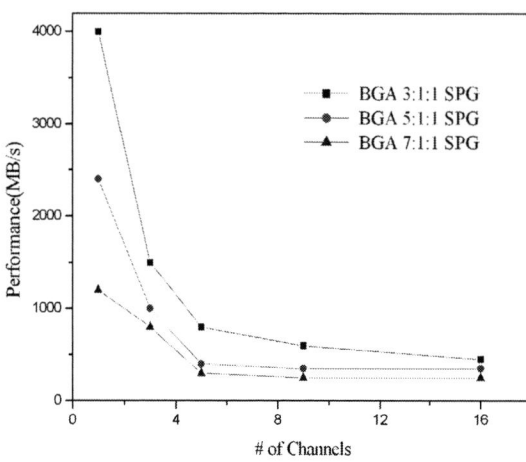

Fig. 3 Per-Pin Datarate for a BGA Package

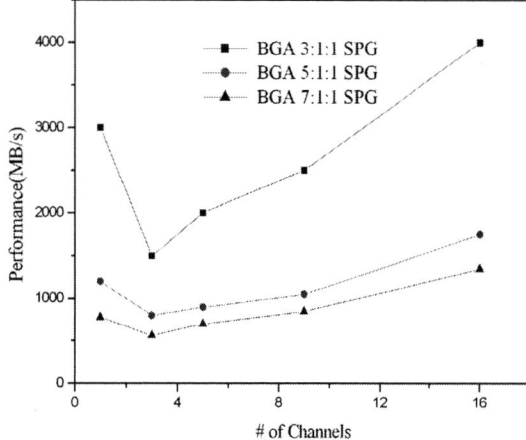

Fig. 4 Throughput for a BGA Package

4. Conclusions

This thesis has presented a comprehensive look at the noise problems within BGA packaging. The parasitics of the package can cause supply bounce, signal-to-signal coupling, capacitive bandwidth limiting, and impedance discontinuities, which were discussed in detail. To verify the models, simulations were performed. It was found that a lower SPG results in a higher per-pin datarate.

References

1. Wang, T.K, Chen, S.T, Wu, S.M, "Modeling noise coupling between package and PCB power/ground planes with an efficient 2-D FDTD/lumped element method", IEEE Transactions on Advanced Packaging, v 30, n 4, November, 2007, pp. 864-871
2. LaMeres, B.J, "Novel Design Techniques to Reduce Simultaneous Switching Noise in VLSI Packaging," Ph.D thesis, 2005.
3. Hsieh, B, Chiang, K, Wang, Y.P, "DDRII memory packages electrical performance comparison of COSBGA, TFBGA and standard TSOPII," Proceedings of the ASME/Pacific Rim Technical Conference and Exhibition on Integration and Packaging of MEMS, NEMS, and Electronic Systems: Advances in Electronic Packaging 2005, v PART B, pp. 1063-1068
4. Yazdani, Farhang, "Signal integrity characterization of microwave XFP ASIC BGA package realized on low-K liquid crystal polymer (LCP) substrateSource," IEEE Transactions on Advanced Packaging, v 29, n 2, May, 2006, pp. 359-363
5. Tarusawa,Y, Ohshita,K, Suzuki,Y, "Experimental estimation of EMI from cellular base-station antennas on implantable cardiac pacemakers," IEEE Trans. Electromagn. Compat.,vol. 47, no. 4, Nov. 2005, pp. 938-950.
6. Seungbae, Ramji,D, Rahul,J , "Comparative analysis of BGA deformations and strains using digital image correlation and moire interferometry," Proceedings of the 2005 SEM Annual Conference and Exposition on Experimental and Applied Mechanics, 2005, pp. 1873-1880
7. Futatsumori S, Hikage,T, Nojima,T, "In vitro experiments to assess electromagnetic fields exposure effects from RFID reader/writer for pacemaker patients," in Proc. BIOLOGICAL EFFECTS of EMF 4th International Workshop, Oct. 2006, pp. 494-500.

A Lowpass Filter with An Embedded Capacitor for Wideband Noise Suppression in Multi-GHz PCBs

Wei Gao, Lixi Wan, Jun Li
Institute of Microelectronics of Chinese Academy of Sciences
A402,3#,BEITUCHENGWest, CHAOYANG District, Beijing, 100029, China
Telephone: 86-010-82995601 86-010-82995591 Fax: 86-010-62021601
Email: galaxyvenus@126.com lixi.wan@gmail.com zhiying19811123@hotmail.com

Abstract

This paper presents a novel lowpass filter with an embedded capacitor which can substitute conventional filter network and suppress switching noise with a large bandwidth in Multi-GHz PCBs. A new design methodology of the low pass filter network is proposed. A new SPICE model of embedded capacitor based on Transmission Line Theory is built and a comparison between H-spice and GTLE is used to illuminate its validity. With the model and Finite Elements Method, the feature of the embedded capacitor, in which different frequency modes were excited by special exciting positions, was studied and the design procedure of an ECF was developed. As an application example, a typical power supply filter network and its replacement, an embedded capacitor with a 100uF SMD capacitor, were studied.

Introduction

With the development of semiconductor industry, higher level of integration, lower cost, and more functions in a system are demanded intensively. As an indispensable part of most electronic products, low pass filter networks are widely used in electronic systems for noise suppression, especially in high frequency/speed systems. They are normally combined with discrete capacitors, resistors and inductors. As ICs work on higher speed/frequency bands, especially in Multi-GHz systems, the design and integration of filter networks has become more complicated due to parasitic parameters in commercial discrete capacitors or inductors play critical part and leads the working frequency band below to a two hundreds MHz. The reason of low frequency feature of the capacitors is that the equivalent serial inductance (ESL) and resistance (ESR) is too large. Usually, ESL is about hundreds picohenry to nanohenry [1].On account of parasitic parameters, embedded filter studies have been reported in several papers, e.g. [2]-[5].

In [2], J. Uei-Ming has designed and implemented an embedded band pass filter by using a special organic material which has a high dielectric constant ε_r >40. Other than J. Uei-Ming, Gye-An Lee [4] and Mohamadou Baba [5] have designed and implemented their embedded band pass filters

This work was supported by High-Technology Research and Development Program of China (863). Contract number: 2006AA01Z236, 2007AA01Z200. National Natural Science Foundation of China, Contract number: No. 90607006.

by using special structures of inductors. Until now, most works focus on the material and structure of band pass filter with embedded inductors, few ones work on embedded low pass filter, especially, with capacitors only. Because it is difficult to make the capacitance of embedded capacitor large enough to have an excellent low frequency characteristic in minimized PCBs. Furthermore, the conventional filter network depends on adding orders to improve filtering performance [6], which uses many discrete passive components and takes a lot of area, leads to increasing the cost and the size of the PCBs. In the opposite, embedded technology has shown a good choice for high density integration. Meanwhile, the embedded capacitors show a significant improvement in power system noise decoupling and change supply to the IC versus existing surface mount solutions due to small parasitic inductance, which is about from a couple of to teens picohenry [7]. Besides, embedded capacitors have more advantages in: saving room for layout, compatible with processing, better reliability, and lower cost [8-11]. There have been a lot of studies on embedded capacitors, but most concentrate on its material, processing and application. As authors' knowledge, embedded capacitor used as a filter network has not been proposed yet. In the paper, we designed a lowpass filter network which can work on wideband up to Multi-GHz based on embedded capacitor technology.

The paper is organized as following: section 2 describes the design methodology. Section 3 shows SPICE model of embedded capacitor. In section 4, the simulation based on the developed methodology is presented. Following section4, design procedure is given in section 5. After that, section 6 is an application example of ECF. Finally, conclusions are drawn in the last section.

Design Methodology

As we known, the design of filters consists of four steps: design synthesis, physical design, fabrication and testing. Nevertheless there is no obvious difference between this design methodology and others in the 1st, 3st and 4st steps, the specificity of this design methodology focuses on the step2, which will be discussed thoroughly.

Based on the synthesized LC-ladder network which is satisfied the filter specifications, an embedded capacitor with two or more connections on an electrode plane is used as a two or Multi-ports low pass filter network to substitute the LC-ladder network. For an example, an embedded capacitor with two connections, which could be a filter with two ports: one serves as an input, while another as a output – just as the conventional low pass filter network composed of passive

978-1-4244-2739-0/08/$25.00 ©2008 IEEE   211

components does. As seen in Figure 1, the low pass filter can be called as Embedded Capacitor Filter (ECF).

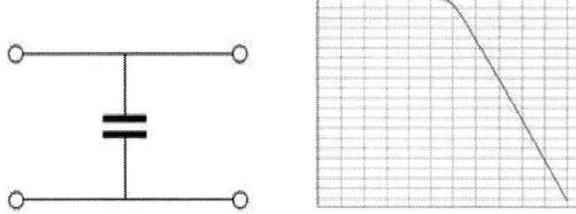

Fig.1 ECF (Embedded Capacitor Filter)

The core of this new design methodology is that the capacitor with special exciting points will has different frequency modes. This feature of capacitor can be explained by Transmission Line Theory: the planar capacitor can be divided into many sections and each section can be approximated using L, C, R and G. At different points, the different values of these four parameters lead to different resonant frequencies which make the capacitor show this special feature. Because the values of parasitic inductors contained connection and distributed inductances are very small, the resonant frequency of ECF can be up to Multi-GHz. However, as it is difficult to make the capacitance of embedded capacitor large enough to provide an excellent low frequency characteristic, a SMD capacitor is used. A cross-section of an ECF in a package/PCB system is shown in Figure 2.

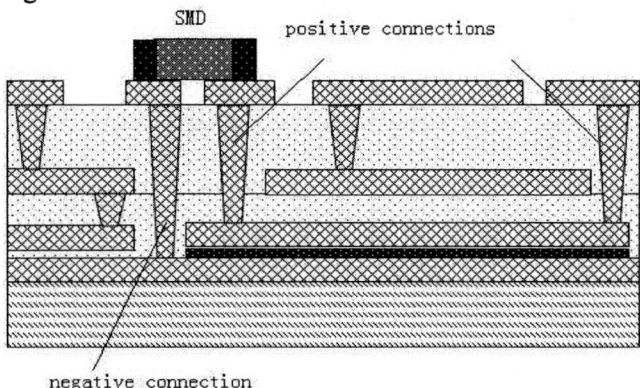

Fig. 2 A cross-section of an ECF in a package/PCB system

SPICE Model of Embedded Capacitor

In order to study ECF with H-spice, an equivalent circuit model of embedded capacitor based on Transmission Line Theory [12] was built. Figure 3 shows the equivalent circuit model of a cell.

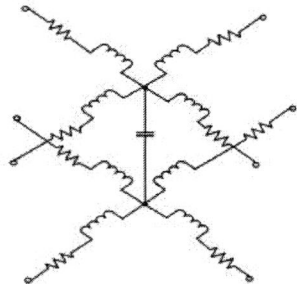

Fig.3 equivalent circuit model of a cell

Because the space between embedded capacitor two electrodes is much less than the width and the length of the conductor plates, it allows the fringe effect of the electrodes to be ignored. The parameters of the cell can be found from below equations. The capacitance can be calculated by equation $C = \varepsilon_r \varepsilon_0 \dfrac{A}{h}$, where A is the area of a cell and d is the thickness of the dielectric material. As we don't care the area of the boundary, we assume that current flows through the electrode uniformly. Equation: $L = \mu_0 \dfrac{hl}{w}$ can be used, where l and w are the length and width of the electrodes, and h is the separation. The resistance of a cell can be calculated using $R = \rho \dfrac{l}{s}$, where ρ is resistivity, l is length of the cell electrode and s is the cross-section of the electrode. We validated this model by example 1. A comparison of waveforms is shown in the figure 4, in which the curves from both H-spice and GTLE [13] match each other very well.

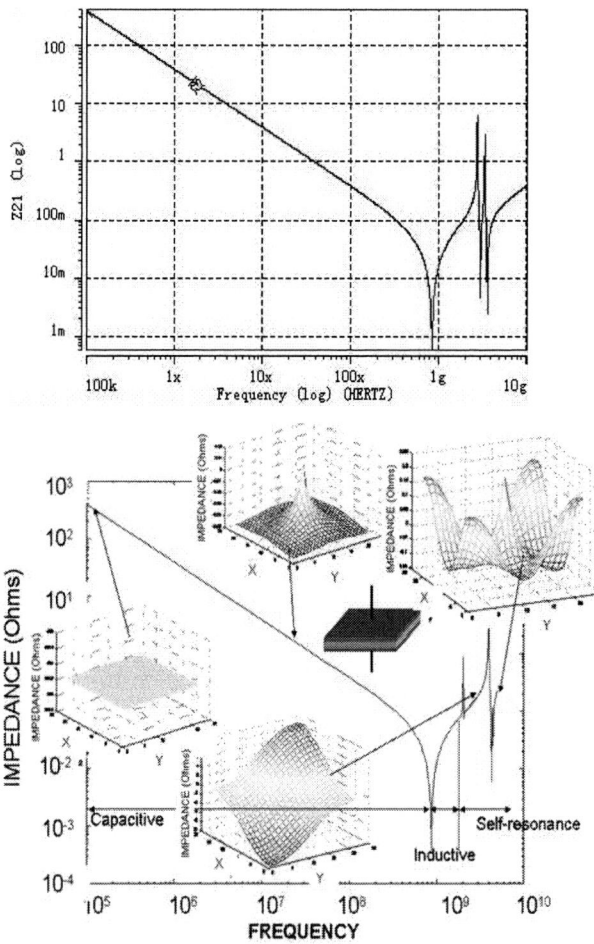

Fig.4 comparison of the waveforms by H-spice (above) and GTLE (below)

Example 1:

In this example, the embedded capacitor used in this study is a planar capacitor of 2×2 cm2 in area, thicknesses of dielectric material and metal are 12 um and 20 um respectively. Dielectric constant is 14. Loss tangent is 0.0001. Center connecting, and neglect the metal loss. The metal is copper with resistivity of 1.69×10-8. Frequency region is from 100 KHz to 5 GHz.

With the same physical dimensions and material parameters, H-spice using this model produces the same results to the one of GTLE which had been proved correct [13] that illuminates the model was built properly.

Simulation

With this model and Finite Elements Method [14], an embedded capacitor filter model is built. Figure 5 is 1/4 square embedded capacitor filter which has been divided into 16 cells. Symmetrically, there are 6 different points: point 152, point 42, point 143, point 33, point 142, and point 32. A simulation by H-spice was run to study the performance of embedded capacitor filter that how the positions of connection affect the range of operating frequency. As Corner or side connection on electrode will result larger inductance than center connection [7], we make one connection at center point 33. In Figure 6, a formula is found that the further the point leaves away from central point 33, the better the performance will be in high frequency.

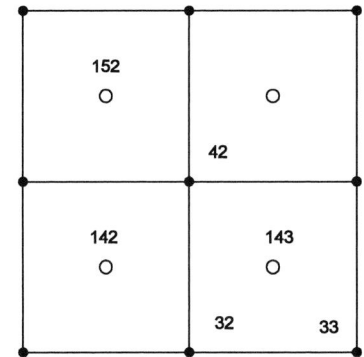

Fig.5 1/4 square embedded Capacitor and 6 different points

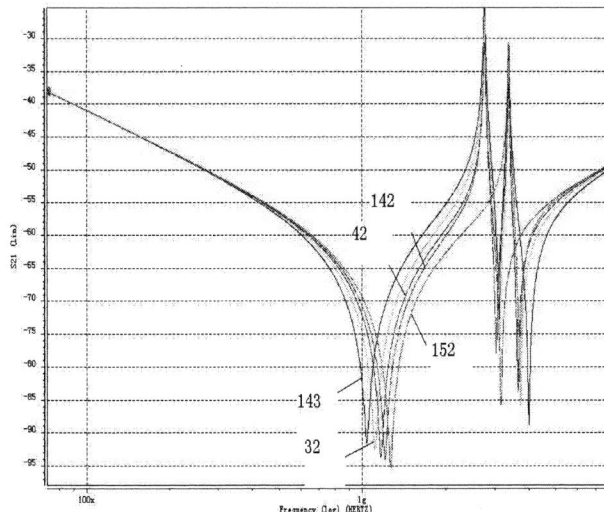

Fig.6 different frequency responses to different two connections

This is because that the current flows from centre to edges in radial path. As the radius increases from centre to edges, the current density decreases which can reduce the distribution inductor contribution. According to

equation $f = \dfrac{1}{2\pi\sqrt{LC}}$, when C is a constant, a smaller inductor will have a higher resonance frequency.

Design Procedure

From above simulation results, it is obvious that an embedded capacitor with two or more connections has an excellent filtering performance in high frequency, but its demerit of awful low frequency filtering also sticks out a mile. To solve this problem, a parallel SMD capacitor which is good at low frequency is easy to use in practice. Now, the ECF composed of an embedded capacitor with two or more connections and a parallel SMD capacitor can work better than a low pass filter network. The design procedure can be generalized as: (1) Calculate the capacitance of embedded capacitor by the highest filtering frequency with equation $f = \dfrac{1}{2\pi\sqrt{LC}}$; (2) Calculate the design physical dimensions of embedded capacitor by thicknesses of dielectric material and material's dielectric constant with equation $C = \varepsilon_0\varepsilon_r\dfrac{A}{d}$; (3) The positions of connection are decided by the highest frequency with the biggest reduction; (4) Choose the parallel SMD capacitor by the lowest filtering frequency with the equation $f = \dfrac{1}{2\pi\sqrt{LC}}$.

Application

ECF finds application in high speed / frequency bands PCBs that not only filters the high frequency noise but also reduces the area and cost of PCBs，which make it possible to achieve high density packaging with smaller dimensions and low cost. In Figure 7, it is a power supply filter circuit of a commercial optical Transimpedance Amplifier Array [15]. We can replace it with an embedded capacitor and a 100uF SMD capacitor. Following above design steps, the parameters of the embedded capacitor used in this ECF are: thickness of dielectric = 14 um; dielectric constant = 16; thickness of metal = 35 um (copper); size: 2X2 cm2. The port positions of this ECF are: port 1 at the center; port 2 at 1.0605 cm to a corner diagonally away from the center. Interesting frequency region is from 1Hz to 10 GHz.

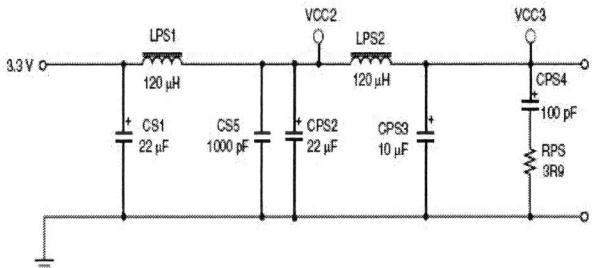

Fig.7 Power Supply Filtering Circuit

In Figure 8, it is the 1st order equivalent circuit of the power supply filtering circuit whose parameters were obtained from the component datasheets.

213

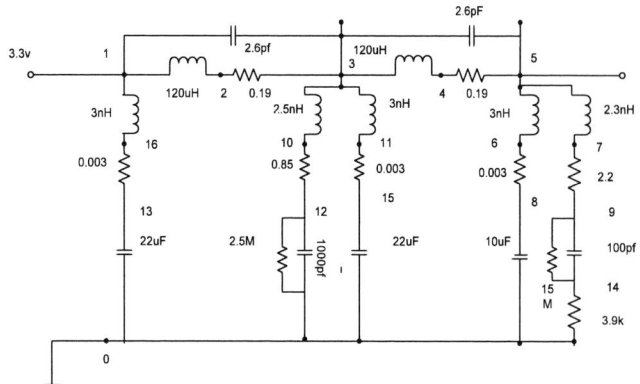

Fig. 8 1st order equivalent circuit of the power supply filtering

In Figure 9, the comparison of filtering performance from conventional filter network and ECF are shown. Smooth response on wideband up to several GHz from ECF was observed. Also, one more advantage of the ECF is that only one SMD was used in the filter, and saved 8 SMDs from the original one. It's obvious that ECF has much higher filtering performance with less area and lower cost with a large bandwidth from 50Hz to Multi-GHz.

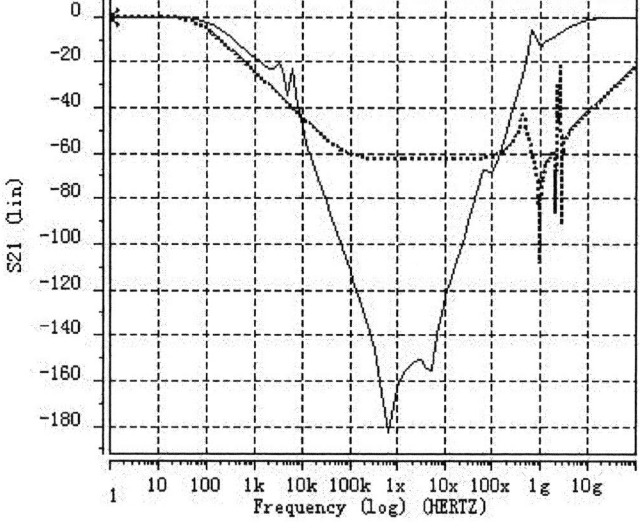

Fig.9 S21 comparison between conventional filter network (solid line) and ECF (dot line)

Experimental Verification

The ECF design in section Application was fabricated as shown in Figure 10. To investigate the ECF performance of suppressing high frequency noise, a two-port vector network analyzer (VNA) and microprobes were used. The insertion loss of the ECF was measured less than -20db from 100MHz to 4GHz. The result comparison between simulation and measurement is shown in Figure 11. Although the results from both measurement and simulation were similar, the differences are obvious. The first resonance frequency from measurement is lower than that from simulation, and it is not as smooth as the simulated one. The reason of the low resonant frequency may come from the larger parasitic of the SMD capacitor than that in the simulation. And the roughness of copper surface could introduce more errors in the transmission line model in the Spice. Even though, the ECF still shows us its performance for suppressing high frequency noise.

Fig.10 ECF in PCBs with port1 and port2

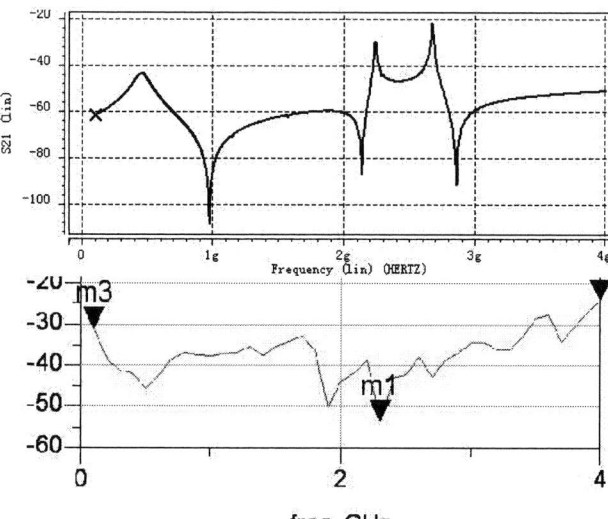

Fig.11 comparison between simulation (above curve) and testing (below curve)

Conclusions

In the paper an Embedded Capacitor Filter design, simulation and application was studied. A new SPICE model of embedded capacitor was built, and a new design methodology of low pass filter network with an embedded capacitor was proposed. Simulation results showed the filter could cover higher frequency band than conventional one does. The new filter can be used in Multi-GHz PCBs for saving space, cheaper in cost, and high density in layout.

References

1. Lixi Wan, etc. "Frequency Limitations of Embedded Decoupling Capacitors in High Speed Circuits and the Need for Nanocapacitors", ECTC, 2004.

2. J. Uei-Ming et al., "Functional embedded RF circuits on multi-layer printed wiring board (PWB) process", 2005 IEEE ECTC Conference, 31 May-3 June 2005, vol. 2, pp. 1634-1641.

3. Li Li et al., "Embedded passives in organic substrate for bluetooth transceiver module", 2003 IEEE ECTC Conference, May 27-30, 2003, pp. 464-469.

4. Gye-An Lee et al.; "Low-cost compact spiral inductor resonator filters for system-in-a-package", 2005 IEEE

Transactions on Advanced Packaging, vol. 28, no. 4, Nov.2005, pp. 761–771.

5. Mohamadou Baba, Stephan Guttowski, Herbert Reichl, "An Efficient Methodology for Design and Implementation of Embedded Bandpass Filters for RF/Wireless Applications", EPTC, 2007.

6. Sgufman J C. "A graphic method for the analysis and synthesis of electromagnetic interference filters", IEEE Trans.On EMC, 1965,7(3): 297-318

7. Lixi Wan, etc. "Design, Simulation and Measurement Techniques for Embedded Decoupling Capacitors in Multi-GHz Packages/PCBs", ECTC, 2005.

8. Robert Croswell, etc. "Embedded Mezzanine Capacitor Technology for Printed Wiring Boards", Circuitree, 08/01/2002

9. William J. Borland and Saul Ferguson, "Embedded Passive Components in Printed Wiring Boards: A Technology Review," CircuiTree, March 2001.

10. Joel Peiffer, "A Novel Embedded Cap Material," PC Fab, February 2001.

11. Lee Patch, et. al., Embedded Decoupling Capacitance Project Final Report, National Center for Manufacturing Sciences Report 0091RE00, December, 2000.

12. David M. Pozar: "Microwave Engineering, Third Edition", 2005, John Wiley& Sons, Inc.

13. Lixi Wan. "Simulation of Switching Noise in Multi-layer Structures Using GeneralizedTransmission Line Equation Method", IEEE, 2002.

14. Dietrich Braess "Finite Elements: Theory, Fast Solvers, and Applications in Solid Mechanics", March, 2007.

15. Vitesse Semiconductor Corporation:"VSC7651 Datasheet-4X3.125Gps Optical Trans-impedance Amplifier Array", August, 2005.

A Simplified Thermal Resistance Network Model for High Power LED Street Lamp

Xiaobing Luo [1,2], Wei Xiong[1], Sheng Liu[2,3] *

[1] School of Energy and Power Engineering, Huazhong University of Science & Technology, Wuhan, 430074, China
[2] Wuhan National Lab for Optoelectronics, Huazhong University of Science & Technology, Wuhan, 430074, China
[3] Institutes of Microsystems, Huazhong University of Science & Technology, Wuhan, 430074, China
*Corresponding author: Sheng Liu, Telephone: 86-13871251668, Fax number: 86-27-87557074, Email:
victor_liu63@126.com

Abstract

Light emitting diode (LED) street lamp heavily relies on successful thermal management, which strongly affects the optical extraction and the reliability/durability of the LED lamp. In this study, a thermal resistance network model was presented to estimate the maximum heat sink temperature of the street lamp, which could be utilized to further evaluate the thermal performance of the street lamp. Two high power LED street lamps, an 114 watts and an 80 watts LED street lamp were used to evaluate the present model. Their heat sink temperatures were calculated by the model, the results showed that the maximum heat sink temperature was about 61°C at the environment temperature of 25°C for the 114 watts LED street lamp. For the 80 watts LED street lamp, the maximum heat sink temperature was about 42.5°C at the environment temperature of 11°C. To prove the model feasibility, experimental investigations on the 114 watts and 80 watts LED street lamp were conducted. The results demonstrated that the heat sink temperature of the 114 watts LED street lamp remained to be stable at about 60°C after several hours' lighting at the room temperature of 25°C. The heat sink temperature of the 80 watts LED street lamp remained to be stable at about 42°C at the room temperature of 11°C. Comparing the results achieved by the thermal resistance model with the experimental results, it was found that the proposed thermal resistance model could be used for temperature estimation and thermal evaluation for the high power street lamp.

Introduction

Theoretically, light emitting diode (LED) has many distinctive advantages such as high efficiency, good reliability, long life, variable color and low power consumption. Recently, LED has begun to play an important role in many applications [1]. Typical applications include back lighting for cell phones and other LCD displays, interior and exterior automotive lighting including headlights, large signs and displays, signals and illumination. LED will soon be used in general lighting, which consumes about 15 percent of the total energy in all over the world. An expectation about high power LED is that it will be the dominant lighting technology by 2025 [2]. Should the goal come to fruition, then up to 40 giga watts per year could be saved in the USA alone. It is generally believed that the LED can be widely used for general lighting in USA. However, in China, with the push of the government for more energy saving, the LED may be used earlier than this time. In China, the estimation by Chinese authorities is that if LED dominates generate lighting

market in 2010, one third of the present power consumption will be saved, which will greatly ameliorate the energy crisis situation in China.

One typical general lighting product of LED is LED street lamp, which is emerging in market, in particular in China. For modern LED street lamps, both optical extraction and thermal management are two critical factors for their high performance. In general, most of the electronic power of street lamp is converted into heat, which greatly reduces the chips' luminosity. In addition, the high junction temperature of LED chips in the lamp will shift the peak wavelength, which will change the color of light. Narendran and Gu [3] have experimentally demonstrated that the life of LEDs decreases with the increase of the junction temperature in an exponential manner. Therefore, a low operation temperature is essential for LED chips in the LED street lamp. Since the market demands that LED street lamp have high power and small size, there is a contradiction between the power density and the operation temperature, especially when applications require LED street lamp to operate at high power to obtain the desired brightness.

In terms of thermal management of LED street lamp, to the authors' best knowledge, there have been no reports or published papers, partially due to the fact that general lighting including street lighting is believed to be a few years away, or due to the concern that this field is highly proprietary as the market for the street lamp is huge. Although there are no reports directly related to the LED streetlamp, there have been some references to introduce the thermal management of high power LED packaging. Wilcoxon and Cornelius [4] described the thermal management approach to a light engine and presented the results of their finite element modeling. The feasibility of the modeling was proven by the experimental data and was used to assess various design aspects of the light engine to understand their effects on the overall thermal resistance. Their results out of the finite element modeling indicated that the junction temperature of the LEDs in this light engine would be close to their maximum values under the application conditions of high environment temperature. Through the use of expensive materials such as diamond/aluminum composites, the LED temperatures could be significantly reduced below the values obtained in this testing. Kim et al. [5] investigated the performance of thermal management system for LED light source in a rear projection TV. Their results showed that decreasing thermal resistance between LEDs and substrate was the most effective way to dissipate heat and the applicable limit of thermal resistance existed for various heat-dissipating conditions of LEDs. They also suggested applying the heat transport system in red, green

978-1-4244-2739-0/08/$25.00 ©2008 IEEE

and blue LED light system to ensure the product quality. Liu's group [6, 7, 8] studied a micro jet array cooling system for the thermal management of a high power LED lighting source. Experimental and numerical investigations on such an active cooling system were conducted. An infrared thermometer and thermocouples were used to conduct on-line temperature measurement and evaluate the cooling performance of the proposed system respectively. The experimental and numerical results demonstrated that the microjet-based cooling system has super cooling performance. Petroski [9] developed an LED-based spot module heat sink in a free convective cooling system. A cylindrical tube longitudinal fin (CTLF) heat sink was used to solve the orientation problem of LEDs. Chen et al. [10] presented a silicon-based thermoelectric (TE) for cooling of high power LEDs. The test results showed that their TE device could effectively reduce the operation temperature of high power LEDs. Acikalin et al. [11] used piezoelectric fans to cool LEDs. Their results showed that the fans could reduce the heat source temperature by as much as 37.4℃. Piezoelectric fans have been shown to be a viable solution to the thermal management of electronic components and LEDs. Treurniet and Lammens [12] presented a thermal design method of a multi-chip LED module that was able to handle an increasing thermal load up to 20 Watts. In addition, they proposed a compact model to estimate the junction temperature of the different dice at an arbitrary load. Arik and Weaver [13] carried out a numerical study to understand the chip temperature profile due to bump defects. Finite element techniques were utilized to evaluate the effects of localized hot spots at the active layer of chip. Tan, Liuxi et al. [14] studied the effects of various defects in terms of voids, cracks, delaminations on the thermal and optical performance of LEDs subjected to both powering and moisture loadings.

For the high power LED street lamp, the present authors [15] already conducted a thermal analysis. A numerical model for an 80 watts LED street lamp was built, the comparison of the simulation with the experiment demonstrated that the numerical model was feasible. Although numerical model can effectively predict the temperature distribution, it is still not convinent for engineering applications in the early design stage. It is necessary to find some closed-form or semi-empirical equations to estimate the temperature.

In this research, a thermal resistance network model was proposed to estimate the maximum heat sink temperature of the street lamp. To prove the model feasibility, experimental studies on an 114 watts street lamp and an 80 watts LED street lamp were conducted. Through the comparison between the experimental results and the thermal resistance modeling results, it was found that the present thermal resistance model would be helpful for the temperature estimation on the high power LED street lamp.

Typical Structure of High Power LED Street Lamp

Figure 1 and Figure 2 are the schematic diagrams of 80 watts and 114 watts LED street lamps. They represent one typical structure of high power LED street lamps to make full use of design freedom of LEDs. The kind of lamp is mainly composed of three parts: high power LED modules, a mechanical frame for both the heat dissipation and support of the LED modules, and four slim PCBs for the power input of LEDs. The lamp frame consists of aluminum base and fins, which are made as one integrated design for saving fabrication cost and decreasing thermal resistance.

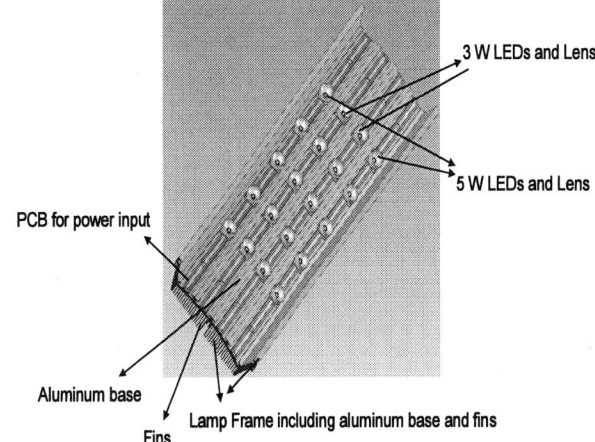

Figure 1. Schematic diagram of the 80-W LED street lamp.

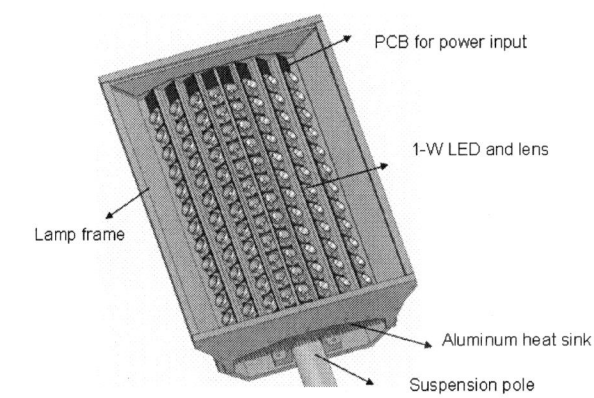

Figure 2. Schematic diagram of the 114-W LED street lamp.

As for the 80 watts LED street lamp shown in Figure 1, twenty high power LED modules are directly bonded on the aluminum base for reducing thermal resistance. They are distributed on the aluminum base in four rows. The modules in the central two rows are 3 watts LED packages, each of which includes three 1 watt power chips. The other two rows consist of 5 watts LED modules, each of the 5 watts packages includes four chips, and they are supplied with 5 watts power. Four slim PCBs are located on the aluminum base and used for providing power for the four rows of LEDs.

For the 114 watts LED street lamp shown in Figure 2, ninety six high power LED modules are bonded onto the heat sink. They are distributed on the heat sink base in eight rows. The ninety six LED modules are the same, their default input powers are 1 watt, however, here they are supplied with 1.188 watts power, therefore, the total input power for this lamp is about 114 watts.

For the above two lamps, when the electronic power is supplied, LEDs generate light and also heat. The heat is dissipated out into the environment through the aluminum base and fins on the base.

Thermal Resistance Model and Its Calculations

Good optical and thermal performances are the design keys to a successful LED street lamp. The optical design of the street lamp is to achieve good light quality and make the road bright and comfortable to the passengers. However, the optical characteristics of the present LED street lamp will not be discussed in this paper, its temperature distribution and thermal performance will be the main concern.

For the street lamps shown in Figure 1 and Figure 2, low thermal resistance is highly appreciated for achieving good thermal and optical performance. The thermal resistance for every LED module in suck kind of street lamp includes four parts, which are demonstrated in Figure 3. The first part is the packaging thermal resistance (R_{cp}) of the high power LED, which is related to the chip packaging technology. The second part is the bonding thermal resistance (R_{bd}) between LEDs substrate and the aluminum base with fins, which is mainly determined by the bonding material and thickness. The third part is the thermal spreading resistance (R_{sp}) between LEDs and the aluminum base with fins, which is affected by many geometrical sizes such as chip substrate size, the size and material of the aluminum base and so on. The last part is the thermal convection resistance (R_{conv}) between the fins and the environment, which is influenced by many factors such as fin structure, area and environment wind speed. In Figure 3, T_j is the maximum junction temperature of the LED chips, T_{c-c} is the maximum temperature in the aluminum base, and T_a is the ambient temperature.

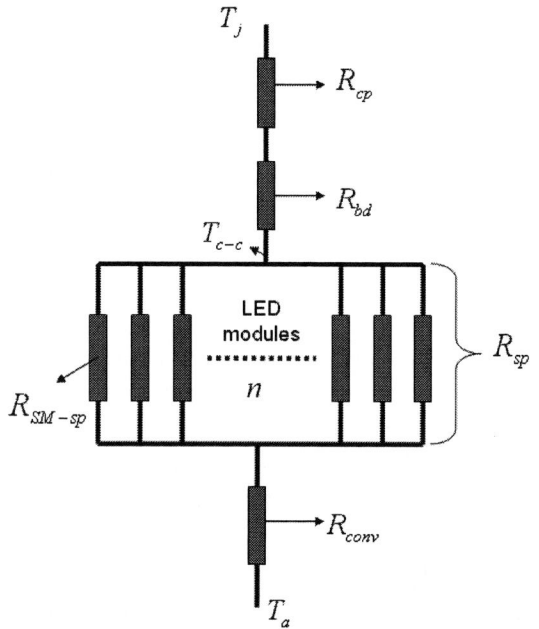

Figure 3. Thermal resistance model of LED module in the high power LED street lamp.

Actually, the bulk material resistance (R_{bk}) exists in the thermal resistance network, which is usually included in the calculation of the thermal spreading resistance(R_{sp}). For the present high power LED lamps, the bulk material resistance is very small so that it can be neglected [15].

In the four thermal resistances, thermal spreading resistance is the most important and difficult to be calculated. For the high power LED street lamps discussed in this paper, many LED modules are distributed on the heat sink and each LED module creates heat, so there are many heat sources distributed on the heat sink. The heat produced by each LED module is transferred to the heat sink and finally is dissipated into the environment. During the heat transfer process from each LED module to the heat sink, one thermal spreading resistance exists because heat transfers from small area such as LED module to the larger area such as heat sink. Therefore, as shown in Figure 3, for one high power LED street lamp, there are many thermal spreading resistances since it consists of many LED modules. These thermal spreading resistances(R_{SM-sp}) are connected in parallel. For the high power LED street lamps shown in Figure 1 and Figure 2, since the size of each LED chip module is the same, also their sizes are very small compared with the heat sink size, although the position for every heat source is different in the base plate, the thermal spreading resistances (R_{SM-sp}) for each LED module in Figure 3 still can be regarded as the same. Therefore, it is very easy to simplify and calculate the final thermal spreading resistance (R_{sp}) of the LED street lamp, which can be expressed as,

$$R_{sp} = \frac{R_{SM-sp}}{n} \qquad (1)$$

where n is the LED module number for the high power LED street lamp, as shown in Figure 3. For the 80 watts street lamp shown in Figure 1, n is 20. For the 114 watts street lamp shown in Figure 2, n is 96.

As to the thermal spreading resistance of the single LED module R_{SM-sp}, its maximum value is defined as:

$$R_{max-SM-sp} = \frac{T_{max} - \overline{T_b}}{Q} \qquad (2)$$

where T_{max} is the maximum temperature in the contacting area between the LED module and heat sink, $\overline{T_b}$ is the average temperatures over the LED module and base areas. Q is the heat transfer rate of each LED module.

About this thermal resistance, a closed-form equation has been proposed by the Lee et al [16,17]. Based on these references,

$$R_{max-SM-sp} = \frac{\psi_{max}}{k \cdot a \cdot \sqrt{\pi}} \qquad (3)$$

where a is the equivalent radius of the LED module, k is the thermal conductivity of the heat sink material.

218

$$\psi_{max} = \frac{\varepsilon \cdot \tau}{\sqrt{\pi}} + \frac{1}{\sqrt{\pi}} \cdot (1-\varepsilon) \cdot \Phi_c \quad (4)$$

In equation (4),

$$\Phi_c = \frac{\tanh(\lambda_c \cdot \tau) + \dfrac{\lambda_c}{Bi}}{1 + \dfrac{\lambda_c}{Bi} \cdot \tanh(\lambda_c \cdot \tau)} \quad (5)$$

where

$$\lambda_c = \pi + \frac{1}{\sqrt{\pi \cdot \varepsilon}} \quad (6)$$

In the above several equations, b is the equivalent radius of the heat sink of the street lamp, t is averaging thickness of the heat sink. R_{conv} is the thermal convection resistance of the heat sink. ε is dimensionless heat source radius, it can be expressed as,

$$\varepsilon = \frac{a}{b} \quad (7)$$

τ is dimensionless heat sink thickness, which is defined as,

$$\tau = \frac{t}{b} \quad (8)$$

Bi is effective Biot Number, it can be expressed as,

$$Bi = \frac{h_{eff} \cdot b}{k} \quad (9)$$

where h_{eff} is the equivalent natural convection efficient for the heat sink of the LED street lamp.

Combining equations (3) to (9) into equation (2), the maximum thermal spreading resistance of the single LED module $R_{max-SM-sp}$ can be calculated. Since this value is acquired, according to equation (2), the maximum temperature in the LED module area T_{max} can be obtained, which is nearly same as the value of T_{c-c}.

Based on equation (1), finally, the total thermal spreading resistance for the LED street lamp can be obtained.

For the other thermal resistance in the model shown in Figure (3), R_{conv} can be expressed as,

$$R_{conv} = \frac{1}{hA} \quad (10)$$

where h is averaging natural convection coefficient in all heat sink area, it is different from h_{eff}. In the original references [15,16], the heat sink connecting with the heat source is a plate, however, for the heat sink of the LED street lamp, it has many fins, to apply the situation into the equations, the heat sink with many fins should be equal to a plate in heat transfer area. Therefore, h is the real heat transfer coefficient based on the heat sink area including fin

area. h_{eff} is an equivalent natural convection efficient when the heat sink of the street lamp is regarded as a plate without its fins. Obviously, h_{eff} is much larger than h. The relation between h and h_{eff} can be described as,

$$Ah = h_{eff} A_{base} \quad (11)$$

In equations (10) and (11), A is the total heat sink area that is exposed to the environment. A_{base} is the heat sink base area, and it does not include the fin area.

The value of h can be calculated by using the methods proposed by reference [18]. Prof.Bar-Cohen in reference [18] gave the detailed analysis and discussions about the methods. It should be noted that the orientation of the fin heat sink of the present LED street lamp is horizontal, therefore, the heat transfer equations for both heat sink base and fins are a little different from those in reference [18]. In this calculation, computation iterations should be used for obtaining the results because of the coupled parameters in the equations.

For the bonding thermal resistance R_{bd} and module packaging thermal resistance R_{cp}, they are easy to be obtained since they are usually provided by the thermal adhesive vendors and LED module vendors.

According to the above discussions, it is noted that all the thermal resistances shown in Figure 3 can be achieved, therefore, the temperatures such as T_{c-c}, T_j in the model can be calculated and estimated.

Thermal Resistance and Temperature Calculations on the 80-W and 114-W LED Street Lamp

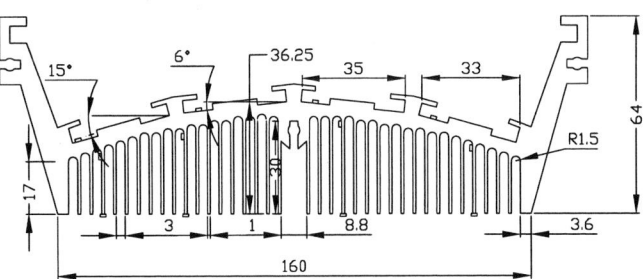

Figure 4. Details of heat sink used in the 80-W lamp. (All dimensions are in mm.)

For the 80 watts and the 114 watts LED street lamps shown in Figure 1 and Figure 2, the sizes of the heat sinks are shown in Figures 4 and 5 in details. The lengths of both heat sinks are 600mm and 550mm respectively. The length directions in Figures 4 and 5 are vertical to the paper.

Based on the above mentioned model, by using equations (1) to (11), and supplying as inputs the parameters such as power, ambient temperature, sizes and physical properties of air and fin, the different values in the models for 80 watts and 114 watts LED street lamps can be obtained and are listed in the following Table 1.

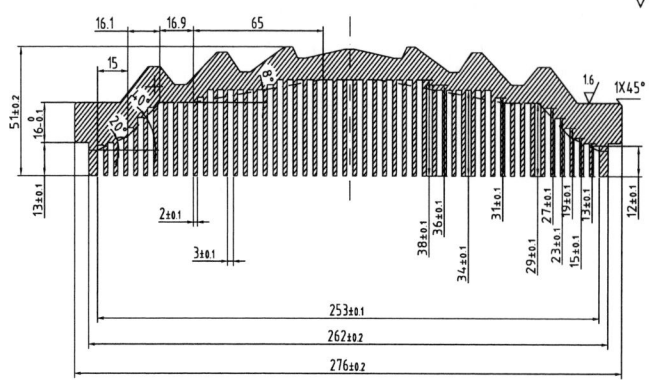

Figure 5. Details of heat sink used in the 114-W lamp. (All dimensions are in mm.)

Table 1. Calculation results for 80 watts and 114 watts LED street lamps

Parameters LED Lamp type	80 watts		114 watts
Averaging heat transfer coefficients. of fins and base (h) (W/m.K)	1.87		1.5
Thermal spreading resistance of the LED street lamp (R_{sp})(K/W)	0.1		0.008
Thermal resistance of thermal interface material (R_{bd})(K/W)	2.5		2.5
Thermal resistances of chip packaging (R_{cp})(K/W)	3-W module	4.1	10
	5-W module	2.42	
Thermal convection resistance (R_{conv})(K/W)	0.863		0.37
Averaging temperature of fins and base (℃)	40.33		60.92
Maximum temperature of fins and base (℃)	45.41		61.7
Averaging junction temperature (℃)	3-W module	68.13	74.332
	5-W module	72.93	

Experimental Study

1. System and temperature measurement

To prove the feasibility of the above thermal resistance network model, the temperature distributions of the aluminum base and fins of the 80 watts and 114 watts LED street lamp were measured by thermocouples in the experiments.

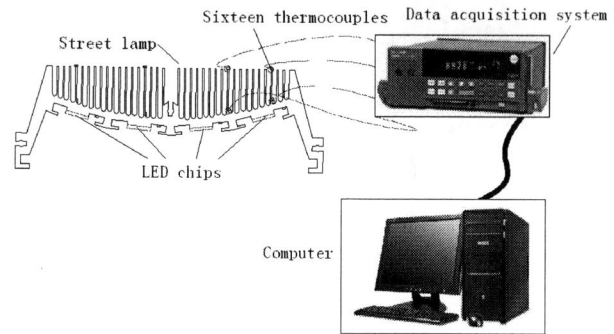

Figure 6. Experimental setup.

Figure 6 shows the experimental setup. The orientation of the heat sink and the system were set up as the application conditions. The fin base was placed blow, the fin tips were on the top. The tests were conducted at a natural environment. For 80 watts and 114 watts LED street lamp tests, the ambient temperature was about 11℃ and 25 ℃ respectively. Several thermocouples were placed at different positions of the aluminum base and fins. The temperature data obtained by the thermocouples was transferred to the data acquisition system and displayed on the PC monitor. The model of the data acquisition system in the experiment was Keithley 2700 multimeter and control unit 7700.

2. Accuracy analysis

In the experiments, the temperature was the main parameter for system evaluation, and it was directly measured by thermocouples. Since there were no other indirectly measured parameters, the errors associated with this experiment mainly included measurement error of the thermocouples and reading error of the digital multimeter.

Standard T-type thermocouples (Cu-CuNi) were used in the experiments. During the temperature range from -30℃ to 150℃, their measurement error was about 0.2℃. The data acquisition system had a reading error of 1℃ since the cold junctions of the thermocouples used the default setup supplied by the system, not the ice bath with constant 0℃. Therefore, the total error of the temperature measurement for the experiments was about 1.2℃.

Experimental Results and Analysis

Figure 7 shows the variation of the heat sink temperature with the operation time for the 80 watts LED street lamp. In the experiments, as described above, the room temperature was about 11℃ and there were sixteen thermocouples to measure the temperatures at sixteen different positions, much data was obtained and it was difficult to display them in one figure synchronously. Thus the temperatures obtained by four thermocouples numbered as 2, 3, 11 and 13 were used for description. In Figure 6, it can be seen that the fin temperature increased as time extended, initially, the fin temperature was nearly the same as the room temperature. After the lamp was activated, its temperature increased. Several hours later, it reached steady situation and the temperature remained to be stable to be nearly 42℃. It is also noted from Figure 6 that the temperatures achieved by all thermocouples showed the same change trend. The temperature differences among the thermocouples were very small, considering the reality that the measurement error was about 1.2℃, it cannot differentiate the temperature difference demonstrated in Figure 6.

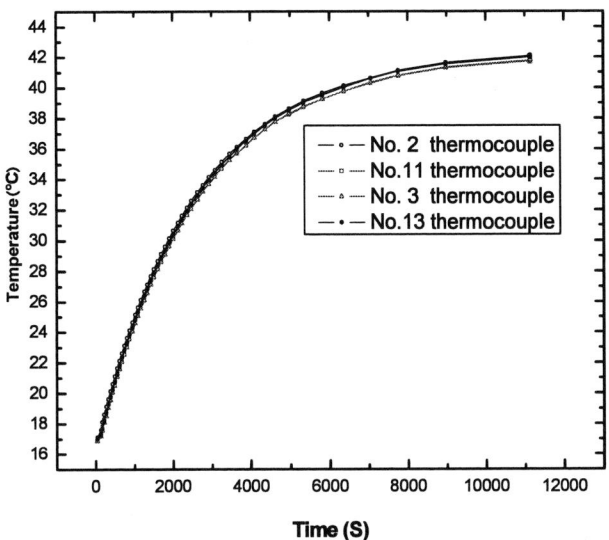

Figure 7. Variation of the heat sink temperature with the operation time for 80-W LED street lamp

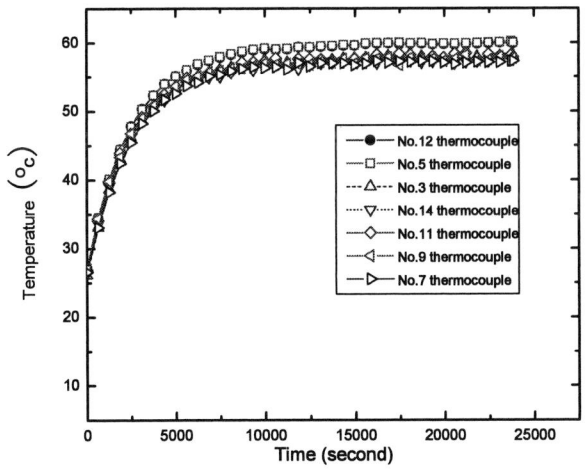

Figure 8. Variation of the heat sink temperature with the operation time for 114-W street lamp.

Figure 8 shows the variation of the heat sink temperature with the operation time for the 114 watts LED street lamp. In the experiments, the room temperature was about 25℃. In Figure 8, it can be seen that the heat sink temperature increased as time extended, initially, the fin temperature was nearly the same as the room temperature. After the lamp was activated, its temperature increased. Several hours later, it reached steady situation and the maximum temperature remained to be stable to be nearly 61℃.

Comparison between Experimental Results and Modeling Results

As discussed in the above sections, the modeling results demonstrate the temperature distribution in different parts of the street lamp, including the junction temperature. However, in the experimental results, the junction temperature could not be achieved because of the measurement difficulty, the surface temperature of heat sink was measured, therefore, in the comparison, the heat sink temperature will be the only parameter to evaluate the modeling result. The comparison results are shown the following table.

Street lamp type	Ambient temperature	Experim ental result	Thermal resistance modeling result
114 watts LED	25℃	61℃	60.92℃
80 watts LED	11℃	42℃	40.33℃

According to Table 1, it is clear that the modeling results are close to the experimental ones, the temperature difference achieved by the two methods for 114 watts LED street lamp was just 0.08℃. For 80 watts LED, it was 1.67℃. The comparison proves the feasibility of the model.

Summary

In this paper, a simplified thermal resistance network model for LED street lamp was proposed and model was validated by experiments for two prototype street lamps. It is expected that the simplified model can be used for rapid design of thermal management for LED street lamps and other similar high power LED applications.

Acknowledgments

We acknowledge the financial support from Key Technology R&D Program of Hubei Province, China. (2006AA103A04) and GuangDong RealFaith Optoelectronics Inc., GuangDong, China.

References

1   Alan, M.: 'Solid state lighting-a world of expanding opportunities at LED 2002', III-Vs Review, 2003, 16, (1), pp.30-33

2   Alan, M.: 'Lighting: The progress & promise of LEDs', III-Vs Review, 2004, 17, (4), pp. 39-41

3   Narendran, N., and Gu, Y. M.: 'Life of LED-Based white light sources', IEEE Journal of Display Technology, 2005, 1, (1), pp.167-171

4   Wilcoxon, R. and Cornelius, D.: 'Thermal Management of an LED Light Engine for Airborne Applications'. Proc. 22th IEEE Semiconductor Thermal Measurement and Management Symposium, Dallas TX, USA, March 2006, pp.178–185

5 Kim, S. K., Kim, S. Y., and Choi, Y. D.: 'Thermal performance of cooling system for red, green and blue LED light source for rear projection TV', Proceeding of The Tenth Intersociety Conference on Thermal and Thermomechanical Phenomena in Electronics Systems, San Diego, CA, USA, May 2006, pp.377- 379

6 Liu, S., Lin, T., Luo X.B., Chen, M.X., and Jiang, X.P.: 'A microjet array cooling system for thermal management of active radars and high-brightness LEDs', Proc. Fifty-Sixth Electronic Components & Technology Conference, San Diego, CA, USA, May 2006, pp.1634-1638

7 Luo, X.B., Chen W., Sun R.X., and Liu S.: 'Experimental and Numerical Investigation of a Microjet Based Cooling System for High Power LEDs', Heat Transfer Engineering, In press

8 Luo, X.B., Liu S.: 'A Microjet Array Cooling System for Thermal Management of High-Brightness LEDs', IEEE Journal of Advanced Packaging, 2007, 30, (3), pp. 475-484

9 Petroski, J.: 'Understanding longitudinal fin heat sink orientation sensitivity for Light Emitting Diode (LED) lighting applications', Proc. International Electronic Packaging Technical Conference and Exhibition, Maui, Hawaii, USA, July 2003, pp. 111-117

10 Chen, J.H., Liu, C.K., Chao, Y.L., and Tain, R.M.: 'Cooling performance of silicon-based thermoelectric device on high power LED', Proc. 24th International Conference on Thermoelectrics, Clemson, South Carolina, USA, June 2005, pp.53- 56

11 Acikalin, T., Garimella, S.V., Petroski, J., and Arvind, R.: 'Optimal design of miniature piezoelectric fans for cooling light emitting diodes', Proc. Ninth Intersociety Conference on Thermal and Thermomechanical Phenomena in Electronic Systems, Las Vegas, Nevada, USA , June 2004, pp. 663-671

12 Treurniet, T., and Lammens, V., : 'Thermal management in color variable multi-chip LED modules', Proc. 26th IEEE Semiconductor Thermal Measurement And Management Symposium, SEMI-THERM 2006, Dallas, Texas, USA, March 2006, pp. 173-177

13 Arik, M., and Weaver, S., : 'Chip scale thermal management of high brightness LED packages', Procceedings of SPIE v 5530, 4th International Conference on Solid State Lighting, Denver, Colorado, USA, August 2004, pp. 214-223

14 Tan, L.X., Li J., Liu, Z.Y., Wang, K., Wang, P., Gan, Z.Y., and Liu, S., : 'A Light Emitting Diode's Chip Structure withLow Stress and High Light Extraction Efficiency', Proc. Fifty-Eigth Electronic Components & Technology Conference, USA, May 2008, pp.783-788

15 Luo, X.B., Cheng, T., Xiong, W., Gan, Z.Y., Liu, S., Thermal Analysis on an 80W LED Street Lamp, IET Optoelectronics, 2007, 1, (5), pp.191-196

16 S. Song, S. Lee, V. Au, Closed-form equation for thermal constriction/spreading resistances with variable resistance boundary condition, in: Proceedings of the 1994 International Electronics Packaging Conference, Atlanta, Georgia, 1994, pp. 111–121

17 S. Lee, S. Song, V. Au, K.P. Moran, Constriction/spreading resistance model for electronics packaging, in: ASME/JSME Thermal Engineering Conference, vol. 4, Maui, Hawaii, 1995, pp. 199–206

18 Bar-Cohen, A., Iyengar, M, and Kraus, A.D., Design of Optimum Plate Fin Natural Convection Heat Sinks, ASME Transactions - Journal of Electronic Packaging, 2003, Vol 125, Number 2, pp 208-216

Improvement of Power Integrity with novel Segmented Power Bus Structures in RF/Digital SOP

Jun Li[1,2], Lixi Wan[2], Wei Gao[2], Cheng Liao[1]

[1] Institute of Electromagnetics, Southwest Jiaotong University, Chengdu, 610031, China
[2] Institute of Microelectronics, Chinese Academy of Sciences，Beijing, 100029, China

Abstract

With the voltage decreasing in power distribution network (PDN) in system-on-packages (SOPs), power integrity will be a critical issue. The cavity resonance modes between power and ground planes can be excited by simultaneous switching noise (SSN) or ground bounce noise (GBN) [1], [2]. To cut down the noise from susceptible devices, isolation techniques are necessary. In this paper, two novel structures combine segmented method and embedded capacitor provided isolation performances from 0.7GHz to 10GHz below -40dB. The novel structures with a bridge for suppressing noise in high frequency and a thin high K dielectric substrate for decreasing the SSN in the entire frequency band. And they were better in performance and simpler in configuration than Electromagnetic Band Gap (EBG) and other isolation structures. Moreover, the analytical process could be a guidance to find better structures for improving the noise isolation.

Introduction

In high-speed mixed signal packages and printed circuit boards (PCBs), the design of low-impedance and low-noise power distribution network (PDN) has become a major challenge due to simultaneous switching noise (SSN) or ground bounce noise (GBN). The SSN causes the voltage level fluctuation and can be coupled through interconnection structure to disturb nearby circuits and other sensitive devices. To ensure reliable circuit, device, and system operation, the noise must be controlled efficiently. To satisfy the working condition of digital circuit and the RF-IC with high sensitivity, the PDN must supply low impedances, that is high degree of $|S_{21}|$, for a wide frequency band.

In previous works, several methods for noise isolation have been studied. Adding decoupling capacitors is limited by the equivalent serial inductance (ESL) above 600MHz [3]. A concept of isolation noise using electromagnetic band gap (EBG) in place of the power plane has been introduced [1]-[4]. The EBG can omnidirectionally reduce noise with a band behavior. However, the EBG has to occupy vast areas, cost and get reversely poor performance in low frequencies [4]. Segmented power bus, in which conducting or perfect electric conductor (PEC) bridges provide a common voltage, is the simplest and cost-effective structure and often used for noise suppression [5], but usually provide relative poor isolation for a wide band. Embedded capacitor with thin film and high dielectric constant is used to store electric charges and decouple noise in the PDN recently [6]. The method of combining embedded capacitor and segmented power bus can achieve good isolation performance. For this notion, two novel segmented power bus structures are shown in this paper.

In section II, a splited ground plane structure was proposed to suppress the noise. Both the simulation and measurement results showed the good isolation performance. Section III displayed a simpler thin film segmented structure to reduce the SSN. The key parameters of this structure were discussed to show the design rules. The novel structures were better in performance and simpler in configuration than EBG and other isolation structures. Conclusions were drawn in section IV

Splited ground planes structure

A mixed-signal system test board of 10mm by 10mm in size is shown in Fig. 1. There are three metallic layers, ground plane, additive ground plane and power plane. Two dielectric substrates are FR4 and the thin high permittivity substrate (DK=16) with the thickness is 100um and 14um separately. The whole structure is cut into two parts, one of them serves for RF-IC while another for digital IC. The key of this novel structure is the continuous ground plane and the separated additional ground plane to enhance the isolation effect. A long strip bridge provides a common voltage for both ICs in power layer. The relevant parameters are denoted as (width, gap), where width=0.5mm is the strip width; gap=0.5mm is the spacing between separated areas. To simulate the practical situation, each IC with five connection vias is played out in $4mm^2$. Between the port1 and port2, the $|S_{21}|$ is calculated using full-wave electromagnetic analysis software. As you seen in Fig. 2, the traditional EBG structure is implemented in 150um FR4 substrate of 10mm*10mm. The comparison of splited ground plane and the traditional EBG structures as shown in Fig. 3. It is shown that the performance of the novel structure, which $|S_{21}|$ below -40dB from 0.65GHz to 10GHz, is much better than that of the traditional EBG structure, which has poor performance in low frequency. Moreover, to achieve a good isolation effect, the EBG must need a long period, and the vias will disturb the effect easily. So, the novel structure in configuration is much simpler.

The work has been supported by National Natural Scien Foundation of China, No. 90607006 and Hi-tech Research a Development Program of China (863 Program) N 2006AA01Z236 and No.2007AA01Z2a6.

978-1-4244-2739-0/08/$25.00 ©2008 IEEE

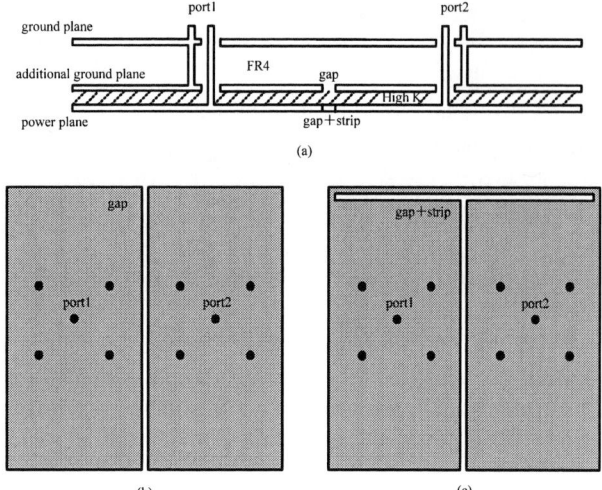

Fig. 1 Splited ground planes structure. (a) Cross section. (b) Top view of additive ground plane. (c) Top view of power plane

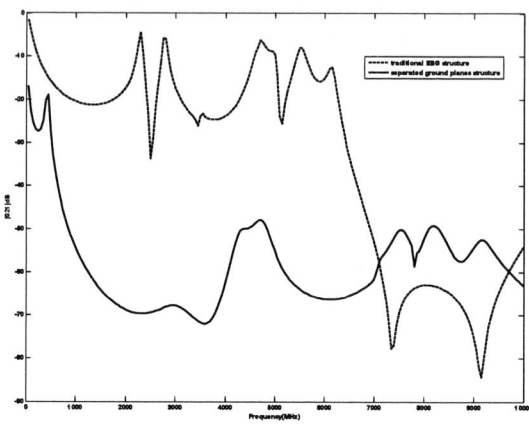

Fig. 3 Comparison of splited ground plane and traditional EBG structure

Fig. 2 The traditional EBG structure. (a) Cross section. (c) Top view of power plane.

Thin film segmented structure

As we know, a lossless continuous rectangular power-return plane pair separated by a thin dielectric substrate can be modeled as 2-D TMz cavity [7]. The transfer impedance between two ports is given by

$$Z_{ij} = jw\mu h \sum_{m=0}^{\infty}\sum_{n=0}^{\infty} \frac{\chi_{mn}^2}{ab(k_{xm}^2 + k_{yn}^2 - k^2)} \cos(k_{yn}y_i)\cos(k_{xm}x_i)\sin c\left(\frac{k_{yn}dy_i}{2}\right) \quad (1)$$

$$\times \sin c\left(\frac{k_{xm}dx_i}{2}\right)\cos(k_{yn}y_j)\cos(k_{xm}x_j)\sin c\left(\frac{k_{yn}dy_j}{2}\right)\sin c\left(\frac{k_{xm}dx_j}{2}\right)$$

Where $k_{xm} = m\pi/a$, $k_{yn} = n\pi/b$, $k = w\sqrt{\varepsilon_0\varepsilon_r\mu}$. $\chi_{mn}^2 = 1$ for $m = n = 0$; $\chi_{mn}^2 = 2$ for $m = 0$ or $n = 0$; $\chi_{mn}^2 = 4$ for $m \neq 0$, $n \neq 0$.

Equation (1) shows the parameter h is in direct proportion to Z_{ij}. For considering the losses, the conductive loss pays an important role in the quality factor. For very thin dielectric layers, an approximate formula for the quality factor due to conductive loss is given by

$$Q_c \approx h\sqrt{\pi f\mu\sigma} \quad (2)$$

So, in this paper using thin film dielectric substrate can reduce the SSN efficiently.

Fig. 4 displays the rule of isolation performance with different relative permittivity of continuous power/ground planes. The continuous power bus can be considered as a capacitor, the lower DK, the lower capacitance and the higher first resonant frequency which marked with number as shown in Fig. 4. In the entire work band, all of the resonance frequencies move to higher frequencies in the lower DK.

To more reduce the thickness of the PDN, a novel thin film segmented structure is proposed as shown in Fig. 5. Power and ground planes with the same size as section II are separated by a thin film high K material with DK=16.The thickness is h=14um. The ground plane remains continuous, and the power plane is divided into two parts with a long strip bridge for connection. The thin high K dielectric substrate is for decreasing the impedances in the entire frequency and the long strip bridge is for suppressing noise in high frequency. The simulation result of thin film segmented structure is displayed in Fig. 6. It is shown that the $|S_{21}|$ are below -40dB from 0.7GHz to 10GHz. So, we can also get good noise isolation performance using the simpler thin film segmented structure.

Experimental Verification

For mechanical consideration, both splited ground planes structure and thin film segmented structure must need attached on a core layer, such as shown in Fig.7. Fig.8 is the measurement result of the thin film segmented structure with a core layer. And the insertion loss of is measured less than -30dB from 500MHz to 4GHz. Although the results from both measurement and simulation are similar, the difference is obvious. The main reason of the differences may be come from the effect of core layer, such as the higher inductance, the couple noise through gaps between the two cavities, and so on. Even though, the results still shows us the performance for suppressing high frequency noise.

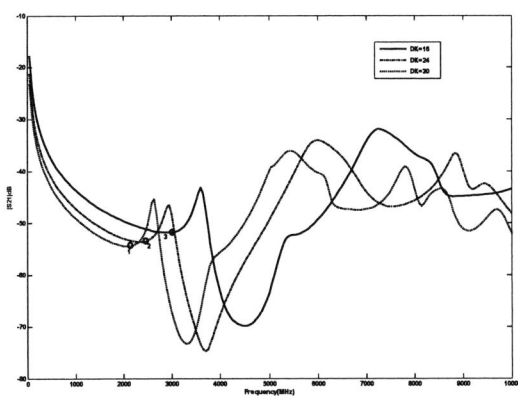

Fig. 4 Isolation performance with different relative permittivity of continuous rectangular power-return plane pair.

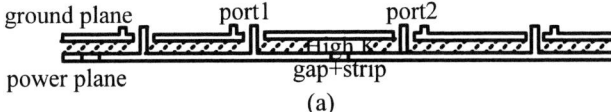

(a)

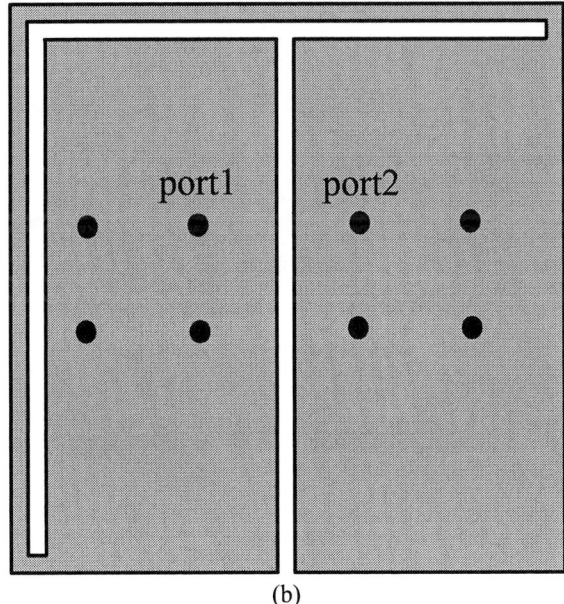

(b)

Fig. 5 Thin film segmented structure. (a) Cross section. (c) Top view of power plane.

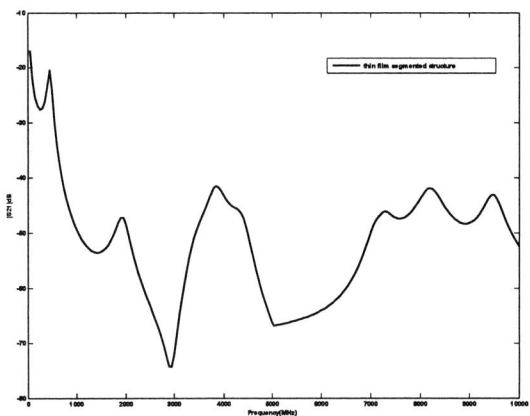

Fig. 6 the |S21| of thin film segmented structure.

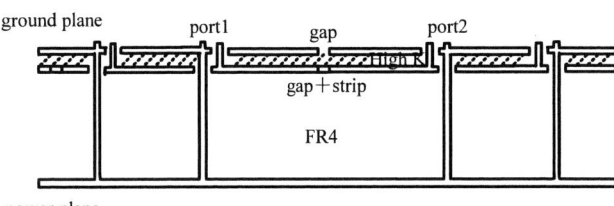

Fig. 7 the cross section of thin film segmented structure with core layer.

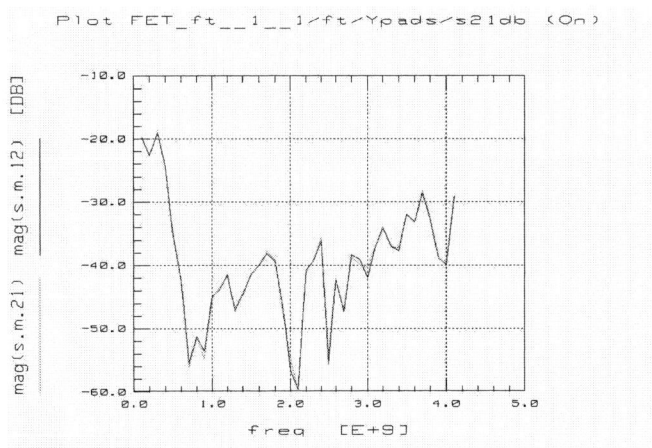

Fig. 8 the measurement of thin film segmented structure with core layer.

Conclusions

In this paper, two novel structures were used in the PDN to improve the power integrity in mixed system effectively. The key of the novel structures is combing segmented method and embedded capacitor. To satisfy the working condition of digital circuit itself and the high sensitivity of RF-IC, the PDN must supply high isolation degree for a wide frequency band. The proposed structures are splited ground planes structure and thin film segmented structure. Both of them can get a good performance from 0.7GHz to 10GHz with a strip bridge for suppressing noise in high frequency and a thin high K dielectric substrate for decreasing noise in the entire frequency band.

225

Through the simulation model and measurement results, the proposed structures in this work have a high practicality for good performances and simply configurations compare with EBG or other structures. The analytical of the thickness h and DK will be a guidance to find better structures for improving the isolation effect.

References

[1] T. L. Wu, Y. H. Lin, T. K. Wang, C. C. Wang, and S. T. Chen, "Electromagnetic Bandgap Power/Ground Planes for wideband suppression of Ground Bounce and Radiated Emission in High-Speed Circuit," IEEE Trans. Micro. Theory and Tech., vol. 53, pp. 2935-2942, Sep. 2005.

[2] T. Kamgaing and O. M. Ramahi, "Design and modeling of High-Impedance Electromagnetic Surfaces for Switching Noise Suppression in Power Planes," IEEE Trans. Electromagn. Compat., vol. 47, pp. 479-489, Aug. 2005.

[3] Tzong-Lin Wu, Chien-Chung Wang, Yen-Hui Lin, Ting-Kuang Wang, and George Chang, "A Novel Power Plane With Super-Wideband Elimination of Ground Bounce Noise on High Speed Circuits," IEEE Microw. Wireless Comon. Lett., vol. 15, pp. 174-176, Mar. 2005.

[4] Xiao-Hua Wang, Bing-Zhong Wang, Yw-Hai Bi, and Wei Shao, "A novel Uniplanar Conpact Photonic Bandgap Power Plane With Ultra-Broadband Suppression of Ground Bounce Noise," IEEE Microw. Wireless Compon. Lett., vol. 16, pp. 276-277, May 2006.

[5] W. Cui, J. Fan, H. Shi, and j. l. Drewniak, "DC Power Bus Isolation With Power-Plane Segmentation," IEEE Trans. Elctromgn. Compat., vol. 45, pp. 436-442, May 2003.

[6] Lixi Wan, etc. "Design, Simulation and Measurement Techniques for Embedded Decoupling Capacitors in Multi-GHz Packages/PCBs", ECTC, 2005.

[7] Minjia Xu, and Todd H. Hubing, "Estimating the Power Bus Impedance of Printed Circuit Boards With Embedded Capacitance," IEEE Trans. Adv. Pack., vol. 25, pp. 424-432, Aug. 2002.

A Study of RF Front-End Filters with Embedded Capacitor Technology

Yunfeng Wang[12], Lixi Wan[12], Lei Li[1]

[1] Shenzhen institute of advance technology, Chinese Academy of Sciences/the Chinese University of Hong Kong,
Shenzhen 518067, China
[2] Institute of Microelectronics, Chinese Academy of Sciences
Beijing, 100029, China
yf.wang@siat.ac.cn

Abstract

As the trend in wireless communication system is toward multi-functionality and higher miniaturization at higher frequency and lower cost, to design a RF front-end modules is a challenge. Filters, baluns, Bluetooth modules, and power amplifier modules on organic substrate have been studied widely [1-5]. Among the modules, the RF filters play a most significant role. Various design and integration approaches for filters have been reported in many publications. But the conventional methods using SMDs can not meet the requirements for better performance and lower cost and size. The embedded passives are promising solutions due to its low parasitic parameters and small size, so embedded filters have been widely studied and also reported in several papers. Base on mature filter theory, better performance needs more passive components, and could not be achieved by reducing circuit orders. For example, band pass filter could be obtained by many orders of combinations of low pass and high pass filters or by many orders of LC resonators. Most researchers focused on designing different topologies and configurations of layout to reach the electrical specifications [6-8]. In this paper, the author firstly proposed that the resonators could be implemented by using only embedded capacitors by designing inductor geometries with rational design of structure, size and shape on one electrode of the embedded capacitors to realize LC resonators, so the bandpass filter (BPF) could be obtained by capacitive-coupled resonator BPF. This paper presents the physical model simulated by HFSS, and discusses the factors influencing the characteristics of the resonator and filter, such as the width, core area, etc. The most important element affecting the multi-order resonator filter is the capacitance of the coupling capacitor whose value could be changed along with the area of the top and bottom electrodes of embedded coupled capacitor. Last, a narrow band pass filter (NBPF) and a wide band pass filter (WBPF) are designed and simulated by HFSS, and the performances are figured subsequently.

Introduction

The wireless communication systems are developing towards the direction of low profile, light weight, low cost, excellent performance and multi-functionality. SOP/SIP (system-on-package/system-in-package), especially the

This work was supported by National Natural Science Foundation of China, No. 90607006 and by Hi-tech Research and Development Program of China (863 Program) No. 2006AA01Z236 and No. 2007AA01Z2a6.

embedded passives technology, is considered to be one of the most challenges and exciting technology to realize the advanced micro-systems, because the passive components take up a large real estate of a PCB, and the electrical performance and reliability are reduced by the longer interconnect and more solder joints. The embedded passives have drawn attractions to the RF front-end circuits and modules due to its low parasitic parameters, small size and high performance and reliability. A bandpass filter is by using embedded passives technology is studied in the work.

In the paper, the author firstly proposed the RF front-end bandpass filter designed with only embedded capacitor. The bandpass filters are composed of 2-order LC resonators coupled by the capacitors. The resonator is implemented by designing spiral inductor geometries on the top electrode of embedded capacitor material, and one end of the resonator is connected to the bottom electrode by a micro-via. NBPF has a size of $2*1.9mm^2$. It exhibits a 3 dB bandwidth of 140MHz, an insertion loss of 2.68dB and return loss of 31.8dB. WBPF has the same size with NBPF，but has a different coupling capacitance between two resonators. It exhibits an insertion loss of 15.1dB, and band width of 500MHz, and a center frequency of 2.4GHz. All the filters' performance are simulated by using 3D EM simulator.

Design

a) Resonator

The resonator of the filter is implemented by a panar spiral inductor geometry on the top electrode to form an inductor, and one end of the inductor is connected to the bottom electrode through a via. Fig.1 (a) shows the typical resonator layout, (b) shows the cross section of the embedded capacitor, and the material specification are listed in Table 1, Metal layers are assumed perfect conductor material, which is PEC for simplicity.

Table1. Specifications of embedded capacitor

Structure	dimensions	material
Layer1	18um	PEC
Dielectric1	14um	Dk = 16 Df = 0.005
Layer2	18um	PEC

As shown from the fig.1, w is the inductor width, s is the space between the winding, and d is the inner diameter of the spiral inductor. These planar spiral inductors are designed with rectangular shapes for finding out optimal geometry. Width and core area of the inductors varies for finding out the most optimal performance. The inductance and the capacitance between the spiral inductor and the bottom electrode comprise of the inductance and capacitance of the resonator, and the mutual inductance and parasitic capacitance among the inductor turns could be ignored because of weak electromagnetic coupling.

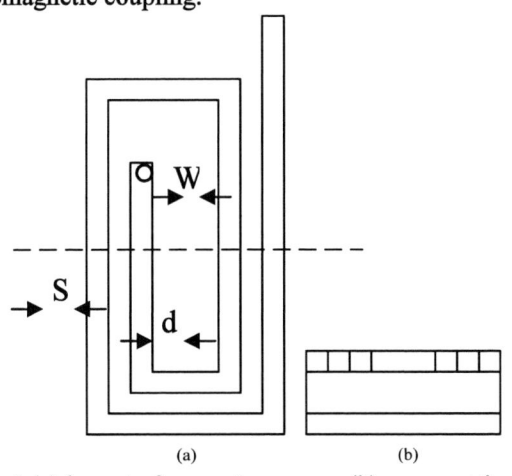

Fig.1 (a) layout of resonator   (b) cross section

b) Multi-order resonator

As shown in fig.3, the bandpass performance of one resonator is not good enough to meet the requirements of bandpass filter, therefore two or more resonators should be coupled by the embedded capacitors to obtain the filter property. The layout and structure of a 2-order resonator filter is shown in fig.2. The resonators are coupled by embedded capacitor. As shown in fig.2, the top electrode of coupled capacitor is connected to the first resonator, and the bottom one is connected to the second resonator through micro-via. Fig.4 shows the insertion loss and return loss of a 2-order resonator filter. The bandpass performance has been improved greatly.

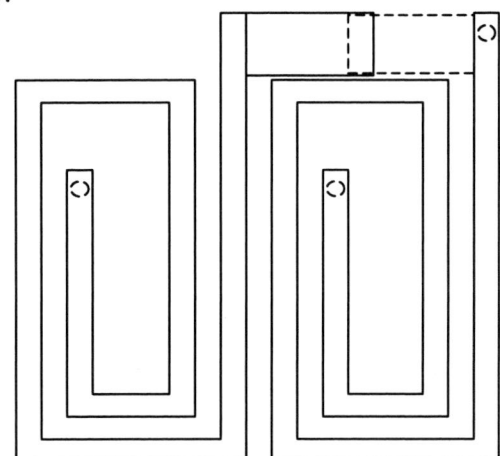

Fig.2 layout of 2-order resonator filter

The factor impacting on bandpass performance of the filter most significant is the value of the embedded coupling capacitor. Whether the value is higher or lower, the performance of filter will go to the bad owing to over coupling or under coupling.

c) NBPF and WBPF

NBPF and WBPF have been designed using 2-order resonators coupled by embedded capacitors. The dimension of each resonator is 2 by 0.9mm², and the one of filter is 2 by 1.9 mm². The electrical characteristic of resonator and filter are simulated by Ansoft 3-D full wave electromagnetic software HFSS. The insertion loss and the return loss of these filters are specified in fig 8 and fig 9. The embedded coupling capacitances are 1.41pF and 1.5pF, respectively.

Width, thickness, and spacing of the conductor lines are designed with 100um, 18um, and 100um to meet the requirements of conventional PCB standard process.

Results and discussion

The electrical characteristic of resonator and filter are simulated by Ansoft 3-D full wave electromagnetic software HFSS.

Fig. 3 shows the insertion loss and return loss of single resonator. The resonator exhibits bandpass feature, but the curve out-of-band drops too slowly to be used for BPF, so the BPF must be composed by more than one resonator.

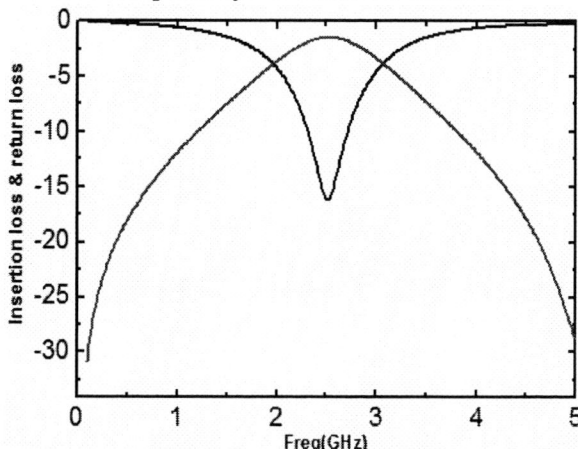

Fig .3 Insertion loss and return loss of single resonator

Fig.4 shows the performance of 2-order resonator. The insertion loss gets higher, and the return loss lower.

Fig. 5 shows a comparison of insertion loss and return loss of one resonator with 1.5 turn and 1.3 by 0.3mm² core area spiral inductor. These inductors have different width varying from 60 to 120 micrometers. As shown in fig. 6, the width is larger, the insertion loss is lower and the return loss is higher.

Fig. 6 shows a comparison of insertion loss and return loss of one resonator with 1.5 turn and 0.1 mm width spiral inductor. These inductors have different core areas varying from 1.3 by 0.2 mm² to 1.3 by 0.5 mm². As shown in fig. 6, these resonators with different core areas, nearly have the same maximal insertion loss and minimum return loss. Nevertheless, resonance frequency decreases along with the increase of the core areas. That is because the larger core areas enhance the capacitance of the resonator, and the resonance frequency descend.

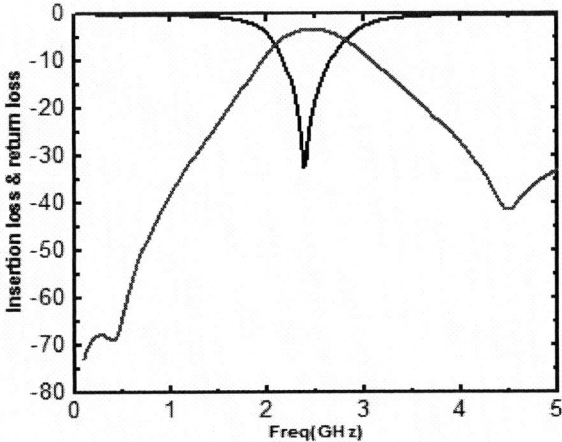

Fig.4 Insertion loss and return loss of 2-order resonator

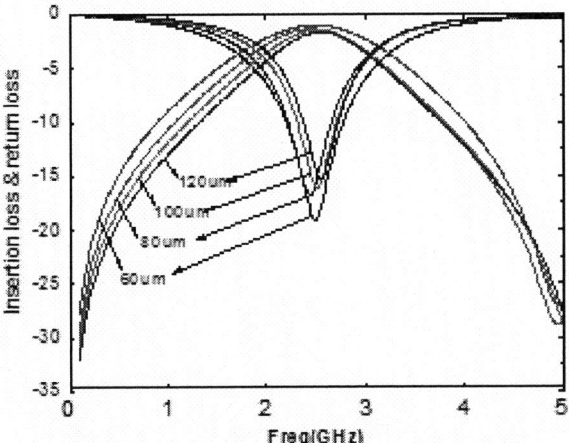

Fig.5 comparison of insertion loss and return loss of a single resonator with different widths

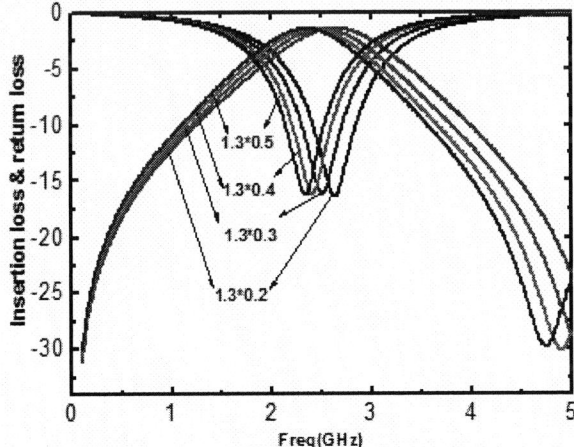

Fig.6 comparison of insertion loss and return loss of a single resonator with different core area

Fig. 7 shows a comparison of insertion loss and return loss of 2-order resonator filter with different coupling capacitor values. The different capacitor values are obtained by changing the areas of the top and bottom electrodes. As shown in fig. 7, if the capacitance is larger, there will be over coupled phenomenon, just like 1.5 pF, and if the capacitance is smaller, there will be under coupled phenomenon, just like 0.3 pF. So the appropriate capacitance is essential for good bandpass filter performance.

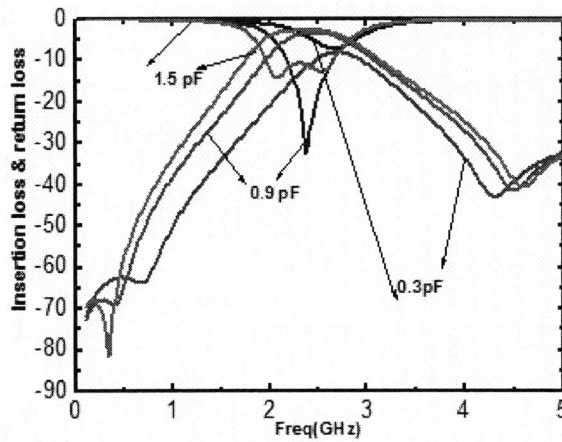

Fig.7 comparison of insertion loss and return loss of 2-order resonator filter with different coupling capacitance

Fig. 8 shows simulated performance characteristics of NBPF. The BPF has insertion loss of 2.68dB, and return loss of 31.8dB at the frequencies ranged from 2.14 GHz to 2.64 GHz. In the low frequency band of 0.1GHz~0.82GHz, the insertion loss is below 40dB.

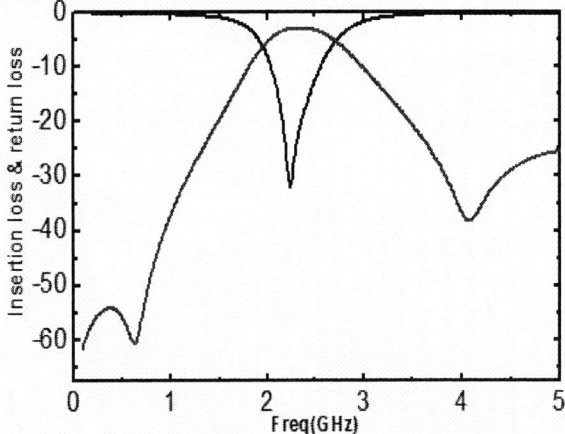

Fig.8 insertion loss and return loss of NBPF

Fig.9 shows simulated performance characteristics of WBPF. It exhibits a center frequency of 2.4GHz, a bandwidth of 500MHz, and an insertion loss of 15.1 dB.

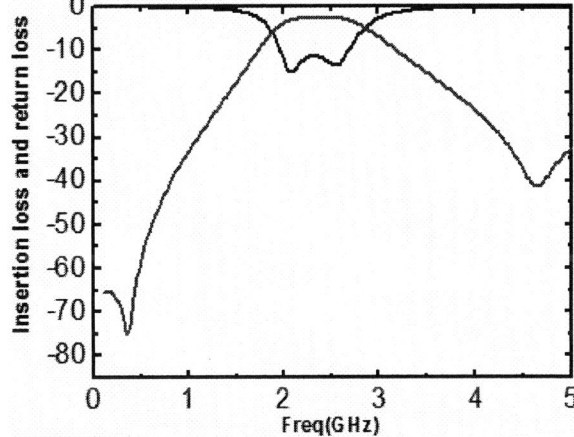

Fig.9 insertion loss and return loss of WBPF

Conclusions

The bandpass filter using only embedded capacitor has proposed, designed and simulated by HFSS. Through comparison and analysis of resonators, it has been found that width and core area of resonator are important design parameters to affect performance and characteristics of the resonator, and the value of coupled embedded capacitor is a significant parameter for the multi-order resonator filter. WLAN and WBPF are designed using 2-order resonators, and these filters are useful for future RF front-end modules with multi-functionalities, low profile and smaller size.

Acknowledgments

The author would like to thank Prof. Lixi Wan and Dr. Lei Li for their technical supports and discussions, and also thank Wei Gao and Jun Li for their comments and discussions.

References

1. Chang-Sheng Chen., "Embedded Capacitors Technology in 2.4 GHz Power Amplifier with Multi-layer Printed Wiring Board (PWB) Process", int'l symposium on Electronic Material and Packaging, 2002.

2. Ching-Liang Weng., "Embedded Passive Technology for Bluetooth Application in Multi-layer Printed Wiring Board (PWB)", Electronic Components and Technology Conference, 2004.

3. Mekita F. Davis., "Integrated RF Architectures in Fully-Organic SOP Technology", IEEE Transacton on Advanced Packaging.

4. Dalmia,S., "Design of inductors in organic substrate for 1-3GHz wireless applications" IEEE MTT-S International,2002

5. Seung J.Lee., " Fully embedded High Q Passives and Band Pass Filters for Low Cost Organic RF SOP Applications". Electronic Components and Technology Conference. 2007.

6. Greg Brzezina., "A Miniature LTCC Bandpass Filter Using Novel Resonators for GPS Applications", Proceedings of the 37th European Microwave Conference, 2007.

7. Mohamadou Baba., "An Efficient Methodology for Design and Implementation of Embedded Bandpass filters for RF/Wireless Applications", EPTC , 2007.

8. Joong Keun Lee., "Design of Bandpass Filter for 900MHz ZigBee Application Using LTCC High Q Inductor ", APMC, 2005.

The Design of the Cache Crossbar based on OpenSPRAC Architecture

Xi-chuan WANG, Bin-feng QIAN

Microelectronic R&D Center, Shanghai University

Key Laboratory of Advanced Display & System Applications, Ministry of Education, Shanghai University

No.149 Yanchang Rd, Shanghai, 200072, PRC

TEL: 021-56331206-114   FAX: 021-56331272

E-mail: wxchuan.keytech@shu.edu.cn

Abstract

Multi-core processor is widely used on the server and desktop computer nowadays. This paper describes the structure of a cache crossbar which used in the multi-core processor SPARC T2. The cores can use the cache crossbar to exchange the data in the L2 cache banks. The multi cores can communicate among each other core by sharing the data in the L2 cache banks. And with the analysis of the CCX, this paper provides a protocol for conneting multi cores and cache banks. The cache crossbar is implemented in SMIC 0.13μm with Design Compiler and can run at 200 MHz.

Keywords: Cache crossbar, OpenSPARC, Multi-core processor

1 Introduction

Multi-core processor is a trend of the development of CPU. AMD and Intel have supplied their own dual-core desktop processor since 2005. Nowadays, AMD and Intel's main desktop processors have integrated with four cores and have the trend to integrate more cores according to the roadmap of these two companies.

UltraSPARC T1 is a multi-core processor introduced by Sun as a server processor in 2005 [1]. In 2006, Sun announced the RTL code and the architecture of the UltraSPARC T1 put it as an open source processor which named as OpenSPARC T1. In the end year of 2007, Sun announced the open source processor OpenSPARC T2. Both of these two multi-core processors are integrated with eight processor cores which exchange the data with L2 cache by cache crossbar. Using cache crossbar to connect multi cores is a popular way.

This paper introduces a cache crossbar that can be used in the multi-core processor which connects the cores and the L2 cache banks. The design is based on the OpenSPARC architecture. The function of the CCX is similar with the CCX in SPARC T2.

2 The CCX of the SPARC T2

The SPARC T2 processor is widely used on the servers. It integrated eight SPARC physical cores on a single chip. Each SPARC physical core has a 16 KB, 8-way associative instruction cache (32-byte lines), 8 Kbytes, 4-way associative data cache (16-byte lines). The eight SPARC physical cores are connected through a crossbar to an on-chip unified 4 Mbyte, 16-way associative L2 cache (64-byte lines). The L2 cache is banked eight ways to provide sufficient bandwidth for the eight SPARC physical cores [2].

The cache crossbar (CCX) connects the 8 SPARC cores to the 8 banks of the L2 cache. An additional port connects the SPARC cores to the IO bridge. A maximum of 8 load/store requests from the cores and 8 data returns/acks/invalidations from the L2 cache can be processed simultaneously. Fig. 1 shows the CCX of the SPARC T2 which connects the cores and the caches.

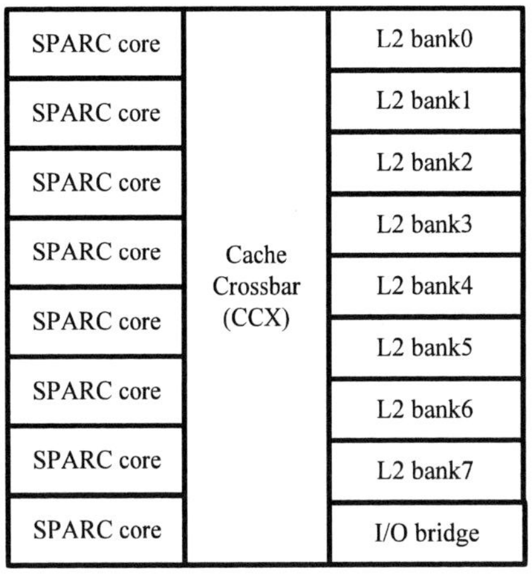

Fig. 1   The CCX of the SPARC T2

3 Structure of the CCX

As the data exchange in the CCX has two directions, the CCX is divided into two separate pieces, the processor to cache crossbar (PCX) and the cache to processor crossbar (CPX). The processor sends the request to the cache through the PCX while the cache gives the data to the processor through the CPX. PCX has eight sources (SPARC cores) and nine targets (cache banks + I/O). The following paragraph describes the detail structure of the PCX and the structure of CPX is similar except the number of the sources and targets.

The PCX is also divided into two parts, the arbiter and the data path slice. Since multiple sources can request access to the same target, arbitration within the crossbar is required. The arbiter issues grants in age order, with oldest having highest priority. There are nine arbiters in the PCX and each arbiter connects to the eight sources. With the arbitration of the arbiters, all the targets can arbitrate independently and simultaneously. The data path slice also has nine slices. Each slice connects to one cache bank and has eight data packet inputs from the processor. The data packet come through the PCX includes the address, the data, the control signal and the destination ID etc. and is used by the L2 cache bank controller. Fig. 2 shows the structure of the CCX.

978-1-4244-2739-0/08/$25.00 ©2008 IEEE   231

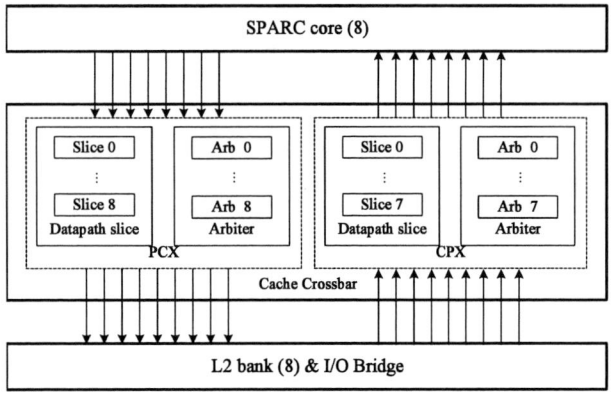

Fig. 2   The Structure of CCX

3.1 Arbiter

The arbitration requirements of the PCX and CPX are identical. The arbiter decides which source can get data in the target. And the arbiter is designed for reuse by CPX and PCX.

The arbiter performs the following functions:

Queue transactions from each source to a depth of two.

Issue grants in age order, with oldest having highest priority.

Resolve requests of the same age without persistent bias to any one source.

Have the ability to stall grants based on input from the target.

Stall the source if the queue is full.

Handle two packet transactions atomically.

Each target has its own arbiter. All targets can arbitrate independently and simultaneously.

3.1.1 Timing

The Arbiter uses a pipeline to perform the arbitration. The basic cycles of the pipeline are PQ, PA and PX which are shown as Tab. 1. And there will has some additional cycles to send the data packet depend on the request type.

Tab. 1   Pipeline stages

PQ	PA	PX
SPARC cores issue requests	Arbitration for target Send the grant to the muxes	Transmit grant to SPARC core Perform data muxing

A source requires access to a target and it asserts a request in the PQ cycle. In the cycle immediately following the request, the source sends its data packet to the crossbar. The arbiters arbitrate among the sources when there have requests in the PA cycle and send a grant back to the source that wins arbitration in the PX cycle. The crossbar sends the packet to the target after the data ready signal which is given in the PX cycle.

The Timing diagram is shown as Fig. 3. And Fig. 3(a) shows an atom transaction with an atom request while Fig. 3(b) shows a normal transaction. As there is an atom request in the Fig. 3(a), the high level of the grant will stay for two cycles. An atom request means there are two requests, and these two requests should be sent continuously [3].

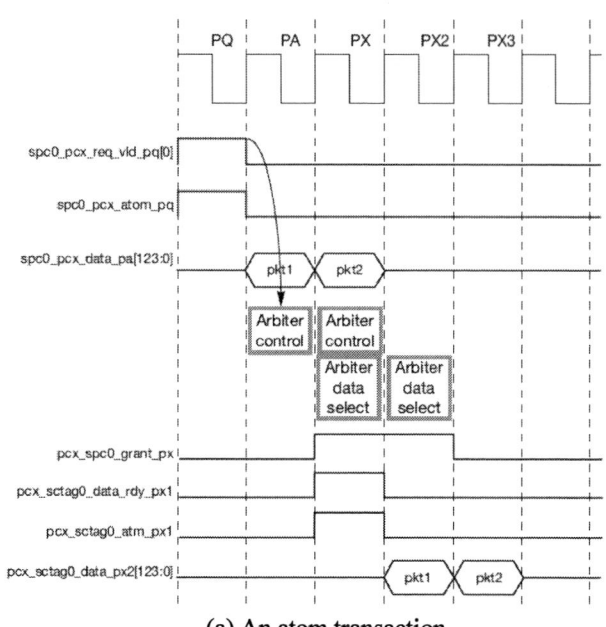

(a) An atom transaction

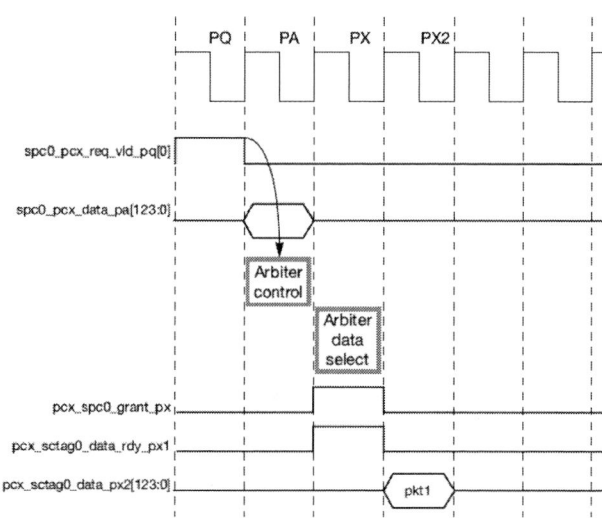

(b) A normal transaction

Fig. 3   Arbiter Timing Diagram

3.1.2 Structure

The structure of the arbiter is shown as Fig. 4. All requests will be sent into the input register in the arbiter first. The input register is used as the PA cycle of the pipeline while the request register is used as the PX cycle of the pipeline. The request will send to the request register next cycle if there is no previous request under processed, or the request will send into the FIFO. The FIFO is used to store the request when previous request is under processed. The logic on the left side of Fig. 4 performs arbitration alternating between the ascending and descending priority encoders. An 8-bit grant signal will send to the eight sources after the arbitration with one source corresponds to one bit of the grant signal.

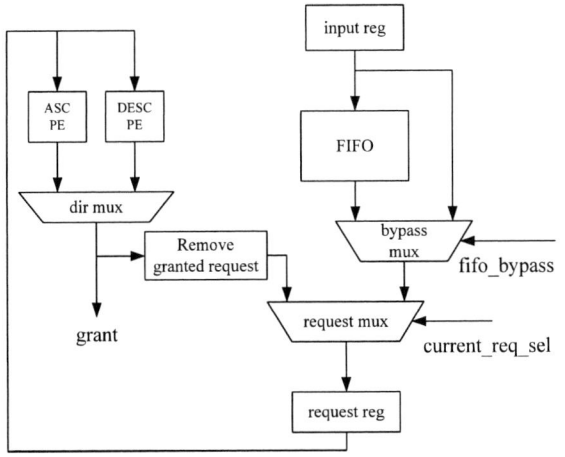

Fig. 4   Arbiter Structure

3.2 Data path slice

Each target is associated with a data path slice. A data path slice steers data from one of M (M=8) sources to one target. It does so based on the grant signals from the arbiters. While the function of a data path slice is essentially a M:1 mux, bringing together M busses, each of over 100 bits, to a common point for muxing is not physically practical. Therefore, the slices are composed of 2:1 and 3:1 muxes that pick data from a "current" source or a neighboring source. The Fig. 5 shows the structure of the PCX data path slice.

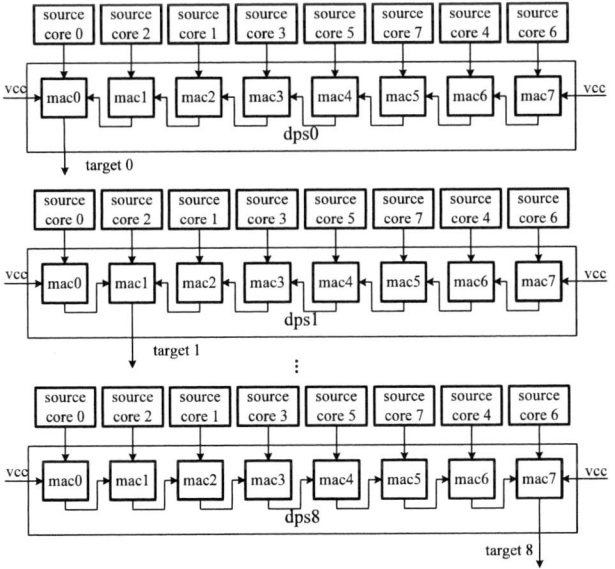

Fig. 5   Data Path Slice Structure

Every data path slice has M (M=8) macs. The mac has two parts, data buffer and the mux. The data buffer is a fifo to store the data send by the core. Each core has its own data buffer, in order to store the data simultaneously. The structure of the mac is shown as Fig. 6. Each mac has one input from the core and one or two inputs from his neighboring macs. In the Fig. 5, the mac1 in the dps1 has two neighbors, so it has two neighbor's inputs. And other macs in the dps1 only have one neighbor's input.

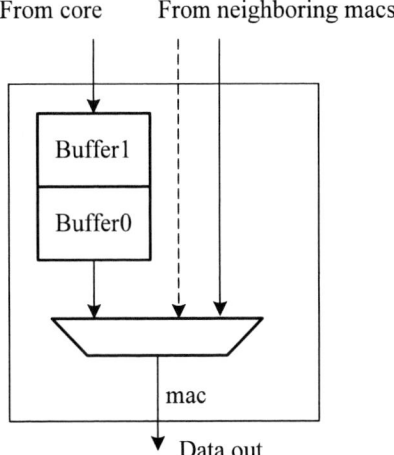

Fig. 6   Mac Structure

4 The Protocol

The protocol defines an input/output ports interface. Using the protocol, other cores, not only the SPARC cores can use the crossbar to share the data in the L2 cache banks and exchange data with each other cores through the L2 caches. The protocol is described as follow, and is divided into two parts.

(1) Ports from core to CCX

Data port1: It is used to send the data packet. And the width of this port depends on the complexity of the design.

Core request port: when the core wants to send a data packet, the level of this port will be pull up.

Atom request port: If it is an atom transaction, this port will become high level.

(2) Ports from CCX to core

Data port2: This port returns the data from the cache.

Data ready port: The level of this port will become high when the data on the data port2 is ready.

Atom data port: It indicates that it's an atom transaction, and the core should get the data twice.

Full queue port: As there have FIFOs in the crossbar, this port indicates that the FIFO is full, and can't receive any data packets.

Empty queue port: When its level is high, it means the FIFO is not empty, or it means the FIFO is empty. It's important that an atom transaction would happen only when the FIFO is empty.

The ports of the protocol are simple, and the cores don't need to know how L2 cache banks are. It just sends the signal to the port, and gets the data from the cache.

5 Simulations

The CCX is described in Verilog HDL and synthesized by the Synopsys Design Compiler with the stand cell library

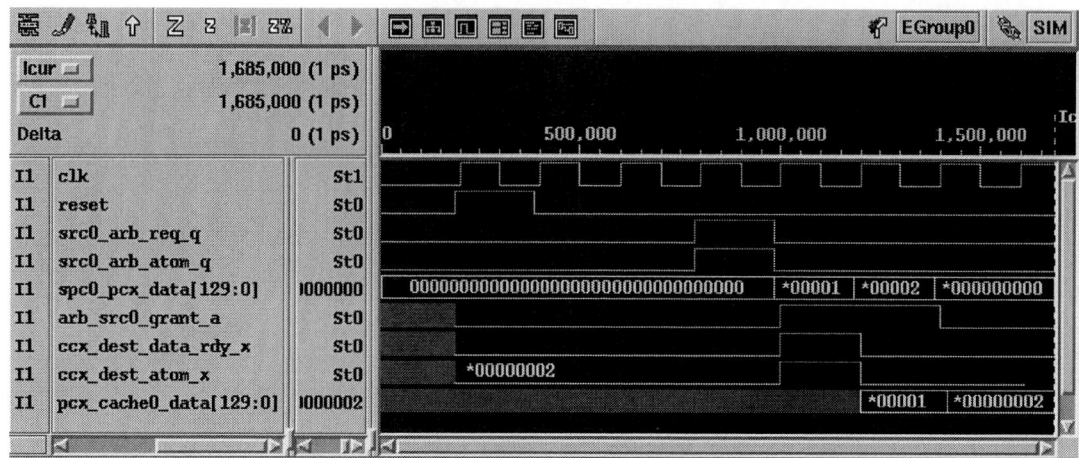

Fig. 7   The Simulation of the netlist

SMIC13. And the result of the synthesis shows the CCX can run at 200 MHz.

The netlist of the CCX is used to run the simulation in the VCS-MX. Fig. 7 shows the result of the simulation. In the Fig. 7 at time 800ns there is an atom request sent by the core. And in the next two cycles, the core sends two data packets to the CCX. And two cycles after the request, the data packet is sent to the cache banks. And the figure is similar with the Fig. 3(a).

6 Conclusions

This paper analyzes the architecture of the crossbar in SPARC T2 and the communication that the eight cores used through the crossbar. Implement the crossbar using the Verilog code and synthesis the design with the SMIC technology library by synopsys Design Compiler. The design simulates with the synopsys VCS tool. With the crossbar the multi-core processor can load/store the data from the L2 cache and the data in the L2 cache can use by all the cores. That means the multi cores can communicate with the other cores by using the CCX. A protocol is given for other type cores to connect with each other with the CCX.

References

1. J.L. Hennessy, D.A. Patterson. Computer Architecture: A Quantitative Approach, Fourth Edition [M]. Morgan Kaufmann Press, 2006.
2. Sun Microsystems, Inc. OpenSPARC™ T2 Core Microarchitecture Specification [s]. 2007.
3. Sun Microsystems, Inc. OpenSPARC™ T1 Core Microarchitecture Specification [s]. 2008.

Interpolating Algorithm Optimization and FPGA Implementation in Image Scaling Engine

RAN Feng[1], LIU Jing[2], XU Meihua[2]

1. Microelectronic Research and Development Center, Shanghai University

2. School of Mechatronical Engineering and Automation, Shanghai University

Campus P.O.B. 110, 149 Yanchang Rd., Shanghai 200072, P.R.China

ranfeng@mail.shu.edu.cn

Abstract

Bi-cubic interpolation algorithm is commonly used in image scaling, but traditional cubic interpolation has its own shortcomings such as complicated computation, long computational time and so on. For these problems, the paper studies traditional cubic kernel function and proposes an optimized algorithm with adjustable coefficients. This algorithm utilizes an modifying coefficient λ to amend the coefficients in the kernel function, which helps the scaling system choose best algorithm with different images. Then, the superiority is verified by MATLAB simulation and the optimized algorithm is applied to the image scaling engine called scaler through the verification in FPGA.

1 Introduction

Image scaling refers to the transformation of image resolution, which is commonly used in the field of image processing[1]. It is the key technology of video image processing in FPD (flat panel display), mainly used in image scaling engine called scaler, which transforms images in different modes into those with fixed resolution by interpolation and makes them displayed on screen. In the processing of scaling, especially in the enlarging operation, there will be the phenomenon of image distortion, which is caused by the pixel positions not exist in the image before transformation. Therefore, it is necessary to adopt corresponding interpolation operation to identify additional pixel values in output images. The nearest neighborhood, bilinear interpolation method and bi-cubic interpolation method are the most common interpolation algorithm. Among them, the nearest neighborhood algorithm is the simplest one with minimum amount of computation, but it has a significant shortcoming which will cause mosaic phenomenon. The pixel values will be continuous in the image scaled by bilinear interpolation, but due to its nature of low-pass filter, the high frequency component of the image will be lost, making the image contour blurred to a certain extent. Bi-cubic interpolation has strong ability in the image detail performance as well as generating smooth edge and the contrast, brightness and color can maintain the original image information well. But as its high computational complexity and long computational time[4][6], traditional bi-cubic interpolation is not able to achieve the best scaling result on every image in application. Dealing with the problems above, how to obtain the best results with bi-cubic interpolation algorithm is studied and the interpolation kernel function is amended in this paper. The optimized algorithm is verified in MATLAB simulation and applied in scaler through the verification in FPGA.

2 Optimization of Interpolation Algorithm

2.1 Bi-cubic Interpolation Algorithm

In numerical analysis, interpolation algorithm can be generally expressed as

$$g(x) = \sum_{k=0}^{n-1} C_k \times s(x - x_k) \qquad (1)$$

Where, $s(x-x_k)$ is the kernel of the interpolation function, C_k is the kth value of original function, and the discrete function is converted into continuous one by the operation of convolution. The distinction among different interpolation algorithms mainly lies in the difference of kernel function and n, the number of interpolation points. If the kernel function has third power as its maximum power and both the first and second derivative are continuous, the algorithm can be called to cubic interpolation algorithm. As image is two-dimension signal and the use of algorithm in image interpolation is its application in two-dimension, it is called bi-cubic interpolation[5].

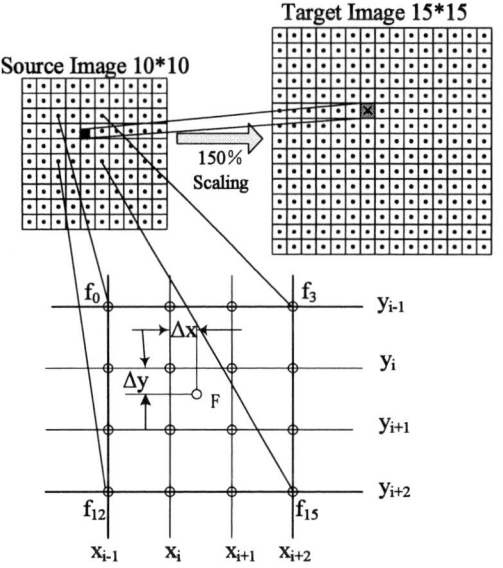

Figure.1 Process of image scaling using bi-cubic interpolation

Figure.1 shows the processing of enlarging the source image with resolution of 10×10 to 1.5 times by bi-cubic interpolation algorithm. The algorithm gets a 4×4 field to form a square surface (the length is 3 units) from the original image. Assuming that F is the interpolation point in calculation and f0 to f15 are the nearest 16 points in the original image, temporary interpolations F1 to F4 are calculated first by four cubic interpolation in the horizontal direction. Then with these four temporary points as objects, one interpolation in vertical direction with the same principle is operated and the value got from this is the pixel value of the new point. The pixel value is calculated by

$$f(x,y) = A g B g C \qquad (2)$$

Where,

$$A = [S(1+Vx) \quad S(Vx) \quad S(1-Vx) \quad S(2-Vx)]$$

$$B = \begin{bmatrix} f(i-1,j-1) & f(i-1,j+0) & f(i-1,j+1) & f(i-1,j+2) \\ f(i+0,j-1) & f(i+0,j+0) & f(i+0,j+1) & f(i+0,j+2) \\ f(i+1,j-1) & f(i+1,j+0) & f(i+1,j+1) & f(i+1,j+2) \\ f(i+2,j-1) & f(i+2,j+0) & f(i+2,j+1) & f(i+2,j+2) \end{bmatrix}$$

$$C = [S(1+Vy) \quad S(Vy) \quad S(1-Vy) \quad S(2-Vy)]^T$$

(3)

Compared with bilinear interpolation algorithm, bi-cubic interpolation method has its shortcomings such as large amount of computation, occupying more computing time. But its scaling result is better than bilinear interpolation as each new pixel is derived from associated 16 points in source image.

2.2 Interpolation Kernel Function

In bi-cubic interpolation algorithm, it differs in calculating amounts, scaling effects and implementation when interpolation kernel function varies. Cubic interpolation kernel function from Keys[2] is commonly used. Reference [2] shows the derivation of it, which is defined on the subinterval (-2,-1), (-1,0), (0,1) and (1,2). Outside the interval (-2,2), the interpolation kernel is zero. Besides, the function should be symmetric. Therefore, the general form can be written as follows,

$$s(\omega) = \begin{cases} A_1|\omega|^3 + B_1|\omega|^2 + C_1|\omega| + D_1, 0 < |\omega| < 1 \\ A_2|\omega|^3 + B_2|\omega|^2 + C_2|\omega| + D_2, 1 < |\omega| < 2 \\ 0, |\omega| > 2 \end{cases} \qquad (4)$$

According to the above constraints of the function, a group of functions related to the parameters is obtained, which is showed in formulas (5).

$$\begin{cases} s(0) = D_1 = 1 \\ s(1^-) = A_1 + B_1 + C_1 + D_1 = 0 \\ s(1^+) = A_2 + B_2 + C_2 + D_2 = 0 \\ s(2^-) = 8A_2 + 4B_2 + 2C_2 + D_2 = 0 \\ -C_1 = s'(0^-) = s'(0^+) = C_1 \\ 3A_1 + 2B_1 + C_1 = s'(1^-) = s'(1^+) = 3A_2 + 2B_2 + C_2 \\ 12A_2 + 4B_2 + C_2 = s'(2^-) = s'(2^+) = 0 \end{cases} \qquad (5)$$

The paper makes A=1[2][3], and obtains the cubic interpolation kernel function from the equations above. The function is written as formula (6) and its function curve is shown in figure.2.

$$s(\omega) = \begin{cases} 1 - 2|\omega|^2 + |\omega|^3, |w| < 1 \\ 4 - 8|\omega| + 5|\omega|^2 - |\omega|^3, 1 < |w| < 2 \\ 0, |\omega| > 2 \end{cases} \qquad (6)$$

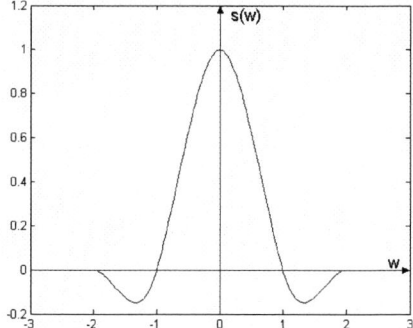

Figure.2 Function curve of s(w)

Scaling the same image for certain times by these two kernel functions under MATLAB environment, and calculating computational time in the whole operation, it is found that it takes more time by Keys's kernel function, which shows that it has more computational complexity and is not able to enhancing the scaling result notably by Keys's kernel function.

2.3 Optimization of Interpolation Kernel Function

Figure.1 shows that, pixel value of the new point F to be interpolated is related with the nearest 16 points in original image, including 4 direct neighbor-points (f5, f6, f9, f10) and 12 indirect neighbor-points. Among them, the influence of direct neighbor-points is the most important, while the indirect ones influence points to be interpolated by influencing direct neighbor-points. For different image sources, how to balance the influence of direct and indirect neighbor-points is one of the important factors which are related to the image scaling result.

When (3) is substituted into (6), the following equations are obtained.

$$\begin{cases} s(1+Vx) = -|Vx|^3 + 2|Vx|^2 - |Vx| \\ s(Vx) = |Vx|^3 - 2|Vx|^2 + 1 \end{cases} \qquad (7)$$

Let , it is obtained that,

$$s(1+Vx) = -m, s(Vx) = m + (1 - |Vx|) \qquad (8)$$

If m=0, cubic interpolation degrades to linear interpolation, and the point to be interpolated is influenced only by the 4 direct neighbor-points. Therefore, bi-cubic interpolation is divided into 2 parts.

(1) Linear component $(1-|\triangle x|)$ shows the impact of direct neighbor-points.

(2) m shows the impact from indirect neighbor-points.

The influence from the former is the most important, while the latter impacts on the pixel value of interpolated points through the influence to direct neighbor-points, which is mainly manifested in the change rate of pixel value between neighbor-points. For the images with flat picture, it is able to recover the source information making full consideration to direct neighbor-points and changing rate of pixel values between neighbor-points. But for complex images, it is necessary to reduce the affect of indirect neighbor-points as the pixel value of neighbor-points changes rapidly. In order to achieve the best result, an amendment to interpolation kernel function is needed, and in this paper, an amendment coefficient λ ($\lambda \in [0,1]$) is added to formula (8), which revises formula (8) in the form of

$$s(1+\nabla x) = -\lambda m, s(\nabla x) = \lambda m + (1 - |\nabla x|) \qquad (9)$$

For λ=0, formulas (9) expresses bilinear interpolation. For λ=1, it expresses bi-cubic interpolation without optimization. The larger λ is, the more influence changing rate between neighbor-points takes to new points, while the smaller λ is, the more influence direct neighbor-points takes to points to be interpolated. For one image with rich detail and high frequency component, λ should be smaller to reduce the impact of indirect neighbor-points and increase that of direct ones due to the acute change between neighborhoods.

3 Experiment Results

3.1 MATLAB Simulations and Results Analysis

The paper utilizes the texture image shown in figure.3 (a) as source image. Enlarging it into 3 times with bilinear and bi-cubic interpolation method with different values of λ, the paper makes comparison among different methods through the upper left part of images after scaling. Figure.3 shows that dealing with images with rich details, the edge is still relatively vague and the source information is not able to be well reverted by bilinear interpolation method, while bi-cubic interpolation is able to resolve those problems. But when λ differs, the difference between images after scaling is not so clear by human eyes. Therefore, PSNR (peak signal to noise ratio) is also calculated to measure the quality of image scaling with different λ. The larger λ is, the better the scaling result is. For the texture image in figure.3 and the Lena image which is commonly used in digital image processing, the paper calculates the PSNR values with different method, showed in table.1. Besides, the processing time with different λ is also calculated. From table.1, it is shown that, for the texture image with rich high frequency component, scaling result achieves the best when λ is 0.3, while for Lena picture, scaling result achieves the best when λ is 0.5.

3.2 FPGA Implementation in Scaler

In hardware implement, λ can be fixed according to occasions and consumer's needs to get better scaling effect with calculation and complexity as little as possible. In computer and consumption display, it is not so strict in image quality as in speed and cost, the paper uses λ =0.5 in the design of scaler. Figure.4 shows main blocks of hard architecture.

Table.1 PSNR and simulation time in MATLAB simulation

	Bilinear	Bi-cubic	Bi-cubic	Bi-cubic
λ	0	1	0.5	0.3
PSNR(Fig.3)	23.8816	25.2308	25.7419	25.8704
Time/s	11.9840	32.7660	31.3280	32.8910
PSNR Lena)	27.5226	28.1363	28.3594	28.2229

(a)Source image

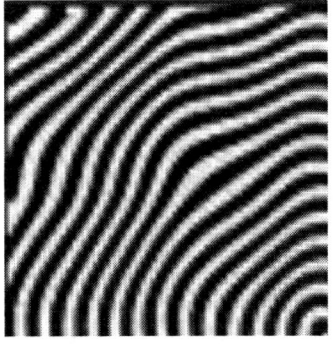

(b)Bilinear interpolation

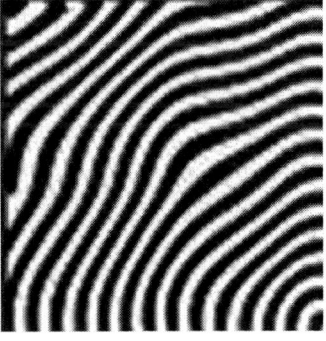

(c)Bi-cubic interpolation $\lambda = 1$

(d) Bi-cubic interpolation $\lambda = 0.3$

Figure.3 Results of enlarging to 3 times

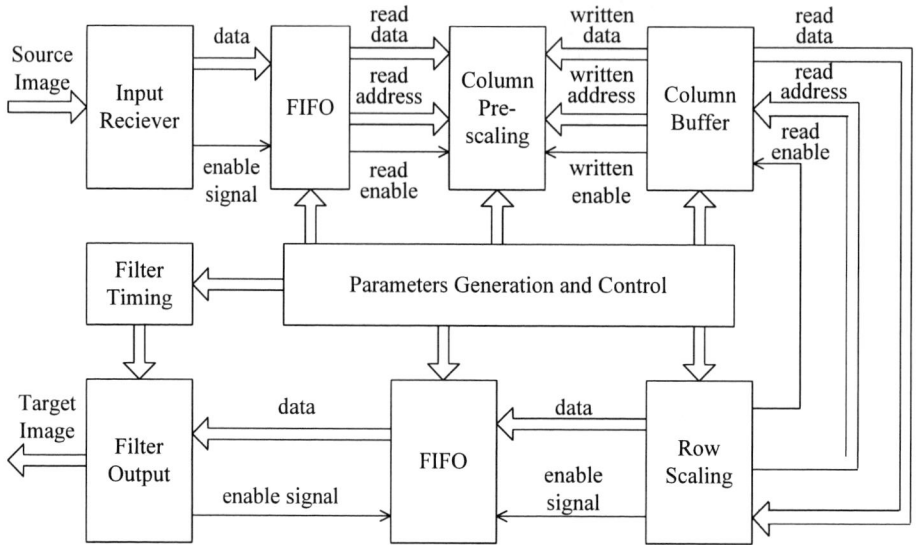

Figure 4 Main blocks of hard architecture by bi-cubic interpolation

The design divides the scaling into 2 units, column and row scaling, and each unit only needs 4 associated points in the interpolation operation, which saves hardware resources and makes FPGA implementation much easier. In the scaling processing, image dataflow is scaled in column pre-scaling and then in row scaling. Finally, with the appropriate timing control unit, the output image data is displayed in FPD. FIFO and buffer in figure.4 are used to store image data. Due to the different directions between column and row unit when reading image data before scaling, the system employs different data buffers in order to satisfy value requirement.

4 Conclusions

By MATLAB simulation, the paper verifies that compared with bilinear interpolation, bi-cubic interpolation algorithm has advantages in detail performance and recovery of contrast and brightness. It also amends coefficients of traditional bi-cubic interpolation algorithm, optimized the algorithm and its implementation. FPGA verification results show that it is able to achieve the desired scaling result by optimized bi-cubic interpolation algorithm in scaler.

Acknowledgments

The authors would like to acknowledge the National Natural Science Foundation of China for providing financial support for this work under Grant No. 60773081 and Grant No. 60777018, and also to acknowledge the financial support by the Shanghai Municipal Committee under Grant No.AZ028 and Grant No.06DZ22013

References

1. Yuming Xu, Shuangchun Wen. Analysis and Implementation of Interpolation Algorithm in Digital Image[J]. Fujian Computer, No.1, 2007, pp.91-94
2. Robert G. Keys. Cubic Convolution Interpolation for Digital Image Processing[J]. IEEE Transactions on Acoustics, Speech, and Signal Processing, Vol.Assp-29, No.6, December 1981, pp.1053-1160
3. R. Bernstein. Digital Image Processing of Earth Observation Sensor Data[J]. IBM J. Res. Develop, Vol.20, 1976, pp.40-57
4. Macro Aurelio Nuno-Maganda, Miguel O. Real-Time FPGA-Based Architecture for Bicubic Interpolation: An Application for Digital Image Scaling[C]. Proceeding of the 2005 International Conference on Reconfiguration Computing and FPGAs (ReConFig 2005)
5. Azhen ZHANG, Zhenglin Zhang, Xuecheng Zou, Zuquan Xiang. Design of Image Scaling Engine Based on Bicubic Interpolation Algorithm[J]. Microelectronic and Computer, Vol.24, No.1, 2007, pp.49-51
6. Hu Yi, Jun Yu. Design of Image Scaling Block in Digital Video Based on FPGA[J]. Digital Television, No.5, 2005, pp.41-43

The Research of the Inclusive Cache used in Multi-Core Processor

Bin-feng QIAN, Li-min YAN

Key Laboratory of Advanced Display & System Applications, Ministry of Education, Shanghai University
Microelectronic R&D Center, Shanghai University
Campus P.O.B.211, 149 Yanchang Road, Shanghai 200072, PRC
TEL: 021-56382144   FAX: 021-56331272
E-mail: Kevin.Qian@msn.com

Abstract

Multi-core processor is becoming popular today. As the number of the core increase, the communications among cores also become complex and difficult. Caches are used in multi-core processors for sharing data and increasing performance. It becomes a channel for cores to communicate with each other. Intel's next generation multi-core processor Nehalem which using an inclusive L3 cache to enhances the performances. This paper describes the function of the inclusive cache in the Nehalem and analyzes advantage of the MESIF cache coherence protocol by comparing with the standard MESI protocol. This paper also gives a structure of the cache that can be used to implement. The control flow is analyzed in order to ensure the operation of read/write cache will accord with the MESIF protocol.

Key words: Inclusive cache, Nehalem, Multi-core processor

Introduction

As the frequency of the processor is higher and higher, the power consume of the processor becomes more and more. A very fast single core chip means it needs a very effective cool down system to reduce the temperature in order to let the chip work correctly. Of course the faster core, the better cool system, and the more expansive cost. The performance of the processor that the user need is higher and higher year by year. Some years ago the processor manufacturer let the chip run faster to satisfy the demand of the user. Recent years, the processor manufacturers produce the multi-core processor to deal with the heat and power consume problem. Multi-core processor becomes a necessary choice for the company who want to develop the performance of the processor. It still has problems in a multi-core processor, as there are more than one cores on a single chip, and every cores need to be provided data to process. Sometimes maybe the data core A need is used by the core B, so share the data with each other cores is a key point for multi-core processor to improve the performance.

As the main memory (DRAM) has much more latency of the access time than the cache (SRAM), the speed to access the main memory becomes a performance-limiting bottleneck for the processor [1]. Using cache (SRAM) as a template store is a solution for the latency of the access time for DRAM. In the multi-core processor architecture cache is not only used to reduce the latency but also used to share the data and improve the performance.

Intel's next generation microarchitecture Nehalem is a multi-core processor which enhances the performances by adding an inclusive shared L3 (last-level) cache that can be up to 8MB in size. In addition to this cache being shared across all cores, the inclusive shared L3 cache can increase performance while reducing traffic to the processor cores. Some architectures use exclusive L3 cache, which contains data not stored in other caches. Thus, if a data request misses on the L3 cache, each processor core must still be searched, or snooped, in case their individual caches might contain the requested data. This can increase latency and snoop traffic between cores. With next generation microarchitecture, a miss of its inclusive shared L3 cache guarantees the data is outside the processor and thus is designed to eliminate unnecessary core snoops to reduce latency and improve performance [2].

Inclusive Cache of Nehalem

An inclusive cache means the lower cache has all the date that available in the higher cache. In Nehalem, it appears as the L3 cache has all the data in the L1/L2 cache in the four cores.

The characteristic of the inclusive cache is indicated by the operation [3]. There is a data request send to L3 cache when one of the four cores has a miss，for example if the data is not in the L1 and L2 cache of the core A which will cause a miss and send the request to L3 cache. The L3 cache will reply a hit, when the requested data is in the L3 cache, or it will give a miss. If it's a miss, as there has 3 other cores, the inclusive L3 cache can guarantee the data is not in other cores, as all the data on the chip has a copy in the inclusive L3 cache. It means the data should load from the main memory. If it's a hit, the data also could be in another core except L3 cache. And the inclusive cache uses a set of "core valid" bits to ensure which core has the data. Fig. 1 and Fig. 2 show how the inclusive cache works.

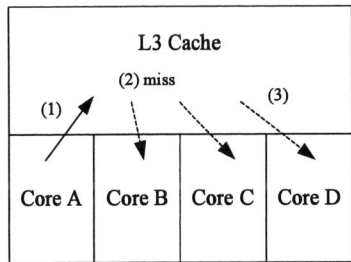

Fig. 1   **A miss in L3 Cache**

(1) A data request sends from the core A to L3 cache
(2) The data is not in L3 cache and cause a miss
(3) Guarantee the data is not in other cores, don't need to send request

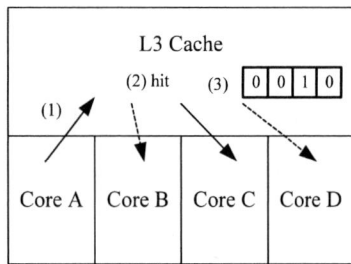

Fig. 2 A hit in L3 Cache

(1) A data request sends from the core A to L3 cache

(2) The data is in L3 cache and there is a hit

(3) The L3 cache query the "core valid" bits and a request sends to core C

Coherence Protocol in Nehalem

As there are four cores on the chip, and different cores could read/write the same address, it will cause the cache coherence problem. The common way to solve the problem is using the MESI (Modified, Exclusive, Shared, Invalid) protocol. The MESI protocol has four states:

Modified: The cache line resides exclusively in this cache only, and the content is modified relative to memory

Exclusive: The cache line resides exclusively in this cache only, and the content is same as memory

Shared: The cache line resides in this cache is shared with other caches. And content is same as memory

Invalid: The cache line contains no valid memory copy.

A modified/exclusive line is owned by the cache. The content in the line can be changed without telling other caches. If a local cache attempts to write a non-owned line (Shared), a message should be broadcasted to other remote caches. If a local cache wants to read the content which owned by another caches, the owner should give up ownership. Table 1 shows the protocol of the MESI.

Table 1   MESI protocol and states

	Clean /Dirty	Unique?	Can Write?	Can Forward?	Comments
Modified	Dirty	Yes	Yes	Yes	Must write back to share or replace
Exclusive	Clean	Yes	Yes	Yes	Transition to M on write
Shared	Clean	No	No	Yes	Implies clean, can forward
Invalid	NA	NA	NA	NA	Can not read

Intel's multi-core processor Nehalem uses an advanced way called MESIF. They adapted the standard MESI protocol to include an additional state, the Forwarding (F) state, and changed the role of the Shared (S) state. In the MESIF protocol, only a single instance of a cache line may be in the F state and that instance is the only one that may be duplicated [4]. Other caches may hold the data, but it will be in the shared state and cannot be copied. In other words, the cache line in the F state is used to respond to any read requests, while the S state cache lines are now silent. This makes the line in the F state a first amongst equals, when responding to snoop requests. By designating a single cache line to respond to

requests, coherency traffic is substantially reduced when multiple copies of the data exist.

When a cache line in the F state is copied, the F state migrates to the newer copy, while the older one drops back to S state [5]. Table 2 shows the MESIF protocol.

Table 2   MESIF protocol and states

	Clean /Dirty	Unique?	Can Write?	Can Forward?	Comments
Modified	Dirty	Yes	Yes	Yes	Must write back to share or replace
Exclusive	Clean	Yes	Yes	Yes	Transition to M on write
Shared	Clean	No	No	No	Does not forward
Invalid	NA	NA	NA	NA	Can not read
Forwarding	Clean	Yes	No	Yes	Must invalid other copies to write

The Fig. 3 demonstrates the advantages of MESIF over the traditional MESI protocol, reducing two responses to a single response.

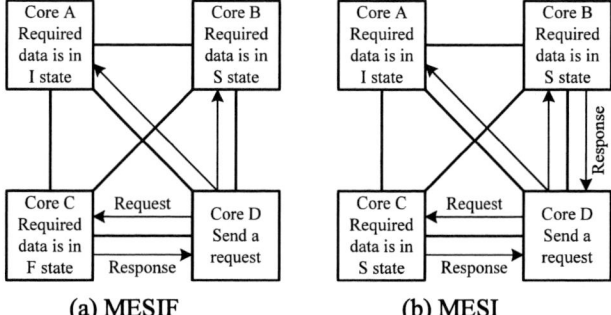

(a) MESIF   (b) MESI

Fig. 3   MESIF versus MESI Protocol

As there is an F state in the MESIF protocol, the response number is reduce to one compared with the MESI protocol. This reduces the traffic.

Structure of the cache

Although Intel has not discussed the detail physical design of Nehalem, the structure of the cache can be concluded form the function that previous paragraph has mentioned. The structure in this paper doesn't include the L1 cache and the main memory, in order to simplify the problem.

Each core of the design has a 256KB L2 cache which is 8 way associative. The tag section of the L2 cache contains the MESIF state bits. As other cores could require the data in the local cache, the snoop unit is used to response the request sent by the remote cache. And there are three L2 cache interfaces in the snoop unit. The snoop unit also has a tag interface which inquires the tag value when the remote cache sends the required the address. And the tag interface can change the tag value in the tag memory as the request from the remote caches may change the MESIF state of the cache line.

Each L2 cache has a cache control unit which controls the read/write of the L2 cache. It contains three parts, the L1 cache interface, the L2 cache interface and the L3 cache interface. The L1 cache send a miss signal to the L2 cache through the L1 cache interface. The control logic is control the access to the local L2 cache. And if there is a miss in the local L2 cache, it will use the L3 cache interface to inquire the L3 cache core valid bits. It will read the data by the L2 cache interface as there has a bit of the core valid is set. The structure of the L2 cache is shown as Fig. 4.

The L3 cache is an inclusive cache which is 8MB 16 way associative. The tag section of the L3 cache contains the core valid bits. The control unit is used to control the read/write of the L3 cache. It contains four L2 cache interface and one arbitration logic. As a miss signal will send to the L3 cache from the L2 cache. There are four interfaces to connect the L2 cache. Because multi source can request the L3 cache, arbitration logic is needed in order to arbitrate the request.

The Fig. 5 shows the structure of the L3 cache.

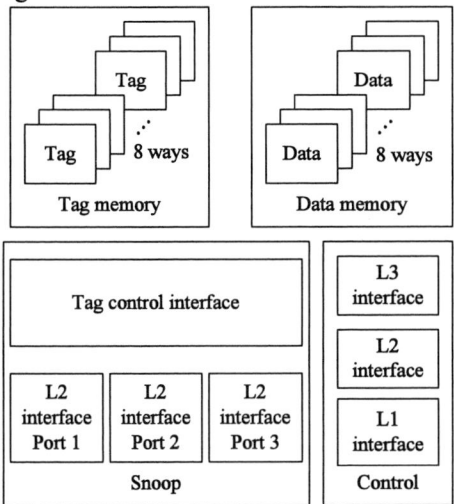

Fig. 4   **Structure of L2 cache**

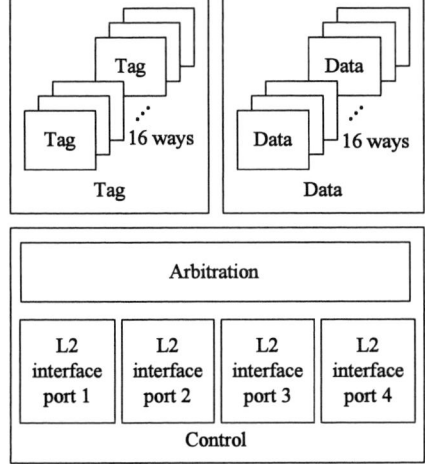

Fig. 5   **Structure of L3 cache**

Arbitration unit in L3 cache

The arbitration unit will scan the four ports one by one. As there is a request on the port, the unit will query the L3 cache first to ensure whether the data is on the chip. If the data is on the chip there is a hit and the core valid bits are all zero, the

L3 cache will send the data to L2. The operation when there is a miss or no all the bits are zero is shown as Fig. 6. The figure only shows the arbitration flow of the port 1.

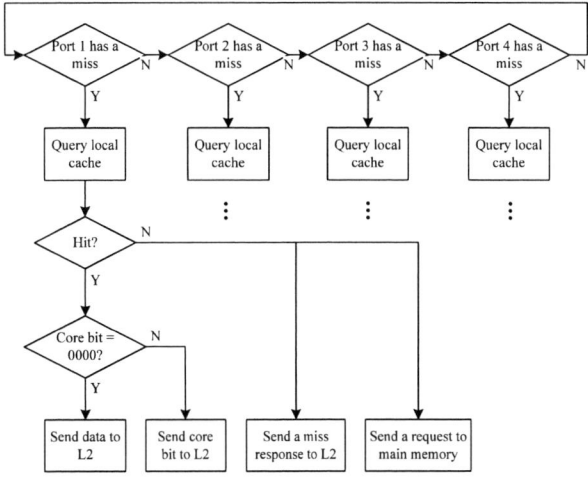

Fig. 6   **Arbitration flow**

Snoop unit in L2 cache

The snoop unit is snooped the request from the remote L2 cache. It snoops the three ports alternately. When there is a request, it will look up the data in the local cache. It will return the data to the remote cache and change the MESIF state in the local cache depend on the current state of the MESIF bits. The Snoop flow is shown as Fig. 7. The operation for the por2 and port3 is not shown as it is similar with the port1.

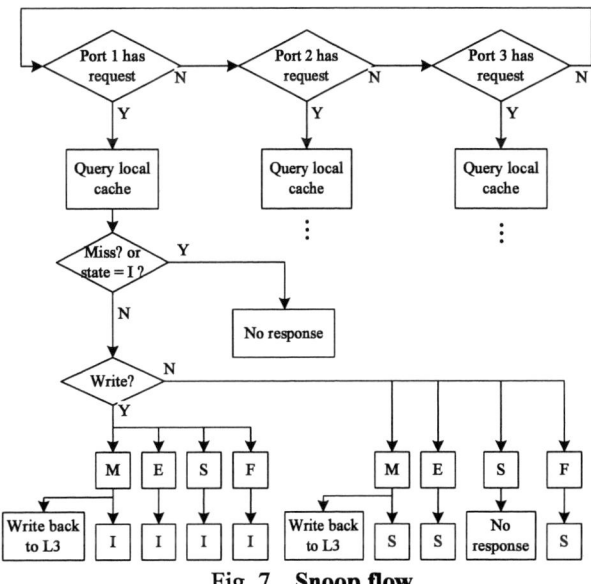

Fig. 7   **Snoop flow**

Control unit in L2 cache

The function of the control unit in the L2 cache is controls read/write the cache. And it is in idle state when there is no miss signal in L1 cache interface. Whenever there is a miss request in L1, it begins the control logic.

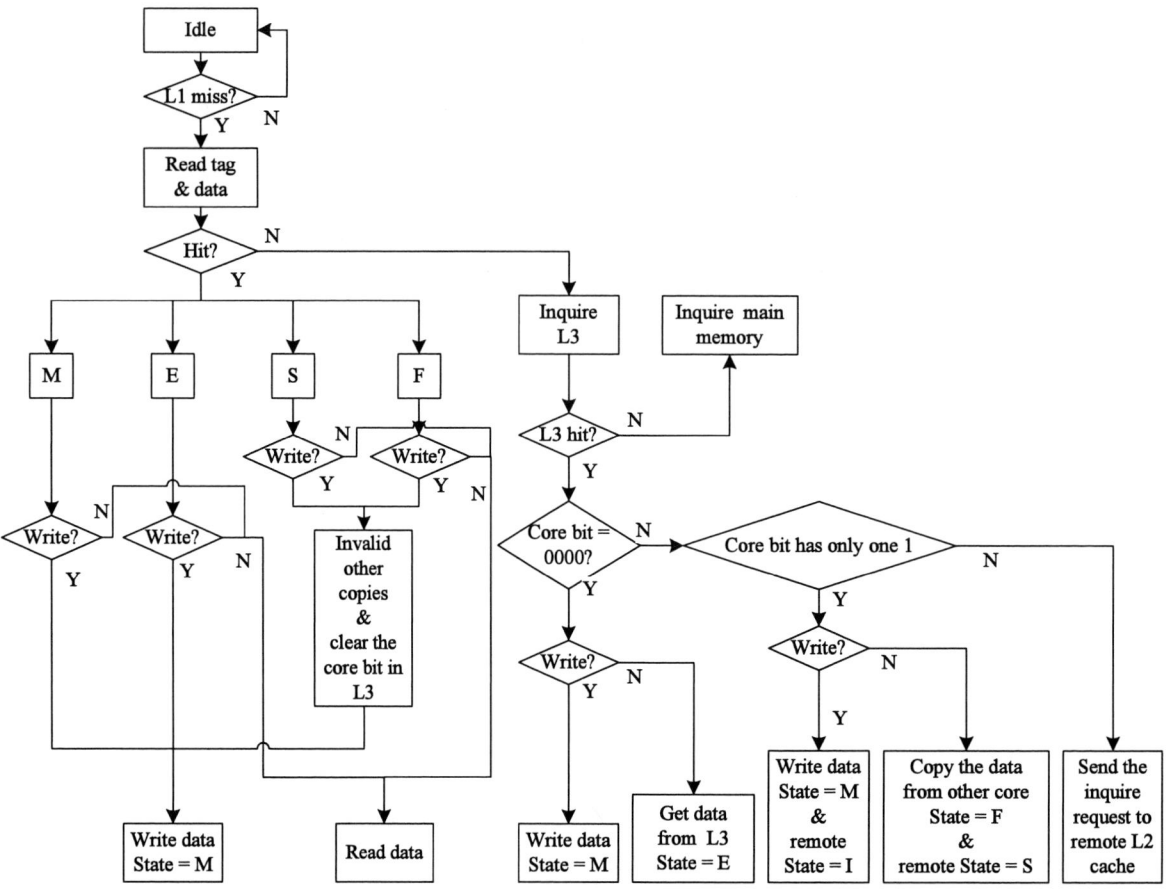

Fig. 8   **Control flow in L2 cache**

It'll ensure whether the request data is in the L2 cache first. If there is a hit, it controls the operation depend on the MESIF state. And if there is a miss, it will first inquire the L3 cache to find whether the request data is on the chip. If the data is on the chip, it controls to get the data either from L2 cache or L3 cache depend on the core valid bits. The Fig. 8 shows the control flow of control unit.

Conclusions

The trend of the processor development is multi-core and the effect that the cores communicate with each other determines the performance of the processor. The advantage of an inclusive cache is that it can handle almost all coherency traffic without disturbing the private caches for each individual-core. The L3 cache insulates each of the cores from as much coherency traffic as possible by using the MESIF cache coherency protocol. This improves the performance of a multi-core processor.

References

1. Mark Balch. Complete Digital Design: A Comprehensive Guide to Digital Electronics and Computer System Architecture [M]. Beijing: Tsinghua Press, 2004.
2. White Paper Intel next generation microarchitecture Nehalem.
3. Intel Developer Forum. Inside Intel® Next Generation Nehalem Microarchitecture [C]. 2008.
4. Hum, H. et al. Forward State for use in Cache Coherency in a Multiprocessor System [P]. US Patent No. 6,922,756 B2. July 26, 2005.
5. Coherency Leaps Forward at Intel[EB/OL]. http://www.realworldtech.com/page.cfm?ArticleID=RW T082807020032&p=5

Effects of Phosphor's Location on LED Packaging Performance

Zongyuan Liu[12], Sheng Liu[123]*, Kai Wang[23], Xiaobing Luo[24]

[1]Institute for Microsystems, School of Mechanical Science and Engineering, Huazhong University of Science & Technology, Wuhan, China, 430074

[2]Division of MOEMS, Wuhan National Laboratory for Optoelectronics, Wuhan, China, 430074

[3]School of Optoelectronics Science and Engineering, Huazhong University of Science & Technology, Wuhan, China, 430074

[4]School of Energy and Power Engineering, Huazhong University of Science & Technology, Wuhan, China, 430074

Email: victor_liu63@126.com, Telephone: 86-27-87542604

Abstract

High power LED packaging is crucial for the development of solid state lighting. Different optical structures with various states of phosphor will greatly influence the packaging performance such as luminous efficiency, quality of white light and color uniformity. Through the analysis of five different optical structures, it is found that the location of phosphor layer has low impact on the light extraction efficiency, except structures in which phosphor is directly coated on the chip. However, the correlated color temperature (CCT) and luminous efficacy greatly depends on the structure and location of phosphor layer. The light extraction efficiency and CCT of curved phosphor layer is normally higher than those of plane shape. However, considering the variations of luminous efficacy, it's suggested that the structure with plane and remote phosphor may be suitable for obtaining high performance LED modules.

Introduction

Light emitting diodes (LEDs) has made remarkable progress in the past four decades with the rapid development of compound semiconductor technology. Since the first red LEDs that was invented by Holonyak and Bevacqua in 1962[1], great efforts have been put into the study to obtain brighter LEDs. In early 1990s, Nakamura successfully grew the blue and green LEDs on GaN substrate, which has a profound impact on the progress of LED technology[2, 3]. Nowadays there are three basic methods to obtain white light LEDs, one is the monochromatic RGB LEDs, another is phosphor converted LEDs (PC LEDs), the third is based on UV LEDs. The most commercially available white LEDs is dichromatic PC LEDs[4], which generates the white light by mixing the blue light from LED chip with the broadband yellow light excited by phosphor[5, 6]. Since the white LEDs has superior characteristics such as high efficiency, small size, long life, dependable, low power consumption, high reliability and mechanically rugged[7], therefore, a new concept of illumination named solid state lighting (SSL) in terms of high power LEDs is proposed[8]. With the accelerated advancement of LED technology, the input electrical power of LEDs was jumped to 1W or more by increasing the chip size from 350 microns to 1mm. The luminous efficiency of high power LEDs was also increased from 25 lm/W in 1999 to more than 100 lm/W in 2007 for a driving current of 350mA[9]. Consequently, the market for high power LEDs is growing rapidly in various applications such as large size flat panel backlighting, street lighting, vehicle forward lamp, museum illumination and residential illumination[10, 11]. It has been widely accepted that SSL will be the fourth illumination source to substitute incandescent lamp, fluorescent lamp and HID lamp[12]. The broad application prospect of SSL has attracted great attention on the study of LEDs[13].

The main functions of LED packaging are to protect the LED chip, to enhance the light extraction and to provide a path for dissipating the generated heat. Packaging is the critical bridge between LED chips and applications for end-users. Now LED packaging technology has been improved much to make full of the potentials of rapidly developing high power LED chips[14, 15]. Products of corporations such as the Lumileds's Luxeon K2 series, the Cree's XLamp series, the Osram's Golden Dragon series, offer high performance LED modules in terms of various packaging structures, in which the phosphor is dispersed on the chip surface. To improve the packaging performance, researcher from the lighting research center proposed the scattered photon extraction method (SPE) by moving phosphor layer to remote location[16, 17]. Results showed that there was over 60 percent improvement in light output and efficacy compared to commercial white LEDs. Based on the SPE concept, Allen et al. applied the role of internal reflection on the packaging components to improve the light extraction, and the increase of efficiency was at least 26%[18]. By dispersing the TiO2 nano-particles into to epoxy, Mont et al. found that the refractive index of epoxy can be improved to 1.68, and the gradient-index epoxy layer can enhance the optical transmittance[19]. Using wafer level technology and silicon process, Lim et al. packaged the LED chips on silicon substrate with a high sag rectangular lens[20]. Tsou and Jeung et al. also developed the packaging platform based on silicon to improve the light extraction, reliability and thermal fatigue[21]. Recently, Osram proposed a new wafer level packaging technology by spin coating phosphor on GaN wafer. By controlling the thickness of phosphor, this feasible method could improve the color distribution of LEDs[22]. To optimize the packaging structure, optical design is also important. Researchers tried to set up the precise optical model and present novel structures[23, 24, 25, 26]. However, now all these works have not systematically analyzed the effects of basic elements such as phosphor layer on the packaging performance.

Phosphor plays an important role in LED packaging. There are energy conversion and re-emission, scattering and absorption in phosphor layer. Different optical structures with various states of phosphor in terms of location, thickness, concentration, shape, etc. will greatly influence the performance such as luminous efficiency, quality of white light and color uniformity. Past studies simplified the parameters of phosphor with scattering and absorption coefficient by Monte Carlo ray tracing, and achieved some useful conclusions[27, 28, 29, 30]. To verify the correction of these coefficients, Zhu *et al.* firstly achieved the scattering and absorption parameters of phosphor by two-integrating-spheres test platform[31]. However, these works are not enough to understand the influences of phosphor on LED packaging.

To further investigate the focused issues in optical design, this paper mainly concerns the effects of phosphor's location on LED packaging performance. In addition to prove that the results are reliable and implementable for packaging production, five different optical structures are discussed by changing phosphor layer from close to chip to being remote from chip. A non-sequential ray trace method was applied, which was based on the Monte-Carlo theory. The power ratio of yellow light to blue light is used to illustrate the correlated color temperature (CCT). The parameters of the model were validated and the total ray trace number was finally 800000 to obtain converged results.

Problem Statement

The basic optical elements in one LED module are chip, phosphor layer and lens. Some modules may have a reflector in each of them. Although the performance of LED module greatly depends on the technology of LED chip, packaging elements such as the phosphor layer are also very important in the process of trying to achieve high quality white LEDs.

When alternating the phosphor layer from being close to chip to being remote from chip, the propagation path and energy of light under and up phosphor layer will be affected in terms of the scattering and absorption of phosphor layer, the reflection of reflector, the absorption of chip, the refraction of lens, etc.. The absorption of phosphor and chip will influence the output power of blue light and yellow light. The scattering of phosphor will disorder the light propagation. The directions of light rays could be converged to central angles by the reflection of reflector and the refraction of lens or be changed to side angles. It is difficult to evaluate the effects of these factors on the performance of packaging experimentally, since a LED module is so compact that each element should be precisely controlled. Taking the fabrication of phosphor layer and the dispersion of silicone as an example, the thickness of phosphor layer is around 100 microns, and the flow of silicone is still uncontrollable because of the complicated visco-elasticity. Therefore, to fabricate the phosphor layer and silicone with desired shape such as a free form surface, further research and advanced techniques are needed.

It is feasible to analyze the novel packaging design theoretically before the prototyping. Since the fabrication of phosphor layer is so difficult and the state of phosphor is the primary factor affecting the performance of LEDs, it is important to analyze the basic influences of phosphor with some simple numerical models. As the first step to realize the influence, this paper is focused on issues such as the relationship between LED's performance and phosphor layer's location. The objectives of this paper are: 1) to investigate the effects of phosphor's location on the light extraction efficiency of LEDs; 2) to investigate the effects of phosphor's location on the variation of CCT; and 3) to provide some suggestions on the packaging design.

Numerical Model and Simulation

This paper is mainly concerned about five different optical structures that are depicted in Figure 1. In order to evaluate the impact of reflector on the packaging performance, three numerical models in terms of Types III, IV, V have a reflector on each to compare other two non-reflector models of Types I and II. In order to minimize the effects of lens' size on light propagation, the radius of lens of all types is large enough to be 4 mm. The base diameter of the reflector cup is 3 mm, and the height is 2 mm. The optical parameters of surface on the board and reflector are 85% perfect reflection, 5% scattering and 10% absorption.

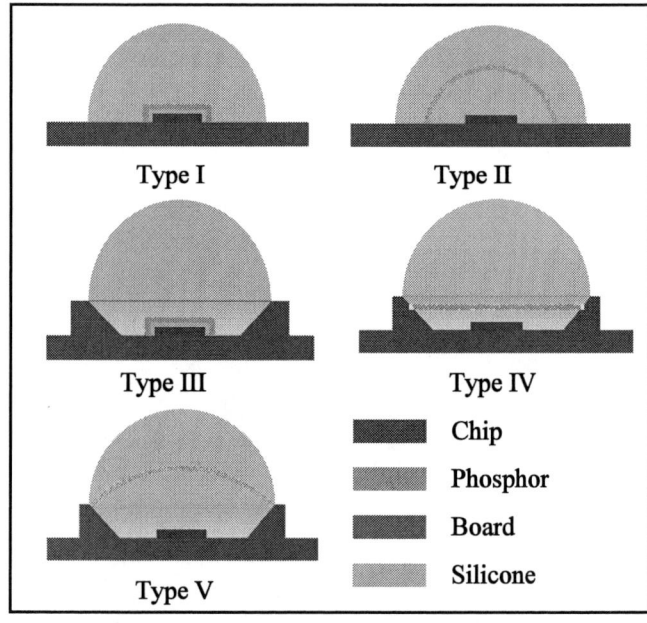

Figure 1. Optical structures for the numerical simulation.

In Types I and III, the phosphor layer is conformally coated with shape of square. Simulation model alters the distance of phosphor layer to chip surface from 0 mm to 0.1mm. In Types II and V, model changes the location by increasing the radius of phosphor's surface. In Type IV, the height of phosphor layer is increased from 0.2 mm to 1.9 mm. The thickness of phosphor layer is 0.1mm in all cases.

The LED chip model is depicted in Figure 2, which is a simplified Cree LED chip with a size of 1mm×1mm. The top surface of Silicon substrate is coated with Au to reflect the back emitted light. We considered that the light emits uniformly from MQW layer, since the direction of photons that are excited by the combination of electrons and holes in MQW layer is arbitrary. Considering that the area of top and down surfaces is much larger than that of the side surfaces, the model defines the two plane surfaces as the light sources by ignoring the side emitting lights. The optical properties of chip are illustrated in Table 1 [32].

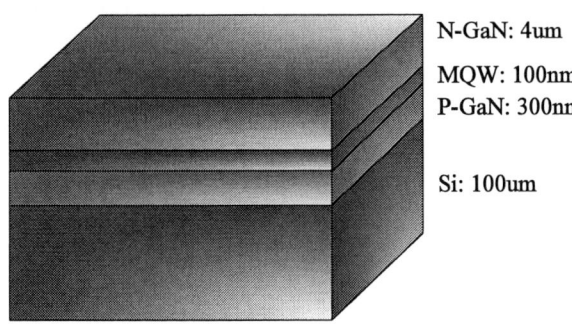

N-GaN: 4um

MQW: 100nm
P-GaN: 300nm

Si: 100um

Figure 2. LED chip model for numerical simulation.

Table 1. Optical properties of LED chip.

	N-GaN	MQW	P-GaN
Refractive Index	2.42	2.54	2.45
Absorption Coefficient (mm^{-1})	8	8	8

Definition of the optical property for phosphor layer is very important, since it relates to the correction and precision of simulation results. When one beam of blue light passes through one phosphor particle, phenomenon such as the transmission, refraction, scattering and absorption will occur simultaneously, and thus affects the light propagation and weakens the blue light energy. The absorbed blue light will re-emit broad-band yellow light, which should be affected by other phosphor particles. The average radius of phosphor particle is 5~8 um as shown in Figure 3, and generally the number of particles is 10000~100000 per mm^3 in the mixture of phosphor silicone. Since there may be millions of phosphor particles in the phosphor layer, light will encounter many particles in the propagation path and thus be scattered many times before it transmits through phosphor layer.

Figure 3. SEM picture of the phosphor particles.

Considering the phosphor layer as a bulk scattering material, it is an easy and effective method that applies the scattering and absorption coefficient to represent the total effects of phosphor particles on light propagation. Therefore, when a beam of rays passes through phosphor layer, the transmitted light energy can be expressed by Equation 1.

$$I(x) = I_0 e^{-(\mu_\alpha + \mu_s)x} \qquad (1)$$

where I_0 is the initial power of light and $I(x)$ is the residual power; μ_α and μ_s are the absorption coefficient and scattering coefficient of phosphor; and x is the thickness of phosphor layer.

The scattering distribution function of phosphor layer, which is used to generate the random directions of the scattered light rays, is based on the Henyey-Greenstein model. It is indicated in Equation 2.

$$p(\theta) = \frac{1 - g^2}{4\pi(1 + g^2 - 2g\cos\theta)^{3/2}} \qquad (2)$$

where g is called the anisotropy factor. Since the scattering of light by phosphor particles is isotropy, g is zero.

The numerical model assumes the absorption and scattering coefficient of phosphor is 8 mm^{-1} and 11.85 mm^{-1} for blue light. It is considered to be transparent when the incident rays are yellow light; therefore, the absorption coefficient is set as 0 mm^{-1}. Because of the non-absorption, the yellow light will be scattered more times than the blue light, which induces that the yellow light may be exhausted later by phosphor particles in the propagation process. Therefore, in the numerical model the scattering coefficient of phosphor layer is 16.25 mm^{-1} for yellow light.

Since phosphor layer is the mixture of phosphor particles and silicone, the refractive index can be expressed by Equation 3.

$$n_{p-s} = \alpha_p n_p + \alpha_s n_s \qquad (3)$$

where n_{p-s}, n_p, n_s are the refractive index of phosphor layer, phosphor crystal and silicone, respectively. α_p, α_s are the volume ratio of phosphor particles and silicone in the phosphor layer. This paper defines that the refractive index of the phosphor layer is 1.7, which does not consider the influences of wavelength on the refractive index.

Since the transmittance of silicone materials is over 95% for visible light, it is generally considered that the silicone materials are transparent to visible light. Therefore, the absorption coefficient is 0 mm^{-1} in the numerical simulation. The refractive index of silicone is 1.5, which can be purchased from corporations such as DowCornig.

Based on the above settings, the simulation steps for the numerical model are:

1) Defining the material properties of each layer;
2) Defining the top and down surface of MQW layer as light source, and then inputting the light power;
3) Ray tracing and collecting the simulation data;
4) Re-defining the material properties of each layer;
5) Re-defining the top and down surface of phosphor layer as light source, the power of which is calculated from the absorbed blue light; and the conversion efficiency of phosphor is 80%;
6) Ray tracing and collecting the simulation data; and
7) Calculating the simulation results of each case.

For simplicity, the energy of light is calculated by optical power in watts. In this simulation procedure, the blue light output and yellow light output are computed separately to easily investigate the impact of the phosphor's location on the color temperature. The yellow blue ratio, which is the value of yellow light power divided by blue light power, is used to represent CCT.

In the simulation results, total efficiency is the ratio of total output power to initial optical power, which is emitted from chip. Packaging efficiency is the ratio of total output power to the power that is calculated by subtracting the absorbed power of chip from the initial power. These are illustrated in the Equations 4, 5 and 6.

$$\eta_{Total} = \frac{P_{Out}}{P_{Total}} \qquad (4)$$

$$\eta_{Packaging} = \frac{P_{Out}}{P_{Total} - P_{Chip}} \qquad (5)$$

$$P_{Total} = P_{Out} + P_{Chip} + P_{Phosphor} + P_{Interface} + P_{Lost} \qquad (6)$$

where P_{Chip} is the absorbed power by chip; $P_{Phosphor}$ is the conversion loss by phosphor; $P_{Interface}$ is the sum of reflection loss on the surface of board and Fresnel loss in the interface; and P_{Lost} is the exhausted power in light propagation path.

Simulation Results and Discussions

The simulation results of five numerical models are shown in the following figures.

In the numerical simulation of Type I, which is shown in Figure 4, it is clear that packaging efficiency and total efficiency are increased slightly when the phosphor layer is changed to remote place from 0.01 mm to 0.1mm. However, there is a sudden variation if the phosphor is dispersed on chip surface directly, which means the distance is 0 mm. The variation is strange since the total efficiency is significantly higher than those cases with a small distance but the packaging efficiency is lower.

Compared to Type I, the efficiency of Type II is obviously higher. This is contributed to the curved surface of phosphor layer. In Figure 5, the fluctuation of efficiency is also smaller than Type I, which is 0.56% for total efficiency and 1.79% for packaging efficiency. The maximum value of the efficiency lies around the radius of 2 mm, which is the half of the lens's radius.

It can be found that the trend and value of efficiency in the third numerical model are similar to Type I, which is shown in Figure 6. This may be caused by the size of reflector. Since the angle of cone is 38.7°, this may induce most of the light rays to be directly emitted out without being reflected. However, if the angle of cone is large enough and the height of reflector is bigger, the affection on light propagation may be significant and thus make the the light extraction efficiency to be greatly varied.

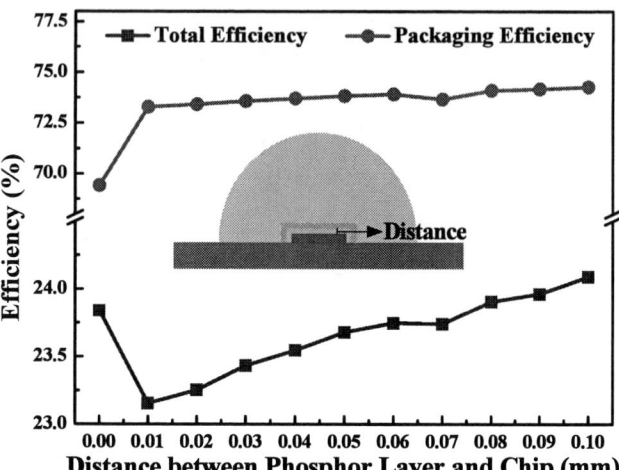

Figure 4. Effects of phosphor's location on the performance of type I.

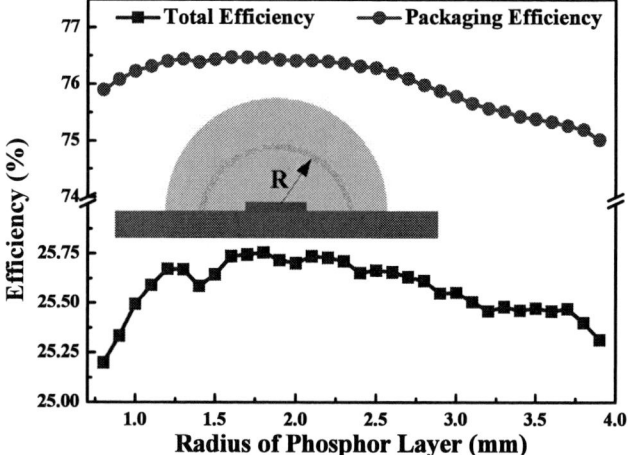

Figure 5. Effects of phosphor's location on the performance of type II.

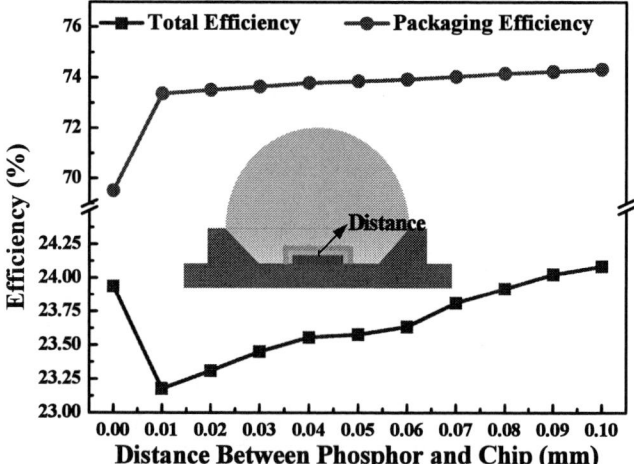

Figure 6. Effects of phosphor's location on the performance of type III.

In LED modules with a reflector, it is another usual approach that firstly coated the chip with silicone layer and then dispersing the phosphor on previously cured silicone. In Figure 7, simulation results show that the maximum value of the efficiency lies on the height of 1mm, when changing the height of phosphor layer from 0.2 mm to 1 mm. However, the

246

performance is still not greatly improved, which is around 66.5% for packaging efficiency and 22.3% for total efficiency.

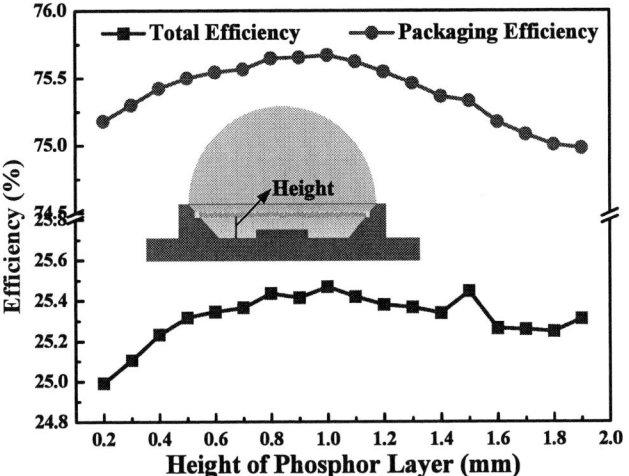

Figure 7. Effects of phosphor's location on the performance of type IV.

There also exists the maximum value of efficiency in the fifth numerical model. However, compared to the above situations, the fluctuation of the efficiency is so small that the influence of radius can be neglect. This is shown in Figure 8.

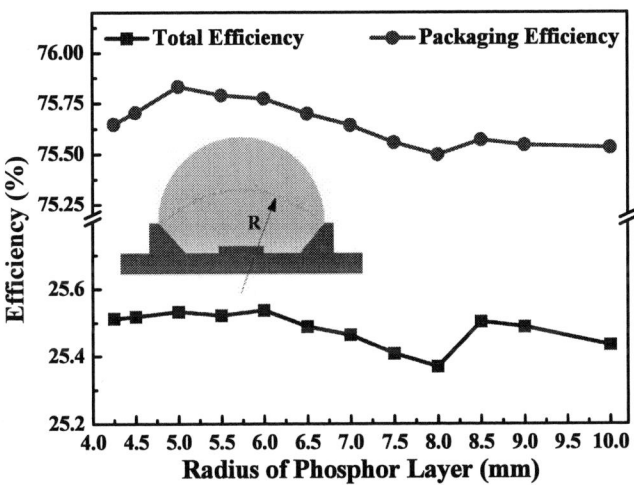

Figure 8. Effects of phosphor's location on the performance of type V.

The effects of phosphor's location on the color temperature are depicted in Figure 9. Obviously, the color temperature of LEDs tends to be warm white when the location of the phosphor is farther. The color temperature of Type IV, which has a reflector and a plane phosphor layer, is usually higher than the other types. The minimum yellow blue ratio is appeared in Type II, in which the phosphor layer is hemispherical. Results of these figures also indicate that the effects of reflector on the color temperature are small.

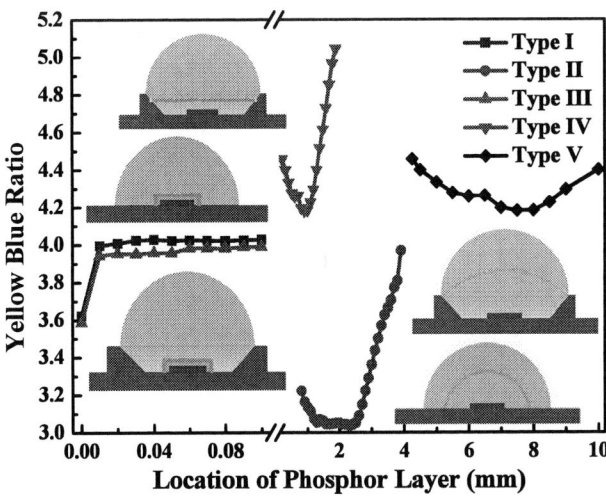

Figure 9. Effects of phosphor's location on CCT.

Attention should be paid to the fact that when the location of the phosphor layer is farther from the chip, the size and surface of the phosphor layer will be bigger. Therefore, the light emitted from phosphor layer will have less chance to be absorbed by chip. This is the main reason that why the remote phosphor case could have low CCT. However, due to the absorption and isotropy scattering of phosphor layer, LED modules with remote phosphor don't always have high performance. In the second, fourth and fifth models, at first the yellow blue ratio is descended and then ascended with the increase of the phosphor's location, while the efficiencies have almost inverse trends.

It can be found that the light extraction and CCT of phosphor layer with a curved surface is generally higher than those with a plane surface. This is because the curved surface could improve the critical angle for side emitting light and provide more chances to allow the rays to be extracted from the surface by reducing the time of multi- scattering. When the chip is located on the center of curved surface, most energy of blue light could directly enter the phosphor layer and thus reduce the backscattered light.

It should be remembered that the efficiency of all simulations is calculated by optical power, which is not the luminous efficiency. Although the optical power efficiency is changed slightly, the luminous efficiency can be changed significatn because of the variation of CCT. Therefore, considering all the influences of the phosphor's location on the performance, the structure of Type IV is predicted to have the highest luminous efficiency and better quality white light.

Conclusions

Based on the Monte-Carlo ray tracing method, five numerical models are studied in this paper. Simulation results show that the location of phosphor layer has low impact on the light extraction, except structures that the phosphor is directly coated on chip. However, CCT greatly depends on the structure and location of phosphor layer and thus affect the luminous efficiency. Normally the efficiency fluctuations are in the range of ±0.5% for total efficiency and ±1% for packaging efficiency, which are mainly due to the isotropy scattering and absorption of phosphor since they affect the blue and yellow light propagation. Through the simulation, it is found that the light extraction efficiency and CCT of curved phosphor layer is higher than those of plane shape. In one word, a plane and remote phosphor layer may be a suitable optical structure, especially when considering the thermal effects.

Acknowledgment

The support from Guangdong Real Faith Optoelectronic Inc. is appreciated.

References

1. Holonyak, J. N. and Bevacqua, S. F., "Coherent (Visible) Light Emission from Ga(As$_{1-x}$P$_x$) Junctions", *Applied Physics Letters*, Vol. 1, No. 4 (1962), pp. 82-83.
2. Nakamura, S., Mukai, T. and Senoh, M., "High-Power GaN PN Junction Blue-Light-Emitting Diodes", *Japanese Journal of Applied Physics*, Vol. 30, No. 12A (1991), pp. L1998-L2001.
3. Nakamura, S., Senoh, M. and Mukai, T., "High-Power InGaN/GaN Double-Heterostructure Violet Light Emitting Diodes", *Applied Physics Letters*, Vol. 62, No. 19 (1993), pp. 2390-2392.
4. Yam, F. K. and Hassan, Z., "Innovative Advances in LED Technology", *Microelectronics Journal*, Vol. 36, No. 2 (2005), pp. 129-137.
5. Schlotter, P., Schmidt, R. and Schneider, J., "Luminescence Conversion of Blue Light Emitting Diodes", *Applied Physics A: Materials Science & Processing*, Vol. 64, No. 4 (1997), pp. 417-418.
6. Nakamura, S. and Fasol, G., The Blue Laser Diode.: GaN Based Light Emitters and Lasers, Springer-Verlag Berlin and Heidelberg GmbH & Co. K, 1997.
7. Evans, D. L., "High-Luminance LEDs Replace Incandescent Lamps in New Applications", *Light-Emitting Diodes: Research, Manufacturing, and Applications*, SPIE, 1997, pp. 142-153.
8. Zukauskas, A., Shur, M. S. and Gaska, R., Introduction to Solid-State Lighting, John Wiley & Sons New York, 2002.
9. Krames, M. R., Krames, M. R., Shchekin, O. B. et al., "Status and Future of High-Power Light-Emitting Diodes for Solid-State Lighting", *Display Technology, Journal of*, Vol. 3, No. 2 (2007), pp. 160-175.
10. Steranka, F.M. Bhat, J., Collins, D. et al., "High Power LEDs - Technology Status and Market Applications", *Physica Status Solidi (a)*, Vol. 194, No. 2 (2002), pp. 380-388.
11. Craford, M. G., "LEDs for Solid State Lighting and other Emerging Applications: Status, Trends, and Challenges", *Fifth International Conference on Solid State Lighting*, SPIE, 2005, pp. 594101-594110.
12. OIDA, "Light Emitting Diodes (leds) for General illumination, An OIDA Technology Roadmap Update 2002," 2002.
13. Pelka, D. G. and Patel, K., "An Overview of LED Applications for General Illumination", *Design of Efficient Illumination Systems*, SPIE, 2003, pp. 15-26.
14. Ya-Ju, L., Ya-Ju, L., Tien-Chang, L. et al., "High Brightness GaN-Based Light-Emitting Diodes", *Journal of Display Technology*, Vol. 3, No. 2 (2007), pp. 118-125.

15. Haque, S., Steigerwald, D., Rudaz, S. et al., "Packaging Challenges of High-Power LEDs for Solid State Lighting", *www.lumileds.com/pdfs/techpaperspres/manuscript_IMAPS_2003.PDF*, (2000), pp.
16. Narendran, N., "Improved Performance White LED", *Fifth International Conference on Solid State Lighting*, SPIE, 2005, pp. 594108-594106.
17. Narendran, N., Gu, Y., Freyssinier-Nova, J. P. and Zhu, Y., "Extracting Phosphor-Scattered Photons to Improve White LED Efficiency", *Physica Status Solidi*, Vol. 202, No. 6 (2005), pp. R60-R62.
18. Allen, S. C., Allen, S. C. and Steckl, A. J., "Elixir-Solid-State Luminaire with Enhanced Light Extraction by Internal Reflection", *Display Technology, Journal of*, Vol. 3, No. 2 (2007), pp. 155-159.
19. Mont, F. W., Kim, J. K., Schubert, M. F. et al., "High Refractive Index Nanoparticle-Loaded Encapsulants for Light-Emitting Diodes", *Light-Emitting Diodes: Research, Manufacturing, and Applications XI*, SPIE, 2007, pp. 64861C-64868.
20. Lim, C.-H., Jeung, W.-K. and Choi, S.-M., "LED Packaging using High Sag Rectangular Microlens array", *Micro-Optics, VCSELs, and Photonic Interconnects II: Fabrication, Packaging, and Integration*, SPIE, 2006, pp. 618516-618517.
21. Tsou, C. F. and Huang, Y. S., "Silicon-Based Packaging Platform for Light-Emitting Diode", *Advanced Packaging, IEEE Transactions on [see also Components, Packaging and Manufacturing Technology, Part B: Advanced Packaging, IEEE Transactions on]*, Vol. 29, No. 3 (2006), pp. 607-614.
22. Braune, B., Petersen, K., Strauss, J. et al., "A New Wafer Level Coating Technique to Reduce the Color Distribution of LEDs", *Light-Emitting Diodes: Research, Manufacturing, and Applications XI*, SPIE, 2007, pp. 64860X-64811.
23. Borbely, A. and Johnson, S. G., "Prediction of Light Extraction Efficiency of LEDs by Ray Trace Simulation", *Third International Conference on Solid State Lighting*, SPIE, 2004, pp. 301-308.
24. Lee, S. J., "Light-Emitting Diode Lamp Design by Monte Carlo Photon Simulation", *Light-Emitting Diodes: Research, Manufacturing, and Applications V*, SPIE, 2001, pp. 99-108.
25. Sun, C.-C., Lee, T.-X., Ma, S.-H. et al., "Optical Modeling for LED in Mid-Field Region", *International Optical Design Conference 2006*, SPIE, 2007, pp. 634217-634217.
26. Chi, W. and George, N., "Light-Emitting Diode Illumination Design with A Condensing Sphere", *J. Opt. Soc. Am. A*, Vol. 23, No. 9 (2006), pp. 2295-2298.
27. Borbely, A. and Johnson, S. G., "Performance of Phosphor-Coated LED Optics in Ray Trace Simulations", *Fourth International Conference on Solid State Lighting*, SPIE, 2004, pp. 266-273.
28. Falicoff, W., Chaves, J. and Parkyn, B., "PC-LED Luminance Enhancement due to Phosphor Scattering", *Nonimaging Optics and Efficient Illumination Systems II*, SPIE, 2005, pp. 59420N-59415.

29. Kim, J. K., Luo, H., Schubert, E. F. et al., "Strongly Enhanced Phosphor Efficiency in GaInN White Light-Emitting Diodes using Remote Phosphor Configuration and Diffuse Reflector Cup", *Japanese Journal of Applied Physics*, Vol. 44 (2005), pp.

30. Luo, H., Kim, J. K., Schubert, E. F. et al., "Analysis of High-Power Packages for Phosphor-Based White-Light-Emitting Diodes", *Applied Physics Letters*, Vol. 86, No. 24 (2005), pp. 243505-243503.

31. Zhu, Y., Narendran, N. and Gu, Y., "Investigation of the Optical Properties of YAG:Ce Phosphor", *Sixth International Conference on Solid State Lighting*, SPIE, 2006, pp. 63370S-63378.

32. Lee, T.-X., Gao, K.-F., Chien, W.-T. and Sun, C.-C., "Light Extraction Analysis of GaN-Based Light-Emitting Diodes with Surface Texture and/or Patterned Substrate", *Opt. Express*, Vol. 15, No. 11 (2007), pp. 6670-6676.

Computer Simulation of Crack Propagation in Power Electronics Module Solder Joints

Hua Lu[a]†, Steve Ridout[a], Chris Bailey[a], Wei Sun Loh[b], Agyakwa Pearl[c], and Mark Johnson[c]

[a] School of Computing and Mathematical Sciences, University of Greenwich, 30 Park Row, London SE10 9LS, UK

[b] Department of Electronic and Electrical Engineering, University of Sheffield, Mappin Street, Sheffield, S1 3JD, U.K

[c] University of Nottingham, School of Electrical and Electronic Engineering, University of Nottingham, Park, Nottingham, NG7 2RD, U.K.

†Telephone: +44(0)2083318536, Email address: h.lu@gre.ac.uk

Abstract

A numerical modelling method for the analysis of solder joint damage and crack propagation has been described in this paper. The method is based on the disturbed state concept. Under cyclic thermal-mechanical loading conditions, the level of damage that occurs in solder joints is assumed to be a simple monotonic scalar function of the accumulated equivalent plastic strain. The increase of damage leads to crack initiation and propagation. By tracking the evolution of the damage level in solder joints, crack propagation path and rate can be simulated using Finite Element Analysis method. The discussions are focused on issues in the implementation of the method. The technique of speeding up the simulation and the mesh dependency issues are analysed. As an example of the application of this method, crack propagation in solder joints of power electronics modules under cyclic thermal-mechanical loading conditions has been analyzed and the predicted cracked area size after 3000 loading cycles is consistent with experimental results.

Introduction

Power electronic modules (PEM) are widely used in the industry and consumer products for the control and conversion of electric power. A typical power module consists of several layers of insulator, conductor and semiconductor plus some metal wires, encapsulations, electric terminals and the casing components [1]. The materials in power modules are assembled together in the packaging process to form power electronic circuits and the mechanical structure. Since PEMs contain different materials, they are highly inhomogeneous devices and there are many interfaces in them. Furthermore, PEMs are often used in extreme conditions and this poses a great challenge for the designers of highly reliable PEMs.

Of the several possible failure mechanisms of a PEM, solder joint fatigue is one of the most important one and it is this mechanism that will be discussed in this work. In particular, the focus of this paper will be on the substrate solder interconnect which is the solder layer connecting the isolation substrate to the baseplate (see Fig. 1). Under cyclic thermal mechanical loading, materials in PEM expand and contract at different rate and this causes the solder joint to deform and suffer from damage over time. Fatigue cracks initially appear at the most stressed locations and then propagate. In Fig. 2, the scanning acoustic micrograph shows that after 8000 cycles under a cyclic thermal loading condition, much of the solder interconnect of a substrate solder joint has cracked. In this experiment, four dies are

soldered to a ceramic substrate. In the picture the light part in the middle of each solder joint is the intact area.

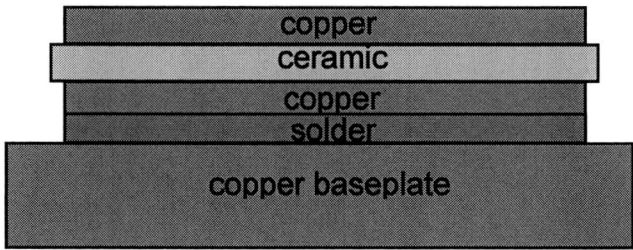

Figure 1. The cross section of the substrate mountdown solder interconnect.

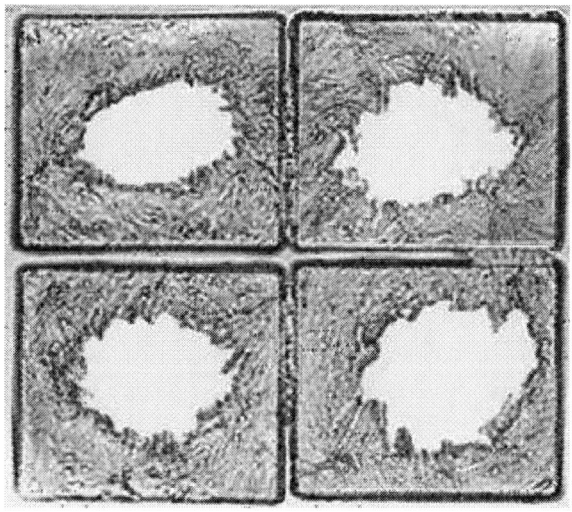

Figure 2. Scanning Acoustic Micrograph images of the isolation substrate solder joints after 7000 cycles.

The solder interconnect functions as the mechanical support of the substrate as well as the heat conduction path. In service conditions, if the cracked area is too large compared to the total solder-substrate interfacial area, or if the cracked area is directly under the active die which is the heat source the thermal resistance may increase significantly and the temperature of the components on the substrate will be too high so that the semiconductor device cannot work reliably. Therefore, in order to design a highly reliable PEM, it is important to be able to predict the growth of solder joint crack propagation and the time to failure of the solder joint. In the

978-1-4244-2739-0/08/$25.00 ©2008 IEEE

following sections, methods of computer modelling of solder joint fatigue crack will be discussed and an example of using a damage accumulation method to predict crack area in PEM substrate solder joint will be given.

Computer modelling of solder joint fatigue crack

Solder joint fatigue crack and failure can be modelled in different ways. The most practical method is to treat solder as a homogeneous material in which damage indicators such as the accumulated inelastic strain or plastic work density over a thermal loading cycle can be calculated using Finite Element Analysis (FEA) method [2, 3]. These damage indicators can then be used in an empirical lifetime model to obtain the lifetime of the solder joint. In this type of simulation crack is not included in the modelling and the damage indicators are usually calculated as the average values over the predicted crack path. For large solder joints this may cause problem because the average values cannot represent the damage in solder joint well and lifetime prediction will not be accurate unless in some particular cases when the damage distribution in solder joint is relatively simple [4]. In this work, a different approach based on the disturbed state concept [5] is used. It is assumed in this method that the accumulation of damage is a continuous process and the apparent material properties change over time. The damage in solder joint is represented by a damage parameter D.

Fig. 3 shows the concept of the damage parameter D. A material under cyclic loading is divided into two continuous parts. One is the intact part and other is the damaged which is represented by voids in the figure. As the damaged part increases in volume over time, D increases in value and the average material properties changes. When the whole volume is damaged D reaches its maximum value of 1.

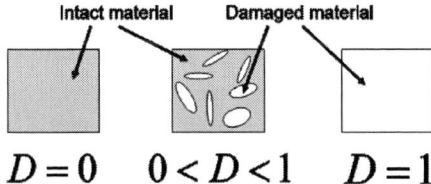

Figure 3. Damage D and the microstructure of solder material.

Mathematically, D is defined as the fraction of damaged material and $1-D$ is the fraction of intact material. The intact fraction has the mechanical material properties of the original solder joint. The damaged fraction has a Young's Modulus of 0, meaning it cannot resist any stress. The average Young's modulus is given by linear interpolation between the damaged and intact properties:

$$E_{avg} = DE_{damaged} + (1-D)E_{intact} \qquad (1)$$

But because $E_{damage} = 0$,

$$E_{avg} = (1-D)E_{intact} \qquad (2)$$

When the damage parameter D or the Young's modulus in any part of a solder joint reaches a certain preset value, that part of the solder joint is considered to have cracked. By tracking the cracked volume in a solder joint the crack area at any time can be predicted. If the failure of the solder joint is defined as the cracked area reaching certain value, the lifetime of the solder joint and therefore the PEM can be predicted. Alternatively, other parameters such as the electric resistance across the solder joint can be sued as a failure criterion,

Using with Stress-based Creep Law

An FEA simulation using E_{avg} will provide the average stress over both intact and damaged fractions. But because only the intact fraction has resistance to deformation, the stress within the intact fraction is greater than the average stress. Assuming the strain is the same within both the intact and damaged fractions, the relationship is as follows.

$$\sigma_{intact} = \frac{\sigma_{average}}{1-D} \qquad (3)$$

The intact stress σ_{intact} is fed into the creep law to calculate the creep strain rate which in turn is used for the calculation of the damage.

Damage Calculation

Following the work by other researchers including Towashiraporn [7] the damage is a function of the accumulated effective creep strain φ_{acc}:

$$D = 1 - e^{-B\varphi_{acc}} \qquad (4)$$

This is the simplest form for a damage law, containing only one material constant B.

Time discretisation

In this work the damage law is implemented using the following explicit time discretisation:

$$D^i = 1 - e^{-B\varphi_{acc}^{i-1}} \qquad (5)$$

where i is the time step. This results in optimistic predictions (less damage) rather than the conservative predictions an implicit discretisation would predict. In a typical thermal/mechanical cycling simulation, the damage builds up gradually over many cycles, therefore the damage increase per time step is usually quite small and the difference between an implicit and explicit scheme negligible.

Volume Averaging

A problem with the above law is its mesh dependence. This occurs because a local concentration of stress leads to a local concentration of damage, and a crack whose width is determined by the mesh spacing. Following similar work by Desai et al. [6], the mesh dependence is overcome by volume averaging the accumulated creep strain φ_{acc}. This work uses a Gaussian distribution to perform the weighted average:

$$\varphi_{acc}^{avg} = \frac{\int_V \varphi_{acc} e^{-r^2/2\omega^2} \, dV}{\int_V e^{-r^2/2\omega^2} \, dV} \qquad (6)$$

251

where r is the distance from the centre of the element and ω is the standard deviation, which can be thought of as a material property representing the scale of the region over which micro-crack interaction occurs, therefore we refer to it as the damage length scale. Since the influence of φ_{acc} over a unit volume at distances greater than 3ω is only 1.1% or less compared to the influence at the centre, these are ignored to save computational resources.

Determining Lifetime

To predict lifetime using the above damage law a failure criterion is required. As it has been mentioned above this can be the cracked area size. But other methods can also be used. For example, as the damaged volume increases the electric resistance across the solder joint also increases. Since the resistance can be measured more easily, in this work the increase in this resistance is used. It works by assuming the electrical conductivity is a linear function of damage just like the Young's Modulus:

$$k = k_{\text{intact}}(1 - D) \qquad (7)$$

A fixed potential difference is applied across the joint and the current I is predicted. The resistance is given by Ohm's law: V=IR. The resistance of a completely intact joint is calculated as R_0. The failure condition is defined as the time at which the relative resistance R/R_0 has increased beyond an arbitrary critical value.

Test case: Displacement-controlled cyclic loading

The mesh dependence of the above damage law will now be investigated for the simple test case shown in Fig. 4. The bottom surface is kept fixed and the top surface is cyclically displaced between -12μm and +12μm with 5 min ramps and 5 min dwells.

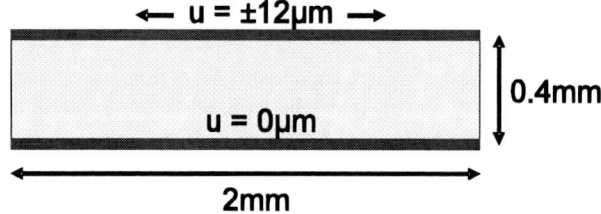

Figure 4. The dimensions and boundary conditions of the displacement controlled cyclic loading test case. (u = displacement)

The number of cycles to failure N_f was predicted using a failure criterion of $R/R_0 > 10$. Using a mesh spacing of 40μm and $\omega = 40$μm, the damage contours at the point of failure are as shown in Fig. 5.

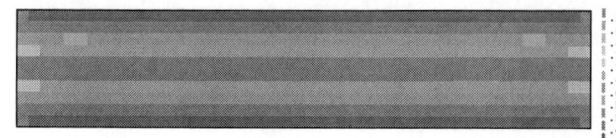

Figure 5. The damage contours after 12 displacement controlled cycles (just before the resistance increases by 10 times)

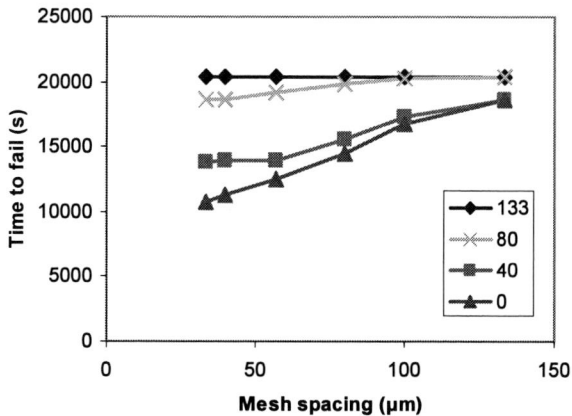

Figure 6. The effect of mesh density and crack length scale on the time to fail under displacement controlled loading. (The graph legend shows the crack length scale in μm)

The effect of changing both the mesh density and the length scale parameter is shown in Fig. 6. This shows that when a length scale of 0 is used N_f is highly mesh dependent. The reason is as follows:

- From the start of the simulation, the damage is not uniformly distributed.
- At an area of peak damage, the material is weaker than the surrounding area and so will subsequently experience a greater elastic strain.
- The greater elastic strain means that $\sigma_{\text{eff,intact}}$ will be even greater than before, leading to even greater creep strain and damage accumulation on the next time step.
- Positive feedback causes the damage to accumulate in a narrow band limited in thickness by the mesh and crack length scale.
- The finer the damaged band, the greater the strain within the band and the quicker failure will occur.

With a length scale of $\omega = 133$μm is used, N_f is completely independent of mesh density over the range investigated because even for the coarsest mesh, the large ω is the limiting factor regarding crack thickness. The smaller length scales exhibit progressively greater mesh dependence.

Test case: Force-controlled cyclic loading

This test uses the same geometry as above, but with a cyclic force boundary condition as shown in Fig. 7. The damage contours at failure are very similar to those predicted for the displacement controlled test in Fig. 5, but the effect of changing the mesh density and crack length scale is different.

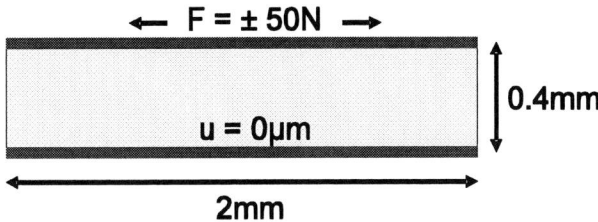

Figure 7. The dimensions and boundary conditions of the force controlled cyclic loading test case. (F = force, u = displacement)

As seen in Fig. 8, the mesh dependence in this case is far less than for the displacement controlled test, having an 18% effect over the range investigated as opposed to 73%. This is because unlike the displacement controlled test, a thinner damaged band does not necessarily experience a greater strain and $\sigma_{\text{eff,intact}}$. The average stress throughout the mesh depends on the applied force and is independent of the crack thickness. For this reason, increasing ω doesn't affect the mesh dependence; it just increases the time to failure slightly.

The reason for mesh dependence is that a finer mesh means more concentrated stress hot spots. Even if average stress in a region stays the same, a greater variation (i.e. more pronounced peaks and troughs) leads to a higher average creep strain rate and therefore greater overall damage build up. This effect is not a result of the damage law, and would be a present even when running a standard non-damage based analysis.

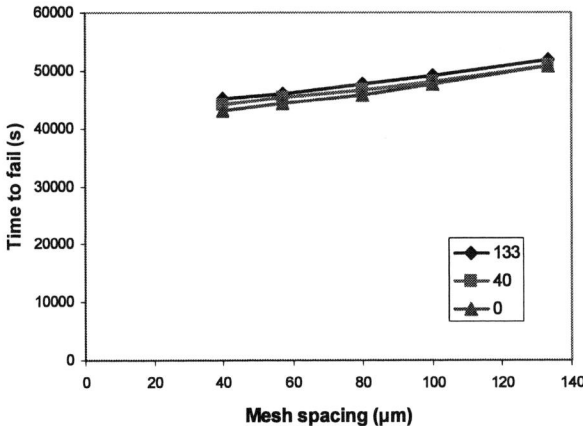

Figure 8. The effect of mesh density and crack length scale on the time to fail under displacement controlled loading. (The graph legend shows the crack length scale in μm)

Speeding up computation

A drawback of using the damage law to predict N_f is that thousands of cycles are typically required to destroy a joint using accelerated testing. To model this using FEA can be computationally expensive so a method has been presented here to solve this problem.

For the displacement-controlled and force-controlled problems described above, simulations have been run using different B constants and the results shown in Fig. 9 and Fig. 10.

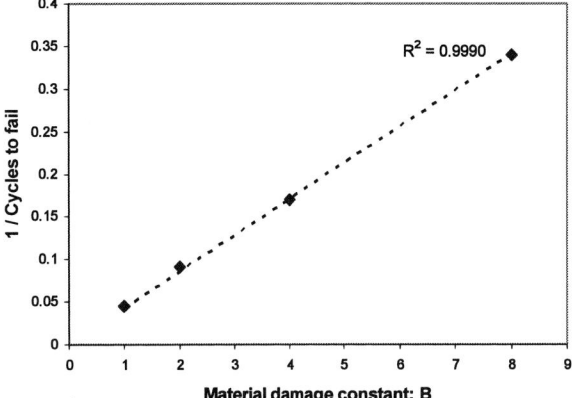

Figure 9. The effect of the damage parameter B on N_f under displacement controlled loading

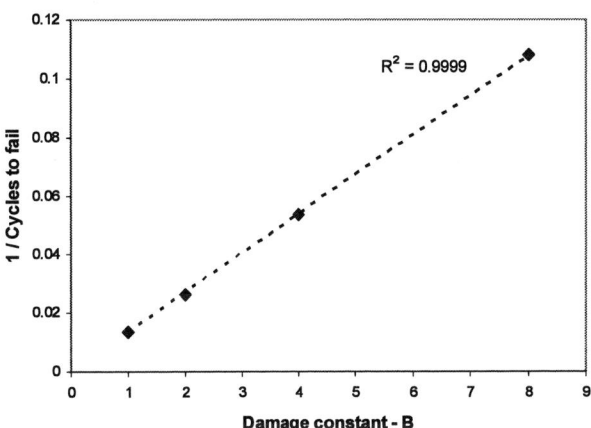

Figure 10. The effect of the damage parameter B on N_f under force controlled loading

The damage constant B is shown to be inversely proportional to N_f under both displacement-controlled and force-controlled tests. Thermal cycling simulations of chip resistor solder joints have also shown the same trend. Assuming the relationship holds under all conditions, this makes it possible to shorten the computation time by using the damage law with an unrealistically high B constant and finding the actual N_f as follows:

If the real B value is B_{real} and B_{sim} is used for the simulation yielding $N_{f\text{:sim}}$, the real cycles to fail $N_{f\text{:real}}$ can be found with:

$$N_{f:\text{real}} = N_{f:\text{sim}} \frac{B_{\text{sim}}}{B_{\text{real}}} \qquad (8)$$

It is important not to make B_{sim} so large that a disproportionate amount of damage is contributed by part of one cycle. For example, $B_{\text{real}} = 0.1$ and using $B_{\text{sim}} = 10$ gives

$N_{f:sim}$ = 1.5 cycles, and all of the damage occurs in the first half of each cycle. In this case using Eq. 8 to obtain $N_{f:real}$ = 150 is inaccurate, in fact $N_{f:real}$ should be 200 cycles.

The upper bound on $N_{f:real}$ due to the above error can be estimated by rounding $N_{f:sim}$ up to the nearest integer before using Eq. 8 and similarly the lower bound can be obtained by rounding $N_{f:sim}$ down before converting to $N_{f:real}$. As long as $N_{f:sim}$ is kept to a reasonably high number the error in $N_{f:real}$ should be small.

This method allows the damage law to be used to model any number of real world cycles in a reasonable time.

Crack propagation in PEM solder interconnect

To demonstrate the use of the method described earlier in this paper, the crack propagation in a PEM substrate solder joint has been carried. The basic structure is shown in Fig. 1 and a 2D mesh of the cross section is shown in Fig. 11. The dimensions of the isolation substrate that has been analyzed in this work are 58×50mm². The thickness of the ceramic, the copper and the baseplate are 0.38, 0.3 and 5 mm respectively. The mechanical properties of the materials are listed in Table I.

Table I. Mechanical properties of the substrate mount-down component. The temperature T is in Celsius.

	E(GPa)	ν	CTE(ppm/°C)
Alumina	270	0.22	7.4
Cu	103	0.3	17.3
SnAg	54.05-0.193T	0.4	21.85+0.02039T

In this simulation nonlinear material properties of the SnAg solder are used. The visco-plastic/creep constitutive equation for SnAg alloy is,

$$\dot{\varepsilon}_{cr} = A \times \sinh^n(\alpha\sigma_e)\exp(\frac{-Q}{RT}) \qquad (9)$$

where R is the gas constant, T is the temperature in Kelvin, σ_e is the von Mises equivalent stress, A, n, α, Q are material constants and their values are listed in Table II [9].

Table II: Creep parameters for solder materials.

	A(s)	n	α(1/MPa)	Q/R
SnAg	9.00E+05	5.5	0.06527	8690

The cyclic temperature load has a minimum temperature of -40°C and a maximum temperature of 120°C. Fig. 12 shows the 3D FEA model of the substrate and the damage parameter D distribution in the solder layer. In this example the value of B_{sim} is 15 and the number of cycles is 6.

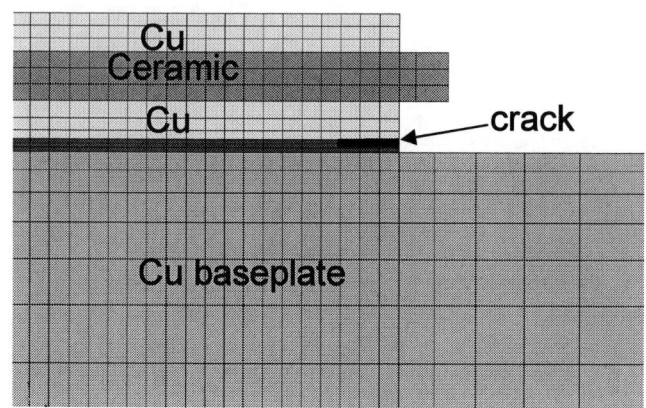

Figure 11. A 2D FEA mesh of PEM substrate solder joint. Crack initiate at the edge.

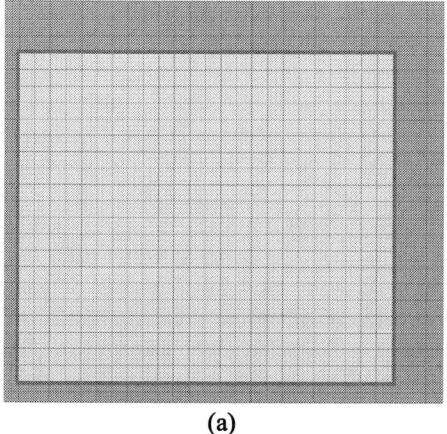

(a)

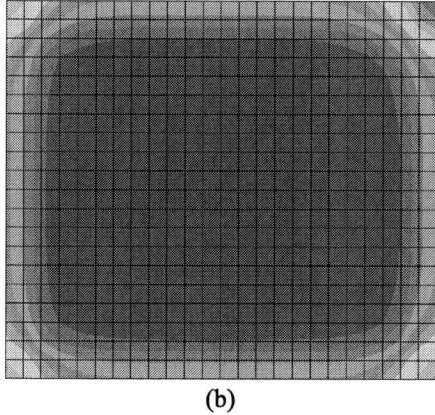

(b)

Figure 12. (a) A model of the substrate solder interconnect, and (b) the damage accumulated in solder joint.

Experimental results have shown that the crack propagation rates for the four corners are only slightly different. In order to further reduce the simulation time, the tiles are assumed to be symmetrical so that the small area highlighted in Fig. 13 has been modelled. The crack propagation rate obtained using this model can be regarded as the average of the cracked areas of the four corners.

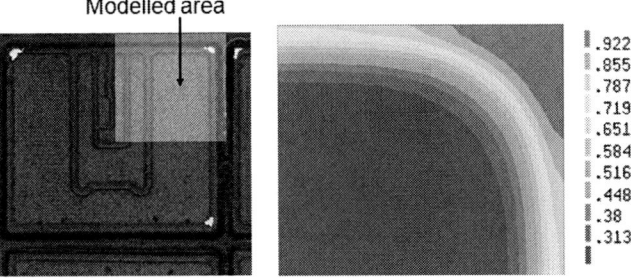

Figure 13. Symmetric model and damage distribution. $B=15$, $N_{f:sim}=9$.

Based on past experiences, the value of B for SnAg solder joint is estimated at 0.05. This means that the equivalent real number of cycles is

$$N_{f:real} = N_{f:sim} \frac{B_{sim}}{B_{real}} = 9 \times \frac{15}{0.05} = 2700$$

Compared to the experimental results shown in Fig. 14, the simulated results are similar to the observed cracked area at about 3000 cycles.

Figure 14. Cracked area in substrate solder joints. N=3000.

Conclusions

A modelling method based on the disturbed state concept has been analyzed. Highly accelerated simulation has proved possible. The method has been to model the crack propagation in power electronics module substrate solder joints. The simulation results and experimental observation show good agreement. Further work will be carried out to determine the damage law parameter.

Acknowledgments

The authors wish to acknowledge the support of the Innovative electronics Manufacturing Research Centre (IeMRC) and the United Kingdom Technology Strategy Board for the project 'Modelling of Power Modules for Lifetime, Accelerated Testing, Reliability and Risk'. The authors would like to thank project partners Semelab, Ltd, Dynex Semiconductor Ltd., Goodrich Engine Control SR Drives Ltd., Areva T&D Ltd and Rolls Royce Plc for their contribution to the project. S. Ridout would also like to thank the Materials Processing Metrology Programme of the UK Department of Trade and Industry, EPSRC (Engineering and Physical Sciences Research Council), Prime Faraday and the NPL (National Physical Laboratory) in the UK for their support.

References

1. Sheng, W.W. and Colino, R.P., Power Electronic Modules, CRC Press (2005)
2. Syed, A., "Accumulated creep strain and energy density based thermal fatigue life prediction models for SnAgCu solder joints", *Proceedings of the 54th Electronic Components and Technology Conference*, pp.737-746, (2004)
3. Schubert, A., Dudek, R., Auerswald, E., Gollhardt, A., Michel, B., Reichl, H., "Fatigue life models for SnAgCu and SnPb solder joints evaluated by experiments and simulation", *53rd Electronic Components & Technology Conference*, 2003 Proceedings : pp.603-610, (2003)
4. Lu, H., Tilford, T., Bailey, C. and Newcombe, D., "Lifetime Prediction for Power Electronics Module Substrate Mount-down Solder Interconnect", HDP07 Proceedings, (2007)
5. Chandra S. Desai and Russell Whitenack, Review of models and the disturbed state concept for thermomechanical analysis in electronic packaging. Journal of Electronic Packaging, vol.123 pp.19-33, (2001)
6. Chandra S. Desai, Cemal Basaran, and Wu Zhang. Numerical algorithms and mesh dependence in the disturbed state concept. International Journal for Numerical Methods in Engineering, vol. 40, pp.3059-3083, (1997)
7. Towashiraporn, P., Subbarayan, G. and Desai, C. S., A hybrid model for computationally efficient fatigue fracture simulations at microelectronic assembly interfaces. International Journal of Solids and Structures, vol.42, pp.4468-4483, (2005)
8. Lau, J.H. (editor), Ball Grid Array Technology, McGraw-Hill (1995), p.396

Process Simulation of DRIE and its Application in Tapered TSV Fabrication

Min Miao[1,2 *], Hongguang Liao[1,3], Xin Wan[1], Liwei Zhao[1], Yunxia Guo[1], Yufeng Jin[1]

[1]. National Key Laboratory on Micro/Nano Fabrication Technology, Peking University, 100871, Beijing, China

[2]. Inst. of Information Microsystem, & Dept. of Telecomm. Engineer, Beijing Information Science and Technology University, 100101, Beijing, China

[3]. Shenzhen Graduate School of Peking University, Shenzhen 518055, China

* Email: miaomin@ime.pku.edu.cn, miaomin@bistu.edu.cn; Tel: +86-10-62752536

Abstract

TSV (Through Silicon Via) has been widely welcomed as an enabling technology for three-dimensional integration in a package with high density. The developing of a drilling method for TSVs with tapered sections by experimental methodology can be a tedious task. The authors thus explore the possibility to simulate the process conditions effectively with an in-house developed simulator which utilize hybrid line and cell evolution algorithms. The micro-fabrication using the parameters obtained by the simulation has demonstrated TSVs with tapered sectional profiles and filled with electroplated coppers as expected, which validates the effectiveness of the simulated results.

Introduction

In recent years, TSV (Through Silicon Via) has been widely accepted as the one of the major enabling techniques for the effective and highly dense three-dimensional integration of hetero- or homogeneous ICs and micromachined functional structures, such as sensors, actuators and antennas, in a package. This type of vertical (z-axial) interconnects may provide a short and low resistance/impedance inter-layer signal path for these stacked chips or modules, so that the comparatively high signal loss, temporal delay and noise pickup associated with the long interconnects in the in-plane integration methodologies, can be mostly eliminated [1]. In addition, TSV may provide the flexibility in 3D interconnect wiring and layout for both digital and RF/microwave circuitry. Then, the formation, i.e. the drilling, of the via, is one of the critical process steps, for the success in the TSV micro-fabrication. Among the technical options, the DRIE (Deep Reactive Ion Etching) is considered as the ideal one for the batch drilling of deep vias into Si substrates[2].

The tapering in the DRIE via drilling can be effective in realizing a smooth and relatively conformal deposition of Si oxide, and uninterrupted barrier/seed layers for a successful bottom-up via electroplating afterwards. As almost all the DRIE fabrication recipes are developed for vertical sidewalls for now, it can be tedious and time-killing to experimentally explore the optimal condition for fabricating vias with tapered section. Furthermore, there are barely any simulators publicly available for the design and verification of the DRIE via drilling.

The authors thus used an in-house developed process simulator for DRIE to explore conditions for the drilling of tapered TSV. And initial results obtained with experimental micro-fabrication, have not only demonstrated the effectiveness of the DRIE simulators, but also provide a practical way to drill a tapered via.

Principle and Simulation

Si DRIE etching is capable of batch fabrication, as compared with laser drilling methodology. The most frequently used process can be considered as a alternation between the passivation of the sidewall, and the reactive ion etching (RIE) of the Si and passivation (protection) film at the via bottom (namely, the Bosch process [2]). Though developed for holes with highly vertical sidewalls and high aspect ratio, e.g. for the comb drives with large electrostatical force, the process may drill out sections with different shape by combining various process conditions. However, it is quite a tedious task to experimentally explore the adequate condition for drilling TSV with the tapered section as desired.

The authors thus utilize an in-house developed simulator （Fig. 1） to explore various process parameters for drilling tapered vias, which has become a component in a commercial design software toolkit for microsystem design and simulation, i.e. part of the "RECIPE " module of the Intellisuite[TM] from IntelliSense company (www.intellisense. com).

The development of this simulator on the feature scale, started with a detailed analysis of the dynamics and distribution of the plasmas constituents involved in the ICP (Inductively coupled plasma) Bosch DRIE process, and its effects on the surfacial reaction on the Si wafers. Then possible mechanisms on the plasma-substrate interface in etching and deposition (passivation), were theoretically and experimentally investigated to provide .the necessary groundwork for the setup of ICP model. In the model, the Bosch process was rebuilt on physical level; the etching and passivation mechanism was confirmed by theoretic analysis and categorization, and was further verified with specially designed experiments. On the basis of the results obtained, a 2D string-cell hybrid graphic model, along with the related data structure, was developed. The algorithms based on hybrid string and cell evolution mechanisms and shadow detection algorithm were built up.

The self-unconsistance, void and fragment problems in the coupling of string and cell structures were solved. The evolution algorithm of the string model is improved and the validity is verified by simulation test. The detailed program flow (as shown in Fig.2) and the rules for the string and cell evolution as in the algorithms are explained in detail in Ref. [3, 4]. By modifying the string and cell structures, fragments and voids may be further eliminated. Currently, a three dimensional simulator with evolution, shadow detection and special angle calculation capability and with OpenGL 3D display modules are under development and will be issued soon.

To demonstrate the effectiveness of the simulation tool, the formation of the micro scallops on the sidewall and the process is simulated and the simulation result is compared with experimental result for validation, as in Fig 3. Then the Lag effect is adopted as a testbed and again the effectiveness is verified.

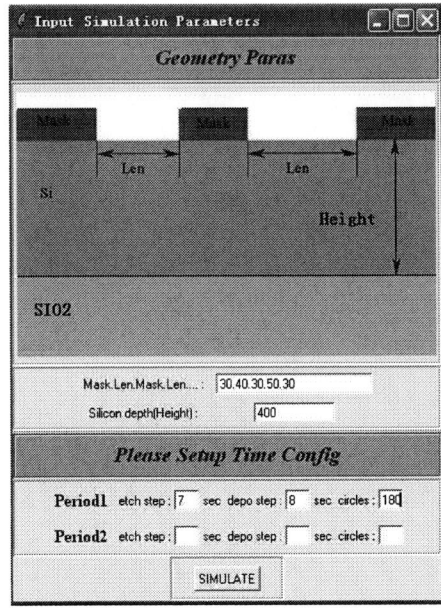

(a)

(b)

Figure. 1 Main and parameter input interface (a), and initial geometry definition interface of the Intellisuite RECIPE module (b).

Driven by the demand from TSV micro-fabrication process, the tunability of the sectional profile of the sidewall

was simulated by the authors, during which vertical shape, U-shape and V-shape are included. The Fig.3 displays an example of the notching at the via opening and the scalloping on the sidewall, which is frequently found in DRIE [2] and may endanger the integrity and continuity of isolation/ barrier/seed layers for TSV, since the process used for the deposition, such as PECVD and sputtering, are not capable of adequately conformal deposition. A ratio of 10:1 can often be found between the thickness of the barrier/seed layers on the opening and that of the layers at the via bottom, which may deteriorate the electrical distribution uniformity in the vias immerged in the copper electroplating solutions, for the via filling. To ensure the full coverage of the barrier/ seed layer on the sidewall and the bottom, thicker layers than reported for the multilayered Cu interconnect process for ICs [5], e.g. up to 1 micron, may be required.

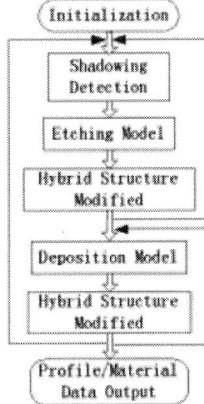

Figure 2. A brief simulation flow [3, 4].

To circumvent the issues mentioned above, starting from a recipe simulation for standard vertical sidewall vias and trenches (the process parameters is shown in Fig. 1, and the deposition/ etching time is 5s/6s), the authors explore various possible combinations of conditions, and thus by enlarging the deposition/etching time ratio, obtain an optimal one for vias with a maximum tapered angle of 12° (Fig.4, complement angle of θ) and a minimum sidewalls smoothness (scallops) of less than 2 μm The deposition/ etching time in a cycle is taken as 8s/7s. The simulated results are displayed in Fig.4 and the distribution factors for simulation are listed in Table. 1.

Micro-fabrication Verification and Results

The fabrication of the vias with vertical sidewall has been firstly used to partially verified the effectiveness of the module (Fig. 5 and Fig.6). The etching/deposition times in the cycle are set to be 5s/6s. The simulator has been shown to be able to reveal the etching reaction dynamics of ICP-based Si DRIE.

Vias with tapered sectional profile is fabricated (drilled), and the results and final electroplated samples are demonstrated in Fig. 7, which validate the effectiveness of the simulator and the anticipation in the DRIE conditions selection. The rounded tapering may be due to the ion reflection at the bottom of the via. SEM images obtained before the electroplating proves the comparatively conformal

257

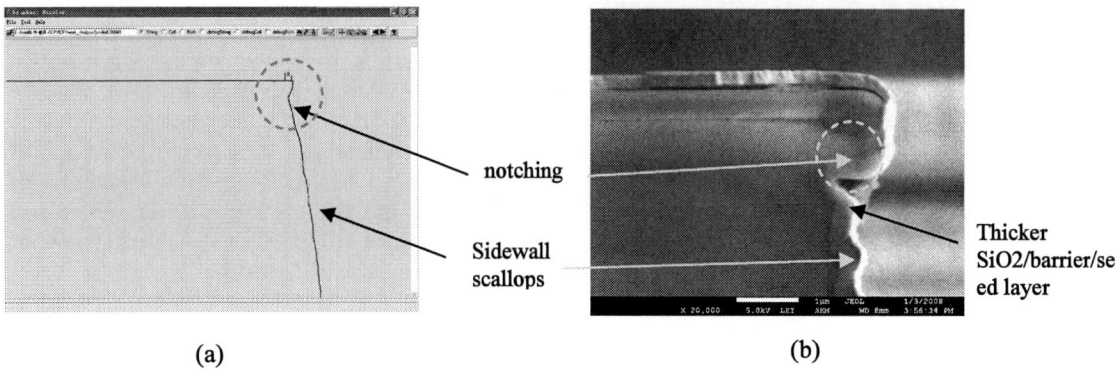

(a)   (b)

Figure. 3 The notching (undercut at the opening of the via) : (a) simulated; (b) fabricated, with a thicker and continuous barrier/seed layer coated.

Table1   The parameter of a physical model for the simulation of the etching of Vias with vertical sidewalls

Si （μm/s）		Polymer （μm/s）			Ion distribution factor
C_{uni}	C_i	C_{uni}	C_i	C_d	σ
0.023	0.04	0	0.06	0.0044	0.02

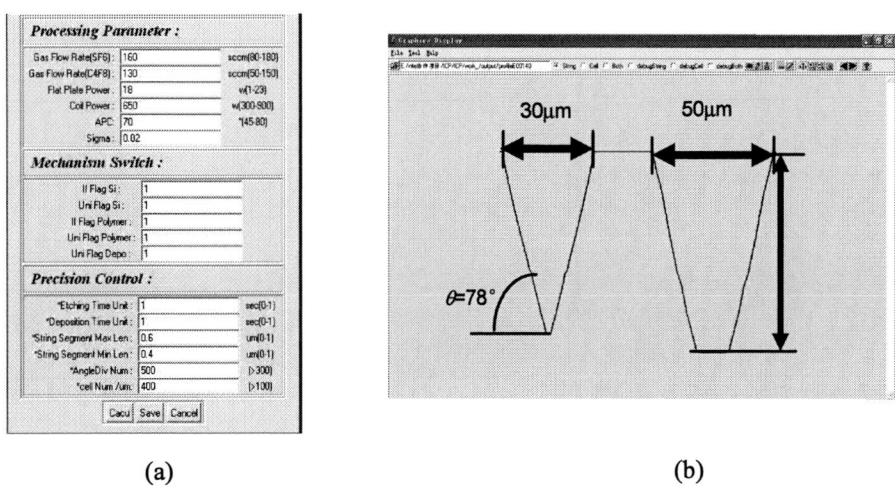

(a)   (b)

Figure. 4   Simulation for the tapered via: (a) Simulation input; (b) simulation results.

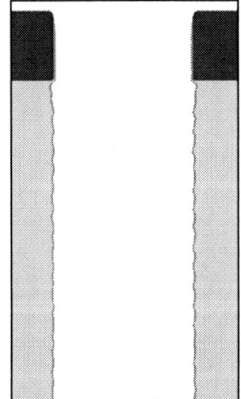

Figure.5   The simulation of a via with vertical sidewalls based on the parameters listed in Figure.1 and Table 1.

deposition of barrier/seed layer both on the sidewall and the bottom.

In all, the DRIE can benefit the TSV process development, in the following aspects: to give the initial process parameters for tapered via etching, to shorten the development cycle, to lower the associated costs, and to reveal the dynamics and potential problems of the process.

Conclusions

Considered as a key enabling technology for three-dimensional heterogeneous and homogeneous integration with high density in a package, TSV has been gaining wider attentions. Among the technological options, Si DRIE is widely considered as the most promising one due to its batch fabrication capabilities and compatibility with Si-based IC and micromachining processes. The drilling methodology for tapered TSV sections can be essential for an adequately conformal coverage of barrier/seed layer. However, the experimental exploration for an optimal process condition for DRIE drilling may be too tedious to implement.

The authors thus explore the possibility to simulate effectively the process conditions with an in-house developed

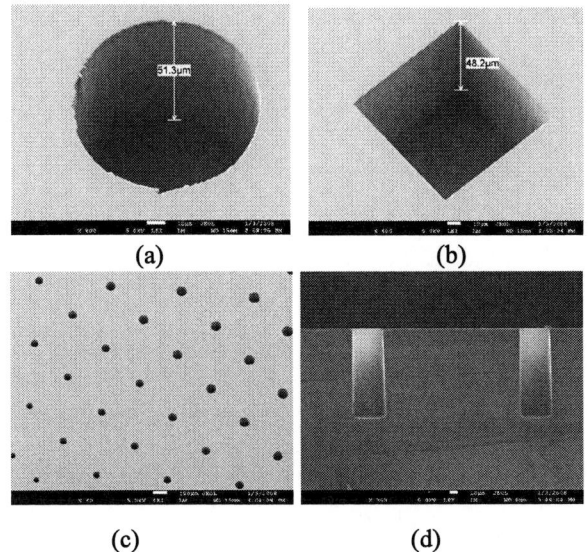

(a)   (b)

(c)   (d)

Figure.6 Fabricated results for verifying the effectiveness of the simulator:(a) a rounded via; (b) a square via; (c) rounded via arrays; (d) vertical shape of the via sidewall, partially demonstrating the effectiveness of the simulation.

Figure. 7 Fabricated TSV with tapered sectional profile and copper electroplated.

simulator utilizing hybrid line and cell evolving algorithms. The micro-fabrication has demonstrated TSVs with tapered sections and filled with electroplated coppers, which validates the effectiveness of the simulated results. Therefore, the DRIE tool can be of great help in tuning the sectional profile for satisfying TSV formation. Currently, the authors are further exploring the optimal conditions for optimal section profiles with the tool discussed above.

Acknowledgments

The work has been funded by CAST foundations (Project No. CAST200705).

References

1. Pozder S.; Chatterjee R., Jain A., et al, "Progress of 3D Integration Technologies and 3D Interconnects", IEEE *International Interconnect Technology Conference*, Volume A , June 2007, pp. 213 – 215
2. Robert Bosch Gmbh, "Challenges, developments and applications of silicon deep reactive ion etching", *Microelectronic Engineering*, Vol: 67 – 68 (2003), pp. 349 – 355.
3. Zhou R., "The modelling, simulation and experimental verification of high aspect ratio silicon etching process techniques", Ph.D dissertation, Peking University, China, 2005.
4. Zhou R., Zhang H., Hao Y. and Wang Y., "The simulation of the Bosch process with string-cell hybrid Method", *Journal of Micromechanics and Microengineering*, volume 14, issue 7 (2004), pages 851 - 858.
5. Ochi S.; Echigo F., "A study of Advanced ALIVH interconnection Technology". *8 th IEEE international symposium on advanced packaging materials*, 2002, pp. 356- 361.
6. Puech M., Thevenoud JM, Gruffat JM., Launay N., et al, "Cost Effective DRIE for Mass Production of MEMS and 3D Interconnections", *Equipment for Electronic Products Manufacturing*, Vol. 144, No.6, (2006), pp.37-44.

Simulation Study on the Warpage Behavior and Board-level Temperature Cycling Reliability of PoP Potentially for High-speed Memory Packaging

Wei Sun, W.H. Zhu, Kriangsak Sae Le and H.B. Tan
United Test & Assembly Center Ltd (UTAC)
Packaging Analysis & Design Center
5 Serangoon North Ave 5, Singapore, 554916
Email: Sun_Wei@sg.utacgroup.com Tel: +65-65511345

Abstract

PoP is a potential solution to high-speed memory packaging. For PoP package, warpage is known as a concern over package stacking and SMT yield [1]. The PoP package under current study has these features such as fine pitch which is 0.5mm for both top and bottom, small ball size and that most solder balls are located at the package's two longer edges. Therefore the solder joint reliability (SJR) in Temperature Cycling on Board (TCoB) test may also pose a concern..

The current paper talks about the systematic simulation and optimization of warpage and TCoB SJR for DRAM PoP package. For warpage study, 3D finite element analysis (FEA) was performed. Not only room temperature warpage, but also reflow temperature warpage was investigated. Full factorial DOE analysis with approximation model determination was conducted for both material selection and structural optimization. Based on this study, material selection and layout design guidelines were quickly derived to optimize the warpage performance of this package. In SJR simulation study, various package and stacking configurations were proposed and simulated in an effort to improve the SJR in TCoB test. Suggestions for improvements were made based on those simulation results.

Key words: FEA, warpage, wCSP, PoP, memory packaging, DOE, board-level temperature cycling test, solder joint reliability

1. Introduction

There are growing demands for high-density, high-speed and small form-factor memory packages. High-speed DRAM, an obvious example, transitioned from TSOP to FBGA and wCSP (or wBGA) in the past few year. Much work is still needed to increase the density and speed of DRAM package in a cost-effective manner, both at the die level and package level. At the package level, stacking more than one chips vertically, by wirebonding currently, is a natural way. Usually DRAM die has most of the bond pads located at centerline of active surface. Therefore, in a stack-die DRAM package, either the bond pads are routed through Redistribution Layer (RDL) to the edges of the die (thus wirebonded to the substrate) or long loop wires are used to connect the center located bond pads to the substrate. Figure 1 illustrates these two forms of typical stack-die DRAM packaging.

It is observed that both packaging forms have long electrical paths between bond pads and substrate. Therefore, their electrical performance is expected to be compromised. A summary of the pros and cons of these two forms of stack-die DRAM packaging is given in Table 1.

PoP, known for its good flexibility and testability, also presents a potential solution to high-density and high-speed DRAM packaging. A schematic cross-section picture of such PoP is shown in Figure 2. Unlike the normal PoP package where all the four sides of package are populated with solder balls, the DRAM wCSP PoP only has solder balls at the longer two sides as shown in Figure 3, mainly to shorten the conductive path between top package and PCB.

Three basic wCSP PoP stacking configurations were proposed with the company as test vehicles for high-speed DRAM packaging as shown in Figure 4. In Figure 4 (a), top wCSP is joined to the pre-deposited solder balls on bottom wCSP. Figure 5 shows the side views on the physical sample of this Pre-solder configuration. Due to the stoppage of the mold cap of bottom wCSP, the solder joints between top and bottom exhibit slender profile as shown in the cross-section picture in Figure 6. Castellation configuration in Figure 4 (b) is similar to Pre-solder. However, mold cap in Castellation allows the use of panel molding. The Castellation design also works as a test vehicle to study the process feasibility so that future PoP bottom package can adopt this design to avoid direct gate molding. Pre-bump configuration in Figure 4 (c) adopts substrate with pre-bumped copper studs as shown in Figure 7 (a). As such, the packaging process can also use panel molding. Figure 7 (b) illustrates the copper studs after panel molding. It should be noted that the solder joints between top and bottom packages in Pre-bump configuration are no longer of slender profile as no pre-deposited solder balls are needed anymore.

For PoP package, warpage is known as a concern over package stacking and SMT yield [1]. Furthermore, because of these features of wCSP PoP configurations such as fine pitch, small ball and that most solder balls are located at the package two longer edges, SJR under TCoB test could also become a concern.

The current paper focuses on the simulation and optimization of warpage and TCoB SJR for wCSP PoP packaging of DRAM chips. For warpage study, 3D FEA was performed. Not only room temperature warpage, but also reflow temperature warpage was investigated. Simulation-based full factorial DOE analysis with approximation model determination was conducted for both material selection and structural optimization at the single wCSP package level. Based on this study, material selection and layout design guidelines were quickly derived to optimize the warpage performance of this package. In solder joint fatigue simulation for TCoB performance evaluation, apart from the three basic

978-1-4244-2739-0/08/$25.00 ©2008 IEEE

configurations other stacking configurations were proposed based on design and process feasibility and simulated in an effort to find out the significant factors that impact TCoB reliability and to provide guidelines on design and material selection for TCoB reliability improvement.

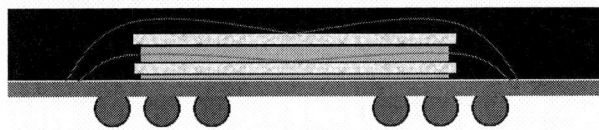

a: Long wire version

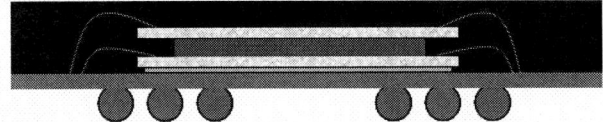

b: RDL version

Figure 1: Two forms of two-die stacking in DRAM packaging

Table 1: Summary of pros and cons of two formats of stack-die DRAM packaging

Long wire	Pros	1. Lower cost because no RDL process is need in the front end.
	Cons	1. Compromised electrical performance. 2. Assembly difficulty such as wiresweep and paste on wire or film on wire 3. Failure in one die may cause entire package scrapped.
RDL	Pros	1. Relatively easy assembly process.
	Cons	1. Compromised electrical performance. 2. Wafer cost is high.

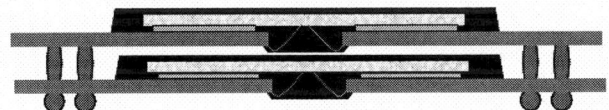

Figure 2: Cross-section picture of wCSP PoP

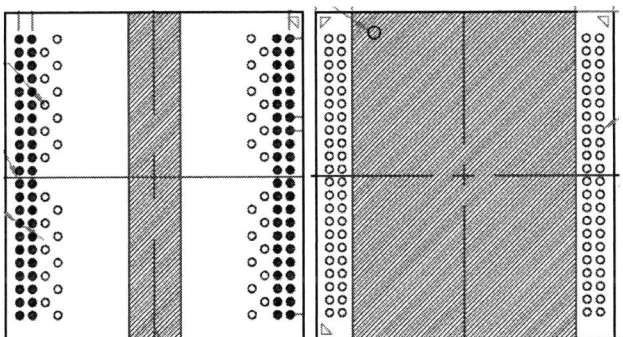

Figure 3: Planar view of wCSP ballout for PoP

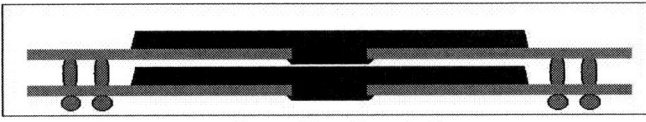

a: Pre-solder (Leg 1-1)

b: Castellation (Leg 2-1)

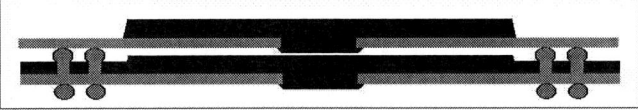

c: Pre-bump (Leg 3-1)

Figure 4: Three basic configurations proposed for wCSP PoP

a: View on the shorter side

b: View on the longer side

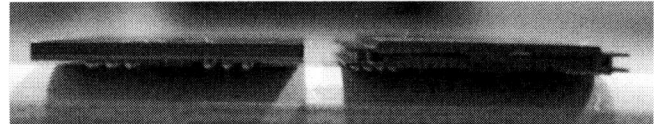

c: Side view of a normal wCSP and a wCSP PoP

Figure 5: wCSP PoP engineering sample

Figure 6: Cross-section picture of solder joint between top and bottom packages in pre-solder configuration of wCSP PoP

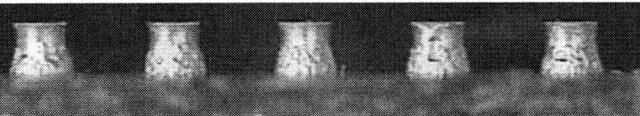

a: copper studs on bare substrate

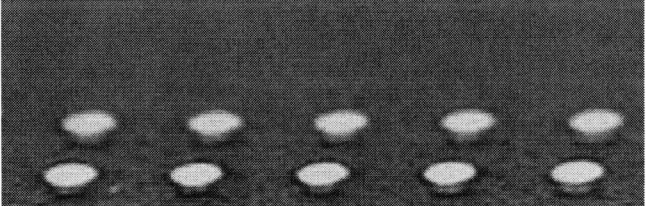

b: copper studs after panel molding

Figure 7: Pictures of physical Pre-bump wCSP PoP

2. Simulation-based DOE Analysis of Warpage

Warpage is not only a concern for automation of ball mounting and singulation processes, it is also a factor of yield loss in package stacking and SMT of PoP [1]. Validated FEA simulation, with the use of DOE methodology, provides a systematic approach of analyzing the impact of a series of design and material parameters on output responses such as warpage [2-14]. The combination of ANSYS/Mechanical, EXCEL and JMP proves a powerful and low-cost tool for simulation-based DOE analysis [14]. With the help of simulation based DOE analysis, design and material selection guidelines can be quickly obtained. An approximation model can be also be efficiently derived to replace further simulation efforts.

In this study, the approach in [14] was adopted to investigate the impact of design and material on the warpage of single package as shown in Figure 8. It is obviously seen from Figure 8 that the warpage is significant because of the un-optimized design/material and large package size.

Figure 8: Single package before PoP stacking

2.1 Effect of Material Properties on Warpage

The purpose of this study is to provide guidelines for material engineers to screen the material candidates for further evaluation. The current analyses used the package dimensional information highlighted in bold in Table 2. Temperature cooling down from 175°C to 25°C and ramping up from 175°C to 260°C with stress-free state assumed at 175°C were simulated. The average CTE for simulation input was calculated using Equation 1 when Tg falls in the simulated temperature range. Figure 9 illustrates how the warpage mode and value are defined in this paper.

Simulation was firstly performed to confirm the correlation between simulation results and actual measurement on feasibility build. It is shown in Figure 10 that good agreement was achieved between simulation and measurement. The analysis was then moved on to simulation-based DOE study. A 9-parameter and 2-level material DOE matrix was defined in Table 3 and 4 for room and reflow temperature warpage study respectively. A full factorial DOE approach was used.

Table 2: Nominal dimensions of the wCSP PoP under study

wCSP for PoP	Dimension (mm)	
	Size	Thickness
Die	11.94x7.68 **19.49x7.68**	0.110
Die Attach	Calculated	0.040
Bond Window	12.54x0.9 **20.09x0.9**	N.A.
EMC	12.54x8.68	0.350
Solder Mask	12.54x11.58 **20.09x11.58**	0.025
Core		0.060
Cu		0.012

$$CTE_{ave} = \frac{\alpha_1(T_g - T_{final}) + \alpha_2(T_{ref} - T_g)}{T_{ref} - T_{final}} \quad \text{(Equation 1)}$$

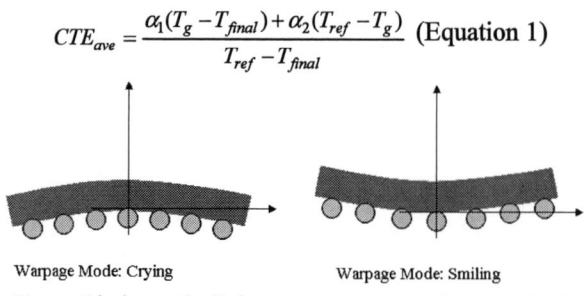

Warpage Mode: Crying

Warpage Value has negative (-) sign

Warpage Mode: Smiling

Warpage Value has positive (+) sign

Figure 9: Definition of warpage mode and value

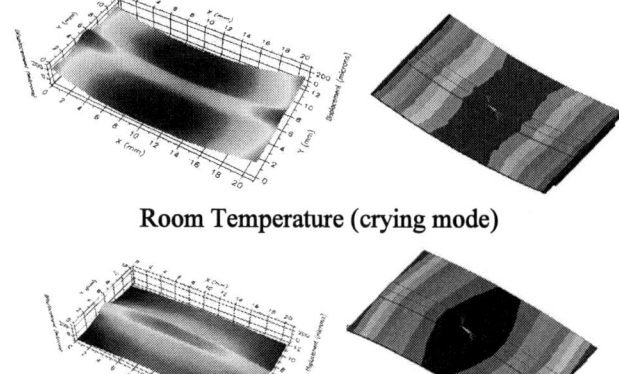

Room Temperature (crying mode)

Reflow Temperature (smiling mode)

Figure 10: Agreement between simulation and measurement on samples from feasibility build

Table 3: Material DOE matrix (room temperature)

Material	Room Temp Properties			
EMC	E (MPa)	14000	CTE (ppm/K)	15
		32000		35
Die Attach	E (MPa)	500	CTE (ppm/K)	150
		2500		350
Solder Mask	E (MPa)	2000	CTE (ppm/K)	80
		4000		120
Core	E (MPa)	20000	CTE (ppm/K)	XY: 12/Z: 20
		30000		XY: 16/Z: 60

Table 4: Material DOE matrix (reflow temperature)

Material	Reflow Temp Properties			
EMC	E (MPa)	300	CTE (ppm/K)	30
		2000		60
Die Attach	E (MPa)	1	CTE (ppm/K)	150
		10		350
Solder Mask	E (MPa)	100	CTE (ppm/K)	100
		500		150
Core	E (MPa)	7000	CTE (ppm/K)	XY: 6/Z: 100
		15000		XY: 12/Z: 200

Figure 11 shows that good fit achieved between simulation results and the approximation model by regression analysis of DOE results. Figure 12 gives the Pareto plots showing the effect of material properties on warpage at both room and reflow temperature. The current selectable materials include EMC and substrate core materials as listed in Table 5 and 6 respectively. It is clearly seen from the Figure 12 that higher average CTE and larger modulus EMC are desirable and have large impact on room temperature warpage. CTE of substrate core also plays some role. Low CTE substrate core material is desirable at room temperature. The materials properties listed in Table 5 and Table 6 were plugged into the DOE derived approximation model. Warpage of each set of material combination was quickly calculated. Figure 13 (a) and (b) show the normalized warpage performance at room and reflow temperature respectively. It is noticed that same as the feasibility build the simulated warpage of current material selection has crying mode at room temperature and crying mode at reflow temperature. It is clear seen from Figure 13 (b) that the selection of material has insignificant impact for reflow temperature warpage. However, at room temperature EMC E, F and low CTE core substrate are highly desirable as illustrated by Pareto plots in Figure 12. Due to the high cost of EMC_F, eventually material combination of EMC_E and low CTE core substrate were chosen for the next stage of study.

Table 5: EMC candidates

EMC	E@ 25 (MPa)	E@ 260 (MPa)	α1 E-6	α2 E-6	Tg (C)	Average CTE E-6
EMC_A	19600	300	14	55	135	24.9
EMC_B	23000	800	10	39	123	20.1
EMC_C	20500	300	10	41	110	23.4
EMC_D	25500	1000	8	31	125	15.7
EMC_E	21000	650	12	45	110	26.3
EMC_F	15000	250	18	62	120	34.1

Table 6: Substrate core materials candidates

Core	E@25 (MPa)	α1 E-6	α2 E-6	Tg (C)
Standard	27000	XY:15 Z: 32	XY: 11 Z: 142	185
Low CTE	29000	XY: 11.5 Z: 22	XY: 6.5 Z: 115	180

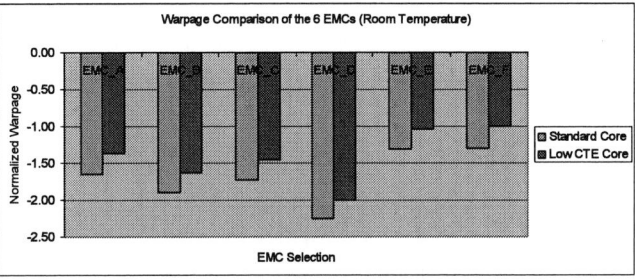

a: Warpage at room temperature

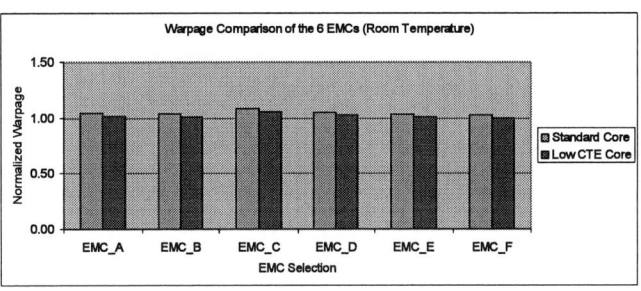

b: Warpge at reflow temperature

Figure 13: Warpage of the material candidates at room and reflow temperature

2.2 Effect of Structural Dimensions

The purpose of the structural DOE analysis is to study how the change of structural dimensions such as die and substrate core thickness impacts the warpage. The structural DOE matrix is listed in Table 6. Again a full factorial DOE was performed and good fit achieved between approximation model calculation and simulation results. It is shown in Figure 14 that thinner die, bond line thickness (BLT) and thicker mold cap (or larger mold coverage) can help to bend up the room temperature warpage that is in crying mode. It is also observed that although the core thickness does not play a significantly role in room temperature warpage, thicker core thickness can help to reduce the reflow temperature warpage which has smiling mode.

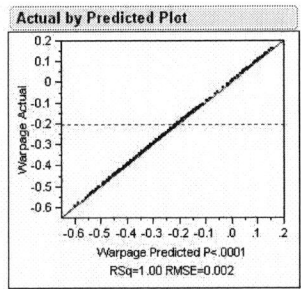
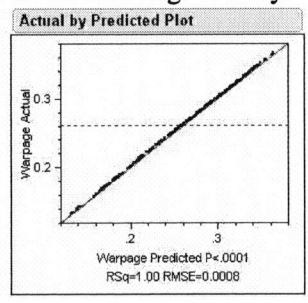

Figure 11: Simulated vs. Approximation (left: room temperature warpage, right: reflow temperature warpage)

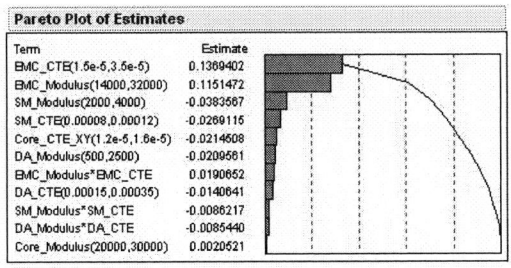

a: for room temperature warpage

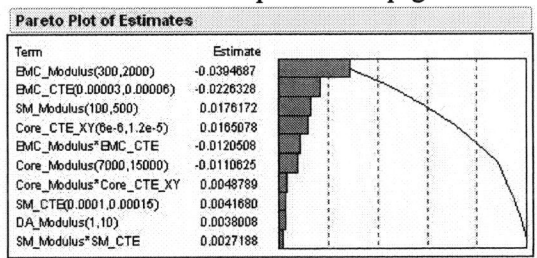

b: for reflow temperature warpage

Figure 12: Pareto plots for material optimization

Table 6: Structural DOE matrix under study

Geometry	Dimensions	
Mold Cap Thickness	Lower Bound	0.25mm
	Higher Bound	0.28mm
Die Thickness	Lower Bound	75μm
	Higher Bound	110μm
BLT (Bond Line Thickness)	Lower Bound	30μm
	Higher Bound	60μm
Substrate Core Thickness	Lower Bound	60μm
	Higher Bound	110μm

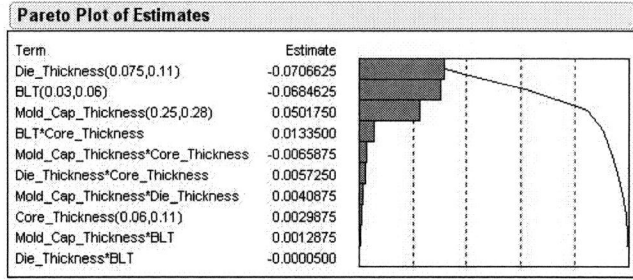

a: for room temperature warpage

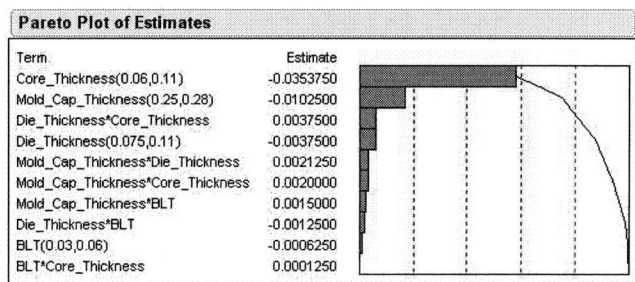

b: for reflow temperature warpage

Figure 14: Pareto plot for structural optimization study

2.3 Summary and Conclusions

In this study, both material and structural DOE studies were performed. Given by material engineers the available material list, it was found that high average-CTE and high modulus EMC are highly desirable for warpage at both room and reflow temperature. Low CTE substrate core material can also help to reduce the warpage at both room and reflow temperature. Several structural changes, namely increase in mold cap thickness (or increase the mold coverage) and reduction in die thickness and BLT, all lead to increase of mold compound material and help to reduce the room temperature warpage that is very important when product transfers to mass production mode. Based on DOE analysis, approximation models based on regression analysis were derived and ready for material and design engineers to use. Those approximations models can replace further simulation analyses once new material or new design need to be evaluated.

3. TCoB Reliability Evaluations and Optimization

For SJR prediction in TCoB test, it was found in [15] that Schubert's hyperbolic sine constitutive model plus his creep strain energy density based fatigue correlation model gives the best overall prediction accuracy for typical lead-free solders like SAC305/405. The current analysis used this approach for the solder joint fatigue simulation and reliability prediction.

The SAC305/405 solder constitutive equation and fatigue model are shown in Table 7 and 8 respectively. Other material properties involved and dimensional information is listed in Table 2 and 9. The current analyses used a smaller package size due to computational efficiency. The materials used did not adopt the warpage-optimized material selection either. However, the trend derived from current study should apply to the larger package size and optimized material selection in previous section. The FEA used 3D quarter symmetric models typically as shown in Figure 15. The TCoB testing condition for simulation input is -40°C to 125°C with 15mins dwell/ramp on 1.1mm PCB board that is typically used in DIMM module.

Table 7: Constitutive equations for SAC305/405

Solder composition	Constitutive equation
Sn3.8Ag0.7Cu Sn3.5Ag0.75Cu Sn3.5Ag0.5Cu (Schubert et al. [16])	$\dot{\varepsilon} = 277984[\sinh(0.02447\sigma)]^{6.41} \times \exp\left(\dfrac{-6500}{T\left(^oK\right)}\right)$
E (MPa)	61251-58.5T (degree K) (Schubert et al. [16])
υ	0.36
CTE (ppm/K)	20.0

Table 8: Schubert's energy-based fatigue correlation model for his hyperbolic sine constitutive equation [16]

Creep Energy Density	$N_{characteristic} = 345 w_{acc}^{(-1.02)}$

Table 9: Material properties used in simulation

	Material Properties for Simulation			
	E@25C (MPa)	α1 E-6	α2 E-6	Tg (C)
Silicon	131000	2.6	-	-
Die Attach	1890@-65C 658@25C 16@100C 5.3@175C	70	350	40
EMC	23000 **20000**	9 **14**	38 **55**	133 **135**
Solder Mask	2400	60	130	100
Core	28500	XY: 14 Z: 30	XY: 7 Z: 180	185
Cu	117000	17.3	-	
PCB	23000	XY: 17 Z: 80		

264

Figure 15: FE mesh for the Pre-bump configuration

3.1 Simulation of Basic Configurations

The SJR evaluation is firstly done for the three basic stacking configurations as shown in Figure 4. The results are plotted in Figure 16. Note that in the current and subsequent discussions all the predicted characteristic lives are normalized against bottom package of Leg 1-1 Pre-solder whose characteristic life is 820 cycles predicted by simulation.

It is shown in Figure 16 that Pre-solder design gives the best TCoB reliability for both top and bottom packages. We also notice that in Pre-solder and Castellation designs, solder joint reliability of top package is much better than that of bottom. The better solder joint reliability of top package probably attributes to its slender and compliant solder joint profile which can better withstand the cyclic loading due to CTE mismatch. In Castellation design, due to the extra mold coverage on the two edges, has less solder joint reliability for both top and bottom packages compared with the similar Pre-solder design. In Pre-bump design, it seems that the existence of copper studs and extra mold materials has significant negative impact on the SJR of bottom package. However, for top package, it is expected to have similar solder joint reliability as bottom package because both have similar solder joint shape.

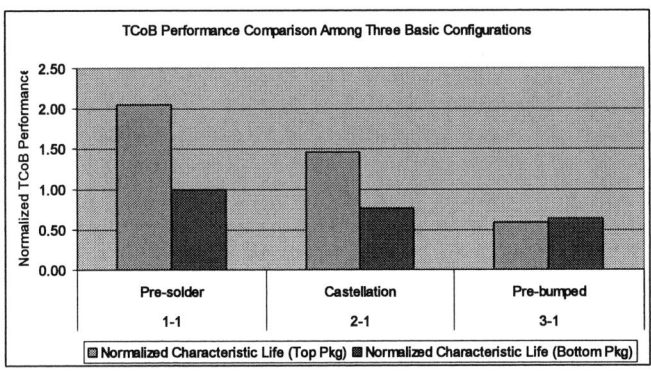

Figure 16: TCoB performance comparison among the three basic stacking configurations

3.2 Effect of Alternative EMC (Epoxy Molding Compound)

TCoB test is typically conducted under temperature range below the Tg of EMC. Therefore α1 CTE of EMC plays an important role in package's SJR as it effectively influences the CTE mismatch between package and PCB. Below Figure 17 shows the effect of using alternative EMC on solder joint reliability. The alternative EMC, highlighted in bold in Table

9, has α1 CTE of 14ppm/k. It is clearly seen from Figure 17 that by using a EMC whose α1 is closer to that of substrate and PCB the solder joint reliability can be significantly enhanced. It must be noted that the preference of higher α1 CTE EMC in TCoB SJR does not conflict with the high average CTE EMC requirement in warpage optimization. An EMC with both high α1 and average CTE is achievable. It is also interesting to notice that use of alternative EMC has different level of impact on SJR. A highlight is that the alternative EMC can enhance the top package's solder joint reliability of Castellation by almost 100%, making it even better than that of Pre-solder.

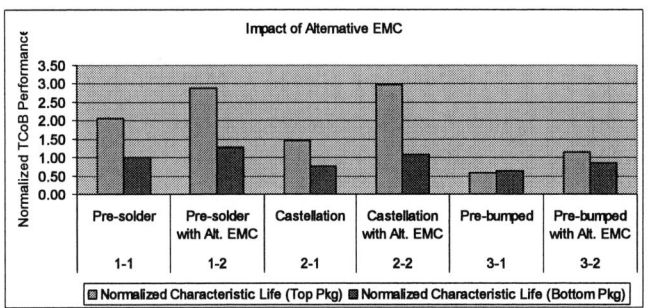

Figure 17: Effect of α1 CTE of EMC on solder joint reliability

3.3 Effect of Mold Coverage on Top Package

It could be more economical to adopt a top package which is fully covered by EMC as shown in Figure 18 and 19. Simulation results in Figure 20 demonstrates that this adoption can significantly improve the SJR of both top and bottom packages. However, it is hard to explain explicitly the reason of this phenomenon as the CTE mismatch among those different materials of the package is hard to judge before a simulation is really done.

Figure 18: Leg 1-6

Figure 19: Leg 2-4

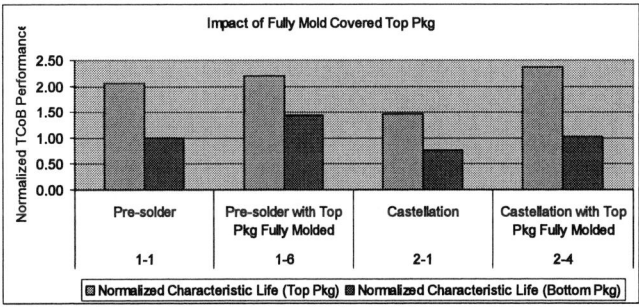

Figure 20: Effect of mold coverage on top package

3.4 Effect of Applying Dummy Solder Balls

Earlier simulation results show that SJR of bottom package, which interfaces with PCB, is critical in determining the overall reliability performance. However, the small number of solder joints and small solder joint size make the SJR performance not so satisfactory. Application of dummy solder joints as shown in Figure 3 and 21 was proposed to reinforce the bottom package's SJR. Figure 22 shows that the use of dummy solder balls can significantly enhance the reliability of bottom package and bring both packages to similar number of fatigue cycles.

Figure 21: Leg 1-7

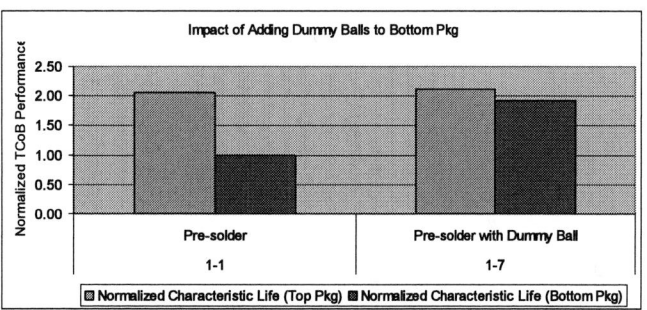

Figure 22: Effect of using dummy solder joints for PoP bottom package

3.5 Effect of Using JEDEC Ballout on Bottom Package

Because the requirement of high speed, the original ballout only puts the solder balls at the two longer edges of the package, which aims to shorten the conductive path between top package and PCB. However, the ballout is not a JEDEC compliant one and earlier study also shows that SJR is possibly a concern without applying dummy solder balls. The study in this part explores the SJR of moving back to JEDEC ballout as shown in Figure 23. It is shown in Figure 24 that JEDEC ballout, similarly effective as adding dummy solder balls, can also significantly increase the TCoB reliability. This is because JEDEC ballout uses 0.8mm pitch, larger solder ball size and larger solder mask opening.

Figure 23: Leg 1-8

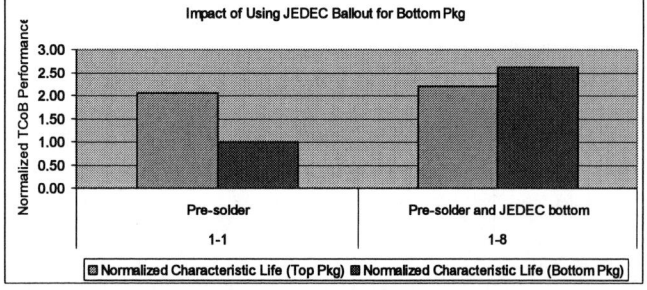

Figure 24: Impact of using JEDEC ballout for PoP bottom package

3.6 Effect of Package Size

The package size is also expected to play a role in affecting SJR. In this study, a larger PoP, whose dimensions were used in earlier warpage DOE study 2, was simulated and compared with nominal design. As shown in Figure 25, the larger package size will reduce the TCoB reliability, probably because of the larger die size used and thus increased CTE mismatch.

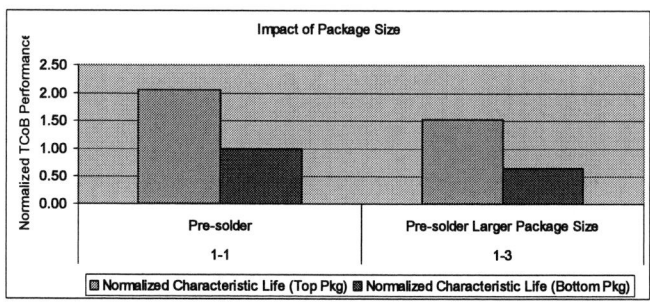

Figure 25: Effect of package size on solder joint reliability

3.7 Summary and Conclusions

The initial stacking configurations may see SJR concern in TCoB test based on simulation results. It was found out that a high α1 CTE EMC, increase the EMC coverage on top package, application of dummy solder ball on bottom package and use of JEDEC ballout and ball size on bottom package can help to enhance the overall solder joint reliability. Larger package size, together with larger die size, will tend to reduce the SJR compared with a smaller package.

4. Overall Summary and Conclusions

Thermo-mechanical simulation was performed in the early stage of wCSP PoP development to aid the design and material selection. Warpage simulation was performed in first stage as significant warpage would interrupt the automation of several processes and shed some concerns over the yield of package stacking and SMT. Simulation-based DOE analysis was performed for both material and structural optimization. Material selection and design guidelines were quickly obtained to mostly reduce the warpage at both room and reflow temperature. In TCoB simulation, it was found from early simulation that the SJR of the initial wCSP PoP package design might be unable to meet the reliability requirements. Several design and material selection proposals were simulated and evaluated. From the simulation results, development engineers can quickly identify the feasible solutions to overcome the weakness of initial design.

Acknowledgments

The authors would like to thank UTAC R&D management team for their support. The assembly work done by process group for test vehicles preparation is greatly appreciated.

References

1. Akito Yoshida et al, "A Study on Package Stacking Process for Package-on-Package (PoP)", Proceedings of ECTC2006, pp.825-830.
2. A. Mertol, "Application of the Taguchi Method on the Robust Design of Molded 225 Plastic Ball Grid Array

Packages", IEEE Trans. Comp. Packag., Manufact. Technol. B, vol. 18, pp. 734-743, Nov. 1995

3. A. Mertol, "Optimization of High Pin Count Cavity-up Enhanced Plastic Ball Grid Array (EPBGA) Packages for Robust Design", IEEE Trans. Comp. Packag., Manufact. Technol. B, vol. 20, pp. 376-388, Nov. 1997

4. A. Mertol, "Application of the Taguchi Method to Chip Scale Package (CSP) Design", IEEE Trans. Adv. Packag., vol. 23, 2000, pp. 266-276

5. A. Dasgupta, M. G. Pecht, B. Mathieu, "Design-of-experiment Methods for Computational Parametric Studies in Electronic Packaging, Finite Element in Analysis and Design", 30 (1998), pp.125-146

6. S. Stoyanov, C. Balley and M. Cross, "Optimization Modeling for Flip Chip Solder Joint Reliability, Soldering & Surface Mount Technology", 14/1, 2002, pp. 49-58

7. B. Vandevelde, E. Beyne, G. Q. Zhang, Jo F. J. M. Caers, D. Vandepitte and M. Baelmans, "Solder Parameter Sensitivity for CSP Life-Time Prediction Using Simulation-Based Optimization Method", IEEE trans. on Electronics Packag. Manufact., vol. 25, 4, October 2002, pp. 318-325

8. B. Vandevelde, E. Beyne, G. Q. Zhang, Jo Caers, D. Vandepitte, M. Baelmans, "Parameterized Modeling of Thermo-mechanical Reliability for CSP Assemblies", Journal of Electronic Packaging, vol. 125, December 2003, pp. 498-505

9. W. D. van Driel, G. Q. Zhang, J. H. J. Janssen, L. J. Ernst, "Reponse Surface Modeling for Nonlinear Packaging Stresses", Journal of Electronic Packaging, vol. 125, December 2003, pp. 490-497

10. W. D. van Driel, G. Q. Zhang, J. W. C. de Vries, M. Jansen, L. J. Ernst, "Virtual Prototyping and Qualification of Board Level Assembly", Proceedings of EPTC2003, December 2003, pp. 772-775

11. S. G. Jagarkal, M. M. Hossain, D. Agonafer, M. Lulu, S. Reh, "Design Optimization and Reliability of PWB Level Electronic Package", Proceedings of ITHERM'04, June, 2004 Las Vegas, USA, pp. 368-376

12. C. C. Lee, S. M. Chang, K. N. Chiang, "Design of Double Layer WLCSP Using DOE with Factorial Analysis Technology", Proceedings of EPTC2004, December 2004, pp. 776-781

13. A. A. Giunta, "Use of Data Sampling, Surrogate Models, and Numerical Optimization in Engineering Design", AIAA 2002-0538, American Institute of Aeronautics and Astronautics

14. Wei Sun et al, "Warpage Simulation and DOE Analysis with Application in Package-on-Package Development", Proceedings of EuroSimE2008, pp.244-251

15. Wei Sun et al, "Experimental and Numerical Assessment of Board-level Temperature Cycling Performance for PBGA, FBGA and CSP", Proceedings of EPTC2006, pp.121-126

16. A. Schubert et al., "Fatigue Life Models for SnAgCu and SnPb Solder Joints Evaluated by Experiments and Simulation", Proceedings of ECTC2003, pp.603-610

Miniaturization Design of Backside-Via Structures Underneath Collector-Up HBTs Using A 3-D Finite-Element Model

H. C. Tseng*, P. H. Lee, and J. H. Chou
*Nanotechnology R & D Center, Kun Shan University, Tainan 71003, Taiwan, China
Department of Engineering Science, National Cheng Kung University, Tainan 70101, Taiwan, China
e-mail:hctseng2@giga.net.tw

Abstract

To carry out the miniaturization design of backside-via packaging structures underneath collector-up HBTs, a 3-D finite-element model has been developed for analyzing temperature-distribution phenomena within the configurations. The results are demonstrated on the three-finger InGaP/GaAs collector-up HBT. Compared to previous reports, backside-via structures can be further reduced by 42% while maintaining the same heat-dissipation performance.

1. Introduction

High-power amplifiers (HPAs) for wireless communication systems have shown significant progress in recent years and the interests for HPA-based dual-band systems are growing rapidly [1, 2]. Miniaturizing HPAs is crucial for downsizing multi-band cellular phones. Due to the high-power density and high-linearity characteristics [3], the HBT is promising for HPA applications. The miniaturization of HPAs depends mainly on thermal management of HBTs. Conventional HPAs employing emitter-up HBTs suffer from large collector capacitance which limits RF performance. Besides, their thermal resistance is large because heat dissipates mostly through emitter wiring. Thus, special thermal designs, such as ballast resistors [4], thermal shunts [5], and wide finger pitches, have been required to improve thermal conduction. On the other hand, HPAs composed of collector-up HBTs, whose collector capacitance is one third that of emitter-up HBTs, exhibit better RF performance [6]. Furthermore, in the collector-up HBT with a backside-via structure, heat dissipates directly from the backside via, so that collector finger pitches can be minimized.

We present a 3-D finite-element model (FEM) to analyze thermal performance in collector-up HBTs and to conduct the miniaturization design of backside-via packaging structures underneath collector fingers. It is shown that numerical methods are most appropriate for developing heat-dissipation structures in HBTs [7, 8]. For this advanced packaging-technology analysis, the FEM methodology was used to calculate temperature distributions around the transistor by adjusting the thickness of plated heat sink (PHS) layer. Also the maximum operation temperature in the three-finger collector-up HBT with different finger pitches was examined. In this Letter, the results are demonstrated on the GaInP collector-up HBT, which is an attractive candidate for used in advanced HPAs because of its low-power supply voltages and high-power handling capability [9].

2. Finite-Element Model

A three-finger collector-up HBT with a backside-via packaging structure, which is right under HBT fingers, is depicted in Fig. 1. In the original configuration, the thicknesses of collector wiring, collector, base, emitter, PHS layer, and substrate are 1.3, 0.8, 0.1, 0.2, 12, and 30 μm, respectively. The collector area is 4.5 μm × 40 μm.

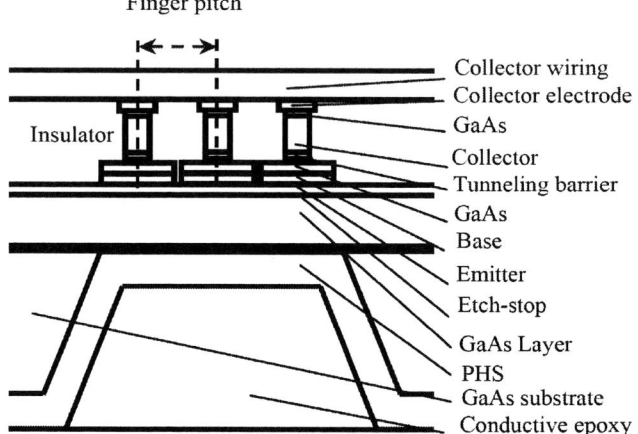

Fig. 1. Schematic diagram of the InGaP collector-up HBT with a backside-via packaging structure.

In this work, the thermal resistance is defined as the difference between the maximum temperature of the HBT and the bottom surface temperature, so we build up a 3-D thermal FEM, as shown in Fig. 2, to evaluate the junction temperature and temperature distributions. The materials and their properties used are as described in [10]. The main heat source is at the collector layer, and a steady-state heat transfer condition is assumed for this efficient simulation. The boundary conditions for the surfaces in contact with surrounding air at room temperature are as follows: convection at the bottom surface and adiabatic for the rest surfaces. Some of the chosen elements are specified by convections or heat fluxes as surface input loads at the element faces. The nodal temperatures are the solution output. Unstructured meshes are adopted instead of a free mesh so that regions with large temperature gradients can be treated by much smaller finite-element meshes.

A 2-D simulation can save lots of time and resources in most practical conditions. Usually, it is the desired mode of analysis if it works. However, a real electronic device is three dimensional. In building a 3-D model, the number of the element is always 100 times larger than that needed for a 2-D model. To reduce the model size and be more efficient, a quarter-symmetry model is established by taking advantage of symmetry of the device. Within the studied device, which is principally made of GaAs, the thermal resistance between interfaces of each layer is negligible, and the thermal as well as mechanical properties of each layer are treated as constant.

978-1-4244-2739-0/08/$25.00 ©2008 IEEE

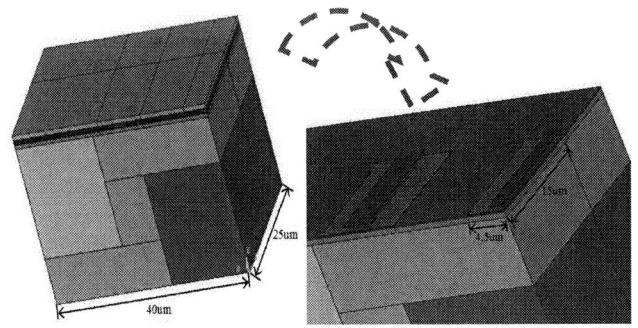

Fig. 2. A 3-D thermal model built up in this work.

3. Results and Discussions

A cross-sectional view of the model is illustrated in Fig. 3(a), and the temperature distribution of the collector-up HBT with 15 μm finger pitch is displayed. It is found that heat transfers from the collector layer, through the emitter, GaAs and PHS layers, to the isothermal bottom of the model. The simulation result of the 3-D model is presented in Fig. 3(b).

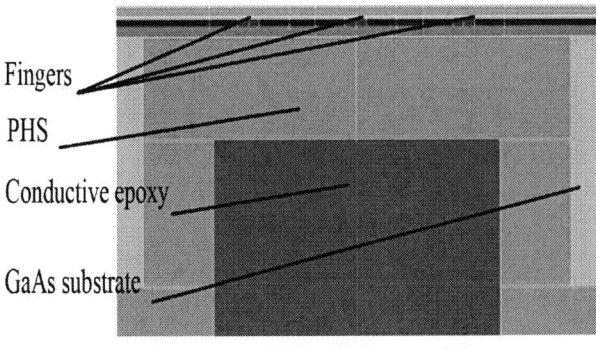

(a) Simulation model

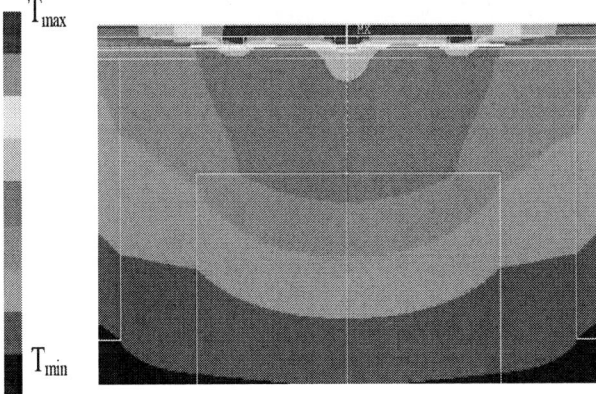

(b) Temperature distribution within the HBT

Fig. 3. Cross-sectional view of the three-finger transistor with 15 μm finger pitch.

From the temperature distribution demonstrated, it can be seen that the maximum temperature occurs at the collector layer where heat dissipation takes place, and most of the heat spreads from the PHS layer. It is worth mentioning that the thermal performance observed here is comparable to those reported in [9, 10], but with a much thinner configuration.

According to the 3-D FEM analysis, it is obvious that the PHS layer and the conductive epoxy constitute the major bulky portion of the backside-via structure. Considering the key advantage of using collector-up configuration is the

shrinkage of device dimensions, it is very important that the thickness of bulk part must be reduced without sacrificing the thermal performance. From this point of view, the thickness of the heat-dissipation structure underneath HBT fingers has to be thinned.

Contrary to the thicker backside-via structure [9], for which the maximum temperature was reduced by increasing the thickness of PHS layer, we reduce the maximum temperature by decreasing the thickness of PHS layer, as shown in Fig. 4, and the thickness of GaAs substrate is thinned to less than 20 μm. The 2-D analysis, which overestimates maximum temperatures within the device, is included in Fig. 4 for comparison. Noticeably, the overall thickness can be reduced by 33 and 42% for finger pitches of 10 and 18 μm, respectively.

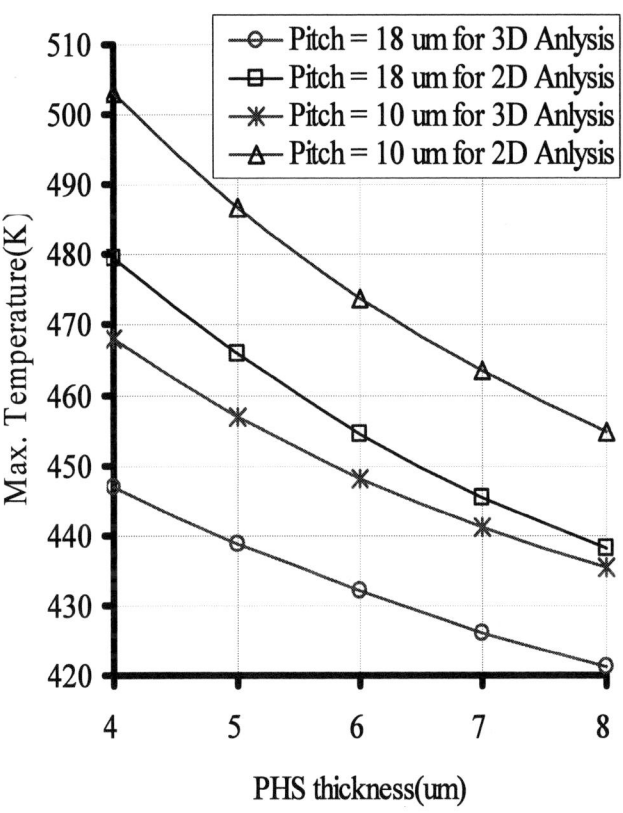

Fig. 4. Dependence of the maximum temperature on the thickness of PHS layer.

4. Conclusions

The temperature distribution of backside-via structures has been analyzed using the 3-D thermal FEM. Results on multifinger InGaP collector-up HBTs demonstrate that the thickness of backside-via packaging structures can be further reduced by 42%, and the achieved heat-dissipation performance will not be deteriorated. From this efficient analysis, it is believed that miniaturization of collector-up HBTs with backside-via structures should be a valuable aid in designing small-scale HPAs.

References

1. Chen, P., *et a.l*,"Application of GaInP/GaAs DHBT's to power amplifiers for wireless communications", *IEEE*

Trans. Microwave Theory Tech., Vol. 47 (1999), pp. 1433-1438.

2. Megahed, M., Egomepe, E., Ayvazian, M., and Glasbrener, M., "Design considerations for single package radio TM (SPR) solution for EGSM/DCS dual band cellular phones", *IEEE MTT-S Int. Microwave Symp. Dig.*, 2003, pp. 1707-1710.

3. Liu, W., Kim, T., Ikalainen, P., and Khatibzadeh, A.,"High linearity power X-band GaInP/GaAs heterojunction bipolar transistor", *IEEE Electron Device Lett.*, Vol. 15 (1994), pp. 191-192.

4. Liu, W., Khatibzadeh, A., Sweder, J., and Chau, H.F., "The use of base ballasting to prevent the collapse of current gain in AlGaAs/GaAs heterojunction bipolar transistor", *IEEE Trans. Electron Devices*, Vol. 43 (1996), pp. 245-251.

5. Fraysse, J.P., Vendier, O., Soulard, M., and Auxemery, P., "2W Ku-band coplanar MMIC HPA using HBT for flip-chip assembly", *IEEE MTT-S Int. Microwave Symp. Dig.*, 2002, pp. 441-444.

6. Kroemer, H., "Heterostructure bipolar transistors and integrated circuits", *Proc. IEEE*, Vol. 70 (1982), pp. 13-25.

7. Gao, G.B., Wang, M.Z., Gui, X., and Morkoc, H., "Thermal design studies of high-power heterojunction bipolar transistors", *IEEE Trans. Electron Devices*, Vol. 36 (1989), pp. 854-862.

8. Kim, C.W., *et al.*, "3-V operation power HBTs for digital cellular phones", *IEICE Trans. Electron.*, Vol. E79-C (1996), pp. 617-621.

9. Osone, Y., Mochizuki, K., and Tanaka, K., "Thermal performance of collector-up HBTs for small high-power amplifiers with a novel thermal via structure underneath the HBT fingers", *IEEE Trans. Compon. Packag. Technol.*, Vol. 28 (2005), pp.34-38.

10. Tseng, H.C., Lee, P.H., and Chou, J.H., "Thermal performance analysis for packaging configuration design of GaInP collector-up HBTs as small high-power amplifiers", *Proc. IEEE Int. Conf. Electron. Packag. Technol. (ICEPT'07)*, 2007, pp. c-28.

Modeling Ion Transport through Molding Compounds and its Relation to Product Reliability

M. van Soestbergen,[1,*] R.T.H. Rongen,[2] L.J. Ernst[3] and G.Q. Zhang[3,4]

[1]Materials Innovation Institute, [2]NXP Semiconductors, Quality and Analytical Services, [3]Delft University of Technology, fundamentals of Microsystems Engineering, [4]NXP Semiconductors, Strategy and Business Development, * Mekelweg 2, 2628 CD Delft, the Netherlands, m.vansoestbergen@tudelft.nl

Abstract

Nowadays highly filled epoxy molding compounds are used as material for encapsulation of microelectronic devices. These molding compounds always contain a very low concentration of ionic impurity. In addition ionic species can originate from chemical processes inside the encapsulation. In the presence of an electrical field ions will migrate through the encapsulation, which might eventually result in failures, such as corrosion or dendrite growth. Although these failures are well-known they still lack a knowledge based description of their failure mechanism. Therefore a model describing the transport of ions might be useful to give more insight into these failures. However, calculating the transport of ions is numerically very challenging since it requires a multi-physics model on a multi-time and length-scale. Besides, the notion of a maximum ion concentration due to volume constraints opposed by the molding compound increases the complexity of the mathematical framework even further and results in a model that is very difficult to solve. In this paper we discuss several simplified models for the transport of ionic species that might be used to model their corresponding failure mechanisms. Further, we show the conductivity of molding compounds as a function of temperature and discuss how this accelerates the transport of ions.

Introduction

From an engineering point of view the reliability of a product can be defined as the probability that a product will perform its intended function during a specified period of time under stated conditions. To determine the reliability of products, test are performed at high external stimuli in order to accelerate their time to failure. Such an acceleration is necessary because the time to market of microelectronics is much shorter than their lifetime. Consequently, the acceleration factor between test and operating conditions needs to be known to calculate the lifetime of a product from the test results. These acceleration factors are usually determined by extrapolation of the test results over a wide range of the external stimuli. Besides, an activation energy based, i.e. Arrhenius, function is frequently used to extrapolate the test results to operating temperature. Furthermore, the assumption of mathematical separable variables is commonly used from a practical point of view, i.e. the external stimuli do not depend upon each other. However, this assumption becomes ambiguous if e.g. the product heats up due to power dissipation as a consequence of an applied electrical potential.

A more accurate acceleration factor will thus automatically result in a better prediction of the lifetime of products. To accurately determine the acceleration factor a profound knowledge of the failure mechanism is necessary.

Where we define the failure mechanism as the physical phenomena causing the failure and, in connection, the failure mode as the observation of the failure.

Bond pad corrosion, electrochemical migration and charge creep are examples of failure mechanisms that have a basis in the transport of ionic species through the molding compound. Sources of ions are, amongst others, the fabrication processes of the die, leaching of the die attach glue at high temperatures, or the synthesis of the molding compound itself. The ions in the molding compound will migrate through the encapsulation in the presence of a electrical field and might accumulate in a region where they can cause failure of the product.

Modeling ion transport requires a highly non-linear set of equations which is numerically very challenging to solve due to the formation of the so-called diffuse or electrical double layer at interfaces. In this layer we have an ionic concentration profile with a large gradient, where its thickness is typically in the order of nanometers for molding compounds. Using a numerical method to solve these transport equations, such as the finite element method, will result in extremely fine element distribution near interfaces. Thus, the degrees of freedom blow up to an extremely large number for models at the product size, even if only two dimensions are considered.

In this paper we will first present an overview of the literature on the failure mechanisms given above. Next, we will discuss (simplified) models for ion transport and show how they are related to the previously described failure mechanisms. As important input parameter for modeling the transport of ions we show volume resisivity data for molding compounds at high temperatures in the last part of this work.

Failure Mechanisms
Charge creep [1,2]

When a product is subjected to a high electrical field, applied either inside or outside the encapsulation, ions will migrate through the molding compound. Due to the migration of ions charge can accumulate at the die-to-molding compound interface (Fig. 1). The accumulated charge might result in a large electrical field in the die. In most cases this field is completely screened out by the metal of the back-end stack. However, there is a possibility that at some regions of the die surface there is no metal in the back-end, such that the underlying electrical circuit is subjected to a high electrical field. At these open regions the field can be sufficiently high to invert the silicon between two neighboring transistors. Consequently, a leakage path is created, which might lead to anomalous electrical behavior of the product and thus device failure. One of the main characteristic of this mechanism is the recovery of failures. When the applied electrical field vanishes, the ions will eventually spread out over the encapsulation due to diffusion such that the accumulated charge and thus the failure vanishes as well.

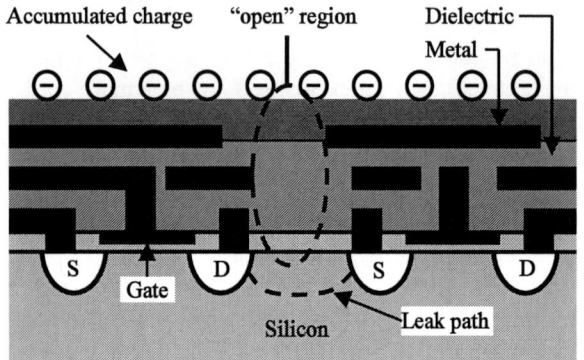

Fig. 1 schematics of a back-end cross-section where charges accumulate on top of the die surface, resulting in a high electrical field at the open region and consequently a leak path between the two neighboring transistors.

The effect of charge creep was studied experimentally using a test chip by van der Pol et al. and Bruggers et al. [1,2] These authors used a specially designed gateless transistor which opened when charge accumulated on the die surface. Using a calibration method they were capable of determining the electrical potential on top of the die surface from the measured current of the gateless transistors. They showed that the charging time of the die surface as a function of temperature is determined by the temperature dependent volume resistivity of the molding compound.

Recently failures, as described above, have been observed during qualification of a product. [3] The product was subjected to the thermally stimulated parasitic gate leakage test as described in the AEC Q100 standard. During this test products are exposed to mobile ions from a corona discharge at a tungsten needle placed above the products. After 2 minutes of exposure at 150° C the products are rapidly cooled down to room temperature. Due to the decrease in temperature the mobility of the ions decreases and the accumulated charge at die surface is 'frozen'. After exposure an unacceptably large increase in sleep current was noted (Fig. 2). This failure was addressed to leaking paths between neighboring transistors due to charge creep. Again the volume resistivity was the predominant factor for failure as a function of temperature.

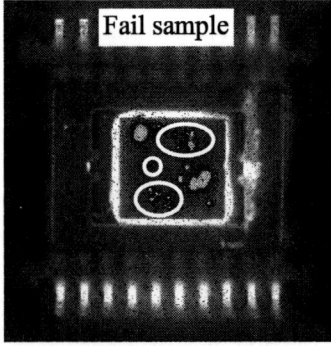 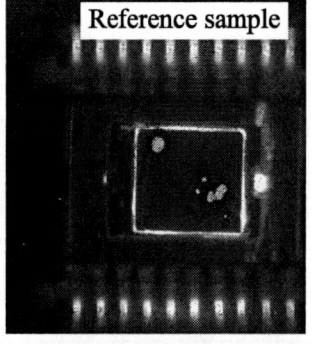

Fig. 2 Photon Emission analysis of a die diffused with a three metal SOI technology, assembled in a Heatsink Thin Shrink

Small Outline Package (HTSSOP20) showing an increasing in sleep current (circles) after exposure to an ion source.

Electrochemical migration

Electrochemical migration is a process that might occur when two metals with different electrical potentials are in contact with an (aqueous) electrolyte [4-7]. In the most simple case an oxidation process takes place at the anode forming metal ions. Under the presence of the electrical field the metal ions migrate toward the cathode where they deposit either in a dendrite or a broader plated form. Dendrites will be formed when a small nucleation site at the cathode is electrochemically favored to form a dendrite tip [4]. The formation of dendrites will results in leakage currents between leads, even resulting in a short circuit when the deposited metal bridges the distance between the anode and cathode completely.

Copper, lead and silver are most susceptible to suffer from electrochemical migration. However, halogen ionic contaminates, such as chloride, can even cause gold to migrate. Here, due to multiple chemical reactions secondary complex cations are formed. These cations migrate to the cathode where they precipitate. Harsányi reports reaction schemes for different forms of electrochemical migration [6].

Several authors [4,5,7] report that the solubility product of the metal ions in water is the rate limiting factor for migration. Harsányi [5] examined the electrochemical migration of 0.5 mm wide metal lines with a spacing of 0.2 mm on alumina substrates. The lines were not covered with a dielectric material. Test were performed at 95% relative humidity, 40° C and a electrical potential difference between the lines of 10 Volt. From Table 1 it can readily be seen that mean time to failure (MTTF) correlates well with the solubility product of the metal. Besides, it was noted that for silver palladium and platinum alloys the migration was almost completely eliminated. Using an identical experimental procedure as above, Manepalli et al. [7] applied benzocyclobutene (BCB) between the silver lines and found a threshold potential for migration greater than 3 Volt for lines separates by 0.1 mm.

Table 1 Mean time to failure of metal tracks on a substrate after [5], width of the tracks 0.5 mm, gap 0.2 mm, RH=95%, T=40° C and V=10 Volt.

Metal	MTTF [hours]	Solubility -log[Me(OH)$_n$]
Silver	~100	7.7
Lead	~700	14.92
Copper	~900	19.63
Tin	~2000	25.3

According to the theory described above electrochemical migration can only occur under moist conditions. However, Tan and Ong [8] observed copper dendrites after 100 hour of dynamic operational life test in the adhesive part of the tape that stabilizes the leads and argued that this is a form of 'dry' migration. We observed similar failures for quad flat packages (QFP) where the tapes showed a vast amount of voids (Fig. 3). However, the dendrite formation could be suppressed by a dry bake of the product before encapsulation. We therefore believe that the presence of water is necessary for electrochemical migration, however, the concentration can be very low.

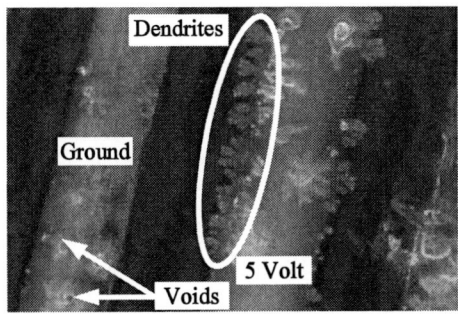

Fig. 3 Dendrites on the adhesive tape at a positively charged copper lead, the tape contains an extensive amount of voids.

Bond pad corrosion

Corrosion in microelectronics (Fig. 4) is based on essentially the same mechanisms as corrosion of large scale objects, such as bridges or cars [9]. During the process the following processes can be identified.

Water vapor absorbed by the molding compound adsorbs at the die surface. Several authors report that the adsorption of water on the die surface is accurately described by the Brunauer-Emmet-Teller (BET) equation [9-11]. The water can adsorb at the die surface in the micro voids of the encapsulation and at places where the material is (partly) delaminated [10]. Further the water can ingress along the leads of the package. Koelmans [11] report that the inactivation of adsorption sites on the die surfaces is the main function of the encapsulation to protect against corrosion.

Ionic contamination dissolves in the thin adsorbed water film resulting in a local electrolyte. The ionic contamination can either already be present on the die surface due to fabricating of the die or can be transported from another location in the presence of an electrical field or ion concentration gradient [12-14]. Amongst others, ionic sources can be the low level of ionic impurity always present in the molding compound [15], a decomposition of the die attach glue or an environmental source such as the solder flux [13].

In case of aluminium bond pads, a 20 to 100 Å thick native oxide grows rapidly on the bare material when exposed to air [14,16]. The native oxide protects the aluminium against corrosion in pure water. In basic or acid solutions the oxide can loose its protective function dependent on the electrical potential of the metal and the pH of the solution. The thermodynamically possible chemical reactions are then represented in a Pourbiax diagram [12,17]. Besides, ionic species, such as chloride or bromide, can attack the native oxide layer as well [14]. These species are often catalysts such that a small concentration can result in catastrophic amount of corrosion [16,18,19].

When metal corrodes we have an oxidation reaction at the anode and an accompanying reduction reaction at the cathode. The oxidation and reduction reactions will generally lead to an acidic and basic environment at their corresponding electrodes, respectively [12]. As a results of the chemical reactions there is a charge transfer between the anode and cathode. Due to the condition of a neutral charge balance the amount of charge flowing, i.e. electrical current, is equal for both the anode and cathode such that the corrosion rate kinetics depends on anode to cathode surface ratio [20].

The rate of corrosion is either determined by the reaction kinetics of the corroding metal or by the transport of charges between the electrodes [12,20]. For microelectronics the latter process is rate limiting for corrosion [12]. In case of galvanic corrosion of a gold-aluminium system, commonly used in microelectronics, the hydrogen evolution at the more noble metal, i.e. gold, determines the charge transfer rate between the two metals [20]. Such that the transport of hydroxyl ions is rate limiting for the corrosion kinetics.

Koelmans [11] reports that corrosion of aluminium metallization is governed by the surface resistivity of the die-to-molding compound interface. His work lacks an acceleration factor for corrosion. However, it is used by Dunn and McPherson [13] who determined an acceleration factor based on a temperature and humidity dependent mobility of corrosive contaminants. They calculated the dependency of the mobility from the time-to-failure of products. Consequently, it is ambiguous whether they have really determined the temperature and humidity dependency of the ionic mobility or just of the time-to-failure. Peck [21] reports a model based on an empirical fit of data over a temperature and humidity range. His acceleration function is defined as the product of an Arrhenius function and a power law for the temperature and humidity dependency, respectively.

The authors of both accelerating factors are concerned with the rigorousness of applying their factors above the glass transition temperature of the encapsulation [13,21]. The current trend for new molding compounds is to decrease their glass transition temperature to values as low as 110° C. This is far below the temperature of the Highly Accelerated Stress Test (HAST), which is 130° C. Therefore, despite the fact that they are adopted by the JEDEC standards [22], the above mentioned accelerating factors can not be used in this case. This shows the necessity of a better understanding of the failure mechanism and thus a better description of the acceleration factor.

Fig. 4 Examples of a corroded aluminium bond pad, the bond wire did not show any corrosion.

Modeling Ion Transport
Full model

The transport of ionic species i through a polymeric electrolytes is given by the Nernst-Planck equation [23]

$$J_i = -D_i\left(\nabla c_i - \frac{z_i c_i}{V_T}\nabla V\right) \tag{1}$$

where J_i is the flux (numbers of ions per unit area per unit time), D_i the diffusion coefficient of species i, c_i its concentration, z_i its valence, V the local electrostatic potential, and V_T the thermal voltage ($=k_B T/e \sim 25.6$ mV at room temperature). Combining eq. (1) with the conservation law

$$\dot{c}_i = -\mathrm{div}(J_i) \tag{2}$$

and Gauss' law for the electrostatic field

$$\varepsilon_r \varepsilon_0 \mathrm{div\,grad}V = -e\sum_i z_i c_i \tag{3}$$

where ε_r is the relative dielectric permittivity of the polymer and ε_0 the permittivity of vacuum, yields a self-consistent model for the transport of ions.

Solving eqs. (1)-(3) results in a large ion concentration gradient near charged interfaces, the so-called double layer with a thickness in the order of the Debye length

$$\lambda_D = \sqrt{\frac{\varepsilon_r \varepsilon_0 k_B T}{2e^2 c_\infty}} \tag{4}$$

where c_∞ denotes the ion concentration in the bulk. The Debye length for commercial molding compounds is in the order of several nanometres [23,24]. In connection, we can define the Debye time [25]

$$\tau_D = \frac{\lambda_D{}^2}{D} \tag{5}$$

for the local dynamics of the double layer. The Debye time is in the order of seconds or smaller, while full equilibrium of the layer is reached at a time scale of hours for realistic models [23]. This results in a model with a length scale ranging from nanometres to millimetres and a time scale ranging from seconds to hours for a complete encapsulation. It is a delicate, almost impossible, task to solve such models.

Besides these numerical problems, the Nernst-Plack equation predicts unrealistic high ion concentrations at high electrode potentials. Already at a potential of 1 Volt it predicts an equilibrium ion concentration of $\sim 10^{17}$ times its bulk concentration [23-25], which is physically impossible due to volume constraint opposed by the molding compound. To include a maximum ion concentration in the mathematical framework described above we can replace eq. (1) by [24,25]

$$J_i = -D_i\left(\left[1 + \frac{c^* c_i}{1 - c_i c^*}\right]\nabla c_i - \frac{z_i c_i}{V_T}\nabla V\right) \tag{6}$$

where c^* denotes the reciprocal of the empirical maximum ion concentration.

We have solved eqs. (2), (3) and (6) for an one-dimensional model with a length $L=20$ nm and different values of c^*. At $x=0$ we have applied symmetry boundary conditions and at $x=L$ we have a blocking electrode, i.e. $J_i=0$, and an electrical potential of 0.5 Volt. The initial ion concentration, c_∞, is 4 mM, i.e. 100 ppm [23], and the Debye length is 1 nm. The results (Fig. 5) clearly show a plateau value equal to product of c_i and c^*. The ion concentration at the electrode for $c^*=0$ is ~250 times its bulk concentration. Here the maximum concentration is bounded by the small model size, i.e. the ion concentration is limited due to the

depletion of ions from the bulk. Nevertheless, the electrode concentration is rather high when compared to the solubility of NaCl in water, which is ~6 M. When c^* increases the thickness of the layer where the ion concentration is at its maximum increases as well. We will refer to this layer as the condensed layer.

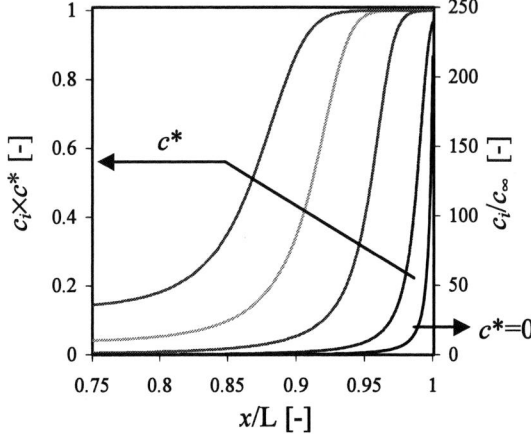

Fig. 5 Ion concentration profiles at an electrode, $V(x=L)=0.5$ Volt, $L=20$ nm, $D=1$ nm, $T=293$ K, $c_\infty=4$ mM, $c^*=0.004$, 0.0135, 0.032, 0.0625 (colored lines) and $c^*=0$ (black line).

In the condensed layer we have

$$\lim_{c_i c^* \to 1}\left\{\frac{c^* c_i}{1 - c_i c^*}\right\} = \infty \tag{7}$$

which will cause for a large repulsive force when the ion concentration, c_i, tends to its maximum value, $1/c^*$. A large potential difference between the electrodes will result in a large attractive force working on the ions. Consequently, high electrode potentials will result in numerical problems and instability of the model. Due to this, and the previously mentioned problems without a maximum ion concentration, simplified models for the transport of ions are useful.

Circuit models

The size of the model discussed above is rather small. As a result ions are depleted from the bulk material. When the model is sufficiently long no ion depletion will occur, such that concentration gradients in the bulk are absent. Hence it can be seen from eq. (6) that the ionic current equals

$$I = e\sum_i z_i J_i = \frac{2eDc_\infty}{V_T}\nabla V = \frac{\varepsilon D}{\lambda_D{}^2}\nabla V = \frac{1}{\rho}\nabla V \tag{8}$$

where is the volume resistivity [23]. Consequently, the bulk material can be modeled as a resistor, $R=\rho L$.

In case of thin double layers having a fast equilibration time compared to the remaining system we can use the equilibrium double layer properties to describe the dynamics of the system [26]. The capacitance of the equilibrium double layer including volume constraints for a $z{:}z$ electrolyte is [27]

$$C_D = \frac{\frac{\varepsilon\sqrt{c^* c_\infty}}{\lambda_D}\sinh\left(\frac{zV_{DL}}{V_T}\right)}{\left[1 + 4c^* c_\infty \sinh^2\left(\frac{zV_{DL}}{2V_T}\right)\right]\sqrt{\ln\left(1 + 4c^* c_\infty \sinh^2\left(\frac{zV_{DL}}{2V_T}\right)\right)}} \tag{9}$$

where V_{DL} is the potential drop across the double layer. We now have a resistor and capacitor in series. The dynamics of this system is described by

$$\frac{dV_{DL}}{dt} = \frac{V^0 - V_{DL}}{RC(V_{DL})} \tag{10}$$

where V^0 is the applied electrical potential across the system. Note that the product RC is proportionally dependent on $\lambda_D L/D$, which is known as the relaxation time [28]. Obviously, the current in the systems equals

$$I = \frac{V^0 - V_{DL}}{R} \tag{11}$$

We have solved eqs. (2),(3) and (6) for the input parameters given above. Only now we kept the volume constraint, c^*, constant at 0.004 mM^{-1} and varied the length of the model ranging form 20 nm to 2 μm. The results are presented in Fig. 6, where we have normalized the current with the initial current, I_0. For the smallest models the decay in current is dominated by the depletion of ions from the bulk. The largest model, L=2 μm, coincides perfectly with the RC-model of eqs. (9)-(11). This indicates that for this model there is a negligible amount of ion depletion in the bulk. Consequently, the volume resistivity of the bulk remains constant whereas it increases for the smaller models. As a result of the constant resistivity the decay in current is completely dominated by the formation of the double layer. This explains the peculiar behaviour of the current at prolonged time for L=2 μm.

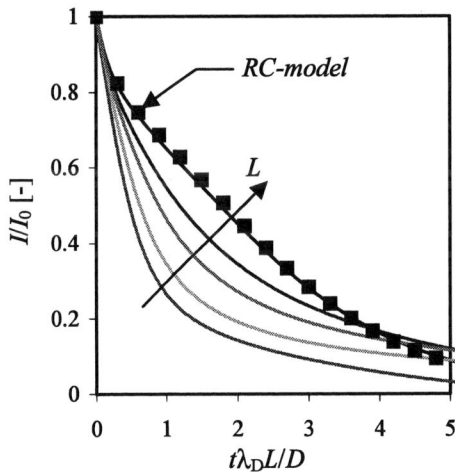

Fig. 6 *Current decay between two parallel electrodes,* $V(x=L)$=0.5 Volt, T=293 K, c_∞=4 mM, c^*=0.004 mM^{-1}, *lines; eqs. (2),(3),(6) for L=20, 40, 200, 2000 nm, squares; eq.(11).*

Simplified finite element theory

We can extend the RC-model presented above, valid in only in one dimension, to two or three dimensions using the finite element method [23]. When there is no decrease in ion concentration in the bulk there will be no charge separation, such that Gauss' law -eq. (3)- becomes

$$\varepsilon_r \varepsilon_0 \text{divgrad} V = 0 \tag{12}$$

If we assume the double layer negligibly small compared to the bulk size we can replace it by a mathematical description.

At interfaces we have charge accumulation according to

$$\frac{d\sigma}{dt} = -I \cdot \mathbf{n} = \frac{1}{\rho} \nabla V \cdot \mathbf{n} \tag{13}$$

where σ is the total accumulated charge in the double layer and $\mathbf{n}$ is the outward normal on the interface. When charge accumulates at an interface with an isolator we have two boundary conditions. First we have

$$\varepsilon_{r,i} \nabla V_i \cdot \mathbf{n} + \sigma = \varepsilon_{r,e} \nabla V_e \tag{14}$$

where subscript i and e denote the isolator and electrolyte, respectively. Secondly, there is an electrical potential difference across the double layer due to the accumulated charge. If we again assume that the double layers are in quasi-equilibrium we can use the equilibrium properties of the double layer to describe this potential difference

$$V_{DL} = \text{sgn}(Q)V_T \left\{ 2\text{asinh}(0.5|Q|) + c * c_\infty Q^2 \right\} \tag{15}$$

where Q is a dimensionless surface charge

$$Q = \frac{\sigma \lambda_D}{\varepsilon_0 \varepsilon_r V_T} \tag{16}$$

When charge accumulates at an conductor interface we only need to consider the potential difference across the double layer according to eq. (15).

Using a finite element model as outlined above has the following advantages over a full model. First, in the bulk we only have one degree of freedom, the electrical potential, which leads to an enormous reduction of the system matrix. Next, we do not need to calculate the exact structure of the diffuse layer. This will lead to a much coarser mesh at interfaces. However, the model is limited to systems without any ion depletion of the bulk.

Moving boundary

As discussed previously, the term containing c^* in eq. (6) causes numerical difficulties at high electrode potentials due to the formation of the condensed layer. We can however replace the condensed layer by a mathematical description for the ion concentration and electrical potential. The ion concentration in this layer is straightforward, namely $1/c^*$. The electrical potential at the head of the condensed layer directly follows from Gauss' law –eq. (3)-

$$V_{con} = \frac{ze l_c^2}{2c * \varepsilon_0 \varepsilon_r} + \left. \frac{\partial V}{\partial x} \right|_{x=lc} l_c + V^{elec} \tag{17}$$

where l_c is the condensed layer thickness and V^{elec} is the applied potential on the electrode.

The only unknown parameter in eq. (17) is the thickness of the condensed layer. We can determine the thickness of this layer from its dynamics. Note that the total amount of ions per unit electrode area in the condensed layer equals, $c_{tot}=l_c/c^*$. The kinetics of the condensed layer is bounded by the conservation of mass, i.e. the ions can redistribute throughout the model, but their total amount cannot change. Therefore we can assume that the increase in ions in the condensed layer is proportionally dependent on the ion concentration at head of the layer. Hence we obtain a condition for a moving boundary

$$\frac{dl_c}{dt} = H\left(\sum_i c * c_i - 1\right)\beta c * \frac{dc}{dt} \qquad (18)$$

where H denotes the Heaviside step function and is the proportionality factor mentioned above.

Application of eq. (18) to the electrode of a model will cause it to move into the material after reaching an ion concentration equal to $1/c*$ with a velocity equal to the formation of the condensed layer. The electrical potential of the electrode then decreases according to eq. (17).

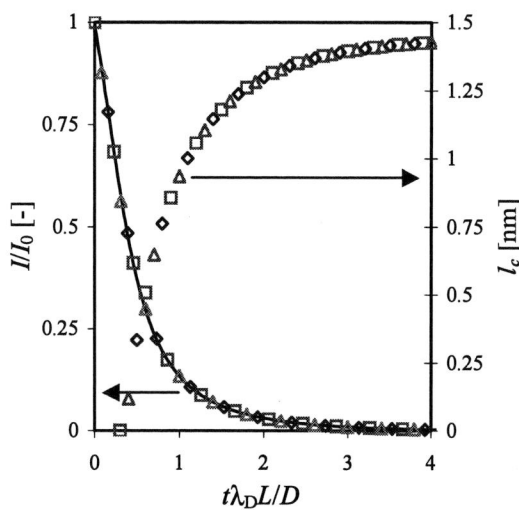

Fig. 7 Ionic current decay between two parallel electrodes including a moving boundary, V(x=L)=0.5 Volt, L=20 nm, λ_D=1 nm, c_∞=4 mM, c*=0.0625 mM^{-1}., =293 K, black line; full model, squares; =1, triangles; =5, diamonds; =10.

We now return to the previously used model including eqs. (2),(3) and (6) and compare the result of this full model with the model including a moving boundary according to eqs. (17) and (18). We varied the proportionality factor, , from 1 to 10. The results coincide perfectly for all values of . Fig. 7), this shows that the velocity of the condensed layer is indeed bounded by the conservation of mass as indicated above. Further, the current decay of the model including the moving boundary coincides with that of the full model. Besides, the thickness of the condensed layer, l_c, agrees very well with the equilibrium concentration profile presented in Fig. 5. This shows that a moving boundary according to eqs. (17) and (18) can be used to simplify the model.

Failure mechanism modeling

The models presented in the previous section apply when charge is equally distributed in the package. We have used these models in previous work to study failures due to charge creep through the encapsulation. A RC-network approach was used to model the failures that occurred after the thermal stimulated gate leakage test (Fig. 2). From this analysis we were capable of determining the influence of different molding compounds on the failures [3]. Where the volume resistivity at high temperatures was different for the various compounds.

Further we used the simplified finite element model to explain the experimental results of Bruggers et al. [2]. The model results were in fair agreement with the experimental determined charge accumulation on top of the die surface [23,29]. Besides, the model predicts that the accumulated charge, and thus the failures, should vanish at prolonged time. However, the time-scale of the experimental data was not sufficiently long to underpin this prediction.

In case of corrosion we need to model the ionic current originating from the chemical reactions at the metal surfaces. To determine the charge transfer rate per unit area at the metal-electrolyte interfaces the Butler-Volmer equation including the Frumkin correction for the diffuse layers might be useful [30]

$$I = k\left\{c_R \exp\left(\frac{-\alpha_O \eta}{V_T}\right) - c_O \exp\left(\frac{\alpha_R \eta}{V_T}\right)\right\} \qquad (19)$$

where c_O and c_R are the ion concentrations in the oxidized and reduced state, respectively, k is a kinetic constant, α_O and α_R are transfer coefficients and is an overpotential according

$$\eta = V^{elec} - V_{rp} - V^{eq} \qquad (20)$$

where V^{eq} is a thermodynamic number for the chemical reaction and V_{rp} the potential at the reaction, or Helmholtz plane. This plane is located near the electrode into the electrolyte and indicates the plane where the chemical reactions occur. Eq. (19) is a boundary condition for the ionic current and can thus be coupled to eq. (3).

In aqueous environments the Frumkin correction is not generally used since here the formation of the diffuse layers is relatively fast compared to corrosion. However, in molding compounds the formation is rather slow due to the low conductivity of the medium. Without the Frumkin correction the concentrations c_O and c_R can be simply lumped into the overpotential.

Now consider as an example the conversion of gaseous oxygen with electrons and water into hydroxyl ions

$$O_2(g) + 4e^- + 2H_2O \leftrightarrow 4OH^- \qquad (21)$$

which might occur at the cathode. The values for α_O and α_R are typically equal to 0.5 such that Eq. (19) becomes

$$I = k_c\left\{\exp\left(\frac{-\eta}{2V_T}\right) - c_{OH,r}\exp\left(\frac{\eta}{2V_T}\right)\right\} \qquad (22)$$

where k_c is a kinetic constant dependent on the water concentration and subscript r indicates the concentration at the reaction plane. Note that we considere the concentration of water and oxygen as a constant at the reaction plane. In this case the kinetic constant, k_c, determines the influence of water concentration on the reaction rate.

Volume resistivity

Until now we tactically circumvent using the diffusion coefficient by scaling time for all the results of the previous sections with the relaxation time, $\lambda_D L/D$. However, we can calculate the diffusion coefficient from volume resistivity data using eq. (8). Fig. 8 shows volume resistivity data of four commercially available molding compounds at elevated temperatures. When we assume an ion concentration of 1 mM the diffusion coefficient varies from ~10^{-21} m^2/s at 50° C to

~10^{-16} m^2/s at 200° C. The typical size between two adjacent bond pads is in the order of tenths of micrometers and the value for the Debye length is in the order of nanometers. This results in a relaxation time in the order of 10^2 to 10^7 seconds.

Fig. 8 shows a discontinuity in the volume resistivity at 150° C (1000/T=2.36 1/K). At higher temperatures the activation energy, E_a, increases from 0.56 eV to 2.5 eV. This will thus increase the accelerating factor for ion transport related failures by a large extent.

There is no data available for the volume resistivity as function of humidity. However, this data is necessary to interpret the results of humidity enhanced reliability tests.

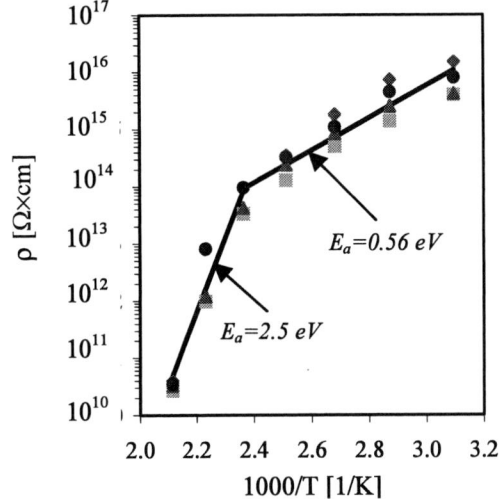

Fig. 8 Arrhenius plot of the volume resistivity for four different molding compounds (suppliers data).

Conclusions

We have presented a simplified mathematical framework for the transport of ions through the molding compound that encapsulates microelectronics. Numerical problems were circumvented by (partly) replacing the double layer by mathematical equations. This framework can be used to study the effect of ion transport on their related failures mechanisms, namely, charge creep, dendrite growth and corrosion. The former can be modeled without any chemical reactions showing a fair agreement with experimental result. The latter two failure mechanisms do involve chemical reactions in the encapsulation.

To in incorporate these reactions we require a description of the charge transfer rate at the electrodes. Although the Butler-Volmer equation is amongst the most fundamental relations in electrochemistry, this equation cannot be used for the charge transfer rate since corrosion in encapsulations is transport limited. Therefore, we suggest to use the Butler-Volmer equation including the Frumkin correction. This correction accounts for the formation kinetics of the double layer. Based on experimental observations, the formation of the double layer can be on a time-scale comparable to the corrosion time-scale.

We believe that water absorption by the molding compound increases the mobility of ions and thus decreases its volume restistivity. However, no experimental data is available to show this effect. Further, water can increase the charge transfer rate at electrodes. Consequently, there will be a significant effect of the (local) water concentration on the corrosion rate.

According to JEDEC standards the acceleration factor for corrosion is an unknown function of the applied electrical potential for all failure models. Solving the mathematical framework presented here will eventually lead to an electrical potential dependent corrosion rate.

To solve the complete mathematical framework presented here for realistic values of the input parameters is the topic of the ongoing research by the authors.

Acknowledgments

This research was carried out under project number MC3.05236 in the framework of the Research Program of the Materials Innovation Institute M2i (www.m2i.nl), the former Netherlands Institute for Metals Research.

References

1. van der Pol, J.A. *et al.*, "Model and Design Rules for Eliminating Surface Potential Induced Failures in High Voltage Integrated Circuits", *Microelectronics Reliability* Vol. 40, (2000), pp. 1267.
2. Bruggers, H.J. *et al.*, "Reliability Problems due to Ionic Conductivity of IC Encapsulation Materials under High Voltage Conditions", *Proc Int. Symp. on Power Semiconductors and ICs,* 1999, pp. 197.
3. van Soestbergen, M. *et al.*, "Electrical Characterization of Plastic Encapsulations Using an Alternative Gate Leakage Test Method", *Proc. Int. Reliability Physics Symposium,* Phoenix, 2008, pp. 462.
4. Steppan, J.J. *et al.*, "A Review of Corrosion Failure Mechanisms during Accelerated Tests", *Journal of the electrochemical society* Vol. 134, (1987), pp. 175.
5. Harsanyi, G., Inzelt, G. "Comparing Migratory Resistive Short Formation Abilities of Conductor Systems Applied in Advanced Interconnection Systems", *Microelectronics Reliability* Vol. 41, (2001), pp. 229.
6. Harsanyi, G., "Electrochemical Processes Resulting in Migrated Short Failures in Microcircuits", *IEEE CPMT* Vol. 18, (1995), pp. 602.
7. Manepalli, R., "Silver Metallization for Advanced Interconnects", *IEEE Trans. Adv. Packaging* Vol. 22, (1999), pp. 4.
8. Tan, S.H., Ong, S.H., "A Dry Migration? Copper Dendrite Growth in Adhesive Tape During Burn-in", *Proc. Int. Symp. Physical and Failure Analysis of Ics,* Singapore, 2001, pp. 178.
9. Comizolli, R.B. *et al.*, "Corrosion of Electronic Materials and Devices", *Science* Vol. 234, (1986), pp. 340.
10. Hoge, C.E., "Corrosion Criteria for Electronic Packaging Part I- A Framework for Corrosion of Integrated Circuits", *IEEE Trans. Comp. Hybrids Man. Tech.,* Vol 13, (1990), pp. 1090.
11. Koelmans, H., "Metallization Corrosion in Silicon Devices by Moisture-Induced Electrolysis", *Proc. Int. Reliability Physics Symposium,* 1974, pp.168.

12. Osenbach, J.W., "Corrosion-induced Degradation of Microelectronic Devices", *Semicond. Sci. Technol.*, Vol. 11, (1996) pp. 155.

13. Dunn, C.F., McPherson, J.W., "Recent observations on VLSI Bond Pad Corrosion Kinetics", *J. Electrochemical Soc.*, Vol. 135, (1988), pp. 661.

14. Iannuzzi, M., "Reliability and Failure Mechanisms of Nonhermetic Aluminum SIC's: Literature Review and Bias Humidity Performance", *IEEE Trans. Comp. Hybrids Man. Tech.*, Vol 6, (1983), pp. 181.

15. Lantz, L., Pecht M.G., "Ion Transport in Encapsulants Used in Microcircuit Packaging", IEEE Trans. Comp. Pack. Tech., Vol. 26, (2003), pp. 199.

16. van Gestel, R., Reliability related research on plastic IC-packages: A test chip approach, Delft university press (Delft, 1994).

17. Pourbaix, M., Atlas of electrochemical equilibria in aqueous solutions, NACE international (Houston, 1974).

18. Paulson, W.M., Lorigan, R.P. "The Effect of Impurities on the Corrosion of Aluminum", *Proc. Int. Reliability Physics Symposium*, 1976, pp.42.

19. Pecht, M., "A Model for Moisture Induced Corrosion Failures in Microelectronics Packages", *IEEE Trans. Comp. Hybrids Man. Tech.*, Vol 13, (1990), pp. 383.

20. Gellings, P.J., Introduction to corrosion preventiom and control for engineers Delft university press (Delft, 1976).

21. Peck, D.S., "Comprehensive Model for Humidity Testing Correlation", *Proc. Int. Reliability Physics Symposium*, 1986, pp.44.

22. JEDEC standard JEP122C "Failure Mechanisms and Models for Semiconductor Devices" (available at http://www.jedec.org/Catalog/catalog.cfm)

23. van Soestbergen, M. *et al.*, "Modified Poisson-Nernst-Planck Theory for Ion Transport in Polymeric Electrolytes", *J. Electrostatics*, (2008), in press.

24. van Soestberegen, M. *et al.*, "Transport of Corrosive Constituents in Epoxy Moulding Compounds", *Proc. Eurosime*, London, 2007, pp. 726.

25. Kilic, M.S. *et al.*, "Steric Effects in the Dynamics of Electrolytes at Large Applied Voltages II. Modified Poisson-Nernst-Planck Equations", *Physical Review E*, Vol. 75, (2007), pp. 021503.

26. Kilic, M.S. *et al.*, "Steric Effects in the Dynamics of Electrolytes at Large Applied Voltages II. Double Layer Charging", *Physical Review E*, Vol. 75, (2007), pp. 021502.

27. Freise, V., "Zur Theorie der Diffusen Doppelschicht", *Zeitachrift fur Elektrochemie*, Vol. 56, (1952), pp. 882.

28. Bazant, M.Z. *et al.*, "Diffuse-charge Dynamics in Electrochemical Systems", *Physical Review E*, Vol. 70, (2004), pp. 021506.

29. Biesheuvel, P.M. *et al.*, "Ionic Polarisation Layers in Polymer Electrolytes" *Proc. Eurosime*, Freiburg, 2008.

30. Biesheuvel, P.M. *et al.*, "Corrosion Modeling in Low-conductivity Solid Electrolytes: Influence of Dynamic Formation of Polarization Layers on Corrosion Rate Using a Frumkin-corrected Butler-Volmer Model", In preparation.

Reliability Analysis of Copper Interconnections of System-in-Packaging Structure using Finite Element Method

Shih-Ying Chiang, Shin-Yueh Yang, Chan-Yen Chou, Ming-Chih Yew, and Kuo-Ning Chiang
Advanced Microsystem Packaging and Nano-Mechanics Research Lab PME
Advanced Packaging Research Center
National Tsing Hua University, HsinChu, Taiwan 300, China
Phone：886-3-5742925   Fax：886-3-5745377
E-mail：knchiang@pme.nthu.edu.tw

Abstract

The system-in-package (SiP) is among the popular designs which meet the trend of integrated circuit (IC) development. The SiP structure investigated in this study includes seven sub-chips attached to the chip carrier, and polymer was applied around the chips. The polymer is an exceptional stress buffer layer reducing the maximum shear stress in the solder joints, but it also affects the copper interconnection which suffers from significant stress/strain concentration under thermal loading due to coefficient of thermal expansion (CTE) mismatch. In this paper, several parameter studies for a radio frequency front end module (RF-REM) incorporated with the novel wafer level chip scale package (WLCSP) technology is proposed to reduce the stress concentration behavior both in the package-level structure and the board-level structure in order to enhance reliability. In investigating the physical phenomenon of SiP structure, 2D and 3D finite element analysis (FEA) were both used. The analysis indicated that the stress concentration behavior was aggravated, especially in the vias at the chip edge. Finally, the compromised optimal location of the vias and the thickness of the adhesive are determined to minimize the stress concentration, which is due to the expansion of the filler polymer.

Introduction

The SiP is among the most popular designs which meet the trend of IC development, featuring higher working frequency and increased complexity. It is known for its small size, light weight, and multiple functionalities. For a radio frequency device in the wireless communication system, the use of SiP technology would not only reduce the cost and size of the RF package but would also enhance product performance. Another advantage of using the SiP in a wireless system design would be related to saving development time. Moreover, advanced packaging technology recently focuses on the WLCSP which completes the packaging operation directly on the wafer with the chip scaled size, and then dices the whole packaged wafer for the assembly in a flop chip manner. The packaging and testing of the separated dice are replaced by whole wafer fabrication and wafer level testing. In order to reach the further objective of miniature packages, which reduce costs, the use of the WLCSP is a suitable solution for a wireless module design.

However, due to the CTE mismatch between the silicon chip and the organic PCB, the solder joints in the WLCSP suffered shear stress, resulting to the package being absorbed by the junction. As the die size increases, the peripheral solder joints on the chip incur the maximum shear strain upon thermal loading, making the solder joints the weakest portions in the WLCSP. The reliability issues surrounding the WLCSP with a large die size are considered as serious problems. Many improvements have been developed to release the shear stress in the solder joints due to the large CTE mismatch. Among the improvements to the WLCSP is setting the soft stress-buffer-layer (SBL) structure under the solder bumps, which then releases the shear stress in the solder joints. [1-3] The stress-buffer-layer could decrease the shear strain on the solder joints, relieving the shear stress. Yuan et al. [4] and Wei et al. [5] have introduced a novel WLCSP with a fan-out redistribution layer having stress-release capability. In the fan-out WLCSP, the chip is first placed on an 8-inch chip carrier which has the same as an 8-inch wafer. The trenches between the chips are then filled with filler polymer. As shown in Fig. 1, the polymer is an exceptional stress buffer layer reducing the maximum shear strain in the novel fan-out WLCSP.

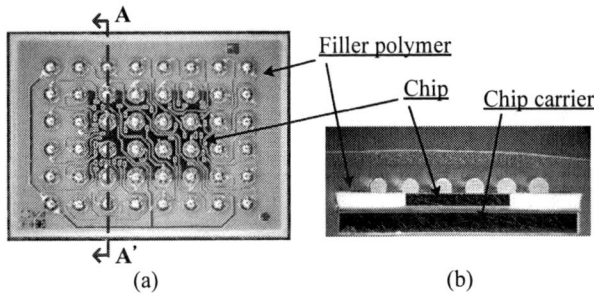

Fig. 1: (a) The testing sample of the fan-out WLCSP; (b) The cross-sectional view of A-A'

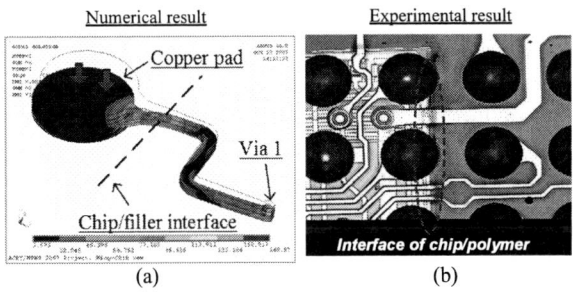

Fig. 2: The comparison between the result of finite element analysis (pattern of von Mises stress) with the experiments at trace [6]

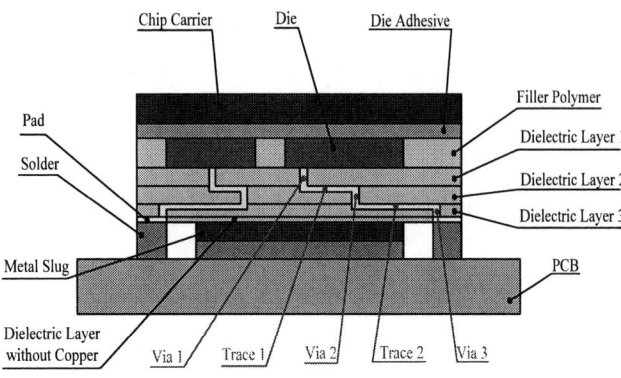

Fig. 3: The sketch of RF SiP structure.

Yew et al. [6] have indicated that the redistribution traces in the fan-out WLCSP may suffer significant stress/strain, making it crack under thermal loading due to the CTE mismatch between the chip and the filler polymer. Shown in Fig. 2 is the comparison between the FEA with the experiments at trace. The maximum von Mises stress in the FE model happened in the aforementioned region, and a crack of the trace was discovered in the experiment's real sample. By considering the equivalent stress in the FE model, the most critical region in the fan-out WLCSP through thermal loading can be achieved.

In this paper, an RF-FEM incorporated with the novel WLCSP technology is investigated. As shown in Fig. 3, the chips were mounted on the chip carrier by die adhesive material, and filler polymer was applied around the chips. Moreover, the redistribute interconnects were laminated and covered with the chips and filler polymer. The module forms, the metal slug, and the pads were then finally attached to the PCB test board by the solder paste layer. Due to the metal traces buried in high CTE material, the reliability of the interconnect trace in this structure becomes an important design issue. In this research, package-level and board-level reliability on trace have been discussed using both 2D and 3D FEA. Comparison of the two different package design criteria, a suitable solution with the best position, and a parameter of geometry could be attained.

Finite Element Model Description

The RF-FEM structure in this paper is composed of seven chips, as shown in Fig. 4. This SiP structure is composed of seven chips including power amplifier (PA), switches, diplexer, regulator, and low noise amplifier (LNA). We used the finite element analysis software ANSYS® to investigate the trace physical behaviors of the SiP structure.

Firstly, we use 2D FEA to investigate the trace physical behaviors in order to save the computational time and to quickly determine the major cause which influenced package reliability. In this assumption, a 2D plane strain element was used, but actual behaviors that we can't consider in this model. One of these is practical trace width in this simplified model, regarding it as infinity shell along the direction out of the 2D plane. The stiffness of the trace in 2D assumption will be higher than that in the real condition. However, the stress concentration caused by the CTE mismatch between the traces and the dielectric material was still the same. Consequently,

we could get reasonable results by observing the physical behaviors on traces and vias. Since our goal in this study is to look into how dielectric material influences the metal trace in SiP structure, it is necessary to consider all trace connective components. The 2D FE model needs to include multiple chips and solder bumps to connect with traces, and the line B-B' in Fig. 4 is the cross-section we chose.

Table 1: The material properties of RF-SiP.

	Young's modulus (GPa)	Poisson ratio	CTE (ppm/°C)
Chip Carrier	148	0.3	5
Adhesive	0.05	0.4	167
Chip	112.4	0.28	2.62
Dielectric	0.09	0.41	150
Copper	Nonlinear	0.35	17
Filler Polymer	0.05	0.4	167
PCB	18.2	0.3	16
Solder	Temp. dependent	0.35	23.9

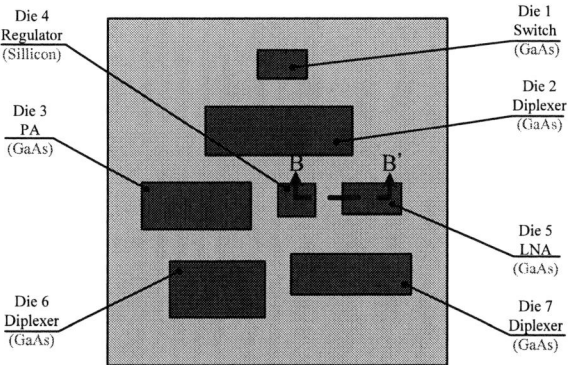

Fig. 4: The sketch of position of seven chips

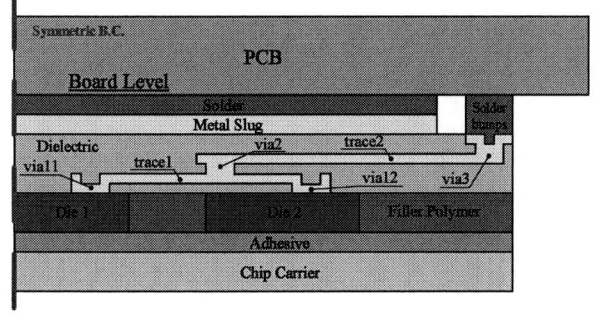

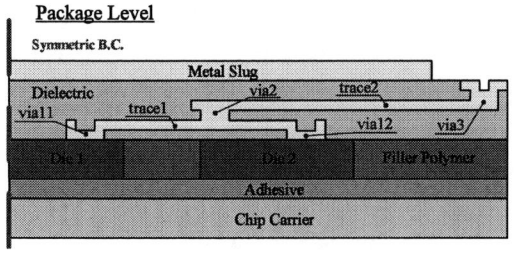

Fig. 5: The sketch of 2D FE model of board-level and package-level

Shown in Fig. 5 is the sketch of the 2D FE model. The figure shows that the board-level structure in this study and

280

the package-level are similar with the same design without the PCB and solder materials. The maximum plastic strain can be obtained by applying thermal cycle loading or uniformly arising thermal loading, so we could perform the latter process to be able to save computational time. In this study, reliability analysis in the package-level and board-level structures was conducted, and thermal rising from room temperature 25℃ to 125℃ was applied. The filler polymer and dielectric material have the advantage of being the stress buffer layers; however, the property of the large CTE is presumed to be the main reason affecting package reliability. Reliability may be affected by five parameters, as shown in Fig. 6. The five parameters will be described as follows: (a) the position of via11 located on die1; (b) the position of via12 located on die2; (c) the thickness of the adhesive material; (d) the thickness of the chip carrier; and (e) the thickness of the dielectirc3. In figuring out how the major driving force would affect the stress-concentrated regions, a discussion of several parameters will be done in the following section.

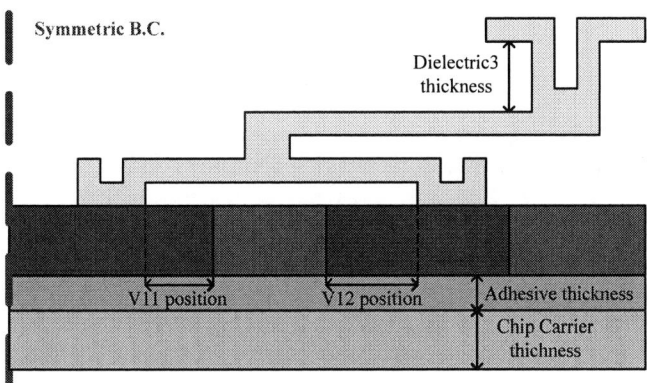

Fig. 6: Five different parameters study in 2D FE model

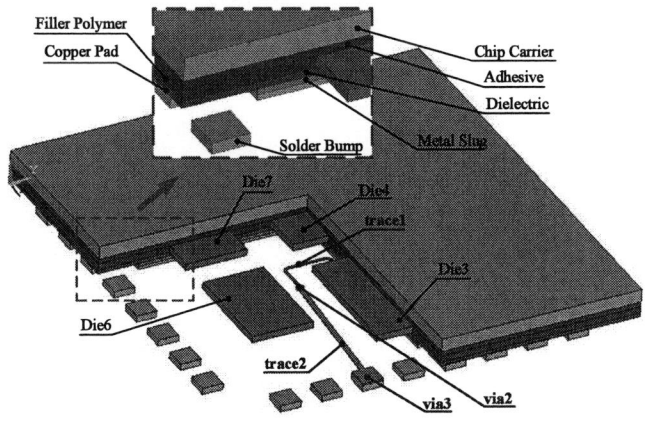

Fig. 7: The sketch of 3D FE model

In the 3D model, die3 (PA) is the major heat source which is a generated form chip. It can be assumed that traces between die3 and the filler polymer will fail easier than the others. With the previous assumption, we modeled a trace between die3 and die6, as shown in Fig 7. Similar with the 2D FE model, via1 and via3 are located on the die and the solder bump respectively, and via2 is the interconnection between via1 and via2. In the 3D FEA, we focused on the parameters

which influence the stress concentration behavior in the 2D FE model to a large extent.

Parameters study with 2D FEA

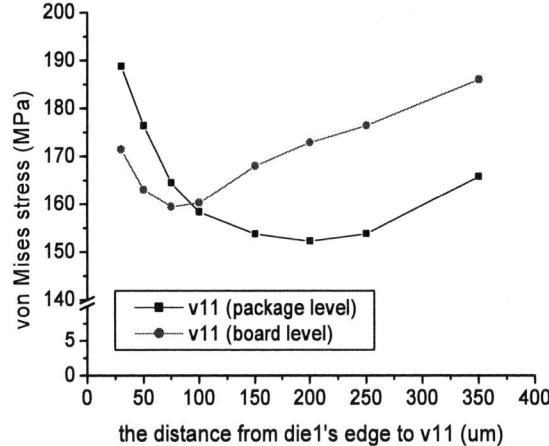

Fig. 8: Different distance from the right side of die1 to via11

In the package-level and board-level structures, via11 and via12 have large stress concentrations during thermal loading due to their adherence to chips that could be regarded as fixed ends. The region above the chip/polymer boundary in trace1 produced stress concentration due to the expansion of the filler polymer. When the via becomes closer to the chip edge, it will produce a high von Mises stress. In order to find an optimal rule about the via position, we could compare both board-level and package-level structures with the same condition. The distance between the die1 edge and via11 was changed from 30μm to 350μm, and the results are shown in Fig. 8. The results reveal that via1 neither comes close to, nor moves apart from, the chip/polymer edge, thus making it an unreliable solution in this specific FE model. The via was located a short distance away, about 50μm to 150μm in the package-level structure, and was judged to be in a better position.

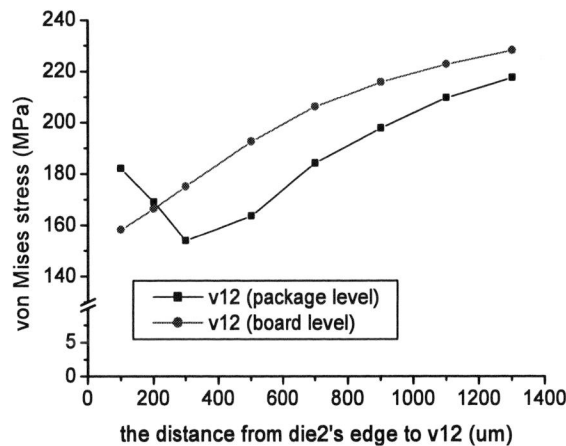

Fig. 9: The influence of changing the distance from left edge of die2 to via2

As Fig.9 shows, the location of via12 in the package-level condition is similar with that of via11. The trend reveals that the via needs to be placed away from the chip edge in order to reduce the von Mises stress. However, the difference observed in the board-level is "the greater the distance between this corner and the left edge of die2, the higher the von Mises stress that occurred in the sensitive region." The results show that the expansion of the high CTE material between trace1 and the chips would be restrained from this enclosed region. The region will push the via's corner during the thermal loading.

Die1 and die2 could be regarded as one body embedded in soft material, so their mobility might influence the stress in the vias. The adhesive under die1 and die2 is a buffer layer. When the adhesive becomes thicker, the mobility of die1 and die2 also increases. As shown in Fig. 10, the thickness of the adhesive was modified from 20μm to 150μm. Increasing the thickness of the adhesive layer could reduce the stress on the vias. In addition, the researchers also altered the thickness of the chip carrier, as shown in Fig. 11. The result indicates that the thickness of the chip carrier is not an important design variable because the change of the von Mises stress is not distinct.

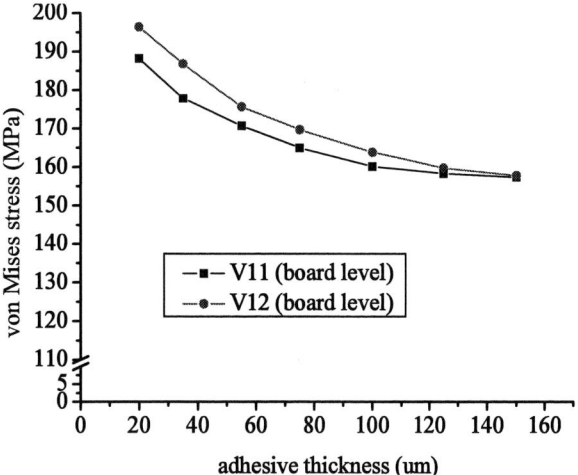

Fig.10: The influence of changing adhesive thickness.

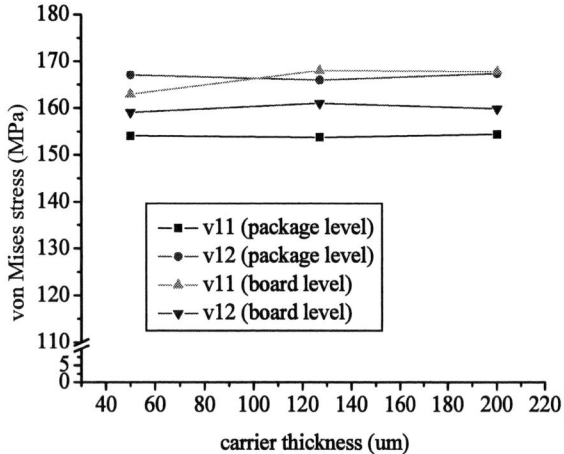

Fig.11: The influence of changing chip carrier thickness.

Another sensitive region, which might crack under thermal loading, is the corner between via3 and trace2. The corner stress-concentrated effect in the board-level structure is more critical than that of the package-level structure due to via3 being at a fixed-end with lower mobility. Due to this, the researchers only discussed the board-level structure and altered the thickness of dielectric3 from 40μm to 80μm, as shown in Fig. 12. The result shows that when the thickness of dielectric3 increases, the stress concentration in the corner becomes more serious in via3.

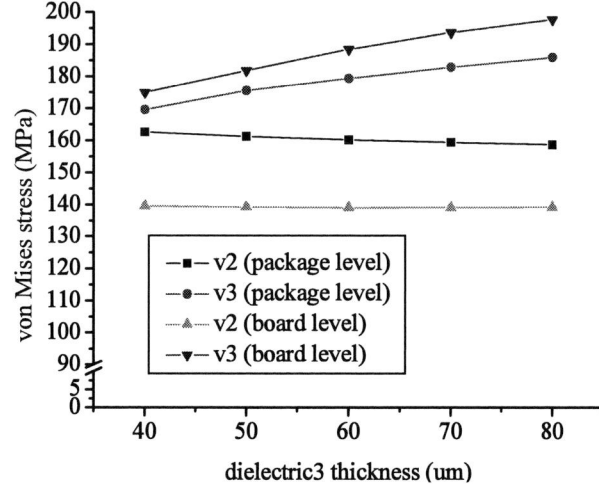

Fig.12: The influence of changing dielectric3 thickness.

Parameters study with 3D FEA

In generalizing the phenomenon of forward discussion of which the most important design variable is the position of the vias and the thickness of the adhesive, we investigated the stress concentration behavior in the 3D FE model. The sketch in Fig. 13 shows the different parameters studied in the 3D model, and these are described as follows: (a) the position of the via which moves along the X-axis; (b) the position of the via which moves along the Y-axis and is only located on the chip edge; (c) and the thickness of the adhesive material.

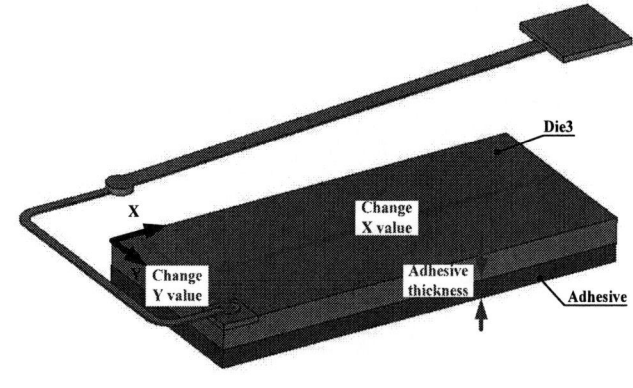

Fig. 13: Three different parameters study in 3D FE model

We located the via on the midline of the Y-axis, which is Y=500μm and was modified from 100μm to 1800μm along the X-axis, as shown in Fig.14. In the package-level structure, the via near the chip/polymer boundary condition induces stress concentration due to the expansion of the filler polymer. When the via moves along the X direction toward the other chip edge, a higher von Mises stress is produced both in the package-level and the board-level structures. Moreover, the von Mises stress in the board level is higher than the stress in the package level. In this study, the trend of 3D FEA is similar with the 2D results.

Second, we located the via on the place of X=100μm and changed the location of the via from 100μm to 900μm along the Y-axis, as shown in Fig.15. The result indicates that the stress concentration behavior on the corner of the chip is more serious than that of the others. This is due to the shear stress that will be induced by the CTE mismatch between the chip and the dielectric material when the distance to neutral point (DNP) is larger. Thus, we can infer that the worst position of the via is the location on the corner of the chip, which has worse reliability both in the board-level and in the package-level structures.

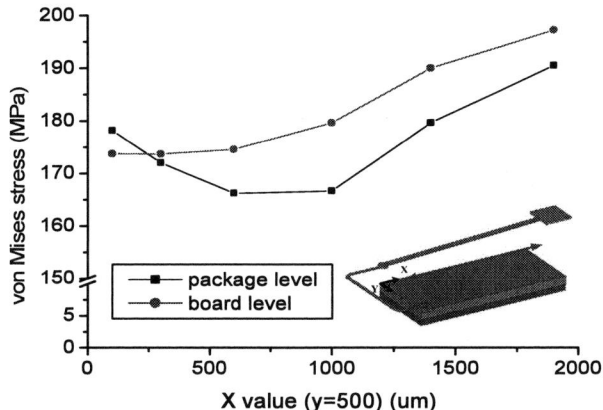

Fig.14: The influence of changing position in X direction in 3D FE model.

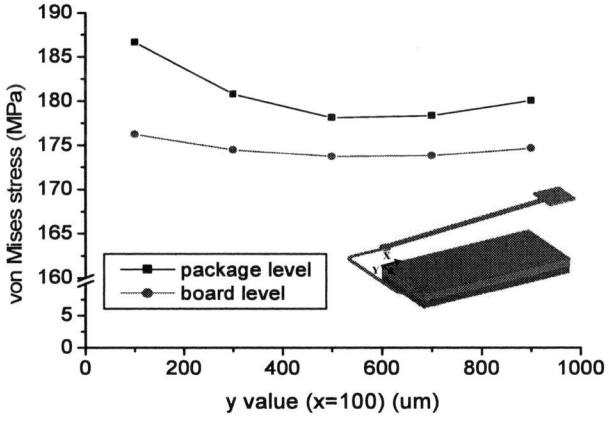

Fig.15: The influence of changing position in Y direction in 3D FE model.

Finally, we considered the influence of changing the adhesive thickness. We placed the via on X=100μm and Y=100μm, and then increased the thickness from 30μm to 150μm. As shown in Fig. 16, stress does not vary conspicuously when thickness increases. This is much different with the result of the 2D FEA in which increasing the thickness of the adhesive layer could reduce the stress on the vias. The reason why this occurs would be the element we used in the 2D model, which is in plane strain condition regarded as an infinity shell along the direction out of the 2D

plane. The stiffness in 2D assumption will be higher than that in the real one, confining the expansion of the filler polymer out of the 2D plane. The mobility of the chip in the 3D model will not be affected by the filler polymer..

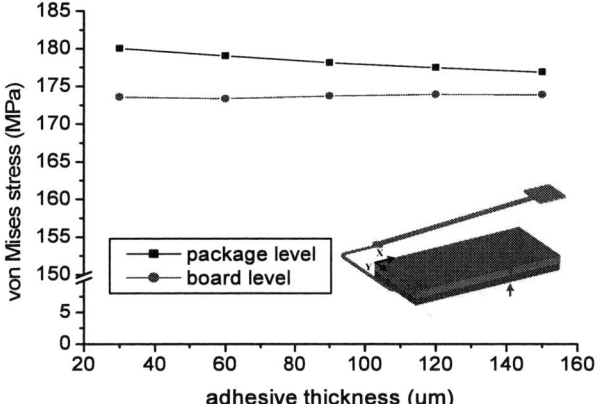

Fig.16: The influence of changing adhesive thickness in 3D FE model.

Conclusions

In this paper, we expanded the application of the novel WLCSP technology to the SiP structure. The reliability of the interconnection metal traces embedded in a large CTE material is an important issue for this specific package design. Both board-level and package-level FE models were discussed in this study, and the results indicate that there were two major driving forces for failure. One is due to the CTE mismatch among the copper interconnections, the filler polymer, and the chips. The other is due to the expansion of the filler polymer which will pull the copper interconnection, aggravating the stress concentration behavior especially at the chip/polymer edge both in 2D and 3D FEA. Results show that the mobility of the via will influence the concentrated stress laid upon it. Therefore, the fixed via might have a larger stress concentration. Although 2D FEA can quickly determine the major cause which influenced the package reliability, the plane strain condition used in the 2D model may not completely reveal the phenomenon of the real one. Thus, 3D FEA is not only needed to confirm the 2D results but it is also required that the real physical behaviors be investigated.

The utilization of WLCSP technology into a SiP structure could reduce the total cost involved in package processing and package size. However, the reliability of metal interconnection must be carefully considered. In this study, we presented the preliminary simulated results of this specific structure, but comparison with results of experimental data in further analysis is encouraged.

Acknowledgments

The authors would like to thank the National Science Council, Taiwan (Project NSC96-2221-E-007-070).

References

1. A. Badihi, "Ultrathin wafer level chip size package," *IEEE Transactions on Advanced Packaging*, Vol. 23, No.2, pp.212-214, 2000.

2. C. C. Lee, H. C. Liu and K. N. Chiang, "3D Structure Design and Reliability Analysis of Wafer Level Package with Bubble-Like Stress Buffer Layer," *Proceedings of 9th Intersociety Conference on Thermal and Thermomechanical Phenomena in Electronic Systems*, Las Vegas, USA, pp. 317-324, 2004.

3. T. Kawahara, "SuperCSP™," *IEEE Transactions on Advanced Packaging*, Vol. 23, No. 2, pp. 215-219, 2000.

4. Cadmus Yuan, G. Q. Zhang, C. S. Huang, C. H. Yu, C. C. Yang, W. K. Yang, M. C. Yew, C. N. Han, and K. N. Chiang, "Design and analysis of a novel fan-out WLCSP Structure," *A Proceedings of the 7th International conference on Thermal, Mechanical and Multi-Physics Simulation and Experiments in Micro-Electronics and Micro-Systems*, EuroSIME2006, Como (Milano), Italy, April 2006, pp.297-304.

5. H. P. Wei, M. C. Yew, W. K. Yang, K. N. Chiang, "Reliability Analysis of a Package-on-Package Structure using the Novel WLCSP Technology with Fan-Out Capability," *The 8th International Conference on Electronics Materials and Packaging*, EMAP 2006, 11-14 December 2006, Hong Kong.

6. M. C. Yew, C. J. Wu, K. N. Chiang, "Trace Line Failure Analysis and Characterization of the Panel Base Package (PBP™) Technology with Fan-Out Capability," *Proceedings of the 11th Intersociety Conference on thermal and thermomechanical Phenomena in Electronic Systems*, ITherm2008, Florida, USA, May 2008, pp. 862-869.

A Multi-scale Interfacial Delamination Model of Cu-SAM-Epoxy Systems

H. B. FAN, Cell K.Y. Wong, Matthew M.F. YUEN
Department of Mechanical Engineering,
Hong Kong University of Science and Technology
Clear Water Bay, Kowloon, Hong Kong SAR, China

Abstract

Interfacial delamination, due to the presence of dissimilar material systems, is one of the primary concerns in electronic package design. The mismatch in coefficient of thermal expansion between the different layers in the packages can generate high interfacial stresses due to thermal loading during fabrication and assembly.

More and more functional materials at the nano scale are, such as self-assembly monolayer (SAM) and CNT, used in electronic packaging for the improvement of the interfacial performance, traditional continuum model without considering these nano materials are obviously not suitable to study performance of electronic packages. In this study, a multi-scale model was built to investigate interfacial failure between EMC and SAM coated copper substrate. The interfacial material behavior was derived from the molecular dynamics simulation. The constitutive relation for the EMC-SAM-Cu interface under tensile load was derived from MD simulation. Tapered double cantilever beam tests (TDCB) were conducted on the laminated specimens to quantify the load during delamination propagation along the EMC-Cu interface with SAM and without SAM. Finite element models of the DCB test were built using ANSYS with interfacial element at the Cu-EMC interface. The constitutive relations from MD simulations in the form of a traction-displacement plot were introduced into the cohesive zone model to study the constitutive response of the EMC-Cu interface under the tensile loading, which is traversed across the length scale from nanoscale to macroscale. and assigned to the continuum model. The critical loading forces for the EMC/Cu interface with SAM and without SAM were obtained from the multi-scale model. It was found that interfacial strength between EMC and Cu substrate could be improved by SAM. Based on the proposed method, the predicted results were found to be comparable with those from experimental measurement.

Introduction

Interfacial delamination at material interface is one of the key failure modes in electronic packages. This is particularly prevalent at interface with materials having a significant mismatch in coefficient of thermal expansion. Thermal loading during various stages of the fabrication, assembly and qualification will induce high stresses in the electronic package leading to delamination.

A number of approaches have been developed to investigate interfacial delamination. It is generally accepted that the propagation of interfacial delamination can be evaluated by the fracture failure criteria. The crack growth along bi-material interface is strongly dependent on the mode mixity of the opening mode (mode I) and shearing mode

(mode II) [1]. Evans and Hutchinson [2] summarized the existing knowledge in the interfacial fracture of layered materials. However, more and more functional materials at the nano scale are used in electronic packaging for the improvement of the interfacial performance, traditional continuum model without considering these nano scale material properties are obviously not apt for the performance study of electronic packages.

Due to the difficulty of measuring nano-materials behavior, Molecular dynamics (MD) simulation is a well-established tool for modeling the material performance at an atomistic level including modulus, adhesion, thermal conductivity, solubility, diffusion and reactivity [3-10]. With the limitation of MD method, it is impossible to build the full package model using MD technique due to long calculation times and costly calculations. Several methodologies on how to couple nano-scale models and continuum models for studying material performance of composites have been established, which includes hand-shaking method [11], coarse-grained molecular dynamics (CGMD) method [12] and virtual internal bond method [13-16]. However, there are still lots of issues in using these methods, including difficulty in implement computational technique and mismatch of time scale. The Virtual Internal Bond (VIB) model has been well developed to investigate failures in bi-material systems on both macro- and micro-levee [12-14]. But, it is suitable for the bulk materials rather than description of atomic interaction along a prescribed interface. Moreover, VIB based on the simplified atomic potentials without considering bond torsion, bending and electronic static force is not enough to describe the complicated reality of the material at nanoscale.

In this study, an interfacial MD model was built to study atomistic behavior of the interface under tensile loading. The model contains epoxy resin polymer connecting a Cu substrate by SAM. MD simulations were conducted to derive the constitutive response of the interfacial MD model under tensile strain applied on the system. Finite element model of DCB test was built and the constitutive relation derived from the interfacial MD model was assigned to the TDCB model representing the behavior of the SAM at the EMC/Cu interface. Based on the multi-scale model, the material behavior at nanoscale was passed onto the continuum model under tensile loading condition.

Molecular Dynamics Simulation

The Materials Studio software (Accelrys Inc.) was used for the amorphous model construction and the atomic simulations. The condensed-phase optimized molecular potential for atomistic simulation studies (COMPASS) force field was used in the simulation. COMPASS force field enables accurate and simultaneous prediction of structural, conformational, vibrational, and thermophysical properties for

978-1-4244-2739-0/08/$25.00 ©2008 IEEE   285

a broad range of molecules in isolation and in condense phases under a wide range of conditions of temperature and pressure. The advantage of the COMPASS force field is its high accuracy in predicting the molecular properties of polymers, metals and their interfaces.

Delamination, especially that occurring between copper (Cu) and Epoxy Molding Compound (EMC) in plastic packages, is one of the major problems in electronic packaging design. Self-assembly monolayer (SAM) has been suggested as adhesion promoter of EMC-Cu system. Having an organic end, thiol can largely enhance adhesion in Cu-EMC system through strong covalent linking (C-S-Cu). In this study, one SAM candidate was involved and their chemical structure is shown in Figure 1.

$$NH_2 \!-\!\!\!\bigcirc\!\!\!- SH_2$$

Figure 1: Chemical structures of SAM

In this study, the bulk materials, EMC and the Cu substrate were connected by the SAM. SAM solvent develops a thin monolayer on the Cu surface that allows direct linkage between Cu and epoxy resin. The fully cured epoxy network is composed of Diglycidyl Ether of Bisphenol-A (DGEBA) epoxy resin reacted with Methylene Diamine Dianilene (MDA) curing agent. A method was presented by Wong et al. [9] on how SAM was adsorbed on the Cu substrate and connected the EMC and the Cu substrate. Based on same method, the fully cured epoxy network was layered with a copper surface cleaved from a crystal copper structure, corresponding to the (001) plane. To avoid surface effect, periodic boundary conditions in all directions were used. All the chains could move freely. Interface roughness affects many physical and mechanical properties at different length scales, even atomic scale roughness. In this study, copper substrate with roughness of 3nm was considered in the simulation. Two kinds of MD models were built: One is model with SAM treated on the copper substrate, the other is model without SAM on the copper substrate, and the others are. Figure 2 shows the morphological configurations of the systems with minimum potential energy.

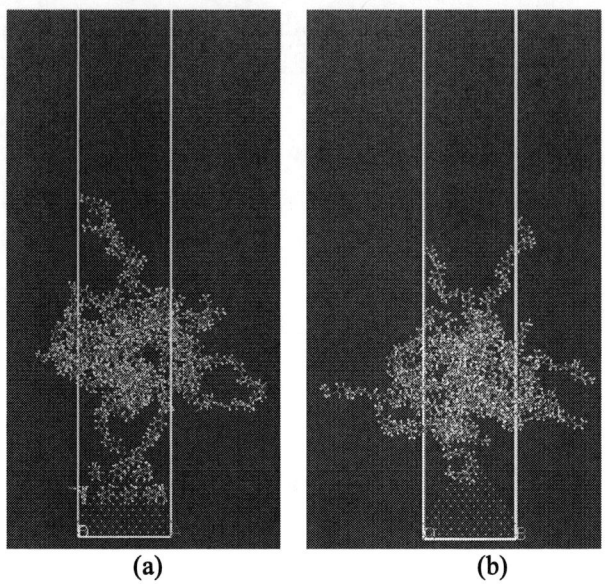

(a)   (b)

Figure 2: MD model of the EMC-Cu system (a) with SAM and (b) without SAM

A tensile displacement was applied on the system in a single simulation step and energy equilibration was conducted to relax the whole system with the maintained displacement. Next same displacement was applied on the system and the procedure was repeated until interfacial bonds were broken. In each simulation, the whole system was equilibrated for about 100ps at a temperature of 25°C using the ensemble of the constant number of particles, constant-volume and constant temperature (NVT). Interfacial stresses can be obtained for the applied displacement from MD simulations. The interfacial stress is simply related to the interfacial energy by the following relation:

$$\sigma = (E_{tot} - (E_{EMC} + E_{Cu})/(A \cdot \Delta d) \qquad (1)$$

where σ is the interfacial tensile stresses. A is the interfacial area and Δd is the displacement applied on the system in each simulation step. E_{tot} is the total energy of the whole system, E_{EMC} is the energy of the EMC without the Cu substrate, E_{Cu} is the energy of the Cu substrate without the EMC.

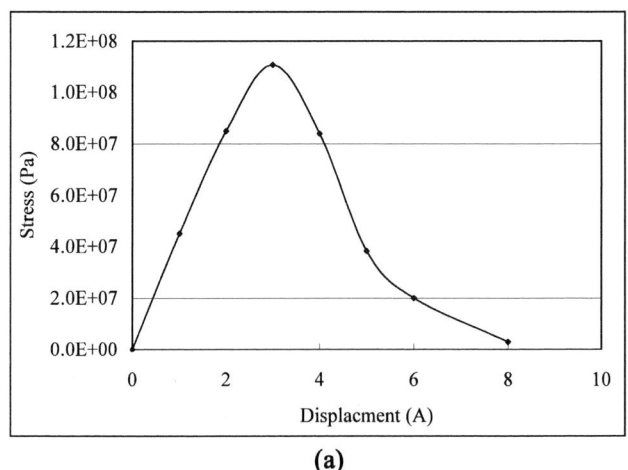

(a)

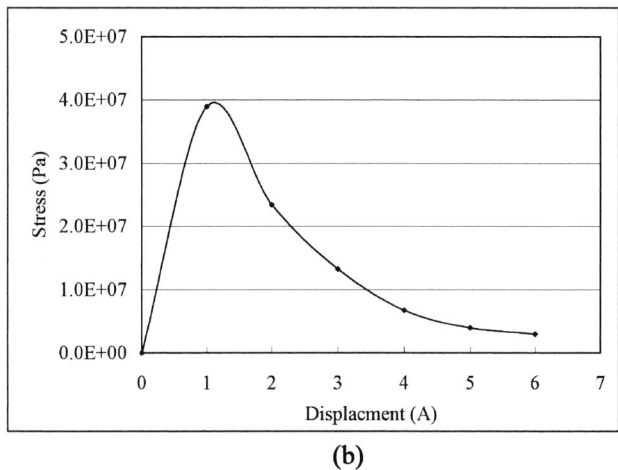

(b)

Figure 3: Stress-displacement relation for the EMC-Cu system (a) with SAM and (b) without SAM

Figure 3 shows the stress-displacement relation obtained from MD simulations. The relation for both the tensile stress and shear stress show the nonlinear behavior of the interface, increasing stress first and then decreasing stress with the increasing displacement. In each curve there is a bifurcation point A, where the continuous failure mode will transfer to a discontinuous failure mode. At the bifurcation point, the interaction force between the atoms will be at the highest value where the macroscopic response of the interface is the initiation of delamination. After the bifurcation point, the discontinuous failure mode takes over and the stress-strain curve describes the de-adhesion process, such as the propagation of cracks.

Multi-scale Model of Tapered Double Cantilever Beam Test

Tapered double cantilever beam (TDCB) test was carried out to evaluate fracture toughness the tensile adhesion between EMC and copper substrate under tensile mode. The advantage of using TDCB specimen is that fracture toughness can be directly obtained by measuring the critical loads for delamination initiation. In this study, two kinds of samples were prepared for TDCB test: one is sample with SAM treated at the EMC/Cu interface, the other is sample without any treatment at the interface (control sample).

In order to study the interfacial delamination of EMC and Cu substrate under mechanical loading, it is necessary to perform finite element analysis to extract some useful information. A more realistic multi-scale model is shown in Fig. 4. The mesh was refined at the interface between the EMC and copper to capture the steep stress gradients expected. A pre-crack with length of 39mm was made at the interface as shown in Figure, and cohesive element was used at the EMC/Cu interface to describe the behavior of SAM and the initial thickness of the element is set to zero. Both EMC and Cu materials are assumed to be linear elastic, homogeneous and isotropic. The constitutive relations derived from the interfacial MD model as shown in Fig. 3 is assigned to the cohesive zone elements as the description of the atomic interaction between the EMC and copper substrate. Boundary conditions and loading were applied to the model as those in the TDCB test. The displacements of nodes at the surface of the hole in the bottom block were constrained and tensile displacement was applied on the surface of the top hole.

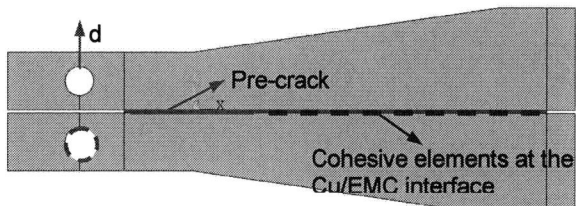

Figure 4: The diagram of bi-material system assembled with EMC and copper leadframe

Results and Discussion

The tensile forces were calculated for the multiscale model under the tensile displacement and plotted against the displacement for three kinds of models respectively, as shown in Fig. 5. All tensile force increased with the increased displacement and reached the maximum value where delamination initiated, and then decreased to zero for the remainder of the displacement. The higher maximum tensile force means the higher adhesion between EMC and Cu substrate. Predicted result from TDCB test simulations showed that the adhesion force between EMC and SAM treated Cu substrate was higher than that for the control sample.

Experimental results were also shown in Figure 5. It can be seen that all predicted results from simulations were a little bit higher that those experimental results. The difference between the simulation and experimental values can be attributed to more complicated crosslink density of EMC, SAM structure on the Cu substrate and bonds between EMC and SAM. Moreover, voids or impurity inside the real samples can also degrade the interfacial properties.

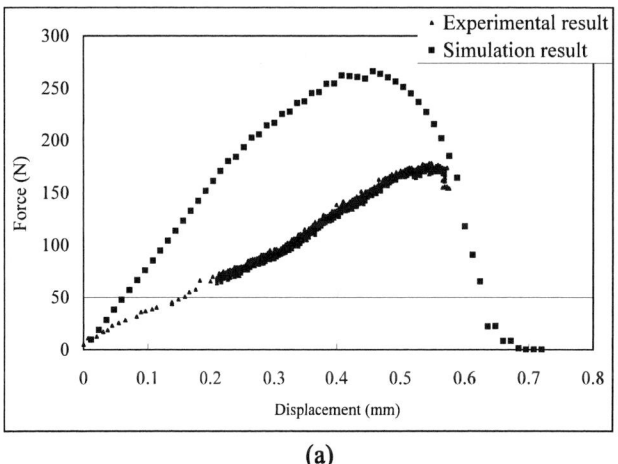

(a)

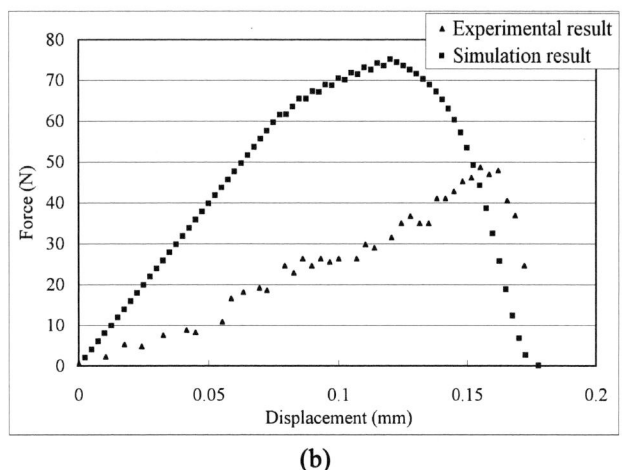

(b)

Figure 5: Force-displacement curve for the EMC-Cu system
(a) with SAM and (b) without SAM

The mulitscale model presented in this study has led to some interesting results providing a method on how the material behavior at nanoscale was passed on to the continuum model. The model is suitable for the explanation of the phenomena in the TDCB test. In contrast to other multi-scale methods, the method presented in this study has significant advantages.

Compared to other multi-scale methods, the model in this study combines different spatial and temporal scales together and the interfacial material property as the key part of the model is directly derived from the MD simulation. The constitutive relation derived from MD simulation is based on simple calculations, which avoids the complicated numerical equations to derive the constitutive relation. Complicated numerical equations to solve the overlapping domain involved in the method for coupled continuum models with molecular models are not needed in the multiscale model. The performance of nanostructure can be directly inlaid in the finite element model by using the cohesive element helps to adopt the molecular based material constant into continuum mechanics element.

Considering the difficulty on experimental measurement of the behavior of nano materials, such as SAM in this study, at the interface, MD simulations are used to model not only deformation of SAM between the EMC and Cu substrate but also defects initiation and defect prorogation of the system under external loading. Based on the method presented in this study, the atomistic information including deformation, void nucleation and interfacial debonding were extracted and represented by the constitutive relation. Therefore, a general framework of multi-scale model was developed to study the macroscopic behavior of the system consisting of nano scale materials.

Conclusions

The paper is focused on failure of EMC/copper interface using the TDCB as a testing platform. A new methodology using a multi-scale model to predict failure force at the interface is proposed. The interfacial constitutive relation is derived from MD simulations. The methodology

has shown the promising result indicating the method can lead to a meaningful method for prediction of materials properties of system containing nano scale materials in electronic packages.

Acknowledgment

The author would also like to thank Prof. Ping Gao for providing the software for the computer analysis.

References

[1] C. F. Shih, "Cracks on Bimaterial Interfaces: Elasticity and Plasticity Aspects," *Materi. Sci. Eng.*, vol. A143, pp. 77-90, 1991.

[2] A. G. Evans, and J. W. Hutchinson, "Mixed Mode Cracking in layered Materials," *Adv. Appl. Mech.*, vol. 29, pp. 63-201, 1991.

[3] Tanaka G. and Goettler L. A., "Predicting the Bonding Energy for Nylon 6,6/Clay nanocomposites by Molecular Modelling," Polymer, 43, pp.541-553, 2002.

[4] Gou J., Minaie B., Wang B., Liang Z. Y. and Zhang C., "Computational and Experimental Study of Interfacial Bonding of Single-Walled Nanotube Reinforced Composites," Comput. Mater. Sci., 31(3-4), pp. 225-236, 2004

[5] N. Iwamoto, L. Moro, B. Bedwell, P. Apen, "Understanding Modulus Trends in Ultra Low k Dielectric Materials Through the Use of Molecular Modeling," Proceedings of the 52nd Electronic Components and Technology Conference, May 28-31, San Diego, CA pp 1318-1322, 2002.

[6] H. B. Fan, Edward, K. L. Chan, Cell, K.Y. Wong and M. F. F. Yuen, "Investigation of Moisture Diffusion in Electronic Packaging by Molecular Dynamic Simulation," J. Adhesion Sci. Technol. 20:1937-1947, 2006.

[7] H. B. Fan, Edward, K. L. Chan, Cell, K.Y. Wong, and Matthew, M. F. Yuen "Molecular Dynamic Simulation of Thermal Cycling Test in Electronic Packaging," ASME Journal of Electronic Packaging, 129:35-40, 2007.

[8] H. B. Fan, and M. F. F. Yuen, "Material properties of the Cross-linked Epoxy Resin Compound Predicted by Molecular Dynamics Simulation," Polymer 48:2174-2178, 2007.

[9] Cell, K.Y. Wong, H. B. Fan, and M. F. F. Yuen, "Interfacial adhesion study for SAM induced covalent bonded Copper-EMC interface by Molecular Dynamics Simulation," IEEE Transactions on Components and Packaging Technologies, Vol. 31, pp.297-308, 2008.

[10] H. B. Fan, K. Zhang, M M F Yuen, "Effect of defects on thermal performance of carbon nanotube investigated by molecular dynamics simulation." In: Proc. EMAP, Hong Kong, pp 451-454, 2006.

[11] E. Lidorikis, M. E. Bachlechner, R. K. Kalia, A. Nakano, P. Vashishta, and J. Voyiadjis, ''Coupling Length Scales for Multiscale Atomistics-Continuum Simulations: Atomistically Induced Stress Distributions in Si/Si3N4 Nanopixels,'' Phys. Rev. Lett., 87, p. 086104, 2001.

[12] R. E. Rudd, and J. Q. Broughton, ''Concurrent Coupling of Length Scales in Solid State Systems,'' Phys. Status Solidi B, 217, pp. 251–291, 2000.

[13] H. Gao and P. Klein, "Numerical Simulation of Crack growth in an Isotropic Solid with Randomized Internal Cohesive Bonds," J. Mech. Phys. Solids, 46, 187-218, 1998.

[14] P. Klein and H. Gao, "Crack Nucleation and Growth as Strain localization in a Virtual-bond Continuum," Engng. Fracture Mech. 61, 21-48, 1998.

[15] B. Ji and H. Gao, "A Study of fracture mechanisms in Biological Nano-composites via the Virtual Interbal Bond Model," Mater. Sci. Eng. A, 366, 96-103, 2004.

[16] H. Gao and B. Ji, "Modeling Fracture in nanomaterials via A Virtual Internal Bond method," Engng. Fracture Mech., 70, 1777-1791, 2003.

Three Dimensional Corner Delamination Analysis for Fan-Out Chip Scale Package

Yu-Ren Chen[1], G. S. Shen[1]
Hung-Chun Yang[2], Huang-Chun Lin[2], Tz-Cheng Chiu[2],
1. ChipMOS Technologies Ltd, Tainan, Taiwan, China
2. Department of Mechanical Engineering, National Cheng Kung University, Tainan, Taiwan, China
E-mail: stoke_chen@chipmos.com

Abstract

The problem of a corner delamination in a fan-out chip scale package subjected to thermomechanical load is investigated. The fracture mechanics parameters, including the stress intensity factors, the strain energy release rate, and phase angles, for a quarter-circular corner delamination between silicon die and fan-out redistribution polyimide layer are obtained by using numerical finite element approach with 3D virtual crack closure technique (VCCT). Results of the analysis indicated that contact between crack faces occurs under temperature cycling condition. The delamination driving forces are modes-II and –III dominant. In addition, the strain energy release rate is highest near the location where the delamination crack front intersects die-to-molding compound interface. Therefore, the delamination is expected to propagate first around the periphery of die surface. Parametric study is also conducted to investigate the effects of material properties and geometrical dimensions on the delamination driving forces. The calculated fracture mechanics parameters may be combined with data on the resistance to interface fracture for predicting reliability of the package.

Keywords: Delamination, Fan-Out Package, Virtual Crack Closure Technique (VCCT)

I. Introduction

Driven by the need for electrical performance improvement and system integration, the interconnection between silicon die and its package has become one key area for packaging technology development. One of the interconnection schemes received a lot of attention lately is die surface fan-out redistribution interconnect [1, 2]. In a fan-out package the silicon die is first encapsulated in molding compound except its top surface. A metal-polymer redistribution layer is then fabricated on top for interconnection between die-level bond pads and BGA pads. This type of interconnect involves structures in thin-layered configuration and materials having high coefficients of thermal expansion. As a result, the structure is susceptible to interfacial delamination during package assembly or under service condition. It is therefore important to understand the interaction of materials and geometry and their effects on delamination driving forces to ensure reliability of the fan-out package.

Debonding occurring on the interfaces of dissimilar materials is typically initiated from corner or edge where stress concentration is the highest. Growth of the interface defect, therefore, is a 3D problem in nature. Since the analytical solutions for 3D interface crack problem are limited to cases with simplified geometry, numerical finite element approach is employed to investigate the silicon-polyimide interfacial delamination initiated from the corner of die in a fan-out package. The fracture driving forces, including strain energy release rate, stress intensity factors and phase angles are obtained by using collapsed quarter-point singular crack-tip finite element with the 3-D virtual crack closure technique (VCCT) [3,4]. In this paper, a brief description on the fracture mechanics parameters for interface fracture and the corresponding 3-D VCCT technique applied to calculate these parameters are first given. To validate the proposed approach, the problem of a quarter circular corner crack subjected to remote mechanical load is investigated by using the 3-D VCCT and compared to solutions from other approaches. The approach is then applied to determine the fracture mechanics parameters along a realistic curvilinear corner delamination crack front in a fan-out package under uniform cooling from 125°C to -40°C. The effects of material properties and geometrical dimensions on the delamination driving forces are also investigated. The calculated fracture mechanics parameters may be combined with data on the resistance to interface fracture for predicting reliability of the package.

II. Fracture Mechanics Parameters

The asymptotic stress field at the crack tip of an interface crack between two dissimilar isotropic materials can be expressed as [5]

$$
\sigma_{ij} = \frac{1}{\sqrt{2\pi r}} \left\{ \mathrm{Re}[Kr^{i\varepsilon}] \, \tilde{\sigma}_{ij}^{\mathrm{I}}(\theta,\varepsilon) \right. \\
\left. + \mathrm{Im}[Kr^{i\varepsilon}] \, \tilde{\sigma}_{ij}^{\mathrm{II}}(\theta,\varepsilon) + K_{\mathrm{III}} \, \tilde{\sigma}_{ij}^{\mathrm{III}}(\theta) \right\},
$$

(1)

where r and θ are the in-plane coordinates of the plane normal to the crack front, K is the complex stress intensity factor, $K = K_1 + iK_2$, and are the angular variations of stress components for each mode. The oscillatory index is defined as

$$
\varepsilon = \frac{1}{2\pi} \ln \left[\frac{\mu_1 + \kappa_1 \mu_2}{\mu_2 + \kappa_2 \mu_1} \right],
$$

(2)

where $\kappa = 3 - 4\nu$ for plane strain, μ is the shear modulus, is the Poisson's ratio and the subscripts = 1 and 2 refer to the materials above and below the crack plane, respectively. It may be seen from (1) that the complex stress intensity factor K has a unit of $(\text{stress}) \times (\text{length})^{1/2-i}$ and varies with r near the interface crack tip. This oscillating nature makes characterizing the stress intensity factors difficult. To overcome this issue, Rice [6] suggested an alternative definition of stress intensity factors for the interface crack

$$\hat{K} = K_I + iK_{II} = KL^{i\varepsilon}, \tag{3}$$

where L is a certain material length, invariant to crack length or other geometric dimensions in different applications. It may be seen that $\hat{K}$ has a unit of (stress)$\times$(length), which is the same as that for the classical stress intensity factor. The corresponding mode-mixities or phase angles are given by

$$\psi = \tan^{-1}\left(\frac{K_{II}}{K_I}\right), \quad \varphi = \tan^{-1}\left(\frac{K_{III}}{K_I}\right). \tag{4}$$

The in-plane and anti-plane crack opening displacements at a small distance ρ behind the crack tip are given, respectively, by

$$\delta_2 + i\delta_1 = \frac{c_1 + c_2}{2\sqrt{2\pi}\,(1 + 2i\varepsilon)\cosh(\pi\varepsilon)}\hat{K}\rho^{1/2}\left(\frac{\rho}{L}\right)^{i\varepsilon},$$

$$\delta_3 = \sqrt{\frac{2}{\pi}}\left(\frac{1}{\mu_1} + \frac{1}{\mu_2}\right)K_{III}\rho^{1/2}, \tag{5}$$

$$c_1 = \frac{\kappa_1 + 1}{\mu_1}, \quad c_2 = \frac{\kappa_2 + 1}{\mu_2}.$$

The virtual crack closure technique is based on the hypothesis that crack growth is self-similar and the released strain energy during crack growth is the same as the amount of work required to close the crack. Therefore, the strain energy release rate for a crack to extend by an infinitesimal distance may be expressed as

$$G = G_I + G_{II} + G_{III},$$

$$G_I = \frac{1}{2\Delta}\int_0^\Delta \sigma_{22}(r,0)\delta_2(\Delta - r)\mathrm{d}r,$$

$$G_{II} = \frac{1}{2\Delta}\int_0^\Delta \sigma_{12}(r,0)\delta_1(\Delta - r)\mathrm{d}r, \tag{6}$$

$$G_{III} = \frac{1}{2\Delta}\int_0^\Delta \sigma_{13}(r,0)\delta_3(\Delta - r)\mathrm{d}r.$$

Unlike the crack problem in homogeneous medium, the mode I and II stress intensity factors K_I and K_{II} can not be determined from G_I and G_{II}, respectively. Instead, an additional integral, which is given by

$$G_{I\text{-}II} = \frac{1}{2\Delta}\int_0^\Delta [\sigma_{22}(r,0)\delta_1(\Delta - r) + \sigma_{12}(r,0)\delta_2(\Delta - r)]\mathrm{d}r, \tag{7}$$

is introduced [3]. From (1), (5-7) it may be shown that, for an "opening" crack tip, the stress intensity factors can be determined from the following relationship:

$$K_I^2 = \frac{8\cosh^2(\pi\varepsilon)}{c_1 + c_2}\{(G_I + G_{II}) + \mathrm{Re}[A_1](G_I - G_{II}) - \mathrm{Im}[A_1]G_{I\text{-}II}\}, \tag{8a}$$

$$K_{II}^2 = \frac{8\cosh^2(\pi\varepsilon)}{c_1 + c_2}\{(G_I + G_{II}) - \mathrm{Re}[A_1](G_I - G_{II}) + \mathrm{Im}[A_1]G_{I\text{-}II}\}, \tag{8b}$$

$$K_{III}^2 = \frac{4\mu_1\mu_2}{\mu_1 + \mu_2}G_{III},$$

where

$$A_1 = \frac{\pi\left(\frac{1}{2} + i\varepsilon\right)}{\cosh(\pi\varepsilon)\mathrm{B}\left(\frac{1}{2} + i\varepsilon, \frac{3}{2} + i\varepsilon\right)}, \tag{9}$$

B being the Beta function. Note that the signs of the stress intensity factors can be determined by inspecting the signs of the stress components at the crack tip.

When large-scale contact occurs between crack faces, the crack opening displacement δ_2 equals to zero for the closed crack tip. From (5), it can be seen that K_I is zero when the crack tip is closed. The relationship between the mode-II stress intensity factor and G_{II} is given by

$$K_{II}^2 = \frac{16}{(c_1 + c_2)(1 - \beta^2)}G_{II},$$

$$\beta = \frac{\mu_1(\kappa_2 - 1) - \mu_2(\kappa_1 - 1)}{\mu_2(\kappa_1 + 1) + \mu_1(\kappa_2 + 1)}. \tag{10}$$

The relationship between KIII and GIII described in (17) is also valid for closed crack tip, since the anti-plane mode is not coupled to the in-plane modes.

By following Roeck and Abdel Wahab [4], finite element discretization for calculating (6) and (7) is implemented for collapsed quarter-point singular crack-tip elements. For the schematic of quarter-point finite element shown in Fig. 1 (crack faces are on the negative-x,z plane, while the crack front is along z-axis),

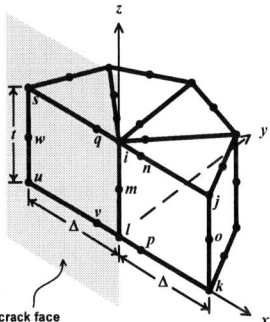

Fig. 1: Collapsed quarter-point singular elements.

the strain energy release rate G_I is given by

$$G_{\mathrm{I}} = \frac{1}{2t\Delta}\left[F_y^i\left(C_1\delta_y^q + C_2\delta_y^w - 2C_2\delta_y^s - \frac{C_2}{2}\delta_y^u \right) \right.$$
$$+ F_y^l\left(C_2\delta_y^w - \frac{C_2}{2}\delta_y^s - 2C_2\delta_y^u + C_1\delta_y^v \right)$$
$$+ F_y^m\left(\frac{C_1}{2}\delta_y^q + C_3\delta_y^w + C_4\delta_y^s + C_4\delta_y^u + \frac{C_1}{2}\delta_y^v \right) \quad (11)$$
$$\left. + F_y^n\left(\delta_y^q + \frac{1}{2}\delta_y^s \right) + F_y^p\left(\frac{1}{2}\delta_y^u + \delta_y^v \right) \right],$$
$$C_1 = 6\pi - 20, \quad C_2 = \pi - 4,$$
$$C_3 = \pi - 2, \quad C_4 = -\frac{5}{4}\pi + 4,$$

where F is the nodal force, δ is the crack opening displacement, and the superscript denotes nodal location shown in Fig. 1. Similarly, G_{II} and G_{III} can be calculated by replacing (F_y, δ_y) in (11) with (F_x, δ_x) and (F_z, δ_z), respectively. The finite element discretization for $G_{\mathrm{I\text{-}II}}$ is given by

$$G_{\mathrm{I\text{-}II}} = \frac{1}{2t\Delta}\left[F_x^i\left(C_1\delta_y^q + C_2\delta_y^w - 2C_2\delta_y^s - \frac{C_2}{2}\delta_y^u \right) \right.$$
$$+ F_x^l\left(C_2\delta_y^w - \frac{C_2}{2}\delta_y^s - 2C_2\delta_y^u + C_1\delta_y^v \right)$$
$$+ F_x^m\left(\frac{C_1}{2}\delta_y^q + C_3\delta_y^w + C_4\delta_y^s + C_4\delta_y^u + \frac{C_1}{2}\delta_y^v \right)$$
$$+ F_x^n\left(\delta_y^q + \frac{1}{2}\delta_y^s \right) + F_x^p\left(\frac{1}{2}\delta_y^u + \delta_y^v \right)$$
$$+ F_y^i\left(C_1\delta_x^q + C_2\delta_x^w - 2C_2\delta_x^s - \frac{C_2}{2}\delta_x^u \right) \quad (12)$$
$$+ F_y^l\left(C_2\delta_x^w - \frac{C_2}{2}\delta_x^s - 2C_2\delta_x^u + C_1\delta_x^v \right)$$
$$+ F_y^m\left(\frac{C_1}{2}\delta_x^q + C_3\delta_x^w + C_4\delta_x^s + C_4\delta_x^u + \frac{C_1}{2}\delta_x^v \right)$$
$$\left. + F_y^n\left(\delta_x^q + \frac{1}{2}\delta_x^s \right) + F_y^p\left(\frac{1}{2}\delta_x^u + \delta_x^v \right) \right].$$

III. Validation of the 3-D VCCT

The problem of a quarter-circular corner crack as shown in Fig. 2 is considered as a validation of the approach described in Section II.

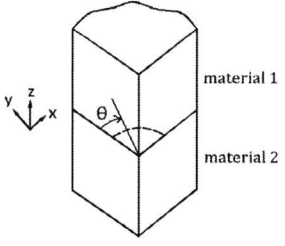

Fig. 2: Quarter circular corner crack (materials 1 & 2 are identical).

Figure 3 shows the finite element mesh around the crack tip.

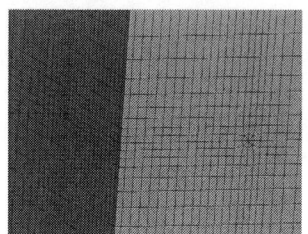

(a) Mesh around crack tip

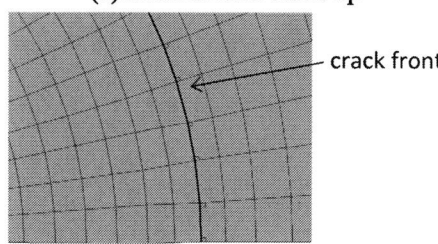

(b) Mesh along the crack front

Fig. 3: Finite element mesh for the corner crack problem.

It may be seen from Fig. 3 that, in order to satisfy the self-similar crack growth assumption, singular elements with identical dimensions are used in-front and behind of the crack tip. Numerical results for the crack problem are obtained by post-processing the finite element solutions calculated using ANSYS. Figure 4 shows the stress intensity factor K_{I} for the crack shown in Fig. 2 (assuming material 1 is the same as material 2) under uniform remote tensile loading $\sigma_z = \sigma_0$.

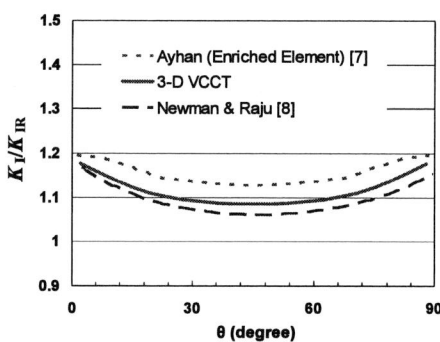

Fig. 4: Mode I stress intensity factors for the quarter circular crack, $K_{\mathrm{IR}} = 2\sigma_0(a/\pi)^{1/2}$.

It can be seen from Fig. 4 that K_{I} values obtained from the 3-D VCCT is very close to those obtained by enriched element approach [7] and by nodal force method [8].

IV. Corner Delamination in Fan-Out Package

In this section, a corner delamination on the interface of Si die and redistribution polyimide layer in a fan-out package subjected to uniform cooling from 125°C to -40°C is considered by using the 3-D VCCT. The fan-out package with dimensions depicted in Fig. 5 is considered by using a quarter model with symmetric boundary conditions.

292

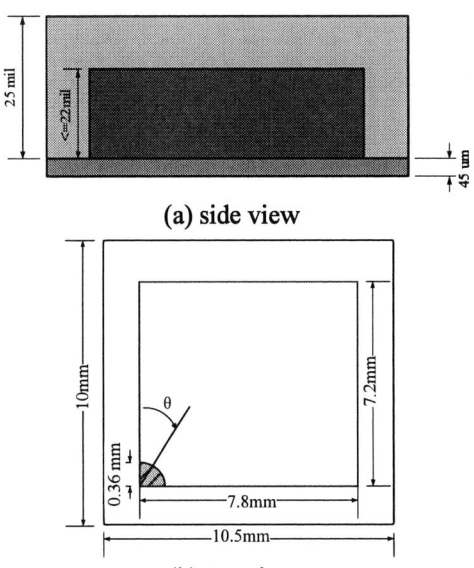

(a) side view

(b) top view

Fig. 5: Schematics of a fan-out package containing an interface corner crack between die and the redistribution layer.

Finite element mesh for the model is shown in Fig. 6.

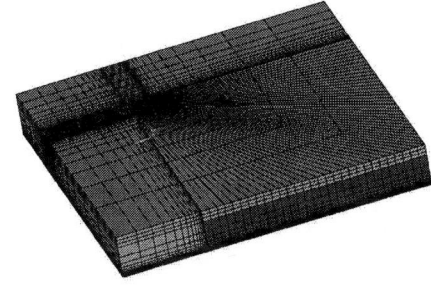

Fig. 6: Finite element mesh for the fan-out package containing an interface crack.

Material properties of the fan-out package are given in Table 1.

Table 1: Material properties for the fan-out package.

Material	E (GPa)		. (ppm/°C)
Si	150	0.17	2.9
mold compound MC1	23(-60°C) 17(40°C) 12.8(146°C)	0.33	8(-30°C) 11(90°C) 17(135°C)
mold compound MC2	27.4(-50°C) 25.4(10°C) 23.3(120°C) 19.9(150°C)	0.33	5.5(-40°C) 7.2(135°C)
polyimide	3.3	0.3	55

Results of the finite element analysis are shown in Figs. 7-9.

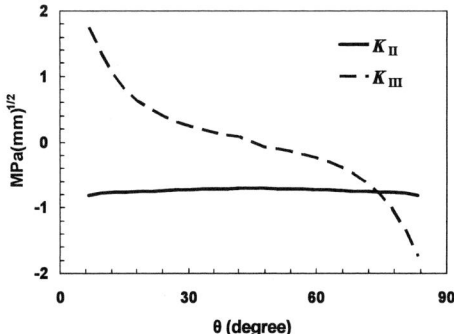

Fig. 7: Stress intensity factors for the interface corner crack in fan-out package with MC1 compound, die thickness: 22mil, L = 1mm.

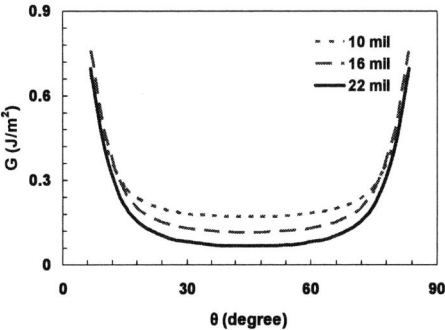

Fig. 8: Strain energy release rate for the interface corner crack in the fan-out package with MC1 compound.

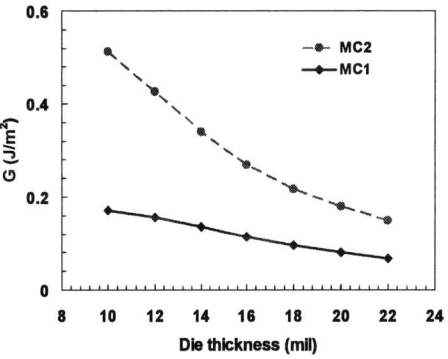

Fig. 9: Strain energy release rate at θ = 45°for the interface corner crack in the fan-out package.

It is observed from the analysis that the crack faces of the corner delamination are in contact at -40°C. Therefore, it can be concluded that the delamination driving force does not have mode-I component. From Fig. 7 it can be seen that the mode-II stress intensity factor remains relatively constant for all angles. On the other hand, the mode-III stress intensity factor is most significant near the periphery of die (θ = 0 or 90°). This is reflected in the plot of strain energy release rate vs. crack front location, which is shown in Fig. 8. It may be seen from Fig. 8 that the strain energy release rate is higher near the periphery of die. This is indicating the likelihood of faster delamination growth around die periphery.

It is observed from Figs. 8 and 9 that, as die thickness increases, the strain energy release rates around the interior part of the crack front reduce. Consequently, the risk of delamination growth is expected to be lower for package with thinner die. Fig. 9 also shows the comparison between two types of mold compound on the delamination driving force. It can be seen that the strain energy release rate for package with MC1 compound is about 50% lower than that for package with MC2 compound. The lower G value for the case with MC1 may be attributed to the more compliant nature of the compound. Therefore, the interfacial fracture reliability for package with MC1 compound is expected to be better than than that with MC2 compound.

V.   Conclusions

In this paper, the interfacial fracture problem is analyzed by using numerical finite element approach with 3-D VCCT for extracting the fracture mechanics parameters. The methodology is applied to study the corner delamination in a fan-out package. Fracture mechanics parameters including the strain energy release rate and the stress intensity factors are obtained along the curvilinear crack front. Effects of dimension and material on the trend of fracture reliability are discussed. In addition, the fracture parameters may be combined with experimental data on the fracture resistance to quantitatively predicting fracture reliability for the interface of interest.

Reference

1. K. Beth et al, "The Redistributed Chip Package: A Breakthrough for Advanced Packaging," Proc. 57th Electronic Components and Technology Conf., Reno, NV, May 2007, pp. 286-291, 2007.
2. M. Brunnbauer et al. "An Embedded Device Technology Based on a Molded Reconfigured Wafer" Proc. 56th Electronic Components and Packaging Technology Conf., San Diego, CA, May 2006, pp. 547-551, 2006.
3. W. T. Chow and S. N. Atluri, "Finite element calculation of stress intensity factors for interfacial crack using virtual crack closure integral," Computational Mech., 16, 417-425, 1995.
4. G. De Roeck and M. M. Abdel Wahab, "Strain energy release rate formulae for 3D finite element," Engng. Frac. Mech., 50, 569-580, 1995.
5. T. Nakamura, "Three-dimensional stress fields of elastic interface cracks," J. Appl. Mech., 58, 939-946, 1991.
6. J. R. Rice, "Elastic fracture mechanics concepts for interfacial cracks," J. Appl. Mech., 55, 98-103, 1988.
7. A. O. Ayhan, Finite element analysis of nonlinear deformation mechanisms in semiconductor packages, Dissertation, Lehigh Univ., 1999.
8. J. C. Newman Jr. and I. S. Raju, "Stress-intensity factor equations for cracks in three-dimensional finite bodies subjected to tension and bending loads," In Computational Methods in The Mechanics of Fracture, S. Atluri (ed.), Elsevier Science Publishers, Amsterdam, 311–334, 1986.

Frequency Dielectric Constant and Loss Tangent Extracting of Organic Material Using Multi-length Microstrip

Sung-Mao Wu , Chi-Chang Lai, *Hung-Hsiang Cheng, Yu-Che Tai, *Chen-Chao Wang
Department of Electrical Engineering, National University of Kaohsiung, Kaohsiung, Taiwan, China
Tel: 886-7-5919436, Fax: 886-7-5919374, E-mail: sungmao@nuk.edu.tw
Corporate R&D Center, ASE Electronics, Inc., Kaoksiung, China
Tel: 886-7-3617131 ext 15291 , Fax: 886-7-3613094, E-mail: Alexcc_wang@aseglobal.com

Abstract

Organic material using for packaging substrate is selected and multi-length microstrip lines in same trace width are designed and performed on it. Novel formulas deliver to extract dielectric constant and loss tangent varying with frequency for selected organic materials will be shown in this paper. Performances of microstrip lines are measured by Agilent Vector Network Analyzer up to 20GHz and SOLT calibration used to get two-port S-parameters. Then, novel formulas are used to extract material parameters in ADS software by measurement date.

Introduction

For multi-chip and system integrated in one packaging, analysis techniques of substrate circuit for System-in-Packaging (SiP) design are demanded violently in recent years. Electrical parameters of packaging substrate materials affect mainly system performance, for this reason, the issue about properties of packaging material is important. For active or passive devices in microwave/RF application, incorrect electrical parameters of packaging substrate materials will bring on confused design during design stage, therefore, accurate electrical parameters of substrate materials varying with frequency lead correct electrical performance design and fast time to market during design stage.

In Simulation setup, the substrate vendors always provide the packaging material parameters at 1MHz, 1GHz or 10GHz. However, it courses mismatch results comparing with actual measurement case by the difference between provided parameters and real. One method to extract parameters of organic material is using an organic material and designed multi-length lines in the same trace width.

Extracting dielectric constant and loss tangent has finished by the theory of microstrip line phase delay and power loss, then comparisons and discussions of microstrip line frequency responses and real case measurement are presented, EM solver extracting with unit material parameters setting, and EM solver extracting with frequency varying material parameters setting.

Experiment and Theory

In our study, material parameters of the selected material are shown in table1 which is material composition with BT/glass fiber, shown dielectric constant is 4.05 at 10GHz and loss tangent (Dissipation factor) is 0.0078 at 10GHz. Multi-length microstrip lines named TL1, TL2, TL3 and TL4, are designed and performed on our selected material, real design dimension and two layers substrate cross-section are shown in Figure 1. The thickness of Cu layers are 20 um, dielectric layer is 200 um. The trace lengths we design and measure are 3150um for TL1, 4700um for TL2, 6300um for TL3 and 12600um for TL4 respectively.

Table 1: Material Parameters

Organic Material			
parameters	value	parameter	value
Composition	BT/Glass Fiber	Dielectric constant	4.05 @10GHz
Tg (℃)	205 (DMA)	Dissipation Factor	0.0078@ 10GHz
Peel Strength (kN/m)	≥ 0.95 (12umCu)	CTE(PPM/ ℃) α 1	14~16 (X,Y) 40~60 (Z)
Water Absorption (%)	0.4 (85C/85%RHx16 8hrs)		

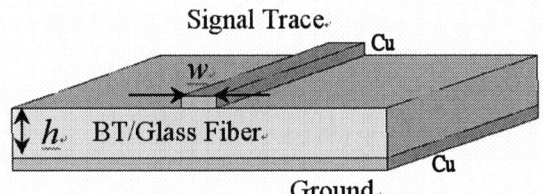

Fig.1 (a) 3D structure for microstrip line and thickness of Cu is 20 um, dielectric (h) is 200 um

Fig.1 (b) Real sample for microstrip line and length are 3150um, 4700um, 6300um, 12600um respectively

Performances of microstrip lines are measured by Agilent Vector Network Analyzer up to 20GHz and SOLT calibration used to get two-port S-parameters. According to measurement results, several electrical parameters of organic material like as effective dielectric constant (ε_{eff}), dielectric constant (ε_r) and loss tangent ($\tan\delta$) are extracted.

1.dielectric constant extracting

Supposed that the electromagnetic wave propagates forward to $+z$ direction, and L is the length of microstrip line. For extracting the effective dielectric constant, a formula is derived by theorem of microstrip line phase delay and shown like: [1]

$$\varepsilon_{\text{eff}} = \left(\frac{\theta \cdot C}{360 \cdot f \cdot L}\right)^2 \tag{1}$$

Where, θ can replace by $S_{21}(\text{phase})$.

When $w/h \leq 1$, the relationship between effective dielectric constant and dielectric constant is: [2] [3]

$$\varepsilon_{\text{eff}} = \frac{\varepsilon_r+1}{2} + \frac{\varepsilon_r-1}{2} \cdot [(1+12h/w)^{-(1/2)} + 0.04(1-w/h)^2] \tag{2}$$

That h is thickness of dielectric and w is the width of top copper, as Figure 1 (a). Therefore, we can derive effective dielectric constant (ε_{eff}) and dielectric constant (ε_r) value from equation (1) and equation (2).

2 loss tangent extracting

Consider the loss of microstrip line as

$$\alpha = \alpha_c + \alpha_d \tag{3}$$

Where, α_c is conductor (or ohmic) losses in strip conductor and ground plane; α_d is dielectric losses in substrate; α is total losses including α_c and α_d [4]. Total loss α can be gotten, as below:

$$|S_{11}|^2 + |S_{21}|^2 + \text{Power loss} = 1$$

$$\Rightarrow \text{Power loss} = 1 - \left(|S_{11}|^2 + |S_{21}|^2\right)$$

$$\Rightarrow \text{Power loss (in dB)} = -20\log e^{-\alpha \cdot l} \text{dB}$$

$$= 0\text{dB} - 10\log\left(|S_{11}|^2 + |S_{21}|^2\right)\text{dB}$$

$$\Rightarrow -20\log e^{-\alpha \cdot l}\text{dB} = -10\log\left(e^{-\alpha \cdot l}\right)^2 \text{dB}$$

$$= -10\log\left(|S_{11}|^2 + |S_{21}|^2\right)\text{dB}$$

$$\Rightarrow \left(e^{-\alpha \cdot l}\right)^2 = \left(|S_{11}|^2 + |S_{21}|^2\right) \Rightarrow \left(e^{-\alpha \cdot l}\right) = \sqrt{\left(|S_{11}|^2 + |S_{21}|^2\right)}$$

$$\Rightarrow -\alpha \cdot l = \ln\sqrt{\left(|S_{11}|^2 + |S_{21}|^2\right)}$$

$$\Rightarrow \alpha = -\frac{1}{l} \cdot \ln\sqrt{\left(|S_{11}|^2 + |S_{21}|^2\right)} \text{, (Np/m)} \tag{4}$$

where, l is the length of microstrip line.

Conductor loss α_c, When $w/h \leq 1$, can be gotten by [5] [4], as below:

$$\alpha_c = \frac{10R_s}{\pi\ln 10} \frac{\left(\frac{8h}{w} - \frac{w}{4h}\right)\left(1 + \frac{h}{w} + \frac{h}{w}\frac{\partial w}{\partial t}\right)}{hZ_0(\varepsilon_r)\exp\left(\frac{Z_0(\varepsilon_r)}{60}\right)} \tag{5}$$

where, $Z_0(\varepsilon_r) = \frac{Z_0(\varepsilon_r = 1)}{\sqrt{\varepsilon_{\text{eff}}}}$, $\frac{\partial w}{\partial t} = \frac{1}{\pi}\ln(\frac{2h}{t})$ (6)

, for $\frac{w}{h} \geq \frac{1}{2\pi}$

h, w and t are thickness of dielectric, width of strip and thickness of strip, respectively. Loss tangent $\tan\delta$ can be gotten, as below:

$$\tan\delta = \frac{\sigma}{\omega\varepsilon} = \frac{2\alpha_d}{\beta} \tag{7}$$

Where, we can shown α_d from equation (3)

$$\alpha_d = \alpha - \alpha_c \tag{8}$$

Thus we got loss tangent $\tan\delta$ from equation (7), equation (8), equation (4), and equation (5), (6).

Results and Discussion

Using above formulas, these electrical parameter such as ε_{eff} (effective dielectric constant), ε_r (dielectric constant) and $\tan\delta$ (loss tangent) are extracted in ADS software, and shown in Figure 2.

The key application of material parameters varying with frequency is to improve accuracy of EM simulation result during substrate in design stage. Compare and verify extracting result is necessary and to make EM solver more powerful in complex substrate circuit design. Shown in figure 3, "HFSS" means that simulated using constant parameters; "Meas" which shows measurement results;"HFSS by Extracted" which shows simulation data by using varying parameters with frequency. We compared the results of measurement by VNA with SOLT calibration, EM simulate with fixed dielectric constant and loss tangent by HFSS, and also EM simulate with frequency varying material parameters by HFSS.

According to the result of comparison between different EM simulation setting and measurement, when EM solver is used to evaluate circuit performance, simulation with frequency-varying material parameters setting coincides with measurement data. Noting that obvious amplitude difference of S21 in dB after 10 GHz in figure 3 (d) caused by some unavoidable reasons, like the copper oxidation in air and variables of fabrication.

shown in this study too. The simulation by extracting parameters is matching with measurement.

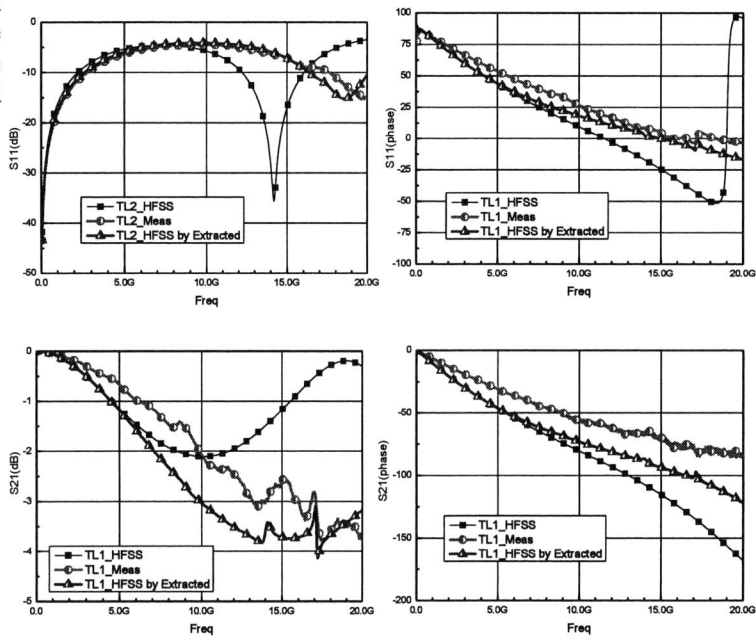

Fig.3 (a)   The comparison of S11 and S21 in dB and Phase with both measurement and simulation. The length of TL1 is 3150 um.

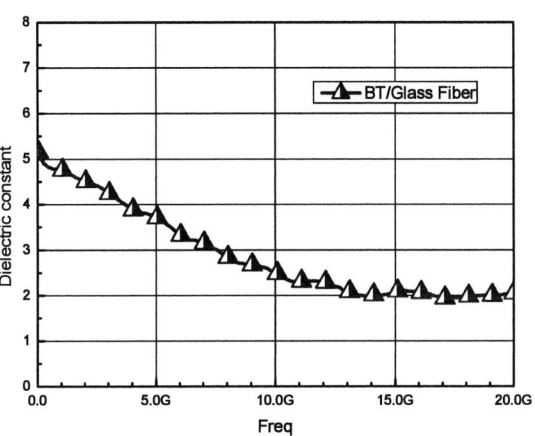

Fig.2 (a) dielectric constant

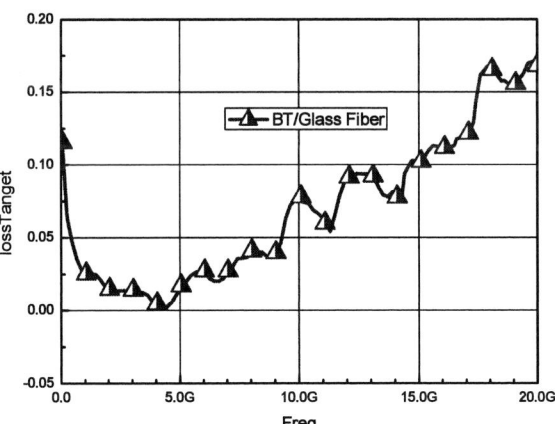

Fig.2 (b) loss tangent

Fig.2   extracting parameter varying with frequency, dielectric constant and loss tangent

Conclusion

In this paper, we presented the extracting technology of dielectric constant and loss tangent varying with frequency for advanced package substrate design by simple multi-length microstrip lines. Different material parameters setting used in EM simulation and compare with measurement data are

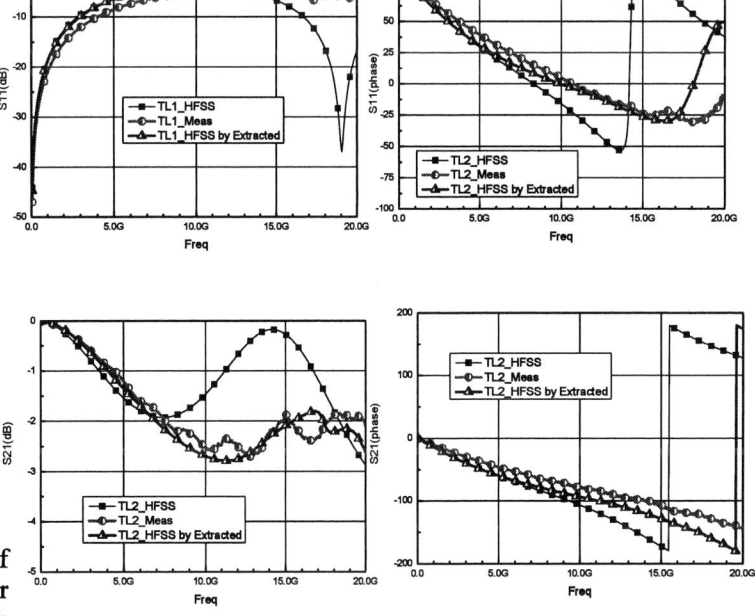

Fig.3 (b) The comparison of S11 and S21 in dB and Phase with both measurement and simulation. The length of TL1 is 4700 um

References

[1] Jingook Kim, Junho Lee, Namhoon Kim, Junwoo Lee, H.Y.Bang, Y.H. Chung, and Joungho Kim, "Microwave Frequency Dielectric Constant and Loss Tangent Measurements of PCB Materials Using Strip-line Structure", *APACK 2001 Conference on Advances in Packaging*.

[2] Schneider, M.V., "Microstrip Lines for Microwave Integrated Circuit", *Bell System Technical Journal*, Vo.1 48, 1969, pp.1421-1444.

[3] Hammerstad, E. O., and F. Bekkadal, *Microstrip Handbook ELAB Report*, STF 44 A74169, University of Trondheim, Norway, 1975

[4] Günter Kompa, *Practical Microstrip Design and Applications*, artech house,2005.

[5] Wheeler, H. A., "Transmission-line Properties of Parallel Wide Strips by Conformal Mapping Approximation," *IEEE Trans.*, Vo1. MTT-12, 1964, pp. 280-289.

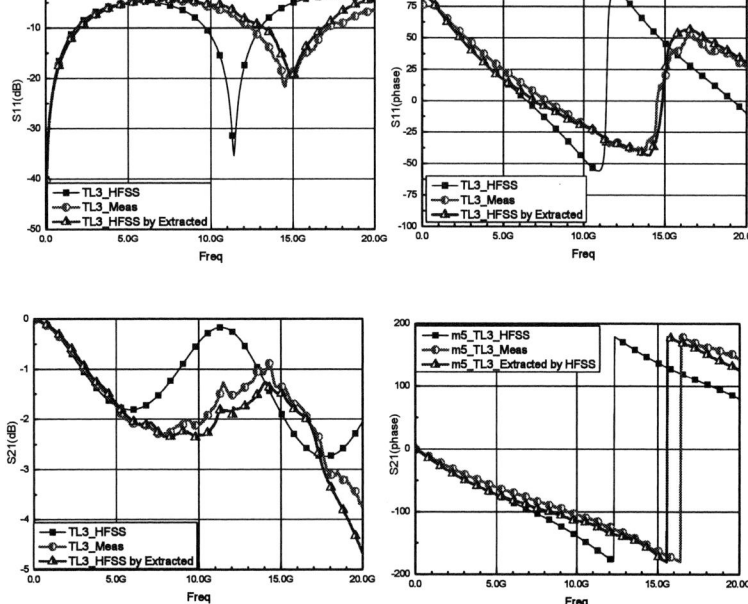

Fig.3 (b) The comparison of S11 and S21 in dB and Phase with both measurement and simulation. The length of TL1 is 6300 um

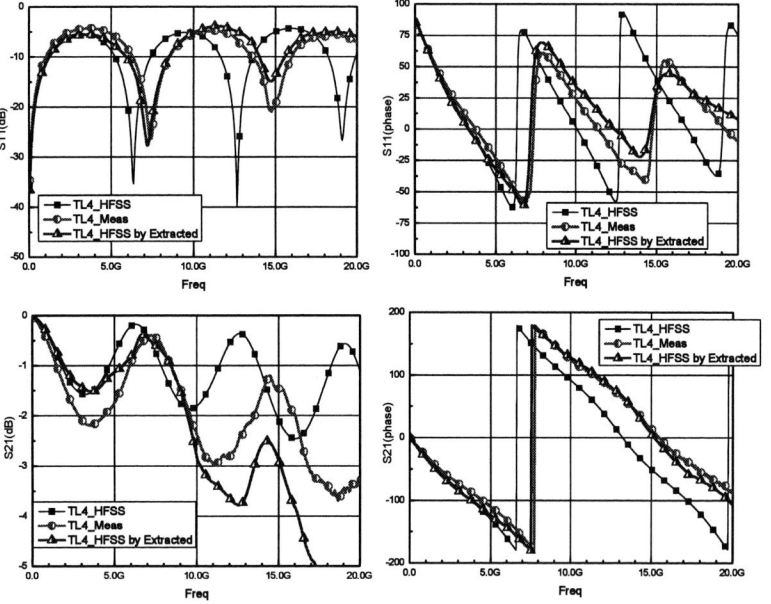

Fig.3 (b) The comparison of S11 and S21 in dB and Phase with both measurement and simulation. The length of TL1 is 12600 um

Thermo-mechanical Design of Large Die Fine Pitch Copper/Low-k FCBGA and Lead-free Interconnections

Kalyan Biswas *, Shiguo Liu*, Xiaowu Zhang**, TC Chai**

* IBIDEN Singapore Pte Ltd, 31 Kaki Bukit Road 3, #06-22 Techlink, Singapore 417818
Tel: (65) 67406137; Fax: (65) 62962170; E-mail: kbiswas.isp@ibiden.com
** Institute of Microelectronics (IME), A*STAR (Agency for Science, Technology and Research), Singapore
11 Science Park Road, Singapore Science Park II, Singapore 117685

Abstract

Device speed and functionality requirements are quickly forcing the semiconductor industry to incorporate copper and low-k dielectric materials. Compared to the commonly used aluminum metallization scheme on the traditional silicon dioxide and silicon nitride passivation, a Cu/low-k combination offers higher on-chip communication speed and a lower overall device cost. However, the low-k materials have intrinsically lower modulus and poorer adhesion compared to the commonly used dielectric materials. Thus, thermo-mechanical failure is one of the major challenges for development of a Cu/low-K large die flip chip package. In this paper, a two-dimensional plane strain analysis is performed on the diagonal cross-section of the package. A series of parametric study is performed to study the effect of bevel cut at die corner, effect of bevel cut depth, effect of filling materials of bevel cut, effect of Cu post bumps vs. lead-free bumps, effect of Cu post height and effect of Cu post diameter. Findings of these simulations could be used as a design consideration for the design of the Cu/low-k larger die flip chip package

1. Introduction

The added functionality, performance & demand for cost reduction for electronics devices has driven the semiconductor industry to incorporate copper and low-k dielectric materials. Compared to the commonly used aluminum metallization scheme on the traditional silicon dioxide and silicon nitride passivation, a Cu/low-k combination offers higher on-chip communication speed and a lower overall device cost [1-5]. However, the low-k materials have intrinsically lower modulus and poorer adhesion compared to the commonly used dielectric materials. Because of the lower adhesion strength of the low-k layer, the potential failure mode, moves from the die-underfill interface to the interface between the low-k layer and the silicon. So, this is very important to reduce the stresses within the flip-chip package to prevent the delamination between the low-k and the silicon interface.

Coefficient of thermal expansion (CTE) mismatch between the die and the substrate and the local CTE mismatch between the solder and the adjacent materials has been a problem in the solder reliability. Due to higher reflow temperature for lead-free solder and lower standoff of fine pitch bumps, a chip size of 20mmx20mm has resulted in extremely high shear strain at the solder joint between the silicon chip and the organic buildup substrate. An underfill encapsulant was used to fill the gap between the chip and substrate around the solder joints to improve the reliability of the flip chip interconnects system and a proper selection of underfills is important to reduce the stress on low-k die and to improve solder joint reliability [6-8]. But, then also some concern remains on mechanical reliability of a very large die

and fine pitch FCBGA. Therefore, Copper post bumps with lead-free solder (SnAg) tip have been employed in this study to enhance interconnection standoff height. For comparison, a lead-free solder (SnAg) bump has been analyzed. In this study, Polymer Encapsulated Dicing Lane Technology (PEDL) developed by IME has been used to address the dicing challenges of Cu/low-k and simultaneously improve the reliability of the packages by reducing the corner stress of the Cu/low-k [13-14].

Effects of different crucial package dimensions which play an important role in reducing the stress in low k layer and improve solder fatigue life was described in our earlier work [15]. In this paper, we have introduced a series of parametric study to analyze the effect of bevel cut at die corner, effect of bevel cut depth, effect of filling materials of bevel cut, effect of Cu post bumps vs. lead-free bumps, effect of Cu post height and Cu post diameter.

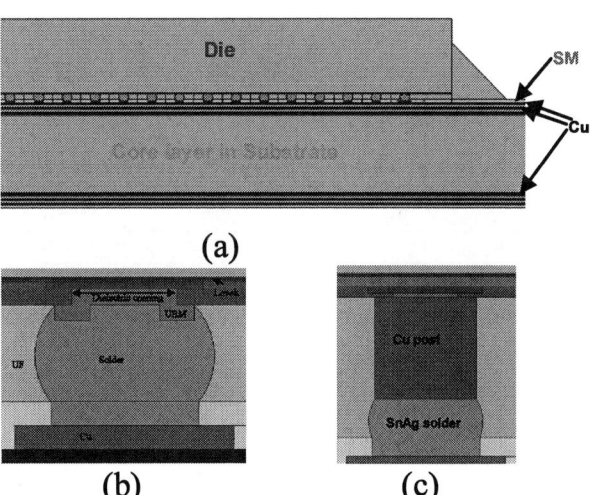

Fig. 1: (a) Diagonal cross section of the package (b) The detailed structure of lead-free solder bump (c) The detailed structure of Cu post bump

2. Package Description

A large die flip chip package with 20 mm x 20 mm die size and fine solder bump pitch (1st level solder pitch 150 μm) is considered in this analysis. The die thickness is 600 micron. The 45 mm x 45 mm substrate is consisted of 2-2-2 buildup configuration. The substrate core thickness was 0.8 mm. Underfill is used between chip and the substrate. Some dimensions of the package for the base model could be found in Table I.

Table I: Dimensions of the base model

Design Factors	Unit (mm)
Die Thickness	0.6
Core thickness of Substrate	0.8
solder bump pitch	0.15
solder ball height	0.08
Cu post diameter	0.08
Cu post height	0.08

3. Finite Element Model

A two-dimensional plane strain analysis is performed on the diagonal cross-section of the package. Low-k material is modeled using one layer of elements instead of a detailed Cu/low-k structure. Fig. 1(a) shows the cross section of the model and Fig. 1(c) shows the detailed structure of the Cu post interconnections. Stress distribution in low-k layer was determined when the package is cooled down from temperature of 260 °C to -40 °C. Regarding solder bump reliability, a two-dimensional nonlinear analysis was performed to determine the maximum equivalent inelastic strain range per TC (Δε) in solder bumps under temperature cycling (TC). The condition of TC test was set between -40°C and 125°C with 15 min hold time and 15 min ramp time. The maximum equivalent inelastic strain range per TC (Δε) in solder was compared for optimization study.

Table II: Material properties used

Materials	CTE (ppm/C)	Young's Modulus (GPa)	Poisons Ratio
Silicon	2.7	131	0.28
Low-k	23	7.7	0.3
USG	0.57	66	0.18
RDL	17	110	0.34
Dielectric	57	5.5 (-65°C), 4.0 (25°C), 1.2 (150°C)	0.35
Cu	17	110	0.34
Solder	22	44.4 (25°C), 18.8 (125°C)	0.4
Core in Substrate	16	24.5	0.22
Build Up material	95	3	0.41
Solder mask	52	4	0.4

The material properties used in simulation are listed in Table II [9-12]. Viscoplastic Constants for the lead-free solders were obtained from literature [9-10].

4. Simulation Results and Analysis

After simulation, the maximum shear stres was found at the outermost edge of die. But, as low-k layer under UBM of outermost solder bump is critical for package reliability, shear stress at this area as well as outermost edge of die was extracted to compare. The shear stress distribution at the die corner and underneath outermost solder bump is shown in Fig. 2.

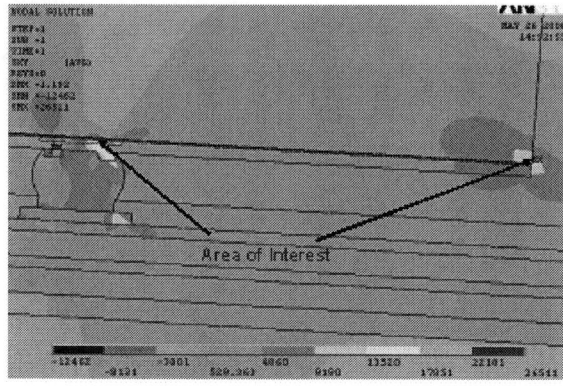

Fig. 2: Shear stress at die corner and underneath outermost interconnect

The maximum equivalent inelastic strain range per TC (Δε) in solder bumps was compared for optimization study. The smaller the value of Δε is, the better solder bump reliability is. A typical equivalent inelastic strain contour of solder bump is shown in Fig.3.

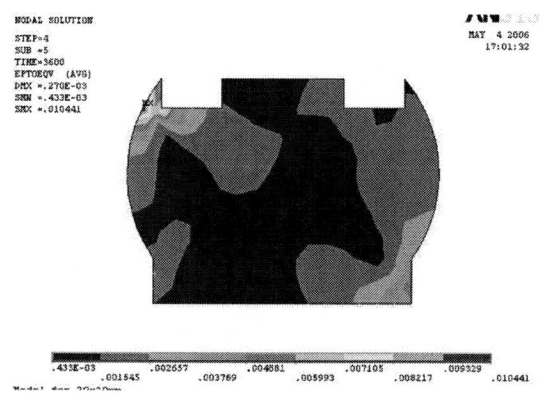

Fig. 3: Equivalent inelastic strain contour after one cycle

4.1 Effects of Bevel Cut

Polymer Encapsulated Dicing Lane Technology (PEDL) was developed by Institute of Microelectronics (IME) to address the dicing challenges of Cu/low-k and simultaneously improve the reliability of the packages. The details of this process can be found in literature [13]. In this process a bevel cutting is carried out in the scribed line with depth about 30 ~ 50um. Effect of bevel cut on low-k stress and solder bump reliability was studied by mechanical simulation. Shear stress distribution near the die edge with bevel cut is shown in Fig. 4. Fig 5 shows the effect of bevel cut on low-k stress near die corner. Both shear stress (Sxy) and peel stress (Sy) was compared. Almost, 33% decrease in the max. shear stress at corner when bevel cut is used. Peel stress (Sy) also decrease when bevel cut is used. Effect of bevel cut underneath the outermost solder bump is small. Only 4% decrease in shear stress of low k layer underneath the outermost bump when bevel cut is used.

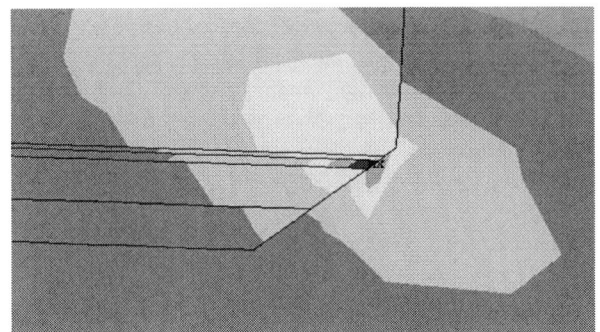

Fig. 4: Shear stress contour near the die edge with bevel cut.

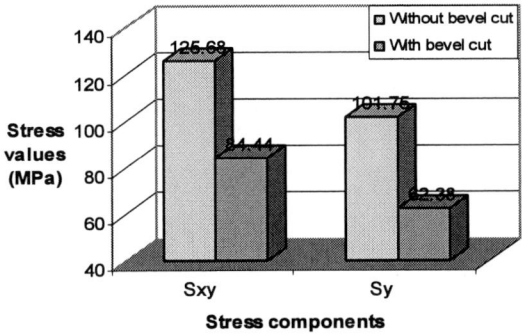

Fig. 5: Effect of bevel cut on low-k stress near die corner

Table III: Effect on Bevel cut on solder joint reliability

	The Max. inelastic strain range per TC cycle (Δε) in solder bump
Without bevel cut	0.8047%
With bevel cut	0.7558%

The max. Δε in solder bump decreases when bevel cut is used (Table III). Thus, PEDL approach gives potential advantages to overcome the weakness of low-k materials on the flipchip package and their reliability performance.

4.1.1 Effect of bevel cut deptht

Bevel cut depth has also an impact on die corner stress. Table IV shows the effect of bevel cut on stress near die corner. From simulations, it is observed that almost 11% decrease in the max. shear stress at corner when bevel cut depth increases from 20um to 40um. Bevel Cut depth has almost no effect on stress in low-k layer underneath the outermost solder bump. Δε in solder bump increases very little when bevel cut depth increases.

Table IV: Effect on Bevel cut depth on stress in low k layer

Cut Depth	Sxy	Sy
20um	84.44	-62.38
40um	75.11	-56.95
60um	76.63	-56.89
80um	72.93	-57.11

4.1.2 Effect of filling material

Effect of filling material of bevel cut on stress in low-k layer at die corner was also studied [Fig 7]. From simulation, it is noticed that nearly 22% decrease in the max. shear stress in low-k layer at die corner when bevel cut is filled by PI (Polyimide) instead of underfill when bevel cut depth is 20um [Fig.6]. The shear stress at low-k layer underneath the outermost bump does not change when bevel cut is filled by PI instead of underfill. Also, Δε does not change when bevel cut is filled by PI instead of underfill.

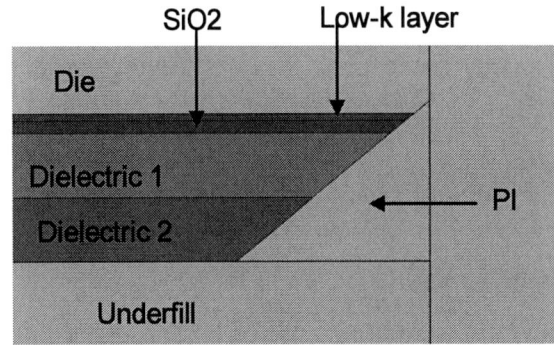

Fig. 6: Bevel cut filling material near die corner

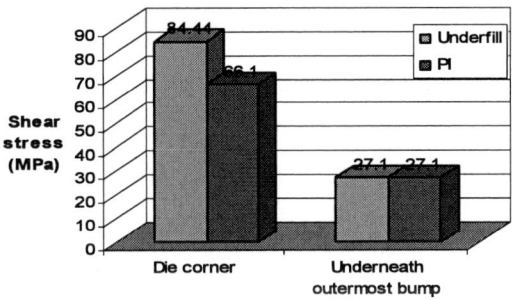

Fig.7: Effect of filling material on low-k stress near die corner

4.2 Effect of Cu Post Bumps

Cu-post (refer to Fig. 1) interconnects are used to increase the standoff height of the interconnection. Effects of Cu-post interconnects were evaluated by FEA. Keeping bevel cut depth 20um and filling by underfill in bevel cut, maximum shear stress at the die edge and underneath the outermost solder bump were obtained.

Table V: Effect of Cu post on stress in low k near die corner

Model	Sxy (MPa)	Sy (MPa)
Solder bumps with bevel cut	84.44	-62.38
Cu post with bevel cut	80.57	-92.66

301

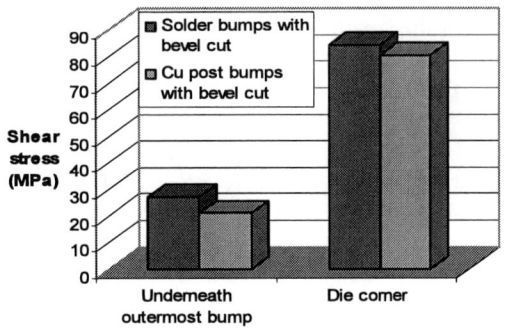

Fig.8: Effect of Cu post bump on low-k stress (Sxy)

Table 5 shows the effect of Cu post bumps on stress in low-k layer near die corner. Shear stress at die corner and underneath the outermost solder bump decreases when we use Cu post bumps instead of solder bumps [Fig. 8]. However peel stress (Sy) in low-k layer increase if we use Cu post bump.

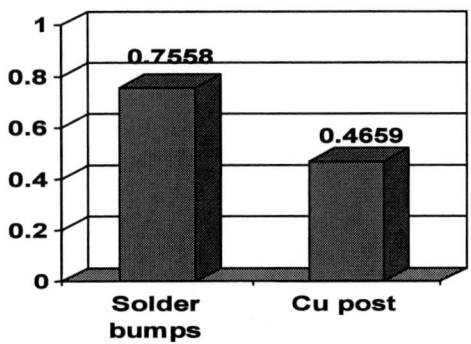

Fig.9: Effect of Cu post bump on solder bump reliability

The max. $\Delta\varepsilon$ in solder bump also decreases when we use Cu post bumps [Fig. 9]. So, Cu post bump is good to reduce the shear stress at the low-k layer and to reduce $\Delta\varepsilon$ in solder bumps.

4.2.1 Effect of Cu post height

Three different Cu post height (40 micron, 80 micron and 120 micron) were used in this study. The post diameter was fixed (80 micron). From simulation, it is noticed that the stresses in low-k layer near die corner and underneath outermost bump decrease when Cu post height increases [Fig. 10]. 11% decrease in the shear stress in low-k layer underneath outermost bump when Cu post height increases from 80um to 120um.

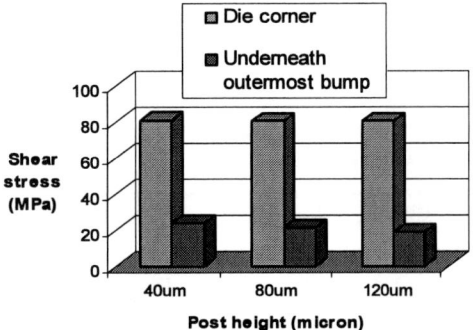

Fig.10: Effect of Cu post bump height on low k stress

4.2.2 Effect of Cu post diameter

Keeping the post height 80 micron, three different Cu post diameter were studied in this paper. Effect of Cu post diameter on stress in low k layer is shown in Table VI. From results, it is noticed that there is very little effect of Cu post diameter on low-k stress. Only, 1.9% increase in the shear stress in low-k layer underneath outermost bump when Cu post diameter increases from 80um to 100um

Table VI: Effect of Cu post diameter on stress in low-k

Post diameter (micron)	Die corner	Underneath outermost bump
60	80.81	20.75
80	80.57	21.03
100	80.22	21.44

5. Conclusions

We have presented a thermo-mechanical analysis of Cu/Low-K large die flip chip package with emphasis on thermally induced stress in low-k layer and the 1st level solder bump reliability. Effect of bevel cut at die corner, bevel cut depth, filling materials of bevel cut portion, Cu post bumps vs. lead-free solder bumps, Cu post height and Cu post diameter were simulated. Some of the conclusions from our simulation are summarized as follows.

(i) The max. shear stress at die corner (Fig. 4) and $\Delta\varepsilon$ in solder bumps decrease when bevel cut is used.

(ii) Significant decrease in the max. shear stress at die corner when bevel cut depth increases. However, very little effect is seen in $\Delta\varepsilon$ when bevel cut depth increases.

(iii) Bevel cut at die corner can largely reduce the max stress at the low-k corner when bevel cut is filled by PI instead of underfill. However, $\Delta\varepsilon$ does not change when bevel cut is filled by PI instead of underfill.

(iv) Cu post bump is good to reduce the max shear stress at the low-k layer and to reduce $\Delta\varepsilon$ in solder tips; however, Cu post increases Sy (peel stress) at the low-k layer.

(v) Cu post bump is good to improve bump reliability to reduce $\Delta\varepsilon$ when we use Cu post bumps instead of leadfree solder bumps.

(vi) Significant decrease in the shear stress in low-k layer underneath the outermost bump when Cu post height increases.

These findings could be used as a design guideline for the design of the Cu/low-k large die flip chip package.

Acknowledgments

This work is the result of a project initiated by the 8th IME Electronic Packaging Research Consortium (EPRC VIII), the members of which are Advanced Micro Devices (Singapore) Pte Ltd, BOC Gases Pte Limited, Ibiden Singapore Pte Ltd, Infineon Technologies Asia Pacific Pte Ltd, NEPES Corporation, United Test and Assembly Center Ltd, MicroCircuit Technology (2002) Pte Ltd, Cookson Semiconductor Packaging Materials A Division of Cookson Singapore Pte Ltd, Institute of Microelectronics, Singapore Institute of Manufacturing Technology, Institute of High

Performance Computing and Institute of Materials Research & Engineering. The authors are grateful to members of EPRC VIII Project 1 as well as IME staffs who had contributed and made this work possible.

References

1. S. Chungpaiboonpatana and F. G. Shi, "Packaging of Copper/Low-k IC Devices: A Novel Direct Fine Pitch Gold Wirebond Ball Interconnects Onto Copper/low-k Terminal Pads," IEEE Transactions on Advanced Packaging, Vol. 27, No. 3, August 2004, pp. 476-489

2. L. L. Mercado, S-M Kuo, C. Goldberg and D. Frear, "Impact of Flip-Chip Packaging on Copper/Low-k Structures," IEEE Transactions on Advanced Packaging, Vol. 26, No. 4, November 2003, pp. 433-440.

3. Guotao Wang, Steven Groothuis and Paul S. Ho, "Effect of Packaging on Interfacial Cracking in Cu/ Low k Damascene Structures" Proc. of ECTC, 2003, pp. 727-732.

4. Chih-Tang Peng, Chang-Ming Liu, Ji-Cheng Lin, Hsien-Chie Cheng, and Kuo-Ning Chiang, "Reliability Analysis and Design for the Fine-Pitch Flip Chip BGA Packaging" IEEE Trans. on Components and Packaging Technologies, Vol. 27, No. 4, December, 2004, pp. 684-693.

5. Eiji Hayashi, Shinji Baba, Shiori Idaka, Akira Maeda, Mitsuru Satoh, Michitaka Kimura, "Realization of Pb-free FC-BGA Technology on Low-k Device", Proc. of 55thECTC, USA, 2005, pp. 9-13

6. Sarathy Rajagopalan, Kishor Desai, "Underfill for Low-K Silicon Technology", Proc. of IEEE/SEMI Int'l Electronics Manufacturing Technology Symposium, 2004.

7. Xiaowu Zhang, D. Pinjala, Mahadevan K. Iyer "Comprehensive Analysis of A Larger Die, Copper Pillar Bump Flip Chip Package with No-Flow Underfill" Proc. of EPTC, 2005, pp. 575-578.

8. Tong Hong Wang, Yi-Shao Lai, Meng-Jen Wang, "Underfill Selection for Reducing Cu/low-K Delamination Risk of Flip-chip Assembly", 2006 Electronics Packaging Technology Conference. pp. 233-236.

9. HS Ng, TY Tee, et al, "Absolute and Relative Fatigue Life Prediction Methodology for Virtual Qualification and Design Enhancement of Lead-free BGA," Proc. of 55thECTC, USA, 2005, pp. 1282-1291.

10. TO Reinikainen, et al, "Deformation Characteristics and Microstructural Evolution of SnAgCu Solder Joints," Proc. EuroSime Conference, Germany, Apr. 2005.

11. M. G. Pecht, et al, "Electronic Packaging Materials and Their Properties" CRC Press LLC, Boca Raton, 1998.

12. http://www.inemi.org

13. Seung Wook Yoon, David Wirtasa, Samuel Lim, V. Ganesh, Akella Viswanath, Vaidyanathan Kripesh and Mahadevan K. Iyer. "PEDL (Polymer Encapsulated Dicing Line) Technology for Copper/Low-k Dielectrics Interconnect," 2005 Electronics Packaging Technology Conference. pp. 711-715.

14. Akella G.K.Viswanath, Wang Fang , Tai-Chong Chai, Navas Khan, Srinivasamurthy Sampath, "Structural optimization of fine pitch, large die flip chip package", Proc. of EPTC, 2004, pp. 105-108.

15. Kalyan Biswas , Shiguo Liu, Xiaowu Zhang, TC Chai, Ser-Choong Chong, "Structural Design for Cu/Low-K

Larger Die Flip Chip Package", Proc. of EPTC, 2006, pp. 237-242.

Study on Non-Uniformity of Through-Mask Electroplated Ni Thin-Film

Jun Tang, Hong Wang, Rui Liu, Shengping Mao, Xiaolin Zhao, Guifu Ding

Research Institute of Micro /Nano Science and Technology, Shanghai Jiao Tong University, Shanghai 200240, China

Tel: +86-021-34206689, Jun Tang, Email: tjmnri@sjtu.edu.cn

Abstract

Through-Mask Electroplating has been widely used in the fabrication of chips, BGA substrates and PCBs etc. The uniformity of plating thin-film is the major factor contributed to the reliability of the products. Currently, it is usually by setting optimum plating parameters and adopting electrochemical method to achieve the uniformity of plating. However, the problem of non-uniform distribution of electric field, which is the major cause of the non-uniformity of the plating thin-film, has not been solved. In this paper, Finite Element Method (FEM) was developed to analyze the non-uniform distribution of electric field under different conditions in the process of electroplating. The results show that different thickness of photo-resist and size of electroplating cell are two major factors contribute to the uniformity of plating thin-film. The uniform of electroplating cell can be improved by adding in-chip auxiliary electrode. Also better uniformity of the plating film in radial direction can achieved by setting a shield in the proper position of the plating solution and annular out-chip auxiliary electrode (Cu) around the wafer. The simulation results were consistent with experimental results, which proved that Finite Element Method is an effect way to simulate the electroplating process.

Index terms--Through-Mask Electroplating, Finite Element Method (FEM), Uniformity of Thickness

Introduction

With the development of integrated circuits technology toward high-density, low-cost, and miniaturization, through-mask electroplating has been widely used in the fabrication of chips, BGA substrates and PCBs etc. The uniformity of thickness of the electroplating thin-film is one of the major factors in the process of through-mask electroplating. The non-uniformity of the film leads to the thickness non-uniformity of microstructures and the non-uniformity of the plating film between the edge and the centre of the wafer. The latter is also known as "terminal effect" [1-2]. Traditional method to solve the problem of non-uniformity problem is by setting optimized parameters in the process, such as current density, temperature, stirring speed and pH value of solution, etc [3]. The uniformity in diameter direction can also be improved by adding different shape shield between the anode and the wafer. However, there is no ultimate approach to resolve the primary non-uniform distribution of the electric field, the main reason for the non-uniform plating, which is concerned with geometry of the plating system. In order to avoid these two kinds of non-uniformities mentioned above, in-chip auxiliary electrode was introduced to solve the former problem in this paper, meanwhile, a shield in the proper position of the plating solution and annular copper out-chip auxiliary electrode around the wafer are developed to achieve better uniformity of the plating film in radial direction.

There are several methods to simulate the electroplating process: numerical [4], analytical [5] and finite element method [6]. Finite element method has its verified unique advantages in electric field simulation for different geometries. In this paper, Finite element method is used to simulate the primary distribution of electric field in different electroplating systems to obtain the optimized parameters.

FEM Modeling of Electroplating Process and Experimental Method

According to Faraday's law, the relation between electricity and mass of depositing material is described as follows:

$$m = \frac{QM}{zF}$$

In the above equation, m is the mass of the deposited mental material, Q is the total charge used in the deposition, M is Moore mass, z is the number of electrons taking part in the reaction, F is the Faraday constant.

Then, the thickness of deposited thin-film D can be obtained as follows:

$$D = \frac{M}{z\rho F} Jt$$

Where ρ represents the density of the deposited material, J is current density, t represents the plating time. Theoretically, the thickness of electroplating thin-film is proportional to the current density and deposited time. Among other factors, electric field, which determines J, is the major factor contributed to the uniformity of thickness. Consequently, the distribution of electric field in electroplating system is derived to learn the uniformity of the plating thin-film.

The 2-D FEM model is shown in Fig.1. The geometry size of the model is the same with the electroplating system in order to compare the simulation results with experimental one. The model in Fig.1 is used to analyze the effects of the plating cell size and thickness of photo-resist on uniformity, as well as to achieve microstructure uniformity by obtaining optimized parameters of in-chip auxiliary electrode. The Cathode (seed layer) was set to 0V, and anode was set to 2V as boundary conditions.

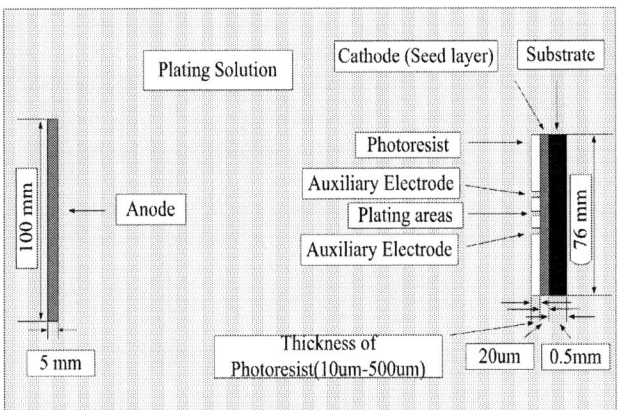

Fig.1. Finite Element Method Modeling of the electroplating process.

In this paper, shield in the proper position of the plating solution and annular copper out-chip auxiliary electrode around the wafer are developed to achieve better uniformity of the plating thin-film in radial direction. In order to obtain their optimized parameters, another 2-D model is shown in Fig.2. Part of the mesh results are showed in the Fig too. The parameters of this electroplating system includes the diameter of the aperture of the shield, distance between aperture and wafer, the width of the annular out-chip auxiliary electrode and the distance between out-chip auxiliary electrode and wafer. Material properties of all the models above are shown in Table I. The cathode including wafer and out-chip auxiliary electrode was set to 0V, and anode was set to 2V as boundary conditions. All simulations were carried out by using ANSYS software.

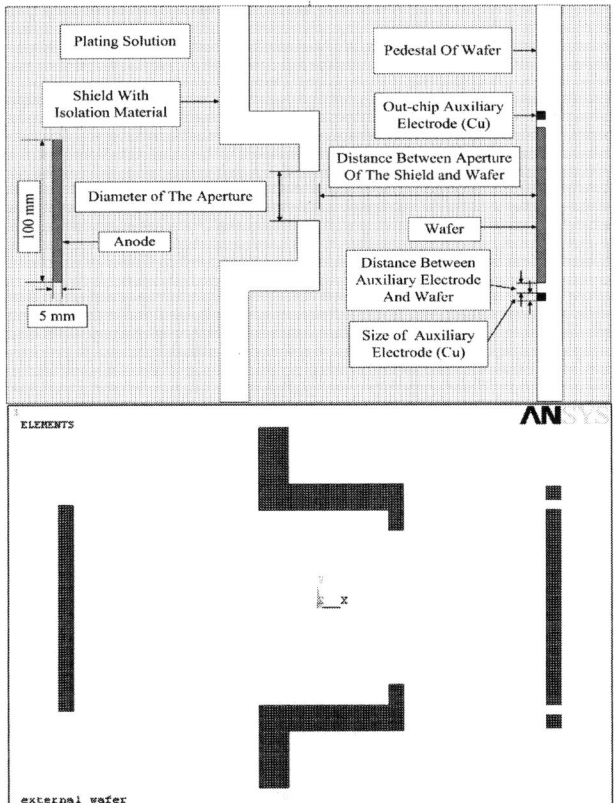

Fig.2.Another Finite Element Method Modeling of the electroplating process and part of the mesh results.

TABLE I
Material Properties of FEM Simulation

Material	Ni	Cu	Solution	Si	Isolation
Resistivity Ω*M	68.4*10⁻⁹	16.7*10⁻⁹	0.66	2.3*10²	10¹²

Experiments for comparing with the simulation results were carried out in following conditions: Plating temperature is about 18℃, current density is about 10mA/CM² and the pH value of plating solution is 3.8 during plating. The thickness of plated Ni film is about 10um in our experiments. After removing the photo-resist, species were measured by a Dektak profiler.

In order to quantify the variation of the non-uniformity ERROR upon the process conditions, we define the non-uniformity as follows:

$$ERROR = \frac{(H_{edge} - H_{middle})}{H_{middle}} \times 100\%$$

Here H_{edge} is the thickness at the edge of a microstructure, and H_{middle} is the thickness of a microstructure in the center. A negative non-uniformity ERROR means that the thickness of the center is smaller than the edge.

Results and Discussion

A. Efforts of Thickness of Photo-resist and Width of Electroplating Cell on Uniformity.

The distribution of thickness across an electroplating cell has been studied as a function of its width [3].FEM simulation results also proved the above fact, as shown in Fig.3. It shows that Electric field crowding occurs at the edge of a wide microstructure, and the center of a narrow microstructure, leading a higher plating rate and thicker film in these areas. So the thickness distribution of a wide electroplating cell is rabbit-ears profile, and cap-like thickness distribution can be found in narrow electroplating cell.

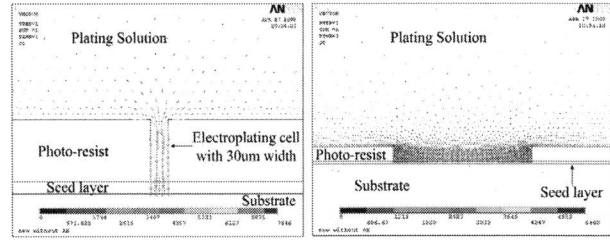

Fig.3. Distribution of the electric field of electroplating cell with different widths. Electric field crowding occurs at the edge of a wide microstructure, and center of a narrow microstructure, leading a higher plating rate and thicker film in these areas.

The effect of the thickness of the photo-resist on uniformity of thickness of the thin-film was also studied in our model as shown in Fig.4. It shows that non-uniformity decreases with the increasing thickness of photo-resist and the decreasing width of the electroplating cell.

Experimental results compared with simulation results are shown in Fig.5. Best uniformity could be achieved, when width of electroplating cell is 100um at certain plating conditions. Simulation results are proved to be correct. However absolute values of the experiments are larger than the simulation one. It is because that only the primary distribution of electric field was concerned in our simulation, but secondary electrochemical distribution of electric field also infects the uniformity of Ni thin film.

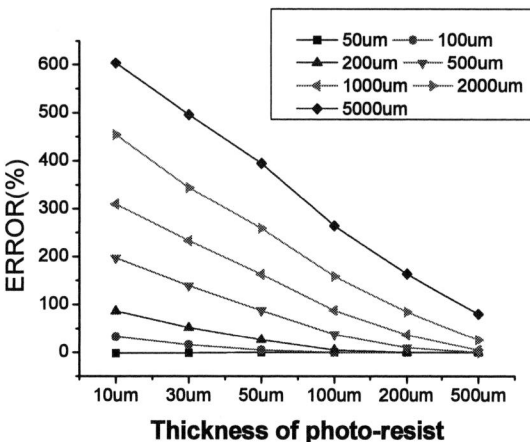

Fig.4. The non-uniformity of Ni film as a function of the thickness of photo-resist and width of the electroplating cell. Non-uniformity decreases with increasing thickness of photo-resist and decreasing width of the electroplating cell.

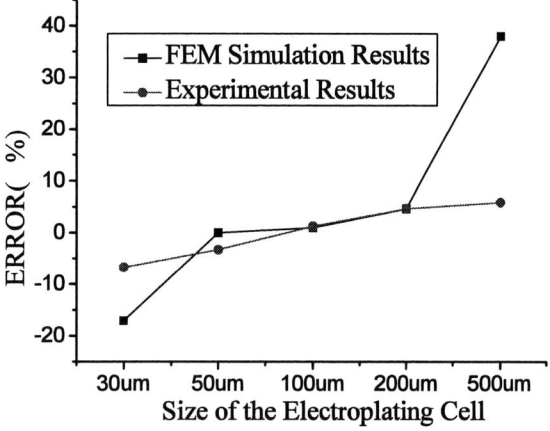

Fig.5. Experimental and FEM simulation ERROR results as a function of width of the electroplating cell at a fixed thickness of the photo-resist of 100um. Absolute value of non-uniformity decreases to 0 with width of the electroplating cell up to nearly 100um, and then it increases with increasing width of the electroplating cell.

B. Efforts of In-Chip Auxiliary Electrode on Uniformity.

In-chip auxiliary electrode near the electroplating cell was introduced to improve the uniformity of the microstructure, which is becoming more and more important especially in MEMS devices. FEM Simulation results show that electric field crowding at the edge could be removed by adding the in-chip auxiliary electrode.

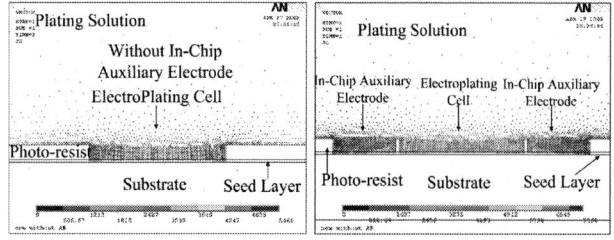

Fig.6. Distribution of electric field of electroplating cell without and with in-

chip auxiliary electrode. Electric field crowding at the edge obviously alleviate after adding the auxiliary electrode.

The distance between electroplating cell and in-chip auxiliary electrode and the width of auxiliary electrode are two major factors by means of adding auxiliary electrode. Non-uniformity of Ni film as a function of two factors mentioned above is shown in Fig.7 and Fig8.It was also experimentally proved correctly as shown in Fig.7 and Fig8. According to the results, the distance should be set as smaller as possible, and width should be set as larger as possible.

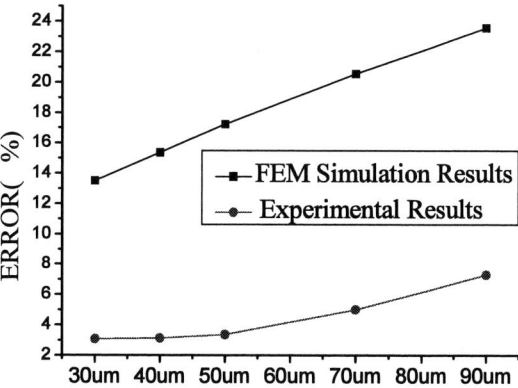

Fig.7. Non-uniformity of Ni film as a function of distance between plating cell and auxiliary electrode at a fixed width of the plating cell of 1000um. Non-uniformity increases with increasing distance.

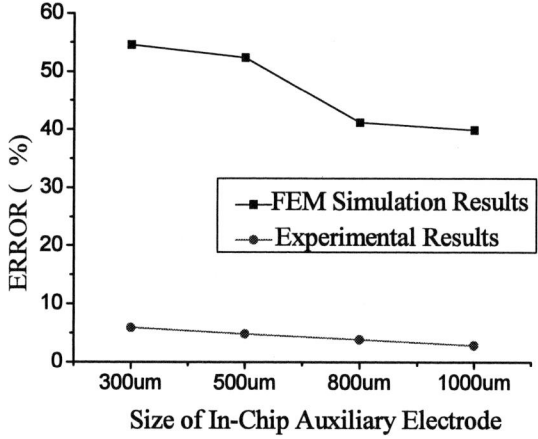

Fig.8. Non-uniformity of Ni film as a function of width of the in-chip auxiliary electrode at a fixed distance of 1000um. Non-uniformity increases with increasing width.

C. Efforts of shield and Out-Chip Auxiliary Electrode on Uniformity in Radial direction.

Orthogonal test table $L_{16}(4^4)$ is used for study the effect of diameter of the aperture, distance between aperture of the shield and wafer, width of the out-chip auxiliary electrode (Cu) and distance between auxiliary electrode and wafer on uniformity of the wafer. In order to quantify the variation of the non-uniformity across the wafer, we define non-uniformity η as follows:

$$\eta = \frac{h_{max} - h_{min}}{(h_{max} + h_{min})/2}$$

Here h_{max} is the max thickness across the wafer, h_{min} is the min thickness across the wafer. The statistic results are shown in Table II. Here A is Diameter of the aperture, B is distance between aperture of the shield and wafer, C is width of out-chip auxiliary electrode and D is distance between out-chip auxiliary electrode and wafer. According to the orthogonal test results, theoretical optimized conditions are $A_4B_3C_4D_1$ and the simulation result is 12.041%. Without all the auxiliary facility mentioned above, simulation result is 155.96%. Experimental results are 9.92% and 57.53% respectively corresponding to the simulation results. Simulation results are larger than experimental results. It is because electrochemical factors improved the uniformity, which were not concerned in our modeling. The method by adding a shield in the proper position of the plating solution and annular copper out-chip auxiliary electrode around the wafer has been proved could remarkably improve the uniformity of Ni Thin-film in radial direction.

Table II
Orthogonal Test $L_{16}(4^4)$ Table

Factor Number	A (cm)	B (cm)	C (cm)	D (cm)	η
1	3	1	0.1	0	184.33
2	3	3	0.3	0.2	85.16
3	3	5	0.5	0.4	111.69
4	3	7	0.7	0.6	125.42
5	4	1	0.3	0.4	170.74
6	4	3	0.1	0.6	118.96
7	4	5	0.7	0	12.05
8	4	7	0.5	0.2	102.07
9	5	1	0.5	0.6	140.23
10	5	3	0.7	0.4	99.18
11	5	5	0.1	0.2	119.66
12	5	7	0.3	0	54.79
13	6	1	0.7	0.4	97.47
14	6	3	0.5	0.6	112.82
15	6	5	0.3	0	41.98
16	6	7	0.1	0.2	131.48
I	506.60	592.78	554.43	290.31	
II	400.97	416.12	352.67	438.37	
III	413.87	282.54	466.82	479.09	
IV	383.75	413.76	331.27	497.43	
R	122.85	308.82	220.31	204.28	

Conclusions

According to simulations and experimental results above, the following conclusions can be drawn:

1. Thicker photo-resist could obtain better uniformity of thin-film of electroplating cell. There is a critical width of electroplating cell could achieve best uniformity corresponding to different plating conditions.

2. Adding in-chip auxiliary electrode can improve uniformity of thin-film of electroplating cell. Smaller distance between electroplating cell and in-chip auxiliary electrode and wider in-chip auxiliary electrode can obtain better uniformity.

3. The method by adding a shield in the proper position of the plating solution and annular copper out-chip auxiliary electrode around the wafer can remarkably improve the uniformity of Ni Thin-film in radial direction.

4. FEM can be successfully used to simulate the distribution of the electric field in electroplating process, and guide the design.

Acknowledgments

This project was sponsored by national high technology research and development program "863", subsidized project No. 2006AA4Z326.

References

1. Chen, KW. *et al*, "Investigation of overpotential and seed thickness on damascene copper electroplating," *Surface & Coatings Technology*, Vol.200, (2006), pp. 3112-3116.

2. K.M.Takahashi, "Electroplating copper onto resistive barrier films," *Journal of the Electrochemical Society*, Vol.147, (2000), pp.1414-1417.

3. Luo, J.K. *et al*, "Uniformity Control of Ni Thin-Film Microstructures Deposited by Though-Mask Plating," *Journal of The Electrochemical Society*, Vol. 152, No. 1 (2005), pp. C36-C41.

4. Toshikazu Okubo, *et al*, "Patterned Copper Plating Layer Thickness Made Uniform by Placement of Auxiliary Grid Electrode about Ball Grid Arrays," *Chemical Engineering Communications*, Vol. 193, (2006), pp. 1503-1513.

5. Yu.V.Litovka, *et al*, "Numerical Calculation of the Electric Field in an Electroplating Bath With Bipolar Electrodes," *Theoretical Foundations of Chemical Engineering*, Vol. 40, No. 3 (2006), pp. 305-310.

6. F. Druesne1, "Electroplating simulation and design tool," *Journal of Engineering Manufacture*, Vol.217, No.5 (2003), pp. 705-707.

Simulation and Analysis for Backward Compatibility of Solder Joints under Thermal Cycle

Ning Ye-xiang, Pan kai-lin, Li Ni

School of Mechanical & Electronical Engineering, Guilin University of Electronic Technology

Guilin，China，541004

ningyx666@gmail.com

Abstract:

In this paper, the backward compatibility solder joints were chosen in simulation of perimeter PBGA272 assembly. A double-symmetric plane FE model of a PBGA272 was established using the software ANSYS. Based on the maximum von Mises stress and von Mises strain, the position of the most danger solder joints were obtained under thermal cycle with the temperature condition from -40℃ to 125℃ (JESD22-A104-B Condition G) , Viz. the inner (1#) solder joint and the outside (6#) solder joint are the two key solder joints which are the easiest to failure. On the basis of above analysis, the geometry parameters of the chosen assembly are optimized by design of experiment (DOE). The factors included PCB size, PCB thickness, chip size, chip thickness, substrate size, substrate thickness, solder height and solder radius. The simulating results have shown that substrate thickness (factor F), solder radius (factor H) and solder height (factor G) performed the main factors. The optimal scheme is $F_3H_2G_1C_2D_1E_2B_3A_2$(substrate thickness 0.7mm, solder radius 0.38mm, solder height 0.4mm, chip size 2.54mm, chip thickness 0.4mm, substrate size 13.5mm, PCB thickness 1.8mm and PCB size 15mm) by comprehensively considered with every factor's effect.

1. Introduction

In response to the European Union (EU) Restriction of Hazardous Substances (RoHS) and other countries' impending lead-free directives, the electronics industry is moving toward lead-free soldering. Transitioning tin-lead (SnPb) soldering to totally lead-free soldering is a complex issue and involves movement of the whole electronics industry supply chain. In reality, there is a transition period. Forward compatibility (FC) refers to the compatibility of lead-free solder paste with lead-containing component finish, and backward compatibility (BC) refers to the compatibility of eutectic Sn-Pb solder with lead-free component finish [1].However, in some areas such as medical equipment, military and aerospace are exempt from the RoHS, because the reliability of lead-free solder joints for these high reliability applications is still unknown, these areas are still use SnPb solder paste while the component manufactures didn't produce leaded components, so the BC situation is inevitable. On the other hand, in the early transition, components manufacturers were slow in responding to the lead-free transition or there was insufficient demand initially, consumer electronics manufacturers were impossibly buy all the lead-free components, so the FC situation will be exist for some time. Lastly, the complete transition to lead-free may not be possible also due to temperature limitations of the device. [2-4].

The reliability of mixed solder joints has attracted attention in resent years. For FC solder joints, its reliability is equivalent or better than the reliability of the SnPb solder joints under lead-free reflow profiles[5-6], although there will be more voids comparison to BC solder joints[7], So this paper only discuss the BC solder joint. For BC solder joints, the key in BGA/CSP backward compatibility assemblies is to use the proper reflow profile to achieve complete mixing of SnPb paste with lead-free components [8-9], and one of the assumption in this paper is that the lead-free BGA component and the SnPb paste are fully mixed.

2. Finite Element (FE) Modeling

The modeling is a lead-free PBGA272(Fig. 1) assembled with SnPb paste (BC assembly) on FR-4 board, and a double-symmetric cross section view of a PBGA272 and its size is shown in Fig.2, solder ball radius is 0.38mm. However, Cu pad and solder mask which only slightly affect package internal stress and strain, are neglected. Generally speaking, The most realistic model currently practicable is a double-symmetric 3D model containing each thermal (inner) ball and peripheral ball, Rainer Dudek [10] comprised various modeling assumptions and simplifications and found that a slice or a plane model can provide valuable information on the stress or strain range to be expected and is useful in parametric study , so this paper use a double-symmetric plane modeling, as shown in Fig.3. The material properties were incorporated in the FE model as displayed in Table 1[11]. The Anand plasticity data table was activated for the BC solder ball material and incorporated the constants as given in Table 2[12], note that the BC solder ball material properties in Ref.[12] is a approximation, because it was linear adding by the volume ratio of lead-free component and SnPb paste. Solder material were meshed in ANSYS using VISCO108 elements, whereas all other components materials were meshed using PLANE82. Thermal cycling loading followed JESD22-A104-B Condition G[13] with -40℃/125℃ temperature range, as shown in Fig4, after 4 cycles, the loading is finished.

Fig. 1 PBGA272 layout and solder distribution pattern

Table1 Model Material Properties of package components

Material	Elastic Modulus (MP)	Poisson's Ratio	CTE	
			a_x & a_y (10^{-6}/℃)	a_z (10^{-6}/℃)
PCB (FR4)	18200	0.25	15	50
Substrate	22268.7	0.3	15.5	52.5
Solder joint	33.5e3	0.35	24.9	
Chip	131000	0.3	2.8	
Molding resin	16000	0.25	15	

Table2 Anand material parameters of BC solder joint

$A(s^{-1})$	$Q/R(K)$	ξ	m	$\hat{s}$ (MP)	n
9.47e6	8258.42	3.89	0.14	63.02	0.016

	h_0(MP)	a	s_0(MP)
	8251.98	1.47	45.15

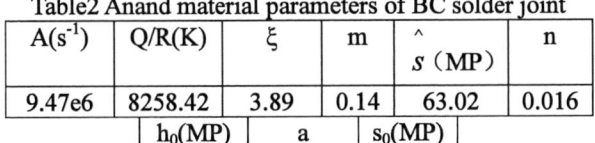
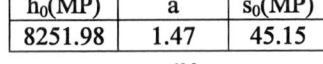

Fig.2 Double-symmetric cross section view of a PBGA272 and geometry size

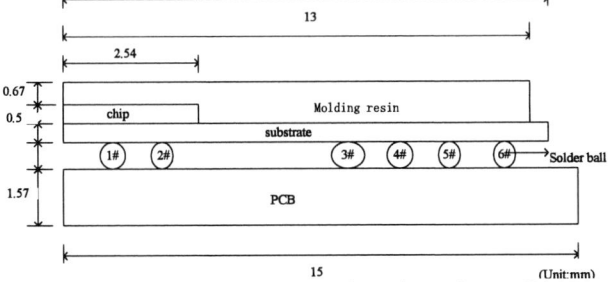
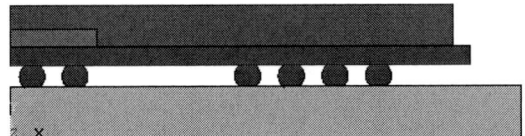

Fig.3 Double-symmetric plane geometry modeling

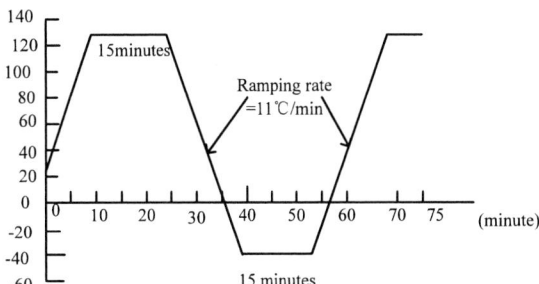

Fig.4 Temperature loading

3. Result and Discussion

Fig.5 and Fig.6 show the von Mises stress and von Mises plastic strain of the assembly after 4 cycles. As can be seen from Fig.5, the maximum von Mises stress lies in the top of left corner of the inner (1#) solder joint, the possible reason was that the hardness of the chip is nearly 10 times higher than that of substrate while the thermal expansion coefficient of the chip is nearly 10 times smaller than that of substrate, so this will lead to large extrusion to solder joints. But in Fig.5, the maximum von Mises plastic strain lies in the top of right corner of the outside (6#) solder joint. We can predict the location of crack propagation according to the maximum stress or maximum plastic strain [14], so 1# and 6# are the two key solder joints which are the easiest to failure.

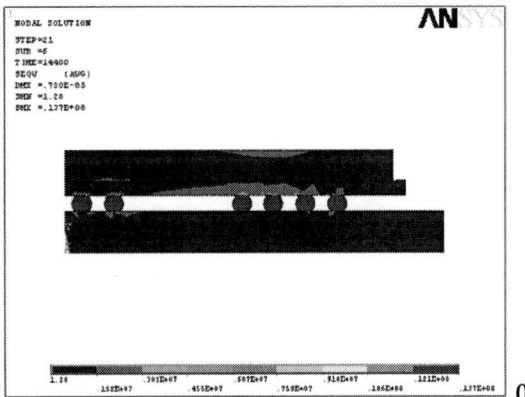

Fig.5 Von Misex stress

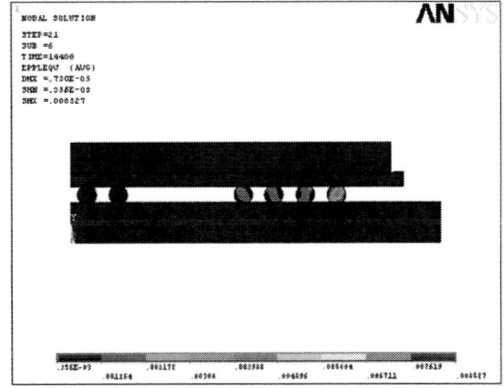

Fig.6 Von Mises plastic strain

4. DOE Simulation and Range Analysis

The purpose of DOE is to reduce the maximum von Mises plastic strain, the smaller of the maximum von Mises plastic strain, the harder of the solder joints to failure. The factors which will affect the von Mises plastic strain include environment, material properties and geometry, as shown in Fig.7 [15]. In this study, we only consider the geometry factors including PCB size, PCB thickness, chip size, chip thickness, substrate size, substrate thickness, solder height and solder radius. PCB size has two levers, and the rest have three levels, shown in Table 3. A mixed orthogonal test is designed, and the orthogonal table $L_{18}(2^1 \times 3^7)$ is adopted (Table4).

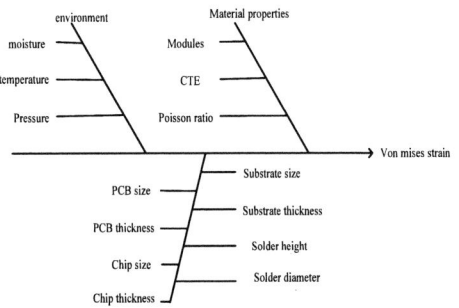

Fig.7 Fishbone diagram of effecting on von Mises strain

Table 3 Factors and levels

The software ANSYS is adopted to do the simulation, the loading condition (JESD22-A104-B Condition G) and the material Properties are the same with in section 2.

Range analysis is then used to deal with the DOE simulation results by using von Mises plastic strain as evaluation index, shown in Table 5. In Table 5, where K_i is the sum of the results which behalf of No. i level of arbitrary

column, $k_i = \dfrac{K_i}{s}$, where s is the number of times that the level appears, $R = Max.\{K_1, K_2, K_3\} - \min\{K_1, K_2, K_3\}$ or $R = Max.\{k_1, k_2, k_3\} - \min\{k_1, k_2, k_3\}$, for the first column,
$K_1 = 0.000479 + 0.008527 + 0.007815 + \cdots + 0.009365 = 0.06665$
$k_1 = K_1/9 = 0.00741$
$R = Max.\{k_1, k_2\} - \min\{k_1, k_2\} = 0.00020$

Table 3 Factors and levels

Factor	Levels		
	Level-1	Level-2	Level-3
A: PCB size (mm)	18	15	—
B: PCB thickness (mm)	1.4	1.57	1.8
C: Chip size (mm)	3	2.54	3.5
D: Chip thickness (mm)	0.4	0.5	0.6
E: Substrate size (mm)	13	13.5	14
F: Substrate thickness(mm)	0.4	0.56	0.7
G：Solder height (mm)	0.4	0.6	0.5
H：Solder radius (mm)	0.45	0.35	0.5

As can be seen from the range analysis (Table 5), based on the value of the range (R), the factors which effect the maximum von Mises plastic strain of the solder joint could be ranked as follows: F > H > G > C > D > E > B > A, the main effect plot (Fig.8) also indicates that substrate thickness (factor F), solder radius (factor H) and solder height (factor G) performed the main factors. The optimal scheme is $F_3H_2G_1C_2D_1E_2B_3A_2$(Viz. substrate thickness 0.7mm, solder radius 0.38mm, solder height 0.4mm, chip size 2.54mm, chip thickness 0.4mm, substrate size 13.5mm, PCB thickness 1.8mm and PCB size 15mm) by comprehensively considered with every factor's effect

Table 4 Orthogonal table $L_{18}(2^1 \times 3^7)$

Factors Test No.	PCB size	PCB thickness	Chip size	Chip thickness	Substrate size	Substrate thickness	Solder height	Solder radius
1	1	1	1	1	1	1	1	1
2	1	1	2	2	2	2	2	2
3	1	1	3	3	3	3	3	3
4	1	2	1	1	2	2	3	3
5	1	2	2	2	3	3	1	1
6	1	2	3	3	1	1	2	2
7	1	3	1	2	1	3	2	3
8	1	3	2	3	2	1	3	1
9	1	3	3	1	3	2	1	2
10	2	1	1	3	3	2	2	1
11	2	1	2	1	1	3	3	2
12	2	1	3	2	2	1	1	3
13	2	2	1	2	3	1	3	2
14	2	2	2	3	1	2	1	3
15	2	2	3	1	2	3	2	1
16	2	3	1	3	2	3	1	2
17	2	3	2	1	3	1	2	3
18	2	3	3	2	1	2	3	1

Table 5 Simulation results of DOE and range analysis results

Run order	PCB size (A)	PCB thickness (B)	Chip size (C)	Chip thickness (D)	Substrate size (E)	Substrate thickness (F)	Solder height (G)	Solder radius (H)	ξ_x
1	1	1	1	1	1	1	1	1	0.00479
2	1	1	2	2	2	2	2	2	0.008527
3	1	1	3	3	3	3	3	3	0.007815
4	1	2	1	1	2	2	3	3	0.006871
5	1	2	2	2	3	3	1	1	0.010843
6	1	2	3	3	1	1	2	2	0.004715
7	1	3	1	2	1	3	2	3	0.008414
8	1	3	2	3	2	1	3	1	0.005313
9	1	3	3	1	3	2	1	2	0.009365
10	2	1	1	3	3	2	2	1	0.006334
11	2	1	2	1	1	3	3	2	0.012472
12	2	1	3	2	2	1	1	3	0.004332
Table 5 continued									
13	2	2	1	2	3	1	3	2	0.005232
14	2	2	2	3	1	2	1	3	0.006913
15	2	2	3	1	2	3	2	1	0.009451
16	2	3	1	3	2	3	1	2	0.012415
17	2	3	2	1	3	1	2	3	0.00439
18	2	3	3	2	1	2	3	1	0.006979

K_1	0.06665	0.04427	0.04406	0.04734	0.04428	0.02878	0.04866	0.04371
K_2	0.06847	0.04403	0.04846	0.04433	0.04691	0.04499	0.04183	0.05273
K_3	—	0.04688	0.04266	0.04351	0.04398	0.06141	0.04468	0.03874
k_1	0.00741	0.00738	0.00734	0.00789	0.00738	0.00480	0.00811	0.00729
k_2	0.00761	0.00734	0.00808	0.00739	0.00782	0.00750	0.00697	0.00879
k_3	—	0.00781	0.00711	0.00725	0.00733	0.01024	0.00745	0.00646
R	0.00020	0.00047	0.00097	0.00064	0.00049	0.00544	0.00114	0.00233
Factor: main → secondary	FHGCDEBA							
Optimum scheme	$F_3H_2G_1C_2D_1E_2B_3A_2$							

Note: ξ_x is the Maximum von-Mises plastic strain

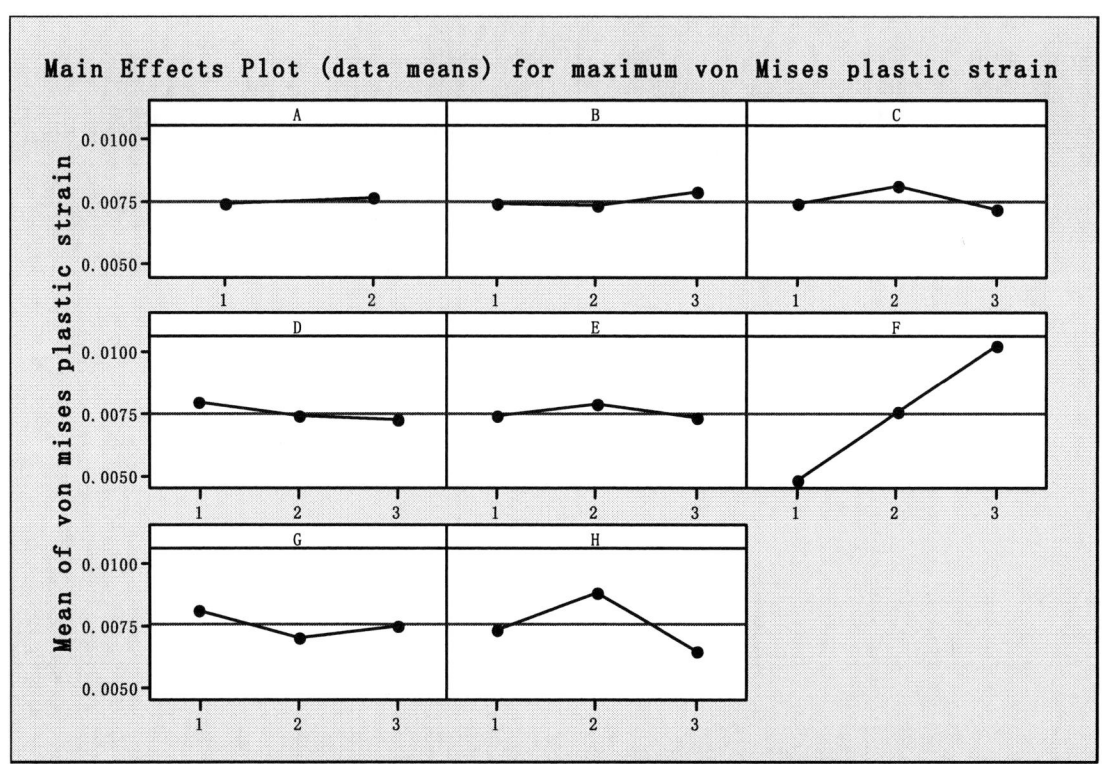

Fig.8 Main effect plot-maximum von Mises plastic strain

5. Further Study

It is noted that the material properties of BC solder in Ref. [12] is a approximation, further study should focus on the material properties of mixed solder; On the other hand, one of the assumption in this study is that the lead-free BGA component and the SnPb paste are fully mixed, so the proper reflow profile is needed to study to achieve the fully mixed solder joints; Verification tests which can verify the FE results also need to be done in the future.

6. Conclusions

In this work, the process of the thermal cycle was simulated on the temperature condition from -40℃ to 125℃ (JESD22-A104-BConditionG), the inner (1#) solder joint and the outside (6#) solder joint are two key solder joints which are the easiest to failure .

From the DOE simulation and range analysis, the factors which effect the maximum von Mises plastic strain of the solder joint could be ranked as follows: F > H > G > C > D > E > B > A; The optimal scheme is $F_3H_2G_1C_2D_1E_2B_3A_2$ (Viz. substrate thickness 0.7mm,solder radius 0.38mm, solder height 0.4mm, chip size 2.54mm, chip thickness 0.4mm, substrate size 13.5mm, PCB thickness 1.8mm and PCB size 15mm) by comprehensively considered with every factor's effect.

References

1. Anand Kannabiran, Elavarasan T. Pannerselvam, et al, "Investigation of the Forward and Backward Compatibility of Solder Alloys with Component Finishes for Hash and OSP PCB Finish," SMTA International Conference Proceedings, September, 2006.
2. Jianbiao Pan, "Estimation of Liquidus Temperature When SnAgCu BGA/CSP Components Are Soldered with SnPb Paste," 7th International Conference on Electronics Packaging Technology, Shanghai, China, August, 2006, pp. 225-230.
3. Jasbir Bath, Lead-free Soldering, Springer US (New York, 2007), pp. 173-P197.
4. Anand Kannabiran, Student Member, "Forward and Backward Compatibility of Solder Alloys with Component and Board Finishes," IEEE transactions on electronics packaging manufacturing, Vol.30, No.2, (2007), pp. 138-146.
5. Nurmi ST, Ristolainen, "Reliability of Tin-lead Balled BGAs Soldered with Lead-free Solder Paste," Soldering and Surface Mount Technology, Vol.14, No.2, (2002), pp. 35-39.
6. Lau J, Hoo N, Horsley R, Smetana J, et al, "Reliability Testing and Data Analysis of Lead-free Solder Joints for High-density Packages," Soldering and Surface Mount Technology,Vol.16 , No.2, (2004), pp. 46-68.
7. Smetana J, Horsley R, Lau J, Snowdon K, et al, "Design, materials and process for lead-free assembly of high-density packages," Soldering & Surface Mount Technology, Vol.16, No.1, (2004), pp. 53-62.
8. Hua F, Aspandiar R, Rothman T, Anderson C, et al, "Solder Joint Reliability of Sn-Ag-Cu BGA Components Attached with Eutectic Pb-Sn Solder Pate," Journal of

Surface Mount Technology,Vol.16 , No.1, (2003), PP. 34-42.

9. Hua F, Aspandiar R, Anderson C, Clemons G, et al, "Solder Joint Reliability Assessment of Sn-Ag-Cu BGA Components Attached with Eutectic Pb-Sn Solder," Proceedings of the SMTA international conference ,Chicago, IL, September, 2003, pp. 246-252.

10. Rainer Dudek, "Reliability Prediction of Area Array Solder Joints," Journal of electronic packaging, Vol.125, (2003), pp. 562-568.

11. B.A.Zahn. "Finite Element Based Solder Joint Fatigue Life Predict Same Die Size-stacked-chip Scale Ball Grid Array Package," Proceedings 27th International Electronics Manufacturing Technology Symposium, San Jose, CA, United States, 2002, pp. 274-284.

12. Jiang Ting-biao, Du Chao, Xu Long-hui, "Finite Element Analysis and Fatigue Life Prediction of BGA Mixed Solder Joints," Proceedings of HDP'07 International symposium, Shanghai, June, 2007.

13. JEDEC standard, JESD22-A104B, Temperature Cycling, Jul. 2000.

14. Ma Xin, Qian Yi-yu , Liu Fa, "Application of Numerical Simulation of Stress - Strain Field Distribution in the Solder Joints," Electronics Process Technology, Vol.22, No.2, (2001), pp. 51-55.

15. Liu Yuzhi, "The Study of Maximum Equivalent Strain of PBGA Package under Thermal cycles", Master Degree thesis of National Cheng Kung University, 2005, pp. 74-75.

First-Principles Study on the Elastic Anisotropy of Au–Sn Intermetallic Compounds

Rong An, Chunqing Wang, Yanhong Tian

Department of Electronics Packaging Technology, School of Materials Science and Engineering, Harbin Institute of Technology, 92 Xidazhi Street, Harbin 150001, P.R. China
anyieng@yahoo.com.cn, wangcq@hit.edu.cn, +86-451-86418359

Abstract

Independent elastic constants of single-crystal Au_5Sn and AuSn were determined through first-principles calculations to characterize their elastic anisotropy. The ideal elastic moduli and Poisson's ratio of Au_5Sn were determined using the Voigt–Reuss–Hill method and were very close to the range of experimental results; but the ideal Young's modulus of AuSn was much smaller than the experimentally obtained values. This unusual discrepancy in the Young's modulus values of AuSn is probably attributed to its extraordinarily high anisotropy in Young's modulus with a 95 GPa difference between its maximum and minimum values. Au_5Sn exhibits relatively low anisotropy in Young's modulus with the maximum-minimum difference of 38 GPa.

1 Introduction

Au–Sn solder is widely used as preforms, coatings, or bumps for flip chip bonding in electronic/optoelectronic packaging because of its environmental friendliness, superior creep resistance, high thermal and electrical conductivities, reduced oxide formation, and good corrosion resistance.[1,2] It contains two intermetallic compounds, Au_5Sn (ζ') and AuSn (δ).[2] Although the vast majority of research[1-5] has focused on the microstructure of the solder and on its evolution during interfacial reaction between the solder and a pad to bridge the gap between the microstructure and mechanical property of the solder joints, surprisingly few studies of the mechanical properties of these Au–Sn compounds themselves exist in the literature (Table I)[6-8]. This is attributable mainly to the difficulty in preparing the single-phase samples and the limitations of the available experimental techniques.

The miniaturization of microelectronic packages requires the use of more fine solder joints; as the solder joints become increasingly small (less than 100 μm in diameter) and contain only a few grains, their mechanical property, however, cannot be determined from conventional mechanical tests with bulk samples, and there may be a considerable variation in mechanical behavior from joint to joint because of the anisotropy of mechanical property.[3,4] Thus, in addition to characterizing the morphology and crystallographic orientation of the grains in joints, there is a strong need to determine the elastic constants of these intermetallics to represent their elastic anisotropy.

First-principles calculation is being widely utilized for the prediction of material properties by virtue of the increase of computing power and the development of the density functional theory (DFT). Previous researches[9-12] provided evidence on the applicability of DFT calculations with the pseudopotential method for Sn-based intermetallic compounds. It can be seen from these examples that the experimental data on single-crystal elastic constants can be enriched using the Voigt–Reuss–Hill method, and bounds can be placed on polycrystalline elastic moduli, so that they can be compared with existing polycrystalline data. In our study, we computed the full set of elastic constants by performing the first-principles pseudopotential total energy calculations on both Au_5Sn (ζ') and AuSn (δ). Our objects are (i) to determine the ideal elastic properties of the ζ' and δ phases and (ii) to investigate their elastic anisotropy.

2 Methodology

The Au_5Sn (ζ') crystallizes in the space group $R3$ (No. 146) belonging to trigonal crystal structure (Fig. 1a), which can be described by two cell parameters a and c.[13] The Sn atom occupies the 3a site (0, 0, 0), and there are three Au sites: Au1 locates at 3a positions (0, 0, 0.3307), Au2 locates at 3a positions (0, 0, 0.6693), and Au3 resides at 9b positions (0.3333, 0.3403, 0.1667). The crystal structure of AuSn (δ) is depicted in Fig. 1b. The compound crystallizes in a hexagonal lattice with the $P6_3/mmc$ space group (No. 194).[14] It has four atoms per unit cell; there is one position of Au at 2a site and one position of Sn at 2c site. We used their conventional cells for all the first-principles calculations.

The CASTEP code[15,16] was used in the present calculations, wherein the Vanderbilt-type ultrasoft pseudopotential,[17] the PW91 form[18] of the generalized gradient approximation (GGA), and a plane-wave basis set were employed to describe electron–ion interactions, to take into account exchange–correlation effects, and to represent electronic wavefunctions, respectively. The cut-off energy of plane-wave basis sets was 540 eV for all the calculations on either Au_5Sn or AuSn. An 8×8×8 Monkhorst–Pack[19] k-point mesh was employed to approximate definite integral over the Brillouin zone of Au_5Sn, and a 9×9×6 k-point mesh was used to AuSn. For Au_5Sn, increasing the plane-wave cut-off energy to 640 eV and the k-point mesh to 10×10×10 changed the total energy by less than 0.01 eV/atom and lattice constants by less than 0.2%; and for AuSn, increasing the cut-off energy to 640 eV and the mesh to 11×11×8 changed the total energy by less than 0.007 eV/atom and lattice constants by less than 0.1%. Therefore, the present computations were precise enough to reproduce the ground state properties.

978-1-4244-2739-0/08/$25.00 ©2008 IEEE

Table I. Experimental Young's Moduli of Au₅Sn and AuSn

Material	Specimen/technique	v	E (GPa)
Au₅Sn (ζ')	Bulk, resonance	0.4	62[6]
	Diffusion couples, nanoindentation		76[8]
AuSn (δ)	Bulk, resonance	0.3	71[6]
	Thick film, microindentation		101[7]
	Bulk, nanoindentation		87[8]

Table II. Calculated Elastic Stiffness (C_{ij}) and Compliance (S_{ij}) Constants of Single-Crystal Au₅Sn and AuSn.

				11	33	44	12	13	14	15
Au₅Sn	540 eV, 8 × 8 × 8 k points	PW91	C_{ij} (GPa)	160	181	30	117	102	-1.0	0.6
			S_{ij} (10^{-3}/GPa)	15	9.5	33	-8.4	-3.5	0.7	-0.4
	640 eV, 10 × 10 × 10 k points	PW91	C_{ij} (GPa)	165	173	34	121	104	-1.6	1.4
			S_{ij} (10^{-3}/GPa)	15	10	30	-8.3	-3.7	1.1	-0.9
AuSn	540 eV, 9 × 9 × 6 k points	PW91	C_{ij} (GPa)	103	164	14	88	62	-	-
			S_{ij} (10^{-3}/GPa)	36	8.1	69	-29	-2.7	-	-
		PBE	C_{ij} (GPa)	115	165	15	86	61	-	-
			S_{ij} (10^{-3}/GPa)	21	7.8	65	-14	-2.4	-	-
		LDA	C_{ij} (GPa)	126	205	24	103	72	-	-
			S_{ij} (10^{-3}/GPa)	25	6.3	42	-19	-2.0	-	-
	640 eV, 11 × 11 × 8 k points	PW91	C_{ij} (GPa)	108	158	17	89	66	-	-
			S_{ij} (10^{-3}/GPa)	30	8.8	59	-23	-2.9	-	-

11, 22, etc. are the tensor subscripts.

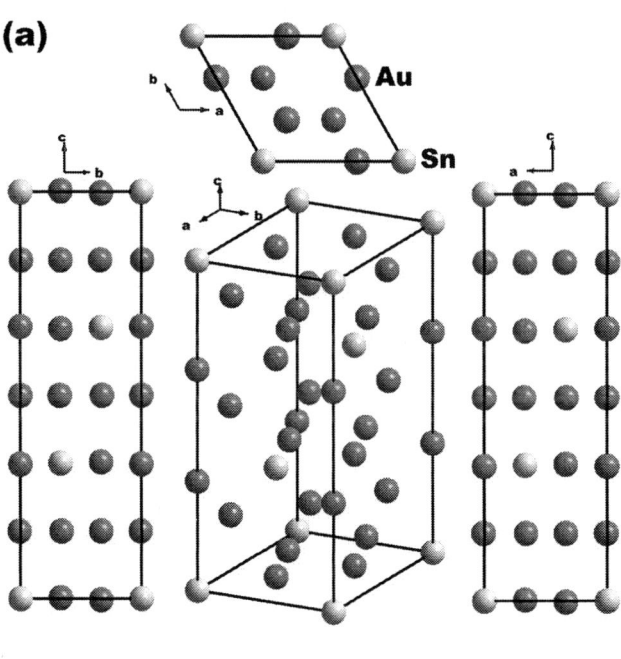

(a)

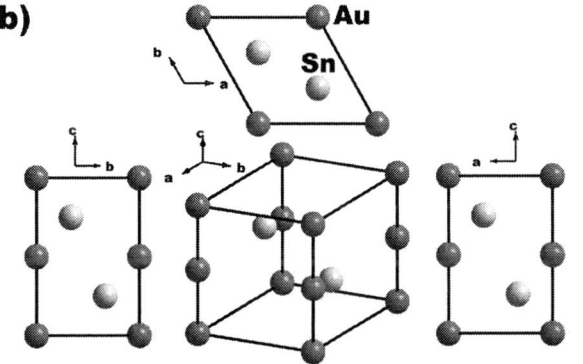

(b)

Fig. 1. The crystal structures of (a) Au₅Sn and (b) AuSn.

Table III. Bounds on the Elastic Properties of Polycrystalline Au₅Sn and AuSn.

	Bound	K (GPa)	G (GPa)	E (GPa)	v
Au₅Sn	Voigt	127.21	28.48	79.51	0.3958
	HS upper	127.19	27.97	78.18	0.3976
	HS lower	127.19	27.81	77.76	0.3981
	Reuss	127.17	27.16	76.07	0.4003
AuSn	Voigt	88.34	17.80	50.04	0.4056
	HS upper	87.86	15.59	44.17	0.4162
	HS lower	87.51	13.50	38.51	0.4266
	Reuss	87.31	11.87	34.06	0.4350

Unit cell was fully optimized in order to obtain the equilibrium crystal structure. Lattice parameters and internal atomic coordinates were independently modified to minimize the total energy and interatomic forces. The Broyden–Fletcher–Goldfarb–Shanno (BFGS) minimization scheme[20] was used for geometry optimization. The tolerances for geometry optimization were selected as follows: difference in total energy within 5×10^{-6} eV/atom, maximum ionic Hellmann-Feynman force within 0.01 eV/Å, maximum ionic displacement within 5×10^{-4} Å, and maximum stress within 0.02 GPa.

We set the strain of the optimized equilibrium cells to a finite value by applying a given homogeneous deformation to optimize the internal atomic coordinates and calculate the resulting stress; each of the second-order elastic constants was determined by means of a least-squares linear fit of stress against strain. To calculate the elastic coefficients precisely, the stricter criteria for convergences were selected to optimize the internal atomic coordinates: difference in total energy within 1×10^{-6} eV/atom, ionic Hellmann–Feynman forces within 0.002 eV/Å, and maximum ionic displacement within 1×10^{-4} Å. For the Au₅Sn compound with $R3$ symmetry, its independent elastic constants are c_{11}, c_{33}, c_{44}, c_{12}, c_{13}, c_{14}, and

c_{15}; for the AuSn with $P6_3/mmc$ symmetry, its independent elastic constants are c_{11}, c_{33}, c_{44}, c_{12}, and c_{13}. Two strain patterns (the first with nonzero xx component, and the second with nonzero zz and yz components) were used to generate the stresses related to all seven independent elastic constants of Au$_5$Sn; similarly, two other strain patterns (the first with nonzero zz component, and the second with nonzero xx and yz components) were employed to yield all five elastic coefficients of AuSn that are independent with each other. Three positive and three negative amplitudes were applied for each strain component with maximum strain value of 0.6% in these computations. We demonstrated the convergence of the calculations by increasing the energy cutoff to 640 eV and the k-point mesh to 10×10×10 for Au$_5$Sn and by increasing the energy cutoff to 640 eV and the k-point mesh to 11×11×8 for AuSn, after which there was no difference in the calculated elastic stiffness constants of Au$_5$Sn within 8 GPa and in the calculated elastic stiffness constants of AuSn within 6 GPa (Table II). Polycrystalline elastic parameters, such as bulk modulus and Young's modulus, were estimated from the compliance tensor components using the Voigt method and the Reuss method,[21] and the Hashin–Shtrikman (HS) bounds[22] was also used to place tighter bounds within Voigt and Reuss bounds. In addition, the Voigt–Reuss–Hill (VRH) average[21] was employed to determine the theoretical polycrystalline elastic property.

3 Results and discussions

The elastic stiffness determines the response of a crystal to an imposed strain (or stress) and provides information about bonding characteristics near the equilibrium state. Investigation of elastic stiffness is essentially the first step to understand the mechanical properties of a solid. Table II includes the full set of theoretical second-order elastic constants of Au$_5$Sn and AuSn. Since their second-order elastic constants have not been reported in the literature, it is impossible to compare the present theoretical elastic constants with others; it is however noted that Au$_5$Sn has great elastic stiffness constants relative to AuSn. For instance, the elastic constants representing stiffness against uniaxial strains, c_{11} and c_{33}, of Au$_5$Sn, are 160 and 181 GPa, respectively, which are slightly higher than those of AuSn (103 and 164 GPa, respectively). The c_{44} of Au$_5$Sn, which correspond to the resistance to shear deformation, are 30 GPa, and are over twice that of AuSn (14 GPa). These results agree with the hardness of these two Au–Sn intermetallic compounds reported in Ref. 8 that the hardness value of Au5Sn (2.5 GPa) is much larger than that of AuSn (1.1 GPa).

Table III lists the isotropic bulk (K) and shear (G) moduli calculated from the corresponding single-crystal data using the VRH approximation. The Young's modulus E and Poisson's ratio v were calculated from the K and G using the interrelationship of these four elastic parameters based on the isotropic elasticity of the materials, and the results are also listed in the table. The VRH Young's modulus of Au$_5$Sn is 78 GPa, which is larger than that (62 GPa) obtained using resonance[6] and very close to that (76 GPa) determined by Chromik et al. using nanoindentation[8] (Table I). The VRH

average on the Poisson's ratio of Au$_5$Sn (0.398) is fairly consistent with the reported experimental value (0.4). Lee et al.[9,10] reported similar results about Cu$_6$Sn$_5$, Ag$_3$Sn, and Ni$_3$Sn$_4$ that the Young's moduli determined through first-principles calculations are close to those obtained using nanoindentation, but larger than those values measured with bulk samples using resonance or tensile loading technique. This difference can be attributed to the following possible sources: first, the DFT computation typically deduces the elastic constants and the mechanical parameters at absolute zero, and Young's modulus decreases with increasing temperature; second, defects in material are not taken into account in the overall calculation. In contrast with Au$_5$Sn and other Sn-based intermetallics, however, the VRH average of AuSn on Young's modulus (42 GPa) is surprisingly far less than any of the values determined through resonance, microindentation, or nanoindentation. To clarify that the elastic constants calculated with the PW91 are not in error,[23] we examined the predictions from other GGA formulations, i.e., Perdew-Burke-Enzerhof (PBE), and also the local density approximation (LDA); the calculation results are listed in Table II. For the GGA-PBE, no difference is observed in the calculated elastic constants within 10%, and the VRH Young's modulus of 54 GPa is still much less than the experimental results (71~101 GPa). The elastic constants calculated with the LDA are typically larger than that predicted from the PW91 or PBE, but the VRH Young's modulus of 62 GPa is also less than the experimental results. Thus, this unusual minus deviation from the experimental values can be derived not from the calculation error but from the elastic anisotropy, which will be discussed later in this paper.

Either of Au$_5$Sn and AuSn is a non-cubic crystal; their elastic anisotropy can be measured using three dimensionless quantities A_G, A_K, and A_E, defined as $A_G = (G_V-G_R)/(G_V+G_R)$, $A_K = (K_V-K_R)/(K_V+K_R)$, and $A_E = (E_V-E_R)/(E_V+E_R)$,[24,25] respectively. The subscripts V and R here designate the Voigt and the Reuss averaging schemes and they represent the maximum and minimum limits of the true polycrystalline elastic moduli. The A gives the relative magnitude of the elastic anisotropy presented in crystals. It is always positive and is zero for crystals which are elastically isotropic. For Au$_5$Sn, the anisotropy factors, viz., A_G, A_K, and A_E (in percent), are 2.4%, 0.016%, and 2.2%; and for AuSn, these factors are up to 20%, 0.59%, and 19%, respectively. The results indicate that AuSn is much more anisotropic than Au$_5$Sn; these are supported by the fact that the residual indents of AuSn in nanoindentation testing exhibited significant asymmetric pileup different from Au$_5$Sn, which is due to the plastic anisotropy of AuSn.[8] Furthermore, these results are also in accordance with the data on the orientation-dependent Young's moduli in these crystals. A three-dimensional surface representation of elastic anisotropy is an illustrative way of showing the variation of elastic modulus with crystallographic direction. The directional dependence of Young's modulus for trigonal crystals can be given by[26]

$$E = \left[\begin{array}{l} \left(1-l_3^2\right)^2 s_{11} + l_3^4 s_{33} + l_3^2\left(1-l_3^2\right)\left(2s_{13}+s_{44}\right) \\ + 2l_2 l_3 \left(3l_1^2 - l_2^2\right)s_{14} + 2l_1 l_3 \left(3l_2^2 - l_1^2\right)s_{15} \end{array} \right]^{-1} , \quad (1)$$

and for hexagonal crystals,

$$E = \left[\left(1-l_3^2\right)^2 s_{11} + l_3^4 s_{33} + l_3^2\left(1-l_3^2\right)\left(2s_{13}+s_{44}\right) \right]^{-1} , \quad (2)$$

where s_{ij} are the single-crystal elastic compliance constants in the two suffix notation, and l1, l2, and l3 are the direction cosines to the *a*, *b*, and *c* axes, respectively. In this representation, an isotropic system would have a spherical shape, and so the degree of deviation of the geometry from a sphere indicates the degree of anisotropy in a specific property of system. The representations on the directional dependence of the Young's moduli of both Au₅Sn and AuSn show clear deviations from a spherical shape as shown in Fig. 2 and Fig. 3, but the more deviation occurs in AuSn. Therefore, the two intermetallics have a high degree of anisotropy in Young's modulus, but the degree in AuSn is much higher than that in Au₅Sn.

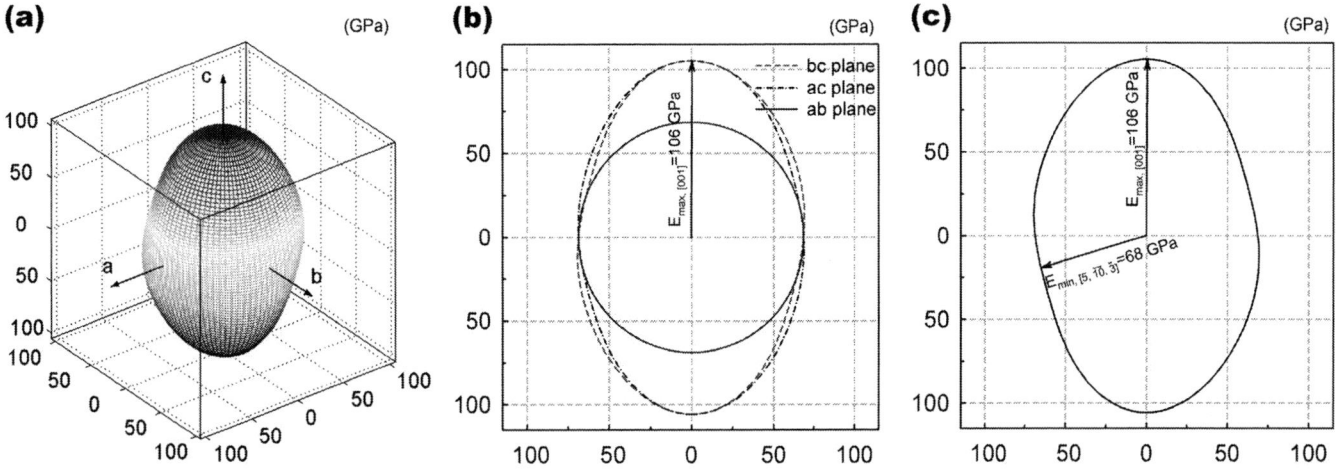

Fig. 2. (a) Directional dependence of Young's modulus in Au₅Sn. Plane projections of the directional dependence of Young's modulus on (b) the (100), (010), (001), and (c) (120) planes.

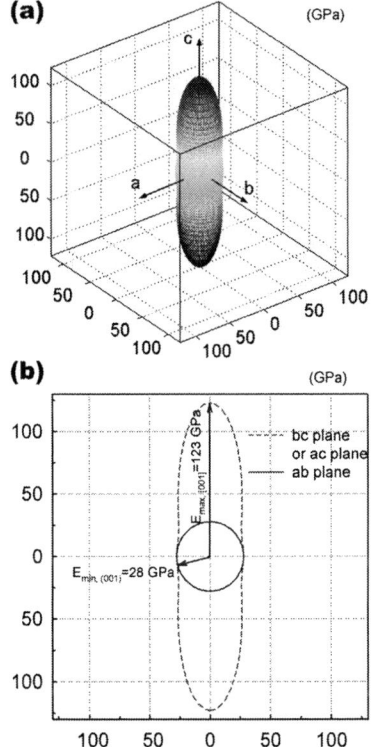

Fig. 3. (a) Directional dependence of Young's modulus in AuSn. (b) Plane projections of the directional dependence of Young's modulus on the (100), (010), and (001) planes.

The anisotropic Young's moduli of both the compounds in the (100), (010), and (001) planes are depicted in Fig. 2b and Fig. 3b. For either Au₅Sn or AuSn, significant in-plane elastic anisotropy appears in the (100) and (010) planes, but no elastic anisotropy in the (001) plane, which is consistent with the atomic arrangements in itself. The maximum Young's modulus of Au₅Sn is 106 GPa, which is in the [001] direction; the minimum is 68 GPa, which is along the [$\bar{5}$, 10, 3] direction (Fig. 2c). For AuSn, the maximum modulus of 123 GPa is reached in the [001] crystallographic orientation; the minimum of 28 GPa is obtained along any orientation in the (001) plane, as shown in Fig. 3b. The difference between the maximum and minimum moduli of Au₅Sn is 38 GPa, which is 49% of the VRH Young's modulus; this large difference implies that the distinct disagreement in its experimentally obtained Young's modulus can be attributed to the strong elastic anisotropy plus the differences induced during sample preparation and the restriction imposed by limited experimental techniques available. This explanation is partially supported by making the comparisons between the maximum-minimum difference of Young's modulus and the bound of the reported experimental results. Examples include the cases of Cu₆Sn₅ and Ag₃Sn; these two Sn-based intermetallics fall within orthorhombic crystal structure. The difference between the maximum and minimum moduli of Cu₆Sn₅ (calculated using the elastic constants in Ref. 9) is more than 20 GPa, and the experimental Young's moduli summarized in Ref. 9 show a wide range from 85 GPa to 125 GPa. Here again, the calculated difference of Ag₃Sn (using the data in Ref. 10) is 34 GPa, and the reported experimental values are scattered within the large range of 70 to 94 GPa. For AuSn, the difference of Young's modulus between the maximum and minimum is up to an unusual large value of 95 GPa, which is 2.3 times itself VRH Young's modulus. This is graphically illustrated in the surface representation of its elastic anisotropy. The representation of AuSn has a needle-like shape, meaning that a slight degree of deviation of crystallographic orientation from the [001] direction in AuSn will produce a significant decrease in Young's modulus, as depicted in Fig. 3a. Thus, when the test samples primarily have the [001] texture, it is obvious that the experimentally obtained Young's moduli will be much larger than the VRH Young's modulus. So far it has been impossible to quantitatively compare the modulus values obtained with samples that possess different textures because the textures of samples are not readily accessible. However, because of such distinct anisotropy of elasticity, it is clear that, the fewer grains there are in the intermetallic compounds of the increasingly small solder joints, the greater the difference between the Young's modulus value of the bulk samples and that measured with the intermetallics in the solder joints.

Conclusions

The seven elastic constants of single-crystal Au_5Sn and the five elastic constants of single-crystal AuSn, have been determined from first-principles calculations using the pseudopotential plane-wave method. The ideal bulk, shear, and Young's moduli, as well as Poisson's ratios, of these two polycrystalline intermetallic compounds were calculated using the VRH and HS method. Au_5Sn exhibits quite distinct anisotropy in Young's modulus, with a difference of 38 GPa between its maximum and minimum modulus; AuSn shows a much higher degree of anisotropy than Au_5Sn, that is, the maximum-minimum difference of the Young's moduli of AuSn is 95 GPa. In addition to the metallurgical imperfections of test samples and the restrictions imposed by the limited experimental techniques available, high elastic anisotropy is one of the causes of the discrepancy in experimentally and theoretically obtained elastic modulus values. It is believed that with time the elastic anisotropy of these intermetallics will become a more important consideration in the accurate characterization of the mechanical behavior of the small solder joints.

Acknowledgments

This work is financially supported by the National Natural Science Foundation of China under grant No. 50675047/E052105 and Joint Project between Samsung Electronics Co. Ltd. (Korea) and Harbin Institute of Technology (HIT). The authors thank J. Cheng and J.C. Zhu, Harbin Institute of Technology, for their valuable help with the execution of CASTEP code, and the Shanghai Supercomputer Center (SSC) for supercomputing resources provided.

References

1. G.S. Matijasevic, C.C. Lee, C.Y. Wang, Thin Solid Films 223, 276 (1993).
2. D.G. Ivey, Micron 29, 281 (1998).
3. H. Song, J. Morris, M. McCormack, J. Electron. Mater. 29, 1038 (2000).
4. H. Song, J. Ahn, J. Morris, J. Electron. Mater. 30, 1083 (2001).
5. J.-W. Yoon, H.-S. Chun, S.-B. Jung, J. Mater. Res. 22, 1219 (2007).
6. F.G. Yost, M.M. Karnowsky, W.D. Drotning, J.H. Gieske, Metall. Mater. Trans. A 21, 1885 (1990).
7. A. Vicenzo, M. Rea, L. Vonella, M. Bestetti, P.L. Cavallotti, J. Solid State Electrochem. 8, 159 (2004).
8. R.R. Chromik, D.-N. Wang, A. Shugar, L. Limata, M.R. Notis, R.P. Vinci, J. Mater. Res. 20, 2161 (2005).
9. N.T.S. Lee, V.B.C. Tan, K.M. Lim, Appl. Phys. Lett. 88, 031913 (2006).
10. N.T.S. Lee, V.B.C. Tan, K.M. Lim, Appl. Phys. Lett. 89, 141908 (2006).
11. R. An, C. Wang, Y. Tian, H. Wu, J. Electron. Mater. 37, 477 (2008).
12. R. An, C. Wang, Y. Tian, J. Electron. Mater. 37, 968 (2008).
13. K. Osada, S. Yamaguchi, M. Hirabayashi, Trans. Jpn. Inst. Met. 15, 256 (1974).
14. J.-P. Jan, W.B. Pearson, A. Kjekshus, S.B. Woods, Can. J. Phys. 41, 2252 (1963).
15. M.C. Payne, M.P. Teter, D.C. Allan, T.A. Arias, J.D. Joannopoulos, Rev. Mod. Phys. 64, 1045 (1992).
16. M.D. Segall, P.J.D. Lindan, M.J. Probert, C.J. Pickard, P.J. Hasnip, S.J. Clark, M.C. Payne, J. Phys.: Condens. Matter 14, 2717 (2002).
17. D. Vanderbilt, Phys. Rev. B 41, 7892 (1990).
18. J.P. Perdew, Y. Wang, Phys. Rev. B 45, 13244 (1992).
19. H.J. Monkhorst, J.D. Pack, Phys. Rev. B 13, 5188 (1976).
20. T.H. Fischer, J. Almlöf, J. Phys. Chem. 96, 9768 (1992).
21. R. Hill, Proc. Phys. Soc. A 65, 349 (1952).
22. J.P. Watt, L. Peselnick, J. Appl. Phys. 51, 1525 (1980).
23. A.E. Mattsson, P.A. Schultz, M.P. Desjarlais, T.R. Mattsson, K. Leung, Modell. Simul. Mater. Sci. Eng. 13, R1 (2005).
24. D.H. Chung, W.R. Buessem, J. Appl. Phys. 38, 2010 (1967).
25. D.H. Chung, W.R. Buessem, J. Appl. Phys. 39, 2777 (1968).
26. J.F. Nye, Physical Properties of Crystals (Oxford: Oxford University Press, 1985).

Numerical Analysis of Interfacial Delamination in Thin Array Plastic Package

Ke Xue[1,2], Jingshen Wu[1,2], Haibin Chen[1,2], Yongqiao Sun[1,2], Kenneth Kwan[3] ,John Yuen[3] , Angus Lam[3]

1. Department of Mechanical Engineering, The Hong Kong University of Science and Technology, Clear Water Bay, Kowloon, Hong Kong
2. Center for Engineering Materials and Reliability, Fok Ying Tung Graduate School, The Hong Kong University of Science and Technology, Clear Water Bay, Kowloon, Hong Kong
3. ASAT Holdings Limited, Dongguan, China

Abstract

It is well-known that high thermal stresses induced interfacial delamination is a typical failure mode in multi-layered micro-electronic components. Therefore, it is a key factor in package design to minimize interfacial stresses to enhance package reliability. In this paper, two designs of TAPP with different die attach pastes were manufactured. It was found that some test vehicles had cracks near the lead-frame paddle/mold compound side interface after 1000 temperature cycles (-65°C to150°C) although the electrical test passed. The failed units were analyzed and the physical location and shape of the failure was determined. Meanwhile finite element analysis was performed to find the locations of highest stress, and the expected modes of failure. A quarter three-dimensional (3-D) FEM model was constructed using commercial software ANSYS 11.0. The distribution of thermal stresses in the package was calculated and mapped, which were used to predict the most possible delamination origin and to describe the entire failure history of the package subjected to different thermal histories. It also identified material properties of the as the primary factors in this failure mechanism.

Introduction

With the increasing need for higher input/output (I/O) counts and miniaturization, novel electronic packages such as the thin array plastic package (TAPP) are continuously being developed. The TAPP technology provides a platform that integrates digital, analog, RF integrated circuits, along with high-performance passive components for system-in-package implementation. [1]

Despite these advantages, thermo -mechanical reliability of these emerging packages is still one of the major concerns in the electronic industry. Based on previous study it is found that these reliability problems are often triggered by loadings associated with packages manufacturing and testing processes. [2] Studies have shown that delamination and cracking are common in all types of electronic packages where the interfaces pose possible sites of fracture. [3]-[5] Failures like delaminations, package cracking, and metal shift occur due to the build-up of residual stress and warpage in the packages because of the TCE mismatch between the package materials as the package suffering temperature changes.

Finite element method is widely used in packaging failure analysis. It can predict the internal package stress distribution and help explain the stress transfer mechanism between the die, die paddle, and plastic after molding. [6] In the past decades, extensive research has been done in failure analysis by using FEM. For example, Moore and Jarvis [7] have used the finite element method to identify a root cause of reliability failures in small multichip BGAs. Lin et al. [8] have implemented finite element analyses to address the stress distribution in the stacked die leaded package and verified by the scanning acoustic microscope. Mei et al. [9] have developed a nonlinear finite element model for predicting the deformation, stress, and fracture behavior of delaminated plastic packages induced by mechanical and hygro-thermal loads. Zhao et al. [10] have investigated stress distributions leading to delamination at the chip/underfill and the underfill/substrate interfaces in flip-chip assemblies.

But there is little research effort found in open literature regarding the thermo-mechanical reliability of TAPP since it is a relatively new package solution. In this study, a detail of finite element analysis of thermal stress of TAPP was operated, combined with physical inspections including C-SAM, cross-section and SEM photos, to elicit the clear explanation of the failure mechanism.

Physical Description of the Package

TAPP is a leadless, Pb-free and multi-row packaging solution. The very thin, fine-pitch package with an exposed die attach pad allows for optimal thermal performance and a power/ground ring option for enhanced electrical characteristics. The package has a similar form factor to a quad flat no lead (QFN) package; however, it can have up to three rows of peripheral I/O pads allowing for higher density. A schematic cross-section view of TAPP is shown in Fig.1.

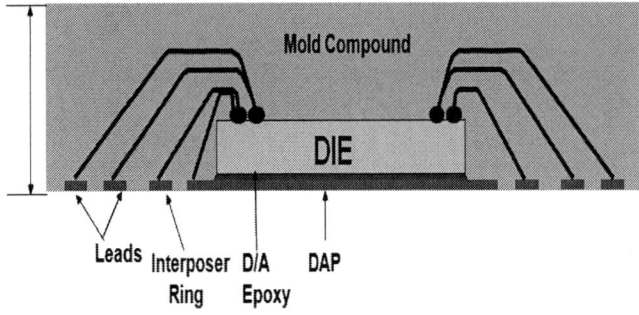

Fig.1. Cross section of TAPP

As can be seen from the figure, TAPP has an exposed pad with a die sitting on the top. Molding compound (MC) covers only one side of the lead-frame, leading to an imbalanced structure with a MC-rich top and a multi-layer (die/die-attach/pad) bottom. Due to the presence of dissimilar materials in the top and bottom of the package, the mismatch in the coefficient of thermal expansion between the different layers will result in high interfacial stresses upon heating or cooling of the structure during fabrication, testing, assembly, or in field use.

For the TAPP of interest in this study, the package is 11 mm square in size and 0.9mm high. The centered die attach pad is 5.4mm square in size and its thickness is 0.064mm. The 148 I/O pads are gathered in two rows along the periphery and the pad pitch is 0.2mm. Each I/O pad has a size of 0.4mm by 0.3mm. The die sits on the top of DAP with 4.8mm square and 0.28 thickness.

Observations of the Failure Modes

Reliability evaluation and SAM were performed on two groups of TAPP after 1000 temperature circles, which used ABLEBOND® A2200 and ABLEBOND® 84-1LMIS as die attach materials respectively. The difference between these two die attach materials will explain later. The results showed that the ABLEBOND® A2200 group has samples failed at the side interface between DAP and molding compound but the ABLEBOND® 84-1LMIS group did not have any failures. The A-scan mode can directly distinguish the failure spots. It was validated by scanning electronic microscopy (SEM) results, as shown in Figs 2, and 3, respectively.

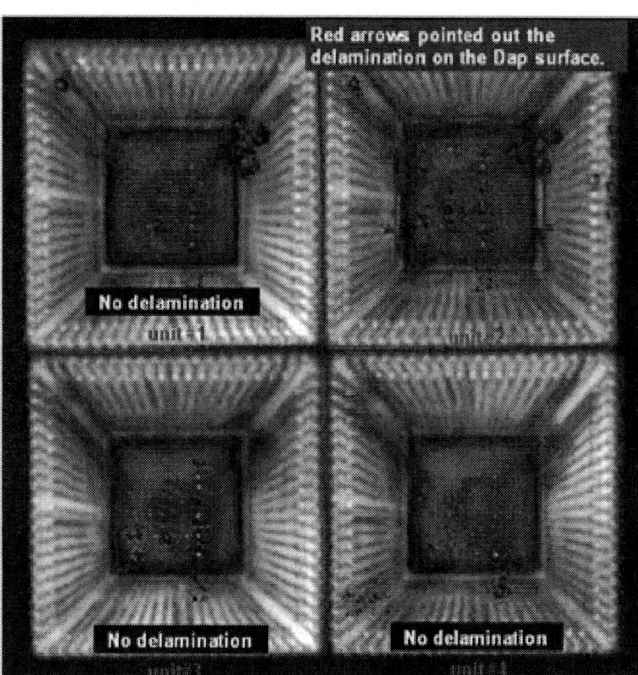

Fig.2. SAM pictures of TAPP (unit #1 and unit #2 used ABLEBOND® A2200, unit #3 and unit #4 used ABLEBOND® 84-1LMIS)

The failure means DAP lift up from mold compound when externally inspection. The failed unit had been cross section and it was found that the DAP edge had separated from molding compound and the crack propagated to mold compound.

Finite Element Simulation

The objective of this investigation is to evaluate the stress level inside the package body. Results will be used to justify the possible failures. For taking advantage of geometry symmetry a 3-D quarter finite element model was built by

using commercial FEA code ANSYS®11.0. Because the geometry difference is not significant between DAP and die, a detailed 3-D die attach fillet was also constructed into this model for the sake of getting more accurate results. Figs 4 and 5 showed FEA model for TAPP.

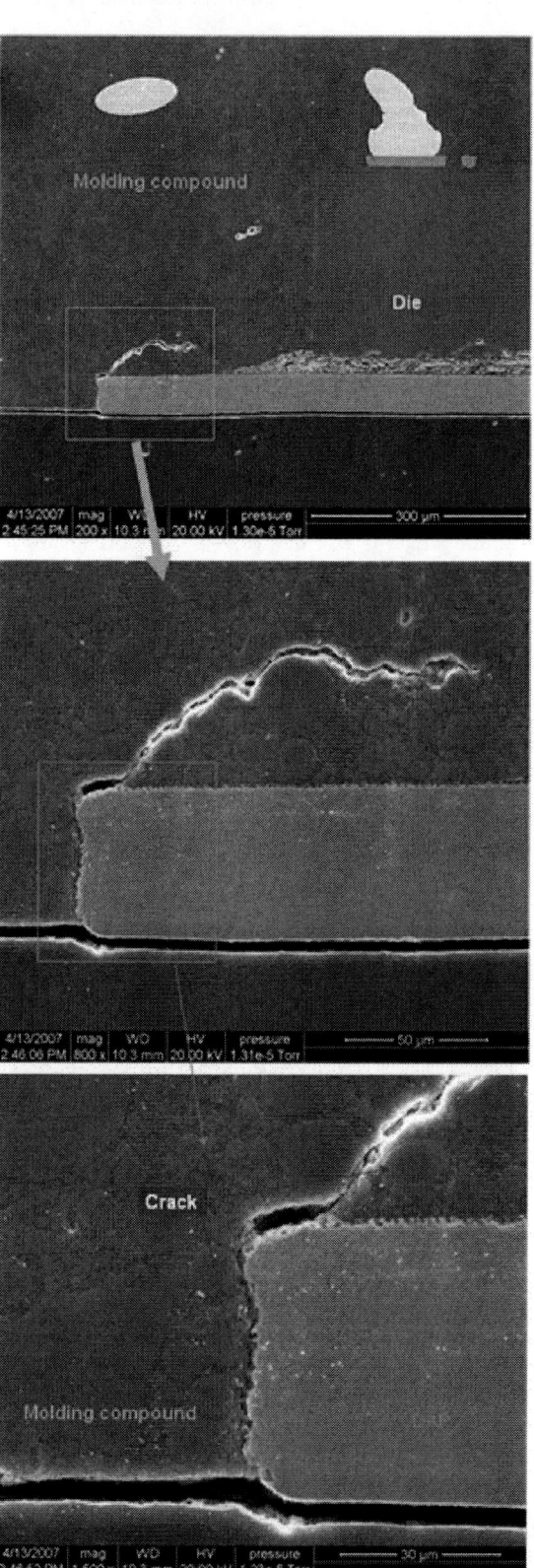

Fig.3. The failure description (interface delamination and cracking kink into molding compound) in TAPP

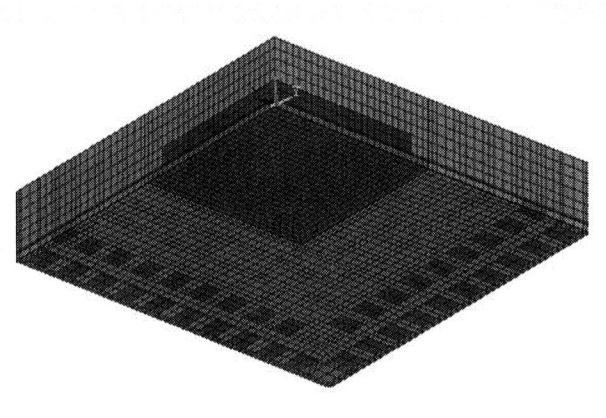

Fig.4. Three-dimensional FEA model of TAPP

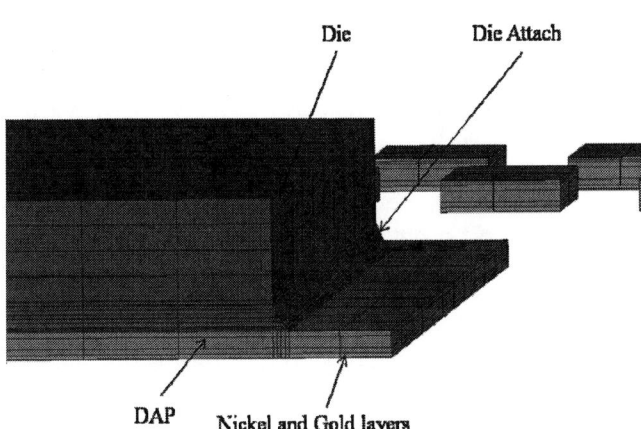

Fig.5. Three-dimensional FEA model of TAPP (encapsulant not shown)

In this study, a temperature-dependent linear-elastic formulation was used for the material models. It is possible that the molding compound material model may exhibit visco-elastic behavior. However, the temperature-dependent, linear-elastic model is justified because the extent of viscoelasticity exhibited by molding compound was found to be negligible over the temperature range of interest. The die-attach adhesive is a very thin layer with a low stiffness relative to other materials in the package, and hence was also modeled as elastic material. The properties for these materials were listed in Table 1.

The current TAPP production design use SUMITOMO BAKELITE EME-G770C encapsulant and Ablestik A2200 and 84-1LMIS die-attach adhesives.

Table 1 Packaging material properties ([11] and vendor-provided data)

	Modulus(MPa)	Poisson ratio	CTE (ppm)
EMC(EME-G770C)	27000	0.3	8, below T_g (130°C); 37, above T_g (130°C);
Die attach I (*ABLEBOND® A2200*)	170, at 25°C ; 81, at 150°C	0.3	73.4, below T_g (-43°C); 195, above T_g (-43°C);
Die attach II (*ABLEBOND® 84-1LMIS*)	3940, at 25°C ; 1960, at 150°C	0.3	41.1, below T_g (53°C); 142.4, above T_g (53°C);
Die (Silicon)	161000	0.21	2.6, at 25°C ; 3.2, at 170°C
Copper	117000	0.34	17
Gold	51500	0.4	14.2
Nickel	206843	0.3	13.4

The model was assumed stress free at 135 °C (the molding temperature) and the temperature of the entire model was reduced to -65 °C.

The deformation caused by the uniform temperature drop is shown in Fig. 6. The four corners of the package contracted not only horizontally but also vertically. The DAP warped like a "crying face" due to the squeezing effect by the surrounding molding compound.

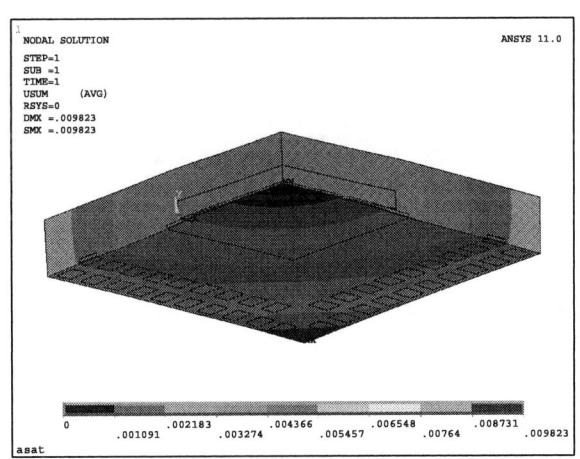

Fig.6. Deformation of TAPP under uniform temperature load

In investigating the cracking of the packaging body earlier, attention was focused on any high tensile or shear interface stresses since these could initiate delamination, and provide a site for crack growth. The regions of highest maximum principal stress and highest in-plane shear over the whole model were found to be near the DAP and molding compound interface.

Fig.7 shows a contour plot of σ1 for the DAP edge region. Fig.8 shows a contour plot of in-plane stress σxy for the same region. Both of the highest values occurred at the top corner of the DAP. The use of sharp corners creates singular points at these locations with unbounded stresses.

321

After demonstrating the highest stress occurred in the top edge of the DAP, the stress distribution along this edge was analyzed to find the maximum stress point, so that the most possible crack initiation origin could be predicted. Firstly a path was defined in the package model, the start point A was at the corner of DAP edge and end point B at the midpoint of DAP edge. Then the nodal stresses along this path were collected and plotted, shown in Figs.9 and 10.

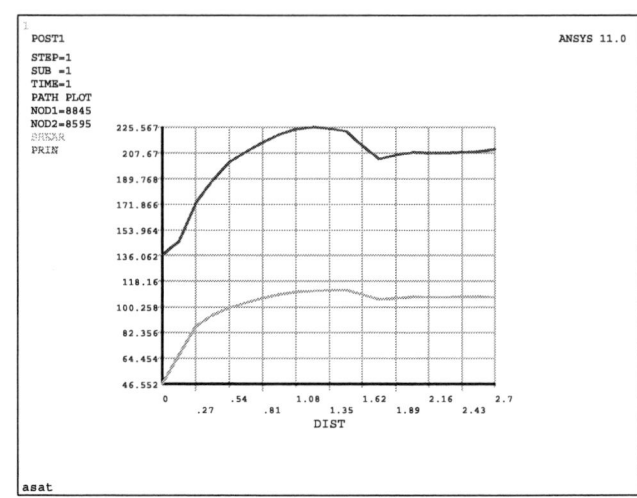

Fig.10. σ_1 and σ_{xy} stress distribution in the path

The results predict that the maximum peeling stress ( ) and in-plane shear stress ( ) at the DAP-MC interface would be at the point approximate 1.35mm away from the DAP corner (the length of the DAP is 5.4mm), where DAP-MC interfacial delamination most likely initiates. The interfacial crack would propagate downwards to the bottom surface of DAP after initiation, and on the other side propagate into molding compound where stress highly concentrated. This stress concentration was induced by inorganic fillers in the molding compound. The predictions matched well with experimental observation under a SEM.

The two set of TAPPs using different die attach materials were compared here. The simulation results showed that the 84-1LMIS group has better performance than A2200 from a stress point of view, although it has a little bigger in-plane shear stress than A2200, in Figs 11 and 12. It was also matched what we had known in the inspection process.

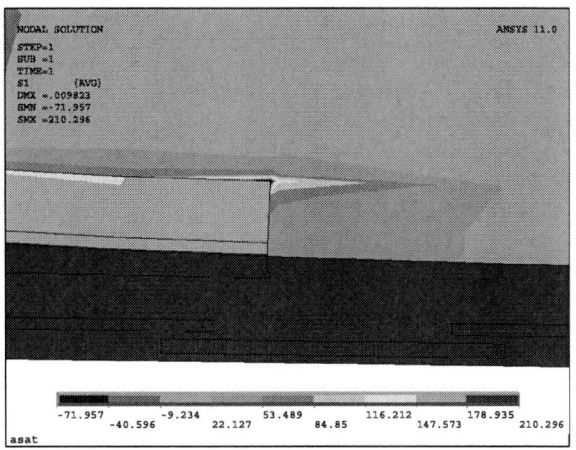

Fig.7. σ_1 principal stress at the DAP edge

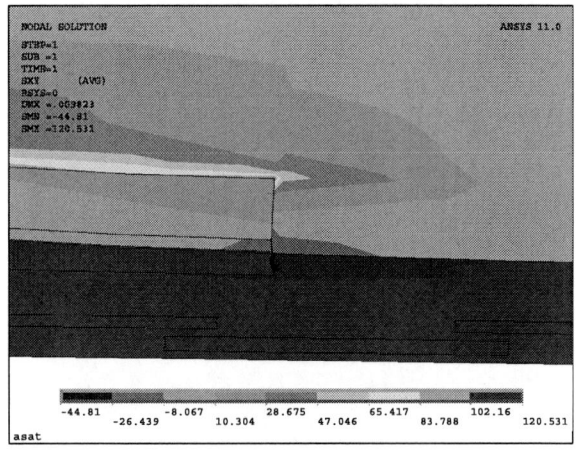

Fig.8. σ_{xy} shear stress at the DAP edge

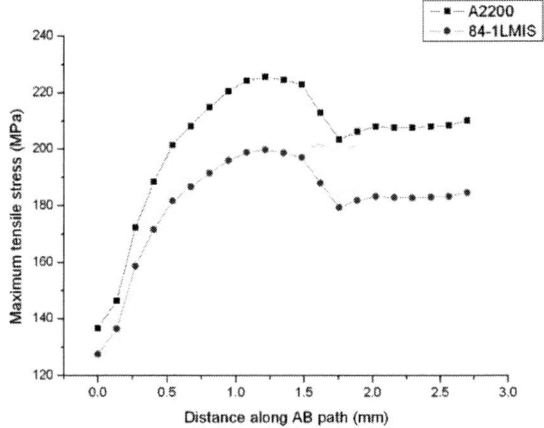

Fig.11. Maximum tensile stress σ_1 comparison between A2200 and 84-1LMIS

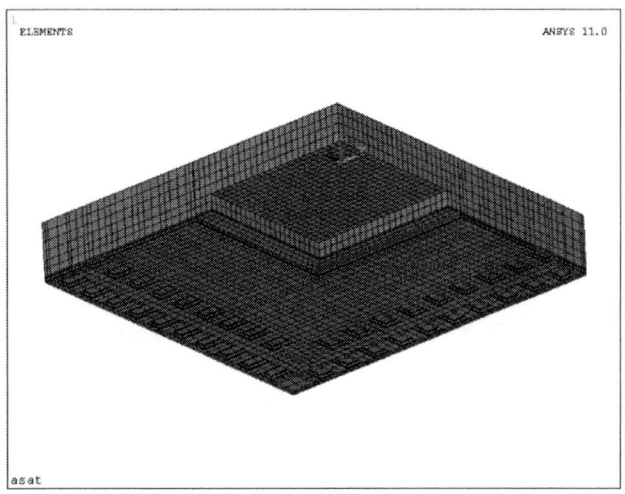

Fig.9. Stress path plot in molding compound model

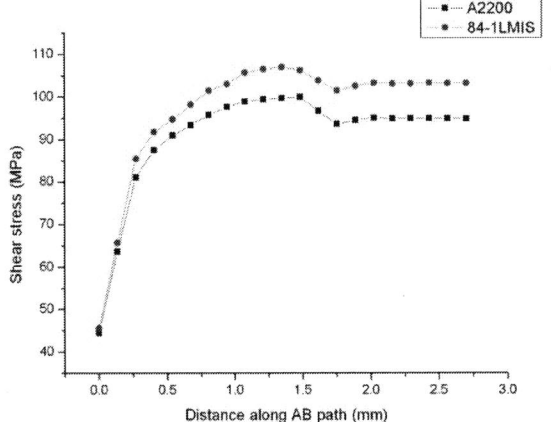

Fig.12 In-plane shear stress σ_{xy} comparison between A2200 and 84-1LMIS

Conclusions

The finite element analysis demonstrated that high stress levels exist in the interface between die attach pad and molding compound. It explained how the cracks found during testing came to exist. Under a large temperature drop, a very localized tensile stress arises at the DAP-MC interface, exactly at the point approximate 1.35mm away from the DAP corner. This stress is high enough to cause of the molding compound around the copper DAP. The crack can propagate both downwards to the bottom surface of DAP and upwards into the molding compound due to localized stress concentration around inorganic fillers of MC.

Recommendations

As the failure mechanisms has been clarified, parametric study based on finite element method could be conducted for the goal of achieving optimal design of TAPP. The correlations between thermo-mechanical properties and the growth of DAP-MC interfacial crack could be established according to the theory of fracture mechanics.

References

1. Kelvin. J. Chen, K. W. Chan. *et al*, "High-Performance Large-Inductance Embedded Inductors in Thin Array Plastic Packaging (TAPP) for RF System-in-Package Applications," *IEEE MICROWAVE AND WIRELESS COMPONENTS LETTERS*, Vol. 14, No. 9 (2004), pp. 449-451.

2. W.D. van Driel, M.A.J. van Gils, R.B.R. van Silfhout and G.Q. Zhang, "Prediction of Delamination Related IC & Packaging Reliability Problems," *Microelectronics Reliability*, Vol. 45 (2005), pp. 1633-1638.

3. R. J. Harries and S. K. Sitaraman, "Numerical Modeling of Interfacial Delamination Propagation in a Novel Peripheral Array Package," *IEEE Trans. Comp. Packag. Technol.*, Vol. 24, No. 2 (2001), pp. 256-264.

4. C. LeGall, D. L. McDowell, and J. Qu, "Delamination cracking in encapsulated flip chips," in *Proc. 46th Electron. Comp. Technol. Conf.*,1996, pp. 430–434.

5. J. Wu, "Reliability assessment of flip-chip and CSP," in *Proc. SEMICON'98: Test, Assembly, Packag.*, Singapore, 1998, pp.994–1000.

6. G. Kelly, <u>The Simulation of Thermomechanically Induced Stress in Plastic Encapsulated IC Packages</u>, Dordrecht, The Netherlands:Kluwer, 1999.

7. T.D. Moore and J. L. Jarvis, "Failure Analysis and Stress Simulation in Small Multichip BGAs," *IEEE Trans. Adv. Packag*, Vol. 24, No. 2 (2001), pp. 216-223.

8. T. Y. Lin. *et al*, "Failure Analysis of Full Delamination on the Stacked Die Leaded Packages," *J. Elec. Packag*, Vol. 125 (2003), pp. 392-399.

9. Y. H. Mei, S. Liu, and E. Suhir, "Parametric study of a VLSI plastic package subjected to encapsulation, moisture absorption and solder reflow process," in *Structural Analysis in Microelectronic and Fiber Optic Systems*. New York: ASME, 1995, vol. 12, pp. 159–174.

10. J. H. Zhao, X. Dai, and P. S. Ho, "Analysis and modeling verification for thermal-mechanical deformation in flip-chip packages," in *Proc. 48th Electron. Comp. Technol. Conf.*, 1998, pp. 336–344.

11. *Metals Handbook*, 9th ed., vol. 2, American Society of Materials, Materials Park, OH, 1994.

Design and Implementation of LED Daylight Lamp Lighting System

Rongfeng Guan, Dalei Tian, Xing Wang

Institute of Materials Science and Engineering, Henan Polytechnic University

No. 2001, Century Avenue, Jiaozuo City, Henan Province, China

Abstract

LEDs have becoming the most suitable candidate replacing traditional fluorescent lamps because of its energy-efficient, the introduction of high brightness LEDs with white light and monochromatic colors have led to a movement towards general illumination. This revolutionizes the optoelectronics market, enabling engineers to use LEDs for general lighting applications as well as medical, indoor lighting and automotive solutions. So variable LED array modules were developed, they are making great strides in terms of lumen performance and reliability, however the barrier to widespread use in general illumination still remains the cost or luminous efficiency, special requirements concerning optical properties and optomechanical layout have to be met.

In order to meet the requirements of indoor illumination, a LED daylight lamp model was designed, it can replace traditional fluorescent lamp without insteading additional power supply establishment. The optical properties of the model were simulated using optical analysis software, its luminous efficiency is about 41 lm/W, the illuminance is about 50 lux when the distance is 1.5 m between the center of the model and measured spot, With the theoretically-optimized design of the LED model, experiments based on the results of the optimal simulation in the laboratory were conducted to verify the performance of the proposed LED model, it reaches a power factor of about 0.8 at 11 W. Results of the simulation are very similar with the measured values, it was testified that simulative method is one of the effective tools for LED lighting optical design.

Introduction

LEDs are a revolution in lighting. They allow us to do things with light that were previously impossible. With LEDs you're free to create any lighting effect or installation you can imagine. LEDs offer many advantages over traditional lighting sources. Exactly which ones are important will depend on the specific application, they include but not limited to: very long lifetimes (50,000 hours); lower maintenance costs; more efficient than incandescent and halogen lamps; light up instantly; fully dimmable without color variation; directly emit colored light without filters; complete spectrum of colors; dynamic color control and tunable white point; total design freedom with hidden light; directed light for more efficient systems; vibration-proof lighting; no mercury; no IR or UV radiation in visible light. The market for LED solutions is expanding rapidly with the significant efficiency increase of LED chips. This revolutionizes the optoelectronics market, enabling engineers to use LEDs for general lighting applications as well as medical, specialty lighting and automotive solutions, where previously less efficient technologies had to be used[1].

It is well known that fluorescent lamps have been widely used in homes because of their energy saving, but they have the serious shortcoming of mercury pollution which restricts their applications. So LED has becoming the most suitable candidate replacing fluorescent lamps [2]. YAN Jun simulated a traditional LED's light distribution through carrying out simulation of LED's optical encapsulation structure by means of Monte Carlo method, the simulation results fit the experimentation results well in the light distribution[3]. YANG Yi designed a novel uniform illuminance system with LED source to create a illuminated circular region with desired size in a screen at a prescribed place, by using ray-tracing software based on Monte-Carlo method, the simulation results show that in the illuminated region the luminous uniformity is better than 85%, and the efficiency of whole system is more than 82%[4]. A LED daylight lamp was designed in this paper, its optical properties were simulated using illumination analysis software, sample was fabricated based on the results of the simulation, and results of the simulation are very similar with the measured values. The properties of the LED daylight lamp are perfect: 11W LED daylight lamp can be comparable with traditional 36 W fluorescent lamp, but the former is more energy-efficient.

Model for the simulation

LED array model has been shown in Figure 1. The transparent tube is used to protect the PCB and LEDs, the type of the LEDs are high-power 1 W. The dimension of the PCB is 580×25×1mm, the distance between two LEDs is 50mm, we assume a LED array consisting of 10 LEDs installing on the PCB.

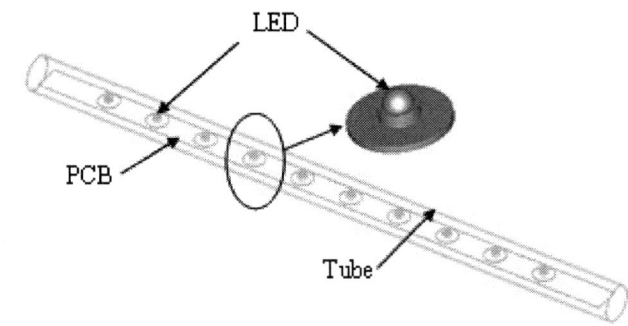

Fig.1. LED array model

Optical simulations

In the optimization algorithm process, a software was utilized to compute light distribution and illuminance of the model with certain combination of the following design and control parameters considered. Two assumptions should be made about the optical properties of this LED model. First, the surface of the PCB is a perfect reflector without any losses. Second, each LED is a perfect reflective diffuser. In the model, some parameters should be set first, each LED emits 60 lumen in a Lambertian pattern.

A number of 10000 rays of each LED were traced in the model to get a rough idea of performance. Figure 2 represents a polar candela distribution plot of rays, it can be seen from the figure, total luminous flux is 453.87 lm, so its luminous efficiency is 41 lm/W. Figure 3 indicates a rectangular candela distribution plot of the model, its has a narrow beam angle, and the intensity is fairly uniform with a scattering angle of about 120°.

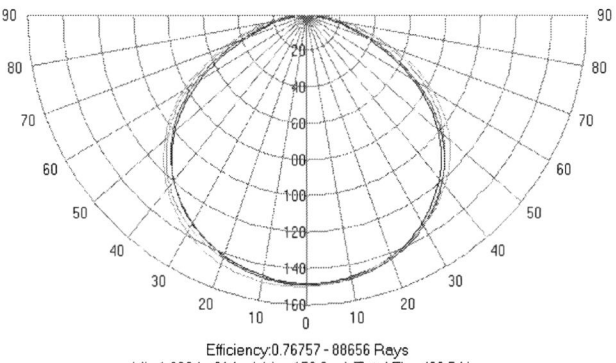

Efficiency:0.76757 - 88656 Rays
Min:1.6864e-014 cd, Max:150.2 cd, Total Flux:460.54 lm

Fig.2. Polar candela distribution plot of the model

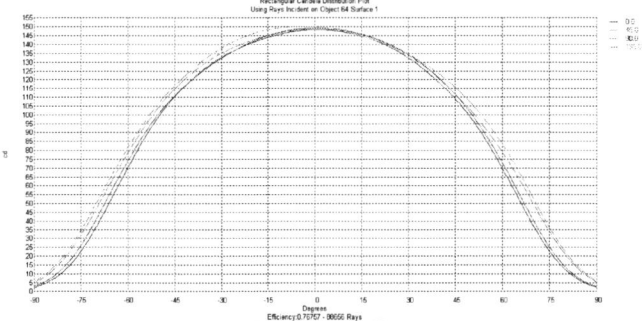

Fig.3. Rectangular candela distribution plot of the model

To get the illuminance distribution, a screen was placed before the model to receive rays. The distance between the center of the LED model to the screen is 1.5m, which meets the requirements of general lighting. Figure 4 shows a illuminance map of the model, 100,000 rays were traced, it can be seen that the illuminance distribution is not homogeneous, it has a gradient of about 6 lux from center to 0.5 m range, the maximal illuminance is 49.072 lux, total luminous flux is 453.87 lm. Simulations were performed with different distance in addition, they were compared with measured values in following experiment.

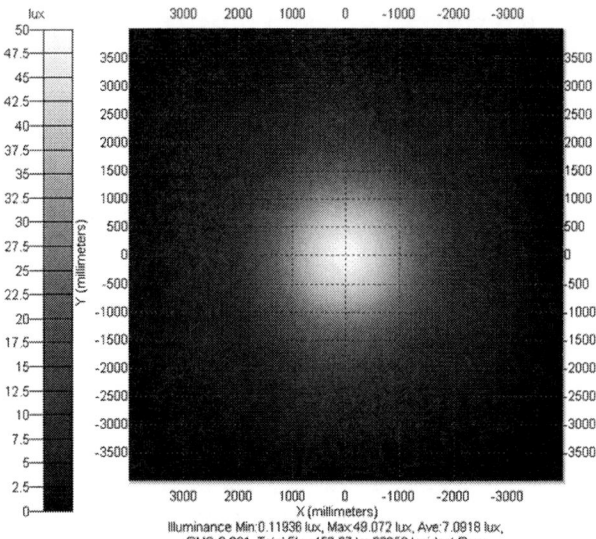

Illuminance Min:0.11936 lux, Max:49.072 lux, Ave:7.0918 lux, RMS:9.861, Total Flux:453.87 lm 88656 Incident Rays

Fig.4. Illuminance map of the model

Experiment and test

With the theoretically-optimized design of the LED model in hand, experiments based on the results of the optimal simulation in the laboratory were conducted to verify the performance of the proposed LED model. Figure 5 shows the sample lighting in an illuminant state, it reaches a power factor of about 0.8 at 11 W.

Fig.5. The lighting picture of the sample

The illuminance of the simulation and experiment were compared. Figure 6 describes the results of the simulative and measured values of the illuminance, the distance represents the optical path from the center of the LED daylight lamp to measured spot, it can be seen that results of the simulation are similar with the measured values, the simulative results fit the experimental results to a certain extent in the illuminance distribution, it also testified that simulative method is one of the effective tools for LED lighting optical design. There are some differences between simulation and experiment, because there are still some differences in conditions, such as materials and suppose in the simulation.

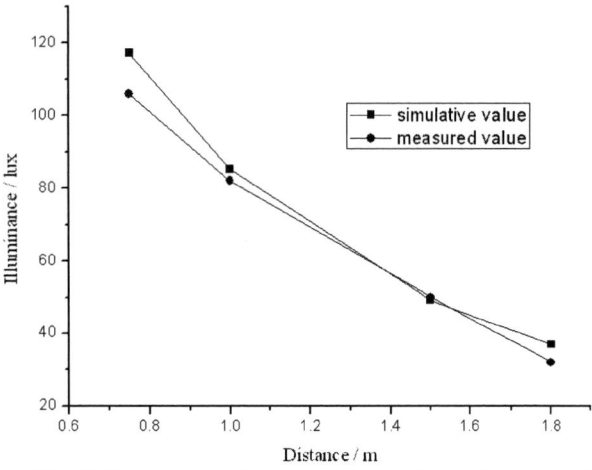

Fig.6. The compare between simulative and measured values of the illuminance

Conclusions

A LED daylight lamp model was designed, the optical properties of the model were simulated using illumination analysis software, its luminous efficiency is 41 lm/W according to simulative results, the illuminance is about 50 lux when the distance is 1.5 m between the center of the model and measured spot. Experiments based on the results of the optimal simulation in the laboratory were conducted to verify the performance of the proposed LED model, results of the simulation are similar with the measured values, it was testified that simulative method is one of the effective tools for LED lighting optical design. The proposed LED daylight lamp can be used in tunnel, living room, replacing traditional fluorescent lamp.

Acknowledgments

The authors acknowledge the following projects for financial support: Key Scientific and Technological Research Projects of Henan Province(No. 072102240027), Dr Fund of Henan Polytechnic University(No. 648602), Postgraduate Degree Thesis Innovation Fund of Henan Polytechnic University(No. 644005) .

References

1. Kuckmann, O., "High power LED arrays special requirements on packaging technology," *Proc. of SPIE,* San Jose, CA, Jan. 2006, pp. 613404.
2. Gatzeit, F. M., Sidorov, A. M, "*Distribution of illuminance from led modules,*" *Actual Problems of Electron Devices Engineering*, No. 15-16 (2004), pp. 310-315.
3. YAN J., YU Y., "LED's optical encapsulation structure design based on Monte Carlo simulation method," *Chinese Journal of Luminescence*, Vol. 25, No. 1 (2004), pp. 90-94.
4. YANG Y., QIAN K. Y., LUO Y., "A novel LED uniform illuminance system based on nonimaging optics," *Optical Technique*, Vol. 33, No. 1 (2007), pp. 110-115.

Effect of Bonding on the Packaged RF MEMS Switch

LE YANG, XIAO-PING LIAO*, JING SONG

Key Laboratory of MEMS of Ministry of Education, Southeast University,
Nanjing 210096, China
Email: xpliao@seu.edu.cn

Abstract

Thermal mismatch induced by the die bonding structure will have significant effect on the reliability and performance of RF MEMS switch. In this paper, a Cell Library Method (CLM) is introduced as an alternative against conventional FEM simulation to simplify the study on packaging of RF MEMS for X-band operation. The calculations derived from CLM are in good agreement with the results simulated by ANSYS. Strain caused by thermal stress which is generated during the die bonding process obviously increases the pull-in voltage. According to the research, the pull-in voltage of the RF MEMS switch with no stress on the beam will varied from 25V to 43V after bonding process (a 400-micron-long beam is taken for example). If the excitation voltage remains unchanged, the switching time of silicon-based fixed-fixed polysilicon surface-micromachined beam will extend to be about 2 times as that of before bonding (a 200-micron-long beam is taken for example).

Key words: RF MS packaging; die bonding; cell library; switching time

1. Introduction

In recent years, the urgent demands in wireless applications and communication systems require expanding development for the application of RF MEMS devices such as switch, varactor and filter since those offers low power consumption, low loss, high linearity and high Q factor compared with conventional communication components. To accelerate the commercialization of RF MEMS products, development for packaging technology is one of the most critical issues should be solved in advance. RF MEMS switch is highly sensitive to the stress induced by the packaging process. This stress leads to significant packaging effects on switch's performance and reliability. Die bonding process is one of the important steps in RF MEMS packaging. Thermal mismatch caused by the change of the temperature in the die boding process will introduce additional thermo-elastic strain and geometric deformation into the mechanical structure. Thus, RF MEMS switch's performance and reliability will significantly influenced by this process.

In order to predict and minimize the packaging effects on the switch's performance, package-device co-design is preferred to the traditional design method since the traditional way usually concern about the separated devices rather than the overall structure. However most of the previous works associated with the co-design idea are based on the conventional FEM simulation, there is no theoretical model accounting for the mechanical interactions between the concerned structures. Rabinovich [1, 2] applied the co-design conception to FEM simulations by separating the device element and the substrate element for the first time. Kobrinksy [3] extracted the anchor element from the device structure in his study on the buckling behavior of a fixed-fixed beam fabricated with the surface micro-machined process. These works have already involved the idea of the Cell Library Method (CLM) although they are still not enough for a comprehensive study of the entire package-device system.

This paper represents a complete Cell Library model of the packaged RF MEMS switch and theoretically solves the package-device co-design problem. This method is used to investigate the effects of the die bonding process on the pull-in voltage and the switching time of a fixed-fixed polysilicon surface-micromachined beam. An equivalent spring constant approach is used to obtain the closed-form expression of the pull-in voltage. The results based on the CLM demonstrate extraordinary agreement with the results derived from ANSYS. The deviation of the switching time of the fixed-fixed beam after packaging process is also researched in this paper. The variations of the switching time are agreed well with the results simulated by ANSYS.

2. Modeling of RF MEMS Switch

Fig.1 shows the 2D model of the die bonding structure with an adhesive interlayer between the top chip layer and the bottom package substrate layer. Since the structure are assumed undergo a small, elastic deformation without reliability failures, the first step is to solve the distribution of the strain along the surface of the chip layer as the temperature drops from the packaging temperature to the environment temperature.

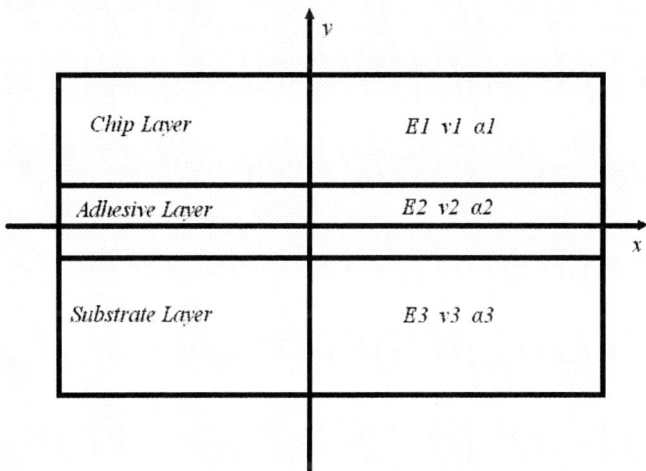

Fig.1 2D Model of Die Bonding Structure

In the planar coordinate system in Figure 1, some fundamental equations are established by treating the chip and the substrate as classical beams, and the adhesive layer as an elastic one. The contribution of the axial stress to the bending stiffness is neglected because the rotary inertia of the bonding structure is relatively large.

The equilibrium equations of the force and moment are expressed as follows,

978-1-4244-2739-0/08/$25.00 ©2008 IEEE   327

$$E_1' \frac{d^2 \overline{u_1}(x)}{dx^2} = \frac{dF_1(x)}{h_1 dx} = \frac{\tau_0(x)}{h_1} \qquad (1)$$

$$E_2' \frac{d^2 \overline{u_2}(x)}{dx^2} = \frac{dF_2(x)}{dx} = -\frac{\tau_0(x)}{h_2} \qquad (2)$$

$$\frac{dQ_1(x)}{dx} = -\frac{dQ_2(x)}{dx} = \sigma_0(x) \qquad (3)$$

$$-E_i' I_i \frac{d^3 w_i(x)}{dx^3} = \frac{dM_i(x)}{dx} = Q_1(x) - \tau_0(x)\frac{h_1 + h_2}{2} \qquad (4)$$

The horizontal strains at the interface of each layer are as follows,

$$\frac{du_1(x)}{dx} = \frac{d\overline{u_1}(x)}{dx} + \frac{d^2 w_1(x)}{dx^2}\frac{h_1}{2} + \alpha_1 \Delta T \qquad (5)$$

$$\frac{du_2(x)}{dx} = \frac{d\overline{u_2}(x)}{dx} - \frac{d^2 w_2(x)}{dx^2}\frac{h_2}{2} + \alpha_2 \Delta T \qquad (6)$$

If the thermal expansion of the adhesive layer is neglected, the continuity of the interfacial displacement is described as,

$$u_i(x) - u_i(x) = k_x \tau_0(x) = \frac{h_0}{G_0}\tau_0(x) \qquad (7)$$

$$w_1(x) - w_2(x) = k_z \sigma_0(x) = \frac{h_0}{G_0}\sigma_0(x) \qquad (8)$$

The boundary conditions at the free edge are

$$F_i(l) = 0 \quad M_i(l) = 0 \quad Q_i(l) = 0 \qquad (9)$$

where $E_i' = \frac{E_i}{1-v_i^2}$, $G_0 = \frac{E_0}{2(1+v_0)}$, E_i is the Young's modulus, h_i is the thickness of the beam, α_i is the thermal expansion coefficient, u_i is the horizontal displacement at the interface, $\bar{u}_i$ is the axial displacement, w_i is the vertical displacement, F_i is the horizontal force, Q_i is the shear force and M_i is the bending moment. Subscript i=0, 1, 2 corresponds to the case of the adhesive layer, chip layer and package substrate layer, respectively. k_x represents the horizontal interface compliance and k_z stands for vertical interface compliance. τ_0 and σ_0 are the interfacial shear stress and peeling stress, respectively.

Utilize Equations (1) to (9) into a seventh-order ordinary differential equation of shear stress τ_0 so that an analytical solution can be obtained upon the boundary and symmetric conditions. The other concerned results can be derived from this solution including the peeling stress, the two-dimensional displacements and strain distribution along the chip surface which is denoted by,

$$\varepsilon_{top}(x) = \frac{du_{top}}{dx} = \frac{d\overline{u_1}}{dx} - \frac{d^2 w_1}{dx^2}\frac{h_1}{2} \qquad (10)$$

If the thermal expansion of the adhesive layer is taken into account, the equations mentioned above can be transformed into non-zero boundary conditions of the beam layers as,

$$F_i(l) = -\frac{F_0}{2} \qquad M_i(l) = \pm\frac{F_0}{2}\frac{h_1 + h_2}{2}$$

(+ for i=1,-for i=2)
$$\qquad (11)$$

where $F_0 (\frac{1}{E_1' h_1} + \frac{1}{E_2' h_2} + \frac{1}{E_0' h_0}) = \frac{\alpha_1 + \alpha_2 - 2\alpha_0}{2}\Delta T$.

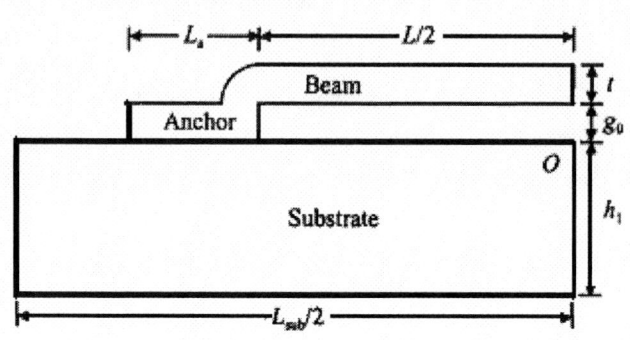

Fig.2 Substrate-Anchor-Device Joint Model

The average strain of the substrate can be expressed as $\varepsilon_0 = \frac{1}{L}\int_{x_0-\frac{L}{2}}^{x_0+\frac{L}{2}}\varepsilon_{top}(x)dx$ and the average curvature of the substrate can be expressed as $\phi = \frac{1}{L}\int_{x_0-\frac{L}{2}}^{x_0+\frac{L}{2}}\frac{d^2 w_1(x)}{dx^2}dx$, respectively. According to Equation (10) and (4), the two parameters mentioned above can be estimated. x_0 is the midpoint location of the beam along the axial direction of the chip. Fig.2 shows the schematic structure of the joint model for FEM investigations by ANSYS. Only the half of the model is established due to symmetric condition. Some typical geometric values are as follows, the substrate length L_{sub}=2mm; the substrate thickness h_1=400μm; the beam length L=50~800μm; the beam thickness t=2μm; the gap height g_0=2μm; the Young's modulus E_1=170MPa; the Poisson ratio v=0.28 and the length of the anchor L_a=8μm. Using small deformation theory the model can be described as follows,

$$\begin{Bmatrix} u_s \\ \phi \end{Bmatrix} = \frac{1}{E't}\begin{bmatrix} S_1 & S_2 \\ S_2 & S_3 \end{bmatrix}\begin{Bmatrix} P \\ M \end{Bmatrix} =$$

$$\begin{Bmatrix} \varepsilon_r \frac{L}{2} - \int_{-\frac{L}{2}}^{0}\left[\frac{P}{E't} - \frac{1}{2}\left(\frac{dv(x)}{dx}\right)^2\right]dx \\ -\frac{qL}{2P} + \frac{1}{k_0}(\frac{q}{P} - \frac{12M}{E't^3})\tanh(k_0\frac{L}{2}) \end{Bmatrix} + \begin{Bmatrix} \varepsilon_0\frac{L}{2} + (g_0+\frac{t}{2})\phi_0 \\ \phi_0 \end{Bmatrix} \qquad (12)$$

S_1~S_3 are the undetermined coefficients. And the Least Square Method is then adopted to extract the fitting parameters S_1~S_3 from the calculated data, with S_1=18.9763μm, S_2=7.8688 and S_3=6.5922μm^{-1}. The diagonal symmetry of the anchor's nodal matrix follows the Elastic Reciprocity Theory. P and M are the force and moment of the joint node, respectively. ε_r is the initial strain of the beam. u_s and φ are the translational and rotational degree of freedom of the Substrate-Anchor-Device joint node , respectively. $k_0 = \sqrt{\frac{12P}{E't^3}}$ and $\phi_0 = -k_0\frac{L}{2}$ represent the initial slope of the anchor under the substrate bending, ε_r here represents residual strain of the beam induced the bonding process, and $v(x)$ is the vertical deflection of the beam, formulated by:

328

$$v(x) = \frac{qL^2}{8P} - \frac{1}{k_0^2}(\frac{q}{P} - \frac{12M}{Et^3}) - \frac{qx^2}{2P} +$$

$$\frac{1}{k_0^2}(\frac{q}{P} - \frac{12M}{Et^3})\frac{\cosh(k_0 x)}{\cosh(k_0 \frac{L}{2})} \tag{13}$$

Iterative computation is carried out between Equations (12) and (13) until the converged solutions of M, P and $v(x)$ are obtained. The second brace in Equation (12) indicates the effects of the substrate deformation, and the first brace includes the effect of axial stress on the bending stiffness and the effect of stress-stiffening, which are indispensable for a precise prediction of the beam's nonlinear behavior under the relatively large deformation in the pull-in dynamic.

3. Effect of Bonding on the Switch

The structure of the fixed-fixed beam is widely used in RF MEMS switch recently. With the development of the research on RF MEMS, more and more attentions are focused on the performance of that. The pull-in voltage and the switching time are important parameters of RF MEMS switch. Effects of die bonding on the pull-in voltage and switching time will therefore be studied next as a complete demonstration for the application of the CLM in the package-device co-design.

Estimation approach for the pull-in voltage V_p using the substrate-anchor-device joint model is as follows. An equivalent of initial stress σ_{eff} is obtained by solving the stress value in the packaged beam near the pull-in location v_p at the first. The pull-in location for the fixed-fixed beam should be lower than that of the ideal parallel capacitance structure. Cheng [4] reports the pull-in location is about 40~42% of the gap height for an ideal fixed-fixed beam, and in this paper an approximation of $v_p = 0.4\, g_0 = 0.8\mu m$ is adopted. Secondly, an equivalent spring constant K_{eff} is extracted numerically by substituting the equivalent initial stress into the basic equations and solving the load-deflection relationship within the linear, small deformation range. The obtained equivalent spring constant is then substituted into a closed-form expression to compute the pull-in voltage after the die-bonding process. This quasi-static pull-in voltage can be expressed as

$$V_p = \sqrt{\frac{2.95 K_{eff} g_0^3}{\varepsilon_0 (1 + 0.42 g_0/w)}} \tag{14}$$

Where ε_0 is the dielectric constant and w is the width of the beam. If the strain on the switch was treated as plane strain, the bracket term will approach to a limit of 1. Equation (14) modifies the Osterberg's pull-in voltage formula for an ideally fixed-fixed beam [5] by introducing the equivalent spring constant and revising the fitting coefficient. The modified formula expands the previous assumptions of rigid substrate, rigid anchor and small deformation, and is more applicable for the actual fixed-fixed beam.

The epoxy-adhering technique on the FR4 substrate is often adopted in the laboratory research. A fixed-fixed beam on the FR4 substrate is taken as an example. Both the mechanical response and the pull-in voltage are simulated by ANSYS to verify the joint model. The maximum deflection $v(x)$ and the pull-in voltage are represented in Tab.1.

Tab.1 Comparison of mechanical response and pull-in voltage between calculations and ANSYS

L (μm)	σ (MPa)	$v(x)$ (e-2 μm)		$V_p(V)$			
				Quasi-static		Dynamic	
		CLM	FEM	CLM	FEM	CLM	FEM
200	0	0	0	98	103	90	100
200	36.9	-4.5	-4.3	116	128	106	128
300	0	0	0	45	46	41.3	46
300	36.9	-5.4	-5.1	63	68	58	67
400	0	0	0	25	26	23	26
400	36.9	-5.7	-5.3	43	46	40	45

According to the data in Tab.1, the maximum deflection $v(x)$ caused by the thermo-elastic strain is about 2.5% of the gap height for the ideal fixed-fixed beam. So that the variations of capacitance between the beam and the center line of the CPW after die bonding process can be neglected. It also can be seen from Tab1.1 that the quasi-static pull-in voltages calculated in CLM are agreed well with the results derived from ANSYS. That is means the CLM is as precise as the FEM in the research of the quasi-static pull-in voltage.

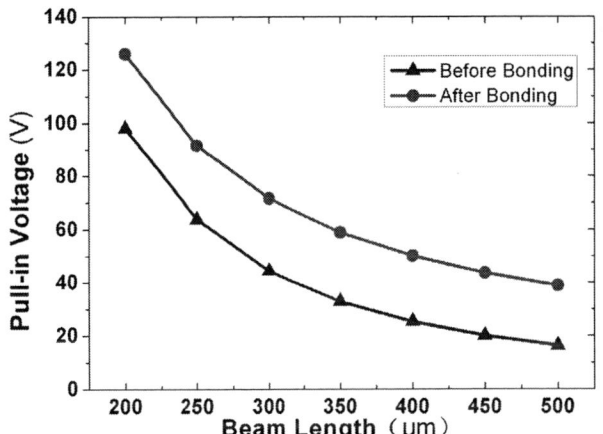

Fig.3 Effects on the quasi-static pull-in voltage of RF MEMS switch

Fig.3 illustrates that the quasi-static pull-in voltage is dramatically diverted after packaging process. As the packaging process is the indispensable and ultimate process to RF MEMS switch, the designed pull-in voltage must be reevaluated after packaging.

Nielson [6] reports that the relationship between the quasi-static pull-in voltage and the dynamic one is obtained by

$$\frac{V_p'}{V_p} = \sqrt{\frac{27}{32}} \approx 0.919 \tag{15}$$

The dynamic pull-in voltages in Tab.1 demonstrate that CLM is not as precise as FEM in that case. The reason for that is the fixed-fixed beam is simplified as a parallel-plate capacitor by Nielson so that the fringe capacitance and curvature of the beam are not taken into consideration.

When a voltage V_s is applied to the RF MEMS switch, the fixed-fixed beam will move to contact with the dielectric layer. The response of the RF MEMS switch is nonlinear due to the inherent nonlinearity of electrostatic forces. The dynamic equation of the RF MEMS switch is derived using the Newtonian dynamic equation of motion and formulated by [7]

$$m\frac{d^2x}{dt^2}+b\frac{dx}{dt}+kx=F(V_s) \tag{16}$$

$$F(V_s)=-\frac{1}{2}\frac{\partial C}{\partial x}V_s^2 \tag{17}$$

where m is the mass of the beam, x is the displacement from the up-state position, b is the damping coefficient and is dominated by the squeeze-film damping under the bridge., C is the capacitance of the switch and V_s is the voltage applied on the switch. Generally speaking, the damping coefficient in the fixed-fixed beam is quite small. In another words, b in Equation (16) can be ignored to obtain an approximate solution. And a close-form expression of switching time can be obtained. It is approximated as

$$t_s \approx 3.67\frac{V_p'}{V_s\omega_0} \tag{18}$$

with V_p' is the dynamic pull-in voltage and ω_0 is the resonant frequency of the beam.

Tab.2 Comparison of the switching time between calculations and ANSYS

Length (μm)	Stress (MPa)	V_s (V)	V_p' (V)	ω_0 (Hz)	Switching Time (μs)	
					CLM	FEM
200	0	128	100	287230	9.98	10
200	36.9	128	128	181546	20.21	18

A 200-micron-long beam is taken for example. According to the results in Tab.2, if the excitation voltage remains unchanged, the switching time will extends to about 2 times of that before bonding.

It is relevant to mention that the epoxy-adhering technique on the FR4 substrate which is often adopted in the laboratory research has actually a larger and more ambiguous effect on the device response due to greater thermal mismatch and smaller bending stiffness of the coupled package-device structure.

It is clearly seen from the research mentioned above that the effect of the die bonding process on the switch is not only significant, but also the combination of multiple factors. Omitting these factors will reduce the accuracy, repeatability and predictability of the designed performance; make it more difficult for the testing and calibration of the packaged devices.

4. Conclusions

A joint model based on the Cell Library Method is presented in this paper to solve the MEMS package-device co-design problem. Effects of die bonding process on the pull-in voltage of a fixed-fixed beam are studied in this method. Both anchor compliance and substrate deformations are taken into consideration. Results show that the packaging effects on RF MEMS switch are both significant and complicated. The pull-in voltage and the switching time of RF MEMS switch vary obviously after the die bonding process. Neglecting these effects in the device design will decrease the repeatability and predictability of the responses of packaged devices.

ACKNOWLEDGMENTS

This work is supported by National Natural Science Foundation of China (60676043) and The National High Technology Research and Development Program of China (863 Program, 2007AA04Z328).

References

1. Bart S.F, Zhang. S, Rabinnovich.V.L, Cunningham S, Coupled package-device modeling for MEMS, Technical Proceedings of the 1999 International Conference on Modeling and Simulation of Microsystems, 1999, Chapter 7: 232-262.
2. Rabinovich V.L, Kuijk J, Zhang S, Bart S.F, Gillbert J.R, Extraction of compact models for MEMS / MOEMS package-device co-design, SPIE Proceedings, 3680: 114-119.
3. Kobrinsky M.J, Deutsch E.R, Senturia S.D, Effect of support compliance and residual stress on the shape of fixed-fixed surface-micromachined beams, J. MEMS, 2000, 9: 361-369.
4. Cheng J, Zhe J, Wu X, Analytical and finite element model pull-in study of rigid and deformable electrostatic microactuators, J. Micromech. Microeng, 2004, 14: 57-68.
5. Osterberg P.M, Senturia S.D, M-test: a test chip for MEMS material property measurement using electro-statically actuated test structures, J. MEMS, 1997, 6: 107-118.
6. Gregory N. Nielson, George Barbastathis, Dynamic Pull-In of Parallel-Plate and Torsional Electrostatic MEMS Actuators, J.MEMS, 2006, 8791-21
7. Rebeiz Gabriel M. RF MEMS theory, design, and technology. Hoboken, NJ: Wiley & Sons, 2003.

Genetic Algorithm Optimization in TFT-LCD Drive System

Zhi-Jie Tang[1] Ran Feng[1] Mei-Hua Xu[2]

[1]Key Laboratory of Advanced Displays and System Application,
Ministry of Education, Shanghai University, Shanghai 200072, China
[2]Microelectronic Research and Development Center, Shanghai University, Shanghai 200072, China
[1]Email: joviertang@163.com

Abstract

This paper describes a novel TFT-LCD drive method for high gray scale display based on Genetic Algorithm optimization. The method can optimize scanning drive topology structure and improve the efficiency of scanning. This method uses the traditional DAC to drive the low-bit gray scale while the high-bit gray scale is generated by applying the new topology structure. This method is applied to a 1280*1024 dots RGB TFT-LCD panel which can display 256-level gray scale with 6-bit DAC source driver.

1 Introduction

In recent years, the thin film transistor liquid crystal displays (TFT-LCDs) have been developed giving flat panel displays with high quality.

The color depth of TFT-LCD is decided by the number of colors which the device can represent, and the number of colors is related to the number of bit of gray scale. The higher display gray scale can get higher display quality [1, 2].

Genetic algorithms are optimization methods which use the ideas of the evolution of the nature. Simple as genetic algorithms are, they are efficient [3, 4].

Now, most medium or large size LCD display devices have 6- or 8-bit gray scale system. In order to improve gray scale, this paper reports a design for a novel drive method with lower accurate voltage DACs and lower scan frequency. It uses digital advanced sub-frame modulation (ASFM) cooperating with the voltage DAC to generate the high gray scale.[5] We use genetic algorithms to optimize the ASFM scan drive sequence to get the best efficiency and the lowest request for system resources.

The driving system is shown in Fig.1.

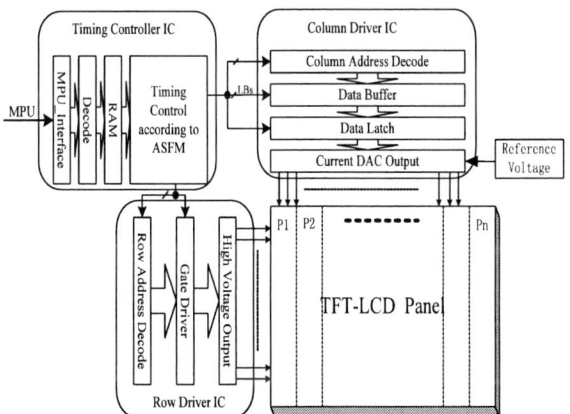

Fig.1 the TFT-LCD driving system of ASFM

From the Fig.1, we can see that the entire display panel is divided to S_N sub-partitions (P1, P2, ---, Pn), which are scanned by ASFM. In the system, there are three ICs (timing controller, column driver and row driver) and TFT-LCD panel. The display data are transported from micro processor unit (MPU) to timing controller IC. The timing controller IC generates the driving control timing, address and data of the column driver IC and row driver IC according to the ASFM. The column address is decoded and the corresponding low-bits of the pixel data are latched and voltage DAC converted in the column driver IC. The row address is decoded to drive the corresponding row of the panel.

2 Mathematics Indication of ASFM

In traditional TFT-LCD drive system, the drive sequence of the frame modulation is shown in Tab.1.

Tab.1 Traditional 8-level gray-scale drive sequence

S	1	2	3	4	5	6	7
GB	X_0	X_1	*	X_2	*	*	*
0	0	0	0	0	0	0	0
1	1	0	0	0	0	0	0
2	0	1	1	0	0	0	0
3	1	1	1	0	0	0	0
4	0	0	0	1	1	1	1
5	1	0	0	1	1	1	1
6	0	1	1	1	1	1	1
7	1	1	1	1	1	1	1

In the Tab.1, S means sequence; GB means bit power of gray data. For example, if the gray data is 4, then the valid drive is "X0=0, X1=0, X2=1".

From the table, we can see all the data drive only need three valid drive outs (X0, X1, X2), whose sequences are 1,2 and 4,while the other drive sequences are waiting sequences. We can get drive efficiency is

$$Eff = \frac{T}{T_{all}} = \frac{n}{2^n - 1} = \frac{3}{7} = 42.86\% \quad (1)$$

T means the number of the valid drive sequence; T_{all} means the number of the all drive sequence. n means the number of gray-scale level bits.

Then the ASFM divides the entire panel to many sub-partitions. The ASFM drives one sub-partition while the other sub-partitions are in waiting sequences. The ASFM drive sequence is shown in Tab.2.

From the table, we can see the drive efficiency is

$$Eff = \frac{T}{T_{all}} = \frac{p * n}{2^n - 1} = \frac{2 * 3}{7} = 85.71\% \quad (2)$$

P means the number of the sub-partitions.
Then, we can see the efficiency is enhanced.

Tab.2 ASFM 8-level gray-scale drive sequence

扫描序号	P1	P2
1	X_2	*
2	*	X_0
3	*	X_2
4	*	*
5	X_0	*
6	X_1	*
7	*	X_1

We find a matrix $X_{l\times m}$, which is according to the limits:

(1) $x \in \{0,1\}$, x is the element of the $X_{l\times m}$.

(2) $A \times X = \begin{bmatrix} n & n & \cdots & n & n \end{bmatrix}_{1\times m}$,

while $A = \begin{bmatrix} 1 & 1 & 1 & \cdots & 1 & 1 \end{bmatrix}_{1\times l}$

(3) $P_{l\times 1} = X \times B$ $\quad and \quad$ $p_i \leq 1$, $\quad i = 1,2,3,\cdots l$,

while $B = \begin{bmatrix} 1 & 1 & 1 & \cdots & 1 & 1 \end{bmatrix}^T_{1\times m}$.

(4)The spaces between adjacent two elements which are equal to 1 are $2^{n-1}-1, 2^{n-2}-1, \ldots, 2, 1, 0$.

Then we define that the matrix $X_{l\times m}$ is the ASFM drive sequence matrix. The number of the sub-partitions is m.

3 Genetic algorithm optimization

According to the ASFM drive sequence matrix definition, we need to find the maxim sub-partitions number to enhance the drive efficiency using genetic algorithms.

Firstly, we design the coding method:

According to the limit (1), we use binary code system. The gene is made up of column m, row 1 and element x of $X_{l\times m}$, shown in Fig.2.

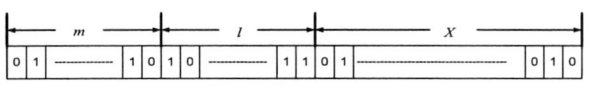

Fig.2 Gene structure

Then we design the fitness function. Firstly, we design the penalty functions because of the many limits to the matrix $X_{l\times m}$.

According to the limit (2), we set the penalty function $f_\alpha(H)$. We calculate the matrix $A \times X$,

$$g_i(p) = \begin{cases} 1 & (A \times X)_i = n \\ \alpha & (A \times X)_i <> n \end{cases} \quad i=1,2,\cdots,m \quad (3)$$

α is the penalty factor, normally set to 0.1~0.3.

Then, $f_\alpha(H) = \dfrac{\sum_{i=1}^{l} g_i(p)}{m}$ (4).

According to the limit (3), we set the penalty function $f_\beta(H)$. We calculate the matrix $X \times B$,

$$k_i(x) = \begin{cases} 1 & \sum_{j=1}^{m} x_{i,j} \cdot b_j \leq 1 \\ \beta & \sum_{j=1}^{m} x_{i,j} \cdot b_j > 1 \end{cases} \quad i=1,2,\cdots,l \quad (5)$$

β is the penalty factor, normally set to 0.1~0.3.

Then, $f_\beta(H) = \dfrac{\sum_{i=1}^{l} k_i(x)}{l}$ (6)

According to the limit (4), we set the penalty function $f_\gamma(H)$. We calculate the space number $s(H)$,

Because of the maximum of $s(H)$ is $n \times m$,

We get the $f_\gamma(H) = \dfrac{s(H)}{n \times m}$ (7)

Then we get the fitness function:

$$Fit(H) = f_\alpha(H) \cdot f_\beta(H) \cdot f_\gamma(H) \quad (8)$$

We use Matlab genetic tool to run the optimization. then we amend the fitness function:

$$Fit(H)' = 1 - Fit(H) = 1 - f_\alpha(H) \cdot f_\beta(H) \cdot f_\gamma(H) \quad (9)$$

We use stochastic uniform selection function, scattered crossover function and uniform mutation function. We use proportional fitness scaling function. We set the mutation rate to 0.1.

Then we get the optimization result shown in Fig.3.

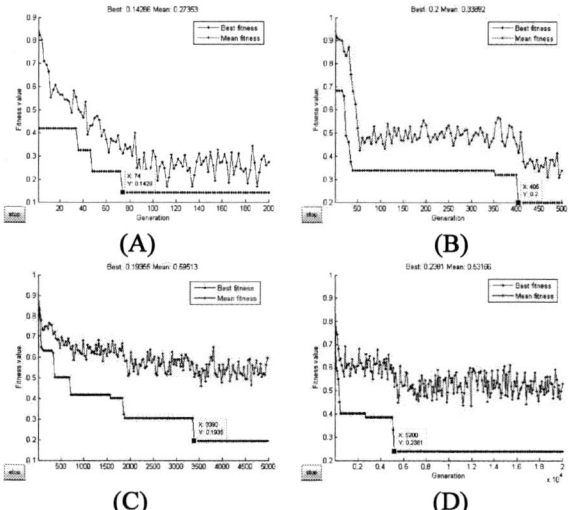

Fig.3 Optimization result
(A) 8-level gray-scale; (B) 16-level gray-scale;
(C) 32-level gray-scale; (D) 64-level gray-scale.

The optimization result data are shown in Tab.3.

Tab.3 optimization result data

Index	Gray Level	Efficiency	Part umber
1	8	85.71%	2
2	16	80.00%	3
3	32	80.65%	5
4	64	76.19%	8
5	128	71.65%	13
6	256	81.57%	26
7	512	75.73%	43
8	1024	77.22%	79

The efficiency comparison between genetic ASFM (GASFM) with traditional frame modulation (TFM) is shown in Fig.4.

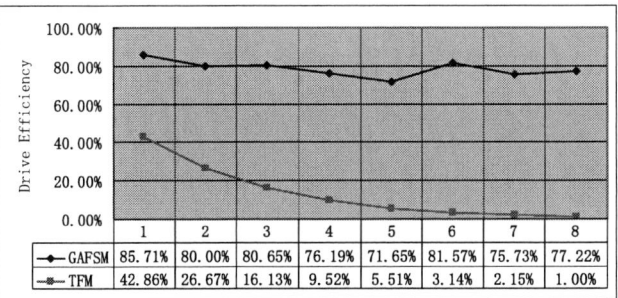

Fig.4 Efficiency comparison

	1	2	3	4	5	6	7	8
GAFSM	85.71%	80.00%	80.65%	76.19%	71.65%	81.57%	75.73%	77.22%
TFM	42.86%	26.67%	16.13%	9.52%	5.51%	3.14%	2.15%	1.00%

From the figure, we can see that the drive efficiency in enhanced enormously.

Then, data trans frequency:

$$Clk_{sys} = \frac{Rows \times Cols \times (2^n - 1) \times Clk_{fresh}}{DataWidth} \quad (10)$$

While, $Rows$ means TFT-LCD rows, $Cols$ means TFT-LCD columns,n means gray bits，Clk_{fresh} means TFT-LCD fresh frequency, $DataWidth$ means data trans width.

Then, to a 1280*1024 TFT-LCD panel , $DataWidth$ set to 8, Clk_{fresh} set to 60Hz, we can get the data trans frequency shown in Tab.4.

Tab.4 Data trans frequency comparison

Gray level	GASFM Data Frequency (MHz)	TFM Data Frequency. (MHz)
8	34.41	68.81
16	49.15	147.46
32	60.95	304.74
64	77.41	619.32
128	96.04	1248.46
256	96.41	2506.75
512	116.82	5023.33
1024	127.30	10056.50

From the table, we can realize higher gray-scale level drive using lower data trans frequency by GASFM.

4 System realization

With the proposed method, we designed the drive system for a 1280*1024 dots RGB TFT-LCD whose source drive DAC is 6-bit.

We get the 8-bit RGB data from DVI, and then we store the data to SDRAM. Secondly, we read the data from SDRAM according to GASFM to drive TFT-LCD panel. We can get 256-level gray scale display effect. The system realization picture is shown in Fig.5.

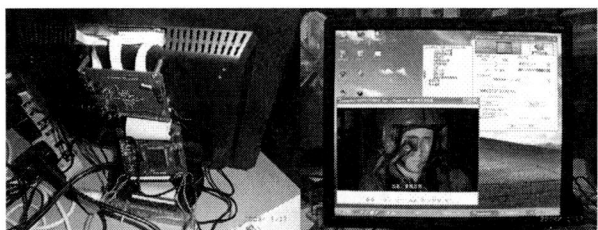

Fig.5 The picture of system realization.

Conclusions

We use Genetic algorithms to optimize TFT-LCD drive system successfully. We realized 256-level gray-scale 1280*1024 TFT-LCD drive It proves this method is feasible and effective.

Acknowledgments

The authors would like to acknowledge Shanghai Science and Technology Department Foundation for providing financial support for this work under grant No.06DZ22013 and 077062008, and also to acknowledge the financial support from National Natural Science Fund grant No.60773081 and No.60777018.

References

1. D.McCartney, "Tuning LCDs to Tune TV," Information Display, Vol. 20, No. 10, 2004, pp.14-17.
2. J.Virginia, "The Future Looks Bright for LCD TV," Information Display, Vol.19, 2003, pp.10-14.
3. Holland,J.H, "Outline for a logical theory of adaptive systems,"Journal of the Association for Computing Machinery,Vol.9,1962,pp.297-314.
4. Goldberg,D.E.,"Genetic algorithm in search, optimization and machine learning," Addison-Wesley Publishing,NewYork,1989.
5. H.Mano *et al.* Multicolor Display Control method for TFT-LCD, SID'91 Digest, 1991, pp.547-550.

The design of the Ku band Dielectric Resonator Oscillator

Guoguang Yan

Motorola Company

Motorola R&D center, No. 68, Dongxin road, Binjiang district, Hangzhou city, China, 310053.

tj2003wakx@163.com

Abstract

Phase noise is considered as a significant source of performance degradation of irreducible error rate in millimeter-wave and microwave systems, in particular, for applications employing low cost and moderate bit rate systems. Many new ways to improve oscillator phase noise haw been proposed. There was the new developments using dielectric resonators assembled on monolithic microwave integrated circuits. The paper provides a new approach to accurately simulate the Dielectric Resonator and design the GaAs MESFET Dielectric Resonator Oscillator (DRO) in the 12.75GHz by Negative Resistance theory and Harmonic Balance theory with use of two EDA (Electronic Design Automation) tool (CST&ADS). Passive microwave & RF component design is traditionally seen as CST's core competence. The soft of CST will be used in accurately simulate the Dielectric Resonator. The soft of Advanced Design Systerm of the Agilent will be used in the nolinerity analyses and optimization design of the DRO . After the EDA tool simulate and optimization, The unprecedented performance of DRO was found . At @12.75GHz, output power exceed 13dBm, Phase noise less -104dBc.

Introduction

Dielectric Resonator Oscillator (DRO) have potential advantages for integrated oscillators in microwave frequencies. Today, millimetre-wave and microwave systems require oscillators with low phase noise and excellent frequency stability. Low phase noise and excellent frequency stability resonator have been used as the frequency stability element of communication system. Dielectric Resonator Oscillator (DRO) have essentially advantages compared with other resonators, because of good temperature stability and assembiling.

Many new ways to improve Dielectric Resonator Oscillator phase noise have been proposed. There was the new developments using dielectric resonators assembled on monolithic microwave integrated circuits.

The article introduces a new approach to accurately simulate the Dielectric Resonator and design the GaAs MESFET Dielectric Resonator Oscillator (DRO) in the 12.75GHz by Negative Resistance theory and Harmonic Balance theory. Passive microwave & RF component design is traditionally seen as CST's core competence. CST will be used in accurately the Dielectric Resonator. The soft of Advanced Design Systerm of the Agilent will be used in the nolinerity analyses and optimization design of the DRO(12.75GHz).

Dielectric Resonator Oscillator (DRO) design process

A feedback type oscillator circuit was used in order to miniaturize the middle power oscillator. It was easy to debug

the circuit of oscillator to proper results. The block diagram of the oscillator shown in figure1&2.*The active component is GaAs MESFET. A matching circuits was used to minimize the frequency fluctuation by change of the load impedance.*

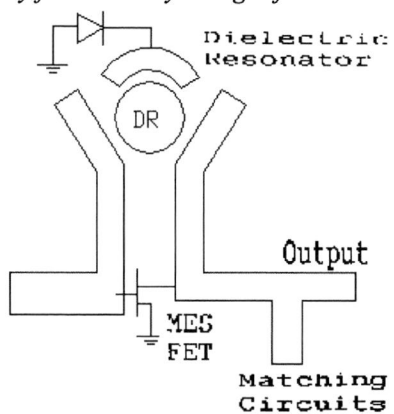

Figure1. Block diagram of the FET DRO

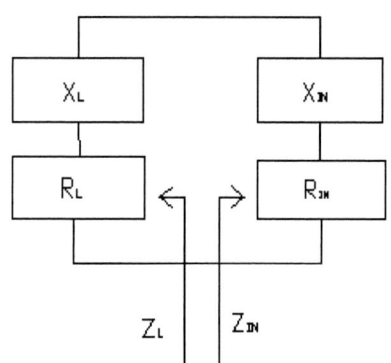

Figure2. Block diagram of the DRO

Process shown:

1. choice the appropriate GaAs MESFET and circuit such as common drain or common source circuit. Positive feedback network or *negative feedback network was up to gate* $\Gamma_{IN}>1$ in GaAs MESFET, FET is instability states .

2. debug output match circuit for stability states : $R_{in}+R_L=0$, $X_{in}+X_L=0$ 。

3. DR simulate by CST.

4. big signal simulate and optimize by ADS

Dielectric Resonator(DR) design and result

DR(*Dielectric Resonator*) is very important part of DRO . Columnar dielectric is high Q *resonator. It has high* R_0, resonator frequency ， $F_o = \dfrac{1}{2\pi\sqrt{LC}}$. DR coupling

978-1-4244-2739-0/08/$25.00 ©2008 IEEE 334

configuration and equivalent circuit shown in figure3、figure4.

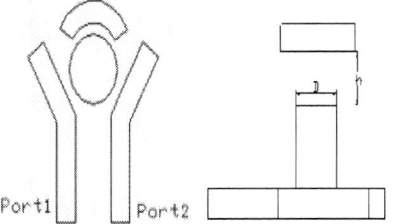

Figure3. DR coupling configuration

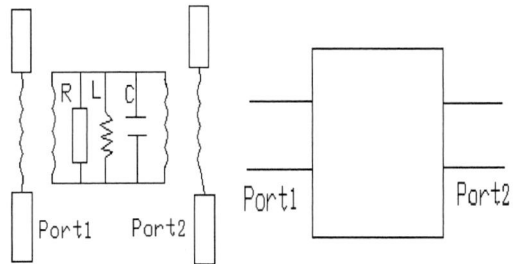

Figure4. DR equivalent circuit and model

Use columnar dielectric：D = 4.2mm，H=2.4mm, h=8mm, ε_r=30 in CST . First, in CST, set up the model of DR coupling configuration and design two port. Second, take S- parameter of model , after CST electromagnetic simulate .shown in figure 5,6.

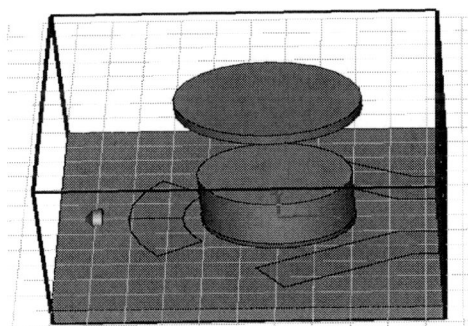

Figure5. DR model in CST

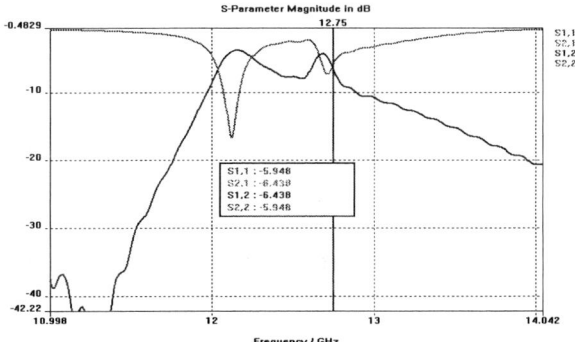

Figure6. S- parameter of model

Dielectric Resonator Oscillator (DRO) design and result

Use GaAs MESFET ATF26884 to do signal simulate and optimize by ADS, shown in figure 7、8、9.

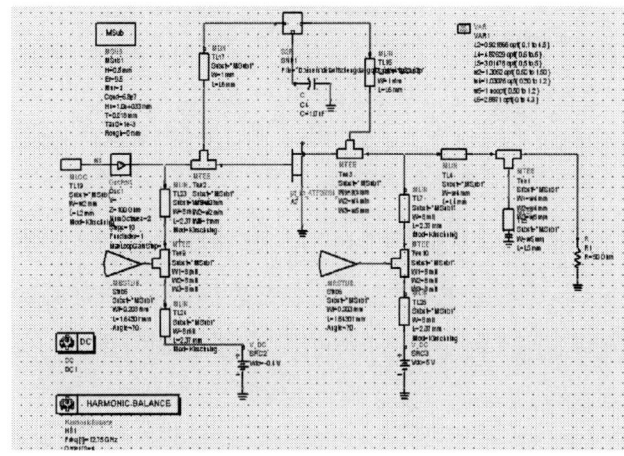

Figure7. DRO model in ADS

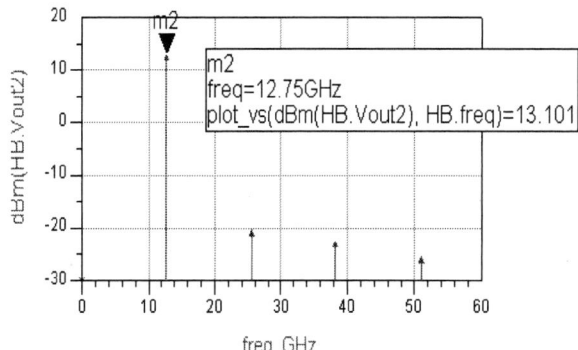

Figure8. Output frequency and power

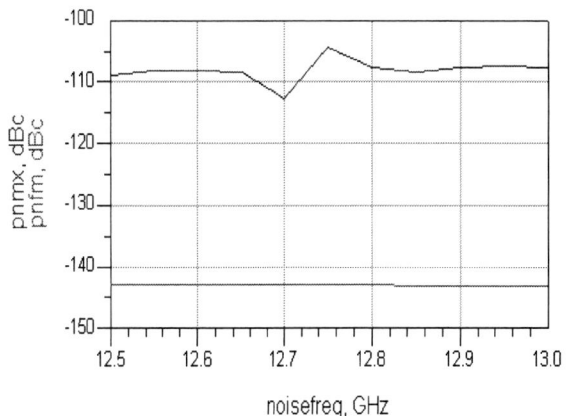

Figure9. Output phase noise

Conclusions

The unprecedented performance of DRO was found , @12.75GHz, output power exceed 13dBm, Phase noise less -104dBc.

References

1. Oliver Bernard，"Simulate and build a Ku — band DRO"，Microwave&RF，MAY2000。
2. Best R E. Phase-locked loops theory, design, and applications [M] . 3rd ed. McGraw-Hill Inc, 1995. 251289
3. Jones, R ，"Low phase noise dielectric resonator oscillator "，Frenquecy Control , 1990

Simulation of Multi-bit Digital Delta-Sigma Modulator

Wen-Rong Yang*, Yuan-Yuan Cheng, Jiong-ming Wang

Key Laboratory of Advanced Display and System Applications (Shanghai University), Ministry of Education

Microelectronic Research & Development Center

Shanghai University, Shanghai 200072, P. R. China

*Email: yang_w_r@sina.com

Abstract

The paper describes a four-order delta-sigma modulation (DSM) with 15 levels quantizer which is used in a 24-bit 44.1-kHz sample-rate audio digital-to-analog converter (DAC). An odd level quantizer has been chosen instead of an even level to reduce quantization noise. The noise transfer function (NTF) is designed to have the zeros optimally in order to increase DR. The peak SNR of the DSM is about 130dB, which is enough for an audio DAC designed with a 0.35 um CMOS technology.

1. Introduction

The development of multimedia systems has increased the demand for an audio digital-to-analog converter (DAC) and the demand for low-cost, wide dynamic range and high linearity of a DAC. Because of its inherent benefit Delta-sigma modulation (DSM) is the most suitable DAC topology to satisfy these requirements. DSM can reduce the bits of input digital signal at the cost of increasing sample rate and a great many of digital circuits, but the area and the complexity of the analog part can be greatly reduced and the implement of digital circuit is very convenient by using developed digital signal process and CMOS technique. Therefore the cost of a DAC can be decreased and that's why the topology of DSM is popular now.

In actual application the tradeoff among power, stability and DR must be made. Large DR needs high over sample rate and multi-bit quantizer, which will increase power and the difficulty of the implement of the circuit. For stability multi-bit quantizer is needed for high order DSM but it will increase the difficulty of the internal design of DAC. So the over sample rate and the number of the bits of the quantizer must be chosen seriously in order to achieve those anticipated goals.

The paper describes a four-order multi-bit delta-sigma modulation (DSM) which operates in a 24-bit 44.1-kHz sample-rate digital-to-analog converter in 0.35um CMOS technology. To improve the performance of DSM, an odd level quantizer and optimal zeros have been used. The DSM can achieve 130dB SNR.

2. DSM Architecture

The proposed DAC adopts the architecture shown in Fig.1. The key choice for the overall performance is the definition of the characteristics of the DSM, which not only affects the performance of the analog filter but also fixes the SC filter sampling frequency by its output data rate.

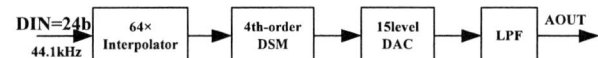

Fig.1 Signal flow diagram of the multi-bit delta-sigma DAC

The DSM DR depends on three main parameters: over sample ratio, noise-shaping order, and the bits number of its quantizer. The quantizer resolution and DSM's order selection involve four key considerations. The first is to make the in-band quantization noise as low as possible, even negligible in the overall noise budget. [1] Analog circuit noise (thermal and 1/f noise) is directly related to device size and power dissipation. So it is advantageous to allocate the most noise budget to the analog components. The second is to make out-of-band signal-to-noise ratio as low as possible, because it decides the amount and complexity of out-of-band filtering.[2] The third is to reduce the effect of limit-cycle oscillation, which disturbs noise shaping in DAC and creates the superfluous signal (called an idle tone). [3] So multi-bit quantization must use an odd level quantizer instead of an even level to reduce the effect of quantization noise. Because an odd-level quantizer just has one level at midscale, the output code will be mainly zeros with occasional "+/-1's" for a small input signal. This minimizes the generation of high-frequency idle tones. The last is clock jitter immunity. Modulation by jitter will fold down the high frequency quantization noise and/or raise the noise spectra at side-bands of the fundamental signal.[5] Therefore, the amount of high-frequency quantization noise affects the jitter sensitivity. In order to reduce high-frequency noise, the quantizer resolution must be increased.

The working frequency of the DSM is 64 times Fs (2.8224MHz) obtained by using the interpolator filter to increase the sampling frequency of the audio signal from the Nyquist frequency (Fs=44.1 KHz) to 64Fs. To achieve the overall performance of the DAC, the specifications of the DSM are: SNR more than 120dB in the audio band; signal attenuation less than 0.001dB in the audio band, and the noise gain less than -90dB.

To meet the above specifications, a 4th-order DSM with a 15 levels quantizer was chosen. The noise transfer function (NTF) has been optimized in consideration of both in-band and out-of-band performance. The noise transfer function (NTF) chooses a high-pass transfer function given by the Chebyshev response and is designed to have two zeros at dc and two conjugate zeros at about 20 kHz, which is optimally spreaded in the audio band and provides 6dB more DR than the case all zeros at dc. [4] The DSM architecture is shown in Fig.2. The basic configuration is a single-loop with multiple feed-forward paths. The g1 feedback loop generates two conjugate zeros at about 18 kHz.

The DSM signal transfer function (STF) and NTF are given in the Equation (1) and (2), which has been shown at the end of the paper. The values of the coefficients (a1, a2, a3, a4, b1, c1, c2, c3, c4, g1) are indicated in Table I and refer to the exact MATLAB model studied for the NTF and STF design. But in the practical implementation of the digital circuit, these coefficients must have been approximated to reduce the complexity of the circuit and die area by avoiding using multiplier. This approximation will not reduce the DR and SNR significantly, and if the DSM performances become worse these coefficients must be adjusted.

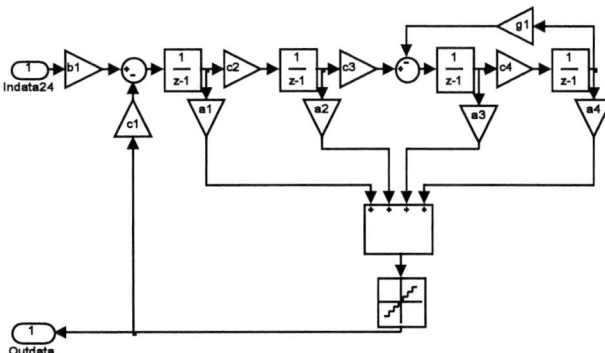

Fig.2 4th-order DSM

TABLE I
Coefficients in the DSM

Coefficients	exact value	Approximated value as power of 2 sum
a1	1.2630	$2^0+2^{-2}=1.25$
a2	0.7344	$2^{-1}+2^{-2}=0.75$
a3	0.2243	$2^{-2}=0.25$
a4	0.0289	$2^{-6}+2^{-7}=0.0234$
c1=c2=c3=c4	1	$2^0=1$
b1	1	$2^0=1$
g1	0.0017	$2^{-10}+2^{-11}=0.0015$

3. Matlab Simulation

Fig.3 shows the NTF and STF magnitude response. In-band signal attenuation is about zero and noise gain is less than -90dB. Fig.4 shows the curve of the SNR versus input signal amplitude. The peak SNR is larger than 130dB, so the performance goals can be achieved. Fig.5 shows the curve of the input and output signal frequency spectrum. In audio band the output signal frequency spectrum is equal to the input and the noise frequency spectra is out of the band, so the noise won't influence the input signal, that is, the noise is shaped and pushed to higher frequencies.

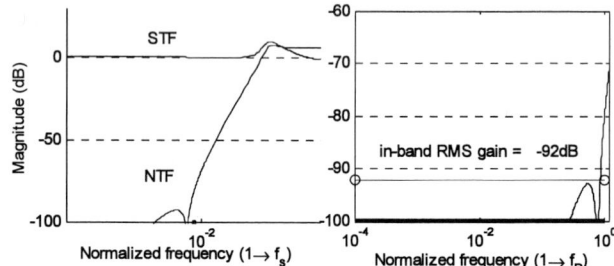

Fig.3 STF and NTF magnitude response

4. Circuit Implement

According to the structure and the coefficients of the DSM, the data flow graph is shown in Fig.6, in which 'Z-1' stands for delay unit, 'S' means shift and '+' means addition. It can be seen that there are 14 addition or shift operations between two input data. These operations can be classified into 4 groups: (1)addition (2,3,5,7); (2)shift and addition(8,9,10,11); (3)feedback addition(4,6); (4)result addition (1,12,13,14). So in the circuit only 4 adders are needed.

In the circuit five kinds of registers have been used: Ureg, Mreg, Yreg, Sreg, Vreg. According to the need of the circuit, Mreg and Ureg is 24 bits while others are 12 bits. Fig.7 shows the circuit structure of the DSM. The circuit can be described by Verilog program and produced by a 0.35um CMOS technology.

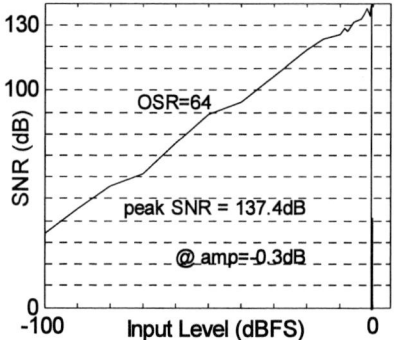

Fig.4 DSM SNR versus input signal

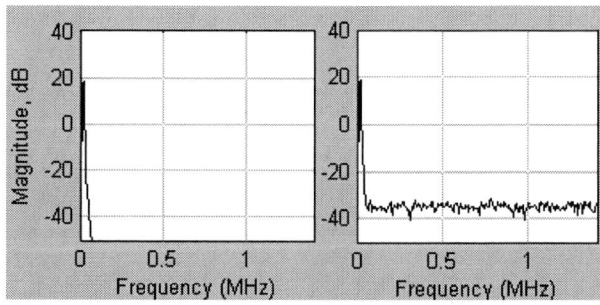

Fig.5 DSM input and output signal frequency spectra

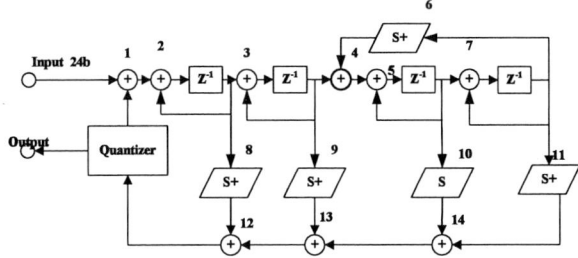

Fig.6 Data flow graph of DSM

5. Conclusions

In the paper, a high DR 4th-order multi-bit DSM has been presented using 0.35um CMOS technology. An odd level quantizer has been chosen reduce the amount of quantization noise. The noise transfer function (NTF) is designed to have the zeros optimally to increase DR. According to the results of the simulation, the DSM is suitable for audio DAC that needs high performances, whose SNR can achieve to 130dB.

References

1. Fujimori and T. Sugimoto "A 1.5-V 4.1-mW dual-channel audio delta-sigma D/A converter" IEEE J. Solid-State Circuits, vol. 33, no. 12, pp. 1879-1886, Dec. 1998.

2. Vittorio Colonna, Marzia Annovazzi, Gianluigi Boarin, Gandolfi, Fabrizio Stefani, and Andrea Baschirotto, "A 0.22-mm2 7.25-mW per-Channel Audio Stereo-DAC With 97-dB DR and 39-dB SNRout," IEEE J. Solid-State Circuits, vol. 40, no. 7, pp. 1491-1498, July 2005.

3. Yasuyuki Matsuya, Yuuichi Kado, Jun Terada, Takashi Eguchi, Maoko Tamura, and Hajime Fujiwara, "A 4th-order local-feedforward D/A converter that prevents limit-cycle oscillation," Symposium On VISL Circuits Digest of Technical Papers, pp. 330-335, 2002.

4. Marzia Annovazzi, Vittorio Colonna, Gabriele Gandolfi, Fabrizio Stefani, and Andrea Baschirotto, "A low-power 98-dB multibit audio DAC in a standard 3.3-v 0.35um CMOS technology," IEEE journal of solid-state circuit, vol.37, No.7, pp. 825-834,July 2002.

5. Ichiro Fujimori, Akihiko Nogi and Tetsuro Sugimoto, "A multibit delta-sigma audio DAC with 120-dB dynamic range," IEEE journal of solid-state circuit, vol.35, No.8, pp. 1066-1073, Augus

$$H(z) = \frac{a_1z^3 + (a_2c_2 - 3a_1)z^2 + (3a_1 - 2a_2c_2 + a_3c_2c_3 + a_1c_4g_1)z + (-a_1 + a_2c_2 - a_3c_2c_3 + a_4c_2c_3c_4 - a_1c_4g_1 + a_2c_2c_4g_1)}{z^4 + (-4)z^3 + (6 + c_4g_1)z^2 + (-4 - 2c_2g_1 - 2c_4g_1)z + (1 + c_4g_1)}$$

$$NTF(z) = \frac{1}{1 + c_1H(z)} \quad \ldots\ldots(1) \qquad STF(z) = \frac{b_1H(z)}{1 + c_1H(z)} \quad \ldots\ldots(2)$$

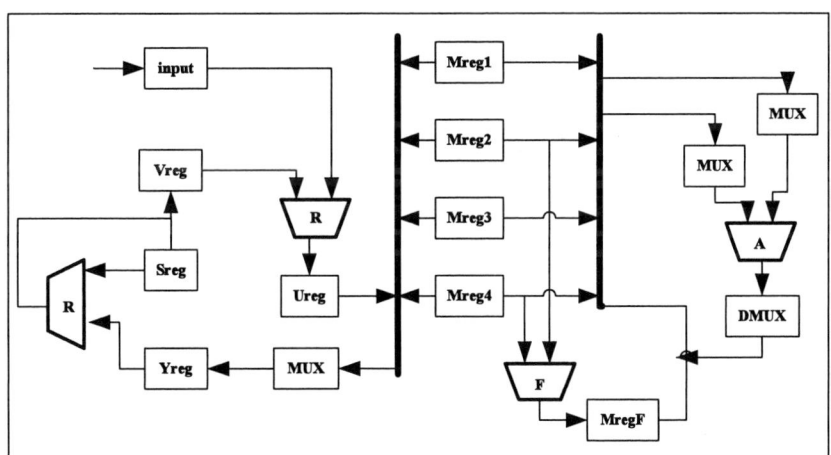

Fig.7 the circuit structure of DSM

The Research of Sub-space Partition Strategy and Bit Scanning Control Method Based on Human's Vision Nonlinearity Rule

Li-min Yan, Shen-nan Qiu

Key Laboratory of Advanced Display & System Applications, Ministry of Education, Shanghai University
Microelectronic R&D Center, Shanghai University
Campus P.O.B. 221, 149 Yanchang Road, Shanghai 200072, PRC
TEL: 86-021-56331323-117 FAX: 86-021-56331272
E-mail: yanlm@shu.edu.cn

Abstract

Flat Panel Display (FPD) technology is having the trend to replace CRT displayers both in industry and family usage, and becoming more and more important all over the world.

Gray scale is one of the key parameters of display quality. The main method to control gray scale levels is Pulse Width Modulation (PWM). It contains much time redundancy in this process, and requests higher scan frequency. On the other hand, to solve the problem based on human's vision nonlinearity rule, we had to add more gray scales to satisfy human vision.

In order to those problems, this paper proposes a sub-space partition strategy and bit scanning control method based on human's vision nonlinearity rule, to optimize the scanning time redundancy problem and reduce the scan frequency.

Key Word: FPD, gray scale, sub-space, vision nonlinearity

1. Introduction

Since the FPD was developed, the display quality is improved significantly, due to the maturation of PDP and TFT-LCD techniques, the increase of the rates of finished products; CRT displayers are facing the threat of being replaced.

As one of the most important judgments parameters, the gray scale describes the layers of image's luminosity. Normally the more the gray scales are, the better the display quality is. In digital system, the control ability of gray scale is measured by the number of it [1].

Unfortunately, human eyes' feeling are not linear to the luminance, with different strength levels, the eyes would have different reaction. Concretely, when the luminance is low, human eyes feel more than the quantity it increased, on the contrary when the luminance is high, the eyes need more stimulation to get the sense of increase on the light. So, a better rectification is needed to satisfy the human's vision nonlinearity [2].

2. Gray Scale rectification

Nowadays, the widely used rectification methods are called inserting gray scale rectification and extracting gray scale rectification. Inserting gray scale rectification can rectify the data without any incensement in gray scale levels. The method is as follow.

$$P_i = Q\left[\left(\frac{S_i}{S_L}\right)^{\gamma_{inv}} \cdot n\right] \qquad (1)$$

The parameter P_i represents the levels after rectified, $Q[]$ is an integer function, S_i is the concrete display scale, and S_L is the maximum of the scale($1 \leq i \leq L$), $\gamma_{inv}=1/0.45$, parameter n is the number of the maximum scale level, so the scale number of a display panel is n+1.

Take 16 scale levels as an example, the rectification result is shown by Table 1. In this table we can find insufficiency and turbulence of gray scales, especially on the low and middle levels which are the key field on display, the relative luminance of levels fewer than 4 are almost 0, so the scale warp is serious.

Table 1 the inserting rectification with 16 levels

Original	Rectified	Integer (optimize)
0	0	0
1	0.0389	0
2	0.1824	0
3	0.448	0
4	0.848	0 (1)
5	1.392	1
6	2.088	2
7	2.9424	2 (3)
8	3.9584	3 (4)
9	5.1424	5
10	6.4992	6
11	8.032	8
12	9.744	9
13	11.6416	11
14	13.7264	13
15	15	15

To reduce the drawback, extracting gray scale rectification is introduced. This method can improve the display ability by added the levels of gray scale and profundity of rectification.

The method is based on the previous one, what we need to do is adding levels until every degree can be clearly show off. The formula is as follow.

$$P_i = Q\left[\left(\frac{S_i}{S_L}\right)^{\gamma_{inv}} \cdot n \cdot 2^k\right] \qquad (k > 0) \qquad (2)$$

In the expression, n is the number of maximums gray scale, after we finished the rectification all the effective scale can be shown in the display. The characteristic curve of the display has matched human's eyes, they are all nonlinearity. Parameter k is an integer, the sum of k and pre-expanding scale value is the value after-expanding.

Table 2 shows the expended 16 levels scales value when k is 2/3/4 respectively. The essential of the table is establishing a relationship between 16 levels scale and 64/128/256 levels scale. After the expanding, gray scale of 16 levels can satisfy human's eyes. Of course the more levels we expand, the better display quality we will get.

Table 2 The extracting rectification with 64/128/256 levels

4bits	6bits rectified	6bits adjusted	7bits rectified	7bits adjusted	8bits rectified	8bits adjusted
0	0	0	0	0	0	0
1	0.146089	1	0.292178	1	0.584356	1
2	0.681668	2	1.363335	2	2.726670	2
3	1.678418	3	3.356835	3	6.713671	6
4	3.180819	4	6.36163	6	12.723276	12
5	5.222553	5	10.445105	10	20.890211	20
6	7.831429	7	15.662857	15	31.325715	31
7	11.130914	11	22.261828	22	44.523656	44
8	14.841656	14	29.683312	29	59.366624	59
9	19.282108	19	38.564216	38	77.128433	77
10	24.369001	24	48.738003	48	97.476005	97
11	30.117858	30	60.235716	60	120.471431	120
12	36.542423	36	73.084847	73	146.169693	146
13	43.656224	43	87.312448	87	174.624896	174
14	51.471518	51	102.943036	102	205.886071	205
15	60	60	120	120	240	240

This method can solve the distortion problem; unfortunately another drawback is introduced, which is the enhancement of gray scale levels. Take 16 levels scale as an example, to get the optimum result, the gray scale levels is as follow:

$$n = \left(\frac{S_{16}}{S_2}\right)^{\gamma_{inv}} = \left(\frac{15}{1}\right)^{\gamma_{inv}} \approx 411 \qquad (3)$$

Obviously, conventional scan strategy will induce time redundancy and the drive circuits must be designed rigorously [3]. With the levels is becoming larger, high frequency will be the bottleneck of FPD technology. So a better strategy which cut the whole display area into subspaces is needed to reduce the spending of hardware resource.

3. Linear bit scanning and sub-space partition

The direct scan can be seen as an extremeness of sub-space partition, it cuts the area into several sub-spaces whose number equals the levels of gray scale, and each sub-space has the same weight. We also take 16 levels as an example; different levels are shown in Fig. 1.

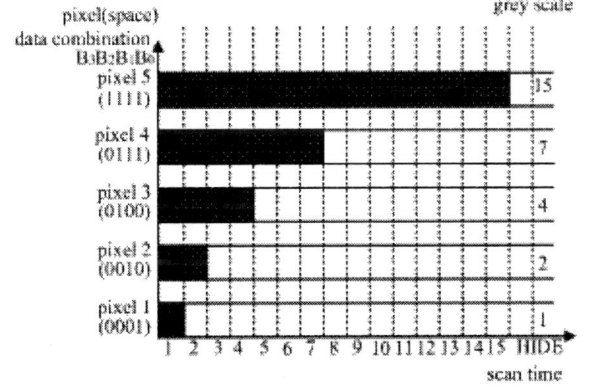

Fig. 1 Direct scan sequence

Fig.1 shows different levels of 1, 2, 4, 7 and 15. And we can find how the direct scan works; first, it divides the whole scan time into several time units N_{base}, at each N_{base} a serial of data would be send to the displayer, if the levels is N, the drive circuits should send the data N times with clearance of N_{base}. So the number of scan is proportional to the gray scale levels. Naturally the bit scanning is introduced.

The basic concept of linear bit scanning is as follow. Suppose every pixel in the image needs n bits to describe the gray scale level ($N=2^n$), to get the required level, the controller can send the lowest bit to the display system, after the time delay defined by the weight of the bit, the controller sends the lower bit .Until all the bits are finished and the relevant times delay are passed by, the pixel shows out the gray scale level. In the whole process, controller only needs to send the data n times, and every time interval is corresponding to the bit weight, so we call the strategy bit scanning.

The different gray scale levels got from 4 bits weight can display 16 gray scale levels.

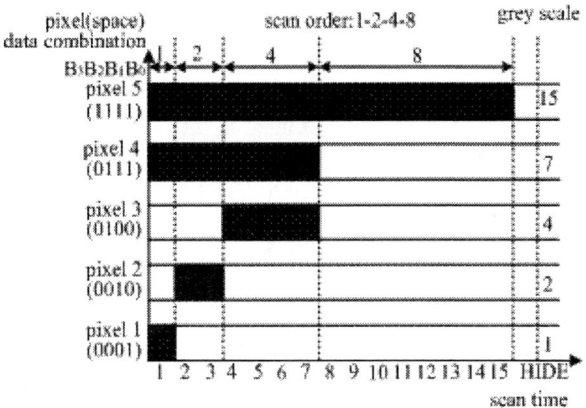

Fig. 2 Linear bit scan sequence

In Fig. 2 the controller scans by the weight order of 1-2-4-8, which represent by byte $B_0B_1B_2B_3$. Through different combination of weights, we can get all the 16 gray scale levels. Column j stores the data 0001B which represents the scan time or weight that corresponds to the gray scale in level 1; column j+1 stores the data 0100B which represents the weight that corresponds to the gray scale in level 4; the data 1001B in column j+3 describes the gray scale in level, 9 which turns on the display in weight 8 and 1; at last 1111B in the column j+4 shows the level 15, that means the display works in the whole scan times. To scan order, it has the same effect that by the weight order of 1-2-4-8, or by 8-1-4-2, or by any other orders. Because from the human's eyes vision, the only important thing is the total scan time, rather than how to range its order [2]. This kind of irrelativeness can also be found between different pixels, if we turn on some pixels by one order, others by another, the image will be just the same as normal. And the irrelativeness laid a good foundation for the partition of sub-space.

Sub-space is proposed on base of those thoughts. When we use linear bit scanning, the total scan times is 4, and every scan time, the weight is equal to n ($n \geq 0$) times power of 2. So if the duty cycle is same, the bit scanning would reduce

75% scan times compare to the direct scanning. On the direct scanning side the scan efficiency is just 5/16=31.25% in count of hide scan, the 11 scan times redundancy can be reused in sending data of gray scale levels to other sub-spaces [1]. If the time redundancy of two sub-spaces is staggered, the efficiency can be improved. What we need to do is optimize every weight order in every sub-space, in the most ideal condition; we can even fill all the redundancy time [2]. And this is the bit scanning for sub-spaces partition.

Let the display area be a space, the area with the same weight scan order be in one sub-space, we can scan by rows in every sub-space, and jump between sub-spaces. Also, we take 16 levels as an example, we divide the whole area into two sub-space, which are scanned by different weight order, the details can be seen in table 3.

Table 3 The scan sequence of 2 sub-spaces

Scan time	0	1	2	3	4	5	6	7	8	9	10	11	12	13	14	15	16
Sub_space0	X_3	*	*	*	*	*	*	*	X_0	X_1	*	X_2	*	*	*	X_x	
Sub_space1		X_0	X_1	*	X_3	*	*	*	*	*	*	*	X_2	*	*	*	X_x

The scan time begins from 0, X_3, X_2, X_1, X_0 and X_x represent 4 times of weight scan and 1 hide scan. The first sub-space corresponds to the scan time from 0 to 15, its weight order is 8-1-2-4; the second sub-space begins from scan time 1 to 16, and weight order is 1-2-4-8. Both of two sub-spaces can generate 16 kinds of different levels of gray scale. In detail the controller turns on the first sub-space during five time-blocks, that are 0,8,9,11,15, while turns on the second sub-space when 1,2,4,12,16. Because during every time-block the controller can just scan one sub-space, the total effective scan time is 10(0,1,2,4,8,9,11,12,15,16). We can find out the efficiency of scan equals 10/17=59%, that's twice the common method would get.

So when using bit scanning strategy, the more the sub-spaces are partitioned, the higher the scan efficiency would be. To get better, we can also use the redundant time left in the two sub-spaces, to scan the third or the forth sub-space. Now the key problem turns out to be how to divide every sub-space to minimize the time redundancy. Nowadays there are many strategies, the demands and conditions are preferentially considered. Here we propose a better strategy.

Suppose all the sub-spaces are scan by sequence, the number of them is m, gray scale level is 2^n, so we can work out the total number of scan time unit T_{base} is 2^n+m-1 (including one hide scan), and controller should $m \times (n+1)$ times(including m times of hide scan). The relationship between m and n is as follow [2].

$$2^n+m-1 \geq m \times (n+1)$$

$$m \leq \frac{2^n-1}{n} \qquad (4)$$

So the maximum value of m is $Q\left[\dfrac{2^n-1}{n}\right]$ (integer function) [3]. This is just a theoretical value, it is hard to realize, and normally the value is less than m.

We take 16 levels of gray scale as an example. In this situation n equals 6, $Q\left[\dfrac{2^n-1}{n}\right]$=10, so m equals 10. We use C

language to develop an algorithm that can exhaustive all the possible combinations of the weight order in every sub-space. We display one example below in Table 4, the gray scale level is 64.

Table 4 The scan sequence of 10 sub-spaces (time point)

Sub_space1	0	32	33	35	39	47	63
Sub_space 2	1	17	49	51	52	56	64
Sub_space 3	2	34	42	43	45	61	65
Sub_space 4	3	11	12	44	46	62	66
Sub_space 5	4	20	24	25	27	59	67
Sub_space 6	5	21	53	57	58	60	68
Sub_space 7	6	22	26	28	29	37	69
Sub_space 8	7	15	16	48	50	54	70
Sub_space 9	8	10	18	19	23	55	71
Sub_space 10	9	13	14	30	38	40	72

According to this table, the utilization rate of scan is 70/73=95.89%, while the original is 7/64=10.94%. That increases almost 9 times.

The scan sequence of sub-spaces is as following Table 5.

Table 5 Scan points distribution of 10 sub-spaces with 64 levels

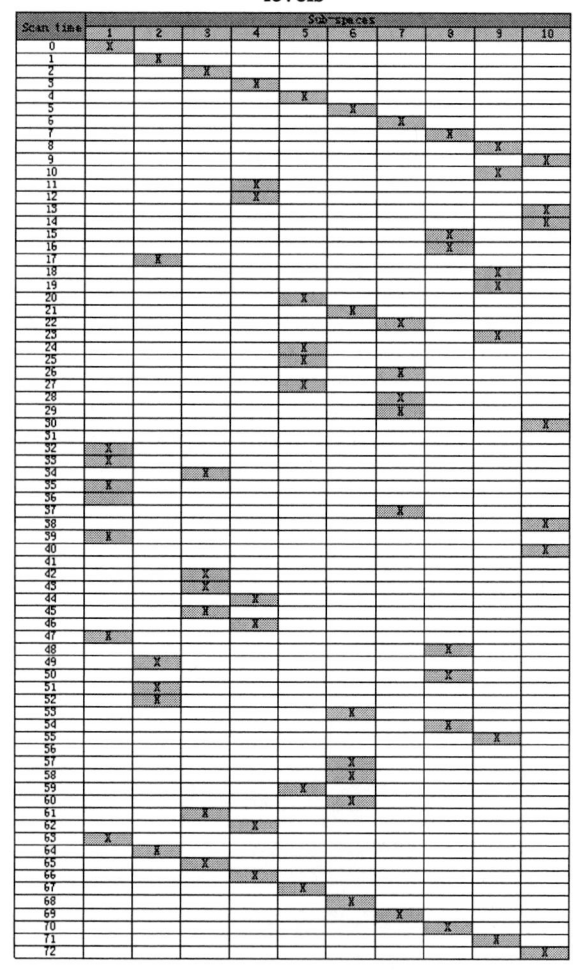

4. Implementation and verification

To verify our design, we program a complete controller for generating scan sequence and weights order by using Verilog Hardware Description Language.

The design is based on 16 gray scale levels, pixels in every frame is 128*96, the data bus's width is 16 bits, so we divide the display into 8 sub-spaces.

The controller system contains three main function blocks, the gray scale level expending block, the sub-spaces generating block, and the RAM. In the expanding block, a RAM is added to store relevant gray scale levels that had been calculated before, through look up process, we can accomplish the gray scale rectification.

At last we also use the EDA tool - Quartus II to simulate the result, the waves forms are shown as following Fig. 3.

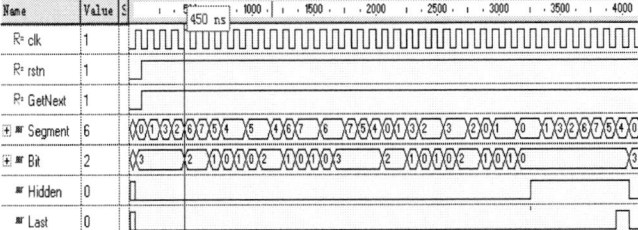

Fig. 3 The wave form of sub-spaces' generation

In Fig. 3, Segment is the signal of the selection of sub-spaces, Bit represents the weight of every gray scale level, and Hidden is the signal of hide scan. Through the wave's forms, we can find out every sub-space's weight is just match the sequence of sub-spaces, which means the design works.

5. Conclusion

Through our job above mentioned, we established a model of linear bit-scanning with sub-spaces partition for FPD. The strategy improves the scan efficiency significantly and reduces the time redundancy; these advantages can release the harsh demands for drive circuits. On the other hand, the partition strategy has large room to be enhanced, new methods such as fractal theory can handle the conflict between weights and sequence better.

Obviously, after adding RAM controller, data controller, the whole system is suitable to be embed in modern display.

References

1. K Inuka H Kimura, M Mizukami, et al. 4.0-inch TFT-OLED Displays and A Novel Digital Driving Method. SID, 00 [C], 2000. 924-927.

2. Ki-Duck Cho, Heung-Sik Tae, Sung-Il Chien. Improvement of low gray scale linearity using multi-luminance-level subfield method in plasma display panel [J]. IEEE Transactions on Consumer Electronics, Vol.48, Issue: 3, 2002, 377-381

3. V. P. Felix, R. Narasimman, and R. C. Agarwal. Perceptually Optimum Gray-Scale Transformation of Multibin Time History Data [J]. IEEE Journal of Oceanic Engineering, VOL. 22, NO.1, January 1997.

Layout Optimization and Modeling of an ESD-protection n-MOSFET in 0.13um Silicide CMOS Technology

Jiang Yuxi, Li Jiao, Ran Feng and Dian Yang

Key Laboratory of Advanced Display and System Application (Shanghai University), Ministry of Education
Microelectronic Research & Development Center of Shanghai University
Address: No.149, Yan Chang Rd., Shanghai, 200072, China
Mail: jiangyuxi@shu.edu.cn   Tel: 86-021-56331323-112   Fax: 86-021-56331272

Abstract

In this paper, a lot of CMOS devices with different device dimensions, spacings, and clearances have been drawn and fabricated to find the optimized layout rules for electrostatic discharge (ESD) protection in 0.13um Silicide CMOS Technology. The dependences of layout parameters on ESD protection ability of GGNMOS are investigated by using the TLP (transmission line pulsing) measurement technique. A DC model for modeling ESD NMOS snapback characteristics is also presented in this paper.

1. Introduction

The gate-grounded NMOS (GGNMOS) devices become the most vulnerable element to an ESD event due to the thin gate oxide and low drain breakdown voltage in submicron CMOS technology. The ESD protection mechanism of this device is based on snapback characteristics [1,2]. The devices under ESD stress are operated in unconventional regions, such as high operation voltage and current. Conventional compact models such as the Simulation Program with Integrated Circuit Emphasis (SPICE) are not suitable for ESD modeling. In addition, for the expensive cost, the ESD protection circuits are desired to sustain the highest ESD stress voltage in limited layout area. The circuit designers often confuse on how to predict and optimize the ESD protection devices. To optimize the layout design of the gate-grounded NMOS (GGNMOS) devices, some layout parameters which can affect ESD robustness will be investigated.

As shown in Fig. 1, a schematic I-V characteristic of a typical NMOS is given. The I-V curve can be divided into four regions. Region 1 and region2 are the linear and saturation regions respectively, described by the standard MOS equations. Region 3 is the avalanche breakdown region while region 4 is the snapback region. The standard MOS equations are no longer valid in these two regions. The NMOS ESD protection devices operated in region 1 and region 2 are under normal conditions and go to region 3 and region 4 under ESD stress.

Fig.1 also shows a schematic I-V curve of a typical GGNMOS transistor obtained by transmission line pulse (TLP) technique. The GGNMOS snapback parameters in the curve, such as V_{t1}, V_h, I_{t2}, are critical for measuring the ESD failure threshold voltage (ESDV) level of ESD protection devices. (I_{t1},V_{t1}) is the trigger point, which decides when the ESD protection device turns on. (I_{t2},V_{t2}) is the second breakdown point, ESDV level is represented by the second breakdown current I_{t2}. (I_H,V_H) is the holding point, V_H shall

ensure proper voltage clamping, low V_H provides a low-impedance discharge path.

The ESD protection device should be designed in the way that the operating voltage in the high current regime remain smaller than the thin gate oxide breakdown voltage (BV$_{ox}$). However, the voltage should be larger than the supply voltage (VDD) with a safety margin to avoid any unintentional triggering of the ESD protection device due to noise or voltage overshoot.

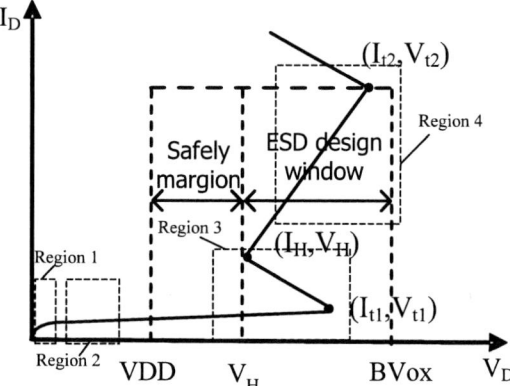

Figure 1: Schematic I-V curve of a GGNMOS transistor

In this paper, a lot of CMOS devices with different device dimensions, spacings, and clearances have been drawn and fabricated to find the optimized layout rules for electrostatic discharge (ESD) protection in 0.13um Silicide CMOS Technology. The dependences of layout parameters on ESD protection ability of GGNMOS are investigated by using the TLP (transmission line pulsing) measurement technique. A DC model for modeling ESD NMOS snapback characteristics is also presented in this paper.

2. Layout optimization design and TLP results of single finger GGNMOS

The main layout parameters which affect ESD robustness of CMOS devices are channel length (L), channel width (W), the spacing from drain contact to poly-gate edge (DCGS), silicide block layer width (Wsb) and the spacing from source contact to substrate diffusion (BS) as shown in Fig.2.

The TLP (transmission line pulsing) technique is widely used to measure the snapback behavior of GGNMOS during the high ESD stress. These TLP measured results and relations will be shown and analyzed in the following. When the relations between GGNMOS snapback parameters and layout parameters are well understood, the circuit designer

978-1-4244-2739-0/08/$25.00 ©2008 IEEE   344

can predict the behaviors of the GGNMOS devices under high ESD current stress, and design area-efficient ESD protection circuits to sustain the required ESD level [3, 4].

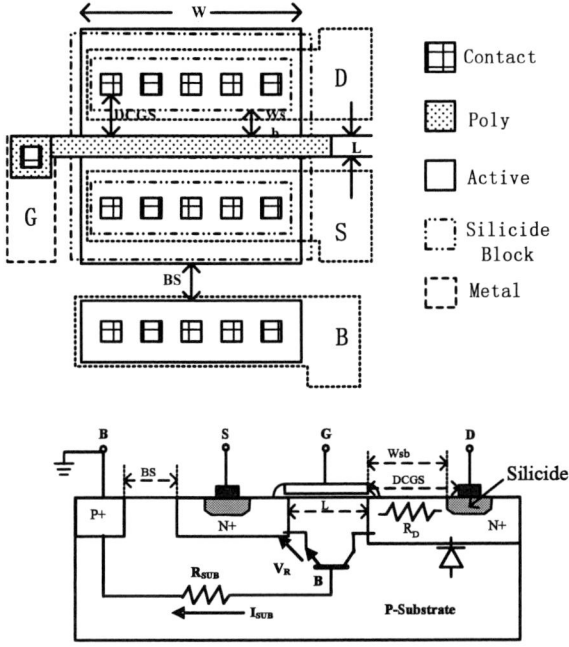

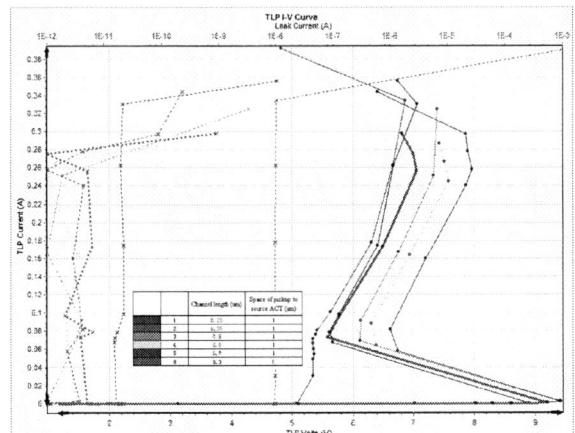

Fig. 2. Layout of the single-finger NMOS transistor.

2. 1 Channel Length (L) of single finger GGNMOS

The TLP measure data of GGNMOS devices with different channel length is shown in Fig.3. According to these results, the holding voltage V_h, is proportional with the gate length. The changes of snapback voltage V_{t1} is small when the gate length increases.

Fig.3. The TLP measurement data of gate-grounded NMOS devices with different channel length (L=0.28, 0.3, 0.35, 0.5, 0.6, 0.8um)

As the cross section shown in Fig.2, the gate length is decreasing so as to decreases the equivalent base width of the parasitic bipolar transistor, and then improves the current gain β. So, it has always been assumed that the minimum gate length device shows the best ESD-performance. But as shown in Fig.4 , the second breakdown current I_{t2} drops a little bit   when the gate length   is the minimum

length(L=0.13um). The reason of this reverse effect is attributed to the trade off between power dissipation and melts volume, which is probably too small in the shortest gate length devices [5]. The optimum value for gate length is about 0.5um from the TLP measurement results.

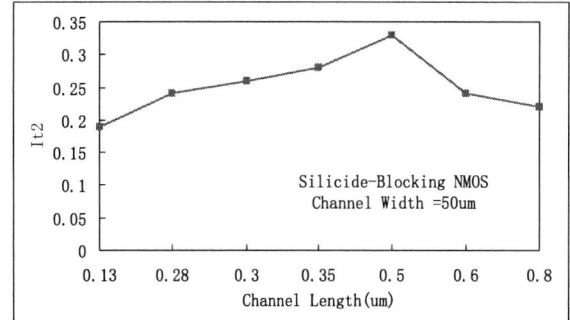

Fig.4. Gate length dependence of It2 for silicided-blocking devices

2.2 Channel Width (W) of single finger GGNMOS

According to the TLP data shown in Fig.5, The single finger GGNMOS shows an obvious width dependence of It2. In Fig.5, the It2 of the NMOS device is increased when the device channel width is increased. Redraw the relationship between channel width and It2 when It2 measured in mA/um, as shown in Fig.6. When channel width is increased, although the It2 of the NMOS is increased, the current sustainable ability per channel width of the NMOS is decreased. This is due to the nonuniform turn-on effect among the long channel width of the large dimension devices [6]. According to the data of Fig.6, for single finger GGNMOS, if the finger width Wf is larger than Wmax=50um, the non-uniform turn-on effect will be obvious. So, in order to decrease the non-uniform turn-on effect, more fingers are drawn and connected in parallel to form the large-dimension NMOS devices.

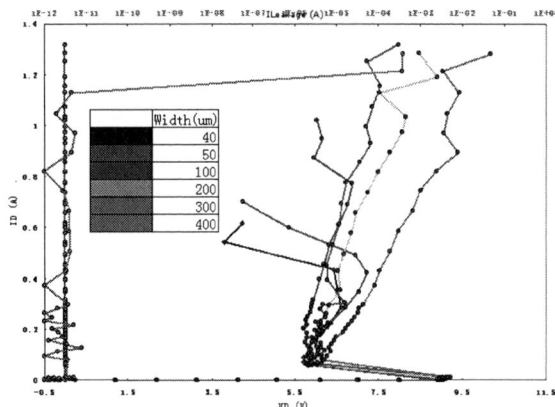

Fig.5. The TLP measurement data of single finger gate-grounded NMOS devices with different channel width (W=40, 50, 100, 200, 300, 400um; L=0.3um)

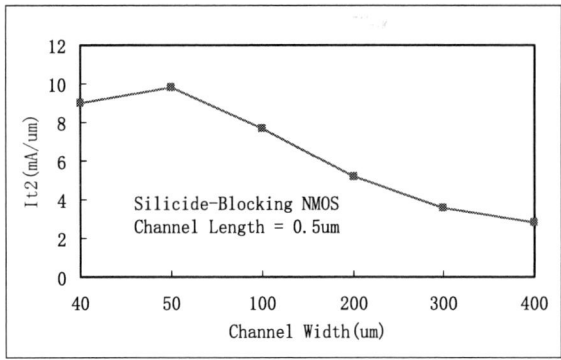

Fig.6. Dependence of It2 (mA/um) on the channel width

2.3 The spacing from drain contact to poly-gate edge (DCGS) and the silicide block layer width (Wsb)

As shown in Fig.2, the spacing from drain contact to poly-gate edge (DCGS) is equal to the spacing of SAB to contact plus the silicided block layer width (Wsb). So, increase Wsb is equal to increase DCGS. According to the TLP data shown in Fig.7 and Fig.8, the increase in the Wsb improves the ESD robustness of NMOS devices by increasing the parasitic ballast resistance.

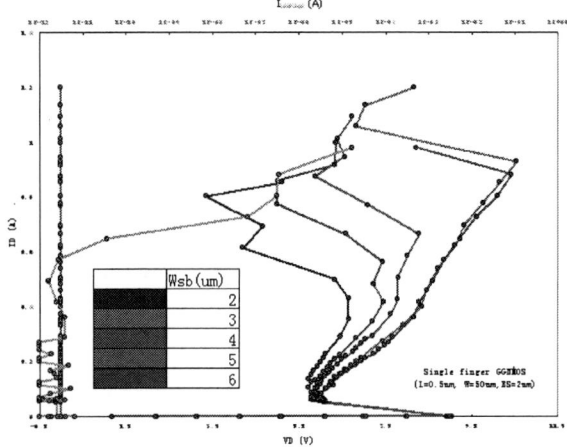

Fig.7 The TLP measurement data of single finger gate-grounded NMOS devices with different silicide block width (Wsb=2, 3, 4, 5, 6,0um;W=50um,L=0.5um,BS=2um)

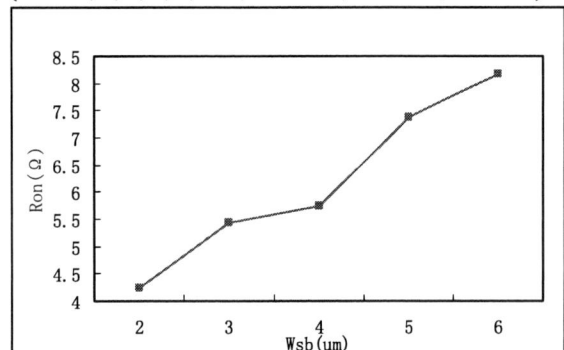

Fig.8 The Ron dependence of the silicided block layer width (Wsb)

2.4 The spacing from the source active to the substrate diffusion (BS) of single finger GGNMOS

The spacing from the source active to the substrate diffusion has been illustrated in Fig.2 and marked as "BS".

As shown in Fig.9, the incensement of ESD capability is not obvious when BS increasing.

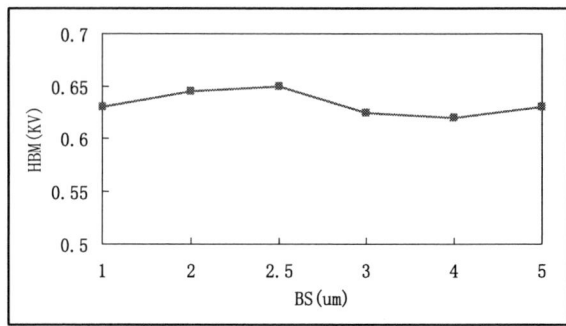

Fig.9 Dependence of HBM robustness on the BS spacing of single finger GGNMOS (W=50um, L=0.5um, Wsb=2um)

3. Layout optimization design and TLP results of multi-finger GGNMOS

To sustain the required ESD protection level, GGNMOS devices are always drawn with large device width. Such GGNMOS devices with large dimensions are often achieved by multi-finger to reduce the total layout area and the nonuniform turn-on effect. The layout of multi-finger GGNMOS is shown as Fig.10.

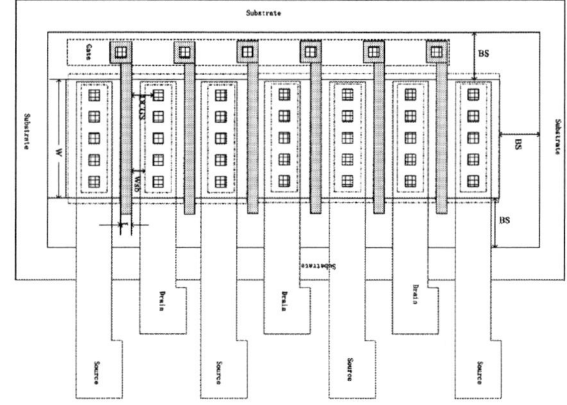

Fig.10 The layout view of multi-finger GGNMOS (M=6, Wf=50um,L=0.5um Wsb=2um)

3.1 Multi-finger NMOS Layout Floorplan

Different kinds of layout floorplan of multi-finer GGNMOS have been shown in Fig.11 [7, 8]. The total channel widths of these multi-finger GGNMOS devices are 200um. In Fig.11 (a), the multi-finger GGNMOS has 2 parallel fingers, and every finger is drawn with a finger width of 100um. In Fig. 11(b) and Fig.11(c), there are the typical layout structures of multi-finger MOSFET without any additional pick-up guard ring inserted into the source region. The difference between the layout of Fig.11 (b) and Fig. 11(c) is the layout floorplan of the source and drain. In Fig. 11(d), there are two additional pick-ups inserted into the central source region. In Fig. 11(e), there is one additional pick-up inserted into the central source region.

Each multi-finger GGNMOS device has 4 parallel fingers in Fig.11 (b), Fig .11(c), Fig.11 (d), Fig.11 (e), and every finger is drawn with a unit-finger width of 50um. So, the total channel width for each multi-finger GGNMOS device is 200um.

Table.1. the dependence of HBM level with different layout Floorplan

	Total Width(um)	Finger Width(um)	Finger Number	Layout Floorplan	Ron(Ω)	HBM(KV)
1	200	100	2	SDS	3.23	2
2	200	50	4	DSDSD	1.21	1.8
3	200	50	4	SDSDS	1.77	2.21
4	200	50	4	DSBSBDSBD	5.83	0.435
5	200	50	4	SDSBSDS	0.989	2.36

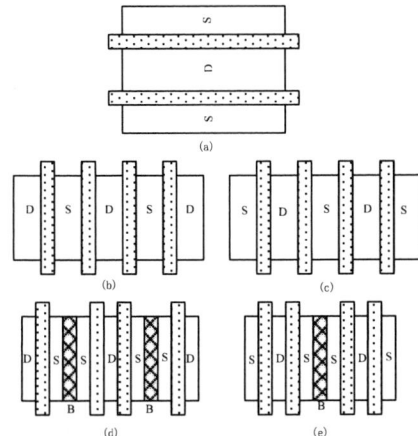

Fig.11 The layout floorplan of the multi-finger GGNMOS devices.

The TLP measurement data of multi-finger gate-grounded NMOS devices with different layout floorplan have been shown in Fig.12. According to the data of Table 1, the layout floorplan (SDSBSDS) of the GGNMOS (M=4, Wf=50um, L=0.5um, Wsb=2um, BS=1um) can archive the best HBM ESD capability.

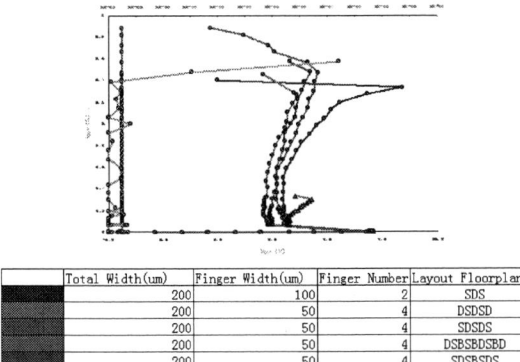

	Total Width(um)	Finger Width(um)	Finger Number	Layout Floorplan
	200	100	2	SDS
	200	50	4	DSDSD
	200	50	4	SDSDS
	200	50	4	DSBSBDSBD
	200	50	4	SDSBSDS

Fig.12.The TLP measurement data of multi-finger GGNMOS devices with different layout Floorplan.

3.2. The spacing from the source active to the substrate diffusion (BS) of Multi-finger GGNMOS device

According to the data shown in Fig.13, the wider spacing BS has higher HBM ESD level. This is due to that wider BS contributes the larger parasitic well resistor, which will turn on the parasitic lateral npn BJT in the NMOS device under ESD stress quickly and uniformly.

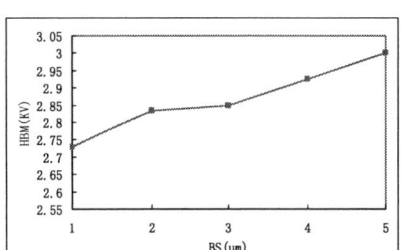

Fig.13. Dependence of HBM robustness on the BS spacing of Multi-finger GGNMOS (M=6, Wf=50um, L=0.5um, Wsb=2um)

4. DC model for ESD NMOS snapback characteristic

A DC model explains the relationship between the layout parameter and the ESD robustness of NMOS, as shown in Fig.14. This model consists a MOS transistor extracted from BSIM 3V3, a parasitic bipolar transistor modeled by Verilog-A, a substrate resistance and series resistance [9-12]. Equations for modeling the high current behavior of NMOS transistor have been developed. Simulation results and comparison to the TLP data for 0.13um NMOS device are presented as shown in Fig.15.

4.1 MOS transistor modeling

Equations for I_{DS} are listed below.

$$V_{DSAT} = \frac{(V_{GS}-V_T)L\varepsilon_{SAT}}{V_{GS}-V_T+mL\varepsilon_{SAT}} \qquad (1)$$

$$I_{DSAT} = \frac{WC_{OX}v_{SAT}(V_{GS}-V_T)^2}{V_{GS}-V_T+mL\varepsilon_{SAT}} \qquad (2)$$

For $V_{GS}>V_T$
if $V_{DS} \leqslant V_{DSAT}$

$$I_{DS} = \frac{WC_{OX}u_{eff}}{L}(m-1)\left(\frac{kT}{q}\right)^2 e^{\frac{V_{GS}-V_T}{mkT/q}}(1-e^{\frac{-V_{DS}}{kT/q}}) \qquad (3)$$

if $V_{DS}>V_{DSAT}$

$$I_{DS} = I_{DSAT}\left(1+\frac{V_{DS}-V_{DSAT}}{V_A}\right) \qquad (4)$$

347

For $V_{GS} \leq V_T$

$$I_{DS} = \frac{WC_{OX}u_{eff}}{L}[(V_{GS} - V_T)V_{DS} - \frac{m}{2}V_{DS}^2]\frac{1}{1 + \frac{V_{DS}}{L\varepsilon_{SAT}}} \quad (5)$$

4.2 Parasitic bipolar modeling

Equations for I_C and I_B are given by [13]

$$I_C = I_{oe}(e^{\frac{V_{bS}}{kT/q}} - e^{\frac{V_{bD}}{kT/q}}) \quad (6)$$

$$I_B = I_{oc}(e^{\frac{V_{bS}}{kT/q}} - 1) \quad (7)$$

Here I_{oc} and I_{oe} are reverse saturation current due to diffusion of holes in the collector and emitter of the NPN.

$$I_{oc} = \frac{I_D/M - I_{DS}}{\exp(I_{SUB} \cdot \frac{R_{SUB}}{kT/q})} \quad (8)$$

$$I_{oe} = \frac{I_D(M-1)/M - I_{DS}}{\exp(I_{SUB} \cdot \frac{R_{SUB}}{kT/q})} \quad (9)$$

4.3 High current behavior modeling

Take M as the avalanche multiplication factor, avalanche generation current I_{gen} can be given by [14]

$$I_{gen} = (M - 1)I_i \quad (10)$$

Here I_i is an incident current.

When the parasitic bipolar transistor turned on,

$$I_i = I_{DS} + I_C \quad (11)$$

Therefore

$$I_{gen} = (M - 1)(I_{DS} + I_C) \quad (12)$$

Generation current is formed with base-emitter current I_B and substrate current I_{SUB}, thus

$$I_{gen} = I_{SUB} + I_B \quad (13)$$

Therefore,

$$I_{SUB} = (M - 1)(I_{DS} + I_C) - I_B \quad (14)$$

4.4 Avalanche multiplication factor M

M can be extracted from a single I-V measurement. Before bipolar turn-on take place, all the generation current acted as substrate current and M can be determined by:

$$M = \frac{I_D}{I_D - I_{SUB}} \quad (15)$$

After the bipolar turns on, M can no longer be extracted from measurements after snapback, since some of the generation current flows into the emitter. Here, M is given by miller formula:

$$M = \frac{1}{1 - K1 \times \exp(-\frac{K2}{Vd - Vdsat})} \quad (16)$$

K1,K2 are the parameters related to drain depletion width and impact ionization coefficients.

Take the peak value of M as obtained by Equation as a function of V_D, with a plot of *(1-1/M)* vs. *1/(V_D-V_{DSAT})*, *K1* and *K2* can be extracted from the intercept and the slope [14].

4.5 Parasitic resistor modeling

The drain resistance can be modeled by an open-collector/emitter method for the parasitic BJT based on Ebers-Moll equivalent model combining a measurement of the emitter-collector voltage, developed by H.T. Kim.[15]

$$V_{EC} = \frac{kT}{q}\ln[\frac{I_B + I_E(1 - \alpha_R)}{\alpha_R[I_B - I_E(1 - \alpha_R)/\alpha_F]}] + R_C(I_B + I_E) + R_E I_E \quad (17)$$

Here α_F and α_R are the large-signal forward and reverse common-base current gains, respectively.

The substrate resistance R_{SUB} of ESD MOSFET is not a constant. It can be described as a dynamic substrate resistance, modeled by a current controlled voltage source V_{SUB} [16]

$$V_{SUB} = R_{SUB} \cdot I_{SUB} - R_{DRAIN} \cdot (I_{DS} - I_S) \quad (18)$$

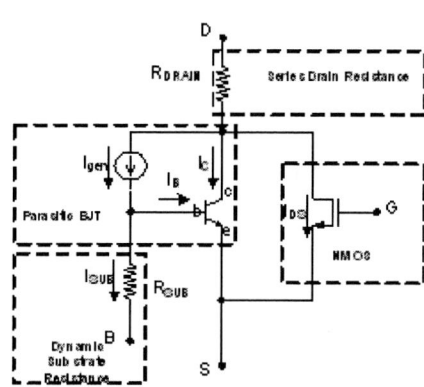

Fig.14. Equivalent circuit of NMOS DC model including the parasitic bipolar transistor

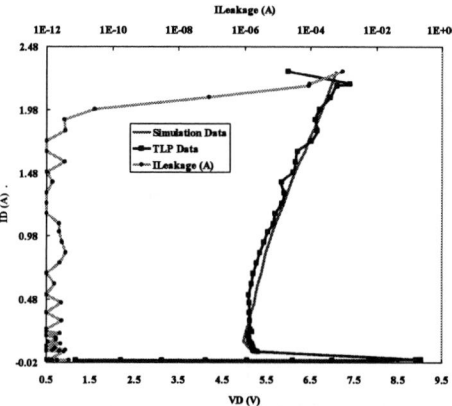

Fig.15. TLP measurement data and simulation results for 0.13um GGNMOS device, Wf=50um, L=0.5um, M=6, t_{rise}=10ns.

Conclusion

This paper presents the layout optimization design to improve the ESD robustness of NMOS in 0.13um silicide CMOS process. The optimized layout parameters have been verified to effectively improve ESD robustness of CMOS devices. A DC model for modeling ESD NMOS snapback is also presented in this paper. Simulation results and comparison to the TLP data for 0.13um NMOS device are presented.

Acknowledgments

This work was supported by Shanghai Leading Academic Discipline Project. (Project Number: J50104)

Reference

1. A. Amerasekera, C. Duvvnury, "ESD in Silicon Integrated Circuit", Second Edition, John Wiley, Chichester, (England, 2002).

2. Albert Z.H. Wang, "On-chip ESD protection for integrated circuits," Kluwer Academic Publishers (Boston, 2002), pp. 54-55.

3. Tung-Yang Chen, Ming-Dou Ker, "Analysis on the Dependence of Layout Parameters on ESD Robustness of CMOS Devices for Manufacturing in Deep-Submicron CMOS Process" *IEEE Trans. on semiconductor manufacturing,* vol.16, no.3(2003), pp.486-500.

4. A. Amerasekera, S. Ramaswamy,M.-C. Chang, and C. Duvvury, "Modeling MOS snapback and parasitic bipolar action for circuit-level ESD and high current simulations," *Proc 34th Int. Reliability Physics Symp*, Dallas, TX, 1996, pp. 318–326.

5. K. Bock, B. Keppens, V. De Heyn, G. Groeseneken, L.Y. Ching, A. Naem, "Influence of gate length on ESD-performance for deep sub micron CMOS technology," *Proc. Electrical Overstress/Electrostatic Discharge (EOS/ESD) Symp,* 1999, pp.95-104.

6. Kwang-Hoon Oh, Charvaka Duvvury, "Analysis of Nonuniform ESD Current Distribution in Deep Submicron NMOS Transistors", *IEEE Trans. Electron Devices,* vol. 49, No. 12(2002), pp.2171-2182.

7. Thomas L. Polgreen, Amitava Chatterjee, "Improving the ESD failure threshold of silicided n-MOS output transistors by ensuring uniform current flow" , IEEE Trans. on Electron Devices, VOL. 39, NO. 2, (1992), pp. 379-389.

8. Ming-Dou Ker, Che-Hao Chuang, and Wen-Yu Lo, "Layout Design on Multi-finger Mosfet for On-Chip ESD Protection Circuits in a 0.18um Salicided CMOS Process", Proc. of the 8th IEEE International Conference on Electronics, Circuits and Systems (ICECS 2001), 2001, Vol. 1, pp. 113-116.

9. M. Mergens,W.Wilkening, S. Mettler, H.Wolf, and W. Fichtner, "Modular approach of a high current MOS compact model for circuit-level ESD simulation including transient gate coupling behavior," *Proc 37th Int. Reliability Physics Symp*, San Diego, CA, 1999, pp. 167–178.

10. C. Russ, K. Bock, P. Roussel, G. Grosenken, H. Maes, and K .Verhage, "A compact model for the ground-gated nMOS behavior under CDM ESD stress," *Proc. Electrical Overstress/Electrostatic Discharge (EOS/ESD) Symp*, Orlando, FL, 1996, pp.302-315.

11. Junjun Li, "Compact Modeling of On-Chip ESD Protection Devices Using Verilog-A," *IEEE Trans. Computer-Aided Design of Integrated Circuits and Systems*, vol.25, No. 6 (2006), pp. 1047-1063.

12. Chen J.Z., Amerasekera, E.A. and Duvvury C, "Design methodology and optimization of gate-driven NMOS ESD protection circuits in submicron CMOS processes," *IEEE Trans. Electron Devices*, vol.45, No. 12 (1998), pp. 2448-2456.

13. R.S.Muller, Device Electronic for Integrated Circuits, New York, Wiley(1986).

14. A. Amerasekera, S. Ramaswamy,M.-C. Chang, and C. Duvvury, "Modeling MOS snapback and parasitic bipolar action for circuit-level ESD and high current simulations," *Proc 34th Int. Reliability Physics Symp*, Dallas, TX, 1996, pp. 318–326.

15. H.T. Kim, I.C. Name and K.S. Kim, "Extraction of source and drain resistances in MOSFETs using parasitic bipolar junction transistor ", *Electronics Letters*, Vol.41 , No. 13(2005), pp. 772-774.

16. S.Ramaswamy, A.Amerasekera, and M-C. Chang, "A Unified Substrate Current Model for Weak and Strong Impact Ionization in sub-0.25um NMOS device", *IEDM 97*, 1997, pp. 885-888.

A Verification Method Based On A Mixed-Signal System For MV06

HuYue-li1,2, ZhangYi-chi2, XuanXiang-guang3
(1.School of Mechatronics and Automation, 2.School of Microelectronic R&D Center
Shanghai University, Shanghai 200072, China
3. Technology Center, Shanghai Feilo Co.,Ltd. Shanghai 200050, China)
Key Laboratory of Advanced Displays and system Application, Ministry of Education, Shanghai University
Campus P.O.B. 25, 149 Yanchang Rd., Shanghai 200072, China
E-mail:huyueli@shu.edu.cn

Abstract

With the rapid development of mixed-signal system-on-chip (SOC), the verification before tape-out is a critical phase during the design flow in order to guarantee the products yield. A design flow based on mixed-signal verification is proposed. Exemplified with the design of a mixed MCU (MV06), this paper proposes the verification precept and introduces the simulation principle and method of mixed signals in the Synopsys simulation environment, and then describes the simulation process with the Discovery AMS and presents the verification result. This methodology is implemented in our tape-out, the feasibility and efficiency of the method has been verified.

1. Introduction

According to the Moore Law, the dimension and function of an SOC has rapidly expanded. Due to the large scale of both digital and analog part of the SOC, the interface for analog/digital signal (AMS) is becoming more and more complicated, which has become a bottle-neck for design verification.

Traditional simulation method extracts the RC information of the final placement and simulates using Spice which has two disadvantages. First, it increases design period, because the placement has to be redesigned if the simulation result is not right. Second, the simulation rate is very slow, which is unbearable for a large scale IC[0]. A new resolution, which simulates both the digital part and the analog part described by HDL with a digital circuit simulator in its prophase, and simulate the digital part and the analog part with respective digital simulator and analog simulator (such as Hspice 、 Specter 、 NanoSim) in its anaphase, can provide high simulation rate and precision. In its prophase, the simulation rate is pretty much the same as a digital circuit, and its simulation precision is related with the description scheme of the analog part, which is relatively low because of the irrelevance between the actual analog circuit and its HDL simulation model.

Exemplified with the design of a mixed-signal MCU (MV06), this paper introduces the method for mixed-signal verification in the Synopsys simulation environment with Discovery AMS (the platform for mixed-signal synchronal simulation) and presents the verification result.

2. The platforms for mixed-signal verification

In order to fully utilize the respective merit of the digital simulator and the analog simulator and to address the mixed-signal synchronal simulation problem, many EDA suppliers are providing a co-simulation resolution, which integrates the digital simulator and the analog simulator with a "platform". The digital part is simulated with digital simulator and the analog part is simulated with analog simulator, and the interface signals of the two parts are synchronized through a "platform". This resolution not only increases the simulation rate but also realized the simulation of a whole system.

2.1 Introduction of the platforms

Most EDA suppliers have offered platforms for mixed-signal verification, such as Discovery AMS by Synopsys, Incisive AMS in Incisive by Cadence and ADVance MS by Mentor Graphics.

Mixed-signal verification tools of the day integrate the circuit simulator and the logic simulator via various interface modules or signal conversion mechanisms, and the analog and digital simulation tasks are completed with the circuit simulator and the logic simulator respectively. For example, Virtuoso AMS Simulator by Cadence integrates Spectre (FastSpice, Hspice, cdsSpice) and NC-verilog via two interface module such as A2D and D2A; Discovery AMS by Synopsys integrates NanoSim and VCS via special signal conversion mechanism; and ADVance MS by Mentor Graphics integrates Eldo, ModelSim and Mach into one environment.

2.2 Discovery AMS

Discovery AMS platform by Synopsys is employed in this paper to verify the mixed-signal design. Discovery AMS integrates leading EDA software such as NanoSim, Hspice and VCS. LUT based simulation software NanoSim not only greatly increases simulation rate but maintains certain simulation precision, Hspice is the domain criterion of the transistor-level simulation software, and VCS is the Verilog HDL simulator with highest market possession rate. Discovery AMS, which closely integrates NanoSim and VCS, is capable of flexible control between gate circuit rate and transistor-level precision, as is depicted in Fig. 1[0].

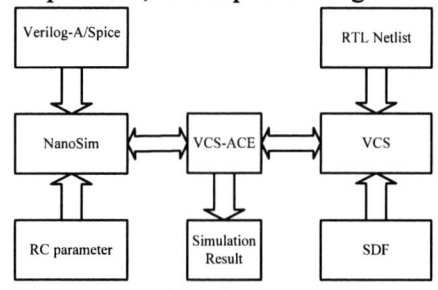

Figure 1. Discovery AMS

One thing that matters a lot is the signal conversion between digital field and analog field[0]. The integration of NanoSim and VCS features unique mechanism that completes the logic to voltage conversion or voltage to logic conversion on the digital/analog borderline. It employs resistance mapping table to match the driving resistance for digital/analog conversion, and the MOSFET threshold resistance is used to match the intensity of the digital signal in logic to voltage conversion. The user may create a customized resistance mapping table, or just adopt the default one. The mapping principle is depicted in Fig. 2. Conversion between digital and analog signals is carried out by means of a resistance mapping table. Different digital driven intensities can be mapped into pull-up or pull-down resistances with different values when digital signals are transferred into analog signals on the borderline. And as for the analog to digital conversion, analog output impedance is mapped into different driven intensities in Verilog. Analog to digital conversion and digital to analog conversion can use either different conversion rules (non-bidirectional mapping) or same rules (bidirectional mapping), and both the threshold of low/high level and rising/falling time can be set according to certain conditions.

The simulation environment of NanoSim-VCS, which quite resembles the simple digital simulation environment, can be easily integrated into different phases of the design process. Mixed-signal simulation can be carried out in the prophase, in the metaphase, or in the anaphase of the design process. In our project, mixed-signal simulation is performed in the anaphase for function and timing verification of the design after placement and routing.

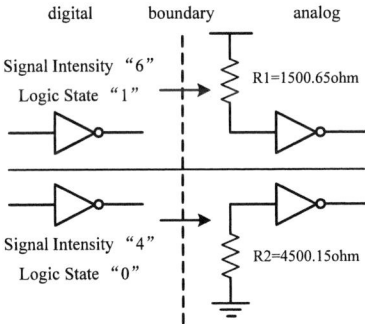

Figure 2. The principle of signal conversion

3. Introduction of MV06

An 8-bit Successive Approximation AD converter is integrated into the MCU, which widens the application extension. In some applications which require an ADC, using the ADC-integrated MCU will reduce the cost.

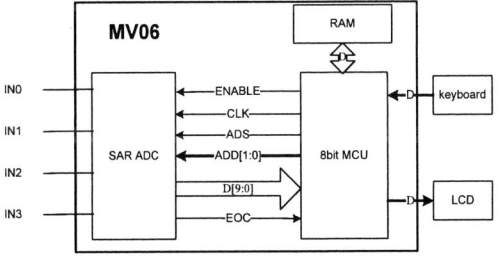

Figure 3. System structure

Fig. 3 is the block diagram of the MV06 system. The system functions as follows: the ADC converts the analog input to 10-bit digital output, which is stored to the SFR of the MCU directly by the ADC control module. The ADC, which is based on CHRT0.35um2P4M technology, employs Successive Approximation AD converters to complete conversion of 4 channel analog inputs to 10-bit digital outputs at the speed of 1Msps.

The ADC conversion timing is depicted in Fig. 4. ENABLE is the enable signal of the ADC which is valid with high level and the tsp time must be longer than 300ns to ensure the circuit stabilization. Set-out time for initial signal ADS must be longer than 3ns and the ADS pulse width should at lest be one clock cycle. The SAR （Successive Approximation Register） structure requires 10 clock cycles for the ADC to perform one conversion. When EOC (end of conversion) is low, the ADC is in the state of conversion and the output D[9:0] is in the process of successive approximation. When EOC is high, AD conversion is done and output D[9:0] is stable.

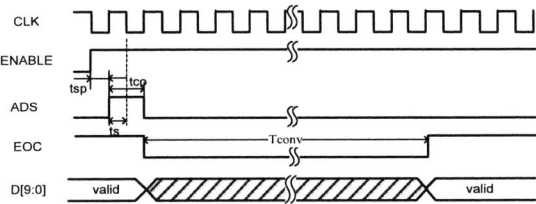

Figure 4. Timing diagram of SAR ADC

4. Mixed-Signal (AMS) Verification

The mixed-signals of the ADC-integrated MCU are simulated via the proposed mixed-signal simulation platform. Logic circuit netlist and sdf delay information of the MCU core and the ADC control circuit are simulated in VCS, and the SPICE netlist of the ADC is simulated in NanoSim.

4.1 Procedures of Verification

Detailed procedures of NanoSim-VCS simulation are described as follows[0]:

(1) Extract the size of the MOSFET from Spice netlist. Run Gentech and select the simulation technical angle, temperature and voltage according to the size and Spice simulation model offered by agent supplier. Invoke HSPICE or Spectre simulator and generate the LUT-based technology file which is used by NanoSim. The NanoSim precision is guaranteed because the technology file, which reflects technical related effects such as short- channel effect, speed- saturation effect and Substrate bias voltage effect, is directly generated by Spice simulator according to the simulation model of specific technology[0].

(2) Before setting up the mixed-signal simulation platform, the circuit model to be simulated should be divided into two parts at first. The digital part is designated to VCS for simulation while the analog part is designated to NanoSim for simulation. vcsAD.init is the configuration file which is described as follows:

use_spice -cell ADC;

choose nanosim -n ./ADC.net -o 01 -C cfg - p ./ADC035.tt_25_3p3.tech;

351

```
set bus_format _%d;
set rmap resis_map;
```

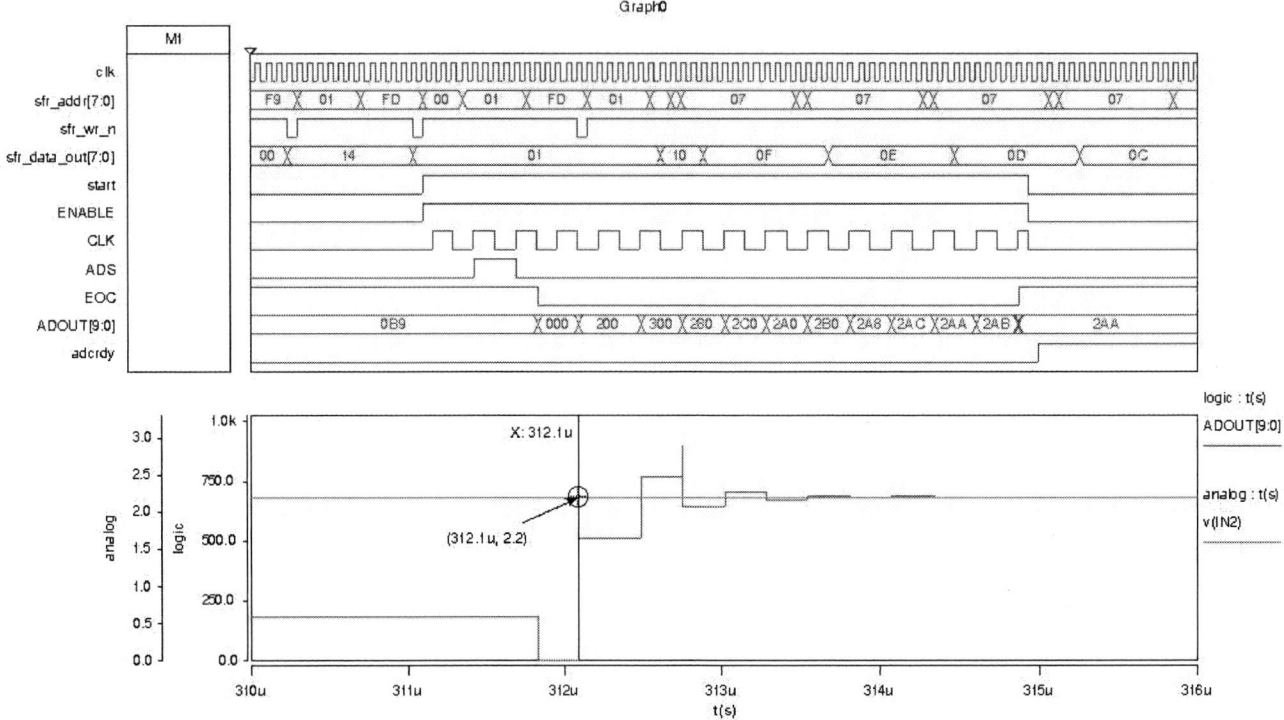

Figure 5. Simulation wave of mixed-signals triggered by software

Line one suggests that the ADC is an analog module. Line two chooses NanoSim as the simulator. Option -n,-o and C respectively specifies the path of the netlist, the prefix of the output file and the path of the configure file. Line three defines the bus format in Spice netlist. Line four specifies the resistance mapping file which corresponds the range of the transistor driving resistance with the Verilog logic intensity. Such correspondence can be neither unidirectional nor bidirectional. In this paper, an bidirectional resistance mapping table is employed to perform the conversion between analog and digital signals.

(3) The NanoSim configure file sets up the mixed simulation environment and tells NanoSim how to carry out the simulation process. The configure file may include information as follows:

Nodes whose waveforms are to be saved and the format of the output waveforms;

Power consumption and timing analysis which are to be executed;

Modified information about the circuit;

Precision control of the simulation;

(4) Compile the Verilog and Spice netlists by the main simulator VCS and perform co-simulation. The following line is the operation command:

>vcs +ad -f run.f +v2k -l comp.log +define+sdfapr

Different from the simulation command of a mere digital circuit, there is a "+da" option which specifies the configure file of VCS mixed-signal simulation. A simv file will be generated after the execution of the command, and a

*_uod.out file, which contains the waveform of the entire simulation result, will be generated after the execution of the simv file.

(5) Load and inspect the output simulation file into CosmosScope to analysis the simulation result.

CosmosScope is a graphical wave analyzer and supports various formats of wave files. It can display the digital and analog wave at the same time and possesses powerful computational and analytical functions of the waveform. Simulation debugging of mixed signals as well as observation of the conversion of A/D interface signals can be easily performed by means of CosmosScope.

4.2 Verification Result

Entire-chip mixed-signal Simulation is verified with the ADC works in the software-trigged mode. The NanoSim Version is z-2007.03-SP2, the VCS is Y-2006.06-SP1-9, and the working platform is Solaris 10. The simulation result shows that the ADC works normally and meets the design request. The analog part selects typical technology corner at 25° with the working voltage of 3.3V during simulation, and the input system clock is 15MHz.

The simulation wave of mixed signals triggered by software is shown in Fig. 5. Analog input is switched to channel 2 with 2.2V direct voltage, as shown at the bottom of Fig. 5 (v（IN2）). The conversion clock is a quarter frequency divider of the system clock. Digital signal is presented in the upper wave window of CosmosScope and analog signal is presented in the lower window. ADOUT[9:0] in the upper window of Fig. 5 is the 10-bit output of the ADC, which is displayed with hexadecimal form in digital view, and ADOUT[9:0] in the lower window is the decimal step analog output wave which is converted from the hexadecimal digital signal by CosmosScope. These two ADOUT[9:0]s correspond with each other. The full scale of analog ADOUT[9:0] output is 3.3V, which corresponds to digital code of 0 to 1023. It is obvious that the conversion regulation happens to agree with the successive approximation process of an SAR ADC. During the sampling time of the ADC, v（IN2）is 2.2V and the obtained conversion result is 10'h2AA, which basically matches the theoretical result.

5. Conclusion

Existing mixed-signal simulation platform of some important EDA companies is discussed in this paper. The NanoSim-VCS method for mixed-signal simulation has made a successful verification of the entire ADC-integrated MCU chip. The method compromises between simulation speed and precision, which not only largely reduces simulation time but also guarantees certain precision. In this way, difficulty for verification is reduced and precision and efficiency of simulation is enhanced. The ADC-integrated system-level chip MV06 verified with the proposed method has achieved a successful tape-out at the first time, indicating the validity and efficiency of this method.

Acknowledgments

The authors would like to thank National Natrural Secience Foundation of China (Foundation Number: 60773081，60777018) and Postgraduate Innovation Fund of Shanghai University for financial support.

References

1. PRAKASH R, PETER P, LEENA S. System-on-a-chip verification methodology and techniques [M]. Kluwer Academic Publishers, 2002: 129-151.
2. Discovery AMS: A Comprehensive Mixed-Signal Verification Solution [Z]. Synopsys, 2003.
3. Raul Salvi. A Verification Methodology for Large Mixed-Signal SoC Designs Using NanoSim-VCS [Z]. Synopsys, 2003.
4. Synopsys. Discovery AMS: NanoSim-VCS User Guide [Z], Version Z-2007.03.
5. Synopsys. Circuit Simulation and Analysis Tools Reference Guide [Z], Version Z-2007.03:195-278.

Effects of Phosphor's Thickness and Concentration on Performance of White LEDs

Zongyuan Liu[12], Sheng Liu[123]*, Kai Wang[23], Xiaobing Luo[24]

[1]Institute for Microsystems, School of Mechanical Science and Engineering, Huazhong University of Science & Technology, Wuhan, China, 430074

[2]Division of MOEMS, Wuhan National Laboratory for Optoelectronics, Wuhan, China, 430074

[3]School of Optoelectronics Science and Engineering, Huazhong University of Science & Technology, Wuhan, China, 430074

[4]School of Energy and Power Engineering, Huazhong University of Science & Technology, Wuhan, China, 430074

Email: victor_liu63@126.com, Telephone: 86-27-87542604

Abstract

The state of phosphor will greatly influence the packaging performance such as luminous efficiency, quality of white light and color uniformity. The analysis presents that the small variations of thickness and concentration could significantly influence the light extraction efficiency and the correlated color temperature (CCT). With the increase of thickness and concentration, the light extraction efficiency is reduced, and the yellow blue ratio is increased, which means the color will tend to be warm white or yellow light. The reflector has slight influence on the performance, and the remote phosphor location is a better choice in packaging. When the thickness and concentration are determined, the manufacture tolerance for the variation are in the range of ± 0.02 mm for thickness and ± 5 mm^{-1} for concentration.

Introduction

Light emitting diodes (LEDs) has made remarkable progress in the past four decades with the rapid development of compound semiconductor technology. Since the first red LEDs that was invented by Holonyak and Bevacqua in 1962[1], great efforts have been put into the study to obtain brighter LEDs. In early 1990s, Nakamura successfully grew the blue and green LEDs on GaN substrate, which has a profound impact on the progress of LED technology[2, 3]. Nowadays there are three basic methods to obtain white light LEDs, one is the monochromatic RGB LEDs, another is phosphor converted LEDs (PC LEDs), the third is based on UV LEDs. The most commercially available white LEDs is dichromatic PC LEDs[4], which generates the white light by mixing the blue light from LED chip with the broadband yellow light excited by phosphor[5, 6]. Since the white light LEDs has superior characteristics such as high efficiency, small size, long life, dependable, low power consumption, high reliability and mechanically rugged[7], therefore, a new concept of illumination named solid state lighting (SSL) in terms of high power LEDs is proposed[8]. With the accelerated advancement of LED technology, the input electrical power of LEDs was jumped to 1W or more by increasing the chip size from 350 microns to 1mm. The luminous efficiency of high power LEDs was also increased from 25 lm/W in 1999 to more than 100 lm/W in 2007 for a driving current of 350mA[9]. Consequently, the market for high power LEDs is growing rapidly in various applications such as large size flat panel backlighting, street lighting, vehicle forward lamp, museum illumination and residential illumination[10, 11]. It has been widely accepted that SSL will be the fourth gerneration illumination source to substitute incandescent lamp, fluorescent lamp and HID lamp[12]. The broad application prospect of SSL has attracted great attention on the study of LEDs[13].

The main functions of LED packaging are to protect the LED chip, to enhance the light extraction and to provide a path for dissipating the generated heat. Packaging is the critical bridge between LED chips and applications for end-users. Now the LED packaging technology has been improved much to make full of the potentials of rapidly developing high power LED chips[14, 15]. Products of corporations such as the Lumileds's Luxeon K2 series, the Cree's XLamp series, the Osram's Golden Dragon series, offer high performance LED modules in terms of various packaging structures, in which the phosphor is dispersed on the surface of chip. To improve the packaging performance such as the light extraction efficiency, researchers have made remarkable efforts and achieved many valuable results[16, 17, 18, 19], for example, the scattered photon extraction method (SPE) [20, 21], the silicone based packaging platform [22], wafer level packaging [23], et al..

Phosphor plays an important role in the LED packaging. There are energy conversion and re-emission, scattering and absorption in the phosphor layer. Different optical structures with various states of phosphor in terms of location, thickness, concentration, shape, etc. will greatly influence the performance such as luminous efficiency, quality of the white light and color uniformity. Past studies simplified the parameters of phosphor with scattering and absorption coefficient by Monte Carlo ray tracing, and achieved some useful conclusions[24, 25, 26, 27]. To verify the correction of the coefficients, Zhu et al. firstly achieved the scattering and absorption parameters of phosphor by two-integrating-spheres test platform[28]. But these works are not enough to understand the influence of phosphor on LED packaging.

To further investigate the focused issues in optical design, this paper mainly concerns the effects of phosphor's thickness and concentration on the performance of white LEDs. Two cases are considered in this paper. The first case changes the thickness, and the second case changes the concentration. Each case considers five different optical structures to prove that the results are reliable. A non-sequential ray trace method was applied, which was based on Monte-Carlo theory. The power ratio of yellow light to blue light is used to illustrate the correlated color temperature (CCT). The parameters of these models were validated and the total ray trace number was finally 800000 to obtain converged results.

978-1-4244-2739-0/08/$25.00 ©2008 IEEE

Problem Statement

In the LEDs packaging process, dispersion of the phosphor silicones is a key step. It is a general method through altering the thickness and concentration of phosphor silicone to improve the quality of white LEDs. However, constrained by the material properties and packaging technique, it is difficult to fabricate the phosphor layer with desirable thickness and uniform concentration distribution.

Normally increasing the thickness or concentration of phosphor layer will enhance the effects of scattering and absorption. This should decrease the blue light output and increase the yellow light output. Therefore, CCT and luminous efficiency may be varied. However, considering the cost and technology of the production, it is not a suitable choice that the phosphor layer is too thick but with low concentration or too thin but with high concentration. Differences of the influencing mechanism for thickness and concentration are that the thickness changes the total distance for all scattering, while the concentration changes the average distance for one scattering.

As a basic step to realize the differences, it is necessary to investigate the effects of thickness and concentration on the performance of LEDs theoretically. Two cases are discussed in this paper. One case is to keep the concentration to be the same, but change the thickness. The other one is opposite, assuming the thickness unchanged but increasing the concentration. Five optical structures are discussed including the hemispherical phosphor layer and remote location of phosphor layer. The objectives of this paper are: 1) to analyze the effects of phosphor's thickness and concentration on the light extraction efficiency of LEDs; 2) to analyze the effects of phosphor's thickness and concentration on the variations of CCT; and 3) to provide some suggestions on the improvement of LED packaging.

Numerical Model and Simulation

The five optical structures are schematically shown in Figure 1. In the first optical structure, the phosphor is directly coated on chip. In the second and fourth models, the chip is firstly packaged with a thin silicone film, and then coated with the phosphor layer. The thickness of film is 0.05 mm. The difference of two types is that the fourth model has a reflector. The phosphor layer is a hemispherical membrane with the radius of 2 mm in the third model. The fifth optical structure firstly disperses silicone in the reflector with height of 1.9 mm, and then coats phosphor silicone uniformly on the cured silicone. In order to minimize the effects of lens' size on the light propagation, the radius of lens in all types is large enough to be 4 mm. The base and top radius of reflector is 1.5 mm and 4 mm, and the height is 2 mm. Optical parameters of the surface on the board and reflector are 85% perfect reflection, 5% scattering and 10%. For the first case, the thickness of phosphor layer is increased from 0.04mm to 0.2 mm. In the second case, the thickness is kept to be 0.1 mm thick.

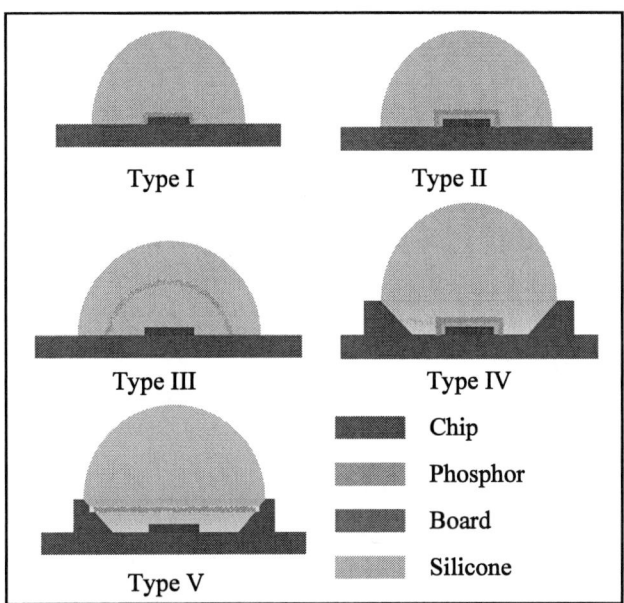

Figure 1. Optical structures for the numerical simulation.

The LED chip model is depicted in Figure 2, which is a simplified Cree LED chip with a size of 1mm × 1mm.. The top surface of the Silicon substrate is coated with Au to reflect the back emitted light. We considered that the light emits uniformly from MQW layer, since the direction of photons that are excited by the combination of electrons and holes in MQW layer is arbitrary. Considering that the area of top and down surfaces is much larger than that of the side surfaces, the model defines the two plane surfaces as the light sources by ignoring the side emitting lights. The optical properties of the chip are illustrated in Table 1[29].

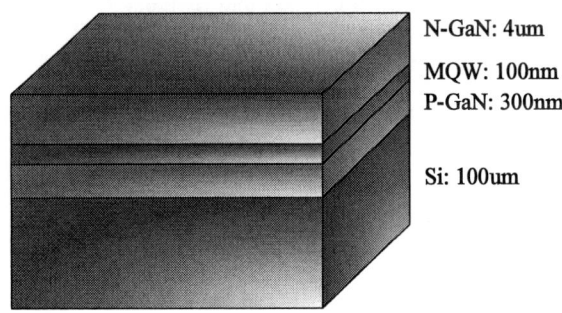

N-GaN: 4um

MQW: 100nm
P-GaN: 300nm

Si: 100um

Figure 2. LED chip model for the numerical simulation.

Table 2. Optical properties of LED chip.

	N-GaN	MQW	P-GaN
Refractive Index	2.42	2.54	2.45
Absorption Coefficient (mm^{-1})	8	8	8

Before defining the optical properties of phosphor, we should discuss the differences of the light propagation in the phosphor layer for two cases. The average radius of phosphor particle is 5~8 microns in Figure 3, and generally the number of particles is 10000~100000 per mm^3 in the mixture of the phosphor silicone. Since there may be millions of phosphor particles in the phosphor layer, the light will encounter many particles in the propagation path and thus be scattered many times before it transmits through the phosphor layer. This paper applies the mean free length to illustrate the scattering characteristic of phosphor. The mean free length is the average distance that a ray of light travels through the material without striking a phosphor particle. It can be calculated with Equation 1.

$$l = \frac{1}{n\sigma} \tag{1}$$

where n is the density of phosphor particles, and σ is the cross-section area of a particle.

It is obvious that the concentration could change the mean free length and thus affect the total scattering time. However, the thickness enhances the scattering characteristics by increasing the sum of the mean free length.

Figure 3. SEM picture of the phosphor particles.

Considering the phosphor layer as a bulk scattering material, it is an easy and effective method that applies the scattering and absorption coefficient to represent the total effects of phosphor particles on the light propagation. Therefore, when a beam of rays passes through the phosphor layer, the transmitted light energy can be expressed by Equation 2.

$$I(x) = I_0 e^{-(\mu_\alpha + \mu_s)x} \tag{2}$$

where I_0 is the initial power of light and $I(x)$ is the residual power; μ_α and μ_s are the absorption coefficient and scattering coefficient of phosphor; and x is the thickness of phosphor layer. The variation of concentration will also influence μ_α and μ_s. Therefore, this paper utilizes the sum of scattering and absorption coefficient to represent the concentration in the following discussions.

The scattering distribution function of phosphor layer, which is used to generate the random direction of scattered light, is based on the Henyey-Greenstein model. It is indicated in Equation 3.

$$p(\theta) = \frac{1-g^2}{4\pi(1+g^2-2g\cos\theta)^{3/2}} \tag{3}$$

where g is called the anisotropy factor. In the phosphor layer, since the scattering of light by the particles is isotropy, g is zero.

In the first case, it assumes that the absorption and scattering coefficient of phosphor is 8 mm^{-1} and 11.85 mm^{-1} for blue light, and 0 mm^{-1} and 16.25 mm^{-1} for yellow light. In the second case, the variations of the concentration are listed in Table 2. The refractive index of the phosphor layer is defined as 1.7, which doesn't consider the influences of wavelength.

Table 2. Optical properties of phosphor for the third case.

	Absorption Coefficient (mm^{-1})		Scattering Coefficient (mm^{-1})	
	Blue Light	Yellow Light	Blue Light	Yellow Light
1	20	0	29.5	40.45
2	13.3	0	19.8	26.95
3	10	0	14.8	20.33
4	8	0	11.85	16.25
5	6.65	0	9.88	13.48
6	5.72	0	8.465	11.62
7	5	0	7.4	10.173
8	4.44	0	6.57	8.99
9	4	0	5.92	8.09

Since the transmittance of silicone materials is over 95% for visible light, it is generally considered that the silicone materials are transparent to visible light. Therefore, the absorption coefficient is 0 mm^{-1} in the numerical simulation. The refractive index of silicone is 1.5, which can be purchased from corporations such as DowCornig.

Based on the above settings, the simulation steps of the numerical model are:

1) Defining the material properties of each layer;
2) Defining the top and down surface of the MQW layer as the light source, and then inputting the light power;
3) Ray tracing and collecting the simulation data;
4) Re-defining the material properties of each layer;
5) Re-defining the top and down surface of the phosphor layer as the light source, the power of which is calculated from the absorbed blue light; and the conversion efficiency of the phosphor is 80%;
6) Ray tracing and collecting the simulation data; and
7) Calculating the simulation results of each case.

For simplicity, the energy of the light is calculated by optical power in terms of watts, since the lumens (lm) of luminous needs to be converted by the visual sensitivity function from the optical power. In this simulation procedure, the blue light output and yellow light output are computed separately to easily investigate the impact of phosphor's location on the color temperature. The yellow blue ratio, which is the value of yellow light power divided by blue light power, is used to represent the color temperature.

In the simulation results, total efficiency is the ratio of total output power to initial optical power, which is emitted from the chip. The packaging efficiency is the ratio of total output power to the power that is calculated by subtracting the absorbed power by the chip from the initial power. It is illustrated in Equations 4, 5 and 6.

$$\eta_{Total} = \frac{P_{Out}}{P_{Total}} \qquad (4)$$

$$\eta_{Packaging} = \frac{P_{Out}}{P_{Total} - P_{Chip}} \qquad (5)$$

$$P_{Total} = P_{Out} + P_{Chip} + P_{Phosphor} + P_{Interface} + P_{Lost} \qquad (6)$$

where P_{Chip} is the absorbed power by chip; $P_{Phosphor}$ is the conversion loss by phosphor; $P_{Interface}$ is the sum of reflection loss on the surface of board and Fresnel loss in the interface; and P_{Lost} is the exhausted power in the light propagation path.

Simulation Results and Discussions

The simulation results of the five numerical models are shown in the following figures.

From Figures 4, 5 and 6, it is clear that the thickness has great impact on the performance of LEDs. With the increase of the thickness, the total efficiency and packaging efficiency of the five structures are all reduced, while the yellow blue ratio are increased.

When the phosphor layer is thicker, more blue light must be absorbed and thus enhance the yellow light output. Therefore, the curve of yellow blue ratio is ascended gradually. However, it does not mean that thicker layer is better. Because there is 20% conversion loss by the phosphor, more energy will be lost for thicker layer and thus reduce the light extraction efficiencies. This is confirmed by the simulation results in Figures 4 and 5.

From the curves of the performance, it can be found that the tendencies of type II and IV are almost the same, which indicates that the reflector has small impact on the light extraction. This may be due to the fact that the dimensions of reflector could not affect the light propagation effectively. The cone angle of the reflector is near 52°, only small light rays with high intensity could be reflected.

Another important finding is that the variation of the thickness has smaller influence on the light extraction efficiencies for cases with remote phosphor location than other cases with proximate phosphor location. However, except the third structure, the variations of CCT in the structures with remote phosphor location are all higher than that in the first structure. This indicates that the curved surface of phosphor presents smaller impact on the light propagation.

The purpose of adjusting the thickness is to obtain high luminous efficacy and high quality white light LEDs. Simulation results indicate that the thickness should be carefully controlled in the range of ±0.02 mm. Otherwise,

there will be color shift and great variation of the luminous efficacy.

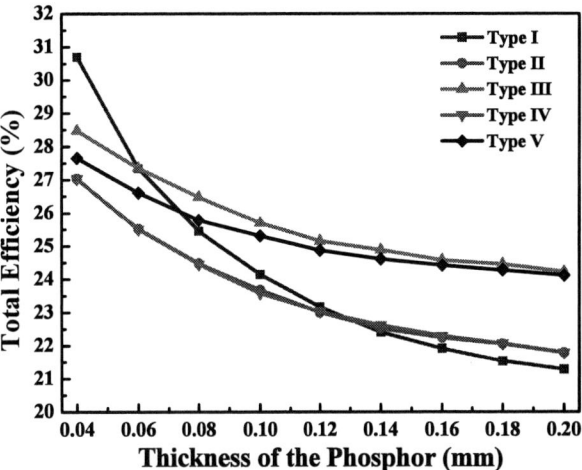

Figure 4. Effects of phosphor's thickness on the total efficiency.

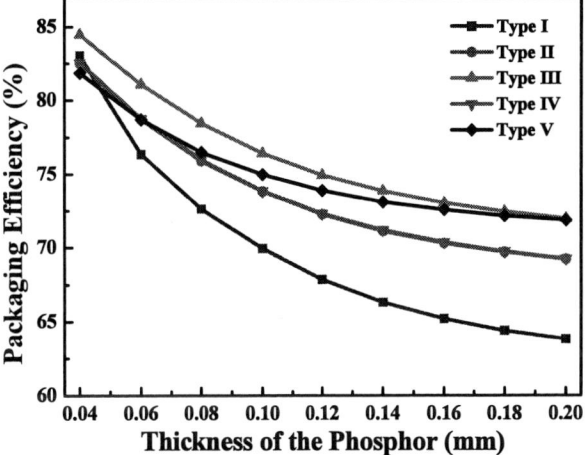

Figure 5. Effects of phosphor's thickness on the packaging efficiency.

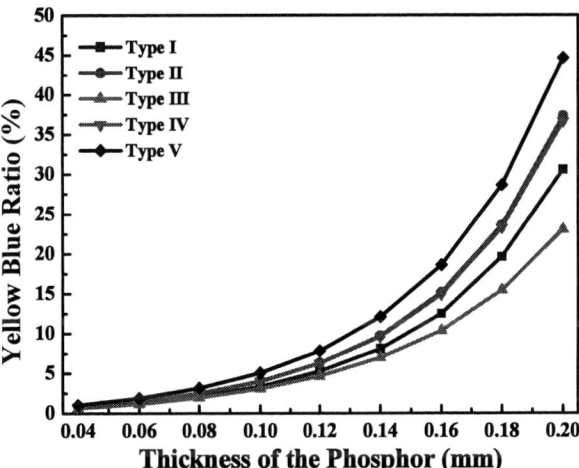

Figure 6. Effects of phosphor's thickness on CCT.

The concentration of the phosphor also has great impact on the performance of LEDs as shown in Figures 7, 8 and 9. With the increase of the concentration, the light extraction efficiencies are reduced more significantly than those of thickness. The variation of the CCT is also remarkable, in which the maximum value of the yellow blue ratio even reaches 150.

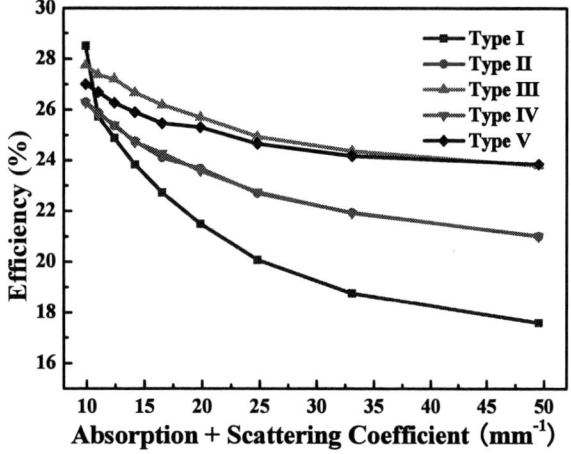

Figure 7. Effects of phosphor's concentration on the total efficiency.

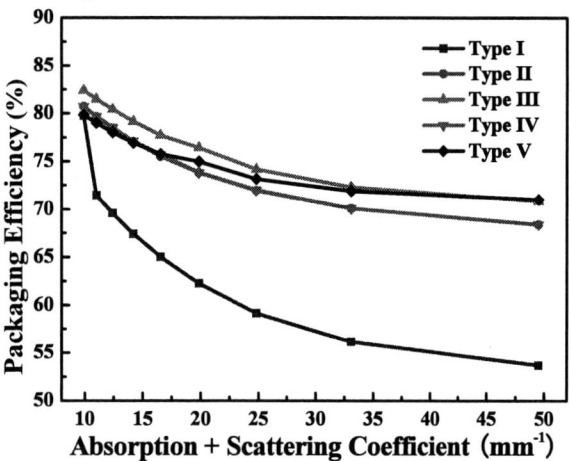

Figure 8. Effects of phosphor's concentration on the packaging efficiency.

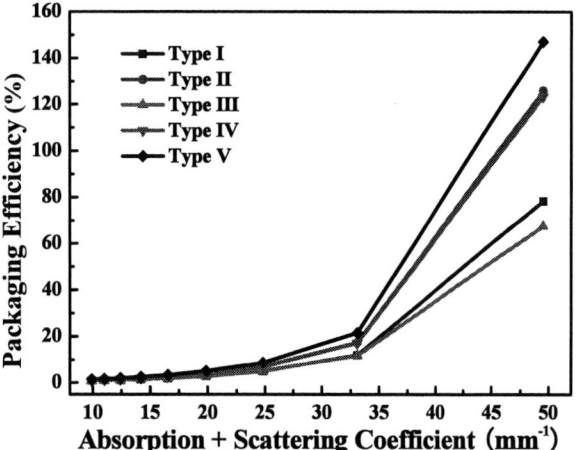

Figure 9. Effects of phosphor's concentration on the yellow blue ratio.

It can be found that tendencies of all structures are almost the same as the effects of thickness. The reflector also influences the performance slightly. The variations of light

extraction efficiency for structures with remote phosphor location are lower than those with proximate phosphor location. This confirms that the variation of thickness and concentration could have the same effects on the characteristics of phosphor. The difference is that concentration could influence the performance more effectively. From the simulation results, concentration should be controlled in the range of ±5 mm^{-1} to generate the desired high performance LEDs.

Since small variation of the thickness and concentration can influence the performance significantly, the thickness and concentration should be precisely controlled during the production process. Generally we expect that there is more light extracted from a LED module, however, the evaluation criteria of LED performance are luminous efficacy, color rendering index (CRI) and CCT. Thin or low concentration film obviously shows high light extraction efficiency, but because most of the extracted light is blue light, this will greatly reduce the luminous efficacy and induce the color of the light tending to be blue color. Thick or high concentration layer has relatively better white light, but the light extraction efficiency could not satisfy the requirement of high luminous efficacy. Therefore, it is important to find the balance point between the thickness and concentration to obtain high performance white light LEDs.

Conclusions

The effects of phosphor's thickness and concentration on the performance of LEDs are investigated. Simulation results show that small variations of thickness and concentration could significantly influence the light extraction efficiency and CCT. With the increase of thickness and concentration, the light extraction efficiency is reduced, and the yellow blue ratio is increased, which means the color will tend to be warm white or yellow. The reflector has slight influence on the performance, and the remote phosphor location is a better choice in packaging. When the thickness and concentration are determined, the manufacture tolerance for the variation are in the range of ±0.02 mm for thickness and ±5 mm^{-1} for concentration.

Acknowledgement

The support from Guangdong Real Faith Optoelectronic Inc. is appreciated.

References

1. Holonyak, J. N. and Bevacqua, S. F., "Coherent (Visible) Light Emission from Ga(As$_{1-x}$P$_x$) Junctions", *Applied Physics Letters*, Vol. 1, No. 4 (1962), pp. 82-83.
2. Nakamura, S., Mukai, T. and Senoh, M., "High-Power GaN PN Junction Blue-Light-Emitting Diodes", *Japanese Journal of Applied Physics*, Vol. 30, No. 12A (1991), pp. L1998-L2001.
3. Nakamura, S., Senoh, M. and Mukai, T., "High-Power InGaN/GaN Double-Heterostructure Violet Light Emitting Diodes", *Applied Physics Letters*, Vol. 62, No. 19 (1993), pp. 2390-2392.
4. Yam, F. K. and Hassan, Z., "Innovative Advances in LED Technology", *Microelectronics Journal*, Vol. 36, No. 2 (2005), pp. 129-137.

5. Schlotter, P., Schmidt, R. and Schneider, J., "Luminescence Conversion of Blue Light Emitting Diodes", *Applied Physics A: Materials Science & Processing*, Vol. 64, No. 4 (1997), pp. 417-418.

6. Nakamura, S. and Fasol, G., The Blue Laser Diode.: GaN Based Light Emitters and Lasers, Springer-Verlag Berlin and Heidelberg GmbH & Co. K, 1997.

7. Evans, D. L., "High-Luminance LEDs Replace Incandescent Lamps in New Applications", *Light-Emitting Diodes: Research, Manufacturing, and Applications*, SPIE, 1997, pp. 142-153.

8. Zukauskas, A., Shur, M. S. and Gaska, R., Introduction to Solid-State Lighting, John Wiley & Sons New York, 2002.

9. Krames, M. R., Krames, M. R., Shchekin, O. B. et al., "Status and Future of High-Power Light-Emitting Diodes for Solid-State Lighting", *Display Technology, Journal of*, Vol. 3, No. 2 (2007), pp. 160-175.

10. Steranka, F.M., Bhat, J., Collins, D. et al., "High Power LEDs - Technology Status and Market Applications", *Physica Status Solidi (a)*, Vol. 194, No. 2 (2002), pp. 380-388.

11. Craford, M. G., "LEDs for Solid State Lighting and other Emerging Applications: Status, Trends, and Challenges", *Fifth International Conference on Solid State Lighting*, SPIE, 2005, pp. 594101-594110.

12. OIDA, "Light Emitting Diodes (LEDs) for General Illumination, An OIDA Technology Roadmap Update 2002," 2002.

13. Pelka, D. G. and Patel, K., "An Overview of LED Applications for General Illumination", *Design of Efficient Illumination Systems*, SPIE, 2003, pp. 15-26.

14. Ya-Ju, L., Ya-Ju, L., Tien-Chang, L. et al., "High Brightness GaN-Based Light-Emitting Diodes", *Journal of Display Technology*, Vol. 3, No. 2 (2007), pp. 118-125.

15. Haque, S., Steigerwald, D., Rudaz, S. et al., "Packaging Challenges of High-Power LEDs for Solid State Lighting", *www.lumileds.com/pdfs/techpaperspres/manuscript_IMAPS_2003.PDF*, (2000), pp.

16. Borbely, A. and Johnson, S. G., "Prediction of Light Extraction Efficiency of LEDs by Ray Trace Simulation", *Third International Conference on Solid State Lighting*, SPIE, 2004, pp. 301-308.

17. Lee, S. J., "Light-Emitting Diode Lamp Design by Monte Carlo Photon Simulation", *Light-Emitting Diodes: Research, Manufacturing, and Applications V*, SPIE, 2001, pp. 99-108.

18. Sun, C. C., Lee, T. X., Ma, S. H. et al., "Optical Modeling for LED in Mid-Field Region", *International Optical Design Conference 2006*, SPIE, 2007, pp. 634217-634217.

19. Chi, W. and George, N., "Light-Emitting Diode Illumination Design with A Condensing Sphere", *J. Opt. Soc. Am. A*, Vol. 23, No. 9 (2006), pp. 2295-2298.

20. Narendran, N., "Improved Performance White LED", *Fifth International Conference on Solid State Lighting*, SPIE, 2005, pp. 594108-594106.

21. Narendran, N. Gu, Y., Freyssinier-Nova, J. P. and Zhu, Y., "Extracting Phosphor-Scattered Photons to Improve White LED Efficiency", *Physica Status Solidi*, Vol. 202, No. 6 (2005), pp. R60-R62.

22. Tsou, C. F. and Huang, Y. S., "Silicon-Based Packaging Platform for Light-Emitting Diode", *Advanced Packaging, IEEE Transactions on [see also Components, Packaging and Manufacturing Technology, Part B: Advanced Packaging, IEEE Transactions on]*, Vol. 29, No. 3 (2006), pp. 607-614.

23. Braune, B., Petersen, K., Strauss, J. et al., "A New Wafer Level Coating Technique to Reduce the Color Distribution of LEDs", *Light-Emitting Diodes: Research, Manufacturing, and Applications XI*, SPIE, 2007, pp. 64860X-64811.

24. Borbely, A. and Johnson, S. G., "Performance of Phosphor-Coated LED Optics in Ray Trace Simulations", *Fourth International Conference on Solid State Lighting*, SPIE, 2004, pp. 266-273.

25. Falicoff, W., Chaves, J. and Parkyn, B., "PC-LED Luminance Enhancement due to Phosphor Scattering", *Nonimaging Optics and Efficient Illumination Systems II*, SPIE, 2005, pp. 59420N-59415.

26. Kim, J. K., Luo, H., Schubert, E. F. et al., "Strongly Enhanced Phosphor Efficiency in GaInN Shite Light-Emitting Diodes using Remote Phosphor Configuration and Diffuse Reflector Cup", *Japanese Journal of Applied Physics*, Vol. 44 (2005), pp.

27. Luo, H., Kim, J. K., Schubert, E. F. et al., "Analysis of High-Power Packages for Phosphor-Based White-Light-Emitting Diodes", *Applied Physics Letters*, Vol. 86, No. 24 (2005), pp. 243505-243503.

28. Zhu, Y., Narendran, N. and Gu, Y., "Investigation of the Optical Properties of YAG:Ce Phosphor", *Sixth International Conference on Solid State Lighting*, SPIE, 2006, pp. 63370S-63378.

29. Lee, T.-X., Gao, K.-F., Chien, W.-T. and Sun, C.-C., "Light Extraction Analysis of GaN-Based Light-Emitting Diodes with Surface Texture and/or Patterned Substrate", *Opt. Express*, Vol. 15, No. 11 (2007), pp. 6670-6676.

Digital Dimmable Controller in CCFL Module based on Variable Frequency Technique

RAN Feng[1][2], LI Tiezhu[1], XU Meihua[1][2], WU Jian[1]

1. Key Laboratory of Advanced Displays and system Application, Ministry of Education, Shanghai University
2. Microelectronic Research and Development Center, Shanghai University
Campus P.O.B. 110, 149 Yanchang Rd., Shanghai 200072, China
ranfeng@mail.shu.edu.cn

Abstract:

A novel simple control method to improve the ignition behavior of cold cathode fluorescent lamp (CCFL) in digital-dimming mode is proposed in this paper. To extend the lamp life, we designed a digital-dimming controller (DDC) in cold cathode fluorescent lamp based on variable frequency technique. The design was simulated by Cadence Spectre. It is effective to eliminate the ignition current spike and reduce the high ignition voltage.

Key words:

Backlight, CCFL, DDC, frequency-shift

0. Introduction

Flat and thin-shape are both the current trend in display industry. Backlight module is a crucial component for driving light source in flat display panel technologies, and its performance will determine the display quality. The LCD with CCFL satisfies the demand on display performance, size, and efficiency. In consideration of the cost, efficiency, and uniformity ratio of illumination, the CCFL is still the best choice [1]. The inverter that drives the lamp is powered by 6~11 Vdc from computer's batteries or an adaptor. The CCFL requires 1~2 kV to fire and inverter efficiency and size are extremely critical. CCFL typically operates with sinusoidal voltage of 700~800Vrms and current of 5~6 mArms. The operation frequency is normally recommended between 25~85kHz, and a sinusoidal lamp voltage waveform is preferable. Dimming control capability is a very desirable feature for applications. These formidable requirements demand a highly efficient conversion topology and max circuit integration.

Recently, three familiar methods, which are duty-ratio, frequency, and voltage controls, are used to dim the gas-discharge lamp [2-3]. The duty-ratio control is easy when used to regulate the lamp current, but the asymmetrical lamp current and poor lamp crest factor will result in discoloration of the lamp [4]. The frequency control is the most common technique in regulating the lamp current. However, the dimming range is restricted by the switching frequency range, and the soft-switching is not easy to achieve the entire switching frequency range. Although the voltage control has a good dimming performance, the circuit is too complicated to practically use for low-power applications. The digital dimming control method, which is capable of widening the dimming range, is the latest technique used to modulate the brightness of CCFL. The main detrimental effect resulting from this method is a substantially reduced lamp lifespan due to a restriking phenomenon caused by current spikes. According to some manufacturers' test reports, an abnormal operation current, such as one with current spikes, can substantially shorten the usage lifespan of CCFL.

To extend the lamp life, a novel simple control strategy to restrain the current spike is proposed in this paper. We design a digital-dimming controller based on variable frequency technique to reduce the high ignition voltage and eliminate the ignition current spike. Simulation is by using Cadence Spectre, and results are close to the theoretical prediction.

1. Conventional Dimming Controller

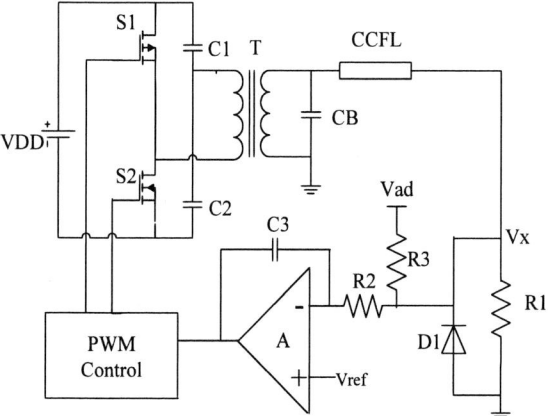

Fig 1 conventional analog dimmer

In the conventional analog dimming circuit, shown in Fig 1, a feedback loop consists of a commute circuit and an error amplifier A. Due to the amplifier's characteristic, when V_{ad} changes, both of the amplifier's input voltages are still equal. So the current through R1 and R2 changes, and the voltage of V_x and the current through CCFL change too. If V_{ad} is high, the current through CCFL is low, and accordingly the dimming effect is realized. The dimmer's drawbacks lie at its nonlinearity and its low power efficiency, particularly under low-level luminance conditions.

Low frequency digital dimming circuit is modified on the analog dimming circuit slightly. The R3 and V_{ad} are replaced by diode and pulse signal LFD respectively. When LFD is high voltage level, due to the input voltage over modulating V_x voltage, the lamp current drops, closer to zero; when LFD is low voltage level, due to the diode's insulating function, the lamp is in full illuminant. This approach can cause a substantial reduction in the lamp's lifespan because CCFL is repeatedly struck by spikes during the dimming operation. In addition, the boost transformer will generate annoying hum due to the high spikes resulted from the restriking behavior. To eliminate the aforementioned high ignition voltage and high lamp current spike, a novel control strategy and its realization

978-1-4244-2739-0/08/$25.00 ©2008 IEEE

circuit are proposed in this paper.

2. Digital-dimming Controller with Variable Frequency Oscillator

We design a digital dimming controller based on variable frequency technique. The soft-start is adopted to reduce the high ignition voltage and to eliminate the ignition current spike.

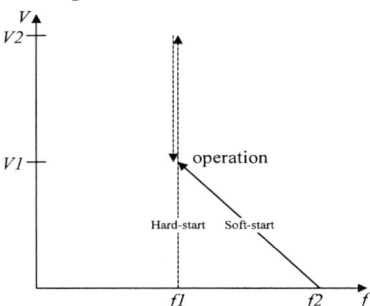

Fig 2 a comparison between hard-start and soft-start

Fig 2 compares the ignition characteristics of two different methods for CCFL, where V represents lamp voltage, and f represents lamp frequency. The hard-start (dashed) is a conventional igniting technique with constant frequency drives and during the whole start-up lamp's voltage rises from zero to $V2$, then drops back to $V1$. The start-up voltage is very high; However the soft-start (real line) is proposed method and during the whole start-up the lamp's voltage rises from zero to $V1$ with the change of frequency. The soft-start substantially reduces ignition voltage.

2.1 Digital Dimming Controller based on Variable Frequency Technique

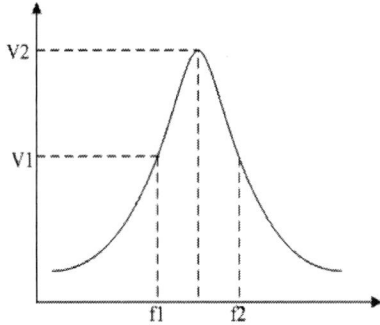

Fig 3 the relation of startup voltage and frequency

Besides, the relation of startup voltage and frequency in the adopted resonant inverter are illustrated in Fig 3. It is apparent that the selection of the operating frequency is determined by the frequency responses of the employed resonant circuit. In this paper, we can preset $f2$ as the original frequency. Another operating frequency $f1$ ($f1 < f2$) is adopted to drive CCFL as the operation frequency during the ON state of CCFL.

Fig 4 is the circuit of the proposed variable frequency digital dimming controller, in which slew-rate limiter, voltage-controlled current source, variable frequency oscillator, and MOSFET gate driver are included.

In this circuit, a slew-rate limiter receives the standard LFD signal and outputs a trapezoidal shape signal V_{LFD} to regulate the voltage-controlled current source I_a. The feedback signal V_t inspects the change of the lamp's current to control the frequency of variable

frequency oscillator. In addition we need to choose the best parameter of DDC to eliminate the current spike.

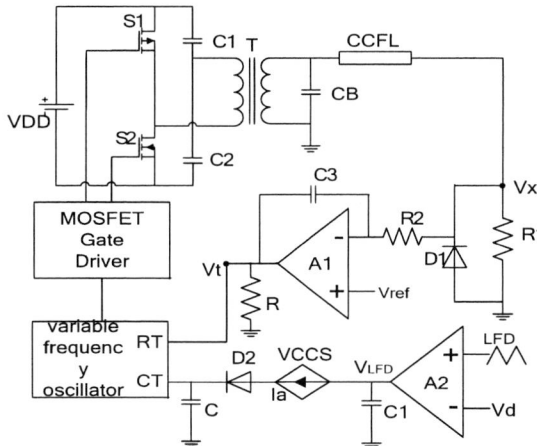

Fig 4 the equivalent circuit of DDC

2.2 Variable Frequency Oscillator

RC oscillator is quite popular in micro controller and other ASIC applications where the accuracy is not very important and the frequency is not very high. The common architecture of RC oscillator is shown in fig 5. Here, a current inversely proportional to resistor R generated by V-to-I and a reference voltage V_{ref}. The current is used to charge and discharge the capacitor C under the CK control. It is between two reference voltages, V_H and V_L ($V_H > V_L$).Two comparators, RS trigger, are used to generate the square wave oscillator output, CK. The output CT sawtooth wave will be achieved. However, there are limitations with the conventional RC oscillators as far as accuracy and frequency are concerned.

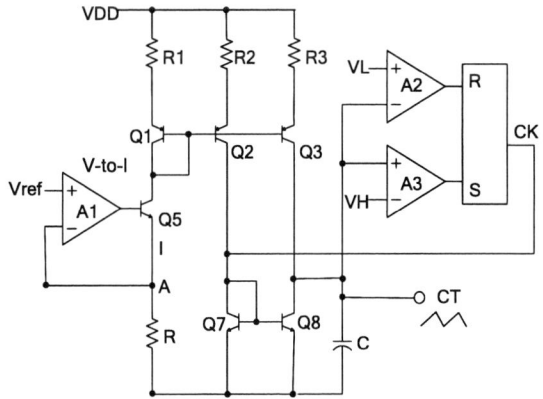

Fig 5 conventional RC oscillator

Fig 6 proposes a programable variable frequency oscillator via the improvement of conventional RC oscillator. The circuit adds 2 switches Q16 and Q17, which is controlled by RS trigger output /CK and CK respectively. The auxiliary current I_a is decided by R_t、V_t、V_A, and the part in the dashed frame is the programable one [5].

361

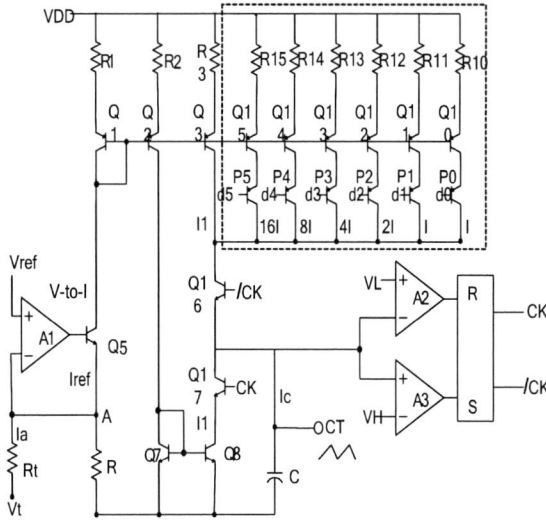

Fig 6 programmed RC oscillator

First, without regard to the change of exterior voltage V_t, we assume $V_t=V_A$, then the auxiliary current $I_a=(V_A-V_B)/R_t=0$. The current in the circuit $I_{ref}=V_{ref}/R$ is generated in the V-to-I section of the circuit, consists of R1, Q1, Q5, R_t, R and the differential amplifier A1. The current I flowing through Q3 and Q8, is proportional in Q1. The charge and discharge current of capacitor C is controlled by Q16 and Q17. When /CK signal retains high voltage level, Q16 is on and Q17 is off, then capacitor C is charged through Q3, the charge current is I_c, the charge time is $T_1=V_{ref}*C/I_c$; When CK signal retains high voltage level, Q17 is on and Q16 is off, then capacitor C is discharged through Q8, the charge current is I_c, the charge time is $T_2=V_{ref}*C/I_c$; So the cycle of the oscillator is $T=T_1+T_2=2*V_{ref}*C/I_c$.

Considering improving the clock's precision, we intercalate six switches P5-P0 by controlling their on-off, we can inching the charge current of capacitor C, and change the charge time, thereby complete mending the oscillator frequency. Every controlled spur track is adopted current mirrors design and mirrors the identical current, their magnitude rise progressively in double. When d5-d0 are 111111, P5-P0 are off, then the charge current of capacitor I_c is the min charge current I_1, the cycle of the oscillator is the min ; when d5-d0 are 000000, P5-P0 are on, then the charge current of capacitor I_c is the max charge current I_1+32I, the cycle of the oscillator is the max; When d5-d0 are 011111, the charge current of capacitor I_c is the mid charge current I_1+16I, the cycle of the oscillator is the mid . by setting the value of d5-d0, we can itching the charge current I_c, thereby the oscillator's precision can be adjusted in allowable error range.
The maximum frequency of oscillation is:

$$F_{max}=\frac{1}{T}=\frac{I_1+32I}{2*V_{REF}*C} \qquad (1)$$

The middle frequency of oscillation is:

$$F_{mid}=\frac{1}{T}=\frac{I_1+16I}{2*V_{REF}*C} \qquad (2)$$

The minimum frequency of oscillation is:

$$F_{min}=\frac{1}{T}=\frac{I_1}{2*V_{REF}*C} \qquad (3)$$

If we suppose that the frequency of oscillation is F_{mid}, so the frequency is adjusted between F_{max} and F_{min} the adjustable frequency range is

$$\Delta F=\frac{\pm16I}{2*V_{REF}*C} \qquad (4)$$

The minimum adjustable current is I, so the adjustable frequency precision is

$$\Delta F_{min}=\frac{\pm I}{2*V_{REF}*C} \qquad (5)$$

The frequency of oscillation is influenced by source and temperature. The bias is range 1% and 10%. So we order $\Delta F=\pm10\%$, the bias can be amended in range of $\Delta F_{min}=\Delta F/16=\pm0.625\%$ .In other words, the precision of the output CT wave may be assured between $\pm0.625\%$.

3. Simulated and Measured Results

From the over formula, the resistance R and R_t in the circuit are adjusted to control the spur track current. The C, R, R_t are assembled to choose the charge and discharge time of capacitance, changing V_t can change the output frequency. With 0.6um BiCMOS technics, we use Cadence Spectre to simulate it, and the oscillator's sinulation waveform reflects the CCFL in the corresponding four states.

State 1:

In this state, V_t retains low voltage level, and the additional current $I_a=V_A/R_t$ is max. The current $I_{ref}=V_{ref}/R+V_{ref}/R_t$. The current to capacitance C is max. The frequency of the output CT sawtooth wave is max. Then the operating point is the original one, closer to 90kHz, and CCFL is not turned on in fig 7.

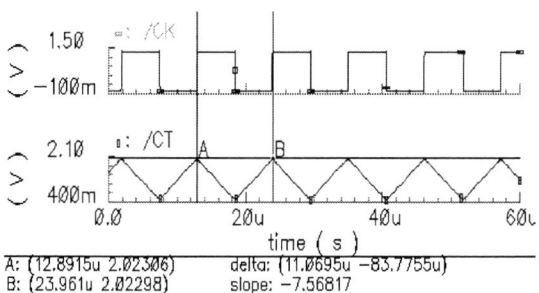

Fig 7 output waveform in V_t retains low voltage level

State 2:

As V_t signal goes from low to high, the auxiliary current $I_a=(V_A-V_t)/R_t$ drops progressively to zero. The current to capacitance C decreases from the max to the min. The frequency of the output CT sawtooth wave decreases progressively, ultimately reaches f=60 kHz, as shown in Fig 8. Then the operating point will move back gradually toward the original one, and CCFL will be turned on. Therefore, the current of the exterior capacitor C is modulated to generate the mentioned ignition behavior with continuously variable frequency.

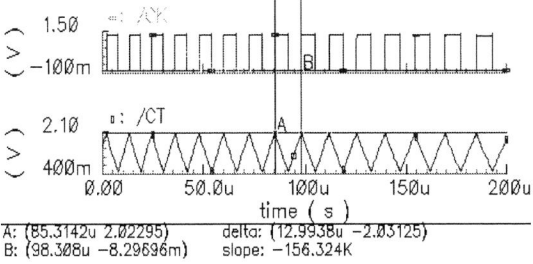

A: (85.3142u 2.02295)   delta: (12.9938u −2.03125)
B: (98.308u −8.29696m)   slope: −156.324K

Fig 8 output waveform as V_t goes from low to high

In other words, we employ continuously variable frequency technique to achieve soft-starting, then to eliminate the spikes of lamp current and lamp voltage completely. The change of V_t signal must be controlled properly. For instance, the dimming range is restricted when it is too large. However, if it is too small, soft starting will not be achieved due to the insufficient amount of frequency shift.

State 3:

In this state, V_t signal retains high voltage level, the auxiliary current $I_a=0$ and keeps constant minimum current to capacitance C. Hence, the operating frequency is still min. In other words, the frequency of the output CT sawtooth wave keeps constant, closer to 60kHz, as shown in Fig9. CCFL is in normal state, when the average brightness of CCFL can be controlled by the digital dimming approach.

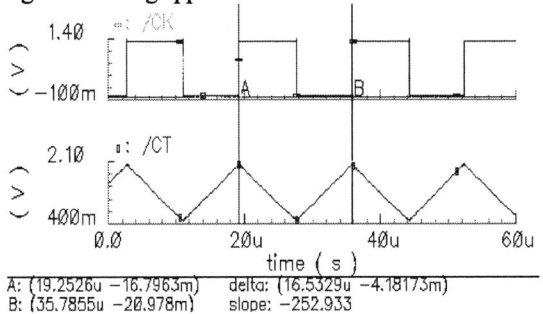

A: (19.2526u −16.7963m)   delta: (16.5329u −4.18173m)
B: (35.7855u −20.978m)   slope: −252.933

Fig 9 output waveform as V_t retains high voltage level

State 4:

When V_t signal goes from high to low, the auxiliary current $I_a=(V_A-V_t)/R_t$ rises progressively up to max. The frequency of the output CT sawtooth wave rises progressively, ultimately reaches f=90kHz, as shown in Fig 10. In other words, with the uninterrupted variable frequency, the current in the exterior capacitance C is adjusted and shifted away from the normal operating point so as to turn off CCFL.

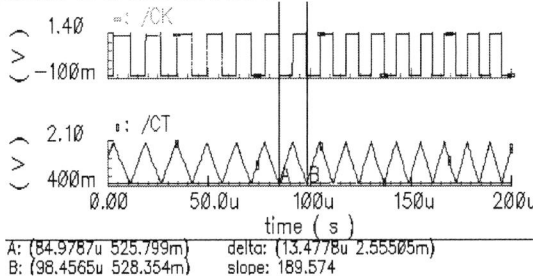

A: (84.9787u 525.799m)   delta: (13.4778u 2.55505m)
B: (98.4565u 528.354m)   slope: 189.574

Fig 10 output waveform as V_t goes from high to low

4. Conclusion

In this paper, variable-frequency oscillator structure is proposed and by simulation the frequency change in different controller states is achieved. In addition digital-dimming based on variable frequency technique is realized to reduce the high ignition voltage and eliminate the ignition current spike.

Acknowledgment

The authors would like to acknowledge the National Natural Science Foundation of China for providing financial support for this work under Grant No. 60773081 and Grant No. 60777018, and also to acknowledge the financial support by the Shanghai Municipal Committee of Science and Technology under Grant No. 06DZ22013.

References

[1] Y. L. Lin, and A. F. Witulski, "Analysis and design of current-fed push-pull resonant inverters cold cathode fluorescent lamp drivers," IEEE-IAS Conference Record, 1996, pp. 2149-2152

[2] R. Redl, and K. Arakawa, " A low-cost control IC for single-transistor zvs cold-cathode fluorescent lamp inverters and dc/dc converters, " IEEE APEC'97, vol. 2, 1997, pp. 1042-1049.

[3] G. H. Kweon, Y. C. Lim, and S. H. Yang, " An analysis of the backlight inverter by topology, " IEEE ISIE'01, vol. 1, 2001, pp. 896-900.

[4] M. S. Lin, W. J. Ho, F. Y. Shih, D. Y. Chen, and Y. P. Wu, " A cold-cathode fluorescent lamp driver circuit with synchronous primary-side dimming control," IEEE Trans. Ind. Electron., Vol. 45, No. 2, Apr ,1998, pp. 249-255.

[5] Bala F, Nandy T. Programmable High Frequency RC Oscillator[C].IEEE VLSI Design, 2005, pp 511-515.

Design of a Low Voltage Band-gap Reference Circuit for OLED-On-Silicon

XU Meihua[1][2][3], WU Jian[1], RAN Feng[1][2][3], LI Tiezhu[2]

1. Microelectronic Research and Development Center, Shanghai University
2. School of Mechatronical Engineering and Automation, Shanghai University
3. Key Laboratory of Advanced Displays and system Application, Ministry of Education, Shanghai University
Campus P.O.B. 25, 149 Yanchang Rd., Shanghai 200072, China
mhxu@staff.shu.edu.cn

Abstract

This paper presents a design of low voltage band-gap reference circuit for OLED-On-Silicon. In order to make the op-amp working in the high-gain area, the boost technique is used in the amplifier to increase the gate drive ability. The reference source uses first-order temperature compensation design to eliminate the temperature influence to voltage source. Simultaneously, the power dissipation is greatly decrease because the amplifier designed is working in the weak inversion layer. The simulation is conducted in chartered 0.35um 2-poly 4-metal 3.3V/18V high voltage process, and the results show that the proposed design meets the scheduled requirement and realizes the application of source voltage under 1.8V.

1. Introduction

OLED-On-Silicon technique is one branch of AM-OLED technique. The combination of OLED technique and CMOS technique prompts the development of a new generation micro displayer. OLED integrated into single crystal silicon, combining video sub systems, display-driven circuit, control circuit, PC interface and other functional components, constitute a micro display system with low total cost.

Accurate voltage reference plays nontrivial role in the OLED-On-Silicon chip and ensure the normal operation of chip internal blocks, for instance, voltage-current converter, contrast regulation DAC, oscillator and so on. The operation voltage of the OLED-On-Silicon driving chip proposed in the paper varies between 2.7V and 3.6V while operation temperature varies from minus 10 degree to 75 degree. Since power supply can be decreased with the development of CMOS process, the system can work with a power supply less than 1V. The design adapts pure CMOS process. As to the voltage reference insensitive to temperature and voltage of power supply, it considers not only the working environment of the chip, but the case of low working voltage also.

2. Principle of band-gap voltage reference

Current mirror configuration with high accuracy or voltage regulation circuit is needed in band-gap structure. A cascade current mirror or voltage pre-stabilization circuit can reduce errors. However, the voltage of power supply will inevitably be increased so that it is inapplicable to low voltage. As for OLED on Silicon chip, the requirement of operation temperature is not very rigorous and it can be satisfied through first-order compensation in the specific application. First-order temperature compensation is to offset the T's first-order term of V_{BE} with the negative temperature coefficient V_T.

The band-gap reference circuit with first-order temperature compensation is shown in Fig.1. There're two current sources, which are proportional to BJT voltage V_{BE} and $\triangle V_{BE}$. The amplifier serves as feedback loop and the

voltages of the two input node N1 and N2 of the amplifier keep equal all the time. The respective W/L and current I of the upper three PMOS are the same. The resistance R_{2a1} and R_{2a2} are equivalent to R_{2b1} and R_{2b2} respectively. Thus there's the following equation.

$$I = \frac{V_{EB2}}{R_2} + \frac{V_T \ln(N)}{R_1} \quad (1)$$

Where N represents the area ratio of the emission, V_T represents the thermal voltage, and R_2, $R_{2a1}+R_{2a2}$ and $R_{2b1}+R_{2b2}$ keep equal. Since the current flowing into R3 is also I, the reference voltage is eq.(2).

$$V_{REF} = \frac{R_3}{R_2}[V_{BE2} + (\frac{R_2}{R_1}\ln N)V_T] \quad (2)$$

The formula in the bracket of the above equation is the common voltage of band-gap voltage reference. The satisfying temperature characteristic can be obtained through adjusting the scale between R2 and R1. In addition, reference voltage with a specified range can be achieved through regulating the scale between R3 and R2.

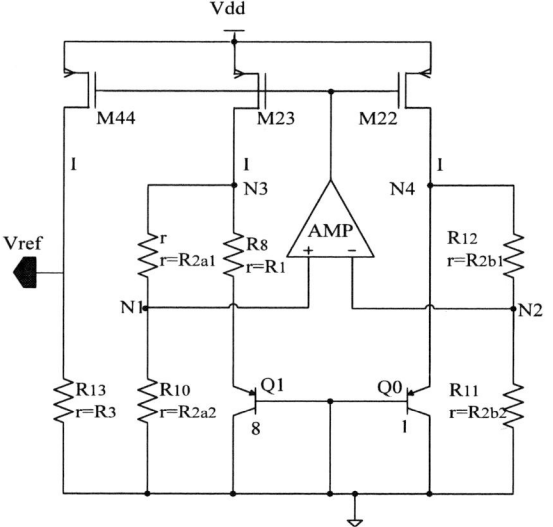

Fig.1 band-gap voltage source

3. Design of low voltage band-gap circuit

3.1 The structure of the band-gap circuit

When the band-gap circuit is designed with a low power supply, the common-mode input voltage range of the amplifier must be taken into account first. A pair of NMOS is used here as the input differential stage of the amplifier, and the lowest common-mode input voltage is $V_{thn}+2V_{DS(sat)}$, which must be less than $V_{EB(on)}$ in order to make sure the amplifier work properly. For example, assumed that $V_{EB(on)}$=0.7V and $V_{DS(sat)}$=50mV, then V_{thn} must be less than 0.6V. This can be satisfied by using most common technologies. However, the influence of temperature variation on the base-emitter and the threshold voltage of the BJT can not be neglected. The temperature coefficient of V_{BE} TC equals approximately to -2mV/K while the temperature coefficient of NMOS transistor threshold voltage TC is larger than -2mV/K. As the temperature becomes high, the value of $V_{EB(on)}$ drops down faster than the V_{thn} does. That's the reason why $V_{thn}+2V_{DS(sat)}$ will be larger than $V_{EB(on)}$ when the circuit is working at high temperature. In this condition, the amplifier actually does not work in the high-gain linear region so that the restraining function against the environment variation of the feedback will be weakened and the band-gap voltage will vary a lot. Therefore, in this paper the amplifier with a pair of NMOS as the input differential stage is not accepted in the design of low-voltage circumstances. However, the maximum value of the common-mode input voltage of the amplifier with a pair of PMOS as the differential input stage equals to V_{DD}-$2|V_{DS(sat)}|$ $+|V_{thp}|$, while the minimum one is $V_{DD}+2|V_{DS(sat)}|+|V_{thp}|$. Let $|V_{DS(sat)}|$=50mV,$|V_{thp}|$=0.2V, then the design of a band-gap with about 1V reference voltage output can be achieved.

With the CMOS technology of 3.3V and 5V supply voltage, the PMOS input differential stage of the amplifier will be effectively on so that the amplifier will work in the linear region and will have a high gain. Besides, the whole feedback loop circuit can obtain a very satisfied result. Considering the design of 1V output voltage, the structure of the band-gap reference circuit is redesigned.

When the supply voltage approaches to 1V, the maximum common-mode input voltage equals to V_{DD}-$2|V_{DS(sat)}|$+ $|V_{thp}|$. $V_{BE}{\approx}0.7V$ will probably exceed the maximum common mode input voltage, thus will make the amplifier work in the nonlinear region along with the decrease of the gain. The difference between the typical band-gap voltage reference and the one we designed in the Fig. 2 is the input-ports of the amplifier in the typical band-gap voltage reference are connected to node N3 and N4, where the voltage is equals to the base-emitter voltage of the BJT V_{BE}. However, in this paper, they are connected to N1 and N2. Because the sum of the voltage of R_{2b1} and R_{2b2} is equals to V_{EB2}, the voltage of the node N1 and N2 can be calculated by the following eq.(3).

$$V_{in+} = (\frac{R_{2b2}}{R_{2b1} + R_{2b2}}) V_{EB2} \qquad (3)$$

In this way, the minimum supply voltage can be expressed as eq.(4).

$$V_{DD(min)} = (\frac{R_{2b2}}{R_{2b1} + R_{2b2}}) V_{EB2} + |V_{thp}| + 2|V_{DS(sat)}| \qquad (4)$$

From the equation (4), it can be seen that the proportion of power supply voltage has been reduced. This is applicable to design process with all the CMOS technologies.

3.2 Design of amplifier

The design of amplifier is critical to the implementation of the band-gap reference. The power supply rejection ratio (PSRR) of the band-gap reference is determined by the feed-back effect of the amplifier loop.

It employs voltage-boosting technology to improve the amplifier performance. The voltage-boosting technology improves the driving ability of the gate through keeping the

V_{BS} of the PMOS positive, and makes the amplifier work in the high gain area. The positive-biased V_{BS} will decrease the threshold of the PMOS.

$$|V_{th}| = |V_{t0}| + \gamma(\sqrt{2|\Phi_f| - V_{sb}} - \sqrt{2|\Phi_f|}) \qquad (5)$$

Where, $|V_{t0}|$ is the value of $|V_{th}|$ when V_{sb} equals to 0. γ is the body effect parameter. $|\Phi_f|$ is the Fermi potential of the substrate. The positive bias increases the V_{sb} and reduces the $|V_{th}|$, hence V_{DD}-$|V_{thp}|$ increases. So, the amplifier works in the high gain area.

The diagram of amplifier is shown in Fig.2. It is designed as a micro power dissipation amplifier, and its bias current is provided by the core circuit of the band-gap-reference. The amplifier primarily works in the weak inversion region, and it needs small current and low voltage. The load-currents mirror each other so that no additional bias-circuits are needed, that dramatically cut-down the power dissipation. The amplifier has two poles: the dominant pole results from R_{out} and $C_{load,}$ while the second pole results from conductance of the load transistors of the input-stage and internal capacity. The ratio of the poles is mainly determined by the load capacitance and the ratio of the currents at two stages. By decreasing the C_{load} and the ratio of the current enough phase-margin can be attained. Normally, the ratio is approximate to 1. The differential gain A_{V01} of the first stage is obtained.

$$A_{V01} = \frac{g_{m2}}{g_{m4}} \qquad (6)$$

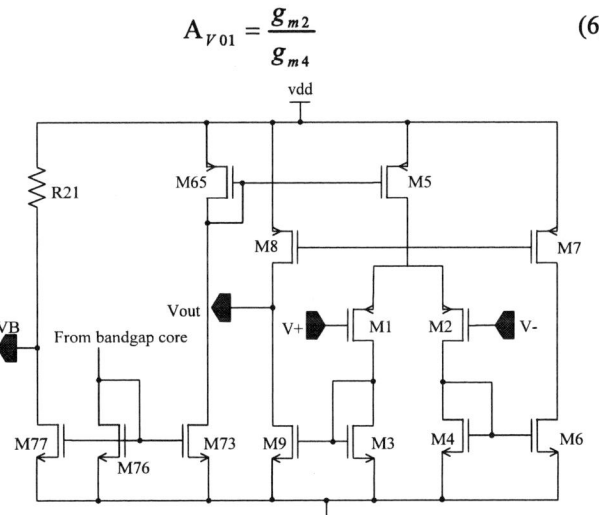

Fig.2 Micro power dissipation amplifier

The gain is very small and approximates to 1 at the first stage, but the second stage will provide enough gain. The total gain of the amplifier is given by eq.(7).

$$A_{V0} = \frac{g_{m1}(S_6 / S_4)}{(g_{ds6} + g_{ds6})} = \frac{(S_6 / S_4)}{(\lambda_6 + \lambda_7)n_1 V_t} \qquad (7)$$

At room temperature, the gain will reach 60dB while the typical device parameter adopted. The bandwidth is shown as eq.(8).

$$GB = \frac{g_{m1}}{C}(S_6 / S_4) \qquad (8)$$

The simulation result is shown in Fig.3. The gain is 72 dB. The phase margin is 25K and the power dissipation is 1.1uW.

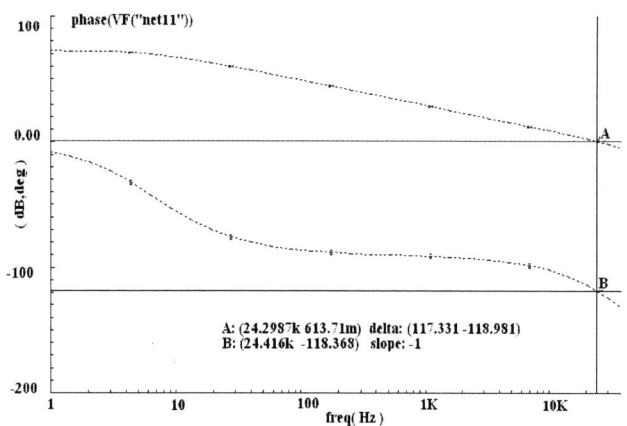

A: (24.2987k 613.71m) delta: (117.331 -118.981)
B: (24.416k -118.368) slope: -1

Fig.3 Amplifier gain and phase response

3.3 DC level shift current mirror

The employment of DC level shift current-mirror is try to make the amplifier work properly, especially in some technologies where $V_{thn}>|V_{thp}|$. The structure of the DC level shift current mirror is given in Fig.4(b). From Fig.4(a), we can get the drain-source voltage of transistor MA08 where $V_{in+}+|V_{GS8}|-V_{GS10}$ is approximately equals to V_{in+}. Assumed that $|V_{thp}|=V_{thn}$, then $|V_{GS8}|=V_{GS10}$. If $|V_{thp}|<V_{thn}$, the drain-source voltage of MA08 will be lower than V_{in+}. If V_{in+} is low, then the drain-source voltage of MA08 may be less than the saturation voltage. $V_{in+}=[R_{2b2}/(R_{2b1}+R_{2b2})]\cdot V_{EB2}$ is defined and MA08 can indeed work in the triode region when the Vdd is 1V. That will reduce the gain of the amplifier, as shown in Fig.4(b), the drain-source voltage of MA08 equals to $V_{in+}+|V_{GS8}|-V_{GS10}+V_{EB16}$ with the addition of the parasitic longitudinal BJT working as the DC level shift current source. In this way, even though the V_{in+} is 0V, MA08 will work in the saturation region due to the V_{thn} will be certainly 0.6V lower than $|V_{thp}|$.

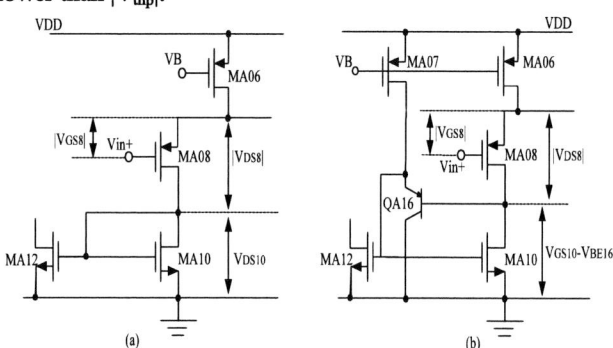

Fig.4 DC level shift current mirror

3.4 Start-up circuit

The start-up circuit of band-gap reference is comprised of M1，M2，M3，M4, shown in Fig.5. The M1 and M2 work as an inventor. The W/L ratio of M1 and M2 is less than 1 and W/L ratio of M1 equals to the W/L ratio of PMOS current mirror. It is for the purpose of removing the threshold voltages errors which may be made by geometry influence of layout that we do like this. When the circuit's working current is 0 ampere, the gate voltage of PMOS current mirror is the same as the gate voltage of M1, which will be pulled up near to power supply voltage. The drain voltage of M1，M2 will

be pulled down to turn on M3、M4, making the current stream into to the core circuit of band-gap and the amplifier. It will increase the output current of PMOS current mirror and lower the gate voltage as well. Meanwhile, the gate voltage of M1 decreases, while the drain voltage of M1 and M2 rises, and M3 and M4 will be shutted finally.

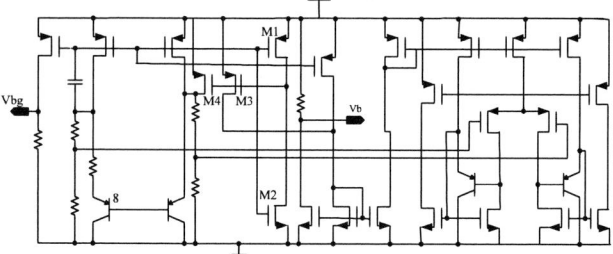

Fig.5 Band-gap voltage reference schematic circuit diagram

The pivot of the design is the W/L ratio of M2. The band-gap circuit doesn't work well if the M3 and M4 haven't been switched off completely when the whole circuit is turned on. Therefore, the dimension of W/L ratio to M2 will be selected at the break over voltage for maximum power voltage and maximum working temperature. Moreover, in order to avoid the influence of process error, it had better select the long channel device.

3.5 Offset and noise

The error of band-gap reference is mainly caused by the geometry mismatch of BJT, the threshold value mismatch between PMOS current source geometry and threshold and the offset of amplifier. The PMOS current source mismatch's influence is represented by N. The influence of amplifier's offset can be expressed as eq.(9).

$$V_{ref} = \frac{R_3}{R_2}[V_{EB2} + \frac{R_2}{R_1}(V_T \ln N + \frac{R_2}{R_{2a2}}V_{OS})] \qquad (9)$$

The amplifier mismatch is magnified to (R_2/R_{2a2}) times. **The proper way to reduce this effect is by increasing N and reducing R2/R1.** The offset voltage is made up of two parts. One is the system offset caused by the current mirror's channel-length modulation, and the other is the random offset caused by device's threshold value and geometry mismatch. The system offset can use the method of reducing transistor size and the proportion of bias current to cancellation. The random offset can be adjusted by the layout, device symmetry and compact location.

The influence of noise is similar to the offset and it is magnified to (R_2/R_{2a2}) times also. However, most of the noise is broadband thermal noise. The capacitance which is cascade between the import and output port works as a low-pass filter. Furthermore, the positive and negative feedback loop of band-gap reference requires frequency compensation to stabilize. Frequency compensation always adopt the method of capacitance compensation, which effects as Miller capacitance compensation. The higher of the capacitance, the more stable it can afford, but it may increase the start-up time.

4. Simulations and implementation

Simulations of band-gap voltage reference are conducted in chartered process which employs 0.35um 2-poly 4-metal

3.3V/18V high voltage process. The circuit layout is shown in Fig.6. Its area is 100um².

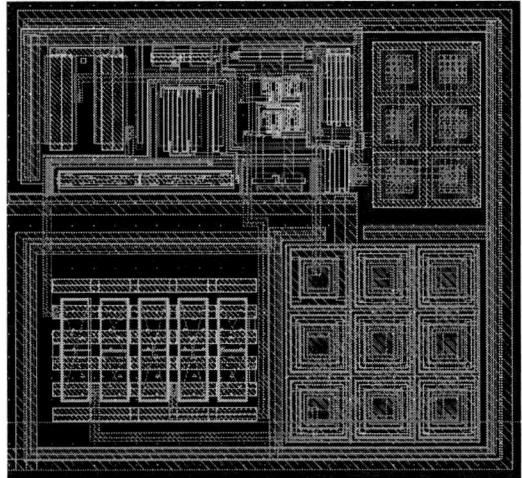

Fig.6 Band-gap voltage reference layout

Simulation is illustrated in Fig.7 when source voltage is 1V. Band-gap voltage reference is 389mV and the error of which will not exceed to 0.6mV when temperature varies from 0°C to 100°C. TC equals to 10ppm/K in room temperature (27°C).

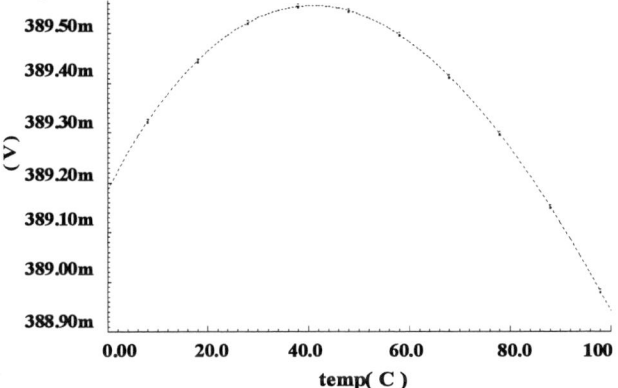

Fig.7 Voltage vs temperature curve when Vdd=1V

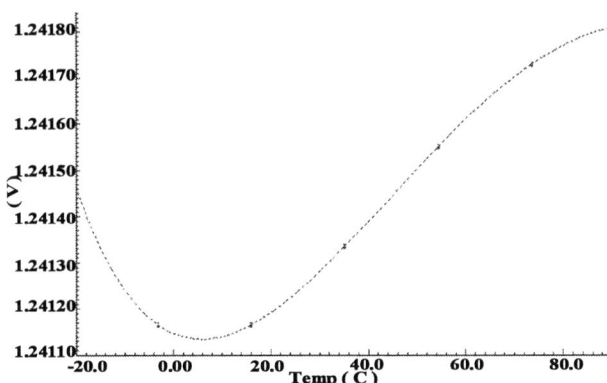

Fig.8 Voltage vs temperature curve when Vdd=2V

When the source voltage is 2V, the simulation is shown in Fig.8. Band-gap voltage reference is 1.241V and the error of which will not exceed to 0.7mV when temperature varies from -20°C to 100°C. TC equals to 10ppm/K in room temperature (27°C).

The suppression rate of output voltage vs. source is shown in Fig.9 when source voltage varies from 0.9V to 3.6V. The PSRR decreases from 20dB when source voltage is 0.9V. It still keeps 45dB when source voltage is 1V, and it reaches 70dB when source voltage is 3V. It indicates that source suppression capacity of band-gap output source is enhanced as the source voltage increasing.

It's shown in simulations that the proposed design of band-gap voltage reference meets the design requirement and it realizes the application of source voltage under 1.8V when low voltage is fully considered.

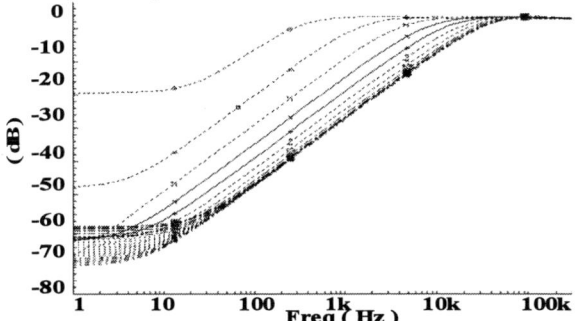

Fig.9 Power suppression ratio for output voltage

Acknowledgments

The authors would like to acknowledge the National Natural Science Foundation of China for providing financial support for this work under Grant No. 60773081 and Grant No. 60777018, and also to acknowledge the financial support by the Shanghai Municipal Education Committee under Grant No.AZ028.

References

1. Gray P R, Meyer R G, <u>Analysis and Design of Analogy Integrated Circuits</u>(Beijing, Publishing House of Electronics Industry，2002), pp.323-327.
2. Phillip E.Allen and Douhlars R.Holberg, <u>CMOS Analog Circuit Design</u>(Second Edition), Oxford University Press, Inc2002, pp.153-159.
3. Banba H, Shiga H, Umezawa A, et al. "A CMOS band-gap reference circuit with sub-1-V operation". IEEE J. Solid-State Circuits. May 1999, 34, pp.670~674.
4. Vittoz E, Fellrath J. "CMOS Analog Integrated Circuits Based on Weak Inversion Operation". IEEE JSSC, 1977, SC-12(3), pp.224~231.
5. Boni Andrea. "Op-Amps and startup circuits for CMOS band-gap references with 1-V supply". IEEE Journal of Solid-state Circuits, 2002, 37(10), pp.1339~1342.

Effects of Cu on the electromigration behavior of Al interconnect by using First-Principles method

Chun Yu, Hao Lu

School of Materials Science and Engineering, Shanghai Jiaotong University,
800 Dongchuan Road, Shanghai 200240, P.R.China
Email address: yuchun1980@hotmail.com

Abstract

Electromigration resistance of Al could be improved through adding a small amount of Cu elements. In this paper, Al31Cu supercell was constructed to calculate the effects of the solute elements on the properties of face central cubic (FCC) Al, including the diffusion activation energy, electronic structure etc, to explain why Cu can suppress the EM process occurred in pure Al interconnect, by employing the density functional theory based on the first-principles method. The calculated diffusion activation energy for pure Al with vacancy-mediated mechanism is -1.29 eV, which is comparable with the experimental result, -1.48 eV. The addition of Cu atom results in the increase of diffusion energy of Al atom near the solute atom, it is -1.57 eV. Since EM failure is greatly related with the diffusion activation energy, EM resistance of Al interconnect is expected to be enhanced by Cu solute. Likewise, the value of the density of states of the systems at the Fermi level $(N(E_F))$, as well as that of the Al atoms at the specific sites could also be reduced slightly by Cu. These two results indicate that the nearest neighbor Al atoms could be stabilized by the above elements. Our calculations are in good agreement with the previous calculations and experimental phenomenon at some extent.

1 Introduction

Naturally, Electromigration (EM) is also a kind of diffusion process, but it is directional and accelerated. Under big current stressing, a large number of metal atoms or ions migrate from cathode side to anode side, this process could induce irreversible damage to the integrated circuits [1,2]. Since Al is a kind of common electronic material, the EM damage occurred in Al interconnect still attracts a lot of attentions. After a large number of experiments, researchers found that a few weight percent of Cu addition could effectively improve the EM resistance of Al interconnect [3], therefore, Al(Cu) has been become the most popular interconnect material for electronic packaging. However, up to data, the exact suppressing mechanism is not fully understood. Since EM process is related with atomistic motion and interaction between atom and atom, atom and defect, there exists strong motivation to understand the improvement mechanism from a fundamental, microscopic theory. The results can be also expected to guide designing of electronic materials.

It is observed experimentally that Al atomic flux was dramatically reduced as the concentration of Cu exceeded a certain threshold. Only after Cu was depleted, did the Al atoms start drifting [3,4]. The atomic flux induced by EM can be described as [5]

$$J = CDF/kT \qquad (1)$$

where C is concentration of the migrating species; $D=D_0exp(-Q/kT)$ is the diffusion rate, and in which D_0 presents vibration frequency, Q is diffusion activation energy, k is Boltzmann's constant and T is absolute temperature; $F=Z^*eE$ is the driving force for EM, in which e is electron charge, Z^* is effective valence and E is electric field strength.

Based on equation (1) and the aforementioned reports, we can conclude that there are two factors contributing to the improvement of the EM resistance of Al alloy interconnect, namely, the effects of moving electrons on host atoms are scattered by solute atoms, or the host Al atoms are stabilized. The former prediction was checked by Dekker et al [6]. However, there were few reports on the investigation of the impacts of solute atoms on the lattice diffusion and stability of host atoms in order to explain why Cu can improve the EM lifetime of Al interconnects.

In this paper, we have investigated the electronic structures of the dilute Al(Cu) system, as well as the dynamic parameters of atoms, by using ab initio calculations based on the density functional theory (DFT) [7]. These calculations are expected to achieve a more qualitative understanding of the effects of alloying species in EM process.

2 Methodologies

To calculate the diffusion activation energy of the atom in bulk Al, a 32-atom supercell based on FCC Al (a=4.0495 Å) was constructed, the point defect (solute or vacancy) was formed by substituting or subtracting one host atom at a specific site. And then, solution enthalpy, vacancy formation energy, vacancy-solute binding energy, and the barrier energy were estimated firstly. The solution enthalpy (E_{Cu}^f) of a solute atom in a 32-site Al supercell is calculated by [8]

$$E_X^f = [E(Al_{n-1}, Cu) - \frac{(n-1)}{n} E(Al_n) - E(Cu)] \qquad (2)$$

where $E(Al_{n-1}, Cu)$ is the total energy of the system containing (n-1) Al atoms and one solute atom, $E(Al_n)$ is the total energy of the perfect system containing n Al atoms, and $E(Cu)$ is the single point energy of Cu solute. Also, the heat of formation for a vacancy (E_V^f, V is short for vacancy) can be described as [9]

$$E_V^f = E(Al_{n-1}, V) - \frac{n-1}{n} E(Al_n) \qquad (3)$$

where, $E(Al_{n-1}, V)$ is the energy of the system containing one vacancy defect. The formation energy of vacancy-solute pair (E_{CuV}^f), as well as the binding energy (E_{CuV}^b) between solute atom and vacancy, are calculated by [10]

$$E_{CuV}^f = E(Al_{n-2}, Cu, V) - E(Al_{n-1}, Cu) + \frac{1}{n} E(Al_n) \qquad (4)$$

$$E_{CuV}^b = [E(Al_{n-1}, Cu) + E(Al_{n-1}, V) - E(Al_{n-2}, Cu, V) - E(Al_n)] \qquad (5)$$

where, $E(Al_{n-2}, Cu, V)$ is the total energy of the supercell system containing a solute-vacancy pair.

The migration barrier energy, E_b, is calculated by [10]

$$E_b = E(Al_{n-2}, Cu, V) - E_{sadd}(Al_{n-2}, Cu, V) \qquad (6)$$

where $E_{sadd}(Al_{n-2}, Cu, V)$ is the energy for a transition state. And therefore, the diffusion activation energy can be expressed as

$$Q = E_V^f + E_{CuV}^b + E_b \qquad (7)$$

Our calculations were performed by employing the DMOL³ code [11]. Exchange and correlation potentials were described by the Perdew-Zunger functional [12], adding a non-local correction in the form of the generalized gradient approximation (GGA) of Perdew and Wang [13]. Brillouin zone (BZ) integrations were performed with the special k-point method over a $10 \times 10 \times 10$ Monkhorst-Pack mesh [14]. And the orbital cutoff is 5.1 Å. All the calculations were done using the supercell approach with periodic boundary conditions. The defect calculations were done at constant volume thus relaxing only the atomic positions. Convergence tests were performed. The tolerances of energy, gradient, and displacement convergence are 2×10^{-5} Ha, 4×10^{-3} Ha/Å, and 5×10^{-3} Å, respectively. The maximum gradient for most of the optimized structures is less than 2×10^{-3} Ha/ Å. For diffusion barrier energy calculation, we use the Linear Synchronous Transit (LST) method [15].

3 Results and discussions
3.1 activation energy

In many systems, the addition of solute element has a profound effect on the diffusion rate of the host atoms. EM is also a directional accelerated diffusion process. Though many works have been done on the diffusion of the solute atom in FCC Al, however, little work cared how the solute element impacts the diffusion energy of the host. In this section, the diffusion energies of the solute atoms in FCC Al were calculated to verify our calculation method, and then the barrier energy of the host Al atom which is closest to the solute atom was calculated. Here, we focus our attention on the vacancy-mediated diffusion.

The total energies for the 32-site supercell structures, namely, perfect Al, vacancy-containing Al, solute-containing Al, and solute-vacancy pair containing systems were calculated based on DFT. According to equation (2), the solution energy for Cu was calculated as -0.098 eV. Our result is very close to that calculated by Wolverton et al by employing VASP code [16], it is -0.085 eV. It is said the more negative the solution energy, the much stronger the Al-X interactions, and therefore the lower solubility of the solute in a solvent. Since Cu has weakly negative solution energies in Al, they are expected to have relatively higher solubility. This result can also be verified through the binary phase diagram and experiment.

In a one vacancy-containing 32-site pure Al supercell system, the formation energy of a vacancy is about 0.63 eV. In addition, the calculated vacancy-mediated diffusion barrier energy with GGA for an Al atom in a vacancy-containing supercell is 0.66 eV. Therefore, the self-diffusion activation energy in Al is about 1.29 eV with vacancy-mediated mechanism, regardless of the Al-vacancy bonding energy. This value is a little lower than the reported activation energy for Al self-diffusion in bulk Al, 1.48 eV [17]. Cu solute enhances the diffusion barrier of the nearest Al atom to 0.94 eV, the diffusion energy of the nearest neighbor Al atom is therefore enhanced to 1.57 eV. Hence, we can find that the Cu solute can improve the diffusion activation energy of Al. According to equation (1), the EM resistance of Al interconnect is also improved. This result is qualitatively comparable to the experimental phenomenon.

Moreover, the calculated vacancy-mediated activation energy of Cu in bulk Al is 1.43 eV. This result is also within the range, 1.23-1.47 eV [18]. Since Al has a higher activation energy than Cu, we can predict that in Al(Cu) interconnects, Cu will diffuse first, while the NN Al diffusion is retarded due to a higher diffusion activation energy. This conclusion is also in good agreement with the experimental phenomenon [19]. Liu et al [20] also found that the activation energy of Al is slightly larger than that of Cu in the grain boundary.

3.3 electronic properties

The stability of a system depends on the location of the Fermi level and the value of $N(E_F)$ [21]. Generally, a high $N(E_F)$ results in an unstable structure relate to the original one; and vice verse. We calculated the density of states (DOS) of the doped systems, and therefore $N(E_F)$. Fig.1 shows the DOS curves of Al32 and Al31Cu. It is seen that the DOS curves of these three systems have almost the same shape. However, at the energy of -0.13 Ha, there forms a strong bonding peak due to the hybrid between Cu d and Al s and p states. This peak has a positive effect on the stability of the doped system. Meanwhile, at the Fermi level (which located at zero eV, denoted by the dashed line perpendicular to the energy coordination), the DOS value, namely, $N(E_F)$, changed slightly. They are 337.56, and 332.11 states/Ha for Al32, Al31Cu, respectively. Therefore, from the point of view of electronic structure, Cu solute can be beneficial in improving the stability of Al system.

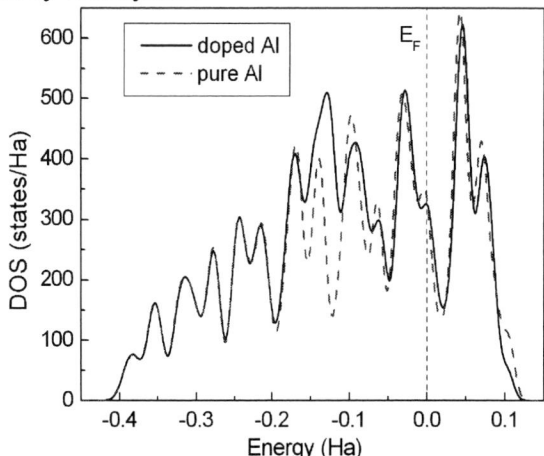

Fig.1 DOS of Al supercell before and after doping

The results calculated by Dekker et al indicate that the dopant has bigger effect on the nearest neighbor (NN) host atoms, relative to the other farther host atoms. Fig.2 shows the partial DOS (PDOS) of the Al atom at pure Al system, and the NN and the second NN host atoms at doped system. It is seen that the $N(E_F)$ of NN and SNN Al atoms is slightly lower than that of the Al in pure Al system. Moreover, that of the NN Al atom is the smallest one. This result demonstrates that the solute atom has the biggest effect on the NN atoms. This is also in good agreement with the conclusion drawn by Dekker. The decrease of $N(E_F)$ also indicates that the stability of Al atom was strengthened, and therefore the EM resistance.

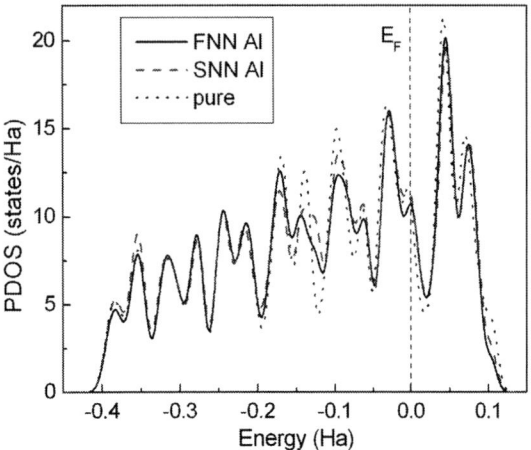

Fig.2 site-projected density of states curves of Al before and after addition of Cu (FNN and SNN present the first nearest and second nearest neighbor Al atoms, pure Al is Al atom at perfect Al supercell, and E_F is the Fermi level)

Conclusions

In this paper, first-principles calculations were performed to try to explain why Cu can improve the EM resistance of Al interconnects. A 32-atom supercell based on FCC Al was constructed, which included Al atoms, one solute atom, and/or one vacancy. The formation energies, diffusion energies, electronic properties of the doped systems were calculated. The results show that Cu solute can improve the diffusion n energy of neighbor host atoms. Meantime, due to a lower activation energy of Cu relative to neighbor Al atoms, Cu solute would shift first in EM process, and then Al atoms follow. Hence, the EM lifetime of Al interconnect is elongated. This is in good agreement with experimental phenomenon. In addition, according to the electronic structure, the N(EF) of Al atoms around Cu solute are decreased slightly, which indicates that these host atoms are stabilized. Moreover, much closer to Cu, more stable the host atoms are.

References

1. Hu, C.K., Rodbell, K.P., Sullivan, T.D., etc, "Electromigration and stress-induced voiding in fine Al and Al-alloy thin-filmed lines", *IBM J.Res.Develop.*, 39 (1995), pp. 465-497.

2. Tu, K.N., "Recent advances on electromigration in very-large-scale-integration of interconnects", *Journal of Applied physics*, 94 (2003), pp. 5451-5473.

3. Hu, C.K., Small, M.B., Ho, P.S., "Electromigration in Al(Cu) two-level structures: effect of Cu and kinetics of damage formation", *J. Appl. Phys.*, 74 (1993), pp. 969-978.

4. Lloyd, J.R., Clement, J.J., "Electromigration damage due to copper depletion in Al/Cu alloy conductors", *Appl. Phys. Lett.*, 69 (1997), pp. 2486-2488.

5. Huntington, H.B., Grone, A.R., "Current-induced marker motion in gold wires", *Journal of Physics and Chemical of Solids*, 20(1961), pp. 76-87..

6. Dekker, J.P., Gumbsch, P., Arzt, E., etc, "Calculation of the electromigration wind force in Al alloys", *Phys. Rev. B*, 59 (1999), pp. 7451-7457.

7. Kohn, W., Sham, L.J., "Self-consistent equations including exchange and correlation effects", *Phys. Rev. A*, 136 (1965), pp. A1133-A1138.

8. Domain, C., "Ab initio modelling of defect properties with substitutional and interstitials elements in steels and Zr alloys", *Journal of Nuclear Materials*, 351 (2006), pp. 1-19.

9. Satta, A., Willaime, F., deGironcoli, S., "Vacancy self-diffusion parameters in tungsten: Finite electron-temperature LDA calculations", *Phys. Rev. B*, 57 (1998), pp. 11184-11192.

10. Vincent, E., Becquart, C. S., Domain, C., "Ab initio calculations of vacancy interactions with solute atoms in bcc Fe", *Nuclear Instruments and Methods in Physics Research B*, 228 (2005), pp. 137-141.

11. Delley, B., "An all-electron numerical method for solving the local density functional for polyatomic molecules", *J. Chem. Phys.*, 92 (1990), pp. 508-517.

12. Perdew, J.P., Zunger, A., "Self-interaction correction to density-functional approximations for many-electron systems", *Phys. Rev. B*, 23 (1981), pp. 5048-5079.

13. White, J.A., Bird, D.M., "Implementation of gradient-corrected exchange-correlation potentials in Car-Parrinello total-energy calculations", *Phys. Rev. B*, 50 (1994), pp. 4954-4957.

14. Monkhorst, H.J., Pack, J. D., "Special points for Brillouin-zone integrations", *Phys. Rev. B*, 13 (1976), pp. 5188--5192.

15. Halgren, T.A., Lipscomb, W.N., "The synchronous-transit method for determining reaction pathways and locating molecular transition states", *Chem. Phys. Lett.*, 49 (1977), pp. 225-232.

16. Wolverton, C., Ozoliņš, V., "First-principles aluminum database: Energetics of binary Al alloys and compounds", *Phys. Rev. B*, 73 (2006), no. 144104.

17. Lundy, T.S., Murdock, J.F., "Diffusion of Al[26] and Mn[54] in Aluminum", *J. Appl. Phys.*, 33 (1962), pp. 1671-1673.

18. Du, Y., Chang, Y.A., Huang, B.Y., etc, "Diffusion coefficients of some solutes in fcc and liquid Al: critical evaluation and correlation", *Materials Science and Engineering A*, 363 (2003), pp. 140-151.

19. Hu, C.K., Small, M.B., Ho, P.S., "Electromigration in Al(Cu) two-level structures: Effects of Cu and kinetics of damage formation", *J. Appl. Phys.*, 74 (1993), pp. 969-977.

20. Liu, X.Y., Xu, W., Foiles, S.M., etc, "Atomistic studies of segregation and diffusion in Al-Cu grain boundaries", *Appl. Phys. Lett.*, 72 (1998), pp. 1578-1580.

21. Yu, C., Liu, J. Y., Lu, H., etc, "First-principles investigation of the structural and electronic properties of $Cu_{6-x}Ni_xSn_5$ (x=0,1,2) intermetallic compounds", *Intermetallics*, 15 (2007), pp. 1471-1478.

A Mixed- Signal Physical Design and Its Verification

Hu Yue-li, Yan Ke

Key Laboratory of Advanced Display and System Applications (Shanghai University), Ministry of Education
Microelectronics Research & Development Center Shanghai University
Campus P.O.B.221, 149 Yanchang Rd, Shanghai 200072, China
E-mail: huyueli@shu.edu.cn

Abstract

More and more analog and mixed-signal (AMS) blocks are integrated into SoC (system-on-chip) platform due to intense market competition. A mixed signal /mixed power SoC design, using Synopsys Libraries, based on Chartered 0.35um Salicide 2P4M CMOS mixed signal process is introduced in this paper. The method of supply the core with different powers, isolated the digital part and the analog part and splitting the padring is also proposed. The physical layout design and its verification are implemented using Astro and Calibre.

1. Introduction

SoC technology is a new trend for VLSI design. It provides unprecedented market prospects and development opportunities for IC industry. Today's high performance of manufacturing processes of integrated circuits in CMOS technology allows and even forces due to cost issues the developers to add more and more functionality in one circuit. The trend is to place especially analog and even RF parts besides digital circuits on the same die. The most common building blocks added in this way are e.g. PLL's, A/D and D/A converters.

In this paper, a new method of mixed-signal physical design SHU-MV06, an 8-bit MCU based on Intel 8051 architecture and integrating a 10-bit SAR ADC is proposed. The core and the ADC are supplied with independent powers. The padring are spitted by deleting the fillers between the digital and the analog pads. This feature is very useful to isolate noisy outputs from the reminder of the padring. The reminder could be analog or digital.

2. Issues in Mixed signal design

A mixed signal / mixed power design introduces problems which do not exist in a straight forward digital design. Four key problems are described as follows:

2.1 Cross-couple noise

Since most analog signals are sensitive to the switching noise of digital signals, a simple way in which this noise can be limited is by keeping all analog circuitry well separated from the digital sections

2.2 Delay

Speed is limited by switching and propagation delays along the interconnections. With deep submicron technologies, the switching speed depends more and more critically on the interconnection capacitances, which are becoming the dominant loading factors. Hence, the longer the interconnections the higher the delays. By increasing the interconnection widths, propagation delays decrease in some measure, but switching delays get worse, and area is wasted.

Interconnection length remains the fundamental parameter used to control delays.

2.3 Substrate noise

Substrate noise is injected by digital switching into the substrate of integrated circuits, where it propagates to other sections of the circuit. Noise injection is getting particularly significant with deep-submicron technologies, where short channel lengths and high speed produce concentrations of high energy charge carriers in the channel region of MOS devices. Noise power is transferred by capacitive coupling into the interconnections, and can disrupt analog signals and high impedance digital signals. The techniques used to prevent or limit substrate noise depend on the technological process used.

2.4 ESD Protection

ESD protection has not only to be considered in the design of the protection circuit in the I/O libraries; it is mandatory to search for full-chip ESD solution. The overall methodology to design the chip especially the padring with the IO's and the power routing determines the level of ESD protection. Every ESD event on a pad generates current spikes which will find their way through the chip. If no dedicated current path for these spikes is foreseen, the current will find its own way. It will break down the weakest link in the design and will use a current path which is not necessarily designed for such high currents. Damages on the chips are inevitable.

The better way for the design is to provide ESD current paths between any possible pin to pin combinations. Every pad must have dedicated breakdown devices which conduct the ESD current to wide metal lines (normally the power rails) whereas the wide metal lines have also to be connected together by dedicated break down devices.

3. An Example of mixed signal /mixed Power Design

3.1 ADC Milkyway Reference Library Preparation

The ADC implemented with Chartered 0.35um Salicide 2P4M CMOS Analog/RF Single Gate (3.3V) process is a hard IP .We can access cells in another library by making a reference library of the library we are working in. The premise of using the ADC hard IP in Astro is that we must create an ADC Milkyway reference library which contains the physical cells .We use ADC as a reference library of the main design and it is a macro in the layout during P&R using Astro.

In addition to GDS, the Library Exchange Format (LEF) is another method to provide library information from a third-party database to Milkyway. LEF defines the elements of an IC process technology and associated library of cell models. LEF contains library information for a class of designs. This library data includes layer, VIA, placement site type, and

978-1-4244-2739-0/08/$25.00 ©2008 IEEE   371

macro cell definitions. The flow of importing the ADC LEF data to create a Milkyway reference library is shows in Figure 1.

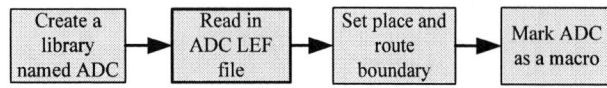

Fig.1 Flow for Creating a Milkyway Library from LEF

3.2 Core and ADC Powers Separation

The SHU-MV06 is a mixed signal /mixed power design .We choose independent power to supply the core and the ADC ,meaning that with the same or different potential level but isolated from each other for noise sensitive digital and analog parts.

3.2.1 Powers Declaration in the Gate Level Netlist

Generally, the gate level netlist generated by DC doesn't include the power and ground declaration ,so we must add the power and ground supply in the netlist. The declaration of power and ground supply is as follow:

//supply for Astro

supply1 VDD;

supply1 AVDD;

supply0 VSS;

supply0 AGND;

VDD and AVDD is the power net, and VSS and AGND is the ground net .VDD and VSS supply the core and AVDD AGND supply the ADC.

3.2.2 Power Pad Selection

 The libraries that Synopsys provides allow the user to use different power buses for the IO ring. There are 3 power buses:

- vdd : standard power bus.

- vddq: the quiet power bus.

- vddo: the I/O power bus (noisy).

The I/Os of the Synopsys libraries use these buses in a novel way. Furthermore there is a patent pending on the IO design that allows to reduce the noise on the noisy power bus.

Synopsys provides in the IO libraries power pads that allow the power buses to be used in four different ways: "Optimal solution", "Basic solution"," Intermediate solution (1)" and "Intermediate solution (2)".We choose "Basic solution" in the design.

With the "Basic solution" ,user can use the following combination of pad : pvdf, pv0f. This combination shorts all the power buses. All the power ring is supplied with the same power supply. The advantage of choosing this solution is that this combination gives the minimum number of power pads. We should manually add the power pads in the gate level netlist. The codes are as follow:

pvdf pad_vdd0 (.PAD(VDD));

pv0f pad_gnd0 (.PAD(VSS));

pvdf pad_vdd1 (.PAD(VDD));

pv0f pad_gnd1 (.PAD(VSS));

pvdf pad_avdd (.PAD(AVDD));

pv0f pad_agnd (.PAD(AGND));

We choose two couple powers to supply the core to guarantee that the core can be supplied well and prevent the IR drop and EM phenomenon. But they are in the different location on the chip.

3.3 Padring Splitting

Several applications need to isolate a part of the padring from the rest of the padring to guarantee that the digital parts and the analog parts can be supplied independently.

Synopsys supports two cells to cut the padring : PxVCE cuts the VDDO, VDDQ and VSSO, VSSQ rails but maintains VDDi and VSSi. ESD protection in Synopsys library is based on the current path theory and that it still exists a connection between two split padrings over VDDi and VSSi, makes this cut cell suitable for good ESD protection. PxVCF cuts all power rails except VSSi.

Placing n cut cells in the padring generates n padring segments which are isolated from each other. This generates n different VDDO, VDDQ, VSSO and VSSQ rails. All of them have to be connected properly that means every VDDO, VDDQ, VSSO and VSSQ rail has to be connected to at least one power pad to get ESD protection. Splitting the padring could also generate different VSSi and VDDi segments depending on the cut cell used. Using only PxVCE as cut cell do not generate additional VDDi or VSSi. If several VSSi and VDDi are generated they must be also connected to at least one power pad.

Although the pad PxVCE can well split the padring, but we can't find it in Synopsys Chartered 0.35 IO Library. It only exists in 0.25,0.18 and 0.13 process. At last we decide to delete the filler cells between the digital pads and the analog pads to split the padring. The layout splitting the padring in Astro is shown in figure 2.

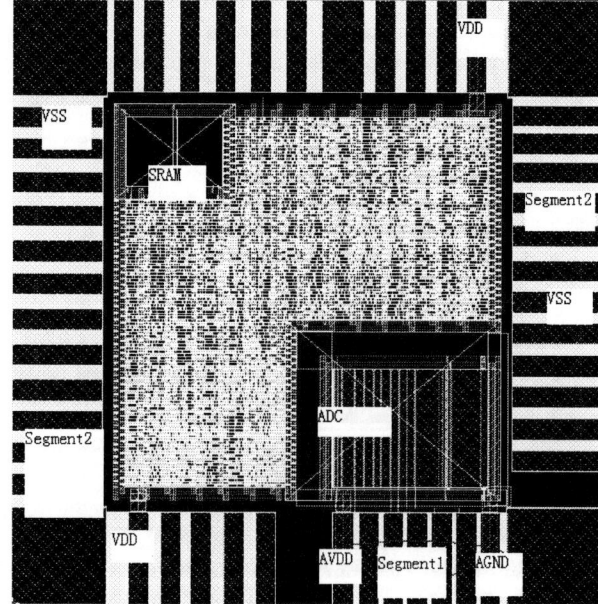

Fig.2 The layout splitting the padring in Astro

The above design consists of two padring segments separated by deleting the pad filler cells. Segment 1 serves as analog part of the padring. The second segment (Segment 2)

372

supplies the main digital circuit with power VDD and VSS. In order to reduce the noise we leave a gap between the ADC and the digital parts.

4. Physical verification

We use Mentor Graphics Inc's EDA tool Calibre to do the physical verification.

4.1 GDS and Netlist Modifying

The physical verification in this design encompasses Design Rule Check (DRC) and Layout Versus Schematic Comparison(LVS). DRC verification for this design passed easily but LVS verification encountered some difficulty. So we just introduce LVS verification. We use ADC as a macro during P&R using Astro and the GDS exported by Astro doesn't contain the ADC's GDS ,so we should combine the two parts together by Cadence Inc's EDA tool Virtuoso through the functionality of "Stream in". We embedded the Calibre tool into the Virtuoso tool and GDS can be simultaneously exported during DRC verification. The final layout of the design is shown in Figure 3. The netlist extracted after P&R using Astro is in verilog format and we must use "v2lvs" command to covert the verilog netlist into spice netlist which is available for Calibre.

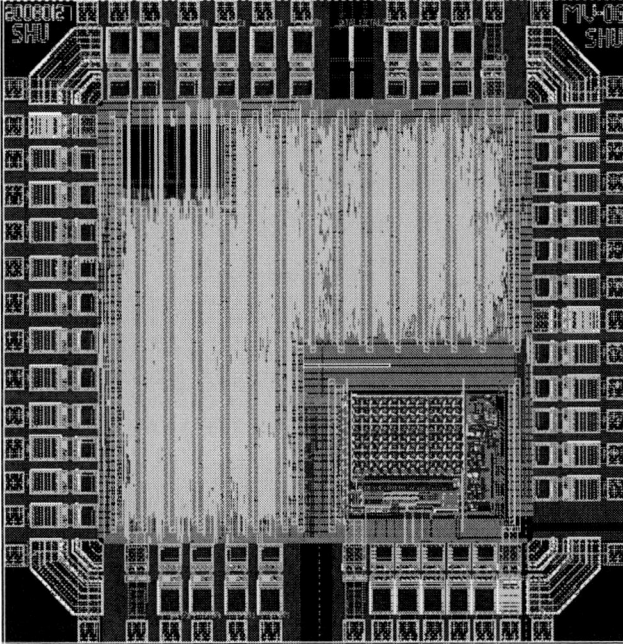

Fig.3 The final layout in Virtuoso

The power pads we use in this design are pvdf and pv0f. The pvxf short all power buses, then all power rings of IO will be shorted. There exists only one power net in the IO, so before deleting the pad fillers we must add to the CDL netlist the following:

    *.CONNECT VDD VDDQ
    *.CONNECT VDD VDDO
    *.CONNECT AVDD VDD
    *.CONNECT AVDD vdd !

    *.CONNECT VSS VSSQ
    *.CONNECT VSS VSSO
    *.CONNECT AGND VSS

    *.CONNECT AGND gnd!

vdd! and gnd! are the power ports of the ADC. From the above codes we can infer that all the power nets including the digital parts and the analog parts are joined together, but we must separate the two parts and isolate the digital pads and the analog pads.

In the layout ,ADC's power pads are named as pvdfa and pv0fa and the analog input and output pad pc3d00 is named as pc3d00a. Generally,pvdfa,pv0fa and pc3d00a are just duplicate copies of pvdf,pv0f and pc3d00 and they are same in the architecture. We just distinguish the analog and digital pads in layout and CDL netlist for LVS verification. In the CDL we add pvdfa ,pv0fa and pc3d00a imitating the CDL of pvdf ,pv0f and pc3d00 in order to match the layout. In the CDL of pvdfa,pv0fa and pc3d00a , the power buses VDDQ and VDDO are replaced as AVDDQ and AGNDO. By far the GDS and the CDL of pvdfa,pv0fa and pc3d00a are ready.

After deleting the pad fillers, the digital and the analog power nets are spitted in the layout ,so we must modify the CDL as follows:

    *.CONNECT VDD VDDQ
    *.CONNECT VDD VDDO

    *.CONNECT VSS VSSQ
    *.CONNECT VSS VSSO

    *.CONNECT AVDD AVDDQ
    *.CONNECT AVDD AVDDO
    *.CONNECT AVDD vdd!

    *.CONNECT AGND AGNDQ
    *.CONNECT AGND AGNDO
    *.CONNECT AGND gnd!

By default, LVS does not separate substrates in different potentials. So The net VSS and AGND are both recognized to stamp to the substrate by LVS and SCONNECT conflicts error will occur. A method of solve this problem is to define Nwell ring or/and LVS_PSUB2 ring as substrate separation .In the LVS rule file, using the $SUB_SEPARATION environmental variable can realize the aim put forward above. An example is as follow:

    setenv SUB_SEPARATION NW_PSUB2_RING # [NO || NW_RING || NW_PSUB2_RING]

Note: LVS_PSUB2 ring functions being as a pseudo nwell ring does not force any physical separation between substrates as it is not a real physical process layer. Using this LVS technique will be on one's own risk.

4.2 Verification Results

After the data preparation is ready tell the detail path of the GDS and the CDL to the tool, and then run it .The reports of the verification is as follow:

    ######################################
             CALIBRE SYSTEM REPORT
    ######################################
    REPORT FILE NAME: mv06.lvs.rep
    LAYOUT NAME: netlist ('mv06')

SOURCENAME:
/disk3/home/tr03/JWY/calibre/v2lvs/mv06_apr.spi ('mv06')

RULE FILE: runset.cal

RULE FILE TITLE: Mentor Calibre LVS Runset for 0.35um Dualgate (3.3V/5.0V) & SiGe Process

HCELL FILE: list

CREATION TIME: Tue Mar 4 20:36:55 2008

CURRENT   DIRECTORY: /disk3/home/tr03/JWY/calibre/lvs_sg

USER NAME: tr03

CALIBRE VERSION: v2007.3_36.25   Fri Oct 12 13:06:51 PDT 2007

OVERALL COMPARISON RESULTS

```
 #    ##################       _  _
 #    #                  #    * *
# #   #   CORRECT        #     |
##    #                  #    \__/
 #    ##################
```

```
**********************************************
               CELL  SUMMARY
**********************************************
```

Result	Layout	Source
CORRECT	pc3d00a	pc3d00a
CORRECT	pc3d01d	pc3d01d
CORRECT	pc3o02	pc3o02
CORRECT	pc3x12	pc3x12
CORRECT	pv0f	pv0f
CORRECT	pvdf	pvdf
CORRECT	pvdfa	pvdfa
CORRECT	ADC_4	ADC
CORRECT	RS4	RS4
CORRECT	mv06	mv06

According to the reports, a conclusion can be drawn that the GDS is matched with the spice netlist. The LVS verification has passed. In the CELL SUMMARY section a few of the cells are listed. The cells pc3d00a ,pvdfa and pv0fa we creating are all correct.

5. Conclusion

The ADC IP is used as a macro during place and route in Astro and the ADC's GDS is finally embedded into the whole design by the Virtuoso EDA tool. In mixed-signal design the digital and the analog parts are noise sensitive ,so they must be isolated. In order to supply the digital and the analog parts well independent powers are needed. For mixed powers design LVS does not separate substrates in different potentials and we should draw an nwell ring or PSUB2 ring around the analog block in the layout to pass the LVS verification.

Acknowledgments

The authors would like to thank National Natural Science Foundation of China (Foundation Number: 60773081 , 60777018) and IC Special Foundation of Shanghai Municipal Commission of Science and Technology   (Grant No. 077062008).

References

1. Yueli Hu, Wenyi Jing, Ying Liu, "Integrating ADC IP in a SoC and its Verification", *2007 IEEE International Symposium on High Density packaging and Microsystem Integration(HDP '07),* Shanghai, China, June 26-28 ,2007:pp.376-379.

2. Enrico Malavasi and William H. Kao, "Current lssues in a Const raint-Driven Mixed Signal Physical Design Flow", *1997 IEEE International Symposium on Circuits and Systems,* Hong Kong, June 9-12,1997:133-136.

3. B. R. Stanisic, N. K. Verghese, D. J. Allstot, R. A. Rutenbar and L. R. Carley, "Addressing Substrate Coupling in Mixed-Mode ICs: Simulation and Power Distribution Synthesis",*IEEE Journal of Solid State Circuits,* March 1994.

4. Jing Wenyi "Design and System Integration of Analog-to-Digital Converter IP"[D] ,   Master Dissertation , Shanghai University，Shanghai，China，（2008），pp.39-43.

5. Documentation for Chartered0.35 Rev1.0, 2004,12, Synopsys Inc.

6. Standard Verification Rule Format(SVRF)Manual ,Calibre v2004.1_1 ICverify v8.9 February 2004, Mentor Graphics Corporation 2004.

7. Synopsys Physical Implementation Online Help, Version V-2004.06, September 2004 ,Synopsys, Inc.

Hardware /Software Co-design for Viterbi Decoder

Ming Li*, Tao Wen
Microelectronics Research and Development Center,
Key Laboratory of Advanced Display and System Applications, Ministry of Education,
Shanghai University, China
*E-mail: eeliming@163.com   Tel: 86-21-56331323

Abstract

Convolutional code and Viterbi decoding is one of the methods used for channel coding. Convolutional coding is widely used in many aspects because of its excellent performance. However, both the coding algorithm and hardware implementation are very complex. Along with the continuous deepening of SOC design, hardware-software co-design technology has become more and more important as one part of SOC design. A hardware-software platform used for the design of the Viterbi decoder is proposed in this paper. The platform includes embedded software, hardware acceleration, and peripheral interface. Good practicality and powerful verification functions are proved, and optimum design for the Viterbi decoder can be achieved by using the platform.

1. Introduction

Convolutional coding was proposed by Elias in 1955 [1]. The n-k check-up elements are related to not only the information units of the current group but also the information units inputted to the encoder in the past groups. So it does need not only extracting decoding information from the current receiving code group but also extracting related information from the code groups received in the past and in the coming time during the decoding process of convolutional codes. As a result, it has not found an effective mathematic tool in the process of analyzing the convolutional codes so that it is hard to analyze the performance.

Viterbi algorithm, proposed by Viterbi in 1967, is a maximum likelihood decoding algorithm based on the trellis graph of code. Since the VB algorithm has been proposed, it has developed rapidly both in theory and in the practical application. It is widely used in various data transmission systems. However, the decoding process with Viterbi algorithm is very complex. So, it is hard to design the decoder with hardware method only. This paper proposes a hardware-software co-design and co-verification method, which mainly establishes a platform based on the ARM embedded system and FPGA. ARM embedded system generates test data, which can verify the performance of Viterbi decoder with software and hardware. The platform carries out with the reconfigurable FPGA through downloading the comprehensive nets to the FPGA. The drivers of bus and peripherals can be generated by software console.

The method will accelerate design and save resources. Moreover, it can carry out interactive verification between hardware and software, and discover the errors in the design and correct them timely.

2. Hardware and software co-design
2.1 Hardware and software co-design theory

For a hardware and software system, when the system hardware increases, proportion of area increases and delay decreases. When system software increases, the situation is contrary. The so-called "optimal system cost" means: the design meets minimum system latency under the restraints of the area, or meets the minimum system area or maximum cost of system under delay constraints [The speed (performance) contrast the area (prices)], that is, the pursuit $[(T*S)_{min}]$. Hardware and software co-design methodology, in essence, is to give you an algorithm, which automatically searches for the best hardware and software constraints compromise. Finally, system generates effective system architecture [6].

Co-design of software and hardware involves in the contents: HW-SW co-design process, HW-SW division, HW-SW parallel synthesis, and HW-SW parallel simulation.

2.2 The Design Flow of the collaborative design with software and hardware

The design flow of the co-design with software and hardware usually can be divided to following steps:

Step 1: describe the system with the HDL and C language, then simulate it using software and verify its behavioral function;

Step 2: according to the function of the system, partition it to the achievement of the sections using software and hardware, respectively. Each section is designed firstly by using VerilogHDL, and then is synthesized to make sure the behavior and performance forecast correct.

If no problem is found, detailed design is carried out in step 3.

The final step deals with the test of the system. A typical design flow is showed in Fig. 1.

978-1-4244-2739-0/08/$25.00 ©2008 IEEE

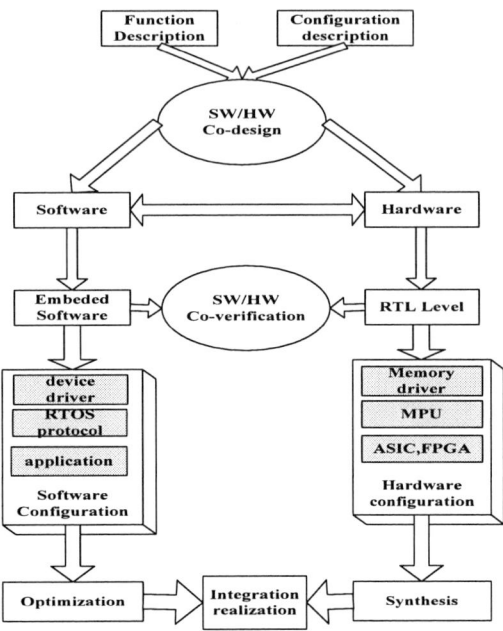

Fig. 1 The flow of co-design

The hardware and software's partition is an important subject in the SOC design. During the process, some basic principles should be followed: the section which requires high speed and low power dissipation is better to be achieved by hardware and the one which owns the character of low-volume and high-variety is done by software. Besides, in order to increase the processing speed and reduce the power dissipation, the process unit and special hardware are recommended to run concurrently. Parallel synthesis of hardware and software method adds more constraints and limits compared to the synthesis of IC design, especially for the embedded system and the synthesis of interfaces. The big problem faced now is the synthesis problem of the interfaces' connections between SOC and varieties of IP core.

3. Architecture and SW/HW partitioning results of the Viterbi decoder

3.1 The top-level architecture of Viterbi decoder.

As shown in Fig.2, the Viterbi decoder includes four modules as follows:

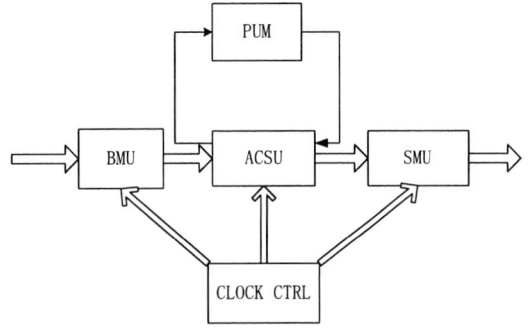

Fig.2 Viterbi decoder system architecture

(1) BMU(branch metric generation unit)

BMU is transition metric unit and its function is to generate corresponding merit of skip branch according to the input code sequence. The generated sequence merit is delivered to ACS unit, completing the process as bit calculation. Toward to the Channel of AWGN, the Euclidean distance is often used to calculate the merit of skip branch. For instance, a Viterbi decoder ,which will decode the input code as the pattern of (n,k,m), has an output sequence like $\{y_0, y_1 \cdots y_{n-1}\}$ corresponding to skip branch, but the input coded signal is $\{x_0 x_1 \cdots x_{n-1}\}$. In that way, the merit of branch B is $|y_0-x_0|+|y_1-x_1|+\ldots+|y_n-x_n|$.

(2)ACS (add，compare，select）unit

ACS is in the most basic computing unit in the Viterbi decoder, because Viterbi decoding algorithm searches of the shortest path algorithm。 The ACS module not only receives the code sequence from the branch merit module, but needs the path merit of last state and information related to state shift. It is necessary to calculate the sum of branch merit and path merit firstly, then select the smallest path merit.

(3)PMU (Path metric storage unit)

Every time of every state has a survival path because the process of adding, comparing and selecting are executed all the time. Therefore, it is crucial how to save the survival path merit. In some universal designs, it is common that the designs adopt the number of flip-flop or latch twice as the number of the design's state in order to decrease to the number of the flip-flop in the design.

(4)SMU (Survivor path memory unit)

The arbitral bit which is generated from the ACS module will be saved in the survival path memory and will be used to search the survival path. Generally, the method of searching survival path can be classified into two basic modes. One is named register-exchange (RE) and the other is called Trace-Back (TB).

3.2 System hardware/software partition

HW/SW partition is an important subject of the design of SOC [11]. It should follow the major principle: the tasks with low complexity and high speed should be implemented through hardware while the tasks with high complexity and low speed through software. The flexibility of software design and the efficiency of hardware design are incompatible, thus the purposes of reasonable HW/SW partitions are to find a relatively most optimal processing solution for the sake of obtaining best system performance. The factors such as target architecture and the cost of software and hardware implementation should be considered during the hardware and software partition.

According to the previous analysis, in the modules of the Viterbi decoder, the most appropriate one for software implementation is add-compare-select unit, also called butterfly unit. The butterfly unit refers to Redix-4 algorithm and is of great complexity, which just accord with the partition principle, thus the add-compare-select unit can be implemented with software. The add-compare-select unit is one part of Viterbi decoder. Software implementation can't be integrated into the accelerator. In the embedded system, the part described with software can be converted to ELF files. The following is the conversion command from ELF file to Verilog HDL memory format:

%fromelf –vhx -16x2 xxx.elf -output rom0.dat

The algorithms in other parts are not complicated and can be implemented with hardware. In the early period, the algorithms can be designed with FPGA, not only shorting development time, but also finding the design errors timely.

4. System architecture to achieve and Hardware/ Software co-verification for Viterbi decoder
4.1 Viterbi decoder system architecture

After hardware and software partitioned, tasks of the system will conduct a detailed division of labor. When hardware and software access their tasks, the system carries on collaborative design, it can seen from the Fig.3 that software system is made up of embedded software, ARM kernel and linux operating system. An embedded operating system can be encapsulated in ARM, to accomplish complex algorithms and to finish a variety of tasks instead of personal computers. We design software system with C program.

The hardware platform is composed of FPGA accelerator. FPGA simulation system includes on-chip bus and hardware acceleration units. Hardware acceleration modules make use of reusable IP cores to meet the different requirements of the simulation and system constraints through a variety of configurations for IP cores. Design of IP core is described by the Verilog HDL language.

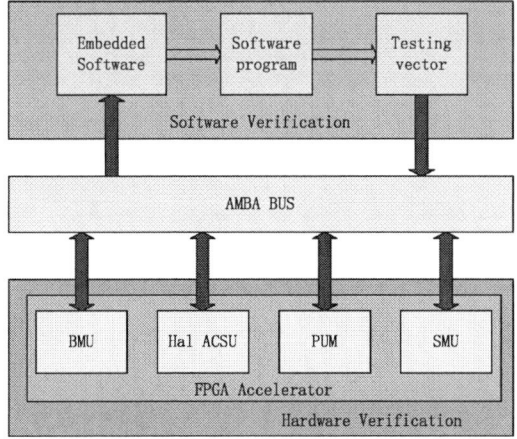

Fig.3 System platform for co-design of Viterbi decode

4.2 Hardware and software co-verification of system

After system architecture designed completely, we also completed the TOP level of Viterbi decoder. The next question is how to verify TOP. Verification of the design process occupy most of the time, so the growing gap becomes distinct between engineer's design capability and verification capability. In the design of the hardware and software co-verification, behavioral models are described by the high-level language in embedded console, and PLI (Program Language Interface) is used between behavioral modules and real RTL descriptions.

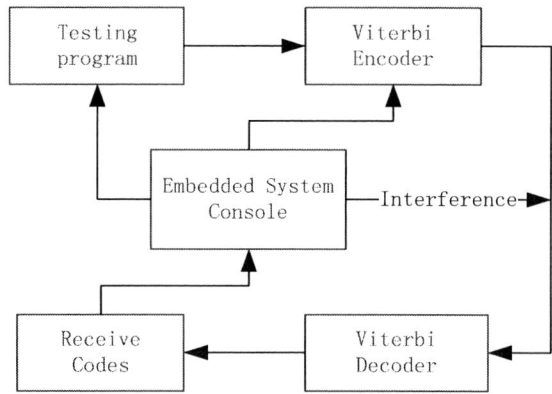

Fig.4 Co-verification for system

In the process of hardware verification, the software is also being verified. The key of software design is to produce transplantable programs, which can use not only the Viterbi decoder testing platform, but also on other platforms. For example, the interface between hardware and software is implemented through AMBA bus and we have to drive the bus for communication between Embedded Platform and FPGA. The following driver is a section of the code:

```
struct AMBA_device {
    struct device       dev;
    struct resource     res;
    u64                 dma_mask;
    unsigned int        periphid;
    unsigned int        irq[AMBA_NR_IRQS];
};
```

5. Testing results and analysis

The main testing procedures are written by high-level language. In testing process, we first make coding for testing signals, which are transmitted to Viterbi decoder. We inject interference signals artificially and judge the output codes of Viterbi decoder correct or not.

The Viterbi decoder is verified in the proposed platform. When inputting codes, we get the survivors path and TB data. Viterbi decoder can find wrong codes and output the correct code sequences.

The simulation results of Viterbi decoder are shown in Fig.5.

377

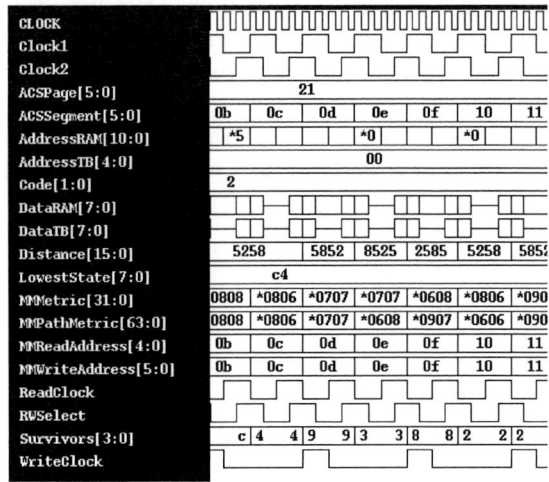

Fig. 5 The simulation waveform of Viterbi decoder

6. Conclusions

The paper discusses a co-design and co-verification of hardware and software Viterbi decoder. The main tasks of the paper emphasize importance on partition of hardware/software and analyze the results of partition. We process the co-design for Viterbi decoder using the results of partition. The results show that the performance of Viterbi decoder fully achieves the requirements of design. The application of hardware and software co-design methods can speed up the design process and reduce errors in design, also increase the fabrication successful rate of tapout.

Acknowledgments

The authors will express their sincere gratefulness to the Key Laboratory of Advanced Display and System Applications in which all the experiments are completed.

References

1. Xinmei wan, Guozhen Xiao. Error Correcting Code-Principles and Methods, Publishing House of Xi'an University of Electronic Science and Technology , 2001

2. P.J.Black and T.H.Meng. A 140-Mb/s, 32-state, radix-4 Viterbi decoder. IEEE journal of Solid-state Circuits, vol. 27, pp. 1877-1885, Dec. 1992.

3. Ming Li*, Jing-sai Jiang. "A test platform for Turbo encoder
&decoder based on S3C2410". IEEE journal of HDP, pp. 400-403,2007

4. K.Hu,M.D.Caldell,W.W.Lin, "A Viterbi decoder memory management system using forward traceback and all-path traceback ", Consumer Electronic,1999.ICCE.Internationgal Conference on,1999, pp.68-69.

5. K. Chadha, J.R. Cavallaro, "A reconfigurable Viterbi decoder architecture," Signals, Systems and Computers, 2001. Conference Record of the Thirty-Fifth Asilomar Conference on, Volume: 1 , 4-7 Nov. 2001, pp. 66 – 71, vol.1

6. Giovanni de Micheli and Rajesh K.Gupta, "Hardware/Soft-

ware Co-Design", in Proceedings of the IEEE, VoZ. 85, No. 3, March 1997, pp. 349-365.

7. F. Balann et al., "Hardware/Software Co-design of Embedded Systems: The POLIS Approach", Kluwer Academic Publishers, 1997

8. T. Z. Sun, W. J. Yuan, and H. F. Zhang, "Manual for Embedded Design and Linux Driver Development: Based on ARM9 Microprocessors", Beijing: Publishing House of Electronics Industry, 2005.

9. Wayne H. Wolf, "Hardware-Software Co-Design of Embedded Systems", Proceeding of the IEEE, Vol 82, No. 7, Pages 967-988, July 1994.

10. http://www.codingtechnologies.com/

11. Aiping Wen, Zhen Yang. "Key technologies and methods of SOC design." Microelectronic technology, 2001.5

Attenuators Using Thin Film Resistors for RF Application

Yiqin Sun[1,2]， Lei Li[1], Han Lin[1], Zhiyuan Yu[2], Mian Huang[1], Lixi Wan[3]

[1] Shenzhen Institute of Advanced Technology, Chinese Academy of Sciences/the Chinese University of Hong Kong,
Shenzhen 518067,China
[2] School of Physical Electronics, Univ.of Electro. Sci.& and Tech, China 610054, ChengDu,
[3] Institute of Microelectronics of Chinese Academy of Sciences, Beijing 10029, China
Yq_sun@yahoo.cn, lei.li@siat.ac.cn

Abstract

This paper presents a simple methodology for deducing the resistors of attenuators based on the relationship of S parameters and ABCD metric. Single attenuators composed of embedded thin film resisitors simulate up to 20 GHz. For 30 dB attenuators is difficult to implement for wide bandwidth, cascade attenuators can solve the problem.

Key words: thin film resistor, attenuators, cascade attenators

Introduction

Embedding passive components (capacitors, resistors, and inductors) within printed wiring boards (PWB) is one of a series of technology advances enabling performance increases, size and weight reductions, and potentially economic advantages in electronic systems. With the development of the thin film material, embedding technology has been realized. A precision integral resistor process has been successfully developed using resistive materials such as tantalum nitride (TaN), nickel chromium (NiCr) and nickel phosphorus (NiP) deposited on substrate[1]–[3]. Low cost methodologies of resistor fabrications are developed, a sheet of copper foil with an integrated NiCrAlSi resistive layer is laminated to the substrate[4]. Integrating the thin film onto rolls of copper foil can be outsourcing from a third party company. Tolerances are directly related to the etch capability, and a 10% tolerance is achievable using standard processes.

Resistors have several applications in high frequency circuits such as attenuators, terminations, wilkinson power divider, stabilising and feedback elements in transistor amplifiers et .al. As frequency increasing, the parasitic effects of embedded risistor in multi-layer circuits become complication. Models based on EM effects become necessary. Furthermore, EM oriented models with physical/geometrical parameter information allow statistical analysis and yield optimization taking into account process variations and manufacturing tolerances.

Attenuators are passive resistive elements, they are used in a wide variety of applications and can satisfy almost any requirement where a reduction in power is needed. Attenuators are used to extend the dynamic range of devices such as power meters and amplifiers, reduce signal levels and match circuits to prevent any reflections back. The thin film Resistor attenuators have several good performance such as Miniature size, high reliability, stable performance, Easy installation and Design flexibility and so on. There are five common attenuator topologies used in microwave circuits, the tee, the pi, the bridged tee, the reflection attenuator and the balanced attenuator. The tee, pi and bridged tee each require two different resistor values, while the reflection and balanced attenuators need only a matched pair of resistors. Recently, majority of designers use the attenuator of thin film resistors and their combination which are transition from Tee-type and Pi-type attenuator, make the instrumentation and electronic equipment have light weight, small size, and better performance.

Thin Film Resistor Conductance

In fabrication of thin film resistors, apart from resistor material, the substrate also plays an important role. Thin film resistors should be prepared on a suitable substrate for high frequency microelectronics applications. The basic requirements for a substrate to be used in thin film resistor are low dielectric loss and high thermal conductivity. Sheet resistance is inversely proportional relationship with the film thickness. An optimal sheet resistance can be realized by controlling the the film thickness. According to Resistance's law, we obtain the following equations:

$$R = \rho \frac{L}{S} = \rho \frac{L}{W \cdot h} = \frac{\rho}{h} \cdot \frac{L}{W} \qquad (1)$$

$$R_\square = \frac{\rho}{h} \qquad (2)$$

$$\sigma = \frac{1}{\rho} \qquad (3)$$

Where S、W、L、h、σ and ρ is the surface area、width、length、thickness、conductance and resistivity of the resistance film, respectively. $R_\square$ is the sheet resistance with units $\Omega / \square$. the resistors can be calculated in (1) using the standard DC technique, which is determined by the ratio of length to width on a planar layer, and have nothing to do with the area size; the resistivity of the resistor film can be calculated from the sheet resistance and the thichness of the thin film in (2), then the conductance can be obtained in (3). As microwave resistor, the width of resistor is allowed to adjust to be equal to that of transmission line, So the process tolerance should be controlled.

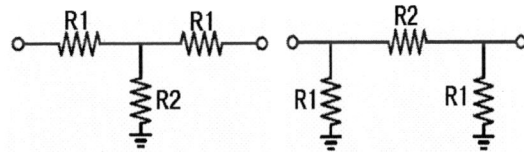

Fig.1. (a) Structure of Tee-network Attenuator

(b) Structure of Pi-network Attenuator

RF Resistor attenuators calculation

In general, Impedance convertor isn't needed in the input and output port of RF attenuator circuit, for the input and output impedance is equal to characteristic impedance. The structure of T-network and Pi-network attenuators, combination of three resistors R1 and two R2, is shown in Fig.1.(a). S21 is the magnitude of attenuation, Z_0 is characteristic impedance (Z_0 =50Ohm). From the symmetrical structure, there is $S_{21} = S_{12}$ and $S_{11} = S_{22}$, and the attenuator circuit match the impedance of system, there is $S_{11} = 0$. Two resistor R1 and R2 can be deduced from S-parameter and ABCD metric, which is as follows:

T-type's S parameter: $\begin{bmatrix} S_{11} & S_{12} \\ S_{21} & S_{22} \end{bmatrix} = \begin{bmatrix} 0 & S_{12} \\ S_{21} & 0 \end{bmatrix}$.

T-type's ABCD metric :

$$\begin{bmatrix} A & B \\ C & D \end{bmatrix} = \begin{bmatrix} 1+\dfrac{R_1}{R_2} & 2R_1+\dfrac{R_1^2}{R_2} \\ \dfrac{1}{R_2} & 1+\dfrac{R_1}{R_2} \end{bmatrix}$$

According to the relationship of the S-parameters and ABCD metric, we can obtain the following equations:

$$A = 1+\frac{R_1}{R_2} = \frac{1+S_{21}^2}{2S_{21}} \quad (4)$$

$$B = 2R_1+\frac{R_1^2}{R_2} = Z_0 \cdot \frac{1-S_{21}^2}{2S_{21}} \quad (5)$$

$$C = \frac{1}{R_2} = \frac{1}{Z_0} \cdot \frac{1-S_{21}^2}{2S_{21}} \quad (6)$$

$$D = 1+\frac{R_1}{R_2} = \frac{1+S_{21}^2}{2S_{21}} \quad (7)$$

From (4) and (6), the resistance of R1 and R2, as shown in Fig. 1, is calculated through the following equations.

$$R_1 = Z_0 \frac{1-S_{21}}{1+S_{21}}$$

$$R_2 = 2Z_0 \frac{S_{21}}{1-S_{21}^2}$$

Similarly, the electrical resistance of resistors R1 and R2 of the Pi-type attenuators is deduction as follows.

$$R_1 = Z_0 \frac{1+S_{21}}{1-S_{21}}$$

$$R_2 = Z_0 \frac{1-S_{21}^2}{2S_{21}}$$

If the attenuation dB is known, then the magnitude of attenuation is $S_{21} = 10^{-\frac{dB}{20}}$.

From the above calculation process, we can conclude that this method is simple and practical. The resistors of any form attenuator can be deduced from S-parameter and ABCD metric.

Simulation and Analyse

A field solver program uses Maxwell's equations along with specified boundary conditions to create field solutions in space and time using ODE and PDE numerical techniques. These numeric solutions find approximations to differential equations at specified points that form a mesh throughout the model. Electromagnetic Simulator was used to understand the discrepancies between the measurements and the circuit based model.

Table 1: Attenuator Resistor Values

Type	dB	R1/Ω	R2/Ω
Pi	10	96.3	71.1
	30	292.5	17.6
Tee	10	26	35.1
	30	46.9	3.2

In the field solver, Both approaches can be used to define the thin film resistors with appropriate sheet resistance: using an ideal impedance boundary and defining resistor's conductance of material properties. while the transmission line portions are represented by copper. The layout of R1 and R2 are shown in Figure 1 for both the Tee-network and Pi-network topologies. Table 1 gives the value for attenuators of a 1/10 and 1/1000 reduction in power respectively.

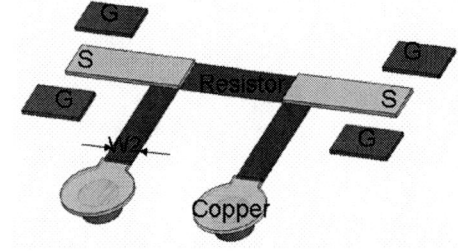

Figure 2. EM model of Pi-network attenuator

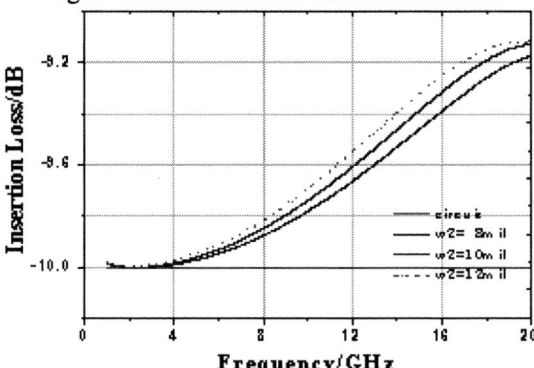

Figure 3. 10 dB attenuation

The above table 1 shows that resistor is moderate when attenuation is 10 dB, take 10 dB Pi-network atteuator for example, The simulation layout done in HFSS can be seen in Figure 2. The thickness of the film is 0.4um, sheet resistance is 25 $\Omega/\square$. The foil is laminated to 7mil thick FR4 ($\varepsilon_r = 4.0$, $\tan\delta = 0.012$) A comparation of the simulated to the circuit response can be seen in figure 3. With frequency increasing, the size of the thin film resistor can be compared with wavelength, thin film resistor become a lossy transmission line and no longer is DC resistor. When using longer resistors on low sheet resistance foils, there are tree primary ingredients: standard transmission line losses (conductor losses, dielectric losses, and radiation losses), skin effect and non-uniform lateral currents [4]. Figure 3 shows the same resistance of different size have different attenuation, the smaller size of resistor is, the better attenuation is.

Table 2: Cascade Attenuator Resistor Values

Type	dB	R1/Ω	R2/Ω	R3/Ω
Tee	30	30	19	25
	15	35	19	

When attenuation is 30 dB, resistor is more longer (292.5 Ω) or smaller (3.2 Ω) for lower sheet resistance or high sheet resistance. Embedded thin film resistors have an operating range as a result of capacitance to ground [5], very long resistors tend to show larger parasitic effects, so the thin film resistor is too longer to adopt the single-network structure of the attenuators. Casecade atteuator in Figure 4 should be work and could be have a ultra wide band. Table 2 gives the value for two cascade attenuators. Figure 5 and 6 show the insertion loss and vswr of the both anttenuator. The fluctuation of attenuation is not more than 2.5 within 16GHz.

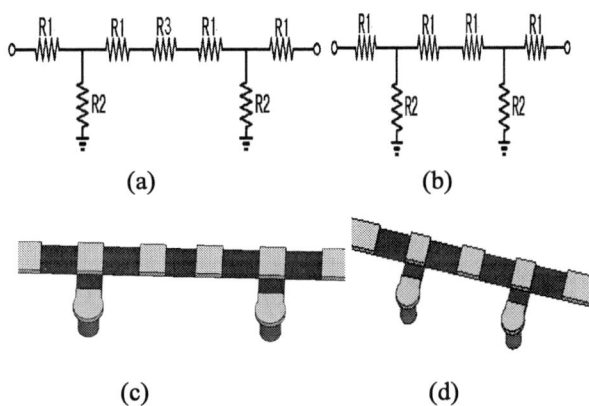

(a)   (b)

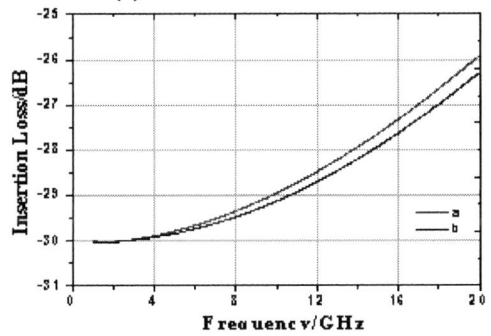

(c)   (d)

Figure 4. (a) One resistor and two Tee-network cascade attenuator. (b)Two Tee-network cascade attenuator (c) EM model of a circuit (d) EM model of b circuit

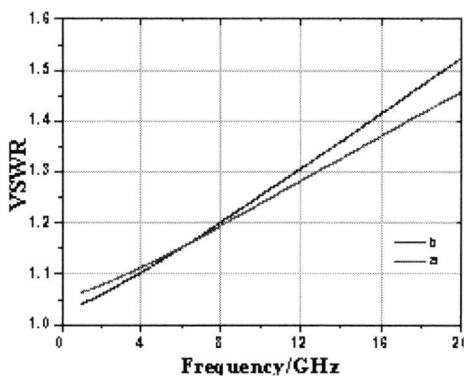

Figure 6. VSWR of 30 dB cascade Attenutor

Conclusions

This paper presented RF attenuators calculation and model simulation using resistive boundary conditions or material property in a field solver program using the finite element method. Adopting cascade attenuators can obtain ultra wide band. Using control accuracy technique of thin film resistors can receive desired results.

References

1. Renu Sharma, Seema Vinayak. et al, "RF Parameter Extraction of MMIC Nichrome Resistors," *Microwave and Optical Technology Letters,* Vol.39, No.5 (2003), pp. 409-412.

2. J. T. Wang and S. Clouser, "Thin film embedded resistors," *IPC Rev.* Jun. 2001, pp. 7–12.

3. B. P. Mahler, "Integral resistors in high frequency printed wiring boards," *Microw. J.* vol. 43, no. 2(2000), pp. 108.

4. Stephen Horst, Swapan Bhattacharya. et al, "Modeling and Characterization of Thin Film Broacband Resistors on LCP for RF Applications," *Proc 56th Electronic Components and Technology Conf,* June (2006) 1751-1755.

5. F.johsnndmann,R.Henderson. *et al*, "Parameterized RF Models of Embedded Resistor Components Using EM Simulation in LTCC Substrates," *13th European Microelectronics and Packaging Conference & Exhibition, Proceedings* · Strasbourg, May.2001, pp. 1210-1213.

Figure 5. 30 dB attenuation

These attenuators are simple and if feeder length is small, perform very well at high millimeter wave frequencies.

Task scheduling and management in single-chip multi-processor system

HuYue-li[1,2], WangYao-ming[2], XuanXiang-guang[3]

1.School of Mechatronics and Automation,

2.School of Microelectronic R&D Center Shanghai University, Shanghai 200072, China

3. Technology Center, Shanghai Feilo Co.,Ltd. Shanghai 200050, China

Key Laboratory of Advanced Displays and system Application, Ministry of Education, Shanghai University

Campus P.O.B. 25, 149 Yanchang Rd., Shanghai 200072, China

E-mail:huyueli@shu.edu.cn

Abstract

The single-chip multi-processor system, which uses the technique of parallel processing, can well improve the performance and processing ability of the CPU. This paper proposes a method of parallel processing based on the task library, and presents a strategy about task distribution and scheduling. It has been implemented by hardware and the verification about task loading and task scheduling has made, based on a single chip multi-processor system, comprised of 4 sub-processors which are all based on the architecture of Intel 8051. The verification result indicates that this method can well improve the working efficiency and parallel computing ability of SCMP system.

1. Introduction

The microprocessor has shown exceptional growth with the development of microelectronics technologies in the past 10years. The working frequency of the microprocessor has bumped up from MHz level to GHz level. Meanwhile, the processor can executes more than one instruction in the same clock cycle. The processor's capabilities have been well improved as the Instruction Level Parallelism (ILP) technique such as instruction pipeline processing; superscalar executing and Very Long Instruction Word (VLIW) are in development. However, it will become very complex in processor's hardware implementation with the increase of instruction's parallelism and complexity [1-2]. Moreover, the improvement of the processor's performance is heavily dependent on the growth of its working frequency. It will lead to higher power consumption when increases the chip's clock speed blindly [3]. This power density not only presents packaging and cooling challenges, but can pose problems for reliability as well. This kind of chip can't be used in the hand-held equipment. It is for this reason that we have to design the task parallel processing mechanism based on the architecture of single-chip multi-processor in order to resolve the problem as above-mentioned. Based on the architecture of single-chip multi-processor (SCMP),this paper not only proposes a method for parallel processing, but presents a strategy about task loading and task scheduling also. It has been implemented by hardware. The verification about task loading and task scheduling has made, base on a single chip multi-processor system ,comprised of 4 sub-processors which are all based on the architecture of Intel 8051. The results of the verification indicate that this method is feasible and it will well improve the ability of parallel processing in SCMP system.

2. Task parallel processing

2.1 The advantage of task parallel processing

We have found a shortcut for task parallel processing, depending on the development of micro-electronics and character of single-chip multi-processor's architecture. We have shift from instruction level parallelism (ILP) to task level parallelism (TLP) for the purpose of improving the performance of the processor. The parallelism of the program has shifted from coarse-grained to fine-grained also. Three are three factors for realizing the parallelism. First, the program exist the parallelism; second, parallelism can be extracted; third, the hardware must support the parallelism [4]. The architecture of single-chip multi-processor and parallel programming based on task library can meet the requirements of above-mentioned, which are necessary for parallelism. The programmer can call the tasks which lie in the task library, utilizing the idea of modular programming. The single-chip multi-processor makes the tasks to be executed parallel in the sub-processors, realizing the parallel processing.

2.2 Parallel programming based on task library

For simplifying the procedures of designing the compiler and parallel program programming, parallel programming based on task library can well satisfy this demand [5-6]. The tasks can be either some simple arithmetic and logical operation or some complex control algorithms. The tasks have interface parameter and can be located by task number. The programmer could call these tasks through task name, then the compiler generates the task number flow, using these task names. The task scheduling and management module could assign and manages the tasks according to the task number flow. Figure 1 illustrates the architecture diagram of parallel programming system.

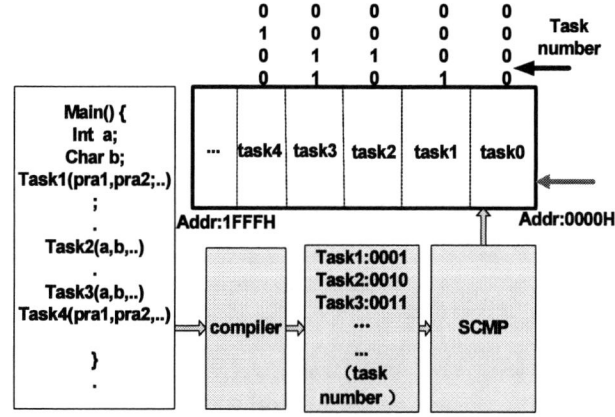

Fig.1 Programmer system based on task library

2.3 The task scheduling and management system in the SCMP

The task scheduling and management system may be considered to be an operating system which is running in the SCMP system, but it is implemented in hardware. It works like a software operating system which is characterized for task scheduling, task management and memory management. The user program is a master process, to the task scheduling and management system, which contains many child processes and subtasks. This hardware operating system can assign the tasks to the sub-processor automatically to achieve the task parallel processing, according to working conditions of the sub-processors and memory usage of the whole system. This task scheduling and management system could pre-fetch instruction which can be considered as pre-fetch the task number. The task number which has been fetched will be put into the task pool. The volume of the task pool is always proportionate to the number of sub-processors.

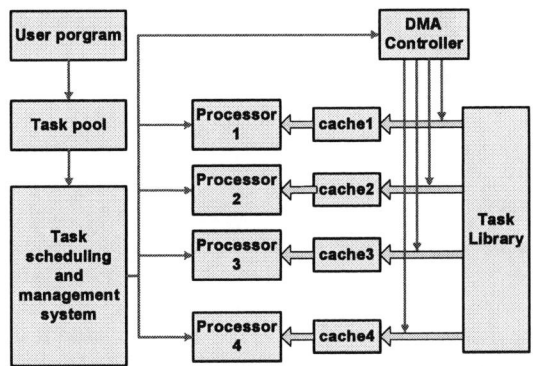

Fig.2 System Architecture of single chip multi-processor

The task scheduling and management system directly assigns the tasks to the sub-processors which are free. Figure 2 shows the architecture of single-chip multi-processor system which contains the task scheduling and management module.

3. Task scheduling and management

3.1 Task loading

It is not only that the executable code, which will be executed in the sub-processor, can be found in the task library according to the task number, but the task's physical address that lies in the task library will be found also. The sub-processor has its local high-speed cache. The task scheduling and management system assigns the task to the sub-processor and completes a direct memory access (DMA) operation at the same time. The DMA operation will load one of the task's executable code that is stored at the task library into the sub-processor's local cache. It will ensure that all the code that the sub-processor requires lies at the local cache and make the Cache Hit Rate reach to 100%. When the DMA operation has been accomplished, the dormant sub-processor will be awakened to process the new task. Although it takes some time in loading the task's executable code into the sub-processor local cache, it will make the Cache Hit Rate reach to 100% and the sub-processor never has to read the instruction and data from the task library. It is impossible that the read conflict happens among the sub-processors. The new system would improve the processor's processing ability consumedly. This study adopt 100% task loading and can be called coarse-grained parallel processing.

3.2 Relevant tasks and irrelevant tasks

The tasks can be classified into relevant tasks and irrelevant tasks. The tasks that have links among them can be called relevant tasks such as producer process and consumer process. Relevant tasks need data communication and data generated by prior task will be used in the later. The tasks that have no links among them can be called irrelevant tasks. They don't need message interacting. To the relevant tasks, it is necessary to design task directive statements to point out the correlation between two tasks when you are using task library for programming and compiler developing so that the compiler can generate correct task number flow quickly, using these directive statements. To the irrelevant tasks, however, they have no use for task directive statements. The compiler just needs to map the task name to task number. Existing architectures of SCMP are Shared Memory structure and Message Routing structure. This study adopts the first one and messages of relevant tasks are passed using public variables in the shared memory.

3.3 Task scheduling and management

In single-chip multi-processor system, the working efficiency of each sub-processor directly affects that of the entire system. Therefore, it is crucial to develop a reasonable strategy for task scheduling and distribution in order to enhance the working efficiency of the single-chip multi-processor system. In our proposed system, four sub-processors are employed and a forward inquiry-distribution strategy is adopted. The forward inquiry-distribution strategy, which cyclically inquires the status of each sub-processor in sequence (P1-P2-P3-P4-P1) will immediately execute a DMA task loading ,awake the certain sub-processor and set its working status as "busy" when it detects the idle status of that sub-processor. When the sub-processors are ALL BUSY, the process of task distribution is halted and tasks are prevented from streaming into the task library. And when any one of the sub-processor turns into idleness, task is immediately distributed to that idle one and new task is fetched from the task library to the task pool. Figure 3 is the Schematic plot of forward inquiry-distribution strategy.

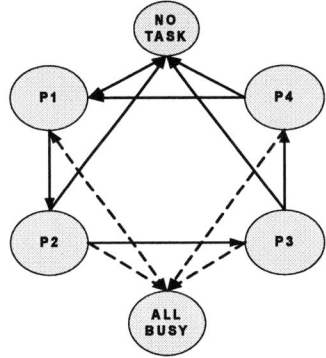

Fig.3 Schematic plot of forward inquiry-assign strategy

In our proposed system, it has been defined a task register used to indicate the number of tasks in the task pool. When a new task flows into the task pool, a right shift operation of task register will be done and the upper bit will be set. On the contrary, when a task flows out of the task pool, a right shift operation of task register will be done also and the upper bit will be reset conversely. Thus, the number of the tasks in the task pool can be counted by the number of set bits in the task register. Figure 4 illustrates the modifying operation of task register.

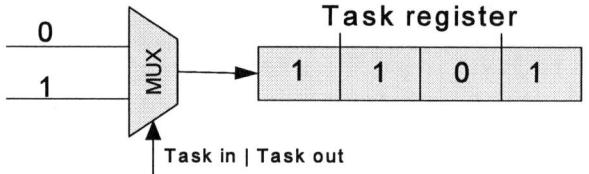

Fig.4 Schematic plot of task register modification

We also define a sub-processor state register in our proposed system. Every bit of this register indicates the working status of a certain sub-processor. When a new task has been assigned, the corresponding bit will be set and indicates that the sub-processor is busy. When the sub-processor has accomplished the assigned task, it will reset the corresponding bit to indicate this processor is free. Task scheduling and management system monitors the value of these two registers continuously and achieves task scheduling and management.

4. Verification of task scheduling and management system

In this proposed system, task pool has a capacity of 4 tasks and there are 4 tasks to be distributed. The MCU will output two PWM signals with Duty ratio of 50% and 25% when task1 has been executed. Task2 is an operation of consecutive multiplication and task3 is an operation of consecutive addition. Task4 is a program which will let the MCU output the ASCI II character of Shanghai University. There are four MCUs which are compatible with Intel 8051 instruction set in our system. Task library is generated by memory synthesis tool named Memory Compiler based on SMIC0.18 CMOS process. It is an ROM with storage capacity of 4K and bit-width of 8.

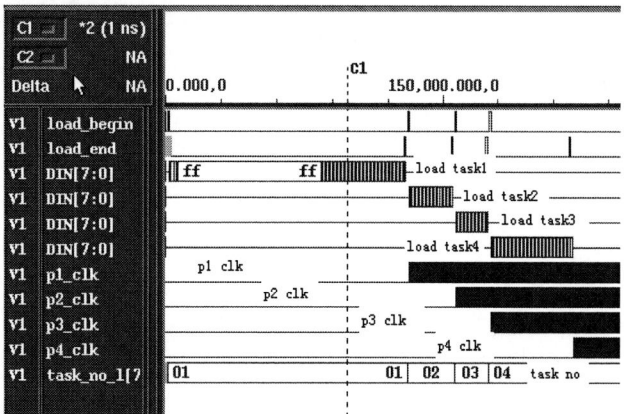

Fig.5 Simulation result of task loading and assignment

We assume that the task flow sequence is defined as follows: 0001->0010->0011->0100 and these tasks have been filled into the task pool preliminarily. Figure 5 illustrates the simulation result of distribution operation. As shown in Figure 5, load_begin and load_end present task loading begin signal and task loading end signal, respectively. DIN stands for the name of data bus which connects local cache to task library.

Clocks of sub-processors are presented by signal p1_clk, p2_clk, p3_clk and p4_clk, respectively. It can be seen from Figure 5 that the clock of a sub-processor won't be valid until certain task has been assigned and task scheduling and management module distributes four tasks to sub-processors automatically, following the task flow:0001->0010->011->0100. Figure 6 shows simulation result of DMA operation. In figure, DMA module works correctly when it control the read operation to sub-processor's local cache.

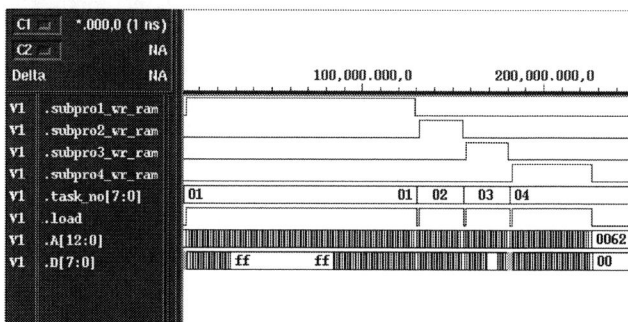

Fig.6 Simulation result of DMA module

For instance, when task_no equals to 0001, task scheduling and management system direct writes the executable code of task1 into sub-processor's local cache. During the time of DMA operation, write enable signal named subpro_wr_ram keeps valid till one task has been loaded into local cache completely ensuring all of the task's machine code can be loaded. The different size of each task results in different DMA operation time and different holding time of write enable signal to each task. Because task1 is larger than other tasks, it costs most time for DMA operation as shown in Figure 6. It can be seen in Figure 7 that subprocessor1 outputs two PWM signals with desired Duty ratio, subprocessor2 performs consecutive multiplication, subproessor3 performs addition all the time and subprocessor4 continuously outputs the ASCI II character of

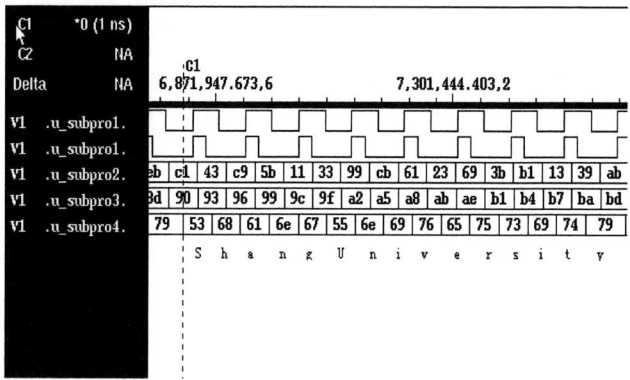

Fig.7 Simulation result of task processing in sub-processors

Shanghai University correctly. All the processors work correctly according to task flow. We can make a conclusion from simulation result that task management module, task distribution module, task execution module and DMA operation module are all in normal operation.

5. Conclusion

This paper mainly introduces task parallel processing in single-chip multi-processor system and proposes a task scheduling and management system. After analyzing the features of task parallel processing in SCMP system, a task parallel processing method is proposed and task loading and scheduling strategy is presented. It has been implemented by hardware. The verification about task loading and task scheduling has made, base on a single chip multi-processor system ,comprised of 4 sub-processors which are all based on the architecture of Intel 8051. The results of the verification indicate that this method is feasible and it will well improve the ability of parallel processing in SCMP system. When this method is adopted, the multi-tasks parallel processing can be easily realized.

Acknowledgments

The authors would like to thank National Natrural Secience Foundation of China (Foudation Number: 60773081，60777018) and Postgraduate Innovation Fund of Shanghai University for financial support.

References

1. XiangHuifang, HuYueli, "Design of single-chip multi-processor Architecture Based on SDZX-MV-02 MCU IP CORE," Computer easurement&control, Vol.14, No. 7(2006), pp. 942-945.

2. GuoLamei, HuYueli,Design, "A kind of MCU Bus Architecture," Computer Measurement &Control, Vol.13, No.7(2005), pp. 715-717.

3. Michael Keating, David Flynn, Rob Aitken, Alan Gibbons, Kaijian Shi, Low Power Methodology Manual: For System-on-Chip Design, Springer, pp. 50-125.

4. ZhangMingxuan,WangYongwen, High-Performance Micr-
oprocessors:Techniques and Architecture, National University of Defense Technology Press, (Changsha, 2004), pp. 98-120.

5. Anath Grama, Anshul Gupta, George Karypis, Vipin Kumar, Introduction to parallel computing (Second Edition), Pearson Education Limited(New York, 2003), pp. 25-9.

6. Hesham EI-Rewini, Mostafa Abd-EI-Barr. Advanced Computer Architecture Processing, John Wily & Sons, Inc(USA, 2005), pp. 100-120.

Analysis of Factors Affecting Color Distribution of White LEDs

Zongyuan Liu[12], Sheng Liu[123]*, Kai Wang[23], Xiaobing Luo[24]

[1]Institute for Microsystems, School of Mechanical Science and Engineering, Huazhong University of Science & Technology, Wuhan, China, 430074

[2]Division of MOEMS, Wuhan National Laboratory for Optoelectronics, Wuhan, China, 430074

[3]School of Optoelectronics Science and Engineering, Huazhong University of Science & Technology, Wuhan, China, 430074

[4]School of Energy and Power Engineering, Huazhong University of Science & Technology, Wuhan, China, 430074

Email: victor_liu63@126.com, Telephone: 86-27-87542604

Abstract

The color uniformity is a critical index in the evaluation of high quality white light emitting diodes (LEDs). The main factor affecting the color distribution is the state of the phosphor. The secondary factor is the optical structure. This paper analyzes two parameters of the phosphor layer (thickness and concentration) and six optical structures. Results indicate that the structures with reflector have lower color uniformity. The hemispherical shape of phosphor layer could improve the performance of color distribution and make the variation smaller. Between the two parameters of phosphor, the concentration has the greatest impact on the color uniformity, the second one is the thickness. Through the analysis, it is suggested that the thickness and concentration should be precisely controlled. Otherwise, the color uniformity will be varied significant and may generate serious yellow rings.

Introduction

Light emitting diodes has made remarkable progress on the past four decades with the rapid development of compound semiconductor technology. Since the first red LEDs that was invented by Holonyak and Bevacqua in 1962[1], great efforts have been put into study to obtain brighter LEDs. In early 1990s, Nakamura successfully grew the blue and green LEDs on GaN substrate, which has a profound impact on the progress of LED technology[2, 3]. Nowadays there are three basic methods to obtain white LEDs, one is the monochromatic RGB LEDs, another is phosphor converted LEDs (PC LEDs), the third is based on UV LEDs. The most commercially available white LEDs is dichromatic PC LEDs[4], which generates the white light by mixing the blue light from LED chip with the broadband yellow light excited by phosphor[5, 6]. Since the white light LEDs has superior characteristics such as high efficiency, small size, long life, dependable, low power consumption, high reliability and mechanically rugged[7], therefore, a new concept of illumination named solid state lighting (SSL) in terms of high power LEDs is proposed[8]. With the accelerated advancement of LED technology, the input electrical power of LEDs was jumped to 1W or more by increasing the chip size from 350 microns to 1mm. The luminous efficiency of high power LEDs was also increased from 25 lm/W in 1999 to more than 100 lm/W in 2007 for a driving current of 350mA[9]. Consequently, the market for high power LEDs is growing rapidly in various applications such as large size flat panel backlighting, street lighting, vehicle forward lamp, museum illumination and residential illumination[10, 11]. It has been widely accepted that SSL will be the fourth illumination source to substitute incandescent lamp, fluorescent lamp and HID lamp[12]. The broad application prospect of SSL has attracted great attention on the study of LEDs[13].

The main function of LED packaging is to protect the LED chip, enhance the light extraction and provide a path for dissipating the generated heat. The packaging is the critical bridge between the LED chips and applications for end-users. Now the LED packaging technology has been improved much to make full of the potentials of rapidly developing high power LED chips[14, 15]. Products of corporations such as the Lumileds's Luxeon K2 series, the Cree's XLamp series, the Osram's Golden Dragon series, offer high performance LED modules in terms of various packaging structures, in which the phosphor is dispersed on the surface of chip. To improve the packaging performance such as the light extraction efficiency, researcher have made remarkable efforts and achieved many valuable results[16, 17, 18, 19], for example, the scattered photon extraction method (SPE) [20, 21], the silicone based packaging platform [22], wafer level packaging [23], et al..

High luminous efficacy, in general, is the main purpose of optical design to improve light extraction, but the color uniformity should not be neglected in the efforts to obtain high quality white light LEDs. It is another key evaluation index like the correlated color temperature and color rendering index, but nowadays there doesn't attach enough importance on the color distribution. Non-optimized or un-suitable structure and process could induce unexpected phenomenon such as the yellow ring. We found this situation in many corporations' packaging modules as shown in Figure 1. This color ununiformity (adj not a noun!) will influence the real performance of illumination, thus leading to discomfort for the eye's vision.

Figure 1. The yellow ring in some LED modules.

Fundamentally, the main factor affecting the color distribution is the state of the phosphor. The secondary factor is the structure that could change the light propagation and intensity distribution in the space. There are energy conversion and re-emission, scattering and absorption in the phosphor layer. Both the blue light and converted yellow light must be sufficiently scattered by the phosphor particles to generate uniform white light. Different optical structures with various states of phosphors in terms of location, thickness, concentration, shape, etc. will greatly influence the performance of LEDs, especially the color uniformity. Lumileds proposed a new phosphor coating technology named conformal coating, which dispersed a uniform phosphor film on the surface of the chip. This can improve that the variation of the CCT drops to ~80K[24].

validated and the total ray trace number was finally 800000 to obtain converged results.

Problem Statement

The LED module generally consists of the heat slug, chip, phosphor, silicone packages and lens. Reflector is also necessary for some modules, for example, the Golden Dragon of Osrams. It is well known that the performance of LED module greatly depends on the technology of chip, but the packaging elements are also critical in the process of trying to achieve high quality white light LEDs. In the optical design for the packaging, the primary considerations are the phosphor, lens and reflector.

Table 1. The optical structures and variable parameters for the numerical simulation.

I		Location (mm)	0	0
		Thickness (mm)	**0.04~0.2**	0.1
		Concentration (mm⁻¹)	19.85	**9.92~49.5**
II		Location (mm)	0.05	0.05
		Thickness (mm)	**0.04~0.2**	0.1
		Concentration (mm⁻¹)	19.85	**9.92~49.5**
III		Location (mm)	2	2
		Thickness (mm)	**0.04~0.2**	0.1
		Concentration (mm⁻¹)	19.85	**9.92~49.5**
IV		Location (mm)	0.05	0.05
		Thickness (mm)	**0.04~0.2**	0.1
		Concentration (mm⁻¹)	19.85	**9.92~49.5**
V		Location (mm)	1.9	1.9
		Thickness (mm)	**0.04~0.2**	0.1
		Concentration (mm⁻¹)	19.85	**9.92~49.5**

To further investigate the focused issues in the efforts of trying to obtain high color uniformity for white light LEDs, this paper discusses various parameters of phosphor layer on the performance of LEDs. The parameters include the thickness, concentration. In addition to prove that the results are reliable and implantable for packaging production, six different optical structures are studied. A non-sequential ray trace method was applied, which was based on the Monte-Carlo theory. The power ratio of yellow light to blue light, which is named yellow blue ratio, is used to illustrate the correlated color temperature (CCT). The curve of color distribution was fitted by non-linear least square method. The uniformity of color distribution is the ratio of the max value to the min value in the range of ±65° of the emitted angle, in which the light intensity is considered effective for the design of illumination lamps. The parameters of the models were

The phosphor layer is the mixture of phosphor particles and silicone. It absorbs blue light and re-emits yellow light. When alternating the thickness and concentration of the phosphor layer, there will be different optical behaviors for blue light and converted yellow light. The variations of these behaviors will affect the initial energy of extracted blue light and yellow light.

The lens and reflector are normally used to enhance the light extraction and change the light distribution by the refraction of materials and reflection on the interface. They could affect the final energy of extracted blue light and yellow light. Now the commonly used packaging materials

However, there still are not enough studies on these elements to better understand the design guidelines for nowadays used packaging structures. As the basic steps to realize these factors, the numerical simulation and analysis are focused on the issues about the relationship between the color uniformity and the characteristics of phosphor layer and the packaging structure. The effects of phosphor layer are the primary considerations in this paper. Obviously it is difficult to explain the influencing mechanism only by these simple models, but this investigation could make clear of what the optical design should avoid. Therefore, the objectives of this paper are 1)To simulate different structures with various states of phosphor layer; 2) To explain the simulation results and trying to find the principles about these factors; and 3)To provide some suggestions for the optimized optical design.

Numerical Model and Simulation

This paper mainly concerns six different optical structures that are depicted in Table 1. For each structure, the thickness and concentration of phosphor are changed to investigate the effects of variation. To evaluate the impact of the reflector on the color uniformity, three numerical models in terms of type IV, V and VI have the reflector to compare the front three non-reflector models of type I, II and III. The base and top diameter of the reflector cup is 3 mm and 8mm, and the height is 2 mm. The optical parameters of the surface on the board and reflector are assumed to have 85% perfect reflection, 5% scattering and 10% absorption. The radius of the lens for all types is large enough to 4 mm to minimize the effects of lens's size on the light propagation. Since this paper mainly discusses the effects of phosphor, the dimensions of the reflector and the lens are assumed to be the same.

In the first model, the phosphor is directly coated on the chip. The variable parameters are the thickness and concentration. In the second and fourth structures, the chip is firstly packaged with a thin silicone film, and then coated with the phosphor layer. Both types increase the thickness of the silicone film to change the distance between the phosphor layer and the chip. When the variable parameters are the thickness and concentration, the distance is kept to be 0.05mm. In the third structure, the phosphor layer is a hemispherical film with increasing the radius to change the location. The fourth structure firstly disperses the silicone in the reflector, and then disperses the phosphor silicone uniformly to make a thin phosphor film. Through increasing the volume of the inner silicone, the height of the film could be changed from 0.2 mm to 1.9 mm. The sixth structure only discusses the effects of location, which increases the radius of the spherical phosphor layer. For all types, when one of the three variable parameters is changed, the another two parameters are kept un-changed. For example, when the location is changed, the thickness and concentration are 0.1 mm and 19.85 mm^{-1} in these models. This paper applies the sum of the absorption and scattering coefficient to represent the concentration. This will be explained in the following discussions.

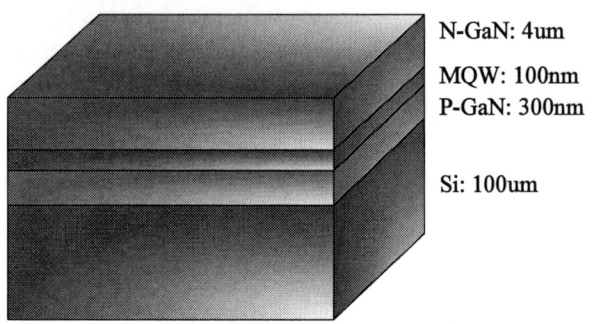

Figure 2. The LED chip model for the numerical simulation.

The LED chip model is depicted in Figure 2, which is a simplified 1mm×1mm Cree LED chip. The top surface of the Si is coated with Au to reflect the back light. We considered that the light emits uniformly from the MQW layer, since the direction of the photons that are excited by the combination of the electrons and holes in the MQW layer is arbitrary. Considering that the area of the top and down surface is much larger than that of the side surface, the model defines the two surfaces as the light sources by ignoring the side emitting lights. The optical properties are illustrated in Table 2[25].

Table 2. Optical properties of the LED chip.

	N-GaN	MQW	P-GaN
Refractive Index	2.42	2.54	2.45
Absorption Coefficient (mm^{-1})	8	8	8

The definition of the optical properties for the phosphor layer is very important, since it relates to the correction and precision of the simulation results. When one beam of blue light passes through one phosphor particle, the phenomenon such as the transmission, refraction, scattering and absorption will occur simultaneously, and thus affects the light propagation and weakens the blue light energy. The absorbed blue light will re-emit broad-band yellow light, which should be affected by other phosphor particles. The average radius of phosphor particle is 5~8 microns as shown in Figure 3, and generally the number of particles is 10000~100000 per mm^3 in the phosphor silicone. Since there may be millions of phosphor particles in the phosphor layer, the light will encounter many particles in the propagation path and thus be scattered many times before it transmits through the phosphor layer. During the multi-scattering process, the blue light energy will be weakened gradually by absorption, but the converted yellow light energy will be increased for each scattering. The scattering will also influence the intensity distribution of the light rays. It's well known that the white light is generated by the combination of the yellow light and blue light. If the power ratio of the yellow light to the blue light is too high or too low in partial zones of the space, the color of the light will tend to become yellow or cool white. This is the main reason affecting the color uniformity.

Figure 3. SEM picture of the phosphor particles.

Since the optical behavior of the light in the phosphor layer is so complicated, the simplification of the optical properties for the phosphor is necessary to make the simulation practicable. Considering the phosphor layer as a bulk scattering material, it is an easy and effective method that applies the scattering and absorption coefficient to represent the totally effects of the phosphor particles on the light propagation [26, 27, 28, 29]. Therefore, when a beam of rays passes through the phosphor layer, the transmitted light energy can be expressed by Equation 1.

$$I(x) = I_0 e^{-(\mu_\alpha + \mu_s)x} \tag{1}$$

where I_0 is the initial power of the light and $I(x)$ is the residual power; μ_α and μ_s are the absorption coefficient and scattering coefficient of the phosphor; and x is the thickness of the phosphor layer. Because the variations of the concentration could influence μ_α and μ_s, this paper applies the sum of μ_α and μ_s to represent the concentration. High concentration indicates that the sum is bigger.

The scattering distribution function of the phosphor layer, which is used to generate the random direction of the scattered light, is based on the Henyey-Greenstein model. It is indicated in equation 2.

$$p(\theta) = \frac{1 - g^2}{4\pi(1 + g^2 - 2g\cos\theta)^{\frac{3}{2}}} \tag{2}$$

where g is called the anisotropy factor. In the phosphor layer, since the scattering of the light by the particles is isotropy, g is zero.

Since there are two parameters for the phosphor layer, the definition for the optical properties should be specific for each case. When the thickness is changed, the concentration of the phosphor is unchanged in the first case. Therefore, the optical parameters keeps the same. This paper assumes that the absorption and scattering coefficient is 8 mm-1 and 11.85 mm-1 for blue light. However, it is considered to be transparent when the incident ray is yellow light. Therefore, the absorption coefficient is set as 0 mm^{-1}. Because of the non-absorption, the yellow light will be scattered more times than the blue light, which induces that the yellow light should be exhausted later by the phosphor particles in the propagation process. Therefore, the scattering coefficient is 16.25 mm^{-1} for yellow light.

For the second case, since the concentration is changed, the variation of the optical parameters is listed in Table 3.

Table 3. Optical parameters of phosphor in the second case.

	Absorption Coefficient (mm^{-1})		Scattering Coefficient (mm^{-1})	
	Blue light	Yellow Light	Blue Light	Yellow Light
1	20	0	29.5	40.45
2	13.3	0	19.8	26.95
3	10	0	14.8	20.33
4	8	0	11.85	16.25
5	6.65	0	9.88	13.48
6	5.72	0	8.465	11.62
7	5	0	7.4	10.173
8	4.44	0	6.57	8.99
9	4	0	5.92	8.09

Since the transmittance of the silicone materials is over 95% for visible light, it is generally considered that the silicone materials are transparent to visible light. Therefore, the absorption coefficient is 0 mm^{-1} in the numerical simulation. The refractive index of silicone is 1.5, which can be bought in corporations such as DowCornig.

Based on the above settings, the simulation steps of the numerical model are:

1) Defining the material properties of each layer;
2) Defining the top and down surface of the MQW layer as the light source, and then inputting the light power;
3) Ray tracing and collecting the simulation data;
4) Re-defining the material properties of each layer;
5) Re-defining the top and down surface of the phosphor layer as the light source, the power of which is calculated from the absorbed blue light; and the conversion efficiency of the phosphor is 80%;
6) Ray tracing and collecting the simulation data; and
7) Calculating the simulation results of each case.

In this simulation procedure, the blue light output and yellow light output are computed separately to easily investigate the impaction of the phosphor's states on the color temperature. The yellow blue ratio, which is the value of the yellow light power divided by the blue light power, is used to represent the correlated color temperature.

Therefore, the color distribution of the LEDs is the variation of the yellow blue ratio in the space. High yellow blue ratio means that the CCT is lower and the light tends to be warm white. Oppositely, low yellow blue ratio means the CCT is higher and the light tends to be cool white.

The color uniformity is the ratio of the maximum value to the minimum value in the effective angels. In the following discussions, to calculate the color uniformity, this paper only analyzes the data in the range of ±65° of the emitted angle. This is due to the fact that most of the light energy is in this range, and the intensity of the side angle is obviously small and may be undetectable by eyes.

Simulation Results and Discussions

The simulation results are shown in the following figures. Each figure contains the color distribution curve, the color uniformity and the illustration for the specific case.

The effects of phosphor's thickness on the color distribution are depicted in Figures 4-8. It can be found that the yellow blue ratio curve is ascended when the thickness is increased. That means the color tends to be warm white or yellow light. The rising tendency of the side angles are more significant than that of central angles, especially in the first, second and fourth structures. That means there are fewer flat zones but more fluctuant zones in the cure. Therefore, the color uniformity in those cases is reduced gradually and the phenomenon of yellow ring is more serious. Although we could find the same variation of the curve in the third and fifth structures from 0° to 180°, the variation in the range of ±65° is not so significant. The color uniformity is even slightly increased.

The reflector also has great impact on the color uniformity when the thickness is changed. For example, the color uniformity in the second structure is around 0.5~0.6, but in the fourth structure it is aroud 0.3~0.4. The maximum value points in the curves are also obvious higher than those without reflector.

The remote phosphor location could make the variation of color uniformity smaller. In the third and fifth structure, the color uniformities are even increased slightly.

It is a better choice to utilize the thinner phosphor layer to improve the color uniformity, but attention is paid to the color trends of the LEDs. Because the concentration is not changed, the thinner phosphor layer in all cases has high CCT or even blue color in some zones. Therefore, in the progress to achieve high color uniformity LEDs, the thickness is not the unique approach.

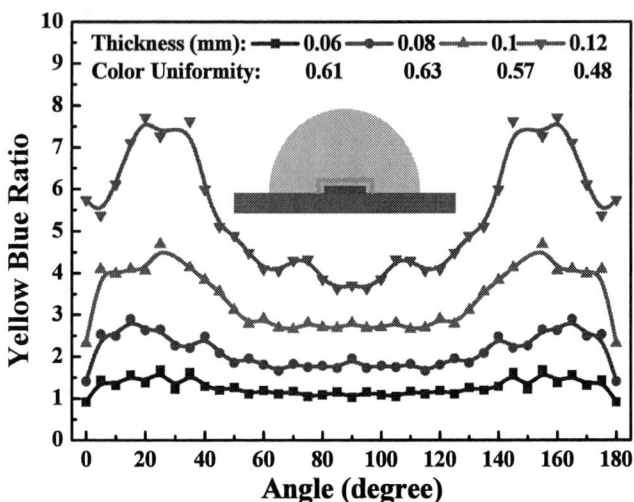

Figure 5. Effects of phosphor's thickness on the color distribution for the second structure.

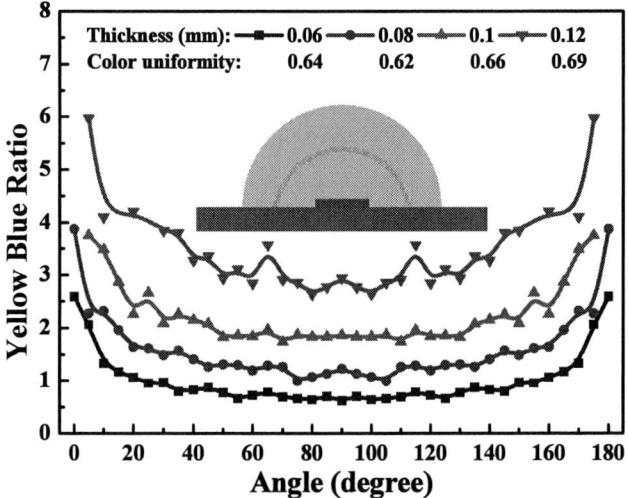

Figure 6. Effects of phosphor's thickness on the color distribution for the third structure.

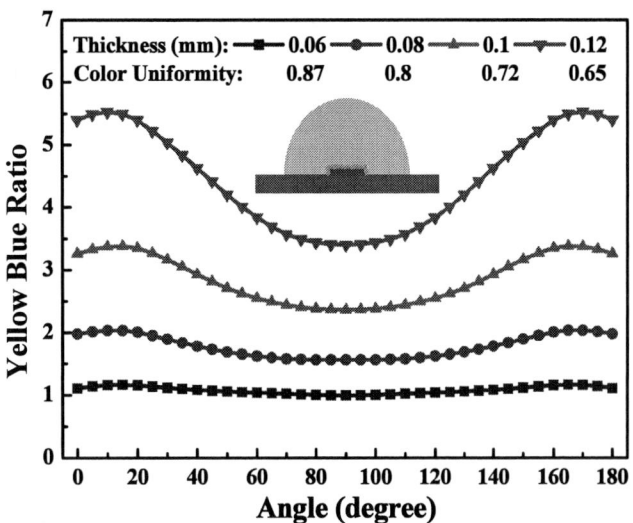

Figure 4. Effects of phosphor's thickness on the color distribution for the first structure.

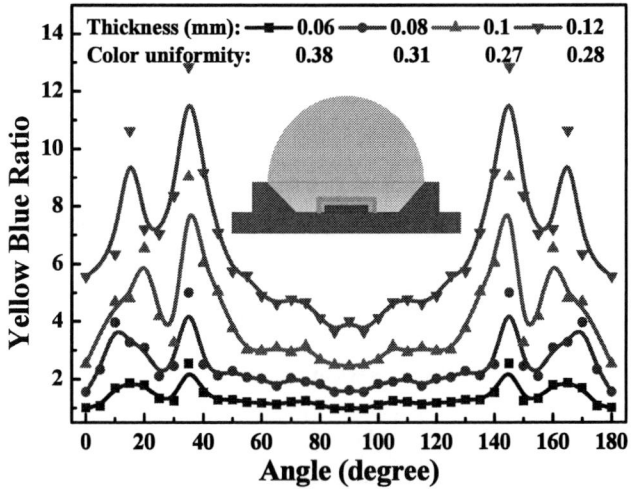

Figure 7. Effects of phosphor's thickness on the color distribution for the fourth structure.

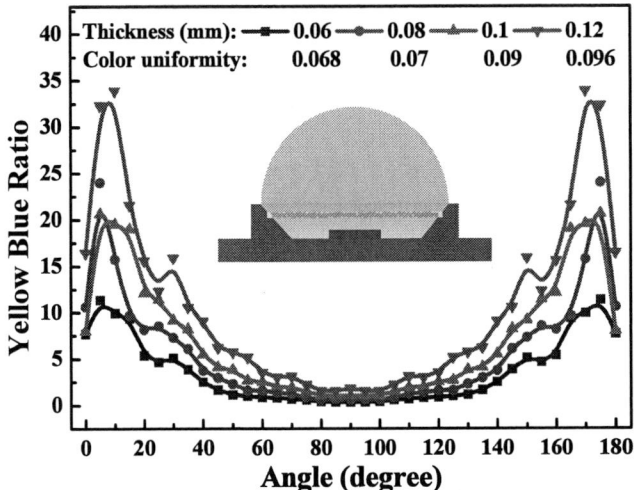

Figure 8. Effects of phosphor's thickness on the color distribution for the fifth structure.

The effects of phosphor's concentration on the color distribution are depicted in Figures 9-13. The results show that the concentration has almost the same influence on the color distribution. When the concentration increased, the yellow blue ratio curve is ascended. The rising tendency of the side angles is also more significant than that of central angles. The yellow ring will also appear and be more serious with the increase of concentration. The color uniformity for structures with reflector is also higher than other structures without reflector. However, there still are some different variations of the color distribution, which indicate that the concentration could affect the color distribution through different manner.

Firstly, once the concentration is more than the 19.85 mm^{-1}, the yellow blue ratio will jump up to be higher than 10. This color is almost yellow light and is not what the LED packaging desires. But if the concentration is lower than 19.85 mm^{-1}, the color of the LEDs is generally fluctuated in the acceptable ranges. Therefore, when considering the deposition of the phosphor particles in the silicone, the concentration should be paid special attention in the production to avoid generating partial high concentration areas.

Secondly, the variation of the color uniformity is not linearly reduced with the increase of concentration. In the second, third and fourth structures, the maximum value is among the middle concentration. In the fifth structure, the color uniformity is even linearly increased as that of thickness.

Thirdly, with the increase of the concentration, the variation of the color uniformity is bigger than that of thickness. For the structures with remote phosphor location, the variation is also significant. For example, in the third and fifth structures, the fluctuations are 21.7% and 45.5%. In the first case, the fluctuations are 10.1% and 29.2% with the increase of thickness. This indicates that the concentration could change the color distribution more effectively. This mainly dues to the enhanced scattering and absorption effects of phosphor.

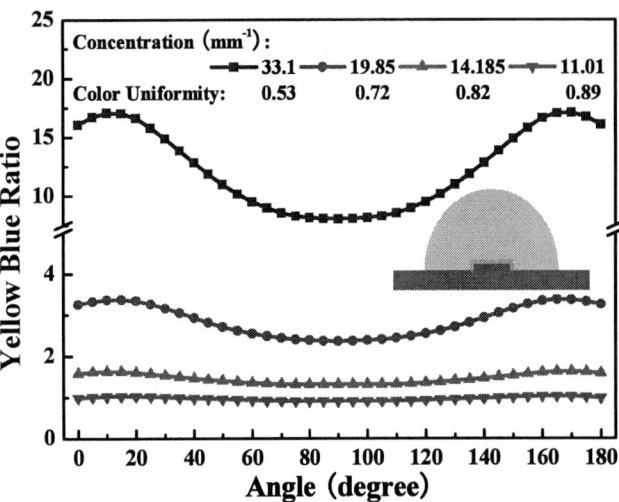

Figure 9. Effects of phosphor's concentration on the color distribution for the first structure.

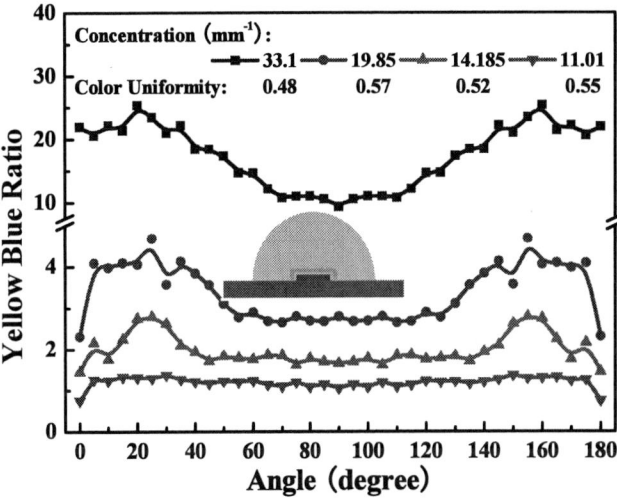

Figure 10. Effects of phosphor's concentration on the color distribution for the second structure.

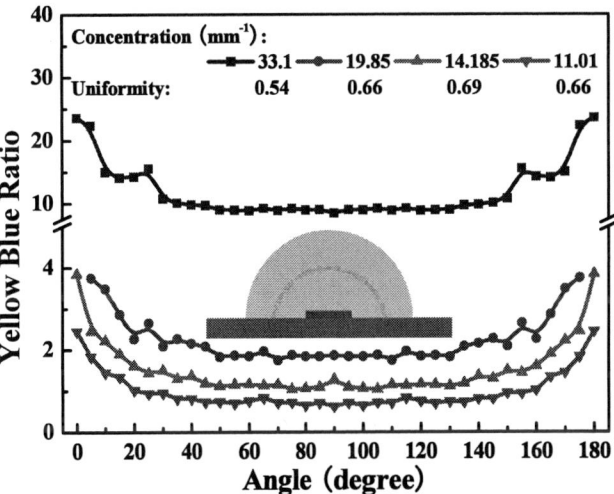

Figure 11. Effects of phosphor's concentration on the color distribution for the third structure.

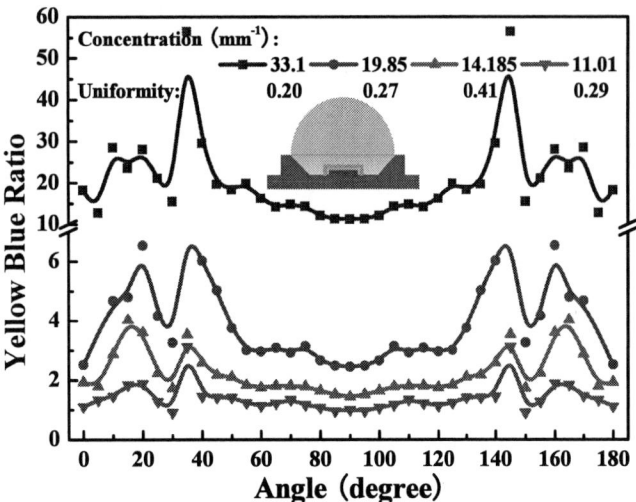

Figure 12. Effects of phosphor's concentration on the color distribution for the fourth structure.

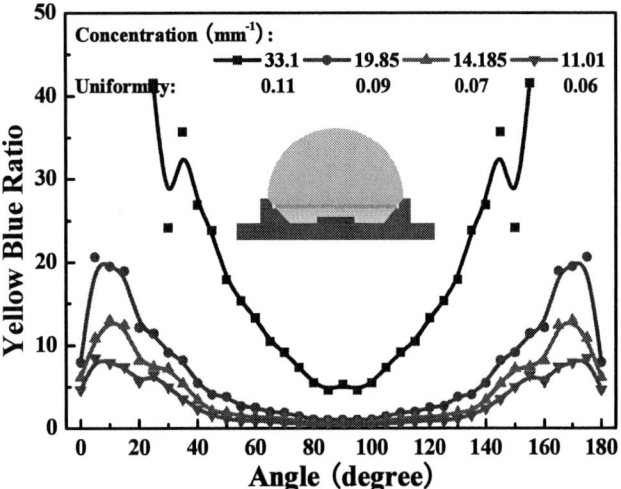

Figure 13. Effects of phosphor's concentration on the color distribution for the fifth structure.

Through the comparison of the effects of the two parameters in the phosphor layer, it can be conclude that the concentration has the biggest influence on the color distribution, the next factor is the thickness. Both the thickness and concentration could greatly affect the value of yellow blue ratio and the color distribution curve simultaneously. However, the concentration could influence the fundamental characteristics of phosphor more significantly. When the thickness and concentration are changed, the total effects of the absorption and scattering for the phosphor layer are also changed at the same time. Thicker or higher concentration layer could absorb and scatter more light rays, and thus increase the extracted yellow blue ratio. Therefore, small variation of the thickness and concentration will influence the color remarkably.

The another important factor affecting the color distribution is the optical structure. There are enormous differences whether the structure contains reflector and the shape of the phosphor layer is curved or not. The reflector reflects the side light rays and changes the rays to different directions. Since the light emitted from the chip is lambertian, most of the blue light could be extracted out without reflection. After the convergence of the lens, there is only a little blue light energy in the side angles. But the light emitted from the phosphor layer is arbitrary, part of the side yellow light could be reflected to the central angles while other parts could be reflected to the side angles. Therefore, the yellow blue ratio is reduced in central angles and increased in side angles. If there is not reflector, both of the blue light and yellow light could be converged by the lens without other disturbances. Although the emitted patterns for chip and phosphor are different, without the enhancement of the yellow light in the central and side angles, the differences are not as large as those cases with reflector. Therefore, the color uniformity is higher.

It can be found that the color uniformity of the third structure is more stable than other structures. This is due to the surface of the phosphor layer is curved. The curved surface could improve the critical angle for the light and provide more chance to let the rays extracted from the surface. In the third structure, the surface is hemispherical and the chip is located in the center. Therefore, the blue light emitted from the chip could keep the intensity distribution still like lambertian after passing through the phosphor layer. Since the phosphor layer and the lens are homocentric, the yellow light emitted from the phosphor layer could also easier emit out and keep the radiation pattern like what it is excited. That means the intensity distribution of the yellow light is near lambertian. The variations of the location, thickness and concentration could not change the fundamental effects of phosphor layer on the light propagation. Therefore, the color uniformity of curved surface is high and more stable.

Conclusions

Two parameters of the phosphor layer and six optical structures are discussed in this paper to analyze the factors affecting the color distribution of white light LEDs. Results show that the optical structure is the primary factor that deserves special attention. Reflector could reflect the side emitting light rays and change the rays to different directions. This shows that the color uniformity for the structures with reflector is lower. Another important factor is the shape of the phosphor layer. Hemispherical phosphor could improve the color uniformity and make the variation more stable. In the two parameters of the phosphor, the concentration has the biggest influence on the color distribution, the next factor is the thickness. Since small variation of the thickness and concentration could induce significant changes of the color, it should precisely control the steps of the fabrication technique of the phosphor to avoid generating serious yellow ring. In one word, it is difficult to improve the color uniformity only adjusting one or two factors. There needs more efforts on the studying of the influencing mechanism for the color distribution.

Acknowledgement

The support from Guangdong Real Faith Optoelectronic Inc. is appreciated.

References

1. Holonyak, J. N. and Bevacqua, S. F., "Coherent (visible) light emission from ga(as$_{1-x}$p$_x$) junctions", *Applied Physics Letters*, Vol. 1, No. 4 (1962), pp. 82-83.
2. Nakamura, S., Mukai, T. and Senoh, M., "High-power gan pn junction blue-light-emitting diodes", *Japanese Journal of Applied Physics*, Vol. 30, No. 12A (1991), pp. L1998-L2001.
3. Nakamura, S., Senoh, M. and Mukai, T., "High-power ingan/gan double-heterostructure violet light emitting diodes", *Applied Physics Letters*, Vol. 62, No. 19 (1993), pp. 2390-2392.
4. Yam, F. K. and Hassan, Z., "Innovative advances in led technology", *Microelectronics Journal*, Vol. 36, No. 2 (2005), pp. 129-137.
5. Schlotter, P., Schmidt, R. and Schneider, J., "Luminescence conversion of blue light emitting diodes", *Applied Physics A: Materials Science & Processing*, Vol. 64, No. 4 (1997), pp. 417-418.
6. Nakamura, S. and Fasol, G., The blue laser diode.: Gan based light emitters and lasers, Springer-Verlag Berlin and Heidelberg GmbH & Co. K, 1997.
7. Evans, D. L., "High-luminance leds replace incandescent lamps in new applications", *Light-Emitting Diodes: Research, Manufacturing, and Applications*, SPIE, 1997, pp. 142-153.
8. Zukauskas, A., Shur, M. S. and Gaska, R., Introduction to solid-state lighting, John Wiley & Sons New York, 2002.
9. Krames, M. R., Krames, M. R., Shchekin, O. B. et al., "Status and future of high-power light-emitting diodes for solid-state lighting", *Display Technology, Journal of*, Vol. 3, No. 2 (2007), pp. 160-175.
10. F.M. Steranka, J. Bhat, D. Collins et al., "High power leds - technology status and market applications", *Physica Status Solidi (a)*, Vol. 194, No. 2 (2002), pp. 380-388.
11. Craford, M. G., "Leds for solid state lighting and other emerging applications: Status, trends, and challenges", *Fifth International Conference on Solid State Lighting*, SPIE, 2005, pp. 594101-594110.
12. OIDA, "Light emitting diodes (leds) for general illumination, an oida technology roadmap update 2002," 2002.
13. Pelka, D. G. and Patel, K., "An overview of led applications for general illumination", *Design of Efficient Illumination Systems*, SPIE, 2003, pp. 15-26.
14. Ya-Ju, L., Ya-Ju, L., Tien-Chang, L. et al., "High brightness gan-based light-emitting diodes", *Journal of Display Technology*, Vol. 3, No. 2 (2007), pp. 118-125.
15. Haque, S., Steigerwald, D., Rudaz, S. et al., "Packaging challenges of high-power leds for solid state lighting", *www.lumileds.com/pdfs/techpaperspres/manuscript_IMAPS_2003.PDF*, (2000), pp.
16. Borbely, A. and Johnson, S. G., "Prediction of light extraction efficiency of leds by ray trace simulation",

Third International Conference on Solid State Lighting, SPIE, 2004, pp. 301-308.

17. Lee, S. J., "Light-emitting diode lamp design by monte carlo photon simulation", *Light-Emitting Diodes: Research, Manufacturing, and Applications V*, SPIE, 2001, pp. 99-108.
18. Sun, C.-C., Lee, T.-X., Ma, S.-H. et al., "Optical modeling for led in mid-field region", *International Optical Design Conference 2006*, SPIE, 2007, pp. 634217-634217.
19. Chi, W. and George, N., "Light-emitting diode illumination design with a condensing sphere", *J. Opt. Soc. Am. A*, Vol. 23, No. 9 (2006), pp. 2295-2298.
20. Narendran, N., "Improved performance white led", *Fifth International Conference on Solid State Lighting*, SPIE, 2005, pp. 594108-594106.
21. N. Narendran, Y. Gu, J. P. Freyssinier-Nova and Zhu, Y., "Extracting phosphor-scattered photons to improve white led efficiency", *Physica Status Solidi*, Vol. 202, No. 6 (2005), pp. R60-R62.
22. Chingfu, T. and Yu-Sheng, H., "Silicon-based packaging platform for light-emitting diode", *Advanced Packaging, IEEE Transactions on [see also Components, Packaging and Manufacturing Technology, Part B: Advanced Packaging, IEEE Transactions on]*, Vol. 29, No. 3 (2006), pp. 607-614.
23. Braune, B., Petersen, K., Strauss, J. et al., "A new wafer level coating technique to reduce the color distribution of leds", *Light-Emitting Diodes: Research, Manufacturing, and Applications XI*, SPIE, 2007, pp. 64860X-64811.
24. Steigerwald, D. A., Steigerwald, D. A., Bhat, J. C. et al., "Illumination with solid state lighting technology", *Selected Topics in Quantum Electronics, IEEE Journal of*, Vol. 8, No. 2 (2002), pp. 310-320.
25. Lee, T.-X., Gao, K.-F., Chien, W.-T. and Sun, C.-C., "Light extraction analysis of gan-based light-emitting diodes with surface texture and/or patterned substrate", *Opt. Express*, Vol. 15, No. 11 (2007), pp. 6670-6676.
26. Borbely, A. and Johnson, S. G., "Performance of phosphor-coated led optics in ray trace simulations", *Fourth International Conference on Solid State Lighting*, SPIE, 2004, pp. 266-273.
27. Falicoff, W., Chaves, J. and Parkyn, B., "Pc-led luminance enhancement due to phosphor scattering", Nonimaging Optics and Efficient Illumination Systems II, SPIE, 2005, pp. 59420N-59415.
28. Kim, J. K., Luo, H., Schubert, E. F. et al., "Strongly enhanced phosphor efficiency in gainn white light-emitting diodes using remote phosphor configuration and diffuse reflector cup", Japanese Journal of Applied Physics, Vol. 44 (2005), pp.
29. Luo, H., Kim, J. K., Schubert, E. F. et al., "Analysis of high-power packages for phosphor-based white-light-emitting diodes", Applied Physics Letters, Vol. 86, No. 24 (2005), pp. 243505-243503.

Development of OLED demonstration system based on SD card

Chen Zhangjin, Jin Chen, Wang Hao
Micro-electronic Research & Development Center, Shanghai University
Campus P.O.B. 221, 149 Yanchang Road, Shanghai 200072, PRC
E-mail:haidaideer@yahoo.com.cn

Abstract

In this paper, it proposed a feasibility project to the realization of portable AM-OLED display. For the request of the portability of OLED display, we developed a demonstration system based on Altera's FPGA using Nios II soft-core and SD Card. At first, it introduces some backgrounds about OLED display and SD card. Then it shows the AM-OLED driving method. After that, it also briefly imports some information about the SD card and its internal FAT file system structure, finally analyzes the entire system.

1 Introduction

With the development and popularity of the portable multimedia electronic products, the electronic products must have serial interfaces to connect a variety of mobile and portable storage devices. SD card not only has advanced built-in copyright protection, but also has a compact size to suit all kinds of portable digital products. Because of these advantages, a growing number of electronic products provide expanded SD card interface.

In this paper, for the request of the portability of OLED display, we analysis the AM-OLED display driver and SD card internal structure, then determine the entire structure of demonstration system, finally integrate the two parts into the entire system.

OLED is considered as one of the most promising next generation flat panel displays have been widely studied in the past decades. In the second half of the 1990s, panels of passive panel (PM-OLED) had batch process at first. But the next round of industry trends will be triggered by the AM-OLED [1].

With the development and popularity of portable multimedia electronic products, the requirement of portable devices are dramatic increased, such as the mass storage on audio and video. The design of electronic products must have a rich interface in order to connect to a variety of mobile, portable storage devices, such as CF (Compact Flash) Card, Memory Stick and SD (Secure Digital) Memory Card [2].

Gartner pointed out that the SD card meets in size, performance, security and other aspects of all requirements. The miniSD Memory Card used in cell phone has also played a key role on the market growth. SD card's format will become the main specification in memory cards.

The FAT (File Allocation Table) file system has its origins in the late 1970s and early1980s and was the file system supported by the Microsoft and MS-DOS operating system. In SD card, its internal file system is according to FAT file system [3].

2 Display driving of AM-OLED

OLED (organic light-emitting diode), also called OELD (organic electro luminescent devices), with many advantages such like self-luminous, fast response, wide angle, low power, low cost. It is regarded as the most optimistic alternative after the LCD display. According to OLED driving mode, it can be divided into active (AM) and passive (PM). Though the currently products available in the market focus on passive OLED, the active OLED is the real successor of the wide-screen LCD market.

The resolution of AM-OLED panel, which used by the system, is 240×RGB(H)×320(V) QVGA. Each scan completes eight-pixel points, so a line needs to be scanned 30 times to complete. SCK is the clock signal of horizontal scanning and GCK is for vertical scanning.

During data transmission, GCK should be detected whether overturned at first. If overturned, it is said that the horizontal scanning can start. Then the data can transfer at rising edge of SCK when the horizontal scanning control signal SSP pulse is activated.

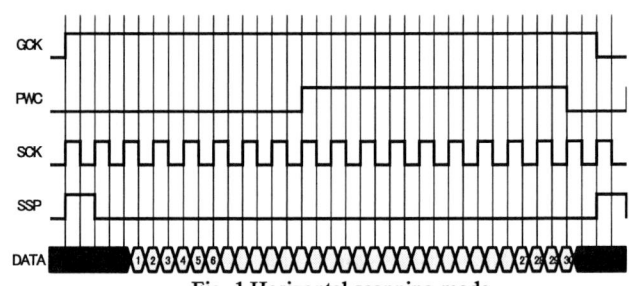
Fig. 1 Horizontal scanning mode

G1SP and G2SP are both columns scanning driving pulse signal. G2SP is a pulse signal appears between the adjacent G1SP signals. It controls when the pixels are off.

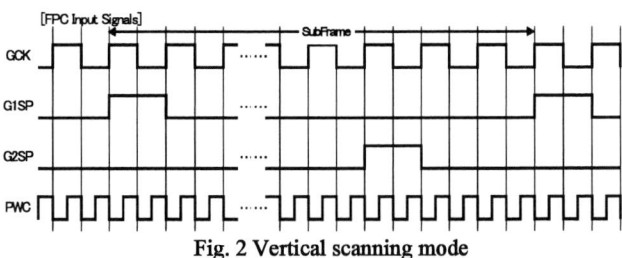
Fig. 2 Vertical scanning mode

In the whole driving time, each scanning line sequential postpones a PWC clock cycle. The following diagrams are the AMOLED driving mode.

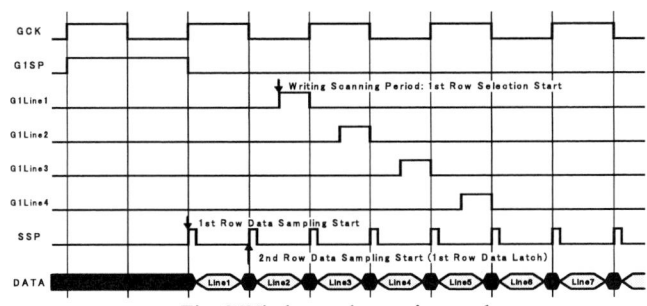

Fig. 3 Whole panel-scanning mode

3 The internal structure of SD card

SD Card is short for Secure Digital Card. SD Card standard is established by Japanese Toshiba, Panasonic and SanDisk. SD card has total of nine lines, of which six communication lines, three power lines. SD card can support two operating modes: SD card mode and SPI mode.

In SPI mode, it only allows a DAT data transmission line. Its data transmission speed seems much more slowly compared to the SD card mode using four data lines. As the back-end of the OLED display driver has a relatively large data processing, we used SD card mode to alleviate the pressure on the back-end in this paper.

The SD card transfer timing clock is given by the host to the SD card, can verify from 0 ~ 25MHz. CMD and DAT are both serial transferred based on clock, input is active at rising edge of the clock while the output is active at the falling edge of the clock.

The work status of SD card can be divided into three types: non-activity mode, card verification mode and data transfer mode. After put the SD card into the slot and power up, it step into non-active mode.

Then we send CMD0 to send it into idle, at the same time, it will be sent to card verification mode. The ACMD41 is sent to detect and confirm the working voltage whether it is match. Now, SD Card is ready for transfer. CMD2 is sent to confirm address and CMD3 returns the SD card's relevant address, and then SD card will change modes from verification mode into transfer mode. Now, this time the SD card has been on standby. The following chart is the whole process of SD card's initializing.

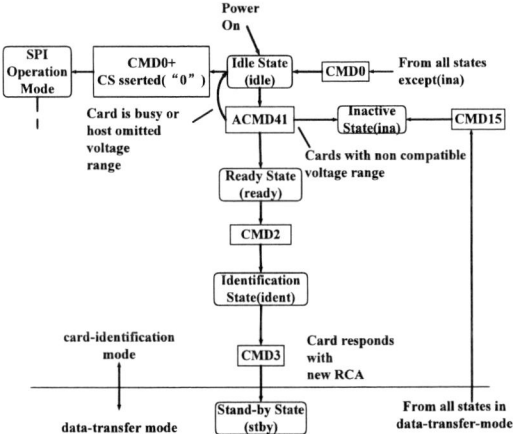

Fig. 4 SD card initial and identification mode

After entering the transfer mode, we can send a series of CMD to SD cards to read and write or even programming. In this paper, we only need read SD card operation, the operation of writing is not in consideration. In transmission mode, firstly we sent CMD9 get some special card information, such as card's block size, capacity, and so on. CMD7 is sent the card into the transfer status, CMD17 and CMD18 can order the SD cards transfer data in single or multi-block operation, see below.

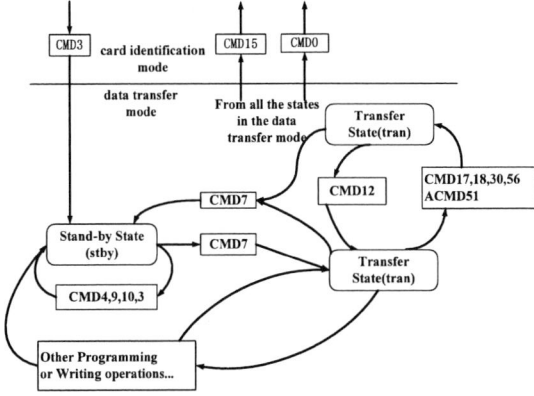

Fig. 5 SD card transfer mode

When SD card in transmission state, the host sent command to SD card to choose block size and single or multi-block transmission type. After SD card receive the command, it puts the data as host request type on the DAT transmission line. Certainly, response returned to the host is sent on the CMD line as usual, back to SD card. In the process of multi-block transmission, host CMD can send suspension command on the CMD line to stop multi-block transmission immediately.

4 FAT file system

SD card's file system is reference to ISO / IEX 9293 standards. The logical structure use FAT file system. According to its internal data areas, it can be divided into Partition Area and Regular Area. In the Regular Area, it is divided into System Area and User Area. The Partition Area contains Master Boot Record and Partition Table. System Area includes Partition Boot Sector, File Allocation Table and Root Directory. The user area is the real data storage area for the user.

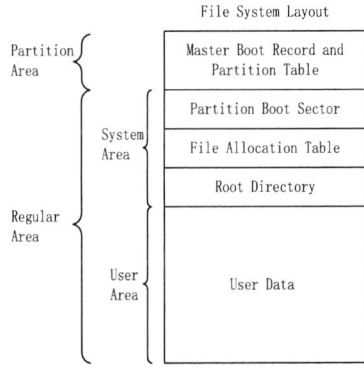

Fig. 6 SD card file system

Master Boot Record is located in 0 cylinder 0 head and 1 sector of the disk, used by disk boot. It is generally a hidden sector. In the Master Boot Record, it contains some executable codes and has a length of 446 bytes. The content of this part is not specified by SD card. In the following partition table, it includes not only the start and the end and its total number of the head, sectors and tracks, but also contains the information such as used sectors and the type of the partitions.

From Partition Boot Sector in the system area, we can know the size of sector, the total number of sectors, clusters per sector and root directories, and so on. Follow the Partition Boot Sector is FAT Table. There also has a copy, two FAT tables are exactly the same. After the two FAT tables, there is a description table about the root directory, called Root Directory.

The final part of the user area is the real user data storage area, use cluster as a unit. The size of cluster will be verified according to SD card capacity, but different sector of the clusters should start at 2. Data transmission capacity in this area of SD card is accordance with the entire sector or the minimum transmission size defined in physical layer. As mentioned in the previous section, we will use large-capacity transmission type, prepare for the OLED display driver.

5. The overall framework of the demonstration

The hardware structures are showed in Fig.7. The entire demonstration system is based on Altera's Nios II and SOPC system. The initial and read works to the SD card are on the Nios II soft-core platform. When program is running, it reads the SD card's contents then put pixels data into SRAM. After that, it access control module through SRAM, in accordance with the specific method of scanning mode to show the pixels data on the OLED screen.

SOPC (System On Programmable Chip) is a flexible and efficient SOC solution proposed by Altera Corporation. It integrates many functional modules need by system design such as processor, memory, I / O system design into a single programmable device, then constitute a programmable system-on-chip.

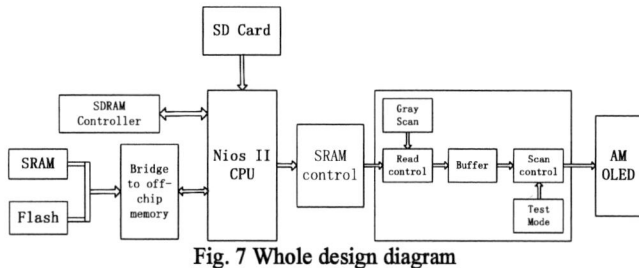

Fig. 7 Whole design diagram

Figure 8 shows the entire demonstration systems in SOPC part. It's including the Nios II soft-core processor provided by Altera Corporation, the analyzer and read work to SD card is completed by this soft-core. SOPC integration also includes a number of SDRAM, SRAM, Flash controller, FIFO, SD Card I / O port and the LED display, and so on.

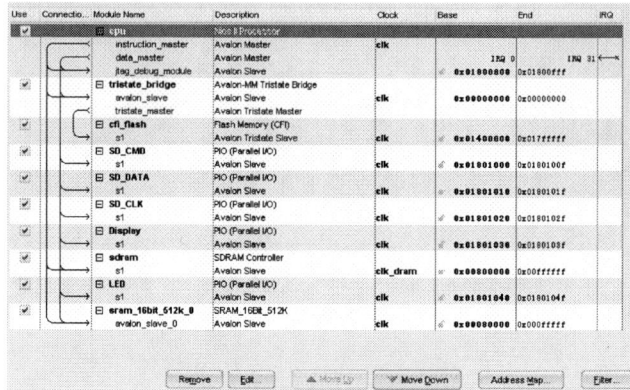

Fig.8 SOPC part design diagram

Following diagram shows the software code debugging environment -- Nios II IDE. It is the basic software development tool of Nios II family of embedded processors. All software development tasks can be completed under the Nios II IDE, including editing, building and debugging. Nios II IDE provides a unified development platform for all Nios II processor systems.

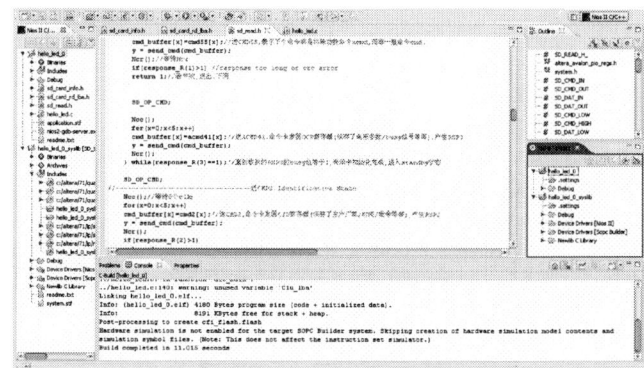

Fig. 9 Nios II part design diagram

In Nios II processor, we divide the operations of SD card reader and analytical into sd_initial, sd_info, sd_cmd, sd_res and main functions. The initial part completed the initialization of the SD card. As the process referred in the third part of this paper, the whole initial process send CMD0, CMD1, CMD3 to receive the corresponding response and has successfully completed the initialization of power-on. Sd_cmd and sd_res completed sending CMD and analyzing the response from the SD card.

After the completion of initialization process, SD card is standby. We can still send a series of command to do read or some other operations to SD card. Sd_info module is to isolated useful information from the data read from SD card, such as SD card's capacity, format, and some important information in FAT table.

According to data separated from the read data from SD card, we can caculate the beginning and the end of address of the data stored in the SD card, according to the FAT table format mentioned in the forth part of this paper.

We will put the data read from SD card in accordance with the length and width of the picture at its appropriate address in SRAM, and then the SRAM controller can easily read data from the SRAM as the picture's format to send the RGB data to scanning module, in accordance with the form of AM-OLED drive to drive AM-OLED display. Since then, we complete the entire demonstration system.

Conclusions

In this paper, we introduce the development of AM-OLED demonstration system based on SD card. After anaylzing the functions and the realization of any part in the demonstration, we complete to constructe the entire system. It proposed a feasibility project to the realization of portable AM-OLED display.

This demo system currently can only show BMP format picture, because Nios II has not supported the compression format picture and not added extract function. However, Nios II and SOPC are all open platform.If we want to add the function algorithm or hardware interfaces, it will be very convenient. For the subsequent escalation of this, it provided a good basis. In future, we only need add the corresponding algorithm into Nios II soft-core, it will support a variety of image formats, or even play videos.

Acknowledgments

The authors would like to acknowledge the National Natural Science Foundation of China for providing financial support for this work under Grant No. 60773081 and Grant No. 60777018, and also to acknowledge the financial support by the Shanghai Municipal Committee under Grant No.AZ028 and Grant No.06DZ22013

References

1. Gao Xinyu, "Next round of industry trends -- triggered by AM-OLED," *Electrical appliances,* No.03(2008), pp21-23.
2. MEI, SanDisk, Toshiba, <u>SD Memory Card Specifications</u>, SD Group(April 2001), pp.1-83.
3. Microsoft Company, <u>Microsoft Extensible Firmware Initiative FAT32 File System Specification</u>, Microsoft (December 6, 2000), pp.7-25.
4. Arnold M., <u>Verilog Digital Computer Design: Algorithms to Hardware</u>, Prentice-Hall(NJ, 1998).
5. Altera Corp., <u>Quartus II Version 7.1 Handbook</u>, Altera(2007), pp. 2159-2355.
6. Altera Corp., <u>Nios II Software Developer's Handbook</u>, Altera(2007).

Combinational Test Generation for Transition Faults in Acyclic Sequential Circuits

Shi Hui[1,2], Ran Feng[1,2], Zhang Jinyi[1,3]

1. Key Laboratory of Advanced Displays and system Application, Ministry of Education, Shanghai University
2. Microelectronic Research & Development Centre, Shanghai University
3. Key Laboratory of Special Fiber Optics and Optical Access Networks, Ministry of Education, Shanghai University
No.149 Yanchang Road, Shanghai 200072, P.R.China
Email: shihui.keytech@shu.edu.cn   Tel: 086-021-56331323-197   Fax: 086-021-56331272

Abstract

This paper presents a combinational test generation method for transition faults in acyclic sequential circuits. In this method, to generate test sequences for transition faults in a given acyclic sequential circuit is performed on its extend time-expansion model. The model is composed of two copies of time-expansion model of the given circuit and extends in the close two sequences to generate 2 vectors for the transition faults with some restrictions. Experimental results show the method can achieve the higher fault efficiency with the lower test generation times than conventional method.

1 Introduction

As the speed of modern VLSI circuits reaches the gigahertz range, delay testing is becoming more important. Until now, many delay fault models have been investigated [1]. If propagation time of rising or falling signal transition through the segment of circuit exceeds a specified limit, the segment of circuit has a delay fault. So Transition fault is a kind of delay fault. Test for the transition faults can effectively find the delay fault.

For any delay fault model, a two-pattern test is obligatory. However, test generation for a sequential circuit is always a hard problem just under simple fault models such as the single stuck-at fault model and may be unsolvable in a large sequential circuit. Therefore, design for testability (DFT) becomes an important approach to reduce the test generation effort for sequential circuits. The circuits used full scan become combinational circuits and can easily achieve about 100% fault coverage with the present algorithms and programs. This is the main reason for the widespread acceptance of the full-scan design. However, the problem of hardware overhead, delay and test time make designer to explore partial-scan techniques instead of full-scan.

Partial-scan techniques based combinational test generation using balanced structure has been proposed by Gupta [2]. A circuit is called a balanced circuit if it is acyclic and all paths between any pair of nodes have the same number of FFs, where a node can be a primary input (PI), gate, FF or primary output (PO). Strongly balanced circuit has been proposed by Balakrishnan and Chakradhar and internally balanced circuit has also been proposed by Fujiwara [3] with changing some property of the balanced structure. The test patterns of stuck-at faults in a balanced (resp. strongly balanced and internally balanced) sequential circuit can be obtained by using its combinational equivalent circuits and combinational test pattern generation algorithms.

Cheng and Agrawal have proposed another partial scan technique that breaks feedback to make the sequential

structure acyclic in the test mode. The sequential circuits become feedback free called acyclic sequential circuits, which are super class of balanced sequential circuits. With time-expansion model (TEM) [4], test patterns generation can also using combinational test pattern generation algorithms.

Just like the scan technique for stuck-at fault, an enhanced scan technique [5] is given for delay fault, which replaces flip-flop (FF) by an enhanced scan FF (ESFF). An ESFF can store two bits and apply two consecutive vectors to implement two-pattern test for delay fault. However, hardware overhead caused by extra memory elements of ESFFs is very high. It can only be alleviated by using partial scan techniques. With a partially enhanced scan technique, a sequential circuit becomes an acyclic sequential circuit. TEM can used to be the test generation model of the acyclic sequential circuit for the single stuck-at fault. But to detect transition faults need a two-pattern test, there are still some challenging problems in the test generation for transition faults. This paper presents a test generation method for transition faults in acyclic sequential circuits. Using this method, test generation for transition faults becomes more efficient than those methods with sequential ATPG.

2 Preliminaries

2.1 Circuit Model

A sequential circuit is considered as a connection of combinational logic blocks (CLB) and FFs. A CLB is a combination of combinational logic gates between two nodes. The nodes represent primary inputs (PI), primary outputs (PO) or FFs. In this paper, we assume that FFs are of D-type. This assumption does not impose restriction on circuit representation because any FF of other types can be modeled by a D-type FF and some logic gates. The target circuits, i.e. acyclic sequential circuits have no circle. For example, Figure 1(a) shows an acyclic sequential circuit S. Node a, b, c and o

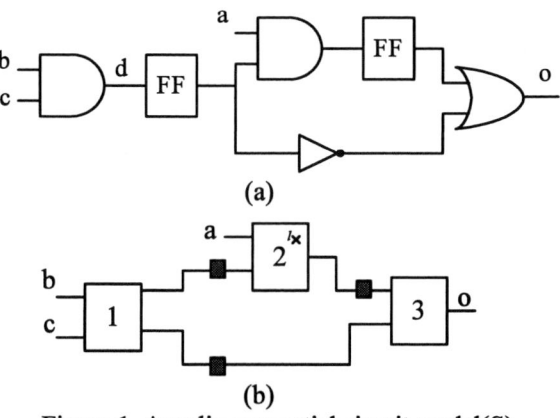

Figure 1. Acyclic sequential circuit model(S)

are primary inputs and a primary output, respectively and FF are registers of D-type. Figure 1(b) is the block model corresponding to Figure1 (a) by using CLBs to denote combinational circuit and shadows to denote FF. Graph G=(V,A,w) [6] is also used to denote the acyclic sequential circuit.

2.2 Fault Model

Target faults discussed in this paper are transition faults in acyclic sequential circuits. There are two transition faults associated with each line in an acyclic sequential circuit: a slow-to-rise fault and a slow-to-fall fault. Under the transition fault model, the extra delay caused by a transition fault is assumed to be large enough to prevent the transition through the faulty site from reaching any FF or any primary output within a specified time. In this paper, it is assumed that transition faults in acyclic sequential circuits are tested in the slow- fast-slow testing manner [1]. In the slow test circle, the transition faults is initialized and in the fast test circle the transition faults is active and propagate to the PO. Under this assumption, we can consider the internal time between a fast test circle and next slow test circle is too long to transition faulty occurred in initialization. Similarly as stuck faults, it is assumed that a circuit has only one transition fault at same time.

It is proved that if a transition fault is testable under the at-speed testing manner, the fault is also testable under the slow-fast-slow testing manner. Hence, the slow-fast-slow testing never misses any testable transition fault in the at-speed testing.

2.3 Time-Expansion Model

In test generation for acyclic sequential circuits, time-expansion model (TEM) is used. In this model, a given acyclic sequential circuit S is transformed into a TEM circuit C(S) [4]. Figure 2 shows a time-expansion model (TEM) C(S) for S in Figure 1. TEM C(S) is a combinational circuit derived by connecting CLBs according to their sequential depths. A sequential depth between two CLBs is defined by the number of FFs on a path between the two CLBs. (e.g. the sequential depth between CLB2 and CLB3 is one.) If a CLB has several different sequential depths from another CLB in S, the CLB with different sequential depths should be duplicated in accordance with its sequential depths in C(S) (e.g., CLB1 has two different sequential depths to CLB3 in S, one and two, and hence it is double in C(S)).

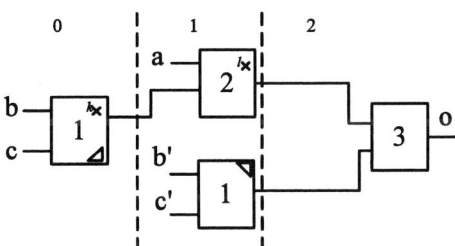

Figure 2. Time-expansion model: C(S)

A single stuck fault in the CLB duplicated is also duplicated and become a multiple stuck fault. A multiple to single transform is used for the multiple stuck faults in TEM. However, in transition fault model a two pattern test is used

under the slow-fast-slow testing manner. In this manner, no transition fault occurs in initialization. So a multiple transition fault can be regarded as multiple respective single transition faults and we can handle it one by one.

For an acyclic sequential circuit, test patterns gained in a TEM are transformed into test sequence for its original circuit. For transition faults a two-pattern test is used and the two test patterns are transformed into one test sequence. We use (u,v) and (o_u,o_v) denote the two vectors and the desired outputs of the circuit C(S) and giving Table 1 to show the relation between vector and sequence.

Table 1 PIN sequence

sequence	0	1	2	3
a	X	u_a	v_a	X
b	u_b	$u_{b'}=v_b$	$v_{b'}$	X
c	u_c	$u_{c'}=v_c$	$v_{c'}$	X
o	X	X	o_u	o_v

As the sequence depth interval of b and b' (c and c') is one, the test vector (u,v) need to satisfy the condition $u_{b'}=v_b$ ($u_{c'}=v_c$). Thus, it is not sufficient to only use TEM to generate test sequences for an acyclic sequential circuit.

3 Method of extend time-expansion model

3.1 Extend time-expansion model (ETEM)

For transition fault, a two pattern test is used under the slow-fast-slow testing manner. The first vector u is used in the first slow test circle for initialization and the second vector v is in the fast test circle used to detect the fault. The two test pattern u and v has the sequence interval of one. As Table 1 show if a CLB connected with PI (e.g. CLB1 in Figure 2) is duplicated in the two sequence depth between which the interval is one, the first vector u of these PI in the late sequence depth should be equal to the second vector v of these PI in the early sequence depth (e.g. $u_{b'}=v_b$ and $u_{c'}=v_c$ in Table 1).

In order to satisfy these condition and generate transition tests for an acyclic sequential circuit, we define a extend time-expansion model (ETEM). In this method, to generate test sequences for transition faults in a given acyclic sequential circuit are performed on its extend time-expansion model (ETEM). ETEM is composed of two copies of a TEM of the given circuit and extends in the close two sequences to generate 2 vectors for the transition faults. For a given TEM circuit C(S), ETEM need make two copies of C(S), named C1(S) and C2(S). C1(S) begins with sequence depth 0 and C2(S) begins with sequence depth 1, i.e. the CLB in C2(S) is one sequence after the corresponding CLB in C1(S). Figure 3 show the ETEM corresponding to the circuit in Figure 1.

A CLB connected PIs may be duplicated in TEM. In ETEM circuit, if a CLB connected PIs appears both in C1(S) and C2(S) in the same sequence depth. These PIs of the CLB in C1(S) and C2(S) should be connected correspondingly for the test generation. In Figure 3 the CLB1 appears in C1(S) and C2(S) in the sequence depth 1. So $b1$ and b' ($c1$ and c') are connected, the test generation of ETEM can satisfy $u_{b'}=v_b$ ($u_{c'}=v_c$).

399

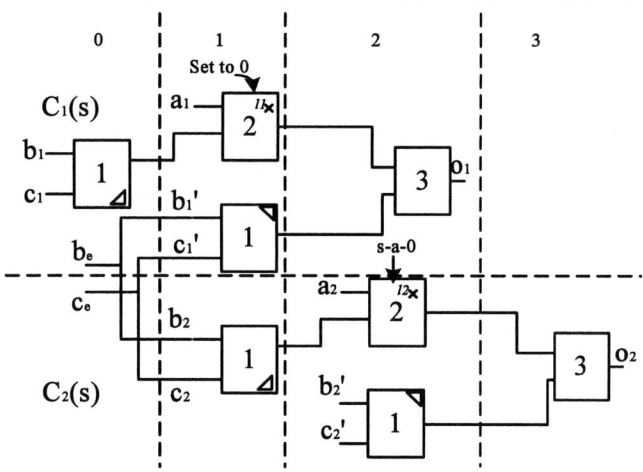

Figure 3. Extend time-expansion model

A two-pattern test (u,v) for the slow-to-rise(slow-to-fall) fault associated with a line l in a combinational circuit with the following two properties:

a. The value of 0 (resp. 1) is justified to l by u.

b. The stuck-at 0 (resp. 1) fault associated with l is detected by v.

To test generation for transition faults in ETEM is similar to stuck-at test generation in the light of following restrictions. If a slow-to-rise(slow-to-fall) fault associated with a line l in C(S) ,the corresponding line $l1$ in C1(S) should be set to 0 (resp. 1) and the corresponding line $l2$ in C2(S) has a stuck-at-0 (resp. 1) fault. The vector for the stuck fault propagated to primary output must make an output in C1(S) different from the corresponding output in C2(S).

If line l in CLB2 has a slow-to-rise fault, with ETEM, set $l1$ in CLB2 of C1(S) to 0 and $l2$ in CLB2 of C2(S) has a stuck-at-0 (as Figure 3 shown). The gained vector must make o_1 different from o_2. A two input XOR gate can be add to o_1 and o_2 for convenient test generation by software with a constraint of output of XOR gate equal to 1.

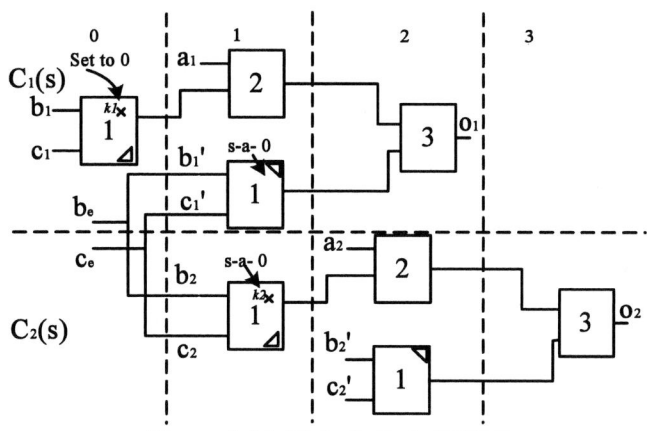

Figure 4. Multiple faults in ETEM

Another problem of ETEM, line k in the CLB1 of TEM circuit C(S) has a slow-to-rise (slow-to-fall) transition fault. Using ETEM circuit to test generation for the fault, line $k1$ in CLB1 in sequence depth 0 of C1(S) should be set to 0(resp. 1) and line $k2$ in CLB1 in sequence depth 1 of C2(S) has a s-a-0(resp. 1) fault according to the restrictions. The CLB1 of C1(S) in sequence depth 1 is the same CLB with one sequence depth after the CLB1 of C1(S) in sequence depth 0.

When line k in CLB1 of TEM has a slow-to-rise (slow-to-fall) and line $k1$ in CLB1 of C1(S) in sequence depth 0 is set to 0(resp. 1), the CLB1 of C1(S) in sequence depth 1 is also has a s-a-0(resp. 1) fault. Show in Figure 4.

So we get a conclusion that for ETEM circuit, if a CLB appears both in C1(S) and C2(S) in the same sequence depth, in which the CLB of C2(S) has a stuck–at fault, the stuck–at fault in the CLB of the sequence depth occurs simultaneously both in C1(S) and C2(S). So a Multiple-to-Single transformation must be use to deal with these fault. The model shows in Figure 5.

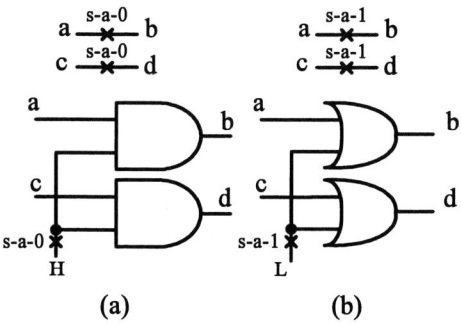

Figure 5 Fault Multiple-to-Single model

If node d in the original sequence circuit of Figure 1 has a slow-to-rise (slow-to-fall) fault, according to the restrictions set d_1 to 0 (resp. 1) and d_2 has a s-a-0(resp. 1) fault in ETEM. The d_1' in ETEM is the same gate output in the original sequence circuit after one sequence and also should have a s-a-0(resp. 1) fault. So a Multiple-to-Single transformation must be use to the stuck fault in d_1' and d_2. The ETEM of Figure 1 for slow-to-rise fault after Multiple-to-Single transformation is shown in Figure 6. For slow-to-fall fault, Multiple-to-Single transformation for ETEM also can use the combination of low level and OR gate (Figure 5 b).

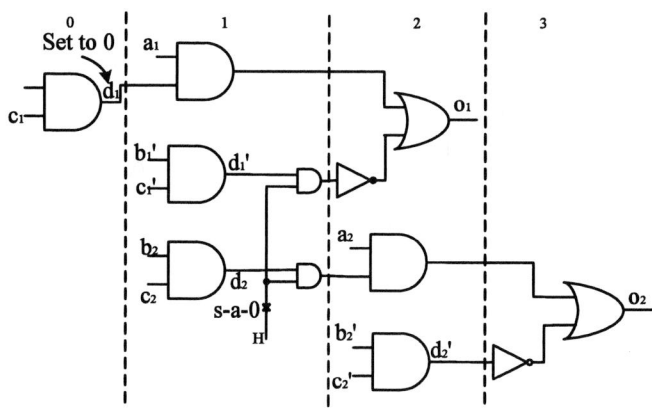

Figure 6. ETEM after Multiple-to-Single

3.2 Test Generation Procedure

The test generation method using ETEM is also proposed. This method can generate test sequences for all the testable transition faults and can identify almost all the untestable transition faults in a given acyclic sequential circuit. Main Procedure show as follows.

Main Procedure:

S1: Create a transition fault list F of S.

S2: Construct a TEM C(S) of S with DAG [7] algorithms

S3: Construct ETEM of S based on TEM C(S). Use

Multiple-to-Single transformation for slow-to-rise and slow-to-fall fault respectively, if it is needed.

S4: Find a remaining fault f in F and find the corresponding fault in ETEM.

S5: Perform test generation based on restrictions that set to 0 (resp. 1) and has the stuck-at 0 (resp. 1) fault respectively in corresponding line in C1(S) and C2(S) of ETEM.

S6: Convert a test pattern generated in ETEM into A two-pattern test (u,v) for C(S).

S7: Add (u,v) to test patterns list T and remove f from fault list F.

S8: If the fault list F is not empty, repeat perform procedures from S4 to S7.

4 Experimental Results

The proposed ETEM is applicable to all classes of acyclic sequential circuits. If the given circuit is not acyclic, a partial enhanced scan procedure can be used to make it acyclic. In this section, we consider selecting those ISCAS'89 benchmark circuits that are either acyclic or having at least one FF remaining after making them acyclic by partial enhanced scan. According to these conditions, several test circuits are selected from ISCAS'89 benchmark circuits. In Table 2, the column "Circuit" lists the circuits for our experiment. The column "Total FF" denotes the total number of Flip-flops in the circuit. Four columns under "Acyclic sequential circuits" namely "Scanned FF", "Scan (%)" "PIs" and "POs" give the number of scan FFs used to make the circuit acyclic, the percentage of scan FFs and the PIs and POs (include PPIs and PPOs) of the acyclic sequential circuit after scanned, respectively. The columns "Scanned FF" and "Scan (%)" under "Balanced sequential circuits" show the number of scan FFs used to make the circuit balanced and the percentage of scan FFs.

Compare with the method in balanced sequential circuits [6], the method of ETEM is in acyclic sequential circuits and can effectively reduce the hardware overhead of enhanced scan in the most tested circuits. In the circuit S1423, most each FF has a self feedback; therefore, the hardware overhead is almost same in acyclic sequential circuit and balanced sequential circuit.

Next we give the test generation experimental result for the circuits in the Table2. Table 3 lists the test generation results ETEM for transition faults, respectively. The columns

Table 2 Circuit Statistics

Circuit	Total FF	Acyclic sequential circuits				Balanced sequential circuits	
		Scanned FF	Scan (%)	PIs	POs	Scanned FF	Scan (%)
S1196	18	0	0	14	14	16	88.9
S1238	18	0	0	14	14	16	88.9
S1423	74	71	95.9	88	76	72	97.3
S5378	179	30	16.8	65	79	124	69.3
S9234	228	152	55.7	171	174	210	92.1

"fault", "detect", "untestable" and "aborted" denote the number of target transition faults, detected faults, identified untestable faults and aborted faults in the test circuit, respectively. The Column " TGT [s] " denotes test generation time which includes fault simulation time. The Column "FE [%]" is the fault efficiency, obtained with the formula (detect + untestable)/fault. From Table 3, we can see that method of ETEM using combinational test generation is more efficiency in test generation time than the sequential test generation. In fault efficiency, method of ETEM reduces the number of aborted faults and changes them into detected faults and untestable faults. So using our method can gain high fault efficiency than using Sequential ATPG.

5 Conclusions

In this paper, we present an ETEM and the test generation method based on ETEM for transition faults. This method for transition faults in acyclic sequential circuit is complete and can effectively low the hardware overhead of enhanced scan. The method also uses combinational test generation. Experimental results show the method can gain the higher fault efficiency with the lower test generation times than using Sequential ATPG.

Acknowledgements

Thank you for the support from Shanghai Science and Technology Department Foundation for providing financial support for this work under grant No.06DZ22013 and 077062008, Leading Academic Discipline Project of Shanghai Educational Committee grant No.J50104 and also to acknowledge the financial support from National Natural Science Fund grant No.60773081 and No.60777018.

Table 3 Test generation results

Circuit	method	fault	detect	untestable	aborted	FE [%]	TGT [s]
S1196	ETEM	2124	2090	31	3	99.8	0.56
	Sequential		2042	21	61	98.1	3.82
S1238	ETEM	2102	2068	27	7	99.7	0.53
	Sequential		2033	14	55	97.3	3.88.
S1423	ETEM	2326	2310	16	0	100	0.16
	Sequential		2302	0	24	98.9	0.59
S5378	ETEM	5080	4810	231	39	99.2	25.62
	Sequential		4633	228	219	95.6	3271.23
S9234	ETEM	3756	3634	105	17	99.5	15.14
	Sequential		3535	94	127	96.6	3368.54

REFERENCES

1. A. Krsti'c and K.-T. Cheng, Delay fault testing for VLSI circuits,Boston: *Kluwer Academic Publishers*, 1998.

2. R. Gupta and M. A. Breuer, "Testability properties of acyclic structures and applications to partial scan design," *Proc. IEEE VLSI Test Symp.*,Apr. 1992, pp. 49 – 54.

3. H. Fujiwara, "A new class of sequential circuits with combinational test generation complexity," *IEEE Trans. Comput.*, vol. 49, no. 9, pp. 895 – 905, Sep. 2000.

4. T. Inoue, T. Hosokawa, T. Mihara, and H. Fujiwara, "An optimal time expansion model based on combinational ATPG for RT level circuits," *Proc. ATS*, pp.190-197, Dec. 1998.

5. I. Dervisoglu and G. E. Stong, "Design for testability:Using scanpath techniques for path-delay test and measurement," *Proc. Int. Test Conf.*, pp. 365–374, 1991.

6. S. Ohtake, S. Miwa and H. Fujiwara, "A method of test generation for path delay faults in balanced sequential circuits," *Proc. 20th IEEE VLSI Test Symp.*, pp. 321–327, 2002.

7. YongChangKim, Vishwani D. Agrawal, and Kewal K. Saluja, "Combinational Automatic Test Pattern Generationfor Acyclic Sequential Circuits" *Computer-Aided Design Of Integrated Circuits And Systems*, VOL. 24, NO. 6, JUNE 2005

8. T. Inoue, T. Hosokawa, T. Mihara, and H. Fujiwara, "An optimaltime expansion model based on combinational ATPG for RT level circuits," *Proc. ATS*, pp.190-197, Dec. 1998.

9. X. Liu, M. S. Hsiao, S. Chakravarty and P. J. Thadikaran, "Efficient transition fault ATPG algorithms based on stuck at test vectors," *Journal of Electronic Testing: Theory and Applications*, Vol. 19, No. 4, pp. 437–445, Aug. 2003.

10. T. Iwagaki, S. Ohtake and H. Fujiwara, "Reducibility of sequential test generation to combinational test generation for several delay fault models," *Proc. IEEE 12th Asian Test Symp.*, pp. 58–63, 2003.

Design of a CMOS Charge Pump for high-performance phase-locked loop

XUAN Xiangguang[1], RAN Feng [2], XU Meihua [2]
1. Technology Center of Shanghai FeiLe Limited Company
2. Microelectronic Research and Development Center, Shanghai University
Campus P.O.B. 110, 149 Yanchang Rd., Shanghai 200072, China
xuanxiangguang@feilo.com.cn
ranfeng@mail.shu.edu.cn

Abstract

In conventional CMOS charge pump circuits, there are some current mismatching characteristics which result in a phase offset in phase-locked loop circuits. This paper presents a new charge pump circuit after detailed analysis of the current mismatch problem. It combines an error amplifier with reference current sources to achieve good current matching characteristics and lower phase noises, and at the same time it can eliminate charge sharing by using charge removal transistors. The circuit was designed by chartered 0.35um CMOS technology and simulated by Spectre tools. Simulation results show that the good current matching characteristics can be obtained with the proposed designing scheme.

1 Introduction

Since the conception of phase locking was proposed in the Thirties of the 20th Century, it has been widely applied in electronics and communication fields[1], especially used in large scale digital circuits. CP-PLLs (Charge Pump Phase-Locked Loop) are mainly used to generate signals and renew the clock pulses during the data transmission with high speed[5][6]. As a key model, the charge pump plays an important role in assuring PLLs stability. It converts the digital signals in PFDs (Phase Frequency Detector) into analog signals of VCOs (Voltage Controlled Oscillator). When the phase-locked loop was locked in a certain frequency, the output voltage of charge pump is demanded to be a fixed value, and any tiny change of which will result in apparent frequency offset. Therefore, it is very important to design a charge pump circuit which can send a stable output voltage in CP-PLLs plan.

2 Principle Analysis

Fig. 1 shows the circuit diagram of a conventional charge pump. In the charge pump, the digital output signals (UP and DN) of the PFD control the two circuit sources (I_{UP} and I_{DN}), and charge the capacitance C_L via two switches which generally substituted by two MOSFET to obtain the DC level V_{ctrl} needed by the Voltage Controlled Oscillator.

In Fig. 1 the I_{UP} and I_{DN} should be completely equal in theory, but there are many nonideal effects which will result in their mismatching in practice[4]. Another ubiquitous problem is the charge sharing in conventional charge pumps which results from the parasitic capacitance of node A and B. The level in node A will be charged to V_{DD}, and in node B be discharged to GND when the signal UP and DN are invalid, whereas the node A level will be falling and the node B level will be rising when the signal UP and DN are valid. The difference between V_{ctrl} and node A will not be uniform to the difference between V_{ctrl} and node B, thus bring on the charge

redistribution among C_L, A and B. Because the I_{ds} will change with V_{ds}, current source I_{UP} or I_{DN} will share the charge. It will result in current mismatch which make V_{ctrl} jittering, and influence the circuit performance.

When the phase-locked loop is working, however, there is phase error; even the phase can't be locked because of the charge and discharge current mismatch in charge pump. After the loop comes into the locked status, the periodic up, down reset signals will still enter the charge pump due to a no dead zone PHD is used. The output V_{ctrl} would be held if the charge and discharge current are well matched. Generally, the net current generated by the charge pump is not equal to zero because of the current mismatching, it will make the V_{ctrl} increase a fixed value in every phase comparing time. The control voltage V_{ctrl} should be held in an average value to maintain the loop in a locked status (as shown in Fig. 2(a)), then the phase-locked loop shall bring on phase error which make the net current of the charge pump be zero in every period, as shown in Fig. 2(b) where A1 and A2 have the equal area.

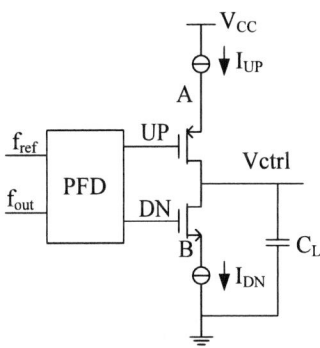

Fig. 1 Conventional charge pump schematic

The phase error resulted from the current mismatch can be expressed as the following formula where Δt_{on}, I_{cp} and $|I_{UP}-I_{DN}|$ respectively represent the dead zone time of the PFD, the period of the reference clock, and the offset between CP current and charge, discharge current.

$$\Phi_\varepsilon = 2\pi \cdot \frac{\Delta t_{on}}{T_{ref}} \cdot \frac{|I_{UP} - I_{DN}|}{I_{cp}} \qquad (1)$$

978-1-4244-2739-0/08/$25.00 ©2008 IEEE   403

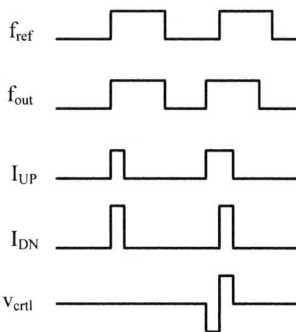

(a) Charge and discharge current mismatch

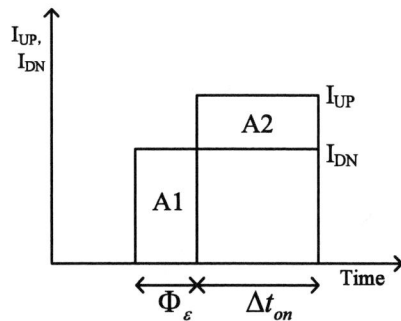

(b) Conventional charge pump current mismatch

Fig. 2 Phase offset causation

The eq.(1) indicates that, to lower the phase error, the dead zone time Δt_{on} and mismatch current $|I_{UP}-I_{DN}|$ should be reduced, but the CP current I_{cp} should be increased while the reference clock period is fixed. Holding a definite dead zone time will be propitious to overcoming the PFD dead zone, and the higher current I_{cp} will increase the power consumption and noise, so lessening the mismatch current $|I_{UP}-I_{DN}|$ is the key to lower the CP phase error.

To solve above problem, this paper presents a novel charge pump circuit which use error amplifier and reference current source to obtain the improved characteristic for current match, and reduce the PLL's phase noise. Simultaneously, the charge sharing is effectively restrained by using charge removal transistors. So the circuit possesses good current match and high working speed.

3 Circuit Design

There are different designing modes[2][3] to implement the schematic in Fig. 1, as shown in Fig. 3. The linear switch of digital PFD's output control set in current source directly connects with the control voltage in Fig. 3(a). This structure will lead to the control level unsteady while switching on and off, and also make the current source transfer from linear to saturation one, which result in the current mismatch. Generally, the steady control voltage must be obtained by additional voltage follower and complementally control switches. In Fig. 3(b) the switch directly controls the grid of the current source which is assured to work in saturation, but in this structure, the transconductance of M1 and M2 influences the switch time constant which can not be too small. The charge pump widely used now is illustrated in Fig. 3(c) where the switch joints to the source of the current source, thus the transconductance of M1 and M2 is independent of the switch time. In this structure, we can use

the low bias current to get the high output current of the charge pump, and moreover the switch only joins to one transistor with low parasitic capacitance, so its working speed is higher than one in Fig. 3(b).

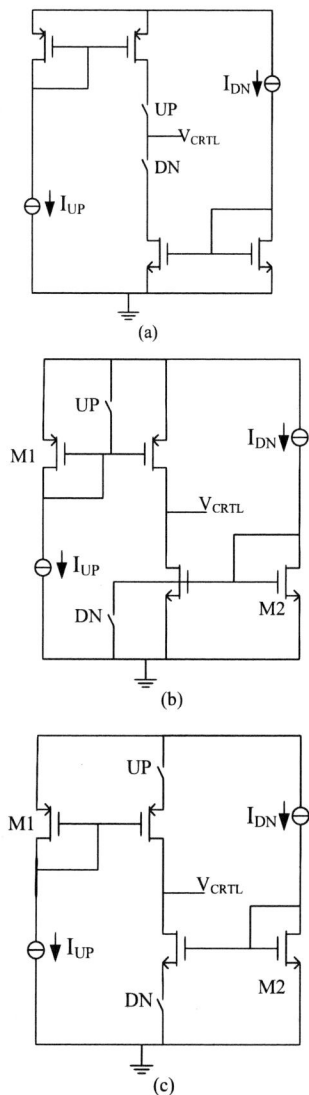

Fig. 3 Charge pump structure

Based on the above analysis, this paper presented a novel charge pump circuit which summarized the advantage of Fig. 3(c) structure. Its schematic is shown in Fig. 4. The proposed CP circuit greatly improved the circuit performance by enhancing the current match and lessening the charge sharing, and at the same time, possessed the characteristics of high operating speed and low power consumption.

In Fig. 4, capacitance C1, C2 fill the role of stabilizing the node E and F's voltage to avoid instantaneous grid voltage fluctuation of the current source. The voltage V_c in node C will change with V_{ctrl} by inserting an error amplifier which has the gain high enough to make Vc equal V_{ctrl}. Moreover, M11 is designed to equal M9, M8 equal M10, M3 equal M5, and M4 equal M6, so the current I4 will be the same as I3, I1 when UP, DN level is holding high, and I2 will be the same as I1, I3 when UP, DN level is holding low. Finally the current

I4 equals I2, which lead up to the result of almost perfect drain-source current matching.

M7 and M12 are named as charge eliminating transistor. When the transistors transfer from saturation to cutoff, the charge resting on the channel will be emitted to the source, and that the drain will not be impacted. When the UP and DN is low, the spare charge will be removed from node A and B, so that the charge sharing can be successfully restrained.

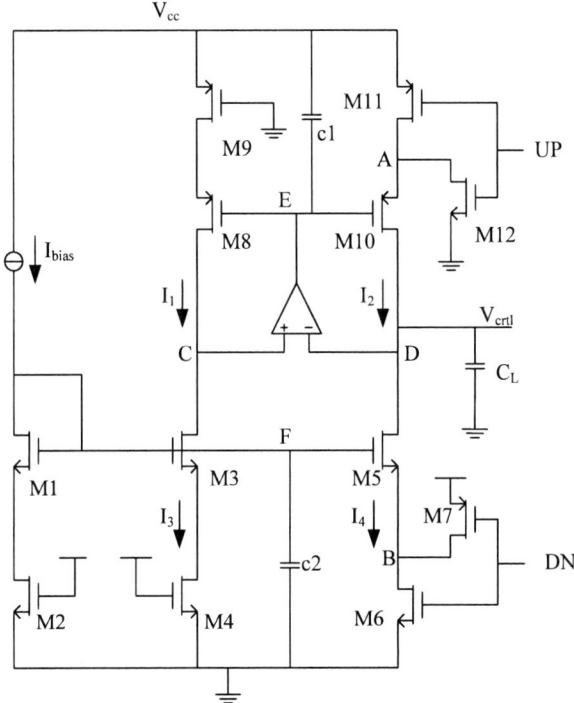

Fig. 4 Improved charge pump circuit

The shortcoming of the structure is that it will confine the dynamic range, but it is not important in most situations. While V_{ctrl} less than $Vg5-V_{tn}$ and DN invalid, the current will move to the output via M7. The V_{ctrl} will not less than V_{tn} during NMOS used as VCO current control, and forcing the Vg5 less than $2V_{tn}$ will be easy to implement. Because the V_{ctrl} is up to $Vg10+|V_{tp}|$, it restricts the V_{ctrl} range. Transistor M7 and M12 also improve the switch speed of current, and supply DC level for node A, B while switch is turned off, which will prevent the pending nodes influencing the control voltage.

4 Simulation Results

Based on the chartered 0.35uCMOS technique, the proposed charge pump circuit, at the 3.3V power source, is simulated with Spectre tools. The simulation results are shown in Fig. 5. Fig. 5(a) shows the simulation result of CP charge circuit, and Fig. 5(b) shows the CP discharge result. In the diagram the V_{ctrl} ascends or drops linearly without sunken or bulgy undulation, and the simulation graph accords with the requirement.

Fig. 5(c) presents the relationship of charge/discharge current and output node voltage. From Fig. 5(c), we can see that the node voltage range from 1V to 3V, while output current is not change in the main, therefore it can fully meet the requirement for the charge pump.

This paper studied the current mismatch causation of current CMOS charge pumps, and then proposed a novel CP circuit which use error amplifier and reference current source to get good match characteristic, and use charge removal transistor to restrain the charge sharing. All the design has been simulated with Spectre tools. The proposed circuit has also been successfully applied in the PLL clock frequency multiplier for the LVDS drivers, and its phase error is efficiently restrained.

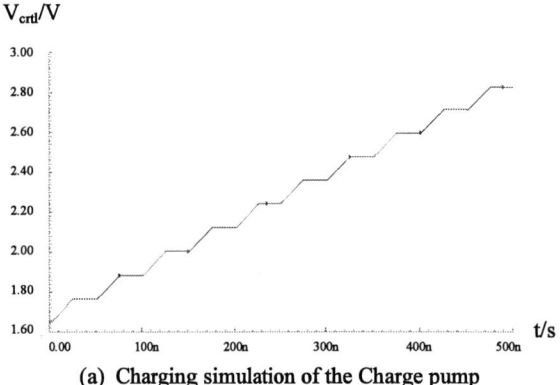

(a) Charging simulation of the Charge pump

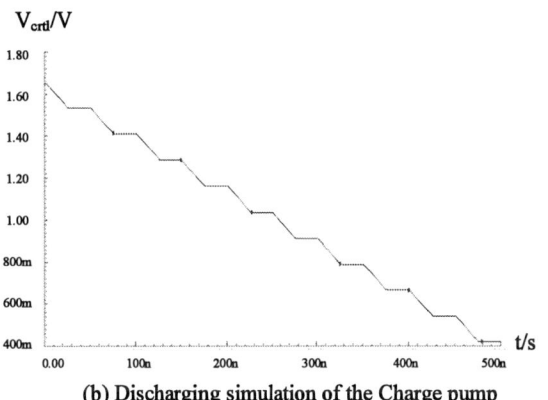

(b) Discharging simulation of the Charge pump

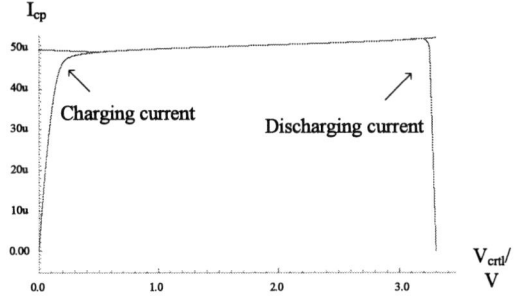

(c) Relationship between current and output voltage of charge pump

Fig. 5. Simulation results

Acknowledgments

The authors would like to acknowledge the National Natural Science Foundation of China for providing financial support for this work under Grant No. 60773081 and Grant No. 60777018, and also to acknowledge the financial support by the Shanghai Municipal Committee under Grant No. AZ028 and Grant No. 06DZ22013

References

1. Behzad Razavi, "Integrated Circuit Design of Analog CMOS, " XIAN, XIAN JIAOTONG University Publishing Company, 2002, pp. 432-470

2. W. Rhee, "Design of high-performance CMOS charge pumps in phase-locked loops," *IEEE Int. Symp. Circuits and Systems,* Vol. 1, 1999, pp. 545-548

3. P. Larsson, "A 2-1600 MHz CMOS Clock Recovery PLL with Low-Vdd Capability," *IEEE Journal of Solid-State Circuits,* Vol. 34, No. 12 (1999), pp. 1951-1959.

4. Kyung-Soo Ha and Lee-Sup Kim, "Charge-pump reducing current mismatch in DLLs and PLLs," *ISCAS 2006 IEEE International Symposium on Circuits and Systems,* May. 2006, pp. 21-24

5. Mark Van Paemel, "Analysis of a Charge-pump PLL: A New Model," *IEEE Journal of Solid-State Circuits,* Vol. 42, No. 7 (1994), pp. 2490-2498

6. Qu qiang, Zeng lieguang, "A dead zone Phase Frequency Detector used in Phase-locked Loop with high speed," *Micro Computer Information*, Vol. 12, No. 2 (2006), pp. 235-237

Performance Simulation for EVPD with Equivalent Circuit Models

Zhihua Li, Lixi Wan

Institute of Microelectronics, Chinese Academy of Sciences, Beijing 100029, China
lzhandzy@ime.ac.cn，86+10-82995593

Abstract

Edge-view photodetector (EVPD) is a new semiconductor photodetector with 3-D structure for simple optical assembly and packaging. In this paper, the equivalent R-L-C circuit model of EVPD was built and the R-L-C values were obtained by fitting the model S parameters to the measurement. A commercial P-I-N photodetector was also studied with the same approach and the R-L-C values were compared with those of EVPD. As a result, the series resistance, capacitance and inductance of EVPD were much higher than that of the PIN PD. It may be a reason of the poorer frequency properties of EVPD. To improve the EVPD performance, the material growth of I layer, Ohm contact process and anode process were suggested to be optimized in the future work.

1. Introduction

Optical interconnects in or between electronic systems can provide a solution for high-speed interconnects to solve the inherent limits of electrical interconnects in operating speed and distance, channel density and power dissipation. But it must undergo an integration of many functions into a small package. At the same time, semiconductor industry demands for higher levels of integration, lower costs, and a growing awareness of complete system configuration have continued to drive system in package (SiP) solutions. So optical interconnects in or between systems should have simpler structure, higher integration and lower cost before its extensive commercial application.

Edge-view photodetector(EVPD), which has an unique 3-D structure to allow directly coupling with plane waveguide, is a new photoelectric device developed in our group [1, 2]. For its potential to simplifier optical interconnects structure and reduce the cost of optoelectronic integration, EVPD has attracted much attention[3, 4]. But the performance of EVPD needs to be improved before it can be used in optical interconnects. In this paper, for the purpose of electric properties studying and guiding for performance improvement, the equivalent R-L-C circuit models of EVPD and a commercial P-I-N PD were built and simulated by using Agilent ADS software. The R-L-C parameters of EVPD and P-I-N PD were compared and analyzed to provide a reference for EVPD optimization.

This work was supported by National Natural Science Foundation of China, No. 90607006 and by Hi-tech Research and Development Program of China (863 Program) No. 2006AA01Z236 and No. 2007AA01Z2a6.

2. EVPD modelling and simulation

The EVPD structure is shown in figure 1. In fact, EVPD is similar to conventional P-I-N PD except that the light entry facet is on the slope. Figure 1 shows that the active area is partly on the top plane and bottom plane. It results from the unperfect process of window definition. The equivalent R-L-C circuit model based on the EVPD structure is shown in figure 2. This model consists of three parts: high mesa part,

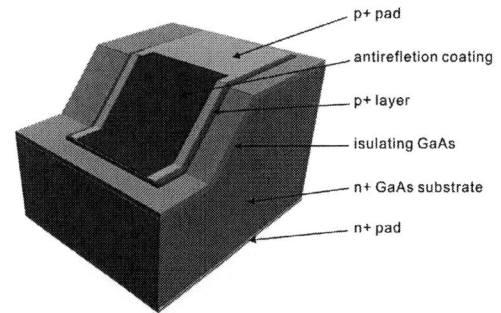

Figure 1. Schematic structure of EVPD

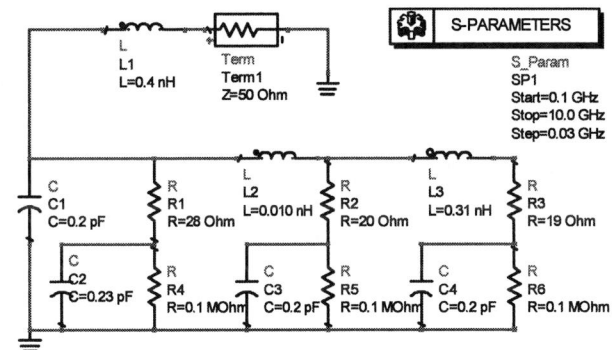

Figure 2. Equivalent R-L-C circuit model of EVPD

slope part and low mesa part. As figure 2 shows, R1, R2 and R3 are the parasitic series resistances of the three parts. R4, R5 and R6 are the parallel resistances of I layer in the three parts. C1 is the pad capacitance and C2, C3, C4 are the junction capacitances of the P-I-N structure of the three parts. L1 is the lead wire inductance between the p$^+$ pad of EVPD and its slide. L2 and L3 are the parasitic inductances between the three parts. Term is the port for S parameter simulation. Agilent ADS software was used to simulate the single-ended S parameters whose curves could be changed from the change of the R-L-C values in the model. The actual S parameters are measured by using a vector network analyzer HP8510C. The simulated S parameter curves can be fitted to the measured ones by changing the R-L-C values in the model. When the R-L-C values are as figure 2 shows, the simulated curves are consistent well with the measured ones, shown in figure 3. There are a real part and image part of the S parameter in the figure, in which the data of both parts from simulation should be fit the measured ones synchronously, which can unveil the real R-L-C values.

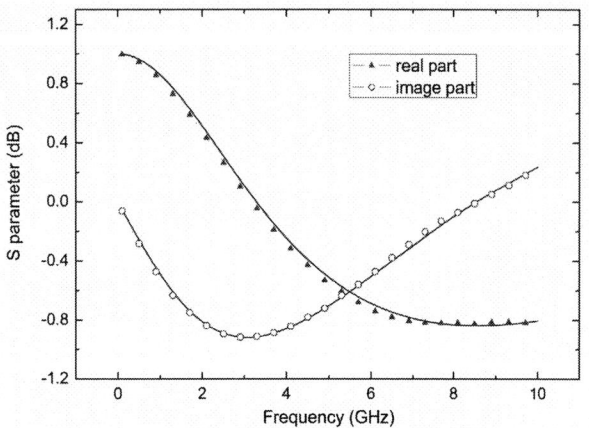

Figure 3. S parameters fitting of EVPD, the symbols of triangles and rotundities are the measured data, the curves are the simulations.

3. Planar P-I-N PD model and simulation

For comparison, we studied a commercial P-I-N PD with the same method. Figure 4 shows the structure of the P-I-N PD. It has round active area and round anode pad. The cathode pad is on the back substrate, which is the same as EVPD. The equivalent R-L-C circuit model is shown in figure 5.

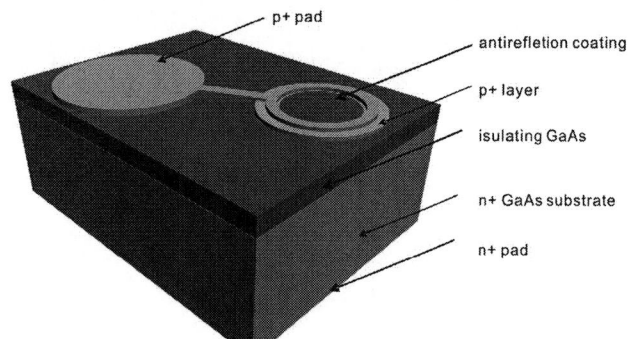

Figure 4. Schematic structure of P-I-N PD

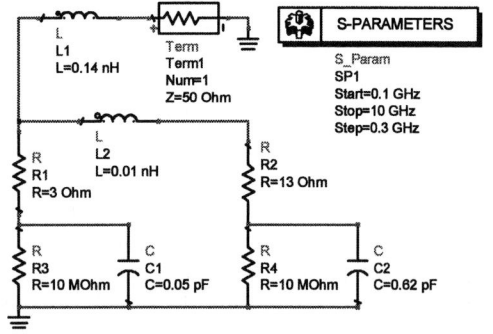

Figure 5. Equivalent R-L-C circuit model of P-I-N PD

As figure 5 shows, R1 and R2 are the parasitic series resistances of the pad area and active area. R3 and R4 are the parallel resistances of I layer under the pad area and the light entry area. C1 and C2 are the pad capacitance and junction

capacitance respectively. L1 is the lead wire inductance between the anode pad of the P-I-N PD and its slide. L2 is the parasitic inductance of the metal stripe wire connecting anode pad with active area. Term is the port for S parameter simulation. If the R-L-C values are shown in figure 5, the simulated S parameter data are perfectly fitted to the measured curves, as figure 6 shows.

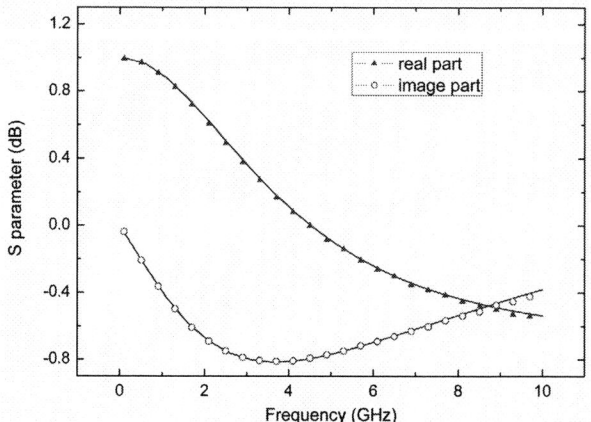

Figure 6. S parameters fitting of P-I-N PD, the symbols of triangles and rotundities are the measured data, the curves are the simulations.

4. Discussion

In fitting process we found that the parallel resistances such as R4, R5, R6 in EVPD model and R3, R4 in P-I-N PD model have little influence on the curves shape when their values are higher than 1MΩ. So the values of these resistances are inaccurate. But we can obtain the approximate value by calculating the slope of I-V curve. Figure7 shows the I-V curve of EVPD. We can calculate that the holistic parallel resistance of EVPD is about 57KΩ from figure 7(a), and the holistic series resistance is about 10Ω from figure 7(b), which can be as the initial values in the fitting process.

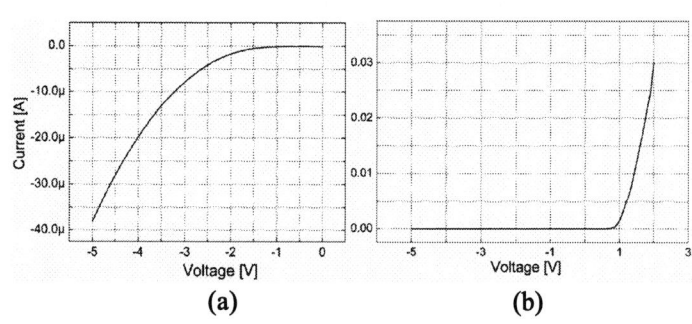

(a)   (b)

Figure 7. The I-V curve of EVPD

For the simplicity of comparison, we consider the holistic R-L-C values of EVPD and P-I-N PD. It is supposed that the values of holistic series resistance, parallel resistance and capacitance are the parallel results of all corresponding components in the equivalent circuit, and the holistic inductance is the sum of inductance units in series. As a result, the R-L-C values of EVPD and P-I-N PD are shown in table 1.

408

Table 1. The R-L-C values of EVPD and P-I-N PD

	EVPD	P-I-N PD
Series resistance (Ω)	7.23	2.44
Parallel resistance (Ω)	$\sim 5 \times 10^4$	$> 10^8$
Capacitance (μF)	0.83	0.67
Inductance (nH)	0.72	0.15

It is well known, a PD with P-I-N structure is expected to have low series resistance, low capacitance, low inductance and high parallel resistance. Table 1 shows all of the parameters of the P-I-N PD are better than that of EVPD. It indicates that the performance of EVPD is poorer than P-I-N PD which showed in our previous photoelectric measurement. The measurement detail: an 850nm vertical cavity surface emitting laser (VCSEL) with a multimode fiber pigtail was used to constructed the optical link and the responsivity and S(21) parameters are measured in HP8510C system. As a result, the responsibility of EVPD is 0.13A/W, much less than 0.6A/W for P-I-N PD. Figure 8 shows S(21) parameters of EVPD and P-I-N PD. The 3dB bandwidth of EVPD is less than 500MHz while it is about 2.5GHz for P-I-N PD.

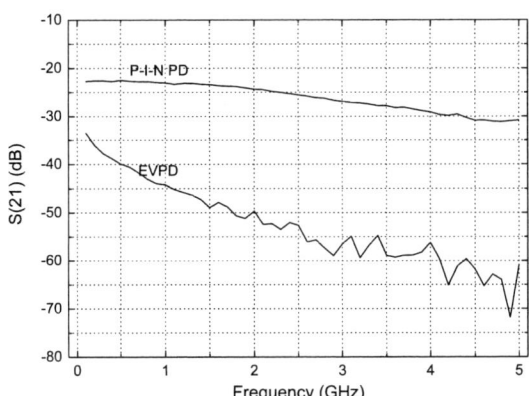

Figure 8. S(21) parameter of EVPD and P-I-N PD

Based on the model and simulation results which could be translated into the physic requirements in each layer, some processes should be optimized to improve the EVPD performance. Firstly, the parallel resistance is sensitive to the crystal quality of I layer. The material growth of I layer should be optimized to improve the insulating properties, which may enhance the responsibility of EVPD. Secondly, appropriate ohm contact process can efficiently reduce the parasitic series resistance and improve the frequency properties of EVPD. Thirdly, the capacitance and inductance are mainly influenced by the anode area and shape, especially the metal stripe line which may lead to high parasitic inductance. The anode pad should be re-designed.

Conclusion

In this paper, we built equivalent R-L-C circuit models of EVPD and commercial P-I-N PD. The R-L-C values are obtained by fitting the single-ended S parameters measured. This may be a helpful method analyzing the structure of semiconductor devices. Based on the comparison of the R-L-

C values of EVPD and P-I-N PD, it is indicated that the material growth of I layer, Ohm contact process and anode process of EVPD should be optimized in the future work.

References

1. Z.H. Li, H.J. Shen, C.Y. Yang, B.X. Li, L.X. Wan, and D. Guidotti, "Edge-view photodetector for optical interconnects," Optics Letters **32**, 2096-2098 (2007).
2. Z.H. Li, H.J. Shen, C.Y. Yang, B.X. Li, L.X. Wan, "A New Photodetector for Chip-to-Chip Optical Interconnects," in *2008 Science Development Report*, J. Hou, Y. Guo, and S. Hu, eds. (Science Press, Beijing, 2008), pp. 136-139.
3. B. Hitz, "850-nm Photodetector Can Be Integrated into Waveguide," photonics spectra **41**, 91 (2007).
4. "Research Highlights," Nature Photonics **1**, 608-609 (2007).

A March-CL Test for Interconnection Faults of SOC

Zhang Jinyi[1, 3], Yang Xiaodong[2], Yang Yi[3], Zhang Dong[3], Dong hui[3]

1. Key Laboratory of Advanced Displays and system Application, Ministry of Education, Shanghai University
2. Microelectronic Research & Development Centre, Shanghai University
3. Key Laboratory of Special Fiber Optics and Optical Access Networks, Ministry of Education, Shanghai University
No.149 Yanchang Road, Shanghai 200072, P.R.China
Email: zhangjinyi@staff.shu.edu.cn   Tel: 086-021-56331323-113   Fax: 086-021-56331272

Abstract

Shrinks of feature size, high working frequency, and rising number of the IP cores integrated in SOC make the problem with interconnection test critics. A March-CL test for interconnection faults of SOC is proposed in this article. According to the method, eight test patterns are used to detect stuck and delay faults of interconnection between IP cores. The IP connected by interconnection under test (IUT) is wrapped and complied with IEEE1500. Short test time and low area overhead are achieved with the method. Moreover, modified wrapper cell structure with simple control logic is adopted for detecting delay in March-CL test. Finally, March-CL test is applied to ITC'02 bench, and result proves that the method covers 100% of stuck, bridge and delay faults in synchronous interconnection test.

1 Introduction

Interconnection between core used to be ignored, because of its simple architecture and less test time than that for internal-core [1]. However, continuously growing number of IP cores integrated in SOC causes interconnection between IP more and more. Furthermore shrinks of feature size, especially when contemporary deep submicron technology (DSM) and working frequency up to multiple GHzs, make crosstalk and delay of interconnection be a major issue [2]. Therefore, interconnection test cannot be neglected as before. Signal integrity (SI) problem caused by cross-coupling capacitance and inductance between interconnections, include excessive signal delay, overshoot, undershoot, glitches, oscillation, and even signal speedup [3].Moreover, delay of SI problem, stuck and bridge occupy up to 90% of interconnection faults, which are addressed by the March-CL test proposed in this article.

Many literatures attempt to provide solution to addressing interconnection test. M. H. Tehranipour et al. proposed a test mechanism complied with IEEE1149.1 standard, which includes modified driving cell to generate test patterns and observation cell for multiple transition (MT) fault model [4]. Qiang Xu et al. proposed a compacted patterns method for SI test and algorithms for test architecture optimization.[1], and present a new IEEE Std. 1500-compliant wrapper designs and overshoot detector inside the wrapper[5].in [6], a method to detect crosstalk faults using a periodic square wave test signal was proposed.

March-CL is also complied with IEEE Std.1500, which is proposed to enhance controllability and observablility of IP cores in SOC. As well known, huge scale of SOC leads to difficulty in testing IP cores, which is addressed through adding wrapper cell to interconnection between IP cores in IEEE1500. Storage ability of wrapper cell makes it similar to testing memory and interconnection. Due to the similar physical characteristics of IP interconnection and memory cell, this article proposes to adopt March algorithm used in memory test to generate patterns, which are able to detect stuck faults, bridge faults and delay fault. Meanwhile, IEEE1500 can potentially support SI interconnection test if two-pattern at-speed signal transition capability is provided at the core test wrapper of the interconnection driving side [5]. This article adopts a modified wrapper cell under simple control to detect delay fault, the test mechanism is complied with IEEE1500, and able to test signal integrity and non-signal integrity in unified mode, which is detailed in the section 3.

Final part of this article gives experiment in ITC'02. Simulation result by Modelsim and Pspice proves that March-CL test is able to test interconnections at both sides of IP cores synchronously and covers 100% percent of stuck, bridge and delay faults with comprehensive comparison to all the eight patterns.

2 Interconnection faults and March
2.1 Fault model

As growing number of IP cores in SOC, interconnections are suffered many types of faults caused by material defects, technological process flows or design imperfection [2]. As shown in Fig, 1, main types are stuck, bridge and delay faults. The former two types are static, while delay is caused by crosstalk between interconnection. Thus, another model is required to describe crosstalk, rather than traditional stuck model.

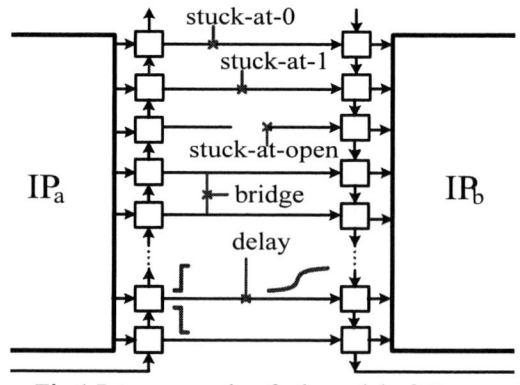

Fig.1 Interconnection fault model of IP cores

Many researchers use maximum aggressor (MA) fault model to describe and analyze crosstalk and detect fault types of interconnections. MA is provided in [7] by M. Cuviello et al. The model takes specific interconnections as a victim line, and interconnections at the both sides of victim as aggressor line, which affect the victim through passive components, such as capacity. In general, the crosstalk result could be noise (causing ringing and functional error) and delay (causing performance degradation, as shown in Fig, 1). MA model shows that all aggressors at both of sides simultaneously invert to the same direction, while the victim line is kept quiescent (for maximal ringing) or makes an opposite transition (for maximal delay)[4]. Table 1 shows the test patterns for detecting faults according to MA model.

Table 1. MA fault model and test patterns [4]

	P_{g0}	P_{g1}	N_{g1}	N_{g0}	d_r	d_f
A	↓	↓	↑	↑	↓	↑
V	0	1	1	0	↑	↓
A	↑	↑	↑	↑	↑	↑
Vector 1: AVA	000	010	111	101	010	101
Vector 2: AVA	101	111	010	000	101	010

2.2 Principle of March

March algorithms is widely used in memory test, which write and read memory cells according to up or down address order and analyze the faults of cell with readout response. To address memory fault test, March algorithm has been proved an effective method, which includes a variety of branches, such as March A, March B, March Y, and so on[8]. All of them have complexity of O(N), and have different targets and characteristics. Some are used to reduce test application time, and some to high fault coverage. Regularly, practical applications of March algorithm integrate several branches for higher fault coverage.

Because of the interconnection between two IP cores shows similar physical characteristics with memory, this article proposes an IEEE1500-complied March test to generate patterns for interconnection test, which are able to detect stuck faults, bridge faults and delay fault. As shown in Fig.2, an SOC wrapper contains several wrapper cells, which represent as flip-flop. Therefore, interconnection fault model can be simply formulated in memory fault model.

Differences still exist between interconnection and memory test. As the layout of IP Interconnections being simpler than memory, the number of fault modules necessary to consider is less than memory. Moreover, the two tests mechanisms are different, which will be discussed in following section. Thereby, this article presents a new March-CL (March for Connection Line) test, which is improved to suit for Interconnection test.

3 March-CL test and wrapper cell for delay test

March-CL test is a method to detect stuck, bridge and delay faults of interconnections between IP cores complied to IEEE1500. The detail application of March-CL is introduced in part 3.1. For achieving to delay detecting with March-CL, part 3.2 proposes a modified wrapper cell structure.

3.1 March-CL test

In despite of comparability of March test in memory and interconnections between IP cores, there are still several differences between two kinds of test mechanisms, main aspects of which are described as follow:

a. The read/write of memory is processed to single cell. If number of cells under test is n, it is 2n steps to take for completing the read and write to all of cells. In contrast, read and write to interconnection can be regarded as approximately simultaneous, which means that test of n independent interconnections needs merely two steps to load patterns and extract response. Unfortunately, wrappers complied with IEEE1500 around interconnections makes the test more time-consuming.

b. Choices of cell or IUT are different. In memory test, cell under test assigned to is determined by address decoder, and the loading of specific pattern chooses IUT. Moreover, time to load patterns is sum of whole scan chain length. The signal loading and extract are to the same direction along the scan chain (as Fig. 2), unlike the accordance to up or down address order of memory test.

c. March test for memory changes the state of a single cell once a time, which increases the complexity of couple fault test procedure. Thanks to the simultaneous transition of all interconnections, steps to detect same fault of interconnections is significantly reduced.

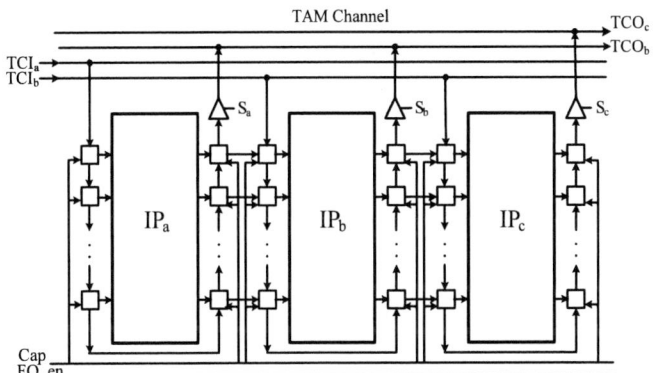

Fig. 2 Interconnection fault test structure

For above reasons, notations of March-CL are modified to adapt for interconnection test. Test patterns that are able to cover MA model is expressed as follow:

{wr0, wr01, w10, wr01, wr0, wr1, wr10, wr1}

The above notation "wr" is writing and reading process to an interconnection. and "0"("1", "01", "10") is data ground, where "wr0(1)" means 0(1) is written into and read out from all of interconnections, "wr01" means writing and reading 0 to IUT, and "1" to other interconnections, "10" means opposite. For instance of three interconnection, the third step "wr10" represents data ground is 101, where second interconnection is current IUT.

The test flow (see Fig.2) is:

a. Enabling the wrapper cells of IP$_a$ (WC$_a$) to load test patterns when wrapper cells of IP$_b$ (WC$_b$) are waiting for patterns arriving at right location.

b. WC$_b$ are enabled to capture the test patterns and WC$_a$ is hold to current situation through this clock cycle.

c. WC$_a$ transports next test pattern, while response of former pattern is extracted through WC$_b$.

Therefore, test patterns are loaded to WOC (Wrapper Output Cell) through WIC (Wrapper Input Cell) for IP$_a$, and response is seized by WIC and extracted through WOC for IP_b. Test time for one interconnection is expressed as

$$Ti = (max \{SC_a, SC_b\} +1)* P+min \{SC_a, SC_b\} \quad (1)$$

Where SC is the shift time through scan chain, the subscript a (b) means the scan chain is belong to IP_a (IP_b), and P is the number of pattern, which is eight in March-CL test. Both SC_a and SC_b are proportional to scan chain length of IP cores, respectively. Noted that formula (1) is according to test for interconnection at one side of core. To avoid scheduling conflict, test for interconnection at the other side will take Ti again, which leads to waste in time, in fact. Therefore, synchronous test for both sides interconnection is adopted in the experiment. More detail will be given in next part.

Test for other interconnections is similar as described above. Therefore, taking 2k aggressor interconnections at both sides into account, it takes (k+1) phases to complete test for all of interconnections. Application time in all is (k+1)*Ti.

Taking four interconnections as a group, current IUT are the first and third interconnections in first phase, which are both affected by two neighboring interconnections. Test pattern according to March-CL is as Fig3.

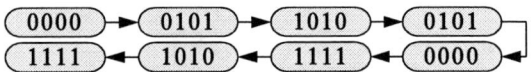

Fig. 3 test patterns in March-CL

Analyzing of response is divided into two aspects:

First, detecting stuck faults and bridge faults is performed by analyzing the response truth table (as table2), where "And" means wired-and short of bridge faults, and "Or" means wired-or. For instance, interconnection "A" and "B" is neighboring, and "A" is IUT suffered interference from "B". Through analyzing responses of four inputs combinations, one of five states is got to diagnose the situation of "A".

Table2. Response truth table

A	B	Response of A					
		Normal	Sa0	Sa1	So	Bridge	
						And	Or
0	0	0	0	1	Hiz	0	0
0	1	0	0	1	Hiz	0	1
1	0	1	0	1	Hiz	0	1
1	1	1	0	1	Hiz	1	1

Second, delay is another main issue in DSM. Serious interference in MA model is achieved through inverts of interconnections according to March-CL patterns. For example, invert form 0101 to 1010 leads to maximal rising delay of third interconnection suffered from interference of second and fourth one. However, standard IEEE1500 wrapper is not sufficient to observe delay fault. In next part, a modified wrapper cell is present to improve the observation.

3.2 Wrapper cell for delay test

A modified observation and driving cell is proposed in [4], which are controlled by complicated signals and works at two modes: SI test and non-SI test. During test, interconnection can be set to aggressor, victim or non-SI states, which increase the steps of test procedure. Moreover, some of test patterns for SI test mode are able to detect static faults, which are reloaded in another mode. To avoid the time waste and complexity of control architecture, a simply modified wrapper cell is adopted in this article for delay test at concerned work frequency.

However, in need to construct scan chain, traditional wrapper structure is not suit for March-CL test. March-CL request interconnection maintain the pattern loaded in it, so as to achieve adjacent interconnection flip simultaneously, which leads to intensive disturbance according to MA model. Taking second pattern for instance, interconnections are suggested to stand at situation as 0101 until third pattern (1010) is loaded in the right wrappers. Traditional wrapper will flip in whole pattern loading process, which is ought to be avoid. A modified wrapper structure is adopted in this article. As shown in Fig.4. ,a D flip-flop is added to original "do" port in wrapper, and controlled by "FO_en" signal, which will set high level to enable original do pass, and low level to prohibit current "do" invert. Both Dffs are driven by rising edge of "CLK". "Cap" is control signal to enable wrapper capture data pass through "di" port. "Tsel" chooses wrapper work mode, which is including normal function and test modes. In test process, "Tsel" is kept at 1 in test mode.

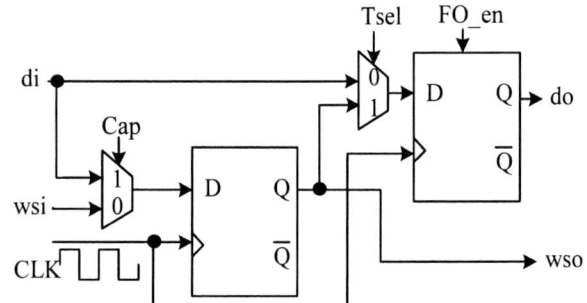

Fig. 4 An improved wrapper structure

The wrapper work process is: "Cap" is set 0 to transport test patterns to corresponding wrappers linked to interconnection, while "FO_en" is set 0 to maintain interconnection situation; when pattern is loaded, scan chain hold for one clock to ensure sample completed. In the moment, "FO_en" and "Cap" are set 1 to transport "di" signal thorough interconnection. Noted that clock of receive wrapper is later than that of stimuli wrapper (as shown in Fig. 5). The phase difference between clocks ("CLK_1", "CLK_2" in Fig. 5) of wrappers at both end of interconnection is acceptable delay region. (ADR), which is defined as the time interval that signals loaded in one end of interconnection must be transport to another end. If signal arrives after ADR, first sample of DFF will be the same as former sample, and is difference from second sample, which is right stable signal after 32 clock cycles, if no static fault. Therefore, comparison of two samples is able to detect delay of concerned ADR, which is convenient to adjust by modified phase difference between clocks of stimuli and receive wrappers.

Noted that delay is not detectable with this structure, if any Stuck-at fault exists. However, bridge does not affect delay

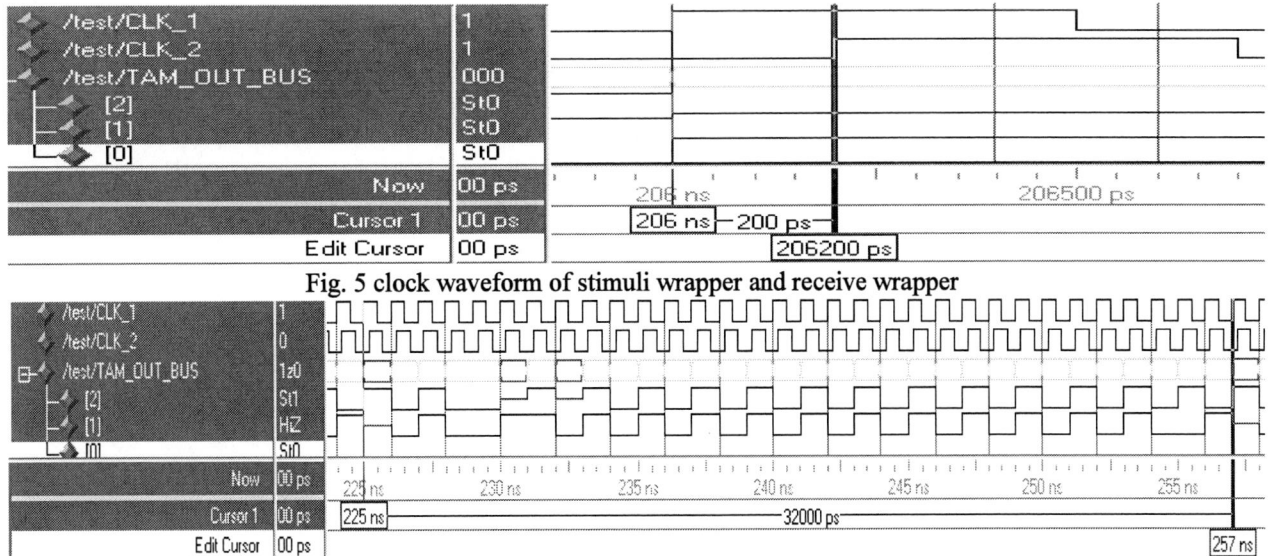

Fig. 5 clock waveform of stimuli wrapper and receive wrapper

Fig.6 response waveform of interconnection test

test effectiveness, for transition is able to be observed in receive wrapper.

Moreover, the proposed structure liberates scan chain in synchronous test for interconnection at both sides, for that next pattern could be loaded while samples are transported. The pattern loading of one side is independent on situation of the other side, which realizes full use of shift time.

As described in previous part, formula (1) is not applicable in synchronous test, so a modified formula is as follow:

$$Ti = (LSC*3+1)* P+LSC \qquad (2)$$

where LSC is largest length of scan chains constructed by WIC or WOC of all the IP cores under test. Taking one c6288 of ITC'02 bench for instance, whose number of input and output port with wrappers is 32 and number of scan chain is 1, LSC is 32, either. Ti of interconnection at both sides is ((64+1)*8+1)*2 with formula (1) in serial test, while in synchronous test, Ti is (32*3+1)*8+32, which means the former test method reduce application time by nearly 40%. It's obvious that the numbers of WIC and WOC should better be close to each other.

4 Experiment results

Experiment is carried out in ITC'02 benchmark circuit d695. To simulate at-speed state of interconnection; frequency of all cores under test is set to at 1 GHz. Software to simulate test scheduling is Modelsim. Interconnection at both sides of core (as IP_b shown in Fig.2) is under test simultaneously.

Taking c6288 for instance again, there is one scan chain around the middle c6288. Fig.6 shows response of receive wrapper to second pattern, input sequence of which is 0101... Stuck-open fault shown as red part of wave, presents high resistance, and is noted by two blue cursors at interval of 32 clock cycles, sampled by Dff twice. Moreover, two samples in front of 225ns and four followed are different from pattern, which are sufficiently used to detect faults in the corresponding interconnection, but not to diagnose fault

types. To achieve that, eight patterns are necessary to comprehensive comparison.

According to delay detecting, Pspice is applied to simulate sample result of added Dffs in wrapper. Fig. 7 shows situation of wrapper working with clock at frequency of 1 GHz in

lumped RC model. Simulation object is three adjacent interconnections. The middle one is victim inverting from 0 to1, while ones at two sides invert from 1 to 0 simultaneously. The dot crude wave in the figure is input voltage(Vi) of victim interconnection, and thin is output voltage(Vo), which is later than Vi up to 0.7ns, and there is a unstable period after Vo reaching high level, obviously. "T_CLK" is sample clock, frequency and phase of which results in the precision and ADR as discussed above. DFFA, straight line in figure, is sample result. Taking care of Dff's own delay, grey line means Vo is at level between valid 0 and 1 level, and sample by second rising edge presents 0 different from Vi, which proves delay occurs.

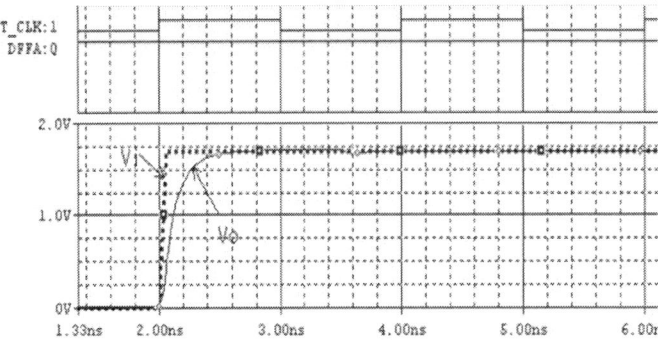

Fig. 7 delay detecting simulation by Pspice

Conclusions

The article presents a March-CL test for interconnection fault between IP cores, which is able to detect stuck faults, bridge faults and delay through a simple modified wrapper cell complied with IEEE 1500. Synchronous test for interconnection at both sides of one core is achieved in ITC'02 and wrapper with another Dff is proved to successful in detecting delay by Pspice model. The coverage of 100% faults under control of simple logic is provided through experiment.

However, frequent transitions of wrapper cell lead to great power, especially in considering more interconnections under test. To addressing the issue, patterns order should be optimized to reduce transitions of wrapper. In addition, because test time significantly depends on scan chain length, breaking a long scan chain into shorter ones will shorten application time at the cost of occupying more TAM wires.

Acknowledgments

Thank you for the support from Leading Academic Discipline Project of Shanghai Educational Committee (Project No.J50104).

References

1. Qiang Xu, Yubin Zhang, Krishnendu Chakrabarty. "SOC Test Architecture Optimization for Signal Integrity Faults on Core-External Interconnections," *Design Automation Conference*, June 2007, pp.676-681.

2. T. Garbolino, K. Gucwa, M. Kopec, A. Hlawiczka. "Crosstalk-Insensitive Method for Testing of Delay Faults in Interconnections Between Cores in SoCs," *Proceedings of Mixed Design of Integrated Circuits and Systems, 2007. 14th Int. Conf.* , June 2007, pp. 496-500.

3. S. Kundu, et al, "On Modeling Crosstalk Faults," *in IEEE Trans. Computer-Aided Design of Integrated Circuits and Systems*, Dec. 2005, 24(12):1909–1915.

4. M. H. Tehranipour, N. Ahmed, and M. Nourani. "Testing SoC Interconnections for Signal Integrity Using Extended JTAG Architecture," *IEEE Transactions on Computer- Aided Design*, May 2004, 23(5):800–811.

5. Qiang Xu, Yubin Zhang, Krishnendu Chakrabarty. "Test-wrapper designs for the detection of signal-integrity faults on core-external interconnections of SoCs," *Int. Test Conf.*, Oct 2007, pp.1-9.

6. Ming Shae Wu, Chung Len Lee. "Using a periodic square wave test signal to detect crosstalk faults," *IEEE Design & Test Computers*, March 2005, 22(2):160-169.

7. M. Cuviello, S. Dey, X. Bai, Y. Zhao. "Fault Modeling and Simulation for Crosstalk in System-On-Chip Interconnections," *Proceedings of the International Conference on Computer-Aided Design*, 1999, pp. 297-303

8. CF. Wu, CT. Huang, KL. Cheng, CW. Wang, and CW. Wu. "Simulation-based test algorithm generation and port scheduling for multi-port memories," *in Proc. IEEE/ACM Design Automation Conf.*, June 2001, pp.301-306.

Multi-clock SOC Test schedule based on TWC&S

Zhang Jinyi [1,3], Jiang Yanhui [2], Lin Feng [2], Wang Jia [3], Sun yan [3]

1. Key Laboratory of Advanced Displays and system Application, Ministry of Education, Shanghai University
2. Microelectronic Research & Development Centre, Shanghai University
3. Key Laboratory of Special Fiber Optics and Optical Access Networks, Ministry of Education, Shanghai University
No.149 Yanchang Road, Shanghai 200072, P.R.China
Email: zhangjinyi@staff.shu.edu.cn   Tel: 086-021-56331323-113   Fax: 086-021-56331272

Abstract

Both wrapper and TAM are important components in SOC test architecture, and wrapper scan chain optimization and TAM optimization are NP-hard problems. Addressing wrapper scan chain optimization or TAM optimization separately leads to SOC test time sub-optimal. This paper presents a TWC&S (TAM/wrapper co-optimization and scheduling) algorithm for multi-clock SOC after combining the advantages of wrapper scan chain optimization and TAM optimization, and it aims to decrease test time of multi-clock SOC extremely. To demonstrate the validity of the proposed algorithm, experiment is performed on the multi-clock SOC $MCDS_2$. The results show that there exist inflexion of test time, and the optimal test time is 23.2% when the widths of test bus are 20 vs. the least one. The phenomenon, decreasing degree of test time decreases slowly with test bus widths increase, demonstrates that trade-off will be achieved properly among test time, test area and design complexity.

Key words: multi-clock SOC; wrapper optimization; TAM optimization; co-optimization; scheduling

1 Introduction

With the development of semiconductor technology and the improvement of design ability of integrated circuit (IC), the complexity of system design increases and many Intellect Property (IP) cores are integrated in a single chip, such as microprocessor, analogy IP cores, digital IP cores, and memory. The reused IP cores become much welcomed by system integrators because it will greatly reduce the design time of System-on-Chip (SOC). However, as IP cores deeply embedded in a single chip, it is more difficult to test IP cores from primary input and output ports of SOC. Therefore, the SOC test becomes a challenging problem. In order to solve the SOC test problem, the international test standard IEEE1500 based on the test for embedded cores was proposed in 2005. The standard presents a unified specification for the IP cores' wrap for the first time. For reducing the test complexity of SOC and then reducing the test time, the test source/test sink, test access mechanism (TAM), and wrapper [1] are proposed in SOC research area; this is the work of the SOC DFT (design-for-test).

It has proved in some literatures that wrapper scan chain optimization, TAM optimization, and test schedule are NP-hard problems [2, 3, 4]. In order to solve those problems and reduce the test time of SOC, then the test cost can be reduced, many methods for TAM optimization, wrapper scan chain optimization, and test schedule were proposed in literatures. Such as the BFD algorithm [5] that deals with wrapper scan chain optimization, the ILP [6] method that deals with TAM optimization, test schedule concurrently; and other

enumerated heuristic method [7]. The test schedule method [8] that considers the power, source, and process constraints was also proposed in the literature.

However, current SOC usually involves multi clocks, especially the SOC used in communication systems; it increases not only the difficulty of SOC design, but also the difficulty of SOC DFT. In order to furthest reduce test time of multi-clock SOC under test bus widths constraint; this paper proposes a TWC&S algorithm for multi-clock SOC. In order to prove the efficiency of the proposed algorithm, the simulations are performed on the $MCDS_2$ which is one of the typical multi-clock SOCs. Simulations show that there exist an important inflexion, and the optimal test time is 23.2% when the widths of test bus are 20 vs. the least one. The phenomenon, decreasing degree of test time decreases slowly with test bus width increase, demonstrates that it unnecessary to increase test bus widths extremely to obtain the least test time. This is helpful to constraint the test area and test complexity. Thereby trade-off will be achieved properly among test time, test area and design complexity according to actual necessity.

2 Multi-clock SOC

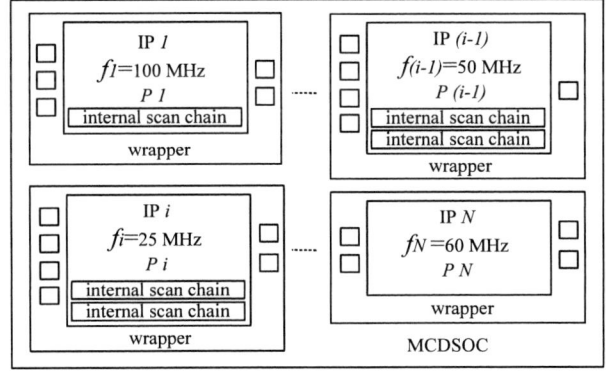

Fig.1 multi-clock SOC

In this section, the multi-clock SOC for test optimization will be described. Multi clocks exist in the SOC, the individual IP cores in SOC are tested in certain clock frequency, and however, there is only one test clock for one IP core. The IEEE standard wrapper is introduced to wrap each IP core, by which the test stimulus are transported from TAMs to each IP core, and test response are transported from the each IP core to TAMs, finally collected by auto-test equipment (ATE). It's assumed that the amount of IP cores is N in multi-clock SOC, and the test frequency of each IP core is different. $\{I_1、I_2……I_N\}$ are the numbers of input ports about each IP core. $\{O_1、O_2……O_N\}$ are the numbers of output ports about each IP core. $\{P_1、P_2……P_N\}$ are numbers of test pattern. $\{f_1、$

978-1-4244-2739-0/08/$25.00 ©2008 IEEE   415

f_2 ......f_N} are the test frequency. The multi-clock SOC is shown in Fig.1. The wrapper scan chain optimization, TAM optimization, and test schedule in SOC top level are necessary for reducing test time.

3 The TWC&S algorithm for multi-clock SOC

3.1 Test problems of multi-clock SOC

Current SOC usually involves multi clocks. In order to reduce the test complexity of multi-clock SOC, the wrapper scan chain optimization is performed on IP core level. However, the test time of multi-clock SOC decreases with increasing of the TAM widths. Thereby, it is necessary to optimize the TAM on SOC level. Therefore, only the sub-optimal test time not the optimal one will be obtained when the optimizations of wrapper scan chain and TAM are performed separately.

In order to furthest reduce test time of multi-clock SOC under test bus widths constraint, problems of the TAM /wrapper co-optimization and scheduling of multi-clock SOC will be described firstly. It's assumed that the amount of IP cores with different test frequency is N in a multi-clock SOC. In order to get the optimal test time for multi-clock SOC, it's necessary to obtain the numbers of TAMs, the partition for the TAM widths, the test schedule for each IP core with different test frequencies, and optimization for wrapper scan chain in each IP core under widths of test bus constraint.

3.2 Test time model of multi-clock SOC

Wrapper scan chain Optimization, TAM Optimization and test schedule are NP-hard problems, in order to solve those problems, the test time for each IP core must be modeled firstly. If S_i is the clock period for test pattern scanned in, S_o is the clock period for test pattern scanned out, and the P_i is the amount of test pattern. The model of test time for each IP core (t_i) can be formulated from the model [9] of test clock numbers when the test frequency is considered; it's shown in equation 1 as follows.

$$t_i=((1+\max (s_i,s_o))P_i + \min (s_i,s_o))/f_i \qquad (1)$$

Where, for IP core i, S_i/S_o is its scan in/out amount of clock period, P_i is its amount of test pattern, and f_i is its test frequency.

3.3 The TWC&S algorithm

In this section, the TWC&S algorithm will be presented for multi-clock SOC, and it aims to obtain the optimal test time.

Width of test bus: W_{max}
Numbers of TAM: N_{tam}
Numbers of IP cores in multi-clock SOC: N
TAM partition: W_1, W_2... W_{Ntam}
The test time for multi-clock SOC: T_{mcsoc}

Tab.1 Basic parameters

The algorithm flow of TWC&S algorithm for multi-clock SOC is shown in Fig.2. The basic parameters related to the algorithm are shown in Tab.1. The TAMs are optimized firstly when the widths of test bus, the numbers of TAMs, and the numbers of IP cores are achieved. The steps for TAM optimization is shown in following.

To begin with, the algorithm is initialized to confirm the test time of each IP core using the least test bus, the test time of each IP core is checked and compared, and each IP core is reordered in descending order, it's shown from step 1 to step 2 in Fig.2.

Then, some IP cores will be cycled and located according to test time of current IP cores. The IP cores, whose test time are twice or more when compared to the current IP cores, are searched and compared iteratively, then those IP cores are collected to one group, and the TAM partition strategies are determined in each group, the process is shown from step 3 to step 7 in Fig.2. Because the test time of IP cores are closed related to the TAM widths, the test time of IP cores are directly influenced by the TAM partition, and the test time of SOC is influenced finally. Therefore, the certain strategies for TAM partition are explored according to the differences of test time among the current IP cores; it aims to extremely close the test time of IP cores and reach the balance of test time in each group, then the redundant test time will be reduce extremely and the test time of multi-clock SOC is least. The strategies used for partition are shown as follows. The strategy A, increase the TAM widths for the IP cores groups which need much more test time, is introduced when test time of each IP cores group exist great gap. The strategy B, increase TAM widths for each IP cores group, will be introduced when the test time of each IP core is closed together. The process is shown from step 8 to step 13 in Fig.2.

Finally, the test time of each IP core is compared iteratively after the certain strategy performed. The IP cores group and the strategy for TAM partition will be reconfirmed, and then the minimum test time on each TAM will be got, the process is shown in step 14 in Fig.2.

In order to obtain the optimal test time of IP cores, the BFD algorithm [5] is adopted to optimize the wrapper scan chain when the TAM widths are determined, the process is shown in step 15 in Fig.2, the basic idea is shown in following. To begin with, for each internal scan chain l, assign l to wrapper scan chain S, such that {(S_{max})-(length(S) + length(l))} is minimum, where S_{max} is wrapper scan chain with current maximum length, S_{min} is wrapper scan chain with current minimum length. Finally, add the primary inputs/ outputs to the wrapper scan chains.

The TAM optimization and wrapper scan chain optimization are iterative process, it is shown from step 5 to step 15 in Fig.2. The wrapper scan chains are necessary to be optimized to get the best balance scan chains in each IP core when the TAM is optimized and the best partition is achieved; the TAM partition is necessary to be done after wrapper scan chains optimized, the process will be terminated when the total TAM widths exceed to test bus widths.

The IP cores are scheduled to test parallelly to get the optimal test time for multi-clock SOC when wrapper scan chain and TAM are optimized. The assignment process of IP cores to TAMs is shown from step 16 to step 21 in Fig.2. For IP core i, the total test time in current TAMs is compared and the TAM$_j$ that costs the least time is found, then, the IP core i is assigned to TAM$_j$.

The process of WTC&S algorithm for multi-clock SOC

1. Calculate the test time of core i when using the least TAM width
2. Compare the test time of all core. Sort them in descending order
3. Find the current first core: IP_{s1}, and return it test time: T_{IPs1}
4. Find the current $(N/2+1)$ core: IP_{s2}, and return it test time: T_{IPs2}
5. Compare T_{IPs1} and T_{IPs2}
      if $T_{IPs1} >= 2 * T_{IPs2}$
6.   turn to step 8, adopt strategy A for TAM optimization
      else
7.   turn to step 11, adopt strategy B for TAM optimization
8. strategy A: Increase TAM width (W_1) for IP_{s1}, judge whether

       if $\sum_i^N W_i \leq W_{max}$

9.   transfer to BFD algorithm to recalculate test time of core
       $(IP_{s1}...IP_{s(N/2)})$, and return it, then turn to step 2
      else
10.   cancel width increasing, and turn to step 16 for core assignment
11. strategy B: Increase TAM width (W_i+2) for each TAM, judge whether

        if $\sum_i^N W_i \leq W_{max}$

12.   transfer to BFD algorithm to recalculate test time of core
        $(IP_{s1}...IP_{sN})$, and return it, then turn to step 2
       else
13.   cancel width increasing, and turn to step 16 for core assignment
14. Compare the total test time on each TAM, find the max test time
15. BFD algorithm for Wrapper scan chain optimization
16. determinate the TAM number
17. Sort core in descending order after TAM/Wrapper cooptimization
18. For core i
19.   Compare the total test time on each TAM
20.   Find TAM_j, such that the test time on TAM_j is minimum
21.   Assign core i to TAM_j

Fig.2 The pseudo-code of the TWC&S algorithm

4 Verification

In order to demonstrate the validity of the proposed algorithm for multi-clock SOC test, some IP cores in multi-clock SOC $MCDS_1$[10] is referenced. The information of $MCDS_1$ is shown in Tab. 2 as follows. The SOC consists of 14 cores. First 10 cores are from "d695" in ITC'02 SOC benchmarks. The column 1 is the order of IP each core. The column 2 is the at-speed test requirement of each IP core. The "flexible (≥ 2)" in column "wrapper list" denotes that wrapper with any number of test pins can be designed. The last column is the test frequency of each IP core.

the IP cores from 1st to 10th, 13th, and 14th except the 11th and 12th which need at-speed test in multi-clock SOC $MCDS_1$ are referenced as multi-clock SOC $MCDS_2$. The basic information about $MCDS_2$ is shown in Tab.3; there are twelve IP cores with different test clocks. The numbers of IP cores are shown in column 1. The test frequency (MHz) of IP cores are shown in column 2. The input/output numbers are shown in column 3/4. The test pattern of IP cores are shown in column 5. The numbers of scan chain in IP cores are shown in column 6.

Tab.2 the information of $MCDS_1$

core	at-speed requirement	wrapper list (pins)	test freq. (MHz)
1	no	flexible (≥ 2)	50
2	no	flexible (≥ 2)	50
3	no	flexible (≥ 2)	50
4	no	flexible (≥ 2)	50
5	no	flexible (≥ 2)	50
6	no	flexible (≥ 2)	50
7	no	flexible (≥ 2)	50
8	no	flexible (≥ 2)	50
9	no	flexible (≥ 2)	50
10	no	flexible (≥ 2)	50
11	yes	fixed (64)	100
12	yes	fixed (32)	200
13	no	flexible (≥ 2)	20
14	no	flexible (≥ 2)	25

Tab.3 Basic information of $MCDS_2$

IP	test frequency	input numbers	output numbers	Test pattern	scan chain numbers
1	50	32	32	12	0
2	50	207	108	73	0
3	50	34	1	75	1
4	50	36	39	105	4
5	50	38	304	110	32
6	50	62	152	234	16
7	50	77	150	95	16
8	50	35	49	97	4
9	50	35	320	12	32
10	50	28	106	68	32
11	20	77	150	95	16
12	25	38	304	110	32

The proposed algorithm for multi-clock SOC is verified in $MCDS_2$ when the TAM numbers are two, and the experiment results are shown in Tab.4. The widths of test bus are shown in column 1. The TAM partition results are shown in column 2. The best assignment for each IP core into TAM 1 (or TAM 2) are shown in column 3; where 1(2) represents IP core is assigned to TAM1 (TAM2). The optimal test time of multi-clock SOC is shown in column 4. The results with different TAM widths will also be obtained, but the results are similar [5] when compared add the TAM numbers to add the widths of TAM not the numbers of TAM in a certain widths of test bus, therefore, two TAMs for optimization are reasonable.

It's shown in Tab.4 that the optimal test time is 2817.7us when the widths of test bus are 20, the test bus are divided into 12 and 8, the 1st, 8th, 9th, 10th, and 12th IP cores are assigned to TAM 1, and the 2nd, 3rd, 4th, 5th, 6th, 7th, and 11th IP cores are assigned to TAM 2. The optimal test time is 23.2% when the widths of test bus are 20 vs. the least one. It's clearly that decreasing degree of test time decrease slowly with test bus widths increase when the widths of test bus exceed 20, thus, there is an important inflexion of test time.

Tab.4 Optimization Results when $N_{tam}=2$

Test bus widths	TAM partition	Assignment of IP cores	Test time $(10^3 us)$
4	2+2	(2,2,1,1,2,1,2,1,2,1,2,2)	12.147
12	8+4	(1,1,2,1,2,2,1,2,1,2,2,1)	5.0666
20	12+8	(1,2,2,2,2,2,2,1,1,1,2,1)	2.8177
26	14+12	(1,2,2,2,2,2,1,2,1,1,2,1)	2.2277
30	16+14	(2,1,1,1,2,2,2,1,2,1,2,1)	1.9076
34	18+16	(2,2,1,2,2,2,1,1,2,1,2,1)	1.6101
40	22+18	(2,1,2,1,2,2,2,1,2,1,2,1)	1.5184
46	24+22	(2,1,2,1,1,2,2,1,2,2,2,1)	1.3334
56	30+26	(2,1,2,1,1,2,2,1,2,2,2,1)	1.3080

The test time of multi-clock SOC varies with the widths of test bus when the TAM numbers are 2. It's shown in Fig.3, the test time is shown in dotted line, and the decreasing degree of test time is shown in asterisk line. It's clearly that the test time decrease greatly with the widths of test bus increase when the widths of test bus are small enough, but decreasing degree of test time decreases slowly with the widths of test bus increase when the widths of test bus are too large. Therefore, it's unnecessary to obtain the least test time through adding the widths of test bus extremely, then the test area and design complexity will not increase extremely, and the design process will be accelerated.

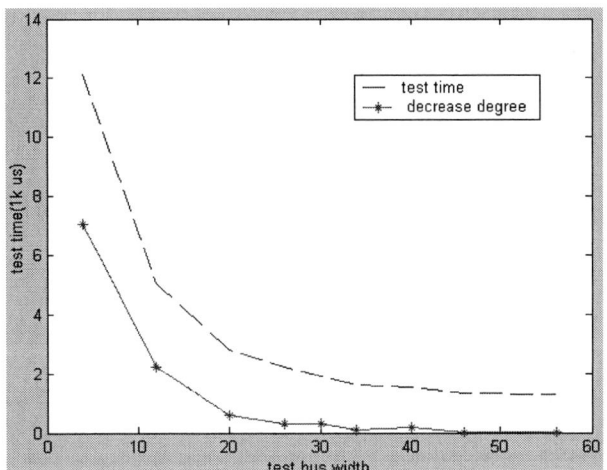

Fig.3 Decreasing degree of test time

5 Conclusions

This paper presents the TWC&S algorithm for multi-clock SOC after combining the advantages of wrapper scan chain optimization and TAM optimization, and it aims to extremely decrease test time of multi-clock SOC under the widths of test bus constraint. Experiment is performed on multi-clock SOC $MCDS_2$ to prove the efficiency of the proposed algorithm; and the results show that the proposed algorithm can reduce test

time of multi-clock SOC efficiently. It is also found that there exist inflexion of test time, and the optimal test time is 23.2% when the widths of test bus are 20 vs. the least one. The phenomenon, decreasing degree of test time decreases slowly with test bus widths increase, demonstrates that trade-off will be achieved properly among test time, test area and design complexity.

Acknowledgements

Thank you for the support from Leading Academic Discipline Project of Shanghai Educational Committee (Project No.J50104).

References

1. Y.Zorian, E.J.Marinissen, S.Dey. Testing embedded- core based system chips[C]. *In Proc. ITC,* 1998: pages130-143.

2. Iyengar V., Chakrabarty K., Marinissen E.J. On using rectangle packing for SOC wrapper/TAM co-optimization, *Proc 20th IEEE, VLSI Test Symposium 2002 (VTS 2002),* 2002, pages: 253-258.

3. Tomokazu Yoneda, Masahiro Imanishi, Hideo Fujiwara. An SoC Test Scheduling Algorithm using Reconfigurable Union Wrappers [C]. *Design, Automation and Test in Europe (DATE'07),* Apr. 2007, pages 231-236.

4. S. K. Goel, Erik Jan Marinissen. Cluster-based test architecture design for system-on-chip[C]. *Proceeding of the 20th IEEE VLSI Test Symposium,* 2002, 259~264.

5. Iyengar, Chakrabarty, Marinissen. Test wrapper and test access mechanism co-optimization for system-on-chip[J]. *Journal of electronic testing: Theory and Applications 18,* 2002: pages 1023-1032.

6. K.Chakrabarty. Test Scheduling for Core-Based Systems Using Mixed-Integer Linear Programming [J]. *IEEE Transactions on Computer-Aided Design,* 19(10), 2000: pages1163–1174.

7. V.Iyengar, K.Chakrabarty, E.J. Marinissen. Efficient test access mechanism optimization for system-on-chip[J]. *IEEE Transactions on Computer-Aided Design of Integrated Circuits and Systems,* VOL. 22, NO. 5, 2003:pages635-643.

8. E.Larsson, Z.Peng. Test Scheduling and Scan-Chain Division Under Power Constraint[C]. *In Proceedings IEEE Asian Test Symposium (ATS),* 2001: pages 259–264.

9. S. Goel, E. Marinissen. TAM architecture and their implication on test application time[C]. *In Proceedings of the International Workshop on Test Embedded Core-based Systems,* 2001:pages3.3-1-10.

10. Yoneda T, Masuda K, Fujiwara H. Power-Constrained Test Scheduling for Multi-Clock Domain SoCs [C]. *In Proc. DATE '06,* 2006: Pages1-6.

Research on the Characteristics Theory of Reverse SoC TAM Design Based on Dual-Balanced Strategy

Zhang Jinyi [1,3], Wang Jia [3], Lin Feng [2], Jiang Yanhui [2], Zhou yi kai [3]

1. Key Laboratory of Advanced Displays and system Application, Ministry of Education, Shanghai University
2. Microelectronic Research & Development Centre, Shanghai University
3. Key Laboratory of Special Fiber Optics and Optical Access Networks, Ministry of Education, Shanghai University
No.149 Yanchang Road, Shanghai 200072, P.R.China
Email: zhangjinyi@staff.shu.edu.cn   Tel: 086-021-56331323-113   Fax: 086-021-56331272

Abstract

The paper presents a reverse SoC TAM design based dual-balanced strategy, which is on the basis of IEEE1500. Firstly test scheduling is executed according to the conceptual TAM architecture that is physically realizable, and then the real TAM architecture can be reversely established according to the test scheduling result. Since the test scheduling is not limited by TAM architecture, the test scheduling optimization can involve both top level and IP level and obtain the cross-level combined optimization between these two levels. The experimental results on the ITC'02 show the better availability and reliability and the performance improvement on test time of the proposed method, compared with several other representative approaches. Particularly, the method of this paper is based on the practical test cost, thus is of greatly practical value.

1 Introduction

The SoC design methodology based on reusable IP cores is widely popular in integrated circuit designer, due to its short development period. With the increasing size and higher and higher complexity of SoC, the test for SoC also becomes more and more challanging and has become the development bottleneck of SoC design methodology. To solve the SoC test problem, the international test standard IEEE1500 based on the test for embedded cores was proposed in 2005. The standard present an unified specification for the IP cores' wrapper for the first time. The academic usually devides the SoC test issues into three components: wrapper, Test access mechanism (TAM) and test scheduling. There is only wrapper under technology sepecification constraint while the other two parts are sitll left to SoC designer to further explore.

Different wrappers can result in different scan chain architectures of an IP core, each of which directly determines the test time of the core. Test efficiency of wrapper is related with the balance degree between the scan chains, thus besides the physical realizability of TAM, an effective balancing algorithm of scan chains is required to optimize wrapper. The algorithm should generate a scan chain architecture which responds to the least test time under the certain TAM width specified by the top level. Test scheduling needs to consider when and how to test each IP core to make the test cost of SoC minimum. During the optimization for test cost of SoC, the test scheduling not only considers the test cost of one IP core, but also the impact of other IP cores on the test scheduling. The eventually available test scheduling solution is obtained under the constraint of test time, the scale of test circuit and other parameters.

Test scheduling and TAM design are two important factors for the reusable core-based SoC DFT. They directly determine the eventual test efficiency and test cost of SoC. The traditional DFT approach takes advantage of the line flow, which firstly determines the TAM architecture and then implements test scheduling optimization. Iyengar proposed ILP based on multiple TAM design [1] and Geol presented cluster scheduling algorithm on TestRail. But these test scheduling are restricted by the existing TAM architectures and the complexity of the practical IP cores, and can only be optimized on top level. Such First-TAM-Then-Scheduling method can achieve limited test efficiency improvement.

The paper presents a reverse SoC TAM design based dual-balanced strategy, which avoids the disadvantage of the traditional DFT design flow for SoC and is based on IEEE1500. Firstly test scheduling is executed according to the conceptual TAM architecture that is physically realizable, and then the real TAM architecture can be decided according to the test scheduling result. Both kinds of design flow are shown in Fig. 1. Since the test scheduling is not limited by TAM architecture, the test scheduling optimization can involve both top level and IP level and obtain the cross-level combined optimization between the two levels. The experimental results on the ITC'02 show the better availability and reliability and the performance improvement on test time of the proposed method, compared with several other representative approaches. Particularly, the method of this paper is based on the practical test cost, thus is of greatly practical value.

Traditional Design Flow

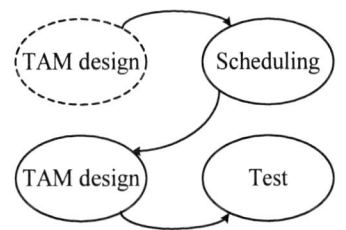

Duel-Balanced Reverse Design Flow

Fig.1 Design flow

2 Modeling of IP level

The IP core which is tested with scan method is connected to the TAM architecture of top level with test chains. Test chains consist of internal scan chains of the IP core and the wrapper cells recommended by IEEE1500. They deliver the test patterns of the IP core. The test time of the IP core is related to the test chains lengths and can be expressed as follows:

$$T = [1 + \max(s_i, s_o)] \cdot p + \min(s_i, s_o) \qquad (1)$$

where S_i and S_o represent the maximum input and output scan chain length, respectively. P is the number of test pattern. In the premise of certain number of test pattern, the test time can be reduced only by decreasing S_i and S_o simultaneously.

The test chain architecture can be denoted as a relationship matrix, as shown in Fig.2. S_x represents an internal scan chain or a wrapper cell while C_x a test chain. Any element a_{ij} in the matrix can be valued with 0 or 1, corresponding to the open or connection state. On the basis of relationship matrix, the test chain architecture on IP level can be described as an optimization model. The differences between the practical input and output test chains lengths and theoretically most optimal test chains lengths are viewed as basic elements while their sum is the target function. The mathematical model is shown in expression (2), where the variable I and O represents the number of input ports and output ports, respectively and $l(s_x)$ represents the length of internal scan chain S_x. For any TAM width m, there always exists a relation matrix A_m to minimize the result of the target function. The resulting matrix corresponds to the best connection solution of test chains for the IP core under the circumstance of TAM width m. For a specific IP core, there can be a set of relation matrixes under different test widths which are used during the test scheduling of top level.

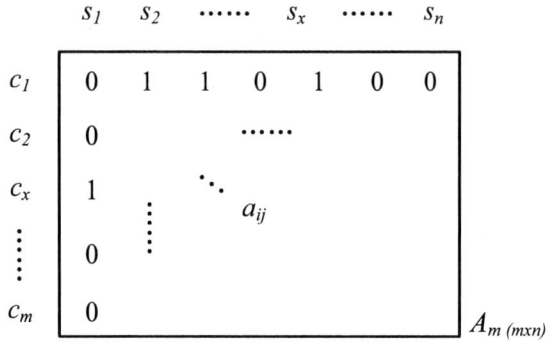

Fig.2 The relation matrix

3 Modeling of top level

In hardware, top level needs to provide TAM architecture for the test chains of IP level to achieve testability; in software, top level needs to solve the test scheduling of multiple IP cores to accelerate test speed and fully use test resource. Thus, modeling of top level should be partitioned into two above parts. Since it's almost impossible to simultaneously achieve the most optimal software and hardware designs, the paper presents a cost function as the measurement guidance to get the trade-off solution. The software and hardware cost are unified in order to determine the best solution of top level testability by the total cost.

The test cost of SoC comprises many parts, among which the most predominant are test time and area overhead. The following will introduce the construction of the two cost components.

$$
\begin{aligned}
\min \quad & \sum_{1 \le i \le m} \left(|L_i - \beta| + |V_i - \gamma| \right) \\
s.t. \quad & L_i = \sum_{j=1}^{n} a_{ij} l(s_j) + I_i \\
& V_i = \sum_{j=1}^{n} a_{ij} l(s_j) + O_i \\
& \beta = \frac{\sum_{j=1}^{n} l(s_j) + I}{m} \\
& \gamma = \frac{\sum_{j=1}^{n} l(s_j) + O}{m} \\
& \sum_{i=1}^{m} I_i = I, \sum_{i=1}^{m} O_i = O, \sum_{i=1}^{m} a_{ij} = 1 \\
& I_i \in Z^+, O_i \in Z^+, i = 1, \cdots m, j = 1, \cdots n
\end{aligned}
\qquad (2)
$$

3.1 Test time cost

Test time cost is mainly from the charge of test equipment, manpower programming and so on. Manpower and material cost will vary with different test equipment. According to the quoted price from the relative companies, the general and linear cost function of test time can be summarized as the following:

$$\Psi(T) = \alpha \cdot T + k \qquad (3)$$

In (3), T represents the test time of SoC, α represents the unit test cost for the time and k represents the fixed cost of every test. Different test equipments correspond to different α and k which are deduced by the real quoted price. The test time of SoC T is related to the balance degree between IP level and top level. Appropriate balance optimization solution can greatly decrease test time T, thus can effectively reduce the time cost of test cost.

3.2 Areas overhead cost

The test circuit will inevitably increase the area of SoC. To make the modeling of top level economically practical, the area overhead cost caused by test circuit should be considered. Since it's difficult to calculate the accurate area of test circuit, the paper makes an assumption to simplify the problem. The assumption is that the pads are tightly placed in all the four sides of SoC and the only way to add pad for test is to exten current area, as is shown in Fig.3. On the premise of the above assumption, the increased area of test circuit can be measured with the number of increased pad and calculated with formula (4).

$$\Delta S(B)=\begin{cases}\left[\dfrac{B^2}{4}+\dfrac{B}{2}(a+b)\right]\cdot l^2 & \text{B is even}\\[2mm]\left[\dfrac{B^2}{4}+\dfrac{B}{2}(a+b)+\dfrac{b}{2}-\dfrac{a}{2}-\dfrac{1}{4}\right]\cdot l^2 & \text{B is odd}\end{cases}\quad(4)$$

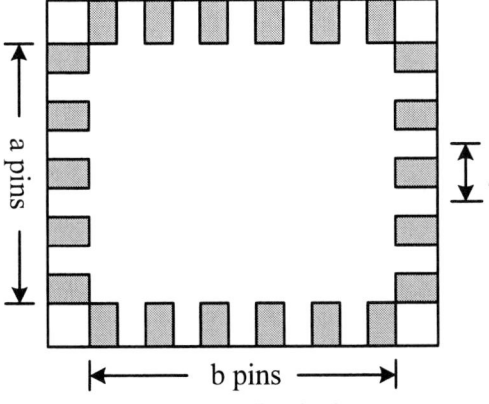

Fig.3 Pad distribution

In the formula, l represents the width of each pad; B represents the added test channels, each of which contains one input port and one output port to deliver test patterns. a and b represents the original vertical and horizontal pad number, respectively. Finally the area overhead cost can be obtained according to the area increment, as is shown in the following:

$$\Phi(B)=\gamma\cdot\Delta S(B)\quad(5)$$

where γ represents the unit manufacture cost of area.

3.3 Test cost function

When the test time cost and test area overhead cost are determined, the test cost function can be expressed as follows:

$$COST=\Phi(B)\cdot(1-\delta)+\Psi(T)\cdot\delta$$

$$s.t.\quad B\cdot T\geq\sum_{i=1}^{n}N_i\quad(6)$$

where δ is the weighted factor to adjust the weight of test time cost and area overhead in the total cost. The constraint shows that the information capacity of the test scheduling must exceeds the summation of test data of each IP core so that the lost of test information can be avoid.

According to the above test cost function, the most optimal test time T and test channel B that minimize the total test cost can always be found. They together determine the theoretical modeling of top level. Only the practical information capacity (without considering idle bits between test chains) is viewed as constraint during the construction of test cost function. Under the case of certain test channel width, the practical test time usually exceeds that of theoretical modeling. Thus the algorithm of top level should generate the test scheduling solution whose test time exceeds that of theoretical one least.

4 The hardware realization of dual-balance strategy

The paper mainly focuses on the impact of dual-balanced strategy on the test performance of SoC without any specified TAM architecture. The design of TAM architectures can be referred in [3], [4] and [5]. To realize the proposed test scheduling solution, and avoid causing too much hardware overhead, the serial scheduling mechanism that controls the test state of each IP core with shift registers is used in the paper. The control mechanism is shown in Fig.4.

The architecture consisting of shift register array and update register array, which is similar to the JTAG technique, can effectively control the test sequence of the IP cores. Meanwhile it can also reduce the number of control pins. The last bits of the shift register array controls the read/write operation of the update register array. The output of the update register array provides the enable signals for the IP cores. Once some IP cores are needed to be tested, the specified bits are assigned to the active state, and then the relative IP cores are connected electronically to TAM.

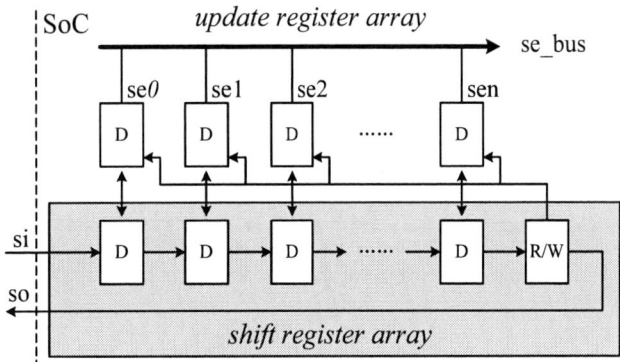

Fig.4 Test control mechanism

5 Experimental results

To validate the availability and effectiveness of the proposed algorithm, the experiments results based on ITC'02 are present in the following.

The objective function and constraint of modeling of IP level have been given in (2). With the modeling, the optimized results of IP level are generated by professional programming software called Lingo. One of the experimental results on IP level is shown below. This result is achieved on the benchmark circuit called d695. Details about d695 are given in table1. Table2 and table3 show the results of core 2 and core 10 on IP level. The test time here is measured by clock cycle.

Table 1 The test information of d695

IP name	test pattern	chain number	longest chain	Input number	output number
c6288	12	-	-	32	32
c7552	73	-	-	207	108
s838	75	1	32	34	1
s9234	105	4	54	36	39
s38584	110	32	45	38	304
s13207	234	16	41	62	152
s15850	95	16	34	77	150
s5378	97	4	46	35	49
s35932	12	32	54	35	320
s38417	68	32	55	28	106

The core 2 is a purely combinational core, containing no scan chains while the core 10 is a sequential core. Both cores adopt the scan-based test methodology. For the sequential core, the largest TAM width is determined by the internal scan chains of the core. When the test time can not be reduced by adding TAM width, the upper bound of TAM width for the core is established. Considering the practical operationability for the combinational core, an upper bound 20 is specified as a constraint so that the software can handle the situation. There is obvious trend that the test time stepladder decreases with the increase of test channel width. The effectiveness of IP core is higher when the test widths are in the Pareto optimal point, which should be used in the test scheduling. The test rectangle which can be eliminated of the sequential core is far more than that of combinational core, as can be seen from table2 and table3. For instance, the test time of core 10 keeps constant when width is between 18 and 31, thus the test rectangles whose widths range from 19 to 31 can be eliminated. Moreover there are not so many test rectangles that can be compressed in the combinational cores. The reason exists in that the chaining for the combinational core is not constrained by the unbalanced length of internal scan chains and can more randomly generate the test chains that are connected to top TAM architecture with the shortest length.

experiments are implemented on these common test channel widths.

The test time obtained with the proposed algorithm is shown in the last column of table4. For comparison, the test time of previous ILP [1], heuristic method [4] (BFD algorithm for the IP level optimization) and cluster method [2] are listed in column 2, 3 and 4, respectively. The test time is measured by clock cycle. As is shown in table 4, when the test channel widths are small, the test time obtained with the proposed algorithm is evidently less than that of other methods. But when the test channel widths gradually become larger, the test time is more than that of the other methods, since there are some bottleneck IP cores whose individual test time is so high that the total test time can not be reduced. Fig.5 shows one of the details about the test scheduling solution with the test channel width 64. It can be seen that the core 7 is the bottleneck core. The proposed method considers the impact of test time and area overhead on the test cost when modeling on top level, and obtains the test design solution which minimizes the channel number and test time in the practical manufacturing. Considering the practical situation that low test channel widths which are always less than 30 are usually adapted to reduce cost, the proposed method is of high application value.

Table 2 The optimal data for the core 2 of d695

Width	Time(cc)	Width	Time(cc)	Width	Time(cc)
1	15292	7	2279	13	1250
2	7719	8	1985	14	1176
3	5146	9	1764	15	1103
4	3896	10	1617	16~17	1029
5	3161	11	1470	18	955
6	2646	12	1396	19~20	882

Table 3 The optimal data for the core 10 of d695

Width	Time(cc)	Width	Time(cc)	Width	Time(cc)
1	120188	6	21182	12~15	10625
2	60128	7	17663	16	7586
3	40137	8	15100	17	7174
4	30132	9~10	14144	18~31	7106
5	24701	11	11033	32	3863

Table 4 The test time for d695

Channel	T_{ILP}	T_{heu}	T_{clus}	T_{dbs}
16	42568	43723	44330	35680
24	28292	30317	30021	23303
32	21566	23021	23488	18236
40	17901	18459	19034	18184
48	16975	15698	16194	16325
56	13207	13415	13479	13605
64	12941	11604	11033	13213

Test scheduling is an NP-hard problem. To realize the test scheduling, the heuristic algorithm proposed by Iyengar[4] is used in the paper to generate the test scheduling solution. In the algorithm, the parameters p and d should be preferentially determined. The range of p and d are $1 \leq p \leq 10$, $0 \leq d \leq 4$, as that of [5]. The test scheduling should be implemented essentially on the basis of the test channel widths decided by the test cost function, and then the obtained results are compared with that of other approaches. Since there are only several test channel widths in most of current papers, the

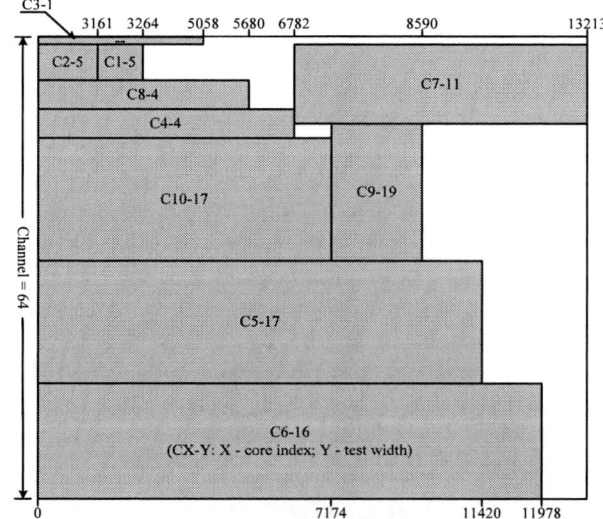

Fig.5 The test scheduling for d695 under the test channel width 64

6 Conclusions

The paper presents a reverse SoC TAM design based dual-balanced strategy, which constructs the accurate optimization modeling on IP level and top level, respectively. In this way, the DFT problem of SoC can be explored from the mathematic view and the theoretically best optimal test solution can be achieved. Since the test scheduling is prior to TAM architecture, the test scheduling optimization can involve both top level and IP level and obtain the dual-balanced optimization between the two levels and prominently improve the test efficiency of SoC DFT. Comparing with the result of some existing approaches under several test channel widths, the result of the proposed method is better when the channel width is less than 32. Considering the requirement of practical application cost, the proposed test method is of high practical value.

Acknowledgements

Thank you for the support from Leading Academic Discipline Project of Shanghai Educational Committee (Project No.J50104).

References

1. Iyengar V., Chakrabarty K.. "Test Wrapper and Test Access Mechanism Co-Optimization For System-on-Chip". *Journal of Electronic Testing: Theory and Applications*, Vol. 18(2002), pp. 213-230.

2. Goel S.K., Marinissen E.J., "Effective and efficient test architecture design for SOCs"; *Proceedings International Test Conference 2002*, 7-10 Oct. 2002, pp. 529 – 538.

3. Pouget J., Larsson E., Peng, Z.; An efficient approach to SoC wrapper design, TAM configuration and test scheduling; *Proc 8th IEEE European Test Workshop*, 25-28 May. 2003, pp. 51 – 56.

4. Aerts J. Marinissen E.J, "Scan Chain Design for Test Time Reduction in Core-Based ICs". *Proceedings IEEE International Test Conference(ITC)*, Washington, DC, October 1998, pp. 448 – 457.

5. Iyengar V., Chakrabarty K., Marinissen E.J. "On using rectangle packing for SOC wrapper/TAM co-optimization", *Proc 20th IEEE,VLSI Test Symposium 2002 (VTS 2002)*, 2002, pp. 253-258.

Optimization of hierarchical SOC test time based on genetic algorithm

Li Jiao[1,2,3], Zhang Jinyi[2,4], Shi Hui[3], Luo xiao wei[4]

1. College of Sciences, Shanghai University
2. Key Laboratory Advanced Display and System Applications (Shanghai University), Ministry of Education
3. Microelectronic Research & Development Center, Shanghai University
4. Key Laboratory of Special Fiber Optics and Optical Access Networks (Shanghai University), Ministry of Education
Box 221, No.149 Yanchang Road, Shanghai 200072, China
0086-21-56331323-123, lijiao@staff.shu.edu.cn

Abstract

Test time optimization is necessary for modular testing of hierarchical system-on-chip (SOC) that contain embedded IP core. In this paper, we consider the case of non-interactive design transfer between IP core vendor and IC integrator. We proposes a method based on genetic algorithm which can efficiently optimize the test time of hierarchical SOC. Utilizing international reference circuit provided by International Test conference 2002(ITC'02), we execute the experiment and results suggest that this method is superior than recently proposes methods for hierarchical SOC test time.

1 Introduction

Embedded cores are now increasingly being used in large system-on-chip (SOC) designs. In a SOC, different types of Intelligence Property (IP) cores are usually incorporated into single SOC design. These changes have several benefits including higher performance, lower power consumption, and smaller size, and so on [1]. However, testing of SOC has been changed to be a costly step in the manufacturing process due to huge test data volume that requires large automatic test equipment (ATE) memory and long test application times.

Several techniques [2-4] have been proposed to reduce the cost of SOC testing. In these techniques, test architecture design and test scheduling techniques that reduce the test processing times often have been used. Though concurrent scheduling of test reduces test application time, there exist many resource constraints, such as test access mechanism (TAM) width, power consumption, precedence relationship, resource conflicts, etc. TAMs transport the test stimulus from the source to the core-under-test (CUT) and responses from CUT to the sink. So, TAM width limitation is the most essential problem in test scheduling [5].

In general, every IP core in the SOC are considered as the same level in test mode, but this assumption is not always in realistic. Multilevel IP core often been contained in a SOC and formed a hierarchical SOC, an example is shown in figure 1(a). It contains the eight cores: core A ~ E and core B1~ B3, and core B1~ B3 have been contained by core B. Figure 1(b) shows the corresponding design hierarchy tree, consisting of three levels. The SOC level is named "level 0" and a core (it can be named sub-IP core) embedded in a core of "level n" has "level n+1". A core of "level n" can be named hierarchical core.

The TAM within the hierarchical core may have been designed by core vendor, thus assigning extra TAM wires to the sub-IP cores is not feasible. A hierarchical core itself may

be an SOC，hence it has its own test constraints such as TAM, power, etc. A complete scheduling should consider the constraints of both the top level SOC and hierarchical embedded cores at the same time.

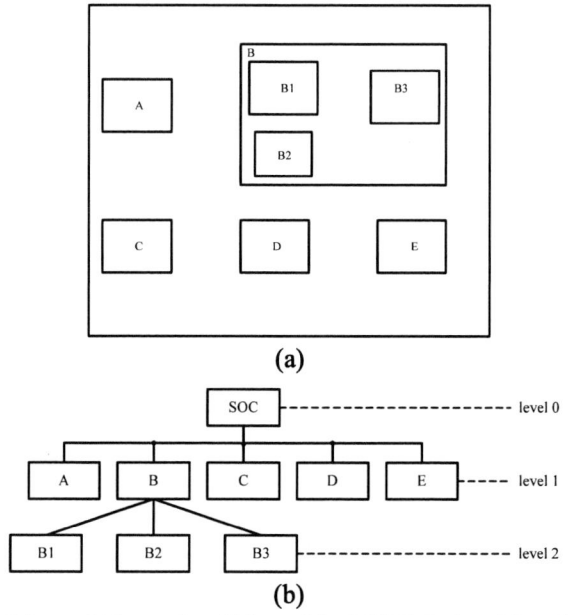

Figure 1 Example of hierarchical SOC design (a) and the corresponding design hierarchy tree (b)

Generally, there are two scheduling models often to be used optimize test time of hierarchical SOC. Those are integer linear programming (ILP) and rectangular bin packing [5, 6]. The ILP (Integer Linear Programming) algorithm, is actually defined a restrictive conditions which withdraw from the cycle. Because the optimal solution of NP-Hard problem only through listed all possible solution, and then to determine the optimal solution, therefore, the restrictive conditions prescribed by the ILP is to tell the algorithm under what conditions can jump out of circulation in the search process. Then obtain the optimal solution. But the computation of this algorithm is larger when SOC size is large. For Bin-packing algorithm, currently there are mainly two-dimensional (2-D) and three-dimensional (3-D) bin packing model. On 2-D packing model, the height of each rectangle corresponds to a different TAM width and the width of the rectangle represents the core test application time for anyone fixed TAM width. In the space which formed by limited TAM width, select the best Scheduling strategy so that the space of all IP rectangle occupy the shortest length, which will be equivalent to minimize the test of time. If co-consider the power factors, Bin-packing algorithm can be extended to 3-D model. Bin-

978-1-4244-2739-0/08/$25.00 ©2008 IEEE

packing algorithm can be more use of graphic image analysis methods to solve the test scheduling problem, but for larger changes in the composition of SOC, it is very difficult form of regularity scheduling strategy.

In this paper, a test scheduling mode which named MBGA (mode based on genetic algorithm) for hierarchical SOC based on genetic algorithm (GA) has been proposed, which can efficiently optimize the SOC test time.

The rest of this paper is organized as follows: a brief review on TAM architecture for hierarchical SOC is presented in section 2, while section 3 presents the proposed MBGA mode for optimize test time of hierarchical SOC. In section 4 the effectiveness of the proposed mode is evaluated with experimental results. And last conclusions are given in section 5.

2 TAM Architecture for hierarchical SOC

In hierarchical SOC, multilevel TAM design is necessary. An example is shown in figure 2. In which the TAM architecture consists of two sub-TAMs (TAM1 and TAM2) with widths of W1 and W2, respectively. Core A and core B are connected to TAM1, and they can only serial test. In core B, TAM1 divide into two sub-TAMs (TAW$_{B1}$ and TAW$_{B2}$) with widths of W_{B1} and W_{B2} ($W_{B1} + W_{B2}$ =W1), sub-IP core B1 and sub-IP core B2 are connected to TAW$_{B1}$, sub-IP core B3 is connected to TAW$_{B2}$, core B1 and core B2 can only serial test, but they can parallel test with core B3. Core C, core D and core E are connected to TAM2. Similarly, core C, core D and core E only can serial test through TAM2.

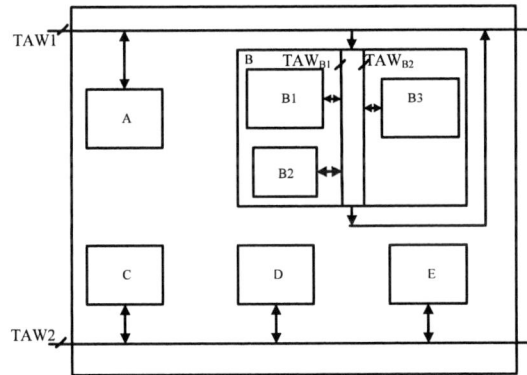

Figure 2 TAM Architecture for hierarchical core

Based on the above discussions, it can be concluded that the IP cores which connected the same TAM (include sub-TAM) can serial test, but the IP cores which connected the different TAM (include sub-TAM) can only be parallel test. So, the test time of SOC can be decided by the maximum of the times taken among the TAMs.

Generally, four problems [3] are stated in order of increasing complexity to the design of efficient wrapper and TAM co-optimization.

P_W: Design a wrapper for a given core, such that the core testing time is minimized, and the TAM width required for the core is minimized.

P_{AW}: Given N cores and B TAMs of test widths w_1, w_2, w_B, determine an assignment of cores to the TAMs and a wrapper design for each core, such that the testing time is minimized.

P_{PAW}: Given an SOC having N cores and B TAMs of total width W, determine a partition of W among the B TAMs , an assignment of cores to TAMs, and a wrapper design for each core, such that the total testing time is minimized.

P_{NPAW}: Given an SOC having N cores and a total TAM width W, determine the number of TAMs, a partition of W among the TAMs, an assignment of cores to TAMs, and a wrapper design for each core, such that the total testing time is minimized.

But, these four problems don't care of multilevel TAM design in hierarchical core. For example, in figure 2 the sub-TAMs width of core B1, core B2 and core B3 have been fixed by the core vendor, therefore, the SOC integrator optimizes the top-TAM widths to minimize the testing time must be under the constraints of these sub-TAMs width. From above, we known that Core B1 and core B2 cannot be tested in parallel, however, if we don't care of the constraints of these sub-TAMs, a test-planning will evaluate a TAM architecture in which core B1 and core B2 are places on different TAM partition and thereby tested in parallel. If this design yields a lower SOC testing time than a design in realistic, the test-planning will incorrectly select the former TAM architecture. Thus, it is important to formulate the problem correctly and provide the right set of constraints to an optimization for hierarchical SOC.

3 MBGA for optimize test time of hierarchical SOC

The TAM of hierarchical SOC optimization problem has often been stated as follows [7]:

Given the test set parameters for the top-level cores and the total TAM width W for the SOC, determine a wrapper design for each core, and a partition of W among the cores in the test schedule, such that the SOC testing time is minimized under the constrains that: 1) W is not exceeded at any time and 2) the cores which contain some sub-cores receive at least their pre-specified TAM widths.

Wrappers design for each core don't been consider by our. So, the problem that will resolve can be address as it: Given the total TAM width of the SOC, we determine an optimal number of TAMs, an optimal partition of the total TAM width among the TAMs, an assignment of cores to each TAM , such that the overall testing time is minimized.

3.1 Genetic Algorithms

A special case of our problem is equivalent P_{NPAW}, if the cores in SOC not contain some sub-cores. Because the problem of P_{NPAW} is a NP-hard problem, therefore, the TAM of hierarchical SOC optimization problem is NP-hard.

Genetic algorithms (GA) are considered to be a way which can effectively solve the NP-hard problem. So we selected it to solve the problem.

GA is stochastic optimization search algorithms based on the mechanics of natural selection and natural genetics. It consists of population of solutions called chromosomes. Here the chromosome is an encoding of the solution to a given problem. During each generation, a new population of individuals is created from the old, by applying three genetic operators.

a. Selection: This operator selects the individuals from the old generation. The individuals with a better fitness possess higher chance of getting selected.

b. Crossover: This operator generated two new chromosomes from the couple of selected chromosomes. A random point on the chromosome also known as cross-site is selected.

c. Mutation: This operator chooses a random chromosome and modifies it to form the new chromosome.

3.2 MBGA formulation

The main formulation of MBGA will be stated, these include: the proper encoding of the chromosomes, crossover operator decide, mutation operators decide and the fitness measure.

a. Chromosomes coding form: In order to optimize the expression of the method solutions, we formulate an appropriate coding form of the chromosomes. In this form, contains three parts as shown as Fig 3. Part 1 is a binary string, part 2 and part 3 is a real number string. (1) The number of TAMs. Since the total TAM width W can be encoded with a binary string of ($\log_2^W$+1) bits, the partition part is an array of size ($\log_2^W$+1). (2) Every TAMs width. This array size is equal to the decimal value of the first part; (3) Assignment of IP cores to TAMs. The array size is equal to the total number of cores in the SOC and every bit corresponding to the number of TAMs. It represents which TAM assigned to the every IP core.

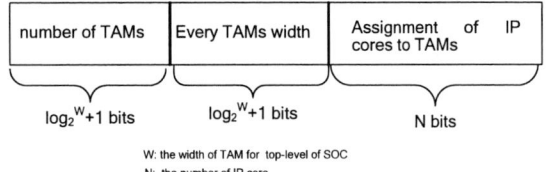

W: the width of TAM for top-level of SOC
N: the number of IP core

Figure 3 Chromosome coding form

b. Crossover operators: Because of the character of part 1 differed with part2 and part 3, we adopt different methods to determine operators about them and make more research space for solutions. For part 1, apply a single point crossover. But apply arithmetic crossover to part 2 and part 3, it can ensure $\sum_{j=1}^{B} w_j > W$. Where, w_j is the width of the TAM j. B is the number of TAMs. The operation of three parts of coding was synchronized, but each other constraints mutually.

c. Mutation operators: The mutation operator brings effectiveness into the chromosomes introducing newer search options. We apply uniform mutation to the chromosomes.

d. Fitness measure: Fitness of the chromosome is measured in terms of the time required to test all cores in the SOC, which is explained as follow. If the core i is assigned to TAM j, then the testing time for core i is given by [8]

$$T_i(w_j) = (1 + \max\{S_i, S_o\}) \times P_i + \min\{S_i, S_o\} \quad (1)$$

Where P_i= Number of test patterns for core i

S_i = Length of the longest wrapper scan-in chain

S_o = Length of the longest wrapper scan-out chain

w_j = Width of the TAM j

For the cores which included some sub-cores, if the TAM width specified by core vendor is w and corresponding test time for the core is T_w, then

$$T_i(w_j) = \begin{cases} \infty, for\ all\ w_j < w \\ T_w, for\ all\ w_j \geq w \end{cases} \quad (2)$$

Let x_{ij} is a 0-1 variable defined as follows:

$$x_{ij} = \begin{cases} 1, if\ core\ i\ is\ assigned\ to\ TAMj \\ 0, otherwise \end{cases} \quad (3)$$

$$(\text{Where } 1 \leq i \leq N, 1 \leq j \leq B)$$

Now, the test time for testing all cores on TAM j is given by $\sum_{i=1}^{N} T_i(w_j) \times x_{ij}$. Since all the TAMs can be used simultaneously for testing, so the total time needed to test all the cores in the SOC is $\max_j \left\{ \sum_{i=1}^{N} T_i(w_j) \times x_{ij} \right\}$, subject to $\sum_{j=1}^{B} w_j = W, 1 \leq j \leq B$, i.e., the sum of all sub-TAM widths is W.

3.3 Evolution Process

We first generate the 300 chromosomes as initial population. Then select 60 chromosomes which have less test time taken as "best class chromosomes". These chromosomes have been copied to the next generation. The crossover rate has been set up 0.6 and the mutation rate is 0.03, then start crossover operation and mutation operation to generate next generation. Repeat this process up to certain predefined number of generations. For our experimentation we took 200 generations to obtain the results.

4 Experimental results

In order to verify the proposed algorithm in this paper, an experiment has been performed on one ITC'02 benchmarks circuits – p93791. It is a largest SOC among ITC'02, it contains 32 cores, of which 18 are combinational circuits and 14 are sequential circuits. And it is appropriate for the experiments because it is hierarchical, containing multiple levels of IP cores.

Table 1 compared the experimental result used the proposed MBGA mode for optimize hierarchical SOC with 2-D bin packing model presented in [6]. TAM widths supplied to each hierarchical core were fixed at 12 bits prior to top-level TAM design.

The test time obtained from [9] is presented in table 1 also. Results exhibit that this result is better than our experimental result sometimes. This is because of that every IP core in the SOC are considered as the same level in [9].

Table 1 Experimental Result for p93791

W	B	TAM Partitions	Assignment of IP cores	Test Time (clock cycles)		
				Method of [9]	2-D Bin Packing Model	MBGA Model
24	2	12,12	2,2,1,1,1,1,1,1,1,1,2,2,2,1,1,1, 1,1,1,2,2,2,1,2,1,1,2,1,2,1,1,1	1206532	1248795	1219200
32	2	20,12	1,2,2,2,1,1,2,2,1,2,2,1,2,2,2,1, 2,1,1,1,2,1,1,2,2,2,1,1,2,2,2,2	878971	975016	896195
48	3	12,24,12	2,1,2,2,1,2,2,2,1,1,3,1,1,2,1,2, 3,2,2,2,1,2,3,1,1,1,3,3,1,3,2,1	723400	627934	614631
56	3	20,12,24	3,3,1,1,1,3,2,3,1,2,2,1,2,2,3,2, 3,1,1,1,3,2,1,2,3,1,1,1,3,2,3,3	511001	568436	523875

5 Conclusions

In this work the MBGA mode has been used to optimize the test time of hierarchical SOC. Experimental results for p93791 among ITC'02 achieved improvement over 2-D bin packing model. At the same time, experimental results proved that if the constraints of sub-TAMs width haven't been considered, then the test-planning will select the incorrectly TAM architecture.

Acknowledgments

Thank you for the support from Leading Academic Discipline Project of Shanghai Educational Committee (Project No. J50104).

References

1. DRajesh K, Gupta and Yervant Zorian, "Introduction Core-Based System Design," *IEEE Design & Test of computers.* 14(4):15-25, December 1997.

2. E. J. Marinissen, S. K. Goel, and M. Lousberg, "Wrapper Design for Embedded Core Test," *Proceedings of International Test Conference*, pp. 911-920, 2000.

3. V. Iyengar, K. Chakrabarty, and E. J. Marinissen, "Test Wrapper and Test Access Mechanism Co-Optimization for System-on-Chip," *Proceedings of International Test Conference (ITC)*, pp. 1023–1032, 2001

4. U. Ingelsson, S. K. Goel, E. Larsson, and E. J. Marinissen, "Test Scheduling for Modular SOCs in an Abort-on-Fail Environment," *Proceedings of IEEE European Test Symposium (ETS)*, pp. 8–13, 2005.

5. Tai-ping Wang, Cheng-Yu Tsai, Ming-Der Shich, and Kucn-Jone Lee,. "Efficient Test Scheduling for Hierarchical core Based Design," *IEEE VLSI-TSA International Symposium*, April.2005, pp.200-203

6. Vikram Iyengar, Krishnendu Chakrabarty and Erik Jan Marinissen, "On Using Rectangle Packing for SOC Wrapper /TAM Co-Optimization," *IEEE VLSI Test Symposium*, 2002, pp.200-203

7. K. Chakrabarty, V. Iyengar and M. D. Krasniewski,. "Test planning for Modular Testing of Hierarchical SOCs," *IEEE Transaction in CAD of Integrated Circuits and Systems*, March 2005, pp.435-448

8. Marinissen. E. J, Goel. S. K, Lousberg. M, "Wrapper design for embedded core test," *International Test Conference*, Oct. 2000, pp.911-920

9. Chattopadhyay, S.; Reddy, K.S.; "Genetic algorithm based test scheduling and test access mechanism design for system-on-chips," *Proceeding of VLSI Design Conference*, Jan. 2003 PP:341 - 346

Effective Dielectric Constant Method for Trace Impedance Control

Te-Chun Wang and Yin-Guang Zheng
ASE Assembly & Test (Shanghai) Limited.
No.669, Guoshoujing Rd., Pudong, Shanghai 201203, China.
Email: Ted_Wang@aseglobal.com, JV_Cheng@aseglobal.com,
Phone: +86-21-50801060 ext.6550

Abstract

Impedance mismatch induces signal reflections along with degradations in transmission lines. Those distortions of signals can result in electrical function failures in high frequency or high speed devices. Impedance control thus has long been developed in PCB and IC packaging substrate manufacturing for better interconnect signal integrity. During our PBGA substrate impedance controlled production, an intrinsic gap between measured impedance and impedance from post-simulation with measured geometry was found. An engineering approach to define "effective dielectric constants" and to bridge the gap between the post-simulations and the measurements has been reported. This engineering procedure has been proved to be convenient and useful in practices. The statistics of the effective Dk distribution is better than direct impedance compensation. The deviations were found to be approximately the same for different stacks and transmission line types. The processing-related factors, for example, the conductor-dielectric interface roughness deviated from the ideal 2D simulation geometry, have been discussed as the possible causes.

Introduction

Signal integrity has long been widely studied in all kinds of high-speed and high frequency applications. Among other design-related factors to degrade the signal integrity, such as crosstalk between traces, characteristic impedance discontinuities of interconnects affecting the signal integrity can be attributed to PCB or substrate processing as well as designing. [1][2] Impedance mismatch between adjacent parts of a transmission line path can result in distortion of signals and fail the electrical functions of the components. The characteristic impedance control procedure has been considered one of the most necessary tools in high electrical performance interconnects production.

During the impedance control procedure, the deviation of the measured characteristic impedance from those of designed can be categorized into two main causes. First one is the mismatch between the designed geometry and produced geometry, e.g., conducting trace width or dielectric layer thickness. This should be considered as the production capability for process engineers to improve. The second cause of the deviation between the design and the final product comes from within the design loop, namely the error between impedance simulation and impedance measurement. Even the produced cross-sectional geometry is measured to meet all the designed values, the measured impedance still deviate from the targeted simulated impedance value.

In this report, we present an engineering method to define the deviation between the simulation and measurement of characteristic impedance for IC packaging organic substrates. A self-consistent simulation procedure can be obtained for first pass production of impedance-controlled samples.

Experimental

Ball grid array substrate samples with sizes ~30mm*30mm and signal trace length from ~10mm to above 15mm were prepared by industrial organic substrate processes for 4 and 6 layers stacks. A cross-section schematics for microstrip and stripline transmission line stacks under test are shown in Fig. 1. The characteristic impedance simulation was performed on Ansoft 2D simulator before and after the production for transmission line models.

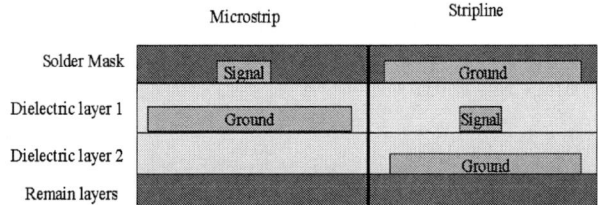

Fig. 1. Schematic cross sections of microstrip and stripline transmission lines. The theoretical Characteristic impedance can be obtained by 2D field solver (Ansoft) simulation for these models.

The characteristic impedance was taken from produced samples. Afterwards the samples were cross-sectioned and the 2D geometry was measured under optical microscopes. As we post-simulated the impedance with the measured geometry of produced samples we can compare the measured impedance to a more realistic impedance value. For example, even if the width of the produced trace was read to be much wider than the originally designed value and the measured impedance was lower than the designed impedance, the post-simulated impedance should still equal to the measured impedance, if other geometrical parameters were exactly produced as designed values. However, we found this straightforward assumption was in fact not true in our productions. We are going to discuss in detail this impedance gap between the post-simulations and the measurements.

The measurements of characteristic impedance were conducted by using Tektronix TDS8000 sampling scope with 80E04 TDR (Time Domain Reflection) module and TDA software. [3][4] In all our samples under test, high frequency probes were placed on substrate ball pads on the ball sides. The traces under control were on or near the chip sides of the substrates, connected to the probes through via holes. The configuration of TDR measurement is shown in Fig. 2.

978-1-4244-2739-0/08/$25.00 ©2008 IEEE

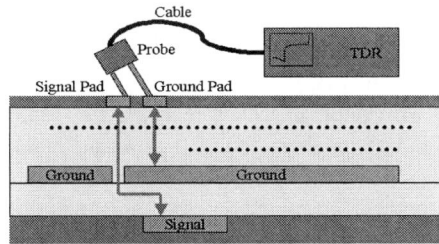

Fig. 2. The Time Domain Reflection (TDR) measurement configuration of a single-ended microstrip transmission line in 2D side view of the substrate.

A typical TDR impedance profile is shown in Fig. 3. The impedance value of the trace under test can be obtained from the plateau of the curve, as shown between green vertical lines. The dip at the beginning of the impedance curve can be attributed to the parasitic effect, mostly capacitive, of probed pads and the through via holes before the signal traveling through the main trace part on the other side of the substrate.

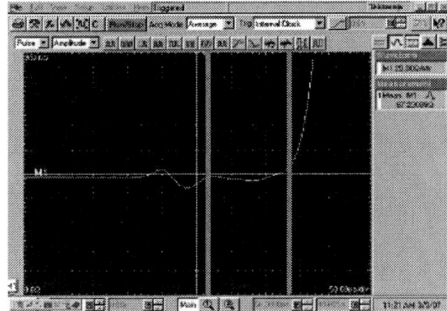

Fig. 3. A TDR impedance profile example of ball side probed configuration. The impedance value of trace under test can be obtained from the plateau region of the curve, as marked by the two green lines in the figure.

Results and Discussions

During the impedance control production, the impedance monitored samples were further cross-sectioned and measured. It was found that there was always a significant gap between the TDR-measured impedance and the post-simulated impedance based on the measured geometry. An example of the gap between the post-simulation and the measurement for a stripline design is demonstrated in Table 1.

Table 1. An example with the Dk optimization procedure for impedance control.

Sample	1st-designed	1st-produced	2nd-designed	2nd-produced
Transmission line Type	Stripline	Stripline	Stripline	Stripline
Dielectric thickness (above Cu)	34	20	20	17
Cu Thickness	15	19.5	19.5	21
Trace Width (Space)	40	36.5	36.5	38
Dielectric Thickness	34	31	31	30.5
Pre-Simulation Dk	3.6		2.3	
Pre-Simulation Z0	39.5		40.5	
Measured Z0 (Ohms)		40		38
Post-Simulation Z0_Dk 3.6		32.3		29.2
Post-Simulation Z0 (effective Dk)		40.5 (2.3)		36.5(2.3)

The post-simulation of the first lot showed a lower impedance than the measured value. A lower Dk value for the target 40 Ohms with the actual geometry was found and the second lot was designed with the lower Dk value and measured geometry of the first lot.

The first sample showed a lower post-simulated impedance, 32.3 Ohms, than the measured 40 Ohms. Although the measured impedance was close to the target value, however

compared to the designed value, a thinner first dielectric layer above copper trace surface was produced. The gap between the post-simulation and the measurement of impedance cannot be explained by this thickness deviation.

Trying to compensate this gap, the simulation was modified by varing Dk input. The Dk value for the target 40 Ohms with the actual geometry was found to be as low as 2.3. The design of second lot was modified by simulation with the optimized "effective Dk". As we post-simulated with the measured geometry of the second lot the impedance 36.5 Ohms appeared close to the measured 38 Ohms, while the deviation was still significant by using Dk 3.6.

This Dk value modification procedure was applied to a series of impedance controlled productions with similar processing. The results were summarized in table 2. The average effective Dk is 2.56, with standard deviation 0.16 for 8 samples. The "effective Dk" was significantly lower than the material Dk 3.6. This result implied a substantial and repeatable gap between the ideal 2D simulation and the measurement of impedance.

Table 2. Examples of post-simulation measurement gap analysis.

	Sample	1	2	3	4	5	6	7	8
	Type	Microstrip	Stripline	Microstrip	CPW	Stripline	Em-CPW	Microstrip	Microstrip
Layer 1	H	20		20	20			24	22.5
	T1	19.5		20	20			15.5	17.5
	W1(S1)	36.5		46	65(80)			35	25
	H1	31	20.5	28	27	32	32	31	32.5
Layer 2	T2		19			15	15		
	W2(S2)		30			23	40		
	H2		32			32	32		
	Z0	40	42	52	43	56	44	59	66.5
	Dk_E	2.3	2.6	2.4	2.5	2.6	2.7	2.8	2.6

Average Dk ~2.56 with standard deviation ~0.16. Notations: H: Solder mask thickness (above 1st Cu layer), T1: 1st Cu layer thickness, W1(S1): 1st layer trace width (space), H1: Dielectric thickness (above 2nd Cu layer), T2: 2nd Cu layer thickness, W2(S2): 2nd layer trace width (space), H2: Inner dielectric thickness. Z0: Measured impedance, Dk_E: Post-Simulation effective Dk of dielectric layers, CPW: Co-planar waveguide, Em-CPW: Embedded co-planar waveguide.

To evaluate the effectiveness of the "Dk compensation" approach, we alternatively calculated the impedance error percentage to define the gap between the post-simulation and the measurement. The post-simulated gaps were expressed as ratios of percentage with respect to the values of measurement for all cases:

Err%= 100%*[Z0_measured-Z0_post-simulated]/Z0_measured

The impedance errors along with effective Dk of all cases are listed in table 3.

Table 3. Statistical data comparison of "effective Dk" and "Z0 Err%" of produced samples.

Sample	1	2	3	4	5	6	7	8	AVE	STDV	STDV/AVE
Effective Dk	2.3	2.6	2.4	2.5	2.6	2.7	2.8	2.6	2.563	0.1598	0.06236058
Z0 Err %	19.5	15.5	13.7	12.8	14.5	13.6	7.97	9.77	13.42	3.5047	0.26120197

Effective Dk represents the modified Dk values needed to eliminate the impedance gap between post-simulation and measurement. The Z0 Err% represents the percentage ratio between impedance gaps and the measured values.

In our impedance controlled substrate preparation, the effective Dk approach provided a convenient and effective method to obtain targeted characteristic impedance for all designs. For the Dk method the standard deviation was 0.16 or 6.23 % with respect to material Dk 3.6 of ABF dielectrics. While for the impedance error percentage presentation, the standard deviation was 3.5 or 26% with respect to the average value of 13.42. Compared to the impedance percentage method the effective Dk method appeared to be more stable. This can be concluded from the deviation statistics of the cases.

It was noted that for various designs and stacks, the simulation-measurement gap existed and the "effective Dk" distributed in a relatively narrow range. The root cause of the gap seemed to be independent on the designing for different stacks. The deviation of the cross section geometry of actual samples from ideal models in simulations appeared more likely to be the root cause. As we examine the cross sections, we noticed the serious interface roughness between conductor and dielectric layers. This processing induced microscopic defect might play important roles for these fine pitch designs.

(a) A stripline sample.   (b) A microstrip sample.

Fig. 4. Typical cross section measurement photos from samples with ~30 um trace width. The interface roughness was serious, as high as 5 micrometers, especially at interfaces above dielectrics and beneath the copper layers.

As examples, Fig. 4 shows cross section photos of samples with 30 um width designs. The geometry appeared non-ideal compared to the simulation input and serious roughness can be observed especially at above laminated dielectric layers. The Rz was estimated to be as high as 5 micrometers. It was found that if the inner edges of the rough conductors were taken for the geometrical reading, we should get data with thinner and narrower traces with higher dielectric thickness. And 2~3 micrometers, about half of the roughness, compensation was found to obtain equivalent results as effective Dk method.

Most of conductor roughness researches were focused on frequency domain analysis on skin effect losses. [5] Although our results showed only an engineering issue in impedance control, still maybe we can go deeper into the time domain physics of roughness in the future.

Conclusions

During the trace impedance control production of PBGA packaging substrate, we found that a gap existed for all our samples between post-simulation and measurement. To compensate the gap and find a practical method to target the demanded impedance, an "effective Dk method" was developed. To match the TDR measured impedance values, the post-simulations were found to demand much lower Dk input values. While the modified Dk values are too low to be in physical nature, the effectiveness of this method was shown by the narrow statistical distribution, compared to the direct linear error impedance presentation. The copper-dielectric interface roughness related non-perfection of cross section geometry was suggested to play certain role for this gap. Further investigation is needed to clarify the interface roughness effect on this gap.

Acknowledgement

We appreciate inspiring technical discussions with C. T. Chiu of ASE Kaoshiung Lab. and S. M. Wu of National Kaohsiung University.

References

1. Johnson, Howard et. al, High Speed Signal Propagation: Advanced Black Magic, Prentice-Hall PTR (2003) .
2. Bogatin, Eric. Signal Integrity – Simplified, Prentice Hall PTR (2003).
3. "TDR Theory", Hewlett-Packard Application Note 1304-2, November 1998
4. "TDR Impedance Measurements: A Foundation for Signal integrity", Textronix application note (2006).
5. Deutsch, A. Krabbenhoft, R.S. Surovic, C.W. Winkle, T.-M. Schuster, C. Kwark, Y.H. and Klink, E. , "Practical Considerations in the Modeling and Characterization of Printed-Circuit Board Wiring", IEEE Workshop on Signal Propagation on Interconnects (2006) pp. 119 - 122.

Study on No-fillet SMT Solder Joint Reliability Based on Solder Joint Shape CAD

Wu ZhaoHua[1]   Zhou DeJian[2]   Huang ChunYue[1]

1.   School of Mechanical & Electrial Engineering of Guilin University of Electronic Technology
2.   Guangxi University of Technology Guilin Guangxi 541004, China
Email: emezdj@guet.edu.cn,   phone: +86-773-5601392

Abstract

Four process parameters including, the pad length, the pad width, the stencil thickness and the stand-off, are chosen as four control factors, and by using an $L_{25}(5^6)$ orthogonal array, the no-fillet chip component solder joints shapes with 25 different process parameters combinations are established. And then all the shape prediction models of the 25 solder joints are built through the Surface Evolver soft, after that, the finite element analysis models are set up by converting the above ones. Afterwards, a non-linear finite element analysis on the no-fillet chip component solder joints under thermal cycles are performed by using ANSYS soft, and the thermal fatigue lifes of them are achieved through Coffin-Manson equation. Finally the variance analysis was performed based on the above lifes figures. The research show that with 95% confidence the stand-off has a significant effect on the reliability of the no-fillet chip component solder joints, whereas the pad length, the pad width and the stencil thickness have little effects on the reliability of the ones.

Key words

Solder joint shape; No-fillet; Process parameters; Variance analysis; Reliability

0. Introduction

The SMT (surface mount technology) as a newly emerging mount technology has got a wide-ranging application and a rapid development with advantages including high density, fine pitch, high accuracy and high reliability[1]. The high density assembly technology has the feature the average spot numbers of the solder joints are bigger than 30 /cm^2 per unit area on the PCB surface of the circuit module product. Its characteristic is to utilize the assembly space adequately and to reduce the area and volume occupied by the circuit module as possible, so that the electronic equipments can be more lighter and tinier. In order to improve the assembly density, the following methods can be used: high-density packaging, more I/O, fine-pitch and microminiaturization components such as BGA or CSP. Moreover, the miniaturization tendency of the chip component have been developed from 1005 to 0402, 0201. The 0201 component have been used now. For enhancing the package density of chip components, the method of reducing the solder pad size also can be used, besides using smaller

component. The no-fillet solder joint of chip component appearance is just to improve the mount density.

The no-fillet solders take dual duties of the electric connection and mechanical connection, so their qualities affect the performance and reliability of the product directly. The research have indicated that: the solder quality have direct correlation with the generalized solder joint shape parameters involving the pad size, the solder quantities, the solder process parameters and the solder joint physical dimensions after shaping and so on.[2] Therefore, the reasonable design of correlative solder parameters is important to improve the quality of the no-fillet solder joint before mounting. In this paper, the typical SMT solder joint— 0402 chip component solder joint is studied as an analysis object based on the shape CAD technology, with establishing its prediction model to predict the no-fillet solder joint shapes under different process parameters combinations. The design of experiment (DOE) method is applied to study the relationship between the process parameters and the solder joint reliability of no-fillet solder joints. The results can be used as references for carrying optimal design of process parameters in practically about no-fillet solder.

1. SMT solder joint shape CAD technology [3]

Usually, solder joint shape refers to the geometry size which melting solder can reach along the moist metal surface where solder components foot and printed circuit board were welded, the Metal surface contact angle and the solder fillet shape. In short, it is the appearance structure of solid joint after shaping. Research shows that the solder joint shape and the reliability of solder joint have a direct relation. This paper reveals the relations which stress distribution and the solder joint shape, deformation, and with above results forecasting and controlling the solder joint shape, and optimizing the solder joint for high reliability are feasible.

Fig1 is the schematic diagram of SMT solders joint shape CAD technology based on stress analysis. The basic idea is: Firstly , establishing the model of solder joint shape with the original design parameters of the solder joint using the basic theory of SMT solder joint shaping and CAD method, simultaneously getting an elementary solder joint shape; Secondly, calculating, analyzing and evaluating the practical stress distribution and the weakness points of the solder joint by using the established analytical model of solder joint shape; Lastly, making the feedback and shape correction according the adjusted data by analyzing and evaluating, and forming a new solder joint shape. Again and again, until a reasonable solder joint shape is obtained.

Foundation Project: advanced project(A New Method of Electrical Interconnection)
Introduction of author: Wu Zhaohua, female, born in Nanchang, Jiangxi. professor of GUET, engaging in mechatronics and SMT research

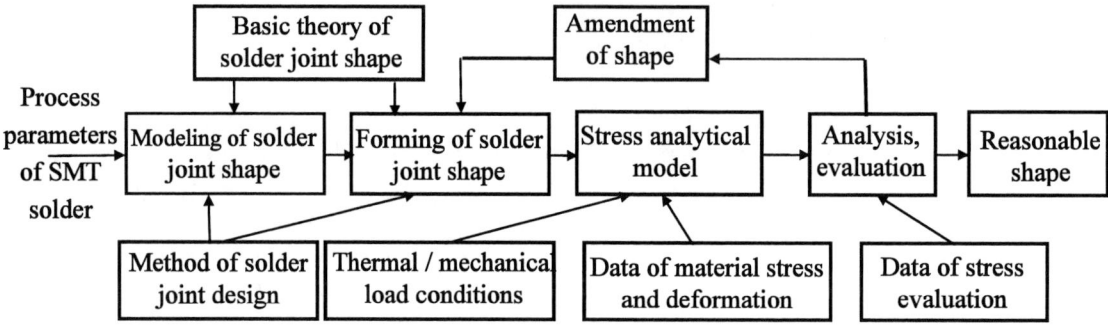

Fig1 SMT solders joint shape CAD theory based on a stress analysis method

2．Modeling and forecasting of no-fillet chip components solder joint shape

The solder joint assemble plots of the fillet chip component and the no-fillet chip component are showed in Fig2. Seen from the two charts, no-fillet solder joint is shaped while at the lengthwise section the pad size decreasing from a+b to b. The decrescence of pad length brings the pad area into decreasing, resulting in the improvement of the package density of circuit board.

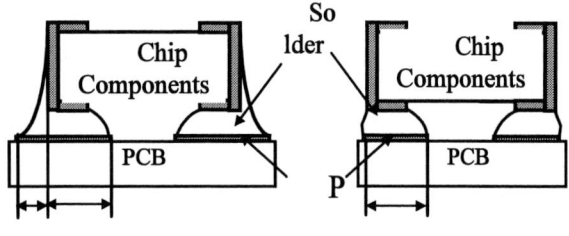

（a）fillet solder　　　（b）no-fillet solder

Fig2 Assembly diagram of Chip components solder joint

2.1 Orthogonal design of process parameters

By using the orthogonal design for the test with process parameters combinations of no-fillet SMT solder joints, it can reduce the number of solder joint shape prediction models and the FEA models, and it can get all the information impacting the model performance, with part of combinations among all ones to set up the prediction and analysis model of the solder joints.

The study object of this paper is chip components (0402). Based on the practical assembling process parameters, four factors, which affect the reliability of no-fillet chip components solder joints, are selected as critical factors: length of bonding pad, width of bonding pad, height of stand-off and stencil thickness. The stand-off is defined as the height measured from the lowest point on the solder joint of no-fillet components to PCB. Each critical factor has five levels, which are shown in Table1. Since there are 4 five-level factors, an $L_{25}(5^6)$ orthogonal array with the first 4 columns can be used to arrange the combinations of the factors, which are shown in Table2.[4]

Table1　Factors and levels

Factors	Levels				
	1	2	3	4	5
pad length L (mm)	0.28	0.30	0.33	0.35	0.40
pad width W (mm)	0.45	0.48	0.51	0.53	0.55
stand-off H (mm)	0.15	0.13	0.11	0.17	0.10
stencil thickness S (mm)	0.10	0.115	0.125	0.135	0.15

2.2 3D shape prediction of no-fillet chip component solder joint

No-fillet chip component solder joint shape model is set up by Surface Evolver software based on the least energy principle and finite element numerical analysis method. The key of setting up the No-fillet chip component solder joint prediction model lies in building the three-phase system energy control equation of the No-fillet chip component solder joint successfully while using the Surface Evolver software.

The total energy of the no-fillet chip component solder joint system E consists of three major energy portions: the surface tension energy E_s, the gravitational energy E_g, and the external energy E_f. While the molten solder expands along the metal surface, the total energy of the solder joint three-phase system can be defined as:

$$E = E_S + E_G + E_f \qquad (1)$$

The partial correlation energy governing equations are deduced as follows:

(1) The surface tension energy E_s

The surface tension energy E_s consists of gas-liquid surface energy E_{s1} and liquid-solid surface energy E_{s2}. The E_{s1} can be given by

$$\text{T\_z} = -\text{TENS} \cos(Z_{angle} \cdot \pi / 180)$$

$$\text{T\_x} = -\text{TENS} \cos(X_{angle} \cdot \pi / 180)$$

Table2 Orthogonal test designs with process parameters combinations of joints

Number	Number of row					
	L	W	H	S	Strain Range $\Delta\varepsilon$	Fatigue life (cycles)
No.1	1	1	2	4	0.1090737	3883.37
No.2	1	2	5	5	0.183554	1039.79
No.3	1	3	4	1	0.121361	2963.77
No.4	1	4	1	3	0.110501	3757.64
No.5	1	5	3	2	0.096391	5310.27
No.6	2	1	3	3	0.113878	3481.91
No.7	2	2	2	2	0.087223	6839.18
No.8	2	3	5	4	0.13115	2435.31
No.9	2	4	4	5	0.151804	1681.74
No.10	2	5	1	1	0.1102934	3775.57
No.11	3	1	1	5	0.137518	2159.86
No.12	3	2	3	1	0.135089	2259.54
No.13	3	3	2	3	0.143709	1932.01
No.14	3	4	5	2	0.133735	2317.90
No.15	3	5	4	4	0.153838	1626.01
No.16	4	1	4	2	0.12997	2491.68
No.17	4	2	1	4	0.1053985	4235.39
No.18	4	3	3	5	0.160859	1452.30
No.19	4	4	2	1	0.11425	3453.28
No.20	4	5	5	3	0.13382	2314.18
No.21	5	1	5	1	0.102619	4531.87
No.22	5	2	4	3	0.125358	2730.34
No.23	5	3	1	2	0.105437	4231.48
No.24	5	4	3	4	0.1029835	4491.38
No.25	5	5	2	5	0.153721	2151.81

In the formulas: TENS is the surface tension of melting filler metal, T_z is the equivalent surface tension of melting filler metal and the pad surface, Z_{angle} is the angle of melting filler metal, T_x is melting filler metal, X_{angle} is the angle of melting filler metal on the metal surface.

1) The surface potential energy of solder pad and melting filler metal can be described:

$$Es = \iint T\_z(-\bar{k}) \cdot d\bar{A} = \oint \bar{w} \cdot d\bar{s} \tag{2}$$

$\nabla \times \bar{w} = T\_z \cdot \bar{k}$, ∇ is Laplace operator

thus
$$Es = \oint T\_z(-y)\bar{i} \cdot d\bar{s} \tag{3}$$

2) The surface potential energy of component solder

vertical plane (X=0) and melting filler metal can be described:

$$Es = \iint -T\_x(-\bar{i}) \cdot d\bar{A} = \oint \bar{w} \cdot d\bar{s} \tag{4}$$

thus
$$Es = \oint T\_x(-y)\bar{j} \cdot d\bar{s} \tag{5}$$

For the sideline fixed, the corresponding surface potential energy from metallized surface of components base and liquid state solder interface can be ignored.

(2) The gravitational energy E_g,

Gravitational energy may also change into the contact surface or the sideline integral equivalent value, using the vector product revolution per minute.

$$Eg = \iiint_V \rho gz dV = \oiint_{\partial V} \bar{F} \cdot d\bar{A} \tag{6}$$

where $\nabla \cdot \bar{F} = \rho gz$, $\bar{F} = \rho gzx\bar{i}$, thus

$$Eg = \oiint_{\partial V} \rho gzx\bar{i} \cdot d\bar{A} \tag{7}$$

(3) Volume constraints

Solder joint volume is constant, the vector form is:

$$V = \iiint_V dV = \oiint_{\partial s}(x\bar{i} + y\bar{j} + z\bar{k}) \cdot d\bar{s}/3 \tag{8}$$

Thus, the equivalent volume of base pad (z=0) and solder joint volume is 0, the equivalent surface volume can be written as:

$$V = \iint z\bar{k} \cdot d\bar{A}/3 = -z\oint y\bar{i} \cdot d\bar{s}/3 \tag{9}$$

In terms of the energy control equations above and the constraint conditions, the no-fillet chip component solder joint shape forming model can be set up, as shown in Fig3. For the symmetry of the geometry, one of the no-fillet chip component solder joint model is set up as Fig1. Through the shape prediction model, with inputting different process parameters, the no-fillet chip component solder joint 3D shape can be predicted, as shown in Fig4.

Parts of prediction results are shown in Fig4, which the

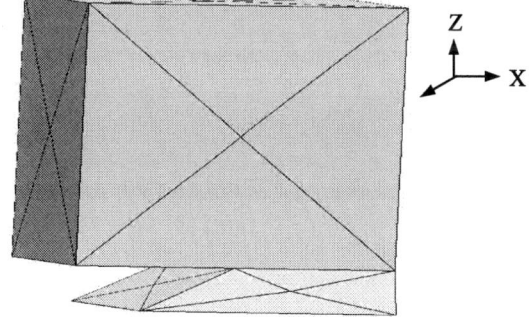

Fig3 3D shape prediction of no-fillet chip component solders joint prediction

combinations of the process parameters are corresponding to RUN2, RUN 13, RUN 18 and RUN 25 shown in Table2. It can be seen from Fig2, with the change of process parameter combinations, the shape of the No-fillet chip component solder joint is different, and the different shape of solder joint will result in the different fatigue lifes. Utilizing the no-fillet chip component solder joint shape prediction results, the solder joint surface data under sorts of process parameters combinations can be obtained, and then the finite element analysis (FEA) models can be set up.

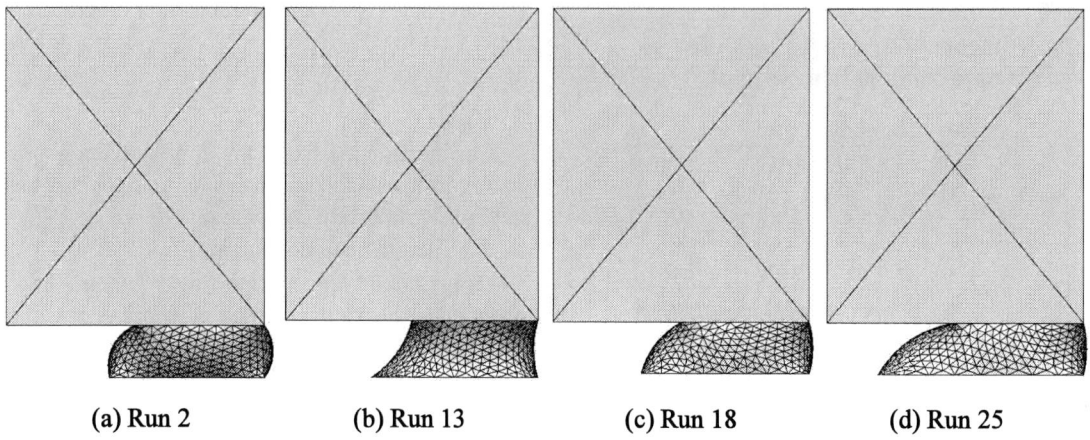

| (a) Run 2 | (b) Run 13 | (c) Run 18 | (d) Run 25 |

Fig4 Shape prediction results of no-fillet solder joint

3. Finite element analysis and life prediction of no-fillet chip component solder joints

3.1 Finite element analysis models of no-fillet chip component solder joints

For the no-fillet chip component solder joints under thermal cycle load, the coefficients mismatch of thermal expansion (CTE) between the ceramic package, solder joints and the PCB can induce solder joint fatigue failure. By the nonlinear finite element method, the thermal stress/strain analysis of the no-fillet chip component solder joints can be performed, and the maximal plastic strain scope is also calculated. And then the corresponding fatigue life of the no-

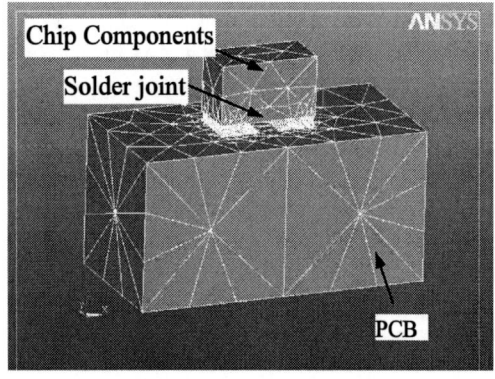

Fig5 Finite element model of the no-fillet solder joints

fillet chip component solder joints can be acquired through the Coffin-Manson equation.

With the before-mentioned surface coordinate, the geometry models and the relevant finite element analysis (FEA) models can be set up, the 3D FEA model of no-fillet chip component solder joints just shown as Fig5. In this case, the combination of process parameters is the stand-off of 0.10mm, the stencil thickness of 0.115mm, the pad length of 0.55mm and the pad width of 0.4mm. The finite element model consists of 6874 elements with 3607 nodes. The imposed displacement boundary condition is: at the bottom of PCB four corner points can not move in every direction. The temperature cycle profile using in this study is a cyclic temperature load changing between -55℃ and +125℃ with 10 minutes dwell time, 5 minutes ramp time and a period of 30 minutes (see Fig6). Detailed material properties of the no-fillet chip component package are listed in Table3.

3.2 Finite element analysis results of no-fillet chip component solder joints.

The FEM analysis result of the no-fillet chip component solder joints is shown in Fig7, which shows the stress

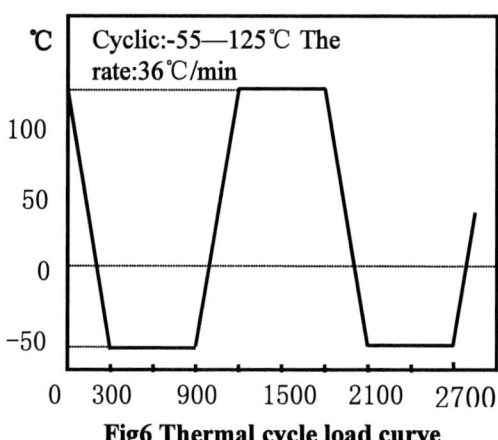

Fig6 Thermal cycle load curve

distribution within the whole no-fillet chip component solder joints at the third cycle. Fig8 shows the strain distribution within the solder joints at the same cycle. Fig9 and Fig10 show the stress and strain distributions in the interior of the solder joints. As shown in Fig7 and Fig8, the maximum stress and strain occur at the solder joints within the whole no-fillet chip component. It can be seen from Fig9 and Fig10 that: (1) the stress and strain distribution in the solder joint are not uniform; (2) the interface between the solder joint and the chip component is the high stress/strain region, the maximum stress/stain occur at this region. Thus, the solder joint cracks should occur firstly at the interface between the solder joint and the chip component, and then extend along this region, finally extend in the whole interface between the solder joint and the chip component. The FEM analysis results of no-fillet chip component solders joints under the other 24 kinds of process parameters combinations in Table2 are similar with above.

3.3 Thermal fatigue life of no-fillet chip component solder joints

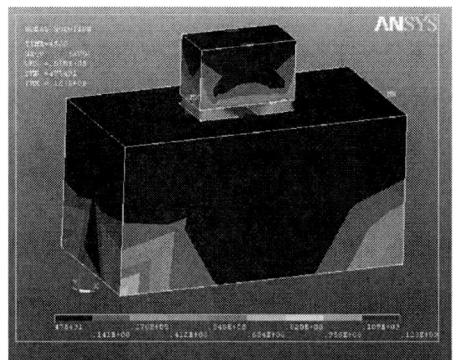

Fig7 Stress distribution of the no-fillet solder joints

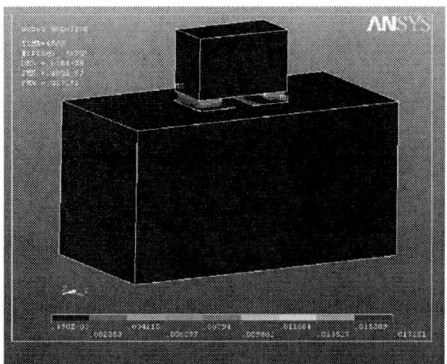

Fig8 Strain distribution of the no-fillet solder joints

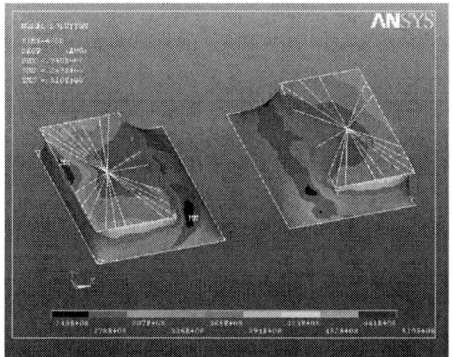

Fig9 Stress distribution of the outermost solder joints

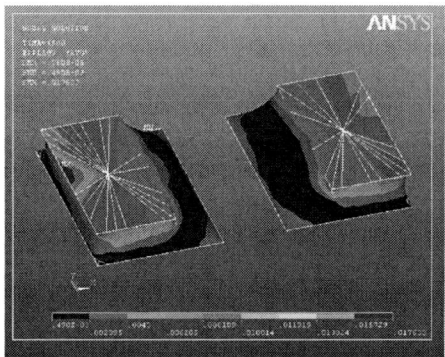

Fig10 Strain distribution of the outermost solder joints

The thermal fatigue life of the no-fillet chip component solder joint can be estimated based on the plastic strain and Coffin-Manson equation[5][6], which is shown in equation(10).[5][6]

Table3 Material properties

Material	Elastic Modulus (Mpa)	Poisson Ratio	CTE (10^{-6}/℃)
Chip Components	3.79×10^5	0.21	6.7
FR4 PCB	1.1×10^4	0.28	15
Solder 63Sn/37Pb	1.0×10^4	0.4	21

$$N_f = \frac{1}{2}(\frac{\Delta\gamma}{2\varepsilon_f'})^{(1/C)} \qquad (10)$$

where N_f is the number of cycles to failure, $\Delta\gamma$ is shear strain range, $\Delta\gamma = \sqrt{3}\Delta\varepsilon$, $\Delta\varepsilon$ is strain range that can be obtained from FEA generated hysteresis curve according to Fig11. It can be seen from Fig11 that the hysteresis curve can stable within a few thermal cycles.

In calculation, $\Delta\varepsilon$ is the strain range of the third cycle, ε_f' is fatigue ductility coefficient, C is fatigue ductility exponent.

$$C = -0.442 - 6 \times 10^{-4} Tm + 1.74 \times 10^{-2} \ln(1 + f)$$

$$Tm = \frac{1}{2}(T_{\max} + T_{\min}) = \frac{1}{2}(125 - 55) = 35$$

Where, Tm is the average temperature of thermal cycle, f is the thermal cycle frequency, f=48(cycle/day), thus, C=-0.395.

In this case, the strain range is 0.153721 and consequently, the no-fillet chip component solder joint with the process parameters combination of *L5W5H2S5*, should have the fatigue life of 2151.81 cycles according to Coffin-Manson equation. The fatigue life of the other 24 no-fillet chip component solder joints is shown in Table2.

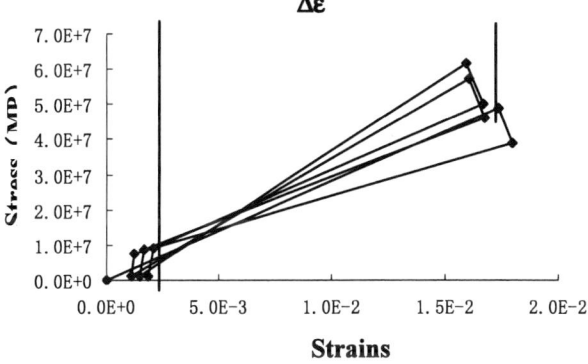

Fig11 Stress and strain hysteresis curve of no-fillet solder joint

Table5 Results of thermal fatigue life variance analysis (2)

Source	warp sum of squares	degrees of freedom	variance sum of	Ratio of F	$F_{0.01}$	$F_{0.05}$	significance
pad width	9186899.14	4	2296724.785	2.10	5.41	3.26	not significant
stencil	8232913.91	4	2058228.4775	1.88	5.41	3.26	not significant
stand-off	17361555.14	4	4340388.785	3.97	5.41	3.26	significant
error	13126178.81	12	1093848.23				
summation	147496018.07	24					

Table4 Results of thermal fatigue life variance analysis (1)

Source	warp sum of squares	degrees of	variance sum of squares	Ratio of F	$F_{0.01}$	$F_{0.05}$	significance
pad length	1998664.38	4	499666.095	0.36	7.01	3.84	not significant
pad width	9186899.14	4	2296724.79	1.65	7.01	3.84	not significant
stencil thickness	8232913.91	4	2058228.48	1.48	7.01	3.84	not significant
stand-off	17361555.14	4	4340388.79	3.12	7.01	3.84	not significant
error	11127514.43	8	1390939.30				
summation	44388689.45	24					

4. Variance analysis of thermal fatigue life of no-fillet chip component solder joints

As a method to analyze experimental data[7], the analysis of variance can be used to determined whether the factor has a significant effect on the result of experiment. By analyzing the fatigue life data of no-fillet solder joint with the variance analysis, the effect significance can be obtained which the four factors imposes on the solder joint reliability including pad length, pad width, stencil thickness and gap height. And according to the significance, the reliability of no-fillet solder joints can be improved by controlling the relevant process parameter to the factor.

According to the theory of orthogonal variance analysis, the sum of deviation square, degree of freedom, variance estimation and variance ratio(F value) of each factor can be worked out. The sum of deviation square of each column in orthogonal table can be got by the following formula:

$$SS_j = \frac{\sum T_j^2}{n} - \frac{T^2}{N} \qquad (11)$$

In the formula (11), T_j is the sum of Experimental data under the same level in the j column; n is the repetitive of each level; N is the total number of experiments; T is the sum of experimental results.

The sum of deviation square can be calculated by formula (12):

$$SS_{sum} = \sum y_j^2 - \frac{T^2}{N} \qquad (12)$$

The formula of variance is formula (13):

$$MS_j = \frac{SS_j}{DF} \qquad (13)$$

Where DF is degrees of freedom, equal to the level number of corresponding factor minus 1. Variance ratio is as formula(14):

$$F_j = \frac{MS_j}{MS_{error}} \sim F(DF_j, DF_{error}) \qquad (14)$$

According to the above formula, the sum of deviation square, degrees of freedom, variance estimation and Ratio of F can be calculated (see Table4). Form the table 4, It shows that all the four factors which influence the reliability of no-fillet solder joint are not significant. The possible reasons are: the error is big, the freedom degree of error is less and the testing sensitivity is not high. For improving the testing sensitivity, the sum of deviation square of pad length whose significance is minimal is merged into the sum of error square and the corresponding freedom degree is also merged into the error freedom degree. Then the recalculation analysis results are shown in Table5.

From the data in Table5, the following conclusion can be drawn. The ratio of F of the stand-off is 4.98, which is bigger than the critical value $F_{0.05}(4,12)$, but smaller than $F_{0.01}(4,12)$. So when significance level α =0.05, the stand-off has a significant effect on the reliability of no-fillet solder joints. The ratio of F of pad width and stencil thickness are all smaller than the critical value $F_{0.05}(4,12)$. It shows that these two factors are not remarkable. According to the ratio of F as shown in Table4 and Table5, the decreasing order of the factor impacting fatigue life is the gap height, the pad width, the stencil thickness and the pad length. Based on above analysis, the factor of the stand-off height has a significant effect on the reliability of solder joints. Consequently, controlling the stand-off height as a critical process parameter for higher solder reliability is essential while designing the process parameters of the no-fillet chip component in practice.

5. Conclusions

With the above analysis, some conclusions can be drawn as follows:

(1) With the change of process parameters combinations, the shapes of no-fillet solder joints will change; consequently the solder joint fatigue lifes will change too.

(2) Under the thermal cycle load condition, the stress and strain distribution in the no-fillet solder joints is not uniform. The interface between the solder joint and the Chip Component is the high stress/strain region. The maximum stress/stain occurs at this region. Thus, the solder joint cracks occur firstly at the interface between the solder joint and the Chip Component, and then extend along the interface to the whole bonding plane, resulting in the failure of the solder joints.

(3) The results of variance analysis on the thermal fatigue lifes show that: with 95% confidence the stand-off has a significant effect on the reliability of no-fillet solder joints; The sequence of significance effect of the factors on the reliability in a descending order is stand-off, pad width, stencil thickness, pad length.

References

1. Ii Chunquanm, Zhou Dejian, WU Zhaohua, "Study on SMT Solder Joint Quality Fuzzy Diagnosis Technology Based on the Theory of Solder Joint Shape," CHINA MECHANICAL ENGINEERING, Vol. 15, No. 21 (2004), pp. 1967-1970.

2. Zhou Dejian, Pan Kanlin, WU Zhaohua, "Prediction Three-Dimensional Solder Joint Shapes of SMT Based on Minimal Energy Principle," ACTA ELECTRONICA SINICA, Vol. 29, No. 5 (1999), pp. 66-68.

3. Zhou Dejian, , "Theory of SMT solder joint shape and CAD Technology," Zhejiang University (1998),.

4. He Shaohua, Wen Zhuqing, Lou Qing, Experimental design and data processing, National Defense Scientific and Technical University Press (Changsha China, 2002), pp. 65-67.

5. Manson S.S, Thermal stress and low cycle fatigue, McGraw-Hill (New York, 1966), pp. 155-187.

6. Solomon H.D,"Fatigue of 60/40 solder," IEEE Transaction on Components, Hybrids, and Manufacturing Technology, , Vol. 4, No. 19 (1986), pp. 423-432.

7. Ren Luqun, Test Technologies, Machinery Industry Press (Beijing China, 1987), pp. 68-94.

Simulation and Experiment Study of Dispensing Patterns Influence on Underfill Filling Process

XIE Jinghua   DENG Guiling   GENG Fei   Wang Junquan

College of Mechanical & Electrical Engineering, Central South University, Changsha, 410083, China
E-mail: shoot_cry@hotmail.com

Abstract

Dispensing pattern is important for flip-chips underfill filling quality. This paper uses three-dimensional flow model to simulate underfill capillary flow and filling process for L and I dispensing patterns. The model reckons in influences of surface tension, contact angle and radius of curvature, especially contact angle and radius of curvature at wall of gap between chip and substrate, side wall of chip and surface of sold bump joints. In order to track the moving interface, the VOF model is applied. On the other hand, an experimental device has been constructed to research underfill filling process of the two dispensing patterns. Both simulation results and experiment results show that dispensing patterns have great influences on underfill filling velocity and flow front profile. In L filling pattern, underfill flow velocity is faster in diagonal direction. Comparing with experimental results and simulation results, it can be concluded the 3-D simulation model established can simulate underfill filling process well for the both L type and I type dispensing pattern.

Keywords: dispensing pattern; underfill, capillary flow; 3-D numerical simulation; surface tension.

Introduction

The basic target of flip-chips application is to improve filling velocity, decrease solidifying time and prolong the service life of flip-chips [1]. Filling velocity and solidifying time mainly have relationships with the characteristic of undefill material and work environment, meanwhile the flip chip size, bumps dimension and dispensing pattern of also have rather influences [2-4].

There are many types of dispensing pattern: I pattern, L pattern and U pattern, as shown in Fig 1. Although the dispensing pattern along more than one edge can increase flow rate, it will increase the probability of gas caught and void formation [5-8].

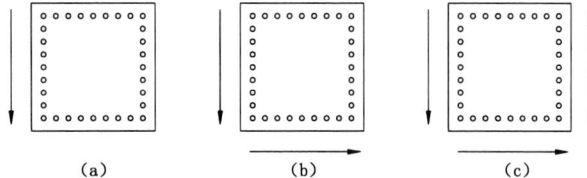

Figure 1. Schematic of dispensing pattern:
(a) I pattern; (b) L pattern; (c) U pattern.

The service lives of flip-chips depend on the filling quality between chips and substrates to a large extent. If the filling result is unsatisfactory, there will be air bubbles, fissures and air interstices etc [9]. At present international researches on the process mainly through experiments and two-dimensional numerical simulation mainly. Because the two-dimensional model can't consider properly the influence of surface tension in the direction of height, so the simulation results compared with experimental results have obvious deviations [3].

In order to make simulation results reflect the real filling situation between chip and substrate, this work reckons in the influences of surface tension, contact angle and radius of curvature in the computation model. We have established a three-dimensional transient model to simulate the filling process, especially considering the influence of contact angle and radius of curvature at the rims of flip-chips, substrates and solder bumps. In order to verify the numerical simulation results, an experimental system of underfill filling has been established to observe and record the flow situation in the process of underfill filling. Through comparison of simulation results and experimental results, the model can simulate exactly the phenomena of wave in flow front edge and faster filling velocity at rim in underfill capillary flow front process and dispensing patterns influences on filling velocity.

Mathematics model and numerical simulation

Simulation object

The gap between the flip chip and substrate are about $75 \sim 300um$ through solder bumps realizing them connect each other, the underfill is dispensed after interconnection. According to capillary flow principle, surface tension draws the underfill flow into the gap, as Figure 2 shown.

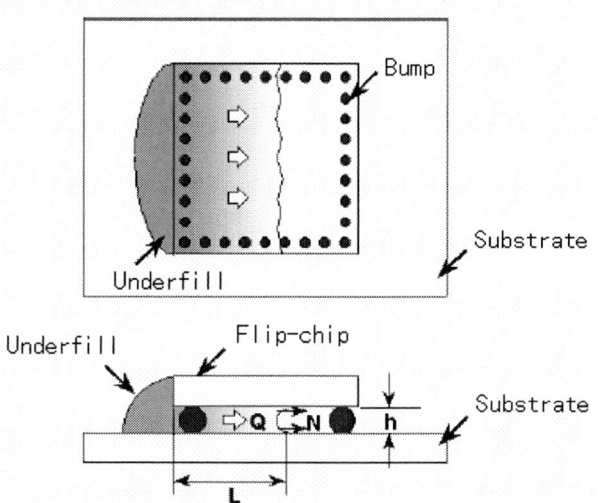

Figure 2. Schematic of underfill flow

Underfill filling process is affected by many parameters. These parameters not only include characteristic parameters of underfill, but also include geometric parameters such as solder bump dimension, flip-chip dimension and the distance between chip and substrate etc. The filling quality of underfill also has close relationship with dispensing patter. In addition, filling process is affected by temperature, dispensing amount of material, dispensing needle speed, and etc, but this paper neglects their influences.

978-1-4244-2739-0/08/$25.00 ©2008 IEEE   438

Table I Some parameters of the underfill			
Viscosity	Density	Contact angle	Surface tension
2kg/m.s	1200kg/m$_3$	30º	0.03

Table I shows the characteristic parameters of underfill used in the simulation. Underfill is non-Newtonian flow. Because the Reynolds number of underfill (order of magnitude is 10^{-5}) is less than the critical Reynolds number, so flow characteristic is laminar flow [6].

Mathematical model

Because the gap dimension between chip and substrate is an important influence parameter in filling process. The dimension in height direction can't be neglected, therefore, two-dimensional model can't explain the influence of surface tension in filling process, so this simulation adopts three-dimensional transient model simulating this process.

The process of underfill filling can be looked as gas-liquid two-phase flow, and considering that both the flow and gas are all continuums and have no pervasion with each other. Under the condition of continuously flowing, the acting forces of underfill locate in balanced state. The two-phase flow only interacts with external objects at interface[10]. The acting force between two-phase flow is internal force, which has equal value, reverse direction, and momentum interchange, but has no contribute to the equation of momentum of whole system. This simulation adopt universal VOF model (namely volume of flow). VOF model simulates two kinds of flow by solving equation of momentum and handling the volume fraction of each flow which flowing through the domain.

The VOF formulation relies on the fact that two phases are not interpenetrating. In each control body, the sum of volume fraction of all phases is 1. The mesh which does not contain the object flow named "empty" mesh, the mesh which overflow the flow named "full" mesh, the mesh which contains interface named "half" mesh. All the variables and the domain of their attribute shared by each phase represent the average of volume.

The equation of volume fraction:

$$\frac{\partial \alpha_L}{\partial t} + \vec{v} \cdot \nabla \alpha_L = \frac{S_L}{\rho_L} \qquad (1)$$

where α_L is the volume fraction of flow phase, ρ_L is the density of flow phase, S_L shows the quality source of flow phase. While supposing the quality source as 1, also namely the quality of the flow phase is limitless.

The equation of volume fraction is used for control the change of volume fraction of two phases, but the continuity equation and momentum equation are the basic governing equation in the process of simulate underfill filling. The process of undefill filling can be regarded as transient flow process of incompressible non-Newtonian flow, the continuity equation of this process is as the following:

$$\frac{\partial \rho}{\partial t} + \frac{\partial (\rho u)}{\partial x} + \frac{\partial (\rho v)}{\partial y} + \frac{\partial (\rho w)}{\partial z} = 0 \qquad (2)$$

where ρ is the density of the filling flow; u, v, w is the velocity component of the velocity in the direction of X, Y, Z respectively.

The momentum equations are as the following:

$$\begin{cases} \rho \left(\frac{\partial u}{\partial t} + u\frac{\partial u}{\partial x} + v\frac{\partial u}{\partial y} + w\frac{\partial u}{\partial z} \right) = -\frac{\partial P}{\partial x} + \rho g_x + \mu \nabla^2 u \\ \rho \left(\frac{\partial v}{\partial t} + u\frac{\partial v}{\partial x} + v\frac{\partial v}{\partial y} + w\frac{\partial v}{\partial z} \right) = -\frac{\partial P}{\partial y} + \rho g_y + \mu \nabla^2 v \\ \rho \left(\frac{\partial w}{\partial t} + u\frac{\partial w}{\partial x} + v\frac{\partial w}{\partial y} + w\frac{\partial w}{\partial z} \right) = -\frac{\partial P}{\partial z} + \rho g_z + \mu \nabla^2 w \end{cases} \qquad (3)$$

where μ is the dynamic viscosity; g is the gravitational acceleration; ∇^2 is the Laplace operator

Model and mesh division

The geometrical model of flip-chip is 5.4×5.4×1(mm), and the gap between chip and substrate is 0.1mm. Considering the convenience of computing and the consumption of computing time, now the whole flow domain is simplified as a cube of 5.4×5.4×0.1(mm). The shape of solder bump is simplified as a cylinder with the height 0.1 mm and the diameter 0.15 mm, the pitch of solder bump is 0.6 mm, and the distance between peripheral solder bumps and edges is 0.6 mm, as shown in Figure 3. The purpose of the simplification is to make the model more regular, the mesh division more convenient, and the calculation time less.

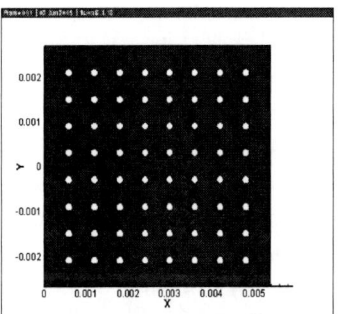

Figure 3. Distribution maps of solder bumps

In order to ensure the quality of mesh division, the whole domain is divided into six groups, and then each group is divided mesh respectively. And then all groups of mesh are merged. Because the model relative to the middle section geometric is symmetry, considered the convenience of computing and the consumption of computing time, so the computation only accounts to half of the reality model and puts the middle section as the symmetry interface.

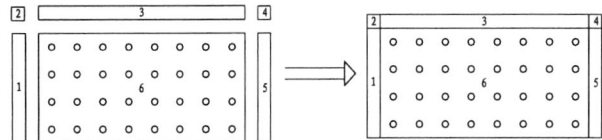

Figure 4. Schematics of mesh grouping: the external dimensions of the first group and the fifth group is 2.4×0.3mm; the external dimensions of the second group and the fourth group is 0.3×0.3mm; the external dimensions of the third group is 4.8×0.3mm; the external dimensions of the sixth group is 2.4×2.4mm.

The local mesh division of model, as Figure5 shows:

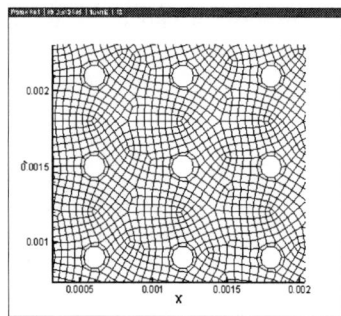

Figure 5. Local mesh division charts

The volume of computing field is 2.810mm³, the minimum mesh volume is 4.341625e-05mm³, the maximum mesh volume is 2.248620e-04mm³, and the total meshes are 23128.

The boundary conditions and initial conditions

The commercial software FLUENT is used in the simulation of underfill filling process of two types dispensing pattern: I pattern and L pattern. In order to reflect the influence of surface tension, contact angle and radius of curvature, wall adherence model and surface tension model are adopted. The wall adhesion model is used to adjust the surface normal in cells near the wall. The dynamic boundary condition results in the adjustment of the curvature of the surface near the wall [14].

If θ_w is the contact angle at the wall, then the surface normal at the cell next to the wall is:

$$\hat{n} = \hat{n}_w \cos\theta_w + \hat{t}_w \sin\theta_w \qquad (4)$$

where $\hat{n}_w$ and $\hat{t}_w$ are the unit vectors normal and tangential to the wall respectively. The combination of this contact angle with the normally calculated surface normal one cell away from the wall determine the local curvature of the surface, and this curvature is used to adjust the filling velocity in the surface tension calculation [14].

The driving force of material is influenced by surface tension, contact angle and radius of curvature in filling process. This paper establishes three-dimensional transient model which reckons in the surface tension influences on the rims of flip-chips, substrates and solder bumps

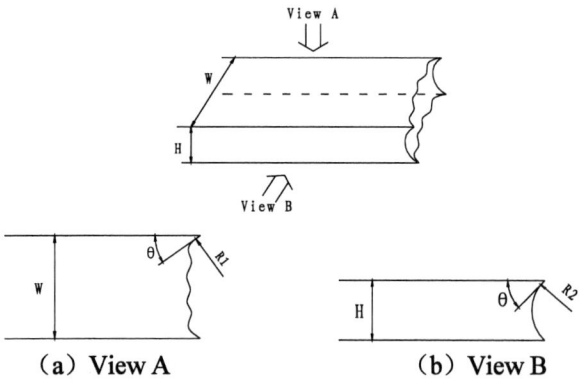

(a) View A   (b) View B

Figure 6. Schematics of flow front shape in a surface tension-driven underfill flow: (a) shape of flow front; (b) shape in width direction; (c) shape in height direction.

1.1.1 Boundary condition on flow front

The flowing situation in the flow front is shown as Figure 6:

According to Young—Laplace equation:

$$\Delta p = \sigma\left(\frac{1}{R_1} + \frac{1}{R_2}\right) \qquad (5)$$

where Δp is the pressure difference between the two sides of flow front. It means the pressure drop from the flow entrance to the current location of the melt front, σ is the surface tension coefficient, R_1 and R_2 is the radius of curvature in the direction of width and height in flow front respectively.

As shown in Figure 6（b）:

$$R_2 = \frac{H}{2\cos\theta} \qquad (6)$$

where H is the height of filling domain.

Because H is not change, so R_2 is a fixed value in height direction.

At the side wall edge, because of influence of wall, the radius of curvature there changes. That makes the underfill filling velocity at the side wall edge change.

At the edge of wave shape, different waves have different radius of curvature. That makes the filling velocity at the filling front non-uniform.

1.1.2 Other boundary conditions

The change of the driving force not only happens at the wall edge of the chip and substrate, but also happens at the edges solder bumps because of the existence of them. The flowing situation of underfill filling between solder bumps as shown in Figure 7:

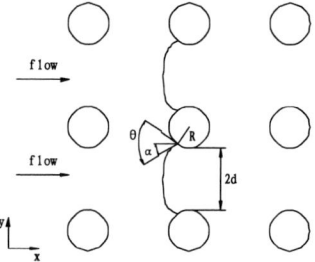

Figure 7. Schematic of underfill filling between solder bumps

At the rims of solder bumps, angle α changes with underfill filling in flowing direction. The radius of curvature exposes to change in solder bump joints, which causes the surface tension of solder bumps to change. This causes the underfill filling velocity among solder bumps to change. The non-uniform of filling velocity will cause the underfill flow front to appear wave shape.

1.1.3 Initial condition

The surface tension is the main driving force of underfill filling in flowing process. The ambient pressure has very small function of flowing process. So the initial pressure is set up to: P_{inlet}=0atm, P_{outlet}=0atm. The initial velocity is set up to V_{inlet}=0m/s. The pressure balancing model is adopted here. Air is defined as the primary-phase and underfill as the secondary-phase. The initial volume fraction of air is 1and the initial volume fraction of underfill is 0. Unsteady discrete solution controls and PRESTO method were adopted in the model. The momentum equation adopted First Order Upwind. The two phases have not mixed each other.

Analysis results

Simulation results and discussion

Because the calculation consumed time in model is long. In order to realize the balance of simulating time and precision of calculation, according to simulation experiences, the time step size is set up to 0.001s and the max iteration per time step is set up to 20.

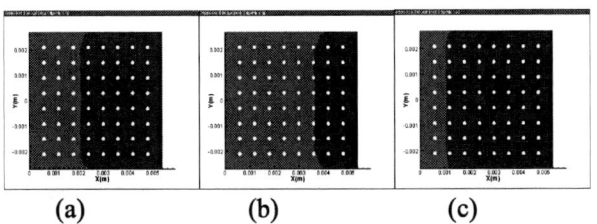

| (a) | (b) | (c) |

Figure 8. Filling position in I pattern: (a) filling time is 2s; (b) filling time is 6s; (c) filling time is 8s.

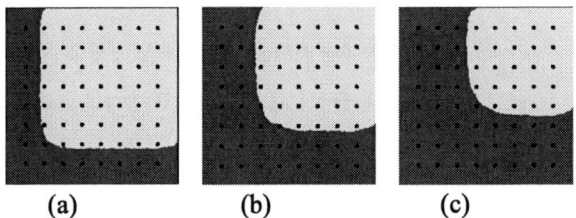

| (a) | (b) | (c) |

Figure 9. Filling position in L pattern: (a) filling time is 2s; (b) filling time is 6s; (c) filling time is 8s.

From the simulation results, the two types of filling pattern are similar at first. The two types all show that flow velocity at chip edges is obviously higher than the velocity inside chips. And then flow rate in L pattern becomes higher than in I pattern, as shown in Fig 10.

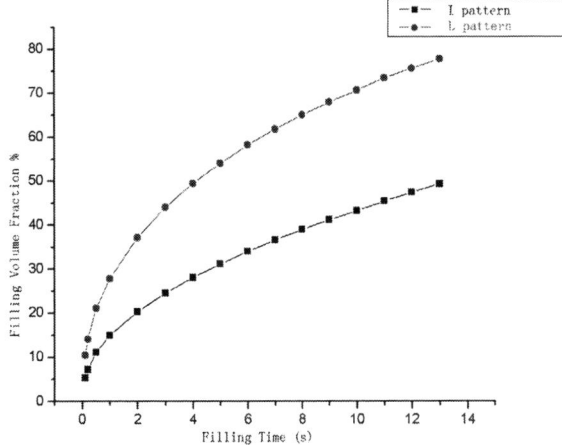

Figure 10. Schematics of filling volume fraction in two types of filling pattern

The simulation result shown in Fig. 11 (a) has reflected the influence of surface tension in the rim of chip and substrate. The underfill flow front appears obviously concave. This flow front is compatible with that illustrated in Fig. 6. Fig. 11 (b) shows that the pressure distribution from model

inlet to flow front presents the linear decrease trend. In this process the average velocity of underfill filling maintains a relatively stable value, seeing Fig. 11 (c).

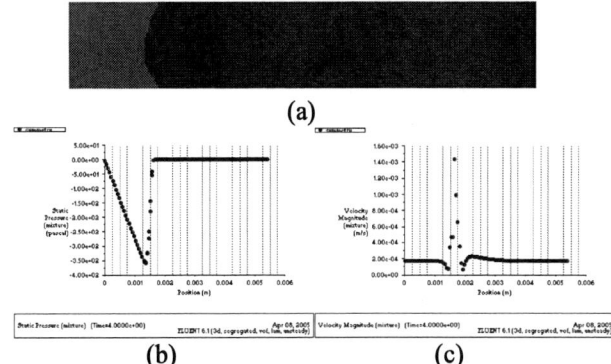

Figure 11. Filling position, pressure distribution curve and velocity distribution curve in middle section in gap width direction when filling time is 4s: (a) filling position; (b) pressure distribution curve; (c) velocity distribution curve.

In this process, inlet pressure is set up to 0. The internal pressure presents the linear decrease downward trend with flow front moving. According to the flow front pressure formula (8) of Sejin Han[12].

$$p_{\text{front}} = p_{inlet} - \Delta p \tag{7}$$

$$\Delta p = \frac{12\eta \left(\dfrac{dX}{dt}\right) X}{h^2} = \frac{12\eta VX}{h^2} \tag{8}$$

$$p_{\text{front}} = p_{inlet} - \frac{12\eta VX}{h^2} \tag{9}$$

where $\triangle p$ is the pressure drop from the flow inlet to the current location of the flow front(X), η is the viscosity of underfill, V is the average velocity glue of liquid among them, show glue average velocity that liquid advance, show glue liquid advance distance, express the height of the interval.

From formula (9), with the increase of filling displacement, values of p_{front} will appear linear decrease trend. This change is compatible with the pressure illustrated in Fig. 11 (a). With increasing of pressure, the average velocity maintains a stable value [12].

The forming process of air bubble was caught in the numerical simulation for the first time, as shown in Fig 12.

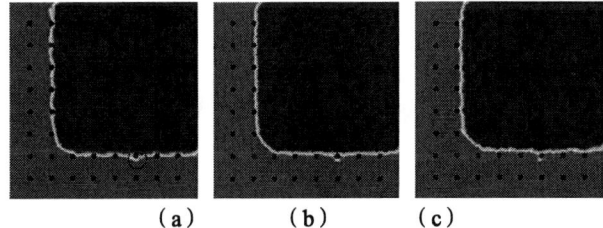

| (a) | (b) | (c) |

Figure 12. Air bubble forming process:
(a) t=2.6s; (b) t=2.8s; (c) t=3.0s

Fig.12 shows that air bubble shape and position changes at different time.

Simulation results and discussion

In order to observe and analyse the real flowing situation, the underfill filling experimental system is established. The internal diameters of the syringe needle is 0.84mm, external diameter is 1.24mm; The journey of the syringe needle is 5.4mm; The height of solder bump is 0.1mm and syringe needle moves along the chip; the maximum velocity is 4mm/s and the most acceleration is 1000mm/s^2. The glass plate used has been polished fine. Its geometric dimension is 5. 4 mm×5. 4 mm.

The experimental process is as follows:

To assure the environment temperature, dispensing volume and dispensing velocity fixed values, we can observe the underfill filling process and the formation of wave shape.

Experimental results:

From Fig.13, underfill velocity at the rim of model is obviously higher than the velocity inside the model and the obvious wave shape appears at flow front. the flow rate in L pattern is higher than in I pattern. Underfill flow velocity is faster in diagonal direction in L pattern. The experimental results are compatible with the simulation results well.

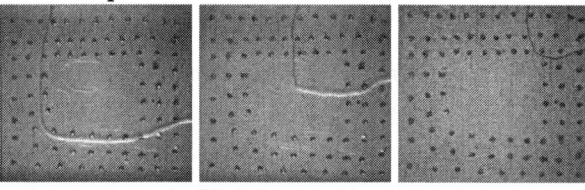

Figure 13. Underfill filling position (experiment) in L pattern

Conclusion

The three-dimensional transient model has been established for the different filling pattern. The model reckons in the influence of surface tension, contact angle and radius of curvature in the computation model. From the simulation results, at the rims of chip and solder bumps the velocities are obviously faster. The flow front appears wave shape. The flow rate in L pattern is higher than in I pattern. Underfill flow velocity is faster in diagonal direction in L pattern.

The underfill filling experimental system has been set up. The change of filling velocity at the rims of chip, substrate and solder bump is compatible with the simulation result. From the experimental results, the flow rate in L pattern is higher than in I pattern obviously.

Acknowledgements

The paper is supported by the foundation item of the national natural science foundation of China, major project (50390064) and general project (50475138).

References

[1] STEVEN J. ADAMSON. Review of CSP and Flip Chip Underfill Process and When to Use the Right Dispensing Tools for Efficient Manufacturing. Globaltronics Technology Conference, Singapore, 2002.

[2] G. HI, M.H. GORDON, W.F. SCHMIDT, R.P. SELVAM. Flow Properties of Liquid Underfill Encapsulations and Underfill Process Considerations. Electronic Components and Technology Conference, 2001.101-108.

[3] FENNY LIU, Y. P WANG, KEVIN CHAi, T. D. Her. Characterization of Molded Underfill Material for Flip Chip Ball Grid Array Packages. Electronic Components and Technology Conference, 2001.

[4] YEWCHOON CHIA, HONGSEE YAM, S. H. LIM, K. S. CHIAN, SUNG YI, WILLIAM T. CHEN. An Optimization Study of Underfill Dispensing Volume. IEEE Transactions on Components, Packaging, and Manufacturing Technology. 2003, Vol. 26, No.3: 205-210.

[5] TIE WANG, T. H. CHEW, Y. X. CHEW, P. MIAO, L. FOO. Assessment of Flip Chip Assembly and Reliability Reflowable Underfill. Electronic Components and Technology Conference, 2001

[6] WEN-BIN YOUNG, WEN-LIN YANG. Underfill Viscous Flow between Parallel Plates and Solder Bumps. IEEE Transactions on Components and Packaging Technology. 2002, Vol. 25, No.4: 695-700.

[7] L.NGUYEN, C. QUENTIN, P. FINE, B. COBB, S. BAYYUK, H. YANG, S. A. BIDSTRUP-ALLEN. Underfill of Flip Chip on Laminates: Simulation and Validation. IEEE Transactions on Components, Packaging, and Manufacturing Technology. 1999, Vol. 22, No.2: 168-176.

[8] Y. GUO, G. L. LENMANN, T. DRISCOLL AND E. J. COTT. A Model of the Underfill Flow Process: Particle Distribution Effects. Electronic Components and Technology Conference, 1999. 71-76.

[9] DAVID MILNER, DANIEL F. BALDWIN,PH. D. Effects of substrate design on underfill voiding using the low cost, high throughput flip chip assembly process. International Symposium on Advanced Packaging Materials, 2001.51-56.

[10] LIU RUXUN, SHU QIWANG. Certain New Methods of CFD. Science Publisher, 2003

[11] MATTHEW K. SCHWIEBERT, WILLIAM H. Leong. Underfill Flow as Viscous Flow between Parallel Plates Driven by Capillary Action. IEEE Transactions on Components, Packaging, and Manufacturing Technology —Part C. 1996,Vol. 19, No.2:133-137

[12] SEJIN HAN, K. K. WANG. Analysis of the Flow of Encapsulant during Underfill Encapsulation of Flip-Chips. IEEE Transactions on Components, Packaging, and Manufacturing Technology —Part B. 1997, Vol. 20, No.4: 424-433.

[13] STEVEN J. ADAMSON, JAMES J. KLOCKE, LARS NIELSON. Evaluation of Printed Circuit Board layout for Chip Scale Packages that Require Underfill and the Effect of Adjacent Passive

[14] J. U. BRACKBILL, D. B. KOTHE, C. ZEMACH. A Continuum Method for Modeling Surface Tension. J. Comput, 1992. 335-354.

Study on Thermal Simulation Technology for SMA in Lead-free Reflow Soldering

Wu Zhaohua [1]   Zhou Dejian [2]

1. School of Mechanical & Electrical Engineering of Guilin University of Electronic Technology
2. Guangxi University of Technology

Guilin Guangxi, China 541004

Email: emezdj@guet.edu.cn,   phone: +86-773-5601392

Abstract

The thermal mathematical model under intensive convection in lead-free reflow soldering process is constructed for surface mounted assemblies, and the heat conduction governing equation, initial condition and boundary condition are made. A thermal mathematical model of SMA is transformed into solid model which can be analyzed by the thermal analysis software FLOTHERM. Then by defining the reasonable boundary condition and model simulating analysis, the thermograph contour map and profile for the SMA are obtained. Using the data relevant to simulation reflow profile and orthogonal experiment design to set up the factor effect table, the degree of the critical parameters of lead-free reflow soldering process affecting the key index of thermal profile would be obtained. The results show that the heating rate is mainly affected by the conveyer belt speed, temperature of heating zone No.6 and No.2; the maximum temperature is mainly affected by the conveyer belt speed, temperature of heating zone No.6 and No.7; and the soldering time of extension melting point temperature is mostly affected by conveyer belt speed.

Keywords

Surface Mounted Assembly; Lead-free Reflow Soldering; Orthogonal Experiment Design; Thermal Analysis

1. Introduction

With more attention paid to the health and environment problem recently, the Pb harmfulness in conventional Sn-Pb solder paste used in the reflow soldering process is gradually paid attention to by public. Pb-free electronic products are a tendency and the conventional reflow soldering process replaced by lead-free reflow soldering process is also inevitable. In comparison with the conventional reflow soldering process, the peak temperature in the lead-free one is 30~40^{0}C higher and the available process window is greatly reduced, which makes it difficult to set the thermal profile of lead-free reflow soldering affecting the product assembly quality. However, the traditional method of setting thermal profile through repeated tests and adjustment cannot meet the request of high renewal speed and increasingly strong competition of

electronic products. Just under this background, simulation and prediction research of lead-free reflow soldering process can greatly shorten the process preparation time, cut experiment cost, improve reflow soldering quality, decrease reflow soldering defects and strengthen industry's competition ability.

In this paper, the thermal numerical model of SMA under forced hot air reflow soldering is constructed, and the micro-unit temperature-controlling equations for the SMA, initial conditions and boundary conditions are established. And a numerical model for heat field of SMA is transformed into a solid model by the thermal analysis software FLOTHERM with reasonable boundary conditions. Through the simulation and analysis on the above model, the temperature profile and temperature distribution contour of each heat zone for the SMA are obtained in the lead-free reflow soldering. With the orthogonal experiment design method, collecting the data from the simulated temperature profile, setting up the factor effect table using the averaging method and analyzing the correlative factors ,the degree of influence on the lead-free reflow soldering process of the critical process factors can be obtained. The research results have certain reference value for carrying out optimization design of lead-free reflow soldering process, reducing the setting time of lead-free reflow soldering heat profile and increasing the production efficiency[1][2].

2. Foundation of the simulation model for lead-free reflow soldering SMA

In forced hot air lead-free reflow soldering process, the main way of SMA heat transfer is heat convection. The heat conduction governing equation[3] can be obtained as formula 1, with taking no account of the heat exchange between SMA and PCB and supposing the temperature is consistent in the same heating zone.

$$\rho c V . \partial T / \partial t = 9.3 U air^{1/2} . A(T_w - T_f) \qquad (1)$$

With no inner heat resource being in SMA micro-unit in lead-free reflow soldering, supposing all the materials are isotropic, the heat field governing equation[4][5] of either SMA micro unit is:

$$K(\partial^2 T/\partial x^2 + \partial^2 T/\partial y^2 + \partial^2 T/\partial z^2) = \rho C \partial T / \partial t \qquad (2)$$

When calculating the SMA transient temperature that changes over time, it often assumes that the temperature is uniform and equal to the ambient temperature. If the initial temperature is set to 25 ℃, the initial condition is:

$$T |_{t=t0} = T(x,y,z,t_0) = T_0 \qquad (3)$$

Foundation Project: advanced project(A New Method of Electrical Interconnection)

Introduction of author: Wu Zhaohua, female, born in Nanchang, Jiangxi. professor of GUET, engaging in mechatronics and   SMT research

978-1-4244-2739-0/08/$25.00 ©2008 IEEE   443

And the convection condition is known in every moment in forced hot air lead-free reflow soldering, through Newton formula getting the heat density from surrounding medium, and referring to the Fourier assumption that the heat intensity is in direct proportion to the temperature gradient, the convection boundary condition is acquired as follows:

$$-k_n \partial T / \partial n \mid_{M=sc} = q_{Sc}(M,t) = h(Te-Ts_c) \quad (4)$$

Because SMA temperature changing is uneven, according to Fourier assumption, radiation boundary condition can be described as:

$$k_n \partial T / \partial n \mid_{M=sr} = q_{sr}(M,t) = \varepsilon F \sigma [(T_r+273)^4 - (T_{sr}+273)^4] \quad (5)$$

Then, the heat boundary condition can be obtained from the equation (4) and equation (5).

$$k_n \partial T / \partial n + h(Te-Ts_c) + q_w + \varepsilon F \sigma [(T_r+273)^2 + (T_{sr}+273)^2][(T_r+273)+(T_{sr}+273)](Tr-Tsr) = 0 \quad (6)$$

As SMA structure is complex, it should be made a reasonable simplification when simulated by FLOTHERM. However, the chip component can be considered a mass point because its size and heat capacity are small, its temperature change is broadly in line with the PCB's and it has little effect on the temperature distribution of SMA. For the bigger component with regular shape, its pins can be ignored when calculating and replaced by regular geometric shape, and the heat conduction relation between it and PCB can be calculated in accordance with the following formula [6].

$$G = \frac{N k_p A_p}{l_p} + \frac{k_a A}{t} \quad (7)$$

Where, G is the heat conductivity, N is the pin number, k_p is the pin's heat conductivity, A_p is the pin's cross sectional area, l_p is pin's length, k_a is air's heat conductivity, A is the air area underneath SMA, t is air gap thickness between SMA and PCB.

The solder joints that have small heat absorption capacity can be ignored when simplifying the model. And for the anisotropic PCB, its material parameters can be dealt with the average method. Then the SMA simulation model established by FLOTHERM is shown in Figure 1 and the model consists of two QFPs, four SOICs, four SOPs and one PLCC.

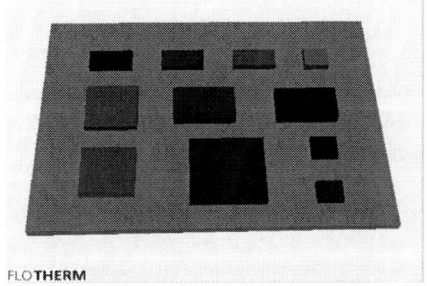

Fig1 SMA Simulation Model

3. Thermal analysis and heat profile simulation for lead-free reflow soldering SMA

For the analysis on the lead-free reflow soldering SMA, it is often done that the heat convection and heat radiation are exerted on the simulation model as boundary conditions, and it is assumed that the hot air temperature does not change over time, heating zones have different temperature and the temperature load changes over time. To express the lode changing over time on the simulation, each inflection point is considered as a load-step in the load-time curve, and the corresponding SMA temperature load curve is shown in Figure 2.

For the temperature field simulation, a 10-zone forced hot air reflow stove of Muliticore LF300 is chosen, and the data of every temperature zone is shown in table 1. The SMA is heated in every heating zone for 30 seconds and cooled by the forced convection air of 25 ℃ with the convection coefficient 50 W/m² ℃. According to the above parameters and convective boundary condition loaded on the simulation model using FLOTHERM, the thermograph contour maps are obtained and shown in Fig.3(a)~(j).

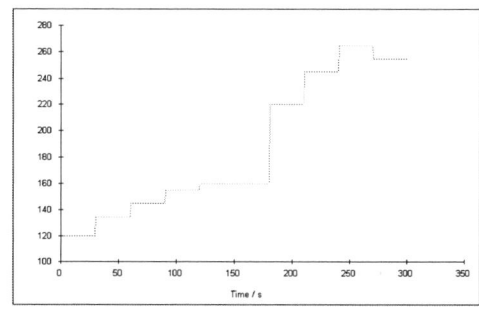

Fig2 SMA Temperature Load Curve

Table1 Temperature Setting in Heating Zones

Heating zone	1	2	3	4	5
Temperature(℃)	120	135	145	155	160
Heating zone	6	7	8	9	10
Temperature(℃)	160	220	245	265	255

Seen from the figures, the temperature distribution is uneven, especially in the preheating zones from one to four because the materials of components and PCB are different. In the preheating stage, the SMA's temperature increases rapidly from room temperature to a higher temperature, resulting in the great difference in temperature in materials with different heat capacity ,just shown in Fig.3(c). While, the temperature distribution becomes uniform in the 5th, 6th and 7th heating zones in the heat preservation stage with a long heating time, shown in Fig.3 (g). For the larger device, it should be set a higher temperature in the 9th heating zone, which makes the device that can not arrive to the reflow temperature in the 8th heating zone reach the reflow temperature, and the solder paste has been reflowed in the 8th ,9th ,10th heating zones in reflow stage.

After obtaining the thermograph contour maps through

simulation and using the FLOTHERM transient thermal analysis module, reflow soldering thermal profile changing over time for any point in SMA is obtained. The thermal profile of typical point in SMA simulation model is shown in Fig.4.

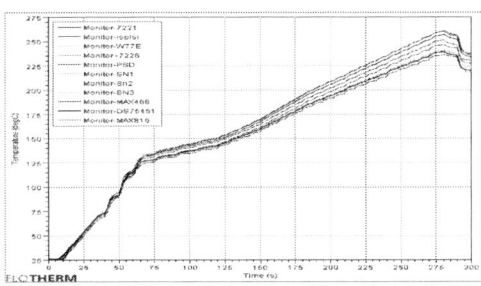

Fig4 Thermal Profile of Typical Point in SMA Model

4. Effect analysis of lead-free reflow soldering process parameters based on orthogonal design

The influence critical reflow soldering process parameters on the reflow soldering temperature profile is obtained through the simulation involving three critical parameters: the maximal heating ratio RS, the reflow soldering peak temperature PT, the heating time of temperature over melting point TAL. In this paper, a 10-zone reflow stove for simulation is in action. Because in the preheating stage the temperature has a slight variation, it is supposed the preheating temperature is constant. Thus eight variable parameters are set consisting of the conveyor belt speed and the temperature of the latter seven heating zones.

Table2 Effect Factor Level Table

factor	Level 1	Level 2	Level 3
S	70	80	
T_1	150	155	160
T_2	155	160	165
T_3	155	160	165
T_4	215	220	225
T_5	235	240	245
T_6	255	260	265
T_7	245	250	255

Note: S is the conveyer belt speed （cm·min-1）, T1～T7 are temperature from the fourth temperature zone to the seventh（0C）.

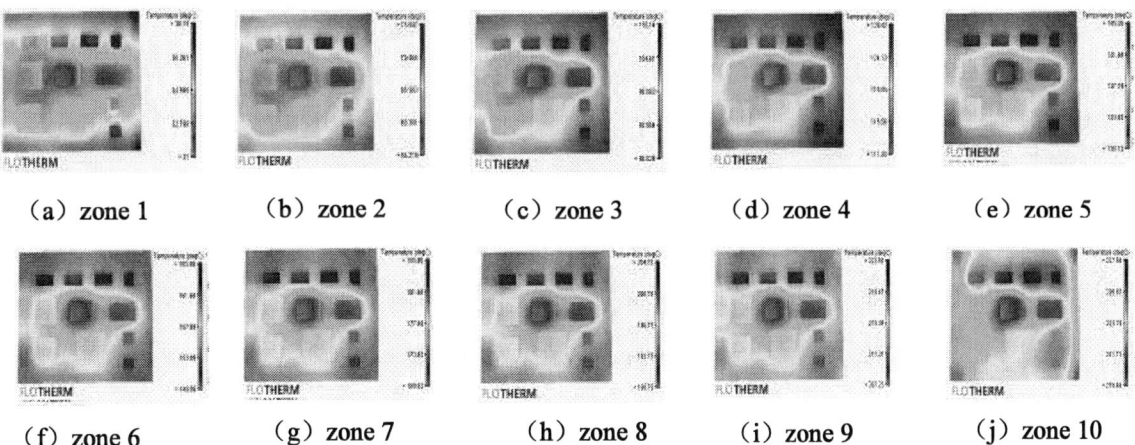

（a）zone 1　　（b）zone 2　　（c）zone 3　　（d）zone 4　　（e）zone 5

（f）zone 6　　（g）zone 7　　（h）zone 8　　（i）zone 9　　（j）zone 10

Fig3 Thermograph Contour Maps in Heating Zones

Table3 Orthogonal Test Data

Test	S/cm .min⁻	T_1/°C	T_2/°C	T_3/°C	T_4/°C	T_5/°C	T_6/°C	T_7/°C	R_s/°C.s⁻¹	P_T/S	T_{AL}/S
1	70	150	155	165	220	240	255	250	0.64	242.14	70.50
2	70	155	155	155	215	235	260	245	0.65	241.84	62.62
3	70	160	155	160	225	245	265	255	0.70	247.38	78.00
4	70	150	160	160	215	240	265	245	0.69	245.87	70.50
5	70	155	160	165	225	235	255	255	0.65	246.32	65.25
6	70	160	160	155	220	245	260	250	0.64	246.58	73.50
7	70	150	165	155	225	235	265	250	0.66	246.24	65.25
8	70	155	165	160	220	245	255	245	0.57	238.61	73.50
9	70	160	165	165	215	240	260	255	0.62	241.90	67.12
10	80	150	155	155	215	245	255	255	0.79	244.59	59.37
11	80	160	155	160	225	240	260	250	0.79	241.69	59.37
12	80	165	155	165	220	235	265	245	0.88	244.72	49.69
13	80	150	160	165	225	245	260	245	0.81	240.85	62.19
14	80	155	160	155	220	240	265	255	0.73	246.11	55.94
15	80	160	160	160	215	235	255	250	0.73	240.57	49.69
16	80	150	165	160	220	235	260	255	0.80	245.10	49.37
17	80	155	165	165	215	245	265	250	0.81	245.19	49.06
18	80	160	165	155	225	240	255	245	0.72	237.25	49.37

Table4 the Factor Effect Table of R_S(°C/s)

	S	T1	T2	T3	T4	T5	T6	T7
Level 1	0.647	0.732	0.742	0.708	0.715	0.728	0.683	0.720
Level 2	0.791	0.710	0.718	0.713	0.720	0.708	0.718	0.712
Level 3		0.715	0.697	0.735	0.722	0.720	0.755	0.725

Table5 the Factor Effect Table of P_T(°C)

	S	T_1	T_2	T_3	T_4	T_5	T_6	T_7
Level1	244.10	244.13	243.73	243.77	243.33	244.13	241.58	241.52
Level2	242.90	243.29	244.38	243.2	243.88	242.49	242.99	243.74
Level3		243.07	242.38	243.52	243.29	243.87	245.92	245.23

Table6 the Factor Effect Table of $T_{AL}(s)$

	S	T_1	T_2	T_3	T_4	T_5	T_6	T_7
Level1	69.58	62.86	63.26	61.01	59.73	56.98	61.28	61.31
Level2	53.78	60.96	62.85	63.41	61.54	62.13	62.36	61.23
Level3		61.23	58.95	60.64	63.24	65.94	61.41	62.51

By using the orthogonal design, it only needs to choose parts of process parameters to get the desired experiment data and result and to get the factor affecting the quality most. In this paper, a eight-factor L_{18} ($2^1 \times 3^7$) orthogonal array[7] with one two-level factor and seven three-level factors are used ,shown in Table 2, and 18 times experiments are carried out. Taking the monitory point in the middle component of SMA simulation model as study object, the orthogonal test data are obtained shown in Table 3.

With an average value analysis using factor effect table, it shows the influence of factors on the indicators. As each factor being in different level occurs at the same frequency, the effect value of each indicator for each factor in level 1 can be gained through getting the average date from all the date, which will more directly reflect the degree of influence on one index for each factor changing from level 1 to level 3. In the factor effect table, the variation quantity and direction of effect value for each factor in deferent level can reflect the influence degree and influence direction of each factor on the index.

(1) The factor effect table for R_S (table 4)

Seen from the table 4, it shows that the factors affecting R_S greatly is the conveyer belt speed S, the temperature T_6 of the 6th heating zone and the temperature T_2 of the second one in order.

Considering the values, when the conveyer belt speed S changes from 70cm/min to 80cm/min, the heating rate changes from $0.647^0C/s$ to $0.791^0C/s$，increased by $0.144^0C/s$;the set temperature T_6 of the sixth heating zone changes from 255^0C to 265^0C，the heating rate R_S is increased by $0.072^0C/s$;and the set temperature T_2 of the second heating zone changes from 155^0C to 165^0C，the heating rate R_S is increased by $0.045^0C/s$. So the effect degree of the conveyer belt speed on the temperature ascending rate has more than twice the one of the set temperature in the heating zone. Moreover, increasing the conveyer belt speed will raise the heating rate, but increasing the set temperature in the heating zones will reduce the heating rate.

(2) The factor effect table for the reflow peak temperature P_T (table 5)

Seen from the table 5, it shows that the factors affecting the reflow peak temperature P_T is the set temperature T_6 of the 6th heating zone, the set

temperature T_7, of the 7th heating zone the conveyer belt speed S in sequence.

(3) The factor effect table for the time T_{AL} of the temperature over the melting point (table 6)

Seen from the table 6 , it shows that the factors affecting T_{AL} is the conveyer belt speed S, the set temperature T_5 of the 5th heating zone, the set temperature T_4 of the 4th heating zone one by one.

5. Conclusions

Through the rounded analysis of the experiment data derived from the orthogonal experiment, it is known that the effects of the reflow soldering process parameters on the key index of the temperature profile are concluded as follows:

①The factors affecting the heating rate R_S obviously are the conveyer belt speed S the set temperature T_6 of the 6th heating zone, the set temperature T_7 of the 7th heating zone in sequence ,and the influence degree by S is the biggest.

②The factors affecting the reflow peak temperature P_T are the set temperature T_6 of the 6th heating zone, the set temperature T_7 of the 7th heating zone , the conveyer belt speed S in sequence, and the influence degree by T_6 is the biggest.

③The factors affecting the time T_{AL} of the temperature over melting point are the conveyer belt speed S,the set temperature T_5 of the 5th heating zone, the set temperature T_4 of the 4th heating zone in sequence, and the influence degree by S is the biggest.

References

1. Ma Xin, Dong Benxia, "Develpement of Lead-free Solder," Electronic Process Technology, Vol.23, No.2 (2002), pp.47-52.

2. Bie Zhengye, "the Present Situations and Applications of Lead-free Soldering Technology," Motor and Electric Appliances Technology, No.6 (2002), pp.12-14.

3. Sarvar F, "Effective Transient Process Modeling of the Reflow Soldering of Surface Mount Assemblies," IEEE /ASME-THERM V Conf, Orlando, FL, 2006, pp.195-202.

4. Qian Zuoqin, "Preliminary Application of Finite Volume Scheme in Solving Temperature Field of Heat Transfer," Wuhan Shipbuilding, No.6 (1998), pp.7-10.

5. S.S Lauer, Finite Element Method in Engineering,

Science Press (Beijing, 1991), pp.125-130.

6. Free J A, "Recent Advances in the Thermal/ flow Simulation: Integrating Thermal Analyses into the Mechanical Design Process," CA, No.2 (1995), pp.15-19.

7. Yangde, Experimental Design and Analysis, China Agriculture Press (Beijing, 2002), pp.90-97.

A Novel High Effective Envelope-tracking Amplifier for OFDM Systems

Yingliang Li[1]; Jide-Zhao[2]

1Motorola Hangzhou R&D center, Hangzhou, 310053, China

2 Institute of physical & electric, Ludong University, Yantai, 264025, China

Abstract

OFDM (Orthogonal Frequency Division Multiplex) mobile communication system is so popular in recent because it has so many virtues as effective frequency spectrum, wideband, high speed rate, etc. But it has so many challenges in technical field, such as the high efficiency of PA (Power amplifier) and linearity. RF (radio frequency) amplifier with ET (Envelop-Tracing) structure is described for OFDM systems in this paper. The first study the ET amplifier model, and then a high-efficiency OFDM amplifier is presented with GaN HEMT (Heterostrucutre Field-Effect Transistors), and includes the theoretic analysis and circuit design. Finally, the design circuit is simulated by ADS (Advanced design systems) software, and gets the average DE (Drains Efficiency) of the amplifier is as high as 50% for an OFDM (Wimax) modulated signal with EVM of 2.94% at an average output power of 50 W and gain of 13.0 dB from simulating result. The design has good efficiency and linearity. All of these performances are satisfied with the OFDM system's requirement of RF. This design may be applied for the OFDM systems compare with the standard of OFDM.

Keywords- Amplifier; Envelop tracing, OFDM; Wimax.

1 Introduction

Cost, efficiency, and distortion are the main concerns in selecting the bias scheme for a PA. The traditional approach to linearly amplify the nonconstant envelope modulated signal is to "back off" the linear Class-A or Class-AB PA's output power until the distortion level is within acceptable limits. Unfortunately, this lowers efficiency significantly, especially for high PAR signals. Thus, there is an inherent tradeoff between linearity and efficiency in PA design. In addition to excellent linearity, high efficiency is essential for low-cost high reliability OFDM (Orthogonal Frequency Division Multiplex) base-station power amplifiers. However, in order to obtain better linearity and efficiency for new wireless base stations, i.e. OFDM systems. For example, the Wimax specifications for the transmitter design are summarized in Table I [1].

Recently GaN HFETs can provide higher voltage operation and higher power density at microwave frequencies than other high power devices and thus are attractive for application to commercial high-power base stations. The use of this GaN power amplifier within an envelope-tracking (ET) architecture was recently reported in many papers, and it has so many virtues, especially the efficiency and linearity. OFDM systems is wide band modulating systems, with modern wireless communication systems evolving to more spectrally efficient and higher data-rate modulation formats, highly linear PA's are required to avoid the out-of-channel interference [e.g., adjacent channel power ratio (ACPR)] and distortion [e.g., EVM]. OFDM systems standard employs 64-QAM modulation and 128 carriers at a 54-Mb/s data rate.

This modulation format has a high envelope peak-average ratio (PAR) of 8~10 dB [2]. This paper presents performance characteristics of the amplifier with ET structures, which employs a dynamic supply voltage for efficiency enhancement. The amplifier is used the GaN HFET. The high-efficiency performance of the RF power amplifier is analyzed by ADS. Simulations shows class AB with OFDM single input and ET amplifier has high-efficiency and good EVM.

Table I: 802.11 e Wimax transmitter specifications

Frequency	2.1-10GHz
Numbers of Carriers	1,285,121,024
Channel bandwidth	3.5,5,7,10MHz
Carrier type	OFDM
Modulation	64QAM,BPSK,etc.
EVM	-30dB
Spectrum mask	-20dBc@11MHZ offset; -20dBc@11MHZ offset -20dBc@11MHZ offset

2 High effective PA of ET using GaN HEMT

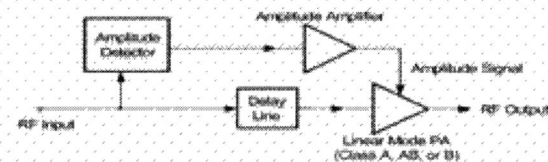

Figure.1. Block diagram of envelope tracking amplifier

ET PA system is figure 1, and it is composed of amplitude detector, amplitude amplifier, and delay line and RF amplifier. It has tow ways of RF amplifier and ET amplifier. RF amplifier permit the largely distortion of amplitude for improving the RF effective; ET signal will distinguish the ET information of amplitude, and then amplify the signal to control the power of PA, and power voltage will variety with the ET signal variation, and the RF way need delay line for synchronizing the ET signal. The amplifiers are always working at saturation area. Delay line is realized DPD

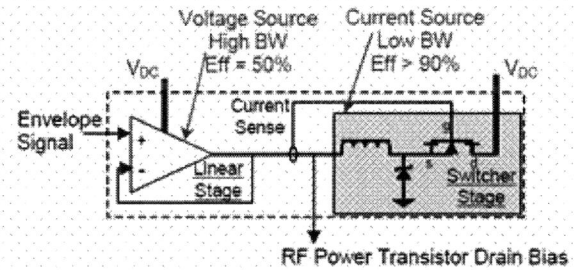

Figure. 2. Schematic diagram of high efficiency envelope amplifier

(Digital Perspective Distortion) technical in paper [3].

The envelope amplifier used in this work, shown schematically in Fig. 2, comprises a linear stage to provide a wideband voltage source and, in parallel, a switching stage to provide an efficient current supply. The output voltage of the envelope amplifier follows the input envelope signal with help of an operational amplifier. The current is supplied to the RF amplifier drain from both the linear stage and the switching stage through a current feedback network which senses the current flowing out of the linear stages and turns on/off the switch [5].

A RF3821 GaN HFET transistor [4] and the device is output 50W of max power. It was used to implement the Class-AB RF PA,as shown in Fig. 3. Quarterwave transmission lines are used at the input and output to short the even harmonics. For a 2.5GHz single-tone test, the measured gain is 13 dB and the DE is 52% at an output power of 50W by the datasheet of RF3821.

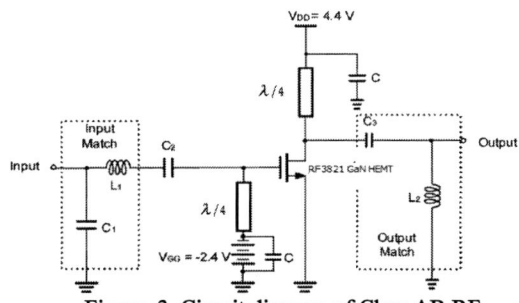

Figure. 3. Circuit diagram of Class AB RF amplifier output stage with GaN HEMT

3 Simulated and results

The simulation ET system includes a wide-band high-efficiency envelope amplifier and the Class-AB RF PA (Power Amplifier). Assuming the Wimax of 2.5GHz RF OFDM signal input the PA. In fact, it is required upconvert the baseband signal to IF of 50MHz. at a sample rate of 110MHz and then up convert to 2.5GHz by FPGA device. Because this part is so complex, we don't describe detail and detail may get from paper [5]. In the simulation, we use the DC power of 27V by RF3821 requirement, and input the visual signal of 2.5GHz with band 5M by ADS mode of generic OFDM models.Fig. 4 show curves of gain, DE and Pout. When the input is 40dBm and DE may arrive about 50%.This design improve the DE compare with traditional design method. Figure 5 shows the EVM performance with 3% input power more than 50W.

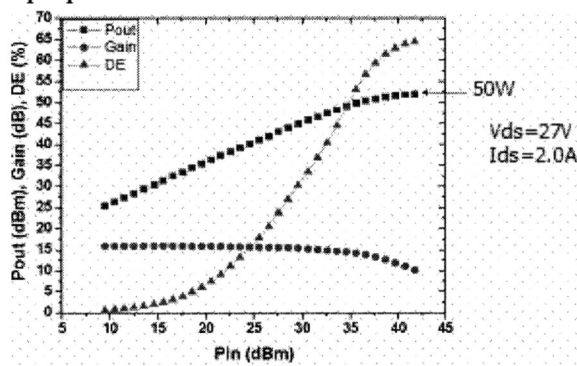

Figure. 4. ET amplifier. Performance when f=2.5GHz by ADS simulated

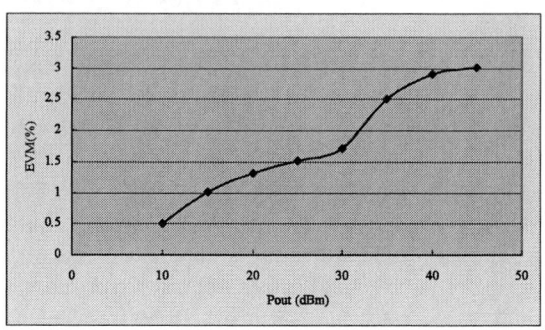

Figure. 5. OFDM signal EVM when f=2.5GHz by ADS simulated

4 Conclusions

In this paper, an OFDM power amplifier using RF3821 GaN HEMT transistor was presented with very high DE of 50.7%, together with average output power of 50W and gain of 13.0 dB. The high-efficiency is gotten by GaN HEMT device with ET structure. It also shows good linearity corresponding to EVM of 2.94%. This design solve the challenging for OFDM systems

Certainly, this paper only analyzes the ET structure and simulation, if need realize the ET systems for OFDM, we have so many working need do in the future as following:
(1)Study the OFDM technical.
(2)Study the PAR (peak-average ratio), because OFMD is multi-carriers systems, PAR is easy to become so large.
(3)Study the manufacture technical due to OFDM system is so complex, and it has so many differences compare with the traditional manufacture.

References

[1] Part11: WIMAX Medium Access Control (MAC) and Physical Layer Specifications, IEEE Standard 802.16e/f/g 2005/2007.

[2] Feipeng Wang, Annie Hueiching Yang, Design of Wide-Bandwidth Envelope-Tracking Power Amplifiers for OFDM Applications, IEEE TRANSACTIONS ON MICROWAVE THEORY AND TECHNIQUES, VOL. 53, NO. 4, APRIL 2005, pp 1244~1255.

[3]P. Draxler, J. Deng, D. Kimball, I. Langmore, P.M. Asbeck, "Memory Effect Evaluation and Predistortion of Power Amplifiers," *2005 IEEE MTT-s Dig.*, TH2B, 2005

[4] Chen Yuquan, Microwave and RF, China Academic Journal publishing, Beijing,China,2007, Vol.46 (11) 15~18

[5] F.Wang, A. Ojo, D. Kimball, P. M. Asbeck, and L. E. Larson, "Envelope tracking power amplifier with pre-distortion for WLAN 802.11 g," in IEEE MTT-S Int. Microwave Symp. Dig., 2004, pp. 1543–1546.

[6]T. Marra, D. Kimball, J. Archambault, W. Haley, and J. Thoreback, " Envelope tracking efficiency enhancement for CDMA base station high power amplifier," IEEE Topical Workshop on Power Amplifiers for Wireless Communications, 2002.pp:891~901

Biographies

LI Ying-liang (1979-), male, he receive the M.S degree on 2004 from Chongqing University of Posts &

Telecommunications ,currently, he is with senior RF engineer, Motorola Hangzhou R&D center, His research focuses on micro-electronics, mobile communication RF etc.

A Dual-band MEMS PA Study for Mobile Communication Systems

ZHAO Ji-de1, LI Ying-liang 2, WEI De-fang 1

1. Institute of physical & electric, Ludong University, Yantai, 264025, China;
2. Hangzhou R&D center, Motorola, Hangzhou, 310053, China

Abstract

This paper described a dual-band PA with MEMS capacity switch, and it is composed of the series MEMS capacity switch, cascode amplifier and other circuit, and gets the dual-band of 2.1GHz /2.3GHz PA by the switch capacity variation. The first, a MEMS capacity switch is designed, analyzed by EDA of HFSS, and obtain the capacitive curve with 2.1GHz and 2.3GHz, insert loss is about 2dB and isolation is about 50dB.The second, the MEMS switch is applied in the PA circuit, and the circuit model is designed basing on the CMOS technology. The performance is obtained by EDA of ADS simulation, and the gain is 12.4dB and 11.5dB with 2.1GHz and 2.3GHz and PAE is about 50% when output 25dBm at two band. MEMS switches are used to implement a variable filter and matching network that allows the PA to realize the dual-band and high efficiency. It has good performance, and it may be applied for mobile communications systems, such as 3.5G (Generation), 4G mobile communications systems.

Key words: **Dual-band; MEMS switch; PA; PAE.**

1 Introduction

Mobile ubiquitous world will require band-free RF (Radio-Frequency) circuits that can support the use of a variety of frequency bands due to so many mobile communication systems in the global. Mobile communication is developing quickly with many kinds of type methods, such as 3G(Generations) communications systems, there have four methods WCDMA, CDMA 2000 ,TD-SCDMA and WIMAX etc. These mobile communications services will promote the conversion to broadband communications and a ubiquitous communications environment in which all kinds of devices and objects are interconnected and real space and virtual space interact in the future. These standards work with difference frequencies in difference standards, even if it is the same standard, it may work the different frequency in different countries. All of these are required the device of broadband for satisfying with different frequencies. However, it's very difficult that realize the RF broadband due to the many limitations, for instance, the linearity, distortion.

A RF circuit is an essential element of an MS (Mobile device) and BTS (Base transmitter station). RF circuit can be applied to individual circuits like the Power Amplifier (PA) and transmit-receive duplexer that handle high frequency signals or to all of these circuits combined. It is difficult for a PA, a key device of an RF circuit, to operate at multiple frequency bands with adequate performance. It would therefore be of great benefit to develop technology that could solve this problem. As we know, performance of PA is very critical in mobile communications, and it has many challenges as PAE (Power added efficiency), multi-band and linearity, especially the linearity and efficiency are very critical. There have many kinds of PA, for example the Class A, B, AB, E, and so on. However, PA's linearity and efficacy is the relation of inverse ratio, so need balance them for getting the difference applications. So it is very difficult to get broadband, high linearity and high efficacy devices. MEMS (Micro-Electric-Machine-Systems) are multi-across subject technology with many virtues, such as miniaturization, lower power and high linearity. If MEMS is applied to the PA of mobile communication systems, it will get good performance of linearity and efficiency. It is gotten broadband by configuration the multi-switches [1]. If a PA port has two ways signal input, MEMS switch may realize the signal switch with good linearity and lower loss.

This article outlines the configuration and features of a high efficiency PA equipped with dual -band-switch, and describe a prototype 2.1GHz/2.3GHz dual-band PA using MEMS switches based on the proposed configuration. The first, the MEMS switch of capacitance is described, and then use it to design the dual-band PA. The PA is class A/E, and it has good performance, and it may be applied for mobile communications systems, such as 3.5G (Generation), 4G mobile communications systems.

2 MEMS capacitive switch

There are many different types of MEMS switches in both series and shunt configurations [2]. These switches can be DC contact switches, where there is direct metal contact between the two plates or capacitive switches as discussed following. As a rule, DC power is the mainly supply power in PA systems, and the switch power is fit for the total circuit of PA and get good topologies if use the MEMS switch configurations. So it is a lower power PA system with MEMS switch configuration, because high power needs high voltage to supply.

To maximize the potential of the proposed PA configuration, switches that can exhibit both low loss and high isolation across a wide band are required. One such switch now being researched and developed is the MEMS capacitive switch that uses micromachining technology in many papers [3]. MEMS capacitive series switch is the DC power, and the function of switch is realized by metal bridge two states which are Down-state and Up-state. It is used CPW (Coplanar Wave) structure for getting the high frequency (GHz). MEMS capacitive switches are fabricated with a metal dielectric-air gap-metal cross section as shown in Figure 1.The upper metal plate (also known as a bridge) can be actuated from an up state to a down state. In the up state, the plate is relaxed and the air gap is present between the dielectric and the upper plate, and the capacity between the plates with the signal line is so little, the resistance is not affected by the capacity, the signal is transferred. In the down state, an electrostatic force is applied by an external control voltage to the upper plate causing it to collapse and eliminating the air gap between the dielectric and upper plate; the capacity between the plates with the signal line is so large, the resistance is affected by the capacity, and the signal is truncated. A cavity of substrate is etched for costing the loss of radiation as figure 1.

The actuation voltage of MEMS switch is determined by equating the electrostatic force of the applied voltage to the mechanical restoring force of the beam. This is dependent on the spring constant, k, for a fixed-fixed bridge as shown in Figure 1

$$K = \frac{32Et^3w}{l^3} \qquad (1)$$

Where E is Young's modulus, t is the bridge thickness, w is the bridge width, and l is the bridge length over which the electrostatic force is applied. In the case of the switch shown in Figure 1, the electrostatic force is applied over the two ground planes on either side of the center conductor, denoted by $l/2$. The actuation voltage is given by [4]

$$V_{sw} = \sqrt{\frac{2k}{\varepsilon_0 A} g^2 (g_0 - g)} \qquad (2)$$

Where $g0$ is the bridge height in the up state (at 0V bias), g is the current beam height, and $A = wl$ is the area of the bridge that overlaps the lower-ground plane. Equation (2) shows that there are two possible voltages for every bridge height. This is due to the instability that occurs when the electrostatic force exceeds the restoring force. At this point, the bridge pulls down. This instability occurs at a bridge height of [4].

$$g = \frac{2}{3}g_0 \qquad (3)$$

Knowing the bridge height, and substituting for k and A, the pull-down voltage can be determined [4]:

$$V_p = V_{sw|(2/3)g_0} = \sqrt{\frac{256Eg_0^3t^3}{27\varepsilon_0 l^4}} \qquad (4)$$

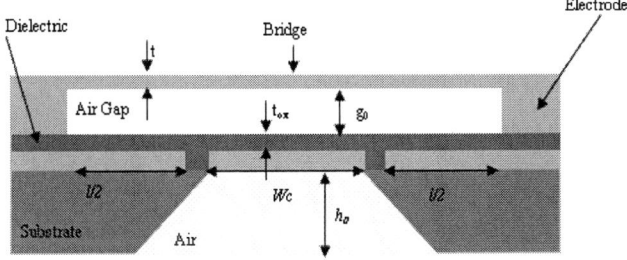

Figure 1: MEMS capacitive switch cross

The pull-down voltage must not exceed the breakdown voltage of the dielectric layer, therefore limiting the thickness and type of the dielectric material. MEMS switches are based on a process developed at Carleton University and built in Carleton's Microelectronics Fabrication Lab [5]. Dimensions of the MEMS switch are listed in Table 1. The trapezium cavity is the 1000μm×500μm×120μm

Table 1: Dimensions of the MEMS switch

Parameter	Dimension	Parameter	Dimension
Thickness of bridge, t	2μm	Dielectric thickness, t_{ox}	120μm
Width of bridge, W	290μm	Signal conductor width, W_c	120μm
Length of bridge, l	400μm	Thickness of substrate, h_0	500μm
Bridge height at 0V bias, g_0	3μm	Bottom border length of trapezium	3000μm

The pull down voltage is 3.3V by formula (1) ~ (4), if bridge material is golden. The switch is realized by validation

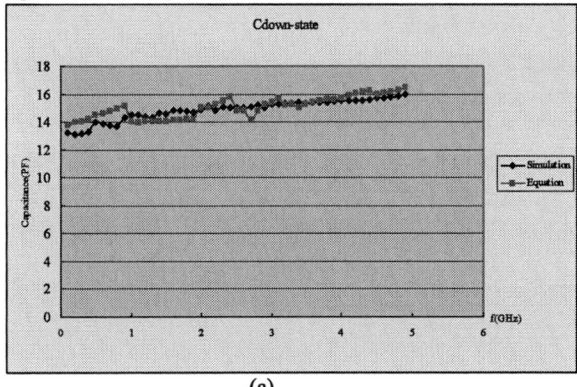

(a)

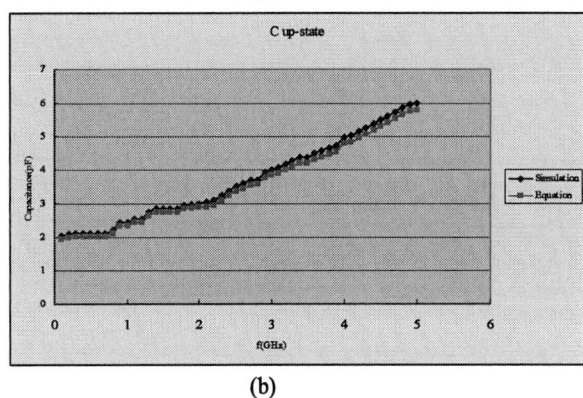

(b)

Figure 2: MEMS switch capacitance calculate and simulation; (a) Down-state; (b) Up-state

of capacitance, when down state, the capacitance value is large, signal is cut off; contrarily, it's transferred. The capacitance value is got by formula in paper [6], as following figure2 with simulation of EDA HFSS software and equation result. The insert loss is 2dB and isolation is 50dB by simulation.

3 Dual-band MEMS capacitive switch PA design

In this paper, the main goal of this PA is to increase efficiency at lower power. Other methods for increasing efficiency at low power include dynamically adjusting the power supply voltage [7, 8] or using parallel amplifiers (such as the Doherty configuration) [9].

Dual-band PA is realized by MEMS capacitive switch, because the switch has two states corresponding to two capacitances, it is realized dual-band filter using it. If choice wide-band amplifier CMOS device, and dual-band amplifier is gotten with configuration MEMS switch. Design circuit shows as figure 3, use the MEMS switch as dual-band filter on the RF input front-end. The circuit model come from the paper [4], the input transistor, M1, is matched with a lowpass network formed by Lg and Cg to limit harmonic content. The source bond wire inductance is modeled with Lbondwire. This is actually formed of several bond wires in parallel to lower the inductance and to provide heat sinking to the die. DC blocking is provided by the bypass capacitors, C_{DC}. The π network formed by C_{MEMS1}, C_{MEMS2}, and L_{match} is a variable matching network that adjusts the 50Ω load, RL, to the optimal load impedance. DC VDD is 3.3V, and it supplies the MEMS switch and transistor, adds the dual-band filter to the front-end of RF for realizing the dual-band in this paper. Choice the L=20nH, and the switch Up-state, the capacitance is about 2.87PF, when down state, C=14PF. So we can get the resonator frequency f1=2.1GHz (Up) and f2=1.9GHz (down), and get the dual-band signal input and amplifier.

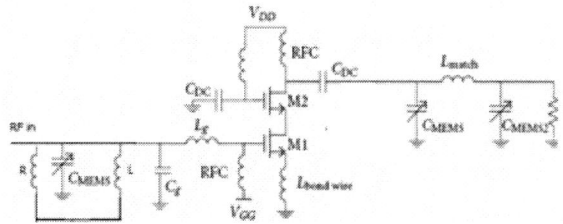

Figure 3: Dual band MEMS
PA circuit

It's so import for choice the amplifier device M1 and M2, because need realized the wide-band amplifier, need choice the wide-band amplifier of covering 1.9~2.1GHz. This paper choice the An MWT-871 GaAs MESFET transistor [10] was used to implement the Class-AB RF PA. The circuit is simulated by ADS software as figure 4. The gain is 12.4dB and 11.5dB with 2.1GHz and 2.3GHz and Output is about 25dBm. The amplifier characteristic and

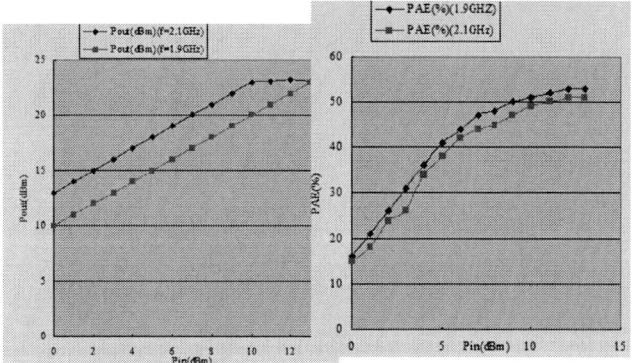

Figure 4: Dual band MEMS PA Simulation
for Gain and PAE

PAE is figure 4.

Simulations compare the performance of the class AB PA with a fixed and variable matching network. The Pout graph in Figure 4 shows minimal difference frequency band. The design goal of 25dBm output power is achieved at an input power of 12 dBm for a gain of 13 dB at 1.9GHz and 2.1GHz. The power added efficiency (PAE) graph shows the expected increase in efficiency. At Pin = 0 dBm, there is a 15%

increase in PAE Even at Pin =10 dBm, there is 52%. The high efficiency and dual-band are realized.

4 Conclusions

A design procedure for a MEMS capacitive switch, emphasizing integration in a standard IC design, has been presented at first. Then use a filter structure with capacitive MEMS switch to realize the dual-band function of PA. Then the PA is simulated by ADS software. The PA is designed in 0.18μm CMOS to operate at 1.9GHz and 2.1GHz with forward gain of about 12 dB by simulation result. Its PAE is 50% when input 10dBm.But this paper only design the circuit and simulation, need more working to do next if require realize the structure application:

(1) Study the MEMS switch technical of machining.

(2)MEMS integrated with other circuit.

(3)Improving the PA's efficiency for changed the new circuit topologies.

Reference

[1] John Danson, Calvin Plett, and Niall Tait. Using MEMS Capacitive Switches in Tunable RF Amplifiers [J]. EURASIP Journal on Wireless Communications and Networking, Vol.27 (12):1~9.

[2] G. M. Rebeiz and J. B. Muldavin, RF MEMS switches and switch circuits, IEEE Microwave Magazine, 2001,Vol. 2(4) pp. 59~71.

[3] L.Dussopt and G.M. Rebeiz, Intermodulation distortion and power handling in RF MEMS switches, varactors, and tunable filters, IEEE Transactions on Microwave Theory and Techniques,2003, Vol. 51, No. 4, part 1:1247~1256.

[4] John Danson, Calvin Plett, and Niall Tait, Using MEMS Capacitive Switches in Tunable RF Amplifiers, EURASIP Journal onWireless Communications and Networking Volume 2006, Hindawi Publishing Corporation., P:1~9,

[5] J. Rose, L. Roy, and N. Tait, Development of a MEMS microwave switch and application to adaptive integrated antennas,in Proceedings of the IEEE Canadian Conference on Electrical and Computer Engineering (CCECE '03), Montreal, Canada, May 2003, Vol.3:1901~1904.

[6] Dong Qiaohua , Liao Xiaoping. Analysis of Pull In Voltage of RF MEMS Switches, Journal of Semiconductors, Vol.29 (1) :165~167.

[7] L.-H. Lu, H.-H. Hsieh, and Y.-S. Wang, A compact 2.4/5.2-GHz CMOS dual-band low-noise amplifier, IEEE Microwave Wireless Components Letters, 2005, Vol.15, No.10:685~687.

[8] G. Hanington, P.-F. Chen, P. M. Asbeck, and L. E. Larson, "High-efficiency power amplifier using dynamic powersupply voltage for CDMA applications," IEEE Transactions on Microwave Theory and Techniques, 1999,Vol. 47, No.8:1471~1476.

[9] B. Sahu and G. A. Rincon-Mora, A high-efficiency linear RF power amplifier with a power-tracking dynamically adaptive buck-boost supply, IEEE Transactions on Microwave Theory and Techniques, 2004, Vol. 52:112~120.

[10] F.Wang, A. Ojo, D. Kimball, P. M. Asbeck, and L. E. Larson, Envelope tracking power amplifier with pre-

distortion for WLAN 802.11 g, in IEEE MTT-S Int. Microwave Symp. Dig., 2004: 1543~1546.

Biographies

ZHAO Ji-de (1951-), male, currently, he is with the Department of Physics and Electrical Engineering, Ludong University, China, where he is a professor and director. His research focuses on micro-electronics, MEMS etc.

LI Ying-liang(1979-), male, currently, he is with the Hangzhou Motorola R&D center, Hanzhou city, China, where he is a senior RF engineer. His research focuses on mobile communications.

Optical Analysis of A 3W Light-Emitting Diode (LED) MR16 Lamp

Kai Wang, Sheng Liu*, Xiaobing Luo, Zhongyuan Liu, and Fei Chen
Division of MOEMS, Wuhan National Laboratory for Optoelectronics
No. 1037, Luo Yu Road, Wuhan City, Hubei Province, 430074，China
*Corresponding author, Email: victor_liu63@126.com, Telephone: 86-13871251668

Abstract

Optical analysis is critical to the evaluation of an LED MR16 lamp, especially when the lamp is still in its early stage of development and applications and when optimization is needed for making use of unique characteristics of LEDs. In this study, optical research of a 3W LED MR16 lamp was conducted. The experimental results demonstrated that the LED MR16 lamp's angle of full width at half intensity was about 30 degrees and the luminous efficiency reached as high as 53.5lm/W, which was accepted for the illumination requirements. Numerical simulation was also conducted. The feasibility of the numerical model was proven by comparison of the simulations with the experimental data. Through the simulation and the corresponding analysis, it was found that tested 3W LED MR16 had high optical efficiency (90.2%) into all angles, but had poor beam pattern design, especially for the low uniformity of the beam pattern and low optical efficiency of light into main beam (less than 60%). To improve the illumination performance, a modified non-imaging optical design method was suggested to enhance the uniformity of beam pattern, to enhance the optical efficiency into main beam and to reduce the optical system volume. A novel high power LED MR16 with high beam pattern uniformity and high optical efficiency was designed as an example.

1. Introduction

Theoretically, the light-emitting diode (LED) has many advantages, such as good reliability, long life, variable color and low power consumption. As the LED's lumen efficiency increases rapidly in recent years, LED has begun to play an important role in many applications, such as backlighting for cell phones and other LCD display, street lighting, and interior and exterior automotive lighting [1]. One typical general lighting product of LED is the LED MR16 lamp. MR16s are typically used as outdoor spotlights or accent lighting for part of a room, in restaurants, museums, or retail displays [2]. As shown in Figure 1, compared to the conventional halogen lamp based MR16 in use which count on the multifaceted reflector (MR) to project light onto the target surface, LED chips are unique in terms of their small size, their nearly two-dimensional shapes, and the possibility of using large numbers in an array. The resulting direct lighting characteristics, together with the possibility of various designs of reflectors and lens systems, could significantly enhance MR16 lamp's optical performance.

In order to promote the wide applications of the LED general lighting in general and MR16 lighting in particular, it is essential to conduct an analysis of the optical performance of some newly developed LED MR16s. This paper focuses on the optical analysis for a 3W LED MR16 existing in the market. It presents the problem statement, physical and numerical modeling, experimental work, and comparisons. The results are discussed in terms of the advantages and drawbacks of this design. To improve the illumination performance, a modified non-imaging optical design method is suggested and a novel high power LED MR16 with high beam pattern uniformity and high optical efficiency is designed as an example.

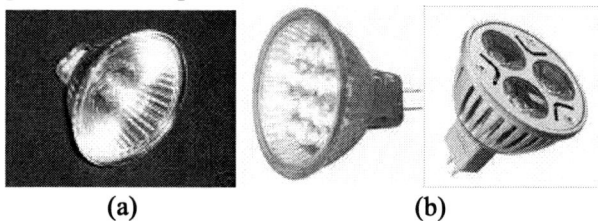

(a)   (b)

Figure 1. (a) A conventional halogen MR16; and (b) different kinds of LED MR16s.

2. Problem Statement

Shown in Figure 2 is a decomposed diagram of the 3W LED MR16 lamp, with power consumption 3 W and with LED chips distributed on the surface of the substrate and the heat sink. The lamp is mainly composed of five parts: 6 LEDs (0.5W for each), convergent lens array, MCPCB substrate, a lamp frame with heat sink and fins, and driving circuit for the power input of the LEDs. The objectives of this research are (1) to provide an optical analysis of the developed design, (2) to find out the possible flaws in design, and (3) to come up with possible improvements.

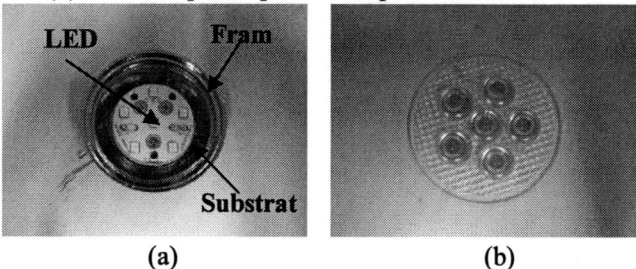

(a)   (b)

Figure 2. (a) A 3W LED MR16 lamp without lens; and (b) lens array of the 3W LED MR16 lamp.

3. Optics Experiments and Results

To analyze the optical performance of the 3W LED MR16 lamp, the optical metrics used to characterize the lamp with and without lens, such as total luminous flux, luminous efficiency, correlated color temperature, as well as color-rendering index were measured by UV-VIS-near IR spectrum photo colorimeter measurements, integrating sphere and illuminometer. As shown in Table 1, the experimental results demonstrate that the LED MR16 lamp's luminous efficiency and color-rendering index reach as high as 53.5lm/W and 65.9 respectively, which are accepted for the illumination

978-1-4244-2739-0/08/$25.00 ©2008 IEEE   456

Table 1. Optical performance of the 3W LED MR16

	Total Luminous Flux (lm)	Luminous Efficiency (lm/W)	Correlated Color Temperature (K)	Color-Rendering Index	Lens Efficiency
3W LED MR16 without lens	190.3	59.0	5227	66.1	—
3W LED MR16 with lens	171.6	53.5	5152	65.9	90.2%

requirements. From comparing the experimental results of the two test conditions, the 3W MR16 lamp with and without lens, we can find that total luminous flux and luminous efficiency decrease obviously when adding lens array above the LEDs, which is probably caused by total internal reflector loss and Fresnel loss at the lens's surface, and lens material absorption loss. Although different energy loss occurs during light's propagation in the lens, the lens efficiency reaches as high as 90.2%, which is much higher than that of some other LED illumination products existing in the market. The light intensity distribution curve of the MR16 lamp was also measured (Figure 3), and the angle of full width at half intensity is about 30 degrees, which is useful when validating the numerical model mentioned later in this paper.

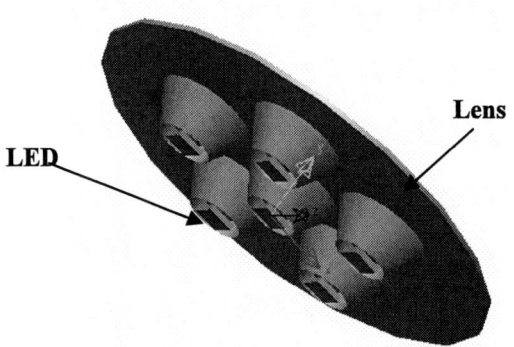

Figure 4. Optical model for the 3W LED MR16 lamp.

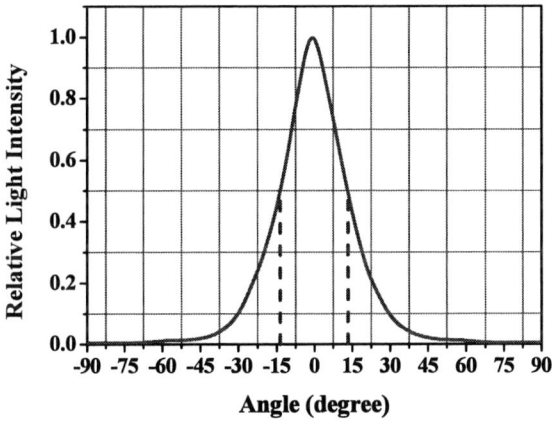

Figure 3. Light intensity distribution curve of the 3W LED MR16 lamp.

$$NCC = \frac{\sum_x (A_x - \overline{A})(B_x - \overline{B})}{\left[\sum_x (A_x - \overline{A})^2 \sum_x (B_x - \overline{B})^2 \right]^{1/2}}$$

where A_x (B_x) is the simulation intensity or irradiance (the experimental) value of the and $\overline{A}$ ($\overline{B}$) is the mean value of A (B), which is different from each other with the changing angle value along the x axis.

The modeling algorithm is summarized in Figure 5, where the threshold value T of the NCC may vary from one case to another, depending on the applications [4]. If the NCC is below a threshold value, the parameters, such as the lens material's index and surface scattering characteristics, can be adjusted.

As shown in Figure 6, according to the modeling algorithm, we changed the lens's scattering characteristic and the value of the index, and adjusted them many times until the NCC was as large as 99.5%. Thus, the precise optical modeling for single LED module is finished, and we can do further simulation for the whole LED MR16 lamp.

4. Optical Simulation and Analysis of Results

4.1 Precise Optical Modeling for the LED MR16 lamp

We simulated the LED MR16 numerically by the widely used Monte Carlo ray tracing method. Firstly, we made some simple measurements to determine the geometrical parameters of the LED MR16 optical module and estimated the original material characteristic of the lens (Figure 4). We obtained the emitted rays by using Monte Carlo ray tracing from the LED source. Then one million rays were emitted from the LED source, and the light intensity distribution was obtained. To quantify the similarity between the simulation light intensity distribution and the experimental measurement, the normalized cross correlation (NCC) is applied [3]. The NCC is written as:

4.2 Optical Modeling and Analysis of the Lens Design

The whole 3W LED MR16 lamp was also simulated. Figure 8(a) shows the simulation illuminance distribution on a test area 3 meters wide and 3 meters high that is 2 meters away from the lamp. As shown in Figure 9, the illuminance at the center of the beam pattern is much higher than that of the rest beam pattern and the illuminance decreases fast from the center of the beam pattern to the edge, which results in low uniformity of the beam pattern. The poor beam pattern distribution on the test target also results in a lot of lights irradiating out of the desired space angle. From Figure 8(a) we can find that the optical efficiency of light into main beam is also very low (less than 60%), which means more than 40% of the light exiting from the lens surface doesn't irradiate on the desired receive target and is wasted. This poor illumination performance is mainly caused by the poor design of the lens, which couldn't control irradiation direction of lights effectively. Most lights irradiate from the LED, especially for the lights generate at the center of LED chip, should be accurately designed to irradiate at the desired points on the target plane. Therefore, a modified lens design method is needed to control lights much more accurately.

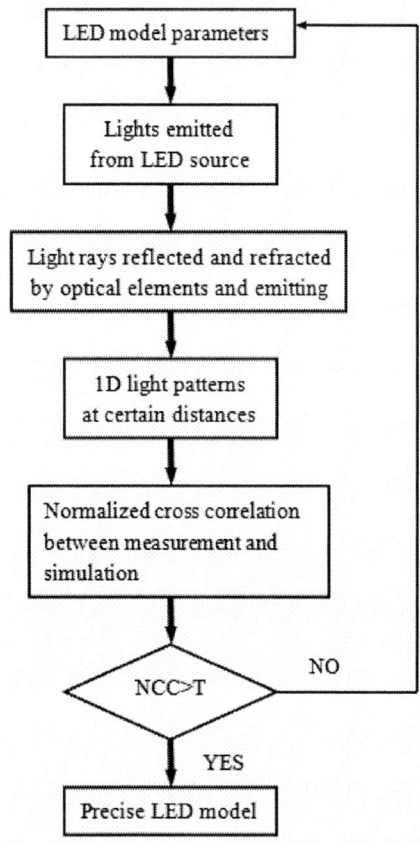

Figure 5. Modeling algorithm for a LED model.

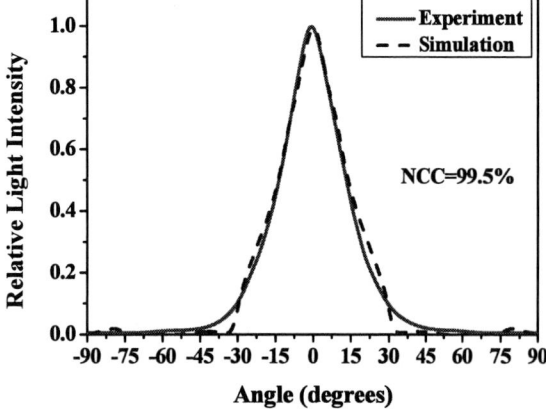

Figure 6. Simulation light distribution curve versus experimental measurement for the 3W MR16 LED lamp.

5. Improvements

5.1 A Modified Non-imaging Optical Design Method

To improve the illumination performance, a modified non-imaging optical design method is suggested to enhance the uniformity of beam pattern, to enhance the optical efficiency into main beam and to reduce the optical system volume. The method is briefly described as follows: Firstly, we divide space distribution of light energy of LED source and illumination target into several energy cells, and build the energy mapping relationship between these cells [5]. Secondly, calculate the coordinates and normal vector of each point on the freeform surface according to the energy mapping relationship, Edge Ray principle and Snell law. Thirdly, we construct freeform optics using the points obtained in the second step. Finally, we validate freeform lens or surface design by ray trace simulation. Based on this method, random shape uniform beam pattern, such as circularity, rectangle, etc., could be designed, especially for the small size light source design cases.

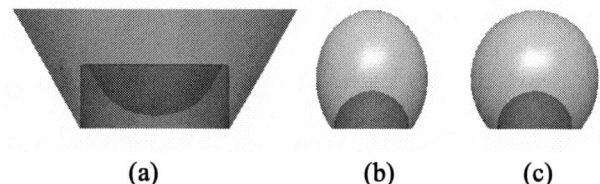

(a) (b) (c)

Figure 7. (a) Single lens of the 3W LED MR16 lamp; (b) novel 30 degrees LED MR16 lamp 's lens; and (c) novel 60 degrees LED MR16 lamp's lens.

5.2 A Novel LED MR16 Lamp Design and Simulation Results Analysis

A novel 3W LED MR16 lamp was designed as an example. To compare with the 3W LED MR16 lamp analyzed above, we used the same kinds of LEDs as the light source and the same degree of lamp's light distribution angle as the design target. As shown in Figure 7(b), the volume of the novel 30 degrees LED MR16 lamp's single lens is only about one third of that of the 3W LED MR16 lamp mentioned above, which could provide much more space for heat dissipation design in the size confined MR16 lamp, and makes it possible for ultra high power LED MR16 integrated by multi-LEDs. From comparing Figure 8(a) with Figure 8(b), we can find that the novel LED MR16 lamp lens has much better light control ability than the conventional lens. More than 85% lights exiting from the lens surface irradiate within the desired receive target, which is much higher than 60% of the MR16 lamp mentioned above, and this optical efficiency could be enhanced further more by optimization. As shown in Figure 9, the novel MR16 lamp has a higher uniformity illuminance distribution across the target, especially for the central illumination area and the MR16 lamps with large distribution angles.

However, obvious light rings still exist at the edge of the beam pattern, which mainly caused by two main reasons as follows: 1) Some lights, especially for those with large mergence angles, couldn't irradiate to the desired points on the target only by one refracted optical surface and diverge, which mainly caused by the limits of the largest deflection

458

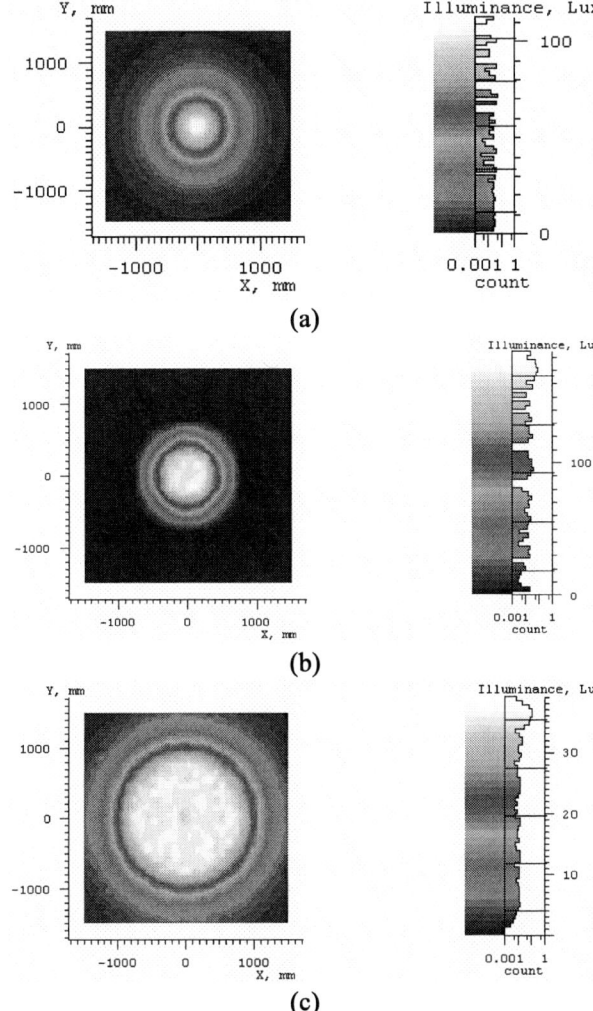

(a)

(b)

(c)

Figure 8. Simulation illumination performance of different LED MR16 lamps on a 3m×3m target in 2 meters away: (a) the 3W LED MR16 lamp; (b) novel 30 degrees LED MR16 lamp; and (c) novel 60 degrees LED MR16 lamp.

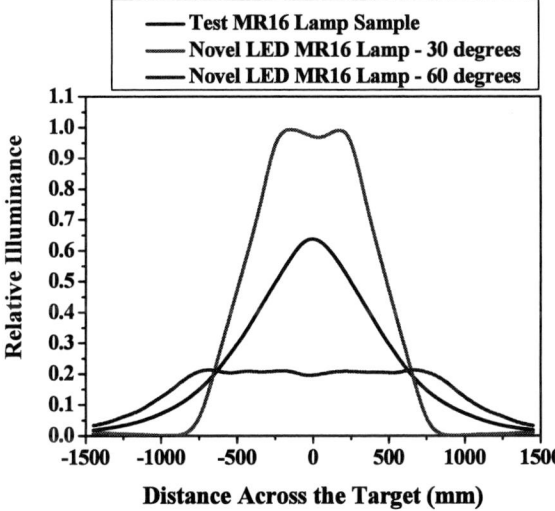

Figure 9. Different kinds of LED MR16 lamps' relative illuminance distributions on the target.

angles of lights refracted by the optical surface. To control large mergence angle lights more effectively, two or more optical surfaces need be designed simultaneously in the future work. 2) The LED light source is simplified to point light source during the freeform optics calculation. However, actually, LED is an extended source with the length of 0.6mm and width of 0.6mm, and lights generate at the edge of the LED couldn't irradiate at the desired points on the target, which also results in light rings. Considering more than one point light source distributed at different positions of LED simultaneously during calculation is one of the possible solutions to the extended light source problem.

6. Conclusions

In this study, optical analysis, including both models and experiments, of a 3W LED MR16 lamp was conducted. Experiment results demonstrated that the illumination design of this lamp was acceptable for the illumination requirements. Numerical simulation was also conducted. The feasibility of the numerical model was proven by comparison of the simulations with the experimental data. Through the simulation and the corresponding analysis, it was found that tested 3W LED MR16 had high optical efficiency (90.2%) into all angles, but had poor beam pattern design, especially for the low uniformity of the beam pattern and low optical efficiency of light into main beam (less than 60%). To improve the illumination performance, a modified non-imaging optical design method was suggested. A novel high power LED MR16 with high beam pattern uniformity was designed as an example. The novel LED MR16 lamp lens's volume was only about one third of that of the 3W LED MR16 lamp and the optical efficiency into main beam reached as high as 85%, which were much better than the tested LED MR16 sample. However, obvious light rings still existed at the edge of the beam pattern, which probably could be reduced by designing two or more optical surfaces and point light sources simultaneously in the future work.

Acknowledgments

The authors would like to acknowledge the support of Science and Technology Department of Guangdong Province and Guangdong Real Faith Enterprises Group Co., Ltd.

References

1. Alan, M., "Solid state lighting-a world of expanding opportunities at LED 2002", *III-Vs Review*, Vol. 16, No.1 (2003), pp.30-33.
2. Brown, D., Nicol, and D., Ferguson, I., "Investigation of the spectral properties of LED-based MR16 bulbs for general illumination", *Optical Engineering*, Vol. 44 (11) (2005).
3. Sun, C. C., Lee, T. X., Ma, S. H., Lee, Y. L., and Huand, S. M., "Precise optical modeling for LED lighting verified by cross correlation in the midfield region", *Optics Letters*, Vol. 31, No. 14 (2006), pp. 2193-2195.
4. Wang, K., Luo, X. B., Liu, Z. Y., Zhou, B., Gan, Z. Y., and Liu, S., "Optical analysis of an 80-W light-emitting-diode street lamp", *Optical Engineering*, Vol. 47 (1) (2008).

5. Parkyn, W. A., "Illumination lenses designed by extrinsic differential geometry", *Proceedings of SPIE*, Vol. 3428 (1998), pp. 389-396.

Research on the Cascaded Inverters Based on Simplex DC Power Source

XIN yibo[1][2],   CHEN wenqing[2] , FANG huajing[1]

(1. HuaZhong University of Science and Technology , Wuhan 430074, China

2. Luoyang Institute of Science and Technology, Luoyang 471003, China)

Abstract

Presents a multi-level cascaded inverter with on single DC power source and give a generalized modulation method in this paper. Simulation and experiment shown this cascaded inverter needs less numbers of full-bridge unit than conventional cascaded topology to get same output levels. This cascaded converter can save a number of switching devices and separated powers source. The step modulation method makes harmonic distortion of output voltage being less than conventional cascaded converter's.

Key words

cascaded inverter；single DC power source；step modulation

1. INTRODUCTION

Multi-level inverter has been widely researched in high power level application with high voltage output. Power energy with characteristic of high capacity high quality can be achieved by this type of inverter, in which relatively small-capability low-voltage switches are adopted. So this technique has been widely concentrated in such application as medium-high voltage transducer and neatly AC transmission system[1].

The general function of the multilevel inverter is to synthesize a desired AC voltage from several levels of DC voltages. A multi-level converter was presented in which the two separate DC power sources were the secondary of two transformers coupled to the utility AC power[2]. Only one power source is used without the use of transformers in this paper. The interest here is interfacing a simplex DC source with a cascade multi-level inverter where the other power sources are capacitors. Each phase of a cascade multi-level inverter's amount is n for 2n+1 levels in applications that involve real power transfer. In this paper, a plan is proposed that premits the using a simplex DC source with the remaining n-1 DC sources being capacitors. It is shown that a simplex DC source can maintain the DC voltage level of the capacitors and make choice of a fundamental frequency switching model to produce a resembled sine curve wave.

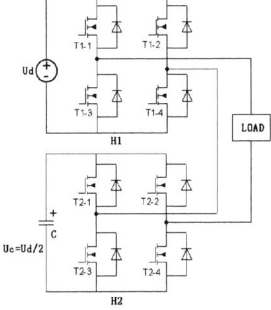

Fig.1 Topology of the multi-level
cascaded H-bridges inverter

2. MODELING AND EXAMPLED

To up build a multi-level cascade inverter with a simplex DC source, to use capacitors as the DC sources for others but the first one is indispensable . Discuss a simple cascade multi-level inverter with two H-bridges as shown in Fig.1. The DC source for the first H-bridge (H_1) is a DC source output voltage V_{dc}, while the second H-bridge (H_2) is a capacitor voltage to be charge up V_{dc} /2. The output voltage of the H_1 is denoted by V_1 and the H_2 is denoted by V_2 , so that the output of this two DC source multi-level cascade inverter is $v(t) = v_1(t)+v_2(t)$. By opening and closing the switches of H_1 appropriately, the output voltage V_1 can be made equal to $V_{dc},0,-V_{dc}$, while the output voltage of H_2 can be made equal to V_{dc} /2,0,-V_{dc} /2 by opening /closing its switches appropriately. Therefore, the output voltage of the inverter can keep the values $3V_{dc}/2,V_{dc}$, $V_{dc}/2,0,-V_{dc}/2,-V_{dc}$,-$3V_{dc}/2$, which is seven-levels portray in Fig. 2.

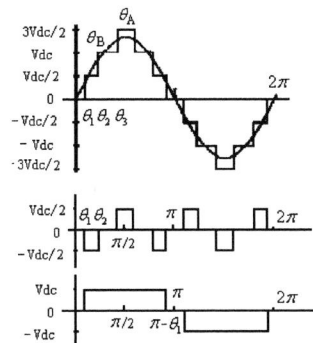

Fig. 2 seven-level cascaded staircase
waveforms of inverter

We can see in Fig. 2, for$\theta_1 \leq \theta \leq \theta_2$, $V_1=0$ and $V_2=V_{dc}$ /2 is chosen. The fact that the output voltage level V_{dc} /2 can be achieved in two different ways is exploited to keep the capacitor voltage regulated. Specifically, one measures the capacitor voltage V_c and the inverter current i [3]. If $V_c< V_{dc}$ /2 and $i>0$, one sets$V_1=V_{dc}$ and $V_2=V_{dc}$ /2 then capacitor is being charged.   Table 1 shows how a waveform can be generated using the topology of Topology of the multi-level cascaded H-bridges inverter.

Table 1   Output voltages of a
seven-level converting circuit

θ	System State	V_1	V_2	V_1+V_2
$0 \leq \theta \leq \theta_1$	$V_c<V_{dc}/2\ i>0$	0	0	0
$\theta_2 \leq \theta \leq \theta_2$	$3V_c<V_{dc}\ i<0$	0	V_{dc} /2	$V_{dc}/2$
$\theta_1 \leq \theta \leq \theta_2$	$V_c>V_{dc}/2\ i>0$	V_{dc}	0	V_{dc}
$\theta_1 \leq \theta \leq \pi/2$	$V_c>V_{dc}/2\ i<0$	V_{dc}	V_{dc} /2	$3V_{dc}/2$

By choosing the charge value of the capacitor voltage to be one half that of the DC source, the nominal values of the levels are equally spaced. However, this is not required. The criteria required for this capacitor regulating scheme is that (1) the desired capacitor voltage is less than the power source voltage;(2) the capacitance value is chosen large enough so that the variation is small;(3) the capacitor charging cycle is greater than the capacitor discharge cycle[4].

3. CONTROL STRATEGIES

Approach sine curve wave with the stairs wave, give or get an electric shock more many even, the frequency chart characteristic is more well[5]. This kind of strategy the biggest advantage is in brief a realization, the switch frequency is lowest, the switch exhausts small, but this for big power situation is importance. The step wave makes creation exportation wave form principle, such as shown in Fig.3.

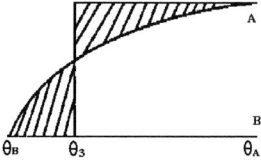

Fig. 3. The step wave produces principle

Fig. 2 medium the calculation of $\theta_1, \theta_2, \theta_3$ carry on according to low harmonic minimal principle, so request figure 2 exportation wave form areas to be equal to sine wave area, namely Fig. 2 medium the shadow area is equal. Because Fig. 2 medium θ_A and θ_B can establish an amount to compute according to the H bridge mold piece, so θ_3 can compute to get with the equation (1).

$$\int_{\theta_B}^{\theta_3}(\sin\theta - B)d\theta - \int_{\theta_3}^{\theta_A}(A - \sin\theta)d\theta = 0 \qquad (1)$$

solve above equation, we have:

$$\theta_3 = \frac{A\theta_A - B\theta_B + \cos\theta_A - \cos\theta_B}{A - B} \qquad (2)$$

Take the $\theta_A = arcsinA$, $\theta_B = arcsinB$ into(2), we have Take the θA=arcsinA, θB=arcsinB into(2), we have:

$$\theta_n = n\theta_A - (n-1)\theta_B + N(\cos\theta_A - \cos\theta_B) \qquad (3)$$

where, $n=1,2,3$, $A=n/N$, $B=n-1/N$. N is an amount of the H bridge, here $N=2$.

4. EXPERIMENTAL RESULTS

If the nominal capacitor voltage is chosen as $V_{dc}/2$, then one can compute the switching angles θ_1, θ_2, andθ_3. Following the development, the Fourier series expansion of the output voltage waveform of the multilevel inverter as shown in Fig. 2 is[6]:

$$\frac{U_{AN}}{V_d/2} = 2NM\cos(\omega_i t) +$$

$$\sum_{n=1}^{\infty}(\frac{4}{n\pi})\cos(\frac{n\pi}{2})\sin[\frac{Mn\pi}{2}\cos(\omega_1 t)]$$

$$[\cos(n\theta_{c1}) + \cdots + \cos(n\theta_{ck})]\cos(n\omega_c t) \qquad (4)$$

$$+ \sum_{n=1}^{\infty}(\frac{4}{n\pi})\cos(\frac{n\pi}{2})\sin[\frac{Mn\pi}{2}\cos(\omega_1 t)]$$

$$[\sin(n\theta_{c1}) + \cdots + \sin(n\theta_{ck})]\sin(n\omega_c t)$$

Equation(4) expresses the relationship between n and harmonic.

when $n=1$、3、5、…,

$$\cos(\frac{n\pi}{2}) = 0, \frac{U_{AN}}{V_d/2} = 2NM\cos(\omega_1 t) \qquad (5)$$

harmonic content is zero.

$$\frac{U_{AN}}{V_d/2} = 2NM\cos(\omega_i t) +$$

$$\sum_{n=1}^{\infty}(\frac{4}{n\pi})\cos(\frac{n\pi}{2})\sum_{n=1}^{\infty}(-1)^{\frac{k-1}{2}}J_k(\frac{Mn\pi}{2}) \qquad (6)$$

$$[A\cos(nk_c \pm k)\omega_1 t + B\sin(nk_c \pm k_{c1})\omega_c t]$$

where, $A=cos(n\theta_1)+cos(n\theta_2)+ \dots +cos(n\theta_N)$, $B=sin(n\theta_1)+sin(n\theta_2)+ \dots +sin(n\theta_N)$.

Because $n=2$、4、6、…,when $B=0$ and $n<2N$,then $A=0$. The output voltage U_{AN} not any longer contains harmonic $2Nkc\pm1$order or lower, and contains $2Nkc\pm1$ order or higher just only.

Figure 4 shows the output voltage waveforms of seven-level cascaded staircase with sinusoidal reference. f_0=50Hz, the modulation ratio M=0.9 and the carrier/fundamental frequency ratio k=30.As the carrier frequency can be expressed as $f_c= Kf_o$= 1500Hz, the equivalent switching frequency of the seven-level inverter is[7]: $f_{eq}=2nf_c=2\times7\times1500$= 21KHz, the dominant harmonic should be centered around this frequency. Lower order harmonic component are cancelled with the method. From the harmonic spectrum we can see the amplitude about 1%~2% of some harmonic appearing at 385^{Th}~415^{Th} harmonic (see Fig. 5). The corresponding frequency is 19.75KHz~21.25KHz, just around the equivalent switching frequency. The simulated results agree with the theoretical analysis. As the reactance of the motor winding is in proportion to frequency, then these high order voltage harmonic can hardly generate current harmonic.The amplitude of current harmonic are absolutely negligible, and hence the current THD is within 1%.The inductance of the winding here works as low pass filter.

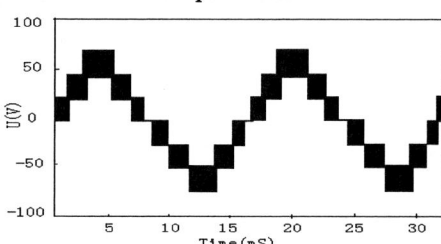

Fig. 4 Output voltage waveforms of the inverter

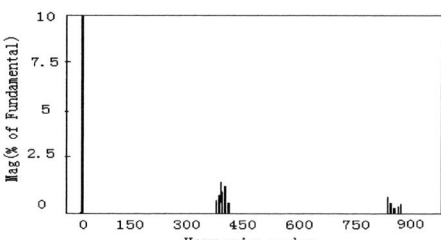

Fig.5 Output voltage harmonic analysis

5. CONCLUSIONS

Multilevel converters, due to advantages of high power quality waveforms, low electromagnetic interference and high voltage capability, are the research hot points, especially in their corresponding topology and control. A multi-level cascade inverter topology has been proposed that requires only a simplex DC power source. The output waveform is simply dictated by the particular set of switching angles computed for that value of modulation index which give the smallest THD. Experimental results were in close agreement with the predicted results. Subject to specified constraints, it was shown that the voltage level of the capacitors can be controlled while at the same time choosing the switching angles to achieve a specified modulation index and eliminate harmonics in the output waveform. Experimental results indicate that the proposed technique is effective the reducing harmonics in multilevel cascaded inverters. With the development of the further research on multi-level technology and the development of DSP skills, high-capacity and high-voltage switch skills, optical communication and isolation technology, and etc., the multi-level inverter will be further widely used.

References

1. CHEN Guo-cheng , Novel Power Electrnic Converting Technology, China Electric Power Press (Beijng, 2004)，pp96~104.

2. Lai J S ,“ MuliL-lever lnverter a Survey of Topologies Controls and Applicaions，” *IEEE Trams on Ind Elect*,Vol.49 No.4(2002),pp.724-738.

3. Dongsheng Zhou, D.G Rouaud "Experimental comparisons of Neutral-Point Balancing Strategies for Three-Level Topology," *IEEE Trans. On Power Electronics*, Vol.16, No.6(2001), pp. 872-879.

4. Kouki Matsuse et al, "DC Voltage Control Strategy for a Five-level Converter," *IEEE Proc.of PESC'99*, pp. 521-527.

5. M.N.Ratnayake, Y.Murai,T.Watanabe "Novel PWM Scheme to Control Neutral Point Voltage Variation in Three-level Voltage Source Inverter," *IEEE Industry Application Con. Of 1999*, Vol.3, pp .1950-1955.

6. Qiang Song, Wenhua Liu, Qingguang Yu et all,“A Neutral-Point Potential Balancing Algorithm for Three-Level NPC Inverters Using Analytically Injuected Zero-Sequence Voltage," *Applied Power Electronics Con. Of APEC'2003*, Vol.l,(2003),pp.228-233.

7. Tolbert L M, Peng F Z, Cunningham T, "Charge balance control schemes for cascade multilevel converter in hybrid electric vehicles," *IEEE Trans. on Ind. Electron*, 49(4) (2002),pp.1058-1064.

8. A.R.beig,V T.Ranganathan, "Space Vector Based Bus Clamped PWM Algorithms for Three-Level Inverters: Implementation, Performance Analysis and Application Consideration,"*IEEE Proc.of APEC*'03,Vol.1(2003), pp.569-575.

Simulation on Thermal Characteristics of LED Chips for Design Optimization

Ting Cheng[1], Xiaobing Luo [1,2]*, Suyi Huang[1], Zhiyin Gan[2,3], Sheng Liu[2,3]

[1] School of Energy and Power Engineering, Huazhong University of Science & Technology, Wuhan, China, 430074
[2] Wuhan National Lab for Optoelectronics, Huazhong University of Science and Technology, Wuhan, China, 430074
[3] Institute for Microsystems, Huazhong University of Science and Technology, Wuhan, China, 430074
*Corresponding author: Xiaobing Luo, Telephone: 86-13971460283, Fax number: 86-27-87557074, Email:
luoxb@mail.hust.edu.cn

Abstract

As thermal performance is importance for high-power LED devices, there exists a need to build a validated model to clarify the thermal transfer mechanisms in the LED chip in terms of the chip materials and structures. High-power LED was numerically investigated using the finite element method. A series of substrate materials with different thermal conductivity and thicknesses were studied. The impact of varying cooling capacities on chip junction temperature is also presented.

Introduction

Light-emitting diode (LED), comparing with conventional light sources, provides a direct transfer of electrical energy into light. LED technology has flourished for the past few decades. It has many distinct advantages, such as high efficiency, reliability, rugged construction, low power consumption and durability. The ramifications are significant and indeed the LED has been foreseen as an "ultimate lamp" for the future[1][2]. Typical applications include back lighting for cell phones and other liquid crystal display (LCD) displays, interior and exterior automotive lighting including headlamps, large signs and displays, signals and illumination. As the next generation lighting source, LED has attracted increasing attention in commercial and scientific research communities.

Junction temperature is an important indicative parameter about the whole LED device performance. Elevated junction temperature would cause light output reduction, accelerated chip degradation, increased thermal stress in its package, changed light color and decreased forward voltage.

Electric current goes through active layer. The inner quantum conversion is only about 15 %, 85% of the input power transfer into heat, then a great of heat generates in the active layer. Heat needs go through the bulk chip and reach heat spreading packaging part. As the chip size is only 1*1mm2 and its thickness is 1mm, heat flux has much difficulty to go through layers. At present time, chip driving current could be up to 750mA and heat flux has increased to $180 W / m^2$. Huge heat causes a big problem for the thermal management of the LED devices. To obtain a lower junction temperature, more and more efforts have been devoted to the LED chip design in recent years. Then different chip configurations and various substrate materials have been used to improve heat dissipation conditions. Nowadays there are mainly three structure types of chip configurations: p-side- up

chip (Fig. 1), vertical chip (Fig. 2) and flip chip (Fig. 3). Sapphire, Si, and SiC are used for the chip substrate to obtain better thermal conductivity. No matter which type of LED chip is, heat flux nearly reaches $5*10^6 W / m^2$. To satisfy the demand of harsh environments and long life, junction temperature should be below 120℃. It needs to spread heat out the chip quickly and effectively.

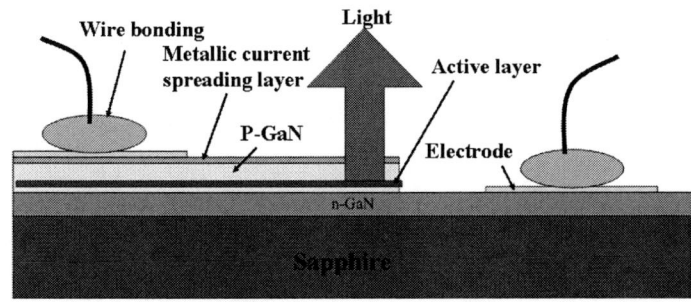

Fig 1. Schematic of P-side-up chip.

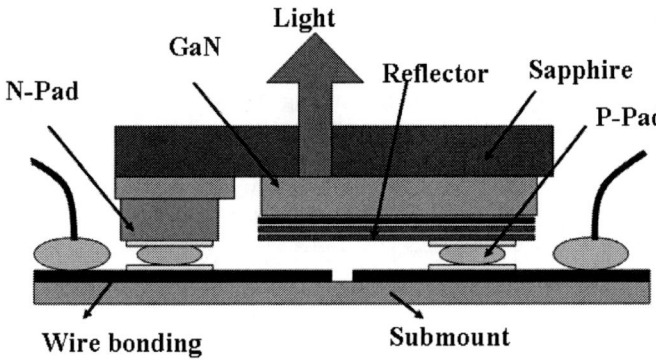

Fig 2. Schematic of flip chip.

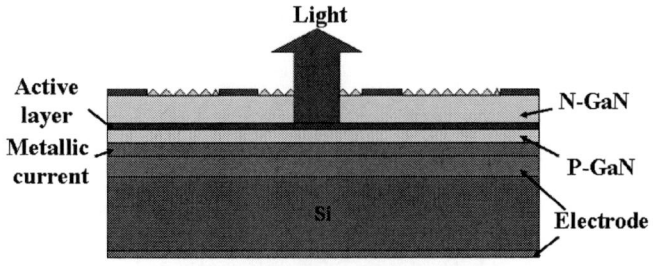

Fig 3. Schematic of vertical chip

Nowadays, sapphire, Si and SiC, are used as chip substrate materials. Substrate layer is the main part in chip to strengthen chip, which is about $90\,\mu m$. It is always the thickest part in the chip. Thermal resistance of this part is large. As its materials usually have poor heat transfer performance. Therefore the substrate is the main part to improve the chip thermal resistance. To clarify the heat transfer in chip, especially the influence of the chip substrate on heat transfer procession, and make sure the potentiality and the optimal effect of this part, some research results are needed.

Table 1. Thermal characteristic of substrate materials

Substrate Material	Thermal conductivity($W / (m\ k)$)
Sapphire	57
Si	157
SiC	400

Besides the chip itself, LED device cooling is another key point for heat transfer. Packaging technology and its materials contribute significantly to the cooling capability. There are various thermal designs for LED packaging modules, such as plate fin heat sink, heat pipe, micro-jet[4][5][6], micro-channel, pyroelectricity cooling. Its packaging substrate materials include Si, AlN, ceramics, Direct Bonded Copper (DBC), AlSiC Metal Matrix Composite (MMC), and low temperature co-fired ceramic (LTCC) [6]. Good packaging could lead to an excellent heat transfer performance and obtain a lower junction temperature. Considering the available packaging size and cost, there is still space limitation of cooling part. If we make the relationship between the heat transfer capability and the chip junction temperature clear, we will be able to provide relatively optimum design of LED chip and packaging..

Problem statement

The research focuses on the chip part and finite element numerical method is used for various simulations. Figure 4 presents the schematic structure in the simulation model. In simulation model, chips, interface material (TIM), sub-mount copper are all included. Chip configuration uses p-side-up chip which is more commonly used. We consider four layers in the structure: GaN (p-layer, active layer, n-layer), substrate layer, interface material layer (TIM), and sub-mount copper layer. Other parts of the whole LED module are simplified, and their thermal influence is considered in the bottom face.

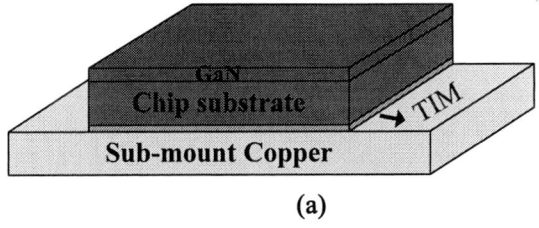

(a)

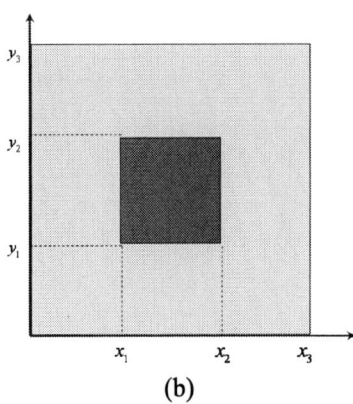

(b)

Fig 4. Schematic of the model.

The governing equation for the model used in the simulation model is Laplace's equation expressed as follows,

$$\nabla^2 T = \frac{\partial^2 T}{\partial x^2} + \frac{\partial^2 T}{\partial y^2} + \frac{\partial^2 T}{\partial z^2} = 0$$

which is subjected to a uniform flux distribution within the heat source area A_s. Outside the heat source area, and a convective or mixed boundary condition on the bottom surface along the edges of the plate, the following conditions are also required,

$$\left.\frac{\partial T}{\partial z}\right|_{z=0} = -\frac{Q / A_s}{k}$$

$$\left.\frac{\partial T}{\partial z}\right|_{z=t} = -\frac{h}{k}[T(x,y,t) - T_f]$$

$$\left.\frac{\partial T}{\partial x}\right|_{x=0,x_1,x_2,x_3} = 0$$

and

$$\left.\frac{\partial T}{\partial y}\right|_{y=0,y_1,y_2,y_3} = 0$$

Results and discussion

In section I, the relationship between LED junction temperature of the chip and thermal conductivity of substrate is the simulation purpose. Junction temperature is collected and the thermal conductivity of the substrate material is increased $50\,W / k$ each time. According to the data line in Figure 5, when the thermal conductivity is less than $200\,W / k$, thermal resistance has an obvious decrease. While the numerical data shows there is no obvious improvement to chip thermal resistance when the thermal conductivity of chip substrate is above $200\,W / k$. Nowadays, substrate materials of the LED chip include mainly three types: sapphire, Si, and SiC. As the thermal conductivity of SiC is close to $490\,W / (m\ k)$, which is much larger than $200\,W / (m\ k)$. Considering that the manufacture of the light emitting chip is complicated and costly, it is not necessary to pursue another higher thermal conductivity material for the chip substrate.

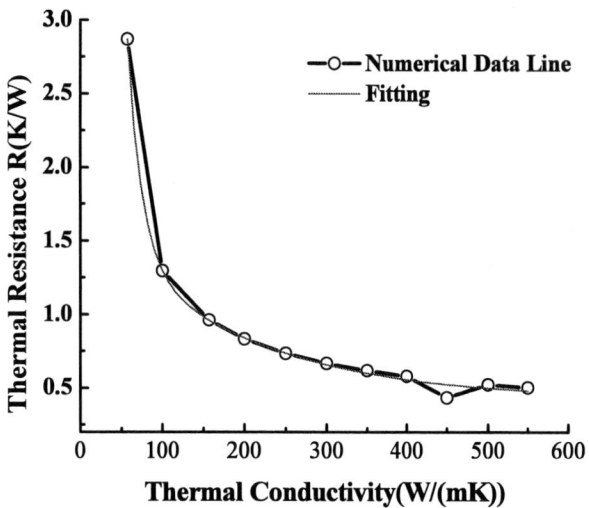

Fig 5. Relation between substrate material thermal conductivity and thermal resistance.

In section II, Sapphire and SiC are studied as the chip substrate material respectively. To analyze the relation ship between chip substrate thickness and the junction temperature, substrate thickness decreases with 10 μm each time.

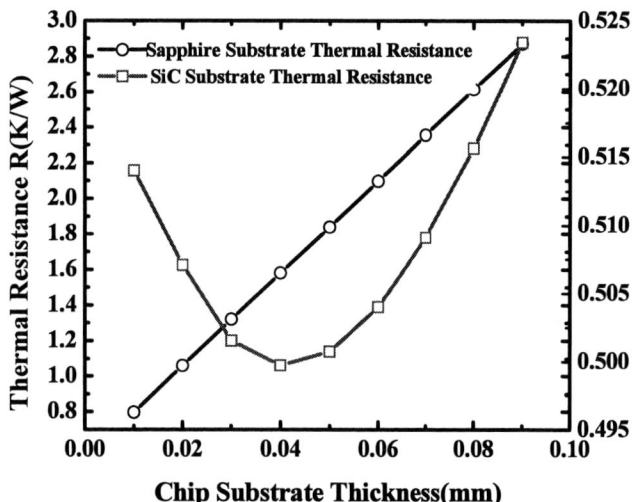

Fig 6. Relation between thermal conductivity and thermal resistance.

There are different results of the sapphire and the SiC substrate. As to sapphire substrate, with the thickness decreasing, thermal resistance of the module goes down linearly. It is confirmed the following relationship exists:

$$R = \frac{\delta}{\lambda}.$$

where δ is the thickness of layer, λ is the thermal conductivity of material. This formula shows that thermal resistance is linearly dependent on the material thickness.

As to SiC substrate, results are very different The relationship between the SiC substrate and the thermal resistance is nearly quadratic in function. There exists an extreme value of thermal resistance, when the substrate thickness is about 0.0439mm. As the thermal conductivity of

SiC is about $400 W / k$. There is much heat energy stored in the substrate. Therefore, its influence to heat transfer is much more significant than the substrate thickness change. The extreme value is caused by the thermal conduction and the thickness.

Junction temperature depends on the heat transfer capability of the packaging board and the cooling device. Equivalent heat coefficient on the bottom face in the module is an equivalent parameter, which counts for in the heat transfer of packaging interface, sub-mount material, and cooling part. In section III, to find the relationship between the coefficients and junction temperature, this coefficient on the bottom face is changed in simulation model.

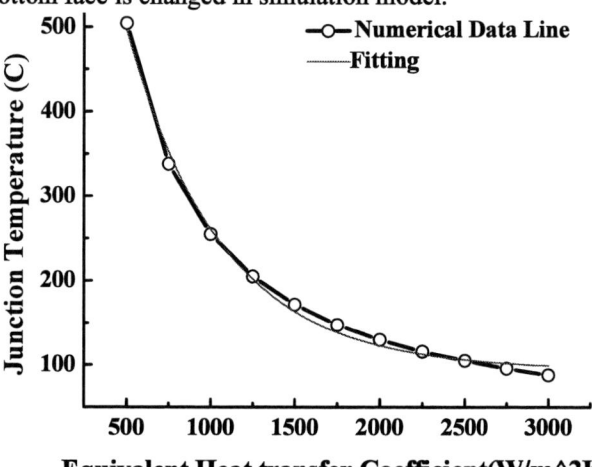

Fig 7. Relation between equivalent heat transfer coefficient and junction temperature.

In this model, the area of the bottom face is 4 mm2. When this coefficient is less than $2000 W / m^2 K$, the data decrease fast, therefore the cooling improvement is significant. Then the change in data became much smaller with the equivalent heat transfer coefficients growing larger, the improvement for chip junction is minimum. According to the analysis fitting line, an optimal coefficient does exist. Therefore there is no need to seek a material with a much better heat transfer capability when considering engineering applications.

Conclusions

During the design process, there is a universal thought that the larger the thermal conductivity of the substrate and the thinner the substrate is, and the lower the junction temperatures of the LED chip is. Therefore, there is an obvious trend in the LED chip design that researchers use the substrate with high thermal conductivity and thin thickness, and also prefer the chip packaging part with powerful cooling capability by using different board substrates and cooling methods. In this study, the simulations demonstrate:

a) Increasing thermal conductivity of the chip substrate can improve junction temperature. But there is no need to pursue much better thermal conductivity material, which has higher thermal conductivity than SiC ($490 W / mK$).

b) For different substrate materials, decreasing the thickness of the chip substrate has different influence on the junction temperature.

c) The capability of the cooling part is important to improve the chip junction temperature. While it needs to find a balance between the cost, manufacture and junction temperature when choosing the packaging design.

References

1. Holonyak, N., "Is the light emitting diode (LED) an ultimate lamp," Am. J. Phys., Vol.68, (2000), p.864-866.

2. Craaford, M .G., Holonyyak. M, and Kish. F.A., "In pursuit of the ultimate lamp," Scientific Amer, (2001) pp.83-88.

3. Lumileds Lighting, http://www.lumileds.com/technology (2004).

4. Hayes, D.J., Cox, W.R., "Micro-jet printing of polymers for electronics manufacturing", In, Adhesive Joining and Coating Technology in Electronics Manufacturing, Proceedings of 3rd International Conference, (1998), pp.168-173.

5. Luo, X., and Liu, S., "A Closed Micro Jet Cooling System for High Power LEDs," Electronic Packaging Technology, 2006. ICEPT'06. 7th International Conference on, (2006), pp.1-7.

6. Liu, S., Yang, J., Gan, Z., Luo, X., "Structural optimization of a micro-jet based cooling system for high power LEDs", International Journal of Thermal Sciences, 47: (2008), pp.1086-1095.

7. Wang, C. B., Kao, S., Lin, Y., and Cheng, J., "Low temperature co-fired ceramic (LTCC) tape compositions, light-Emitting diode (LED) modules, lighting devices and methods of forming thereof," (EP Patent 1,760,784, 2007).

Application of BP Neural Network in FBG Sensing System Performance Improvement

Zhang Jian[a], Zhao Hong[b] ,Rong Xian-wei[a]

a. Harbin Normal University, Harbin, 150080,China

b.Harbin University of Science and Technology, Harbin, 150040, China

Email:Jian.9626@gmail.google.com, Cell phone: 13163682975

Abstract

In order to improve measuring precision, a neural network application in non-linear compensation of FBG sensing system, which is based on coarse wavelength division multiplexer (CWDM), was studied in this paper. Simulation results show that a perfect linearization was realized with the system characteristics after neural network compensation.

1. Introduction

Fiber Bragg Grating (FBG) sensing element is the hot spot in the sensor research area home and abroad. The key technique of FBG sensing is how to demodulate the wavelength shift $\Delta\lambda$ effectively on time. The edge demodulation technique is one of the most dominant dynamic measuring demodulation techniques. But the linearity of edges is not ideal in general. That is the main obstacle for precision and span promotion of the whole measuring system. There is an automatic non-linearity calibration function, which can eliminate non-linear error, in microprocessor based sensing system. There are three methods of realizing non-linearity calibration. That is look-up table method, curve fitting method and neural network method generally. In fact, in instrument calibrations, the curve fitting or approximate straightway is adopted to determine the relationship between input and output traditionally. Take the system as a linear system factitiously and cause bigger error. Curve fitting method is through solving matrix equation to get fitting curve's coefficients. As noise exists, there may be morbid matrix occurred and cause trouble in matrix equation solution. In order to overcome this defect, BP neural network method is adopted to calibrate the non-linearity existed in FBG dynamic sensing demodulation system which is realized by making use of coarse wavelength division multiplexer (CWDM) in this paper. The instrumental indication precision is improved. Simulation experiment shows that system linearity can be improved tremendously after adopting BP neural network method to do the calibration.

Artificial Neural Network, ANN is called as Neural Network for short. It work is in the way of simulating brain. A non-linear system is constructed by a great deal and simple neural cells. It is a kind of pure mathematical model based on modern neurosciences research achievements. It is a non-linear function approaching method without selecting a basic function series beforehand. It has the advantages of self-learning, self-organizing and self-adopting, connatural parallel contracture and parallel processing, distributed knowledge storing, fault tolerant etc. It can approach any non-linear continuous function with the precision at will. So it is fit to solve the non-linear problems that can not be described exactly. It is a new kind of black-box operation method. It is approved that a corresponding BP neural network can be found to approach any continuous function defined in a closed interval. So BP neural network has learning ability from measuring system input and output relationship through non-linear mapping. It can be applied for non-linear calibration of measuring system. We make use of BP neural network to realize non-linearity calibration for FBG dynamic sensing demodulation system which is constructed with CWDM.

2. Principle

CWDM is a three-port device applied widely in photo-communication. That is input port, transmission port and reflection port. The outline and photo-characteristic is showed in Fig.1. The basic principle of FBG dynamic sensing wavelength detection based on CWDM is shown in Fig.2. ASE is wideband light source, PD is photoelectricity detector, IMG is matching liquid. The reflect signal of FBG sensor is divided into two parts by a 3dB coupler, and are accepted by two detectors. The relationships between transmission and wavelength, reflectivity and wavelength of CWDM are functions which is difficult to express in explicit formulation. But BP neural network is suit to solve this problem to the

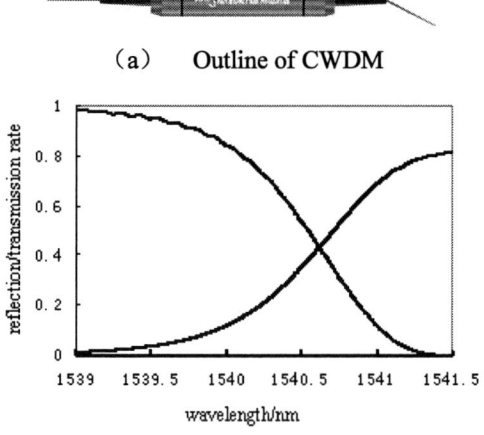

（a） Outline of CWDM

（b）Photo-characteristic of CWDM

Fig.1 Outline and Photo-characteristic of CWDM Fiber Coupler

moment. The nonlinear calibration model by BP neural network is shown in Fig.3.

978-1-4244-2739-0/08/$25.00 ©2008 IEEE

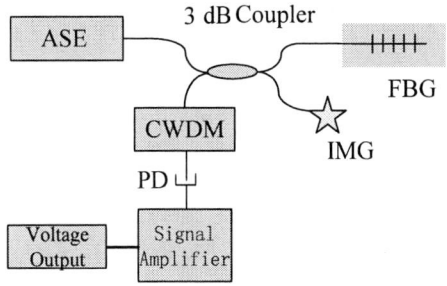

Fig.2 FBG Dynamic Sensing Demodulation Schematic
Diagram Based on CWDM

Back Propagation (BP) model is an error back propagation learning arithmetic applied in forward multi-layer neural network. The BP neural network adopts parallel structure,

then

$$a1 = \tan sig(W1p + b1), \quad a2 = purelin(W2a1 + b2).$$

The input p thereinto is corresponding to the circuit voltage output in Fig.2, output a2 is corresponding to wavelength λ.

Now, for the CD-ROM version of your paper, we prefer to have color figures and graphics, where appropriate; these will be viewable in color through the CD-ROM's browser (Acrobat Reader), but you should assure that they will be understandable when printed on a B&W printer. Be cautious especially of colors such as yellow, light blue, etc.

Your text should flow completely to the foot of the page. On 8.5x11" sheets, the top of your title (and the top line on each succeeding page) should be 0.67" from the top (you may need to adjust slightly, to match the Format Specification sheet), and the columns should continue to within 0.67" of

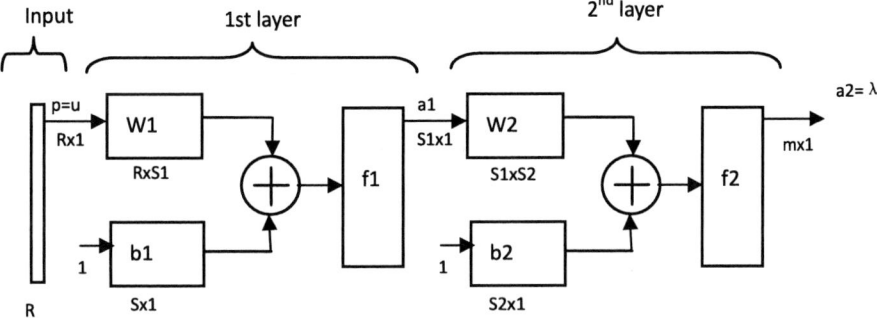

Fig.3 BP Neural Network Sketch

includes input layer, implicit layer and output layer. Implicit node passes to output node through process function and gives out the output finally. Learning process of the arithmetic is constituted with information forward propagation and error back propagation. In the forward propagation process, input information is processed from input layer through implicit layers step by step and passes to output layer. The first layer cell status has effects on the next layer cell status only. If the expected result did not get from the output layer, then switch to back propagation. Error signal (the difference of target and network output) goes back along the original connection and to minimize the mean square deviation by modifying the weights of layer cells. It has proved that neural network has powerful abilities on non-linear mapping and generalization. A two-layer network, the first layer transfer function is curvilinear function, the second layer transfer function is linear function, can simulate any function with limited discontinuity points. Refer to Fig.3, there are R entries for input and S1 cells, R×S1 orders weight matrix W1; There are S2 cells. The first layer output is the second layer input. There are S1 entries for input and S2 cells and S1×S2 orders weight matrix W2 in layer 2. inputs for layer 2 are a1 and outputs are a2. Let

$$f1 = \tan sig(x) \qquad f2 = purelin(x),$$

the bottom.

If you are using A4 paper, then you may need to adjust the borders in this template file. In "Page Setup", select "paper size" of A4; then select "Margins" and set left and right to 13.8 mm, and top and bottom to 25.2 mm. The gutter between columns should already be about 5 mm; it can be set under Format/Columns.

See the Format Specification sheet for additional details.

3. Simulating calibration with Matlab

Matlab is a set of high-powered value computation and visual software. It integrated value analysis, matrix computation, signal processing and graphical display with a convenient and friendly interface environment for user. There is a special tool box for designing and training of neural network in Matlab. This tool box provides abundant computations and operations which is involved in neural network arithmetic. That saves us mass of works on non-linear calibration for FBG wavelength detection by applying Matlab to simulate BP neural network. The main calibration procedure is as follows:

(1) Voltage and wavelength calibration: we need to measure some groups of wavelength and voltage and store the data in

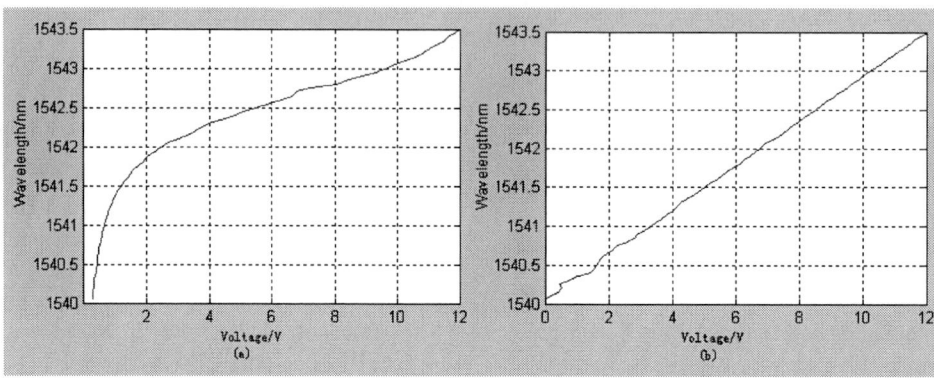

(a) Before calibration   (b) After calibration

Fig.4 Relationship Before and After Calibration between Wavelength and Voltage

a text file for training the neural network to be established.

(2) Neural network establishment: net=newff(minmax(p),[10,1],{'tansig','purelin'},'trainlm') ; newff() is the BP neural network function established. minmax(p) is the span of network input p of preconditioned sample data, [10, 1] represents that implicit node number is 10, output node number is 1. {'tansig','purelin'} means that the transfer function in implicit layer is tansig. The purelin function is adopted in output layer. Trainlm expresses the learning arithmetic. Basic BP arithmetic adopts gradient decent method to make mean square error (mse) run to minimum till up to requirement. But in practice, there are defects of low convergence speed and local extremum point. There are more than ten quick learning arithmetic methods provided in Matlab neural network tool box. One of them is excitation learning method, for example, traingdm arithmetic in which momentum factor is introduced, learning speed

(4) Simulation: a= sim(net,p) neural network output is simulated according to well trained network and input vector.

4. Conclusion and expectation

We made use of some groups of wavelength/voltage data training the neural network established, and use the well-trained neural network to simulate. The relationship between before and after calibration with the neural network is shown in Fig.4. Linearity is improved tremendously after calibration. For neural network can be calibrated at any moment with input and output data, this guarantees its precision and interchangeability.

variation arithmetic traingda, elastic learning arithmetic trainrp; the other category adopts value optimization arithmetic, for example, conjugate gradient learning arithmetic traincgf etc, Quasi-Newton arithmetic trainbgf etc, Levenberg-Marguardt arithmetic trainlm. Thereinto Levenberg-Marguardt value optimization arithmetic is suitable for middle or small scale neural network and with the most fast learning speed. The arithmetic adopted in this paper is trainlm.

(3) Learning: [net,tr]=train(net,p,t); t is target vector. According to the network learning error back propagation arithmetic, the new network weights and bias can be gotten after training from network weights and bias got from the former training till the net gotten with minimum error. The net is called the best neural network after training.

Non-linear calibration discussed in this paper is realized with a personal computer equipped with Matlab software and data acquisition card with bulk volume and more costly. In order to miniaturize the system and make it more cost/effective, we consider constructing a neural network with DSP to realize system non-linear calibration. Its diagram shows in Fig.5.

By adopting the kind of BP neural network mentioned above to calibrate the non-linearity of FBG wavelength detection system, FBG sensing demodulation system can be made small in size, light in weight, high precision, higher in linearity and interchangeability.

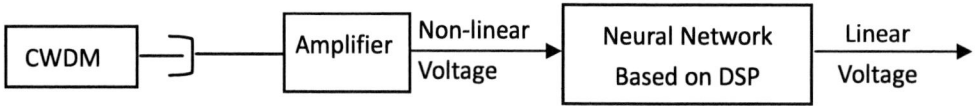

Fig.5 Neural Network Calibration System Based on DSP

5. Acknowledgments

This work is Supported by Scientific Research Fund of Heilongjiang Provincial Education Department(NO : 11531242).

References

1. Thompson M.L. A modeling chemical process using knowledge and neural networks，AICHEJ, 2001, 40 (8): 1328-1340

2. Tan Chao, Li Jiang-yuan. Neural network method for virtual instrument non-linear calibration, Modern Science Instruments, 2002, 6: 34~36

3. Cheng Xiao-huai, Zhang Yong-bin, Zhu Zhong-kui. Neurao network application in measuring instrument calibration, Chinese Journal of Scientific Instrument, 2001,22 (4): 52~53

4. A.S. Paterno, J.C.C.Silva, M.S. Milczewski, L V R Arruda and H J Kalinowski. Radial-basis function network for the approximation of FBG sensor spectra with distorted peaks, Meas.Sci.Technol,2006,17:1039-1045

5. E.Rivera and D.J. Thomson. Accurate strain measurements with fiber Bragg sensors and wavelength references, Smart Mater. Struct., 2006,15: 325-330.

Numerical Simulation of Solder Spreading and Solidification during Solder Jet Bumping Process

Dewen Tian, Chunqing Wang, Yanhong Tian
Department of electronic packaging technology, school of material science and engineering,
Harbin Institute of Technology, Harbin, 150001
Email: tiandw@hit.edu.cn Phone: 86-451-86418359 Fax: 86-451-86416186

Abstract

A VOF model is developed to simulate the solder spreading and solidification during solder jet bumping process. This model is based on fixed mesh method, and accounts for the surface tension, wetting effects, and heat transfer with solidification. The visualizations of the transient impact processes are employed in order to compare and validate the numerical model presented. Results show the spreading and recoiling process coupled with the solidification leads to a final cone-shaped solder bump. The variation of gravitational potential energy in the impingement is too small to be neglected. The simulated results are in excellent agreement with the photographic images.

1. Introduction

Joint shape is closely related with the reliability during the service. As an emerging technology[1,2], solder jet bumping process is different from other bumping processes that the bump formation happened in a very short time. The final joint shape is dependent on the solder spread and solidification. Better control of the process requires a fundamental understanding of the impingement physics.

In recent years, various theoretical models, such as the truncated sphere model, the force-balanced solution and the energy-based method, have been developed to predict solder geometry in electronic packaging[3,4]. Among these models, the surface evolver based on the minimum energy is most frequently used to evaluate and optimize the packaging processing[5,6]. However, these models are based on the static or quasi static theories, and can only predict the final bump shape rather than the solder shape evolution. Furthermore, they can only give an accurate prediction of the final bump shape formed under equilibrium or close to equilibrium states, such as the hot-air reflow bumping, which is long enough for the solder to reach its wetting balance on the substrate. For the solder jet bumping process, the solder will solidify as soon as it approaches the substrate. To achieve accurate prediction, the solder spread and solidification must be incorporated into the model. To date, very little information has been reported in the literature on numerical prediction of the solder bump formation in solder jet process[7-10]. There is still a lack of sufficient and clear understanding of the bump formation in solder jet process.

The aim of the present study is to develop a VOF model to predict the solder bump formation during the solder jet process, which will provide detailed information of fluid flow, solder solidification, and reveal the bump formation physics.

2. Numerical Model

2.1 Fluid flow and free surface tracking

The droplet impingement is solved using one-field volume of fluid tracking method, which can model the motion of multiple fluid phases. The mass and momentum conservation equations for Newtonian fluid under laminar flow conditions are given by

$$\nabla \cdot (\rho \vec{u}) = 0 \tag{1}$$

$$\frac{\partial}{\partial t}(\rho \vec{u}) + \nabla \cdot \left(\rho \vec{u} \vec{u} \right) = -\nabla p$$
$$+ \nabla \cdot \left[\mu (\nabla \vec{u} + \nabla \vec{u}^T) \right] + \rho \vec{g} + F_{VOF} + S_y \tag{2}$$

where ρ is the density, t is the time, $\vec{u}$ is the velocity vector, p is the pressure, μ is the viscosity, $\vec{g}$ is the gravitational acceleration vector, F_{VOF} is the continuum surface force vector, and S_y is the moment source term.

This single set of flow equations is used throughout the domain and mixture properties as defined blow was used. The mixture properties are calculated as

$$\phi = \alpha \phi_l + (1-\alpha) \phi_g \tag{3}$$

where ϕ is property variable; the subscript l and g represents liquid and gas phase respectively; α is the fraction of liquid phase.

When in a particular computational cell:

$\alpha = 0$: the cell is empty of the liquid;

$\alpha = 1$: the cell is full of the liquid;

$0 < \alpha < 1$: the cell contains the interface between the liquid and gas

The tracking of the interface between the liquid and gas is accomplished by the solution of a continuity equation for the volume fraction of liquid.

$$\frac{\partial \alpha}{\partial t} + \vec{u}_l \cdot \nabla \alpha = 0 \tag{4}$$

Surface tension is modeled as a smooth variation of capillary pressure across the interface. Following Brackbill et al.[11], it is represented as a continuum surface force (CSF) and is specified as a volumetric source term in the momentum equation as

$$F_{VOF} = \sigma \frac{\rho \kappa \nabla \alpha}{1/2(\rho_l + \rho_g)} \tag{5}$$

where σ is the surface tension coefficient, κ is the curvature and defined in terms of the divergence of the unit normal $\hat{n}$ as

$$\kappa = \nabla \cdot \hat{n} = \frac{\nabla \alpha}{|\nabla \alpha|} \tag{6}$$

A piecewise-linear scheme from the work of Youngs[12] is used to the interface reconstruction. It assumes that the interface between two fluids has a linear slope within each cell, and uses this linear shape for calculation of the advection of fluid through the cell faces.

The source term S_y is used to modify the momentum equation in a mushy zone (a region in which the liquid fraction lies between 0 and 1). The velocity also decreases to zero, when the mush becomes completely solid as represented by

$$S_y = -Au \qquad (7)$$

The form of A is derived from the Darcy law:

$$A = \frac{-C(1-\lambda)^2}{\lambda^3 + \eta} \qquad (8)$$

The value of C depends on the morphology of the porous media and measures the amplitude of the damping.

2.2 Solidification and heat transfer

An enthalpy porosity technique is used for modeling the solidification process. The enthalpy of the material is computed as the sum of the sensible enthalpy, h and the latent heat, ΔH, as follows:

$$H = h + \Delta H \qquad (9)$$

where

$$h = h_{\text{ref}} + \int_{T_{\text{ref}}}^{T} c_p dT \qquad (10)$$

Here, h_{ref} is the reference enthalpy, T_{ref} is the reference temperature, c_p

The liquid fraction, β is defined as

$$\begin{aligned}
\beta &= 0 && \text{if } T < T_s \\
\beta &= 1 && \text{if } T > T_l \\
\beta &= \frac{T - T_s}{T_l - T_s} && \text{if } T_s < T < T_l
\end{aligned} \qquad (11)$$

where T_s is the solidus temperature, T_l is the liquidus temperature.

The latent heat content is written in terms of the latent heat of freezing, L

$$\Delta H = \beta L \qquad (12)$$

Heat transfer in the droplet is modeled by solving the following energy equation

$$\frac{\partial}{\partial t}(\rho H) + \nabla \cdot (\rho \vec{u} H) = \nabla \cdot (k \nabla T) + S_h \qquad (13)$$

where k is the thermal conductivity, S_h is the source term and can be written as

$$S_h = \frac{\partial \rho \Delta H}{\partial t} + \nabla(\rho u \Delta H) \qquad (14)$$

In a mushy zone, this source term must be included.

2.3 Numerical solution

The droplet impingement model is solved with and the computational domain mesh and the boundary conditions shown in Fig.1. Only half of the model is actually computed with a symmetric boundary. The constant pressure boundary condition was used in the simulation for that the velocity profiles at these planes of the solution domain are not known.

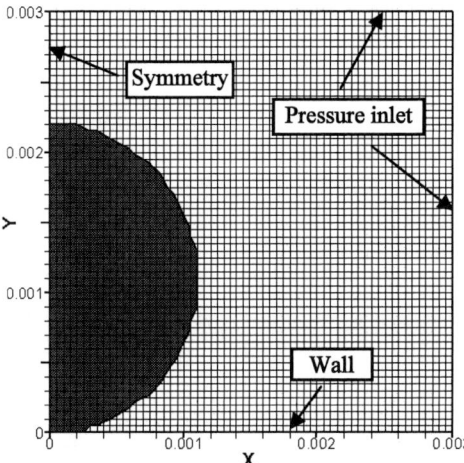

Fig. 1 Computational domain mesh and boundary conditions

Treatment of wall adhesion and movement of the contact line deserves special attention. Rather than impose this boundary condition at the wall itself, a contact angle is used to adjust the surface normal in cells near the wall. If θ is the contact angle at the wall, then the surface normal at the live cell next to the wall is

$$\hat{\mathbf{n}} = \hat{\mathbf{n}}_w \cos(\theta) + \hat{\mathbf{t}}_w \sin(\theta) \qquad (15)$$

where n_w and t_w are the unit vectors normal and tangential to the wall, respectively. The calculated surface normal one cell away from the wall determines the local curvature of the surface, and then this curvature is used to adjust the body force term in the surface tension calculation. The contact angle used in the current model is $140°$, and later results show excellent agreement between the predictions and experimental results.

A convective heat transfer was applied on the wall by specifying a heat transfer coefficient and a heat-sink temperature as

$$q = h_f(T_w - T_f) = h_{ext}(T_{ext} - T_w) \qquad (16)$$

where h_f is fluid-side local heat transfer coefficient, T_w is wall surface temperature, T_f is the local fluid temperature, h_{ext} is external heat transfer coefficient, T_{ext} is external heat-sink temperature, in the present model, $T_{ext} = 298K$, $h_{ext} = 500000$.

The fluid-side heat transfer coefficient is computed based on the local flow-field conditions as

$$q = k_f \left(\frac{\partial T}{\partial n} \right)_{wall} \qquad (17)$$

where k_f is the fluid conductivity, n is the local coordinate normal to the wall.

The initial temperature of droplet and the ambient gas is 523K and 298K, respectively. The properties used in the present model are listed in table 1.

Table 1 Properties used in the model

Properties	Solder	Gas
Density (kg/m^3)	7500	1.225
Viscosity (Pa·s)	0.002	0.0018
Thermal conductivity (W/mK)	73	0.0242
Specific heat (J/kgK)	250	1006.43
Latent heat (J/kg)	64762	
Solidus temperature (K)	489	
Liquidus temperature (K)	494	
Surface tension coefficient (N/m)		0.431

3. Experiment

In order to examine the validity of the simulation, a high speed videography system was designed to capture the solder motion at every moment. The main components of the apparatus are a translation stage, a solder droplet generator, a gas chamber, an x-y precision work stage, and a high speed camera with data acquisition system. A schematic drawing of the experimental setup is shown in Fig. 2. The solder droplet chamber is fitted with the translation stage on the top of the gas chamber. The solder droplet generator utilizes a temperature controlled unit to heat the Sn3.0Ag0.5Cu solder to a molten state and a pressure-driven unit to squeeze a molten solder droplet with the diameter 2.2mm out of the nozzle. The droplet would fall in the chamber filled with ambient gas, Ar, for a distance before it impacts on the substrate. The substrate (coller clad plate) is placed on the x-y work stage to reach the intended location precisely. The falling height can be adjusted to achieve an intended impinging velocity. In this study, the impact velocity is taken 0.5m/s. The high speed camera (DALSA 0256) is capable of recording 955 frames per second and were fitted with a Japan AVENIR CCTV 16mm lens to magnify the small solder droplet throughout the spreading process. The rapid motion involved in the solder impingement was captured with EPIX video acquisition system. The XCAP™ image analysis system was used to control the beginning of the capture and save the images.

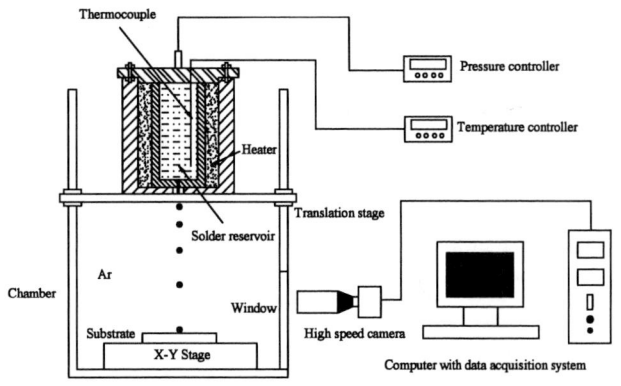

Fig.2 Schematic diagram of experimental setup

4. Results and Discussion

4.1 Solder shape evolution during droplet impingement

In order to investigate the solder geometry quantitatively, three characteristic geometry variables (ζ, ξ, η,), as shown in Fig.3. It is noted that ζ is defined as the maximum spread; whereas, ξ is defined as the position of the contact line. η is the droplet height. The three variables are made dimensionless with the initial droplet diameter d_0, as $\zeta^* = \zeta / d_0$, $\xi^* = \xi / d_0$, and $\eta = \eta / d_0$. The achieved postprocessing pictures used in the subsequent analysis include much information, such as the pressure distribution (left), velocity distribution (right) and the freezing front (right), which are all clearly denoted in Fig.3.

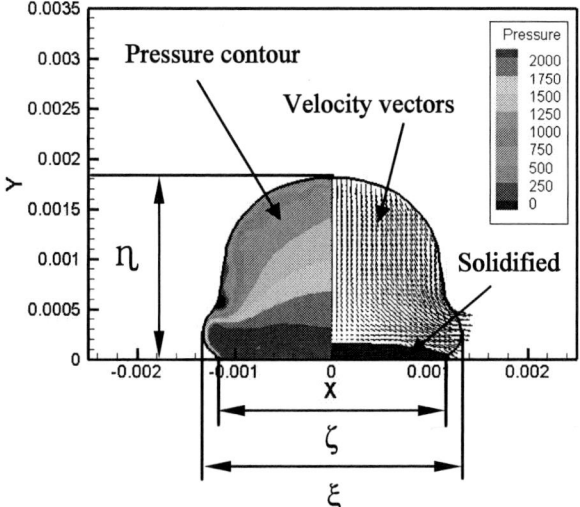

Fig.3 Definition of geometry variables and postprocessing illustration

Fig. 4 shows the solder shape evolution during the droplet impingement. When the droplet approaches on the substrate, the droplet spreads outward driven by the pressure gradient induced by impingement. The region of droplet in contact with the substrate has a relative high pressure. The outflow molten solder will accumulate at the periphery to form a toroidal rim. As a result of directional heat removal, the lower layer of the droplet solidifies at a planar mode. The solidification leads to a liftup of the periphery of the splat. The solder reaches its maximum spread at the moment t=4.2ms. Subsequently, the accumulated molten solder at the periphery will recede and rise from the center. The molten solder will oscillate and continue to solidify upward. The speed of the freeze front decreases because the rate of heat removal via the substrate decreases. The total solidification time of the droplet in the case is 15.8ms and the final bump is approximately cone-shaped. It is found that the predictions are in good agreement with the photographic images.

Fig.5 shows the dimensionless geometry evolution during droplet impingement. ξ^* experiences a nonmonotonic variation that it increases with time first, and then decreases to a constant value. However, ζ^* increases with time monotonically. This is because that the contact line could not move as long as the solder near it solidified, whereas, the solder above the bottom layer is still in molten state and can deform. The profile of η^* indicates that the droplet height decreases first, and then experiences a slight oscillation to reach a constant value.

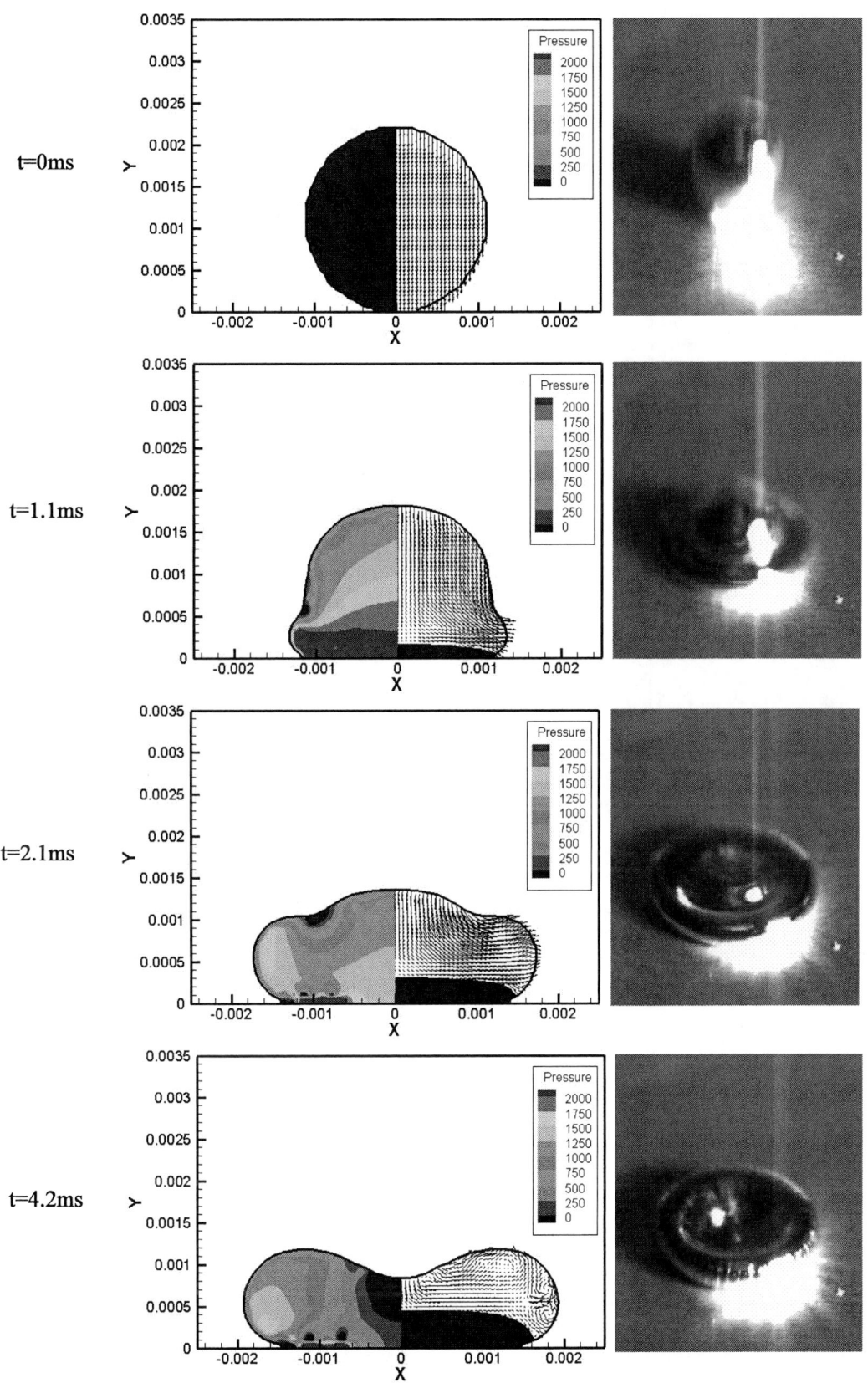

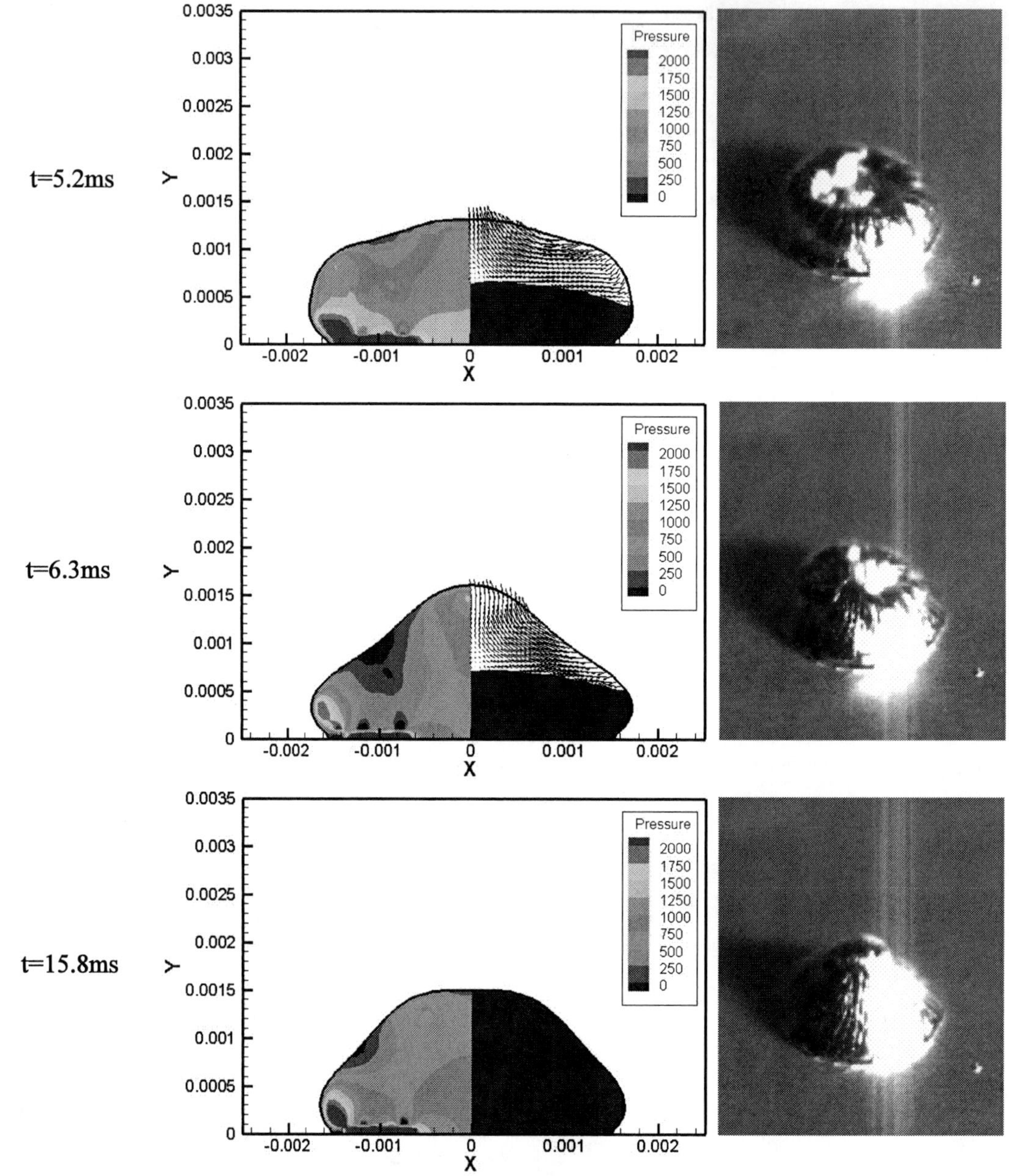

t=5.2ms

t=6.3ms

t=15.8ms

Fig.4 Solder shape, pressure distribution, velocity distribution, and freeze front as a function of time and comparison between simulation and photographic images

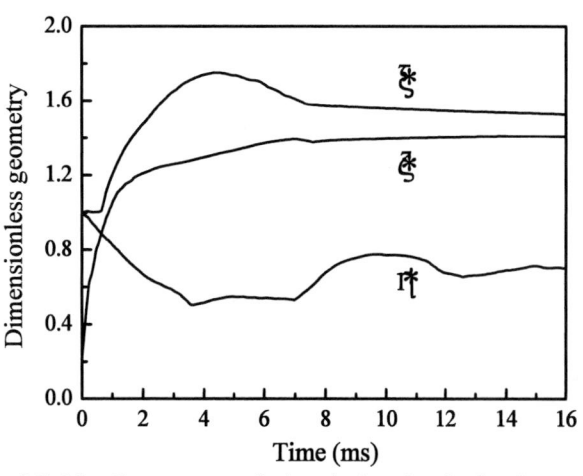

Fig.5 Solder Geometry evolution during droplet impingement

4.2 Energy analysis during droplet impingement

From the energy point of view, the impingement is actually the process of transformations of energies, such as gravitational energy, kinetic energy, surface energy and viscous dissipation. The energy components can be calculated from the modeling results. The surface energy SE can be written as

$$SE = S_{gl}\sigma_{gl} + S_{ls}(\sigma_{ls} - \sigma_{gs}) \qquad (18)$$

where S_{gl} and S_{ls} are the liquid-gas and liquid-solid interface area respectively; σ_{ls}、 σ_{gs} and σ_{gl} are the interfacial tension of the liquid-solid, gas-solid, and gas-liquid interfaces.

σ_{ls} is usually difficult to be determined, but it can be achieved by the Young's equation

$$\sigma_{gs} = \sigma_{ls} + \sigma_{gl}\cos\theta \qquad (19)$$

Substituting equation (19) in to equation (18), we can get

$$SE = \sigma_{gl}(S_{gl} - S_{ls}\cos\theta) + S_0\sigma_{sg} \qquad (20)$$

where S_0 is the substrate area. The last term in equation (20) is remains constant during droplet impingement. It does not make contribution for the energy variation, so it is omitted, and the surface energy is rewritten as

$$SE = \sigma_{gl}(S_{gl} - S_{ls}\cos\theta) \qquad (21)$$

The gravitational potential energy GE and the kinetic energy KE can be achieved using integral over the whole computational domain D as

$$GE = \iiint\limits_{D}(\rho_l gz\alpha)dV \qquad (22)$$

$$KE = \iiint\limits_{D}\left(\frac{1}{2}\rho_l Ve^2\alpha\right)dV \qquad (23)$$

where $Ve = \sqrt{u_x^2 + u_y^2 + u_z^2}$, u_x , u_y , and u_z are the velocity components at x , y , and z directions respectively.

The viscous dissipation VD is calculated using the energy balance method as

$$SE^0 + GE^0 + KE^0 = SE + GE + KE + VD \qquad (24)$$

where the superscript 0 denotes quantities evaluated at the initial condition (t=0).

Fig. 6 is the energy components evolution during droplet impingement. The decrease of gravitational potential energy during the whole impingement process is very small, and can be neglected. At first, the kinetic and surface energy decreases, whereas the viscous dissipation increases. At the moment t=2ms, the surface energy reaches its minimum value. When most of the kinetic energy are converted into surface energy and dissipated by the viscous force, the splat reaches its maximum spread. Meanwhile, the surface energy also reaches its first peak. And then the surface energy is released, the molten solder begins to recede and the kinetic energy begins to increase. In the subsequent stage, the surface energy and the kinetic energy experience an oscillation, and then remain constant. During the whole impingement, the viscous dissipation increases monotonically, for that it is non-reservable.

5. Conclusions

A numerical model has been developed to investigate the solder spreading and solidification during solder jet bumping process. The following conclusions can be drawn:

(1) Solder solidifies at a planar mode and the final bump shape is approximately cone-shaped.

(2) With the increase of time, ξ^* increases first, and then decreases to a constant value; ζ^* increases monotonically; η^* decreases first, and then experience an oscillation to reach a constant value.

(3) The decrease of gravitational potential energy is very small during the impingement and can be neglected. The surface energy and the kinetic energy oscillate during the

droplet impingement. The viscous dissipation increases monotonically with the increase of time.

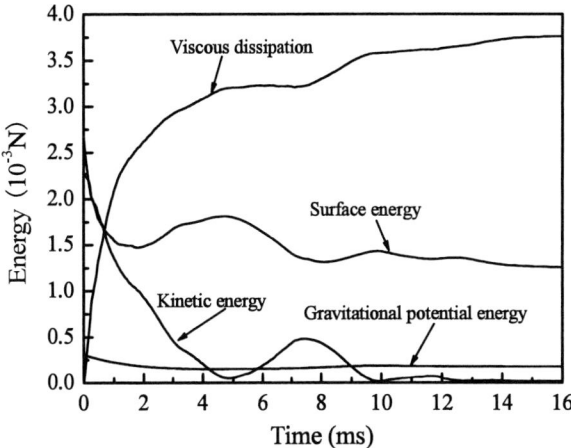

Fig.6 Energy components evolution during droplet impingement

Acknowledgments

This research is supported by the National Natural Science Foundation. (Grant No. 50675047/E052105)

References

1. Gallagher, C., et al., "Solder jet technology for advanced packaging," Proceeding of SPIE on Optoelectronics, Photonic Devices, and Optical Networks Dublin, Ireland, 2005, pp. 615-621.
2. Liu, Q., et al., "High precision solder droplet printing technology and the state-of-the-art," Journal of Materials Processing Technology, Vol. 115, No. 3 (2001), pp. 271-283.
3. Heinrich, S. M., et al., "Solder Geometry Prediction in Electronic Packaging: An Overview," Advances in Electronic Packaging, Vol. 19, No. 2 (1997), pp. 1371-1376.
4. Chiang, K. N., et al., "An overview of solder bump shape prediction algorithms with validations," IEEE Transactions on Advanced Packaging, Vol. 24, No. 2 (2001), pp. 158-162.
5. Yeung, B. H., et al., "Evaluation and optimization of package processing and design through solder joint profile prediction," IEEE Transactions on Electronics Packaging Manufacturing, Vol. 26, No. 1 (2003), pp. 68-74.
6. Chen, W. H., et al., "Predicting the Liquid Formation for the Solder Joints in Flip Chip Technology," Journal of Electronic Packaging, Vol. 128, No. 4 (2006), pp. 331-338.
7. Dietzel, M., et al., "Marangoni and variable viscosity phenomena in picoliter size solder droplet deposition," Journal of Heat Transfer, Vol. 125, No. 2 (2003), pp. 365-376.
8. Waldvogel, J. M., et al., "Solidification phenomena in picoliter size solder droplet deposition on a composite substrate," International Journal of Heat and Mass Transfer, Vol. 40, No. 2 (1997), pp. 295-309.

9. Wang, W., et al., "Prediction of solder bump formation in solder jet packaging process," IEEE Transactions on Components and Packaging Technologies, Vol. 29, No. 3 (2006), pp. 486-493.

10. Young-Soo, Y., et al., "Spreading and solidification of a molten microdrop in the solder jet bumping process," IEEE Transactions on Components and Packaging Technologies, Vol. 26, No. 1 (2003), pp. 215-221.

11. Brackbill, J. U., et al., "A continuum method for modeling surface tension," Journal of Computational Physics, Vol. 100, No. 2 (1992), pp. 335-354.

12. Youngs, D. L., "Time-dependent multi-material flow with large fluid distortion," Numerical Methods for Fluid Dynamics, Vol. 1, No. 1 (1982), pp. 41-51.

Research of Design-for-Testability of CMOS Image Sensor

Zhaohui Ou[1], Feng Lin[2]

Microelectronics R&D center，Shanghai University

Key Laboratory of Advanced Display and System Applications (Shanghai University) Ministry of Education

Shanghai 200072, P.R.China

zhaohuiou@shu.edu.cn[1]

lobol@shu.edu.cn[2]

Abstract

CMOS image sensor has experienced explosive growth in recent years. As increasing of number of pixel scale and complexities of circuit, testability of image sensor chip has become an important problem that must be dealt with by both design and test engineers. A systematic approach to handle testability of CMOS image sensor circuits is urgently needed, because current test methods less address this domain. In this paper, a uniform and systematic approach is explored to the testability problem of CMOS image sensor, and it covers the image sensor defect analysis, fault model definition and test system design. The experimental data shows the fault coverage can be up to 99.3%.

Introduction

CMOS image sensor is becoming more and more popular in recent years and has become a significant silicon technology driver. With development of CMOS image sensor manufacture process and circuit design technology, the pixel number of one chip has increased from hundreds of thousands to several millions and more. An SOC system includes image sensor, ADC, image processor, external interface etc generally, which have become the image sensor system develop trend.

Most prior researches on design-for-testability focused on digital, analog and mixed-signal SOC [1, 2, 3, 4, 5], and also some study explored the test cost optimization [6, 7]. However, since the large-scale or high frame rate CMOS image sensor system is used in more and more application, there is a need for efficient test methodologies that can handle the testability of CMOS image sensor. Figure 1 shows image sensor system basic diagram, which includes pixel array model, PGA/readout, ADC, memory RAM, and/or image DSP block. For PGA, READOUT, ADC, memory and DSP blocks, as well as SOC test method, many studies have already discussed it. To address the test need of image sensor, the paper is going to discuss the CMOS image sensor defect type, fault model, test pattern generation and test system design.

This paper focuses on design-for-testability of CMOS image sensor system, and is organized as follow: section 2 presents a detailed analysis of CMOS image sensor pixel cell and relative component defect cases, and thus analyze fault mode case; section 3 discusses the test pattern generator method; section 4 presents research of test system for image sensor system with test bench and ATE and additional controller; section 5 shows fault coverage result by an example design test and simulation result. Finally, Section 6 summarizes the conclusions of the work.

Fault Model

CMOS image sensor is a optic-electric device which have special physical structure, circuit cell and manufacture process. Its special characteristic is discussed in [8, 9]. The defect models are studied detailed.

Figure 2 depicts the cross-section of basic components of image sensor photo-diode [8]; and figure3 is the basic circuit of CMOS image sensor pixel with APS structure [8,9]. The Image sensor comprises a two-dimensional array of pixels converting the light incident at its surface into an array of electrical signals. To perform color imaging for the color application image system, a color filter array (CFA) is typically deposited in a certain pattern on the top of the image sensor pixel array. The basic photo-diode is made of CMOS P-N silicon material. The photo-diode converts the optical signal to electronic signals response to the light strength. The electronic signals are very weak and need a sequential analog PGA or amplifier to scale up and an ADC sample to convert to digital signals finally.

For analyzing defect case, the photo-electric transfer equation in silicon can be simply expressed as equation 1:

$$Y = a \cdot Q \cdot C \cdot \int_0^T f(x)dx + N_1 + N_2 \qquad (1)$$

Where

Y the value of ADC output point that converted from photo input to electronic output of each pixel. Y also represents the photo strength of one pixel

Q the photo strength of pixel surface.

C the photo to electronic transfer factor

α the transfer efficiency factor of silicon photodiode

N_1 the summary noise effect of photodiode, including the thermal, 1/f, dark current etc effect factors.

N_2 the summary noise effect of PGA/REDOUT/ADC Analog component

$f(x)$ the transfer curve of photo to electronic

T the exposure time length

978-1-4244-2739-0/08/$25.00 ©2008 IEEE

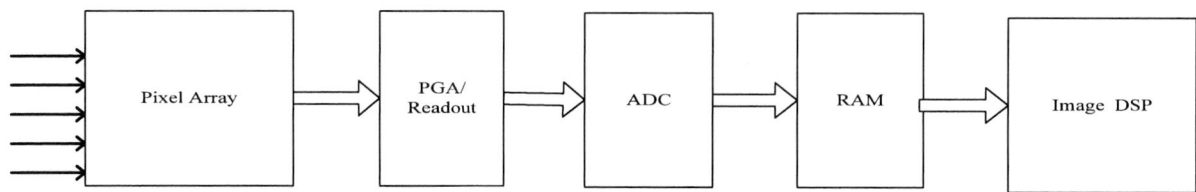

Figure 1 CMOS image sensor system diagram

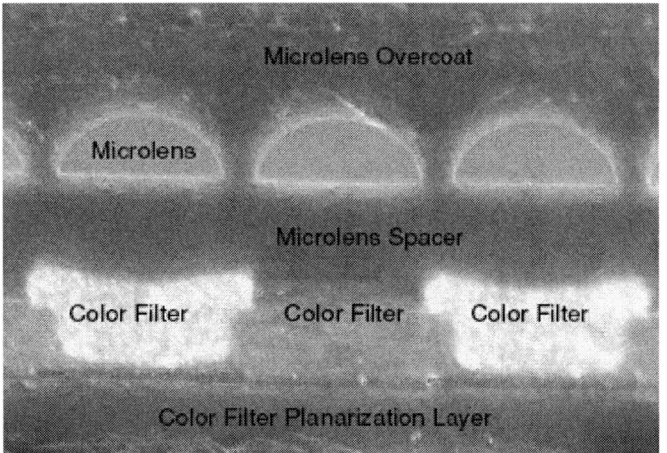

Figure 2 cross–section photograph of CFA photo-diode

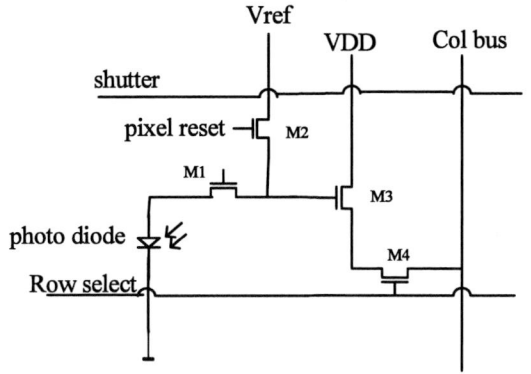

Figure 3 circuit of CMOS image pixel

The photodiode is the core device in photo to electric transfer system and is implemented by conjoining P and N type material. The exposure time T is positive scale with output Y, which is expressed as f(x). The factor α represents silicon photodiode transfer efficiency for considering the various factors which affect the photodiode work performance.

Temporal noise sets the fundamental limit on image sensor performance [10]. In a CMOS image sensor several additional sources contribute to temporal noise. These include the pixel reset, follower, and access transistor thermal, shot, and 1/f noise and the column amplifier thermal and 1/f noise. This paper classifies all noise into N_1 and N_2. N_1 represents sum of noise of pixel cells and N_2 represents sum of noise of other sequence process devices.

According to the above analysis, the CMOS image sensor defects cases can be expressed as follow:

a) photodiode defects caused by impurity inleakage photodiode in manufacture stage, short, open bridge etc

connection defects which result in totally dark, totally light or loss partial energy (i.e. abnormal α value).

b) analog device defects, the short, open and bridge etc defects that occur in the pixel cell and sequence PGA,ADC etc analog circuit

c) digital relative block defects. There are many digital logic embedded in the image sensor system, such as sensor expose controller, ADC timing controller, address decoder etc.

d) package defects. The image sensor is different form other digital/analog/mixed chip. It must be placed in dark space with accurate aperture. Even small size package deviation will lead the system to fail.

Now, the fault model of CMOS image sensor is given, as shown in the table 1. There're four types of fault model: stuck-at-light, stuck-at-dark, stuck-at-performance and stuck-at-package:

Table 1 CMOS image sensor fault model definition

NO	Fault model type	Equivalent defect case
1	Stuck-at-light	Device/metal short to power, open, bridge etc, digital gates defect
2	Stuck-at-dark	Device/metal short to GND, open, bridge etc, digital gates defect
3	Stuck-at-performance	Photodiode error, noise effect, dark-current, relative analog device defect
4	Stuck-at-package	Package defect, optical position excursion, impurity pollute

Test Pattern

A low-cost and high coverage test pattern is an important issue in the DFT design. As we known, there are many test pattern generator algorithms [11,12] and EDA implementation tools, such as D algorithm used for detecting single fault of digital combinational circuits, TetraMax ATPG for generating test vectors and BIST solution for memory block etc, but the test pattern for image sensor system is very special since they are optical photo vectors and not electronic signals. As the equation (1) shown, the output result Y is realated to three factors, exterior photo strength Q, expose time T and the photodiode transfer characterization. The photodiode transfer characteristic is determined in manufactured process, but the Q and T two factors are decided by some external controllers. In an practical system, the optical photo test patterns are generated in a dark-box and are controlled by ATE with optical LED source.

480

Test patterns number which meets the photo strength levels requirement for full test coverage is correlatively with the sensor pixel resolution and is presented as equation 2:

$$P = 2^R \cdot N \tag{2}$$

Where
P   the required number of test pattern for full test coverage, i.e. the photo pattern number.

R   resolution of the sensor pixel designed, i.e. the resolution of ADC located in image sensor system.

N   number of color filter. In general case, it is 1 for gray image application or 3 for color image application.

The Table 2 presents a example for explaining the relation of between P and R, where N is assumed to be 1.

Table 2 .example of test pattern and pixel resolution

R(resolution)	P(pattern)
4-bit	16
6-bit	64
8-bit	256

Test System

Figure 3 presents the internal architecture diagram of image sensor system for design-for-testability. In this system, TAM (Test Access Mechanism), memory BIST and DSP digital logic scan chain are all known technologies [13,14,15], and the image BIST is new which used for bypass-streaming out the image pixels test data and generate relative control signals in test status. The image pixel test data are exported without any judgment and process; on the final stage, an external test judger will be employed.

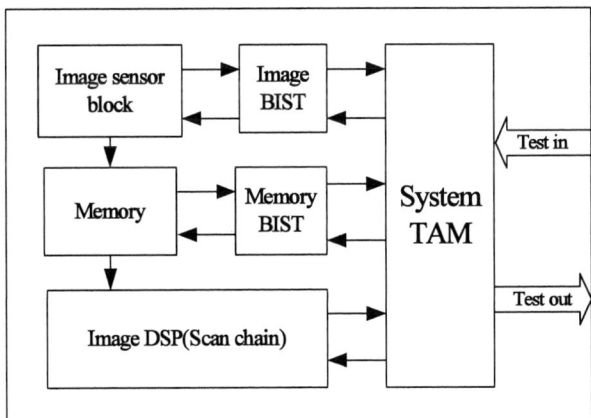

Figure 3 DFT structure diagram of image sensor

Whole test system for image sensor chip consists of four main build blocks, as is shown in figure 4, test pattern generator, DUT, test controller and ATE.

Test pattern generator generates the optical test photo-pattern. Which consists of a reflect surface and a led controlled and be placed in the dark-box. DUT is the Design under Test, CMOS image sensor chip. Since build-in image sensor BIST is very simple for saving chip area cost, an external test controller is used to generate the controller signals and receive the bulk test data and compare them with expectation value for image sensor block test. ATE test handle and relative controller bench, used to test result bin-

out and test worker relative process same as general digital/analog/SOC ATE.

A system test bus is employed to transfer the test data and receive the test control command. For saving the test cost, most the test IO should be shared with other function IO and communicate with ATE and/or other test controller serially.

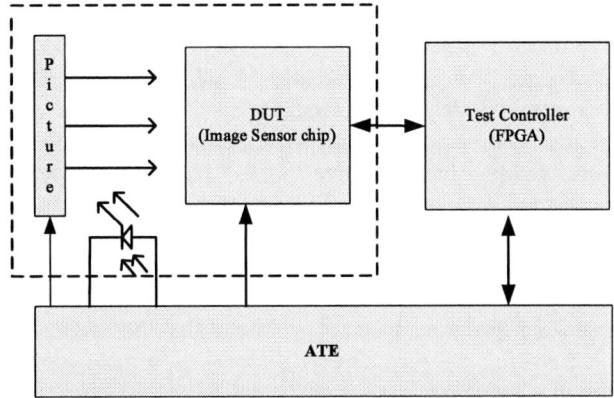

Figure 4 CMOS image sensor test system

Experimental Results

The fault coverage [16] is the most important factor for evaluating the performance of a DFT solution. The fault coverage can be expressed approximately as equation (3):

$$FC \approx N/M \tag{3}$$

Where FC is the fault coverage rate, N is the number of detected faults, M is the total number of faults.

The proposed solution is verified on an practical chip platform. The basic characteristic of example chip is shown in Table 3. And Table 4 presents the summary of the fault coverage got in the final practical testing and fault simulation result.

Table 3 example product characteristics

Technology	CMOS 0.25um,4-layer metal,1 layer poly, n-well
Gray level of pixel	8bits,256 level
Number of pixels	64*64
Pixel size	30*30(um)
Transistors per pixel	APS, 3 transistors
Frames per second	2000~5000
Pixels interconnect	Metal1 and poly
Photodiode	N-well/p-sub diode
Package	12pins

Table 4 experimental fault coverage test data

Fault type	Fault Coverage
Stuck-at-light	100%
Stuck-at-dark	100%
Stuck-at-performance	Bin-out process, about 98.9
Stuck-at-package	99.5%
Total average fault coverage	99.3%

Table 4 shows FC experimental result rate can reach up to 99.3%, which means more the 99% fault can be detected. In general case, the rate of good product of the CMOS image sensor manufactured by the final process factory is large than 95%. Thus, the final rate of good product is 99.965‰ (100% - 5%*99.3%), and the fault chips that are miss-judged in test stage is not more than 0.0035‰.

Conclusions

The DFT design for image sensor system is blank area of IC design. This paper presented an effective and practical technique for design-for-testability of CMOS image sensor systems composed of high performance image pixels. The work firstly studies image pixels and optical-electronic transform characteristic, and then gives some further analysis on defect case and fault type. Finally, a test schedule is found for testing optical sensor system with a special ATE dark-box environment. On applying the test strategy to some example image sensor-based systems, we found that the average area overhead is very low. The verification on a mass-produced design example indicates that the test solution is efficient and the fault coverage can be up to 99.3%. The design-for-testability of CMOS image sensor is a new issue. The future work includes developing a methodology for reducing time cost and enhancing the robustness.

Reference

[1] Fernandes J. M. , Santos M. B. etc, "DFT and Probabilistic Testability Analysis at RTL", *High-Level Design Validation and Test Workshop, 2006. Eleventh Annual IEEE International* Nov. 2006 Page(s):41 - 47

[2] Sehgal A., Ozev S., and Chakrabarty K.; "Test Infrastructure Design for Mixed-Signal SOCs With Wrapped Analog Cores"; *Very Large Scale Integration (VLSI) Systems, IEEE Transactions* on Volume 14, Issue 3, March 2006 Page(s):292 - 304

[3] Yamamoto T. et al., "A mixed-signal 0.18-um CMOS SoC for DVD systems with 432-MSample/s PRML read channel and 16-Mb embedded DRAM," *IEEE J. Solid-State Circuits*, vol. 36, no. 4, Apr. 2001, pp. 1785–1794.

[4] Kundert H., et al., "Design of mixed-signal systems-on-a chip," *IEEE Trans. Computer.-Aided Des. Integr. Circuits Syst.*, vol. 19, no. 12, Dec. 2000, pp. 1561–1571.

[5] Iyengar V., K. Chakrabarty, and Marinissen E. J., "Test access mechanism optimization, test scheduling and tester data volume reduction for system-on-chip," *IEEE Trans. Comput.-Aided Des. Integr. Circuits Syst.*, vol. 22, no. 5, May 2003, pp. 593–604.

[6] Calvano J. V., Lubaszewski M. S.; "Efficient Test Access Mechanism Optimization for System- on-Chip"; *Computer-Aided Design of Integrated Circuits and Systems*, IEEE Transactions on Volume 22, Issue 5, May 2003 Page(s):635 - 643

[7] Huang Y., Cheng W-T., Tsai C.-C., Mukherjee N., Samman O., Zaidan Y., and Reddy S. M., "Resource allocation and test scheduling for concurrent test of core-based SOC design," in *Proc. Asian Test Symp.*, 2001, pp. 265–270.

[8] Gamal A. E., and Eltoukhy H., "CMOS IMAGE SENSOR", *IEEE CIRCUITS & DEVICES MAGAZINE* MAY/JUNE 2005, pp.7-18

[9] Chapinal G., "128x128 CMOS Image sensor with analog memory for synchronous image capture",*IEEE Sensors Journal*,2,n2, april 2002, pp 120-127.

[10] Singh K., "Noise analysis of a fully integrated CMOS image sensor," in *Proc. SPIE*, vol. 3650, San Jose, CA, Jan. 1999, pp. 44–51

[11] Agerbo E. and Cornils A., "How to preserve the benefits of design patterns". *Proceedings of OOPSLA'98*, Vancouver, BC, Canada, October 1998, pp. 134-43.

[12] Guennec A. L., SunyéG. and Jézéquel J.-M., "Precise Modeling of Design Patterns". *In proceedings of UML'00*, October 2000, pp. 482-496.

[13] Iyengar V., Chakrabarty K., and Marinissen E. J., "Efficient Test Access Mechanism Optimization for System-on-Chip", *IEEE TRANSACTIONS ON COMPUTER-AIDED DESIGN OF INTEGRATED CIRCUITS AND SYSTEMS*, VOL. 22, NO. 5, MAY 2003, pp.635-643.

[14] Aerts J. and Marinissen E. J., "Scan chain design for test time reduction in core-based IC's," *Proc. Int. Test Conf.*, 1998, pp. 448–457.

[15] Iyengar V., Chakrabarty K., and Marinissen E. J., "Test wrapper and test access mechanism co-optimization for system-on-chip," *J. Electron. Testing: Theory Applicat.*, vol. 18, 2002, pp. 213–230.

[16] Hashempour H., Meyer F.J., Lombardi F., "Analysis and measurement of fault coverage in a combined ATE and BIST environment Instrumentation and Measurement", *IEEE Transactions* on Volume 53, Issue 2, April 2004 Page(s):300 - 307

A 10-bit 40MSPS Pipeline Analog-to-Digital Converter

Cai Jun, Ran Feng, Xu Meihua

Key Laboratory of Advanced Display and System Application (Shanghai University), Ministry of Education

Microelectronic Research & Development Center of Shanghai University

No.149, Yan Chang Rd, Shanghai, 200072, China

Abstract

A 40MSample/s, 10-bit, 3.3V pipeline ADC is presented. In order to achieve very low power consumption, it employs a high bandwidth low-power amplifiers technique and a low power low offset dynamic comparators technique. The ADC is designed in $0.35 \mu m$ CMOS technology and occupies $1.2*0.8mm^2$.

1. Introduction

Analog-to-digital (A/D) conversion and digital-to-analog (D/A) conversion are critical interfaces in mixed-signal processing systems. The pipelined analog-to-digital converter (ADC) has become the most popular ADC architecture with improved characters in speed, resolution, dynamic performance, and low power. The pipeline ADC is suited to CMOS implementation with potential applications include video imaging, wireless LAN, personal communication systems, and disk drive read channels. Many of these applications make use of 10-bit analog-to-digital converters sampling at around 40MHz. In such applications, it is highly desirable for both power and cost [1,2,3,4].

A 40MSample/s, 10-bit, 3.3V pipeline ADC is designed which features a low power consumption. In order to achieve very low power consumption, high bandwidth low-power amplifiers technique and low power low offset dynamic comparators technique are employed. With the digital correction, the comparator accuracy requirements can be relaxed, and static power dissipation of precision comparators can be eliminated.

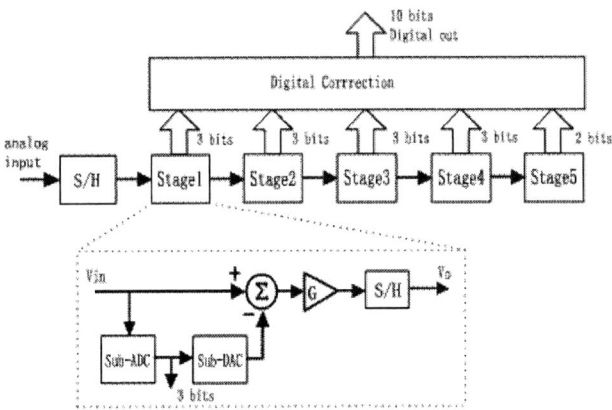

Figure 1 Block Diagram of the System.

2. ADC architecture

To achieve high integration in today's integrated circuit design, A/D converter is becoming a circuit block in a large digital signal-processing chip, which shares the same supply voltage with the digital circuit. Pipeline ADCs provide a balance of size, speed, resolution, power dissipation, and analog design effort. Pipeline ADCs consist of numerous consecutive stages, each containing a track/hold (T/H), a low-resolution ADC and DAC, and a summing circuit that includes an inter-stage amplifier to provide gain. Separate track-and-hold (T/H) amplifiers for each stage release each previous T/H to process the next incoming sample, enabling conversion of multiple samples simultaneously in different stages of the pipeline [5,6].

Figure 1 shows a block diagram of the system. The sample and hold circuit is followed by a five stages pipelined ADC. In this schematic, the analog input, VIN, is first sampled and held steady by a sample-and-hold (S&H). The flash ADC in stage one quantizes the analog input to three bits. The 3-bit digital output is then fed to a 3-bit DAC to generate an analog output that is subtracted from the input of the stage. This "residue" is then gained up by a factor of four and fed to the next stage (Stage 2). This gained-up residue continues through the pipeline, providing three bits per stage until it reaches the 2-bit flash ADC, which resolves the last 2LSB bits.

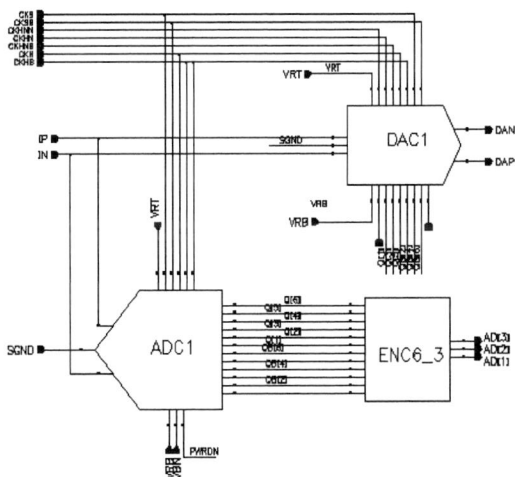

Figure 2 Structure of the First Stage

Figure 2 shows the structure of the first stage which includes a 3-bit flash ADC, a 3-bit DAC and a encode circuit. Although each stage generates three raw bits, because the inter-stage gain is only 4, each stage (Stages 1 to 4) effectively resolves only two bits. In fact, the 3-bit flash ADCs in Stages 1 through 4 require only about four bits of accuracy. The extra bit is simply to reduce the size of the residue by one half, allowing extra range in the next 3-bit ADC for digital error correction. The effective number of bits of the entire ADC is therefore 10 bits.

All the output of each stage is applied to digital correction circuit. Because the bits from each stage are determined at different points in time, all the bits corresponding to the same sample are time-aligned with shift registers before being fed to the digital-error-correction logic. When a stage finishes processing a sample, determining the bits, and passing the residue to the next stage, it can then start processing the next sample received from the sample-and-hold embedded within each stage. As the input signal is effectively amplified further down the pipeline, the resolution requirements for the later amplifiers are significantly relaxed. Therefore the performances of the first and second stages are the most critical.

3. Sample and Hold Architecture

As an important part in the pipelined ADC, the switched capacitor sample and hold circuits are optimized to minimize power dissipation. The basic structure of the sample and hold circuit used in the pipelined ADC is shown in figure 3.

This circuit is the standard switched capacitor sample and hold circuit used in charge redistribution analog to digital converters. The sample and hold circuit has two phases of operation: the sampling phase and the hold phase.

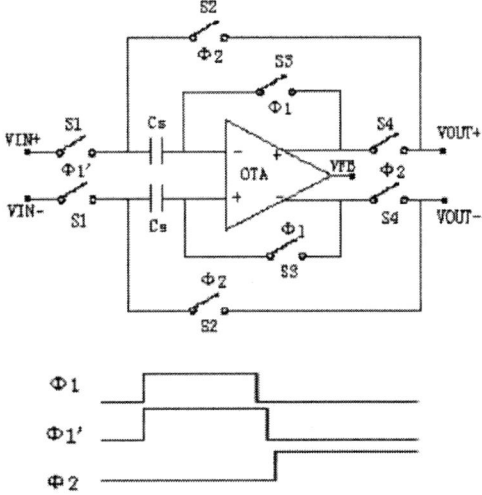

Figure 3. Architecture of Sample and Hold

During the sampling phase, $\Phi 1$ is high and the amplifier output nodes are tied to the input nodes. The capacitor bottom plates are all tied to the appropriate input voltage, either VIN+ or VIN-. At the end of the sampling phase, the switches connecting the input signal to the capacitor bottom plates and the switches connecting the amplifier inputs and outputs are turned off. During the hold phase of operation, the amplifier output nodes and the capacitor bottom plates are connected. The amplifier output VOUT+ and VOUT- will be the same value as the amplifier inputs voltage. Differential input signal path is helpful to reduce the clock feed-through influence, and the error can be cancelled by taking the input differentially as long as it is present on both signal inputs.

4. Comparator Architecture

The comparator is a fundamental part of an analog to digital converter. Its function is to make a comparison between two input signals and produces an output indicating which of the two inputs is larger. A comparator may be thought of as doing this by subtracting the two inputs and generating a binary output of 1 if the difference is positive and 0 if the difference is negative. Comparator nonidealities affect circuit performance in important ways. One of the most important nonideal characteristics is the offset of the comparator. When the comparator computes the difference of between the two input signals, an internal offset voltage is added to this difference. Thus, when the two inputs are close together, the comparator may make a wrong decision. Some techniques for reducing comparator offsets have been developed and digital error correction technique is widely used.

The circuit diagram for the comparator used in this project is shown in figure 4. This comparator circuit consists of a latch, pre-charge transistors, input transistors, cascade transistors, bias transistors, input sampling switches, and reference sampling switches.

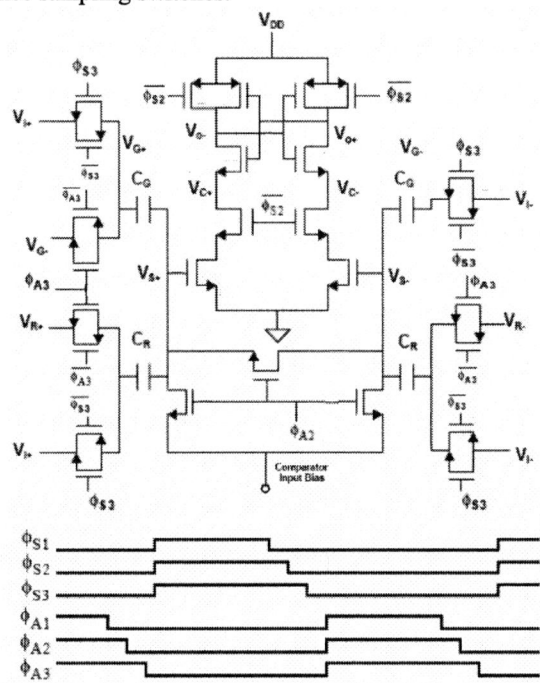

Figure 4 Circuit Diagram for the Comparator

The operation of this circuit is as follows. While clocks are high, the charges at nodes V_{S+} and V_{S-} are initialized to values determined by the reference voltages V_{R+} and V_{R-} and the bias voltage. While clocks are high, the output nodes V_{O+} and Vo- are pre-charged to V_{DD}. V_{C+} and V_{C-} are also pre-charged to V_{DD} at this time. V_{A+} and V_{A-} are pre-charged to ground at this time. During this time, V_{S+} and V_{S-} are floating and follow the input voltages V_{I+} and V_{I-} through the capacitive coupling provided by C_S and C_R. When goes low, latching begins while the input remains connected. Both V_{O+} and Vo- begin to discharge. The differential input voltage produces a differential voltage on nodes V_{S+} and V_{S-}. This differential voltage produces a mismatch in impedance that causes the nodes Vo+ and Vo- to discharge at unequal rates. The slow node is charged back to V_{DD} while the fast node continues to discharge to ground. For $V_{S+} - V_{S-} < 0$, the output is high. Otherwise the output is low. Thus a decision has been made.

Figure5 shows a simulation result of the comparator. It is clearly that the latency of the comparators is small enough for high-speed application.

484

In stage 1, six of these comparators are used together to form a sub-ADC. In order to construct a useful ADC, a set of comparators with a range of input threshold voltages is required. This is accomplished by appropriately choosing the value for the ratio of resister. This ratio determines the amount of charge initially stored at the summing nodes V_{S+} and V_S-. When the decision has been made by sub-ADC, it will be passed on to the encode circuit to generate a 3-bit output.

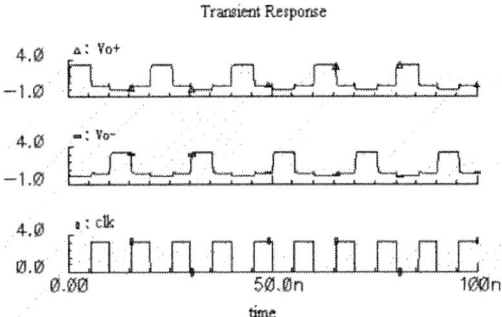

Figure 5 Simulation Waveform of the Comparator

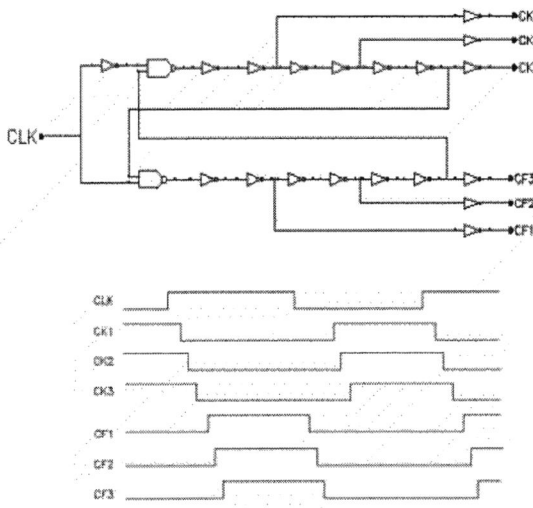

Figure 6 Clock Circuits and Timing Diagram

5. Clock Generation

In order to control the timing of operation, it is necessary to generate a set of clock signals with two non-overlapping phases. It is essential that the sampling phase and hold phase do not overlap. Overlapping of the phases would result in signal charge leaking away during the hold phase. The circuit used for generating these non-overlapping clocks is shown in figure 6. A timing diagram declaring how these clocks are generated is also shown.

6. Power and Noise considerations

For the switched-capacitor design techniques that are thermal noise limited by kT/C, the required power consumption for a given dynamic range, DR, is

$$P \propto (DR \cdot V_{DD}) / (V_{DD} - 2V_{DSAT})^2 \qquad (5-1)$$

It shows that analog power increases as VDD decreases [5]. Since the 0.35 μm process available for this project does not have a low Vt option, the typical supply voltage is chosen to be 3.3 V for better power efficiency in the analog circuitry. Switched capacitor circuits can provide multiple functions while dissipating no static power. The sample and hold, the feedback network for the inter-stage amplifier, and the subtraction of the DAC output from the input are all implemented with a simple switched capacitor scheme.

To reduce overall noise, one technique is to design the circuit in such a way that supply noise is not coupled into the signal. One very effective way of accomplishing this is to use fully differential circuit architectures. Fully differential circuit architecture has both positive and negative input and output terminals. Therefore, the signal never has to be referenced to either the positive or negative supply. To first order, the supply noise couples equally onto both the positive and negative signal. When the difference is taken, this supply noise is cancelled out. The fully differential architecture also has the advantage that variations in current are reduced because increases in current in a positive branch of the circuit tend to be balanced by reductions in a negative branch. This technique is very effective, but it has the disadvantage that extra hardware and complexity is required in comparison to a single ended circuit.

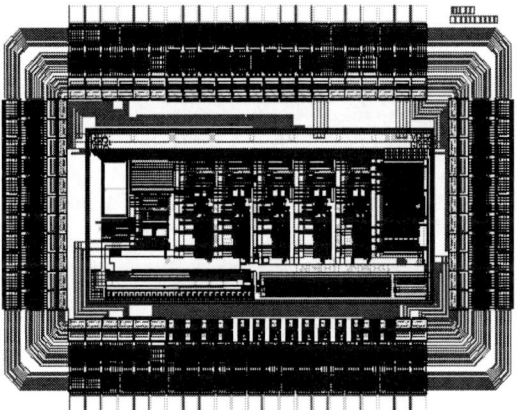

Figure 7 Layout of Pipeline ADC

7. Layout Considerations

When perform the layout of the pipiline ADC, much care was taken in the layout to isolate sensitive nodes in the circuit from sources of interference. For example, the sample and hold amplifier input is the most critical node of the integrated circuit, grounded dummy capacitors are used to shield the sampling capacitors from interference and to improve matching by making the edge effects similar. The signal lines are kept close together and surrounded on four sides by grounded conductors to minimize capacitive coupling of sources of interference onto the signal.

8. Conclusions

This paper presented a 40MSample/s，10-bit，3.3V pipeline ADC which employs a high bandwidth low-power amplifiers technique and a low power low offset dynamic comparators technique to achieve low power consumption. The pipeline ADC is designed in 0.35 μ m CMOS technology and occupies $1.2*0.8mm^2$. Figure 7 shows the layout of the pipeline ADC.

9. Acknowledgments

The authors would like to acknowledge the National Natural Science Foundation of China for providing financial support for this work under Grant No. 60773081 and Grant No. 60777018, and also to acknowledge the financial support by the Shanghai Municipal Committee under Grant No.AZ028 and Grant No.06DZ22013.

References

1. Shan-feng Cheng, "A Low Power 33MSample/ s, 10 Bit Pipel ine Architecture ADC". Jouarnl of Fudan University (Natural Science) Vol. 40 No. 3 Jun. 2001. pp.335-341.

2. David William Cline, "Noise, Speed, and Power Trade-offs in Pipelined Analog to Digital Converters", Ph.D. Dissertation, University of California, Berkeley, pp.320-323.

3. T. Byunghak Cho, P. R. Gray, "A 10 b, 20Msample/s, 35mW Pipeline A/D Converter," IEEE Journal of Solid-state Circuits, Vol. 30(1995), pp. 166- 172.

4. Lewis S H, Fetterman H S, Gross G F, et al, "A 10-b 20Msample/s analog-to-digital converter", IEEE J Solid-state Circuits, 1992, 27(3): 351

5. A. Baschirotto, "Switched-capacitor circuits," in Trade-Offs in Analog Circuit Design: The Designer's Companion, C. Toumazou et al, Eds. Netherlands: Kluwer, 2002, pp. 443459.

6. Dong-Young Chang, "A 1.4-V 10-bit 25-MS/s pipelined ADC using opamp-reset switching technique", IEEE J Solid-state Circuits, Vol. 38, No.8 (2003), pp. 1401- 1404.

Analytical Analysis on the Effect of Time Duration of Acceleration Pulse to a JEDEC Board in Drop Test

Jiang Zhou
Department of Mechanical Engineering
Lamar University, PO Box 10028, Beaumont, TX 77710, USA
jenny.zhou@lamar.edu

Abstract

The objective of this paper is to simulate a JEDEC test board dynamic response with the use of a simplified analytical model. A block-diagram based Matlab/Simulink model was also built to perform the parametric study. It is found that the desirable predominated mode and no-ringing dynamic response can not be achieved for the test board and input profile with current JEDEC standard. Time durations of the input acceleration plays an important role in the dynamic response. The system response can be designed by carefully choosing the acceleration time duration. It is very meaningful in the design of board level drop test in the electronic industries. With the time duration adjusting to 1.5 times of the system period for the standard JEDEC test board, no-ringing response occurs. A closed-form theoretical solution was obtained for half-sine acceleration pulse input. The analytical simulation was confirmed by the theoretical results.

1. Introduction

More and more people have been using portable telecommunication devices, such as mobile phones, personal digital assistances, laptop PCs, etc. It is not uncommon for those portable electronic products to be accidentally dropped onto the ground. Vulnerable elements inside such products may experience very high accelerations and dynamic stresses. This ultimately causes failures in solder joints, intermetallic layers at solder-pad interface, or board via cracking. The impacts and shocks thus can lead to the failure of electronic packages and the malfunction of the products. Manufacturers usually determine the fragility of such products by conducting experimental drop tests. JEDEC provided the standardized board level drop test method [1]. Usually, experiments are not only expensive but also time consuming. An alternative approach is to develop analytical dynamic models and to perform numerical simulations. Although a full-scale finite element analysis with complicated geometry can be carried out, it is very time-consuming, and the problem can be so complicated that it is difficult to capture some most important affected factors in engineering design. Therefore, it is necessary to develop an efficient analytical analysis for the problem [2-12].

In this paper, a predicative analytical model was developed to simulate the JEDEC board level drop test. Block-diagram based SIMULINK model was also developed to perform the parametric study. The FEA modal analysis was performed for the JEDEC test board. The foundamental frequency based on the modal analysis was used to find the the stiffness of the test board. The developed Matlab/Simulink model was used to check the dynamic response of a JEDEC test board. It is found that the desirable predominated mode and no-ringing

dynamic response can not be achieved for the test board and input profile with current JEDEC standard. Time durations of the input acceleration plays an important role in the dynamic response. In the other words, the system response can be designed by carefully choosing the impact time duration. It is very meaningful in the design of board level drop test in the electronic industries. With the time duration adjusting to 1.5 times of the system period, i.e., 7.5 milliseconds for the standard JEDEC test board, no-ringing response occurs. A closed-form theoretical solution was obtained for half-sine acceleration pulse input. The analytical simulation was confirmed by the theoretical results.

2. JEDEC Board Level Drop Impact Test Method

JEDEC provided the standardized board level drop test method. It recommends mounting 15 components on the test board in 3 rows of 5 components each. The test board size and layout is shown in Fig. 1. Figure 2 is the typical drop apparatus and mounting scheme. The test board is mounted on a base plate by 4 screws at the corners. This base plate is then mounted on a drop table. The drop table, guided by guide rods, is allowed to strike on a rigid base from some specified height. A half sine-impulse is produced when the table strikes the rigid base. A layer of felt is used on the strike surface to obtain the desired load conditions.

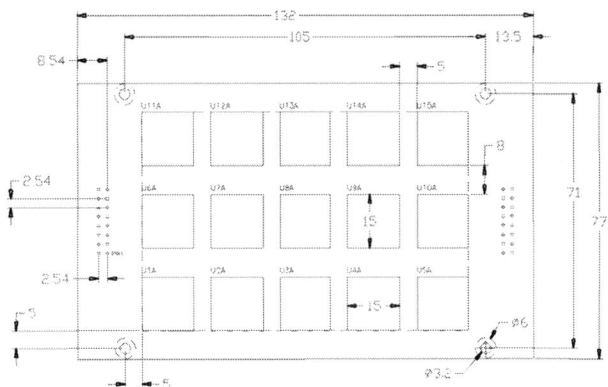

Figure 1. Test board size and layout [1]

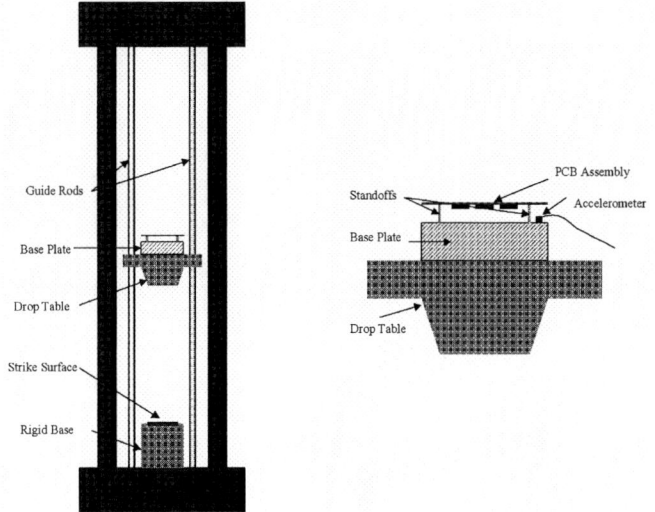

Figure 2. Typical drop apparatus and mounting scheme [1].

3. Modal Analysis with Finite Element Method

Modal analysis is performed by finite element method for the PCB board in the drop test. Commercial software ANSYS is used to determine the natural frequencies and mode shapes. Standard JEDEC test board full model was built with length of 133mm, width of 77mm, and the height of 1 mm. 15 component or chips of same type in a 3 rows by 5 columns format are mounted on the board. Material properties used in the calculation are listed in Table 1. The first five mode shapes are shown in Figure 3. The first five natural frequencies are around 200Hz, 535Hz, 901Hz, 1235Hz, and 2100Hz, respectfully, with a certain chip size. Different chip sizes from 6mm to 15mm were simulated. It was found that the fundamental frequencies are in the range of 200Hz to 240Hz.

Table1. Material properties used in FEA simulation

	Modulus (MPa)	Poisson Ratio	Density x 10^{-3} (gm/mm^3)
PCB	22000	0.25	2.1
Component	130000	0.278	2.5

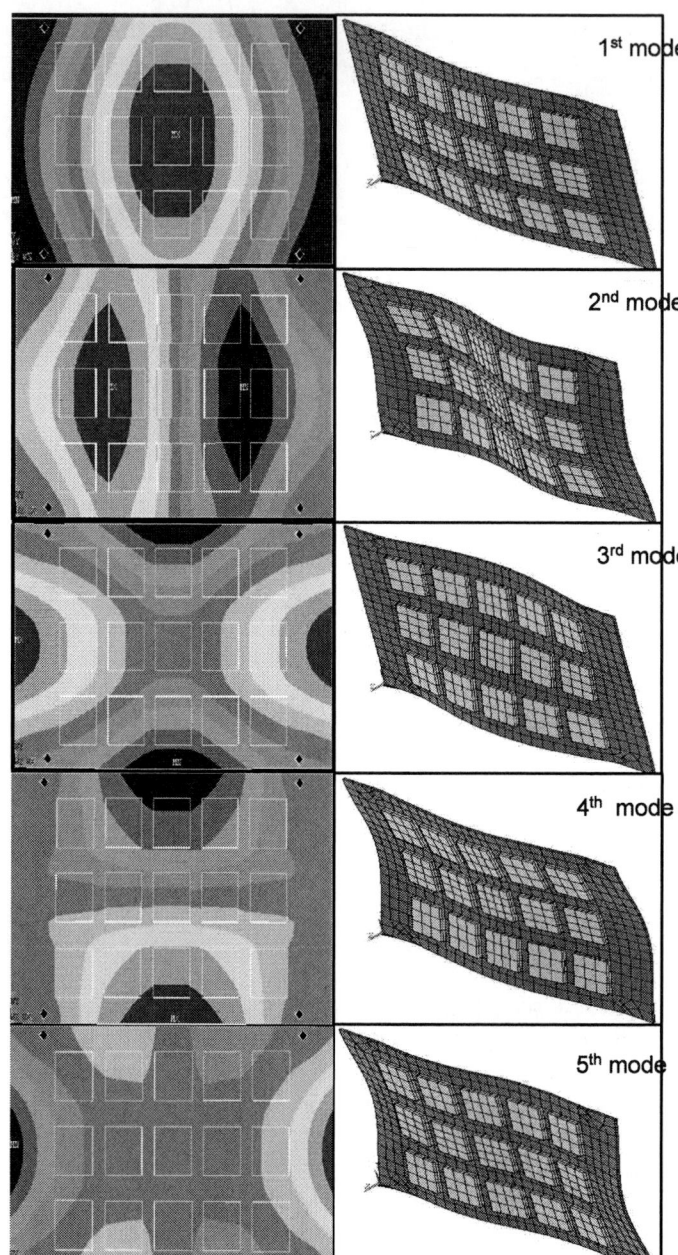

Figure 3. Model analysis results of the first five modes

4. Analytical Model for the JEDEC Board in Drop Test

The test board in the drop test can be simplified as the one-degree-of-freedom (1DOF) mass-spring-damper system. The equation of motion is

$$M\ddot{x}(t) + B\dot{x}(t) + Kx(t) = f(t) \qquad (1)$$

where M is the mass of the system, B is the damping coefficient, and K is the spring constant, x(t) is the maximum displacement of the test board of the system, and f(t) is the applied impact impulse.

JEDEC suggested that the test board is subjected to the half-sine pulse acceleration shown in Fig. 4. The mathematical expression is

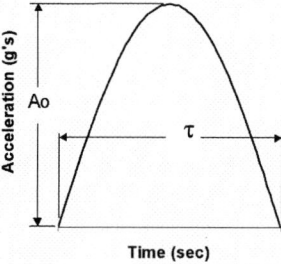

Figure 4. Typical shock test half-sine pulse, where A_0 is peak acceleration and t_w is time duration [1]

$$A(t) = A_0 \sin(\frac{\pi t}{\tau}) \qquad (2)$$

where, where A_0 is peak acceleration and t_w is time duration. Thus, the impact pulse $f(t)$ in Eq. (1) is assumed to be

$$f(t) = MA_0 \sin(\frac{\pi t}{\tau}) = G_P \sin(\frac{\pi t}{\tau}). \qquad (3)$$

In order to perform the parameter study easily, A MATLAB/SIMULINK model is developed to determine the system's dynamic response. SIMULINK is block-diagram-based model to analyze and design dynamic systems. It is an interactive, block-diagram-based tool for modeling and analyzing dynamic systems, and is tightly coupled with MATLAB and supported by blocksets and extensions. Using such a tool, the relationship between input and output can be obtained and visualized easily and quickly with selected system parameters. The block diagram of the system represented by Eq. (1) is sketched and the corresponding SIMULINK models are shown in Fig. 5. The input half-sine pulse is generated in Matlab workspace before the execution of the SIMULINK model.

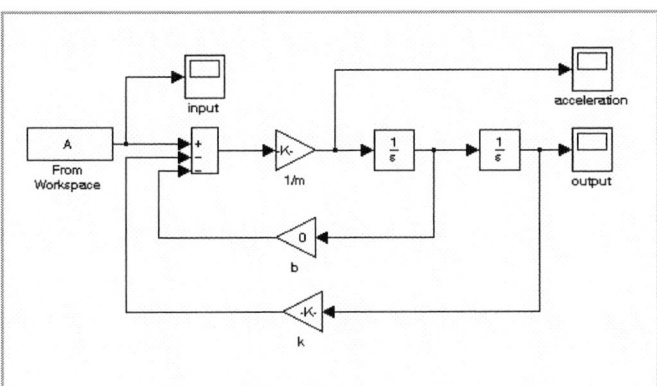

Figure 5. Block diagram and SIMULINK model

5. JEDEC Test Board Simulation

As described earlier, JEDEC test board is of the dimension as 133mmX77mmX1mm. The density of the PCB is given in Table 1. The weight of the components is negligible, compared with that of the test board. Therefore, the mass M is obtained as about 25 gram. Based on the FEA simulation results in Section 3, we assume the fundamental frequency of the test board is around 200 Hz. The system stiffness K can be calculated as ~40 N/mm, followed by the following equation.

$$\omega_n = \frac{1}{2\pi}\sqrt{\frac{K}{M}} \quad \text{Hz} \qquad (4)$$

For the first step analysis, the damping effect is neglected. The input half-sine function is defined by JEDEC standard as peak acceleration A_0 is 1500Gs and time duration τ is 0.5 milliseconds in Eq. (3).

With all the parameters defined above in the Matlab/SIMULINK model in Figure 5, the input pulse and system dynamic response for the output displacement and acceleration are obtained in Figures 6-8. The output

displacement is oscillating up and down with the same peak value.

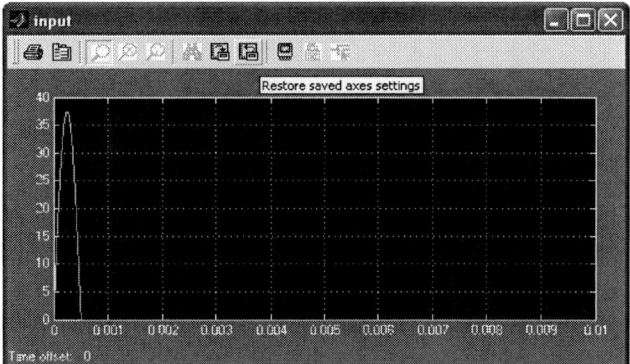

Figure 6. Input excitation to the JEDEC test board with standard input (1500Gs and 0.5 ms). The horizontal axis is time (seconds) and the vertical axis is the input excitation corresponding to Gs.

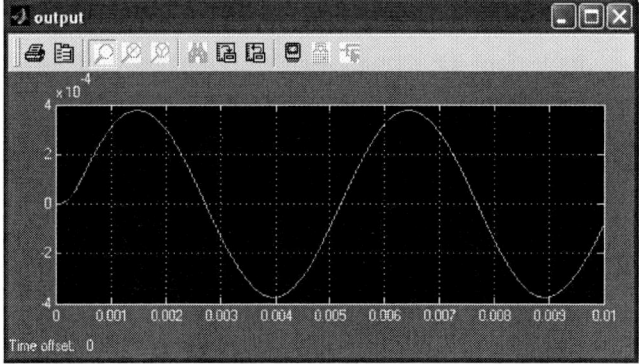

Figure 7. Output displacement of the JEDEC test board with standard input (1500Gs and 0.5 ms). The horizontal axis is time (seconds) and the vertical axis is the displacement (m).

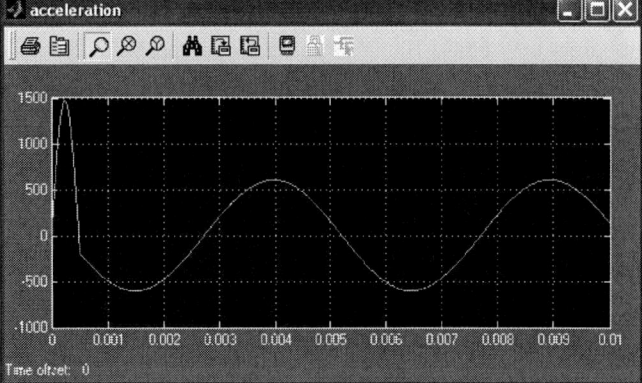

Figure 8. System response of acceleration of the JEDEC test board with standard input (1500Gs and 0.5 ms).. The horizontal axis is time (seconds) and the vertical axis is the acceleration (Gs).

In order to compare the system's input and output, the input half-sine acceleration and output acceleration is plotted in the same figure, as demonstrated in Fig. 9. The time duration of the input pulse is 0.5 milliseconds. During the first 0.5

milliseconds, the input and output are coincident with peak acceleration of 1500 Gs. Then the output acceleration drops to ~ 600 Gs, which is about 40% of the original value.

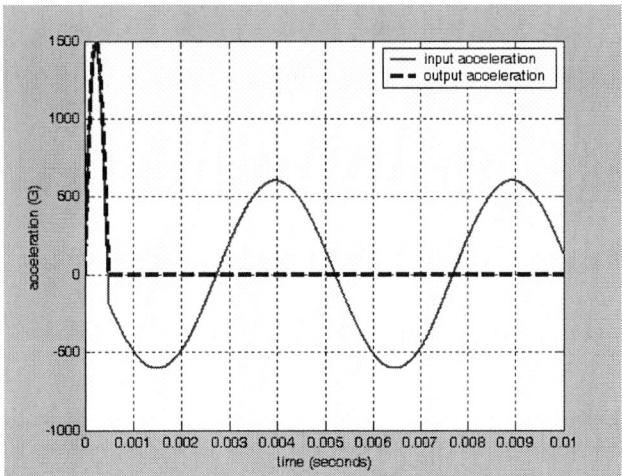

Figure 9. The input half-sine acceleration and output acceleration.

6. Discussion - Predominated Mode Response Condition

It is possible and also desirable to have the system response with one predominated mode, and the rest of the response is close to zero. Such kind of response helps greatly to identify and isolate the failure root cause only from the drop shock, and easily extract the solder joint stain life data from the experiment. It is obviously that the output displacement of the JEDEC standard test board in Fig. 7 is not such the case.

We know that the time period of the system is 5 milliseconds, since system's frequency was assumed to be 200 Hz. Now we change the input time duration to be 1.5 times system period, i.e., 7.5 milliseconds. The other system parameters keep the same. With the Matlab/SIMULINK model, the relationship between the input and output can be obtained and visualized easily and quickly with selected system parameters. The input pulse and system dynamic response for the output displacement and acceleration are obtained in Figures 10-12. In order to compare the system's input and output, the input half-sine acceleration and output acceleration is plotted in the same figure, as demonstrated in Fig. 13.

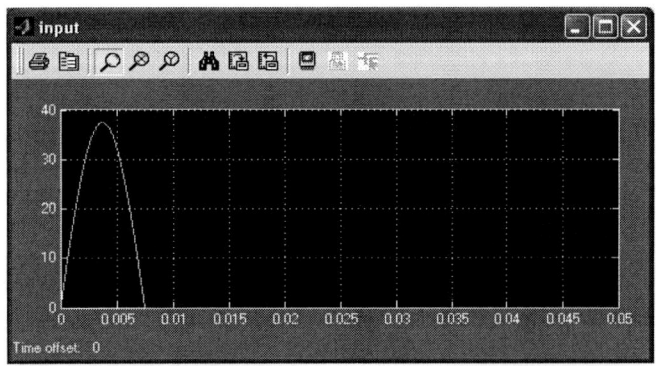

Figure 10. Input excitation to the JEDEC test board with input time duration of 7.5 milliseconds. The horizontal axis is time (seconds) and the vertical axis is the input excitation corresponding to Gs.

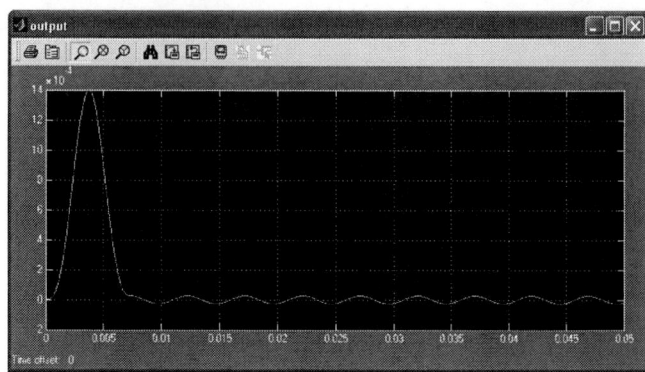

Figure 11. Output displacement of the JEDEC test board with input time duration of 7.5 milliseconds. The horizontal axis is time (seconds) and the vertical axis is the displacement (m).

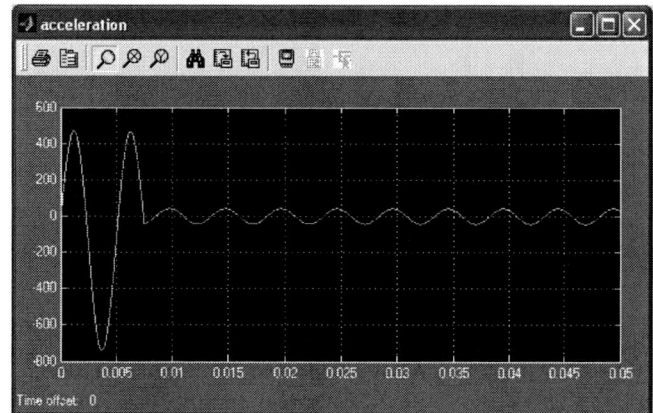

Figure 12. System response of acceleration of the JEDEC test board with input time duration of 7.5 milliseconds. The horizontal axis is time (seconds) and the vertical axis is the acceleration (Gs).

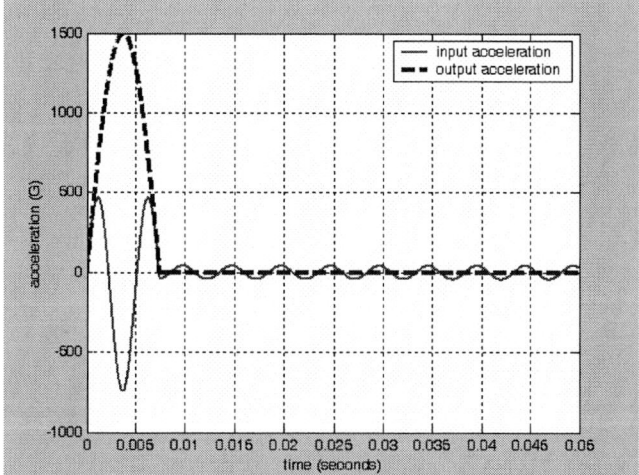

Figure 13. The input half-sine acceleration and output acceleration.

We are delighted to see the displacement response in Fig. 11 has a predominated mode, and the other modes are close to zero with almost no ringing. We can see the no-ringing phenomena better when we compare the Fig. 7 with Fig. 11. The output acceleration is first coincident with the input to a

490

point of ~ 1/3 of the original value, and then goes out of phase, and then returns back to the last 1/3 of input curve. The acceleration value is close to zero after 7.5 milliseconds.

7. Theoretical Closed-form Results

The theoretical closed form solution can explain the reason of the predominated mode and no ringing phenomena [2]. With the half-sine input of magnitude G_p and duration τ, the equations and initial conditions can be expressed piecewise by Eqs. (13) and (14).

$$\begin{cases} m\ddot{x} + kx = G_p \sin(\pi t / \tau) \\ x(0) = \dot{x}(0) = 0 \end{cases} \quad \text{when } t \leq \tau \quad (5)$$

and

$$\begin{cases} m\ddot{x} + kx = 0 \\ \text{ICs } x(\tau), \dot{x}(\tau) \text{ obtained from Eq (13)} \end{cases} \quad (6)$$

$$\text{when } t > \tau .$$

The displacement response of tested board subjected to a half-sine pulse can be expressed by the following two equations, i.e., Eqs (7) and (8).

$$x(t) = \frac{G_p}{m} \frac{1}{\omega_n^2 - (\pi/\tau)^2} \left[\sin\left(\frac{\pi t}{\tau}\right) - \frac{\pi}{\tau\omega_n} \sin(\omega_n t) \right], \text{ when } t \leq \tau$$

$$(7)$$

and

$$x(t - \tau) = -\frac{G_p \pi}{m\tau\omega_n \left(\omega_n^2 - (\pi/\tau)^2\right)} \left\{ \sin(\omega_n t) + \sin[\omega_n(t - \tau)] \right\}$$

$$\text{when } t > \tau \quad (8)$$

It is found, from Eq. 8, that when $\tau = (i + \frac{1}{2})/\omega_n$, where i = 1, 2, 3, ..., the no ringing occurs. This is the reason that we picked time duration 1.5 times system period, i.e., 7.5 milliseconds.

6. Conclusions

The following conclusions are drawn from the analyses of this paper.
- A predicative analytical model was developed to simulate the JEDEC board level drop test. Block-diagram based SIMULINK model was also developed to perform the parametric study.
- The FEA modal analysis was performed for the JEDEC test board. The first five modes were obtained. The foundamental frequency is around 200 -240 Hz.
- The desirable predominated mode and no-ringing dynamic response can not be achieved for the test board and input profile with current JEDEC standard.
- With the time duration adjusting to 1.5 times of the system period, 7.5 milliseconds for the standard JEDEC test board, no-ringing response occurs.

- A closed-form theoretical solution was obtained for half-sine acceleration pulse input. The analytical simulation was confirmed by the theoretical results.
- It is found that certain input pulse time results in no ringing response. For the half-sine input, the no ringing conditions are $\tau = (i + \frac{1}{2})/\omega_n$, where i = 1, 2, 3, ...
- Time durations of the input profiles play an important role in the dynamic response. In the other words, the system response can be designed by carefully choosing the impact time duration. It is very meaningful in the design of board level drop test in the electronic industries.
- The developed Matlab/SIMULINK models include the damping constants, while the results here were obtained by ignoring the damping. Further research can be performed by the models and theoretical analysis for the effect of damping.
- SIMULINK is a very powerful and useful tool in the design of the portable electronic products to quickly extract the system response and analyze the stability. The visual interface of the developed models presents results in a way that one can immediately identify the effects of changing inputs and system parameters.
- The developed models can be applied not only to the design of the portable electronic products, but to other engineering applications where dynamic response analysis is needed.

References

1. JEDEC Standard JESD22-B111, Board Level Drop Test Method of Components for Handheld Electronic Products.
2. Zhou J., "Analysis on the Effect of Input Impact Profiles in Drop Test," Procedings of IMECE 2007, ASME International Mechanical Engineering Congress and Exposition, November, 2007, Seattle, Washington
3. Tee T.Y., Ng H.S., Lim C.T., Pek E., and Zhong Z., "Board Level Drop Test and Simulation of TFBGA Packages for Telecommunication Applications," *The Proc. of the 53rd IEEE/EIA Electronic Components and Technology Conference*, 2003.
4. Tee, T. Y., Ng, H. S., Lim, C. T., Pet, E., and Zhong, Z. W., "Impact Life Prediction Modeling of TFBGA Packages under Board Level Drop Test", *Microelectronics Reliability Journal*, Vol. 44(7), pp.1131-1142. 2004.
5. Suhir E., "Could Shock Tests Adequately Replace Drop Tests?" *52nd Electronic Components and Technology Conference*, pp 563-73, June 2002.
6. Steinberg, D. S., Vibration Analysis for Electronic Equipment. 2nd Ed., John Wiley & Sons, New York, NY, 1988
7. Zhou, J., Kallolimath K., and Lahoti, S., "Analytical and numerical analysis of drop impact behavior for a portable electronic device," *IEEE International Conference on Thermal, Mechanical and Multiphysics Simulation and Experiments in Micro-*

Electronics and Micro-Systems (EuroSimE), April 17-20, 2006, Milan, Italy, pp 138 – 144

8. Zhou, J., Corder, P., "Block-diagram based SIMULINK analysis for the drop impact response of a mobile electronic system," IEEE International Conference on Thermal, Mechanical and Multiphysics Simulation and Experiments in Micro-Electronics and Micro-Systems (EuroSimE), April, 2007, London, England

9. Suhir E., "Dynamic Response of a Rectangular Plate to a Shock Load, with application to Portable Electronic Product," IEEE Transactions on Components, Packaging and Manufacturing Technology, Vol.17, No. 3, 1994.

10. Suhir E., "Is The Maximum Acceleration An Adequate Criterion Of The Dynamic Strength of a Structural Element in an Electronic Product?" *IEEE Transactions on Components, Packaging, and Manufacturing Technology*, Vol. 20, No. 4, December 1997.

11. Lall, P., Panchagade, D., Liu, Y., Johnson, W., Suhling, J., "Models for Reliability Prediction of Fine-Pitch BGAs and CSPs in Shock and Drop-Impact", *54th Electronic Components and Technology Conference*, June 2004.

12. Luan J., and Tee T. Y., "Effect of Impact pulse parameters on Consistency of the Board level Drop Test and Dynamic Responses," *Electronic Components Technology Conference*, 2005.

Thermal Management of A Multi-core Master Processing Unit (MPU) for An Ultrascalable Computing Platform

Ting Cheng[1], Wei Xiong[1], Xiaobing Luo[13], Suyi Huang[1], Zhiyin Gan[23], Sheng Liu[23*]

[1] School of Power and Energy Engineering, Huazhong University of Science & Technology, Wuhan, China, 430074
[2] School of Mechanical Science and Engineering, Huazhong University of Science & Technology, Wuhan, China, 430074
[3] Division of MOEMS, Wuhan National Laboratory for Optoelectronics, Wuhan, China, 430074
*Corresponding Author: Sheng Liu, Email: victor_liu63@126.com, Telephone: 86-13871251668, Fax: 86-27-87557074

Abstract

Thermal management of a super computer is one key point in the whole system design. The working temperature of electronic device in the super computer is the important parameter to show the cooling performance. Thermal design was conducted for one super computer. The experiment was conducted to test the temperature of an ultrascalable computing platform in order to evaluate the cooling performance, the results demonstrate that the thermal management by such a simple method is good.

Introduction

With the increasing clock rate of microprocessors and high density electronic components in super computer[1], power dissipation is becoming a critical problem of super computer design. As we know, electronic devices are especially sensitive of the temperature, their performances descend due to the increasing temperature[2], so thermal issues are becoming especially critical for high-performance super computer. Thermal management of the equipment is becoming a challenging task for designers [3].

In this paper, one simple thermal management method for a super computer was proposed, the experimental study on the thermal management of one MPU was conducted. The experimental test data shows the cooling performance of a pin-fin sink and fan thermal management system is good.

Experimental procedures

The study goal is to test the cooling system of a multi-core master processing unit (MPU) for an ultra-scalable computing platform. Figure 1 shows the components of this super computer, it consists of several MPU to realize the calculation function. The structure of a MPU is clearly demonstrated in Figure 2, it contains sixteen calculation flashboards as shown in Figure 3, several communication boards and power modules. For all boards and modules, they generate much heat when computer works. To lower the temperature inside the MPU, several fans were distributed inside the MPU as shown in Figure 2. One row of five fans were placed on the cabinet left wall, another row of five fans were placed between the communication boards and the calculation boards in the middle of the whole cabinet. As for each board and module, the heat sink was used for cooling core component. The power of a CPU was nearly 50W when it is full loaded; FPGA has a full power of 30W. To test the cooling effect, thermocouples were placed on the bottom of fin heat sinks.

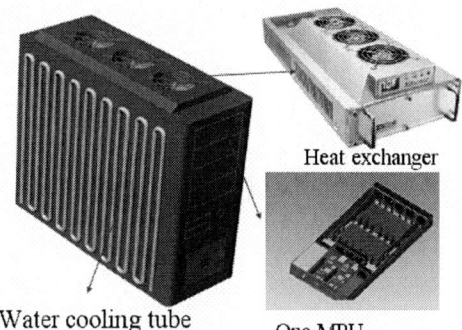

Figure 1. The picture of a super computer.

Figure 2. The picture of one MPU module.

Figure 3. The picture of one calculation flashboard.

Ten thermocouples were attached at ten different heat sinks on calculation flashboards and two thermocouples were attached at two different heat sinks on one communication module. Figure 4 demonstrates the thermocouple distribution on different heat sinks, in Figure 4, six thermocouples were located on the top of CPU heat sinks. They were marked as number 1, 3, 8, 11, 17 and19, six thermocouples were located on the top of FPGA heat sinks. They were marked as number 13, 7, 9, 10, 18 and 20. Figure 5 shows the experimental setup, the temperature data obtained by the thermocouples was transferred to the data acquisition system and reported on the PC monitor, the model of the data acquisition system in the experiment are Keithley 2700 multimeter and control unit 7700.

Result and analysis

Since different usage conditions of the super computer will result in different heat load, several cases were tested. The test cases in which the super computer works in three different loads were as follows, 1) all the CPUs were load free; 2) the calculation CPUs worked at between 67% and 69% of full load, the communication CPUs worked at between 73% and 75.5% of full load; 3) the calculation CPUs and the communication CPUs both worked at 99% of full load. The ambient temperature was about 19℃.

The results show that when all the CPU load free, the temperature of communication module CPU heat sink was as high as 51℃, all the heat sinks had a short balance time, other temperatures achieved by different thermocouples were shown as Figure 6. When the calculation CPUs worked at between 67% and 69% of full load and the communication CPUs worked at between 73% and 75.5% of full load, the temperature of communication module CPU heat sink was 49℃, all the heat sinks temperature changed little . Other temperatures were showed as Figure 7, when the calculation CPUs and the communication CPUs both worked at 99% of full load, the temperatures of heat sinks were different at first, but several minutes later they balanced at the same degree after the load was free. All these test results demonstrate that the thermal management solution based on the fan and fin could be used.

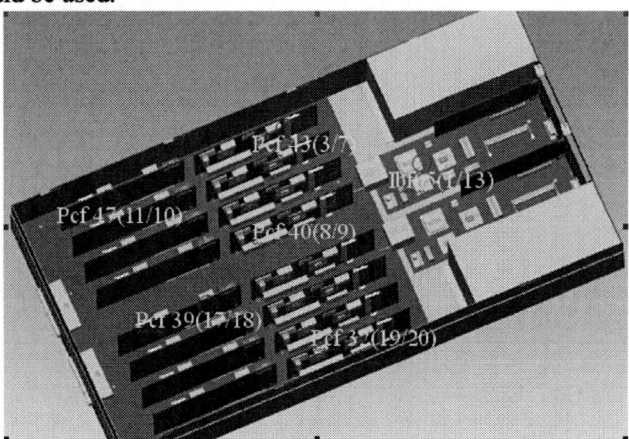

Figure 4. Thermocouple distribution at various positions of the MPU module.

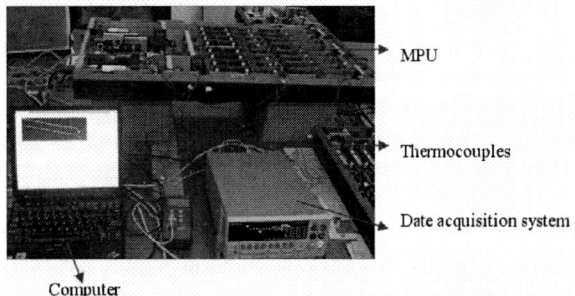

Figure 5. Experimental setup.

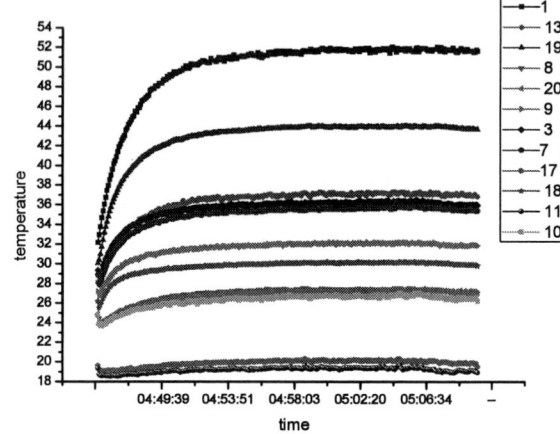

Figure 6. Variation of temperature of the CPU heat sinks with the operation time for all of them worked at empty load.

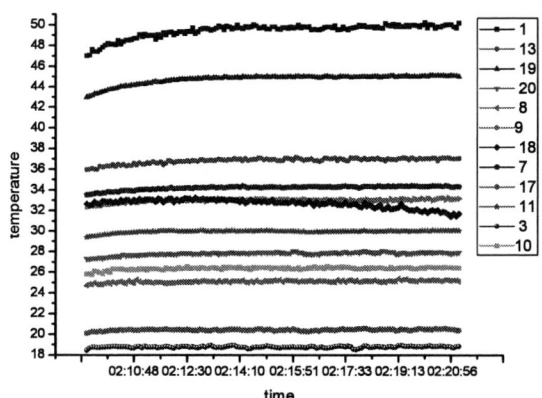

Figure 7. Variation of temperature of the CPU heat sinks with the operation time for the calculation CPUs worked at between 67% and 69% of full load, the communication CPUs worked at between 73% and 75.5% of full load.

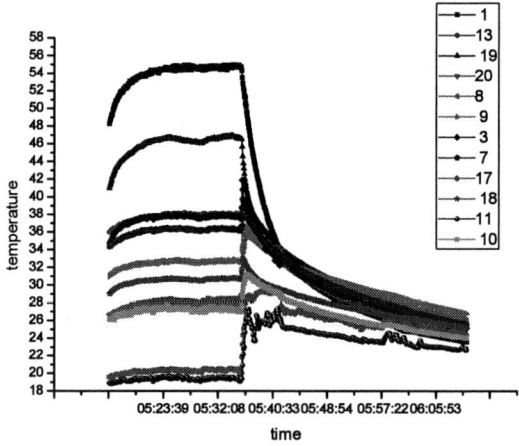

Figure 8. Variation of temperature of the CPU heat sinks with the operation time for the calculation CPUs and the communication CPUs loads change.

According to the data shown in Figure 8, there is an obvious working temperature difference for different boards at different loads. The calculation program can maintain the electronic devices running at stabilized output. Different working temperatures are caused by the air flow in the box. The spatial arrangement of the boards also affects the temperatures. The air flow blowing direction from the fans is stochastic. Air flow in the cabinet is non-form, especially when it is near cabinet. The flow from two rows fans was interacted. In addition, electronic elements in the MPU disturbed the airflow. All the above factors led to different temperatures.

Conclusions

Powerful fans, plate or fin heat sinks are still effectively used in computer cooling and they are less cost, easy maintenance. In this paper, heat sink with fans were used for thermal management of super computer, experimental tests were also conducted. The temperature distribution at different locations and different loads were achieved. The results demonstrate the present cooling system works well.

Acknowledgments

The authors would like to thank Shanghai Redneurons Co. Ltd for providing experimental test environment.

References

1. Bhave.Ninad, Okamoto.Nicole, "Modeling non-coplanarity effects on thermal performance of computer chips", ISAPM. 4419927; 3-5 Oct. (2007), pp.47 – 52

2. Brooks.D, Martonosi,.M., "Dynamic thermal management for high-performance micro-processors", HPCA, (2001), 903261; 19-24 Jan, pp.171 - 182

3. Nakanishi.M, Nakayama.W, Behnia.M, Soodphakdee.D," A new approach to the design of complex heat transfer systems: notebook-size computer design", THERM 2002. The Eighth Intersociety Conference, pp.595 - 599

Investigation of Thermal Performance of Various Power-Device Packages

Xuejun Fan[1,2]

[1]Department of Engineering Mechanics South China University of Technology Guangzhou 510640, China
[2]Department of Mechanical Engineering Lamar University, P.O. Box 10028 Beaumont, Texas 77710, USA
E-mail: xuejun.fan@lamar.edu

Abstract

Continuing trends of miniaturization, rising switching frequencies and increasing packaging densities require increased current handling capability of packaged devices in applications related to power conversion. Traditionally, these ever-increasing demands are met by improvements in silicon efficiency. Nevertheless, with silicon efficiency pushed to the limit, major semiconductor power-device manufacturers are now looking for innovative packaging options for power devices to achieve the next level of breakthroughs in electrical and thermal performance. This paper presents a comprehensive study of thermal behaviors of various power-device packages. CFD-based FLOTHERM has been applied to calculate the junction-to-ambient thermal resistance with the industry standard-specified board attachment. Fundamental cooling mechanisms associated with different packaging technologies, including wire-bond, strap bonding, flip chip and ball grid array (BGA), and wafer-level packaging are investigated. The impact of internal package design on the thermal performance of various packages is discussed in detail. A thermal analysis of multichip module for leadless and BGA technologies is also presented.

Introduction

Applications demanding high-power conversion such as voltage regulator module for microprocessors, automotive electronics and telecommunications, have introduced a trend for achieving higher power densities at lower cost [1,2]. Over the past decades, this trend has been successfully met by increasing silicon efficiency; however, future requirements dictate further improvement in overall system efficiency, which can only be achieved through innovations in packaging [3-5]. Accordingly, in recent years, semiconductor industry has taken aggressive steps towards achieving small form factor power packages with significant improvements in electrical and thermal performance. From traditional plastic injection-molded and wire-bonded package, power packaging has come a long way where state of the art IC packaging techniques such as ball-grid-array, chip-scale packaging and leadless, and wafer-level packaging are being used [6-9].

Leaded packages such as TO-220s and axial leaded devices had been the packaging configurations of power devices for the longest time. However, as miniaturization and functional integration became the dominant drivers for electronics components and modules, new technologies emerged[2]. The DPAK package was introduced first in mid-1980, which caused a major paradigm shift in the package design arena. At the same time, an alternative package with SOT-223 came along, which offered smaller outline than DPAK yet used the footprint and pin-outs. Then came D2PAK in the early 1990s, which offered usage of bigger die size in a package, thus increasing current handling capability

significantly and reducing thermal resistance of the package to some extent. The SO-8 packages were introduced in the mid-1990, which allowed significant size reduction compared to DPAK packages and resulted in fewer packages required for assembly while reducing board space. This improvement was made possible since the die size reduction and more cells per inch of silicon enabled designers to achieve the same type of RDS(on) performance in SO-8, which was previously only available in TO-220 and DPAK configurations [8].

However, continued miniaturization drive demanded even smaller and more efficient package than SO-8 packages to meet future thermal and electrical requirements. This new surge of design revolution has two specific market demands – higher current requirements in microprocessors and low RDS(on) in a smaller area for power management. Accordingly, a new stream of packaging configurations emerged for packaging discrete power devices as well as multi-chip modules since the beginning of 2000 [7]. Figure 1 presents a snapshot of the evolution of power-device packaging from the leadframe based to flip chip and ball grid array technologies.

Liu et al [10, 11] introduces the general methodology of the simulation interface and platform. This simulation tool has been applied to MLP package family. The results agree well with those from the classic ANSYS and measurement. Tounsi et al. [12] developed specific thermal simulation tools to perform electro-thermal simulation of power devices or circuits. 3D and transient flow spreading effects in multilayered substrates commonly used in power component packaging as well as in hybrid power circuits were considered. Kasem [13] investigated the influence of the design and physical limitations on the performance of thin quad flat packages (TQFPs) b using a 3-D finite element scheme. A methodology for low profile 48-lead TQFPs was outlined. Ganesa-Pillai and Chen [14] presented a finite-element thermal analysis of a boost converter module, which integrated all the semiconductor devices and the snubber circuits of a boost converter on a ceramic substrate. The effects of different substrates and use of multiple current sharing components were examined. Katsis and Van Wyk [15] compared the thermal impedance of modules with varying void area at a constant power dissipation level in order to develop a relationship between thermal impedance and void area. The effect of aging on thermal transient behavior was correlated to finite element thermal simulations. Chiriac and Lee [16] performed a detailed thermal analysis for the wirebonded GaAs devices by using numerical simulations. The main focus was on the impact of die attach thermal conductivity, substrate's top metal layer thickness, and via wall thickness on the overall thermal performance of GaAs IC device . Arik Garg, and Bar-Cohen [17] explored the thermal challenges in advanced system-on-package (SOP)

978-1-4244-2739-0/08/$25.00 ©2008 IEEE

electronic structures, as well as candidate thermal solutions for these highly demanding cooling needs. Detailed three-dimensional (3-D) finite-element simulations were used to study the temperature distributions in a typical SOP package, and to provide guidance for the development and implementation of "compact thermal models". Direct liquid cooling by immersion of the components in inert, nontoxic, high dielectric strength perfluorocarbon liquids was seen effective over a range of anticipated SOP power dissipations. Chiriac and Lee [18] performed a detailed numerical study to examine the thermal characteristics of a chip set at the system level. The chip set included the Power Amplifier (PA) module, power management and base-band packages, front-end receiver package and memory. Detailed solid modeling was applied to the PA module with the GaAs (Gallium Arsenide) device bonded to a multi-layer ceramic substrate. Frank [19] discussed two methods of defining the thermal junction-ambient resistance and the commonly used wave solder assembly technology. The test setup and the results of tests done with various packages and transistors were also described. Kandasamy and Subramanyam [20] numerically evaluated the performance of the package different die sizes and apoxy molding compounds at different power levels. The use of heat slug was investigated to identify its effect on heat dissipation for IC generation.

In this paper, a comprehensive study to investigate thermal behaviors of various types of power-device packages was presented. The packages under investigation include traditional DPAK, D²PAK, leadless package, SO, flip chip, BGA, and wafer-lever CSP packages. We have analyzed the motivations for small form factor packaging of power devices from a power conversion point of view. Current and future design requirements of voltage regulator modules for microprocessors are also analyzed. The extensive thermal simulations have been performed to understand the thermal limitations with regard to each technology and the possible options to improve the thermal performance.

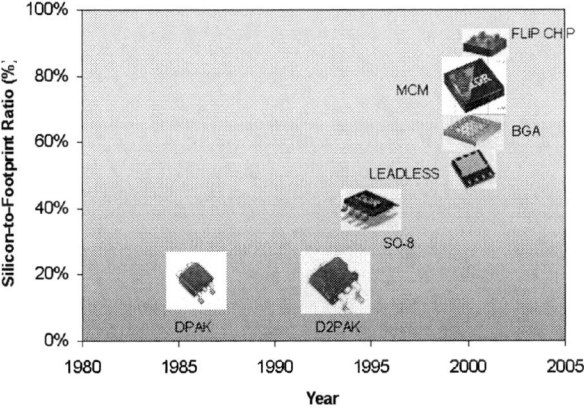

Fig. 1 Evolution of discrete and multi-chip packaging configurations of power devices

Background

The junction temperature of power device depends on many factors, including the packaging design (interconnect and structure), board selection (material, structure and interface), heat sink attachment (heat sink design and interface materials), ambient conditions (ambient temperature and air flow speed), as well as the system integration. The package design plays a very important role in overall thermal management because it provides the first 'gate' for the heat dissipation from the silicon chips. The demands of the continuing miniaturization of the system with maximized thermal performance require innovative package designs.

The thermal resistance is defined as

$$R = \frac{T_{j,\max} - T_{\text{ref}}}{P} \qquad (1)$$

where $T_{j,\max}$ is the maximum junction temperature, T_{ref} is the reference temperature and P is the power dissipation of the package. There are different selections of the reference temperature, such as following:

R_{j-a}: junction-to-ambient thermal resistance, where T_{ref} is taken as the ambient temperature;

R_{j-c}: junction-to-case thermal resistance, where T_{ref} is taken as the maximum case temperature on the top of the package;

R_{j-b}: junction-to-board thermal resistance, where T_{ref} is taken as the maximum board temperature that the package is attached to;

R_{j-l}: junction-to-lead thermal resistance, where T_{ref} is taken as the lead package for leaded packages.

Thermal resistance such as R_{j-c}, R_{j-b}, or R_{j-l} is a local measure for the thermal performance in a particular heat dissipation path. For example, R_{j-b} represents the thermal resistance between the junction and the board only and therefore, does not fully represent the thermal performance of the overall package. Therefore, in our investigations, the thermal resistance of a package is defined as the junction-to-ambient thermal resistance in steady-state, i.e., R_{j-a}.

Unless otherwise stated, the package is mounted on 1 in² of 2 oz copper on FR4 in our following investigations according to the industry standards. The ambient temperature (temperature in chassis) is assumed to be 50°C under the natural convection condition. The effects of the ambient temperature and the air speed (forced convection) on the thermal resistance are also briefly addressed. The FLOTHERM simulation tool has been used for the analysis.

Modeling

For the leadframe-based packages, the lead fingers are modeled discretely for DPAK, D²PAK, and SO-8 packages. When the leadframe has a great number of leads, the detailed modeling will lead to the huge amount of grid cells in the numerical analysis. Therefore, a lumped cuboid with an orthotropic (i.e., directionally dependent) conductivity is used. This means that the conductivity in the direction along the leads would be obtained by an assumption of parallel resistance, whereas the conductivity in the transverse direction would be obtained by an assumption of series resistance. This would avoid spurious spreading of the lumped leadframe cuboid in the transverse direction in the model. The results show that the lumped model presents an excellent agreement (1% difference in junction temperature) with detailed model. For the flip-chip/BGA packages, the solder balls are lumped and replaced by an equivalent volume of same material. The gaps between the solder balls, when underfill is not applied, are filled with air. Considering that the thickness of gaps is rather small (~ 100μm), the air inside is almost in still state, and is modeled in conduction mode.

Most of studies do not consider the impact of radiation effect. This may be true in forced convection condition, in which convection becomes dominant in heat dissipation. However, under the natural convection condition, the radiation effect may be comparable to the convection. In Fig. 2, the differences of thermal resistance due to the radiation effect are shown for a DPAK package.

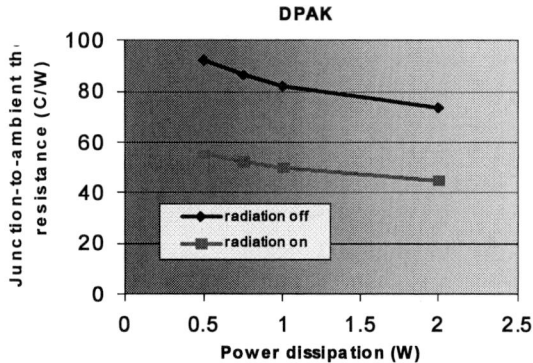

Fig. 2 Radiation effect on thermal resistance for DPAK packages in natural convection

Radiation effects play a very significant role in heat dissipation under natural convection. Almost 50% of heat is dissipated through the radiation. This suggests that in thermal modeling of a package under natural convection, the radiation effect must be turned on. Therefore, all the thermal models developed in this paper will consider the radiation effect.

DPAK and D²PAK Packages

DPAK or D²PAK packaged power devices have excellent current handling and thermal dissipation capabilities. Table 1 lists the typical dimensions of the DPAK and D2PAK packages, respectively. The package dimension in the table refers to the foot-print-area. It can be seen that D²PAK is almost twice as large as DPAK in terms of die area and package thickness

Table 1 Dimensions of DPAK and D²PAK (unit: mm)

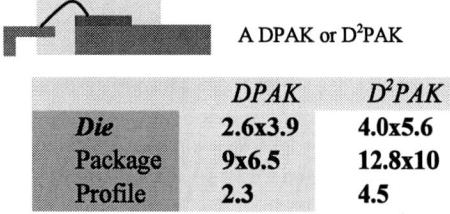

A DPAK or D²PAK

	DPAK	*D²PAK*
Die	2.6x3.9	4.0x5.6
Package	9x6.5	12.8x10
Profile	2.3	4.5

Fig. 3 plots the thermal resistance as function of power dissipation for DPAK and D²PAK respectively. The differences in thermal resistance are not significant (less than 5°C/W), in spite of obvious differences in physical sizes between DPAK and D²PAK. This is primarily because the package is mounted on 1 in² of 2 oz copper on FR4. The uniformly-distributed copper on the top of FR4 makes the board highly thermal conductive. As an extreme opposite situation where the package is mounted onto 1 in² of bare FR4 only with 0.3 W/m²K thermal conductivity, Table 2 list the

results of the thermal resistance for both DPAK and D²PAK packages. It clearly shows that the board material has significant influence on the package thermal performance. This means that the better thermal performance may be achieved even with smaller packages by effective board design. Nevertheless, without loss of generality, in our following studies all thermal resistance data refer to the 2 oz copper on FR4 according to the standards. This makes it easier to compare the simulated data with the published data.

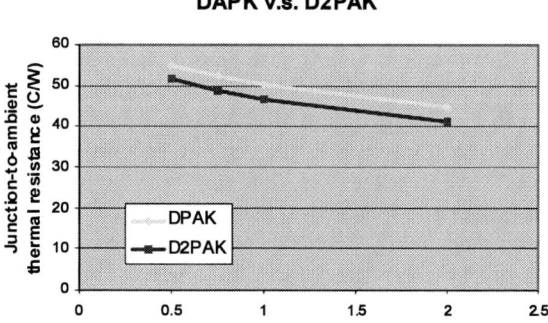

Fig. 3 Thermal resistance as function of power dissipation for DPAK and D²PAK packages

Table 2 Thermal Resistance of DPAK and D²PAK Packages

	DPAK	*D²PAK*
Bare FR4	186.6	104.5
2oz copper on FR4	50.2	46.6

Standard Outline (SO) Packages

The original standard IC SO-8 package uses wires on the top of chip to connect all leads (see. Fig.4a). The standard power MOSFET SO-8 package uses wires to connect the source and gate to the leads, but for the drain sides, the leads are directly connected with the die pad, as shown in Fig. 4b. Thermal behaviors between these two packages can be very different, which will be discussed later. In order to improve the thermal and electrical performance of SO-8 packages, a solid copper strap that covers the surface of the die is adopted to replace the wirebonds connecting source to the leadframe (Fig. 4c). This provides highly conductive path in addition to the existing path through the die pad connected to the leads.

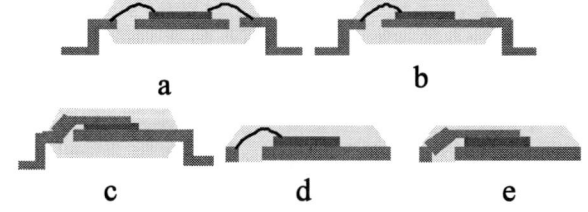

Fig. 4 Various types of SO-8 packages a: standard IC; b: standard power MOSFET; c: strap-bonded; d: leadless; e: strap-bonded with leadless

Thermal performance can be further improved by providing a direct path from the backside of the copper die attach pad to the board (Fig. 4d). The main thermal path is through the large copper pad exposed on the bottom of the package, which would improve thermal resistance dramatically. This package is usually called leadless package or micro-leadframe package.

Fig. 5 plots the thermal resistance of different types of SO-8 packages with same die size and package dimension. We notice a huge difference in thermal resistance between the standard IC SO-8 and standard power MOSFET SO-8. The direct connection of copper die pad to the leads in standard power MOSFET SO-8 provides a path for heat dissipation to reduce the junction temperature dramatically. Reduction of thermal resistance by using strap bonding is about 15.2% over the standard power SO-8 package, as shown in Fig.5. When the leadless package format is adopted, the thermal resistance can be reduced to 52°C/W, comparable to the thermal resistance of DPAK packages, but with much smaller package dimensions and lower profile. The thickness of DPAK and leadless packages are 2.3 mm and 1.1 mm, respectively. Fig. 5 suggests that even though the package dimension and chip size remain unchanged, thermal performance can be significantly improved by an appropriate package design and improved interconnect. For example, with an ambient temperature of 50°C and 1 Watt power dissipation for the chip, the junction temperature with standard IC SO-8 package is as high as 201.8°C. However, with leadless package, the junction temperature drops to 102.3°C.

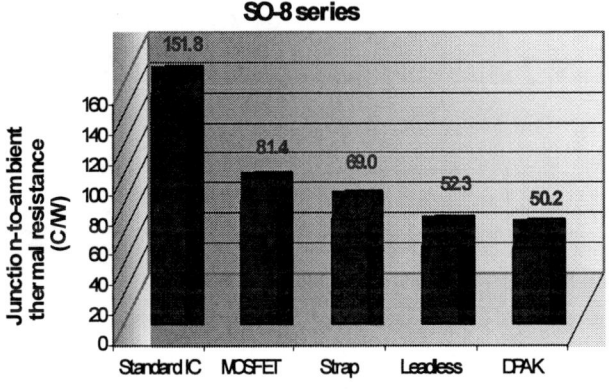

Fig. 5 Comparison of thermal resistance for various SO-8 packages

The question remains if any further thermal benefit could be gained with the strap bonding applied to a SO-8 leadless package, as shown in Fig. 4e. Fig. 6 plots the simulation results compared to other types of power SO-8 packages. The improvement is almost negligible in comparison to leadless packages from a thermal point of view. The heat dissipation by the direct contact of copper pad to the board in leadless package is so dominant that heat dissipation through strap bonding is minimal.

Fig. 5 and Fig. 6 present a clear picture how the package design affects the thermal behaviors. We notice that in Fig. 4, three of these packages, i.e., standard power SO-8, strap bonded, and leadless packages, have applications in power electronics. However, the comparisons of these packages to the standard IC SO-8 and strap boned leadless packages,

reveal the fundamental cooling mechanism associated with the package design. The package thermal performance can be enhanced by providing a direct path for the heat dissipation between chip and board. When a package is designed such as leadless, in which the heat dissipation is maximized in one path, the improvement by additional heat path would be insignificant.

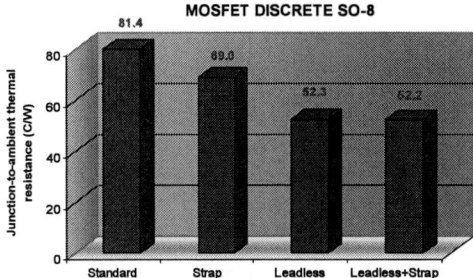

Fig.6 Comparison of strap bonding applied to leadless SO-8 with other types of SO-8 packages

Flip Chip (FC)/Ball Grid Array (BGA) Packages

Flip-chip-on-board or wafer-level power package uses the chip-scale packaging technology to bring all of the terminals on a single side of the die (see Fig. 7). Since the die itself is the package, a flip chip on board package has a nearly 100% silicon-to-footprint ratio. The main heat dissipation path will be the solder balls and underfill (if any). Although the chip is directly attached to the board through the solder balls, which gives the direct path of heat dissipation, questions remain how the solder ball geometry, layout and number of solder balls would affect the thermal performance. Would it be necessary to use the underfill from a thermal point of view? In order to address these questions, thermal models are developed here with chip size same as that used for SO-8 package. A single solder ball diameter is assumed to be 250μm and the pitch between the solder balls is 800μm. Three cases for solder ball layouts, 2x2, 3x4, and 4x5 arrays, are considered, as shown in Fig. 7.

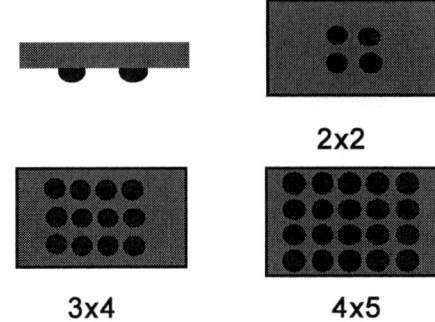

Fig. 7 Flip chip on board packages with different numbers of solder balls

In Fig. 8 the thermal resistance is shown for a flip-chip-on-board package with different combinations of solder balls when underfill is not applied. The thermal resistance results in the presence of underfill are given in Fig. 9. The results have shown that underfill has major contributions to the thermal dissipation despite of its relative low thermal conductivity (~ 0.9 W/Km).

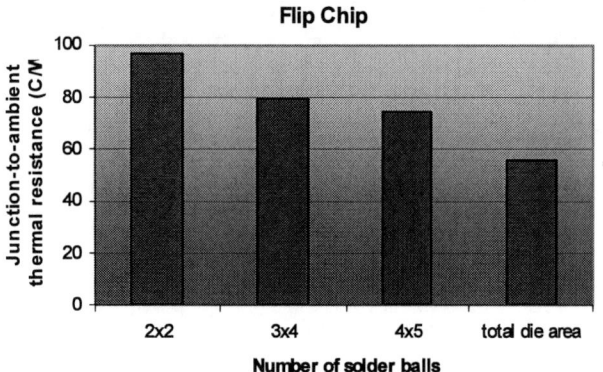

Fig. 8 Thermal resistance of a flip-chip-on-board package with different numbers of solder balls

When underfill is applied, the number of solder ball on thermal resistance has insignificant effect. However, in the absence of underfill, thermal resistance can be significantly reduced with increasing of the number of solder balls. As an extreme case, where the entire contact area between board and chip is solder contact, Fig. 8 shows that the minimum thermal resistance for this package is about 55.7°C/W. The difference between the extreme case and the case with 2x2 solder balls with underfill is not significant. These results imply that in flip chip applications, underfill is necessary to improve both thermal and reliability performances. The number of solder balls with underfill virtually gives no significant influence on thermal behaviors.

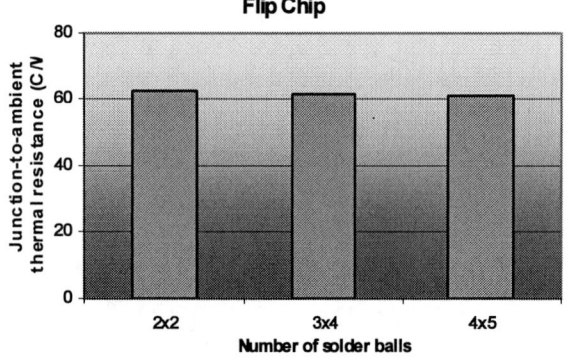

Fig. 9 Thermal behaviors of flip chip on board packages with underfill

Another version of flip chip application in power devices is the ball grid array package shown in Fig. 10a. The drain side is connected to the solder ball through the conduction layer, which is encapsulated by the epoxy-based materials. In this package the die size is same as flip chip on board package. A full array of solder balls with 6x5 is assumed over the package. The heat can be dissipated through the solder balls directly beneath the chip and those connected to the drain side. Fig. 11 investigates the effect of the conduction layer on the thermal dissipation. It is interesting to note that, unlike the strap bonds used in SO-8 packages (Fig. 4c), where the copper strap improves thermal performance significantly, the conduction layer (copper) used in BGA as shown in Fig. 10 has negligible effect on thermal resistance. "No conduction layer" in Fig. 11 means that the conduction layer has very low

thermal conductivity (~ 0.9 W/Km) which, of course, is not realistic. The results imply that the heat dissipation is dominant through the path of solder balls under the chip. Fig. 11 also shows the effect of underfill on thermal resistance. The improvement is about 12% reduction over the same package without underfill.

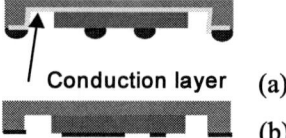

Fig. 10 a): ball grid array (BGA) MOSFET package; b): large contact interconnect MOSFET package

A ball grid array approach, even with multiple balls per connection has a limited contact area with a printed circuit board and hence the thermal performance junction to board and conduction efficiency cannot be maximized. Therefore an underfill material is required in the above applications. An alternate interconnection methodology that addresses this issue has been developed using a large area solder-contact technique. Fig. 12 presents the results of thermal resistance for large contact interconnect compared to the BGA package discussed before. We notice a 10% reduction in thermal resistance over the BGA package.

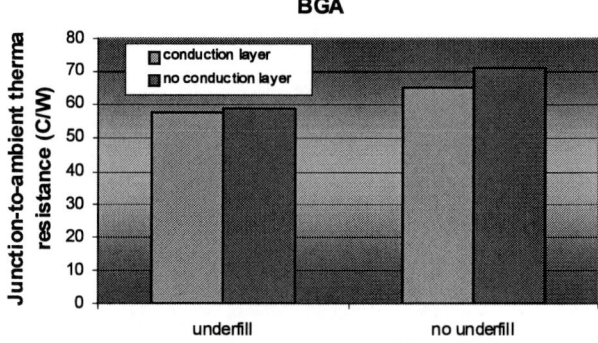

Fig. 11 Thermal resistance of full array BGA MOSFET packages

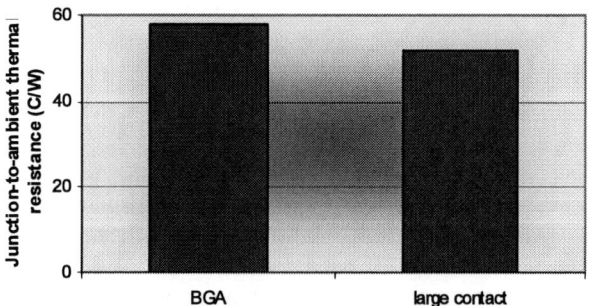

Fig. 12 Comparison of thermal resistance between BGA and large contact area power device package

Multichip Module (MCM)

A multichip module, which contains a control IC and two power MOSFETs was studied. The leadless package technology was selected and the package profile is below

1mm. A similar module with a ball grid array (BGA) is also studied, as shown in Fig. 13. Two MOSFETs and an IC chip are assumed to dissipate power 0.4, 0.4 and 0.2 Watt, respectively. Fig. 14 shows that leadless (so-called 'PIP') has better thermal performance, with 16.9% lower thermal resistance, than 'BGA'. In Fig.15 the maximum junction temperatures for each chip in the module are given for both packages under steady-state condition. The difference of the junction temperature in different chips is very small in spite of the non-uniformity in power dissipation.

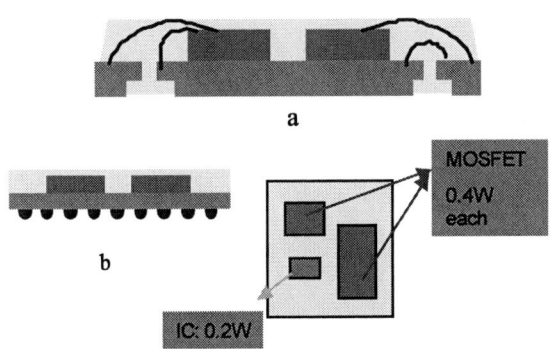

Fig. 13 Multichip module: a: PIP; b: BGA

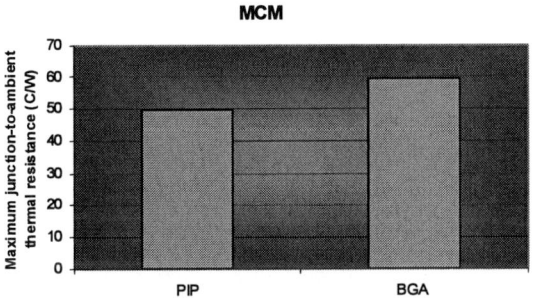

Fig. 14 Comparison of thermal performance between 'PIP' and 'BGA' of multichip module

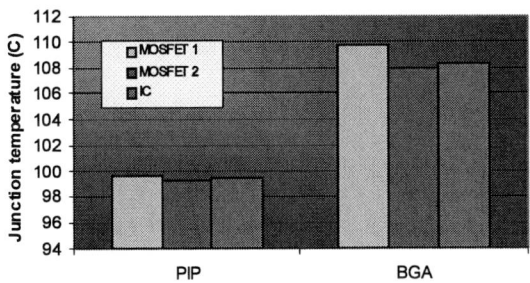

Fig. 15 junction temperature for each chip in MCM

Conclusions

Thermal models have been developed for various power-device packages. The CFD-based FLOTHERM simulation tool has been applied to predict the junction-to-air thermal resistance of different packages. It has been found that the difference in thermal resistance between DPAK and D2PAK under the same power dissipation is not significant (less than

5°C/W), in spite of the large differences in die size and package dimensions. This is primary because the package is mounted onto 2oz copper on FR4, which makes the board highly thermal conductive. Due to the large contact area between the copper pad and board, DPAK package displays the least thermal resistance (~50°C/W) and thus sets a baseline to evaluate other types of packages.

SO-8 power packages present a wide range of thermal resistance (50-80°C/W) when different interconnect technology and package format are applied. Strap bonding can improve thermal performance about 15% over the traditional SO-8 power MOSFET. The leadless or micro leadframe package further reduces the thermal resistance to the level comparable to the DPAK, with smaller package dimensions.

Flip-chip-on-board packages or ball grid array (BGA) packages have relatively good thermal performance (50-70°C/W) due to the direct solder interconnect to the board. When underfill is applied, the size and number of solder balls do not matter. However, the thermal performance has strong dependence on the number of solder balls if underfill is not used. The large area contact technology developed maximizes the thermal performance of the flip chip packages to the level of traditional DPAK.

For multichip modules, leadless module achieves better thermal performance (16.9% reduction of thermal resistance) than a similar module using BGA version.

A superior advantage of flip chip packages over wire-bonded packages is the realization of double-sided cooling mechanism. When a heat sink is attached, the thermal behaviors may be completely different from the behaviors shown above. Care must be taken to use the above results in an actual system application either from the predicted thermal resistance or the data from the data sheets, because thermal resistance depends not only on the package design and interconnect, but also on the ambient conditions, heat sink attachment and the board selection.

References

1. Stephen H. Gunther, Frank Binns, Douglas M. Carmean, and Jonathan C. Hall, Managing the impact of increasing microprocessor power consumption, Intel Technology Journal, Q1, 2001
2. A. Gentchev, Designing high-current VRM-compliant CPU power supplies, Electronic Design, October, pp.155-158, 2000
3. P. Mannion, MOSFET's break out of shackles of wire bonding, Planet EE, March 1999
4. Mike Speed and Wharton McDaniel, New power MOSFET packages cut DC-DC converter size, PCIM, June 2001
5. R. Aschenbrenner, J. Gwiasda, J. Eldring, E. Zakal, H. Reichel, Flip-Chip Attachment Using Non-Conductive Adhesives and Gold Bumps, International Journal of Microcircuits and Electronic Packaging, Vol. 18, No. 2, pp. 154-161, 1995
6. Xuejun Fan, "Combined thermal and thermomechanical modeling for a multi-chip QFN package with metal-core printed circuit board", Proceedings of the Intersociety Conference, v 2, ITherm 2004 - Ninth Intersociety

Conference on Thermal and Thermomechanical Phenomena in Electronic Systems, p 377-382, 2004

7. Xuejun Fan, "Development, validation, and application of thermal modeling for a MCM power package", *IEEE Semiconductor Thermal Measurement and Management Symposium*, p 144-150, 2003

8. Xuejun Fan and Shatil Haque, "Emerging MOSFET packaging technologies and their thermal evaluation", *Thermomechanical Phenomena in Electronic Systems - Proceedings of the Intersociety Conference*, p 1102-1108, 2002

9. Xuejun Fan et al, "Multi-physics modeling in virtual prototyping of electronic packages - combined thermal, thermo-mechanical and vapor pressure modeling", *Journal of Microelectronics Reliability*, 44 (12): 1967-1976, 2004

10. Scott Irving, Yong Liu, "Duane Connerny, Timwah Luk, "SOI die heat transfer analysis from device to assembly package", EuroSimE 2006, 2006

11. Yuan Xiang Zhang, Lihua Liang and Yangjian Xia, " Highly efficiency modeling automation for electronic package thermal analysis", ECTC57, Reno, pp1931-1935, 2007

12. Tounsi, P., Dorkei, J.-M.; Leturcq, P., "Thermal modeling for electrothermal simulation of power devices or circuits," 5th European Conference on Power Electronics and Applications, Brighton, England, Sep. 13-16, 1993

13. Kasem, M., "Thermal management and design aspects of high performance plastic quad flat packages for smart-power ICs," IEEE 8th International Symposium on Power Semiconductor Devices and ICs, Maui, HI, USA, May 20-23, 1996.

14. Ganesa-Pillai, M. and Chen, Q., "Thermal management of an integrated power module with multiple power devices," 22nd International Telecomunications Energy Conference, Phoenix, AZ, USA, Sep. 10-14, 2000.

15. Katsis, D.C. and Van Wyk, J.D., "Experimental measurement and simulation of thermal performance due to aging in power semiconductor devices," 37th IAS Annual Meeting and World Conference on Industrial Applications of Electrical Energy, Pittsburgh, PA, United States, Oct. 13-18, 2002.

16. Chiriac, V. A. and Lee, T-Y T., "Impact of die attach material and substrate design on RF GaAs power amplifier devices thermal performance," ASME Journal of Electronic Packaging, v. 125, pp. 589-596, 2003

17. Arik, M., Garg, J. and Bar-Cohen, A., "Thermal modeling and performance of high heat flux SOP packages," *IEEE Transactions on Advanced Packaging*, v. 27, pp. 398-412, 2004

18. Chiriac, V. and Lee, T-Y T., "Thermal evaluation of power amplifier modules and RF packages in a handheld communicator system," ITherm 2004 - Ninth Intersociety Conference on Thermal and Thermomechanical Phenomena in Electronic Systems, Las Vegas, NV, Jun . 1-4, 2004

19. Frank, W., "Comparison of simulation and measurement of thermal resistance of surface mounted power packages for high power density ballasts," 2005 European Conference on Power Electronics and Applications, Dresden, Germany, Sep. 11-14, 2005.

20. Kandasamy, R. and Subramanyam, S., "Application of computational fluid dynamics simulation tools for thermal characterization of electronic packages," International Journal of Numerical Methods for Heat and Fluid Flow, v 15, pp. 61-72, 2005

Wafer Level LED Packaging with Integrated DRIE Trenches for Encapsulation

Rong Zhang and S. W. Ricky Lee
Department of Mechanical Engineering
Hong Kong University of Science and Technology
Clear Water Bay, Kowloon, Hong Kong, China

Abstract

A novel encapsulation process for wafer level LED arrays is presented. In this process, 4 inch P-type single crystal silicon wafers served as the substrates for flip-chip mountable LED chips. The wafer substrates were fabricated by wafer level lithography and plating process. An UV curable epoxy was applied as the encapsulant. The encapsulation process takes advantage of square trenches fabricated by deep reaction ion etching (DRIE) process as barriers to limit the spread of the epoxy encapsulant, and can adjust the geometry of the encapsulation via controlling the volume of the epoxy and the dimension of the trenches. The packaging and encapsulation process of LED arrays were completed on wafer level. LED packages can be directly obtained after wafer dicing.

Introduction

Light emitting diodes (LEDs) have been widely used since its invention in 1962. LEDs can only emit red light in its early age. The power of LED chips is usually below 0.1 W. Therefore, the conventional low power LEDs are normally used for decoration, signal indicators, and message display panels. They are not suitable for general illumination applications. Afterwards, AlInGaN blue and green semiconductors technologies were developed. It becomes feasible to use LEDs to generate white light [1, 2]. In recent years, high power LED technology has enjoyed a rapid progress. The power of single LED chip can reach to more than 1W, which is 10 times higher than conventional LEDs. These developments greatly expand the applications of LEDs. Due to the properties such as long life, low power consumption, compact size, and fast response to input signals, LEDs become a promising candidate of solid-state lighting (SSL) [3-6].

LEDs now play more and more roles in mobile phones, automobiles, display panels, traffic lights, and general illuminations. Since the power limit and the concentrating light emitting profile, LEDs are mostly used in the form of LED arrays in these applications. However, currently most LEDs are packaged on individual component basis. Encapsulation molds are required to encapsulate LED packages. The component level packaging process has a relatively low throughput and high cost, which is obstructive to the propagation of LEDs.

People are trying to apply wafer level packaging (WLP) process to fabricate LED packages. C. T. Sou, Y. S. Huang and G. W. Lin demonstrated a silicon-based packaging platform for a LED array [9]. K. H. Lee and S. W. R. LEE presented a wafer level printing process for yellow phosphor coating for white light LEDs [10]. Chang-Hyun Lim, Won-Kyu Jeung and Seog-Moon Choi demonstrated a wafer level process to fabricate LED lens in batches [11].

This study is to develop a novel encapsulation process for LED arrays. This approach is an SMT-compatible batch process and is able to encapsulate LED arrays directly on wafer without molds, so a wafer level packaging (WLP) process can be performed. Fig. 1 shows a schematic pattern of this wafer level encapsulation process. A 4-inch wafer serves as the substrate for the LED arrays. The fabrication of the bond pad patterns on wafer is performed by wafer level plating process.

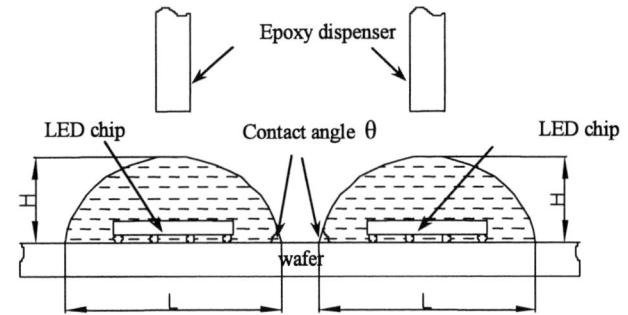

Fig. 1 A schematic drawing of wafer level encapsulations

Experimental Processes

Dummy Encapsulation Process on Bare Wafer

Before encapsulating a real functional LED package, the dummy encapsulation process was performed. A 4 inch P-type single crystal silicon wafer was used as the substrate. The wafer was deposited with a 3μm thick low-temperature-oxide (LTO) layer. An UV curable epoxy Loctite352 is used as the encapsulant. The viscosity of the epoxy is 15000-26000 mPa.s at 25℃. The specific gravity of the epoxy is 1.06 at 25℃.

One straightforward method to make dome-shape encapsulations on wafer as shown in Fig.1 is to dispense epoxy droplets on wafer, then cure the epoxy droplets to form solid encapsulations. In this situation, the wafer surface does not provide constraints in the horizontal plane. The epoxy droplet can spread freely on the surface; therefore the geometry of the encapsulation is completely determined by the contact angle between the encapsulation and the wafer surface. In this experiment, the contact angle values of epoxy droplets on wafer surface under free spread and UV curing condition were investigated. A syringe was used to dispense epoxy droplets onto wafers. The syringe is attached to a micrometer, so it can control the volume of the droplets. The minimum volume the syringe can control is 0.33 mm3. The wafer surface was cleaned by acetone and then air dried. An UV lamp of 75 mW/cm2 power density was used to cure the

978-1-4244-2739-0/08/$25.00 ©2008 IEEE

epoxy droplets. The experimental condition is shown in Fig. 2 (the UV lamp is not shown).

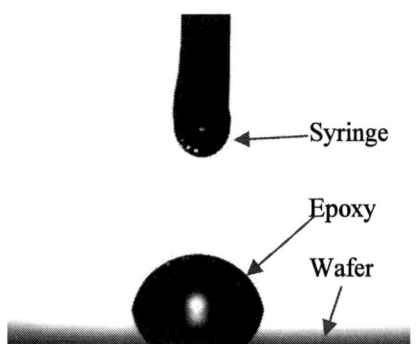

Fig. 2 An epoxy droplet on wafer surface

In order to obtaining better constraints to limit the epoxy droplets within a certain region to control the geometry of the encapsulation, double-line trenches were cut on wafer by a dicing machine using the blade "Disco NBC-ZH 2050-SE 27HEED". The cutting speed is 10 mm/sec. The parameters of a trench are shown in Fig. 3. Due to the thickness of the dicing blade, the width of the cutting grooves is fixed at around 35 μm. The depth of trenches D is fixed at 200 μm. The pitch between the inner and outter trenches P is fixed at 35 μm. The dimensions of the trenches are shown in Fig. 3. In this pattern, the region of the wafer surrounded by the square trench is regarded as one "substrate". For example, in Fig. 3, the side length of the square substrate is L. After the cutting of the trenches, epoxy droplets were dispensed onto the substrates, and then went through the UV curing to form solid encapsulations.

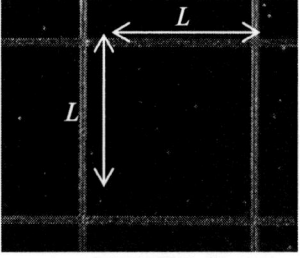 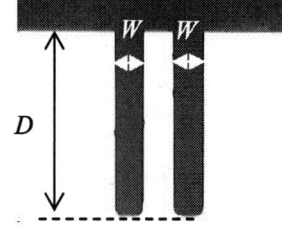

a) Top view   b) Cross-section view

Fig. 3 Dimensions of a single-line trench substrate

Fig. 4 shows an encapsulation that was formed on a substrate. In such a square substrate, the contact angles around the edges vary everywhere; therefore the contact angle value cannot be used to represent the geometry. Here, the ratio H/L is used to represent the geometry; the ratio $k = \dfrac{V_d}{V_s}$ is used to represent the substrate's capacity for holding epoxy droplets within it, where V_d denotes the volume of the actual droplet and can be controlled by the operation, Vs denotes the volume of the virtual semi-sphere as shown in Fig. 4 and can be calculated by $V_s = \dfrac{1}{2}\dfrac{4\pi(L/2)^3}{3}$. For a given substrate, an epoxy droplet can be held within the substrate without overflowing as long as the k value is smaller than a certain critical value k_c. By this definition, the higher the k_c, the wider

range the ratio H/L can be adjusted. In this experiment, the k_c–L curve was obtained when the range of L is from 3 to 5 mm.

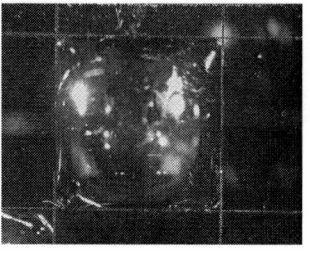 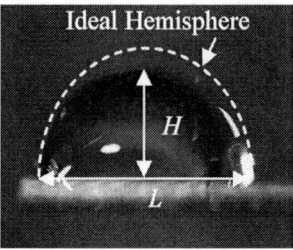

a) Top view   b) Side view

Fig. 4 An epoxy encapsulation on substrate

Wafer Level Encapsulation Process for LED Array

In the present process a 4-inch P-type wafer serves as the substrate for LED arrays. The fabrication of the wafer substrate is performed by micro fabrication and wafer level plating process.

The wafer is firstly deposited with 3 μm silicon low-temperature-oxide (LTO) layer. The LTO layer serves as the mask for patterning in the subsequent deep-reaction-ion-etching (DRIE) process due to its high resistance to plasma etching comparing to silicon (etching rate by DRIE is SiO2:Si = 1:50). Subsequently, 50 μm deep square trenches similar to the trenches in the dummy encapsulation process were fabricated by DRIE process (Fig. 6a). The side length of the trenches L = 3 mm. The trenches have 4 openings in the corners as shown in Fig. 5 for the subsequent Al traces for electric paths. The pitch between the outer and inner trench P is 35 μm. The layout of the square trenches is shown in Fig. 5.

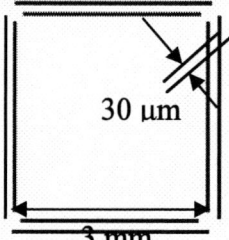

Fig. 5 Layout of the trenches

After the trenches were etched, a pure Al layer of 2.5 μm thick is sputtered on wafer and then be patterned by dry etching using Cl2 and BCl3 gas (Fig. 6b). Subsequently, a SiO2 layer which serves as the passivation layer of the bond pads for the LED flip-chip mounting is deposited onto the wafer using PECVD method, and then be patterned by dry etching(Fig. 6c). 500Å thick TiW and 5000Å thick Cu seed layer for the subsequent electroplating process are then sputtered onto the whole wafer for the subsequent electroplating process (Fig. 6d). A 31 μm thick photo resist (PR) is coated on wafer and patterned, and then 6 μm thick Cu and 30 μm thick SnPb solder layers are electroplated (Fig. 6e). After the plating, the photoresist, the sputtering Cu and TiW layers are stripped; and then the whole wafer went through reflow to form solder bumps (Fig. 6f).

504

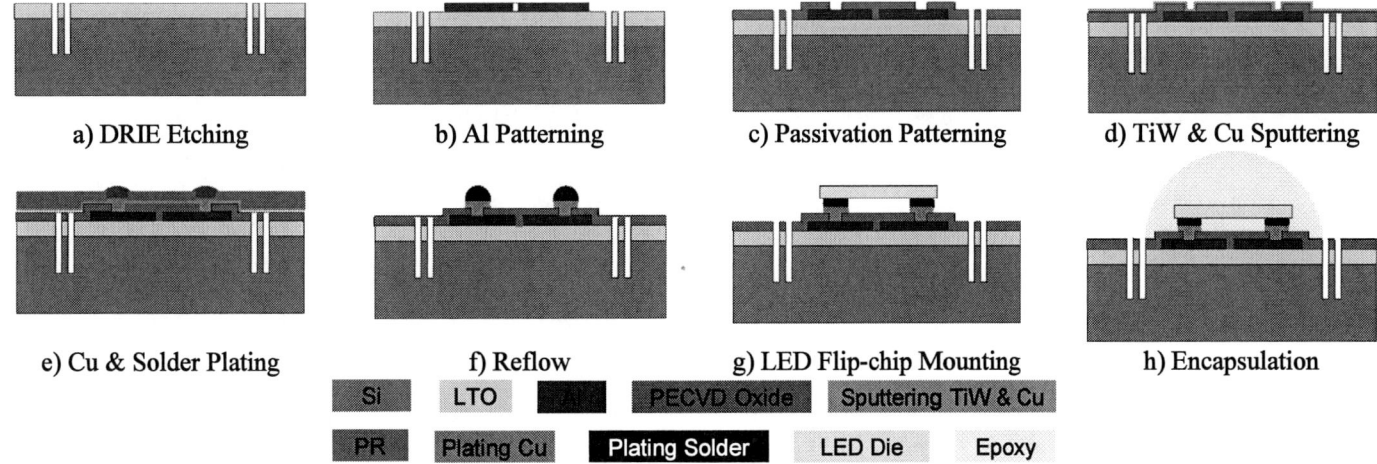

a) DRIE Etching   b) Al Patterning   c) Passivation Patterning   d) TiW & Cu Sputtering

e) Cu & Solder Plating   f) Reflow   g) LED Flip-chip Mounting   h) Encapsulation

| Si | LTO | Al | PECVD Oxide | Sputtering TiW & Cu |
| PR | Plating Cu | Plating Solder | LED Die | Epoxy |

Fig. 6 The wafer level packaging process for LED arrays

The fabrication of the wafer substrate is completed. Surface-mountable LED chips are then flip-chip mounted onto the substrates on wafer. Finally, an UV curable epoxy droplet was dispensed onto the substrate, and was cured to form the encapsulation (Fig.6g). The wafer level encapsulation process for LED array is completed until now. LED arrays are encapsulated on wafer level. After the wafer dicing, functional LED units are obtained.

Results and Discussion

Dummy Encapsulation Process on Bare Wafer

The contact angle curve of the epoxy on wafer is shown in Fig. 7. Due to the high viscosity, the epoxy droplets spread very slowly on wafer. It took more than 30 minutes to complete the spread process. The final contact angle of a free spread droplet is less than 20°. Referring to Fig. 1, for a free spread droplet, the ratio H/L is completely determined by the contact angle θ. The encapsulation is too flat with such small contact angle. The black color dot shows the final contact angles of epoxy droplet under UV curing. It is much larger than the final contact angle under free spread condition. However, geometry of the droplet is still too flat.

This experiment implies that an encapsulation with a large contact angle is unfeasible to obtain on a flat surface under the experimental condition. Normally, the encapsulations of LEDs are semi-sphere. To form an encapsulation of approximate semi-sphere, the ratio H/L should be closed to 1/2, and the contact angle should be 80° or even larger. To obtain such a high contact angle under free spread condition, the wettability of the surface to the epoxy should be very low. A large contact angle can be also obtained by curing the epoxy as soon as possible, or using an epoxy of a very high viscosity. However, the UV light power and the viscosity of the epoxy should be so high that it is difficult to obtain such an ideal result. Therefore, extra constraints are required to limit the spread of the epoxy droplets. The double-line trenches shown in Fig. 3 are to provide such constraints. The epoxy droplets were dispensed onto the substrates surrounded by the cutting trenches as defined above, and then went through the UV curing to form solid encapsulations.

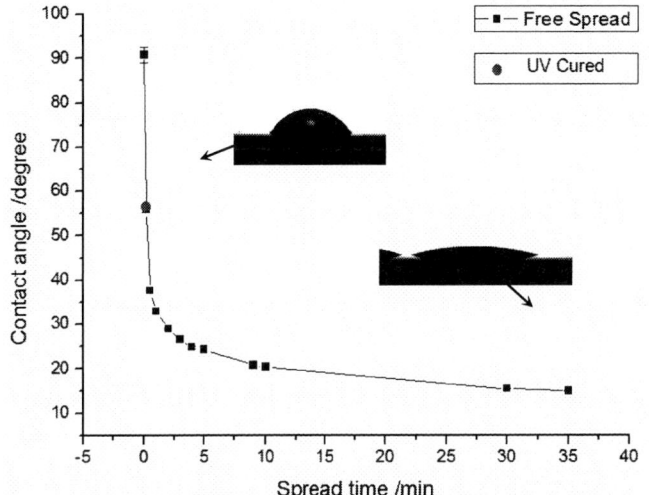

Fig. 7 The contact angle curve of epoxy droplets on wafer

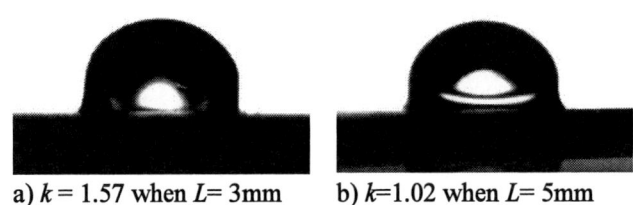

a) $k = 1.57$ when $L= 3mm$ b) $k=1.02$ when $L= 5mm$

Fig. 8 The geometry of dummy encapsulations

The results are shown in Fig. 8. The epoxy droplet can be held within the area surrounded by the trenches, the geometry is improved.

Fig. 9 shows the trends of the critical k value kc and H/L value with respect to the dimension of the substrates. The circular dots denote that the epoxy droplets can be held within the substrates without overflow, while the cross dots denoting that the epoxy droplet overflowed outside the substrate. The triangle dots denote the critical k values. The rectangular dots denote the H/L values at critical k values.

505

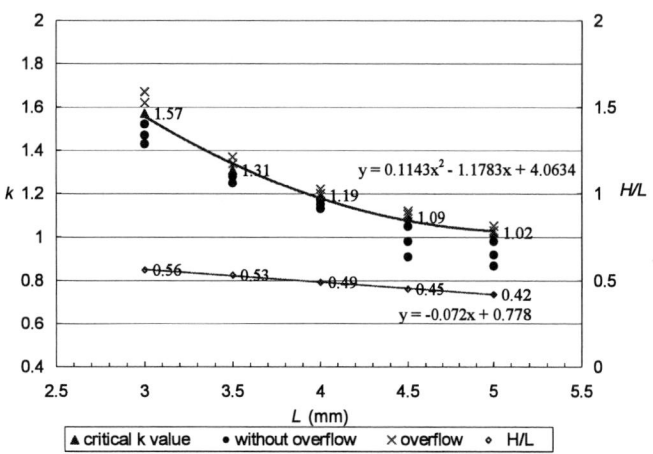

Fig.9 Trends of k_c and H/L on double-line trench substrates

It can be seen from the figure that, the H/L values improved significantly comparing with those on bare wafer surface as shown in Fig. 10. When L = 3 mm, the H/L value increases more than 30%. With the increase of the dimension of the substrates L, the critical V value and H/L value decrease. The he critical k value can be higher than 1, providing an encapsulation of a dome-shape geometry. The critical k value decreases with the increase of the dimension of the substrates. When the dimension of the substrates increases to 5 mm × 5 mm, the critical k value drops to 1.02.

The mechanism of the constraint effect can be explained as follow: After the droplet is dispensed, it spreads freely within the area surrounded by the inner trench until it encounters the inner trench and then be stopped, as shown in Fig.10a. If the volume of the epoxy does not exceed a certain value, then the epoxy can be constrained by the inner trench and will not overflow it. If the volume of the epoxy exceeds a certain value, the constraint of the inner trench will be broken though. The epoxy will flow down into the inner trench, and then overflow it, and finally be stopped by the outer trench, as shown in Fig. 10b. Once the volume of the epoxy is too large, the droplet can break through the constraint of the outer trench and overflow outside the substrate (shown in Fig. 10c). The spread process is illustrated in Fig.10d.

Fig. 11 shows a dummy encapsulation array on wafer when L = 3.5 mm, k = 1. The epoxy droplets can be constrained within the substrates without overflow, even though the pitch between substrates is 35 μm. The constraint provided by the trenches is effective. This constraint effect can be utilized to realize the wafer level encapsulation process for LED arrays packaging.

Wafer Level Encapsulation Process for LED Array

The dispensing process presented above was performed to encapsulate real LED packages. In this process, the trenches are fabricated by DRIE process instead of mechanical cutting. The trenches have 4 openings for electric paths in the four corners as shown in Fig. 5. The openings are for electric paths. Fig. 12 shows a corner of the encapsulation. It can be seen that the constraint effect provided by the trenches still works. The epoxy is constrained within the outer trench. Fig.

13 shows the cross-section view of an LED package. The LED chip is fully embedded in the encapsulation of k = 1 (volume = 7.07 mm3). The gap between the chip and the substrate is well filled.

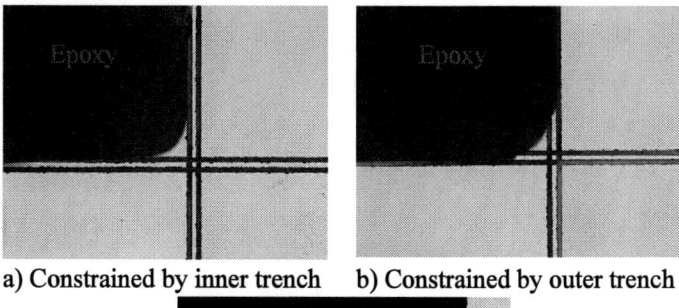

a) Constrained by inner trench   b) Constrained by outer trench

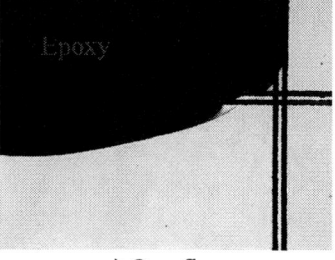

c) Overflow

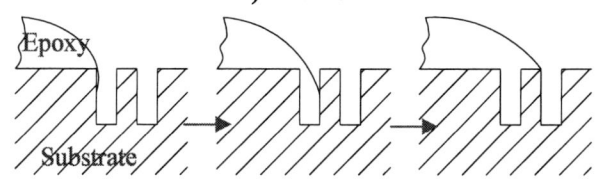

d) Spread process of the droplet
Fig. 10 The constraint effect of the trenches

Fig. 11 A dummy encapsulation array on wafer

Comparing to the dummy encapsulation, the LED chip occupies an extra volume in the encapsulation, however the influence caused by this extra volume to the dispensing process is small. This influence can be evaluated through a simple calculation. The dimension of the chip is $1.1mm \times 1.1mm \times 0.08mm$. The volume ratio of the chip with respect to the encapsulation can be estimated as $\frac{1.1mm \times 1.1mm \times 0.08mm}{7.07mm^3} \times 100\% = 1.4\%$, which is small and can be neglected.

506

Fig. 12 The corner of a LED encapsulation

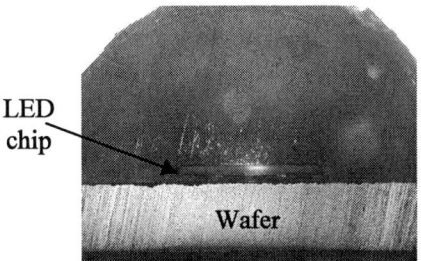

Fig. 13 The cross-section of an LED encapsulation

The influence of the volume of the chip can be neglected, however, the LED chip can interferes with the spread of the epoxy droplet on the substrate. The chip becomes a small "substrate" during the epoxy dispensing, and can stop the spread of the epoxy droplet on the substrate. According to the trend of kc in Fig. 9, the smaller the substrate, the larger the volume of the droplet the substrate can hold. Since the dimension of the LED chip is merely 1.1 mm×1.1 mm, it can hold a drop of considerable volume, and defeat the dispensing process, as shown in Fig. 14.

To eliminate this effect, the tip of the epoxy dispenser (the syringe) should be closed to the chip to provide extra pressure to assist the droplet to envelop the chip to ensure a smooth spread on the substrate, as shown in Fig. 16. Fig. 17 shows the encapsulated LED arrays on wafer. After dicing, functional LED packages as shown in Fig. 18 can be directly obtained.

 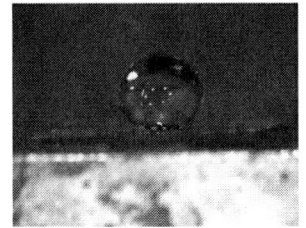

a) Top view   b)Side view

Fig.14 Droplet stands on the top of the chip without wetting the side walls.

Fig. 16 The dispensing of the epoxy encapsulant

Fig. 17 Encapsulated LED arrays on wafer

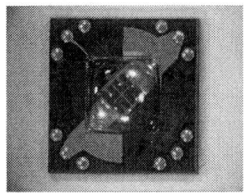 

a) Top view of an LED package   b) Side view of an LED package

c) LED package lighted up

Fig. 18 A functional LED package with desired encapsulation

Conclusions

The wafer level encapsulation process for LED arrays packaging has been presented. The double-line trenches can significantly improve the geometry of the epoxy encapsulations for both dummy encapsulations and real LED packages. The k value and H/L value were used to represent the geometry of the encapsulation. High k and H/L are desired. The critical k value and H/L value decrease with the increase of the dimension of the substrates. The volume of the LED chips brings little influence to the encapsulation process. However, the the LED chip can interferes with the spread of the epoxy droplet on the substrate. The LED chip can become a small substrate during the dispensing of the epoxy droplet and stop the spread of the droplet thus defeat the dispensing process. The tip of the syringe should be closed to the LED chip to provide extra pressure to assist the dispensing of the epoxy. Under proper operations, a wafer level LED packaging is successfully performed.

Acknowledgments

The LED chips were provided by the Advanced Packaging Technology (APT) Ltd. The authors would like to acknowledge APT's support to this study.

References

1. D. A. Steigerwald, J. C. Bhat, D. Collins, R. M. Fletcher, M. O. Holcomb and M. J. Ludowise, "Illumination with Solid State Lighting Technology," *IEEE Journal on Selected Topics in Quantum Electronics*, Vol. 8, No. 2, Mar-Apr 2002.

2. S. Muthu, F. J. P. Schuurmans and M. D. Pashley, "Red, Green, and Blue LEDs for White Light Illumination," IEEE Journal on Selected Topics Quantum Electronics, Vol. 8, Mar–Apr 2002, pp. 333–338.

3. P. C. H. Chan, "Electronic Packaging for Solid-state Lighting," *Proc. of 6th International Conference on Electronic Packaging Technology*, 2005.

4. M. G. Craford, "LEDs Challenge the Incandescents," *IEEE Circuits and Devices Magazine*, Vol. 8, Sep 1992, pp. 24-29.

5. N. Narendran and Y. Gu, "Life of LED-Based White Light Sources," *IEEE/OSA Journal of Display Technology*, Vol. 1, No. 1, Sep 2005.

6. J. Y. Tsao, "Solid-state Lighting: Lamps, Chips and Materials for Tomorrow," *IEEE Circuit and Devices Magazine*, Vol. 20, May-Jun 2004, pp. 28-37.

7. M. S. Shur and R. Zukauskas, "Solid-state Lighting: Toward Superior Illumination," *Proceedings of the IEEE*, Vol. 93, Oct 2005, pp. 1691-1703.

8. R. F. Karlicek, Jr., "High Power LED Packaging," *Proc. of Conference on Lasers & Electro-Optics (CLEO)*, 2005.

9. C. Tsou, Y. S. Huang and G. W. Lin, "Silicon-based Packaging Platform for Light Emitting Diode," *Proc. of 6th International Conference on Electronic Packaging Technology (ICEPT)*, 2005.

10. K. H. Lee and S. W. R. LEE, "Process Development for Yellow Phosphor Coating on Blue Light Emitting Diodes (LEDs) for White Light Illumination," *Proc. of 8th Electronics Packaging Technology Conference*, Dec 2006, pp. 379-384.

11. C-H. Lim, W-K. Jeung and S-M. Choi, "LED Packaging using High Sag Rectangular Microlens Array," *SPIE Proc. of Micro-Optics, VCSELs, & Photonic Interconnects II : Fabrication, Packaging, & Integration*, Vol. 6185.

Impact of Assembly Process Technologies on Electronic Packaging Materials

T. Tilford [1], C. Bailey [1], A.K. Parrott [1], J. Rizvi [1], C. Yin [1], K.I. Sinclair [2] and M.P.Y. Desmulliez [2]

1) Computational Mechanics and Reliability Group, University of Greenwich, Greenwich,
London, SE10 9LS, United Kingdom

2) MicroSystems Engineering Centre (MISEC), School of Engineering & Physical Science, Heriot Watt University,
Edinburgh, EH14 4AS, United Kingdom

Abstract

Assembly processes used to bond components to printed circuit boards can have a significant impact on these boards and the final packaged component. Traditional approaches to bonding components to printed circuit boards results in heat being applied across the whole board assembly. This can lead to board warpage and possibly high residual stresses.

Another approach discussed in this paper is to use Variable Frequency Microwave (VFM) heating to cure adhesives and underfills and bond components to printed circuit boards. In terms of energy considerations the use of VFM technology is much more cost effective compared to convection/radiation heating.

This paper will discuss the impact of traditional reflow based processes on flexible substrates and it will demonstrate the possible advantages of using localised variable frequency microwave heating to cure materials in an electronic package.

Flexible substrate subjected to Lead-free soldering process

Environmental legislation such as Reduction of Hazardous Substances (RoHS) means that electronics manufacturing companies have to adopt lead-free soldering practices and technologies. This means that printed circuit boards will be subjected to higher assembly temperatures during the reflow process because the melting point of lead-free solders is in general higher than tin-lead solder. With the requirement to adopt lead-free soldering practices there are concerns that:

higher processing temperatures will affect the behavior of the substrate during the reflow assembly process

adoption of lead-free solders will affect the reliability of the interconnects between the substrate and the components

Flexible substrates are made from organic materials where adhesives and copper are used to establish the circuit tracks on the substrate. Typical base materials for flexible substrates include polyimide (PI), Polyethylene Terephthalate (PET), Polyethylene Naphthalate (PEN), Polyvinyl Chloride (PVC) or Liquid Crystal Polymer (LCP). A typical polyimide flexible substrate used in this study is shown in Figure 1.

Figure 1: Typical flexible substrate (Courtesy: Budapest University of Technology and Economics).

Flexible substrates can be prone to moisture uptake and this can affect their behavior during the reflow process and during reliability testing. Together with cure shrinkage this can change the physical appearance of a flexible substrate after thermal processing. For example, a flexible substrate made from polyvinyl chloride (PVC) could be distorted severely if not handled carefully as seen in Figure 2.

Figure 2: Distortion of PVC flexible substrate after thermal processing (Courtesy: TWI Ltd, UK).

Numerical Model

Figure 3 illustrates a portion of a flexible substrate containing the base carrier film material, dielectric and copper tracks. Both the copper tracks and the dielectric are bonded to the base material using suitable adhesives. Three types of material, namely, PI, LCP and PEN are discussed in this paper as the base carrier film materials.

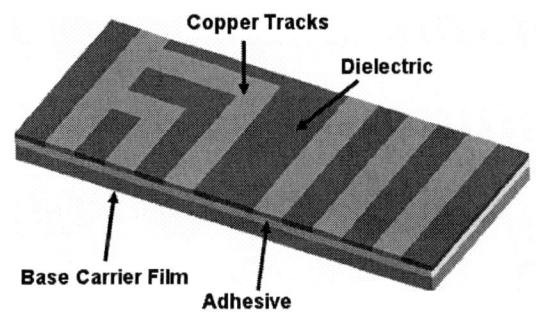

Figure 3: Illustration of a section of board with a dielectric

Finite element simulations have been undertaken to predict the behavior of a flexible substrate as it passes through a lead-free reflow temperature profile. The reflow temperature is applied on the top and bottom of the substrate with a heat transfer coefficient of 100 W/m^2-K. The mechanical material

978-1-4244-2739-0/08/$25.00 ©2008 IEEE

properties used in this simulation are shown in Table-1. The computer simulations were carried out using the modeling software PHYSICA [1]. The amount of deformation that can occur in a 14cm * 12cm substrate has been modeled where the substrate had different base carrier films and thickness of copper.

Table-1: Material properties used in computer modelling

Materials	Properties		
	Young's Modulus (MPa)	Poisson's Ratio	CTE (10-6/K)
Insulator	4400	0.22	50
Copper	132400	0.34	16.7
Adhesive	300	0.3	150
Polyimide	2500	0.34	30
LCP	340000	0.4	17
PEN	6100	0.45	20

Figure 4 illustrates the upward displacement of a flexible substrate made of PI, PEN and LCP. In this calculations standard base carrier layer thicknesses of 50μ, 125μ and 100μ are used for PI, PEN and LCP respectively. Clearly we can see that the thinner PI material shows much greater amounts of displacement than the others. This is of course aided with the higher CTE for this material.

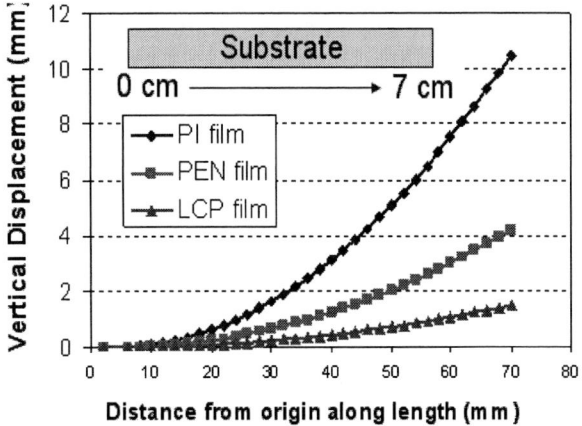

Figure 4: Upward displacement of substrate at peak temperature in a reflow process.

Another design variable that has been investigated is the dimensions of the copper tracks in the substrate and the impact this can have in stresses at the copper/adhesive interface. Figure 5 illustrates this region of a flexible substrate.

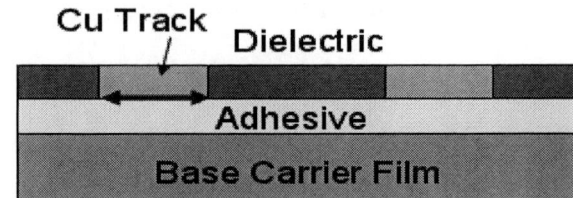

Figure 5: Arrow in the illustration show the region of interests where delamination can occur.

As seen in the material data provided in Table-1, the dielectric insulator has comparatively higher coefficient of thermal expansion (CTE) value than copper. Therefore, this insulator will try to expand more than copper when heated and push from both sides onto the copper conductors. Moreover, the adhesive possesses a very high value of CTE that will cause higher expansion of the adhesive layer. In that case the possible delamination can occur either in adhesive-substrate or in copper-adhesive interfaces.

Shear stress along the copper track-adhesive interface is illustrated in Figure 6. Cleary we can see significant changes in the shear stress at the copper-adhesive interface near the dielectric material.

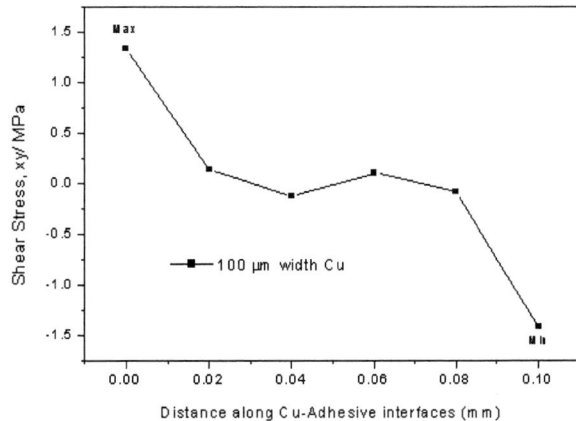

Figure 6: Shear stress (σ_{xy}) at peak reflow temperature.

The effect of copper track width has been investigated, for the layout illustrated in figure 3, is shown in figure 7. The impact of copper track width on the stress is also shown where a wider track reduces the stress. The decreasing tendency is almost similar to all the three carrier films.

It is important to note that increase in the copper track width eventually increases the volume percentage of copper in the circuit boards. This higher amount of copper in fact determines the behavior of flexible substrate and reduces the distortion during lead-free reflow soldering. Clearly we can see that the maximum stress at the copper-adhesive-dielectric interface is given when PI is used as the base carrier film. Therefore, it is evident that the type of carrier film also determines the failure criteria in flexible substrate.

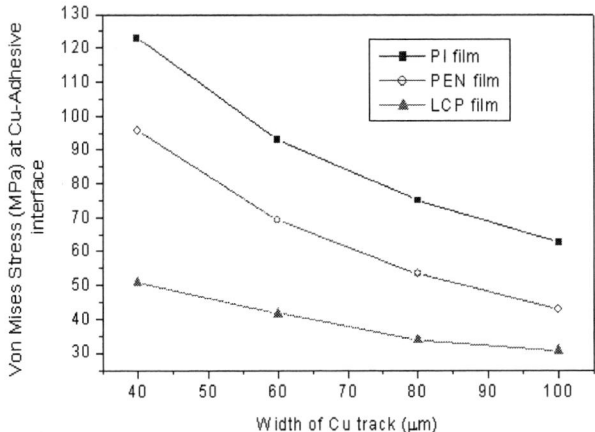

Figure 7: Stresses along copper-adhesive interface as a function of copper track width.

Figure 8 represents the comparative shear stress, σ_{xy} along the copper adhesive interfaces. The stress is plotted at peak reflow temperature of 250 °C. It is obvious from this figure that using dielectric material as an insulator in between the copper conductors can reduce the shear stress generation along adhesive-copper interfaces. It is seen that when dielectric material is used shear stress is almost zero along the interfaces except the edges whereas a significant increase in shear stress values is noticed for the flexible substrate without dielectric material. The increased shear stress values in the adhesion interface could cause a lift of copper tracks from the substrate leaving the circuit's electrical performance at risk.

In conclusion, the base carrier material used in a flexible substrate governs the magnitude of warpage in the seen during the reflow process. Higher volume of copper in the flexible substrate reduces warpage during reflow. A dielectric material in between the copper tracks can reduce the shear stresses along the copper/adhesive interface.

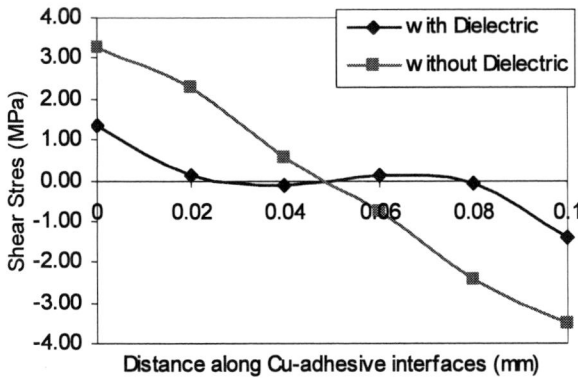

Figure 8: Shear stress distribution along the copper-adhesive interfaces.

The FAMOBS system

The FAMOBS system was proposed by Sinclair et al in 2006 [2]. It consists of a rectangular metal cavity which is partially filled with a dielectric material, such as a ceramic. The remaining section of the oven, referred to as the 'load section', is open ended and filled with air.

The composition of the FAMOBS oven is depicted schematically in Figure 9. The oven is intended to be mounted on the arm of a precision placement machine, enabling the load section to be placed over a single component on a printed circuit board. The ability to perform localised heating and curing enables the process to be optimised for and individual component as opposed to optimising the process for the entire board which is required in convection heating.

The design of the system allows electric fields to build up internally but prevents electromagnetic energy propagating from the open end of the oven. The ability for a wave to propagate through an enclosed structure, such as the FAMOBS oven, is dependant upon the operating frequency of the system. Propagation is only possible above a critical 'cut-off' frequency, which depends on the dimensions of the cavity and the electrical properties of the material contained within it. The cut-off frequency of the ceramic filled section FAMOBS system therefore differs from that of the air filled load section.

The system is operated at frequencies at which wave propagation occurs in the ceramic section but not in the air section. The fields in the load section are excited by the fields in the ceramic section but, as the wave cannot propagate, their amplitude decreases exponentially with distance from the ceramic-air boundary, a phenomena known as evanescent decay. Figure 10 shows the system arrangement with diagrammatic illustration of wave amplitude through the oven centreline.

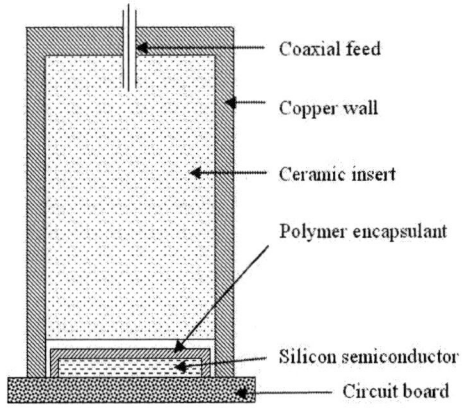

Figure 9: Schematic of prototype system

Heating patterns generated in the load are therefore dependant upon the electric field distribution which, in turn, depends on the operating frequency. The system can be operated at a large number of discrete harmonic resonant frequencies each resulting in a differing modal structure within the ceramic. The heating pattern within the load can therefore be controlled through the choice of operating frequency.

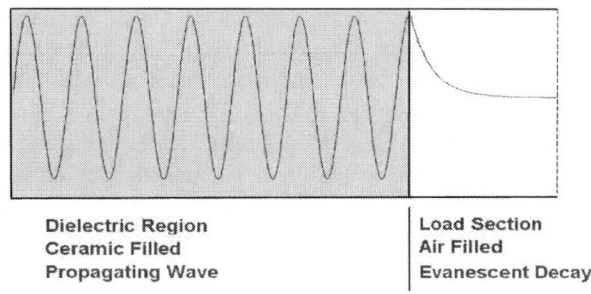

Dielectric Region
Ceramic Filled
Propagating Wave

Load Section
Air Filled
Evanescent Decay

Figure 10: Schematic of field arrangement within FAMOBS dual-section oven system.

Assessment of the performance of the FAMOBS system has been carried out using both experimental and numerical approaches. Experimental studies of the system have been conducted using a prototype FAMOBS system, with initial results demonstrating that typical thermosetting encapsulant materials can be heated and cured very rapidly. The material can be heated to temperature in excess of 200 °C in less than 20 seconds with a significant (but as yet unquantified) degree of cure being achieved during this time.

Numerical Model

A numerical model describing the system has been developed to expedite the design process and aid understanding of the system processes. In order to accurately model the process of microwave curing a holistic approach must be taken. The process cannot be considered to be a sequence of discrete steps, but must be considered as a complex coupled system.

Significant inter-linking between processes exists. The dielectric properties of the polymer material are influenced by temperature and degree of cure, leading to substantial alteration the electromagnetic field distribution during the process. The thermosetting reaction is often exothermal leading to further coupling between cure and temperature. Thus, despite the very different thermal and electromagnetic timescales, there is a requirement to closely couple the computational electromagnetic, thermal, curing and stress analyses. The process coupling is illustrated in figure 11 below.

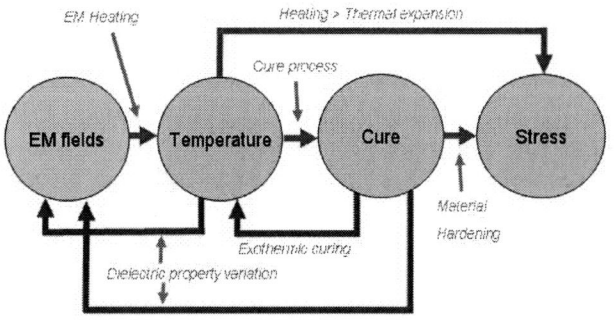

Fig. 11: Processes coupling in microwave curing

The numerical model utilised in this contribution is an extension of the approach developed by the authors for food processing applications [3]. The model comprises a Yee scheme Finite Difference Time Domain (FDTD)

electromagnetic solver [4] coupled with an unstructured finite volume method (FVM) multi-physics package [1], both of which were co-developed by the authors. Electromagnetic and thermophysical solutions are solved within independent numerical domains, with coupling implemented through an inter-domain cross mapping process. The thermophysical analysis mesh is confined to the load material while the electromagnetic mesh occupies the entire oven domain (encompassing the thermophysical domain). A cross mapping algorithm transfers required data between the two domains based on spatial coordinate sampling approach. The model is more completely described in previous publications [5, 6].

The numerical model has been used to analyse heating curing and thermal stress generation within an idealised microelectronics geometry. The geometry consists of a silicon block of dimensions 19.0mm x 19.0mm x 2.0mm, mounted to a polymer block of dimensions 19.0mm x 19.0mm x 1.0mm. The geometry is depicted in figure 12.

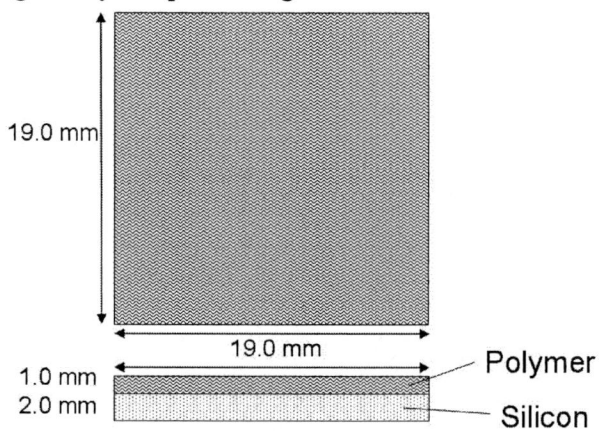

Figure 12: Idealised microelectronics load

The load has been placed inside a FAMOBS oven of internal dimensions 25.5mm x 25.5mm x 114.8mm. The oven is filled with a dielectric material of dimensions 25.5mm x 25.5mm x 104.8mm, with a dielectric constant of 6.0 and zero loss tangent. The idealised load is placed centrally within the oven cross section, 0.5 mm from the air-dielectric interface and with the polymer side of the load facing the dielectric material.

Variable frequency microwave heating of the load was modelled for a period of 40 seconds, with a total of 80 different modes being excited in turn. Each mode was excited for 0.5 seconds, requiring 80 simulation timesteps. The modes were excited in ascending frequency between 8.135 GHz (a $TM_{3,1,5.5}$ mode) and 13.70 GHz. (a $TM_{4, 4, 3.5}$ mode). The total power deposited into the load was scaled to 1.0 Watts.

Figures 13 to 16 show the distribution of simulation variables over the external surface of the idealised microelectronics load after the final solution timestep (40 seconds). Figure 13 shows the distribution of electric field intensity, with the modal '4 by 4' pattern resulting from the excited TM4, 4, 3.5 clearly evident. As the load does not completely fill the load section of the cavity (which is 25.5mm square with the load being 19.0 mm square) the modal pattern is cropped somewhat with field peaks on the edges of the load – a situation likely to result in edge

512

overheating issues. Figure 14 shows the variation of temperature over the outer surface of the load. It is evident from this plot that the variable frequency operation of the system results in a relatively even temperature distribution over the polymer material. Figures 15 and 16 respectively show the rate of cure and degree of cure of the material at the surface of the load.

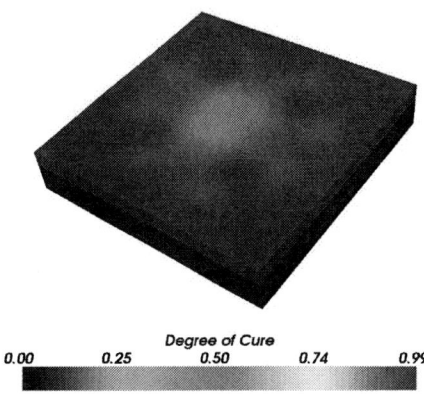

Figure 16: Distribution of degree of cure over idealised microelectronics load at t = 40 seconds

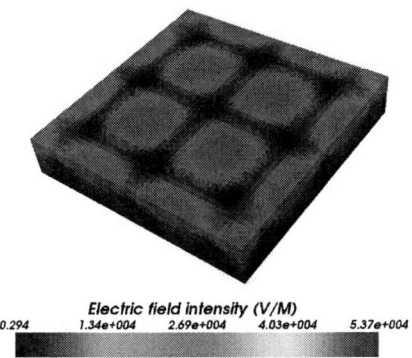

Figure 13: Distribution of Electric field intensity over idealised microelectronics load at t = 40 seconds

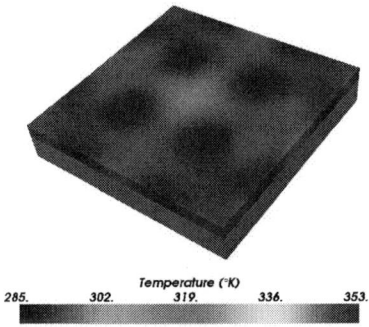

Figure 14: Temperature distribution over idealised microelectronics load at t = 40 seconds

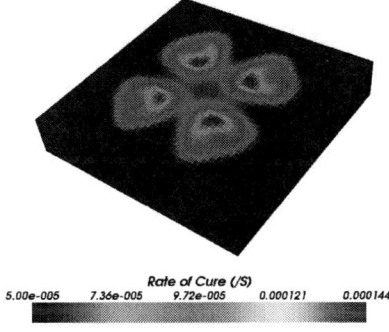

Figure 15: Distribution of rate of cure over idealised microelectronics load at t = 40 seconds

Conclusions

The impact of lead-free soldering on the thermo-mechanical behavior of flexible circuit boards and their solder joints has been discussed. Based on the above results, and the assumptions used in the models, the following conclusions can be provided:

The base carrier material used in a flexible substrate will govern the magnitude of warpage in the seen during the reflow process.

Models detailed above give reasonable predictions of warpage at peak reflow temperature as observed in experiments

Higher vol % of copper in the flexible substrate will reduce warpage during reflow.

A dielectric material in between the copper tracks can reduce the shear stresses along the copper/adhesive interface.

The use of Microwave Heating is showing very promising results. Excellent heating rates can be achieved and the ability to selectively and locally heat materials could provide significant advantages to this approach.

Acknowledgements

The authors would like to acknowledge the financial support provided by the European Union (EU) through the project FLEXNOLEAD. This project is collaboration between the following organizations: TWI Ltd, GTS Flexible Materials Ltd, Eldos Co. Ltd, Flexible Technology Ltd, Flex-ability Ltd, Kmed Manufacturing, Epigem Ltd, International Consulting Bureau, ICI Belgium N.V., KIC International Sales Inc., Freudenberg Forschungsdienste KG, Budapest University of Technology and Electronics and the University of Greenwich. The project is co-ordinated and managed by TWI Ltd and is partly funded by the EC under the CRAFT programme, Project Reference: COOP-CT-2004-513163.

Acknowledgement is also given to the IeMRC for supporting the FAMOBS project to investigate the use of microwave heating for assembly processes. We would also like to acknowledge the contributions of the industrial partners to the FAMOBS project which includes Renishaw, GE-Aviation, National Physical Laboratory, and Henkel.

References

[1] PHYSICA (1996-2007). Physica Ltd, 3 Rowan Drive, Witney, Oxon, United Kingdom, http://www.physica.co.uk.

[2] Sinclair, K.I., Desmulliez, M.P.Y. & Sangster, A.J., 2006, "A novel RF-curing technology for microelectronics and optoelectronics packaging", *Proc. IEEE Electronics Systemintegration Technology Conference (ESTC 2006)* . 2006, Vol. 2, pp 1149-1157.

[3] T. Tilford, E. Baginski, J. Kelder, A. K. Parrott and K. A. Pericleous, (2007) "Microwave Modelling and Validation in Food Thawing Applications', *Journal of Microwave Power and Electromagnetic Energy*, 41(4), pp. 30-45.

[4] Yee, K., 1966. "Numerical solution of initial boundary value problems involving Maxwell's equations in isotropic media". IEEE Transactions on Antennas and Propagation, 14: 302–307

[5] T. Tilford, K.I. Sinclair, C. Bailey, M.P.Y. Desmulliez, G. Goussettis, A.K. Parrott and A.J. Sangster, "Multiphysics Simulation of Microwave Curing in Micro-Electronics Packaging Applications", *Journal of Soldering and Surface Mount Technology*, Volume 19, Issue 3, 2007, pp. 26-33.

[6] T. Tilford, K. I. Sinclair, G. Goussetis, C. Bailey, M. Y. P. Desmulliez, K. Parrott and A. J. Sangster, "Numerical Simulation of Encapsulant Curing within a Variable Frequency Microwave Processing System', *Proc. EuroSimE 2008*, pp.447-454.

Board Level Assembly of Inertial MEMS Devices Based on Surface Mounting Technology

Kejia Li[1,2], Young Sun[3], Hongguang Liao[1,2] Yufeng Jin[2]

1. Graduate School of Peking University，Beijing 100871，China;
2. National Key Lab of Micro/ Nanometer Fabrication Technology, Institute of Microelectronics,
Peking University, Beijing 100871, China
3. Nokia Telecommunications Ltd., Beijing 100176, China
jinyf@ime.pku.edu.cn, 86-10-62752536

Abstract

In this paper, the board level assembly of inertial MEMS devices using Surface Mounting Technology (SMT) in mass production has been studied. MEMS microphone is taken as a representative device. Meanwhile, assembly precision and strength are also analyzed in the experiments. A further analyze in assembly strength has been done using theoretical simulation. The result indicates the full compatibility of MEMS devices with SMT. Design principles of the application of inertial MEMS devices in SMT has also been discussed.

Key words: board level assembly, inertial MEMS devices, Surface Mounting Technology, assembly precision, assembly strength

1. Introduction

The MEMS fabrication technologies have been developing rapidly in recent years. Novel methods of mass production are highly desired for MEMS devices with various structures. Featured with high density, high reliability, small scale and low cost, Surface Mounting Technology (SMT) proves to be a promising candidate, which has been used extensively in board level assembly. Design principles of MEMS devices application in SMT need to be discussed, especially for the inertial devices which dominate MEMS market [1].

In the MEMS devices SMT packaging, assembly precision, assembly strength and thermal forces are three key parameters. To the best of our knowledge, few studies have been reported on assembly precision. In the studies of assembly strength, the researches are mainly focused on the effects of Ag content on fracture resistance of Sn-Ag-Cu lead-free solders [2]. The influence of solders' number, arrangement on assembly strength also needs further discussion. Meanwhile, lots of work has been done with respect to the thermal forces, especially the effect of void formation on thermal fatigue reliability of lead-free solder joints, thermal cycling damage and stress of CSP and BGA's solders [3-5].

In this paper, we will focus on MEMS packaging using SMT in mass production, assembly precision and assembly strength of MEMS devices packaging. MEMS microphone is taken as a representative device in SMT mass production. A further analyze in assembly strength has been done using theoretical simulation.. The result indicates the full compatibility of inertial MEMS devices with SMT. Moreover, we will shed some light on the design principles of the application inertial MEMS devices in SMT.

2. MEMS Devices Packaging Design

It's well known that MEMS devices contain various fragile movable mechanical structures, like cantilever and thin films. Necessary cautions need to be taken to protect those movable actuators in the SMT.

Fig.1 MEMS Microphone

Figure 1 shows the typical structure of MEMS microphone. It has extensive applications in cell phones, MP3 and other electronic products. In this part, MEMS microphone is taken as a representative device to draw up some packaging design principles.

The fragile vibration film of MEMS microphone is vulnerable to high pressure. In modern SMT, devices are placed from material package to board mainly through vacuum absorber. Therefore, special adjustment should be made in the aspect of absorption position, action and velocity. Custom-absorber with special shape should be made for special material sometimes. As shown in figure 2, SMT absorber should keep away from the sound inlet of MEMS microphone when picking it up. The negative pressure in the vacuum absorber may destroy the vibration film. It's also necessary to keep a relatively low mounting speed.

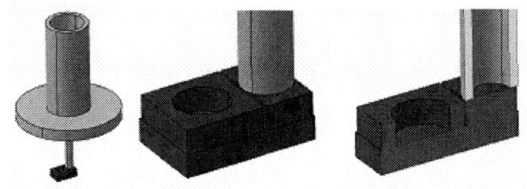

Fig.2 MEMS Microphone and Vacuum Absorber

Another consideration of MEMS microphone in SMT process is the board choice. It is not recommended to use flexible board, which is inclinable to distortion and circuit interconnects failure would deteriorate the device performance.

Each kind of MEMS device's packaging should be carefully considered according to its own special structure. In the meanwhile, some common design rules should also be followed. The thermal stress is a very important factor to induce performance failures of MEMS device. In the SMT, devices will go through 255℃ reflowing process. The thermal stress generated in board, solder, packaging material and chip

978-1-4244-2739-0/08/$25.00 ©2008 IEEE   515

may lead to deterioration of device performance. Many reports indicate that the non-uniform thermal stress accounts for the device thermal fatigue in the SMT [3-6]. Meanwhile, void formation, diameter and height of solder joints also affect thermal reliability. Therefore, those factors should be carefully considered in MEMS packaging design.

3. Experiments and Results

In this investigation of MEMS SMT assembly, assembly precision and assembly strength are mainly discussed, which play very important role in the SMT assembly process.

Assembly precision is the precision requirement on the X, Y axis and angle directions in the packaging of board level. It plays an important role in many MEMS applications. For example, the performance of the most commonly used inertial MEMS devices - accelerometer and gyroscope - are highly dependent on the precision X, Y axis or angle directions. The devices could not work well with low assembly precision, even causing system failure. Besides, it is the foundation of upper level assemble and final adjustment.

In order to analyze the assembly precision, High-speed Modular Mounter NXT made by Fuji Corporation and HO1 Mount head ($\pm$0.03mm at 3 sigma precision) are used in the experiment. Thirty-two devices are also used. Each one is picked up from material boxes and then mounts on the board after rotating a specific angle. The angles are 0°, 90°, 180°and 270°.

Tab.1 Assembly Precision at Different Angles

Degree	Average			3sigma		
	X[μm]	Y[μm]	Q[deg]	X[μm]	Y[μm]	Q[deg]
0	-2	3	-0.003	13	11	0.042
90	-2	2	-0.003	19	9	0.032
180	-6	1	-0.008	28	31	0.054
270	-5	0	-0.003	13	28	0.039

Table 1 shows the experiment results which can be summarized in table 2. It reveals that the displacements in X and Y directions are less than $\pm$0.03mm in the 3 sigma precision. This is also the highest accuracy that the equipment can reach.

Tab. 2 Summarization of Assembly Precision

Total	X (μm)	Y (μm)	Q (deg)
Average	-4	1	-0.004
3sigma	20	22	0.043
Maximum	6	13	0.024
Minimum	-25	-20	-0.029

For inertial MEMS devices, another important parameter is the angle precision. From table 2, the result shows the average deviation is 0.043°. From the accelerometer datasheets of Analog Device and MEMS IC Corporation, the tolerance of angle error of accelerometer is $\pm$1°. That means assembly precision is totally in the acceptance range.

Assembly Strength is defined as the solder strength between device and board. It equals to force applied on device divided by solder-to-board contacting area. If assembly strength is not strong enough to bear mechanical impact,

collision and vibration, the integrated electronic system can't work even if the anti-strike strength of MEMS device itself is high.

In order to analyze assembly strength, LF300 soldering tin made by Henkel Loctite Corporation, Multiple layer non-halogen electronics circuit board and Dage 4000-TP10 bond tester are used. The solder mask consists of selective Ni/Au layers and OSP layers. The measure precision of bond tester is 0.01g and the push force ranges from 0kg to 100kg. The moving speed is about 262.1μm/s in the experiment. Table 3 shows the different sizes of components. Considering the difference of inertial MEMS devices, A, B represent the packaging of gyroscopes and C, D represent the accelerometers' packaging.

Tab. 3 Different sizes of components

No.	packaging	size (mm)	Solder	
			number	Diameter(mm)/height(mm)
A	BGA	12×12×0.9	289	0.3/0.25
B	BGA	11×10×1.0	133	0.3/0.22
C	CSP	0.96×1.02×0.65	4	0.32/0.24
D	CSP	1.67×1.67×0.41	8	0.32/0.24

The results are shown in table 4. It reveals that the assembly strength is almost the same despite the different average push force. The error is 5.3% ((38-36)/38) and the assembly strength has little relationship to solder's size, distribution and height.

Tab. 4 Assembly strength of different component

Components	Average push force (kgf)	area (mm²)	Assembly strength(MPa)
A（BGA）	74.721	20.418	36
B（BGA）	36.061	9.396	38
C（CSP）	1.169	0.322	36
D（CSP）	2.483	0.643	38

The average of maximum acceleration of ordinary accelerometer chips is about 10,000g in current market (Analog Devices8, MEMSIC7, STMicroelectronics9, Freescale10). If it is converted to pressure, the pressure is about 71 Pa which is far less than 37MPa. That means if MEMS devices can't work due to collision, the reliability of device itself is the limiting factor rather than that of SMT assemble.

In order to know the force distribution in the device-to-board separation, finite element software ANSYS is used to simulate the whole process.

Component C is taken as an example model. Simplified structures are shown in figure 3. The properties of different layers and materials are shown in table 5. According to the Von Mises Yield Criterion, material response is assumed to be elastic prior to yield. Therefore, elastic linear analysis should be used at first to get distribution and changes of each node's

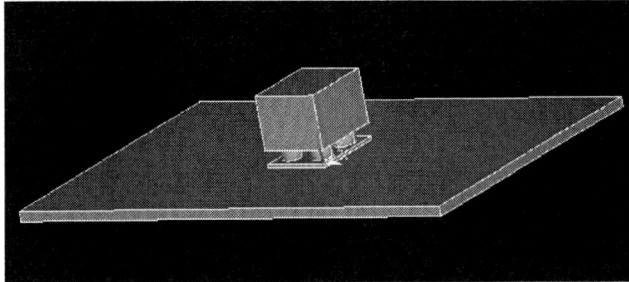

(a) three-dimension structure

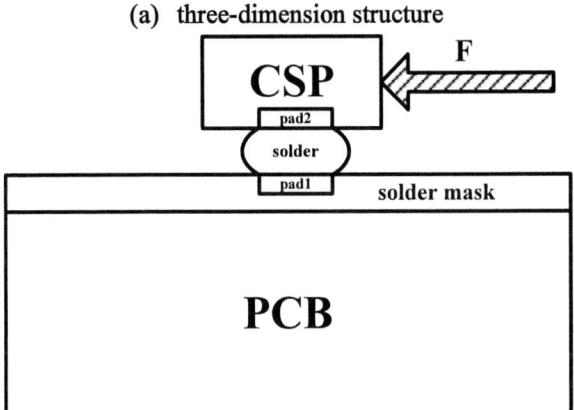

(b) front-section view

Fig.3 The structure of the model

stress. Then structural nonlinear analysis would be applied to get the approach tear force at which device falls from PCB. In this part, only elastic linear analysis is discussed. The boundary conditions are set as bellow. In the direction of X, Y and Z, the displacement of each PCB face is zero. The displacement of area with applied force is zero in X and Z directions. In the simulation, 0.1MPa, 0.5MPa, 1MPa, 2MPa, 3MPa, 4MPa, 5MPa and 6MPa pressures are applied on the model separately.

Tab. 5 Size and Material Properties of Each Layer

Layer No.	Material	Young's Modulus (GPa)	Poission's ratio	size
PCB[12]	FR4	18.2	0.19	7×4×0.12
Solder mask[12]		6.87	0.35	0.96×1.02×0.04
Pad1[12]	Cu	68.9	0.31	R=0.14;h=0.02
Solder[3,12]	Sn(95%) Ag(30%) Cu(0.05%)	51	0.31	R=0.16;h=0.16
Pad2[3]	Ni	200	0.31	R=0.14;h=0.02
CSP[11]	rosin	26	0.3	0.96×1.02×0.65

Figure 4 shows the relationship between Von Mises stress and applied pressure. It derives from the node with maximum Von Mises stress. It reveals that as the increase of the applied pressure, Von Mises stress increases. The fitting equation can be drawn as follows.

$$V=27.73P-0.01 \qquad\qquad eq. 1$$

Where P means pressure and V stands for Von Mises stress. Figure 5 shows the areas with maximum stress- the edges and the center of solders.

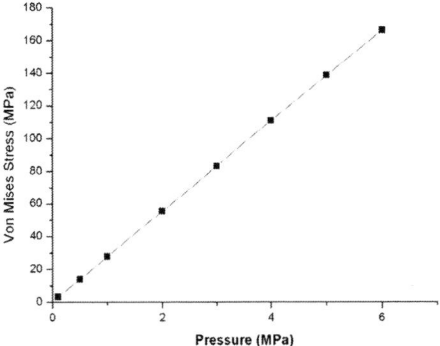

Fig. 4 The relationship between Pressure and Von Mises Stress

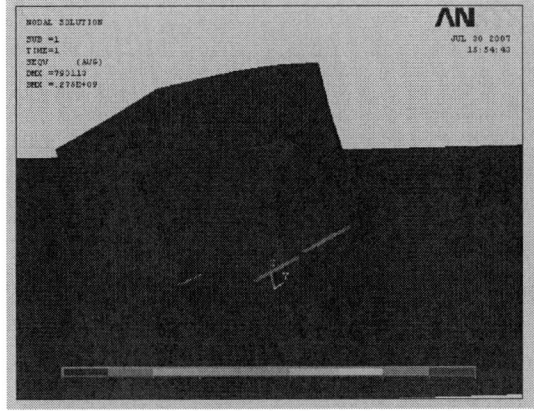

(a) Von Mises stress distribution- from 3D view

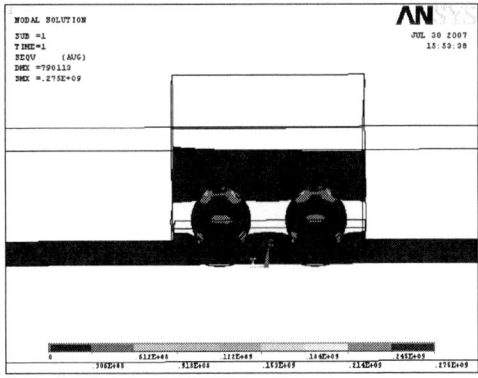

(b) Von Mises stress distribution- from xy plane

Fig.5 Von Mises stress distribution when CSP will fall

4. Conclusions

In this paper, MEMS microphone is taken as the sample of MEMS devices. According to the special mechanical structure of MEMS microphone, assembly actions and board choice in SMT are mainly discussed. Other common MEMS packaging problems in SMT and design principles are also mentioned.

Assembly precision is analyzed in the experiments. It reveals that the displacements in X and Y directions are less than ±30μm in the 3 sigma precision, and average deviation is 0.043° in the angle precision. Compared with the datasheets of MEMS products in the current market, those assembly precisions are totally in the acceptance range.

Assembly strength is also studied both from experiment and simulation. The experiment indicates that the assembly strength is almost the same despite different average push force. The error is 5.3% ((38-36)/38) and the assembly strength has little relationship with solder's size, distribution and height. Meanwhile, stress distribution in the device-to-board separation and process of nearly break-off situation are presented by finite element simulation tool ANSYS. The fracture is prone to happen in the middle and edges of solders, which provide certain hints in further strength studies.

In conclusion, those results have proved the full compatibility of MEMS devices with SMT. In the future, packaging materials and processes should be given more attention in MEMS devices packaging.

References

1. Yufeng Jin, et al,. Introduction of MEMS Packaging, Science Press (Beijing, 2006).
2. Dawoong Suh, Dong W. Kim, Pilin Liu, "Effects of Ag Content on Fracture Resistance of Sn-Ag-Cu Lead-free Solders under High-Strain Rate Conditions ". Materials Science and Engineering A, (2007), 460-461:595-603.
3. Lianxi Shen, Sung Yi, Caers, "A damage parameter based on fracture surface for fatigue life prediction of CSP solder joints," International Conferences on Electronic Materials and Packaging, 2001, Jeju Island, South Korea, Nov., 2001,pp. 412 – 416.
4. Hu Yong-fang, Xue Song-bai, Wu Yu-xiu, "FEM analysis of stress and strain and evaluation on reliability of soldered CBGA joints under thermal cycling,". Trans. Nonferrous Met. Soc. China, Vol. 11, No. 15 (2005), pp. 317-322.
5. Do-Seop Kim, Qiang Yu, Shibutani, T. et al., "Effect of void formation on thermal fatigue reliability of lead-free solder joints," *The Ninth Intersociety Conference on Thermal and Thermomechanical Phenomena in Electronic Systems,* Las Vegas, Nevada, June 2004, pp. 325 – 329.
6. Chang-Chun Lee，Hsing-Chih Liu and Kuo-NIng chiang, "3-D structure design and reliability analysis of wafer level package with stress buffer mechanism," *Components and Packaging Technologies*, Vol.30, No.1 (2007), pp. 110-118.
7. MEMSIC. MXR9500G/M Datasheet [EB/OL/. http://www.memsic.com.cn/memsic/data/products/MXR 9500G&M/MXR9500G&M.pdf. 2007-7-28
8. Analog Devices. ADXL320 Datasheet [EB/OL]. http://www.analog.com/UploadedFiles/Data_Sheets/AD XL320.pdf. 2007-7-28
9. STMicroelectronics. LIS3LV02DL Datasheet [EB/OL]. http://www.st.com/stonline/products/literature/ds/12094. pdf. 2007-7-28
10. Freescale. MMA7260QT Datasheet [EB/OL]. http://www.freescale.com/files/sensors/doc/data_sheet/M MA7260QT.pdf. 2007-6-03
11. Jinsong Xie,Jiaqi Zhong, Chuan Li et al. "Thermal Force Analysis of CSP chips," *Specific Equipment of Electronics Industry*，Vol 123 (2005)，pp. 32-34,42.
12. Chang-Chun Lee, Hsing-Chih Liu, and Kuo-Ning Chiang et al. "3-D Structure Design and Reliability Analysis of Wafer Level Package With Stress Buffer Mechanism." *Components and Packaging Technologies*, Vol. 30, No.1 (2007), pp. 110-118.

Encapsulation of Organic Light-emitting Devices for the Application of Display

C Y Li[1,2], B Wei[1] and J H Zhang[1,2]

1 Key Laboratory of Advanced Display and System Applications (Shanghai University), Ministry of Education
2 The School of Mechanical & Electronic Engineering and Automation, Shanghai University, Shanghai, 200072, China
Email: jhzhang@staff.shu.edu.cn; Tel: 86-21-56331976

Abstract:

Organic light-emitting diodes (OLED) is being considered as one of the most potential technology for the application of flat panel display (FPD) due to the low driving voltage, high luminance and high efficiency. Furthermore, OLED displays with plastic substrate also have many attractive features, such as ultra-thin and light in substrate.

However, short operation-time hinders the process of OLED technology to industrialization due to its extremely sensitive to the operation environment. The key issues in achieving such displays are how to protect flexible OLED (FOLED) from moisture and oxygen.

In this paper, some recent progresses in OLED packaging have been described and compared, such as the characteristics of oxygen and water permeability in different species of plastic substrates, the merits and demerits of chemical, physical and hybrid encapsulation methods. Finally, effects of the barrier film onto a plastic substrate and passivation film on the performance (efficiency and lifetime) OLED device have been discussed. In addition, in order to observe the roughness of passivated film, the oxygen and water permeability, and the fabrication method of FOLED , some widely-used measurement technology have also been reviewed.

Key words: FOLED; passivation film, oxygen/water permeability

1 Introduction

Organic light emitting diodes (OLED) have been studied extensively since the high-performance green emission was demonstrated for a bilayer OLED by Tang and VanSlyk in 1987[1]. Promising prospect in flat full-color panel displays [2, 3] has attracted the interest of study in the efficiency and operation time of OLED. With the advancement in the technology of OLED, the efficiency and lifetime have been improved dramatically [4, 5], and several kinds of OLED displays have been commercialized.

The OLED have many attractive features for display applications, such as high brightness, high efficiency, a wide viewing angle and fast response time. Moreover, OLED have high visibility by self-light emission, and no backlighting is required. Most of the above, OLED can be fabricated into lightweight and thin flexible displays. Therefore, OLEDs are considered potential to replace liquid crystal displays which are dominant flat panel displays nowadays [6-8].

To date, a lot of efforts have been focused on fabricating OLEDs on various flexible substrates [9]. However, the plastic substrates often meet stricter encapsulation problem than glass substrate. The barrier properties of these substrates are not sufficient to protect the electroluminescent polymeric or organic layers in OLEDs due to the penetration of the chemically reactive oxygen and water molecules into active layers of the devices, especially during the process of voltage applying on the device. The OLED structures and active electroluminescent materials are studied to degrade rapidly in the presence of oxygen and moisture [10].

Currently, OLEDs are being fabricated on a glass substrate, and then encapsulated with metal or glass lids. An OLED display on a plastic substrate is being expected because it is thinner and lighter compared with that on a glass substrate. Moreover, by utilizing the flexibility of the substrate, displays of various forms can be fabricated including so-called flexible display [11]. However the short operation lifetime limits the process of flexible OLED technology to industrialization. The key issue in fabricating OLEDs on a plastic substrate is to prevent from moisture and oxygen. It is because OLEDs are very sensitive to both [12]. TFE technology is the key technique to meet the requirements of small-size electronic products.

In the paper, the characteristics of oxygen and water permeability in different species of plastic substrates have been described. Next, we compared the merits and demerits of three encapsulation methods, which are chemical, physical and hybrid packaging technology. Finally, effects of the barrier film onto a plastic substrate and passivation film on the performance (efficiency and lifetime) OLED device have been discussed. In addition, in order to observe the roughness of passivated film, the oxygen and water permeability, and the fabrication method of flexible OLED (FOLED), some widely-used measurement technology have been reviewed.

2. OLED requirements to water and oxygen permeation
2.1 Degradation mechanism of OLED

1) H_2O and Oxygen

Permeation barriers for OLEDs are required because the devices degrade in the presence of atmospheric gases, primarily moisture and oxygen. The degradation mechanism is illustrated in Fig. 1[13]. Accordingly, high-performance TFE should be fabricated to realize long-time driving OLED display. The high-quality TFE should meet good diffusion barrier to water and oxygen penetration, high flexibility or mechanically robust to avoid cracking during bending. Furthermore, the TFE should have good heat diffusion effect. This is because the device may degrade when the current is turn on, resulting from the electro-chemical process. The heat accumulation will promote the growth of pinhole in the cathode area and the formation of dark spot in the emissive layer (Fig. 1). Finally, we should assure a lower effect of damage when depositing passivation layer on device, since the organic materials cannot withstand high process temperature and also require an inert processing environment. Figure 2 compared the lifetime of OLEDs without treatment and with treatment with two ways.

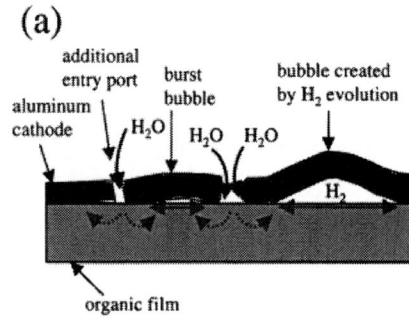

(a)

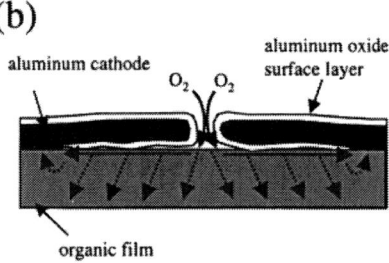

(b)

Fig. 1 Formation of dark-spot of OLED under operation resulted from the electro-chemical process of oxygen and water

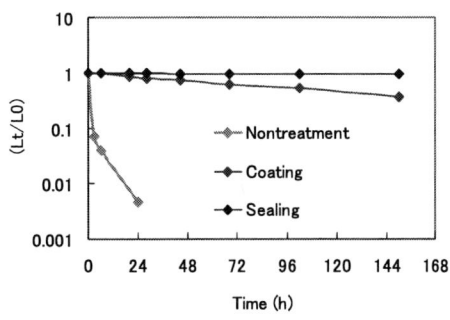

Fig. 2 Lifetime of OLED encapsulated by different treatment, no treatment, coating by organic/metal layer, and sealing with glass

In addition, the size of OLED also plays an important role in the degradation at the application of external voltage. The Joule heating produced by the application of external voltage can be transferred more rapidly away from a device with smaller size of active region [14]. The voltage-evolution is different for two different sizes of OLEDS when the voltage goes up to a certain value. For the thermal inhibition effect, the emission intensity of big-size of OLEDs might increase slower than that of small-size of OLEDs.

2.2 Permeation requirements

To date, three species of substrates have been used for fabricating OLED. However, the permeability of oxygen and water, and flexibility are varied, as shown in Table 1. Glass substrate-based OLED is of buffering capability, but can not used for flexible OLED. Metal thin film-based OLED has been used for FOLED fabricated by Universal Company; however, the cost is relatively high. FOLED with plastic substrate is flexible, but poor permeability of oxygen and

water. Therefore, the approaches to fabricating high-quality passivition film are being widely investigated.

Table 1 Characteristics of several species of substrates

Substrate	Glass	Metal thin film	Plastic		
			PET	PP	PE
O_2	●	◎	○	○	○
Water	●	◎	○	○	○
Flexibility	○	◎	●	●	●

● excellent   ◎ good   ○ poor

Some researches have found that for OLEDs to have reliable performance lifetime, exceeding 10,000 h, oxygen and moisture transmission rates must be below 10-6 g/m2 day [15]. A barrier film must be added on the plastics substrate to protect OLEDs from moisture and oxygen. It has reported that the level of moisture impermeability for OLEDs must be less than 10-5g/m2 /day [16].

3. Encapsulation methods

3.1 Physical packaging technology

Figure 3 is one widely used encapsulation technology for OLED device, in which an UV curing epoxy adhesive (2) sealing the sheet-shape glass (3) with desiccators (4) onto the OLED (1). The total thickness for such frame is about 1.4 mm, depending on the selected glass cover. Auch et al. [17] proposed ultra-thin glass as the substrate and encapsulating cover for flexible OLED. But ultra-thin glass is different in toughness and brittle. Present ultra-thin glass cutting technology is easy to cause the edge micro fissure flaw. Using this type of encapsulation, the common method is to seal display device in the dry inert gas environment

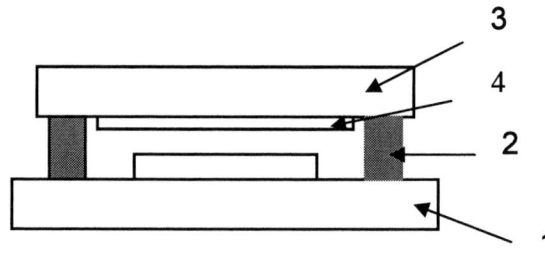

Fig. 3 Schematic of sealed OLED

1 glass or plastic 2 adhesive, 3 sheet-shape, 4 desiccators

For achieving thinner, simpler and low-cost encapsulation, some reported the metal back electrode embeded in a thick metal coating [18]. However, pinhole deflect is inevitable during the process of the metal evaporation, providing space for the invasion of oxygen and moisture. Therefore this method is not a very effective method to enhance the lifetime of OLED.

Rendu et al. [19] have investigated the behavior of two polymeric materials used as a protecting layer: cellulose acetate and polyvinylidene chloride. This type of method only has some effect on luminescent material packaging, but can not spread its application widely. Moreover, its effect needs further experiment test.

3.2 Chemical Vapor Deposition

Jeong have studied UV–visible and infrared transmission of three sorts of poly (p-xylylene) films obtained by vapor-deposition polymerization [20]. This polymer thin film has the outstanding impediment to water and oxygen [21], and its light transmittance rate can reach 95% in visible light and near infrared area. In addition, Huang et al. [22] recently proposed Plasma Enhanced Chemical Vapor Deposition (PECVD) to prepare silicon nitride (SiNx) thin film for straightly sealing OLED. The SiNx thin film prepared by PECVD was developed in the 1970s. Because it has the excellent properties against humidity and ion corrode; therefore, it has been utilized in micro electron package. The disadvantage of preparing SiNx by PECVD needs 180 ℃, far beyond the bearing temperature of organic materials for OLED.

Therefore Huang et al. [22] has optimized PECVD deposition parameter. The film temperatures deposited at the substrate from 20℃ to 180℃ and the RF powers from 10 W to 30 W were investigated. It was suggested that SiNx films deposited at low temperature could be applied in OLED packaging effectively.

3. 3 Hybrid packaging technology

Hybrid packaging technology is also so-called multi-layer barrier film, had been considered as one of most potential method. It would compensate the disadvantages resulted from single inorganic or organic package. The FOLED often uses two approaches to thin film encapsulation, which are single-layer and multi-layer thin films. Single-layer thin film passivation employs PECVD or thermal evaporation deposition, to fabricate one buffer layer in the substrate to prevent the permeability from water and oxygen. Some researchers [23-25], have proposed novel encapsulation using multi-layer composed of alternating oxide and polyacrylate layers. American Vitex Corporation develops the Barix technology which polymer layer and inorganic layer alternately, has realized the OLED package effectively [26]. Compared to traditional packaging method, it is lower cost, thinner and longer OLED life. Moreover, this method can eliminate interaction between various protective layers.

Sang-Hee [27] et al. used atomic layer deposition (ALD) to encapsulate FOLED for the first time. The merit of ALD is to form thin film simultaneously at both sides of substrate, therefore may reduce stress caused by different thermal diffusion coefficient between substrate and dielectric medium thin film. However, due to slow growth rate for thin film during the process of ALD, water vapor and air possibly have influence on the device.

In addition, H. Lifka et al. [28] proposed a novel multilayer stack of silicon nitride - silicon oxide - silicon nitride -silicon oxide - silicon nitride as thin film encapsulation by PECVD.

4. Measurement of permeation rates

The surface morphology of the flexible substrate was examined using atomic force microscopy. Its optical transmission was often tested by absorption spectrometry.

It is worth to mention that one OLED degradation measurement system has been developed by shanghai university. The system is simple and in spots. It uses photodiode to measure the relative brightness of device during operation. The current density is set to be 10mA/cm2. Data of brightness is recorded by computer with an interval of one hour.

One of the most important properties of a good encapsulation layer is a low permeation rate for water and oxygen. However, to date, none of the commercially available systems meet the sensitivity requirements for the low permeation rates required for OLEDs. To be able to measure these required low permeation rates Nisato et al. developed the so-called calcium test [29, 30]. Using this method, the bulk permeation and defect-based permeation could be discriminated by observing spot percentage on the Ca film. However, the permeation produced from oxygen or water permeation could not be identified. Clearly, it is important to develop advanced measurement method or techniques for understanding multilayer permeation barriers, particularly the permeation from oxygen.

5. Conclusions

The various encapsulation techniques for OLEDs have been discussed. Barrier requirements and permeation measurement techniques were also investigated. The realization of high-quality FOLED displays requires further study in the technology of thin-film permeation barriers.

6. Acknowledgments

This work was supported by the National Nature Science Foundation of China (50575229), the Education Commission of Shanghai (2006ZZ04), the Innovation Fund of Shanghai University （A10-0109-07-014）and Program for New Century Excellent Talents in University under the grant number 07-0535.

The corresponding author, J.H Zhang, would also thank the financial support provided by the Shanghai Shuguang Program under the grant number 05SG042 and Shanghai Leading Academic Discipline Project under the grant number Y0102.

References

1. C.W.Tang and S.A.VanSlyke, "Organic electroluminescent diodes," *Appl. Phys. Lett.,* Vol. 51, No. 3 (1987), pp. 913–915.
2. J. R. Sheats, H. Antoniadis, M. Hueschen, W. Leonard, J. Miller, R. Moon, D. Roitman, and A Stocking, "Organic Electroluminescent Devices," *Science,* Vol. 273, No. 5277, (1996) pp.884-888
3. S.Forero, P.H.Nguyen, W.Brutting, and M.Schwoerer, "Charge carrier transport in poly (p-phenylenevinylene) light-emitting devices," *Chem. Phys.*, Vol.1, No.8, (1999), pp.1769 - 1776.

4. M.A. Baldo, S. Lamansky, P. E. Burrows, M. E. Thompson, and S.R. Forrest, "Very high-efficiency green organic light-emitting devices based on electrophosphorescence," Appl. *Phys. Lett.*, Vol. 75, No. 1, (1999), pp.4–6.

5. T.Tsuji, S.Kawami, Y.Fukuda, and S.Miyaguchi, "Improvement of operating lifetime red phosphorescent organic light emitting devices," *In Proc. 8th Int. Display Workshops*, 2001, pp.1423–1426.

6. L. S. Hung and C. H. Chen, "Recent progress of molecular organic electroluminescent materials and devices," *Mater. Sci. Eng.*, R 39,(2002), pp. 143–222.

7. R. H. Friend, R. W. Gymer, A. B. Holmes, J. H. Burroughes, R. N.Marks, C. Taliani, D. D. C. Bradley, D.A. Dos Santos, J. L. Brédas, M. Lögdlund, and W. R. Salaneck, "Electroluminescence in conjugated polymer," *Nature*, Vol. 397, (1999), pp. 121–128.

8. S. R. Forrest, "The road to high efficiency organic light emitting devices," *Organic Electron.*, Vol. 4, (2003), pp. 45–48.

9. C. Fou, O. Onitsuka, M. Ferreira, M. F. Rubner, and B. R. Hsieh, "Fabrication and properties of light-emitting diodes based on selfassembled multilayers of poly(phenylene vinylene)," *J. Appl. Phys.*, Vol. 79, (1996), pp. 7501–7509.

10. A. B. Chwang, M. R. Rothman, S. Y. Mao, R. H. Hewitt, M. S.Weaver, J. A. Silvermail, K. Rajan, M. Hack, J. J. Brown, X. Chu, L. Moro, T. Krajewski, and N. Rutherford, "Thin film encapsulated flexible organic electroluminescent displays," *Appl. Phys. Lett.,* Vol.83, (2003), pp. 413–415.

11. G. Gu, P. E. Burrows, S. Venkatesh, and S. R. Forrest, "Vacuum deposited, nonpolymeric flexible organic light emitting device," *Opt. Lett.*, Vol. 22, (1997), pp. 172–174.

12. J. K. Mahon, J. J. Brown, T. X. Zhou, P. E. Burrows, and S. R. Forrest, "Requirements of flexible substrate for organic emitting devices in flat panel display applications," *Soc. Vacuum Coaters,* Vol. 505, (1999), pp.456–459.

13. Michel Schaer, Frank Nüesch, Detlef Berner, William Leo, and Libero Zuppiroli, "Water Vapor and Oxygen Degradation Mechanisms in Organic Light Emitting Diodes," *Adv. Funct. Mater.,* Vol.11, No. 2, April, (2001), pp.116-121

14. B. Wei, M Ichikawa, K Furukawa, T Koyama, Y Taniguchi, "High Peak Luminance of Molecularly Dye-Doped Organic Light-Emitting Diodes Under Pulse Voltages," *Journal of Applied Physics*, 98, 0445061 (2005).

15. Van Nostrand Reinhold, Microelectronics Packaging Handbook, (New York, 1989), ch. 10.

16. P. E. Burrows, G. L. Graff, M. E. Gross, P. M. Martin, M. Hall, E.Mast, C. Bonham, W. Bennet, L. Michalski, M. S. Weaver, J. J. Brown, D. Fogarty, and L. S. Sapochak, "Gas permeation and lifetime tests on polymer-based barrier coatings," *Proc. SPIE*, Vol. 4105,(2000), pp. 75-83.

17. Auch M D J, Soo O K, Ewald G, et al, "Ultra-thin glass for flexible OLED application," *Thin Solid Films,* Vol.417, No.122, (2002), pp.47-50.

18. Lim S F, Wang W, Chua S J, "Degradation of organic light-emitting devices due to formation and growth of dark spots,"*Mat.Sci.&Engi.*B,Vol.85, No.223, (2001),pp.154-159.

19. Rendu P L, Nguyen T P, Carrosis L, "Cellulose acetate and PVDC used as protective layers for organic diodes," *Synth. Met*. Vol. 138, No.122, (2003), pp.285-288.

20. Jeong Y S, Ratier B , Moliton A , et al., "UV-visible and infrared characterization of poly (p-xylylene) films for waveguide applications and OLED encapsulation," *Synth. Met.,* Vol.127, No.123, (2002), pp.189-193.

21. Beach W F, Lee C, Bassett C R, et al., "Xylylene Polymers: in Encyclopedia of Polymer Science and Engineering," New York: Wiley, 1989.

22. Huang W D, Wang X H, Sheng M , et al, "Low temperature PECVD SiNx films applied in OLED packaging," *Journal of Functional Materials and Devices,* Vol. 98, No. 3, (2003), pp.180-184.

23. T. B. Harvey, S. Q. Shi, and F.So, "Passivated organic devices," U.S. Patent 5686360, 1997.

24. D. Affinito, "Environmental barrier material for organic light emitting device and method for making," U.S. Patent 6268695, 2001.

25. G.L.Graff, M.E.Gross, J.D. Affinito, M.K.Shi, M. Hall, and E.Mast, "Environmental barrier material for organic light emitting device and method for making," U.S. Patent 6522067, 2003.

26. Vitex Systems, Inc., San Jose, CA.

27. Sang-Hee Ko Park, Jiyoung Oh, Chi-Sun Hwang, et al., " Ultra thin film encapsulation of organic light emitting diode on a plastic substrate," *ETRI Journal*, Vol.27, No. 5 , (2005), pp.545-550.

28. H. Lifka, H. A. van Esch, J. J. W. M. Rosink, "Thin Film Encapsulation of OLED Displays with a NONON Stack", *SID 04 DIGEST*, pp.1384-1387

29. G. Nisato et al., "Evaluating high performance diffusion barriers: the Calcium test," *Proceedings of the IDW-Asia Display 2001*, 2001.

30. G. Nisato et al., "Thin Film Encapsulation for OLEDs: Evaluation of Multi-layer Barriers using the Ca Test," *Proceedings of the SID 2003 International Symposium*, 2003, P88.

A Study of Fluid Coolant with Carbon Nanotube Suspension for Microchannel Coolers

Yi Fan[1], Yifeng Fu[2,3], Yan Zhang[1], Teng Wang[2], Xiaojing Wang[1], Zhaonian Cheng[2] and Johan Liu[1,2]

[1]Key Laboratory of Advanced Display & System Applications , SMIT Center and School of Automation and Mechanical Engineering, Box 282, No 149 Yan Chang Road, Yan Chang Campus, Shanghai University, 200072, China

[2]SMIT Center and Bionano Systems Laboratory, Department of Microtechnology and Nanosciences (MC2)
Chalmers University of Technology, Se 412 96 Gothenburg, Sweden

[3]SHT Smart High Tech AB, Nordgårdsvägen 19, 428 34 Kållered, Sweden

jliu@chalmers.se

Abstract

In this work, silicon microchannel coolers were made using the Deep Ion Reactive Etching (DIRE) technique. Stable and homogeneous Carbon NanoTube (CNT) suspension was also prepared. Meanwhile, a closed-loop cooling test system was developed to investigate the heat removal of the silicon microchannel cooler with different coolants. The experimental setup included a test module, a minipump for providing controllable flow, and a fan system for cooling the circular fluid. Beside the inlet and outlet of the test module, two thermocouples and pressure gauges were set up to measure the temperature and pressure of the fluids. The heat removal of the silicon microchannel cooler using different CNT volume fraction of suspension coolant was studied. The results show that the microchannel cooler with CNT suspension as coolant could strengthen the heat removal capability of microchannel cooler. In addition to heat transfer enhancement, the microchannel cooler with CNT suspension coolant did not produce extra pressure drop in the present study.

1. Introduction

The resistance of the flow of electrical current through leads, poly-silicon layers and transistors results in significant internal heat generation with an operating microelectronic component. In digital applications, a typical very-large-scale-integrated (VLSI) chip containing 106 transistors can dissipate up to 100W, and more complex circuits integrating of high power components may reach a heating power as high as 400W. In the absence of proper cooling, the temperature of such operating circuit would rise until it reaches a value at which either operation ceases or components lose their physical integrity. So providing effective and compact heat removal solutions is an essential element of the electronics packaging approach and its importance increases as the trend in the electronics industry moves towards higher packaging density and more severe operation conditions.

Many new thermal removal techniques have been developed in response to this situation, among which microchannel liquid cooling has been considered as one of the very promising cooling solutions [1-3]. A typical microchannel cooler configuration is a finned structure, which is cooled by forced convection. The power is dissipated on the circuit side and the heat is conducted through the substrate to the fins where it is transferred to the coolant. The major advantage is the small volume, the high heat transfer coefficient, small power and possible wafer-level integration with chips.

Besides the microchannel structure, the coolant adopted also plays an important role in the heat transfer of the microchannel coolers. And the heat dissipation depends mainly on the mass transport of the coolant. To improve further more heat dissipation, another method is to increase the heat transfer coefficient. The most frequently used coolant in the microchannel heat sink is pure water. Better results may be achieved with other fluids. For example, fluids with particles of nanometer dimension suspended, namely nanofluids, are particularly attractive as a new type of coolant [4]. Although the mechanisms of nanofluids have not been fully revealed, several experimental studies have shown that the effective thermal conductivity of the fluids can significantly be increased by adding very small volume fraction of nanoparticles [5,6].

Recently, Choi et al. [7] reported high thermal conductivity of Carbon NanoTube (CNT) suspensions in a synthetic poly (a-olefin) oil. They reported anomalous enhancement of the thermal conductivity with a small amount of CNTs added, which suggests a strong potential of CNT suspension as an enhanced heat transfer medium.

In this work, stable and homogeneous CNT suspension was achieved by attaching hydrophilic functional groups onto the surfaces of CNTs. Using the DIRE process, silicon microchannel cooler was also prepared. The heat removal capability of silicon microchannel cooler with CNT suspension as coolant was investigated.

2. Cooler and coolants preparation

2. 1 Microchannel cooler

The silicon microchannel coolers were fabricated in this work. The DIRE process was chosen to achieve channels with high aspect ratio. The parallel channels etched are 100μm wide and about 200μm deep. The wafer was then diced into chips of a rectangular shape 10×8mm. The geometrical dimensions of the coolers are given in Fig. 1. On the other side of the silicon chip, a 1μm thick serpentined copper trace was integrated onto the chip to simulate as the heat source. And a 500μm thick aluminum plate was glued on silicon microchannel chip as a lid to construct the final microchannel cooler.

2.2 CNT suspension coolant

It is evident that pristine CNTs (PCNTs) are not only aggregated, but entangled [8], which means they are not ready to form stable suspensions in most heat transfer fluids. Even with the use of surfactant like oleylamine, they can not be dispersed into nonpolar liquid. Therefore, more studies should be done not only for measurements of thermal conductivities of CNT suspensions in different kinds of base fluids, but also for the methods to produce stable and homogeneous CNT suspensions.

In the present work, the commercial CNTs purchased were produced by chemical vapor deposition method. A chemical treatment method was adopted to produce stable and homogeneous suspensions of CNTs in pure water. This method includes two steps. Firstly, samples of CNTs (1g) were suspended in 40ml of concentrated nitric acid for 2 hours and refluxed at 140°C. Second, the samples of CNTs were washed with water until the supernatant attained a pH around 7 and dried.

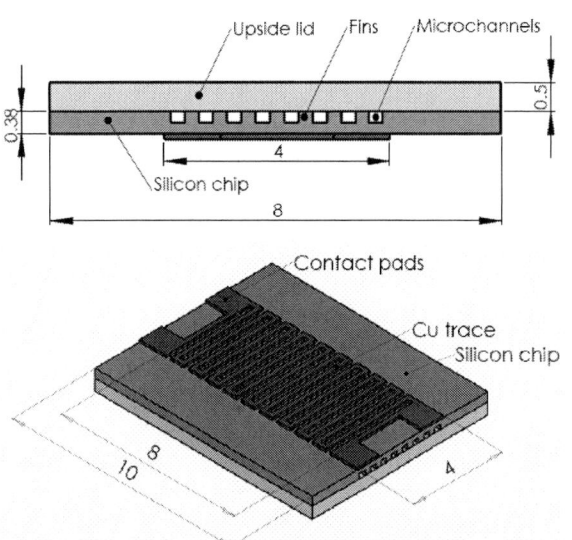

Fig. 1 Geometrical dimensions of the microchannel cooler with integrated heating resistor. All units are in millimeters.

3. Heat removal measurement setup
3.1 Closed-loop cooling system

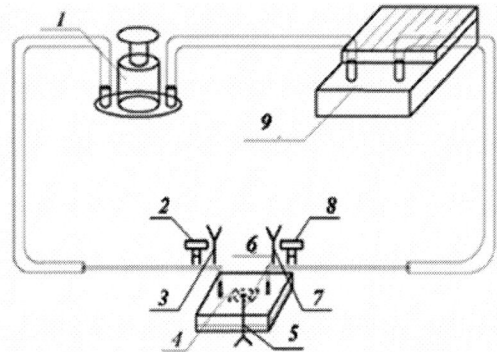

Fig. 2 Schematic view of the experimental facility: 1-minipump, 2-inlet pressure gauge, 3-inlet thermocouple, 4,5,6-test module, 7-outlet thermocouple, 8-outlet pressure gauge, 9-fan system.

In this study, we investigated the heat removal of CNT suspension coolants. The experimental facility was designed as a close loop cooling system and constructed as illustrated schematically in Fig. 2. We can see from Fig. 2 that the experimental setup mainly consisted of a test module including silicon microchannel cooler and a supporting frame, a minipump

for providing a continuous, stable and controllable flow, and a fan system for cooling the circular fluid. Beside the inlet and outlet of the module, thermocouples and pressure gauge were set up to measure the temperature and pressure of fluids. The experiments were performed in a close loop, and the inlet temperature was controlled to be the same in all the cases studied.

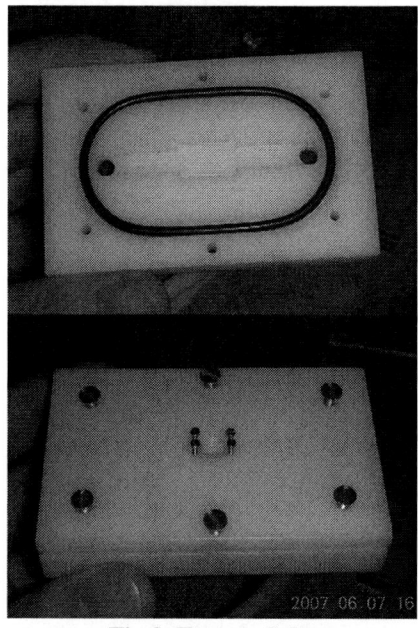

Fig.3. Test module.

3.2 Test module

Fig.3 gives the pictures of the test module. The test module consisted of a test silicon microchannel cooler, a base plate and an upside lid. The silicon microchannel cooler was fixed in the middle rectangular cavity of the base plate, and the base plate and the upside lid distributed the fluid flow through a test silicon microchannel cooler. When the coolants flow into the microchannel in the test module, the power from the heating resistor was dissipated by the microchannel cooler. The heat was conducted through the substrate to the silicon fins in microchannel cooler, and then transferred to the coolant. An O-ring was used to prevent the fluid leakage. Four screws were set in the base plate to act as the electrical connection parts between the Cu trace and the power supply. A thermocouple was inserted into a hole in the center and placed on the surface of the heating resistor to measure the temperature.

4. Measurement results
4.1 Effect of heating power.

In the first case study, the effects of heating power Wchip on the temperature of chip Tchip, the outlet temperature Toutlet and the pressure drop between inlet and outlet were tested. The CNT volume fraction of suspension coolant was 1%, the fluid flow was 27 ml/min, the room temperatures Troom=20°C, and inlet temperature Tinlet = 12.7°C.

524

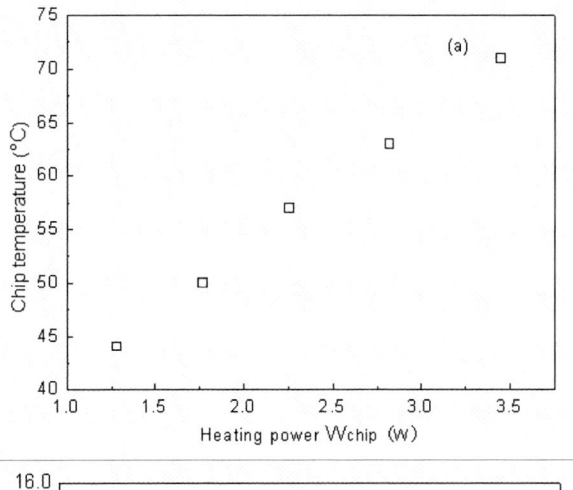

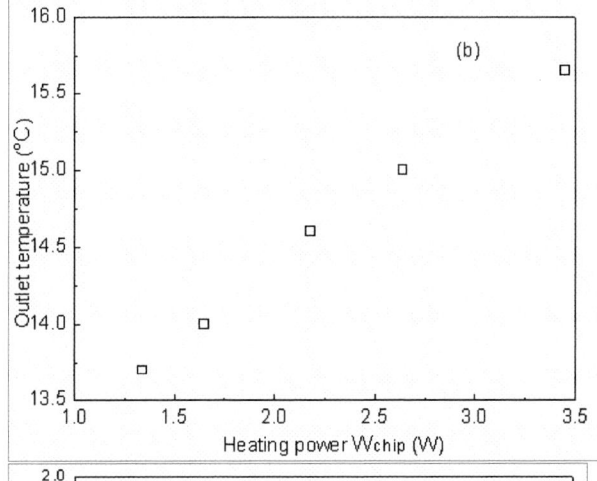

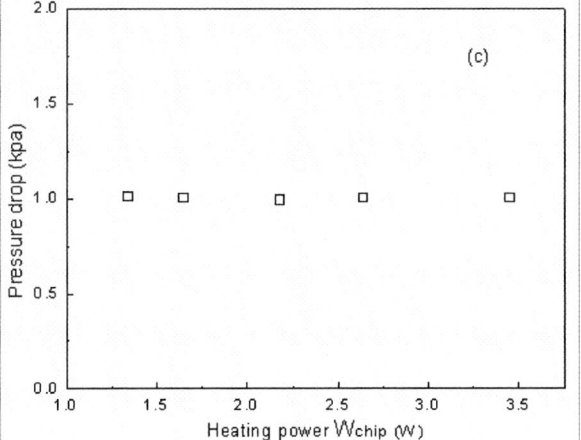

Fig.4. Measurement results for different heating power

    (a) Variation of temperature of chip T_{chip} with heating power.

    (b) Variation of outlet temperature T_{outlet} with heating power.

    (c) Variation of pressure drop between inlet and outlet with heating power.

Fig.4 (a) shows the dependence of temperature of chip Tchip on the heating power. It can be found that there was a linear relationship between Tchip and heating power. This linear relationship is corresponding to a constant thermal resistance of cooling system. That means the heating power is an external factor for coolers, which can not change the heat transfer

properties of cooler and influence the cooling efficiency of the coolers.

Fig.4 (b) shows the dependence of outlet temperature Toutlet on the heating power. It can be seen that the outlet temperature goes up while the heating power increases. Under the same inlet temperature, higher outlet temperature means more heat was transferred into the coolant.

As shown in Fig.4 (c), the pressure drop in the cooling assembly nearly keeps constant while the heating power is increasing, which means the pressure drop is independent of the heating power

4.2 Effects of CNT volume fraction of suspension coolants

In the second case study, the effects of CNT volume fraction of suspension coolants on the temperature of chip T_{chip}, the outlet temperature T_{outlet} and the pressure drop between inlet and outlet were obtained. The heat power was 3.45W, the fluid flow was 27 ml/min, the room temperatures $T_{room}=20°C$, and inlet temperature $T_{inlet}= 12.7°C$.

Fig.5 (a) shows the dependence of temperature of chip T_{chip} on CNT volume fraction of suspension coolants. It can be seen that the T_{chip} decrease while the CNT volume fraction of suspension coolants increases because of higher thermal conductivity in higher CNT volume fraction of the suspension coolants.

In this case study, the outlet temperature T_{outlet} is found independent of CNT volume fraction, as shown in Figure.5 (b). In our previous work, the simulation results for different thermal conductivity of coolants indicated that the outlet temperature keeps constant with increasing thermal conductivity of coolant, which illustrates that the outlet temperature is independent on the thermal conductivity of fluid, due to the energy conservation [9]. The measurement data from this study is coincident well with our previous simulation work.

Figure.5.(c) also shows the pressure drop between inlet and outlet nearly keeps constant while the CNT volume fraction is increasing. This means the CNT suspension coolant did neither produce any extra pressure drop.

4.3 Compare between DW and CNT suspension coolant

Table 1 shows the measurement results of silicon microchannel cooler with Distilled Water (DW) as coolant and with 1% CNT suspension as coolant.

In this case study, the flow rate was controlled at 27ml/min, and the room temperatures were 20° C. It can clearly be seen from Table 1 that the temperature of silicon chip grew while the power increased. Under the same power, the temperatures of silicon chip with 1vol% CNT suspension as coolant were lower than those with DW as coolant. The heat removal efficiency of the same silicon cooler by using 1vol% CNT suspension as coolant was about 7% higher than that by using DW as coolant. Meanwhile, the measurement results of pressure drop between the inlet and outlet indicated that compared with DW, the CNT suspension coolant did not produce extra pressure drop, which is also a benefit for this application.

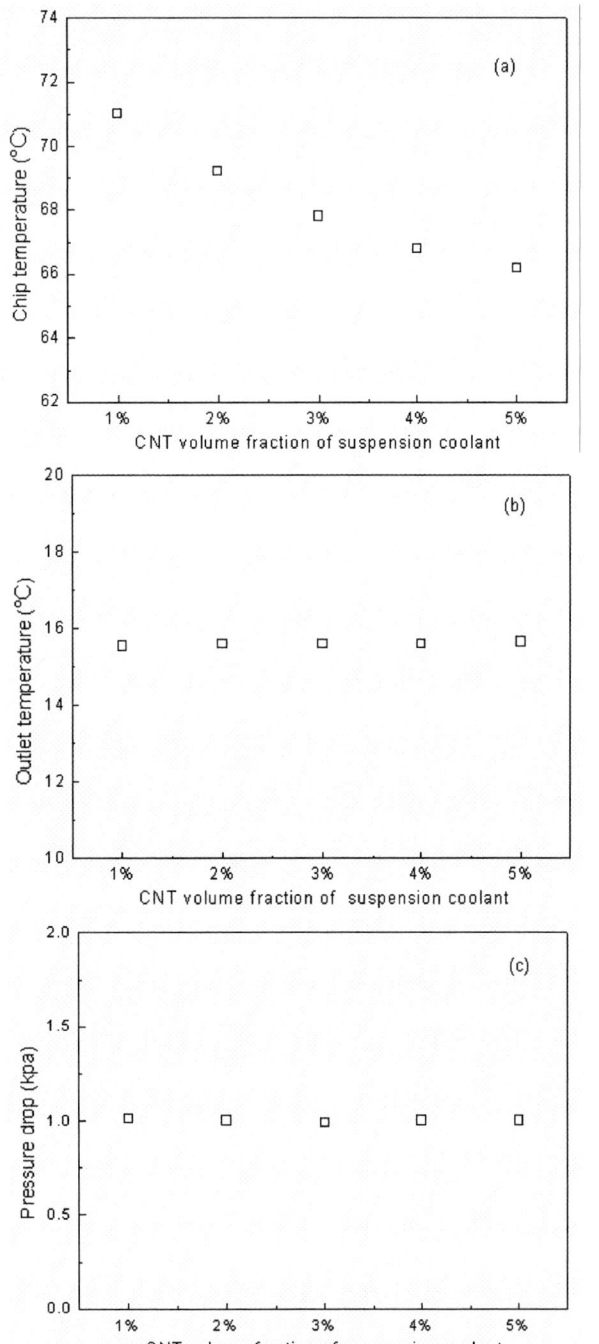

Fig.5. Measurement results for different volume fraction of suspension coolants

(a) **Variation of temperature of chip T_{chip} with CNT volume fraction of suspension coolant.**

(b) **Variation of temperature of outlet T_{outlet} with CNT volume fraction of suspension coolant.**

(c) **Variation of pressure drop between inlet and outlet with CNT volume fraction of suspension coolant.**

Tab.1 Measurement results of silicon microchannel cooler with DW coolant and with 1% CNT suspension coolant

		DW as coolant			1vol% CNT suspension as coolant		
T_{room} (°C)		20			20		
Flow rate (ml/min)		27			27		
P_{inlet} (kPa)		3.0			3.0		
P_{outlet} (kPa)		2.0			2.0		
Temp. (°C)		T_{chip}	T_{inlet}	T_{outlet}	T_{chip}	T_{inlet}	T_{outlet}
Power (W)	1.77	51.6	12.7	13.9	50.4	12.7	14.1
	3.45	74.4	12.7	15.4	71	12.7	15.5

Conclusions

In the present work, the stable and homogeneous CNT suspension was prepared and the silicon microchannel coolers were prepared by using the DIRE technique. Meanwhile, a close loop cooling test system was developed to investigate the heat removal of silicon microchannel cooler with CNT suspension coolant. The measurement results of heat removal show that CNT suspension coolant could strengthen the heat removal capability of microchannel coolers. In addition to heat transfer enhancement, the CNT suspension coolant did not produce any extra pressure drop.

Acknowledgements

Johan Liu, Yifeng Fu and Teng Wang's contribution was partly financially supported by the Seventh Framework Programme of the European Union (Project name: Nano Packaging Technology for Interconnection and Heat Dissipation under the contract No: 216 176) and partly by the EU EC-GEPRO project. The work at SMIT Center, Shanghai University is partly sponsored by the Henkel Shanghai Region Electronics Materials Research and Failure Analysis Center.

References

1. D.B. Tuckerman & Pease R.F.W., "High-performance Heat Sinking for VLSI," IEEE Electron Device Lett., Vol. EDL-2, (1981), pp. 126-129.
2. Z. MO, R. Morjan, J. Anderson, et al., "Integrated Nanotubes Microcooler for Microelectronics Applications", Proceedings of 2005 IEEE Electronic Components and Technology Conference, pp.51-54.
3. Teng Wang, Martin Jönsson, Elisabeth Nyström, Zhimin Mo, Eleanor E. B. Campbell and Johan Liu, "Development and Characterization of Microcoolers using Carbon Nanotubes," Proceedings of the 1st IEEE CPMT Electronics Systemintegration Technology Conference (ESTC2006), September 5-7, 2006, Dresden, Germany, pp.881-885.
4. X. Wang, X. Xu, S.U.S. Choi, "Thermal conductivity of nanoparticle-fluid mixture", J. Thermophys. Heat Transfer 13 (1999) , pp.474–480.
5. S. Lee, S.U.S. Choi, S. Li, J.A. Eastman, "Measuring thermal conductivity of fluids containing oxide

nanoparticles," J. Heat Transfer 121 (1999), pp. 280–289.

6. B.X. Wang, L.P. Zhou, X.F. Peng, "A fractal model for predicting the effective thermal conductivity of liquid with suspension of nanoparticles," Int. J. Heat Mass Transfer 46 (2003), pp. 2665–2672.

7. S. U.S. Choia, Z. G. Zhang, W. Yu, F. E. Lockwood, E. A. Grulke, "Anomalous thermal conductivity enhancement in nanotube suspensions", Appl. Phys. Lett. Vol. 79, No 14 (2001), pp. 2252-2254 .

8. Huaqing Xie, Hohyun Lee, Wonjin Youn, and Mansoo Choi, "Nanofluids containing multiwalled carbon nanotubes and their enhanced thermal conductivities," Appl. Phys. Vol. 94, No 8 (2003), pp. 4967-4971.

9. Xiaolong Zhong, Teng Wang, Johan Liu, Yan Zhang, Zhaonian Cheng, "Computational Fluid Dynamics Simulation for On-chip Cooling with Carbon Nanotube Micro-fin Architectures," Proceeding of The IEEE 8th International Conference on Electronic Materials and Packaging, Hong Kong, Dec. 11-14, 2006, pp.117-123

Novel Pore-sealing Technology in the Preparation of Low-k Underfill Materials for RF Applications

Kuo-Yuan Hsu and Jihperng Leu*

Department of Materials Science and Engineering, National Chiao-Tung University
1001 University Road, Hsinchu, Taiwan 30049, China.
jimleu@mail.nctu.edu.tw, 886-3-5131420

Abstract

Underfill materials had been widely employed in the flip-chip packaging to fill the gaps of solder bumps connecting IC chip and organic substrate in order to prevent failure of the solder joints. For radio-frequency (RF) device applications, underfill materials should possess low dielectric constant to alleviate power loss at high-frequency, in addition to good thermal and mechanical properties. In this study, a novel approach of incorporating porosity through porous silica filler was attempted to develop low-k underfill materials. An inorganic, sacrificial material, hexamethylcyclotrisiloxane (D_3), was used to temporarily seal the interconnected pores in the porous silica at temperature < 95 °C by thermal and solvent pretreatment methods, and was later removed thermally at 125-165 °C during the crosslinking reaction of underfill materials. For underfill materials with 15% filler content, a 7.8% reduction in dielectric constant has been successfully demonstrated and achieved by pore sealing of porous silica (60% porosity) using solvent pretreatment, while maintaining the mechanical strength of porous silica, 2.6 GPa. Moreover, the adhesion between epoxy and porous silica in the underfull materials was found to critical in preserving its mechanical strength when pore sealing pretreatment was applied.

Introduction

Silicon-based integrated circuit (IC) devices continues scaling toward 45 nm node, while the products becomes smaller and faster with more functionalities and complexity. Flip-chip technology, which utilizes an area array of solder bumps to connect IC chip and organic or ceramic substrate, has been widely adopted in the packaging of microprocessors, graphic chips, and DSP chips due to its advantages such as high I/O density and short interconnects. In the reliability of flip-chip packaging, the most challenging problem is the solder joint fatigue which mainly arises from thermal mechanical stress induced during temperature cycle due to coefficient of thermal expansion (CTE) mismatch between silicon chip and substrate. [1] The solder joints reliability can be alleviated by using underfill material, which consists of epoxy resins, hardener, catalyst and silica filler, and offers mechanical properties ranging from rigid to compliant depending on the requirements dictated by the solder type and the low-k dielectrics in the copper interconnect. [2]

In addition to the thermo-mechanical properties, the electric properties, especially the dielectric constant of underfill materials used in RF application must deal with the RC delay and power dissipation in the flip-chip packaging, which can be illustrated by Eqs. (1)-(3) [3-5]:

$$C = \varepsilon \, L \, T / W \qquad (1)$$

$$RC = 2 \, \rho \, \varepsilon \left(\frac{4L^2}{W^2} + \frac{L^2}{T^2} \right) \qquad (2)$$

$$P \infty 2\pi \cdot fV^2 \varepsilon \cdot \tan \delta \qquad (3)$$

where C is capacitance, ε is dielectric constant, L is line length, T is thickness, W is pitch, ρ is resistivity, and P is power consumption. As the development toward smaller T and W of solder bumps, but still in 60-100 m range in the next 2-3 technology nodes, the increase of RC delay may not be an urgent issue. However, high operating frequency would cause an increase of power consumption and crosstalk noise if ε is fixed. With the explosive growth in cell phones and wireless products, radio-frequency (RF) devices (up to 100 GHz) have become the mainstream products in microelectronics industry. Such RF devices may suffer large loss of RF energy when operated at high-frequency. [6] Therefore, underfill materials shall possess low dielectric constant to alleviate power loss, and excellent mechanical properties to support solder joints to ensure solder joints reliability for high-frequency applications.

In this paper, novel low dielectric constant (low-k) underfill materials was developed by incorporating porosity (kvacuum = 1) through porous silica fillers and pore sealing treatment. In particular, an inorganic, sacrificial material, hexamethylcyclotrisiloxane (D3), was employed to temporarily seal the interconnected pores in porous silica at temperature < 95 °C, and was then removed thermally at 125-165 °C during the crosslinking reaction of underfill materials. The pore sealing pretreatment methods and curing profile were investigated by differential scanning calorimetry (DSC) and rheometer. The temperature profile for pore-sealing pretreatment and the outgassing of sacrificial material was first investigated by rheometer and differential scanning calorimetry (DSC). Then the morphology and pore size/volume of sacrificial materials/porous silica were characterized, leading a proposed mechanism of pore sealing. Finally, the impact of pore sealing pretreatment on the dielectric constants and moduli of underfill materials were examined and discussed.

Experimental

Two formulation systems involving different underfill materials and pore-sealing pretreatment were investigated in this study. The underfill materials consisted of 85% organic components and 15% filler to illustrate our approach. The organic components contained epoxy resins, a medium viscosity, bisphenol-A based Epikote 828 and a low viscosity bisphenol-F based epoxy resin Epikote 862, a hardener, methyl hexahydrophthalic anhydride (MeHHPA,), and a catalyst, 2-ethyl-4-methylimidazole (2E4MI). For the fillers,

978-1-4244-2739-0/08/$25.00 ©2008 IEEE

conventional solid silica and porous silica were employed for comparative study. In particular, we prepared two different underfill systems (A and B) in which different pretreatment were applied onto porous silica fillers. The formulations for conventional underfill (A) and five low-k underfill materials with different pretreatment (B1, B2, B3, B4 and B5) were summarized in Table 1.

The porous silica here possessed irregular shape with an average pore size of 6 nm. Since epoxy resin possessed fluidity prior to curing reaction, it could easily back-fill the pores in porous silica fillers by capillary force when underfill formulation was prepared. Thus, in order to reduce the dielectric constant of underfill material, the pore volume of porous silica was pretreated and occupied by an organic or inorganic sacrificial material. Specifically, an inorganic sacrificial material, hexamethylcyclotrisiloxane (D3), with melting temperature of 64.5 ℃ and boiling temperature 134 ℃ was chosen as a sacrificial material for pore sealing in this study.

Tab. 1 Formulations of conventional and low-K underfill materials

	Organic part (85%)	Filler (15%)	Pretreatment (D₃ : Porous Silica)
A	Epikote862, Epikote828, MeHHPA and 2E4MI	Solid silica	None
B1	Epikote862, Epikote828, MeHHPA and 2E4MI	Porous Silica	None
B2	Epikote862, Epikote828, MeHHPA and 2E4MI	Porous Silica	Thermal Melting (1 : 1)
B3	Epikote862, Epikote828, MeHHPA and 2E4MI	Porous Silica	Thermal Melting (3 : 1)
B4	Epikote862, Epikote828, MeHHPA and 2E4MI	Porous Silica	Solvent (1 : 1)
B5	Epikote862, Epikote828, MeHHPA and 2E4MI	Porous Silica	Solvent (3 : 1)

In addition, two pretreatment methods namely thermal treatment and solvent treatment were investigated regarding their efficiency in the pore-sealing. In the thermal treatment method, the sacrificial material was first mixed with porous silica at desired weight ratio, either 1:1 or 3:1. Then, the mixture was heated at 95℃ isothermally until the sacrificial

material was melted to complete the pore-sealing treatment and then cooled down to room temperature. In the solvent treatment, the sacrificial material was dissolved in diethyl ether until the solution became limpid. Then, the porous silica filler was added into sacrificial material/diethyl ether solution at the desired weight ratio, either 1:1 or 3:1 and stirred the solution until uniform. Finally, diethyl ether was removed at room temperature to avoid any loss of sacrificial material. Afterward, the pore volume, specific area, porosity and pore-size distribution of porous silica with and without pretreatment were characterized by Brunauer-Emmett-Teller (BET) (Quantachrome NOVA-1000A). Then this mixture was combined with epoxy resins, hardener, and catalyst at room temperature to form an underfill formulation. The sacrificial material was then removed at the temperature such that the viscosity of epoxy resin increased drastically to prevent its reflow into the pores; thus a low-k underfill was achieved.

For underfill materials based on conventional silica (A) and porous silica (B1), the curing profile was to ramp from 30 ℃ to 170 ℃ at 5 ℃/min rate, then stay at 170 ℃ for 2 hours. In contrast, for the underfill formulations (B2, B3, B4 and B5) with pretreatment, the curing profile for removing the sacrificial materials and crosslinking epoxy resins simultaneously was to ramp from 30 ℃ to 135 ℃ at 5 ℃/min, then stay at 135 ℃ for 30 minutes, later ramp from 135 ℃ to 170 ℃ at 5 ℃/min, and lastly cure isothermally at 170 ℃ for 2 hours.

Differential scanning calorimetry (DSC) (PerkinElmer Diamond) and rheometer (TA Rheomatric 1000) were used to characterize the thermal reactivity and viscosity behavior of underfill systems. In the study dielectric constants and moduli of cured underfill materials were measured by RF impedance-material analyzer (HP 4291B) and dynamic mechanical analyzer (DMA) (PerkinElmer DMA 7), respectively. In addition, scanning electron microscopy (SEM) (JEOL JSM-6700F) was used to examine the surface morphology of porous silica and its mixture with sacrificial material.

Results and discussion

Characterization of thermal reactivity and viscosity

DSC and rheometer were first employed to examine the curing profile and the viscosity in underfill system. Figure 1 showed the thermal reaction of underfill formulation B1 by DSC from 30℃ to 250℃ at a heating rate of 5 ℃/min. It was found that the underfill system B1 started crosslinking at 125 ℃ and reached the full acutest degree of curing reaction at 165 ℃. From rheometer data of underfill formulation B1 as shown in Figure 2, the viscosity decreased slightly at 125 ℃, then increased drastically starting 165 ℃. Based on DSC and rheometer data, we learned that even though underfill starts crosslinking at 125 ℃, the epoxy resins still possessed higher mobility due to its lower viscosity. As curing reaction continued, epoxy resins formed more and more network structure which inhibited their movement and increased the viscosity substantially at T > 165 ℃.

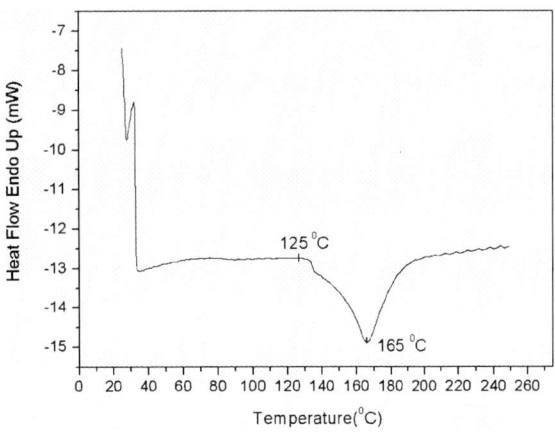

Fig. 1 DSC curve of underfill system B1

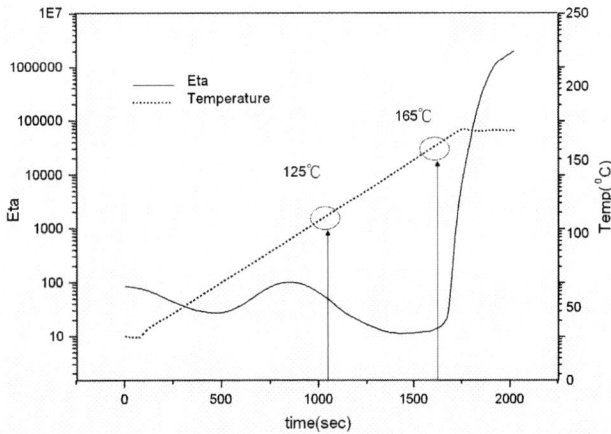

Fig.2 Viscosity as a function of temperature for underfill formulation B1 as measured by rheometer

Based on the relationship between viscosity and temperature of the underfill system, B1, an inorganic material with boiling temperature at 134℃ could serve as a sacrificial materials to seal the pores of porous silica fillers and hinder the reflow of epoxy resins back to the pores by its vapor outgassed from the sealed pore until the completion of curing.

Characterization of pore-sealing treatment

Afterward, SEM was used to investigate the morphology of porous silica fillers with and without pore sealing pretreatment. Figures 3(a) and 3(b) illustrated the morphology of sacrificial material and pure porous silica by SEM, respectively. Sacrificial material showed smooth surface as illustrated by Figure 3(a). In contrast, pure porous silica showed a sponge-like structure which was presumably composed of large number of pores randomly dispersed and interconnected in silica matrix. Such porous silica fillers were expected to have excellent capability of absorbing sacrificial materials in liquid state due to its porosity.

Figures 4(a) and 4(b) showed the SEM morphology of porous silica after different pore-sealing treatment for B2 and B3, respectively. For 1:1 sacrificial material ratio/silica, it was hard to find sacrificial material on the silica surface, as illustrated Figure 4(a), indicating that sacrificial material was

well absorbed by pores into the inner of silica fillers. As more sacrificial material was added to its mixture with porous silica as 3:1 sacrificial material ratio/silica, sacrificial material overflowed onto outer surface and left uniform coating outside the porous silica particles as illustrated in Figure 4(b). It appeared that during pore-sealing treatment the sacrificial material in liquid phase was absorbed into the inner pores of silica particles. With more and more amount absorbed, sacrificial material overflowed onto exterior surface.

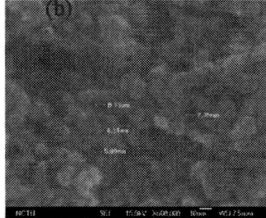

Fig.3 SEM photos of (a) sacrificial material and (b) porous silica without pretreatment

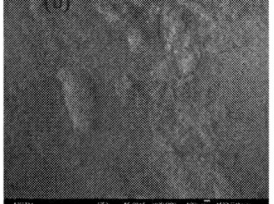

Fig.4 Morphology of porous silica after pore-sealing treatment with sacrificial material ratio/silica: (a) 1:1 and (b) 3:1

Subsequently, Brunauer-Emmett-Teller (BET) was utilized to measure pore volume, specific surface area, porosity and pore size distribution for porous silica filler with and without pore-sealing pretreatment. Since conventional solid silica filler exhibited extremely low level of porosity, BET measurement and analysis were primarily applied to B underfill systems with porous silica. Table 2 summarized the pore volume, specific surface area, and porosity of porous silica of B1-B5 underfill materials. Pure porous silica (B1) possessed 306 m2/g surface area and 0.64 cm3/g pore volume with 60% porosity, which was relatively high in order to provide high absorbing capability. However, the surface area and pore volume of B2-B5 were reduced to < 50% of original value in B1 because of the pore-sealing by sacrificial material pretreatment. Moreover, B5 system exhibited the lowest surface area and pore volume when high weight ratio of sacrificial material to porous silica was used. Also, the decreasing rate of surface area was larger than that of pore volume as listed in Table 2. It implied that sacrificial material could seal the small-size pores more readily than big ones due to the size difference of pores. In terms of pretreatment method, the much reduced surface area and pore volume in B5 as compared to B3 indicated that solvent treatment was more efficient in pore-sealing than thermal pretreatment. The difference was more pronounced in high D3/porous silica ratio. It could be inferred that the solvent method supplied good mobility for sacrificial materials to fill-in and seal the pores within porous silica.

Tab. 2 BET data of porous silica in B1-B5

	Surface area (m²/g)	Pore volume (cm³/g)	Porosity (%)
B1	306	0.64	60
B2	143	0.31	40
B3	85	0.23	33
B4	139	0.3	40
B5	42	0.13	22

Figure 5 showed the pore size distributions of porous silica with five different pretreatment ranging from B1 to B5. For as-received porous silica (B1), pore size ranged up to ~30 nm with an average of 6.6 nm. For B2 and B3 porous silica with pretreatment, the average pore size was reduced from 6.6 nm in B1 to ~5 nm due to pore sealing by the sacrificial material, while pore size distribution curves of B2 and B3 were shifted down, showing that their volume fractions of larger pores were also significantly reduced. In contrast, B4 and B5 showed excellent pore-sealing efficiency. The pore size distribution curves were shifted down to lower than ~2 nm. The average pore size was reduced to 1.86 nm and 1.73 nm for B4 and B5, respectively. Moreover, the pore size distribution curvature of B4 and B5 was sharper than the distribution curves of B2 and B3.

This showed that D3 in melted state by thermal pretreatment could not fill in the pores of porous silica effectively due to its high viscosity inhibiting its capillary movement inside the interconnected pores. D3 material may fill the pore in limited scale such as large pores with partial impregnation, and also reside onto the exterior surface of porous silica. In contrast, solvent treatment was more efficient in sealing the pores effectively and uniformly due to the much reduced viscosity of D3 in the solvent in the capillary movement of pore sealing.

Based on BET data and SEM images, a simple model was proposed and schematically illustrated in Figure 6. For pore sizes below a threshold dimension, sacrificial material sealed the pores completely, making the open pores into a solid phase. In the second mode, for pore size above the critical dimension, the sacrificial material sealed the large pores partially by a coating layer on the sidewall of pores, making the open pores into smaller, open pores.

Dielectric constants and moduli of underfill materials

The dielectric constants and moduli of various cured underfill systems, measured by RF impedance-material analyzer and DMA, were summarized in Table 3. The B1 underfill system had the same dielectric constant with system A. This indicated that the pores in the silica of B1 system were all back-filled with epoxy resins and thus showed no reduction in the dielectric constant. In contrast, underfill

systems such as B2-B5 with pretreatment in porous silica showed a reduction of the dielectric constant, from 7.5% to 10%, respectively.

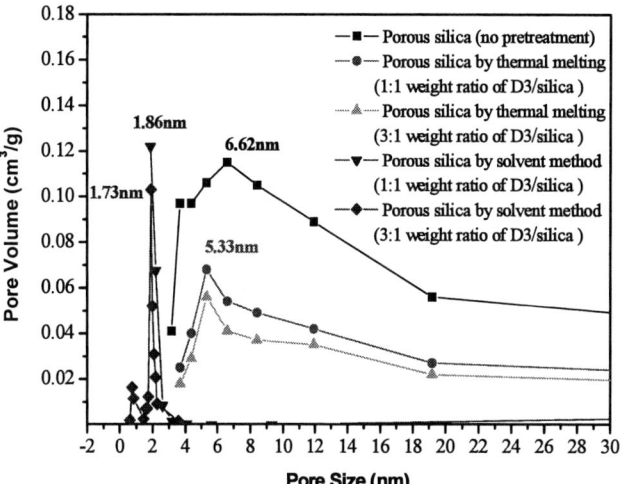

Fig. 5 Pore size distributions of porous silica fillers in B1-B5 from BET

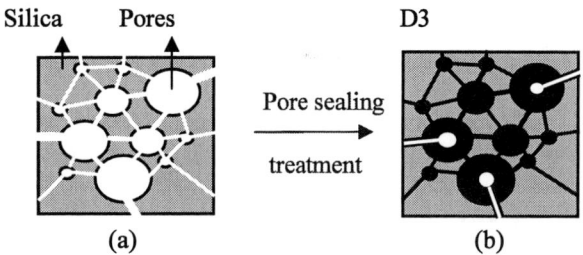

Fig.6 2-D pore-sealing model: (a) prior to pore-sealing pretreatment and (b) after pore-sealing pretreatment

Table 3 Dielectric constants and moduli of A and B1-B5 underfill materials

	Dielectric constant	Modulus (GPa)	Tanδ At room temperature
A	3.2	3.0	0.02
B1	3.2	2.6	0.024
B2	2.96	1.5	0.04
B3	2.88	1.8	0.06
B4	2.95	2.6	0.026
B5	2.90	1.9	0.05

In composite material, its dielectric constant could be predicted by Equation (4) [7]:

$$Logk = \sum Vi \cdot Logki \qquad (4)$$

where k are the dielectric constant of composite materials, Vi and ki are the volume fraction and dielectric constant of i-th component, respectively. There were two assumptions when applying this equation; namely: (1) The low-k underfill material was an uniform mixture of epoxy resins, silica and pores, and (2) The volume faction retained by sacrificial materials during pretreatment was completely removed after curing and left as pores with k=1. Using Equation (4), the predicted dielectric constants of (B2, B4) and (B3, B5) systems were 2.87 and 2.82, respectively, as compared to the measured dielectric constants, (2.96, 2.95) and (2.88, 2.90) whose discrepancy was within 3%. The result indicated that the pore-sealing treatment in (B2, B4) and (B3, B5) systems indeed delivered the expected effect in retaining the porosity and thus reducing the dielectric constant. The successful construction of low-k underfill materials could be attributed to two factors: (1) Pores were sealed by sacrificial material, therefore raw epoxy resins could not enter the pores at room temperature, and (2) the sacrificial material boiled at 130~140 ℃ during curing reaction, which ougassed out of the pores and hindered the backfill of epoxy resins into the pores with a positive, outward pressure until the completion of curing reaction.

In addition, the moduli of low-k underfill materials were characterized to study any effect of pretreatment on the mechanical strength. From Table 3, underfill material A had the highest modulus due to the dense structure of solid silica. For underfill system B1 without pretreatment, its modulus drops 13.3% to 2.6 GPa due to the loose structure of porous silica, although its dielectric constant was the same as that of A with solid silica. Compared to the control sample with porous silica, B1, the moduli of underfill materials B2 to B5 dropped 42%, 30%, 0% and 26.9%, respectively. The degradation of moduli in B2 and B3 was more severe than the reduction of pore volume retained by sacrificial material. In addition, large discrepancy (~42%) in modulus was found between B2 and B4 at the same D3/porous silica ratio. Similar variation was found between B3 and B5 at 3:1 D3/porous silica ratio. Figure 7 showed the theoretical modulus value E and experimental data of underfill materials, B1-B5. E is the theoretical moduli of composite materials calculated as 2.6 GPa according to Eq. (5), where Vi and Ei are the volume fraction and modulus for i-th component assuming perfect adhesion at various interfaces. [8]

$$\frac{1}{E(L)} = \sum \frac{Vi}{Ei} \qquad (5)$$

From Figure 7, the theoretic modulus remained at 2.6 GPa regardless the ratio of sacrificial materials. The modulus of B1 was measured as 2.6 GPa which was the same as the theoretic value. For thermal pretreatment, the modulus degraded to 1.5-1.8 GPa as more D3 was added. The degradation of modulus may be caused by the void defect within epoxy and the poor adhesion between epoxy and silica due to the overflow of excess D3 onto exterior surface of silica. In contrast, the modulus of underfill materials prepared by solvent pretreatment showed theoretic value, 2.6 GPa in B4 (1:1 D3/silica ratio), but 1.9 GPa in B5 as more D3 was added. Moreover, large discrepancy (42%) was observed between B2 and B4. This clearly showed that porosity can be retained by solvent pretreatment at 1:1 D3/silica ratio to delivered the

reduction of dielectric constant (B4) while maintaining the mechanical strength as that of porous silica (2.6 GPa). It was believed that solvent method provided D3 a better mobility in filling the pores uniformly without aggregation or partial coverage on the exterior surface at 1:1 D3/silica. When more D3 was added, aggregation of D3 or partial coverage on the exterior silica surface may degrade the adhesion between epoxy and silica, thereby deteriorate its mechanical strength.

This hypothesis was further confirmed by the loss factor, obtained from DMA measurement, tanδ, of various low-k underfill materials as summarized in Table 3. A high tanδ indicated high internal energy dissipation due to poor adhesion. [9] For different pretreatment methods under the same D3/porous silica (1:1) ratio, B4 showed much lower loss factor (0.026) compared to B2 (0.04). At higher D3/porous silica (3:1) ratio, the loss factor in both B3 and B5 was higher than the rest of group. This implied that D3 could be impregnated into pore structure effectively, thus sealed the porous silica without leaving residue on the exterior surface of silica. Therefore, B4 possessed the same modulus as control B1 because there was no loss of adhesion based on its extremely low tanδ. As D3 portion was increased or thermal pretreatment was employed in the pore sealing pretreatment, high tanδ indicated poor adhesion between epoxy and silica because of the overflow of D3 or D3 residue on the exterior of silica surface. The degradation of adhesion between epoxy and silica weakened the mechanical strength of the underfill prepared at high D3/ silica ratio. [9]

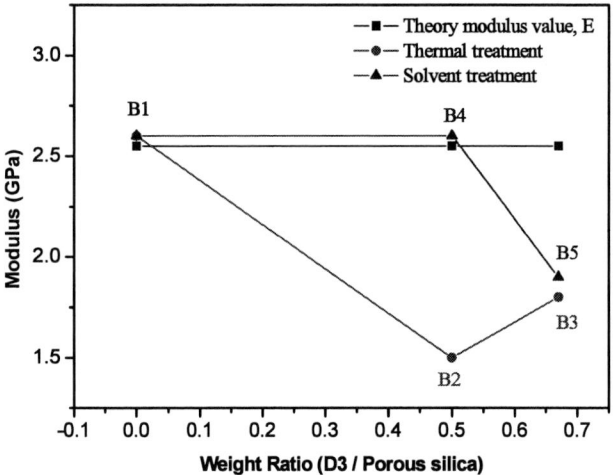

Fig.7 Moduli of underfill materials, B1-B5, as a function of weight ratios

Conclusions

A novel approach by incorporating porosity (kvacuum = 1) through porous silica filler was employed in this study to develop low-k underfill materials using an inorganic, sacrificial material, hexamethylcyclotrisiloxane (D3), with melting temperature of 64.5 ℃ and boiling temperature 134 ℃ for pore-sealing pretreatment prior to the underfill formulation. The temperature-dependent viscosity and the curing reaction were found to be important in the selection of sacrificial material, curing profile, and pretreatment methods for pore sealing . The mechanisms for retaining porosity in the underfill system were summed up in two stages; namely (1) The sacrificial material when heated at 95 ℃ or dissolved in ether, easily flowed into pores and sealed the pores partially or completely such that the starting epoxy resins could not enter these sealed pores at room temperature, and (2) The sacrificial material outgased at boiling point, 134 ℃, at which large gas pressure out of pores hindered the reflow of the epoxy resins during the curing reaction, in addition to the high viscosity of resins induced by crosslinking. Compared to thermal method, solvent pretreatment method was efficient and uinform in the capillary pore sealing due to its much reduced viscosity. In summary, we successfully developed an easy and convenient pore sealing by solvent method to achieve a 7.8% reduction in dielectric constant of underfill materials with 15% filler content (porosity of porous silica: 60%), while maintaining the mechanical strength of porous silica, 2.6 GPa. Moreover, the adhesion between epoxy and porous silica in the underfull materials was found to critical in preserving its mechanical strength when pore sealing pretreatment was applied. The mechanical strength of low-k underfill materials could be further improvement when the interfacial adhesion was increased by optimization of curing profile and selection of sacrificial materials.

Acknowledgments

The authors wish to thank the instrumentation support by Industrial Technology Research Institute (ITRI) and D3 materials provided by Shin-Etsu Silicone Taiwan Corp.

References

1. Z. Zhang and C. P. Wong, "Recent Advances in Flip-Chip Underfill: Materials, Process, and Reliability," *IEEE T. ADV. Packaging*, Vol. 27, No. 3 (2004), pp. 515-524.
2. H.-Y. Chen, *et al*, "Thermal Behavior Analysis of Lead-free Flip-Chip Ball Grid Array Packages with Different Underfill Material Properties," 8th *International Conference on Electronic Packaging Technology and High Density Packaging (ICEPT-HDP) Conf*, Shanghai, China, July. 2008, pp. 28
3. P. S. Ho and J. Leu, *et al*, Low Dielectric Constant Materials for IC Applications, Springer (Berlin, 2003).
4. S. P. Murarka, *et al*, Interlayer Dielectrics for Semiconductor Technologies, Elsevier/Academic Press (Boston, 2003).
5. Z. J. Yang, D. Z. Zhang, and H. Z. Zheng, "Process for Cu matel and low dielectric materials," *NDL Communication*, Vol. 7, No. 4 (2001), pp. 40-46.
6. Z. Feng, W. Zhang, B. Su, K. C. Gupta, and Y. C. Lee, "RF and mechanical characterization of flip-chip interconnects in CPW circuits with underfill," *IEEE T. Microw. Theory*, Vol. 46, No. 12 (1998), pp. 2269-2275.
7. Y. Rao, J. Qu, T. Marinis, and C. P. Wang, " A precise numerical prediction of effective dielectric constant for polymer-ceramic composite based on effective-medium theory", *IEEE T. Compon. Pack.*, Vol. 23, No. 4 (2000).
8. W. D and J. Callister, Materials and Science and Engineering an Introduction, Wiley (New York, 2003).
9. T. Murayama, *et al*, Dynamic Mechanical Analysis of Polymeric Material, Elsevier (New York, 1978).

A Reliability Investigation of MEMS Transducers with Comb Structures

Ping AN, Yandong HE, Yufeng JIN, Yilong HAO*

National Key Laboratory of Micro/Nano Fabrication Technology / Institute of Microelectronics,
Peking University, 100871, P.R.China
anping@ime.pku.edu.cn, jinyf@ime.pku.edu.cn, ylhao@ime.pku.edu.cn
Telephone: 86-010-62752561   Fax: 86-010-62752789

Abstract

An overview of reliability investigation of micro electro mechanical system (MEMS) transducers with comb structures was presented. With the development of the MEMS industrialization, the reliability is underway to meet the need of market. To build a reliability research and develop platform, MEMS center of Peking University made its roadmap of a general developing plan. In this paper, the roadmap was presented and the short-term roadmap about failure analysis (FA) was illustrated. In addition, FA techniques of localization, characterization and sample preparation for MEMS were summarized. Finally, some FA cases of the center of PKU were discussed and a destructive physical analysis (DPA) case about the particle impact noise detection (PIND) was presented.

Introduction

Reliability researches about MEMS are particularly important to break the bottle-neck of MEMS applications. With the advent of MEMS applications in application in automobile industry, consumer electronics, space vehicles etc., the MEMS sensors and actuators will take a more important role for their lower weight, lower capacity and higher reliability than normal transducers. Moreover, electric consumers and car electronics need the products with more reliable, more stable and longer life MEMS devices but low cost. To speed up the application of MEMS devices in these fields, it is necessary to develop the reliability all-round.

However, the general reliability problems have not been discovered sufficiently and systematically yet for the problems do be numerous and complicated. The problems would involve in the material, physics, devices and even system. Furthermore, different samples generated different reliability issues. Radio frequency (RF) switches often relate to creep and lifetime period. But the microengines with gear often involve in the friction and the wear. Moreover, it is very hard to perform systematical experiments for the time consuming and the cost. Only one experiment to determine the life of the devices often takes several months. While the study of MEMS reliability problems was started more than ten years ago in the world, most related researches were pursued separately or specially in different programs. The general studies of MEMS reliability was carried out by Sandia Laboratory of USA in the end of the last century. The groups in Sandia studied on the poly-Si structures which made by surface processes and took a series of experiments on the special micro-engine with gears and combs. [1] The results involved in failure mechanisms, such as the adhesion, the wear and the friction, and FA such like the influence to the lifetime by the humidity etc. French Space Agency (CNES) performed the general reliability studies five years ago. Their achievements were mainly on the environmental tests and some basic mechanism researches like thermal deformation. [2] In China, many special reliability problems have been studied. However, the general research on MEMS reliability has just started and seldom issued some useable mechanism of failure and some results about lifetime model. Achievements of systematic reliability study had not been issued yet. Therefore, it is very essential to study the MEMS reliability generally and systematically. In this paper, we present a roadmap of how to perform the systematic research in PKU (Peking University).

In addition, Micro comb structure made by bulk-Si undertakes a standard sample in MEMS or MOEMS for it takes great applications in MEMS as a senor or an actuator normally. The sensor with comb structures like resonators, accelerometers and gyros can be easily produced by standard bulk-Si processes and can realize a 2D or 3D direction sensing with a high level sensitivity, a high reliability and an expandable ability. The comb actuator also has the most popular application in RF MEMS as well for its integratability with integrated circuit (IC) in a die. Therefore, the reliability investigation was discussed by using MEMS devices with comb structures as samples.

In the present paper, the roadmap of reliability investigation was introduced first. Then the FA as the basic problems was studied and the failure investigation was presented in details by introducing some novel FA techniques and some results of a MEMS structure with comb structures.

1 Roadmap

The roadmap of how to study the complex reliability problems in MEMS center of PKU follows four parts as showed in Fig. 1: FA, test structure, lifetime and harsh environment. FA involves in investigating failure modes, failure analysis and building failure mechanism. These will be performed in three levels in turns of die level, module level and system level. The test structure part relates the studies of designing test structure like stress gauge to extract the parameters of the material and processes etc. These studies will be carried out to establish a database of the parameters. The lifetime is the studies about mean time to failure (MTTF), Mean time before failure (MTBF), storage lifetime and accelerated stress test and modeling etc. The harsh environment will be explored to find some novel method to anti-harsh stresses and promote the quality of products. As the most important one and the most urgent one, the FA of MEMS was studied first.

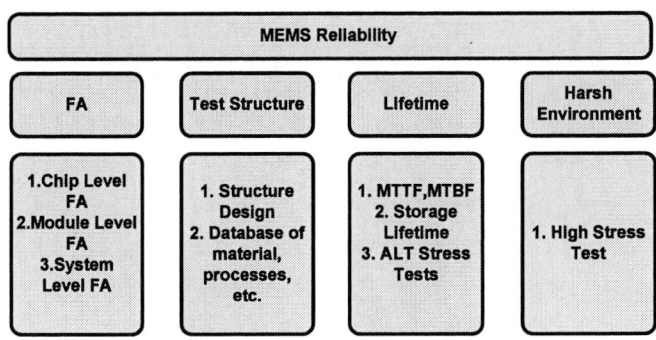

Fig. 1 Subjects in MEMS reliability.

2 FA

FA is a procedure involving failure sites localization, root cause investigation and failure model establishment. As the first step of FA, the failure sites localization was pursued by using a series of techniques in three levels of production: die level, module level and system level (Fig. 2). While many FA techniques are based on IC techniques, some of the MEMS FA techniques are quiet different with the ones of IC for FA of MEMS involves in not only electrical analysis, but also mechanical, material and circumstance analysis. In this section, the techniques for MEMS failure analysis were discussed in three parts: 1) Failure localization techniques; 2) Characterization techniques; 3) Subsidiary techniques. In addition, some FA cases by using parts of these techniques were presented in turns of Fig. 2 as: 1) Yield rate investigation; 2) Failures localization in die level; 3) Failure analysis in die level; 4) Failure investigation in module level; 5) Exploratory experiment of PIND test.

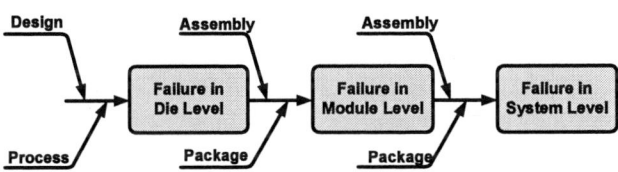

Fig. 2 Failure localization in production

2.1 FA Techniques

In MEMS failure studies, the main FA techniques can be parted into failure modes localization, failure characterization and other subsidiary techniques for preparing samples.

In first step of FA, the failures were localized in macroscopic analysis and microscopic analysis first. The techniques for localization were used in MEMS FA and were listed in Table 1. In which, optic microscopy and SEM are both popular in MEMS FA to detect the surface failures, and the confocal IR image, X-Ray inspection, ICT and SAM are used in localizing the defects under the surface layer.

In the following steps, the spotted failures were determined to be which kind of failure modes and found the root causes by characterization techniques (Table 2). Where, AFM, Raman Spectroscopy, X-Ray Diffraction, Nano Indenter, Optical Scanning Profilometer, and Full Field Optical Profilometer can be used in detecting the deformation of the structure to calculate the residual stress, and the material characterization techniques like TEM, EDXS,

EELS, AES and SIMS can be used to determine the contaminations or other special failures related to the material.

During the analysis, some subsidiary techniques were used to prepare specimens or test plane for special uses (Table 3).

Table 1 Techniques for failure modes localization

Technique	Applications in MEMS FA	Key Features
Optic Microscopy	Surface inspection: wear, broken, contamination etc.	XY Resolution: 0.25μm Magnification: 1000X
Scanning Electron Microscopy (SEM)	Surface inspection: wear, broken, contamination etc.	XY Resolution(3keV): 20～30Å Magnification: 300000X
Confocal IR Laser Microscopy	Delamination, failures under the surface	Limited depth of field, but perfect in focus
X-Ray Inspection	Metal contaminations, package	XY Resolution: 1μm
Industrial Computerized Tomography (ICT)	Metal contaminations, package	XY Resolution: 10μm
Scanning Acoustic Microscope (SAM)	Delamination	XY Scanning range: 1.3mm² ～ 76mm², XYZ Resolution: 5μm

Table 2 Techniques for characterization

Technique	Applications in FA	Key Index
Atomic Force Microscopy (AFM)	Surface topography and roughness, friction	Resolution: 10 pm Surface: conductive or insulate
Raman Spectroscopy	Stress	Depth: 1μm XY Resolution: 1μm Spectrum resolution: 0.2 cm⁻¹
X-Ray Diffraction	Stress	Spot size: Φ50μm
Nano Indenter	Stress, fracture	
Optical Scanning Profilometer	3D topography / roughness /deformation / stress	Z Resolution:1μm

535

Full Field Optical Profilometer/ (Vibrometer)	3D topography /stress Dynamic deformation	XY Resolution: < 1μm , Z Resolution: <10nm, （Stroboscope）
Transmission Electron Microscopy (TEM)	Wear/ contaminations /defects in atomic scale	Resolution: 2 Å
Energy Dispersion X-ray Spectroscopy (EDXS)	Wear / contaminations	Range: 1um
Electron Energy Loss Spectrum (EELS)	Wear / contaminations	Range: 2 nm
Auger Electron Spectroscopy (AES)	Wear / contaminations	Resolution: <12nm
Second Ion Mass Spectrometry (SIMS)	Contaminations /delamination / doping	Spatial resolution : 0.5um，depth resolution: 100A

Table 3 Techniques for sample preparation

Technique	Applications in FA	Key Index
Focus Ion Beam	High precision dicing	Precision：Å No induced contaminations Range: < 50×30μm²
Parallel Polishing	Layer removing	Precision:1μm～ 1nm
Lift-off	Layer or parts removing	
Laser Ablation		Velocity: several picoseconds

2.2 Failure investigation

In this sub-section, some cases were presented in turns to investigate all the failure modes of a kind of MEMS transducer with comb structures in the die level and the module level.

2.2.1 Yield rate investigation

In the large scale production, failures of the comb structure were investigated to improve the reliability. The static capacitance test and the surface check by microscope and SEM were carried out to screen the unqualified structures. In the screening, almost all of the failure modes were found as: over thin finger, hogging finger, structure disjunction, etch pits and some intact structures. The former three failures turned out to be over etched and the broken fingers normally were induced by

random shock operating or man made errors. However, about 19% structures in die level after dry etching release showed an over lager static capacitance (Fig. 3). For the structures looked like qualified ones with a perfect surface, the failures needed to be investigated and analyzed in details.

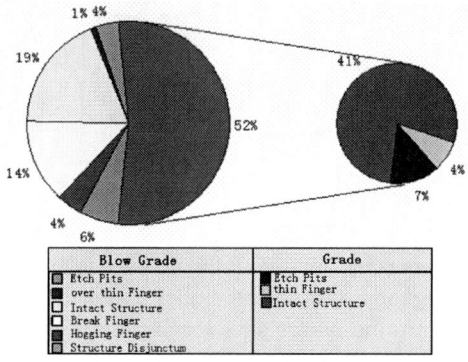

Fig.3 The yield map-graph of comb devices

2.2.2 Failures localization

By surface inspection, failures like contaminations, etch pits, and the failures by over etching were recognized. Fig. 4 showed some contaminations by SEM/EDX in 10 kV, where (a) was a SEM photo of a contamination and (a'), an EDX curve with a Si peak, shows that the contamination was a broken comb. So did the (b)~(e): (b') with C peak shows us that the contamination is a kind of organics. The organic turned out to be photosensitive resist (PR); (c') with Si and O peaks represented that it could be some dust of SiO2 etc; (d') with peaks of C, Na, Cl etc. shows that the contamination must be a furfur from operator; And (e') with peaks of Fe, Ni and Cr represents that it is a metal particle which could come from a stainless tweezers. The failure modes with over etched structures were showed in Fig. 5, where (a), (b), (c) were caused by over etch and broken fingers (d) were caused randomly by manmade. The etch pits (e) were induced by the wet etching. The last one (f) showed connection combs under the structure layer. Its root cause turned out to be the un-uniformity of the dry etching.

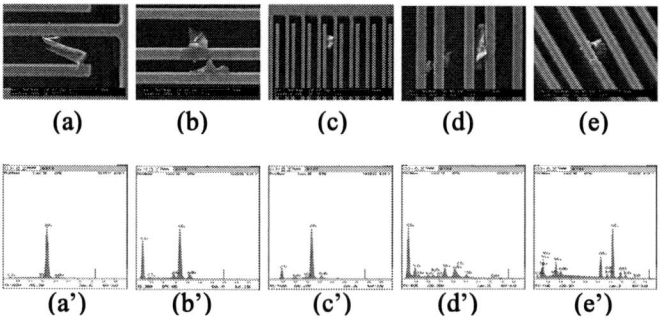

Fig. 4 SEM photos of contaminations and material analysis by EDX. (a) and (a') SEM photo and EDX analysis of a broken comb respectively. (b) and (b') Some PR points. (c) and (c') A particle of dust. (d) and (d') A furfur from operator. (e) and (e') A metal particle from tweeter.

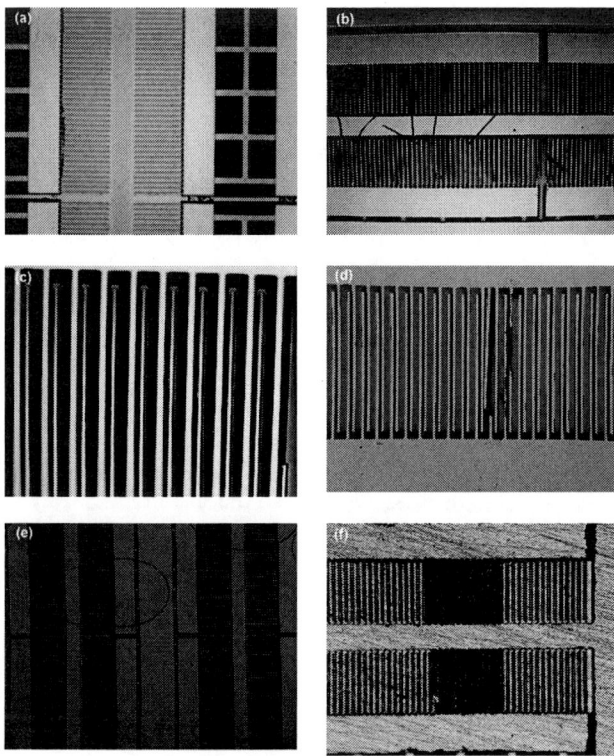

Fig. 5 Micrographs of failure modes
(a) Structure disjunction. (b) Hogging Finger. (c) Thin Finger. (d) Break Finger. (e)Etch Pits. (f)Connection combs with perfect surface

After investigation of a 3000 samples, a classifier was built for automatic classifying the potential failure modes. The classifier was built by an improved Back-Propagation (BP) neural-net and can be used with combining auto pattern recognition to failure modes. To build the classifier, a series of investigations of failure modes of devices were carried out. As an application, the classifier was built by investigating a kind of comb capacitance micro-engine. The result of training indicates that the classifier is effective with a very low error which is less than one chip in 300 chips in one wafer. The result of testification shows that the classifier fits to the wafers from different batches (Fig. 6). [1,3-6]

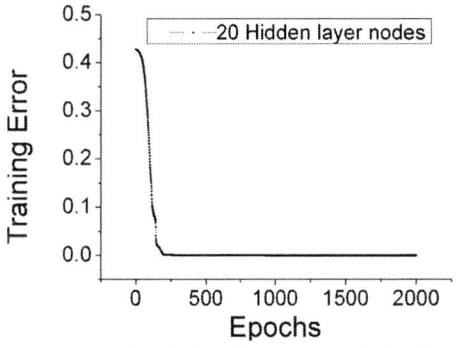

Fig. 6 Training errors of the classifier

2.2.3 FA in die level

In failure investigation, the 19% failed structures with the perfect surface needed the further analysis to find the root cause of the failure. To determine that there were some metal particles ling under the combs, X-Ray inspection with a lower than 1um precision was carried out as in Fig. 7. However, the figure showed that there were not any metal particles which made the capacitance abnormal. To find what had happened under the combs, the technique of sample preparation, parallel polishing, needed to be performed. [7-9] The combs were filled in epoxy and were polished physically after the epoxy solidified. The precision of the polishing was less than 1μm. After removing the structure layer, the pattern under the combs was obtained as in Fig. 8. The black field turned out to be the connection of the combs which was testified by EDX. By comparing the different fields in line (a) and line (b) by EDX, the black field was determined to be Si which was same as the combs. By further analyzing the processes, we found that the connection of the combs was caused by un-uniformity of the dry etching to the combs with high aspect ratio.

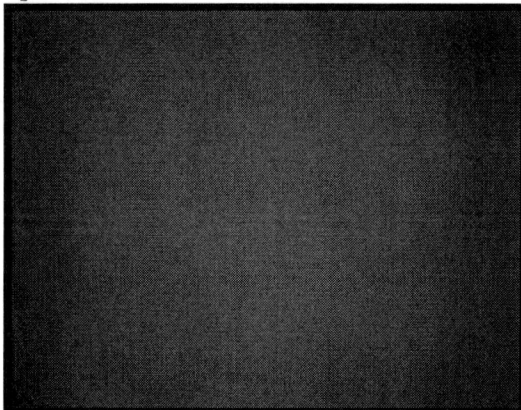

Fig.7 X-Ray transmission inspection showed that there were no metal particles located between structure and substrate

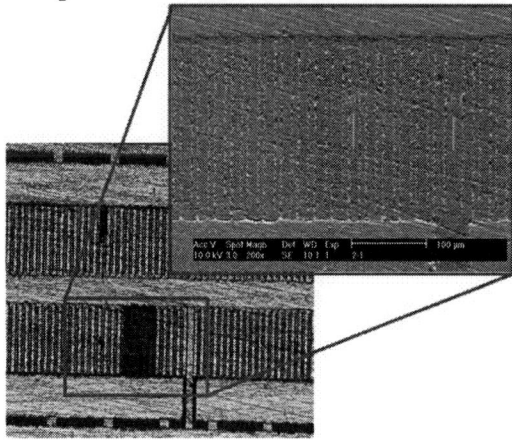

Fig. 8 The optical and Close-up SEM micrograph of the structure showed some abnormal phenomena occurred in the bottom

In die level, some issues about stress were focused for it caused the deformation of the combs and further induced the static capacitance drifting with the temperature outside. One of the techniques to study the residual stress in micro field is Raman spectroscope. The stress was calculated by the shifts of the spectrum. From Fig.9, we found in the area of the bonding, the shift of spectrum was obvious as the line in 10um which was the bonding area.

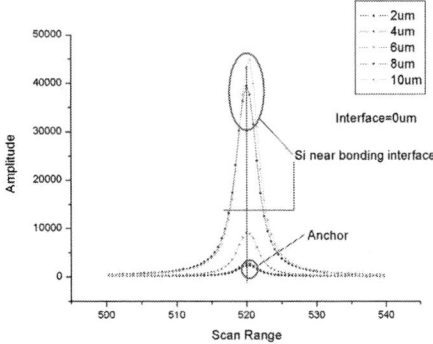

Fig.9 The spectrum shift due to the residual stress in the interface of bonding

2.2.4 FA in module level

In module level, MEMS reliability was related to the assembling and packaging. As to the assembling, the adhesive strength, the adhesive uniformity will determine the quality of the assembling. In addition, the un-uniform adhesion would cause a different stress to the MEMS die which could be very important to the structure's reliability. Therefore, a series of experiments were carried out to study the failures of the assembling. One of the studies used the SAM to detect the adhesive uniformity as Fig. 10. After investigated the acoustic reflect wave of the materials of the die substrate and the package shell, the interface of the die and the shell was detected. In Fig.10, the interface was spotted with some gaps especially in the corners of the die. The further examinations showed it was caused by surface tension during the epoxy solidifying. As to the package, a FA case was about gas leakage. After the plat bonding, some modules showed unqualified in atmosphere test for their over large vapor concentration. The interface of the bonding board was checked by X-Ray inspection and the reason turned out to be some holes lied on the sealing board (Fig. 11). The holes kept some gas which released after the sealing.

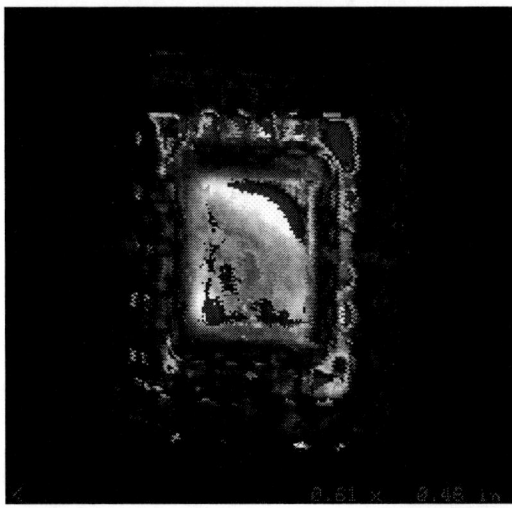

Fig. 10 The acoustic microscopic figure of failed assembly

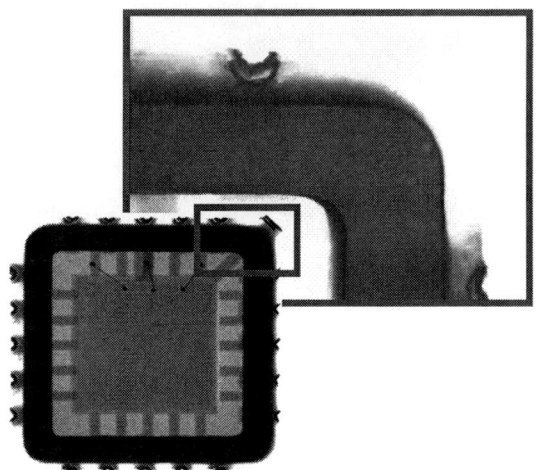

Fig.11 Air leakage points of the sealing by X-Ray inspection

2.2.5 Exploratory experiment of the PIND test

With the development of MEMS into the market and into the space applications, some standard should be built. However, only a very few standard for MEMS have been built until now. In fact, most products of MEMS were test and screened by using the methodologies and standards of IC or other traditional discrete components. But many of that can not be used in MEMS for many MEMS modules have movable parts. One of these tests is PIND which is used to detect the movable particles concealed in the module normally. As some parts of the MEMS transducers with comb structures are movable, whether the PIND test is still usable or not should be re-testified. In one of the experiments, the traditional PIND (4511L) was used to detect a module with some broken combs. The result of the test showed not any particles in the module. It meant that the PIND can not detect the broken comb for the comb weight only about 3×10^{-10}g, which was much lower than the normal particles in other modules which were more than 2×10^{-6} g. Therefore, the PIND should be developed to detect much lighter and smaller particles to meet the MEMS requirements.

Another experiment was performed to find if the PIND could destroy the combs or not. The results showed that the combs can not destroy under a very low frequency of (25~250Hz).

Therefore, traditional PIND can be used in MEMS DPA. However, it needs to be developed to detect the super lighter particles and the frequency of the equipment needs to be tunable. Moreover, the further test method to build a new standard needs to be testified.

Conclusions

MEMS reliability is very important and essential to help the products into the space applications and consumer electrics market. To build a reliability research platform and realize the design reliability, the roadmap was built to subsumption the complex and difficulty problems. The short-term roadmap was focused on FA and the study logic was presented by steps of different production status. In MEMS FA, many conventional FA techniques were used in localization and characterization. Some subsidiary techniques were used in the sample preparation too. In addition, some

FA cases of a MEMS transducer with comb structures were presented in turns of FA: yield investigation, failures localization and root causes analysis in die level and some failure modes in module level etc. The last case was an exploratory experiment of PIND used in MEMS DPA. MEMS center of Peking University developed a know-how in MEMS FA and related techniques to promote the reliability of the products with comb structures.

Acknowledgments

The authors would like to thank Mr. Wei Cai and Mrs. Aibin Xu of CEPREI Certification Body for their technique support.

References

1. www.sandia.gov
2. Lafontan X. *et al*, "The advent of MEMS in space", Microelectronics Reliability, Vol. 43 (2003), pp. 1061–1083.
3. Graupe and Daniel 1997 Principles of Artificial Neural Networks Advanced Series on Circuits and Systems (Singapore, River Edge, NJ World Scientific Publishing Co)   3 31-109
4. Sebastian P 2000 Reliability Modeling of MEMS Using Neural Networks (NASA/TP¾ - 210192) 14-30
5. Rumelhart D E and Hinton G E and Williams R J 1986 Learning representations of back-propagation errors Nature (London) 323 533-536
6. Ki-Dong L 2001 Pattern Classification and Clustering Algorithms with Supervised and Unsupervised Neural Networks in Financial Applications (Dissertation: Bell & Howell Informaiton and Learning Company)
7. Y. N. Hua, S. P. Zhao, L. H. An, Z. R. Guo & Shailesh Redkar, Failure analysis and investigation of bondpad peeling problem during bonding, Electrochemical society proceedings, Volume 99-35, 2001, pp. 161-168.
8. Guenther Benstettera, Michael W. Ruprecht and Douglas B. Hunt, A review of ULSI failure analysis techniques for DRAMs 1. Defect localization and verification, Microelectronics Reliability 42, 2002, pp. 307–316.
9. K. A. Peterson, P. Tangyunyong and D. L. Barton, Failure Analysis for Micro-Electrical-MechanicalSystems (MEMS), Proceeding of the 23rdInternational Symposium for Testing and FailureAnalysis, Santa Clara, October 27-31, 1997, pp.133 – 142.
10. Walraven J 2005 A Failure Analysis Issues in Microelectromechanical Systems (MEMS) Microelectronics Reliability 45 1750–1757
11. Merlijn W and Spengen V 2003 MEMS reliability from a failure mechanisms perspective Microelectronics Reliability 43 1049-1060
12. Ingrid D W 2003 MEMS reliability Guest Editorial Microelectronics Reliability 43 1047-1048
13. Spearing S M 2000 Materials Issues in Microeletromechanical System(MEMS) Acta Mater. 48 179-196
14. Clifford F 2005 Industry Study on Issues of MEMS Reliability and Accelerated Lifetime Testing (43d Annual International Reliability Physics Symposium, San Jose) 312-316
15. Jeremy A W and Edward I C J and Lynn R S 2001 Failure Analysis of Radio Frequency (RF)
16. Microelectromechanical Systems (MEMS) (Reliability, Testing, and Characterization of MEMS/MOEMS, Rajeshuni Ramesham, Editor,254 Proceedings of SPIE) 4558 254-259

Low Temperature Sinterable (Zn,Mg)TiO₃ Microwave Dielectrics

Lih-Shan Chen[1], Ming-Liang Hsieh[1], Hsiang-Chen Hsu[2] and Shen-Li Fu[1]
[1]Department of Electronic Engineering, I-Shou University, Kaohsiung, Taiwan, China
[2]Department of Mechanical Engineering, I-Shou University, Kaohsiung, Taiwan, China

Abstract

In this work, low temperature sintering of microwave dielectric $(Zn_{1-x}Mg_x)TiO_3$ were studied. Sintering temperature of $(Zn_{1-x}Mg_x)TiO_3$ system was lowered by the addition of dopants that could lower the sintering temperature while maintaining the dielectric properties. $(Zn,Mg)TiO_3$ can be well densified at low temperatures for 4wt% Bi_2O_3 + 1wt% V_2O_5 -doped specimens. Dielectric properties of 4wt% Bi_2O_3+ 1wt% V_2O_5 -doped $(Zn_{0.9}Mg_{0.1})TiO_3$ ceramics sintered at 850 °C for 1 h were ε_r=26.85, Q×f =42351GHz and τ_f= 4.92 ppm/℃.

Introduction

The rapid progress in wireless communication system has increased demand for the development of dielectric resonators for microwave frequencies. Basic requirements of microwave dielectric resonators are high performance with low loss and stable temperature coefficient of resonance frequency (τf). High dielectric constant enables the dielectric resonator to be smaller since the dielectric constant is inversely proportional to the size. Low dielectric loss gives resonators less power loss. Stable temperature coefficient of resonance frequency assured that the resonators can be used in any environment without detrimental effects on the signal.

Many microwave dielectrics have been developed according to the needs of specific applications. Owing to the compatibility with low temperature co-fired ceramic (LTCC) process, low temperature sinterable dielectrics for microwave applications recently has received much attention. Highly conductive internal electrode metals, such as silver, copper, and their alloys can be applied for LTCC. The purpose of this work is to investigate low temperature sinterable dielectrics with distinguished dielectric properties in the microwave range. The microwave dielectrics were prepared by using conventional mixed oxide route, and identify sintering additives that could lower the sintering temperature while maintaining the dielectric properties.

$(Zn,Mg)TiO_3$ and their modified systems, having high dielectric constant and near zero temperature coefficient of resonant frequency, are promising materials for microwave applications [1-4] and are selected as microwave dielectrics in this study. However, sintering temperature of $(Zn,Mg)TiO_3$ is greater than 1200℃ and is incompatible with LTCC. Various kinds of dopants were added as low temperature sintering promoters to reduce sintering temperature of $(Zn,Mg)TiO_3$ in this study. Crystalline structures and microstructure of the materials were observed and dielectric properties of the sintered specimens were examined.

Experimental Procedures

Microwave dielectrics $(Zn,Mg)TiO_3$ were prepared by conventional solid-state reaction method. Reagent grade raw materials ZnO, MgO and TiO_2 with high purity (>99%) were used. The starting materials were weighed, according to the composition $(Zn_{1-x}Mg_x)TiO_3$ where x = 0, 0.1 and 0.2, and were ball milled in distilled water for 24 h with zirconia balls. Mixtures were dried and calcined at 800℃ for 4 h. The phases of calcined powders were analyzed by X-ray diffractometer (ModelX1, Scintag, USA) using Cu Kα radiation of 2θ from 20º to 60º for phase identification. The calcined powders were ground and then mixed with various ratio of high purity (>99%) oxides V_2O_5 and Bi_2O_3. The added powders were ball milled in distilled water for 24 h with zirconia balls and dried. After drying, the mixed powders were pressed into pellets under a pressure of 98 MPa and then sintered at temperatures of 800–950℃ for 4 h in air and cooled to room temperature in furnace.

Crystalline structures of the sintered specimens were analyzed by using X-ray diffraction patterns using Cu Kα radiation of 2θ from 20º to 60º for phase identification. Microstructures of the specimen were observed by scanning electron microscopy (SEM). The bulk densities of the sintered pellets were measured by the Archimedes methods. The relative permittivity and quality factors were measured by Hakki–Coleman dielectric resonator method and a network analyzer (Model HPE8364A, Hewlett-Packard, Palo Alto, CA) was employed in the measurement. Dielectric properties were measured in the frequency range between 6 and 10 GHz. The temperature coefficient of resonant frequency (τ f) was calculated from the following equation:

$$\tau_f = \frac{\Delta f_o}{f_o \Delta T}$$

where fo is the resonant frequency at 25 ℃, △fo is the difference between the resonant frequency at 25 and 75℃, and △T is the temperature difference between 25 and 75 ℃.

Results and Discussion

Figure 1 shows densities of $(Zn_{1-x}Mg_x) TiO_3$ (x = 0, 0.1 and 0.2) with 4wt% Bi_2O3 + 1wt% $V2O5$ sintered at different temperatures for 1 h. The Bi2O3+V2O5 doped $(Zn,Mg)TiO3$ ceramics can be well densified below 1000oC, as shown in Fig.1. When x=0.1, the bulk density of $(Zn_{1-x}Mg_x)TiO_2$ ceramics sintered at 850℃ is 4.77 g/cm³. The density decreased with the amounts of the MgO increased, which is because that density of $MgTiO_3$ and $ZnTiO_3$ is 3.89 g/cm³ and 5.16 g/cm³, respectively.

978-1-4244-2739-0/08/$25.00 ©2008 IEEE

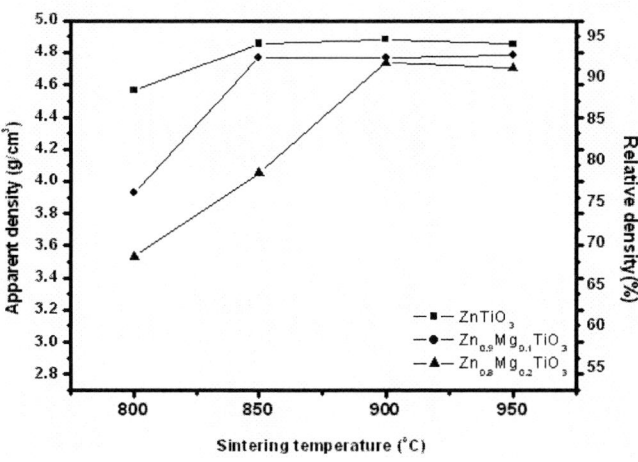

Fig.1 Densities of (Zn$_{1-x}$Mgx) TiO$_3$ (x = 0, 0.1 and 0.2) with 4wt% Bi$_2$O$_3$+1wt% V$_2$O$_5$ sintered at different temperatures for 1 h.

Figure 2 shows microstructures of (Zn$_{0.9}$Mg$_{0.1}$)TiO$_3$ dielectrics added with 4wt% Bi$_2$O3+1wt% V$_2$O$_5$ sintered at 800-900℃ for 1 h. Coalescence of grains and elimination of pores are observed as sintering temperature increased from 800 to 900 ℃, as shown in Fig.2. Grain size of (Zn$_{0.9}$Mg$_{0.1}$)TiO$_3$ ceramics increased with the increase of sintering temperature.

(a)

(b)

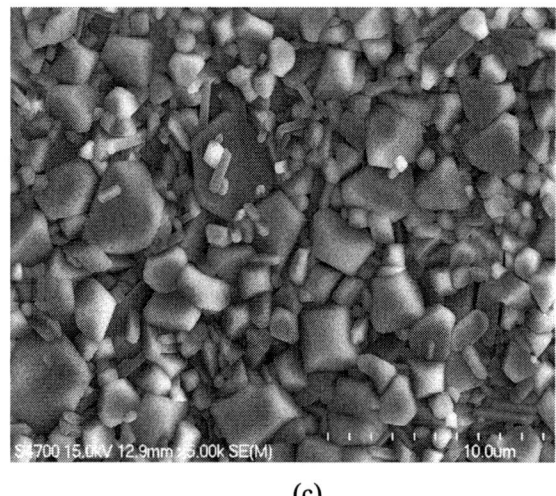

(c)

Fig.2 SEM micrographs of the 4wt% Bi$_2$O$_3$ + 1wt% V$_2$O$_5$ doped (Zn$_{0.9}$Mg$_{0.1}$)TiO$_3$ ceramics sinter at (a)800 ℃ (b)850℃ (c)900℃ for 1h.

X-ray diffraction patterns of (Zn$_{1-x}$Mg$_x$)TiO$_3$ (x=0-0.2) with 4wt% Bi$_2$O$_3$ +1wt% V$_2$O$_5$ sintered at 800-900℃ for 1 h are shown in Figs.3-5. When the sintering temperature is 800 ℃, it was found that the specimens contained three phases: ZnTiO$_3$, Zn$_2$TiO$_4$, and TiO$_2$. ZnTiO$_3$ was decomposed into Zn$_2$TiO$_4$ and TiO$_2$ during sintering at temperatures above 800 ℃.

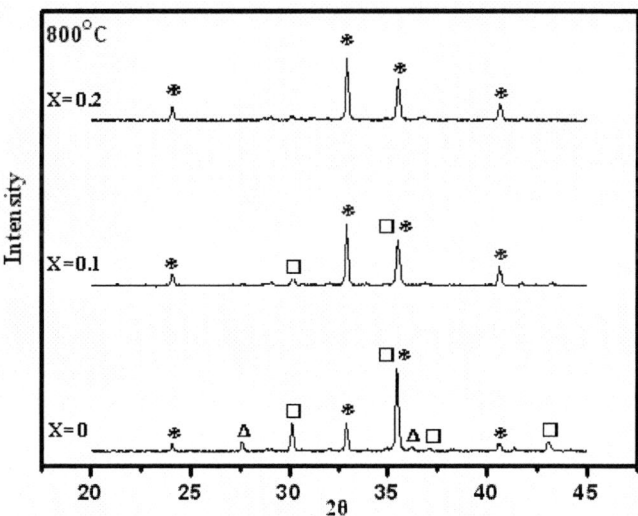

Fig.3 XRD of (Zn$_{1-x}$Mg$_x$)TiO$_3$ (x=0-0.2) specimens sintered at 800℃ (＊:ZnTiO$_3$ (hexagonal) □:Zn$_2$TiO$_4$ (cubic) Δ:TiO$_2$).

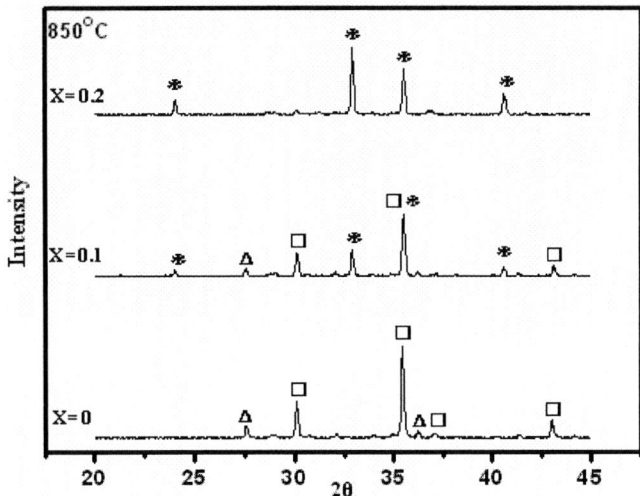

Fig.4 XRD of $(Zn_{1-x}Mg_x)TiO_3$ (x=0-0.2) specimens sintered at 850℃ (＊:ZnTiO₃ (hexagonal) □:Zn₂TiO₄ (cubic) Δ:TiO₂).

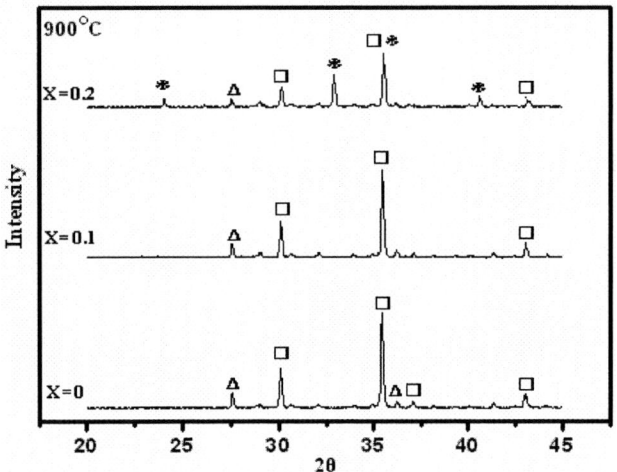

Fig.5 XRD of $(Zn_{1-x}Mg_x)TiO_3$ (x=0-0.2) specimens sintered at 900℃ (＊:ZnTiO₃ (hexagonal) □:Zn₂TiO₄ (cubic) Δ:TiO₂).

Microwave dielectric properties of $4wt\%Bi_2O_3+1wt\%V_2O_5$ doped $(Zn_{1-x}Mg_x)TiO_3$ (x=0-0.2) ceramics sintered at different temperatures are shown in Figs. 6-8. It was found that the added content of MgO played an important role in determining dielectric properties of the sintered specimens. Relative permittivity of (Zn,Mg)TiO₃ decreased as the amount of magnesium content increased, as shown in Fig.6. The εr values increased from 10.5 for $(Zn_{0.8}Mg_{0.2})TiO_3$ sintered at 800℃ to 30.3 for ZnTiO₃ sintered at 900℃.

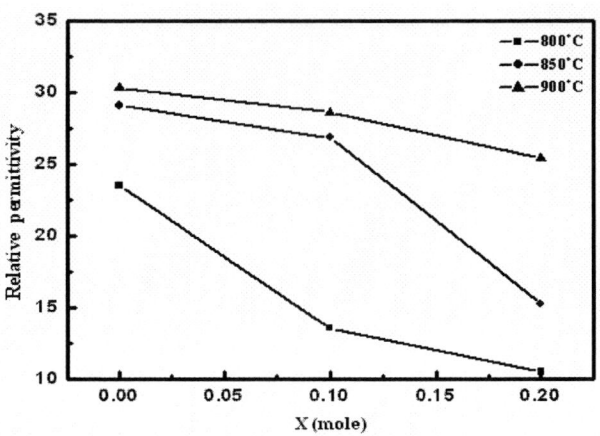

Fig.6 Dielectric constant of 4wt% Bi₂O₃+1wt% V₂O₅ doped $(Zn_{1-x}Mg_x)TiO_3$ (x=0-0.2) sintered at different temperatures.

Contrary to the relative permittivity, the Q×f values of $(Zn_{1-x}Mg_x)TiO_3$ increased as the amount of magnesium increased, as shown in Fig.7. This result arose because MgTiO₃ has a higher quality factor than ZnTiO₃ in the microwave region. The MgTiO₃ dielectric has a quality factor Q of 21,800 at 8 GHz, whereas that of ZnTiO₃ is 3000 at 10 GHz.

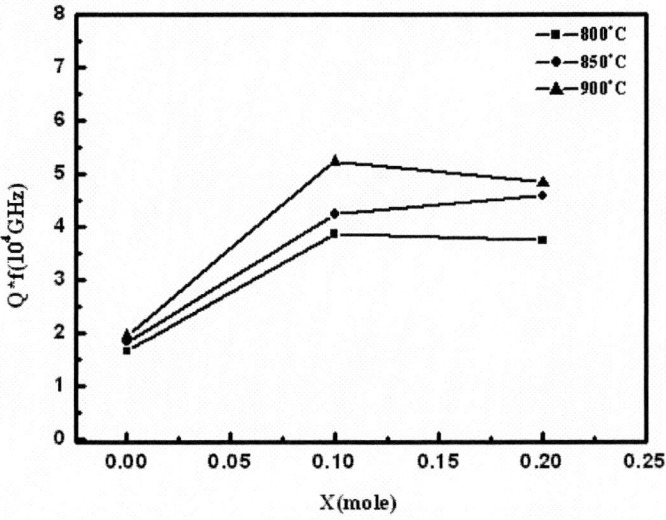

Fig.7 Q×f values of 4wt% Bi₂O₃+1wt% V₂O₅ doped $(Zn_{1-x}Mg_x)TiO_3$ (x=0-0.2) sintered at different temperatures.

The τ f values of $(Zn_{1-x}Mg_x)TiO_3$ ceramics depend on the crystalline structure. The τ f values of ZnTiO₃, Zn₂TiO₄ and TiO₂ are -55, -70 and +450 ppm/℃, respectively [5]. The positive value of τ f of ZnTiO₃ dielectric sintered at 800–900℃ was attributed to the rutile phase from the phase decomposition of ZnTiO₃. The τ f values of $(Zn_{1-x}Mg_x)TiO_3$ shifted to more negative values as x increased from 0 to 0.2, as shown in Fig.8. This result was attributed to the decrease in the amount of the rutile phase.

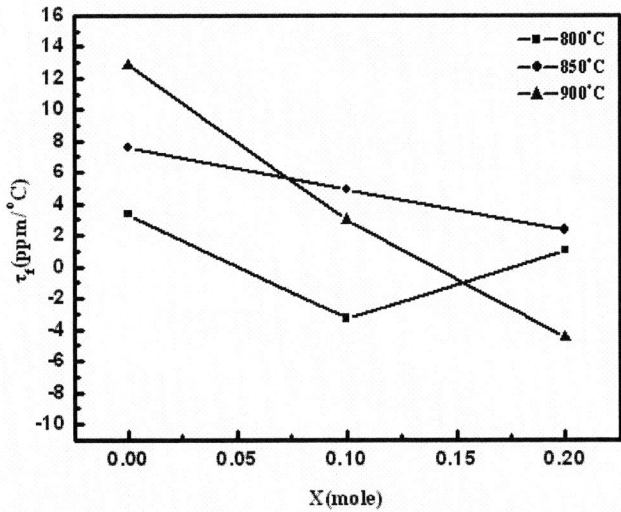

Fig.8 τf values of 4wt% Bi_2O_3+1wt% V_2O_5 doped $(Zn_{1-x}Mg_x)TiO_3$ (x=0-0.2) sintered at different temperatures.

Conclusions

Microwave dielectric properties of $(Zn,Mg)TiO_3$ ceramics depend on the crystalline structure. $(Zn_{0.9}Mg_{0.1})TiO_3$ doped with 4wt% Bi_2O_3+1wt% V_2O_5 has been successfully sintered at 850℃ for 1 h had an average density of 94.41% with a dielectric constant of 26.85, Qxf of 42351 and τ= 4.92 ppm/℃.

Acknowledgements

The authors are grateful to the National Science Council of R.O.C. and I-Shou University for financial support under project No. NSC 96-2221-E-214-001 and ISU 96-01-06, respectively.

References

1. F. H. Dulin and D. E. Rase, "Phase Equilibria in the System ZnO–TiO2" J.Am. Ceram. Soc., Vol.43, No.3 (1960), pp.125–31.
2. H. T. Kim, S. Nahm, and J. D. Byun, "Low-Fired $(Zn,Mg)TiO_3$ Microwave Dielectrics" J. Am. Ceram. Soc., Vol.82, No.12 (1999), pp. 3476–80.
3. O. Yamaguchi, M. Morimi, H. Kawabata, and K. Shimizu, "Formation and Transformation of $ZnTiO_3$" J. Am. Ceram. Soc., Vol.70 No.5 (1987), pp. C-97–C-98.
4. H. T. Kim, Y. H. Kim, and J. D. Byun, "Phase Transformation and Thermal Stability in Zinc Magnesium Titanates" J. Korean Phys. Soc., Vol.32, No.2 (1998), pp. S159–S161.
5. H. Jantunen, A. Uusimaki, R. Rantiaho, S. Leppavuori, "Temperature Coefficient of Microwave Resonance Frequency of a Low-Temperature Cofired Ceramic (LTCC) System" J. Am. Ceram. Soc. Vol. 85, No.3 (2002), pp. 697-699.

Vibration and Buckling of a Carbon Nanotube Inserted with a Carbon Chain

Zhili Hu[1], X.M.Guo[2], Johan Liu[3]

1 Key Laboratory of Advanced Display and System Applications, Ministry of Education & SMIT Center, School of Mechatronics Engineering and Automation, Shanghai University, P.O.Box 282, Shanghai 200072, China

2 Shanghai Institute of Applied Mathematics and Mechanics, Shanghai University

3SMIT Center & Bionano Systems Laboratory, Department of Microtechnology and Nanoscience, Chalmers University of Technology, SE-412 96 Gothenburg, Sweden

xmguo@shu.edu.cn

Abstract

An elastic string-elastic shell model is developed to study vibration behaviors of a carbon nanowire. The present model predicts that non-coaxial vibration between the C-chain and the innermost tube does not occur due to negligible bending rigidity of the C-chain. In addition, it is found that the C-chain has most significant effect on the lowest frequency associated with radial vibration mode for circumferential wave-number 2 (n=2). In particular, the effect of the C-chain on axisymmetric radial breathing frequencies (n=0) predicted by the present model is found to be in reasonable agreement with known experimental and modeling results available in the literature.

To study the buckling of a carbon nanowire made of the C-chain inserted inside a CNT, not only the elastic string-elastic shell model that used above but also the molecular dynamics (MD) simulation are adopted.

For radial buckling, both continuum model and MD simulation predict that C-chain increases buckling stress by more than 40% to (2, 8), (0, 9), (3, 7) and (5, 5) tube. The continuum model predicts that circumferential wave number n=2 at buckling. Related with this, MD simulation shows that the cross-section of CNT is an ellipse at buckling. In particular, the nonlinear effects of vdW interaction have significant influence on the critical buckling loading.

For the buckling under axial compression, two different kinds of buckling have been observed and investigated, namely the C-chain's unconditional buckling, and the tube's wall buckling, respectively. After the unconditional buckling, all displacements on the C-chain are in the same plane, and, the wave amplitudes' localization phenomenon appears. To the tube's wall buckling, the C-chain could either increase or decrease axial buckling strain depending on specific CNT chirality and temperature.

Introduction

Recent findings of linear carbon-atom chains in the core of multiwalled carbon nanotubes (MWNTs) have stimulated considerable interest from broad range of researchers [1-6]. Typically, these carbon nanowires are made of a linear carbon chain (i.e. C-chain) inserted in singlewalled nanotube (SWNT) or MWNT of innermost diameter about 0.7nm. Such nanostructures offer a new type of carbon nanowires and have potential application in nanoelectronics, nanomechanics and nanomaterials. Therefore, it is relevant to study basic physical behavior of this new type of nanowires. The present work is motivated by these experimental findings, and aims to study mechanical vibration of such carbon nanowires, with particular emphasis on the effect of the inserted C-chain on vibrational frequencies and critical buckling loading.

It is known that elastic shell model have been effectively used to study vibration and buckling of unfilled hollow SWNTs or MWNTs [7]. In addition, although an isolated linear C-chain is known to be highly unstable and cannot keep its straight-line shape [8], an inserted linear C-chain can keep its perfectly straight shape due to vdW interaction with the surrounding (innermost) CNT (typically of diameter about 0.7nm) [1-6]. Thus, for a carbon nanowire made of a linear C-chain inserted inside an N-walled CNT, the inserted C-chain can be modeled appropriately as a straight elastic string with negligible bending rigidity. In particular, deflection of the elastic string would be coupled with deflection of the innermost CNT through the vdW forces between the C-chain and the innermost CNT. In doing so, vibration of the nanowire is governed by elastic string equations for the inserted C-chain coupled with elastic shell equations for the surrounding CNTs. To the best of our knowledge, such coupled vibration of CNTs filled with the C-chain has not been studied in the literature. As will be seen below, the inserted C-chain does cause some interesting new vibration phenomena as compared to vibration and buckling of unfilled hollow CNTs.

Results show that the c-chain has the most significant effect associated with the innermost tube, rather than with other outer tubes. Thus, to understand essential effect of the c-chain, let us first consider a singlewalled CNT (N=1) of diameter 0.7nm.

Continuum Model

Let us consider a long linear C-chain inserted inside the core of an N-walled CNT of length L, as shown in Fig.1. With the elastic shell model for CNTs [7], each of the N concentric CNTs is defined by two coordinates, longitudinal coordinate x ($0 \leq x \leq L$) and circumferential angular coordinate θ ($0 \leq \theta \leq 2\pi$), and the N CNTs are labeled with subscript k (k=1, 2,...,N), from the innermost tube (k=1) to the outermost tube (k=N). In particular, the radius of tube-k is R_k (k=1, 2,...,N). Let that longitudinal displacement, circumferential displacement and radial displacement (inward positive) of tube-k be $u_k(x,\theta)$, $v_k(x,\theta)$ and $w_k(x,\theta)$ (k=1, 2,...,N), respectively. On the other hand, the C-chain is defined by the longitudinal coordinate x ($0 \leq x \leq L$). Thus, the displacements of any point x of the C-chain are defined by longitudinal component u(x), and two components perpendicular to the C-chain, X(x) (along the direction θ =0) and Y(x) (along the direction $\theta = \pi/2$), as shown in Fig.1. For the buckling section, because only single-walled nanotube will be considered, the subscript k will be omitted.

Thus, at any point (x, θ) of the innermost tube (k=1), the distance change between the innermost tube and the C-chain after deformation is $[-w_1 - X\cos\theta - Y\sin\theta]$. Therefore, the attractive vdW force per unit area, acting on the point (x, θ) of the innermost tube (k=1) due to the vdW interaction with the C-chain, is given by

$$-c_0[w_1 + X\cos\theta + Y\sin\theta] + P_{ext} - P_{int} \quad (1)$$

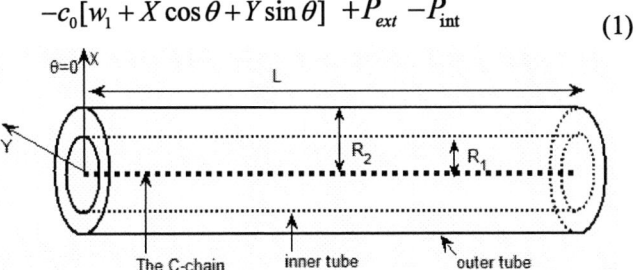

Fig.1 An elastic model for a nanowire made of a linear C-chain inserted inside a double-walled CNT.

where c0 is the vdW interaction coefficient between the innermost tube and the C-chain. It is c0=27.8GP/nm here. In connection with this, we noticed that the interaction coefficient between two parallel graphene sheets at their equilibrium separation is about 100 GPa/nm [10].

Now, three equations of motion for the C-chain, modeled as an elastic string, are given by

$$A\rho u_{tt} = EAu_{xx} \quad (2)$$

$$A\rho X_{tt} = F_x + FX_{xx} \quad (3)$$

$$A\rho Y_{tt} = F_y + FY_{xx} \quad (4)$$

where ρ = mass of the C-chain per unit axial length, A is the cross section area of C-chain (defined consistently with graphene sheets), E is Young's modulus of the C-chain, and F is an initial axial force acting on the C-chain (it will be set zero in free vibration discussed in the present paper).

For vibration problem, an N-wall CNT is governed by the following three equations [7]

$$R_k^2 \frac{\partial^2 u_k}{\partial x^2} + \frac{1}{2}(1-v)\frac{\partial^2 u_k}{\partial \theta^2} + \frac{R_k}{2}(1+v)\frac{\partial^2 v_k}{\partial x \partial \theta} - vR_k\frac{\partial w_k}{\partial x} +$$
$$(1-v^2)\frac{D}{EhR_k^2}[\frac{1}{2}(1-v)\frac{\partial^2 u_k}{\partial \theta^2} + R_k^3\frac{\partial^3 w_k}{\partial x^3} - \frac{R_k}{2}(1-v)\frac{\partial^3 w_k}{\partial x \partial \theta^2}]$$
$$= \frac{\rho h}{Eh}(1-v^2)R_k^2\frac{\partial^2 u_k}{\partial t^2}$$

$$\frac{R_k}{2}(1+v)\frac{\partial^2 u_k}{\partial x \partial \theta} + \frac{R_k^2}{2}(1-v)\frac{\partial^2 v_k}{\partial x^2} + \frac{\partial^2 v_k}{\partial \theta^2} - \frac{\partial w_k}{\partial \theta}$$
$$+(1-v^2)\frac{D}{EhR_k^2}[\frac{3R_k^2}{2}(1-v)\frac{\partial^2 v_k}{\partial x^2} + \frac{R_k^2}{2}(3-v)\frac{\partial^3 w_k}{\partial x^2 \partial \theta}]$$
$$= \frac{\rho h}{Eh}(1-v^2)R_k^2\frac{\partial^2 v_k}{\partial t^2}$$

$$vR_k\frac{\partial u_k}{\partial x} + \frac{\partial v_k}{\partial \theta} - w_k - (1-v^2)\frac{D}{Eh}R_k^2\nabla^4 w_k + (1-v^2)\frac{D}{EhR_k^2}$$

$$[-R_k^3\frac{\partial^3 u_k}{\partial x^3} + \frac{R_k}{2}(1-v)\frac{\partial^3 u_k}{\partial x \partial \theta^2} - \frac{R_k^3}{2}(3-v)\frac{\partial^3 v_k}{\partial x^2 \partial \theta} - w_k - 2\frac{\partial^2 w_k}{\partial \theta^2}]$$

$$= \frac{1}{Eh}(1-v^2)R_k^2[\rho h\frac{\partial^2 w_k}{\partial t^2} - p_k] \quad (5)$$

For the buckling problem, the governing eq. will be take as [9]

$$D\nabla^4(\nabla^2 + \frac{1}{R^2})^2 w + \frac{2D(1-v)}{R^6}(\frac{\partial^6 w}{\partial x^2 \partial \theta^4} + \frac{\partial^4 w}{\partial x^2 \partial \theta^2} - R^4\frac{\partial^6 w}{\partial x^6}) = \nabla^4 p(x,\theta)$$
$$+F_x\frac{\partial^2}{\partial x^2}\nabla^4 w + 2\frac{F_{x\theta}}{R}\frac{\partial^2}{\partial x \partial \theta}(\nabla^4 w + \frac{\nabla^2 w}{R^2}) + \frac{F_\theta}{R^2}\nabla^4(\frac{\partial^2 w}{\partial \theta^2} + w) - \frac{Eh}{R^2}\frac{\partial^4 w}{\partial^4 x} \quad (6)$$

where D is effective bending stiffness of a SWNT, Eh is in-plane stiffness of a SWNT, ρh is the mass density per unit (lateral) area of each tube, and v is Poisson's ratio of SWNTs. In particular, for a SWNT, D, Eh, and ρh are three independent material constants, irrespective of the specific definition of thickness h. The effective values for these constants are D=2eV, Eh = 360 J/m2, and ρh = (2.27g/cm3) × 0.34 nm [7]. In other words, for different definitions of the thickness h, the constants E, and ρ are defined in such way that the products Eh and ρh remain unchanged for different definitions of h.

Molecular Dynamics Simulation

Radial buckling of both filled and unfilled (2, 8), (0, 9), (3, 7) and (5, 5) SWCNT with length of 2~20 Å will be studied by MD simulation. During the simulation, several hundred atoms are put in Berendsen thermostat to keep the system in a stable temperature. The instantaneous total energy that consists of covalent potential (i.e. the C-C bond energy) and vdW potential on the unrigged atoms are monitored. Time step is adopted as 0.5fs such that it is neither too large for accuracy nor too small for computation feasibility. The Brenner's reactive empirical bond-order (REBO) potential will be adopted in this paper for the short-range (covalent) interactions [12], while Lennard-Jones (L-J) potential is applied to calculate vdW potential, with a cutoff distance of 2 Å. For two atoms with a distance x apart, the vdW potential has form

$$U(x) = -\frac{A^*}{x^6} + \frac{B^*}{x^{12}} \quad (7)$$

Where A* is taken as 15.2 eV×Å^6, and B* is 24.13×10^3 eV×Å^{12} [11].

For radial buckling, external forces inwardly radial directions are applied to every atom in the tube to imitate the radial pressure. Temperature is fixed to 300K. Force intensity acting on every atom is the same, which is gradually increasing with a speed of 107N/s. The statistical data reveal that the shirking speed of tube surface towards the axis is about 1.3m/s before buckling. This increasing speed of force is taken because no prominent difference would occur if lower speed is in use.

For axial buckling, three rings at each ends are constrained and move closer at a speed of 6m/s, which is chosen for similar reason mentioned above. Both temperatures of 3K and 300K are used because under higher or lower temperature, buckling behaviors especially the role of the C-chain differ a lot according to the author's observation.

Both potential energy and average radius of the tube has been recorded to find out the buckling time. For convenience, the obtained values for radial buckling will be given in terms of the external radial pressure, which can be obtained as force intensity per atom / area of an atom. In addition, redundant simulations have been taken to obtain statistical average.

Results on Vibration

Let us first consider axisymmetric vibration of the N-wall CNT characterized by $v_k = 0$ and $w_k(x,t) = w_k(x + \pi, t)$ (k=1, 2,...,N) and consider situation with n=0. In this case, it is seen from eqs. that X=Y=0 and thus the C-chain remain at the initial central line of the CNT. It can be verified that the second equation of (5) for each tube is always met, then we have 2N equations for $u_k(x,t)$ and $w_k(x,t)$ (k=1, 2,...,N). Furthermore, it can be verified that the effect of the C-chain on longitudinal and circumferential motions of the CNT is negligible. Therefore, we shall focus on the effect of the C-chain on axisymmetric radial vibration, especially radial breathing modes (RBMs) characterized by $u_k = 0$ and $w_k = w_k(t)$.

It is seen that the last term on the right-hand side is the effect of the C-chain on the radial breathing mode of a SWNT. For example, let us take Eh=0.34 TPa nm, ν=0.2, R1=3.6 Å, we find that RBM's frequency of a SWNT of diameter 0.7 nm increases from 9.777×10^{12} (unfilled) to 9.823×10^{12} (filled by a C-chain). In other words, the C-chain changes the radial breathing frequency of a SWNT by +0.476 %.

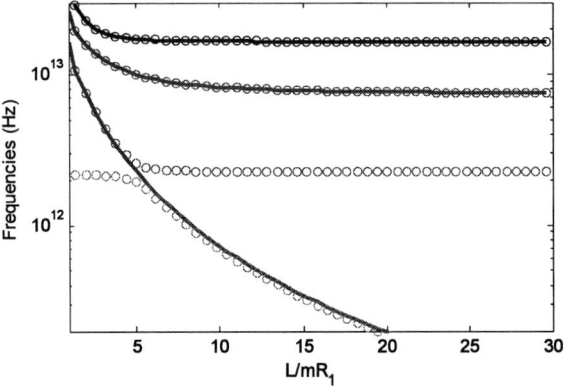

Fig.2 Four frequencies of a SWNT (of diameter 0.7nm) filled with a linear C-chain (circlets) and three frequencies of a SWNT (of diameter 0.7nm) without a linear C-chain (solid lines) when n=1.

In an attempt to compare our results with any known data, we noticed that density functional calculation [3] predicts an increase of the RBM frequency of a SWNT of diameter

0.69nm from 370 cm-1 to 374 cm-1, which leads to a relative increase of about +1%. In addition, experimental data reported in [5] indicated that the highest RBM's frequency of a C-chain filled MWNT (of innermost diameter 0.6nm) increases from 387 cm-1 to 390 cm-1, with a relative increase about +0.8%. Therefore, our results appear to be in reasonable agreement with these known data [3, 5]. The small difference between our results and the data reported in [3, 5] could be attributed to the fact that the interaction coefficient c0 used in the present paper has been underestimated due to the neglected curvature effect of the innermost tube of small diameter (0.7nm). It is expected that the curvature effect would lead to a larger interaction coefficient c0 and a larger relative increase of RBM frequencies due to the inserted C-chain.

To the Coupled rod-like vibration (n=1), free vibration of the coupled c-chain and the singlewalled CNT is governed by 4 algebraic equations [7].

For each given (L/(mR1)), there are four frequencies for the filled singlewalled CNT. The four frequencies are plotted in Fig.2 for L/(R1m) ranging from 1 to 30, with comparison to three frequencies of the unfilled singlewalled CNT without the inserted c-chain, denoted by dashed lines.

It is seen from Fig.2 that this new added frequency is about 2.457 THz , much lower than the longitudinal frequency (6.184THz) and the torsional frequency (13.799THz) of the unfilled SWCNT of diameter 0.7nm. In fact, this new frequency can be estimated from the free vibration of the c-chain within a rigid CNT (whose deflection w1 of the innermost tube is negligible). And, it can be verified that the frequency of the c-chain vibrating inside a rigid CNT is about 2.365 THz, which is very close to the above-mentioned second frequency (2.457 THz) for sufficiently large ratio L/(mR1) (say, >5). This confirms the idea that the second lowest frequency ω_2 is essentially associated with free vibration of the c-chain insider the almost rigid SWCNT.

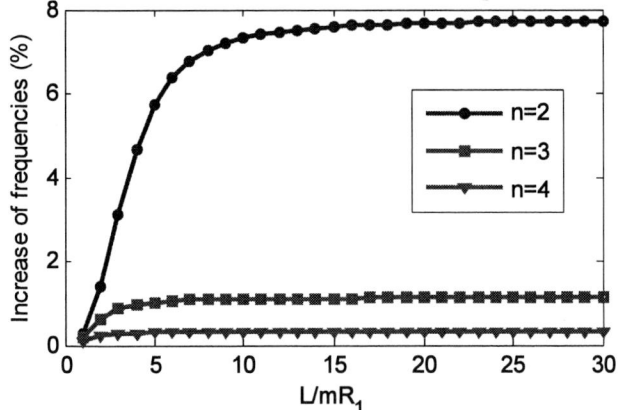

Fig.3 Relative increase of the lowest frequency of a SWNT (of diameter 0.7nm) due to the inserted C-chain for various circumferential wave-number n>1.

Now, let us consider vibration of higher-order modes with circumferential wave-number $n \geq 2$. In this case, X=Y=0, and one has 3N equations (5) for uk, vk and wk (k=1,2...N), where the net pressure acting on each tube is given in [7].

To understand essential effect of the c-chain, let us first consider a singlewalled CNT (N=1) of diameter 0.7nm. In this case, eq. (5) leads to three algebraic equations.

Three frequencies of the singlewalled CNT for each wave-number $n \geq 2$ can be calculated as the eigen-roots. The effect of the c-chain on the lowest frequency (radial mode) is plotted in Fig.3 for n=2, 3, and 4, respectively. It is seen that, the c-chain increases the lowest frequency as much as 7.7%, 1.1%, and 0.3% for sufficiently large L/(mR1), when n=2, 3 and 4, respectively. Especially, we noticed that the effect on the lowest frequency when n=2 is as high as 7.7%.

Here it is noticeable that, if real CNT e.g. (5,5) SWNT, not the ideal CNT with innermost redius 0.7nm is considered, the C-chain's influence will be much larger.

Results on Redial Buckling

Now, let us consider radial buckling of the CNT filled with C-chain under radial pressure in the linear premise, which means the influence of the pressure-induced diameter reduction on the vdW interaction coefficient c is neglected and the latter is treated to be pressure-independent with

$$F = 0, F_x = 0 \text{ and } F_{x\theta} = 0 \tag{8}$$

As mentioned above, buckling of the inserted C-chain is couple with the CNT only if circumferential wave number n=1. For radial buckling, we have $n \geq 2$ and X=Y=0. Also, here $P_{int} = c[R_0 - R]$. Taking this into Eq (1) the attractive vdW force per unit area, acting on the point (x, θ) of the tube due to the interaction with the C-chain can be simplified as

$$p = -cw - c[R_0 - R] + P_{ext} \tag{9}$$

Substituting Eqs. (9, 8) and solution form into Eqs. (6), gives an algebraic equation for A, which has form

$$M(m, n)A = 0 \tag{10}$$

Thus, the existence condition of a non-zero solution of A is

$$M = 0 \tag{11}$$

This condition determines a relationship between the applied circumferential strain ε_θ and m, n. Thus, solving the equation obtains the critical buckling (circumferential membrane) force for each given (L/(mR)) and n=2,3,4, and therefore obtains critical buckling (circumferential) strain via constitution law.

Both filled SWCNT and unfilled SWCNT are calculated, according to which, the lowest strain value occurs with n=2 when $L/mR \rightarrow +\infty$.

Results shows that with D=0.85 eV, the C-chain raises the critical buckling strain by 39.2% to the ideal CNT of radius 3.630 Å, or 46.6% for (2, 8) SWCNT of radius 3.588 Å, or 59.8% for (0, 9) SWCNT of radius 3.523 Å, or 70.5% for (3, 7) SWCNT of radius 3.479 Å, or 97.1% for (5, 5) SWCNT of radius 3.390 Å, respectively.

In the above linear analysis, the vdW interaction coefficient c is assumed to be a constant, independently of the applied pressure. However, the actual interaction coefficient c is very sensitive to small change of the CNT's radius due to the applied pressure, and therefore the coefficient c would vary considerably during the pre-buckling procedure. To take this nonlinear vdW effect into account, the radius of the tube should not be always taken as its initial value R, but a changing value r. Also, the coefficient c should be taken as a function of changing radius r, i.e. c=c(r), which has an effect on the predicted critical buckling strain. At last, it has

$$\varepsilon_c(c(r_c)) = 1 - r_c / r_0 \tag{12}$$

where r_0 is the initial radius of the tube, r_c is radius at the buckling state, and ε_c is the critical buckling strain. This equation is nonlinear because c=c(r) is a nonlinear function. Notably, according to Eq. (7), $\varepsilon_c = \varepsilon_c(c)$ is nearly a linear function. Also, Eq.(14) should be rewritten as

$$M\mid_{c=c(r)} = 0 \tag{13}$$

Solving these equations (12, 13) together with relation (6) via any nonlinear numeric algorithm obtains radius r_c and thus the critical buckling strain ε_c of CNT with a C-chain.

Results from MD simulation are preferred to be expressed in terms of the external radial pressure. To make comparison, the critical external radial pressure at buckling is presented and thus the strains should be transformed into external radial pressure, which can be obtained from

$$p = \frac{F_\theta}{R} \tag{14}$$

Taking this into Eq. (1) and notice X=Y=0 and w=0, the critical external radial pressure at buckling has form

$$P_{ext} = \frac{F_\theta}{R} + P_{int} = \frac{Eh}{R}\frac{\varepsilon_c}{1 - v^2} + P_{int} \tag{15}$$

Here in view of the nonlinear vdW effect, Pint should not be taken as $c[R_0 - R]$ as in Eq.(9), and should be directly calculated from the first derivative of the energy-interlayer spacing relation between the innermost tube and the C-chain. This applied internal pressure is caused by the vdW force between the CNT and the C-chain, and is dependant on CNT's radius. As a result, the C-chain would raise critical external radial pressure more than for critical buckling strain, although according to our calculation (see Table 1), this influence is not as significant as that from Eq. (12).

Results from both the continuum model (with the nonlinear dependence of the vdW interaction on the reduced radius) and the MD simulation are summarized in Table 1. Results from the continuum model based on linear analysis are not listed because these values are too low compared with results from MD simulation.

With a lot of analysis, consequently, if the non-linear dependence of the vdW interaction on the reduced radius, and the chirality-dependent D has been considered, we think that the continuum model agrees well with MD simulation, in that the critical pressures and the increments of the critical pressures given by one method are reasonably consistent with those by another. And, in particular, MD simulation shows

Tab.1 Comparison on critical external radial pressure of radial buckling of filled or unfilled CNT via MD simulation or continuum model. The C-chain increases critical external radial pressure by >50% to all real kinds of SWCNTs with D=0.85eV.

CNT type	Final internal pressure /GPa	Initial c/c0	critical external radial pressure of CNT/GPa					
			MD simulation			Continuum model (with the non-linear dependence of the vdW interaction), D=0.85eV		
			unfilled	filled	raise	unfilled	filled	raise
Ideal R=3.630 Å	0.3	1				8.5	12.9	51.3%
(2,8) R=3.588 Å	0.5	1.24	10.9	17.6	60.9%	8.8	14.4	66.4%
(0,9) R=3.523 Å	1.0	1.71	11.8	19.6	66.9%	9.3	17.4	86.2%
(3,7) R=3.479 Å	1.0	2.13	11.4	21.0	83.8%	9.7	19.9	105%
(5,5) R=3.390 Å	3.0	3.24	11.1	28.7	157%	10.5	27.3	160%

Tab.2 Critical external radial pressure of CNT/GPa via continuum model (with the non-linear dependence of the vdW interaction on the reduced radius) with D of values other than 0.85eV.

CNT	unfilled	filled	raise
D=1.0 eV			
Ideal	10.0	14.6	45.1%
(2,8)	10.4	16.2	55.6%
(0,9)	11.0	19.3	75.6%
(3,7)	11.4	22.0	92.8%
(5,5)	12.3	29.7	141%
D=1.1 eV			
Ideal	11.0	15.7	41.9%
(2,8)	11.4	17.3	51.6%
(0,9)	12.1	20.6	70.2%
(3,7)	12.5	23.4	86%
(5,5)	13.6	31.4	131%
D=1.2 eV			
Ideal	12.4	16.8	39.2%
(2,8)	12.5	18.5	48.3%
(0,9)	13.2	21.8	65.8%
(3,7)	13.7	24.7	80.8%
(5,5)	14.7	33.0	123%

that the originally circular cross-section becomes an ellipse after buckling, which corresponds to a circumferential wave number n=2, as predicted by the continuum model.

Results on Axial Buckling

In this case, it supposes that the C-chain and the tube are compressed simultaneously such that these they are of the same axial strain load. And, we focus on short (say, length<10 nm) CNTs.

Now we consider buckling of the CNT inserted with a C-chain under axial compression, which indicates

$$F_\theta = 0 \text{ and } F_{x\theta} = 0 \qquad (16)$$

As that mentioned above, when $n \geq 2$, the C-chain will be uncoupled with the tube, which means

$$F_x = F_y = 0. \qquad (17)$$

For buckling modes with circumferential wave-number $n \geq 2$, there should be a correlated equation for the tube, i.e. an equation for $w(x,\theta)$. For buckling modes with circumferential wave-number $n=1$, it is necessary that for either the tube or the C-chain there are correlated equations, then we have one equation for $w(x,\theta)$ and two equations for $X(x)$ and $Y(x)$. Thus, when n=1, for a simply-supported CNT, solutions for $X(x)$ and $Y(x)$ are,

$$X(x) = P\sin\frac{m\pi x}{L}$$

$$Y(x) = Q\sin\frac{m\pi x}{L} \qquad (18)$$

where P and Q are two constants. Here both $X(x)$ and $Y(x)$ are expressed as $\sin{m\pi x}/{L}$, in order to make consistency with Eqs. (6), which implies at buckling, all displacements on the C-chain are within a plane.

Nevertheless, because here the C-chain shares the same axial strain with the tube, it is necessary to examine the pre-strain of C-chain. There is

$$\varepsilon_x = \frac{F_x}{K}, K = \frac{Eh}{1-v^2} \qquad (19)$$

where F_x is pre-stress on axial-direction. ε_x is pre-strain on axial-direction for both the tube and C-chain. Substituting Eq. (19) into Eq. (2) gets

$$-F = EA\varepsilon_x = \frac{EAF_x}{K} \qquad (20)$$

Here the value of EA is taken as 742 nN/nm.

Solving equations obtained above for either $n \geq 2$ or $n=1$, there gets critical buckling (axial) strains for each given L/(mR) and results are put in Fig.4, which suggest that buckling of the nanostructure comprised of CNT with a C-chain inside could happen at any pressure. Thus, this result implies the unconditional buckling of C-chain. And in a certain perspective, buckling of tube is in fact a post-buckling problem which will be dealt with through MD simulation below.

Examined results from the MD simulation, the compression can mainly induce two different kinds of buckling, the first of which is buckling of the C-chain since the beginning of compression, another of which is wall buckling of CNT when there is relatively large axial strain.

Before wall buckling of the tube in MD simulation, through MD's graphics (see Fig.5), the buckling of C-chain happens almost since the very beginning of axial compression and no pre-buckling behavior has been observed by the author yet, which seems consisted with the unconditional buckling predicted with continuum model above. After buckling of C-chain, wave is formed on the C-chain, with wave length observed as 8~9 C-C bond lengths. During this process, all displacement of the C-chain are in the same plane (see Fig.5.d), which is consist with analysis above (see analysis for Eq. (18)).

It is interesting to mention the wave amplitudes' concentration phenomenon here. Soon after the C-chain's buckling, sine wave is formed on the C-chain (see Fig.5.a). Later, with the increase of compression, one or two wave crests begin to have larger amplitudes (see Fig.5.b). At last just before the tube's buckling, almost all the displacement concentrate into the one or two wave peaks, and other crests seem vanished (see Fig.5.c).

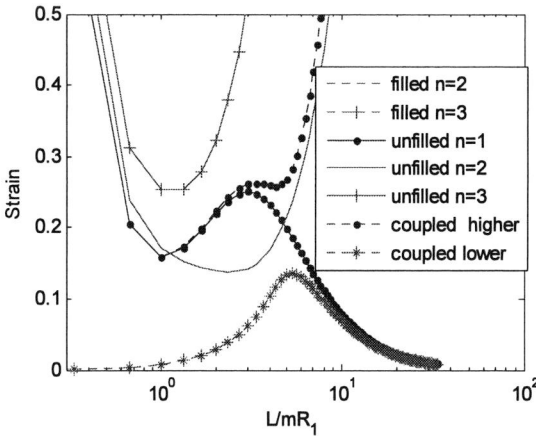

Fig.4 Buckling strains of filled or unfilled CNT under axial compression. In this figure, data related to CNT filled with C-chain is plotted as dotted line, while data related to unfilled CNT is plotted as solid line. Both terms of coupled higher and coupled lower refer to the situation that the tube is coupled with the C-chain when n=1.

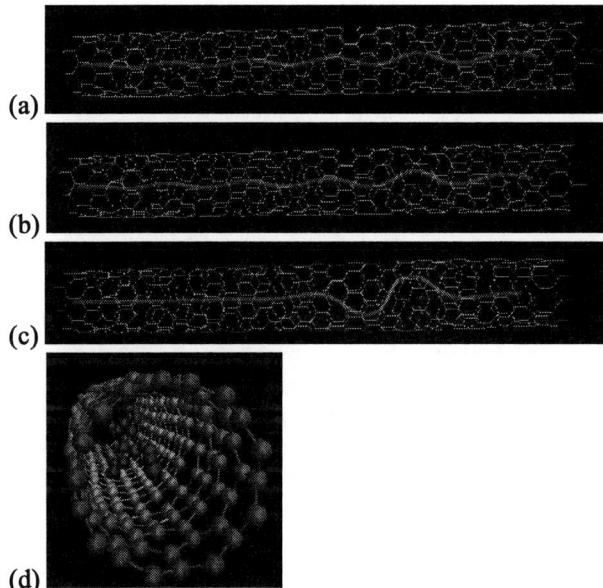

Fig.5 Buckling of C-chain before buckling of CNT and the wave amplitudes' concentration phenomenon from (a) to (c). And, (d) shows that all displacements on the C-chain are in the same plane.

At last let's consider wall buckling of the tube. Strain energies of filled or unfilled CNTs are calculated during the MD simulation. The critical buckling strains are determined via monitored energies. Although column buckling also happens in some case, the "sudden change point" on the energy curve relates to only wall buckling, according to the author's observation. The result suggests that the C-chain prominently influence the buckling behaviors of the nanostructure. Moreover, the influence of C-chain depends on specific CNT chirality and temperature, and buckling strain of filled CNTs can be either larger or lower than that of empty CNT.

Conclusions

An elastic string-elastic shell model is developed to study vibration behaviors of a carbon nanowire. It is shown that vibration of the inserted C-chain is coupled with vibration of the CNT only for vibration modes with circumferential wave-number n=1. In other cases, such as axisymmetric modes (n=0) or higher-order vibration modes with n ≥ 2, total resultant van der Waals (vdW) force acting on the C-chain due to the innermost tube always vanishes and therefore vibration of the CNT does not cause vibration of the inserted C-chain, although the existence of the C-chain does have an effect on vibration of the CNT through the chain-CNT vdW forces acting on the innermost tube. The present model predicts that non-coaxial vibration between the C-chain and the innermost tube does not occur due to negligible bending rigidity of the C-chain. In addition, it is found that the C-chain has most significant effect on the lowest frequency associated with radial vibration mode for circumferential wave-number 2 (n=2). In particular, the effect of the C-chain on axisymmetric radial breathing frequencies (n=0) predicted by the present model is found to be in reasonable agreement with known experimental and modeling results available in the literature.

Meanwhile, to study the buckling of a carbon nanowire made of the C-chain inserted inside a CNT, not only the elastic string-elastic shell model that used above but also the molecular dynamics (MD) simulation are adopted.

For radial buckling, both continuum model and MD simulation predict that C-chain increases buckling stress by more than 40% to (2, 8), (0, 9), (3, 7) and (5, 5) tube. The continuum model predicts that circumferential wave number n=2 at buckling. Related with this, MD simulation shows that the cross-section of CNT is an ellipse at buckling. In particular, the nonlinear effects of vdW interaction have significant influence on the critical buckling loading.

For the buckling under axial compression, two different kinds of buckling have been observed and investigated, namely the C-chain's unconditional buckling, and the tube's wall buckling, respectively. After the unconditional buckling, all displacements on the C-chain are in the same plane, and, the wave amplitudes' localization phenomenon appears. To the tube's wall buckling, the C-chain could either increase or decrease axial buckling strain depending on specific CNT chirality and temperature.

References

1. Z. Wang, X. Ke, Z. Zhu, F. Zhang, M. Ruan, and J. Yang, Phys. Rev. B 61, R2472 (2000).
2. X.L. Zhao, Y. Ando, Y. Liu, M. Jinno, and T. Suzuki, Phys. Rev. Lett. 90, 187401 (2003).
3. Á. Rusznyák, V. Zólyomi, J. Kürti, S. Yang, and M. Kertesz, Phys. Rev. B 72, 155420 (2005).
4. M. Jinno, Y. Ando, S. Bandow, J. Fan, M. Yudasaka, and S. Iijima, Chem. Phys. Lett. 418, 109 (2006).
5. T. Hayashi, H. Muramatsu, A.K. Yoong, H. Kajitani, S. Imai, H. Kawakami, M. Kobayashi, T. Matoba, M. Endo, M.S. Dresselhaus, Carbon 44, 1130 (2006).
6. C. Fantini, E. Cruz, A. Jorio, M. Terrones, H. Terrones, G. Van Lier, J.-C. Charlier, M. S. Dresselhaus, R. Saito, Y. A. Kim, T. Hayashi, H. Muramatsu, M. Endo, and M. A. Pimenta, Phys. Rev. B 73, 193408 (2006).
7. Z.L. Hu, X.M. Guo and C.Q. Ru, Nanotechnology 18 485712 (2007).
8. R.H. Baughman, Science 312, 1009 (2006).
9. Z.L. Hu, X.M. Guo and C.Q. Ru, Nanotechnology 19 305703 (2008).
10. C.Y. Wang, C.Q. Ru, and A. Mioduchowski J. Appl. Phy. 97, 114323 (2005).
11. L.A. Girifalco, M. Hodak, and R.S. Lee, Phys. Rev. B 62, 13104 (2000).
12. D. W. Brenner, Phys. Rev. B 42, 9458(1990)

Author Index

A

An, Bing 129, 914, 964, 534
An, Rong 314
Andersson, Cristina 600
Aschenbrenner, Rolf 84
Azarian, Michael H. 1

B

Bai, Shuo 563
Bailey, C. 509
Bailey, Chris 250
Bao, Shengxiang 1059
Bei, Liu 826
Bian, Lifei 703
Bin, Liu 183
Biswas, Kalyan 299
Bo, Tao 885
Buchwalter, S. 577
Busby, J. 577

C

Cai, Jian 121
Cai, Miao 856
Cai, Xiong-Hui 129
Chai, Tc 299
Changshun, Jiang 1055
Chen, Fei 456
Chen, Guoqin 748, 751
Chen, Haibin 319
Chen, Hsin-Yuan 1040
Chen, Jin 394
Chen, Jing 660
Chen, Lih-Shan 540
Chen, Mingxiang 152
Chen, Quayle 1086
Chen, Su 748
Chen, Wei 804
Chen, Wenqing 461
Chen, Xi 740, 744
Chen, Xin 788
Chen, Xuefan 889
Chen, Y. R. 968
Chen, Yu-Ren 290
Chen, Zhengrong 944
Chen, Zhi 558
Chen, Zhuo 117
Chena, Qian 125
Cheng, Dongfang 1017, 1077
Cheng, Hung-Hsiang 157, 204, 295
Cheng, Ting 464
Cheng, Ting 493
Cheng, Xingming 679
Cheng, Yuan-Yuan 337
Cheng, Zhaonian 192, 523, 600, 725

Cheng, Zhong 774
Chiang, Kuo-Ning 23, 279
Chiang, Shih-Ying 279
Chiu, Chi-Tsung 157
Chou, Chan-Yen 279
Chou, J. H. 268
Chu, Huabin 626
Chuan, Tang 720
Chuan-Pei, Xu 794
Chung, T.F. 134
Chung, Tom 28
Chunling, Ren 1055
Chunyue, Huang 1092
Chunyue, ... 431
Conway, Paul P. 57
Cui, C.Q. 134
Cui, Shan-Ling 804

D

Dan, Huang 713
Dang, B. 577
Deng, Guiling 438
Deng, Zongquan 751
Desmulliez, M.P.Y. 509
Dhiman, Harpreet S. 994
Ding, Dongyan 558, 563
Ding, Guifu 304, 699
Dong, Zhang 410
Dou, Xinyu 121
Du, Bin 964
Du, Chao 1021
Du, Ruxu 605
Du, Xinyu 679
Du, Zhibo 1059
Dudek, Rainer 145
Duh, Jenq Gong 873, 927

E

Ernst, L. J. 937, 271, 636

F

Fan, H. B. 285
Fan, Jing-Yu 192
Fan, Pingyue 905
Fan, Xuejun 496, 931, 994
Fan, Yanhua 609
Fan, Yi 523
Fang, Huajing 461
Fang, Qiang 1052
Fei, Qin 171, 178, 183, 613
Feng, Lin 415, 419
Feng, Ran 331, 344, 398, 483
Feng, Tao 121
Frémont, Hélène 636, 974

A-1

Author Index

Fu, Ran..707
Fu, Shen-Li...540
Fu, Xingming...949
Fu, Yifeng......................................523, 641
Fu, Yonggao..765

G

Gan, Zhiyin.............................464, 493, 667
Gao, Guohua..187
Gao, Jian...788
Gao, Jie..735
Gao, Wei...................................50, 211, 223
Gao, Xiang..558
Gao, Yulai...600
Gao, Ziyang...28
Garant, E. Perfecto J...............................577
Gautier, Christian.............................636, 974
Geng, Fei...438, 111, 165
Gruber, P...577
Gu, Xiaolong...758
Guan, Rongfeng......................................324
Gui-Ling, Deng......................................807
Guiling, Deng..813
Guo, J. J...848
Guo, J.D..863
Guo, Jianjun..758
Guo, X.M..544
Guo, Yunxia...256
Guoliang, Wang..72
Guoyuan, Li..720

H

Haihua, Jiang...................................1008, 729
Han, Ding...885
Han, Lei...106, 778
Han, Lei..798
Han, Qiang...931
Hao, Hu...807
Hao, Wang.......................................394, 774
Hao, Yilong..534
He, Jie..717
He, Yandong...534
He, Yanping...558
Hochstenbach, P......................................651
Hong, Zhao..468
Hongjun, Liu......................................72, 75
Hou, H. N...1047
Hou, Zhezhe..589
Hsieh, Ming-Liang....................................540
Hsu, C. L...968
Hsu, Hsiang-Chen.....................................540
Hsu, Kuo-Yuan....................................1040, 528
Hu, Anmin.......................................735, 740
Hu, Anming......................................117, 744
Hu, Dyi-Chung...23

Hu, Jun...626
Hu, L...1047
Hu, W. G...632
Hu, Zhili...544
Huang, Dejian..431
Huang, Liugang......................................1082
Huang, Louie......................................39, 572
Huang, Mian.............................379, 605, 609
Huang, Qing-An....................................80, 99
Huang, Qiuping.......................................111
Huang, Suyi.....................................464, 493
Hughlett, E...577
Hui, Dong...410
Hui, Shi..398, 424
Hung, Chih-Pin.......................................157
Hung, Mike.......................................39, 572

I

Inoue, Masahiro.......................................655

J

Jansen, K M. B...................................937, 636
Ji, Hongjun...592
Ji, Li-Na..867
Jia, Wang..415, 419
Jian, Zhang...468
Jian, Zhou..713
Jiang, Sijia...655
Jiang, Ting-Biao....................................1021
Jianqiangwang..589
Jiao, Li..344, 424
Jiao, Qinghua...754
Jide-Zhao,..449
Jie, Yingliang..626
Jillek, W...1026
Jin, Li..774
Jin, Ling...626
Jin, Xing...836
Jin, Yufeng.....................................256, 515, 534
Jinyi, Zhang...............398, 410, 415, 419, 424
Johan, Liu...75
Johnson, Mark...250
Jun, Cai..483
Jun, Hu...103
Junquan, Wang..813

K

Kai, Qiao...140
Kai, Zhou Yi...419
Kai-Lin, Pan.....................................308, 832
Kang, S...577
Kaulfersch, Eberhard.................................145
Ke, Li..161
Ke, Yan...371

Author Index

Kelly, Robert 788
Kettner, Paul 43
Kim, Bioh 43
Kim, Daewon 592
Kim, Hongbae 584
Kim, Jong-Hoon 841
Kim, Jongmyung 584, 592
Kim, Keun-Soo 841
Knauf, Benedikt J. 57
Knickerbocker, J. 577
Knickerbocker, S. 577
Kong, Lingwen 763
Kui, Dai 794
Kuo, C. Y. 968
Kwan, Kenneth 319
Kwon, Henri Hk. 121

L

Lai, Chi-Chang 295
Lai, Xiao-Wei 129
Lai, Yi-Shao 984, 988
Lam, Angus 319
Lauro, Paul 956
Le, Kriangsak Sae 260
Lee, Aching 39
Lee, Jaisung 567, 944, 949
Lee, P. H. 268
Lee, S. W. Ricky 503
Lee, Yong-Won 841
Lei, Chen 1063, 1067
Leng, Yi 551
Leu, Jihperng 1040, 528
Leung, Vincent Chi-Kuen 93
Li, Baoxia 50
Li, C Y 519
Li, Dou-Xing 617
Li, Fenghui 769
Li, Gongke 909
Li, Han 103
Li, Jun 211, 223
Li, Junhui 106
Li, Kejia 515
Li, Lei 227, 379, 605, 609
Li, Lin-Kai 717
Li, Ming 117, 375, 558, 563, 735, 740, 744, 836, 852
Li, Mingyu 584, 592
Li, Peng 896
Li, Qingxia 551
Li, Quan-Yong 856
Li, Ren 720
Li, Song 208
Li, Tiezhu 360, 364
Li, Xiaohui 1017, 1036, 1077
Li, Xiaoyan 1070, 1073, 765, 769
Li, Xueping 699

Li, Yan.. 563
Li, Yingliang.. 449
Li, Ying-Liang 452
Li, Yuanxun 703
Li, Zhihua 407
Li, Zhihua 50
Liang, Le... 597, 684
Liang, Lihua 197, 889
Liao, Cheng 223
Liao, Hongguang 256, 515
Liao, Xiao-Ping 327
Lijun, Yang.. 826
Lin, Bryan... 572
Lin, C. Y. 968
Lin, Changyong 152
Lin, Feng... 479
Lin, Han 379
Lin, Penrong 684
Lin, Tsung-Shu 1040
Lin, Tzu-Chih 157
Lin, Xiaoqin 1031
Lin, Y.C. 1082
Lina, Pengrong 597
Ling, Jin 103, 183
Ling, Li 826
Liu, Bin 852
Liu, Changqing 57
Liu, D. S. 968
Liu, Deming 707
Liu, Jianhui 763
Liu, Jing 235
Liu, Johan 192, 523, 544, 600, 641, 655, 725, 754
Liu, Junwen 99
Liu, Li 754
Liu, Lilin 707
Liu, Na 1070, 1073
Liu, Rui... 304, 699
Liu, Sheng152, 216, 243, 354, 386, 456, 464, 493, 551, 667
Liu, Shiguo 299
Liu, Wenming 152
Liu, Y. 632
Liu, Yang 783
Liu, Yingli 703
Liu, Zhen 717
Liu, Zhi-Quan 617, 629, 664
Liu, Zhongyuan 456
Liu, Zongyuan 243, 354, 386
Liusang, ... 1013
Loh, Wei Sun 250
Lou, Minyi 914, 944
Lu, Dafu 703
Lu, Dechun 1059
Lu, Hao 368
Lu, Hua 250
Lu, M. 577
Lu, Minfei 763

Author Index

Lu, Minhua 956
Lu, Sheng 660
Lu, Xiu Zhen.. 725
Lui, Tung-Chin 28
Luo, Fei 660
Luo, Le... 1031, 111, 165
Luo, Xiaobing........ 216, 243, 354, 386, 456, 464, 493, 667
Luo, Yi..34
Lv, Lei 778
Lv, Yao 46, 50

M

Ma, Lili 1059
Ma, Vivian Wei 93
Ma, Xiaosong... 636
Ma, Xiao-Song.. 729
Mao, Dali 117, 558, 563, 735, 740, 744, 852
Mao, Jiwang.. 740, 852
Mao, Shengping...304, 699
Maslyk, Dan...63
Mcelroy, J. B. 648
Mclellan, Neil 1003
Mei, Hujie 679
Meihua, Xu 161, 483
Meng, Songhe 748
Miao, Min 256
Michel, Bernd 145
Ming-Shan, Yang 717
Minliang, Zhang 72, 75
Muc, Xiu 667

N

Na, Chen 171, 613
Ni, Li... 308, 832
Ning, Chen.. 1008, 729
Nong, Hong-Mi.. 1021

O

Ou, Zhaohui... 479

P

Pan, Kailin 140
Pan, Kai-Lin... 896
Pan, Yingfeng... 551
Pang, Xue-Yong 629, 664
Pape, H.. 937
Pargfrieder, Stefan 43
Parrott, A.K... 509
Pearl, Agyakwa.. 250
Pecht, Michael 1
Pei, Li Pei 713
Peng, Chung-Nan... 927
Peng, Jin.. 882

Peng, Li 832
Peng, Yafang 1052
Pfahl, R. C. 648
Poh, Edith S. W. 671
Poirier, Patrick 974
Pun, Kelvin 134

Q

Qi, Fangjuan 589
Qi, Quan 66
Qian, Bin-Feng 231, 239
Qiang, Huang 1055
Qiao, Jiaping 819
Qin, Cha... 589
Qin, Liancheng 909
Qin, Ming 80
Qin, Suqiong.. 679
Qiu, Shen-Nan 340

R

Ran, Feng... 235, 360, 364, 403
Regard, Charles 636, 974
Ridout, Steve 250
Rizvi, J.. 509
Röllig, Mike... 145
Rongen, R.T.H... 271
Rongzheng, Ding 1055
Ruhmer, R. Weisman K... 577
Rzepka, Sven 145

S

Salo, Antti... 1086
Schlottig, G... 937
Schmitt, E. 1026
Semkow, K. 577
Sham, Man-Lung 28
Shang, J. K. 848, 863, 878
Shang, Jian-Ku 617, 629, 664
Shang, Jintang..99
Shang, Pan-Ju 617, 664
Shen, G. S. 290, 968
Shen, Guang-Ping 80
Shen, Xuwei 783
Shen, Yidong 744
Shi, Daniel Xun-Qing 93
Shi, Daniel 86
Shi, Ying.. 754
Shih, D.-Y. 577
Shih, Da-Yuan 956
Sinclair, K.I. 509
Song, Jing 327, 900, 99
Suganuma, Katsuaki 841
Sun, F. L 632, 651
Sun, Fenglian 783

A-4

Author Index

Sun, Li 589
Sun, Peng 86, 93
Sun, Rong 605, 609
Sun, Wei 260, 671
Sun, Yiqin 379
Sun, Yongqiao 319
Sun, Young 515
Sundlof, B. 577
Sun-Fujiang, ...1013

T

Tai, Yu-Che 295
Tan, H.B. 260, 671
Tan, Wei 679
Tang, Jia-Jie 165
Tang, Jieying 900, 99
Tang, Jun 304, 699
Tang, Zhi-Jie 331
Tao, Wenkai 600
Tao, Yuan 125
Te Gan, Richard 671
Teng, Hui 688
Tian, Dalei 324
Tian, Dewen 472
Tian, Xiao-Wei 621
Tian, Yanhong 314, 472, 684, 597
Tilford, T. 509
Tong, An 171, 178, 183, 613
Tsai, Mars 23
Tseng, Andy 572
Tseng, H. C. 268
Tsui, Anthony C. 688
Tu-Yunhua, ...1013

V

Van Der Sluis, O. 636, 980
Van Driel, W. D. 651, 980, 636
Van Soestbergen, M. 271

W

Wan, Lixi 211, 223, 227, 379, 407, 46, 50, 54, 605, 609
Wan, Xin 256
Wang, Baohua 660
Wang, Bo 964
Wang, Chen-Chao 157, 295
Wang, Ching-Chun 988
Wang, Chunqing 314, 472, 592, 684
Wang, Fuliang 819
Wang, Hong 304, 699
Wang, Honghui 187
Wang, Jiaji 905
Wang, Jiongming 1017, 1036, 1077, 337
Wang, Junquan 438
Wang, Kai Jheng 873

Wang, Kai 243, 354, 386, 456, 667
Wang, Lei 567
Wang, Li-Ding 34
Wang, Lifeng 783
Wang, Ling 765
Wang, Qian 121, 567, 944, 949
Wang, Shao-Qing 629
Wang, Te-Chun 428
Wang, Teng 523, 641
Wang, Vicky 63
Wang, Weiqiang 1
Wang, X.J. 878
Wang, Xiao-Dong 34
Wang, Xiaojing 523
Wang, Xi-Chuan 231
Wang, Xing 324
Wang, Yu 852
Wang, Yunfeng 227
Wang, Z. G. 848, 878
Wanga, Chunqing 597
Wangyao-Ming, ...382
Webb, D. Patrick 57
Wei, B. 519
Wei, De-Fang 452
Wei, Luo Xiao 424
Wen, Long 944, 949
Wen, Tao 375
Wen, Zhang 75
Weng, Xue-Tao 208
Wong, Cell K Y 692, 285
Wu, Dejian 43
Wu, Feng-Shun 125, 129, 914, 964
Wu, Gaohui 748, 751
Wu, Jian 360, 364
Wu, Jian 80
Wu, Jingshen 1003, 319
Wu, Li-Ming 804
Wu, Sung-Mao 295
Wu, Yanan 754
Wu, Yanhong 111
Wu, Yi-Ping 125, 129
Wu, Yiping 914, 964
Wunderle, B. 937

X

Xi, Hongjia 914
Xi, Yanyan 152
Xia, Y.H. 1026
Xia, Yangjian 197
Xiangguang, Xuan 161
Xiangzhao, ...1013
Xian-Wei, Rong 468
Xiao, A. 937
Xiaodong, Yang 410
Xiaojing, Wang 72, 75

Author Index

Xiao-Song, Ma 1008
Xie, Bin 86, 93
Xie, Guangchao 679
Xie, Jinghua 438
Xie, Xiaoqiang 567
Xin, Yibo 461
Xiong, Wei 216, 493
Xiu, Ziyang 748, 751
Xu, Gaowei 111
Xu, Hongbo 584
Xu, Leon 1086
Xu, Li 725
Xu, Liyuan 900
Xu, Meihua 235, 360, 364, 403
Xu, Mei-Hua 331
Xu, Po 117
Xuan, Xiangguang 403
Xuanxiang-Guang, ... 350, 382
Xue, Ke 319
Xuefeng, Shu 920, 923

Y

Yan, Guoguang 334
Yan, Li-Min 239, 340
Yan, Sun 415
Yan, T. L. 632
Yan, Yongchang 1070
Yan, Yong-Chang 1073
Yang, Chengyue 50
Yang, D. G. 909
Yang, Daoguo 140
Yang, Dao-Guo 856
Yang, Dian 344
Yang, Guoji 187
Yang, Hongbo 688
Yang, Le 327
Yang, Lianfa 140
Yang, Shihua 684
Yang, Shin-Yueh 279
Yang, Wang 826
Yang, Wen-Kung 23
Yang, Wenrong 1036
Yang, Wen-Rong 337
Yang, Xiaohua 769
Yang, Zhen-Guo 621, 867
Yang, Zhijun 788
Yang, Zhiqin 763
Yang, Zongbao 703
Yanga, Shihua 597
Yanhui, Jiang 415, 419
Yeh, Chang-Lin 204, 984, 988
Yew, Ming-Chih 23, 279
Ye-Xiang, Ning 308, 896
Ye-Yuming, ... 1013
Yi, Fuxi 909

Yi, Yang 410
Yiming, Jiang 774
Yin, C. 509
Yin, Guangfu 605
Yin, Yansheng 609
Yiping, Wu 885
Yngve, Wang 183
Yu, Chun 368
Yu, Chun-Fai 23
Yu, Hk 1052
Yu, Roy 13
Yu, Shuhui 605, 609
Yu, Zhang 923
Yu, Zhiyuan 379
Yuan, C. 980
Yue, Cong 192, 725
Yue-Li, Hu 371, 350, 382
Yuen, John 319
Yuen, Matthew M F. 692, 285
Yule, Fan 161
Yunxin, Wu 823
Yuxi, Jiang 344

Z

Zbrzezny, Adam R. 1003
Zeng, Lu 765
Zeng, Qiulian 758
Zhai, Qijie 600
Zhang, G. Q. 651, 980, 271, 636
Zhang, Hua 80
Zhang, Huaiwu 703
Zhang, Huiliang 688
Zhang, J H 519
Zhang, Jue 1017, 1036, 1077
Zhang, Jun 1082
Zhang, Kouchi 140
Zhang, Lili 600
Zhang, Li-Na 798
Zhang, Liqing 584
Zhang, Qin 667
Zhang, Rong 503
Zhang, Shu-Qiang 157, 204
Zhang, Tong-Yi 707
Zhang, X.F. 863
Zhang, Xiaowu 299
Zhang, Xu 50
Zhang, Yan 192, 523, 600, 725
Zhang, Yuanxiang 197
Zhang, Z. F. 848
Zhang, Zheming 1003
Zhang, Zhen-Qiang 34
Zhang, Zhikun 655, 754
Zhang, Zong-Bo 34
Zhangjin, Chen 394
Zhangyi-Chi, ... 350

Author Index

Zhanhui, Li	823
Zhao, Ji-De	452
Zhao, Liwei	256
Zhao, Wu	431
Zhao, Xiaolin	304
Zhao, Xinbing	758
Zhao, Zhenqing	567
Zhaohua, Zhou	443
Zhe, Wang	826
Zhendong, Zhao	923
Zheng, Xu	605
Zheng, Yin-Guang	204, 428
Zhigang, Li	920, 923
Zhili, Long	823
Zhong, Jue	106
Zhong, Li-Jun	856
Zhou, Hua	431
Zhou, Jian	111
Zhou, Jiang	487
Zhou, Jianwei	944, 949
Zhou, Ming	688
Zhou, Tiao	994
Zhouping, Yin	885
Zhu, Haiqing	187
Zhu, Q. S.	848
Zhu, Swen	43
Zhu, W.H.	260, 671
Zhu, Zhi-Jun	125
Zou, Changhui	819

A-8

9781424427390

2008 International Conference on Electronic Packaging Technology & High Density Packaging

Shanghai, China
28-31 July 2008

IEEE Catalog Number: CFP08553-POD
ISBN: 978-1-42442-739-0

2008 International Conference on Electronic Packaging Technology & High Density Packaging

Shanghai, China
28-31 July 2008

Pages 551-1095

IEEE Catalog Number: CFP08553-PRT
ISBN 13:   978-1-4244-2739-0

Copyright © 2008 by The Institute of Electrical and Electronics Engineers, Inc.
All Rights Reserved

Copyright and Reprint Permissions: Abstracting is permitted with credit to the source. Libraries are permitted to photocopy beyond the limit of U.S. copyright law for private use of patrons those articles in this volume that carry a code at the bottom of the first page, provided the per-copy fee indicated in the code is paid through Copyright Clearance Center, 222 Rosewood Drive, Danvers, MA 01923.

For other copying, reprint or republications permission, write to IEEE Copyrights Manager, IEEE Operations Center, 445 Hoes Lane, Piscataway, New Jersey USA 08854. All rights reserved.

IEEE Catalog Number: CFP08553-PRT

ISBN 13:   978-1-4244-2739-0

LOC: 2008906231

Additional Copies of This Publication Are Available from:

IEEE Service Center
445 Hoes Lane
Piscataway, NJ 08854
Phone: (800) 678-IEEE
    (732) 981-1393
Fax: (732) 981-9667
E-mail: customer-service@ieee.org

Directed by
China Institute of Electronics, China
Science & Technology Department, Ministry of Education, China
Foreign Affairs Office of Shanghai Municipality, China
Department of High and New Technology, Development & Industrialization, Ministry of Science and Technology, China
Ministry of Industry and Information Technology, China
China International Culture Exchange Center

Sponsored by
China Electronics Packaging Society, of China Institute of Electronics, China (CEPS)
Fudan University, China (Fudan University)
The Component, Packaging, and Manufacturing Technology Society of IEEE (IEEE-CPMT)
The International Microelectronics and Packaging Society (IMAPS)

Organized by
China Electronics Packaging Society of China Institute of Electronics (CEPS)
Shanghai Integrated Circuit Industry Association (SCIA), China
Fudan University, China (Fudan University)
Beijing Faith Consulting Co., Ltd.

Co-organized by
Hong Kong Applied Science & Technology Research Institute (ASTRI), China
Shanghai Zhangjiang Group Co., Ltd, China
Shanghai Zhangjiang Hi-Tech Park Management Committee
Shanghai Pudong New Area Science & Technology Association, China

Editors
Keyun Bi
China Semiconductor Industrial Association

Fei Xiao
Department of Materials Science, Fudan University, Shanghai, 200433, China

Table of Contents

Qualification for Product Development 1
Weiqiang Wang, Michael H. Azarian and Michael Pecht

High Density 3D Integration 13
Roy Yu

A Study of Thermal Performance for the Panel Base Package (PBPTM) Technology 23
Ming-Chih Yew, Chun-Fai Yu, Mars Tsai, Dyi-Chung Hu, Wen-Kung Yang and Kuo-Ning Chiang

FMEA of System-in-Package (SiP) -based Tire Pressure Monitoring System 28
Man-Lung Sham, Tung-Chin Lui, Ziyang Gao, Tom Chung

Ultrasonic Bonding of Polymer Microfluidic Chips 34
Zong-bo Zhang, Yi Luo, Xiao-Dong Wang,, Zhen-Qiang Zhang, Li-Ding Wang

Mold Array Package for POP Applications 39
Aching Lee, Louie Huang, Mike Hung

New Technologies for advanced high density 3D packaging by using TSV process 43
Paul Kettner, Bioh Kim, Stefan Pargfrieder, Swen Zhu

Recent Progress of Ohmic Contact on ZnO 46
Yao Lv, Lixi Wan

Low-cost High-efficiency 4 Channel Pluggable Parallel Optical Transceiver Using Optoelectronic MCM Packaging Technologies 50
Baoxia Li, Lixi Wan, Chengyue Yang, Wei Gao, Yao Lv, Zhihua Li, and Xu Zhang

SiP/SoP Technology and Its Implementation 54
Lixi Wan

Packaging of Polymer Based Microfluidic Systems Using Low Frequency Induction Heating (LFIH) 57
Benedikt J. Knauf, D. Patrick Webb, Changqing Liu, Paul P. Conway

Analysis of the Reliability of Package-on-Package Devices Manufactured Using Various Underfill Methods 63
Vicky Wang and Dan Maslyk

Meeting Thermal Performance and Reliability Challenges for a Thermally Enhanced Ball Grid Array Package (TEBGA 66
Quan Qi

Numerical Simulation of the Micro-channel Heat Sink on Non-uniform Heat Source 72
Zhang Minliang, Wang Xiaojing, Liu Hongjun, Wang Guoliang

Heat Transfer Simulation of Nanofluids in Micro Channel Cooler 75
Liu Hongjun, Wang Xiaojing, Zhang Minliang, Zhang Wen, Liu Johan

Flip-chip on Board packaging of a Thermal Wind Sensor 80
Guang-ping Shen, Ming Qin, Qing-An Huang, Hua Zhang, Jian Wu

System-in-Package Solutions with Embedded Active and Passive Components 84
Rolf Aschenbrenner

Design Advisor for Package-on-Package (PoP) Manufacturing 86
Bin Xie, Peng Sun and Daniel Shi

Warpage Reduction of Package-on-Package (PoP) Module by Material Selection & Process Optimization 93
Peng Sun, Vincent Chi-Kuen Leung, Bin Xie, Vivian Wei Ma, Daniel Xun-Qing Shi

A New Process to Fabricate Cavities in Pyrex7740 Glass for High Density Packaging of Micro- System 99
Junwen Liu, Qing-an Huang, Jintang Shang, Jing Song, and Jieying Tang

DPA Tests on SiP Device 103
Jin Ling, Hu Jun, Han Li

Table of Contents

Ultrasonic Features in Wire Bonding and Thermosonic Flip Chip..106
Junhui Li, Lei Han, Jue Zhong

Development of Three-dimensional Multichip Module Based on Embedded Substrate with Multiple Interconnections...111
Gaowei Xu, Yanhong Wu, Fei Geng, Qiuping Huang, Jian Zhou, Le Luo

Novel Low-Temperature Micro-insert Bonding Technology for 3D Package.....................................117
Po Xu, Anming Hu, Zhuo Chen, Ming Li, Dali Mao

High-Q On-Chip Inductors Embedded in Wafer-Level Package for RFIC Applications......................121
Tao Feng, Jian Cai, Henri HK Kwon, Qian Wang, Xinyu Dou

Thermo and Mechanical Study on Integrated High-density Packaging...125
Qian Chen, Yi-ping Wu, Feng-shun Wu, Zhi-jun Zhu, Yuan Tao

Research on the Contact Resistance and Reliability of Flexible RFID Tag Inlays Packaged by Anisotropic Conductive Paste..129
Xiong-hui Cai, Bing An, Yi-ping Wu, Feng-shun Wu, Xiao-wei Lai

Ultra-Fine Via Pitch on Flexible Substrate for High Density Interconnect (HDI).............................134
Kelvin Pun, C.Q. Cui, T.F. Chung

Introduction of Microelectronics Manufacturing Engineering into Professional Education: a Joint Effort among Industry, Government and Universities...140
Kailin Pan, Daoguo Yang, Qiao Kai, Kouchi Zhang, Lianfa Yang

FEA Based Reliability Prediction for Different Sn-Based Solders Subjected to Fast Shear and Fatigue Loadings..145
Rainer Dudek, Eberhard Kaulfersch, Sven Rzepka, Mike Röllig, Bernd Michel

Thermo-Mechanical Analysis of a Wafer Level Packaging by Induction Heating.............................152
Wenming Liu, Mingxiang Chen, Yanyan Xi, Changyong Lin, Sheng Liu,

Synthesizing SPICE-Compatible Models of Power Delivery Networks with Resonance Effect by Time-domain Waveforms...157
Chen-Chao Wang, Shu-Qiang Zhang, Hung-Hsiang Cheng, Tzu-Chih Lin, Chi-Tsung Chiu, and Chih-Pin Hung

Optimization of CAVLC Algorithm and Its FPGA Implementation...161
Xu Meihua, Li Ke, Xuan Xiangguang, Fan Yule

Thermal Management and testing of MCM with embedded chip in Silicon Substrate.......................165
Fei Geng, Jia-jie Tang, Le Luo

Strain rate effect and Johnson-Cook models of lead-free solder alloys..171
Qin Fei, An Tong, Chen Na

A Scale Reduced Computation Scheme for Peeling Stress of Solder Joints under Drop Impact.........178
An Tong, Qin Fei

Dynamic Bending Tests and Numerical Simulation of Board Level Electronic Package...................183
Qin Fei, Wang Yngve, Liu Bin, An Tong, Jin Ling

A New Method for The Investigation of Strip Warpage of MAP-QFN...187
Guohua Gao, Honghui Wang, Guoji Yang, Haiqing Zhu

Numerical Investigation on the Effect of Filler Distribution on Effective Thermal Conductivity of Thermal Interface Material...192
Cong Yue, Yan Zhang, Johan Liu,, Zhaonian Cheng, Jing-yu Fan

Development of Moisture Automation Analysis System for Microelectronic Packaging Structures......197
Yangjian Xia, Yuanxiang Zhang, Lihua Liang

High Speed Package Design and Electrical Performance Analysis..204
Shu-Qiang Zhang, Hung-Hsiang Cheng, Yin-Guang Zheng, Chang-Lin Yeh

Table of Contents

Methodology for Modeling Simultaneous Switching Noise in BGA Packaging 208
Song Li, Xue-tao Weng

A Lowpass Filter with An Embedded Capacitor for Wideband Noise Suppression in Multi-GHz PCBs 211
Wei Gao, Lixi Wan, Jun Li

A Simplified Thermal Resistance Network Model for High Power LED Street Lamp 216
Xiaobing Luo, Wei Xiong, Sheng Liu

Improvement of Power Integrity with novel Segmented Power Bus Structures in RF/Digital SOP 223
Jun Li, Lixi Wan, Wei Gao, Cheng Liao

Study of RF Front-End Filters with Embedded Capacitor Technology 227
Yunfeng Wang, Lixi Wan, Lei Li

The Design of the Cache Crossbar based on OpenSPRAC Architecture 231
Xi-chuan Wang, Bin-Feng Qian

Interpolating Algorithm Optimization and FPGA Implementation in Image Scaling Engine 235
Feng Ran, Jing Liu, Meihua Xu

The Research of the Inclusive Cache used in Multi-Core Processor 239
Bin-feng Qian, Li-Min Yan

Effects of Phosphor's Location on LED Packaging Performance 243
Zongyuan Liu, Sheng Liu, Kai Wang, Xiaobing Luo

Computer Simulation of Crack Propagation in Power Electronics Module Solder Joints 250
Hua Lu, Steve Ridout, Chris Bailey, Wei Sun Loh, Agyakwa Pearl, and Mark Johnson

Process Simulation of DRIE and its Application in Tapered TSV Fabrication 256
Min Miao, Hongguang Liao, Xin Wan, Liwei Zhao, Yunxia Guo, Yufeng Jin

**Simulation Study on the Warpage Behavior and Board-level Temperature Cycling Reliability of PoP
Potentially for High-speed Memory Packaging** 260
Wei Sun, W.H. Zhu, Kriangsak Sae Le and H.B. Tan

**Miniaturization Design of Backside-Via Structures Underneath Collector-Up HBTs Using A 3-D Finite-
Element Model** 268
H. C. Tseng, P. H. Lee, and J. H. Chou

Modeling Ion Transport through Molding Compounds and its Relation to Product Reliability 271
M. van Soestbergen, R.T.H. Rongen, L.J. Ernst and G.Q. Zhang,

**Reliability Analysis of Copper Interconnections of System-in-Packaging Structure using Finite Element
Method** 279
Shih-Ying Chiang, Shin-Yueh Yang, Chan-Yen Chou, Ming-Chih Yew, and Kuo-Ning Chiang

A Multi-scale Interfacial Delamination Model of Cu-SAM-Epoxy Systems 285
H. B. Fan, Cell K.Y. Wong, Matthew M.F. Yuen

Three Dimensional Corner Delamination Analysis for Fan-Out Chip Scale Package 290
Yu-Ren Chen, G. S. Shen

**Frequency Dielectric Constant and Loss Tangent Extracting of Organic Material Using Multi-length
Microstrip** 295
Sung-Mao Wu, Chi-Chang Lai, Hung-Hsiang Cheng, Yu-Che Tai, Chen-Chao Wang

**Thermo-mechanical Design of Large Die Fine Pitch Copper/Low-k FCBGA and Lead-free
Interconnections** 299
Kalyan Biswas, Shiguo Liu, Xiaowu Zhang, TC Chai

Study on Non-Uniformity of Through-Mask Electroplated Ni Thin-Film 304
Jun Tang, Hong Wang, Rui Liu, Shengping Mao, Xiaolin Zhao, Guifu Ding

vi

Table of Contents

Simulation and Analysis for Backward Compatibility of Solder Joints under Thermal Cycle 308
Ning Ye-xiang, Pan kai-lin, Li Ni

First-Principles Study on the Elastic Anisotropy of Au-Sn Intermetallic Compounds 314
Rong An, Chunqing Wang, Yanhong Tian

Numerical Analysis of Interfacial Delamination in Thin Array Plastic Package 319
Ke Xue, Jingshen Wu, Haibin Chen, Yongqiao Sun, Kenneth Kwan ,John Yuen, Angus Lam

Design and Implementation of LED Daylight Lamp Lighting System 324
Rongfeng Guan, Dalei Tian, Xing Wang

Effect of Bonding on the Packaged RF MEMS Switch 327
Le Yang, Xiao-Ping Liao, Jing Song

Genetic Algorithm Optimization in TFT-LCD Drive System 331
Zhi-Jie Tang Ran Feng Mei-Hua Xu

The design of the Ku band Dielectric Resonator Oscillator 334
Guoguang Yan

Simulation of Multi-bit Digital Delta-Sigma Modulator 337
Wen-Rong Yang, Yuan-Yuan Cheng, Jiong-ming Wang

The Research of Sub-space Partition Strategy and Bit Scanning Control Method Based on Human's Vision Nonlinearity Rule 340
Li-min Yan, Shen-nan Qiu

Layout Optimization and Modeling of an ESD-protection n-MOSFET in 0.13um Silicide CMOS Technology 344
Jiang Yuxi, Li Jiao, Ran Feng and Dian Yang

A Verification Method Based On A Mixed-Signal System For MV06 350
HuYue-li, ZhangYi-chi, XuanXiang-guang

Effects of Phosphor's Thickness and Concentration on Performance of White LEDs 354
Zongyuan Liu, Sheng Liu, Kai Wang, Xiaobing Luo

Digital Dimmable Controller in CCFL Module based on Variable Frequency Technique 360
Feng Ran, Tiezhu Li, Meihua Xu, Jian Wu

Design of a Low Voltage Band-gap Reference Circuit for OLED-On-Silicon 364
Meihua Xu, Jian Wu, Feng Ran, Tiezhu Li

Effects of Cu on the electromigration behavior of Al interconnect by using First-Principles method 368
Chun Yu, Hao Lu

A Mixed- Signal Physical Design and Its Verification 371
Hu Yue-li, Yan Ke

Hardware /Software Co-design for Viterbi Decoder 375
Ming Li, Tao Wen

Attenuators Using Thin Film Resistors for RF Application 379
Yiqin Sun,Lei Li, Han Lin, Zhiyuan Yu, Mian Huang, Lixi Wan

Task scheduling and management in single-chip multi-processor system 382
HuYue-li, WangYao-ming, XuanXiang-guang

Analysis of Factors Affecting Color Distribution of White LEDs 386
Zongyuan Liu, Sheng Liu, Kai Wang, Xiaobing Luo

Development of OLED demonstration system based on SD card 394
Chen Zhangjin, Jin Chen, Wang Hao

Table of Contents

Combinational Test Generation for Transition Faults in Acyclic Sequential Circuits.................................398
Shi Hui, Ran Feng, Zhang Jinyi

Design of a CMOS Charge Pump for high-performance phase-locked loop..................................403
Xiangguang Xuan, Feng Ran, Meihua Xu

Performance Simulation for EVPD with Equivalent Circuit Models 407
Zhihua Li, Lixi Wan

A March-CL Test for Interconnection Faults of SOC..................................410
Zhang Jinyi, Yang Xiaodong, Yang Yi, Zhang Dong, Dong hui

Multi-clock SOC Test schedule based on TWC&S..................................415
Zhang Jinyi, Jiang Yanhui, Lin Feng, Wang Jia, Sun yan

Research on the Characteristics Theory of Reverse SoC TAM Design Based on Dual-Balanced Strategy..................419
Zhang Jinyi, Wang Jia, Lin Feng, Jiang Yanhui, Zhou yi kai

Optimization of hierarchical SOC test time based on genetic algorithm..................................424
Li Jiao, Zhang Jinyi, Shi Hui, Luo xiao wei

Effective Dielectric Constant Method for Trace Impedance Control 428
Te-Chun Wang and Yin-Guang Zheng

Study on No-fillet SMT Solder Joint Reliability Based on Solder Joint Shape CAD 431
Wu Zhao, Hua Zhou, DeJian Huang, ChunYue

Simulation and Experiment Study of Dispensing Patterns Influence on Underfill Filling Process..................438
Jinghua Xie, Guiling Deng, Fei Geng, Junquan Wang

Study on Thermal Simulation Technology for SMA in Lead-free Reflow Soldering 443
Wu Zhaohua Zhou Dejian

A Novel High Effective Envelope-tracking Amplifier for OFDM Systems..................................449
Yingliang Li; Jide-Zhao

A Dual-band MEMS PA Study for Mobile Communication Systems..................................452
Ji-de Zhao, Ying-liang Li, De-fang Wei

Optical Analysis of A 3W Light-Emitting Diode (LED) MR16 Lamp..................................456
Kai Wang, Sheng Liu, Xiaobing Luo, Zhongyuan Liu, and Fei Chen

Research on the Cascaded Inverters Based on Simplex DC Power Source..................................461
Yibo Xin, wenqing Chen, huajing Fang

Simulation on Thermal Characteristics of LED Chips for Design Optimization 464
Ting Cheng, Xiaobing Luo,, Suyi Huang, Zhiyin Gan, Sheng Liu

Application of BP Neural Network in FBG Sensing System Performance Improvement..................................468
Zhang Jiana, Zhao Hongb ,Rong Xian-weia

Numerical Simulation of Solder Spreading and Solidification during Solder Jet Bumping Process 472
Dewen Tian, Chunqing Wang, Yanhong Tian

Research of Design-for-Testability of CMOS Image Sensor..................................479
Zhaohui Ou, Feng Lin

A 10-bit 40MSPS Pipeline Analog-to-Digital Converter 483
Cai Jun, Ran Feng, Xu Meihua

Analytical Analysis on the Effect of Time Duration of Acceleration Pulse to a JEDEC Board in Drop Test 487
Jiang Zhou

Thermal Management of A Multi-core Master Processing Unit (MPU) for An Ultrascalable Computing Platform..................................493
Ting Cheng, Wei Xiong, Xiaobing Luo, Suyi Huang, Zhiyin Gan, Sheng Liu

viii

Table of Contents

Investigation of Thermal Performance of Various Power-Device Packages 496
Xuejun Fan,

Wafer Level LED Packaging with Integrated DRIE Trenches for Encapsulation 503
Rong Zhang and S. W. Ricky Lee

Impact of Assembly Process Technologies on Electronic Packaging Materials 509
T. Tilford, C. Bailey, A.K. Parrott, J. Rizvi, C. Yin, K.I. Sinclair and M.P.Y. Desmulliez

Board Level Assembly of Inertial MEMS Devices Based on Surface Mounting Technology 515
Kejia Li, Young Sun, Hongguang Liao, Yufeng Jin

Encapsulation of Organic Light-emitting Devices for the Application of Display 519
C Y Li, B Wei and J H Zhang,

A Study of Fluid Coolant with Carbon Nanotube Suspension for Microchannel Coolers 523
Yi Fan, Yifeng Fu, Yan Zhang, Teng Wang, Xiaojing Wang, Zhaonian Cheng and Johan Liu,

Novel Pore-sealing Technology in the Preparation of Low-k Underfill Materials for RF Applications 528
Kuo-Yuan Hsu and Jihperng Leu

A Reliability Investigation of MEMS Transducers with Comb Structures 534
Ping An, Yandong He, Yufeng Jin, Yilong Hao

Low Temperature Sinterable (Zn,Mg)TiO3 Microwave Dielectrics 540
Lih-Shan Chen, Ming-Liang Hsieh, Hsiang-Chen Hsu and Shen-Li Fu

Vibration and Buckling of a Carbon Nanotube Inserted with a Carbon Chain 544
Zhili Hu, X.M.Guo, Johan Liu

Analysis of Electromagnetic Wave Propagation Characteristics in Rotating Environments 551
Yi Leng, Yingfeng Pan, Qingxia Li, Sheng Liu

Electrodeposition of Palladium Films on Ni-Co Coatings 558
Yanping He, Dongyan Ding, Xiang Gao, Zhi Chen, Ming Li, Dali Mao

Titania Nanostructures Fabricated Through Anodization of Ti6Al4V Alloy 563
Yan Li, Dongyan Ding, Shuo Bai, Ming Li, Dali Mao

The Influence of Low level Doping of Ni on the Microstructure and Reliability of SAC Solder Joint 567
Zhenqing Zhao, Lei Wang, Xiaoqiang Xie, Qian Wang, Jaisung Lee

BGA Assembly Process Development for 45nm ELK CUP Devices 572
Andy Tseng, Bryan Lin, Louie Huang, Mike Hung

C4NP for Pb-Free Solder Wafer Bumping and 3D Fine-Pitch Applications 577
D.-Y. Shih, B. Dang, P. Gruber, M. Lu, S. Kang, S. Buchwalter, J. Knickerbocker, E. Perfecto J. Garant,
S. Knickerbocker, K. Semkow, B. Sundlof, J. Busby, R. Weisman K. Ruhmer and E. Hughlett

Manufacture of Hourglass-shaped Solder Joint by Induction Heating Reflow 584
Hongbo Xu, Mingyu Li, Liqing Zhang, Jongmyung Kim and Hongbae Kim

The Effects of Ni Nanoparticles Addition on Shear Behavior and Microstructure of Sn-Ag Lead-free Solder 589
Fangjuan Qi, Li Sun, Zhezhe Hou, JianqiangWang, Cha Qin

Nanoscale Analysis of Ultrasonic Wedge Bond Interface by Using High-Resolution Transmission Electron Microscopy 592
Hongjun Ji, Mingyu Li, Chunqing Wang, Jongmyung Kim and Daewon Kim

Limited ß-Sn grain number of miniaturized Sn-Ag-Cu solder joints 597
Shihua Yang, Chunqing Wang, Yanhong Tian, Pengrong Lin, Le Liang

Table of Contents

Manufacture, Microstructure and Microhardness Analysis of Sn-Bi Lead-Free Solder Reinforced with Sn-Ag-Cu Nano-particles 600
Lili Zhang, Wenkai Tao, Johan Liu, Yan Zhang, Zhaonian Cheng, Cristina Andersson, Yulai Gao and Qijie Zhai

A comparison study of two different methods to synthesize magnetic slurry for the fabrication of magnetic films 605
Xu Zheng, Rong Sun, Shuhui Yu, Lei Li, Mian Huang, Guangfu Yin, Ruxu Du, Lixi Wan

Microstructure and Properties of Barium Strontium Titanate Thin Films Prepared on Copper Foils via Addition of PEG to the Sol Precursor 609
Yanhua Fan, Shuhui Yu, Rong Sun, Lei Li, Mian Huang, Yansheng Yin, Lixi Wan

Dynamic Behavior Tests of Lead-free Solders at High Strain Rates by the SHPB Technique 613
Qin Fei, An Tong, Chen Na

Formation of Double-layer Cu3Sn in Solid-State Aging Process at the Interface of Eutectic SnBi Solder and (100) Single Crystal Cu 617
Pan-Ju Shang, Zhi-Quan Liu, Dou-Xing Li, and Jian-Ku Shang

Synthesis of High Purity O-cresol Novolac Epoxy Resins 621
Xiao-Wei Tian, Zhen-Guo Yang

Defect Analysis of Copper Ball Bonding 626
Huabin Chu, Jun Hu, Ling Jin, Yingliang Jie

Fundamental Influence of Segregated Bi on the Mechanical Properties of Interconnect of Bismuth-containing Solder and Copper 629
Xue-Yong Pang, Zhi-Quan Liu, Shao-Qing Wang, and Jian-Ku Shang

Effects of Bi and Ni Addition on Wettability and Melting Point of Sn-0.3Ag-0.7Cu Low-Ag Pb-free Solder 632
Y. Liu, F. L Sun, T. L.Yan, W. G. Hu

Moisture Diffusion Model Verification of Packaging Materials 636
Xiaosong Ma, K.M.B. Jansen, L.J. Ernst, W.D van Driel, O. van der Sluis, G.Q.Zhang, Charles Regard, Christian Gautier, Hélène Frémont

Recent Progress of Carbon Nanotubes as Cooling Fins in Electronic Packaging 641
Johan Liu, Yifeng Fu, and Teng Wang

Environmentally Friendly Electronics for High Reliability 648
J. B. McElroy, R. C. Pfahl

Morphology, Evolution and Performance of IMC in SAC105 Solder/UBM (Ni (P)-Au) 651
F. L. Sun, P Hochstenbach, W. D. Van Driel , G. Q. Zhang

Development of High Temperature Stable Isotropic Conductive Adhesives 655
Zhikun Zhang, Sijia Jiang, Johan Liu, and Masahiro Inoue

Microstructural and physical characteristics of Sn-Ag-Cu-Mg Lead-free Solders 660
Sheng Lu, Fei Luo, Jing Chen, Baohua Wang

Microstructural Investigation on the Interfacial Evolution of SnBi/Cu Interconnect during Reflow and Solid-State Aging 664
Zhi-Quan Liu, Pan-Ju Shang, Xue-Yong Pang, and Jian-Ku Shang

Dynamic Mechanical Properties of the Transparent Silicone Resin for High Power LED Packaging 667
Qin Zhang, Xiu Mu, Kai Wang, Zhiyin Gan, Xiaobing Luo and Sheng Liua

Study of Five Substrate Pad Finishes for the Co-design of Solder Joint Reliability under Board-level Drop and Temperature Cycling Test Conditions 671
Wei Sun, W.H. Zhu, Edith S. W. Poh, H.B. Tan and Richard Te Gan

The Influence of Heat Treatment on the Adhesion between Molding Compound and Lead Frame 679
Wei Tan, Xinyu Du, Guangchao Xie, Suqiong Qin, Hujie Mei, XingMing Cheng

Table of Contents

Shear Fracture Behavior of Sn3.0Ag0.5Cu Solder joints on Cu Pads with Different Solder Volumes..........................684
Yanhong Tian, Chunqing Wang, Shihua Yang, Penrong Lin, Le Liang

Effect of Moisture and Temperature on Al-Cu Interfacial Strength 688
Hui Teng, Huiliang Zhang, Hongbo Yang, Ming Zhou, Anthony C. Tsui

Thiol based chemical treatment as adhesion promoter for Cu-epoxy interface..........................692
Cell K Y Wong and Matthew M F Yuen

Design of Testing Chip for Measuring Mechanical Properties of Thin Films..........................699
Rui Liu, Hong Wang, Xueping Li, Jun Tang, Shengping Mao, Guifu Ding

The Design and Fabrication of RF Band Pass Filter by LTCC Technology 703
Yuanxun Li, Yingli Liu, Huaiwu Zhang, Dafu Lu, Lifei Bian, Zongbao Yang

Cu Out-Diffusion Kinetics in Pre-plated Cu-alloy Leadframes Investigated by a Developed EDX-based Oxidation Test..........................707
Lilin Liu, Ran Fu, Deming Liu, Tong-Yi Zhang

Corrosion Performance of Pb-free Sn-Zn Solders in Salt Spray..........................713
Huang Dan, Zhou Jian, Li Pei Pei

The Synthesis and Curing Kinetics of the Silicon-containing Epoxy Resin with Environmentally Friendship Flame Retardance 717
Ming-shan, Yang, Jie He, Lin-kai Li, Zhen Liu

Nonlinear Optical Fluorinated Polyimide/Inorganic Composites for Optical Interconnections and Devices..........................720
Li Guoyuan, Tang Chuan and Ren Li

Nano-Thermal Interface Material with CNT Nano-Particles For Heat Dissipation Application..........................725
Li Xu, Cong Yue, Johan Liu, Yan Zhang, Xiu Zhen Lu and Zhaonian Cheng

The Humidity and Thermal Characteristics of Die-Attach (DA) and its Impact on the Package Reliability 729
Chen Ning, Ma Xiao-song, Jiang Haihua

Influence of Crystal Orientation on the Oxidation Failure of Copper for IC Package 735
Jie Gao, Anmin Hu, Ming Li, Dali Mao

The Influence of Pre-heat Treatment on Peeling Resistance of Oxide Film of Copper Alloy Lead Frames..........................740
Jiwang Mao, Xi, Chen, Anmin Hu, Ming Li, Dali Mao

Interfacial Reactions and Reliability of Sn-Zn-Bi-XCr Solder Joints with Cu Pads 744
Yidong Shen, Anming Hu, Xi Chen, Ming Li, Dali Mao

The Thermal and Electric Conduction Properties of High Dense TiB2P/Cu Composites Fabricated by Squeeze Casting Technology..........................748
Guoqin Chen, Ziyang Xiu, Songhe Meng, Gaohui Wu, Su Chen

Microstructure and Properties of Environmental-friendly Sip/1199Al Composites Used for Electronic Packaging 751
Ziyang Xiu, Guoqin Chen, Zongquan Deng, Gaohui Wu

Preparation of Polysulfoneamide Electrospinning Nanofibers..........................754
Li Liu, Ying Shi, Qinghua Jiao, Yanan Wu, Zhikun Zhang, Johan Liu

Liquid-State Interfacial Reaction of Sn-10Sb-5Cu High Temperature Lead-free Solder and Cu Substrate 758
Qiulian Zeng, Jianjun Guo, Xiaolong Gu, Xinbing Zhao

Application of Embedded Components in Package Substrate..........................763
Lingwen Kong, Zhiqin Yang, Jianhui Liu, Minfei Lu

Roadmap of Reliability: Methodology and Application in Guangdong's Lead-Free Technology..........................765
Xiaoyan Li, Ling Wang, Yonggao Fu, Lu Zeng

Table of Contents

Effect of Isothermal Aging on Interfacial IMC Growth and Fracture Behavior of SnAgCu/Cu Soldered Joints 769
Xiaoyan Li, Xiaohua Yang, Fenghui Li

Electrochemical Corrosion Behaviors of ITO Films at Anodic and Cathodic Polarization in Sodium Hydroxide Solution 774
Wang Hao, Zhong Cheng, Li Jin, Jiang Yiming

Effects of Bonding Pressure on Nonlinear Dynamic Characteristic of the Ultrasonic Wire Bonding System 778
Lei Lv, Lei Han

Effects of Ni addition on Microstructure and the Shear Strength of Sn-3.0Ag-0.5Cu/Cu Solders Joints 783
Lifeng Wang, Xuwei Shen, Fenglian Sun, Yang Liu

An Investigation of Capillary Vibration during Wire Bonding Process 788
Jian Gao, Robert Kelly, Zhijun Yang, Xin Chen

The Optimization of Hierarchical SOC Test Architecture to Reduce Test Time 794
Xu Chuan-pei, Dai Kui

The Influence of Heating Temperature on Alignment Precision in Thermosonic Flip-chip Bonding 798
Li-na Zhang, Lei Han

A High Speed Image Preprocessing Method for IC Wafer Inspection 804
Wei Chen, Li-Ming Wu, Shan-Ling Cui

The Influence Discipline of Temperature of High Viscosity Fluid Jetting 807
Hu Hao, Deng Gui-ling

The Influence of Structural Parameters of Electromagnetic Fluid Jetting Dispenser 813
Wang junquan, Deng Guiling

Dynamic Phase-frequency Characteristic of Thermosonic Wire Bonder Transducer 819
Fuliang Wang, Changhui Zou, Jiaping Qiao

Study of Prepress Force on Piezoelectric Transducer of Wire Bonding 823
Li zhanhui, Wu yunxin, Long zhili

Simulation and Experimental Research on Water-jet guided Laser Cutting Silicon Wafer 826
Wang Yang, Li Ling, Yang Lijun, Liu Bei, Wang Zhe

Research on Solder Joint Intelligent Optical Inspection Analysis 832
Li Ni, Pan Kai-lin, Li Peng

Productivity Improvement of Stack Package Line through Die Bonding Process & Scheme Optimization 836
Xing Jin, Ming Li

Developing the Stencil Printing Process for 01005 Lead-Free Assemblies 841
Yong-Won Lee, Keun-Soo Kim, Katsuaki Suganuma and Jong-Hoon Kim

Shear of Sn-3.8Ag-0.7Cu Solder Balls on Electrodeposited FeNi Layer 848
Q. S. Zhu, J. J. Guo, Z. G. Wang, Z. F. Zhang and J. K. Shang

IMC Formation between Electroless Ni/Pd/Au Surface Finish and SnAgCu Solder 852
Jiwang Mao, Bin Liu, Ming Li, Yu Wang, Dali Mao

Optimization of the Fatigue Life of Epoxy Molding Compounds based on BP Neural Network Prediction Model 856
Miao Cai, Dao-guo Yang, Quan-yong Li, Li-jun Zhong

Electromigration Behavior of the Ni/SnZn/Cu Solder Interconnect 863
X.F.Zhang, J.D.Guo, J.K.Shang,

Analysis on Cracking Blind Vias of PCB for Mobile Phones 867
Li-Na Ji, and Zhen-Guo Yang

xii

Table of Contents

Interfacial Reaction and Failure Mechanism for SnAgCu Solder Bump with Ni(V)/Cu Under Bump Metallization During Aging 873
Kai Jheng Wang and Jenq Gong Duh

Effect of displacement rate on lap shear test of SAC solder ball joints 878
X.J. Wang, Z.G. Wang, J.K. Shang

Capability Study on the Destructive Pull Test of 1 Mil Gold Wire Bond and Its Asymmetric Distribution 882
Jin Peng

Mechanical Reliability Estimation for μBGA Solder Joints Based on Heating Factor Q. 885
Tao Bo, Yin Zhouping, Ding Han and Wu Yiping

Electromigration Simulation with Consideration of the Atomic Concentration Gradient 889
Xuefan Chen, Lihua Liang

Finite Element Analysis of Reliability on Compliant Wafer Level Packaging With Compliant Layer 896
Peng Li, Kai-lin Pan, Ye-xiang Ning

On the Study of the In-Use Stability of a DCA Assembled MEMS Device 900
Liyuan Xu, Jing Song, Jieying Tang

IC Chip Crack Issues due to Mounting Process For Ultra-thin IC Smart Card Module 905
Pingyue Fan, Jiaji Wang

Optimization of Interface Strength for SCSP Based on Uniform Experimental Design 909
Gongke Li, D. G. Yang, Liancheng Qin, Fuxi Yi

Evaluate Anti-Shock Property of Solder Bumps by Impact Test 914
Hongjia Xi, Minyi Lou, Bing An, Fengshun Wu, Yiping Wu

Cavitation instability in Valanis-Landel hyperelastic IC packaging material 920
Li Zhigang, Shu Xuefeng

Moisture Absorption and Void Growing Effects on Failure of Electronic Packaging 923
Zhao Zhendong, Li Zhigang, Zhang Yu, Shu Xuefeng

Mechanical Test after Temperature Cycling on Lead-free Sn-3Ag-0.5Cu Solder Joint 927
Chung-Nan Peng, Jenq-Gong Duh

Reliability Challenges and Design Considerations for Wafer-Level Packages 931
Xuejun Fan, Qiang Han

Mixed Mode Interface Characterization Considering Thermal Residual Stress 937
A. Xiao, G. Schlottig, H. Pape, B. Wunderle, K M. B. Jansen, L. J. Ernst

The Effect of Strain Rate and Strain Range on Bending Fatigue Test 944
Minyi Lou, Long Wen, Zhengrong Chen, Jianwei Zhou, Qian Wang, Jaisung Lee

Dynamic Properties Testing of Solders and Modeling of Electronic Packages Subjected to Drop Impact 949
Long Wen, Xingming Fu, Jianwei Zhou, Qian Wang, Jaisung Lee

Electromigration in Pb-free Solders 956
Minhua Lu, Da-Yuan Shih, Paul Lauro

Effect of Stand-off Height on the Microstructure and Fracture Mode of Cu/Sn-9Zn/Cu Solder Joint under Tensile Test 964
Bo Wang, Fengshun Wu, Bin Du, Bing An, Yiping Wu

Failure Mode Analysis of Lead-free Solder Joints under Differential Reflow Profiles by High Speed Impact Testing 968
C. Y. Lin, Y. R. Chen, and G.S. Shen D. S. Liu, C. Y. Kuo, And C. L. Hsu

Influence of Underfill Methods on the Solder Joint Fatigue of Wafer Level Packaging 974
Charles Regard, Christian Gautier, Hélène Fremont, Patrick Poirier

xiii

Table of Contents

The Role of the Molecular Simulation Approach for IC-backend Developments 980
C. Yuan, O. van der Sluis, W. D. van Driel, G. Q. Zhang

Strain-rate and Impact Velocity Effects on Joint Adhesion Strength 984
Chang-Lin Yeh, Yi-Shao Lai

Parametric Study on Board-level Electronic Test Device Subjected to JEDEC Vibration Loads 988
Chang-Lin Yeh,, Yi-Shao Lai, Ching-Chun Wang

Modeling Techniques for Board Level Drop Test for a Wafer-Level Package 994
Harpreet S. Dhiman, Xuejun Fan, Tiao Zhou

Effect of Shear Rate on Lead Free Solder Joint Strength 1003
Zheming Zhang, Jingshen Wu, Adam R. Zbrzezny, Neil Mclellan

Analysis and Comparison of Thermal Stress and Hygrothermal Stress of SiP Device By QFN Packaging 1008
Jiang Haihua, Ma Xiao-song, Chen Ning

Crack Growth Analysis of Ball Grid Array Resistor's Solder Joint Subjected to Thermal Cycling and 4 Point Cycling Bending 1013
Xiangzhao, Ye-Yuming, Sun-Fujiang, Tu-Yunhua, Liusang

A Dual-Output Voltage Reference for High-Accuracy Pipelined ADC 1017
Dongfang Cheng, Xiaohui Li, Jue Zhang, Jiongming Wang

Study of Interface Reliability in QFN Device under Hygro-Thermal Environment 1021
Ting-biao Jiang, Hong-mi Nong, Chao Du

In-situ Observation on Electrochemical Migration of Lead-free Solder Joints under Water Drop Test 1026
Y.H. Xia, W. Jillek, E. Schmitt

The Reliability Study of Sub 100 Microns SnAg Flip Chip Solder Bump on FR4 Substrate under Thermal Cycling 1031
Xiaoqin Lin, Le Luo

Analysis and Solving of the EMI effect on LC-VCO in mixed-signal ICs 1036
Wenrong Yang, Jiongming Wang, Jue zhang, Xiaohui Li

Thermal Behavior Analysis of Lead-free Flip-Chip Ball Grid Array Packages with Different Underfill Material Properties 1040
Hsin-yuan Chena, Kuo-yuan Hsua, Tsung-shu Linb and Jihperng Leua

Enrichment and Removal of Heavy Metals Contained in PCB Boards by Multiwalled Carbon Nanotubes for WEEE Directive 1047
L. Hua, H. N. Hou

A Study on Application of N&K Analyzer in OLED Failure Analysis 1052
Qiang Fang, Yafang Peng, HK Yu

Failure Analysis of the First Wire's Bond 1055
Ren Chunling, Huang Qiang, Ding Rongzheng, Jiang Changshun

Investigation of Electromigration in Copper Interconnection of ULSI 1059
Dechun Lu, Shengxiang Bao, Lili Ma, Zhibo Du

Study on MCM Interconnect Test Generation using Ant Algorithm and Particle Swarm Optimization Algorithm 1063
Chen Lei

MCM Interconnect Test Scheme based on Adaptive Genetic Algorithm 1067
Chen Lei

Investigation on Fatigue-Creep Interaction Damage Model for Solder 1070
Na Liu, Xiaoyan Li, Yongchang Yan

xiv

Table of Contents

Study of Plasticity Damage Mechanics Constitutive Model for SnAgCu Solder Joint.....................................1073
Xiao-yan Li, Yong-chang Yan,Na Liu

A Design for Increasing the Immunity to RFI of Protection IC of Lithium-ion Battery.................................1077
Dongfang Cheng Jue Zhang Xiaohui Li Jiongming Wang

The Effect of the Different Teflon Films on Anisotropic Conductive Adhesive Film (ACF) Bonding.......................1082
Jun Zhang, Y.C. LIN, Liugang Huang

Reliability Study of Flexible Display Module by Experiments 1086
Quayle Chen, Leon Xu, Antti Salo

Thermal Fatigue Life Analysis and Forecast of PBGA Solder Joints On the Flexible PCB Based on Finite Element Analysis...1092
Huang Chunyue

xvi

Analysis of Electromagnetic Wave Propagation Characteristics in Rotating Environments

Yi Leng[2,3], Yingfeng Pan[2], Qingxia Li [1], Sheng Liu[1]*

1. National Laboratory for Optoelectronics, No.1037, Luoyu Road, Wuhan, Hubei Province, 430074, P. R. China
2. AFRA, Wuhan, 430019, P. R. China; 3. FineMEMS Inc., Shanghai, 201203, P. R. China
victor_liu63@126.com, 86-13871251668

Abstract

During the parameter acquisition of rotating components, wireless data transmission is an efficient way to resolve data transmission problem in rotating environments, and the design of wireless data transmission system is based on EMWP (Electromagnetic wave propagation). In this paper, EMWP characteristics in rotating environments are preliminarily studied in order to provide a strong theoretical support for developing high-performance rotating component radio monitoring system. In particular, the Doppler frequency shift and the EMW polarization characteristics in rotating movement are fundamentally analyzed. Finally, a test bench of rotating component is built to test EMW propagation characteristics of both stationary transmitter and rotating transmitter.

I. Introduction to Radio Monitoring System in Rotating Environments

Radio data transmission is a kind of data communication technique using electromagnetic wave propagation (EMWP). During the parameter acquisition of rotating components, radio monitoring system is a measuring system of radio data transmission in rotating environments, which is widely used in the fields of aerospace, industry, automotive, etc. because of big bandwidth, strong penetration ability and long transmission distance. The typical radio monitoring systems of rotating components include car tire pressure monitoring system [1 , 2], truck tire pressure monitoring system, measurement instrumentation of measuring and recording data from aero-engine rotors, aircraft propellers, helicopter rotors, landing gears, etc. [3].

In general, radio monitoring system is comprised of the rotating transmitter mounted on the rotating components and nearby fixed receiver. The collected parameters and process signals from the rotating components are required to be transmitted wirelessly to fixed receiver reliably and accurately in rotating environments [4,5]. The measurement is made by such transducers as the resistance strain gage, thermocouple, thermistor, pressure transducer, accelerometer, etc.. The transducer converts the measured quantity to an electrical signal which is subsequently transmitted without a conductor to a fixed receiver. The typical radio monitoring system is shown in block diagram form in Figure 1. The rotating transmitter consists of power supply module, sensor unit, signal conditioning module, wireless transmission module and transmitting antenna. The rotating transmitter is generally encapsulated as a whole reliably. Various kinds of sensor units acquire raw data from the rotating components. The acquired parameters are processed by signal conditioned module and transferred to wireless transmission module. The fixed receiver consists of receiving antenna, wireless receiving module, data process module, data storage module, control and display module. A receiving antenna captures the FM signal and demodulates it by wireless receiving module. The demodulated signal will be sent to control and display module through by certain bus or stored into data storage module temporarily.

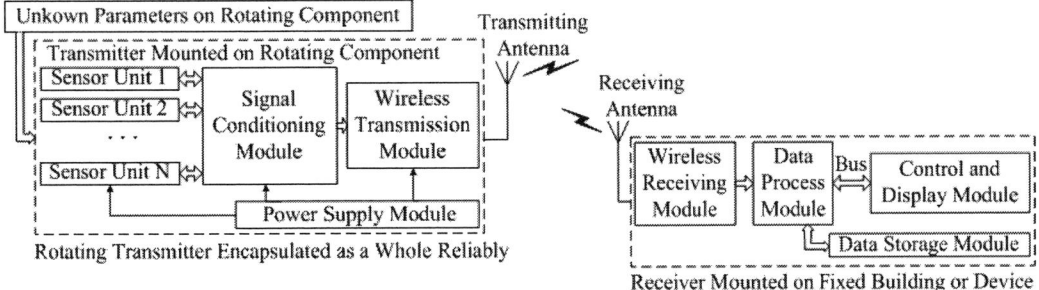

Fig.1 Block diagram of radio monitoring system for rotating components.

Most of the past efforts have been made to predict the EMW propagation characteristics between the stationary or translating wireless transceiver system [6-10]. Thus the primary purpose of this paper is to preliminarily analyze EMWP characteristics in rotating environments, and to set up a test bench of rotating component for measuring EMW propagation characteristics of both stationary transmitter and rotating transmitter.

II. Analysis

A. T-R Geometry in Rotating Environments

According to radio monitoring system for rotating components, the transmitter is rotating and the fixed receiving antenna is stationary or translating.

Assuming that observed electromagnetic wave is uniform plane wave in the receiving side. As shown in Figure 2, the transmitter is in the xOy rotating plane. S_T is the position of the rotating transmitter. S_R is the position of the fixed receiver. S_R' is the projection of S_R upon the xOy plane.

978-1-4244-2739-0/08/$25.00 ©2008 IEEE   551

For a rotating component, the transmitter rotating linear velocity is v_T, the rotating radius is R_T, the angular velocity is ω_n, the transmitter rotating velocity is f_n. In xOy plane mounted with rotating transmitter, $2\pi N$ ($N = 0,1,2,...$) represents angles at x-axis, $2\pi N + \pi/2$ represents angles at y-axis.

Let's assume time variable $t = 0$, the initial angle $\alpha_T(0) = \alpha_0$. Thus, the rotating angle of the transmitter around the axis is

$$\alpha_T(t) = v_T t / R_T + \alpha_0 \qquad (1)$$

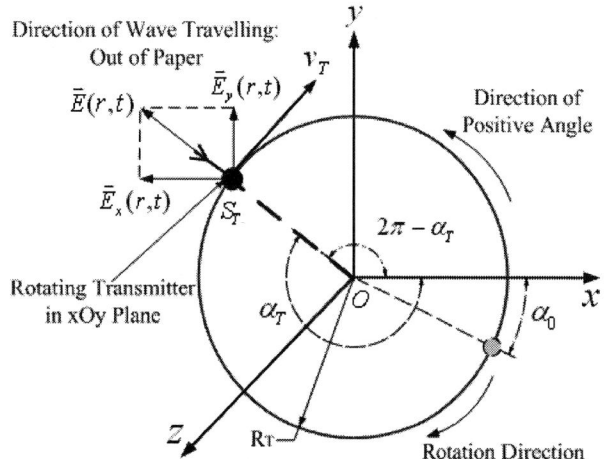

Fig.2 Illustration of transmitter rotating angle in xOy plane.

As shown in Figure 3, the new rectangular coordinate system $x_T y_T z_T$ is built with its origin at S_T. The direction of the rotating linear velocity v_T is $S_T x_T$-axis, the direction of OS_T is $S_T y_T$-axis and $S_T z_T$-axis is parallel with Oz-axis. The coordinate system $x_T y_T z_T$ is rotating around relatively stationary Oz-axis in the coordinate system $x_T y_T z_T$.

It is assumed that φ_{TR} is azimuth angle of EMW incident direction in $x_T S_T y_T$ plane, θ_{TR} is elevation angle in $x_T y_T z_T$, α_{TR} is space angle between transmitter rotating velocity direction and incident wave direction in the receiving antenna side. It is also assumed that the distance from the receiving antenna to the rotating axis is d_S and the height from the receiving antenna to its projection upon xOy plane is h_P.

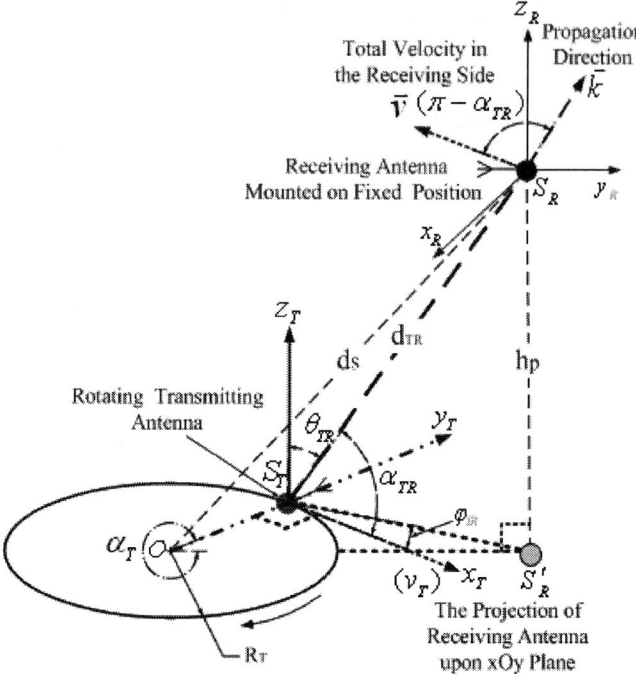

Fig.3 Position and angle relationships between rotating transmitting antenna and receiving antenna

From triangle $\triangle S_T S_R' S_R$, we have

$$\sin\theta_{TR} = \frac{\sqrt{d_s^2 - h_p^2 + R_T^2 - 2R_T\sqrt{d_s^2 - h_p^2}\cos\alpha_T(t)}}{\sqrt{d_s^2 + R_T^2 - 2R_T\sqrt{d_0^2 - h^2}\cos\alpha_T(t)}}$$

(2)

Therefore, the azimuth angle is

$$\varphi_{TR}$$
$$= \pm\left(\arccos\frac{R_T - \sqrt{d_s^2 - h_p^2}\cos\alpha_T(t)}{\sqrt{d_s^2 - h_p^2 + R_T^2 - 2R_T\sqrt{d_s^2 - h_p^2}\cos\alpha_T(t)}} - \frac{\pi}{2}\right)$$

(3)

Then, according to (2) and (3), we have

$$\cos\alpha_{TR} = \sin\alpha_T(t)\sqrt{\frac{(d_0^2 - h^2)}{d_0^2 + R_T^2 - 2R_T\sqrt{d_0^2 - h^2}\cos\alpha_T(t)}}$$

(4)

It can be seen from (4) that the T-R distance d_{TR} and space angle α_{TR} can be obtained at any time in rotating environments when the line-of-sight between transmitter and receiver is un-obstructed. Furthermore, the T-R distance and space angle are related to rotating radius, rotating linear velocity, rotating angle, projection height, etc.

B. Doppler Shift in Rotating Environments

As illustrated in Figure 3, we consider the simple case in which the coordinate axes of observers S_T and S_R are parallel, with their origins coinciding at time $t = 0$. The receiving antenna S_R moves with velocity $\vec{v}(t)$ relative to the transmitting antenna S_T.

The receiver with HF receiving antenna moves with translating velocity v_R and receives radio signal from the rotating transmitter, whose rotating velocity is v_T and translating velocity is also v_R. Therefore, the total velocity in the receiving side is

$$\vec{v}(t) = v_T \cos \alpha_T(t) \tag{5}$$

The Lorentz transformation of space-time coordinates between moving transmitter and receiver, with the use of dyadic notation, is given by [6]

$$ct' = \gamma ct - \gamma \vec{\beta} \cdot \vec{r} \tag{6.a}$$

$$\vec{r}' = \bar{\bar{\alpha}} \cdot \vec{r} - \gamma \vec{\beta} ct \tag{6.b}$$

where $\bar{\bar{\alpha}} = \bar{\bar{I}} + (\gamma - 1)\dfrac{\vec{\beta}\vec{\beta}}{\beta^2}$, $\vec{\beta} = \dfrac{\vec{v}}{c}$, $\gamma = \dfrac{1}{\sqrt{1-\beta^2}}$,

$\beta^2 = \vec{\beta} \cdot \vec{\beta}$, and $c = 3 \times 10^8$ m/s is the velocity of light in vacuum space. The position vector is $\vec{r} = \hat{x}x + \hat{y}y + \hat{z}z$. In matrix notation, the unit dyad $\bar{\bar{I}}$ is a diagonal matrix.

Consider the receiver and transmitter in relative motion, and assume that the receiver receives a plane wave from the transmitting antenna. According to the rotating transmitter on the wheel, the plane wave is expressed as

$$\begin{bmatrix} \vec{E}(\vec{r},t) \\ c\vec{B}(\vec{r},t) \end{bmatrix} = \begin{bmatrix} \vec{E}_0 \\ c\vec{B}_0 \end{bmatrix} \cos(\vec{k} \cdot \vec{r} - \omega t) \tag{7-a}$$

According to the receiver S_R, the plane wave is expressed as

$$\begin{bmatrix} \vec{E}'(r,t) \\ c\vec{B}'(r,t) \end{bmatrix} = \begin{bmatrix} \vec{E}'_0 \\ c\vec{B}'_0 \end{bmatrix} \cos(\vec{k}' \cdot \vec{r}' - \omega t') \tag{7-b}$$

where wave vector $\vec{k} = \hat{x}k_x + \hat{y}k_y + \hat{z}k_z$,

$k^2 = k_x^2 + k_y^2 + k_z^2 = \omega^2 \mu \varepsilon$.

According to the Lorentz transformation formulas and comparing with (7), the phase factor in (7) becomes

$$\vec{k}' = \left(\vec{k} \cdot \bar{\alpha} - \gamma \vec{\beta} \frac{\omega}{c} \right) \tag{8.a}$$

$$\frac{\omega'}{c} = \gamma \left(\frac{\omega}{c} - \vec{\beta} \cdot \vec{k} \right) \tag{8.b}$$

The Doppler effect is a consequence of (8). Using the dispersion relation for isotropic media and letting the angle between $\vec{k}$ and $\vec{\beta}$ be α_{TR}, we find that

$$\omega' = \gamma \omega (1 - n\beta \cos \alpha_{TR}) \tag{9}$$

As can be seen, $\vec{k}$ and $\vec{\beta}$ are in the same direction when the receiver is receiving radio signal from the rotating transmitter. The frequency is shifted downward. Otherwise,

the frequency is shifted upward when the rotating transmitter is approaching the receiver. When the receiver happens to be moving perpendicularly to $\vec{k}$, we have the transverse Doppler shift $\omega' = \gamma \omega$, which is purely relativistic effect. In an isotropic medium, $k = n\omega/c$, where n is refraction ratio.

Substitution of (4) in (9) yields

$$f'_c = \gamma f_c \left(1 - n\beta \sin \alpha_T(t)\right)$$

$$\square \sqrt{\frac{d_s^2 - h_p^2}{d_s^2 + R_T^2 - 2R_T \sqrt{d_s^2 - h_p^2} \cos \alpha_T(t)}} \tag{10}$$

where $f_c = \omega_c/(2\pi)$ is carrier frequency.

The rotation of transmitting antenna leads to Doppler effect in receiving antenna side. The typical selected parameters are configured as follows: the carrier frequency $f_c = 5.8$GHz, the rotating radius $R_T = 0.6$m, the transmitter rotating velocity $f_n = 300 r/s$, the rotating initial angle $\alpha_T(0) = 0$ rad, the distance from the receiving antenna to the rotating axis $d_s = 1000$m, and the height from the receiving antenna to its projection upon xOy plane is $h_p = 10$m. Thus, typical Doppler frequency variation is shown in Figure 4.

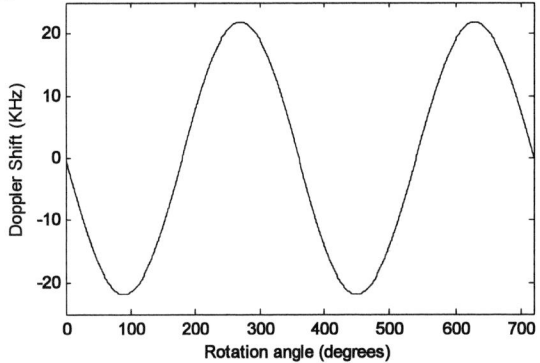

Fig.4 Typical Doppler frequency variation over rotation period.

As can be seen from the above figure, Doppler frequency curve is similar to the periodicity and fluctuation of cosine function in rotating environments. The fluctuation range is up to tens of KHz. The frequency shift is minimal at 0 and 180 degrees. The frequency shift is negatively maximal at 90 degrees and positively maximal at 270 degrees, respectively.

Doppler frequency variation is related to many other factors.

(1) The relationship between Doppler frequency shift and working frequency in rotating environments

Doppler frequency variations at different carrier frequencies are shown in Figure 5.

As can be seen from the above figure, the higher the working frequency is, the bigger the Doppler frequency shift fluctuation range is. The frequency shift variation range is only several KHz at 315MHz, while the frequency shift variation range is more than 40 KHz at 5.8GHz. However, their curve outlines are both similar to the periodicity and fluctuation of cosine function.

553

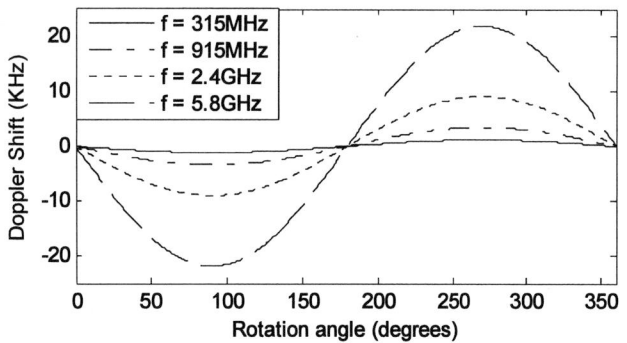

Fig.5 The relationship between Doppler frequency shift and working frequency.

(2) The relationship between Doppler frequency shift and rotating velocity in rotating environments

Doppler frequency variations at different rotating velocities are shown in Figure 6.

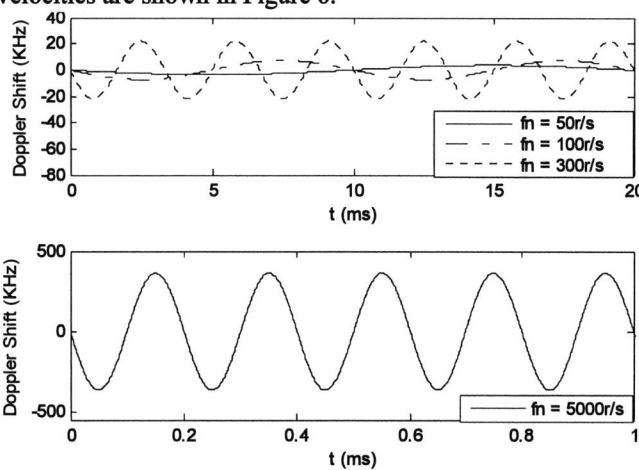

Fig.6 The relationship between Doppler frequency shift and rotating velocity.

As can be seen from the above figure, the higher the rotating velocity is, the bigger the Doppler frequency shift is. The Doppler frequency shift is even up to several hundred KHz, which is related to rotating velocity and angles, and results in neighboring carrier interference in the receiving antenna side. Furthermore, the higher the rotating velocity is, the faster the Doppler frequency shift change velocity is. The duration of implementing a frequency shift change period is from hundreds of microseconds to tens of milliseconds.

(3) The relationship between Doppler frequency shift and rotating radius in rotating environments

Doppler frequency variations at different rotating radii are shown in Figure 7.

As can be seen from the above figure, the higher the rotating radius is, the bigger the Doppler frequency shift is, which is up to hundreds of KHz, even up to several MHz.

(4) The relationship between Doppler frequency shift and the height from receiving antenna to its projection upon rotating plane

Doppler frequency variations at different projection heights are shown in Figure 8.

As can be seen from the above figure, the bigger the projection height is, the less the Doppler frequency shift is.

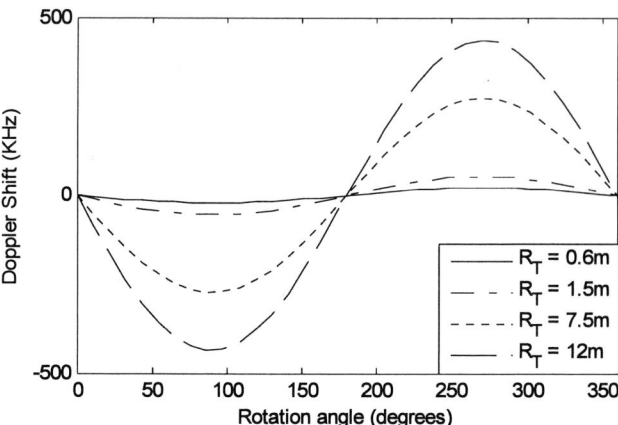

Fig.7 The relationship between Doppler frequency shift and rotating radii.

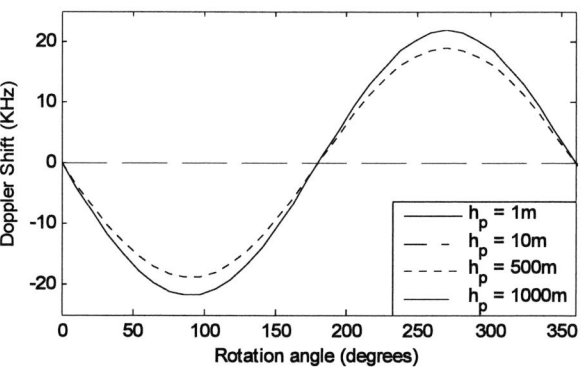

Fig.8 The relationship between Doppler frequency shift and projection height.

$h_p \ \Box \ d_s$, the frequency shift is almost changeless while receiving antenna is in the direction of rotating axis, i.e. $h_p = d_s$, the frequency shift is equal to zero.

C. Polarization response in Rotating Environments

The polarization of a wave is the direction of the electric field. We handle all polarization problems by using vector operations on a two-dimensional space using the far-field radial vector as the normal to the plane.

As shown in Figure 9, a rotating transmitting antenna's electric field is supposed to be normal to the wave propagating direction and the direction of the electric field is changeless, which means that it is linear polarization. $\vec{E}(r,t) = E_{R_T}\hat{R}_T$, $\hat{R}_T$ is the changeless direction of the electric field. E_{R_T} is the phasor in the direction of the unit vector $\hat{R}_T$.

When the receiver is shifting and the transmitter is rotating, the direction of the electric field is rotating around the tire shaft. Thus, we can express the direction of the electric field in terms of a plane wave propagating along the positive xOy plane.

$$\vec{E}(r,t) = E_x(r,t)\hat{x} + E_y(r,t)\hat{y} \qquad (11)$$

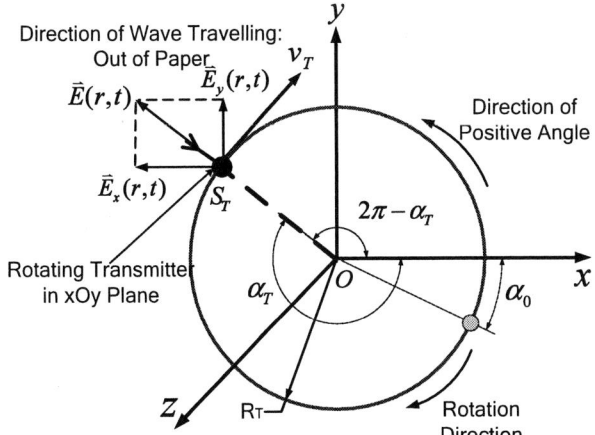

Fig.9 Polarization direction of rotating transmitting antenna.

where $E_x(r,t)$, $E_y(r,t)$ are the phasor components in the direction of the unit vectors $\hat{x}$ and $\hat{y}$.

For transmitter rotating clockwise (CW), the direction of wave traveling is out of paper:

$$E_x(r,t) = \left|\vec{E}(r,t)\right|\sin\left(\frac{v_T}{R_T}t + \alpha_0 + \frac{\pi}{2}\right) \quad (12.a)$$

$$\vec{E}_y(r,t) = \left|\vec{E}(r,t)\right|\sin\left(\frac{v_T}{R_T}t + \alpha_0 + \frac{\pi}{2} + \frac{\pi}{2}\right) \quad (12.b)$$

Therefore, $\hat{\rho}_L = e^{j\pi/2}$

where $\hat{\rho}_L$ is the linear polarization ratio, a complex constant.

For transmitter rotating counter clockwise (CCW), the direction of wave traveling is negative z direction, and $\hat{\rho}_L = e^{-j\pi/2}$.

The above equations shows that the tip of the electric field traces over time in space and appears as a circle with the electric field rotating either clockwise or counter clockwise at the rate $v_T t / R_T$ for wave propagation perpendicular to the paper. When $\hat{\rho}_L = e^{j\pi/2}$, the electric field is constant in magnitude but rotates CW (left hand), which forms left-handed circular polarized wave. When $\hat{\rho}_L = e^{-j\pi/2}$, the electric field is constant in magnitude but rotates CCW (right hand), which forms right-handed circular polarized wave. As a result of transmitter rotation, the linear polarizations are converted into the circular polarizations, including especially left-handed and right-handed circular polarizations.

For transmitter and receiver, polarization mismatch adds an extra loss. We determine polarization efficiency by applying the scalar product between normalized polarization vectors. The transmitting antenna propagating in the $\hat{z}$-direction has the components

$$\vec{E}(r,t) = E_x(r,t)\left[\hat{x} + \hat{\rho}_t\hat{y}\right]$$

The incident wave on the receiving antenna is given by

$$\vec{E}_i(r,t) = E_{yr}(r,t)\left[\hat{x} + \hat{\rho}_r\hat{y}\right]$$

where the wave is expressed in the coordinates of the transmitting antenna. $\vec{E}_i(r,t)$ is electric field on the receiving antenna. $\hat{\rho}_r$ is the linear polarization ratio of the receiving antenna.

We obtain the polarization efficiency

$$p_r(t) = \frac{1 + |\hat{\rho}_t|^2|\hat{\rho}_r|^2 + 2|\hat{\rho}_t||\hat{\rho}_r|\cos(\delta_1 - \delta_2)}{(1 + |\hat{\rho}_t|^2)(1 + |\hat{\rho}_r|^2)} \quad (13)$$

where δ_1 and δ_2 are the phases of the polarization ratios of the transmitting antenna and receiving antenna, respectively.

As expressed in terms of the polarization ratio of the receiving antenna, the above formula is determined by both the amplitudes and phases change. Therefore, a formula using circular polarization ratios would be more useful, because only phase changes under rotation.

Two arbitrary polarizations are orthogonal only if

$$|\hat{\rho}_t| = \frac{1}{|\hat{\rho}_r|} \text{ and } \delta_1 - \delta_2 = \pm\pi \quad (14)$$

We can use (13) with circular polarizations whose polarization ratio magnitudes are constant with rotation of the transmitting antennas. The maximum and minimum polarization efficiencies occur when $\delta_1 - \delta_2$ equals 0 or $\pm\pi$, respectively. The maximum or minimum polarization efficiency becomes

$$P_{\text{max/min}} = \frac{(1 \pm |\hat{\rho}_t||\hat{\rho}_r|)^2}{(1 + |\hat{\rho}_t|^2)(1 + |\hat{\rho}_r|^2)} \quad (15)$$

Thus, we can design specific receiving antenna so that the optimum polarization response and least polarization loss are obtained.

III. Test Results

To measure EMW propagation characteristics in rotating environments, a Continuous Wave (CW) signal at a frequency of 315MHz was transmitted from the rotating transmitter in the rotating measurement bench and it was received by the fixed receiver at a frequency of 315MHz.

An open typical outdoor ground is selected to set up measurement bench. The rotating transmitter and fixed receiver have a clear, unobstructed line-of-sight path. The typical parameters of the measurement bench are chosen as follows: f_c =315MHz, R_T =0.28m, d_s =5m, h_p =1m, α_0 =0rad, relative dielectric constant ε_r =15, the height from bottom of the rotating component to ground plane is 1m, the height from the receiving antenna to ground plane is 2m.

The measurement bench of rotating component is shown in Figure 10. The specific rotating transmitter is modified from TPMS (Tire Pressure Monitoring System) wheel sensor, which mainly consists of pressure and temperature sensor, tire valve set for fixing transmitter, button battery and transmitting antenna. The input power is +5dBm at a frequency of 315MHz. A standard omni-directional antenna is used as receiving antenna and connected to spectrum analyzer by RF (Radio frequency) cable, which is subsequently recorded and analyzed by specific application software. The receiver integrated with clock module is also recorded and analyzed by specific application software.

Start up testing motor and configure rotating velocity as f_n =5r/s, i.e. $T_n = 1/f_n$ =200ms. Wireless transmission rate is 9.8kbps and the transmitting interval is approximately 5ms. Each data frame's length is 5ms with a cumulative counter for counting transmitting frame number triggered by centrifugal switch.

(a) Specific transmitter　(b)Testing motor and fixing devices

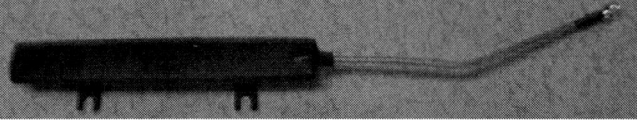

（c）Receiving antenna

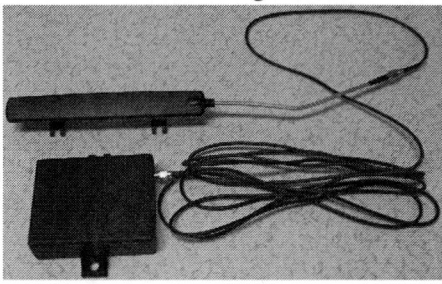

（d）Receiver integrated with clock module

Fig.10 Measurement setup of rotating component.

The specific application software based on LabView Environment is used for recording power value and reading time. The receiver is used to record data frame number and corresponding receiving time. During one rotating period, data are recorded in 30 degrees step with total of twelve received power values. Thus, received signal strength curve of rotating transmitter is gained by analyzing recording power, number and time, which is shown in Figure 11.

On the other hand, single antenna used as transmitting antenna is placed on the axis of rotating component and is fed CW signal by signal generator. Received signal strength of stationary transmitter is shown in Figure 11.

It can be seen from the above figure, received power variation range in rotating environment is up to 14dB while received power in stationary environment is constantly 34dB. For rotating transmitter, received power is minimally 30dB at 0 degree while received power is 44dB at 360 degrees.

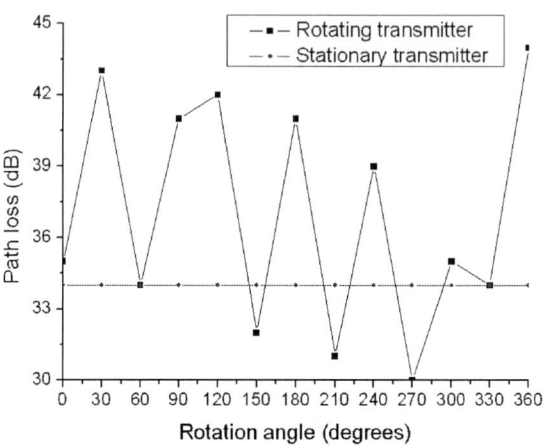

Fig.11 Path loss of both rotating transmitter and stationary transmitter

Comparing with stationary transmitter without rotating Doppler effect and polarization conversion, Figure 11 shows that rotating environment leads to big received power variation range and fast fading wireless channel. Besides, received power varies periodically.

IV. Conclusions

The mathematical relationships, including the distance between rotating transmitter and fixed receiver as well as the unobstructed line-of-sight space angle at any time, in rotating environments, are derived, which are also related to the parameters of rotation radius, angle, velocity, space wave vector, etc. The relationships between the rotation Doppler frequency shift and working frequency, rotation radius, velocity, etc. are established. The results show that the Doppler frequency shift in rotation environments varies periodically like cosine function curves. The higher the rotating velocity and the rotating radius are, the greater the frequency shift becomes, which can reach up to several KHz, even up to several MHz. Furthermore, the transmitter's rotation changes the transmitting linear polarization wave into circular polarization wave, and polarization mismatch causes excess power losses.

In addition, comparing with stationary transmitter without rotating Doppler effect and polarization conversion, test results show that rotating environment leads to big received power variation range and fast fading in wireless channel, as well as periodically varying received power.

Acknowledgments

The authors would like to thank Huazhong University of Science & Technology, Shanghai FineMEMS Inc., Jiujiang Baohua FineMEMS Inc. for the provision of TPMS samples and releasing some technical data. Second, I would like to thank my fellows, especially Mr. Xiaoping Wang, Mr. Wan Cao, Mr. Jun Yang for their support. Finally, I want to express my appreciation to the financial support of this research by China 863 High-Tech Program with the contract number of 2004AA404270, and financial support of this research by Shanghai Science and Technology Commission's Program with the contract number of 057062013.

References

1. Sun B. H., Li J. F., Liu Q. Zh., "Polarisation-diversity antenna for TPMS application [J]," Electronics Letters, 2007, 43(11):603-605.
2. Leng Yi., Li Qing xia, Liu Sheng, "Wheel Antenna of Wireless Sensors in Automotive Tire Pressure Monitoring System[C]," Wireless Communications, Networking and Mobile Computing, International Conference on, Sept. 2007:2755-2758.
3. Russell G. D., "Wireless Telemetry for Gas-Turbine Applications [R]," TM-2000-209815, ARL-MR-474, NASA and U.S. Army Research Laboratory, Glenn Research Center, Cleveland, Ohio, 2000.
4. Leng Yi, Li Qingxia, Liu Sheng, et al, "A Direct Tire Pressure Monitoring System Based on Wireless Sensor and CAN Bus," Chinese Journal of Scientific Instrument, 2008, 4.
5. Li Qingxia, Zhang Hongxia, and Leng Yi, "Study on Dynamic Antenna in Tire Pressure Monitor System, Automotive Engineering (Chinese)," 2007, No.6.
6. Kong J.A., Electromagnetic Wave Theory [M]. Beijing:Higher Education Press, 2002:889-903.

Electrodeposition of Palladium Films on Ni-Co Coatings

Yanping He[1], Dongyan Ding[1], Xiang Gao[1], Zhi Chen[2], Ming Li[1], Dali Mao[1]

[1] Lab of Microelectronic Materials & Technology, School of Materials Science and Engineering, Shanghai Jiao Tong University, Shanghai 200240, China
[2] Department of Electrical and Computer Engineering, University of Kentucky, Lexington, KY 40506, USA
dyding@sjtu.edu.cn, 86-21-34202741

Abstract

Co-Ni alloys have the properties of high hardness, good wear and corrosion resistance. A transition layer of Co-Ni coating will help enhance the hydrogen sensing stability of Pd films. In this work, Pd films were electrodeposited on Co-Ni coated copper substrate and silicon wafers. The influence of deposition parameters on the microstructure of Co-Ni coatings and Pd films were investigated. Experimental results indicated that scallop shell-like Co-Ni alloys could be fabricated on copper wafers. The tendency to form the shell-like deposits increased with increase of deposition time. While on silicon wafers, scallop shell-like Co-Ni alloys could not be fabricated. SEM and AFM analyses indicated that both composite films have a large surface area. Results showed that Pd films could be shaped by the prime films and thus maintain a large surface area.

1. Introduction

Hydrogen is one of the most important reducing gases used in many conventional chemical processes and semiconductor industry. However, the presence of 4% hydrogen gas in air will cause explosion and in many applications there is a demand to monitor the hydrogen atmosphere. Accordingly, hydrogen sensors are needed for process control and hydrogen detection for the sake of safety. Palladium, as an ideal hydrogen sensing material that can absorb up to 900 times its volume of hydrogen and form palladium hydride, has attracted much attention [1–4]. Pd Films change their resistance after absorption of hydrogen because electrical resistance of the hydride is greater than the Pd's resistance [5–6]. Although commercial Pd film sensors have been available in the market, several drawbacks still hinder their widespread applications. For example, the response of Pd film sensors is slow and the response speed is acceptable only at higher H2 concentrations. Furthermore, adhesion of the Pd film to the substrate is generally not so strong to afford considerable phase transition or volume expansion of Pd film.

Co-Ni alloys have been shown to have the properties of high hardness, good wear and corrosion resistance. A transition of Co-Ni coating will help enhance the hydrogen sensing stability of Pd films. L.M. Chang [7] reported that Ni-Co composite coating deposited under ultrasonic power becomes uniform, compact and the grains become very fine. In this work, Pd films were electrodeposited on Co-Ni coated copper substrate and silicon wafers. Through controlling the deposition conditions, various film morphologies and thicknesses could be obtained. The influence of deposition parameters on the microstructure of Co-Ni coatings and Pd films were investigated.

2. Experimental

Copper substrates used in this work have a size of 1 cm × 1 cm. They were cleaned through electrolytic degreasing, activated with 10%HCl solution, and then washed with deionized water before electrodeposition.

P-type Si (100) wafers with a resistivity of $0.15\Omega\cdot cm$ is also used in this work. These wafer samples were ultrasonically washed with absolute alcohol and then through RCA1 and RCA2 cleaning to prevent pollution. After that they were dried with nitrogen gas for electrodeposition.

All chemicals for the electroplating solutions were analytical grade. Table 1 shows composition and parameters of the plating solution for the Co-Ni alloy electroplating. Bath solutions were adjusted to a pH value of 4.0 using 10%NaOH and 10%HCl solutions.

Tab. 1 Composition and parameters of the plating solution for the Co-Ni alloy electroplation.

Parameters	Coatings	
	Co-Ni alloys	Pd
Solution Concentrations	$NiCl_2$ 200 g/L Crystallization regulater 200 g/L $CoCl_2$ 100 g/L Boric acid 40 g/L	$PdCl_2$ 3 g/L
pH	4.0	8.0
Temperature	60℃	45℃
Current Density	2A/dm^2	0.2 A/dm^2
Deposition Time	30s ~ 8min	10s ~ 40s

3. Results and Discussion

3.1 Co -Ni coatings on copper substrate

Fig. 1 presents SEM images of Co-Ni coatings fabricated with different deposition time at 60℃ with a current density of 2.0A/dm2 and pH 4.0. As shown in Fig 1(a), a lot of irregular micro-crystals began to format on the surface of deposition layer after electrodeposition for 30s. The particles kept on growing up with increase of the deposition time. Some of the crystals grew abnormally, forming scallop shell-like depositions. It can be indicated from Fig. 1(d) that these abnormally grown scallop shell-like Co-Ni crystals continuously grew up, "eating up" small particles along the growth orientation. The length of some big scallop shells even grew up to a length of about 0.5μm. A significant vertical

978-1-4244-2739-0/08/$25.00 ©2008 IEEE 558

dividing line can be seen in the center of scallops with symmetric parallel lines on each side. The grain' preferred growth at crystal orientation during electrodeposition process

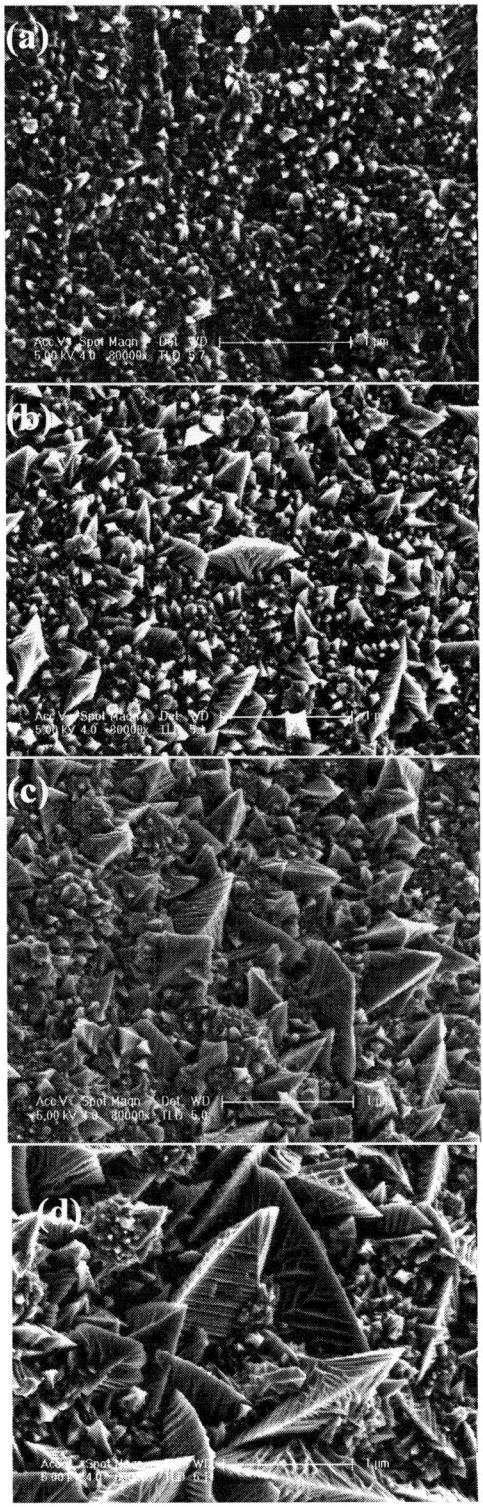

may be one reason to form the above abnormal growth phenomenon. Moreover, since the scallops were irregular, their growth was unparallel and the scallops might come across with each other, which may hinder their further growth.

Table 2 shows the height differences and the root mean square of roughness of the samples. It can be indicated from Tab. 2 that irregularity of surface roughness becomes more significant with increase of deposition time. It suggests that the tendency of scallops' abnormal-growth become more and more obvious.

Tab. 2 Influence of the deposition time on the surface roughness of the Co-Ni alloys.

Parameters	Deposition Time			
	30s	2min	4min	8min
Z range （nm）	307.54	575.74	846.61	1379.2
RMS （nm）	41.890	82.160	121.63	230.31

Furthermore, when we use EDX analysis to check the compositions of depositions corresponding to deposition time of 2min, 4min and 8min it can be found that mass content of Co-Ni of the coating keep basically same (Tab. 3).

Tab. 3 Composition of the Co-Ni alloys with different deposition time

Element	Deposition time		
	2min	4min	8min
Ni at%	14.66	17.19	17.11
Co at%	85.34	82.81	82.89

To analyze the influence of Pd films on morphologies of Co-Ni coatings, this work chooses the sample of Co-Ni alloys deposited for 4min (shown on Fig. 2). Pd film was plated onto the sample for 15s and the sample were examined with SEM and AFM (Fig. 3). Compared with the sample before Pd deposition, the surface morphology of Co-Ni alloys kept consistent with the one without Pd films deposition although the length of scallops increased to a certain level. Pd films were evenly coated on the scallops. Since the scallops grew vertically to the surface, the sides had big surface areas. If the growth of scallops could be controlled to such that more regular and small sized scallops could be fabricated, large specific surface areas would be obtained for better hydrogen sensing.

Fig. 1 SEM morphologies of Co-Ni depositions fabricated with different deposition time. (a) 30s, (b) 2min, (c) 4min, (d) 8min.

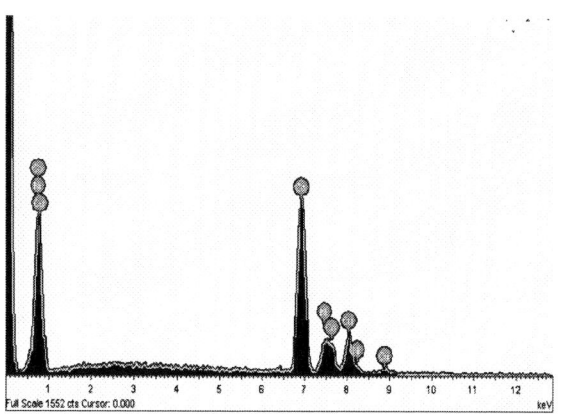

Fig. 2 Typical EDXA pattern of Co-Ni alloys deposited on copper substrate.

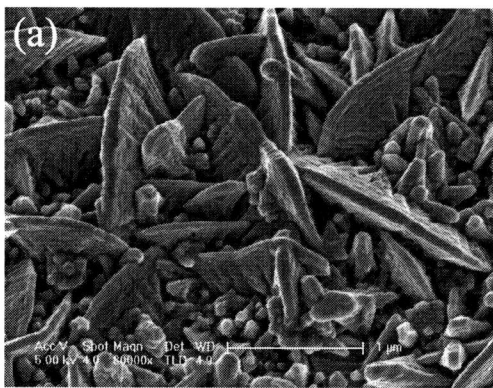

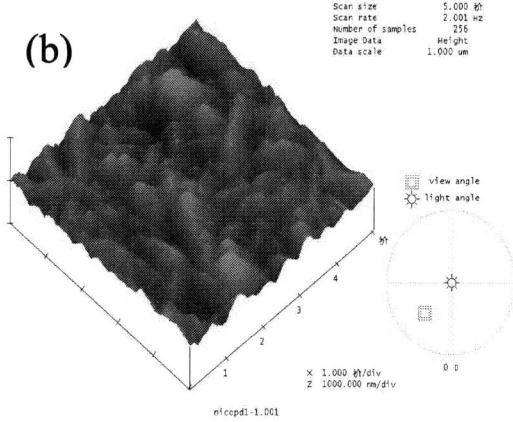

Fig. 3 SEM and AFM morphologies of Pd film deposited for 15s.

3.2 Co-Ni coatings on silicon wafer

To deposit Co-Ni alloys on silicon wafers, control groups with current density (1.0A/dm2, 2.0A/dm2) and deposition time (20s, 1min, 2min) were chosen. Fig. 4 shows SEM morphologies of Co-Ni deposition on polishing surface of silicon wafers. Compared with Co-Ni deposition on copper substrate, quite different morphologies were found. It shows that nucleation on silicon wafers is much more difficult. We can find that deposition particles grew up with increase of time. After a deposition for 2minutes, the samples under either current density were almost fully covered with Co-Ni alloys. The size and shape of particles deposited under different current density were quite different. The particles under a higher current density usually had a bigger size. Furthermore, surfaces of particles under a lower current density were much smooth, they presented sphere at the beginning of deposition. With time going on, the particles grew up, colliding with each other. And their shape also became irregular. On the other hand, particles deposited under a higher current density were also spherical. However, there were many acupuncture-like structures on their surfaces, especially for the sample with a deposition time of 20 seconds, both smooth spherical particles and acupuncture-like depositions could be seen. With increasing the deposition time, all of the surfaces of deposited particles could have acupuncture-like structures.

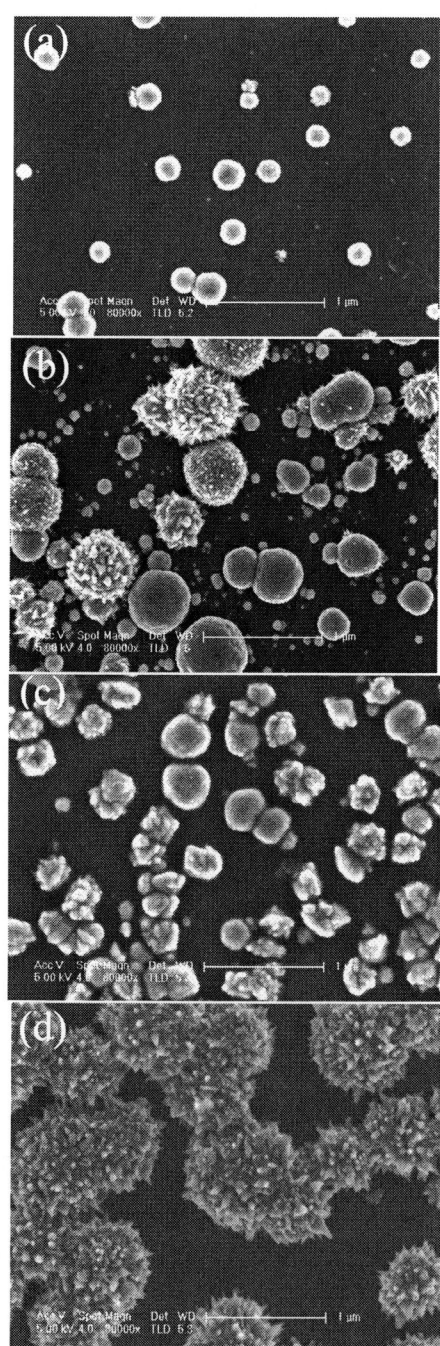

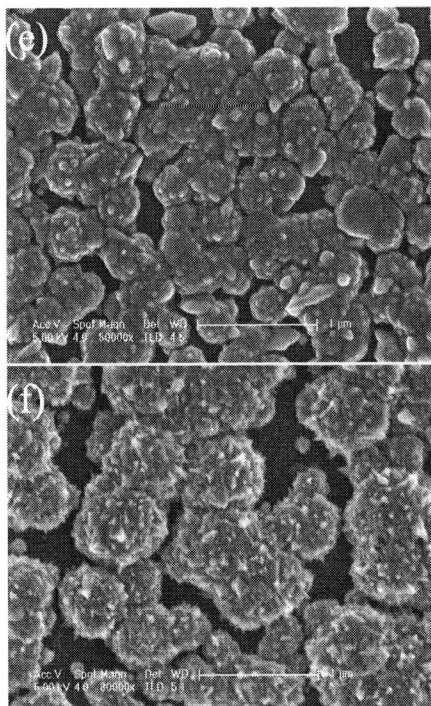

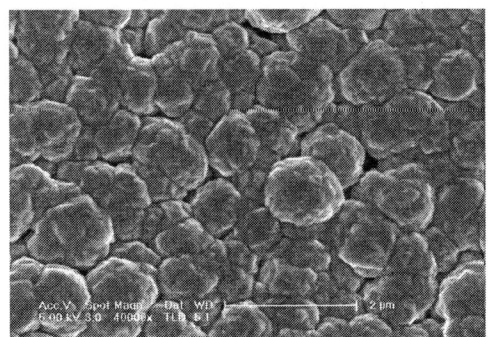

Fig. 5 SEM morphology of the Pd film deposited on the surface of Co-Ni coating (electrodeposition time is 15s).

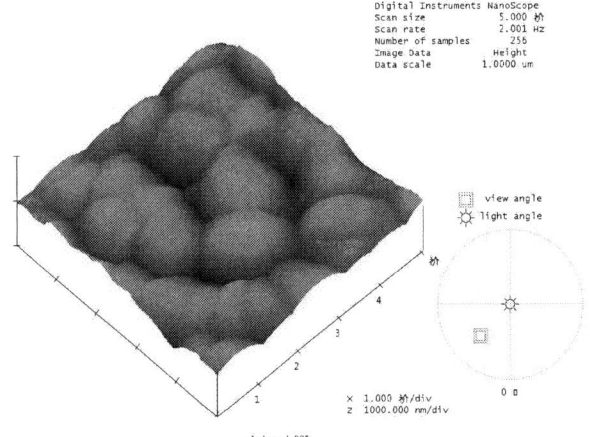

Fig. 6 AFM morphology of the Pd film deposited on the surface of Co-Ni coating (electrodeposition time is 15s).

Tab. 5 Surface roughness of the Pd film deposited on the surface of Co-Ni coating.

Parameters	Data
Z range (nm)	630.15
RMS (nm)	99.008

Fig. 4 SEM morphologies of Co-Ni depositions fabricated with different time and currents. (a) 1.0A/dm2 20s, b) 2.0A/dm2 20s, c) 1.0A/dm2 1min, d) 2.0A/dm2 1min, e) 1.0A/dm2 2min, f) 2.0A/dm2 2min.

The samples of Co-Ni depositions fabricated with current density of 1.0A/dm2 and 2.0A/dm2 for 2 minutes were choosen for EDX analyses. Tab. 4 shows composition of the Co-Ni films. Ni accounted for a large proportion when current density was low. While under high current density, the proportion of Co became larger. A sample of Co-Ni alloys deposited for 2min under the current density of 2.0A/dm2 was chosen to for Pd deposition for 20s. SEM and AFM observations were conducted (Figs. 5 and 6). It shows that Pd film was evenly coated on the surfaces. Since the size of acupuncture-like structures were quite small, these structures were completely covered by the Pd film. As a result, the surface after Pd deposition was much more smooth. Tab. 5 shows the surface roughness of Pd film deposited on the surface of Co-Ni coating.

Tab. 4 Composition of the Co-Ni films fabricated with different current density.

Element	Current Density	
	1.0A/dm^2	2.0A/dm^2
Ni at%	95.36	22.57
Co at%	4.64	77.43

Conclusions

In summary, the electrodeposition technique is a useful method for depositing Co-Ni particles on the surface of copper substrate and silicon wafer. Scallop shell-like Co-Ni alloys could be fabricated on copper wafers. While on silicon wafers, scallop shell-like Co-Ni alloys could not be fabricated. Analyses indicated that both composite films have a large surface area. Results showed that Pd films could be shaped by the prime films and thus maintain a large surface area.

Acknowledgments

We thank the supports by National Natural Science Foundation of China (No. 60641004) and Shanghai Pujiang Program (No. 07pj14047).

References

1. Christofides, C., Mandelis, A., "Solid-state Sensors for Trace Hydrogen Gas Detection," *J. Appl. Phys*, 68 (1990), pp. R1-R30.

2. Rahimi, F., Iraji, A., Razi, F., "Characterization of Porous Poly-silicon Impregnated with Pd as a Hydrogen Sensor, " *J. Phys. D*: Appl. Phys. 38 (2005), pp. 36-40.

3. Xu, T., Zach, M. P., Xiao, Z. L., Rosenmann, D., Welp, U., Kwok, W. K., and Crabtree, G. W., "Self-assembled Monolayer-enhanced Hydrogen Sensing with Ultrathin Palladium Films," *Appl. Phys. Lett.*, vol. 86 (2005), pp. 203104-1–203104-3.

4. Villatoro, J., Monzón-hernández, D., "Fast Detection of Hydrogen with Nano Fiber Tapers Coated with Ultra Thin Palladium Layers," *Optics Express,* vol. 27, (2005), pp. 5087-5092.

5. Walter, EC, Favior, F, Penner, M. *Anal Chem*, vol. 27,(2002), pp.1508-1514.

6. Abyaneh, M Y, Fleischmann M. "The Electrocrystallisation of Nickel". *Electroanal Chem*, 1981, 119:187.

7. Chang, L.M., Guo H.F., An M.Z., "Electrodeposition of Ni–Co/Al2O3 Composite Coating by Pulse Reverse Method Under Ultrasonic Condition," *Materials Letters*, Volume 62, Issue 19, 15 July 2008, pp.3313-3315.

Titania Nanostructures Fabricated Through Anodization of Ti6Al4V Alloy

Yan Li, Dongyan Ding, Shuo Bai, Ming Li, Dali Mao

Lab of Microelectronic Materials & Technology, School of Materials Science and Engineering, Shanghai Jiao Tong University,
Shanghai 200240, China
dyding@sjtu.edu.cn, 86-21-34202741

Abstract

This work reports on the fabrication and thermal stability of self-organized titania nanostructures on Ti6Al4V alloy. Ti6Al4V sheets were anodized in 1M NaH_2PO_4 containing 0.5 wt% HF. And the anodized sheets were heat-treated at different temperatures to test their thermal stability. SEM observations revealed that, for the two-phase Ti6Al4V alloy, there were two different kinds of nanostructures (nanotubes grown at α-phase region and inhomogeneous nanopores grown at β-phase region) formed on the substrate surface. The nanotubes can withstand a high temperature of 650 °C without collapsing but sinter to densification at 675-700 °C.

1. Introduction

Recently reports on the synthesis of titania nanostructures have aroused considerable scientific interest due to wide applications of the oxide materials. Titania can be used in heterojunction solar cells, water photolysis, fuel cells, molecular filtration, tissue engineering[1,2], photocatalytic devices[3], and lithium-ion batteries[4], exhibiting highly variable and tunable wetting behavior[5,6]. Titanium oxides can also be used as an important material for chemical gas sensor in harsh industrial environments and also at low temperature. Titania-based devices are used not only as oxygen sensors, but they can be used as excellent hydrogen sensors. For applications where the use of vertically oriented titania nanotubes has been studied, these advantages have manifested themselves in an extraordinary enhancement of the extant TiO2 properties. For example, hydrogen sensors based on the use of titania nanotube arrays of single-micron length exhibit an unprecedented 50 000 000 000% change (about 8.7orders of magnitude) in electrical resistivity upon exposure to alternating atmospheres of nitrogen containing 1000 ppm hydrogen and air at room temperature[7]. Consequently, it is important to fabricate high quality oxide structures.

It has been reported how to fabricate titania nanostructures through Pure Ti and its alloy. However, the important roles of alloying elements in the process of fabricating nanostructures have not been studied. Our work focuses on how different elements act in the anodization process. As one of the most important biomaterials, Ti6Al4V has been used for many years. Fabricating a thick and stable titania nanostructures on the surface of Ti6Al4V will improve the bioactivity. Alloying Al and V in Ti leads to two different phases, which show different nanostructures, after anodization. This phenomenon provides us a good chance to investigate the roles of different elements in the anodization process. Besides, titania nanotubes could be applied in the industry as excellent gas sensor, the thermal stability of titania nanotubes is important for a practical use. Thus, investigating the nanotubes' thermal stability at different temperatures is another important part of this paper.

2. Experimental

The Samples used were Ti6Al4V alloy sheets. The sheets were ground with emery papers up to #2000, and mirror-finished with diamond paste, and ultrasonically cleaned with absolute alcohol. Finally they were rinsed with deionized water and dried in N2 stream. Electrochemical anodization was carried out with a DC Voltage Stabilizer as illustrated in Fig. 1. All of the samples were fabricated at 15V (for one and a half hours) in the electrolytes of 1M NaH2PO4 containing 0.5 wt% HF. After the anodization, the samples were rinsed with deionized water and dried in N2 stream. Some of these anodized samples were then heat-treated at different temperatures including 650°C, 675 °C and 700 °C.

The anodized and heat-treated samples were characterized with Scanning Electron Microscope (FEI SIRION 200). The composition of the surface structures was investigated by Energy Dispersive X-Ray analysis (OXFORD INCA).

3. Results and Discussion

As shown in Fig. 2, Ti6Al4V alloy studied here consists of two different phases (α and β). The image was acquired after mechanical polishing and chemical etching in 0.5% HF.

After the anodization, two different nanostructures formed on the surface of the Ti6Al4V sheet. One was self-organized nanotubes formed at the α-phase region; the other was inhomogeneous nanopores formed at the β-phase region. Fig. 3 shows SEM images of different titania nanostructures at the α-phase and β-phase regions, respectively. The nanotubes have an inner pore diameter of about 60 nm. The wall thickness and the length of the nanotubes are approximately 20nm and 260 nm, respectively.

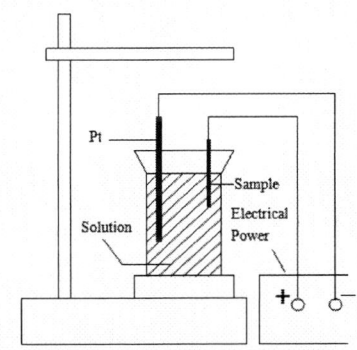

Fig.1 Schematic illustration of the anodization equipment.

978-1-4244-2739-0/08/$25.00 ©2008 IEEE

Fig. 2 Optical image of two-phase structures in Ti6Al4V alloy.

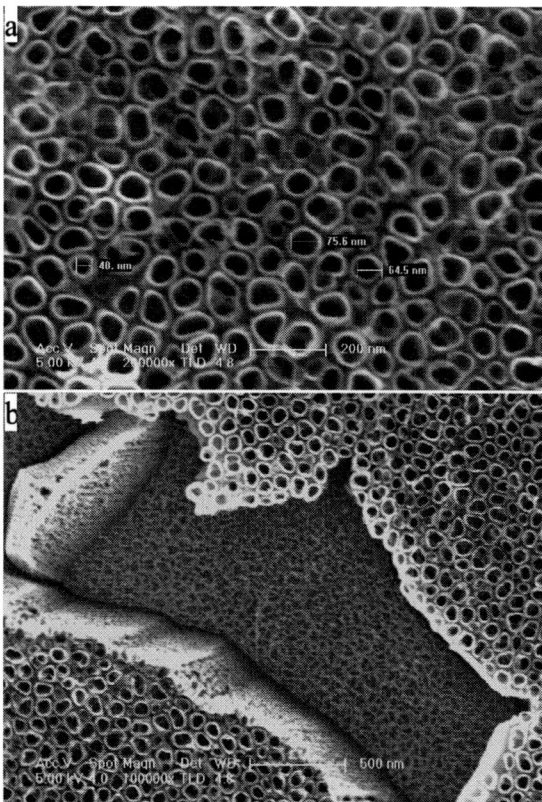

Fig. 3 SEM images of Ti6Al4V nanostructures showing (a) nanotubes on α-phase region, and (b) nanopores on β-phase region.

Table 1 shows the composition at α- and β-phase regions after anodization. From the data in the table, it is obvious that Al element is rich in α-phase region and V element is rich in β-phase region. For the two-phase alloy Ti6Al4V, the two different phases behave differently during the anodization process, which results in different nanostructures. Self-organized nanotubes formed at the α-phase region. In contrast, nanopores formed at the β-phase region.

The whole anodization process can be concluded as two stages. For the first stage, a compact oxide layer formed on the surface, and then randomly distributed localized dissolution events occur over the entire surface leading to pores growth under the top oxide layer [8]. During the final

stage, the pore growth morphology gradually changes to a homogeneous and self-organized morphology. During the process a competition between formation and dissolution of the oxide tubes takes place [9-12]. Finally, the pore growth and dissolution of oxide layers occur at an equal rate.

In order to test the thermal stability of the titania nanostructures fabricated from Ti6Al4V. Several of these anodized Ti6Al4V alloy sheets were heat-treated at elevated temperatures. Fig. 4 shows top-view image of nanotubes heat-treated at 650°C and 675 °C. Fig. 5 shows top-view image of nanopores heat-treated at 650°C and 675 °C. And Fig. 6 shows the surface of the anodized sheet heat-treated at 700°C.

Table 1 Composition of α- and β-phase regions after anodization.

Tested Areas	Elements (wt%)			
	Ti	Al	V	O
α-phase region	79.00	4.72	3.67	12.61
β- phase region	81.33	3.81	5.07	9.79

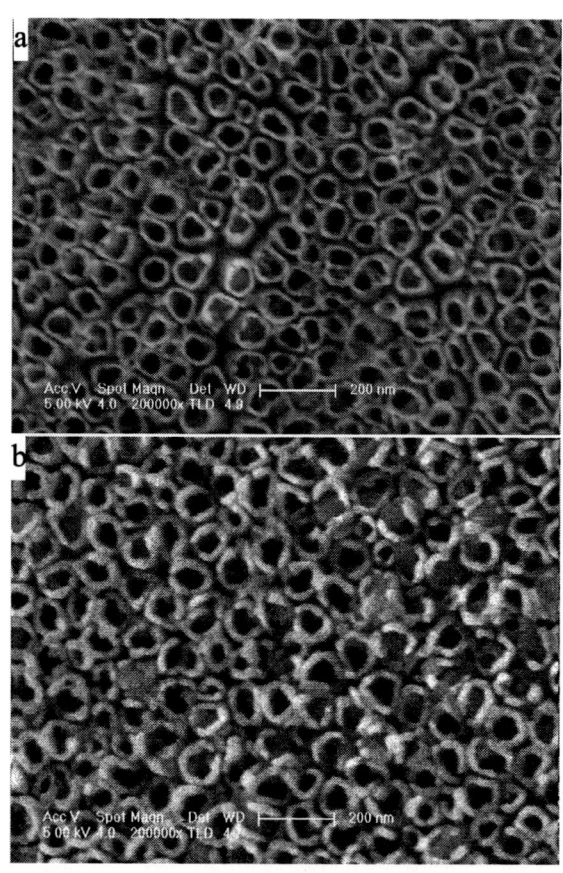

Fig. 4 SEM images of nanotubes on α-phase region heat-treated at different temperatures. (a) 650°C, (b) 675 °C.

Fig. 4 indicates that nanotubes (formed at α-phase region) could withstand 650°C without any change in nanotube morphology, and some nanotubes began to collapse a little when the temperature increases to 675 °C. Fig. 5 shows that nanopores (formed at the β-phase region) sintered into dense coating at 650°C, and the coating became even denser when the temperature rose to 675°C. The nanotubes (formed at α-

564

phase region) were much more stable than the nanopores (formed at β-phase region) at high temperatures. When the temperature increased to 700ºC, sintering densification took place on all of the nanostructures formed on the Ti6Al4V alloy sheet (Fig. 6). For pure TiO2 nanotubes heat-treated at 650ºC, a substantial collapse of the tubular morphology has been reported [13]. The thermal stability of nanotubes formed on Ti6Al4V alloy improveda little compared with pure TiO2 nanotubes fabricated from Ti sheet.

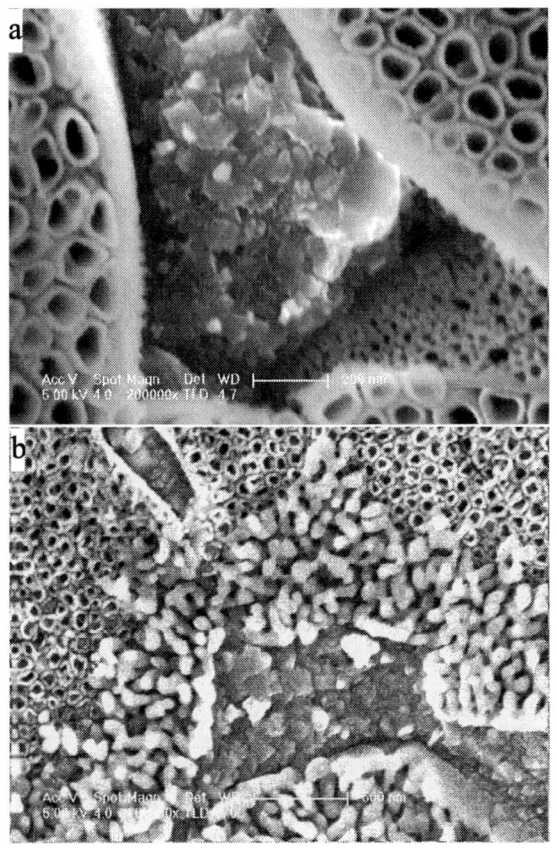

Fig. 5 SEM images of nanopores on β-phase region heat-treated at different temperatures. (a) 650ºC, (b) 675 ºC.

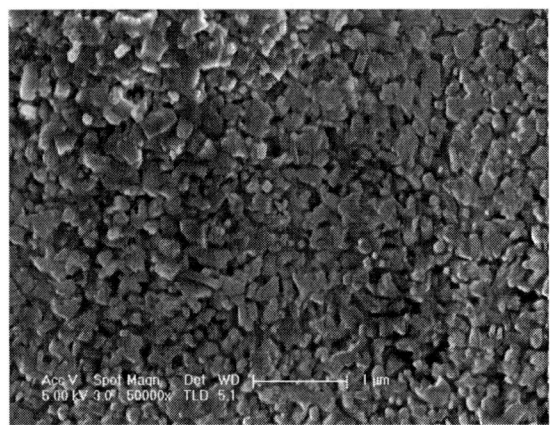

Fig. 6 SEM image of nanostructures on the surface of Ti6Al4V heat-treated 700ºC.

Conclusions

In summary, titania nanostructures can be easily fabricated through anodization of Ti6Al4V alloy. For the two-phase material of Ti6Al4V the nanostructure growth behavior is clearly different for each kind of phase. The α-phase regions could lead to formation of self-organized nanotubes with an average diameter of about 60 nm and length of about 260nm. Whereas, β-phase regions could only lead to formation of nanopores. It is found that the nanotubes have a tendency to collapse with increase of the temperature to above 650ºC, and nanotubes are more stable than nanopores. The nanotubes fabricated through anodization of Ti6Al4V can withstand 675ºC without big change. These nanotubes are better than pure TiO2 nanotubes that collapse at 650ºC.

Acknowledgments

We thank the support by Shanghai Pujiang Program (No. 07pj14047) and National Natural Science Foundation of China (No. 60641004).

References

1. Hueso, L., Mathur, N., "Dreams of a hollow future," *Nature*, Vol. 427, (2004), pp. 301
2. Tenne, R., Rao, C. N. R., "Inorganic nanotubes," *Philos. Trans. R. Soc. Ser. A*, Vol. 362, (2004), pp. 2099
3. Fujishima, A., Honda, K., "Electrochemical photocatalysis of water at a semiconductor electrode," *Nature*, Vol. 238, (1972), pp. 37-38
4. Kavan, L., Gratzel, M., Gilbert, S. E., Klemenz, C., Scheel, H. J., "Electrochemical and Photoelectrochemical Investigation of Single-Crystal Anatase," *J. Am. Chem. Soc.*, Vol. 118, (1996) , pp. 6716-6723
5. Wang, R., Hashimoto, K., Fujishima, A., Chikuni, M., Kojima, E., Kitamura, A., Shimoshigoshi, M., Watanabe, T., "Light induced amphiphilic surface," *Nature* ,Vol. 388, (1997), pp. 431-432
6. Feng, X., Zhai, J., Jiang, L., "The fabrication and switchable superhydrophobicity of TiO₂ nanorod films," *Angew. Chem. Inc. Ed.*, Vol. 44, (2005), pp. 7463-7465
7. Paulose, M., Varghese, O. K., Mor, G. K., Grimes, C. A., Ong, K. G., "Unprecedented ultra-high hydrogen gas sensitivity in undoped titania nanotubes," *Nanotechnology* , Vol.17, (2006), pp. 398-402
8. Taveira, L. V., Macak, J. M., Tsuchiya, H. L., Dick, F. P., Schmuki, P., "Initiation and growth of self-organized TiO2 nanotubes anodically formed in NH₄F/(NH₄)₂SO₄-electrolytes," *J Electrochem Soc.*, Vol. 152, (2005), pp. 405-410
9. Beranek, R. Hildebrand, H., Schmuki, P., "Self-organized porous titanium oxide prepared in H₂SO₄/HF electrolytes," *Electrochem Solid State Lett*, Vol. 6, (2003), pp. B12-B14
10. Macak, J. M., Tsuchiya, H., Schmuki, P., "High-aspect-ratio TiO₂ nanotubes by anodization of titanium," *Angew. Chem. Inc. Ed.*, Vol. 44, (2005), pp. 2100-2102
11. Ghicov, A., Tsuchiya, H., Macak, J. M., Schmuki, P., "Titanium oxide nanotubes prepared in phosphate

electrolytes," *Electrochem Commun*, Vol. 50, (2005), pp. 3679-3684

12. Macak, J. M., Sirotna, K., Schmuki, P., "Self-organized porous titanium oxide prepared in Na_2SO_4/NaF electrolytes," *Electrochim Acta*, Vol. 7, (2005), pp. 49-52

13. Varghese, O. K., Gong, D., Paulose, M., Grimes, C. A., Dickey, E. C.,"Crystallization and high-temperature structural stability of titanium oxide nanotube arrays," *J. Mater. Res.*, Vol. 18, (2003), pp. 156-165

The Influence of Low level Doping of Ni on the Microstructure and Reliability of SAC Solder Joint

Zhenqing Zhao, Lei Wang, Xiaoqiang Xie, Qian Wang, Jaisung Lee
Samsung Semiconductor China R&D CO. LTD.
Science Plaza 7F, International Science & Technology Park,
Suzhou Industrial Park, Jiangsu Province, China 215021
Tel: 86512-62888288-8821, Fax: 86512-62888388, E-mail: zq.zhao@samsung.com

Abstract

In this paper, the behavior of BGA solder joints microstructures was studied as a function of Ni doping in SAC solder. Three kinds of solder compositions were selected including Sn3.0Ag0.5Cu, Sn1.0Ag0.5Cu and Sn1.0Ag0.5Cu0.02Ni to value the influence the effect of Ni doping, OSP and Au/Ni pad was employed on the PCB side. Emphasis was placed on studying the effect of low level doping with Ni on the joint microstructure and subsequent reliability. Both solder composition and PCB surface finish had a notable effect on the interfacial microstructure, the Ni addition can give rise to needle like NiCuSn IMC formation and reduce the grain size locally at solder/NiAu pad interface after one time reflow according to top-view interface analysis, and had no obvious effect on the IMC evolution of solder/OSP pad, the phenomena was investigated from the perspective of metallurgy.

Bending and drop tests were conducted to evaluate the effect of solder composition and pad finish on the joint reliability. It was found that the decrease of Ag concentration and Ni addition in SAC solder could significantly improve the drop test performance when NiAu pad was used. In bending test, OSP pad show better performance than Au/Ni pad. The correlation between joint microstructure and joint reliability was discussed in detail. The work can give some directions on the solder alloy design and choice of pad finish in electronic packaging.

1 Introduction

Due to the consideration of environmental protection, the adoption of lead-free solders has become an inevitable trend in the electronic packaging industry. Nevertheless, compared to the lead-containing solders, the lead-free solders are in general stiffer and more brittle. This makes the lead-free solder joints more fragile to dynamic loads, which are frequently encountered for portable electronic devices during normal or excessive use conditions.

In long-term industry trends, greater integration of functions and features in devices is needed. This implies that a greater number of interconnects are required to facilitate these new features. Mobile computing also exerts down pressure on device size, consequently the solder joints in these packages shrink with every new generation of electronic products [1-5]. These two trends combined to constantly reduce the size of the second level interconnect, thus making each individual interconnects weaker.

In array packages, the Ni-based metal and OSP pad are often used as the surface finish or Under Bump Metallurgy (UBM), when Ni reacts with commercial Cu-bearing SAC

solder, the reaction depends strongly on the Cu concentration [6-8], the composition of solder will play a very important role on the joint microstructure and subsequent joint reliability, the choice of solder composition and pad finish are main factors to determine the joint reliability, especially when the joint dimension decreased. Dopants addition in solder is a common way to increase the solder properties and joint reliability in solder alloy design. So in this paper, the solder composition, pad finish and their influence on solder joint reliability were comprehensively investigated in fine picth BGA interconnection.

2 Experiment prcedures

2.1 Materials, Process and analysis

The package test vehicle was BGA products with 88 solder interconnects. The solder ball samples included Sn3.0Ag0.5Cu, Sn1.0Ag0.5Cu, Sn1.0Ag0.5Cu0.02Ni. The diameter and pitch of the solder interconnects were 0.3 mm and 0.5 mm, respectively. The pad finish on package side is Ni/Au, two types of pad finish including OSP pad and Ni/Au pad were used on PCB side to study the effect of Ni doping and pad finish on joint microstructure and reliability. The identical fluxes was employed in the lead free ball mount process at 250 ℃ peak temperature. In the surface mount process, the Sn-3Ag-0.5Cu solder paste was employed for assembly. the surface mount temperature profile had a plateau at around 150℃ for 82 s and a stage over 220℃ for 35 s with a peak temperature at around 260℃.

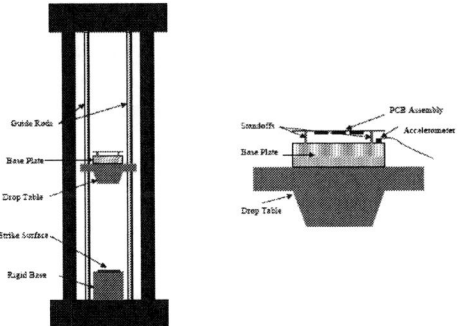

Figure 1 Drop test schematic

The commonly used cross sectional analysis after reflow and reliability test was conducted. The specimens were polished by 150C, 600C, 1200C, 2000C silicon carbide abrasive paper consequently. And then they were fine polished by 1μm diamond polishing solution and 0.05um silica gel polishing solution consequently. Finally, the cross-sections were etched by etchant to observe the metallographic phases. The etchant constitution: 96.5%vol. Methanol, 1%vol. HNO3 and 2.5%vol. HCl. Top view analysis of IMC was also performed to observe the size and morphology of IMC grains. The solder joints were splited from middle and then polished by 2000C silicon carbide abrasive paper until quite near to IMC layers. The specimens were then dip into corrosive to get rid of residual solder matrix and reveal IMC after cleaning by ultrasonic for thirty minutes. The etchant used for top view analysis is 90%vol. Methanol, 6%vol. HNO3 and 4%vol. HCl.

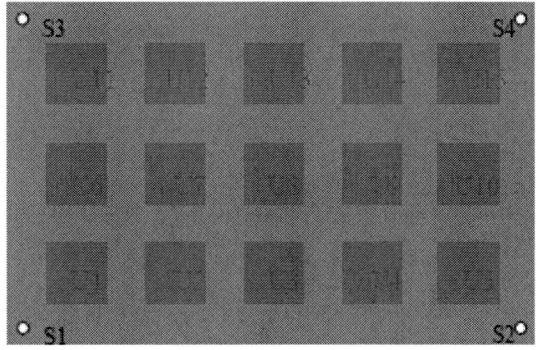

Figure 2 Components locations and grouping on board as per JEDEC-B111

2.2 Drop and bending test

The drop tests were performed according to JEDEC standard JESD22-B111 with the acceleration of 1500G and failure criteria of 100Ω(Show in Fig.1). Seven packages (U1, U2, U3, U8, U13, U14, U15) were mounted on a 132 x 77 x 1 mm3 drop test board in a layout regulated by JEDEC, show in Fig. 2, in which the mounted packages were individually numbered. The test board and each of the mounted packages were daisy-chain designed so that the overall electrical resistance of daisy-chained solder joints could be individually measured for each mounted package during drop test. After drop tests, the solder joints were cross-sectioned, polished and etched to observe the fractured microstructure.

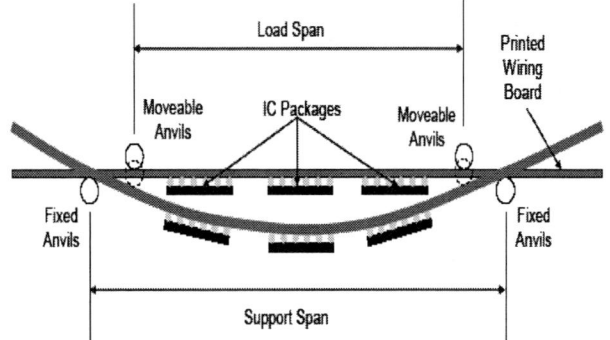

Figure 3 Schematic showing 4-Point bend setup

The bending test conditions of PCB and assembly method are all according to the 'Board level Cyclic Bend test method for Interconnect Reliability Characterization of Components for Handheld Electronic Products', JEDEC JESD22B113 (see Fig.3). In bending test the span for support Anvils and load Anvils are 110mm and 75mm, respectively. Load Anvil vertical displacements are 2mm for Ni/Au pad and 3mm for OSP pad. The load profile is Sinusoidal and Cyclic Frequency is 3Hz.

3.Results and discussions

3.1 Interfacial IMC morphology

Fig.4 shows that the interfacial IMC of different solder compositions in top-view IMC observation between solder and Ni/Au substrate side, it can be seen that there is no obvious difference when various solder alloy was employed.

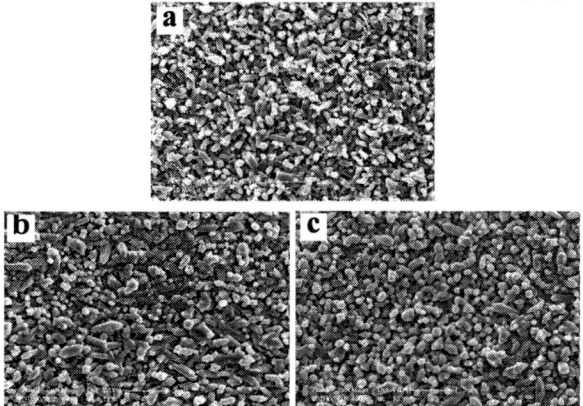

Figure 4. Top view IMC of solder/NiAu substrate pad
(a) Sn3.0Ag0.5Cu (b) Sn1.0Ag0.5Cu (c) Sn1.0Ag0.5Cu 0.02Ni

(b)

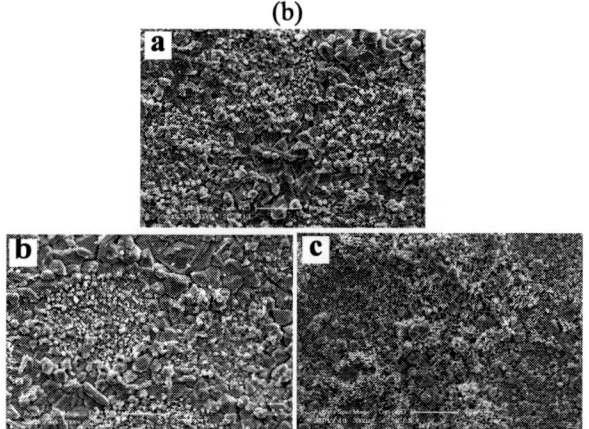

Figure5. Top view of solder and NiAu PCB side interface (a) Sn3.0Ag0.5Cu, (b) Sn1.0Ag0.5Cu, (c) Sn1.0Ag0.5Cu0.02Ni

Fig.5 showed IMC compound morphology soldered with three solders in the case of NiAu pad finish, as can be seen from the micrograph, the morphology is different when different solders were used. According to interfacial IMC EDX analysis, in the case of Sn3.0Ag0.5Cu and Sn1.0Ag0.5Cu, nearly all the interfacial IMC are (CuNi)6Sn5, but there are some particles deposited on large dimension (CuNi)6Sn5 grains. But for Sn1.0Ag0.5Cu0.02Ni, it is quite different in the IMC constitution, almost all interfacial IMC are (NiCu)3Sn4, only a small portion interfacial IMC are(CuNi)6Sn5(the larger grain in Fig.5c). There is a significant difference between no nickel and nano nickel particles-included Sn1.0Ag0.5Cu in the IMC grain size. The IMC size of Sn1.0Ag0.5Cu is diameter of 0.512um with a length of 2.15um, and the IMC size of Sn1.0Sg0.5Cu0.02Ni is diameter of 0.475um with a length of 1.62um.(shown in Fig.6) that is to say, the addition of Ni in SAC solder change the phase structure of interfacial IMC. In particular, 0.02wt%

Ni reduced the grain size of intermetallic compounds compared to no nickel included Sn1.0Ag0.5Cu. This phenomena is probably because the added nano-Ni particle can help the nuclear of (NiCu)3Sn4.

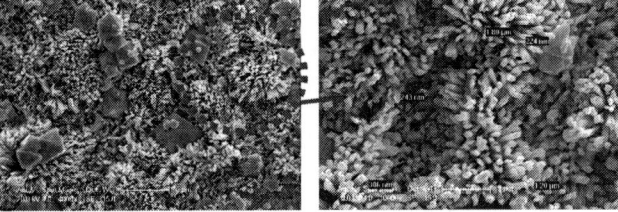

Figure6. Fine (NiCu)3Sn4 IMC formation at Sn1.0Ag0.5Cu0.02Ni and PCB NiAu interface

From high magnification top-view SEM observation of IMC, a interesting phenomena was found, when the Ag content in SAC solder is 3.0, some small particles deposited on Cu6Sn5, shown in Fig.7, EDX analysis indicated that the small particles are Ag3Sn. According to phase diagram analysis, this is probably because the primary phase formed in the solidification is Ag3Sn. There are no such phenomena when the Sn1.0Ag0.5Cu and Sn1.0Ag0.5Cu0.02Ni solder was used. Few studies has focused on the formation mechanism of the particle-like Ag3Sn, some recent publications reported the formation of Ag3Sn can decrease the interfacial energy and hamper the growth of IMC, but if the formation of Ag3Sn can degrade the interfacial properties is still unknown and not studied in this paper.

Figure7. Ag3Sn particles morphology on CuNiSn IMC

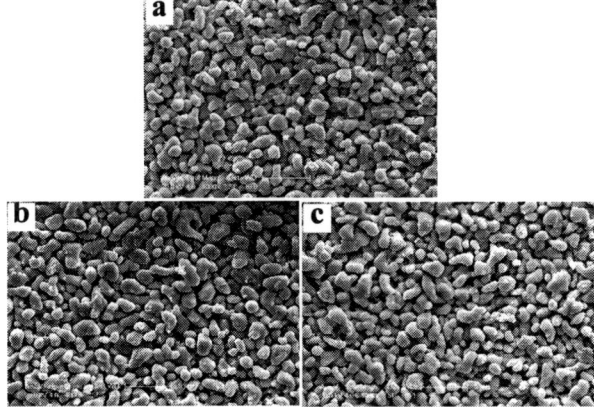

Figure8. Top view of solder and OSP PCB side interface (a) Sn3.0Ag0.5Cu, (b) Sn1.0Ag0.5Cu, (c) Sn1.0Ag0.5Cu0.02Ni

Fig.8 illustrate the interfacial IMC of different solder compositions in top-view IMC observation when OSP pad was used on the PCB side, EDX results showed the IMC is Cu6Sn5, the results showed that the Ni addition in solder has no influence on the SnCu IMC microstructure. According to our observation, it can also be seen that the concentration of Ag in SAC solder has no obvious influence on the shape and size of SnCu intermetallic compounds with OSP pad was used on PCB side.

3.2 Drop test performance

According to previous study [9], the solder joint reliability (SJR) is governed by two properties: the bulk property of solder itself and the property of the interface formed between the solder and base metal or pad. The strain rate experienced by solder joints or the boards during drop/shock testing is estimated to be ~102/sec., which belong to dynamic-to-impact loading conditions. Under these conditions, the behavior of the metallic materials is dominated by elasticity. In other words, plasticity is suppressed under these high strain rates. Therefore, elastic compliance is becoming a key material property for shock performance. A high-compliance solder is expected to be favorable for shock performance because it tends to lower stress transferred to vulnerable joint regions. Since compliance is not very sensitive to microstructure, the constituent phase need to be optimized for higher bulk compliance. Among all constituent phases in SAC, the primary Sn phase has the highest compliance and need to be optimized in the solder alloy design. There are two alloying elements in SAC: Ag and Cu, interfacial reaction and resultant interface characteristics especially on Ni are known to be very sensitive to Cu content. Ag, which does not participate in interfacial reactions, is therefore selected for bulk optimization. According to the Sn-rich region of Sn-Ag-Cu ternary diagram [10], the lower Ag content give rise to a more primary Sn phase and therefore is expected to result in higher compliance. Fig.9 (a) and (b) shows the solder bulk microstructure of Sn3.0Ag0.5Cu and Sn1.0Ag0.1Cu, respectively, compared Fig.9 (a) and (b), it can be seen that the fraction of primary Sn content in Sn1.0Ag0.5Cu is higher than Sn1.0Ag0.1Cu, according to the analysis above, solder joint with Sn1.0Ag0.1Cu should show better drop test performance than Sn3.0Ag0.5Cu.

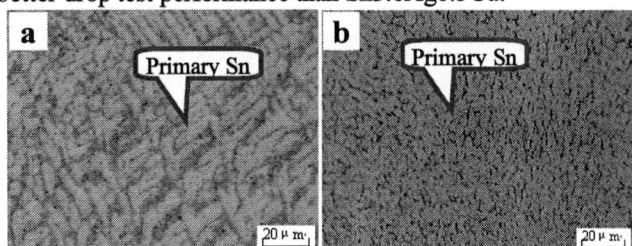

Figure9. Optical photo of solder bulk in solder joint (a) Sn3.0Ag0.5Cu (b) Sn1.0Ag0.5Cu

Fig.10 shows a weibull plot of drop test failure analysis of three solder alloys, it can be seen that the Sn1.0Ag0.5Cu0.02Ni solder showed better drop test performance than the other two alloys when NiAu pad was used on PCB pad. According to our analysis above, the needle like (NiCu)3Sn4 formation is the main influence factor for the small grain size IMC can inhabit the crack propagation after crack formation in drop test. But for OSP pad on PCB, the Ni doping has no influence on the formation of interfacial IMC, shown in Fig.8. The Ni addition in solder alloy doesn't have any effect on the drop performance improvement when OSP pad was used (shown in Fig.10(b). the lifetime of SAC305 is even a little higher than Sn1.0Ag0.5Cu and Sn1.0Ag0.5Cu0.02Ni. According to the analysis above, the Ni doping take effect only when NiAu pad finish on PCB side, so the influence factors are so complicated, the solder composition (including Ni addition), pad finish must be considered synthetically.

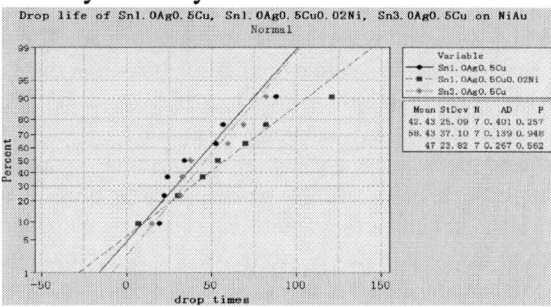

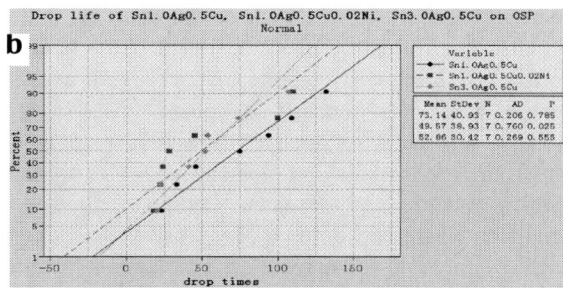

Figure10. Weibull plot for drop test

The failure mode of solder joint during drop test is illustrated in Fig.11. All the failure occurred on the PCB pad. There are totally two kinds of failure mode: pad crack (or called pad lift) as shown in Fig.11 (a) and (b), and crack between interfacial IMC and pad finish. For Au/Ni/Cu pad solder joint, the crack is between (Cu,Ni)6Sn5 and Ni finish, as shown in Fig.11(b). For OSP (Cu) pad solder joint, the crack propagates along the interface of Cu6Sn5 and Cu pad, as shown in Fig.11(c). the addition of Ni has no influence on the fracture mode in our study.

Interfacial IMC is a chemical reaction product during interfacial reaction. Interfacial reaction of solder is highly dependent on a base metal and solder constitution. For example, interfacial reaction between the SAC solder and Cu board /substrate pad leads to two types of IMC: Scallop shaped Cu6Sn5 next to solder and planar Cu3Sn next to Cu, the interfacial strength or joint integrity inversely scales with the overall thickness of IMC layers on base metal, Cu3Sn is particularly more vulnerable to shock fracture than Cu6Sn5. In our research, from fracture mechanism analysis in drop test, if the assemblies broken in solder joints, almost all fractured at the IMC, the fracture mode is shown in Fig.11.According to the analysis above, both the solder bulk and interfacial microstructure has a great effect on the joint, their influences are complicated and need to be synthetically considered.

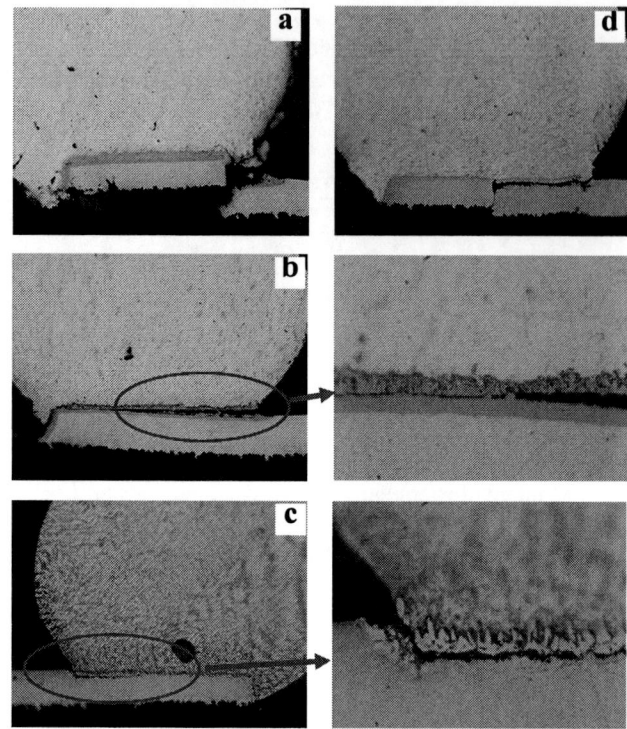

Figure.11. Cross-sectional SEM micrograph of Crack formation and propagation between solder and PCB in drop test

3.3 Bending test

Fig.12 showed the weibull plot of drop test failure for three types of solder alloys when the pad finish on PCB side is OSP, as can be seen in the Weibull plot. Sn3.0Ag0.5Cu and Sn3.8Ag0.7Cu showed better drop test performance. In the case of NiAu pad, Sn1.0Ag0.1Cu showed better bending test performance. The test results indicated that if different pad finish was employed, bending test performance is different for NiAu and OSP.

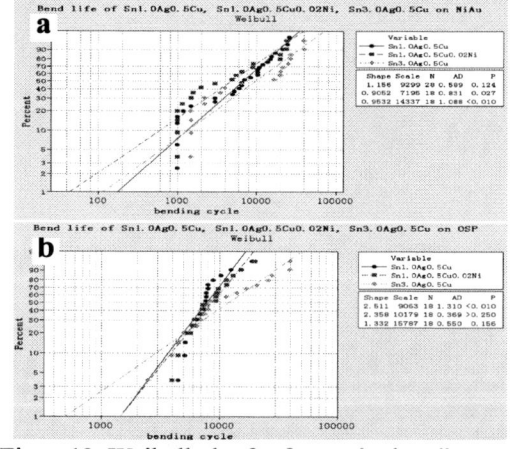

Figure12. Weibull plot for four point bending test
(a) NiAu pad, (b) OSP pad

Failure mode was studied to find the root cause of phenomena. Four kinds of failure modes were observed in the analysis. The first one is the test board pad de-laminates from the laminate core. A small amount of epoxy resin usually remains with the pad. Once the pad delaminates, it is free to move up and down as the test board is bent. This causes eventual fatigue failure of the test board trace. The second Failure mode is due to test board trace failure. However, in this case, the pad does not delaminate from the laminate core. The trace fatigues and cracks near the region of the solder mask opening. This type of failure mode is most difficult to detect. Fig.13 showed this kind of failure in bending test of our study. The third Failure mode is due to solder fatigue failure near the test board pad. The forth failure mode is due to solder fatigue near the component interface. This failure mode was observed only rarely in the present study, because in most cases the component pad was somewhat larger than the test board side. The result is consistent with previous publication.

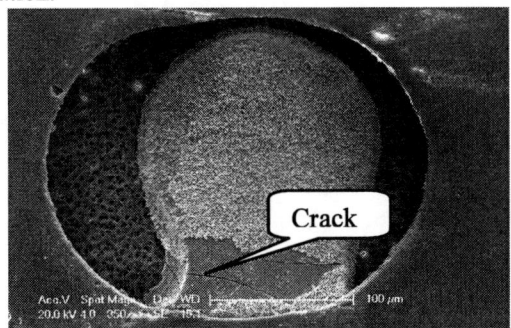

Figure13. Crack formation on PCB NiAu pad in bending test

In the present study, OSP and NiAu finish were used on the test boards. For OSP pad, there were no interface failures observed. For test boards with Ni/Au finish, it is possible to observe interface failures at the test board pad interface. The test boards with NiAu pad on PCB side showed poor bending test performance under same conditions is probably due to that the failure mechanisms are different for OSP and NiAu pad in bending test.

4.Conclusion

In this paper, the interfacial IMC morphology with Ni doping in SAC alloys was investigated. When Ni/Au pad was used on the PCB side, the morphology of IMC was sensitive Ni addition in contrast to OSP pad. Fine (NiCu)3Sn4 structure IMC formed at interface. In drop test, the Ni addition can improve the drop performance when NiAu pad used on PCB, but has no influence in the case of OSP. This is because that the fine grain can inhabit the crack propagation according to our analysis. Microstructure analysis in drop test results showed that the influence of bulk solder and interface IMC microstructure on the drop test performance is complicated and must be considered synthetically. In bending test, OSP pad showed a little better bending test performance than NiAu pad, the doping of Ni doesn't have obvious effect on bending test performance either OSP or NiAu pad used on PCB. Fracture mode analysis showed that NiAu pad on PCB side is more fragile to interface failure in bending test.

References:

1. J.W. Jang, D.R. Frear, T.Y. Lee and K.N. Tu, Morphology of interfacial reaction between lead-free solders and electroless Ni-P under bump metallization, J. Appl. Phys. 88(11), 2000, 6359-6363

2. C.M. Tsai, W.C. Luo, C.W. Chang, Y.C. Shieh, and C.R. Kao, Cross-interaction of under-bump metallurgy and surface finish in flip-chip solder joints, J.Electr. Mater. 33 (12) (2004) 1424-1428

3. M.N. Islam, A. Sharif, and Y.C. Chan, Effect of volume in interfacial reaction between eutectic Sn-3.5%Ag-0.5%Cu solder and Cu metallization in microelectronic packaging, J. Electr. Mater. 34 (2), 2005, 143-149

4. Z. Chen, M. He, and G.J. Q, Morphology and kinetic study of the interfacial reaction between the Sn-3.5Ag solder and electroless Ni-P metallization, J. Electr. Mater. 33 (12) (2004) 1465-1472

5. C.E. Ho, R.Y. Tsai, Y.L. Lin, and C.R. Kao, Effect of Cu concentration on the reactions between Sn-Ag-Cu solders and Ni, J. Electr. Mater. 31(6), 2002, 584-590

6. C.E. Ho, Y.W. Lin, S.C. Yang, and C.R. Kao, Volume effect on the solder reaction between SnAgCu solders and Ni, IEEE Proceedings of International Symposium on Advanced Packaging Materials: Processes, Properties and Interfaces, 2005. 39-44

7. Yanghua Xia, Xiaoming Xie, Chuanyan Lu, Junling Chang, Coupling effects at Cu(Ni)-SnAgCu-Cu(Ni) sandwich solder joint during isothermal aging. Journal of Alloys and Compounds, 417 (2006), 143-149

8. S.J. Wang, C.Y. Liu. Study of interaction between Cu-Sn and Ni-Sn interfacial reactions by Ni-Sn3.5AgCu sandwich Structure.

9. D.Q. Yu, L. Wang, C.M.L. Wu, C.M.T. Law. The formation of nano-Ag3Sn particles on the intermetallic compounds during wetting reaction. Journal of Alloys and Compounds, 389 (2005), 153-158

10. K.S. Kim , S.H. Huh, K. Suganuma, Effects of intermetallic compounds on properties of Sn–Ag–Cu lead-free soldered joints. Journal of Alloys and Compounds, 352 (2003), 226-236

BGA Assembly Process Development for 45nm ELK CUP Devices

Andy Tseng[1], andy.tseng@aseus.com, Bryan Lin[2] ,bryan_lin@aseglobal.com,
Louie Huang[2], louie_huang@aseglobal.com, Mike Hung[2], mike_hung@aseglobal.com,

[1]ASE US, 3590 Peterson Way, Santa Clara, CA95054, USA, TEL: 1-408-986-6502,

[2] ASE Group, 26. Chin 3rd RD., Nantze Export Processing Zone, Kaohsiung, Taiwan 811,China, (TEL) +886-7-361-7131.

Abstract

The object of this study is to develop a set of optimized assembly process parameters for BGA package using 45nm ELK (extreme Low-K) and CUP (circuit under pad) wafer which is driven by high speed and high I/O requested. Due to chip size shrinkage with electrical performance improvement, most of 0.13μm and 90nm wafer process technology are moving toward 65nm and even 45nm now. The ELK dielectric material for Inter-Level Dielectric (ILD) with the CUP has been designed to get more space for active circuit layout. But the poor mechanical properties of the low-k dielectric and the CUP structure circuit pad design make packaging assembly more challenges. The impacts of IC packaging assembly processes are including the wafer sawing, wire bonding, and molding process. For mass production purpose, the most effective parameters for 45nm ELK CUP wafer have been studied such as sawing blade type, sawing speed and sawing feeding rate for different wafer thickness, the wire bond time, bond power and bond force. To solve bond wire sweep and mold void issues, the properties of different molding compounds have been studied and assembly process parameters have been optimized. In the end, a real functional die of 45nm ELK with CUP design has been assembled into package level for reliability test using optimized process parameters.

Introduction

In the last decade, the electronics devices are requested to be smaller size, lighter weight, faster speed and lower cost. In the mean while, the I/O numbers increases for more functionality. Therefore, the wafer process technology is moving fast from 0.18um, 0.15um and 0.13um Aluminum standard wafer down to 90nm and 65nm Copper Low-K wafer for fast speed applications and more dice per wafer. Recently, the 45nm ELK wafer process technology is introduced and completely developed. The copper is used for inner layer circuits to replace Aluminum metal and the lower dielectric material (ELK) is used for inter-Level dielectric (ILD) to reduce RC circuit delay [1]. As wafer technologies shrinking rapidly faster than the size of wire bond pad, it becomes that bond pad occupy the circuit layout area in a silicon chip. If the active circuit of a chip can be placed under bond pads, the die cost could be reduced obviously and the design flexibility would also be improved. For those reasons, the ELK wafers [2] with circuit under pad (CUP) design are getting more and more popular.[3]

Figure 1. An Bond wire on ELK CUP bond pad structure. As the soft property of low-K material, the bonding process windows are squeezed: Too little force will cause bond fail and too much force create cracks or peeling.

The mechanical performance of low-K material such as brittleness, porosity, inferior adhesion and poor intrinsic mechanical strength do impact the adhesion between the interfaces of materials inside the wafer. Many of studies and papers are talked about the low-K wafer or in the last couple years [4][5][6][7]. Figure 1 show the ball shape bonded on ELK CUP wafer. The most defects of assembly process for ELK CUP wafer are the sawing kerf metal-peeling, wafer topside/backside chipping and bonding pad metal-peeling. Figure 2.1, 2.2 and 2.3 are shown the defect in wafer saw process.

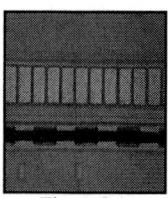 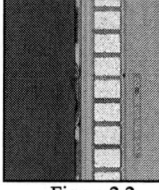 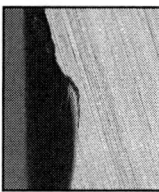

| Figure 2.1 | Figure 2.2 | Figure 2.3 |

Figure 2.1. Sawing street and test pad residue
Figure 2.2 Topside chipping
Figure 2.3. Backside Chipping.

Figure 3.1 shows the bond pad peeling after wire bond process. The figure 3.2 is an ILD delamination during temperature cycling caused by molding process Those defects are known from poor mechanical properties. The new assembly processes need be studied and be optimized, especially for mass production purpose.

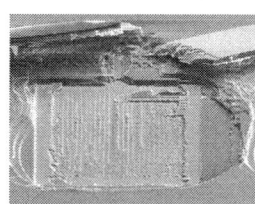 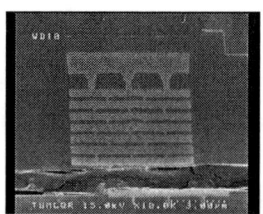

Figure 3.1. Bond pad metal peeling

Figure 3.2, Metal peeling after molding

978-1-4244-2739-0/08/$25.00 ©2008 IEEE

With previous experiences on 0.13um, 90nm and 65nm Low-K wafer [8][9][10] and CUP pad [11][12][13], the mechanical strengths, thermal compatibility, CTE (Constant of Temperature Expansion) of materials have been studied and learned. The sawing blade types, the cutting speed, the feeding rate, and the bonding parameters have been refereed to this study. Molding compounds selection for lower stress and smaller particle filler for poor intrinsic mechanical strength have been performed to reduce wire sweep and delamination issues. In the end, a real functional wafer has been assembled into packages and been qualified for mass production.

Process Experiment:

45nm Cu ELK CUP 300mm wafer has been used for this study and also for assembly process optimization.

Part 1: Wafer Sawing Process:

The challenges of top side peeling or chipping and backside chipping always happened in wafer sawing process due to brittle and lower hardness ELK wafer with CUP design. The cracking size after sawing is monitored and measured. Table 1 is sawing experiment matrix. The spindle and sawing speed are discussed. Three speeds (high, middle and low) are set for each item (Spindle and feed speed).

Table 1: Experiment matrix for sawing process optimization

		Feed (mm/s)		
		Low	Middle	High
Spindle (rpm)	Low	X	X	X
	Middle	X	X	X
	High	X	X	X

Part 2: Wire Bonding Process.

Bond power, bond time and bond force are the key factors of wire bonding process. The bond energy and bond force form the IMC (intermettallic Compound). The relationship of these two parameters is critical and inseparable. Increasing force beyond an optimal level will cause bond diameter to exceed diameter requirement at the desired squash height. If the force level is too low for a given amount of ultrasonic energy, the silicon can fracture below pad structure and causing cratering. An experiment matrix is set in able 2.

Table 2: wire bonding experiment matrix

1st bond	Bond Current	Bond Time	Bond Force
Cell 1	Low	Low	Low
Cell 2	Low	Middle	Middle
Cell 3	Low	High	High
Cell 4	Middle	Low	Middle
Cell 5	Middle	Middle	High
Cell 6	Middle	High	Low
Cell 7	High	Low	High
Cell 8	High	Middle	Low
Cell 9	High	High	Middle

The data of ball size, ball shear and wire pull are for bondability evaluation and process parameters optimization. The experience of CUP bond pad structures in the metal layers and the via arrangement have been evaluated and

discussed in papers [11][12][13] and it will be referred to this study. The test die has bond pad pitch 40um and bond pad opening 35x80mm. Gold wire diameter is 0.7 mills.

Part 3: Molding Process.

Thermal mismatch between various materials within a package is the main cause of the package reliabilty failure. i.e., the CTE of molding compound contacted with die surface directly in plastic package is significantly larger than the silicon die and is one of the major sources for the die surface stress. When selecting a molding compound, the properties including Tg, (glass transition temperature), moisture absorption rate, flexural modulus/strength, CTE, thermal conductivity, and adhesion properties need be taken consideration. After several molding compounds with different material properties have been studied and the less stress and lower modulus mold compound were selected for ELK and CUP wafer. Wire sweep and delamination are the most defects were found after molding process. Wire sweep will affect electrical performance by changing the mutual inductance of adjacent wires, NNS (simultaneous switching noise), and it will fail the function by circuit short if wires actually touch. Delamination happened when thermal mismatch between compound and low-K die within a package. Two molding compounds have been selected for final evaluation (brand A and brand B) shown in table 3.

Table 3. Mold compound properties for evaluation

Item		Unit	Compound A	Compound B
Spiral flow		cm	150	105
Flow Viscosity		Pa-s	17	10
Specific Gravity			2.2	2
Tg		°C	150	145
CTE	α1	ppm/°C	7	9
	α2	ppm/°C	27	43
Flexural modulus		GPa	26	17.5
Mold shrinkage		%	0.08	0.15

Experiment Result and Discuss.
Part 1: Wafer Sawing:

By applying different speed level of spindle and feeding rate in wafer sawing process, the chipping measurement data are shown in table 4 and 5. The chipping length is not touching the seal ring and within QA criteria which is shown in figure 4.

Table 4. Topside chipping data.

Top side chipping (max/min) um		Feed (mm/s)		
		Low	Middle	High
Spindle (rpm)	Low	22/19	21/18	21/19
	Middle	20/16	21/17	22/18
	High	21/19	22/18	21/17

Table 5. Backside chipping data.

Back side chipping (max/min) um		Feed (mm/s)		
		Low	Middle	High
Spindle (rpm)	Low	60/41	93/74	101/87
	Middle	76/45	100/85	98/65
	High	111/83	111/89	112/99

Spindle (rpm)	Feed Rate (mm/s)	Die Corner	Test Pad	Metal Layer
Low	Low			
Middle	Middle			
High	High			

Figure 4. Topside chipping images at different locations.

Part 2 Bonding Process:

The Bondability study are shown in table 6.1, 6.2 and 6.3. Table 6.1 is wire pull data. Table 6.2 is ball shear data and table 6.3 is ball size data.

Cell 1 has two lifting issues and the image is shown in figure 5.1 which wire pull force is 1.8 gram and 1.9 gram. The most common root cause of ball lifting will be incorrect wire bond parameter settings such as improper bond energy and bond force, bond pad corrosion or contamination, excessive bond pad probing, Kirkendall voiding, excessive thermal stress resulting in excessive intermetallic formation, cratering and so on. After failure analysis, it is improper bonding parameters setting. The other cells have no wire peeling, ball lifting and NSOP (Non stitch on pad) was found and the wire-pull force is greater than 2.0 gram which is beyond product specification.

Table 6.1. Wire pull data.

| | First Bond | | | Wire Pull | | | |
	Current	Time	Force	Ave	Peeling	lifting	NSOP
1	L	L	L	5.1	0	2	0
2	L	M	M	5.3	0	0	0
3	L	H	H	5.4	0	0	0
4	M	L	M	5.4	0	0	0
5	M	M	H	5.5	0	0	0
6	M	H	L	5.6	0	0	0
7	H	L	H	5.5	0	0	0
8	H	M	L	5.6	0	0	0
9	H	H	M	5.7	0	0	0

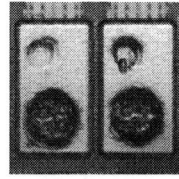

Figure 5.1.. Two Ball lifting images.

The wire ball shear data is shown in table 6.2 and all of them of each cell is passed 5.0 gram of product specification. No cratering was found. The ball shape image was shown in figure 5.2.

Table 6.2. Ball shear.

| | First Bond | | | Ball Shear | | | Cratering |
	Current	Time	Force	Max	Min	Ave	
1	L	L	L	8.9	7.3	8.1	0
2	L	M	M	9.7	7.1	8.3	0
3	L	H	H	9.2	6.4	8.2	0
4	M	L	M	11.7	10.4	10.9	0
5	M	M	H	13.1	9.9	11.7	0
6	M	H	L	10.3	7.9	9.2	0
7	H	L	H	13.9	11.7	13.1	0
8	H	M	L	9.8	8.5	9.3	0
9	H	H	M	11.2	9.4	10.3	0

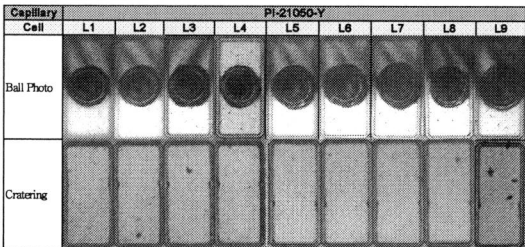

Figure 5.2. No crating was found.

Table 6.3 shows wire ball size measurement.. Cell 1 has larger thickness due to lower bond force and bond time.The measure point X, Y and thickness are shown in figure 5.3,

Table 6.3. Ball shape and size.

| | First Bond | | | Ball Size | | |
	Current	Time	Force	X	Y	Thickness
1	L	L	L	31.6	31.8	7.86
2	L	M	M	32.4	32.3	7.18
3	L	H	H	32.6	33.1	6.79
4	M	L	M	32.6	32.8	6.21
5	M	M	H	33.2	33.3	5.56
6	M	H	L	33.5	34.1	5.48
7	H	L	H	33.5	33.6	5.25
8	H	M	L	32.7	32.9	6.11
9	H	H	M	33.6	34.5	5.51

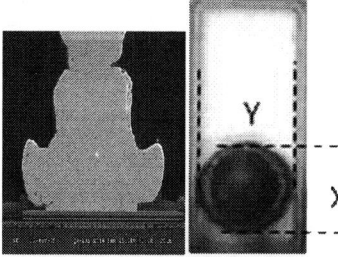

Figure 5.3. Ball shape and size.

Part 3: Molding Process:

X-Ray and SAT have been used to measure the wire sweep, delamination and void inside molding compound. Delamination happened inside ELK layer also can be detected by SAT. Figure 6.1 shows no abnormal was found in compound A. Figure 6.2 shows delamination happened at die corner in compound B. Figure 7.1 is compound A wire sweep 6.5% and figure 7.2 is compound B wire sweep 7.9%.

Compound A will be used for package qualification based on the measurement data.

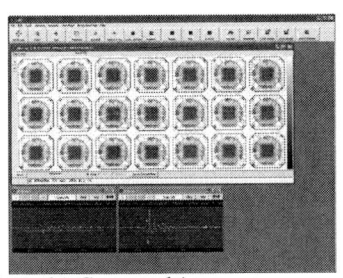

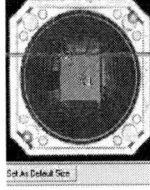

Figure 6.1, Compound A. No delaminationl was found.

Figure 6.2 Compound B. Delaminationl was found at die coner.

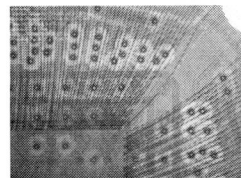

Figure 7.1, Compound A wire sweep 6.5%

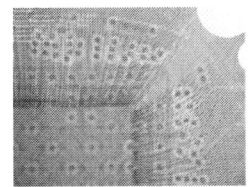

Figure 7.2, Compound B wire sweep 7.9%

Package Level Qualification:

Table 8. The key process data

um	Item	Max	Min	Ave	
Wafer Saw	Kerf Width	39	23	33	PASS
	Topside chipping	No touch seal ring			PASS
	Backside chipping	102	49	67	PASS
Die Attach	Die shear (Kg)	15.05	15.03	15.04	PASS
	Epoxy void	None			PASS
	Epoxy thickness	30	25	28	PASS
Wire Bond	Wire pull (>2g)	5.3	4.1	4.6	PASS
	loop height	239	231	235	PASS
	Ball size	36	33	34	PASS
	Ball shear(> 5g)	11.14	7.39	9.22	PASS
	Cratering	None			PASS
Mold ing	Wire sweep (<10%)	5.6%	5%	5.2%	PASS
	Mold void	None			PASS
Ball Mount	Ball shear (> 0.6Kg)	1.55	1.34	1.44	PASS
SAT	Delamination	None			PASS
PKG	Co-planarity (Max 152um)	102	69	86	PASS

31x31mm HSBGA 899 balls with 1.0 ball pitch and 0.6mm ball size has been assembled for qualification. 4 layers green substrate with 0.36mm thickness has been used for this package. The wafer is 300mm size 45nm ELK CUP wafer and has been ground and sawn into 10x10mm die size with

12 mil thickness. 4N 0.7mil gold wire, liquid type epoxy and compound A are adopted. The solder ball composition is SAC305. Optimized parameters of assembly process of wafer sawing, wire bonding and molding are performed for packaging assembly.

The key process data are shown in table 7. Topside /backside chipping have been measured and no chipping touched the sealing ring. Figure 8 shows the sawing results. Bond line thickness and epoxy bleeding length have been measured after die attached. The data of wire pull, ball shear and ball shape is passed specification and no cratering was found. The maximum wire sweep 5.6% after molding process. The minimum solder ball shear is 1.34kg beyond specification. The final check is packaging outline dimension and the coplanarity is within 152um, all the check point and data are passed QA inspection and process criteria.

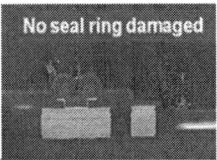

Figure 8, Die corner top view After sawing and it is pass criteria by no chipping to damage seal ring.

MSL (Moisture sensitivity level) 3 for pre-condition, and the reliability test items are PCT (Pressure Cooker Test), TCT (Temperature Cycling Test), HTST (High Temperature Storage Test), HAST (High Accelerated Stress Test) and THT (Temperature Humidity Test).

Figure 9.1 is an IMC structure at time zero. And figure 9.2 is IMC cross-section after HTST 1000 hours. Both has no abnormal were found.

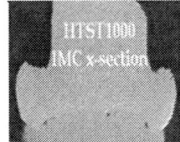

Figure 9.1   Figure 9.2

Figure 10. An X-Ray image to show the wire sweep the within packaging qualification specification.

Figure 10. X-Ray Image : No Wire short happened and wire seep is 5.6%.

Table 8 shows the reliability data that 31x31mm HSBGA passed MSL3 TCT, HTST and HAST test. The SAT picture showns in figure 11 has no any delamination happened.

Table 8. 31x31mm HSBGA 899 balls passed MSL 3 TCT, HAST, and HTST test for The 45nm ELK CUP wafer.

Testing item	MSL 3 260C	TCT – C 1000cycle	HTST 1000hr	HAST 168hr
Sample size	0/135	0/45	0/45	0/45

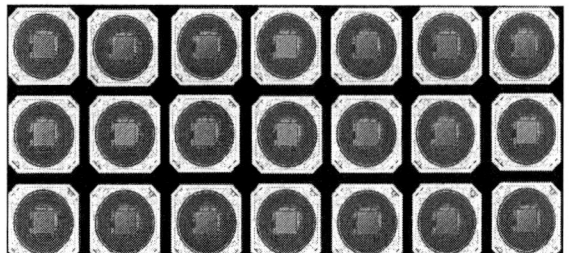

Figure 11. SAT Image shows no delamination happened.

Conclusion

While the wafer technology is moving toward nanometer and the IC chip size becomes smaller and smaller, the performance and the speed of electronics device is getting higher and getting faster, and also, the I/O count is increasing for more functionality. The semiconductor manufacturers tend to use Cu circuit and ELK dielectric material with CUP bond pad design to improve electrical performances and design flexibility. The IC packaging becomes challenge due to the material properties changes. The parameters of wafer sawing process, wire bonding process and molding process have been developed for 45nm Cu ELK CUP wafer and 31x31mm HSBGA 899 balls packaging has been qualified for Cu ELK 45nm wafer with CUP by using optimized process.

Acknowledgment

The authors would like to thank R&D group of Advanced Semiconductor Engineering, Kaohsiung, Taiwan for their support in data collection and consulting.

Reference:

[1] KY Chou, MJ Chen, CC Lin, YS Su, CS Hou, TC Ong, "Die cracking evaluation and improvement in ULSI plastic package", IEEE, int. conf. on Microelectronic test structure", 2001 pp 239-244.

[2]. WR Anderson, WM Gonzalez, SS Knecht, W Fomler, "ESD protection under wire bonding pads:, in proc. Od EOS/ESD symp, 1999, pp88-94.

[3]. Jonathan Tan, Zhao Wei Zhong, Hong Meng Ho, "Wire bond process development for Low-K material,

[4]. Bob Chylak, et al, "Wafer probe, wire bond, and packaging issues for bonding over Cu/low-K dielectric materials", K&S Willow Grove, PA, SEMICON, China, Technology Symposium, IEMT Symposium, 2003.

[5]. Y.F. Yao, et al, "Assembly process development of 50um fine pitch wire bonded devices", ECTC 2004.

[6]. Yoon-Joo Kim et al, "Low-K wire bonding", Amkor Inc, ECTC 2006.

[7]. Chang-Lin Yeh, et al, "Pull test simulation and optimization for wire bond on Cu/Low-K wafers", ASE,

[8]. Andrew F. Hmiel, Claire Ruitiser, "Wire bond short reduction by encapsulation", K&S, SEMICON, International Electronics Manufacturing Technology Symposium, IEMT, 2003

[9]. Min-Shuoh Liang, Louie Huang, "Impact of assembly process to the Cu/Low-K chip integrity", ASE.

[10]. Inderjit J.Y. On, L. Levine, "Enhancing fine pitch, high I/O devices with copper ball bonding", ECTC, 2003.

[11]. Ming-Dou Ker, et al, "Active device under bond pad to save I/O layout for high-pin-count SOC", ISQED, 2003.

[12]. Ken Tamala, et al. "Resolution of a fine pitch wire bonding reliability problem", SEMICON, Singapore, 2006.

[13]. Kevin J. Hess, et al, "Reliability of bond over active pad structure for 0.13um CMOS Technology", Motorola, ECTC 2003

C4NP for Pb-Free Solder Wafer Bumping and 3D Fine-Pitch Applications

D.-Y. Shih, B. Dang, P. Gruber, M. Lu, S. Kang, S. Buchwalter, J. Knickerbocker, E. Perfecto* J. Garant*, S. Knickerbocker*, K. Semkow*, B. Sundlof*, J. Busby*, R. Weisman* K. Ruhmer** and E. Hughlett**

IBM T. J. Watson Research Center 1101 Kitchawan Rd., Yorktown Heights, N.Y. 10598

* IBM System and Technology Group

**SUSS MicroTec, Inc., Waterbury, VT, USA

Email: dys@us.ibm.com

Abstract

Controlled Collapse Chip Connection – New Process (C4NP) technology is a novel solder bumping technology developed by IBM to address the limitations of existing bumping technologies. Through continuous improvements in processes, materials and defect control, C4NP technology has been successfully implemented at IBM in the manufacturing of all 300mm Pb-free solder bumped wafers. Both 200 μm and 150 μm pitch products have been qualified and are currently ramping up volume production.

Extendibility of C4NP to 50 μm ultra-fine pitch microbump application has been successfully demonstrated with the existing C4NP manufacturing tools. Targeted applications for microbumps are three-dimensional (3D) chip integration and the conversion of memory wafers from wirebonding (WB) to C4 bumping. The metrology data on solder volume, bump height, defect and yield have been characterized by RVSI inspection. This paper reviews the C4NP processes from mold manufacturing, solder fill and solder transfer onto 300 mm wafers, along with defect and yield analysis. Reliability challenges as well as solutions in the development and qualification of flip chip Pb-free solder joint are also reviewed. In addition to a suitable under bump metallurgy (UBM), a robust lead-free solder alloy with precisely controlled composition and special alloy doping is needed to enhance performance and reliability.

Introduction

Demand for the flip chip interconnect structure has grown rapidly since it was first introduced in IBM's System/360 computer over 40 years ago. [1] As feature size continues to scale down, the number of transistors and interconnects on a chip has increased rapidly. As a result, the number of chip-to-package input/output (I/O) interconnects have also increased significantly in the past decades [2]. On the one hand, flip-chip I/O pitch is being reduced continuously to meet the requirement of I/O counts in high-performance and high-bandwidth applications. Fine-pitch wire bond interconnection which is still popularly used in memory wafers needs to be replaced by fine-pitch area interconnects due to the performance limitation of wire-bond technology in the high-frequency regime. Furthermore, fine pitch interconnection is in high demand for 3D integration of semiconductor chips because of the benefits in power distribution, signal latency, small form factor as well as chip to chip communication bandwidth, etc. Therefore, manufacturing of fine pitch C4 (≤50μm) interconnects is required for future packaging structures. A number of solder bumping technologies have been developed and used in manufacturing production. These include evaporation [1], paste screening [3], electroplating [4], and the direct attach of preformed solder spheres. [5]

However, not all of these C4 bumping technologies are extendable to fine-pitch applications for volume production at low cost. In addition, as the microelectronic packaging industry is moving to the "green" Pb-free solder as mandated by the European Union (EU) RoHS legislation, the high Sn-content solder has presented major challenges in meeting reliability requirements. The selected bumping technology is preferred to have alloy flexibility to precisely control composition and allow alloy element "doping" to form a multi-component solder alloy to deliver enhanced performance and reliability. To address these major issues associated with Pb-free solder, C4NP technology has been developed [6], along with IBM's commercialization partner, SUSS MicroTec. [7] The newly developed manufacturing tools for 300mm wafers have backward 200mm compatibility. Thus, the high volume manufacturing (HVM) toolset that has been delivered to IBM routinely bumps 300mm wafers, but can seamlessly process 200mm whenever the need arises.

Reliability Challenges for Pb-free Solders

Compared with leaded solders, Pb-free solders, such as the popularly used SnCu, SnAg and SnAgCu, all contain more than 95 w.t.% of Sn, which not only have higher melting temperature than eutectic PbSn but are highly reactive with UBM and substrate pad. This necessitates a thicker and/or more robust reaction barrier layer to survive the multiple reflows and various reliability tests without being totally consumed. In addition, two critical reliability challenges were encountered during qualification. One is commonly referred to as ILD (Interlayer Dielectric) delamination [8, 9] due primarily to the use of fragile low-k dielectric layer which is aggravated by the use of high yield strength Pb-free solders, and on large chips due to high DNP (distance from neutral point) issues. During cooling down of the chip/solder/laminate assembled module, the differential thermal expansion mismatch (CTE) between chip and laminate, and the inability of the high yield strength Pb-free solder to deform has often caused cracking in the ILD layer in the back-end- of-line (BEOL) structure, shown in Fig. 1.

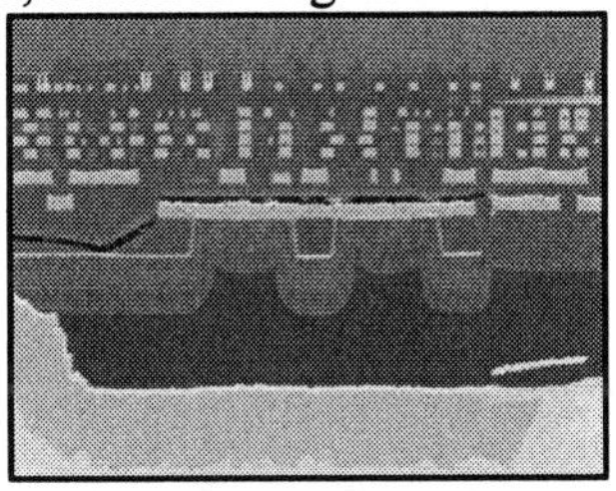

Fig. 1. SEM image of cracked BEOL.

The open circuit can be detected by using acoustic scan imaging and identified as a "white bump", shown in Fig. 2.

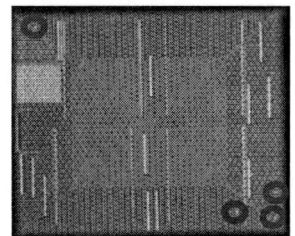

Fig. 2. Acoustic scan image showing "white bumps".

Reducing the cooling rate, which reduces manufacturing throughput, is less practical. Optimizing solder composition along with improving the adhesion strength in the BEOL structure were both shown to effectively mitigate ILD cracking. With the continued movement to ultra low-k materials in the 32 and 22 nm nodes of silicon technology the chip package interaction (CPI) problem becomes even more challenging.

To solve the ILD cracking problem, C4NP bumping technology was readily used to facilitate quick debug, optimization and, finally, identification of the solder solution among a comprehensive list of solder candidates. Table 1 correlates solder composition with hardness which agrees with laminate warpage data [9]. Low Ag and Cu content solders have lower hardness and more compliant mechanical properties which is needed to absorb the stresses before they can be transmitted to the BEOL ILD layer to cause cracking. Fig. 3 shows a module level microhardness indentation measurement on the solder joint.

Table 1: Module level solder joint microhardness and indentation measurements.

wt% Sn	wt% Ag	wt% Cu	Hardness, HV Mean (std dev)
97.6	2.2	0.2	**16.0 (0.6)**
98.5	1.3	0.2	**14.5 (0.8)**
98.5	0.9	0.6	**14.0 (0.0)**
98.6	1.2	0.2	**14.0 (0.6)**
99.5	0.3	0.2	**12.0 (0.9)**
99.3	0	0.7	**11.5 (0.5)**

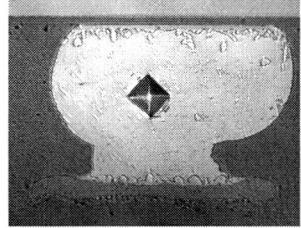

Figure 3. A typical module level micro-hardness indentation measurement on solder joint.

Electromigration (EM) is another major challenge of using high Sn solder due to its highly anisotropic crystal structure [10], high diffusivities [11], and highly reactive property [12]. Compounded by high current density and local Joule heating, the solder joint can fail prematurely with a wide spread of failure times. To enhance the EM life time, various solutions have been reported, including the use of improved UBM

structure and thickness, solders with optimized compositions and specially doped alloys [13]. The latter solution is shown in Fig. 4 on the EM performance of SAC alloy. The increase in resistance is plotted as a function of test time when the solder joints were stressed at 5.2×10^3 A/cm^2 and 150 °C for 1100 hrs. For SAC alloy, shown in Fig. 4(a), some samples showed early failures due to resistance increases that exceeded failure criteria. In comparison SAC solder doped with a minor alloying element showed significantly enhanced performance by eliminating the early failures, as shown in Fig.4(b). The specially doped quaternary solder alloy can be readily used by C4NP technology.

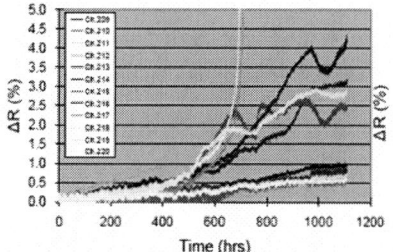

Fig. 4(a) SnAgCu solder joint stressed at 5.2×10^3 A/cm^2, at 150 °C for 1100 hrs. Some samples had early failures.

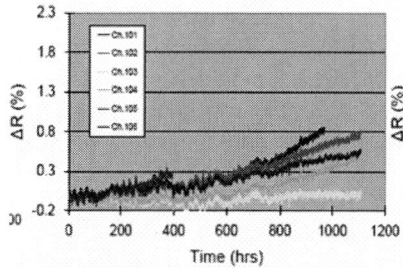

Fig. 4(b). A specially doped SnAgCu solder joint stressed under the same condition. Early failures have been eliminated.

Sn pest, the allotropic transformation of β-Sn (body centered tetragonal) into α-Sn at temperatures below 13 °C, is a long term reliability threat. It has been observed in Pb-free solders at low temperatures, as shown in Fig. 5(a) and Fig.5(b). [14,15] The transformation normally takes long time to happen. It is accompanied by an increase in volume by 26% and solder joint could practically disintegrate. The presence of residual stress in solder joint accelerates the transformation process. Doping the solder with a small amount of Bi and Sb was reported to suppress Sn pest formation. [16]

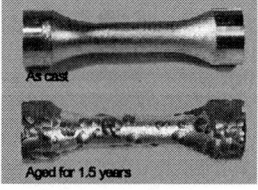

Fig. 5(a). Top view Sn0.5Cu samples aged for 1.5 years at 8 °C, compared to an as machined sample.

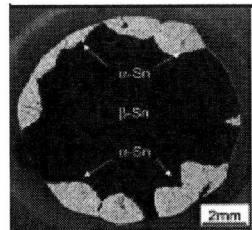

Fig. 5(b) Cross-section view shows Sn pest transformation starts from the surface in the highly stressed grip area.

Good drop reliability is critically important for hand-held devices where Pb-free solders are not as good as leaded solders. When doped with a small amount of Ni, Ce or Ti, the drop reliability of low Ag ($\leq$ 1%) SAC BGA joint was shown to improve significantly. [17,18]

Kirkendall voids are another reliability issue when solders, both Pb-containing and Pb-free, are joined to Cu and annealed for a prolonged time. Fig. 6(a) shows a SAC alloy joined to Cu and annealed at 150 °C for 1000 hrs. The voids form an almost continuous layer and seriously impact the reliability of BGA joints. However, when doped with a small amount of Zn the voids are eliminated as shown in Fig. 6(b). [19]

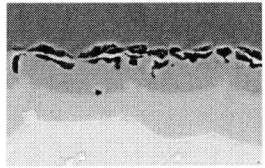 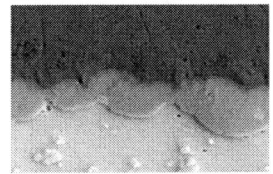

(a)   (b)
Fig. 6 (a). SAC , (b) doped SAC, after annealing

For C4NP technology changing solder alloys is simple. It is accomplished by changing the fill head in the mold fill tool, which is done in less than an hour. The process temperatures of the solder reservoir and the mold can be adjusted to accommodate a particular solder alloy. This flexibility allows C4NP to be backward compatible with existing solder alloys and UBM stacks as well as to enable the use of any multi-component new solder materials and UBM stacks. As previously discussed, "dopants" can be added to the solder to improve EM performance, suppress Kirdendall voiding, reduce copper pad consumption, suppress Sn pest, etc. The ability to maintain alloy flexibility with precisely controlled solder composition and alloy doping is critically important to deliver enhanced performance and reliability.

In the following, we will discuss C4NP processes, qualification and yield data for 200 and 150 µm pitch applications and metrology data for micro-bump for 3D chip stacking.

C4NP Process Flow

C4NP process starts with a glass mold in which the UBM I/O pads of an entire wafer are replicated in a mirror image with tiny cavities etched into the glass plate. These cavities are filled with solder as the mold is scanned below a fill head. The fill head contains a reservoir of molten solder and a slot through which the solder is injected into the mold cavities. The cavity depth and diameter determine the volume of the solder bumps that will be subsequently transferred to the wafer. The

filled mold is inspected automatically and then aligned below a wafer with exposed UBM pads facing the mold. Mold and wafer are heated above the solder melting point, vapor flux is applied to 'scrub' the pads and solder surface and then brought into close proximity/contact. The solder forms spherical balls which transfer from the mold to the UBM pads on the wafer, where they wet and solidify. Wafer and mold are separated, and the mold is cleaned for reuse. Fig. 7 describes this process flow.

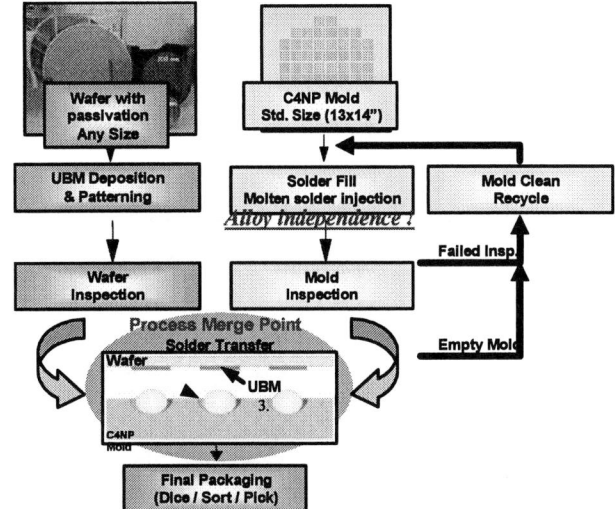

Fig. 7. C4NP process flow.

Mold Fabrication and Solder Transfer to Wafer

C4NP mold uses borofloat glass, which has a CTE closely matching that of silicon. A photolithographically defined pattern and wet etching were used to create the cavities. The molds are scanned beneath a solder injection head which fills the cavities with liquid solder precisely to the surface of the mold. Therefore, the solder volume transferred to the wafer at contact is directly a function of the glass cavity volume. The solder fill, auto inspection, solder transfer and cleaning processes are performed using Mold Fill, Mold Inspection, Solder Transfer and Mold Cleaning tools to provide the desired product C4 pattern on the wafers. Figure 8 shows the SEM image of typical wet chemically etched cavities in the C4NP glass mold. A typical glass mold exhibits three major types of deviations from perfect planarity. As shown in Fig. 9, type 1 is related to localized waviness and can not be compensated by the compression force during the solder transfer operation. Type 2 is the global waviness and can be accommodated by deformation under an increased compression force. Type 3 is from the non-uniformity in the glass mold thickness, which can be overcome through wedge compensation in transfer tool.

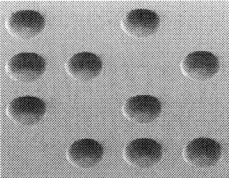

Figure 8. C4NP glass mold with etched cavities.

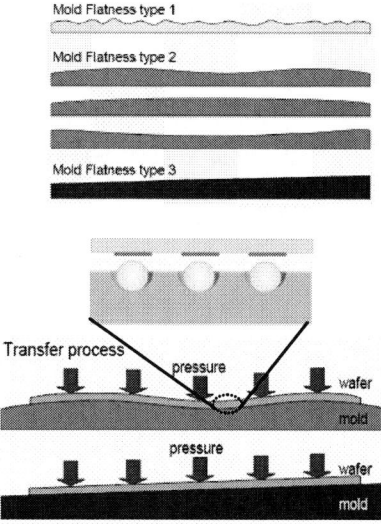

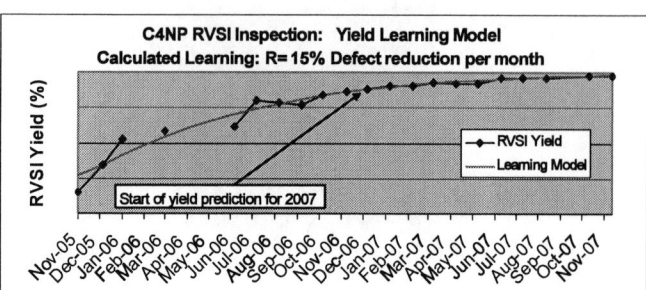

Fig. 10. C4NP Yield learning improvements

Fig. 9 describes three types of glass surface waviness (1) local, (20 global and (3) Wedge.

For the 200, 150 and down to 50 µm pitch applications, as received glass molds have been successfully used for mold fill and wafer transfer. For further extension to very fine pitch applications where solder volume becomes extremely small, to ensure a successful solder transfer, the flatness of the glass surface needs to be improved, along with techniques to increase the depth and sidewall angle of mold cavities to increase the solder stand-off height above the glass surface that overcomes the local non-flatness, as shown in Fig. 9.

Wafer Bumping Yield Improvements

C4NP bumping yield has improved significantly as process matures and the manufacturing tools were installed. Defect root cause analysis has resulted in process improvements in patterning the UBM pads, as well as in the mold and mold fill areas which contributed to impressive yield improvements. Figs. 10 illustrate the significant yield improvement over the last two years. [20, 21] The yield data is derived from RVSI inspection of the 200 µm pitch product wafers. Yield learning model showed a 15% defect reduction per month since the start of the C4NP program. The model had a correlation of actual vs predicted of 0.88. Initially, the major yield detractor was contamination, mainly caused by residual resist not completed stripped and contaminants generated from the manually operated low volume manufacturing tools. With the installation of a new resist stripper and the movement of manufacturing to high volume manufacturing (HVM) tools with FOUP to FOUP automatic handling and tightened cleanliness control, this type of defects was mostly eliminated. Also, as the quality of glass molds and mold fill processes improved, a significant reduction in missing C4s, along with improvements in volume uniformity and co-planarity, has contributed to improvements in transfer yield.

All these improvements have resulted in excellent bumping yield for both the 200 and 150 µm pitch applications. Best wafers have consistently achieved 100% yield. The feasibility for 50 µm pitch wafer bumping was successfully demonstrated in the same manufacturing environment using the same set of C4NP tools for mold fill and wafer transfer. [22] The mold inspection tool (MIT) which works well for the 150 to 200 µm pitch was unable to handle the high density bumps (~11,000 bumps per chip at 50 µm pitch) due to lack of pixel density. Using an improved inspection tool developed by RVSI, very high bumping yield was demonstrated. The good results were largely attributed to improvements in the volume uniformity of mold cavities, the elimination of oxides debris by performing mold fill in pure nitrogen ambient and high transfer yield. Four molds have been fabricated and characterized, as shown in Table 2, where molds 1 and 2 are defect free and molds 3 and 4 each contains < 1ppm defect among a total of ~ 9 million etched cavities for each mold.

Table 2. Micro bump mold cavity dimension, cavity volume uniformity and defect have been characterized.

Mold ID	Depth (µm)	Top Dia. (µm)	Solder Vol. (x10³ µm³)	3σ volume uniformity
1	13.6	39.9	11.9	8.0%
2	14.0	40.1	12.3	4.6%
3	13.9	40.0	12.1	4.8%
4	8.8	32.6	5.4	3.7%

Figure 11 compares mold filling under N_2/O_2 ambient (Fig. 11(a)) to that under pure N_2 (Fig.11(b)). More solder bridging between adjacent cavities is observed under N_2/O_2 mixture environment, while bridging is mostly eliminated in N_2 ambient. Solder bridging for standard C4s (pitch ≥150µm) does not require special care because the spacing between adjacent cavities is longer. However, bridging is much more sensitive to micro-bumps because of shorter spacing between cavities.

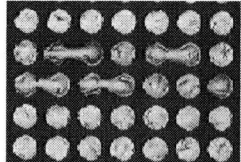

Fig. 11(a) Mold fill in O₂/N₂ mixture gas.

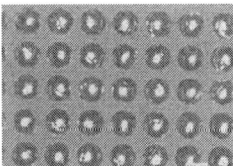

Fig. 11(b) Mold filled in Pure N$_2$.

With the high yield filled molds, micro-bumps were successfully transferred from glass molds to both 200 and 300mm wafers which have been patterned with a three-layer UBM, as shown in Figs. 12 and 13. With the aid of formic acid vapor flux, excellent wetting was achieved for the SnAg solder micro-bumps. The UBM pads are ~ 28µm in diameter. An additional reflow was performed to reshape the micro-bumps and uniform bump heights were obtained. Excellent height uniformity has been measured for the transferred microbumps (co-planarity < 2 µm). The preliminary results suggest that C4NP technology can be readily scaled down to 50 µm to meet the increasing demand on I/O density.

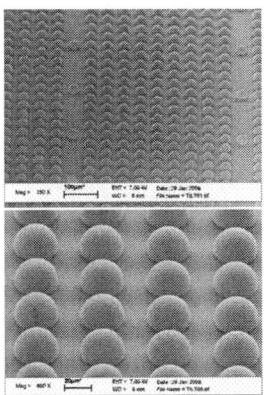

Fig. 12. SEM images of 50 µm microbumps.

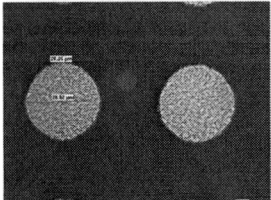

Fig. 13. Optical image of the sputtered UBM.

Technology Applications

Solder bumping is commonly used for chip packaging and, depending on applications, are often characterized by bump size and pitch. Bump sizes range from ≤ 250µm used in BGA, CSP down to around 25µm micro-bump applications. Advanced 3D integration pushes further toward the lower limit. 3D micro-bumping feasibility with 25µm size bumps on 50µm pitch using C4NP has been demonstrated. Figure 14 provides an overview of the various applications and associated wafer bumping technologies. C4NP covers the entire range of solder bumping applications where bump size and pitch are defined by the glass molds.

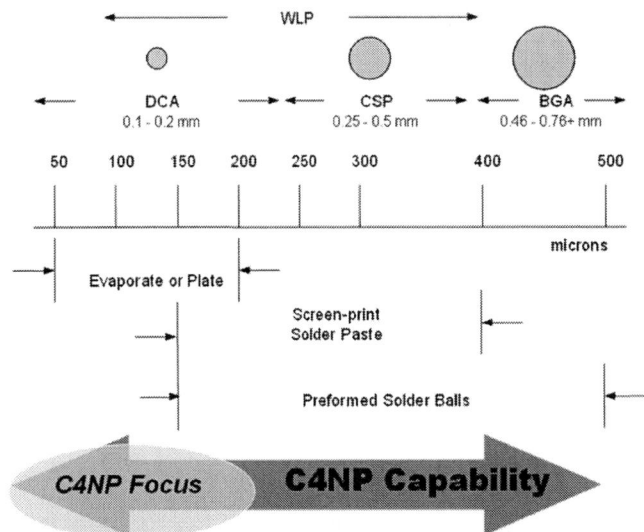

Fig. 14. Wafer bumping applications and technology.

UBM for Pb-free Flip Chip Bumping

UBM is the direct interface between solder and chip, an integral part of solder bumping technology. Typical UBM serves as a low resistance contact interface by providing wettability to solder, adhesion to BEOL metallization and passivation and a hermetic seal between UBM and IC pad. For PbSn solder, commonly used UBM stacks are Cr/CrCu/Cu/Au (original C4 from IBM), Ti/Cu, TiW/Cu, Ti/NiV, Cr/CrCu/Cu and Al/NiV/Cu. Usually, these UBMs are sequentially deposited by sputtering or plating. Intermetallic compounds (IMCs) are formed between Sn and Cu or Ni UBM that provides the required adhesion between bump and chip pad. Because of the high Sn content and highly reactive nature of Pb-free solders, a robust reaction barrier layer is needed in UBM.

The barrier layer commonly used for electroplated Pb-free solders is electroplated Ni. C4NP provides the opportunity to eliminate electroplating of solder. If electroplating of UBM can also be avoided, the entire infrastructure required for electroplating such as chemistry procurement, analysis, mixing, pumping and waste treatment can be avoided. A suitable method to avoid electroplating is the implementation of an electroless Ni/immersion Au (ENIG) UBM process. ENIG is a well known and commonly practiced UBM for Pb-free solders. It offers a significant cost advantage over other UBM technologies by avoiding the use of a photolithography or vacuum process to form a "capture pad". The ENIG metallurgy is simply applied directly on the pad metallurgy in the passivation openings, where solder is applied with C4NP. Other low cost UBM processes such as a sputtered UBM (TiW/Ni) can also eliminate the need of electro-plating. An overview of the cost comparison of different UBMs and solder bumping choices is summarized in Figure 15.

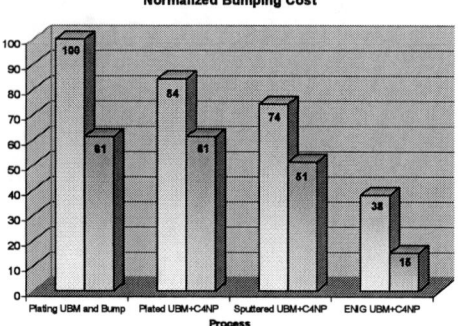

Fig.15. Cost comparison between different bumping technologies and UBMs.

3D Applications

Integrated circuit (IC) system performance is significantly enhanced by 3D integration of chip and packages because of the benefits of high bandwidth, low latency, low power, and small form factor for a variety of applications. Today, chip stacking with wire bonding has already been widely used to reduce the overall form factor and thickness of products in memory and handheld applications.[23] As the demand for higher performance and higher bandwidth continue to increase, chip stacking with high-density thru-silicon-vias (TSV) interconnection is being developed and receiving more attention.[24] As a "dry" bumping technology, C4NP is evaluated for 3D chip stacking such that the issues associated with the TSV, such as trapping of moisture or contaminants in TSVs can be significantly reduced. TSV chips with C4NP bumps have been stacked using various methods. As shown in Figure 16, chips can be joined on a substrate with a sequential reflow process. In sequential reflow, the bottom chip is joined first onto the substrate, followed by subsequent chips.

A sequential process can avoid relative displacement between chips as each subsequent chip is joined into the stack. Figure 16 also shows examples of 2-layer and 3-layer stacks of thinned TSV chips, utilizing sequential reflow of C4 interconnections. A major drawback of a sequential reflow process is that multiple reflows are necessary to complete the stack assembly. Multiple reflows require more processing time and lead to more dissolution of UBM, especially for the C4s contained in the lowest level of the stack, a concern for high reliability and high-performance applications. An alternative, parallel reflow process has also been demonstrated. A tacky flux is used to hold the stacked chips in place before the reflow process. Up to 4 layers of thinned TSV chips have been successfully formed with a single reflow step, shown in Fig. 17. With the self-centering effect of C4 bumps, a small amount of displacement between the chips is well compensated during reflow. To meet the demand for high I/O counts in high-performance and high-bandwidth applications, flip-chip I/O pitch needs to be reduced continuously. According to the International Technology Roadmap for Semiconductors (ITRS), the area-array flip chip I/O bump (C4) pitch will be less than 70 μm for high-performance applications by 2018. [2]

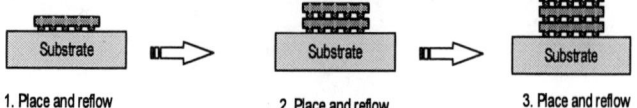

Figure 16. Sequential reflow process and photos of 2-Layer and 3-Layer stacks of thinned TSV Chips.

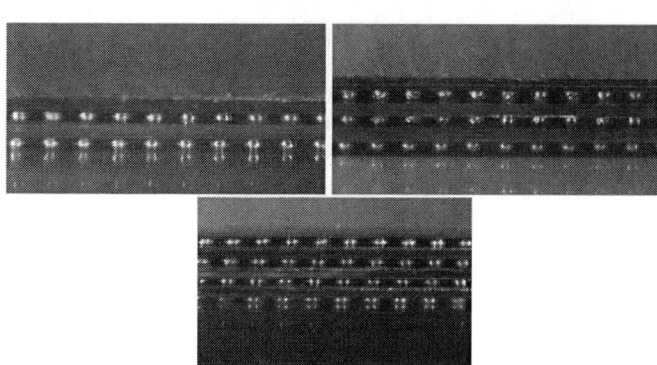

Figure 17. Multi-layer stacking through parallel reflow: 4-layer stack of TSV chips after hot-removal.

Table 3. Reliability test vehicle

Chip	4 on 8 Test Vehicle	3 on 6 Test Vehicle
Chip Technology	Low K CVD/BEOL	Low K CVD/BEOL
Chip size (mm)	14.7 X 14.7	15.6 x 11.2
UBM a Pad (um)	TiW/Cu/Ni/Cu /110	TiW/Cu/Ni/Cu /85
#. of C4s	4699 C4s/Chip	4985 C4s/Chip
Pitch (um)	200	150
C4 Solder	SnAg	SnAg
Laminate Design	4-2-4	3-2-3
Chip Size (mm)	42.5 x 42.5	42.5 x 42.5
RO pad Dia. (um)	130 (SRO = 100 um)	115 (SRO = 80um)
Lid	1 mm thick	2 mm thick
BSM	Sn3.0Ag0.5Cu BGAs on Cu OSP	Sn3.0Ag0.5Cu BGAs on Cu OSP
JEDEC Precon.	Level 3 30C/60%RH/192 Hrs	Special 30C/70%RH/120 Hrs
Precon. T	245 C / 3 X	245 C / 3 X

Reliability Results

The test vehicles for C4NP technology development were specifically designed to evaluate all aspects of C4 reliability, including solder fatigue, metal migration, high temperature stability, chip back end of line structures, and EM. A comparison of the 4 on 8 test vehicle and 3 on 6 test vehicle is summarized in Table 3.

Multiple processing lots of these test vehicles were evaluated to assess process control and repeatability. Reliability evaluations encompassed standard JEDEC stress conditions (TC, THB, HAST, HTS, LTS) and some new stress conditions designed specifically to evaluate Pb-free solders such as C4 electromigration. Prior to stressing, all 4 on 8 test 8 modules were subjected to JEDEC pre-conditioning (Level-3). The 3 on 6 test modules were subjected to a customer defined preconditioning and tests which is shown in Table 4.

Table 4. Reliability Data for 3 on 6 SnAg solder.

Test	Conditions	Duration	Results
DTC	-55/125C	1250 cyc	Pass
DTC	-40/125C	1250 cyc	Pass
ATC	0/100C	1500 cyc	Pass
HAST	130C/85%RH	96 Hrs	Pass
THB	85C/85%/3.6V	1000 hrs	Pass
HTS	150C	1000 hrs	Pass
C4 EM	110C/130C/150C-0.5A/0.7A	2000hrs	Data supports up to 275watts (assume 2.5V) at 100C, 100KPOH
Wettability	Initial Join Wettability	----	Good pull strength, no nonwetts

Conclusions

C4NP has demonstrated the ability to meet critical manufacturing and reliability requirements for C4 interconnections. C4NP processes have been reviewed from mold manufacturing to lead-free solder transfer onto 300 mm wafers. To enhance the performance and reliability of Pb-free solder joint the ability to maintain alloy flexibility is critically important. Applications including flip chip interconnects and 3D integration were described. C4NP micro bumping results in support of 3D packaging, and early manufacturing yield results from 300 mm wafer development and manufacturing were provided. Lastly, the most recent lead-free reliability data for both 200μm & 150μm C4 pitch was summarized

Acknowledgments

The authors would like to thank all our co-workers at IBM & Suss MicroTec for their contributions that made C4NP possible. We are also grateful to Gary Dawson at RVSI Inc. for his technical support on wafer inspection.

References

1. L.F. Miller, "Controlled Collapse Reflow Chip Joining," *IBM J. Res. Develop.*, 13, p.239, 1969.
2. International Technology Roadmap for Semiconductors, Assembly and Packaging, 2006.
3. V. Kripesh, et al., "Ultra-Fine Pitch Pb-free &Eutectic Solder Bumping with Fine Particle Size Solder Paste for Nano Packaging," *Proc. 53th ECTC Conf*, 2003.
4. H. Gan, et al., "Pb-free Micro-joints (50 μm pitch) for the Next Generation Micro-systems: the Fabrication,

Assembly and Characterization," *Proc. 56th ECTC Conf*, 2006.
5. K. Tatsumi et al., "An Application of Micro-ball Wafer Bumping for Flip Chip Interconnection" *Proc. 55th ECTC Conf*, p.855, 2005.
6. P. Gruber et al., "Low Cost Wafer Bumping", *IBM J. Res. & Dev.* Vol. 49 No. 4/5, 2005.
7. E. Laine et al., "C4NP Technology for Lead Free Solder Bumping," *Proc. 57th Electronic Components and Technology Conf*, 2007.
8. M. Uchida et al., "Low-Stress Interconnection for Flip Chip BGA Employing Lead-Free Solder Bump", *Proc. 57th ECTC Conf*, P.885, 2007.
9. J. Sylvestre, A. Blander, V. Oberson, E. Perfecto, K. Srivastava, "The Impact of Process Parameters on the Fracture of Device Structure", *Proc. 58th ECTC Conf*, p.82, 2008.
10. M. Lu et al., "Effect of Sn Grain Orientation on electromigration degradation mechanism in high Sn-based Pb-free Solders," *Appl. Phys. Lett.*, Vol. 92, No. 21, p.211909, 2008.
11. M. Lu et al., "Comparison of Electromigration Performance for Pb-free solders and Surface Finishes with Ni UBM," *Proc. 58th Electronic Components and Technology Conf*, p.360, 2008.
12. S.K. Kang and A.K. Sarkehl, "Lead (Pb)-free Solders for Electronic Packaging," *J. Electron. Mater.*, Vol. 23, No. 8 (1994), pp. 701-707.
13. M. Lu, unpublished data
14. Y. Kariya et al., "Tin pest in lead-free solders", *Soldering & Surface Mount Technol.* P.39, 2000.
15. S.K. Kang Kang et al., "Formation of Ag₃Sn Plates in Sn-Ag-Cu Alloys and Optimization of their Alloy Composition", *Proc. 53th Electronic Components and Technology Conf*, p.64, 2003.
16. D. Henderson et al., US Patent 6,805,974.
17. H. Kim et al., "Improved drop Reliability Performance with Lead Free Solders of Low Ag Content and Their Failure Modes", *Proc. 57th ECTC Conf.*, p.962, 2007.
18. W. Liu et al., "The Superior Drop Test Performance of SAC-Ti Solders and Its Mechanism", *Proc. 58th ECTC Conf.*, p.452, 2008.
19. S.K. Kang et al., "Interfacial Reactions of Sn-Ag-Cu Solders Modified with Minor Zn Alloying Addition", *J. Electronic Materials*. P.479, 2006.
20. E. Perfecto et al., "C4NP Technology: Manufacturability, Yields and Reliability", *Proc. 58th ECTC Conf.*, 2008.
21. J. Busby et al., "C4NP Lead Free Solder Bumping and 3D Micro Bumping", *ASMC conf*. 2008.
22. B. Dang et al., "50μm Pitch Pb-Free Micro-bumps by C4NP Technology", *Proc. 58th Electronic Components and Technology Conf.*, 2008.
23. M. Karnezos, "3D packaging: where all technologies come together," Proc. IEEE 29th IEMT Symposium, pp. 64- 67, 2004.
24. K. Takahashi, et al., "Process integration of 3D chip stack with vertical interconnection," *Proc. 54th ECTC Conference, vo. 1, p. 601. 2004.*

Manufacture of Hourglass-shaped Solder Joint by Induction Heating Reflow

Hongbo Xu[1], Mingyu Li[1], Liqing Zhang[1], Jongmyung Kim[2] and Hongbae Kim[2]

1. Shenzhen Graduate School, Harbin Institute of Technology

HIT Campus, Shenzhen University Town, Xili,Shenzhen 518055, P. R. China

myli@hit.edu.cn, 0755-26033463

2. Jeonnam Provincial College

Jeonnam, Korea, 517-802

Abstract

Induction heating reflow method, which can achieve the solder bumping and interconnecting process in a simple way, was studied to control the height and shape of solder interconnects employed in electronic packaging application. A solder joint model was built to investigate the temperature distribution in the joint during the whole induction heating process by ANSYS software. Based on the simulation, the localized melting phenomenon is defined and identified by scanning electron microscope (SEM) observation. In this work, the barrel-shaped solder joints with high heights and the hourglass-shaped solder joints can be obtained which is useful to increase the solder joint lifetime. The mechanism of solder joint height and shape control, which can be explained by the local melt phenomenon, is discussed and demonstrated by the different morphologies of Ag_3Sn intermetallic compound (IMC). The findings of this paper will help to provide an understanding of the whole solder interconnecting process during induction heating reflow and the effects of electromagnetic field on solder interconnect shape controlling.

Introduction

Solder joint geometry plays an important role among the factors affecting solder joint fatigue performance. [1] Some modeling and calculation work has shown that compared with the barrel-shaped solder joints, the hourglass-shaped solder joints would have a higher standoff height and lower plastic strain and stress during the temperature cycles, thus a longer lifetime. [2, 3] A stacked solder bumping technology has been developed to fabricate triple-stacked hourglass-shaped solder joints with a high-melting-point solid solder ball inside. [1, 4] The hourglass shape of solder joint improve fatigue lifetime by about 60% over the conventional barrel-shaped solder joint. Several other approaches have been employed to control the shape of solder joint, such as double-bump technology [5] or using single copper base [6], multi-copper column [7] and ceramic column in the joint [8]. The basic scheme is adding non-melting standoffs to hold the weight of the chip during the reflow process. However, these technologies either are complex to be implemented, or have high cost-effectiveness.

Based on these above facts, when only the surfaces of solder balls melt, and the melted layer is sufficient to form the interconnects, the solid core inside can act as a standoff and will maintain the height of the solder joint. Considering the skin effect for alternating electromagnetic field, induction heating at a suitable frequency may achieve this kind of heating process. Li et al. has used induction heating power supply achieve the selective heating of solder bumps with a

high heating/cooling rate. [9] This work presents a solder reflow utilizing induction heating to control the geometry of solder joints. In this work, the temperature distribution during the process will be discussed by using FEM simulating. Through the simulation results, a localized melting phenomenon was defined, and the hourglass-shaped solder joints can be formed by using this phenomenon, which was respected to extend the fatigue lifetime of solder joints.

Experimental procedure

Figure 1 (a) shows a schematic diagram of the induction heating process. In the experiments a FR4 PCB board (thickness 0.4 mm) was selected and the solder balls, 760 μm in diameter, of 96.5wt%Sn3.5wt%Ag eutectic alloy were manually placed on pads. The microstructure of the solder ball is shown in Fig. 1 (b). The Under Bump Metallization (UBM) on the PCB board consisted of a bottom Cu layer (35 μm thick), an electroplated Ni layer (12 μm thick) and a top Au layer (0.05 μm in thickness). The diameters of the Cu pad and the passivation opening were 800 μm and 760 μm, respectively. The induction heating apparatus was operated with a maximum operating power of 6 kW and frequency of 0.3 MHz. The coil used was a 15 mm outer diameter, two-turn, copper inductor with a pitch of 4.4 mm. The solder balls and the FR4 board were laid and heated in the center of the coil to fabricate solder bumps. Afterwards, the board with the solder bumps was flipped on to a new PCB board without solder bumps, and heated to form solder joints.

In comparison with the induction heating reflow, a hot-air reflow test was carried using a SODRTEK® ST 325 Digital Convective Soldering/Desoldering System by PACE. The diameters of the Cu pad and passivation opening used here were 800 μm and 600 μm, respectively. The same solder balls were used as the previous experiment. The reflow profile of the soldering reflow process is given as curve 1 in Fig. 1 (c). The temperature was measured by a thermocouple in real time.

An infrared pyrometer, Metis MP25, with a temperature resolution of 0.1 ℃ was introduced into the induction heating experiment instead of a thermocouple which may itself generate considerable heat in the presence of a strong alternating electromagnetic field. The emissivity of the infrared thermometer was set to 0.20, which had been calibrated by using the thermocouple to test the same spot under the hot air condition. In the temperature test the measurement point was located on the top surface of the solder bumps. A typical temperature profile is shown in Fig. 1(c). Curve 2 is the temperature profile of a solder bump heated by induction heating. It is noticed that the

978-1-4244-2739-0/08/$25.00 ©2008 IEEE

heating/cooling rates are much higher than those for hot air reflow as seen in curve 1.

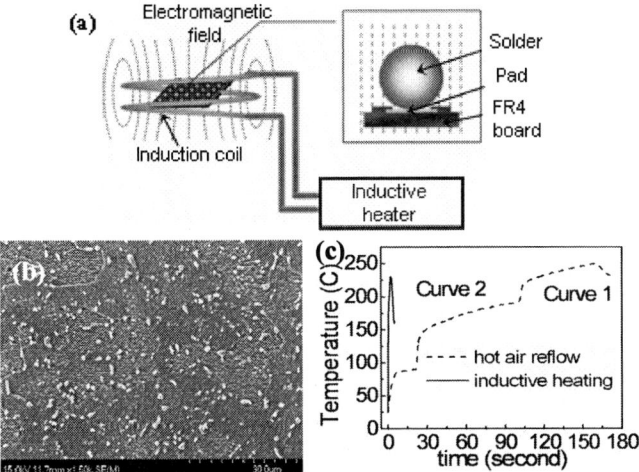

Fig.1 Experimental setup for solder joint shape control via inductive heating. (a) Schematic overview of the solder ball, PCB board and inductive heating coil. (b) Original microstructure of Sn3.5Ag solder ball. (c) Temperature profiles for an interconnecting process by hot air, and a solder bumping process by inductive heating at $I_{coil} = 29.0$ A.

Finite element modeling

Finite element model

A three-dimensional finite element model for thermal simulation was developed using the nonlinear FEM analysis program ANSYS. The model includes all the key elements during the induction heating process, such as an induction coil, a FR4 substrate, pads, and solder balls. In this model, the diameter of a solder ball is 0.76 mm with 1 mm pitch; the thickness of BT is 0.4 mm. Since it is impossible to include every geometrical detail in the model, some simplifications have to be assumed. Au metallization were ignored in this model, and computational analyses were also made under an assumption that contact between material pairs was perfect. Figure 2 shows finite element meshes of the solder bump.

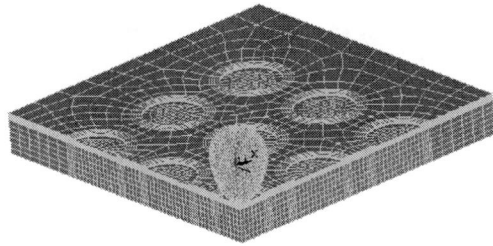

Fig. 2 Finite element of the solder bump.

Boundary conditions

Vector wave equations derived from Maxwell's equations, which can be used to calculate the electromagnetic field in the system, are as follows:

$$\nabla \times (\frac{1}{\mu}\nabla \times E) - \omega^2 \varepsilon_c E = -j\omega J_i \tag{1}$$

$$\nabla \times (\frac{1}{\varepsilon_c}\nabla \times H) - \omega^2 \mu H = \nabla \times (\frac{1}{\varepsilon_c}J_i) \tag{2}$$

where μ is the permeability, E is the electric field strength, H is the magnetic field intensity, ω is the angular frequency, ε is

the dielectric constant, σ is the electrical conductivity, ε_c is the combined contribution of the induction current and the displacement current, which can be calculated as $\varepsilon_c = (\varepsilon - j\sigma/\omega)$, and J_i is the impressed current or resource current.

The boundary conditions of the electromagnetic field are as flowed:

Dirichlet boundary condition

$$A|_\Gamma = g(\Gamma) \tag{3}$$

Neumann boundary condition

$$\frac{\partial A}{\partial n}\Big|_\Gamma + f(\Gamma)A|_\Gamma = h(\Gamma) \tag{4}$$

where A is the considered variable, n is the outside normal vector of the boundary Γ, $g(\Gamma)$ is the function of the location, and both $f(\Gamma)$ and $h(\Gamma)$ are regular functions.

Here, we consider that (1) convective heat exchange within the molten solder balls may be ignored due to its small effect on temperature distribution; (2) the latent heat of phase change of the solder balls is considered in this study; (3) all the materials are isotropic. So the mathematical formulation of the energy equation for calculating the temperature field of the package is as follows:

$$\rho c_p \frac{\alpha T}{\alpha t} = \lambda(\nabla^2 T) + \dot{q} \tag{5}$$

where ρ is the density of the material; C_p is the specific heat; λ is the heat conductivity; $\dot{q}$ is the energy density of heat resource, which is generated by induction eddy current.

For the surfaces of the package, the boundary condition is

$$q = \beta(T - T_0) \tag{6}$$

where q is the density of heat flow rate; and β is the equivalent coefficient of heat transfer considering the solder flux volatilization, the values of which used in this paper are listed in Table 1; T and T_0 are surface and ambient temperature, respectively.

TABLE 1 VALUES OF HEAT TRANSFER COEFFICIENT β CONSIDERING FLUX VOLATILIZATION.

Temperature T K	Heat transfer coefficient β W(m²K)⁻¹
293	50
393	150
394	450
773	550

Since the computational analyses were made under an assumption that the contact between material pairs is perfect, both heat flux and temperature at the material interface are continuous, and the interfacial condition is

$$\lambda_1 \frac{\partial T_1}{\partial n} = \lambda_2 \frac{\partial T_2}{\partial n} \tag{7}$$

$$T_1 = T_2 \tag{8}$$

where T_1 and T_2 are the interface temperatures of different materials, n is the normal direction of the interface.

The latent heat H can be simplified as the equivalent specific heat capacity C^*:

$$C = C_0 + C^* = C_0 + \frac{H}{T_s - T_l} \qquad (9)$$

where C is the equivalent specific heat capacity, C_0 is the actual specific heat capacity, T_l and T_s are the beginning temperature and ending temperature of the solid-liquid coexistence period respectively. In this paper, for Sn3.5Ag solder H is 59 520 J Kg^{-1}, T_l is 219℃, T_s is 225℃, and C^* is 9 920 J Kg^{-1} K^{-1} correspondingly.

The physical parameter of the materials of Sn3.5Ag solder, Ni metallization, Cu pad, FR4 Board and air are listed in Table 2.

TABLE 2 PHYSICAL PARAMETER OF MATERIALS.

		SnAg	Ni	Cu	FR4	Air
Density ρ Kg m^{-3}		7360	8900	8960	1200	—
Relative permeability μ		0.99998	49.886	0.999999	1	1
Heat conductivity λ W m^{-1} K^{-1}		33	92	393	0.23	—
Resistivity ρ Ω m	293 K	10.8e-7	6.2e-8	1.58e-8	--	--
	393 K	15.768e-8	10.416e-8	2.2594e-8		
	493 K	20.763e-8	14.632e-8	2.9388e-8		
	693 K	30.672e-8	23.064e-8	4.2976e-8		
specific heat capacity C J Kg^{-1} K^{-1}	293 K	222	435	385	700	--
	373 K	239	477	389		
	473 K	260	528	402		
	505 K	250	477	404		
	573 K	242	543	410		
	673 K	241	519	418		
	773 K	240	535	427		

Analysis and discussions

Results of thermal simulation

Thermal fields and temperature profiles of the solder joint during induction heating reflow soldering were analyzed using the three-dimensional model described earlier.

In order to validate the veracity of the FEM model, the temperature profiles of the solder ball being heated by the induction heating were measured by infrared thermometer at the top of the solder ball. In Fig. 3 the solid line shows the measured temperature profile of a solder ball being heated by electromagnetic field, and the dotted line shows the simulated temperature profile. The predicted temperature profile matches very well with the measured one, which indicates that the three-dimensional FEM model of thermal simulation of induction heating reflow bumping is reliable.

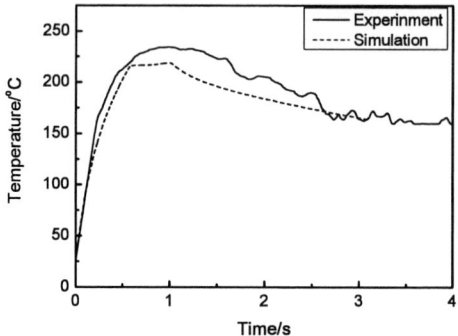

Fig. 3 Temperature profiles test by experiment and calculated by simulation

Figure 4 shows typical temperature distributions of the solder ball subjected to induction heating for 0.05 sec, 0.1 sec, 0.34 sec and 1 sec, respectively. Fig. 4 shows that there was a temperature gradient inside the solder ball. In Fig. 4 (a), the peak temperature appears near the interface, because the pad has a smaller mass and higher eddy current, thus a higher temperature, which made the pad transfer the heat to the solder ball. Later, the heat generated by the solder ball itself became the major heat resource, and the temperature around the equator of the solder ball is the highest in the solder bulk, as shown in the Fig. 4 (b) and (c). After that, the generated heat and the dissipated heat were nearly balanced, and the input heat from the pad became the major factor to effect the temperature distribution, as shown in Fig. 4 (d).

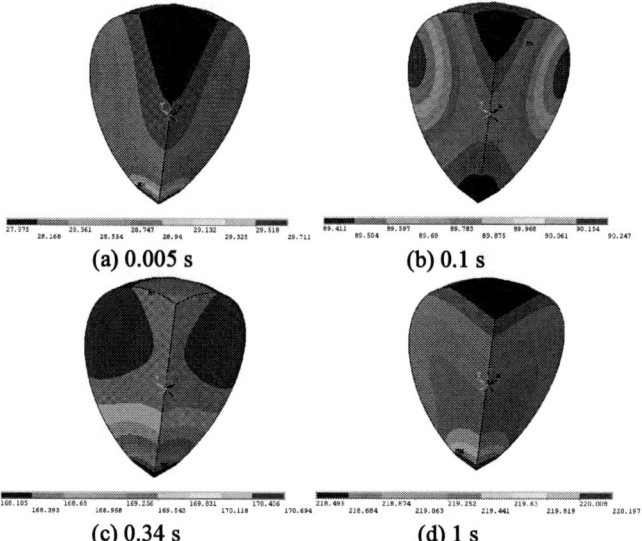

Fig. 4 The thermal field of the solder ball heated by (a) 0.005 s, (b) 0.1 s, (c) 0.34 s, (d) 1 s.

Figure 5 shows the temperature profiles of 6 different nodes on the surface of the solder joint and board. To distinguish these curves, only the part of the temperature profile between 0.68 sec and 1.0 sec was attached here, which was also the most important part for the final thermal field. It can be seen that the temperature at the edge of the pad was the highest, and the temperature at the interface was a little lower. At the same time, the temperature on the surface of the solder ball was lower than the interface. The top of the solder ball has the lowest temperature. This trend lasted until 1.0 sec, the end of the induction heating.

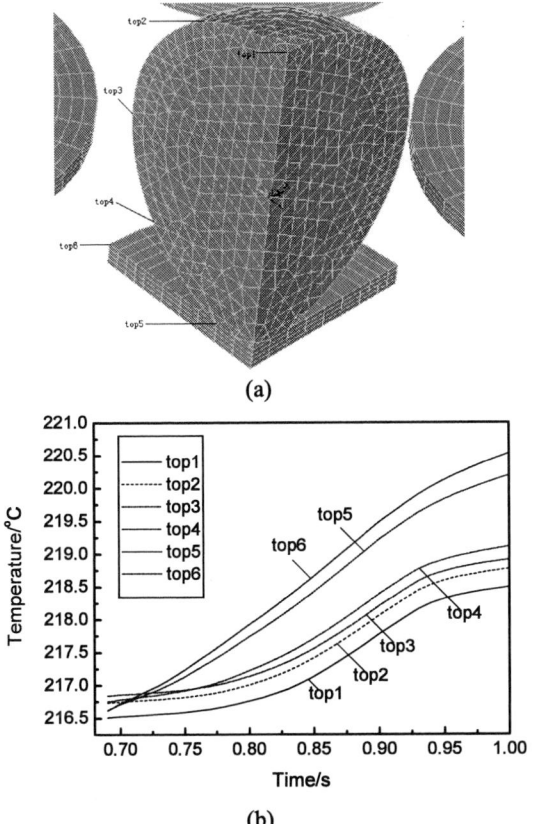

(a)

(b)

Fig. 5 Temperature profiles of 6 different nodes on the surface of the solder joint and board, (a) nodes locations, (b)temperature profiles between 0.68 s~1.0 s.

Localized melting phenomenon

The non-uniform temperature distribution in the solder joint implies that the high temperature region can be concentrated to the thin layer around the interface between the solder ball and the pad, leaving a big enough solid core on the top. That is defined as the localized melting phenomenon, and could also be used to control the solder joint geometry during the induction heating reflow process.

Solder bumping

Figure 6 (a) shows the cross-section of a solder bump formed by induction heating. The solder bump has the bullet shape which is different compared to the traditional oblate spheroid shape obtained by hot air reflow. Normally, the typical standoff height h of the oblate-spheroid-shaped solder bump is around 650 μm after hot air reflow. In contrast, h for the bullet-shaped solder bump is approximately 735 μm, which is an increase of over 13%. In order to demonstrate the effect of the localized melting phenomenon as mentioned above, the microstructure of the solder bump was studied. Fig. 6 (b), (c) and (d) are the microstructure images of regions B, C and D of Fig. 6 (a) respectively. The Ag$_3$Sn IMC morphology and the grain size of the β-Sn phase are evidently different. From Fig. 2 (b) to (d), the morphology of the Ag$_3$Sn IMC changes from small particles dispersed evenly inside the β-Sn grain to needle-like ones located mainly along grain boundaries. Also the β-Sn grains coarsen. The microstructure in Fig. 6 (b) shows the similar morphology to the original microstructure of the solder ball in Fig. 1 (b). Time and temperature above the solder liquidus during the

heating stage of the reflow process has been found to play a significant role in determining the IMC morphology by Lee et al. [10] Accordingly, the morphology of Ag$_3$Sn can be used as the criterion in this study to decide whether the solder melt. It is considered that the morphologies of the Ag$_3$Sn and β-Sn grains in different positions within the solder bumps indicate their dissimilar melting and solidification processes. Because the Ag$_3$Sn and β-Sn phases in the center do not change, so the center of the solder remains in solid state during the whole reflow process. The morphology of the outer layer and interfacial solder alloy indicates the re-melting process that has occurred. In other words, there is a solid core in the center of bullet-shaped solder bumps during induction heating, which can maintain its bullet shape.

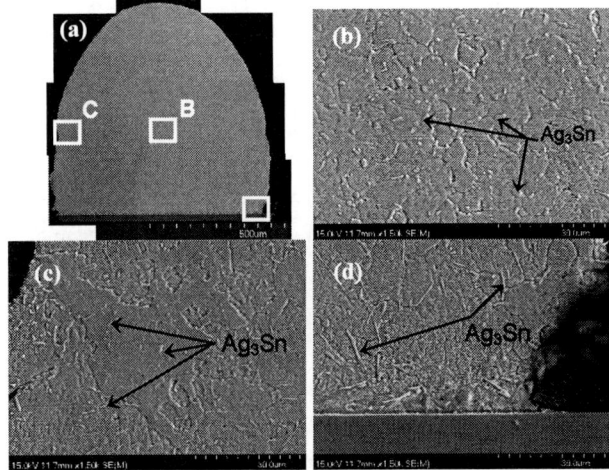

Fig.6 (a) Bullet-shaped solder bumps fabricated by inductive heating. The SEM image of solder bump microstructures (b) in region B with small particle Ag$_3$Sn and β-Sn grain size, (c) in region C with bigger β-Sn grain size and (d) in region D with needle-like Ag$_3$Sn.

Solder interconnections

Figures 7 (a) and (b) are the SEM images of solder interconnections reflowed by hot air and induction heating, respectively. In normal hot air reflow processes only the barrel-shaped solder joint can be formed, however, in this work the induction heating method can create hourglass-shaped joints, without any other complex structures or processes compared to other reflow methods. As mentioned above, the hourglass-shaped solder joint has a greater lifetime for both the increase of the standoff height h and the change of the shape which can be described by the midpoint diameter d_m. In Fig. 7 (a), h and d_m of the barrel-shaped solder joint is 547.4 and 827.6 μm respectively. In contrast, in Fig. 7 (b), h and d_m of the hourglass-shaped solder joint is about 732.0 and 597.6 μm respectively, an increase of nearly 40% in h and a reduction of 27.8% in d_m compared with the barrel-shaped solder joint. As mentioned in a previous work, when the standoff height increases by 20% with the solder volume increasing by 14%, the lifetime of solder joint will increase by over 50%. [11] The effect of standoff height on improving the lifetime of solder joints is obvious. Furthermore, the observation of microstructure was also performed, which helps to form the hourglass-shaped solder joint. The dotted line in Fig. 7 (b) is drawn so as to distinguish the edges of regions with clearly different microstructures. The microstructure at the interface, shown magnified in the circle in Fig. 7 (c), has the original morphology in Fig. 1 (b) and Fig. 6 (b); the microstructure outside the circle at the upper right corner is shown in Fig. 7 (d), which includes the needle-like Ag3Sn, is similar to those of Fig. 2 (d), and the eutectic region with the Sn phase and Ag3Sn particles, similar to those in Fig. 3 (a). It is clear that there is also a solid core in the center of the hourglass-shaped solder joint formed by induction heating, similar to the bullet-shaped solder bump.

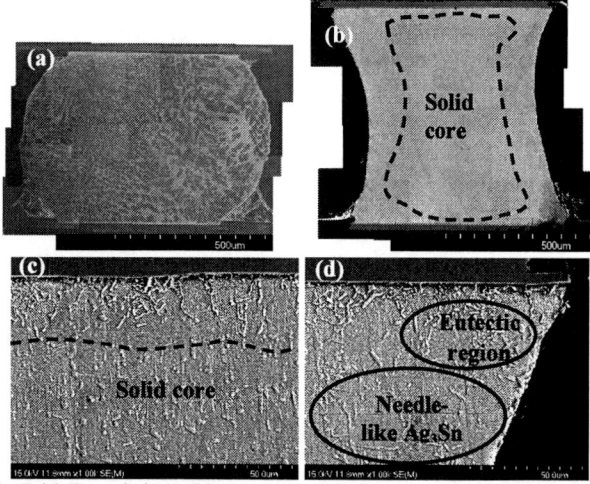

Fig.7 (a) Barrel shaped interconnection fabricated by hot air reflow. (b) Hourglass-shaped interconnection made by inductive heating reflow. (c) The interfacial microstructure including the solid core and the melting part near the pad. (d) The microstructure at the upper right corner consisted by the eutectic region and needle-like Ag3Sn region.

The essential factor for the formation of both bullet-shaped solder bumps and hourglass-shaped solder joints is a structure with a solid core in the center surrounded by liquid solder during the heating stage of the induction heating reflow process. This demonstrated that the localized melting phenomenon discussed above was the main reason for the formation of both the bullet-shaped solder bump and the hourglass-shaped solder joints.

Conclusions

In summary, this paper introduced a unique method of soldering through induction heating in a high-frequency electromagnetic field. A solder joint model was built to investigate the temperature distribution in the joint during the whole induction heating process through ANSYS. The localized melt phenomenon is defined and identified by scanning electron microscope (SEM) observation. Shape controlled solder joints can be formed through this technique, such as hourglass shape, which can apparently increase the lifetime of the solder joint. The formation of hourglass-shaped

solder joints can be explained by the localized melting phenomenon.

Acknowledgments

The authors would like to express their gratitude to the National Natural Science Foundation of China for supporting this work under grant No. 50405010. Additional support from the NURI project offered by the government of the Republic of Korea is also acknowledged.

References

1. Liu, X et al, "Effects of Solder Joint Shape and Height on Thermal Fatigue Lifetime," IEEE T. Compon. Pack. T., Vol. 26, No. 2 (2003), pp. 455-465.
2. Yu, Q et al, "A study of the effects of BGA solder geometry on fatigue life and reliability assessment," Proc. Thermal Thermomech. Phenom. Electron. Syst., Seattle, USA, 1998, pp. 229-235.
3. Satoh, R et al, "Thermal fatigue life of Pb-Sn alloy interconnections," IEEE Trans. Comp. Hybrids Manufact. Technol. Vol.14, No. 1 (1991), pp. 224-232.
4. Liu, X et al, "Stacked solder bumping technology for improved solder joint reliability," J. Microelectron. Rel. Vol. 41 (2001), pp. 1979-1992.
5. Yeung, B et al, "Evaluation and Optimization of Package Processing and Design Through Solder Joint Profile Prodiction," IEEE T. Electron. Pack. Vol. 26, No. 1 (2003), pp. 68-74.
6. Kawahara, T, "Super CSP™," IEEE T. Adv. Packaging Vol. 23, No.2 (2000), pp. 215-219.
7. Liao, E et al, "Fatigue and Bridging Study of High-Aspect-Ratio Multicopper-Column Flip-Chip Interconnects Through Solder Joint Shape Modeling," IEEE T. Compon. Pack. T. Vol.29, No. 3 (2006). pp. 560-569.
8. Hong, B et al, "Ceramic column grid array technology with coated solder columns," Proc. IEEE 50th Electron. Comp. Technol. Conf., Las Vegas, NV, USA, 2000, pp. 1347-1353.
9. Li, M et al, "Eddy Current Induced Heating for the Solder Reflow of Area Array Packages," IEEE Trans. Adv. Packaging, Vol. 31, No. 2 (2008), pp. 399-403
10. Lee, J et al, "Formation and Growth of Intermetallics around Metallic Particles in Eutectic Sn-Ag Solder," J. Electron. Mater. Vol. 32, No. 11 (2003), pp. 1240-1248.
11. Liu, C et al, "Enhancing the Reliability of Wafer Level Packaging by Using Solder Joints Layout Design," IEEE T. Compon. Pack. T. Vol. 29, No. 4 (2006), pp. 877-885.

The Effects of Ni Nanoparticles Addition on Shear Behavior and Microstructure of Sn-Ag Lead-free Solder

Fangjuan Qi, Li Sun，Zhezhe Hou，JianqiangWang, Cha Qin

Shijiazhuang Railway Institute, shijiazhuang, Hhebei, 050043, China

E-mail address: sun-li-sun-li@sohu.com

Tel: +86-0311-87936726

Abstract

In this article, the effects of Ni nanoparticles addition on shear property and microstructure of Sn-3.5Ag Lead-free solder joint was studied. The nickel nano-composite Sn-3.5Ag solder was prepared by adding dispersant to the dry nanoparticles and mechanically stirred Ni nanoparticles into the Sn-3.5Ag Lead-free solder paste. The shear force of the Sn-3.5Ag solder, 0.5 and 1.0 wt% nickel nano-composite solder was tested respectively at reflow 120s and 240s. The result shows that adding nickel nanoparticles can improve the shear performance of the soldered joint; the shear force of the soldered joint is highest when adding 0.5wt% Ni nanoparticles at reflow 240s. The SEM observations shows that the hexagonal Cu_6Sn_5 IMC（intermetallic compound）in the inside solder is disappears gradually and the morphsa of the IMC that on the interface of the solder joint becomes planar after adding Ni nanoparticles into solder.

Keywords：Lead-free solder, Ni nanoparticles, shear performance, microstructure, IMC

Introduction

Environmentally conscious manufacturing is becoming a very important objective for the electronics industry. It is highly desirable to minimize the environmental impact of electronic manufacturing processes. The electronics industry is being forced to eliminate lead from products, due to the undeniable evidence of lead toxicity. Strict legislation to ban the use of lead-based solders have provided an inevitable driving force for the development of lead-free solder alloys[1]. These lead-free solders are mostly based on Sn-containing binary and ternary alloys. Among them, Sn–Ag system is one of the earliest commercially available lead-free solders and has been recommended for general-purpose use as substitutes for Sn–Pb eutectic solder.

Hence, many researchers have concentrated on the addition of other elements into Sn–Ag solder so as to obtain better solder alloys exhibiting refined microstructure. The most attractive progress was achieved by adding minor rare earth elements into the Sn–Ag alloys and thus leading to improve the microstructure stability of those solders. For the rare earth elements are surface active and almost insoluble in many metals, such as Sn, Ag, Bi, etc., they tend to be enriched at grain boundaries and/or phase boundaries for their relative large atomic radius and thus improve the microstructure stability of the solder.

In recent years, a composite solder using nickel particle reinforcement, prepared by mechanically dispersing 15.0% of Ni particles into eutectic Sn3.5Ag solder paste, has been developed and has shown improved mechanical properties over the Sn-3.5Ag. The average size of the Ni particles used was approximately 5μm. It has also been reported that nickel particle reinforced composite solder has comparable wetting characteristics to the Sn-3.5Ag solder, and creep resistance was significantly improved at all test temperatures [2-4]. So the effects of Ni nanoparticles addition on shear behavior and microstructure of Ag-3.5Sn Lead-free soldered joint was studied in this paper in order to enhance the shear strength of the Lead-free solder.

1. Experimental Procedures

1.1 Preparation of nano-composite solder pastes

The Sn-3.5Ag solder paste was produced by Shenzhen Earlysun Technology Co., LTD. The nano-composite solder was prepared by blending pre-weighed solder paste with different weight percentages of pure nickel nanoparticles. Pure nickel nanoparticles having an average size of 50-70 nm were produced by NACHEN S&T LTD (China). Two weight percentages were selected for nickel nanoparticles: 0.5 and 1.0 wt.%.

Since almost all dry nanoparticles are in the form of agglomerates due to their surface properties, the nature of manufacturing process, handling, and storage, dispersant was added to the nickel nanoparticles to improving nickel nanoparticles dispersing property under the condition of the stable supersonic power. The best supersonic time is 5 minutes [5].

The nickel nanopowder that was dispersed was mixed with Sn-Ag solder paste. The resultant mixture was mechanically stirred in a glass bottle for a full 30 min, so as to ensure a homogeneous distribution of the reinforcing nanoparticles in the Sn-Ag solder paste.

1.2 Shear test and microstructure observation

The Sn-3.5Ag solder paste and 0.5 and 1.0 wt.% nickel nano-composite solder pastes were printed on the PCB (print circuit board) with electronic components and soldered at peak temperature of 250℃. Reflow times of all solders were 120 seconds and 240 seconds respectively.

The shear force of specimens as-soldered was testing by using shear testing apparatus. Fig.1 shows the structure of the shear testing apparatus.

978-1-4244-2739-0/08/$25.00 ©2008 IEEE

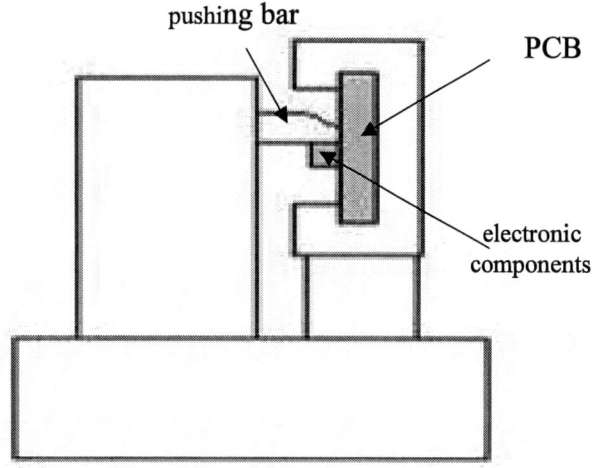

Fig.1 the diagram of shear testing apparatus

Subsequently, the specimens were cross-sectioned, mounted at room temperature epoxy resin and polished for examination of the microstructure using scanning electron microscopy (SEM) equipped with an energy dispersive x-ray (EDS) attachment.

2. Results and Discussion

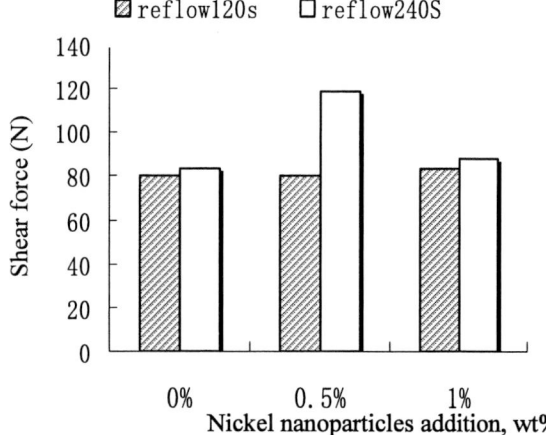

Fig.2 The comparison of different solders at reflows 120s and 240s

Fig.2 shows the shear force of soldered joint that adding 0wt%, 0.5wt% or 1wt% nickel nanoparticles at reflow 120s is lower than that of adding the same percentage nickel nanoparticles at 240s. With increasing of the amount of Ni nanoparticles addition, the increasing of the shear force of solder joint is little at reflow 120s. The reason probably is the addition of a small quantity of Ni nanoparticles could decrease the amount of the thick hexagonal Cu_5Sn_6 IMC in the solder. The thick hexagonal Cu_5Sn_6 IMC could decease the shear force of the solder[6]. The shape of the thick hexagonal Cu_6Sn_5 IMC in the solder shows in Fig. 3.

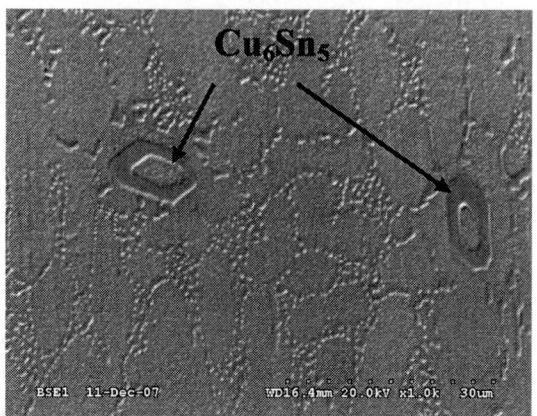

Fig.3 the shape of Cu_6Sn_5

When adding 0.5wt% Ni nanoparticles at reflow 240s, the shear force has a very significant increase. It is probably because the disappearing of the hexagonal Cu_6Sn_5 IMC from solder with the extension of the reflow time. The other reason is the morphsa of the IMC of 0.5wt% Ni reflow 240s on the interface of the solder joint is planar. The morphsa of the IMC of Sn-3.5Ag solder reflow 240s and the on the interface of the solder joint is scallop. The planar IMC obtained better shear property than scallop IMC[6]. Fig. 4 shows the comparison of the morphsa of the IMC on the interface of soldered joint.

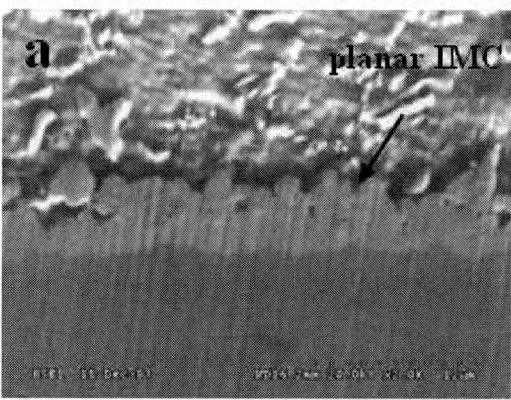

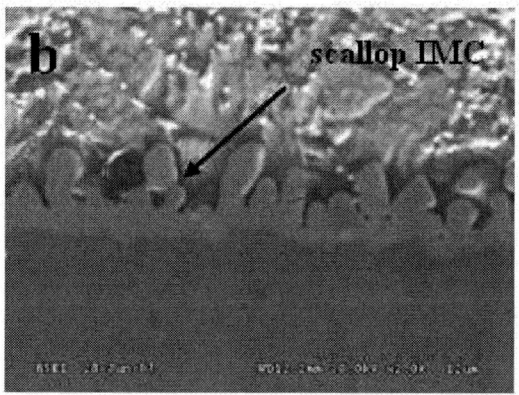

Fig.4 The comparison of the morphsa of the IMC (a) 0.5wt% Ni reflow 240s (b) Sn-3.5Ag reflow 240s

Fig.5 shows the comparison of the IMC on the interface of solder joints when adding 0.5wt% Ni nanoparticles at reflow

590

120s and 240s. With the increase of the reflow time, the morphsa of the IMC that on the interface of the solder joint becomes more planar and the thickness of the intermetalic compound layers increase. The thickness of the intermetalic compound layer at reflow 120s is 2.7~3.96μm, at 240s is 3.68~6.78μm. The changes of the morphsa of the IMC and the thickness of the intermetalic compound layer maybe lead to the increase of the shear force.

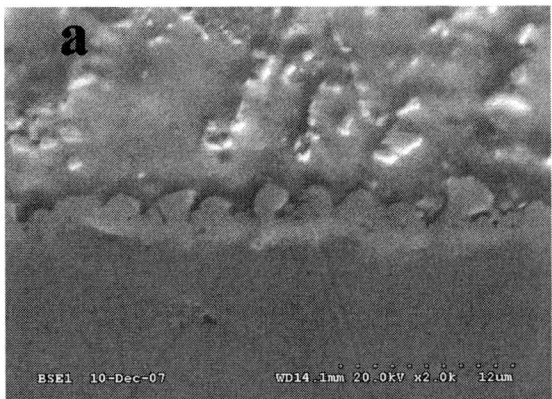

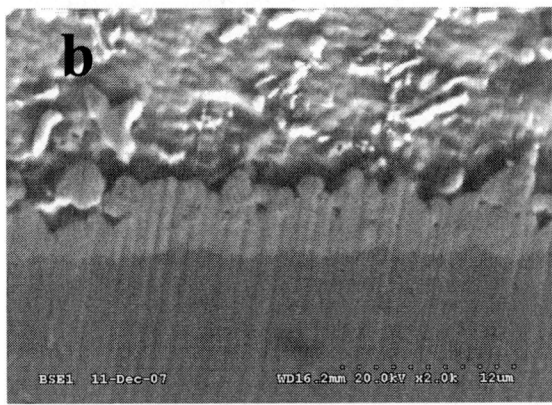

Fig.5 The comparison of the IMC (a) 0.5wt% Ni reflow 120s (b) 0.5wt% Ni reflow 240s

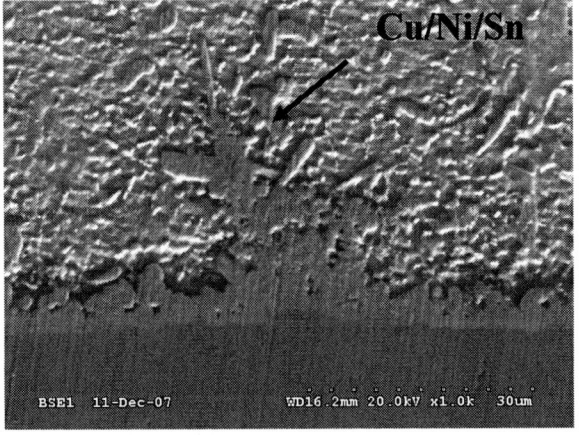

Fig.6 The morphsa of the Cu/Ni/Sn IMC

The shear force of 1wt% Ni is lower than 0.5wt% Ni. It is maybe because there are too many lumpish Cu/Ni/Sn IMC

formed in the inside of the solder and the solder joints. The shear performance probably is influenced because of the formation of the lumpish Cu/Ni/Sn IMC. Fig.6 shows the morphsa of the lumpish Cu/Ni/Sn IMC.

Conclusions

(1) The shear performance of soldered joint was enhanced after adding nickel nanoparticles into the solder.

(2)The shear force of the solder joint is highest when adding 0.5wt% Ni nanoparticles at reflow time 240s during reflowing process.

(3) The Cu/Sn IMC on the interface of the solder joints changes into Cu/Ni/Sn with the additional of Ni nanoparticles.

(4) The hexagonal Cu_6Sn_5 IMC in the inside solder is disappears gradually by adding Ni nanoparticles.

（5）When adding Ni nanoparticles, the morphsa of the IMC that on the interface of the solder joint change from scallop to planar.

Acknowledgments

This work was supported by Natural Science Foundation of Hebei Province (No.E2006000382) and the Education Department of Hebei Province (No.2005206).

References

1. Arulvanan Periannan, Zhong Zhaowei, Shi Xunqing. "Effects of process conditions on reliability, microstructure evolution and failure modes of SnAgCu solder joints". Microelecton Reliab, Vol.64 (2006), pp. 432–439

2. Fu Guo, Lead-free Soldering Technology and Application, Science Press (2006), pp.195-199

3. Liu Zhi Jie,, "Effects of Ni Element on the Property of Vacuum Brazing Joint 0f Al-Si-Cu Base Alloy," Aluminum fabrication, Vol. 164, No. 5 (2005), pp. 6-8

4. G. Martin, "An overview of the current status of lead-free assembly and related issues," Circuit World, Vol. 29, No. 4(2005), pp. 23-27

5　Lu Chengjie, Zhang Zhenzhong，Zhou Jianqiu，Zhang Shaoming, " Influence of Different Dispersants on the Dispersion Stabilities of Nickel Nanopowders in Ethanol”, Materials Review, Vol. 21, No. 8(2007), pp.165-166

6　D. Ma, W. D. Wang, and S. K. Lahiri, "Scallop formation and dissolution of Cu–Sn intermetallic compound during solder reflow". Journal Of Applied Physics (2002), Vol. 91, No. 5, pp, 3312-3317

Nanoscale Analysis of Ultrasonic Wedge Bond Interface by Using High-Resolution Transmission Electron Microscopy

Hongjun Ji[1,2], Mingyu Li[1], Chunqing Wang[2], Jongmyung Kim[3] and Daewon Kim[3]

[1] Harbin Institute of Technology Shenzhen Graduate School, HIT Campus, Shenzhen University Town, Xil, Shenzhen 518055 P.R.China

[2] State Key Laboratory of Advanced Welding Production Technology, Harbin Institute of Technology 92, Xidazhi Street, Nangang, Harbin 150001 P.R.China

Jeonnam Provincial College, Jeonnam 517-802, Korea

E-mail: jhj7005@hit.edu.cn; myli@hit.edu.cn; wangcq@hit.edu.cn

Abstract

In this paper, the bond interface of Al-1wt.%Si wire bonded on Au/Ni/Cu pad at atmosphere temperature was analyzed by using high resolution transmission electron microscopy. Nano-scale characteristics at bond interface indicated that elemental aluminum diffusing into gold layer was with the feature of step-level periodicity. Due to exceeding solid solubility limit, intermediate Au_8Al_3 phase penetrated among the Al-Au solid solutions. The diffusion distance was not more than 100nm analyzed by energy x-ray dispersive spectrum. This process controlled by solid diffusion reaction was realized by the rapid and periodic ultrasonic vibration according to the theoretical calculation based on the Fick's Law and the observation of the deformation twins among the bond wire and within the interfacial diffusion layer.

Introduction

Ultrasonic wedge bonding is one of the most important methods to realize signal and power interconnection. However, one crucial problem encountered is little knowledge about ultrasonic effects on bonding process, especially on bond interface due to the slightness of interface reaction and the limitations of observing methods, resulting in instability of bond quality and hindrance to the technique innovation.

Ultrasonic softening effect and its enhanced plastic deformation of aluminum wire have been certified. Dislocations, point defects etc. preferentially absorbed acoustic energy, which freed the defects from their pinned positions and consequently allowed metal to deform heavily under relatively low compressive load [1]. Recently, the reactant between thermosonic gold ball and aluminum pad was identified as intermetallic phase (IP) Au_4Al, presented by Li J. H. etc [2], and they ascribed it to atomic diffusion because of the activation of ultrasonic and thermal energy. However, Au_8Al_3 formed when it came to the bond interface between Al-Si wire and gold pad during ultrasonic wedge/wedge bonding, as reported by Geißler U. etc. [3]. They demonstrated that recrystallization of bonded wire occurred near the interface and further showed that IP layer grew thicker with the increase of ultrasonic power. However, all of above are absent of atomic scale identification to the bond interface, especially the cause of these phases was still debated [4].

Therefore, a thorough, accurate nano-scale observation and analysis is necessary. In this study, the joining behavior of Al-1%Si wedge bonding on Au/Ni/Cu pad was investigated using high resolution transmission electron microscope (HRTEM) and the bonding mechanism was discussed experimentally and theoretically.

Experimental Procedures

In the experiment, the bonder used was AW121Z Au/Al wire bonding machine. The wedge was flat 30COB series 2025-L, and the ultrasonic wedge/wedge method was used to bond the Al-1%Si wire of a diameter of 25μm on Au/Ni/Cu pad. The parameters were as follows: 150mW ultrasonic power (resonant frequency ~60kHz), 60gf (1gf=9.8mN) bonding force and 20ms bonding time at room temperature after technique window was optimized by the deformation standard (in the range 20%~50%) and by pull test (pull force was above 6gf).

At the beginning, in order to analyze total features of bond interface, the bond footprint on pad was observed under JSM-6460LV scanning electron microscopy (SEM) at 20kV acceleration voltage after etching aluminum wire away using 5% NaOH solution. Since the standard sample for TEM is a disc of a diameter of 3mm, and their thickness should be thinned enough to be electron-transparent, TEM samples were prepared with "H-bar" method by FEI DB235 FIB equipment. All TEM lamellae were cut perpendicular to the wire axial direction (ultrasonic vibration direction). Then, the bond interface was analyzed with FEI Tecnai G2 F30 field emission TEM at 300kV accelerated voltage. Convergent beam electron diffraction (CBED) enables the identification of the type and the structure of IPs at bond interface.

Results and Discussion

In order to investigate the distribution of actual joining area, we initially observed characteristics of bond footprints on the pad after chemically etching the bond wire away (first bond), as shown in Fig. 1. The footprint was typical of wedge bonding and characteristically exhibited a torus-shaped peripheral zone within which was the bond zone. Joint marks (corrugations) dispersed mainly at bond periphery with directional and periodic features evidently. These marks and their distributions were attributed to the plastic flow of bond wire during bonding process [5]. Because the reactants were too slight to be identified quantitatively using energy dispersive x-ray analysis equipped on the SEM, FIB-HRTEM experiments were performed. All TEM lamellae were cut perpendicular to the vibration direction of ultrasonic as shown in Fig. 1 black line.

978-1-4244-2739-0/08/$25.00 ©2008 IEEE

Fig. 2 shows the transmission electron microscopy (TEM) images of the bond wire (Fig. 2(a)) and the bond interface

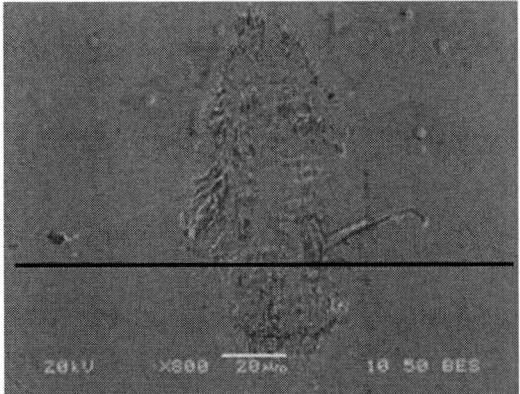

Fig. 1 SEM image of the bond footprint on the pad after chemically etching the bond wire away indicating the actual joining area (micro-joining marks) at the bond interface. The black line shows the location and direction of the TEM sample preparation.

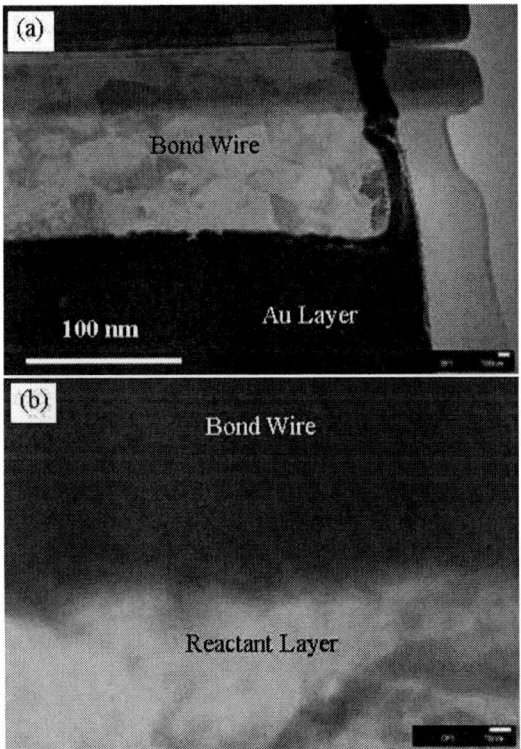

Fig. 2 (a) TEM bright field image of the grains morphology on the bond wire (b) TEM dark field image of the interfacial reactant.

(Fig. 2(b)). On the cross-section of the bond wire, no evident dislocations or other defects were observed, but the grain size was rather diversity, ranging from 100nm to 1μm. The microstructures of the bond wire adjacent to the interface were fragmentized. In Fig. 2(b), the darker area indicated where more aluminum distributed and brighter area presented more gold. Interdiffusion between aluminum and gold was evident. Bond interface was joined perfectly, and a reactant layer was observed.

The reactant at bond interface was identified by CBED as Au_8Al_3, as shown in Fig. 3. Its crystal structure is rhombohedral, belonging to space group R 3 c, and the lattice parameters are a=7.724Å and c=42.083Å. The zone axis was [7 21 5̱]. No other kinds of IP were found during electron diffraction (ED) analysis.

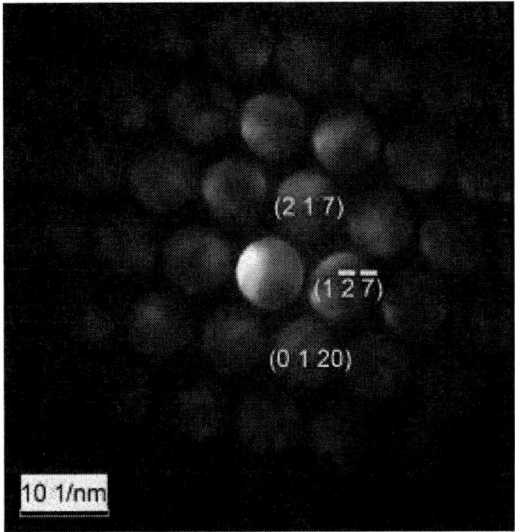

Fig. 3 Convergent beam electron diffraction pattern of the Au_8Al_3 intermetallic phase.

Associated with Fig. 1, content of the joint marks on the bond footprint could be these Au_8Al_3 phase uncovered after etching the bond wire away. Obviously, morphology and distribution of the reactant intermetallic phase had important influence on the bond qualities (such as electrical resistance) and reliability. However, due to the traditional concept of fusion welding or soldering, the layered structure of the reactant was assumed as a matter of course. Such as in thermosonic ball bonding (TSB) of gold wire on aluminum pad, which the bonding area must be preheated to 150 or 200℃, a layered interfacial reactant of Au_4Al and/or Au_8Al_3 was identified under TEM or energy x-ray dispersive spectrum (EDS). Regarding ultrasonic wedge bonding (USB) at room temperature, a layered reactant of Au_8Al_3 was also reported based on the CBED though it was absent of the EDS analyses and high resolution TEM (HRTEM) images. The analyses accuracy of CBED is about several nanometers, which can not give any dimensional information about the reactant. Therefore, a direct analyses method, such as EDS line scan, HRTEM should be used to identify its dimension.

During our previous study, the discrete joint marks were firstly found, which was an important phenomena indicating the microstructure and distribution of the interfacial reactant. Further, the bond resistance varied with bonding parameters fluctuently, which also implied the joining process at bond interface was periodic [6]. So, in order to make sure whether the reactant layer was continuous or not, HRTEM was used to observe the atomic-scale joining characteristics at bond interface, including EDS line scan and fast Fourier transformation (FFT).

Fig. 4 shows the high resolution images of the reactant layer and the FFT and reverse FFT images (magnified images) of their rectangular parts. The diffusion phenomena with the features of the interference patterns were shown in Fig. 4(a). In some parts, novel intermetallic phase formed according to the diffraction patterns analyzed and the interplanar distances measured in the high resolution images. It was also identified as Au_8Al_3. From the reverse FFT image, the twin crystals were obvious, as shown in Fig. 4(a) D. In other area, as seen in Fig. 4(b), the Au_8Al_3 phase was also found. No other IP was found. It should be referred that these intermetallic phase was rather small and dispersed among the interface individually.

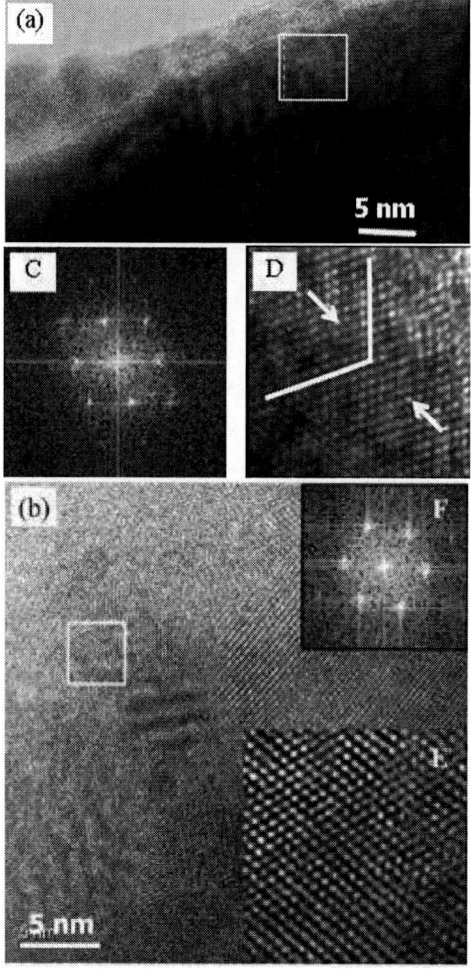

Fig. 4. (a) and (b) High resolution TEM (HREM) images of the interdiffusion characteristics of the bond interface. Images C and F are FFT diffraction pattern of the rectangular parts in (a) and (b), respectively, their magnified parts are shown in D and in E. Some twin crystals in area D were present.

Furthermore, the distribution of elemental Al and Au analyzed by EDS line scan appeared no obvious flat step, but indeed, it indicated that the diffusion of elemental Al into Au layer occurred during bonding process and the diffusion distance was about 100nm, as shown in Fig. 5. Evidently, based on the above images and analyses, the interfacial reactant Au_8Al_3 was dispersed in the Au-Al solid solutions as precipitating particles. This foundation is rather different from the previous reports. It is found for the first time that the constitutes of the wedge bond interface is a composite structures, neither solid solutions only formed by atomic diffusion nor layered IPs only. The structure of the wedge bond interface can be illustrated in Fig. 6.

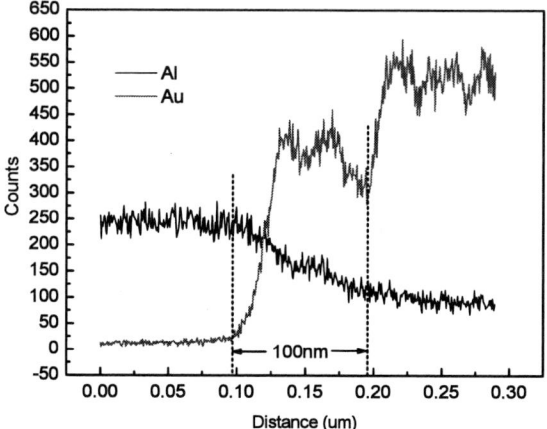

Fig. 5 EDS line scan showing the distribution of elemental aluminum and gold at the interface. No obvious flat step was found, indicating no large dimensional layered phase precipitation.

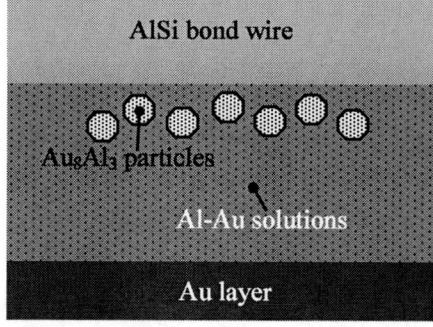

Fig. 6 Schematic of the constitutes of the wedge bond interface showing the Al-Au solutions added with Au_8Al_3 particles.

Evidently, these features were resulted from the diffusion and reaction processes during ultrasonic bonding. It was known that the relative motion at the interface played a very important role in ultrasonic bonding. Surface contaminants and oxides were broken down. So, pure contact interfaces partially exposed and contacted, especially at bond periphery [7-9]. However, at ambient temperature, there was another more important issue for the IP formation, activation energy.

Interfacial friction and the rapid, cyclic and large deformation of the bond wire produce a certain amount of heat energy. But the flash temperature was hardly measured accurately due to small bond size and short bonding time. Experimental results showed it was not more than 300℃ [10] even below 80℃ [11]. By theoretical calculation based on flash temperature model, the temperature rise during bonding process was not more than 321℃ [12].

Since Au_8Al_3 was formed only in the gold pad, calculation of the time needed for the formation of IP region by diffusion at various temperatures can be done using the semi-infinite rode approximation in the solid state, as shown in equation 1.

$$C_x = C_0 \left(1 - \text{erf} \left(\frac{x}{2\sqrt{Dt}} \right) \right) \qquad (1)$$

The finite atomic concentration of the Al in the Au pad was set to $C_x = 0.3$ (at $x = 0$, $C_0 = 1$). The diffusion time t can be defined as the bonding time. The relationship between diffusion coefficient D and temperature is calculated from equation 2 [13]:

$$D = 0.025 \exp(-\frac{164.3 \times 10^3}{RT}) + 0.83 \exp(-\frac{212.3 \times 10^3}{RT}) \qquad (2)$$

Therefore, diffusion distance x can be estimated under different temperatures, as seen in Table 1. According to these results, the interfacial temperature should be above 500℃ if the maximal thickness of the reactant layer was 100nm.

Table 1 the diffusion distance for the time of 20ms containing 30 at.% Al inside the solid Au, as a function of the temperature.

$T(℃)$	D (cm²/sec)	X (nm)
80	1.22×10^{-26}	2.28×10^{-5}
165	6.36×10^{-22}	5.21×10^{-3}
175	1.74×10^{-21}	8.61×10^{-3}
300	2.63×10^{-17}	7.68
400	4.45×10^{-15}	13.77
500	2.01×10^{-13}	92.57
600	3.85×10^{-12}	405.13

However, the temperature measured in experiments and calculated in theories was not more than 350℃, which was too low to activate such a long distance. As seen in Tab 1, under 400℃, the diffusion distance was only 13.77nm.

Another more important mechanism for the formation of the metallurgical wedge bond was short-circuit diffusion reported by Li et al [2]. Compared with the fast diffusion path provided by the increasing dislocations among the aluminum pad in thermosonic gold ball bonding, in our study, twin crystals were observed evidently not only in the bond wire adjoining to the interface but also among the reactant as seen in Fig. 4(a) D. No obvious dislocations were observed. These plane defects, twin crystals, can also provide the fast diffusion path and make the diffusion process occur under relatively lower temperature. Thus, the reaction diffusion along the twin boundaries promoted faster diffusion of elemental Al into Au and intermetallic phase Au_8Al_3 was precipitated when exceeded its solid solubility.

Generally, the slipping deformation (perfect dislocation) is the mainly mechanism for the plastic deformation of the face-centered cubic (FCC) metals. Another more uncommon mechanism for the plastic deformation is twinning (partial dislocation) when the deformation proceeds under lower temperature or high strain rate. During USB process, the strain rate of the bond wire in height can be calculated based on the condition: bonding time was 20ms. Deformation rate of the cross section of the bond wire can reach 50%. It indicated that the stain rate was about 2500% per second. Furthermore, USB experiments were carried out at ambient temperature. Therefore, the deformation mechanism of the bond wire may be changed from the slipping to the twinning. This found is rather novel because it occurred in the coarse grains of the aluminum-silicon alloy. This work will be done further in order to certify the effects of ultrasonic in the bonding/processing on the metals.

Conclusions

According to our experimental observation in atomic-scale and theoretical analyses, the wedge bond interface of ultrasonic Al-1%wt.Si wire/Au pad possesses the evident effects of rapid, cyclic ultrasonic vibration, such as the joint marks, large amounts of twin-crystals. It was found for the first time that the bond interface was composed of Al-Au solid solutions added with Au_8Al_3 particles, completely different from the previous concept that the interfacial reactant was a layered solid solutions or a layered intermetallic compounds.

Thermal energy can hardly activate such longer diffusion distance. Twin crystals distributing in the bond wire adjoining to the interface and among the reactant provided the fast diffusion path for the formation of the metallurgical bonds at ambient temperature.

The deformation mechanism of the bond wire during the bonding process may be twining due to the rapid ultrasonic vibration effects under relatively lower temperature.

Acknowledgments

This work was financially supported by NURI Project.

References

1. Langenecker B., "Effects of Ultrasound on Deformation Characteristics of Metals," *IEEE Trans. Sonics Ultrasonics*, Vol. 13 (1966), pp. 1-8.
2. Li J. H., Han L., Zhong J., "Microstructural Characteristics of Au/Al Bonded Interfaces," *Mater. Charat.*, Vol. 58 (2007), pp. 103-107.
3. Geiβler U., Ramelow M. S., Lang K., Reichl H., "Investigation of Microstructure Processes during Ultrasonic Wedge/Wedge Bonding of AlSi1 Wires," *J. Electron. Mater.*, Vol. 35 (2006), pp. 173-180.
4. Qi J., Huang N. C., Li M., Liu D. M., "Effects of Process Parameters on Bondability in Ultrasonic Ball Bonding," *Scripta Mater.*, Vol. 54 (2006), pp. 293-297.
5. Ji H. J., Li M. Y., Wang C.Q., et al, "Evolution of the Bond Interface during Ultrasonic Al-Si Wire Wedge Bonding Process," *J. Mater. Processing Technology*, Vol. 182 (2007), pp. 202-206.
6. Ji H. J., Li M. Y., Wang C. Q., et al, "In Situ Measuement of Bond Resistance Varying with Process Parameters during Ultrasonic Wedge Bonding," *J. Mater. Processing Technology*, in press.
7. Li M. Y., Ji H. J., Wang C.Q., et al, "Irregular Characteristics of Bond Interface Formation in Ultrasonic Wire Wedge Bonding," *J. Mater. Sci. Technol.*, Vol. 22, No. 4 (2007), pp. 483-486.
8. Li M. Y., Ji H. J., Wang C. Q., Bang H.S., Bang H. S.,

"Interdiffusion of Al–Ni System Enhanced by Ultrasonic Vibration at Ambient Temperature," *Ultrasonics*, Vol. 45 (2006), pp. 61-65.

9. Ji H. J., Li M. Y., Wang C. Q., Bang H.S., Bang H. S., "Comparison of Interface Evolution of Ultrasonic Aluminum and Gold Wire Wedge Bonds during Thermal Aging," *Mater. Sci. Eng. A-Struct.*, Vol. 447 (2007), pp. 111-118.

10. Joshi K. C., "The Formation of Ultrasonic Bonds between Metals," *Welding J.*, Vol. 50 (1971), pp. 840-848.

11. Harman G. G., Leedy K. O., "An Experimental Modal of the Microelectronic Wire Bonding Mechanism," *10th Annu. Pro. Reliability Physics Symp.*, Las Vegas, NV USA; 1972, pp. 49-56.

12. Jeng Yeau-Ren, Horng Jeng-Haur, "A Microcontact Approach for Ultrasonic Wire Bonding in Microelectronics," *ASME J. Tribol.*, Vol. 123 (2001), pp. 725-731.

13. Smithells C.J. Smithells Metals Reference Book, 7th edition, Butterworth-Heinemann, (Oxford, 1992).

Limited β-Sn grain number of miniaturized Sn-Ag-Cu solder joints

Shihua Yang[a], Chunqing Wang[a], Yanhong Tian[a], Pengrong Lin[a], Le Liang [b]

[a] Department of Electronics Packaging Science and Engineering, School of Materials and Engineering
Harbin Institute of Technology, P.O. Box 436, 92, Xidazhi Street, Nangang, Harbin, 150001, China
Email: yangshihua@hit.edu.cn

[b] Interconnect Product and Technology Center SAMSUNG SEMICONDUCTOR (CHINA) R&D CO., LTD
Science Plaza 7F, International Science & Technology Park, Suzhou Industrial Park, Suzhou, 215021, China
Email: le.liang@samsung.com

Abstract

As-reflowed Sn-Ag-Cu solder joints in various diameters were found to contain only several β-Sn crystal grains. With the solder joints increasingly miniaturized, there is no obvious change in the grain number in a solder joint. The aged Sn-Ag-Cu solder joints are composed of very limited number of β-Sn crystal grains as well. It appears that the solder joint size and thermal aging have less influence on the growth of β-Sn grains.

Introduction

Sn-Ag-Cu (SAC) solder alloys have been widely used as the electronics industry standard lead-free solder to replace the traditional near-eutectic Sn-Pb solders. [1] Several SAC alloys with compositions lie near a ternary eutectic point of approximately Sn-3.5Ag-0.9Cu are commercially available and the investigations on the optimization of SAC alloy compositions have been well demonstrated in the literature. [2-6] The researches showed that Ag content less than 3wt.% was suitable to minimize the formation of large Ag_3Sn plates and Cu content no higher than 0.7~0.9wt.% to reduce the difference between the liquidus and the solidus temperature of the SAC alloy and the formation of large Cu_6Sn_5 rods. [2,7] So the Sn-3.0Ag-0.5Cu alloy was recommended and used in the subsequent study.

There is a common characteristic in the SAC alloys that near eutectic SAC alloys are more than 95 atomic percent Sn. The nature of solidification in SAC alloys results in solder joints comprised of highly oriented large Sn grains. [8,9] LaLonde et al. quantitated the β-Sn grains in Sn-3.8Ag-0.7Cu solder balls via electron backscatter diffraction (EBSD) and polarized light microscopy (PLM). [10] Based on their results, it was estimated that a single 900μm-diameter solder ball contained on average eight individual β-Sn grains, independent of cooling rate (0.35~3.0℃/s). Henderson et al. found that typical SAC solder ball grid array joints, 900μm in diameter, were composed of 1 to 12 different Sn grains and the average was estimated at approximately 8 grains. [11] Lehman et al. also illustrated that both SAC solder balls and solder joints contained few β-Sn grains or only one β-Sn grain. [12] It is well known β-Sn is highly anisotropic in thermal expansion and elastic properties.[13] Therefore, the thermomechanical response of SAC solder joints composed of several β-Sn grains depends significantly on the size and orientation of their β-Sn grains. It has been confirmed that the anisotropy of β-Sn is correlated with the formation of fatigue cracks in SAC solder joints during temperature cycling that is different from typical observation in Sn-Pb solder joints. [14,15]

However, to meet the urgent demands of future electronic packaging, the solder joints have to become increasingly miniaturized. The researches on the analysis of β-Sn grain number during the miniaturization of solder joints are insufficient. The change in grain number will have direct influence on assessing the reliability of SnAgCu solder joints. The main aim in this paper is to investigate whether a correlation between the grain number evolution and the change in the joint size indeed occurs.

Experimental

Commercial Sn-3.0Ag-0.5Cu BGA solder balls in different diameters ranging from 200 to 600μm were employed. The solder balls were bonded to the Cu pads prefluxing RMA flux in a reflow process with a maximum temperature of 245℃ for 25s. The proportion between the diameter of a solder ball and the corresponding pad on FR-4 board was 1.2. After the first reflow, some of the samples were aged isothermally at 150℃ for 36days. Samples were subsequently mounted in epoxy and metallographically cross sectioned using standard polishing procedures. The polishing had to be performed with great care using adequately low compression force to have the surface free from any polishing damage.

β-Sn is birefringent in reflecting light from a polished surface because of its tetragonal crystal structure. Therefore, a polarized light microscope with a nearly crossed polarizer/analyzer pair was used to identify the different β-Sn grains. By viewing the reflected light through an analyzing polarizer, nearly or partially crossed with respect to the incident polarized light, contrast can be seen because the different crystallographic orientations exhibit a distinguishable contrast. Each sample was rotated over a wide variety of angles to get high contrast and then the contrast differences were used to discern and quantitate differently orientated grains of a solder joint.

Results and Discussion

SAC Solder joints are nearly comprised of eight or few β-Sn grains per joint. [10-12,16] Fig. 1 shows typical polarized light micrographs of the as-reflowed SAC solder joints in diameter of 400μm. The solder joint in Fig. 1a just has only one β-Sn grain and Fig. 1b four β-Sn grains. It was clearly observed that the SAC solder joints contains few crystal grains. Base on the PLM analysis, all of the as-reflowed SAC solder joints contain no more than six β-Sn grains of each joint and the number of β-Sn grains per cross section of each

978-1-4244-2739-0/08/$25.00 ©2008 IEEE

solder ball as a function of solder size is plotted in Fig. 2. The results suggest that the average number of β-Sn grains per solder ball cross section is independent of solder joint size. This is a result of solidifying characteristic in SAC solder joints. The solidification of the β-Sn phase is rapid because of the significant undercooling and because of the intrinsically rapid growth kinetics of β-Sn. The expeditious solidification and the concomitant release of latent heat of fusion results in the increase in temperature of solders and thus the nucleation of other β-Sn grains is restrained. It has been reported that the degree of undercooling increases monotonically with decreasing sample size, and the growth rate of β-Sn increases in proportion to the undercooling of β-Sn. [8] But the growth rate of β-Sn grains is so extremely fast that the solder joints with the diameters less than 600μm can completely solidify in the split second. Therefore, the decrease in sample size produces negligible effect on the growth of Sn. And the larger and smaller SAC solder joints have almost the same average number of β-Sn grains.

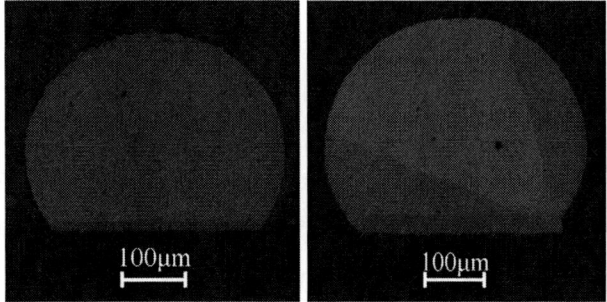

Fig. 1 Cross-polarized light micrographs of SAC solder joint in diameter of 400μm

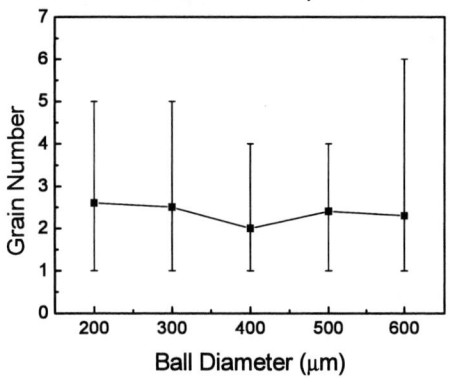

Fig. 2 The number of β-Sn grains per solder joint cross section versus solder ball diameter

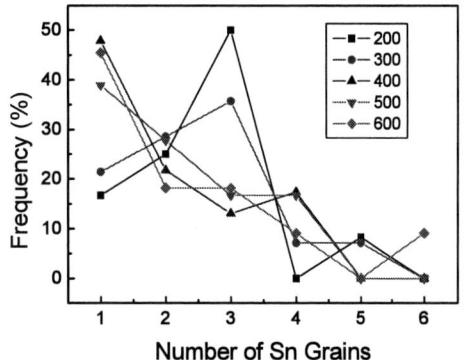

Fig. 3 The frequency versus number of β-Sn grains of each solder joint

Fig. 3 illustrates the frequency of β-Sn grains per cross section in the as-reflowed solder joints as a function of the number of β-Sn grains per cross section determined by PLM. It can be seen that more than 80% of SAC solder joints have no more than 3 β-Sn grains. There is a further possibility that the bigger solder joints are composed of only one β-Sn grains, and the smaller joints three β-Sn grains.

The change in the number of β-Sn grains as a function of the size of solder joints aged for 36 days was shown in Fig. 4. No marked difference in grain number between the aged solder joints and as-reflowed ones were observed as well. This indicates that aging can not give rise to the visible change in the crystal orientation of β-Sn grains in SAC solder joints. It seems that the long time aging has little influence on the β-Sn grain growth.

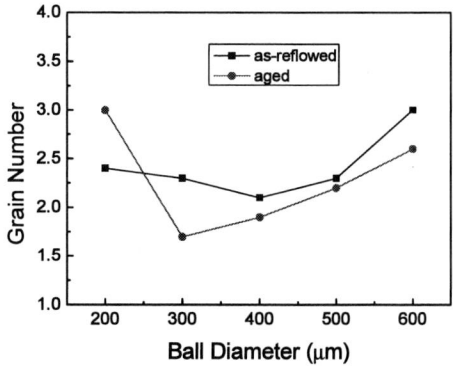

Fig, 4 The β-Sn grain number of SAC solder joints in different size

Due to the limited grain phenomenon of SAC solder joints and the high anisotropic properties of β-Sn grains, further in-depth researches on the orientation distribution of β-Sn grains and the distribution of misorientation angles between neighbouring β-Sn grains should be conducted to observe the evolution of the orientation in the solder joints in different diameters.

Conclusions

Sn-3.0Ag-0.5Cu solder joints in different diameters were found to contain very limited β-Sn grains. The results showed that more than 80% of SAC solder joints have no more than 3 β-Sn grains as determined by PLM. No marked influence of SAC solder joint size on the β-Sn grain number was observed.

The influence of aging time on the evolution of the number of β-Sn grains per cross section in the solder joints was investigated. Aging time was also found to have little influence on the β-Sn grain growth.

Acknowledgments

This work is financially supported by the National Natural Science Foundation of China under grant No. 50675047/ E052105 and Joint Project between Samsung Electronics Co. Ltd. (Korea) and Harbin Institute of Technology (HIT).

References

1. C.M. Miller, I.E. Anderson, and J.F. Smith, "A Viable Tin-Lead Solder Substitute - Sn-Ag-Cu," *J. Electron. Mater.*, Vol. 23, No. 7 (1994), pp. 595-601.

2. S.K. Kang, P.A. Lauro, D.-Y. Shih, D.W. Henderson, and K.J. Puttlitz, "Microstructure and Mechanical Properties

of Lead-free Solders and Solder Joints Used in Microelectronic Applications," *IBM J. Res. & Dev.*, Vol. 49, No. 4-5 (2005), pp. 607-620.

3. K.W. Moon, W.J. Boettinger, U.R. Kattner, F.S. Biancaniello, and C.A. Handwerker, "Experimental and Thermodynamic Assessment of Sn-Ag-Cu Solder Alloys," *J. Electron. Mater.*, Vol. 29, No. 10 (2000), pp. 1122-1136.

4. K. W. Moon, and W. J. Boettinger, "Accurately Determining Eutectic Compositions: The Sn-Ag-Cu Ternary Eutectic," *JOM*, Vol. 56, No. 4, (2004), pp. 22-27.

5. S.K. Kang, W.K. Choi, D.-Y. Shih, D.W. Henderson,T. Gosselin, A. Sarkhel, C. Goldsmith, and K.J. Puttlitz, "Ag₃Sn Plate Formation in the Solidification of Near-ternary Eutectic Sn-Ag-Cu," *JOM*, Vol. 55, No. 6 (2003), pp. 61-65.

6. I.E. Anderson, "Development of Sn-Ag-Cu and Sn-Ag-Cu-X Alloys for Pb-free Electronic Solder Applications," *J Mater Sci: Mater Electron*, Vol. 18, No. 1-3 (2007), pp. 55-76.

7. K.S. Kim, S.H. Huh, and K. Suganuma, "Effects of Intermetallic Compounds on Properties of Sn-Ag-Cu Lead-free Soldered Joints" *J. Alloys Compd.*, Vol. 352, No. 1-2 (2003), pp. 226-236.

8. R. Kinyanjui, L.P. Lehman, L. Zavalij, and E. Cotts, "Effect of Sample Size on the Solidification Temperature and Microstructure of SnAgCu Near Eutectic Alloys," *J. Mater. Res.*, Vol. 20, No.11 (2005), pp. 2914-2918.

9. D. Swenson, "The Effects of Suppressed beta Tin Nucleation on the Microstructural Evolution of Lead-free Solder Joints," *J Mater Sci: Mater Electron*, Vol. 18, No. 1-3 (2007), pp. 39-54.

10. A. LaLonde, D. Emelander, J. Jeannette, C. Larson, W. Rietz, D. Swenson, and D.W. Henderson, "Quantitative Metallography of beta-Sn Dendrites in Sn-3.8Ag-0.7Cu Ball Grid Array Solder Balls via Electron Backscatter Diffraction and Polarized Light Microscopy," *J. Electron. Mater.*, Vol. 33, No. 12 (2004), pp. 1545-1549.

11. D.W. Henderson, J.J. Woods, T.A. Gosselin, J. Bartelo, D.E. King, T.M. Korhonen, M.A. Korhonen, L.P. Lehman, S.K. Kang, P. Lauro, D.Y. Shih, C. Goldsmith, and K.J. Puttlitz, "The Microstructure of Sn in Near-eutectic Sn-Ag-Cu Alloy Solder Joints and Its Role in Thermomechanical Fatigue," *J. Mater. Res.*, Vol. 19, No. 6 (2004), pp. 1608-1612.

12. L.P. Lehman, S.N. Athavale, T.Z. Fullem, A.C. Giamis, R.K. Kinyanjui, M. Lowenstein, K. Mather, R. Patel, D. Rae, J. Wang, Y. Xing, L. Zavalij, P. Borgesen, and E.J. Cotts, "Growth of Sn and Intermetallic Compounds in Sn-Ag-Cu Solder," *J. Electron. Mater.*, Vol. 33, No. 12 (2004), pp. 1429-1439.

13. K.N. Subramanian, and J.G. Lee, "Effect of Anisotropy of Tin on Thermomechanical Behavior of Solder Joints," *J Mater Sci: Mater Electron*, Vol. 15, No. 4 (2004), pp. 235-240.

14. M.A. Matin, E.W.C. Coenen, W.P. Vellinga, and M.G.D. Geers, "Correlation between Thermal Fatigue and Thermal Anisotropy in a Pb-free Solder Alloy," *Scripta Mater.*, Vol. 53, No. 8 (2005), pp.927-932.

15. M.A. Matin, W.P. Vellinga, M.G.D Geers, "Thermomechanical Fatigue Damage Evolution in SAC Solder Joints," *Mater. Sci. and Eng.* A, Vol. 445, (2007), pp. 73-85.

16. A. U. Telang, and T. R. Bieler, "The Orientation Imaging Microscopy of Lead-free Sn-Ag Solder Joints," *JOM*, Vol. 57, No. 6 (2005), pp. 44-49

Manufacture, Microstructure and Microhardness Analysis of Sn-Bi Lead-Free Solder Reinforced with Sn-Ag-Cu Nano-particles

Lili Zhang[1], Wenkai Tao[1], Johan Liu[1,2], Yan Zhang[1], Zhaonian Cheng[2], Cristina Andersson[2], Yulai Gao[3] and Qijie Zhai[3]

1. Key Laboratory of Advanced Display and System Applications & SMIT Center, School of Automation and Mechanical Engineering, Shanghai University, No.149 Yanchang Road, Shanghai 200072, P.R. China

2.SMIT Center & Bionano Systems Laboratory, Department of Microtechnology and Nanoscience, Chalmers University of Technology, SE-412 96 Goteborg, Sweden

3. School of Materials Science and Engineering, Shanghai University, No.149 Yanchang Road, Shanghai 200072, P.R. China

Corresponding author: E-mail: jliu@chalmers.se

Abstract

This paper investigates a composite solders obtained by adding Sn-3.0Ag-0.5Cu (SAC) nano-particles into conventional eutectic Sn-58Bi solder paste. The microstructure analysis and the measurement of the Vickers microhardness have been carried out. Utilizing the self-developed Consumable-electrode Direct Current Arc (CDCA) technique, the Sn-3.0Ag-0.5Cu nano-particles with an average particle size between 20 and 80nm are prepared.

The reinforced lead-free Sn-Bi solder was prepared by thoroughly blending the nanometer-sized SAC particles into the eutectic Sn-Bi solder paste. The SAC reinforced Sn-Bi composite solder paste was printed onto ENIG/Cu metalized substrate and reflowed in a conventional reflow oven. After reflow, the morphology of the as-solidified reinforced composite solder was observed by means of SEM and TEM. The Vickers microhardness measurements indicated that the addition of SAC nano-particles enhances the overall strength of the eutectic solder, and the results agree well with the theory of dispersion strengthening.

1. Introduction

Tin–lead solders have widely been used in electronics manufacturing for many decades. This solder has many benefits, such as ease of handling, low melting temperature, good workability, ductility, and excellent wetting on Cu and its alloys. At present, soldering technology has become indispensable for the interconnection and packaging of virtually all electronic devices and circuits [1]. However, due to the toxicity of lead, many countries are going to ban the use of lead and its compounds. In Europe, the move towards lead-free solder alternatives has stepped-up significantly since Japanese industry declared their going lead-free time schedule and also as a result of legislation within the European Union. The European Union has approved the directives on waste from electrical and electronic equipment (WEEE) and restriction of hazardous substances (RoHS) [2,3].

For surface mount technology, solders are demanded high reliability and dimensional stability to improved mechanical properties [4,5]. A potential method for enhancing the mechanical properties of a solder alloy and consequently of solder joints has been reported [6] is to add second-phase particles to a solder matrix in order to form a composite solder. These particles can either be metallic or intermetallic. Generally, composite solders contain fine second-phase reinforcing particles dispersed uniformly in the solder matrix.

There are two possible ways to manufacture the composite solders. One is in-site reaction and the other is mechanical mixing. The presence, formation, and growth of a second phase have been studied as the potential mechanisms controlling solderability and reliability of solder joints [6-8]. The nano-particles obstruct movement of dislocation and pin grain boundaries so as to prevent the solder matrix from plastic deformation.

With the development of nano technology, various nano-sized particles were selected as reinforcements in manufacturing composite solders. Masazumi Amagai [9] studied the different kinds of nano-particles in Sn-Ag based lead free solders, and the present paper concentrates on the intermetallic compounds growth after four time reflow processes and thermal aging. Also, these nano- particles were studied on whether they can reduce the occurrence frequency of intermetallic compound fractures in high impact pull tests and affect solder hardness and displacement in drop tests. Moreover, the nano-sized metallic particles (Ni, Cu, and Mo) reinforced composite solders were developed, and the results showed that the mechanical properties, such as microhardness and creep resistance were improved [10-12].

The purpose of this paper is to investigate the microstructure analysis and the measurement of the Vickers microhardness of composite solders, where the solder is obtained by adding Sn-3.0Ag-0.5Cu (SAC) nano-particles into conventional eutectic Sn-58Bi solder paste.

2. Experimental procedure

2.1 Manufacture of nano-composite solder paste

The nano-particles of Sn-3.0Ag-0.5Cu were prepared by using the Consumable-electrode Direct Current Arc (CDCA) technique and preserved in an inert gas atmosphere to prevent oxidation and agglomeration. The nano-particles had an average particle size between 20 and 80nm. Then the nano-composite solder paste was prepared by thoroughly blending the nanometer-sized Sn-3.0Ag-0.5Cu particles into the eutectic Sn-58Bi solder paste. The matrix materials used for this study were solder paste with an average diameter of about 38-53 microns of Sn-58Bi particles. The SEM image of the nano-particles prepared is shown in Fig.1 [13]. The details of manufacturing of nano-particles are given in [13,19]. From Fig. 1, we can see that the SAC nano-particles have a spherical morphology and an average size about 20~80nm. The SAC nano-particles are less than that of the Sn-Bi

978-1-4244-2739-0/08/$25.00 ©2008 IEEE

particles in diameter about 3 orders, and can therefore be used as the reinforcements.

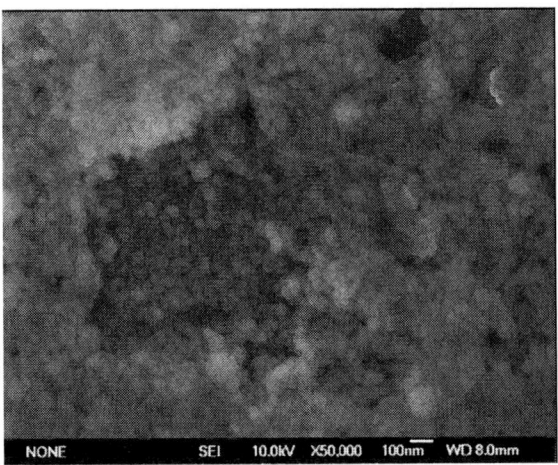

Fig. 1 Morphology of Sn-3.0Ag-0.5Cu alloy nano-particles measured by SEM [13]

Different weight percentages of nano-particles were added into the Sn-Bi solder paste. Then, the nano-composite solder paste was stirred for 30 min so as to ensure a homogeneous distribution of the reinforcement particles in the solder matrix. Finally, the pastes were immediately sealed and preserved in a refrigerator at -10℃.

2.2 Microstructure characterization and indentation testing

The sample preparing of solder joint follows a traditional SMT process. The copper pad on the FR-4 substrate is 50μm in height with 4*4mm and deposited with Ni/Au(ENIG) surface finish. The thicknesses of the Au and Ni layers are 0.1μm and 8–10μm, respectively. The soldering temperature of 180℃ was used in this work, while the reflow time was 10min. At 180 ℃, the SAC nano-particles did not melt because its melting point is higher than the soldering temperature, while Sn-Bi solder with a lower melting point was melted sufficiently. Four different weight percentages, 1wt.% , 2wt.%, 4wt. % and 5 wt.% SAC nano particles, were tested in the present work.

After reflow, regular mechanical grinding, polishing and etching were used to study the morphology and the growth of IMCs. All the samples in this work were cut, cold mounted in epoxy, polished with 0.25μm diamond paste until a mirror surface was obtained. Then, the samples were etched with a 94% C2H5OH-4% HNO3-2% HCl (in vol.%) solution. The polished and etched samples were observed by means of SEM and TEM with the target of observing the morphology, size and distribution of the second-phase nano-particles in the eutectic microstructure.

The Vickers microhardness (Hv) of the composite solder joint samples was measured using a MH-3 Microhardness tester. Higher loads can cause load dependence of hardness, and promote the occurrence of localized microscopic cracking. In order to minimize uncertainties from measurement and cracking of the as-consolidated composite

solder samples, an indentation load of 10 g for a dwell time of 10 s was chosen.

3. Results and discussion

3.1 Microstructure observations

Fig. 2 shows the microstructure of the Sn–58Bi solders and the composite Sn-Bi-SAC solder in the as-cast condition. The features observed are quite typical of a eutectic Sn-Bi solder and consist of fine alternating lamellae of the two constituent phases. In the figures from scanning electron micrograph, the light regions are the tin-rich phase, while the dark regions represent the bismuth-rich phase. The SEM micrograph of the eutectic Sn-Bi solder joint and the morphology of the composite solder joint are shown in Fig. 2(a), (c) and Fig. 2(b). (d), respectively. From Fig. 2(a), it can be seen that the distribution of the bismuth-rich phase is continuous and regular, while it is snatchy in Fig. 2(b). In the enlarged figures of Fig. 2(c) and (d), both Fig. 2(c) and (d) reveal that the nano-sized grain dispersed evenly in the tin-rich phase, but there lies more nano-sized grain in Fig. 2(d) than that in Fig. 2 (c).

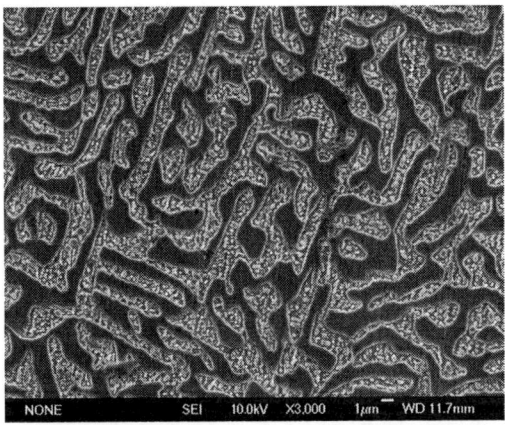

(a) Morphology of the eutectic Sn-Bi solder joint

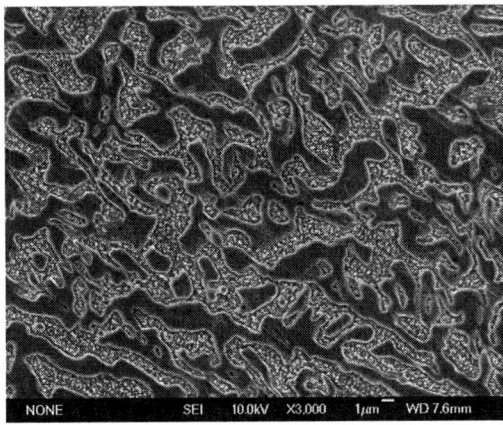

(b) Morphology of the Sn-Bi-1%SAC composite solder

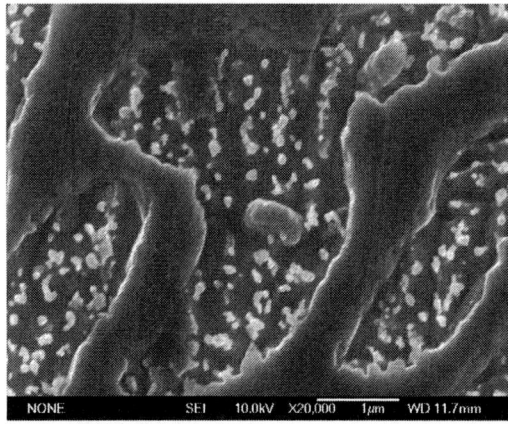

(c) Morphology of the eutectic Sn-Bi solder joint

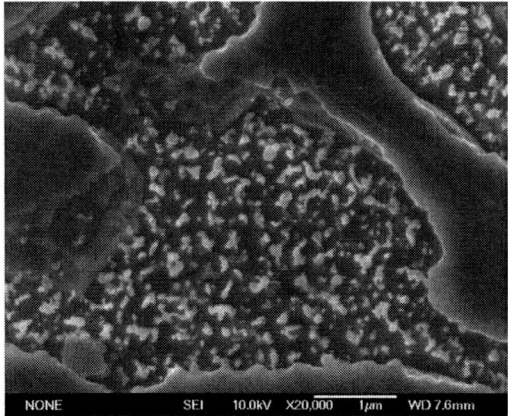

(d) Morphology of the Sn-Bi-1%SAC composite solder

Fig. 2. SEM micrograph of the eutectic Sn-Bi solder joint and the Sn-Bi-SAC composite solder joint

From Fig. 2(a)-(d), it can be seen that the addition of small amount of Sn-3.5Ag-0.5Cu nano-particles to the eutectic Sn-Bi solder marginally alter the as-solidified microstructure. Comparing with a pure eutectic solder, the addition of nano-particles is resort to form a large number of nano-sized grain with uniform distribution. The SAC nano-particles dispersed through the composite microstructure obstruct movement of dislocation and pin grain boundaries so as to prevent the grain from growth [14,15]. As a result, the grains were refined.

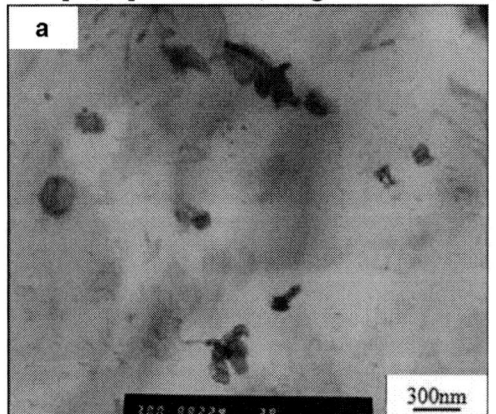

(a) TEM image of the composite solder joint (generally view)

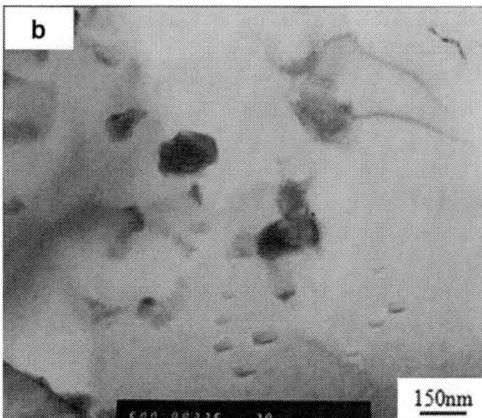

(b) TEM image of the solder composite joint (enlarged view for nano-particles)

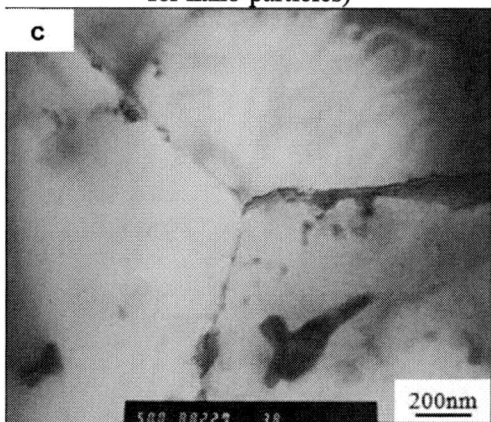

(c) TEM image of the solder composite joint (local view for grain boundaries)

Fig. 3 TEM images of the Sn-Bi-1%SAC composite solder joint

The morphology of the Sn-Bi-1%SAC composite solder joint was observed by transmission electron microscopy as shown in Fig. 3. From Figure 3(a), it can be seen that the nano-particles distributed uniformly generally. Although small amount of the nano-particles were agglomerated, the reunited nano-particles just contained several grains, as shown in Fig. 3(b). It is deduced that the SAC nano-particles pinned the movement of dislocation [8] and this could explain the fact that the grains were refined with nano-particles addition to the eutectic solder.

We can also see from Fig. 3(c) that the nano-particles were easy to gather into crystal grain boundaries or near the grain boundaries. Because of these agglomerated particles, the grain boundaries would be pinned. It indicates that the dispersed nano-particles will result in grain refining.

3.2 Microhardness measurements

The value of hardness provides a measure of the resistance of the material to deformation, densification, and cracking [16]. The influence of SAC nano-particle reinforcement on microhardness is shown in Fig. 4. It can be observed that the microhardness of the composite solder increased with increasing in nano-particles addition percentage into the eutectic Sn-Bi solder paste.

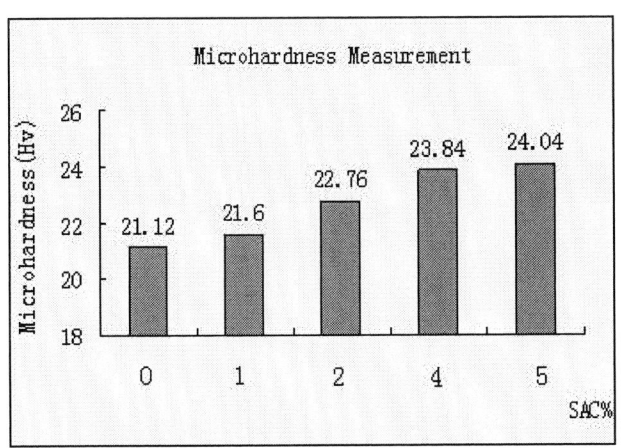

Fig. 4. Bar graph showing the influence of SAC nano-particles on microhardness of the eutectic Sn-58Bi

The microhardness of the solidified composite solders increases 14% for 5.0 wt.% addition of SAC nano-particles, comparing to the eutectic Sn-Bi solder. The measured Vickers microhardness of the composite solder is 24.04Hv, which is higher than the microhardness value of Sn-Bi eutectic solder reported in [17]. The observed increase in microhardness of the composite solder is attributed to the presence of second-phases in the microstructure. Microhardness measurements indicates that the addition of Sn-3.0Ag-0.5Cu nano-particles enhance the strength of the eutectic Sn-Bi solder.

From the dispersion-strengthening theory [18], the microhardness of the composite solder will increase with increase in nano-particles content, and the results in our work agree well with the theory.

4. Conclusions

This paper presents some summary of the work on manufacture, microstructure and microhardness analysis of the Sn-58Bi lead-free solder reinforced with Sn-3.0Ag-0.5Cu (SAC) nano-particles. The addition of SAC nano-particles into eutectic Sn-Bi solder resulted in a refined crystal grain structure. It is due to the fact that the SAC nano-particles obstruct movement of dislocation and pin grain boundaries to prevent the grain from growth. The Vickers microhardness variation with SAC nano-particles content has also been studied. It was observed that the microhardness of the composite solder increased with an increase in nano-particles addition in the eutectic microscale Sn-Bi solder paste. The results from the present work agree well with the theory of dispersion strengthening.

Acknowledgments

We acknowledge the final support from the Ministry of Science and Technology within the 863 program on nanosolder paste (no: 2006AA03Z339). This work is also supported by the National Natural Science Foundation of China (grant no. 10702037). Supported by the Shanghai Rising-Star Program (no: 06QA14020) is also appreciated.

References

1. Katsuaki Suganuma, "Advances in lead-free electronics soldering," *Current Opinion in Solid State and Materials Science,* No.5(2001), pp.55–64.
2. Young-Sun Kim, Keun-Soo Kim, Chi-Won Hwang, Katsuaki Suganuma, "Effect of composition and cooling rate on microstructure and tensile properties of Sn–Zn–Bi alloys," *Journal of Alloys and Compounds*, Vol. 352, No. (2003), pp. 237–245.
3. C. Andersson, Z. Lai, J. Liu, H. Jiang, Y. Yu, "Comparison of isothermal mechanical fatigue properties of lead-free solder joints and bulk solders," *Materials Science and Engineering* A, Vol. 394, No. (2005), pp. 20–27.
4. H. Mavoori, Dimensionally stable solders for optoelectronics and microelectronics, *J. Met. June*, 2000, pp. 30–32.
5. Y. Wu, J.A. Sees, C. Pouraghabagher, L.A. Foster, J.L. Marshall, E.G. Jacobs, R.F. Pinizzotto, "The formation and growth of intermetallics in composite solder," *J. Electron. Mater.* Vol. 22, No. 7 (1993), pp. 769–777.
6. F. Guo, J. Lee, S. Choi, J.P. Lucas, T.R. Bieler, *J. Electron. Mater* 30, 1073 (2001), pp.758-762.
7. I. Dutta, B.S. Majumdar, D. Pan, W.S. Horton, W. Wright, Z.X. Wang, *J. Electron. Mater*, Vol. 33, No. 258 (2004), pp. 637-639.
8. J.P. Liu, F. Guo, Y.F. Yan, W.B. Wang and Y.W. Shi, "Development of Creep-Resistant, Nanosized Ag Particle-Reinforced Sn-Pb Composite Solders," *Journal of Electronic Material*, Vol. 33, No. 9(2004), pp. 959-963.
9. Masazumi Amagai, "A study of nanoparticles in Sn－Ag based lead free solders," *Microelectron Reliab* (2007), pp. 1-16.
10. K.Mohankumar, A.A.O.Tay, in *Proceedings of 6th Electronics Packaging Technology Conference, IEEE*, 2004, pp. 455–461.
11. J.P. Liu, F. Guo, Y.F. Yan, W.B. Wang, Y.W. Shi, *J. Electron. Mat.* Vol. 33, No.9(2004), pp. 958.
12. D.C. Lin, C.Y. Kuo, T.S. Srivatsan, G.X. Wang, in Proceedings of the ASME Heat Transfer Division—2003 vol 3: Heat Transfer Equipment, Heat Transfer in Manufacturing and Materials Processing, Visualization of Heat Transfer, (2003), pp.253–258.
13. Xinzhi Xia, Changdong Zou, Yulai Gao, Johan Liu, Qijie Zhai, "Preparation Techniques and Characterization for Sn-3.0Ag-0.5Cu Nanopowders," *Proceeding of 2007 International Symposium on High Density Packaging and Microsystem Integration*, Shanghai, 2007, pp.302-304.
14. D.C. Lin, S. Liu, T.M. Guo, G.-X. Wang, T.S. Srivatsan, M. Petraroli, "An investigation of nanoparticles addition on solidification kinetics and microstructure development of tin-lead solder," *Materials Science and Engineering A*, Vol. 360 (2003), pp. 285-292.
15. D.C. Lin, G.X. Wang, T.S. Srivatsan, Meslet Al-Hajri, M. Petraroli, "Influence of titanium dioxide nanopowder addition on microstructural development and hardness of tin–lead solder," *Materials Letters*, Vol. 57 (2003), pp. 3193– 3198.
16. D. Lin, G.X. Wang, T.S. Srivatsan, Meslet Al-Hajri, M. Petraroli, "The influence of copper nanopowders on

microstructure and hardness of lead–tin solder," *Materials Letters,* Vol.53, No. (2002), pp. 333– 338.

17. Yasuyuki Miyazawa, Tadashi Ariga, "Microstructure Change and Hardness of Lead Free Solder Alloys," *Dept. of Metallurgical Engineering*, Tokai University, pp. 616-619.

18. J. Shen, Y.C. Liu, Y.J. Han, Y.M. Tian, and H.X. Gao, "Strengthening Effects of ZrO_2 Nanoparticles on the Microstructure and Microhardness of Sn-3.5Ag Lead-Free Solder," *Journal of Electronic Materials*, Vol. 35, No. 8(2006), pp. 1672-1679.

19. Wanbing Guan, Suresh Chand Verma,Yulai Gao, Cristina Andersson, Qijie Zhai1 and Johan Liu, "Characterization of Nanoparticles of Lead Free Solder Alloys," *Proceedings of the 1st IEEE CPMT Electronics Systemintegration Technology Conference (ESTC 2006)*, Germany, Dresden, 2006, pp.7-12.

A comparison study of two different methods to synthesize magnetic slurry for the fabrication of magnetic films

Xu Zheng[1,2], Rong Sun[2], Shuhui Yu[2], Lei Li[2], Mian Huang[2], Guangfu Yin[1], Ruxu Du[2], Lixi Wan[3]

1. College of Materials Science and Engineering, Sichuan University, Sichuan, 610065, China
2. Shenzhen Institute of Advanced Technology, Chinese Academy of Sciences, Shenzhen, 518067, China
3. Institute of Microelectronics of Chinese Academy of Sciences, Beijing, 10029, China
Email 1: yingf@scu.edu.cn, Phone 1: 86-28-85413003, Fax 1:86-28-85416050
Email 2: rong.sun@siat.ac.cn, Phone 2: 86-755-26803554, Fax 2: 86-755-26803554

Abstract

A new spiral thin-film inductor structure is designed, in which the insulating magnetic thin film takes the places of both the insulating layer and the magnetic layer in the traditional structure. For the fabrication of the insulating magnetic film, a magnetic slurry which is a soft magnetic composite composed of organic polymer and nano Fe_3O_4 particles was synthesized. The magnetic slurry was synthesized by chemical co-precipitation method and microemulsions, respectively. Fe_3O_4 particles with a grain size of 20nm tends to agglomerate and is constrained by the network of silicon gel in the first slurry prepared via co-precipitation. While in the microemulsions, water-soluble organic material disperses the nanoFe_3O_4 whose size lays in the range of 6-15nm. The two kinds of magnetic slurries have different magnetic susceptibilities, which is 110cm^3/mol, and 344cm^3/mol, respectively. The surface of the film fabricated with the microemulsions slurry is smooth. In contrast, there are a lot of flaws for the film derived from the co-precipitated slurry. Therefore, the microemulsions method is more suitable to synthesize magnetic slurry for the fabrication of magnetic film.

1. Introduction

With the development of Electronic Packaging Technology, the thin-film device featured by "small, thin and precise" plays a more and more important part. The thin-film inductor employed as the elementary electronic device has aroused many people's interest, such as improving its structure and process to be used for the embedded system [1]. The thin-film inductor with a spiral shape guides the development of thin-film inductors [2,3]. At present, the magnetic layer of the spiral thin-film inductor is mainly metal magnetic film, including Fe, FeCoTaN, NiFe and so on, which not only requires complicated process such as magnetron sputtering but also needs additional insulating barriers [4,5,6].

In this paper, a new concept of making magnetic slurry and film was put forward, and a new structure of spiral thin-film inductor was designed (Fig.1.). The new structure is composed of spiral, base plate and insulating magnetic layer, which takes the places of the metal magnetic layer, underlayer and insulating layer such as polyimide and silica dioxide assembling the traditional spiral thin-film inductor [2]. The fundamental part in the new design is to spin coating or dip coating the magnetic slurry which is a soft magnetic composite composed of organics and ferrite. The soft magnetic composite with high permeability, low coercive force and high electrical resistivity is a chief orientation of

magnetic core, and it can fully develop these merits when a film is fabricated [7, 8]. Therefore, the fine magnetic material is the key part of the new spiral thin-film inductor. Our work is focused on the synthesis of magnetic slurry and the comparison of the magnetic films derived from different slurries.

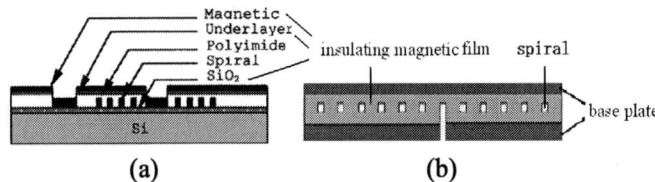

Fig.1. The comparison of the structures (cross section) between the traditional and new spiral thin-film inductors. (a) is the traditional spiral thin-film inductor; (b) is the new spiral thin-film inductor, which is designed in this paper, without additional insulating barrier)

2. Experiment

2.1. Chemicals

Tetraethylorthosilicatewere (TEOS), Iron (III) chloride hexahydrate ($FeCl_3 \cdot 6H_2O$) and iron (II) chloride tetrahydrate ($FeCl_2 \cdot 4H_2O$) were purchased from Tianjin Damao Chemical reagent Factory. Ammonia liquor (25%), span-80 and oleic acid were purchased from Tianjin Kaitong Chemical reagent Co. Polyethylene glycol (average molecular weight 1000), n-butyl alcohol and cyclohexane were purchased from Tianjin Guangfu Fine Chemical reagent institute. All reagents and solvents were of analytical grade and used without further purification.

2.2. Preparation of magnetic slurry

The synthesis of magnetic slurry adopted the current existing magnetic fluid technology, which has been well developed already, and the common methods include chemical co-precipitation, mechanical porphyrization, microemulsions, thermal decomposition and so on [9,10,11,12]. This paper compared the slurries prepared with chemical co-precipatation and microemulsions. The films derived by dip coating were investigated.

The functional component of magnetic slurry is nano Fe_3O_4 particles, which were synthesized by chemical co-precipitation [9, 13 , 14] and microemulsions [12] respectively in this paper. In the first experiment, 3.65g $FeCl_3 \cdot 6H_2O$ and 2.68g $FeCl_2 \cdot 4H_2O$ were dissolved in 50mL distilled water. The solution was stirred for 10min and filtered to form an orange solution. Consequently, 25mL of $NH_3 \cdot H_2O$ was added to the solution drop by drop,

978-1-4244-2739-0/08/$25.00 ©2008 IEEE

maintaining vigorous stir at 50°C for 1h. The precipitate was rinsed by distilled water and ethanol twice, and then 200mL ethanol and 3 drip oleic acid were added, followed by the disperse under stirring and ultrasound for 20 min. The suspension was poured into three-necked bottle equipped with a mechanical stirrer, and maintained in 40 °C and vigorous stir. Then $NH_3·H_2O$ 20ml and tetraethylorthosilicate 2.4ml were added in the bottle, and stirred for 3h the colloid magnetic material was isolated by magnet. Finally the magnetic material and 5g polyethylene glycol (1000) were dissolved in 50 ml distilled water to form magnetic slurry [10, 11].

In the second method, the mixture of the n-butyl alcohol 13ml, span-80 3ml and cyclohexane15ml were poured in the three-necked bottle under 50°C and vigorous stirring. After 10 min 16.7 ml of the ferric salt solution ($FeCl_3·6H_2O$ and $FeCl_2·4H_2O$) was added drop by drop, and the concentration of iron salt was 0.27M. Consequently, the $NH_3·H_2O$ 6mL was added dropwise at 50°C and in vigorous stirring for 1h, at last a black magnetic slurry was obtained.

2.3. Fabrication of magnetic film

Dip coating was used to fabricate the magnetic film, which are simpler than the process of metal magnetic film deposited with magnetron sputtering. We got 30ml magnetic slurry in 50ml beaker, then dipped the glass slide by the vertical sliding equipment TL0.01 with a speed of 30mm/min. Then the damp slide was baked at 220°C for 30min by electric muffle furnace.

2.4 Charaterization

By using transmission electron microscope (TEM, CM-200, PHILIPS), Gouy magnetic balance (CTP-II, Nanjing Sangli Electronics, Ltd.), scanning electron microscopy (SEM, S-4700 , HITACHI) and optical microscope(DMC, Shenzhen Hailiang Precision Instrument Co. Ltd.), we studied the magnetic slurry and magnetic film.

3. Result and discussion

3.1. The component state of magnetic slurry

The microstructure of magnetic slurry is studied by transmission electron microscope (TEM) and its macroscopical state with and without the magnet is showed in Fig.2.

The magnetic slurry synthesized by chemical co-precipitation method is in metastable state, whose microstructure is shown in Fig.2 (a) (1). The size of the nanoparticles is about 20 nm and they suspend in the network of the silicon gel, which easily form agglomerated particles and the dispersibility is poor. The TEOS hydrolyzed in alkali environment to form a silicon gel and a network configuration like polymer [13, 14], which separates some nano Fe_3O_4 particles aggregation so as to keep the magnetic slurry temporary stable (Fig.2 (a) (2)). As a result, the magnetic slurry forms precipitate when it is near the magnet (Fig.2 (a) (3)).

The magnetic slurry synthesized by microemulsions method shows good stabilization, whose microstructure is shown in Fig.2 (b) (1). The size of nanoparticles is in the range of 6-15 nm, and there are lots of colloid which coats these nanoparticles and disperses them. We think that the colloid comes from surfactant. As soon as the water-soluble

colloid appears, it will coat the nano Fe_3O_4. There is no deposit when the slurry is centrifugated in 8000 r/min for 10min. The magnetic slurry does not form precipitate under the magnet, but the whole magnetic slurry is picked up (Fig.2. (b) (3)).

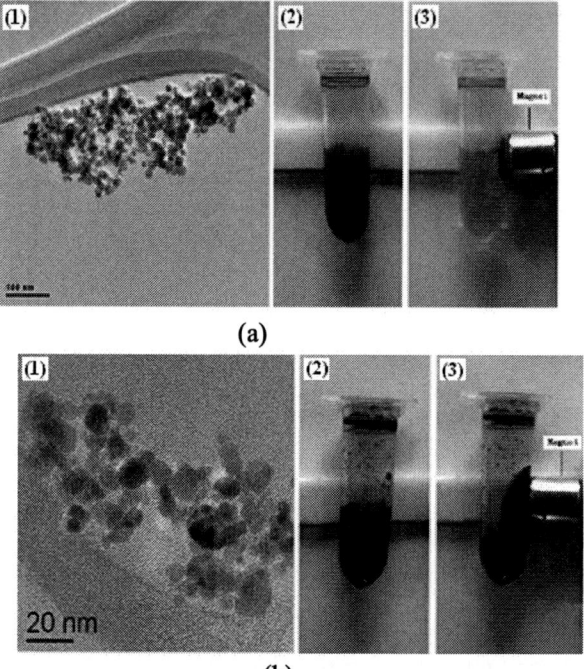

Fig.2. The TEM and macroscopical state of two magnetic slurries. (a), the magnetic slurry synthesized by chemical co-precipitation method: (1),TEM of magnetic slurry; (2), the slurry keeps stable when it is laid naturally; (3), magnet dissociated the magnetic particles from the slurry. (b), the magnetic slurry synthesized by microemulsions method: (1), TEM of magnetic slurry; (2), the slurry maintains stable when it is laid naturally; (3), the magnetic slurry is picked up by magnet

3.2. The magnetic performance of magnetic slurry

We study the magnetic performance by Gouy magnetic balance [15], and Fig.3 shows that the weight of the samples increases with the magnetic density's enhancing. The relation of magnetic density and sample weight indicates the sample is of magnetic performance, which includes magnetic susceptibility and magnetic induction intensity. And molar susceptibility is calculated by the formula below.

$$X_M = \frac{2(\Delta W_{tube+sample} - \Delta W_{tube})ghM_{sample}}{H^2 M_{sample}}$$

Using these data of Fig.3 we know that average molar susceptibility of the magnetic slurry (a) is 110cm³/mol. By the same way we get the average molar susceptibility of slurry (b), 344cm³/mol, which is higher than the former. The different magnetic material component of magnetic slurry results in their change in magnetic susceptibility. The magnetic slurry synthesized by chemical co-precipitation method has less magnetic material than the other because there are many agglomerated particles in the network configuration of silicon gel. In contrast, the magnetic slurry synthesized by microemulsions method contains abundant magnetic material due to good dispersibility.

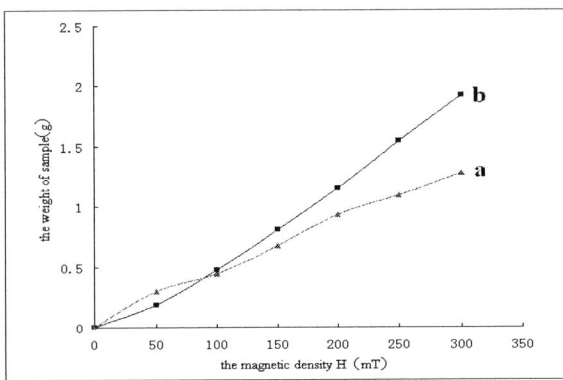

Fig.3. The magnetic performance comparison of magnetic slurry. (a) is the magnetic slurry synthesized by chemical co-precipitation method; (b) is the magnetic slurry synthesized by microemulsions method

3.3. The appearance of magnetic thin film

By optical microscope we compare the different films. The film prepared with chemical co-precipitation method is very rough with lots of big flaws (Fig.4. (a)). we think the dark part of the surface is the ferrite and the other part is the organic material, which means the ferrite cannot be dispersed in the organic material and the ferrite easily forms into big blocks. The reason of forming this structure is that the slurry contains the agglomerated particles covered by the organic material, and many aggregates shrink to form some big blocks when the slurry is baked. And the film has poor mechanical property and magnetic performance. In contrast, the optical microscope photo of the film from microemulsions is smooth and without big flaws except some pollutants which are from the air. The SEM photo shows that the film is homogeneous and maybe contains nanoparticles which can't be observed in SEM. The magnetic character of the film will be studied in future work. These data and figures indicate that microemulsions method is a better way than chemical co-precipitation method to fabricate a useful magnetic film.

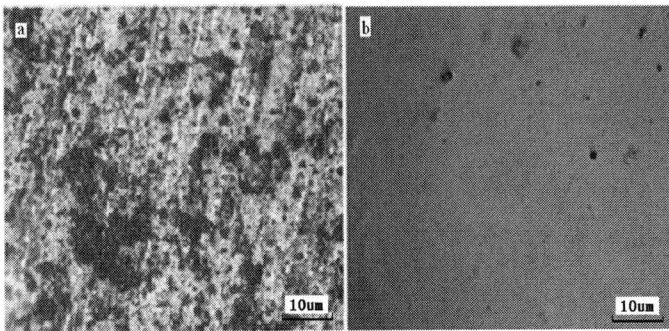

Fig.4. The optical microscope comparison of two kinds of film. (a), the film fabricated by chemical co-precipitation magnetic slurry is very rough with lots of big flaws; (b), the film fabricated by microemulsions magnetic slurry, is smooth and with few flaws)

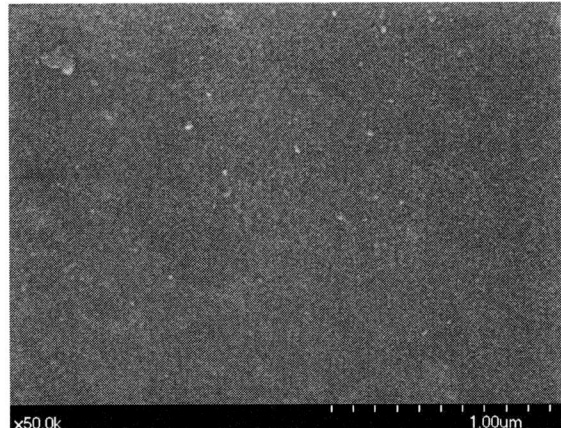

Fig.5. SEM image of the film fabricated by microemulsions magnetic slurry. The appearance of film is homogeneous and no big particles can be seen when the magnification is 50 thousand.

4. Conclusion

Different magnetic slurries had been synthesized by chemical co-precipitation method and microemulsions method, which are different in the component state and magnetic performance.The size of the nano Fe_3O_4 particles in the magnetic slurry synthesized with microemulsions,is in the range of 6-15nm and is covered with water-soluble colloid. As a result, the slurry does not deposit either centrifugated in 8000 r/min or under the magnet adsorption. Meanwhile the average molar susceptibility is $344cm^3/mol$, which is much higher than the slurry prepared with co-precipitation. Consequently, the film fabricated with microemulsions magnetic slurry is much better than the other method, with smooth surface and few flaws observed under the microscope, and no big particles are observed in SEM with the magnification of 50 thousand. The microemulsions ensure the synthesis of fine magnetic slurry and smooth film. With this technology, we can simplify many process flows, cut down energy consumption and promote the rapid development of thin-film inductor.

References

1. Stephen O'Reily, Maeve Duffy, et al "Characterisation of embedded filters in advanced printed wirring boards," *Microelectronics Reliability*, Vol. 41, (2001), pp. 781-788.

2. M. Yamaguchi, K. Suezawa, *et al* "Magnetic thin-film inductors for RF-integrated circuits," *Journal of Magnetism and Magnetic Materials*, Vol. 216, (200), pp. 807-810.

3. Xiao-Li Tang , H.W. Zhang, et al "Study of the impact of winding form and film thickness on thin-film inductors," *Microelectronic Engineering*, Vol. 81, (2005), pp.212-216.

4. Chong Leia, Yong Zhou, *et al* "Fabrication of a solenoid-type inductor with Fe-based soft magnetic core," *Journal of Magnetism and Magnetic Materials*, Vol. 308, (2007), pp. 284–288.

5. K. Seemann, H. Leiste, C. Ziebert, "Soft magnetic FeCoTaN film cores for new high-frequency CMOS compatible micro-inductors," *Journal of Magnetism*

and Magnetic Materials, Vol. 316, (2007), pp. 879–888.

6. Y. Zhuang, M. Vroubel, B. Rejaei, J. N. Burghartz, "Integrated RF inductors with micro-patterned NiFe core," *Solid-State Electronics*, Vol. 51, (2007), pp. 405-413.

7. L.K. Varga, "Soft magnetic nanocomposites for high-frequency and high-temperature applications," *Journal of Magnetism and Magnetic Materials*, Vol. 316, (2007), pp. 442-447

8. M. Anhalt, "Systematic investigation of particle size dependence of magnetic properties in soft magnetic composites," *Journal of Magnetism and Magnetic Materials*, Vol. 320, (2008), pp. 366-369.

9. Gou Maling, Xie Zhen , Wang Hui, et al "Preparation of A Magnetic Fluid by Chemical Co - deposition Method," *West China Medical Journal*, Vol.22, No.2 (2007), pp. 343-344.

10. He QiangFang, Li Guoming, et al "Preparation of Water-based Fe_3O_4 Magnetic Fluid," *Chinese Journal of Applied Chemistry*, Vol.22, No. 6 (2005) pp.665-668.

11. Xiong Longrong, et al "Synthesis and Stability of PEG-covered Fe_3O_4 Magnetic Fluid," *Material Review*, Vol.21, (2007), pp.195-197.

12. Zhang Wenju, Chai Bo, et al "Study on Preparation and Properties of Fe_3O_4 Magnetic Fluid by One-step Method with a Microemulsion Reactor," *Lubrication Engineering*, Vol. 2, (2005), pp.57-59..

13. Ziyang Lu, Gang Wang, Jiaqi Zhuang, Wensheng Yang, "Effects of the concentration of tetramethylammonium hydroxide peptizer on the synthesis of Fe_3O_4/SiO_2 core/shell nanoparticles," *Colloids and Surfaces A*, Vol. 278, (2006), pp. 140-143.

14. Chen Lingyun , Li Fengsheng, et al "Preparation and Characterization of Magnetic Fe_3O_4/SiO_2 Composite Nanoparticles," *Journal of Materials Science and Engineering*, Vol.23, No.5 (2005), pp. 556-560.

15. Luo Xin, Fang Yuxun, "Gouy Magnetic Balance Measuring Ferrofluids," Magnetism and Stability" *Journal of East China Geological Institute*, Vol. 20, No. 2 (1997), pp.120-125.

Microstructure and Properties of Barium Strontium Titanate Thin Films Prepared on Copper Foils via Addition of PEG to the Sol Precursor

Yanhua Fan[1,2], Shuhui Yu[1], Rong Sun[1], Lei Li[1], Mian Huang[1], Yansheng Yin[2], Lixi Wan[3]

[1] Shenzhen institute of advance technology, Chinese Academy of Sciences/the Chinese University of Hong Kong, Shenzhen 518067, China

[2] Institute of Materials Science and Engineering, Ocean University of China, Qingdao 266100, China

[3] Institute of Microelectronics of Chinese Academy of Sciences, Beijing, 10029, China

Email: yuushu@gmail.com, Phone: 86-755-26803624, Fax: 86-755-26803589

Abstract

$Ba_{0.7}Sr_{0.3}TiO_3$ thin films have been deposited on copper foils via sol-gel method. The films have been processed in almost inert atmosphere so that the substrate oxidation is avoided while allowing the perovskite phase to crystallize. Polyethylene glycol (PEG) with the molecular weight of 200 was employed to modify the BST precursor solutions. The effect of PEG on the microstructure and dielectric properties of BST thin films has been investigated. Leakage current for BST thin film made from the sol with PEG modification is reported.

Key words: BST thin film, dielectric constant, Leakage current, dielectric loss

I Introduction

Dielectric thin film is the key material for the development of System in Package (SiP) technology. Barium strontium titanate (BST) with perovskite structure exhibit excellent electrical and optical properties which show a variation over a broad range, depending on the ratio of Ba/Sr [1,2]. Therefore, BST is one of the candidates for embedded capacitors in SiP.

As the competition of the electronics market become stronger, substrate requires inexpensive materials and a flexible material that will facilitate lamination into printed wiring boards. Copper foils are the optimal choice because of its excellent properties. However, acquiring a useful BST thin film on the flexible copper foils is a great challenge, due to the ease of copper oxidization, cracking, poor crystallinity, and poor adhesion of the films to the flexible copper foil. Some groups[3-7] have reported Barium zirconate titanate (BZT),Lead zirconate titanate (PZT) thin films deposited on copper foils. However, little data is available to identify the microstructure of the thin film, such as SEM images, probably due to the difficulty to obtain a thin film with good quality.

S.Yu et al have found that polyethylene glycol (PEG) addition to the sol precursor is helpful to eliminate the cracks and to promote the formation of the perovskite phase in the lead-contained thin films prepared with sol-gel method. The analysis of the mechanism was based on the interaction between PEG and Pb ions. [8,9] In order to investigate the effects of PEG on the lead-free dielectric thin films, a systematic study on the effects of PEG addition amounts on the BST thin films is conducted in this paper..

BST thin film with the ratio of Ba/Sr equal to 7:3, was prepared with sol-gel process on Cu foils with La_2O_3 buffer layer. The La_2O_3 was selected owing to its high electrical breakdown field strength and high dielectric constant (k=27)[10]. PEG was directly added to the sol precursor. The relationship between the PEG addition amount and the microstructure and dielectric properties of the derived BST thin films is discussed.

II Experimental Procedure

Barium acetate $Ba(CH_3COO)_2$ (purity ≥ 99%, Shanghai Chemical Reagent Co.), strontium acetate $Sr(CH_3COO)_2 \cdot 1/2H_2O$ (purity ≥ 99.5%, Shanghai Chemical Reagent Co.) and titanium butoxide $Ti(OC_4H_9)_4$ (purity ≥ 98%, Shanghai Chemical Reagent Co.) were used as starting materials. Glacial acetic acid was selected as solvents. To completely eliminate the water associated with strontium acetate, $Sr(CH_3COO)_2 \cdot 1/2H_2O$ was dried at about 70°C for 24 h. The barium acetate and the strontium acetate were dissolved in glacial acetic acid at 60°C with a molar ratio of 7:3, respectively. The solutions were then mixed and stirred. The Ti $(OC_4H_9)_4$ (in proper molar ratio) was added into the mixture to get a precursor solution. The solution was diluted to 0.3 M with glacial acetic acid. The mount of 20wt%, 30wt%, 40wt%, 50wt%, 60wt%, 70wt% of PEG with molecular weights of 200, based on the BST weight in the solutions, was added to the BST solution respectively. Then, the mixtures were completely stirred to make the sol homogeneous.

BST solution was deposited by spin coating at 3000 rpm for 30s on top of the La_2O_3/Cu foil. The La_2O_3 buffer layer with the thickness 10 nm was prepared by sol-gel method. The wet BST films were dried on a hotplate at 105°C for 5 min to remove the solvent, followed by a pyrolysis at 450°C for 10 min and a pre-annealing at 600°C for 10 min in RTP-500 (Rapid Thermal Processor, Beijing Institute of Applied Physics, China) in almost inert atmosphere which was achieved by flowing the high purified argon at a rate of 1 L/min, to avoid the oxidation of the Cu foil while allowing the film to crystallize. The above process was repeated to achieve the desired film thickness. Finally, the films with a thickness of 600 nm were obtained. At last, the samples were annealed at 700°C, 750°C, 800°C and 850 °C respectively, for 10 min in almost inert atmosphere.

The crystalline structure of the film was analyzed by X-ray diffractometer with Cu Kα radiation (XRD, Bruker, D8 Advance,Germany). The surface morphology was examined with Field-emission scanning electron microscope (SUPRA[TM] 25, Germany). Au top electrode having a diameter of 1mm was deposited by direct current sputtering to measure electrical properties. The dielectric properties of $Ba_{0.7}Sr_{0.3}TiO_3$ thin film were measured at room temperature using an oscillation voltage of 500 mV with the frequencies

978-1-4244-2739-0/08/$25.00 ©2008 IEEE

ranging from 40 Hz to 1 MHz, using an HP4294A impedance analyzer (Agilent, USA). Leakage current was measured with TFANALYZER2000 (aix ACCT, Germany).

III Result and Discussion

3.1 Crystallization and microstructures

Fig. 1 shows an X-ray diffraction pattern for the BST thin film prepared on copper foil with La_2O_3 buffer layer. Dominant perovskite phase has formed in all the samples. The films annealed at 700 C and 850 C show a preferred orientation in the (100) planes. However, further study is needed to clarify the reason. In Fig.1, it can also been found that the BST diffraction peak around 45° split into (002) and (200) peaks with the increase of the annealing temperature from 700℃ to 850℃, and at 850℃, the (200) peak is stronger and tends to be as the dominant peak. This phenomenon indicates that the tetragonal phase formed in the film and became dominant with the increase of annealing temperature Other researchers also reported that as the extent of splitting increased, tetragonal phase became much stronger [11]. The peak around 29° of La_2O_3 phase disappears at 850℃. The reason is possibly that La^{3+} entered into the lattice of BST unit cell at 850℃, which in return acted as dopant to the BST system.

Fig.2 shows the SEM images of the surface morphology of the $Ba_{0.7}Sr_{0.3}TiO_3$ thin film with La_2O_3 buffer layer. From the figure, it can be readily seen that the BST thin film are well crystallized and the grain size is tens of nanometers.

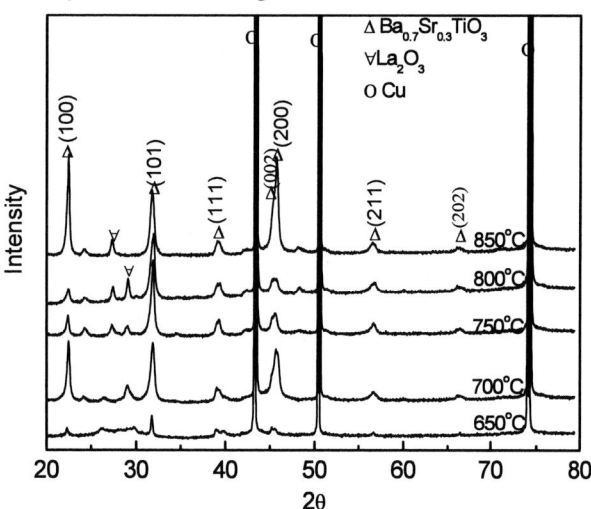

Fig.1. X-ray diffraction patterns of $Ba_{0.7}Sr_{0.3}TiO_3$ thin films with a thickness of 600 nm, and annealed at different temperatures.

The influence of the PEG additive on the film surface texture was investigated by SEM, as shown in Fig.3. From the Fig.3, it can be seen that the film prepared from the BST solution without PEG cracked seriously. In contrast, the thin films made from the PEG-modified solution were crack-free. The improvement of texture is probably resulted from the dispersion effect of the thermal stress in the thin film by the additive of PEG.

Fig.2 SEM of $Ba_{0.7}Sr_{0.3}TiO_3$ thin film annealed at 850℃

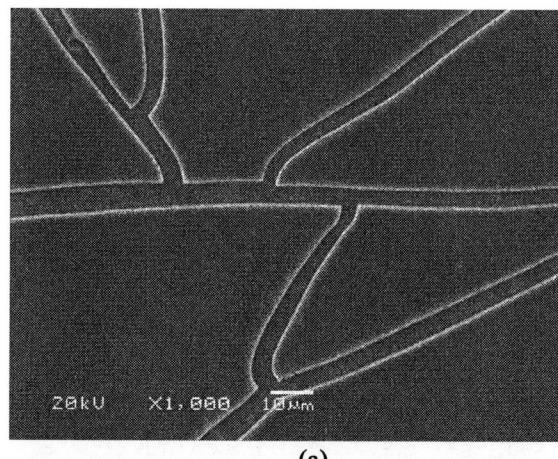

(a)

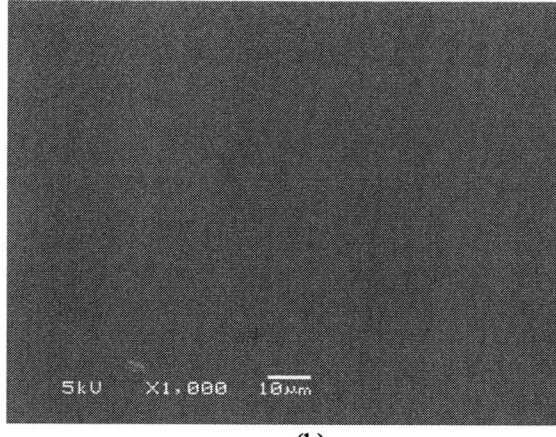

(b)

Fig.3 SEM of $Ba_{0.7}Sr_{0.3}TiO_3$ thin film annealed at 800℃, a: without PEG, b: with 60wt%PEG.

In order to analyze the effect of PEG on the crystallization of BST film, average grain sizes of BST thin film were calculated from Scherrer equation [12], based on the (101) plane:

$$D=K\lambda/FW(S)\cos\theta$$

Where K is Scherrer constants (0.89); D is the size of the grain; FW(S) is the half width of the X-ray diffraction peaks. Fig.4 shows a plot of the average calculated grain size as a function of the amount of PEG additive. The effect of amount of PEG additive on the ultimate grain size clearly demonstrates that grain size increased with increasing the amount of PEG from 20wt% to 30wt%, and arrived at 190 A° when the addition amount is 30wt%, then switched to reduce with increasing the amount. It is likely that in the first stage of the formation of BST, a large amount of PEG attached to the surface of BST grain, reducing the surface energy of crystal grain, which decreased the growth rate of crystal. Another reason is possibly that a large amount of PEG separated the BST crystal, which restrained the secondary crystal of BST crystal grain.

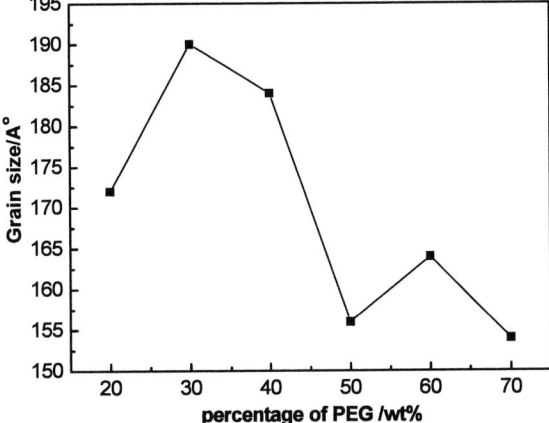

Fig.4 Average grain size as a function of amount of PEG additive for a series of $Ba_{0.7}Sr_{0.3}TiO_3$ thin films prepared on copper foil, annealed at 800℃.

3.2. Electrical Properties

Fig.5 shows the frequency dependence of dielectric constant (ε_r) and dielectric loss (tanδ) of the $Ba_{0.7}Sr_{0.3}TiO_3$ thin film grown on the copper foil with La_2O_3 buffer layer. As shown in Fig.5 (a), the dielectric constant decreased with the frequency. Such a monotonous decrease in dielectric constant with frequency can be attributed to hopping conduction, which could be resulted from oxygen vacancies existing in the film, or the interfacial layer formed between thin film and electrode [13]. The dielectric constant of the thin film made from the BST solution modified by 20wt%, 30wt %, 40wt %, 50wt %, 60wt %, 70wt %PEG, were 529, 460, 625, 446, 241, 375 respectively at the frequency of 1 kHz. The maximal value for the dielectric constant at 1kHz occurs in the thin film derived from the BST solution modified with 40wt% PEG. When the frequency increased to 1MHz, the thin film made from the BST solution modified by 40wt% still keep the maximal dielectric constant which is 422.

The curves in Fig.5 (b) show a variation in the dielectric loss of the films with frequency. The dielectric loss of the BST thin film were much higher, which could be attributed to a large amount of oxygen vacancies existing in the thin films and the interfacial layer formed between thin film and electrode. On the whole, the dielectric loss of the thin films prepared from the BST solution modified by 40wt% and 50wt% are relatively low, which is around at 0.1.

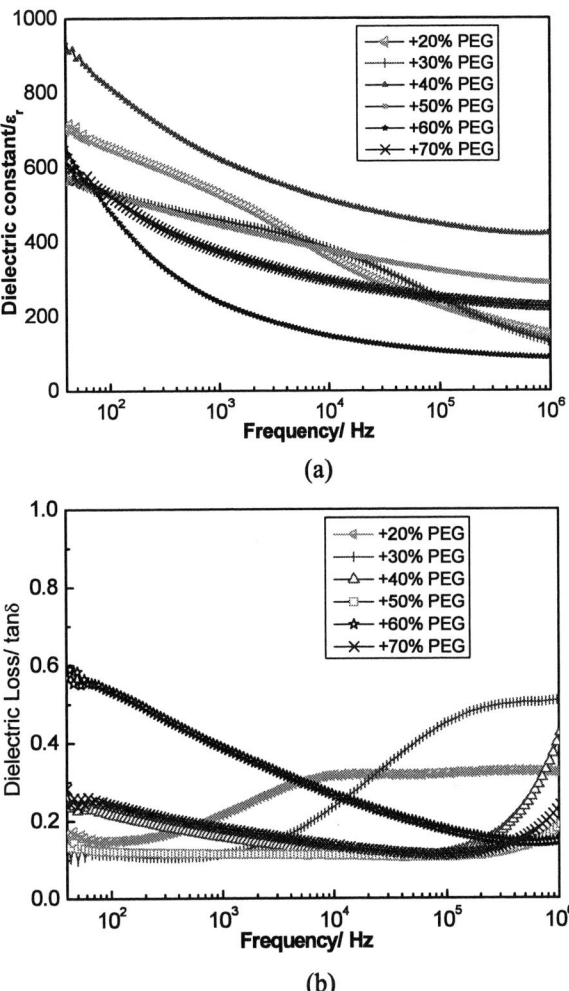

(a)

(b)

Fig.5. Frequency dependence of dielectric constant (a) and dielectric loss (b) for $Ba_{0.7}Sr_{0.3}TiO_3$ thin films prepared on copper foils, with a thickness of 600 nm, and annealed at different temperatures.

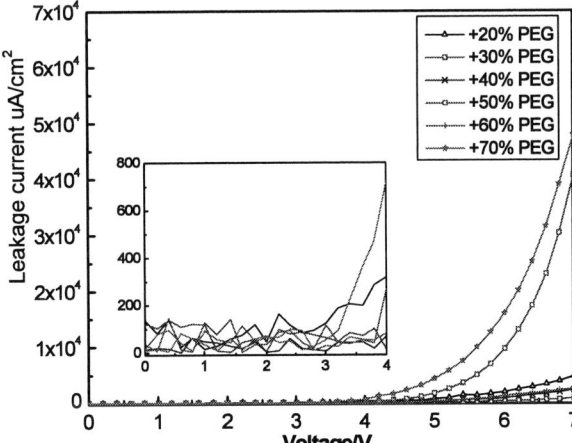

Fig.6 Leakage current as a function of voltage for $Ba_{0.7}Sr_{0.3}TiO_3$ thin films prepared on copper foil, with 600nm thickness.

The leakage current density of the BST films as a function of applied electric field is plotted in Fig. 6. As shown in the figure, when the applied electric field was below 50 kV.cm^{-1} (3 voltage), leakage current density had no much difference for all the thin films derived from the BST solution modified by PEG. It should be indicated that when the applied voltage was above 50 kV.cm^{-1}, both of the thin films prepared from the BST solution modified by 50wt% and 70wt% of PEG had much high-leakage current density. Whereas, the leakage current density of the thin film prepared from the BST solution modified by 40wt% and 30wt% were much lower. While the applied electric field was around 116kV.cm^{-1} (7 voltage), the leakage current density of the thin films derived from the BST solution modified by 40wt% and 30wt% were lower eight times than that of the thin films prepared from the BST solution modified by 50wt% and 70wt%.

As a result, appropriate amount (40wt%) of PEG additive not only eliminate the cracks, but also promote the growth of crystal grain in the BST thin film. Excess PEG additive could weaken the compactness of the BST thin film, which was detrimental to the electrical properties of the BST thin film.

3.2. Conclusion

Ba$_{0.7}$Sr$_{0.3}$TiO$_3$ thin films were prepared on the copper foil via a sol-gel method. X-ray diffraction analysis and SEM images reveal that the Ba$_{0.7}$Sr$_{0.3}$TiO$_3$ thin film with a buffer layer was well crystallized at 850℃. A certain amount of PEG200 additive can avoid the formation of the surface cracks of the BST thin film significantly. 30-40wt% of PEG200 additive can increase the grain size of the BST thin film. The dielectric constant of the thin film derived from the BST solution modified by 40wt% was 422 at 1 MHz, and its dielectric loss was around at 0.1. While the applied electric field was around 116kV.cm^{-1} (7 V), the leakage current density of the thin films derived from the BST solution modified by 40wt% and 30wt% were lower eight times than that of the thin films prepared from the BST solution modified by 50% and 70%. Therefore, the optimal amount additive for PEG200 was around 40%.

Reference

[1] X.F. Chen, W.Q. Lu, W.G. Zhu, S.Y. Lim, S.A. Akbar, "Structural and thermal analyses on phase evolution of sol–gel (Ba,Sr)TiO$_3$ thin films," Surface and Coatings Technology, Vol. 167 (2003), pp. 203206.

[2] Z.Q. Wei, H.P. Xu, M. Noda, M. Okuyama, "Preparation of Ba$_x$Sr$_{1-x}$TiO$_3$ thin films with seeding layer by a sol–gel method," Journal of Crystal Growth, Vol. 237-239 (2002), pp. 443-447.

[3] T. Kim, J.N. Hanson, A. Gruverman, A.I. Kingon, "Ferroelectric behavior in nominally relaxor lead lanthanum zirconate titanate thin films prepared by chemical solution deposition on copper foil," Applied Physics Lettters, Vol. 88 (2006), pp. 262907.

[4] J.F. Ihlefeld, W. Borland, J-P. Maria, "Synthesis and Properties of Barium Titanate Thin Films on Copper Substrates," Materials Research Society Symposium Proceedings, Vol.902E (2006), pp. 0902-T02-03.1.

[5] J.F. Ihlefeld, J-P. Maria, "Dielectric and microstructural properties of barium titanate zirconate thin films on copper substrates," Journal of Materials Research, Vol. 20, No. 10 (2005), pp. 2838-1844.

[6] J.T. Dawley, P.G. Clem, "Dielectric properties of random and <100> oriented SrTiO3 and (Ba,Sr)TiO3 thin films fabricated on <100> nickel tapes," Applied Physics Lettters, Vol. 81, No.16 (2002), pp. 3028-3030.

[7] M.D. Losego, L.H. Jimison, J.F. Ihlefeld, J-P. Maria, "Ferroelectric response from lead zirconate titanate thin films prepared directly on low-resistivity copper substrates," Applied Physics Lettters, Vol. 86 (2005), pp. 172906.

[8] S.H. Yu, K. Yao, S. Shannigrahi, F.T.E. Hock, "Effects of poly(ethylene glycol) additive molecular weight on the microstructure and properties of sol-gel-derived lead zirconate titanate thin films," Journal of Materials Research, Vol. 18, No. 3 (2003), pp. 737-741.

[9] S.H. Yu, K. Yao, F.T.E. Hock, "Structure and Properties of (1-x)PZN-xPT Thin Films with Perovskite Phase Promoted by Polyethylene Glycol," Chemistry of Materials, Vol.18 (2003), pp. 5343.

[10] N.K. Park, D. K. Kang, B-H. Kim, "Electrical properties of La$_2$O$_3$ thin films grown on TiN/Si substrates via atomic layer deposition," Applied Surface Science, Vol. 252 (2006), pp. 8506-8509.

[11] X.T. QU, L. Li, "Preparation and XRD Analysis of Barium Titanate (BaTiO$_3$) Nano-powers," Journal of Shanxi University (Nat.Sci.ED.), Vol.30, No. 1 (2007), pp. 61-63.

[12] J.W. Huang, "Handbook of MDI Jade (revision),"University of Zhongnan, 2006, pp. 36.

[13] S. Lahiry, A. Mansingh, "Dielectric properties of sol–gel derived barium strontium titanate thin films," Thin solid Films, Vol. 516 (2008), pp.1656-1662.

Dynamic Behavior Tests of Lead-free Solders at High Strain Rates by the SHPB Technique

Qin Fei, An Tong, Chen Na

College of Mechanical Engineering and Applied Electronics Technology,
Beijing University of Technology, Beijing 100124, China
Email:qfei@bjut.edu.cn, Phone: +86-10-67392173

Abstract

Dynamic behavior of solder joints in microelectronic packages is key issue for drop/impact reliability design of mobile electronic products. The dynamic mechanical behaviors of 63Sn37Pb, 96.5Sn3.5Ag and 96.5Sn3.0Ag0.5Cu solder alloys at high strain rate were investigated by using the split Hopkinson pressure/tension bar testing technique. Stress-strain curves of the three materials were obtained at strain rate $600 \ s^{-1}$, $1200 \ s^{-1}$ and $2200 \ s^{-1}$, respectively. The experimental results show that all the three materials are strongly strain rate dependent. Among them 96.5Sn3.5Ag is the most sensitive to strain rate, while 96.5Sn3.0Ag0.5Cu has the greatest yield stress and tensile strength. Relations of the tensile strength, fracture strain and yield stress of the materials with strain rate were proposed by fitting the experimental data.

1 Introduction

When portable electronic products drop or work in the circumstances of high acceleration, the solder joints, which act as mechanical supports and electrical interconnection[1], can easily reach strain rate of $10^3 \ s^{-1}$ [2]. However, there is not enough understanding currently to the mechanical behavior of lead-free solders at high strain rate, which is urgently needed for reliability design of mobile electronic products.

Most research works on mechanical behavior of solder materials focus on quasi-static or creep properties used for thermal fatigue life analysis. Mechanical behavior of Pb and Pb-free solders at low strain rates of $10^{-6} \sim 10^{-1} s^{-1}$ were investigated by Wilde et al.[3], Amagai et al.[4] and Chen et al.[5] and other researchers. Dynamic behavior of Pb solder was investigated by Dai et al.[6] and Lee et al.[7] using the split Hopkinson torsing bar technique. Wang and Yi [2] obtained the stress-strain curves of 63Sn37Pb solder at high strain rates by using the split Hopkinson pressure bar test (SHPB). More recently, the SHPB technique was used by Siviour et al. [8] to investigate stress-strain behavior of Pb and Pb-free solders at high strain rates and in various temperature conditions.

For solder joints under drop/impact loadings, cracking is often found along the interface of the Printed Circuit Board (PCB) and the solder joints or in the intermetallic compound layer (IMC) [9]. This failure is caused by dynamic tensile stress or peeling stress in the solder[10]. Furthermore, as pointed by Wang and Yi [2], for solder alloy its tensile behavior is significantly different from its compressive behavior. Therefore there is necessity to investigate tensile behavior of solder materials.

In this paper the dynamic mechanical behaviors of 63Sn37Pb, 96.5Sn3.5Ag and 96.5Sn3.0Ag0.5Cu at high strain rate were investigated by using the split Hopkinson pressure/tension bar technique. Effect of strain rate on mechanical behaviors of the three solder alloys was discussed.

2 Experimental Procedures

Commercial ingots of 63Sn37Pb (the SnPb), 96.5Sn3.5Ag (the SnAg) and 96.5Sn3.0Ag0.5Cu (the SnAgCu) were melt in a crucible and casted into rods. The process was intentionally slowed down to make sure that there were no voids formed in the rods. Then the cast rods were treated by quickly immerging them into cold water and were aged for 15 days. The rods then were machined into specimens in specific shape and dimensions. Specimens for quasi-static tension and compression tests are in size of Φ8x56 mm and Φ12x12 mm, respectively. For the split Hopkinson pressure bar tests the specimens have size of Φ12x6 mm. For the tensile bar test, the specimen is of Φ6x8 mm but with additional two screw ends it has a total length of 40 mm.

The quasi-static tension/pressure tests were conducted with a MTS-809 material tester at strain rate of 0.001 s^{-1}. The split Hopkinson pressure/tension bar setup was designed to be capable to test at maximum strain rates of 2400 s^{-1}. All the tests were carried out at room temperature.

For the quasi-static tests, three specimens for each kind of solders were tested and then to average the three specimens' data as the final results. While for the split Hopkinson tension/pressure bar tests, there were five specimens at one strain rate for each kind of solders. The final data were calculated by averaging available data of the five specimens.

3 Results and Discussion

Quasi-static behavior For the quasi-static tests, the mechanical properties of the three materials are listed in Table 1. Both in tension and compression, the ultimate stress of the SnAg is almost the same as that of the SnPb, but it's yield stress is slightly less than the SnPb. The SnAgCu has the greatest ultimate stress, but its fracture strain is the same as the SnPb. Under quasi-static loadings, linear hardening can be observed for the three materials. It suggests that an bilinear or tri-linear elastic-plastic material model can be used to model their quasi-static behaviors in numerical simulations.

Compressive behavior at high strain rate Figs 1~3 show the stress-strain curves recorded by the split Hopkinson pressure bar tests at three strain rates. All the materials have strain rate effect, and their dynamic yield stresses are almost three times greater than that of static for all the three materials. For example, the quasi-static yield stress of the SnAg is 29 MPa but it increases to 77 MPa when the strain rate is 2200s^{-1}. Using 2% true strain as a reference point to define the flow stress[8], the relations between the flow stress y and strain rate x for the SnPb, SnAg and SnAgCu can be fitted as

978-1-4244-2739-0/08/$25.00 ©2008 IEEE

Table 1 Quasi-static Mechanical Properties

Solders	Young's Modulus (GPa)	Yield Stress (MPa)	Ultimate Stress (MPa)	Fracture Strain
SnPb	30	35	43	0.16
SnAg	45	29	42	0.20
SnAgCu	54	38	51	0.16

$y=0.0032x+67.7$, $y=0.0122x+82.2$ and $y=0.0101x+95.4$, respectively, as shown in Fig.4. Slopes of the three lines are $m_1=0.0032$, $m_2=0.0122$ and $m_3=0.0101$, which can be interpreted as the strain rate sensitivities. This shows that the SnAg has the greatest sensitivity, 0.0122. Wang and Yi [2] proposed a relationship between the flow stress and strain rate of the SnPb, similarly, based on the test data we derived relationship between dynamic yield stress and strain rate for the two lead-free solders as

$$\frac{\sigma}{\sigma_0} = 1 + \left(\frac{\dot{\varepsilon}}{D}\right)^{\frac{1}{q}} \qquad (1)$$

Where σ_0 is the quasi-static yield stress, $D=175.1s^{-1}$, $1/q=0.200$ for the SnAg and $D=754.6s^{-1}$, $1/q=0.261$ for the SnAgCu, respectively.

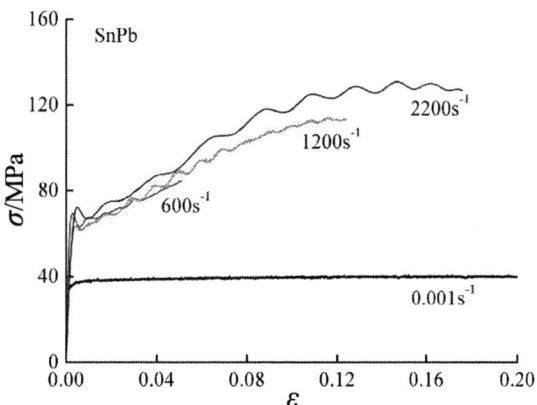

Fig. 1 SnPb stress-strain curves in compression

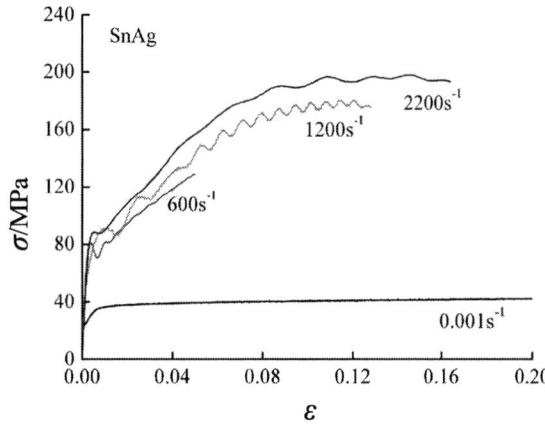

Fig. 2 SnAg stress-strain curves in compression

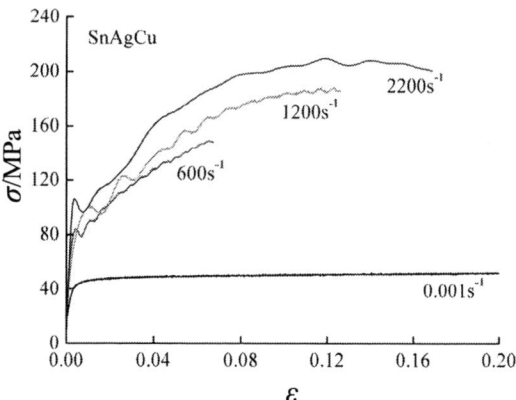

Fig. 3 SnAgCu stress-strain curves in compression

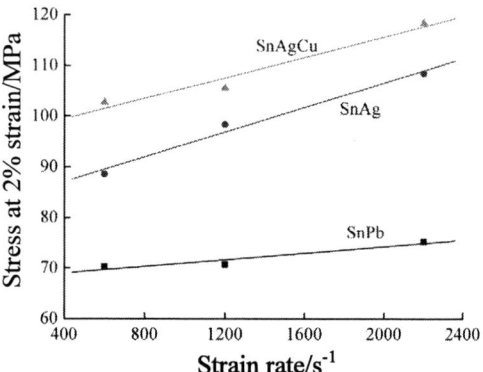

Fig. 4 Flow stress-strain rate curves in compression

Tensile behavior at high strain rate Figs 5~7 show the stress-strain curves recorded from the split Hopkinson tension bar tests at different strain rates. Comparison was made in Table 2 to show the difference of tensile strength σ_b and fracture strain ε_f for the three materials. In quasi-static conditions the tensile strength of the SnAg is almost the same as that of the SnPb. However, in dynamic the SnAg has much higher strength than the SnPb because of the strain rate effect. Generally the tensile strength of the three materials increases significantly as increasing of strain rate. Let σ_b is the dynamic tensile strength and σ_b^0 is the strength in quasi-static, the test data can be fitted by

$$\frac{\sigma_b}{\sigma_b^0} = 1 + \left(\frac{\dot{\varepsilon}}{\dot{\varepsilon}_0}\right)^{\rho m} \qquad (2)$$

Where $\dot{\varepsilon}_0 = 0.001s^{-1}$, is the reference strain rate, m is the strain rate sensitivity as discussed in the preceding section, ρ was material constant identified by the test data as 22.76, 6.035 and 5.57 for the SnPb, SnAg and SnAgCu, respectively.

Table 2 shows that the tensile strength increases as strain rate increases but the fracture strain decreases. This indicates that the materials tend to brittle failure at high strain rate. Fig.8 presents linear relations fitted between the fracture strain ε_f and the strain rate for three materials, and equations of $y=-0.000012x+0.163$, $y=-0.00006x+0.197$, and $y=-$

0.00004x+0.157 are applicable to the SnPb, SnAg and SnAgCu respectively. Obviously, the two lead-free solders are easier to be brittle than Pb solder at high strain rates. A fractured SnAg specimen after the split Hopkinson tension bar test conducted at strain rate 1800 s^{-1} was presented in Fig.9.

For the quasi-static loadings, linear strain hardening can be observed for all the three materials both under compression and tension. However, for the high strain rate tests, strain hardening is not linear and strain softening was observed and the softening becomes more significant as the strain rate is higher. This might be caused by the temperature rising induced by the heat transformed from plastic work during the dynamic deformation of the specimen, which can be regarded as an adiabatic process.

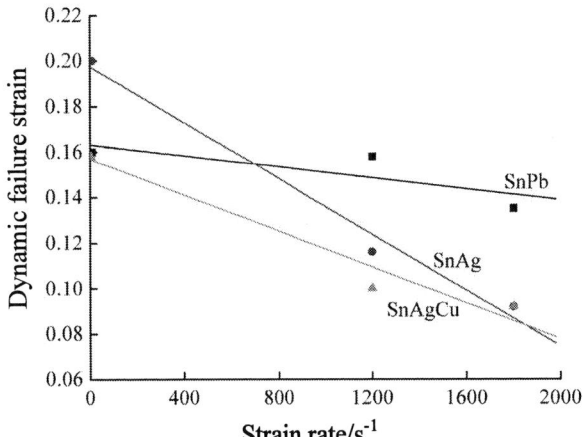

Fig. 8 Failure strain-strain rate curves

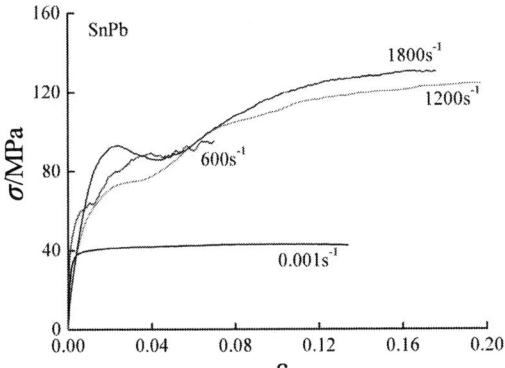

Fig. 5 SnPb stress-strain curves in tension

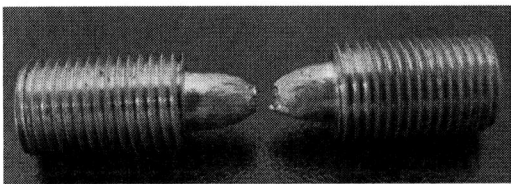

Fig. 9 Fractured specimen of SnAg at tensile strain rate of 1800s^{-1}

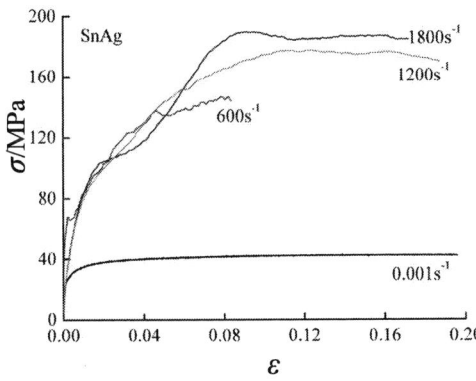

Fig. 6 SnAg stress-strain curves in tension

Table 2 Mechanical Properties

Strain Rate (s^{-1})	SnPb		SnAg		SnAgCu	
	σ_s/σ_b (MPa)	ε_f	σ_s/σ_b (MPa)	ε_f	σ_s/σ_b (MPa)	ε_f
0.001	35/43	0.160	29/ 42	0.200	38/51	0.160
600	63/99	-	66/146	-	73/154	-
1200	58/103	0.158	72/158	0.116	83/162	0.100
1800	64/111	0.135	77/173	0.092	87/172	0.092

4 Conclusions

The dynamic mechanical behaviors of the SnPb, SnAg and SnAgCu were investigated at high strain rate by using the split Hopkinson pressure/tension bar testing technique. Some observations can be made as:

1) Stress-strain behaviors of the three materials under quasi-static loadings are featured as linear strain hardening, which can be modeled by bilinear or tri-linear material models. Significant nonlinear strain hardening and softening were observed under high strain rate dynamic loadings.

2) All the three materials are sensitive to strain rate and their flow stresses increase obviously as strain rate increases. The lead-free solders are more strain rate sensitive than Pb solder, especially for the SnAg.

3) The tensile strength increases as strain rate increases, but failure strain decrease. The SnAgCu has higher yield

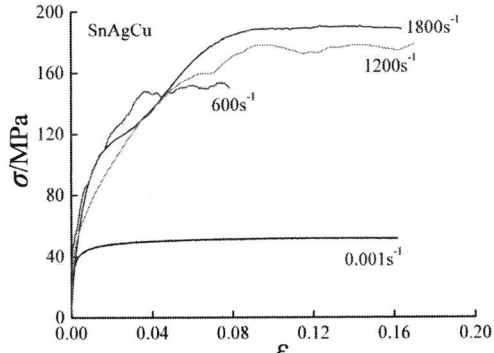

Fig. 7 SnAgCu stress-strain curves in tension

stress and tensile strength than SnAg but lower fracture strain.

Acknowledgments

The authors would like to thank the financial support from the National Natural Science Foundation of China (NSFC) under the Grant No.10572010 and the Science & Technology Development Project of Beijing Education Committee under Contract No.KM200610005013.

References

1. Rao R. Tummala, *et al*. Microelectronics Packaging Handbook (Second edition). 1997, Chapman & Hall, New York

2. B. Wang, S. Yi. Dynamic plastic behavior of 63wt%Sn37wt%Pb eutectic solder under high strain rate. *Material Science Letters*, 2002, 21: 697-698

3. Jurgen Wilde, Klaus Becker, Markus Thoben, et al. Rate dependent constitutive relations based on Anand model for 92.5Pb5Sn2.5Ag solder. *IEEE Transactions on Advanced Packaging*, 2000, 23(3): 408-414

4. Masazumi Amagai, Masako Watanabe, et al. Mechanical characterization of Sn–Ag-based lead-free solders. *Microelectronics Reliability*, 2002, 42: 951-966

5. Xu Chen, Gang Chen, Masao Sakane. Prediction of stress-strain relationship with an improved Anand constitutive Model For lead-free solder Sn-3.5Ag. *IEEE Transactions on Components and Packaging Technologies*, 2005, 28(1): 111-116

6. Dai L H, Lee S W R. Strain rate-dependent punch shear behavior of 63Sn-37Pb solder joints, *Proc Inter-PACK'01*, Kauai, Hawai, July, 2001, 2001:1-7

7. Lee S W R, Dai L H. Characterization of strain rate-dependent behavior of 63Sn-37Pb solder using Split Hopkinson Torsional Bars(SHTB). *Proc 13th Symposium on Mechanics of SMT & Photonic Structures*, New York, USA, Nov., 2001:1-6

8. Siviour C R, Walley S M, Proud W G, J E Field. Mechanical properties of SnPb and lead-free solders at high rates of strain. *Physics D: Applied Physics*, 2005, 38: 4131-4139

9. Chang-lin Yeh, Yi-shao Lai and Chin-lin Kao. Evaluation of board-level reliability of electronic packages under consecutive drops. *Microelectronics Reliability*, 2006, 46: 1172-82

10. Jin-en Luan, Tong Yan Tee, Kim Yong Goh and Hun Shen Ng. Drop impact life prediction model for lead-free BGA packages and modules. *Proc EuroSIME Conference*, Germany, April,2005

Formation of Double-layer Cu₃Sn in Solid-State Aging Process at the Interface of Eutectic SnBi Solder and (100) Single Crystal Cu

Pan-Ju Shang,[a] Zhi-Quan Liu,[a] Dou-Xing Li,[a] and Jian-Ku Shang [a,b]

[a] Shenyang National Laboratory for Materials Science, Institute of Metal Research, Chinese Academy of Sciences, Shenyang 110016, China, (Email: zqliu@imr.ac.cn)

[b] Department of Materials Science and Engineering, University of Illinois at Urbana–Champaign, Urbana, IL 61801, USA, (Email: jkshang@uiuc.edu)

Abstract

Transmission electron microscopy (TEM) observations were carried out to investigate the microstructural evolution of SnBi/single crystal (100) Cu interface during reflow and solid-state aging process. It was found that there were two kinds of IMCs, Cu_6Sn_5 and Cu_3Sn, at the interface after reflow. The Cu_3Sn grains grew along [100] direction of Cu with columnar morphology. During aging, new triangle Cu_3Sn grains were found at the triple junction sites of Cu/Cu₃Sn interface. With the prolonging of solid-state aging, two distinct Cu_3Sn layers formed between Cu and Cu_6Sn_5 layer. Bi segregation was detected at the Cu/Cu₃Sn interface after the sample was solid-state aged for 360 hrs at 393K.

Introduction

The flip chip technology has been developed to satisfy the miniaturization and multifunction of electronic devices. Because it is unlike the conventional wire bonding interconnection, the input/output connections are made with solder bumps which can be distributed over the surface of the chip, not just along the periphery. The chips can be joined to the substrate through these solder bumps [1-3]. As the diameter of solder bumps scale down to about 50 m [4], the intermetallic compounds (IMCs) as the reaction products of solder and under bump metallization (UBM) after reflow will take up more volume in the whole solder balls. As we know, the formation of IMCs is inevitable for the metallic bonding between solder and UBM. However, due to the brittle nature and the tendency to produce defects, thick IMC layer at the solders/substrates interface will degrade the reliability of solder joints [5]. Therefore, understanding the growth mechanism of IMCs is important for improving the reliability of electronic packaging and devices.

In the last few decades, due to the toxicity of Pb to environment and human health, many attentions have been paid to lead-free solders. Until now, almost all of the lead-free solders are tin-based alloys. For most tin-containing solder joints on Cu, a uniform scallop-type Cu_6Sn_5 forms firstly between Sn and Cu and grows up quickly. This Cu_6Sn_5 layer is often much thicker than Cu_3Sn layer that has a layer-type morphology in wetting and solid-state reactions. On the other hand, despite its slower development, Cu_3Sn layer plays a critical role in determining the reliability of a solder/Cu interface as the Cu_3Sn or its interface with Cu is prone to voiding either by the Kirkendall effect or by solute segregation [6, 7]. Since voids are formed as Cu_3Sn phase grows, understanding the growth mechanism of Cu_3Sn phase

is necessary to addressing reliability concerns on interfacial voiding.

There have been many studies reported on the growth mechanism of IMCs [8-11], but they were mainly focused on Cu_6Sn_5. The growth of scallop-type Cu_6Sn_5 has been analyzed by a nonconservative ripening mechanism, which proceeds with a constant surface area while the total volume increases with time [12]. The discussion of the growth mechanism for Cu_3Sn phase has been rather scarce because of the limitation of analytical instruments. Although other studies have analyzed the thermodynamics and kinetics of Cu_3Sn formation, the detailed growth mechanism still remains unclear. In this work, TEM investigations were carried out to analyze the microstructure of solder/substrate interface during reflow and solid-state aging process. Efforts were made to clarify the growth mechanism of Cu_3Sn on single crystal Cu substrate.

Experimental Procedure

The single crystal Cu (100) sheets with the size of 10 mm × 2.5 mm × 2 mm were cut as the substrate in this experiment. The single crystal Cu was first ground mechanically and carefully surface polished by 0.5 m diamond paste, and then cleaned in acetone, ethanol and distilled water in an ultrasonic bath in order. Upon air drying, a commercial eutectic Sn-Bi solder paste was dispersed on copper surfaces, and then two copper surfaces were placed together face to face, between which several fine brass wires (the diameter is less than 20 m) were put in to control the thickness of solder. The reflowing process was carried out on a heating plate until the solder was completely melted for about 3 seconds, and then cooled in air as reflowed samples. Aging of samples were carried out in silicone oil bath at 393K for different hours.

To prepare TEM sample, a cross-sectional specimen in thickness of 300 m was cut from the solder joint by spark erosion, and then mechanically ground to a final thickness of 40 m. TEM foils were produced by ion milling (Gatan model 691 PIPS) at 5.0keV and 5 A with low milling angle (< 6°) at room temperature. The TEM observation was carried out in a FEI Tecnai F30 TEM operating at 300kv and equipping with energy dispersive X-ray spectroscopy (EDS).

Results and Discussions

Fig.1 shows the interfacial microstructure of SnBi/(100) single crystal Cu interface after reflow. It can be seen that there are two kinds of IMCs, Cu_6Sn_5 and Cu_3Sn, at the interface. For clarity, the Cu/Cu₃Sn and Cu₃Sn/Cu₆Sn₅ interfaces are indicated by black and white arrows, respectively. The grain size of Cu_6Sn_5 is larger than that of

978-1-4244-2739-0/08/$25.00 ©2008 IEEE

Cu_3Sn. Moreover, the thickness of Cu_3Sn layer is uniform and the $Cu/Cu_3Sn/Cu_6Sn_5$ interfaces are smooth generally. Almost all of the Cu_3Sn grains are columnar-type and their long axes perpendicular to the Cu/Cu_3Sn interface.

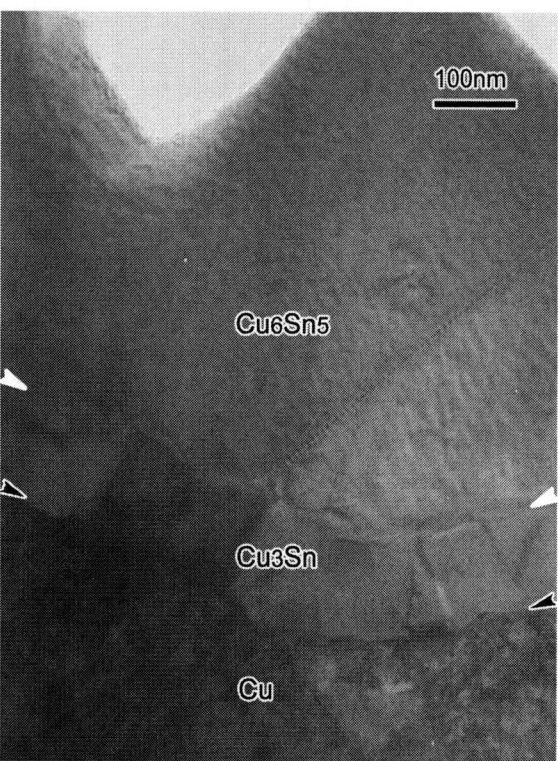

Fig.1 The interfacial microstructural images of eutectic SnBi/(100) single crystal Cu solder joints after reflowing.

Fig. 2 Bright field (BF) image (a) showing the interfacial microstructure after solid-state aged for 24 hrs at 393K,and diffraction pattern (b) showing orientation relationship between (100) single crystal Cu and columnar Cu_3Sn.

After solid-state aging for 24 hrs at 393K, the interface changed slightly.Fig. 2(a) shows the interfacial microstructure after the sample was aged for 24 hrs. The Cu_3Sn grains still remained columnar except that the grain size grew. The orientation relationship between (100) single crystal Cu and columnar Cu was verified by selected area electron diffraction (SAED), as shown in Fig. 2(b). The diffraction spots from Cu

substrate were marked and indexed in white, while those from Cu_3Sn in black.The spots of $(2\text{-}10)_{Cu3Sn}$ and $(\text{-}402)_{Cu}$ coincide, indicating an orientation relationship of

$$(2\text{-}10)_{Cu3Sn}//(\text{-}402)_{Cu}, [122]_{Cu3Sn}//[010]_{Cu} \quad (1)$$

As we known, in wetting reactions, the Cu_6Sn_5 phase first forms at the solder/Cu interface, Cu_3Sn forms later between Cu and Cu_6Sn_5 phase from solid-state reaction, because the melting point of Cu_3Sn and Cu_6Sn_5 are higher than the reflowing temperature (443K). The columnar growth of Cu_3Sn should result from the following favourable conditions. Firstly, the reacting species can diffuse fast in the way normal to Cu_6Sn_5/Cu interface in the early stage of the reaction, because the diffusion distance is shortest in this direction. Secondly, there are more multiple nucleation sites at the Cu_6Sn_5/Cu interface, and few other heterogeneous nucleation sites in the early part of the growth process. Moreover, there is a large driving force for Cu_3Sn phase to nucleate according to Sn-Cu binary phase diagram. Once nucleated, the Cu_3Sn crystals could grow rapidly without nucleation of new grains. Thirdly, during growth, Cu_3Sn crystals maintain a special orientation relationship with Cu as shown in eqn. (1). So far, the relationship in eqn. (1) is the only one being observed, and its uniqueness still remains a question. However, the preference of the $[122]_{Cu3Sn}//[010]_{Cu}$ significantly reduces the nucleation and growth of Cu_3Sn crystals along other orientations.

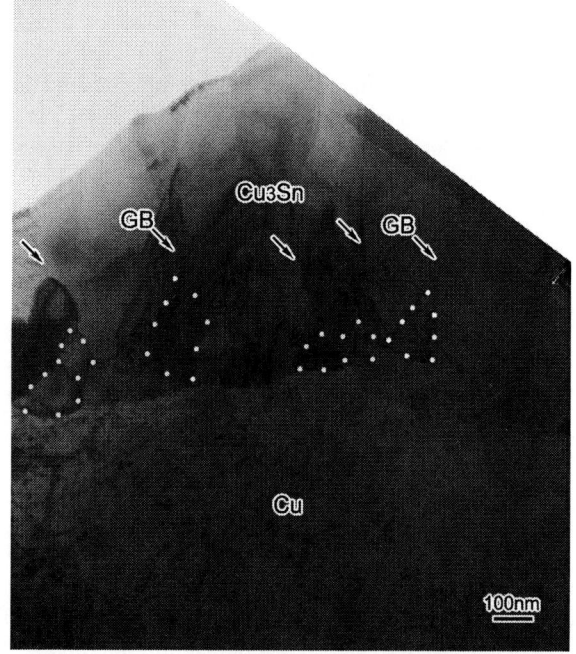

Fig.3 TEM images of the sample aged for 48 hours showing nucleation and growth of triangle Cu_3Sn grains at the triple junction sites between Cu/Cu_3Sn interface and columnar Cu_3Sn grains.

As the solid-state aging goes on, the triangle Cu_3Sn grains began to nucleate at the triple junction sites of Cu_3Sn grain boundaries. Fig. 3 shows the interfacial microstructure after the sample was solid-state aged for 48hrs at 393K. In Fig. 3, the grain boundaries of reflowed columnar Cu_3Sn and small triangle Cu_3Sn grains were indicated with arrows and white dots, respectively. It can be seen clearly the new Cu_3Sn grains nucleated at the Cu_3Sn grain boundaries.

With the prolonging of solid-state aging, more and more Cu_3Sn grains nucleated, grew and aggregated to form a second Cu_3Sn layer, which was made up of equiaxial grains developed from triangular grains. Figure 4 shows the interfacial microstructure of the solder joint aged for 168 hours. On Cu substrate there were two distinct Cu_3Sn layers: an initial columnar layer (layer 1) and a newly-formed equiaxed layer (layer 2), separated by the interface marked with white dots. The appearance of the equiaxed Cu_3Sn grains marked the end of the directional growth process for Cu_3Sn.

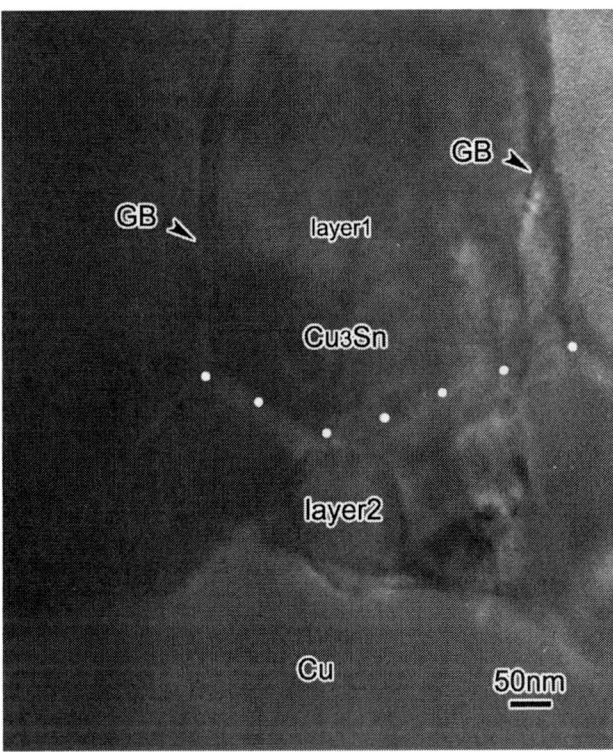

Fig. 4 TEM image of the sample aged for 168 hours showing the formation of equiaxed Cu_3Sn layer (layer 2) between columnar Cu_3Sn layer (layer 1) and Cu substrate. The grain boundaries between columnar Cu_3Sn grains were indicated by arrows.

The triple junction nucleation and the formation of new equiaxed layer of Cu_3Sn are closely related to the diffusion of Sn during solid-state aging. In Sn-Cu diffusion couple, Paul et al [13] verified that Cu diffused faster than Sn in Cu_3Sn phase through their marker experiment. It means that Cu dominates the reaction during the growth of Cu_3Sn. However, it does not mean that the diffusion of Sn can be neglected. By contrary, in the early stage of the growth process, Cu_3Sn layer is very thin so that the diffusion distances for reacting species are very short. As Cu_3Sn grew thicker, diffusion of Sn through the long column becomes increasingly more difficult, especially in comparison with the diffusion along the column boundaries, which should still be very fast. Therefore, the grain boundaries in columnar Cu_3Sn layer, which are perpendicular to Cu_3Sn/Cu interface, can act as fast channels for supplying reacting species. Once a triangle grain is

precipitated, the original columnar grain boundary is split into two grain boundaries between triangle/columnar Cu_3Sn grains. This enables new nucleation on these new triple junction sites at triangle/columnar grain boundaries, whose number is multiplied due to splitting. At last, newly-formed triangle grain would grow to equiaxial grain and aggregate to form an equiaxed layer. The protruding of equiaxial Cu_3Sn grains (in layer 2) into columnar Cu_3Sn layer (layer 1) at the grain boundaries, indicated by arrows in Fig. 3, provides a direct evidence to support the explanation above.

In our previous work, it was found that on polycrystalline Cu, Bi segregated to the Cu/Cu_3Sn interface after the sample was solid-state aged for 24 hrs at 393K [14]. On single crystal Cu, Bi segregation was also detected by elemental mapping in STEM. Fig. 5 is elemental mappings of the interface showing the Bi segregation in eutectic SnBi/ (001) single crystal Cu solder joint after the sample was aged for 360 hours at 393K. It can be seen clearly that two Bi particles about 50nm in size were formed at the Cu/Cu_3Sn interface. Around these particles there were also interfacial voids. This phenomenon is the same as that on polycrystalline Cu, where the voids formed at the side of Bi particles. Compared to that on polycrystalline Cu, interfacial segregation of Bi took longer aging time on single crystal Cu. Moreover, when Bi segregation took place, the thickness of Cu_3Sn layer was about twice larger than that on polycrystalline Cu. it was supposed that the Bi segregation at the Cu/Cu_3Sn interface was affected by not only the thickness of Cu_3Sn but also by the structures of Cu_3Sn layer formed during reflow and subsequent solid-state aging.

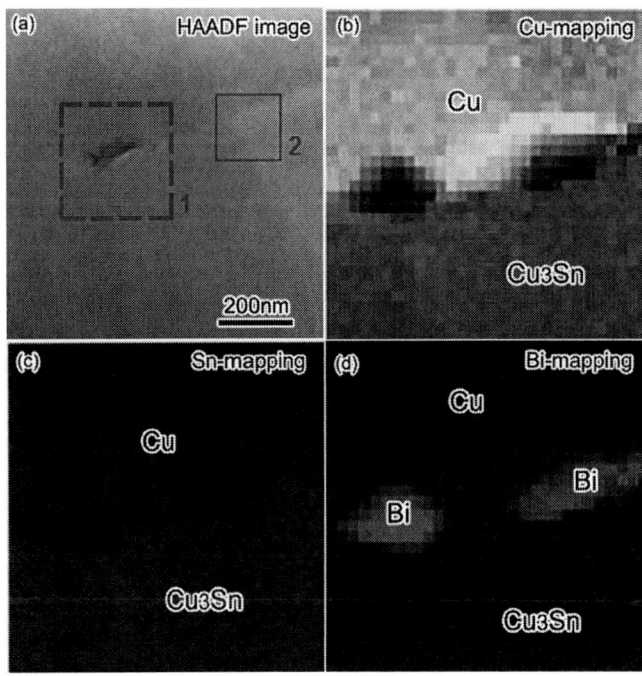

Fig. 5 Elemental mapping of the interface showing the segregation of Bi in eutectic SnBi/ (100) single crystal Cu solder joint aged for 360 hours at 393K: (a) HAADF image; (b) Cu mapping; (c) Sn mapping; (d) Bi mapping.

Conclusions

During the reflow and solid-state aging process of eutectic SnBi/(100) single crystal Cu solder joint, the growth behavior of Cu_3Sn was investigated by TEM. Columnar-type Cu_3Sn grains grew along the Cu [100] direction with a preferred orientation relationship of $(2-10)_{Cu3Sn}//(-402)_{Cu}$, $[122]_{Cu3Sn}//[010]_{Cu}$. In the subsequent solid-state aging process, new triangle Cu_3Sn grain nucleated at the triple junction sites of Cu_3Sn and substrate. This process was dominated by the diffusion of Sn along the Cu_3Sn grain boundaries. Upon further growth, the equiaxed grains formed a second layer of Cu_3Sn between copper and the columnar Cu_3Sn, creating a double layer Cu_3Sn structure. Bi was also found at the $Cu_3Sn/(100)$ single crystal Cu interface, although it took longer aging time than that on polycrystalline Cu.

Acknowledgments

The authors gratefully acknowledge the financial support from the National Natural Science Foundation of China under Grant No. 50228101 and 50571100, and the National Basic Research Program of China under Grant No. 2004CB619306.

References

1. Lee, T.Y., Tu, K.N., Frear, D.R., "Electromigration of eutectic SnPb and SnAg3.8Cu0.7 flip chip solder bumps and under-bump metallization," *J. Appl. Phys.,* Vol. 90, No. 9 (2001), pp.4502-4508.

2. Li, D.Z., Liu, C.Q., Conway, P.P,"Characteristics of intermetallics and micromechanical properties during thermal ageing of Sn–Ag–Cu flip-chip solder interconnects," *Mater. Sci. Eng., A,* Vol. 391 (2005), pp.95-103.

3. Liao, C.N., Wei, C.T., " Effect of Intermetallic Compound Formation on Electrical Properties of Cu/Sn Interface during Thermal Treatment," *J. Electron. Mater.,* Vol. 33, No.10 (2004), pp.1137-1143.

4. Agarwal, R., Ou, S.E., Tu, K.N., "Electromigration and critical product in eutectic SnPb solder lines at 100 °C," *J. Appl. Phys.,* Vol. 100 (2006), pp.024909.

5. Choi, S., Bieler, T.R., Lucas, J.P., Subramanian, K.N., " Characterization of the Growth of Intermetallic Interface Layers of Sn-Ag and Sn-Pb Eutectic Solders and Their Composite Solders On Cu Substrate During Isothermal Long-Term Aging," *J. Electron. Mater.,*Vol. 28, No. 11(1999), pp.1209-1215.

6. Tu, K.N., Zeng, K., "Tin-Lead (SnPb) Solder Reaction in Flip Chip Technology," *Mater. Sci. Eng., R* Vol. 34 (2001), pp. 1-58.

7. Liu, P.L., Shang, J.K., "Segregant-induced Cavitation of Sn/Cu Reactive Interface," *Scripta Mater.,* Vol. 53 (2005), pp. 631-634.

8. Lee, J.H., Kim, Y.S., " Kinetics of Intermetallic Formation at Sn-37Pb/Cu Interface During reflow Soldering," *J. Electron. Mater.,*Vol. 31, No. 6(2002), pp.576-583.

9. Rönkä, K.J., Van Loo, F.J.J., Kivilahti, J.K., " A Diffusion-Kinetic Model for Predicting Solder/Conductor Interactions in High Density Interconnections," *Metall. Mater. Trans. A,* Vol. 29A (1998), pp.2951-2956.

10. Bader, S., Gust, W., Hieber, H., "Rapid Formation of Intermetallic Compounds by Interdiffusion in the Cu-Sn and Ni-Sn Systems," *Acta Metall. Mater.,*Vol. 43, No. 1 (1995), pp. 329-337.

11. Mei, Z., Sunwoo, A.J., Morris, Jr. J.W., "Analysis of Low-Temperature Intermetallic Growth in Copper-Tin Diffusion Couples," *Metall. Trans.,* Vol. 23A (1992), pp.857-864.

12. Gusak, A.M., Tu, K.N., "Kinetic theory of flux-driven ripening," *Phys. Rev. B,* Vol.66 (2002), pp.115403.

13. Paul, A., Kodentsov, A.A., van Loo, F.J.J., " Intermetallic Growth and Kirkendall Effect Manifestations in Cu/Sn and Au/Sn diffusion Couples," *Z. Metallkd.,* Vol. 95 (2004), pp.913-920.

14. Shang, P.J., Liu, Z.Q., Li, D.X., Shang, J.K., " Bi-Induced Voids at the Cu_3Sn/Cu Interface in Eutectic SnBi/Cu Solder Joints," *Scripta Mater.,* Vol. 58 (2008), pp. 409-412.

Synthesis of High Purity O-cresol Novolac Epoxy Resins

Xiao-Wei Tian; Zhen-Guo Yang[*]

Department of Materials Science; Fudan University; Shanghai; China

220 Handan Rd, Shanghai, 200433, China

Email: zgyang@fudan.edu.cn; Phone: 86-21-65642523

Jiang-yan Sun; Zheng Ji

Shanghai Sinyang Semiconductor Materials Co., Ltd.; Shanghai 201616; China

Abstract: O-cresol novolac epoxy resins are used widely as electronic encapsulating materials. When the pitch wires in electronic systems are more slender, there are pressing demands for high purity o-cresol novolac epoxy resin which are with little content of hydrolyzable chloride. In this paper, o-cresol novolac resin was synthesized from para-formaldehyde and o-cresol in the presence of a mixed catalyst (oxalic acid and another co-catalyst). Then, the resultant resin was further reacted with epichlorohydrin to prepare o-cresol novolac epoxy resin. The performances and structure of o-cresol novolac resin and o-cresol novolac epoxy resin were characterized by epoxy equivalent weight, softening point, the content of hydrolyzable chloride and inorganic chlorine, FTIR, ^{13}CNMR. And the relationships between the performances and the reaction conditions were also studied. The results indicate that, o-cresol novolac epoxy resins were synthesized with certain performances through the adjustment of the reaction conditions, which were suited for the pressing demands for high purity applications.

Key Words: o-cresol novolac resin; o-cresol novolac epoxy resin; high purity; hydrolyzable halide

1. Introduction

Epoxy resins, especially the o-cresol novolac epoxy resin (ECN) are widely used for semiconductor device encapsulation in the microelectronic industry, for protecting components from chemicals and mechanical stress, ensuring a good electrical insulation, and offering a good thermal conductivity. The major criteria for measuring quality of these resins are the content of hydrolyzable chloride and the epoxy equivalent weight. Since the existent of hydrolyzable chloride can erode the pitch wires and make electronic systems invalidate. When the pitch wires in electronic systems are more slender, there are exigent demands for high purity o-cresol novolac epoxy resin with little content of hydrolyzable chloride. The epoxy resins have become the main factor which have a great influence on the reliability of an encapsulated semiconductor device. The types of halide in ECN were classified into two kinds: organic halide and inorganic chlorine[4,7,10]. The main component of organic halide is the hydrolyzable chloride which is the important factor to measure the quality of ECN. There are many researches which had been done to reduce the content of hydrolyzable chloride in ECN, and have developed into many different methods.

With the researches of the relationship between the content of organic chloride and the polycondensation reaction of ECN, many researchers can produce high-purity ECN by control the amount of alkali metal hydroxide[1]. Further, some researchers use different inert polar solvents to restrain theside reaction[2], such as dimethyl sulfoxide, dimethyl sulfone. A common method to obtain high purity ECN is using a phase transfer catalyst (PTC) during the polycondensation, such as benzyl triethyl ammonium chloride (BTEAC), so that the synthesis can be taken under mild condition. Some researchers also develop different methods from another thoughts, such as a post treatment process for reducing the undesirable hydrolyzable chloride[3], but the molecular weight content distribution became broader after this post-treatment process which is also undesirable.

The helpful method to reduce the undesirable hydrolyzable chloride is to understand the nature of the reaction, and reduce the side reaction. In this paper, based on this thought, a high purity o-cresol novolac epoxy resin is synthesized successfully, which meets the demands for the developments of microelectronic industry.

2. Experiments

2.1. Materials

Ortho-cresol (CP), para-formaldehyde (powder, >95%), epichlorohydrin (ECH, CP), sodium hydroxide (CP) and oxalic acid (powder, CP) was purchased from Sinopharm Chemical Reagent Co., Ltd., China. All solvent used were commercial products and used without further purification.

2.2. Preparation of ECN

2.2.1. Preparation of OCN

1 mol o-cresol and 0.85mol para-formaldehyde were placed in a four-necked round-bottom flask with a thermometer, a stirrer, and a reflux condenser was fitted to allow for reflux of the water that formed. At first, half of oxalic acid and another co-catalyst were added. They are reacted for 3hours under $100\pm5\,^{\circ}\text{C}$ and continuously stir. Then the other half of oxalic acid and another co-catalyst were added, and another 3 hours' reaction was taken under $100\pm5\,^{\circ}\text{C}$ and continuously stir. Nitrogen was also used to keep reaction under inert conditions. After reaction, the appropriate amount of methyl isobutyl ketone (MIBK) was added into the flask to dissolve the reaction mixture. So the reaction mixture was washed by deionized water till that the water phase's pH was 7. After water washed, MIBK was desolventized under the condition of $200\,^{\circ}\text{C}$ and 0.1MPa minus pressure for a long time. Thus, the o-cresol novolac resin (OCN) was obtained.

978-1-4244-2739-0/08/$25.00 ©2008 IEEE

The FTIR spectrum (KBr) of OCN exhibited absorption at 3424cm^{-1} (the stretching vibration of Ar-OH); 1605, 1505cm^{-1} (the stretching vibration of Ar); 2914cm^{-1} (the stretching vibration of CH$_2$); 1445, 1379cm^{-1} (the shearing vibration of CH$_2$); 3011, 2853cm^{-1} (the stretching vibration of CH$_3$).

The ^{13}C NMR data for OCN were assigned with reference to the chloroform solvent peak at a chemical shift of 77ppm. These are listed below.

No.	d (ppm)	Carbon atom
1	15	CH$_3$
2	32	Ar-CH$_2$-Ar (o-o)
3	37	Ar-CH$_2$-Ar (o-p)
4	41	Ar-CH$_2$-Ar (p-p)
5	115~135	Ar C
6	149~153	Ar C-OH

And the structure of OCN was predicted as below.

2.2.2. Preparation of ECN

To a four-neck reactor equipped with a thermometer, a stirrer, and a device for condensing co-distillate mixture of water, epichlorohydrin (ECH) and separating them into an organic phase and an aqueous phase, were added 1 mol of o-cresol novolac resin, which hydroxyl equivalent was 118, 7.5 mol of ECH, and a phase transfer catalyst[8]. The mixture was stirred to form a homogeneous solution under nitrogen and then heated to 65~70℃ under 0.08Mpa minus pressure. After reaching equilibrium of the pressure and the temperature, to the mixture was added 1.1 mol of 50wt% aqueous sodium hydroxide solution at a constant rate over five hours while water contained in the reaction system was azeotropically distilled and condensed. The condensed azeotrope was separated into an organic phase and an aqueous phase, the organic phase, mainly ECH was sequentially recycled into the reaction system and the aqueous phase was discarded. After the reaction was completed, unreacted ECH was distilled off under reduced pressure which was about 0.1MPa minus pressure. Then the appropriate amount of MIBK was added into the flask to dissolve the reaction mixture. And 10wt% sodium hydroxide solution was added in the reaction mixture to refine, and the mixture's temperature was 85~85℃ for 2 hours. So the resultant crude epoxy resin was washed off by deionized water till that the water phase's pH was 7. MIBK was distillated from the resulting mixture under the condition of 200℃ and 0.1MPa minus pressure for a long time. Thus, the o-cresol novolac epoxy resin (ECN) was obtained.

The FTIR spectrum (KBr) of ECN exhibited absorption at 1602, 1500cm^{-1} (the stretching vibration of Ar); 2920cm^{-1} (the stretching vibration of CH$_2$); 1446, 1379cm^{-1} (the shearing vibration of CH$_2$); 2998, 2868cm^{-1} (the stretching vibration of CH$_3$); 910cm^{-1} (the stretching vibration of epoxy group).

The ^{13}C NMR data for ECN were assigned with reference to the chloroform solvent peak at a chemical shift of 77ppm. These are listed below.

No.	d (ppm)	Carbon atom
1	42	Ar-CH$_2$-Ar (o-o)
2	36	Ar-CH$_2$-Ar (o-p)
3	42	Ar-CH$_2$-Ar (p-p)
4	120-135	Ar C
5	153	Ar C-O
6	44	CH$_2$ of Epoxy group
7	50	CH of Epoxy group
8	73	O-CH$_2$-Epoxy group
9	16	CH$_3$
10	131	Ar C-CH$_3$

And the structure of ECN was predicted as below.

2.4. Characterizations

The hydroxyl equivalent of OCN was measured by acetylation reaction according to GB 2895-82. The epoxy equivalent weights (EEWs) of ECN were determined by hydrochloric acid /acetone titration method. The softening points of OCN and ECN were measured by Ball and Ring method according to GB 12007.6-89. The content of inorganic chlorine was measured by GB4613-84. And the content of hydrolyzable chloride was measured by GB4618-84[5].

^{13}C nuclear magnetic resonance (^{13}C NMR) characterization were carried out by DMX500 NMR spectrometer using chloroform -d$_6$ as the solvent and tetramethylsilane (TMS) as internal standard. FTIR spectra were recorded on a Nexus 670 infrared spectrometer.

3. Results and discussion

3.1. Synthesis of OCN

3.1.1. Influence of the molar ratio of HCHO to O-cresol

For the sake of obtaining o-cresol novolac epoxy resin, the o-cresol novolac resin was synthesized by excessive amount of o-cresol with formaldehyde under acid catalyst according to the chemical reactions outlined in Scheme 1.

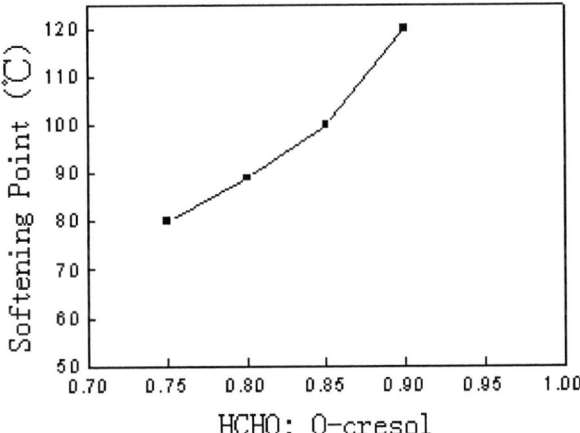

Scheme 1 Schematic diagram of the synthesis of o-cresol novolac resin

The results depicted in Figure 3.1 indicate that the softening point increases with the increase in the HCHO: *o*-cresol molar ratio from 0.75 to 0.90 at 0.05 increments. And it could influence the softening point and epoxy equivalent weight of ECN. As a polycondensation reaction, the reaction was expected to occur since there was more HCHO for polymer formation, and this would necessarily lead to higher molecular weight polymers.

Therefore, it was important to select a suitable molar ratio of HCHO to o-cresol.

Fig 3.1 Softening point values at different molar ratio of HCHO to *o*-cresol

3.1.2. Influence of co-catalyst

Different acids are used as catalysts when OCN was synthesized, and the polycondensation reaction of o-cresol and HCHO was so exothermic that its catalyst should use some weak acid, such as oxalic acid. In this research the amount of oxalic acid was 2.5wt% of o-cresol. After the determination of the molar ratio of formaldehyde to *o*-cresol, another catalyst was used as co-catalyst to obtain the exact control of the softening point of OCN. The results were listed in Figure 3.2 below.

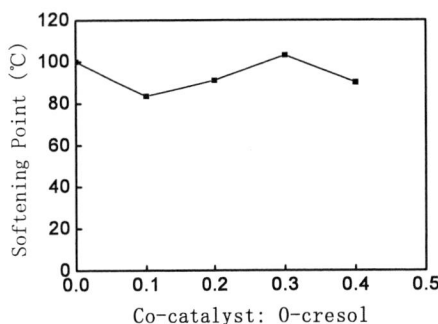

Fig 3.2 Softening point values at different amount of co-catalyst

From these results, it was found that there was an appropriate amount of the co-catalyst, which was used in the mass batch production to stabilize the product's quality. Table 3.1 was the typical properties of o-cresol novolac resin while the molar ratio of co-catalyst to o-cresol was 0.3.

Table 3.1 Analytical results obtained for prepared OCN resins

No.	Hydroxyl equivalent (g/mol)	Hydroxyl equivalent (g/mol)	Excessive o-cresol (%)
1	118	103	0.2
2	117.2	102	0.1
3	120.3	103	0.2
4	118.6	103	0.2
5	118.6	101	0.2

3.2. Synthesis of ECN

3.2.1. Influence of the molar ratio of ECH to OCN

In the synthesis of epoxy resin, the amount of epichlorohydrin (ECH) is often excessive to promote the reaction as well as the conversion of phenolic hydroxyl in OCN, and the excessive ECH could be seen as a good solvent to dissolve the o-cresol novoalc resin and the resultant o-cresol novolac epoxy resin. But the amount of ECH was too excessive to reduce the reaction rate, and the reaction would be not economical. Thus the influence of the ratio of ECH to OCN was researched by experiments. The results were shown in Figure 3.3 and Figure 3.4.

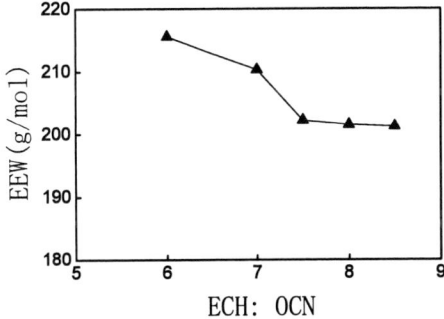

Fig 3.3 EEW values at different molar ratio of ECH to OCN

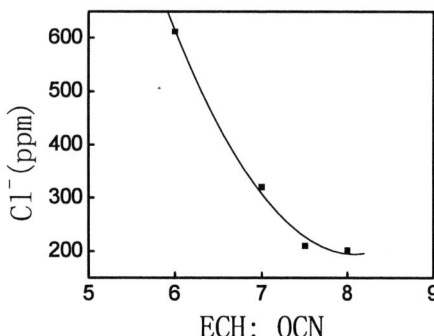

Fig 3.4 The content of hydrolyzable chloride values at different molar ratio of ECH to OCN

The more excessive the amount of ECH was, the lower the epoxy equivalent weight and the content of hydrolyzable chloride of ECN was. The reason was that with the increasing amount of ECH, the reaction was easier taken on, and the less side reaction was happened. But when the amount of ECH was more, the influence to ECN was not distinct, so for economical reason, the molar ratio of ECH to OCN was 7.5. Scheme 2 was the reaction process of ECN.

Scheme 2 Schematic diagram of the synthesis of ECN

3.2.2. Influence of NaOH

As it had the abilities of opening epoxy group and closing epoxy group, aqueous sodium hydroxide solution (50wt%) was used as the catalyst. The additive method and amount of NaOH were important to affect the epoxy resin's performances, such as the content of hydrolyzable chloride and the epoxy equivalent weight[9]. In order to obtain an high purity ECN, aqueous sodium hydroxide solution was dripped into the reaction system continuously according to our experiments. The affect of the amount of sodium hydroxide was researched by experiments, and the results were listed in Table 3.2.

Table 3.2 EEWs and the content of hydrolyzable chloride values at different amount of NaOH

NaOH: OCN	0.96	0.98	1.0	1.1
EEW (g/mol)	206.0	204.6	202.8	202.6
Hydrolyzable chloride (ppm)	609	406	321	210

With the amount of sodium hydroxide was increased, the content of hydrolyzable chloride was reduced, and the epoxy equivalent weight almost was not affected.

3.2.3. Influence of reaction temperature

During the synthesis, there was a competition between the normal addition reaction and the abnormal addition reaction of ECH, so the reaction temperature was an important factor to determine the content of hydrolyzable chloride[6]. The reaction was divided into two parts, that the first part was the addition part of sodium hydroxide, and the another part was the reaction part. And the reaction temperature's effect was listed in Table 3.3 below.

Table 3.3 EEWs and the content of hydrolyzable chloride values at different reaction temperature

Reaction temperature (℃)	65+70	70+65	70+70
EEW (g/mol)	210.6	203.5	202.2
Hydrolyzable chloride (ppm)	412	203	210

From these results, it was found that the content of hydrolyzable chloride was decreased with the increase of the temperature of addition part, and increased with the increase of the temperature of reaction part. The reason was that the reaction of abmromal addition of ECH was mainly occurred in the reaction part, and the active energy of abnormal addition was higher than the normal addition. So, the reaction with a two stages of temperature was better to obtain ultra-high purity o-cresol novolac epoxy resin.

Table 3.4 showed the typical performances of ECN including different types. From the data, our product were excelled than the products from Japan, and more suitable for microelectronic industry. That meaning was that the optimized process of synthesis of ECN was very useful and economical.

Table 3.4 Analytical results obtained for different ECN

		EEW (g/mol)	Softening point (℃)	Hydrolyzable chloride (ppm)
EPICLON[11]	N-660	206	66	<600
	N-655	201	58	<400
Our products	1	202.6	62	210
	2	202	64	206
	3	205	62.5	204

4. Conclusions

The high purity o-cresol novolac epoxy resins were successfully synthesized via polycondensation of ortho-cresol and para-formaldehyde, followed by epoxidation with epichlorohydrin. With the optimized reaction conditions, the resultant epoxy resins exhibited the lower content of hydrolyzable halide and inorganic chlorine than other normal resins. These pronounced good properties make it an attractive candidate for electronic encapsulation applications and composite materials.

References

[1]Wang Chunshan, Pham HaQ, Bertram JamesL. Preparation of epoxy resins [P]. US: 4499255, 1985-02-12.

[2]Zeng Kunliao, Wang Chunshan. Concurrent addition process for preparing high purity epoxy resins [P]. US: 5028686, 1991-07-02.

[3]Wang Chunshan, Zeng Kunliao Method for reducing the aliphatic halide content of epoxy resins using a solvent mixture including a polar aprotic solvent [P]. US: 4785061, 1988-11-15.

[4]Dong Gengjiao, Zhu Xiong, Chen Ming, etal. Study On O-cresol Epoxy Resin [J]. Thermosetting Resin. 1995, (4):15-20.

[5]Sun Manling. Principle and technology of Epoxy resins [M]. Beijing: China machine press, 2002.

[6]Shi Zijin, Pan zheng. The preparation of high purity ECN resin used for encapsulation of integrated circuit [J]. Synthetic resin and plastics. 1996, 13 (1): 19-22.

[7]Kuen Yuanhwang, Chen Honghsing, An Bangduh, etal. Process for preparing a high purity epoxy resin [P]. US: 6001873, 1999-12-14.

[8]Meng Hong, Yu Zaizhang, Li Baofang, etal. The kinetics of etherification of o-cresol formaldehyde resin in presence of phase-transfer-catalysts [J]. Synthetic resin and plastics. 1994, 11 (1): 10-14.

[9]Li Baofang, Dong Gengjiao. The synthesis process of high pure o-cresol formaldehyde epoxy resin with low chlorinty [J]. Thermosetting Resin. 1998, (1): 22-26.

[10]Wang Chunshan, Zeng Kunliao. Synthesis of high purity o-cresol formaldehyde novolac epoxy resins [J]. Polymer Bulletin. 1991, 25: 559-565.

[11]Technical data sheet of EPICLON n-600 series. Dainippon ink & chemicals, Japan.

Defect Analysis of Copper Ball Bonding

Huabin Chu, Jun Hu, Ling Jin, Yingliang Jie
Guangdong Yuejing High Tech Co., ltd
No.10, 2nd Nanxiang Road, Guangzhou Science Park, Guangzhou City, 510663, China
huxbill@21cn.com, 86-20-82075328-8074

Abstract

Copper ball bonding has gained more and more popularity compared to gold ball bonding due to its better electrical conductivity, better thermal conductivity, higher mechanical strength, lower cost and better reliability and so on. However copper wire has two well-known disadvantages, one is its greater hardness, the other is being easy to be oxidized when forming FAB (free air ball). These disadvantages damage ball bonding process feasibility and stability and bring many problems that are proving difficult to resolve such as none stick, weak bonding, missing ball, different size and shape bonding ball, wire breaking, aluminum sputter, cratering, short tail and finally reduce product yield. Though new materials and improved equipments are also very important to copper ball bonding, promoting bonding process to decrease weld defects is necessary and sustaining. Copper ball bonding process parameters are various and complicated. Many studies have been conducted on the relationship between bonding process and bondability. However many impact of process on bonding are still unclear and misconception, this paper intends to help understand the relevant between. This paper utilizes DOE (design of experiment) to optimize separately FAB parameters such as fire current, fire time and $0.9N_2/0.1H_2$ gas flow rate and bonding process parameters such as ultrasonic power, bonding force, bonding temperature and bonding time to obtain the feasible and stabile bonding process. The bonding effect under different process parameters were compared and analyzed by 3D digital microscope. This paper also indicated the failure mode and the failure mechanism of ball shear and wire pull by die tester

1 Introduction

Wire bonding is generally considered the most cost-effective and flexible interconnect technology and is used to assemble the vast majority of semiconductor device. The wire is generally made up of gold, copper and aluminum. Gold wire is quite most used and aluminum wire is mainly used in high power device, in recent years copper wire as the replacement of gold wire has gained more and more popularity due to its better electrical conductivity, better thermal conductivity, higher mechanical strength, lower cost and better reliability and so on. Both gold wire and copper wire can use ball bonding process which use a combination of heat, pressure and ultrasonic energy to make a weld joint at each end of wire.

However copper wire has greater hardness and is being easy to be oxidized when forming FAB. Both two great challenge seriously damage bonding process stability and reduce product yield. So research on defect analysis is necessary and it can be a useful feedback for wire bonding process improvement. [1]

According to these background, this paper focuses on internal copper wire bonding transistor where 0.8mil copper wire is used. The study on FAB and bonding performance will help us to understand copper ball bonding and next assembly many high I/O and fine pitch devices, like GPU, MCU etc al, which use no more than 1 mil copper wire. In this paper different defect mode such as cratering, aluminum sputter, none stick, wire breaking, short tail etc under various bonding conditions was discovered and researched by DOE. This work will help to understand how the bonding process parameters affect the weld joint and how to prevent the occurrence of bonding defect. The outer bonding appearance was showed and analyzed and ball shear strength and wire pull strength were tested and compared. Though this work is simple and obvious, it is only a beginning and further work is currently onging in our lab.

2 Experiment

2.1 Materials, Equipment and Sample

Copper ball bonding was performed with automatic bonder ASM [®] with protective forming gas $(0.9N_2/0.1H_2)$. 1mil copper wire was from MK Electron in South Korea, which has a purity of 99.99% up, 4-16% elongation and more than 5 cN breaking load. The chip has a size of $0.44*0.44$ mm^2 and a height of 230μm. The pad of Fe/Ni alloy leadframe is Ag coated by electroplate. The first bonding ball was sheared and copper wire was pulled by die tester DAGE[®]4000. The outer bonding appearance and bonding fracture interface after shear test and pull test were observed by 3D digital microscope KEYENCE[®]VHX-600E.

2.2 DOE

First copper FAB forming process was optimized by DOE, then ball bonding and subsequent wedge bonding process was optimized separately by DOE. For FAB DOE, the variables are fire current, fire time and $0.9N_2/0.1H_2$ flow rate and the corresponding inspection functions are ball size and surface appearance. For bonding DOE, the variables are contact time (CT), contact power (CP), contact force (CF), bonding time (BT), bonding power (BP), bonding force (BF) and bonding temperature (BT*). All above parameters span was shown in table 1, 2 and 3 separately.

Table 1 FAB DOE span

	Fire current (A)	Fire time (μs)	$0.9N_2/0.1H_2$ flow rate (L/min)
Low	300	500	0.3
High	500	1000	0.9

	CT (ms)		CP (Dac)			CF (g)
Low	1	1	0	0	0	55
High	5	5	20	40	45	100

Table 2 Bonding DOE span

	BT (ms)		BP (Dac)			BF (g)	BT* (°C)
Low	2	2	20	30	10	50	200
High	10	10	80	135	50	120	260

Table 3 Bonding DOE span

3 Result and Discussion

DOE unit of copper ball bonding was carried out after FAB parameters were optimized by DOE. Here we only observe and compared the outer appearance and size of FAB through 3D digital microscope. Smooth surface and real sphere of uniform size FAB was as good and accepted. Further work about its hardness, oxidation and residual stress will be studied.

Cratering, aluminum sputter, none stick, short tail, missing ball, wire breaking and weak bonding are the most familiar defects for copper ball bonding and all these problems will decrease product yield. It's found that certain extreme bonding condition would result in above problems.

3.1 Cratering and aluminum sputter

Petal-like is the feature of bonding crater. As shown in Fig.1, it's a serious cratering effect and mainly caused by stronger ultrasonic power and bigger bonding force. Hard copper FAB plus ultrasonic power and bonding force may bring horizontal and vertical destructive impact on aluminum pad layer and Si layer below and further lead to the occurrence of cratering. Fig.2 shows that the first bonding ball was pulled out from the crater when capillary moving and aluminum pad layer around copper ball sputters also can be seen. Aluminum sputter is due to excessive bonding energy and harder copper FAB. Actually many more latent craters are exposed only after thermal aging or long-time normal work and don't be discovered in packaging product line.

Fig. 1 Cratering

Fig. 2 Aluminum sputter

3.2 None stick

To prevent cratering, generally ultrasonic power and bonding force were set lower, however the first bonding ball may be none stick on aluminum pad because these two dissimilar metals are difficult to be metallurgical joint [2]. It's found that none stick on pad happened and a residual hit-pit instead from Fig.3. Fig.4 is a picture of the shifted ball from Fig.3. DOE result indicates that appropriate bonding power and bonding force are effectual to prevent cratering and none stick and certainly are important to promote bonding process stability.

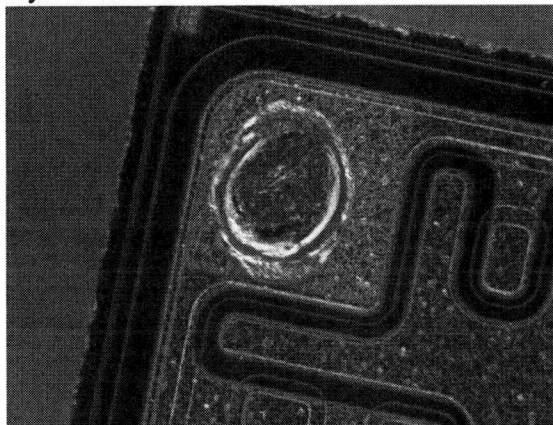

Fig. 3 None stick on pad and a hit-pit instead

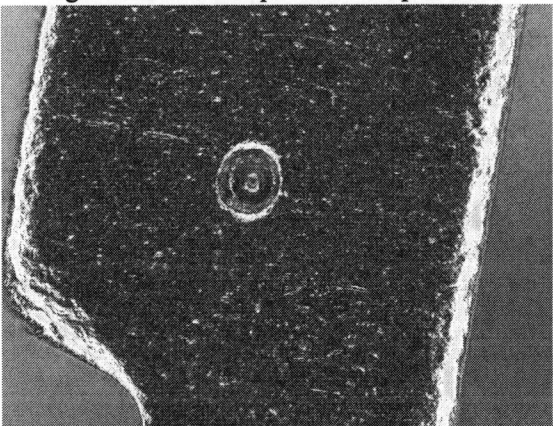

Fig. 4 Shifted ball on leadframe

3.3 Wire breaking and Short tail

For the second wedge bonding joint, short tail and wire breaking are two main defect mode. Weak welding may induce short tail length and short tail will lead to small FAB size or missing ball. Just like the first ball bonding joint, the most difficult challenge for the second bonding is the metallurgy of copper and sliver coating. It's notable that copper-silver can form bulk miscible system (random alloys) [3-4]. Generally the performance of the second wedge bonding is improved through increasing bonding force, ultrasonic power and bonding time and not to worry about damaging leadframe. However such so high bonding energy can result into great stress to copper wire tail. It may accelerate work harden during bonding and finally copper wire is breaking at welding end. Fig.5 indicates wire breaking at end point of wedge bond.

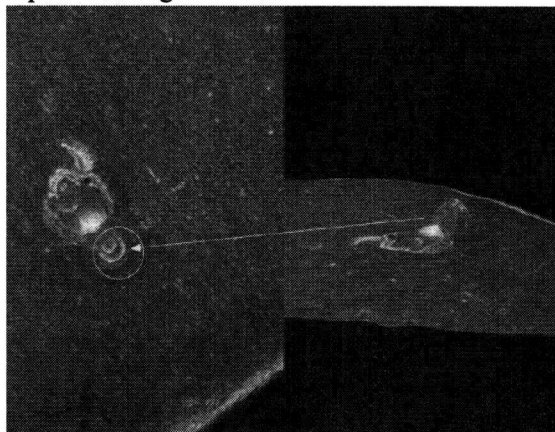

Fig. 5 Wire breaking at end point of wedge bond

3.4 Ball shear and wire pull failure mode

Copper ball shear strength and wire pull strength were measured only on DOE optimized samples. Fig.6 is a picture of the typical fracture layer morphology after ball shear test. It looks there's not residual copper on pad and inversely some aluminum was pulled out and deformation. It can be explained by that copper is harder than aluminum and there's not sufficient IMC (intermetallic compound) formed in order that copper ball is left from pad. Fig.7 presents that copper wire end is breaking and wedge bond is still strong and not floating after wire pull test. In fact it's found that copper wire breaking point is mostly at copper ball neck in wire pull test .In this study, copper ball shear strength was 25.4g and copper wire pull strength was 10.5g.

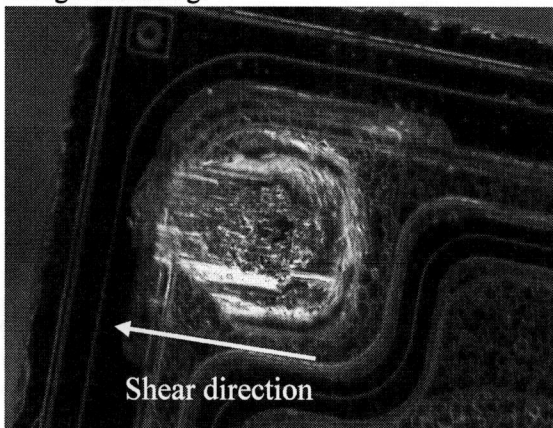

Shear direction

Fig. 6 Ball Shear fracture

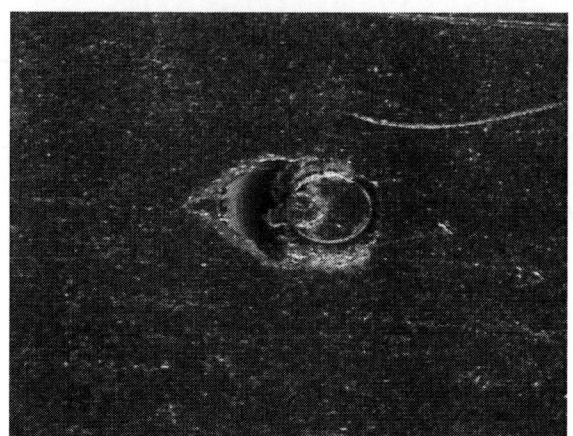

Fig. 7 Copper wire pulled from wedge bond

4 Conclusions

Many main defect mode of copper ball bonding was discovered and studied by DOE in this paper. From this investigation, following conclusions can be drawn: Stronger ultrasonic power and bigger bonding force may induce cratering, however if ultrasonic power and bonding force are set lower, none stick may happen, so appropriate coupling setting of ultrasonic power, bonding force and bonding time is very important to copper ball bonding stability. Short tail and weak welding can be avoided by enforcing the second bonding energy, however larger bonding energy may bring great stress into copper wire end and accelerate wire breaking. It's found copper ball is left from aluminum pad after ball shear test because IMC is not formed enough. In addition it's found that copper wire breaking point is mostly at copper ball neck in wire pull test. For 0.8 mil copper wire, optimized ball shear strength was 25.4g and wire pull strength was 10.5g.

Next, mechanical performance and microstructure of FAB will be researched. For bonding, diffusion and interface reaction of copper-aluminum are research emphases later. All above work is for improving copper ball bonding process stability and product yield, though many huge challenge exists.

Acknowledgements

The authors thanks Dingzu Dai, Haowei Zhang etc colleague and thanks Gongqun Ouyang and Shaowei Zhao from HIT for their assistance.

References

1. Harman, G., Wire Bonding in Microelectronics: Materials, Processes, Reliability, and Yield, McGraw-Hill(USA, 1997), PP. 212-213.

2. Irfan AY, Sare Celik, Ibrahim CELIK, "Comparison of Properties of Friction and Diffusion Welded Joints Made between the Pure Aluminium and Copper Bars", 1997.

3. Haasen P. Physical Metallurgy, Cambridge University Press, 3rd Edn., Ch. 12,1996.

4. John Beleran, Alejandro Turiano, Dodgie R.M. Calpito et al. "Tail Pull Strength of Cu Wire on Gold and Silver-Plated Bonding Leads" *SEMICON,* Singapore, 2005.

Fundamental Influence of Segregated Bi on the Mechanical Properties of Interconnect of Bismuth-containing Solder and Copper

Xue-Yong Pang,[a] Zhi-Quan Liu,[a] Shao-Qing Wang,[a] and Jian-Ku Shang [a,b]

[a] Shenyang National Laboratory for Materials Science, Institute of Metal Research, Chinese Academy of Sciences, Shenyang 110016, China (Email: zqliu@imr.ac.cn)

[b] Department of Materials Science and Engineering, University of Illinois at Urbana-Champaign, IL 61801, USA (Email: jkshang@uiuc.edu)

Abstract

We employed density functional theory to investigate bismuth segregation at Cu/Cu$_3$Sn interface. Firstly, we considered five initial constructions by adopting the adhesion energies criterion. By comparing adhesion energies of them, the so-called "Between-Cu" construction was found to be the most energy-favored (0.98J/m^2). Secondly, based on "Between-Cu" construction, we introduced eight possible segregation sites. Among these eight sites, Site-2 was determined to be the most likely segregation site with adhesion energy as low as 0.53J/m^2, which is almost half of the adhesion energy of the initial construction. Finally, we analyzed atomic structure and electronic density, from which it was found that the interfacial space was enlarged by bismuth segregation and the bond between the first layer and the second layer at copper side was weakened. Our calculated result is qualitatively consistent with reported experimental results and can explain them well.

Introduction

In electronic packaging, interfacial strength between solder and metallic substrate is a critical factor in determining the reliability of a solder joint. When a solder alloy is used to bond copper surfaces by the reflow soldering process, intermetallic compounds (IMC) often form as a result of interfacial reactions [1]. The presence of IMC's creates a complex interfacial system consisting of multiple distinct interfaces between solder alloy and IMC, between different IMC's, and between IMC and copper substrate. Failure of any of these interfaces could cause a total loss of the joint reliability to that it is important to understand atomic mechanism of interfacial failure in solder interconnects [2, 3].

Because of the toxicity of lead on environment and human body, international organizations have made laws to restrict using lead. From this view of points, Bi-containing solders such as eutectic and near-eutectic Sn-Bi and Sn-Ag-Bi have attracted much attention because of their low melting points and better mechanical properties. In interfacial reactions of Bi-containing solders with Cu, first; two intermetallic compounds, Cu$_6$Sn$_5$ and Cu$_3$Sn may form during reflow and aging treatment. Bi does not form intermetallic phases with Cu and has very limited solubility in both Sn and Sn-Cu intermetallics. Given sufficient time, e.g. by isothermal aging in the solid state, Bi tends to segregate onto the interface between Cu$_3$Sn and Cu [1]. The segregate weakens the solder interconnect greatly resulting in brittle fracture. Bi segregation was known to cause intergranular embrittlement in copper. The embrittlement has been explained well by electronic density calculation [4]. However, the reason why segregated Bi greatly weakens the interface strength of solder interconnect is not fully understood yet.

In this study, First-principles calculations were performed to investigate the change of adhesion energy of the Cu/Cu$_3$Sn interface with segregated Bi atoms. In addition, the electronic structure of the interface was calculated to analyze basic mechanics of the failure induced by Bismuth segregation.

Methodology

The plane wave pseudopotential method and generalized gradient approximation (GGA) were employed in our calculations. The self-consistent PW91 density was determined by iterative diagonalization of Kohn-Sham Hamiltonians, coupled with a Pulay mixed scheme [5]. Ground-state atomic geometries were determined by minimizing the Hellman-Feynman forces. The Brillouin zone was sampled with Monkhorst-Pack k-point grid [6]. The plane-wave cut-off energy in our calculations was 350eV. This set of parameters assured a total energy convergence of 0.01eV per atom.

Computation details and results

Experimentally, the orientation relationship between Cu and Cu$_3$Sn has not bee reported yet. To simplify the model, we took Cu(100)($\sqrt{2} \times \sqrt{2}$)/Cu$_3$Sn(010) interface to investigate the intrinsic mechanics of segregation-induced embrittlement. The Cu/Cu$_3$Sn interface was constructed by six layers of Cu$_3$Sn, seven layers of Cu and ten angstrom vacuum. Five constructions were introduced by adopting the adhesion energy criterion. The ideal work of adhesion, W_{ad}, is defined as the reversible work needed to separate an interface into two free surfaces. It can be calculated by the difference in total energies between the interface and its isolated slabs

$$W_{ad} = (E_A^{tot} + E_B^{tot} - E_{A/B}^{tot})/A. \qquad (1)$$

Here, E_A^{tot} and E_B^{tot} are the total energy of the relaxed, isolated Cu$_3$Sn and Cu slabs in the same supercell when one of the slabs is retained and the other is replaced by a vacuum, respectively. $E_{A/B}^{tot}$ is the total energy of the Cu/Cu$_3$Sn interfaces system. A is the total interface area of the unit cell. Generally, the mechanical work needed to separate an interface is larger than the ideal work of adhesion, W_{ad}. As discussed by Finnis [7], this is due to neglecting plastic and diffusion degrees of freedom.

Table 1 shows the W_{ad} values for five interface structures. Among them, the so-called "Between-Cu" construction was found to be the most energy-favored (0.98J/m^2). We took

978-1-4244-2739-0/08/$25.00 ©2008 IEEE

'Between-Cu' as our initial interface for following calculations.

Table 1 Calculations of the relaxed adhesion energies (W_{ad}) for the five Cu/Cu$_3$Sn systems.

	Top-Cu	Top-Sn	Between-Cu	Between-Sn	Bridge
W_{ad} (J/m^2)	0.75	0.74	0.98	0.96	0.94

In order to determine the site preference of Bi at the Cu/Cu$_3$Sn interface, the total energies of the Cu/Cu$_3$Sn interface supercell and the correlated crystal model were calculated. It is well known that the site preference can be deduced from the magnitude of heat of formation of the supercell or cluster. [8] In this work, the following expression was adopted to estimate the heat of formation per atom H at the Cu/Cu$_3$Sn interface:

$$H=[E_i(n,m,l)-n \cdot E_c(Sn)-m \cdot E_c(Cu)-l \cdot E_c(Bi)]/(n+m+l)$$
(2)

where $E_i(n,m,l)$ is the total energy of the Cu/Cu$_3$Sn supercell with n, m and l atoms of Sn, Cu and Bi, respectively. $E_c(Sn)$, $E_c(Cu)$ and $E_c(Bi)$, are the energies per atom in the β-Sn, *fcc*-Cu and base-centered monoclinic Bi unit cell. Due to the large radius of Bi, the segregation mode was replacement. Based on the so-called "Between-Cu" construction, eight possible segregation sites were examined as shown in Fig.1. Table 2 shows the site preference energies and the adhesion energies of different segregated sites. By comparing the interfacial formation energies of these eight sites, Site-2 (replacing Cu atom belonging to Cu slab) was determined to be the most likely segregation sites (-0.269eV/atom) with adhesion energy as low as 0.53J/m^2, which is almost half of the adhesion of the initial construction.

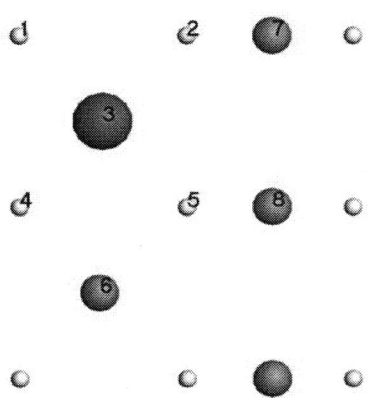

Fig.1. Eight possible segregation sites of Bi. Green and red spheres represent tin and copper atoms respectively, which belong to Cu$_3$Sn slab; white spheres represent copper atoms which belong to Cu slab.

The mechanical strength of the interface is due ultimately to the atomic bonding strength, so it is necessary to investigate the bonding nature of the interface. We used planar-average charge density, the charge density and density of states (DOS) to examine the interfacial electronic structures of the most stable so-called "Between-Cu" interface.

The planar-average charge density is shown in Fig.2. There are some significant differences in the interfacial bonding characteristics with and without Bi. Firstly, it is clear that the interfacial space was enlarged after Bi segregation, which results from the greater radius of Bi atom. Secondly, after Bi segregation, part of charge density was transferred to Cu$_3$Sn sublayer; the decrease of electronic density at interface means that the interfacial bond is weakened. Thirdly, there was severe depletion of charge density between Cu sublayer and interface layer, which indicates that the fracture would likely occur between these two layers.

Table 2 Calculations of the site preference energies and the adhesion energies of different sites.

	H_{form}(eV/atom)	W_{ad}(J/m^2)
Site1	-0.260	0.44
Site2	-0.269	0.53
Site3	-0.266	0.79
Site4	-0.263	0.45
Site5	-0.268	0.52
Site6	-0.251	0.44
Site7	-0.257	0.54
Site8	-0.256	0.53

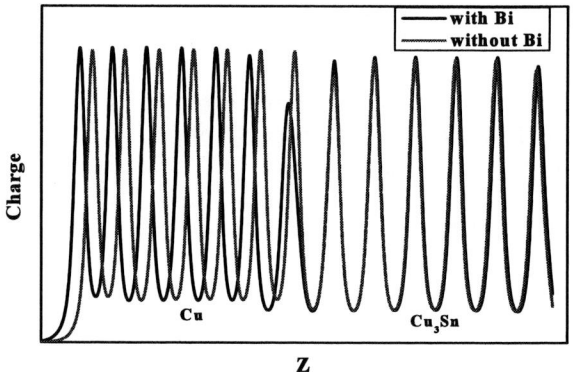

Fig.2. Planar-average charge density.

More intuitionistic pictures emerged from charge density comparison before and after Bi segregation, which are shown in Fig.3. It can be seen that after segregation, Bi atom and interface Cu atoms exhibited large displacements. The structure of interface layer changed from *fcc*-Cu to Cu$_3$Sn structure. Bi atom occupied Sn atom site in Cu$_3$Sn, and the interfacial Bi atom became a continuation part of bulk Cu$_3$Sn. Before segregation, interface Cu atoms showed strong bonding with sublayer Cu and Cu$_3$Sn. After segregation, due to large radius of Bi, interface space grew. Interface layer kept interacting with Cu$_3$Sn but apart from Cu slab, and the bonding between interface layer Cu and sublayer Cu was almost broken. From Fig.3 we can see that Bi atom almost only interacted with interface Cu but did not interact with sublayer Cu. According to planar-average charge density, we can predict that break will occur between interface layer and Cu sublayer. This can explain why Bi segregation induced embrittlement of Cu/Cu$_3$Sn interface [1].

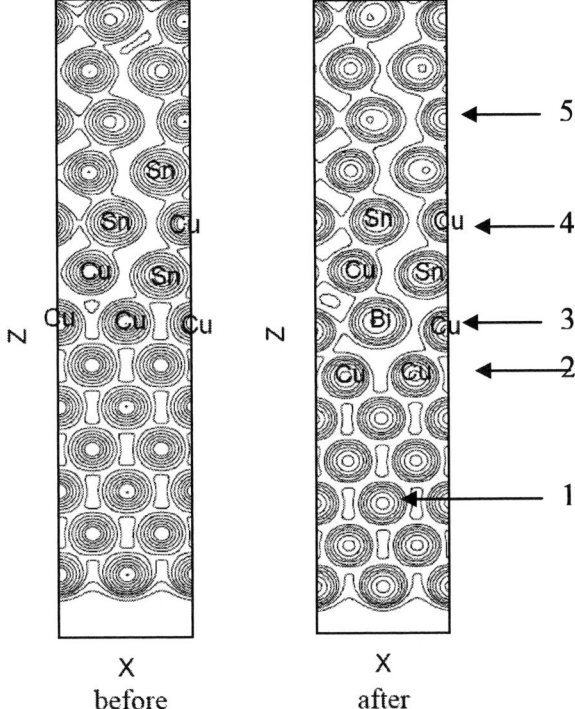

X before   X after

Fig.3. Comparison of projected charge density before and after segregation.

For further understanding of the reason, density of states (DOS) analysis was performed and results are shown in Fig.4. There are some important differences for DOS among five different Cu atoms. There is a peak at -12eV which is contributed by s-orbit of Bi (not present in bulk Cu). The interface Cu has the most intense peak, followed by Cu$_3$Sn sublayer Cu, Cu sublayer Cu and Cu$_3$Sn bulk Cu. Bi has stronger interact with Cu$_3$Sn sublayer Cu than Cu sublayer Cu, which indicates that the interface layer is closer to Cu$_3$Sn. At Cu side, Bi atom reduced Cu s-d hybridization, increased d-orbit which exhibits a closed 3d shell and most importantly an s-band is more than half full. The d-orbit of sublayer Cu atom was almost fully filled and moved to high energy exhibiting bulk properties. At Cu$_3$Sn side, Bi atom reduced Cu-Sn s-p hybridization which almost extends into bulk Cu$_3$Sn. None of these effects is good for fracture toughness.

Conclusions

We have performed the first-principles study of Bi segregated Cu/Cu$_3$Sn interface using density functional theory calculation. The energy analysis in our study indicates that Bi segregation reduced the adhesion energy of Cu/Cu$_3$Sn interface to by approximately half, which caused serious reliability problem of Bi-containing solder system.

To explore the nature of the different interfacial adhesion energies, we analyzed Cu/Cu$_3$Sn interfaces using planar-average charge density, the charge density and DOS. Bi segregation enlarged interfacial space and caused severe lost

of charge density at the interface, and the structure of interface layer changes from *fcc*-Cu to Cu$_3$Sn structure. Bi atom reduced Cu s-d hybridization, Cu-Sn s-p hybridization and increased d-orbit which exhibits a closed 3d shell and s-band is more than half full. These effects result in weakening of the interface.

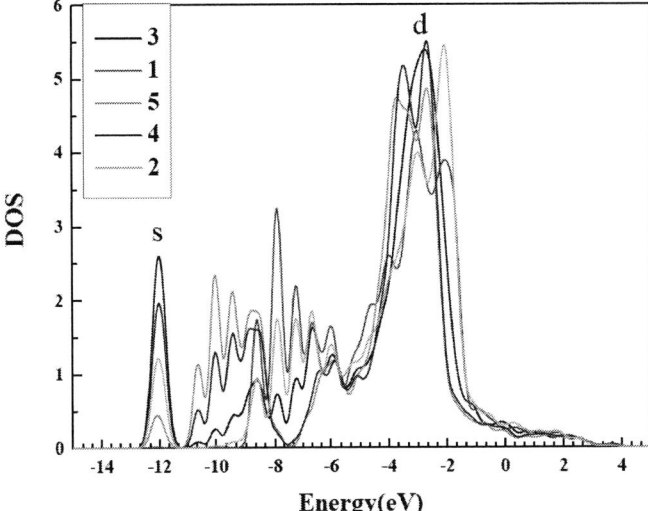

Fig.4. DOS of Cu atoms which were pointed at Fig.3

Acknowledgments

We thank Dr. Lei Zhang for helpful discussions, and gratefully acknowledge the financial support from the National Basic Research Program of China (No.2004CB61936, No.2004CB619300 and No.2006CB605103).

References

1. Liu, P. L. and Shang, J. K, "Interfacial segregation of bismuth in copper/tin-bismuth solder interconnect," Scripta Mater., Vol. 44, No.7 (2001), pp. 1019-1023.

2. Mei, Z. and Morris, J. W, "Characterization of eutectic Sn-Bi solder joints," J. Elecron. Mater., Vol. 21, No.6 (1992), pp. 599-607.

3. Vianco, P.T. *et al*, "Intermetallic compound layer growth by solid state reactions between 58Bi-42Sn solder and copper," J. Elecron. Mater., Vol. 24, No.10 (1995), pp. 1493-1505.

4. Duscher, G. *et al*, "Bismuth-induced embrittlement of copper grain boundaries," Nature Materials, Vol. 3 (2004), pp. 621-626.

5. Boettger, J. C, "Nonconvergence of surface energies obtained from thin-film calculations," Phys. Rev. B, Vol. 49, No. 23 (1994), pp. 16798-16800.

6. Monkhorst, H. J. and Pack, "Special points for Brillouin-zone integrations," J. D, Phys. Rev. B, Vol. 13, No. 12 (1976), pp. 5188-5192.

7. Finnis, M.W, "The theory of metal - ceramic interfaces," J. Phys: Condens. Matter, Vol. 8, No. 32 (1996), pp. 5811-5836.

8. Song, Y. *et al*, "First principles study of site substitution of ternary elements in NiAl," Acta Mater., Vol. 49, No. 9 (2001), pp. 1647-1654.

Effects of Bi and Ni Addition on Wettability and Melting Point of
Sn-0.3Ag-0.7Cu Low-Ag Pb-free Solder

Y. Liu, F. L Sun, T. L.Yan, W. G. Hu

School of Material Science and Engineering, Harbin University of Science and Technology,
Mailbox 317, 4# Linyuan road, Dongli Dist., Harbin, China, 150040,
Email: Yang_Liu@hrbust.edu.cn; Phone: +86-451-89051367; Fax: +86-451-86392522

Abstract

Bi and Ni were added to Sn-0.3Ag-0.7Cu low-Ag solder, to fabricate new low-Ag solders, Sn-0.3Ag-0.7Cu-XBi (X= 1.0, 3.0, 4.5) and Sn-0.3Ag-0.7Cu-XNi (X=0.05, 0.10, 0.15). Melting point tests were carried out with DSC (differential scanning calorimetry) instrument. Wettability tests were conducted on a wetting balance instrument. Test results of the two new solders were compared with that of Sn-0.3Ag-0.7Cu respectively to study the effects of the adding elements on the melting point and wettability of the low-Ag Pb-free solder. It shows that Bi addition has striking positive effects on decreasing the melting point and improving wettability. Ni addition could improve the wettability as well, although not as much as Bi does. And Ni has a negative effect on melting point. With proper adding amount as X=3.0, Bi could significantly improve the wettability and decrease melting point at the same time. However, too much Bi addition could increase the melting range between liquidus and solidus, which may lead to the initiation of solidification crack of the solder joints.

Introduction

Solder is widely used to connect chips to their package substrates in flip chip technology as well as in surface mount technology[1]. The key points of Pb-free solder development are melting point, wettability, mechanic properties (strength, toughness, creep resistance), physical chemistry character (conductivity, oxidation resistance, corrosion resistance). These characters or properties should perform as well as or better than those of Pb-eutectic solder[2].

At present, one of the most commonly used Pb-free solders is near-eutectic Sn-Ag-Cu solder alloy, which perform good wettability, low melting temperature and good comprehensive properties[3]. However, it also appears more brittle and inferior drop resistance than past used eutectic Pb-based solder. And the Sn-Ag-Cu Pb-free solder is much higher in cost than Pb-based solder due to high Ag-content [4].To improve its performance while decrease the cost, low-Ag solders has become an interest in solder alloy research. A new low-Ag solder, Sn-0.3Ag-0.7Cu, has been developed and used in some electronic products. However, in part because of slightly inferior in wettability and in part because of slightly higher melting point, some electronic companies are reluctant to adopt it into their high-end products. It's necessary to improve the performances for its extensive application in the future.

Experiments

Solder alloys used in this study were fabricated by adding Bi and Ni to Sn-0.3Ag-0.7Cu solder. Alloy contents are Sn-0.3Ag-0.7Cu-XBi (X= 1.0, 3.0, 4.5) and Sn-0.3Ag-0.7Cu-XNi (X=0.05, 0.10, 0.15).

Melting point is one of the most important basic performances, which determines the profile of soldering temperature[5]. Melting point test was conducted with Differential Scanning Calorimetry (DSC) instrument. Testing solder amount is 10mg. N_2 was used as a protective gas during DSC test. Temperature range was from 150 to 260. Referring to the Japanese industry standard JIS Z 3198-1-2003, melting temperature was obtained as the following[6]. Prolong the baseline from low temperature side to high temperature side till it meets the tangent of maximal slope point on the low temperature side of the endothermic peak. The temperature at intersection corresponds to the beginning of solder alloy melting.

Wettability tests were on the basis of wetting balance method[7]. The structure of wetting balance instrument is shown in Fig.1. Test solder is placed in a pot and heated to molten state. The test sample is fixed to the clamp over the

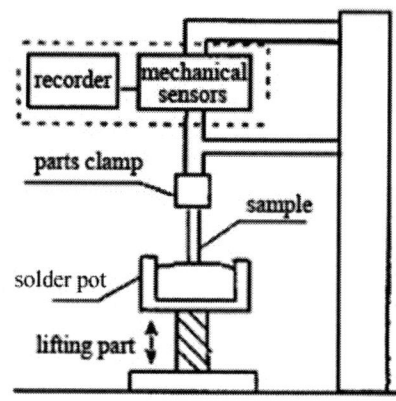

Fig.1 Sketch of testing equipment

solder pot. The equipment is adjusted to balance before starting test. Test sample gets down and immerses into the molten solder alloy at a certain speed. When the sample gets into the preset depth, it will stop and stay in the molten solder for a period of time, then lift upward. During the whole wetting process, the force between the solder and the sample piece can be recorded as a function of time by the mechanical sensor and signal translator. Fig.2 is a typical wetting profile, which is a continuous time-force profile. At the beginning of the wetting process, test sample is subjected to an uplift action. With continued wetting, the surface tension between solder and base metal will offset the uplift action. And the resultant force becomes 0 at t_0. The resultant force continues to increase and reaches $2/3F_{max}$ at t_1. The sum of t_1 and t_0 is

978-1-4244-2739-0/08/$25.00 ©2008 IEEE

called wetting time (t). When resultant force stops increasing, it will reaches the maximal wetting force (F_{max}). Wetting time and maximal wetting force are very important indexes to evaluate the wettability of a solder[8]. They indicate how fast and to what degree could a wetting process progress. So in this research, t and F_{max} of the new produced low-Ag solder are tested and studied.

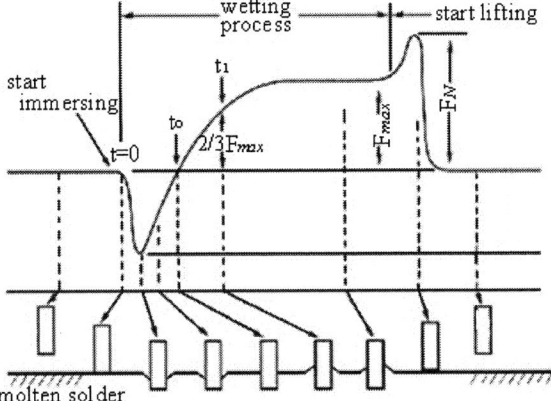

Fig.2 Typical wetting curve

Wettability test in this study was carried out on a SAT-5100 wetting balance instrument. Test sample is pure copper piece without coating in a dimension of 30mm×5mm×0.3mm. Considering the different melting temperatures, wettability tests of the two new fabricated low-Ag solders were conducted under different wetting temperatures. With lower melting temperature, wetting tests of Bi-added low-Ag solders were carried out at 240°C. EC19S-18 soldering flux was used eliminate the surface oxide and activate the interface between solder and base metal. Experimental parameters were shown in table 1

Table 1 Wetting test parameters

Item	Parameter
Immersion depth	3mm
Immersion speed	5mm/s
Staying time in molten solder	15s

Results and discussion

DSC melting endothermic profiles of Sn-0.3Ag-0.7Cu-xBi are shown in Fig.3. Melting temperature is shown in table2. It shows that Bi addition decrease the melting temperature significantly. With the increase of adding amount, melting temperature decreases continuously. Bi addition also make the small Endothermic peaks move towards low temperature side, which indicate the increasing of melting range between liquidus and solidus, which may lead to the initiation of solidification crack of the solder joints. Small Endothermic peaks near the main peak indicate the existence of low melting point eutectic.

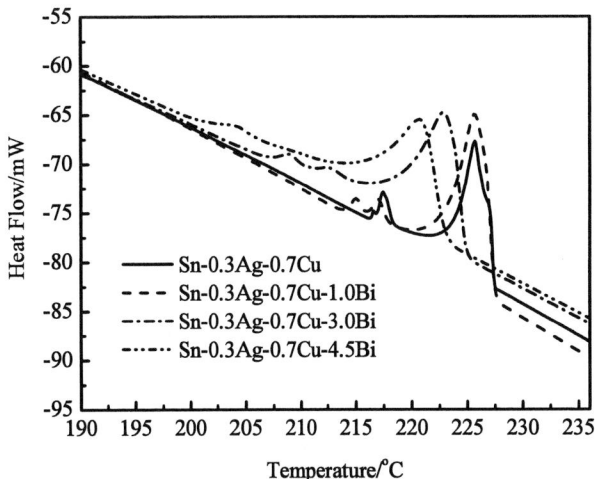

Fig.3 Sn-0.3Ag-0.7Cu-xBi DSC melting profile

Table 2 Melting temperature of Bi adding solder

Solder (wt.%)	Melting temperature(°C)
Sn-0.3Ag-0.7Cu	216.21~222.62
Sn-0.3Ag-0.7Cu-1.0Bi	213.93~220.54
Sn-0.3Ag-0.7Cu-3.0Bi	206.71~217.33
Sn-0.3Ag-0.7Cu-4.5Bi	200.11~213.09

DSC melting endothermic profile of Sn-0.3Ag-0.7Cu-xNi is shown in Fig.4. It shows that both the main peaks and small peaks move towards high temperature side. Melting temperature is shown in table3. Ni addition increases the melting temperature of the low-Ag solder. With the increase of adding amount, melting temperature increases slightly. So considering usual industrial requirements for melting temperature, Ni addition has a negative effect on soldering process.

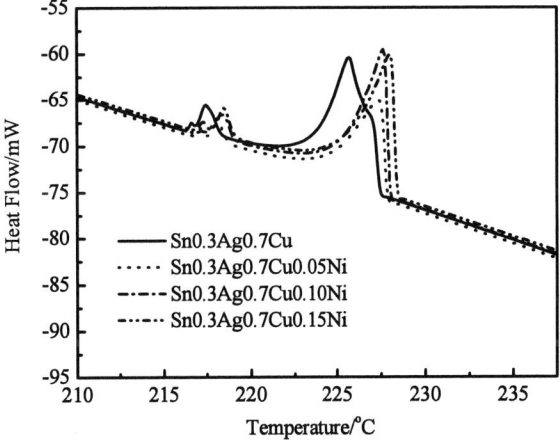

Fig.4 Sn-0.3Ag-0.7Cu-xNi DSC melting profile

Table 2 Melting temperature of Ni adding solder

Solder (wt %)	Melting temperature(°C)
Sn-0.3Ag-0.7Cu	215.31~222.64
Sn-0.3Ag-0.7Cu-0.05Ni	216.74~223.92
Sn-0.3Ag-0.7Cu-0.10Ni	217.26~224.40
Sn-0.3Ag-0.7Cu-0.15Ni	217.25~224.38

Wetting curves of Sn-0.3Ag-0.7Cu-xBi is shown in Fig.5, and wettability results is work out on the basis of wetting curve, see Fig.6. It shows that with the increasing amount of Bi addition, wetting time decreased continuously, while maximal wetting force first increased then decreased. Wetting time decrease from 4.597s to 1.721s with 62.6% drop. The maximal wetting force increase from 2.52mN to 3.21mN with 27.4% rise. So the effect of Bi addition on the wettability is really significant. Sn-0.3Ag-0.7Cu low-Ag solder with Bi addition could fairly meet the industrial production requirement, $t \leq 2.5s$, $F_{max} \geq 3.0mN$.

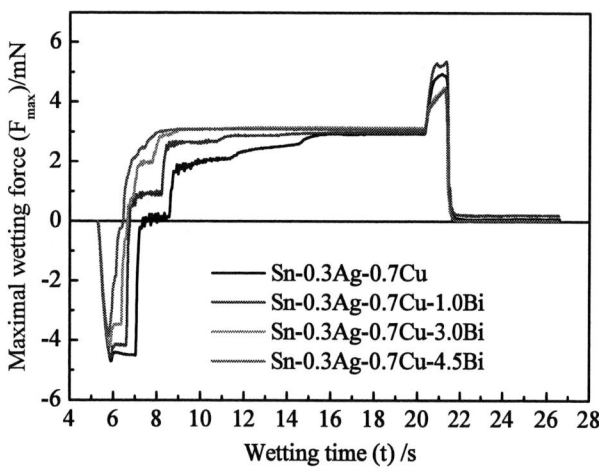

Fig.5 Wetting Curves of Sn-0.3Ag-0.7Cu-xBi at 240℃

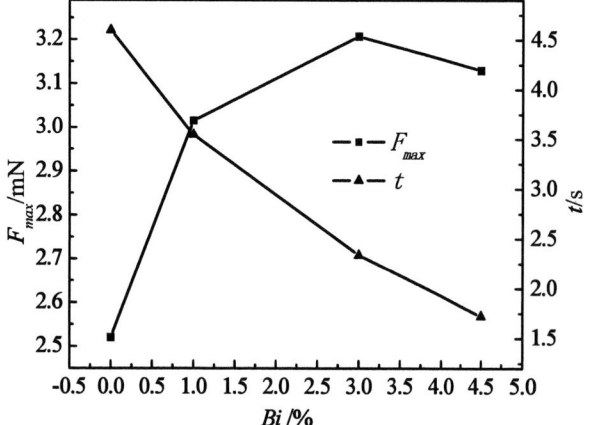

Fig.6 Wetting results of Sn-0.3Ag-0.7Cu-xBi at 240℃

Wetting curves of Sn-0.3Ag-0.7Cu-xNi is shown in Fig.7, and wettability results is work out on the basis of wetting curve, see Fig.8. It shows that Ni has relatively small effects on wettability than that of Bi. It slightly increases the maximal wetting force and keeps the wetting time almost the same.

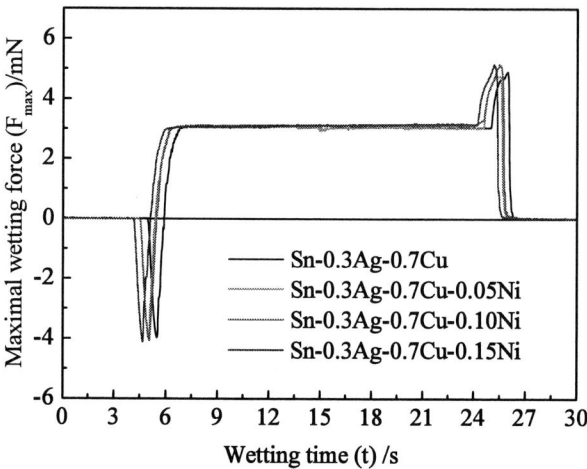

Fig.7 Wetting Curves of Sn-0.3Ag-0.7Cu-xNi at 260℃

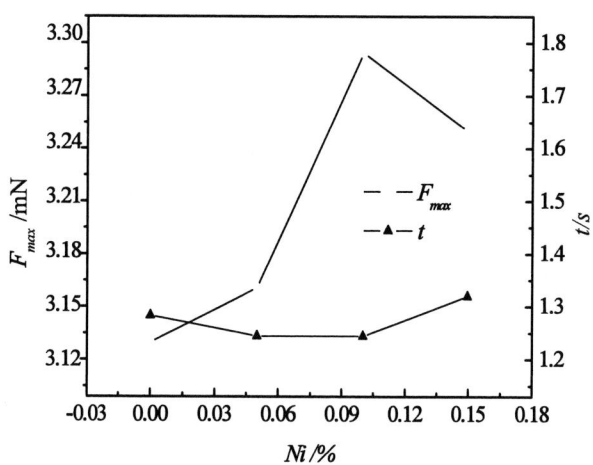

Fig.8 Wetting results of Sn-0.3Ag-0.7Cu-xNi at 260℃

Conclusions

The melting point of Sn-0.3Ag-0.7Cu low-Ag solder alloy is 222.6°C, which is higher than that of NEMI and JEI-TA recommendatory Pb-free solder. Bi addition has a positive effect on decreasing the melting point of Sn-0.3Ag-0.7Cu solder. The melting point of Sn-0.3Ag-0.7Cu-3.0Bi low-Ag solder has decreased to 217.3°C, which is almost equal to that of recommendatory lead free solders. However, Bi addition also increased the melting range between liquidus and solidus, which may lead to the initiation of solidification crack of the solder joints. Ni addition increased the melting point slightly. Ni addition has a negative effect on soldering process.

Bi addition has a striking positive effect on wettablility of Sn-0.3Ag-0.7Cu low-Ag solder. It could not only decrease the wetting time, but also increase the maximal wetting force. With proper adding amount, Bi could improve solder wettability significantly. Among solder compositions in this study, Sn-0.3Ag-0.7Cu-3.0Bi showed best wettability in accordance with the comprehensive consideration of both wetting time and maximal wetting force. With the increasing amount of Ni addition, the maximal wetting force increased and the wetting time remained the same. So, to some extend, Ni addition could also improve wettability of the low-Ag solder, although the improvement is not as significant as that of Bi addition caused.

Future Work

Some tests would be conducted further to investigate the effects of Bi and Ni elements addition on Sn-0.3Ag-0.7Cu low-Ag Pb-free solder joints microstructure and performance such as shock-resistance or aging-resistance under different service conditions.

Acknowledgments

The financial support for this work from the Natural Science Foundation of China (No. 50575060) is gratefully acknowledged.

References

1. Zeng K, Tu K N. "Six cases of reliability study of Pb-free solder joints in electronic packaging technology," *Mater Sci Eng Rep*, No. 38 (2002)，pp. 55-105.

2. Abtew M , Selvadury G. "Lead-free solders in microelectronics," *Materials Science and Engineering* , 2000 ,Vol. 27, No.1 (2000), pp. 95-100.

3. Kisiel R. "Lead-free technologies for electronic equiptment assembly," *Proceedings of SPIE — The Intern-ational Society for Optical Engineering*, 5125 (2002), pp. 348-352.

4. Wang Chunqing, Li Mingyu, Tian Yanhong, Kong Lingchao. "Review of JISZ 3198: Test Method for Lead-free Solders," *Electronics Process Technology,* Vol. 25, No. 2, pp. 25-32

5. Ma Xin, He Peng. lead-free Soldering Technology in Electronic Assembly, Harbin Institute of Technology Press.(Harbin: 2006), pp. 218-231

6. P.M. Sargent, A.C.T.Tang and F.H.Gordon. "An Experimental Study of the Variation of Wettability of SMDs Using the Micro-global Wetting Method." *IEEE/CHMT IEMT Symposium,* (1991), pp. 166-170

7. Zhang Deyun, Zhuang Hongshou. Brazing and soldering manual, Mechanical Industry Press, (Beijing1998): 1-24

8. Liu Dan, The Improvement of the Solder Paste Performance and the Effect of Composition on It. Harbin Institute of Technology.(2006), pp. 25-31

9. DU Chang-hua; CHEN Fang; DU Yun-fei "Investigation for Solderability of Sn-Cu, Sn-Ag-Cu Lead-free Solders," Electronic Components & Materials, No.11, (2004), pp. 33-38

Moisture Diffusion Model Verification of Packaging Materials

Xiaosong Ma[1], K.M.B. Jansen[1], L.J. Ernst[1], W.D van Driel[3], O. van der Sluis[1,2], G.Q.Zhang[1,3]
Charles Regard[4,5,6], Christian Gautier[4,5], Hélène Frémont[6]
[1] Delft University of Technology, Mekelweg 2, 2628 CD Delft, the Netherlands
[2] Philips Applied Technologies, High Tech Campus 7, 5656 AE Eindhoven, the Netherlands
[3] NXP, Gerstweg 2, 6534 AE Nijmegen, the Netherlands
[4] NXP Semiconductors, [5] LaMIPS, Université de Caen, 2, rue de la Girafe, 14000 Caen, France
[6] IMS Bordeaux Université de Bordeaux, 351 cours de la libération, 33405 Talence, France
Phone: +31-(0)15-2782859, Fax: 31-(0)15-2782150 e-mail:X.Ma@tudelft.nl

Abstract

The use of the non-hermetic material for electronic packaging does raise a potential concern, i.e. moisture induced interfacial delamination and pop corning during reflow. Therefore, it is very important we can correctly model the moisture absorption property. In this study, moisture absorption and desorption properties of three kinds of package materials were investigated. Moisture absorption equilibrium weight gain and diffusion coefficient at different temperature and different humidity are characterized. Moisture absorption processes are simulated using a 3D model at conditions according to the moisture sensitivity test levels. Finally moisture absorption is verified by our research carrier.

1.Introduction

Thermosetting polymers are widely used as encapsulants in electronic packaging for their outstanding properties. The mechanical properties of polymers are, however, strongly influenced by various environmental factors. Moisture penetrating into the polymer induces expansion which results in reliability failure, such as stress cracking[1,2], delamination and pop-corning during reflow [3]. Such moisture and temperature induced failure has long been recognized as an important issue for package reliability but there is often a lack of reliable material data. Therefore it is very important to characterize and model the moisture uptake and its effect on the mechanical properties of packaging materials used in the electronic industry. In this paper, three different packaging polymers are studied. The water absorption rate is initially proportional to the square root of time and gradually reaches an equilibrium level. The diffusion coefficients and saturation levels at different conditions were obtained by fitting the equation to the full series of the temperature and humidity steps. And then the moisture absorption equilibrium and diffusion coefficient are determined by the experimental data.

In order to check the accuracy of material moisture models, package moisture absorption tests are done at NXP Nijmegen and Caen. The tests are conducted at different temperature and relative humidity according to the moisture sensitivity test levels. A 3D simulation model is established using MSC.Marc software. Moisture

2.Moisture diffusion test and modeling

2.1 moisture diffusion tests samples

In order to know moisture diffusion properties of our package three kinds of packaging materials' moisture properties are studied. They are molding compound, underfill and die attach. The molding compound is from NXP Nijmegen. It contains about 90% silica and the average filler size is about 15μm in diameter. The material is post-cured at 180°C for 3 hours. Underfill is a commercially available underfill from Namics. It is a high reliability underfill compared to other underfills. Filler content is 55%. The sample is cured at 150°C for 20 min in the oven by using Teflon foil between two glass plates. The die attach material contains about 90% silver particles and the average size of the silver particle is about 5μm in diameter. The sample is cured at 180°C for 15 minutes. To shorten the moisture equilibrium time the samples should be as thin as possible. Therefore the samples are grinded and polished to a thickness of about 0.4mm. For the pure epoxy sample can be polished to the thickness about 0.1mm but for the silica or sillver filled epoxy they just can be polished to the thickness about 0.2 to 0.4 mm because of the filler size.

2.2 Equipment of Moisture Absorption Test

A programmable SGA-100 temperature and relative humidity chamber is used for moisture absorption tests. It can perform two kinds of moisture absorption tests. One is isothermal moisture absorption test and another is temperature and humidity cycle absorption test, which means that you can change the temperature and relative humidity at the same time. It can automatically graph the sample weight change with the time, temperature and relative humidity. And the diffusion coefficient can be obtained by using the software provided by the supplier. Temperature range is from 5 to 90°C and relative humidity range is from 5 to 95 RH%. The weight sensitivity of the equipment is 1μg. If high temperature and high relative humidity are needed at the same time, an extra heater is needed.

2.3 Moisture Absorption Theory

There are many parameters that govern the moisture absorption process. The moisture content gain, M_t was determined by:

$$M_t (wt\%) = \frac{m_t - m_0}{m_0} \times 100\% \qquad (1)$$

where m_t and m_0 are the weights of wet specimen at exposure time, t, and of the dried specimen respectively.

978-1-4244-2739-0/08/$25.00 ©2008 IEEE

M_{sat} is saturated moisture gain in the sample and it is dependent on the temperature. It can be determined by[4]:

$$M_{sat} = k_0 \cdot X_{RH} \cdot e^{(-H/RT)} \qquad (2)$$

where k_0 is the proportional constant, X_{RH} is the relative humidity factor ranging from 0 to 1, i.e. 0RH% to 100RH%. H, is activation energy constant, R is the universal gas constant, 8.314JK^{-1}mol^{-1}, T is the absolute temperature. M_{sat} can be obtained by the moisture absorption test at different temperature and different relative humidity. A linear fit of $\ln M_{sat}$ versus the reciprocal of the absolute temperature $1/T$ is used to determine the activation energy constant. From the intercept with $\ln M_{sat}$-axis, at $1/T = 0$, k_0 can be determined.

The diffusion constant D is supposed to also follow the Arrhenius temperature dependency and is assumed to be independent of the relative humidity

$$D = D_0 \cdot e^{(-E_D/RT)} \qquad (3)$$

It can be rewritten in the following form:

$$\ln D = \ln D_0 - E_D/RT \qquad (4)$$

where D_0 is the constant pre-exponential coefficient; R, the universal gas constant; T, the absolute temperature. From the slope of the data plot of $\ln D$ against $1/T$, E_D/R can be determined. From the intercept with the $\ln D$-axis, at $1/T = 0$, $\ln D_0$ can be determined.

2.4 Moisture Diffusion Tests and Modelling

Here the experimental procedure is described briefly. All samples are pre-dried at 125ºC for 24 hours before they are put in the SGA-100. The temperature is set at 20ºC, 40ºC, 60ºC respectively and a sequence of relative humidity is programmed (0%, 80%, 5%, 60%, 5%, 40%, 5%RH). At every relative humidity, the dwelling time is from 200 to 600 minutes. The time is dependent on the thickness of the sample and the temperature. Generally moisture absorption time is short if the sample is thin in thickness and the temperature is high, references [6, 7] give experimental procedures in detail.

Diffusion coefficients are obtained at each temperature respectively. For M_{sat} three date points are obtained at each temperature and different relative humilities. Using equations (2), (3) and (4) the diffusion coefficient and M_{sat} models are established. Table 1 is model parameters for three kinds of packaging materials in the carrier.

Table 1 Parameters of packaging material models

parameters	MC	Underfill	Die attach
k_0 [-]	10.82	38.43	1.087
H [kJ/mol]	8.63	10.72	3.46
D_0 [mm²/s]	1.2E-09	1.054E-09	7.0E-06
E_d [kJ/mol]	15.76	16.13	37.16

Using these models diffusion coefficients at different temperatures and M_{sat} at different temperatures and different relative humilities can be calculated. In the moisture

absorption and desorption simulation all material diffusion properties are calculated from these models.

3. Moisture diffusion simulations

3.1 Moisture diffusion model

A 3D carrier finite element model is constructed by using MSC.Marc software. In modeling, the transient moisture diffusion equation is analogous, see table 2, to the transient heat conduction equation, and can be described as

$$\frac{\partial C}{\partial t} = D \cdot \left(\frac{\partial^2 C}{\partial x^2} + \frac{\partial^2 C}{\partial y^2} + \frac{\partial^2 C}{\partial z^2} \right)$$

where C is the moisture concentration, x, y, z are spatial coordinates, D is the diffusion coefficient which measures the rate of diffusion and t is the time.

Table 2 FE thermal moisture analogy

Properties	Thermal	Moisture
Field variable	Temperature, T	Wetness, W
Density	ρ (kg/m³)	1
Conductivity	k(W/mºC)	D*C$_{sat}$ (mg/s mm)
Specific capacity	c_p	C$_{sat}$ (mg/mm³)

However, unlike temperature, the moisture concentration is discontinuous along the bi-material interface. Therefore, moisture wetness, w, is introduced as the field variable as it is continuous across the multi-material interfaces. It is defined as

$$w = \frac{C}{C_{sat}}$$

where C_{sat} is the maximum moisture concentration that can be absorbed by the material, the lower limit $w = 0$ means it is dry, and upper limit $w = 1$ means it is fully saturated with moisture. By using the wetness approach, the moisture diffusion problem can now be solved as a typical heat transfer problem. For moisture absorption modeling, the initial condition is $w = 0$ for the whole package, and boundary condition is $w = 1$ at the external surface which are exposed to the ambient moisture. For moisture desorption boundary condition $w = 0$ at external surfaces which are exposed to the dry ambient.

Two 3D half finite element models are as follow, one part is used as moisture diffusion model and the other part is used as hygro-thermal mechanical modeling later.

Using the wetness approach, the moisture diffusion implementation in commercial FE software becomes straight forward with the help of appropriate user subroutines.

3.2 Moisture absorption at different conditions

In order to check the accuracy of our material moisture diffusion model and compared to measurement, moisture diffusion simulations are carried out at following conditions, such as 85ºC/85RH, 85ºC/60RH, 60ºC/60RH and 30ºC/60RH. Here only some simulation results are given and others just given the final results. Fig. 2 is weight change rate at 85ºC/85%RH for 168 hours. Form this graph it shows that moisture almost near saturation in about 48 hours. The final moisture weight change rate is about 0.0945% for SGA-100 moisture absorption dates. This weight change rate result is fit well with the results from Nijmegen.

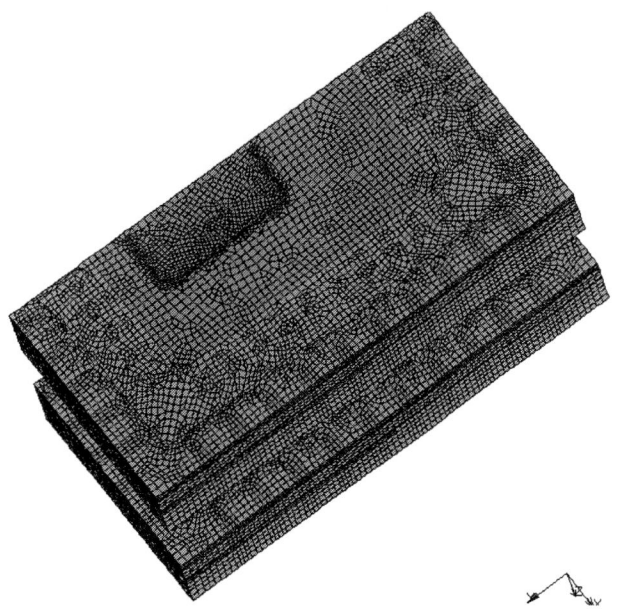

Fig.1 3D moisture absorption FE model

Table 3 is material moisture absorption properties at 85°C/85RH used in the simulation.

Table 3 Material moisture properties at 85°C/85RH

material	D(mm²/s)	C_{sat}(mg/mm³)
MC	6.04e-6	3.765e-3
Underfill	4.67e-6	13.65e-3
Die attach	26.47e-6	11.40e-3

From Table 3 we know that molding compound has the lowest moisture saturation level but underfill has the highest moisture saturation level. It is reasonable because the filler contents of molding compound is 90% and filler contents of underfill is about 55%. Most of moisture exists in the epoxy not in the filler.

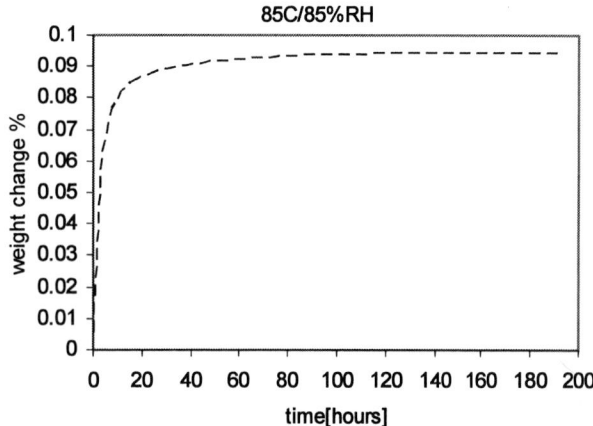

Fig. 2 Weight change with time at 85°C/85RH

Table 4 is material moisture absorption properties at 60°C/60%RH used in the simulation.

Table 4 Material moisture properties at 60°C/60%RH

material	D(mm²/s)	C_{sat}(mg/mm³)
MC	2.31e-6	1.374e-3
Underfill	3.11e-6	7.355e-3
Die attach	10.37e-6	7.38e-3

Fig. 3 is the moisture diffusion simulation result using dates in Table 4

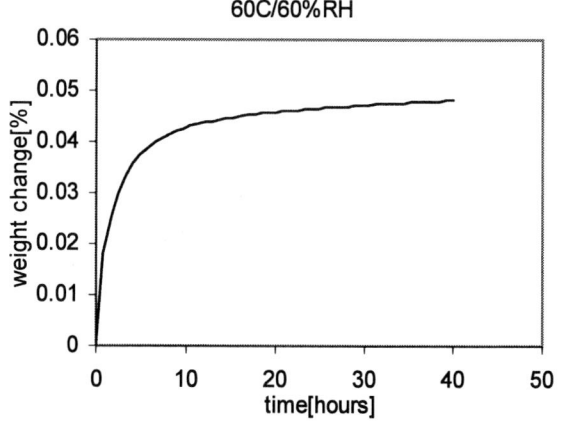

Fig 3 Moisture diffusion simulation at 60°C/60%RH

From this simulation we know that moisture absorption occurs quickly at first then slow down and finally reach equilibrium in very longer time. Moisture absorption simulation at 85°C/60RH and 30°C/60RH weight changes are 0.069% and 0.035% respectively.

4. Package moisture diffusion experimentation

In order to check our material diffusion models two separately moisture diffusion tests are done at NXP Nijmegen, Netherlands and NXP Caen, France.

4.1 Package moisture diffusion experiment in NXP Nijmegen, Netherlands

Moisture absorption tests are done according to the moisture sensitive level standard, MSL. Moisture absorption at NXP Nijmegen only provided the final weight change rate but no more weight change dates with time. All samples are pre dry at 125 °C for 24 hours before they are put into humidity oven. Table 5 is the weight change rate at different pre conditions.

From Table 5 we know that most of the test results are reasonable according to the moisture absorption mechanism. High temperature and high relative humidity results in high weight change. However due to the low weight of sample combined with the accuracy of the balance there are some difference in the table 4, bold and italic dates.

If simulation results are compared with the test results from NXP, generally speaking they fit well. For example at 85°C/85RH, simulation weight change rate is 0.0945% and 3 tests results weight change rate is between 0.08 to 0.089%

Table 5 Package weight change at NXP Nijmegen

MSL	Test	wt%	average
1 85°C/85%RH 168 hours	control lot	0.088%	0.113%
	lot 1	0.089%	
	lot 2	*0.194%*	
	lot 3	0.080%	
2 85°C/60%RH 168 hours	control lot	0.053%	0.051%
	lot 1	*0.071%*	
	lot 2	0.053%	
	lot 3	*0.027%*	
3 30°C/60%RH 192 hours	control lot	*0.080%*	0.053%
	lot 1	*0.061%*	
	lot 2	0.036%	
	lot 3	0.035%	
3acc 60°C/60%RH 40 hours	control lot	*0.045%*	0.031%
	lot 1	*0.018%*	
	lot 2	0.026%	
	lot 3	0.036%	

only one is 0.194%. At 60°C/60%RH, simulation weight change rate is 0.048% and the test results weight change rate is between 0.018 to 0.045% only one is 0.045%. NXP explains the deviation is from low weight of samples combined with the accuracy of balance.

4.2 Package moisture diffusion experiment at NXP Caen, France

Moisture diffusion tests at NXP Caen, France are specially done for checking our material diffusion models.

4.2.1 Equipments

Following equipments are used for experiments. Hot dry oven, Heraeus UT6120 oven is used for drying samples. Temperature range can be regulated between 10°C to 300 °C. Humidity oven, Sapratin 320H30, is used for moisture absorption. It can regulate relative humidity from 10% to 98% and temperature from 10 °C to 90 °C. Balance, Ohaus Adventurer SL Pro As64, is used to perform weight measurement with a 0.1mg precision.

4.2.2 Experimental procedure

All samples are pre dried at 125 °C for 24 hours in the hot dry oven and initial weights are measured after the pre dry. Then samples are put into the humidity oven for moisture absorption at different conditions. Samples are measured in groups because of the low weight of the sample. Samples weights are measured by balance frequently at beginning stage and longer time after moisture absorption weight change reached certain level.

4.2.3 Test results

Following condition moisture absorptions are done, at the condition of 85°C/85%RH and 60°C/60%RH. At 85°C/85%RH, 50 components are measured at the same time in order to have high weight. At 60°C/60%RH, 50 and 100 components are measured at the same time respectively, See Table 6 test results.

Table 6 Package weight change at NXP Caen

condition	wt% Test 1	wt% Test 2	wt% Average
85°C/85%RH	0.0661%		
60°C/60%RH	0.0599%	0.049%	0.0545%

5. Comparison between tests and simulations

Moisture absorption between NXP Nijmegen tests results and simulations are compared in section 3.1 but only final results are shown and compared. Here moisture weight change with time is compared between NXP Caen tests results and simulations.

Fig 4 is test and simulation results comparison at 85°C/85%.

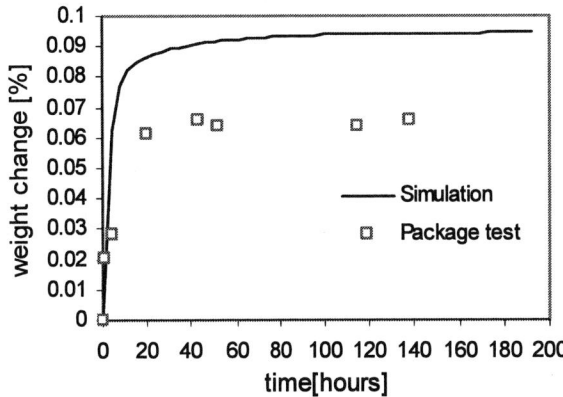

Fig. 4 Test and simulation comparison at 85°C/85%RH

In Fig. 4 full curve is the simulation result and the squares are test results at different time. We know that the difference is about 0.03%. Real package moisture absorption test result is lower than the simulation result but the deviation is not very big.

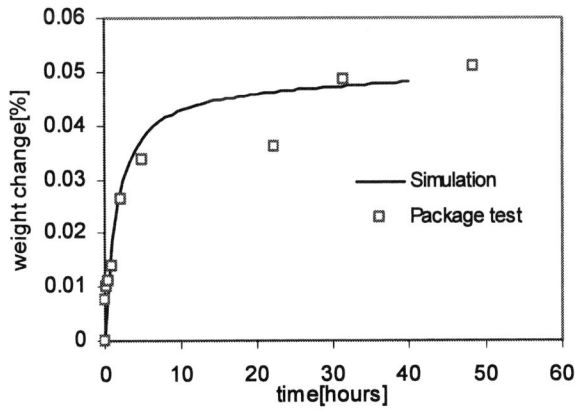

Fig. 5 Test and simulation results comparison at 60°C/60%

In Fig.5 full curve is the simulation result and squares are tests results at different time. Simulation and test results are fit quite well.

From above two tests and simulation comparison we know that experiments and simulation fit reasonably well and can be used for the future simulation. The differences at 85°C/85%RH are mainly from low weight of the samples combined with the accuracy of balance. In order to increase the weight of the samples 50 or 100 components are measured at the same time.

6. Moisture Desorption Simulation and Tests

Before moisture sensitivity level qualification tests all sample are pre dried at 125°C for 24 hours. Our material diffusion models are establish from 20 to 60°C and if they are suitable for desorption are very useful for desorption simulation. Therefore simulation and two desorption are conducted for this purpose.

Two separate moisture desorption tests are done after moisture absorption at 85°C/85%RH for 168 hours. Test 1 and test 2 have 50 components measured at the same time in order to increase the weight of the samples measured.

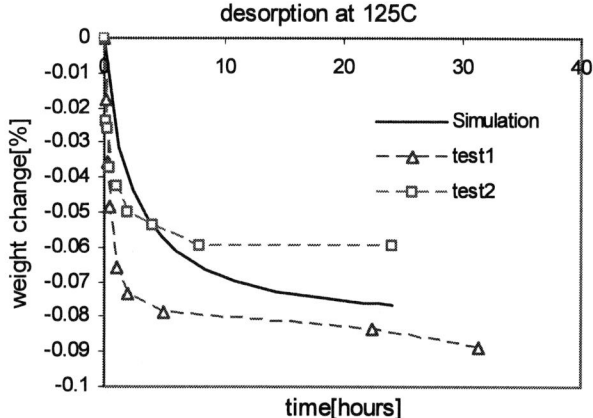

Fig. 6 Moisture desorption comparison

In Fig 6 the full curve is the desorption simulation result and the square and triangle doted curves are two test results. From Fig. 6 it can be seen that the desorption differences between the two tests is about 0.02%. The simulation result is between the two desorption tests. This means that our material moisture models can be used in desorption at about 125°C.

7. Conclusions

From the tests and simulation in this study we find the above method is suitable for modeling moisture absorption and desorption. Our material moisture properties' models are good enough for the simulation after comparison with the different moisture absorption test from NXP Nijmegen and NXP Caen. Desorption simulation and tests results are fit well too. This means that our model can be used for the moisture desorption at about 125°C. There is deviation between the simulation and tests. These deviations mainly from the low weight of the samples combined with the accuracy of the balance in the verification experiments. Our SGA-100 has a accuracy of 1μg but most manfully operated balance has a accuracy of 0.1mg. Another reason maybe the results of delamination in the component such that there is more moisture in the sample delamination interface. Howerver not all weight change are higher than our simulation. More investigations are need to explain these.

References

1. Gils M.A.J.van, Haberts P.J.J.H.A, Zhang G.Q., Driel W.D.van., Schreurs P.J.G. Characterization and modelling of moisture driven interface failures. Microelectronics reliability 2004;44:1317-1322.
2. Tee T.Y., Zhong Zhaowei Integrated vapour pressure, Hygroswelling, and thermo-mechanical stress modelling of QFN package during reflow with interface fracture mechanics analysis. Microelectronics reliability 2004;44:105-114.
3. Lee K.C., Vythilingam A., and Alpern P. A simple moisture diffusion model for the prediction of optimal baking schedules for plastic SMD packages Microelectronics reliability 2005;45:1688-1671.
4. Merdas I., Thominette F., Tcharkhtchi A., Verdu J. Factors governing water absorption by composite matrices Composites Sicience and Technology 2002;62:487-492.
5. Nogueira P. et. al Effect of water sorption on the structure and mechanical properties of an epoxy resin system. J. of Applied Polymer Science 2001;80:71-80.
6. Xiaosong Ma, K.M.B.Jansen, L.J.Ernst, W.D.van Driel, O.van.der.Sluis, and G.Q.Zhang, "Characterization and Modeling of Moisture Absorption of Underfill for IC Packaging"IEEE Proc 8th ICEPT2007, Sahnghai,China, Aug.2007, pp. 380-384.
7. Xiaosong Ma, K.M.B.Jansen, L.J.Ernst, W.D.van Driel, O.van.der.Sluis, and G.Q.Zhang, " Characterization of moisture properties of plolymers for IC packaging," Microelectrics Reliability, Vol. 43(2007), pp. 1695-1699.

Recent Progress of Carbon Nanotubes as Cooling Fins in Electronic Packaging

Johan Liu[1,2], Yifeng Fu[2,3,4] and Teng Wang[1]

[1]SMIT Center & Bionano Systems Laboratory, Department of Microtechnology and Nanoscience, Chalmers University of Technology, SE-412 96 Göteborg, Sweden;

[2]Key Laboratory of Advanced Display and System Applications and SMIT Center, School of Automation and Mechanical Engineering, Box 282, No. 149 Yan Chang Road, Yan Chang Campus, Shanghai University, Shanghai 200072, China

[3]SHT Smart High Tech AB, Nordgårdsvägen 19, 428 34 Kållered, Sweden

[4]FOAB Elektronik AB, S:t Jörgens väg 8, 422 49 Hisingsbacka, Sweden

Corresponding author: jliu@chalmers.se

Abstract

As power density in electronic system is approaching a level that conventional cooling methods can't handle, reliability of microsystems is of large concern. In this paper, novel schemes for heat dissipation on electrical components have been reviewed with a strong focus on using Carbon Nanotubes (CNTs) as cooling fins as a basic approach. With an extraordinary high thermal conductivity, CNT is considered as an ideal material for thermal management in high heat flux microsystems. Fabricated onto a silicon substrate to form microchannels, the CNT based cooling fins show high heat dissipation efficiency. This paper reviews the ongoing research on CNT based microchannel cooler development carried out so far. Both experimental and simulation results are included and presented.

Introduction

The electronic industry has been growing at an amazing rate for nearly half a century. Since Gordon Moore proposed that transistors on chips would double every two years [1], this has become a law which witnessed the fast development of electronics. Nowadays hundreds of thousands of components are integrated on a chip at a millimeter scale. When chip size decreases with the increasing components, the power density is also increased dramatically. In this situation, thermal issues attract big concern in electronic packaging field. According to the estimation of the International Technology Roadmap for Semiconductor 2004 (ITRS2004) Update [2], heat generated in "cost-performance" and "high-performance" single-chip packaging will reach 1.08W/mm^2 and 0.64W/mm^2, respectively. Such high heat flux threatens heavily on the reliability of components as well as microsystems. However, conventional natural or forced air flow can't meet requirements any longer. In order to make devices operate at endurable temperature, researchers are mainly working on the novel cooling schemes: heat pipes, high thermal conductivity thermal interface materials (TIMs) and high efficient heat sinks.

Micro heat pipes can be fabricated in silicon wafer [3-6] and remove big amount of power by being mounted on the electrical devices with a direct or indirect contact. Power density as high as 200W/cm^2 has been demonstrated to be dissipated in microsystems by micro heat pipes [7]. By optimized design and choosing proper working fluid, heat removal effect of heat pipes can be further improved.

Interface contact resistance contributes a large part of the thermal resistance in a whole electrical system; therefore, developing high thermal conductivity interface materials to reduce the contact thermal resistance is another promising solution to the thermal issues currently. Lots of research has been done in this field, and the best available commercial TIMs normally have a thermal conductivity of about $4 \text{W·m}^{-1}\text{·K}^{-1}$. Recently, researchers are adding conductive particles into matrix to make new TIMs [8-13], among which CNTs are the most commonly used fillers due to their high thermal conductivity [14, 15].

Heat sink plays a very important role in the thermal management of electrical systems. It functions by efficiently transferring thermal energy from a high temperature component to a lower temperature object with much bigger heat capacity and dissipation area. So researchers are making great effort to find high heat capacity materials for heat sink and enlarge the heat exchange area. Conventional heat sinks are made of metal (Aluminum and Copper are most commonly used) materials as fins and normally work with the assistance of a fan to enhance the heat convection. In 1981, D. B. Tuckerman and R. F. W. Pease proposed a high-performance forced liquid cooling method [16]. They fabricated microscopic channels in silicon substrate, through which liquid coolant would flow. Since the heat exchange area per unit volume was so great, the heat dissipation effect was dramatically improved to be able to deal with a power density as high as 790W/cm^2.

In D. B. Tuckerman and R. F. W. Pease's work, silicon was used as fin material. As the outstanding thermal properties, people are trying to utilize CNTs as cooling fins in micromachined coolers. This paper will focus on the recent progress of CNTs as cooling fins in microcoolers.

Carbon Nanotube

Carbon nanotubes (CNTs) were discovered in 1991 [17]. They are allotropes of carbon with a nanostructure that can have a length-to-diameter ratio greater than 1000000.

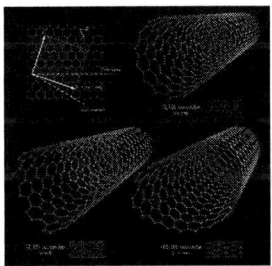

Fig. 1 Structure of single-walled carbon nanotubes

Figure 1 shows the structure of single-walled carbon nanotubes (SWNTs). CNTs made from cylindrical carbon molecules are very special in thermal, electrical and mechanical. A thermal conductivity up to 6600W/m·K has been reported [18]. As a new material, CNTs are attracting more and more attention, and they are very potentially useful in nanotechnology, electronics, optics and aeronautics.

CNTs can be manufactured by Chemical Vapor Deposition (CVD) at suitable temperature. During the CVD process, process gas (such as ammonia, nitrogen, hydrogen, etc.) and carbon-containing gas (such as acetylene, ethylene, ethanol, methane, etc.) are demanded as reactors, and catalyst particles such as nickel, cobalt and iron are also needed.

Microchannel coolers

Nearly all of the microchannel coolers are designed to be two parts: a substrate with channels fabricated and a cover made of metal. Adhesives are used between the interface of the substrate and cover to prevent fluid leakage. Since the first microchannel cooler was made by D. B. Tuckerman and R. F. W. Pease in 1981, as shown in Figure 2, a lot of research has been presented on microchannel flow, microchannel heat transfer and microchannel heat sinks in electronics. Even two-layered bi-directional microchannel heat sink was designed [19]. In 2001, Choondal B. Sobhan and Suresh V. Garimella completed a comparative analysis on heat transfer and fluid flow in microchannels [20]. With the development in the field of micro machine, more complicated structure such as zigzagged microchannels is realizable [21, 22]. Because most of these microchannels were formed by making grooves in silicon substrate, silicon was used as fin material.

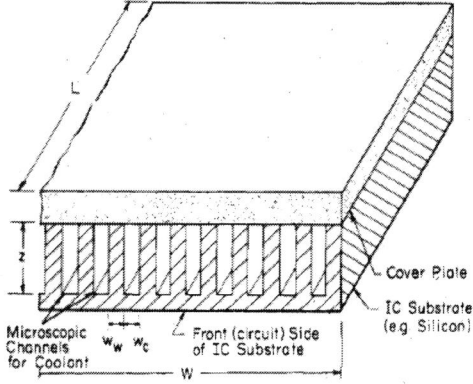

Fig. 2 Schematic view of the microchannel heat sink made by D. B. Tuckerman and R. F. W. Pease.

CNT based microchannel cooler

CNT into cooling systems as fin material was introduced by Liu and his research group in 2004 [23, 24]. By combination of the high thermal conductivity material with the high heat transfer efficiency structure, it is a tempting and promising scheme for thermal issues in electronics.

In [23, 24], a bare silicon chip was chosen as the substrate and. After CVD synthesis, CNTs array was grown from the catalyst as cooling fins. Finally, a lid was bonded to seal the CNTs and form the microchannels. Figure 3 shows the flow chart. Figure 4 and Figure 5 show the SEM pictures of the one-dimensional and two-dimensional CNTs arrays respectively.

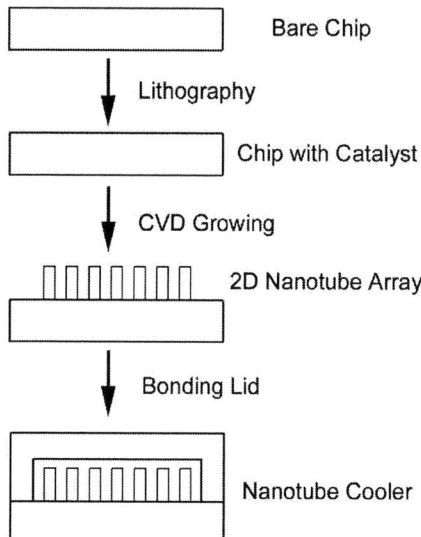

Fig. 3 Manufacturing procedure of CNT based microchannel cooler

Fig. 4 SEM picture of one dimensional CNT fins on silicon chip

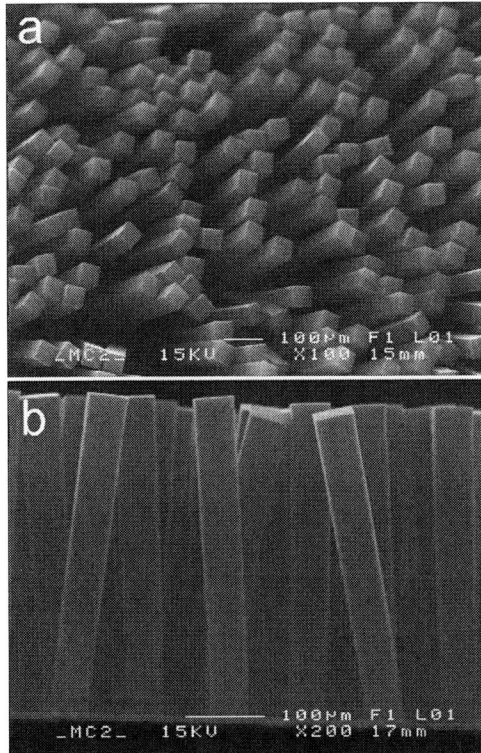

Fig. 5 SEM picture of two dimensional CNT fins on silicon chip. (a) Top view; (b) Side view

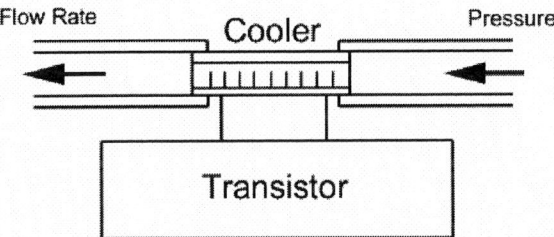

Fig. 6 Cooling experiment setup of the CNT-based microchannel cooler

Tab. 1 Characterization result of the CNT-based microchannel cooler

Cooler with Fins		Cooler without Fins	
RT = 21 °C		RT = 20 °C	
Flow Rate = 6.1 ml/s		Flow Rate = 7.0 ml/s	
P_{input}	$T_{transistor}$	P_{input}	$T_{transistor}$
8.9 w	48 °C	7.2 w	54 °C
15.6 w	72 °C	11.3 w	76 °C
25.7 w	108 °C	19.6 w	119 °C

Afterwards, experiments were performed to test the CNT-based microchannel coolers, and it was also compared with the cooler without CNTs array as cooling fins. The experiment setup was shown as Figure 6, and experiments result was shown as Table 1. Although the CNTs array decreased the flow rate of the fluid by 12%, the CNT-based microchannel cooler showed much higher cooling capacity. With 23% higher input power, the CNT-based microchannel cooler kept the transistor temperature 6°C lower than the reference cooler (the cooler without CNTs fins).

Ekstrand et al and Wang et al further developed Mo's research by simulation and experimental method respectively [25, 26]. Ekstrand et al simulated the heat transfer in the microchannel cooler by FEMLAB when water was used as coolant, then she compared the CNT-based cooler with the reference cooler. Results showed that thermal resistance of the microchannel cooler was reduced from 0.99K/W to 0.43K/W when CNT fins were introduced, although pressure drop between inlet and outlet increased. On the other hand, due to the gap between the bottom of the lid and the top surface of the CNTs array, flow velocity of the water at the bottom of the channels was significantly reduced, which led to a great reduction of heat exchange between water and CNTs. Figure 7 shows the real CNTs array and the geometry in modeling. Wang fabricated three kinds of microcoolers: CNT-based microchannel cooler, silicon-based microchannel cooler and microcooler without any fin. In his experiments, the power transferred from the heater to the cooler cannot be measured precisely because there was an unknown portion of heat dissipated by natural convection. In addition, there was a considerable thermal resistance existing between the heat resistor and the cooler, which made an accurate calculation more difficult. However, because experimental conditions were well controlled, the research was meaningful to reveal the advantages and drawbacks of the CNT-based microchannel cooler. Wang et al's experimental results were shown in Table 2, which indicated that the CNT-based microcooler was about 10-15% better than silicon-based cooler. To cool the resistor to the same temperature (42°C), CNT-based cooler needed a lower water flow rate (35.7ml/min vs. 38.5ml/min) under a higher power (2.13W vs. 1.95W) loaded on heat resistor. Additionally, the bottom plate of the CNT-based cooler was much thicker than that of the silicon-based cooler because CNTs were grown on the plate while silicon-based cooler was made by etching into the plate, which meant the CNT-based cooler had a bigger thermal resistance between heat resistor and cooling fins.

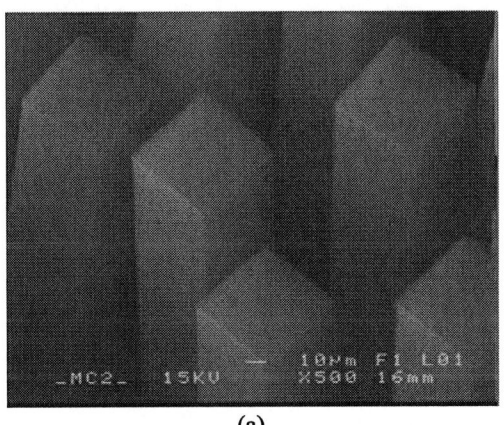

(a)

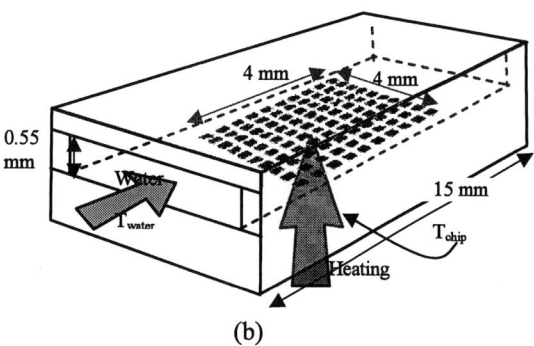

(b)

Fig. 7 (a) Real CNTs array; (b) Modeling geometry

Tab. 2 Performance comparison of microcoolers

Sample	1	2	3
Description	CNT cooler	Silicon Cooler with fins	Silicon cooler without fins
Power of resistor (W)	2.13	1.95	1.95
Flow rate (ml/min)	35.7	38.5	34.5
Temperature of resistor (°C)	42	42	50

Room temperature = 23°C

K. Kordas et al. grew multi-walled carbon nanotubes on Si/SiO$_2$ templates from a xylene/ferrocene precursor by catalytic chemical vapor deposition deposited at 770°C to form a thick CNT film [27]. Then they applied laser to etch the film to get 10×10 fin array blocks, as shown in Figure 8. The structure they designed to test the CNT fins consisted of a thermometer flip chip mounted on a customized silicon substrate, and the array of the fin block itself being soldered onto the backside of the flip chip, as shown in Figure 9.

Fig. 8 Morphology and structure of laser-patterned CNT films. (a) Microstructure of three cooler blocks laser etched next to each other in the CNT film (scale bar: 500 μm). (b) Grooves and a pyramidal fin obelisk of aligned nanotubes (scale bar: 50 μm). (c) Close-up image of the aligned nanotubes (scale bar: 5 μm). The inset in the figure shows an energy filtered electron microscopy image of a well-graphitized nanotube (scale bar: 50 nm)

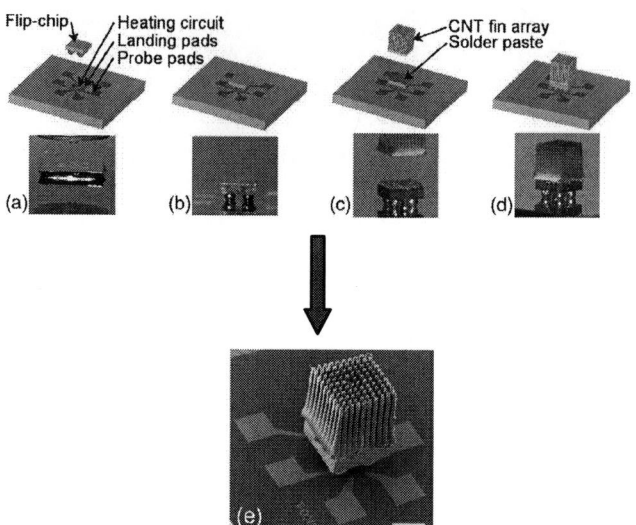

Fig. 9 (a) Positioning and (b) soldering the flip chip on the Cu landing pads of the substrate (this structure also served as a reference). (c) Solder paste dispensing, CNT array positioning, and (d) soldering on the Cu coated backside of the chip. (e) Field emission scanning electron microscopy image of an assembled structure (scale bar: 500 μm).

The assembly was measured on a probe station to record the chip temperature versus heating power characteristics under various thermal loads and cooling gas flow rate with a flow perpendicular to the upper facet of the chip. Reference measurements were also done without the nanotube fins, under the same condition. Results showed that using natural convection, 11% more power was observed to be dissipated from the chip that had the attached nanotube fin structures. While under forced N$_2$ flow, the cooling performance with the fins was improved by 19%. These results meant that applying the nanotube fin structure would allow the dissipation of about 30 and 100 W/cm^2 more power at 100 °C from a hot chip for the cases of natural and forced convections, respectively. Because of the low density of carbon nanotubes, these results also meant an extra power dissipated per weight of added CNT structure can be as high as 1.1 and 3.7kW/g for these two cases respectively. Kordas et al. also compared the cooling performance of the CNT fins with copper fins, results showed that copper fins performed fairly equally to CNT fins. CFD simulation showed the heat transfer coefficient varied from 20 to 500W · m^{-2} · K^{-1} depending on the local flow rate, which agreed with published values very well.

Zhong et al. carried out CFD simulation for a series of CNT microfin cooling architectures based on both one and two dimensional fin array models [28, 29]. They considered the influence of microfin structures, fluid speed, heating powers and effective thermal conductivity on cooling effects, and they also compared the simulation with experiment results. Modeling geometry were shown as Figure 10. The 2D CFD simulation indicated that the heat transfer capability of coolers was considerably dependent on the number of fin rows and the cooling capability of 2D carbon nanotube fin array was more efficient than that of 1D one. Simulation also showed that the fluid speed was the key factor of heat transfer, and the heat generated by the chip was removed mainly by liquid mass flowing in the channels of micro-fin architectures. The pressure drop between the inlet and outlet of the cooling device was an important factor limiting the fluid speed, and the excessive pressure drop may destroy the carbon nanotube fin structures. Maximum temperatures in fin arrays were dependent linearly on the chip heating power, and the linear relationship was corresponding to a constant thermal resistance of cooling system. Finally, they found the effective transverse thermal conductivity of carbon nanotube fins was also a crucial factor for cooling efficiency of the microfin structures, but they needed further experiments for verification.

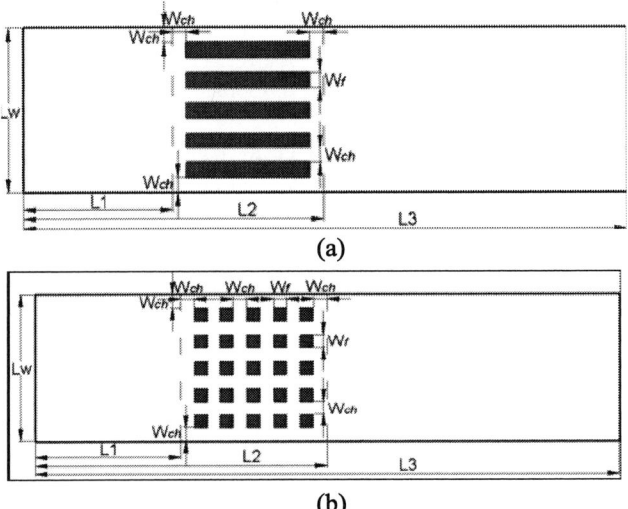

(a)

(b)

Fig. 10 (a) 1D CNT array; (b) 2D CNT array

Ming Hu et al. studied the air flow through carbon nanotube arrays using molecular dynamics simulations [30]. They found that for 1.4nm diameter tubes separated by 25nm, the air flow can be well described by the free molecular flow theory. They estimated for such array, the pressure gradient was about 0.1atm/μm at 1atm air pressure and 5m/s flow velocity, indicating that the flowing air can only pass through an array of no more than about 400 carbon nanotubes in series. The carbon nanotube-air system was shown in Figure 11. They also gave suggestion for CNTs array design for thermal energy exchange with air.

Seok Pil Jang and Stephen U.S. Choi investigated the cooling performance of a microchannel heat sink with nanofluids numerically [31]. The results showed that cooling performance of the microchannel heat sink was enhanced by about 10% when nanofluids containing diamond (1 vol.%, 2nm) was used. They thought nanofluids reduced both the thermal resistance and the temperature difference between the heated microchannel wall and the coolant.

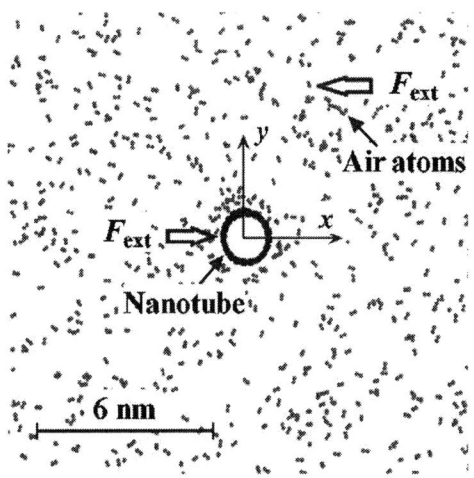

Fig. 11 Schematic of carbon nanotube-air system. The bar indicates mean free path of air molecules between collision

Summary

Due to the unusually high thermal conductivity of carbon nanotubes, the CNT-based microchannel cooler is very promising, which brings hope to the thermal engineers in electronics. The previous studies are very encouraging. Both experiments and simulations have demonstrated the feasibility to utilize CNTs as cooling fins. However, it's still a very long way ahead. One of the biggest obstacles is the weak adhesion between CNTs and substrate. Since CNTs are attached to the substrate by Van der Waals force, they are easily separated by external force, such as air or water flow. Fortunately, progress has been achieved to transfer CNTs for stronger adhesion [32]. Another challenge is the optimization of the fin geometry including CNTs' length and density, fin's width and channel's width. Working fluids is also a crucial factor and some related work is published in another paper in this conference. Proper fluids not only can improve heat dissipation, but also protect the cooling fins.

Acknowledgements

The authors at Chalmers University of Technology would like acknowledge the financial support by the Seventh Framework Programme of the European Union (Project name: Nano Packaging Technology for Interconnection and Heat Dissipation under the contract No: 216 176).

References

1. Excerpts from A Conversation with Gordon Moore: Moore's Law. Intel, 2005.
2. International Technology Roadmap for Semiconductors, 2004 Update, Assembly and Packaging, Jointly sponsored by European Semiconductor Industry Association, Japan Electronics and Information Technology Industries Association, Korea Semiconductor Industry Association, Taiwan Semiconductor Industry Association, and the Semiconductor Industry Association, 2004.
3. G. P. Peterson, A. B. Duncon, and M. H. Weichold, "Experimental Investigation of Micro Heat Pipes Fabricated in Silicon Wafers," Journal of Heat Transfer, Vol. 115 (1993), pp. 751-756.
4. M Le Berre, S Launay, V Sartre and M Lallemand, "Fabrication and experimental investigation of silicon

micro heat pipes for cooling electronics," Journal of Micromechanics and Microengineering, Vol. 13, No. 3 (2003), pp. 436-441.

5. D. A. Benson, R. T. Mitchell, M. R. Tuck, D. R. Adkins, and D. W. Palmer, "Micro-Machined Heat Pipes in Silicon MCM Substrates," Proceedings of the 1996 IEEE Multi-Chip Module Conference, pp. 127-129.

6. Amab K. Mallik, G. P. Peterson, and Mark H. Weichold, "Fabrication of Vapor-Deposited Micro Heat Pipe Arrays as an Integral Part of Semiconductor Devices," JOURNAL OF MICROELECTROMECHANICAL SYSTEMS, Vol. 4, No. 3, (1995), pp. 119-131.

7. Z. Jon Zuo, Mark T. North, and Kevin L. Vert, "High Heat Flux Heat Pipe Mechanism for Cooling of Electronics," IEEE TRANSACTIONS ON COMPONENTS AND PACKAGING TECHNOLOGIES, Vol. 24, No. 2, (2001), pp. 220-225.

8. Jun Xu, Timothy S. Fisher, "Enhancement of thermal interface materials with carbon nanotube arrays," International Journal of Heat and Mass Transfer, 49 (2006), pp. 1658–1666.

1. 9. Johan Liu, Michael Olugbenga Olorunyomi, Xiuzhen Lu, Wen Xuan Wang, Tomas Aronsson and Dongkai Shangguan, "New Nano-Thermal Interface Material for Heat Removal in Electronics Packaging," 2006 Electronics Systemintegration Technology Conference, Dresden, Germany, pp.1-6.

10. Kai Zhang et al., "Study on Thermal Interface Material with Carbon Nanotubes and Carbon Black in High-Brightness LED Packaging with Flip-Chip Technology," 2005 Electronic Components and Technology Conference, pp. 60-65.

11. Farhad Sarvar, David C. Whalley and Paul P. Conway, "Thermal Interface Materials - A Review of the State of the Art," 2006 Electronics Systemintegration technology conference, Dresden, Germany, pp. 1292-1302.

12. Ravi S. Prasher, Jim Shipley, Suzana Prstic et al., "Thermal Resistance of Particle Laden Polymeric Thermal Interface Materials," Journal of Heat Transfer, Vol. 125, (2003), pp. 1170-1177.

13. Ravi S. Prasher, Paul Koning, James Shipley and Amit Devpura, "Dependence of Thermal Conductivity and Mechanical Rigidity of Particle-Laden Polymeric Thermal Interface Material on Particle Volume Fraction," Journal of Electronic Packaging, Vol. 125, (2003), pp. 386-391.

14. Hone J., "Carbon Nanotubes: Thermal properties," Dekker Encyclopedia of Nanoscience and nanotechnology, pp. 603-610.

15. Savas Berber, Young-Kyun Kwon and David Tománek, "Unusually High Thermal Conductivity of Carbon Nanotubes," PHYSICAL REVIEW LETTERS, Vol. 84, No. 20, (2000), pp. 4613-4616.

16. D. B. TUCKERMAN AND R. F. W. PEASE, "High-Performance Heat Sinking for VLSI," IEEE ELECTRON DEVICE LETTERS, Vol. 2, NO. 5, (1981), pp. 126-129.

17. Sumio Iijima, "Helical microtubules of graphitic carbon," natue, Vol. 354, (1991), pp. 56-58.

18. Savas Berber, Young-Kyun Kwon and David Tománek, "Unusually High Thermal Conductivity of Carbon Nanotubes," PHYSICAL REVIEW LETTERS, Vol 84, No. 20, (2000), pp. 4613-4616

19. Kambiz Vafai and Lu Zhu, "Analysis of two layered microchannel heat sink concept in electronic cooling," International Journal of Heat and Mass Transfer, 42 (1999), pp. 2287-2297.

20. Choondal B. Sobhan and Suresh V. Garimella, "A comparative analysis of studies on heat transfer and fluid flow in microchannels," Microscale Thermophysical Engineering, 5 (2001), pp. 293–311.

21. E. G. Colgan, et al., "A Practical Implementation of Silicon Microchannel Coolers for High Power Chips," 21st IEEE SEMI-THERM Symposium, 2005, pp. 218-225.

22. E. G. Colgan, et al., "High Performance and Subambient Silicon Microchannel Cooling," Journal of Heat Transfer, Vol. 129, (2007), pp. 1046-1051.

23. Zhimin MO, Johan Anderson and Johan Liu, "INTEGRATING NAN0 CARBONTUBES WITH MICROCHANNEL COOLER," Proceeding of HDP'04, (2004), pp. 373-376.

24. Zhimin Mo , Raluca Morjan, Johan Anderson, Eleanor E.B. Campbell and Johan Liu, "Integrated Nanotube Microcooler for Microelectronics Applications," 2005 Electronic Components and Technology Conference, pp. 51-54.

25. Lisa Ekstrand, Zhimin Mo, Yan Zhang and Johan Liu, "Modelling of Carbon Nanotubes as Heat Sink Fins in Microchannels for Microelectronics Cooling," 5th International Conference on Polymers and Adhesives in Microelectronics and Photonics, (2005), pp. 185-187.

26. Teng Wang, Martin Jönsson, Elisabeth Nyström, Zhimin Mo, Eleanor E.B. Campbell and Johan Liu, "Development and Characterization of Microcoolers using Carbon Nanotubes," 2006 Electronics Systemintegration Technology Conference, (2006), pp. 881-885.

27. K. Kordas et al., "Chip cooling with integrated carbon nanotube microfin architectures," APPLIED PHYSICS LETTERS 90, 2007.

28. Xiaolong Zhong, Yan Zhang, Teng Wang, Zhaonian Cheng, Johan Liu, "Computational Fluid Dynamics Simulation for On-chip Cooling with Carbon Nanotube Micro-fin Architectures , Proceedings of the 8th IEEE International Conference on Electronic Materials and Packaging (EMAP), December 11–14, 2006, Hong Kong, pp117-123.

29. Zhong, Xiaolong, Yi Fan, Johan Liu, Yan Zhang, Teng Wang, and Zhaonian Cheng, "A Study of CFD Simulation for On-chip Cooling with 2D CNT Micro-fin Array," Proceedings of the 2007 International Symposium on High Density Packaging and Microsystem Integration (HDP'07), Shanghai Everbright Convention & Exhibition Center, Shanghai, China, June 26-28, 2007, pp.442-447.

30. Ming Hu, Sergei Shenogin and Pawel Keblinskib, "Air flow through carbon nanotube arrays, "APPLIED PHYSICS LETTERS 91, 2007.
31. Seok Pil Jang and Stephen U.S. Choi, "Cooling performance of a microchannel heat sink with nanofluids," Applied Thermal Engineering, 26 (2006), pp. 2457-2463.
32. T. Wang, B. Carlberg, M. Jönsson, G.-H. Jeong, E. E.B. Campbell and J. Liu, "Low temperature transfer and formation of carbon nanotube arrays by imprinted conductive adhesive," APPLIED PHYSICS LETTERS 91, 2007.

Environmentally Friendly Electronics for High Reliability

J. B. McElroy, Sr. and R. C. Pfahl, Jr.
International Electronics Manufacturing Initiative (iNEMI)
2214 Rock Hill Road, Suite 110, Herndon, Virginia 20170, U.S.A.
jmcelroy@inemi.org; bob.pfahl@inemi.org

Abstract

In 2006 when the European Union's RoHS Regulation went into effect, a number of global firms who produce high-reliability products such as servers and telecommunication equipment had decided to take the exemption allowed for Pb containing solders in these applications. As a result they had not completed the tests necessary to prove the reliability of Tin Silver Copper (SAC) alloys in these applications. In 2007 it became apparent to many of these firms that they could no longer procure components with traditional SnPb surface finishes, and thus they faced an unknown reliability risk. In 2006 iNEMI had begun a study to evaluate the reliability of "Pb-Free BGAs in SnPb Assemblies." This paper will report on the results of this initial study and will then report on several studies currently under way to evaluate the reliability of new green-materials in high-reliability applications.

Introduction

Since 1998, iNEMI's collaborative agenda has been increasingly dominated by projects that address infrastructure gaps facing the electronics industry as it makes the transition to be compliant with materials restrictions and end-of-life regulations. While the original emphasis was on materials and process-related issues, the scope has broadened over time to include supply chain and business process issues. More recently, our efforts have been dominated by the issues faced by the high-reliability product sectors.

In 2006 when the European Union's RoHS Regulation went into effect, a number of global firms who produce high-reliability products such as servers and telecommunication equipment had decided to take the exemption allowed for Pb containing solders in these applications. The primary justification for these exemptions was that they had not completed the evaluations necessary to prove the reliability of Tin Silver Copper (SAC) alloys in these long life time/mission critical applications. In 2007 it became apparent to many of these firms that they could no longer procure components with traditional SnPb surface finishes, and thus they faced either an unknown reliability risk or a source of supply risk.

Many members of the high-reliability community now recognize that they quickly must follow the consumer market in evaluating new "green materials" if they are to take advantage of the high volume low-cost components produced for the consumer market. These same firms now recognize the cost and time required to evaluate the reliability of new materials in each of their applications. Consequently they are evaluating the feasibility of introducing Pb-free solders and surface finishes at the same time they introduce bromine flame retardant free (BFR-free) Printed Circuit Boards (PCBs). This paper will describe a number of projects that the iNEMI consortium has established to determine the reliability of Pb-free and BFR-free components in electronic assemblies.

High-Reliability Electronics

While Pb-free assembly is now well established in the consumer segment, a number of challenges remain before this technology can meet the stringent reliability requirements of mission-critical products such as medical electronics, telecommunication infrastructure equipment and high-end computer servers. Many of these products enjoy exemptions from, or are out of scope of, the European Union's RoHS; however, the long-term outlook indicates that conversion to Pb-free assembly is inevitable.

Many iNEMI members are designing and manufacturing high-reliability electronics. Many of these firms are taking advantage of Pb exemptions offered under the EU RoHS Directive and, therefore, expect to continue to use SnPb assembly for some time (i.e., several years). Over the longer term, however, most of these same companies are planning for Pb-free conversion. Due to the supply chain conversion to Pb-free, this posture is creating a number of challenges/risks:

- The vast majority of component suppliers are converting to tin finishes for I/O terminals. Tin whiskers remain a concern for high-reliability applications, although mitigation practices are now established that can substantially reduce the risk.
- Most Pb-free BGAs are not compatible with SnPb assembly processes.
- Either need to have ongoing availability of SnPb-compatible BGAs, or
- Need to develop a robust process for mixed assembly.

Many high-reliability applications make use of large, complex PWBs. Once this segment does convert to Pb-free, there will be several issues to resolve:

- Highly complex boards stress materials and processes used for Pb-free assembly (e.g., laminates).
- Component survivability due to higher processing temperatures (especially for wave solder).
- Greater temperature variability over these complex assemblies.
- Robust repair processes are still in development.

iNEMI High-Reliability RoHS Task Force

In order to address some of these issues, iNEMI formed the High-Reliability RoHS Task Force to work with the supply base to ensure that high-reliability product needs are met. The group has now completed four position papers and recommendations to help provide guidance for industry.

The first paper addressed the availability of SnPb-compatible BGAs for the high-reliability segment. It also communicates clear requirements for tin whisker mitigation and testing practices. This position paper is available at: http://thor.inemi.org/webdownload/projects/ese/High-Rel_RoHS_recommends.pdf

The second document published by the group covered manufacturing issues associated with thermally complex assemblies. This document is available at: http://thor.inemi.org/webdownload/projects/ese/High-Reliability_RoHS/High_Rel_position_021606.pdf

The third document dealt with the qualification and use of RoHS 5 and RoHS 6 subassemblies in high-reliability products. It is available at: http://thor.inemi.org/webdownload/projects/ese/High-Reliability_RoHS/High_Rel_position_061206.pdf

The fourth document "Recommendations for Managing Pb-Free Alloy Alternatives," addressees the proliferation of alloy alternatives that are being created.

iNEMI's High-Reliability Task Force is working to develop an industry view and communicate results to the supply base. Proposed tasks include:
- Define list of key knowledge gaps for high-reliability Pb-free:
  - Reliability.
  - Manufacturing.
- Create matrix of work underway to close gaps:
  - Consider existing sources of data.
  - Consider all industry/university cooperative efforts.
- Create timeline for completion.
- Coordinate work and communicate results.

The BGA Challenge

For companies that participate in industries that require high-reliability/long-service-life products, the rationale for staying with SnPb assembly was very clear: in many cases Pb-free technology had not been able to demonstrate the required level of reliable performance for long-life, mission-critical applications. This was further compounded by the fact that in some cases the ability to move to Pb-free assembly was impeded by the non-availability of Pb-free parts for older, low-volume or out-of-production BGA components. While it is increasingly obvious that virtually all electronics products will be Pb-free over time, there are a number of knowledge gaps that must be closed before Pb-free reliability can be predicted with the same certainty that SnPb assembly can deliver. The risks and potential long-term costs associated with this uncertainty to users and producers of these products are too great to ignore.

High-reliability hardware companies were relieved to understand that exemptions would be granted for continued use of SnPb in their products; however, the rapid transition of the supply chain to meet the high-volume needs of consumer electronics is causing problems with the availability of critical components (new and existing) in their traditional surface finish configurations. Many suppliers have elected to convert all of their components over to Pb-free connection finishes. In a number of cases, this presents compatibility issues (especially with BGAs) for those users who are allowed to continue to use SnPb assembly processes.

The strategy of staying with SnPb assembly is starting to create supply continuity issues for those who thought that exemptions (or being out of scope) would temporarily insulate them from the movement to Pb-free material systems that industry has much less experience with.

iNEMI SnPb BGA Availability Task Group

iNEMI has formed the SnPb BGA Availability Task Group, made up of firms that are either taking the Pb exemption or are out of scope for EU RoHS. The efforts of this group are focusing on ways to work with the BGA component supply base (integrated circuit as well as packaging firms) to support SnPb-compatible BGAs, assist with questions of long-term reliability, and/or other solutions to address concerns. Activities to date include:
- Developed a list of critical BGA component families.
- Estimated Total Available Market (TAM) for these devices in a SnPb assembly version.
- Worked with the supply base to come up with alternatives that meet this critical market need.
- Organized a workshop with BGA supply base to help ensure balanced solutions. Event was held on March 1, 2007, at Hewlett-Packard in Cupertino, California.

Results from BGA Workshop

The following conclusions were reached from this workshop:
- So much emphasis has been put on RoHS conversion that the supply base still seems to be somewhat surprised/misinformed that there are significant ongoing needs for SnPb BGAs.
- More education is definitely needed on the unique characteristics of high-reliability product lifecycles (high development costs, long product availability window, decades of support, etc.) and how the RoHS transition has exacerbated the situation.
- More consensus and education is needed on remaining knowledge gaps and what must be done to reduce risk. It is clear that the supply base is somewhat skeptical of the level of risk/severity of gaps remaining.
- The high-reliability users have enjoyed the technology, availability, and low cost of consumer-driven components and would like to continue to do so. Staying with SnPb BGAs is moving them away from this paradigm (during the extended transition period).
- Suppliers are treating SnPb BGAs as custom parts with all the implications that this brings (limited competition, limited availability, less favorable pricing).
- Perhaps iNEMI should develop a business case for accelerated investment to close knowledge gaps (based on cost and availability issues of custom vs. standard parts).

- The proliferation of Pb-free metallurgies is becoming a significant issue for industry. While it is inevitable that change will occur as we learn more about Pb-free, it is counterproductive to diverge to so many alloys. This could be a good opportunity to combine forces to drive convergence.
- The position being taken by military and aerospace (SnPb forever!) is especially problematic and would drive the re-establishment of dedicated military parts lines if it is to be achieved.

BGA Business Potential

While total unit volume is relatively small, the packaged semiconductor revenue generated from high-reliability products is significant (18% of BGA and 12% of CSP value). More importantly, these critical devices leverage over 25% of the total hardware revenues for electronics! It is believed that this analysis should help to justify the critical need for these components during this lengthy transition period.

Results from the iNEMI Pb-Free BGAs in SnPb Assemblies Project

iNEMI is in the process of completing a four-year project on this subject. Highlights from this effort will be presented at the ICEPT Conference.

New High-Reliability Projects

Based on recommendations from the iNEMI High-Reliability RoHS Task Force and Results from BGA Workshop, iNEMI has started a new round of projects addressing the needs of high-reliability applications. Included are three new projects:

- *Evaluation of Pb-Free Component & Board Finish Reliability.* This project will evaluate the effects of alternative surface finishes for circuit boards and package substrates on Pb-free solder joint reliability during mechanical and thermal stress testing.
- *Pb-Free Early Failure Project.* This new iNEMI effort will be working to determine whether a large sample size will reveal Pb-free early failures in accelerated thermal fatigue testing of Pb-free solder joints. All components for the test vehicle are on order and many have been received. The test vehicle board design is complete.
- *BFR-Free High-Reliability PCB.* This project is a follow-on to iNEMI's current BFR-Free PCB Project. Proposed project objectives include:
  - Build on industry knowledge and capability and on results of the iNEMI BFR-Free Project: Test Phase (Phase II).
  - Consider unique market segment requirements.
  - Identify technology readiness and gaps.
  - Stimulate supply capability.
  - Determine BFR-free board-level reliability for various components.

Conclusions

This paper highlights some of the collaborative work undertaken to address the specific needs of the high-reliability product segments as they continue to work towards the Pb-free transition. While this change is inevitable, there are still a number of risks that must be mitigated before mission-critical applications can again enjoy the benefits of full utilization of consumer-driven components in a Pb-free assembly environment. The Pb-free conversion experience has revealed that the high-reliability community must respond to material changes being introduced in consumer products if they wish to take full advantage of the cost reductions high volume manufacturing provides.

Acknowledgments

The authors wish to thank the many individuals who have participated in the iNEMI roadmapping process and the iNEMI members who have conducted the projects that we have summarized.

Morphology, Evolution and Performance of IMC in SAC105 Solder/UBM (Ni (P)-Au)

F. L. Sun[1], P Hochstenbach[2], W. D. Van Driel[2,3], G. Q. Zhang[2,3]

[1] Department of Mat. Sci. and Eng., Harbin University of Sci. and Tech.,
52# Xuefu road, Harbin, 150080, China, Email: sunflian@163.com
[2] NXP Semiconductors, the Netherlands.
[3] Delft University of Technology, the Netherlands

Abstract

To enhance the ability of lead-free solder joint to resist failures induced by mechanical impact and shock, some researchers have introduced low-Ag lead-free solder. In this study, the formation and evolution of IMC, the fracture morphology and performance of solder joint between SAC 105 solder and under bump Metallization (UBM) have been studied after different temperature storage aging and multi-reflow for WLCSP. The morphology and fracture surface of IMC have been analyzed by SEM, EDX and deeply etching techniques. The behavior of IMC and correlation with fracture mode of joint at high speed ball pull (HSBP) test has been investigated. It has been found that high temperature storage (HTS) aging results in IMC growing from one layer that is (Cu_6Sn_5) IMC to two different layers that are (($Cu,Ni)_6 Sn_5$ and $(Cu,Ni)_3Sn_4$). Multi-reflow results in the IMC layer thickness increasing a little and some needle-like IMC grain spalling into the solder.

Introduction

With the development and promotion of portable electronic devices, the ability of electronic packages and assemblies to resist the sudden shock impact is becoming a growing concern. Some new kinds of solders have been introduced to improve the brittle fracture–resistance of solders joint (1-4). It has been reported that the solder joint between Low-Ag solder and under ball metallurgy (UBM) appears higher resistance to drop impact than high-Ag solder (4-8). But the fracture morphology and mechanism in Low-Ag SAC Solder/UBM for wafer-level chip-scale packages (WLCSP) are unclear, typically, under high speed pull condition. The formation and evolution mechanism of intermetallic compound (IMC) and their relation with joint strength need to be investigated.

In this study, the formation and evolution of IMC, the fracture morphology and performance of solder joint between SAC 105 solder and UBM have been studied after different temperature storage aging and multi-reflow for WLCSP. The morphology and fracture surface of IMC have been analyzed by SEM, EDX and deeply etching techniques. Also, the difference of IMC in interface has been compared with high-Ag solder. The behavior of IMC and correlation with fracture mode of joint at high speed ball pull (HSBP) test has been investigated.

Experimental Procedure

Solder alloy: SAC105 (Sn-1Ag-0.5Cu),
UBM: Electroless Nickel Immersion Gold (ENIG).

Accelerated aging methods: two sorts of high temperature storage (HTS) aging process are as followed:
1) HTS aging at 150 ℃ for 500Hrs, 1000hrs, 2000hrs.
2) HTS aging at 200 ℃ for 20hrs, 80hrs, 160hrs, 280hrs.
Multiple reflow profile is as Fig. 1

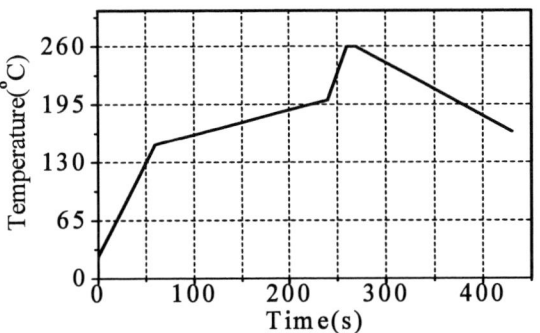

Fig. 1 Reflow soldering profile

Solder joint strength assessment: HSBP test is down in Dage series 4000HS machine. Pulling velocity is 1200mm/s.
Sample preparation for microanalysis:
Cross sections of solder joint were prepared with standard metallographic methods. Some were slightly etched and some were deeply etched to clean up the solder materials and made the IMC visible. Fracture surface analysis was performed by SEM, EDS and deeply etched technology.

Results and Discussions

HTS Affects the Morphology and Evolution of IMC

Some experiments proved that HTS affects the Morphology and Evolution of IMC in interface significantly.

Fig. 2 shows the evolution of IMC in SAC105/UBM in room temperature aging and after HTS aging at 200℃ for 280h respectively.

Fig. 2 a) and c) are the top view interfacial IMC grains after removing the solders by deep-etching technology. Fig. 2 a) shows the morphology of IMC grains at room temperature. And Fig. 2 b) shows the morphology of cross section. The composition of IMC grains are similar to $(Cu,Ni)_6Sn_5$ which is identified by EDS and the composition of point "o "is shown in table 1.

Fig. 2c) and d) show two different layers IMC $(Cu,Ni)_6Sn_5$ and $(Cu,Ni)_3Sn_4$ after aging. Also, the composition of point A and point B are shown in table 1. From Fig. 2c), only $(Cu,Ni)_6Sn_5$ layer can be seen but from Fig. 2d), $(Cu,Ni)_3Sn_4$ is visible.

978-1-4244-2739-0/08/$25.00 ©2008 IEEE   651

Table 1 The composition of IMC before and after aging

areas	Sn	Cu	Ni	Sort of IMC
O	54.25	23.19	22.56	$(Cu, Ni)_6Sn_5$
A	45.90	20.27	33.80	$(Cu, Ni)_6Sn_5$
B	53.23	3.70	43.37	$(Cu,Ni)_3Sn_4$

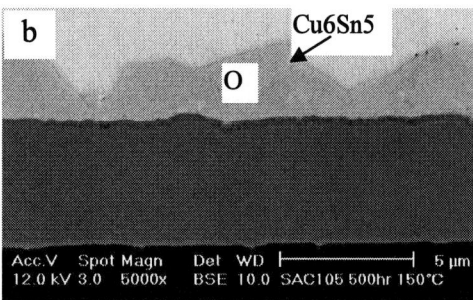

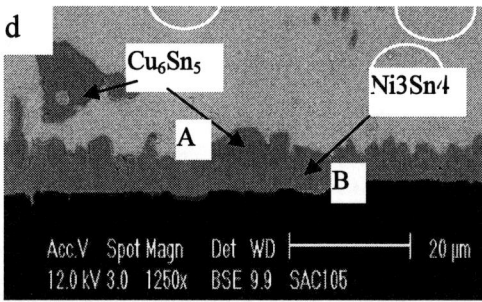

Fig. 2 Evolution of interfacial IMC

In Fig. 1 c), it can be found that the grains (in the circle) of $(Cu,Ni)_6Sn_5$ have grown longer than before aging. So, it could be deduced that HTS aging can change the IMC shape from one to another.

Above mentioned results of formation and evolution of IMC are similar to the results of reference (6). Both SAC105 used in this paper and SAC405 used only in reference (6) appears two layers IMC which are stone-like $(Cu, Ni)_6Sn_5$ and needle-like $(Cu,Ni)_3Sn_4$. But the aging conditions are different.

The only difference is that HTS were carry out at 200 ℃ for 280hrs for SAC 105, but in the paper (6) reflow were carry out at above 300 ℃ for 1min for SAC 405. So, it can be considered that the Ag content does not effect the performance of IMC nearly, especially the composition of $(Cu, Ni)_6Sn_5$ and $(Cu,Ni)_3Sn_4$. The key factor to lead IMC to growing and developing is Cu content in solder and the thickness of Ni layer of UBM.

It should be pay attention to that there is no $(Cu,Ni)_3Sn_4$ appearance nearly in the interface after aging at 150℃ for 2000hrs and it appears only after at 180℃ for 2000hrs. This indicates that SAC 105 appears a better ability to resist-aging.

Multiple Reflow Affects the Morphology and Evolution of IMC

The reflow profile is as Fig. 1 and the cross section morphology of solder joint is as Fig.3. Reflow affects the morphology of IMC, but the composition of IMC is no evident change and both are similar to $(Cu,Ni)_6Sn_5$. Fig. 3 a) b) show the cross section morphology of IMC after one time reflow and 6 times reflow respectively. After one time reflow as in Fig. 3a), some IMC grains grow as a stone-like attaching to the surface of UBM discontinuously. The average grain size is about 3.0μm. But, after 6 times reflow, the IMC grains were broken and spalled into solder. The thickness of IMC area is about 20μm.

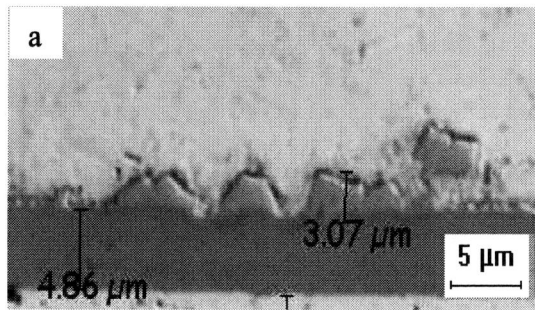

Fig. 3 IMC morphology after multiple reflow

According to the result of solder joint strength assessment by HSBP test as in Fig. 4, multiple reflow process no decrease the joint strength evidently. Every numerical value in the diagram is the medial value of 10 times measurement results in case error.

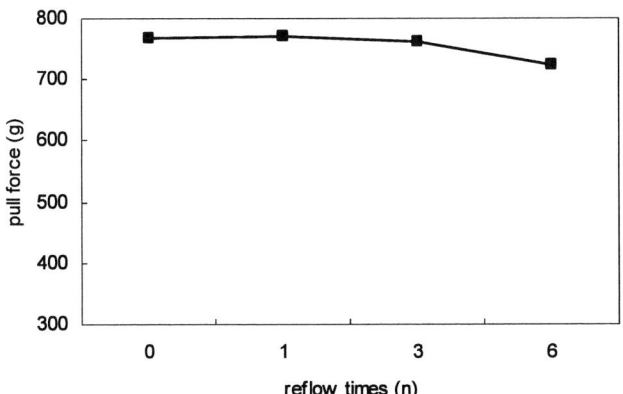

Fig. 4 Reflow affects HSPS of joint

Multiple reflow (peak temperature 260 ℃) makes the IMC grains broken and spreaded into larger region in solder near the interface than before. The hardness of IMC area would be decreased because the IMC was broken. So the brittleness of IMC layer and stress concentration of solder joint would not be increased after multiple reflow. This is why the HSBP not to be decreased evidently after multiple reflow. But the thickness of IMC layer would increase if raising the peak temperature (above 300℃) and extending the keeping-time which has been proved by paper (6). This is the key factor to affect the performance of IMC.

The correlation between fracture mode of joint and IMC

After HTS aging, the IMC in SAC105/UBM appears two layers: one is $(Cu,Ni)_6 Sn_5$, the other layer near Ni-P is Ni3Sn4. Fig 5 shows the different fracture surface of IMC. According to some test results that fracture often occurs between the two layers (see Fig 5 a). In Fig. 5a), the lift area is uper layer and the right is lower layer. Also fracture occurs between Ni3Sn4 and UBM (see Fig 5 b). Occasionally the numeral value of HSBP is lower if fracture occurs between $(Cu,Ni)_6 Sn_5$ and $(Cu,Ni)_3Sn_4$. Fig. 5c) is the sketch diagram to describe the fracture mode.

Discussion

In the Fig. 5) lift area, some nuclear grains appear and their composition is similar to $(Cu,Ni)_6Sn_5$. Based on the phenomenon, it could be deduced that $(Cu,Ni)_6Sn_5$ grain grow again because of the growth of new $(Cu,Ni)_3Sn_4$ layer. As pointed out by (9-11), Grain boundary diffusion of Cu atoms through the interfacial IMC layer is the controlling mechanism of the dissolution of Cu into the molten solders. Cu diffusion from in IMC $(Cu,Ni)_6Sn_5$ to from $(Cu,Ni)_3Sn_4$ may be possible.

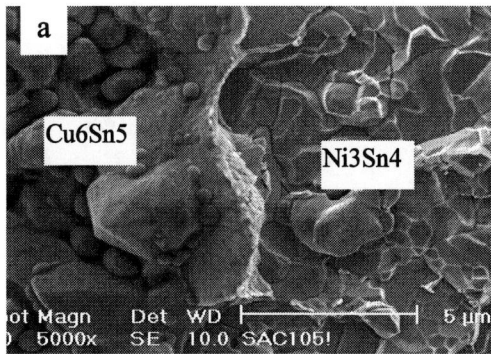

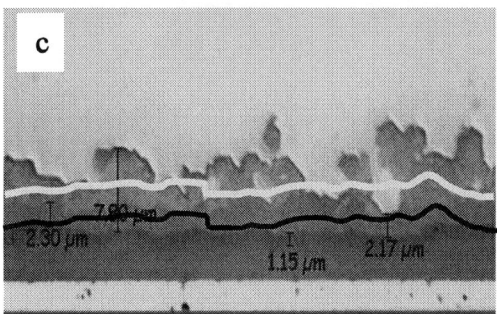

Fig. 5 Fracture surface of IMC

Conclusions

At room temperature storage, $(CuNi)_6 Sn_5$ exists in a stone-like in the interface of solder/UBM in SAC105. HTS aging results in IMC growing from one layer $(CuNi)_6 Sn_5$ IMC to two different layers IMC, and upper layer is $(Cu,Ni)_6 Sn_5$ and lower layer near Ni-p layer is $(Cu,Ni)_3Sn_4$. Moreover, HTS aging results in $(CuNi)_6 Sn_5$ grains growing toward solder ball direction.

By HSBP, the solder joint fracture often occurs at the interface of two IMC layers and the joint strength is lower. Sometimes fracture occurs in the interface Ni3Sn4 layer and UBM which lead to higher joint strength.

Acknowledgments

This work was supported by the National Natural Science Foundation of China (No. 50575060).

The authors are particularly grateful to Daoguo Yang and Ludo Krassenburg for their contributions and support towards this study.

References

1. Ranjit Pandher, Monnir Boureghda, "Identification Of Brittle Solder Joints Using High Strain Rate Testing Of BGA Solder Joints" IEEE *45 th Annual International Reliability, Physics Symposium,* Phoenix, 2007, pp.107-112

2. X. J. Zhao , J.F.J.M.Caers , J.W.C. de Vries etc, "A Component Level Test Method for Evaluating the Resistance of Pb-free BGA Solder Joints to Brittle Fracture under Shock Impact" *2007 Electronic Components and Technology Conference*, pp. 1522-1529

3. Ranjit Pandher and Monnir Boureghda, "High speed ball pull- A predictor of brittle solder joint" *2006 electronics packaging technicalogy conference,* pp.701-707

4. Daewoong Suha, Dong W. Kimb, Pilin Liua, etc., "Effects of Ag content on fracture resistance of Sn–Ag–Cu lead-free solders under high-strain rate conditions" *Materials Science and Engineering: A*, Vol. 460-461, No.15, July (2007), pp595-603.

5. Shinichi Terashima, Yoshiharu Kariya, Takuya Hosoi and Masamoto Tanaka, "Effect of silver content on thermal fatigue life of Sn-xAg-0.5Cu flip-chip interconnects" *Journal of Electronic Materials*, Vol. 32, No. 12, Dec., (2003), pp. 1527-1533

6. Y. Jeon, S. Nieland, et al, "A study on Interfacial Reaction between Electroless Ni-P Under Bump Metallization and 95.5Sn-4.0Ag-0.5Cu Alloy" *Journal of electronic materials,Vol.32, No. 6, (2003),* pp. 548-557.

7. Huann-Wu Chiang, Kenndy Chang, Jun-Yuan Chen , "The effect of Ag content on the formation of Ag3Sn plates in Sn-Ag-Cu lead-free solder '*Journal of Electronic Materials'*,Vol.35, No. 12 Dec. (2006), *pp.* 2074-2080.

8. C. Brizer,er al "Drop test Reliability improvement of lead-free fine pitch BGA using different solder ball composition," *Proc. 55th electronic components and technology conference*, June (2005),pp 1194-1200

9. M. L. Huang, et al "Role of Cu in dissolution kinetics of Cu metallization in molten Sn-based solders." *Applied physics letters 86,181908 (2005).*

10. Jeong-Won Yoon et al "Interfacial reaction and mechanical properties of eutectic Sn–0.7Cu/Ni BGA solder joints during isothermal long-term aging," *Journal of Alloys and Compounds,* Vol. 461, Issues 1-2, pp.L1-L34.

11. Y. C. Chan, et al "The Effect of Cooling Rate on the Growth of Cu-Sn Intermetallics in Annealed PBGA Solder Joints," *Journal of Electronic Packaging, Vol. 125,(2003) pp.153*

Development of High Temperature Stable Isotropic Conductive Adhesives

Zhikun Zhang[1], Sijia Jiang[2], Johan Liu[1,2] and Masahiro Inoue[2]

[1]Key State Laboratory of New Displays and System Applications and SMIT Center, School of Automation and Mechanical Engineering, Shanghai University, Box 282, 149 Yanchang Rd., Shanghai, 200072, China

[2]SMIT Center & Bionano Systems Laboratory, Department of Microtechnology and Nanoscience, Chalmers University of Technology, SE-412 96 Gothenburg, Sweden

Corresponding author: jliu@chalmers.se

Abstract

In this paper, a novel isotropic conductive adhesive containing different fillers, which shows better properties at high temperature compared with traditional ones, was developed and the properties of various fillers were investigated. Silver flakes as the main filler and secondary materials such as silver nanoparticles, carbon nanotubes (CNT) with different volume ratios are combined to form bi-modal or tri-modal systems for electrical or thermal conduction. The curing behavior of the adhesive was investigated by Differential Scanning Calorimeter (DSC). The properties such as glass transition temperature (Tg), storage modulus were detected by Dynamic Mechanical Analyzer (DMA). Thermogravimetric Analysis (TGA) was used to determine the decomposition behavior. The distribution of the carbon nanotubes used in the adhesive was also characterized by SEM. The glass transition temperature of the formulated adhesives is above 200℃. The in-plane bulk electrical resistivity (lower than $5.2*10^{-5} \Omega*cm$)as well as the shear strength (up to 15 MPa) of the adhesives with different volume percentages of various fillers were also obtained.

1. Introduction

As an environment friendly replacement of solders, the isotropic conductive adhesives are used more and more in the electronic packaging field. It has many advantages comparing with traditional solders, such as low processing temperatures and less environment contamination and so on [1-2]. With the development of the microelectronic systems involving more and more functionality and complexity, the heat produced by microelectronic components is much more difficult to dissipate.

Therefore isotropic conductive adhesives used should be applicable under high temperature. This means that the glass transition temperature (Tg) must be as high as possible. Therefore new kinds of isotropic conductive adhesive must be developed. Usually, polymer-based isotropic conductive adhesives are mainly composed of epoxy resin, curing agent and conductive fillers. The polymer resin provides mechanical interconnection and the conductive filler provides electrical conductivity [3]. Generally speaking, the fillers include silver flakes and other conductive particles, such as carbon nanotubes and silver nanoparticles. Epoxy resin contains a series of polymer resins with different properties, which basically determines the application temperature range of adhesives. We have developed a modified epoxy based resin system to elevate the glass transition temperature and enhance simultaneously the mechanical properties of the system.

2. Experiment
2.1 Preparation of Adhesives

The ICAs used in the paper are prepared by a modified epoxy matrix named AF2, silver flakes and silver spherical nanoparticles for bi-modal, as well as Multi-Wall Carbon NanoTubes (MWCNT) for tri-modal adhesives. Silver flakes, of which the size is ranging from 3μm to 10μm and the silver nano particles ranging from 500nm to 1000 nm are used. The MWCNT are with the outer diameter of 20-40 nm and with a length of 5-20μm.

To improve the dispersity of CNT in the epoxy matrix, the CNTs are chemically treated in 98% sulfuric acid and 65% nitric acid (volumetric ratio 3:1) solution at 50℃ for 16 hours with ultrasonic vibration. Then CNTs are washed by de-ionized water until the PH value becomes 7 and dried in the oven at 50~70℃. And the influence of strong acid treatment is observed by SEM, and the SEM samples are prepared by mixing 2% volume ratio of non-treated and treated CNT into the matrix AF2 and cured at 150 ℃ for 30 minutes in the oven [4]. Then small pieces of as-cured adhesives are taken and gold sputtered for the following SEM observation.

The bimodal adhesives are prepared by first mixing manually silver nanoparticles of 1%-5% (volume) into the matrix AF2 following by adding a specific volume ratio of silver flakes, keeping the volume ratio of fillers as 39.238% as constant. When preparing the tri-modal samples, the volume ratio of silver nanoparticles is kept as 1 vol % and that of treated CNTs are 0.1 vol %, 0.3 vol % and 0.5 vol % respectively, both of which are mixed into the matrix firstly and then silver flakes are introduced to keep the volume ratio of fillers as 39.238%. All the samples are stored under -20 ℃. Table 1 shows the different ratios of fillers in detail.

Table 1 the volumetric composition of samples

Sample ID	Matrix(%)	Ag Flakes (%)	Ag nano (%)	CNT (%)
AgF201	60.762	38.238	1	-
AgF202	60.762	37.238	2	-
AgF203	60.762	36.238	3	-
AgF204	60.762	35.238	4	-
AgF205	60.762	34.238	5	-
AgF206	60.762	39.238	0	-
3F201	60.762	38.138	1	0.1
3F202	60.762	37.938	1	0.3
3F203	60.762	37.738	1	0.5

978-1-4244-2739-0/08/$25.00 ©2008 IEEE

2.2 DSC Measurement

The Differential Scanning Calorimetry (DSC, Perkin Elmer Pyris 1) is used to characterize the curing behaviors of the isotropic conductive adhesives. Dynamic scanning is utilized on samples of appropriate weight, usually heating from 50°C to 300°C with the heating rate of 5°C /min. The samples are placed in hermetic aluminum DSC pans and scanned in a nitrogen purge gas (20ml/min) atmosphere. A series of samples of bimodal conductive adhesives with different volume ratios of silver nanoparticles are prepared and compared.

2.3 DMA Characterization

DMA (Perkin Elmer DMA 7e), which stands for dynamic mechanical analyzer, characterizes the storage modulus, loss modulus and tangent α of the adhesives and also shows the behavior of glass transition. A 3-point bending measuring system is utilized with heating rate 5°C/min, temperature ranging from 30 to 300 °C under force frequency of 1Hz.

The free standing samples of adhesives are prepared by curing for 30 minutes at 150 °C in the oven. Firstly, the PTFE tape is attached onto the glass plate to make a rectangle duct-like mold with 3mm width and 0.7mm thickness. Then five different adhesives are printed manually to ensure the flatness of the samples. After curing and cooling down to the room temperature in the oven, these adhesives are detached from the PTFE tape.

2.4 TGA Measurement

The decomposition behavior of bimodal ICAs is characterized by the Thermogravimetic Analysis (TGA, Perkin Elmer Pyris 1). Several samples (cured at 150°C for 30 minutes) with around 5mg weight are prepared and compared with temperature scanning from 50°C to 900°C at the heating rate of 20°C/min.

2.5 Bulk Resistivity Measurement

In order to indicate the relationship between the electrical resistivity and the loading of fillers, especially the effects of secondary nano-fillers in the adhesives, the in-plane bulk resistivity is measured by the 4-probe method. The as-prepared adhesives are printed manually onto the glass substrates with a stripe in-between two tapes onto the glass to make samples of 50mm length, 3mm width and 0.05mm thickness. Then, all the samples are cured at 150°C for 30 minutes in the oven and kept cooling down to room temperature. According to the 4-probe method, the constant current of 0.2 A is applied to the samples by DC Power Supply (Velleman PS 613), and the voltage between the end points of 40mm long of samples is measured by a multimeter (FLUKE 79 III TRUE RMS) [5]. The measuring procedure is shown in Figure 1. The length l, the width w and the thickness h of the tested stripe of adhesives are all measured again after cured and cooled due to the possible change of dimension during curing. Thus, the in-plane resistivity ρcan be calculated by the formula

$$\rho = \frac{U \times w \times h}{I \times l} \qquad (1)$$

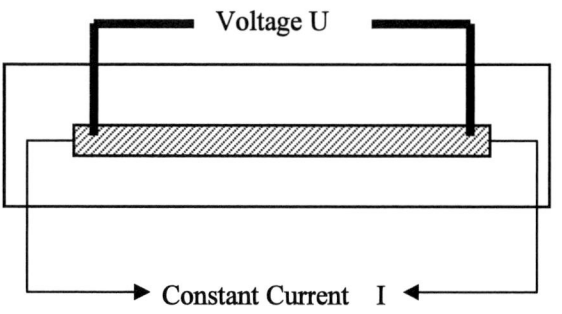

Figure 1 Schematic of 4-probe method

2.6 Lap shear strength Measurement

Compared with traditional Sn-Pb solders, the big shortcoming of adhesives is lower shear strength, which is affected by many factors such as epoxy matrix, metallization of pads, loading of fillers, curing conditions and even the height of the joint [6~7]. In this paper, the lap shear strength of bimodal adhesives is tested by using the Instron 5548 Microtester. A shear rate of 10^{-3}/s is used for all the shear tests. Copper metallization pads of the 2mm diameter on the FR4 substrate (16 mm wide, 50 mm long and 1.6 mm thick) was used in order to mimic the practical application. All the samples are cured at 150°C for 30 minutes in the oven. Figure 2 suggests the testing structure in detail.

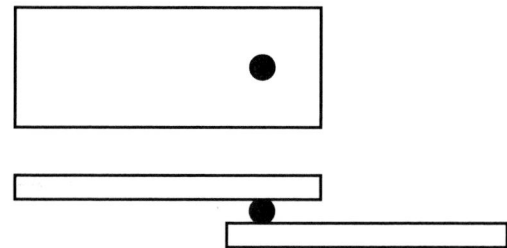

Figure 2 Schematic structure for lap shear test the black dot in the above is a pad on the FR4 substrate and in the below, the dot suggests the adhesive joint

3. Results and Discussion
3.1 SEM and dispersity of CNTs

The strong acid treatment used in this work is the most commonly used method to modify the surface of CNTs. It is believed that after treatment in the sulfuric and nitric acid solution for 16 hours at 50°C, the –OH and –COOH groups are introduced onto the outer walls [4], which can strengthen the dispersity of CNT in epoxy matrix.

As shown in Figure 3, two typical SEM pictures are selected among some randomly taken pictures to indicate the improvement of dispersity of CNTs in the matrix. In picture (b), the carbon tubes, which are the white slimlines, are more uniformly distributed in the matrix.

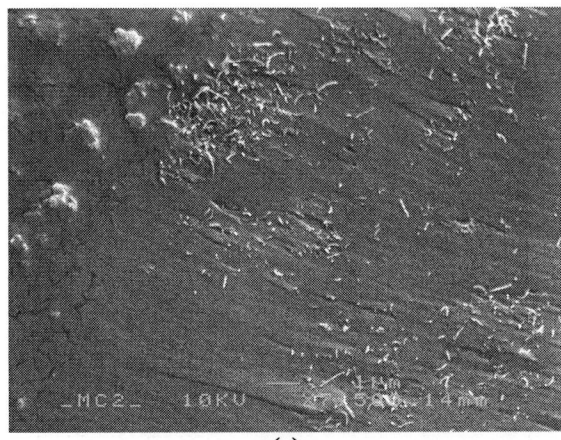

(a)

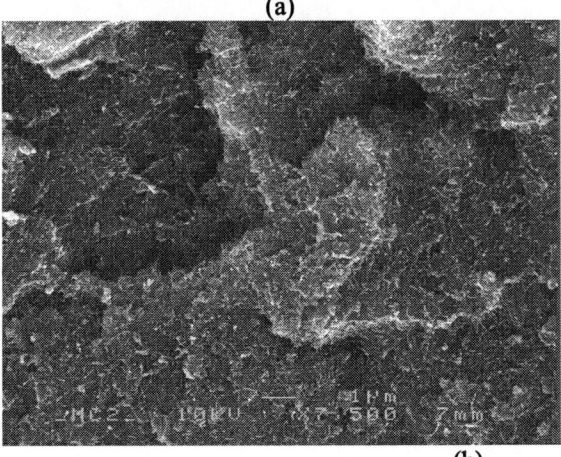

(b)

Figure 3 (a) SEM of non-treated CNTs in AF2
(b) SEM of treated CNTs in AF2

3.2 Behavior of curing and decomposition

From the curve of the DSC information such as the onset temperature and the peak temperature of curing can be gathered and the processing conditions can be decided. In Figure 4, it can be seen that all the samples of AgF201~AgF205 start to polymerize around 130~135℃, reach the peaks at about 150℃ and almost complete after 170℃, which suggests that the processing temperature can range from 130℃ to 180℃ with different curing times. Though there are slight shift among curves of five samples which accounts more for operation and machine, the difference of fraction of silver nanoparticles has no apparent influence on the curing behavior, which dominantly depends on the composition of the epoxy matrix.

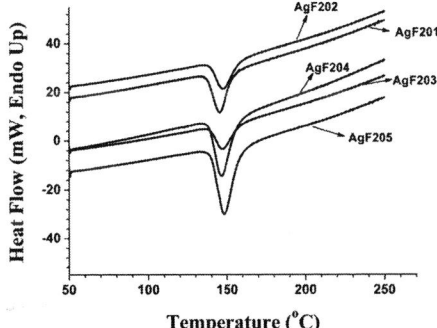

Figure 4 DSC curves of the five bimodal ICAs

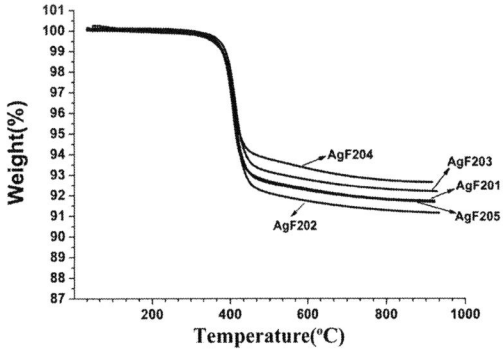

Figure 5 TGA curves of five bimodal ICAs

In Figure 5, the decomposition behavior of five samples is shown by TGA curves. All of them start to lose their weight quickly above 350℃ which indicates the decomposition temperature of the polymers in the cured adhesives is around 350℃.

3.3 Thermo-dynamic properties

In Figure 7, the DMA result of AgF206 with only silver flakes inside is shown. The storage modulus is decreasing gradually during temperature scanning from 50℃ to 300℃ with heating rate 5℃/min. During 190~230℃, the tan δ starts to increase sharply after a plateau, which indicates that the glass transition is around 200℃. Also, there is a relatively abrupt decease of the storage modulus around 160℃ probably since the defects in the cured adhesives like imperfect cross-linking cause the local molecular blocks to move.

In Figure 8, the storage modulus, loss modulus and tan δ of the samples AgF201~AgF205 are shown respectively. There is a turning point in all the storage modulus of the five samples around 250℃ and the curves of storage modulus of AgF203 and AgF204 are almost overlapping though different ratio of silver nanoparticles. All the curves of loss modulus and tanδ show similar tendency as AgF206 in Figure 8. Therefore, the behavior of the five samples shows little difference independent on the composition of the silver flakes and the nano particles, which suggests that the ratio of fillers do not influence the thermo-dynamic properties of adhesives such as the glass transition temperature.

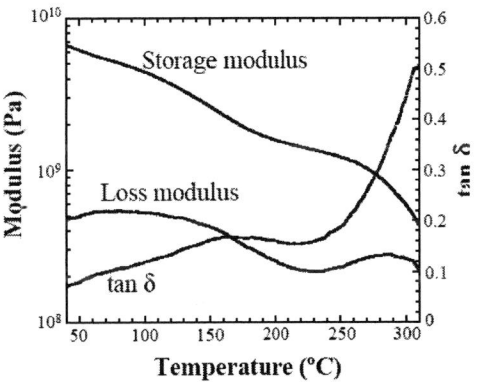

Figure 6   DMA Curve of AgF206

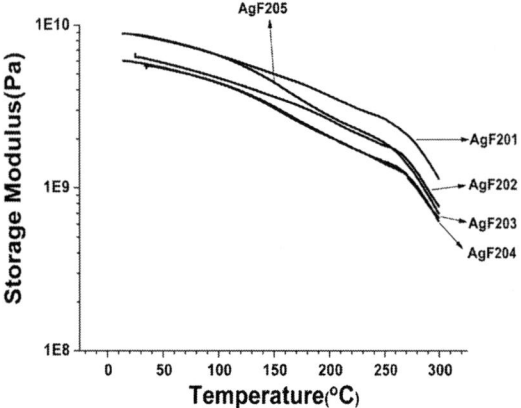

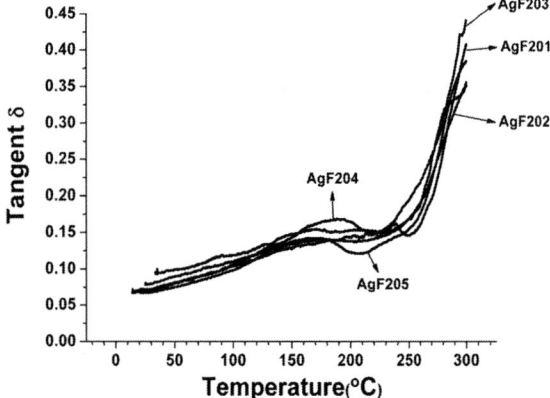

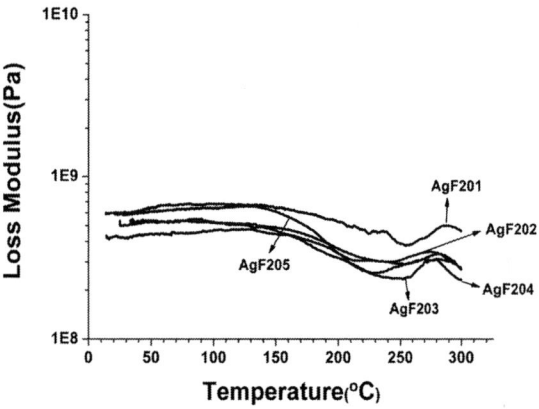

Figure 7 Curves for storage modulus, loss modulus and tanδ of AgF201~AgF205

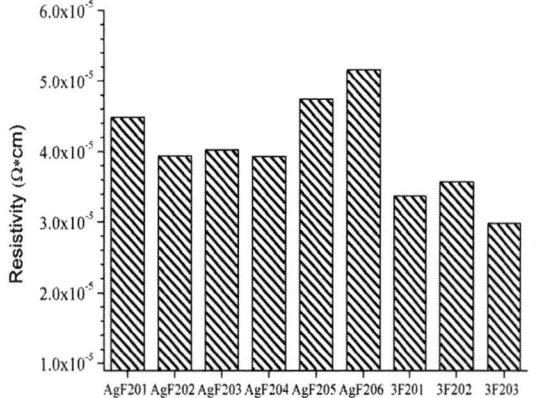

Figure 8 Histogram of resistivity of bimodal and trimodal samples

3.4 In-plane resistivity

Figure 9 shows the resistivity of five bimodal adhesives (AgF201~AgF205) and three tri-modal adhesives (3F201~3F203) as well as one reference adhesive (AgF206) with only silver flakes inside.

Compared with the reference, both bimodal and trimodal samples show lower resistivity, and also all three trimodal samples posses lower resistivity than AgF201, in which there is the same amount of silver nanoparticles as trimodal samples. It is known that ICAs become conductive during curing due to the shrinkage of epoxy matrix. However, micro-size fillers are always difficult to form perfect contact which means that there are many gaps between flakes, underperforming the conductance of adhesives. If appropriate amount of nano-size particles are introduced and mixed uniformly, it is believed that these smaller particles can fill into the gaps in-between flakes, which helps decrease the contact resistance between flakes [8]. However, as mentioned above, nanoparticles are not expected to disperse uniformly in the matrix due to large surface area, which increases the viscosity when mixing. Referring to the sample AgF205 with 5% nano-size silver, the resistivity turns to increase, probably because the increased viscosity inhibits the uniform distribution of nanoparticles.

As to the three trimodal samples, compared with the sample AgF201 with the same amount of nano-size silver, a small amount of CNTs (0.1~0.5 vol%) can decrease the resistivity as low as to $3*10^{-5}\Omega*cm$ with lower viscosity of all samples.

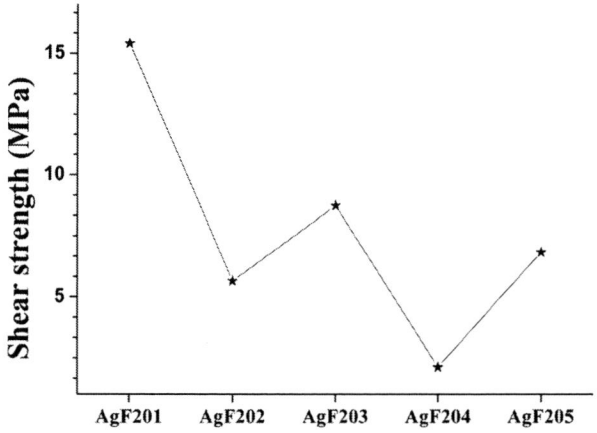

Figure 9 Relationship between shear strength and fraction of Ag nanoparticles

3.5 Lap shear strength

As indicated in Figure 10, there is a tendency that the shear strength decreases with the amount of silver nanoparticles. The most robust joints shown in the figure is made of AgF201 with the lowest ratio of silver nano-particles added with a value of about 15 MPa. And it is observed that most of the samples are broken in an interfacial mode with relatively low shear rate 10-3/s. Maybe the nanoparticles can deteriorate the adhesion between the copper metallization pads when keeping the total loading of fillers the same. However, a more comparably experiment, which can rule out the other factors such as applied pressure and joint height, should be designed to examine the tendency and relationship more accurately.

4. Conclusion and Future work

In this paper, high temperature stable isotropic conductive adhesives are designed, and the properties of the conductive adhesives, both bimodal and trimodal, are also studied. The glass transition temperature of the adhesives are around 200~230 ℃, independent of the fraction of nano-size fillers. The nano-size fillers, silver nanoparticles and treated CNTs strengthen the electrical conductance of the adhesives, bringing the resistivity as low as $3*10^{-5}\Omega*cm$. Strong acid treatment enhances the uniform distribution of CNTs in the matrix, however proper surfactants are needed to be developed in order to increase the amount of added CNTs with low viscosity. More comparable and accurate approach should be designed to confirm the measurement and comparison of shear strength of different adhesives.

Acknowledgments

The work of Inoue was financially supported by the Swedish Foundation for International Collaboration in Research and Higher Education (STINT). Liu and Jiang are supported partly by the Seventh Framework Programme of the European Union (Project name: Nano Packaging Technology for Interconnection and Heat Dissipation under the contract No: 216 176) and partly by the EU EC-GEPRO project.

References

[1] A.O. Ogunjimi, et al. A Review of the Impact of Conductive Adhesive Technology on the Interconnection, J. of Electronics Manufacturing, 1992; 2(3): 109

[2] D.D. Chang, et al. An Overview and Evaluation of Anisotropically Conductive Adhesive Films for Fine Pitch Electronic Assembly, IEEE Transactions on Components, Hybrids and Manufacturing Technology, 1993;16(8): 828

[3] Daoqiang Lu, C.P.Wong, High Performance Conductive Adhesives IEEE Transactions on Electronics Packaging Manufacturing, Vol. 22, No.4, October 1999:324-329

[4] H.P. Wu, et al. Properties Investigation on Isotropic Conductive Adhesives Filled with Silver Coated Carbon Nanotubes, Composites Science and Technology 67 (2007):1182-1186

[5] J. Liu, Conductive Adhesives for Electronics Packaging (Electrochemical Publications Ltd. Isle of Man, British Isles, 1999), Chapter 1

[6] SG Prolongo and A.Ureña, Effect of surface pre-treatment on the adhesive strength of epoxy–aluminium joints. Int J Adhesion and Adhesives (2008), doi:10.1016/j.ijadhadh.2008.01.001

[7] R. Kahraman, Mehmet Sunar, Bekir Yilbas, Influence of adhesive thickness and filler content on the mechanical performance of aluminum single-lap joints bonded with aluminum powder filled epoxy adhesive, J. Mater. Process. Tech. (2008):183-189

[8] Hongjin Jiang, Moon.Kyoung-Sik, Yi Li, C.P.Wang. Ultra high conductivity of isotropic conductive adhesives, Electronic Components and Technology Conference (2006): 485-490.

Microstructural and physical characteristics of Sn-Ag-Cu-Mg Lead-free Solders

Sheng LU[a*], Fei LUO[a], Jing CHEN[b], Baohua WANG[a]

a.Department of Welding and Materials Forming, b. Department of Mechanical Engineering
Jiangsu University of Science and Technology
2 Mengxi Road, Zhenjiang, Jiangsu, 212003, PRC
E-mail: Lusheng119@yahoo.com.cn, Tel: 86-511-4407569

Abstract

Sn-Ag-Cu-Mg lead-free solders based on a eutectic Sn-Ag-Cu alloy were prepared by means of orthogonal experimental design. The melting temperature and the wettability of Sn-Ag-Cu-Mg solders were tested by differential thermal analysis (DTA) method and SAT solder checker respectively. Combined with the microstructure analysis, the influence of trace addition of Mg on the melting points and wettability was investigated. The results indicate that Mg, combined with the other alloying elements Ag and Cu, plays a key role in modifying the melting characteristics of the alloy solders. And with the added Mg, the wettability of solders is depressed.

Keywords: lead-free solders; Sn-Ag-Cu-Mg; melting temperature; wettability; microstructure

1 Introduction

Because the Sn-Pb alloy solder has many advantages such as low melting point, rich reserves, cheap price, it is a widely used joining material applying in the modern manufacturing industry, especially the domestic electric appliances, the electronic communication, computers and such related electronic industries. However, the element lead and the lead compounds are highly toxic[1~2]. Many developed countries have limited and forbid its application in the electronic industries gradually through the legislation [3~5]. Therefore, it is an imminent mission to research and develop new lead-free solder to substitute the Sn-Pb solder[6].

The constant development of the electronic industry calls for new solders with better performance. As a result of the poor creep resistance of the traditional Sn-Pb solders [7], they have not been able to satisfy the reliability requirements of the electronic industry. Recently, the domestic and foreign existing research results have indicated that the alloy solders composed by the elements Sn、Ag、Cu have a good application prospect. They have good mechanical properties, better weld performance, and they are easy to control in the manufacture and preparation[8]. However, there are also some disadvantages about the Sn-Ag-Cu solders, such as high melting points, coarse microstructure, and worse weldabilities than Sn-Pb solders. The objective of the present study was to design and prepare Sn-Ag-Cu-Mg alloys on basis of orthogonal design. Further efforts were

focused on clarifying the relationship between the concentration of Mg and the melting temperature, wettability, and other physical properties for the Sn-Ag-Cu-Mg lead-free solders.

A statement of the problem or situation, and the approach that is taken to resolve it. The second paragraph in the Section should start with a "line break" (using the enter/return key), but do not add an extra blank line. The slight indent will clearly define the paragraphs, and the "ICEPT Text" Style includes a slight spacing (1 point) between paragraphs within the Section.

This first Section may also contain a summary of the past developments and background of what is already known, and published elsewhere. This is best summarized in your own paper, with references to other publications containing more-extensive discussions of this background information. [1] The references are placed at the end of the paper. [2]

Remember that you should not re-state material that is readily available in the archival literature; simply summarize it, then add a reference or two. Many peer-reviewed papers in our fields have been published in the IEEE Transactions on Components and Packaging Technologies, and on Advanced Packaging – see www.cpmt.org/trans/.

Table 1 Factors and levels of the OED

Levels	Factors		
	Ag wt.%	Cu /wt.%	Mg/wt. %
1	3.20	0.50	0.1
2	3.50	0.65	0.4
3	3.80	0.80	0.7

Table 2 Arrangement and composition of SACM solders based on the OED

No.	Composition /wt. %			
	Ag	Cu	Mg	Sn
1	3.2	0.5	0.1	balance
2	3.2	0.65	0.4	balance
3	3.2	0.8	0.7	balance
4	3.5	0.5	0.4	balance
5	3.5	0.65	0.7	balance
6	3.5	0.8	0.1	balance
7	3.8	0.5	0.7	balance
8	3.8	0.65	0.1	balance
9	3.8	0.8	0.4	balance

2 Experiment procedure

The experimental Sn-Ag-Cu-Mg (SACM)solder alloys was designed by the orthogonal experimental design (OED) with a L9(4³) form shown in table 1. Their composition are among the range (wt%) of Ag 3.2~3.8, Cu 0.5~0.8, Mg 0.1~0.7 and Sn the balance. The corresponding experiment arrangement is listed in table 2. For the purpose of comparison, a Sn-4.0Ag-0.50Cu (SAC) alloy was among the experiment.

All experimental alloys mentioned above were melted in a resistance furnace with pure metals of 99.99wt. % Sn, 99.90wt.%Ag, 99.90wt. %Cu and 99.50wt.% Mg, 99.90wt.% . After melted, the ingot was kept at 750℃for 10 minutes and stirred for the uniformity. The solder ingots were re-melted at 400℃ twice for more homogeneous composition.

A differential thermal analysis (DTA) device, Perkineler instrument Pyris Diamond was adopted to determine the melting points and the melting rang of solders. The experiment was carried out in a helium atmosphere with the sample weight around 15mg and a heating rate of 5℃/min.

A SAT 5100 solder tester was used to evaluate the wettability of experimental solders according to the standard of IEC 60068-2-54, soldering solderability testing by the wetting balance method. The parameters of the experiment are given as below: immersed depth: 1 mm, immersed time: 5 second, immersed velocity: 2mm/sec,immersed angle: vertical, temperature: 260℃, flux: ROL1. The parameters to be measured include T_0, the specimen touches the surface of molten solder;T_1, the specimen begins wetting by molten solder, corresponding to a certain point where the curve starts rising;T_2,the buoyancy force corresponds with the recorded force; F_{max}, the max wetting force.

To observe the microstructures, the specimens were polished up with 0.3μm Al2O3 and etched with FeCl3 solution. Microstructure observation was finished by means of a ZEISS optical microscopy (OM).

3 Experimental results and discussion

3.1 Melting temperature

The results of the OED are shown in table3.

Table 3 The melting temperatures of solders

No.	Onset/ ℃	End / ℃	Melting range/℃
SACM -1	217.0	237.2	20.2
SACM -2	216.4	234.7	18.3
SACM -3	211.8	233.2	21.4
SACM-4	210.9	237.6	26.7
SACM -5	211.2	236.9	25.7
SACM -6	216.5	234.2	17.7
SACM -7	215.2	245.9	30.7
SACM -8	217.8	237.2	19.4
SACM -9	211.3	231.7	20.4
SAC	218	225	7

Compared with the solder of Sn-4.0Ag-0.50Cu, it can be seen obviously that the trace addition of Mg into Sn-Ag-Cu

solders can modifies the melting characteristics. It was reported that Sn-Ag-Cu ternary alloy has a eutectic temperature around 217℃ or 218℃. The results of table 3 indicate that that with the addition of Mg, the melting temperature of the SACM solders decreases in some extent. Some specimens even have large scale reduction. However, it does not show a simplex trend that the melting temperatures decrease along with the addition of Mg element. The total amount of alloying elements Ag, Cu and Mg has complex influence on the melting temperature. Further analysis in melting range shows that the more the addition of Mg is, the broader the melting range may reach. It also seems that the solders with lower melting temperatures exhibit wider melting ranges.

The variance analysis about the OED is summarized in table 3 where Ⅰ,Ⅱ and Ⅲ stand for the average values of melting temperature or melting range corresponding to counterpart level of every factor while R is the variance, the difference of maximum and minimum for Ⅰ,Ⅱ and Ⅲ.

Table 4.Variance analysis on the OED

Factors	Melting temperatureTo/℃			
	Ⅰ	Ⅱ	Ⅲ	R
Ag	215.1	212.9	214.8	2.2
Cu	214.3	215.1	213.2	1.9
Mg	217.1	212.9	212.7	4.4
	Melting range/℃			
	Ⅰ	Ⅱ	Ⅲ	R
Ag	19.97	23.37	23.50 3	3.5
Cu	25.87	21.13	19.83 4	6.0
Mg	19.10	21.8	25.93 3	6.8

Table 4 The wettability test results of the SACM solders

Alloy No.	T0	T1	Fmax(mN)
SACM-1	1.56 s	0.34 s	0.63
SACM-2	1.77 s	0.59 s	0.81
SACM-3	/	/	/
SACM-4	3.31 s	0.89 s	0.26
SACM-5	2.79 s	0.72 s	0.7
SACM-6	/	/	/
SACM-7	/	/	/
SACM-8	/	/	/
SACM-9	/	/	/
SAC	0.90 s	0.18 s	0.7

It is well known that the value of variance could make a clear understanding of how a factor influences a certain target. According to the R value of table 3, some conclusion could be achieved. To the melting temperature, Mg is the major factor while Ag the second important factor. The influencing order is Mg>Ag>Cu. As for the melting range, Mg and Cu play the same import role with such order Mg> Cu>Ag. Obviously, the element Mg plays a key role in modifying the melting characteristics of the SACM solders. It is believed that a narrow melting range and a low melting point are positive aspects for good lead-free solder alloys. To make a Sn-Ag-Cu-Mg solder with better melting characteristic, it is necessary to consider the influence of not only the trace addition of Mg, but also the mix addition of Ag and Cu.

3.2 Wettability

The results of wettability test are shown in table 5.It can be seen that the wettabilities of SACM-1,2 are best among the SACM specimens. According to the evaluation standard, both test results are passed. Though the results of specimens SACM-4,5 are not satisfied by the standard, inferior, the are superior to the rests, whose evaluation parameters can't be measured because their zero cross time surpassed the test time.

Combined with the consideration of compositions of each solder alloys, the solders that perform well in wettability have relatively fewer total addition of alloying elements Ag,Cu,Mg, especially no more than 0.4wt.% Mg .

Moreover, compared with the wettability test result of the Sn-3.5Ag-0.75Cu solder, it can be discovered that although the addition of Mg reduces the melting temperatures, but it also worsens the wettability of the solders dramatically. It is considered that Mg is prone to oxidize, and the oxide film produced in the thermal process increases the surface tension of the liquid solder, thus prevents the solder from spreading over the copper wire surface. Therefore, the addition of Mg has negative influence on the wettability.

3.3 Microstructure

The optical micrographs and SEM images of SACM alloys are shown in fig.1 to fig.5. It is obviously that with different trace addition of Mg, and other addition amount of Ag and Cu, the microstructure of every Sn-Ag-Cu-XMg (x=0.1, 0.4, 0.7) are distinguished from each other remarkably. As seen in fig.1 and fig3, SACM-1 and SACM-2 exhibit the typical microstructure of near eutectic Sn-Ag-Cu solder with dendritic primary β-Sn and fine eutectic mixture of Ag$_3$Sn and Cu$_6$Sn$_5$ in the Sn-rich matrix.

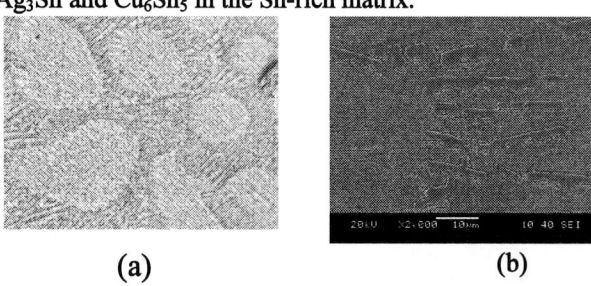

(a) (b)

Fig.1. Microstructures of SCAM-1,
Sn-3.2Ag-0.50Cu-0.1Mg (a) OM, (b) SEM

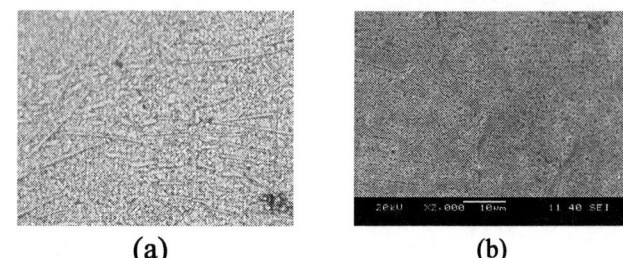

(a) (b)

Fig.2. Microstructures of SCAM-8, Sn-3.8Ag-0.65Cu-0.1Mg
(a) OM, (b) SEM

But with the same amount of Mg 0.1 and 0.4 respectively, and more addition of Ag and Cu, as shown in fig.2 and fig.4, SACM-8 and SACM-9 almost lose the typical microstructure of near eutectic Sn-Ag-Cu. The eutectic phase becomes more coarsened while its fraction decreases more considerably.

As shown in fig.5, with as more as 0.7wt % Mg, none typical fine eutectic phase are observed in and Mg is found existing in Ag3Sn platelet phase and it suggests that little Mg dissolve in Ag3Sn to form (AgMg)3Sn eutectic phase. Besides, new bulk-like SnAgCuMg-rich phases are found in SACM-5 and SCAM-3.

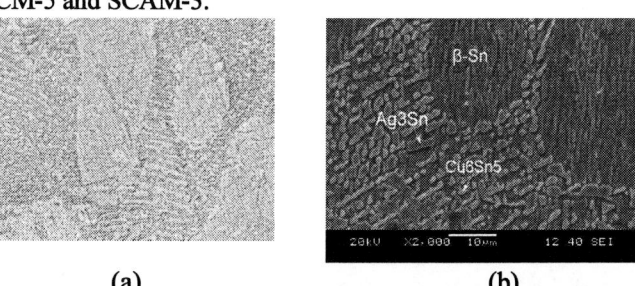

(a) (b)

Fig.3. Microstructures of SCAM-4, Sn-3.2Ag-0.65Cu-0.4Mg
(a) OM, (b) SEM

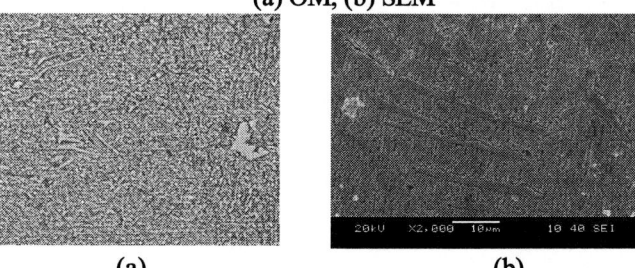

(a) (b)

Fig.4. Microstructures of SCAM-9, Sn-3.8Ag-0.80u-0.4Mg
(a) OM, (b) SEM

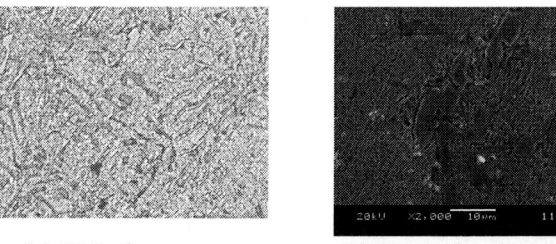

(a) OM of Sn-3.5Ag- (b) SEM of Sn-3.2Ag-0.80u-
0.65Cu-0.7Mg 0.7Mg

Fig.5. Microstructures of (a) SCAM-5, (b) SCAM-3

4 Conclusions

Experimental Sn-Ag-Cu-Mg lead-free solders were prepared by means of orthogonal experimental design. Properties of these SACM solders are evaluated. It is concluded that, with trace addition of Mg into ternary Sn-Ag-Cu, the melting temperatures decrease, and the melting ranges are wider for those solders with lower melting points. The element Mg combined with the other alloying elements Ag and Cu, plays a key role in modifying the melting characteristics of the SACM solders. With the addition of Mg, the eutectic network and phases become coarsened and disorder, and the dendritic primary phase disappears gradually. For SACM solder with 0.7% Mg, the bulk-like phase was discovered. The addition of Mg has negative influence on the wettability.

Acknowledgments

The present work was supported by the high and new technology industrial development program of jiangsu, grant No. JHB05-40, and the key laboratory of advanced welding technology of Jiangsu Province, P.R.C, grant No. JSAW 06-08.

References

1. Zhitian Hu, Qianjin He, Daorong Xu. [J]. Welding technology，2005，34(3)：4-8. Henry Y. Lu , Haluk Balkan , K.Y. Simon Ng, "Effect of Ag content on the microstructure development of Sn-Ag-Cu interconnects", J Mater Sci: Mater Electron, 17 (2006),pp. 171–188

2. Daquan Yu, Jie Zhao, Lai Wang. "Effects of rare-earth elements on wettability of Sn29Zn alloy"[J] Non-ferrous metal journal of China,2003,13(4) ,pp. 1001-1004.

3. RICHARD B, Lead-free soldering--world's apart, National Physical Laboratory, NPL News, UK, http://www. lead - free. org, 2004.

4. Grum S. Components materials re-examined for lead-free compatibility [J]. Electron Packg Prod, 2000, 4(6), pp. 13-15

5. Frear D R. Trends and issues in Pb-free soldering for electronic packing [J]. Electron Tech, 2001, 118(2), pp. 81–86.

6. Yanxiang Kuang. "Micro-electronic package technology of the new century"[J] Electron processing technology, 2001(1) , pp. 1-6.

7. K.S. Kim, S.H. Huh, K. Suganuma, "Effects of fourth alloying additive on microstructures and tensile properties of Sn–Ag–Cu alloy and joints with Cu", Microelectronics Reliability, 43 (2003) ,pp.259–267

8. Martin Goosey, "Soldering considerations for lead-free printed circuit board assembly–an Environment Guide.", Circuit World, Vol.31, No. 3 (2005), pp.40-44.

Microstructural Investigation on the Interfacial Evolution of SnBi/Cu Interconnect during Reflow and Solid-State Aging

Zhi-Quan Liu,[a] Pan-Ju Shang,[a] Xue-Yong Pang,[a] and Jian-Ku Shang [a,b]

[a] Shenyang National Laboratory for Materials Science, Institute of Metal Research, Chinese Academy of Sciences, Shenyang 110016, China, (Email: zqliu@imr.ac.cn)

[b] Department of Materials Science and Engineering, University of Illinois at Urbana–Champaign, Urbana, IL 61801, USA

Abstract

The microstructures of SnBi/Cu interconnect after reflow and its evolution during solid-state aging at 393K were investigated using transmission electron microscopy (TEM). It was found that after reflow there were two kinds of intermetallic compounds (IMCs) - Cu_6Sn_5 and Cu_3Sn in solder joint. Above these IMC layers there was a Bi-rich layer which consisted of discontinuous Bi particles. During solid state aging, the Cu_3Sn phase grew faster into Cu substrate than into Cu_6Sn_5, and the Bi-rich layer did not change much. However, new Bi particles were observed at the Cu_3Sn/Cu interface, which introduced interfacial voids around Bi particles. First principles calculation revealed that Bi segregation reduced the adhesion energy of Cu/Cu_3Sn interface and caused serious reliability problem.

Introduction

In microelectronic industry, solder interconnects have been extensively used for many years to provide electrical connections as well as mechanical and physical connections [1,2]. With the trend of miniaturization and functionality in modern consumer electronics, a large number of input/output (I/O) connections are needed to improve the packaging density on high circuit density device chips. As the density of solder bumps rises , the individual solder joints are scaled to smaller dimensions. Therefore, the microstructural defects in the small solder joints, such as the impurity atoms and voids at the interfacial region, become increasingly more important for the component reliability between solder and metallization [3].

Due to its lower melting temperature and better mechanical properties, eutectic SnBi alloy is a potential candidate to replace eutectic SnPb solder in some applications. Since Bi does not form any compound with Cu, interfacial reactions between SnBi solder and Cu substrate result in two kinds of Intermetallic compounds (IMCs) – Cu_6Sn_5 and Cu_3Sn. During solid state aging, segregation of Bi element will happen at the Cu_3Sn/Cu interface followed by voiding, which is detrimental to the reliability of the interconnect [4]. Although this phenomenon has been observed and reported before, the detailed process and mechanism remains unclear, partially due to the low resolution of scanning electron microscopy (SEM) and scanning Augur microprobe (SAM) used in these studies.

In this work, transmission electron microscopy (TEM) was used to investigate the interface of the eutectic SnBi/Cu interconnects, to clarify the microstructural evolution during wetting and solid-state reactions. The effects of Bi segregation were discussed with first principles calculations.

Experimental Procedure

Both oxygen-free high conductivity (OFHC) Cu and single crystal Cu sheets were used as substrate. The Cu sheets with the size of 10 mm × 2.5 mm × 2 mm were cut first, and then ground mechanically and carefully surface polished by 0.5μm diamond paste. After that they were cleaned in an ultrasonic bath with acetone, ethanol and distilled water in order. Upon air drying, a commercial eutectic Sn-Bi solder paste was dispersed on copper surfaces. Two copper surfaces were placed together face to face, between which several fine brass wires (the diameter is less than 20μm) were put in for controlling the thickness of solder. The reflowing process was carried out on a heating plate at 423K until the solder was completely melted for about 3 seconds, and then cooled in air as reflowed samples. Aging of samples were carried out in silicone oil bath at 393K for different hours.

To prepare TEM sample, a cross-sectional specimen in thickness of 300μm was cut from the solder joint by spark erosion, and then mechanically ground to a final thickness of 40μm. TEM foils were produced by ion milling (Gatan model 691 PIPS) at 5.0keV and 5μA with low milling angle (< 6°) at room temperature. Microstructural observations were carried out to investigate the interfaces of the solder joint.

First principles calculation was conducted on a model of $Cu(100)(\sqrt{2}\times\sqrt{2})$/$Cu_3Sn(010)$ interface to investigate the intrinsic mechanics of segregation-induced embrittlement. The Cu/Cu_3Sn interface was constructed by six layers of Cu_3Sn, seven layers of Cu and ten angstrom vacuum.

Results and Discussions

In the wetting process, liquid SnBi solder reacted with solid Cu substrate to form solid Cu_6Sn_5 firstly. With the consumption of Sn from SnBi solder, Bi was left at the Cu_6Sn_5/solder interface (the frontier of the wetting reactions) to form a Bi-rich layer. Fig. 1 shows the microstructure of the interface after reflowing, where the Bi-rich layer consists of discontinuous Bi particles ranging from several hundreds to two thousands nanometers. The beneath Cu_6Sn_5 layer is about 1000 nm thick on Cu substrate. The typical scallop-like morphology of Cu_6Sn_5 is shown in Fig. 2a, whose frontier was marked with black dots. Its grain size is not homogeneous, which should be due to the blocking of Sn diffusion by Bi particles. The interface between Cu_6Sn_5 and Cu was enlarged and shown in Fig. 2b, where Cu_3Sn layer about 150 nm thick was observed on single crystal Cu. The reflowed grain size and thickness of Cu_3Sn layer were smaller on single crystal Cu than on polycrystalline Cu substrate.

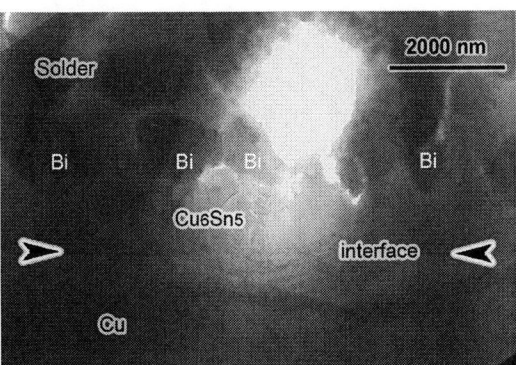

Fig.1 TEM image of the interface between SnBi solder and single crystal Cu after reflow.

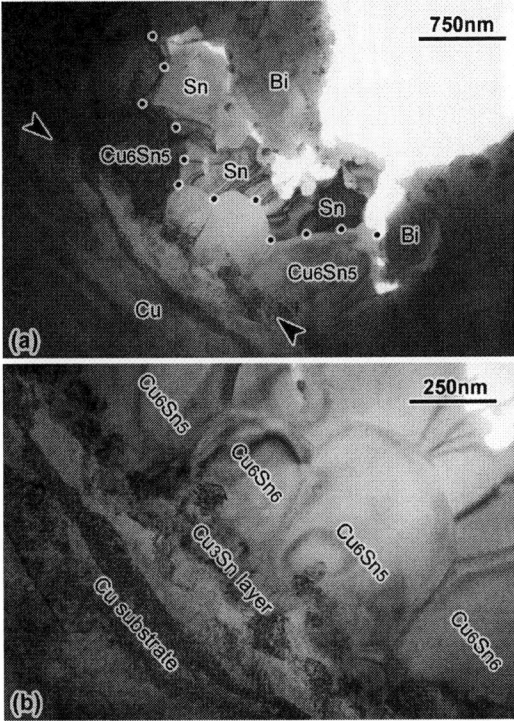

Fig.2 The interfacial microstructure showing the morphology of Cu_6Sn_5 (a) and Cu_3Sn (b) after reflow.

In the following solid state aging, Cu_6Sn_5 layer kept growing into solder as Cu diffused to its frontier to react with Sn. For the Bi-rich layer above Cu_6Sn_5 layer, there were no obvious changes on its position and morphology. On the other hand, Cu_3Sn layer grew up quickly resulting from solid reactions between Cu and Cu_6Sn_5. After three days aging at 393K, the interface in Fig.2a has changed as shown in Fig.3. The original position of Cu_3Sn layer after reflow was indicated with white arrows and dots, while after aging the Cu/Cu_3Sn and Cu_3Sn/Cu_6Sn_5 interfaces were marked with black lines. It was observed that after aging the thickness of Cu_3Sn layer increased from 150nm (see Fig.2) to about 950nm (between broken black lines in Fig.3). According to its reflowed position, Cu_3Sn layer was thickened on both sides. It consumed Cu_6Sn_5 layer for about 340nm, and expanded into Cu substrate for about 610nm. That is to say, Cu_3Sn layer grew faster into Cu side than into Cu_6Sn_5 side. This implies

that in this experiment Cu is the main diffusion species for the growth of Cu_3Sn phase [5,6].

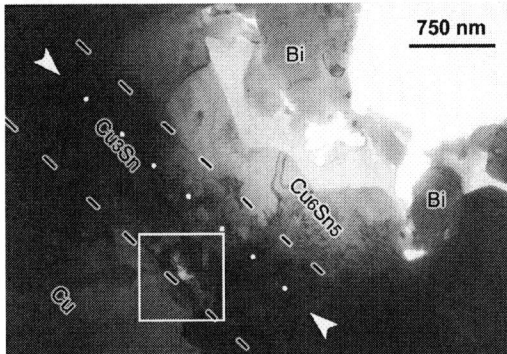

Fig.3 TEM image of the interface showing the growth of Cu_6Sn_5 and Cu_3Sn during solid-state aging.

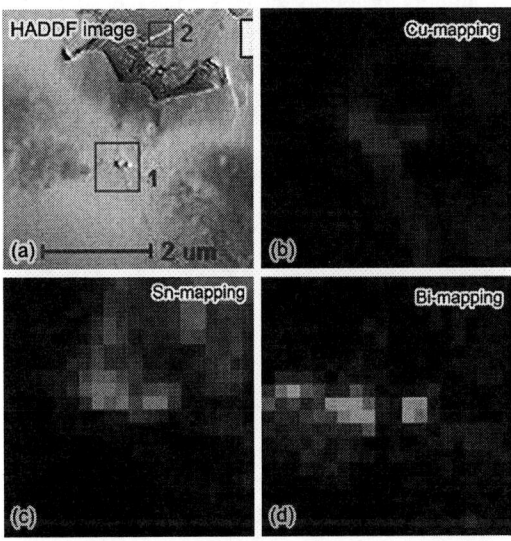

Fig.4 Different elemental mapping showing segregation of Bi at Cu/Cu_3Sn interface during solid-state aging.

The Cu/Cu_3Sn interface was further studied by scanning transmission electron microscopy. Fig.4a shows a high-angle annular dark-field (HAADF) image of the area in Fig.3. Elemental mapping was conducted on the outlined area of the Cu/Cu_3Sn interface in Fig.3 and Fig.4a. Precipitation of Bi particles took place as shown in Fig.4d. It was found that Bi atoms segregated to the interface of Cu_3Sn/Cu to form Bi particles. At last, voids were formed at this interface, locating mainly at Cu side [7]. As Cu_3Sn/Cu_6Sn_5 interface was Bi free and Cu_3Sn grew faster into Cu, it is believed that segregation of Bi atom is more related to the growth of Cu_3Sn.

The segregation of Bi atoms at $Cu(100)/Cu_3Sn(010)$ interface was analyzed using first principles calculations. The "Between-Cu" construction was found to be the most energy-favored. Segregation of Bi on the most favorable site on this construction could decrease the initial adhesion energy to about half. According to the further investigations on atomic structure and electronic density, it was revealed that the interfacial space between Cu and Cu_3Sn was enlarged by bismuth segregation. The bond between the first layer and the second layer at copper side was weakened [8]. These calculated results are qualitatively consistent with reported experimental results and can explain them well.

Conclusions

Two IMC layers, Cu_6Sn_5 and Cu_3Sn, were formed in eutectic SnBi/Cu solder joint during reflow at 423K for a short time. The Bi-rich layer appeared above the Cu_6Sn_5 layer and was made up of discontinuous Bi particles ranging from several hundreds nanometers to two micrometers. During solid-state aging, the position and morphology of this Bi-rich layer did not change. However, as a product of solid state reaction between Cu and Cu_6Sn_5, Cu_3Sn grew up quickly into Cu substrate, and segregation of Bi took place at the Cu/Cu_3Sn interface. Theoretical analysis showed that Bi segregation reduced the interfacial adhesion energy to by approximately half, which induced voids around these Bi particles.

Acknowledgments

The authors gratefully acknowledge the financial support from the National Natural Science Foundation of China under Grant No. 50228101 and 50571100, and the National Basic Research Program of China under Grant No. 2004CB619306.

References

1. Kang, S.K., Sarkhel, A.K., "Lead(Pb)-free solders for electronic packaging," *J. Electron Mater.*, Vol. 23 (1994), pp. 701-707.
2. Lau, J.H., Wong, C.P., Prince, J.L., Nakayama, W., "Electronic packaging – design, materials, process, and reliability," McGraw-Hill, New York, 1998.
3. Zeng, K., Stierman, R., Chiu, T.C., Edwards, D., Ano, K., Tu, K.N., "Kirkendall void formation in eutectic SnPb solder joints on bare Cu and its effect on joint reliability," *J. Appl. Phys.*, Vol. 97 (2005), pp. 024508 (8 pages).
4. Liu, P.L., Shang, J.K., "Segregant-induced Cavitation of Sn/Cu Reactive Interface," *Scripta Mater.*, Vol. 53 (2005), pp. 631-634.
5. Paul, A., Kodentsov, A.A., van Loo, F.J.J., " Intermetallic Growth and Kirkendall Effect Manifestations in Cu/Sn and Au/Sn diffusion Couples," *Z. Metallkd.*, Vol. 95 (2004), pp.913-920.
6. Vuorinen, V., Laurila, T., Mattila, T., Heikinheimo, E., Kivilahti, J.K., "Solid-state reactions between Cu(Ni) alloys and Sn," *J. Electron. Mater.*, Vol. 36 (2007), pp. 1355-1362.
7. Shang, P.J., Liu, Z.Q., Li, D.X., Shang, J.K., "Bi-Induced Voids at the Cu_3Sn/Cu Interface in Eutectic SnBi/Cu Solder Joints," *Scripta Mater.*, Vol. 58 (2008), pp. 409-412.
8. Pang, X.Y., Liu, Z.Q., Wang, S.Q., Shang, J.K., "First-principles investigation on Bi segregation at solder interface", to be submitted.

Dynamic Mechanical Properties of the Transparent Silicone Resin for High Power LED Packaging

Qin Zhang[a, b], Xiu Mu[c], Kai Wang[b, d], Zhiyin Gan[a, b], Xiaobing Luo[a, e], and Sheng Liu[a, b, c]*

[a] Institute for Microsystems, School of Mechanical Engineering, Huazhong University of Science & Technology, Wuhan, China, 430074

[b] Division of MOEMS, Wuhan National Lab for Optoelectronics, Wuhan, China, 430074

[c] Shanghai FINE MEMS INC., Wuhan Branch, Wuhan, China, 430074

[d] School of Optoelectronics Science and Engineering, Huazhong University of Science & Technology, Wuhan, China, 430074

[e] School of Energy and Power Engineering, Huazhong University of Science & Technology, Wuhan, China, 430074

* Corresponding author, Email: victor_liu63@126.com,
Telephone: 86-13871251668, Fax: 86-27-87557074

Abstract

Aiming to study dynamic mechanical properties of the transparent silicone resins for LEDs packaging and the effects of dynamic mechanical properties on the reliability of high power LEDs, dynamic mechanical analysis (DMA) has been adopted to study one silicone resin used for packaging high power white light LEDs. The viscoelastic behavior of silicone resin was obtained from multi-frequencies dynamic mechanical temperature spectra measured by dynamic mechanical analysis instrument in compression mode. The order of storage modulus was 10^6 whether in the glassy state or rubbery state, storage modulus in glassy state was about 50 times more than that in rubber state. Glass transition was found at 40℃ and the frequency had increased accompanied by growing of temperature of glass transition. The activation energy of glass transition was also calculated. A master curve of storage modulus was generalized according to time-temperature superposition principle and WLF equation. Furthermore, characteristics of storage modulus, loss factor, glass transition, etc. were used for exploring how to increase the reliability of LEDs and the reasons which led to the failure of LEDs.

Introduction

As an important encapsulation gel for high power LED (HP-LED) packaging, silicone resin can provide physical protection of the chip and inter-connecting wires, and enhance the optical efficiency. Typical structure of the HP-LED is shown in Figure 1. Molding compounds are plastic, resin, silicone gel, etc. Compared with epoxide resin, silicone resin has excellent thermal and light resistance characteristics [1]. It has found from industrial practices that silicone gels eliminate the yellowing issue of conventional epoxy type encapsulates. The properties of silicone resin affect seriously the life and reliability of LEDs. Thermo-mechanical properties are essential for the fundamental understanding of LED packaging behaviors.

Dynamic mechanical spectrum is one of the most important properties for viscoelastic materials. The material's dynamic mechanical properties are usually studied by dynamic mechanical analysis (DMA) instrument [2, 3]. The DMA technique can measure the glass transition temperature more precisely and sensitively than DSC or TMA. The principle of dynamic mechanical analysis is measuring the amplitude and phase lag of the oscillating mechanical stress within wide temperature and frequency ranges. Then the storage modulus, loss modulus and tangent delta are calculated by using the resultant displacement and phase lag [4]. DMA can also study the structure, glass transition, chains' motion, etc.

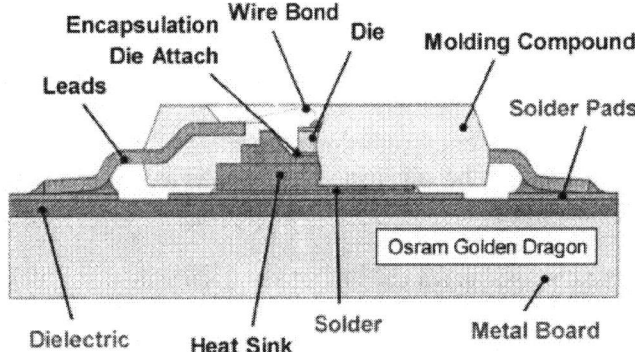

Figure 1. Schematic of LED structure [1].

The time-temperature superposition principle (frequency-temperature superposition), which accounts for the equivalence of time and temperature, is important in investigating viscoelastic behavior of materials [5]. Based on the hypothesis that the mechanical behavior at long time is analogous to the behavior at high temperature, the mechanical behavior at long time or low frequency could be derived from the mechanical behavior at high temperature or high frequency.

The paper is organized as follows. 1) Preparation of specimens and DMA experiment conditions were stated; 2) The experiment results, multi-frequency dynamic mechanical temperature spectroscopy, were listed, the temperature of glass transition as a function of frequency was obtained; and . 3) The correlation between one parameter and others was discussed and the reliability questions associated these parameters were analyzed.

Experiments

Specimen Preparation. Two parts of transparent silicone resin for LED packaging was well mixed, then removed air bubbles thoroughly under vacuum condition before being injected it into moulds. Curing temperature was 150℃ and the curing time was 1 hour. The rate of temperature rise during heating process was about 1.5℃/s. Diameter 15mm circular

978-1-4244-2739-0/08/$25.00 ©2008 IEEE

samples were cut from the 2mm thick sheets of the silicone resin.

Dynamic Mechanical Analysis (DMA) testing. Because the cured silicone resin was very soft polymer material, the best clamping modes were shear or compression. Compression mode was adopted. The dynamic mechanical properties of silicone resin samples for LED packaging was investigated by using dynamic mechanical analyzer, diamond DMA, PerkinElmer Instruments. The vibratory frequencies applied were 0.5, 1, 2, 5, and 10Hz within the temperature range from -50℃ to 180℃. The heating rate is 2℃/min. Gas environment is just un-flowing air. The static force was 9.8N.

The viscoelastic motions with various frequencies within the whole temperature range were obtained, with cyclic loading on measured sample. It could be written as:

$$E = E' + iE'' \tag{1}$$

where E' is storage modulus which values the rigidity of materials, E'' is loss modulus.

$$\tan \delta = \frac{E'}{E''} \tag{2}$$

where $\tan \delta$ is loss factor which values the viscidity of materials.

Results

The multi-frequency dynamic mechanical temperature spectroscopy of the silicone resin is shown in Figure 2. From the picture, it can be found that the glassy moduli (dynamic moduli under $T = T_g$) and rubbery moduli (dynamic moduli under $T = T_g$) are fairly constant. The orders of storage modulus and loss modulus both are 10^6 whether in the glassy state or in the rubbery state. Storage modulus in glassy state is about 50 times more than that in rubbery state. In the curves of $\tan \delta$, the peak of the loss factor shifts to a high temperature that accompanied by an increase of T_g values. It is because higher frequency induces more elastic-like behavior [6].

Specially，there is a abnormal peak of storage modulus within temperature range from 0℃ to 30℃. Following the increasing of frequency, the peak shows the same trend, increasing continuously. This special phenomenon might be caused by crystallizing at about 25℃.

Choosing the maximum of the transition in the loss factor curve as the glass transition temperature, the T_g of the silicon resin are acquired. The relationship between the mobility of polymer chains and temperature is characterized by T_g. Actually, the higher temperature (or lower frequency) is, the lower glass transition temperature is. The temperatures of glass transition for silicone resin are listed in Table 1.

Table 1. T_g for silicone resin

Frequency (Hz)	0.5	1	2	5	10
T_g (℃)	37.6	40.4	43.4	47.9	50.7

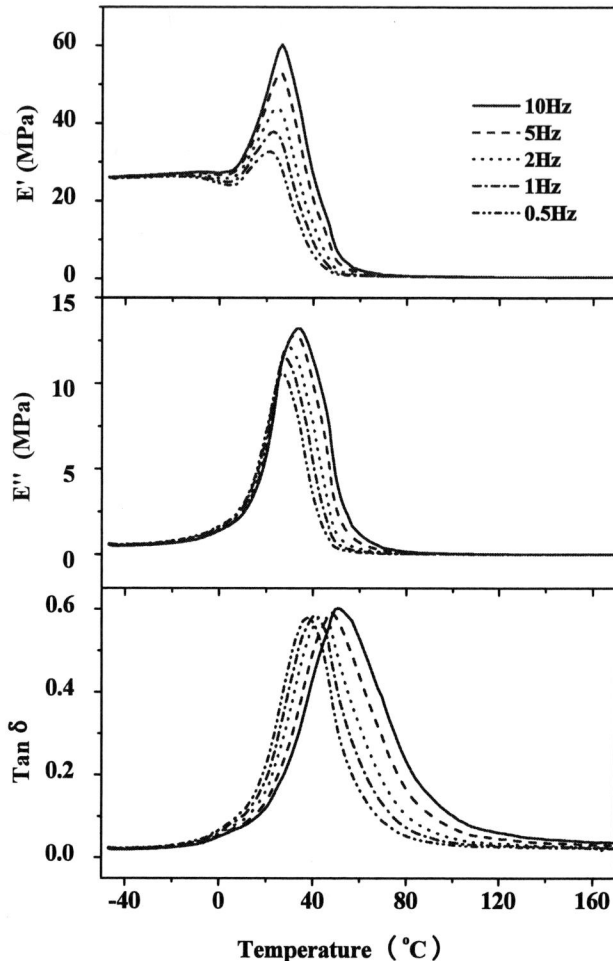

Figure 2. Storage modulus (E'), loss modulus (E''), and loss factor ($\tan \delta$) as functions of temperature for the silicone resin at 0.5, 1, 2, 5, and 10 Hz frequencies.

The relationship between the shift of glass transition temperature and frequency can be characterized by activation energy (ΔE). Furthermore, ΔE represents the energy barrier of glass transition relaxation. The activation energy for the glass transition of silicone resin can be calculated based on the modified Arrhenius equation [6, 7]:

$$\Delta E = R \frac{d \ln \omega}{d(1/T)}$$

$$= 2.303R \frac{d \lg \omega}{d(1/T)} \tag{3}$$

where ω is testing frequency (in Hz), R is universal gas constant ($8.314 J \cdot mol^{-1} \cdot K^{-1}$), and T is absolute temperature (in K). Figure 3 plots $1/T_g$ versus $\ln \omega$ for silicone resin. Activation energy of $192.11 KJ \cdot mol^{-1}$ for the glass transition of silicone resin is obtained.

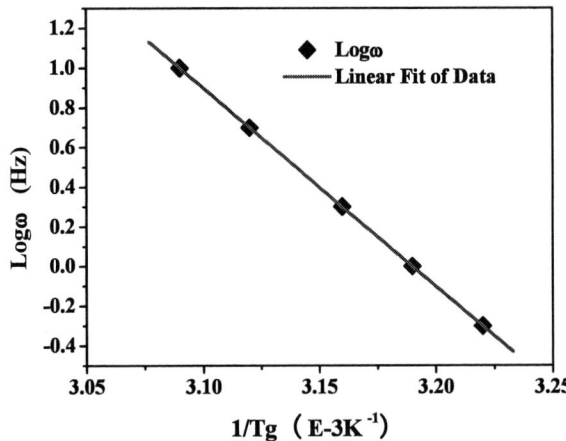

Figure 3. $1/T_g$ versus $\ln \omega$ for silicone resin.

Viscoelastic analysis

Based on the time-temperature superposition principle (TTS), master curves of dynamic modulus could be constructed by isothermal data. The construction method is level shifting isothermal curves on the frequency scale to overlap at a reference temperature. The overlapped curve is called master curve. Thus, the frequency range enhanced at this reference temperature. The shift factor is supposed to satisfy Williams-Landel-Ferry (WLF) equation:

$$\log(\alpha_T) = -\frac{C_1\left(T - T_r\right)}{C_2 + \left(T - T_r\right)} \qquad (4)$$

where α_T is the shift factor, C_1 and C_2 are constants which value the viscoelastic coefficients at T_g, T_r is the reference temperature (in K), and T is the measurement temperature (in K). The temperature scope of application of WLF equation ranges from T_g to $T_g + 100\,℃$.

The temperature of glass transition (T_g), $40\,℃$, is adopted as reference temperature T_r. The storage modulus isothermal curves are shown in figure 4. The master curve of storage modulus is generated by shifting and overlapping these isothermals as shown in Figure 5. The α_T is deduced from master curve of storage modulus. As shown in Figure 6, there is linear relation between variable of temperature and variable of $\log(\alpha_T)$.

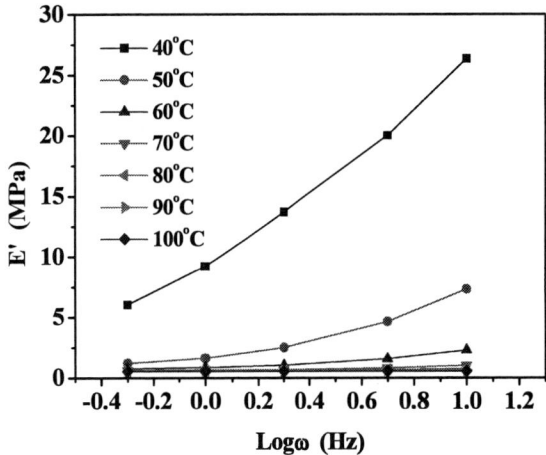

Figure 4. Storage modulus (E') isotherms as a function of frequency.

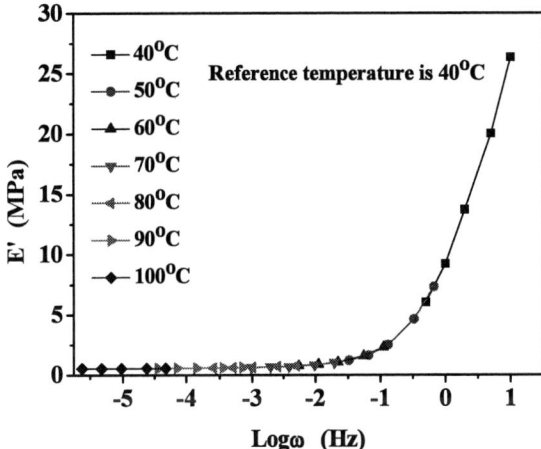

Figure 5. Master curve of storage modulus.

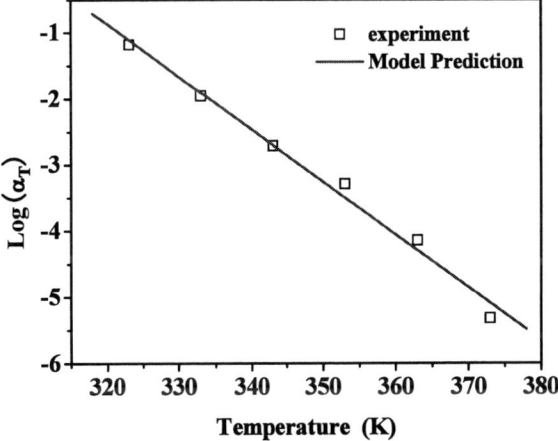

Figure 6. Experiment data and prediction of shift factor.

Reliability analysis of LEDs based on viscoelastic behavior of the silicone resin

With the input power increasing, packaging becomes a critical issue for HP-LEDs. Material challenges for the packaging of HP-LEDs include light extraction, thermal yellowing, UV yellowing, stress/delamination, lifetime, etc. [8]. For choosing an encapsulation material, the following considerations are needed: (1) material property measurement, (2) failure mode and failure mechanism analysis, and (3) key factors in failure.

Based on viscoelastic behavior of the silicone resin, three types of failure mechanisms can be analyzed. First, the coefficient of thermal expansion (CTE) mismatch leads to deformation, stress, and maybe delamination. As shown in Figure 1 and Table 1, T_g (40℃) is higher than usual ambient temperature. Because CTE changes significantly when glass transition occurs, the CTE mismatch between the silicone resin and other packaging materials is inevitable in thermal cycle. It will create stress and may lead to poor mechanical deformation and stresses along interfaces, separation or delamination [1]; Second, chemical aging and physical aging may occur. As far as polymer aging is concerned, the chemical aging would occur when the temperature is higher than T_g, whereas physical aging occurs; Thirdly, reliability degradation at vibration mode may occur. According to TTS principle, the mechanical behavior at high frequency is analogous to the behavior at low temperature. With the frequency increasing, storage modulus increases. High storage modulus would bring out poor shock absorption capacity of silicone resin, because storage modulus represents rigidity of materials. Then the separation or wire-bond breakage might occur.

We can see that T_g is a key in terms of reliability for HP-LEDs. Notably, the value of T_g obviously changes with changing frequency. The value of T_g as a function of frequency can be obtained from multi-frequency dynamic mechanical temperature spectroscopy or dynamic mechanical frequency spectroscopy.

Conclusions

The viscoelastic properties of transparent silicone resin for HP-LED packaging have been investigated by dynamic mechanical analyzer. The multi-frequency dynamic mechanical temperature spectroscopy of the silicone resin was obtained and used to study the characteristics of storage modulus and loss factor. The important parameters, activation energy and shift factor was calculated. A further viscoelastic analysis has allowed us to gain dynamic modulus within much wider frequency range. Finally, the relation of the reliability for HP-LEDs and the encapsulating silicone resin has been analyzed. It is shown that the viscoelastic properties of packaging materials for HP-LEDs are very important in analysis and enhancing the reliability of HP-LEDs.

Acknowledgments

Financial support for this work was provided by China Hubei Provincial Science & Technology Department (2006AA103A04). We wish to thank Mr. Yixiang Qian, adviser of thermo analysis instruments, SII NanoTechnology (Shanghai) Inc, and engineer Ms. Ling Wang, Analytical and Testing Center, Huazhong University of Science and Technology, for their kind assistance.

References

1. Robert F. Karlicek, Jr., "High Power Packaging", *2005 Conference on Lasers & Electro-Optics (CLEO)*, Baltimore, Maryland, USA, May. 2005, pp. 337-339.

2. Diez-Gutierrez, *et al.*, "Dynamic Mechanical Analysis of Injection-mould Discs of Polypropylene and Untreated and Silane-treated Talc-filled Polypropylene Composites," *Polymer*, Vol. 40, Issues 21 (1999), pp. 5345-5353.

3. D.J.T. Hill, *et al.*, "Dynamic Mechanical Properties of Networls Prepared from Siloxane Modified Divinyl Benzene Pre-polymers," *Polymer*, Vol. 41, No. 26 (2000), pp. 9131-9137.

4. Lecon Woo, *et al.*, "Relating Dynamic Mechanical Data to Flexible PVC Low Temperature Performance," *Thermochimica Acta*, Vol. 284, No. 1 (1996), pp. 57-66.

5. MEhdi Hojjati. *et al.*, "Viscoelastic Behavior of Cytec FM73 Adhensive During Cure", *Journal of Applied Polymer Sience*, Vol. 91, No. 4 (2004), pp. 2548-2557.

6. D.Z. Chen, *et al.*, "Dynamic Mechanical Properties and in Vitro Bioactivity of PHBHV/HA Nanocomposite," *Composites Science and Technology*, Vol. 67, Issues 7-8 (2007), pp. 1617 – 1626.

7. Vistasp M. Karbhari, *et al.*, "Multi-frequency Dynamic Mechanical Thermal Analysis of Moisture Uptake in E-glass/vinylester Composites," *Composites: Part B*, Vol. 35, Issue 4 (2004), pp. 299–304.

8. Yuan-Chang Lin, *et al.*, "Materials Challenges and Solutions for the Packaging of High Power LEDs", *International Microsystems, Packaging, Assembly Conference Taiwan*,Taiwan, China, October, 2006, pp. 1-4.

Study of Five Substrate Pad Finishes for the Co-design of Solder Joint Reliability under Board-level Drop and Temperature Cycling Test Conditions

Wei Sun, W.H. Zhu, Edith S. W. Poh, H.B. Tan and Richard Te Gan
United Test & Assembly Center Ltd (UTAC)
Packaging Analysis & Design Center
5 Serangoon North Ave 5, Singapore, 554916
Email: Sun_Wei@sg.utacgroup.com Tel: +65-65511345

Abstract

The current paper evaluates the impact on solder joint reliability (SJR), using board-level drop test, temperature cycling on board (TCoB) test, and component-level high-speed solder ball shear (HSSBS) test, of five existing and under-development substrate pad finishes including Electrolytic Nickel-Gold (NiAu), Organic Solderability Preservative (OSP), stencil printed Solder-on-Pad (SoP), Immersion Tin (ImSn) and Electroless Nickel-Electroless Palladium-Immersion Gold (NiPdAu). The board-level drop test was performed following standard JEDEC conditions with 1500G/0.5ms. The TCoB test was performed following IPC-9701 with -40°C~125°C and 15 minutes dwell/ramp. The HSSBS test was performed using Dage 4000-HS machine with 500mm/s shearing speed and 30μm shearing height.

It is seen from drop test results that NiAu leg gives the poorest SJR. All of the other four pad finishes give very satisfactory drop test durability. In TCoB test results, however, it is observed that NiAu gives the best SJR. OSP gives very poor SJR as compared with the other four counterparts whose 2-parameter Weibull characteristic lives exceed 2000 temperature cycles with significant margin. Subsequent process improvements were made in an effort to improve the SJR of OSP in TCoB test. However, even with these improvements the characteristic lives of two improved OSP testing legs only marginally pass the SJR requirement of TCoB. HSSBS test was also performed to understand the resistance to brittle intermetallic compound (IMC) failure of the five finishes under time zero and ageing condition. It is observed that although SoP sees very good SJR in both drop and TCoB tests, it shows bad resistance to brittle IMC failure in HSSBS test. Encouragingly, NiPdAu, which shows good SJR in both drop and TCoB tests, is also found to give excellent resistance to brittle IMC failure in HSSBS test even after ageing at 125°C for 1000 hours.

Key words: pad finish, NiAu, OSP, SoP, Immersion Tin, NiPdAu, solder joint reliability, IMC, drop test, temperature cycling test, high-speed solder ball shear test

1. Introduction

Electronic package development is going extreme in miniaturization. However, the stringent and even increasing reliability requirements are never relaxed. Meeting one reliability requirement may compromise another. So the overall reliability of a package is always a trade-off result after considering every aspect of the reliability requirements. The board-level drop test and TCoB test are currently two most prevailing board-level reliability tests. Board-level SJR of a package is influenced by many factors including both design and material. However, the trend of package miniaturization does not leave much space for design to play around with to improve board-level reliability. Change of material selection is a viable way, but may require large DOE matrix to make sure certain material set can meet all the reliability requirements, at least, with minimum allowable margin.

Pad finish is the direct interface between package and solder or between solder and PCB. It plays a significant role in determining the type and characteristics of IMC formed at those interfaces and even the microstructure of bulk solder joint. In both TCoB and drop tests it is understood that crack near or at the substrate/solder interface is the most common failure mode. Therefore, board-level SJR can be greatly enhanced by selecting the appropriate pad finish material. It should be noted that change of substrate solder pad finish material, however, will not impact other gauges of package performance such as warpage. It is therefore an excellent choice when board-level SJR needs to be improved without impacting the overall package design and other reliability aspects.

The focus of study in this paper is given to the impact of substrate pad finish on board-level SJR of CSP packages. Several substrate pad finishes exist in production or R&D. NiAu, be it processed by electrolytic or electroless/immersion, has been used for long on BGA packages for good wirebondability, solderability and ability to sustain the high temperature during assembly. With the trend that more and more packages are going into mobile applications, recent attention was paid to OSP pad finish, known for its enhancement of board-level drop reliability [1-2]. Although not widely in mass production use, some other pad finishes are under development. Those pad finishes include SoP, Sn and NiPdAu. Similar as wafer level bumping, SoP can be formed by stencil printing or plating at the substrate level. Sn, as a pad finish material, can be formed by electroplating or immersion process. Initial study in [2] found that its drop test performance is even better than OSP, but the TCoB test performance is unknown in that study. NiPdAu is usually processed by electroless Ni, electroless Pd and Immersion Au [3-5]. So more popularly it is called ENEPIG. In this finish Ni layer bonds securely to the copper metallization, while the Pd layer serves as a diffusion barrier for reducing or preventing Ni from out diffusing during thermal processes. The final Au layer adheres to Pd layer and readily receives solder or bonding wires. Recently some research has been done for NiPdAu

978-1-4244-2739-0/08/$25.00 ©2008 IEEE

pad finish for laminate packages. It was demonstrated in [3, 6] that NiPdAu is potentially an excellent pad finish material because it has both good solderability with lead-free solders and wirebondability using all of the three popular wire metals, Au, Cu and Al. In the study in [3, 5, 6] it was found that Au wirebonding on NiPdAu has good reliability even after thermal ageing and cycling. Cu wirebonding was also found to perform similarly well on NiPdAu [6]. The thickness of deposited Pd was found to have significant impact and less than 0.2μm was recommended in [3] and only 0.06μm was used in [5]. Otherwise, brittle interface could be easily formed and lead to brittle cracking. NiPdAu as BGA pad finish was also studied and evaluated mainly using ball shear/pull test in [3, 5, 7]. In those literatures, however, no studies of the pad finish impact on SJR in both drop and TCoB tests were conducted. This paper aims to fill up the vacancy and share our evaluation results on the board-level reliability of the five finish materials mentioned earlier. Board-level drop test and TCoB tests are the major board-level tests required by most package qualifications. These two tests are adopted in this study to gauge the performance of the five pad finishes. HSSBS test, which simulates impact loading on solder ball, is also performed to understand the resistance to brittle IMC cracking for the five finishes.

2. Drop Test

The study of substrate pad finish impact in this paper employed an 11x10.5mm wCSP test vehicle whose ball layout is shown in Figure 1. The test vehicles were assembled with low-Ag lead-free solder composed of 98.48%-Sn/1%-Ag/0.5%-Cu/0.02%-Ni, which is known to be soft and can improve board-level drop reliability. Drop tests were performed following JESD22-B111 board-level drop test standard. The JEDEC drop test setup and testing conditions are shown in Figure 2 and 3 respectively. The drop test matrix is listed in Table 1 with all the five pad finishes mentioned in previous section. Table 2 shows us the cross-section pictures of the five pad finishes. Those cross-section pictures were taken at the solder pad region of as-received substrates. It should be noted that the SoP pad finish in this study was formed by stencil printing of SAC305 solder paste followed by reflow process. Therefore a dome consisting of solder SAC305 is seen in the cross-section picture.

Examination of repeatability of drop testing conditions was firstly performed to ensure consistency. Good repeatability was achieved with high Cpk value as shown in Figure 3. Then 300 drops were performed for each of the five legs with four pieces of PCB assembly in each leg. Failure analysis found that the solder joint crack, occurred at the IMC layer (Figure 4), is the primary failure mode. A comprehensive data analysis was performed using a variety of methods including number of drops up to first failure, Weibull analysis of all the components on each PCB, Weibull analysis of group E&F on each PCB, arithmetic mean of Group E&F as well the number of survived units. The results comparison based on those data analysis methods is shown in Figure 5 and 6. It is noticed that NiAu leg encountered very early failures. During the test electrical discontinuity was detected even after two drops in three PCB

assemblies. Although these very early failures could be attributed to experimental errors that induced during ball mount process, SMT assembly process, or by possible test board quality issue, those five legs were subject to the same assembly process and mounted onto the same batch of PCB test board. The other four legs showed overall good drop test reliability, although one of the components in OSP leg failed as early as 11 drops. The best drop test performance is seen on ImSn leg. Failure for this leg did not occur until 48 drops. It is observed that OSP, ImSn and SoP show similar drop test reliability. The similar drop test performance could be due to the fact all the three pad finishes produced the same type of Cu-Sn IMC at the solder/pad interface. This is proved by subsequent SEM analysis on IMC microstructure as shown in Figure 8b to 8d. Overall drop performance of NiPdAu, although seems not as good as that of OSP, ImSn and SoP, is highly satisfactory. There was no failure occurrence to NiPdAu leg before 31 drops, which is much better than NiAu and even better than OSP in terms of first failure occurrence. The good drop test reliability, together with the excellent wirebondability and solderability, makes NiPdAu potentially an excellent solution for BGA packages targeted at mobile applications.

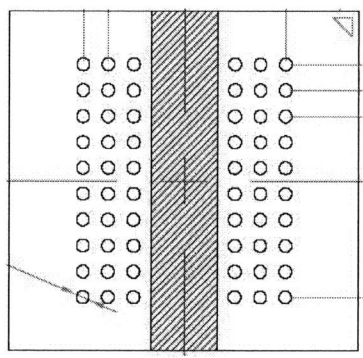

Figure 1: View of test vehicle from solder ball side

Figure 2: Drop test apparatus

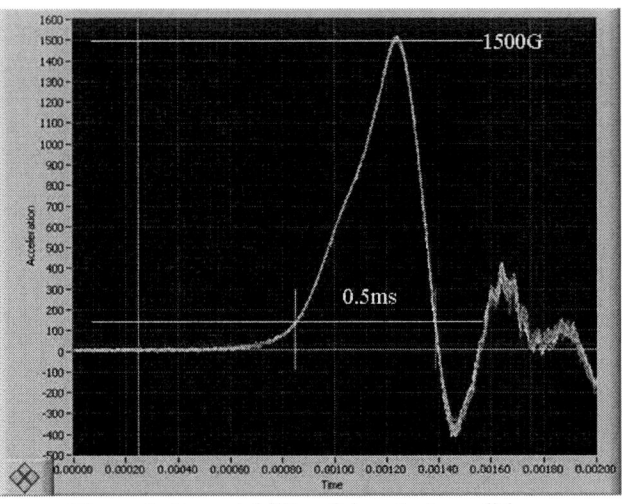

Figure 3: Repeatability of drop test conditions

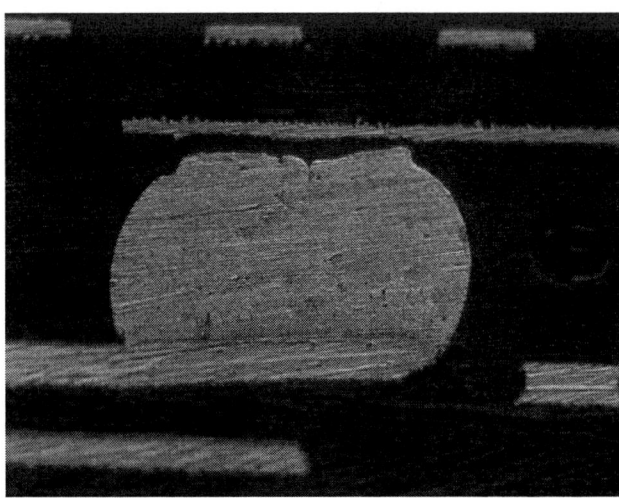

Figure 4: IMC crack in drop test (Leg 3: SoP)

Table 1: Drop test matrix

Leg #	Leg Name	Substrate Pad Finish	Process Technology
1	NiAu	NiAu	Electrolytic
2	OSP	OSP	N.A.
3	SoP	SoP	Printing (SAC305)
4	ImSn	Sn	Immersion
5	NiPdAu	NiPdAu	ENEPIG

Table 2: Cross-section picture of the 5 solder ball pad finishes

Leg #	Leg Name	Cross-section view of pad finish
1	NiAu	
2	OSP	
3	SoP	
4	ImSn	
5	NiPdAu	

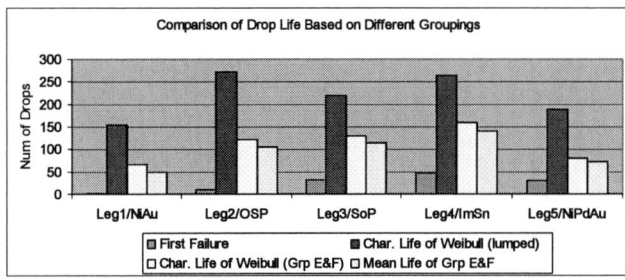

Figure 5: Statistical analysis of drop test results

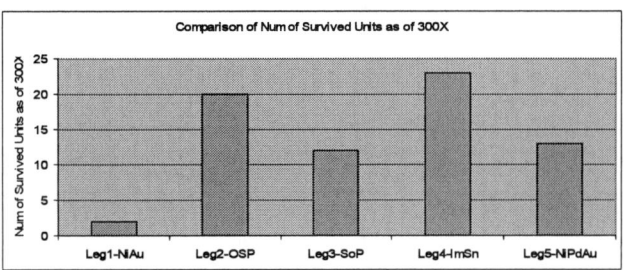

Figure 6: Drop test performance comparison based on number of survived units up to 300 drops

3. Temperature Cycling on Board Test

The TCoB test for pad finish impact study employed similar testing matrix of the drop test with some added legs as shown in Table 4. In Table 4, a larger CSP package with 5 more rows of solder balls was added to the original drop test matrix. Furthermore, two "RE" legs were also added later to improve the SJR of OSP in TCoB test. The "RE" legs were improved in some process parameters based on the recommendations of packaging engineers.

The TCoB test was conducted under -40°C-125°C and 15 minutes dwell/ramp. The PCB test board used was 1.1mm following PCB board thickness usually used by DIMM module.

The TCoB test was conducted until enough data was generated for statistical analysis. Failure analysis on failed samples shows that the failure mode is very typical solder joint crack near the substrate side as shown in Figure 7. But unlike the drop test, the crack in TCoB test initiates and propagates inside the bulk solder material. Microstructure analysis of IMC was also performed for both time zero and after-TCoB-test assemblies. Figure 8 (a-e) shows the comparison of IMC microstructure for each leg before and after TCoB test. In NiAu leg, the initial needle-like Cu-Ni-Sn IMC in Figure 8 (a) has small grain size and dense distribution. Literature survey in [8] shows that this needle-like Cu-Ni-Sn IMC is $(Ni,Cu)_3Sn_4$. During TCoB test, some chunky shape $(Ni,Cu)_6Sn_5$ IMC is formed at the expense of the needle-like $(Ni,Cu)_3Sn_4$ [8]. However such chunky shape IMC only appears in selective locations and the major needle-like IMC still has fine grain size and dense distribution. OSP, ImSn and SoP all enable direct solder attachment to copper pad. Therefore, the IMC formed at these finishes are the same as shown in Figure 8 (b-d). Comparing Figure 8 (b-d) and findings in literature [9], it is understood that the scallop-like IMC at time zero in Figure 8 (b-d) is Cu_6Sn_5. Because there is no Ni layer to prevent the inter-diffusion between copper and solder, the IMC growth rate is reported to be much higher in these three pad finishes than that in NiAu [10]. It is therefore observed from Figure 8 (b-d) that in OSP, ImSn and SoP legs IMC size was greatly increased due to thermal effect and stress induced in temperature cycling test [11]. For NiPdAu, it was reported in [5, 12] that similar as the NiAu the IMC formed with Sn-Au-Cu solder is also Cu-Ni-Sn. However, the current IMC morphology as shown in Figure 8 (e) is quite different from that given in [5, 12]. In [5], the IMC formed between NiPdAu and SnAg3.5Cu0.5 solder is small in grain size and much smaller than the grain size of IMC formed on OSP. It is also shown in the study of [12] that the Cu-Ni-Sn IMC formed between NiPdAu and SAC305 is needle-like. However, the current study found that the IMC morphology as shown in Figure 8 (e) is almost is pyramid-like. Furthermore, after TCoB test, the IMC grain size seems unchanged. Further study will be performed to analyze the IMC.

The 2-paramter Weibull plots for all the eight legs are given in Figure 9 and the Weibull parameters compiled in Table 5 based on 90% confidence level. From Figure 9 it is seen that NiAu, OSP, SoP, ImSn and NiPdAu have characteristic lives of 2912X, 1124X, 2283X, 2287X and 2266X respectively. The results clearly show that although NiAu gives the worst drop test reliability, it shows superior performance in the TCoB test. Therefore, in applications where temperature cycling is the major concern over SJR, NiAu should still be the primary choice for pad finish material. SoP, ImSn and NiPdAu also produced excellent SJR in TCoB test with all the three giving similar performance. The similarity of SJR of SoP and ImSn is probably due to the their direct solder attachment to copper with no influence of other material, unlike the situation in OSP leg where OSP material might disperse into the solder and impact the reliability. Together with the drop test results shown earlier, it is evidenced that SoP, ImSn and NiPdAu can meet both strict drop reliability and TCoB test reliability requirements. Especially for NiPdAu, in view of the good wirebondability and solderability [3-7], and good board-level SJR in both drop and TCoB test, it may become the ideal choice when all of these reliability requirements are to be

met. In the TCoB testing of OSP pad finish, it performed worst although it had overall good drop test reliability. Leg 2 and 6, which were the initially tested legs for OSP finish, saw many failures before 1000 cycles or even 500 cycles. It is known that OSP is an organic material and will be removed from copper during reflow soldering. Therefore it was postulated that OSP, ImSn and SoP legs should give similar SJR in TCoB test since each enables direct solder attachment to copper. This similarity was observed in the drop test where solder joint crack in the IMC is the mode of failure. Such similarity of SJR was also observed in SoP and ImSn in TCoB test as mentioned earlier. However, the current experimental results of TCoB test clearly show significantly inferior SJR of OSP to Im-Sn and SoP. It is observed that solder joint crack in TCoB test usually occurs in the solder neck area that is some distance away from the IMC and is actually inside the bulk solder as shown in Figure 7. It is also understood that OSP thickness control is a challenge in the supplier side and the thickness is very difficult to measure for quality control purpose in the user side. Therefore we hypothesized that the OSP layer in Leg 2 and 6 substrates could be too thick to be completely removed in reflow soldering such that the remaining OSP or its resultant residues dispersed into the microstructure of the solder and somehow made it more prone to fatigue failure. Two process changes aimed at removing OSP completely were proposed. The first change was to increase the peak temperature of ball mounting reflow process by 5 to 10 degree C, while the second change was to increase the dwell time above 217C by 5 seconds in package SMT onto PCB. The first and second changes are name 6-RE1 and 6-RE3 in Table 4. Assemblies of Leg 6-RE1 and RE3 were subject to TCoB test again. This time no failure was logged before 1000 cycles. From the Weibull plot and parameters in Figure 9 and Table 5, it is seen that both improvements give characteristic lives much more than the original OSP legs. With the later first failure and improved characteristic life, the solder joint reliability of OSP finish can now meet the requirements of most customers. However, it should be noticed that even with these improvements, SJR of OSP is still not as good as that of SoP and ImSn. IMC analysis was also performed on RE1 and RE3. It is shown in Figure 8 (f) that no difference in IMC morphology is found between the original OSP (Figure 8 (b)) and OSP based on RE1 and RE3 changes. This probably indicates that IMC might not be the cause of TCoB reliability improvement. It is probably the microstructure of the bulk solder that improved the SJR in TCoB test. Future study needs to be carried out to understand this.

Table 4: Additional legs of TCoB test for pad finish study

Leg #	Pad Finish	Remark
6	OSP	Large package size 11x12.5mm More solder ball: 90B (15x6)
6-RE1	OSP	On top of Leg 6, increase the peak temperature by 5 - 10C in ball mount reflow
6-RE3	OSP	On top of Leg 6-RE1, increase dwell time above 217C by 5 -10 sec during SMT reflow

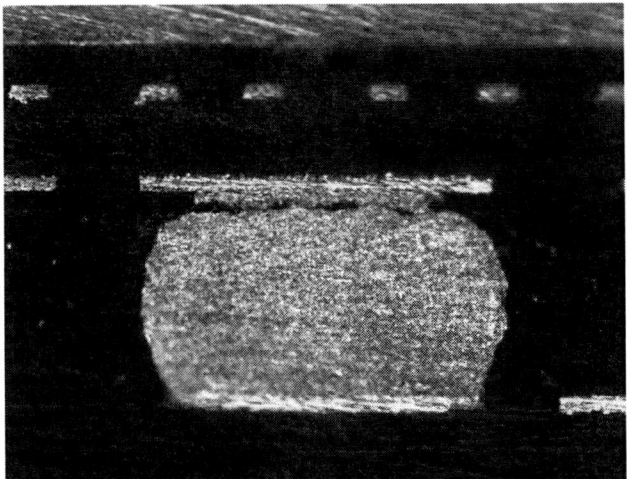

Figure 7: Solder joint crack in TCoB (Leg 4: ImSn)

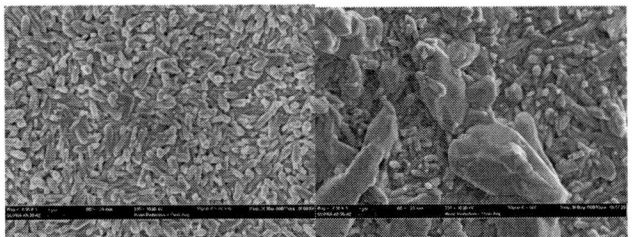

a: Leg 1-NiAu (Left: Time Zero, Right: After TCoB)

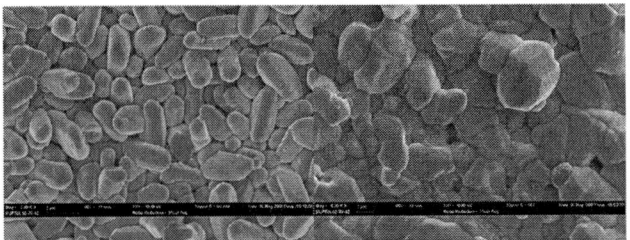

b: Leg 2/6-OSP (Left: Time Zero, Right: After TCoB)

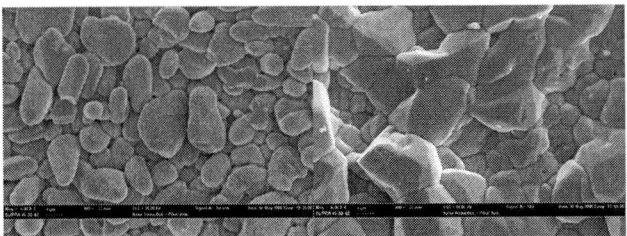

c: Leg 3-SoP (Left: Time Zero, Right: After TCoB)

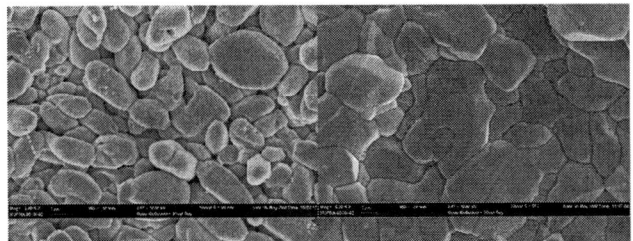

d: Leg 4-ImSn (Left: Time Zero, Right: After TCoB)

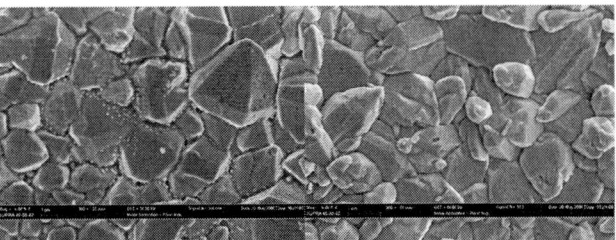

e: Leg 5-NiPdAu (Left: Time Zero, Right: After TCoB)

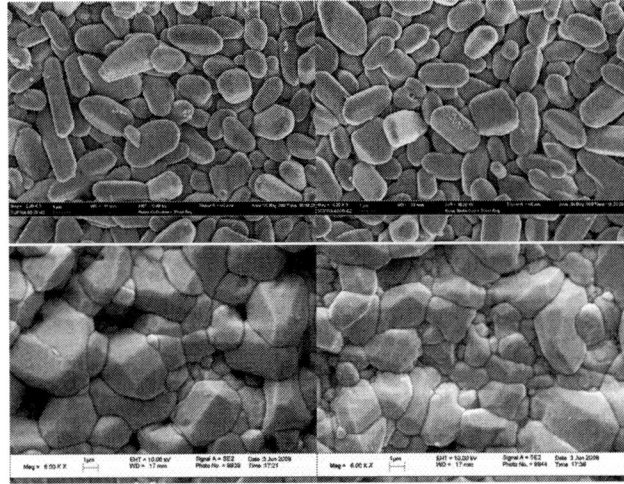

f: Leg 6 OSP RE1 (left) and RE3 (right)
time zero (top) and after TCoB (bottom)

Figure 8: SEM scan (6K Magnification) of IMC

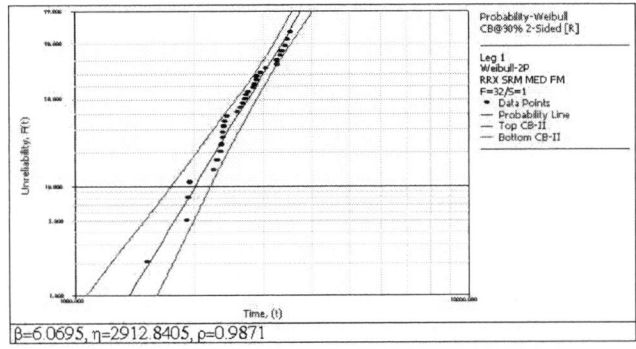

Leg 1: NiAu, 11x10.5mm wCSP with 60 balls

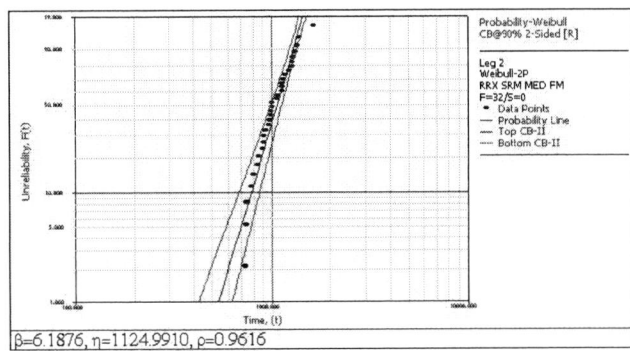

Leg 2: OSP, 11x10.5mm wCSP with 90 balls

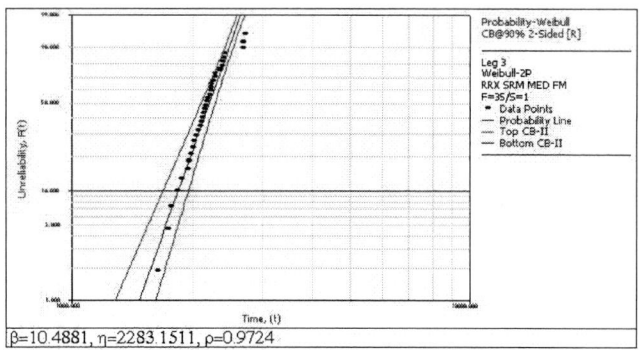

Leg 3: SoP, 11x10.5mm wCSP with 60 balls

β=10.4881, η=2283.1511, ρ=0.9724

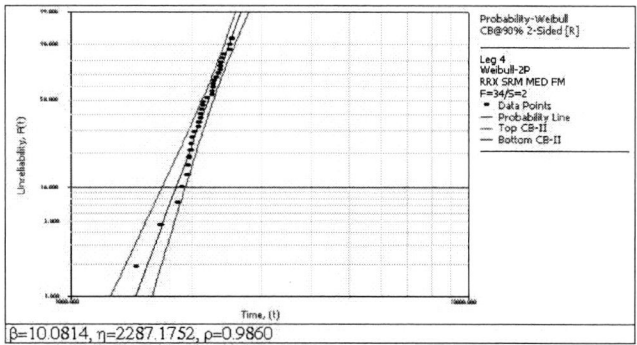

Leg 4: ImSn, 11x10.5mm wCSP with 60 balls

β=10.0814, η=2287.1752, ρ=0.9860

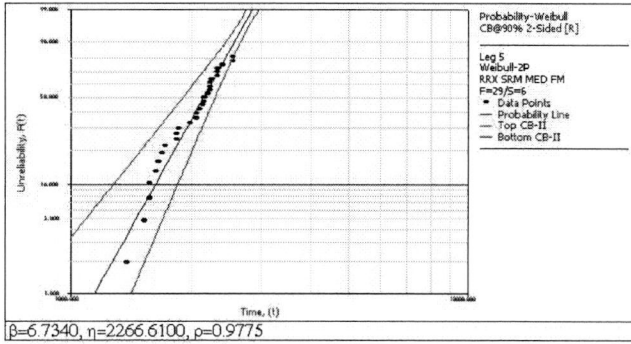

Leg 5: NiPdAu, 11x10.5mm wCSP with 60 balls

β=6.7340, η=2266.6100, ρ=0.9775

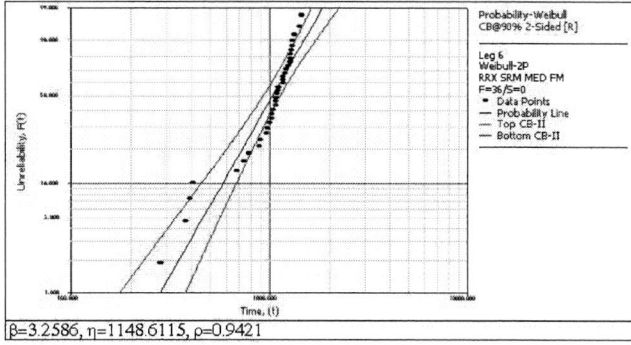

Leg 6: OSP, 11x12.5mm wCSP with 60 balls

β=3.2586, η=1148.6115, ρ=0.9421

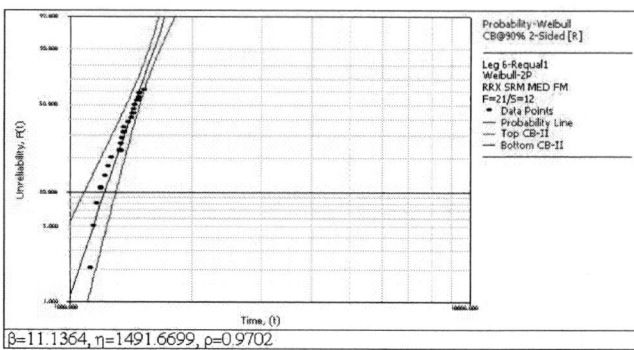

Leg 6 RE1: OSP, 11x12.5mm wCSP with 90 balls

β=11.1364, η=1491.6699, ρ=0.9702

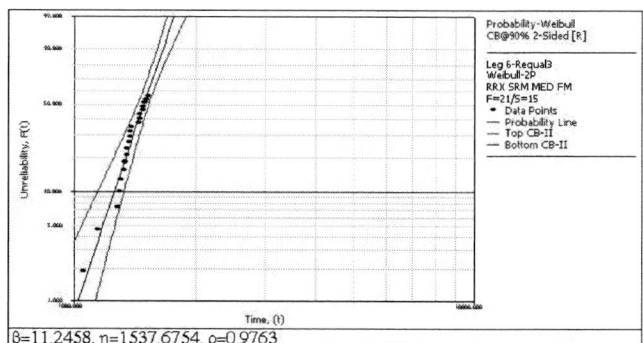

Leg 6 RE3: OSP, 11x12.5mm wCSP with 90 balls

Figure 9: 2-parameter Weibull plots of the 8 tested legs

β=11.2458, η=1537.6754, ρ=0.9763

Table 5: Compiled results of Weibull plots for all the 8 legs

Leg #	Confidence Level	η	β
1	90%	2770<η<3062	4.8<β<7.6
2	90%	1065<η<1187	5.1<β<7.5
3	90%	2207<η<2361	8.6<β<12.7
4	90%	2222<η<2354	8.1<β<12.6
5	90%	2115<η<2428	4.9<β<9.2
6	90%	1047<η<1259	2.5<β<4.2
6-RE1	90%	1447<η<1536	8.1<β<15.3
6-RE3	90%	1494<η<1582	8.3<β<15.3

4. High-speed Solder Ball Shear Test

HSSBS test was performed using Dage 4000-HS machine with 500mm/s shearing speed and 30μm shearing height as shown in Figure 10. The reason of using high-speed shearing is to simulate the strain rate when solder is subject to sudden loading such as the case in drop test. Because of the high strain rate and strain-rate hardening effect of solder, load is effectively transferred to the solder/pad interface and IMC failure, the primary failure mode in drop test, can be observed [10, 13-14]. Therefore HSSBS provides a quick evaluation tool with which IMC strength under high strain-rate can be quickly evaluated. It is also found in [10, 13-14] that the occurrence of IMC failure mode in HSSBS test has some degree of correlation with the drop test results.

The same testing matrix as used in drop test was used for the HSSBS test. For each leg 24 solder balls from 4 different packages were sheared. Two groups of testing were performed with the first group coming from time zero or as-received packages and the second group aged at 125°C for 1000 hours. Three failure modes, namely bulk solder failure, IMC failure and partial-IMC-partial-bulk-solder failure, were observed. Typical failure pictures of bulk solder and IMC are shown in Figure 11. The testing results are shown in Figure 12.

From Figure 12 it is observed by looking at the time zero data that NiAu finish is relatively good in terms of IMC failure occurrence although it gave poorest SJR in drop test. Interestingly SoP pad finish, which showed very good drop test reliability, was found to give the most IMC failures at time zero in HSSBS test. While several IMC failures were found in OSP leg at time zero test, no such failure mode was observed in ImSn and NiPdAu legs that showed good resistance to brittle failure under high-speed shear test at time zero.

Aging the samples for 1000 hours at 125°C is expected to increase the IMC thickness similarly as temperature cycling [11] and thus generates more brittle failures in HSSBS test. It is observed from Figure 11 that all the pad finishes did generate more IMC failures except NiPdAu. Same as the results from time zero samples, aged NiPdAu still did not give any brittle IMC failure, which shows very good resistance to impact loading even after thermal ageing. The robustness of NiPdAu found in HSSBS study may have some correlation with the IMC analysis in TCoB test where we did not observe a significant growth in IMC for NiPdAu pad finish. SoP finish in HSSBS study after ageing is still the worst, giving half of the failure mode as IMC failure. NiAu, OSP and ImSn also saw increase in IMC failure modes after ageing with NiAu seeing only slight increase.

The trend observed in the current HSSBS test for both time zero and aged samples is consistent with the findings in [5, 7, 12]. However, the observation is that the ranking of the five pad finishes in HSSBS test does not show a good correlation with that in board-level drop test. This probably indicates that using HSSBS test to gage the drop test performance of different pad finishes might not provide a good insight into their real SJR in drop test

Figure 10: Apparatus of high-speed ball shear test

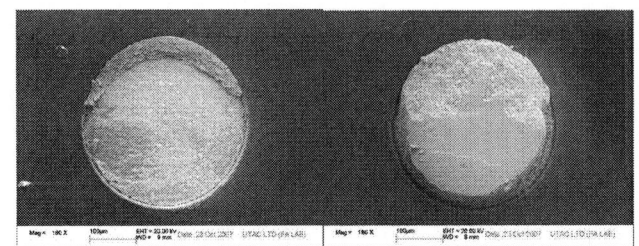

Figure 11: Failure mode under HSSBS test
(left: brittle IMC failure, right: bulk solder failure)

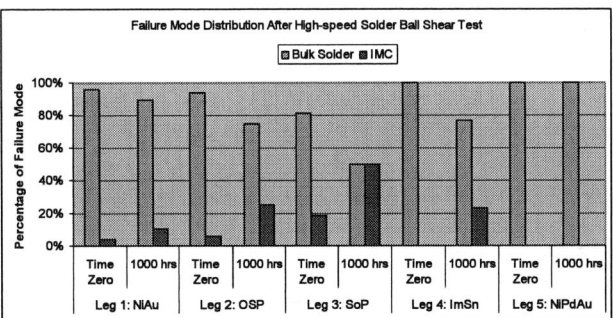

Figure 12: HSSBS testing results on 5 pad finishes

5. Summary and Conclusions

Board-level drop test, TCoB test and HSSBS test were performed to evaluate the impact of substrate pad finishes on SJR. It is found in the current study that:

1. NiAu gave worst SJR in board-level drop test. OSP, SoP, ImSn and NiPdAu showed overall good performance in drop test.

2. NiAu enabled the best SJR in TCoB test. SoP, ImSn and NiPdAu performed similarly with good SJR.

3. SJR of OSP in TCoB test showed strong dependency on reflow soldering conditions. However, even with improved process parameters, OSP still resulted in much less reliability than the other four.

4. NiAu should still be the primary choice when temperature cycling load is the major concern in board level and no impact loading like drop is likely to happen in the field application.

5. SoP, ImSn and NiPdAu can meet SJR requirements from both board-level drop test and TCoB test.

6. Failure analysis on time zero and TCoB tested samples showed that NiPdAu resulted in minimum IMC growth. While NiAu saw IMC grown in selective IMC grains, OSP, SOP and ImSn saw significantly growth in IMC.

7. IMC sees strong growth in OSP, SoP and ImSn under temperature cycling or thermal ageing because no diffusion barrier such as Ni is there between copper and solder. Devices using these pad finishes can have good drop test performance at time zero but might see compromised drop reliability after some use time.

8. HSSBS testing results shows that NiPdAu performed best in resistance to high strain-rate impact loading with no IMC failure for both time zero and 1000 hours aged samples. NiAu ranked second to NiPdAu. SoP saw most brittle IMC failures in both time zero and thermal aged samples.

9. It is observed from this HSSBS test that no direct correlation from failure mode percentage and drop test results

is found. However, the HSSBS test does help to us to evaluate the ageing effect on failure mode. Mostly importantly, NiPdAu is found in this study to show excellent resistance to high temperature ageing.

Acknowledgments

The authors would like to thank UTAC R&D management for their support. The test vehicle preparation work done Assembly Engineering Department, failure analysis work done by industrial attachment students and useful discussions with Annie of Material Group of R&D are greatly appreciated.

References

1. Y. S. Lai et al., "Experimental studies of board-level reliability of chip-scale packages subjected to JEDEC drop test condition", Microelectronics Reliability, Vol 46 (2006), pp.645-650

2. Christian Birzer et al., "Drop Test Reliability Improvement of Lead-free Fine Pitch BGA using Different Solder Ball Composition", Proceedings of EPTC2005, pp.255-261

3. Kuldip Johal et al., "Electroless Nickel / Electroless Palladium / Immersion Gold Process For Multi-Purpose Assembly Technology", SMTA 2004 USA

4. Kudip Johal, et al., "Electroless Nickel / Electroless Palladium / Immersion Gold Platingi Process for Gold- and Aluminium-Wire Bonding Designed for High-Temperature Applications", Pan Pacific 2004 USA

5. Tsukada, K., "Development of new surface finishing technology for PKG substrate with high bondability", Advanced Packaging Materials: Processes, Properties and Interfaces, 2005. Proceedings. International Symposium on, pp.110-114

6. Petar Ratchev et al., "Mechanical Reliability of Au and Cu wirebonds to Al, Ni/Au and Ni/Pd/Au Capped Cu Bond Pads", Microelectronics Reliability, Vol 46 (2006), pp.1315-1325

7. Yee, D.K.W., "Is electroless nickel / electroless palladium / immersion gold (ENEPIG) the solution of lead free soldering on PCB and IC packaging applications?", Microsystems, Packaging, Assembly and Circuits Technology, 2007. IMPACT 2007. International, pp.208-218

8. Sven Lamprecht et al., "Impacts of Bulk Phosphorous Content of Electroless Nickel Layers to Solder Joint Integrity and Their Use as Gold- and Aluminum-Wire Bond Surfaces", www.atotech.com

9 S. W. Ricky et al., "Impact of IMC Thickness on Lead-free Solder Joint Reliability under Thermal Aging: Ball Shear Tests vs. Cold Bump Bull Tests", IMPACT 2006. International, pp.1-4

10 Fubin Song et al., "High-speed Solder Ball Shear and Pull Tests vs. Board Level Mechanical Drop Tests: Correlation of Failure Mode and Loading Speed", Proceedings of EPTC2007, pp.1504-1513

11 Luhua Xu et al., "Isothermal and Thermal cycling Ageing on IMC Growth Rate in Lead-Free and Lead-Based Solder Interface", IEEE Trasactions on Components and Packaging Technologies, Vol. 28, NO. 3, September 2005, pp.408-414

12 Chun Hsien Fu et al., "Investigation of IMC growth and solder joint reliability on new surface finish-ENEPIG", IMPACT 2006. International, pp.331-334

13. X. J. Zhao et al., "A Component Level Test Method for Evaluating the Resistance of Pb-free BGA Solder Joints to Brittle Fracture under Shock Impact", Proceedings of EPTC2007, pp.1522-1529

14. Michael E. Johnson et al., "Using High Speed Shear and Cold Ball Pull to Characterize Lead Free Solder Alloys and Predict Board-level Drop Test Performance", Proceedings of EPTC2007, pp.536-542

The Influence of Heat Treatment on the Adhesion between Molding Compound and Lead Frame

Wei Tan, Xinyu Du, Guangchao Xie, Suqiong Qin, Hujie Mei, XingMing Cheng
Henkel Huawei Electronics Co., Ltd.
Songtiao Industrial Park, Lianyungang, Jiangsu 222006, China
Wei.tan@cn.henkel.com

Abstract

Surface Chemistry and surface energy of copper leadframe after thermal treatment at 175C were studied by XPS and Contact angle. The adhesion after JEDEC level 3 and reflow between epoxy molding compound and leadframe after thermal treatment was measured too. The results shows that the adhesion between epoxy molding compound and heated lead frame increased with heat the leadframe up to 3 minutes then start to decreased. The adhesion force becomes consistent after 7 minutes heating. Form XPS analysis it was suggested that the major chemical of leadframe surface is cupric hydroxide and cuprous oxide after heating the leadframe for short period. With increasing of heating time, cupric oxide was observed on leadframe surface. Copper can't be found no matter how long the leadframe was heated. Compare with adhesion test result, it can be concluded that cupric hydroxide and cuprous oxide has great contribution on the adhesion improvement. The existence of cupric oxide will decrease the adhesion significantly. Surface energy measurement shows that higher adhesion is related to higher surface energy.

1 Introduction

Currently copper is the most common leadframe material in the electronic packaging industry because of their excellent thermal and electronic properties, solder joint reliability and low cost. However, it is susceptible to thermal oxidation during the packaging process such as die bonding, wire bonding and transfer molding because of the high affinity with oxygen. The oxidation of copper can decrease the adhesion between leadframe and molding compound, which can lead to delamination and popcorn[1,2,3]. However, some literature reported an initial increase in adhesion of Cu leadframe with oxidation for thin oxides. For instance, Chong's[1] work showed an initial increase in adhesion strength of an OCN mold compound with oxidation times less than or equal to 120 min, 5 min and 1 min for the temperatures of 150, 200 and 300 C, respectively. At all three temperatures, maximum adhesion strength occurred at bake times that resulted in an oxide layer 20 to 30 nm thick. The results from this study also shows an increase in adhesion strength for exposure times less than 7 minutes at 200 C. They also found that the failure mechanism after oxidation caused by the void between the oxidation layer and the copper leadframe. Ohsuga[4] reported that the adhesion force is depended on the oxidation degree and the type of copper alloy. Berriche[5] found that for the non-oxidated copper leadframe, the failure is inside of molding compound. But for oxidated copper leadframe, the failure is in the oxidation layer. Kang[6] 's work shows different copper alloy has different failure mechanism.

In this paper, the adhesion between epoxy molding compound and Cu leadframe after JEDEC level 3 and 260C reflow was studied with different heating time of Cu leadframe under 175C. Surface chemistry and surface energy of lead frame was studied by XPS and Contact angle. The correlation between adhesion force after reflow and surface chemistry was discussed.

2 Experiments

Copper leadframe was purchased from QPL Limited. The leadframe was heated at 175C in air oven with different time before using. All leadframe was kept in desiccator after cooling.

XPS spectra of the copper leadframe surface were collected by the XPS spectrometer (ESCALAB 250, Thermol Electron Co., USA) equipped with Al Ka radiation source (1486.6 eV). The analysis was done with a take-off angle of 90°, the pass energy of 20.0 eV and the spot size of 500 μm. The binding energy was calibrated by the lowest component of C1s as 284.6 eV.

The transfer molding machine (Hornby Electronic Co., Ltd.) was used for preparation of the Tap pull samples. KL-G730-2(Henkel Huawei Electronics Co., Ltd.) epoxy molding compound was used for this study. A tap pull sample can be taken as two pieces of leadframes connected by EMC, shown as Figure 1. The enclosed area is encapsulated by EMC and the shadowed area on the leadframe surface is attached by EMC. After transfer-molding, the tap pull samples were post-cured in the oven at 175 °C for 6 hours. The tensile test was done with a Universal testing machine (AGS-5kNA, Shimadzu Corporation). The tensile force was reported as the indication of adhesion force between epoxy molding compound and leadframe.

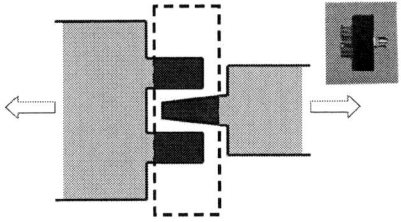

Figure 1. The schematic graph of a tap pull sample for the tensile test. Two pieces of leadframes are connected by EMC. The enclosed area is encapsulated by EMC and the shadowed area on the leadframe surface is attached by EMC. The right graph is a tap pull sample photo taken by camera.

The surface energy of leadframe were measured by contact angle tester (DSA100,Kruss GmbH). Pure water and pure diiodo-methane were used to measure the contact angle of leadframe surface. Owens, Wendt, Rabel and Kaelble method was used to calculate the surface energy.

3 Results and discussion

The adhesion between molding compound and thermal treated leadframe was shown in figure 2. It can be found that the adhesion increased significantly after heat the leadframe less than 3 minutes. Maximum adhesion value was achieved after 3 minutes heat. The adhesion force dropped down quickly with further increasing heating time. The adhesion decreased to the common level after 30 minutes heating, which is similar to the adhesion without any heating. Prolong heat time doesn't show any obvious effect on the adhesion.

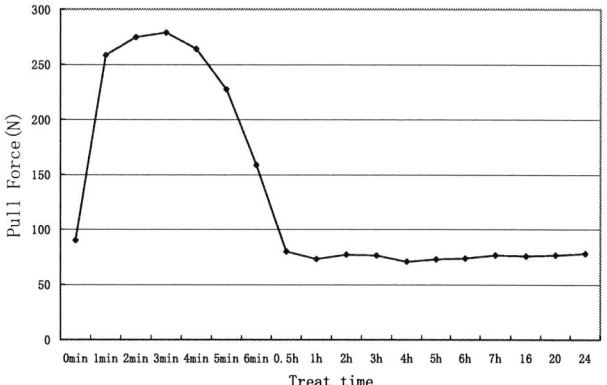

Figure 2 the pull force between molding compound and treated leadframes at different treat time

The influence of heating time on surface energy was shown in figure 3. The surface energy was increased with heating then decrease after 3 minutes. After 30 minutes heating, the surface energy doesn't change much and become stable, which is similar to adhesion strength. Form this result it can be said that higher surface energy can increase the adhesion strength after reflow between molding compound and Cu lead frame

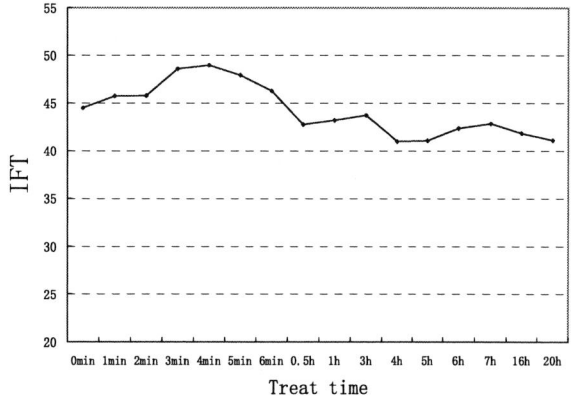

Figure 3 the relationship of treat time and surface energy

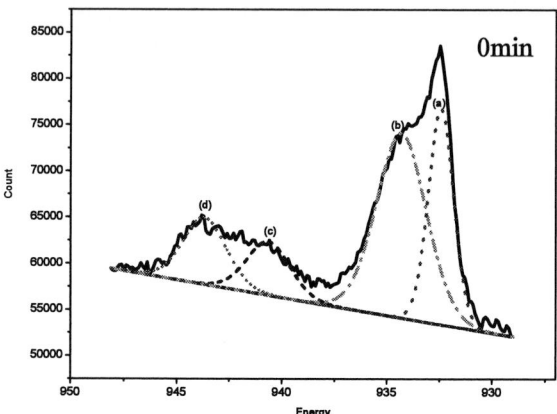

Figure4-1The Cu $2P_{3/2}$ XPS lines for heat 0 minutes.

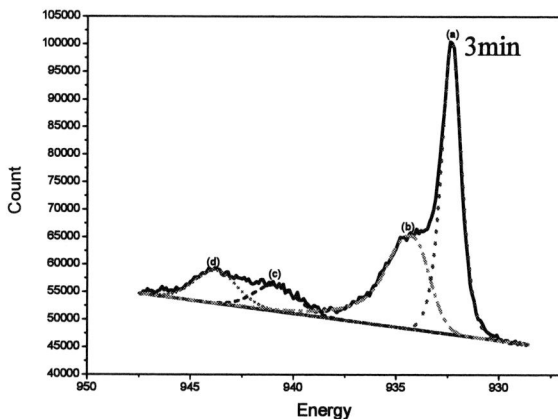

Figure 4-2 The Cu $2P_{3/2}$ XPS lines for heat 3 minutes

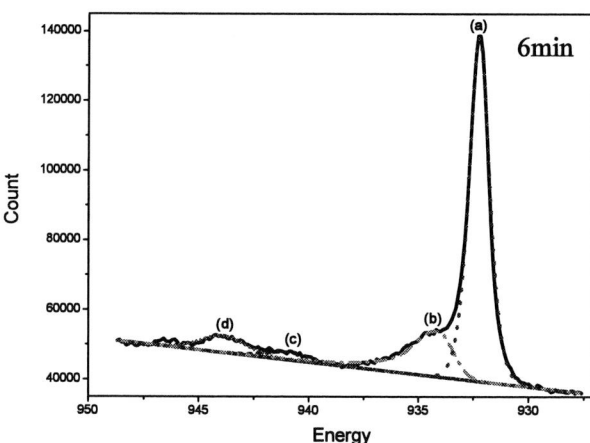

Figure 4-3 The Cu $2P_{3/2}$ XPS lines for heat 6 minutes

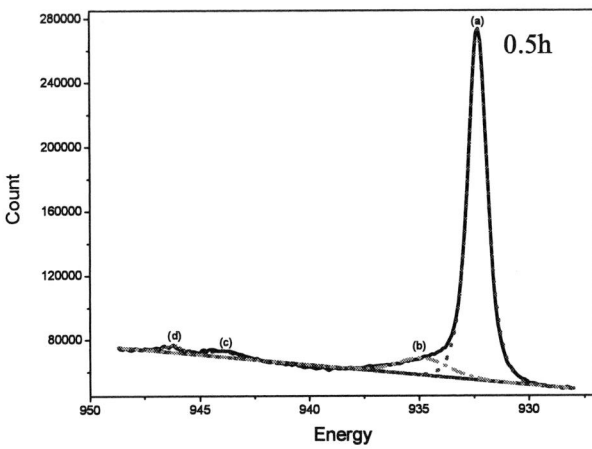

Figure 4-3 The Cu 2P$_{3/2}$ XPS lines for heat 0.5h

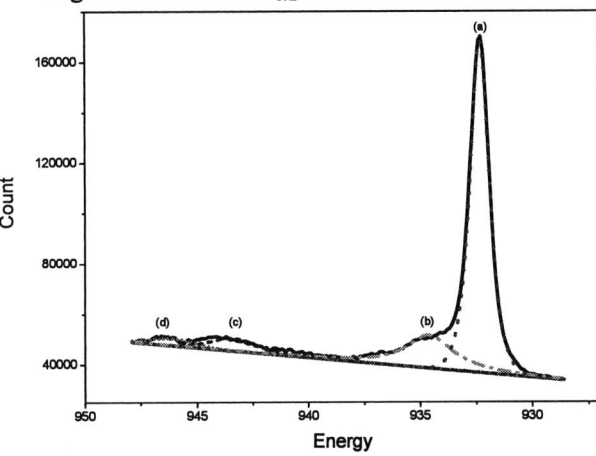

Figure 4-4 The Cu 2P$_{3/2}$ XPS lines for heat 16h

Cu 2p3/2 XPS lines for different heat time of copper leadframe at 175C were shown of Figure 4. Each line was deconvolved to four peaks which location and energy show in table 1. It can be found that the peak of 941.2eV disappeared when heating time reach to 30min, and the 946.2eV peak appeared at the same time. The energy of peak 932.4eV increased with heating time until 30min then reduced with further heating. Conversely, the energy of 934.6eV and 943.9eV were decrease firstly and then increase.

Table 1 the location and energy of the peaks of Cu 2P$_{3/2}$ XPS lines with different treat time

Peak	932.4	934.6	940.7	943.9	946.2
0min	0.02	0.03	0.01	0.01	0
3min	0.03	0.02	0.01	0.01	0
6min	0.06	0.02	0	0.01	0
0.5h	0.1345	0.0156	0	0.0029	0.0024
16h	0.08	0.014	0	0.01	0.002

Based on the previous XPS studies on Copper and its compounds[5,6,7], the peak at 932.4 eV can be assigned to the metal copper or the cuprous state; the other 3 peaks at 934.0 eV,941.2 eV and 943.9 eV are assigned to the cupric state, namely, cupric hydroxide or cupric oxide, which are difficult to be differentiated. In a previous report on the two cupric satellites[8](940.7eV and 943.9eV in this paper), a high left satellite and a low right satellite means that the cupric

hydroxide is dominant while a low left satellite and a high right satellite means that the cupric oxide is dominant. On the other hand, the 946.2eV is the shake up peak of cupric state. According to this analysis, the main composition of leadframe surface are cupric hydroxide and cuprous oxide after heating till to 6 minutes. However, the main composition on leadframe surface change to cupric oxide, cuprous oxide and very little cupric hydroxide after 30min heat treatment.

From table 1, it can be found that the content of cuprous oxide increase along with the heating time, then decrease after heat 16 hours. Cupric oxide can be found from XPS spectrum after 30 min heating and cupric hydroxide decrease when prolong the heat time.

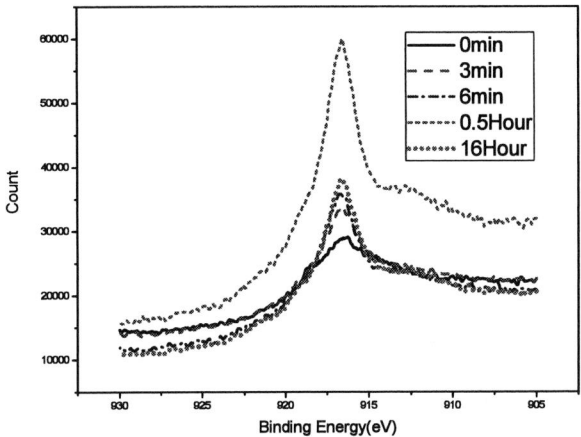

Figure 5 the Cu(LMM) Auger lines of samples with different heat times

Figure 5 shows a group of Cu(LMM) Auger lines for the samples which heated at different times. It should be noted that the metal copper and the cuprous state can not be distinguished from the Cu 2p3/2 XPS spectra[5]. But the metal copper can be qualitatively distinguished from AES spectra[8,9]. The main peak at 916.0 eV is attributed to cuprous oxide and cupric hydroxide, which is difficult to be further separated. For all the samples, the copper metal peak around 918.4eV[8-11] is undiscernable, which suggests that pure copper hardly exists on any surface of samples.

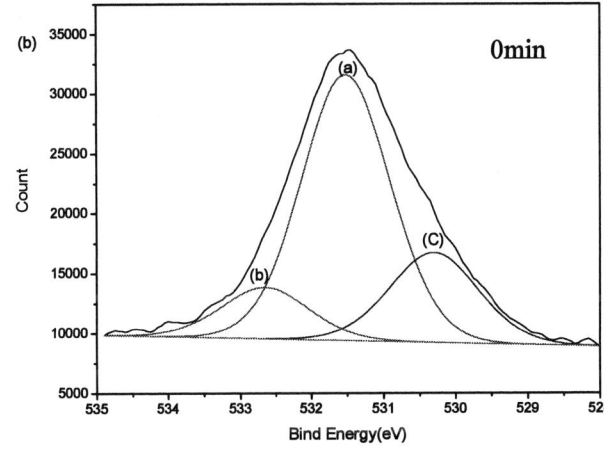

Figure6-1 The O 1s XPS lines for heat 0 minutes.

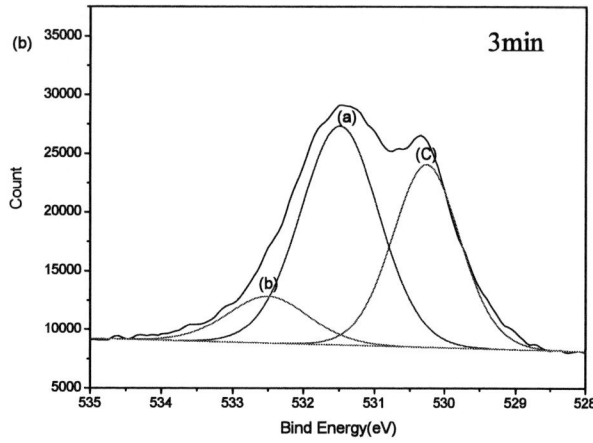

Figure6-2 The O 1s XPS lines for heat 3 minutes.

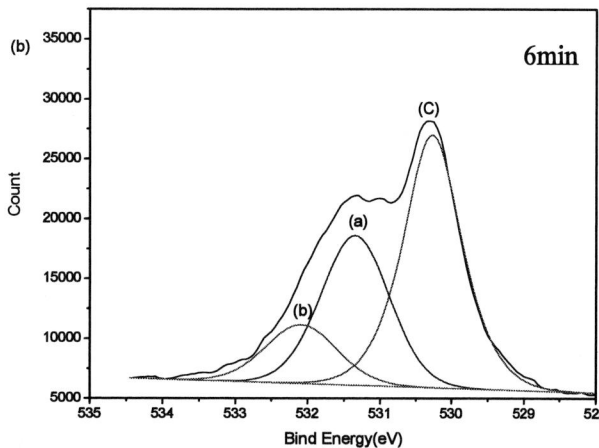

Figure6-3 The O 1s XPS lines for heat 6 minutes.

Table 2 the location and energy of the peaks of O1s XPS lines with different treat time

Peak	530.36	531.5	532.4
0min	0.03	0.08	0.02
3min	0.04	0.06	0.01
6min	0.05	0.03	0.01
0.5h	0.0696	0.0334	0.019
16h	0.05	0.03	0.02

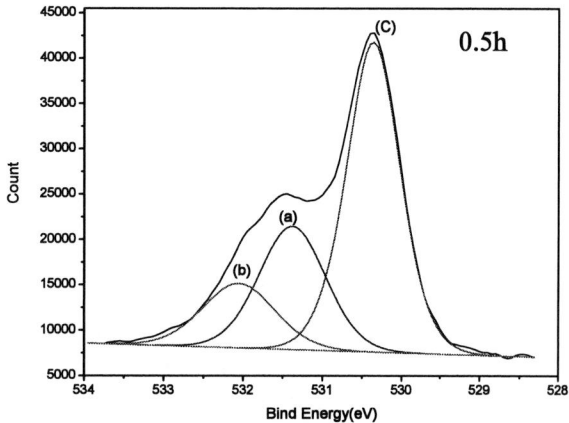

Figure6-4 The O 1s XPS lines for heat 0.5 hours.

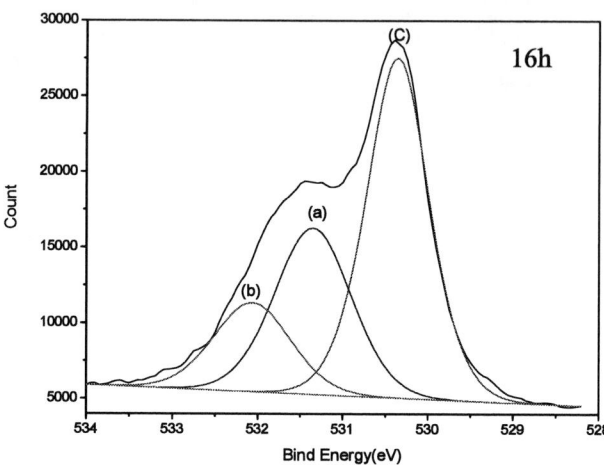

Figure6-4 The O 1s XPS lines for heat 16hours.

More information about the surface of the leadframes with different heat time can be obtained from O1s XPS lines which are showed from Figure 6-1 to figure 6-4, each lines was deconvolved to three peaks around 530.36eV , 531.5eVand 532.4eV which energy are showed in table 2.

The peak at 530.36 eV is assigned to cuprous oxide20, 22, 26. The peak at 531.5 eV may be attributed to the mixed effect of cupric hydroxide[8,11,12] and carbonyl group[13, 14].The peak at 531.5 eV may be attribute to the mixed effect of carbon-oxygen single bond (C-O-C)[14] and water[15]. The surface contamination is believed belong to carbonyl-oxygen content on the leadframe surfaces. From the figure6-1 to 6-4 and table 2, it can be concluded that the main composition of the surface are cuprous oxide and cupric hydroxide. The trace of water on the surface can be attributed to the moisture absorbed from atmosphere.

According to above analysis, the higher adhesion force after 3 minutes heating time can be assigned to the higher content of cupric hydroxide and cuprous oxide on leadframe surface. The reduction of adhesion with further heating is due to the existence of cupric oxide. This different in adhesion could be due to different failure mechanism between molding compound to cuprous oxide and to cupric oxide. When cupric oxide existence on leadframe surface, the failure will take place between the cupric oxide layer and copper. However, when cuprous oxide is the major chemical on leadframe surface, the failure may take place at between molding compound and cuprous oxide layer. Further study need to do conducted to confirm this.

Conclusions

1) Adhesion between epoxy molding compound and Cu leadframe was measured after JEDEC level 3 condition and 260C reflow. Leadframe with pre-treated with different time under 175C. It was found that the adhesion force increased with thermal treatment at beginning then reduced after 3 minutes heating.

2) From XPS it was found that cupric hydroxide and cuprous oxide is the major chemical on leadframe surface with heating less than 6 minutes. Further heating of leadframe will decrease the concentration of cupric hydroxide and begin to see the increasing of cupric oxide until 30 minutes. No pure metal copper can be found on leadframe surface by XPS.

3) The high adhesion after few minutes heating can be attributed to high concentration of cupric hydroxide and cuprous oxide on leadframe surface. With cupric oxide start to appear on leadframe surface, the adhesion force was reduced. .

4) Higher surface energy of pre-treated leadframe will increase adhesion to molding compound after reflow.

References

1. G.S. Ganesan, G.Lewis, A.Woosley, W.Lindsay and H. Berg,"Level I crack Free Plastic Packaging Technology",IEEE 45th ECT Conference,450,1995

2. E.Takano,T.Mino,K.Takahashi,K.Sawada,S-Y.Shimizu, H.Y. Yoo,"the oxidation control of copper Leadframe Package for prevention of Popcorn Cracking",IEEE 47th ECT Conference,78,1997

3. T.Tubbs and P.Procter," Relationship of Delamination, mold compound formulation and device failure modes", Technical Paper,Dexter Electronic Division, The Dexter Corporation,1994.

4. C.T. Chong,A. Leslie,L.T.Beng,and C.Lee, "Investigation on the effect of copper leadframe oxidation on package delamination," in Proc. 45th ECTC,1995,pp.463-469

5. H.Ohsuga, H.Suzuki, T.Aihara, and T.Hamano, development of molding compounds suited for copper leadframes," in Proc.44th ECTC,1994,pp.141-146

6. R.Berriche,S.C.Vahey and B.A.Gillett, "Effect of oxidation on mold compound-copper Leadframe adhesion", 1999 international symposium on advanced packaging materials, pp.77-82

4. Teck-Gyu,Kang,Ik-Seong Park,Jong-Heon Kim and Kwang-Seong Choi."Characterization of Oxidized copper leadframes and copper/Epoxy molding compound Interface adhesion in Plastic Package",IEEE,1998,106-111

5. Wu, C.-K.; Yin, M.; O'Brien, S.; Koberstein, J. T. Chem. Mater. 2006, 18, 6054-6058.

6. Yin, M.; Wu, C.-K.; Lou, Y.; Burda, C.; Koberstein, J. T.; Zhu, Y.; O'Brien, S. J. Am. Chem. Soc.2005, 127, 9506-9511.

7. Ghijsen, J.; Tjeng, L. H.; van Elp, J.; Eskes, H.; Westerink, J.; Sawatzky, G. A. Phys. Rev. B1988, 38, 11322-11330.

8. Mclntyre, N. S.; Sunder, S.; Shoesmith, D. W.; Stanchell, F. W. J. Vac. Sci. Technol. 1981, 18,714-721.

9. Chusuei, C. C.; Brookshier, M. A.; Goodman, D. W. Langmuir 1999, 15, 2806-2808.

10. Poulston, S.; Parlett, P. M.; Stone, P.; Bowker, M. Surf. Interface Anal. 1996, 24, 811-820.

11. Cano, E.; Lopez, M. F.; Simancas, J.; Bastidas, J. M. J. Electrochem. Soc. 2001, 148, E26-E30.

12. Chavez, K. L.; Hess, D. W. J. Electrochem. Soc. 2001, 148, G640-G643.

13. Ektessabi, A. M.; Hakamata S. Thin Solid Films 2000, 377-378, 621-625.

14. Lopez, G. P.; Castner, D. G.; Ratner, B. D. Surf. Interface. Anal. 1991, 17, 267-272.

15 Sinapi, F.; Delhalle, J.; Mekhalif. Z. Mater. Sci. Eng. C 2002, 22, 345-353.

Shear Fracture Behavior of Sn3.0Ag0.5Cu Solder joints on Cu Pads with Different Solder Volumes

Yanhong Tian[1], Chunqing Wang[1], Shihua Yang[1], Penrong Lin[1], Le Liang[2]

1-State Key Lab of Advanced Welding Production Technology, Harbin Institute of Technology, Harbin 150001, China,
2-Samsung Semiconductor(China) R&D Co.,Ltd, Suzhou, China 215021
tianyh@hit.edu.cn, 86-451-86418359

Abstract

To meet the urgent demands of future electronic packages, the solder joints have to become increasingly miniaturized. Comparied to the large solder joints, mechanics behavior for the samll solder joints is very different, resulting in a series of reliability issues. Therefore, it is very important to understand the mechanics behavior of the small solder joints.

In this paper, the shear test of the as-reflowed and aged Sn-3.0Ag-0.5Cu solder joints on Cu pads with the diameters of 200μm to 600μm were conducted, and fracture behavior was observed using SEM. The results show that shear strength of the solder joint increases with the decreasing of the solder joints volumes. For the large volume solder joints, the fracture occurs close to the interface, and the solder joint shows strong brittleness. Whereas, for the small solder joints, the fracture occurs within the bulk solder, and the solder joint shows ductile property. The Ag_3Sn and Cu_6Sn_5 intermetallic compounds (IMCs) at the interface region have a prominent effect on the shear property and the propagation of the fracture.

1 Introduction

The solder joints have been greatly miniaturized due to the demands for high-density assembly, as a results of this, the diameters of solder joints in array-array packages can range from 760μm (BGA joints) to 75μm (flip-chip joints). The variation in diameter leads to a great volume difference. At the same time, the micro-interconnection might be reduced down to the sub-micron range and at such a small scale the solder joints might consist of only limited numbers of crystal grains. The future micro-joint will have distinct mechanical behavior compared to larger joints and bring reliability issues.

The solder joints are mainly subjected to the shear loading in real applications due to the mismatch of coefficient of thermal expansion. Shear loading is more important situation in real solder joints. Meanwhile, shear test is easy to be carried out and understood. Therefore, ball shear test is a widely adopted method to determine the ability of area array solder joints to withstand mechanical shear force.

The mechanical properties of the solder joints depend on the microstructure [31,37,38 40], therefore, it is necessary to investigate the effects of the microstructure on the shear strength. Some literatures[1,2] reported the shear strength of the 760μm Sn-Ag-Cu solder joints under the conditions of multi-reflow and thermal aging, it was found that the shear strength of the solder joints didn't change much after multi-reflow and aging, and the microstructure was not changed significantly, and all the fractures occurred close to the interface. D.-G.Kim et al [32] investigated the shear behavior of the Sn-37Pb/Au-Ni subjected to the isothermal aging. They concluded that the decrease of the aged solder joints is owing to the smoothness of the IMCs, and the decrease of the contact surface between the IMCs and the solders. A. Kar et al[33] reported that the thickness of the brittle IMCs had an important effect on the interconnect strength. Lee et al [34,35] reported that the decrease of the solder strength is caused by the growth of the IMCs and the increase of the roughness between the solder and the Cu_6Sn_5 layer. Kirkendal voids at the interface was also an important factor which decreases the solder joint strength.

In this paper, shear test of the Sn-3.0Ag-0.5Cu/Cu solder joints with different solder volumes was performed, and the fracture behavior was investigated. The effect of IMC morphology on the shear strength and the fracture behavior of the solder joints was also studied.

2. Experimental Materials and Methods

The Sn-3.0Ag-0.5Cu lead-free solder balls with the diameters of 200μm,300μm,400μm,500μm,600μm were used. The ratio of the diameters of the solder balls to that of the corresponding copper pads on the FR-4 substrates was approximate 1.2:1. The solder balls were soldered to the pads with prefluxing RMA flux in a reflow oven with a maximum temperature at 250℃ for 30s. The as-reflowed solder joints were then followed the isothermal aging at 150℃ for 4, 12, 96, 384, 864, and 1536 hours respectively.

Shear tests were performed on the as-reflowed and aged solder joints using a DAGE 4000 bond tester. The shear height was fixed at 15μm from the substrate surface, and the shear speed was 60μm/s. For each sets of shear tests, the shear force was an average of at least 30 measurements. Fig.1 shows the schematic drawing of the shear test for the solder joints. Microstructure and fractography analyses were conducted using scanning electron microscopy (SEM) in secondary electron mode equipped with energy dispersive x-ray (EDX).

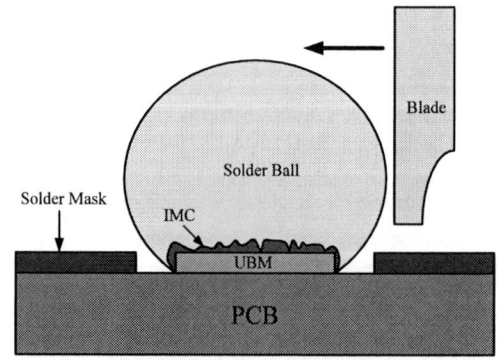

Fig.1 Schematic drawing of the shear test

3 Results and Discussion

3.1 Shear strength of the solder joints

Fig.2 shows the curves of the relationship between the shear strength and the solder joints with different solder volumes. It indicates that the shear strength decreases remarkable with the increasing of the solder volumes, which shows an apparent volume effect. Fig.1 also shows that the strength of the solder joints decrease after isothermal aging process.

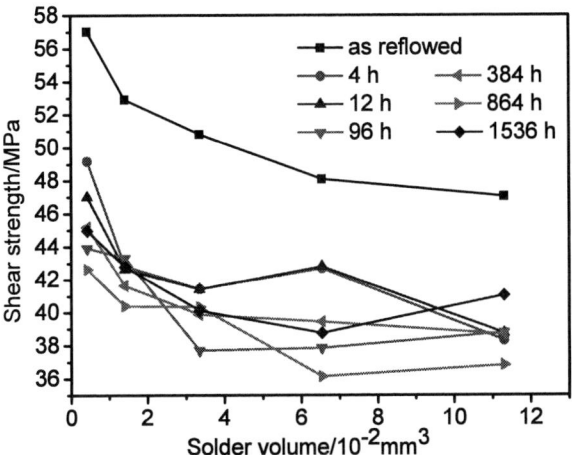

Fig.2 Shear strength variations with increasing of solder volumes after reflow and aging

3.2 Shear fracture behavior of the solder joints

Fig.3 shows the cross section of the solder joints with different solder volume subjected to the shear test.

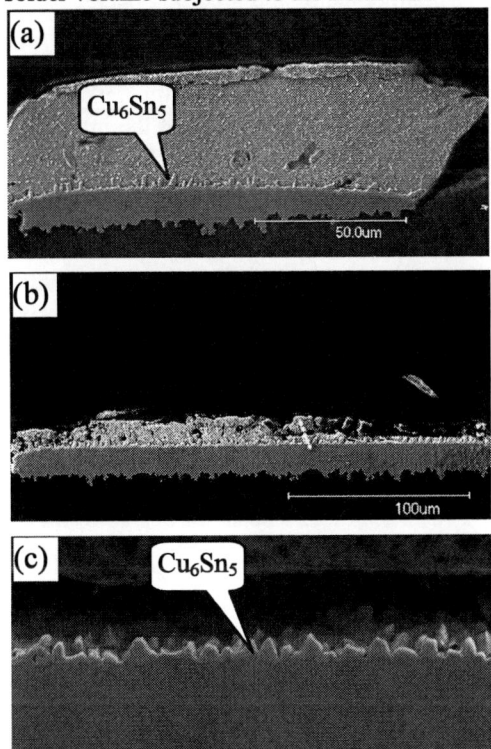

Fig.3 Cross section of the as-reflowed solder joints after shear test with the diameters of (a)200μm, (b) 400μm, and (c) 600μm.

It can be seen that the fracture location was very different even if the hight of the shear blade was same for the solder joints. For the 200μm as-reflowed solder joints, as seen from the Fig.3(a), the fracture occurred inside the solder joint and far away from the interface, which belongs the typical ductile fracture. With the increase of the solder volume, the fracture happened near the interface, as shown in Fig.3(b). Brittle fracture occurred for parts of the 600μm solder joints, and the fracture between the bulk solder and Cu_6Sn_5 IMC was observed, as shown in Fig.3(c). Therefore, it can be concluded that the fracture behavior of the solder joints shows an obvious volume effect, that is, with the increase of the solder volume, the position of the fracture occurs changes from the inner solder joint to the interface of the solder joint. The fracture position of the as-reflowed solder joints indicates that the decrease of the shear strength for the large solder joint is caused by its brittle property.

After the thermal aging, the fracture behavior of the solder joints changed into a mixing way. For the 200μm solder joints aged at 150℃ for 36 days, part of the fracture occurred at the interface between the solder and the Cu_6Sn_5 IMC, part of the fracture happened inside the bumk solder, showing a mixing fracture mode, as seen from the Fig.4(a). For the 600μm solder joints aged at 150℃ for 36 days, the fracture occurred near the interface of the Cu_6Sn_5 IMC and the solder, as shown in Fig.4(b). The fracture behvior of the solder joints after aging proves that the reason of the decrease of the shear strength after aging was the growth of the IMC layer and the coarsen of the IMC particles.

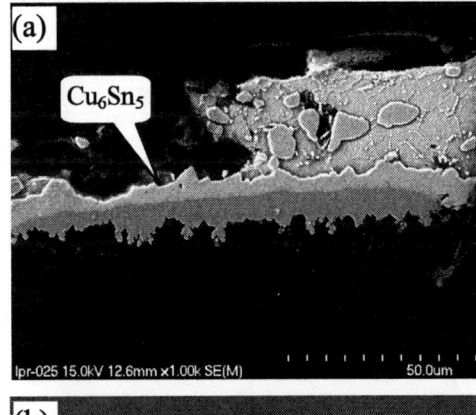

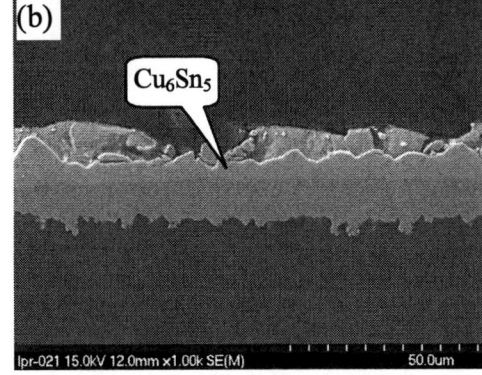

Fig.4 Cross section of the aged solder joints after shear test with the diameters of (a)200μm and (b) 400μm.

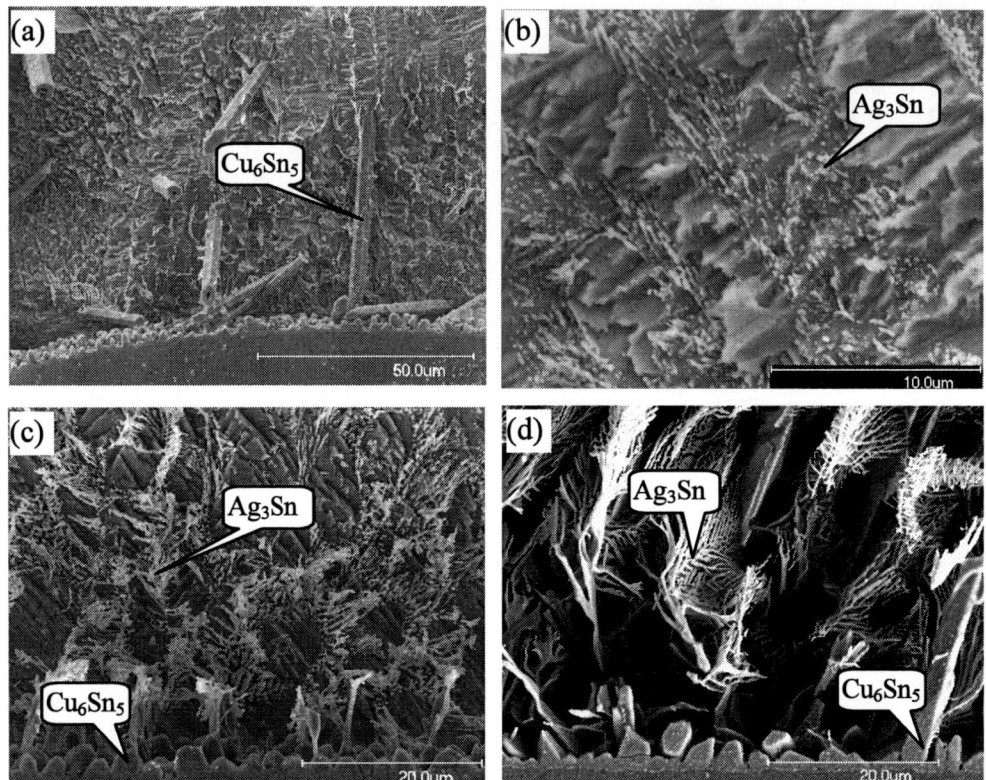

Fig.5 SEM images of the microstructure of the as-reflowed solder joints with the diameter of (a) 200μm, interface, (b) 200μm, bulk sulder (c) 400μm and (d) 600μm

3.3Microstructure of the solder joints

The morphology and distribution of the IMCs inside the solder joints has a significant effect on the shear strength and fracture behavior. Fig.5 shows the microstructure of the solder joints with different volumes. A layer of scalloped Cu_6Sn_5 intermetallic compound was found at all the interface of the solder joints. For the small solder joints, the scallop of the Cu_6Sn_5 was finer than that of the large solder joints, as shown in Fig.5(a)-(c).The finer scallop would increase the contact area between the IMC and the bulk solder, which would increase the shear strength of the solder joints. The morphology of Ag_3Sn phases inside the solder joint shows an obvious volume effect. The Ag_3Sn phases inside the 200μm solder joint show very fine particles in nanoscale, and these dispersive nano-particles could increase the strength and ductility of the solder joints. With the increase of the volume of the solder joints, the morphology of the Ag_3Sn phases changed into net-dendritic or featheriness, as shown in Fig.5 (c) and (d).

It can be considered that the main reason of the change of shear strength and fracture behavior of the Sn-Ag-Cu solder joints with different volumes is the siginificant morphology variations of the Ag_3Sn particles. For the large volume solder joints, the net-dendritic shape of the Ag_3Sn leads to the ductile property, and small plastic deformation during the shearing occurs. Furthermore,the strength of the bulk solder is stronger than that of the interface interconnection, resulting in the brittle fracture. For the small volume solder joints, the dispersive Ag_3Sn particles strengthen the plasticity of the solder joint, and large plastic deformation of the solder joint happens during the shear test, leading to the ductile fracture within the bulk solder.

After the thermal aging, the scallopped Cu_6Sn_5 phases at the interface change into the plain layer, as shown in the Fig.6 (a) and (c). The plain layer decreases the interconnection strength between the bulk solder and the IMC, therfore, the fracture occurs at the interface during the shear test after thermal aging. It should be noted that the nanoscale Ag_3Sn particles inside the samll solder joints coarsen after aging, as seen in the Fig.6(b), which will decrease the strength significantly. However, the net-dendritic or featheriness Ag_3Sn phases within the large solder joints globurize and aggregate, which will also decrease the the strength of the solder joint, at the same time , the plasticity of the solder joint will be increased. That's why the fracture occurs close to the interface after aging not at the interface like the as-reflowed solder joints.

4 Conclusions

(1) The shear fracture behavior of micro solder joints had significantly volume effect. Shear strength increased obviously with the decrease of solder volume after hot air reflow and aging.

(2) The effect of the IMC morphology was the dominant factor that leads to the mechanical properties variation of different volume micro solder joints. The morphology of Ag_3Sn IMCs within the as-reflowed solder joints changed from net-dendritic or featheriness to fine granular with the decrease of the soder volume. The morphology of Ag_3Sn IMCs becomes bulky granular after aging process.

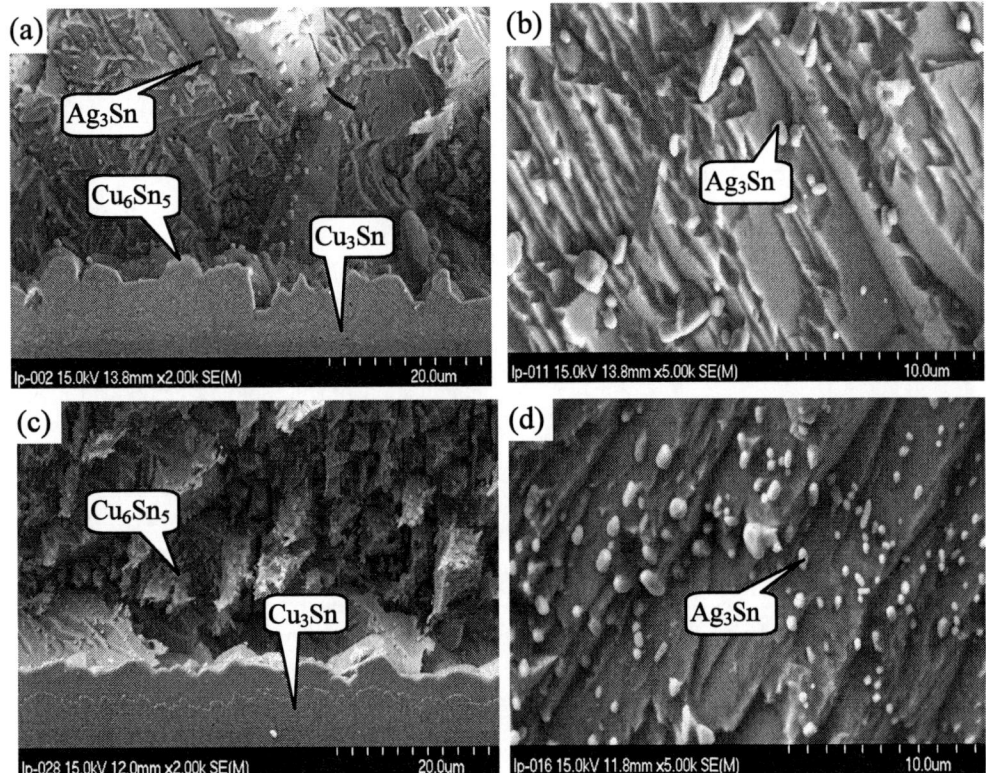

Fig.6 SEM images of microstructure of the solder joints aged at 150℃ for 36 days (a) 200μm, interface, (b) 200μm, bulk solder,(c) 600μm, interface, (d) 600μm, bulk solder.

Acknowledgement

This work was financially supported by the National Natural Science Foundation of China (No. 50675047/8052105) and Joint Project between Samsung Electronics Co. Ltd.(Korea) and Harbin Institute of Technology (HIT)

References

1. M.A. Matin, J.G.A. Theeven, et al. Correlation between localized strain and damage in shear-loaded Pb-free solders. Microelectronics Reliability. 2007, 47:1262~1272.
2. K.S. Kim, S.H. Huh, K. Suganuma. Effects of cooling speed on microstructure and tensile properties of Sn-Ag-Cu alloys. Materials Science and Engineering A. 2002, 333:106~114.
3. K.S. Kim, S.H. Huh, K. Suganuma. Effects of intermetallic compounds on properties of Sn-Ag-Cu lead-free soldered joints. Journal of Alloys and Compounds. 2003, 352:226~236.
4. J. J. Sundelin, S. T. Nurmi, et al. Mechanical and microstructural properties of SnAgCu solder joints. Materials Science and Engineering A. 2006, 420: 55~62.

5. Po-Cheng Shih, Kwang-Lung Lin. Correlation between interfacial microstructure and shear behavior of Sn-Ag-Cu solder ball joined with Sn-Zn-Bi paste. J Mater Sci. 2007(42):2574~2581.
6. Jeong-Won Yoon, Seung-Boo Jung. Effect of immersion Ag surface finish on interfacial reaction and mechanical reliability of Sn-3.5Ag-0.7Cu solder joint. J. Alloys Compd. doi:10.1016/j.jallcom.2007.04.014.
7. D. -G. Kim, H. -S. Jang, J. -W. Kim, et al. Correlation between the interfacial reaction and mechanical joint strength of the flip chip solder bump during isothermal aging. Journal of Materials Science: Materials in Electronics. 2005, 16:603~609.
8. A. Kar, Ma. Ghosh, A. K. Ray, R. N. Ghosh. Effect of copper addition on the microstructure and mechanical properties of lead free solder alloy. Materials Science and Engineering A. 2007, 459:69~74.
9. Hwa-Teng Lee, Ming-Hung Chen, et al. Influence of interfacial intermetallic compound on fracture behavior of solder joints. Materials and Engineering A. 2003, 358:134~141.
10. Hwa-Teng Lee, Ming-Hung Chen. Influence of intermetallic compound on the adhesive strength of solder joints. Materials and Engineering A. 2002, 333:24~34.

Effect of Moisture and Temperature on Al-Cu Interfacial Strength

Hui Teng, Huiliang Zhang, Hongbo Yang, Ming Zhou, Anthony C. Tsui
GEM Electronics(Shanghai) Co., Ltd.
438# Zhaoxian Road, Jiading Development Industrial Park, Jiading, Shanghai 201821, China
teng.hui@gemservices.com; huiliang.zhang@gemservices.com;
hongbo.yang@gemservices.com; ming.zhou@gemservices.com; tony.tsui@gemservices.com
Telephone: 086-021-59167500

Abstract

In power management semiconductor industry, the most comprehensively used metals in microelectronic packages include Copper, Aluminum, Nickel, Gold and Silver. When different metals contact to each other, Intermetallic Compound (IMC) will form at the interface. Under different conditions, IMC may vary to show very complicated characteristic and composition, some Moderate IMC will increase interfacial strength and form reliable bond, but as the brittle intermetallic compound layer thickness increases, the joint contact resistance will increase and the bonding strength will decrease. Excessive IMC growth acts as a major cause for bonding failure such as discontinuous connection, even crack or bond lift. Especially when a package undergoes reliability test of moisture and high temperature, IMC growth will accelerate much faster than in the ambient environment.

This paper mainly studies the IMC growth and interfacial adhesion of Al wire bond on Cu based leadframe with Bare Cu surface, Ni plated surface and Ag plated surface. The interfacial adhesion force is measured by bond shear and wire pull test at initial condition, after Temperature Cycling (TC, -65°C to 150°C), Autoclave (AC, Ta=121°C, relative humidity =100%, Pressure=15 psi), High Temperature Storage (HTS, 150 °C), simulating several typical package storage and usage conditions. The failure mechanism is studied from the fracture surface characteristics and cross section analysis after different aging time. In Autoclave environment with moisture and temperature effect, Al and Cu form an electrochemical couple. Because Al has low electrode potential and is more active, Al acts as anode to lose electrons and corroded in the galvanic corrosion. Further more, in HAST (130C/85%RH, 80% of rated BV) test, the effect of electrical bias to galvanic corrosion is analyzed.

Key Words: Interfacial Adhesion, Moisture, High Temperature, IMC, Galvanic Corrosion

1. Introduction

Plastic packages encapsulated with epoxy molding compound are popular in the consumable and industrial microelectronic packages because of their relatively low cost. However, because of their high permeability, plastic packages offer less resistance to moisture ingress than hermetic packages. Although encapsulation is employed to hermetically seal the microelectronic packages to prevent die from moisture outside, moisture permeation still remains one of the critical concerns associated with the reliability of microelectronic devices. Moisture can permeate any molding compound. The rate of permeation will vary with compound thickness and type, and relative humidity. Moreover, due to the coefficient of thermal expansion (CTE) mismatch of epoxy molding compound and Cu leadframe /silicon die, delamination of the molding compound along the leadframe interface or die surface can take place. This delamination can provide a path for moisture and contaminant ingress along the interfaces where the materials are no longer adhering to each other.

In microelectronic packaging, gold wires are the most successful interconnection materials widely used in wire bonding technology. Aluminum and its alloy films are the primary choice as the metal for contacting silicon and for carrying current to and from the bonding wire. However, the reliability problem of gold wire bond is associated with intermetallic compound formation that occurs at the gold wire-aluminum bonding pad interface. It's well known that particular Kirkendall voids $AuAl2$ (purple plague) and $Au5Al2$ (white plague) are brittle and easy to result in bonding reliability failure [1].

Copper wire bonding is an alternative interconnection technology that serves as a viable and cost saving alternative to gold wire bonding for its high thermal conductivity, low electrical resistance, high pull strength, low cost, and better connection of the wire to aluminum pad. Its excellent mechanical and electrical characteristics attract the high-speed, power management devices and fine-pitch applications [2]. To establish a reliable copper ball bonding process, it is important to investigate the reliability of Cu/Al bond in comparison with that of Au/Al bond. The comparison study regarding the interface between the Cu-Al and Au-Al conducted in former study shows that the rate of intermetallic growth of Cu-Al interface is about 1/10 of that of an Au-Al interface at comparable temperature [3, 4]. The Cu-Al interface provides lower electrical resistance, lower heat generation, and longer package life compared to the Au-Al interface.

In previous study, many papers work on the Cu wire bonding on Al pad, while Al wire bonding on Cu is relatively seldom mentioned. While in power management semiconductor, Al wire and Cu based leadframe (such as bare Cu, Ni plating copper, and Ag plating copper) are also commonly used. So it's very necessary and significant to study the Al wire bonding performance on copper interface.

In the experiment, two kinds of leadframe are prepared. One type leadframe is Ni plating on lead and bare Cu on pad. Another type leadframe is Ag plating on lead post and pad. For the Ni plating leadframe, Al wire bond from bare Cu pad to Ni plating lead post, so as to get Al-Cu and Al-Ni interface. For Ag plating on both pad and lead post leadframe, attach die with Al bonding pad on the pad first, then wire bond from die surface to lead post, so as to get Al-Al, Al-Ag interface.

978-1-4244-2739-0/08/$25.00 ©2008 IEEE

To accelerate the moisture and temperature effect to simulate rapid intermetallic growth in the bond interface, the wire bond samples were not encapsulated and exposed directly in reliability environments: Autoclave (AC, Ta=121°C, RH=100%, Pressure=15psi), Temperature Cycling (TC, -65°C to 150°C), High Temperature Storage (HTS, 150°C). At different time points, the samples were taken out to measure wire pull, wedge shear and cross section inspection.

2. Interfacial Bonding Strength

When the bonding interface is reliable, the wire pull break mode is in the center of Al wire, or in the neck of 1st bond, or in the heel of 2nd bond. So wire pull can reflect the bondability indirectly from one aspect.

The wire pull force of Autoclave tested samples is shown in Fig 1. For Al wire bond on Cu and Ni surface, wire pull data drop a lot after AC96hrs, AC192hrs. Failure mode is in the 1st bond on leadframe pad: Al-Cu interface. For Al wire bond on Al and Ag surface, AC96hrs wire pull force reduce a little, but after AC192hrs, wire pull decrease to nearly zero, failure mode is in 2nd bond on Ag plating lead post. The result indicates that Autoclave weakens Al-Cu and Al-Ag interface severely.

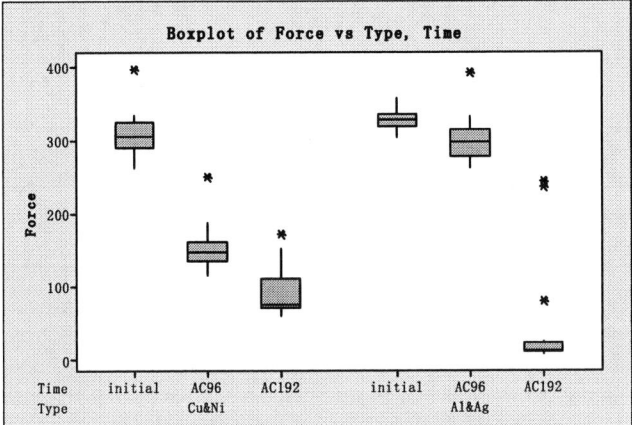

Fig 1 Wire Pull after AC192hours

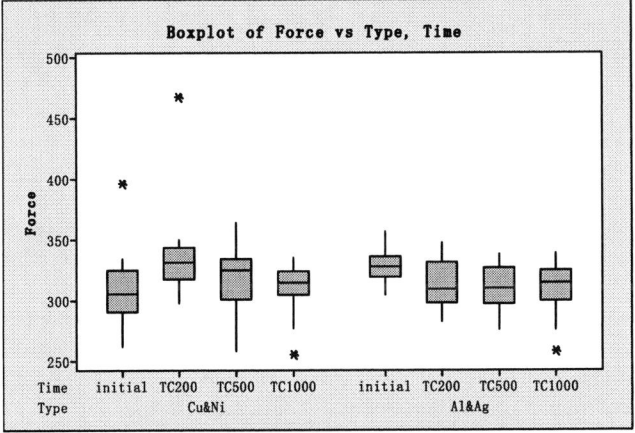

Fig 2 Wire Pull after TC1000cycles

In Temperature Cycle environment (Fig 2), for Al wire bond on Cu and Ni surface, and Al wire bond on Al and Ag

surface, wire pull data drop very slightly. It indicates that TC has no obvious effect to all four interfacial strength degradation.

In High Temperature Storage environment (Fig 3), for Al wire bond on Cu and Ni surface, wire pull data decreases in some degree after HTS500hrs. Failure happens in the 1st bond on leadframe pad. For Al wire bond on Al and Ag surface, wire pull data drop very slightly. It shows that HTS will impair to the bonding interface.

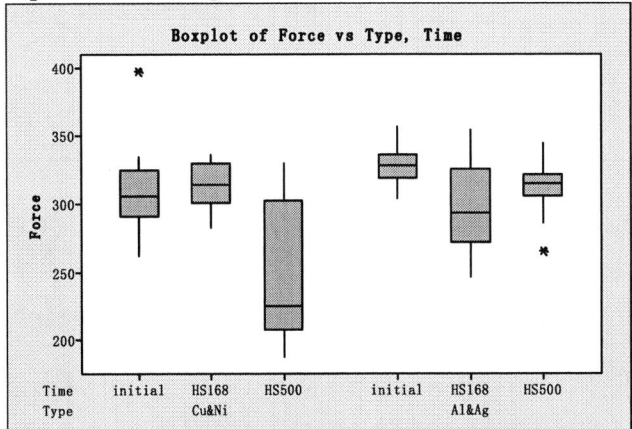

Fig 3 Wire Pull after HTS500hours

Besides wire pull test, wedge shear test is another kind of destructive test that measures the interfacial force of wedge bonding directly [5]. Bonding strength was evaluated by measuring the shear strength before and after AC, TC, HTS aging to investigate the effects of IMC growth on wire bondability.

In AC environment (Fig 4), Al-Cu interfacial force drops a lot after AC96 and AC192hrs. Al-Ni and Al-Al interface force increased with longer aging time. Al-Ag interface suddenly drop to nearly zero after AC192hrs. The result proves the wire pull degradation result that fail mode occurs in Al-Cu and Al-Ag interface.

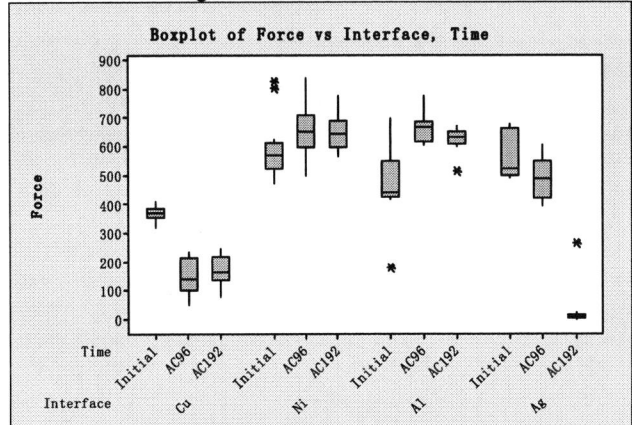

Fig 4 Wedge Shear after AC192hours

In TC environment (Fig 5), all interfacial force has a little enhancement after TC500cycles but start to drop after TC1000cycles.

689

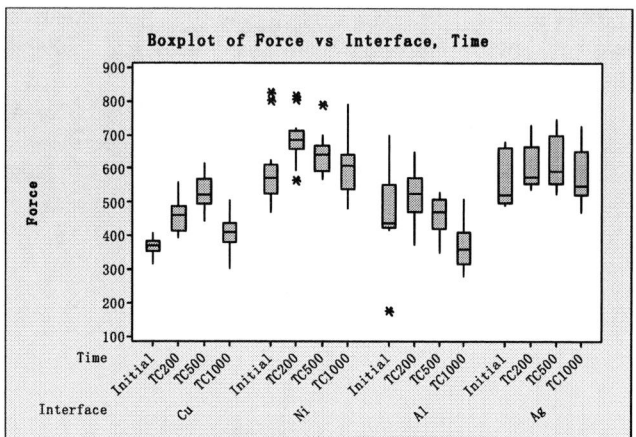

Fig 5 Wedge Shear after TC1000cycles

In HTS environment (Fig 6), all interfacial forces have slight variation after HTS500hours.

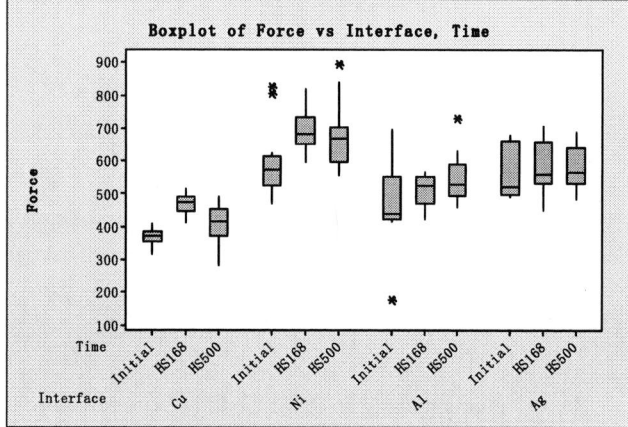

Fig 6 Wedge Shear after HTS500hours

From above experiments, the following conclusions can be drawn:

1) AC is the most severe condition to impair the bonding on Al-Cu and Al-Ag interface. TC and HTS have relatively slight effect to the bonding force.

2) Al wire bonding to Al pad make electrically and mechanically stable mono-metallic bonding interface because Al-Al mono-metallic interface does not form brittle IMC phases, so Al-Al interface is very reliable. And Al-Ni is more reliable than Al-Cu, Al-Ag bonding.

When two different metals connect, at a certain temperature level, the bonding interface will diffuse into each other. Generally, moderate IMC growth increases the bonding strength. However, as the brittle intermetallic compound layer thickness is increased, the joint contact resistance will increase and the bonding strength will decrease. Excessive IMC growth act as a major cause for bonding failure [6, 7].

The Cu-Al bond system has been studied by many former works. Several variables such as the shape of specimen, annealing temperature, and time can affect the formation and the growth of IMC, and thereby change the possible IMC phases. However, CuAl, CuAl2, Cu9Al4 are identified as the main intermetallic compound formed according to the research result [2].

3. Galvanic Corrosion

The experiment shows that AC weakens the bond strength of Al-Cu more severely than TC and HTS. Besides the high temperature effect to intermetallic compound growth, one big difference is AC moisture environment. DI water with some ionic impurities provides the electrolyte environment for corrosion.

To investigate the failure mode at bonding interface, cross section sample was observed using SEM. Al wire corrosion should be the main cause of wedge shear fast degradation. See below Fig 7 of Al wire corrosion.

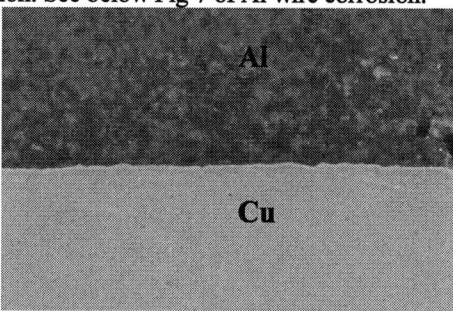

Fig 7 Corrosive Al Wire Bond on Cu

Corrosion induced failure plays an important role in microelectronic device and package failures. The main corrosion related failure mechanisms in microelectronic components are intermetallic compound formation and electrochemical corrosion. Device failures associated with corrosion include openings in wires, peeling, metallic migration, and induced electrical leakage. If sufficient moisture is present inside of a device, high temperatures can cause formation of steam which has the potential to crack the package commonly called "pop-corning" damage [8].

The basic requirements for electrochemical corrosion include electrically conductive anode, cathode, interconnecting electrolyte and the difference of electrode potentials between two dissimilar materials anode and cathode couple. Corrosion rate in microelectronic packaging depends on a number of factors: the package type, the area ratio of anode to cathode, fabrication and assembly processes, and environmental conditions such as moisture level, conductivity of electrolyte, pH value, ionic or organic contaminants, temperature and electrical bias.

Galvanic corrosion is also called dissimilar metal corrosion. Basically it refers to a corrosion phenomenon induced when two different metals are coupled in a corrosive electrolyte. When two dissimilar metals are brought into electrical contact under electrolyte such as water with ions, one of the metals with lower electrode potential acts as anode and corrodes faster than its natural corrosion while the other one with higher electrode potential acts as cathode and corrodes slower than its natural corrosion or even is stopped from corrosion. As Al is the material most often used for die bonding pad metallization, corrosion of Al is one of the most common failure mechanisms in microelectronic devices [9, 10].

While Al wire bonds onto bare copper leadframe, Al-Cu galvanic couple to form an internal short galvanic cell in the AC electrolyte environment. Since the electrode potential of Al is more negative than that of Cu, Al is more active than

690

Cu. So Al acts as the galvanic cell anode and loses electrons to be oxidized and these electrons are transferred to outside circuit: Al - 3e- ->Al^{3+}. Cu acts as the galvanic cell cathode, and gets the electron from external circuit. Hydrogen evolution reaction and oxygen reduction reaction are two-cathode-depolarization reactions. Because the moisture environment in the microelectronic packaging is usually weak acid or neutral, oxygen reduction happens: $2H^2O + O^2 + 4e^-$ -> $4OH^-$. The corrosion product of aluminum is usually believed to be $Al(OH)_3$.

The presence of an electrical field greatly accelerates the kinetics of formation of intermetallic compound phases. In power management microelectronics, for N channel MOSFET device, HAST electrical bias VDS is positive. So in the MOSFET interconnection, electrical current will flow from MOSFET drain to source, and the electron flow from source to drain which is in contrary with the electron flow direction in Al-Cu galvanic cell. So the electrical bias restricts the galvanic reaction and protects the Al layer to avoid corrosion. For P channel MOSFET, HAST electrical bias VDS is negative, the electrical current will flow from MOSFET source to drain, and the electron flow from drain to source which is the same with the electron flow direction in Al-Cu galvanic cell. So the electrical bias promotes the galvanic reaction and accelerates Al layer corrosion.

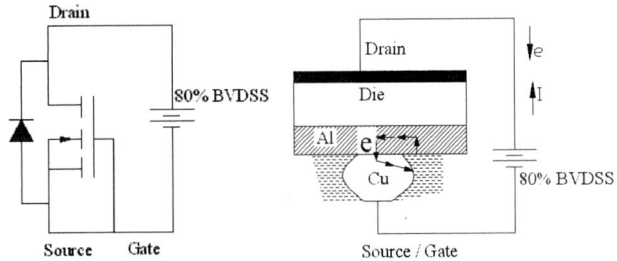

Fig 8 Electrical Bias Effect

4. Conclusions

This paper mainly studied four kinds of interfacial strength variation under different condition aging. Al-Cu interface was weakened severely by Autoclave environment, while TC and HTS have relatively slight effect to the bonding strength. Besides the high temperature effect to intermetallic compound growth, AC moisture environment provides the electrolyte environment for Al wire corrosion. Corrosion mechanism of Al-Cu is analyzed and summarized in this paper. Al-Cu galvanic couple in the electrolyte formed an internal short galvanic cell; Al acts as the galvanic cell anode and loses electrons to be oxidized; Cu acts as the galvanic cell cathode. Electrical bias effect was also analyzed. In HAST condition, electrical bias restricts N-channel galvanic reaction and protects the Al layer to avoid corrosion, and promotes P channel galvanic reaction and accelerates Al layer corrosion.

Temperature and moisture sensitivity level play a significant role in the plastic package reliability. Intermetallic compound growth and corrosion induced failure was and will be a long-term concern in microelectronic packages.

Acknowledgments

The authors wish to express our appreciations to the Reliability Laboratory for providing reliability and cross section analysis. We also acknowledge our colleagues' hard work for samples assembly and data collection.

GEM managements are appreciated for their permission to publish this paper.

References

1. Ying Zeng, "Study of Copper Applications and Effects of Copper Oxidation in Microelectronic Package", May 2003
2. Hyoung-Joon Kim, Joo Yeon Lee, Kyung-Wook Paik, etc, "Effects of Cu/Al Intermetallic Compound (IMC) on Copper Wire and Aluminum Pad Bondability", IEEE TRANSACTIONS ON COMPONENTS AND PACKAGING TECHNOLOGIES, VOL. 26, NO. 2, JUNE 2003
3. JIN ONUKI, MASAHIRO KOIZUMI, AND ISAO ARAKI, "Investigation of the Reliability of Copper Ball Bonds to Aluminum Electrodes", IEEE TRANSACTIONS ON COMPONENTS, HYBRIDS, AND MANUFACTURING TECHNOLOGY, VOL. CHMT-12. N O. 4, DECEMBER 1987
4. Yanhong Tian, Chunjin Hang, Chunqing Wang, Y. Zhou, "Evolution of Cu/Al Intermetallic Compounds in the Copper Bump bonds during Aging Process", IEEE, ICEPT Aug. 2007
5. W. J. TOMLINSON, ROY V. WINKLE, AND LYNNE A. BLACKMORE, "Effect of Heat Treatment on the Shear Strength and Fracture Modes of Copper Wire Thermosonic Ball Bonds to Al-1% Si Device Metallization", IEEE TRANSACTIONS ON COMPONENTS, HYBRIDS, AND MANUFACTURING TECHNOLOGY, VOL. 13, NO. 3, SEPTEMBER 1990
6. M. Braunovic, and N. Alexandrov, "Intermetallic Compounds at Aluminum-to-Copper Electrical Interfaces: Effect of Temperature and Electric Current", IEEE TRANSACTIONS ON COMPONENTS, PACKAGING, AND MANUFACTURING TECHNOLOGY-PART A, VOL. 17, NO, 1. MARCH 1994
7. S. Murali, N. Srikanth, Charles J. Vath III, "An analysis of intermetallics formation of gold and copper ball bonding on thermal aging", Materials Research Bulletin 38 (2003), 637–646
8. MICHAEL PECHT, "A Model for Moisture Induced Corrosion Failures in Microelectronic Packages", IEEE TRANSACTIONS ON COMPONENTS, HYBRIDS, AND MANUFACTURING TECHNOLOGY, VOL 13, NO. 2, JUNE 1990
9. C.W.Tan, A.R.Daud, M.A.Yarmo, "Corrosion Study at Cu-Al Interface in Microelectronics Packaging", Applied Surface Science 191 (2002), 67-73
10. SIMON THOMAS, HOWARD M. BERG, "Micro-Corrosion of Al-Cu Bonding Pads", IEEE TRANSACTIONS ON COMPONENTS, HYBRIDS, AND MANUFACTURING TECHNOLOGY, VOL. CHMT-10, No. 2, JUNE 1987

Thiol based chemical treatment as adhesion promoter for Cu-epoxy interface

Cell K Y Wong and Matthew M F Yuen

Department of Mechanical Engineering

Hong Kong University of Science and Technology, Clear Water Bay, Kowloon, Hong Kong, China

cacell@ust.hk, meymf@ust.hk

(852) 2358 8814

Abstract

The paper focuses on the use of thiol as adhesion promoter in Cu-epoxy interface. A parametric study on the thiol deposition procedure for Cu-epoxy system was conducted. Thiol concentration and deposition time was found to be important parameter in adhesion. Low concentrated solution leads to insufficient morphological modification while prolong treatment time results in fracture happened along weak boundary layer. The optimum condition for surface treatment with the current thiol was 5mM solution for 10 hours. With this thiol treatment, adhesion which was given as fracture toughness has increased from $4.8 Jm^{-2}$ for the untreated sample to $159 Jm^{-2}$. The 30 times adhesion improvement was related to presence of nanostructure upon thiol modification. Both the chemical bonding and morphological changes was found contributes to adhesion.

Introduction

A general trend in electronic industry requires small package size, higher integration and increased system speeds. To align with this trend, the heat dissipation requirement for high density integrated circuit (IC) packages becomes increasingly important during these years [1]. Copper (Cu)-based leadframes is a popular choice for new generation leadframe material for its excellent thermal conductivity. It, however, have a weak point in term of package reliability [2, 3]. The Cu leadframe generally exhibits poor adhesion with polymer encapsulant as compare to Alloy 42 leadframe [4]. In a plastic packages, epoxy compound are usually employed as encapsulating materials for their excellent insulating and mechanical properties. Nevertheless, the poor bondability to epoxy makes the copper leadframe package fails during its service life or even in its manufacturing process. The likelihood of package delamination and cracking hinders the usage of Cu-based leadframes.

Several approaches [5-7] have been taken in an effort to obtain higher adhesion strength between Cu and EMC. The most common approach was copper oxidation. Chemical oxidation through acidic or alkaline solution is preferred over thermal method because the morphological modification obtained from chemical method gives better adhesion [8]. The chemical modification enhances surface roughness of the substrate. It provides more site of attachment which can interlock mechanically with the polymer. Several groups reported the influence of oxide thickness on adhesion [4, 5, 9-11]. They pointed out that there exists an optimum oxide thickness that exhibits the strongest adhesion. The proposed optimum oxide thickness was 20-40nm from Tankano et al [12] and 20-30nm from Cho et al [9]. Choi et al [4] showed

that when CuO/ Cu_2O ratio fell between 0.2 to 0.3, the maximum adhesion strength was achieved, regardless of oxidation time. Lee et al [13] produced the oxide in a hot alkaline solution and reported the maximum adhesion was reached within 10-20min of immersion. They reported the optimum thickness for Cu2O and CuO were 200nm and 1300nm respectively. The discrepancy among these studies came from the variation in the oxide treatment and more importantly the microstructure obtained from different treatment process. According to Cho et al [9], with formation of CuO, the pull out strength of Cu leadframe from EMC increased due to formation of needle like CuO which interlock the epoxy material. Despite this, under excessive oxidation, the adhesion strength dropped. They explained the phenomenon by formation of low density copper oxide. As fracture happened along the weak copper oxide, the interfacial adhesion deteriorated. The results show that copper oxide which result in characteristic morphology plays a dominant role in copper/ EMC interfacial strength. Strictly control of oxide growth is thus essential for optimal adhesion. However, given that frequent thermal process is likely to be encountered by copper leadframe before molding (die bonding: 150-250°C for a few hours, wire bonding: 200-300°C), oxide thickness control is difficult if not impossible. Cu oxide growth is therefore not an applicable solution for the adhesion problem.

Another method for adhesion enhancement adopted in metal-polymer system is the formation of chemical bonding. In a metal-polymer system, covalent bonds play predominate role in interfacial bonding phenomenon. As reviewed by Pauling [14-16], the covalent bond energy is about ten times stronger than energy of the secondary forces. Thus, it is beneficial to enhance the adhesion between epoxy and Cu through chemical bonds formation. Song et al [6, 17] promote the adhesion between Cu and epoxy resin with the used of azole and triazole compounds. They demonstrated an improvement from 54N/m to 408N/m with polybenzotriazole treatment in a peel test. Eight times enhancement was obtained with the polybenzotriazole treatment. Nevertheless, the study also revealed a decline in adhesion strength with prolongs surface treatment time. The peel strength of Cu/epoxy joint has been dropped from 353N/m to 40N/m as treatment time increased from 15s to 10min. Due to the formation of porous Cu-azole complex layer at prolong treatment, failure happen inside the weak complex layer and therefore dwindle the adhesion strength [6]. The study indicates the importance of promoter candidate selection and the treatment procedure.

The use of self assembly monolayer as promoter for Cu-epoxy adhesion has been introduced by Muller et al. [7]. Through the affiliation of the self assembly coupling agent to

978-1-4244-2739-0/08/$25.00 ©2008 IEEE

Cu substrate, chemical bonding can easily realize between the Cu and epoxy. Hydroxylated thiol, disulfide, ethylene diamine and phthalocyaines were used as surface modifiers in Muller's study. About 20% increase in adhesion strength of epoxy molded button sample has been measured when the Cu was treated with disulfide and ethylene diamine. With the use of amino disulfide treatment, our group [18] has reported a 60% increase in button shear strength. Although the adhesion enhancement with the self assembly coupling agent is not as significant as compare to Song's result in the previous study, sulphur containing molecules like thiol and disulfide is advantageous in forming a chemical linkage between the Cu and epoxy compound. Given the low bonding energy in Cu-S bond, copper reacts readily with thiol/disulfide. By choosing a correct functional group in the thiol/ disulfide coupling agent, it is capable to form strong covalent linkage (C-S-Cu) between the Cu and epoxy. Unlike the Cu-azole complex, this Cu-S covalent bond is strong which can prevent failure happening within the coupling agent.

In addition to the treatment method and deposition time, the interfacial adhesion is also affected by concentration of the treatment solution [6, 19, 20]. From Song's study, they reported a change of adhesion from 408N/m to 327N/m to 187N/m as mole ratio of triazole compound decrease from 0.33 to 0.1 to 0.003. These studies disclosed the essence of research on deposition procedure for a surface treatment. Rather than strengthening the interface, prolonged surface treatment can fail the interface by producing a weak boundary layer.

This paper discusses the influence of solution concentration and treatment time to the adhesion of thiol modified copper substrate and epoxy encapsulate. Copper substrate was modified by thiol solution with concentration ranged from 0.1mM to 10mM for 1 to 7 days. The substrate was then bonded with epoxy based underfill encapsulate. The interfacial adhesion was evaluated by measuring fracture toughness of the interfaces form tapered double cantilever beam (TDCB) test. The fracture toughness results were then related to morphological study of the treated surface. To understand the failure mechanism, failure locus of the fractured sample was examined by elemental mapping in an energy dispersive X-ray analysis (EDS).

Experimental

A. Material

To investigate the surface composition and the adhesion of the modified substrate, two types of substrate were used in this study. C194 Cu based leadframe was employed as substrate for surface composition analysis and the morphological study. In the interfacial adhesion study, fracture resistance experiment was conducted using tapered double cantilever jig machined form red brass. The nominal composition of the Cu substrates was evaluated by X-ray fluorescence spectrometer (XRF) model JSX-3201Z from JOEL. The reported composition for C194 leadframe and the red brass jig was Cu-2.2Fe-0.08Si-0.48Zn-0.06Sn (wt%) and Cu-0.05Si-0.38Zn-0.02Sn (wt%) respectively. A proprietary thiol of 125gmol^{-1} molecular weight were purchased from

Aldrich and used as adhesion promoter in the study. Thiol was applied to Cu substrate through solvent adsorption. Absolute ethanol (>96% purity, Aldrich) was chosen as solvent in dissolving the thiol. Epoxy adhesive adopted in fracture test was Hysol® FP4526, a commercially available underfill obtained from Henkel Loctite®. It is a low viscosity, fast flow epoxy based material designed for capillary underfill on flip chip application.

B. Surface preparation

The thiol was used as received. It was dissolved into absolute ethanol and diluted to 0.1mM-10mM solution. All types of the substrate were cleaned with anti-grease solvent followed by Fry 90 flux (Fry Technology) to reduce surface oxide. To further eliminate carbonate and hydroxide on the Cu surface, the samples were immersed into acetic acid for 10min before adsorption. Thiol was adsorbed onto the pre-cleaned Cu surface from a pre-mixed solvent for 10-168 hours. The treated substrate was then removed from the solution and rinsed thoroughly with an ethanol before blow drying with nitrogen.

C. Surface characterization

Upon deposition, the existence and quality of the SAM modified surface has been characterized by numerous of surface analysis techniques. Detailed descriptions are written below.

Scanning Electron Microscope (SEM)

The surface topography of the SAM treated sample was evaluated by a SEM 6700F from JEOL Ltd. The electron microscope was operated under 5-15kV u in secondary electron (SE) mode. Images under 50,000x magnifications were taken to reveal the microstructure, feather size and coverage of the modified layer. 20,000 x magnifications were adopted to investigate morphology in large area.

X-ray Photoelectron Spectroscopy (XPS)

To investigate the surface composition and the bonding of the treated surface, XPS, model 5600 multi-technique system bought from Physical Electronics, was utilized. A monochromatic AlKα X-ray source was run at 1486.6eV. The spot size was kept at 600μm for all sample studied in this paper.

Time-of-flight secondary ion mass spectrometer (ToF-SIMS)

ToF-SIMS is a dedicated technique for chemical bonding identification for thin molecular layer. Since the mass for bonded and non-bonded molecules are different, it can classify a free adsorbed molecule from a chemically bonded one. In our chemical bonding inspection, a mass spectrometer model TOF SIMS V from ION-TOF GmbH was used. The machine was equipped with a Bi^{3+} ion gun and a reflectron ToF mass analyzer as detector. The spot size of the ion gun was controlled in 200x200μm^2 with less than 2nm penetration depth. Both the thiol treated samples and a control sample prepared from neat ethanol was put under investigation. A fragment having characteristic peaks should be observed in the positive ion spectrum if the thiol molecules were bonded to Cu substrate.

D. Tapered Double Cantilever Beam (TDCB) Fracture test

To determine the fracture toughness (G_{IC}) of the modified Cu-epoxy joint, TDCB fracture test was conducted according to ASTM D3433. The specimen was prepared from epoxy underfill sandwiched between a pair of tapered Cu jig. Tapered geometry guarantees a constant moment during the crack growth. It is advantageous in a stable crack experiment. In order to eliminate the roughness effect, the Cu jigs was polished with P4000 sandpaper before surface treatment. To ensure that crack opened at the right interface, a 60mm precrack was prepared in the front end of the testing jig. 0.5mm thick epoxy underfill was cured between the two jigs at 80°C for 6 hours. The width and total length of the jig was 7mm and 102mm respectively. Fig. 1 showed a schematic diagram for the TDCB configuration.

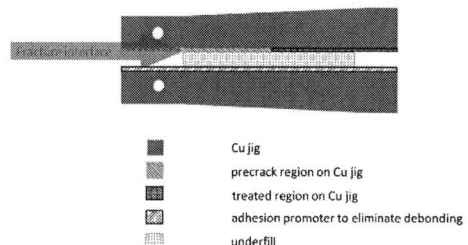

Fig. 1 Schematic diagram for TDCB specimen configuration

The prepared joint was tested with a MTS 858 Universal Testing Machine equipped with a 25kN±5N axial force load cell. Crosshead displacement rate was kept constant as 30μm/min under control displacement mode. It was assumed that the crack developed under linear fracture mechanics condition. The G_{IC} was calculated according to the following equation:

$$G_{IC} = \frac{4P^2[3a^2 + h(a)^2]}{EB^2 h(a)^3} \qquad \text{Eq. 1}$$

where P is the critical load for stable crack opening, B is the width of the specimen, a is the crack length, h is the correcsponding thickness of the jig at the point of crack front.

$$m = \frac{3a^2 + h(a)^2}{h(a)^3} \qquad \text{Eq. 2}$$

where m is a constant for the tapered jig which is determined by the jig geometry

To compare the adhesion of a modified substrate to that of untreated, normalized adhesion was calculated. It was defined as the average G_{IC} for a modified surface divided by that for the control sample. Equation for the calculation is shown below:

$$\text{normalized adhesion} = \frac{G_{IC\ sample}}{G_{IC\ control}} \qquad \text{Eq. 3}$$

E. Failure locus determination

Failure locus of a joint is critical in understanding the adhesion phenomenon. To determine the failure location of the specimen, X-ray microanalysis (EDS) was adopted. The EDS information was obtained from QUANTAX 4010 (manufactured by Bruker AXS) which was installed inside a JEOL JSM-6390 SEM. The EDS system, operated with a LN$_2$-free XFlash® Silicon Drift Detector (SDD), was tailored-made for low energy element detection. This enables sensitive polymer material detection. For our fracture locus analysis, the carbon content measurement is crucial. An accurate analysis can help to locate the failure locus and evaluate the weakest point in a joint. In this EDS experiment, spectra and 2D mapping signals were acquired at 15kV with 1mm^2 beam spot shone on the Cu jig. Acquisition range was set at 0-10keV with at least 10 kc/s detection rate.

Result

A. Surface analysis of thiol adsorbed substrate
Scanning Electron Microscope (SEM)

Fig. 2 showed the surface morphology of the thiol treated Cu surface in 50,000x magnification. Thiols modified the Cu by forming string-like nano-structures on the substrate. The tiny strings were bundled together and covered almost the whole substrate. From the side view picture, the thiol molecules constructed a thin nano scale network layer.

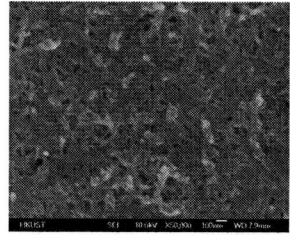

Fig. 2 SEM microscopy of thiol treated Cu: top view (left); side view (right)

X-ray Photoelectron Spectroscopy (XPS)

Fig. 3 illustrated the XPS spectra for a control substrate and a substrate treated with 1mM thiol solution. Table 1 reported the atomic composition of the two samples. In comparing with the control spectrum, significantly higher C, N, S content has been measured from the treated sample. The detection of these element implies thiol molecules were presence on the Cu surface.

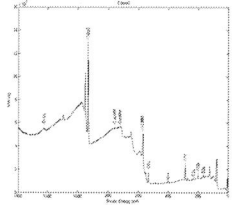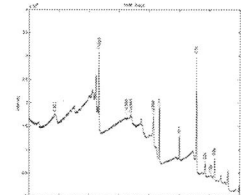

Fig. 3 XPS spectra for a control (left) and a thiol treated sample (right)

Table 1 Atomic percentage of elements on sample surface

Element	control	1mM thiol treated
C1s	40.43	63.70
N1s	1.40	7.86
O1s	11.77	12.03
S2p	2.76	8.26
Cl2p	5.50	0.11
Cu2p3	38.14	8.04

High resolution XPS spectra for Cu2p and S2p peaks are shown in Fig. 4. From the Cu2p spectrum, a 'shake-up satellite' peaks were found in 944eV. This suggested that some Cu (II) compound existed on the sample. In the S2p spectrum, a board peak was observed at ~163.8eV. As reviewed in literature [21, 22], chemisorbed S should give S2p peak at ~162eV. The slightly downward shift of 163.8eV may suggest formation of free unbonded S species on top of the bonded thiol layer.

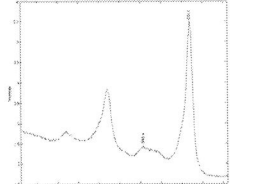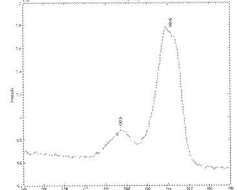

Fig. 4 High resolution spectra for Cu2p (left) and S2p (right) peak on a thiol treated sample

Although XPS is a sensitive tool for surface analysis, identification of Cu-thiol bonding in XPS is not trivial. Since the binding energy for Cu_2S, Cu_2O, CuS and CuO was very close [23] and the oxide formation on Cu was so ready, the differentiation of Cu-S bond from Cu-O bond was almost not possible especially when unknown species might involve. The existence of Cu-S bond should thus be evaluated by other surface technique: ToF-SIMS.

Time-of-flight secondary ion mass spectrometer (ToF-SIMS)

Fig. 5 showed the positive (+ve) ToF-SIMS spectra for a thiol treated sample prepared from 1mM thiol solution and a control sample prepared from ethanol. Both samples were treated with solvent for 2 days before the analysis. From the +ve SIMS spectrum of the control sample, the major mass/charge (m/z) peaks found were 143, 207 and 223. These numbers represent the presence of Cu hydroxide for 143 (Cu_2OH) and Cu oxide for 207 (Cu_3O) and 223 (Cu_3O_2) on the control sample. The fragments indicate that the oxygen dissolved in ethanol can oxidize the Cu substrate during the adsorption process. In comparison to the thiol samples, two distinct peaks were identified. The m/z 124 and 187 were observed on the thiol treated surface only. Table 2 summarized the major peaks obtained from the two samples. As mentioned, the molecular weight of the thiol molecule was 125gmol^{-1}. A characteristic fragment at 124 represents the existence of free unbonded thiol (SR^+). While reacting with Cu surface, 187 peak should be spotted ($CuSR^+$). The identification of the 187 peak revealed strong evidence that the thiol molecules were chemically bonded with the substrate. Another crucial proof for 187 contained Cu was the presence of an isotopic peak at 189. The isotopic peaks originated from $^{63}Cu^+$ and $^{65}Cu^+$ isotopes whose relative abundance was 1:0.45. The area ratio for the 187 and 189 peaks was 1:0.5 which matched the expected value. The surface analysis proves that Cu-S covalent bonds have been realized on a thiol modified copper.

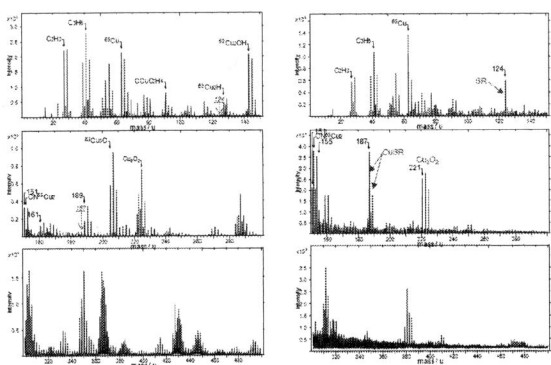

Fig. 5 Positive ToF-SIMS spectra for control (prepared by dipping in ethanol, left) and thiol treated sample (right)

Table 2 The major peaks of treated and control samples obtained from positive ToF-SIMS spectra

Sample	Peak1 (m/z)	Peak2 (m/z)	Peak3 (m/z)
control in ethanol, 2days	124	187 and 189	221, 223, 225
1mM thiol, 2days	143, 145,147	205, 207, 209	221, 223, 225

B.   Tapered Double Cantilever Beam (TDCB) Fracture test
Concentration effect

Fracture toughness (G_{IC}) of the surface modified Cu jig was examined. A control Cu specimen without any treatment was used as benchmark. The fracture toughness was regarded as critical load for stable crack propagation along the sample interface. The average critical load for stable crack propagation for control and 1mM, 2 days specimen was 39±3N and 182±13N respectively. Typical force curves for the control and treated sample were given in Fig. 6.

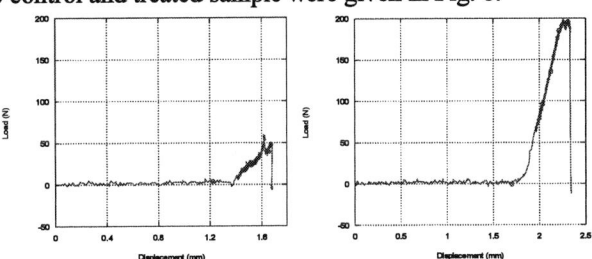

Fig. 6 Typical loading curve for TDCB test on a control (left) and thiol treated specimen (right)

G_{IC} was calculated from the critical load needed for propagating a stable crack according to Eq. 1. Fig. 7 summarizes the G_{IC} results of the control and the thiol modified Cu-epoxy interface in different concentration. The deposition time was fixed at 2 days. The results demonstrate that with sufficient thiol treatment, the interfacial adhesion for Cu-epoxy improves significantly. Fracture toughness of Cu-epoxy is boosted from 4.8±0.8Jm^{-2} to more than 80Jm^{-2} when treated with 0.5mM thiol solution. The trend continues to climb to 104±14Jm^{-2} to 140±17Jm^{-2} for 1mM and 5mM treatment respectively. Despite a positive trend has been revealed by higher concentrated solution, the adhesion decreases as substrate being treated with 10mM thiol. The G_{IC} dropped to 101±7Jm^{-2} with 10mM solution.

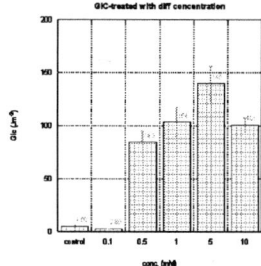

Fig. 7 Fracture toughness of Cu-epoxy interface treated with thiol solution in various concentrations

Deposition time effect

Fig. 8 shows the time effect on G_{IC} for 1mM and 5mM modified interface. With 1mM solution, the recorded G_{IC} is104±14 Jm^{-2} with 48 hours deposition and it reduces to 73±14 Jm^{-2} after 168 hours incubation. A similar trend has been discovered from the 5mM specimen. Maximum G_{IC} is measured as 159±33 Jm^{-2} from 10 hours treated sample. The fracture toughness then drops to 140±17 Jm^{-2} as deposition time increase to 48 hours.

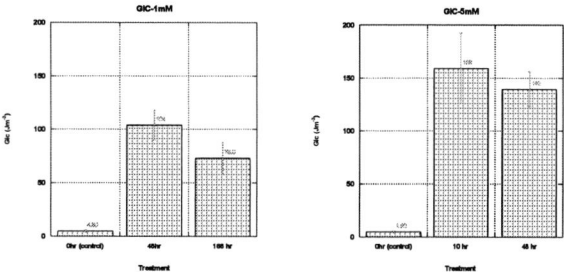

Fig. 8 Fracture toughness (GIC) of Cu-epoxy interface with 1mM thiol (right) and 5mM (left) treatment in treatment time

C. Fracture surface analysis

An optical picture showing the fracture surface for a control and 5mM treated sample is given in Fig. 9. A clean copper surface is observed in the control sample while blue underfill epoxy is found on the 5mM sample. Epoxy residue is found around the jig edge. Some white spot are observed in the center of the thiol treated jig.

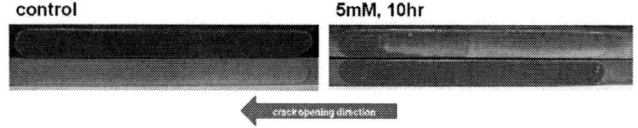

Fig. 9 Optical images of the Cu jig after fractured in an control (left) and 5mM treated (right) specimen

To study the failure mechanism, SEM-EDX was used. 200x magnification SEM picture was taken on the fracture jig (Cu side). The elemental analysis on the Cu jig could roughly determine the failure location. Fig. 10 illustrates a 2D mapping for C, O, Cu and S on a fracture jig from 5mM solution. 2D mapping inside the red square are examined.

Obviously, the white spot (black feature under SEM) consisted of carbon and oxygen mainly. It contains mainly copper in the grey area. This implies that cohesive failure happened at some location in the 5mM specimen. The analysis explains the high fracture resistance in a thiol treated samples.

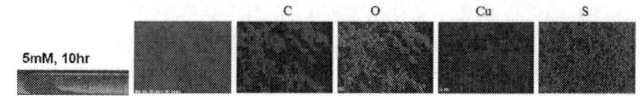

Fig. 10 2D mapping image on a Cu jig after fractured: analysed element included C, O, Cu and S from the left.

Discussion

A. Normalized adhesion

The effect of i) concentration and ii) deposition time was evaluated by the normalized adhesion as calculated from eq. 3. The normalized results are summarized in Table 3 and Table 4 for the i) concentration and ii) deposition time effect respectively.

Table 3 Effect of concentration on interfacial adhesion

treatment	Normalized adhesion
control	1
0.1mM, 48hr	0.6
0.5mM, 48hr	18
1mM, 48hr	22
5mM, 48hr	29
10mM, 48hr	21

Table 4 Effect of deposition time on interfacial adhesion

treatment	Normalized adhesion
control	1
1mM, 48hr	22
1mM, 168hr	15
5mM, 10hr	33
5mM, 48hr	29

B. Morphological analysis

To understand the adhesion phenomenon on the modified surface, surface morphology of the treated specimen were examined under SEM. Pictures were taken under 20,000 x magnifications in order to give topography information in large area.

Concentration effect

Surface morphology and the corresponding normalized adhesion from samples in different concentration are shown in Fig. 11. While comparing with the control sample (time 0 hour), the treated surface is smoother overall. The rolled marks, which were introduced during leadframe manufacturing process, are covered by string like structures upon treatment. Topography on the 1mM, 5mM and 10mM surfaces looked quite similar. Nanoscale strings covered the rolled marks over large area. Nevertheless, little topographical feather was identified on the 0.1mM sample. The rolled marks on the 0.1mM treatment sample are still visible. It seems that, with 2 days dipping, sufficient modification has not been achieved with 0.1mM solution. This explains the extremely low adhesion (0.6x of the control) recorded by the 0.1mM specimen as compared to larger than 20x with high concentrated treatment.

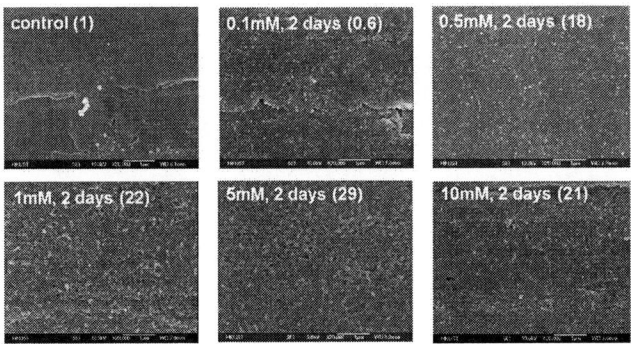

Fig. 11 Morphological image of Cu surface treated with thiol solution in different concentration as taken by SEM, magnification being fixed at 20,000x. Corresponding normalized adhesion were written inside bracket.

Deposition time effect

Fig. 12 and Fig. 13 show the surface morphology of Cu with 1mM and 5mM thiol treatment for different time respectively. String-like features are observed on all these samples, except the 1mM, 168 hours sample. Large holes appear in the sample after 168 hours treatment. The surface appears to be more porous after prolong treatment. This may explain the decline of normalized adhesion from 22 for 48 hours sample to 15 for 168 hours treatment. Interfacial adhesion may deteriorate as failure path shifted to the weak porous layer.

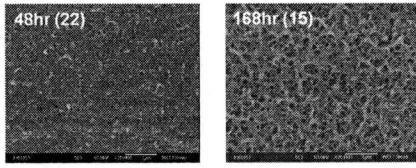

Fig. 12 Morphological image of Cu surface treated with 1mM thiol solution for 48 hours (left) and 168 hours (right), magnification being fixed at 20,000x. Corresponding normalized adhesion were written inside bracket.

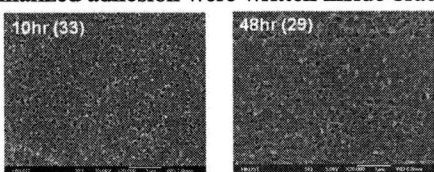

Fig. 13 Morphological image of Cu surface treated with 5mM thiol solution for 10 hours (left) and 48 hours (right), magnification being fixed at 20,000x. Corresponding normalized adhesion were written inside bracket.

The above results reveal that there exists an optimum condition for adhesion. As reported by Bell et al [19] and Plueddemann [24] who studied the coupling effect of silane in metal/glass-polymer system, they illustrated the effect of coupling agent thickness to adhesion. Although the thickness value might have changed with different system, the optimum thickness can also be found in different cases.

C. Failure mechanism study

The 2D mapping data suggested that surface modification can shift the failure path of a joint. Without thiol treatment, failure occurs inside the Cu-epoxy interface. With adequate solution treatment, more than 10 times improvement has been demonstrated. Failure mode has changed from interfacial failure to cohesive failure inside bulk epoxy. It demonstrates the effect of chemical bonding in interfacial adhesion.

Conclusions

This paper focuses on the use thiol treatment for Cu-epoxy interfacial adhesion. Parametric study on the effect of solution concentration and deposition time has been discussed. The findings are summarized in the following:

1. With thiol modification, the interfacial adhesion for Cu-epoxy joint has been improvement for more than 30 times as demonstrated in the currents study. Thiol is thus an effective adhesion promoter for a Cu-epoxy joint.
2. Solution concentration and deposition time are crucial parameters affecting the formation of covalent bonds for adhesion improvement .
3. For prolong time treatment, adhesion may be worsen due to formation of weak structure.

References

[1] L. T. Manzione, *Plastic Packaging of Microelectronic Devices*: Van Nostrand Reinhold, New York, 1990.

[2] S. Kim, "The role of plastic package adhesion in performance." vol. 14, 1991, pp. 809-817.

[3] O. Yoshioka, N. Okabe, S. Nagayama, and R. Yamagishi, "Improvement of moisture resistance in plastic encapsulants MOS-IC by surface finishing copper leadframe," in *39th Electronic Components and Technology Conference*, 1989, pp. 464-471.

[4] K.-S. Choi, T.-G. Kang, I.-S. Park, J.-H. Lee, and K.-B. Cha, "Copper lead frame: an ultimate solution to the reliability of BLP package," *IEEE Transactions on Electronics Packaging Manufacturing,* vol. 23, pp. 32-38, 2000.

[5] H. Y. Lee and J. Yu, "Adhesion strength of leadframe/EMC interfaces," *Journal of Electronic Materials,* vol. 28, pp. 1444-1447, 1999.

[6] S. M. Song, C. E. Park, H. K. Yun, C. S. Hwang, S. Y. Oh, and J. M. Park, "Adhesion improvement of epoxy resin/copper lead frame joints by azole compounds," *Journal of Adhesion Science and Technology,* vol. 12, pp. 541-561, 1998.

[7] R. Muller, K. Heckmann, M. Habermann, T. Paul, and M. Stratmann, "New adhesion promoters for copper leadframes and epoxy resin," *Journal of Adhesion,* vol. 72, pp. 65-83, 2000.

[8] B. Love and P. Packman, "The contributions of morphological and surface chemical modifications to the elevated-temperature ageing of copper-epoxy interfaces," *Journal of Materials Science,* vol. 33, pp. 1359-67, 1998.

[9] S. J. Cho, K. W. Paik, and Y. G. Kim, "The Effect of the Oxidation of Cu-Base Leadframe on the Interface Adhesion Between Cu Metal and Epoxy Molding Compound," *IEEE Transactions on Components, Packaging, and Manufacturing Technology part B,* vol. 20, pp. 167-175, 1997.

[10] P. W. K. Chung, M. M. F. Yuen, P. C. H. Chan, N. K. C. Ho, and D. C. C. Lam, "Effect of copper oxide on the

adhesion behavior of epoxy molding compound-copper interface," in *Electronic Components and Technology Conference*, 2002, pp. 1665-1670 (IEEE cat n 02ch3734-5).

[11] J.-K. Kim, M. Lebbai, J. H. Liu, J. H. Kim, and M. M. F. Yuen, "Interface adhesion between copper lead frame and epoxy moulding compound: effects of surface finish, oxidation and dimples," in *50th Electronic Components and Technology Conference*, 2000, pp. 601-608.

[12] E. Takano, T. Mino, K. Takahashi, K. Sawada, S.-y. Shimizu, and H. Y. Yoo, "Oxidation control of copper leadframe package for prevention of popcorn cracking," in *47th Electronic Components and Technology Conference*, 1997, pp. 78-83.

[13] H. Y. Lee and S. R. Kim, "Pull-out behavior of oxidized copper leadframes from epoxy molding compounds," *J. Adhesion Sci. Technol.,,* vol. 16, pp. 621-51, 2002.

[14] F. M. Fowkes, *Physico-Chemical Aspects of Polymer Surfaces*. New York: Plenum Press, 1983.

[15] R. J. Good, *Treatise on Adhesion and Adhesive*: New York: Marcel Dekker, 1967.

[16] L. Pauling, *The Nature of the Chemical Bond and the Structure of Molecules and Crystals; An Introduction to Modern Structural Chemistry*. New York: Ithaca, N.Y., Cornell University Press; London, H. Milford, Oxford Universty Press, 1940.

[17] S. M. Song, K. Cho, C. E. Park, H. K. Yun, and S. Y. Oh, "Synthesis and characterization of water-soluble polymeric adhesion promoter for epoxy resin/copper joints," *Journal of Applied Polymer Science,* vol. 85, pp. 2202-2210, 2002.

[18] C. K. Y. Wong, H. Gu, B. Xu, and M. M. F. Yuen, "A new approach in measuring Cu-EMC adhesion strength by AFM," *IEEE Transactions on Components and Packaging Technologies,* vol. 29, pp. 543-550, 2006.

[19] J. P. Bell, R. G. Schmidt, A. Malofsky, and D. Mancini, "Controlling Factors in chemical coupling of polymers to metals," in *Silanes and Other coupling Agents*, K. L. Mittal, Ed. The Netherlands: VSP BV, 1992, pp. 49-66.

[20] D. H. Berry and A. Namkanisorn, "Fracture toughness of a silane coupled polymer-metal interface: Silane concentration effects," *Journal of Adhesion,* vol. 81, pp. 347-370, 2005.

[21] T. P. Ang, T. S. A. Wee, and W. S. Chin, "Three-dimensional self-assembled monolayer (3D SAM) of n-alkanethiols on copper nanoclusters," *Journal of Physical Chemistry B,* vol. 108, pp. 11001-11010, 2004.

[22] P. E. Laibinis, G. M. Whitesides, D. L. Allara, V.-T. Tao, A. N. Parikh, and R. G. Nuzzo, "Comparison of the structures and wetting properties of self-assembled monolayers of n-alkanethiols on the coinage metal surfaces, Cu, Ag, Au," *Journal of the American Chemical Society,* vol. 113, pp. 7152-7167, 1991.

[23] C. D. Wagner and G. E. Muilenberg, *Handbook of x-ray photoelectron spectroscopy : a reference book of standard data for use in x-ray photoelectron spectroscopy*. Eden Prairie, Minn.: Physical Electronics Division, Perkin-Elmer Corp., 1979.

[24] E. P. Plueddemann, *Silane coupling agents* Plenum Press, New York 1982.

Design of Testing Chip for Measuring Mechanical Properties of Thin Films

Rui Liu, Hong Wang, Xueping Li, Jun Tang, Shengping Mao, Guifu Ding
Research Institute of Micro /Nano Science and Technology, Shanghai Jiao Tong University, Shanghai 200240, China
Tel: +86-21-34206689; fax: +86-21-34206686; e-mail: lruicxz@yahoo.com.cn

Abstract

Uniaxial tensile test is the most reliable way to measure mechanical properties of thin films. The difficulties of uniaxial tensile test are how to fabricate small and stress-free specimen, align and trip the specimen, generate small forces and measure strain. A novel tensile testing chip to measure thin film specimens with large elongation was proposed in this paper, and it was fabricated by UV-LIGA (Ultraviolet Lithographie GalVanoformung Abformung) technology. This novel testing chip has good alignment and can endure large deformation.

Introduction

With the rapid development of Micro-electro-mechanical systems (MEMS) technology and composite integrated circuit (IC), more thin film materials including silicon, metal and polymers are widely used in these fields. The mechanical properties are necessary for designing devices and evaluating their reliability. However, properties of thin film materials are quite different from those of bulk materials since they are not only affected by materials scale, but also by fabrication processes [1, 2]. So how to measure the properties of thin films becomes an attractive task [3].

Uniaxial tensile test is generally considered as the most reliable way for measuring mechanical properties of thin films. However, the difficulties of uniaxial tensile test are how to fabricate small specimen, align and trip the specimen, generate small forces and measure strain. In order to obtain reliable mechanical properties of thin films, we present a novel uniaxial tensile testing chip to solve these problems. Electrodeposited Ni components are widely used in MEMS and composite IC for its high elasticity and flexibility [4-6]. So the Ni thin films specimen is prepared in this paper.

Design of the Testing Chip

Figure 1 shows the schematic of the testing chip. It consists of a support frame, S-shaped support springs, displacement marks, alignment marks, pin holes, locating holes and the thin film specimen.

The support frame is composed of 50 μm thick electrodeposited Ni metal. Each of the support spring is connected with the electrodeposited Ni frame at one end and the movable table at the other. There is a hole in the end of movable table used for connecting with the force sensor. Moreover, there are four locating holes in the support frame, which can hold the alignment between the testing chip and force sensor. The most important part of the testing chip is the thin film specimen, which is fixed between Ni support frame and the movable table. The typical gauge section of the specimen is 50 μm wide, 100 μm long and 5 μm thick. This testing chip has the several advantages.

(i)   The thin film specimen can be electrodeposited with different thickness
(ii) The thin film specimen has good alignment because both end of it are fixed
(iii) The S-shaped spring can endure large deformation
(iv) The testing chip can be fabricated easily by UV-LIGA technology
(v)   The strain of the thin film specimen can be measured directly

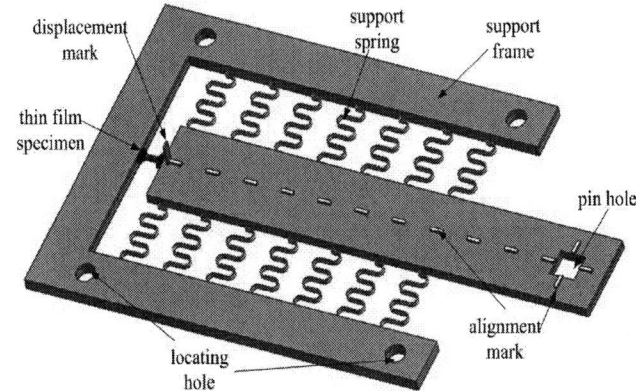

Fig.1 Schematic diagram of the testing chip

The specimen-fixed stage is shown in Fig.2. It is composed of fixed hole, locating pin, absorptive hole and vent pin. The specimen-fixed stage is fixed to adjustable stage by fixed holes, and the vacuum pump is connected with the specimen-fixed stage by vent pin. So the testing chip can be fixed conveniently by this novel specimen-fixed stage. Moreover, the specimen-fixed stage could be used unlimitedly.

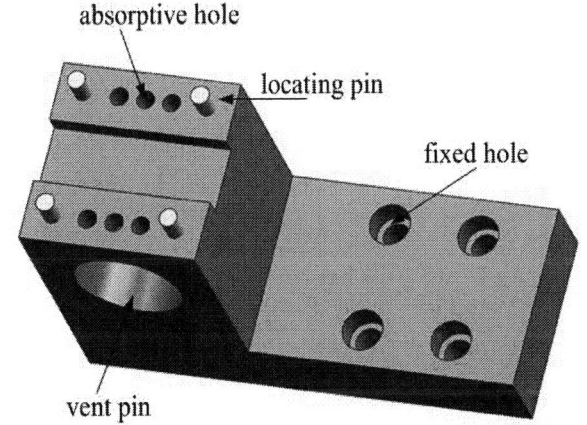

Fig.2 Schematic diagram of the specimen-fixed stage

Fabrication Process of the Testing Chip

978-1-4244-2739-0/08/$25.00 ©2008 IEEE   699

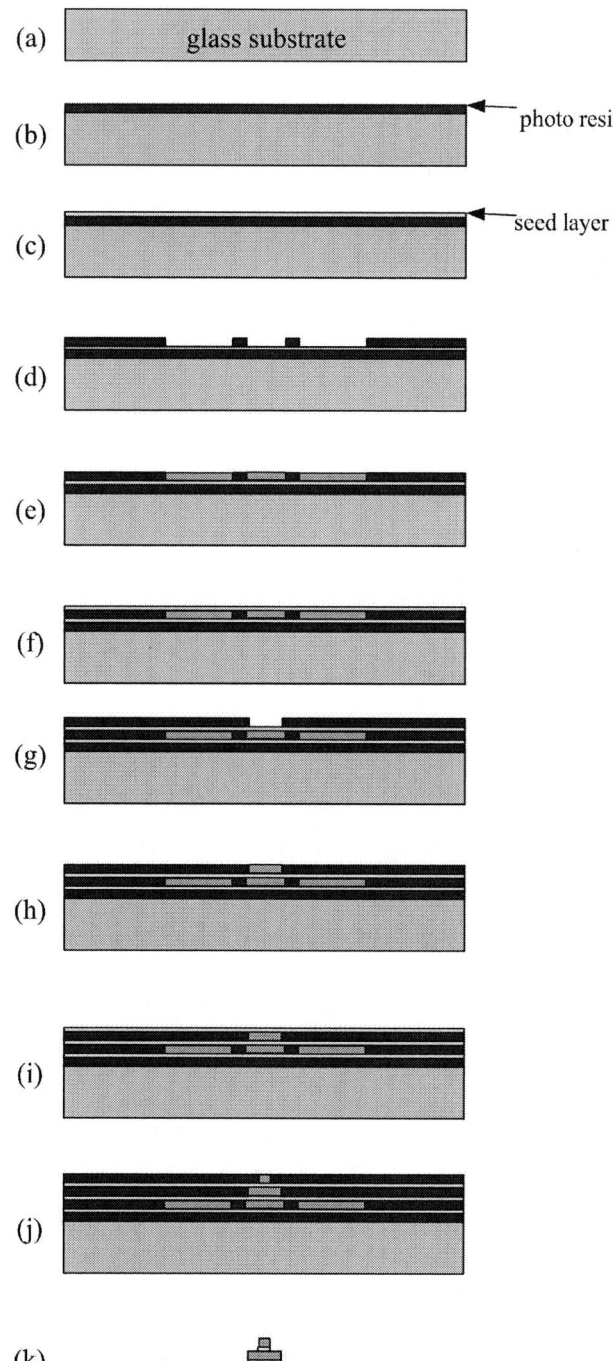

Fig.3 Testing chip fabrication process schematic

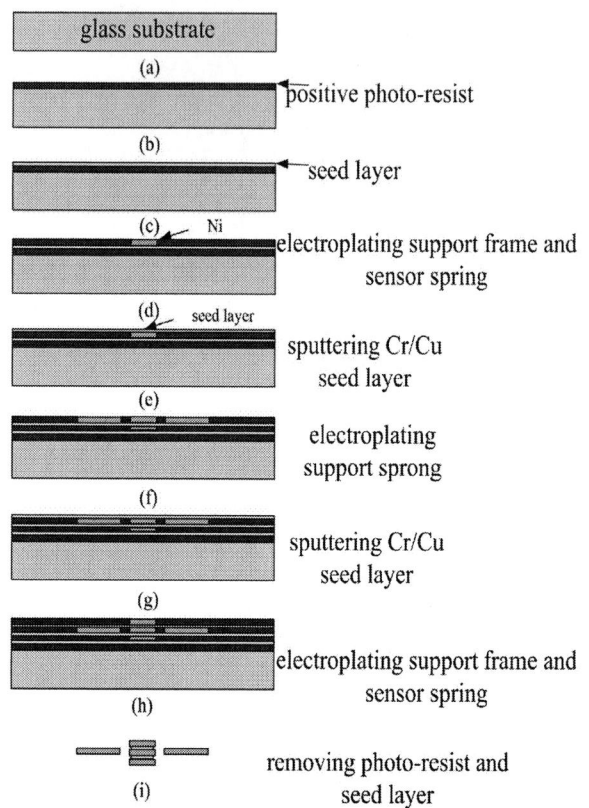

Fig.4 Force sensor fabrication process schematic

Figure 4 is the schematic of force sensor fabrication process. In our testing system, the force sensor is designed according to the testing chip because the force is very slight in measuring properties of thin films [7]. Therefore, the force sensor with accurate resolution is fabricated by UV-LIGA technology.

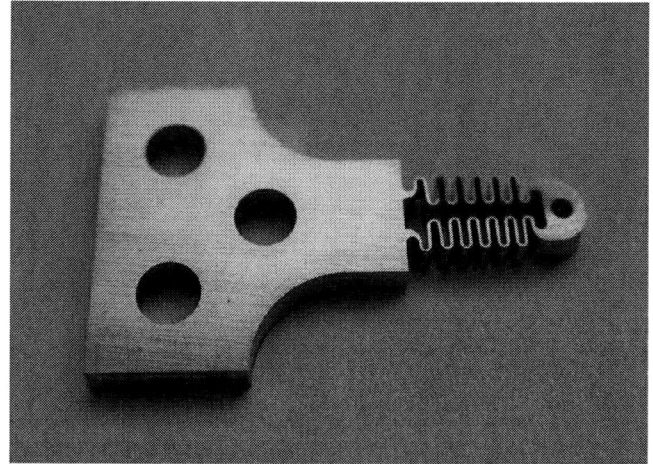

Fig.5 Force sensor fabricated by UV-LIGA technology

The main fabrication process of the testing chip is shown in Fig. 3. Firstly, photo resist (about 30 μm) was spin-coated on the glass substrate as the sacrificial layer. Then the Cr/Cu (about 100 nm thick) was sputtered on the photo resist. Secondly, the photo-resist was spin-coated on the seed layer, and the S-shaped springs and Ni frame were electrodeposited after exposal. The third step was to electrodeposite the thin film specimen. The fourth step was to make the strain marks. Finally, the photo resist and the seed layer were removed, and the testing chip was released.

The force sensor fabricated by UV-LIGA technology is shown in Fig.5. There are two S-shaped springs and a pin hole on one end of the force sensor. It is can be seen that the force sensor has good uniform. Moreover, it has high depth-to-width ratio to ensure accurate horizontality during tensile test.

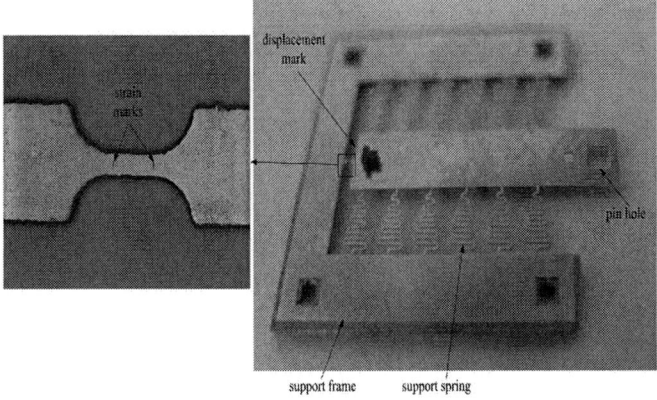

Fig.6 Optical image of the testing chip

Figure 6 shows the optical image of the fabricated testing chip. The testing chip is made up of support frame, spindle shape support springs, movable table, displacement mark and the thin film specimen. It can be seen that two ends of the thin film specimen are connected with the support frame and movable table. So this mode of trip could assure the good alignment of the thin film specimen, and the thin film specimen could keep good state before tensile test. Moreover, the method of fabricated testing chip has high efficiency and mission success rate.

Loading process is realized by connecting the piezoelectric motor to the testing chip through the force sensor. The elastic coefficient of force sensor is calibrated by the specific device. So the tensile force is applied on the thin film specimen by imposing a displacement on one end of the force sensor, while the other end of the force sensor is connected with the movable table of testing chip. The pin holes of the testing chip and the force sensor could be aligned by the electrodeposited marks.

It is very difficult to measure minute displacement during measuring properties of thin films. In our tensile test, the laser displacement sensor is adopted to measure the displacement. Moreover, there is a T-shaped displacement mark fabricated with UV-LIGA technology on one end of the testing chip. The laser displacement sensor in this system consists of an emitter, an acceptor and an indicator. The laser emitted from the laser displacement sensor is reflected by the displacement mark on the testing chip and then received by the acceptor. So the displacement of the film specimen can be measured by this method. Moreover, the tensile force is calculated as $F = k\delta$, where k is the spring constant of the force sensor, and δ is the specimen deflection measured by the laser displacement sensor.

It can be seen that the strain mark of the thin film specimen is formed by hardness gauge in Fig.6. This method is more convenient compared with other methods. Therefore,the strain of the film specimen can be measured through the two strain marks easily.

Fig.7 Optical image of specimen-fixed stage

The specimen-fixed stage fabricated by precision machining technology is shown in Fig.7. The four holes in the end part of the specimen-fixed stage could fix it to the adjustable table. The vacuum pump is connected with the specimen-fixed stage by vent pin, and the testing chip could be held through absorptive holes. Moreover, the testing chip is located accurately by the four locating pins.

Conclusions

A novel uniaxial micro-tensile testing chip to measure micro-scale thin films is presented in this paper. The design of thin film testing chip has good alignment because both end of it are fixed, and testing chip can be fabricated easily by UV-LIGA technology. Moreover, the new specimen-fixed stage can fix the testing chip conveniently, and it could be used unlimitedly.

Acknowledgments

Financial support for this research from the National High Technology Research and Development Program of China (No. 2006AA4Z326) is gratefully acknowledged.

References

1. Kraft O, Freund L B, Phillips R. *et al* "Dislocation plasticity in thin metal films," *MRS. Bull.* Vol. 27, No.1 (2002), pp. 30-36.
2. Taechung Y, Chang J K "Measurement of mechanical properties for MEMS materials," *Meas. Sci. Technol,* No. 10 (1999), pp. 706-716.
3. Haque M A, Saif M T A "A review of MEMS-Based microscale and nanoscale tensile and bending testing," *Exp. Mach.* Vol.44, No.1 (2003), pp. 248-255.
4. Hemker K. J, Last H. "Microsample tensile testing of LIGA nickel for MEMS applications," *Mater. Sci. Eng. A* 319-321 (2001), pp. *882*-886.

5. Mazza E, Abel S, Dual J. "Experimental determination of mechanical properties of Ni and Ni-Fe microbars," *Microsyst.* Tech. Vol.2, No. 1 (1995), pp. 197-202.

6. Kruger C, Bartelink D J, Fritz T, *et al.* "Micro-springs for temporary chip connections," *Sens. Actuators. No.*85 (2000), pp. 371-376.

7. Liu R, Li X P, Wang H, *et al.* "A micro-tensile method for measuring mechanical properties of MEMS materials," *J Micromech Microeng.* 2008 (Unpublished)

The Design and Fabrication of RF Band Pass Filter by LTCC Technology

Yuanxun Li[1], Yingli Liu[1], Huaiwu Zhang[1], Dafu Lu[1], Lifei Bian[1], Zongbao Yang[2]

[1]State Key Laboratory of Electronic Thin Film and Integrated Devices, University of Electronic Science and Technology of China Chengdu, 610054, Sichuan, China
[2]Integrated microcircuit company of Anhui Province, Hefei, 230088, Anhui, China

Abstract

LTCC (Low Temperature Co-fired Ceramics) has been become the key technology of packaging for the integrated of RF passive components due to its higher performance of thermal sink, reliability and plays an important role in increasing higher frequency, decreasing the loss, minimize the volume, etc. This paper mainly focuses on the design and fabrication of RF band pass filter by LTCC technology. The filter model was established according to the capacitor coupled resonate band-pass filter's circuit structure and the connect type of embedded coupling capacitor was neatly designed to make it be linked between in port an out port, for two translations zeros using two stage resonator can be produced. After tuning up the distances between the strip line inductor and ground to adjust the inductance and quality Q, the circuit's whole performance could be improved. The band-pass filters with center frequency separated at 1.8GHz and 1.3GHz were fabricated using ULF140 material as the dielectrics to further validate the model and the samples were tested with high consistence with the simulated data.

Introduction

With the current explosive growth of communication technologies and a wide variety of applications being found for broadband and high frequency of electronic components, LTCC (Low Temperature Cofired Ceramics) with materials that possess special characteristics have been rapidly developed, due to its greater multifunctionality, higher performances, sub-miniaturization and excellent reliability. And it provides an easy way to embed passive components which can be mounted onto the surface of substrates and achieve circuit boards with the desired performances for high-density packaging. Therefore, LTCCs are regarded as a promising technology for the integration of components and substrates for high frequency applications.

As one of the basic passive components, band-pass filters (BPF) play an import role in RF circuits to realize the function of signal band-passing and are extensively used to construct a variety of RF components such as power amplifier modules, transceiver modules and voltage controlled modules [1-6].

In this paper, the design method for BPFs with center frequency separated at 1.8GHz and 1.3GHz is developed based on the equivalent circuit. By employing the proposed design method, the multilayer chip BPF configuration can be made more compact and flexible. The designed chip-type BPF was realized by implementing the multilayer chip inductors and capacitors. Each lumped element was implemented using ceramic material and Ag metal layers. The dielectric constant and loss tangent was chosen to be 14

and 0.0015, respectively for 3D structure design of the chip BPF. The designed BPFs are fabricated by 33um thickness films and constructed by LTCC technology with excellent consistency between simulation and fabrication and merits of simple structures and low cost for wide applications.

Circuit and Stimulation

A. Circuit Model

The BPFs were constructed by π type coupling capacitance circuit model through ADS designing software. The typical circuit and the parameters for BPFs were shown in Fig.1 which is composed of inductors and capacitors by circuit stimulation. Fig.2 depicts the stimulated results.

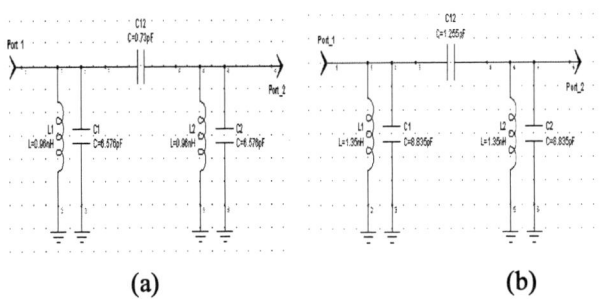

(a)   (b)

Fig.1 The circuit model for band-pass filters (a) center frequency at 1.8GHz, (b) center frequency at 1.3GHz

From Fig.2, it can be seen that the attenuation over 25dB, the center frequency and the band-width about 200MHz could be achieved.

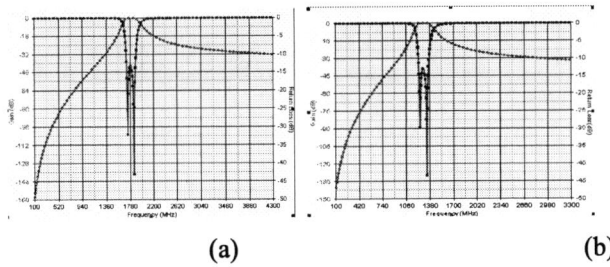

(a)   (b)

Fig.2 The stimulated results of two kind of band-pass filters by ADS software (a) for center frequency at 1.8GHz, (b) for center frequency at 1.3GHz

However, the above results are only used to be referred, because accurate parameters of component model could not be extracted, due to the ADS designing software doesn't take the influences of parasitic parameters and the coupling effect into account during the fabricating process.

978-1-4244-2739-0/08/$25.00 ©2008 IEEE

B. Component Implementation

The HFSS software has offered powerful functions to deal with the three-dimensional electro-magnetic fields' stimulation. According to actual conditions for the fabrication of filters, the equivalent circuit and an EM generated S-parameter database have been investigated with the collection of the electrical response by establishing the proper HFSS model for different available value.

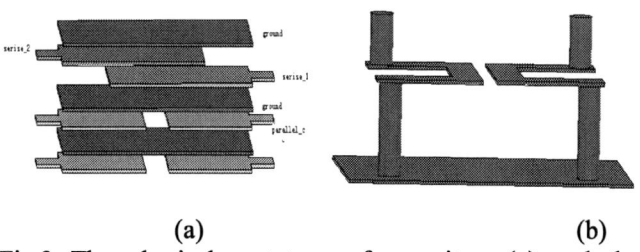

(a)   (b)

Fig.3 The physical prototype of capacitors (a) and the inductors with via-hole (b)

The structures of the capacitors and the inductors embedded in the filters were adopted as shown in Fig.3. To adapt the size of packaging for LTCC chip electronic component, it demands compact routing structure, especially in the Z direction. Therefore, the capacitors were designed to embed between the two ground metals in order to insulate the influences of the outside circuit and overcome the parasitic coupling phenomenon in some degree. From Fig.3. (b), it can be seen that the structure of inductor was made of square spirals structure, which is benefiting for improving Q value. And the via-holes were connected to the capacitors and the ground respectively to construct LC circuit together with the capacitors and to form the shunt resonator units.

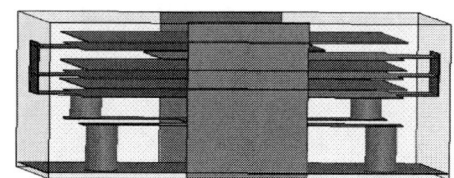

Fig.4 The physical prototype for band-pass filters

The physical prototype for BPFs was shown in Fig.4. The material was Ferro ULF140 with the dielectric constant of 14 and the thickness of each layer was 30um. After optimized, the characteristics of the filters simulated can be achieved and presented in Fig.5.

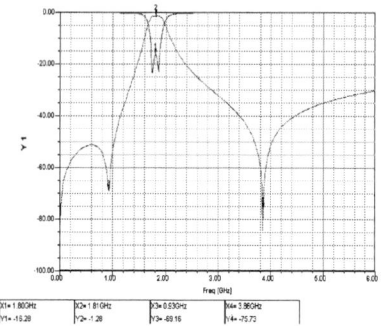

(a)

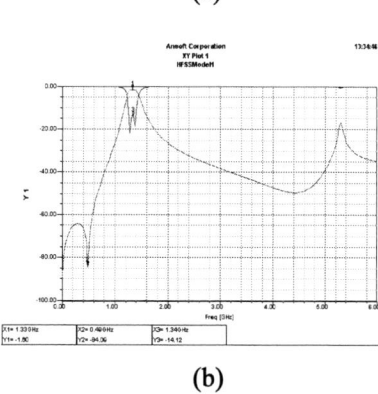

(b)

Fig.5 The stimulated results of two kind of band-pass filters by HFSS software (a) for center frequency at 1.8GHz, (b) for center frequency at 1.3GHz

From the response of three dimension configurations, the center frequency of the BPFs was 1.8GHz and 1.3GHz, the insertion losses were 1.3dB and 1.6dB respectively and the bandwidths were both about 200MHz. The difference from the circuit simulation by ADS software is that there are two translations zeros occurring, which improves the suppression performances for the filters. This can be explained by the theory of intercrossing coupling resonators and filters. So, the designing for the coupling capacitors at the input and output will produce translations zeros distributing outside of the pass-band which do good to improving the performances of filters.

Experimental

For LTCC process technology, the compatibility of system materials with respect to shrinkage, thermal expansion coefficient and chemical compatibility must be considered, different shrinkage rate of the fired specimens during co-firing make the materials distort, which cause deviation of the designed component. While designing silk screen, the pre-cofiring tape must be larger than designed model.

Fig.6. show the graphics of silk screen and relative sizes for one unit which will guarantee the good connection for each lays and formation of capacitors and inductors with the packaging size of 0805.

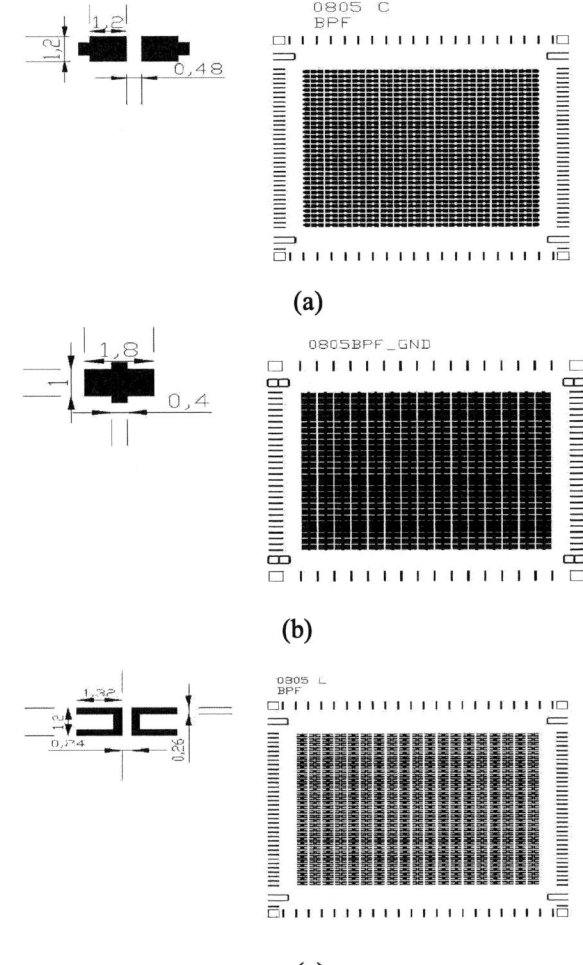

(a)

(b)

(c)

Fig.6. silk screen (a), (b), (c) for the fabrication of BPFs by LTCC technology with the packaging size of 0805

The conventional LTCC technology process was adopted to fabricate the laminated BPFs samples with the outline dimension of 2.0mm×1.2mm×0.9mm shown in Fig.7.

Fig.7. Experimental prototype for the samples

The measurements were carried on by Agilent 8722ES and the collected data was then calibrated to the desired reference plane by the thru-reflect line (TRL) technique through carefully designed calibration standards embedded in the same LTCC tie. The measured responses of the filters are shown Fig.8. From Fig.8, it can be clearly seen that the measured results agree with the simulated data basically. The center frequencies are 1.81 GHz and 1.31 GHz and the insertion loss in the center frequencies are -1.02dB and -

5.88dB with reflection losses -16.4dB and -12.7dB respectively. There are two translations zeros had been observed at 0.91GHz and 3.84GHz for the BPF sample of 1.8GHz. Thus, a good coincidence between simulated and measured data is observed. The differences between the measurement and the stimulation are the center frequencies and the insertion losses are a little larger than the designed. It is believed that the deviation comes from the additional inductive and capacitive parasitic effects caused by the via. On the other hand, the process of LTCC technology also brings many errors which will introduce the manufacturing inaccuracy. Firstly, during the co-firing, the organic solvent in the silver conductor can not dispel all air bubbles, which cause the value of inductance and capacitance smaller and make the self-resonance frequency larger. Secondly, the effective value of the component can not be controlled accurately due to the difficulties of tiny manipulation problems including the printing and laminating.

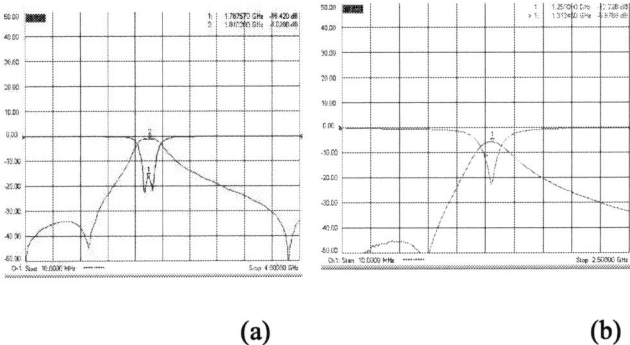

(a)   (b)

Fig.8. (a) The measurement result for 1.8GHz sample
(b) The measurement result for 1.3GHz sample

Conclusions

The structure analysis of LTCC-based passive components is reported for the design of a small multilayer chip BPFs and the filter model was established according to the capacitor coupled resonate band-pass filter's circuit. The BPFs fabricated by LTCC process has small size (2.0mm×1.2mm×0.9mm) using dielectric material ULF140. The center frequencies for BPFs samples are 1.81 GHz and 1.31 GHz, and the insertion loss in the center frequencies are -1.02dB and -5.88dB with reflection losses -16.4dB and -12.7dB respectively. The testing results are in a good agreement with the simulated data which will be helpful for the manufacture of passive components.

Acknowledgement

This work was supported by the Foundation for Innovative Research Groups of the NSFC under Grant No. 60721001, the NSFC Fund of China under Grant No. 60571017, the Youth Fund of University of Electronic Science and Technology of China under Grant No. L08010301JX0725.

References

[1] Tomohiro Seki, Kenjiro Nishikawa, Yasuo Suzuki, et al. 60 GHz Monolithic LTCC Module for Wireless Communication Systems. Proceedings of the 9th European Conference on Wireless Technology, September, 2006, Manchester, UK, 376-379.

[2] Zhenhai Shao and Masayuki Fujise. 60 GHz Narrow Bandpass Filter Based on Circle Patch and LTCC. IEEE, 6th International Conference on ITS Telecommunications Proceedings, 2006, 1173-1174.

[3] Tao Yang, Bo Yan, Shuyi Wang, et al. A compact bandpass filter with two finite transmission zeros using LTCC technology. IEEE 2007 International Symposium on Microwave, Antenna, Propagation, and EMC Technologies For Wireless Communications, 293-296.

[4] Sung-Hun Sim, Chong-Yun Kang, Ji-Won Choi, et al. A compact lumped-element lowpass filter using low temperature cofired ceramic technology. [J] Journal of the European Ceramic Society 23(2003), 2717-2720.

[5]Militaru N., Lojewski G., Banciu M.G. Aperture Couplings in Multilayer Filtering Structures. Signals, Circuits and Systems, 2007, Volume 2, Page(s):1-4

[6]Uysal S. A Double-sided Suspended Substrate Microstrip Lowpass Filter. Microwave Optoeletronics Conference, 2003, Volume 1, Page(s):21-23

Yuan-Xun Li was born in Jiangxi Province, China, in 1979. He received the B.Sc. degree and the M.Sc. degree from the University of Electronic Science and Technology of China (UESTC), Chengdu, in 2004, both in applied chemistry. He is currently pursuing the Ph.D. degree with the Department of Magnetic Materials. His research interests include functional materials and electrical devices.

Ying-Li Liu was born in Sichuan Province, China, in 1968. He received the Ph.D. degree from the University of Electronic Science and Technology of China (UESTC), Chengdu, in 2004, in Microelectronics. He is now a professor of School of Microelectronics and Solid-State Electronics. His research interests include magnetic materials and microelectronic devices.

Huai-Wu Zhang was born in Shanxi Province, China, in 1959. He is now a professor of School of Microelectronics and Solid-State Electronics. His research interests include magnetic materials and microelectronic devices.

Cu Out-Diffusion Kinetics in Pre-plated Cu-alloy Leadframes Investigated by a Developed EDX-based Oxidation Test

Lilin Liu[1*,2], Ran Fu[3], Deming Liu[3], Tong-Yi Zhang[2]

[1]School of Physics and Engineering, Sun Yat-Sen University, Guangzhou 510275, China
[2]Department of Mechanical Engineering, Hong Kong University of Science and Technology,
Clear Water Bay, Kowloon, Hong Kong, China
[3]ASM Assembly Automation Ltd, 4/F Watson Centre, 16 Kung Yip St., Kwai Chung, Hong Kong, China
*Corresponding author: Dr Liu Lilin, email: liullin@mail.sysu.edu.cn, phone: 86-20-84113201

Abstract

The Energy Dispersive X-ray (EDX)-based permeation and oxidation test has been further developed by a novel theoretical analysis, in which the gradient of chemical potential rather than the concentration gradient is employed. The EDX-based permeation and oxidation tests determine Cu flux coefficients in the CuO oxide layers to be 4.17×10^{-26} mol·(m·s·kJ/mol)$^{-1}$exp(-70.30 kJ·mol^{-1}/RT) in temperature range of 250 °C – 400 °C. And the Cu flux coefficients in the Ni layers are 1.72×10^{-32} mol·(m·s·kJ/mol)$^{-1}$ and 5.92×10^{-32} mol·(m·s·kJ/mol)$^{-1}$ at temperatures of 250 °C and 300 °C, respectively.

1. Introduction

A leadframe is the major component in electronic packaging, which provides mechanical support for mounting semiconductor chips and electrical connection between chips and wiring board. The widely used pre-plated leadframe (PPF) comprises a Cu-alloy base plated with Ni, Pd and Au sequentially [1,2]. The Ni layer is to provide the leadframe with oxidation and corrosion resistance by impeding Cu out-diffusion to the leadframe surface because Cu out-diffusion to and oxidation at the leadframe surface is the most severe failure mechanism of PPFs. The ultra-thin Pd and the top Au flash facilitate wire bonding and soldering, and provide protection for the underlying Ni layer from oxidation. To lower the sensitivity of PPFs to Ni layer cracking, the electronic packaging industry calls for a thinner Ni layer with satisfied PPF quality and reliability. Fu et al. [3] introduce a Sn nanolayer between the Cu base and the Ni layer, which leads to the formation of a nanometer-thick Cu-Sn-Ni IMC layer. The IMC nanolayer greatly boosts the oxidation and corrosion resistance of the PPF.

Porosity test is a generally accepted evaluation-index of surface finishes. A major deficiency of this method is the lack of kinetic information. Intensive researches have been carried out on Cu out-diffusion kinetics in diffusion barriers [3-8]. Usually, concentration profiles are measured in diffusion investigations by various characterization means. When a diffusion journey is at the nanometer scale, it will be difficult to measure a concentration profile, especially on an area at the mm^2 scale. In this case, a permeation-like test is appropriate, in which the measured parameter is the total amount of diffusive atoms that diffuse through a membrane. Based on this idea, an Energy Dispersive X-ray (EDX)-based permeation and oxidation testing method has been developed for the investigation of Cu out-diffusion [3]. In the present work, we develop further the EDX-based permeation and oxidation test and use it to systematically investigate the Cu out-diffusion and oxidation behavior through a PPF, which is composed of a nanometer-thick Cu-Sn-Ni IMC layer and Ni/Pd/Au layers.

2. Experimental procedures

Fig. 1(a-c) are schematics of the PPF structures with and without a Sn nanolayer. The base of PPFs is a cold-rolled sheet of Cu-alloy with composition of Fe-2.35wt%, Zn-0.12wt%, P-0.07wt%, and Cu in balance. Two sets of PPF samples were prepared. Set I samples include PPF samples with each having a Sn nanolayer (Sn-PPF) and corresponding control samples (conventional PPFs). Set II samples are designed to measure Cu flux coefficient in Ni layers, which will be described in detail later. So there are no Sn layers in Set II samples, but they have three Ni thicknesses. In a PPF sample containing a Sn nanolayer, Sn, Ni, Pd and Au layers were plated in sequence on the Cu-alloy using the epitaxy electroplating technique. A control PPF sample, as a conventional PPF, was prepared in the same way except without plating the Sn layer. A film thickness measurement system (Micron X of Thermo NORAN) based on the X-Ray Fluorescence technique was used to measure the thicknesses of the Ni, Pd and Au layers after electro-deposition and the measured results were tabulated in Table 1.

The EDX-based permeation and oxidation test is actually a consequent annealing and testing process. This means that a sample is taken out for composition analysis at room temperature after annealing for a period of time at a designed temperature. Afterward, the sample will be annealed again at the designed temperature for another time period, and then taken out for composition analysis again, and so on. The composition analysis is conducted by EDX incorporated in a scanning electron microscope (SEM, JEOL). In the present work, an acceleration voltage of 7.5 kV, a working distance of 10 mm and condenser lens setting of 7 were maintained unchanged during all tests to ensure that the probed volume could be treated as a constant provided that the penetration depth of the electron beam and the gain of the resultant X-ray did not vary significantly with the increasing oxide thickness. Thus, the detected Cu content should be the amount of Cu atoms out-diffusing to the PPF surface. A surface area of 4 mm × 4 mm was scanned to collect X-ray signals from the PPF surface. Experimental data reported here were data averaged over the scanned area.

978-1-4244-2739-0/08/$25.00 ©2008 IEEE

Table 1 Parameters for Set I and Set II of PPF samples:

Samples		Plating thickness(nm)			
		Sn	Ni	Pd	Au
Set I	Conventional PPF (Cu/Ni/Pd/Au)	/	190.0	10.5	1.7
	Sn_PPF (Cu/Sn/Ni/Pd/Au)	~3	182.0	10.7	1.6
Set II	a	/	892.5	19.8	4.5
	b	/	585.0	19.0	4.6
	c	/	228.8	20.8	4.6

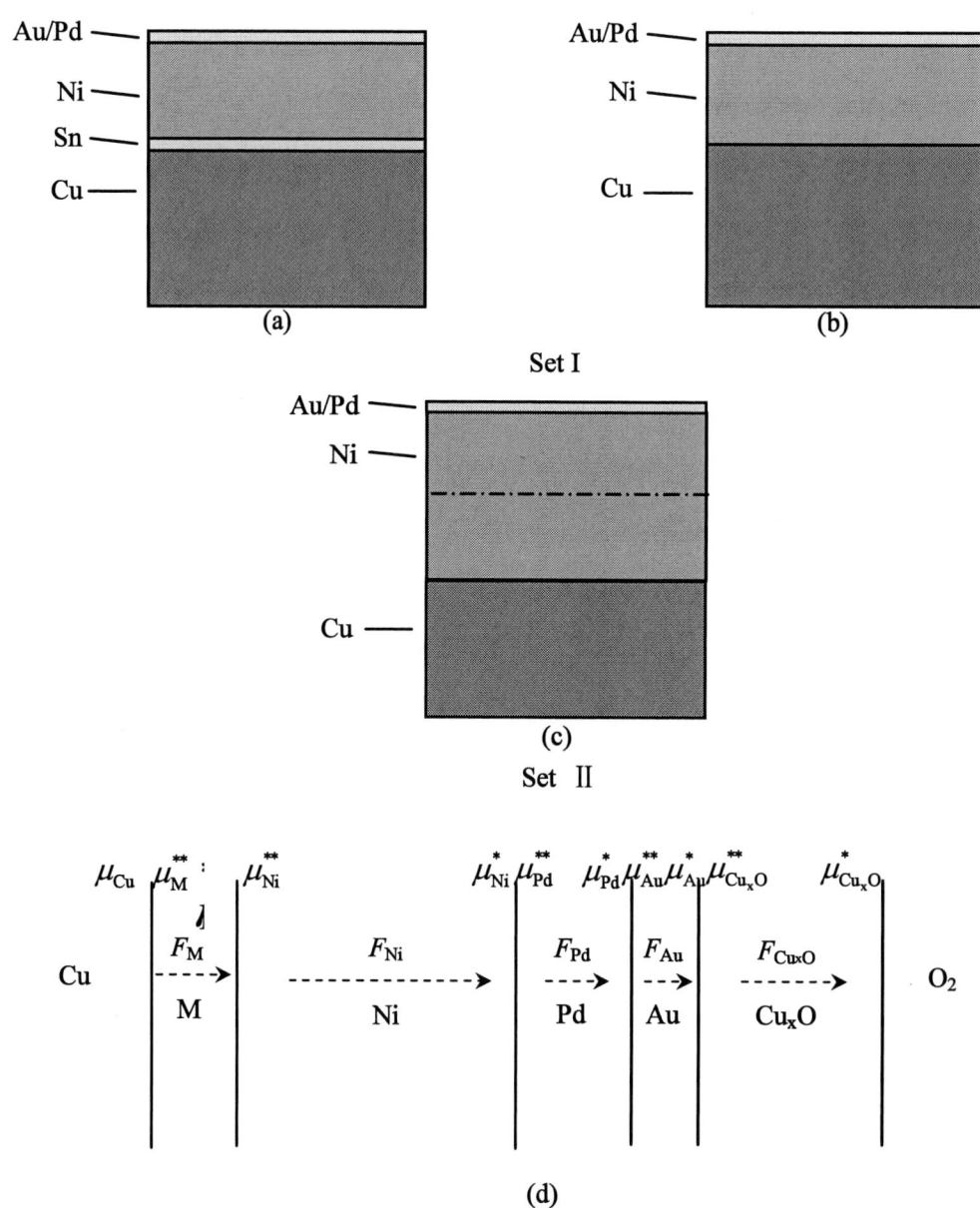

Fig.1 Schematics of the PPT samples. Set I of sample include PPF samples with Sn nanolayer (Sn-PPF) (a) and corresponding control samples (conventional PPF) (b), in which the Ni layer thickness is almost the same. Set II of samples are designed to measure Cu flux coefficients in Ni layers, in which the Ni layers have three thicknesses. The thicknesses of each layer for both

708

Set I and Set II of samples were tabulated in Table 1.

3. Permeation and Oxidation Model

The out-diffusion and oxidation model [3] is further developed here to estimate the Cu flux coefficients in layers of interest. Cu atoms out-diffuse to the PPF surface and oxidize there, which behavior can be treated as a permeation and oxidation process. The schematic drawing of the permeation and oxidation process is plotted in Fig. 1(d), where layer "M" represents the in-situ formed IMC nanolayer or an extra Ni layer. For simplicity, we consider here only a constant permeation flux through all layers although the permeation flux changes with time due to the growth of the oxide layer. We may further assume that the oxidation reaction takes instantly because the oxygen partial pressure in the air is much higher than the equilibrium value, for instance, 9.41×10^{-27} atm at 250 °C, 1.91×10^{-24} atm at 300 °C and 7.64×10^{-21} atm at 400 °C. A much higher oxygen partial pressure provides a considerably stronger driving force for the oxidation reaction and leads to a much larger reaction rate. Since Cu atoms permeate through many layers, it is convenient to express the kinetic behavior in terms of chemical potential gradient rather than concentration gradient because Cu-concentration jumps when crossing an interface in the layers. The chemical potential of Cu atoms at the downward and upward interfaces of the "M", Ni, Pd, Au, and Cu_xO layers are denoted by μ_M^{**} and μ_M^{*}, μ_{Ni}^{**} and μ_{Ni}^{*}, μ_{Pd}^{**} and μ_{Pd}^{*}, μ_{Au}^{**} and μ_{Au}^{*}, and $\mu_{Cu_xO}^{**}$ and $\mu_{Cu_xO}^{*}$, respectively. The Cu chemical potential must be continuous cross an interface, which gives $\mu_{Cu} = \mu_M^{**}$, $\mu_M^{*} = \mu_{Ni}^{**}$, $\mu_{Ni}^{*} = \mu_{Pd}^{**}$, $\mu_{Pd}^{*} = \mu_{Au}^{**}$, and $\mu_{Au}^{*} = \mu_{Cu_xO}^{**}$. The Cu flux from the Cu-alloy base through the "M" layer is then estimated by $F_M = L_M^{Cu}(\mu_M^{**} - \mu_M^{*})/l_M$, where L_M^{Cu} is the flux coefficient of Cu in the "M" layer and l_M is the "M" layer thickness. Similarly, the Cu flux going through each of Ni/Pd/Au barrier layers is estimated by $F_i = h_i(\mu_i^{**} - \mu_i^{*})$, where $h_i = L_i^{Cu}/l_i$ is an effective Cu transport coefficient with L_i^{Cu} and l_i being the flux coefficient of Cu in the "i" layer and the "i" layer thickness, respectively, and the subscript i denotes Ni, Pd, or Au. The Cu flux inside the oxide layer is described by $F_{Cu_xO} = L_{Cu_xO}^{Cu}(\mu_{Cu_xO}^{**} - \mu_{Cu_xO}^{*})/l_{Cu_xo}$, where $L_{Cu_xO}^{Cu}$ is the Cu flux coefficient in the oxide and l_{Cu_xo} denotes the oxide thickness. The same flux condition gives

$$ F = \frac{\mu_{Cu} - \mu_{Cu_xO}^{*}}{l_{Cu_xo}/L_{Cu_xO}^{Cu} + 1/h_{Au} + 1/h_{Pd} + 1/h_{Ni} + l_M/L_M^{Cu}}. \tag{1}$$

The term of μ_{Cu} denotes the Cu chemical potential in the Cu-alloy, while the term of $\mu_{Cu_xO}^{*}$ is the Cu chemical potential at the oxide surface exposing to air. Equation (1)

shows clearly that the Cu out-diffusion and oxidation is driven by the difference between the two Cu chemical potentials.

The flux of Cu atoms links to the growth rate of the oxide layer in the form of $dl_{Cu_xo}/dt = F/N$, where N is a constant related to the volume change from Cu to Cu oxide. Thus, we have an explicit formula to predict the oxide layer thickness as a function of time, i.e.,

$$ \frac{l_{Cu_xo}(t)}{A/2} = \left(1 + \frac{t}{A^2/4B}\right)^{1/2} - 1, \tag{2}$$

where $A \equiv 2L_{Cu_xO}^{Cu}(1/h_{Au} + 1/h_{Pd} + 1/h_{Ni} + l_M/L_M^{Cu})$ and $B \equiv 2L_{Cu_xO}^{Cu}(\mu_{Cu} - \mu_{Cu_xO}^{*})/N$. For the control sample, the thickness of "M" layer equals zero, $l_M = 0$, and A is reduced to $A_0 \equiv 2L_{Cu_xO}^{Cu}(1/h_{Au} + 1/h_{Pd} + 1/h_{Ni})$. Eq. (2) is used to fit the experimental data, oxide layer thickness vs. annealing time, and parameters A_0, A and B can be determined from the fitting. The above analysis also gives

$$ L_M^{Cu} = \frac{N}{\mu_{Cu} - \mu_{Cu_xO}^{*}} \frac{B}{A - A_0} l_M, \tag{3}$$

$$ L_{Cu_xO}^{Cu} = \frac{B}{2} \frac{N}{\mu_{Cu} - \mu_{Cu_xO}^{*}}. \tag{4}$$

When considering the Cu out-diffusion through per unit area, a flux becomes a flux rate. If the flux rate is expressed in terms of atom number (or molar number or atomic fraction) per unit time per unit area, the flux coefficient will be in units of atom number (or molar number or atomic fraction) per unit time per length per molar energy, and $N = x \cdot \rho_{Cu_xO}/M_{Cu_xO}$ here ρ and M_{Cu_xo} denote, respectively, the density and mass number of Cu_xO. Under a constant pressure or stresses, we can calculate the Cu chemical potential at different temperature from its heat capacity

$$ \mu(T) = \mu(T_0) - \int_{T_0}^{T} s(T)dT, \tag{5a}$$

and

$$ s(T) = \int_{T_0}^{T} \frac{C_p}{T} dT. \tag{5b}$$

The general form of heat capacity is given by $C_P = a + bT + cT^{-2}$. Using the general form of heat capacity, we complete the integrations and have

$$ \mu(T) = \mu(T_0) - [\frac{bT^3}{6} + \frac{aT^2}{2} - \frac{bT_0^2 T}{2} + \frac{cT}{T_0} $$
$$ - aT_0 T - c\ln(T) + \frac{bT_0^3}{3} + \frac{aT_0^2}{2} + c\ln(T_0) - c] \tag{5c}$$

Similarly, we can calculate the chemical potentials of the oxide and oxygen gas, from which we are able to determine the Cu chemical potential at the oxide surface exposing to air, $\mu^*_{Cu_xO}$.

4.Results and discussion-Permeation kinetics

High resolution transmission electron microscopy confirmed that a tetragonal DO_{22}-phase IMC nanolayer forms in-situ between Cu substrate and Ni plating and it has a good lattice match with Cu substrate, which will be reported elsewhere along with diffusion kinetic results on the Cu flux in the IMC nanolayers[9]. And the Cu flux coefficient in the Ni plating and in the oxide product layer will be described below as examples.

4.1 Cu flux coefficient in the Ni layers

In the present work, three groups in Set II PPF samples have Ni layer thicknesses of 228.75 nm, 585.0 nm, and 892.5 nm, respectively, and the same Pd/Au top coatings. The permeation and oxidation tests in air were conducted at temperatures 250 °C and 300 °C. When analyzing the EDX-based permeation and oxidation tests on Set II PPF samples, we choose the PPF sample with the thinnest Ni layer as the control sample and take the part of an extra thickness in a thicker Ni layer as the M layer. The EDX analysis of the oxide layers confirms that Cu atoms are oxidized into CuO at the temperatures. We take the Pd layer as a reference to estimate the absolute oxide (CuO) layer thickness from the EDX results. The oxide layer thickness determined from the EDX analysis is consistent with that in the TEM observations. With the group samples with the Ni layer thickness of 228.75 nm as the control samples, Figs. 2(a) and (b) show the oxide layer thickness versus the annealing time at 250 °C and 300 °C, respectively.

We fit the experimental data with Eq. (2) and plot fitting curves in Fig. 2 also. From the least square fitting, we determine parameters A_0, A and B and list them in Table 2. To use Eq. (3) to calculate the Cu flux coefficient in the nickel layer, we must know values of the involved parameters. As mentioned above, the EDX analysis determined the formed oxide to be CuO. Thus, we have $N = \rho_{CuO} / M_{CuO}$ =6.31 (g/cm^3)/79.545 (g/mol)=7.9×10^{-8} mol/m^3. For simplicity, we take the chemical potential at the standard condition, one atmosphere pressure and 298 K, as the reference state. Then, the standard Cu, CuO and O_2 chemical potentials are calculated from their standard enthalpies of formation and the standard entropies. From the literature [10], we have also the values of constants, a, b, and c in the heat capacity for Cu, CuO and O_2. Using the data and Eq. (5c), we have the chemical potentials of Cu, CuO and O_2 and the Cu chemical potential at the oxide surface exposing to air, μ^*_{CuO}. All calculated results are tabulated in Table 3.

With these data and Eq. (3), we calculate the Cu flux coefficients in the nickel layer at temperatures of 250 °C and 300 °C. We have L^{Cu}_{Ni} =6.28×10^{-32} mol·(m·s·kJ/mol)$^{-1}$ and 5.56×10^{-32} mol·(m·s·kJ/mol)$^{-1}$ for the PPF samples with the Ni layer thicknesses of 892.5 nm and 585.0 nm, respectively, at 300 °C, which yields a mean of L^{Cu}_{Ni} =5.92×10^{-32} mol·(m·s·kJ/mol)$^{-1}$. Similarly, we obtain L^{Cu}_{Ni} =1.83×10^{-32} mol·(m·s·kJ/mol)$^{-1}$ and 1.61×10^{-32} mol·(m·s·kJ/mol)$^{-1}$ for the PPF samples with the Ni layer thicknesses of 892.5 nm and 585.0 nm, respectively, at 250 °C with a mean of L^{Cu}_{Ni} =1.72×10^{-32} mol·(m·s·kJ/mol)$^{-1}$. The results list in Table 2 and plot in Fig. 3.

Table 2 Fitting flux parameters by Eq.(2) and flux coefficients calculated by Eq.(3)

sample	T(°C)	A_0 (m)	A (m)	B (m^2/s)	L^{Cu}_M mol·(m·s·kJ/mol)$^{-1}$	$L^{Cu}_{Cu_xO}$ mol·(m·s·kJ/mol)$^{-1}$
Set I	250					3.92×10^{-33}
	300					5.01×10^{-32}
	400					1.12×10^{-31}
					L^{Cu}_{Ni} mol·(m·s·kJ/mol)$^{-1}$	
Set II	250	1.24×10^{-8}	7.25×10^{-8}	4.32×10^{-20}	1.61×10^{-32}	1.36×10^{-33}
		1.24×10^{-8}	1.11×10^{-8}	4.32×10^{-20}	1.83×10^{-32}	1.36×10^{-33}
	300	3.85×10^{-8}	2.03×10^{-7}	5.66×10^{-19}	5.56×10^{-32}	1.28×10^{-32}
		3.85×10^{-8}	3.10×10^{-7}	5.66×10^{-19}	6.28×10^{-32}	1.28×10^{-32}

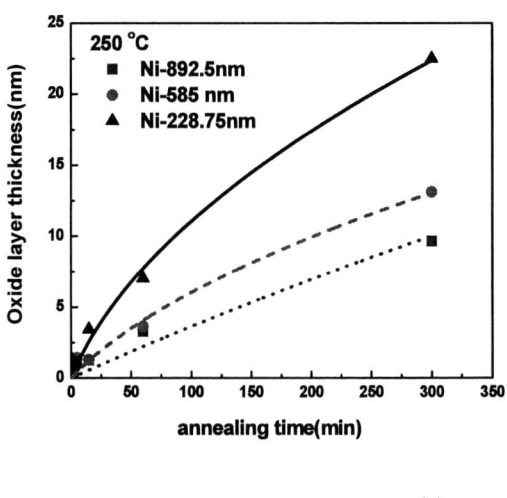 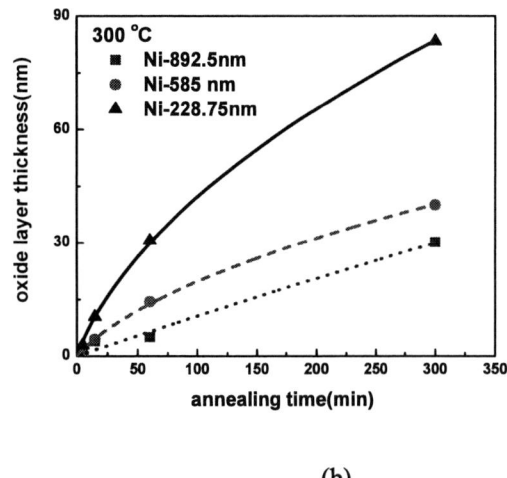

(a)   (b)

Fig. 2 Thicknesses of the CuO layer as a function of annealing time for the set II of PPF samples annealed in air at (a)250^{0}C and (b)300^{0}C.

4.2 Cu flux coefficient in the CuO layers

With Eq. (4), we calculate Cu flux coefficients in the CuO layers at temperatures of 250 °C, 300 °C, and 400 °C for Set I samples and at 250 °C and 300 °C for Set II samples. The values of L_{CuO}^{Cu} at 250 °C are 3.92×10^{-33} mol·(m·s·kJ/mol)$^{-1}$ and 1.36×10^{-33} mol·(m·s·kJ/mol)$^{-1}$ for Sets I and II, respectively, yielding a mean of 2.64×10^{-33} mol·(m·s·kJ/mol)$^{-1}$. The values of L_{CuO}^{Cu} are 5.01×10^{-32} mol·(m·s·kJ/mol)$^{-1}$ and 1.28×10^{-32} mol·(m·s·kJ/mol)$^{-1}$ for Sets I and II, respectively, at 300 °C with a mean of 3.15×10^{-32}

mol·(m·s·kJ/mol)$^{-1}$. The value of L_{CuO}^{Cu} is 1.12×10^{-31} mol·(m·s·kJ/mol)$^{-1}$ for Sets I at 400 °C. The data show clearly that at a given temperature, the values of Cu flux coefficient through the oxide layers obtained from the two sets of samples are more or less the same, which indicates the validity of the EDX-based permeation and oxidation test. From the Arrhenius plot of the Cu-permeability through the oxide layers, shown in Fig. 3 also, we have $L_{CuO}^{Cu} = 4.17 \times 10^{-26}$ mol·(m·s·kJ/mol)$^{-1}$exp(-70.30 kJ·mol^{-1}/RT) at the temperature range of 250 °C - 400 °C.

Table 3 Selected thermodynamic parameters of Cu, O$_2$ and CuO[10]

	$\Delta_f H^0_{298.15}$ (kJ·mol^{-1})	$S^0_{298.15}$ J·K^{-1}·mol^{-1}	a (J·deg^{-1}·mol^{-1})	b (J·mol^{-1})	c (J·deg·mol^{-1})	μ (kJ·mol^{-1})	μ^*_{CuO} (kJ·mol^{-1})
Cu	0	33.15	22.64	6.28×10^{-3}	0	-641(250 °C); -957(300 °C); -1787(400 °C)	
O$_2$	0	49.003	29.96	4.18×10^{-3}	-1.67×10^5	-779(250 °C); -1163(300 °C); -2170(400 °C)	-1900(250 °C); -2698(300 °C); -4814(400 °C)
CuO	-157.168	42.60	38.79	20.08×10^{-3}	0	-1340(250 °C); -1931(300 °C); -3492(400 °C)	

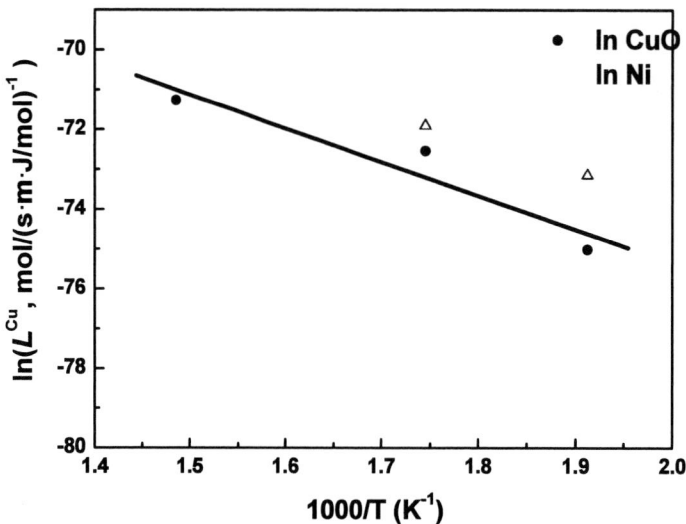

Fig.3 Arrhenius plot for the flux coefficient of Cu in IMC nanolayer formed in Sn_PPF, Cu in Ni layer in Set II of samples and Cu in the growing oxide layer formed in both Set I and Set II of samples.

5.Conclusions and Remarks

The EDX-based permeation and oxidation test has been further developed by a novel theoretical analysis, in which the gradient of chemical potential rather than the concentration gradient is employed. An advantage of using the gradient of chemical potential lies in the continuity of chemical potential of diffusive atoms at an interface between two phases, while the concentration of diffusive atoms may jump when crossing an interface. Another advantage of using the gradient of chemical potential is that accurate measurements of the concentration profile of diffusive element in an IMC, in which the diffusive element is a constituent, could be a challenging task especially when the IMC is exactly stoichiometric. The EDX-based permeation and oxidation tests determine Cu flux coefficients in the CuO oxide layers to be 4.17×10^{-26} mol·(m·s·kJ/mol)$^{-1}$exp(-70.30 kJ·mol^{-1}/RT) in temperature range of 250 °C – 400 °C. Using the Set II PPF samples, we determine Cu flux coefficients in the Ni layers to be 1.72×10^{-32} mol·(m·s·kJ/mol)$^{-1}$ and 5.92×10^{-32} mol·(m·s·kJ/mol)$^{-1}$ at temperatures of 250 °C and 300 °C, respectively.

Acknowledgements:

This work was supported by the Science and Technology Plan Project, JRC-08-GD-001, from the Nansha Science and Technology Bureau and the donated research grant, ASMML04/05.EG01, from ASM Assembly Automation Ltd., Hong Kong, and ASM Semiconductor Materials (Shenzhen) Ltd., China.

References:

[1] Romm, D. Lange, B. and Abbott, D., Application Report No.SZZA026, (Texas Instrument, Austin Texas, 2001).

[2] Liu, L., Liu, D. M., Fu, R., Kwan, Y. F., Yau, C. H., and Zhang, T. Y., *IEEE Trans. Adv. Pack.*, 2006, 29, 683.

[3] Fu, R., Liu, L. L., Liu, D. M., and Zhang, T. Y., *Appl. Phys. Lett.*, 2006, 89, 131911.

[4] Johnson, B. C., Bauer, C. L., and Jordan, A. G., *J. Appl. Phys.*, 1986, 59, 1147.

[5] Hayashi, E., Kurokawa, Y., and Fukai, Y., *Phys. Rev. Lett.*, 1998, 80, 5588.

[6] Meunier, A., Gilles, B., and Verdier, M. *Appl. Surf. Sci.*, 2003, 212, 171.

[7] Rammo, N. N., Makadsi, M. N., and Abdul-Lettif, A. M., *Phys. Stat. Sol.(a)*, 2004, 201, 3102.

[8] Abdul-Lettif A. M., *Phys. B*, 2007, 388, 107.

[9] Liu L.L., Fu R., Liu DM, Huang HY, Zhang TY, Acta Metall., submitted.

[10] Weast R. C, CRC Handbook of Chemistry and Physics, 1st student edition, CRC Press, Inc., Boca Raton, Florida, 1990

Corrosion Performance of Pb-free Sn-Zn Solders in Salt Spray

Huang Dan, Zhou Jian, Li Pei Pei
Jiangsu Key Laboratory of Advanced Metallic Materials
Southeast University, Nanjing, P. R. China, 211189
dd52088@163.com, jethro13@163.com, 025-52090689

Abstract

Environmental regulations around the world have been targeted to eliminate the usage of leaded solders in electronic assemblies. Sn-Zn based alloys are considered to be a very promising candidate for its substantially same melting point as Sn-Pb solder. Fujitsu Limited has developed several lead-free solders with low melting point such as Sn-7Zn-30ppmAl and Sn-9Zn-30ppmAl. Those solders have already launched application to commercial products since 2002. A reliable solder material requires good corrosion resistance in various ambient conditions. In this research, the salt spray corrosion resistances of Sn-7Zn, Sn-9Zn, Sn-7Zn-30ppmAl and Sn-9Zn-30ppmAl alloys were investigated by Salt Spray Tester according to the national standard GB/T2423.17-93. The surface microstructure features before and after corrosion was observed by optical microscopy and scanning electron microscopy (SEM) with energy dispersive analysis of x-ray (EDX) respectively. The results show that comparing with Sn-7Zn-XAl (X=0, 30ppm), the increase of zinc content in Sn-9Zn-XAl (X=0, 30ppm) induces a decrease of corrosion resistance. The main corrosion mode of all solder alloys is intercrystalline corrosion. It was found that 30ppm Al additive to the Sn-7Zn solder improves the corrosion behavior in the salt spray with dissolution rate decreasing. But the addition of 30ppm Al has not too much influence on Sn-9Zn performance of corrosion resistance.

Key words: Lead-free solders; Sn-Zn; Sn-Zn-Al; salt spray corrosion

1. Introduction

Due to environmental and health concerns, regulations around the world have implemented efforts to progressively reduce the use of Sn-Pb solders in electronic assemblies and eventually eliminate lead (Pb) from its products [1, 2]. Sn-Zn based alloys are considered to be a very promising candidate for its closest melting point to that of the Sn-Pb solders and also its low toxicity and cost. Sn-7Zn-30ppmAl and Sn-9Zn-30ppmAl developed by Fujitsu offers superior comprehensive properties and already been used in commercial products since 2002 [3]. Long-term reliability of solder joints requires solder materials must be resistant to corrosion atmosphere, such as air, moisture, air pollutants from industry and oceanic environments [4, 5]. However, there has been little consideration of the corrosion properties of Sn-Zn based lead-free solder up to now [6~8]. In the present study, to make comparisons, the corrosion capability of Sn-7Zn, Sn-9Zn, Sn-7Zn-30ppmAl and Sn-9Zn-30ppmAl alloys in the same salt spray was measured. Although the corrosion environments were different from real environments in which solders are used, but a fundamental understanding of corrosion behavior of the Sn-Zn solders and the influence on the corrosion capability of Al addition.

2. Experimental procedures

The Sn-7Zn, Sn-9Zn, Sn-7Zn-30ppmAl and Sn-9Zn-30ppmAl lead-free solders were made with pure tin, zinc, aluminum (purity of 99.99%). Zn-4.98Al master alloy was firstly prepared and then re-melted with Sn and Zn in a graphite crucible under an argon atmosphere at 450℃. After keeping the temperature for 10min, the melts were cast into a steel mold. The solder alloys used in this investigation were cold rolled to 2 mm thick, and then incised to 50.0 mm×34.5 mm sheets. Prior to the corrosion, the solder plates were polished by # 600 SiC polishing paper and degreased with acetone and rinsed with double distilled water in ultrasonic cleaner. A varnish was applied around the plates after dried so that only the 50.0 mm×34.5 mm surface was exposed to the salt spray.

The corrosion experiments of the plated samples were performed in YWX/Q-250 Salt Spray Tester according to the national standard GB/T2423.17-93 for 16h, 24h, 48h and 96h [9]. After corrosion, the samples were again rinsed with dichromate (50mlH3PO4, +20gCrO3+ 1000ml distilled water) and then distilled water in ultrasonic cleaner for 2 min and dried in air. An electronic microbalance BS110S with 0.1 mg precision was used for measurement of the mass lost of the solders with time during corrosion.

After sanding, polishing and etching treatment, the alloy structure was investigated by Olympus. The surface morphology and deposit composition of the solders after corrosion was observed by FEI-Sirion scanning electron microscopy (SEM) and by GENESIS-60S energy dispersive analysis of x-ray (EDX) respectively.

3. Experimental results

3.1. Microstructure

Fig. 1 shows the microstructures of the Sn-9Zn, Sn-7Zn, Sn-7Zn-30ppmAl and Sn-9Zn-30ppmAl alloys. The microstructure of the Sn-7Zn alloy in Fig. 1(a) shows a typical hypoeutectic structure with the primary β-Sn and the eutectic matrix. Fig. 1(b) of Sn-9Zn shows a typical eutectic structure with β-Sn and zinc particles scattered in β-Sn. Fig. 1(c) also reveals an obvious hypoeutectic structure of Sn-7Zn-30ppmAl alloy but with finer precipitate of the β-Sn phase. Sn-9Zn-30ppmAl also presents eutectic structure in Fig. 1(d).

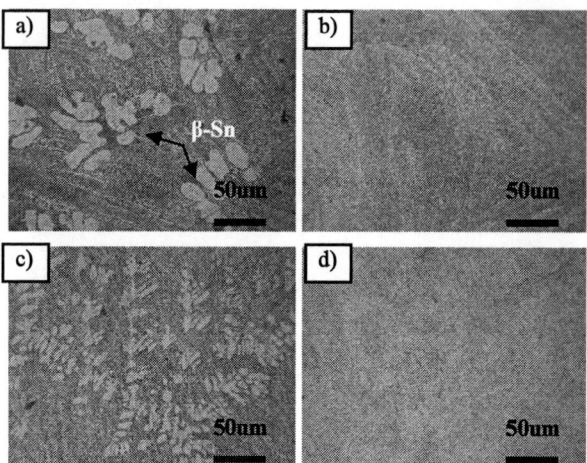

Fig. 1 Optical images of alloys before salt-spray corrosion
(a) Sn-7Zn; (b) Sn-9Zn;
(c) Sn-7Zn-30ppmAl; (d) Sn-9Zn-30ppmAl

3.2. Corrosion rate

The mass loss of the four kind solders with time during the corrosion process in salt spray was shown in Fig. 2. The slope of the curve shows the corrosion rate of the solders. The curve indicates that the corrosion rate is decreasing as corrosion time increased. Sn-9Zn and Sn-9Zn-30ppmAl have more mass loss then Sn-7Zn and Sn-7Zn-30ppmAl. Sn-9Zn and Sn-7Zn almost have the same corrosion rate at the beginning, but after corrosion for 48h, the mass loss rate of Sn-9Zn drops more then Sn-7Zn. Compare with binary compound, Sn-7Zn-30ppmAl exhibits better corrosion ability then Sn-7Zn with much less mass loss and corrosion rate except the first period of 16h. Sn-9Zn-30ppmAl has the same mass loss and corrosion rate with Sn-9Zn, but has inconspicuous corrosion rate drops after corrosion for 48h.

Fig. 2 Salt spray corrosion sum-time curves of the four alloys

3.3. SEM images of corroded alloys

After corrosion tests in salt spay for 16h, 24h, 48h and 96h, the morphology of Sn-9Zn solder is shown in Fig. 3. The

SEM photos of the four kinds solder after corrosion in salt spay for the same time (48h) are presented in Fig. 4. Fig. 5 shows the surface layer structure of the deposit after the salt-spray corrosion.

It is obvious that grain boundaries were the preferred locations of corrosion. A porous tin-rich phase was formed on the surface of Sn-7Zn and Sn-9Zn after corrosion for 48h in Fig. 4. Biggish corrosion hole was found after corrosion for 96h in Fig. 3 (d).

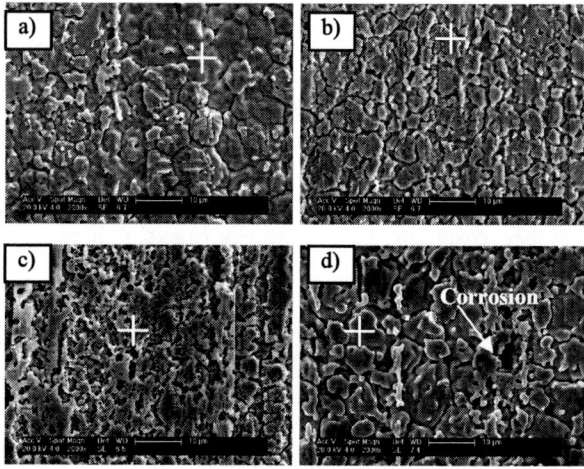

Fig.3 SEM photos of Sn9Zn after salt-spray corrosion
(a) 16h; (b) 24h; (c) 48h; (d) 96h

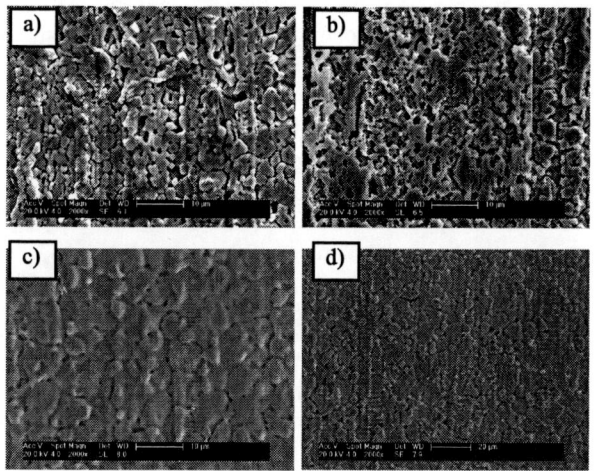

Fig.4 SEM photos of four alloys after 48h salt spray corrosion
(a) Sn-7Zn; (b) Sn-9Zn;
(c) Sn-7Zn-30ppmAl; (d) Sn-9Zn-30ppmAl

714

Fig.5 SEM images of Sn-9Zn-30ppmAl deposit after salt-pray corrosion for 96h

4. Discussion

Differences in the electromotive force (Δemf) between the phases present in the alloy are frequently a good indication of the corrosion potential. The standard emf of Sn and Zn are -0.136 V, -0.763 V respectively [10]. Zn is anodic with respect to Sn, implying a galvanic driving force for the corrosion of Zn, consequently the increase of zinc content in Sn-Zn induces a decrease of corrosion resistance (Fig. 2).

Research results reveal that the corrosion behavior of Sn–13Zn and Sn–2Zn is very similar to that of a pure Zn and Sn respectively [11]. EDX results were presented in Tab.1, the composition of Zn on the remaining solder surface is far less then initially schemed and the amount of Zn on the corroded surface decreasing while corrosion time increased. This suggests that the dissolution of zinc from the Sn-Zn alloy preferably proceeded. Therefore the more content of Zn on the surface of Sn-7Zn after 48h corrosion induced a biggish corrosion rate than the other solder alloys.

Tab.1 The content of Zn (wt %) on the surface of the four alloys after salt-pray corrosion

Corrosion time Solder alloys	16h	24h	48h	96h
Sn-7Zn	/	/	3.29	/
Sn-9Zn	2.21	1.68	1.52	1.41
Sn-7Zn-30ppmAl	/	/	1.72	/
Sn-9Zn-30ppmAl	/	/	1.46	/

During the first period of corrosion time, one observes a rather fast dissolution. This is attributed to the dissolution of zinc, the porous remaining tin phase on a corroded surface can be seen in Fig. 3, but zinc was exhausted shortly in the surface phase. After long time immersion, both alloy components corroded depending on their composition ratios. SEM micrograph in Fig. 5 shows the deposit on the surface of Sn-9Zn-30ppmAl after corrosion for 96h. EDX analysis of the deposit gives the composition as: Sn-12.25wt%, Zn-53.31%, Al-0.54wt%, Cl-4.32%, O-29.58%. This reveals that Sn, Zn and Al may react with chlorine and oxygen ions to form oxide, chloride and oxychloride. These compound and halide cover the surface of the corroded solder alloys to form a passive film, which induced the decreasing of corrosion rate after long time immersion, as can be seen in Fig. 2.

Resistance of the thick passive layer of the different alloys was found to increase with the corrosion time [12]. And it also indicates that the corrosion mechanism of solder alloys changed from charge transfer controlled at the initial stage to diffusion and charge transfer controlled at last.

Corrosion susceptibility of solder alloys depends upon physical and chemical properties of the passive film. It is proposed that decreasing of diffusion and charge transfer is primarily reason for obvious improvement corrosion resistance of Sn–7Zn alloy with 30ppm Al addition. The aluminum atoms might interact with chlorine and oxygen ions and adsorbed on the solder alloy surface to form a compact passive film. These species block the active sites of the solder alloy and inhibit the oxygen and chlorine evolution reactions which are responsible for the anti-corrosion of the passive film.

Fig. 1(c) shows that 30ppm Al addition to the Sn–7Zn alloy markedly induced fine β-Sn phase particles, which should be another reason for the improvement of corrosion resistance. And it may also cause the heaviest mass loss in the initial 16h corrosion time of Sn-7Zn-30ppmAl due to the increase of the galvanic couples area between the β-Sn and zinc phase in the alloy. In conclusion, better corrosion resistance of Sn-7Zn-30ppmAl than Sn-7Zn may attributed to the change in the microstructure, the number of defects and composition of the passive film formed on its surface. Sn-9Zn-30ppmAl almost has the same microstructure; mass loss and corrosion rate with Sn–9Zn indicate 30ppmAl has not too much influence on the corrosion performance of Sn-9Zn.

A comparison of Fig. 3 with Fig. 4 shows that grain boundaries were the preferred corrosive locations of the four solders. The active components and the defects are likely to exist in the grain boundaries during the solidification of alloy. Therefore grain boundary is anodic with respect to grain and the distinction between the area of anode and cathode leading to the formation of active galvanic couples.

5. Conclusions

The increase of zinc content in Sn-Zn based alloys induces a decrease of corrosion resistance. Zinc dissolution is the preferably proceeded and both alloy components corroded depending on their composition ratios after long time immersion. Intercrystalline corrosion is the main corrosion mode of Sn-Zn based solder alloys. It was found that the addition of 30ppm Al improves the corrosion resistance of Sn-7Zn solder alloy in the salt spray solution with dissolution rate decreasing, but has not too much influence on the corrosion performance of Sn-9Zn.

References

1. Shen J, Gao H X, Liu Y C. [J]. Electronics Manufacturing Engineering, 2004, 5(4): 150-153.

2. Yi Li, Kyoung-sik Moon, C. P. Wong. Electronics Without Lead. SCIENCE, [J]. 2005, 308(3): 1419-1420.

3. Kitajima, Masayuki; Shono, Tadaaki. Development of Sn-Zn-Al lead-free solder alloys, [J]. Fujitsu Scientific and Technical Journal, 2005, 41(2): 225-235

4. Sabbar A, Hajjaji S E, Ben B A. [J]. Materials and Corrosion, 2001, (52): 298-301.

5. Masayuki K j. Tadaaki S. [J]. Microelectronics Reliability, 2005, (45):1208-1214.

6. B.Y. Wu et al., Electrochemical corrosion study of Pb-free solders, [J]. Mater. Res. 21 (2006) 62–70.

7. Dezhi Li, Paul P. Conway, Changqing Liu, Corrosion characterization of tin–lead and lead free solders in 3.5 wt.% NaCl solution, [J]. Corrosion Science, 2008, (50): 995–1004.

8. K.L. Lin, T.P. Liu, The electrochemical corrosion behaviour of Pb-free Al–Zn–Sn solders in NaCl solution, [J]. Mater. Chem. Phys. 56(1998) 171–176.

9. Wang L. [J]. Environmental Technology, 1998(1): 14-18.

10. Mulugeta A, Guna S. Lead-free Solders in Microelectronics [J].Materials Science and Engineering, 2000 (27):95-141.

11. Chi-Chang Hu, Chun-Kou Wang. Effects of composition and reflowing on the corrosion behavior of Sn–Zn deposits in brine media [J]. Electrochimica Acta 51 (2006) 4125–4134.

12. Rabab M. El-Sherif, Khaled M. Ismail et al. Effect of Zn and Pb as alloying elements on the electrochemical behavior of brass in NaCl solutions [J]. Electrochimica Acta 49 (2004) 5139–5150.

The Synthesis and Curing Kinetics of the Silicon-containing Epoxy Resin with Environmentally Friendship Flame Retardance

YANG Ming-shan[1,2], HE Jie[2], LI Lin-kai[3], LIU Zhen[1]

1. Department of Material Science and Engineering, Beijing Institute of Petrochemical Technology, Beijing 102617, China;
2. College of Material Science and Engineering, Beijing University of Chemical Technology, Beijing 100029, China;
3. Guangdong Rongtai Industry Co., Ltd., Jieyang, Guangdong 522000, China.

Abstract

Silicon-containing epoxy resin (CNE-Si) was synthesized from diphenylsilandiol (DPSD) and ortho-cresol novolac epoxy resin using SnCl$_2$ as catalyst. The chemical structure of CNE-Si prepared in this paper was characterized by ^{1}H-NMR and FTIR. The thermal stability was analyzed by TGA. The result showed that the –Si- group enhanced the thermal stability of the epoxy resin. The curing kinetics of the system was studied by non-isothermal DSC. The kinetic parameters of the curing reaction including the activation energy were calculated using Kissinger and Ozawa method. The results showed that the system containing CNE-Si has lower curing temperature and more quick curing speed compared to CNE, which offers basic data for the application of CNE-Si in large-scale circuit packaging.

Key words：Silicon-containing epoxy resin; synthesis; curing kinetics

1 Introduction

The higher and higher performances of integrated circuit packaging materials have been demanded with the development of integrated circuit. For example, higher thermal stability of packaging materials is desired because of leaf-free solder. Additionally, because of the put in practice of RoHS and WEEE regulations of EU, halogen-free flame retardance of packaging materials is desired. So the existed packaging materials can not meet the new demands, and it is urgent to develop new materials used in packaging of large-scale integrated circuit.

Organo-silicone resin has been widely used in many fields because of its good thermal stability, anti-oxidation, good weather resistance, good low temperature and electricity properties, environmentally friendship flame retardance. Epoxy resin has good adhesion to metal materials, low cost and high mechanical strength, and has also been used widely. Therefore, the novel polymer materials with excellent performance can be synthesized by combining epoxy with organo-silicone resin through reacting of epoxy group in epoxy resin with hydroxyl group in organo-silicone resin. The internal stress, thermal performance and flame retardance of cured epoxy resin can be greatly improved by integrating organo-silicone into epoxy resin. The attempt to synthesizing a silicone-contained epoxy resin(CNE-Si) by reacting diphenylsilanediol(DPSD) and ortho-cresol novolac epoxy resin(CNE) using SnCl2 as catalyst, which introduces the organo-silicone into the backbone of epoxy resin, was conducted In this paper. The composite was obtained by mixing CNE-Si resin synthesized in the paper with CNE resin and the curing kinetics of the composite was studied using novolac resin as curing agent and 2-methyl imidazole as

accelerator by non-isothermal DSC method, which provides the basic data for the application of CNE-Si resin in large-scale integrated circuit packaging.

2 Experimental

2.1 Materials

Diphenylsilanediol (DPSD), Tin dichloride, analysis regent, was supplied by J&K Chemical Co., Ltd. Ortho-cresol novolac epoxy resin(CNE, JF-45) was supplied by Wuxi Synthesis Resin Factory, China. The novolac phenol-formaldehyde resin with less than 0.5wt% of concentration of dissociated phenol was supplied by Changshu South-east Plastics Co., Ltd., China. 2-methyl imidazole was supplied by Guangzhou Chuanjin Electronic Materials Co., China.

2.2 Synthesis of silicone-containing epoxy

CNE and DPSD (molar ratio is 2:1) were added into the reactor with condenser, stirrer, temperature controller and indicator, stirring for 30min, and Tin dichloride (SnCl2) (the total concentration in the system is 500ppm) was added into the reactor and stirred for 10min. The reactor was heated to the temperature of 140-150℃ under stirring. The reaction maintained for 2hrs at the temperature of 140-150℃ under stirring, and then the reaction product was poured into breaker and cooled to room temperature. The reaction formulation was as following.

2.3 Manufacturing of CNE-Si and CNE composite

30g CNE-Si, 60g CNE, 8.5g novolac phenol formaldehyde resin, 1.5g 2-methylimidazole and 200ml THF were added into 500ml flask and dissolve for 30min under stirring. The product was poured onto glass plate, placed it into vacuum oven and dried for 4hrs at the temperature of 40-50℃ and under the pressure of 10Pa. Then the composite was peeled off and crushed into powder.

2.4 Measurements

FTIR spectrum measurement (KBr pellet method) was conducted with NICOLET 55 infrared spectrometer (NICOLET Co., USA). 1H NMR measurement (300MHa, CDCl3 solvent) was conducted with DMX-300 nuclear magnetic resonance (Bruker Spectrospin Co., Switzerland). TGA measurement (temperature range RT-700℃, heating rate 10℃/min, N₂ atmosphere) was conducted with TGA-DSC 131 thermo-analyzer(SETARAM Co., France). DSC measurement (temperature range:RT-350℃, heating rate: 5, 10,15,20℃/min, respectively; N₂ atmosphere) was conducted with TGA-DSC 131 thermo-analyzer(SETARAM Co., France).

3 Results and Discussion

3.1 The characterization of CNE-Si

The FTIR spectra of CNE-Si were showed in Figure 1.

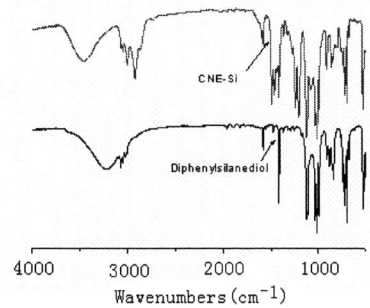

Fig.1 TT-IR of the DPSD and CNE-Si

Figure 1 indicated that the Si-OH absorption peak at 3220cm-1 occurred in the spectrum of DPSD, but the peak disappeared in the spectrum of CNE-Si and a new absorption peak at 3461cm-1 which belongs to C-OH group absorption occurred. In addition, the absorption peaks at 911cm-1 and 1014cm-1 assigned to epoxy group and Si-O-C bond respectively occurred in the spectrum of CNE-Si resin, which explains that the Si-OH group in DPSD molecule reacted with the partly epoxy group in CNE resin and the organo-silicone was successfully integrated into novolac epoxy molecules. Therefore, the silicone-containing epoxy resin has been obtained.

¹H NMR spectrum of CNE-Si was showed in Figure 2.

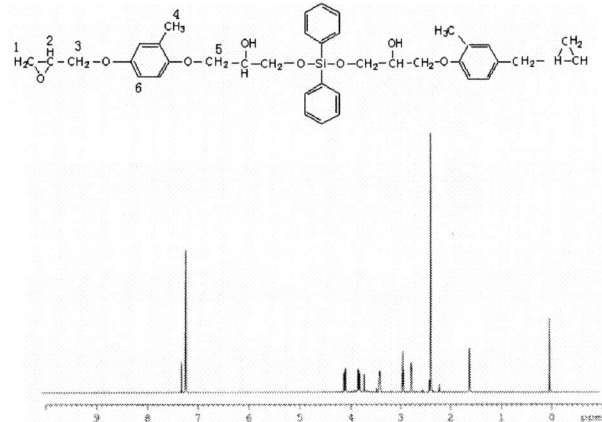

Fig.2 ¹H-NMR spectra of CNE-Si resin

From Figure 2, the peaks assignments can be obtained in Table 1.

Table 1 The chemical shift of hydrogen of the product

Hydrogen	1	2	3	4	5	6
Chemical shift of hydrogen	2.73~2.91	3.38	3.77~4.08	2.36	2.72	7.17
Integral ratio	5.74	2.86	6.06	18.04	6.06	5.74
Area ratio	2	1	2	6	2	2

From Figure 2 and Table 1, it has been shown that the chemical shift of hydrogen of CNE-Si synthesized in this paper were almost equal to the theoretical values, which shows that the synthesized product in this paper is CNE-Si resin and has high productivity gain.

3.2 The thermal stability of CNE-Si

The thermal stability of CNE-Si resin was investigated by TGA and the comparison with CNE was conducted. The results were seen in Figure 3 and Table 2.

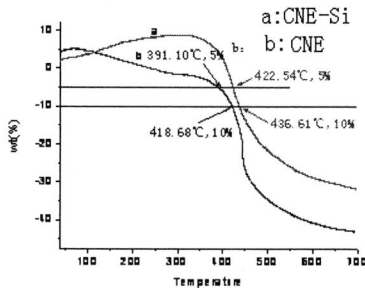

Fig.3 Thermal gravimetric analysis curves of CNE-Si and CNE curing systems

Tab.2 TGA comparison of CNE-Si and CNE curing systems

Weight loss / Type of epoxy resin	temperature/℃	
	5% weight loss	10% weight loss
CNE-Si	422.54	436.61
CNE	391.10	418.68

Figure 3 and Table 2 indicated that the temperatures of weight loss 5% of CNE-Si and CNE were 422.54℃ and 391.10℃, respectively, which shows that the thermal stability of CNE-Si is better than CNE. The reason is that the bond energy of Si-O is 422.5kJ/mol and higher than C-O bond energy (344.4kJ/mol).

3.3 The curing kinetics of CNE-Si

3.3.1 The determination of characterization curing temperature of the composite

The DSC curves of the composite at varied heating rates(Φ) of 5℃/min,10℃/min,15℃/min and 20℃/min were seen in Figure 4.

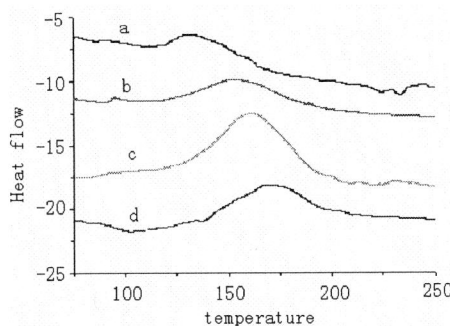

Fig.4 DSC curves of CNE-Si / linear phenolic resin at different heating rates
a)5℃/min;b)10℃/min;c)15℃/min;d)20℃/min

From Figure 4, the initial curing temperature (Ti), peak curing temperature (Tp) and end curing temperature (Tf) were obtained in Table 3.

Table 3 DSC Results of CNE200-Si/ECN at different heating rates

Φ/(℃/min)	T_i/℃	T_p/℃	T_f/℃
5	117.92	131.15	184.76
10	123.38	152.49	188.93
15	128.63	161.12	198.73
20	139.50	169.97	218.70

From Table 3, the T-Φ curves can be obtained in Figure 5.

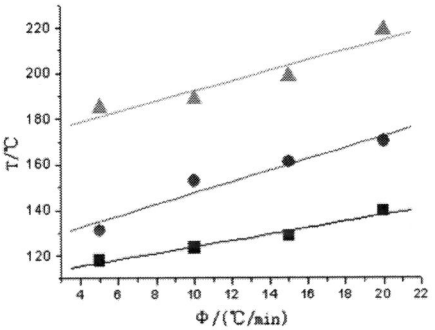

Fig.5 The relationship of heating rates and temperature

According to Figure 5, the beelines were extended to the point at Φ=0 and the characterization curing temperatures can been obtained (see Table 4) which is the important data for the application of composite.

Table 4 The curing temperature of the system when Φ=0

Φ/(℃/min)	T_i/℃	T_p/℃	T_f/℃
0	109.86	122.41	169.88

3.3.2 The determination of curing activation energy and curing reaction constant of the composite

The curing activation energy can be calculated out by formulas of Kissinger and Ozawa as the following:

$$d \ln\left(\frac{\phi}{T_p^2}\right) / d\left(\frac{1}{T_p}\right) = -\Delta E / R \qquad \text{Kissinger}$$

$$\lg \phi = -0.4567 \frac{\Delta E}{RT} + \left[\lg \frac{A\Delta E}{R} - \lg F(x) - 2.315\right] \qquad \text{Ozawa}$$

According to Kissinger formula, the beeline of $\ln(\Phi/T_p^2)$-$1/Tp$ was obtained (see Figure 6). The slope of the beeline is the activation energy. Thus the Kissinger activation energy was 46.05kJ/mol. Similarly, according to Ozawa formula, the beeline of $\ln\Phi$-$1/Tp$ was obtained and the activation energy was 50.43kJ/mol. The average activation energy was 48.24kJ/mol and less than that of CNE (67.11kJ/mol), which indicates that the curing reaction of CNE-Si is easier to conduct than CNE.

4 Conclusions

The silicone-containing epoxy resin has been synthesized and characterized by FTIR and 1H NMR. The results have shown that the product synthesized in this paper is CNE-Si resin. By mixing CNE-Si resin with CNE resin, the composite of epoxy resins has been prepared and its thermal stability and curing kinetics have been investigated systematically by TGA and DSC. The results have shown that the thermal stability of CNE-Si resin is better than CNE resin. The activation energy of the composite was obtained by Kissinger and Ozawa formulas, which indicates that the curing reaction of the composite is easier than CNE resin.

Acknowledgments

The work was sponsored by Beijing Natural Science Foundation (No. 2072007), by National Electronics and Information Technology Foundation of China (No. 292-2007), by Beijing Key Laboratory of Preparation and Processing of New Polymer Materials and by Beijing Key Laboratory of Printing and Packaging Materials and Technology.

References

1　XU J, SUN G C, WANG W S, "Study on Organosilicon Modified Epoxy", Shandong Science, Vol.10, No.4(1997), pp.57.

2　JIA X W, LIU Z G, "Research Progress on Silicon-based Flame Retardants", Chemical Industry and Engineering Progress, Vol.22, No.8(2003), pp.818-821.

3　Mercado L A, Galia M, Reina J A, "Silicon-containing flame retardant epoxy resins: Synthesis, characterization and properties", Polymer Degradation and Stability, Vol.91(2006), pp.2588-2594.

4　Ananda K S, Denchev Z, Alagar M, "Synthesis and thermal characterization of phosphorus containing siliconized epoxy resins", J. European Polymer, Vol.6, No.10(2006), pp.1-2.

5　Wang W J, Perng L H, Hsiue G. H, Chang F C, "Characterization and properties of new silicone-containing epoxy resin", Polymer, Vol.41(2000), pp.6113-6114.

6　ZHANG B, LIU W Q, "The synthesis and Properties of Organic Silicon-Modified Epoxy Resin", Guangzhou Chemisty, Vol.27, No.1(2002), pp.6-7.

7　ZHOU H F, "Study on curing kinetic parameter for polyimide resin", Thermosetting Resin，Vol.20, No.4(2005), pp.11-13.

Nonlinear Optical Fluorinated Polyimide/Inorganic Composites for Optical Interconnections and Devices

[1]Li Guoyuan, [2]Tang Chuan and [3]Ren Li

[1]School of Electronic and Information Engineering, South China University of Technology, Guangzhou China
[2]Department of Computer Science and Technology, Guangdong University of Finance, Guangzhou, China
[3] Institute of Materials Science and Engineering, South China University of Technology, Guangzhou, China

Abstract: A nonlinear optical (NLO) active alkoxysilane chromophore (SGDR1) was successfully synthesized. A fluorinated polyimide/SGDR1 composite was prepared to improve the poor temporal stability of nonlinearity at elevated temperature for the reported poled sol-gel film. The poled composite film was characterized by FTIR, DSC, TGA and UV-Vis. Results show that the composite displays good hydrophobic properties, high glass transition temperature (T_g: 266 °C), and high decomposition temperature (T_{d5}: 433 °C) although the composites were cured at lower temperature (200 °C). UV results show that the λ_{max} of composite film is 293.5 nm. The composites film also exhibits a high contact angle of 90.04°, a low surface energy of 18.17mN/m, and a low water absorption of 0.082%. The second harmonic coefficient d_{33} of the composite was measured to be 18.39 pm/V by using maker fringe technique. The new NLO composite exhibits 85 % of the original d_{33} over 720 hr at 100 °C and possesses much better stability than the reported sol-gel film. It is found that the silicon crosslinked network can significantly improve the properties of the composites and the fluorine and fluorine position in the chromophore side chain of PI can significantly affect properties of composites.
Key words: nonlinear optical material, fluorinated polyimide, chromphore, silicon, composite

1 Introduction

NLO materials have attracted steady attention for many years due to their potential application in integrated optical devices such as optical switching and high-speed electro-optic devices[1-2]. The device-quality NLO materials must possess large optical nonlinearity, high thermal stability and low optical loss [3]. NLO sol-gel materials are promising materials because they have excellent optical quality, refractive index control of films and ease of device fabrication [4-5]. However, one of its major disadvantages is the low temporal stability of nonlinearity at elevated temperature [6]. Recently, organic-inorganic sol-gel materials have received significant attention because of higher thermal stability [7-8]. By incorporation of a high performance organic polymer into an inorganic sol-gel material, the composite may display the desired properties of both components, which may be a potential approach to enhance long-term NLO stability. Aromatic polyimide, especially fluorinated polyimide, have been widely used in the advanced microelectronic and optoelectronic devices manufacturing because they possess a combination of properties, such as high thermal stability, low dielectric constants, high optical transparency and good hydrophobic property [9]. It has also been found that the internal production of water due to the curing of precursor of polyimide, poly(amic acid), aids the hydrolysis of the alkoxysilane, and the carboxylic acid group

of the poly(amic acid), being a Bronsted acid, might have a catalytic effect on hydrolysis and condensation of the alkoxysilane [10]. This will lead to no addition of catalyst and water, and could decrease phase separation between the inorganic and organic phase. Therefore, the incorporation of NLO silicon crosslinked network is a potential approach to decrease the cure temperature and enhance long-term NLO stability. In this study, the fluorinated polyimide/NLO alkoxysilane dye composites were prepared and characterized in order to obtain a NLO materials with better comprehensive properties especially higher stability.

2 Experimental

2.1 Materials
In this work, 1,3-Phenylenediamine (mPDA) was purified by distillation. Disperse red 1 (DR1) was purified by recrystallization from ethanol. Tetrahydrofuran (THF) and Hexane were purified by distillation over sodium chips and benzophenone ketal. 4,4'-(Hexafluoroisopropylidene) diphthalic anhydride (6FDA) was dried under a vacuum at 150 °C overnight before use. The other materials were purchased from Aldrich, Merk and Clariant and used without further purification.

2.2 Synthesis of alkoxysilane chromophore (SGDR1)

The schematic of synthesis of SGDR1 is shown in Fig. 1. At a two-necked round-bottomed Schlenk flask equipped with a magnetic stirrer, a nitrogen inlet and a reflux condenser, the ICTES (2.2 ml 8.08 mmol) and DR1 (1.27 g 4.04 mmol) were dissolved in freshly dried THF (30 ml). The mixture was then stirred at room temperature for 10mins. After that, it was refluxed for 4 hours with a catalytic amount (25 μL) of dibutyltin dilaurate (DBTDL). The solution was poured in distilled hexane and the resulting red precipitated powder was collected and washed again with hexane/THF until no DR1 was detected. The product (SGDR1) was finally dried under vacuum at 60 °C for 24 hour in 80% yield and then stored in a vacuum desiccator.

^{1}H NMR (DMSO) δ (ppm) 8.37 (d, J=8Hz, 2H), 7.94 (d, J=8Hz, 2H), 7.84 (d, J=12Hz, 2H), 7.30 (t, 1H), 6.92 (d, J=12Hz, 2H), 4.15 (t, 2H), 3.73 (q, 6H), 3.67 (t, 2H), 3.53 (q, 2H), 2.96 (q, 2H), 1.45 (m, 2H), 1.15 (m. 12H), 0.53 (t, 2H). Anal. Calcd. for $C_{26}H_{39}N_5O_7Si$: C, 55.57; H, 7.00; N, 12.47, O, 19.95 %. Found: 55.77; H, 7.01; N, 12.38; O, 19.88 %.

Synthesis of Fluorinated Poly(amic acid): m-PDA (2.0 mmol) was dissolved in 10 ml anhydrous NMP at 0 °C, followed by the addition of 6FDA (2.08 mmol) at once. After stirring at a temperature between 0 and 5 °C for 4 hours, the solution was then stirred at room temperature for 20 hours. The resulting fluorinated poly(amic aicd) shown in Fig. 1 was obtained.

2.3 Preparation of NLO Fluorinated Polyimide/SGDR1 Composites

The solution of SGDR1 (0.15g) in THF (1 ml) and NMP (1ml) was added into 4 ml of fluorinated poly(amic acid) solution. The mixed solution was then stirred at room temperature for more than 4 hours to be homogeneous. The resulting solution was filtered through a 0.45 μm Teflon membrane and was then spin-coated on a substrate (glass slide or indium-tin oxide glass) at various speeds ranging from 500 rpm to 4000 rpm for different durations from 5 to 30 seconds under clean-room conditions. After dried in a vacuum oven at 65 °C for 12hr to remove residual solvent, the films were subjected to concurrent poling and imidization. Composite films were poled in a temperature-controlled oven under a corona discharge field using 75 μm diameter tungsten wire in a wire-to-plane geometry of 1.5 cm above a Teflon protected, grounded copper block. A two-step corona poling process was employed to orient the NLO chromophores. The sample was heated to 200 °C at a heating rate of 5 °C/min and successively held at this temperature for 1 hr, while a 6 kV of poling voltage was applied to a corona wire simultaneously. Finally, the sample was cooled to room temperature in the presence of the poling field. The composite films had thicknesses ranging from 1 to 4 μm.

2.4 Characterization

^{1}H NMR spectrum was measured in dimethyl sulfoxide-d_6 (DMSO-d_6) using Bruker 400MHz instrument. Elemental analysis was performed by EURO EA elemental analyzer. FTIR spectroscopy was carried out with Perkin Elmer System 2000 FTIR. Glass transition temperatures (T_g) of the polyimide and composite were determined through DSC analysis that was performed by a TA Instruments 2010 at a heating rate of 10 °C/min in nitrogen atmosphere (50 cc/min). The UV-Vis spectra were measured with a Shimadzu model UV-2501PC spectrophotometer. The contact angle was measured using First Ten Angstroms system with automatic gain control camera with deionized water as reference liquid. The surface energy was automatically computed through Girifalco-Good-Fowkes-Young Model. In water absorption test, the polyimide films on glass slide were cut into 20×20 mm samples. The samples were then dried at 105 °C for 1 hour, weighted, and kept in deionized water at 25 °C for 24 hours and weighted again. SHG measurements were carried out with a p-polarized Q-switched Nd:YAG pulse laser operating with 10 Hz repetition rate and a 4-5 ns pulse width at 1064 nm. A Y-cut quartz crystal was used as the reference.

3 Results and Discussion

The ^{1}H NMR spectrum and the spectrum peak assignment of SGDR1 were shown in Fig. 2. Referred to literature [11], it can be found that the peaks at 8.37, 7.94, 7.84 and 6.92 ppm are attributed to the H atoms of phenylene H_1, H_2, H_3 and H_4, respectively. The peak at 7.20ppm was assigned to the proton

in carbamate, H_9. The peaks at 4.15, 3.73, 3.67, 3.53, 2.96, 1.45 and 0.53 ppm are assigned to the H atom in methylene H_8, H_{13}, H_7, H_5, H_{10}, H_{11} and H_{12}, respectively. The H atoms in methyl H_6 and H_{14} were overlapped at the peak 1.15 ppm. The FTIR spectra of DR1, ICTES and SGDR1 were shown in Fig. 3. The stretching vibration of the hydroxyl group in DR1 is observed at around 3273 cm^{-1}. For ICTES, the strong absorption peak corresponding to N=C=O stretching was observed at 2272 cm^{-1}. In spectrum of SGDR1, the absorption peaks of N=C=O and OH vanished and the new absorption peaks at 3327, 1690 and 1542 cm^{-1} emerged, which contributed by the NH stretching, carbonyl (C=O) stretching and NH bending, respectively. In addition, asymmetric stretching and symmetric stretching of NO_2 were observed at 1509

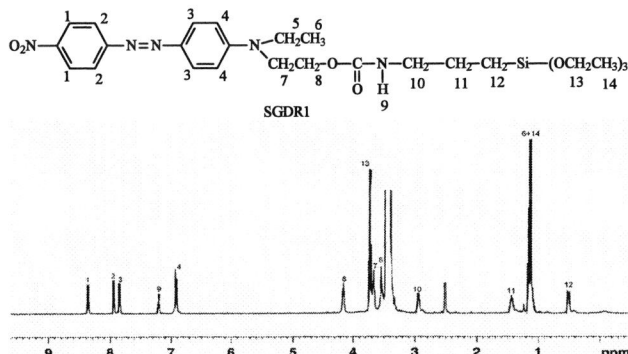

Fig. 1 Synthetic route of SGDR1 and chemical structure of fluorinated poly(amic acid)

and 1343 cm^{-1}, respectively. The experimental element analysis results are agreement with the theoretical one. Based on ^{1}H NMR, FTIR and element analysis results, structure of SGDR1 was confirmed as expected.

The thermal properties of the SGDR1 were characterized through DSC and TGA. Results shown in Fig.4 and Fig. 5 reveal that its melting point of the SGDR1 is 97 °C and the onset decomposition temperature is 225 °C, which ensures that the chromophore could not decompose during curing at 200°C. The UV-Vis results show that the maximum absorption wavelength λ_{max} of SGDR1 in NMP and THF are 498 and 474 nm, respectively.

Figure 2 ^{1}H NMR spectrum of SGDR1

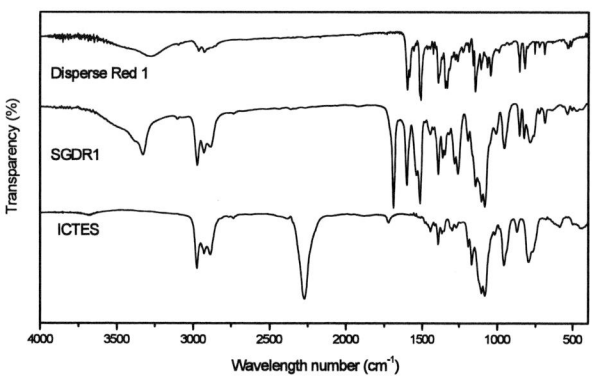

Fig. 3 FTIR spectra of disperse red 1, ICTES and SGDR1

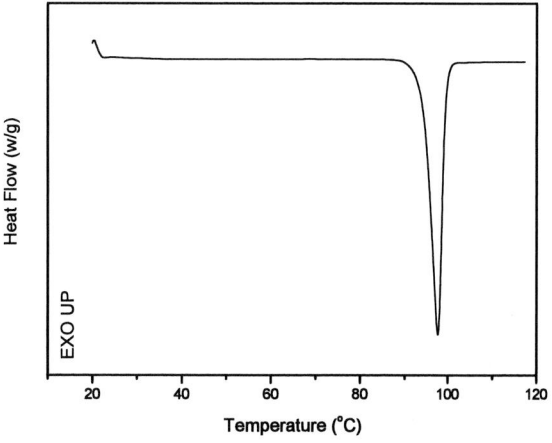

Fig. 4 DSC of SGDR1

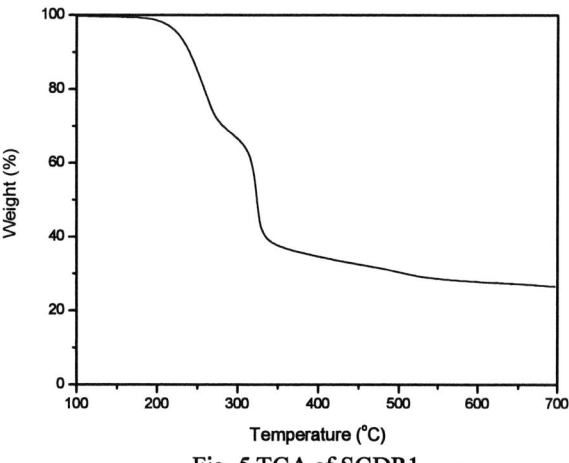

Fig. 5 TGA of SGDR1

FTIR spectra of fluorinated poly(amic acid)/SGDR1composites cured at 65 and 200 °C are shown in Fig. 6. After curing of poly(amic acid)/SGDR1 at 200 °C for 1 hour, the characteristic absorption peak of imide group at 1780 cm⁻¹ was emerged. It is noted that the characteristic absorption peak of Si-O-C₂H₅ around 1070 cm⁻¹ disappeared and the absorption peak around 1100 cm⁻¹ becomes broader after curing. This elucidated that Si-O-Si bond has formed. The negligible change of the symmetric nitro absorption at 1340 cm⁻¹ and the C-N stretching of the aromatic nitro

compound at 860 cm⁻¹ also indicate that the NLO moieties did not decompose during the curing process. This is illustrated by the symmetric nitro absorption at 1340 cm⁻¹ and the C-N stretching of the aromatic nitro compound at 860 cm⁻¹ which showed negligible change after curing. FTIR results suggest that fluorinated polyimide/SGDR1composites were well prepared through curing at 200 °C.

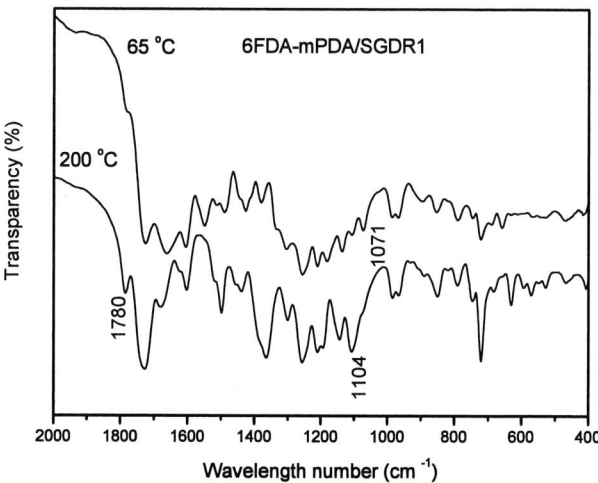

Fig. 6 FTIR spectra of fluorinated poly(amic acid) /SGDR1composites cured at 65 and 200 °C

Thermal properties of fluorinated polyimide/SGDR1 composite were characterized by DSC and TGA and the results are shown in Fig. 7. It can be found that the composite film have high T_g (266 °C) and T_{d5} (433 °C). T_g of the fluorinated polyimide was measured to be 233 °C. Results clearly show that T_g of composite is higher than that of fluorinated polyimide, which is attributed to the formation of the silicon crosslink network.

UV-Vis absorption was used to characterize the transparency of NLO composite film. Fig. 8 shows the UV-Vis spectrum of fluorinated polyimide/SGDR1 composite film. It can be found that λ_{max} is 485 nm. According to the UV results of SGDR1, the absorption peak is attributed to SGDR1.

As well known, water absorption and moisture content are very important to the electro-optical devices. It was reported that incorporation of fluorine or silicon into materials is an effective method to decrease the water absorption [12-13]. The relationship between contact angle and time of the composite film is shown in Fig. 9. The average contact angle, surface energy and water absorption of the composite film are measure to be 90.58°, 18.17 mN·m⁻¹, and 0.08%. Results show that the composite film exhibit excellent hydrophobic properties, which can significantly improve the stability and reliability of the optoelectronic materials and devices.

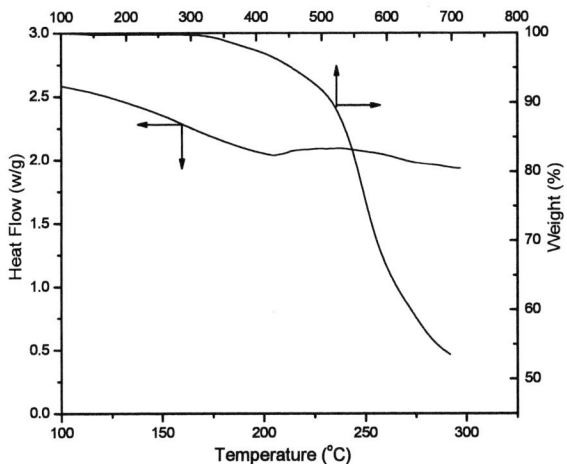

Fig. 7 DSC and TGA of fluorinated polyimide/SGDR1 composite film

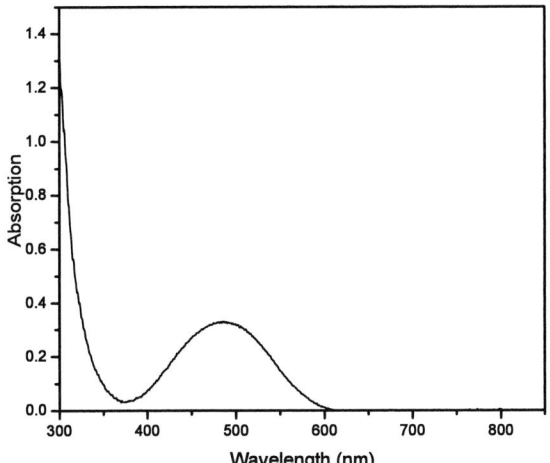

Fig. 8 UV-Vis spectrum of fluorinated polyimide/SGDR1 composite film

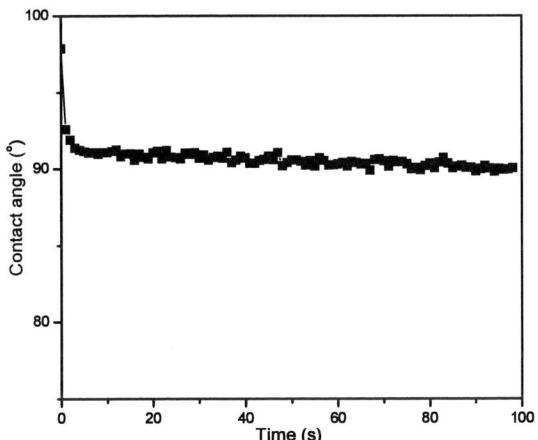

Fig. 9 Contact angles of fluorinated polyimide/SGDR1 composite film

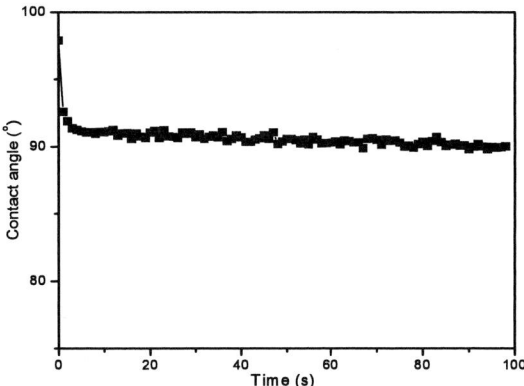

Fig. 10 Temporal stability of d_{33} of NLO fluorinated polyimide /SGDR1 composites at 100 °C

SHG results show that the NLO composite film possess d_{33} value of 18.39 pm/V. And its temporal stabilities of d_{33} value were studied at both room temperature and 100 °C, respectively. Results show that there is no decay of d_{33} value at room temperature after 720 hours. Fig. 10 shows its temporal stability of d_{33} value at 100 °C. It can be found that the composite film remained 85% of the original d_{33} value after 720 hours at 100 °C. Compared with the reported sol-gel films [6], results clearly reveal that fluorinated polyimide/SGDR1 composite exhibited higher temporal stability of d_{33} value. It suggested that the incorporation of fluorinated polyimide into the inorganic materials could lead to better comprehensive properties especially higher stability. The better properties might be attributed to the existence of the fluorine in the side-chain of the polyimide and the formation of the silicon crosslinked network in the composite. The mechanism is still on investigation.

4 Conclusions

A fluorinated polyimide/inorganic composite film were successfully prepared. The effects of silicon crosslinked network on the structures and properties of composites were studied. Results show that the silicon crosslinked network can significantly improve the properties of the composites. Incorporation of fluorinated polyimide into the inorganic material can significantly increase the thermal stability and long-term stability of the d_{33}. The composite film also exhibits better comprehensive properties than the reported inorganic sol-gel materials.

References

[1] S. Dhanuskodi, P. A. A. Mary, Growth and characterization of a new nonlinear optical crystal—ammonium borodilactate, J. Crystal Growth, vol.253 (2003) pp.424.
[2] S. R. Marder, B. Kippelen, A. K.-Y. Jen, N. Peyghambarian, Design and synthesis of chromophores and polymers for elecro-optical and photorefractive applications, Nature, vol.388 (1997), pp 845.
[3] P. Gunter, Nonlinear Optical Effects and Materials, Springer-Verlag, (New York, 2000), pp. 163-168.
[4] J. Ikushima, T. Fujiwara, K. Saito, Silica glass: A material

for photonics, J. Appl. Phys., vol.88 (2000), pp.1201.

[5] R. J. Jeng, Y. M. Chen, J. I. Chen, J. Kumar, S. K. Tripathy, Phenoxysilicon polymer with stable second-order optical nonlinearity, Macromolecules, vol.26 (1993), pp 2530.

[6] D. H. Choi, J. H. Park, J. H. Lee, S. D. Lee, Stability of second-order nonlinear optical properties in sol–gel matrix bearing silylated chalcone and disperse red 1, Thin solid films, vol.360 (2000), pp. 213.

[7] R.-J. Jeng, Ch.-C., Chen Ch.-P. Chang, Ch.-T. Chen, W.-Ch. Su Thermally stable corosslinked NLO materials based on maleimides, Polymer, vol.44 (2003), pp.143.

[8] G. H. Hsiue, R.-H. Lee, R.-J. Jeng, All sol-gel organic-inorganic nonlinear optical materials based on melanines and an alkoxysilane dye, Polymer, vol.40 (1999), pp.6417.

[9] S. Y. Yang, Z. Y. Ge, D. X. Yin, J. G. Liu, Y. F. Li, L. Fan, Synthesis and characterization of novel fluorinated polyimides derived from 4,4-[2,2,2-trifluoro-1-(3-trifluoromethylphenyl) ethylidene] diphthalic anhydride and aromatic diamines, J. Polym. Sci. Part A: Polym. Chem., vol.42 (2004), pp.4143.

[10] M. Palmlof, T. Hjertberg, B. A. Sultan, Crosslinking reactions of ethylene vinyl silane copolymers at prosessing temperature, J. Appl. Polym. Sci., vol.42 (1991), pp.1193.

[11] Y. J. Cui, M. Q. Wang, L. J. Chen, G. D. Qian, Synthesis and spectroscopic characterization of an alkoxysilane dye containing C. I. Disperse Red 1, Dyes and Pigments, vol.62 (2004), pp.43.

[12] L. Ren, F. S. Zeng, P. Ning, Z. Q. Chen, T.-M. Ko, Effect of addition orders on the properties of fluorine-containing copolyimides, J. Appl. Polym. Sci., vol.77 (2000), pp.3252.

[13] Li Ren, G. Y. Li, J. R. Shen, and D. M. Jia, Effects of Monomer Addition Sequences on the Properties of Silicon-Containing Copolyimides, Polym. Int., vol.54, no.7 (2005), pp.1097.

Nano-Thermal Interface Material with CNT Nano-Particles For Heat Dissipation Application

Li Xu[1], Cong Yue[1], Johan Liu[1,2],* Yan Zhang[1], Xiu Zhen Lu[1] and Zhaonian Cheng[2]

[1] Key Laboratory of Advanced Display and System Applications, Ministry of Education & SMIT Center, School of Mechatronics Engineering and Automation, Shanghai University, Shanghai 200072, China

[2]SMIT Center & Bionano Systems Laboratory, Department of Microtechnology and Nanoscience, Chalmers University of Technology, SE-412 96 Gothenburg, Sweden

*Corresponding Author's e-mail: johan.liu@chalmers.se

Abstract

Heat dissipation of electronic packages has become one of the limiting factors to miniaturization. The removal of the heat generated is a critical issue in electronic packaging. With the development of thermal management, thermal interface material (TIM) plays a more and more important role in electronics packaging. A new nano-TIM with nanofibers prepared by using electrospinning has been suggested in recent years. In this experiment study, the carbon nanotube (CNT) nano-particles were added into the polymer solution before the electrospinning to improve the thermal conductivity of nano-TIM. The polymer solution of polyurethane was used for present electrospinning. The effects of a number of process parameters in the electrospinning were studied in this work. Different variables such as the distance between needle tip and collector, the voltage applied, and CNT nano-particles content were studied. The Scanning Electron Microscopy (SEM) was used to characterize nano-TIMs with CNT nano-particles.

Introduction

The integration of electronic chips is increasing every year and the cooling process must increase by the same factor to keep the chip within operating temperatures. In fact, heat dissipation of semiconductor packages has become one of the limiting factors to miniaturization. [1]

Today, typically heat sinks are used as a way to cool devices, which enhance the flow of heat from the device to the atmosphere. However, both the heat sink and the device to removal between the chip and the heat sink.

In this paper, we investigated a nano-TIM preparing process that is related to the embedment of carbon nanotube (CNT) nano-particles into the nano-fibers electrospun. The effects of a number of process parameters in the electrospinning on the nano-TIM with CNT nano-particles were studied in this work. The Scanning Electron Microscopy (SEM) was used to characterize nano-TIMs with CNT nano-particles and the results were also reported in this paper.

Electrospinning

Electrospinning is one of many techniques to make fibers with nanoscale size from polymer solution. The manufactuting process was first reported by Anton Formhals in 1934[6]. In the past years, the electrospinning process has become popular among nanoresearch groups because of its ability to produce fibers in submicron or nanoscale size, with a diameter of around 100 nm. A simple yet powerful electrospinning setup consists of a syringe, a collector, and a high voltage power supply unit. Figure 2 shows the electrospinning setup. [3]

be cooled have microscopic roughness and gaps, which result in that roughly 1% of the material is in physical contact, and the rest is in contact through interstitial air due to the roughness of the two surfaces. (Figure 1)[2]

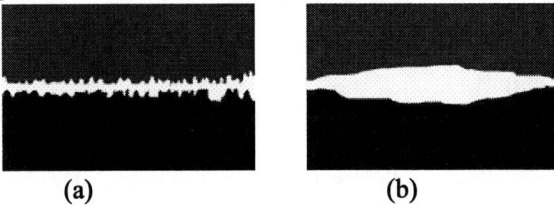

| (a) | (b) |

Figure 1(a): fine heat sink and CPU surface roughness (b) poor heat sink and CPU surface roughness

These rough microscopic gaps need to be filled up with material that will aid in the heat transfer. It doesn't take long for an electronics assembler to realize that a thermal interface material (TIM) is essential when two or more solid surfaces are in the heat path. The thermal interface material is used to fill air pockets and gaps between electronic component and various thermally conducting components.

A new class of Nano-Thermal Interface Materials (Nano-TIMs) with nanofibers prepared by using electrospinning has been suggested in recent years. [2,3,4] And a novel nanostructured polymer-metal composite for thermal interface material application was introduced. [5] Many nanofibers in nano-TIM, which were generated by electrospinning, can fill air pockets and gaps to enhance heat

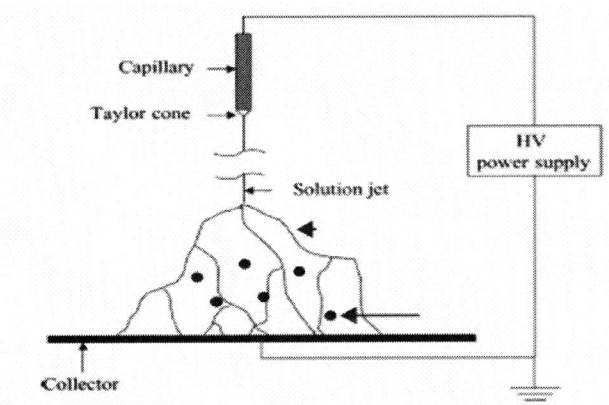

Figure 2: Scheme of electrospinning process with fiber formation and with particles illustrated

In this process, a solid material is dissolved into a solvent to create a solution. Then the polymer solution is added into the equipment that has a nozzle of millimeter diameter. An electronic field of kilovolts is applied between the nozzle (needle tip) and the collector. Under the electrical force, a thin and long fiber is spun by stretching the polymer solution. The diameter of the fiber varies from a few nanometers to a micrometer, and this thin diameter causes an immense length of the fiber, up to kilometers [7,8].

There are several parameters affecting the spinning process: The most important one is the applied voltage of the spinning, which was found that the creation of a Taylor cone was very voltage sensitive. Another important parameter is the distance between the needle tip and the collector. Meanwhile the surrounding environment is also significant which will lead to a great influence on the parameters involved in the electrospinning procedure. The goal of this work is to find the workable process to generate nano-TIM with CNT nano-particles.

Sample Preparation by Electrospinning

1. Materials

CNTs can be thought of being made by rolling up a single atomic layer of graphite to form a seamless cylinder [9,10]. The resulting structure is called single-walled carbon nanotube (SWCNT). If several SWCNTs with varying diameter are nested concentrically inside one another, the resulting structure is called a multi-walled carbon nanotube(MWCNT). In this work, the MWCNT nano-particles were used. The thermal conductivity of nanotubes exceeds that of diamond by a factor of 2 and that of copper by a factor of 15 and is, therefore, ideal for dissipating heat from sensitive active devices.

The CNT nano-particles used in this work is MWCNT powder purchased commercially from Organic Chemistry Corporation of Chengdu of CAS, China and used without further purification. They have a thermal conductivity of about 2000W/mK. The detail of the MWCNT is as given: Percentage pure CNT is over 95%, Diameter is 5-30nm, and Length is 0.5-2μm.

In this work, CNT nano-particles were added into the solution before electrospinning to improve the thermal conductivity of Nano-TIM.

In our experiments, some solutions contained a substance named Tween 80(or Polysorbate 80). Tween 80 is a non-ionic surfactant used to reduce the surface tension of the solution. In the electrospinning process, Tween 80 was found to improve process stability and decrease the amount of droplets appearing on the collector.

2. Three Cases study

The solution used in this work consisted of polyurethane (PU) polymer dissolved in a mixture of THF and DMF (40/60 by weight), and the CNT nano-particles were added to the solution. The first step of electrospinning process is mixing the solution. After weighting the solvent and polymer and mixing the solvent, polymer and CNT nano-particles, it is necessary to put them on the magnetic stirring machine, to let the polymer and the nano-particles fully dissolve into the solvent. This procedure usually took over 4 hours. Sometimes, Tween 80 was also added to the solution before the electrospinning process, which was introduced with the hope that the substance could help to completely disperse the CNT nano-particles.

The electrospinning process involves several processing parameters. A number of parameters have been studied and those with the greater impact on the process seem to be applied voltage, the distance between the needle tip and the collector, and the mass ratio (concentration) of CNT nano-particles relative to PU for nano-TIM with CNT nano-particles.

Three cases study were carried out in order to investigate the influence of the three import process parameters on the nanofibers obtained. In the first case study, the applied voltage was varied under the same mass ratio of CNT nano-particles to polymer resin, and the same distance between the needle tip and the collector (see Table 1). In the second case study, the distance between the needle tip and the collector was varied and the three solutions were electrospun under the same applied voltage and the same mass ratio of CNT nano-particles (see Table 2). In the third case study, four solutions with different mass ratio, in which the mass ratio of CNT nano-particles to polymer resin was 10%, 20%,30%,50% respectively, were prepared to electrospin nano-TIM and other two process parameters were kept constant (see Table 3).

Table 1:Parameters for case study 1

Test	CNT nano-particles (wt%)	Voltage (kv)	Distance (cm)
1	10	20	15
2	10	25	15
3	10	30	15

Table 2:Parameters for case study 2

Test	CNT nano-particles (wt%)	Voltage (kv)	Distance (cm)
1	20	25	15
2	20	25	20
3	20	25	25

Table 3:Parameters for case study 3

Test	CNT nano-particles (wt%)	Voltage (kv)	Distance (cm)
1	10	25	15
2	20	25	15
3	30	25	15
4	50	25	15

Results

In this work, Scanning Electron Microscopy (SEM) was used to study the sample of nano-TIM with CNT nano-particles electrospun. The morphology results showed that the CNT nano-particles were in an agglomerate form both embedding into the nanofibers and attaching onto the surface of nanofibers. Figure 3 and Figure 4 show the SEM picture of the Nano-TIM with CNT nano-particles.

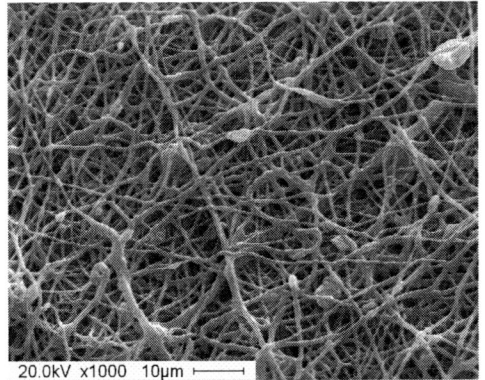

Figure 3: SEM picture of the Nano-TIM with CNT nano-particles (20 wt% CNT nano-particles)

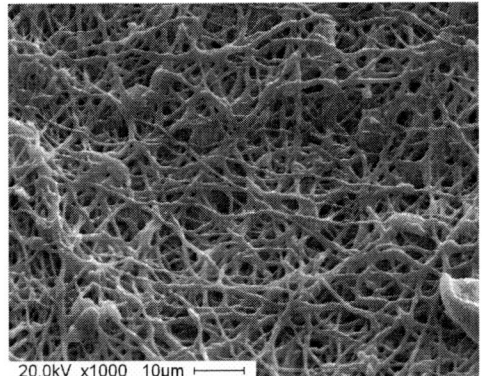

Figure 4: SEM picture of the Nano-TIM with CNT nano-particles (30 wt% CNT nano-particles)

Voltage is a crucial parameter in the electrospinning process. In the first case study, when the voltage less than 20 KV, a number of small droplets were formed at the tip of the needle, and dropped onto the collector. With an increase in voltage, fewer droplets were formed. However, the applied voltage was also limited. One disadvantage of increased voltage was that the circular spot in which fibers were collected on the collector become more ring shaped and a lot of fibers were deposited around the sample, resulting in a waster of material and increased processing time.

From Figure 5, it can be seen that the smaller diameter of nanofibers can be obtained by applying higher voltage. The smallest average diameter of fibers was observed as 700 nm in this work.

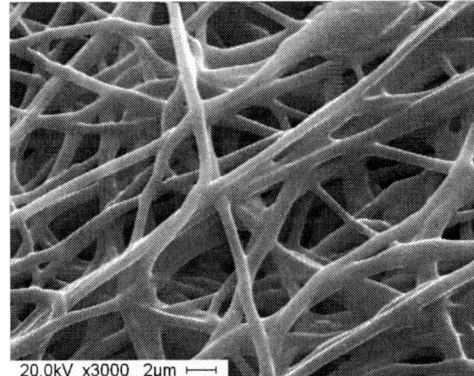

Figure 5: SEM picture of Nano-TIM with CNT nano-particles (10 wt% CNT nano-particles under applied voltage 25 kv)

For the electrospinning setup used in this experiment, the best applied voltage seems to be around 25 KV. Under this voltage, the process was stable with a minimum of droplets formed. The size of the fibers is also under the scale of nanometer.

Varying the distance between the needle tip and the collector will result in a change of electric field strength and the flight time of the fiber generation. A short distance has the same impact as an increased in voltage.

The SEM pictures from the second case study reveal no significant differences in fiber size and quality at different height of the needle tip to collector. However, an increased distance between needle tip and collector led to more droplets formed and unstable fibers, which may be explained by decreased electric field strength (equal to a decreased applied voltage).

The SEM pictures (Figure 6) gives a set of results from the third case study. When adding CNT nano-particles to the polymer solution, the solution changed its characteristic very much. With higher concentration of CNT nano-particles, more droplets formed, which made a useful film by electrospinning to be difficult.

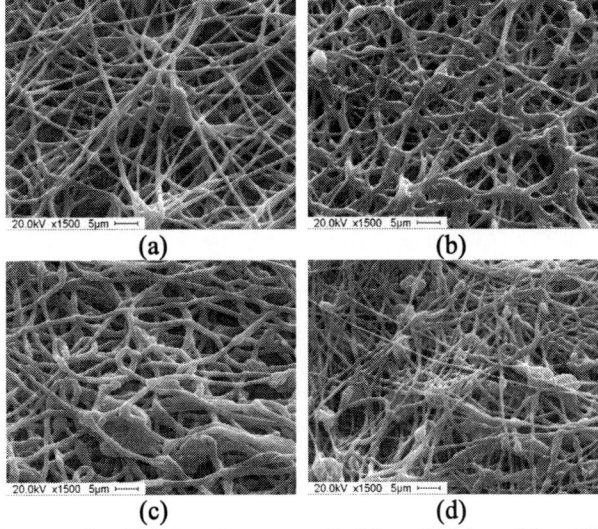

Figure 6: SEM pictures of Nano-TIM with CNT nano-partickes (a) 10 wt% CNT nano-particles (b) 20 wt% CNT nano-particles (c) 30 wt% CNT nano-particles (d) 50 wt% CNT nano-particles

It can also be seen from Figure 6 that the CNT nano-particles were in an agglomerate form both embedding into the nanofibers and attaching onto the surface of nanofibers. The particles should be in nano-size and consequently not visible as spheres on the fibers. Studying the pictures, it is clear that this is not the case. The particles seem to form agglomerate resulting in large lumps attaching onto the fibers. The scale of 785 μm^2 of the SEM picture was measured for this study. Several pictures with different mass ratio of CNT nano-particles but the same scale were chosen to be as the comparison. Average number of agglomerate particles in nano-TIM electrospun with different mass ratio of CNT nano-particles is shown in Figure 7. (A, B, C, D mean the mass ratio of CNT nano-particles to polymer resin was 10%, 20%,30%,50% respectively). It was also indicated that the number of agglomerate particles attaching onto the nanofiber surfaces increased with the increasing of mass ratio of CNT nano-particles. And the size of these agglomerates seems to vary between the experiments.

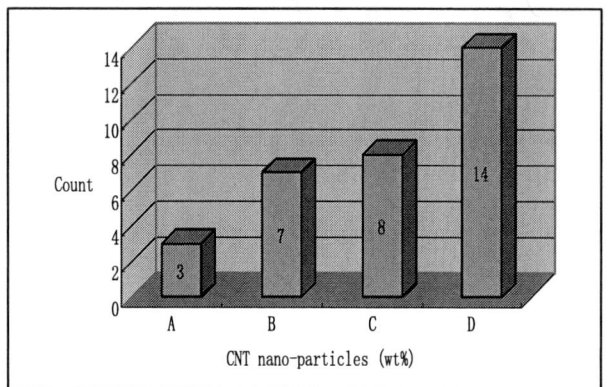

Figure 7: Average number of agglomerate particles in nano-TIM with CNT nano-particles, in area 785 μm^2 from SEM

Conclusion

Nano-TIM samples with various mass ratios of the CNT nano-particles were prepared by using electrospinning in this study for heat dissipation application in electronic packaging. The reasonable processing parameters were obtained. The sample of PU with 50%wt of CNT nano-particles has been obtained under the applied voltage 25 KV and the height 15 cm. The performance of nano-TIM with CNT nano-particles will be investigated in our further work.

Acknowledges

This work was supported by the Ministry of Science and Technology of the People's Republic of China through China-Europe Cooperation project on Science and Technology. This work was also supported by the Shanghai Education Commission.

References

1. Clemens J.M, Lasance, "The Urgent Need for Widely Accepted Test Methods for Thermal Interface Materials," *19th IEEE SEMI-THERM Symposium*, 2003.

2. Johan Liu, Michael Olugbenga Olorunyomi, Xiuzhen Lu, Wenxuan Wang, Tomas Aronsson, Dongkai Shangguan, "New Nano-Thermal Interface Material for Heat Removal in Electronics Packaging", *Proceedings of the 1st IEEE CPMT Electronics Systemintegration Technology Conference (ESTC2006)*, September 5-7, 2006, Dresden, Germany, pp.1-6.

3. WenXuan Wang, Xiuzhen Lu, Johan Liu, Xu Wang, Zhaonian Cheng, Dongkai Shangguan, "Investigation of manufacturing process and effect of processing variables on structure and morphology of Nano-TIM", *Proceedings of the International Symposium on High Density Packaging (HDP)*, Shanghai, July, 2007, pp. 431-435.

4. WenXuan Wang, Lu, X. Z., Liu, J., Olugbenga Olorunyomi, M., Aronsson, T. and Shangguan, D., "New Nano-Thermal Interface Materials (Nano-TIMs) with SiC Nano-Particles Used for Heat Removal in Electronics Packaging Applications", *Proceedings of the 8th International Conference on Electronic Materials and Packaging, Hong Kong*, December, 2006, pp. 631-635.

5. Björn Carlberg, Teng Wang, Yifeng Fu, Johan Liu and Dongkai Shangguan, "Nanostructured Polymer-metal Composite for Thermal Interface Material Applications," *proceedings of the 57th IEEE Electronic Components and Technology Conference (ECTC)*, May 27-May 30, 2008, Orlando, USA, pp 191-197.

6. Formhals A., "Process and apparatus for preparing artificial threads", United States Patent Office, , no. 1, 975, 504, 1934.

7. Seeram Ramakrishna, Kazutoshi Fujihara, et al, An Introduction to Electrospinning and Nanofibers, World Scientific Publishing (Singapore 2005), pp. 91-117.

8. Jie Bai, Yaoxian Li, Meiye Li, Shugang Wang, Chaoqun Zhang, Qingbiao Yang, "Electrospinning method for the preparation of silver chloride nanoparticles in PVP nanofiber," *Applied Surface Science*, Vol.254, No.15, 30 May. 2008, pp.4520-4523.

9. Franz Kreupl, Andrew P. Graham, Maik Liebau, Georg S. Duesberg, Robert Seidel, Eugen Unger, "Carbon nanotubes for Interconnect applications," *Electron Devices Meeting*, 13-15 Dec. 2004, pp. 683-686.

10. Valentine N. Popov, "Carbon nanotubes: properties and application," *Materials Science and Engineering R* Vol.43 (2004), pp. 61-102.

The Humidity and Thermal Characteristics of Die-Attach (DA) and its Impact on the Package Reliability

Chen Ning, MA Xiao-song, Jiang Haihua
Guilin University of Electronic Technology, Guangxi, 541004, China
E-mail:gyver.ning@gmail.com

Abstract

In this paper, A humidity absorption test experiment is carried out under the temperature 20°C, 40°C and 60°C respectively and the environmental humidity was changed in different modes in each case. The experimental sample is from material of DA, and it is 14.5×5.5×0.37mm³ in size, and 94.4 mg in weight. The experiment data are collected to gain the saturated moisture. The moisture diffusion coefficient (D) of the sample can be calculated by fitting the line of data curve and calculating the slope of the line. The constants (Q_D and D_0) which are in the Arrhenius formula are worded out by using the D [1]. In order to analyze the failure model in the interface between DA and Die pad, a new type of System-in-Package (SiP) is modeled. The simulation environments are set as the conditions of stored in the room temperature and processed in the SMT assembly production line. The analysis aim is to calculate the humidity distribution and the hygroscopic and thermal stress in these conditions. The results of FEM analysis shows that the max hygro-mechanical and thermo-mechanical stress appear in the corner of the interface between the Die and the DA, and the corner may be the risk location. The delamination may occur in this place. So it is necessary to analysis the reliability of the interface between the DA and others materials.

1. Introduction:

These years, polymers with filler are used as DA in the electronics package. Usually materials of DA include polymer and thermal conductivity particle such as silver, polyimide, cyanide acid lipid and epoxy resin. [2]. Because organic DA not only can be used conveniently, but also have a low processing temperature, they are the primary materials of DA.

It is common that DA is used for thermal conductivity between Active Die and Die pad in package. The mismatch of hygro-mechanical expansion among different kinds of materials is hoped to be coordinated through using DA. But lots of delamination relater reliability problems are observed in the interface between DA and other materials.

The primary failure mode of micro-electronics package can be explained as follows: different kinds of materials have the different expansion coefficient and thermal expansion coefficient, the decrease of the cohesive strength and the interface delamination may occur between the different kinds of materials' layer [3] .But at present, few hygro-mechanical properties of DA material in micro-electronics package are studied at home and abroad.

In this work, we carry out a humidity absorption experiment to obtain the moisture diffusion coefficient in some temperature, and then calculate Q_D and D_0, finally we model to analysis the humidity distribution and the hygroscopic and thermal stress thought finite element method

(FEM) software. In this paper, DA is investigated as our researching material.

2. Humidity characteristics of DA

Humidity expansion in plastic package electronic devices is a complex process that is affected by lots of factors. First important factor is the existing state of water molecules in the plastic package electronic devices. In a typical storage environment, the state of water molecules may be gaseous, liquid [4], or the state with gaseous and liquid [5]. The humidity diffusion on the face of plastic package electronic devices or into the plastic package electronic devices has the same states as those. But the state of the humidity can change. For example, the humidity absorbed on face of components can agglomerate into liquid water, the moisture in the internal cavities and the interface between different materials and around microcrack can also agglomerate into liquid water.

Because volume of liquid water is smaller than that of water vapor, cavities can absorb more humidity so that the volumes of components in high temperature rapidly expand. But different materials have the different humidity-thermal expansion performance. This condition leads to the reliability of the interface between different materials. The second factor is the path of the humidity expansion. Humidity expand primary through the polymer, but not the interface between different materials. So generally we consider packaged material as the primary diffusion path. The third factor is that there is an interface between polymer and water molecule, water molecule can expand into the component through the interface and hydrogen between water molecule and polymer. Because the factor belongs to chemistry diffusion, we do not take it into account in the analysis FEM. The fourth factor is the thickness of material, it have an influence on the extent and depth of expansion. The expansion shows that polymer can diffuse by moisture absorption, and then shear stress appears [6].

3. Experiment and data processing

The sample is cured between Teflon foil and then is ground and polished to a thickness of about 300μm .The size of the sample is 12.5×4.89×0.385mm³ and 36mg in weight. The sample is made as think as possible so that the moisture absorption time can be shorten. The equipment of moisture absorption test is a programmable SGA-100 temperature and relative humidity chamber.

The humidity absorption test experiment is carried out under the temperature 40°C, 60°C and 80°C, respectively. The environmental humidity was changed in different modes in each case. The experiment data are collected every two minutes so that the increasing weight of experiment sample can be measured. The experiment data can be showed in the

978-1-4244-2739-0/08/$25.00 ©2008 IEEE

coordinate plane. So the saturated moisture can be obtained form the data. By fitting the line of data curve and calculating the slope of the line we can gain the moisture diffusion coefficient (D) (see Fig1). And the moisture saturation concentration (C_{sat}) can be calculated by the relation between temperature, environmental humidity and the mass changed of the sample. Using the D, the constants (Q_D、D_0) which the Arrhenius formula need can be worded out[3].

The Arrhenius formula are showed as follows:

$$D = D_0 \exp(Q_D / RT) \text{----------------------------------- (1)}$$

$$\ln D = \frac{Q_D}{R}\frac{1}{T} + \ln D_0 \text{----------------------------------- (2)}$$

Table1 the diffusion coefficient

Temperature (℃)	Diffusion coefficient(m^2/sec)
20	1.87E-12
40	2.80E-12
60	4.07E-12

Table1 shows the diffusion coefficient in the temperature 20℃.40℃ and 60℃. So the constants Q_D and D_0 are worked out through fitting the data in table1.

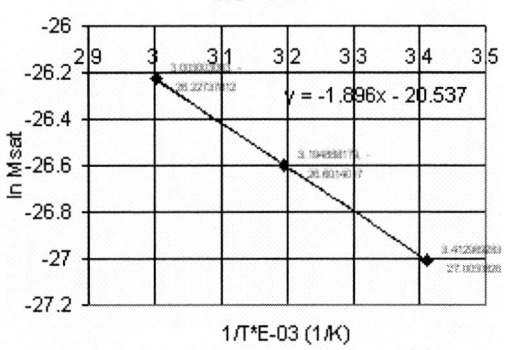

Figure1 The relation between lnD and 1/T

So Q_D and D_0 are calculated from above.

$$Q_D=15.76334kJ/mol$$
$$D_0=1.20E-9m^2/s$$

So the Arrhenius formula is worked out as fellow:

$$D = (1.20E - 9) \times \exp(\frac{15.76334}{8.314} \bullet T) \text{----------------------- (3)}$$

Equation (3) is programmed by FORTRAN embedded in the FEM software to calculate the humidity expansion, hygro-mechanical and thermo-mechanical stress

4. Geometry model

The finite element modeling is an special structure of System-In-Package (SIP) with two flip chips, as showed in Figure1. The size of the component is $6 \times 6 \times 0.85mm^3$.

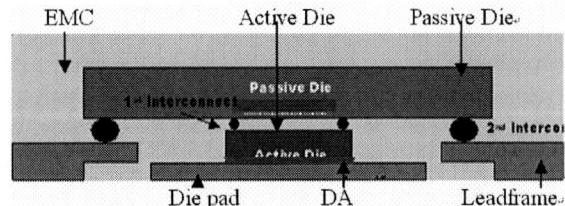

Figure2 Package Construction Model

The figure2 shows the model is composed by DA, Active Die, Passive Die, Die pad, Leadframe and EMC.

As to symmetry of the geometry structure of the model, we select the half component to model. And the research object is the DA between Active Die and Die pad (see Figure3). As is showed in the Figure3, nodes- A-425, B-544,D-554,D-426 are selected to analyze the moisture diffusion and the hygro-mechanic and thermo-mechanic stress.

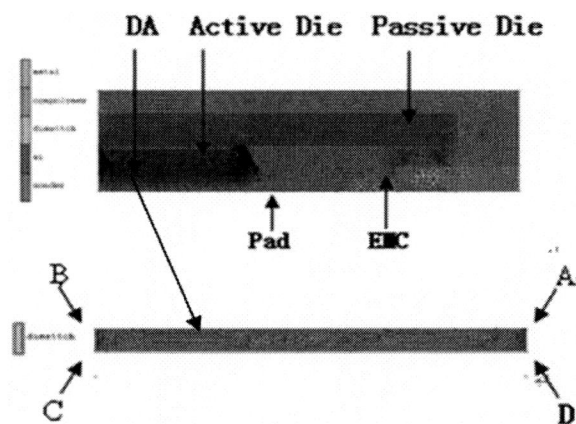

Figure3 SIP and DA model

In the paper two environments are designed to calculate the moisture expansion and moisture-thermal stress distribution. The first environment is set to simulate the storage environmental. Relative humidity is set as 85％,the temperature is set as 298K, location time is set as 168hours. The second environment is set to simulate the reflow heating in SMT process. Relative humidity is set as 0 ％ ,the temperature is set to match the reflow process parameters, the location time is set as 250 seconds.

2.2 Characteristics of Material and Boundary Condition

The Characteristics of all the materials in the SIP are showed in Table2, such as Young's modulus (E), Poisson's ratio (υ), Coefficient thermal expansion (CTE) and Coefficient moisture expansion (CME).

Table2 Characteristics of the material

material	E(GP)	υ	CTE(ppm/℃)	CME (mm^3/mg)
EMC	14470	0.28	34	0.222
Die	131000	0.26	2.3	0
DA	8000	0.25	170	0.52
Die pad	120000	0.3	16.7	0
Solder	43500	0.35	19	0

730

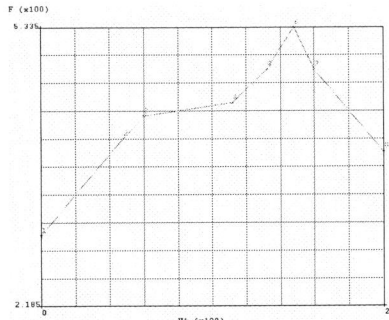

Figure4 The temperature change in flow process

The figure4 shows temperature change in flow process. This is the boundary condition of the finite element modeling. The boundary condition is the environment of reflow heating in SMT process. It is the significant factor for the diffusion expansion, because the diffusion coefficient of DA changes as the temperature changes.

5. Result Analysis

5.1 Analysis of Moisture Expansion

In the paper DA is the research object. Figure7 shows the humidity diffusion of DA changes when the SIP component stored in the room temperature for 168hours

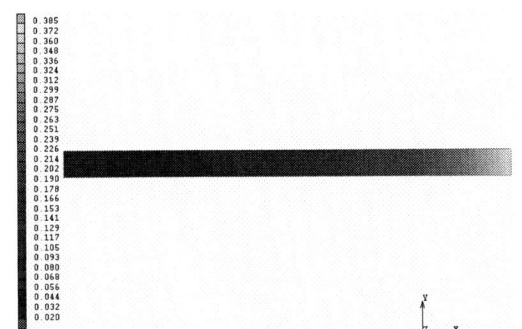

Figure5 Moisture expansion distribution after 168hours

As showed in Figure5, humidity gradually decreases from edge to the middle of the DA.

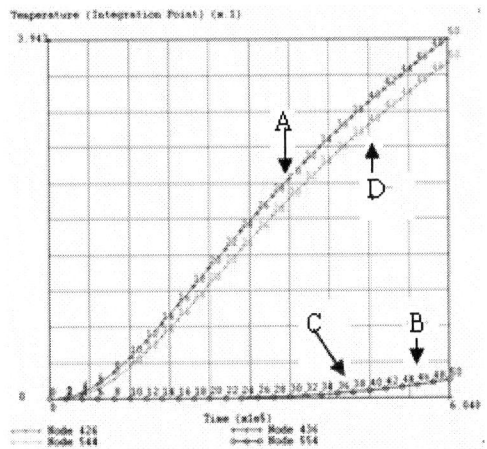

Figure6 the humidity of node A,B,C,D

Figure6 shows Node A that located in the sharp corner of the interface between DA and pad have the maximal humidity and the maximal ratio of the humidity diffusion. Node A is the center of the humidity. And node D has the similar humidity and ratio of the humidity diffusion as Nod A. Because node A has the same lateral scale as node D, they have the similar ability of humidity diffusion. This condition can be showed in nodes B and C. In addition, node B and C have the very low humidity and ratio of the moisture diffusion, and their moisture are too similar to find the distinction of the moisture easily.

Figure7 The contour bands of changing humidity's distribution in reflow heating process

Figure7 shows how the humidity diffusion in DA changes when the SIP component enters the reflow process that experience 250S. We observe the humidity expand form middle to the edge of DA. Because the environmental humidity is set as 0, at the same time, the moisture diffusion coefficient of polymer change with environmental temperature changing incessant,.

Figure8 The contour bands of humidity distribution in reflow process

Figure8 shows the humidity distribution in the end of reflow process. We observe the edge of EMC have low humidity, and the moisture diffuse form middle to the edge of SIP, but the trend of diffusion is not obvious.

Because the humidity of nodes have a obvious change as the lateral scale changes. For observing clarity the changing humidity in DA, the data of nodes A and D, C and B are collected separately. As showed in Figure9,10.

731

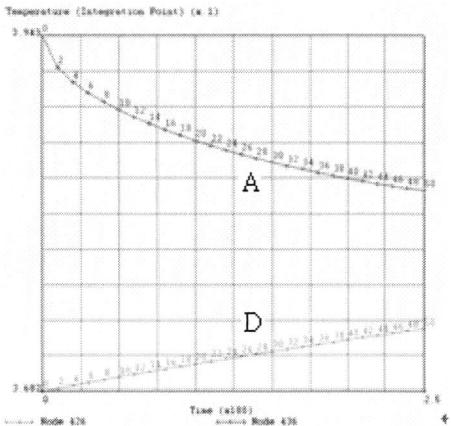

Figure9 The changing humidity of Nodes A,D

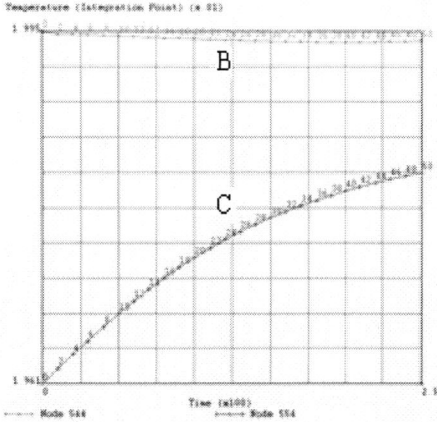

Figure10 The changing humidity of Nodes B, C

Figure9 shows humidity of node A decreases in the flow welding, but the humidity of node D increase

Figure10 shows humidity of node B decreases in the flow welding, but the humidity of node C increase

The special structure of SIP model and the place of the DA lead to the results of calculate. DA locates between the Die pad and Active Die; in addition, it is in the inner of component. So metal don't cover the entire bottom, moisture can diffuse to the air through EMC in the bottom. In the other hand, moisture in the SIP component can diffuse to the air through EMC above. But because Passive Die is on the Active Die, it covers lots of area. The moisture in the DA can't diffuse form lateral face and upper surface easily. So the moisture in the DA diffuse to air primary thought bottom surface. That is the reason why humidity of node A, B decreases in the flow process, but the humidity of node C, D increases.

On the other hand, because the humidity distribution in the DA shows an obviously gradient in transverse and longitudinal, there are different expansion distributions at different places. It leads to the Hygro-mechanical stress in the interface between DA and passive die, DA and Die pad. It also occurs in the region near the interface between different materials.

The result of the moisture diffusion distribution is uploaded to the primary model as the boundary condition so that we can calculate the hygro-mechanical stress in the reflow process.

In addition, we need to set the relative mechanics parameters.

5.2 Analysis of Hygro-mechanical Stress

Hygro-mechanical Stress is calculated by setting the result of moisture expansion as the boundary.

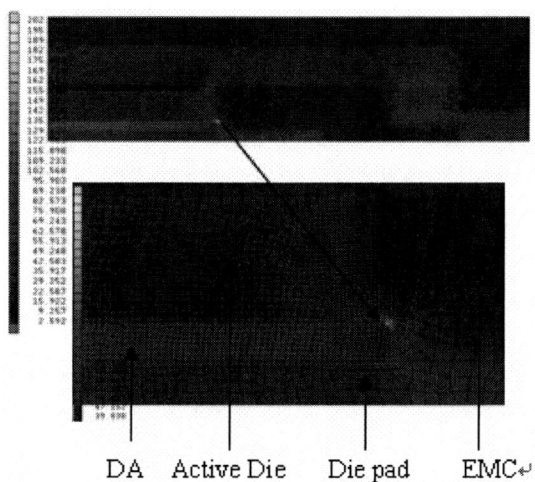

DA   Active Die   Die pad   EMC↵

Figure11 The hygro-mechanical stress distribution in the SIP

Figure11 shows when the SIP component is in the peak value of temperature in the reflow process, the hygro-mechanical stress appears in the Die pad, Die, Solder and DA. Figure11 also shows the max stress appears in the corner of the interface between the Active Die and the DA, and the corner may be the risk location. Its hygro-mechanical stress can reach 202.5MP. As a whole we can observe the stress that appears in the DA is stronger than that in any other place.

Figure12 Hygro-mechanical stress distribution during reflow process

Form Figure12 we observe the hygro-mechanical stress distribution in the peak temperature of reflow process. The hygro-mechanical stress in DA is not uniform. The max stress also appears in the corner of the interface between the Die and the DA.

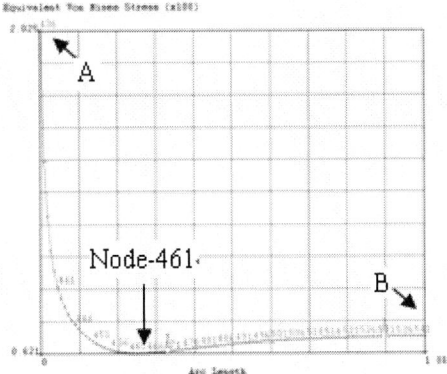

Figure13 The hygro-mechanical stress of path A-B

Figure13 shows the hygro-mechanical stress of path A-B. We observe that stress changes with the changing lateral scale. Node A have the maximal stress 202.5MP, and then the stress decrease quickly as the lateral scale increase, and the node-461 have the lowest stress, but the stress of nodes increase as the lateral scale increase after node-461.

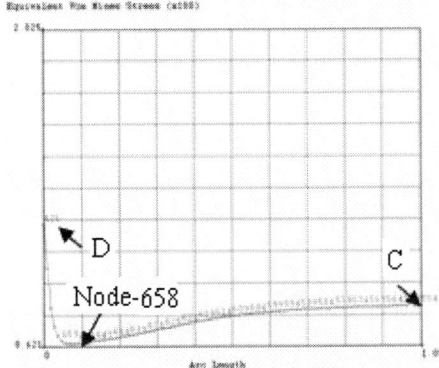

Figure14 The hygro-mechanical Stress of path D-C

Figure14 shows the hygro-mechanical stress of path D-C. We can observe the stress in Fig12 have the similar trend with Fig14. Because the nodes in the path D-C has the same lateral scales as nodes in the path A-B. But the stresses of path D-C are lower than that of path A-B.

5.3 Analysis of thermo-mechanical stress

Thermo-mechanical stress is calculated by setting the temperature change in flow process as the boundary (see Figure4).

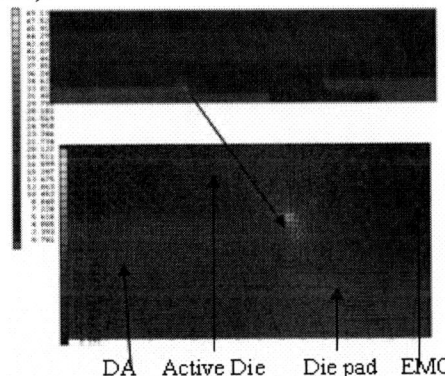

Figure15 The thermo-mechanical stress distribution

Figure15 shows that the max thermal stress also appears in the corner of the interface between the Die and the DA in the peak value of temperature of the reflow process, and the corner may be the risk location. The stress can reach 44.33MP. The thermo-mechanical stress appears in the SIP aggregate in the Die pad, Die and DA. We can observe that the stress that appears in the DA is stronger than that in any other place.

Figure16 thermo-mechanical stress distribution of DA

Form Figure16 we can observe the thermal stress distribution in the peak temperature of reflow process. The thermal stress in DA is not uniform too. The max stress also appears in the corner of the interface between the Die and the DA. As a whole the stress decrease form the edge to the middle of DA. The thermal Stress of path A-B and D-C are showed in Figure17, 18.

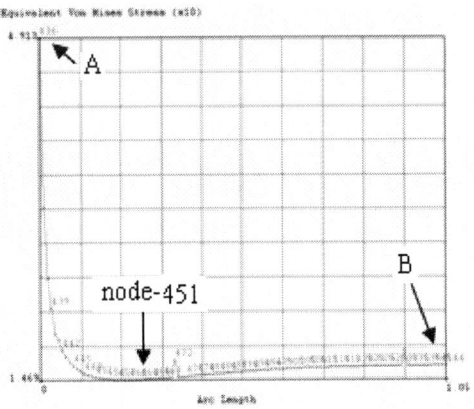

Figure17 The thermo-mechanical stress of path A-B

Figure17 shows the thermal stress of path A-B. we can observe that Stress changes with the changing lateral scale. Node A have the maximal stress 49.13MP, and then the stress decrease quickly as the lateral scale increase, and the node-451 have the lowest stress, but the stress of nodes increase as the lateral scale increase after node-451.

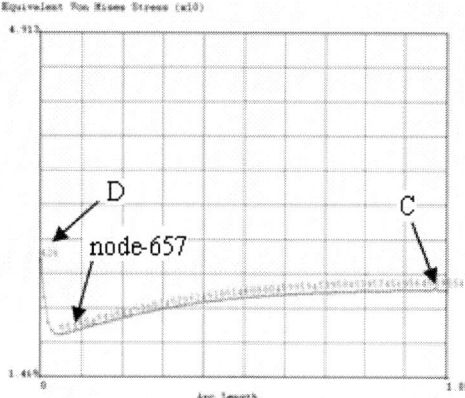

Figure18 the thermo-mechanical stress of path D-C

Figure18 shows the thermo-mechanical stress of path D-C. We can observe the stress in Fig18 have the similar trend with Fig17. Because the nodes in the path D-C has the same lateral scales as nodes in the path A-B. But the stresses of the path D-C are lower than that of path A-B.

6. Conclusions

(1)The results of humidity distribution show that DA have a very good ability of moisture expansion, and in the flow welding process, the moisture can expand into the inner of the SIP component ,even the DA.

(2)The results of FEM analysis shows that the max Hygro-mechanical stress appears in the corner of the interface between the Die and the DA), and the corner may be the risk location. The thermal stress appears in the SIP aggregate in the Pad, Die and DA. And the Hygro-mechanical is stronger than thermal stress.

(3) The results of FEM analysis shows that the max thermal stress appears in the corner of the interface between the Die and the DA too. And risk location may occur in the corner between the active Die and DA. The thermal stress appears in the SIP aggregate in the die Pad and Lead frame, Die and DA. But the thermal stress is smaller than moisture stress.

References

1. KANG Xuejing, The Reliability Analysis of Thermo-Mechanical/Hygro-Thermal Stress in System in Package.(Guilin,2007), pp. 17
2. Charles A.Harper.Electronic Materials and Process Handbook.(Beijing: 2006), pp. 545-546
3. Kailin Liu, Huiqin Ling, Ming Li, Dali Mao"Preparation of Microcones Array Material for Microelectronic Package". 2007 8th International Conference on Electronics Packaging Technology. Shanghai, China, August 14~17, 2007, pp.456
4. M. J. Adamson, Thermal expansion and swelling of cured epoxy resin used in graphite/epoxy composite materials, J. Mater. Sci., vol. 15, pp. 1736–1745
5. X.J. Fan, G.Q. Zhang, L. J. Ernst, A Micro-mechanics Approach for Polymeric Material Failures in Microelectronic Packaging. 3rd. Int. Conf. on Benefiting from Thermal and Mechanical Simulation in Micro-Electronics, EuroSIME2002, pp. 154-164.
6. LUO Haiping, The Study of Thermal and Hygroswelling Stress and Reliability Analysis during Lead-Free Reflow(Guilin,2006), pp.18

Influence of Crystal Orientation on the Oxidation Failure of Copper for IC Package

Jie Gao, Anmin Hu, Ming Li, Dali Mao

Lab of Microelectronic Materials & Technology, School of Material Science and Engineering, Shanghai Jiao Tong University, Shanghai 200240, China

Email: huanmin@sjtu.edu.cn, Tel: +86-21-34202742

Abstract

The effect of crystal orientation on the oxidation failure of pure copper was investigated. XRD results indicated that the oxide film grown on copper surface was mainly composed of Cu_2O. The adhesion strength between Cu(110) and its oxidization film was the highest, whereas, the adhesion strength between Cu(311) and its oxidization film was the lowest. SEM observations revealed that the oxide film grown on Cu(311) delaminated from substrate, while the oxide film grown on Cu(100) and Cu(110) did not reveal such a phenomenon. The oxidation rate was investigated by measuring oxide film thickness using the cathodic reduction method. The thickness of oxide film grown on Cu(100) and Cu(110) was thinner than those on Cu(311) and Cu(111). The activation energy for film growth on Cu(100) was calculated to be the highest while that on Cu(311) was the lowest.

1. Introduction

Since the entry of the information age, the integrated circuit has been gearing towards high density and high performance. Copper and its alloy becomes the most important materials of lead frame because of excellent conductibility and thermal conductivity. But due to its high affinity with oxygen, oxide film will easily form on the surface of lead frame during the storage period or in the heating processes of packaging. The existence of copper oxide will lower the adhesion strength between lead frame and Epoxy Molding Compound (EMC), thus leading to crack and delamination in package[1,2]. Main reason for the low adhesion of lead frame copper alloy with EMC is low oxide film adhesion to copper alloy substrate [3-5].

Many factors have an effect on the bonding strength between oxide film and copper substrate. The optimum oxide thickness to obtain maximum adhesion strength in Cu-base alloy is placed between 20 nm and 30 nm, and it rapidly decrease when the film thickness exceed 20nm[6,7]. High ratio of CuO/Cu_2O will decease the adhesion strength between oxide film and substrate[8]. Kang T.J. [9] found that the adhesion strength became the highest at a CuO/Cu_2O range of 0.2-0.3 regardless of alloy composition and oxide thickness. Adhesion strength also varies with different kinds of copper alloy[8]. It is considered that the alloy elements in the substrate lead to the difference in adhesion strength.

Our laboratory found that copper substrate with different orientations had different oxidation behaviors[10]. In this study, we studied on the oxidation behaviors of single crystal copper with different orientations to investigate the effect of crystal orientation on the oxidation failure of copper.

2. Experimental

The oxidation variation on copper surface with 4 different crystal orientations including Cu(100), Cu(311), Cu(110) and Cu(111), was investigated through researching the composition, thickness, morphology of oxide film grown on the substrate with different orientation and its adhesion strength to copper substrate to find out the effect of the orientation of copper on the oxidation failure of copper.

In order to remove the remaining organic substance and oxide on the sample surface, each sample was firstly cleaned by electrolytic oil removal process for 20 seconds and then pickled by 20% sulfuric acid for 30 seconds. Samples were heated on a flat-bottomed oven with atmosphere in given conditions. Afterwards the following tests were made to characterize some oxidation behaviors of these samples.

To investigate the crystal orientation of copper substrate and composition of oxide film, XRD analysis was employed. The adhesion strength between copper substrate and its oxidization film was investigated by peeling test. The samples were oxidized at different temperature for 10 minutes on electric hot plate, and then taken peeling test. The copper lead frame after oxidation was attached to adhesive tap. Then peel the tap with average forces. We observed the morphology of the oxide on copper surface using the scanning electronic microscope (SEM). The thickness of the oxide layer was measured with CHI660B electrochemical analyzer by using the cathodic reduction method. The cathode was cupric oxide plate; anode was inert platinum electrode, as showed in Fig.1. Oxidized samples were reduced by a constant currency of $0.5mA/cm^2$ in the 0.1mol/L KCl solution and the oxide film thickness was calculated from the quantity of electricity needed to complete the whole reduction reaction.

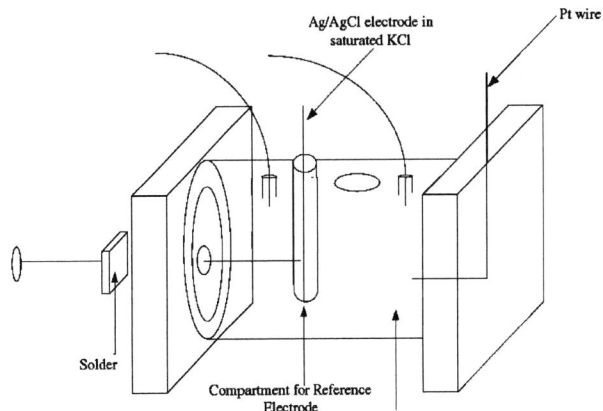

Fig.1 Device for oxide thickness measurement by cathodic reduction method.

3. Results & Discussion

3.1 The orientations of copper with different orientations

The orientations of copper were analyzed by XRD, as showed in Fig.2. The diffraction peak of each sample was strong and the result indicated the orientations of these copper substrate are Cu(100), Cu(311), Cu(110) and Cu(111) respectively.

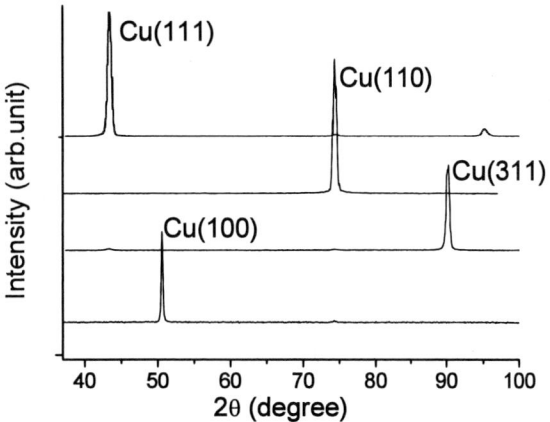

Fig.2 XRD patterns of copper with different orientations.

3.2 Composition of oxide on copper substrate with different orientations

The composition and structure of copper oxide are important factors leading to the oxidation failure of copper. Fig.3 was the XRD result of copper which was oxidized at 280°C for 180 minutes. The diffraction peak of Cu_2O can be found at $2\theta=36.521^\circ$ and 42.423°。 The CuO diffraction peak was observed at $2\theta=38.754^\circ$, but the it was weak and close to the diffraction peak of Cu_2O and Cu so that it was hard to recognize it in the copper oxide film through XRD analysis. As a result, the oxide film of copper of the four orientations was found to be mainly composed of Cu_2O. Early research by Cho, S. J.[8] revealed that the structure of copper oxide was CuO/Cu_2O/Cu, and mainly consisted of Cu_2O. Tomioka, Y. [4] found that there was no CuO in oxide film of pure copper. Our results accord with their research to a large extent.

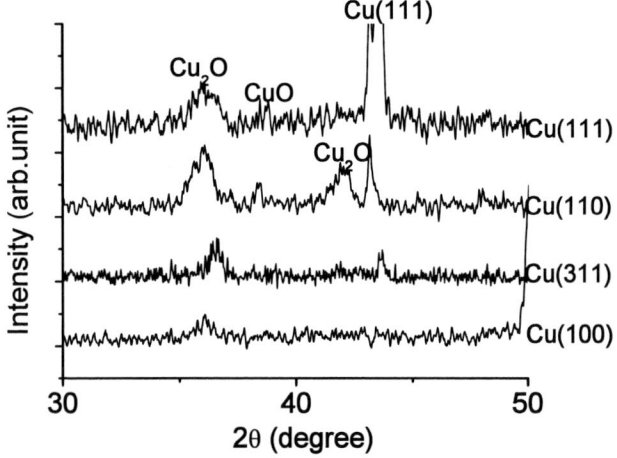

Fig.3 XRD patterns of copper substrate with different orientations oxidized at 280°C for 180 minutes.

3.3 Effect of orientations of copper on adhesion strength of oxide film

All samples were oxidized for 10 minutes from 240°C to 300°C. Fig.4 showed comparison of adhesion strength of substrates with different orientations to their oxidization film. The oxide film grown on all samples began to peeled off at 280°C. The oxide film on Cu(311) peeled off totally both at 280°C and 300°C while that on Cu(110) just peeled off partly at both temperature. The adhesion strength of Cu(100) and Cu(111) to their oxidization film were similar because the oxide film grown on them both began to peeled off totally at 300°C. Therefore, it can be concluded that the adhesion strength between Cu(110) and its oxidization film was the highest, whereas, the adhesion strength between Cu(311) and its oxidization film was the lowest.

	Cu(100)	Cu(311)	Cu(110)	Cu(111)
300℃				
280℃				
260℃				
240℃				

Fig.4 Peeling test for copper with different orientations oxidized at various temperatures for 10 minutes.

3.4 SEM images of oxide film on copper substrate with different orientations

SEM images of oxide film of copper are presented in Fig. 5 and Fig. 6 after oxidized at 280°C for 90 minutes. It can be seen from Fig.5 that the oxide film grown on Cu(100) and Cu (110) surface was smooth and coalescent to the copper substrate. But the oxide film on Cu(311) delaminated from the substrate seriously and the same condition can be found on Cu(111). The delamination of the oxide from copper substrate will lead to oxidation failure of lead frame of copper alloy.

Fig.6 was the morphology of oxide film on Cu(100) and Cu (311) under SEM 50000×. The oxide grains on Cu(100) were smaller and more compact than those on Cu(311). The size of oxide grains on Cu(110) and Cu (111) were between those on Cu(100) and Cu (311). It is obvious that the grains of oxide formed on copper will exert great impact on the adhesion strength between oxide film and copper substrate. Small and the compact oxide grains will account for strong adhesion strength between substrate and oxide film.

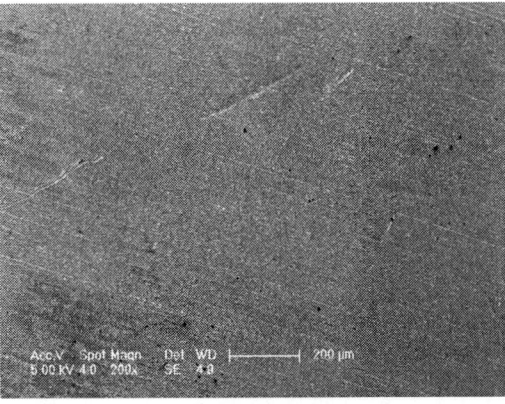

(a) Cu(100)

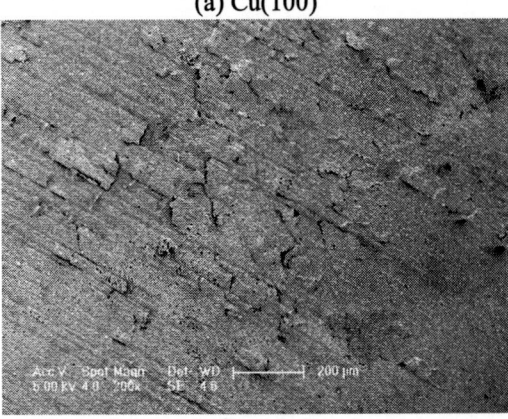

(b) Cu(311)

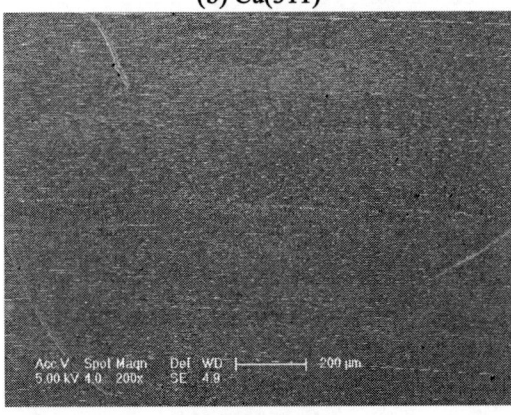

(c) Cu(110)

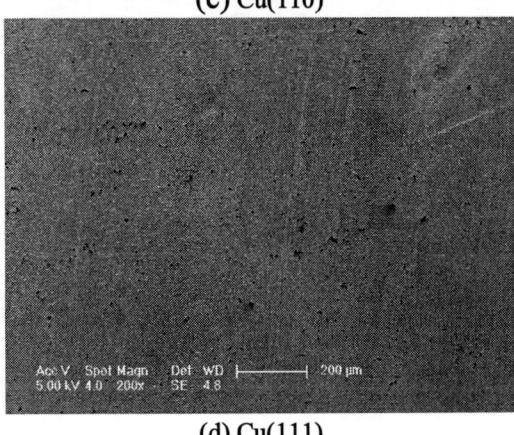

(d) Cu(111)

Fig.5 SEM images of copper substrate with different orientations oxidized at 280°C for 180 minutes.

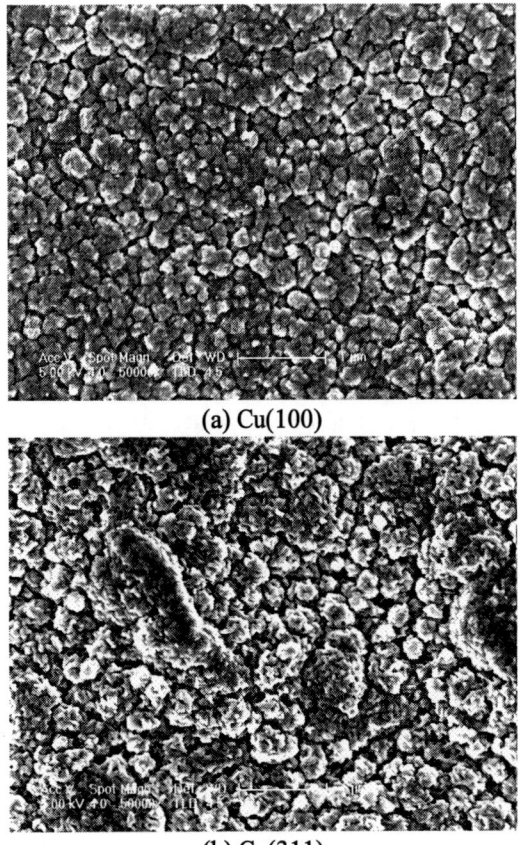

(a) Cu(100)

(b) Cu(311)

Fig.6 SEM images of copper substrate with different orientations oxidized at 280°C for 180 minutes.

3.5 Thickness of copper oxide formed on copper substrate with different orientations

The oxidation rate was investigated by measuring oxide film thickness using the cathodic reduction method. Two groups of copper samples were oxidized at different temperature for 5 minutes and 10 minutes. Fig.7 was the result of samples oxidized for 5 minutes. According to the results, the thickness of the oxide film grown on copper with all orientations rises with the increase of the temperature and the thickness of the oxide film held a linear relation with temperature. The thickness of oxide film grown on Cu(100) and Cu(110) is thinner than those on Cu(311) and Cu(111) through comparing the testing results of the different copper orientations. Another group of samples was oxidized for 10 minutes. The results revealed the similar results.

3.6 Oxidation kinetics of copper

In order to further insight into mechanism of copper leadframe oxidation, the experimental data from various samples were analyzed to obtain activation energy for the rate determining step.

The oxide film thickness growth is expressed by following equation[11],

$$Y=k\times\log(t)+C \qquad (1)$$

Where Y[nm] is oxide film thickness, k is rate constant and t is time. A constant, k generally obeys Arrhenius' equation whose expression is

$$k = A \times \exp(-\frac{E_a}{RT}) \qquad (2)$$

where A is frequency factor, Ea[J/mol] is activation energy, T[K] is absolute temperature and R[J/mol] is the gas constant.

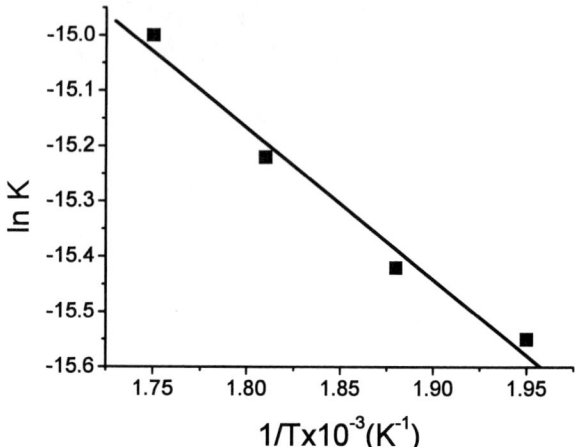

Fig.9 Arrhenius' plot of copper oxide film growth on Cu(100).

Table.1 Oxidation activation energy of copper with different orientations.

	Cu(100)	Cu(311)	Cu(110)	Cu(111)
Ea （kJ/mol）	22.84	9.31	20.47	18.32

Table.1 is the oxidation activation energy of copper with different orientations. The activation energy of Cu(100) is the highest and that of Cu(311) is the lowest. It is known that the higher the activation energy is, the faster the reaction is.

This activation energy is much lower than the activation energy of about 160 or 84 kJ/mol reported for copper oxidation by a diffusion or reaction mechanism, respectively[12]. However, in several other works on copper oxidation, the activation energy for oxygen absorption into copper from the weakly bonded surface state was reported to be about 18kJ/mol[13], which compared quite well with the activation energy determined from the present study.

Acknowledgements

We thank the support from International Cooperation & Communication Plan (No. 20073774) and Shanghai Pujian Programm (No. 05PJ4065). This work is sponsored by International Science and Technology Cooperation Program of China (ISCP) (Contact No. 2008DFA51680)

4. Conclusions

(1) The adhesion strength between Cu(110) and its oxidization film was the highest, whereas, the adhesion strength between Cu(311) and its oxidization film was the lowest.

(2) SEM observations revealed that the oxide film grown on Cu(311) delaminated from substrate, while the oxide film grown on Cu(100) and Cu(110) did not reveal such a phenomenon.

(3) The thickness of oxide film grown on Cu(100) and Cu(110) was thinner than those on Cu(311) and Cu(111). The activation energy for film growth on Cu(100) was calculated to be the highest while that on Cu(311) was the lowest.

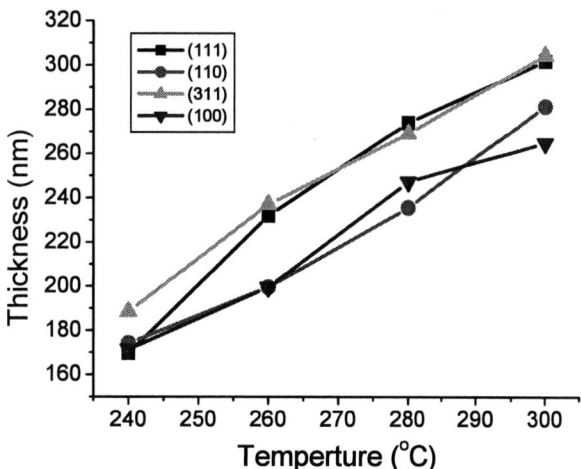

Fig.7 Oxidation thickness of copper substrate with different orientations oxidized at various temperatures for 5 minutes.

Fig.8 showed the results of the growth rate of oxide film grown on Cu(100) in our experiments. For each temperature, the slope of the line was assumed to be equal when we ploted these rate constants by logarithmic scale against inverse of absolute temperature, 1/T.

Fig.9 was an Arrhenius' plot of Cu(100) obtained from our experiment. It showed excellent linearity and activation energy, Ea, can be calculated from the slope of the plots. Similarly, activation energy of Cu(311), Cu(110) and Cu(111) can be obtained in the same way.

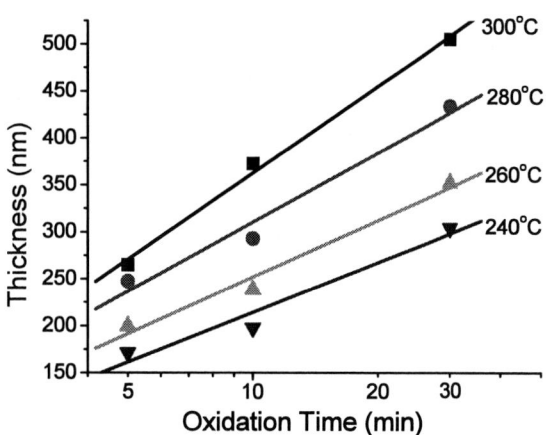

Fig.8 Growth Rate at various temperatures. The oxidation film was formed on Cu(100).

References

1. Yushioka, O., "Improvement of Moisture Resistance in Plastics Encapsulants MOS-IC by Surface Finishing Copper Lead Frame," *Proc 39th ETTC*. 1989, pp. 464-471.

2. Kim, S., "Hybrids and manufacturing technology," *IEEE transactions*, Vol.14, No. 4 (1995), pp. 809-817.

3. Huang, F. X. et al, "The status and development of oxidation of copper alloy for lead frame," *Functional Materials*, Vol.14, No. 4 (2002), pp. 29-32

4. Editorial Committee of School of Production and Technology under Chinese Institute of Electronics Microelectronic Packaging Technology, The Publishing House of University of Science and Technology of China (He Fei, , 2003), pp. 116-117.

5. Qiao, Z. Y. *et al*, "The Research Development and Frontier Questions of Lead-Free Solder," *Rare Metals*, Vol. 20, No. 2 (1996), pp. 139-143.

6. Cho, S. J., "The Effect of The Oxidation of Cu-base Leadframe on The Interface Adhesion between Cu Metal and Epoxy Molding Compound," *IEEE Transactions*, Vol. 20, (1997), pp. 167-175.

7. Takano, E. *et al*, "The Oxidation Control of Copper Leadframe Package for Prevention of Popcorn Cracking ," *Proc 47th Electronic Components and Technology Conf,* 1997, pp. 78-83.

8. Yasuo, T. *et al*, "Oxide Adhesion Characteristic of Lead Frame Copper Alloys," *Proc Electronic Components and Technology Conf,* 1999, pp. 714-720.

9. Choi, K. S. *et al*, "Electronics Packaging Manufacturing," *IEEE Transactions*, Vol. 23, No. 1 (2000), pp. 32-38.

10. Wang, N., Oxidation Failure of Lead Frame Copper Alloys, The Publishing House of Shanghai Jiaotong University (Shanghai, 2006), pp. 22-23.

11. Lahiri, S. K., *et al*, "Kinetics of Oxidation of Copper Alloy Leadframes," *Microelectronics Journal,* Vol.29, No. 6 (1998), pp. 335-341.

12. Scully, J. C., The Fundamentals of Corrosion, Pergamon Press (New York, 1990).

13. Habraken, F. H. P. M., *et al*, "The adsorption and incorporation of oxygen on $Cu(110)$ and its reaction with carbon monoxide," *Surface Science*, Vol. 88 (1979), pp. 285-298.

The Influence of Pre-heat Treatment on Peeling Resistance of Oxide Film of Copper Alloy Lead Frames

Jiwang Mao, Xi, Chen, Anmin Hu, Ming Li, Dali Mao

Lab of Microelectronic Materials & Technology, State Key Lab of Metal Matrix Composites, School of Materials Science and Engineering, Shanghai Jiao Tong University, Shanghai 200240, China

E-mail: huanmin@sjtu.edu.cn, Tel: +86-21-34202542

Abstract

The peeling resistances of oxide film formed on copper alloy lead frames with and without pre-heat treatment were investigated. With the increase of the heat treatment temperature, the peeling resistances of oxide film of C194 decrease, while C7025 show little variances in contrast. Crystal orientation of copper alloy with and without heat treatment was examined by XRD. Compared with peeling test results, it is proved that the surface crystalline orientation of cooper alloy has crucial impact on the cohesive strength between the oxide film and copper substrate. When the surface follow (111) close-packed face, the oxide film forms with better peeling resistance property, otherwise the peeling resistance will deteriorate. Furthermore, peeling resistance of oxide film has correlation with the precipitated phase.

1. Introduction

Copper alloy, with its excellent electrical prosperities and good thermal conductivity, are becoming widely used as lead frame material in high performance packages. However, the high affinity for oxygen of copper and copper alloys will lead to oxidation during packaging process including die attach and wire bond. The oxidation of copper lead frame deteriorates the adhesion between lead frame and epoxy-molding compound (EMC), usually results in delamination and crack [1, 2]. With the development of high-density IC packages and lead-free soldering, this will inevitably aggravate the reliability problem.

Many researchers have studied oxidation behavior of copper alloy lead frames. Lahiri et, al. [3] found that oxidation kinetics of copper alloys obey a logarithmic growth law at temperatures ranging from 200°C to 300°C and the copper substrate played an important role in oxidation. Mino, Toshikazu et al found that the thickness of oxide film of copper alloy should be kept under certain value to ensure the cohesion with EMC [4]. E.Takano showed when thickness of oxide film exceeds 20nm, the cohesive strength between copper alloy and EMC will drop dramatically, and the optimal thickness of oxide film will vary with the composition of alloys and heating temperature.[2]

However, many copper alloys with various compositions and crystal orientation have been developed for use as lead frames. G. Zhou et.al [5, 6] had demonstrated that the oxidation phenomena observed on copper vary with crystalline orientation and showed that the rates of oxidation were different for different crystal orientations. Hong, Shen et. al found that the the oxide layer structure of copper alloy was CuO/Cu$_2$O/Cu, and grain size and segregation of additional elements have great impact on the diffusion of copper atoms.[7]

In this study, two kinds of typical commercial copper alloys commonly used in the industry were investigated. By applying peeling test and measuring thickness of oxide film, the influence of pre-heat treatment on the oxidation behavior of copper alloy are studied.

2. Experimental Procedure

Two kinds of commercial lead frame copper alloys were supplied by Mitsui-High-Tec-Shanghai Co. Table 1 shows their chemical composition. The sample were cut into 25×30 mm and 10 × 10mm size for experiment. Each kind of samples is divided into 4 groups. 3 groups were kept in vacuum at 400°C, 600°C, and 800°C for 5 hours separately. To remove oil and oxide on the surface, all sample sheets with were first cleaned by electrolytic oil removal process for 1 minute, and then pickled in 10% sulfuric acid for 30 seconds, finally cleaned with deionized water and dry up.

These samples were subjected to oxidation process, heated at 5 different temperatures: 220°C, 240°C, 260°C, 280°C, and 300°C separately for 5 minutes on hot plate in air atmosphere.

Table 1 Chemical composition of two kinds of commercial lead frame copper alloys

Sample No.	Alloy Name	Chemical Composition
1	Wieland C194	Cu-2.3%Fe-0.12%Zn-0.07%P
2	Wieland C7025	Cu-3.0%Ni-0.65%Si-.15%Mg

To evaluate the cohesive strength between oxide film and copper alloy lead frame, peeling tests were conducted on 25× 30 mm size samples as follows. Attach the special adhesive tape (ModelE50292-Z) to the sample surface, and then remove the tape with a constant horizontal force. The peeling resistance of oxide film can be evaluated by observing the amount of oxide taken away by the tape.

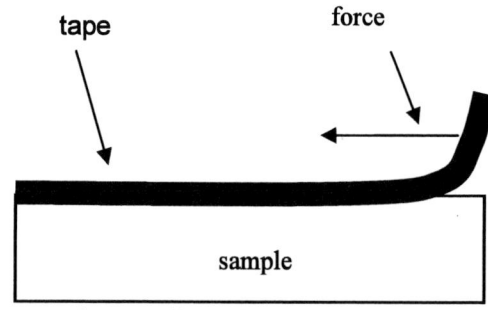

Figure 1 Illustration of Peeling test

The thickness of the oxide film was measured by ChI660B electrochemical analyzer at a constant current density of 0.5mA/cm^2 in 0.1N NaOH solution. Figure 2 shows the devise used in the experiment. The solutions were purged with nitrogen gas for 1 hour before reaction. Appling a constant current, the oxide film at the cathode will be reduced. By recording the cathode potential changed with the reaction time, the thickness of oxide film can be calculated.

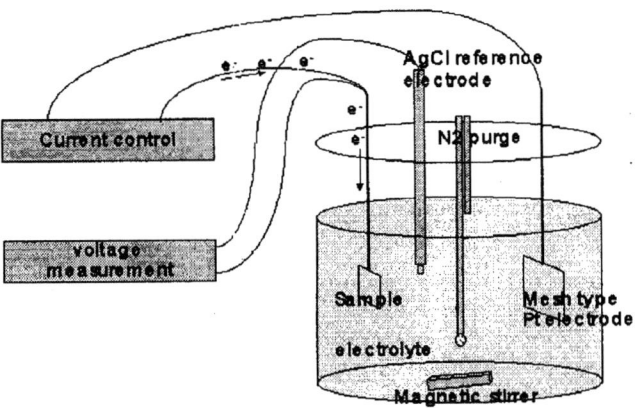

Figure 2 Illustration of cathodic reduction method

The surface morphology of oxide films were examined by Scanning electron microscope. SEM images were obtained on a Philips Sirion-200 scanning electron microscope. C194 and C7025 samples with and without pre-heat treatment were clean by acid and deion water, tested by XRD (Bruker-AXS X-ray diffraction) for crystalline phase identification.

3. Result and Discussion

3.1 Cohesive strength of oxide film forms on cooper alloy with different pre-heat treatment

Peeling tests were applied to test the cohesive strength between oxide film and substrate. Table 2 and 3 demonstrate peeling test results of oxide films formed on C194 and C7025 with different pre-heat treatment at different oxidation temperatures. Dark grey represents absolute peeled off, light grey represents partial peeled off, and white represents no or little peeled off. Cohesive strength can be evaluated by the transition point of non-peel off to peel off.

Table 2 Peeling test results of oxide films formed on C194 alloy with different pre-heat treatment at different oxidation conditions

Peeling test	25°C	400°C heat treatment	600°C heat treatment	800°C heat treatment
300°C oxidation				
280°C oxidation				
260°C oxidation				
240°C oxidation				
220°C oxidation				

Table 3 Peeling test results of oxide films formed on C7025 alloy with different pre-heat treatment at different oxidation conditions

Peeling test	25°C	400°C heat treatment	600°C heat treatment	800°C heat treatment
300°C oxidation				
280°C oxidation				
260°C oxidation				
240°C oxidation				
220°C oxidation				

Table 2 shows with the increase of heat treatment temperature, absolute peeled off phenomena happen at lower oxidation temperature. After high temperature pre-heat treatment, oxide layer formed in the following oxidation process possess poor peeling resistance. We suppose that during high temperature pre-heat treatment, the crystalline structure of C194 copper alloy changed, leading to weak cohesive strength between substrate and oxide films which formed afterwards.

Table 3 shows C7025 presented difference performance from C194. The samples with 400°C pre-heat treatment are tested to have the poorest peeling resistance. While other samples show similar performance.

3.2 Thickness of oxide film formed on copper alloy with different pre-heat treatment

Figure 3 show the oxide film thickness formed on copper alloy with different pre-heat treatment at different oxidation temperatures. The oxide film thickness of C194 and C7025 both increase with the oxidation temperature no matter with or without heat treatment. Under the equivalent pre-heat treatment and oxidation condition, the oxide films of C194 are always thicker than C7025.

As for C7025, under the same oxidation condition, the thickness of oxide film decrease with the increase of pre-heat treatment temperature. That means higher heat treatment temperature is more effective on reducing the thickness of oxide film which formed in the following oxidation process. As for C194, heat treatment at 600°C shows the biggest influence on the thickness reduction, followed by 400°C. However, the 800°C heat treatment has the reverse impact, which results in a remarkable increase of oxide film thickness. Compared with the peeling test results, the peeling resistance property shows slight change after heat treatment. It can demonstrate that the oxide film thickness has no necessary relation with the adhesive strength. .

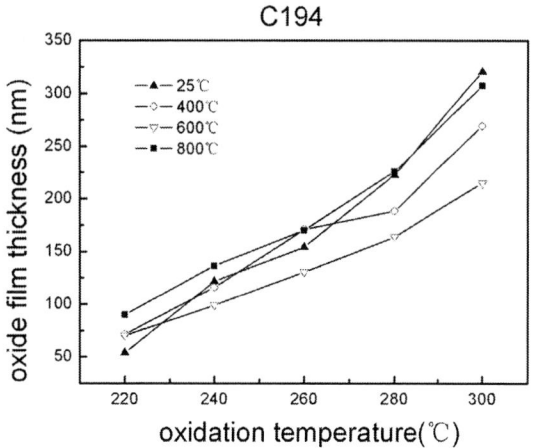

C194

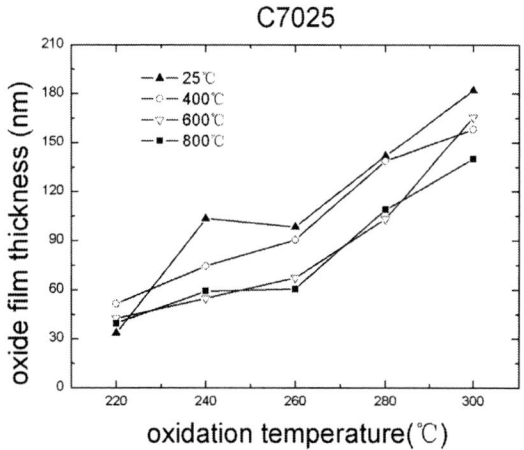

C7025

Figure 3 oxide films thickness of C194 and C7025 with different pre-heat treatment and oxidation conditions

3.3 Comparison of oxide film morphology

Figure 4 shows the surface morphology of C194 after oxide at 280°C. Sample (b) went though pre-heat treatment at 800°C while (a) did not. Without pre-heat treatment, the oxide film surface shows coarse grain size, where no peeled-off phenomena captured. In contrast, with pre-heat treatment, the grain of oxide film become fine and compact, large area of film peeled off from the surface, which can explain well with the result in Table 2.

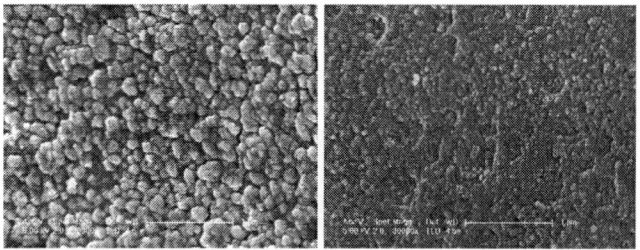

Figure 4 surface morphology of C194 oxidation at 280°C
(a) without heat treatment (b) with 800°C heat treatment

Fiure 5 shows the surface image of C7025 after oxidation at 280°C. (b) went though pre-heat treatment at 800°C while (a) did not. Different from C194, the grain size of the pre-heat

treatment sample slightly changed, which is coherent with the result in Table 3. In particular, large amount of voids were found dispersing on the oxide film of C7025 alloy with pre-heat treatment like Figure 5(b). To some extent, the existing of voids relief the internal stress in the oxide film and is favorable for good peeling resistance.

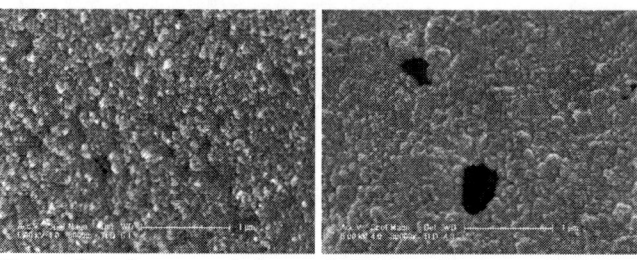

Figure 5 surface morphology of C7025 oxidation at 280°C
(a) without heat treatment (b) with 800°C heat treatment

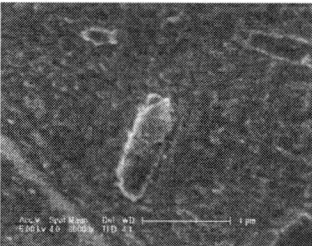

Figure 6 C7025 with 800°C heat treatment oxidation at 280°C

Figure 6 shows the surface of the cooper substrate exposed after peeled off the oxide film. The sample is C7025 oxides at 280°C with 800°C pre-heat treatments. The obvious existence of second phase is observed, which should be the segregation of the trace element in the cooper alloy. When the segregation occurs at the surface, voids will be left at the corresponding spot after oxidation. The relation of voids and segregation is proved to have correspondence.

3.4 Comparison of crystalline structure of copper alloy with different pre-heat treatment

Figure7 shows the XRD result of C194 copper alloy before and after heat treatment. Before heat treatment, there are three obvious copper diffraction peak (111), (200), (311), among which the (311) peak is the most prominent. After 400°C heat treatment, the (200), (311) peak drop significantly to rather low value, while the (111) peak intensified. After 600°C heat treatment, the three peak values are all low. After 800°C heat treatment, the (111), (200) peak nearly faint, at the same time, the (311) peak intensified, the (311) lattice orientation reappear.

Comparing with the oxide film thickness test in Figure3, when the (311) orientation is the dominating orientation in the crystal structure (the sample without heat treatment and go through 800°C heat treatment), the oxide film forms after oxidation process will be thicker. It account for that the oxidation of copper has intrinsic relation with crystal orientation. Copper is FCC structure, among (111), (200), (311) crystal face, (311) face is the least close-packed planes with low atom density, which is favorable for diffusion. Mass data had proved that the oxidation process of copper is via diffusion of cooper atoms. This test result is in accordance with data in literature. [7]

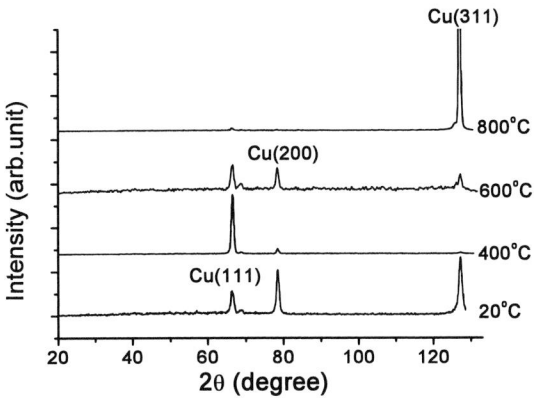

Figure 7 XRD result of C194 alloy with different heat treatment

Figure 8 shows the XRD result of C7025 copper alloy before and after heat treatment. C7025 is quite different from C194. Although peak value decrease as the increase of the heat treatment temperature, (111) orientation dominate before and after heat treatment, which mean the crystal structure and crystal orientation has no big change after heat treatment. After 400°C heat treatment, (200) and (311) orientation occurred. As we can see from Table 3, the peeling resistance of oxide film formed on C7025 alloy with 400°C pre-heat treatment is the poorest, however when orientation is (111), the peeling resistance remains good. Thus the relation between crystal orientation of alloy and peeling resistance of its oxide film is proved again.

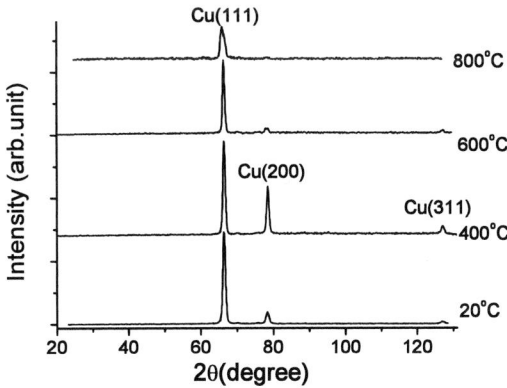

Figure 8 XRD result of C7025 alloy with different heat treatment

Conclusions

Two commercial copper alloy lead frame material Wieland C194 (Cu-Fe-P) and C7025(Cu-Ni-Si) were investigated. The oxidation behavior of the copper alloy before and after heat treatment is examined. The influence of heat treatment temperature, composition of alloy and crystal structure on the peeling resistance property is discussed.

The peeling tests show that the peeling resistance of oxide film on the Cu-Fe-P alloy C194 drop after pre-heat treatment, while the performance of Cu-Ni-Si alloy shows little variance.

Under the equivalent pre-heat treatment and oxidation condition, the oxide films of C194 are always thicker than C7025. The different composition and crystal structure should be the cause of different growth rate of oxide film.

SEM result show that after heat treatment at 800°C, precipitated phase forms in C7025 alloy, result in large amount of voids in the oxide film. To some extent, the existing of voids relief the internal stress in the oxide film and is favorable for good peeling resistance.

XRD result show the C7025 has little crystal structure and orientation change after heat treatment, whereas C194 changed dramatically. Especially after 800°C heat treatment, the dominating (311) peak reveal that the mass texture occurs in the bulk alloy. Since the difference in crystal orientation will lead to different copper atoms diffusion rate, oxide film growth rate and cohesive strength between substrate and oxide film, it could explain the deterioration of peeling resistance of oxide film after heat treatment.

Acknowledgments

The author would like to thank Instrumental Analysis Center of Shanghai Jiaotong University for SEM and XRD analysis. This work is sponsored by International Science and Technology Cooperation Program of China (ISCP) (Contact No. 2008DFA51680).

References

1. Byung Hoon Moon, Hee Yeoul Yoo, et al. "Optimal Oxidation Control for Enhancement of Copper Lead Frame-EMC Adhesion in Packaging". *Pro Electronic Components and Technology Conf,* 1998, pp.1148-1153.
2. Eiji Takano, Toshikazu Mino, et al, "The Oxidation Control of Copper Leadframe Package for Prevention of Popcorn Cracking" .*Proc of Electronic Components and Technology Conf,* 1997, pp.78-83.
3. Lahiri S.K., Singh N.K., Waalib Heng K.W. et al. "Kinetics of Oxidation of Copper Alloy Leadframes" *Microelectronics Journal.* Vol.29, No.6,(1998),pp.335-341.
4. Toshikazu M, Kanako S, Atsushi K, et.al. "Development of Moisture-proof Thin and Large QFP with Copper Lead Frame". *Pro. Electronic Components and Technology Conf.* 1998:1125-1131.
5. Guangwen Zhou , Judith C. Yang, "In situ UHV-TEM investigation of the kinetics of initial stages of oxidation on the roughened Cu(110) surface". *Surface Sci* 559 (2004) ,pp.100－110
6. Guangwen Zhou, Judith C. Yang, "Initial oxidation kinetics of copper (110) film investigated by in situ UHV-TEM," *Surface Sci,* Vol. 531 (2003) ,pp.359－367
7. Hong Shen, et. al, "Oxidation Failure Mechanism of Copper Alloy Lead Frame for IC Package," *Proc. 6th International Conf on Electronic Packaging Technology*, 2005, pp. 314-320

Interfacial Reactions and Reliability of Sn-Zn-Bi-XCr Solder Joints with Cu Pads

Yidong Shen, Anming Hu, Xi Chen, Ming Li, Dali Mao

Lab of Microelectronic Materials & Technology, School of Materials Science and Engineering,
Shanghai Jiao Tong University, Shanghai 200240, China
Email: huanmin@sjtu.edu.cn, +86-21-34202542

Abstract

In this paper, the interfacial reactions of Sn-8Zn-3Bi and Sn-8Zn-3Bi-0.3Cr solders with Cu pads and the growth of IMCs during long-term aging treatment were presented and discussed. Less than 3μm Cu_5Zn_8 IMC layer was found at the Sn-8Zn-3Bi-(0.3)Cr/Cu during the reflow at 230℃ for 1 min. During solid state aging treatment at 150℃ for 4, 9, 16, 25 days, the Cu_5Zn_8 IMC grown up with the extension of aging time, some of the IMCs were scattered into the bulk solders, Zn-poor layer was observed in the Sn-8Zn-3Bi solder joints. In the case of Cr-containing solders, the thickness of Cu_5Zn_8 IMC growth was much thinner than that of in the Sn-8Zn-3Bi solder joints, and no significant Zn-poor layer was found in Cr-containing solders. The growth rate of IMCs at the Sn-8Zn-3Bi-0.3Cr/Cu interface was about 1/2 times than that of in the Sn-8Zn-3Bi solder. During long-term aging, solid phase transition happened in Cr-containing solders, which indirectly and effectively controlled the diffusion of Zn atom to the interface, and thus slowed down the growth rate of Cu-Zn IMCs. It was supposed that Cr-containing Sn-Zn-Bi solders performed better long-term solder joint reliability.

1. Introduction

With the increasing concern on toxicity of lead in eutectic Sn-Pb solders, lead-free soldering has emerged as one of the critical technologies in electronic packing industry. Compared to many other promising lead-free solders, such as Sn-Ag, Sn-Ag-Cu, Sn-Cu, Sn-Bi, the eutectic Sn-Zn solders are widely recommended because of its low melting point, good mechanical properties and low cost [1-4]. However, Zn is easy to be oxidized and decrease wetting properties during reflowing. Recent studies showed that adding a small amount of Bi into Sn-Zn alloys increased the wettability [5]. Nevertheless, it is reported that excessive amount of Bi can degrade the mechanical properties of Sn-Zn solders for the brittleness of Bi. Bi also may decrease the oxidation resistance of Sn-Zn solders [6]. According to previous research, a small amount of Cr could refine the microstructure and enhanced ductibility, meanwhile increased the oxidation resistance of Sn-Zn solders [7].

During the reflow soldering, the solders melts and reacts with pads to form intermetallic compounds (IMCs) at the joint interface. The IMC growth during solid aging in solder joints is of particular interest to the electronic packaging industry. Excessive thickness of the IMC layers at the solder joints can significantly degrade the physical and mechanical properties of the solder joints, particularly in high impact load environment [8-10]. However, the metallurgical behaviors of Cr doped Sn–Zn–Bi solder in joints with different surface finishes and the related joints reliability at package level have not been sufficiently studied yet. In this paper, the study was

carried out to discuss the effect of adding 0.3 wt.% Cr on Sn-9Zn-3Bi solders on the interfacial reactions and reliability during aging treatment with Cu metallization.

2. Experiment procedures

In this study, Sn-8Zn-3Bi, Sn-8Zn-3Bi-0.3Cr were selected as the experimental solders. The alloys were initially melted within induction furnace in low vacuum at 500℃ over 1 h. The compositions of alloys were investigated by Inductively-coupled plasma emission spectrometer (ICP-AES).

Both Sn-8Zn-3Bi and Sn-8Zn-3Bi-0.3Cr solder spheres, weighed 0.1 grams, were placed on the pre-fluxed Cu pads and reflowed at the temperature of 230℃ for 1 min in a convection reflow oven. The bare Cu pads were cleaned in a dilute solution of sulfuric acid and then rinsed with clean water to remove the surface oxide. Then the as-reflowed solders were subjected to isothermal aging treatment at 150℃ for 1, 4, 9, 16, 25 days respectively.

To investigate the microstructures of the solder joints, the as-reflowed and aged samples were cold mounted in epoxy resin, orderly grounded with #320, 800, 1200, 1500, 2000 SiC paper and mechanically polished with 0.5μm diamond powder. After that, the samples were slightly etched with a solution of 2 vol.% nitric acid + 98 vol.% alcohol in order to obtain the cross-sections of the solder/pad interfaces.

The back-scattered electron (BSE) imaging mode of the scanning electron microscope (SEM) was used for characterizing the joint microstructures and IMC growth during reflow soldering and aging treatment. The composition of the IMCs were inspected and identified by the energy dispersive spectroscopy (EDS) analysis.

3. Results and discussion

3.1 Microstructure of as-reflowed solder joints

Fig.1 represents the BSE microstructure of the interface between Sn-8Zn-3Bi-(0.3) Cr and Cu pads after 1 min reflow. Less than 3 μm thickness of IMC layers were found at the interface of Sn-8Zn-3Bi/Cu and Sn-8Zn-3Bi-0.3Cr/Cu joints. EDS analysis confirmed the IMC layer was γ-Cu_5Zn_8. Suganuma et al. [11] also reported that instead of Cu-Sn, the Cu-Zn intermetallics, mainly Cu_5Zn_8, formed at Sn-Zn based solders with Cu pads. Very fine Bi precipitates (extremely white in the BSE image as Fig.1 shows) were dispersed in the solders. Bi didn't take part in the interfacial reactions nor formed any IMCs with Cu [12]. In the case of Cr-containing solders (Fig.1b), some Cr-rich phases were formed in the Sn-8Zn-3Bi-0.3Cr solders, and these particle Cr-rich phases were evenly scattered in the β-Sn matrix.

978-1-4244-2739-0/08/$25.00 ©2008 IEEE

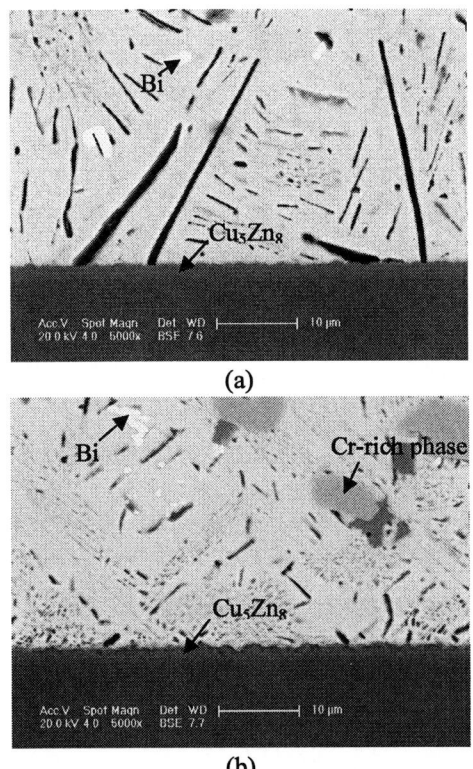

(a)

(b)

Fig.1.Microstructure of as-reflowed solder joints: (a) Sn-8Zn-3Bi/Cu interface, (b) Sn-8Zn-3Bi-0.3Cr/Cu interface.

3.2 Microstructure evolution in solder joints during long-term aging treatment

Fig.2a, 2c, 2e show the microstructure evolution in Sn-8Zn-3Bi solder joints with Cu pads during isothermal aging treatment at 150℃ for 4, 9, 25 days, respectively. Fig.3 shows the EDS line analysis of each element across the Sn-8Zn-3Bi/Cu interface during 25 days' aging. The Cu-Zn IMC at the interface was determined to be Cn_5Zn_8. No Bi was found at the interface which indicated that Bi didn't take part in any interfacial reactions during aging treatment.

After 4 days' aging (Fig.2a), very small size of Cn_5Zn_8 IMCs were found in the bulk solders, and these scattered Cn_5Zn_8 IMCs grown up and the numbers increased with the extension of aging time (Fig.2c, 2e). During long-term aging, Zn atom continuously diffused into Cu sublattice at the interface and a layer of Zn-poor area appeared near the interface in the bulk solders. This Zn-poor layer expanded with the aging time and the thickness of the Zn-poor layer was up to more than 150 μm after 25 days' aging (Fig.2e).

Fig.2b, 2d, 2f show the microstructure evolution in Sn-8Zn-3Bi-0.3Cr solder joints with Cu pads during long-term aging at 150℃ for 4, 9, 25 days, respectively. The thickness of Cn_5Zn_8 IMCs in Cr-containing solder was much thinner than that of in the Sn-8Zn-3Bi solder joints. Some Cr-rich phases were observed in the bulk solders and with the extension of aging time, the number of these Cr-rich phases increased. In addition, no significant Zn-poor layer appeared near the interface in Cr-containing solders.

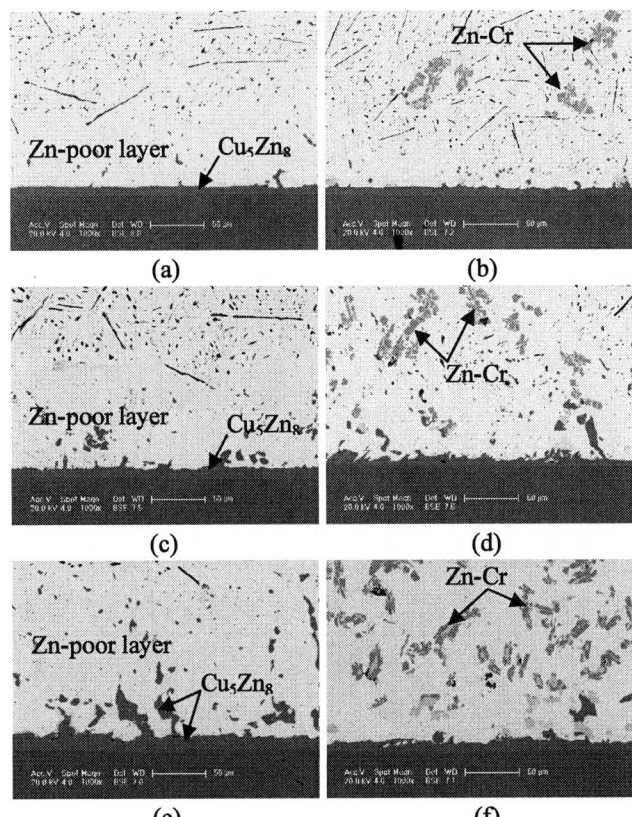

(a)   (b)

(c)   (d)

(e)   (f)

Fig.2. Microstructures evolution in Sn-8Zn-3Bi solder joints with Cu pads during isothermal aging at 150℃ for (a)4, (c)9, (e)25 days and in Sn-8Zn-3Bi-0.3Cr solder joints during isothermal aging at 150℃ for (b)4, (d)9, (f)25 days.

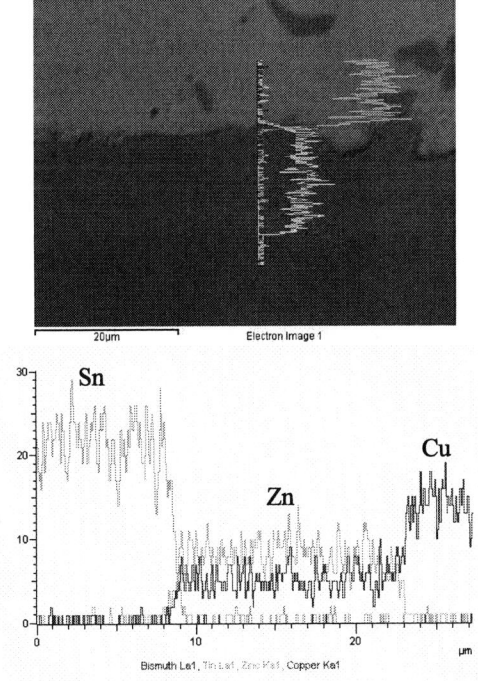

Fig.3. EDS line analysis of Sn-8Zn-3Bi/Cu interface during aging at 150℃ for 25 days.

3.3 Intermetallic growth

The intermetallic growth during aging treatment was observed and measured. Fig.4 represents the Cu-Zn IMCs

745

growth of Sn-8Zn-3Bi-(0.3)Cr solder joints with the extension of aging time. The Cn-Zn IMC at the Sn-8Zn-3Bi/Cu interface was much thicker than that of Cr-containing solder joints with the extension of aging time which indicated Cr can significantly restrain the IMC growth in Sn-Zn-Bi solder joints during long-term aging treatment. After 25 days' aging, the IMC thickness at the Sn-8Zn-3Bi/Cu interface was more than 16.2 μm, compared to less than 11.8 μm in those of Cr-containing solder joints.

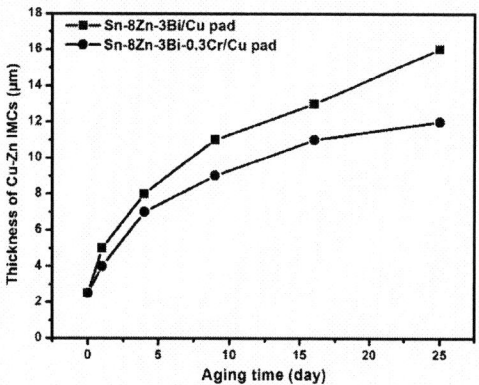

Fig.4. Cu-Zn IMCs growth of Sn-8Zn-3Bi-(0.3)Cr solder joints with the extension of aging time(day).

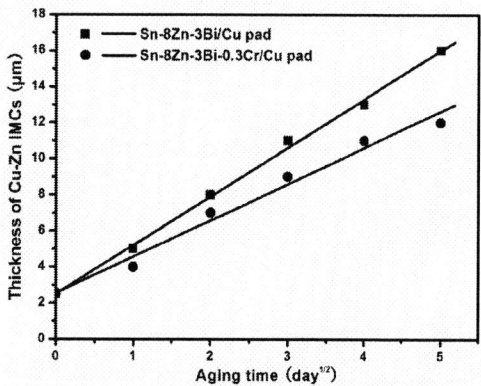

Fig.5. Cu-Zn IMCs growth of Sn-8Zn-3Bi-(0.3)Cr solder joints with the extension of aging time (day$^{1/2}$).

In general, the solid state growth of IMCs can follow a linear growth kinetics. To understand effect of Cr addition on growth rate of IMC layer between Sn-8Zn-3Bi solders and Cu pads, we estimated the IMCs growth rate of Sn-8Zn-3Bi and Sn-8Zn-3Bi-0.3Cr alloy by the following equation [13-14]:

$$X_{(t)} = X_0 + At^n \exp(-Q/RT) \qquad (1)$$

where t is the aging time, $X_{(t)}$ is the IMCs thickness at t time, X_0 is the initial IMCs thickness as-soldered, Q is the activity energy, A is a constant, n is the time constant. It was supposed $n \approx 0.5$, taking the consideration that the interfacial reaction between Sn-8Zn-3Bi solder and Cu pad was controlled by diffusion rate.

So the equation (1) can be simplified as follows:

$$X - X_0 = \sqrt{Kt} \qquad (2)$$

where X is the IMCs thickness at t time, X_0 is the initial IMCs thickness as-soldered, K is the constant of IMC growth rate.

K value can be determined by the slop coefficient in Fig. 5. K values of Sn-8Zn-3Bi and Sn-8Zn-3Bi-0.3Cr solders at 150°C were 1.90×10^{-16} m^2/s and 9.6×10^{-17} m^2/s, respectively.

This result indicated that growth rate of IMCs in Sn-8Zn-3Bi-0.3Cr solder was about 1/2 times than that of Sn-8Zn-3Bi solder.

3.4 Microstructure evolution in Cr-containing solders

In most cases, the thick IMC growth degraded interface integrity and reduced the solder joint reliability, owing to the brittle nature of IMCs and also the mismatches in physical properties such as thermal expansion coefficient and elastic modulus. Thus it was quite reasonable to control the interface microstructure within the optimized conditions. According to the above study, adding a small amount of Cr (about 0.3 wt.%) can significantly restrain the IMC growth at the interface during long-term aging treatment. It was supposed that the reason Cr addition can decrease growth rate of IMCs at the interface of Sn-8Zn-3Bi-0.3Cr/Cu was because of the solid phase transition in Cr-containing solder during long-term aging treatment.

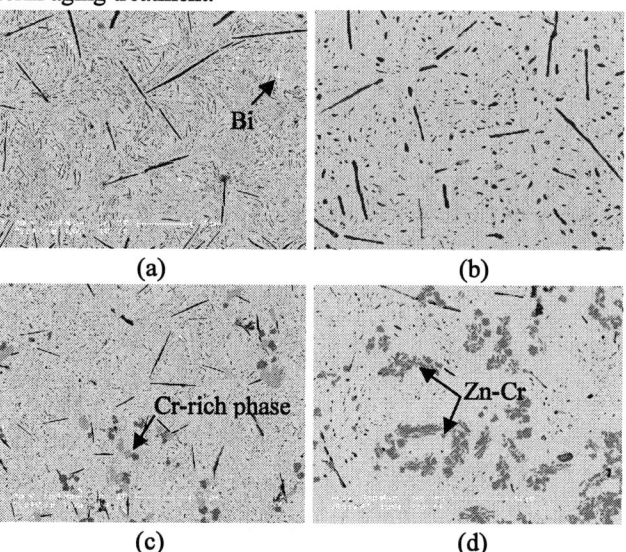

Fig.6. Microstructure in Sn-8Zn-3Bi and Sn-8Zn-3Bi-0.3Cr solders: (a) as-reflowed Sn-8Zn-3Bi solders (b) Sn-8Zn-3Bi solders after aging for 25 days (c) as-reflowed Sn-8Zn-3Bi solders (d) Sn-8Zn-3Bi solders after aging for 25 days.

Fig.6a represents the microstructure in as-reflowed solders and the microstructure of the Sn-8Zn-3Bi solder consists of a typical Sn-Zn eutectic region with a primary Sn-phase, some needle-like Zn-rich phase, and very fine Bi precipitates were dispersed in the β-Sn matrix. After 25 days' aging as shown in Fig.6b, the Zn-rich phase decreased and Bi was not observed in bulk solders, it was supposed that Bi might been completely solid dissolved into β-Sn matrix during long-term aging because the Sn-Bi phase diagram shows that the solubility of Bi in Sn is up to 20 wt.% at 150°C [12].

Fig.6c shows the microstructure in Sn-8Zn-3Bi-0.3Cr solder. Compared to Fig.6a, the Zn-rich phase was much finer which indicated Cr can refine the microstructure in the Sn-Zn based solders. Fig.6d represents the microstructure in Cr-containing solder after aging for 25 days. Compared with the as-reflowed solders (Fig.6c), the microstructures are quite different. Before the reflow, the microstructure was a Sn-Zn eutectic region with a primary Sn-phase, some needle-like Zn-rich phase. Very fine Bi precipitates and some particle Cr-rich phase were evenly dispersed in the β-Sn matrix. Fig.7a shows the EDS analysis of Cr-rich phase in as-reflowed solders, it has a composition of 68-70 Sn and 30-32 Cr (at.%). After aging for 25 days (Fig.6d), the microstructure became a primary β-Sn phase with some petal-like Cr-rich phase and a few of Zn-rich phase. EDS showed this Cr-rich phase had a composition of 20-22 Sn, 66-68 Zn, 10-12 Cr (at.%) (Fig.7b). It implied that during the isothermal aging, Zn involved in the solid phase transition in bulk solders with Cr, transforming Sn-Cr phase in as-reflowed solders into Sn-Zn-Cr phase. It was reported that the growth rate of IMCs should be determined by the diffusion of Zn atom into the Cu sublattice at the interface [11]. This transition had a vital impact on IMC growth for it had indirectly decreased the diffusion of Zn to the interface, thus slowed down the growth rate of Cu-Zn IMCs.

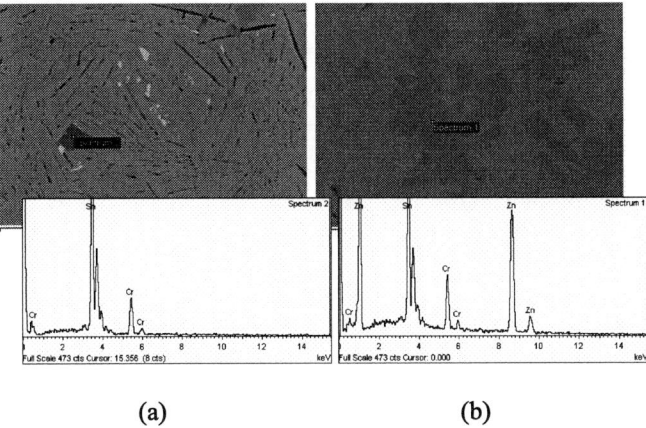

(a)   (b)

Fig.7. EDS analysis of Cr rich phase: (a) as-reflowed, (b) after aging for 25 days.

4. Conclusions

Less than 3μm Cu_5Zn_8 IMC layer was formed at the Sn-8Zn-3Bi-(0.3)Cr/Cu during the reflow at 230℃ for 1 min. During the solid state aging at 150℃, the Cu_5Zn_8 IMCs grown up with the extension of aging time. Some Cu_5Zn_8 IMCs were scattered into the bulk solders, Zn-poor layer was observed in the Sn-8Zn-3Bi solder joints. In Cr-containing solders, the thickness of Cu_5Zn_8 IMC growth was much thinner than that of Sn-8Zn-3Bi solder joints, and no significant Zn-poor layer was found in Cr-containing solders. The growth rate of IMCs at the Sn-8Zn-3Bi-0.3Cr/Cu interface was about 1/2 times than that of Sn-8Zn-3Bi solder. During long-term aging, Cr-rich phase reacted with Zn and transformed Sn-Cr phase into Sn-Zn-Cr phase in the bulk solders. This solid phase transition indirectly controlled the diffusion of Zn atom to the interface, and slowed down the growth rate of Cu-Zn IMCs. It was supposed that this could be the main reason for low IMC growth rate in Cr-containing solders during aging treatment.

Acknowledgments

This work is sponsored by China International Science and Technology Cooperation (Contact No. 20073774). We thank the Instrumental Analysis Center of Shanghai Jiao Tong University, for the use of the SEM equipment. This work is sponsored by International Science and Technology

Cooperation Program of China (ISCP) (Contact No. 2008DFA51680).

References

1. K. Suganuma, "Advances in lead-free electronics soldering", *Curr. Opin. Solid State Mater. Sci.*, Vol. 5 (2001), 55-64.
2. H. J. Lau, C. P. Wong, N. C. Lee, and S. W. Ricky Lee, Electronics Manufacturing with Lead-Free, Halogen-Free & Conductive-Adhesive Materials (New York: McGraw-Hill Handbooks, 2003).
3. T. Ichitsubo, E. Matsubara, K. Fujiwara, M. Yamaguchi, H. Irie, S. Kumanoto, T. Anada, "Control of compound forming reaction at the interface between SnZn solder and Cu substrate", *J. Alloy Comp.*, Vol. 392 (2005), 200-205.
4. S. K. Kang, "Lead (Pb)-free solders for electronic packaging", *J. Electron. Mater.*, Vol. 23 (1994), 701.
5. M. Abtew, G. Selvaduray, "Lead-free solders in microelectronic", *Mater. Sci. Eng.*, Vol. 27 (2000) 94-141.
6. K. S. Kim, J. M. Yang, C. H. Yu, I. O. Jung, "Analysis on interfacial reactions between Sn-Zn solders and the Au/Ni electrolytic-plated Cu pad", *J. Alloys Compd.*, Vol. 379 (2004) 314-318.
7. X. Chen, M. Li, X. X. Ren, A. M. Hu, D. L. Mao, "Effect of small additions of alloying elements on the properties of Sn-Zn eutetic alloy", *J. Electron. Mater.*, Vol. 35, No. 9 (2006) 1734-1739.
8. G. Ghosh, "Interfacial microstructure and the kinetics of interfacial reaction in diffusion couples between Sn–Pb solder and Cu/Ni/Pd metallization", *Acta Mater.*, Vol. 48 (2000), 3719.
9. C. B. Lee, S. B. Jung, Y. E. Shin, and C. C. Shur, "Effect of isothermal aging on ball shear strength in BGA joints with Sn-3.5Ag-0.75Cu solder : Lead-free electronics packaging", *Mater. Trans.*, Vol. 42 (2001), 751.
10. M. McCormack and S. Jin, "The design and properties of new,Pb-Free solder alloy", *JOM*, Vol. 45 (1993), 36.
11. Suganuma K., Niihara K., Shoutoku T., Nakamura Y., "Wetting and interface microstructure between Sn-Zn binary alloys and Cu", *J. Mater. Res.*, Vol. 13 (1998), 2859-2865.
12. D. Frear, H. Morgan, S. Burchett, J. Lau, The Mechanics of Solder Alloy Interconnects, *Van Nostrand Reinhold, New York*, (1994), 7-40.
13. M. Schaefer, R. A. Fournelle, and J. Liang, "Theory for intermetallic phase growth between Cu and liquid Sn-Pb solder based on grain boundary diffusion control", *J. Electron. Mater.*, Vol. 27, No. 11 (1998), 1167-1176.
14. S. Chada, W. Laub, R. A. Fournelle, and D. Shanguan, "An improved numerical method for predicting intermetallic layer thickness developed during the foration of solder joints on Cu substrate", *J. Electron. Mater.*, (1999) 1194-1202.

The Thermal and Electric Conduction Properties of High Dense TiB$_{2P}$/Cu Composites Fabricated by Squeeze Casting Technology

Guoqin Chen[1] Ziyang Xiu[2] Songhe Meng[3] Gaohui Wu[1] Su Chen[1]

1. School of Materials Science and Engineering, Harbin Institute of Technology, Harbin 150001, China
2. Research Academy of Science and Industry Technology, Harbin Institute of Technology, Harbin 150001, China
3.Center for Composite Materials, Harbin Institute of Technology, Harbin 150001, China
chenguoqin@hit.edu.cn, xizy@hit.edu.cn, mengsh@hit.edu.cn, wugh@hope.hit.edu.cn

Abstract

For electronic packaging applications ， TiB$_2$/Cu composites with volume fractions of 50%, 58%, 65% TiB$_2$ content have been fabricated by the patented squeeze-casting technology. The microstructures and thermal and electric conduction properties of the TiB$_2$/Cu composites are investigated. The results show that TiB$_2$ particles are homogeneous and uniformly, and the TiB$_2$-Cu interfaces are clean and free-from interfacial reaction products and amorphous layers; The densifications of the TiB$_2$/Cu composites are higher than 98.2%. The thermal conductivities of TiB$_{2P}$/Cu composites range from 167.3 to 215.4 W/m℃ and decrease with an increase in volume fraction of TiB$_2$ content. The thermal conductivities agree well with predicted values of theoretical models. The electric conductivities of TiB$_2$/Cu composites are between 33.7-43.9 %IACS and decreased with the increasing of TiB$_2$ content. The achievement of higher thermal and electric conduction is attributed to the full densities and high purity TiB$_2$/Cu composites, which are attained through the cost-effective squeeze-casting technology processes.

1. Introduction

The materials for electronic packaging and thermal management applications should have compatible coefficients of thermal expansion (CTEs) with those of semiconductors or ceramic substrates, high thermal and electrical conductivity, and excellent mechanical properties[1-3]. For the cheap, convenient, high thermal and electrical conductivity copper, the copper matrix composites exhibit excellent combination properties[4-5].

Titanium diboride (TiB$_2$) is well known for its high thermal and electrical conductivity, good chemical stability and good thermal shock stability. Thus, the addition of the TiB$_2$ to copper matrix greatly decreases the coefficient of thermal expansion, while reducing the electrical and thermal conductivity much less than the addition of most other ceramic reinforcements[6-9] . Therefore, TiB$_2$ reinforced metal matrix composites have received a great attention recently.The previous researches about the TiB$_{2P}$/Cu composites were mostly focused on the thermal shock resistance properties and fabrication methods[10-12].

In the present study, the TiB$_{2P}$/Cu composites with the content fraction of 50%, 58% and 65% were fabricated by the patent squeeze casting technology, the densification of which were higher than 98.2%, with their microstructures, thermal and electric conduction properties tested and analyzed.

2. Materials and Experimental

2.1 Materials preparation

The reinforcements used in this work were Titanium diboride, TiB2, particles with nominal diameters of 2-3μm, and the reinforcements volume fraction were 50-65 vol.%. The copper matrix was commercially available pure copper(Cu≥99.7 wt.%). This pure copper was chosen for the purpose of high thermal conductivity and low cost. Table1 shows the typical properties of TiB2.

Table1 Properties of TiB$_2$ particle reinforcements

Crystal structure	Hexagonal
Density [g/cm3]	4.4
Melting point [℃]	2930
Hardness [GPa]	30
Young's modulus [GPa]	574
Poission's ratio	0.11
Thermal conductivity [W/(m·℃)]	96

The TiB$_{2P}$/Cu composites were fabricated by squeeze casting technology. The TiB$_2$ preform was first fabricated and preheated. At the same time, copper alloy was melted, after which the molten copper was infiltrated into TiB$_2$ preform under the pressure and held for some time. And then, TiB$_{2P}$/Cu composite was solidified. A flow chart of the above process is shown in Fig. 1. The composites were annealed in vacuum at 700℃ for 1.5 h and furnace cooled in order to release residual stress within the composites.

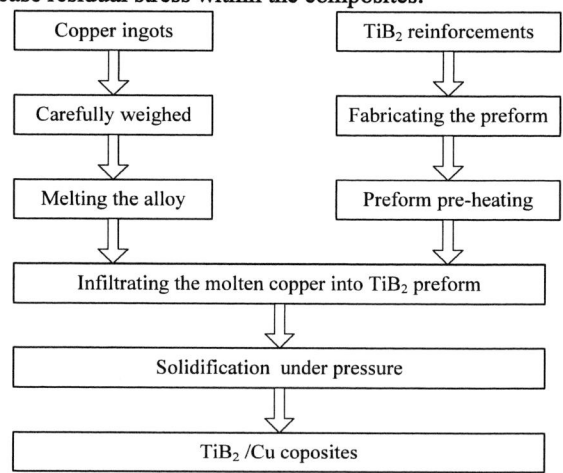

Fig. 1. Flow chart of the squeeze casting technology used for fabricating the TiB$_{2P}$/Cu composite.

2.2 Testing

The microstructure of as-fabricated TiB$_{2P}$/Cu composite were examined using a S-570 scanning electron microscopy(SEM) and Philips CM-12 transmission electron microscope (TEM). The measured density was obtained

978-1-4244-2739-0/08/$25.00 ©2008 IEEE

using the buoyancy (Archimedes) method and compared with the theoretical density to obtain various degree of densification. Thermal conductivity was measured by the laser flash method with the NETZSCH LFA427 thermal constant measuring equipment. The testing sample was cylindrical, 12.7mm in diameter and 3 mm in thickness. Electric conductivity was measured by vortex method with the Forster Sigmatest2.068 electric conductivity measuring equipment. The testing sample was machined into several blocks of 40mm×40mm×2mm.

3. Results and Discussion

3. 1 microstructure examination

Fig.2 reveals the microstructure of as-cast TiB$_{2P}$/Cu composites. The TiB2 particles were observed to be homogeneously distributed in the copper matrix. And the composites were free from common cast defects such as porosity and shrinking cavities because pressure were applied during the solidification of TiB2P/Cu composite.

The interface and the existing of interface effect are the important factors which can affect the properties of composite. Fig.3 illustrates the typical TEM micrographs of the interfaces in TiB$_{2P}$/Cu composites. A large mount of observations indicate that the TiB$_{2P}$/Cu interfaces were clean, smooth and free-from interfacial reaction products and amorphous layers, and no TiB$_2$ particles dissolved observed.

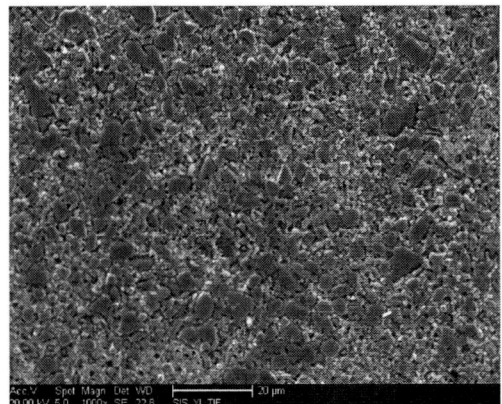

Fig. 2. SEM micrograph of TiB$_{2P}$/Cu composites.

3.2 Densification

The densification of the TiB$_{2P}$/Cu composites are in range of 98.2~99.1％, and decreases with an increase in TiB$_2$ volume fraction under the same processing conditions, which can completely meet the high dense requests for electronic package materials. A dense microstructure was beneficial to electronic packaging applications because of the improvement in mechanical strength and thermal conductivity.

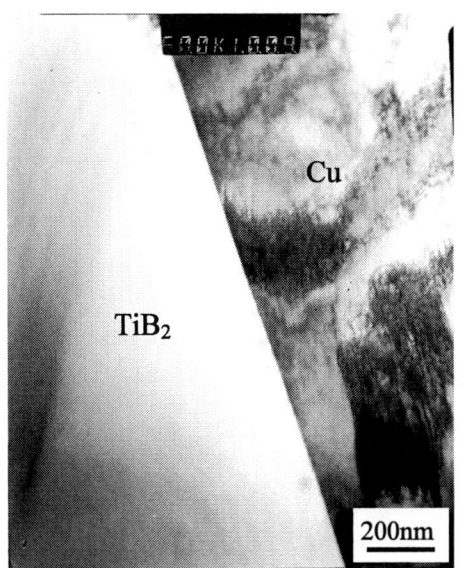

Fig.3 TEM micrographs of the TiB$_2$-Cu interfaces

3.3 Thermal conductivity

The measured thermal conductivities of TiB$_{2P}$/Cu composites were 215.4, 180.5 and 167.3 W/(m·℃) for the 50%, 58% and 65% composites, respectively, which were enough to satisfy the high thermal conductivity requests for electronic package materials. The thermal conductivities of TiB$_{2P}$/Cu composites were increased with the increasing of volume fraction of Cu content. It was attributed to the thermal conductivity of TiB$_2$ which is lower so far than that of copper.

Although the thermal conductivity of PMMCs are mainly decided by the thermal conductivity and content fraction of constituent components, it connects with the densification of materials, interface condition, the figures and distribution of particles. The theoretical predictions model of PMMCs thermal conductivity was the following[14-16]:

(1) Rom:

$$\lambda_c = \lambda_m \times V_m + \lambda_p \times V_p \qquad (1)$$

(2) Maxwel: According to the conductance and the thermal conductivity property of biphase and multiphase, the expression of the thermal conductivity was deduced:

$$\lambda_c = \lambda_m \frac{1 + 2x - 2V_p(x-1)}{1 + 2x + V_p(x-1)} \qquad (2)$$

Where λ is thermal conductivity, V is volume fraction, x equals to λ_m / λ_p, and subscripts c, p and m are composite, reinforcement particle and matrix, respectively.

Generally，the thermal conductivity of matrix Cu is 398 W/(m·K)，the thermal conductivity of TiB$_2$ is 96 W/(m·K). The calculated thermal conductivity were obtained from the above models, and the comparison between predictions and experimental data is shown in table 2.

Table 2 Predicted and experimental thermal conductivities of TiB$_{2P}$/Cu composites

	ROM [W/(m·℃)]	Maxwell [W/(m·℃)]	Experimental [W/(m·℃)]
50% TiB$_{2P}$/Cu	244.0	221.2	215.4,
58% TiB$_{2P}$/Cu	219.4	197.6	180.5
65% TiB$_{2P}$/Cu	197.8	177.9	167.3

The comparison of those with experimental data indicates that the calculated thermal conductivity of Maxwell models is close to that of TiB$_{2P}$/Cu composites. The achievement of higher thermal conduction is attributed the high dense composite fabricated by the patent squeeze casting technology, and the TiB$_2$-Cu interfaces are clean, smooth and free-from interfacial reaction products and amorphous layers.

3.4 Electric conductivity

Fig.4 shows the TiB$_2$ volume fraction effect on electric conductivities for TiB$_{2P}$/Cu composites. the electronic conductivities of three composites are in range of 33.7-43.9 %IACS, and decreased with the TiB$_2$ content increasing. It was attributed to the electronic conductivity of TiB$_2$ which is lower so far than that of copper, and with the increase of TiB$_2$ content, the interfaces within the composites increased as well as the interface electronic resistance.

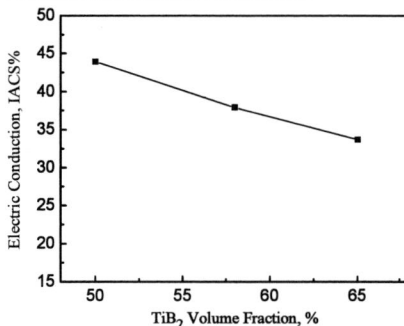

Fig.6 TiB$_2$ volume fraction dependence of electric conductivities for TiB$_{2P}$/Cu composites

The electric conductivity of metal –matrix composites is conceptually similar to the thermal conductivity. For composites containing electric conducting inclusions, the condition of the inclusion/matrix interface (or electric contact resistivity) significantly affects the electric conductivity, since it determines low much current can be carried by the inclusions. The electronic conductivities of materials were determined mostly by electron relaxation time as temperature settled. And the electron relaxation time was affected by crystal lattice style, atom, crystal defects and crystal lattice heat vibration. The factors affecting the electric conduction are very similar to those affecting the thermal conductivity of the composites. So, the achievement of higher thermal and electric conduction was attributed to the full densities and high purity TiB$_{2P}$/Cu composites, which were attained through the cost-effective squeeze-casting technology processes.

4. Conclusions

1) The full densities of the TiB$_{2P}$/Cu composites with volume fractions of 50% ～ 65% TiB$_2$ content have been fabricated by the cost-effective squeeze-casting technology. Then the TiB$_{2P}$/Cu composites were homogenous compound structures of adhesive phase Cu linking TiB$_2$ grains.

2) The thermal conductivities of TiB$_{2P}$/Cu composites at ambient temperature were range from 167.3 to 215.4 W/m℃ and decrease with an increase in volume fraction of TiB$_2$ content., which are agreed with the calculated values of Maxwell model.

3)The electric conductivities of TiB$_{2P}$/Cu composites are between 33.7-43.9 %IACS and decrease with the increasing of TiB$_2$ content.

References

1. M. Robins, "Thermal Management Materials and Designs," *Electronic Packaging and Products*, No. 10(2000), pp. 50~59.
2. C. Zweben, "Advances in composite materials for thermal management in electronic packaging", *JOM,* Vol. 50, No. 6(1998) pp. 47-51.
3. M. Hunt, "Progress in powder metal composites," *SAMPLE Journal*, Vol. 26, No.1(1990), pp. 33-36.
4. C. Zweben, "Advanced Composites and Other Advanced Materials for Electronic Packaging Thermal Management," *2001 IMAPS International Symposium on Advanced Packaging Materials,* Braselton Georgia, 2001, pp. 360~365.
5. E. H. Kerner, "The elastic and thermo-elastic properties of composite media," *Proceedings of the Physical Society*, Vol. 69(1956), pp. 808-815.
6. G. H. Wu, et al, "Properties of High-Reinforcement-Content Aluminum Matrix Composite for Electronic Packages," *Journal of Materials Science-Materials in Electronics*, Vol. 14, No. 1(2003), pp. 9~12.
7. C. Johnston, R. Young, "Advanced thermal management materials," *International Newsletter on Microsystems and MEMs,* Vol. 2, No. 1(2000), pp. 14-15.
8. D. P. H. Hasselman, F. J. Lloyd, "Effective Thermal Conductivity of Composites with Interfacial Thermal Barrier Resistance," *Journal of Composites*, Vol. 21, No. 6(1987), pp. 508~515.
9. Xu, Q. *et al*, "Effect of copper content on the microstructures and properties of TiB$_2$ based cermets by SHS," *Materials Science Forum*, Vol. 475-479, pp.1619-1622.
10. Yih, P. *et al*, "Titanium diboride copper-matrix composites," *Journal of Materials Science*, Vol. 32, No.7, (1997), pp. 1703-1709.
11. Kwon, Y.S. *et al*, "Solid-state synthesis of titanium diboride in copper matrix" *Journal of Metastable and Nanocrystalline Materials*, Vol. 15-16(2003), pp. 253-258.
12. Hong C. Q. *et al*, "Influence of hot pressing on microstructure and mechanical properties of combustion synthesized TiB2–Cu–Ni composite," *Journal of Materials Processing Technology*, Vol. 183 (2007), pp. 445–449.

Microstructure and Properties of Environmental-friendly Sip/1199Al Composites Used for Electronic Packaging

Ziyang Xiu[1], Guoqin Chen[3], Zongquan Deng[2], Gaohui Wu[3]

1 Research Academy of Science and Technology, Harbin Institute of Technology, Harbin 150001, China
2 School of Mechanical Engineering, Harbin Institute of Technology, Harbin 150001, China
3 School of Material Science and Engineering, Harbin Institute of Technology, Harbin 150001, China
xiuzy@hit.edu.cn, chenguoqin@hit.edu.cn, wugh@hope.hit.edu.cn

Abstract

Sip/1199Al composites for electronic packaging applications with high volume fraction of Si particles were fabricated by squeeze-casting technology. The microstructure observation showed that the composites were dense and Si particles distributed uniformly; The linear CTEs of Sip/1199 composites was between $(8.1\sim12) \times 10^{-6}°C^{-1}$, and they were decreased with the increasing content of Si particles as well as annealing treatment; the thermal conductivity can reach 150W/ (m·°C), which was decreased with the increasing contents of Si particles as well as annealing treatment. The composites had excellent mechanical properties and also could be recycled.

1. Introduction

Materials are the basis of the industry, and advanced composites are the major of the current materials industry, especially those used in space and aerospace applications [1-5]. But the traditional metal materials can not meet the demands of the aviation technology development [6,7]. Particulate reinforced aluminum matrix composites, which have low and designable thermal expansion property, excellent thermal conductivity and low density, have been attached much importance [8].

The Sip/Al composites are developed newly, which can inherit the better property of reinforcements and matrix because of no interphase compounds produced in casting process, and are adopted in aerospace, automobile and electronic packaging industry for its low expansion, high thermal conductivity, low density and excellent machining property. Because of the solution ability of the silicon and aluminum, the composites can be recycled [9,10].The 40%Si60/Al composites, CMSHA-40, was produced with PM by Japanese company, but its comprehensive property is not good. The Si/Al composites with excellent properties were produced in America with spray-deposit and liquid metal infiltrating method as well as with spray-deposit and heat isostatic-press method in England Osprey metal company. But these methods require expensive equipments, and thereby increase the costs. The squeeze-casting method has merits of simple equipments, low costs, high density and good uniformity. But the study on Sip/Al composites is on the preliminary stage, the preparation and the basic properties measurement were the major and the further investigation is not well developed [9].

The environmental-friendly Sip/1199 composite with high content (65%) silicon particles were fabricated by squeeze-casting method, and the microstructure, physical property, mechanical property and machining property were studied.

2. Materials and Experimental

2.1 Materials preparation

The reinforcement selected in the study is the high pure silicon particles with nominal diameter of 10um. The matrix alloy is the 1199, and the chemical position is shown in table.1. Firstly, the reinforcements were filled and pressed into a mold to produce a Si perform of 65% volume fraction. Before the casting process, the erform was pre-heated in a tool steel die. After it was heated to 500-600°C, the molten aluminum was poured into the pre-heated die, then the ram was driven downwards and a pressure was applied and maintained until the solidification was complete. In general, the whole infiltration process can be finished within 5 minutes.

2.2 Experimental method

The composites were annealed at 375 °C for 3h and furnace cooled. The microstructures of composites were observed by OLYMPUS PMG3 optical microscope and Philips CM-12 Transmission Electron Microscope. The CTEs of Sip/Al composites were measured on a DIL402C (NETZSCH Corp.) with a heating rate of 5°C/min (20°C~495°C). The thermal conductivity measurement was performed on Thermal Diffusivity NETZSCH LAF 427 Analysis made in Germany. The specimen size was Φ12.7×3mm, and both ends of which were polished. The tested temperature was 25°C. The mechanical properties were measured by three-point bending testing on the Instron5569 universal electron tension system.

3. Results and Discussion

3. 1 Microstructure

3.1.1 Optical microstructure

As seen from Fig.1, the composites were all dense and macroscopically homogeneous and particles clustering were seldom observed. And the composites were free from the micro-pores and obvious defects. A dense microstructure was beneficial to electronic packaging applications because of improvement in the thermal conductivity and mechanical strength as well as elastic modulus, so as to improve the dimension stability and the life cycle of the material.

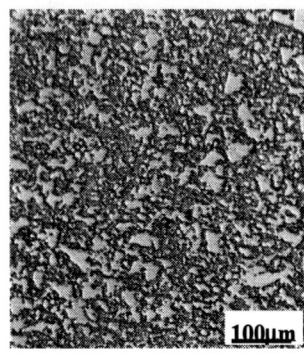

Fig.1 Microstructure of Si/1199 composite

3.2 Physical property

Fig.2 shows the linear CTEs of three Sip/1199 composites at different temperature. It can be seen that the CTEs of composites increase with the temperature. For a metal matrix composite, its CTE is mainly depended on the CTEs of matrix alloy and the influence of interfaces. On the one hand, the CTEs of composites will increase with

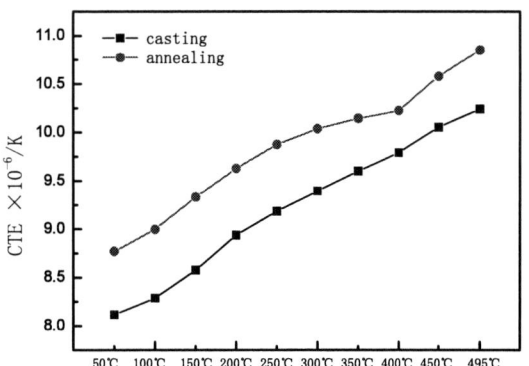

Fig.2 the linear CTEs of three Sip/1199 composites at different temperature

temperature increasing. On the other hand, when the temperature increased, the load-transferring ability of interface was decreased, leading to a decreased restriction against the expansion of the matrix. The Sip/1199Al

composites has a CTE value of $8.1 \times 10^{-6}/°C$ at a range of $20 \sim 50°C$ after annealing. Compared with the value of CMOS chip of Si ($4.1 \times 10^{-6}/°C$), it meets the expansion requirement in electrical package applications. Annealing treatment leads to a lower value of CTE because it lessens the residual thermal stress induced during fabrication process.

In the paper, the thermal conductivities of Sip/1199 composites as-cast and as-annealed were 139.1 W/m·°C and 161.3W/m·°C, respectively. They could meet the demands of electronic package material and were superior to the traditional Kovar alloy (17 W/m·°C) and Invor alloy (10 W/m·°C). Within the particles reinforced metal matrix composites, the heat transferred in metal matrix via free electrons while that transferred in non-metal particles by phonons. Therefore, for thermal conductance in composites, the free electrons and phonons work together. The interfaces between the matrix and particles can scatter the movement of electron and phonon, which hindered the heat conductance. The improvement of reinforcement's volume fraction can bring more and more interfaces, which can hinder the thermal conduction greatly. the thermal conductivities of as-annealed composites were larger than that of the as-cast composites, It was attributed to the annealing treatment can reduce the internal stress within the composites.

3.3 Mechanical property and machining property

The mechanical properties of Sip/1199 composites are shown in table.2. The hardness and bending strength of Sip/1199Al composites as-cast were 306.2 MPa and 316.3 MPa, respectively.

After annealing treatment, the values were decreased slightly. Because of the lower density ($2.4g/m^3$), the special modulus of composites is higher, which is three to four times than Kovar alloy ($13.1GPa·cm^3/g$) and copper ($18.5GPa·cm^3/g$). The high modulus helps to improve the rigidity of package structure, reduce its size and lessen its weight. Meanwhile, the lessening of the apparatus thickness can reduce the thermal resistance and improve the thermal dissipation ability of the apparatus.

Table 1 Mechanical property of the Sip/1199 composites

materials	conditions	Hardness (HB)	Bending strength (MPa)	Special strength (MPa·cm³/g)	Elastic modulus (GPa)	Special mudulus (GPa·cm³/g)
Sip/1199	Casting	306.2	316.3	131.8	121.2	49.69
	Annealing	285.7	293.9	122.5	121.1	49.65

Conclusion

(1). the composites were all dense and macroscopically homogeneous and particles cluster were seldom observed. And the composites were free from the micro-pores and obvious defects. A high density of stacking faults, twin and dislocations are found in silicon particles. The Si-Al interfaces are well-bonded and no interface reactants are observed. The dislocations and eutectic silicon precipitates are found in 1199Al alloy.

(2). the Sip/1199Al composites has a CTE value of $8.1 \times 10^{-6}/°C$ ($20 \sim 50°C$). It is close to the value of CMOS chip of Si ($4.1 \times 10^{-6}/°C$) and meets the expansion requirement in electrical packaging applications. And the annealing treatment can reduce the CTEs.The thermal conductivities of as-cast and as-annealed Sip/1199Al composites are 139.1 W/m·°C and 161.3W/m·°C, respectively. They are superior to the traditional Kovar alloy (17 W/m·°C) and Invor alloy (10

W/m·℃). And, the thermal conductivities are increased after annealing treatment.

(3). the density of composite is lower ($2.4g/m^3$), the special strength and special modulus is higher.

References

1. D. F. Wu, "The mechanical properties of AlN/Al composites manufactured by squeeze casting," *Journal of the European Ceramic Society*, Vol. 22(2002), pp.253-261.

2. H. Fujii, "Application of wetting research to joining and to fabrication of composite materials," *Science and Technology of Welding and Joining*, Vol.4, No. 4 (1999), pp. 187-193.

3. M. R. Ghomashchi, "Fabrication of near net-shaped Al-based intermetallics matrix composites," *Journal of Materials Processing Technology*, Vol.112, No.2 (2001), pp. 227-235.

4. M. Izciler, "Wear behavior of SiC reinforced 2124 Al alloy composite in RWAT system," *Journal of Materials Processing Technology*, Vol.132, No.1-3(2003), pp. 67-72.

5. V. P. McConnell. "High performance in small Packages ," *High Performance Composites*, Vol. 4, No.5(1996) 36-40.

6. V. H. Ozguz, "Materials for 3D packaging of electronic and optoelectronic systems," *MRS Bulletin*, Vol.28, No.1(2003), pp.35-40.

7. C. Zweben, "Advances in composite materials for thermal management in electronic packaging", *JOM*, Vol. 50, No. 6(1998), pp. 47-51.

8. C. Zweben, "Advanced composites and other advanced materials for electronic packaging thermal management," *2001 IMAPS International Symposium on Advanced Packaging Materials*, Braselton Georgia, (2001), pp. 360-365.

9. C. W. Nan, "Effective thermal conductivity of particulate composites with interfacial thermal resistance," *J.Appl.Phys*, Vol.81, No.5(1997), pp. 6692-6699.

10. G. H. Wu, et al, "Properties of high-reinforcement-content aluminum matrix composite for electronic packages," *Journal of Materials Science-Materials in Electronics*, Vol. 14, No. 1(2003), pp. 9-12.

11. M. Robins, "Thermal management materials and designs," *Electronic Packaging and Products*, No. 10(2000), pp. 50-59.

12. W. Zhou, Z. M. Xu. "Casting of SiC reinforced metal matrix composites," *Journal of Materials Processing Technology*, Vol.63(1997), pp.358-363.

13. G. H. Wu, Z.Y. Xiu, China patent, No. 200410043855.9

14. C. W. Chien, "Effects of Si_p size and volume fraction on properties of Al/Si_p composites," *Materials Letters*, Vol.12, No.2 (2002), pp.334-341 .

15. Z. Y. Xiu, "Thermo-physical properties of Si_p/LD11 composites for electronic packaging," *Transactions of Nonferrous Metals Society of Chin*, Vol.15, No.2(2005),pp. 227-230 .

Preparation of Polysulfoneamide Electrospinning Nanofibers

Li Liu, Ying Shi, Qinghua Jiao, Yanan Wu, Zhikun Zhang, Johan Liu

Department of Polymer Science, School of Material Science and Engineering, Shanghai University, Chengzhong Road 20,
Jiading District, Shanghai 201800, China
E-mail:liuli@staff.shu.edu.cn

Abstract

In this paper, we select electrospinning technique to fabricate Polysulfonamide nanofibers. Polysulfonamide with relatively high inherent viscosity was prepared based on 4,4'-diaminodiphenylsulfone and terephthaloyl chloride in the common solvent N,N-Dimethylacetamide (DMAc) by the method of low-temperature solution condensation polymerization. Our research work focused on studying the effect of concentration of solution, processing parameters including voltage and distance from tip to collector on the spinnability of the solution and morphology electrospun Polysulfonamid fibers. The microstructures of the electrospun Polysulfonamid fibers were quantitively investigated by scanning electron microscope (SEM) as a function of processing variables.

Key words: Electrospinning, Polysulfoneamide, Nanofibers

1. Introduction

Electrospinning is not a new investigation. In 1990's, Reneker ansd co-workers, who have demonstrated electrospinning for a wide variety of materials and solutions[1]and have produced a number of different an interesting fiber structers and morphologies[2]. Bulk polyethylene and polyethylene dissolved in paraffin were electrospun by Larrondo and Manley[3,4]. Kim and Lee carried out the electrospinning of poly(ethylene terephthalate), poly(ethylene naphthalate) and their blends[5] in the molten state. Poly(ethylene oxide), which is known as an easily soluble and crystallisable polymer in aqueous solution, has been used for setting the optimum conditions and characterization of fibers [6-8].

Polysulfonamide(PSA) fiber is one of the high-temperature fibers [9-11] and used in many fields because of their characteristics of heat-resistance, flame-retardant, high strength, high modulus and mechanical property. Conventional fiber spinning techniques spun from PSA is wet spinning. Very little research on electrospinning has been performed. The advantage of electrospinning is that the small diameter polymer fibers which are fabricated by it with small pore size and large surface areas have spread their applications range from textile industry to medical science, optics and electronics field[12].

The basic process of electrospinning involves the introduction of electrostatic charge to a stream of polymer melt or solution in the presence of a strong electric field. A high voltage is applied to a metallic capillary, which is connected to reservoir holding polymer solution. With a sufficiently high electrical field, the electrostatic forces can overcome the surface tension of the polymer solution and cause the ejection of a thin jet from the capillary nozzle and then the charged jet undergoes a stretching and whipping process, resulting in the formation of many continuous fibers[13].

Many parameters can influence electrospinning process, including solution properties, governing variables and ambient parameters. In the present work, we study the electrospinning properties of PSA in DMAc. We present the effects of instrument variables such as electrical field and solution variables such as concentration on the fiber forming process by electrospinning. We interpret results of observations made by scanning electron microscope (SEM).

2. Experimental

2.1 Materials

DMAc(CP, purchased from Shanghai Reagent Co., Shanghai, China) used as solvent was dehydrated by molecular sieves (type 4A) prior to use. 4, 4-diaminodiphenyl sulfone and terephthaloyl were purchased from JiangXi LiansDa Co., Ltd., China.

2.2 Synthesis of PSA[14]

With the protection of N_2, certain 4, 4'-diaminodiphenylsulfone was added into DMAc and kept stirring the mixture in the four-necked flask. After the 4, 4'-diaminodiphenylsulfone was dissolved, the same mole terephthaloyl chloride was added at low temperature(below 0 ℃). The reaction was continued for 2~3 hours at 1500rpm at room temperature. The antacid was added at the end, the pH of resulting polymer was adjusted to below 7.

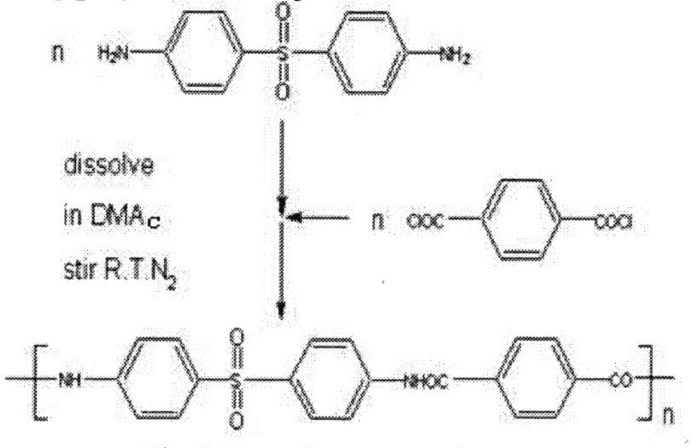

Fig.1 Reaction route of PSA

2.3 Electrospinning setup [15]

The electrical field was provided by a high voltage (HV) power supply, which can generate voltages of up to 50 kV with 200µA direct current. Polymer solution was held in a capillary. The diameters of the needle tip were 1.0mm. The copper circle of the HV generator was hitched to the needle. A grounded aluminum sheet was positioned opposite and perpendicular to the tip of the needle onto which the fibers were deposited. (Fig.1.)

978-1-4244-2739-0/08/$25.00 ©2008 IEEE

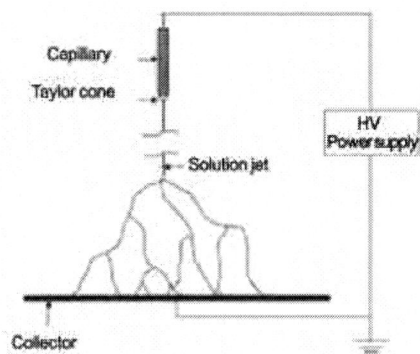

Fig.1. electrospinning equipment

3. Results and Discussion
3.1 Effect of concentrations
3.1.1 Jet forming concentrations and threshold voltage

The concentrations of Polysulfonamide solutions in DMAc，varying in the range 10-20wt% were diluted from reaction solution. It is observed that electrospraying took place where the jet broke into droplets at concentrations below 14wt%. When the concentration was over 18wt%, solution viscosity was so high that surface tension was hard to be overcome and no fibers were formed. Thus the successful concentration range, at which continuous fibers could be generated, was between 14 wt% and 18wt%.

The threshold voltage to start the jet formation is plotted as a function of concentration in **Fig.2.** Higher electrical forces are required to overcome both the surface tension and the viscoelastic force for stretching the fiber when the concentration, or equivalently, the viscosity increases.

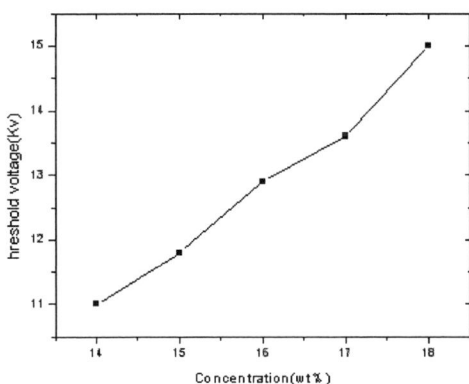

Fig.2. Threshold voltage as a function of concentration.

3.1.2 Effect of concentration on morphology of fibers

Viscosity or the polymer concentration showed an important role in the fiber morphology changes. **Fig. 3** shows a series of Electron micrograph of the fibers obtained from electrospinning of PSA solution concentrations, which are: (a)14wt% (b)16wt% (c)18wt%. An increase in solution concentration results in fibers with larger size. Electrospun fibers were not uniform in diameter and morphology, which varies with the concentration of solution subjected to electrospinning. **Fig. 4** depicts diameter distribution of fibers. Relative low concentration solution was generated to thinner

fibers in a dispersive distribution, whereas it was much more uniform in fibers gained from high concentration

Fig 3. Electron micrograph of fiber morphology of polysulfon amide with different concentration at a voltage 30 kV, collected distance 10cm
(a) 14wt% (b) 16wt% (c) 18wt%

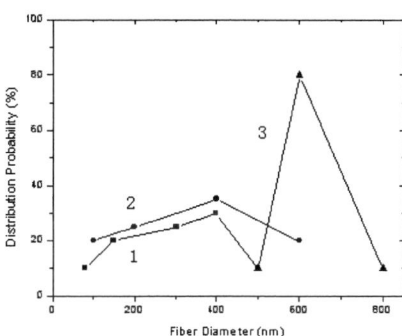

Fig.4. Fiber diameter distribution as a function of PSA weight mass ratio
(1、 14 wt %   2、 16 wt %   3、   18 wt%)

3.2 Effect of spinning voltage

The spinning voltage imposed a significant role on the fiber structure and size. The ultimate goal of our work was to produce nanofibers from PSA solutions. Low electric potential was not enough to overcome the surface tension. As the voltages increased, the force balance between the simultaneous repulsive force and surface tension was able to stretch the fibers and then revealed the spindle shape from a globule mushroom. Thus during the travel of the fibers towards the target, the higher voltage was observed to be more desirable to result in thinner fibers[16]. However, when the potential arrived at a critical value, if the voltage was continued to be increased, there would be more solution jetted so that fibers cannot be split totally before they were collected. Videlicet too high voltage would generate thicker fibers in reverse. Finally spinning would stop when the potential increased to the extent that made spinning speed slower than solution supplying.

The stretching effect of the electrostatic voltage on the fiber thickness was systematically investigated by electrospinning of polymers at different voltage. (**Fig. 5**).

Fig.6 shows fibers average diameter of different voltages. From 15 to 25kv, fibers become more stretched and thinner. Therefore, nanometer PSA fibers were fabricated by regulating voltage, while the other parameters were held at a suitable condition.

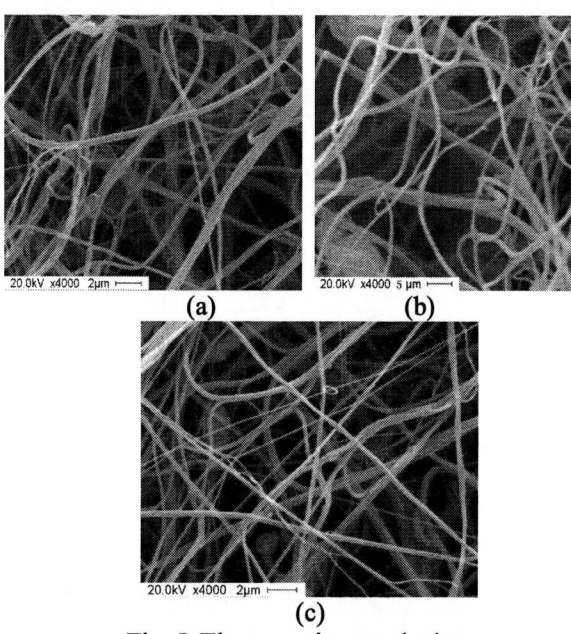

(a)   (b)

(c)

Fig. 5. Electron micrograph picture
of fiber morphology of fibers fabricated with different
voltage at a concentration 14wt%,
collected distance 15cm
(a) 15kv   (b) 20kv   (c)25kv

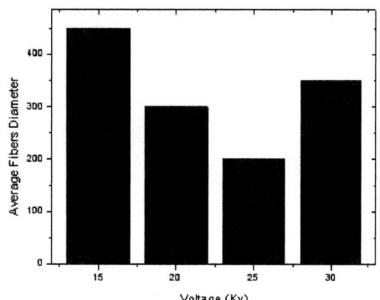

Fig. 6.Average diameter of fibers fabricated at different
voltage

3.3 Effect of collected distance

In the electrospining process, the distance between the spinnerette and grounded collector is another crucial parameter. The longer the distance was, the smaller the electric field strength was required. Thus accept distance and voltage had the opposite effects. On the other hand, longer distance can extend volatile solvent time. From this point of view, fiber diameter would be reduced. In this work, the effect by electrospinning of polymers at 15 and 20cm was compared. (**Fig. 7**) It is seen that fiber gained at 20cm is

thinner than 15cm but the diameter distribution of the latter one is more uniform.

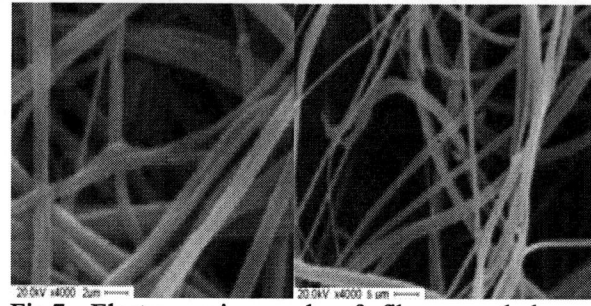

Fig.7. Electron micrograph of fiber morphology of polysulfonamide gained at different collected distance with the concentration of 14wt%, voltage of 30kv (a) 15cm   (b) 20cm

4. Conclusion

In this work, the effects of major process parameters on morphology of electrospun PSA nanofibers were researched. It was found that spinnable concentration range of PSA solution was between 14wt% and 18wt%. Voltages that exert an significant role on fibers diameter had a crucial value. When the concentration was 14wt% and colleted distance was 10cm, the value was 25kv. Lower than it, fibers get thinner with the voltage increased whereas over it, thicker fibers would be generated in reverse. There was a contradiction resulted from collected distance. Extending the distance had the both probability to make fibers thinner or thicker. According to the work, we found fibers gained at the distance of 20cm were thinner than 15cm but the diameter distribution is not that uniform as the latter one. Thus by regulating these parameters, fibers with diameter less than 100nm can be successfully produced.

References

D.H. Reneker, I. Chun "Nanometer diameter fibers of polymer produced by eletrospinning", Nanotechnology, 7(1996), 216-223.

2.   H. Fong, D.H. Reneker, "Electrospinning and the formation of nanofibers. Structure formation in polymeric fibers" 2 (2002), 225-246.

3.   L. Larrondo, RSt. John Manley, "Electrostatic fiber spinning from polymer melts 1. Experimental observations on fiber formation and properties", Polymer Science, 19(1981), 909.

4.   L. Larrondo, RSt. John Manley, "Electrostatic fiber spinning from polymer melts 2. Experimination of the flow field in an electrically driven jet", Polymer Science, 19(1981), 921.

5.   J.S. Kim, D.S. Lee, "Thermal properties of electrospun. Polyesters", Polymer, 2000;32:616

6.   J.M. Deitzel, J. Kleinmeyer, D. Harris, N.C. Beck Tan, "The effect of variables on the morphology of electrospun nanofibers and textiles", Polymer 42(2001), 261

7.   H. Fong, I. Chun, D.H. Reneker, "Beaded nanofibers formed during electrospinning", Polymer, 40(1999), 4585

8.   R. Jaeger, H. SchÖnherr, G.J. Vansco, "Chain Packing in Electrospun Poly(Ethylene Oxide) Visualized by

Atomic Force Microscopy", Macromolecules, 29(1996), 7634

9. H. JIANG, "The development of high temperature fibers ", Fibers and raw materials, 1(2001), 22-23

10. C.F. XIAO, "Polybenzimidazole fiber and its application", Hi-tech Fiber & Application, 28(2003), 3

11. Shanghai Textile Research Institute, "High-temperature resistant polysulfonamide fibers",. Synthetic Fiber, 2(1977), 18-27

12. J.M. DEITZEL et al., J.Polymer 42 (2001), 8163

G.C. Rutledge, Y. Li, S. Fridrikh, "Electrostatic Spinning and Properties of Ultrafine Fibers", National Texile Center Annual Report, November (2001), M01-D22.

14. L. Liu, J. Deng, X. P. Wang, "Preparation of Polysulfonamide/ZnO Nanocomposite by In-Situ Polymerization", Journal of Shanghai Jiatong University (Science), 12(2007), 508

15. Wen Xuan Wang et al, "New Nano-Thermal Interface Materials (Nano-TIMs) with SiC Nano-Particles Used for Heat Removal in Electronics Packaging Applications," Published on the 8th International Conference on Electronic Materials And Packaging (EMAP2006),Hong Kong.

16. M. M. Demir, I. Yilgor, E. Yilgor, B.Erman, "Electrospinning of polyurethane fibers", Polymer, 43(2002), 3303-3309

Liquid-State Interfacial Reaction of Sn-10Sb-5Cu High Temperature Lead-free Solder and Cu Substrate

Qiulian Zeng [1,2], Jianjun Guo[1], Xiaolong Gu[1], Xinbing Zhao[2]

1.Zhejiang Metallurgical Research Institute Co, Ltd.,   2. Zhejiang Universtity

1. Gongsu District, Hangzhou, 310011, China, 2. Xihu District, Hangzhou, 310027, China

Corresponding Arthor: E-mail adress: jjguomail@163.com (J.J.Guo), Tel: 0571-88822773

Abstract

Sn-Sb alloys are potential solders for replacement of high-Pb solders because of their high melting temperature in lead-free solders. However, Cu substrate is extremely dissolved by the Sn-Sb binary alloy during the high temperature soldering process, which will cause serious reliability problem of the solder joint. Based on this critical issue, we designed a new high temperature lead-free Sn-10Sn-5Cu ternary solder to prevent the dissolution of Cu substrate. In this study, liquid-state interfacial reaction between the high temperature lead-free solder and the Cu substrate was investigated. The liquid-state interfacial reaction of the solder on the Cu substrate was carried out at the different temperature of 280 ℃, 320 ℃,360 ℃ and 400 ℃, and the reaction time was 1min, 10mins, 30mins and 60mins, respectively. Microstructure of the Sn-Sb-Cu bulk solder and the solder joint was observed by scattered electron microscope (SEM). The identification of phase composition was determined by Energy Dispersive X-ray Detector (EDX) and electron probe microscopy analysis (EPMA). During the four reaction temperatures, the interfacial reaction products included a scallop Cu_6Sn_5 intermetallic compound (IMC) layer and a flat Cu_3Sn layer adjacent to Cu substrate. IMCs thickness with the reaction time was measured by the area of interface IMCs layer divided by the interface length. The IMCs thickness increased with the reaction temperature and reaction time, and the relationship between IMC thickness and reaction time was linear with square root of time, which signified that the IMC growth dynamics was diffusion controlled. The diffusion coefficient was calculated by the IMC growth rate, which increased with the higher temperature, corresponding to be 2.30×10^{-14} ,6.84×10^{-14} ,1.63×10^{-13}, 1.99×10^{-13} m²/s for the temperatures of 280 ℃, 320 ℃, 360 ℃ and 400 ℃, respectively. And then the diffusion activation energy was determined to be 57.8KJ/mol by fitting the four diffusion coefficients at various temperatures, which indicated that the diffusion mechanism was grain boundary diffusion. Between lower temperature of 280 ℃ and higher temperature of 400 ℃, huge differences existed on the microstructure of IMC in the interior solder of the solder joint.

Introduction

Sn-Sb alloys are potential alternatives replacement for the high-Pb solder in high temperature application of chip interconnection packaging.[1] Most of researches of Sn-Sb alloys focused on the mechanical properties, such as tensile, tensile creep and impression creep behavior of the bulk solder.[2, 3] And it exhibited superior tensile and creep properties over eutectic Sn-Pb solder.[3] The result of El-Daly

et al[4] was that the addition of Ag and Au into Sn-5Sb alloy enhanced the creep resistance greatly, and the creep resistance of addition Au was much higher than that of addition Ag. However, the addition of 3.5%Ag and 1.5%Au decreased the melting point of Sn-5Sb from 240 ℃ to 216 ℃ and 203.5 ℃, respectively.

Cu and Ag substrates are the most common materials in microelectronic interconnection, and the reliability issue of the solder joint due to the interfacial reaction between the substrate and the lead-free solder is one of the major concerns. Chen et al [5] investigate Interfacial reactions between Sn-Sb and Ag or Cu substrate at 200 ℃ and at 250 ℃. Only the Ag_3Sn phase was formed in the Sn-Sb/Ag couples, which was very similar to those in Sn/Ag couples.[6-8] Cu_6Sn_5 and Cu_3Sn phases were formed in the Sn-Sb/Cu couples, and it was concluded that the phases of the Sn-Sb/Cu couples reacting at 260 ℃ were very similar to those of the Sn/Cu system.[9]

From the above literature, it could be concluded that the researches focused on the properties of Sn-Sb binary bulk alloy and Sn-Sb/Ag or Sn-Sb/Cu interfacial reaction at the temperature lower than or equal to 260 ℃. However, the soldering temperature of the Sn-Sb alloys for replacing Pb-rich solder is generally higher than 280 ℃, and Sn-10Sb alloy shows better reliability of the solder joint than the Sn-5Sb in terms of melting point. Additionally, the Sn-Sb binary alloy can dissolve the Ag and Cu substrates greatly.[9] Hence, the third element should be added into the binary to prevent the enormous substrate dissolution. In this study, we designed a new high temperature Sn-10Sn-5Cu ternary alloy, and investigated the liquid-state reaction between the ternary solder and Cu substrate at the temperature higher than 280 ℃.

Experimental Procedures

The high purity Sn and Sb were formulated and melt in the furnace at a temperature of 800 ℃, then Cu element in the form of Sn-20wt.%Cu inter-alloy was added into the binary Sn-Sb alloy. After half an hour the melt was cast into steel mold forming Sn-10Sb-5Cu ternary alloy ingot. The ingot was extruded and then drawn to be wire shape with a diameter of 2mm. The solder wire was cut to segments for the metallurgy examination and for the liquid interfacial reaction.

Liquid-state interfacial reaction between the SnSbCu solder and Cu substrate was carried out on the soldering oven. The sketch diagram of the interfacial reaction between the molten solder and Cu substrate was shown in Fig.1.The solder with a mass of 0.21~0.23g each time was placed on Cu substrate with a size of 10mm×10mm. The reaction surface of Cu was ground firstly and then polished with 0.5μm

978-1-4244-2739-0/08/$25.00 ©2008 IEEE

diamond paste. And some water solution commercial flux was coated on the polishing Cu surface before soldering. The reaction temperature was kept at 280 ℃, 320 ℃, 360 ℃ and 400 ℃, respectively. And reaction holding time varied from 1min to 60mins.

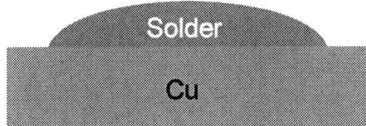

Fig.1 Sketch diagram of the liquid-state interfacial reaction between the Sn-10Sb-5Cu solder and Cu substrate

Microstructure of the Sn-10Sb-5Cu cast alloy and the solder joints were observed by optical microscope with a CCD recorder and JSM-6480 Scanning Electron Microscope with an Energy Dispersive X-ray Detector. And the phase identification in the SnSbCu solder was carried out by electron probe microscopy analysis (EPMA) and EDX. The composition of the ternary cast alloy was determined by Atomic Absorption Spectroscopy (AAS), as shown in table1.

Table 1 The composition of the Sn-10Sb-5Cu solder analyzed by Atomic Absorption Spectroscopy (mass%)

Sb	Cu	Pb	Ni	Zn	Fe	Ag	Cd	Sn
9.96	5.05	0.015	0.001	0.0003	0.0016	0.0009	0.0001	Bal.

Area fraction of the intermetallic compounds (IMCs) in the SnSbCu solder and the thickness of the interface intermetallic compounds were measured by ImageJ software. As we know, the morphology of Cu_6Sn_5 IMC forming at the interface of the solder joint is scallop-like, it is difficult to measure its thickness directly. So, we used the average thickness of IMC to evaluate its growth dynamics, the average thickness of Cu_6Sn_5 was defined by the area of the interface IMC divided by the measured interface length, shown in following equation,[10]

$$z = \frac{S}{L_0} \qquad (1)$$

Where z, S and L_0 are the average thickness of the intermetallic compound layer at the interface, area value of the interface IMCs layer and the selected interface length for measurement, respectively.

Results and discussion

Fig.2 is the optical microstructure of the original Sn-10Sb-5Cu solder, which was composed of near-square or triangle-like SnSb phase with a size of about 50μm and the smaller bright particle Cu_6Sn_5 IMC. The area fraction of SnSb phase and Cu_6Sn_5 IMC was 9.8% and about 10%, respectively.

The SnSb and Cu_6Sn_5 phases were identified by the electron probe microscopy analysis (EPMA) under the backscattered electron mode, shown in Fig.3. The bright coarse phase was SnSb, while the black smaller particles were Cu_6Sn_5, marked by two cycles and numbers, the compositions of which were listed in following table2. According to the equilibrium Sn-Sb binary phase diagram, SnSb IMC formed firstly in the form of Sn_3Sb_2, then decomposing into SnSb solid solution (β phase) and β -Sn solution at the temperature

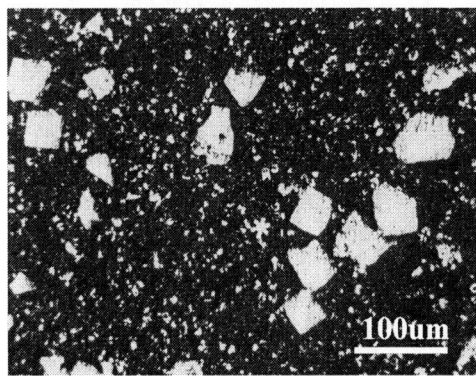

Fig.2 Original microstructure of the Sn-10Sb-5Cu solder

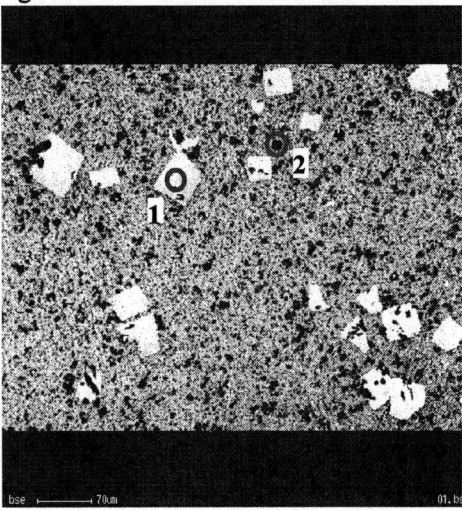

Fig.3 EPMA analyzing the original microstructure of Sn-10Sb-5Cu solder

of 242 ℃, which was approved by the EPMA analysis, shown in Fig.3 and Table2. In SnSb phase, the atoms ratio of Sn and Sb was 56.485%: 43.465% (about 1.33), which lower than that of 1.5 in Sn_3Sb_2 phase, which meant that the decomposition of Sn_3Sb_2 phase occurred. In addition, trace Cu atoms dissolved into the SnSb phase according to the EPMA results (table2). Meanwhile, it was found that some Sb atoms entered the Cu_6Sn_5 phase crystal lattice to form $Cu_6(Sn,Sb)_5$ phase by substituting for the Sn atom positions.

Table 2 EPMA results of SnSb and Cu_6Sn_5 phases

	Element	wt.%	at.%
1 point	Sn	56.485	57.084
	Sb	43.465	42.822
	Cu	0.049	0.093
2 point	Cu	38.909	54.358
	Sn	58.290	43.599
	Sb	2.802	2.043

Fig.4 is the SEM micrographs of the molten SnSbCu solder reacted with the Cu substrate for various reaction time at the temperature of 280 ℃ and 400 ℃. In order to differentiate the SnSb phase from Cu_6Sn_5 IMC in the solder, backscattered electron mode was used here. Fig.4(a)(b) was the case at temperature of 280 ℃, and the corresponding reaction time was 10mins and 60mins, while the static reaction temperature in Fig.4(c)(d) was 400 ℃, and the holding time was 10mins and 60mins, respectively. In the both cases of various temperatures, the interfacial reaction products included a scallop Cu_6Sn_5 IMC layer and a flat Cu_3Sn phase layer between Cu_6Sn_5 phase and Cu substrate, and the total average thickness of the IMCs increased with increasing holding time. At the same reaction time, thickness of Cu_3Sn and Cu_6Sn_5 phases increased with the increasing reaction temperature due to more rapid atom diffusion

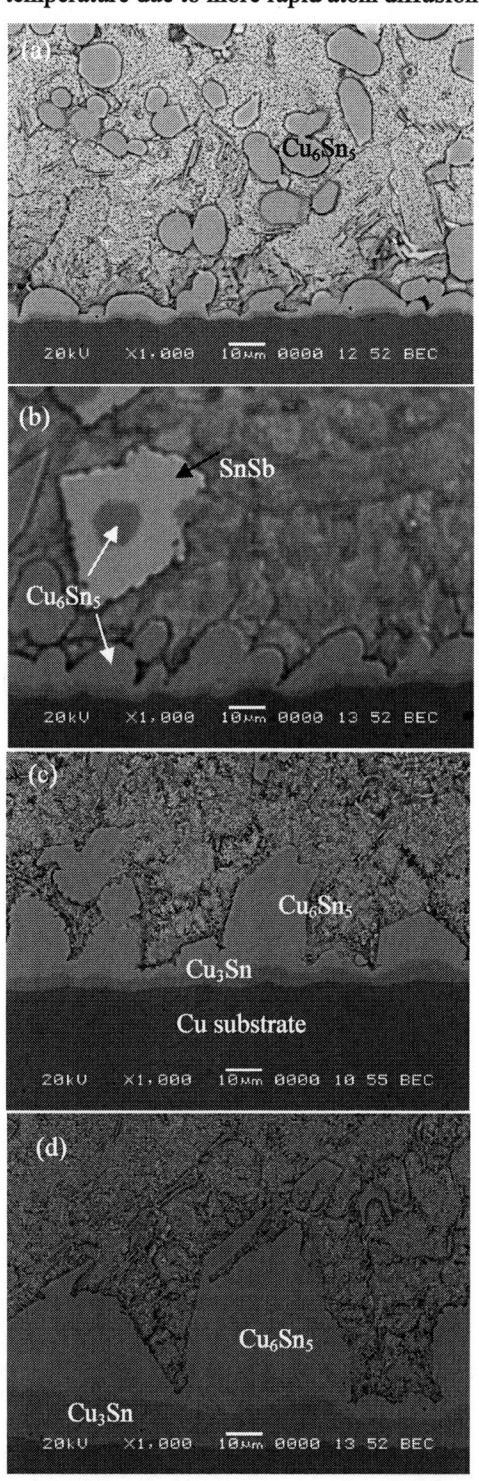

Fig.4 SEM micrographs of the Sn-10Sb-5Cu solder reacted with Cu substrate (a) 280 ℃, 10min (b) 280 ℃, 60min (c) 400 ℃, 10min (d) 400 ℃, 60min

coefficient at higher temperature. Average thickness of the interface IMCs was about 5μm at the temperature of 280 ℃ for a reaction time of 10mins, while in the same reaction time, average thickness of the total IMCs was near 14μm at the temperature of 400 ℃.

The Cu_6Sn_5 IMC morphologies at both of the interface and interior of solder at lower temperature of 280 ℃ were different from that at higher temperature of 400 ℃. At 280 ℃, the scallop Cu_6Sn_5 phase at the interface was smooth and round, while the interface Cu_6Sn_5 phase at the temperature of 400 ℃ was much sharper, that is, the curvature radius of Cu_6Sn_5 IMC was smaller for the former than that for the latter case. In the interior of the solder, the morphologies of Cu_6Sn_5 at the both temperatures were also various. At lower temperature of 280 ℃, most of the Cu_6Sn_5 phase was round with a size of 10μm, while it was nubby and directional strip-like at 400 ℃, which could be observed more clearly in the low magnification SEM micrographs, shown in Fig.5. Such a difference could be explained by larger temperature gradient during the solidification process in the case of higher reaction temperature. Additionally, there was another obvious difference between the both temperatures, the coarse square SnSb phase in the interior solder appeared much more frequently at lower temperature than at higher temperature(Fig4.(b) and Fig.(5)).

Fig.5(a) is the original microstructure of the bulk solder, as discussed above, which comprised the coarse bright SnSb phase and black particle Cu_6Sn_5 IMC. In certain SnSb phase, some Cu_6Sn_5 particles appeared in the interior of SnSb phase, which signified that Cu_6Sn_5 IMC precipitated firstly during the solidification process in the Sn-10Sb-5Cu ternary alloy and SnSb phase nucleated along it. The area fraction of SnSb phase in the solder decreased greatly after the solder remelted to react with Cu substrate (Fig5.(b)). When the temperature of molten solder reached 400 ℃, the SnSb phase can be hardly observed in the interior of the solder (Fig.5(c)). At lower reaction temperature of 280 ℃, many voids existed in the cooling solder joint, while no such defects appeared in the solder joint at the temperature of 400 ℃. Coarse framework type Cu_6Sn_5 IMC formed in the solder at the higher temperature, and little coarse SnSb phase formed in this case. It could be supposed that it was just the framework Cu_6Sn_5 phase inhibited the diffusion and accumulation of Sb atoms, the growth rate of SnSb phase slowing down, then evenly distributed in the solder in the form of small particles. Such results indicated that soldering process of the SnSbCu solder at higher temperature of 400 ℃ was better than that at lower termperature of 280 ℃. However, maybe the most proper soldering temperature should be a value between 280 ℃ and 400 ℃ in order to combine lessening the voids defects with refining the framework Cu_6Sn_5 IMC.

The thickness of the total intermetallic compounds including Cu_3Sn and Cu_6Sn_5 phases increased with longer reaction time at all of the different reaction temperatures, and the relationship between IMC thickness and reaction time was shown in Fig.6. The relationship between IMC thickness and reaction time was expressed by the following empirical Equation,[10]

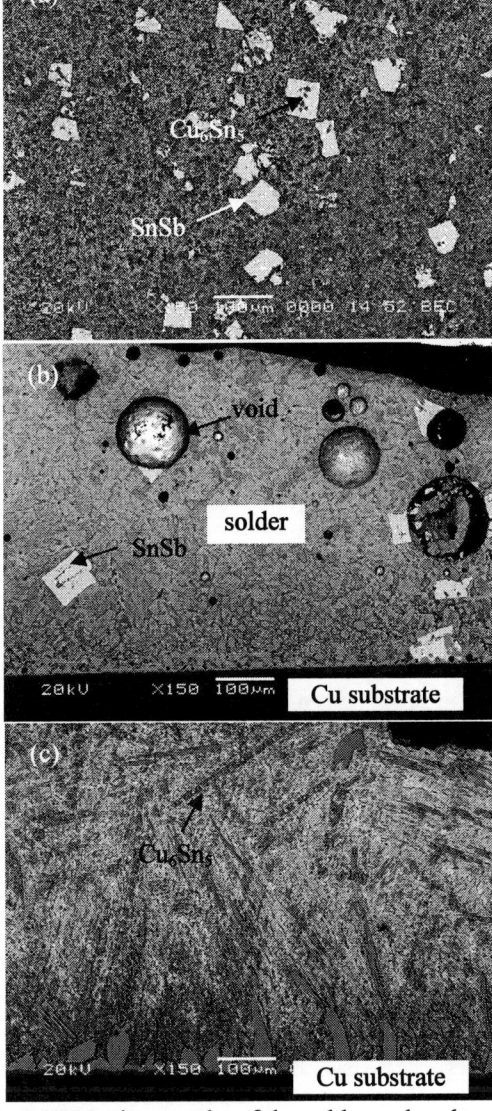

Fig.5 SEM micrographs of the solder and molten solder reacted with Cu substrate for holding time of 30min (a) original microstructure (b) holding temperature of 280 ℃ (c) holding temperature of 400 ℃

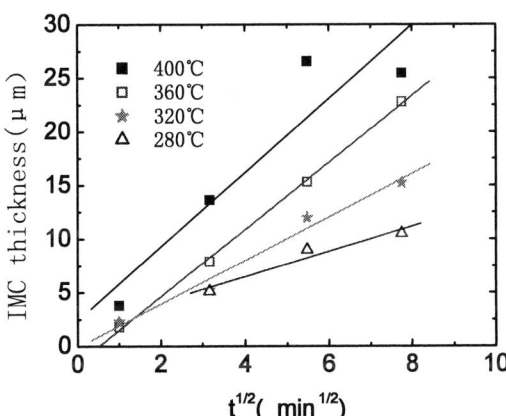

Fig.6 Thickness of the total IMCs at the interface between the Sn-10Sb-5Cu solder and Cu substrate

$$z = D^{1/2}\sqrt{t} + c \qquad (2)$$

Where z, D, t is the average thickness of IMCs, atom diffusion coefficient, reaction time, respectively, and c is a constant.

As we can see from Fig.6, at the four reaction temperatures in this study, the IMC thickness and the square root of time was satisfied well with a linear relationship, which indicated that the growth dynamics of the IMCs was a diffusion-controlled process.

And the diffusion coefficients at various temperatures could be calculated by Eq.2, listed in table3.

Table 3 Diffusion coefficients at various temperatures

Temperature	400 ℃	360 ℃	320 ℃	280 ℃
Diffusion coefficient m^2/s	1.99×10^{-13}	1.63×10^{-13}	6.84×10^{-14}	2.30×10^{-14}

As seen from Fig.6, the IMC at higher temperature was thicker than that at lower temperature in the same holding time. The IMC growth rate could be evaluated by the slope of the curves, i.e. $D^{1/2}$. By comparison between the four cases of different temperatures, the IMC growth rates became larger gradually with the increasing temperature. The diffusion coefficient at 400 ℃ is almost an order of magnitude larger than that at 280 ℃. Additionally, for the same reaction time of 30mins and 60mins, the thickness of Sn-10Sb-5Cu/Cu interface IMC was about 8μm and 10μm at 280 ℃, similar to that of Sn-Sb/Cu (Sb%:3~7%) at 260 ℃.[9] However, it was a half thinner than 16μm and 18μm at 280 ℃ in the Sn-Sb/Ag (Sb%:3~7%) system.[7]

The diffusion coefficients of Cu and Sn atoms at 320 ℃ and 280 ℃ in the current study were similar to that of Sn-9Zn-0.5Ag/Cu ($5.76 \times 10^{-14} m^2/s$) and Sn-9Zn-1.5Ag/Cu ($2.82 \times 10^{-14} m^2/s$) solder joints aging at 250 ℃, reported by Chang et al.[11] However, the values were three orders of magnitude larger than that of Sn-Ag-Cu/Cu joints at solid aging temperature, for example, $7.24 \times 10^{-17} m^2/s$ for aging at 170 ℃,[12] $4.7 \times 10^{-17} m^2/s$ for aging at 125 ℃.[13]

According to Eq.3, the atom diffusion activation energy (Q) could be calculated by fitting the diffusion coefficients at different reaction temperatures, as shown in Fig.7. The diffusion activation energy was calculated to be 57.8KJ/mol, only about a half of the bulk lattice diffusion coefficient, suggesting that the grain boundary diffusion was the predominant diffusion mechanism in the liquid-state interfacial reaction.

$$D = D_0 \exp(-Q/RT) \qquad (3)$$

Where D is the diffusion coefficient, D_0 is a constant, Q is atom diffusion activation energy, R is gas constant, T is absolute temperature.

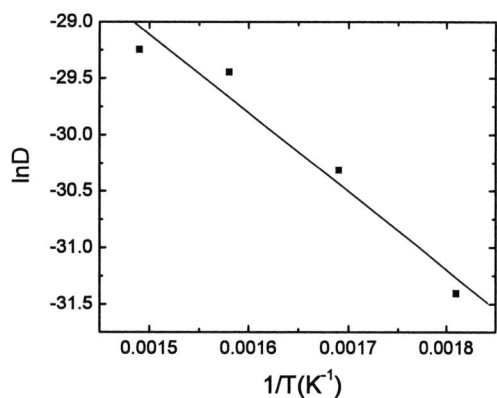

Fig.7 Relationship between the diffusion coefficient (D) and absolute temperature (T)

Conclusions

Liquid-state interfacial reaction between high temperature Sn-10Sb-5Cu lead-free solder and Cu substrate was investigated. The microstructures in the interior solder of solder joint existed distinguished differences between at the reaction temperature of 280 ℃ and 400 ℃.The proportional relationship between IMCs thickness and the square root of reaction time supported the diffusion-controlled growth dynamics. The atom diffusion coefficient increased with the high temperature. And the activation energy indicated that the diffusion mechanism was grain boundary diffusion.

Acknowledgments

The authors would like to acknowledge financial support provided by the Science and technology program (No.D08102) of Zhejiang province.

References

1. Jang, J.W. *et al*, "High-temperature lead-free SnSb solders: Wetting reactions on Cu foils and phased-in Cu-Cr thin films," *Journal of Materials Research*, Vol.14, No.10(1999), pp. 3895-3900.

2. Mahmudi, R., *et al*, "Impression creep behavior of lead-free Sn-5Sb solder alloy," *Materials Science and Engineering A*,Vol. 448, No.1-2 (2007),pp. 287-293.

3. Mathew, M.D., *et al*., "Creep deformation characteristics of tin and tin-based electronic solder alloys," *Metallurgical and Materials Transactions a-Physical Metallurgy and Materials Science*, Vol.36A, No.1(2005), pp. 99-105.

4. El-Daly, A., *et al*, "Creep properties of Sn-Sb based lead-free solder alloys," *Journal of Alloys and Compounds*, (2008), In Press.

5. Chen, S.W., *et al*, "Interfacial reactions in the Sn-Sb/Ag and Sn-Sb/Cu couples," *Materials Chemistry and Physics*, Vol.111, No.1 (2008),pp.17-19.

6. Chen, S.W., *et al*, "Lowering of Sn-Sb alloy melting points caused by substrate dissolution," *Journal of Electronic Materials*,. Vol.35,No.11(2006),pp.1982-1985.

7. Lee, C., *et al*, "The 260 ℃ phase equilibria of the Sn-Sb-Ag ternary system and interfacial reactions at the Sn-Sb/Ag joint," *Journal of Alloys and Compounds*, Vol.458, No.1-2 (2008), pp. 436-445.

8. Lin, C.Y., *et al*, "Phase equilibria of the Sn-Sb-Ag ternary system and interfacial reactions at the Sn-Sb/Ag joints at 400 ℃ and 150 ℃," *Intermetallics*, Vol.16, No.2(2008), pp.230-238.

9. Lee, C., *et al*, "The 260 ℃ phase equilibria of the Sn-Sb-Cu ternary system and interfacial reactions at the Sn-Sb/Cu joints," *Intermetallics*, Vol.15,No8 (2007), pp.1027-1037.

10. Zou, H.F. and Z.F. Zhang, "Solid-state and liquid-state interfacial reactions between Sn-based solders and single crystal Ag substrate," *Journal of Alloys and Compounds*, (2008,)In Press.

11. Chang, T.C., *et al*, "Crystal growth of the intermetallic compounds at the Sn-9Zn-xAg/Cu interface during isothermal aging,"*Journal of Crystal Growth*, Vol. 252, No.1-3 (2003)pp. 401-412.

12. Zhu, Q.S., et al., "Fatigue damage mechanisms of copper single crystal/Sn-Ag-Cu interfaces," *Materials Science and Engineering a-Structural Materials Properties Microstructure and Processing*, Vol.435, (2006), pp. 588-594.

13. Ma, X., *et al*. "Development of Cu-Sn intermetallic compound at Pb-free solder/Cu joint interface," *Materials Letters*, Vol.57,No.22-23(2003), pp. 3361-3365.

Application of Embedded Components in Package Substrate

Lingwen Kong, Zhiqin Yang, Jianhui Liu, Minfei Lu
Shenzhen Shennan Circuits Co.,Ltd.
The CATIC Shahe South Industrial Zone.Shenzhen, China
Yangzq@scc.com.cn, +86 1379-840-9720

Abstract

With the high development of IC industry, the size of IC is reduced and the package density is improved continuously. All of these propose higher requests to package substrate. The development directions of package substrate are miniaturization, ultrathin and multi-layer. Embedded components technology, which includes two types: embedded passive and active components, has applied to package substrate [1,2]. Since passive components account for large quantity, research of embedded passive components in package substrate is got recognition. Embedded capacitance and resistance technology is studied comprehensively. This paper introduced the process of embedded components (capacitance and resistance) with an application example, and investigated a series of potential problems and solutions in the industrialization of embedded passive components.

Introduction

In the past 20~30 years, miniaturization of passive components enormously lagged in the course of high density of IC. For adapting to the demands of CSP, SIP, and other subsequent advanced package forms, miniaturization technology of passive components should be used, and moreover connection technology of chips or installation technology such as exacting automatic assembly must be innovation. Since embedded passive components can solve the problem above very well, this technology is obtained attention and popularization by colleges.

Embedded passive components technology is to form lots of capacitances or resistances at one time by using etching or screen printing on internal layer circuit of multiple laminates whose material is mainly organic insulation material. This technology, which is different from 3D installation technology, doesn't need welding or physical assembling, and it can realize high density, low cost and high reliability at the same time. The figure below shows that the wiring length of embedded capacitances and IC components has shortened to the limit value, which is beneficial to the signal transmission in a high-speed. This new design can reduce self-inductance from current flow in capacitance. Although its application means little in the low frequency scope, it is significant in the high-frequency GHz.

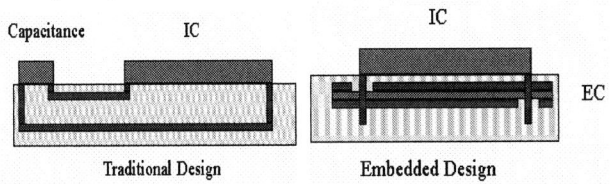

Figure 1 Design of conventional and embedded capacitance

The packaging substrate, as a high-frequency component of the primary package carrier, its function directly affects the infiltration of the function of high-frequency component. Nowadays, along with the development of the thin components, the package substrate which including embedded components has begun to receive attention by designers of the industry.

Introduction of embedded components

According to the material, traditional substrate can be classified as rigid organic package substrate, flexible package substrate, ceramic package substrate, and so on. Although the ceramic package substrate was used in embedded components earliest, due to its exceptive equipments, and high price, its application is limited. Considering the manufacture mobility, the exploitation of embedded components technology in organic substrate receives more attention.

Embedded capacitance material has been produced and listed by Oak — Mitsui, Dupont, 3M and Sanmina-SCI. Currently, organic polymer electrolyte which has the advantage of easy processing is mainly used as electrolyte materiel in embedded capacitances. However, its dielectric constant is only 3-7 usually; therefore the capacity can't exceed the range pF. Such as polyamide film capacitance, which is produced by Dupont (thickness 25 μm and dielectric constant Dk 3.5), the electric capacity is only 122pF/cm². By adding the ferroelectric phase, which has high dielectric constant, such as barium titanate ($BaTiO_3$) into polymer is one of the methods to increase the dielectric constant[4]. For example, with added $BaTiO_3$ particles, the epoxy resin dielectric constant would increase to 23, and the electric capacity would reach 4.65 nF/cm² correspondingly. Theoretical formula of plane capacitance is as followed.

$$C = \xi_r \xi_0 A / t$$

In which, ξ_r is the relative dielectric constant, ξ_0 is vacuum dielectric constant, A is relative capacitance area of upper and lower plane, t is insulation thickness between capacitance layers.

In order to increase the per unit area of the capacitance density, increasing the relative dielectric constant ξ_r or reducing the insulation thickness t are accepted in industry. $BaTiO_3$ is added into the research material by 3M Company, whose minimum insulation thickness reaches 2-micron level in commercial application (when the insulation thickness is too thin, voltage breakdown will be easy to occur).

There are three kinds of processing method, which are screen-printing, compound alloy method, and alloy plating method. Because of the low price and good compatibility, compound alloy has been applied in industry. Resistance material Ohmega Ply of Ohmega Company is very famous. There is a micron-level Ni/P alloy on the surface of copper, which can provide capacitance density with 25~1000 ohm/per unit area.

Processing of embedded components

Embedded components are designed for special units of some high consuming electronic products. The figure below shows an audio unit in a cell phone, which includes embedded capacitances and resistances [3].

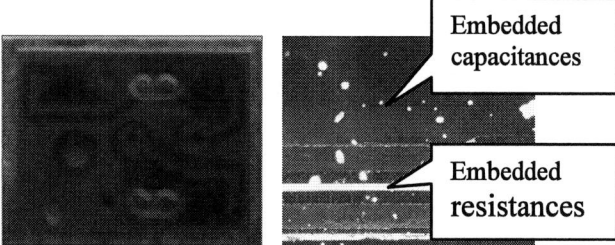

Figure 2 product with embedded capacitances and resistances

Figure 2 shows a normal 4-layer substrate applied embedded capacitances and resistances technology. There is an embedded capacitance in layer 2/3, whose designed capacitance value is 160PF with C-Ply material. And the embedded resistance is in the first layer, whose resistance value is 100 ohm with Ohmega Ply material. Embedded resistance and capacitance values are obtained by testing product sample of different batches.

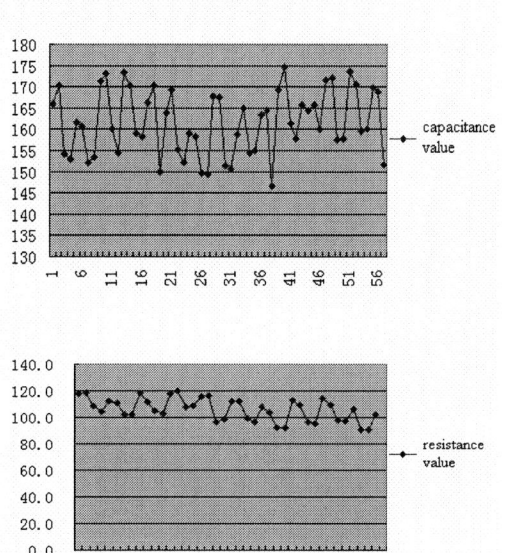

Diagram 1 Values distribution of embedded capacitance and resistance

Diagram 1 indicates that the accuracy of embedded capacitance is controlled within 10%, however, owing to the limits of material and processing, the accuracy of embedded resistance is maintained within 20% lack of secondary

processing such as laser repair, and maintained within 10% applied accurate laser repair.

The capacitance material is very thin. Generally insulation thickness is about 25 microns. Besides of the conventional substrate processing key points, other key points due to the thin substrate should be considered in processing embedded capacitances. For example, level of pressing, horizontal transmission, surface treatment, expansibility and contractibility proportion of the thin substrate.

Twice-graphics, which is general copper lines and resistance graphics, is involved in the processing of embedded resistances. Accurate contraposition is the most important in twice-graphics, such as establishment of etching compensation coefficient, measurement of the panel size, and protection of the thin resistance.

Challenge of substrate with embedded passive components

Design capacity is the bottleneck of the embedded components substrate. Since the main process of this kind of substrate hasn't been identified, there isn't any universal design software. Even if the design software is developed, it will be a long time to be applied popularly.

Embedded components substrate, whose processing is very complex and is as twice as common substrate, demands high processing stability. Its rate of good product is 5% lower than the common substrate.

Conventional test equipments of substrate include flying probe test, universal neilsbed test, and special neilsbed test. Besides, open circuit and short circuit are tested by capacitance and resistance method. However, there isn't a special equipment to test the functional values of designed embedded capacitances and resistances.

Although the substrate with embedded components has low price, but for the application of new technology, materials and processing make the rising cost more obvious. So, embedded components substrate only applied to high-end applications generally.

Conclusions

Embedded passive and active components in substrate are a new technology. Considering the miniaturization and high density of IC, substrate with embedded passive or active components is one of the inevitable choices. In order to popularize and industrialize this technology, material engineers, processing engineers and circuit designers should research together and collaborate.

References

1. Cai Jiqing, The Technology of Embedded Capacitor Substrates, Printed Circuit Information, 05(2004).
2. NIARAKI Asli Rahebeh, MIRZAKUCHAKI Sattar, NAVABI Zainalabedin, RENOVELL Michel, "Test access to deeply embedded analog terminals within an A/MS SoC," Journal of Zhejiang University, 10(2007).
3. Chen Yan, Zeng Shu, "Productive Process of Embedded Capacitor for PCB," Printed Circuit Information, No. 02 (2003).
4. Cai Jiqing, "Epoxy/BaTiO_3 Composite Embedded Capacitor Films and Pastes ," Printed Circuit Information, 07(2006).

Roadmap of Reliability: Methodology and Application in Guangdong's Lead-Free Technology

Xiaoyan Li[a]   Ling Wang[b]   Yonggao Fu[b]   Lu Zeng[c]

[a] School of Materials Science and Engineering, Beijing University of Technology, Beijing, 100022, P. R. China
[b] R & D Center, Guangzhou Electric Apparatus Research Institute, Guangzhou, 510300, P. R. China
[c] Department of Science and Technology, Guangdong Provence, Guangzhou, 510033, P. R. China
xyli@bjut.edu.cn ; xylibjut@yahoo.com.cn

Abstract

Due to the legislations and environment concerns, the applying of lead free technology is the trend in electronic manufacturing industry. Roadmap of Guangdong's lead free technology was established within the framework of industrial roadmapping technology. Reliability is one of the most important features in lead free electronic manufacturing. The corresponding roadmap was formulated in accordingly. This paper presents the methodologies which were employed in the roadmapping of reliability aspect in Guangdong's lead free technology. The employed methodologies include Delphi's method, Brain Storming method and SWOT analytic method. The roadblocks were estimated and the necessary R&D activities were forecasted finally.

Introduction

With the increase of the concerns on environment and health world widely, legislations and technical restrictions on the lead contained electronic devices, such as RoHs, WEEE and the so-called Chinese RoHs, have been formulated and taken effect. The manufacturing and sales of lead contained electronic products are now prohibited in many countries. The electronic industry has been facing the strategic challenge on its long term planning to meets the market's demands. Technology roadmapping approach has been adopted by many firms to schedule their electronic products on this regard. Various roadmaps on lead-free electronics have emerged, such as the roadmaps suggested by Soldertec, JEITA and IPC. Unfortunately, the roadmap on lead-free electronic manufacturing in China is still missing. The establishment of Chinese roadmap for lead-free electronic manufacturing, certainly, is of great importance.

Electronic products manufacturing is one of the most important economic bases in Guangdong's industries. It is estimated that Guangdong's electronic manufacturing holds more than half of the market value in entire China. Roadmap on lead-free technology in Guangdong is of great necessity to be formulated for every reason.

Technology roadmapping, a form of technology planning, is a flexible technique that is widely used within industry to support strategic and long term planning. The approach provides a structured means for exploring and communicating the relationships between evolving and developing markets, products and technologies over time.   It is proposed that the roadmapping technique can help industries survive in competitive environments by providing a focus for scanning the environment and a means of tracking the performance of individual, including potentially disruptive, technologies.

Guangdong's lead-free technology roadmap is deceptively simple in terms of format, but its development poses significant challenges. The principles and methodologies of industrial roadmapping were followed in its formulation. More than 200 experts, either from industries, research institutes or from universities, were involved in its formulation. Guangdong's lead-free technology roadmap covers five sub maps namely maps for lead-free solders, lead-free electronic components, assembly equipments, reliability and standardization respectively. The employed methodologies and the final sub map on reliability were presented here after.

Methodologies

In order to fully take the advantage of lead-free technology developments, Guangdong's electronic industries must have information about the timing of these developments, the resources needed to support the development effort, and the multiple ways in which the technology might be used within the industrial current and future product offerings. For these reason, dynamic near-, mid-, and long-term plans for R&D investments as well as new product and process developments must be developed. The common elements in industrial based roadmaps were shown in Figure 1. Which involves intertwined cycles—those for technologies and those for markets—that are interfaced by lead-free industry's competitive dynamics.

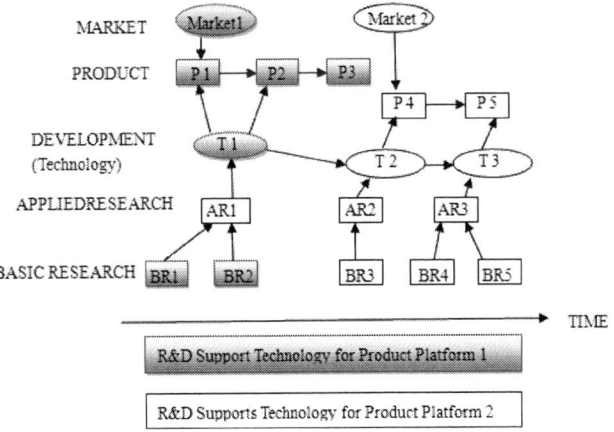

Figure 1   Common elements in an industrial based roadmap [1]

In order to define the details concerning the basic research, applied research, technology development and even product in the market oriented lead-free reliability road map, the methodologies include Delphi's method, brainstorming method and SWOT analytic method were employed [2-5].

Delphi method was used to obtain the most reliable consensus of opinion of a group of experts on lead-free reliability technology by a series of intensive questionnaires interspersed with controlled opinion feedback. Following steps were followed in the Delphi method.

(1)A Delphi team was formed to undertake and to monitor the project.

978-1-4244-2739-0/08/$25.00 ©2008 IEEE

(2)Tow panels were selected to participate in the Delphi exercise. The participants were experts from industries, research institutes and universities.

(3)First round Delphi questionnaire was developed.

(4)The questionnaire was tested for proper wording especially for ambiguities and vagueness.

(5)The first round questionnaires were transmitted to the panelists.

(6)The first round responses were analyzed.

(7)The second round questionnaires were prepared.

(8)The second round questionnaires were transmitted to the panelists.

(9)The second round responses were analyzed.

(10)The report was prepared by the analysis team to present the conclusion on lead-free reliability.

Brainstorming method is a group creativity technique which was adopted to generate a large number of ideas for the solution to the problems related with lead-free reliability technology. The generic process of brainstorming was shown in Figure 2.

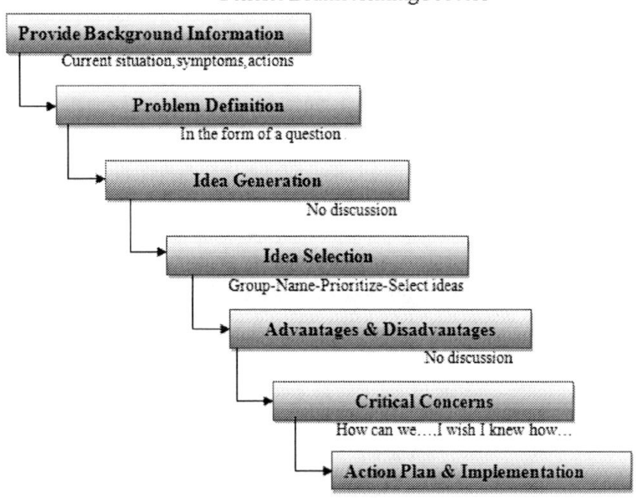

Generic Brainstorming Process

Provide Background Information
Current situation, symptoms, actions

Problem Definition
In the form of a question

Idea Generation
No discussion

Idea Selection
Group-Name-Prioritize-Select ideas

Advantages & Disadvantages
No discussion

Critical Concerns
How can we....I wish I knew how...

Action Plan & Implementation

Figure 2　Brainstorming Process

Following four basic rules were followed in brainstorming process. These are intended to reduce the social inhibitions that occur in groups and therefore stimulate the generation of new ideas. The expected result is a dynamic synergy that will dramatically increase the creativity of the group.

(1) Focus on quantity: This rule is a means of enhancing divergent production, aiming to facilitate problem solving through the maxim, quantity breeds quality. The assumption is that the greater the number of ideas generated, the greater the chance of producing a radical and effective solution.

(2) No criticism: It is often emphasized that in group brainstorming, criticism should be put 'on hold'. Instead of immediately stating what might be wrong with an idea, the participants focus on extending or adding to it, reserving criticism for a later 'critical stage' of the process. By suspending judgment, one creates a supportive atmosphere where participants feel free to generate unusual ideas.

(3) Unusual ideas are welcome: To get a good and long list of ideas, unusual ideas are welcomed. They may open new ways of thinking and provide better solutions than regular ideas.

They can be generated by looking from another perspective or setting aside assumptions.

(4) Combine and improve ideas: Good ideas can be combined to form a single very good idea, as suggested by the slogan "1+1=3". This approach is assumed to lead to better and more complete ideas than merely generating new ideas alone. It is believed to stimulate the building of ideas by a process of association.

SWOT analysis is a strategic planning tool used to evaluate the Strengths, Weaknesses, Opportunities, and Threats involved in lead-free reliability technology. It involves monitoring the environment of the lead-free electronic manufacturing in Guangdong with the aim to identify the key internal and external factors that are important to achieving the sustainable development. It was used to develop a plan that takes into considerations many different factors and maximizes the potential of the strengths and opportunities while minimizing the impact of the weaknesses and threats. A SWOT session is a means of obtaining information from participated experts. It enables participants to take a breath, make a judgment and share their visions on the four pillars mentioned above in order to enrich the collective perception of the way the objectives are pursued. The SWOT sheet used in the roadmapping of lead-free reliability technology was shown in Figure 3.

Roadmap of Reliability

By implement the above roadmapping methodologies, the roadmap concerning reliability aspect in lead-free technology in Guangdong was formulated and shown in Figure 4 [6].

Figure 3　SWOT sheet

Conclusions

By the employment of methodologies such as Delphi's method, Brain Storming method and SWOT analytic method, the roadmap for reliability in Guangdong's lead-free electronic manufacturing was formulated. The near-, mid- and long-term plans for R&D investments were scheduled. However, the updating and modification are necessary due to the present formulated roadmap is the first copy in Guangdong.

References

1. R.N. Kostoff, R.R. Shaller, "Science and technology roadmaps," IEEE Trans. Eng. Manage. 48 (2) (2001), pp.132–143.

2. Irene J. Petrick, Ann E. Echols, "Technology roadmapping in review: A tool for making sustainable

new product development decisions," Technological Forecasting & Social Change, 71 (2004), pp. 81–100.

3. Marie L. Garcia, Olin H. Bray, Fundamentals of Technology Roadmapping, SAND97-0665, Unlimited Release, Distribution Category UC-900, Printed April 1997.

4. Robert Phaal, Clare J.P. Farrukh, David R. Probert, "Technology roadmapping—A planning framework for evolution and revolution," Technological Forecasting & Social Change, 71 (2004), pp.5–26.

5. J. Barrett, "Electronic systems packaging: future reliability challenges," Microelectronics Reliability, 38 (1998), pp. 1277-1286.

6. Roadmap for Guangdong Lead-free Technology, Published by Department of Science and Technology, Guangdong Provence, Jan, 2008.

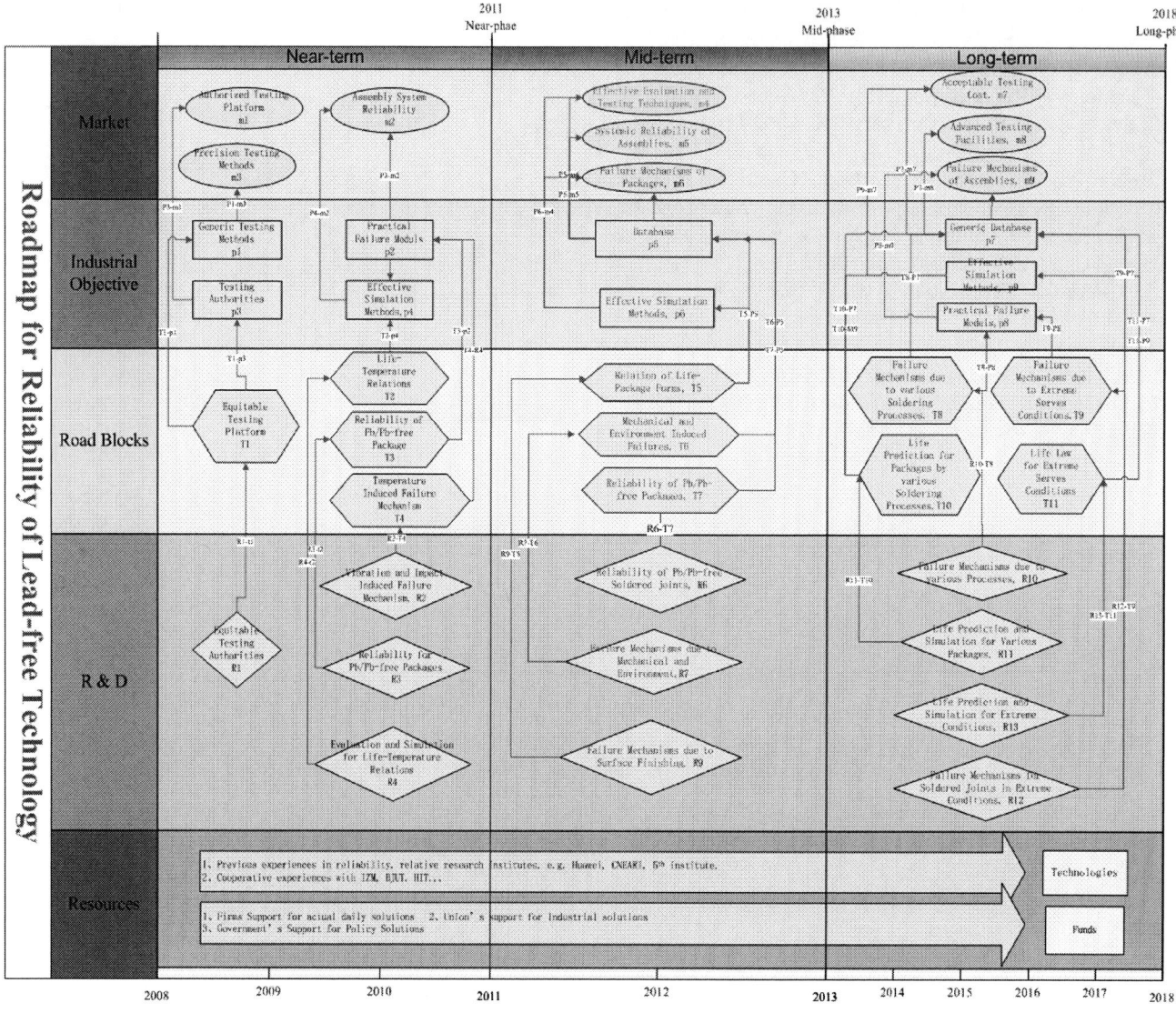

Figure 4 Roadmap for reliability of lead-free manufacturing in Guangdong

Effect of Isothermal Aging on Interfacial IMC Growth and Fracture Behavior of SnAgCu/Cu Soldered Joints

Xiaoyan Li[a]   Xiaohua Yang[b]   Fenghui Li[a]

[a] School of Materials Science and Engineering, Beijing University of Technology, Beijing, 100124, P. R. China
[b] Instrumentation Analysis and Measurement Center, Fuzhou University, Fuzhou, 350002, P. R. China
xyli@bjut.edu.cn ; xylibjut@yahoo.com.cn

Abstract

The reliability of lead free electronic devices depends strongly on the reliability of the soldered joints while the later one was controlled, mainly, by the formation and growth of the interfacial intermetallic compounds (IMCs) between the solder matrix and the substrates. The morphological features, microstructural evolutions and growth kinetics of the IMCs on the interfacial of SnAgCu/Cu soldered joints, under as soldered and isothermal aging condition, were investigated. The three-dimensional IMCs feature was explored by etch the solder matrix out of the SnAgCu/Cu interface. The phases of IMCs were identified by energy dispersive X-ray (EDX). The thickness of the IMCs was measured by element mapping and phase constitution analysis. The SnAgCu/Cu soldered joints were isothermal aged at 125C, 150C and 175C respectively. The corresponding IMCs growth rate was formulated according to the data from various aging time. The growth kinetic of the IMCs was analyzed in the framework of diffusion principles. The tensile strength of the joint was evaluated by in-situ tensile test and the fracture mechanism was analyzed in accordingly. It was found that Cu6Sn5 was formed at the solder and Cu interface during reflowing. With the increase of aging time, the grain size of the interfacial Cu6Sn5 increased and its morphology was changed from scallop-like to needle-like and then to rod-like and finally to particles. The rod-like Ag3Sn phase was formed at the interface of solder and Cu6Sn5 layer with the increase of the aging time. The growth of the IMCs was found follows Arrhenius's diffusion model and the corresponding diffusion factor and active energy were obtained by data fitting. The IMCs growth rate was found increases with the increase of the aging temperature. The fracture site of the soldered joints was changed from the solder matrix to the interfacial Cu6Sn5 layer with the increase of the aging time.

Introduction

In most electronic assembles, solder joints provide both electric connection and mechanical support to the components. The reliability of electric devices depends strongly on the reliability of the soldered joints while the later one was controlled, mainly, by the formation and growth of the interfacial intermetallic compounds (IMCs) between the solder matrix and the substrates. In the past decades, research works were made to explore the formation mechanisms and growth kinetics of the interfacial IMCs with the focus on the morphological features and growth rate under different experimental conditions. Various results were reported by different researchers on that topic, however, the so far proposed mechanisms on the formation and growth of IMCs were not widely accepted, the influences of the IMC's growth on the strength of the soldered joint as well as its reliability

was not well understood. Undoubtedly, further investigation is still of great importance.

Actually, after years of research and development, the transition of the microelectronics industry to lead free soldering has become into practice. Many efforts were made by researchers, in the past decades, either to develop lead free solders, or to modify the soldering profiles in assemblies or to study the reliability of the packages and to estimate the lifetime of the devices. There are indications that soldered structures may be weakened in ways specific to the combination of component, substrate and process[1-3]. Such degradation is often believed to be originated from the formation and evolution of the interfacial intermetallic compounds. On the other hand, the simply implementation of the findings from SnPb soldered joints to Pb-free soldered joints are frequently found misleading[4]. The formation and growth of the IMCs are proven to be controlled by chemical reaction mechanism during reflow process and by diffusion mechanism during thermal aging, however, some researches reported that the kinetics often dose not follow Fick's diffusion law well[5-7]. Current proprietary research is showing the above outstanding problems to be associated with the unique characteristics of the interfacial microstructure evolution of Pb-free soldered joints[8]. With the aims focused on the exploration of the influences of reflow time and isothermal aging temperature on the formation and growth kinetics of IMCs as well as the joint mechanical properties, the present study was performed and the findings on the SnAgCu/Cu interface reported.

Experimental Procedure

Commercial Cu bar with diameter of 3mm was used as the substrates in the present study. A Sn-3.8Ag-0.7Cu Pb-free solder ball was then soldered, by using an infrared reflow oven, in between the two pieces of Cu pads to form a sandwich-like soldered joint specimen. In order to study the effects of the reflow time on the formation and growth of the interfacial IMCs, the specimens were reflowed at peak temperature of 260℃ for 20s and 200s respectively. The preheat temperature and the time, however, were kept the same for all the specimens. The typical temperature profile employed in the reflowing is shown in Figure 1. During the isothermal aging experiment, the specimens were aged at 125℃, 150℃ and 175℃ for 24, 72, 144, 256, 400 and 480 hours so as to evaluate the influence of aging temperature and time on the growth kinetics of the IMCs. For metallogranpic observation of the interfacial microstructure, the specimens were first cross-sectioned perpendicular to the solder-Cu interface of the joints and polished, then etched slightly by using 5% HCl–95%C2H5OH solutions. For the observation of three-dimensional morphology of the IMCs, the solders on the interface were etched deeply by using 10% HCl–90%

C2H5OH solution for 8h followed by etching by 10% HNO3–90%C2H5OH solution for 3h. Scanning electron microscopy (SEM) was used to study the microstructural morphology of the IMCs. X-ray diffraction (XRD) was used to observe the microstructure evolution of the intermetallic grains and to confirm the IMC phases. Energy dispersive x-ray spectrometry (EDS) and XRD were used to characterize the composition of the IMCs and to analysis the elementary distribution. The mean thickness of the IMCs was measured according the elementary mapping and image analysis. Tensile tests the microhardness impressions were performed at different aging time. The fractography of the solder joints, after the tensile tests, were also examined by SEM.

Figure 1 Reflow Temperature Profile

Results and Discussion

The morphologies of the interfacial Cu-Sn intemetallic layers for reflow time 20s and 200s are shown in Fig. 2. The light gray scalloped interfacial regions on the light micrographs are the IMCs. The composition of the IMCs was verified by EDX microprobe analysis. It was found that only the η-phase Cu6Sn5 was formed in as-soldered samples. The intermetallic layer is very rough and irregular. The interface between the η-phase and solder displays a scalloped-like morphology. The formation of η-phase Cu6Sn5 intermetallic layers in soldered joint during the reflow process is believed to be originated from the interfacial reactions between its constituting species, Sn from the solder and Cu from the copper pad. In order to observe the three-dimensional morphology of the interfacial IMC, a deep and selective etching was utilized to remove the solder and expose the intermetallic layer. In this way, the interfacial Cu-Sn intermetallic layer could be viewed in the SEM looking down from the solder side of the interface, as show in Fig.2. The scalloped morphology of the intermetallic layer was clearly visible while the rounded grains and deep channels bwtween the grains could be easily distinguished. It was noted that, the intermetallic layer thickness increases with increasing reflow time. In addition, the size of the scallops increases with reflow time. The increasing sixe of the scallops could be due to some combination of paticle agglomeration and ripening or a competitive grain growth phenomenon according to the literature.

During isothermal aging, the increasing of the thickness of the intermetallic layer and the coarsening of the grains were noted as shown in Figure 3. The Cu6Sn5 η-phase in the layer grows by interdiffusion of Cu and Sn and reaction with each other. In addition, the Cu3Sn ε-phase, verified by EDX, was also found with the increase of the aging time. The Cu3Sn ε-phase forms and grows by reactions between the

Cu substrate and Cu6Sn5 η-phase layer. It was noted that, the solder and Cu6Sn5 η-phase boundary flattens during isothermal aging apart from thickening of the IMC layer, the Cu3Sn ε-phase layer, however, is more uniform in its thickness and is also more planar.

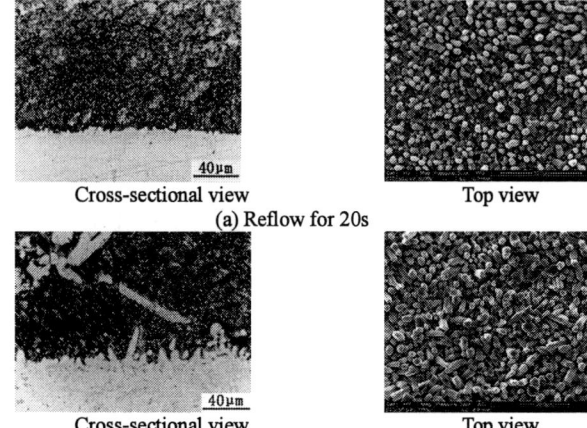

Figure 2 Micrographs of cross-sectional view and top view of interfacial IMCs for different reflow dwell time

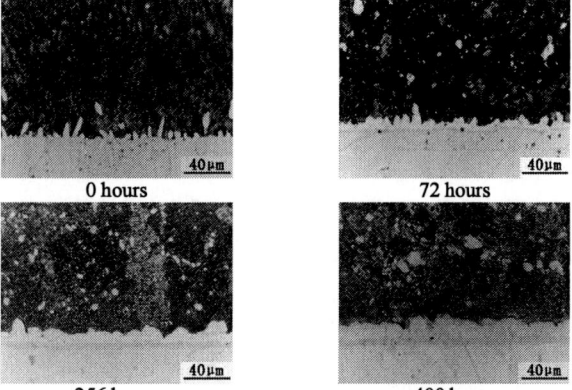

Figure 3 Micrographs of cross-sectional view of interfacial IMCs aged on 175°C for different aging time

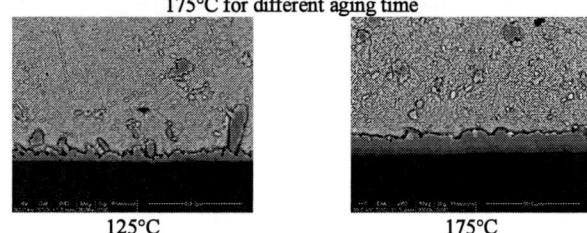

Figure 4 Micrographs of cross-sectional view of interfacial IMCs after aged for 400 Hurs at 125°C and 175°C

With the increase of the aging temperature, the thickness of both the Cu6Sn5 η-phase layer and the Cu3Sn ε-phase layer increases, however, the boundaries of the solder/Cu6Sn5 η-phase layer as well as the Cu6Sn5 η-phase layer/ Cu3Sn ε-phase layer become more planar, as shown in Figure 4.

In order to analysis the element diffusion during isothermal aging, the element mapping, across the IMC lays, was performed at different aging time, as shown in Figure 5. Consequently, the thickness of the Cu6Sn5 η-phase layer and the Cu3Sn ε-phase layer were measured based on the composition and phase constitution analysis.

770

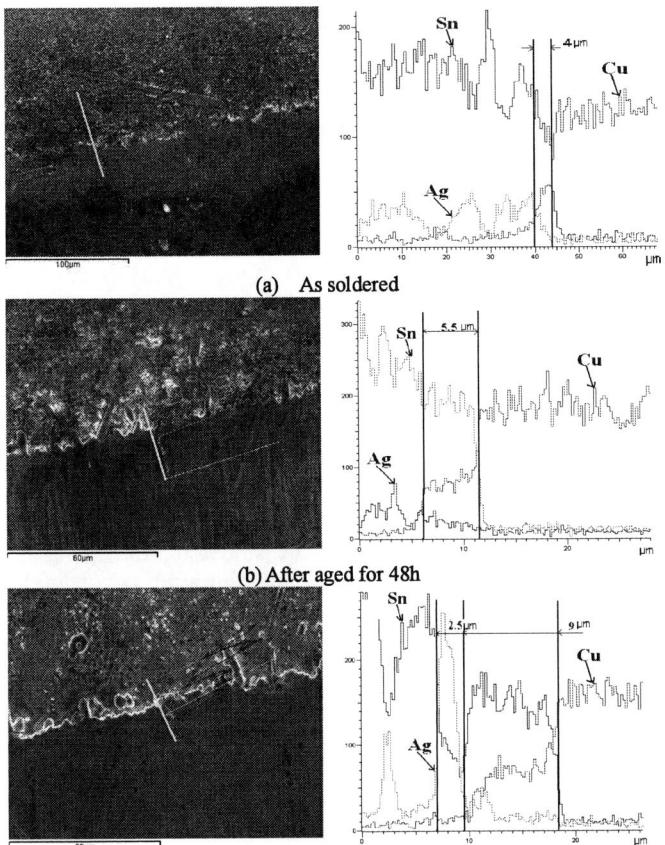

(a) As soldered

(b) After aged for 48h

(c) After aged for 480h

Figure 5   SEM micrographs and EDX spectrums across the interfaces of Sn3.8Ag0.7Cu/Cu joints (150°C)

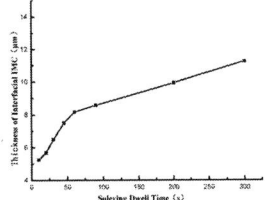
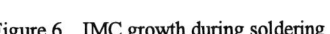

Figure 6   IMC growth during soldering

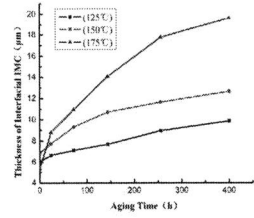

Figure 7   IMC Growth during Isothermal Aging

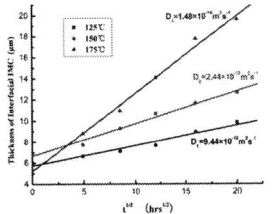

Figure 8   Influence of Isothermal Aging Temperature on Diffusion Factors

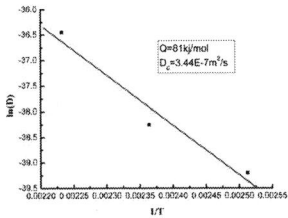

Figure 9 Arrhenius plot for growth of IMC under isothermal aging

The changes of the thickness of IMC layers, during soldering and isothermal aging, was measured in accordance with the above methods, the results were shown in Figure 6 and Figure 7 respectively.   It was found that, the IMC layer thickness increases rapidly in the short soldering time region, the region when the soldering time less than 60s in the present study, while the layer thickness increases slowly when the aging time was further elongated. The variation of the layer thickness with the soldering time may be due to the changes of the mechanisms of IMC formation and growth. Once the solder melts during reflow process, the tin in molten solder immediately reacts with the solder Cu and develops very minute centers of crystalline or nuclei which quickly grow up and form scallop-like islands of the IMC at the interface. This was mainly controlled by the reaction mechanism and the so formed IMC's islands become large enough and connect each other to form a continuous layer of intermetallics. Following the reaction dominated IMC formation process, the grain boundary diffusion and the volume diffusion processes dominated the formation and growth of the IMCs afterward. However, the diffusion of tin or Cu, during the grain boundary diffusion stage and volume diffusion stage, was prohibited by the IMC layer which was formed during reaction stage and finally resulted in the slow down of the growth rate of the IMCs. Nevertheless, the growth rate of the IMCs was found increases with the increase of the aging temperature. This phenomenon, however, could be due to the increase of the diffusion of the tin and Cu at higher temperature.

In order to study quantitatively the growth kinetics of interfacial Cu-Sn intermetallic compound layers in aged surface of soldered joints, the thicknesses of IMC layer in soldered joints aged isothermally at 125, 150 and 175°C for various times were measured and plotted against the square root of aging time, $t^{1/2}$, as shown in Figure 8. It was found that the IMC layer thickness increases linearly with $t^{1/2}$ and the growth rate is faster for higher aging temperature.   Such a $t^{1/2}$ dependence of the growth of interfacial Cu-Sn IMC layer provided a good evidence that the growth was dominated by diffusion process.

In order to have a further understanding of the diffusion process, the activation energy, Q, for the growth of interfacial Cu-Sn intermetallic layer was evaluated by means of an Arrhernius plot of $\ln(D)$ against $1/T$, where D is the interdiffusion coefficient (given by the slope of each line in Figure 8) for the IMC growth and T is the corresponding aging temperature, as shown in Figure 9. From Figure9, the activation energy and the pre-exponential factor, D_0, for interfacial Cu-Sn IMC layer growth could be determined to be 8 KJ/mol and 3.44×10^{-7} $m^2 s^{-1}$ respectively. It should be noted that, however, various activation energies were reported by different researchers. The difference may be due to the difficulty on the individual measurement of the activation energy for Cu6Sn5 growth and Cu3Sn growth on one hand, or the difficulty on the evaluation of the influence of substrate's metallization on the other hand. Undoubtedly, further investigation is needed to reveal the diffusion mechanism of the IMCs growth.

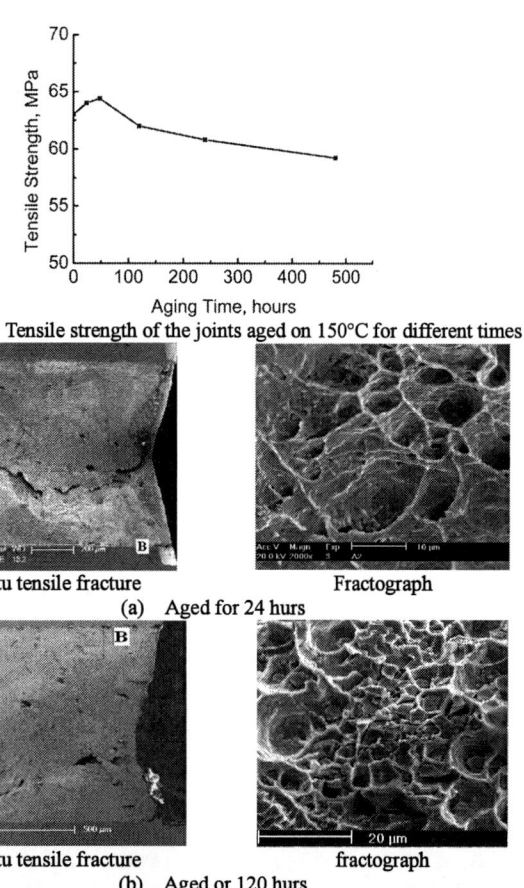

Figure 10　Tensile strength of the joints aged on 150°C for different times

In-situ tensile fracture　　　　　　Fractograph

(a)　　Aged for 24 hurs

In-situ tensile fracture　　　　　　fractograph

(b)　　Aged or 120 hurs

In-situ tensile fracture　　　　　　fractograph

(c)　　Aged for 480 hours

Figure 11　Tensile fractographs of Sn-3.8Ag-0.7Cu/ Cu joint after aged at 150C

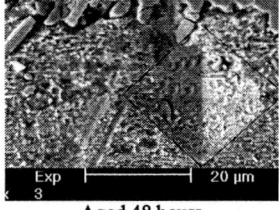
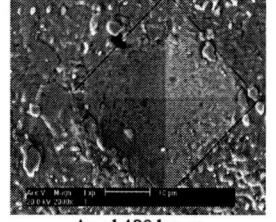

Aged 48 hours　　　　　　Aged 480 hours

Figure 12　Microhardness impressions on the solder matrix

Tensile tests were performed for the sandwich-like soldered specimens on different aging time. Figure 10 shows the changes of the tensile strength of the joints which were aged on 150°C up to 480 hours. It was noted that the joint was strengthen at the early stage of the isothermal aging, however, the tensile strength of the joint decreases with the further aging. This could be due to the influence of the formation and growth of the interfacial IMC layer. At the early stage of the isothermal aging, thin thickness of the IMC layer was formed on the interfacial of the joint The deformation of such IMC layer was constrained, by the

adjacent solder matrix and the Cu substrate, which resulted in the increase of the tensile strength of the joint finally. Nevertheless, with the further growth of the interfacial IMCs, the embrittlement of the IMCs may plays the predominate role in the deformation of the joint which resulted in the decrease of the tensile strength of the joint.

In order to have a further understanding on the fracture behaviour of the soldered joints, in-situ tensile tests were preformed for the joints aged on 150°C for 24 hours, 120 hours and 480 hours respectively. Fractographic examination was also performed with SEM. The results were shown in Figure 11. It was found that, the crack was initiated and propagated mainly along the solder matrix when the aging time was short, i.e. 24 hours in the present case, while the dimples left on the fracture surface indicated a plastic deformation mechanism dominated the fracture. For the joint aged on 150°C for 120 hours, the crack was initiated at the IMC layer but propagated, in the solder matrix, along the solder/IMC interface, as shown in Figure 11 (b). Dimples were also left on the fracture surface. The deviation of the crack path may be due to the difficult deformation of the layer of the IMCs. For the joint aged on 150°C for 480 hours, thicker interfacial IMC layer was formed, the crack was found both initiated and propagated in the layer of the IMCs. In this case, the fracture of the IMC's particles and the cleavage features were observed on the fracture surface of the joint, as shown in Figure 11 (c).

In order to evaluate the influence of isothermal aging on the strength of the solder matrix, microhardness impressions were performed on the specimens aged at 150°C for 48 hours and 480 hours respectively, as shown in Figure 12. It was found that the hardness of the solder matrix decreases with the increase of the aging time. This may implies that the strength of the solder matrix decreases with the increase of the aging time also. However, the changes of the fracture path of the soldered joints, from solder matrix to the layer of IMCs, may indicated that the decrease of the strength of the IMCs was significant compare with that of solder matrix.

Conclusions

Only the η-phase Cu6Sn5 was formed in the reflow process, the formation of the IMC was dominated by reaction mechanism. The growth rate of the IMC was faster at shorter reflow time stage compare with that of longer reflow time stage.

The growth rate of the IMCs increases with the increase of the isothermal aging temperature. The grain boundary diffusion and volume diffusion dominated the IMCs growth during isothermal aging.

The growth of the IMCs follows Arrhenius's diffusion model and the activation energy and the pre-exponential factor, D_0, for interfacial Cu-Sn IMC layer growth in the present study, could be determined to be 8 kj/mol and 3.44×10^{-7} $m^2 s^{-1}$ respectively.

With the increase of the aging time, the fracture paths of the soldered joints changes from the solder matrix to the solder/IMC interface and finally along the IMCs layer. The growth of the IMCs layer and the changes of the mechanical properties of the IMCs and the solder matrix dominated the fracture behavior of the joints.

Acknowledgements

This study was partially supported by National Natural Science Foundation of China （No. 50475043), the Natural Science Foundation of Beijing (No.2052006，No.2082003), which was acknowledged.

Reference

1. Y. G. Lee abd J. G. Duh, "Interfacial Morphology and Concentration Profile in the Unleaded Solder/Cu jont Assambly,"J. Materi. Sci.: Mater. E.ectron., vol.11,2000, pp.33-43.

2. Xiaoyan Li, Zhisheng Wang, "Thermo-fatigue Life Evaluation of SnAgCu/Cu Solder Joints in Flip Chip Assemblies," J. Mater. Process. Tech., vol. 183 , 2007, pp.6–12.

3. Luhua Xu, John H. L. Pang, Kithva H. Prakash and T. H. Low, "Isothermal and Thermal Cycling Aging on IMC Growth Rate in Lead-Free and Lead-Based Solder Interface," IEEE Transaction on Components and Packing Technologies, vol.28, 2005, pp.408-414.

4. L. P. Lehman, R. K. Kinyanjui, J. Wang, Y. Xing, L. Zavalij, P. Borgesen, E. J. Cotts, "Microstructure and Damage Evolution in Sn-Ag-Cu Solder Joints," 2005 Electronic Components and Technology Conference, pp.674-681, May 31- June 3, 2005, Lake Buena Vista, FL, USA.

5. J. H. Lee, J. H. Park, Y. H. Lee, Y. S. Kim and D. H. Shin, "Stability of Changes at a Scalloplicke Cu6Sn5 Layer in Solder Interconnections,"J. Mater. Res. Vol.16, 2001,pp.1227-1230.

6. K. Zeng and K. N. Tu, "Six Cases of Reliability Study of Pb-free Solder Joints in Electronic Packaging Technology," Mater. Sci. Eng.:R:Reports, vol.38, 2002, pp.55-105.

7. J. K. Chen, J. E. Beraun and D. Y. Tzou, "A DualpPhase Lag Diffusion Model for Predicting Intermetallic Compound Layer Growth in Solder Joints," J. Electron. Packing, Vol.123, 2001, pp.53-57.

8. Area Array Consortium Research, Universal Instruments Corporation, Binghamton, NY. USA.

Electrochemical Corrosion Behaviors of ITO Films at Anodic and Cathodic Polarization in Sodium Hydroxide Solution

Wang Hao, Zhong Cheng, Li Jin, Jiang Yiming*
Department of Material Science, Fudan University
220, Handan Rd, Shanghai, People's R China
ymjiang@fudan.edu.cn

Abstract

The electrochemical corrosion behaviors of indium tin oxides (ITO) films were investigated by electrochemical methods in sodium hydroxide solutions. Cyclic voltammetries of ITO films at both anodic and cathodic polarization were carried out. Transmittance spectra, scanning electron microscopy (SEM) and X-ray diffraction (XRD) analysis were used for characterization of the optical transmittance, the corrosion morphology and identification of corrosion product. ITO film remained stable after anodic polarization. In contrast, serious corrosion occurred at cathodic polarization (approximately -1.3V vs. saturated calomel electrode (SCE)). Meanwhile, optical transmittance decreased greatly. The results showed that some of Sn^{4+} in the ITO is reduced to the lower metal state in the form of hydroxides of Sn, which attached to the surface.

1 Introduction

Indium tin oxides (ITO) are widely involved in the application of electrochemical and optoelectronic devices and solar cells because of their high electrical conductivity and their optical transparency to visible radiation [1]. The chemical stability of the ITO layer is of considerable importance determining its use in various application areas. For example, it was noted that ITO is a rather stable photoelectrode in basic solutions but is soluble in acidic solutions [2]. Moreover, ITO is commonly used as substrates for electrodepositing of semi conducting metal oxides [3].

In many of these papers, since the researchers use cathodic potential down to −1.2 V, one of the primary criteria is the electrochemical stability of the underlying ITO substrates under the cathodic stress. However, up to now, majority studies have been focused on the kinetics and mechanisms of the growth [4, 5], modifying procedure of the preparation process [6] and the influence of basic properties by different preparation methods [7-10], without paying much attention to electrochemical stability of ITO. In a general way, oxides are expected to be stable against oxidative processes since the metal cation is at its highest oxidation state. However, some literature results showed that oxide corrosion could occur by through the breaking of surface bonds without change of the metal state. For instance, during the photoelectrochemical corrosion of oxides, such as ZnO or ITO materials, holes are produced which can recombine with electrons from surface bond under illumination. Oxide is dissolved via a multi-step reaction. Another example is given by tin dioxide electrodes applied for organic pollutants treatment in water. The corrosion process occurs owing to the capture of electrons belonging to Sn-O

bonds by highly oxidizing radical species such as OH^- and Cl^- at anodic potentials more positive than 1.0V.

In this work, corrosion behaviors of ITO in NaOH solution were investigated in detail. Both anodic and cathodic polarization tests were carried out on ITO films. Cyclic voltammetries were used to characterize the possible change of electrochemical property of ITO after anodic or cathodic treatment. Furthermore, optical transmittance, X-ray diffraction patterns and SEM were also performed for identification of corrosion product.

2 Experimental

The ITO films for experiments were commercial products (Shenzhen Nanbo Co.Ltd), the thickness of which is 180nm. The cyclic voltammograms of both anodic and cathodic polarization were carried out in a conventional three-electrode cell using an electrochemical workstation model of CHI660B. The saturated calomel electrode (SCE) was used as the reference electrode, and the platinum plate (Pt) was used as the counter electrode. The corrosive medium was a 0.1mol/L NaOH solution, and the scanning rate was 10 mV/s. Transmission spectra were recorded with CVI 240. X-ray diffraction (XRD) analysis was used to obtain the changes of the films' structure after cyclic voltammetry treatment. The XRD experiments were performed on a Rigaku D/max-γB X-ray diffractometer with Cu-Kα radiation and at a 2θ scanning rate of 1.2°/min. And scanning electron microscopy images of the samples were obtained using field emission scanning electronic microscopy (FE-SEM, Philips-XL30).

3 Results and discussions

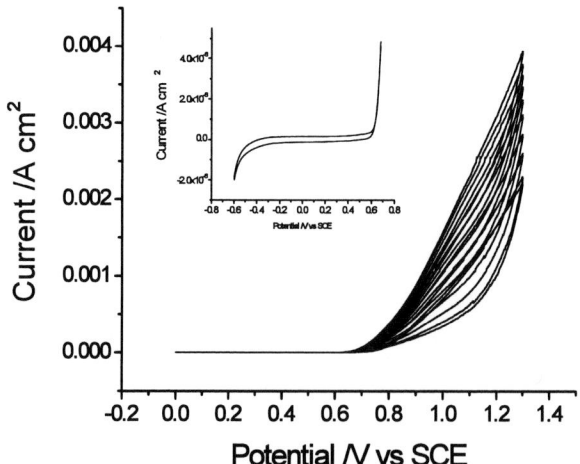

Fig.1 Cyclic vlotammograms of ITO in 0.1 M NaOH between 0 and 1.4 V for the first 10 cycles.

978-1-4244-2739-0/08/$25.00 ©2008 IEEE

The current of Fig. 1, which is related to the oxygen evolution, decreases with increasing cycle number. Thus the overvoltage for oxygen evolution increases during the cycling process. The inset of Fig.1 also shows a typical cyclic voltammogram of ITO between –0.6 and 0.68 V. This range is related to the relatively large "window" between the area of oxygen evolution in the positive direction and hydrogen evolution in the negative direction [11]. The fact that ITO is an oxide in the highest oxidation state of the metal is responsible for the large electrochemical window in the positive range. It is seen that oxygen evolution started at about 0.6 V.

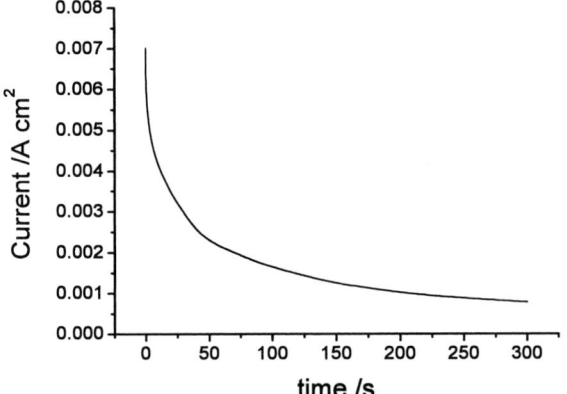

Figure 2 the current-time curve of ITO in 0.1 M NaOH at 1.5 V.

Fig. 2 shows the decrease in the anodic current with time which is in consistent with Fig. 1. After about 5 min the anodic current reaches a nearly constant value.

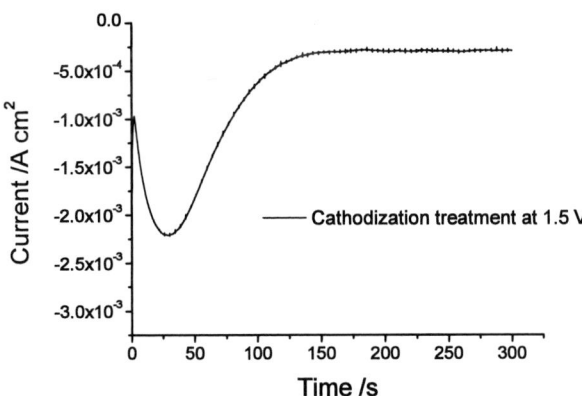

Figure 3 the current-time curve of ITO in 0.1 M NaOH at -1.5 V.

Figure 3 shows a decrease in the cathodic current with time at initial times then followed by an increase. After about 3 min the anodic current attains a nearly constant value. In order to investigate the possible changes in the ITO layer which are responsible for its electrochemical behavior, various properties of ITO (electrochemical characteristics, sheet resistance, optical transmission, surface morphology) before and after anodic (1.5 V for 5 min) and cathodic (-1.5V for 5min) treatment in 0.1M NaOH for 5 min (hereafter called anodization and cathodization) were measured.

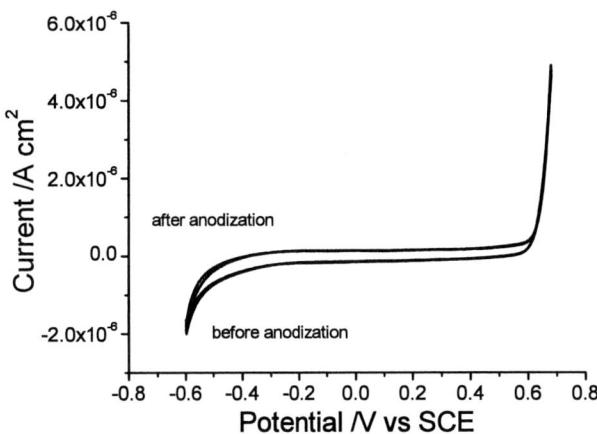

Figure 4 the cyclic voltammograms of ITO in 0.1 M NaOH before and after anodization.

It is clear that no obvious change is observed from cyclic voltammetric i-E curves after anodization at 1.5V for 5min from Fig. 4, indicating that no electrochemical reaction takes place at anodic treatment.

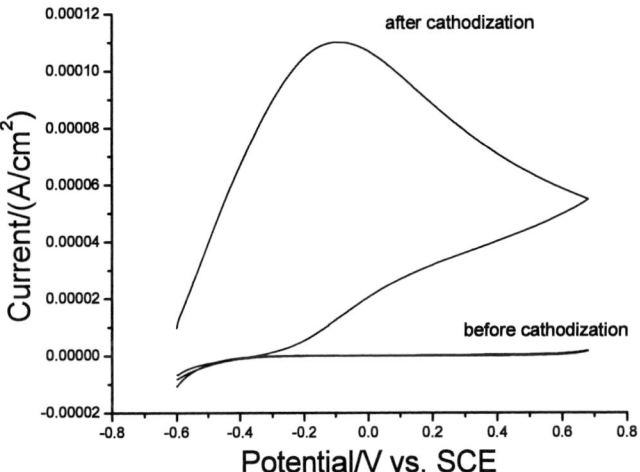

Figure 5 cyclic voltammograms in 0.1 M NaOH before and after cathodization at -1.5V for 5min

For Fig. 5, It is seen that the electrochemical behavior of ITO change dramatically after cycling the electrode potential at −1.5 V for 5 min. The sharp rise in oxygen evolution disappears, a peak at about −0.1 V is found which is related with a deposition/dissolution process. It is also noted that exchange charges (capacity) are substantially greater after cathodization, especially in the left part of cyclic voltammogram. The reason for this could be the larger surface area due to the much greater surface roughness. The results indicate that ITO film is undergoing some changes in the solution during potential cycling in the cathodic potential at −1.5 V. The reaction products accumulating on the film result in much broader voltammetric response and more peak current. The electrochemical reactions, which can occur at this potential range, are obviously the reduction of Sn4+, or In3+ in the ITO film [12] to lower oxidation states. On visual

775

examination, an opaque film was found to form on the surface.

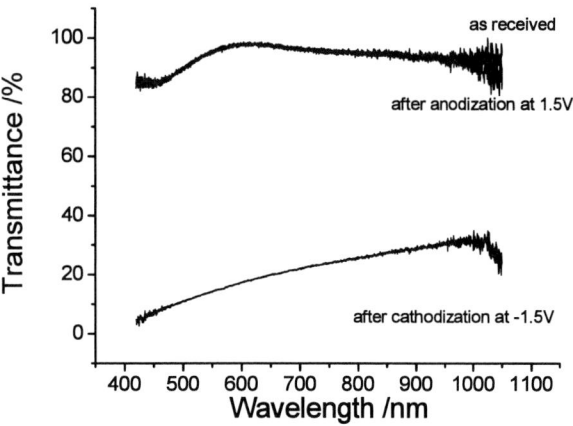

Figure 6 Transmission spectrum change of ITO after anodization and cathodization

Fig. 6 shows the optical transmittance changes after anodic and cathodic treatment. The transmission spectrum remains almost the same after anodization, indicating that there is no considerable dissolution of ITO films or accumulation of reaction products form on the ITO surfaces. This result is in agreement with the cyclic voltammograms results after anodic treatments. It should be noted that there is a dramatic reduction in the light transmittance from 90% to 12% at a wavelength of 500 nm after cathodization, confirming our visual examinations that an opaque film was formed during the cathodic treatment.

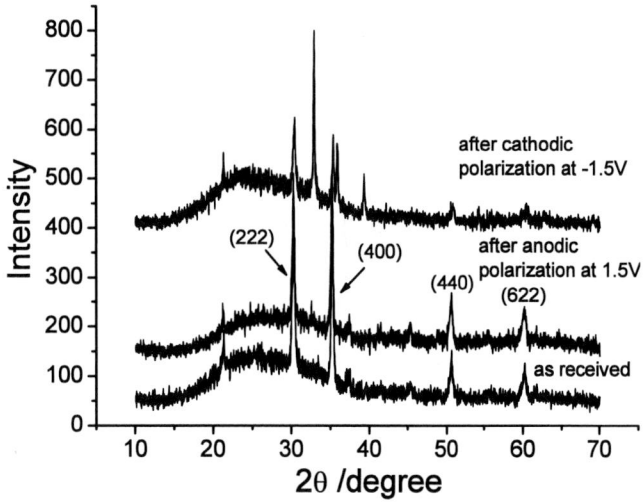

Figure 7 XRD pattern of ITO before and after cathodic polarization

X-ray diffraction patterns of untreated and treated ITO layers both at anodic and cathodic polarizations are presented in the Fig. 7. For the sample untreated, the diffraction peaks correspond to (222), (400), (440), and (622), as which the ones of ITO films at anodic polarization are just the same. It is concluded that the ITO films remain stable in the process of anodic polarization during the oxygen evolution. While the peaks of the sample at cathodic polarization has been changed

dramatically. The 4 diffraction peaks mentioned before has been getting less sharp and there are a few new diffraction peaks. It's possible to estimate that the surface of ITO film was damaged to some extent and some kind of new compound has been created during cathodic polarization, which probably stays on the surface of the film.

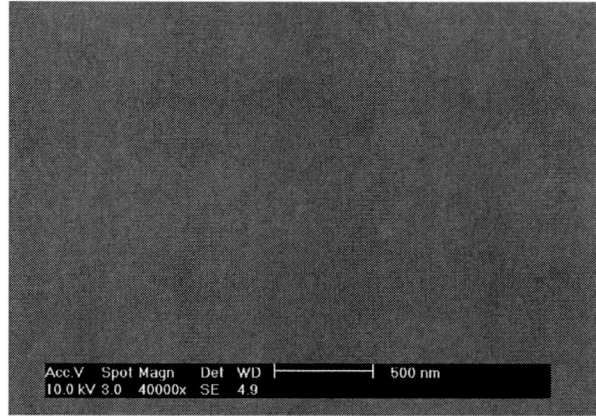

(a) as received

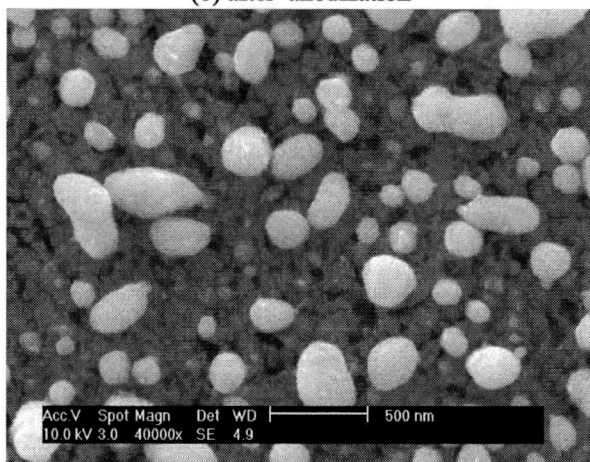

(b) after anodization

(c) after cathodization

Figure 8 SEM images for ITO

In order to confirm the reaction product, SEM images are involved as Fig. 8 shows the results. As can be seen, the surface of the ITO films at anodic polarization has no difference with that of the ITO films untreated, which is in line with XRD results. According to Fig. 8(c), granular products has been found on the surface of ITO film at cathodic polarization. It also shows intergranular corrosion occurs on the surface of ITO film.

The corrosion of ITO exhibits two different behaviors depending on the polarization of different ways. During the anodic polarization, no visible change of electrochemical property (i.e., cyclic voltammograms) has been observed, indicating that no corrosion reaction takes place in ITO film or on the surface. This can also be confirmed by transmission spectrum, XRD and SEM results. Since In and Sn is at its highest metal state in ITO, it is impossible for them to be further oxidation. Besides, the surface of ITO film bubbles up continually in the latter part of cathodic polarization. It is regarded as hydrolysis of oxygen in the reaction. So it is reasonable to figure that the reduction of oxygen in the solution is the cause of the sudden increase of the current. However, it should be noted that during the cathodic polarization, considerable change of electrochemical behavior of ITO has been found. Optical transmittance drops dramatically after cathodic treatment, indicating an opaque corrosion product formed on the surface. Furthermore, new diffraction peaks related to tin hydroxide Sn has been observed on XRD patterns. It is supposed that the corrosion of ITO is associated with the reduction of Sn (IV) oxide to lower metal state, i.e., Sn (II) or Sn. These corrosion products are likely metal hydroxides of Sn during the cathodic polarization. As a consequence, the light throughput in the visible region decreased from about 90% to 12% (see Fig. 6). Our results are in agreement with the previous study [13]. The reduction of Sn (IV) would also cause some roughness and disordering of the crystalline ITO. Evidence for surface roughening is provided in the SEM data.

4 Conclusions

ITO films exhibits two different mechanisms depending on the polarization of different ways in NaOH solution. No obvious changes of cyclic voltammograms and transmission spectrum between untreated ITO film and the ones after anodic polarization, indicating no reaction of ITO occurs at anodic polarization. It can be confirmed with XRD patterns and SEM images. Since In and Sn is at its highest metal state in ITO, it is impossible for them to be further oxidation. In contrast, cyclic voltammograms of ITO film at cathodic polarization shows the reaction that ITO films are involved in occurs. The light throughput in the visible region decreased from about 90% to 12%. It can be seen that granular products has been found and intergranular corrosion occurs on the surface of ITO film in SEM image. It is supposed that some of Sn4+ in the ITO are reduced to the low-hydroxide attached to the surface.

Acknowledgments

This work was supported by National Natural Science Fund, No. 50571027, 50701010 and 10621063; Special Fund of Ministry of Science and Technology, No. 2005DKA104001; Shanghai Leading Academic Discipline Project, No. B113.

References

1. N.R.. Lynam, "Trasparent electronic conductors." Proc. Symp. Electrochromic Materials, Pennington, NJ, 1990, Vol. 90-2, pp. 201-231. 2. A.Nanthakumar and N.R. Armstrong, in H.O. Finklea (ed.), Semiconductor Electrodes, Elsevier (Amsterdam, 1988), pp. 224.
3. C.O. Avellaneda, M.A. Napolitano et al, "Electrodeposition of lead on ITO electrode: influence of copper as an additive," Electrochimica Acta, Volume 50 (2005), Pages 1317-1321.
4. D. Maestre, A. Cremades et el, "Thermal growth and structural and optical characterization of indium tin oxide nanopyramids, nanoislands, and tubes," Journal of APPLIED PHYSICS, Volume 103, 093531 (2008).
5. S.Omanovic, M.Metikos-Hukovic, "A study of the kinetics and mechanisms of electrocrystallization of indium oxide on an in situ prepared metallic indium electrode," Thin Solid Films, Volume 458 (2004), pp. 52–62.
6. Anna Prodi-Schwab, Thomas Luthge et al, "Modified procedure for the sol–gel processing of indium–tin oxide (ITO) films," Journal of Sol-Gel Science and Technology, volume 47 (2008), pp. 68–73.
7. Y.Z. You, Y.S. Kim et al, "Electrical and optical study of ITO films on glass and polymer substrates prepared by DC magnetron sputtering type negative metal ion beam deposition," Materials Chemistry and Physics, Volume 107 (2008), pp.444–448.
8. M.M. Munir, Ferry Iskandar et al, "Optical and electrical properties of indium tin oxide nanofibers prepared by electrospinning," Nanotechnology, Volume 19 (2008), 145603 (6pp)
9. T.J. Stanimirova, P.A. Atanasov et al, "Optical and structural properties of undoped and palladium doped indium tin oxide films grown by pulsed laser deposition," Applied Surface Science, Volume 253 (2007), pp.8206–8209.
10. Y. Zhou, P.J. Kelly, "The properties of tin-doped indium oxide films prepared by pulsed magnetron sputtering from powder targets," Thin Solid Films, Volume 469–470 (2004), pp.18–23
11. Neal R. Armstrong, Albert W. C et al, "Electrochemical and surface characteristics of tin oxide and indium oxide electrodes," Analytical Chemistry, Volume 48 (1974), pp. 741
12. A. Kraft, H. Hennig et al, "Changes in electrochemical and photoelectrochemical properties of tin-doped indium oxide layers after strong anodic polarization," Journal of Electroanalytical Chemistry, Volume 365 (1994), pp.191-196.
13. Jason Stotter, Yoshiyuki Show et al, "Comparison of the Electrical, Optical, and Electrochemical Properties of Diamond and Indium Tin Oxide Thin-Film Electrodes," Chemistry of Materials, Volume 17 (2005), pp. 4880-4888.

Effects of Bonding Pressure on Nonlinear Dynamic Characteristic of the Ultrasonic Wire Bonding System

Lei Lv, Lei Han
College of Mechanical and Electrical Engineering
Central South University, Changsha, China 410083
E-mail: lvleiyhz@yahoo.com.cn

Abstract

Ultrasonic wire bonding is one of the main methods in the package of the chip and substrate of the chip. The vibrations of transducer were tested. Surrogate-data method of phase-randomized based on correlation dimension is proposed to identify the nonlinear chaos of data obtained by ultrasonic wire bonding system. The result indicate there is nonlinear factor in the axial direction of the amplitude transformer terminal when ultrasonic bonding system is loaded different pressure, It is helpful for understanding of transducer and provides the theoretical foundation for further study of the vibration in ultrasonic wire bonding with the nonlinear time series method.

1. Introduction

In the first level IC fabrication industry, wire bonding is the main microelectronics packaging way to make fine-pitch connections between chip and substrate utilizing the ultrasonic vibration and the bonding pressure[1]. With the ULIC developing, more of the lead-number and smaller of the packaging area descend bonding reliability in wire bonding, and more challenges in the packaging field. So further known of the wire bonding packaging process must to be done.

The bonding system is actual a complex nonlinear system, ganging between the amplitude transform and transducer, bonding tool and the amplitude transform coupled, as well as outside noise disturbed, the system shown obviously nonlinear characteristic because of all of these in energy conversion and transmission processing[2]. There will be have many kinds of vibration when bonding, such as the amplitude transformer vibration in 3D dimension, bonding tool transverse direction and tangential direction vibration because of coupled between bonding tool and amplitude transformer, torsion vibration, stray vibration and so on. The complex vibration would bring more and more important effect to micro-connection between the bonding line and substrate because of bonding velocity getting more and more faster and bonding pitch becoming more and more smaller. So whether the complex vibration contains nonlinear factor or not, it would be very important to study wire bonding system and optimize performance of the bonding machine[3]. We make high precision test to the transducer of ultrasonic wire bonding system when bonding by Laser Doppler Vibrometer (LDV), utilize surrogate data algorithm to process the vibration velocity time series measured, and take nonlinearity criteria of test statistics based on correlation dimension to determine whether the vibration contains nonlinear chaotic characteristic.

2 Method of surrogate data

2.1 Principle

The ration distinguishing means of the time series chaotic dynamics usually have two sorts. One is directly recognition time sequence chaotic characteristic, such as counting correlation dimension, Lyapunov exponent, K entropy, complexity and so on.

Although these approaches have made many success, when the data maintain measurement noise or data length is not enough long, we would get false result because of the reliability of these approaches dependent on as much as possible long data[4-6]. So these directly approaches have certain limitation. The other one is that we could test the nonlinear component of the data to indirectly estimate chaos dynamics. In 1992, Theiler and other persons made surrogate data as the approaches to examine the nonlinear element about time series[4,7]. This mean could effectively avoid the limitation of the direct approach, and it would exactly identify nonlinear chaos when we combine it with some special algorithm. Surrogate time series is a stochastic time series, and it is constructed by satisfying a null hypothesis based on some linear and stochastic characteristics. By comparing the nonlinear indices of the original data and surrogate data sets, it is possible to decide whether a linear process is sufficient for describing the original data. If the observed data do not have nonlinear deterministic properties but have linear ones, the estimated properties become almost the same as those of its surrogates. On the other hand, if the original data cannot be described using a linear system, the properties are completely different from those of the surrogates. Because of eliminating the possibility that original data is settled by linear process, so the data must have nonlinear factor[8]. The ultrasonic wire bonding is a process of recombination energy field synthetic action, physical principles very complex. Based on surrogate data algorithm predominance, we take this approach to examine nonlinear element of the ultrasonic wire bonding course.

2.2 Data creation

Surrogate data is produced from original series with data shuffling or phase-randomizing. It has spectral properties similar to the original data, that is, the surrogate data sequence has the same mean, the same variance, the same autocorrelation function, and therefore the same power spectrum as the original sequence, but nonlinear phase relations are destroyed. For testing serial dependence, we often use Random shuffle surrogates.

In this paper, phase-randomizing algorithm from Theiler and Prichar was used[4,9]. The surrogate data are generated by the following steps:

1）We apply the Fourier fast transformed (FFT) to the original time series $x(n)$, and its power spectrum function $X(f)$ will be obtained.

2）Then the stochastic time series is produced by the random function. It will be added to the phase components of the $X(f)$. The phase is randomized, and we get $X'(f)$.

3）The inverse FFT is utilized to $X'(f)$, and then the surrogate time series $x'(n)$ will be gotten.

The surrogate data $x'(n)$ withholds the linear autocorrelation characteristic of the original data, and because of phase randomization, the nonlinear autocorrelation characteristic is wiped off. The imaginary part of surrogate data must be zero. If not, the chaotic characteristic criteria would be valid, because the surrogate data is possibility unreal.

2.3 Nonlinearity criteria

In the paper, correlation dimension was utilized to examine the null hypothesis. The Grallberger-Procaccia algorithm is used to estimate the correlation dimension[11].

We should make some groups of surrogate data from a group of original data, and compute the mean. Then, the confidence of refusing null hypothesis would be improved. In reference [12], the correlation dimension criteria was converted to the S criteria. The following is the definition.

If we set up D_{orig} as the correlation dimension of the original data, $\bar{D}_{surr}$ as the correlation dimension mean of the surrogate data, and σ_{surr} as the standard deviation of the correlation dimension of all the surrogate data, then the S criteria of the $100(1-\alpha)\%$ confidence is shown as the equation (1).

$$S = \left| D_{orig} - \bar{D}_{surr} \right| / \sigma_{surr} \qquad (1)$$

The value of significance level α is made 0.05. If S is more than 1.96, it indicates that there is obvious difference between the original data and the surrogate data, then the original data is nonlinear time series by a 95% degree. When S is less than 1.96, we believe that the original data is random times series by a 95% degree.

2.4 Simulation examination

Duffing oscillator was taken as an example to examine the reliability of the surrogate data algorithm. The following is the equation:

$$\ddot{x} + 0.3\dot{x} - x + x^3 = 0.32\cos 1.2t \qquad (2)$$

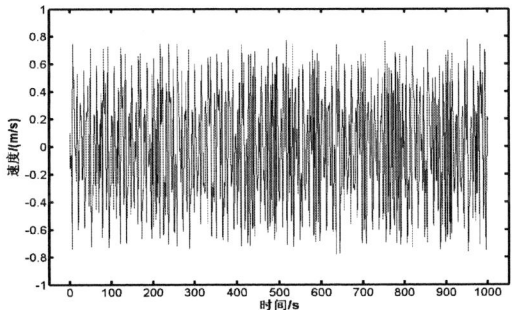

Fig.1 Plot of Duffing oscillator time series

Fig.1 shows the numerical solution and Fig.2 figures the phasor.

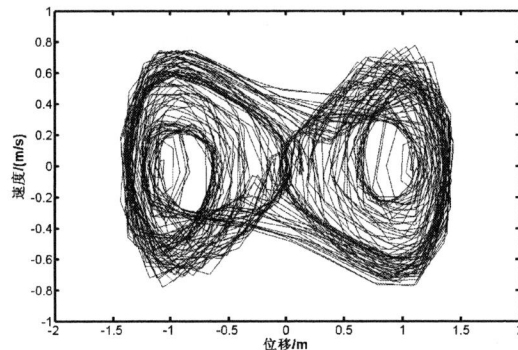

Fig.2 Duffing oscillator phasor

The phasor of Duffing oscillator obviously contents the chaotic characteristic, and this characteristic had been analyzed in [13] about the equation. Then we could get the surrogate data about the vibration time series and the correlation dimension curve of Duffing oscillator by the algorithm. Fig.3 shows the result.

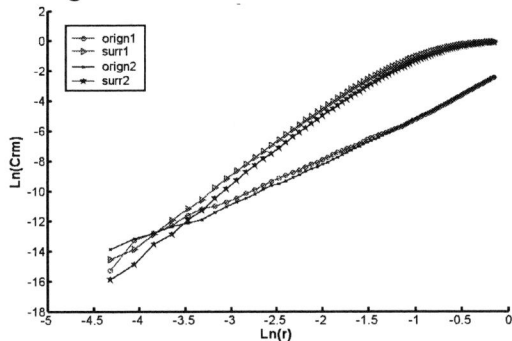

Fig.3 Correlation dimension curve of Duffing oscillator

The value of the standard deviation and correlation dimension are shown as following:

Table 1 The statistical comparison of the Duffing oscillator time series

statistic	original sequence1	surrogate sequence2	original sequence1	surrogate sequence2
sd σ	0.4050	0.1758	0.4054	0.1859
origin D_{orig}	2.5914		2.6040	
surr D_{surr}		3.5956		3.5600

Criteria value of the first group:

$$S = \left| D_{orig} - D_{surr} \right| / \sigma_{surr} = \left| 3.5956 - 2.5914 \right| / 0.1758 = 5.7122$$

Criteria value of the second group:

$$S = \left| D_{orig} - D_{surr} \right| / \sigma_{surr} = \left| 3.5600 - 2.6040 \right| / 0.1859 = 5.1425$$

The null hypothesis is false because the S value of the two group data are both more than 1.96. It indicates that the original data contacts obviously nonlinear chaotic characteristic, coinciding with the academic result. The surrogate data algorithm has upper reliability.

3 Identification of nonlinear chaotic characteristic

3.1 Data acquisition

The vibration signal of the transducer system of a U3000 ultrasonic wedge bonder (Weixun Co., Shenzhen) with the working frequency of approximately 57 kHz was measured by using a Polytec® PSV-400-M2 Laser Doppler Vibrometer (LDV) and software Labview, sampling frequency 512kHz and sampling time 140ms.

A test point is the axial direction of the transducer terminal. The experimental apparatus are shown as Fig.(4).

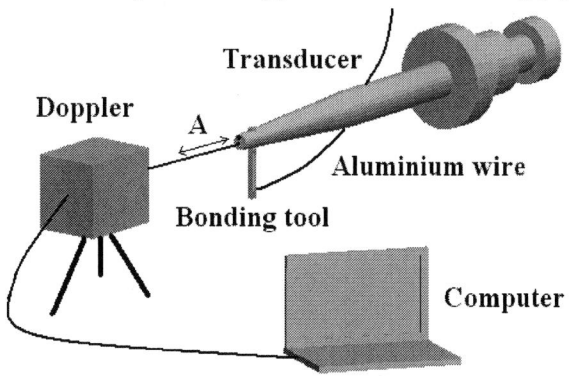

Fig.4 Experimental schematic diagram

The sensing beam of LDV was aligned to the testing point of transducer system. Triggered by the ultrasonic loading signal ,the velocity versus time profiles then was recorded at a sampling rate of 1 MHz. Typical velocity signal measured at the transducer (testing point A) is shown in Fig.5.

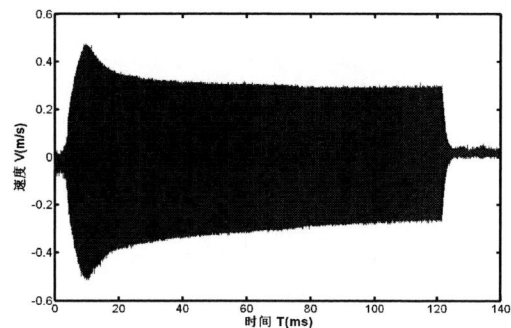

Fig.5 Typical velocity signal at location A by LDV

The value of the bonding strength in different pressures were gotten by Dage Series 4000 Test Equipment. Fig.6 shows the results.

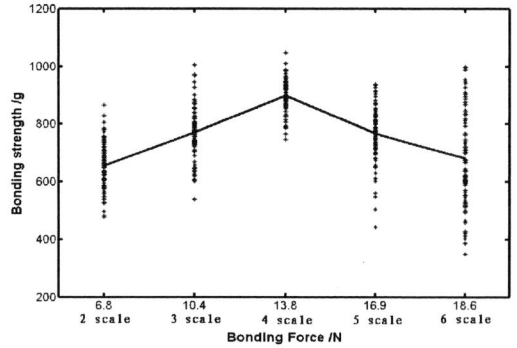

Fig.6 Distribution map of bonding strength

The Fig.6 shows that the wire-bonding strength is changeable unceasing when bonding force changes. In Fig.6, the wire-bonding strength is mainly on 850-950g when the force situations is 13.8N (also pressure 4 scale), and the strength is mainly changes below 800g in other bonding force

situations. As the pressure increasing, the bonding strength turns up an obviously peak in 4 scale, the average boning strength reach the largest value (900.3g); when the pressure continue to increase, the bonding strength appears the tendency of dropping.

3.2 Nonlinear identification of vibration signal

The correlation dimension of the vibration signal from 5ms to 10ms was calculated by the section 2.3, the results as Fig.7.

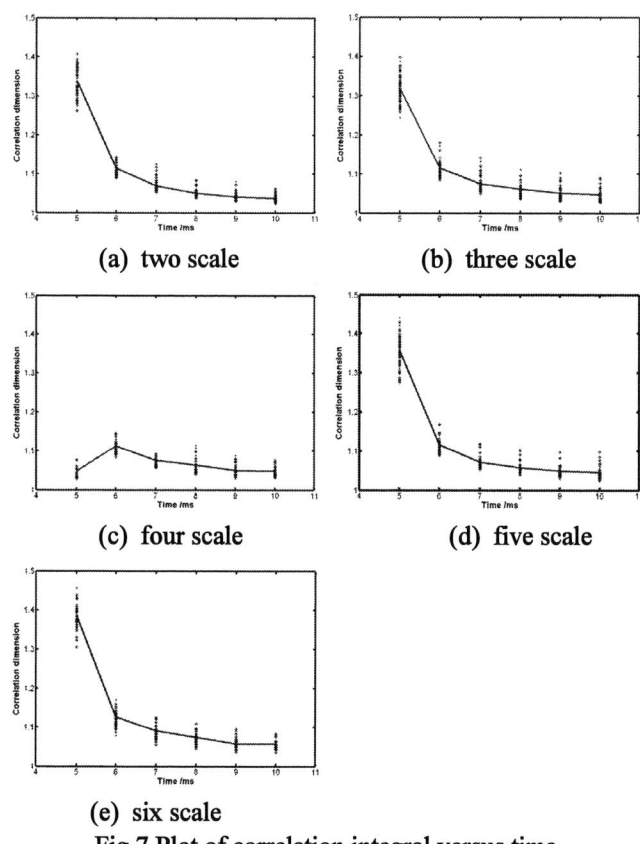

(a) two scale   (b) three scale

(c) four scale   (d) five scale

(e) six scale

Fig.7 Plot of correlation integral versus time in different pressure

In Fig.7, the third map (also Fig.7 (c)) is different from other four maps. The correlation dimension value submits downtrend as the time in Fig.7 (a), (b), (d), (e), but in Fig.7(c), the trend is rising and then dropping, existing a peak which is lower than the maximum of the other four curves .

The curves of correlation dimension about the vibration time series and surrogate data in different bonding pressures are shown in Fig8~Fig12.

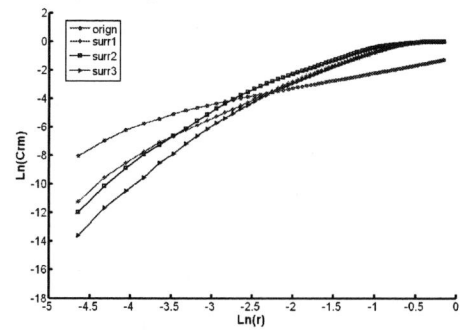

Fig.8 Correlation integral curve of the origin and surrogate data in pressure 2 scale

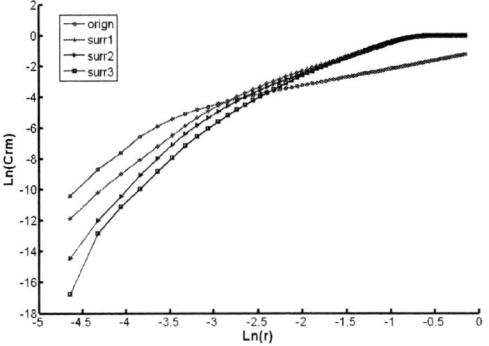

Fig.9 Correlation integral curve of the origin and surrogate data in pressure 3 scale

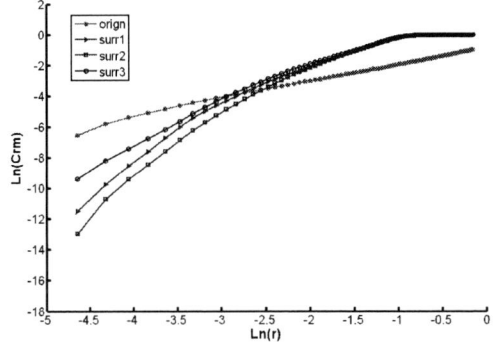

Fig.10 Correlation integral curve of the origin and surrogate data in pressure 4 scale

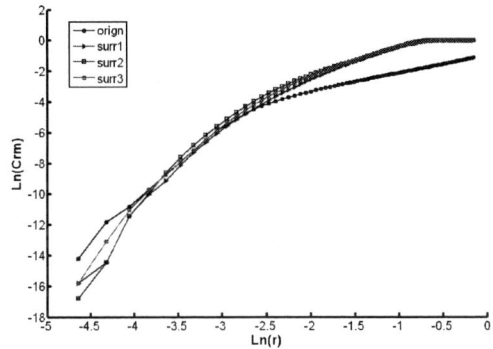

Fig.11 Correlation integral curve of the origin and surrogate data in pressure 5 scale

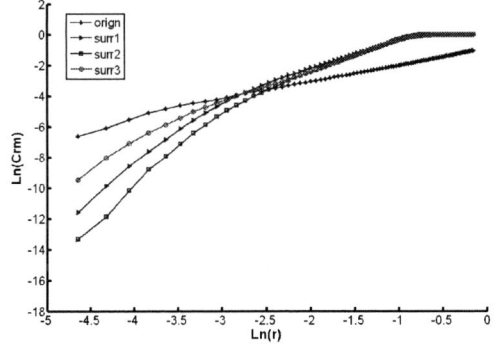

Fig.12 Correlation integral curve of the origin and surrogate data in pressure 6 scale

The approximate straightway of the correlation integral curve was fitted a straight line by the least square method, calculating the slope of the fitting straight line, then we could have the correlation dimension value[14,15].

The Table2~Table6 show the mean, standard deviation and correlation dimension of each group data.

Table 2 The statistical comparison of the vibration time series in pressure 2 scale

statistic	original sequence	surrogate sequence1	surrogate sequence3	surrogate sequence3
mean $\bar{x}$	0.2541	0.2512	0.2510	0.2564
sd σ	0.1208	0.1270	0.1273	0.1160
orign D_{mp}	1.1904			
surr D_{ur}		1.8996	1.8115	1.8392

Table 3 The statistical comparison of the vibration time series in pressure 3 scale

statistic	original sequence	surrogate sequence1	surrogate sequence3	surrogate sequence3
mean $\bar{x}$	0.2439	0.2474	0.2454	0.2460
sd σ	0.1187	0.1112	0.1157	0.1144
orign D_{mp}	1.1455			
surr D_{ur}		1.8866	1.9476	1.8421

Table 4 The statistical comparison of the vibration time series in pressure 4 scale

statistic	original sequence	surrogate sequence1	surrogate sequence3	surrogate sequence3
mean $\bar{x}$	0.2135	0.2143	0.2138	0.2133
sd σ	0.1016	0.1000	0.1010	0.1020
orign D_{mp}	1.0722			
surr D_{ur}		1.9284	1.9910	1.7025

Table 5 The statistical comparison of the vibration time series in pressure 5 scale

statistic	original sequence	surrogate sequence1	surrogate sequence3	surrogate sequence3
mean $\bar{x}$	0.2335	0.2340	0.2350	0.2355
sd σ	0.1136	0.1127	0.1105	0.1095
orign D_{mp}	1.1403			
surr D_{ur}		1.9076	1.7458	1.7960

Table 6 The statistical comparison of the vibration time series in pressure 6 scale

statistic	original sequence	surrogate sequence1	surrogate sequence3	surrogate sequence3
mean $\bar{x}$	0.2208	0.2201	0.2194	0.2190
sd σ	0.1053	0.1067	0.1082	0.1078
orign D_{mp}	1.2174			
surr D_{ur}		1.8169	1.8019	1.8304

The S criteria were calculated by equation (4), following the results in Table 7.

Table 7 Value of S criteria and bonding strength in different pressure scale

pressure scale	2 scale	3 scale	4 scale	5 scale	6 scale
criteria value	5.3447	6.5606	7.9383	6.0971	5.5976

The compared curve of criteria and bonding strength is shown the Fig.13

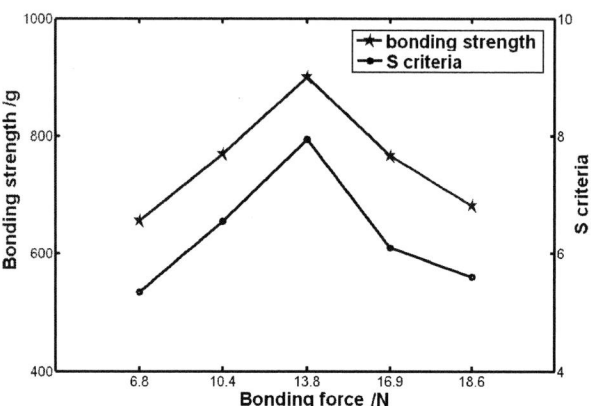

Fig.13 Criteria S and bonding strength compared curve

6 Conclusions

The results tell us the S values of the five are all more than 1.96. The null hypothesis is false, then we could accept the consequence that the vibration signal of the transducer terminal in axial direction contacts the nonlinear chaotic factor when the bonding pressure changes from 2 scale to 6 scale.

In Fig.13, the trend of the bonding strength value changing is the same as the S criteria when the bonding pressure changes. Both of them are the largest in pressure 4 scale (also 13.8N). That makes us preferably known ultrasonic wire bonding system with new visual angle, and be used as a guide for nonlinear dynamics modeling of bonding processing.

Acknowledgements

Project supported by the State Key Development Program for Basic Research of China (Grant No.2003CB716202) and supported by Project supported by the National Natural Science Foundation of China (Grant No.50575230, 50675227).

References

[1] Gao Rong-zhi, Han Lei, "Time-Frequency Analysis of Effects of Bonding Pressure on Ultrasonic Vibration", *Chinese Journal of Machanical Engineering*, Vol . 18, No.15 (2007),pp.1825~1829

[2] Han Lei, Zhong Jue, "Nonlinear Dynamics Behaviors in Flip-Chip Thermosonic Bonding", *Chinese Journal of Semiconductors*, Vol . 27, No.11 (2006),pp.2056~2063

[3] Han lei, Xu Wen-Hu, Li Han-Xiong, "Experiment of Unstable Characteristic on Ultrasonic Bonding System", *Transactions of the China Welding Institution*, Vol . 27, No. 8 (2006),pp.19~22

[4] THEILER J, EUBANK S, LONGTIN A, et al. "Testing for nonlinearity in time series: the method of surrogate data", *Physica D*, 1992, 58: 77~94

[5] Lu Yu, He Guo-Guang, "The identification of Chaos in the Real Traffic Flow Based on the Improved Surrogate-data Technique", *System Engineering*, Vol . 23, No. 6 (2005),pp.21~24

[6] Wang An-Liang, Yang Chun-Xin, "Grassberger-Procaccia Algorithm for Evaluating the Fractal Characteristic of Strange Attractors", *Acta Physica Sinica*, Vol . 51, No. 12 (2002),pp.2719~2729

[7] ENGBERT R. "Testing for nonlinearity: the role of surrogate data", *Chaos,Solitons and Fractals*, 2006, 13: 79~84

[8] Hou Ping-Kui, Gong Yun-Fan, Shi Xi-Zhi, et al. "Detecting Nonlinearity in the Radiated Noise of Underwater Targets", *Acta Acustica*, Vol . 26, No. 2 (2001),pp.135~139

[9] PRICHARD D. "The correlation dimension of differenced data", *Phys Lett A*, 1994, 191: 245~250

[10] Ma Jun-Han, Chen Yu-Shu, Liu Zeng-Rong, "The Influence of the Different Distributed Phase-Randomized on the Experimental Data Obtained in Dynamic Analysis", *Applied Mathematics and Mechanics*, Vol . 19, No.11 (1998), pp.955~964

[11] GRASSBERGER P, PORCACCIA I. "Measuring the strangeness of attractors", *Phys Rev Lett*, 1983, 31: 189~208

[12] ROMBOUTS S, KEUNEN R. "Investigation of nonlinear structure in multichannel EEG". *Phys Lett A*, 1995, 202: 352~358

[13] Jiang Wan-Lu, Zhang Shu-Qing, Wang Yi-Qun, "Numerical Experimental Analysis for Chaotic Motion Characteristic", *Chinese Journal of Mechanical Engineering*, Vol . 36, No. 10 (2000), pp.13~17

[14] ALBAREDA A, PEREZ R, KAYOMBO J H, et al, "Nonlinear mechanical behavior of piezocomposites for ultrasonic transducers". *Ultrasonics*, 2000, 38: 151~155

[15] Liu Hai-Feng, Dai Zheng-Hua, Chen Feng, Gong Xin, et al. "Calculating the Correlation Dimension of Dynamical System with Wavelet Analysis", *Acta Physica Sinica*, Vol .51, No. 6 (2002), pp.1186~1192

Effects of Ni addition on Microstructure and the Shear Strength of Sn-3.0Ag-0.5Cu/Cu Solders Joints

Lifeng Wang, Xuwei Shen, Fenglian Sun,Yang Liu
Harbin University of Science and Technology
317# material college in HUST No.4 Linyuan Road DongLi Zone Harbin, China.
Sxw42816@126.com 13936285643, 0451-81586290

Abstract

The effects of the fourth Ni element on microstructure and monotonic shear strength of Sn-3.0Ag-0.5Cu lead-free solder were investigated. The addition of the Ni element makes the new $(Cu_xNi_{1-x})_6Sn_5$ phase, which improves the solder microstructure. The results indicate that the addition of Ni element increases the shear strength of solder joint. When the content of Ni is 0.10 wt. %, the shear strength of solder joint is the highest by 40.28MPa which is 18% higher than that of Sn-3.0Ag-0.5Cu lead-free solder. The fracture morphology shows that the addition of Ni promotes grain refinement which leads to the increase of shear strength. The shear strength of the solder joints decrease continuously. The fracture microstructure became coarse-rained with the increase of thermal aging time. Dimples and voids appears in the solder joints which lead to the decrease of the shear strength.

Introduction

In order to replace of the past used Sn-37Pb solder, the study of lead-free solder whose solderability is the same as or better than that of Sn-37Pb solder has became the hot spot at the moment. It is gradually recognized that Sn-Ag-Cu is one of the best lead-free solders alloy systems[1], but the coarse of microstructure and eutectic phase affects the application of solder joint of Sn-Ag-Cu alloy systems. The electronic package industry has always been actively searching for proper compositions and effective adding alloys to improve and optimize the performance of Sn-Ag-Cu solder to reach that of the past used eutectic Sn-Pb solder [2].

Some researchers reported that a little amount of other element addition, such as Bi, In, Sb, and Zn could improve solder strength, thermal resistance or fatigue life due to solid solution strengthening effect[3-6]. Previous work reveals that adding a small amount Ni into Sn-Ag-Cu solder can to some extend improve the performance of Sn-Ag-Cu solder joint[7]. The aim of this paper is to investigate the effect of Ni element on the mechanical properties and reliability of solder joint by improving metal structure and refining the eutectic phase.

1 Experimental

1.1 Alloy preparation

The Sn shots (99.9wt% purity), Ni foil (99.9wt% purity) and Sn-3.0Ag-0.5Cu solder were used for alloy preparation. Proper amounts of Sn and Ni elements were weighed and melted together under argon atmosphere. They were melted many times in order to be well-mixed. Sn-Ni alloy and Sn-3.0Ag-0.5Cu alloy were melted to fabricate new alloys Sn-3.0Ag-0.5Cu-XNi (X=0.00, 0.03, 0.05, 0.10 and 10.15) in the resistance furnace at 500°C for hours.

1.2 Preparation of microstructure samples

The metallographic samples of new alloys Sn-3.0Ag-0.5Cu-XNi, which were made with epoxy resin, were grinded with abrasive paper until the surfaces were flat. After that, the samples were grinded by the UNIPOL-820 type grinding polisher with 400#, 800#, 100#, 1500#, 2000# metallographic abrasive paper sequentially. The condition to change metallographic abrasive paper is that the scratches of the sample are covered by the current one. After 2000# was used, the polishing cloth, with diamond polishing spray on its surface, was used to polish until catching the standard. Etching solution (5%HNO$_3$+95% C$_2$H$_5$OH) was used to erode the prepared metallographic samples for about 10 to 20 seconds until the microstructure and compounds could be observed clearly. Then the Olympus multi-function optical metallographic microscope was used to observe and analyze the microstructure. The energy-dispersive x-ray (EDX) was used to check the component of the new produced phase.

1.3 Tensile shear experiment

The tensile shear experiment of solder is under the country standard GB11363-89. 200mg Sn-3.0Ag-0.5Cu-XNi solders were separately weighted and made to tensile shear samples as the structure showed in Fig.1. In order to keep the same clearance 10g vertical load was added at the lap joint during soldering progress. The prepared tensile shear samples were put into aging furnace at 150 °C for different aging times (0, 4, 16, 25days). Tensile shear experiment was carried out on XLD Series liquid crystal screen electronic tester after high temperature storage at load speed of 5mm/min. The max pull was the max shear stress of the solder joint. After the tensile shear experiment, the loading area A of solder joints were measured and noted. After the tensile shear experiment, the scanning electron microscopy (SEM) was used to observe and analyze the microcosmic fracture morphology.

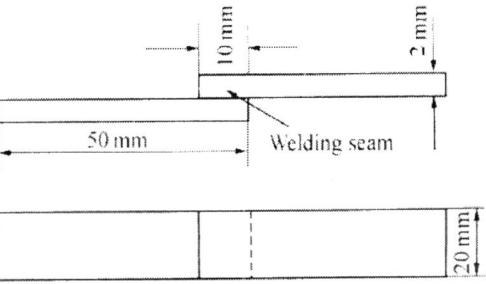

Fig.1 Sample for joint tensile shear test

The shear strength of solder joint is calculated as formula 1-1.

$$\tau = F/A \qquad 1-1$$

Where: τ(MPa) is The shear strength, F (N) is the failure load of solder joint. A (mm²) is the loading area A of solder joint.

2 Results and Discussion

2.1 The microstructure analysis of new solders

The microstructure of alloy determines the capability of alloy. Some specialty can be concluded by observing the change of solder microstructure. The original microstructure is shown in Fig.2.

a) Sn-3.0Ag-0.5Cu

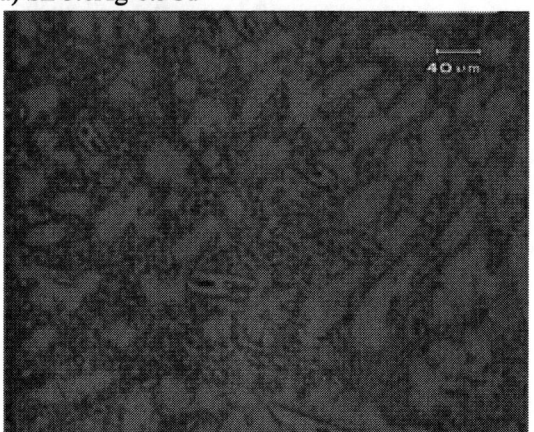

b) Sn-3.0Ag-0.5Cu-0.03Ni

c) Sn-3.0Ag-0.5Cu-0.05Ni

d) Sn-3.0Ag-0.5Cu-0.1Ni

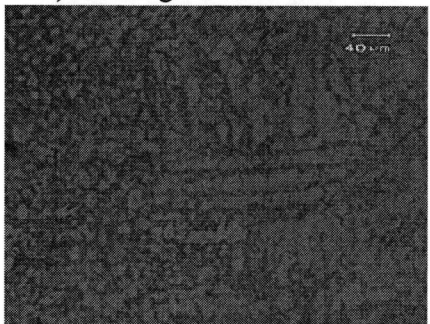

e) Sn-3.0Ag-0.5Cu-0.15Ni

Fig. 2 Microstructure of Sn-0.3Ag-0.7Cu -xNi alloy

Fig.2 a) is the microstructure of Sn-3.0Ag-0.5Cu. It is mainly made of β-Sn phase (white area) and eutectic structure. The eutectic structure is mainly binary eutectic structure, which are granular $Cu_6Sn_5+\beta$-Sn and acicular $Ag_3Sn+\beta$-Sn. The base substance is rich Sn phase and IMC phase distributed dispersively. According to metallography, the dispersively distributed IMC can obviously strengthen the base Sn substance, which can improve the mechanical properties. A few granular new resultant, which dispersively distributes in the microstructure, turn out in Fig.2 b).At the same time, the massive β-Sn phase become smaller than that in Fig.2 a). As the content of Ni element increases, the granular new resultant which dispersively distributes in the microstructure manifolds. When the content of Ni is 0.10wt%, the amount of granular new resultant is the most and the primary phase is fine. At the same time nubbly compound appear in Fig.2 d). Fig.2 e) shows that many compounds appear. Fig.2 shows that the dispersively distributed new resultant appears in primary phase and eutectic structure of Sn-Ag-Cu-Ni solder alloy, which improves the microstructure of solder alloy. The addition of Ni element makes primary phase β-Sn obviously fine, because Ni element provides more nucleation sites to rich Sn phase which separates out first.

The energy-dispersive x-ray (EDX) was used to check the composition of new resultant. Fig 3 shows the target point of energy spectrum analysis. Fig.4 shows the EDX results. It is shows that the new resultant is$(Cu_xNi_{1-x})_6Sn_5$ in the new solder alloys after adding Ni element.

2.2 The change law on the shear strength of solder joint

The shear strength test result is shown in table 1,

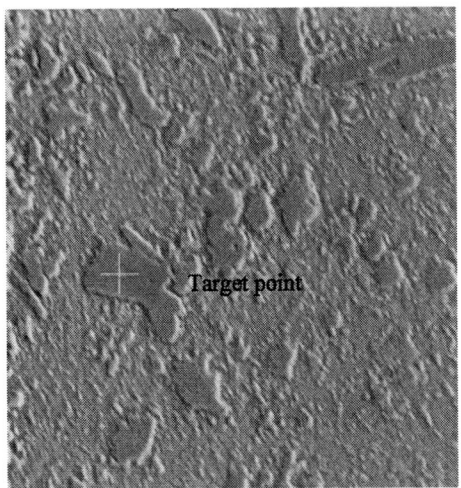

Fig. 3 Energy spectrum target point

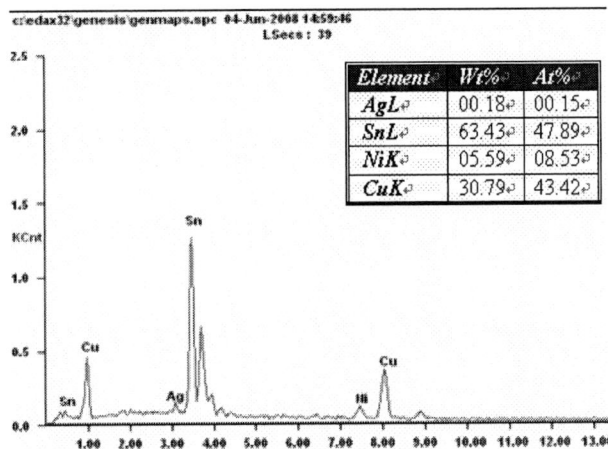

Element	Wt%	At%
AgL	00.18	00.15
SnL	63.43	47.89
NiK	05.59	08.53
CuK	30.79	43.42

Fig.4 EDX of the new resultant

Table 1 Tensile shear experiment data

Solder composition	Ageing time(Days)			
	0	4	16	25
Sn-3.0Ag-0.5Cu-0.15Ni	28.63	26.26	25.77	19.86
Sn-3.0Ag-0.5Cu-0.10Ni	40.28	26.60	27.17	20.42
Sn-3.0Ag-0.5Cu-0.05Ni	38.78	24.48	25.62	20.22
Sn-3.0Ag-0.5Cu-0.03Ni	37.04	26.35	24.28	20.00
Sn-3.0Ag-0.5Cu	32.94	25.52	21.34	19.97

2.2.1 The effect Ni on the shear strength

Fig.5 shows the effect of Ni composition on shear strength. A little amount Ni elemental addition increases the shear strength of solder joint. The shear strength of Sn-3.0Ag-0.5Cu solder is 32.94MPa. It is gradually increased as the increase of Ni content. When Ni content is 0.10wt%, the

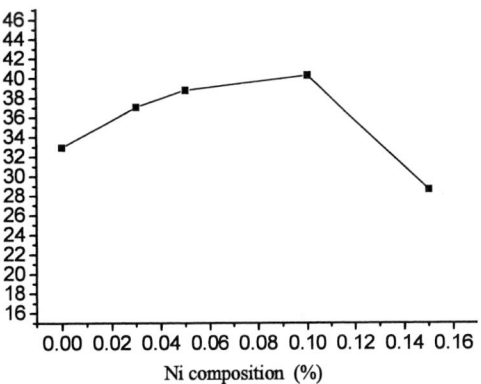

Fig.5 effect of Ni composition on shear strength

shear strength of solder joint reaches the maximal, which is 40.28MPa.The shear strength of Sn-3.0Ag-0.5Cu solder is 32.94MPa. It is gradually increased as the increase of Ni content. When Ni content is 0.10wt%, the shear strength of solder joint, which is 40.28MPa, reaches the highest. It is 18% higher than that of Sn-3.0Ag-0.5Cu lead-free. The reason is that the $(Cu_xNi_{1-x})_6Sn_5$ and $(Cu_xNi_{1-x})_3Sn$ phase can be formed by substituting Ni for Cu atom in binary compounds with Sn. When the Ni content is over 0.10wt%, the strength of solder joint decreases sharply as the content of Ni increased continuously. Large amounts of Ni addition form many compounds which lead to more brittleness and strength decrease of the solder joint.

To study the reason of shear strength deeply decreases. The scanning electron microscope (SEM) was used to observe and analyze the morphology of solder joints fracture. The fracture morphology is shown in Fig.6 in different Ni content.

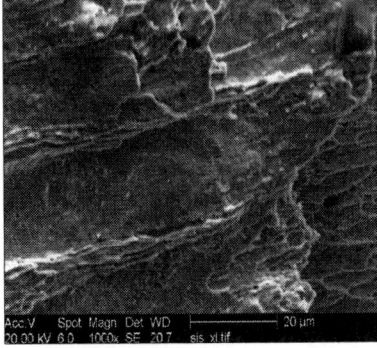

a)Sn-3Ag-0.5Cu

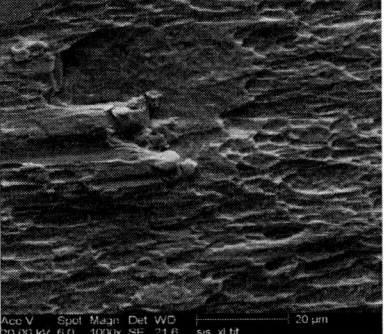

b) Sn-3Ag-0.5Cu-0.03Ni

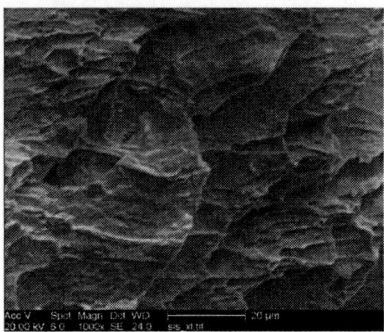

c) Sn-3Ag-0.5Cu-0.05Ni

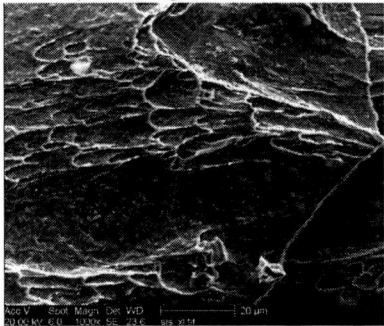

d) Sn-3Ag-0.5Cu-0.10Ni

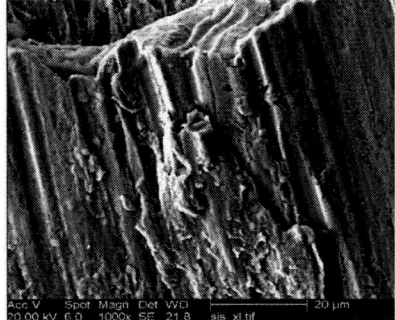

e) Sn-3Ag-0.5Cu-0.15Ni

Fig.6 Fracture surface of Sn-3.0Ag-0.5Cu-XNi/Cu

The fracture morphology of each solder joint with different Ni content are quite different. Fig.6 a) shows that the fracture of Sn-3.0Ag- 0.5Cu appears to be slip-broken or pure-broken. There are large amounts of line shape trace, small amount dimple, and some smooth areas in the fracture. With the addition of Ni, the dimple increases, while the partial smooth area and line shape trace reduce, in the fracture. When Ni addition is 0.10wt%, the partial smooth area and line shape trace cannot be seen anymore. The number of dimple increased and the size of the dimple became smaller, see Fig.6 d). When Ni content is 015wt%, the partial smooth area and line shape trace appear again and the dimple comparatively tails off as shown in Fig.6 e).

On the basis of the above analysis, it shows that the addition of Ni element can increase the strength of solder joint. When Ni content is at 0.10wt%, the shear strength reaches the highest value.

2.2.2 The effect of thermal aging on the shear strength of solder joints and the fracture analysis.

Fig.7 shows the shear strength curves of Sn-3.0Ag-0.5Cu-XNi/Cu. It shows that the shear strength of solder joints

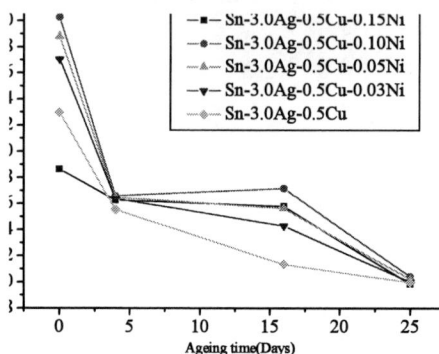

Fig.7 shear strength curves ofSn-3.0Ag-0.5Cu-XNi/Cu.

decrease as the thermal aging time prolonging. They decrease sharply in first 4 days aging then varied relatively smooth in following aging days. Sn-3.0Ag-0.5Cu-0.10Ni solder joints perform the highest shear strength prior to thermal aging. Thus Sn-3.0Ag-0.5Cu-0.10Ni behaves better resistance to thermal degradation.

To find the reason of good shear strength performance, Sn-3.0Ag-0.5Cu-0.10Ni is selected to observe the fracture pattern with different thermal aging times.

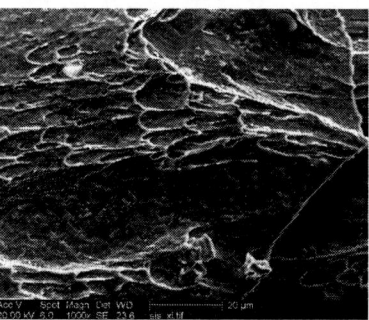

a) before aging

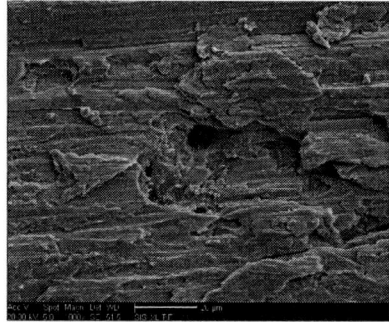

b) 4days

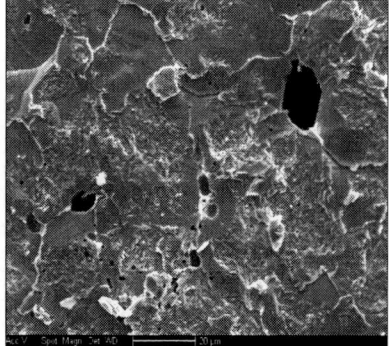

c) 16days

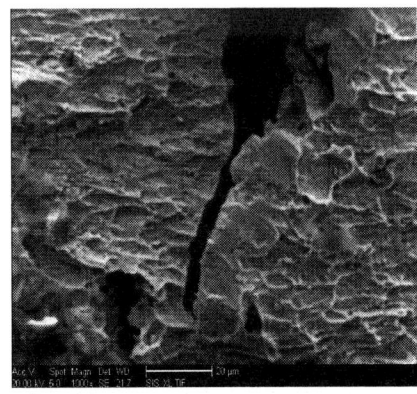

d）25days

Fig.8 Shear fracture surface of Sn-3.0Ag-0.5Cu-0.10Ni/Cu at different ageing time

Fig.8 shows that the fracture of Sn-3.0Ag-0.5Cu-0.10Ni is ductile fracture. Because the typical shape characteristic of tough fracture is dimple and sliding. Large numbers of dimples are observed obviously in Fig.8 a). Microstructure is fine and homogeneous at the same time. After 4 days` aging, dimple becomes coarse and turns out a few voids in the fracture as shown in Fig.8 b). There are plenty of voids after 16 days' aging, and some of the voids joined together and grow bigger. The fractures appear crack after 25 days` aging, as shown in Fig.8 d). Thus, thermal aging has great effects on microstructure of lead-free solder and shear fracture morphology at 150°C. The internal binding force becomes weaker after thermal aging, which causes the crack in the end.

3 Conclusions

In present work, the effects of fourth element additions to Sn-Ag-Cu solder alloys on the microstructure and mechanical properties were studied. The results are summarized as following:

(1) The fine and dispersively distributed $(Cu_xNi_{1-x})_6Sn_5$ phases in primary phase and eutectic structure of Sn-Ag-Cu-Ni solder optimize the microstructure and refine primary phase which improve solder joint mechanical properties.

(2) Ni addition increases the shear strength of solder joint. As the Ni content increases, the improvement of shear strength increases significantly. When the adding amount of Ni is 0.10 wt%, the solder joint perform the highest shear strength by 40.28MPa which is 18% higher than that of Sn-3.0Ag-0.5Cu solder. The fracture surfaces show that the addition of Ni promotes grain refinement which results in the increasing of shear strength.

(3) Thermal aging has great effects on the shear strength and the fracture pattern. Microstructure coursing and voids are observed in the fracture, which result in the decrease of the solder joints shear strength.

4 Acknowledgements

The financial support for this work from the Natural Science Foundation of China (No. 50575060) is gratefully acknowledged

References

1. Bradley Edwin. Lead-free solder assembly: Impact and opportunity. Proceedings-Electronic Components and Technology Conference, 2003,pp: 41-46
2. Bing Lu, Hui Lie, Juan Wang et. "Effect of Er on microstructure and properties of Sn-3.0Ag-0.5Cu lead-free solder alloy," The Chinese Journal of Nonferrous Metals 2007.4.17(4),pp:518-524
3. Y. G. Lee and J. G. Duh, "Interfacial morphology and concentration profile in the unleaded solder/Cu joint assembly," J. Mater Sci Mater Electro., vol. 11, pp: 33–43, 2000.
4. M.E.Loomans, S.Vaynman, G. Ghosh, and M. E. Fine, "Investigation of multi-component lead-free solders," J. Electron Mater. vol. 23, no.8, pp:741–746, 1994.
5. Y. Kariya and M. Otsuka, "Mechanical fatigue characteristics ofSn-3.5Ag-X (x = Bi, Cu, Zn, and In) solder alloys," J. Electron Mater, vol. 27, no. 11, pp:1229–1235, 1998.
6. N. Wade, K. Wu, J. Kunii, S. Yamada, and K. Miyahara, "Effect of Cu,Ag and Sb on the creep-rupture strength of lead-free solder alloys," J Electron Mate., vol. 30, no. 9, pp:1228–1231, 2001.
7. K. Seelig and D. Suraski, "The status of lead-free solder alloys," in Proc.50th Electronic Components and Technology Conf., Las Vegas, NV,May 2000, pp: 1406–1409.

An Investigation of Capillary Vibration during Wire Bonding Process

Jian Gao, Robert Kelly, Zhijun Yang, Xin Chen

Faculty of Electromechanical Engineering, Guangdong University of Technology
729 East Dongfeng Road, Guangzhou, 510090, P.R. China,
gaojian@gdut.edu.cn, +86 20 39322209

Abstract

As the bond pitch size decreases, understanding the behaviour of the capillary tube and monitoring bond quality becomes increasingly important. This paper uses Finite Element Analysis (FEA) and Laser Doppler Interferometer to study the vibration of the capillary tube during the wire bonding process. Using a laser Doppler interferometer, the vibrations were measured along the x, y and z axis under two different conditions, when the capillary was vibrating against a hard surface (low friction) and during wire bonding operations (higher friction). The head of the capillary was shown to describe a circular path when the friction was low and an elliptical path when there was more frictional resistance at the capillary tip. The forces acting on the capillary tube were studied and it was shown that the reason for the vibration in the y-axis was the horizontal reaction due to the frictional resistance when the capillary was not aligned with the x-axis. The path described by the tip of the capillary is discussed. The significance of these results is discussed and using the forces acting on the capillary in the deformed shape, the difference between the circular and elliptical paths is explained. FE analysis is then used to further understand and demonstrate the behaviour of the capillary. The conclusions of the paper show that when the capillary vibrates in a stable condition, the deformed shape does not change, but revolves around the central vertical z-axis.

Key Words: Wire bonding, Laser Doppler Interferometer, FE analysis, Monitoring bond quality.

Introduction

With the increasing complexity of integrated circuits (IC), more wire interconnects are required to transfer this increased functionality to other packages or the main board. There are two problems to be overcome, the space on the surface of the IC is finite so increasingly smaller pitch sizes are used requiring smaller ball bonds, which increases the probability of the failure of the wire bond. Secondly as the number of wires per IC increases, the quality of each bond has to be increased to maintain the same overall chip failure rate. The scope of this project work is to further increase the understanding of the wire bonding process and to establish a technique that can identify bonds that have not been formed correctly, so that remedial action can be taken or the IC component discarded.

There have in recent years been many investigations into the movement of the capillary arm and the vibration modes of the bonding tool itself, that have advanced our understanding of the mechanisms involved in the bonding process. One of the earliest Wilson [1] used holographic interferometry to study the complex motions of the bonding tool in 1972. More recently Hu [2] conducted a detailed study of the vibration of the capillary in free air describing the vibration shape of the capillary and confirming these results with FE analysis. Zhong [3] studied the vibration shape of the capillary, however they were more interested in the design of a new form of capillary and the power transmitted to the bonded area rather than investigating the behaviour involved in the bonding process. Zhang [4] studied the movement on the head and tip of the capillary along the x axis for wedge bonders using a short time Fourier Transform to produce a joint time-frequency analysis. The results of this study show a distinct variation in the power input in the 1st harmonic with the variation of power input and contact force. Lei Han [5] studied the movement on the head and tip of the capillary along the x axis for wedge bonders, then using the process of wavelet analysis, looked for significant variations in the vibration characteristics as the bonding pressure was varied. Separating the signal into different frequency packets, small but significant vibrations at the start of the bonding process in the 256-512 kHz band correlated well with the variation of the bonding pressure. Tsujino [6&7] uses an experimental bonder driven by 2 transducers placed along each axis of the horizontal plane of the perpendicular to the axis of the bonding tool. By slightly varying the frequency of each transducer, circular patterns described by the capillary tip could be produced for equal frequencies and complex square patterns could be produced using slightly different frequencies. Two systems were examined in [6] at frequencies of 160 and 515 kHz, whilst in [7] frequencies of 420 and 980 kHz were studied. The velocity of the capillary was determined by a ring type magnetic vibration velocity detector.

Monitoring the bond quality has become an important issue for the electronic packaging industry, with many different approaches being tried. Two methods using interferometer have already been discussed [4, 5], however work is being carried out in different areas. Rajeswari [8] describes an optical solution for comparing a graphical image of the bond with reference images. Chu [9] suggests that placing piezoelectric sensors on the bonding arm and investigates the best placement of the sensors to produce the signal that can be used to monitor the bonding process. Chiu [10] report a similar process, but investigated the effect of using different types of sensors to monitor the vibration in the bonding arm.

The scope of the work reported in this paper examines how the vibration of the capillary varies under different bonding conditions and how this may be used in future to establish methods that will be able to monitor capillary vibrations and so determine the bond quality.

The methodology of how the interferometer data was acquired is explained, then sample results are presented for

978-1-4244-2739-0/08/$25.00 ©2008 IEEE

the cases of a near frictionless condition and those obtained whilst wire bonding. The significance of these results is discussed and using the forces acting on the capillary in the deformed shape, the difference between the circular and elliptical paths is explained. FE analysis is then used to further understand and demonstrate the behaviour of the capillary. The conclusions of the paper show that when the capillary vibrates in a stable condition, the deformed shape does not change, but revolves around the central vertical z-axis.

Measurement Methodology

The vibrations were monitored along the 3 axes shown in Figure 1. The driving force of the piezoelectric transducer drives the capillary in a cyclic manner along the x-axis. Any vibration recorded along the y and z-axes will be the result of disturbing forces acting on the capillary.

The wire bonder used in this investigation is a Kulicke and Soffa 1488 Gold ball bonder (K&S 1488) with a Small Precision Tool (SPT) capillary UTS-17A-CM-1/16-XL-30°. The vibration was measured by a Polytec Laser Interferometer Vibrometer PSV-400-M2.

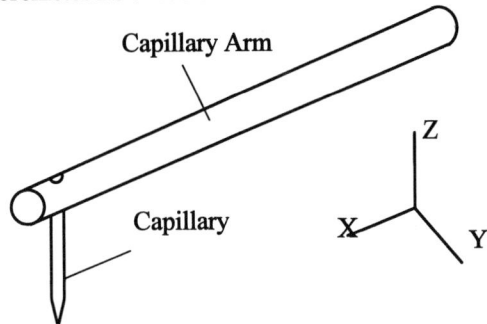

Figure 1 Definition of the axes of measurement

The measurements of the vibration were made by reflecting the laser beam off the capillary arm during the period of time in which the ultrasonic power was driving the capillary. For the x-axis, the beam was reflected off the circular X-section at the end o the capillary arm. The measurement for the y-axis was a little more complex in terms of access, the beam had to be reflected off a mirror placed at 45° to the capillary arm. To avoid any influence of the curvature of the cylindrical shape of the capillary arm and to improve the signal return, a small (1x1mm) metallic mirror was attached to the capillary arm, located just behind the capillary tube. The measurement of the z-axis vibration was made by using 2 mirrors arranged in the manner shown in Figure 2. Locating a third mirror underneath the arm, to ensure a better signal return, proved impossible due to the extremely small clearance between the capillary arm and the second mirror underneath the capillary arm. The laser was triggered to start recording by taking a signal from the ultrasonic generation board of the wire bonding machine and using it as a reference signal. This process enabled a very high data capture frequency of 1.28 MHz to be used.

Experimental Results

a) Without gold wire

Initially the vibration of the capillary arm was measured when the capillary was vibrating against the hard surface of the heat block of the wire bonder to simulate a near

frictionless surface. Examples of these results are shown in Figure 3–6 and summarised in Table 1. The x & y axis data is taken between 6-8ms, z axis data is obtained from the Fourier analysis.

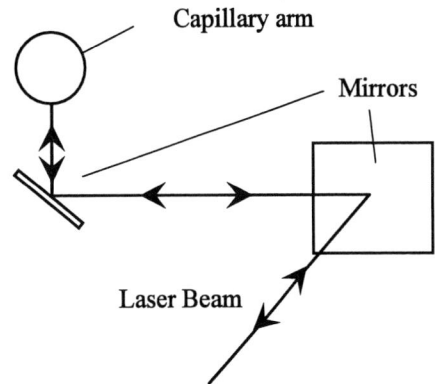

Figure 2 Path of the laser beam to measure z axis vibration

Table 1 Summary of the data from Figures 3-6

	Freque ncy(kHz)	Veloc ity (mm/s)	Ma x (mm/s)	Min (mm/s)
x-axis	60.58	175	181	171
y-axis	60.58	173	180	169
z-axis	60.23	24.0	–	–
,,	120.47	4.75	–	–

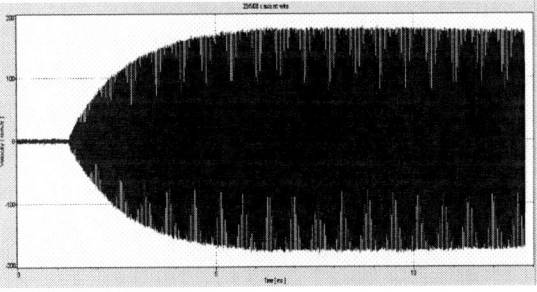

Figure 3 Velocity of vibration x-axis (no wire)

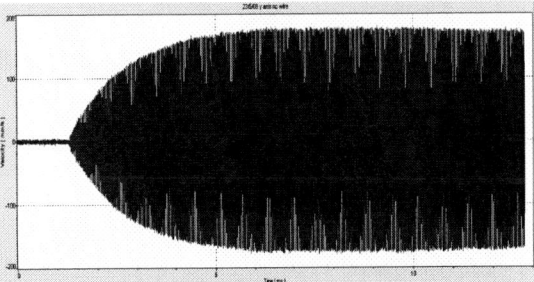

Figure 4 Velocity of vibration y-axis (no wire)

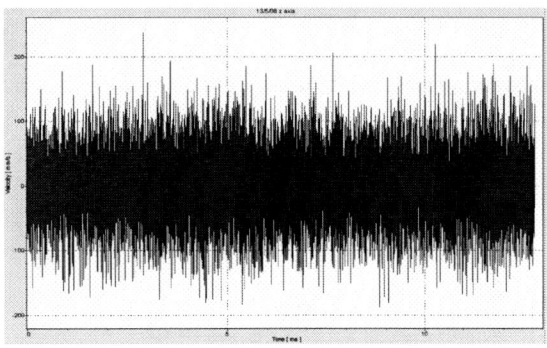

Figure 5 Velocity of vibration z-axis (no wire)

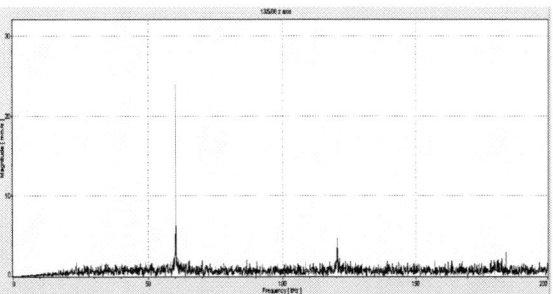

Figure 6 Fourier analysis of the z-axis vibration (no wire)

Generally the data for the y-axis vibration was a little smaller than that for the x-axis. The data for the z-axis is masked by a lot of noise and is only presented as the Fourier analysis of this signal shows distinct peaks at 60 and 120 kHz, which will be discussed later. Figure 7 shows the path of the head of the capillary in the X Y plane. Only having one vibrometer, this path had to be constructed from separate, but similar observations.

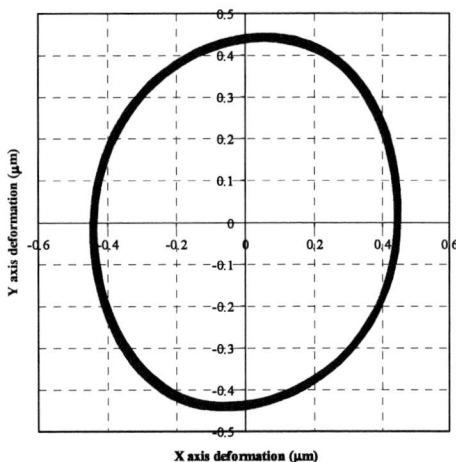

Figure 7 Path of the head of the capillary in contact with a near frictionless surface

b) With Gold wire

The same data was collected for wire bonding on to a bond pad using gold wire (ϕ=25μm), the results of which are shown in Figure 8 & 9 and summarised in Table 2. The presence of the bonding pad and the gold wire have caused a distinct difference in the amplitude of the vibration in the y-axis direction and is only a quarter of the x-axis vibration. The

x & y axis data is taken between 6-8ms, z axis data is obtained from the Fourier analysis.

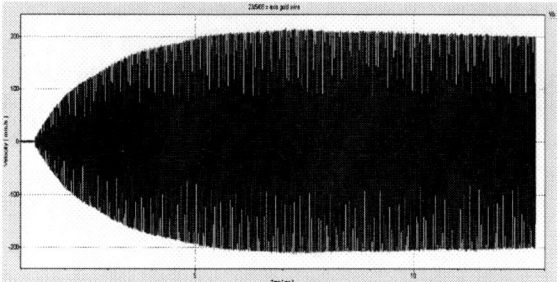

Figure 8 Velocity of vibration x-axis (gold wire)

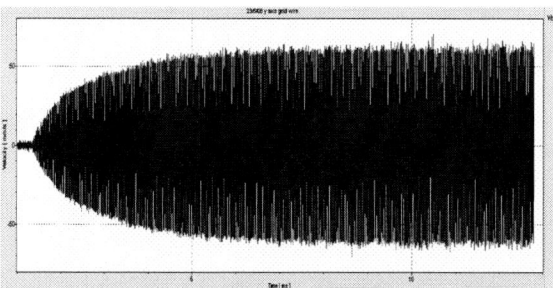

Figure 9 Velocity of vibration y-axis (gold wire)

Table 2 Summary of the data from Figures 8 & 9

	Frequency (kHz)	Velocity (mm/s)	Max (mm/s)	Min (mm/s)
x-axis	59.62	209.9	215.6	204.8
y-axis	59.62	51.0	64.0	57.7

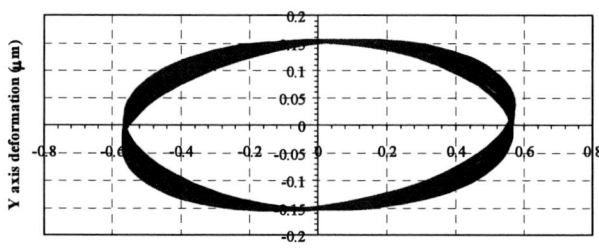

Figure 10 Path of the head of the capillary whilst bonding with gold wire

Discussion

The data collected from the interferometer shows clearly that the capillary arm vibrates in a y-axis direction, (Figure 4 and Figure 9) causing the head of the capillary arm to rotate in a circular or elliptical path.

The reason for this is may be found from considering the deformed shape of the capillary. Figure 11 shows a schematic drawing (not to scale) of the capillary vibrating in air. The tip of the capillary can be seen to describe a vertical arc as it oscillates backwards and forwards. During the situation of wire bonding, when the capillary is pressed against a surface, this motion is not possible and the tip has two choices, either to:

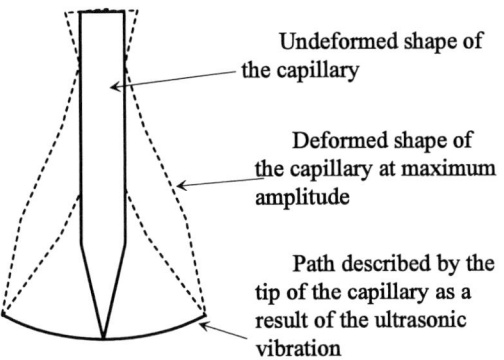

Figure 11 The path described by the tip of the capillary in air

1) Continue to move in the direction of the x-axis in which it is driven and undergo compression of the capillary as it passes through the vertical, or

2) Maintain it deformed shape, and consequently the vertical height, and move in a circular path. There will have to be some sort of disturbing force to cause the tip to move out of the XZ plane, however as soon as this occurs then this circular motion will occur.

Figure 12 shows the external forces acting the capillary in the deformed shape. Assuming a completely frictionless surface at the tip ($R_H = 0$), equilibrium can only be established by the tip moving horizontally until the moment acting at the head of the capillary is equal to the horizontal separation times the reaction force R_V. The capillary tip will move in a circular path about the head of the capillary which will remain stationary. If friction is introduced into the model, then this movement will be resisted by the force R_H. It is this force that acts against the arm of the capillary which causes the vibration in the y-axis and the compressive force acting within the capillary tube that shortens the length of the capillary tube. The effect of this shortening is that the y-axis of the circular path becomes the semi-minor path of an elliptical path.

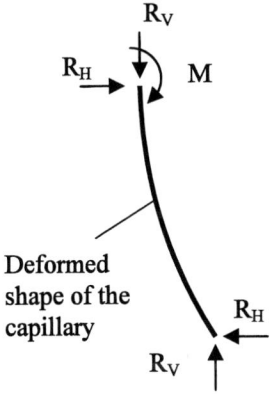

Figure 12 The external forces acting on the capillary

Using this model of the forces acting on the capillary, we can now explain why in the first series of results, when the capillary was vibrating against a hard surface, with virtually no friction, a near perfect circle was described by the head of the capillary arm. Then in the second series of results where gold wire was bonded on to a bonding pad and there is a resistance at the tip of the capillary, the head of capillary then moves in an elliptical manner.

Lei Han [5] reported that in his wavelet analysis, that there was significant high frequency (256-512kHz) vibration in the initial 10ms of the die bonding process that correlated well with the variation in bonding pressure when observing the vibration in the x-axis direction. The largest vibrations occurring with a low bonding pressure (12.2N), and almost indistinct at high bonding pressures (18.8N). Similar, but less distinct peaks can be observed at the lower frequencies of 64-128 and 128-256 kHz. It is suggested from the work in this paper, that these results maybe explained by the path the capillary tip describes in the time before a stable elliptical path is established. When the bond pressure is high, the compression in the capillary tube forces the tip into a stable elliptical path very rapidly. Therefore no high frequency disturbances are seen in the vibration of the capillary. However under conditions of low bonding pressure, with less compression in the capillary tube and correspondingly lower disturbing forces, the path of the capillary tip is unstable. The capillary tip wants to move in the direction in which it is driven, i.e. along the x-axis. However, when describing this path the compression in the capillary increases to a level where the capillary "buckles" and the tip suddenly moves to the elliptical path. It is concluded that this erratic motion may be the cause of the higher frequency vibrations observed by Lei Han and this should be investigated further in future work.

Referring to Figure 11, it can be seen that if the capillary were to move only in xz plane and was in contact with a surface in the xy plane, then a vertical force could be expected to act on the capillary arm as the capillary tube passed through the vertical, undeformed position with a frequency of twice that of the excitation frequency (120kHz). Similar results could also be expected when the head of the capillary describes an elliptical path.

It was for this reason that the vibration in the z direction was observed and whilst the signal was obscured by noise, a distinct peak can be seen in the Fourier analysis of the data at 120kHz. The larger peak at 60kHz however was unexpected and cannot be explained, except to say that the mirror under the capillary arm is extremely close to the arm due to the available space when the capillary is in the bonding position and the movement of the capillary arm may be inducing a vibration in the mirror. Until the methodology of this experiment is improved, this result cannot be taken as significant, however monitoring the disturbance in the z-axis could contribute significant information of the quality of the bond.

FE Analysis

The FE analysis was carried out using ANSYS Version 9.0. The model comprised of 5140 SOLID186 elements (3D 20-node structural solid elements) with each node having 3 DOF. The overall length of the capillary is 11.1mm, however the first 2.2 mm of the capillary tube is constrained in the capillary arm and only the unconstrained length of 8.9 mm is modelled in the analysis. The main diameter of the capillary is 1.585 mm tapering with a 30° angle to a tip diameter of 0.23 mm. The capillary material properties are shown in Figure 13, together with an outline diagram showing the arrangement of elements in the FE model. The loads were applied to the model in the transient analysis by applying physical displacements to the nodes at prescribed time intervals. The model was driven in a harmonic manner at a frequency of 60 kHz in the direction of the x-axis with each cycle requiring 24 separate load steps.

Two harmonic analyses were performed, with the displacements applied to the head of the capillary model in the x-axis direction and the x and y axis directions. The output from the first model was used to input the initial velocities of every node to the undeformed model of the transient analysis so that the correct behaviour is observed in the deformation of the capillary during the harmonic motion. The second harmonic analysis, driven in the direction of the x and y axes in same manner to that observed using the laser vibrometer, when animated showed the motion described in Figure 14. The deformation of the capillary tube remains constant, but revolves about the vertical z-axis. This is assumed to be the stable condition, as constant vibration in the direction of the y-axis is observed by the laser interferometer.

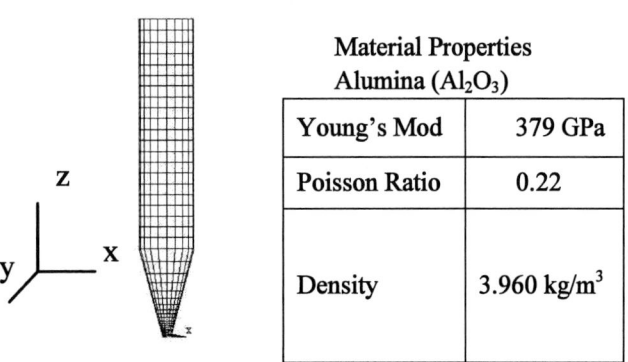

Material Properties Alumina (Al₂O₃)	
Young's Mod	379 GPa
Poisson Ratio	0.22
Density	3.960 kg/m³

Figure 13 The FE Model of the Capillary

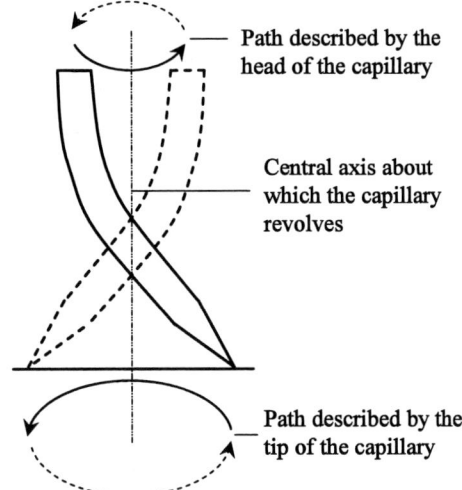

Figure 14 The 3D movement seen in the second harmonic analysis

The purpose of the transient analysis was to show that when the capillary tip is in contact with a surface that a circular motion can be initiated and sustained by applying a small disturbing force on the tip of the capillary and removing

it once the motion has be initiated. This analysis has been problematic due to the overall physical size of the model and the amplitude of the vibration. The vertical displacement when the capillary at the maximum displacement is of the order of 10^{-6} smaller than the length of the model and small residual errors make a large difference to the path described by the capillary tip. This area is the focus of attention in the current research work.

Figure 15 shows the predicted movement of the capillary tip calculated by a finite element transient analysis. The accumulating error can be seen in the path along the x-axis when compared to the predicted path. At 5μm seconds, an attempt is made to correct this growing error, but results in an overcorrection. A mistake is repeated again at 8μm seconds. This error in the FE calculation, relative to the physical size of the model, is very small, but fails to predict accurately the path of the x-axis movement and consequently completely fails to model anything like the y-axis path, which should be identical to the predicted x-axis path but shifted 90° in phase.

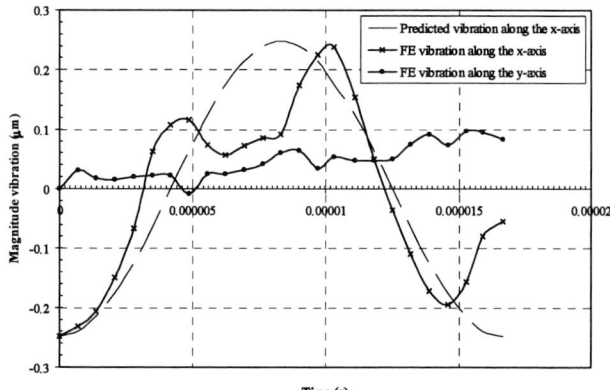

Figure 15 The FE predicted movement of the capillary tip

Conclusions

Using a laser Doppler interferometer, the vibrations were measured along the x, y and z axis under two different conditions, when the capillary was vibrating against a hard surface (low friction) and during wire bonding operations (higher friction). The head of the capillary was shown to describe a circular path when the friction was low and an elliptical path when there was more frictional resistance at the capillary tip. The forces acting on the capillary tube were studied and it was shown that the reason for the vibration in the y-axis was the horizontal reaction due to the frictional resistance when the capillary was not aligned with the x-axis. A possible explanation of the high frequency disturbances noted by Lei Han was suggested and further work in this area may be important in determining the quality of the wire bonding process.

The vibration of the capillary along the z-axis was measured and the Fourier analysis showed that there were measurable vibrations in this direction, although heavily masked by noise.

The harmonic analysis showed that when the capillary oscillates in the same manner of the vibrations observed at the head of the capillary, the deformed shape is maintained and the capillary revolves about the vertical z-axis. Trying to use a transient analysis to show the path described by the capillary tip after initially applying a disturbing force and that a stable path could be established, failed to produce any meaningful results.

Acknowledgments

This work is sponsored by Guangdong Provincial Science and Technology Research Project (No. 2005A10403004 and No.2006A10401003).

References

1 Alan D. Wilson, Bryon D. Martin and Douglas H. Strope. "Holographic Interferometry Applied to Motion studies of Ultrasonic Bonders." *IEEE Transactions on Sonics and Ultrasonics*, Vol SU-19, No 4, (1972), pp 453-461.

2 C.M. Hu, N.Q. Guo, H.J. Du, R.M. Lin, M. Chen. "Vibration Characteristics of the Capillary in ultrasonic wire bonders." *Proc. IMechE Part C: Journal Mechanical Engineering Science*, Vol 221, (2007), pp 897-903.

3 Z. W. Zhong, K.S. Goh. "Investigation of ultrasonic vibrations of wire bonding capillaries." *Microelectronics Journal*, Vol 37, (2006), pp 107-113.

4 Dong Zhang, Ling Shih-Fu. "Monitoring wire bonding via time frequency Analysis of Horn Vibration." *IEEE Transactions on Electronic Packaging Manufact.* Volume 26, No. 3, (2003), pp 216-220.

5 Lei Han, Rongzhi Gao, Jue Zhong, Hanxiong Li. "Wire bonding dynamics monitoring by wavelet analysis." *Sensors and Actuators A*, Vol 137, (2007), pp 41-50.

6 Jiromaru Tsujino, Hiroyuki Yoshihara, Kazuyoshi Kamimoto, Yoshiaki Osada. "Welding Characteristics and temperature rise of high frequency and complex vibration ultrasonic wire bonding." *Ultrasonics*, Vol 36, (1998), pp 59-65.

7 Jiromaru Tsujino, Hiroyuki Yoshihara, Tsutomu Sano, Shigeru Ihara. "High-Frequency ultrasonic wire bonding systems." *Ultrasonics*, Vol 38, (2000), pp 77-80.

8 M. Rajeswari and M.G. Rodd. "Real-time Analysis of an IC Wire-bonding Inspection System." *Real-Time Imaging*, Vol. 5, (1999), pp 409-421.

9 Paul Wing-Po Chu, Chi-Po Chong, Helen Lai-Wa Chan, Kelvin Ming-Wai Ng, Peter Chou-Kee Liu. "P lacement of piezoelectric ceramic sensors in ultrasonic wire-bonding transducers." *Microelectronic Engineering* Vol. 66, (2003), pp 750–759.

10 S.S. Chiu, H.L.W. Chan, S.W. Or, Y.M. Cheung, P.C.K. Liu. "Effect of electrode pattern on the outputs of piezosensors for wirebonding process control." *Materials Science and Engineering B*, Vol. 99, (2003), pp 121-126.

11 Yong Ding, Jang-Kyo Kim, Pin Tong. "Effects of bonding force on contact pressure and frictional energy in wire bonding." *Microelectronics Reliability*, Vol 46, (2006), pp 1101–1112.

12 Daniel T. Rooney, DeePak Nager, David Geiger, Dongkai Shanguan. Evaluation of wire bonding performance, process conditions, and metallurgical integrity of chip on board wire bonds." Microelectronics Reliability, Vol 45, (2005), pp 379–390.

13 Lei Han, Fuliang Wang, Wenhu Xu, Jue Zhong. "Bondabily window and power input for wire bonding." *Journal Microelectronics Reliability* Vol 46 (2006) pp 610-615

The Optimization of Hierarchical SOC Test Architecture to Reduce Test Time

Xu Chuan-pei   Dai Kui

(School of Electronic Engineering, Guilin University of Electronic Technology, Guilin, 541004, China)

Mailing address : Guilin University of Electronic Science and Electronic, Engineering College. Zip code : 541004.

telephone :0773 – 5601344. E-mail : xcp@guet.edu.cn.

Abstract

Modular testing of SOCs mainly concentrates in the SOC test architecture and test structure optimization. At present, most of the studies are based on the assumed flattened SOC. This does not meet the needs of practice SOC, because most actual SOCs are of the structure of layers due to the universal use of the reuse technique. On the basis of IEEE P1500 ring based TAM and test bus based CTW this paper researches on multilevel TAM structures of hierarchical SOC. That two obedience certification of IEEE P1500 and two design processes of hierarchical SOC makes that IP core form providers can be provided in different ways to the SOC integrators. On the basis of the four layer model of SOC test architecture optimization, hierarchical SOC test architecture optimization model is presented according to classifications. This paper adopts black box idea to fuzzy hierarchical sub-core and single-core P_w problems, and puts the flattened SOC test architecture optimization and hierarchical SOC test architecture optimization together in a same test framework. From a macroscopical view, this test framework ignores the structure changing inside of the black box, simplifies the multilevel TAM optimization and has a good scalability to various IP Core; from a microscopical view, finely deals with the different IP sub-core in the black box, and extends downwards by levels, so that achieves the goal of multilevel TAM optimization finally. Based on the test architecture research of hierarchical SOC, the paper applies quantum-inspired evolutionary algorithm to SOC test architecture optimization, and establishes a heuristic process based on a property assumption of searching arithmetic to solve the P_{PAW} . Through the observation of group, QEA can decide the allocation of IP core on TAM and the best individual of the current group, and gradually find the overall best individual by using the updated operation. The heuristic process significantly saves the time of CPU computing.

The paper has partial hierarchical SOC in ITC'02 Test Benchmark as experimentation objects to make the simulation experiments. Experimental results and the results of other algorithms are compared, and the algorithm gets a comparatively short testing time.

1 Introduction

The progress of the deep sub-micron technology allows the design of a complete electronic system on a single chip, which is called the system-on-chip(SOC). As a solution SOC technology has been more widely used and the test for SOC has restricted its design and verification.

Modular testing of the embedded cores in a SOC can simplify the complex problems of test access and application[1]. For modular testing, an embedded core is isolated from surrounding logic using a test wrapper, and a test access mechanism(TAM) is designed to deliver test data from the input/output (I/O) pins of the SOC. In most prior work on TAM design, the SOC hierarchy is assumed to be flattened for the purpose of test. However, this assumption is often unrealistic in practice, especially when older-generation SOCs are used as hard cores in new SOC designs. So multi-level test mechanism optimization is necessary for modular testing of hierarchical SOC. Some important concepts and methods were recently presented in [2,3]. On the basis of the four layer model of SOC test architecture optimization[4], this paper introduces black box treatment to hierarchical IP cores and puts the flattened SOC test architecture optimization and hierarchical SOC test architecture optimization together in a same test framework which adopts more efficient quantum-inspired evolutionary algorithm[5] than other algorithms.

2 Analysis on hierarchical SOC test architecture optimization

2.1 Partition of SOC optimization model

In [4], four problems structured in order of increasing complexity were formulated, such that they serve as stepping stones to the problem of wrapper/TAM co-optimization for assumed flattened SOC. We first review these four problems.

P_w: Design a wrapper for a given core, such that the core testing time is minimized, and the TAM width required for the core is minimized.

P_{AW}: Determine (i) an assignment of cores to TAMs of given widths, and (ii) a wrapper design for each core such that SOC testing time is minimized.

P_{PAW}: Determine (i) a partition of the total TAM width among the given number of TAMs, (ii) an assignment of cores to the TAMs, and (iii) a wrapper design for each core such that SOC testing time is minimized.

P_{NPAW}: Determine (i) the number of TAMs for the SOC, (ii) a partition of the total TAM width among the TAMs, (iii) an assignment of cores to TAMs, and (iv) a wrapper design for each core, such that SOC testing time is minimized.

In an hierarchical SOC, some IP cores could be combined together as a hierarchical IP core which has internal TAM. The top-level TAM must communicate with lower level TAMs within hierarchical-cores and this makes the formation of a multi-level TAM structure. Obviously, hierarchical SOC test architecture optimization could adopt the four layer model of SOC test architecture optimization. The only difference is that the hierarchical core testing time is minimized and the TAM width required for the core is minimized in the P_w problem.

2.2 Translation of the hierarchical SOC optimization

Hierarchical SOC contains a lot of hierarchical IP cores and these IP cores are not in a same flat. The multi-level TAM structure is used in order that the test data could be sent to every IP core. On the basis of the partition of SOC optimization model, this paper adopts black box idea to fuzzy hierarchical sub-core and single-core P_W problems, and puts the flattened SOC test architecture optimization and hierarchical SOC test architecture optimization together in a same test framework. From a macroscopical view, this test framework ignores the structure changing inside of the black box, simplifies the multilevel TAM optimization and has a good scalability to various IP Core; from a microscopical view, finely deals with the different IP sub-core in the black box, and extends downwards by levels, so that achieves the goal of multilevel TAM optimization finally.

2.3 Analysis on hierarchical sub-core

Hierarchical sub-core may be supplied by core vendors in varying degrees of readiness for test integration. For example, the IEEE P1500 proposal on embedded core test defines two compliance levels for core delivery: 1500-wrapped and 1500-unwrapped[6]. Here we describe three other scenarios, based in part on the P1500 compliance levels. We use the term wrapped to denote a core for which a wrapper has been pre-designed, as in and use the term TAM-ed to denote a hierarchical sub-core that contains an internal TAM structure. So the three scenarios are described as: 1) NOT TAM-ed and not wrapped; 2) TAM-ed and wrapped; 3) TAM-ed but not wrapped. For the SOC integration, the third category has the smallest use degree of freedom, because it has identified width of the test, test time

has been determined. The former two types of IP cores can conduct internal testing further structural optimization in SOC integration. The paper intends to make research on test architecture optimization including the second type of IP cores with the complexity between the other two types. The algorithm model can be easily extended to the other two types or the situation of mixed use of several types on the optimization problem.

The second type of hierarchical sub-core contains lower-level TAMs, but it is not delivered in a wrapped form for the functional I/Os and top-level scan chains. Under the test the internal sub-cores must take the first test and under the test for top-level core the hierarchical sub-core must be put in functional mode. The test time is the sum of the test time of sub-cores and top-level cores. Width adaptation can be carried out in the wrapper for the hierarchical sub-core such that a narrow TAM at the SOC-level can be used to access a hierarchical sub-core that has a wider internal TAM. The paper uses a parallel to serial conversion of the test data stream at the inputs of the hierarchical sub-core's internal TAM, and a similar serial-to-parallel conversion at the outputs of the internal TAM. The basic concept is illustrated in Fig.1.

3 Algorithm model based on quantum-inspired evolutionary algorithm

3.1 Solution to $P_W^\cdot$ problem

Consider a SOC that has N IP cores and the test width W which is divided into B TAM partitions of widths {w1, w2, …, wB}. If core i is assigned to TAM j, let the time taken to test core i be given by $T_i(w_j)$. If the core is the signal core, the testing time $T_i(w_j)$ is calculated through the BFD method. And if the core is the hierarchical embedded core, the testing time $T_i(w_j)$ contains two parts: 1) the wrapper1 of the functional I/Os and top-level scan chains, the same calculation as signal core; 2) the wrapper2 of the internal TAM, the method is described as follow:

Suppose hierarchical core i has a top-level internal TAM width of M_i bits and suppose it has B_i TAM partitions of widths $\{m_i(1), m_i(2), \cdots, m_i(B_i)\}$, respectively. Let $T(m_i(j))$ be the total testing time for the embedded cores on the internal TAM partition j for hierarchical core i. For a given value of SOC-level TAM width $M_i^\cdot$ for hierarchical core i, where $M_i^\cdot \le M_i$, we need to determine an appropriate TAM partition $\{m_i^\cdot(1), m_i^\cdot(2), \cdots, m_i^\cdot(B_i)\}$ such that let $\max_{j=1}^{B_i} T^\cdot(m_i^\cdot(j))$ be minimized, where $T^\cdot(m_i^\cdot(j)) = \lceil m_i(j)/m_i^\cdot(j) \rceil \cdot T^\cdot(m_i^\cdot(j))$. To solve this problem, we enumerate all values of $m_i^\cdot(j)$ form 1 to $m_i(j)$, while ensuring that $\sum_{j=1}^{B_i} m_i^\cdot(j) = M_i^\cdot$.

3.2 Solution to P_{AW} problem

The P_{AW} problem is a key part of the whole algorithm model and the paper will apply the quantum-inspired evolutionary algorithm to establish the algorithm model.

Consider a SOC that has N IP cores and the test width W which is divided into B TAM partitions of widths {w1,

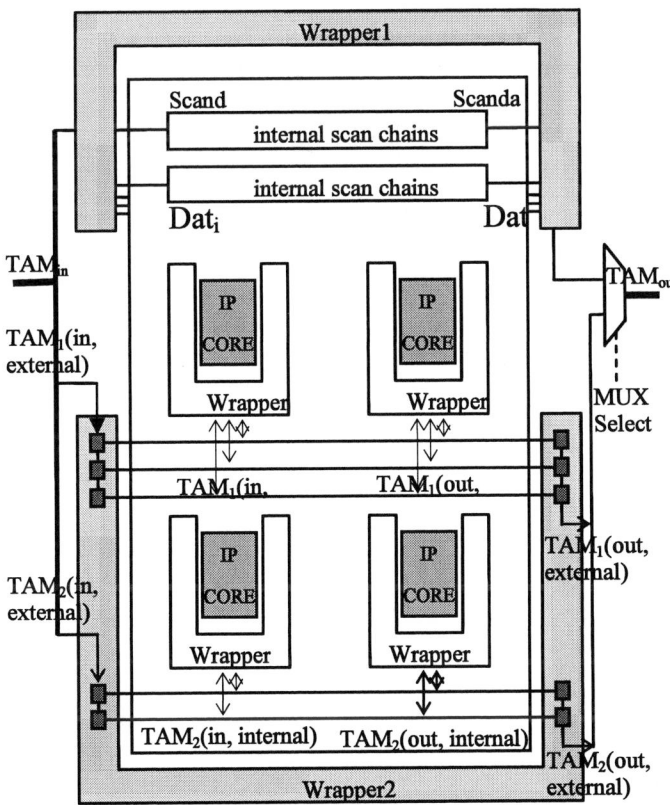

Fig.1 Illustration of the proposed design of Wrapper

$w_2, \ldots, w_B$}. Individual coding of the P_{AW} problem could be described as follow:

$$q^t_j = \begin{bmatrix} \varphi 1_1 & \varphi 1_2 & \cdots & \varphi 1_N \\ \varphi 2_1 & \varphi 2_2 & \cdots & \varphi 2_N \\ \cdots & \cdots & \cdots & \cdots \\ \varphi B_1 & \varphi B_2 & \cdots & \varphi B_N \end{bmatrix}$$

each component of the coding is on behalf of a IP cores, so the length of the individual coding is N. Let $P^t_j = (x^t_1, x^t_2, \cdots x^t_N)$ be the observed values in accordance with the rules of observation, where x^t_i is integer. This shows that if core i is assigned to TAM x^t_i, let the time taken to test core i be given by $T_i(w_{x^t_i})$ and generates a combinational distribution of IP cores on TAM. The paper introduces binary variables $\delta_{x^t_i}$, which are used to determine the assignment of cores to TAMs in the SOC. Let $\delta_{x^t_i}$ be a 0-1 variable defined as follows:

$$\delta = \begin{cases} 1 & , \text{if core i is assigned to TAM x} \\ 0 & , \text{otherwise} \end{cases} \tag{1}$$

The time needed to test all cores on TAM x^t_i is given by $\sum_{i=1}^{N} T_i(w_{x^t_i}) \cdot \delta_{x^t_i}$. All the TAMs can be used simultaneously for testing, so the system testing time can be described as follow:

$$\max_{1 \le x^t_i \le B} \left(\sum_{i=1}^{N} T_i(w_{x^t_i}) \cdot \delta_{x^t_i} \right), (1 \le i \le N, 1 \le x^t_i \le B) \tag{2}$$

Formula (2) could be as fitness formula of the P_{AW} problem.

3.3 Solution to P_{PAW} and P_{NPAW} problem

The purpose of the P_{PAW} problem is in order to achieve the best solution to the scheduling of partition {$w_1, w_2, \ldots, w_B$}. When the partition {$w_1, w_2, \ldots, w_B$} is assured the P_{PAW} problem is translated into the P_{AW} problem. In this paper, the solution was not adopted an enumeration of all the partition[4]. We established a heuristic process based on a property assumption of searching arithmetic. Since the quantum evolutionary algorithm is a search algorithm, it usually contains a corresponding search speed for each of the TAM partition. When a TAM partition is the test solution to minimizing the testing time, the corresponding search time should be fastest in the same iteration times and the best result should be achieved. Based on this idea, the paper randomly generates a certain number of division at first, can also be an enumeration of all the TAM partitions. Let the number of iteration is 100 and then calculate the testing time of each partition. Since the number of iteration is very low the CPU time will be short and this is very effective for the choosing of relatively good TAM partitions. In the above calculation results select the best results of the TAM partitions 5-8 to make the final calculations. Let the number of iteration is 3000 on the basis of repeated tests. This

flexibility in selecting measures ensures the best TAM partition could be chosen, while the number of iterative is enough to overcome the randomness shortcomings of the quantum evolutionary algorithm at a certain extent.

The purpose of the P_{NPAW} problem is in order to achieve the best value of B. When B is assured the P_{NPAW} problem is translated into the P_{PAW} problem. According to the theoretical analysis of the P_{NPAW} problem[4], the paper only calculates under the condition of $2 \le B \le 6$.

4 Experimental Results

The algorithm is working in the environment of the Pentium 4 CPU 1.60GHz, 512M of memory, and VC++6.0. Using standard C programming, simulation results are based on partial hierarchical SOCs in ITC'02 Test Benchmark[7]: p22810, p93791 and p34392. Table 1 presented the experimental results, including TAM partition, the testing time and the improved value. Results of [3] are also presented in the table.

Table 1 Results
(a) P93791

W	B	TAM	Time	Improvement	B	TAM	Time
		Method in the paper				Method in [3]	
16	3	6, 6, 4	1806127	1.28%	4	3,3,5,5	1829530
24	3	16, 8	1220954	0.90%	2	8,16	1232060
32	2	16, 8, 8	914473	0.06%	3	7,9,16	915082
40	3	16, 15, 9	733246	1.50%	3	8,16,16	744410
48	3	16, 16, 16	622644	0.005%	3	16,16,16	622678
56	3	18,15,13,10	556962	2.56%	3	16,16,24	571569
64	4	16,16,16,16	469346	4.78%	3	16,16,32	492899

(b) P22810

W		TAM	Time	Improvement	B	TAM	Time
		Method in the paper				Method in [3]	
16	3	8, 5, 3	457089	8.04%	2	7,9	497079
24	4	9, 8, 5, 2	328853	19.2%	2	8,16	408317
32	4	12, 9, 6, 5	280634	28.4%	2	9,23	392106
40	3	17, 12, 11	280634	22.5%	2	8,32	361895
48	3	17, 16, 15	280634	22.5%	2	10,38	361894
56	3	25, 17, 14	280634	22.5%	2	10,46	361894
64	3	38, 17, 9	280634	22.5%	2	10,54	361894

(c) P34392

W	B	TAM	Time	Improvement	B	TAM	Time
		Method in the paper				Method in [3]	
16	2	8, 8	1146465	3.90%	2	7,9	1192930
24	3	11, 8, 5	833140	-0.39%	3	5,9,10	829938
32	3	16, 10, 6	667943	15.3%	2	16,16	788873
40	3	19, 16, 5	606261	1.99%	3	6,16,18	618597
48	3	22, 17, 9	606261	1.99%	3	6,16,26	618597
56	3	26, 25, 5	606261	1.99%	3	6,16,34	618597
64	3	42, 19, 3	606261	1.99%	3	6,16,42	618597

The paper presents results for p93791 in table 1(a), p22810 in table 1(b) and p34392 in table 1(c). The TAM width of 8 is supplied to each hierarchical sub-core of p22810, and 16 is supplied to p93791 and p34392. The enactment is consistent with [3]. According to the comparison of results, the method generally achieves better results than in the literature [3]. The all improved values are not big, the highest value is 4.78% where $W = 64$ for p93791. The values of p34392 take a 1.99% improvement and the highest value is improved by 15.3% where $W = 40$. The best improved value of p22810 achieves 22.5%. The main reason is that the number of TAM partitions B is more than [3]. That this method in larger value of B has a strong search capabilities, while containing better effect at the same value of B.

5 Conclusions

This paper adopts black box idea to fuzzy hierarchical sub-core and single-core P_w problems, and puts the flattened SOC test architecture optimization and hierarchical SOC test architecture optimization together in a same test framework. Based on the test architecture research of hierarchical SOC, the paper applies quantum-inspired evolutionary algorithm to SOC test architecture optimization and establishes hierarchical SOC test architecture optimization model. Experimental results on partial hierarchical SOCs in ITC'02 Test Benchmark show that the algorithms can achieve the compatible scheduling of all the IP cores and can gets a comparatively short testing time compared to other algorithms.

References

1. Y.Zorian,E.J.Marinissen, and S. Dey, "Testing embedded- core-based system chips", IEEE Comput., vol.32, no. 6, pp. 52-60, Jun. 1999.

2. Vikram Iyengar, Krishnendu Chakrabarty, Mark D. Krasniewski and Gopind N. Kumar. Design and Optimization of Multi-level TAM Architectures for Hierarchical SOCs. Proceedings of the 21st IEEE VLSI Test Symposium (VTS.03).1093-0167. 2003.

3. K.Chakrabarty, V. Iyengar and M. D. Krasniewski. Test planning for Modular Testing of Hierarchical SOCs[J], In IEEE Transactions on CAD of Integrated Circuits and Systems, Vol. 24, No. 3, march 2005.

4. IYENGAR V, CHAKRABARTY K, MARINISSEN E J. Test Wrapper and Test Access Mechanism Co-Optimization for System-on-Chip[C]. ITC International Test Conference. 2001:1023-1032.

5. K.H.Han, J.H.Kim. Quantum-Inspired Evolutionary Algorithm for a Class of Combinatorial Optimization.[C]. IEEE Tran- sactions on Evolutionary computing, 2002:580 -593.

6. E.J. Marinissen et al. On IEEE P1500's standard for embedded core test. *J. Electronic Testing: Theory and Applications*, vol. 18, pp. 365–383, Aug. 2002.

7. E.J. Marinissen, V. Iyengar and K. Chakrabarty. A set of benchmarks for modular testing of SOCs. *Proc. Int. Test Conf.*, pp. 519–528, 2002. (*http://www.extra.research.philips.com/itc02socbenchm*).

（This work is supported by National Natural Science Fund under Grant No. 60766001.）

The Influence of Heating Temperature on Alignment Precision in Thermosonic Flip-chip Bonding

ZHANG Li-na, HAN Lei
Central South University
Central South University, Changsha, 410083, China
E-mail:csu0216@163.com

Abstract

According to the fact that heating will affect the image collection, target identification and track, so it will affect the alignment precision of the chips and substrates in the process of chips package. Sequence images were collected through the thermosonic flip-chip bonder under the different heating temperature, their parameters distribute regulation of translation, scale ,rotation under the influence of the temperature were respectively studied by phase correlation algorithm and 6-parameter affine model hierarchical search algorithm. The distributions of these parameters are highly random and nonlinear. To improve the algorithm speed, a hierarchical search algorithm based on the pyramid was adopted. By Statistical Analysis of the experimental results, it indicated that the higher the temperature, the higher the dispersion degree of translation, scale, rotation. So the temperature effects can not be ignored in thermosonic flip-chip bonding.

1. Introduction

With the improvement of electronic manufacturing process, micromation of organ and high densification of packaging, the requirements of manufacturing equipment's velocity and accuracy are increasingly high in microelectronics region. Among modern electronic packaging equipments, the accurate position of chip and substrate can only ascertained through the vision system [1-3]. In the process of thermosonic flip-chip bonding, the error of alignment precision for chip and substrate should be less than 5um [4]. During the bonding, the temperature of warm table can reach to 150-160oC around, which certainly effects the alignment precision of the chip and substrate.

Heating will cause the air flow, then the flowing air will disturb the radiation from the target, and will brought some aero-optical effects ,such as the intensity distribution diffusion of image points on the imaging focal plane, peak value drop, image blurring , pixel translation and image dithering, etc. That makes the target identification and orientation become difficult [5].

We collected a mass of pictures under different temperature with the thermosonic flip-chip bonder; firstly we adopted phase correlation method to get the translation parameters. This way reduced the influence of light changes between image sequences and increased the computing efficiency and accuracy of the results; then we analyzed zoom and rotation movement of sequence images by 6-parameter affine model. Finally, according to the experimental results, we found the statistical law between temperature and these affine parameters to analyze the effects of temperature for alignment precision in thermosonic flip-chip bonding.

2. Translation motion estimation based on the phase correlation

Phase correlation technique is a non-linear and frequency-domain related technology based on Fourier power spectrum, which often be used to compute the translation between two images, also can be used to compute image rotation and zoom, but the later can only realized when the images are transformed in the logarithm polar coordinate. Literature [6] used this method, but the detection accuracy is not high, which has the relations with the picture size; moreover it is not applicable to horizontal and vertical directions with different scale. But the phase correlation method is unaffected by variable interframe illumination, and it's a robust image-matching method, so there is no need to homogenize brightness before computing the translation.

Supposes f_1 is the reference image, f_2 is the preparative matching image, the relative translation between f_1 and f_2 is (x_0, y_0), we have:

$$f_2(x,y) = f_1(x-x_0, y-y_0) \qquad (1)$$

The Fourier transformation of formula (1) is:
$$F_2(\omega_x,\omega_y) = F_1(\omega_x,\omega_y)\exp\{-i(\omega_x x_0 + \omega_y y_0)\} \qquad (2)$$

The exponential phase deviation factor can be received by computing the cross power spectrum of the two images, just like：

$$\frac{F_1(\omega_x,\omega_y)F_2^*(\omega_x,\omega_y)}{\left|F_1(\omega_x,\omega_y)F_2(\omega_x,\omega_y)\right|} = \exp\{i(\omega_x x_0 + \omega_y y_0)\} \qquad (3)$$

From formula (3), do inverse FFT, then we have:
$$p(x,y) = \delta(x-x_0, y-y_0) \qquad (4)$$

The value of phase correlation function almost is 0 except for the domain of δ function, the coordinates of δ function is the translation (x_0, y_0) .The peak value of δ function in the ideal situation should be 1, but due to the noise and other reasons, often less than 1 .The size of peak value can describe two images match degree.

3. Compute the parameters of scale and rotation

3.1 Image preprocessing

The experimental pictures have tonal distortion, the reasons possibly include: The sensor introduces the entire image signal uniform change and gain; the heating causes reflection and emissivity of the goal change.

If brightness of images is different that means the gray value of picture element is different, and then it will influence the computation of affine parameters and the image clarity evaluation function, which based on the gray value of the images. To reduce the influence of tonal distortion, we must homogenize brightness and balance illumination difference between frames before computing. Literature [7] homogenized the brightness of images directly, but the method we adopted is:

$$I'(x,y) = k\, I(x,y) / \lambda \qquad (5)$$

Where λ is the average gray of an image. The gray value will become lower after divided by λ, which will cause the image visibility fall. To enhance the visibility, each image should be multiplied by the coefficient k in the same group; the choice of k follows the average gray standard deviation is smallest in the same group. According to the above method, the illumination difference between the images eliminated obviously.

3.2 6-parameter affine transformation

Affine transform is the most general form of two-dimensional linear transformation. In the two-dimensional continental space, it can be expressed as follows:

$$\begin{bmatrix} x' \\ y' \end{bmatrix} = \begin{bmatrix} a & b \\ c & d \end{bmatrix}\begin{bmatrix} x \\ y \end{bmatrix} + \begin{bmatrix} e \\ f \end{bmatrix} \qquad (6)$$

Generally speaking, it cannot decompose products of multi-basic transformations.

Affine transformation can use a variety of models to express. Generally speaking, the more parameters of models used, the more accurate to the global movement estimate, but the calculating also become more complexity. Because 4-parameter model is not suitable for the parameter estimation when two directions have the different scale, but 6-parameter model after using the acceleration algorithm its speed may enhance obviously.

This paper according to the 4-parameter affine model hierarchical search algorithm [8] and the hierarchical model-based motion estimation [9] deduced the calculation method of 6-parameter affine model, the process is as follows:

In order to make the calculation easier, we transform formula (6) into:

$$\begin{bmatrix} x' \\ y' \end{bmatrix} = \begin{bmatrix} a'+1 & b \\ c & d'+1 \end{bmatrix}\begin{bmatrix} x \\ y \end{bmatrix} + \begin{bmatrix} e \\ f \end{bmatrix} \qquad (7)$$

The relationship between Image k and Image1 can be expressed as:

$$I_k(x,y) = I_1((a'+1)x + by + e, cx + (d'+1)y + f) \qquad (8)$$

Compute the Taylor series expansion of formula (8) on the domain of (x,y), and reserve the first three sections, we have:

$$I_k(x,y) = I_1(x,y) + (a'x + by + e)g_x + (cx + d'y + f)g_y \qquad (9)$$

Where $g_x = \partial I_1 / \partial x$, $g_y = \partial I_1 / \partial y$. They respectively expresses the horizontal and vertical gradient of an image. According to the least squares principle, the error function can be expressed as:

$$\hat{a'},\hat{b},\hat{c},\hat{d'},\hat{e},\hat{f} = \underset{(a',b,c,d',e,f)}{\arg\min} E(a',b,c,d',e,f) \qquad (10)$$

Choose proper a',b,c,d',e,f to make error E minimum, there E defined as:

$$E(a',b,c,d',e,f) = \sum \begin{bmatrix} I_k(x,y) - I_1(x,y) - (a'x + \\ by + e)g_x - (cx + d'y + f)g_y \end{bmatrix}^2 \qquad (11)$$

Calculate the partial derivatives of formula (11) about a',b,c,d',e,f, then make these formulas equal 0, we have

$$A_k C_k = B_k \qquad (12)$$

where :

$$A = \begin{bmatrix}
\sum M^2 & \sum MN & \sum Mg_x & \sum ML & \sum MO & \sum Mg_y \\
\sum NM & \sum N^2 & \sum Ng_x & \sum NL & \sum NO & \sum Ng_y \\
\sum g_x M & \sum g_x N & \sum g_x^2 & \sum g_x L & \sum g_x O & \sum g_x g_y \\
\sum LM & \sum LN & \sum Lg_x & \sum L^2 & \sum LO & \sum Lg_y \\
\sum OM & \sum ON & \sum Og_x & \sum OL & \sum O^2 & \sum Og_y \\
\sum g_y M & \sum g_y N & \sum g_y g_x & \sum g_y L & \sum g_y O & \sum g_y^2
\end{bmatrix} \qquad (13)$$

$$B = \begin{bmatrix} \sum TM & \sum TN & \sum Tg_x & \sum TL & \sum TO & \sum Tg_y \end{bmatrix} \qquad (14)$$

$$C = [a',b,c,d',e,f] \qquad (15)$$

$$M = xg_x; \quad N = yg_x; \quad L = xg_y;$$
$$O = yg_y; \quad T = I_k - I_1 \qquad (16)$$

3.3 "coarse-to-fine" hierarchical search

To increase computing speed, the paper used "coarse -to - fine" hierarchical search, that is, image Pyramid technology. This technology is usually used for improve computing speed of the image registration and Movement (optical flow) estimate. Regarding the affine transformation model track algorithm, the "coarse- to- fine" hierarchical search conduces to ensure affine parameter convergence correctly through the iterate solving.

There are different pyramid methods, the most common and the simplest is dot sampling to the input images. After sampling, the image is 1 / 4 of the original image, the process is: reducing the original image to its 1/4, we obtain first layer, then reducing the first layer to its 1/4, we obtain second layer, and so on. The layer number is involved with the original image's resolution and size dimension.

Start with a minimum resolution of hierarchical images to calculate the affine transformation parameter and determine rough match position. Then hierarchically search in-order, when we utilize initial parameters to generate the high level of image by subsample, the translation parameters should be multiplied by 2.To simplify procedures, we used fixed iterative number for each layer. Considering the speed and precision of calculating, the iterative number is 4 for each layer. We adopted 960 × 960 pixels picture when we were processing algorithm verification, and found without the influence of tonal distortion, the scale parameters could be precise to four decimal places; the rotation parameters could be precise to two decimal places. If there was tonal distortion, we should eliminate it firstly. Although the precision of scale and rotation would fall, it still can satisfy the computational precision.

3.4 Algorithm verification

Take a 1600 × 1200 pixels high-definition chip image as an example, to save operation time and memory, before computing it will be cut into 960 × 960 pixels as a reference map I0, like Fig.1 (a) shows. Firstly translates reference map I0 8 pixels along the level toward right and 5 pixels along vertical downward, and then enlarge I0 for 1.02, 1.03 times along two directions respectively. Finally rotate it counterclockwise for 0.5°, so we obtain the preparative matching image I', which is 988×998 pixels, like Fig.1 (b) shows. Before computing we enlarge the canvas of picture (a) to the same size of image I', then we use our method to get the registration image(c).

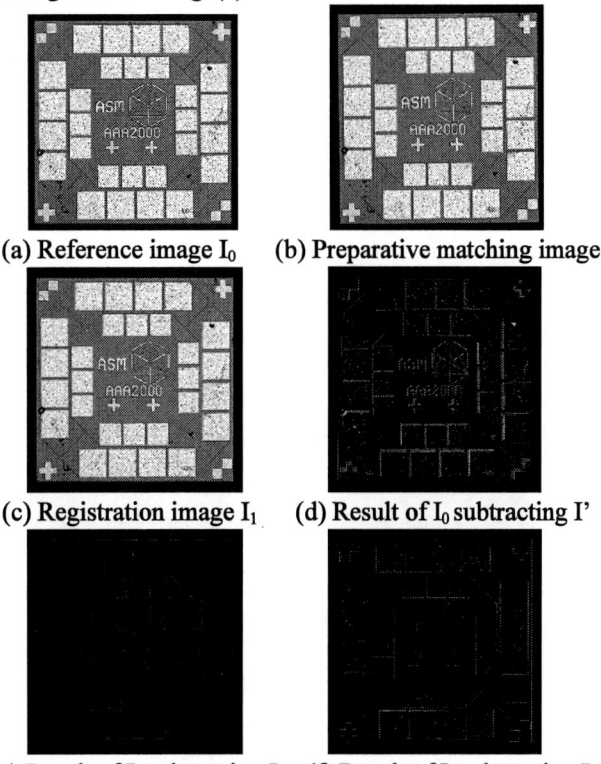

(a) Reference image I_0 (b) Preparative matching image I'

(c) Registration image I_1 (d) Result of I_0 subtracting I'

(e) Result of I_0 subtracting I_1 (f) Result of I_0 subtracting I_1
(6-parameter model)   (4-parameter model)

Fig.1 Image registration

The result of the reference image I0 subtracting the preparative matching image I' like Fig.1 (d) shows, the normalized correlation coefficient was 0.7546; the image (e) is the result of the reference image I0 subtracting the registration image I1. The normalized correlation coefficient of image I0 and image I1 is 09985, the computing time is 26.44s. Compared image (d) with (e), we found the registration precision was satisfying. Concerning the 4-parameter affine model, the computing time is 18.83s, but the correlation coefficient will drop to 0.9360.The result of the registration image subtracting image I0 is the image (f). Obviously the matching precision of 6- parameter model is higher than 4-parameter model's. All calculations were proceeded on: CPU (AMD Sempron(tm), Processor the 2800+), 512M memory, with Matlab7.1.

4 Analysis of the experimental data

The experiment optical system is a 1 time lens, which can't change focus. The lighting system is a circular LED. The image data was collected by black-and-white image collection board (Matrox Meteor-II/Standard). The chip's actual size is 1 mm × 1mm.But owing to the camera optical axis isn't completely vertical to the chip, there is a very small angle, causing the aspect ratio of the chip is not 1:1, the experimental device like Fig.2 shows.

Fig.2 Experimental apparatus

Five groups of chip images were collected through the thermosonic flip-chip bonder under the room temperature (about10oC), 60oC, 90oC, 120 oC, 150 oC respectively. Each group has 250 images; the original size of images is 640 × 480 pixels. To increase computing speed, the images were reduced to the size of 180 × 180 pixels.

Through experiments we found that the higher the temperature, the more blurred images, which is similar to the fuzzy image caused by out of focus, so we adopted the gradient function to evaluate the image definition. Gradient function can expressed as:

$$G = \sum_x \sum_y [I_x^2(x,y) + I_y^2(x,y)] \qquad (17)$$

Where I_x and I_y express the horizontal and the vertical gradient of images separately. Regarding the gradient function, for the good focused image, which has the more incisive edge, should has greater gradient [10]. Before the clarify-evaluation of these images, we should also eliminate the influence of variable interframe illumination. The gradient function value of 250 sequence images collected at 150 °C is showing in Fig.4.

Through the clarify-evaluation, we will delete 100 pieces of fuzzy images from the group under the condition of heating, and then the other 150 images will carry on the statistical analysis.

(a) Clear picture   (b) Fuzzy picture
G=10963154.5   G=5240913

Fig.3 Clear picture and fuzzy picture (150 °C)

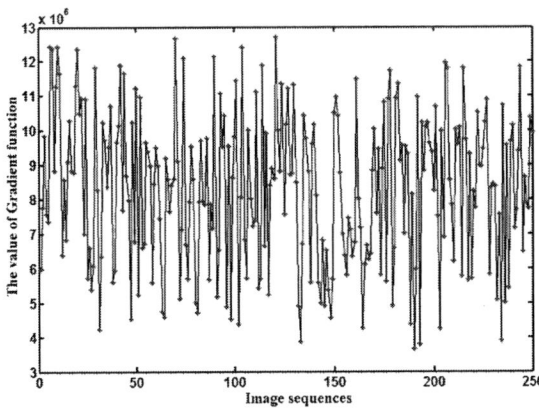

Fig.4 Gradient function value of 250 Images (150 °C)

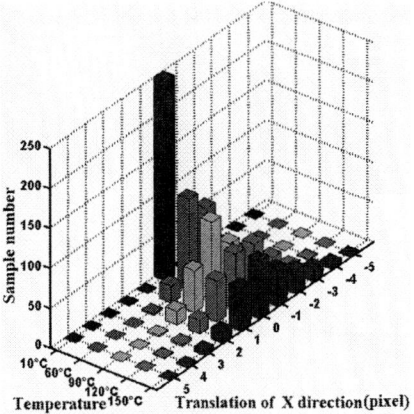

(a) Relation between temperature and X- translation

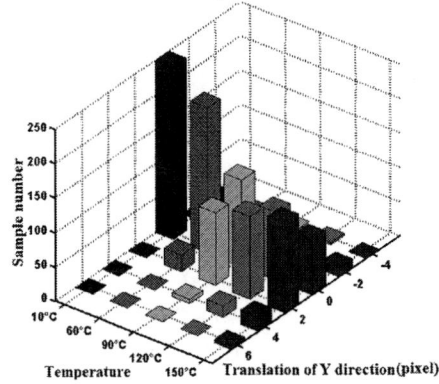

(b) Relation between temperature and Y- translation

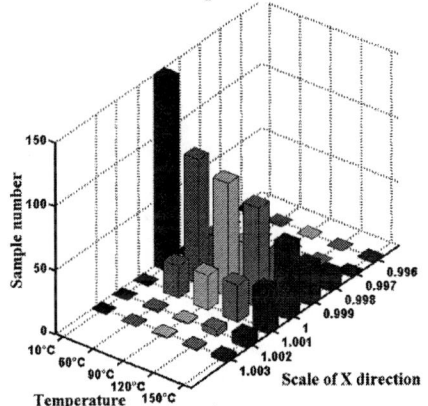

(c)Relation between temperature and X-scale

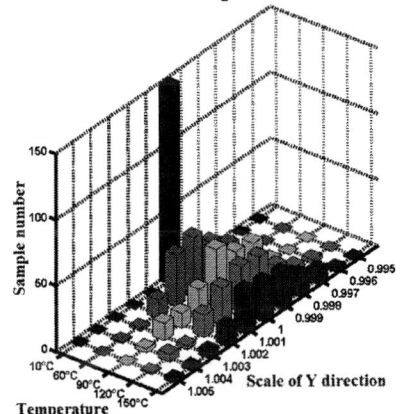

(d) Relation between temperature and Y-scale

The most obvious movement between images is the translation; considering the phase correlation method is unaffected by variable interframe illumination, therefore we adopted it to study the relations between the translation and temperature. But the zoom and rotation between images are not very obvious, to discover the relations between them and temperature, we adopted high accuracy arithmetic: 6-parameter affine model. Took the first image of each group as the reference image to estimate the other images' geometric distortion which caused by heating. Through the affine model's parameter estimation, we obtained the more precise magnification and degrees rotation. Without considering shear transformation, regardless of revolving first or zoom first, there are: $a = a_x \cos(\theta)$, $d = a_y \cos(\theta)$. In view of that the angle θ is small, so we can have $\cos(\theta) = 1$, then we can get $a_x = a$, $a_y = d$. Similarly, we have $c = arcs\,\text{in}(\theta)$, then we can obtain the rotation angles.

Took the first image of each group as the reference image, and then obtained the parameters of translation, rotation and scale under the different temperature separately. Finally compute the mean value of each group parameters, and then took the mean value return to zero. But the operation for translation was especial, the processing was: sought for an image in each group whose translation almost equal or absolutely equal the mean value, then took this image as a template, computed the translation again, finally obtained the relationships between translation of X, Y direction and temperature, like Figure.5 (a), (b) show. The positive translation of X direction expressed that translation toward right along the level, the positive translation of Y direction expressed moves downward along the vertical direction.

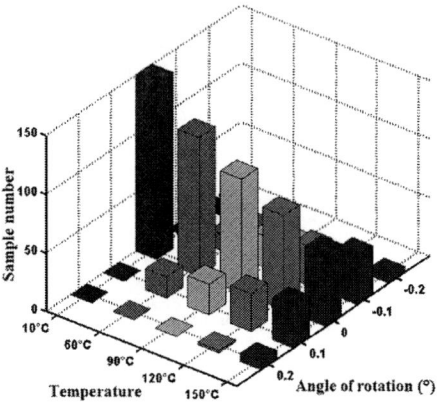

(e) Relation between temperature and rotation

Fig.5 Relations between temperature and the value of translation, scale and rotation

From Fig.5 we found that the temperature influence on translation was more obvious than on rotation and scale. Under the room temperature, there were almost no translation and distortion. But with the increase of temperature, the rotation angle and the scale parameter also increased flexibility, and their distribution approached to the normal distribution, their mean value are almost 0 and 1 respectively. From Fig.5 (b) and (d), we found that the translation and scale of X direction were different with these of Y direction: The Y direction translation was lager than X direction's; its translation distribution didn't approach the normal distribution; compared to X direction, Y direction scale distribution was more scattered under the same temperature. One of the reasons maybe was that there was a small angle between the optical axis and the normal of the chip face，which caused the horizontal parameters not to be bilaterally symmetrical in the object space, and the vertical parameters not to be bilaterally symmetrical in the image space, that is why Y- direction distribution has the dissymmetry.

We calculated the standard deviation of these affine parameters at different temperature respectively. The results like Fig.6 and Fig.7 show. We can find with the temperature increasing, the standard deviation increase, and the discreteness of the affine parameters become higher.

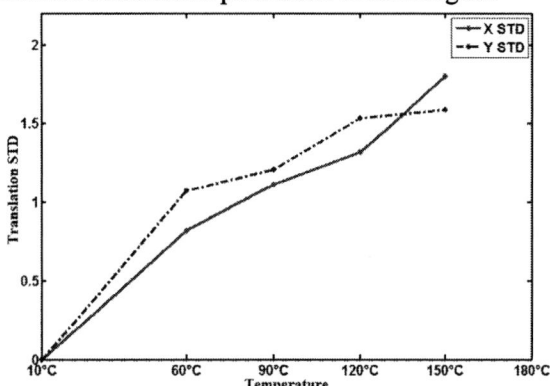

Fig.6 Relationship between displacement standard deviation and temperature

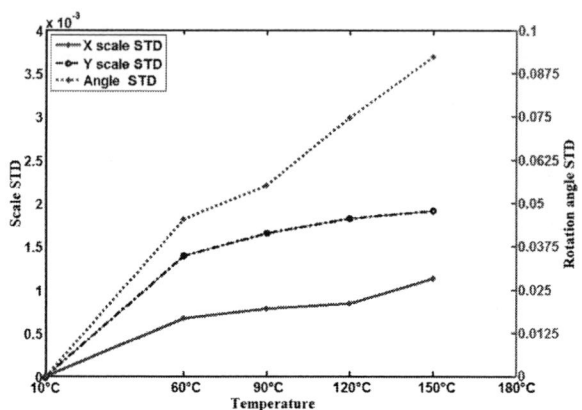

Fig.7 Relationship between temperature and the standard deviation of scale and rotation

5. Conclusion

In this paper, we adopted Phase correlation method and 6-parameter affine model to study the Relationship between temperature and the value of translation, scale and rotation. Owing to the tonal distortion and image blurring, we adopted the clarify-evaluation-function and homogenize brightness to preprocess these images. But because there were so many factors to cause the image change, these interference factors weren't completely eliminated before computing, so which had some influence to the results, but the influence was small to the integral rule. Through the experiment we discovered there was almost no change among the sequence images under the room temperature, but under other temperature, there was obvious translational motion, the higher the temperature, the dithering more intense, and the distributions of affine parameters more discrete, so the influence to alignment precision became bigger. To increase the alignment precision of the chip and substrate in thermosonic flip-chip bonding, the temperature effects can not be ignored.

From above all, we found the influences of temperature on image collection, target identification and track were considerable, and can not be neglected. Under the actual condition, we can analyze the rule of image dithering, distortion and blurring through image data, which has a certain foresight for target identification and track.

References

1. CAO Chang-jiang, ZHANG Chen, LI Zheng-bo, et al. Design of precise position instrument for micro-assemble system [J].Journal of Shanghai Jiao tong University, 2000，34(11):1483-1485.
2. LEI Yuan-zhong, LUO Jian-bin, DING Han, et al. Important academic problem in advanced electronic manufacturing [J]. Bulletin of National Science Foundation of China, 2002, 16(4):204-209.
3. Naren V. Application of microcomputer in submicron level measurements and control of a positioning device[J]. Journal of Microcomputer Applications, 1995, 18(2):149-164.
4. LI Jian-ping,ZOU Zhong-sheng,WANG Fu-liang.Design and realization of machine vision system in thermosoinc flip-chip bonder[J]. Cent. South Univ. (Science and Technology), 2007,38 (1) :116-121.
5. YIN Xing-liang. Air-optic principle. Beijing: Astronaviga

-tion publishing house, 2003.

6. LI Zhong-xin,MAO,Yao-bing,WANG Zhi-quan.A method of image mosaicing using Log-polar coordinate mapping [J].Journal of Image and Graphics,2005,10(1):59-63.

7. LI Qi, FENG Hua-jun, XU Zhi-hai. Image preprocessing techniques for auto focusing[J]. OptoElectronic Engineering, 2004,31(9):66-68

8. LUO Jun, YI Li-ya, HUANG Ben-xiong. A fast hierarchical search matching algorithm [J]. Computer and Information Technology, 2005,13(6):23-27.

9. Bergen, et a1. Hierarehieial Model-Based Motion Estimation[A]. Proceedings of the Second European Conference on Computer Vision[C], 1992,237-252.

10. CAO Mao-yong, SUN Nong-liang, YU Dao-yin. Study on clarity-evaluation-function of out-of-focus blurred image[J]. Chinese Journal of Scientific Instrument , 2001, 22(3): 259-261.

A High Speed Image Preprocessing Method for IC Wafer Inspection

CHEN Wei, WU Li-Ming, CUI Shan-Ling
Faculty of Information Engineering, Guangdong University of Technology
Guangzhou, China, 510006
jkyjs@gdut.edu.cn, 15989249384

Abstract

A lot of vital defects appear in the process of fabrication of IC wafer. Material defect which mainly causes circuits to be connection failure is an important factor of reducing the yield of IC. So the inspection of material defect is one of the most important things to do for guaranteeing the quality of IC. In this paper, a machine visual test system is designed with AOI (Automated Optical Inspection) technology. The surface defects and geometric sizes of the IC wafer can be detected automatically. It introduces the method of image convolution of FPGA, by using assembly pipelining mode to design a real-time median filtering, to apply to the IC chip testing system which has good effect. Defect feature extraction and selection techniques were studied regarding the characteristics of inspection images of IC wafer. This machine vision test system is realized by FPGA based on XILINX Spartan-3A which is a high-speed parallel device.

1. Introduction

On the design of the industrial testing and video monitoring system, real time capability has been the bottleneck of the system design all the time. If we only adopt the traditional MCU, it cannot adapt to the request of real time capability especially in the case of massive data, excessive demands of high speed capability. Although at present, we use the instruction based on the way of pipeline design and DSP which do the processing in high speed, it could be difficulty to achieve the demand due to the corresponding algorithm which is still relied on the serial execution of the instruction. To face with such complex tasks, it can hardly meet the requirement.

The image signal which is collected by CCD, interfered by the various noise sources can make the quality of the image poor. Before carrying out the following processing, it must be pre-processed such as a variety of image filtering, smoothing ,etc.These operations are to improve the quality of the image, to inhibit unnecessary deformation and to strengthen some important features of the image. Because of massive data, simple operational structure of the image pre-processing algorithm of the basic layer, if using general software, it can be a fatal factor to the high real time capability system. For these problems, to design high speed field programmable logical devices, FDGA chip could be a good choice.

FPGA maintain ASIC's high-speed，avoid high cost of development and defects that could not change the internal circuit of the finished product and increase flexibility and adaptability of the design. The regular internal logical block array and varied link-wire source is especially suitable for the fine-grained and high parallelism characteristics of signal processing tasks. The D trigger in FDGA using the structure of multi-level assembly line can get higher processing function and have higher clock frequency. This paper, taking the example of image convolution and median filtering, introduce the hardware design of image pre-processing.

2. System architecture design

（1）Chip selection and architecture design

The preprocess of the video signal selects the chip XC3S400 of Spartan-3 series of Xilinx company, and this series chip is the only FPGA device which possess high-efficiency achievement of DSP function and low cost. By using Spartan-3 series device, it takes a small space in the entire equipment to achieve high performance DSP functions. This device integrates key DSP resources, such as the built-in 18*18 bit multiplier, large memory block (18kb), distributed RAM, and the shift register logic. Besides improving the basic performance, the Spartan-3 series device can enhance the chip utilization. Among them, XC3S400 has internal clock frequency of 326 MHz, I/Os which support 622 Mbps data transfer rate, a high-performance SelectRAM internal memory, up to four digital clock management modules, and eight overall clock multiplexer buffers. This chip is suitable for using as a coprocessor or preprocessor, and the system architecture design show in Fig.1. The CCD acquires the real-time image information, which is digitized by A/D and transferred to FPGA through buffer. Then, through the computation of the image data by FPGA, and the realization of frame buffer by RAM, the data stream finally is transmitted to the PC by PCI bus.

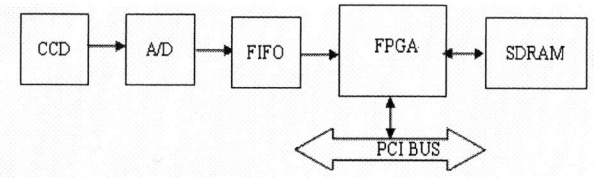

Fig.1 System Architecture

(2) Images of convolution algorithm

In the image pre-processing algorithm, two-dimensional convolution operation is a common operation, such as the image of smooth, each kind of edge detection, filtering ,etc. All above can be achieved by the convolution operation. Although the convolution operation is simple, it's very time-consuming, such as the size of the design template for the $N * N$, an image size for the $M * M$, the output of each point on the need to N^2 multiplications, one division operation, a Image is on the need to $(M - N + 1)^2 * N^2 * N^2$ multiplications and $(M - N + 1)^2$ division operations, but these operations are repetitive. If these operations can be completed by FPGA before entering the follow-up treatment, the overall speed of the system will be substantially improved.

The essence of convolution is a point by point move with template, it usually move template along the row or column, the test data was processed with data that in template. For N*N template, it usually design N-1 cached. The main working manner of the FPGA is parallel computing and Pipeline work, so the high-speed, high-capacity is realized easily. In this paper, the FIFO (First In First Out) is used for data cache, the design of N-1 FIFO to cache data, and when some image data is in the processing, the other data is discarded, and only useful data is saved, so the essence of convolution operation is that the template point is moved point by point, it is usually along the line or out mobile template, each time only data within the template can be processed. So the system can meet the need of real-time requirements, and reduce the waste of resources.

Design 3 * 3 template with two FIFO memories, the width of the FIFO is W (the width of the image), the FIFO access in the adjacent nine pixels A_{ij}, and then we use the template of the corresponding nine Multiplication factors to get value by the operation of addition, shown in Fig.2.

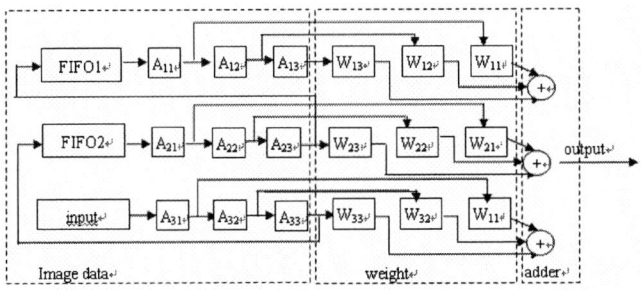

Fig.2   3*3 Template Convolution Filtering Structure

(3) The analysis of Median filtering theory

Median filtering is a non-linear signal processing technology, and it can restrain disturbing pulse and isolated noise very well. Unlike some linear low-pass filter which makes the edge of the picture being blurry while filtering the noise, the median filtering not only filters the noise but also keeps the boundary of object well. Moreover, it can be computed easily and be compatibly implemented by FPGA.

Median filtering is a statistical sort filtering, it sorts the pixels in the fixed window to find the middle gray level value as the value of central pixels in the window. In the two-dimensional data, by slipping in the two-dimensional form, the pixel gray level in the form is sorted to make a monotonous rise or monotonous decline two-dimensional array. The outcome of the median filtering is:

$$G(x,y) = med(F(x-j, y-k), (j,k) \in W)$$

$F(x,y)$ is original image, $G(x,y)$ is image after: processing, W: slipping template.

The windows of the median filtering are various, such as round, crisscross, square, cirque. The magnitude or the pixels in the window is always odd. Different shape and size of window make the effect different. Commonly the size of the window is 3*3 and the shape is rectangle (Fig.3). Only 9 gray levels in the slipping board will be sorted while processing.

$f(x-1,y+1)$	$f(x,y-1)$	$f(x+1,y+1)$
$f(x-1,y)$	$f x,y)$	$f(x+1,y)$
$f(x-1,y-1)$	$f(x,y-1)$	$f(x+1,y-1)$

Fig.3 The window of Median Filter

(4) 3*3 template design

By imposing different weights to the pixels in different location of the template, the convolution operation achieves the computation between adjacent pixels, and finally acquires the pixel values. Because the weight of the points in the median filtering template is 1, it saves a lot of addition, multiplication and division operation units. But with the difference to other convolution operations, median filtering needs to sort large amounts of data, so the design of the hardware module should be considered from various aspects. Although the captured image data is the serial data stream, the FPGA can achieve the template with the parallel pipeline. The median filtering hardware structure shows in Fig.4.

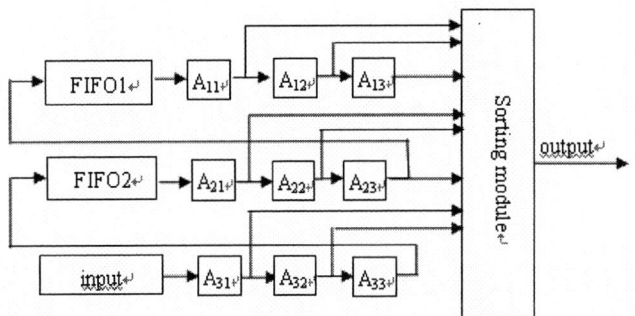

Fig.4 Media Filter Hardware Structure

(5) Sorting based on FPGA

The key of median filtering is the sort of image data. There are many ways to sort, such as the insertion method to sort, and Bubble sort. In this paper, we choose the Bubble sort to sort the 9 data of the 3*3 window and use the Xilinx ISE Simulator to simulate, simulation sequential chart as shown in Fig. 5. We input nine image data(a, b, c, d, e, f, g, h, i), range them from large to small aa, bb, cc, dd, ee, ff, gg, hh , ii, in which the middle ee data is the value of seeking.

3 Experiment result

This design can apply to the visual check of IC chip, by using the image collection and pre-processing technology to process the image data which width is $1024 \times 800 \times 12$ bits per frame. If the system's clock is 50MHz, the max time of processing a frame of image data is 20.5ms, which can meet the request of real-time capability of system completely. One of the filtering results of figure is as follow Fig.6.

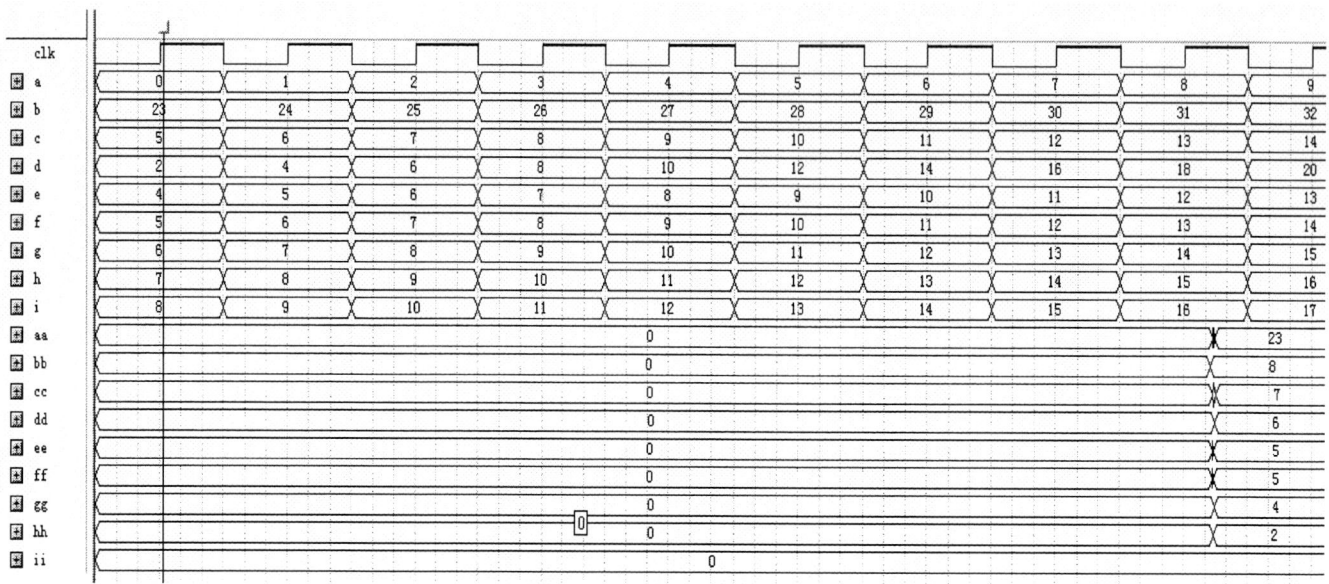

Fig.5 Sequential Chart for Median Filtering

Before filtering   After filtering

Fig.6 Result of Median Filtering

4 Conclusions

We analyse the technology of image preprocessing which is based on FPGA and introduce the method of image convolution, taking median convolution which is used for visual checking in IC chip as an example to carry on experiment for analysis, and prove that the method above can meet the validity of real-time capability of system.

With the development of technology, the cost of FPGA is lower, the capacity of FPGA is higher, the logic units of FPGA are more, and the component which possesses millions of gate level appears.in addition, with the increasing of integration of element, we can design more processing units in a small system. The power consumption is smaller than before. At present, FPGA component is taking place of the primary ASIC to complete the signal processing task. Under the basis of not influencing the post processing result, the image pre-processing system based on FPGA has advantages of improving the quality of picture, reducing the post burden of CPU processing, enhancing real-time capability of system. This paper introduces that the technology of designing pre-processor is very effective in mass data image collection and pre-processing, especially, the design is a good choice for the system of high frequency digital TV preprocessing. .

References

1 DENG Yao-hua, LIU Gui-xiong, WU Li-ming, ZHANG Ying-min, WANG Gui-tang. Adaptive Focus Algorithm of Image Sampling by Lifting Scheme Denoising[J].J.Shanghai Jiaotong Univ.(Sci), 2008, 13(Sup): 127-131

2 ZOU Decai, WU Haitao, LI Yun. Struts Framework Analysis for XILINX FPGA[J]. Aeronautical Computing Technique, 2007, 37(2): 80-83

3 XIAO Wencai, FAN Feng. The FPGA Design of Real-time Video Acquisition System[J]. CHINA DIGITAL CABLE TV. 2006(2): 2104-2108

4 WANG Desheng, XU Wanying, HUANG Xinsheng. The Fast Arithmetic and Simulation of Image Pretreatment Based on FPGA[J]. Computer Simulation, 2007, 24(8): 320-326

5 RAN Xiaoqiang, WEN Desheng, MAN Feng. Design and Implement of Sub-pixel Image Fusion System Based on FPGA[J]. ACTA PHOTONICA SINICA, 2007, 36: 274-277

6 He Wenbo. Research on Infrared Image Pretreatment Algorithm Based on FPGA and DSP. Huazhong University of Science and Technology, 2005.

7 Rabah H, Mathlas H. Linear amy processors with multiple access modes memory for real-time image processing [M]. Proceeding of IEEE, 2002.

8 Shinichi Hirai, Masakazu Zakouji, Tatsuhiko Tsuboi. Implementing Image Processing Algorithms on FPGA-based Realtime Vision System[J]. IEEE Int Conf on Robotics and Automation, 2004,4: 4707- 4712

The Influence Discipline of Temperature of High Viscosity Fluid Jetting

Hu Hao, Deng Gui-ling

(Key Laboratory of Modern Complex Equipment Design and Extreme manufacturing
(Central South University), Ministry of Education Changsha 410083,China)
E-mail:huhaodlu@163.com

Abstract

Viscosity which varies with the temperature remarkably was one of an important factor for fluid jetting. In this paper finite element models of fluid jet-dispenser for thermal analysis and fluid-structure interaction analysis were built through ANSYS. The decided heat flux was loaded to the finite element model for thermal analysis to obtain the temperature field of the fluid field. So the average viscosity of the heated fluid can be worked out. And then the accumulative volume, average velocity and equivalent velocity of two different kinds of fluid between in heating and room temperature condition in fluid-structure interface analysis were compared. The results indicted that the heated fluid was apt to spout. And the losing heat was also worked out in jetting stage which was a reference to next work and designing the heater

Key words: thermal analysis, jetting dispensing, temperature field, high viscosity, fluid-structure interaction analysis

1. Introduction

The technology of the fluid dispensing was widely used in the microelectronic packaging industry. And the conventional contact fluid dispensing mode, including Auger Pump, Time-Pressure and Plunger Pwaston, has always been the mainstream in market. Compared with the conventional contact fluid dispensing technologies, the non-contact jetting dispensing technology has its own prominent merits: higher efficiency, little sensitivity to variations in board height, accurate-jetting with high repeatable, and smaller dot, especially jetting a wide range of adhesive, such as surface mount adhesive, under-fill adhesive, silver epoxy. Being considered as the next generation fluid dispensing technology, jetting dispensing is welcomed in recent years, and is gradually superseding the conventional technologies[1].

At present, some people researched a lot on the structure parameters of fluid jetting dispensing in order to get favorable adhesive dot which is bright surface, suitable form and size, well consistency. And one of important factor is the high quality of the adhesive, such as good mechanical property, and suitable viscosity. With too low viscosity, the initial intensity of the adhesive is no good, and components shift, caved after high temperature; while too high, it is difficult to spout because of the large surface tension. As there is a relationship between temperature and viscosity, controlling the temperature is a good way to control the fluid's viscosity and jetting. This paper researched on how to heat the fluid in order to obtain the suitable viscosity, and what a jetting accumulative volume, average velocity and equivalent velocity of two different kinds of fluid between in heating and room temperature condition in fluid-structure interface

analysis (FSI) through ANSYS simulation. The thermal load was also worked out in jetting stage which was a reference to next work and designing the heater

2. The Temperature-Viscosity Property of Fluid

At present, the kinds of adhesive are extensively used for microelectronic packaging are epoxide resin and acrylic acid, silicon resin and polyurethane are also used. In general, basic resin with silver power can be made conductive adhesive. The dependent relation of viscosity on temperature of different kinds of adhesive, which are usually constituted by basic resin, curing agent, plasticizer, filler, and so on, varies. As high polymer, the temperature-viscosity property of all kinds of adhesive can be described by Arrhenius equation:

$$\eta = A e^{\frac{E}{RT}} \tag{1}$$

Where η is the fluid viscosity, A is viscosity constant, T is temperature, R is gas constant, 8.314 J/mol·K, E is activation energy. Generally, larger is the activation energy, more sensitive is the viscosity dependency on temperature.

As thermosetting resin, its viscosity decrease by heating at first, and increase after curing reaction when temperature reaches some point and as time goes on. At last, the adhesive become solid. In a word, the curve of temperature-viscosity property was like a bathtub. It was obvious that the temperature must be limited to curing reaction during jetting, and heated adhesive should be used up as soon as possible. On the other hand, the temperature for curing reaction varies among different kinds of adhesive. Only accurately control the temperature, right viscosity was available.

In this paper, two kinds of silicone resin were used for simulation whose physical appearance parameters are shown in table 1.

Table 1 physical appearance parameters of silicone resin

T/°C Parameters	20	40	60	80
$\eta 1$ /Pa·s	10.35	7.94	5.94	4.33
$\eta 2$ /Pa·s	33.47	22.69	16.11	11.89
C /J·(Kg·K)$^{-1}$	1550			
ρ/Kg/m^3	972			
K /W/(m·K)	0.35			

Where $\eta 1$ is the first kind fluid, $\eta 2$ is the second kind fluid, ρ is density, C is specific heat, k is coefficient of thermal conductivity.

Considering the temperature range in table 1 the curing reaction hadn't taken into account. Going into the viscosity and temperature data in table1 to carry on the calculation, logarithmic equations were worked out and given by equation (2) and equation (3).

$$\ln \eta = -3.6125 + 14713 / RT \qquad (2)$$

$$\ln \eta = -2.5773 + 14830 / RT \qquad (3)$$

The equation (2) and equation (3) respectively represented the Arrhenius equation of the first and the second fluid. And their specific heat, density and heat conductivity are the same, and vary insignificantly with temperature.

3. Mathematical Model of Jetting Dispenser

Where there were temperature grads in a body or among objects, the heat transfers form high temperature parts to low temperature parts. Here, the heat generated by the heater transferred into the chamber, fluid filed and then ball-needle one by one. And this kind of heat transfer mode was called conduction of heat. The problem of two-dimension nonlinear transient conduction of heat could be described as

$$\rho c \frac{\partial T}{\partial t} = \frac{\partial}{\partial x}(k \frac{\partial T}{\partial x}) + \frac{\partial}{\partial y}(k \frac{\partial T}{\partial y}) + \dot{q} \qquad (4)$$

Where ρ is density, C is specific heat, k is coefficient of thermal conductivity, $\dot{q}$ is inner origin of heat.

The surface of the chamber was surrounded by air and heat transfer because of the temperature difference. As the air flow just because of being heated, this kind of heat transfer mode was called natural convection which was expressed by Newton cooling equation

$$q = h \times \triangle t \qquad (5)$$

Where q is heat flux, h is convection coefficient, $\triangle t$ is temperature difference between the chamber and air.

The heating process of the fluid jet-dispenser was belong to transient heat transfers, and the temperature, the heat flux, the boundary condition and the system internal energy changed obviously in different time. According to the principle of conservation of energy, transient heat balance can be written as (matrix)

$$[C]\{\dot{T}\} + [K]\{T\} = \{Q\} \qquad (6)$$

[K]is transfer matrix including thermal conductivity, convection, radiance and coefficient of form;

[C] is specific heat matrix, considering the increase of system internal energy;

{T}is vector of nodal temperature;

$\{\dot{T}\}$ is equal to $\partial T / \partial t$;

{Q} is vector of nodal heat current, including heat generation.

4. thermal analysis simulation

The process of fluid jetting dispensing was that repeatedly acted by drive force and restoring force, ball-needle moved up and down in the fluid chamber. When the ball-needle stroked down, the fluid around the ball-needle tip was extruded, and the fluid near the nozzle ejected out and formed a dot. It was clear that controlling the viscosity of the fluid, especially the fluid near the nozzle, which influenced the flow of the fluid in the chamber, by controlling the temperature, was a key factor to the quality of the fluid jetting dispensing.

4.1 physical model

The direct resistance heater underneath the chamber was designed to heat the fluid to a suitable temperature (30~60℃ [5]).A part of the fluid jetting dispenser, including the chamber contacting other parts of the fluid jetting dispenser through O ring, ball-needle and adhesive pipeline, was chosen for the thermal analysis, as shown in figure1.

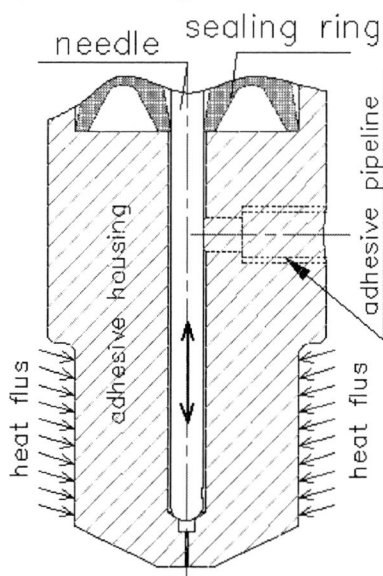

Fig.1 schematic of jetting dispenser for thermal analysis simulation

Clearly, the heat flow affected the uniform of the fluid field temperature, and the fluid jetting next. In order to analyze and solve this problem expediently, some simplification processing was necessary.

1. Considering the axial symmetry of geometry structure, loaded heat flux and boundary condition of the fluid jetting dispenser, it was sensible to choose a half of the dispenser's along one of the longitudinal profiles and build the finite element model.

2. Neglecting solving the flow field in thermal analysis as the fluid not flowed in the process of heating.

3. Neglecting the adhesive pipeline that was small size and it was in favor of meshing finite model.

4. Supposing the uniform heat flow generated by heater flowed into the jetting dispenser totally. Neglecting the energy wasted by heating the heater itself.

5. Heater contacted closely with the chamber taking no account of thermal resistance.

6. Neglecting the thermal radiation because of the low temperature.

4.2 thermal loads

The heater was designed to 15W, and heat got into the fluid jetting dispenser directly, as described by equation (7)

$$-k\frac{\partial T}{\partial n}\Big|_\Gamma = q \qquad (7)$$

Where q is heat flux generated by the heater; T is wall temperature; k is coefficient of thermal conductivity; Γ is the contacting boundary.

According to the formula Q=Aq, thermal power could be transferred to heat flux. A is the surface area of the chamber. As the height of the chamber's bottom is 16mm, diameter is 20mm, the surface area A=1.005×10-3m². The heat flux was worked out to be 14921W/m² and loaded here as shown in fig.1.

4.3 boundary condition

There was thermal transmission when the chamber contacted with the air, and this thermal transmission mode was called convection heat transfer, and belonging to the third boundary condition that can be described by equation (8):

$$-k\frac{\partial T}{\partial n}\Big|_\Gamma = \alpha\left(T - T_f\right)\Big|_\Gamma \qquad (8)$$

Where α is the coefficient of convection heat transfer; T_f is the media temperature; other variables are idem.

The convection boundary condition was to be set up on the boundary of the chamber and air. As the rank of coefficient of the air natural convection was $1\sim10$, and in this paper was set up to 10 W/(m²·℃), the air temperature was 20℃.

The heat transmission mode at the central line of the model was belonging to the adiabatic boundary condition, as shown by equation (9)

$$-k\frac{\partial T}{\partial n}\Big|_\Gamma = 0 \qquad (9)$$

In addition, the most area of the top of the chamber was contacting with the O ring whose coefficient of thermal conductivity was less 2 order than the chamber (carbon steel)'s. The interface which was far away from the heater and the fluid zone was not contacting closely. So it was considered no heat transfer and also set up to adiabatic boundary condition.

4.4 initial condition

Before heating, the temperature of the chamber was considered to be evenly distributed and the value was environmental temperature.

$$T=T_0 \qquad (10)$$

Where T_0 was the environmental temperature.

4.5 simulation of heat transfer

Based on the characteristic of the physical model and the boundary given above, the finite element model was built in ANSYS9.0 as shown in figure2.

The element chosen for thermal analysis was plane thermal element with 4-node: plane55. The material's thermal property including the coefficient of heat transmission, density and special heat of the fluid, chamber and ball-needle, was shown in table1 and table2.

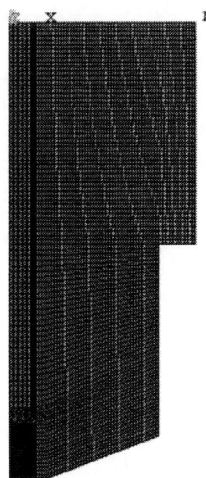

Fig.2. schematic of finite element model in ANSYS

Table2 thermal property parameters of carbon steel (45)

thermal property parameters	value
conductivity /W/(m·K)	50.2
special heat /J·(Kg·K)⁻¹	448
density /Kg/m³	7800

Importing ANSYS the thermal property parameters, loading decided heat flux at suitable part of the chamber, and setting up the boundary condition, it was high time to solve this problem by setting solution time was 10 minutes and initial temperature was 293K(20℃). In order to obtain a homogeneous temperature field, it's a good way to decrease the heat flux and extend the heating time.

4.6 simulation results of heat transmission

With solution over, the result reading from POST1 indicated that the temperature of the fluid under the ball-needle achieved the scheduled temperature （30~60℃） in 4~8 minutes. Temperature fields of this part were shown in figure 3 and figure 4.

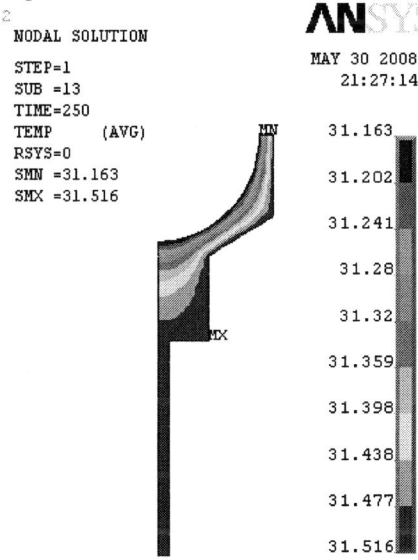

Fig.3. temperature field of fluid under ball-needle in the 4th minute(250s)

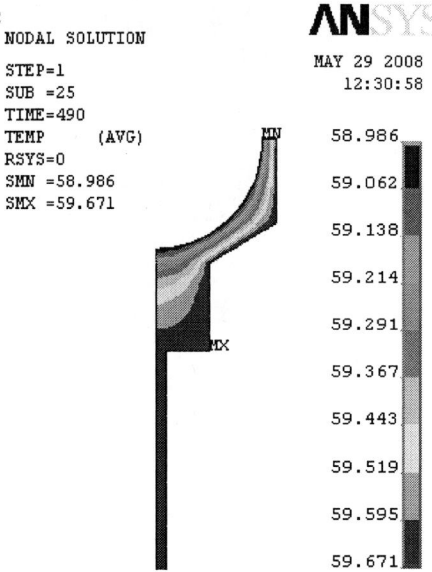

Fig.4. temperature field of fluid under ball-needle
in the 8th minute(490s)

The element temperature of the fluid zone under the ball-needle can be withdrew from the ANSYS database by *VGET command. And the average temperature, maximum temperature, minimum temperature and standard square deviation were obtained shown in table 3 by *VSFUN command.

table3 temperature data in 4th minute and 8th minute

	4th minute	8th minute
average temperature(℃)	31.39	59.42
maximum temperature(℃)	31.52	59.67
minimum temperature(℃)	31.16	58.99
standard square deviation	0.1145	0.2206

Based on the figure 3 and figure 4 and table 3, it was clear that the temperature of fluid field under the ball-needle was uniform complying with the requirement beforehand. And it was convenient for the succeeding FSI analysis. It was also a foundation for the consistency of jetting.

In addition, the average temperature in the fluid field under the ball-needle was worked out by withdrawing data from the ANSYS database and shown in figure 5.

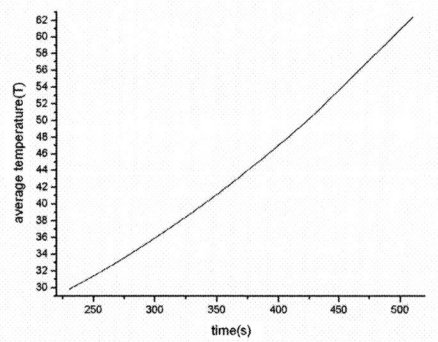

Fig.5. average temperature of fluid field under ball-needle
from 4 to 8 minute

Also, the temperature field of the whole finite element model was given by figure 6.

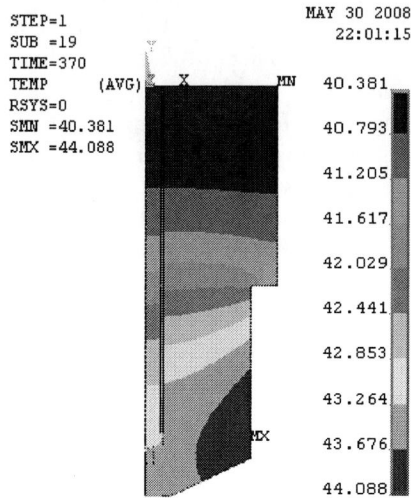

Fig.6. temperature field of the whole model
in the 6th minute(370s)

5. FSI analysis

The jetting dispenser began to work with the ball-needle moving up and down in the fluid field after the temperature reached the scheduled requirement. The initial downward acceleration of the ball-needle was $1600m/s^2$, the stroke was 0.4mm, the gap between the ball-needle and the chamber was 0.2mm, and the radius of the needle was 0.15mm. The acceleration changed every moment as the obstruction of the high viscosity fluid. In addition, it was no need to solve the thermal equation in FSI analysis because the movement time was so short in a stroke.

Fig.8 model of FSI analysis

The simulation results indicated that the main factor to effect fluid jetting was the high viscosity fluid under the ball-needle rather than the fluid between the ball-needle and the chamber. So it was just need to get the part of the tip of the chamber for FSI analysis shown in figure (8). And the average temperature of the fluid field under the ball-needle can be regard as the average temperature of the whole model. This treatment had no much effect on the movement of the ball-needle and was helping to save the solution time.

And then the jetting results of two different kinds of fluid in heating and room temperature condition were compared. Going into the temperature value in the 8th minute to carry on the equation (2) and (3), and the viscosity values of two different kinds of fluid were separately worked out: 5.5776 Pa·s and 16.3866 Pa·s, which were the nominal viscosity in FSI analysis, and the power index was 0.8.

Then, accumulative volume(V), average viscosity($\bar{v}$) and equivalent velocity(v_{eqv}) [3]of the two different kinds of fluid in heating and room temperature condition were worked out in FSI analysis and shown in table4.

Table 4   Accumulative volume (V), average velocity ($\bar{v}$) and equivalent velocity (v_{eqv})

	Fluid1	Fluid1 (heated)	Fluid2	Fluid2 (heated)
V (NL)	221.08	280.22	87.037	172.82
$\bar{v}$ (m/s)	4.468	5.6632	1.9545	3.4926
v_{eqv} (m/s)	8.0857	11.123	2.9722	5.8859

Whether it was in heating condition or in room temperature condition, the accumulative volume, average viscosity and equivalent velocity of the two different kinds of fluid varied with time in the jetting process. For saving ink, only the figures reflecting the accumulative volume and average viscosity in different time were shown in this paper.

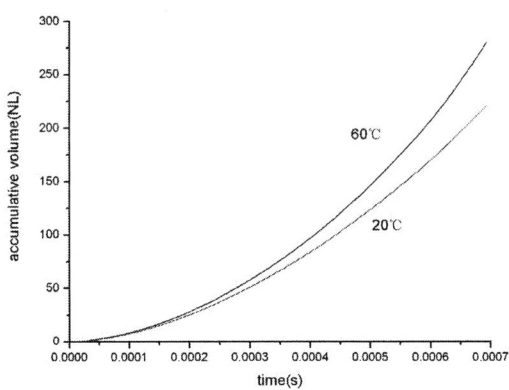

Fig.9 accumulative volume of fluid1 in different time

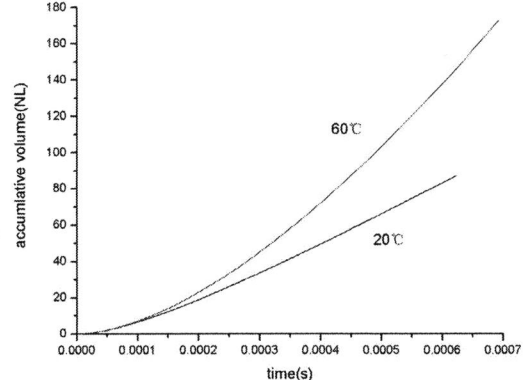

Fig.10 accumulative volume of fluid2 in different time

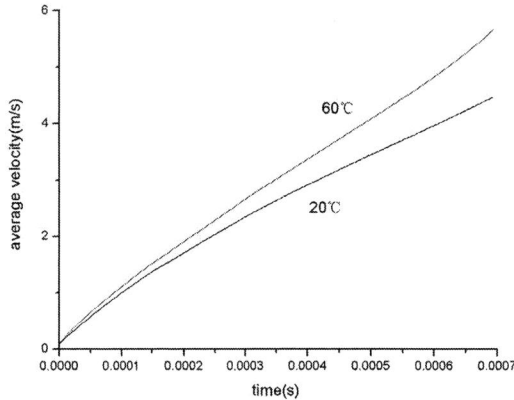

Fig.11 average viscosity of fluid1 in different time

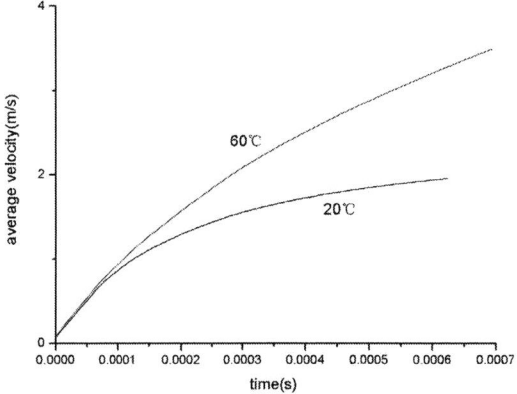

Fig.12 average viscosity of fluid2 in different time

Based on the table 4 and figures 9-12, it was clear that in heating condition the values of accumulative volume, average viscosity and equivalent velocity of the two different kinds of fluid are large than in room temperature condition. It means that in heating condition the fluid was apt to spout by keeping other parameters of the jetting dispenser.

6. Thermal analysis during jetting

The heated fluid spouted and the cool fluid flowed into the chamber through the adhesive pipeline under the air pressure during jetting. So the heat running off with the fluid jetting should be count into the thermal analysis, and negative heat flux should be loaded at the nozzle. At the same time, the room temperature（20℃） should be loaded at the fluid's entrance as the first boundary condition

$$T\big|_{\mathrm{r}} = T_0 \tag{11}$$

Assuming 500 dots were spouted in a second, the losing heat in a second (rate of losing heat) could be worked out according to the equation (12)

$$Q_2 = \rho V C \Delta T \times 500 \tag{12}$$

Where Q_2 —the rate of losing heat, W；ρ—fluid density, Kg/m³；V — volume of a fluid dot which varies with the structure parameters of the jetting dispenser and the fluid viscosity, m³；C—special heat, J·(Kg·℃)$^{-1}$；ΔT—temperature difference of entrance and outlet, ℃.

Going into the data in table1 and table4 to carry on equation (12), the rates of losing heat were separately worked out to be 8.22W and 5.07W for jetting the two different kinds of fluid.

This work offered reference for next work and designing the heater, and the verification should be given by simulation and experiments.

7. Conclusion

(1)In this paper finite element models of fluid jet-dispenser for thermal analysis and fluid-structure interaction analysis were built in ANSYS. And the finite element model's heat flux, boundary condition and initial condition were decided.

(2)The transient thermal analysis was carried on to obtain the temperature field of the fluid flied, and average viscosity of the heated fluid was worked out.

(3)The accumulative volume, average velocity and equivalent velocity of two different kinds of fluid in heating and room temperature condition in fluid-structure interface analysis were compared. The results indicated that the heated fluid was apt to spout.

(4)The losing heat in jetting stage was obtained. This work was reference for next work and designing the heater.

8. Reference

1. H. Quinones, A. Babiarz, C. Deck. Fluid jetting for next generation packages. Pac tech, Berlin, April, 2002.

2. Li Shi-hui. 3—D Numerical Simulation of Adhesive flow in Jet-dispenser and the Design of Structure Parameters of Jet-dispenser[D]. Changsha: College of Mechanical and Electrical Engineering, Central South University,2006.

3. CHEN Kui-yu, DENG Gui-ling, HAN Wei. Numerical Simulation of Cumulative Fluid Volume and Mean Velocity of Jetting Dispensing. Semiconductor Technology, 2007.

4. Chen Kui-yu, Deng Gui-ling. The Influence of actuator Parameters of Fluid Jetting Dispensing. Computer Simulation, 2007.

5. H. Quinones, A. Babiarz, L. Fang. Jetting Technology for Microelectronics. IMAPS Nordic, Stockholm, Sweden, September 2002

6 Steven J. Adamson. Jetting of underfill and encapsulates for high-speed dispensing in tight spaces. http://www.asymtek.com/news/articles/2004_02_jetting_un derfill_tight_spaces.pdf.

The Influence of Structural Parameters of Electromagnetic Fluid Jetting Dispenser

Wang junquan, Deng Guiling

Key Laboratory of Modern Complex Equipment Design and Extreme manufacturing (Central South University), Ministry of
Education Changsha 410083, China
wangjunquannihao@163.com. +8613974847275

Abstract

The fluid flow in the fluid jetting dispenser was analyzed. We present a Numerical study of the jetting process using ANSYS fluid-structure interaction (FSI) analysis tools. The effects of gap between ball-needle and wall, the diameter of fluid chamber, the shape of the needle, the separation angle of dispensing body and the diameter of nozzle etc to mass flow rate are described. The calculation formula of cumulative fluid volume and mean flow velocity are deduced. The distribution of stress of adhesive in the jetting dispenser is obtained. Combining it with data by MATLAB show that the solution is effective, which provides a theoretical basis for choosing appropriate structural parameters of liquid jet-dispenser.

Key words: jet-dispenser; cumulative fluid volume; FSI; mean flow velocity

1 Introduction

As the tendency of electronic system miniaturized and high performance, electronic packaging has the same importance as the chip to the system performance. The technology of fluid dispensing as one of key technologies of the microelectronic packaging also developed rapidly[2,3]. Fluid jetting technology compared conventional fluid dispensing technology has more merits, including jets a wide range of fluids 、 fast jetting eliminates Z-axis motion 、 higher throughput and dispensing ability in tighter space etc, because of these merits fluid jetting technology be used more and more widely ,and traditional fluid jetting dispenser is droved by spring-pneumatic. There are some shortcomings as follows: first, it can't offer great driving force; second, it can't achieve very small dimension needle because of afoul between needle and basement. So, it can't achieve big viscosity and small volume dispensing. Currently, some interrelated product was made, but the study of mechanism is not perfect. There are different aspects on the study of fluid dispenser at home and abroad. Alec Babiarz makes some primary study about the fluid flow in the dispenser. The document presents the hydrokinetics model of dispenser[6]. In the paper, it offer a new drive model spring electromagnetic[1], and probed the structure parameters of the jet-dispenser are the important factor of fluid jetting dispensing based on the theoretical analysis and interaction numerical simulation in the process of dispensing. It present the relations between structure parameter and cumulative fluid volume, fluid mean flow velocity. It also overcomes the drawbacks.

2 the principle of work and the structure parameter of the fluid dispenser

2.1 operating principle of electromagnetic dispenser

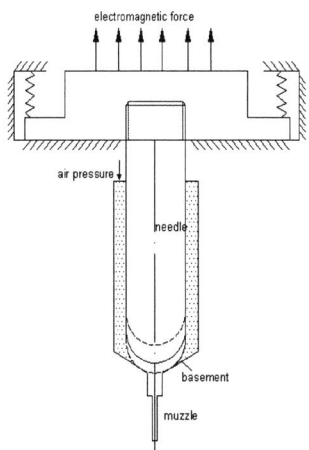

Fig.1 operating principle of electromagnetic dispenser

The fluid dispenser is composed of electromagnetic drive system and fluid dispensing system. Coarse glue filling in the gap between the ball-needle and the wall under the pressure of the air. In the closed position, the armature access to the block. There is a small gap between the base and the needle and avoid the collision with the base. It improves the useful time. In the open position, magnet coils connected to the power .Then, needle upward movement with the electromagnetic force until the needle access to the stroke control knob, the coil lose the power After some seconds ,the needle move down to basement with the spring compress force, separating and ejecting a dot from the fluid, impact the substrate 、 spread out then form the drop. Thus, it accomplishes the whole jetting dispensing. And then the coil connect to the power, the dispenser repeat the last action.

2.2 The key structure parameter of the fluid dispenser

In the motion of fluid, different structure flow model generates different viscous resistance, so the different structural parameters of the jet-dispenser have the different influence to the fluid flow in the process of the fluid jetting. The key structural parameters of the jet-dispenser show as the Fig.2, such as the shape of the needle, the gap between the ball-needle and the wall δ, the diameter of the reserving fluid zone d2,the diameter of the nozzle d1 and the separation angle of dispensing body θ etc.

978-1-4244-2739-0/08/$25.00 ©2008 IEEE   813

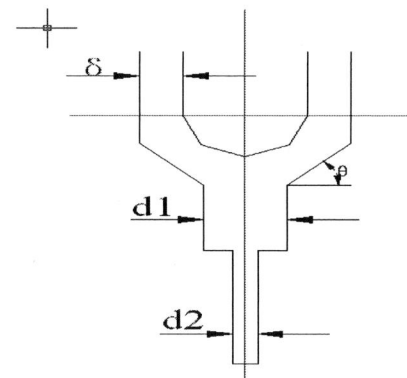

Fig.2 the key structure parameter of the fluid dispenser

The shape of the needle affects the contact area between the coarse glue and the wall. So as to effect the flow of the glue. The gap between the ball needle and the wall affects the flux from the nozzle outlet; the resisting force loss will change if the diameters of the pipe increase but the flow velocity keep constant. The separation angle of dispensing body affects the pressure of the glue so as to the flow of glue to reserving fluid zone. So when different aperture ratio d2/d1 will cause different resisting force loss, then changes the velocity of the flow, affects the flux of the fluid.

3 the calculation formulas of fluid cumulative volume and fluid mean flow velocity

The fluid in the jet-dispenser is non-Newtonian fluid. We suppose that it is power-law fluid in the process of numerical calculation. The basic property of the fluid show as to the table 1.

Formula of cumulative fluid volume:

$$V_{total}' = \frac{s+1}{s+3}\pi a^2 \sum_{i=1}^{n} \frac{1}{2}[v_{max}^i + v_{max}^{i+1}]\Delta t \quad (1)$$

Where s=1/n, n is the power exponent of the Power-law fluid in this paper, , v_{max}^i and v_{max}^{i+1} are the nozzle outlet central velocity of adjacent time step that obtained from the numerical simulation, Δt is the time step-size of simulation computation.

Formula of fluid mean flow velocity:

$$V(T) = \frac{V_{total}(T)}{\pi a^2 T}$$

$$= \frac{1}{2T}\frac{s+1}{s+3}\sum_{i=1}^{n}[v_{max}^i + v_{max}^{i+1}]\Delta t \quad (2)$$

Where $V_{total}(T)$, $V(T)$ is the fluid cumulative volume and fluid mean flow velocity, T is the run duration corresponding the different time step.

Table1 fluid property

name	value
density(kg/m³)	1200
power exponent	0.8
Reference temperature(℃)	22

Table2 parameters of simulation

parameter	value
gap between ball-needle and wall (mm)	0.2
the separation angle of dispensing body (º)	30º
the diameter of fluid chamber (mm)	0.6
the diameter of the nozzle (mm)	0.2
acceleration (m/s²)	1000
stroke(mm)	0.4

4 The model of numerical simulation

4.1 The governing equation of fluid flow

The model of numerical simulation is composing of continuity equation[7], momentum equation and kinetic equation. The flow of the fluid within the flow passage is three-dimensional incompressible flow and to meet the law of conservation of mass and the law of conservation of momentum. Its physical boundary is vertical downward flowing non-cross-section. Thus, the flow velocity and the averaged pressure in the fluid field is axial symmetry. The glue flow control equation:

$$\frac{1}{r}\frac{\partial}{\partial r}(\rho r v_r) + \frac{\partial}{\partial z}(\rho v_z) = 0 \quad (3)$$

$$\frac{\partial v_r}{\partial t} + v_r\frac{\partial v_r}{\partial r} + v_z\frac{\partial v_r}{\partial z} = -\frac{1}{\rho}\frac{\partial p}{\partial r}$$
$$+ \mu\left[\frac{\partial^2 v_r}{\partial r^2} + \frac{\partial^2 v_r}{\partial z^2} + \frac{1}{r}\frac{\partial v_r}{\partial r} - \frac{v_r}{r^2}\right] \quad (4)$$

$$\frac{\partial v_z}{\partial t} + v_r\frac{\partial v_z}{\partial r} + v_z\frac{\partial v_z}{\partial z} = -g - \frac{1}{\rho}\frac{\partial p}{\partial z} +$$
$$\mu\left[\frac{\partial^2 v_z}{\partial r^2} + \frac{\partial^2 v_z}{\partial z^2} + \frac{1}{r}\frac{\partial v_z}{\partial r}\right] \quad (5)$$

$$\mu = \mu_0 D^{n-1} \quad (6)$$

Where v_r, v_z is the fluid velocity of radial and axial, p is the pressure of glue, μ_0 is the titular viscosity, $D = \sqrt{I_2}$, I_2 is the second constant of the strain rate tensor.

$$I_2 = 2(\frac{\partial v_r}{\partial r})^2 + (\frac{\partial v_r}{\partial z} + \frac{\partial v_z}{\partial r})^2 + 2(\frac{\partial v_z}{\partial z})^2 \quad (7)$$

4.2 boundary conditions

In the central line, the change of the radial and axial velocity is zero because of the velocity field and the pressure field is axial symmetry. Namely,

$$v_r(0,z) = \frac{\partial v_z(0,z)}{\partial r} = 0$$

.

In the wall area of the pipe, the radial and axial velocity is zero. Namely, $v_r(R,z) = v_z(R,z) = 0$ .Where R is the radius of the syringe. In the interface between glue and ball-needle, $v_r(r,z) = 0$, $v_z(r,z) = v_{needle}$.

From the document [2], when the needle moves to the bottom and access to the base, the regional grid gets a big deformation. Namely, the simulation process is over when the distance between needle and base is 20μm. The model of numerical simulation show as Fig.3. The basic parameter of simulation show as the table 2.

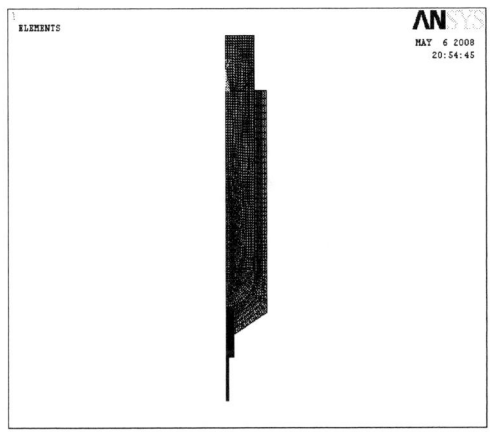

Fig.3 mesh model of ANSY

5 Results and Discussion

5.1 The influence regularity of the shape of the needle

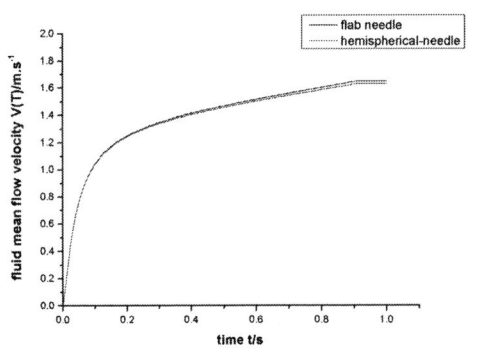

Fig.4 Fluid cumulative volume changed following the shape of the needle

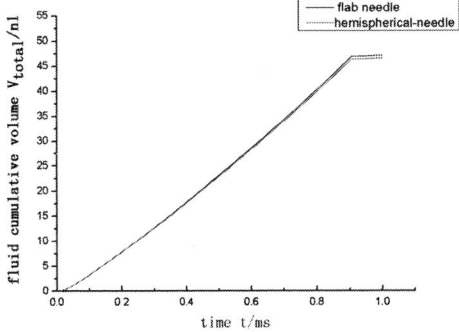

Fig.5 Fluid mean flow velocity changed following the shape of the needle.

Fig.4~Fig5 shows the change regularity of the cumulative volume and fluid mean flow velocity following the time. We draw that the value of the cumulative volume and fluid mean flow velocity of flab-needle is bigger than hemispherical-shaped needle. We can see that the contact area between flab-needle and the base is bigger. Thus, the dispensing glue is even more and faster. Fig.6~Fig.7 shows the shearing stress of different shape needle. The shearing stress of the flab-needle is bigger. So we choose the hemispherical-shaped needle in the micro and accuracy dispensing.

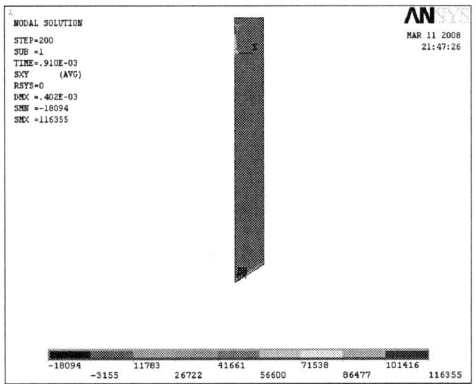

Fig.6 Shearing stress of flab-needle

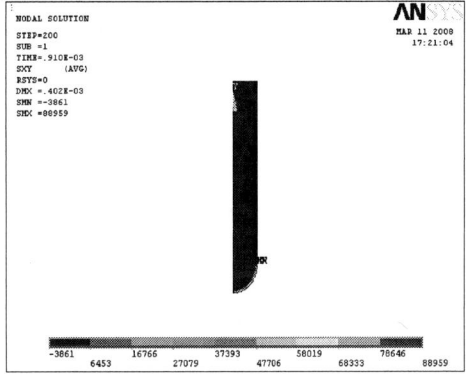

Fig.7 Shearing stress of hemispherical-needle

5.2 The influence regularity of the gap of the ball-needle and the wall

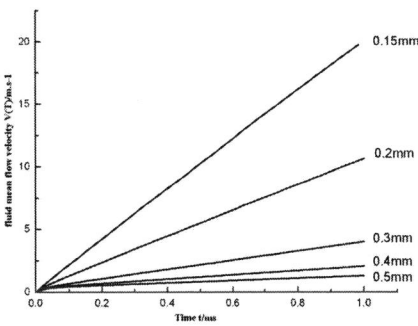

Fig.8 Fluid mean flow velocity changed following the time when the gap between the needle and the wall has different values

815

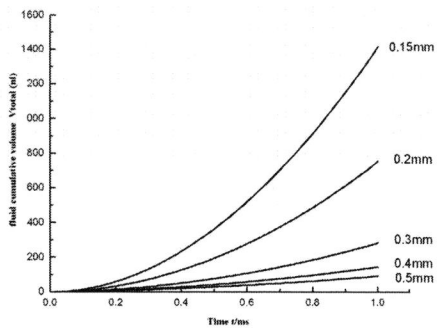

Fig.9 Fluid cumulative volume changed following the time when the gap between the needle and the wall has different values

Fig8~Fig9 shows the change regularity of the cumulative fluid volume an fluid mean flow velocity follow the time ,fluid cumulative volume and mean velocity increase when the gap between the needle and the wall gets small. We know the fluid cumulative volume has direction proportion with the flow velocity of the fluid. The back flow room decreases when the gap between the needle and wall gets small, the pressure increasing causes the raising of the fluid flow velocity, meanwhile causes the increasing of the fluid volume because of the decreasing of the back flow room.

5.3 The influence regularity of the diameter of the nozzle

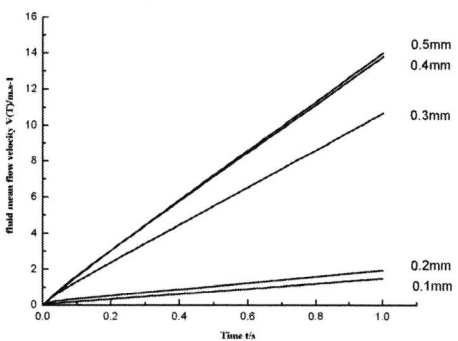

Fig.10 Fluid mean flow velocity changed following the time when the diameter of the nozzle has different values

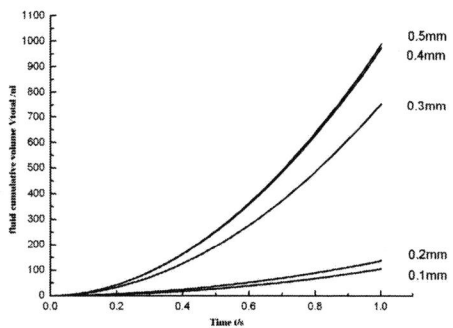

Fig.11 Fluid cumulative volume changed following the time when the diameter of the nozzle has different values

Fig10 ~Fig11 shows the change regularity of the fluid cumulative volume and fluid mean flow velocity following the time. The figures show diameter of the nozzle has great effect to the fluid volume and velocity, although they have the same change regularities, they have great difference at the numerical results. Because of intensify of fluid flow's viscous effect when the diameter of the nozzle decreased, causes the velocity of the fluid decreases, so the cumulative fluid volume also reduces. Meanwhile, the fluid cumulative volume and mean flow velocity increase rapidly because of the pressure of the fluid in nozzle raises rapidly.

5.4 The influence regularity of the diameter of the reserving fluid zone

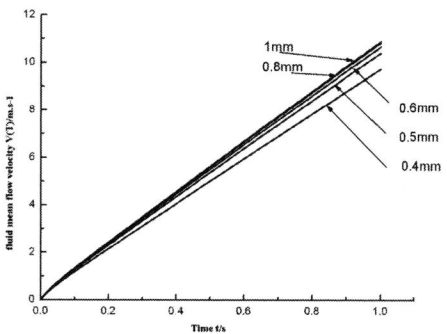

Fig.12 Fluid mean flow velocity changed following the time when the diameter of the reserving fluid zone has different values

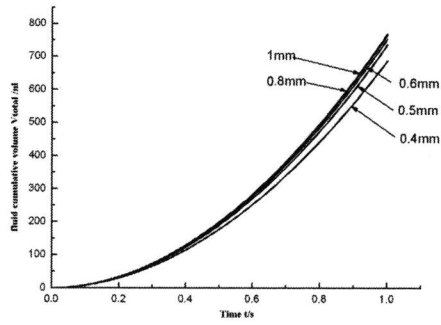

Fig.13 Fluid cumulative volume changed following the time when the diameter of the reserving fluid zone has different values

Fig12~Fig13 shows the change regularity of the fluid cumulative volume and fluid mean flow velocity following the time , fluid cumulative volume and mean velocity increase when the diameter of the reserving fluid zone gets big. But the influence is not obvious , however, when different aperture ratio d2/d1 will cause different revisiting force loss ,the changes the velocity of flow, affects the cumulative volume of fluid.

5.4 The influence regularity of the separation angle of dispensing body

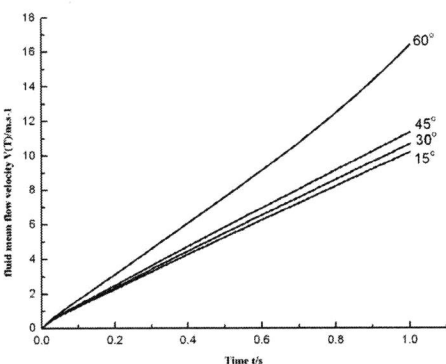

Fig.14 Fluid cumulative volume changed following the separation angle of dispensing body

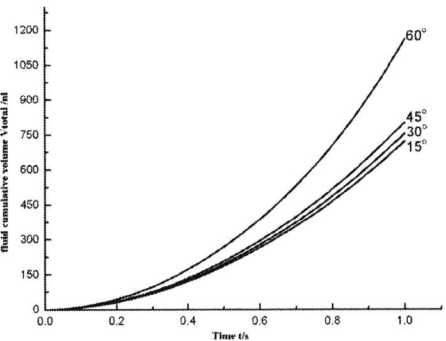

Fig.15 Fluid mean flow velocity changed following the separation angle of dispensing body

Fig14~Fig15 shows the change regularity of the fluid cumulative volume and fluid mean flow velocity following the time， the area of the fluid dispensing and the vertical force increase when the separation angle of dispensing body gets big. Then, the velocity of the glue is faster. So cumulative volume and mean velocity are increasing. The separation angle of dispensing body is limited due to the impact of the structure. The shearing stress increase when the separation angle of dispensing body gets big. Thus, choose the right the separation angle of dispensing body is crucial.

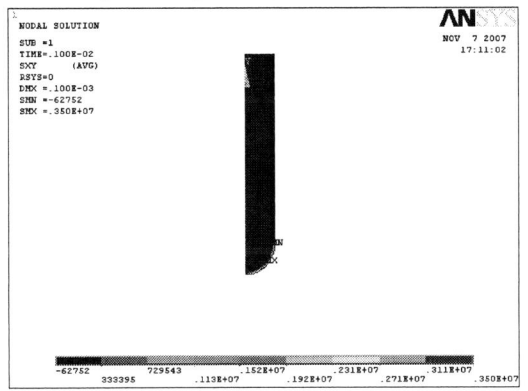

Fig.16 Shearing stress of 15^0 separation angle of dispensing body

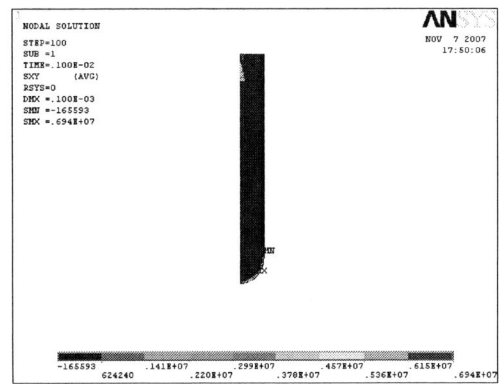

Fig.17 Shearing stress of 60^0 separation angle of dispensing body

6 Conclusions

The fluid cumulative volume and fluid mean flow velocity affected by various factors, and various factors are coherent. From the simulation results, we can obtain:

1）The fluid cumulative volume and the fluid mean flow velocity of flab-needle is bigger than hemispherical-needle. But，the shearing stress is bigger. So, we should select hemispherical-needle in the minim dispensing.

2）The fluid cumulative volume and the fluid mean flow velocity is increasing with the separation angle of dispensing body, however, the shearing stress is decreasing. We should select the appropriate separation angle by physical circumstance.

3）The back flow room decreases when the gap between the needle and wall gets small, the pressure increasing causes the raising of the fluid flow velocity, meanwhile causes the increase of the fluid volume because of the decrease of the back flow room

4）Because of intensify of fluid flow's viscous effect when the diameter of the nozzle decreases, results in the velocity of the fluid decreased, so the cumulative fluid volume reduced.

5）The effect of the fluid cumulative volume and the fluid mean flow velocity with the diameter of the reserving fluid zone is not obvious. But, the aperture ratio d2/d1 will produce great impact, so, we should choose the right size according to the diameter of the nozzle.

In the design and use the fluid dispenser, the effects of parameters are should be considered. And select the appropriate parameters according to the required fluid volume and velocity.

References

1. LaurenceB.Saidman,Duluth,GA.Electrically-operated dispenser.USA.20040230438A1 .OCT.20.2005.
2. Horitio uinones, Alec Babiarz, Christian Deck. Fluid jetting for next generation packages [J], Pac tech, Berlin, April 2002
3. Liu Yanwei, Deng Guiling. The Influence of Fluid Viscosity of Fluid Jetting Dispensing[C], 2007 International symposium on high desity paking and Microsystem integration .June 2007, 273-277.

4. Han wei,Deng Guilin.The influence of actuator parameters of fluid jetting dispensing. semiconductor technique ,2007.4.

5. Deng Guilin,Chen kuiyu. The Influence Regularity of Structural Parameters of Fluid Jetting 2007 International symposium on high desity paking and Microsystem integration .June 2007.

6. Deng Guilin,Zhong jue. Fluid dynamic models for non-Newtonian fluid transferring mechanism in contact dispensing dot process[j] journal of central south university, 2006,37(1):85-90

7. Li Shihui. The three-dimensional simulation of the fluid in the dispenser and the design of structure parameters [D] central south university master's thesis, 2006:23-33.

Dynamic Phase-frequency Characteristic of Thermosonic Wire Bonder Transducer

Fuliang WANG, Changhui ZOU, Jiaping QIAO
Central South University
College of Mechanical and Electronical Engineering, Central South University,
Changsha, 410083, China
wangfuliang@mail.csu.edu.cn, +86-013975145436

Abstract

The resonant frequency of transducer is changed with the state of thermosonic wire bonding interface. The ultrasonic generator need to trace the frequency of transducer, and keep the stimulate frequency consist with the resonant frequency of transducer. The frequency tracking strategy is based on the phase-frequency transducer. The dynamic phase-frequency characteristic was measured with a new FPGA based ultrasonic generator. The dynamic phase-frequency characteristic shows that the stimulation frequency may change by the generator quickly, but the change of phase difference between voltage and current of stimulate electronic signal need 12ms or much longer time, and the final stable phase is much different with the static phase-frequency characteristic measured by HP4191A. This may be one reason for the phase tracking need 3-5ms.

Keywords: phase-frequency characteristic, Thermosonic wire bonding, transducer, ultrasonic generator, frequency tracking.

1 Introduction

Thermosonic wire bonding has been wildly used in semiconductor packaging for several decades for its unique advantages, such as easy process, good bonding strength and excellent metal interconnect between chips and substrates. The transducer is the heart of thermosonic wire bonder; it converts the ultrasonic power into mechanical vibration energy, and welding golden wire with silver pads with the ultrasonic frequency mechanical vibrations and bonding pressure [1-2].

During bonding process, the resonant frequency of transducer is changed with the state of bonding interface, because the damp and rigidity of bonding interface changes with the bonding strength formation. Therefore, the ultrasonic generator need to trace the frequency of transducer, and keep the stimulate frequency consist with the resonant frequency of transducer. The whole bonding process only needs 10-20ms, so the frequency tracking speed is an important factor for ultrasonic generator. It is also related to the stability and consistency of bonding strength. When the stimulate frequency is equal to the transducer resonant frequency, the output vibration amplitude of transducer is best and the bonding process is good, several Hz changing of the stimulate frequency will decrease the output vibration amplitude of transducer quickly and decrease the bonding strength [3].

The frequency tracking strategy is based on the phase-frequency and impedance-frequency character of transducer.

Because the impedance is least and the phase difference is zero when the stimulate frequency is equal to the transducer frequency. In industry technology, only the phase-frequency characteristic is used in the frequency tracking. When the phase difference between voltage and current of stimulate electronic signal is zero, it is considered the transducer in resonance state, otherwise, the stimulate frequency was changed by ultrasonic generator, this process is called as phase lock. Therefore, the phase-frequency characteristic measurement is important for the ultrasonic generator. Currently, the phase-frequency characteristic is measured with the impedance analyzer, such as HP4191A [4-7]. With this method, only the static phase-frequency characteristic can be obtained. With this character, the frequency tracking process needs 3-5ms, this made the bonding process slowly. A better understanding of phase changing process (or dynamic phase-frequency characteristic) is necessary for better frequency tracking strategy.

In this paper, a new type of ultrasonic generator was constructed with a high performance FPGA, and the dynamic phase-frequency characteristic of transducer was measured with this new full digital FPGA based ultrasonic generator.

2 Experiments process
2.1 Static phase-frequency characteristic of transducer

The transducer structure of thermosonic wire bonder used in this paper is show as figure 1. This is a typical PZT transducer; it is consisted by PZT, horn (made of aluminum), bonding tool (made of tungsten carbide steel) and holder.

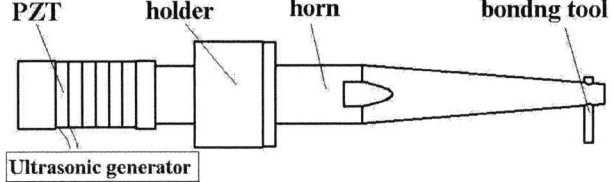

Fig.1 Transducer structure

The ultrasonic generator provide ultrasonic frequency electrical signal to driver the transducer, which made the PZT vibration and the vibration was transferred via horn to bonding tool, and bonding the golden wire to silver pads. The bonding process is about 10-20ms.

With the HP4191A, the static impedance-frequency and phase-frequency characteristic were measured, as shown in figure 2.

978-1-4244-2739-0/08/$25.00 ©2008 IEEE   819

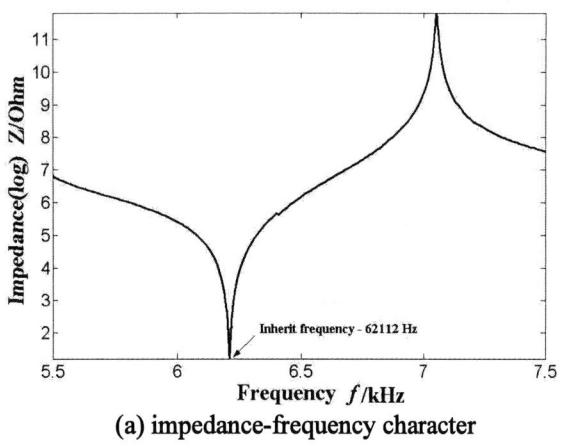

(a) impedance-frequency character

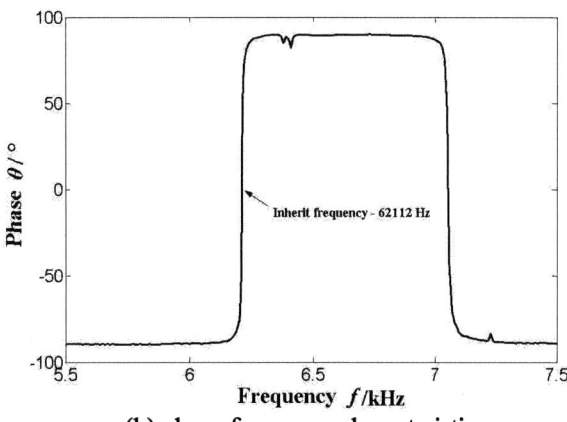

(b) phase-frequency characteristic

Fig. 2 Static character of transducer measured by HP4191A

The measurement results show that the resonant frequency is 62112 Hz and the corresponding impedance is 3.357 Ohm. This is an excellent transducer. However, this phase-frequency characteristic only shows the final phase of transducer in each stimulate frequency state, the difference between begin stimulation and the end of stimulation have not shown in this measurement method.

2.2 FPGA based ultrasonic generator

A FPGA based ultrasonic generator was constructed to driver the transducer. The structure of the new ultrasonic generator is shown in figure 3.

This is a digital phase lock automatic frequency tracking ultrasonic generator. The "phase difference" module compares the phase of sinusoidal signal current and voltage, and out put a digital indicate the phase difference. With the phase difference, the "frequency adjust" changes the output order to DDS (Direct Digital Synthesizer), which changes the stimulate frequency. The phase-frequency characteristic decides the frequency-changing algorithm. If the phase difference is negative, it indicated that the stimulate frequency is too small and the frequency is increased, the opposite happens when the phase difference is positive, those changing process stops when the phase difference equals to zero.

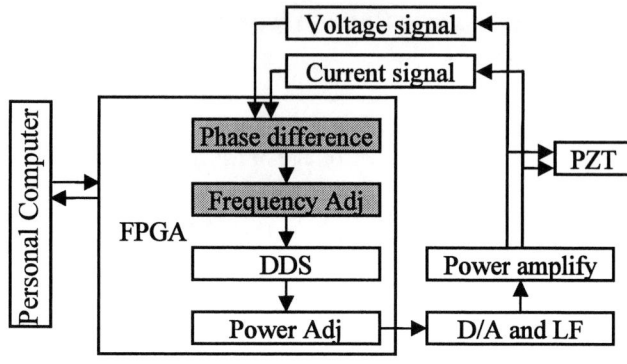

Fig. 3 FPGA-based ultrasonic generator

The FPGA only output the discrete sinusoidal signals, so the D/A converter and low-pass filter (LF) were used to convert the digital signal to analog signal. After the power amplifier, the signals can be used to driver the PZT of transducer.

At the same time, the data of phase changing or frequency changing process were transferred from FPGA to the personal computer according to the instruction given by personal computer. This function can used to measure the phase changing process (or dynamic phase-frequency characteristic) of transducer.

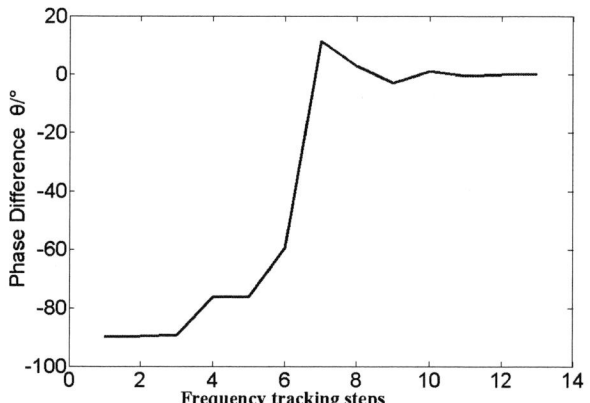

Fig.4 The simulated frequency tracking process with static phase-frequency characteristic

According to the static phase-frequency characteristic shown in figure 2(a), only 16 steps were needed in this frequency tracing process with a simple PID control method, each steps needs 16.7μs, as shown in figure 4. However, the reality frequency tracking time is longer than 3ms. One reason for this kind of error is that the dynamic phase-frequency characteristic has not been considered in figure 4. Actually, the stimulate frequency changing caused phase changing in each steps of figure 4, but the phase changing need time, this dynamic phase-frequency changing process has not been measured with a HP4191A impedance analyzer.

2.3 Dynamic phase-frequency characteristic

With this new FPGA based ultrasonic generator, the dynamic phase-frequency characteristic (phase changing process) of transducer was measured with the constant stimulation frequency or Step-Function changing stimulation frequency.

820

Figure 5 shows the experiment results with the stimulation frequency changed from 0 Hz to 62160 Hz, which happens at the very beginning of bonding starts. At the beginning, the phase is 18 degree, and changed to -10 degree at 1.75ms, the biggest phase is nearly 67 degree at 5.5ms, after 17ms, the final phase tend towards 54 degree. While the measured phase with HP4191A is 56 degree with the stimulation frequency of 62160 Hz. Therefore, this experiments show that the phase changing need a long time (about 12ms or more) to reach a stable state when the stimulation frequency changes, the changing process is similar to the step response. The phase measured by impedance analyzer is the final and stable phase.

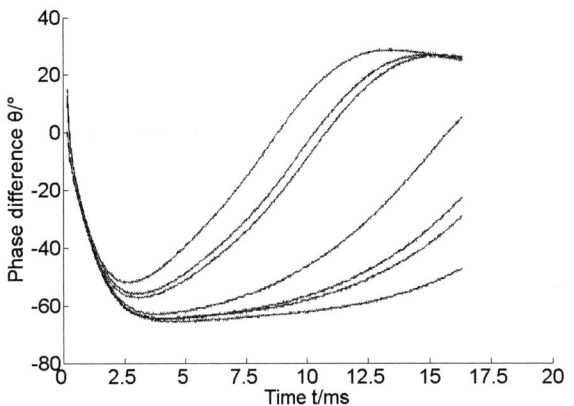

Fig.6 Phase changing process (stimulation frequency is 62130 Hz)

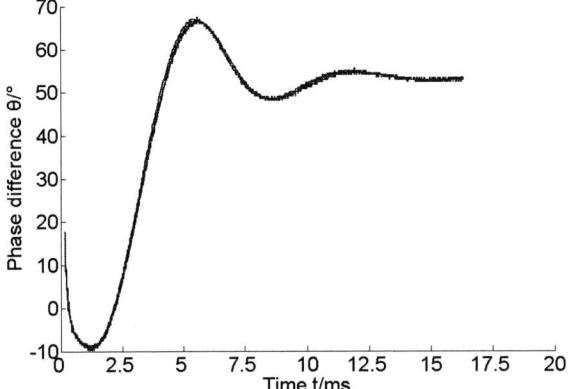

Fig.5 Phase changing process (stimulation frequency is 62160 Hz)

Seven experiment results in figure 5 show that the phase response process is stable. However, when stimulation frequency change to 62125 Hz (more close to the resonant frequency), the phase response process is changed into figure 6. Seven experiments were also carried out in this figure, which shows that the phase response is different although the stimulate frequency are the same and the others experiment conditions have not changed. According the figure 2(b), the final phase should be 17 degree. Only 3 experiments seem to converge at 20 degree, and the other experiments seem to need much longer time to tend to 17 or 20 degree.

When the stimulation frequency changed from f_1 (62058Hz) to a specified frequency f_2 (larger than f_1) at 0ms, which happens during frequency tracking, the experiment results is shown in figure 7. The phase changed from -66 degree to 0 degree and increased to a stable value that decided by the value of f_2. If the f_2 close to f_1, this phase changing process shows much slow, and larger f_2 caused oscillation. According to figure 2(b), the start phase is -66 degree, and the end phase should be 84 degree (f_2 is 62387Hz), but the stable end phase is 60 degree. And all end phase have decreased in different extent, the more close to f_1, the less decrease. The causes for those phenomena are not clear.

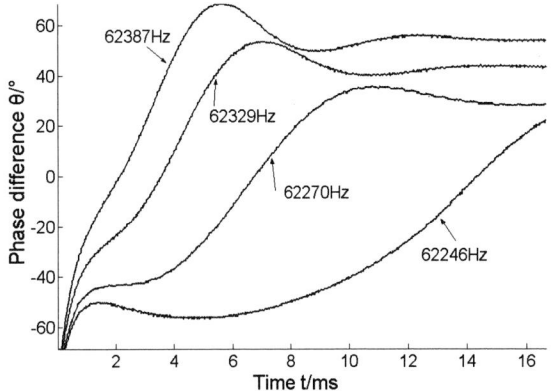

Fig.7 Phase changing process (stimulation frequency changed from 62058Hz to f_2 at 0ms)

When the stimulation frequency changed from f1 (62423Hz) to a specified frequency f2 (less than f1) at 0ms, the experiment results is shown in figure 8. The phase changed from 60 degree to some degree (depend on f2) and increased to a stable phase (depend on f2). If the f2 close to f1, this phase changing process is slow. According to figure 2(b), the starting phase is 85 degree, the end phase should be 84 degree (f2 is 62305Hz), but the start phase and stable phase is 60 and 45 degree. All start and end phase have decreased in different extent. Although the stimulate frequency always larger than the resonant frequency, there is a 0 phase in the changing process.

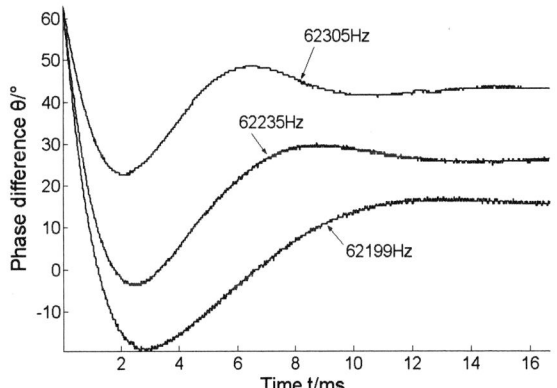

Fig.8 Phase changing process (stimulation frequency changed from 62423Hz to f_2 at 0ms)

3. Conclusions

Experiments show that the HP4191A measured static phase-frequency characteristic of transducer is much different with the dynamic phase-frequency characteristic measured by the new FPGA based ultrasonic generator. The dynamic phase-frequency characteristic shows that the stimulation frequency may change by the generator quickly, but the change of phase difference between voltage and current of stimulate electronic signal need 12ms or much longer time. In addition, the final stable phase much different to the static phase measured by impedance analyzer.

Acknowledgments

The authors are thankful to the supports of The Natural Science Foundation of China (NSFC, Contract No.: 50705098).

References

1. Lopez C, Chai L, Shaikh A, et al. Wire bonding characteristics of gold conductors for low temperature co-fired ceramic applications[J]. Microelectronics Reliability, 2004,44(3):287~294.

2. Rodwell R, Worrall D A. Quality control in ultrasonic wire bonding[J]. Hybrid Circuits, 1985, 34(7):67~72.

3. MAYER M, PAUL O, BOLLIGER D, et al. Integrated Temperature Microsensors for Characterization and Optimization of Thermosonic Ball Bonding Process[J]. IEEE Transaction on Components and Packaging Technology, 2000,23 (2):393-398.

4. Or, S.W. Chan, H.L.W. Lo, V.C. et al. Dynamics of an ultrasonic transducer used for wire bonding. IEEE Transactions on Ultrasonics, Ferroelectrics and Frequency Control, 1998, 45(6): 1453-1460.

5. H. K. Charles, Jr. , , K. J. Mach, S. J. Lehtonen, A. S. Francomacaro, et al. Wirebonding at higher ultrasonic frequencies: reliability and process implications, Microelectronics Reliability, 2003, 43(1):141-153

6. CHU P W, LI H L, CHAN H L W, et al. Smart ultrasonic transducer for wire-bonding applications[J]. Materials Chemistry and Physics, 2002,75(2):95~100.

7. CHIU S S, CHAN H L W, OR S W, et al. Effect of electrode pattern on the outputs of piezosensors for wire bonding process control[J]. Materials Science and Engineering, 2003,99(2):121~126.

Study of Prepress Force on Piezoelectric Transducer of Wire Bonding

Li zhanhui, Wu yunxin, Long zhili
College of Mechanical & Electronical Engineering,
Central South University, Changsha, 410083, China
Lzh-jdx@hotmail.com, +86-731- 8877840

Abstract

The friction contact model of ultrasonic propagation in wire bonding transducer is established. This paper studied piezoelectric driver structure. The relation of tighten moment and apply voltage is deduced. The wire bonding transducer scopes of tighten moment and apply voltage is educed. It provides foundation for design and operating of wire bonding system.

Preface

In first level chip package, there are two method, namely flip chip bonding and wire bonding. The former, which have many advantages, is main technology in next 20 years. [1][2] The later is the popular technology in first level chip package now.

Wire bonding technology becomes mainstream, because of its low costs, and good performance. The study method of wire bonding system including: analysis method, transfer matrix method, equivalent circuit method, finite element method and etc. [3][4][5] All of mentioned methods regard wire bonding system as a whole part, and modeling and resolve. Those methods don't agree with the fact. In the fact, the wire bonding transducer is consisting of many parts. There is a contact interface between each part. The contact interface is different from the integer.

1. Wire bonding system

Wire bonding system configuration (see Fig 1), including Driver, horn, clamping barrel, capillary, chip, workbench, ultrasonic generator and etc. [6~9]

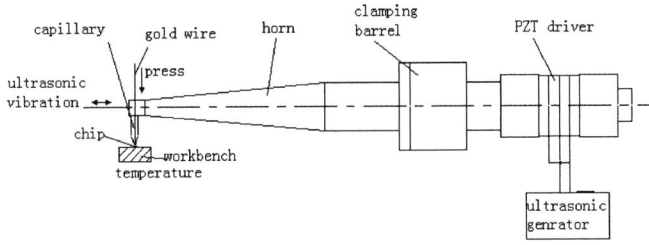

Fig 1 structure wire bonding system

An AC voltage signal is produced by the ultrasonic generator, and is transmited to PZT piezoelectric ceramic transducer, and it is transformed into material particle longitude mechanical vibration because of reversing piezoelectricity principle of piezoelectric material. The vibration is amplified by horn because of the horn's congregate effect. At the end of horn, the particle displacement achieve maximum. It is transformed into transverse mechanical vibration in capillary. The bonding cycle begins with the formation of a ball at the end of a gold wire. The workbench is elavated to a moderate temperature (generally, 100℃ to 160℃) to increase the malleability of the gold. A ball is created by applying an electrically charged want to a length of wire protruding from the capillary of the welding tip that is attached to an ultrasonic transducer. As the ultrasonic transducer moves downward, a ball is pressed onto a chip with a sufficient force to allow even contact between the wire and the chip. The ultrasonic transverse vibration is applied to form a metallurgical ball bonding between gold and chip.

At the position of clamping barrel, the particle extend is zero, namely it is nodal position. Fig 2 is axial vibration profile extend of wire bonding transducer.

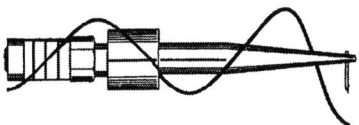

Fig 2 Axial vibration profile of wire bonding transducer

The clamping barrel is fixed on bonding machine. Because the particle vibration displacement is zero at the clamping barrel position, it reduces coupling of transducer and bonding machine, and reduces energy loss.

PZT driver adopt Langevin configuration, which is consist of bolt, cap, PZT, foil and nut. (See to Fig 3) The foil is copper electrode, and connection with the ultrasonic generator. The PZT is anisotropy piezoelectric ceramic. It have higher mechanical quality factor and larger mechanical-electric transform factor. There is a foil between two adjacent piezoelectric ceramic. The AC ultrasonic voltage signal is put on piezoelectric ceramic.

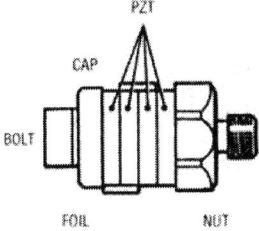

Fig 3 PZT driver

There is no connection amongst piezoelectric ceramic, cap and nut, which are compacted by bolt. Press stress can propagate among those, but pull stress cannot propagate. When applied AC voltage, press stress and pull stress can be produced, contact interface separate. It reduces propagating efficiency, and reduces wire bonding quality. In order to wire bonding quality, it is needed that piezoelectric ceramic, cap and nut in bonding process contact estate.

2 Friction Contact Model

Set there are two anisotropy piezoelectric bodies which contact each other at Z=0 position, and there exits force F and electric field E in the space. The material 1 is in underside,

978-1-4244-2739-0/08/$25.00 ©2008 IEEE   823

and the material 2 is in upside. A longitudinal wave along Z axis perpendicular incidence at contact interface of two bodies from the material 1. At contact interface the incidence wave (n=0 in fig 4) produce reflect wave (n=1 in fig 4) and refraction wave (n=2 in fig4). [10] When incidence wave intensity is enough large, the contact interface can produce adhere and separate two estate.

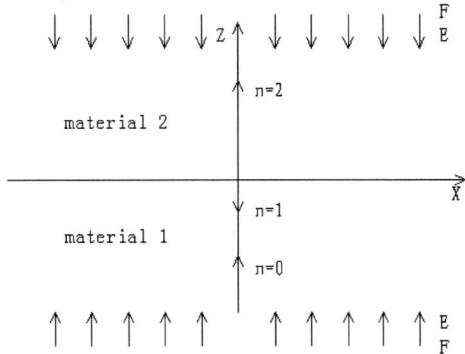

Fig 4 friction contact model

The interface force should continuous at Z=0 interface. When neglect the friction of two contact interface, there are two cases:

(1) Adhere estate: $g = 0, S > 0$

(2) Separate estate: $g > 0, S = 0$

Where: g is relative displacement of two bodies interface; S is two material interface contact force. In addition, electric boundary conditions satisfy the connection electric boundary conditions, namely $D_z^{+} = D_z^{-}$ and $E_z^{+} = E_z^{-}$, D_z^{+}、 D_z^{-}、 E_z^{+}、 E_z^{-} are Z direction electric displace and electric field intensity separately.

Whole problem decompose into adhere estate solution and separate estate solution.

A dimension plane wave's propagation equation:

$$\frac{\partial^2 u}{\partial t^2} = c^2 \frac{\partial^2 u}{\partial z^2}$$

Where: u is particle displacement of body, c is wave propagation velocity.

Set plane wave is round wave number, $k = \omega/c$, ω is angle frequency

So, contact interface force of adhere is:

$S = A\sin(-kct)$

Where: A is extent of contact force

3 Example and analysis

There is a wire bonding transducer system, work frequency is 63 kHz, piezoelectric ceramic material is PZT-4, the main parameters of the material:

$\varepsilon_{33} = 1.15 \times 10^{-8} C/V \cdot m$, $d_{33} = 2.89 \times 10^{-10} C/N$,

$e_{33} = 15.1 C/m^2$, $h_{33} = 2.68 \times 10^9 V/m$,

$\rho = 7500 kg/m^3$. The shape of piezoelectric ceramic is a cirque. Outer diameter is 13 mm, inner diameter is 5 mm, and thickness is 2.3 mm.

From friction contact model, during wire bonding process, guarantee piezoelectric ceramic and cap and nut always contact, the relation of the tighten moment on bolt and voltage on piezoelectric ceramic is resolve. The relation of tighten moment and voltage is gained. (See fig 5)

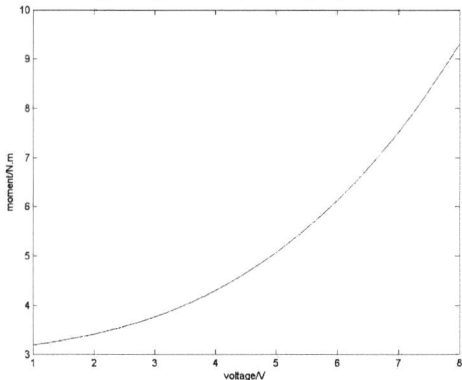

Fig 5 the relation of tighten moment and voltage

In fig5, x-axis denotes the voltage on piezoelectric ceramic, y-axis denotes the tighten moment on bolt. With the increase of loading voltage, the tighten moment on bolt increase.

In order to ensure piezoelectric and cap and nut always contact during wire bonding transducer work, the tighten moment should at upside of curve.

The larger of tighten moment, the better to ultrasonic propagation, from the point of contact interface. But the bolt may break, and ultrasonic energy propagation efficiency is less, if the tighten moment is too large. So the tighten moment should be restrict in a scope.

The bolt's intensity is 10.9 levels, and its yield limit is $1014 MPa$. It is the maximum tighten moment is $8 N \cdot m$. After amend the relation of tighten moment and voltage see fig 6.

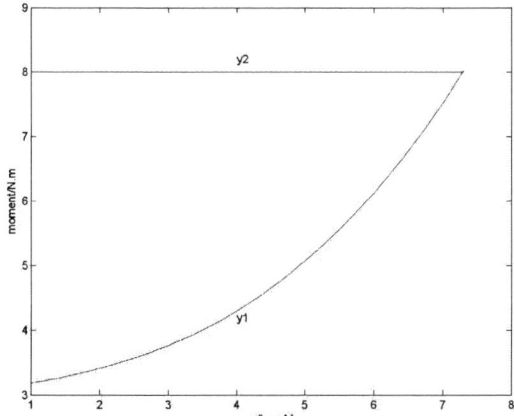

Fig 6 the relation of tighten moment and voltage after amend

From fig 6 the tighten moment should be restrict between curve y1 and curve y2. At the same time, the maximum voltage on piezoelectric ceramic is $7 V$.

Conclusions

The friction contact model of ultrasonic propagation in wire bonding transducer is established, on the base of wire bonding transducer. This paper studied piezoelectric ceramics' connection. The tighten moment load scope of wire bonding transducer and load voltage on piezoelectric ceramic is deduced. The bolt's tighten moment is different at applying different voltage. It provides foundation for design and operating of wire bonding system.

Acknowledgments

The authors gratefully acknowledge research support from The Ph. D. Programs Foundation of Education Ministry of China, No. 20060533068

References

[1]Long zhili, Han lei, Wu yunxin, Zhou hongquan. Effect of different temperature on strength of thermosonic bonding [J]. Welding Transactions, 2005, 26（8）:23~26

[2]WU Yun-xin, LONG Zhi-li, HAN Lei, ZHONG jue. Temperature effect in thermosonic wire bonding[J]. trans Nonferrous Met soc china, 2006,16(3):618~622

[3] Wang fuliang, Li junhui, Han lei. Effect of bonding time on thick aluminum wire wedge bonding strength [J]. Welding Transactions, 2006, 27（5）:47-52

[4]Yeau-Ren Jeng, Jeng-Nan Lin. Study of Interfacial Phenomena Affecting Thermosonic Wire Bonding in Microelectronics[J]. Transactions of the ASME, 2003, 125（6）:576~581

[5]Michael mcbreaty,lee h. kim, nihat m.bilgutay. analysis of impedance loading in ultrasonic transducer systems[J], ultrasonics symposium, 1988, 497-502

[6]Shuyu Lin and Hua Tian.Study on the sandwich piezoelectric ceramic ultrasonic transducer in thickness vibration[J]. Smart material and constructure, 2008, 17(1): 1-9

[7]Em´ılio Carlos Nelli Silva, and Noboru Kikuchi.Design of piezoelectric transducers using topology optimization [J]. Smart Mater. Struct. 1999 (8) ;350–364

[8]ZHOR LERGUET, MEIR SHILLOR, MIRCEA SOFONEA .A FRICTIONAL CONTACT PROBLEM FOR AN ELECTRO-VISCOELASTIC BODY [M]. Electronic Journal of Differential Equations, 2007(170);1–16

[9]P. Schwaller, P. Gro¨ning, A. Schneuwly, P. Boschung, E. Mu¨ ller, M. Blanc, L. Schlapbach. Surface and friction characterization by thermoelectric measurements during ultrasonic friction processes[J]Ultrasonics, 2000 (38) 212–214

[10]Yue Sheng Wang, Gui Lan Yu.Re-polarization of elastic waves at a frictional contact interface II[Incidence of a P or SV wave [J]. International Journal of Solids and tructures,1999(36):4563-86

Simulation and Experimental Research on Water-jet guided Laser Cutting Silicon Wafer

Wang Yang, Li Ling, Yang Lijun, Liu Bei, Wang Zhe
Dept. of Mechanical Manufacturing and Automation, Harbin Institute of Technology,
Harbin 150001, P. R. China
wyyh@hope.hit.edu.cn; liling1@hit.edu.cn, 86413257

Abstract

Silicon wafer cutting is a key process in wafer manufacturing. Traditional cutting methods can not do this work well, but water-jet guided laser processing which is based on the principle of full reflection can do it perfectly. A new simulation model of water-jet guided laser cutting silicon wafer was set up. The model treated the uniform energy input of water-jet guided laser, the cooling effect of the water jet, and the melting and removal of the silicon. The simulation results revealed that the effect of water jet cooling on wafer temperature field is very distinct. Machining silicon wafer, stainless steel1Cr18Ni9Ti and 65Mn were simulated. The results showed silicon wafer had the least heat affected zone of all. Experiments were done using the new water-jet guided laser micromachining system. The results showed that the energy intensity distributed over the water jet cross section nearly homogeneous and the energy decreased little in long working distance. Applying optimized machining parameters to cut silicon wafer, nearly no burrs, no cracks, no heat affected zone were formed, the cutting results were better than diamond saw and gas (Ar) assisted laser cutting.

1. INTRODUCTION

Electronic packaging technology belongs to high-tech domain. The most useful material for this technology is silicon. So cutting out the perfect silicon chips from silicon wafer is very important for electronic packaging. There are some ways of cutting silicon wafer, but as a new development of laser processing, water-jet guided laser processing is a promising method for cutting silicon wafer. Water-jet guided laser processing is based on guiding a laser beam through a fine water jet onto the workpiece [1], as shown in Fig. 1(a), (b), (c). Because of the difference in the reflection coefficient of water and air, the laser beam is fully reflected at the water jet surface. The water jet acts as an optical fiber.

Water-jet guided laser micromachining has some important advantages in comparison with conventional laser processing. (1) Because of no focal point of beam, the long working distance is achieved (2) The water jet guides the laser beam and cools the work piece during the processing, and removes the cut products. (3) The operating costs are in general lower than traditional laser processing. (4) The force on the workpiece is tiny. (5) Fume hazards are much less than laser processing (since most fume is absorbed in the water) and the water jet itself is not dangerous. In this article simulation and experimental researches were done for water-jet guided laser cutting silicon wafer.

2. SIMULATION RESEARCH

Due to the increasing use of lasers in industry, there is significant interest in the development of models that can describe the complex laser-material interactions well enough to guide the selection of process parameters. Since laser processing is a transient process that involves transmission, absorption, conduction, convection, reflection of laser energy, vaporization, melting, fluid flow, moving boundaries, temperature-dependent material properties, and gas dynamics, the selection and implementation model is challenging. A lot of numerical investigations have been carried out on the laser-material interactions [2, 3], but they do not use the water as the cooling method. One of the most important characteristics of water-jet guided laser cutting silicon is the high cooling effect of the water jet.

2.1 Modeling of water-jet guided laser cutting silicon wafer

To evaluate the evolution of the temperature field, the time-dependent heat conduction equation is solved in the work piece domain subject to the appropriate boundary conditions. The governing equation written in terms of temperature T is

$$\rho c \left(\frac{\partial T}{\partial t} + v_x \frac{\partial T}{\partial x} + v_y \frac{\partial T}{\partial y} + v_z \frac{\partial T}{\partial z} \right) =$$

$$\overset{\cdots}{q} + \frac{\partial}{\partial x}\left(k_x \frac{\partial T}{\partial x} \right) + \frac{\partial}{\partial x}\left(k_y \frac{\partial T}{\partial y} \right) + \frac{\partial}{\partial z}\left(k_z \frac{\partial T}{\partial z} \right)$$

$$(1)$$

where ρ is density, c is the specific heat, t is the time, v is velocity, k is the heat conductivity, $\overset{\cdots}{q}$ is the heat generation rate per unit volume.

An explicit solution scheme is used in the simulation, with forward difference approximation for the time derivative. The temperature of a given volume element after a specific time step is therefore found explicitly from the known current temperatures of the element and its neighbors. Some assumptions were made for the simulation. (1)The laser power intensity at the workpiece is uniform throughout the cross-section of the water-jet guided laser. Because of the difference between the index of refraction of water and that of air, the laser beam is totally reflected at the air–water interface. After many reflections, the assumption of uniform laser power intensity at the workpiece is appropriate for water-jet guided laser processing, instead of the assumption of a Gaussian distribution that is widely used for traditional laser processing. (2)The material melting temperature is the maximum temperature reached in the workpiece. The melting

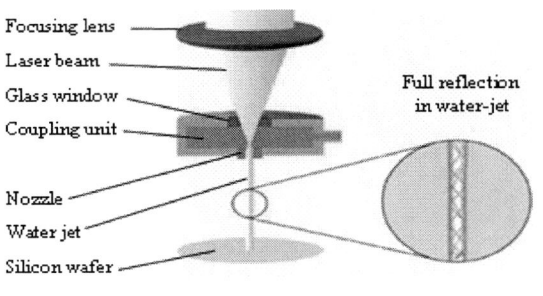  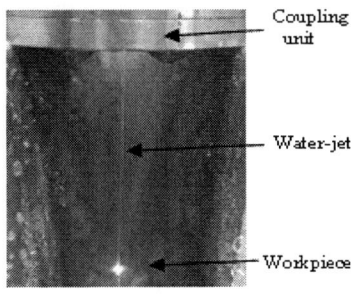

Fig. 1 Silicon wafer cutting by water-jet guided laser (a) principle of cutting (b) infrared CCD photo of cutting process （c） photo

and removal of material occurs element-by-element. After an element of the material has reached the melting temperature and also has absorbed additional incoming energy equal to the latent energy of melting, the element melts and is removed by the action of the water jet.(3)The fluid dynamics of the water jet removal of molten material is not considered.

The general laser heat source $A(z,t)$ can be expressed as:

$$A(z,t) = I(t)\alpha e^{-\alpha z}[\vec{n}\vec{e}_z]$$

(2)

where I is the laser power intensity[4], $I(t) = P(t)/(\pi d^2/4)$, α is the absorption coefficient, z is the distance from the surface, $\vec{n}$ is a unit vector normal to the surface, $\vec{e}_z$ is a unit vector parallel to the z axis. convection heat loss take place at the top and bottom surfaces of the workpiece, the energy losses at these boundaries is calculated by $q_{out} = h(T - T_a)$,where h is the convection coefficient, T_a is the ambient temperature,

The heat transfer coefficient required to model the water jet cooling effect for the current simulation is obtained from empirical Nu formulas reported in a review article by Webb and Ma [4], in which the results of many liquid jet heat transfer experiments were summarized. Webb and Ma observed that the stagnation Nu depends approximately on Pr^n where n varies between 1/2 at a small Pr to 1/3 at a large Pr and recommended the following formulas.

$$Nu = \begin{cases} 0.715 Re^{1/2} Pr^{0.4}, & 0.15 < P_r < 3 \\ 0.797 Re^{1/2} Pr^{1/3}, & P_r > 3 \end{cases}$$

(3)

The heat transfer coefficient associated with the water jet cooling effect is calculated by use of Eqs.(3). For liquid water ranging from 0 to 100 °C, the Pr ranges from about 13.0 down to 1.7.

2.2 Simulation result

2.2.1 Simulations of temperature fields of different water jet convection heat transfer coefficients

A new model was setup to obtain simulation results for water-jet guided laser machining silicon wafer. In this model, the thickness of the silicon wafer is 0.1mm; silicon density is 2330 kg/m³; the thermal conductivity is calculated by $k(T) = 29900/(T - 99)$ (W/m K); the special heat is calculated from $\rho c(T) = (1.4743 + 0.17066T/300) \times 10^6$ (J/m³ K), for T<1683 K and $\rho c(T) = 2.432 \times 10^6$ (J/m³ K), for T>1683 K; the melting temperature is 1683 K; latent heat of melting is 1.79×10^6 J/kg; the simulation end time is 0.2s. The convection heat transfer coefficients (Nus) are 10 W/m² K, 100 W/m² K, and 1000W/m² K.The simulation results show that: the cooling effect of water jet can change the silicon workpiece temperature field. When the cooling effect is small, the heat affected zone is large, as shown in Fig. 2(a); when the cooling effect is large, heat affected zone is small, as shown in Fig.2(b); but when the cooling effect is too large, the workpiece temperature is so low that cannot be machined by laser , as shown in Fig. 2(c). So the conclusion can be drawn: the laser energy and the water jet cooling effect must fit each other well, otherwise the silicon wafer will have large heat affected zone or the silicon wafer cannot be cut.

2.2.2 Simulations of temperature fields of different materials

Three kinds of materials drilling by water-jet guided laser were simulated. They were 1Cr18Ni9Ti, 65Mn, and silicon. Simulation end time is 0.1s, the workpiece thickness is 0.1mm. From the simulation results, as shown in Fig. 3, all three materials had well machining effects. The temperature field of silicon wafer was the smallest of all because the silicon physical quality, the temperature gradient of silicon

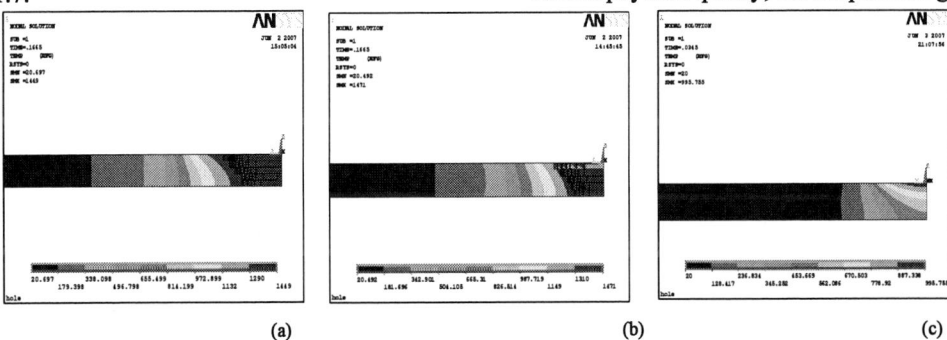

Fig. 2 Simulations of the temperature fields and material remove of different water-jet convection heat transfer coefficients

(a)Nu=10 W/m²K (b)Nu =100 W/m²K (c)Nu=1000W/m²K

827

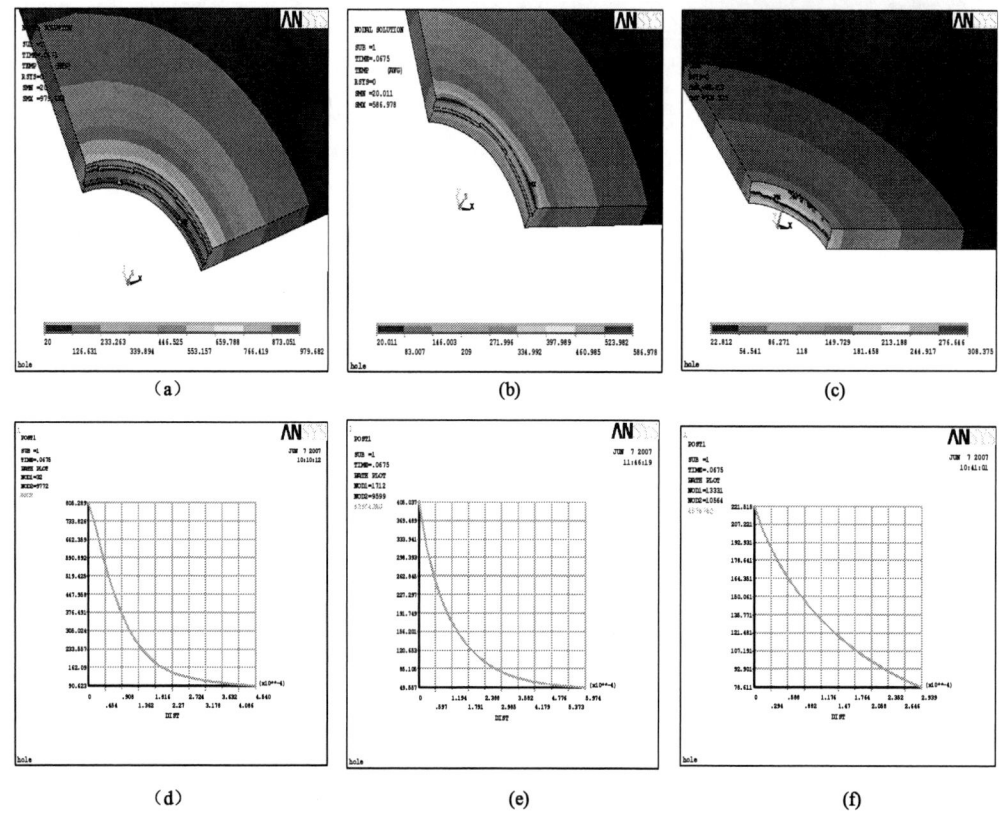

(a)　　　　　　　　　　　(b)　　　　　　　　　　　(c)

(d)　　　　　　　　　　　(e)　　　　　　　　　　　(f)

Fig.3 Simulations of the temperature fields of different materials (a) 1Cr18Ni9Ti (b) 65Mn (c) silicon wafer (d) temperature gradient of 1Cr18Ni9Ti (e) temperature gradient of 65Mn (f) temperature gradient of silicon wafer(the distance is zero at the hole boundary, the distance is calculated from the hole boundary)

wafer was also the smallest of all. This is one reason why water-jet guided laser is suitable for silicon cutting.

3 EXPERIMENTAL RESEARCH

3.1 Experimental setup and procedure

To do the water-jet guided laser cutting silicon wafer experiment, a new water-jet guided laser micromachining

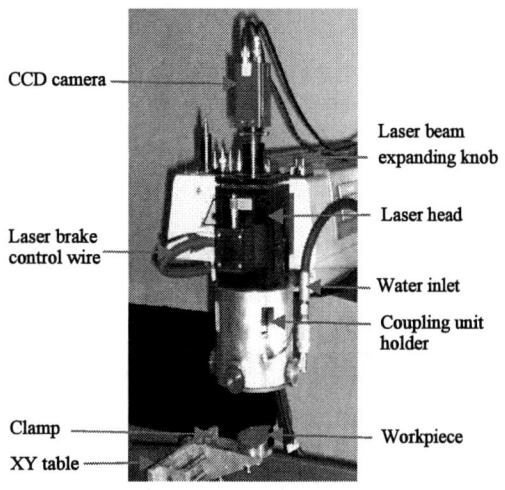

Fig. 4 Water-jet guided laser processing system photo (a)key

system was setup. This complex system includes many parts, as shown in Fig. 4. A Lumonics JK702H Nd:YAG pulsed laser with a maximum average power of 350W was applied in the system [5]. A new key component—the coupling unit was designed. It could form a long, slim, high-pressure and stable water jet; it also made the fluid field in the chamber symmetry. With the coupling unit helping, the coupling quality of the laser beam and the water jet could be easily adjusted and detected by CCD camera. The coupling unit is fed by a continuous water flow provided by a piston pump. Adjustable water pressure ranged from 0 to 10MPa.

The pulsatile pressure could be stabilized by using pumps with several pistons and nitrogen loaded damping elements-accumulator. The water jet sprayed from a sharp-edged nozzle. The particularity of this nozzle is that it can create a

separated water flow. The nozzle length-to-diameter ratio is nearly zero. The diameter of the water jet is smaller than the nozzle diameter because of the vena contract effect [6].

One of the most important things in the water jet guided laser cutting silicon is coupling the focused laser into the water jet fiber. To do this work, a coupling monitoring system was setup, as shown in Fig. 5. By using CCD camera and the developed image software, coupling status can be easily detected and adjusted.

The experiments include 4 groups: (1) Researching the water jet stability, the influence of pressure on max working

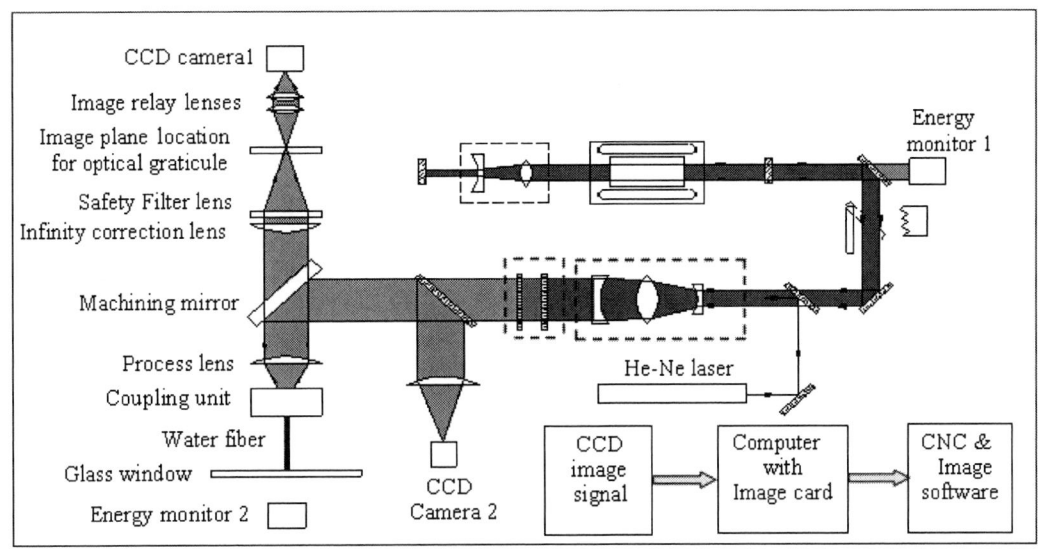

Fig. 5 Coupling monitoring system

distance. (2) The relationship of laser pulse energy and water jet length. (3) Researching the relationship of water-jet guided laser cutting parameters and cutting efficiency. (4) Cutting silicon wafer samples. The proper cutting parameters are shown in table. 1. These parameters were summarized from a lot of other experiments.

Table 1 Cutting parameters employed in cutting silicon wafer

Trial code	Pulse energy [J]	Pulse width [ms]	Pulse rate [Hz]	cutting speed [mm/s]	Nozzle diameter [mm]	Working distance [mm]	Water pressure [MPa]
1	0.2	1.4	35	0.3	0.14	20	3.5

3.2 Experimental results and discussion
3.2.1 Water jet stability and laser pulse energy attenuation

The water pressure and the water jet working length are the most important cutting parameters for water-jet guided laser cutting silicon wafer, because they influence the water jet stability [7] and the laser energy attenuation. The experiment results showed that using the new coupling unit, the stability water pressure is high, as shown in Fig. 6, when the pressure was 5MPa, the stability length was nearly 80mm. When using 0.06J pulse energy, the laser energy decreased little

when the water jet length is shorter than 35mm, but when the water jet length is longer than 55mm, laser energy decreased much, as shown in Fig. 7. This is because when water jet length is long, more laser energy is absorbed by water and impurities in water [8], Raman scattering is also happened [9].

3.2.2 The influence of cutting parameters on slot width

Many water-jet guided laser cutting parameters including pulse energy, scanning speed, nozzle diameter, working distance have influence on cutting efficiency and slot width. The cutting slot width was narrow when nozzle diameter was small, the slot width was shorter than nozzle diameter, as shown in Fig. 8(a). During a long working distance, the slot width changed little, indicating the intensity decreased little in long working distance, as shown in Fig. 8(b). But when the distance was too long, the water jet detached, no energy could be carried on work piece.

The slot width increased as pulse energy increased and scanning speed decreased indicating the increase of laser beam energy density and laser-material reaction time resulted in more heat absorption by the material. For a certain pulse rate, straight slot turned into a series of discrete small holes when the scanning speed increased to a certain extent.

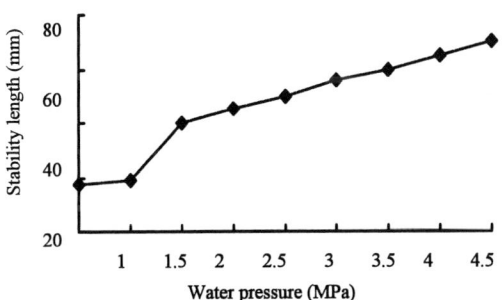

Fig. 6 The influence of water pressure on stability

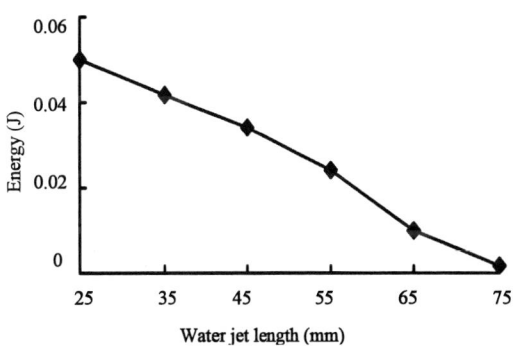

Fig. 7 The influence of water jet length on laser energy

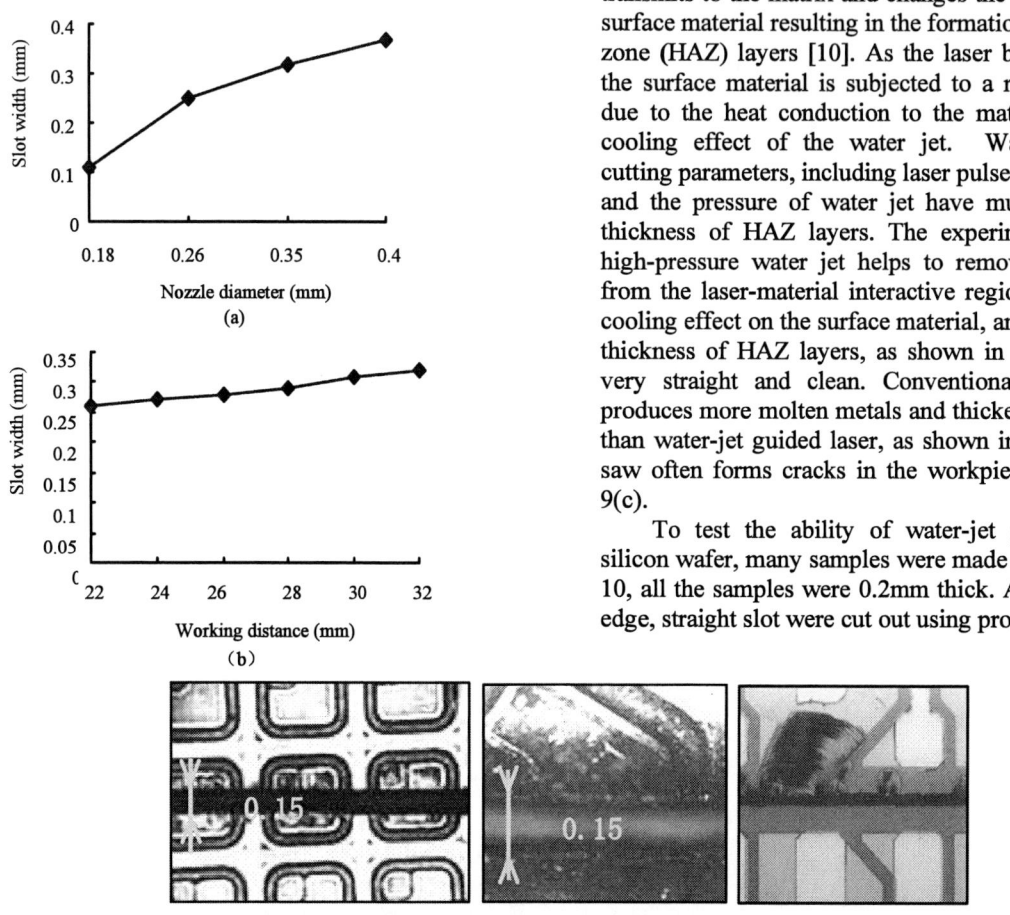

transmits to the matrix and changes the microstructure of the surface material resulting in the formation of heat-affected zone (HAZ) layers [10]. As the laser beam moves forward, the surface material is subjected to a rapid cooling process due to the heat conduction to the matrix material and the cooling effect of the water jet. Water-jet guided laser cutting parameters, including laser pulse parameters, speed, and the pressure of water jet have much influence on the thickness of HAZ layers. The experiment results indicate high-pressure water jet helps to remove the molten metal from the laser-material interactive region, provides a strong cooling effect on the surface material, and helps to reduce the thickness of HAZ layers, as shown in Fig. 9(a), the slot is very straight and clean. Conventional gas assisted laser produces more molten metals and thicker HAZ on workpiece than water-jet guided laser, as shown in Fig. 9(b). Diamond saw often forms cracks in the workpiece, as shown in Fig. 9(c).

To test the ability of water-jet guided laser cutting silicon wafer, many samples were made out, as shown in Fig. 10, all the samples were 0.2mm thick. Arc edge, acute angle edge, straight slot were cut out using proper cutting

Fig. 9 Comparisons of silicon wafers cut by different cutting methods (a) water-jet guided laser (b) gas assisted laser (c) diamond saw

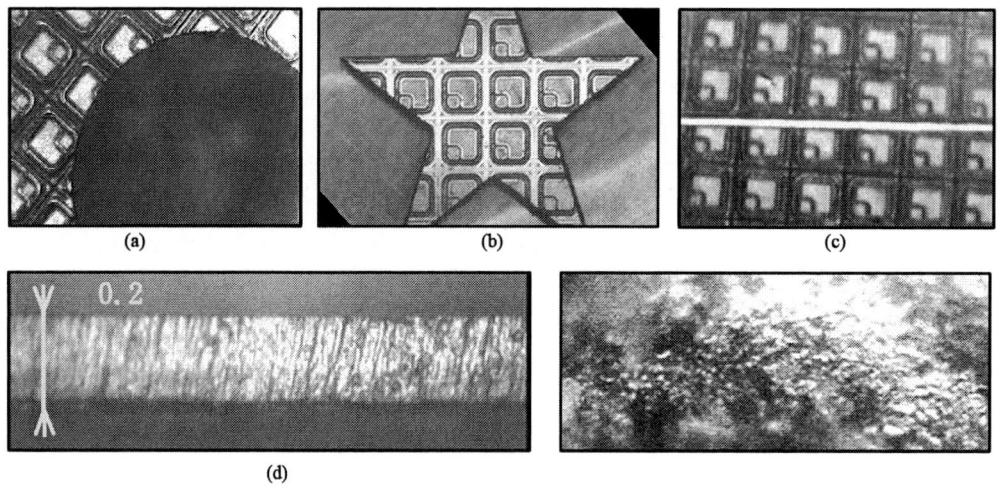

Fig. 10 Silicon wafer cutting samples (a) arc edge (radius is 1.5mm) (b) acute angle edge (circumcircle radius is 2.5mm)
(c) straight slot(the slot is 0.12mm wide) (d) cutting surface(200×) (e) cutting surface(500×)

3.2.3 Comparison of different silicon wafer cutting methods

Water-jet guided laser cutting is a thermal process. Most of the heat melts or vaporizes the material and then the material is washed away by the water jet. The rest of the heat

parameters shown in table. 1, the cut edge shapes are precise, the slot is quite straight. The silicon slot surfaces were smooth and clean, nearly no molten materials ,no heat affected zone, no cracks were found, as shown in Fig. 10(d), (e).

830

Conclusions

Water-jet guided laser processing is a complex thermal process with the water jet cooling. It is a perfect method to cut silicon wafer. A new simulation model was setup, the simulation results showed the water jet cooling makes the convection heat transfer coefficient different from traditional laser, so the temperature field formed by laser and water jet is quite suitable for cutting silicon wafer when choosing proper water pressure and temperature. Because of the high heat conductivity and some other special material properties, simulation result showed that silicon machining temperature gradient is smaller than 65Mn and 1Cr18Ni9Ti. Water-jet guided laser cutting silicon wafer experiments were done using Nd:YAG pulsed laser. The influence of different machining parameters on quality factors including heat affected zone, molten metals and slot widths were investigated. The experiment results showed: 1. Coupling unit could produce stable, slim water jet and could couple laser into the water jet precisely. 2. Long working distance was achieved, the max working distance was more than 100mm. 3. Cutting speed and quality were both well when cutting silicon wafer. 4. The heat affected zone layer and molten material layer produced by Water-jet guided laser were thinner than traditional laser. 5. Using proper cutting parameters, high quality slots could be achieved when cutting silicon wafer.

Acknowledgments

This work was financially supported by the Chinese National Natural Science Funds (project no. 50675053), Weapon equipment pre-research project funds (no. 9140A18030607HT0117), Higher Education doctor Subject Funds (no. 20070213027) and by Development Program for Outstanding Young Teachers in Harbin Institute of Technology (HITQNJS.2007. AUGA18601009).

References

1. B. Richerzhagen, "Industrial Applications of the Water-jet guided Laser," *The Industry Laser User*, Vol. 28(2002), pp. 28-30. [A reference to a journal article …]
2. R.K. Ganesh, W.W. Bowley, *et al*, "A model for laser hole drilling in metals," *Journal of Computational Physics*, Vol. 125(1996), pp. 161-176.[A reference to a journal article …]
3. J.M. Prusa, G. Venkitachalam, *et al*, "Estimation of Heat Conduction Losses in Laser Cutting. International," *Journal of Machine Tools and Manufacture*, Vol. 39(1999), pp. 431-458.[A reference to a journal article …]
4. C.-F.Li, D.B.Johnson, *et al*, "Modeling of Water jet Guided Laser Grooving of Silicon,"*International Journal of Machine Tools & Manufacture*,Vol. 43(2003), pp. 925-936.[A reference to a journal article …]
5. Lumonics. LTD., JK701 & JK702 Nd-YAG Laser Operation & Maintenance Manual. (Lumonics. LTD .,2000), pp. 132-316. [A book reference…]
6. P. Couty, F. Wagner., *et al*, "Laser coupling with a Multimode Water-jet Waveguide," *Optical Engineering*, VOl.44,No.6(2005), pp.1-8.[A reference to a journal article …]
7. F. Wagner, "The Laser Microjet Technology-10 Years of Development," (*Synova SA, Ecublens CH.* 2003), pp.1-5.[A reference to a journal article …]
8. A. Kruusing, "Underwater and Water-assisted Laser Processing: Part 1-General Features, steam cleaning and shock processing," *Optics and Lasers in Engineering*, Vol.41,No.2(2004), pp.307-327.[A reference to a journal article …]
9. Ákos.Spiegel, Nándor.Vágó, *et al*, "High Efficiency Raman Scattering in Micrometer-sized Water Jets," *Optical Engineering*, Vol43(2004), pp.450-454.[A reference to a journal article …]
10. Shanjin. L, Yang. W. "An Investigation of pulsed Laser Cutting of Titanium Alloy Sheet," *Optics and Lasers in Engineering*,Vol.44,No.10(2006), pp. 1067-1077. [A reference to a journal article …]

Research on Solder Joint Intelligent Optical Inspection Analysis

Li Ni, Pan Kai-lin, Li Peng

School of mechanical & electrical Engineering, Guilin University of Electronic Technology

Guilin 541004, China

dylini@gmail.com

Abstract

The existing automatic optical inspection (AOI) equipment can be used to check out the defect of the electronic products, such as solder bridging, tomb stone and so on. Based on the existing AOI, a new intelligent optical inspection analysis (IOIA) technology is brought forward in this paper. The method, which combines off-line analysis and on-line contrast, is employed in the IOIA technology. That is to say, we should combine the conclusions of solder joint shape forecast and reliability analysis with on-line inspection technology of AOI smoothly. The basic principle of this method could be described as following: firstly, the structured light of monocular vision system must be improved to acquire the 3D information of practical solder joint shape. Then the relationships between solder joint shape and its reliability are introduced in this paper, and the 3D shape parameters which have significant effects to the solder joint reliability are extracted based on this relationship. Finally, contrasted the parameters of 3D shape with the data-base which established in off-line, and then the quality evaluation of solder joint could be given immediacy by using the data-base. According to the conclusions we can get the further quality evaluation and classification of the electronic products.

Ⅰ. Introduction

Electronic product is making a continually progress toward higher density and micromation, which bring in great challenge to the surface mount technology (SMT). It was confirmed that the failure of the electronic products is major caused by the solder joint failure. The reliability of solder joint plays a decisive role on the life of the electronic products. Therefore, in the production process of the SMT, the inspection of the products by AOI is very important for insure the reliability of the products.

There are two types AOI system, monocular vision and multi-vision, which classified by the camera number of the system. The classical monocular vision system which using one camera and RGB light source is the product of OMRON. It can acquire limited 3D information by 2D image's color. This 3D information is so limited that it just only assists to judge the solder joints pass or NG, but not to evaluate the solder joints' reliability. The typical multi-vision system is the product of Teradyne. It can get enough 3D information at the cost of inspection speed and image storage space. The biggest disadvantage is the 3D information gained by the system has not used to evaluate the solder joints' reliability.

Aiming at the present research status, based on the existing AOI, a new IOIA technology is brought forward by improve the structured light of monocular vision system and sum up the relationships between solder joint shape and its reliability in this paper. The united of off-line analysis and on-line contrast is employed in the IOIA technology to reduce the

time consumption and to meet the real-time requirements of the system. For the off-line analysis, an analysis result data-base established by adopt experimentation and finite element analysis to analyse the solder joints' reliability under the conditions of various parameters; for the on-line contrast, the result of the data-base called as the IOIA system's real-time conclusions by compare the extracted parameters with the information of the data-base. This method develops a feasible way for the IOIA technology. The established of the data-base needs a long time to maintain and addition, so there is no given the expounding about it. This paper will focus on how to combine the off-line analysis with the on-line contrast, as following.

2. Design of Structured Light

The most important thing for combine the off-line analysis with the on-line contrast is to acquire enough parameters. In the IOIA technology, many parameters are known quantity, such as the material of solder, the size of pad and so on; and the rest one parameter we need is the solder joints' 3D shape parameter which must be acquired by the optical inspection system. Therefore, the primary difficulty is how to obtain the 3D shape parameter.

In computer stereo vision, the monocular vision and the multi-vision have its own advantages and disadvantages. The monocular vision has a simple structure, a small size, but the design of the structured light is difficult. On the other way, the multi-vision has a simple theory which same as the human stereo vision, but the structure of the camera system is too complex, the size is too large, the camera calibration and the image matching are too difficult. The application occasion of the IOIA system is in the real-time detection of the SMT production line, which requires small size and high inspection speed. So we improve the structured light of the classical monocular vision system to acquire the solder joints' 3D shape.

Structured light have three basic forms: pointolite, light sheet and surface light source. The spectrometry[1] proposed by Li Qi, Zhe Jiang University, gained a good effect in measuring 3D object. Based on the spectrometry, a white light combined with the spectrometry is used in this paper, as shown in Figure 1. The first step is using the white ring light to obtain the 2D gray image of the solder joints and components. And then use the spectrometry to get the solder joints' depth information. This paper will focus on obtained solder joints'depth information by improved spectrometry. The white light case will not be discuss, please refer to the relevant literature.

The principle of the spectrometry is based on the active triangle imaging method. In the design of the structured light, the spectral band which consists of multi-light sheet can be seen as a surface light source. There is a one-to-one correspondence between the geometry information (α) of the three-dimensional space and the wave length (λ) of the incident light. According to the correspondence, we could analyse the color information of the 2D color image which acquired by image sensor (CCD). And then we could calculate the depth information of the object based on the geometric relation.[1]

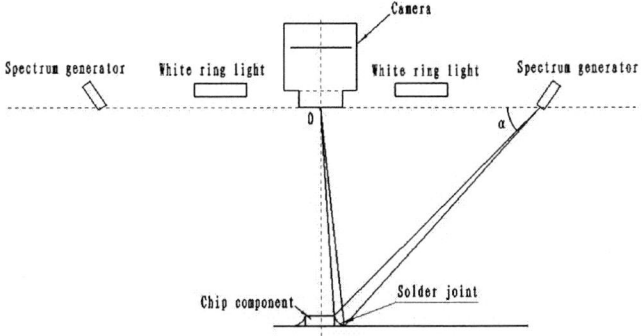

Fig.1 Improved structured light system

In this paper, the depth measurement system is composed with two parallel spectrum generators which are bilateral symmetry about the optical axis. Figure 2 shows the definition of the geometric parameters.

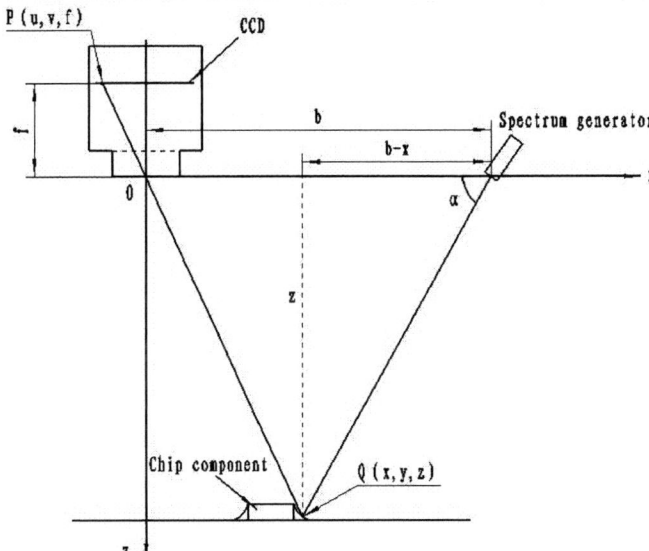

Fig.2 Spectrum measuring system

In Figure 2, take the center of the lens as the origin, the downward of the optical axis as the z-axis, the horizontal towards right as the x-axis, a right hand coordinate OXYZ established as shown. The centers of the light source and the lens are at the same level, and their distance is b. The distance from CCD to the lens center is the focal length of the lens (f).There is a one-to-one correspondence between the object point Q(x, y, z) and image point P(u, v, f). The angle between incidence direction of light and x-axis is α. The coordinates value u, v, f of the image point (P) could be obtained by measuring the image directly. The angle α could be got by the expression as follow:

$$\alpha = \alpha(\lambda) \qquad (1)$$

Where the wave length λ has a one-to-one correspondence to the color, it could be obtained by the color value.

There is similarity between object point triangle and image point triangle, the equation as follow:

$$\frac{x}{u} = \frac{z}{f} \qquad (2)$$

Where the value of the z can be calculated by the triangular relationship, the equation as follow:

$$z = (b-x)\tan\alpha \qquad (3)$$

Take equation (3) into equation (2), we can calculate the x-value of the object point Q, as follow:

$$x = \frac{ub}{u + f \cdot ctg\alpha} \qquad (4)$$

According to the symmetry in turn, we can calculate the y-value and z-value of the object point Q, as follows:

$$y = \frac{vb}{v + f \cdot ctg\alpha} \qquad (5)$$

$$z = \frac{f \cdot b}{u - f \cdot ctg\alpha} \qquad (6)$$

Due to the auxiliary of the structured light system, we can get all the 3D information of the object point Q which can be used for extract 3D shape parameters of the solder joint.

3. Extract Solder Joint Shape Parameters

For the assistant of the spectrum structured light system, we can gain the 3D information for any point, but not all the points have the obvious influence to the reliability of the solder joint. Therefore, considering the efficiency of the system, we only extract the points which have the significant affection to the quality of the solder joint. The 3D information extracted is the shape parameter of the solder joint which we need. It is hard to fix on which point is important because the shape of the solder joints in SMT are various for each other, such as solder joint of chip component and that of QFP. Different solder joint shapes have different shape parameters. We will focus on how to extract the chip component's solder joint shape parameters based on the previous research conclusions in this paper.

2D image sketch map of the chip component's solder joint under spectrum structured light shown as Figure 3.

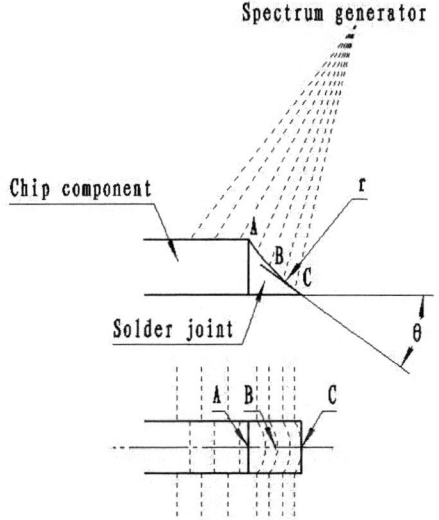

Fig.3 Chip component's solder joint under spectrum structured light

833

In Figure 3, we can see there are distortions of the spectrum structured light on the solder joint's surface. These distortions contain the depth information of the solder joints. Where the r is the curvature radius of the solder joint surface; the θ is the wetting angle of the solder joint; points A, B and C locate on the center profile contour of the solder joint which the B is the middle of the A and C in the x direction. The research[2,3] shows that, the solder joint shape, especially the center profile contour, to be a great influence on the thermal fatigue life of the solder joint. Thus, we only extract the three points A, B and C as the base of the shape parameters. Because these three points in the same Y plane, the values of their y are the same. And the center outline can be seen as a similar arc, so we can take the x-value and z-value of the three points into the standard equation of the round plane for calculate the value of r, as follow:

$$(x - x_0)^2 + (z - z_0)^2 = r^2 \qquad (7)$$

For the wetting angle θ, we can part the solder joint into two types – concave and convex. If the solder joint shape is concave, the θ must less than 45o, usually it will be 15o< θ <45o, so we can get the value of θ by judge the solder joint shape is concave or not; if the solder joint shape is convex, we can introduce an improved Young equation[4] to calculate the value of θ, as follow:

$$\frac{6V}{\pi r^3} = 3\tan\frac{\theta}{2} + \tan^3\frac{\theta}{2} \qquad (8)$$

Where the V is the volume of the solder joint, it can be obtained by previous process; the r is the curvature radius of the solder joint surface, it can be gained by equation (7).

Till then, the shape parameters which have important influence to the solder joint's reliability are all extracted. They are the coordinate value of the three points A, B and C which used for judge the concave and convex of the solder joint shape, the z-value of the point A which used for judge the climbing height, and the wetting angle θ which used for judge the wettability of the solder joint. Based on these shape parameters, we can do the further reliability analysis of the solder joint.

4. Reliability Analysis of the Solder Joint

The analysis to the solder joint's reliability by optical detection is based on the depth research of the internal relations between solder joint shape and its reliability. There are many academic discussions and monographs about the pre-estimate of the solder joint shape and the finite element analysis of the solder joint reliability from Chinese and oversea scholars so far. Wang Guo-zhong[2] et al, who draws the conclusion by experiment is accepted by many people. That is the thermal fatigue life is maximum if the solder joint shape is straight under the conditions of the fixed pad extension length and the fixed stand-off height (SOH) between chip component and substrate. In this paper, we establish an original data-base based on the conclusion firstly, and then extract the solder joint shape by spectrum structured light, finally, give the reliability evaluation of the solder joint directly by simple geometry analyse.

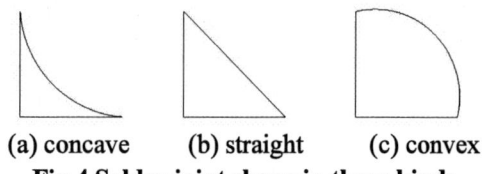

(a) concave   (b) straight   (c) convex

Fig.4 Solder joint shape in three kinds

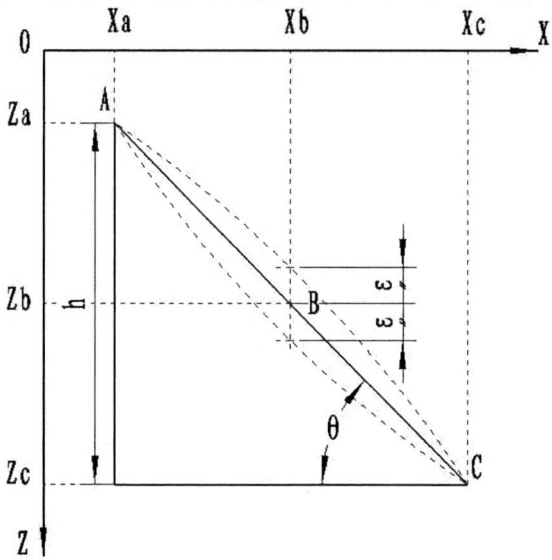

Fig.5 Calculate model

Solder joint shape may be classified into three kinds – concave, straight and convex, shown as Figure 4. We can classify these three kinds solder joint by compare the value of the point A, B and C, the calculate model shown as Figure 5.

In Figure 5, the h is the climbing height of the solder joint; the θ is the wetting angle of the solder joint; ε is the allowance range of the solder joint shape when take the straight as a standard, ε > 0. According to the Figure 5, we can obtain the following conclusions:

1) If $Z_b - Z_{(A+B)/2} > \varepsilon$, the solder joint is over-saturation and the solder paste is excess. At this time, the life of the solder joint can be expression as $N \approx N_H$, where the Z_b is the z-value of point B, the $Z_{(A+B)/2}$ is the mid-point z-value of the A and C, the N_H is the thermal fatigue life of the convex shape solder joint gained by experiment and off-line finite element analysis under the conditions of the fixed pad extension length and the fixed SOH;

2) If $-\varepsilon < Z_b - Z_{(A+B)/2} < \varepsilon$, the shape of the solder joint is reasonable. At the moment, the life of the solder joint can be expression as $N \approx N_O$, where the N_O is the thermal fatigue life of the straight shape solder joint gained by experiment and off-line finite element analysis under the conditions of the fixed pad extension length and the fixed SOH;

3) If $Z_b - Z_{(A+B)/2} < -\varepsilon$, the solder joint is under-saturation and the solder paste is insufficient. And the life of the solder joint can be expression as $N \approx N_L$, where the N_L is the thermal fatigue life of the concave shape solder joint gained by experiment and off-line finite element analysis under the conditions of the fixed pad extension length and the fixed SOH;

4) If $\frac{1}{4}H+G<h<\frac{2}{3}H$, the electronic product could be classified as grade three according to the standard Acceptability of Electronic Assemblies IPC-A-610 D[5] of the electronic products. Where the H is the height of the component, the G is the SOH, the climbing height of the solder joint h is $h=Z_c-Z_a$;

5) If $\frac{2}{3}H<h<H$, the electronic product could be classified as grade one or two according to the standard Acceptability of Electronic Assemblies IPC-A-610 D[5] of the electronic products;

6) If $h<\frac{1}{4}H+G$ or $h>H$, the electronic product is unacceptable according to the visual inspection standard IPC-A-610 D[5] of the electronic products;

7) If $15°<\theta<45°$, the solder joint has a good wetting, otherwise, if the $\theta>90°$, the solder joint has a poor wetting.

Based on the above, we can evaluate and grade the products, pre-estimate the life of the solder joint, get the feedback parameters to the process before.

5. Conclusions

The IOIA technology presented in this paper is an improvement of the AOI. It takes the spectrum structured light to attain the 3D information of the solder joint surface, and then introduces the relationship between the solder joint shape and its reliability to get the basic shape parameters of the solder joint, finally analyses the reliability of the solder joint directly. Because the combine of the off-line analysis and the on-line contrast is adopted in this method, it avoids the slowly finite element analysis holding the precious on-line inspection time. This way is fast and intelligent to satisfy the requirement of the IOIA. At present, the research of the IOIA is at the starting stage. There is needn't the finite element analysis in the   on-line inspection, but it doesn't means that there is no finite element analysis, because we need the conclusions of the finite element analysis and experiment to establish the off-line data-base. Therefore, in the next stage of the research on IOIA, we need to introduce more conclusions of the experiment and finite element analysis to improve the off-line database and advance the IOIA technology.

Acknowledgments

The authors gratefully acknowledge the financial support provided by the Office of Guangxi Education.

References

1. Li Qi, Feng Hua-jun, Xu Zhi-hai et al, "Review of computer stereo vision technique," *Optical Technique*, No. 5 (1999.9)
2. Wang Guo-zhong, Wang Chun-qing, Qian Yi-yu, "An Experimental Study on Effects of Solder Joint Geometry on Thermal Cycle Life in SMT," *Electronics Process Technology*, Vol. 18, No. 5 (1997.9)
3. Zhao Xiu-juan, Wang Chun-qing, Zheng Guan-qun et al, "Section of SMT Soldered Joints and Display of Sectional Profiles," *Transactions of the China Welding Institution*, Vol. 20, No. 2 (1999.6)
4. Wang Da-yong, Feng ji-cai, "Obtaining of Young Equation by Principle of Energy and Establishment of Wetting Angle Model," *Transactions of the China Welding Institution*, Vol. 23, No. 6 (2002.12)
5. IPC-A -610 D, Acceptability of Electronic Assemblies [S].
6. Lin Jian, Lei Yong-ping, Wu Zhong-wei, et al, "Effects of solder joint shape on joint reliability in SMT," *Electronic Components and Materials*, Vol. 27, No. 2 (2008.2)
7. Huang Gui-ping, Li Guang-yun, Wang Bao-feng, et al, "Evolution for Monocular Vision Measurement," *Acta Metrologica Sinica*, Vol. 25, No. 4 (2004.10)
8. Chen Jia-bi, Shu Xian-yu et al, Optical information Technique Principles & Applications, Higher Education Press (Beijing, 2002.7), pp. 321-359.

Productivity Improvement of Stack Package Line through Die Bonding Process & Scheme Optimization

Xing JIN[1,2], Ming LI[1]

1. Lab of Microelectronic Materials & Technology, School of Materials Science and Engineering, Shanghai Jiao Tong University, Shanghai 200240, China;

2. Intel Products (Shanghai) Ltd, Shanghai 200131, China

Abstract

To conform to the ever-emerging market demand, Stacked Memory devices have been more widely utilized. The stacking method also reduces the cost of electronical components through the way that stacking could fully utilize currently on-hand equipment without any new investment. While, starting from late 2003, flash memory manufacturers begin experience capacity degradation, specifically with multiple loop-back workflows induced by Stack CSP devices. By analyzing the process of stacking, the industry's practice considers the control of assembly cost largely depends on the improvement of overall line productivity, specifically the critical bottle-neck area of Die Bonding. This presentation intends to critically describe the methodology and procedures used by Intel's Stack CSP assembly factory, which finally results innovative projects targeting above said productivity improvements. The authors use TRIZ, an inventive problem solving theory and application tool, to analyze and abstract the major contradictions and then sketch out possible solutions. As a result of the applications, the overall Stack CSP assembly factory's productivity increased to a record-high of 340%, far above the industry average, and supports Intel's Stack CSP assembly factory more efficient than benchmarking world-class companies ever since.

The authors of this paper wishes the methodology on productivity and design flexibility at Intel's Stack CSP assembly factory could possibly be proliferated, so as to help achieve productivity and capability maximum output throughout the industry.

1. Background

To conform to the ever-emerging market demand, Stacked CSP devices have been more widely utilized. The method of stacking reduces area size on PCBs consumed by silicon, and thus enables smaller outline of electronic products. The stacking method also reduces the cost of electronical components through the way that stacking could fully utilize currently on-hand equipment without any new investment.

As the demand from portable electronics devices (such as cell phone) calls for more flexible combinations of the chips in the stack, as well as an ever enlarging demand for quantity both in volume and in die count, the bottle neck in productivity emerged through the manufacturers.

Figure.1 below shows the structure difference from single die memory device and stack die device

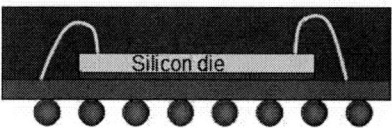

A single die device

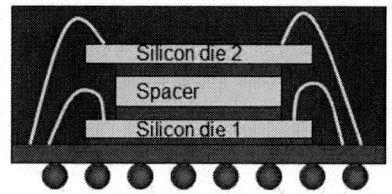

A stacked 2+1 device, 2 active dies and 1 dummy spacer

Figure 1: Structure of normal single die memory and Stacked CSP Device

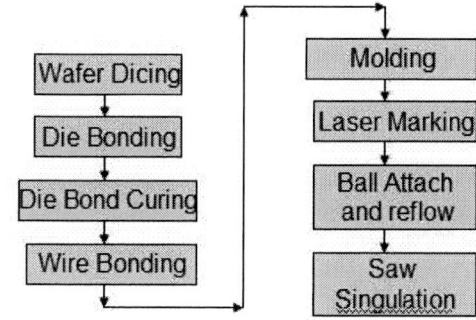

Figure 2: Single die memory product's process flow

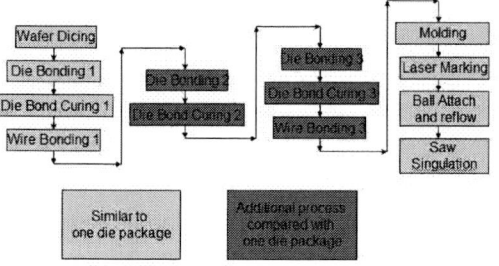

Figure 3: An example of 2 die stack CSP product's process flow

As shown in Figure 2 and 3, as a comparison of two different process flows, simply adding 1 die into the stack would almost double process scheme and complexity, which means halving the productivity. In worse cases, adding 1 die into the stack may introduce adding a spacer (a dummy silicon without electrical function). This brings further degradation to the productivity constraint as if building 3 die stacks. The research from M.T. Zhang et al. also demonstrated the Die Bonding process a constraint to the overall assembly line in stacked memory products. (Reference literal #1).

978-1-4244-2739-0/08/$25.00 ©2008 IEEE

Thus, improvement of the Die Bonding productivity impacts significantly to the productivity of overall assembly line, and thus to the profitability of the companies involved in.

The authors introduced TRIZ as the inventive problem solving theory as a methodology. The authors succeeded in finding solutions of above mentioned problems by using TRIZ as a practical tool as well. In the following segments, the authors attempt to sketch out the profile of the methodology as well as present the practical solutions to the productivity enhancement problems.

2. Introduction of the TRIZ Tool

TRIZ is the acronym of the Russian language "Teoriya Resheniya Izobretatel'skiksh Zadach", namely "the Theory of Inventive Problem Solving". It is a structured methodology developed by G. S. Altshuller, the renowned "Father of TRIZ". The methodology of TRIZ is applied to technical and engineering problems, based on physics laws and a complete summarization and abstraction of the record of innovative patents collected throughout Russia/Soviet history. The tool was previously used (and succeeded) in mechanical industry, while some recent approaches aim at applying this methodology into electrical/electronics industry have been reported.

The TRIZ process could be separated into 2 phases:

Phase 1 is to highly summarize and abstract specific problems into a structured model, to trim-off unrelated information and focus on the key contradictions of the engineering problem. As below Figure 4 shows, the process includes Functionality Analysis, Cause-Effect Chain Analysis, Trimming, and then get to abstracts of the Key Problems. At the end of Phase 1, the Key Problems (also known as contradictions--in engineering or in physics) should be found. This process could either be solved in software-supported environment (such as, TechOptimizer™, etc) or simply by manual works using pencil and paper.

Phase 2 composes of finding a proper "axiomatized" solution for the contradictions. G. S. Altshuller has developed a full scheme of matrix: "39 Engineering Contradictions", maps into "40 Inventive Principles". By using this matrix, engineering contradictions are enabled to get an axiomatized solution. In the Figure 4 below shows, this process has been described as Problem Solving Tools (a.k.a. Classical TRIZ).

As an example, if the TRIZ practitioner points out the "Duration of Action by Moving Object" being the key problem (engineering contradiction), this one could be found as # 15 in the "39 Engineering Contradictions" list. The text of this contradiction #15 points to principle 10, 19, 35, which directs the user go to items 10, 19, 35 in the list of "40 Inventive Principles", cited as below:

Principle 10: Preliminary action (a.k.a. Prior action – "Do it in advance")

Principle 19: Periodic action

Principle 35: Parameter changes (a.k.a. Transformation of physical and chemical states or properties)

At this stage, the TRIZ practitioner could either search through open databases describing historical practices using these principles, or using traditional engineering methodology such as analogy, simulation, intensification, or brainstorm, to get a possible practical solution. With proper verification and qualification procedures, it completes the full cycle of one Inventive Problem Solving worked out by TRIZ.

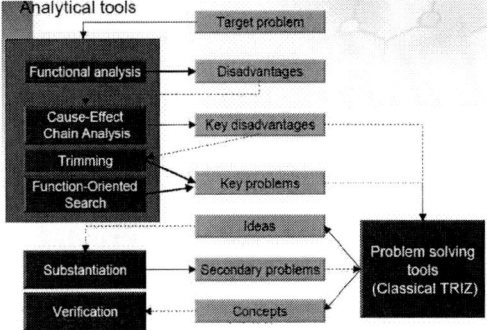

Figure 4: the TRIZ process flow

3. Application of TRIZ in Die Bonding Productivity Projects

To apply TRIZ to solve the productivity problem, the authors first performed boundary condition investigation within the Die Bonding process as a basis of improvement opportunities:

3.1 Boundary Conditions

3.1.1 Manufacturing system

No matter single die or Stacked CSP devices, the products are produced in millions. While each unit of the product should follow certain geometrical specifications during assembly: each die should be ensured to locate within a small range of the designed center; placement or rotation should be monitored and controlled during production;

3.1.2 Incoming Material

As a most efficient method, the Die Bonding process in Stacked CSP devices uses Multi Matrix Arrayed Package (MMAP) and MMAP substrates. In the MMAP substrate, there are up to 10% units carrying flaws—which is inevitable in the production of the MMAP substrate, and are marked out by X-shaped scribes jargonized as "X-outs". Specific to stacked CSP process, the "Assembly Flaws" occurred in the lower level die bonding and wire bonding processes should be picked out as well. Both X-out and Assembly Flaws should be marked and distinguished by Die Bonding process so that valuable silicon dies should be saved from bonding onto;

3.1.3 Design Rule to satisfy Structure and Material needs

As shown on the Figure 5, the stack structure should enable a margin for the bonding wire. This margin must satisfy the wire bonding tool's size, the electrical insulation needs, the tolerance of possible die placement during manufacturing, and possible adhesion agent bleed-out. Thus, a minimum gap from upper die edge to lower die bond pad's interior rim must be created. If this gap could not meet certain value, a spacer has to be used; The spacer needs to be allocated thick enough to nestle the lower die's wire loop, plus additional space to satisfy insulation needs (Refer to Figure 5);

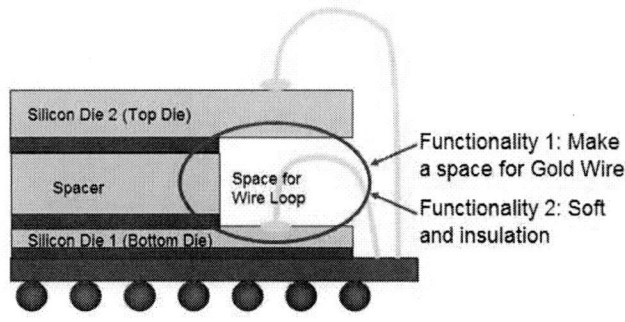

Figure 5: Rationality of a Spacer

Based on above investigation results, two approaches to the productivity problem have been defined. Namely the Manufacturing approach and the Design approach:

3.2 Manufacturing Approach:

A typical Die Bonding process uses liquid epoxy as adhesive. One Dispenser would apply appropriate amount of epoxy onto the substrate (in an intact unit); then the substrate moves to a Bonding location, where a bonder would put a slice of silicon die onto the epoxy, accurately and precisely, to form a good adhesion.

First to perform tests of functionality, cause-effect and practice the trimming work:

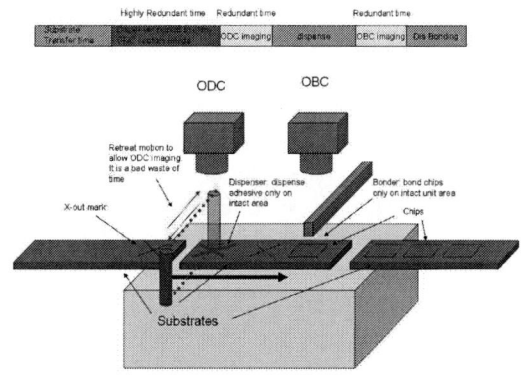

Figure 6: Traditional Die Bonding Schematics

Figure 6 shows the schematics of a traditional Die Bonding process, where two cameras: Optical Dispense Camera (ODC) and Optical Bonding Camera (OBC) are used to determine dispense and bonding locations, to ensure the accuracy which was called for by quality needs. These images also work to distinguish X-out and Assembly Flaw, and if there are such X-outs, the images would instruct the dispenser/bonders to skip the unit.

Because the ODC image captures the exact location of dispensing, the dispenser has one motion to retreat whenever the ODC needs to get an image. Retreat motion moves the entire dispenser (around 25KG) as fast as possible, to make room for the ODC camera. From the Cycle Time Chart shown on top of Figure 6, the dispenser retreat time, as well as ODC/OBC imaging time, are redundant and is better to be removed.

The key problem (also known as engineering contradiction) emerged from these activities: the process must enable dispense/bonding accuracy, as well as to pick out X-outs. This contradicts with a long cycle time of process, which

is not preferred by the die bonding. See Figure 7 for more detail.

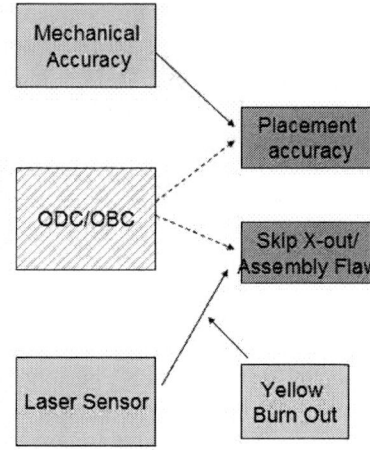

Figure 7: Functionality Test and Trim in Manufacturing Approach of the problem

By functionality test, it is already known dispense/bonding accuracy is part of the reason to use ODC/OBC. As a substitution, the authors noticed the machine's mechanical structure could ensure enough accuracy against the current specification (namely, +/-50um in 3 sigma range), thus the functionality test demonstrated the cameras could be shut off, if the only functionality for them is to maintain accuracy. While for the function of distinguish X-outs, the Classical TRIZ tool should be utilized.

In the TRIZ word, the original problem could be translated as "Duration of Action by Moving Object" (#15) in the 39 Contradictions list. In previous section, the practitioner already found Inventive Principles # 10, 19, 35 and attempting to get solutions.

Principle 10: Preliminary action

Principle 19: Periodic action

Principle 35: Transformation of physical and chemical states or properties

Team researched through literals about TRIZ application in electronics (reference literals #2, #3). Finally the authors developed 2 engineering solutions as following (refer to Figure 8):

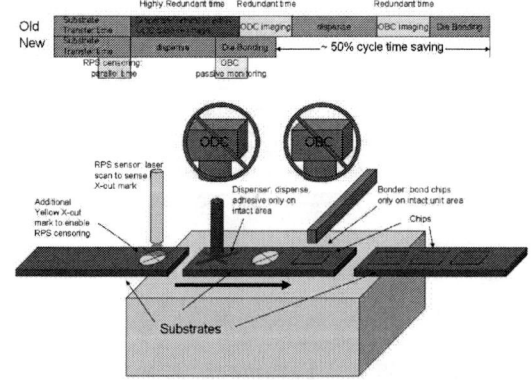

Figure 8: TRIZ improved Die Bonding Schematics

Solution 1 is based on Inventive Principle 10, Preliminary action. The kernel of this solution is to put the X-out recognition prior to dispensing. It uses a reflective laser sensor that could scan through the unit area and detect X-outs. By this activity, the X-out could be detected prior to dispense location, so the ODC could be shut-off, thus dispenser do not need to perform meaningless "retreat motion".

Yet the X-out mark was previously only scribe lines onto the green solder resistant. This was good to human eyes but not enough for laser sensor.

In this regard, the authors turned to Inventive Principle 35, transformation of physical property. We added special burnt onto the substrate surface so that it turns out yellow color comparing to the green solder resistant surface. This yellow color burnt changed the X-out reflection rate, it goes with Principle 35 about the physical property—in light reflection, and makes the laser sensor effective.

3.3 Design Approach

Traditionally, most 2 die Stack products are made with one spacer. The boundary condition analysis above mentioned its rationality.

Again using the TRIZ process to do functionality test, the authors discovered further functionality of the spacer: it should enable space or at least a very soft strength and good insulation to the bottom gold wires, not to bring any damage, deform or electrical leakage to them. See Figure 5 for functionality test of the Spacers.

If a product's structure is composing of 2 live dice and 1 spacer, actually 1/3 of the Die Bonding productivity should be wasted in dummy spacer bonding. From financial or from engineering perspectives, the application of spacer is ineffective. Thus, from the Design Approach, to trim-off the spacer could be an important approach to increase the Die Bonding productivity. The contradiction has been turned to either the space or local material strength in the gold wire 1 area. The authors hypothesized a new material called "spacer-less Material", which could be easily applied as the epoxy or DA Film without extra Die Bonding process needed. The authors tried the TRIZ tool to anticipate spacer-less material's physical property and trying to select material based on the anticipation. See Figure 9, Trimming of the Spacer.

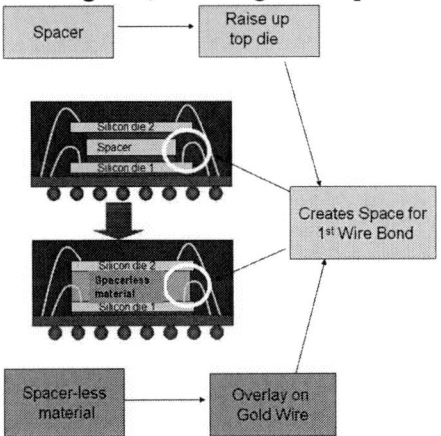

Figure 9: Trimming of spacer

Utilize the Classical TRIZ model to solve the contradictions of: A) Length (or Angle) of Stationary Object; B) Strength.

Using the 39 Contradictions list, the authors found following inventive principles could be utilized to solve the contradictions:

Principle 3: Local quality

Principle 14: Curvature

Principle 15: Dynamization

Principle 17: Another dimension

Principle 35: Transformation of physical and chemical states or properties

Principle 40: Composite materials

Based on data search, the Principles 3, 35 and 40 are considered likely to be useful in solving the contradictions. Two groups of materials are considered as of potential value to solve the contradictions:

First is a group nicknamed Film over Wire (FoW), composed of certain polymer compounds. To these types of polymers, their Young's Modulus before and after the Temperature of Glassification (Tg) differs significantly: they could be very hard in room temperatures but could be very soft if heated above Tg. By installing heater function and properly adjust die bonding temperatures it is easy for Die Bonding module to process FoW materials. This group of material's selection utilizes the Principle 3: Local Quality and Principle 35: Transformation of physical and chemical states or properties.

Table A below shown some candidate FoW materials and their mechanical property.

Table A: Modulus of candidate FoW materials

	Unit	FOW A	FOW B	FOW C
Tg	Deg C	142	191	153
CTE	ppm/K	80/140	51/125	88
modulus	MPa	2500(25C)	3800(25C)	3000(35C)
		73 (250)	170 (250)	70 (250C)

Second is a group of materials uses huge fillers into liquid epoxy, nicknamed Spacer Free Epoxy (SFE). The SFE, like normal electronics adhesive epoxies, uses resin + filler composition. While its filler sizes are huge enough to sustain the epoxy under Die 2, that makes enough room for the 1st Gold Wire loop height. Even in case the epoxy was applied directly over gold wire, the gold wire still got no damage. This group of materials' selection applies the Principle 40: Composite material and Principle 3: local quality. Figure 10 described both 2 groups of materials with different property and innovative principles applied.

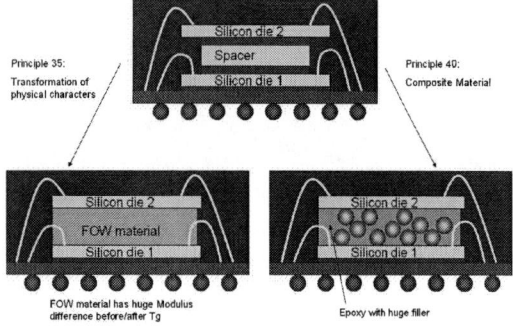

Figure 10: two possible materials by TRIZ method

Conclusions

By now, the implementation of projects through the Manufacturing Approach has already been qualified and in high volume production in one of the assembly facilities of the Stacked CSP memory product assembly line. With the series of improvements, total 340% productivity increase has been achieved. The Design Approach identified projects are also under evaluation phase, by now some limited quantity of FoW material has started application in commercial products.

The authors of this paper are convinced that, the application of TRIZ method in the electronics assembly/packaging industry would be as much successful as it did in the mechanical industry. Related research would be followed up and continuously reported.

References

1. Zhang, M. T. et al., "Dynamic capacity modeling with multiple re-entrant workflows in semiconductor assembly manufacturing," IEEE Int'l Conf on Automation Science and Engineering, 2005
2. SUN, Jianguang et al., "Research on inventive principles classification of electrical design, Principles No 1 to No 20," Journal of Engineering Design, Vol. 14, No. 1 (2007)
3. SUN, Jianguang et al., "Research on inventive principles classification of electrical design, Principles No 21 to No 40," Journal of Engineering Design, Vol. 14, No. 2 (2007)
4. MA, Lihui et al., "Contradiction discovery and solving method based on TRIZ evolution theory and TOC prerequisite tree," Journal of Engineering Design, Vol. 14, No. 3 (2007)
5. TIAN, Qihua et al., "Research on decoupling method of product design based on axiomatic design," Journal of Engineering Design, Vol. 14, No. 6 (2007)

Developing the Stencil Printing Process for 01005 Lead-Free Assemblies

Yong-Won Lee[1], Keun-Soo Kim[2], Katsuaki Suganuma[2] and Jong-Hoon Kim[3]

[1]Mechanical & Manufacturing Technology Center, Samsung Electronics Co., Ltd.
416, Maetan-3Dong, Yeongtong-Gu, Suwon-Si, Gyeonggi-Do, 413-733, Korea
microjoining@naver.com, Tel. 82-31-200-6226

[2]Institute of Science and Industrial Research, Osaka University
Mihogaoka 8-1, Ibaraki, Osaka 567, Japan
kskimm12@sanken.osaka-u.ac.jp, suganuma@sanken.osaka-u.ac.jp, Tel. 81-6-6879-8520

[3]New & Renewable Energy Team, Korea Institute of Industrial Technology
35-3, Honcheon-Ri, Ibchang-Myon, Cheonan-Si, Chungnam, 330-825, Korea
kjhoon@kitech.re.kr, Tel. 82-41-5898-658

Abstract

This paper presents the implementation of 01005 components in a wireless module mass production. Several modifications in processes, material, and equipment were required. A number of manufacturing related issues needed to be addressed. One particularly significant issue for use of 01005 technologies is solder paste printing and the reason is lack of manufacturing experience. Many manufacturers have spent much time and money determining the optimum process conditions for using 01005 components. All of the 01005 manufacturers recommended using an electroform stencil for very small aperture size and fine pitches. This is required a change from the standard manufacturing process that utilizes a laser-cut stencil printing and it is also one of the more expansive stencil fabrication processes. For some manufacturing processes this may be difficult. However, the current state of the art for laser-cut stencil technique of solder paste is very limited for these extremely smallest components due to laser-cutting quality.

In right of this need, we carried out to evaluate if the enhanced laser-cut stencil with post-treatment such as electro-polishing can perform as good as or better than the electroform stencil. The designed experiments are conducted to determine the effect of electro-polishing on the print performance of enhanced stencil for 01005 small apertures. The results showed that the standard laser-cut stencils could be strengthened through optimized electro-polishing treatment after laser-cutting process. This enhanced laser-cut stencil showed acceptable paste release performance and it is low-cost alternative to electroform stencil for 01005 small apertures.

The work described in this paper also examines the release performance of type-3 and type-4 pastes from electroform and enhanced laser-cut stencils. Results of this test showed that paste release performance is significantly better for type-4 comparing to type-3 pastes.

Introduction

Passive components comprise anywhere from 85 to 95% of the components on a common circuit board assembly. Due to size considerations, there has been a constant evolution of passives in order to decrease the overall real estate the passive component takes up. [1] It seems like just a short time ago, the 0201 size was introduced and many manufacturers of semiconductors and consumer products have been interested in using 0201 components. Recently, passive components manufacturers are now introducing and offering low volume quantities of 01005 components, and some early manufacturers are gearing up to develop and qualify assembly processes for their respective applications, while many manufacturers are still developing and optimizing 0201 technologies.[2]

Figure 1. Size comparison for 01005, 0201 and 0402 components.

Ultra miniature 01005 sizes are just 0.4×0.2×0.2mm as compared with the current 0201 sizes (0.6×0.3×0.3mm). Figure 1 shows an optical image of a 01005 component placed next to 0201 component. Although miniature sized components, such as 01005s, enable the effective use of the limited real estate to form complex circuitry, their introduction has posed new challenge during assembly processes. These challenges must be solved in order to gain acceptance as an option for high speed and high yield assembly.

The solder stencil printing process is the crucial point of SMT process. Additional attention needed to be paid to assembly yield and quality related issues. Studies have shown that about 60~70% of defects identified after reflow originated during the solder paste printing process in general, hence improving this process is an important way of increasing assembly yield. [3-5] To achieve a high-yield and high-quality soldering process, a sufficient amount solder paste is needed and variation in paste volume must be minimized. If we print unequal amounts of solder paste we will experience a number of problems including tombstoning. The key consideration is consistently printing the correct amount of solder paste quality onto each pad of the 01005

978-1-4244-2739-0/08/$25.00 ©2008 IEEE   841

components. [4, 6] Since the 01005 components have been introduced but it is still very little, publicly available details on assembling small components and solder stencil printing techniques. Previous experience and earlier studies have shown that electroform stencils with a lower thickness (3mil), and finer powder size (Type-4 or 5) could potentially outperform the described stencil. [1-2], [7-8], [10-11] However, in actual manufacturing applications, electroform stencils are considerably more expansive [9] and longer processing times to fabricate than conventional laser-cut stencils. Moreover, electroform stencil technology thus far are non-standard process in the normal wireless module lines of high-volume manufacturing (HVM). Additional tools such as special squeegee blade for using electroform stencils [10] and standard process changes are required for achieving good release qualities. On the other hand the laser-cutting has the advantage of high-flexibility, high-speed, and is able to cut very small apertures, but the problem with conventional laser-cutting is poor aperture quality. It needs a post-treatment to clean the stencil and remove burrs on the backside. The electro-polishing are used to further smooth surface walls and improve solder paste release. [11] In the electro-polishing, in which the solutions used commonly acide, anodic currents or potentials lead to dissolution and passivation of the metal, promoting the leveling of the metal surface. In the literature, there are some studies comparing the performance of electroform and laser-cut stencils for same aperture size.[7-12] For example, Mattsson *et al*. [9] studied stencil aperture performance, and the 4mil thick laser-cut and electro-polished stainless steel stencil was evaluated as a low-cost option for 01005 volume manufacturing. Other researcher recently have reported the effect of stencil fabrication techniques on solder print qualities such as Iyer *et al*. [12] There findings are very useful data in the stencil selection, the optimum stencil aperture design, solder paste selection, and paste printing process for small apertures such as 01005 and0201 and the factors impacting assembly quality. All the researchers mentioned laser-cut stencil with electro-polishing was shown potentially outperform in solder print quality. However, it still remains to be answered about allows 01005 component while still showing good printing results for 0201 or larger components.

It is motivation of this work to challenge the previous literature results and lesson learned from previous 0201 technology development and manufacturing implementation, and prove or disprove that the electro-polished laser-cut stencil can be performed successfully for 01005 components. In this research work, one of low-cost assembly solution was successfully developed to have mixed 01005, 0201 and 0402 components for SiP module and mobile phone applications. This paper discusses preliminary results from the solder stencil printing process investigations for 01005 small components.

Experimental
Materials and Procedure

A test board with 01005, 0201 and 0402 components was designed for the experiment. Figure 2 is a photograph of the 01005 test vehicle (TV). TV is an 8×8 laminated SiP module, composed of 5×6×2 matrix. A fully populated test vehicle contains 2,160 components for 01005. All 01005 component pads in this design were fixed at 0.175mm(L) × 0.175mm(W) × 0.150mm(S), and all solder pads are solder mask defined pad on three sides.(See Figure 3) The components used in this experiment were 01005, 0201 and 0402 sizes and all provided by Murata Co. The component was fed in tape and reel packaging.

Figure 2. Photograph of 01005 Test Vehicle.

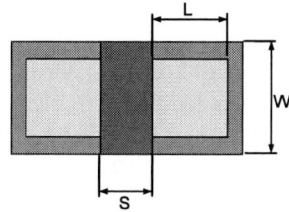

Figure 3. Component pad dimension.

All solder paste printing for the experiment was conducted using 3mil thick stencils with 1:1 opening for 01005 pads. Two types of stencil were manufactured for the evaluations. (Electroform stencil and laser-cut stencil) The electroform stencil was manufactured with the standard electroforming process and it was used for comparing with a laser-cut stencil. For a laser-cut stencil, standard stencil was designed for the aperture experiment. Five different apertures with electro-polishing times were has been tested. For the electro-polishing conditions, the electrolyte was made of chemicals of analytical quality and it consisted of 95wt% phosphoric acid and 5wt% sulphuric acid. This solution was optimized electrolytes from stencil manufacturer and capable of giving a smooth a bright surface finish. A complete stencil consists of a square rigid frame to which the stencil is attached. The frame design is specific to the printer to be used, and is most commonly fabricated from square aluminum tube for the optimum combination of rigidity and lightness, although cast aluminum frames are used for small printer. The frame size is 736mm × 736mm and the metal stencil is 530mm × 530mm. Tensioning is also required in this case to ensure a taut, flat surface. A summary of the stencil fabrication for electro-polishing testing are shown in Table 1. The evaluations by visual inspection of the quality of the surface polish for the aperture wall was checked by scanning electron microscopy (SEM) at magnification of 800X.

Table 1. Laser-Cut Stencil Fabrication Variables.

Laser Cutting Machine	LPKF (Micro Cutting)
Laser-Beam Size	20mm
Laser Type	YAG
E/P Solution	Phosphoric Acid: 95wt% Sulphuric Acid: 5wt%
E/P Time	0, 5, 8, 15 and 20sec
Stencil Mesh Tension	0.55mm – 0.70mm
Stencil Thickness	3mil(80um)

A commercially available solder pastes supplied from Alpha Metal Co., water-soluble (WS609 flux), 95Sn/5Sb solder paste was used throughout the study. The metal content is 90% and the solder melting point was proximately 232~240°C as specified in the supplier's specification.

Since the solder paste is stored at 4°C it has to be brought up to room temperature before being used in the screen printer for the experiments. Screen printing was carried out using Speedline MPM AP Excel printer with metal squeegees angles at 60° were fitted to the printer. The stencil was cleaned before use. A stencil is mounted on the printer frame. A stencil is placed on the PCB substrate, and after vision alignment, a squeegee travels at a certain pressure and speed to push the solder paste through the stencil apertures. The following screen printing parameters were used for all stencil printing: Print speed = 8mm/sec, squeegee type = metal blades, print force = 5.3, balance = 50:50 and print gap = 0(on contact). The solder deposited PCB substrate was inspected in a microscope equipped with a digital camera and documented photographically. The next step in the assembly process is the component pick and place. The pick and place machine from Siemens HS60 was used and programming was carried out with same mounting speeds and forces. Pick-up nozzle is 906 and it is currently in used for 0201 components.

All solder paste reflow was performed in a Heller 1800 convection oven. The reflow system contained 8 heating and 1 cooling zone. The populated PCB substrates were reflowed under nitrogen atmosphere via a Heller convection reflow oven. The nitrogen environment was controlled at less than 400 oxygen ppm level. Peak temperature is 251°C for 38.8 seconds at above 232°C of solder melting point and the belt speed is 16 inch/minutes. Figure 4 is the thermal profile that was used to reflow all boards assembled during the experiments. The final process step was deflux cleaning and a Trek machine was employed for cleaning of flux residues. All boards were visually inspected after stencil printing, after pick and place, and after reflow processes. Component shear test was performed using a Dage-series 4000 shear tester.

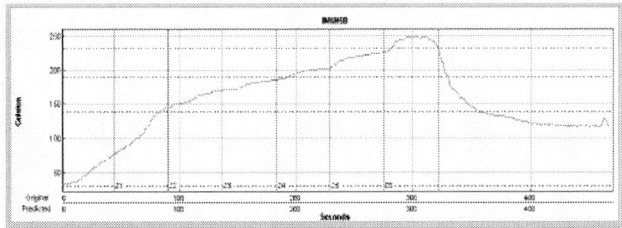

Figure 4. Thermal reflow profile.

Experiment Design

To determine the best condition of electro-polishing for enhanced laser-cut stencil fabrications, test stainless steel sheets were designed with five levels of times; 0, 5, 8, 15 and 20s in this experiments. A stencil was designed based on the results from this experiment. Only one electro-polishing time was selected for the enhanced laser-cut stencil fabrication.

Two variants of stencil types (electroform stencil and enhanced laser-cut stencil) are evaluated for the paste release performance comparison purposes when the aperture volumes to be deposited are kept constant. The variable held constant are stencil thickness (3mil), stencil design (1:1 opening), paste type (type-3 and type-4 particle sizes) and equipments used.

Results and Discussion

Results of the different evaluation steps are described in the following section. The data gathered from the experimentation was arranged and the data was analyzed using JMP ® data statistical software.

Effect of Electro-Polishing on Aperture Quality

In this study, initially the effect of electro-polishing times on the aperture wall quality of the surface polish was investigated. For such, electro-polishing time was varied: 0, 5, 8, 15 and 20sec. As soon as each electro-polished stencil surface was obtained, its polish quality was evaluated by visual inspection. Figure 5 - (a) and (b) shows the SEM images of laser-cut stencil's aperture wall with different electro-polishing times, and electroform stencil's aperture wall for comparison purpose. No electro-polished aperture shows the effects of poor aperture quality as a result of the laser-cutting process. ((a)-①, (b)-①) From the Figure 5 - ②, ③, ④, and ⑤ show the improved aperture quality and an electro-polished aperture significantly has smoother surfaces than no electro-polished aperture (0sec). The best quality of aperture was achieved for electro-polishing times of 8sec and its polish quality was improved with increasing electro-polishing times in general. But, for electro-polishing times of 15 and 20sec, the good aperture quality could not be realized and the corner inside of aperture was more ragged than times of 8sec. It should be pointed out that suitable processing time will improve the surface quality but the excessive processing time will lead to quite lower surface quality. This result therefore that it is desirable to consider process efficiency for high surface quality. Up to now, no optimization of the electro-polishing parameters for many laser-cut stencil manufacturing has been made in industry.

(a)

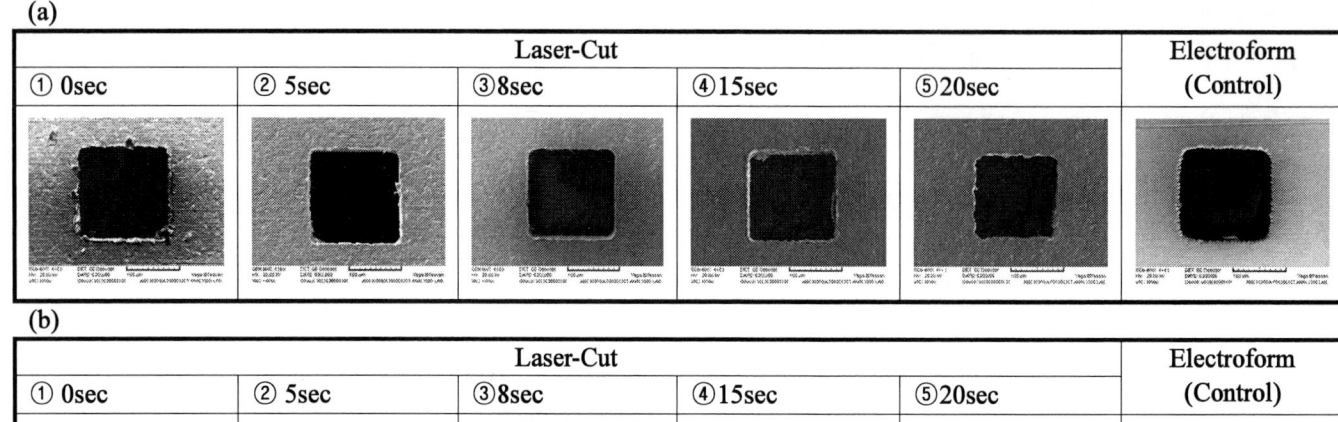

(b)

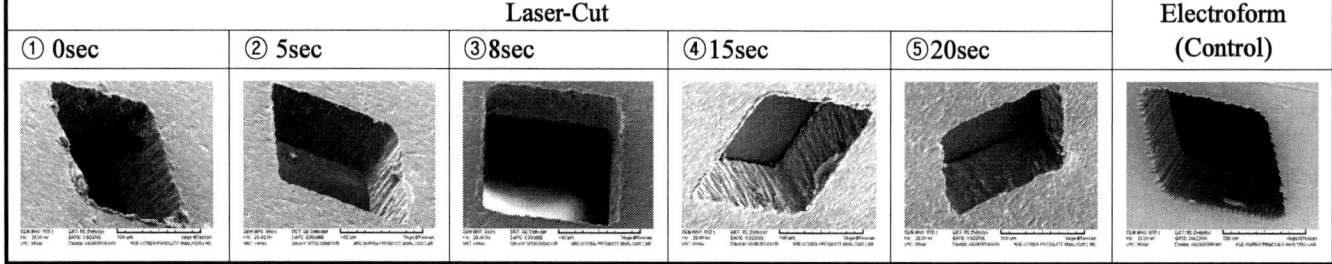

Figure 5. SEM images of laser-cut apertures with different electro-polishing times. (a) Top view at 410X, (b) Angled view at 410X.

However, aperture quality has a direct influence on paste release and because of this, electro-polishing process is necessary to achieve the smoother surface. In addition, the laser-cut aperture size must be adjusted to the electro-polishing employed since the aperture size will change during the processing. [7] For this study, we did not measure the each aperture size after electro-polishing processing at this moment.

Comparison between Different Stencil Fabrication Techniques

Firstly, it is important to evaluate a stencil printing performance for different fabrication techniques. As the most important part of the printing process is the solder volume delivered to the pads. Figure 6 show some representative photographs from solder print deposited for the 175um 01005 aperture arrays. Both stencils were used type-4 solder paste. From the figures, it was observed that there is no significant difference in print qualities between electroform and laser-cut stencils. Therefore, it was assumed that enhanced laser-cut stencils will behave similarly to electroform stencils.

The paste thickness for each stencil type graphically shown in Figure 7 and it was statistically analyzed using a paired t-test. The p-value of less than 0.05 indicates that the factor has statistically significant effect at the 95% confidence level. It shows the result that there is statistically significant difference on solder paste thickness between electroform and laser-cut stencils. From the graph, the values plotted are often called the stencil printed "release" values. Release is defined as the volume of the stencil printed deposit divided by the volume of the apertures. [8] The graph shows us that the enhanced laser-cut stencil demonstrates better paste release for the 01005 solder pads. The better performance of enhanced laser-cut stencils could be due to

improved surface quality of aperture wall by electro-polishing process.

(a) (b)

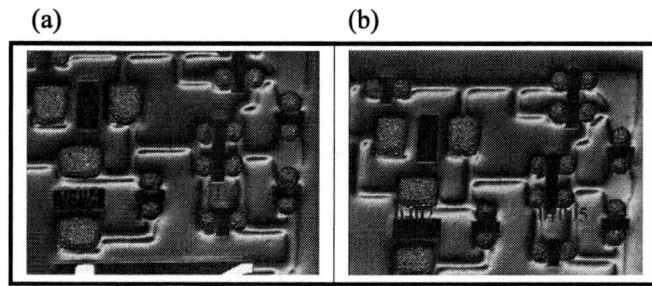

Figure 6. Solder deposition on 01005 and 0402 pads for each of stencil types. Type-4 paste and 3mil thick stencil. (a) Electroform , (b) Laser-cut

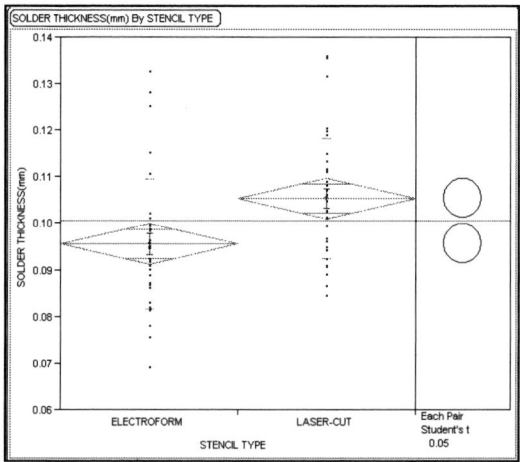

Figure 7. Comparison of electroform vs. enhanced laser-cut stencil. This result was statistically significant. (p-value = 0.0034)

Secondly, the repeat printing experiment was performed on 30 boards for each of the stencil. The objective of this experiment is to determine the print frequency and stencil cleaning interval for each stencil types. Figure 8 and Figure 9 shows paste thickness by repeat printing without wiping for the 0402, 0201, and 01005 components for each of the stencils. From the graphs, it is seen that there was no significant drop in repeat printing for both electroform and laser-cut stencils. However, the paste thickness and print stability for enhanced laser-cut stencil is better than electroform stencil. Visual inspection was performed on 7 boards of the 30 boards printed for each of the stencils. The 1st, 5th, 10th, 15th, 20th, 25th, and 30th print are used in the visual inspection test. There is no significant difference in visual inspection between electroform and enhanced laser-cut stencils.

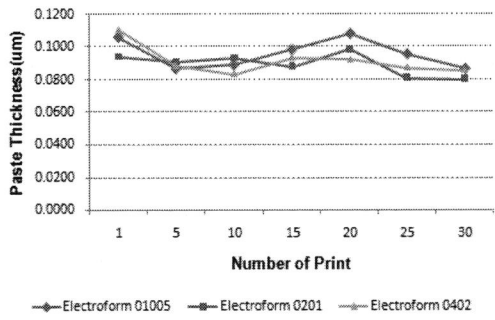

Figure 8. Effect of repeat printing without wiping for electroform stencil. Type-4 paste and 3mil thick stencil.

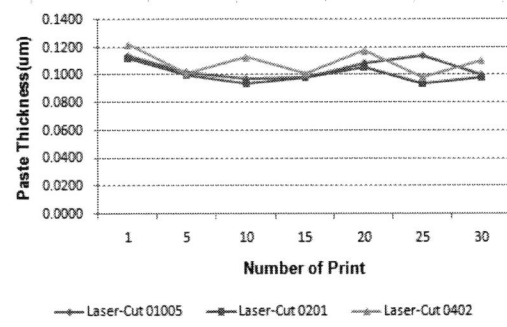

Figure 9. Effect of repeat printing without wiping for enhanced laser-cut stencil. Type-4 paste and 3mil thick stencil.

Effect of Solder Paste Powder Size

Figure 10-(a) shows some representative photographs from solder deposition on 01005 pads with different paste types for electroform stencil. Paste depositions on pads are significantly better for a type-4 paste compare to a type-3 pastes. When a type-3 paste was used, several insufficient solder pastes in pad deposited were observed and paste release was worse for both electroform and enhanced laser-cut stencils. (See Figure 10 (a) - ①) This is in agreement with some article published on 01005 solder printing studies. [2, 8] A type-4 paste has smaller particle size which will give better paste release performance, and will aid in depositing more solder volume in pads. Figure 10-(b) show the photographs from solder deposition on 01005 pads with

different paste types for enhanced laser-cut stencil. Similar print deposition results were observed as when electroform stencil was used. However, enhanced laser-cut stencil showed better paste coverage than the electroform stencil used for a type-4 paste.

(a)

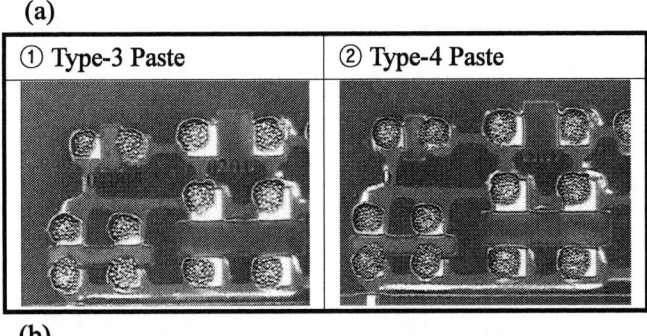

(b)

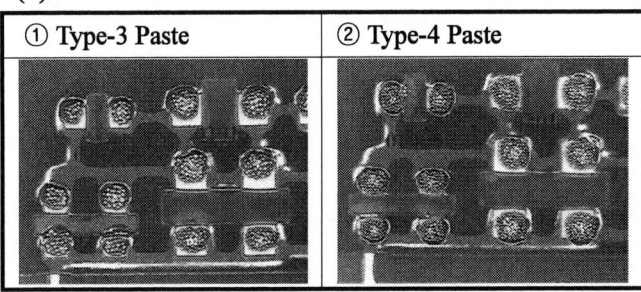

Figure 10. Solder deposition on 01005 and 0201pads with enhanced laser-cut stencil. (a) Electroform, (b) Laser-cut

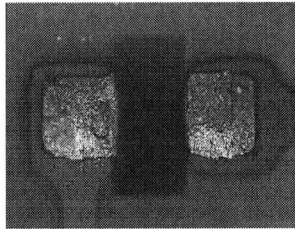

Figure 11. Optical image showing fracture area of 01005 solder joint.

After all boards had completed the reflow soldering, component shear test was performed for each of solder types. Ten components for each type (01005, 0201 and 0402 sizes) on each boards were sheared for each experimental trials. For fracture modes, most of the samples fractured at the solder joint as shown in Figure 11. The result of shear test is shown in Figure 12. A type-3 paste provided an average shear forces of 290gf and 255gf for electroform and enhanced laser-cut stencils respectively. It was a lower shear performance than type-4 paste which conform the notion that finer paste sizes are required for tiny 01005 components. It is clear from the print images that a type-4 paste gives a more consistent, higher volume print than type-3 paste. However, it is interesting to note that enhanced laser-cut stencil has lower shear force than electroform stencil, but highest paste thickness for a type-4 paste as Figure 7. From the graph, enhance laser-cut stencil and type-4 paste combinations had only average shear force of 239gf even though it had average relative paste thickness of 0.105mm. On the other hand the electroform stencil had average shear force of 319gf and average paste thickness of 0.095mm for 01005 components.

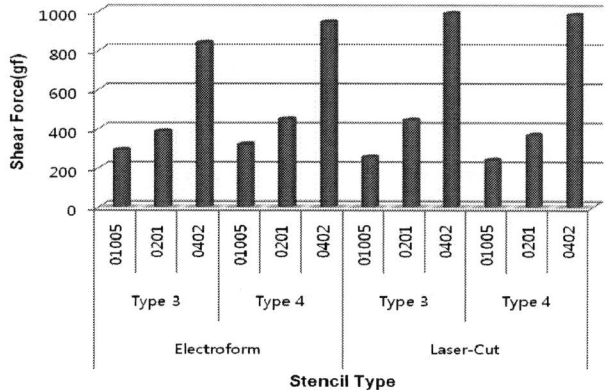

Figure 12. Shear force comparisons: Type-3 vs. Type-4 pastes and Electroform vs. Laser-cut stencils.

The visual inspection of the test vehicles was done manually using a low-power optical micro scope. Defects encountered among the varieties of components (01005, 0201 and 0402 sizes) were analyzed after the production run. The graph in Figure 13 contains process yield information. In general, a comparison between a type-3 and a type-4 pastes shows that type-4 paste have better yield performance for both electroform and laser-cut stencils. The graph note that the electroform stencil / type-4 paste combination's process yield (97.9%) is greater than the enhanced laser-cut stencil / type-4 paste combination's process yield (97.5%). In addition, 0201 and 0402 defects were not encountered during this experiment. For the defect counting, 5 defects including 1 tombstone and 4 misplacements were occurred at the electroform stencil / type 4 paste combinations, and 6 defects including 2 tombstones and 4 misplacements were also observed at the enhanced laser-cut stencil / type-4 paste combinations. It should be noted that only two defect modes were observed during the visual inspection. An example of these defects is presented in Figure 14. Tombstone is caused by differences in initial wetting of the solder between both joints of a component during the reflow soldering, and it is mainly due to unequal amount of printed paste on pads. [13] Preliminary test suggests that paste type (powder size) has an effect on yield and quality of the solder joints after reflow soldering. However, more data are needed to make definitive conclusion for shear force and yield differences between electroform and enhanced laser-cut stencils.

Conclusions

We emphasize that the results discussed in this paper are very preliminary, especially given the relatively small sample size. Due to the limited sample size, we did not attempt to compute defects in terms of parts per million (PPM), which is

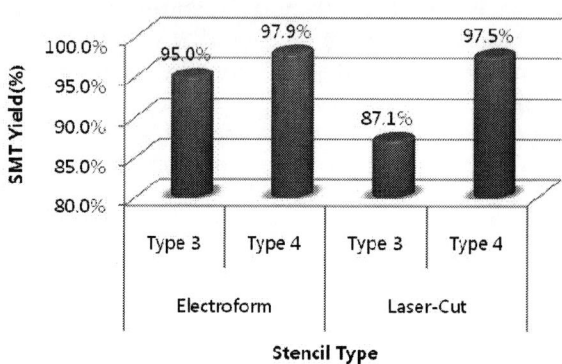

Figure 13.Yield comparisons: Type 3 vs. Type 4 pastes and Electroform vs. Laser-cut stencils.

(a)   (b)

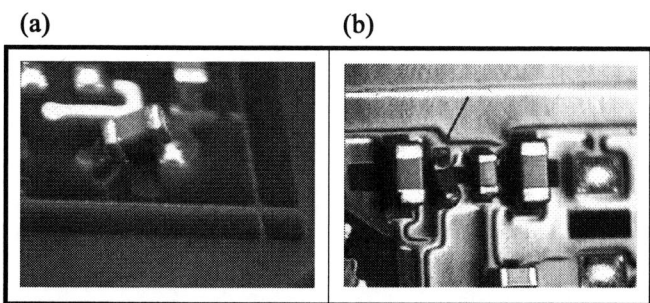

Figure 14. Images of SMT defects. (a) Tombstone, (b) Misplacement

the standard unit used in industry. Further study is needed and it is will be reported in a subsequent paper. The study has also proven that the process can be developed that allows 01005 stencil printing with acceptable yields. The results of the evaluation suggest that recommending the stencil printing technique, along with the electro-polishing post treatment is the significant factor associated with fine-pitch and small aperture such as 01005 components. Some of the critical inferences from experiment are summarized below.

When comparing the five different electro-polishing times for the aperture wall the time of 8sec provided best surface quality of any in the group and it was currently used in the enhanced laser-cut stencil fabrications for our 01005 high volume manufacturing (HVM).

Enhanced laser-cut stencil showed better printing performance and it is low-cost alternative to non standard process such as electroform stencil printing for 01005 small components.

Paste thickness for 01005 components is significantly better for the enhanced laser-cut stencil compared to electroform stencil.

Paste coverage for 01005 pads is better for the enhanced laser-cut stencil when a type-4 paste was used.

Paste depositions on pads are significantly better for a type-4 paste comparing to a type-3 pastes.

Highest shear force was achieved for electroform stencil when a type-4 paste was used.

These some advantages of enhanced laser-cut stencil with electro-polishing technique over the electroform stencil including excellent solder paste release, solder volume increased, durable long stencil life, good manufacturability, fast delivery time and more cost reduction opportunities.

Acknowledgements

All work was performed at ASE Korea Inc. The authors would like to thanks their colleagues and management team at the Packaging Engineering Team of ASE Korea Inc.

References

1. Paul N. Houston. *et al*, "Processing Strategies for High Speed 0201 Implementation", *IEMT2002, Proc 27th Annual IEEE/SEMI International*, San Jose, CA, June. 2002, pp. 166-172

2. Roy Jarvina. *et al*, "01005 SMT Component Assembly for Wireless SiP Modules", *Proc 50th Electronic Components and Technology Conf*, Orlando, FL, May. 2005, pp. 1502-1505

3. T. A. Nguty. *et al*, "Understating the Process Window for Printing Lead-Free Solder Pastes", *Proc 50th Electronic Components and Technology Conf*, Las Vegas, NV, May. 2000, pp. 1426-1434

4. W. Steplewski. *et al*, "Stencil Design for Lead-Free Reflow Process" *Proc 30th International Spring Seminar on Electronics technology*, Cluj-Napoca, Romania, 2007, pp. 330-334

5. Leila J. Ladani. *et al*, "Effect of Selected Process Parameters on Durability and Defects in Surface-Mount Assemblies for Portable Electronics", *IEEE Trans-CPMT*, Vol. 31, No. 1(2008), pp. 51-60

6. Joe Belmonte. *et al*, "Consideration in the Development of the 01005 Component Assembly Process", *Proc SMTA International Conf*, Chicago, IL, Sept. 2005

7. William E. Coleman, "Stencil Print Performance Studies", *Proc SMTA International Conf*, Chicago, IL, Sept. 2005, pp. 94-101

8. Vatsal Shah. *et al*, "Process Development for 01005 Lead-free Passive Assembly; Stencil Printing", *Proc APEX'07*, Los Angeles, CA, Feb. 2007

9. Fredrik Mattsson. *et al*, "Design and Assembly of 01005 Passives Using Pb-free Solder", *Circuit Assembly*, May. 2005

10. Michael R. Burgess. *et al*, "Electroformed vs. Laser-cut: A Stencil Performance Study", *SMT Magazine*, June. 2007

11. William E. Coleman. *et al*, "Choosing a Stencil", *SMT Magazine*, July. 2006

12. Satyanarayan Iyer. *et al*, "Implementing 0201s on High-Density Lead-Free Memory Modules", *IEEE Trans-CPMT*, Vol. 31, No. 1(2008), pp. 41-50

13. Harry Trip, "The persistent Problem of Tombstoning", *Circuit Assembly*, June. 2003, pp.20-22

Shear of Sn-3.8Ag-0.7Cu Solder Balls on Electrodeposited FeNi Layer

Q. S. Zhu[a], J. J. Guo[a], Z. G. Wang[a], Z. F. Zhang[a] and J. K. Shang[a,b]

[a]Shenyang National Laboratory for Materials Science, Institute of Metal Research, Chinese Academy of Sciences, Shenyang, 110016, China

[b]Department of Materials Science and Engineering,University of Illinois at Urbana-Champaign, Urbana, IL 61801, USA

Email: jkshang@imr.ac.cn and qszhu@imr.ac.cn

Abstract

The interfacial reaction between Sn-3.8Ag-0.7Cu solder and FeNi substrate was much slower, resulting in a thin $FeSn_2$ IMC layer at the interface. Interfacial bond between FeNi and Sn-3.8Ag-0.7Cu solder ball was examined by ball shear tests and the shear test results were compared to those on Cu and deposited Ni layers. Ball shear test showed that the Sn-Ag-Cu/FeNi-Cu interface had a comparable strength to that of Sn-Ag-Cu/Cu interface and the shear strength was relatively constant when reflow time increased from 2 minutes to over 15 minutes. The shear fracture occurred within the solder near the interface. Therefore, the electrodeposited FeNi layer may be used as a reliable under ball metallization (UBM).

1. Introduction

For the ball-grid-array (BGA) or flip-chip packaging, one of the major concerns is the reliability of the solder connection. This connection is made in a metallurgical way by the formation of intermetallic compounds (IMCs) between the solder and metallization layer. The related studies[1-3] have indicated that the interfacial intermetallic compounds (IMCs) have a significant effect on the reliability of solder joint. Cu films are widely used as soldering pad or under bump metallurgy (UBM) in solder interconnection for their excellent wetting property and electrical conductivity. When liquid solders react with Cu during reflowing process, the Cu film is consumed over time during reflowing process. This loss of Cu is even more serious for Sn-rich Pb-free solders such as the eutectic Sn-Ag-Cu alloy which requires a reflowing temperature 30-40°C higher than that of traditional Sn-Pb solder. To limit this loss, an electroless Ni-P layer is widely used as the barrier to protect the Cu film [4-7]. However, two major reliability issues arise with electroless Ni-P. One is the formation of the Kirkendall voids near the solder/Ni interface during aging process and the other is the development of a P-rich layer which is more brittle than the IMC and prone to brittle fracture [5-7].

In this paper, the reliability of solder bumps formed on FeNi layer was examined to determine the suitability of FeNi as a potential UBM for Sn-rich Pb-free solders. A layer of 58Fe-42Ni, 2-3μm thick, was electrodeposited on the Cu substrate. After reflowing, submicron meter thick $FeSn_2$ IMC layer with a thickness of 0.2 μm was formed on FeNi while the Cu_6Sn_5 IMC was about 6μm thick on Cu. The maximum shear force for the Sn-3.8Ag-0.7Cu/FeNi connection was comparable to that of the Sn-3.8Ag-0.7Cu/Cu connection after reflowing for different times. Therefore, FeNi layer may be used as a reliable UBM for Sn-rich Pb-free solders.

2. Experimental procedures

A copper sheet, $20 \times 2 \times 0.5mm^3$, was used as substrate. The Cu surfaces were mechanically ground and then carefully electro-polished. Then a layer of about 3 μm Fe-Ni film was electroplated on the Cu sheet. The surface morphology of the FeNi plating is shown in Fig.1.

Fig. 1 Surface morphology of Fe-Ni plating.

The atomic ratio of Fe and Ni elements in the plating was approximately 57:43. The electroplated Fe-Ni surface was carefully polished and then rinsed in water and alcohol. The solder balls of eutectic Sn-3.8Ag-0.7Cu used in this study had a size of 0.8mm. The solder balls, surrounded by solder mask, were placed on the prefluxed substrate. The reflow process was carried out on a BGA&CSP rework system, where the reflowing temperature was controlled at 270 °C. The specimens were reflowed for the different times. The finished specimen for shear test is illustrated in Fig.2. Ball shear tests were performed using a micromechanical testing system. The loading range for the test machine was 50N with a load control accuracy of better than 1% and a displacement resolution of 1 nm. The shear test was performed at a constant displacement control of 0.5mm/min. The microstructures and chemical compositions of the solder/substrate interface were observed with scanning electron microscopy (SEM; LEOsuper35) quipped with energy dispersive X-ray spectroscopy (EDX). After the shear testing, the fracture surfaces were investigated thoroughly by SEM and EDX.

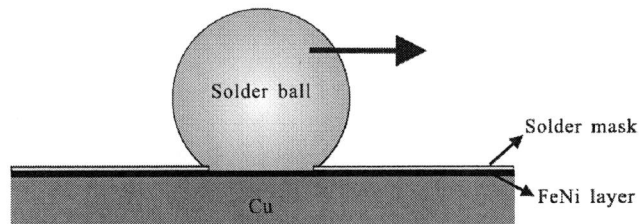

Fig. 2 schematic diagram of the specimen for shear test.

3. Results and discussion

Fig.3 shows cross-sectional SEM images of the interfaces between the Sn-3.8Ag-0.7Cu solder and the Cu, Ni-Cu and FeNi-Cu substrate after reflowing. For the Cu substrate, a scallop morphology IMC layer was formed along the interface, as shown in Fig. 3(a). EDX analysis indicated that the IMC layer was mainly Cu_6Sn_5 phase. The average thickness of the scallop-type Cu_6Sn_5 layer was about 6μm. It is evident that the IMC Cu_6Sn_5 layer at the interface between SnAgCu solder and copper grew rapidly during initial soldering. In the Sn-Ag-Cu/Ni-Cu interface, an IMC layer with irregular morphology was obvious after reflowing for 2 minutes, as shown in Fig.3(b). EDX analysis indicated that the IMC layer was approximtely $(Cu,Ni)_6Sn_5$ phase. The thickness of this IMC layer was above 2 μm. In general, the reaction between molten Sn-based solders without Cu element , such as Sn-Pb solder, and Ni(P) layer resulted in the formation of a Ni_3Sn_4 layer [8]. On the other hand, for the Sn-Ag-Cu solder, a $(Cu,Ni)_6Sn_5$ IMC layer formed at the interface[9]. During reflowing, the Cu atoms in the Sn-Ag-Cu solder moved to the solder/Ni(P) interface, leding to the formation of the Ni-containing Cu_6Sn_5, namely $(Cu,Ni)_6Sn_5$ IMC at the interface between the Ni substrate and the Sn-Ag-Cu solder with high contents ranging from 0.7 to 1 wt.% Cu atoms [10].

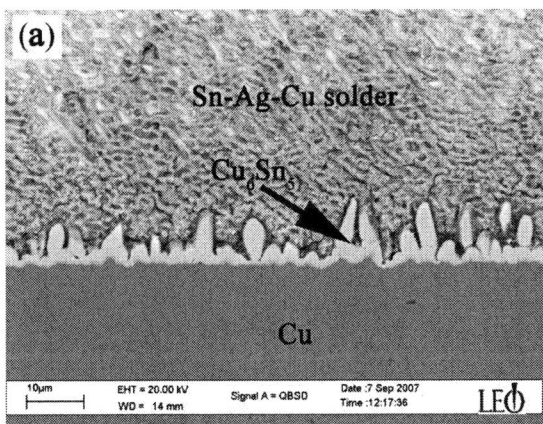

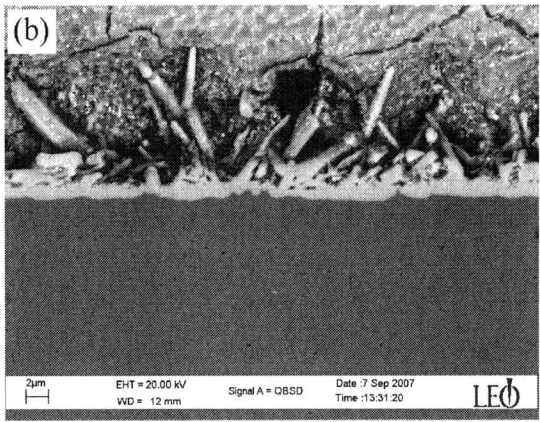

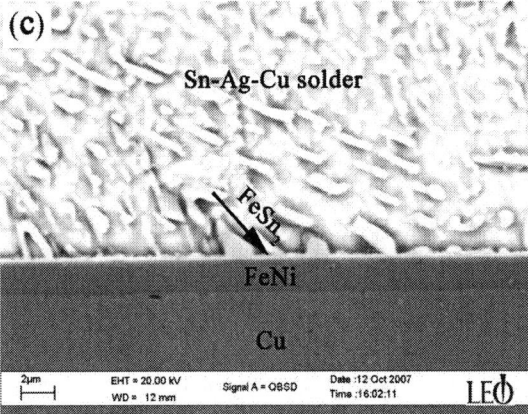

Fig.3 Interfacial IMC between SnAgCu solder and substrate after reflowing for 2 minutes: (a) Cu (b) Ni-Cu (c) FeNi-Cu.

For a comparison, when the liquid Sn-3.8Ag-0.7Cu reacted with the FeNi-Cu substrate, it was observed that a very thin IMC, submicrometer thickness, was formed between the solder and substrate, as indicated by the arrow in Fig.3(c). The average thickness of this plannar IMC layer was about 200 nm. Clearly, the growth rate of this kind of IMC was extremely low during reflow process. TEM analysis indicated that the IMC was $FeSn_2$ phase. Hwang et al. [11] pointed out that Fe reaches a saturated solubility in the liquid Sn quickly and could segregate and nucleate along the interface. On the other hand, the very low solubility of Fe within the liquid Sn would make diffusion become quite slow, leading to an extremely low growth rate of $FeSn_2$ IMC layer[12]. Consequently, compared to the electrodeposited Ni layer, the electrodeposited Fe-Ni layer is a more effective reaction barrier under high temperature reflowing condition.

The shear resistance of the interface are given in Fig.4. For SnAgCu/Ni-Cu connection, the shear force increased with the reflowing time. In contrast, the shear forces of the SnAgCu/FeNi-Cu and SnAgCu/Cu connections were relatively stable. After reflowing for less than 10 min, the shear force of the SnAgCu/FeNi-Cu connection was slightly higher than those of the SnAgCu/Cu and SnAgCu/Ni-Cu connections.

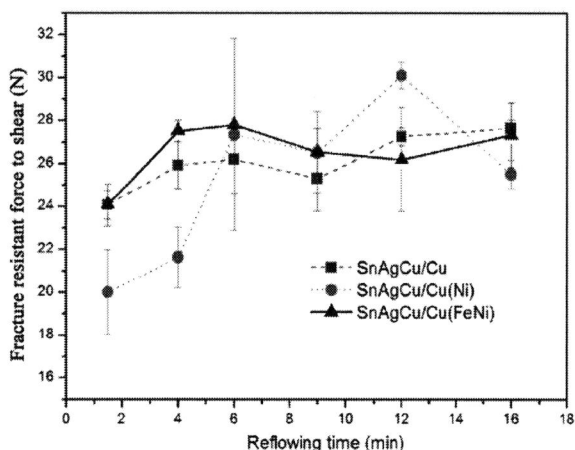

Fig.4 Dependence of the maximum resistant force to shear on the reflowing time for SnAgCu bump/Cu, Ni-Cu, and FeNi-Cu connection.

A careful observation of fracture surface can give a useful explanation for the shear force change. Fig. 5 presents the typical fracture surface images for the SnAgCu/FeNi-Cu connection, where the shear induced ductile dimples occupied most of the surface area. So it is clear that the fracture occurred within the solder near the interface and developed in a ductile manner. From the EDX analysis, the IMCs at the bottom of the ductile dimple were identified as $(Cu,Ni)_6Sn_5$ or Ag_3Sn phase. It is believed that Ni in the Fe-Ni layer dissolved into the liquid solder and combined with Sn and Cu to form the $(Cu,Ni)_6Sn_5$ IMC, while Fe reacted with the Sn to form the $FeSn_2$ IMC at the interface.

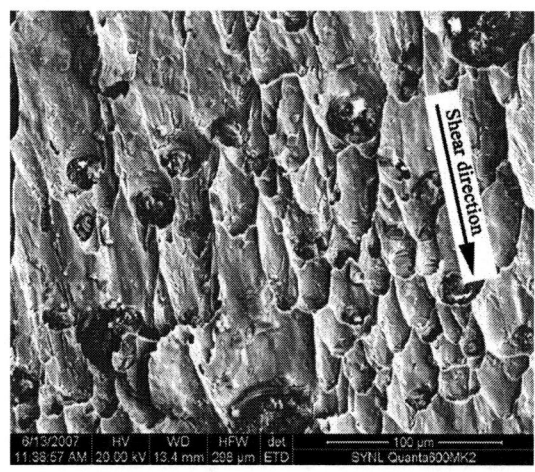

Fig.5 Typical ductile fracture morphology of SnAgCu bump/FeNi-Cu connection.

Similarly, the shear fracture for the SnAgCu/Ni-Cu and SnAgCu/Cu connections mainly occurred within the solder near the interface. Thus, the fracture resistant force to shear would have a strong dependence on the strength of the solder near the interface. At the SnAgCu/Ni-Cu interface, as shown in the Fig.3(b), it is observed that rod or needle-like IMC extended into the solder when the reflowing time was increased. Since the enhanced strength of the solder near the interface can be achieved by such a stiff IMC phase, the force of the SnAgCu/Ni-Cu connection would continually increase with the increasing reflow time. However, the shear force of the SnAgCu/FeNi-Cu connection had no evident dependence on the reflowing time because the interfacial microstructure was highly stable compared to that of the SnAgCu/Ni-Cu interface.

4. Conclusion

In this study, a comparative analysis of the interfacial reactions and shear property of SnAgCu/Cu, SnAgCu/Ni-Cu and SnAgCu/FeNi-Cu connection was conducted. After reflowing, a very thin $FeSn_2$ IMC layer was formed between the Sn-3.8Ag-0.7Cu solder and the FeNi-Cu substrate. The results of shear ball test indicated that the solder/FeNi-Cu connection had comparable mechanical property to those of solder/Cu and solder/Ni-Cu connections. Because of its excellent reaction barrier function and stable mechanical reliability, the electrodeposited FeNi layer appears to be a promising UBM in the electronic packaging.

Acknowledgments

The authors would like to thank H. X. Zhao, J. Y. Min, and H. F. Zou for solder ball preparation and SEM observation. This work was financially supported by National Basic Research Program of China under grant No. 2004CB6193.

References

1. Lee C. B., Jung S. B., Shin Y. E., et al, "Effect of isothermal aging on ball shear strength in BGA joints with Sn-3.5Ag-0.75Cu solder," *Materials Transactions*, Vol. 43, (2002), pp. 1858-1863.
2. Lee H. T., Chen M. H., "Influence of intermetallic compounds on the adhesive strength of solder joints," *Materials Science and Engineering a-Structural Materials Properties Microstructure and Processing*, Vol. 33, (2002), pp. 24-34.
3. Lee H. T., Chen M. H., Jao H. M., et al, "Influence of interfacial intermetallic compound on fracture behavior of solder joints," *Materials Science and Engineering a-Structural Materials Properties Microstructure and Processing*, Vol. 358, (2003), pp. 134-141.
4. Zeng K., Tu K.N., "Six cases of reliability study of Pb-free solder joints in electronic packaging technology," *Materials Science & Engineering Reports*, Vol. 38, (2002), pp. 55.
5. Alam M. O., Chan Y. C., Tu K. N., "Effect of reaction time and P content on mechanical strength of the interface formed between eutectic Sn-Ag solder and Au/electroless Ni(P)/Cu bond pad," *Journal of Applied Physics*, Vol. 94, No. 6 (2003), pp. 4108- 4115.

6. Alam M. O., Chan Y. C., Hung K. C., "Reliability study of the electroless Ni-P layer against solder alloy," *Microelectronics Reliability*, Vol. 42, No. 7 (2002), pp. 1065 - .1073.

7. Chonan Y., Komiyama T., Onuki J., Urao R., et al, "Influence of phosphorus concentration in electroless plated Ni-P alloy film on interfacial structures and strength between Sn-Ag-(-Cu) solder and plated Ni-P alloy film," *Materials Transactions*, Vol. 43, (2002), pp. 1840.

8. Jang J. W., Kim P. G., Tu K. N., et al, "Solder reaction-assisted crystallization of electroless Ni-P under bump metallization in low cost flip chip technology," *Journal of Applied Physics*, Vol. 85, No. 12 (1999), pp. 8456-8463.

9. Komiyama T., Chonan Y., Onuki J., et al, "The influence of phosphorus concentration of electroless plated Ni-P film on interfacial structures in the joints between Sn-Ag solder and Ni-P alloy film," *Materials Transactions,* Vol. 43, (2002), pp. 227-231.

10. Chen W. T., Ho C. E., Kao C. R., "Effect of Cu concentration on the interfacial reactions between Ni and Sn-Cu solders," *Journal of Materials Research,* Vol. 17, (2002), pp. 263-266.

11. Hwang C. W., Suganuma K., Lee J. G., et al, "Interface microstructure between Fe-42Ni alloy and pure Sn," *Journal of Materials Research*, Vol. 18, (2003), pp. 1202-1210.

12. Laurila T., Vuorinen V., Kivilahti J. K., "Interfacial reactions between lead-free solders and common base materials," *Materials Science & Engineering Reports*, Vol. 49, (2005), pp. 1- 60.

IMC Formation between Electroless Ni/Pd/Au Surface Finish and SnAgCu Solder

Jiwang Mao[1], Bin Liu[2], Ming Li[1], Yu Wang[2], Dali Mao[1]

1. Lab of Microelectronic Materials & Technology, Shanghai Jiao Tong University, Shanghai, 200240, China
2. Intel Technology Development (Shanghai), Ltd., Shanghai, 200131, China
Tel: 86-21-34202542, E-mail: mingli90@sjtu.edu.cn

Abstract

The interfacial reaction between OSP, electrolytic Ni/Au, and electroless Ni/Pd/Au surface finish and SnAgCu solder on BGA package were investigated. After multiple reflows, IMC morphology and composition is examined by SEM and EDX. Cu_6Sn_5 formed between LF35 and OSP finish, and Ag_3Sn were found disperse in the solder. Both $(Ni_x,Cu_{1-x})_3Sn_4$ and $(Cu_x,Ni_{1-x})_6Sn_5$ were observed on interface of electrolytic Ni/Au and electroless Ni/Pd/Au surface finish between SAC105.

1. Introduction

Surface finish technology on substrate is experiencing a fast evolution to meet the complexity of new package design, and new assembly technology, including smaller pads, finer pitch, and lighter package. Particularly, new challenges arose by the elimination of lead from electronic devices associate with the RoHs requirement. Surface finish must be not only lead-free itself, but more importantly, to be compatible with the lead-free solder. Thus searching alternative surface finish is on agenda.

There are several surface finishes in use on BGA package pad for lead-free solder as replacement of eutectic SnPb HASL (Hot air solder leveling). These include OSP (organic solderability preservatives); electrolytic Ni/Au; electroless Ni immersion Au, DIG (direct immersion gold). [1] They are widely used and have relatively good reliability record with SnPb solder. The question is how they can perform when soldered with lead-free solder at higher temperature. Each of them has advantages and disadvantages in terms of reliability or cost. DIG forms Cu-Sn IMC at the interface, however, assembly thermal stress can cause Cu immigrate to surface. OSP is easily applied and relatively inexpensive, but processing is complex while shelf time is rather short. In nickel based surface finish, nickel layer acts as a diffusion barrier between solder and substrate, inhibits Cu dissolution result in effective decrease of IMC thickness. Nevertheless, electroless Ni immersion Au finish may have phosphorous enrichment in the nickel layer which leads to a weak solder joint and can not apply in gold wire bond. Therefore, improvement method is studied by researches and industry.

Recently, electroless Ni/electroless Pd/ immersion gold is introduced as a promising alternative surface finish for several advantages. Electroless plating removes plating bars, making substrate design more flexible. The presence of Pd layer improves the solderability and obstructs the diffusion of nickel, leading to a robust solder joint with SnAgCu solder. [2] Meanwhile, Gold layer thickness is significant reduced. Furthermore, eletroless Ni/Pd/Au is a versatile finish which is suitable for soldering, aluminum wire bond as well as gold wire bond. However, the interfacial reaction between eletroless Ni/Pd/Au and lead-free solder is before well known and reliability data is insufficient. Thus, further study is required to understand the IMC formation between electroless Ni/Pd/Au and SnAgCu solder before wide application.

In this paper, three kinds of surface finish in use are applied on SnAgCu solder. LF35 and SAC105 solder are commonly used in industry. OSP and electrolytic Ni/Au finish are used for comparison with electroless Ni/Pd/Au. After ball attach, the BGA packages are subjected to multiple reflows, then IMC morphology and composition are studied.

2. Experimental Procedure

The package test vehicle was BGA package. There are three groups of packages with different solder material and substrate surface finish. Table 1 shows the solder composition and surface finish type of the three groups of samples. In this paper, shortening of the three types of substrate surface finish are used. They are OSP for organic solderability preservatives, NiAu for electrolytic Ni/Au, and NiPdAu for electroless Ni/ electroless Pd/ immersion Au.

Table1 Specimens used in the experiment

Test group	solder material	solder composition	substrate surface finish
1	LF35	Sn1.2Ag0.5Cu with Ni doped	organic solderability preservatives
2	SAC105	Sn1.0Ag0.5Cu	electrolytic Ni/Au
3	SAC105	Sn1.0Ag0.5Cu	electroless Ni/ electroless Pd/ immersion Au

The solder balls were mounted in N_2 atmosphere with peak temperature 250°C. The three groups of sample were subjected to 3 and 10 times of reflow in air. The peak temperature is 260°C, and the time above 217°C is 110 seconds.

The specimens were cold mounted by epoxy, cross-sectioned through a row of solder balls. They were polished by 320C, 600C, 800C, 1200C silicon carbide abrasive papers consequently. Fine polish use 1um diamond polish solution and final polish use 0.05 silica suspension. To observe the metallographic phases, the specimens were etched by 2% HCl, C_2H_5OH solution for 10s. To get top view image of IMC, solder balls on specimens after 10 times of reflow were removed, and residue solder were removed by acid.

The cross-sectional and top view morphology images were observed using scanning electron microscope combined with energy dispersive X-ray (SEM/EDX) analysis. SEM images and EDX results were obtained by FEI SIRION 200.

3. Results and Discussions

3.1 Electoless NiPdAu surface finish

Figure 1 shows the cross sectional view of electroless Ni/ electroless Pd/ immersion Au finish on copper pad after ball attach. The bright thin layer in this SEM image represents the Pd and Au layer. When Pd is introduced, the overall thickness of Pd and Au layer on Ni is much thinner than Au layer in electrolytic Ni/Au finish. This layer is even and no big voids were found.

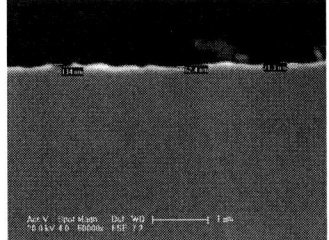

Figure1 electroless NiPdAu surface finish

3.2 IMC thickness after multiple reflows

Figure 2 shows the IMC thickness of different surface finish after multiple reflows. The thickness is defined by the average value of 10 separate measurements by SEM at 5000 magnification. After multiple reflows, the IMC of NiAu and NiPdAu finish is much thinner than OSP finish, which demonstrate that NiAu and NiPdAu finish acted as a diffusion barrier between solder and substrate effectively. After 10 times of reflow, the IMC thickness of NiPdAu finish show little growth compared with NiAu and OSP, it is supposed that the barrier effect of Pd layer will remains after high thermal stress.

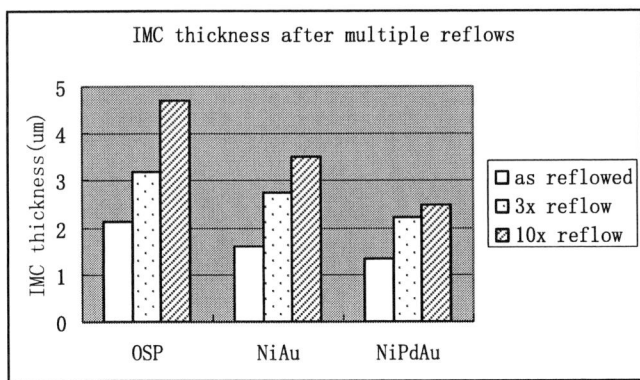

Figure 2 IMC thickness after multiple reflows

3.3 Interfacial morphology

Figure 3 show the IMC morphology of LF35 on OSP finish after multiple reflows. The scallop like IMC grows with the increase of reflow time. The composition of IMC is Cu_6Sn_5, which is reported in literature [3]. After 10 times of reflow , the IMC grow rather thick with big outshoots into the solder.

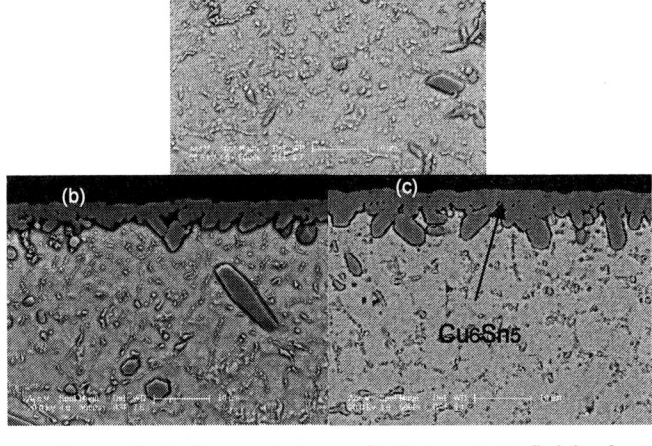

Figure 3   IMC morphology of LF35 on OSP finish after multiple reflows
(a)LF35/OSP as reflowed ; (b)LF35/OSP 3x reflow; (c)LF35/OSP 10x reflow

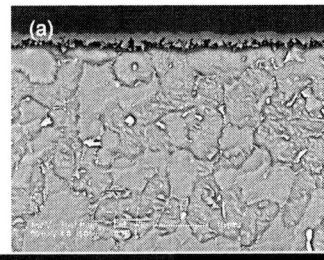

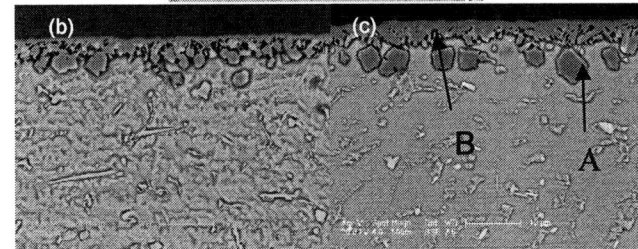

Figure 4 IMC morphology of SAC105 on NiAu finish after multiple reflows
(a) SAC105/NiAu as reflowed; (b) SAC105/NiAu 3x reflow; (c) SAC105/NiAu 10x reflow

Table 2 EDX result of Point A, B, C, D

Atomic %	A	B	C	D
Ni K	22.89	32.49	33.21	21.64
Cu K	29.78	8.91	13.53	31.58
Sn L	47.33	58.60	53.26	46.78

Figure 4 shows the IMC morphology of SAC105 solder and NiAu finish on copper pad. Different from OSP finish, the IMC present two features, the stick-like layer forms first, after 3 times of reflow, polyhedron particles form near IMC surface. After 10 times of reflow, these particles grow and can be found in bulk solder away from the IMC layer. The EDX result show that point A has more Cu content than point B, the composition of the particles is $(Cu_x, Ni_{1-x})_6Sn_5$. Ho. et al has report the same phenomena.[3,4]

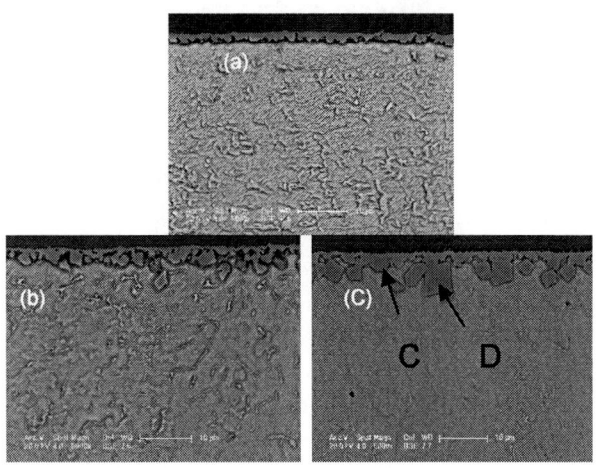

Figure 5 IMC morphology of SAC105 on NiPdAu finish after multiple reflows
(a)SAC105/NiPdAu as reflowed; (b)SAC105/NiPdAu 3x reflow; (c)SAC105/NiPdAu 10x reflow

Figure 5 shows the IMC morphology of SAC105 on NiPdAu finish after multiple reflows. After as reflowed ball attach process, the IMC formed between solder SAC105 and NiPdAu finish is thin and flat in contrast with OSP and NiAu finish. After 3 times of reflow, the interface became uneven, many outshoots occurs and seems to have tendency to come off from the IMC layer. After 10 time of reflow, the polyhedron outshoots grow larger, however the thickness of IMC layer shows little growth. Table2 shows the EDX result of point C and D. It was found that the composition of point C is close to point B while Point D is close to Point A. The IMC growth on NiAu and NiPdAu are similar. The IMC which formed far away Ni layer mores Cu content than IMC formed near Ni layer. It demonstrate that the Cu atom in solder will diffuse to IMC interface where ternary alloy Sn-Ni-Cu form. The phase with higher Cu content will form polyhedron particles and grow rapidly during reflow process.

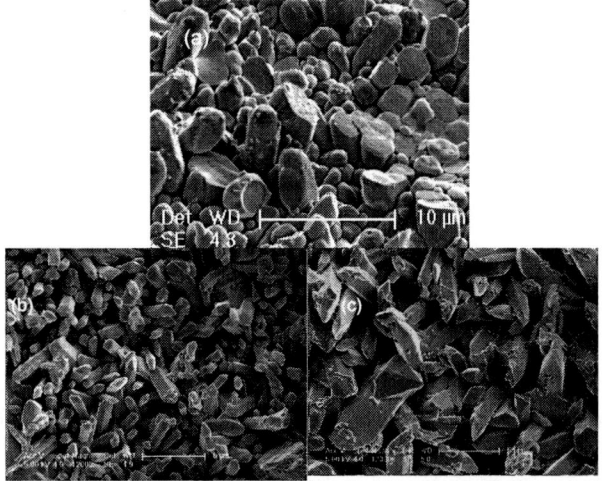

Figure 6 Top-view IMC morphology after 10 times of reflow (etched by acid to remove solder)
(a) OSP (b) NiAu (c) NiPdAu

Figure 6 illustrates the IMC morphology of the three groups of samples after 10 times reflow by top view observation. The Cu_6Sn_5 IMC on OSP finish shows scallop like surface which is the same as the cross-section image. The stick like surface in (b) resemble the B point in Figure 4 while (c) resemble C point in Figure 5. These points in IMC layer combined with less Cu content. The IMC microstructure in NiAu finish is finer and denser than NiPdAu finish.

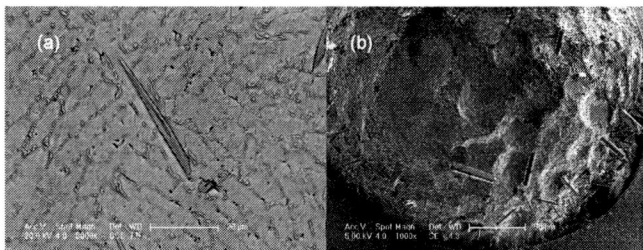

Figure 7(a)cross-section view of LF35 solder after 3 times reflow;(b) LF 35 solder ball etched by acid after 10 times reflow

Another phenomena found in LF35 solder was that after multiple reflows branch-like Ag_3Sn is abundant in solder. Figure 7(a) is a cross section view of Ag_3Sn and (b) is the surface of one solder ball etched by acid. Large amount of Ag_3Sn needles were revealed to the surface. Nano-indentation measurements by Dezhi Li et.al[6]indicated that Ag_3Sn have a higher hardness and brittle than the β-Sn matrix of the SnAgCu solder. The existence of large amount of Ag_3Sn in solder joint will decrease the solder joint reliability. [5, 6]

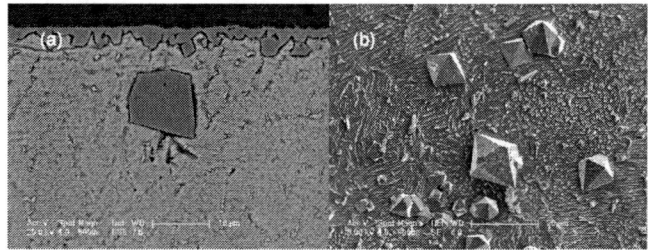

Figure 8 (a) cross- sectional view of NiPdAu 10x reflow; (b) top- view of solder near IMC interface.

Figure 8(b) illustrates the diamond-like dodecahedron forms in the SAC105 solder near the IMC interface. Figure 8(a) is a cross sectional view revealing the position of these particles. The composition of these particle are the same with Point A in Figure 4 and Point D in Figure 5,which has a high Cu content about 30%, while the overall Cu content in the solder material is only 0.5%. The composition is (Cu_x, Ni_{1-x})$_6Sn_5$, which is reported in literature[5,7]. However, the influence of these particles on reliability of solder joint needs further study.

4. Conclusions

After multiple reflows, the IMC between SAC105 and electrolytic Ni/Au and electroless Ni/Pd/Au finish are much thinner than the OSP finish. The composition of IMC on OSP finish is Cu_6Sn_5, which grows relatively fast during reflow. Considering reliability issue, reflow time should be limited for OSP finish. As for electrolytic Ni/Au and electroless Ni/Pd/Au finish, the IMC are demonstrated to have two phases with different morphology. $(Ni_x, Cu_{1-x})_3 Sn_4$ layer with around 10% Cu formed near Ni layer. $(Ni_x, Cu_{1-x})_6 Sn_5$ layer with around 30% Cu formed above and grow to larger size with increase of reflow time. Top view of these particles revealed that they have a regular dodecahedron shape, dispersing near the IMC interface within solder. However, the relationship between IMC and reliability of electroless Ni/Pd/Au surface finish on lead-free BGA package needs further study.

Acknowledgments

The author would like to thank Instrumental Analysis Center of Shanghai Jiaotong University for SEM/EDX analysis. The author would like to thank Yuxiang, Luo in Intel for helpful discussion.

References

1. Kiyotaka Tsukada. "Development of New Surface Finishing Technology for PKG Substrate with High Bondability," *Advanced Packaging Materials: Processes, Properties and Interfaces, 2005. Proc. International Symposium on* 2005, pp.110-114.

2. B. Kobe, N.S. McIntyre. "Investigation of reactions between lead/tin solder and palladium surface finishes,"*Acta Materialia* , Vol.50, (2002), pp. 4667 - 4676.

3. C.E Ho, Y.W.Lin, S.C. Yang, et al, "Effects of Limited Cu Supply on soldering reactions between SnAgCu and Ni". *J. Electron. Mater,* Vol.35, No.5 (2006), pp.1017-1024.

4. C.E Ho, et.al "Volume Effect on the soldering Reaction between SnAgCu and Ni". *Proc. of 10th Inernational Symposium and Advanced Packing Materials,* 2005, pp.39 - 44.

5. Kwang-Lung Lin, Po-Cheng Shih. "IMC formation on BGA package with Sn–Ag–Cu and Sn–Ag–Cu–Ni–Ge solder balls," *J. Alloys Compd,* Vol.452, No. 2 (2008), pp. 291-297.

6. Dezhi Li, et.al, "Characteristics of intermetallics and micromechanical properties during thermal ageing of Sn–Ag–Cu flip-chip solder interconnects" *Mater. Sci. Eng. A ,*Vol.391, (2005), pp. 95 - 103.

7. Pei Yao, Ping Liu, Jim Liu, "Effects of multiple reflows on intermetallic morphology and shear strength of SnAgCu–xNi composite solder joints on electrolytic Ni/Au metallized substrate," *J. Alloys Compd.* Vol. 462, No. 1 (2008), pp. 73 – 79.

Optimization of the Fatigue Life of Epoxy Molding Compounds based on BP Neural Network Prediction Model

CAI Miao, YANG Dao-guo, LI Quan-yong, ZHONG Li-jun
School of Mechanical Engineering
Guilin University of Electronic Technology, Guilin 541004, China, E-mail: caimiao104@qq.com

Abstract

Based on the data from the fatigue test of Epoxy Molding Compound (EMC), firstly with a focus on the application problem of the instability between fitting and prediction error of BP neural network (BPNN), the prediction model of fatigue life for EMC materials is established. In this approach, the network structure is improved with initiative way by reducing input from the perspective of nodes with principal component analysis (PCA). Secondly, in order to deal with the problem of the bottleneck in local flow minimum of BPNN, this study tries to find out the global minimum and improves the convergence performance of the BPNN combining genetic algorithm (GA). The stability and practicality of the GABPNN model is analyzed after training and verifying, and the effect of the input factors on the output factor is studied in turn. Finally, this paper makes use of well-trained GABPNN prediction model to analyze optimum design methods of parameters to predict the fatigue life. The prediction and optimization results show that the well-trained GABPNN model can be used in the forecasting and optimizing design of the fatigue and fracture reliability of the epoxy molding compounds, and is of much practical value.

Key words: optimization; BP neural network (BPNN); fatigue life; principal component analysis (PCA); genetic algorithms (GA); stability.

1. Introduction

When the electronic packaging devices are in service conditions, the packaging materials withstand cyclical temperature changes due to power switching off and on. Because of the thermal expansion mismatch between packaging materials, there is a cycle process of stress and strain in the packaging materials. The device eventually is leaded to failure because of high thermo-mechanical stress level or continuously accumulated inelastic strain. Therefore, it is very necessary to predict the fatigue life of polymer in microelectronics packaging device, and optimize the parameters through the prediction results, in order to improve the reliability of microelectronics packaging devices.

Currently, most of the traditional optimization methods are based on the linear theory (or assumed to be linear systems). However, in reality the implied relationship with regarding to the failure mechanism of the fatigue life of actual material is often a complex nonlinear function. The traditional methods show shortcomings. Therefore, there is an urgent need for a forecasting and optimal design method on the nonlinear problem.

It has been reported that a three-layer feed-forward network can achieve arbitrary precision mapping of the continuous function. Error back-propagation neural network is a multilayer fully-connected feed-forward neural network. At present, the BPNN has been widely used [1~9]. But there are still many aspects that can be improved in using with the BPNN model. Firstly, the key problem is generalization (forecast) capacity for the study of BPNN forecast model [10], and the key issue of the generalization capacity is whether the generalization results is stable. If generalization results are instable, the forecasting model will be no practical value. In the generalization study of BPNN model, most researches are to look for the smallest network structure under the condition by giving the learning samples [4, 10]. However, it is just a "passive" approach to improve the structure. Its essence is to find out the best generalization performance by finding the least number of hidden nodes. There are also many learning algorithms available to obtain appropriate network architecture [10]. However, these learning algorithms have their restrictions when they are used to improve generalization ability for a prediction model. The generalization performance of network model can be improved only when following requirements are meted: 1). A truly learning matrix reflecting the relation between input and output matrix is built. 2). The input dimension of learning matrix is lower. 3). The size of the network structure is smaller. Secondly, we can explore the sensitivity between the input and output factors using the implicit nonlinear relationship and generalization performance of a well-trained BPNN model, so as to realize a purpose of optimal design to the object variables. A preliminary extension to this method has been made recently by [8]. The approach is a brand-new method for optimization design and has large potential for application.

In this paper, based on the experimental data from the fatigue test of Epoxy Molding Compound, the prediction model of fatigue life for EMC materials is established, with a focus on the application problem of the instability between fitting and prediction error of BPNN. The network structure is improved with initiative way by reducing input from the perspective of nodes with PCA. Secondly, to deal with the bottleneck problem in local flow minimum of BPNN, this study tries to find out the global minimum and improves the convergence performance of the BPNN combining with GA. Thirdly the stability and practicality of the GABPNN model is analyzed after training and verifying. Finally, this paper, in turn, makes conversely use of the well-trained GABPNN model to analyze the sensitivity of the input factors on the output factor, and to improve the design methods of parameter optimization to the fatigue life with the prediction model.

978-1-4244-2739-0/08/$25.00 ©2008 IEEE

2. Experiments

Epoxy Molding Compounds are widely used as packaging materials in electronics industry. Fatigue cracking in EMC has frequently observed during temperature cycling test. In order to investigate the fatigue strength of EMC, experiments were conducted [11]. The specimens were made according to ASTM D638ⅣEMC standard, as shown in Fig. 1. Both tensile and fatigue tests were conducted using Instron dynamic material test equipment.

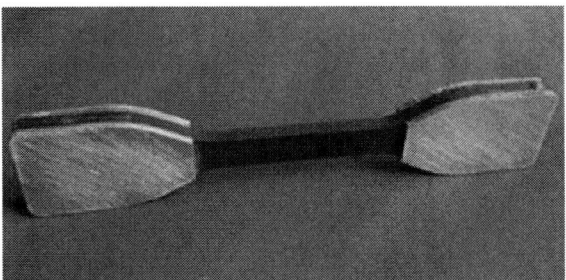

Fig. 1 EMC cake material specimen

Both static and fatigue tests were conducted. Through static tensile experiments, the stress-strain curves mapping the corresponding mechanical properties of the materials can be obtained. Through fatigue experiments, the necessary parameters of fatigue properties can be obtained. The mechanical behavior of EMC material is strongly temperature-dependent. Therefore, the experiments were carried out at different temperatures to investigate the temperature effect. Four temperature levels, i.e. 25°C, 75°C, 125°C , and 150°C, were used in this study. Table 1 lists 26 groups of experimental data, including the mechanical properties, environment parameters, and fatigue performance parameters. In the table, E is the elastic modulus (in MPa), σ_u is the tensile strength (in MPa), ε_u is the tensile fracture strain, q is the stress level, T is the temperature in degree C, and ε_f is the ultimate failure strain which was obtained by converting the fatigue life (N times). These parameters obtained from experiments will be used to train and verify for the BPNN prediction model.

3. Building the GABPNN model

3.1 Modeling theories

1）Back-propagation neural network

Error Back-propagation neural network (BPNN) has become the most extensive application of neural networks [12]. The training process of BPNN can be simply categorized as follows: firstly, weights ω_{ji} and threshold θ_j are initialized; then, training samples are entered into the net. And the error of each sample k can be obtained by calculating its output state. Based on the error, the next step is to adjust repeatedly the weights and thresholds among layers until the network error E1 <ε1 (ε1 is the setting error) with error back-propagation. After training, we can input the tested samples. If the network error E2 <ε2 (ε2 is the test error), it means that the network can be used for the actual forecast.

At present, for the BPNN model some shortcomings exist , such as poor stability between fitting and prediction, being difficult for determining network architecture, et al [4, 10, 13].

Table 1 List of the characteristic parameters of EMC material

NO.	Input					Output
	T	q	E	σ_u	ε_u	ε_f（%）
Factors	X1	X2	X3	X4	X5	Y
1	25	0.55	11710	77.36	0.010248	0.483
2	25	0.55	11710	77.36	0.010248	0.542
3	25	0.55	11710	77.36	0.010248	0.508
4	25	0.65	11710	77.36	0.010248	0.634
5	25	0.65	11710	77.36	0.010248	1.017
6	25	0.65	11710	77.36	0.010248	0.625
7	25	0.75	11710	77.36	0.010248	0.698
8	25	0.75	11710	77.36	0.010248	0.725
9	25	0.75	11710	77.36	0.010248	0.702
10	25	0.85	11710	77.36	0.010248	0.836
11	25	0.85	11710	77.36	0.010248	0.846
12	25	0.85	11710	77.36	0.010248	0.8
13	75	0.65	9691	61.64	0.011841	0.368
14	75	0.65	9691	61.64	0.011841	0.579
15	75	0.65	9691	61.64	0.011841	0.658
16	75	0.75	9691	61.64	0.011841	0.679
17	75	0.75	9691	61.64	0.011841	0.697
18	75	0.75	9691	61.64	0.011841	1.005
19	125	0.65	10490	47.26	0.01154	0.875
20	125	0.75	10490	47.26	0.01154	0.717
21	150	0.65	6854	32.53	0.02327	2.597
22	150	0.65	6854	32.53	0.02327	2.343
23	150	0.65	6854	32.53	0.02327	2.188
24	150	0.75	6854	32.53	0.02327	1.908
25	150	0.75	6854	32.53	0.02327	2.415
26	150	0.75	6854	32.53	0.02327	2.838

2）Principal component analysis [14]

The essence of Principal Component Analysis (PCA) is an exploratory statistical analysis method, which scatters a group of variables in the information on to a certain number of integrated indicators (Principal components). By doing so, principal components (PC) are to use data to describe the internal structure of the data. In fact, it plays a role of reduced-dimension to data.

The relationship of q PCs to p variables is described as following:

$$\left. \begin{aligned} f_1 &= a_{11}X_1 + a_{21}X_2 + \cdots\cdots + a_{p1}X_p \\ f_2 &= a_{12}X_1 + a_{22}X_2 + \cdots\cdots + a_{p2}X_p \\ &\cdots\cdots\cdots \\ f_q &= a_{1q}X_1 + a_{2q}X_2 + \cdots\cdots + a_{pq}X_p \end{aligned} \right\} \quad (1)$$

where $(a_{1i}, a_{2i}, \ldots , a_{pi})'$ are the eigenvectors of q former characteristic roots for the related matrix of variables respectively. The variances from f_1, f_2, ..., f_q are q characteristic roots ($\lambda_1 \geq \lambda_2 \geq ... \geq \lambda_q$) respectively. And (a_{i1}, a_{i2}, ..., a_{iq})' is the load of the ith variable on each PC. In fact, the load often means $(a_{i1}\sqrt{\lambda_1}, a_{i2}\sqrt{\lambda_2}, \cdots a_{q1}\sqrt{\lambda_1})'$, and it is the load of ith variable on each PC standardized. Then standardized PC's scores can be obtained accordingly using the least-squares method.

3）Genetic algorithms

Genetic algorithm (GA) simulates Darwin's natural selection, genetic selection and the process of biological evolution model. It is a search algorithm for global probability of the biological mechanisms based on a genetic variation and a natural selection. It is particularly applicable to solve the nonlinear problems that are complex and difficult for the traditional method [15].

The basic idea for optimizing BPNN with GA is: it is to find the most appropriate linking weight and network structure using the global search feature of GA. These years, many scholars over the world have established prediction models using optimizing neural network with the global search capability of GA [5, 6].

3.2 Modeling theories Establishment of GABPNN model

In this paper, the learning matrix is dealt with by employing PCA for the BPNN model as such to reduce the dimension and to de-noise. It constructs an initiative approach of low-dimensional learning matrix that reflects fully the relationship between prediction factors Xi and predictor Y. It also tries to improve generalization performance of BPNN to make it suitable for the need of predicting fatigue life of EMC.

Considering the coverage of the training samples and the representative of test samples for BPNN, the number of the data from Table 1, as 2, 5, 8, 12, 18, 21, 25, are extracted as the verification samples, and the remaining 19 groups are regarded as training samples. Additionally, to avoid the impact of middleweight among factors, all experimental data are normalized from 0.1 to 0.9 using the equation (2) before using them. And the prediction results can be recovered with the anti-normalization equation (3).

$$X_h = 0.8 * (x_h - x_{min})/(x_{max} - x_{min}) + 0.1 \qquad (2)$$

$$x_h = (X_h - 0.1)(x_{max} - x_{min})/0.8 + x_{min} \qquad (3)$$

where x_h and X_h are the values before and after the normalization, respectively; x_{min}, x_{max} are the minimum and maximum of certain factors prior to primary processing respectively.

The correlation coefficient matrix is shown in Table 2, which was analyzed with SPSS software. It can be seen that the factor X_2 is not related with other factors, but there exists some extent of overlap of information among the factors except X_2. If the original factors are directly input to the network, it is very likely to cause mutual interference among information, which will impact the accuracy and stability of the network prediction.

Table 2: Factor correlation coefficient matrix

	X1	X2	X3	X4	X5
X1	1	0	-0.934	-0.996	0.875
X2	0	1	0	0	0
X3	-0.934	0	1	0.955	-0.952
X4	-0.996	0	0.955	1	-0.914
X5	0.875	0	-0.952	-0.914	1

Through analysis, two PCs, i.e., f1, f2 can be obtained through transforming the five factors with PCA. Table 3 shows the scores corresponding to the various factors Xi.

Table 3: Principal component scores

Factor Xi	PC f1	PC f2
X1	-0.5	0
X2	0	1
X3	0.5	0
X4	0.51	0
X5	-0.49	0

According to Table 3, using equation (1), f1, f2 can be calculated as:

f1= -0.5*X1+0.5*X3+0.51*X4-0.49*X5; f2= X2；

As so an original five-dimensional input has turned into a two-dimensional input. The requirement of the de-noise is meeted.

Therefore, the structure of BPNN model is 2-K-1, where the hidden nodes K will be identified from comparing the effect on the prediction result with different implied nodes identified (see section 5.1). The BPNN topology is shown in Fig. 2. During the learning process of BPNN, the L-M method is used to train network.

By analyzing (see section 5.2), the local minimum problem of BPNN still survives after combining PCA to improve the input data. To solve this problem and to enhance the convergence speed of the network, the GA is employed to optimize the weights of the BPNN model based on PCA, to further enhance the generalization ability and stability of the BPNN model.

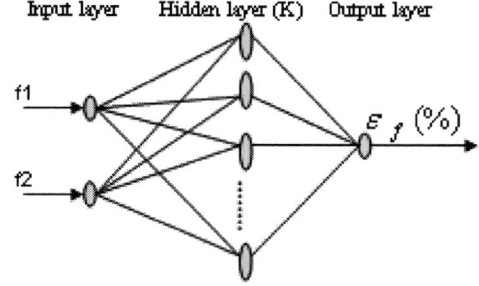

Fig. 2: BPNN topology

4. Approaches of prediction and optimization

After establishing the forecast model (either the traditional method or BPNN), the ultimate goal is to achieve the most reasonable or the best design based on the forecast results. It means to achieve the optimization design of the parameters. After establishing GABPNN prediction model and proving its stability and practicality, we explores the most optimal design of input factors with the implied nonlinear relationship and predictable performance established in the GABPNN model.

Fig. 3 shows the flowchart of analysis in this study. There are three different forecasting models showed as steps ②③④: ② is the simple BPNN model; ③ is the BPNN model combined with PCA; ④ is the GABPNN model after employing PCA. All work before Step ⑤ is prepared to build a stable and reliable prediction model for optimizing the input factors at the last step ⑤.

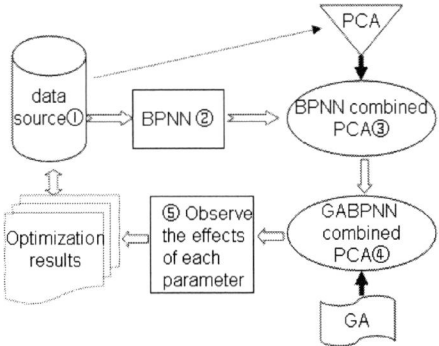

Fig. 3: Flowchart of prediction and optimization

In this work, the optimization method with BPNN model is as follows: by using the predictable performance of the GABPNN model, the change situation of the fatigue life can be observed with following the changing size of input factors among the changes scope of input factors in experimental data. That means it explores the sensitivity of the output factors when the input factors are changing, and seek the optimization way for the output and eventually realize the purpose of optimizing the fatigue life.

5. Results

5.1 BPNN structure

In order to choose a suitable hidden nodes K to help the simulation analysis at steps ③④⑤ in Fig. 3, some relationships (Fig.4), between which are learning fitting mean square error MSE_i and generalization error square SSE_i (i means the number of the hidden nodes), are drawn under different K value in the BPNN model combined with PCA. The relationships, in Fig.4, are drawn in four different initial values of the network for different network structures. Among them, the MSE of each situation are ranged in a descending order   according to MSE.

From Fig. 4, it can be seen that the prediction accuracy changes within a certain range. The small network structure is better for considering the faster convergence speed and better generalization performance of BPNN. As shown in Fig.4 (d) and (e), the prediction results of large network structure are easy to fluctuations although their fitting errors are small and relatively stable. However, it is very stable for the small ones (shown in Fig.4 (a), and (b)). Therefore， 4 (K) nodes for the structure should be chosen after comparing the results between Fig.4 (a) and Fig.4 (b). Finally the definite structure of BPNN model is chosen to be 2-4-1.

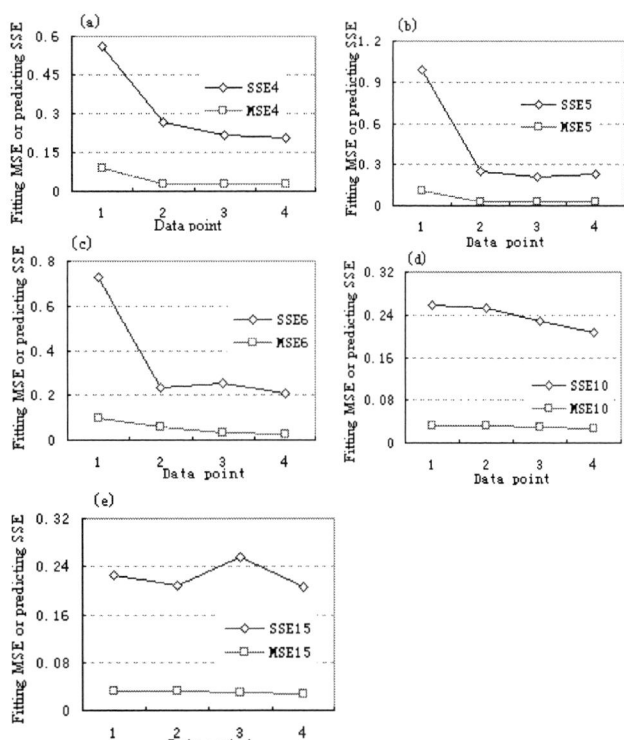

Fig. 4: Relationship between mean square error MSE fitted learning and the prediction error SSE for BPNN: (a) the results of network topology in 2-4-1; (b) the results of network topology in 2-5-1; (c) the results of network topology in   2-6-1; (d) the results of network topology in 2-10-1; (e) the results of network topology in 2-15-1.

5.2 Prediction results

Fig.5 and Fig.6 show the changes of fitness and error square SSE using GA to optimize the weights of BPNN, respectively. It can be seen that the adaptation of the network is basically stable when the generation is 80.

Fig.7 shows the change of MSE during training BPNN only combined with PCA. Behind 100 epochs of training, the training process is terminated due to the reason that the local minimum of the network leads to minimum gradient problems. At this time, the fitting error MSE is 0.0278.

Fig.8 shows the change of MSE during training to GABPNN combined with PCA. It is clearly indicated that the training speed is improved. The fitting error MSE is 0.0028 after 10 epochs. And the fitting accuracy is increased by nearly 10 times than no GA. Furthermore, the accuracy is very close to the requirement of training accuracy (0.001).

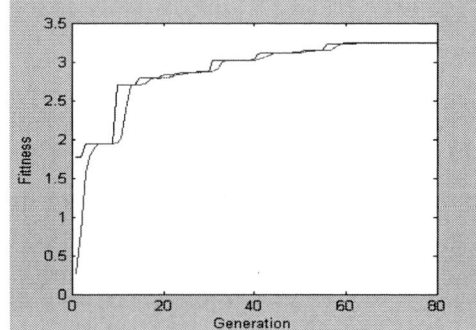

Fig.5: The change of fitness.

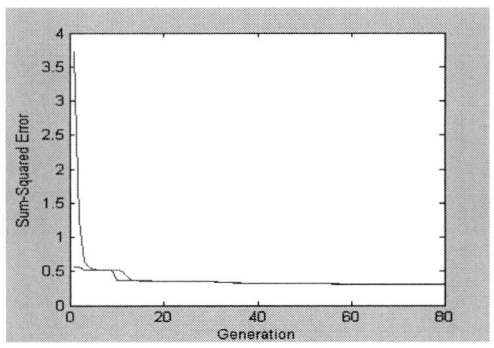

Fig.6: The change of SSE between fitness & settled value.

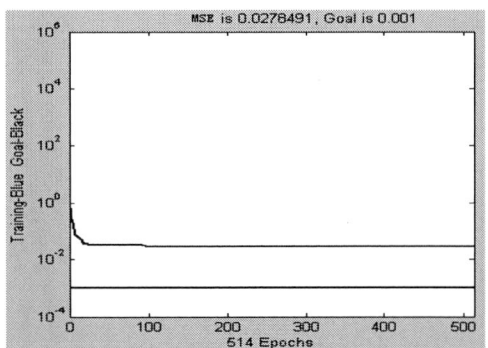

Fig.7: The change of MSE during training to BPNN combined PCA.

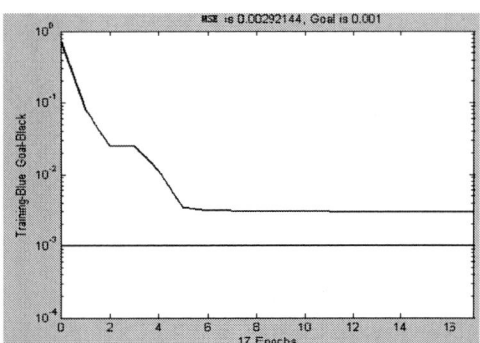

Fig.8: The change of MSE during training to GABPNN combined PCA.

During the analysis process for combining with GA, it shows that only a scope of the optimal solution of BPNN can be found, but not directly the optimal solution.

In order to show the forecasting stability and practicality of the general BPNN model, the BPNN model combined with PCA and the GABPNN model combined with PCA, the relations between the fitting error MSE and its corresponding prediction error SSE for the three different models are presented in Fig. 9 for comparison. As shown, 12 groups of data are given for the general BPNN model after simulatation, and 20 groups are given for the latter two models. The MSE of each situation are ranged in a descending order according to MSE.

Fig.9 shows that with the change of the training fitting error MSE in the general BPNN (not used PCA), the forecast error SSE fluctuates considerably and it has poor stability. So the output results can not be predicted with a new input. It has not practical value.

When PCA is employed into BPNN, the forecast error (SSE) changes in the same direction as that of changing of fitting error (MSE). So the instable problem between the MSE and its corresponding SSE has been solved in the way that avoids overlap interference of information and improves initiatively the BPNN structure. The result shows that the forecast result of BPNN model following this method is nearly close to the need as long as the fitting MSE meets the requirement.

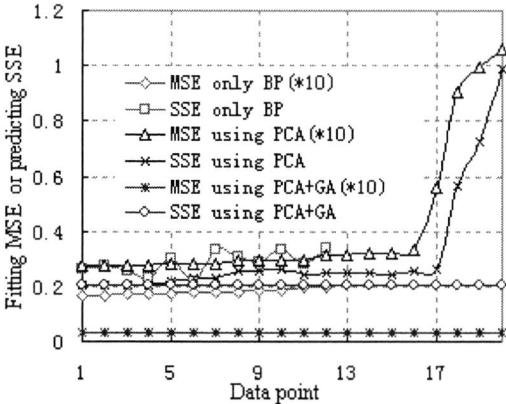

Fig. 9: Relationship between mean square error MSE fitted learning and the prediction error SSE.

Additionally, GA is added into BPNN based on PCA (see PCA+GA in Fig.9). On the one hand, the fitting error MSE of the GABPNN model is much smaller than the previous two cases. On the other hand, the prediction errors of well-trained GABPNN are same for different fitting MSE, and the forecast error SSE is better than that only with PCA. Therefore, it is sure that the GABPNN model based on PCA has a good stability and strong practicability.

In summary, the GABPNN model based on PCA is suitable for the prediction research on fracture reliability of EMC material. The study shows that the GABPNN model employed PCA has a good stability for the prediction results and practical value.

5.3 Optimization results

By using the predictable performance of the well-trained GABPNN, the change process of strain ε_f can be observed following the changing size of input factors (such as T, q, E, σ_u, ε_u) within the change scope of input factors in experimental data (Fig.10 ~ 13).

Fig.10 shows that ε_f is increasing, namely the fatigue life is constantly decreasing with increasing of the temperature and stress level. This prediction matches with the experimental results. It shows that the GABPNN model can effectively predict and picture the sensitivity of the output factor (fatigue life) when the input factors are changing.

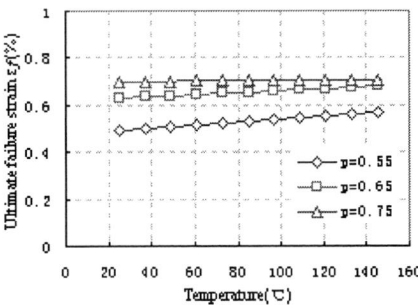

Fig.10: Effect of temperature on ultimate failure strain ε_f (%).

Fig.11 shows the change of the fatigue life at different stress levels and different temperatures. It can be seen that when E is within certain range (7500~8500MPa) the highest fatigue life of the EMC material can be obtained.

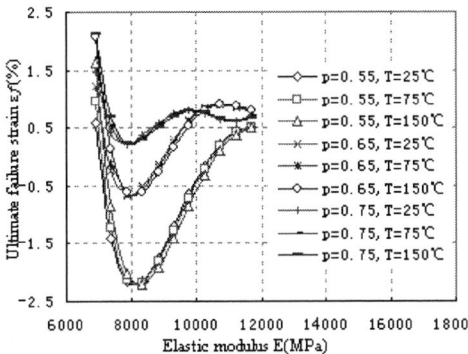

Fig.11: Effect of elastic modulus E on ultimate failure strain ε_f (%).

Fig.12 shows ε_f is slight decreasing with increasing of the tensile strength σ_u, namely the fatigue life is constantly increasing. And the effects of low stress level on fatigue life will be a bit greater by contrast of three levels of stress. But the difference is not big. For example, the influence of σ_u on the fatigue life can be ignored when $p = 0.75$.

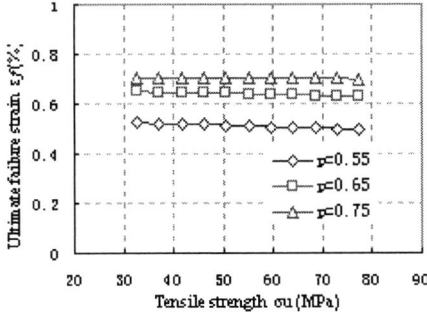

Fig.12: Effect of tensile strength σ_u on ultimate failure strain ε_f (%).

Fig.13 shows that the fatigue life is not affected by changing the tensile strain ε_u. This result not only can be applied to improve the BPNN input, but also is a useful reference in the choice of the packaging material.

Meanwhile, as the tendency of the increasing speed of ε_f by increasing the stress level in Fig.10~13, the speed are all decreasing. It means that the decreasing speed of the fatigue life is gradually become smaller. Therefore, we can expect that there is a certain stress level that will make the fatigue life of the materials reach at the fastest speed. It is also consistent with the actual situation.

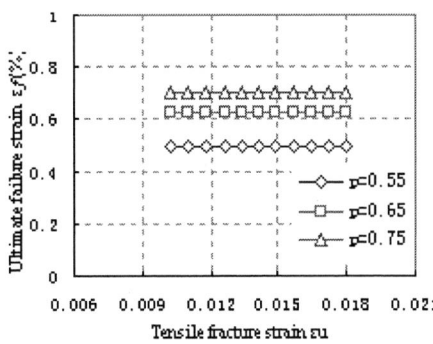

Fig.13: Effect of tensile fracture strain ε_u on ultimate failure strain ε_f (%).

5. Conclusions

In this paper, the BPNN structure is improved initiatively by employing PCA. Then the GA is added to improve BPNN. The forecast result shows that the GABPNN model based on PCA has a good stability and strong practicability.

After establishing GABPNN prediction model and proving its stability and practicality, the sensitivity of the output factors with changing the input factors is explored. A method for parameter optimization of the fatigue life with the model is proposed. The optimization result shows that the GABPNN prediction model can effectively picture the sensitivity of the output factor (fatigue life) with changing the input factors respectively. Based on this study, the optimizing purpose can be achieved by improving the corresponding parameters according to the need in the practical project. In short, the GABPNN model based on PCA is suitable for the prediction and optimization on fracture reliability of EMC material or other material, and has good practical value.

Acknowledgments

The research work in this paper is financially supported by the National Natural Science Foundation of China (NSFC) (Grant No. 60666002).

References

[1] HUANG Ruiyi, YANG Shaohua, LI KunLan,et al. Application of BP Neural Network on the Prediction of non- operation Reliability of Electronic [J]. Electronic Product Reliability and Environmental Testing，2005，5: 7-9.

[2] Yousef Al-Assaf, Hany El Kadi. Fatigue Life Prediction of Composite Materials using Polynomial Classifiers and Recurrent Neural Networks [J]. Composite Structures, 2007, 77: 561–569.

[3] J.A. Lee1, D.P. Almond, B. Harris. The Use of Neural Networks for the Prediction of Fatigue Lives of Composite Materials [J]. Composites: Part A, 1999, 30: 1159–1169.

[4] Jin Long, Kuang Xueyuan,Wang Haihong,et al. Study on the Over Fitting of the Artificial Neural Network Forecasting Model[J]. Acta Meteorologica Sinica, 2004, 62(1): 63–70.

[5] S.H. Mousavi Anijdan, H.R.Madaah-Hosseini,A. Bahrami. Flow stress Optimization for 304 Stainless Steel under Cold and Warm Compression by Artificial

Neural Network and Genetic Algorithm [J]. Materials and Design, 28 (2007) 609–615.

[6] T.S. Li, C.T. Su, T.L.Chiangc. Applying Robust Multi-response Quality Engineering for Parameter Selection using a Novel Neural–genetic Algorithm [J]. Computers in Industry，50 (2003) 113–122.

[7] Xu Liujie，Xing Jiandong, Wei Shizhong, et al.. Optimisation of Chemical Composition of High Speed steel with High Vanadium Content for Abrasive Wear using an Artificial Neural Network [J]. Materials and Design，28 (2007) 1031–1037.

[8] ZHANG Junhong, XIE Anguo, SHEN Fengman, Multi-Objective Optimization and Analysis Model of Sintering Process Based on BP Neural Network [J]. Journal of Iron and Steel Research, International，2007, 14(2): 01-05.

[9] Yaomin Lin, Member, IEEE and Frank G. Shi, Senior Member. Package Design and Materials Selection Optimization for Overmolded Flip Chip Packaging [J]. IEEE, 2006,525-532.

[10] WU Yan,ZHANG Liming, A Survey of Research Work on Neural Network Generalization and Structure Optimization Algorithms [J]. Application Research Computers，2002，19（6: 21-25.

[11] HAO Xiuyun.The Fatigue Failure Analysis on the Microelectronic Packaging Polymer [D]. Guilin University of Electronic Technology，2004. 6.

[12] Shuang Cong. MATLAB Toolbox for the Theory and Application of the Neural Network [M]. HeFei: University of Science and Technology Press China，2003: 45-60.

[13] Hany El Kadi. Modeling the Mechanical Behavior of Fiber-reinforced Polymeric Composite Materials Using Artificial Neural Networks [J]. Composite Structures，2006，73: 1–23.

[14] Xiulin Yu, REN xuesong. Multivariate Statistical Analysis [M]. Beijing: China Statistics Press, 1999.8: P154

[15] Xiaoping Wang, Li-Ming Cao. Genetic Algorithm Theory, Application and Software [M]. XI'AN: Xi'an Jiaotong University Press, 2002.1:18-65.

Electromigration Behavior of the Ni/SnZn/Cu Solder Interconnect

X.F.Zhang[1], J.D.Guo[1], J.K.Shang[1,2]

[1]Shenyang National Laboratory for Materials Science
Institute of Metal Research, Chinese Academy of Sciences, Shenyang 110016, China
[2]Department of Materials Science and Engineering
University of Illinois at Urbana-Champaign, Urbana, IL, 61801, USA
*Corresponding author: jdguo@imr.ac.cn

Abstract

Electromigration in the Ni/SnZn/Cu solder interconnect was studied with an average current density of $4.1 \times 10^4 A/cm^2$ for 168.5h at 150°C. When the electrons flowed from the Ni side to the Cu side, uniform layers of Ni_5Zn_{21} and Cu_5Zn_8 were formed at the Ni/SnZn and Cu/SnZn interfaces. The results are similar to those without passage of an electric current. However, upon reversing the current direction where electron flow was from the Cu side to the Ni side, thicker Cu_6Sn_5 phase replaced Ni_5Zn_{21} phase at the Ni/SnZn interface, whereas at the Cu/SnZn interface, thicker β-CuZn phase replaced Cu_5Zn_8 phase. Meanwhile, Cu-Sn phases also appeared at the Cu/SnZn interface. A kinetic model, based on the Zn and Cu mass transport in the sample, was presented to explain the growth of the intermetallic compound at the anode and cathode.

Introduction

The Ni/Solder/Cu material sequence is a very common material combination in the solder joints of flip chip and ball-grid array (BGA) package. Usually, Cu is used as under bump metallization (UBM) and Ni/Au as a substrate metallization layer in a flip chip package, as shown in Figure 1. During the reflow process, the Au layer is quickly dissolved into the molten solder, leaving the Cu and Ni layers directly exposed to the solder. Thus interfacial reactions in the Ni/Solder/Cu combination are crucial to the reliability of electronic products. Under pure thermal loading, the interfacial reactions are greatly accelerated by the formation of $(Cu,Ni)_6Sn_5$ IMCs [1-3].

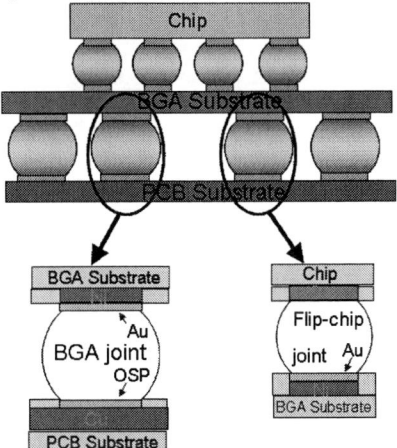

FIG.1 Schematic drawing showing the Ni/Solder/Cu material combination in the solder joints of flip chip and BGA packages

Current-crowding-induced electromigration failure occurs at the Cu side with electrons flowing from the Cu side to the

Ni side, but the joint with electrons flowing from the Ni side to the Cu side survives the current stressing [4-6]. Those results suggest that the failure mechanisms of such joints are highly dependent on current direction. However, there is no report yet on the relationship between IMCs at the interfaces and current direction for Ni/Sn-9Zn/Cu combination.

In this study, Ni/Sn-9Zn/Cu sandwich-structure samples were prepared to avoid asymmetrical Joule heating distribution resulting from current crowding. The coupling effect between SnZn/Ni and SnZn/Cu interfacial reactions under current stressing was investigated and the results showed remarkable differences in the interfacial reaction products when current direction was reversed.

Experimental Procedure

A Ni/Sn-9Zn/Cu sandwich structure was prepared, from the eutectic Sn-9wt.%Zn alloy. Prior to soldering, two copper cubes, 10mm wide, 10mm thick, 10mm long, were ground and carefully polished to form smooth surfaces. Then one of the cleaned substrates was immediately electroless plated with Ni-P. In the as-plated state, the phosphorus content in the electroless Ni-P deposit was about 10at.%, as determined by energy dispersive spectroscopy (EDS). Then the Cu/Ni(P) and Cu cubes were aligned and fixed before the assembly was heated in an oven where the solder was reflowed at 250°C for 90 seconds. After cutting and polishing, the Cu/Ni/SnZn/Cu Blech-structure specimens were finished to be 220 $\mu m \times$ 220 μm in the cross section and 200 μm-length of solder.

Electromigration testing of the solder combination was conducted by applying direct current to both ends of the bar samples using copper wires as the electric leads. The solder joints temperature was monitored by a k-type thermocouple placed next to the solder interconnect. The current density in the samples was $4.1 \times 10^4 A/cm^2$ and the temperature of the samples was about 150±5°C. The time under current stressing was 168.5h. To prevent oxidation, the samples were placed in an N_2 protective atmosphere. After current stressing, the interfaces between the solders and substrates were examined by scanning electron microscopy (SEM). The composition of the precipitates and phases between the solders and substrates was analyzed with EDS. In comparison with the electromigration effect, interfacial reactions in the Ni/SnZn/Cu solder combination, under isothermal aging at 150°C, were also investigated.

Results and Discussion

The as-reflowed interfacial microstructures are shown in Figure 2. On the Cu side consisted of a 3.0-μm-thick, γ-Cu_5Zn_8 compound (63Zn-37Cu in at.% by EDS analysis)

978-1-4244-2739-0/08/$25.00 ©2008 IEEE   863

adhered to the Cu pad and a thinner (0.7- μm in thickness), ε -CuZn$_5$ (83Zn-17Cu in at.%). At the SnZn/Ni interface, a thin Ni$_5$Zn$_{21}$ phase [7] was observed.

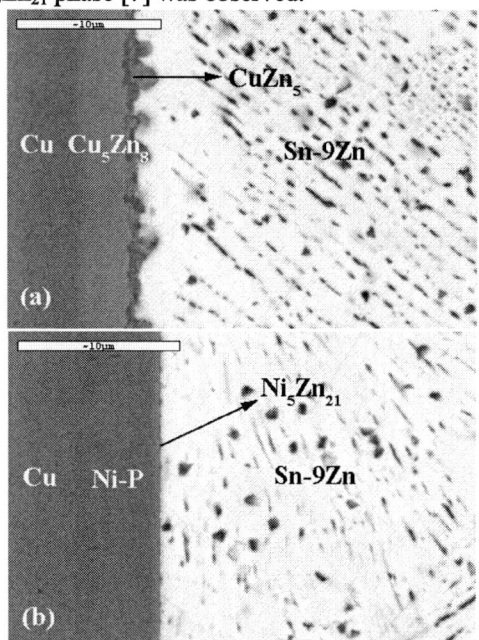

FIG.2 SEM images of the as-reflowed interface microstructure, (a) at the SnZn/Cu interface (b) at the SnZn/Ni interface

Solid state aging at 150°C for 168.5h resulted in growth of the planar Cu$_5$Zn$_8$ phase to approximately 6.1 μm in thickness and disappearance of the Zn-rich ε -CuZn$_5$ IMC at the SnZn/Cu interface, as the ε -CuZn$_5$ may have either coalesced and transformed into Cu$_5$Zn$_8$ compound upon aging [8] or dissolved into the solder. At the SnZn/Ni interface, the Ni$_5$Zn$_{21}$ phase also thickened, but no P-rich layer was observed at the interface. Despite the difference in the Cu or Ni concentration across the Ni and the Cu pads, no coupling effect developed at the interfaces.

Figures 3a shows the observations at the Cu/SnZn (anode) side under current stressing for168.5h with electrons flowing from the Ni side to the Cu side. The Cu$_5$Zn$_8$ IMC layer grew to about 5.7 μm thick, slightly thinner than that found in the aged sample. As with aging, the Zn-rich ε - CuZn$_5$ IMC was no longer found at the interfaces. On the cathode-side (Figure 3b), the Ni$_5$Zn$_{21}$ phase grew to approximately 360nm, slightly thicker than that of the aged condition. Again, no P-rich layer was observed at the interface. Therefore, when the current was directed from the Cu side to the Ni side, interfacial IMCs went through similar evolution to that of thermally aged sample.

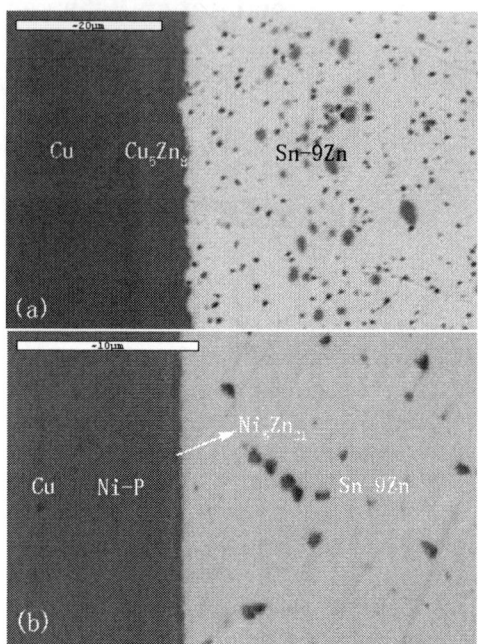

FIG. 3 Interfacial reactions when electrons flow from the Ni side to the Cu side, (a) at the SnZn/Cu interface (anodic) (b) at the SnZn/Ni interface (cathodic)

When the direction of electron flow was reversed, i.e., electrons flowing from the Cu side to the Ni side, the chemistry at the interfaces was completely different. As shown in Figure 4a, at the cathode (the Cu/SnZn interface), a 20.7μm thick β-Cu(Sn$_{1-x}$,Zn$_x$) (about 56Cu-34Zn-10Sn in at.%) layer with a small amount of Sn dissolved in the Zn sublattice was found next to the solder, replacing the Cu$_5$Zn$_8$ IMC layer. A 3.3μm thick Cu$_3$(Sn$_{1-x}$Zn$_x$) (about 73Cu-15Zn-12Sn in at.%) was adhered to the Cu pads. Between the Cu(Sn$_{1-x}$,Zn$_x$) layer and the Cu$_3$(Sn$_{1-x}$Zn$_x$) layer, a 6.5μm thick Cu$_6$(Sn$_{1-x}$Zn$_x$)$_5$ (about 56Cu-12Zn-32Sn in at.%) was formed.

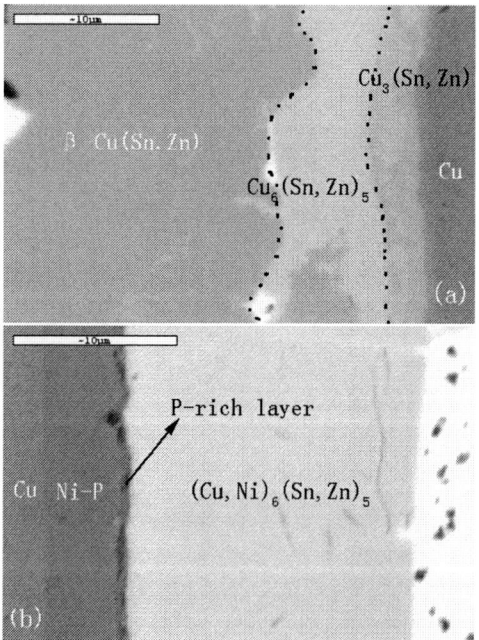

FIG. 4 Interfacial reactions when electrons flow from the Ni side to the Cu side, (a) at the SnZn/Cu side (cathodic) (b) at the SnZn/Ni side (anodic)

Anode-side observation for Ni/SnZn interface, after current stressing for 168.5h, is shown in Figure 4b. A η-(Cu,Ni)$_6$(Sn,Zn)$_5$ phase (about 49Cu-6Ni-37Sn-8Zn in at.%) replaced Ni$_5$Zn$_{21}$ phase and the Cu$_6$Sn$_5$ layer was very thick, approximately 18μm. In addition, a P-rich layer (0.6-μm in thickness) was observed at the interface, which means the Ni consumption was greatly increased under such condition.

From these observations, it is evident that interface reactions were completely different when the current direction was reversed. Such phenomenon has not been reported previously.

The growth of the interfacial region depends upon the mass fluxes flowing in and out. If the inward atomic fluxes are larger than those flowing out, the IMC grows; otherwise, it shrinks. The Zn flux in the sample is the major contributor to the growth of Zn/Cu IMC, since Zn is the dominant diffusing species for the interfacial reaction in Zn/Cu system [10]. In the same way, Cu flux is the major contributor to the growth of Sn/Cu IMC [11]. For Ni/Sn-9Zn/Cu solder combination under current stressing, the IMC growths at the anode and at the cathode are not independent but are related through Zn or Cu fluxes in the solder joint.

In the Sn-9Zn solder, the apparent effective charge of Sn is equal to -18 and that of Zn is -2.5 $\pm$ 0.2 for the electromigration along the c-axis [12], the electromigration driving force of Sn is much larger than that of Zn. So, the back stress induced by electromigration in the solder interconnect is larger than the electromigration driving force of Zn and considered to promote a reversed Zn migration to the cathode side [13]. Therefore, at the anode, the thickness of Cu$_5$Zn$_8$ IMC layer should be a little thinner than that found under annealing. However, at the cathode, the thickness of Ni$_5$Zn$_{21}$ layer should be a little thicker than that found under annealing. This is consistent with the experimental results.

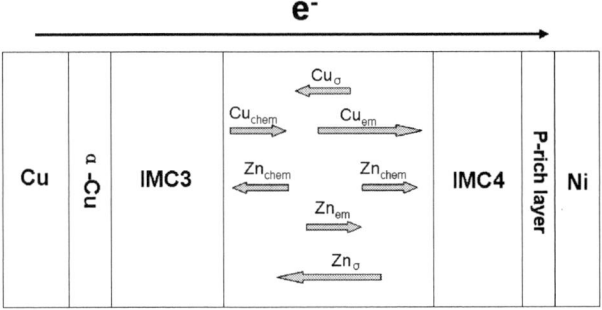

FIG.5 Components of Zn and Cu fluxes in a sample undergoing electromigration when electrons flow from the Cu side to the Ni side

When electrons flow from the Cu side to the Ni side, the atomic fluxes in the sample are shown in Figure 5. Under the combined effect of electron wind force and back stress, the Zn and Cu fluxes in the sample were the major contributor to the growth of intermetallic compounds at the interfaces. The Zn and Cu atomic drift flux is expressed as

$$J'_{cathode/Cu} = J^{Zn} + J^{Cu}$$
$$= (J^{Zn}_{chem'} + J^{Zn}_{em'} - J^{Zn}_{\sigma'}) + (J^{Cu}_{\sigma'} - J^{Cu}_{chem'} - J^{Cu}_{em'})$$

$$J'_{anode/Ni} = J^{Zn} + J^{Cu}$$
$$= (J^{Zn}_{chem'} + J^{Zn}_{em'} - J^{Zn}_{\sigma'}) + (J^{Cu}_{chem'} + J^{Cu}_{em'} - J^{Cu}_{\sigma'}),$$

where $J'_{cathode/Cu}$, the Zn and Cu fluxes flowing to the cathode (the Cu side), $J'_{anode/Ni}$ is the Zn and Cu fluxed flowing to the anode (the Ni side), $J^{Zn}_{chem'}$ and $J^{Cu}_{chem'}$, the Zn and Cu fluxes due to chemical potential gradient, $J^{Zn}_{em'}$ and $J^{Cu}_{em'}$, the Zn and Cu fluxes due to electron wind force , $J^{Zn}_{\sigma'}$ and $J^{Cu}_{\sigma'}$, the Zn and Cu fluxes due to back stress.

As mentioned above, the Zn atomic flux would migrate to the cathode, but the Cu atoms flux would be greatly enhanced and migrate to the anode under the effect of chemical potential gradient and electron wind force. At the anode, as the Cu atoms participated in the interface reaction and the Zn atoms moved out, (Cu,Ni)$_6$(Sn,Zn)$_5$ phase replaced Ni$_5$Zn$_{21}$ phase. Because the growth rate of (Cu,Ni)$_6$(Sn,Zn)$_5$ structure is much faster than that of Ni$_5$Zn$_{21}$, the IMCs growth became faster than that of the annealed state. At the cathode, since many more Cu atoms had migrated in or through the interface under current stressing, the concentration of Cu at the IMCs was increased. Therefore, Cu(Sn$_{1-x}$,Zn$_x$) phase, which has a higher Cu concentration, replaced the original Cu$_5$Zn$_8$ phase. In the same way, Cu$_6$(Sn$_{1-x}$Zn$_x$)$_5$ and Cu$_3$(Sn$_{1-x}$Zn$_x$) layers formed and thickened due to the enhanced fluxes of Cu and Zn atoms under current stressing.

Conclusions

In summary, the electromigration investigation was conducted for the combination of Ni/SnZn/Cu under current stressing of 10^{4}A/cm^2 at 150°C for 168.5h. Based on experimental observations and kinetic analysis, the following conclusions have been reached.

1) When the electrons traveled from the Ni side to the Cu side, the interfacial reactions were similar to those without the passage of electric currents. Ni$_5$Zn$_{21}$ and Cu$_5$Zn$_8$ were formed at the Ni/SnZn and Cu/SnZn interfaces, respectively.

2) When electron flow went from the Cu side to the Ni side, the (Cu,Ni)$_6$(Sn,Zn)$_5$ phase replaced Ni$_5$Zn$_{21}$ phase at the Ni/SnZn interface, while β-CuZn phase replaced Cu$_5$Zn$_8$ phase at the Cu/SnZn interface. The IMCs at the interface became much thicker than those of the annealed state.

Acknowledgments

This work was supported by the National Basic Research Program of China, Grant No. 2004CB619306.

References

1. Wang SJ, Liu CY, "Study of interaction between Cu-Sn interfacial reactions by Ni-Sn3.5Ag-Cu sandwich structure," J. Electron. Mater. Vol. 32, No. 11 (2003), pp.1303-1309
2. Tsai CM, Luo WC, Chang CW, Shieh YC, Kao CR, "Cross-interaction of under-bump metallurgy and surface

finish in flip-chip solder joints," J. Electron. Mater. Vol. 33, No. 12 (2004), pp.1424-1428

3. Zhang F, Li M, Chum CC, Shao ZC, "Effects of substrate metallization on solder/under-bump metallization interfacial reactions in flip-chip packages during multiple reflow cycles," J. Electron. Mater. Vol.32, No. 3 (2003), pp. 123-130

4. Nah JW, Suh JO, Tu KN, "Effect of current crowding and Joule heating on electromigration-induced failure in flip chip composite solder joints tested at room temperature," J. Appl. Phys. Vol. 98, 013725 (2005)

5. Alam MO, Wu BY, Chan YC, Tu KN, "High electric current density-induced interfacial reactions in micro ball grid array (IBGA) solder joints," Acta Mater. Vol. 54 (2006), pp.613-621

6. Lin HY, Hu YC, Tsai CM, Kao CR, Tu KN, "In situ observation of the void formation-and-propagation mechanism in solder joints under current-stressing," Acta Mater. Vol. 53 (2005), pp.2029-2035

7. Kim KS, Ryu KW, Yu CH, Kim JM, "The formation and growth of intermetallic compounds and shear strength at Sn-Zn solder/au-Ni-Cu interfaces," Microelectron. Reliab. Vol. 45 (2005), pp.647-655

8. Huang CW, Lin KL, "Interfacial reactions of lead-free Sn-Zn based solders on Cu and Cu plated electroless Ni-P/Au layer under aging at 150 °C," J. Mater. Res. Vol. 19, No. (2004), pp. 3560-

9. Conrad H, "Effects of electric current on solid state phase transformations in metals," Mater. Sci. Eng. A Vol. 287 (2000), pp. 227-237

10. Smigelkas AD, Kirkendall EO, "Zinc diffusion in alpha brass," Trans. AIME Vol. 171 (1947), pp.130-134

11. Tu KN, Thoimpson RD, "Kinetics of interfacial reaction in bimetallic Cu-Sn thin films"Acta Metall. Vol. 30, No. (1982), pp.947-952

12. Paul S. Ho, Thomas Kwok, "Electromigration in metals," Rep. Prog. Phys. Vol. 52 (1989), pp.301-348

13. Zhang XF, Guo JD, Shang JK, "Abnormal polarity effect of electromigration on intermetallic compound formation in Sn-9Zn solder interconnect," Scripta Mater. Vol. 57 (2007), pp. 513-516

Analysis on Cracking Blind Vias of PCB for Mobile Phones

Li-Na Ji, and Zhen-Guo Yang*
Department of Materials Science; Fudan University
220 Handan Rd., Shanghai 200433, P.R. China
Email: zgyang@fudan.edu.cn, phone: 8621-65642523

Abstract

Microvia interconnection technology, which mainly includes blind via and buried via, has been put forward to meet the trends of "faster, lighter, reliable" in electronic industry. The technology plays a crucial role in the development of low cost and high density printed circuit board (PCB). However, as the cost for raw materials, like copper, keeps increasing, one defect blind via will lead to not only failure of the whole board but great economic loss as well. The defect causes and mechanisms have been diversified due to the complexity of process, which can be traced back to material property, microvia design, manufacturing process and service environment etc..

This paper focuses on systematic failure analysis of blind vias on PCB for novel mobile phones. A series of modern analytical instruments and methods were used to observe defects of the vias. Meanwhile, focused ion beam (FIB) was adopted, for the first time, both at home and abroad, to obtain the microstructure of copper grains in the cracking interface. Characterization techniques such as FT-IR, TGA and SAM were also performed to inspect thermal stability of the bare boards. The results have shown that obvious cracking occurring along the interface of different plating layers is the main cause of open circuit; and that both sulfur impurity on wall of the vias related to incomplete desmear process and inappropriate location of the vias concerned with circuit design predominantly account for blind vias cracking. It has been revealed that the reason of the cracking generation is concerned with the sulfur segregation and the formation of brittle Cu_xS ($1 \leq x \leq 2$). Moreover, the influence of warpage on the reliability of the via has also been mentioned. Based on these defaults, improvement countermeasures and suggestions have been addressed in the paper, which are of significant value for reference to the safe reliability and structural integrity of PCB products during manufacturing and services.

1 Introduction

With the expansion of high density interconnection （HDI）in electronic packaging technology, blind via is a newly developed type of microvias, providing electrical interconnection among different laminates of PCB. Blind vias are comprised of electroless copper on top of buried interconnects formed from various kinds of copper foil and underneath subsequent coatings of electroplated copper. For non-mechanical drilling, blind vias can be constructed by laser ablating the holes in the top layer of the board material. Holes formed are then cleaned and a layer of electroless copper deposited in the hole, followed by electrolytic copper deposited on top of that layer. [1] Therefore, improper circuit and geometry design, unfitting manufacturing process and

ineligible plating quality can directly affect yield and electrical reliability of PCB, among which open circuit is one of the frequent faults that occur during manufacturing. [2]

As blind vias are playing a crucial role in HDI packaging technology for their desirable electrical performance and low reflections, it is critical to evaluate the reliability of these structures. The technologies for manufacturing blind vias have been addressed. [3, 4] The thermomechanical reliability of blind vias is usually evaluated theoretically by experimental tests like FEM simulation and impact tests etc. [5-8] However, comprehensive analyses on failure blind vias after actual services have rarely been reported. M. R. Marks introduced several practical failure cases of flip-chip packaging but skipped the research on blind vias. [9] Also, the chemical mechanism for the cracking generation in the vias has not been well studied.

Based on our former study on failure ball grid array (BGA) solder joints, [10] further analysis on blind vias was carried out. Since visual inspection is the first and important step in failure investigation, [11] a series of modern analytical instruments and methods such as stereomicroscope, optical microscope (OM), digital microscope (3D) and SEM were utilized to observe the geometry design and cracking morphologies of the via and EDS was conducted to attain chemical compositions in the cracking area. Meanwhile, focused ion beam (FIB) was adopted, for the first time, to obtain the microstructure of copper grains in the cracking interface between plating layers. Characterization techniques such as FT-IR, TGA and SAM were also performed to inspect the bare board. The cracking mechanism and the failure causes were definitely revealed. Prevention countermeasures and suggestions were given in the paper as well.

2 Macroscopic and microscopic inspection

Open circuit fault has been detected mainly around BGA solder joints and solder bars of several components and chips by automatic optical inspection (AOI), inducing failure of the whole board. Fig.1 shows the circuit configuration of both sides with components on two types of PCBs, respectively with a dimension of 92 × 37mm × 1.1mm for type A(Fig.1(a)) and 81mm × 38mm × 0.98mm for type B(Fig.1(b)). During macroscopic inspection, circuit configurations of components on type A PCB are principally reasonable, but obvious warpage has been found in the latter one.

978-1-4244-2739-0/08/$25.00 ©2008 IEEE

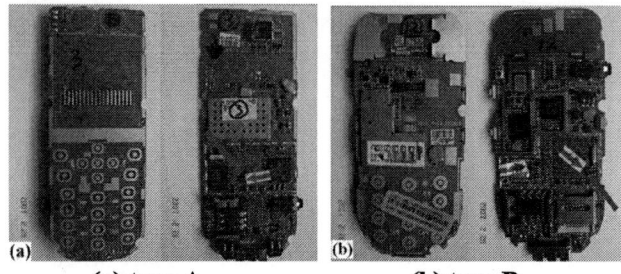

(a) type A   (b) type B

Fig.1 Circuit configuration with components on PCBs

FT-IR results show that the PCBs belong to FR-4 flame resistance substrate and that the raw materials are all bromized epoxy resin. TGA curves in Fig.2 indicate that the materials reaches their thermal decomposing temperature (Td) of about 323℃, while Td of bromized epoxy resin is generally above 300℃ and the soldering peak temperature in surface mounting technology (SMT) of PCB is below 270 ℃. Therefore, it is obvious that the raw materials can endure the instantaneous high temperature and the failure has no relationship with them.

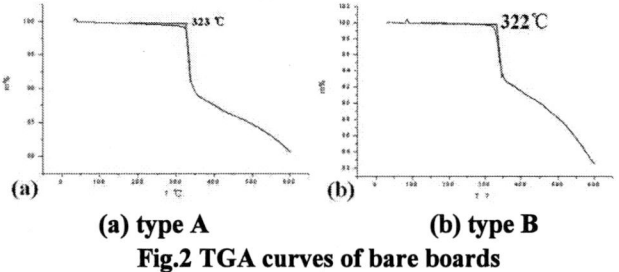

(a) type A   (b) type B

Fig.2 TGA curves of bare boards

Microscopic study was carried out at low magnification using stereomicroscope and at higher magnification using OM and 3D. Fig.3 and Fig.4 respectively present the morphologies of the failure blind vias on both types of PCBs. Distinct cracking can be seen between the copper plating layer and the pad of the blind via from Fig.3(c) and (d). As illustrated in Fig.3(b), the pad is not located right in the centre of the blind via. Consequently, a strong thermal stress will occur around part of the blind via, leading to an asymmetric thermal deformation under conditions of reflow soldering or service. Moreover, the mismatch of coefficient of thermal expansion (CTE) between the plating layer and copper foil will enhance the thermal stress, finally resulting in the fracture from one side of the via. The location of the blind via in Fig.4 indicates that the distance between the via and the BGA solder pad is too close, just like the case shown in Fig.3. An obvious microvoid in the copper foil can be seen from Fig.4(c) and (d), in which there appeared to be some impurity and apparent cracking occurring in the via bottom.

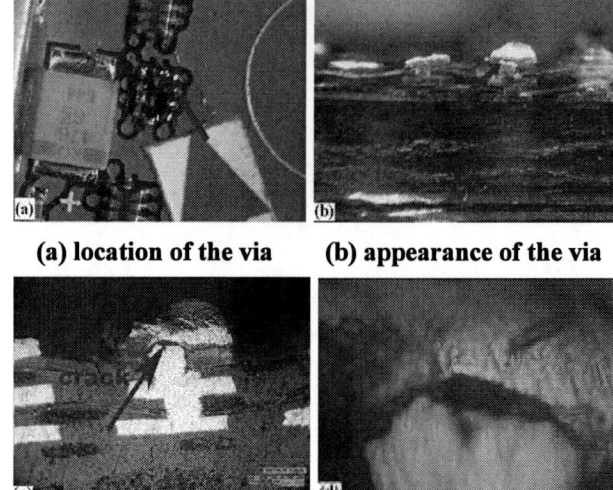

(a) location of the via   (b) appearance of the via

(c) and (d) cracking morphologies
by 3D and OM, 40× and 60×

Fig.3 Morphologies of the failure via on type A PCB

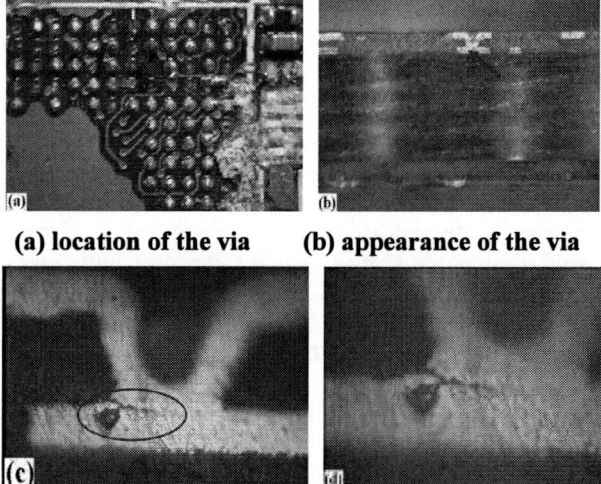

(a) location of the via   (b) appearance of the via

(c) and (d) cracking morphologies, 40× and 60×

Fig.4 Morphologies of the failure via on type B PCB

Cracking morphologies of the blind vias were further observed by SEM and chemical compositions along the fracture line were analyzed by EDS in Fig.5 and Fig.6. Microstructure of the granular near cracking is presented in Fig.7 with part of the deposited copper surface etched by FIB.

SAM with a sensor of 100MHz was used to detect delamination extent of the PCB before and after thermal cycling so as to determine whether delamination influenced the reliability of resin coated copper Foil (RCC). Testing conditions and parameters are listed in Table 1 and the delamination comparison before and after the test are presented in Fig.8. The results show that there is only a slight increase in delamination on the whole board after test and there is almost no noticeable delamination around the failure sections. Therefore, open circuit is not caused by delamination.

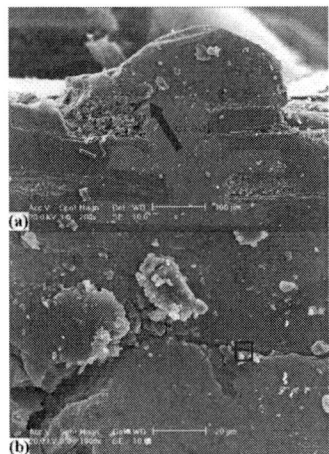

(a) SEM micrograph of blind via (b) cracking morphology

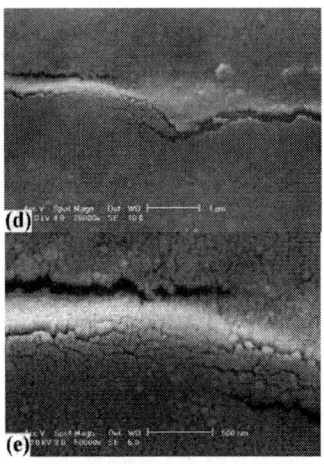

(d) and (e) cracking morphologies
Fig.6 SEM-EDS results of the failure blind via

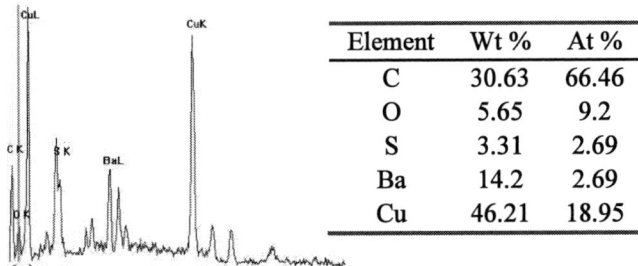

Element	Wt %	At %
C	30.63	66.46
O	5.65	9.2
S	3.31	2.69
Ba	14.2	2.69
Cu	46.21	18.95

(c) EDS analysis of the element compositions in the crack
Fig.5 SEM-EDS results of the failure blind via

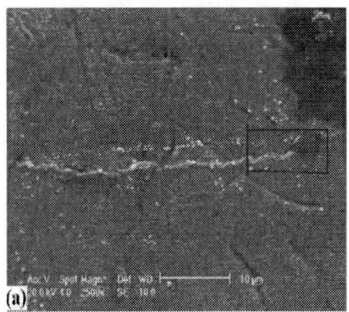

(a) the etching section

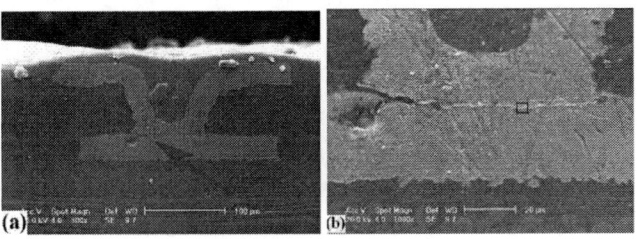

(a) and (b) SEM micrograph of the via

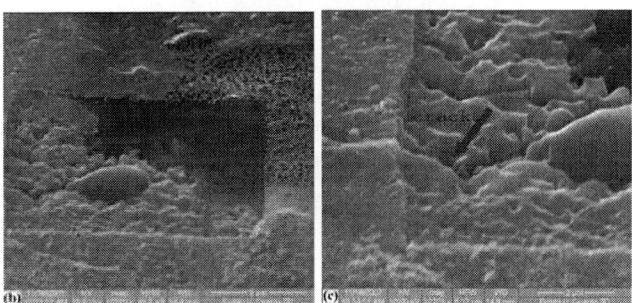

(b) and (c) morphologies of the cracking
Fig.7 Microstructure of the cracking via by FIB

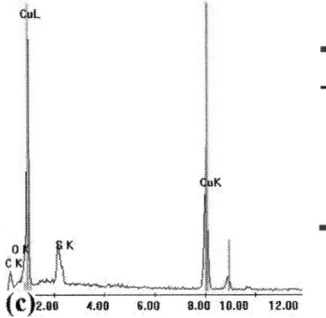

Element	Wt %	At %
C	23.9	58.06
O	4.25	7.76
S	2.65	2.41
Cu	69.19	31.77

(c) EDS analysis of the element compositions in the crack

Table 1 Conditions of thermal cycling on the bare board

range of temperature	-40℃—125℃
minutes of heating-up	13min(up)+20min(remaining)
minutes of cooling-down	17min(down)+20min(remaining)
times for thermal cycling	200

(a) before testing   (b) after testing

Fig.8 SAM inspection of bare board before and after thermal cycling

3 Results and discussion

The primary fault of these two types of PCBs is open circuit. Through chemical compositions and thermal analysis, it can be seen that the material quality of the bare board is acceptable. Meanwhile, the component configuration of type A PCB is reasonable, but warpage has been found in type B PCB through the macroscopic inspection. It can be concluded from the SAM inspection that delamination is not the main cause of failure. The analytical results show that it is the defect of the blind vias that mainly leads to the open circuits.

A big open cracking from one side of the blind via with a result of a separation completely between the pad and the copper-plating layer is shown in Fig.5. An obvious cracking from the bottom of the other blind via that partly separated the via and the copper foil is presented in Fig.6. We can also notice from the location of these two blind vias that the design of position between pads and blind vias are somewhat improper. They are not kept in the same Z-direction and the distance between them is too close. Usually, the region closed packed with deep color dots holds a comparatively faster rate of heating-up. Therefore temperature around the pads covered with solder of dark colour rises up faster than other pad areas without solder during the heating-up segment of reflow soldering; while areas merely with the pads drop faster in temperature than those overlaid with solder in the cooling-down segment of reflow soldering. The mismatch CTE of these two areas results in a strong thermal stress and an asymmetric deformation in the blind vias linked closely to the pad uncovered with solder. Accordingly, the improper circuit design regarding the location of blind vias and pads is one of the main causes for vias cracking.

In addition, the EDS results (Fig.5(c) and Fig.6(c)) demonstrate that a certain amount of sulfur element which usually makes materials embrittlement concentrates at the interface with high energy. Higher magnifications of the

cracking have been shown and many microcracks can be seen in Fig.6(d)and(e). These microcracks are generated along the deposited copper crystal grains, which take on an embrittlement behavior. Microstructure of the grains near cracking has also been obtained by FIB, with a sound characteristic of intercrystalline fracture. It can be deduced that the S impurity in the crack, to some extent, has a deterioration effect on the blind via, inducing a leak bonding force among different layers.

From the point of process view, the blind vias will be plated with several layers of copper after drilling and desmear procedures so as to produce electrical conductivity on the surface of glassfiber and resins in blind vias among the laminates of PCB. As a result, the reliability of them may also be influenced by the process and quality of plating. Electroless plating solution is generally composed of copper sulfate, complex agent, reducing agent, pH conditioning agent and additive containing S and N, etc. A layer of copper is deposited on the wall of the blind via through self-catalyzed oxidation-reduction reaction.

The acquiring of pure copper is, however, confined because of the effect of intergranular embrittlement. The defect in the structure of copper grains is inevitable, and so intergranular cavitation and fracture were observed preferentially at random grain boundaries. [12]

The morphologies of the blind vias cracking exhibit obvious embrittlement deformation behaviors in the fractured copper-plating layer. Generally speaking, S content in pure copper, calculated in weight percentage（%）, is not more than 0.005-0.01%. However, the content of S impurity detected in the fracture interface by EDS is far above the standard, as is shown in the former part. Thus, it can be assumed that this embrittlement fracture is largely affected by segregation of S at grain boundary.

Although it is still unknown that what the structure formed by the segregated S at grain boundaries is, the formation of Cu_xS compounds（$1 \leq x \leq 2$）can naturally be expected, especially Cu_2S, with several crystal structures, according to the equilibrium Cu-S phase diagram. Similar evidence was also found by Boulliard and Sotto [13] that the structure of segregated sulfur on the face of Cu (100) single crystals was 8 S atoms on 17 Cu atoms, with the S atoms arranged in a p(2 x 2) lattice. Note that this coverage approximates the stoichiometry of Cu_2S, phase of which is highly stable. [14] It was also reported that the phase transition from the "high chalcocite" hexagonal phase to the complex "low chalcocite" phase occurs at 103.5℃ for x=2.000 but drops to 90℃ for the slightly more sulfur-rich compound with x=1.990.

Hence, the mechanism of the cracking generation in the blind vias may be established as the following steps. The sulfur impurity that has segregated to the surface along the different layers of the vias probably nucleated as a set of three-dimensional islands of Cu_xS, which would be expected to grow until they coalesced. The nucleation sites may be at dislocation or grain boundary intersections with the surface. [15] However, it is highly undesirable due to several structural transformations of Cu_xS in the temperature interval of interest, which should favor the fatal crack nucleation and propagation at grain boundaries under thermal stress during reflow soldering. Consequently, the conclusion can be reached that the residue of S impurity along copper grain boundary has a significant influence on the reliability of the blind vias, which is the leading cause of the failure. The residue S is mainly concerned with both desmear process and clean degree of the blind vias.

Moreover, through macroscopic observation, apparent warpage can be noticed in type B PCB with a discrepancy of about 1mm between the right and left side of the board. As mentioned above, the type B PCB is only 0.98mm thick, the thickness of which is at least 10% thinner than that of type A PCB. It is both the improper reduction of the thickness and asymmetric configuration of components of the PCB that lead to the excessive warpage and deformation of the board, which will produce a big tensile force or a bending stress around pads linked with blind vias, leading to a cracking in some weak connections of different layers in the vias.

In summary, both the incomplete desmear process before copper-plating and the improper location design of pads and blind vias constitute the main causes of the cracking. The inappropriate dimension design of type B PCB is another potential risk of the failure.

Conclusions

(1) The comprehensive analysis on the two types of PCBs by both macroscopic and microscopic methods shows that it is cracking of the blind vias that mainly causes open circuit fault in the PCBs. The raw materials of the boards are acceptable, and there is also no noticeable delamination around the failure sections.

(2) The residue S impurity appearing in the wall interface of the blind vias has a fatal effect on the structural integrity and deposition quality of the copper-plating with a formation of Cu_xS in the interface of copper grains. The structure of Cu_xS can change with different temperatures, inducing an embrittlement intercrystalline fracture during SMT. Accordingly, the incomplete desmear is the leading factor for failure of the vias.

(3) Improper circuit design concerning locations of pads and blind vias, revealed by OM inspection, leads to strong stresses and asymmetry deformations on part of the vias that will generate cracking in some weak joints once heated. Therefore, the design default is the predominant cause for cracking of the blind vias.

(4) An inappropriate reduction in thickness of type B PCB and together with the asymmetric configuration of components lead to excessive warpage and deformation of the board, which will generate a big tensile force or a bending stress around pads linked with blind vias. However, under some conditions, cracking will occur in those weak connections of the vias. So, the unsuitable dimension design of the PCB is another potential risk of the failure.

Suggestions

(1) It is advised that the quality inspection on the wall of the blind vias should be emphasized before plating so as to provide a clean wall surface for a better copper-plating.

(2) Meanwhile, the design of relative locations among pads and components should be optimized through simulation testing like the combination of finite element methods (FEM) and bench test.

(3) The best thickness of the board should be simulated through FEM and essential property experiments in order to ensure the indispensable stiffness of the substrate.

Acknowledgments

The financial is supported by Shanghai Leading Academic Discipline Project (Project Number: B113).

References

1. R. Soares et al, "High-Density PWB Microvia Reliability for Space Application", Aerosp. Conf., Big Sky, MT, March. 2007, pp. 1-8. [C]
2. P.L Martin. Electronic Failure Analysis Handbook, Sci. Press, (Beijing, 2005), pp. 9. [B]
3. T. Nishiwaki et al, "Comparison of Various Micro Via Technology", Proc ISAPM, Braselton, GA, USA, 2000, pp. 233-237. [C]
4. E.S.W. Leung et al, "A study of microvias produced by laser-assisted seeding mechanism in blind via hole plating of printed circuit board", Int. J. Adv. Manuf. Tech., Vol.24, No. 7-8 (2004), pp. 74-484. [J]
5. T. H. Wang et al, "Stress analysis for fracturing potential of blind via in a build-up substrate", Circ. World, Vol. 32, No. 2 (2006), pp. 39-44. [J]
6. F. Liu et al, "Reliability assessment of microvias in HDI printed circuit boards", Proc 51st Electronic Components and Technology Conf, 2001, pp. 1159-1163. [C]
7. G.Ramakrishna et al, "Role of dielectric material and geometry on the thermo-mechanical reliability of microvias", Proc 52nd Electronic Components and Technology Conf, 2002, pp. 439-445. [C]
8. D. Xie et al, "Impact Performance of Microvia and Buildup Layer Materials and Its Contribution to Drop Test Failures", Proc 57th Electronic Components and Technology Conf, Reno NV, 2007, pp. 391-399. [C]
9. M.R. Marks, "Novel Deprocessing Technique for Failure Analysis of Flip-Chip Integrated Circuit Packages", Pract. Fail. Anal., Vol. 1, No. 6 (2001), pp. 45-52. [J]
10. L.N. Ji et al, "Failure Analysis of BGA Solder Joint in PCB for Novel Mobile Phones", Heat Treat. Met., Vol. 32 suppl.(2007), (with abstracts in both English and Chinese) pp. 373-376. [J]
11. D. Christie, "A Review of the Science and Art of Visual Examination in Failure Analysis", J. Fail. Anal. Prevent., Vol. 6, No. 3 (2006), pp. 1547-7029. [J]
12. M. P. Butron-Guillen et al, Scripta metall., Vol. 24 (1990), pp. 991. [J]
13. J. C. Boulliard et al, Surf. Sci., Vol. 195(1988), pp. 255-269. [J]
14. P. A. Korzhavyi et al. "Theoretical investigation of sulfur solubility in pure copper and dilute copper-based alloys", Acta Materialia, Vol. 47, No. 5 (1999), pp. 1417-1424. [J]

15. R. Kothari *et al*, "Enhanced sulfur segregation in plastically deformed OFHC Cu and the effect of surface segregated sulfur on electric contact resistance", *IEEE Trans-CPMT-A*, Vol. 17, No. 1 (1994), pp. 121-127. [J]

Interfacial Reaction and Failure Mechanism for SnAgCu Solder Bump with Ni(V)/Cu Under Bump Metallization During Aging

Kai Jheng Wang and Jenq Gong Duh[*]

Department of Materials Science and Engineering, National Tsing Hua University, Hsinchu, Taiwan, China
[*]E-mail: jgd@mx.nthu.edu.tw, Fax: 886-3-5712686

Abstract

In flip chip technology, under bump metallization (UBM) composing multi metallic thin-films is usually used to improve the bonding between solder and pad on chip. The Cr/Cu/Au UBM is commonly used to joint with SnPb solder. [1] However, the Sn-Cu intermetallic compounds (IMC) spalled rapidly from interface, due to the quick reaction between Cu and Sn. Therefore the Ni-based UBM, such as Ti/Cu/Ni and Al/Ni(V)/Cu, is advanced to replace Cr/Cu/Au UBM. The primary advantage of Ni-based UBM is slower reaction rate with Sn-Pb solder and Pb-free solder than Cu-based UBM, so the spalling of IMC would be eliminated. Nevertheless, sputtering a pure Ni target is more difficult because of magnetic interference. In order to overcome the magnetic interference, the Delco process which adds 7 at.% V into Ni target is applied.

Only a few literatures were focused on the interface reaction between Al/Ni(V)/Cu UBM and solder. [2-7] Li et al. [3] indicated that $(Cu,Ni)_6Sn_5$ attached well to the Ni(V) in the SnAgCu solder and Al/Ni(V)/Cu UBM system after multiple reflows. However it was revealed that the Ni(V) layer was gradually replaced by the Sn-patch with reflow. Jang and Duh [5] also reported the conformable results for Sn3.0Ag(0.5 or 1.5)Cu solder with Al/Ni(V)/Cu UBM. The Sn-patch was observed in Sn3.0Ag0.5Cu solder system with lower Cu concentration. More Cu was dissolved from Al/Ni(V)/Cu UBM, resulting in the formation of Sn-patch in Ni(V) layer.

Most literatures for Al/Ni(V)/Cu UBM were concerning interfacial reaction, [2, 5-7] yet there was limited data on the mechanical properties. [2, 3] In fact, the mechanical testing methods, such as shear and pull test, would affect the packaging reliability, and the failure mechanism for reliability tests should be better understood. In this study, Sn3.0Ag0.5Cu solder bump was jointed on Ni(V)/Cu UBM and then aged at 125°C for 500-2000h. The heat treatment samples were employed to investigate the interfacial reaction and the mechanical properties between solder and UBM. $(Cu,Ni)_6Sn_5$ IMC formed at the interface between solder and Ni(V) layer after assembly. The thickness of $(Cu,Ni)_6Sn_5$ gradually increased to 4μm during aging at 200°C for 2000h. The rod-like $(Cu,Ni)_6Sn_5$ IMC and fine Ag_3Sn particles were not significantly changed even after 2000h. In addition, the Sn-patch was observed in the Ni(V) layer after 500h, and replaced the original Ni(V) layer.

The mechanical properties for solder joints were estimated with a XYZTEC bond tester. Scanning electron microscope images of the crack surface after testing provided the information for failure modes analysis. The relations between interfacial reaction and mechanical properties would be

discussed. Furthermore, the possible failure mechanism was proposed.

Interfacial reaction of SnAgCu bump with Ni(V)/Cu UBM

The Ni(V) layer was gradually replaced by the Sn-patch as reflow. [3-6] It was supposed that the Sn-patch in the Ni(V) layer would decrease the mechanical property of the solder joint. Since the Sn-patch was the Sn-rich phase, it might hinder the diffusion of Ni in Ni(V) layer. Therefore, IMC would lose the adhesion of the UBM, and the reliability of solder joint became the concern.

In order to study the formation of the Sn-patch between Sn3.0Ag0.5Cu solder and the Ni(V)/Cu UBM, as well as the relationship of the shear testing with Sn-patch, the Ni(V)/Cu thin film was used as the UBM. The thickness was 0.3μm and 0.5μm for sputtered Cu and Ni(V) layers, respectively. The Ni(V)/Cu UBM with the Sn3.0Ag0.5Cu solder was reflowed at 250°C for 90 sec, and then the specimens were treated in 125°C and 200°C, respectively, for 500h, 1000h, and 2000h.

After different heat treatments, the specimens were mounted in G2 resin, and then ionically polished by a cross-section polisher (CP) to avoid the defects from traditional grinding and polishing. The field-emission electron probe microanalyzer (FE-EPMA) was used to observe the interfacial reaction behaviors between solder and UBM, and the compositions of the detected regions were determined.

Figure 1 was the cross-section image of SnAgCu solder with Ni(V)/Cu UMB as reflowed. The diameter of solder joints and pads was about 120μm and 80μm, respectively. The intermetallic compound (IMC) was formed along the interface of solder and UBM after reflowed for 1min. Figure 2(a) was the enlarged image of the interface, and the thickness of IMC was about 2.1μm as shown in Fig. 3. The composition ratio of IMC was 50.54±0.95at.%Cu-3.56±1.22at.%Ni-45.90±0.27at.%Sn, and the IMC was indicated as $(Cu,Ni)_6Sn_5$. There might be another $(Cu,Ni)_3Sn$ IMC adjacent to $(Cu,Ni)_6Sn_5$ and near the Ni(V) layer, yet it was too small to be identified by FE-EPMA.

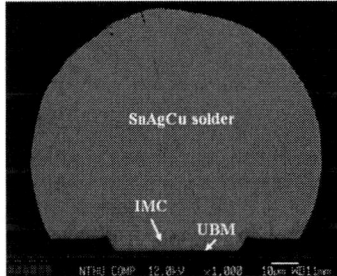

Fig.1 SEM morphology of solder joint as reflow.

The cross section images of the solder joints after heat treatment at 125°C for 500h, 1000h, and 2000h are shown in Figs. 2(b)-(d). The thickness of IMC was increased from

978-1-4244-2739-0/08/$25.00 ©2008 IEEE

2.1μm at as-reflowed to 2.6μm for 2000h aging (Fig. 3). Table 1 lists the composition of IMC. In Fig. 2, a light gray compound was observed, but it was too small to identify the composition even after aging for 2000h. In literatures, the light gray compound was the Sn-rich phase as called "Sn-patch".

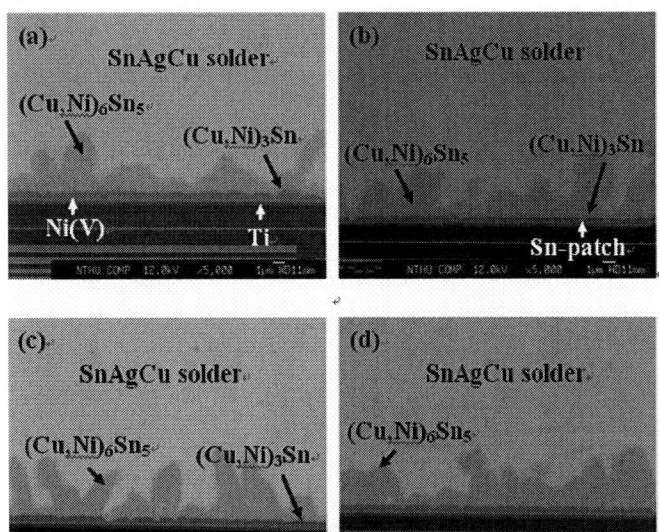

Fig.2 SEM images of solder joint at 125°C for (a) 0h, (b) 500h, (c) 1000h, and (d) 2000h.

Table 1 The compositions of $(Cu,Ni)_6Sn_5$ at 125°C.

	Cu	Ni	Sn
0h	50.54±0.95	3.56±1.22	45.90±0.27
500h	50.45±0.91	3.95±0.85	45.60±0.16
1000h	48.82±1.80	5.75±1.10	45.43±0.92
2000h	50.55±0.59	3.81±0.33	45.64±0.36

Figure 3 shows the cross section images of the solder joints aged at 200°C for different durations. The thickness of IMC was about 3.3μm, and the composition of IMC was $(Cu,Ni)_6Sn_5$. In addition, the Ni concentration in $(Cu,Ni)_6Sn_5$ at 200°C was higher than that at 125°C. Another IMC was formed between $(Cu,Ni)_6Sn_5$ and Ni(V) layer at 200°C for 1000h as shown in Fig 3(b). After identified by FE-EPMA, the composition of the new IMC was 8.56±1.54at.%Cu-34.39±1.70at.%Ni-57.05±0.31at.%Sn, and it was $(Ni,Cu)_3Sn_4$. The Sn-patch replaced the Ni(V) layer completely, and the composition of the Sn-patch was 6.97±1.32at.%Cu-5.39±0.86at.%Ni-68.04±1.23at.%Sn-19.60±1.01at.%V. After aged at 200°C for 2000h, the thickness of $(Cu,Ni)_6Sn_5$ and $(Ni,Cu)_3Sn_4$ was 3.6μm and 0.7μm, respectively, as shown in Fig. 4(b) and Fig. 5. The composition of $(Cu,Ni)_6Sn_5$, $(Ni,Cu)_3Sn_4$, and Sn-patch are listed in Table 2.

From Fig. 3, the thickness of $(Cu,Ni)_6Sn_5$ grew from 2.1μm to 2.7μm after aged at 125°C for 200h. The thickness of IMC was thinner than that of bare Cu substrate reaction

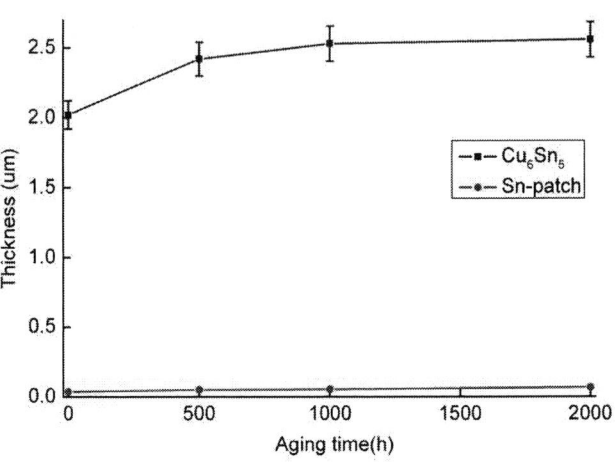

Fig.3 Thickness of IMC and Sn-patch at 125°C

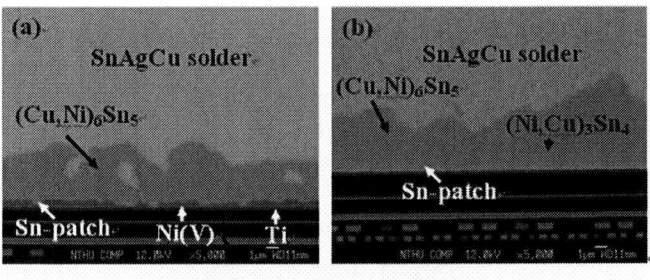

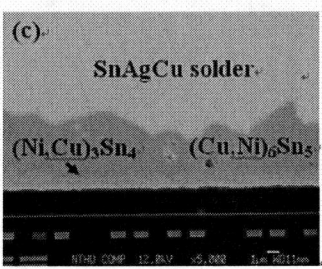

Fig.4 SEM images of solder joint at 200°C for (a) 500h, (b) 1000h and (C) 2000h.

with SnAgCu solder. [8] When SnAgCu solder reacted with either the Ni(V)/Cu or bare Cu substrate, the major IMCs were Cu_6Sn_5 and Cu_3Sn, yet the thickness of Cu_6Sn_5 was different. It was caused by the UBM design. It should be note that the Cu layer onto the Ni(V) UBM was consumed during reflow, and formed $(Cu,Ni)_6Sn_5$ at interface after reflow. The growth of $(Cu,Ni)_6Sn_5$ was resulted from Ni and Sn atoms interactive diffusion during solid state aging. The diffusion rate of Ni was smaller than Cu, and Ni atoms were needed to replace Cu to form $(Cu,Ni)_6Sn_5$. So, $(Cu,Ni)_6Sn_5$ grew very slowly and the Ni(V) layer was also slowly consumed.

The formation of Sn-patch was an interesting phenomenon in the solder joint with Ni(V) layer. From the FE-EPMA quantitative analysis, there was no V in $(Cu,Ni)_6Sn_5$, meaning that V did not diffuse to the IMC and did not react with solder. Therefore, when the Ni diffused to form $(Cu,Ni)_6Sn_5$, the Sn diffused to the Ni(V) layer, and V and Sn would form the Sn-patch. In addition, the Sn-patch was usually observed near the grain boundary of $(Cu,Ni)_6Sn_5$, since the atomic movement was easier at grain boundary than grain. When the

874

Sn-patch was formed, Ni needed to diffuse through the Sn-patch and then reached the IMC. Therefore, the formation of the Sn-patch might decrease the growth of $(Cu,Ni)_6Sn_5$. During heat treatment at $125^{\circ}C$, the amount of Sn-patch was increased, and Ni atoms in the Ni(V) layer needed to diffuse through more Sn-patches, so the growth rate of $(Cu,Ni)_6Sn_5$ was decreased.

Table 2 The compositions of $(Cu,Ni)_6Sn_5$, $(Ni,Cu)_3Sn_4$, and the Sn-patch at $200^{\circ}C$.

		Cu	Ni	Sn	V
$(Cu,Ni)_6Sn_5$	0h	50.54±0.95	3.56±1.22	45.90±0.27	-
	500h	44.60±2.85	9.85±3.10	45.55±0.26	-
	1000h	36.26±2.84	18.04±2.75	45.70±0.23	-
	2000h	37.66±0.98	17.33±1.41	45.01±0.49	-
$(Ni,Cu)_3Sn_4$	0h*	-	-	-	-
	500h*	-	-	-	-
	1000h	8.56±1.54	34.39±1.70	57.05±0.31	-
	2000h	8.96±0.45	34.16±0.61	56.88±0.29	-
Sn-patch	0h**	-	-	-	-
	500h**	-	-	-	-
	1000h	6.97±1.32	5.39±0.86	68.04±1.23	19.60±1.01
	2000h	5.98±0.47	3.55±0.95	71.57±0.40	18.90±0.07

*: $(Ni,Cu)_3Sn_4$ did not form at interface.
**: The Sn-patch was to small to be analyzed.

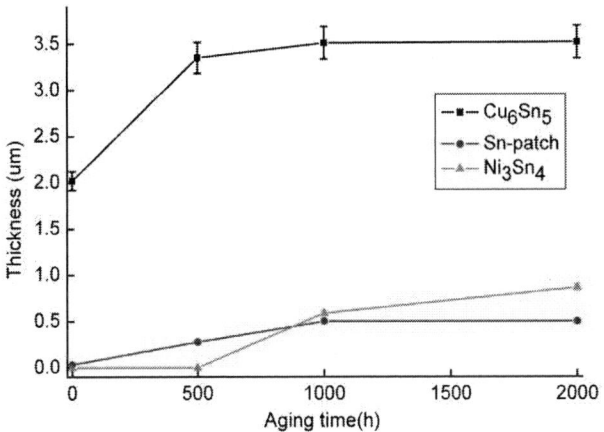

Fig.5 Thickness of IMC and Sn-patch at $200^{\circ}C$.

Luo et al. [9] reported that the Sn3.9Ag0.6Cu solder with Ni layer formed $(Cu,Ni)_6Sn_5$ after reflow, while $(Ni,Cu)_3Sn_4$ formed at interface between $(Cu,Ni)_6Sn_5$ and Ni layer after aged at $180^{\circ}C$ for 150h. From the phase diagram, [10] $(Cu,Ni)_6Sn_5$ is not thermodynamically stable at interface between SnAgCu solder and Ni layer. In this study, $(Ni,Cu)_3Sn_4$ was not observed after aged at $125^{\circ}C$ for 2000h in the interface of SnAgCu solder and Ni(V)/Cu layer. It might be caused by two reasons. One was that $(Ni,Cu)_3Sn_4$ was nucleated at interface , and its growth was too small to be observed at interface. The other reason was that $(Ni,Cu)_3Sn_4$ did not nucleate due to kinetics. It was argued in this study that the possible reason was $(Ni,Cu)_3Sn_4$ did not nucleate. It was noted that the Ni diffusion was rather slow, and the Ni concentration in $(Cu,Ni)_6Sn_5$ was about 3.81at.% even after aged at $125^{\circ}C$ for 2000h. Li et al. [9] reported y could be 0.41 in $(Cu_{1-y},Ni_y)_6Sn_5$ at $240^{\circ}C$, indicating that there could be 22.5at.% Ni in $(Cu,Ni)_6Sn_5$. When the Ni concentration was

more than 22.5at.%, $(Cu,Ni)_6Sn_5$ might transform to $(Ni,Cu)_3Sn_4$. In this study, the Ni concentration in $(Cu,Ni)_6Sn_5$ was only 3.81at.%, and well below the concentration of phase transformation. Therefore, $(Ni,Cu)_3Sn_4$ was not formed at interface between $(Cu,Ni)_6Sn_5$ and Ni(V) layer.

The interfacial reaction between SnAgCu solder and Ni(V)/Cu UBM at $200^{\circ}C$ was different at $125^{\circ}C$. After aged at $200^{\circ}C$ for 1000h, $(Ni,Cu)_3Sn_4$ was formed between $(Cu,Ni)_6Sn_5$ and the Ni(V) layer. The compositions of $(Cu,Ni)_6Sn_5$, $(Ni,Cu)_3Sn_4$, and Sn-patch are listed in Table 2. The Ni concentration in $(Cu,Ni)_6Sn_5$ was 18.04at.% which was higher than that at $125^{\circ}C$, and $(Ni,Cu)_3Sn_4$ easily formed at interface between $(Cu,Ni)_6Sn_5$ and Ni(V) layer.

After aged at $200^{\circ}C$ for 1000h, the Sn-patch replaced the Ni(V) layer. If the Sn-patch was a stable phase at the interface, there would be no source of Ni and Cu in the solder joint to form IMC. From Fig. 5, the thickness of $(Ni,Cu)_3Sn_4$ still increased after aged at $200^{\circ}C$ for 1000h. The growth of $(Ni,Cu)_3Sn_4$ was provided from Ni and Cu atomic diffusions in the Sn-patch and $(Cu,Ni)_6Sn_5$ transformation. However, the thickness of $(Cu,Ni)_6Sn_5$ was not increased, while the concentrations of Ni and Cu in the Sn-patch decreased. This indicated that the growth of $(Ni,Cu)_3Sn_4$ might be supplied by diffusion of Ni and Cu atoms from the Sn-patch. Therefore, the composition of the Sn-patch was not fixed, and the Ni and Cu contents decreased during aging.

Shear testing of SnAgCu Solder Bump with Ni(V)/Cu UBM

In this study, besides morphological observation and composition analysis of the interfacial reactions, the shear strengths of the solder joints with different heat treatments were tested with a XYZTEC bonding tester. The shear speed was $1000\mu m/s$ and the shear height was $10\mu m$. An average shear force of 30 solder bumps was taken for each condition. The fracture surfaces were observed by FE-EPMA.

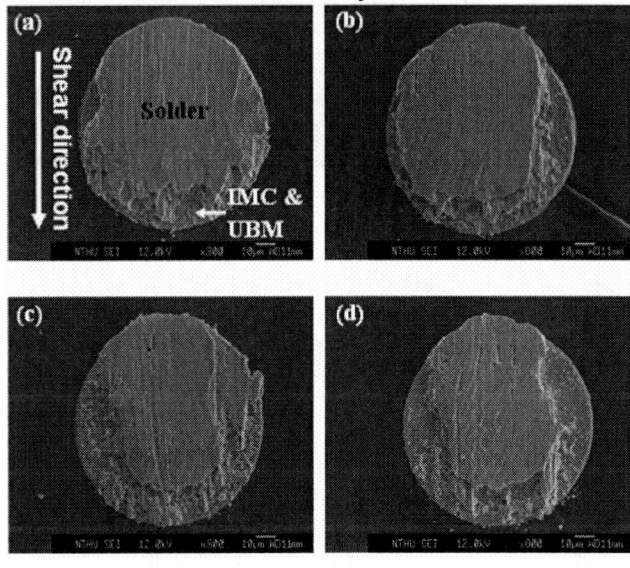

Fig6 Top view images of the failure surface for solder joints after shear testing at $125^{\circ}C$ for (a) 0h, (b) 500h, (c) 1000h, and (d) 2000h.

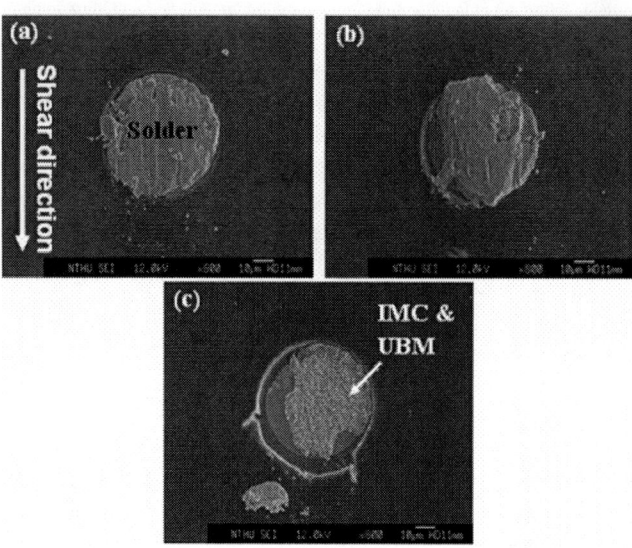

Fig7 Top view images of the failure surface for solder joints after shear testing at 200°C for (a) 500h, (b) 1000h and (C) 2000h.

Figure 6 is the images of fracture surfaces for the solder joints. The fracture surface was observed to be crossed the solder, and IMC. The brittle facture area of the solder joints increased with aging time. Nevertheless, the shear strength of the solder joints remained around 43gf as shown in Fig. 8. It was caused by the slow interfacial reaction, since the thickness of $(Cu,Ni)_6Sn_5$, the amount of Sn-patch, and the consumption of Ni(V) layer did not change significantly, as indicated in Fig. 3. It was noted that the shear strength was 43gf even after aging 2000h at 125°C, implying that the crack behavior of the solder joints was mainly ductile. This is good for the reliability of the solder joints at 125°C for 2000h.

After aged at 200°C, the fracture surfaces of the solder joints (Fig. 7) were different to those at 125°C. The fracture surface was composed of brittle and ductile crack at 200°C for 500h (Fig. 7(a)). The ductile fracture only occurred in the center area of the solder joint, while the brittle fracture was in the peripheral area of the solder joint. The failure behavior of the solder joint for 1000h was similar to that for 500h (Fig. 7(b)). In contrast, after aged at 200°C for 2000h, only brittle fractures were observed in the solder joint. The fracture surface crossed IMC in the center area of the solder joint, and was in the interface of UBM and Si wafer in the peripheral area of the solder joint (Fig. 7(c)). From Fig. 7, the failure of the solder joints in the peripheral area was all along UBM and Si wafer. This was due to the geometry of UBM, for the disc-like UBM exhibited different height between the center and the peripheral area of the solder joint. The thickness of IMC was greater than 3.3μm after aged for 500h, which was about 1/3 of shear height in the peripheral area of the solder joints. However, the shear height was 15μm in the center area of the solder joints. During the shear testing, the brittle IMC did not deform to reduce the shear force immediately, so the UBM broke away with IMC in the peripheral area of the solder joints.

It should be noted that the shear strength of the solder joints was about 36gf after aging 500h at 200°C, and deceased to 31gf after aged for 2000h (Fig. 8). The shear strength at 200°C was smaller than that at 125°C, and the shear strength of the solder joints deceased gradually at 200°C with aging time. If the Ni(V) layer was replaced by the Sn-patch at 200°C, the IMC might lose the adhesion with Sn-patch. However, the shear strength maintained 31gf even after aged 2000h. The possible reason was that the disc-like UBM design eliminated the IMC effect of the solder joint. According to the design of UBM, the crack of the solder joint was usually brittle at edge, yet ductile at center, and the shear strength of solder joint could be improved. Zhang et al. [4] reported that the IMC between the SnAgCu solder and the Ni(V) layer spalled out from the interface after 30 reflows. The shear strength of the solder joints was 60gf, similar to the strength after 1 reflow. The good shear strength of the solder joints was also improved by the UBM design. Therefore, shear testing of solder joint with the disc-like UBM could maintain a good strength even if IMC spalled out from interface or lost adhesion with UBM.

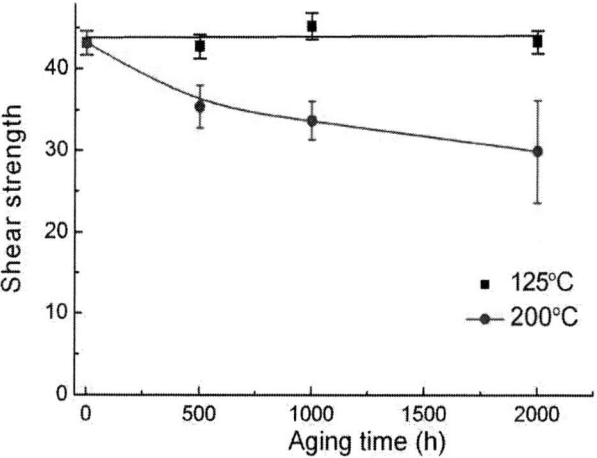

Fig.8 Shear strength of the solder joints after different heat treatment.

Conclusions

In this study, $(Cu,Ni)_6Sn_5$ and $(Cu,Ni)_3Sn$ formed at interface of Sn3.0Ag0.5Cu solder with the Ni(V)/Cu UBM. The thickness of $(Cu,Ni)_6Sn_5$ increased from 2.1μm at as-reflowed to 2.6μm for 2000h at125°C. The shear strength of the solder joints remained around 43gf even after aged for 2000h. However, $(Ni,Cu)_3Sn_4$ formed at interface between $(Cu,Ni)_6Sn_5$ and the Ni(V) layer at 200°C for 1000h. The Ni(V) layer was replaced by the Sn-patch after aged for 1000h, and its composition was 6.97±1.32at.%Cu-5.39±0.86at.%Ni-68.04±1.23at.%Sn-19.60±1.01at.%V. The composition of the Sn-patch was not fixed, with Ni and Cu content decreasing with aging time. Ni and Cu atoms diffused to the IMC to form $(Ni,Cu)_3Sn_4$. Therefore, the thickness of $(Cu,Ni)_6Sn_5$ maintained 3.6μm, and the thickness of $(Ni,Cu)_3Sn_4$ increased to 0.7μm after aged at 200°C for 2000h. The shear strength of the solder joints decreased from 42gf to 31gf at 200°C. It was due to the fact that the thickness of the IMC was thicker at 200°C than that at 125°C. Nevertheless, the fracture surface of the solder joints crossed the IMC and UBM, and the shear strength of the solder joints remained

31gf after aged at 200°C for 2000h. This was resulted from the geometry of the UBM, for the disc-like UBM could improve the shear strength of the solder joints. Therefore, shear testing of the solder joints with the disc-like UBM could maintain a good strength even after aged at 200°C for 2000h.

Acknowledgments

Financial support from the National Science Council, Taiwan, under contract No. NSC- 96-2221-E-007-093-MY3 is acknowledged.

References

1. Liu, A. A. et al, "Spalling of Cu_6Sn_5 spheroids in the soldering reaction of eutectic SnPb on Cr/Cu/Au thin films," *Journal of Applied Physics,* Vol. 80, (1996), pp. 2774.

2. Liu, C. Y. et al, "Electron microscopy study of interfacial reaction between eutectic SnPb and Cu/Ni(V)/Al thin film metallization," *Journal of Applied Physics*, Vol. 87, (2000), pp. 750.

3. Zhang, F. et al, "Failure mechanism of Lead-Free solder joints in flip chip packages," *Journal of Electronic Materials*, Vol. 31, (2002), pp. 1256.

4. Li, M. et al, "Interfacial microstructure evolution between eutectic SnAgCu solder and Al/Ni(V)/Cu thin films," *Journal of Material Researches*, Vol. 17, (2002), pp. 1612.

5. Jang, G. Y, Duh, J. G, "Elemental redistribution and ihterfacial reaction mechanism for the flip chip Sn3.0Ag(0.5 or 1.5)Cu solder bump with Al/Ni(V)/Cu Under bump metallization during aging," *Journal of Electronic Materials*, Vol. 35, (2006), pp. 2061.

6. Tung, C. H, Teo, P. S, Lee, C, "Interface microstructure evolution of lead-free solder on Ni-based under bump metallizations during reflow and high temperature storage," *IEEE Transactions on Device and Materials Reliability*, Vol. 5, (2005), pp. 212.

7. Wu, A. T, Hua, F, "Interfacial stability of eutectic SnPb solder and composite 60Pb40Sn solder on Cu/Ni(V)/Ti under bump metallization," *Journal of Material Researches*, Vol. 22, (2007), pp. 735.

8. Peng, W, Monlevade E, Marques M. E, "Effect of thermal aging on the interfacial structure of SnAgCu solder joints on Cu," *Microelectronics Reliability*, Vol. 47, (2007), pp. 2161.

9. Luo, W. C. et al, "Solid-state reactions between Ni and SnAgCu solders with different Cu concentrations," *Materials Science and Engineering A*, Vol. 396, (2005), pp. 385.

10. Li, C. Y, Chiou, G. J, and Duh, J. G, "Phase distribution and phase analysis in Cu_6Sn_5, Ni_3Sn_4, and the Sn-rich corner in the ternary Sn-Cu-Ni isotherm at 240°C," *Journal of Electronic Materials*, Vol. 35, (2006), pp. 343.

Effect of displacement rate on lap shear test of SAC solder ball joints

X.J. Wang[1], Z.G. Wang[1], J.K. Shang[1, 2]

[1]Shenyang National Laboratory for Materials Science
Institute of Metal Research, Chinese Academy of Sciences, Shenyang 110016, China
[2]Department of Materials Science and Engineering
University of Illinois at Urbana-Champaign, Urbana, IL 61801, USA
Email: jkshang@imr.ac.cn and xjwang@imr.ac.cn

Abstract

Lap-shear tests were combined with elastic-viscoplastic constitutive modeling to study the shear behavior of Sn3.8Ag0.7Cu (SAC) solder ball joints. In the shear rate range of 0.005mm/s to 0.7mm/s, the peak shear force was attained an optimal rate (labeled as v_o), around 0.3mm/s. On either side of v_o, the shear fracture behavior was quite different. Cross-section examinations of the remaining solder joints after fracture showed that the fracture surface moved towards the solder/IMC interface with an increasing shear rate. Two shear modes were evident: below v_o, a cohesive shear fracture controlled by the solder properties and above v_o, solder/IMC interface controlled fracture.

Introduction

Solder joints in microelectronics systems often experience shear loading during their service life[1]. As a general deformation mode to mimic real-life loading configurations of solder joints, the lap-shear method is widely adopted to evaluate the shear, creep, and thermal fatigue behavior of Sn–Pb and Sn-rich solder joints[2-10]. While there is no specific lap-shear test standard documented for solder alloys, ASTM Standards for testing adhesive bonds do exist (e.g., ASTM D1002, D3163, D3164 and D5656). Despite the different test configurations in these standards, the key features are qualitatively the same as that in Fig. 1, using a solder ball in our system instead of the rectangle cross sectioned adhesive[11].

Fig.1 Illustration of single-lap shear test specimen.

The strong dependence of shear strength data on temperature and strain rate has been reported in several papers using the ball shear test method[12-14]. They found that the shear strength increased with shear rate, at shear rates from 0.05mm/s to 88mm/s [13,15]. However, the rate dependent shear behavior for lap joints with solder balls is seldom found in the literature available so far. Because of the different load implementation in lap joints, compared to the method using a shear tool specified in ASTM F1269-89(1991) and JESD22-B117(2000), established by the American Society for Testing and Materials (ASTM) and the Joint Electronic Devices Engineering Council (JEDEC), the particular focus on displacement rate is required.

In this study, the rate dependent shear behavior of lap shear joints made from SAC solder balls was studied by combining lap shear test with viscoplastic modeling. With the change of rates from 0.005mm/s to 0.7mm/s, a strong rate dependence was observed. However, the rate dependent of the shear force was not a monotonic function of loading rate. Instead, the shear strength reached a peak value at an optimal shear rate of 0.3mm/s and decreased in shear rate moved away from the optimal rate.

Experimental

The solder ball used in this study was a Sn3.8Ag0.7Cu sphere with a diameter of 1mm. The specimens for single-lap tests were prepared by joining the solder ball between two copper substrates (see Fig. 1)[16]. Copper plate cut into plates of 15mm×3mm×0.5mm, was used as the substrate. A 0.02mm thick solder mask was made to keep a bare copper opening of 0.9mm in diameter.

To join the solder ball with the two Cu substrates, the substrate was first cleaned and kept dry. The solder balls were dipped in a rosin mildly activated (RMA) flux and mounted on the pads of BGA substrates. Then, the solder balls and the substrates were reflowed at 260℃. After the reflow cycle, the specimens were cleaned using an ultrasonic cleaner with a flux remover.

Upon completion of the reflow process, the single lap tests were conducted at room temperature using a universal tensile testing machine at a constant cross-head speed of 0.005~0.7mm/min. The specimens before and after fracture were mounted in cold epoxy, ground using SiC paper through a series of solder balls and polished with 0.1μm diamond paste. The specimens were observed using a scanning electron microscope (SEM) in back-scattered electron imaging (BEI) mode.

Finite element analysis

The computational model employed in this study is showed in Fig. 2. In order to reduce analysis time, a two dimensional visco-plasticity finite element simulation method was utilized to model the plastic deformation behavior, the possible fracture location during shearing, and the effect of ball shear speed on the shear force in Sn3.8Ag0.7Cu lead-free solder joint. Considering that the focus of our work is on the rate dependent effects, the components of finite element simulation included only the reflowed solder ball and the substrate plates. In the simulation, horizontal displacements were imposed at the far right end of the upper copper plate. The x-direction motion of the far left edge of the lower copper plate is forbidden, while the y-direction movement is allowed. The lower-left corner of the lower copper is completely fixed. The finite element code used in this study was ANSYS.

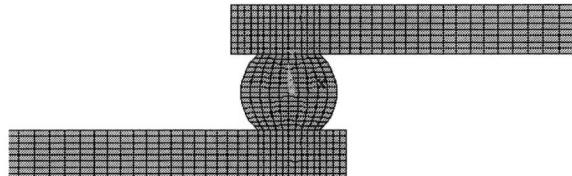

Fig.2 The finite element model for ball shear test

Because the test temperature was in excess of a homologous temperature of 0.5, linear and non-linear, time dependent and independent material properties were incorporated in the finite element model [12, 17, 18].

In the present work, a viscoplastic constitutive model, proposed by Anand (1985) was employed to represent the viscoplastic behavior of solder. The power law equation can be expressed as:

$$\frac{d\varepsilon_p}{dt} = A\exp(-\frac{Q}{RT})[\sinh(\xi\frac{\sigma}{s})]^{1/m} \quad (1)$$

where s is the deformation resistance; $d\varepsilon/dt$ is the inelastic strain rate; A, the pre-exponential factor; Q, the activation energy; R, the gas constant; T, the absolute temperature; ξ, the multiplier of stress; σ, the applied stress; s, a single scalar as an internal variable flow; and m is the strain rate sensitivity of stress.

The evolution equations for s are

$$\frac{ds}{dt} = \left[h_0 \left| 1 - \frac{s}{s^*} \right|^a \left\{ \frac{1-(s/s^*)}{|1-(s/s^*)|} \right\} \right] \frac{d\varepsilon_p}{dt}; \ a>1 \quad (2)$$

With

$$s^* = \hat{s} \left[\frac{d\varepsilon_p/dt}{A} \exp\left(\frac{Q}{RT} \right)^n \right] \quad (3)$$

where h_0 is the hardening/softening constant; s^0, the saturation value of s; a, the strain rate sensitivity of hardening or softening; $\hat{s}$, the coefficient for the saturation deformation resistance; and n is the strain rate sensitivity of saturation. The linear elastic material properties and viscoplastic constants are given in table 1 and 2, respectively.

Table 1 Linear elastic material properties[19, 20]

Material properties	Cu	Solder
Yang's modulus(GPa)	117	49.8
Poisson's ratio	0.34	0.33
Density(g/cm^3)	8.9	7.5

Table 2 Parameters for Anand's model[20, 21]

Description	Specification
$A(\text{sec}^{-1})$	22300
$Q/R(\text{K})$	8900
E	6
m	0.08
$\hat{S}(\text{MPa})$	73.81
n	0.018
$h_0(\text{MPa})$	3321.15
a	1.82
S_0	39.03

3. Results and discussion

The shear load displacement curves of SAC solder joints show typical elastic-plastic response, consisting of a linear elastic region, a rising plastic region before the maximum shear load, and a plastic decay in following and ending in fracture. Fig.3 presented the variation of the maximum shear load with the shear rate. With an increase of shear rate, the shear force initialy increased until the peak value and then decreased rapidly. Such a rate dependence is different from the previous reports of ball shear tests where the shear force increased linearly with the shear rate[15]. In the range of 0.005mm/s to 0.7mm/s, there was an optimal rate (labeled as v_0), around 0.3mm/s, corresponding to the peak shear force. On both sides of v_0, the displacement-load curves, looked quite different. The curve before the maximal shear load value was quite smooth and similar to the tensile curve of bulk SnAgCu material: linear elastic deformation, plastic yield, work hardening, necking down, and material fracture. When the shear rate was over v_0, a flat plateau appeared instead of the plastic flow stage.

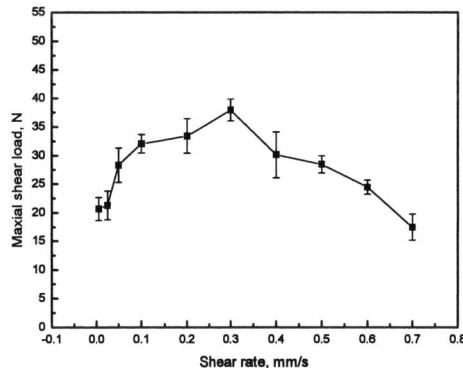

Fig.3 Maximal shear load of solder joints at different shear rate

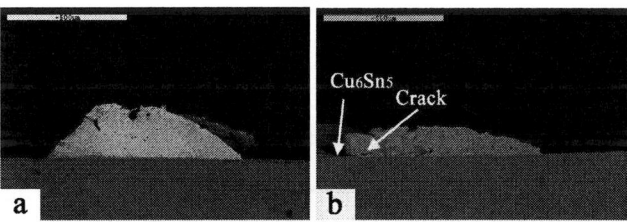

Fig.4 Cross-section views of the fracture bump remaining of the solder joints (a) with a shear speed of 0.05mm/s and (b) 0.3mm/s

The fracture surfaces of the sheared SAC solders are shown in Figs.4. Ductile shear failure in the bulk solder was observed to be the main failure mode in the solder. Micro cracks were also found as the shear rate increased At a shear rate corresponding to the peak shear load, there were both cohesive shear fracture and solder/IMC (Fig.4b) interface fracture. Notice also that the fracture bump remaining was shrinking with the increasing shear rate. Two shear modes were evident: below v_o, cohesive shear fracture controlled by solder properties and above v_o, solder/IMC interface controlled fracture.

Von Mises strain, which reflects the mixed deformation status of a material, was used to identify regions where cracks are more likely to initiate. The von Mises strain is related to the distortional energy, which has been found to control failure of materials. Fig.5 shows plastic zones after a certain displacement is applied (less than the corresponding displacement of the maximum load). It can be observed that the shape of the plastic region is quite similar to the cross-sectional view after shear testing (Fig.4). Also the size of the plastic region decreases and the whole plastic zone moves towards solder/IMC interface as the shear rate increased. It matches well with the change trend of cross-sectional views after the lap shear test (showed by white arrowhead in Fig.5ab).

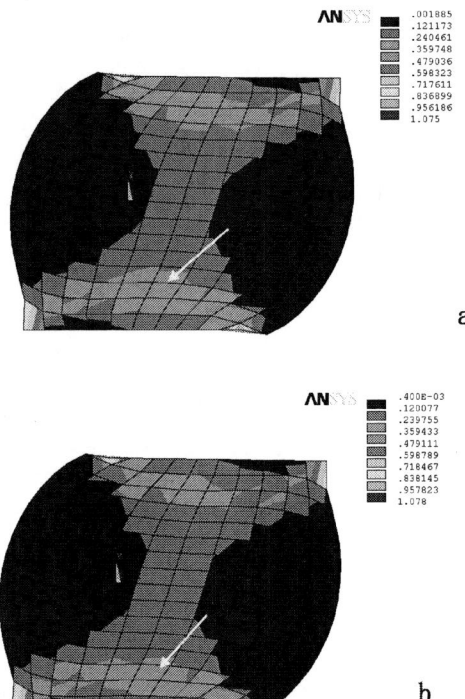

Fig.5 von Mises plastic strain distribution of element result in solder joints at shear rate (a) 0.005mm/s and (b) 0.6mm/s

Conclusions

The numerical modeling results indicated that the maximum von Mises plastic strain zone could be used to rationalize the shear fracture morphology. As the shear rate increased, the contour of the intense plastic deformation zone moved in a similar fashion to the envelope of the shear fracture surface.

Over the shear rate range from 0.005mm/s to 0.7mm/s, an optimal rate, v_o, around 0.3mm/s, corresponded to the peak shear force. On both sides of v_o, the shear responses of the solder joints were quite different. The cross-section examinations of the remaining solder joints after fracture showed that the fracture surface moved toward the solder/IMC interface with an increasing shear rate, in agreement with the modeling analysis.

Acknowledgments

The authors would like to express their appreciation to Mr. and Mrs. Daghfal for editorial revising. Q. S. Zhu and L. Zhang for experimental assistance. This work was supported by the National Basic Research Program of China (Grants Number 2004CB619306).

References

1. K.S.Siow, M.Manoharan, "Mixed mode fracture toughness of lead–tin and tin–silver solder joints with nickel-plated substrate," *Materials Science and Engineering: A*, Vol. 404 (2005), pp. 244-250.

2. W.J. Plumbridge, C.R. Gagg, "Effects of strain rate and temperature on the stress–strain response of solder alloys", *J. Mater. Sci., Mater. Electron.* Vol. 10 (1999), pp. 461–468.

3. J.G. Lee, F. Guo, S. Choi, K.N. Subramanian, J.P. Lucas, T. R. Bieler, "Residual-mechanical behavior of thermomechanically fatigued Sn-Ag based solder joints," *Journal of Electronic Materials*, Vol. 31, No. 9 (2002), pp. 946-952.

4. W. W. Lee, L. T. Nguyen, and G. S. Selvaduray, "Solder joint fatigue models: review and applicability to chip scale packages" *Microelectronics Reliability*, Vol. 40, No. 2 (2000), pp.231-244.

5. J.H.L. Pang, K.H. Tan, X.Q. Shi, Z.P. Wang, "Microstructure and intermetallic growth effects on shear and fatigue of solder joints subjected to thermal cycling aging" *Materials Science and Engineering: A*, Vol. 307 (2001), pp. 42-50.

6. N.M. Poon, C.M.L. Wu, J.K.L. Lai and Y.C. Chan, "Residual shear strength of Sn–Ag and Sn–Bi lead-free SMT joints after thermal shock" *IEEE Transactions on Advanced Packaging*, Vol. 23 , No. 4 (2000), pp. 708-714.

7. C.H. Raeder, L.E. Felton, V.A. Tanzi and D.B. Knorr, "The effect of aging on microstructure, room temperature deformation, and fracture of Sn-Bi/Cu solder joints" *Journal of Electronic Materials*. Vol. 23, No. 7 (1994), pp. 611-617.

8. Y.L. Shen, K. C. R. Abell and S. E. Garrett, "Effects of grain boundary sliding on microstructural evolution and damage accumulation in tin-lead alloys" *International Journal of Damage Mechanics*, Vol. 13 (2004), pp. 225-240.

9. F.S. Shieu, Z.C. Chang, J.G. Sheen, C.F. Chen, "Microstructure and shear strength of a Au-In microjoint" *Intermetallics*, Vol. 8, No. 5 (2000), pp.623-627.

10. W.J. Tomlinson, A. Fullylove, "Strength of tin-based soldered joints" *Journal of Material Science*, Vol. 27 (1992), pp.5777-5781.

11. H.T. Lee, S.Y. Hu, T.F. Hong,Y.F. Chen, "The Shear Strength and Fracture Behavior of Sn-Ag-*x*Sb Solder Joints with Au/Ni-P/Cu UBM" Journal of electronic materials, 2008 in press，DOI: 10.1007/s11664-008-0396-5.

12. R.J. Coyle, P. P. Solan, "The influence of test parameters and package design features on ball shear test requirements" *Proc 26th IEEE/CPMT International Electronics Manufacturing Technology Symposium*, Santa Clara, CA, Oct. 2-3, 2000, pp. 168-177.

13. M.C. Yew, C.Y. Chou, K.N. Chiang, "Reliability assessment for solders with a stress buffer layer using ball shear strength test and board-level finite element analysis" *Microelectronics Reliability*, Vol.47 (2007) 1658–1662.

14. X.J. Huang, S.W.R.Lee, C.C. Yan, S. Hui, "Characterization and analysis on the solder ball shear testing conditions" Soldering and surface mount technology, Vol. 14, No. 1 (2002), pp. 45-48.

15. Julian Yan Hon Chia, B. Cotterell, Tai Chong Chai, "Evaluation of displacement rate effect in shear test of Sn-3Ag-0.5Cu solder bump for flip chip application" *Materials Science and Engineering A*, Vol.417 (2006) pp. 15-20

16. Y.L. Shen, N.Chawla, E.S. Ege, X. Deng, "Deformation analysis of lap-shear testing of solder joints" *Acta Materialia*, Vol. 53 (2005), pp. 2633-2642

17. R. Erich, R.J. Coyle, G.M. Wenger, A. Primavera, "Shear Testing and Failure Mode Analysis for Evaluation of BGA Ball Attachment" *Proc 24th IEEE/CPMT International Electronics Manufacturing Technology Symposium*, Austin, TX,18-19 Oct. 1999, pp. 16-22.

18. R.J. Coyle, D.E.H. Popps., A. Mawer, D.P. Cullen, G.M. Wenger, P.P. Solan, "The effect of modifications to the nickel/gold surface finish on assembly quality and attachment reliability of a plastic ball grid array*" IEEE Transactions on components and packaging technologies*, Vol. 4, No. 26 2003, pp:724-732.

19. Sakai H, Nishimura K, Motegi M, Sakuyama S. "Fatigue life evaluation of lead free solder" *Proc 8th symposium of Microjoining and Assembly Technology in Electronics*, 2002. pp. 443-448.

20. G.Z. Wang, Z.N. Chinese journal of applied mechanics, Vol. 17, No. 3 (2000) pp.133-139.

21. Database for Solder Properties with Emphasis on New Lead-free Solders Release 4.0, on website http://www.boulder.nist.gov/div853/lead_free/props01.ht ml.

Capability Study on the Destructive Pull Test of 1 Mil Gold Wire Bond and Its Asymmetric Distribution

Jin Peng

The Key Laboratory of Integrated Microsystems, Shenzhen Graduate School of Peking University,
Shenzhen university town, Xili, Shenzhen, P.R. China,
Phone: 0755-26032269, Email: Jinpeng@szpku.edu.cn

Abstract

The Statistical Process Control (SPC) of destructive pull test on 1 Au gold wire is studied on two High-Rel devices, namely #1 and #2. It was found that the Cpk of both parts are around 1.2, which have not meet our goal of 1.33 and the ultimately Six Sigma (Cpk= 2). However, this study shows that the traditional method of calculating Cpk has underestimated the capability of process control if considering the tolerance as asymmetric, such as the case for 1mil gold wire bond.

Capability Study

The wire-bond pull test is the universally accepted method to verify the quality of a wire bonding process.[1,2] In the pull test, a small hook is placed in the center of the wire span between the substrate and the lead frame and pulled up in a direction normal to the bonding plane. In the destructive test, the wire is pulled to failure and a pull force value is recorded. The military test standard for microcircuit, MIL-STD-883E [4], specifies the purpose and procedure of destructive pull test. The minimum bond strength requirement (Lower Spec Limit) for 1 mil gold wire bond is 3 gram. Although the military standard is targeting the space and military products, its general application has been extended to high reliability industry, automotive and medical electronics. As for the wire bonding, the most basic interconnecting method, the high yield and reliable bonding requirements against high current, stress and elevated temperature have presented in the 21st century electronics – power LED for solid state lighting, MEMS and SiP etc.

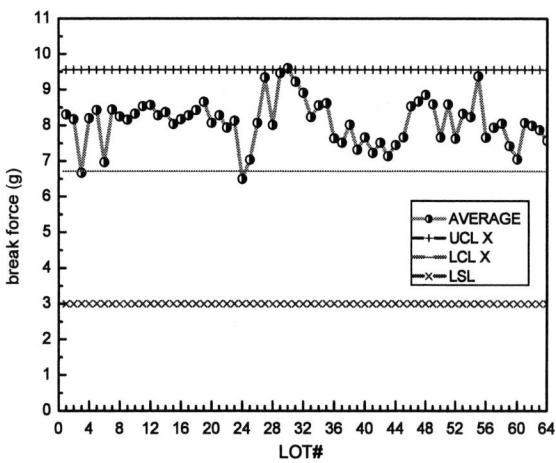

Figure 1. Control chart for #1, 64 lots with 10 wires per lot. Grand average=8.14 g, UCL (upper control limit) = 9.56 g, LCL (lower control limit) = 6.72 g, LSL =3 g, Cpk= 1.14

The data on #1 and #2 were taken from the production run within a year time span. Figure 1&2 are plotted following the standard control chart method [3]. The lower Spec limit (LSL) is 3 gram and no upper spec is defined. The typical destructive pull test reading for a 1 mil gold wire is 7~8 gram.

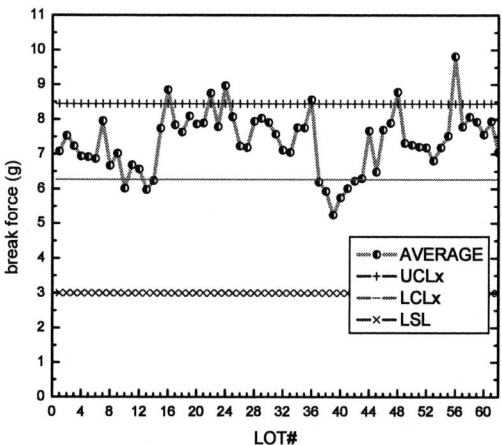

Figure 2. Control chart for #2, 62 lots with 10 wires per lot. Grand average = 8.14 g, UCL = 8.45 g, LCL = 6.28 g, LSL =3 g, Cpk= 1.27

Experimental procedure

K&S Model 4124 Semi-Automatic Gold Ball Bonder was used for bonding, the gold wire is of 1 mil diameter with 99.99 purity and five percent elongation. The work holder temperature was set at 150°C+/- 5°C. All pull test data were collected on Dage Model PC 2400 Wire Pull, Ball Shear and Die Shear Tester. 10 wires were pulled at beginning of each lot following the MIL-PRF-19500 requirement[5].

Table 1. Failure mode recorded in #1 and #2

Codes	Wire breaking mode	# 1	#2
0	operator/machine error	1	0
1	neck break	537	512
2	wire break	101	107
3	bond lifted from die	0	1
4	bond lifted from other than die	1	0
5	die metalization lifted	0	0
6	substrate, package metal lifted	0	0
7	die fracture	0	0
8	substrate fracture	0	0

978-1-4244-2739-0/08/$25.00 ©2008 IEEE

Failure mode

Besides the wire break strength, information regarding the failure mode is collected, see Table 1. For example, wire breaks, lifts etc. are typically categorized and provide important additional information along with the recorded pull strength value. Neck and wire breaking are the main failure modes. Should abnormal breaking mode repeat, corrective actions are required.

Asymmetric distribution

There are many aspects of the wire bonding process that must be considered besides the physical placement of the bonded wire. The wire metallurgy and aging effects, the wire diameter and elongation, surface cleanliness of the bond pad, potential failure mechanisms, the degrading effects of temperature, the materials and morphology of the bond pad metallization can all adversely effect bond quality[1].

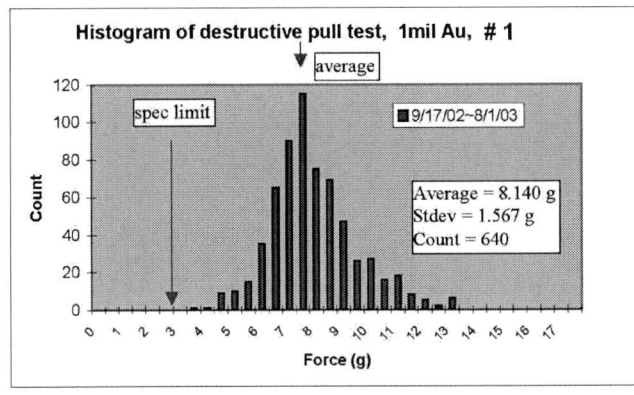

Figure 3. Asymmetrical distribution of breaking force in #1

In Figure 3, among all 640 pull data on #1, the maximum value is 13.4 gram, while minimal value is 3.84 gram and average is 8.14 gram. The range of breaking force variation above the average breaking force (Maximum-Average = 13.5-8.14) is 5.36 gram, while the range below average (Average-Minimum = 8.14 -3.84) is 4.3 gram.

Similar results are also observed in #2, as shown in Figure 4. The maximum force is 7.35 gram, while minimum force is 3.165 gram and average is 7.36 gram. The range of breaking force above the mean is 9.99 gram, while the range below the mean is 4.2 gram. The observed range difference is the result of asymmetrical distribution where breaking force can exceed mean value more than below it.

Because of the apparent asymmetry of wire breaking force distribution, the calculated overall standard deviation using symmetrical Gaussian distribution is larger than the actual spreading near the LCL region. Thus the Cpk rate will predict a higher failure rate (below LCL) than the actual value in the pull test, Figure 5 illustrates this difference. The limitation of Cpk on estimating process capability in with asymmetric tolerance has been studied in recent years. A more accurate index, such as Cpm, Cpmk and C"pmk have been proposed to deal with asymmetric tolerance[6].

However, there is no report on applying those new indexes on wire bond pull test.

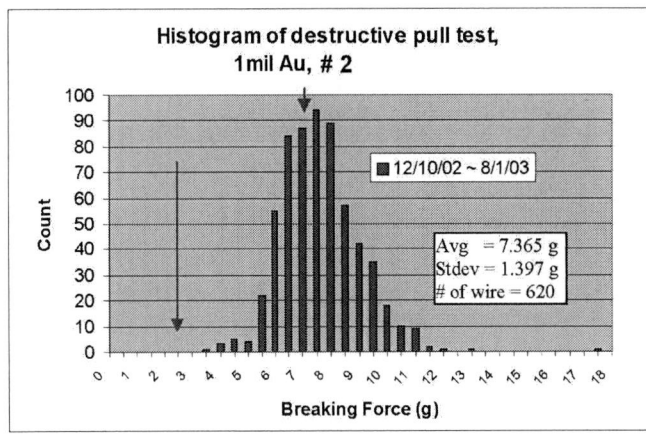

Figure 4. Asymmetrical distribution of breaking force in #2

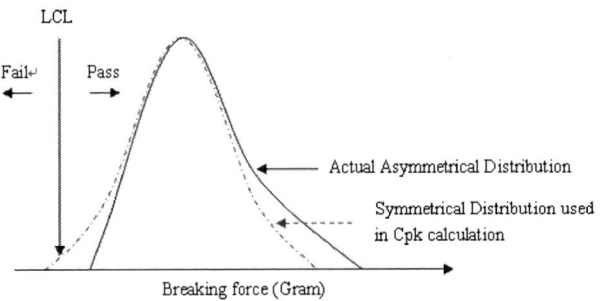

Figure 5. Failure rate schematic, below lower control limit, calculation using symmetrical and asymmetrical distribution.

Conclusions

Previous studies had shown that the wire pull test strength for the gold thermosonic first bond can be fitted with Weibull distribution to extract localization factors [7,8]. In the two-bond (first and second bond) pull test covered in this article, the breaking force is complicated by hook location, pulling angle, wire elongation etc [1]. The exact distribution is yet to be studied, however, it can also be fitted with three-parameter Weibull to characterized its asymmetrical distribution. Given the extend of asymmetrical break force distribution, the Cpk may not be a good index to describe process capability.

To improve the process control of wire bonding process, the failure mode analysis and visual inspection are essential additions to pull test SPC. A well controlled pull testing location relative to the wire will reduce the variation of pull test results. Furthermore, a more uniform wire length and loop height within the device will also reduce breaking force variation, and thus increase the Cpk.

References

1. George G. Harman, " Microelectronic wire bond pull test- How to use it, How to abuse it." IEEE Transactions on components, Hybrids, and Manufacturing Technologies, Vol. CHMT-1, No3, Sept 1978 P203

2. George G. Harman, "Reliability and Yield Problems of Wire Bonding in Microelectronics", published by ISHM (Society for Hybrid Microelectronics).

3. "Statistical process control specification", Internal Spec (unpublished).

4. MIL-STD-883E Mothod 2011.7, "Destructive Bond Pull".

5. MIL-PRF-19500M, "General Performance Specification for Semiconductor devices"

6. W.L.Pearn, et al "Estimating process capability index C pmk for asymmetric tolerance: Distributional properties",. Metrika (2001) 54: 261-279.

7. Cher Ming Tan, et al, "A new Quality Control Parameter in Wafer Fabrication for Wire Bonding Integrity," Proceedings of SPIE Vol. 4229 (2000).

8. S.J. Hu, et al, "Gold wire weakening in the thermosonic bonding of the first bond", IEEE Trans. On Components, Packaging, and Manufacturing Technology, Part A, 18 (1), p230, 1995.

Mechanical Reliability Estimation for μBGA Solder Joints Based on Heating Factor Q_η

Tao Bo, Yin Zhouping, Ding Han and Wu Yiping
National Key Laboratory of Digital Equipment and Technology
Huazhong University of Science and Technology, China, taobo@mail.hust.edu.cn

Abstract.

A novel method of mechanical reliability analysis on vibration fatigue failure of μBGA solder joints, based on the heating factor Q_η, is introduced. Firstly, a two-parameter weibull distribution is used to model the collected data of vibration fatigue lifetime for different Q_η. After that, two explicit functions are deduced in a unified mathematic expression form, which give an intuitionistic description of the MTTF and reliability of solder joints against induced variable Q_η, thus revealing definitely the effect of Q_η on the mechanical fatigue lifetime of solder joints suffering from cyclic vibration loading. Numerical analysis and calculation are performed. The results show that the solder joints reflowed at Q_η near 500 have higher reliable, and those reflowed farther away this optimal process parameter have less reliability.

Keywords. Reliability Estimation, Mechanical Fatigue Failure, Reflow Optimization, Heating Factor

1. Introduction

Solder joints' reliability is one of the most critical issues for the structural integrity of surface mount electronics, especially for the automotive and portable products. For the micro ball grid array (μBGA) package, the mechanical fatigue failure of solder joints is one of the key modes of failure when they suffer vibration, bend, twist, impact, and so on [1-2]. Among them, vibration is one of the important forms of mechanical stress experienced by the printed circuit board (PCB) assemblies using μBGA components in practice. It is reported that vibration failure is the cause of electronics assembly failure in about 20% of all failure modes. Historically, the reliability analysis of eutectic solder joints was focused on the study of the intermetallic compound (IMC) layer forming mechanism and its mean thickness characteristic [3-5]. Many literatures on the formation mechanism of IMC layer and its effect on the reliability of solder joints have been found. However, the measurement of the micro IMC thickness is very inconvenient and complicated, which is usually achieved through the scanning electron microscope (SEM) or energy dispersive X-ray (EDX) analysis of the crossing of solder joints. Evidently, it is hard and even impossible to achieve rapid online quality detection and control for solder joints during reflowing process.

In our previous research [6], a new reflow process parameter of μBGA solder joints, i.e. heating factor Q_η, has been presented to characterize the coupling effects of the reflow temperature and time on the formation of IMC layer. For the eutectic solder, Q_η was defined as the integral of the measured temperature over the melting point temperature with respect to the dwell time above the liquidus. A typical reflow profile characterized by Q_η is shown in Fig.1. It has been found that the Q_η has a dominant influence on the formation of IMC layer. Solder joints could not be formed successfully if the Q_η value is too small, even though they are soldered under a higher temperature or a longer time condition. It is found that in the diffusion-controlled growth phase of IMC layer, there is an approximately linear relationship observed between the mean IMC thickness value δ and the heating factor value Q_η. A linear growth indicates that the formation of the interfacial IMC layer is an interfacial-reaction-controllable process [5]. In other words, if the process parameter Q_η was properly chosen and controlled during reflowing, an ideal IMC layer, which means more reliable solder joints, could be obtained.

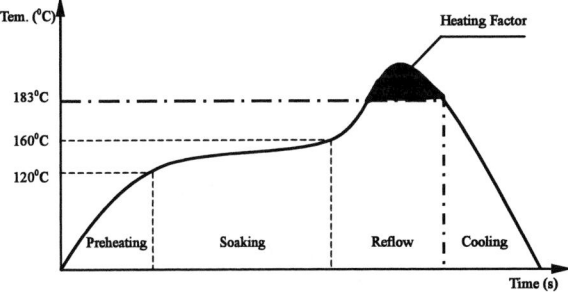

Fig.1 A typical reflow profile characterized by Q_η

This work will continue our previous work to study the influence of the reflow process parameter Q_η on the mechanical reliability of eutectic solder joints to reveal the potential effecting laws of Q_η on the fatigue lifetime and reliability of solder joints. Based on the vibration fatigue lifetime test data of μBGA sold joint, two explicit functions are deduced, which give a more intuitionistic description on the correlation of the Q_η and the MTTF and reliability of solder joints. It will provide a useful and applicable way for the computer-based online solutions to achieve numerical calculation and control for the quantitative reliability estimation during reflow process optimization.

2. Probability Distribution of Solder Joints Lifetime

2.1 Experimental Procedure

Vibration fatigue lifetime data for six groups of μBGA solder joints reflowed using various heating factor had been collected, as listed in the Table 1. In that test, the μBGA CSP46T.75-DC24 balls (diameter 0.3 mm) were mounted on to FR-4 PCBs by a high speed flexible mounter (CASIO YCM-5500V) with a eutectic 63Sn37Pb solder, and a five zone gas convection oven (BTU VIP-70N) was used to reflow the testing assembles under different heating factors, namely 33, 205, 307, 682, 864 and 2004 sºC in the N_2 environment of 20 SCFH, oxygen content of 100ppm. The thickness of the solder paste was 105μm or so. Various heating factor was achieved by controlling reflow time or temperature during the soldering.

978-1-4244-2739-0/08/$25.00 ©2008 IEEE

Table 1 Vibration test result for samples reflowed with different Q_η

No. of failed samples	Vibration time to failure (hour)					
	Q_η=33	Q_η=205	Q_η=307	Q_η=682	Q_η=864	Q_η=2004
1	0.5	0.75	16	33	1.1	0.15
2	1.8	1.8	30.5	50	3.8	0.3
3	4.7	5.1	43	66	9.3	2
4	9.5	10.5	60	79	17	4.3
5	17.8	19	72	92	28	9.2
6	29.8	30	90.5	105	45	17
7	50.5	47	110	118	70	30
8	84.1	70	123	129	102	53
9	148	100	147	143	149	86
10		141				159
Total samples	**11**	**18**	**18**	**24**	**12**	**14**

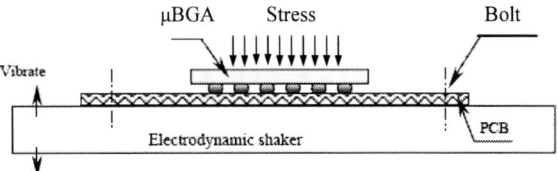

Fig.2 Schematic of vibration test stable

After that, all the assemblies were suffered from cycling vibration to test the fatigue lifetime, which will be used to evaluate the influence of heating factor Q_η on solder joints reliability. A vibration simulator system (K&D 9363) was used to test the vibration fatigue lifetime of the solder joints. The assembled PCBs are fixed to an electrodynamics shaker with four bolts positioned at each corner, as shown in Fig.2. A steel vibration stud of 56g weight was bonded to the top of each µBGA package to accelerate the failure procedure. The shaker was operated with a sinusoidal excitation of RMS acceleration amplitude 9g and frequency 30 Hz. The peak to peak displacement achieved was about 5 mm, and the pressure on solder joint was about 1.54 MPa. The electrical resistance of the solder joint was measured continuously by a computer via analog-to-digital (ADC) cards to capture the fatigue failure of solder joints. Failures were determined by a persistent detected peak in electrical resistance during vibration shock, which is indicative of the crack development of interfacial IMC layer in shear deformation.

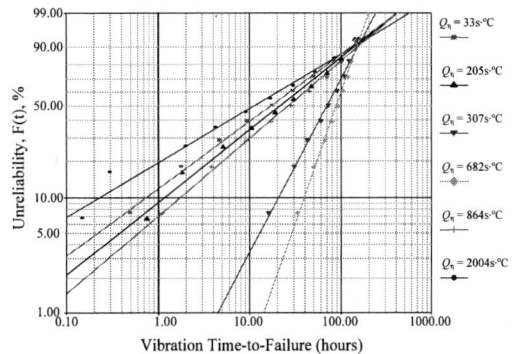

Fig.3 Weibull probability distribution of fatigue lifetime

2.2 Probability Distribution Function of Fatigue Lifetime

The vibration fatigue lifetime of solder joints is appraised with the aid of the well-known two-parameter weibull distribution method. Fig.3 shows the weibull probability plots at various Q_η characteristics. From Fig.3, it can be seen that most of the data fall on the straight-line plots except for several occasional outliers. This suggests that the two-parameter weibull distribution is a reasonable candidate to model the vibration fatigue lifetime of µBGA solder joints for each fixed Q_η, so the probability density function (PDF) of the lifetime can be given by:

$$f(t) = \frac{\beta}{\eta} \times \left(\frac{t}{\eta}\right)^{\beta-1} \times \exp\left(-\left(t/\eta\right)^\beta\right), \qquad (1)$$

where t is the vibration testing time, β is the shape parameter of weibull distribution as a function of the heating factor characteristic Q_η, and η is the distribution scale parameter.

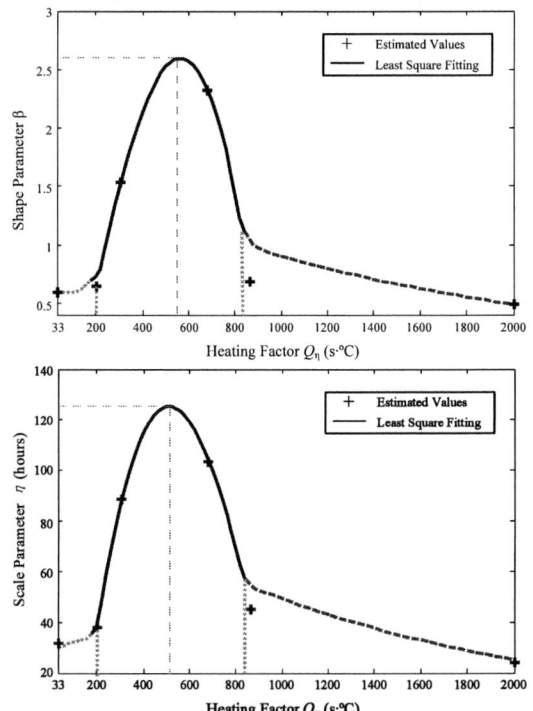

Fig.4 Relationship of β and η as of a function of Q_η

Table 2 Estimated Weibull distribution parameters for samples with different Q_η

Q_η (s°C)	Q_η=33	Q_η=205	Q_η=307	Q_η=682	Q_η=864	Q_η=2004
β	0.5965	0.6443	1.5389	2.3192	0.6889	0.4882
η (hours)	31.8975	38.1421	88.7407	103.4783	45.2106	24.2084

2.3 Estimation of the Distribution Parameters

Applying the principles of least squares and ranking to the testing vibration fatigue lifetime data, the best fitting weibull parameters, namely shape β and scale η, are calculated with respect to various Q_η respectively. They are listed in Table 2 and graphically shown in Fig.4. It is clear that the weibull parameters all vary according to the change of heating factor significantly. The results show that the shape parameter β firstly increases and then decreases with the increment of Q_η. The greatest value occurs near $Q_\eta = 550$s°C. It is also found that the value of β increases slowly before Q_η less than 200s°C and then sharply before $Q_\eta = 550$s°C. After that, it firstly decreases sharply and then asymptotically approximates a constant with the increment of Q_η. The threshold value for Q_η is 840s°C or so. It is interesting that similar law can be found for η except that the greatest value occurs when $Q_\eta = 510$s°C. Piece-to-Piece least square fitting technique is used to find the potential law of β and η against the Q_η respectively, and the resultant piece-to-piece formulae are:

$$\beta(Q_\eta) = \begin{cases} 0.59 \times \exp\left(5.89 \times 10^{-4} Q_\eta\right) & Q_\eta < 200\text{s°C} \\ -1.82 \times 10^{-5} Q_\eta^2 + 1.96 \times 10^{-2} Q_\eta - 2.68 & others \\ 1.67 \times \exp\left(-6.16 \times 10^{-4} Q_\eta\right) & Q_\eta > 840\text{s°C} \end{cases} \quad (2)$$

$$\eta(Q_\eta) = \begin{cases} 29.34 \times \exp\left(8.92 \times 10^{-4} Q_\eta\right) & Q_\eta < 200\text{s°C} \\ -7.35 \times 10^{-4} Q_\eta^2 + 0.789 Q_\eta - 89.28 & others \\ 96.38 \times \exp\left(-6.71 \times 10^{-4} Q_\eta\right) & Q_\eta > 840\text{s°C} \end{cases} \quad (3)$$

3. Reliability Analysis at Various Heating Factor

Although the equation (2) and (3) are drawn from the given test data, it is still reasonable to conclude that both weibull distribution parameters β and η are the functions of induced variable Q_η respectively. So submitting parametric $\beta(Q_\eta)$ and $\eta(Q_\eta)$ into the equation (1) yields the conditional probability density function of the vibration fatigue lifetime at given heating factor Q_η, written as:

$$f\left(t \mid Q_\eta\right) = \frac{\beta(Q_\eta)}{\eta(Q_\eta)} \times \left(\frac{t}{\eta(Q_\eta)}\right)^{\beta(Q_\eta)-1} \times \exp\left(-\left(\frac{t}{\eta(Q_\eta)}\right)^{\beta(Q_\eta)}\right) \quad (4)$$

Typically, $\beta(Q_\eta)$ and $\eta(Q_\eta)$ are given by equation (2) and (3) respectively for the solder paste material of the eutectic 63Sn37Pb alloy. So the mean-time-to-failure denoted by MTTF and reliability denoted by $R(t)$ with respect to Q_η and vibration time t are expressed as:

$$MTTF = \int_0^{+\infty} t \times f\left(t \mid Q_\eta\right) dt = \eta(Q_\eta) \times \Gamma\left(\frac{1}{\beta(Q_\eta)} + 1\right) \text{ hours,} \quad (5)$$

where $\Gamma(\bullet)$ is the Gamma function, and

$$R(t) = \int_t^{+\infty} f\left(x \mid Q_\eta\right) dx = \exp\left(-\left(\frac{t}{\eta(Q_\eta)}\right)^{\beta(Q_\eta)}\right). \quad (6)$$

Herein, two unified expression equations of MTTF and reliability of μBGA solder joints suffering from vibration cycling against the Q_η are deduced respectively. Theoretically speaking, from equation (7) and (8), solder joints' MTTF and reliability could be approximately estimated and predicted from a known reflow profile characterized by certain Q_η value. On the other hand, carefully controlling Q_η of soldering process can achieve specified reliability goal.

3.1 Analysis of MTTF Dependence on Q_η

For the extensively used eutectic 63Sn37Pb solder, the dependence of solder joint MTTF on heating factor can be quantitatively calculated by substituting equation (2) and (3) into equation (5). Clearly, from equation (5) it can be seen that the MTTF only depends on Q_η. Numerical calculations are achieved, and the resultant MTTFs against various heating factor Q_η are graphically shown in Fig.5.

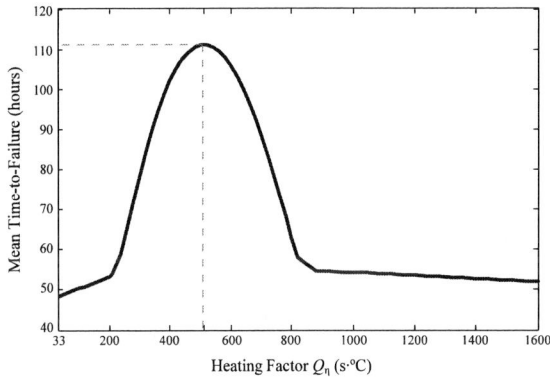

Fig.5 MTTF of solder joint versus Q_η

From Fig.5, it is found that the solder joint MTTF first increases and then decreases with the increment of Q_η. The greatest MTTF occurs at Q_η near 500 s°C. It is also found that the MTTF increases slowly before Q_η less than 200s°C and then sharply till $Q_\eta = 500$s°C. After that, MTTF begins to decrease quickly and then asymptotically approximates a constant with the Q_η increasing. When Q_η is larger than 1000s°C, the decrement of MTTF is indistinctive. The lowest MTTF value of 53 hours or so appears when the Q_η is larger than 1500s°C, and the MTTF change can be negligible with a maximum relative difference less than 5% at an increment interval of ΔQ_η=500s°C. The optimal Q_η for reflow of μBGA assembles is between 300 and 700 s°C. Hence, to achieve more highly reliable solder joint, the heating factor should be controlled in that preferable range during soldering. This is a little difference from the result drawing from a viewpoint of thermal cycling induced failure, where the MTTF monotonously decreases with the increment of Q_η [6]. That means a lower Q_η is controlled during soldering, higher reliable solder joints are obtained in the diffusion-controlled growth phase of IMC layer. That is because that the defective IMC layer of solder joint formed under lower heating factor can further grow during thermal cycling, thus resulting in a longer MTTF.

3.2 Analysis of Reliability Dependence on Q_η

Similarly, substituting equation (2) and (3) into equation (6), the solder joints' reliability dependence on heating factor for the eutectic 63Sn37Pb solder is estimated and calculated. Some typical curves of the reliability $R(t)$ against vibration time t for several heating factor values, i.e. 50, 250, 500, 750 and 1500 s°C are comparatively plotted together, as shown in Fig.6.

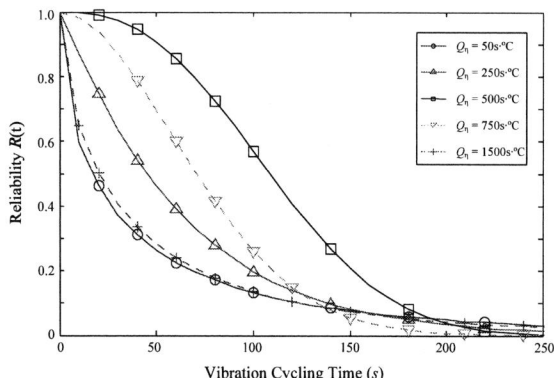

Fig.6 Reliability versus vibration time at various Q_η

From Fig.6, it is found that the reliability of solder joints $R(t)$ versus vibration time t for various heating factor value decreases monotonously while in different ways. This means that all the solder joints reflowed at various heating factor value are degraded by a single damage mechanism when suffering from same vibration cycling. However, for the solder joints tested, those that are soldered under the Q_η at 500s°C have a highest reliability for a given cycling vibration time. For the solder joints reflowed at Q_η away from 500s°C, it is found that the farther away from 500s°C the Q_η is, the lower reliability the solder joints have. The reliability profiles of Q_η=200s°C and Q_η=1500s°C are very close to each other and the variety in terms of vibration time for certain reliability goal is indistinctive. This is much different from the conclusion made from the thermal fatigue testing [6], where the larger the Q_η is chosen and controlled, the more quickly the reliability curve decreases with respect to thermal cycle. Hence, to achieve solder joints with higher reliability, the Q_η should be chosen and controlled near 500s°C during the reflow process.

4. Conclusions

A mechanical reliability analysis method of for μBGA solder joints, based on the eximious reflowing process parameter, i.e. heating factor Q_η, is introduced in this work instead of the usual mean IMC thickness characteristic, by which quantitative estimation and prediction of mechanical reliability can be done. Based on the vibration fatigue lifetime data set, a unified mathematic form is deduced and presented to explicitly reveal the effect of the heating factor on the MTTF and reliability of solder joints suffering form cycling vibration. This is very useful for a computer-based online

solution to achieve an ideal reflow profile to meet certain specified reliability goal by properly controlling Q_η of soldering, or to estimate the lifetime expectancy of solder joints soldered in a known reflow profile. Numerical calculations are achieved, and the result shows that the solder joints reflowed at Q_η near 500s°C are more reliable and have larger MTTF from a statistical viewpoint. For the solder joints reflowed at Q_η away from 500s°C, it is found that the farther away from 500s°C the Q_η is, the lower reliability and less MTTF the solder joints have. What needs to confess is that the validity of the presented numerical results and conclusion drawn based on the given test data needs to be further verified from more comprehensive investigation and test data to avoid drawing incorrect conclusion. For different solder paste materials, they may be invalid. However, the presented quantitative method can be used in any reliability analysis of surface mount solder joints and can find its way in reflow process optimization control.

Acknowledgements

This work is supported by the 863 grand program of China under grant No. 2006AA04A110, National Science Foundation of China under grant No. 50625516 and the China Postdoctoral Fund under grant No. 20070420173.

References

[1] P.L. Tu, Y.C. Chan, K.C. Huang, *et al*. Study of Micro-BGA solder joint reliability. Microelectronics Reliability 2001; 41: 287-293.

[2] Mitchell C. Assembly and Reliability Study for the Micro Ball Grid Array. IEEE/CPMT International Electronics Manufacturing Technology Symposium Proceeding 1997: 344-346.

[3] S. H. Fan et al. The effect of Reflow Condition on the Characteristics of PBGA Solder Joint. IEEE/CPMT International Electronics Manufacturing Technology Symposium Proceeding 1998: 264-268.

[4] Alex C. K. So et al. Reliability Studies of Surface Mount Solder Joints – Effect of Cu-Sn Intermetallic Compounds. IEEE Trans. on Component Packaging and Manufacturing Tech – Part B 1996; 19: 661-668.

[5] Wei Huang et al. Study of the Effect of Reflow Time and Temperature on Cu-Sn Intermetallic Compound Layer Reliability. Microelectronics Reliability 2000; 42: 1119-1234.

[6] Bo Tao, Yiping Wu, Han Ding. A Quantitative Method of Reliability Estimation for Surface Mount Solder Joints Based on Heating Factor, Microelectronics Reliability, vol.46, No.5~6, 2006, pp.864-872.

Electromigration Simulation with Consideration of the Atomic Concentration Gradient

Xuefan Chen, Lihua Liang

Zhejiang University of Technology, Fairchild-ZJUT Joint Lab.. Hangzhou 310014, China

Yong Liu

Fairchild Semiconductor Corp., S. Portland, Maine, 040106, USA

Tel: 86-571-88320294, Email: lianglihua@zjut.edu.cn

Abstract

An enhanced finite element modeling methodology based on commercial software ANSYS Multi-physics and FORTRAN code is developed for the simulation of electromigration. The electronic migration formulation taking into account the effects of the atomic concentration gradient (ACG) has been developed to show the difference in the electromigration (EM) failure mechanisms. An improved algorithm of total atomic flux and corresponding EM formulation are presented with the electro-migration, the thermo-migration, the stress-migration and the atomic concentration gradient migration. The interaction of the various driving forces during the EM process is investigated. Finite element simulations are performed for solder joints under high current density with and without consideration of atomic concentration gradient. Simulations show that atomic concentration gradient migration can't be ignored. It is also found that the atomic concentration distribution majorly depends on the current density, atomic concentration gradient and the stress gradient of the structure.

1 Introduction

EM is a phenomenon of mass transport in metallization structures when the metallization is stressed with high electrical current density. Traditional EM models are based on the solutions of diffusion equations, and the electron wind force is proposed to be the sole driving force. The atomic diffusivity of metal atoms are altered through either interconnect material microstructure or texture in-homogeneity, resulting in atomic flux divergence and hence the formation of voids and hillocks.

In classical electromigration studies, Black equation has been successful in characterizing the operation life of aluminum and copper traces on a chip, but a lot of experimental work and modeling have shown that Black theory is not accurate enough to evaluate the reliability or unable to reasonably simulate the failure due to void formation. One of the reasons is that Black equation only takes into account "electronic wind" by neglecting other driving force factors, so the correctness of the Black equation in predicting the median time to failure of EM is also become questionable. There are lots of researchers focusing on studying the failure mechanism of EM [1-5]. It shows that: metallization under the high electrical current density , the main driving forces of EM diffuse are shown as follows: (1) electron wind force (EWF); (2) stress gradient induced driving force (SGIDF); (3) temperature gradient induced driving force (TGIDF); and (4) atomic concentration gradient induced driving force (ACGIDF). However, the majority of the literatures have ignored the atomic concentration gradient induced driving force, especially for the solder joint structures.

In recent years, the researchers have found the phenomenon of electromigration inside the solder which is adjacent to the under bump metallization (UBM) layer [6-8]. The current density is nonuniform within the solder ball. The nonuniform distribution can be attributed to current crowding in which a large portion of the current enters the solder bump at the nearest corner of UBM and solder ball. The concentration of high current and high temperature at this corner causes the formation of voids. The void propagates across the entire solder/UBM interface and leads to failure.

In this work, atomic concentration distribution based finite element analyses has been performed to show the difference in the EM failure mechanisms at different test conditions for solder structure. An improved algorithm and corresponding EM formulation are presented, taking into account the electromigration, the thermo-migration, the stress-migration and the atomic concentration gradient migration contributions. A sub-model technique is introduced in the multiple physics analysis. A global 3D chip scale package (CSP) with PCB is modeled using relative coarse elements, and a refined mesh sub-model is constructed for the detailed coupled analysis is obtained. Finally, the atomic concentration import to ANSYS, the atomic concentration distribution of the solder joint is gotten, and the comparison with the measurement result is discussed.

2 Migration Formulation

2.1 Practical electromigration formulation

Electromigration is a diffusion process which is controlled by mass transportation. The time dependent evolution equation of the local atomic concentration can be given as

$$\nabla \cdot \vec{q} + \frac{\partial c}{\partial t} = 0 \tag{1}$$

where c is normalized atomic concentration $c=N/N_0$, N is the atomic concentration and N_0 is the initial (equilibrium state) atomic concentration in the absence of a stress field, t is the time; $\vec{q}$ is the total normalized atomic flux. In fact, Eq. (1) is mass conservation equation.

Assuming that driving forces of atomic flux are electrical field forces, thermal gradient, stress gradient, and atomic concentration gradient, respectively, the normalized atomic flux is given by [9-11].

$$\vec{q} = \vec{q}_{Em} + \vec{q}_{Th} + \vec{q}_S + \vec{q}_C \tag{2}$$

where

$$\vec{q}_{Em} = \frac{c}{kT} Z^* e \rho \vec{j} D_0 \exp\left(-\frac{E_a}{kT}\right) \tag{2.a}$$

$$\vec{q}_{Th} = -\frac{c}{kT} Q^* \frac{\nabla T}{T} D_0 \exp\left(-\frac{E_a}{kT}\right) \tag{2.b}$$

$$\vec{q}_s = \frac{c}{kT} \Omega \nabla \sigma_m D_0 \exp\left(-\frac{E_a}{kT}\right) \tag{2.c}$$

$$\vec{q}_c = -D_0 \exp\left(-\frac{E_a}{kT}\right) \nabla c \tag{2.d}$$

Rewrite Eq. (2), we have

$$\vec{q} = \frac{cD}{kT} Z^* e \rho \vec{j} - \frac{cD}{kT} Q^* \frac{\nabla T}{T} + \frac{cD}{kT} \Omega \nabla \sigma_m - D \nabla c$$
$$= c \cdot F(T, \sigma_m, \vec{j}, \cdots) - D \nabla c \tag{3}$$

where

$$F(T, \sigma_m, \vec{j}, \cdots) = \frac{D}{kT}\left(Z^* e \rho \vec{j} - Q^* \frac{\nabla T}{T} + \Omega \nabla \sigma_m\right) \tag{3.a}$$

where k is Boltzmann's constant; e is the electronic charge; Z^* is the effective charge which is determined experimentally; T is the absolute temperature, ρ is the resistivity which is calculated as $\rho = \rho_0 (1 + \alpha(T - T_0))$, α is the temperature coefficient of the metallic material; D is the diffusivity, $D = D_0 \exp\left(-\frac{E_a}{kT}\right)$, E_a is the activation energy, D_0 is the thermally activated diffusion coefficient; $\vec{j}$ is the current density vector; Q^* is the heat of transport; Ω is the atomic volume; σ_m is the local hydrostatic stress.

Assume blocking boundary condition for atomic flux, we have

Boundary conditions:

$$\vec{q} \cdot \vec{n} = 0 \quad on \quad \Gamma \tag{4}$$

Initial conditions (for all nodes):

$$c_0 = 1 \tag{4.a}$$

2.2 Discretization for finite element method implementation based on Galerkin

(1) Spatial Approximation

$$\int w \left(\nabla \cdot \vec{q} + \frac{\partial c}{\partial t}\right) dV = \int w \frac{\partial c}{\partial t} dV + \int w \cdot \nabla \cdot \vec{q} dV$$

$$= \int w \frac{\partial c}{\partial t} dV - \int \frac{\partial w}{\partial \vec{n}} \cdot \vec{q} dV + \int w \cdot (\vec{q} \cdot \vec{n}) d\Gamma \tag{5}$$

$$= \int w \frac{\partial c}{\partial t} dV - \int \frac{\partial w}{\partial \vec{n}} \cdot \vec{q} dV = 0$$

where w is the weight function.

Suppose that $c = \sum_{j=1}^{n} \psi_j c^j$, $\dot{c} = \sum_{j=1}^{n} \psi_j \dot{c}^j$ and $w = \psi_i$, where ψ_i is the shape function. The matrix form of Eq. (5) can be gotten as below

$$[M]\{\dot{c}\} + [K]\{c\} = 0 \tag{6}$$

where the matrix $[M]$ is independent of time, and $[K]$ keep constant in an incremental step.

(2) Time Approximation

Recall the Eq. (1),

$$\nabla(cF - D\nabla c) + \frac{\partial c}{\partial t} = 0, 0 < t < T \quad and \quad c(0) = c_0 = 1 \tag{7}$$

In the finite difference solution of Eq. (7), we replace the derivatives with their difference approximation. The most commonly used scheme for solving above equation is the α-family of implicit approximation in which a weighted average of the time derivatives at two consecutive time steps is approximated by linear interpolation of the values of the variable at two steps:

$$(1-\alpha)\dot{c}_{t_i} + \alpha \dot{c}_{t_{i+1}} = \frac{c_{t_{i+1}} - c_{t_i}}{\Delta t}, \quad for \quad 0 \le \alpha \le 1 \tag{8}$$

When $\alpha = 0$, Eq. (8) gives $\dot{c}_{t_i} = \frac{c_{t_i + \Delta t} - c_{t_i}}{\Delta t}$. Since the value of the function from a step in front is used, it is termed a forward difference approximation. When $\alpha = 1$, we obtain $\dot{c}_{t_i + \Delta t} = \frac{c_{t_i + \Delta t} - c_{t_i}}{\Delta t}$, which is termed the backward difference approximation. In this work, $\alpha = 0.5$ is taken, and it is the Crank-Nicolson scheme which is stable and has the accuracy order of $O((\Delta t)^2)$ [17].

(3) Fully Discretized Equations

From Eq. (6), it can be obtained

$$\left. \begin{array}{l} [M]\{\dot{c}\}_{t_i} + [K]\{c\}_{t_i} = 0 \\ [M]\{\dot{c}\}_{t_{i+1}} + [K]\{c\}_{t_{i+1}} = 0 \end{array} \right\} \tag{9}$$

From Eq. (8), we have

$$(1-\alpha)\Delta t [M]\{\dot{c}_{t_i}\} + \alpha \Delta t [M]\{\dot{c}_{t_{i+1}}\} = [M](\{c_{t_{i+1}}\} - \{c_{t_i}\}) \tag{10}$$

Substitute Eq. (9) into Eq. (10), it can be obtained

$$([M] + \alpha \Delta t [K])\{c_{t_{i+1}}\} = ([M] - (1-\alpha)\Delta t [K])\{c_{t_i}\} \tag{11}$$

Thus, the atomic concentration c in i^{th} step can be obtained based equation (11) where we assume the initial atomic concentration $c_0 = 1$.

2.3 Algorithm Verification

Clement and Lloyd performed the numerical investigations of the 1D electromigration boundary value problem [12]. In the time dependent evolution equation, only electrical field driving force and concentration gradient are considered. A thin film metallic conductor with finite length l stressed by constant current at constant temperature is considered.

Fig.1 shows the normalized atomic concentration at $x=0$ as the normalized time $\tau=a^2Dt$ for various conductor lengths with blocking boundary condition $q(0,t) = q(-l,t) = 0$, where $a = \frac{Z^* e \rho j}{kT}$, j and T are constant. From Fig. 1, the presence of the saturation values at long time indicate a length effect. With the decrease of the conductor length, the electromigration failure becomes immune gradually, which has been observed experimentally. Maybe the different algorithms between this work with weight residual method (Galerkin finite element method) and ref [12] using finite difference techniques for space discretization as well as the different schemes for time discretization induce the difference in Fig. 1.

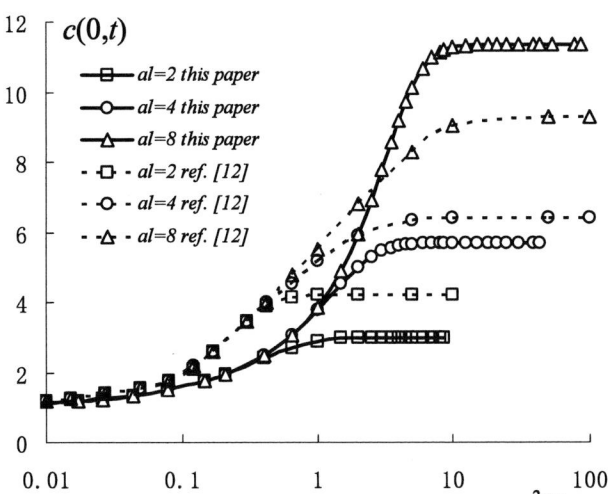

Fig 1 normalized atomic concentration as normalized time $\tau=a^2Dt$ for various conductor lengths with blocking boundary condition.

3 Electromigration Simulation Procedures

Sub-model technique in ANSYS is introduced to get the better response of the electronic migration. The global thermal-electric coupled field model uses Solid-69 element and the global stress model uses Visco-107 element for solder bumps and Solid-45 element for the remaining parts of the model[14-16]. The global structure is modeled using relative coarse elements in the first step as shown in Fig. 3. In the second step, a refined thermal-electric coupled field sub-model and a refined stress sub-model with UBM which considering its Equivalent material parameters(Al/NI(V)/Cu) layer are then constructed as shown in Fig. 4. To examine the effect of the different driving forces of EM, their respective corresponding aotmic concentration needs to be calculated. However, ANSYS post-processing module has the limitation to calculate the aotmic concentration obtained from its solution. In this work, a FORTRAN code is developed to process data from ANSYS migration simulation. That means the EM basic informaton such as atomic fluxes is calculated in ANSYS as mentioned above and its' results are transfered into the self-developed FORTRAN code for the computation of the aotmic concentration due to various EM driving forces based on Eq. (11). After that, the atomic concentration result is transferred back to ANSYS for the post process and further analysis. The flowchart of the simulation and the mathematical implementation is illustrated in Fig.2.

With the operation of the ANSYS APDL routines that have been programmed beforehand the local coordinate, current density, hydrostatic stress and temprature on every node are obtained and saved in an external file. Then the ANSYS reads the external data from the FORTRAN code. Finally after post process,we can plot the atomic concentration distribution.

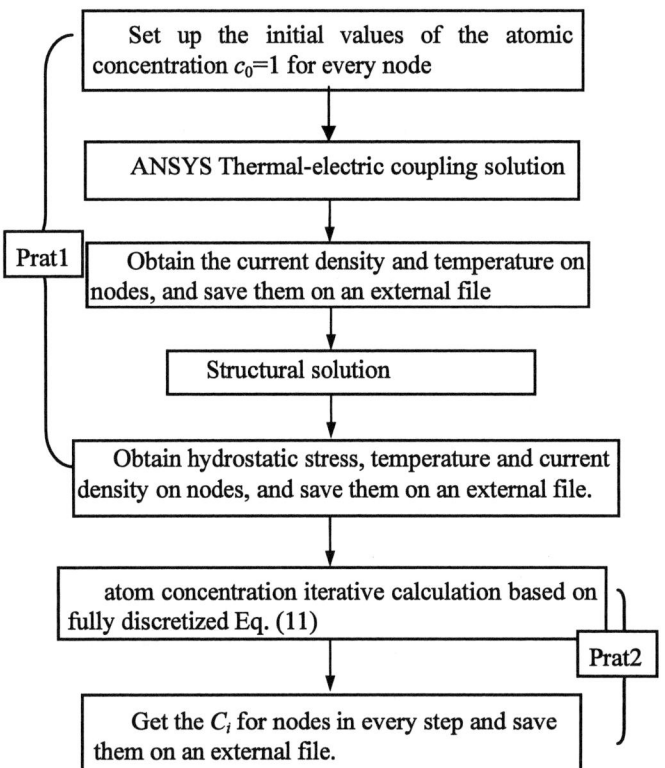

Fig 2 Flow chart of the simulation and the mathematical implementation is illustrated

4 Numerical example

4.1 CSP Structure

To correlate our simulation methodology with the experimental data, the CSP package in reference [6] is modeled, which has 36 bumps with 500 μm pitch. The exterior 20 solder bumps are assumed to connect with each other in a daisy chain as shown in Fig. 3. The schematic cross-sections of the test module and a solder joint are shown in Fig. 3. The under bump metallization (UBM) on the silicon die side is equivalent material parameters(Al/NI(V)/Cu) layer. The diameter of the solder (Sn37Pb63) joint is around 300 μm and the height is about 200 μm. The solder joints are encapsulated in the underfill between the silicon die and PCB.

A global chip scale package (CSP) with PCB is modeled using relative coarse mesh, and a sub-model is built with refined mesh as shown in Figs. 3 and 4.

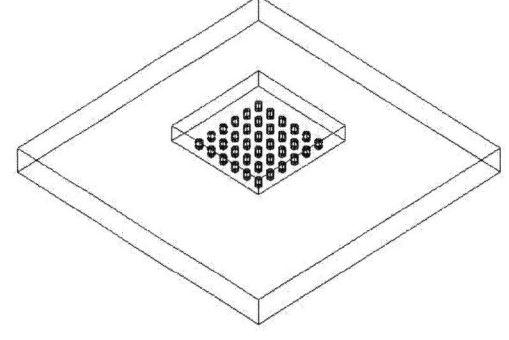

(a) A CSP package structure

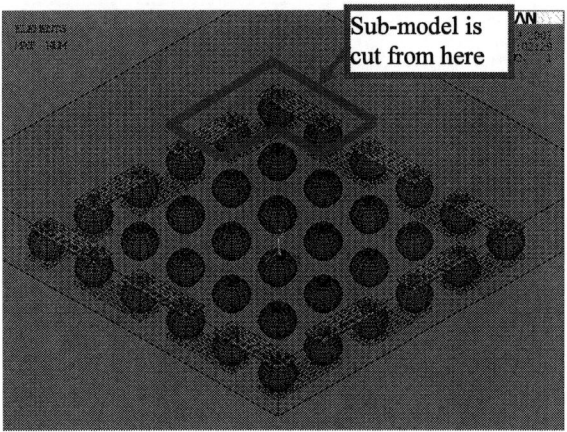

(b) Local view

Fig 3 CSP package model.

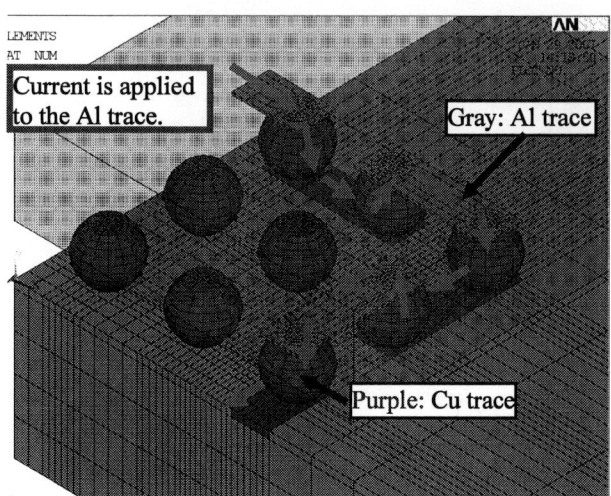

Fig 5 Electron flow in a global model

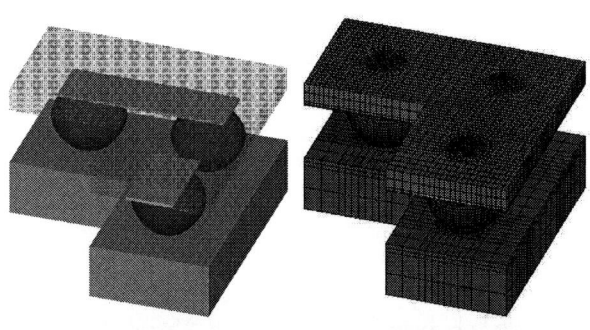

(a) View of the top right of the model

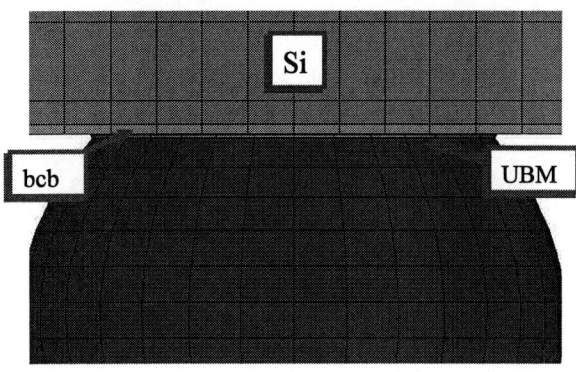

(b) Local view

Fig 4 Sub-model of CSP package

Fig. 5 shows the electron flow for the thermal-electric coupled field whole model with 1.7A current. The free convention boundary condition is applied with 17 W/m²·°C film coefficient and 50°C ambient temperature. Initial condition for all the nodes is that $c_0 = 1$.

Electromigration testing is conducted on flip chip microelectronic package containing eutectic 63Sn37Pb solder joints. Material properties for those are as follows (see Tables 1–2).

Table 1 Material property for numerical model

Material	Density (kg/m³)	Specific Heat (J/(kg*K))	Elastic modulus (GPa)	Poisson's Ratio
63Sn37Pb	8420	150	30.8	0.4
Al	2710	902.1	69	0.33
Cu	8900	385.2	127.7	0.31
Ni	8900	443.8	200	0.31
BCB	1050	2180	2.9	0.34
Die	2300	/	131	0.3
PCB	1900	/	25.4 (x,z), 11 (y) 4.971 (GXY,GYZ) 11.453 (GXZ)	0.39(xy,yz) 0.11(xz)

Material	TEC (/K)	Thermal Conductivity (W/m*K)	Electrical resistivity (Ω-m)
63Sn37Pb	24e-6	50	15.5e-8(1+2.8e-3ΔT)
Al	23e-6	240	2.61e-8(1+4.2e-3ΔT)
Cu	17.1e-6	393	1.58e-8(1+4.3e-3ΔT)
Ni	13.4e-6	91	6.32e-7
BCB	52e-6	0.29	1e17
Die	2.8e-6	150	4.4
PCB	16e-6(X, Z) 84e-6 (Y)	1.7	1e10

The electromigration parameters of 63Sn37Pb solder bump are selected from previous references [6, 13] as shown in the following: activation energy Ea=0.8eV; effective charge number Z*=-33; self-diffusion-coefficient D_0= 9.69E-05 m²/s; heat of transport Q*=0.0094eV; initial electrical receptivity R_0=15.5e-8Ω·m; temperature coefficient resistance α=3.0e-3 /k; atomic volume Ω=2.48e-29 m³/atom.

Table2 ANAND model parameters for 63Sn37Pb

Description	Symbol	63Sn37Pb
Pre-exponential factor	$A(1/s)$	26
Activation energy	$Q/R(K)$	5797
Stress multiplier	ξ	10
Strain rate sensitivity of stress	m	0.256
Coef. for deformation resistance saturation value	$\hat{s}$ (MPa)	83.12
Strain rate sensitivity of saturation value	n	0.043
Hardening coefficient	$h0$(MPa)	92148
Strain rate sensitivity of hardening coeff.	a	1.24
Initial value of s	s_0(MPa)	37.9

4.2 Multiple driving forces mechanism with and without considering atomic concentration gradient

Fig. 6-8 show current density distribution, the temperature distribution, and hydrostatic stress distribution of the corner solder ball under 1.7A. Fig. 9 shows the atomic concentration distribution of the corner ball under the 1.7A after running 500 seconds considering atomic concentration gradient base on the Eq.(3) which consider atomic concentration gradient. Whether the current density distribution or the atomic concentration distribution, it indicates that failure firstly happens in the contact interface between the UBM and solder bump.

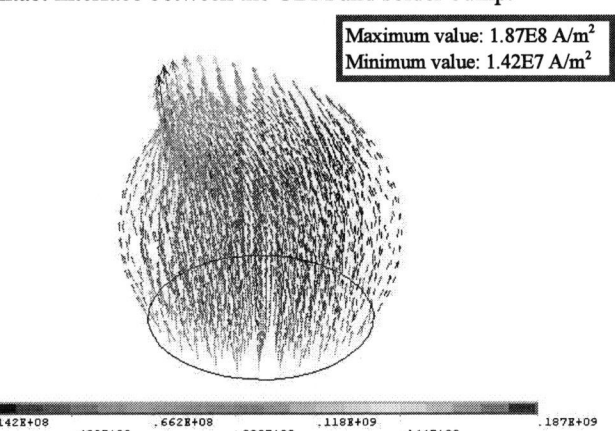

Maximum value: 1.87E8 A/m²
Minimum value: 1.42E7 A/m²

Fig 6 The current density distribution under 1.7A

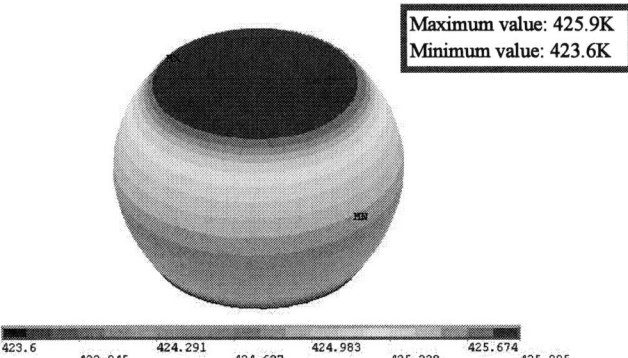

Maximum value: 425.9K
Minimum value: 423.6K

Fig 7 The temperature distribution under 1.7A

Maximum value: 7.04 E +6 MPa
Minimum value: -4.98E+7MPa

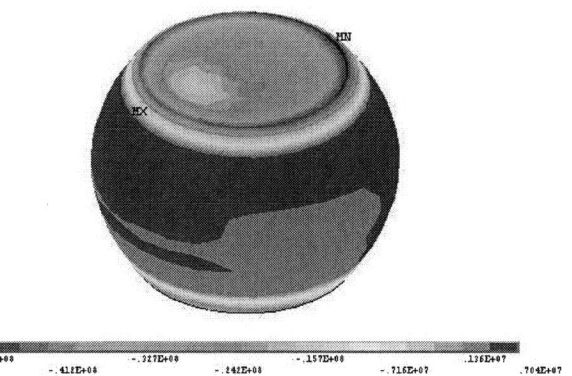

Fig 8 The hydrostatic stress distribution under 1.7A

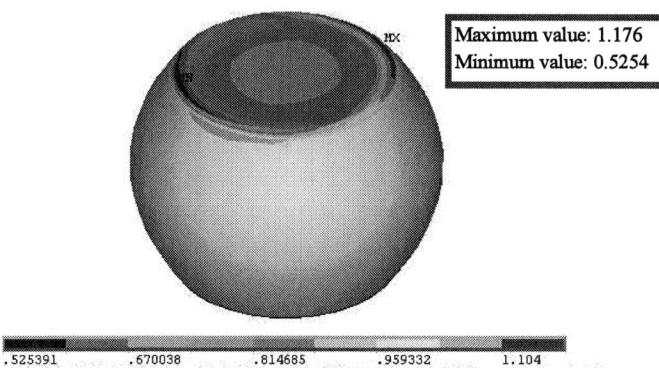

Maximum value: 1.176
Minimum value: 0.5254

Fig 9 Atomic concentration distribution with ACG

Fig 10 shows the atomic concentration distribution of the corner solder ball under the 1.7A after running 500 seconds without considering atomic concentration gradient base on the equation: $\vec{q} = \frac{cD}{kT}Z^* e\rho\vec{j} - \frac{cD}{kT}Q^* \frac{\nabla T}{T} + \frac{cD}{kT}\Omega\nabla\sigma_m$. Fig 12 shows the atomic concentration variation with and without atomic concentration gradient of the designated nodes which are shown in Fig. 11, the solid line means considering the atomic concentration gradient, on the contrary, and the dash line means without considering the atomic concentration gradient. It found that the atomic concentration gradient has a great impact on the atomic concentration distribution and can not be ignored from Fig 12.

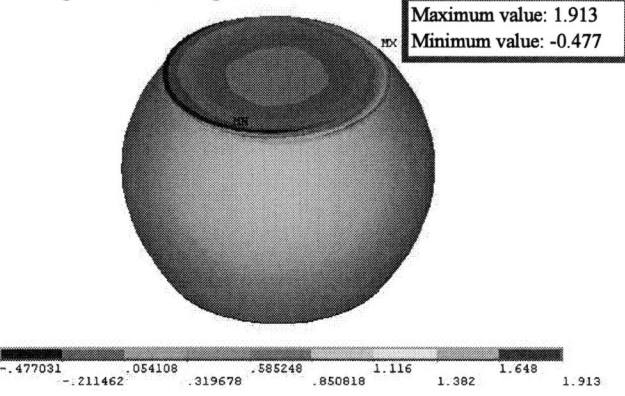

Maximum value: 1.913
Minimum value: -0.477

Fig 10 Atomic concentrations distribution without ACG

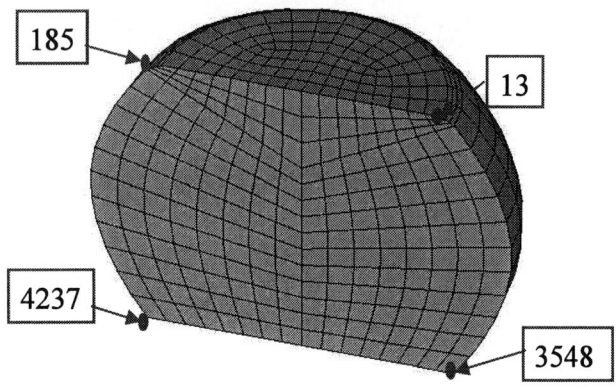

Fig 11 The designated nodes of the corner solder ball

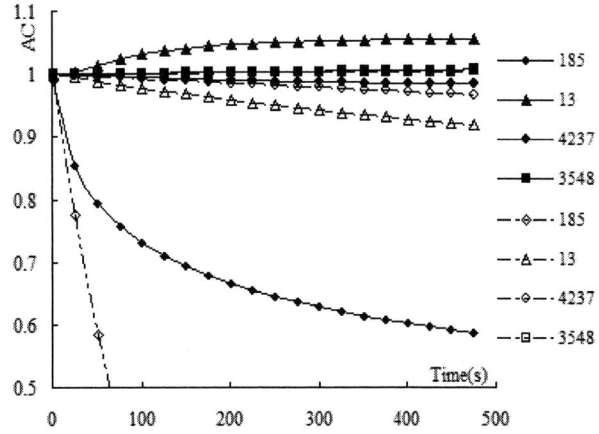

Fig 12 AC variations with and without ACG where solid line: considering atomic concentration gradient; dash line: without considering atomic concentration gradient.

4.3 Multiple driving forces mechanism without stress gradient (SG)

Fig 13 shows the atomic concentration distribution of the corner solder ball under the 1.7A after running 500 seconds without considering stress gradient based on the equation $\vec{q} = \frac{cD}{kT} Z^* e \rho \vec{j} - \frac{cD}{kT} Q^* \frac{\nabla T}{T} - D\nabla C$. Fig 14 shows the AC variation with and without stress gradient, the solid line means considering the stress gradient, on the contrary, and the dash line means without considering stress gradient. It found that the stress gradient also has a great impact on the atomic concentration distribution and can not be ignored from Fig 14.

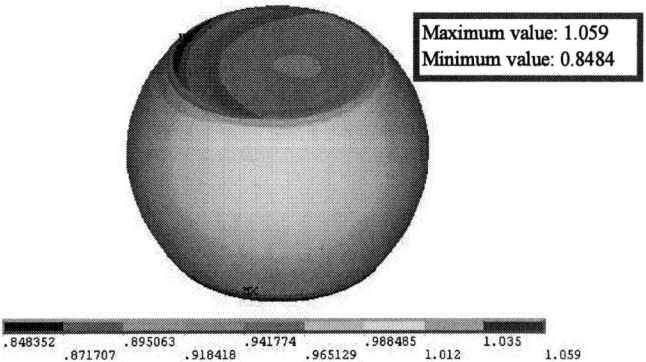

Fig 13 Atomic concentrations distribution without SG

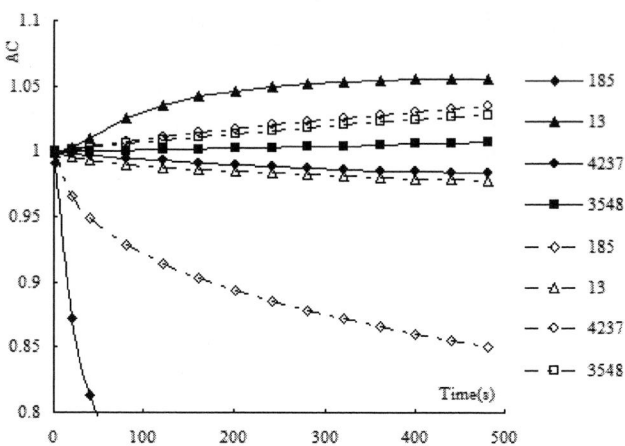

Fig 14 AC variations with and without SG where solid line: considering atomic concentration gradient; dash line: without considering atomic concentration gradient.

5 Conclusions

A new algorithm for EM analysis considering multiple driving forces mechanism based on finite element analyses for simulating the solder joint failure under high current density is presented. The interaction of the various driving forces during the EM process is investigated.

Simulations indicate that atomic concentration gradient migration is an important factor that has a large influence on the damage evolution induced, so it can't be ignored. It is also found that the atom concentration depends on the atomic concentration gradient and the stress gradient of the structure greatly.

Since the ultimate goal of this research project is to predicate the EM time-to-failure of solder joints under high current density stressing, it is important to develop the dynamical propagation algorithm in the further works.

Acknowledgments

The authors many thank Fairchild Semiconductor Corporation for their help. This work is also supported by Zhejiang Science Foundation (No .Y107365).

References

1. Fiks, V. B. *et al*, "On the mechanism of the mobility of ions in metals," *Sov. Phys. Solid State*, (1959), pp:1 - 14.
2. Huntington, H .B., Grone, A. R."Current-induced marker motion in gold wires," *J. Phys. Chem. Solid*, Vol. 20, (1961), pp. 76 - 87.
3. Blech, I. A, *et al*, "Electromigration in thin aluminum films on titanium nitride," *J. Appl. Phys*, Vol. 47, (1976), pp. 1203 - 1208.
4. Korhonen, M. A, Børgesen, P., Tu, K.N, *et al*,"Stress evolution due to electromigration in confined metal lines," *J. of Applied Physics*, Vol. 73, (1993), pp. 3790 - 3799.
5. Tan, C .M and Roy, A."Investigation of the effect of temperature and stress gradients on accelerated EM test for Cu narrow interconnects," *Thin Solid Films*, (2006), pp:288-293.

6. Gee S, Kelkar N, Huang J, Tu K N. "Lead-free and PbSn bump electromigration testing." *Proceedings of InterPACK 2005*, IPACK2005-73417, July 17-22.

7. Brandenburg, S. *et al*, "Electromigration Studies of Flip Chip Bump Solder Joints," S*urface Mount International Conference and Exposition*, SMI 98 Proceedings, 1998.

8. Gupta, D. *et al*, "Interface Diffusion in Eutectic Pb-Sn Solder," *Acta Metallurgica*, Vol. 47, (1998), pp: 5 - 12.

9. Tan, C M and Roy, A. "Investigation of the effect of temperature and stress gradients on accelerated EM test for Cu narrow interconnects", *Thin Solid Films*, Vol. 504, (2006), pp. 288 - 293.

10. Kirchheim,R. "Stress and electromigration in Al-lines of integrated-circuits," *Acta Metall. Mater.*, Vol. 40, No. 2 (1992), pp. 309 - 323.

11. Sarychev, M.E *et al*, "General model for mechanical stress evolution during electromigration," *J. Appl. Phys.*, Vol. 86, No. 6 (1999), pp. 3068 - 3075.

12. Clement, J J and Lloyda,J R. "Numerical investigations of the electromigration boundary value problem", *J. Appl. Phys.*, Vol. 71, No. 4 (1992), pp. 1729 - 1731.

13. Tu,K.n."Rencet advances on electromigration in very large scale integration of interconnects," *J.APPL.Phys*, Vol. 94, No. 9 (2003), pp. 5451 - 5473.

14. Liu, Yong and Scott, Irving, "Power Cycling Simulation of an IC Package: Considering Electro migration,and Thermal-Mechanical Failure," *ECTC2003*, pp. 415 – 421.

15. Liang, L.H. Xu, Y. J. and Liu Y., "Electro-Migration Study in Solder Joint and Interconnects of IC packages," *Proceedings of EuroSIME2006*, Como/Italy, Apr. 2006, pp. 464 - 470.

16. Liang, L.H. and Liu Y., "Reliability study in solder joint under electromigration thermal-mechanical load," *International Conference on electronics packaging tchnology ,ICEPT2006*, pp. 861.

17. Reddy J. N., <u>An introduction to the finite element method (Third Edition)</u>, McGraw-Hill (2006).

Finite Element Analysis of Reliability on Compliant Wafer Level Packaging With Compliant Layer

LI Peng, PAN Kai-lin, Ning Ye-xiang

Guilin University of Electronic Technology, Institute of Mechatronics Engineering, Guilin, 541004, China

Email: lpjr82@gmail.com

Abstract

Along with electronic products developing toward lighter, thinner, and multi-functional integration, Chip Scale Package (CSP) has been widely used in electronic packages. Wafer level packaging (WLP) has become one dominant technology. However, applications of WLP are limited by Solder joint fatigue due to stress generated by the CTE mismatch among different materials. Compliant wafer level packaging (CWLP) technology can be used to enhance thermal fatigue reliability of packages greatly. Structure of CWLP with compliant layer is introduced firstly. Subsequently, ANSYS software is employed, a quarter 3D model is developed based on 128MB DDR SDRAM, and the model is loaded on four thermal cycles from -40℃ to 125℃. Finally, by combining simulation results with FEM results and experimental results in other studies, comparative analyses are performed based on different thickness of compliant layer. FEM results show that, CWLP structure with compliant layer studied is reasonable in relieving the stress generated by CTE mismatch. Parameters, such as thickness of compliant layer and compliant material, are both important factors impact reliability of solder joint greatly. Thermal fatigue reliability can be significantly improved by reasonable selections of these parameters.

Introduction

Chip Scale Package (CSP) has been widely used for microelectronic packages require small form factor and high frequency performance.[1] In order to achieve low cost packaging solution, more and more companies have produced devices processed and packaged in wafer format. Nowadays, Wafer Level Packaging (WLP), which combines advantages of Flip Chip Technology (FCT) and conventional SMT process, has been steadily increased due to the smaller packaging size and lower manufacturing cost. WLP can offer area array arrangement of interconnects and be bumped efficiently at the back end of wafer process. [2, 3]

However, when WLPs are assembled onto PCB, large CTE mismatch between the die, PCB and solder joints can easily result in different thermal expansions and serious thermal stress in the whole package. As a result, solder joint thermal fatigue due to stress generated by the CTE mismatch has limited applications for larger dies greatly.[4,5] In order to overcome this issue faced in engineering practice, Compliant Wafer Level Packaging (CWLP) technology is put forward as a novel concept which addresses packaging and interconnecting needs of ICs by building kinds of compliant structures.

Generally, compliant structure of WLP can approximately be classified to three types: (1) Structure with softer metal material or specific shape for solder joints; (2) Structure with compliant layer; (3) Structure with embedded air-gap. [5, 6, 7] Comparatively, CWLP with compliant layer is much more preferable. As packaging size being increased, during the thermal cycling process, thermal mismatch stress generated by large CTE difference is easily to cause solder joints to crack and fracture. Transitionally, pre-applied underfill layer is used to compensate the stress. This approach can be used to improve the reliability to a certain extent, but process for underfill is not compatible with traditional SMT process, and the production cost will also increase greatly. [4]

In previous papers, 2D and 3D FEM analyses were carried out to research different compliant layer structures of CWLP. Gui-lian Gao et al studied the enhanced thermal fatigue reliability of DRAM package by 2D FEM simulation and built prototype units based on Tessera's compliant WLP technology. [5] Chang-Chun Lee et al used response surface methodology to the solder joint layout design, reliability enhancement of WLP through FEM and experiments. All results show that compliant layer under the solder joint can be effectively used in dissipation of thermal stress; thickness of the die and PCB are important factors impacted reliability of solder joint. [8]

In this paper, a CWLP structure with two compliant layers under the solder joint is studied. Finite element method analysis is employed to develop a 3D FEA model based on 125MB DDR SDRAM die. Then, four thermal cycling is loaded according to JEDEC temperature cycling standard of JESD 22-104C. By combining the FEM results with experimental results of previous studies, comparative analysis is carried out on different thickness of compliant layer.

Structure of CWLP

The cross-sectional view of CWLP structure used in this study is shown in Fig.1. To relief thermal deformation that induced by a larger CTE mismatch of materials during thermal cycling loads, two stress compliant layers between the silicon die and the solder joints are fabricated. In this study, polyimide and BCB are used as two compliant materials, respectively. [8]

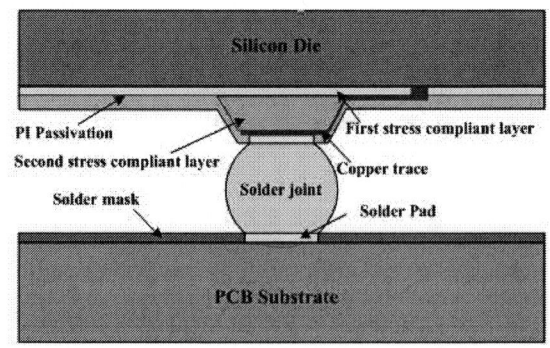

Fig.1 Cross-section of CWLP structure

The first compliant layer with a thickness of 0.01 mm and the second compliant layer with a thickness of 0.075mm are adopted. After redistribution process, solder joints with 0.75 mm×0.75 mm pitch is connected via a copper pad through the electroplated Ni under bump metallurgy. Then, two compliant layers and copper trace are redistributed. Solder mask layer with a thickness of 0.01mm is spin-coated, and lands are opened up by photolithography. Afterwards, solder paste is printed onto the lands, and the solder balls are placed on the solder paste. Finally, reflow process is performed, and the whole wafer is diced into individual packages.

Non-linear CWLP Finite Element Model

CWLP structure with 128 MB DDR DRAM with two dummy joins at each corner is taken as an example to study in this paper. The structure and the size of the die are referenced from the sample supplied by ACE Tech, Inc. This proposed DDR SDRAM is fabricated with a Rambus DRAM layout. Concrete parameters of the die are shown in table 1. [8,9] In order to protect internal solder joints, dummy solder joints, which are designed to be arranged at the corner of the package, can be considered as structural dummy supports with no electrical signals passing through. The die is assembled on the PCB (FR-4 laminate) with a thickness of 1.2 mm.

Numerical simulation of thermo-mechanical properties for DDR SDRAM subjected to thermal cycling loading is performed by 3D finite element modeling. Due to the symmetry of packaging structure, a quarter 3D model with twelve real solder joints and two dummy solder joints on substrate is modeled merely, as shown in Fig. 2. This model type has been widely applied in previous researches for predicting mechanical behavior of such eutectic solder material as well as meeting experimental results. [10, 11]

Table 1 Geometry dimensions of 128MB DDR SDRAM

Die size	7.15 mm×7.65 mm
Die thickness	0.5 mm
Die weight	65.0 dyne
Die-side pad diameter	0.38 mm
Substrate-side pad diameter	0.34 mm
Solder joint pitch	0.75 mm
Solder joint diameter	0.45 mm
Solder volume	0.048 mm3
Numbers of real joints	48
Numbers of dummy joints	8
Die-side contact angle	128.6 degree
Substrate-side pad diameter	136.0 degree
Max width of solder joint	0.48 mm
Standoff height of solder joint	0.308 mm

During meshes generation, the model is mapped under manual-size control. There are 48304 eight-node solid elements and 57129 nodes in the whole model. Based on foregoing simplifications of FEA model, a symmetry boundary condition is placed on the symmetric plane of the package, and the displacement at the center point located in bottom side of the FR-4 substrate is constrained in all degrees of freedom (DOF) to suppose that the package is placed on a rigid basement, as shown in Fig. 3.

According to the temperature cycling standards JESD22-A104C, [12] thermal cycling test condition G and soak mode condition 4 are chosen as thermal loading condition, which is assumed to be within the state-steady condition. The thermal loading applied to CWLP assemblies ranged between -40℃ and 125℃ with 60 min cycling rate and completed four cycles.

In JEDEC standard, typical ramp rate for this situation is 15℃/minute or less for any portion of each cycle, with a preferred rate of 10℃ to 14℃/minute, and the ramp rate of 11℃/minute is used in this study. Loading condition for this FEA simulation is shown in Table3. Representative temperature profile for thermal cycling condition used in this study is shown in Fig.4.

Table 3 Anand model constants of SnAg

Material parameters	Units	96.5Sn3.5Ag
S_0	MPa	39.09
Q/R	K	8900
A	1/s	2.23E4
ξ	-	6
m	-	0.182
H_0	MPa	3321.15
$\hat{s}$	MPa	73.81
n	-	0.018
a	-	1.82

Table 4 Loading conditions for thermal cycle

Temperature range	-40℃ - 125℃
Time/per Cycle	60 min
Time for ramp-up	15 min
Dwelling Time	15 min

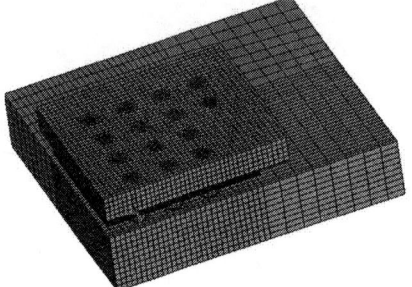

(a) FEA global model

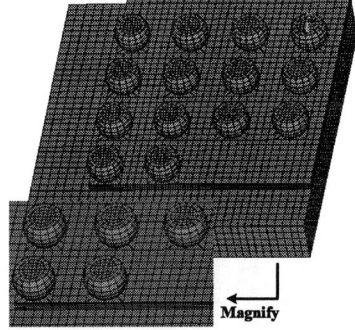

(b) Portions above solder joints

Fig.2 FEA model of the DDR SDRAM structure

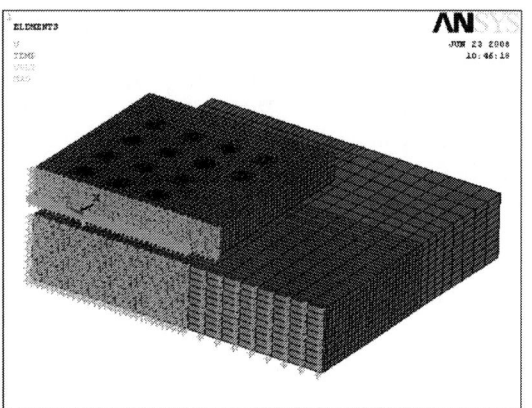

Fig .3 symmetric FEM model and its boundary settings

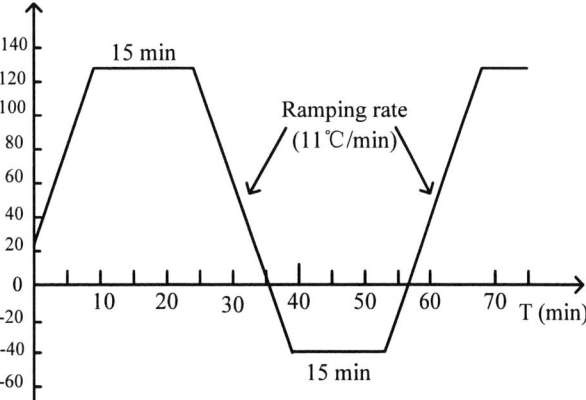

Fig. 4 Representative temperature profile

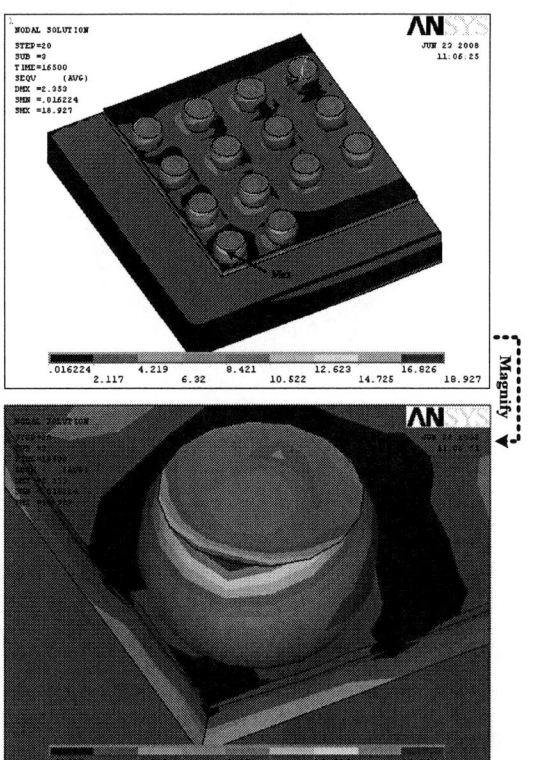

Fig. 5 Distribution of von mises stress of solder joints after four thermal cycles

Result and discussion

After four thermal cycles under the test condition selected, both von mises stress and equivalent plastic strain can be obtained after four thermal cycles, FEA simulation results are shown in Fig. 5 and Fig. 6.

According to FEA results, maximum von mises stress and maximum equivalent plastic strain are appeared at the dummy solder joints on the outer-corner of the die. Due to dummy solder joints arranged on the outer side of the die are only used as a structural support to protect the inner real solder joints, it is of no value to explore their reliability.

FEM simulations are carried out based on a series of compliant layer thicknesses and groups of data are obtained after four thermal cycles. As the comparative result shown in Fig. 7, the thicker the compliant layer thickness, the lower maximum equivalent plastic strain of solders joints. Equivalent plastic strain is significantly reduced by thickness increasing from 0.01mm to 0.06mm. When thickness varies from 0.075mm to 0.1mm, value of strain do not change obviously. So, compliant layer with thickness of 0.07mm to 0.08mm is recommended.

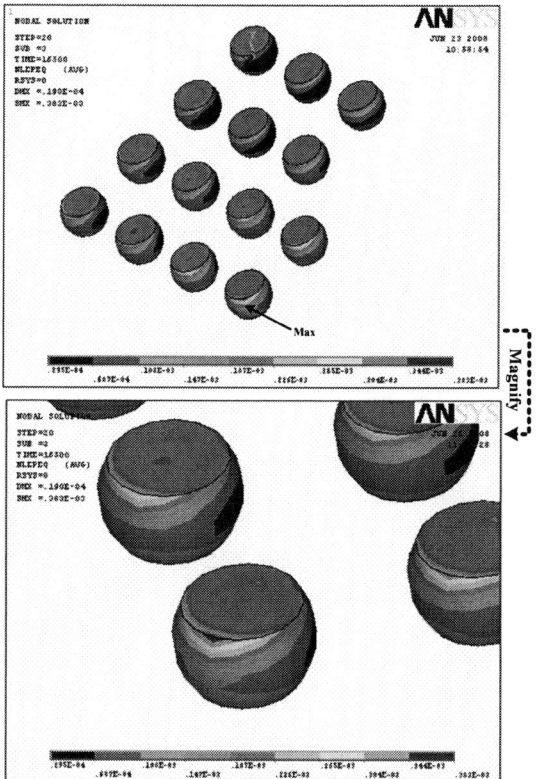

Fig. 6 Distribution of equivalent plastic strain of solder joints after four thermal cycles

Compared with solder joints at die-side, solder joints close at substrate-side exhibit a major fatigue crack in pad-to-solder interface. From a physical viewpoint, what induces this phenomenon is the fact that compliant layers absorb most of thermal stress resulting from CTE mismatch between the die, PCB and solder joints. In that case, the maximum thermal stress of solder joints is transferred into substrate-side. Critical dummy solder joints arranged on the outside of the die are suffered to this stress instead. Simulation results in this study agree with investigated results by Chang-Chun Lee et al.[8, 9] As their results shown, micrograph of the solder joints by ATC testing reveals that, undergoing thermal cycling test, the fatigue failure mainly appears on the outer dummy solder joints close to the substrate-side. Similar result can also be reflected by their FEM analysis.

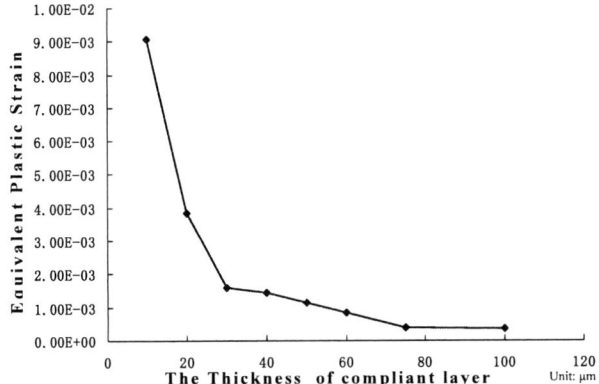

Fig. 7 Maximum equivalent plastic strain of solder joints versus thickness of stress compliant layer

Conclusion

In this study, 128MB DDR SRDAM with CWLP structure is simulated by FEA. Von mises stress and equivalent plastic strain are obtained after four thermal cycle based on the thermal cycling standard JESD22-A104C. Variations of stress and strain generated on by CET mismatch among different materials are investigated and compared with various thickness of compliant layer.

Simulation results show that compliant layer can be used to absorb thermal stress. The maximum thermal stress of solder joints has been transferred into the substrate-side. Dummy solder joints, which is only used as a structural support to protect the inner solder joints exhibit a major fatigue crack in pad to solder interface. As a result, reliability of the inner solder joints can be enhanced in some extent. Thickness of compliant layer is one of important factors to the CWLP structure, the thicker the compliant layer thickness, the lower maximum equivalent plastic strain of solders joints. 0.6mm to 0.8mm is more preferable compliant layer thickness. Reliability of CWLP structure can be enhanced by optimizing the thickness. These findings of this study have some degree of guiding significance in further research on thermal fatigue reliability of CWLP packages.

References

1. Vaidyanathan Kripesh et al, "Design & Development of a Large Die and Fine Pitch Wafer Level Package for Mobile Applications," *Electronic Components and Technology Conference*, 2006 , pp. 570-576.

2. Shu-Ming Chang, et al, "A Novel Design Structure for WLCSP With High Reliability, Low Cost and Ease of Fabrication," *IEEE Transactions On Advanced Packaging*, Vol.17, No.3(2007), pp.377-382.

3. John H. Lau et al, "Effects of Build-Up Printed Circuit Board Thickness on the Solder Joint Reliability of a Wafer Level Chip Scale Package," *IEEE Transactions on Components and Packaging Technologies*, Vol.25, No.1(2002), pp.3-14.

4. Yan-Xiang Kuang, "A Novel Flexible Bumps Technology for FC and WLP," *ELECTRONICS & PACKAGING*, Vol.5, No. 4 (2005), pp. 5-8.

5. Gui-Lian Gao et al, "Compliant Wafer Level Package for Enhanced Reliability," *8th International conference on Electronics Packaging Technology*, Shanghai, China, August, 2007, pp. 217-221.

6. R. Fillion et al, "New Wafer Level Structure for Stress Free Area Array Solder Attach," *SMT & Packaging*, Vol.4, No.3, (2004), pp.16-20.

7. Naoya Watanabe et al, "Wafer-level Compliant Bump for 3D Chip-Stacking, " *2006 Electronic Components and Technology Conference*, 2006, pp.125-130.

8. Chang-Chun Lee et al, "Solder joints layout design and reliability enhancements of wafer level packaging using response surface methodology," *Microelectronics Reliability*, Vol. 47, (2007), pp.196-204.

9. Chang-Ming Liu, et al, "Enhancing the Reliability of Wafer Level Packaging by Using Solder Joints Layout Design," *IEEE Transactions on Components and Packaging Technologies*, Vol.29, No.34 (2006), pp.877-885.

10. Ji-Cheng Lin et al, "Design and Analysis of Wafer-Level CSP with a Double-Pad Structure," *IEEE Transactions on Components and Packaging Technologies*, Vol.28, No.1 (2005), pp.117-126.

11. M.Gonzalez et al, "Finite Element Analysis of an Improved Wafer Level Package using Silicone Under Bump layers," *5th Conference on Micro-electronics and Micro-Systems*, pp.163-168.

12. JEDEC Solid State Technology Association. JESD22-A104C, "Temperature Cycling", May, 2005.

On the Study of the In-Use Stability of a DCA Assembled MEMS Device

Liyuan Xu, Jing Song, Jieying Tang
Key Laboratory of MEMS of Education Ministry, Southeast University, Nanjing, 211189, China
Email: xuliyuanyuan1983@163.com

Abstract

DCA (direct chip attach) is introduced as a main mode for the MEMS (Micro-electronics Mechanical System) devices packaging. However, thermo-elastic coupling occurred in the DCA package will introduce disturbance stresses and deformations, which may give rise to various packaging effects. In this effort, the in-use structural stability of a packaged MEMS device is further investigated. Susceptibility of a DCA assembled microbridge to in-use stiction with respect to packaging parameters is modeled, estimated and measured.

Key words: MEMS package, DCA, strain distribution, in-use stability, peel number

1. Introduction

MEMS miniaturization reduces cost, size, weight, and power consumption. The advantages of surface micromachining technology have made it an inherent choice for many applications in automobile, biomedical, aerospace and defense sectors. But successful commercial applications of MEMS still encountered great technical challenges. Unlike the ICs, the micro-structures usually contain freestanding moving parts resulting from the removal of the sacrificial layer，which are normally fragile and easily damaged. The variety and specialty of MEMS challenge its packaging. One of the challenges is how to provide protection, hermeticity and alignment precision well with accepted cost and keep interconnection at the same time. The packaging technology has become the key point for MEMS applications.

As one of the most commonly used packaging process, DCA has its advantages such as high availability and low cost. However, thermo-elastic coupling occurred in the DCA package will introduce disturbance stresses and deformations. Previous works have characterized their instant influence on the intrinsic static and dynamic properties of a MEMS device[1, 2]. The issue of in-use stiction is mainly addressed which concerns the adhesion between the suspended structure and the substrate induced by external loads or processes. In this paper, we use the microbridge as a typical structure to investigate the in-use stability of a packaged MEMS device. The results and discussion of this experiment provide a useful guidance for failure prediction and reliability design in MEMS.

2. Analysis of model

Two analytical models are combined to estimate the structural stability of a DCA packaged microbridge. One is the substrate model concerning the strain distribution along the top surface of the chip after DCA process. The other is an in-use stiction model of microbridge based on the classical beam theory, which gives the peel bounds of the microbridge with prestress and determines whether the adhesion is sufficient to hold it pinned to the substrate.

2.1. The stiction model of microbridge

The stiction mechanisms, which take an important part in stiction of surface-micromachined structures, are the comprehensive action effect of micro-forces such as capillary forces, electrostatic forces and van der Waals forces. Since microbridge become an important microstructure and has been widely used in RF switches, optical switches, resonators, and sensors, it will be focused of our discussion, which is shown in Fig.1 [3]. Fig. 2[4] is the schematic diagram of a cantilever beam. The micro-machined device fails when the surface force exceeds the restoring force of the bent beam. Mastrangelo and Hsu introduced a peel number N_p to describe the degree of stiction between the micro-structures, defined as the ratio of elastic deformation energy and stiction energy. The microbridge peels completely for $N_p > 1$, which means the energy of elastic deformation is larger than stiction and adhesion will not occur. And microbridge will be pinned for $N_p < 1$, when failure caused by stiction arises. So $N_p = 1$ is the critical state between peel and adhesion. The N_p of micro-structure showed in Fig.1, 2 could be calculated as follows respectively:

$$N_p = \frac{128 E h^2 t^3}{\gamma_s l^4} \quad (1)$$

$$N_p = \frac{3}{8} \frac{E t^3 h^2}{\gamma_s l^4} \quad (2)$$

where E is the Young modulus, γ_s is the interfacial adhesion energy of per unit contact area, t is the thickness of the beam, h is the distance between the two surfaces and l is the length of the beam.

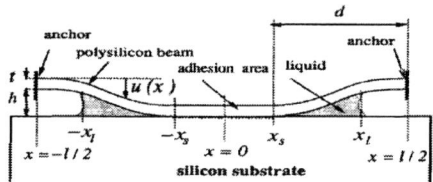

Fig.1. Scheme of microbridge with stiction

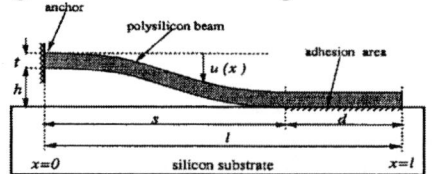

Fig.2. Scheme of cantilever beam with stiction

2.2. Thermally induced packaging effect [2]

As most MEMS devices sensitive to stress essentially, the thermally-induced package effect will influence the performance and reliability of MEMS systems significantly.

The inherent process temperature's change can cause the thermo-mechanical coupling and introduce additional thermo-elastic strain well as geometric deformation, thus influence the properties of MEMS devices directly.

Timoshenko[5] gives the solutions to this kind of problem first. His conclusion is simple and accurate, but is not suitable for solving the strain distribution at the edge of the structure which has low ratio of depth to width. Chen[6] and Suhir[7] based on lap joint analysis and interfacial flexibility analysis respectively to introduce the further development of this issue. Reference [2] further expanded and improved the model of Chen and made it applicable to get the strain of surface. The strain distribution along the top surface of the chip is:

$$\varepsilon_{top}(x) = \frac{du_{top}}{dx} = \frac{d\overline{u}_1}{dx} - \frac{d^2 w_1}{dx^2}\frac{h_1}{2} \quad (3)$$

Where u_1 is the displacement at x direction of the chip, $\overline{u}_1$ is the displacement at x direction of the chip axis, v_1 is the displacement at y direction of the chip and h_1 represent the thickness of the chip.

2.3. The influence of package on the stiction of microbridge

For a microbridge made with surface micromachining technology, the residual stress induced by fabrication process will influence the initial morphology of the microbridge during the "release" of the microstructures from the surrounding sacrificial layer by liquid etchants, which cause two kinds of initial morphology as follows: buckling upward and without buckling. Frequency is used to characterize the performance of both states. For a post-buckling microbridge, its natural frequency of vibration could be calculated as follows[8]:

$$
\begin{vmatrix}
0 & 1 & 0 & 1 & 1 \\
\lambda_1 & 0 & \lambda_2 & 0 & 0 \\
\sin\lambda_1 & \cos\lambda_1 & \sinh\lambda_1 & \cosh\lambda_1 & 1 \\
\lambda_1\cos\lambda_1 & -\lambda_1\sin\lambda_1 & \lambda_2\cosh\lambda_2 & \lambda_2\sinh\lambda_2 & 0 \\
\alpha\lambda_1\frac{2\pi(\cos\lambda_1-1)}{\lambda_1^2-4\pi^2} & -\alpha\lambda_1\frac{2\pi\sin\lambda_1}{\lambda_1^2-4\pi^2} & \alpha\lambda_2\frac{2\pi(1-\cosh\lambda_2)}{\lambda_2^2+4\pi^2} & -\alpha\lambda_2\frac{2\pi\sinh\lambda_2}{\lambda_2^2+4\pi^2} & \beta-\pi\alpha
\end{vmatrix} = 0
$$

(4)

Where $\alpha = -\dfrac{8t\pi^3}{I}\left[\varepsilon\dfrac{l^2}{\pi^2} + 4\dfrac{l}{t}\right]$, $\beta = \dfrac{\rho_l\omega^2 l^4}{E'I}$,

$\lambda_{1,2} = \sqrt{\pm 2\pi^2 + \sqrt{\pi^4 + \dfrac{\rho_l\omega^2 l^4}{E'I}}}$, $E' = E/(1-v^2)$, $\rho_l = \rho h$,

v , I , ρ and ε is Poisson ratio, inertia moment, volume density and strain of the beam. By the application of energy method, the 1st order natural frequency of the beam without buckling is simplified to be[9]:

$$f_1 = 1.028\frac{h}{l^2}\sqrt{\frac{E'}{\rho}\left[1 + 0.295\frac{l^2}{h^2}\varepsilon\right]} \quad (5)$$

After removing the chip from the solution, the morphology of the beam will be changed by micro-forces such as capillary force. The beam may contact to the substrate due to buckling, stiction or the comprehensive action of buckling and stiction. Equation (6) could be used to calculate

the critical length of the beam when it contact to the substrate caused by buckling:

$$\sqrt{\frac{-\varepsilon l^2}{\pi^2} - \frac{t^2}{3}} \bullet (1 - \cos\frac{2\pi x}{l}) = h \quad (6)$$

The critical length of the beam when it contact to substrate owing to stiction is:

$$l_d = (\frac{128Eh^2t^3}{5\gamma_s})^{\frac{1}{4}} \quad (7)$$

And for the beam whose contact is caused by the comprehensive action of buckling and stiction, its critical length could be calculated as:

$$l_d = \sqrt{\frac{256h^2\sigma t}{105\gamma_s} + \frac{8\sqrt{2}\sqrt{256Eh^4\gamma_s t + 512h^4\sigma^2 t^2 + 2205Eh^2\gamma_s t^3}}{105\gamma_s}} \quad (8)$$

Where σ is the stress. Before packaging, it represents residual stress and after packaging, it is the sum of residual stress and the stress introduced by packaging.

We combined the substrate model concerning the strain distribution after DCA process and the in-use stiction model together to determine whether the adhesion is sufficient to hold it pinned to the substrate after package. With the release and packaging process, the morphology of the beam may have changes shown in Table 1.

Table 1. The changes of microbridge's morphology during packaging and release process

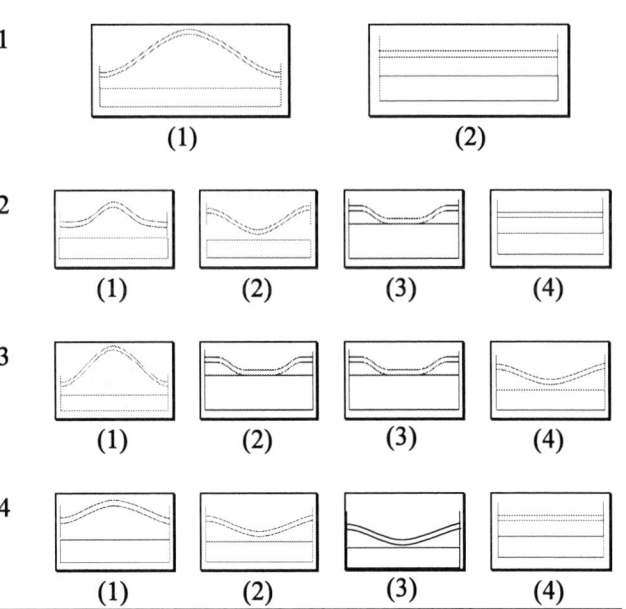

Two initial cases of the microbridges during release are shown in Table 1: buckling upward and straight. After removing from liquid etchants, the morphology of the beam will change because of capillary force. For the beams buckling upward, four morphologies may appear after release, as shown in 2, are buckling upward, buckling downward, stiction and straight respectively. For the straight beam, only the last two situations may appear, that is, stiction and straight. After release, the beam is packaged, which will change the morphology of beam again, shown in 3 and 4. The stress introduced by packaged could be classified as compressive stress and tensile stress. We will discuss below respectively:

1）For the beam buckling upward shown in 2(1), if compressive stress is introduced by packaging, the morphology of beam will change to 3(1). If packaging introduces tensile stress, 4(1) describes the morphology of the beam. Packaging influences the performance of the device such as resonant frequency, which could be calculated by formula (4).

2）For the beam buckling downward shown in 2(2), if compressive stress is introduced by packaging, the beam may contact to the substrate as a result of buckling. As shown in 3(2). The critical length of the beam could use formula (6) to calculate. If packaging introduces tensile stress, the beam will not contact to the substrate, shown in 4(2), just resonant frequency changes, which could be calculated by formula (4).

3）For the stiction beam shown in 2(3), if compressive stress is introduced by packaging, the length of the beam contact to the substrate will increase. If compressive stress is introduced by packaging, it is possible for the beam to release stiction.

4）For the straight beam, if compressive stress is introduced by packaging, the beam will be buckling result in stiction. The critical length can use equation (6) to calculate. If packaging introduces tensile stress, the beam still straight, and resonant frequency could be calculate using (5).

Based on the theory above, we use formula (8) to calculate the influence of packaging on the critical length for the beam without stiction. Both the thickness and the distance between the beam and the substrate is 2μm and v is 0.28. Tensile stress is assumed to be introduced by packaging. The values of material parameter [10] are shown in Table 2.

Table 2. The influence of packaging on the critical length

Material parameters	E （Gpa）	γ_s （ mJm-2)	ε	ld (um)
Before packaging	170	140	-360e-6	159.3
After packaging	170	140	-180e-6	170.3

From the estimate, we can see that after the introduction of tensile stress by packaging, the critical length of the beam increases, that is, more difficult for the beam to contact the substrate after packaging.

3. Design of structure and experiment

In this effort, the microbridge and cantilever beam are selected to evaluate the packaging effect. We use the two polysilicon layer beam structures, which is fabricated of surface-micromachining technology, in Pecking University. A 3000Å silicon dioxide and a 1800Å silicon nitride were grown on a silicon substrate. The thickness of the first polysilicon layer was 0.3μm and the actuating pad and contact pad were lithographed. The polysilicon layers were separated by a 2μm thick sacrificial layer of PSG. A 2μm structural polysilicon layer was deposited, and metal films such as aluminum was sputtered deposition on anchor as interconnections. After the preceding process, silicon wafer were divided into chips by scribing. The following process was begun by releasing the structures in a 25ml HF (40 wt %):40gNF4F:20ml glycerol: 46ml deionized water solution. Once the sacrificial layer was etched, the chips were then removed from the solution to be cleaned by deionized water, acetone and alcohol respectively. After drying the chips by infrared light, we got the chip sample after release. CB602 CRCBOND produced by Fujitsu was used to bonding the chip and the FR4 substrate. According to the curing procedure, we cured it under 120℃ for two minutes. The chip with package is shown in Fig. 3.

Fig.3. The chip with package

3.1 The measurement of resonant frequency

We use LDV (Laser Doppler Vibrometer) to measure the resonant frequency of microbridge before and after package. The LDV is MSV-400M2-2, using 6,400 standard FFT lines, can get the vibration spectrum in the given frequency range of the position where laser spots is. Double-sided adhesive is used to paste the sample in PZT (piezoelectric ceramics), which will drive the entire chip to vibrate, so as to provide the base excitation for structure. The structure response strongly when the exciting is at resonant frequency and the resonant peak is formed. The amplitude-frequency curve could be got by measuring and comparing the amplitude frequency of structure and substrate at the key position, as shown in Fig.4 and Fig.5.

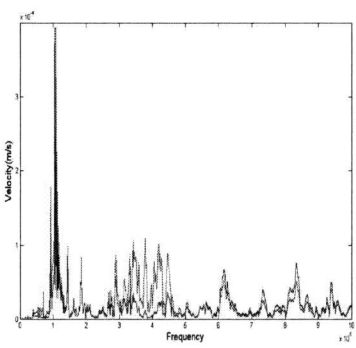

Fig.4. The amplitude-frequency curve of structure (red) and substrate (blue)

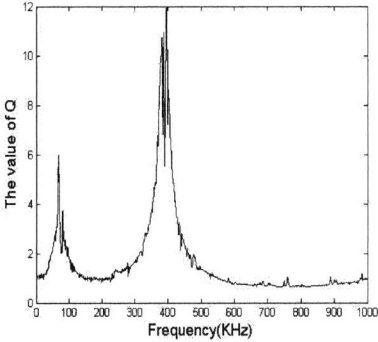

Fig.5. The relationship between Q and frequency

3.2 The measurement of stiction length

LDV is also used to measure the stiction length of the microbridge before and after packaging. PZT drive the entire chip vibration to provide the base excitation for structure. Sweep the beam and judge whether the beam is stiction and how long the stiction length is according to the vibration condition of beam at resonant frequency, shown in Fig.6. We can see that the intermediate part of the microbridge is adhering to the substrate.

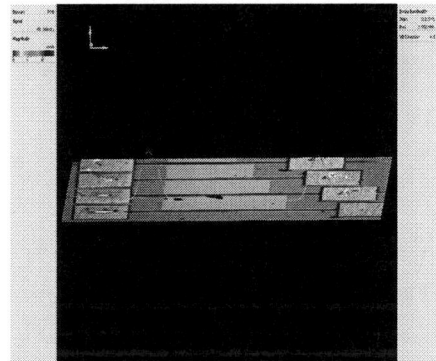

Fig.6. Vibration of microbridge at resonant frequency

4. Results and Discussion

4.1 The influence of packaging on resonant frequency

The experiments were carried out in an environment with a relative humidity of 60% and at room temperature ($15 \pm 5°C$). We used LDV, Fluke and probe station to measure the resonant frequency, resistance and stiction length of microbridges before and after package respectively.

Fig.7 and Fig.8 show the resonant frequency of microbridge and cantilever beam before and after package respectively, which are used to investigate the influence of packaging on the structures without stiction. The theoretical resonante frequency is calculated using formula (4), and the material parameters are shown in Table 2.

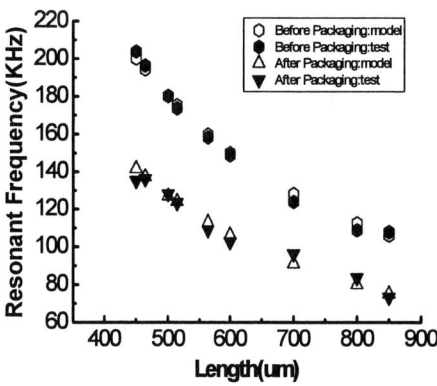

Fig.7. The 1st order resonant frequency of microbridge before and after packaging

The morphology of the microbridge beam before packaging corresponds to 2(1) in Table 1. Comparing the resonant characteristics of amplitude before and after packaging, we can see that the 1st modal frequency decrease uniformly and significantly after packaging, and the decrease amplitude of which is about 25%. It could be seen that tensile stress is introduced by packaging, corresponds to 4 (1) in Table 1. This result verifies that packaging will influence the resonant frequency of beam without stiction. The resonant frequency of cantilever beam before and after packaging shown in Fig.8 does not exhibit obvious change, this is due to the structure of cantilever beam with one end fixed and another movable, which can release the stress caused by packaging and do not affect the resonant frequency.

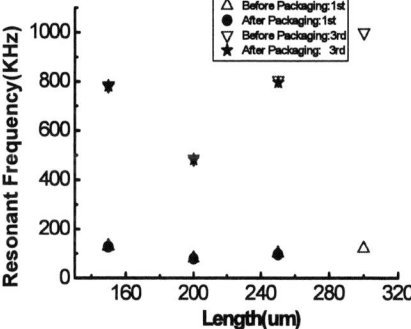

Fig.8. The resonant frequency of cantilever beam before and after packaging

4.2 The influence of packaging on resistance and stiction length

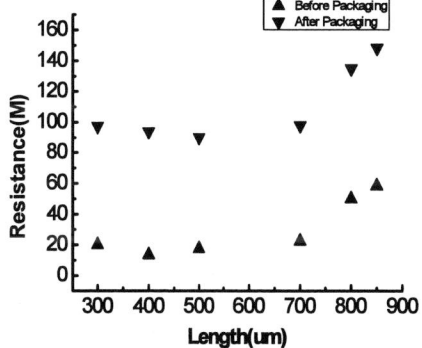

Fig.9. The change of resistance between left anchor and bottom electrode before and after packaging

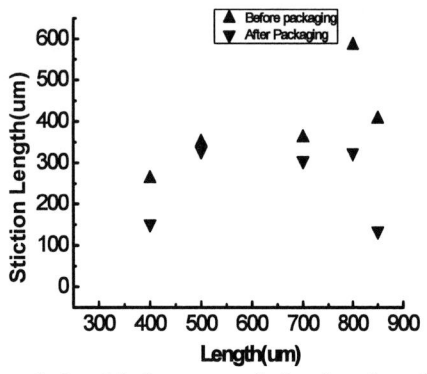

Fig.10. The relationship between stiction length and the length of beam before and after packaging

For stiction beam (corresponds to 2(3) in Table 1), we measure the resistance change between left anchor and bottom electrode before and after packaging to study the influence of packaging on it. The variation of resistance is shown in Fig.9. The resistance after packaging increase significantly due to the decrease of buckling degree, further confirmed that the tensile stress is introduced by packaging (corresponds to 4 (3) in Table 1). The stiction length of doubly-supported beam decreases after packaging, which is also due to the introduction of tensile stress.

5. Conclusions

From the results of experiment, we can see clearly that the thermally induced packaging effect not only influences the performance of MEMS devices, but also exhibits a diversiform effect tendency with the differences of the devices. In this effort, the influence of packaging on the stability of MEMS device is analyzed. We also measured the property change due to thermally induced packaging effect introduced by DCA of both microbridge and cantilever beam. The results show that the coupling between the package and device has a significant influence on the direction and magnitude of the uniaxial stress of the microbridge, and the structural rigidity of structure, which is possible to make the suspended microbridges to be adhered, or the adhered beams to be released. Based on the results, more issues of the in-use stability of a packaged microbridge are still being studied by taking into account the loads of shock and thermal cycling.

References

1. Lishchynska M., O'Mahony C., Slattery O., et al. "Evaluation of packaging effect on MEMS performance: simulation and experimental study". *IEEE Transactions on Advanced Packaging*, Vol.30, No.4 (2007), pp. 629-635.

2. Song J., Tang J. Y., Huang Q. A., "Package level simulation and verification of microsystems". *Proceedings of IEEE Sensors Conference*, 2007, pp. 99-102.

3. Mastrangelo C. H., Hsu C. H., "Mechanical stability and adhesion of microstructures under capillary forces—part II : experiments". *Journal of Microelectromechanical Systems*, Vol.2, No.1 (1993), pp. 44-55.

4. Mastrangelo C. H., Hsu C. H., "A simple experimental technique for the measurement of the work of adhesion of microstructures". *IEEE*, (1992), pp. 208-212.

5. Timoshenko S.P., "Analysis of Bi-Metal Thermostats", *Journal of Optical Society of America*, Vol.11, No.3 (1925), pp. 233-255.

6. Chen W.T., Nelson C.W., "Thermal Stresses in Bonded joints", *IBM Journal of Research and Development*, Vol.23, No.2 (1979), pp. 178-188.

7. Suhir E., "Interfacial Stresses in Bimaterial Thermostats", *Jounal of Applied Mechanics*, Vol.56, (1989), pp. 596 - 600.

8. Nayfeh A. H., Kreider W., Anderson T. J., "An Analytical and Experimental Investigation of the Natural Frequencies and Mode shapes of Buckled Beams", *AIAA Journal*, Vol.33, (1995), pp. 1121-1126.

9. Liu Guangyu, Fan Shangchun. New Technology and Application of Sensors, Beijing University of Aeronautics and Astronautics Publishing House (BeiJing, 1995).

10. Li M., Song J., Huang Q. A., Tang J. Y., "Thermally Induced Packaging Effect on the Resonant Frequencies of a Fixed-Fixed Beam". *Journal of Semiconductors*, Vol.29, No.1 (2008), pp.157-162.

IC Chip Crack Issues due to Mounting Process For Ultra-thin IC Smart Card Module

Pingyue Fan, Jiaji Wang

Department of Material Science, Fudan University, No.220, Handan Road, Shanghai, 200433, China
Tel: 86+21-55664588, Email: 0230004@fudan.edu.cn

Abstract

The reliability problem of thin/ultra-thin die used in IC card becomes the most primary problem which restrains the further application of IC cards. Failure induced by die crack takes over 50% of all failure modes. This paper paid attention to the crack mechanism of thin/ultra-thin die crack problem, especially emphasized on the crack due to mounting process. During the mounting process, the instant pressure made by thimble can reach GPa level, which easily results in the imprint damage. The sliding contact between thimble and die could leave the scuff mark on the die. Photos were taken to describe the figure and size of two kinds of micro damage. LEFM approach was used to analyze the effect of different kinds of damage. According to the results of quantized analysis, the micro damage appeared during mounting process could significantly decrease the intension of die, which finally make the die easy to crack. Packaging process could also bring extra stress to the thin/ultra thin die, which is often neglected during the process because its damage on reliability may not work immediately. Through the temperature circle and mechanical distortion in the daily life, the stress can finally make the die crack and impact the lifetime of the IC smart card. Several methods are put forward to deal with the aforementioned problems, protect the die from crack issues and increase the yield of ultra-thin IC smart card module.

Introduction

IC smart card, as the new type of storage device nowadays, has been more and more widely used at financial, communication, traffic, entertainment, logistic and many other fields because of its big volume, excellent security and convenience to use. The development of IC card is also trend to become versatile, ultra-thin and non-contact, which makes the reliability problem of the core component of IC card, the ultra-thin IC smart card module (less than 0.30mm), has become more and more important.

Depend on the observation of failure IC smart cards, die crack, which occurs at the percentage of 1% as the early failure mode of traditional silicon-based IC application, becomes the most failure mode of ultra thin die, taking over 50% of all failure issues.

The cause of die crack is that the micro damage and residual stress decrease the stress intension of ultra-thin die. So it is necessary to study the crack mechanism of ultra thin die. This paper emphasized on the micro damage due to mounting process, deeply analyzed the mechanism how the micro damage decrease the intension of ultra-thin die, studied the impact of residual stress on chip during package process.

Mounting Process of Ultra-thin IC Cards Module and Its Impact on Chip

IC cards are usually manufactured by integrating the IC smart card module containing NVM or MCU into the plastic card matrix. The process flow of IC card module including: wafer grinding, dicing, transporting to lead frame, wire bonding and encapsulating（fig.1）.

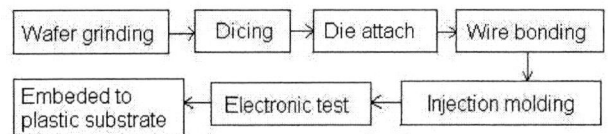

Figure1. Flow chart of process of ultra-thin IC card module

The thickness of IC card module includes the thickness of carrier tape (lead frame), die, adhesive, arc height of gold wire and encapsulated layer. To control the thickness of ultra-thin IC smart card module, it is necessary to use ultra-thin tape (~55 μ m) material, ultra-thin wafer grinding (<150 μ m), low-damage and low-stress dicing, ultra-low arc height (<65 μ m) wire bonding.

During these processes, grinding may leave relative high mechanical residual stress and micro damage on the backside of wafer, dicing may easily yield micro crack around the ultra-thin die because of the effect of dicing bath. Both the above two processes will decrease the intension of die, make the die easy to crack thoroughly after mechanical and thermal stress during package process or daily life. If the die cracked because of this reason, its characteristic is that the cracks on the die should begin at the edge of die and the length of cracks are at random. However, among most failure cases, the final cracks are usually via the center of die or radiate from the center (fig. 2a & 2b). Depend on the truth, we draw the conclusion that the main reason of die crack is the damage on center of die done by thimble step during mounting process.

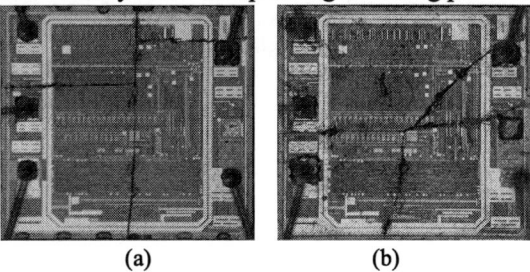

(a)   (b)
Figure2.Typical images of cracked die
(a) crack via the center of the chip; (b) crack radiate from the center

Damage due to thimble step during mounting process

Die bonding during mounting process means using thimble to lift the diced chip from blue membrane, at the same time the vacuum suction picking the chip and adhering it at the proper location on the metal tape（fig.3）.

978-1-4244-2739-0/08/$25.00 ©2008 IEEE

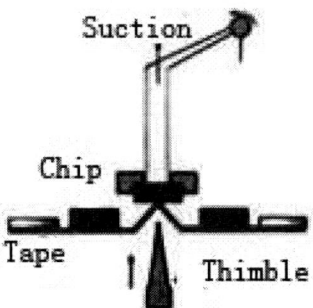

Figure3. Schematic diagram of thimble step

During thimble step, the direct mechanical contact between thimble and chip bring potential risk of die crack. Non-proper operation can make thimble through the blue membrane and leave the micro damage on the chip, which may cause the die crack directly or become the potential reliability problem. In the further backside observation of cracked chips, the cracks caused by thimble are presented in two different modes. Some look like imprint damage, with micro crack around the imprint (fig 4a). Others look like scuff damage because of the relative motion between thimble and chip (fig.4b).

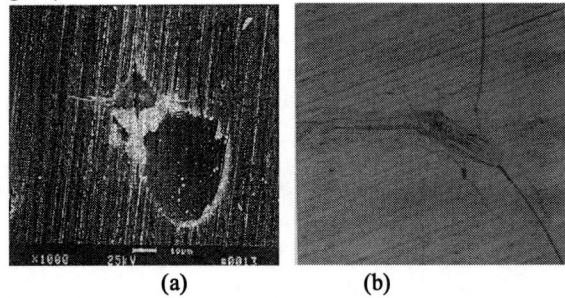

(a)   (b)

Figure4. Typical images of two different kinds of damage
(a) imprint damage; (b) scuff damage

In the final analysis, it is stress that causes the emerging of crack. Silicon belongs to brittle material. When a bulk of silicon has no micro damage, the stress will distribute uniformly outside of the silicon. However, when the micro damage appears, the stress, especially the stress vertical to the surface will concentrate at the crack tip. Once the crack tip stress is bigger than the critical stress intension factor of material, the crack will propagate. In a word, the appearance of micro damage done by thimble could dramatically impact the stress intension of ultra thin die. Aiming to the above two kinds of damage in the observation, we will deeply analyze their impression to the intension of chip.

Mechanism Analysis for Die backside damage Induced Crack

1、Imprint damage

The end of thimble used in the process is round. Besides the relative motion between thimble and chip, the process of thimble and chip could be simplified the round object bring stress vertically on the infinite plane [1] (fig 5.).

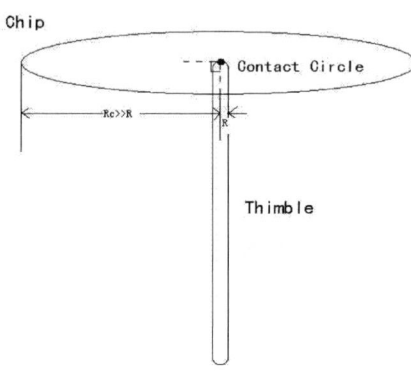

Figure5. Simplified model of thimble step

Because the diameter of thimble is small enough relative to the size of chip, the simplification is reasonable. Then the Vicky Indentater model could be applied to stimulate the process. The contact radius (a) and the vertical stress load (p) satisfy the equation:

$$\left\{ \frac{3}{4} pR(1-v^2)/E + (1-v'^2)/E' \right\}^{\frac{1}{3}} = aP^{\frac{1}{3}}$$

Where R means the radius of thimble, E, v and E', v' mean the Young Modulus and Poisson Ratio of thimble and chip (silicon). At the edge of contact circle, the tensile stress (parallel with the chip) could achieve the maximum:

$$\delta_m = \frac{1}{2}(1-2v)\frac{p}{\pi a^2}$$

Consider that the radius of contact point (a) is very small (the typical curvature radius of thimble is about 1mm), in case of 1N stress is put to chip, the initial tensile stress could achieve GPa level. Once there is delay between thimble and suction, let thimble contact chip too long time, imprint damage is easily to happen at the backside of the chip.

Further more, because of the brittle of silicon, imprint damage always occurs with much micro damage surround the contact point. Meanwhile, under the imprint, there appear extra lateral cracks and radial cracks (fig.6).

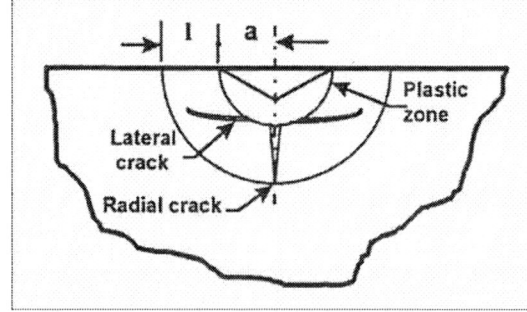

Figure6. Extra cracks surrounding the imprint

Under the condition of forementioned model, the tensile stress away from the contact point satisfies the equation:

$$\delta_r = \delta_m \left(\frac{a}{r}\right)^2$$

Where r means the distance away from the contact point. So in the limited area around the imprint point, the tensile stress on the surface could still make the cracks further extend.

In one word, the imprint damage has considerable impact on the chip, not only because of the imprint damage itself, but also because of its impression on the surrounding area.

2、Scuff Mark Damage

The other main kind of micro damage is scuff damage. After observing several cracked samples, it was found that the scuff marks were appeared in a similar pattern. The typical pattern is just like figure 7, a crack with fornicform at one end of the crack.

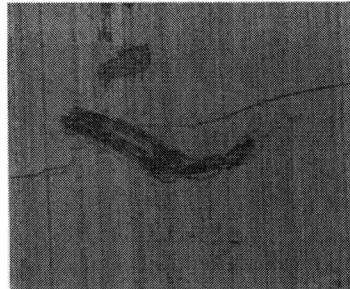

Figure7. The typical pattern of scuff mark

The crack on the surface of the chip will obviously decrease the chip intension, which makes the chip easily to be cracked. Depend on the Griffith theory [2] on crack propagation of the brittle material, the critical stress intension of chip is:

$$\sigma = \sqrt{\left(\frac{2E\gamma}{\pi a}\right)}$$

Where "a" means the length of crack. The Young modulus and Poisson ratio of silicon is 106.9GPa, 3.1J/ m² respectively. So the result is:

$$\sigma = \left(\frac{0.46}{\sqrt{a}}\right) GPa$$

Rely on this equation, we get the relation between crack length and critical intension (fig.8). When the crack length begins from zero to 1 μ m, the intension of chip decrease dramatically. After that, the trend gets smooth.

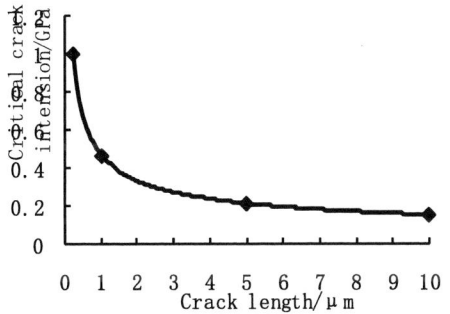

Figure8. Relation between crack length and critical crack intension

Figure9 is the schematic diagram of scuff mark, the data of length and width is the average value of over 20 samples.

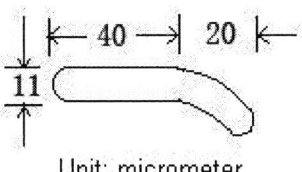

Figure9. Schematic diagram of scuff mark

Use this schematic diagram, we could draw out that:

$$\sigma \approx 56MPa$$

It means even when the stress is less than 1N, the crack would begin to grow. It can be concluded that this kind of micro damage could also make the intension of chip much lower.

After above analysis, the measures can be taken to improve the thimble step include: choosing relative coarse thimble to effectively decrease the stress on chip; better co-operation between thimble and suction to avoid long-time contact between thimble and die; the monitoring of thimble step to reduce the relative motion and changing the thimble in time once it gets aging and wearing. The proper control of adhesive intension between die and bluemenbrane can also benefit to decrease the stress during thimble step.

3、Residual stress and thermal stress on chip during packaging process

Either the imprint damage or scuff damage can hardly lead to direct crack during process. However, there always has residual stress during package process, which will further deteriorate the intension of die and affect the reliability of ultra-thin die, decrease the lifetime of IC smart card.

During the process of injection, curing and encapsulating, the ultra-thin die mainly experience two kind of stress: curing residual stress under high temperature (~180 ℃) and high pressure; thermal stress after the thermal cycling between room temperature and high temperature.

Molding material is a kind of thermosetting material, which can be cured spontaneously. During the process of curing, the molecular chains of material become close. Along with the shrinking of volume, the stiffness of material increase. The curing will limited by other materials, which yield stress. Part of the stress can be released because of the viscoelasticity of molding material. Part of the stress cannot be released, which is so-called curing residual stress.

The non linear finite element analysis software can simulate the spontaneous curing of molding material. The model of ultra-thin IC smart card module is like figure10, including the interaction among molding material, die and metal substrate, neglecting the impact of less critical structure such as gold wire, adhesive layer, etc.

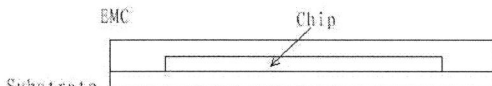

Figure10. Model of ultra-thin IC smart card module

Because of the viscoelasticity of epoxy resin, it is important to choose proper equation to construct the constitutive relation of curing process, which is the key of software simulation.

$$S_i(t) = \int_{-\infty}^{t} C_{ij}(\alpha(\xi),(t-\xi))\left\{\left[\frac{\partial E_j}{\partial \xi}\right]_{\xi} - \left[\frac{\partial E_j^*}{\partial \xi}\right]_{\xi}\right\}$$

In this equation, Si, Ej, Ej* mean stress, strain and initial strain tensor, Cij means the relaxation modulus relation to curing, α (ξ) means the curing parameter relation to curing time, the viscoelastic parameter and strain during chemical shrinkage were measured by DMA. The correct of this model has been proved [3].

The result of simulation shows that the impact of curing residual stress cannot be ignored. The curing residual stress can greatly deteriorate the intension of die. When the stress is over the critical value, it can also lead to micro damage to die, especially at the interface between chip and resin [4].

Besides the curing residual stress, the thermal cycling of package process will bring thermal stress. Because of the thermal expansion coefficient of resin (~45ppm) much bigger than that of silicon (~2.8ppm), the compressive stress is inevitable [5]. The compressive stress can be described as:

$$\sigma = KE\alpha\Delta T$$

Where K is a constant, E means the Young Modulus of molding material, α means the thermal expansion coefficient, ΔT means the differential between T_g of molding material and room temperature.

The above analysis shows that choose the proper molding material with lower curing shrinkage rate, lower Young modulus and Tg will reduce the stress during this process.

4、 Stress during practical use

The encapsulated ultra-thin IC smart card module looks like figure 11.

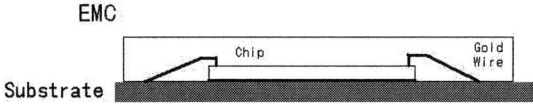

Figure11. Schematic diagram of encapsulated ultra-thin IC smart card module

All of the fore mentioned micro damage including mounting process induced and residual stress induced may not work immediately. The ultra-thin IC smart card module with these damages can also pass the regular electronic test and become the IC card production. Because of the special practical use environment, IC card may receive several kind of distortion in the daily life and the thermal stress from the change of condition temperature. All these stress to card will impact on chip eventually. The micro damages on the surface of die then will further extend, which make the die more and more fragile and finally the IC smart card will lose its function.

Conclusions

Depend on the observation of numbers of cracked chips, the micro damage due to thimble step during mounting is the main reason that causes the die crack. The micro damage appears in two modes: imprint damage because of the improper stress load from thimble; scuff damage because of the relative motion between thimble and die. Both kinds of cracks could dramatically decrease the intension of chip, leaving the hidden trouble inside of the application. During the consequent package process, there exists extra stress on chip including curing residual stress and thermal stress brought by molding material. The mechanical and thermal stress during practical use will impel the growth of micro damage, greatly increase the possibility of die crack. It is necessary to choose the proper thimble and monitoring the so-operation of thimble and suction to avoid the damage by thimble; it is also necessary to use proper molding material to reduce the residual stress due to package process. In this way we can prevent the die from crack risk effectively, improve the reliability of ultra-thin IC smart card.

References

1. Zhu Xiao-kun, Wang Jiaji, "Research of the failure Mechanism of IC Smart Cards", *Journal of Fudan University*, Vol. 44, No. 1 (2005), pp. 149~154

2. Wu Yi-sheng, *Micro Fracture Mechanics*, (1985), pp. 175-188.

3. Liu Shi-long, Qin Lian-cheng, "Influence of cure-residual stress on underfill epoxy reliability in flip-chip assembly under thermal cycling", *Electronics Process Technology*, Vol. 23, No. 6 (2002), pp. 249~252

4. Jiang Ting-biao, Tian Gang-ling, "Finite element simulation analysis on cure-residual stress in PBGA Device", *Electronic Components & Materials*, Vol. 24, No. 2(2005), pp. 3~6

5. Yan Zong-da, *et al*, *Thermal Stress*, (1993), pp. 65-67.

Optimization of Interface Strength for SCSP Based on Uniform Experimental Design

Gongke Li, D. G. Yang, Liancheng Qin, Fuxi Yi

School of Mechanical and Electrical engineering, Guilin University of Electronic Technology, Guilin 541004, China

lgk198110@163.com，86-13788599082

Abstract

There are various interfaces in the SCSP device, such as BT Substrate-adhesive, chip-adhesive, chip-EMC. Under used environment, it is intended to have delamination under hydro-thermal integrated stress to result in failure. Therefore, interface delamination has become the one of major failure modes. On the one hand, in this paper, the whole distribution situations of stress under thermal loading are obtained. After carrying out these analyses, the hazard locations to the essential interface are selected showed as Fig.4. On the other hand, referring to interface delamination failure, some initial cracks are placed in stress concentration interface. The distribution of J-integral is calculated with finite element method. Finally, in order to optimize device's interface strength and enhances the device reliability, the uniform experimental design combining the finite element method is employed in this paper by choosing several structure size as factor of experimental design, such as chip, adhesive, BT Substrate and EMC. The uniform design table with 4 factors 5 levels is used to study the relationship between structure and interface strength J-integral, to optimize interface strength through changing the structure parameters. The SCSP model approximate regression function is obtained by using sample point which produces uniform experimental design and response value is calculated by Finite element method and the optimization solution with the optimization algorithm is carried out. The optimal combination to dimension is obtained under this method. The analyzing results of thermal stress show that maximum thermal stress of device appears at mount tip of the second die. The analyzing results of interface strength show that the maximum value of J-value locates at the interface between BT substrate and adhesive, the crack in this field is unstable state.

1. Introduction

Microelectronic packaging technology is a bridge connecting die and electronic system outside. With the development of the integrated circuit industry and heavy demand on electronic product miniaturization, there is a trend in IC package technology toward miniaturization, high-density, high reliability and 3D package. Stacked die package is a widespread used 3D package technology. Stacked die package not only has high packing density, but also short interconnection at the same time, thereby increasing the running rate of device. Besides, it may also realize the multi-functions through this technology. [1]

There are so many layered structures in stacked die package. Interfaces were formed between the different materials, such as interface between EMC and die or passivation layer, EMC and die, die and adhesive etc. Due to thermal expansion coefficient between moulding compound and other materials are significantly different, it will induce large thermo

mechanical stress in reliability test and reflow process. The delamination occurred in the interface once the interfacial stress reaches the critical stress. Interface delamination will damage the structural integrity of device, even causes failure directly. When stress transfer to the other materials in the device because of interface delamination will induce reliability problem indirectly. So the study about interface delamination is very important for design and manufacture of integrate circuit.[2] This paper take four die SCSP (stacked chip size package) as object of study analyzed the mechanical behavior of device in reflow process through finite element method. Author established the finite element analysis model and calculated the distribution of thermal stress, further calculated the crack tip J-integral value of crack tip in the interface.

The analysis object of this paper is a typical four die stacked SCSP and the diagram of device structure is given in figure 1. There are four die in this package. And the DIE2 is a separate slice and no etching circuit in it. There are two kinds of adhesive between two die and BT substrate. 2025D adhesive between DIE1 and BT substrate and HS-230 adhesive between each die to glue each other. The overall dimensions of package are 11.6mm×8mm×1.4mm.

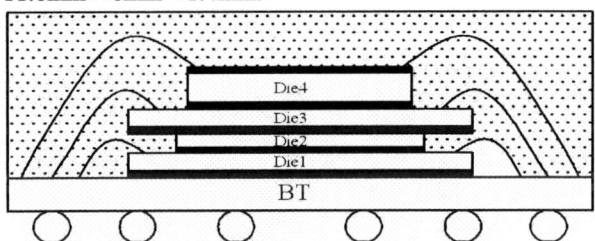

Fig.1 Diagram of device structure

2. Finite analyzing model

Because of the symmetry, only one half of the structure is modeled. The model used 8 nodes plane strain element. There are 4641 elements in the whole model. In order to find the key position where is easiest to induce the interfacial delamination for each bi-material interface, and to calculate the J-integral value, 6 cracks are set in position of stress concentration of the model. The key positions of the finite element model are local mesh refined for calculate the J-integral value of the crack tip in the interface exactly. The position removed from crack tip used the 8 nodes plane strain elements. The crack tip used 1/4 node element in order to stimulate the stress singularity in crack tip exactly. The node numbers of crack tip mesh are 19, 39, 70, 72, 81, and 90. The finite element meshes and position of preset cracks are given in figure 2. Following boundary conditions are applied: nodes along the symmetry axis are fixed in X direction (u=0), and nodes on the bottom of the model are fixed both in X and Y directions (u=v=0). The FEM mesh is presented in fig.2. [3]

978-1-4244-2739-0/08/$25.00 ©2008 IEEE

909

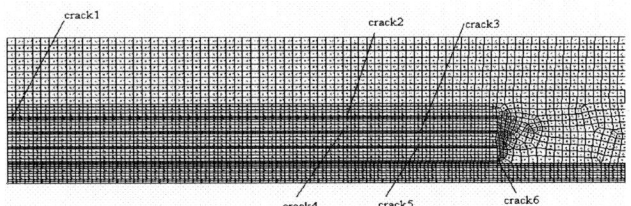

Fig.2 Finite element meshes of SCSP

2.1 materials models

SCSP is composed of die, adhesive of die, adhesive of substrate and EMC. A linear elastic mode is used for silicon die; substrate, their corresponding Young's modulus, Poisson's ratio and coefficient of thermal expansion are presented in table1. The EMC and adhesive will be considered viscoelastic materials. Their response between Young's modulus and temperature were showed in fig 3 and fig 4. [4][5]

Tab.1 Material property parameters of SCSP

Material properties	Young's modulus (Mpa)	Poisson's ratio	CTE (ppm/K)
Si	169000	0. 26	2. 3
adhesive	Fig 3	0. 4	48
EMC	Fig 4	0. 37	24
BT	20000	0.11	13

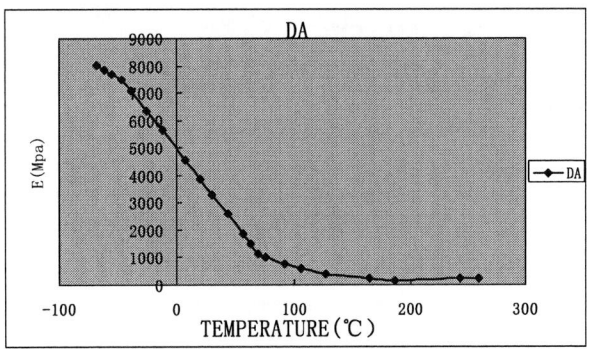

Fig3 Temperature dependent Young's modulus for adhesive

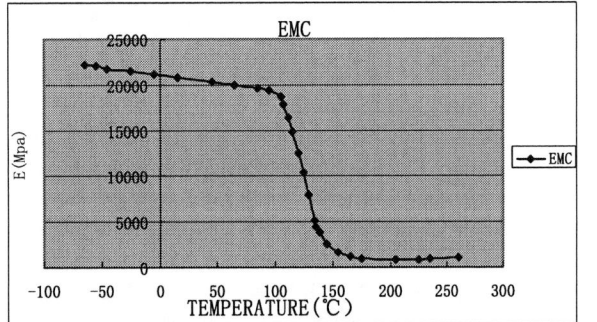

Fig4 Temperature dependent Young's modulus for EMC

Viscoelastic material is a kind of thermosetting polymer. Its Mechanical properties can be described with time-temperature

dependent. Viscoelastic material combines the property of viscosity and elasticity. Its constitutive relation shows below:

$$\sigma_{ij} = \int_0^t 2G(t-\xi)\frac{de_{ij}(\xi)}{d\xi}d\xi + \delta_{ij}\int_0^t K(t-\xi)\frac{d\varepsilon_{kk}(\xi)}{d\xi}d\xi$$

(1)

δ_{ij} : Kronecke symbol

G (t): function of shear relaxation

$K(t)$: function of volume relaxation

The relation between them as follow:

$$K(t) = \lambda(t) + 2G(t)/3 \qquad (2)$$

From formula (1)

$$S_{ij} = \int_0^t 2G(t-\xi)\frac{de_{ij}(\xi)}{d\xi}d\xi \qquad (3)$$

$$\sigma_{kk} = \int_0^t 3K(t-\xi)\frac{d\varepsilon_{kk}(\xi)}{d\xi}d\xi \qquad (4)$$

σ_{kk} :bulk stress

ε_{kk} :bulk strain

S_{ij} : stress deviator

e_{ij} : strain deviator

They can be described by a generalized Maxwell model with limited elements (Prony series) in an approximating form:

$$G(t) = G^\infty + \sum_{n=1}^N G^n \exp(-t/\tau_n) \qquad (5)$$

$$K(t) = K^\infty + \sum_{n=1}^N K^n \exp(-t/\tau_n) \qquad (6)$$

Where N is number of Maxwell elements describing relaxation modulus, τ n is relaxation time.

2.2 Thermal loading

In this paper, the mechanical behavior of SCSP is simulated with thermal loading during the reflow process. Fig.5 is a typical reflow solder temperature curve, which included six stages. The peak temperature is about 260℃.

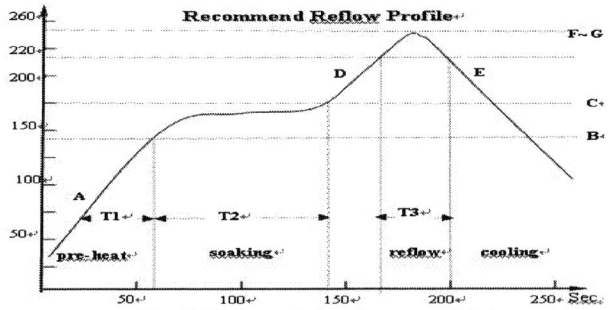

Fig.5 J integral development history

2.3 Results and discussions

Distribution of Equivalent Von Misses stress at peak temperature 260 ℃ was shown in Fig.6. Due to thermal expansion coefficient between EMC, adhesive and die are significantly different, the inhomogeneous distribution is induced in the whole packaging. The effect of this inhomogeneous distribution as following: first of all, the stress concentration appeared in mount part of DIE3, the stress value is the largest about 242Mpa in the whole device. Therefore, the DIE3 will induce product of vertical crack because the die in SCSP were processed by chip thinning technology for control their thickness. The load capacity of this chip in Y direct will decline. Secondly stress concentrate also appeared in the joint part of die, adhesive and EMC. The stress value in this position reaches about 160Mpa. This location is likely to cause the delamination fracture along interface between die and adhesive.

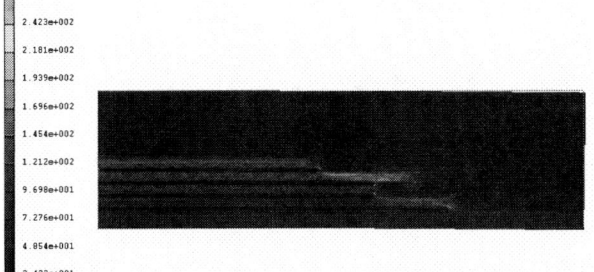

Fig.6 Equivalent Von Mises stress distribution map at 260℃

J-integral history of each crack is shown in Fig.5. According to the Fig.7, it is obvious that the J-integral value of crack6 tip (node19) located in the interface between adhesive and BT substrate is the largest than others. At the same time, it is found that the position also is the section of stress concentrate. The result indicated that the J value will increase gradually with the increase of thermal stress,, and once the J-integral reaches the critical J integral, the crack will be unstable as such so crack growths to fracture directly. Based on above analysis, the crack6 tip is regarded as the key position in this paper.

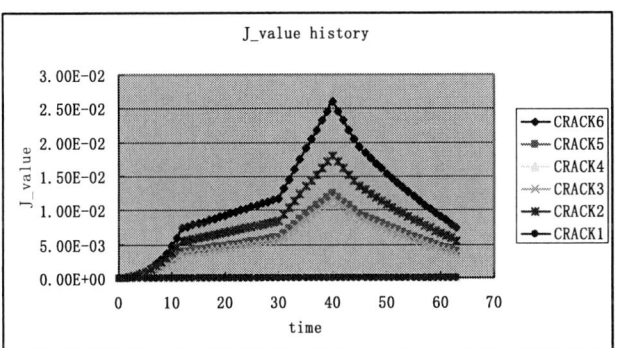

Fig.7 J integral development history

3. Optimization on the interfacial strength for SCSP

The basic thought of uniform experiment design is arrange the experimental scheme reasonably. Through fewer experiments, let experiment data can reflect the characteristics of mathematical model comprehensively. It can assure the experiment point statistical characteristic of uniform distribution. Compare with traditional orthogonal experiment design, it emphasize on uniform distribution of experiment point in order to obtain the more information through the fewer experiments. Therefore, the experiment times significantly fewer than orthogonal experiment design. This method is especially suitable for multi-factors and multi-levels experiments and completely unknown system model. In this paper, the interfacial strength of SCSP through uniform experiment design is optimized by changing the structure of SCSP. All the factors and levels are given in table2. The 4 factors and 5 levels uniform design table is selected to create original simple points of optimization design. [6]

Tab.2 Structure parameter table

factor	level1	level 2	level 3	level 4	level 5	original
XI	3.8	3.9	4	4.1	4.2	3.88
X2	0.1	0.11	0.12	0.13	0.14	0.12
X3	0.15	0.165	0.18	0.195	0.21	0.18
X4	0.015	0.02	0.025	0.03	0.035	0.025

Tab.3 Uniform design table

EXP	X	X2	X3	X4	J-value
EXP 1	4.2	0.1	0.15	0.025	0.0216622
EXP 2	4.2	0.13	0.165	0.035	0.0223572
EXP 3	4	0.11	0.15	0.035	0.0186852
EXP 4	3.9	0.13	0.18	0.02	0.022998
EXP 5	4	0.1	0.195	0.02	0.0263493
EXP 6	3.8	0.1	0.165	0.015	0.0137538
EXP 7	4.1	0.14	0.18	0.03	0.0246149
EXP 8	4.1	0.13	0.195	0.025	0.0274628
EXP 9	3.8	0.14	0.195	0.035	0.0101425
EXP10	4	0.14	0.15	0.015	0.0211928
EXP 11	3.9	0.13	0.165	0.025	0.0206307
EXP 12	4.1	0.12	0.165	0.02	0.024282
EXP 13	3.9	0.12	0.21	0.015	0.0269752
EXP 14	4.2	0.11	0.18	0.015	0.0279798
EXP 15	4.2	0.14	0.21	0.02	0.0318815
EXP 16	3.9	0.11	0.18	0.03	0.0214246
EXP 17	3.8	0.12	0.15	0.03	0.01882255
EXP 18	4.1	0.1	0.21	0.035	0.0260748
EXP 19	4	0.12	0.195	0.03	0.0248677
EXP 20	3.8	0.11	0.21	0.025	0.0130385

Note: X1 is the length of DIE1, X2 is thickness of DIE, X3 is thickness of BT substrate, and X4 is thickness of adhesive

3.1 Regression Analysis

The results of uniform design must be analyzed by regression analysis. The relationship formula between some variables is found out with regression analysis. This paper establishes the approximate function between corresponding structure parameter and maximum J integral value in crack6 tip through regression analysis method. We not only consider the linear term, but also their quadratic term and interaction term. Other parameters are not involved in the function use the original value. The SPSS software is used to process data and regression analysis with backward method. The results as follow (show in table4~6):

Tab.4 ANOVA

	Sum of Squares	Df	Mean Square	F	Sig.
Regression	.001	8	0.000	30.040	0.000(f)
Residual	.000	11	0.000		
Total	.001	19			

Tab.5 Coefficients

Model	Unstandardized Coefficients		Standardized Coefficients	t	Sig.
	B	Std. Error	Beta		
(Constant)	-0.300	0.054		-5.592	0.000
X2	1.983	0.748	4.673	2.652	0.022
X3	1.596	0.658	5.642	2.425	0.034
X1X2	-.355	0.183	-3.555	-1.939	0.079
X1X3	0.247	0.116	3.617	2.120	0.058
X1X4	1.019	0.188	4.848	5.416	0.000
X2X4	-22.296	4.321	-3.481	-5.160	0.000
X3X4	-9.340	2.652	-2.131	-3.522	0.005
X3X3	-6.212	1.276	-7.914	-4.867	0.000

Dependent Variable: J_value

Tab.6 Model summary

Model	R	R Square	Adjusted R Square	Std. Error of the Estimate
	0.978(f)	0.956	0.924	0.001692898

B in table2 is coefficient of regression variable. It is found that X2 has the greatest influence on J-integral value from t-test in table5. On the other hand, X1 and X4 were ruled out because of low t value. Besides, X1X4 and X2X4 also have large influence on J integral value in regression function. The F test value of regression function reached 30.04 large than $F_{0.05}(8,11)=2.95$. This result indicates that the regression function is significant. Regression effect was tested by R test and R2 test value. According to table6, adjusted R2 =0.924

indicated that the degree of correlation is good. Based on the coefficients of regression function in table5, the approximate mathematical model between J-integral and structure parameters can be obtained as below:

$$Y=1.983X_2+1.596X_3-0.355X_1X_2+0.247X_1X_3+1.019X_1X_4-22.296X_2X_4-9.34X_3X_4-6.212X^2-0.3$$

The optimization solution by optimization algorithm can be calculated. The result was shown in table7.

Tab.7 Comparison between initial design and optimum solution

De \ F	X1	X2	X3	X4	J_value
Initial design	3.88	0.12	0.18	0.025	0.0209
Opt solution	3.9	0.14	0.15	0.033	0.0098

3.2 Results and Discussions

Fig5 shows the respond max J-integral value when factors vary within their domain.

Form Fig.5 (a), it shows that J integral value decrease firstly, and then increase when X1 increase. There is a minimum appear in X1=4.Fig.8 (b), (c) shows that when X2 or X3 increases, the J integral value also increase in an approximate linear regular. And the difference of those Figs is slope of curve in Fig.8 (b) larger than another. This result shows the thickness of die has the greatest influence on J-integral than other variable. The result of regression analysis also shows the same thing. Fig.7 (b) shows an inverse ratio relationship between J integral value and X4.

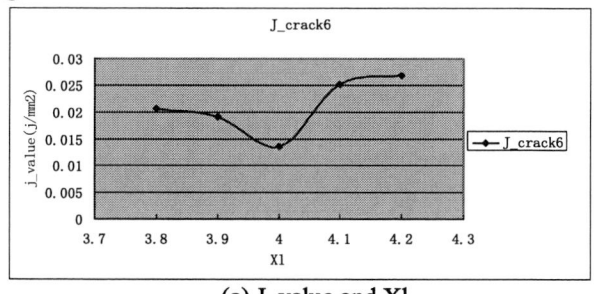

(a) J_value and Xl

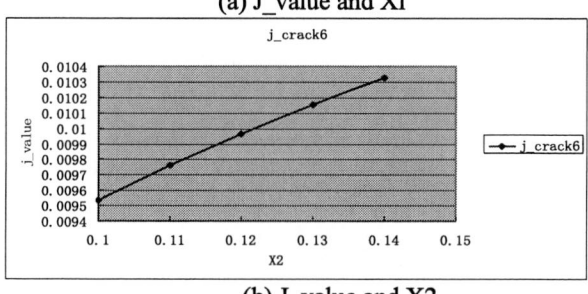

(b) J_value and X2

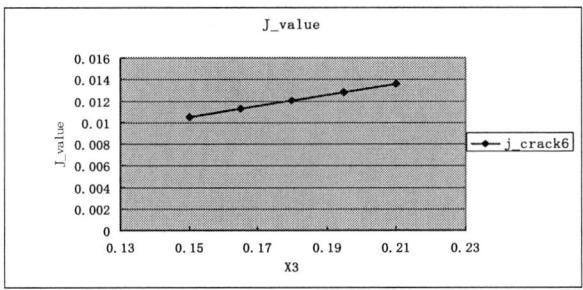

(c) J_value and X3

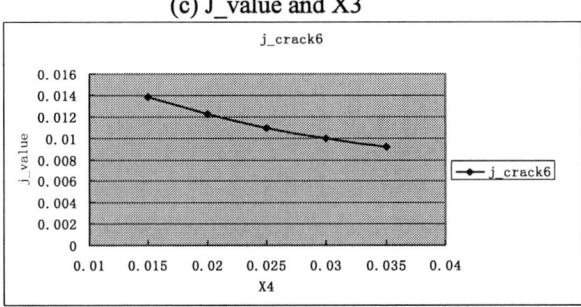

(d) J_value and x4

Fig.8 Curves for relationship between J_value and structure parameters

4. Conclusions

According to the finite element analysis, it is found that the interface between BT substrate and adhesive is the most unstable interface in whole device, and the cracks in this interface easily induce delamination. We established the approximate mathematical model between design variable and J integral value through uniform experiment design and regression analysis method. Variance analysis shows that the significant and correlation of regression function are also good. According to regression function, we obtain the optimization solution through optimization algorithm. The optimization result shows that J integral value is reduced significantly.

Acknowledgments

The research work in this paper is financially supported by the National Natural Science Foundation of China (NSFC) (Grant No. 60666002)

References

1. Andrew A.O.Tay. "Mechanics of interfacial delamination in IC packages undergoing solder reflow," *High Density Microsystem Design and Packaging and Component Failure Analysis*, Jun. 2005, pp. 1-7.

2. Hu Guojun, Andrew A.O.Tay. "Applications of modified virtual crack closure method on delamination analysis in a plastic IC package during lead-free solder reflow," *Electronic Packaging Technology Conference*, Dec. 2005, pp. 7-9.

3. W.D.van Driel, M.A.J.van Gils, G.Q.Zhang. "Prediction of delamination in micro-electronic packages," *Electronic Packaging Technology, Sept. 2005.* pp. 676-68.

4. K.M.B.Jansen, L, Wang, D.G. Yang. "Constitutive modeling of moulding compounds," *IEEE,* 2004, pp. 890~894.

5. J.G.J.Beijer, J.H.J.Janssen, H.J.L.Bressers. "Warpage minimization of the HVQFN map mould," *IEEE*, 2005, pp. 168~174.

6. Fang, K.T.and Ma Changxing. <u>Orthogonal and uniform experiment design</u>, Science press (Beijing, 2001).

Evaluate Anti-Shock Property of Solder Bumps by Impact Test

Hongjia Xi[1], Minyi Lou[2], Bing An[1], Fengshun Wu[1], Yiping Wu[1]

[1] State Key Laboratory of Material Processing and Die & Mould Technology, Huazhong University of Science and Technology, Wuhan National Laboratory for Optoelectronics, Wuhan, 430074, China

[2] Samsung Semiconductor (China) R&D Co., Ltd, Suzhou, 215021, China

Email: ypwu@mail.hust.edu.cn

Abstract

Anti-shock property of lead-free Sn96.5-Ag3.0-Cu0.5 solder bumps was investigated by the high speed impact to explore the relation among the fracture modes and the reflow profile and the microstructure of solder joint. Solder bumps were formed with various reflow profiles and multi-reflow and then subjected to the impact test under a constant speed of 1.8m/s and a shear standoff of 50μm. The results show that the IMC status has a close relation with the impact behavior upon one reflow. When the heating factor increases beyond 800 s-°C, the thickness of IMC layer goes up, and the impact absorbed energy of solder bump raises quickly. Upon multi-reflow using the same profile, the IMC thickness changed a little but the failure modes varied a lot.

Keywords: impact test, solder bump, lead-free solder, intermetallic compound, fracture, failure mode

1 Introduction

Solder joint anti-shock ability has become a issue due to the prevalence of mobile devices which are portable and susceptible to drop conditions [1]. On the other hand, for environment protection concern [2], lead-free solder joints are widely adopted but they exhibit more brittle under impact force, owing to the brittle intermetallic compound (IMC) formed on the bonding interface. Many previously researches [3-5] were done to investigate the anti-shock ability of BGA solder joints, especially in the replacement of board-level drop test which is expensive and time-consuming. It is shown that, in conventional quasi-static strength test, fracture always takes place in the solder bulk, while under high speed impact loading conditions, it is frequently observed that fracturing occurs around the interface between the solder joint and the bonding pads, where intermetallic compounds (IMC) are formed [6].

In the former investigation, several high speed ball shear test apparatus were developed with shear velocity ranging from 0.7m/s up to 4m/s [5, 8, 9]. It was shown that under a strain-rate over 1m/s, brittle fracture of intermetallic compounds on the bonding interface frequently occurred.

In this paper, impact test on the solder bumps is performed to investigate the brittle nature of BGA solder joints by using different reflow profile and reflow times. The microstructure was observed and the failure modes were analyzed.

2 Tester build-up

We developed a high speed shear tester to measure the impact strength and toughness properties of the solder joints, as shown in Figure 1. The solder joint sample was attached on a hammer. When falling along the slick guide rail, it approached a high impact speed, which can be controlled and adjusted by changing the drop height. A quartz piezoelectric sensor and a high sampling data acquisition system were employed to measure the transient structural response of solder bump under dynamic impact loading. The impact force profile was drawn by the computer subsequently, see Figure 1.

(a)

(b)

Fig. 1 (a) High speed shear device and (b) its schematic

Typical profile of impact force vs. time was shown in Figure 2. Shear force increases immediately after the shear tool hits the solder joint and then decreases after fracture within the solder joints initiates. Subsequent oscillates of the profile were due to fixture vibrations, which was determined by the natural frequency of test system. F_{max} is the maximum impact force, which implies that the fracture crack begins to

978-1-4244-2739-0/08/$25.00 ©2008 IEEE

form in the solder bulk or the IMC layer, and the fracture failure of the solder joint will occur at an extremely short time. T is the duration of the first half-sine part of the impact force profile. Total energy absorbed is the area below the first half-sine part, which represents the IMC strength, ductility, toughness of solder joints, respectively [10].

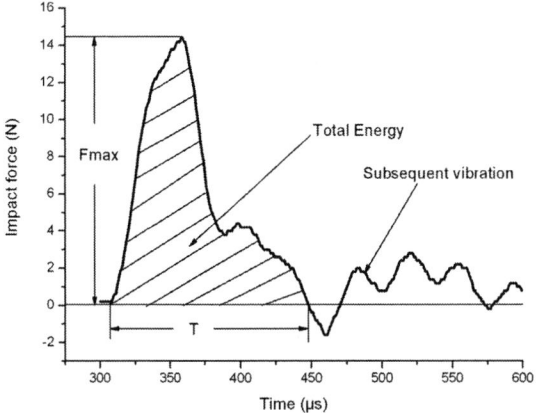

Fig.2 Typical impact force-time profile

3 Experiments
3.1 Sample preparation

The samples were prepared by reflow the Sn96.5-Ag3.0-Cu0.5 solder balls in $\varnothing$450 μm on Cu/Ni/Au pads. The opening diameter of the pad was 400μm. Various heating factors Qη were chosen to evaluate the influence of the reflow process on the anti-shock ability of solder joints. Qη is defined as the integral of the temperature above the liquidus over the dwell time (see Figure 3) [11], i.e. the area of the reflow profile above the liquidus. Apparently, Qη represents the thermal history to affect the formation of interfacial IMC layer, and the latter is regarded to have a large influence on the brittleness of the solder joint.

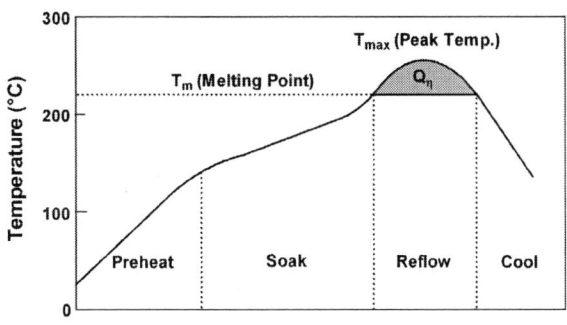

Fig. 3 Typical reflow profile and the definition of the

heating factor Qη

Five reflow profiles were set to approach the different heating factors, ranging from 378 s-°C to 1178 s-°C, as listed in Table 1. After that, we chose profile C5 to perform the multi-reflow process, and the times are 1, 2, 3, 5, 7, respectively. In each sample lot, 30 balls were tested.

Tab. 1 Qηs determined by the reflow profiles

Profile No.	Preheat time (s)	Peak temp(°C)	Reflow time(s)	Heating factor Qη(s-°C)
C1	92	235	42	378
C2	88	239	48	528
C3	93	245	53	742
C4	93	251	59	1003
C5	78	255	62	1178

3.2 Impact test

Impact test was performed on the above samples by the tester. The impact velocity on the tip of the impact tool was set in 1.8m/s. As the structural responses were very sensitive to the impact tool standoff, a constant standoff of 50μm was used, which were recommended by JEDEC standard [12].

3.3 Microstructure and fractography

Solder joint samples with different reflow parameters were mounted in epoxy resin and cross-sectioned, followed by grinding, polishing and etching procedures for microstructure observation. The section of the sample was observed using a scanning electron microscope. The fractography of the samples was photographed with a stereo microscope and their failure modes were determined.

4 Results and discussion
4.1 Microstructure

Sectional Microstructures of the solder joints under different Qηs are shown in Figure 4. IMC appears smooth on the solder side when Qη is below 1000 s-°C (a)-(d), while IMC turned to conchoids on the solder side with growth in thickness (e).

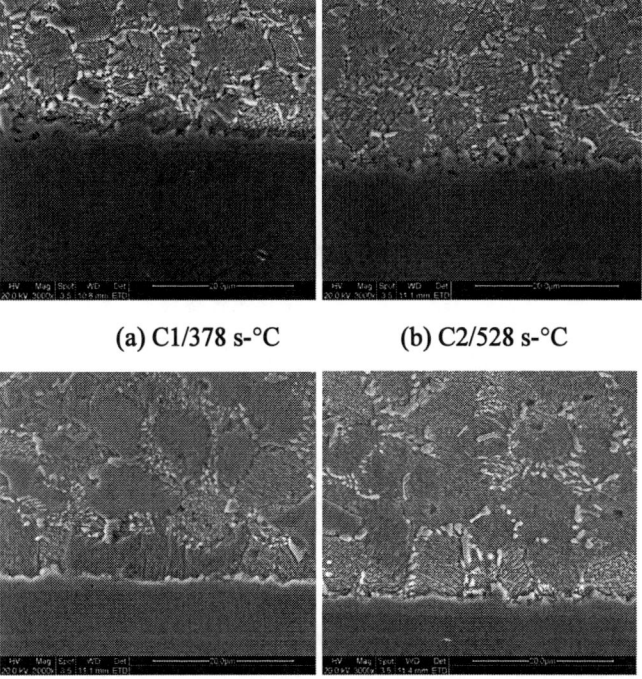

(a) C1/378 s-°C   (b) C2/528 s-°C

(c) C3/742 s-°C   (d) C4/1003 s-°C

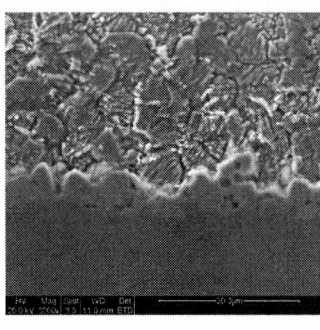

(e) C5/1178 s-°C

Fig. 4 Effect of heating factors on IMC thickness

The sectional structure of solder joints upon multi-reflow were shown in Figure 5. The results showed that, although the thickness of the IMC seems no grow as the numbers of reflow increase, the IMC morphology varies from each other. The possible reason is that, the later reflow process makes the former existed IMC gradually spall and dissolve into the solder body, and continuously the interfacial reaction lasts and new IMC forms on the pad side. Hence, total thickness does not change much due to the balance between IMC formation and dissolution. Nevertheless, the morphology exhibited dissimilarly. The upper surface of IMC in Figure 5 goes flatter with the reflow times increases. This difference may link to macro mechanical properties of solder joints, especially under an impact load.

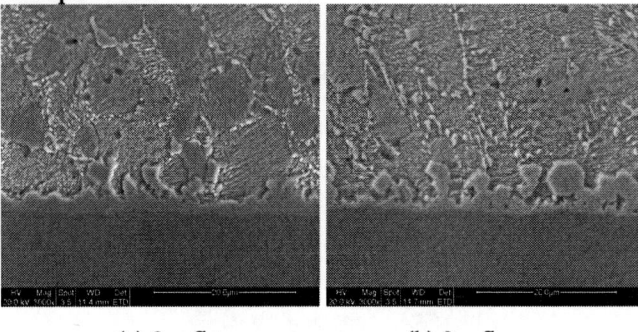

(a) 2 reflows   (b) 3 reflows

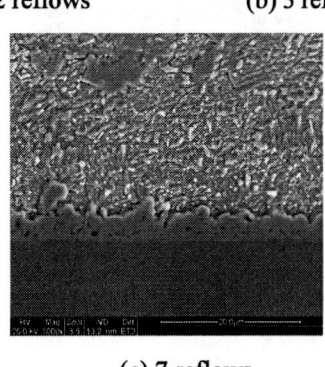

(c) 7 reflows

Fig. 5 Effect of reflow times on IMC microstructure

Figure 6 gives the statistical results of IMC thickness with respect to heating factors. It is obviously that with the increase of heating factor, IMC thickness goes down first and increases

after $Q\eta > 800$ s-°C, but the variation of thickness changes a litter under multi-reflow.

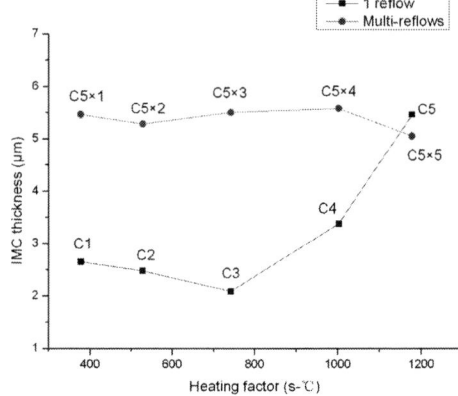

Fig. 6 IMC thickness as a function of heating factor

4.2 Impact test

Figure 7 showed the measured impact profile versus time. A greater IMC strength made a higher impact force and a longer failure time, which implied the better anti-shock ability. As depicted in Figure 8-9, max shear force and total energy varied with the increasing of heat factor. The trend was especially apparent in the energy profile. C1, C2, C3 exhibit similar anti-shock ability while C4 and C5 grow dramatically. Compared Figures 6 and 9, it was obvious that, the anti-shock ability significantly depended on IMC thickness.

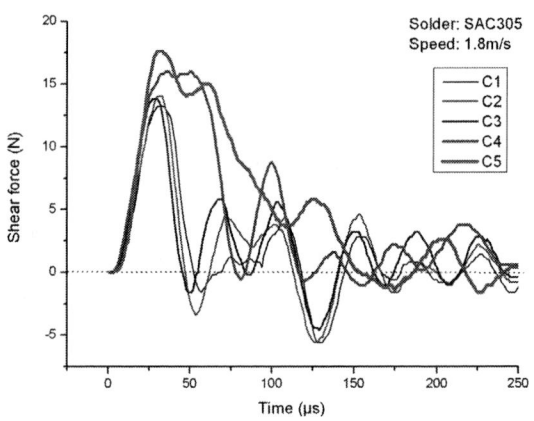

Fig. 7 Dynamic shear force vs. time

In a practical manufacturing process of electronic products, solder joints always undergo multi-reflow conditions which have influences on the reliability of interconnection. As all the samples are prepared under C5 reflow profile, which has the largest heating factor, multi-reflow is conducted to investigate effect of multi-reflow time on the solder ball reliability. Test results shown in Figure 10~12 reveal that anti-shock ability slightly decreases after the second reflow and gradually increases in the following reflows. After 7 times of multi-reflow, characters of mechanical properties reach its peak value. It is interesting that upon multi-reflow the IMC thickness changes a little but impact behavior becomes better. This may attribute to the

916

dissolution of the IMC or the evenness of IMC and need to be explored by the prospective experiments.

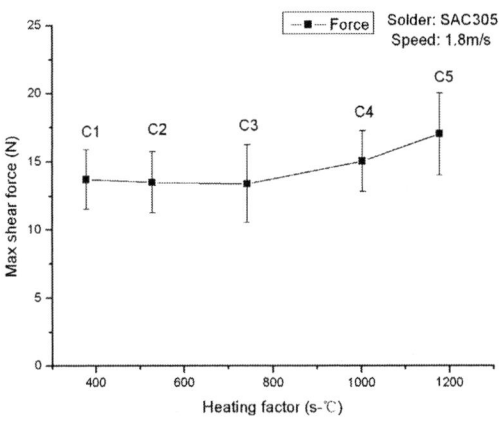

Fig. 8 Max shear force (fracture load) vs. heating factor

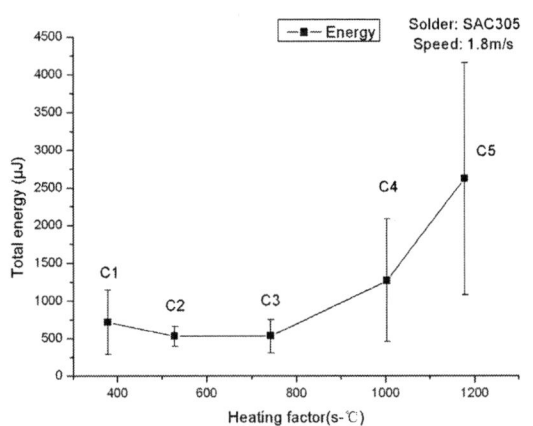

Fig. 9 Total energy absorbed vs. heating factor

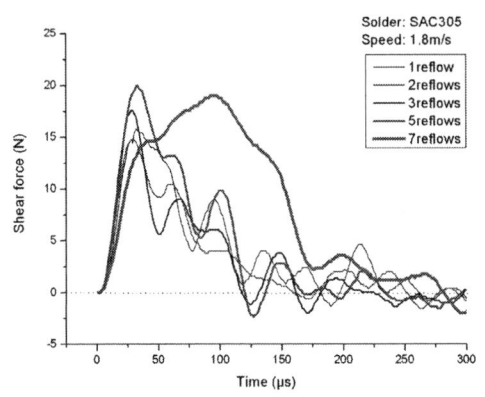

Fig. 10 Dynamic shear force with respect to time

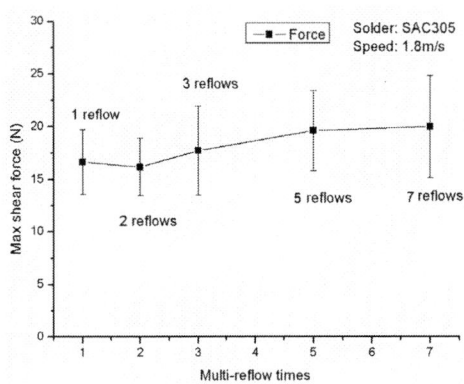

Fig. 11 Max shear force (fracture load) vs. multi-reflow times

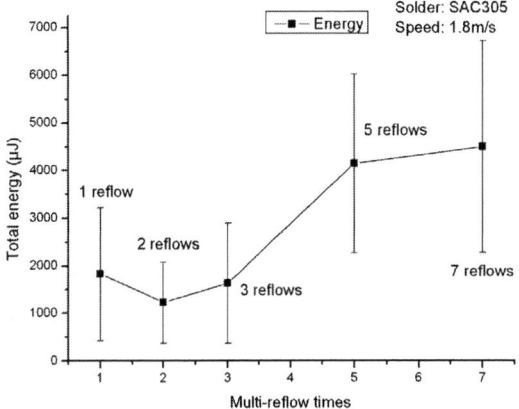

Fig. 12 Total energy absorbed vs. multi-reflow times

4.3 Failure analysis

The failure modes of the fractography upon impact test were sorted into six types as shown in Figure 13, where mode B1 is generally equal to mode 1 in section 2.3, and B2 to B4 is the subdivisions of mode 2. Mode A represents that the solder and soldering strength are higher than that of the bonding strength between copper pad and resin substrate.

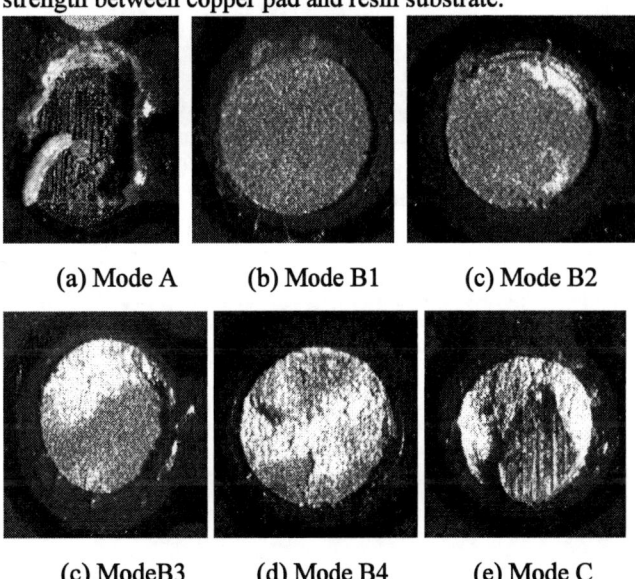

(a) Mode A   (b) Mode B1   (c) Mode B2

(c) ModeB3   (d) Mode B4   (e) Mode C

Fig. 13 Six types of failure modes (180×).

917

(a) Mode A, pad lift, fracture occurs between the Cu pad and substrate.

(b) Mode B1, brittle fracture, with solder remaining on the pad ≤5%;

(c) Mode B2, brittle fracture, with solder remaining on the pad 5%~25%;

(d) Mode B3, brittle fracture, with solder remaining on the pad 25%~75%;

(e) Mode B4, brittle fracture, with solder remaining on the pad 75%~95%;

(f) Mode C, ductile fracture, crack takes place in solder bulk.

In a shear test, usually all of the failures occur in the solder bulk (Mode C in Figure 13), which means that the maximum shear force represents the strength of solder alloy itself. When subjected to a high strain rate loading, the fractography of the samples exhibited different types of failure modes as shown in Figure 13. Under different heating factors, statistical analysis of failure modes showed that all of the samples are brittle failure (Figure 14). Profile C3 had the most brittle fracture modes while C5 was relatively ductile compared to the rest samples.

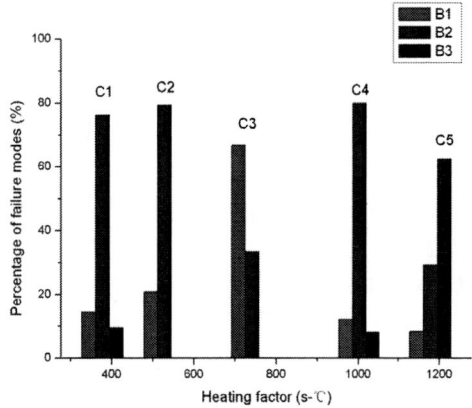

Fig.14 Distribution of failure modes under different heating factors.

Figure 15 gives the percentage of failure modes in impact test with respect to multi-reflow conditions. The dominant failure modes were Mode B1~B4, which corresponded to the bond failure of solder joints. Combined with the mechanical behavior shown in Figures 11 and 12, it is obviously that with the increasing of reflow times, the fracture place turns from the interface to the solder body, indicating that the interfacial strength becomes stronger.

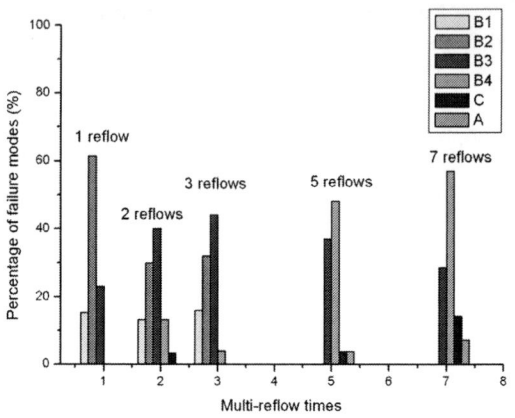

Fig. 15 Failure modes upon various reflow times

6. Conclusions

A desktop solder ball impact tester was developed and the anti-shock ability of solder joints was studied. Solder bumps with composition of Sn96.5-Ag3.0-Cu0.5 was tested under a constant speed of 1.8m/s and shear tool standoff of 50μm. Microstructure analysis indicated that the thickness and shape of IMC layer was affected by heating factor. With the heating factor raises, the mean thickness of IMC grows and also its absorbed energy is increasing. The fact indicates the best heating factor for SAC305 solder was around 1200 s-°C. Upon the multi-reflow test, the thickness of IMC varies a little but the interface between the solder body and IMC becomes flatter, and the anti-shock property turns better with the increasing reflow times.

Acknowledgments

The authors would like to acknowledge the financial support by National Natural Science Foundation of (No. 60776033), National High Technology Research and Development Program of China (863 Program) (No. 2006AA04A110), and Financial Grant from Samsung Inc.

References

1. Goyal, S., Uasani, S, Patel, D. M., "The Role of Case Rigidity in Drop-Tolerance of Portable Products," *IJMEP*, 1999, pp. 175-184.
2. J. Cannis, "Green IC packaging", *Advanced Packaging*, (2001), pp. 33 - 38.
3. M. Date, T. Shoji, M. Fujiyoshi, K. Sato, et al, "Impact Reliabiity of Solder Joints," *The proceedings of the 54th ECTC*, Las Vegas, NV, USA, June 2004, pp. 668-674.
4. Tn-Cheng Chiu, Kejun Zeng, Roger Stierman, et al, "Effect of Thermal Aging on Board Level Drop Reliability for Pb-free BGA Packages", *The proceedings of the 54th ECTC*, Las Vegas, NV, June 2004, pp. 1256-1262.
5. Chang-Lin eh, Yi-Shao Lai, Hsiao-Chuan Chang, et al, "Correlation between Package-level Ball Impact Test and Board-level Drop Test," *Proceedings 7th Electronics Packaging Technology Conference*, Singapore, 2005, pp. 270–275.
6. Shengquan Ou, Yuhuan Xu, K. N. Tu, "Micro-Impact Test on Lead-Free BGA Balls on Au/Electrolytic Ni/Cu

Bond Pad," *Proc.55th Electronic Components and Technology Conf.*, 2005, pp. 467-471.

7. E. Suhir, "Could Shock Tests Adequately Replace Drop Tests," *Electronic Components and Technology Conference*, 2002, pp. 563-573.

8. E.H. Wong, Y-W Mai, R. Rajoo, et al, "Micro Impact Characterization of Solder Joint for Drop Impact Application," *Proc. Electronic Components and Technology Conf.*, 2006, pp. 64-71.

9. Wong EH, Rajoo R, Mai Y-W, et al, "Drop impact: fundamentals and impact characterization of solder joints," *Proceedings of the 55th electronic components and technology conference*, Lake Buena Vista, FL, 2005, pp. 1202-1209.

10. Chang-Lin Yeh, Yi-Shao Lai, "Design Guideline for Ball Impact Test Apparatus", *Journal of Electronic Packaging,* Vol. 129, No. 3 (2007), pp. 98-104.

11. Bo Tao, Yiping Wu, Han Ding, et al, "A quantitative method of reliability estimation for surface mount solder joints based on heating factor $Q\eta$", *Microelectronics Reliability*, Vol. 46, (2006), pp. 864-872.

Cavitation instability in Valanis-Landel hyperelastic IC packaging material

Li Zhigang , Shu Xuefeng

Institute of Applied Mechanics & Biomedical Engineering , Taiyuan University of Technology, Taiyuan 030024, China

Abstract

In this paper, a representative material cell containing a single microvoid is used to investigate void growth under combined vapor pressure and thermal stress. The plastic IC packaging material is assumed to be Valanis-Landel hyperelastic materials. Using the theory of cavity formation and unstable void growth in incompressible hyper-elastic material, we gained an analytical relation between the applied traction (moisture-induced vapor pressure and thermal stress) and void volume fraction in forementioned materials. Numerical analysis showed that the critical stress is existent and the critical stress decreases with increase of the initial porosity.

Key Words: vapor pressure, unstable void growth, hyperelastic material

1. Introduction

Plastic electric packages often crack during the solder-reflow process required to mount them to the board. The problem attracted much interest from researchers. Mostly early studies show that thermal stress induced the popcorn failure. A series of Zhang and Fan's studies indicate that absorbing moistures in uncontrolled humid play an important role. The vapor pressure model based a micromechanics approach was established by Fan and Zhang [1-3]. They have listed four stages which lead to final popcorn failure. In stage 1(preconditioning), the package absorb moisture from the environment, which condenses in micropores in the substrate, solder mask, die attach and along the interfaces. In stage 2, the condensed moisture vaporizes under the high temperature associated with the reflow process, generating high internal vapor pressure which causes microvoids to nucleate, grow rapidly and coalesce. As a result small interfacial delamination zones are initiated. In stage 3, the vapor pressure exerts traction loading on the delaminated interfaces, aggravating the process of delamination and eventually causing the package to bulge. In the final stage, the interface crack (e.g. die/die-attach interface, die-pad/molding compound interface) propagates laterally outwards. When the crack reaches the package exterior, the high-pressure water vapor is suddenly released, producing an audible sound like popcorning. On the other hand Gent and Lindley（1958）[4] observed the sudden formation of voids in hyper- elastic material in their experimental work on rubber cylinders . Ball（1982） set up the theoretical frame for cavitation formation and growth and found a class of hyper-potential functions and also gave the conditions of cavitation formation[5]. Ren and Cheng[6,7] studied the cavitation problem in incompressible or compressible materials too. Guo and Cheng established the theory of the unstable void growth in neo-Hookean plastic IC packaging material[8]. It is interesting to note that inside surface of

voids in plastic electronic packages also suffer the vapor pressure while outside surface of voids suffer thermal load. It is different from dead-load boundary condition of Ball's model. Guo and Chen's research results show that unstable void growth will be produced when the sum of the internal vapor pressure and the remote thermal stress reaches a critical traction, who point out that the critical traction is related to the shear modulus and the initial void volume fraction of these materials. Using Guo's theory and considering the weakening effect on property of material caused by heat and moisture, theoretical explanation of popcorn failure in neo-Hookean plastic IC packaging material was given by Fan.

Considering the variety of plastic IC packaging material, in this paper, a representative material cell containing a single microvoid is used to investigate void growth under combined vapor pressure and thermal stress. The plastic IC packaging material is assumed to be Valanis-Landel hyperelastic materials. Using the theory of cavity formation and unstable void growth in incompressible hyper-elastic material ,we gained a analytical relation between the applied traction (moisture-induced vapor pressure and thermal stress) and void volume fraction in forementioned materials. Numerical analysis showed that the critical stress is existent and the critical stress decreases with increase of the initial porosity.

2. Mechanical Description

Let a thick sphere shell, denoted R in the reference configuration, of external radius R_1 and containing a spherical hole of initial radius R_0, undergoes thermal stress on its external surface and vapor pressure on its internal surface. Denote by r_0 and r_1 the deformed radii in the current configuration. The deformation is assumed to be symmetric.

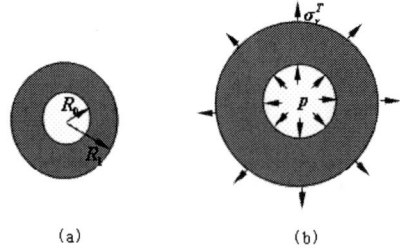

Fig.(1) mechanical model. (a) is the initial configuration of a singe void. (b) is the current configuration of a singe void.

The deformation is assumed to be symmetric. The plastic IC packaging material is assumed to be Valanis-Landel hyperelastic materials. the strain energy function has the form[4]:

$$W = \mu \left[\lambda_R + \lambda_\theta + \lambda_\varphi + \frac{1}{\lambda_R} + \frac{1}{\lambda_\theta} + \frac{1}{\lambda_\varphi} - 6 \right]$$

(1)

Here, μ is the shear modulus of the Valanis-Landel materials. $\lambda_R, \lambda_\theta, \lambda_\varphi$ are the principal stretches of deformation tensor. The principal Cauchy stress σ_i takes the form:

$$\sigma_i = \lambda_i \frac{\partial W}{\partial \lambda_i} - p(R) \qquad i = r, \theta,$$

(2)

In the absence of body forces must satisfy the following equilibrium equation:

$$\frac{d\sigma_r}{dr} + \frac{2}{r}(\sigma_r - \sigma_\theta) = 0$$

(3)

The constant load boundary condition on the surface during the solder-reflow process now requires that

$$\sigma_r(r_1) = \sigma_r^T \quad \sigma_r(r_0) = -p$$

(4)

Thus, the governing equations of the problem of radially symmetric deformation for a thick sphere shell composed of the equilibrium equation (3), the strain-energy function (1), the constitutive equation (2) and the boundary condition (4).

3. Solutions to Governing Equations

Spherically symmetric deformation implies that

$$\lambda_r = \frac{dr}{dR}, \lambda_\theta = \lambda = \frac{r}{R}$$

(5)

By use of the incompressibility condition, one has

$$r^3 - R^3 = r_0^3 - R_0^3$$

(6)

From (5), (6) we obtain:

$$\lambda_r = (1 - \frac{r_0^3 - R_0^3}{r^3})^{\frac{2}{3}},$$

$$\lambda_\theta = \lambda = (1 - \frac{r_0^3 - R_0^3}{r^3})^{-\frac{1}{3}}$$

(7)

Substituting Eq.(1),(2) into the Eq.(3), and using Eq.(7) we have:

$$r\frac{d\sigma_r}{dr} = -2\mu \left[(1 - \frac{r_0^3 - R_0^3}{r^3})^{2/3} - (1 - \frac{r_0^3 - R_0^3}{r^3})^{-2/3} \right.$$
$$\left. - (1 - \frac{r_0^3 - R_0^3}{r^3})^{-1/3} + (1 - \frac{r_0^3 - R_0^3}{r^3})^{1/3} \right]$$

(8)

On integration of Eq.(8) from 0 to R, and using the notation

$$t = (1 - \frac{r_0^3 - R_0^3}{r^3})$$

(9)

We obtain:

$$\frac{\sigma_r(t) - \sigma_r(t_0)}{\mu} = -\frac{2}{3} \int_0^t \frac{t^{2/3} - t^{-2/3} - t^{-1/3} + t^{1/3}}{1 - t} dt$$
$$= t^{2/3} + 2t^{1/3} - t_0^{2/3} - 2t_0^{1/3}$$

(10)

For the spherical shell under consideration, One can define the initial and current void volume fractions f_0 and f

$$f_0 = (\frac{R_0}{R_1})^3, f = (\frac{r_0}{r_1})^3$$

(11)

From the incompressibility condition and the definitions in Eq.(11), it follows

$$t_0 = (\frac{R_0}{r_0})^3 = \frac{f_0}{f} \frac{1-f}{1-f_0},$$

$$t_1 = (\frac{R_1}{r_1})^3 = \frac{f_0}{f} \frac{1-f}{1-f_0}$$

(12)

Substituting Eq.(12) into the Eq.(10) and using Eq.(7), we obtain

$$\frac{\sigma_r^A + p}{\mu} = (\frac{1-f}{1-f_0})^{2/3} + 2(\frac{1-f}{1-f_0})^{1/3} - (\frac{f_0}{f} \frac{1-f}{1-f_0})^{2/3}$$
$$- 2(\frac{f_0}{f} \frac{1-f}{1-f_0})^{1/3}$$

(13)

The formula (13) represents the relationship between the applied traction $\frac{\sigma_r^T + p}{\mu}$ and initial porosities f_0, current porosities f.

4. Unstable void growth and Popcorn Failure in electronic Packages

This section discusses the implications of Eq. (13) on popcorn failure in electronic packages. The right hand side of (13) attains a maximum value when the initial volume f0, the hardening exponent n and the fiber reinforcement parameter α were given. This peak stress value defines the critical traction, say σ_c, which is shown in Fig.2 by the peak of the curve. The critical condition for unstable void growth is given by:

$$\frac{d\bar{\sigma}}{df} = \frac{dF(f_0, f, n, \alpha)}{df} = 0$$

(14)

or equivalently,

$$f_0^{1/3} [f_0^{1/3}(1-f)^{1/3} + f^{1/3}(1-f_0)^{1/3}]$$
$$= f^{5/3} [(1-f)^{1/3} + (1-f_0)^{1/3}]$$

(15)

Figure 2 shows that the critical stress is very sensitive to the initial porosity. We obtained that the critical traction decreases about 1~1.5 times, considering the initial void volume fraction ranges from 1% to 5% (estimated by Fan and Zhang for some polymeric materials commonly used in IC packages). Figure 3 shows the variation of the critical traction σ_c versus initial porosity f_0. It can be seen that

σ_c will become smaller when f_0 becomes bigger. Figure 4 displays the critical void size f_c corresponding to σ_c for different values of initial porosity f_0. The critical tractions and the critical void sizes under the different variety of initial porosity are given by table 1

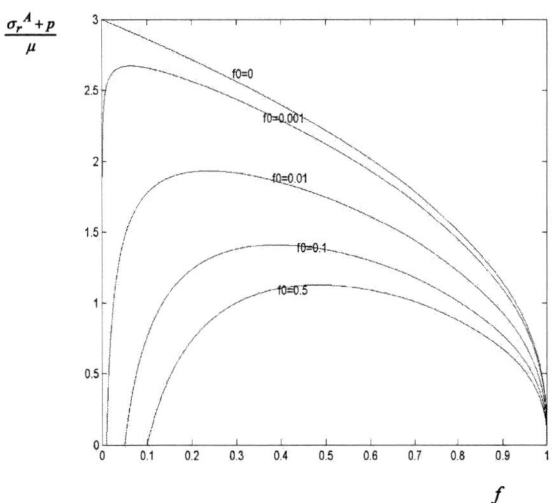

Fig.2 Traction versus void volume fraction

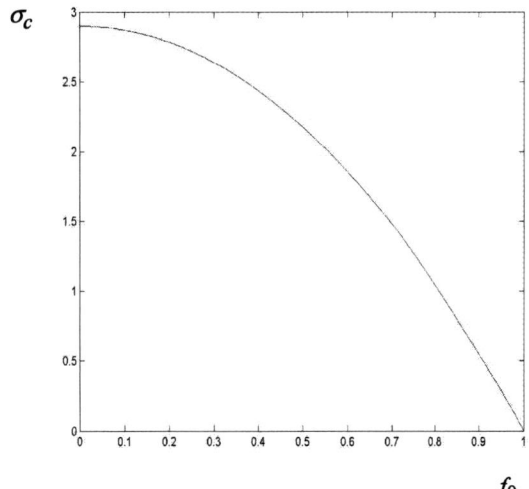

Fig.3 the variation of the critical traction versus initial porosity

Table 1: Critical Values for Typical Initial Void Volume Fractions

f_0	f_c	$(\sigma_r{}^A+p)/\mu$
0	0	3
10^{-4}	0.0729852028	2.6726
10^{-3}	0.1332681804	2.4089
0.01	0.2495773882	1.9336
0.05	0.3980247730	1.4102
0.1	0.4912495866	1.1280
0.2	0.6090529361	0.8136

5.Conclusions

Plastic electronic packages made of Valanis-Landel hyperelastic material will produce popcorn failure when the sum of the vapor pressure and thermal stress attains the critical traction. The critical traction will be closely related to the initial porosity and the shear modulus of this material Furthermore, the critical traction and the critical void size decrease with increase of the initial porosity.

Acknowledgments

The authors would like to thank Dr. X. J. Fan of Institute of Microelectronics, for many valuable discussions. This project was sponsored by the National Natural Science Foundation of People's Republic of China (10672113), the National Natural Science Foundation of People's Republic of China (10672112),the Special PH.D Research Fund for Universities (20050112009), the Natural Science Foundation of Shanxi Province (2007011010) and Shanxi Province Innovation Fund for Graduate Student (19712909536).

Reference

[1] Fan, X.J. , Zhang, G..Q., "Analytical Solution for Moisture-Induced Interface Delamination in Electronic Packaging" Proc. ESIME 2003.733-738.

[2]Fan, X.J. , Zhang, G.Q., and Ernst, L.J., "A Micromecbanics Approach for Polymeric Material Failures in Microelectronic Packaging", Proc. ESIME 2002

[3] Fan, X.J.,& Zhang, S.Y., "Void Behavior Due to Internal Vapor Pressure Induced by Temperature Rise", Journal of Materials Science, 30, 1995.3483-3489.

[4]Ren Jiusheng , Cheng Changjun. Cavitated bifurcation for composed compressible hyperelastic materials. Acta Mechanica Solida Sinica , 2002 , 15 : 208～213

[5]Ren Jiusheng , Cheng Changjun. Cavitated bifurcation for incompressible hyperelastic material. Applied Mathematics and Mechanics , 2002 , 23 : 881～888

[6]Guo. T.F.. & Cheng, L."Thermal and vapor pressure effects on cavitation and void growth. Journal of Materials Science,36(2001). 5871-5879

Moisture Absorption and Void Growing Effects on Failure of Electronic Packaging

Zhao Zhendong, Li Zhigang, Zhang Yu, Shu Xuefeng

Institute of Applied Mechanics & biomedical engineering, Taiyuan University of Technology, Taiyuan 030024, China

Abstract

The purpose of this paper is to study the combined effect of moisture absorption and void growing on the reliability of electronic packaging. Finite element simulation on a plastic PBGA package was carried out for moisture history from the moisture preconditioning (85 ℃ / 85 % RH for 168h) to subsequent exposure to a lead-free soldering process, and the rule of moisture diffusion and the change of stress was found. Then, with the implementation of interface properties into the model study, the critical stress that results in the unstable void growth and the delamination at interface is significantly reduced and comparable to the magnitude of vapor pressure. Finite element results give a good guideline on the underfill material selection, and also give an insight of the failure mechanism associated with moisture absorption.

Keywords: moisture diffusion, moisture absorption, vapor pressure, finite element simulation.

1. Introduction

During solder reflow in surface mounting, the temperature of an IC package is raised up to 230 ℃. The moisture absorbed in the plastic package becomes vaporized and exerts a pressure on the internal surfaces of the package. The induced vapor pressure coupled with the thermal stress could result in a failure mechanism often referred to as "popcorn" package cracking. The moisture-induced failure has long been considered as the consequences of the void initiation, growth and coalescence. Therefore, it is necessary to investigate the problem, in order to have more profound insight into the mechanisms of the delamination failure at interface.

The amount of moisture absorbed by the package prior to reflow had been known to have a strong influence on the susceptibility of the package to popcorn cracking. However, the importance of local moisture concentration on popcorn failure was realized only after Kitano's experiments on moisture sensitivity of electronic package. His work has highlighted the need for accurate estimate of the moisture induced vapor pressure in the package during reflow soldering of board mounting. Cheng and Guo focused on the void growth and interaction in polymeric solids, and established the theory of the unstable void growth in neo-Hookean plastic IC packaging material [1, 2].

The objectives of this study are as follows. First, the finite element analysis on moisture diffusion in microelectronic plastic packages during moisture preconditioning by using commercial FEA software is studied and modeled. Second, the detailed description and derivation for the vapor pressure evolution for three distinct cases are presented to obtain accurate estimate of the moisture induced vapor pressure in the package during reflow soldering. Third, the above vapor pressure model is used together with the equations of deformation and constitutive relations to investigate material behaviors during soldering, and a finite element model of a spherical volume of material containing a spherically shaped

micro-void is postulated and analyzed.

2. Moisture diffusion modeling

2.1. Moisture absorption diffusion

At the temperature and pressure experienced by IC packages during moisture pre-conditioning, the moisture diffusion process can be described rather well with Fick's diffusion equation.

$$\frac{\partial C}{\partial t} = D\left(\frac{\partial^2 C}{\partial x^2} + \frac{\partial^2 C}{\partial y^2} + \frac{\partial^2 C}{\partial x^2}\right) \tag{1}$$

where C is the local moisture concentration, x, y, z are the spatial coordinates, D is the diffusivity which measures the rate of diffusion, and t is the time. However, unlike temperature, which is continuous in nature, moisture diffusion poses a new challenge in that the concentration is discontinuous across bi-material interface. This restricts the use of Fick's diffusion equation with concentration as field variable to homogeneous system only. Therefore, A similar approach is to use moisture wetness, w, as the field variable, which is also continuous across multi-material interface [3]. It is defined as

$$w = \frac{C}{C_{sat}} \qquad 1 \geq w \geq 0 \tag{2}$$

where C_{sat} is the maximum moisture concentration can be absorbed by the material, the lower limit $w = 0$ means it is dry, and the upper limit $w = 1$ means it is fully saturated with moisture. The Eq. (1) can be rewritten as

$$\frac{\partial w}{\partial t} = D\left(\frac{\partial^2 w}{\partial x^2} + \frac{\partial^2 w}{\partial y^2} + \frac{\partial^2 w}{\partial x^2}\right) \tag{3}$$

and it may be solved as a typical heat transfer problem using any commercial FEA software. For moisture absorption modeling, the initial condition is $w = 0$ for the whole package, and the boundary condition is $w = 1$ at the external surfaces which are exposed to the ambient moisture.

The moisture profile of a Plastic Ball Grid Array (PBGA) subjected to JEDEC Level 1 moisture pre-conditioned (85℃ /85RH/168hrs) is depicted in Fig. 1(a), ranging from 0 % to 100 % saturation. The moisture diffuses into the package through mold compound, and gradually spreads into die attach layer. At the end of 168 h of moisture preconditioning under 85℃/85%RH, the package is almost fully saturated with moisture.

978-1-4244-2739-0/08/$25.00 ©2008 IEEE

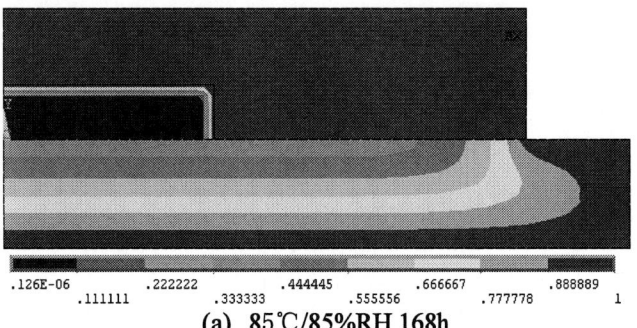

(a) 85℃/85%RH 168h

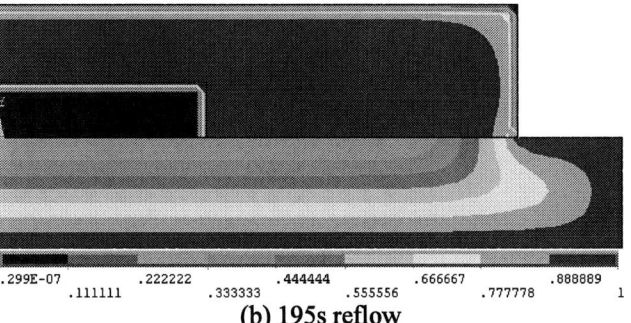

(b) 195s reflow

Fig.1 Transient moisture wetness distribution

2.2. Moisture desorption diffusivity

Moisture desorption diffusivity, especially during reflow, determine the moisture weight loss, and the relationship can be described by the following equation

$$D = D_0 e^{\frac{Q}{RT}} \qquad (4)$$

where D_0 is the diffusion coefficient, Q is the activation energy (ev), R is the Boltzmann constant (8.83e-5eV/K), and T is the absolute temperature (K).

During the 195s solder reflow, external package surface loses a significant amount of moisture due to high moisture desorption rate (see Fig. 1(b)). However, the moisture concentration in the interior of the package, including critical die attach and die/mold compound interfaces, still remains relatively unchanged. The local moisture concentration in these critical interfaces determines the strength of interfacial adhesion and magnitude of internal vapor pressure induced, which partially lead to moisture induced failures, e.g., delamination and popcorning, in the package.

3. Vapor pressure model

Some studies assume that the moisture is always in a single vapor phase throughout the temperature rise, and hence the ideal gas law can be applied for the evolution of the internal vapor pressure inside voids. Since such a vapor pressure model is no linked to the moisture property of the material, it is difficult to estimate the initial or reference vapor pressure at reference temperature.

Because of the inhomogeneous character of a porous material, the element should be established over a finite representative volume, RVE. Therefore, a field quantity, the void volume fraction, f, can be defined as

$$f = \frac{dV_f}{dV} \qquad 1 \ge f \ge 0 \qquad (5)$$

where dV_f is the void volume and dV is the element volume. When $f = 1$, it implies that the delamination occurs at this location.

A useful quantity, the moisture density in the voids, can be described as

$$\rho = \frac{dm}{dV_f} = \frac{dm/dV}{dV_f/dV} = \frac{C}{f} \qquad (6)$$

Let's make a comparison of magnitude of ρ with the ambient moisture density at 85℃/85RH condition. As an illustration, the saturated concentration of moisture in the molding compound, C_{sat}, is typically 0.01g/cm^3 while the saturated vapor density, ρ_g, is 0.0003 g/cm^3 at 85℃/85RH. The fact that $C_{sat} \gg \rho_g$, suggests beyond doubt the condensation of moisture in the molding compound. Such an amount of moisture must condense into the mixed liquid/vapor phase in material.

The state of moisture in the pores along the depth of the solid is depicted in Fig. 2.

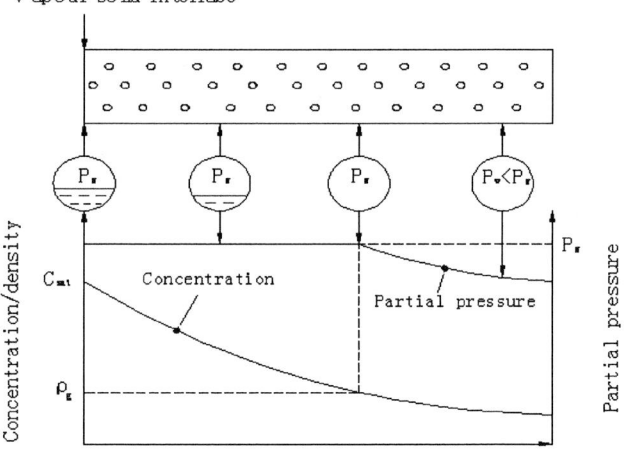

Fig. 2 Moisture concentration and vapor pressure in the solid

Three distinct cases for the vapor pressure evolution have been identified [4]. It is assumed that the water vapor follows the ideal gas law. The vapor pressure can be calculated based on the local moisture concentration after preconditioning which determines the transition temperature, T_1.

The first case is when the moisture density in the voids is low enough such that all the moisture becomes vaporized at preconditioning temperature, T_0

$$p = \frac{Cp_g(T_0)T}{f\rho_g(T_0)T_0} \qquad \text{when } T_0 \ge T_1 \qquad (7)$$

where p is the pressure and p_g is the saturated vapor pressure.

In the second case, the moisture is fully vaporized at a

temperature between preconditioning temperature, T_0, and the peak reflow temperature, T

$$p = \frac{p_g(T_1)T}{T_1} \qquad \text{when} \quad T \geq T_1 \geq T_0 \qquad (8)$$

For the last case, the moisture is not fully vaporized even at reflow temperature, T

$$p = p_g(T) \qquad \text{when} \quad T_1 \geq T \qquad (9)$$

This case uses thermodynamics table, instead of ideal gas assumption, for saturated pressure calculation at particular temperature.

4. Void behavior

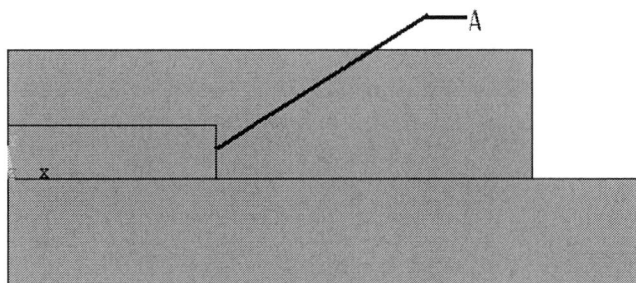

Fig. 3 The position of A

As mentioned before, the above vapor pressure model should be used together with the equations of deformation and constitutive relations to investigate material behaviors during soldering. Although the moisture exists and is evaporated anywhere in polymer materials when the entire package is exposed to the reflow temperature at 220℃, the rupture of the bulk material prior to interface delamination has never been found [5]. For the purpose of analysis a spherical volume of material containing a spherically shaped micro-void in the position A as depicted in Fig. 3 is considered.

Fig. 4(a) shows the finite element results for the cells subjected to thermo-mechanical stress and internal vapor pressures. The integrated Von Mises stress for spherically void are indicated by blue line. The thermo-mechanical stress is only considered in the Fig. 4(b). It can be seen that stresses developed in thermal loading are quite insignificant as compared that of integrated loading. The magnitude of the integrated Von Mises stress is as high as 5.56 MPa, about 4 times greater than normal thermo-mechanical stress. Therefore hygroscopic stresses are the dominant factor for the failure of the PBGA packages. At reflow temperatures, the moisture vaporizes. If the rapidly expanding vapor creates high internal pressure on pre-existing voids in the interface, the simultaneous action of thermal stresses and internal vapor pressure drives pre-existing voids to grow causing material failure.

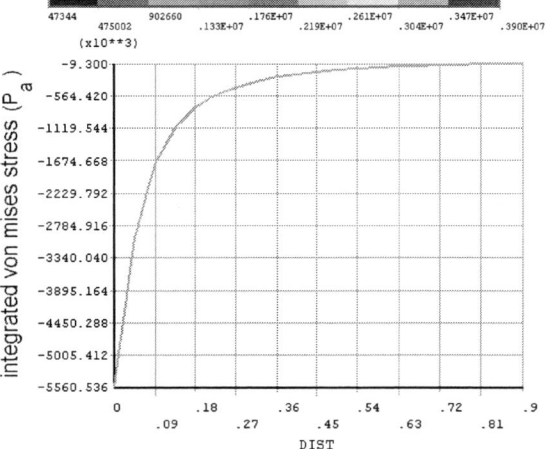

(a) The integrated Von Mises stress

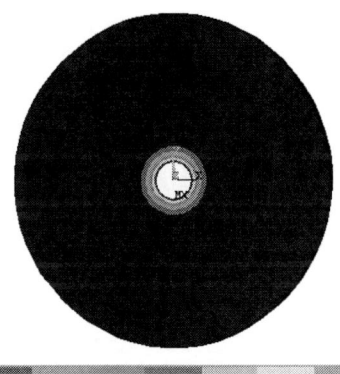

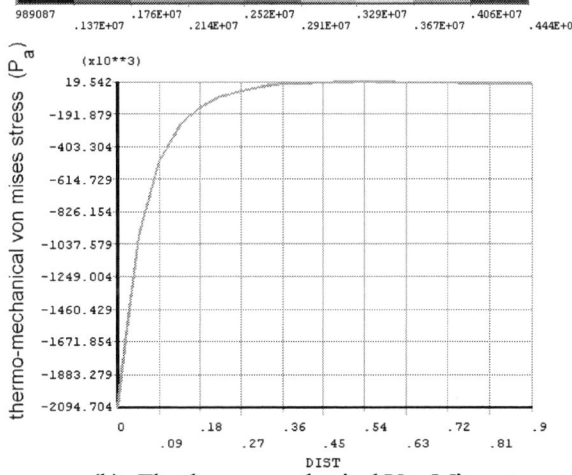

(b) The thermo-mechanical Von Mises stress

Fig. 4 A micro-mechanics model of a singe void in the interface

5. Conclusion

This paper investigated the reliability of a PBGA at reflow temperatures. FEA simulation results revealed the significance of contribution of moisture diffusion, which cause the vapor pressure. Vapor pressure with thermal loading contributes to the interfacial delimitation. Finite element results give an insight of the failure mechanism associated with moisture absorption. The model explains that the interface delamination will unlikely take place without moisture absorption, even though the temperature is increased at a higher level than the soldering temperature.

The moisture induced vapor pressure has been identified as the dominant damage mechanism responsible for popcorn cracking in electronic packages. This reinforces the importance of moisture diffusion and vapor pressure modeling.

Acknowledgments

This project was sponsored by the National Natural Science Foundation of People's Republic of China (10672113), the Special PH.D Research Fund for Universities (20050112009), the Natural Science Foundation of Shanxi Province (2007011010) and Shanxi Province Innovation Fund for Graduate Student (19712909536).

References

[1] T.F. Guo, L. Cheng, "Vapor pressure and void size effects on failure of a constrained ductile film", *Journal of the Mechanics and Physics of Solids*, Vol 51, 2003, pp. 993 – 1014.

[2] Cheng, L., Guo, T.F., Void interaction and coalescence in polymeric materials, *International Journal of Solids and Structures*, Vol 42, August 2006, pp. 735-756.

[3] Ee Hua Wong, Yong Chua Teo, Thiam Beng Lim, "Moisture Diffusion and Vapour Pressure Modeling of IC Packaging", *1998 Electronic Components and Technology Conference*, pp. 1372-1378.

[4] X.J.Fan, J. Zhou, G.Q. Zhang and L.J. Ernst, "A Micromechanics-Based Vapor Pressure Model in Electronic Packages", *Journal of Electronic Packaging*, Vol. 127, September 2005, pp. 262-267.

[5] Xuejun Fan, G.Q. Bang, W. van Driel and L. J. Emst, "Analytical Solution for Moisture-Induced Interface Delamination in Electronic Packaging", *2003 Electronic Components and Technology Conference*, pp. 733-738

Mechanical Test after Temperature Cycling on Lead-free Sn-3Ag-0.5Cu Solder Joint

Chung-Nan Peng, Jenq-Gong Duh [*]

Department of Materials Science and Engineering National Tsing Hua University , Hsinchu , Taiwan, China
[*] Fax: 886-3-5712686, E-mail: jgd@mx.nthu.edu.tw

Abstract

The solder balls on the plastic ball grid array (PBGA) packages provide both electrical input/output and mechanical supports. Mechanical properties and strengths of solder balls significantly influence the PBGA package reliability. Temperature cycling test has been widely used in flip-chip technology (FCT) [1-3] in order to simulate the failure mode due to fatigue and aging under repeatedly heat treatment. In this study, ball shear test was employed for measuring mechanical properties in the joint of the Sn-3Ag-0.5Cu solder ball(φ =100μm) attached to Ti/Ni(V)/Cu UBM. The alloys studied included Sn–3Ag-0.5Cu and baseline eutectic Sn–37Pb. FE-EPMA was used to quantitatively analyze intermetallic compounds and to observe fracture surface. After repeatedly heat treatments, there are some differences on fracture surfaces between Sn-3Ag-0.5Cu and Sn-37Pb joint. The fracture surface of Sn-37Pb joint was ductile after TCT treatment. Only tin rich and lead rich phase were observed in the solder matrix of Sn-37Pb system, and no precipitated compound formed during the heat treatment. The force of shear test was almost the same after 1000 times TCT. In the Sn-3Ag-0.5Cu system, Ag would dissolve in Sn to form Ag_3Sn. There were $(Cu, Ni)_6Sn_5$ and $(Cu, Ni)_3Sn$ IMC formation between Sn-3Ag-0.5Cu solder and Ti/Ni(V)/Cu UBM.[4-6] After TCT treatment, the growth of $(Cu, Ni)_6Sn_5$ IMCs would affect the fracture surface. In this study, the compounds of Ag_3Sn and $(Cu, Ni)_6Sn_5$ in the Sn-3Ag-0.5Cu joint grew slowly. After shear test, the fracture surface appeared brittle. The shear strength of Sn-3Ag-0.5Cu was higher than Sn-37Pb. The relationships between microstructure and strength of the joints as functions of thermal cycling test (TCT) cycle were investigated and discussed.

Keywords: Filp chip, TCT, Sn-37Pb, Sn-3Ag-0.5Cu, shear test, FE-EPMA

Discussion 1

Interface Reaction between the Sn-3Ag-0.5Cu/Sn-37Pb solders and Ti/Ni(v)/Cu UBM.

The application of ball grid array (BGA) packages to electronic equipment is becoming more common, since this type of package has the advantage of having faster electronic transmission, many input/output terminals, and providing better electrical performance. The solder pastes for lead-free joint is Sn-3Ag-0.5Cu and for lead-bearing is Sn-37Pb. The test specimens were fabricated in accordance with industrial standards and provided by United Microelectronics Corporation. The thermal cycling test (TCT) was performed for evaluating the thermal reliability of the package. The test condition was temperature cycle test from -65°C to 150 °C in one cycle for 23min (TCT, -65°C to 150°C, -65°C for 10 min, 25°C for 3min, and 150°C for 10 min).

Cross-sectional image of the Sn-3.0Ag-0.5Cu solder/Ti/Ni(V)/Cu interface after TCT cycle is shown in Fig.1. Only one layer of IMC was found between solder and Ni(V) in all joints after TCT cycle. This IMC was identified as $(Cu, Ni)_6Sn_5$ after FE-EPMA analysis. The thickness of Ni(V) changed from the initial 0.59±0.06 to 0.53±0.03 μ m after TCT1000, while the thickness of $(Cu,Ni)_6Sn_5$ IMC was 2.52±0.06, 2.68±0.13, 2.71±0.07 and 2.80±0.12 μ m for TCT0, 200, 500 and 1000 cycles, respectively. It appeared that $(Cu, Ni)_6Sn_5$ IMC in the joint grew gradually with increasing TCT cycles, as shown in Fig.2.

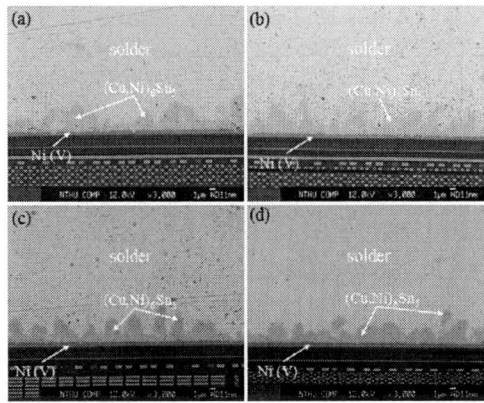

Figure 1 The cross section image of Sn-3Ag-0.5Cu solder/UBM after TCT cycles (a) 0 (b) 200 (c) 500 (d) 1000

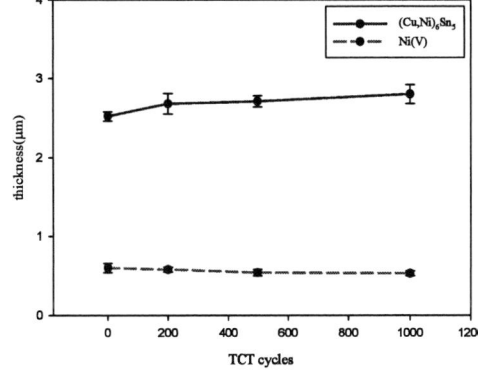

Figure 2 The thickness of $(Cu_{1-x},Ni_x)_6Sn_5$ IMC formed in the Sn-3.0Ag-0.5Cu solder and Ni(V) joints after TCT cycles.

On the other hand, the cross-sectional image of the Sn-37Pb solder/Ti/Ni(V)/Cu interface after TCT cycle is shown in Fig.3.

978-1-4244-2739-0/08/$25.00 ©2008 IEEE

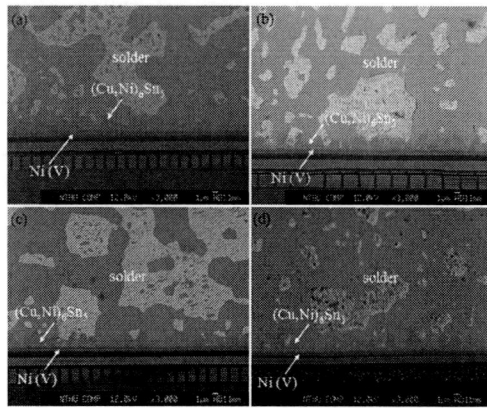

Figure 3 The cross section image of Sn-37Pb solder/UBM after TCT cycles (a) 0 (b) 200 (c) 500 (d) 1000 cycles

The thickness of $(Cu,Ni)_6Sn_5$ IMC was 1.32±0.05, 1.44±0.07, 1.48±0.03, 1.57±0.06μm in the joints after 0, 200, 500, 1000 cycles, respectively, as shown in Fig.4. It is clear that $(Cu,Ni)_6Sn_5$ IMC in the joint grew gradually with increasing TCT cycles. The average thickness of unconsumed Ni(V) in the Sn-37Pb joints during TCT times are 0.63±0.02, 0.61±0.04, 0.61±0.01, 0.57±0.03μm, respectively. The thickness of unconsumed Ni(V) decreased slowly during TCT times for all solder joints. It appeared that the thickness of $(Cu,Ni)_6Sn_5$ in the Sn-3Ag-0.5Cu/UBM was thicker than Sn-37Pb, which was about 1 μm. The main result is that when the Cu layer of Ti/Ni(V)/Cu UBM is fully consumed, the Cu atoms of Sn-3Ag-0.5Cu diffused into the interface between solder and UBM. The cupper supplied by Sn-3Ag-0.5Cu increased with increasing TCT cycles.

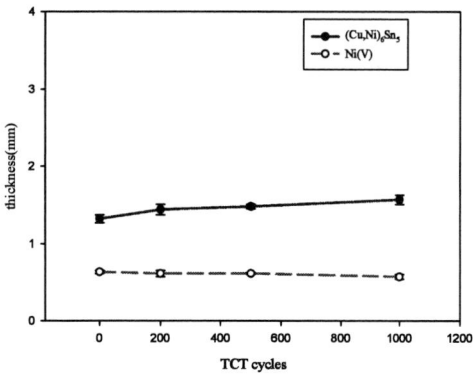

Figure 4 The thickness of $(Cu_{1-x},Ni_x)_6Sn_5$ IMC formed in the Sn-37Pb solder and Ni(V) joints after TCT cycles.

Discussion 2 Ball shear test result

In this study, statistical software was used to analyze the data and optimum parameters were confirmed by bonding 30 balls and by measuring the shear values using XYZTEC series shear tester. The ball shear test has been used to evaluate mechanical properties in the joint of either 63Sn-37Pb solder or the Sn-3Ag-0.5Cu solder ball (φ =100 μm) attached to Ti/Ni(V)/Cu UBM substrates. The thicknesses of Ti, Ni(V), and Cu was of 0.05 μm, 0.65 μm, and 0.8 μm, respectively. The bump strength, the mechanical strength of the solder bumps was evaluated with a shear tester using 1000 μm/s

shear velocity at a shear height of 10 μm.

The shear test samples were evaluated both quantitatively (force and fracture energy) and qualitatively (failure mode). Representative photographs of the failure modes (ductile, brittle) are shown in Figs.5 and 6. The shear mode changed from ductile (failure inside the bulk solder) to partial brittle (failure at the solder/IMC or IMC/UBM interfaces)[7] in the Sn-37Pb and Sn-3Ag-0.5Cu systems with an increase of cycle numbers to 1000. To determine the adhesion between the solder and the UBM, a die shear test was carried out for samples subjected to 0, 200, 500 and 1000 TCT cycles. It was obvious that the fracture mode changed with the number of cycles, and the solder failed at the IMC/UBM interface at the outer ring of the UBM in Sn-3Ag-0.5Cu solder after 1000 cycles, as shown in Fig.6(d). The partial ductile failure in the center could be due to the design of the UBM structure of the test dies in the current study[9]. For the eutectic Sn-37Pb system, most of the solder joints failed inside the solder, showing a ductile failure mode, as indicated in Fig. 5.

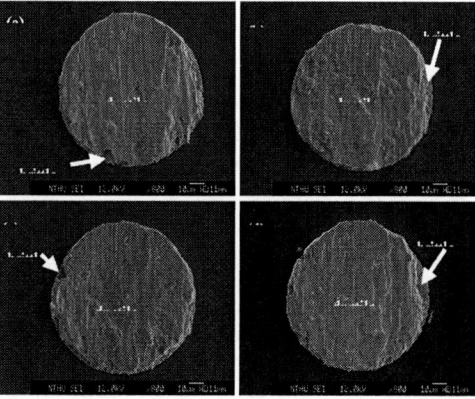

Figure 5 Fracture surface of Sn-37Pb solder joints sheared at a blade height of 10μm after various cycles (a) 0 (b) 200 (c) 500 (d) 1000

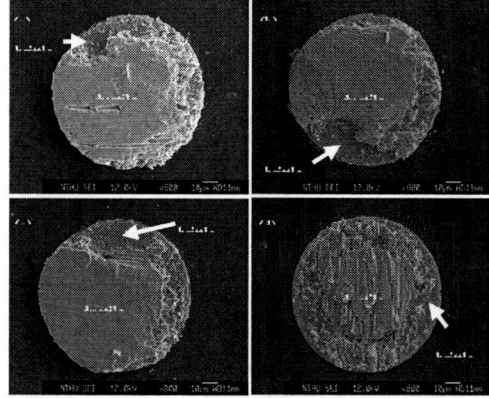

Figure 6 Fracture surface of Sn-3Ag-0.5Cu solder joints sheared at a blade height of 10μm after various cycles (a) 0 (b) 200 (c) 500 (d) 1000

The estimated area percentage of brittle fracture at interface by shear test is presented in Fig. 7. The brittle surface gradually increased with increasing TCT cycles. There was little brittle fracture occurred in the samples of Sn-37Pb system after shear test and the maximum area of brittle fracture surface was about 6.7%. Similar results were reported in literatures [8, 9]. Most of the fracture surface of Sn-37Pb is ductile after shear test. For the as-reflowed Sn-3Ag-0.5Cu system, about 15% of the fractured surface area on the pad side exhibited a brittle nature (Fig. 7). After 1000 TCT cycles, more than 60% percentage of fractured area on the pad side showed features of brittleness.

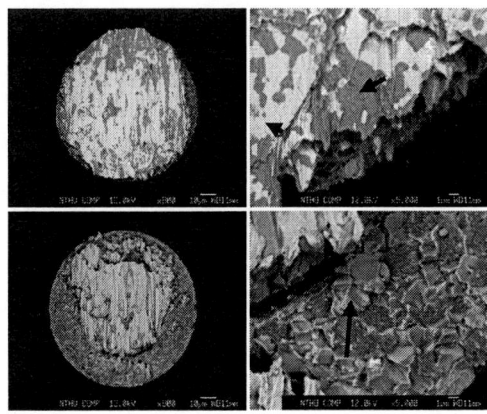

Figure 9 Backscatter electron micrograph (BSE) of shear test fracture surface after 1000 TCT cycles (a) Sn-37Pb solder (b) Sn-37Pb solder (higher magnification) (c) Sn-3Ag-0.5Cu solder (d) Sn-3Ag-0.5Cu (higher magnification)

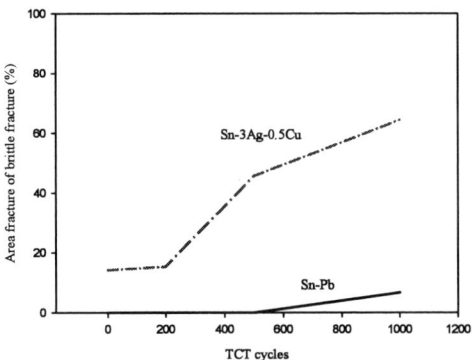

Figure 7 Estimated areas percentage of brittle interfacial fracture by shear test of Sn-3Ag-0.5Cu and Sn-37Pb solder joints as a function of TCT cycle. Each plotted data represented an average value from 30 points in a same sample

Two distinct reflow profiles were selected to represent the eutectic (220°C peak temperature) and lead free (245°C peak temperature) solders. Fig.8 shows the shear strength for both solder joints after TCT cycles. No degradation in the measured shear strength after 1000 TCT cycles is a good indication that the crystal structure of the solder joints are not changed as a result of these heat cycles. Sn-3Ag-0.5Cu solder shows slightly higher shear load on Ti/Ni(V)/Cu as compared to the Pb-Sn one. It could be due to the strengthening effect from the very fine particulate Ag_3Sn phase and flat $(Cu,Ni)_6Sn_5$ in bulk Sn-3Ag-0.5Cu solders[10]. Figs.9 (a) to (d) show the backscatter electron micrographs (BSE) of shear test fracture surface after TCT cycles. The flat $(Cu,Ni)_6Sn_5$ in bulk Sn-3Ag-0.5Cu solder is visible, as indicated in Fig.9(d).

Conclusions

In this study, the formation of intermetallic compound $(Cu,Ni)_6Sn_5$ between the Sn-3Ag-0.5Cu and Ti/Ni(V)/Cu was investigated Temperature cycling was used as criterion for testing fatigue life after repeated heat treatment. The thickness of $(Cu,Ni)_6Sn_5$ IMC grew as the number of TCT cycles increased, while the thickness of Ni(V) decreased with the TCT cycle. The morphology of interface between Sn-3Ag-0.5Cu and Ti/Ni(V)/Cu slightly changed after 1000 TCT cycles, which is similar to that of Sn-37Pb. The thickness of IMC in Sn-3Ag-0.5Cu is 1 μ m thicker than that of Sn-37Pb. There was little brittle fracture in the samples of Sn-37Pb system after shear test and the maximum area of brittle fracture surface on the pad side was about 6.7% after 1000 cycles. For the as-reflowed Sn-3Ag-0.5Cu system, the area of fracture surface area about 15% on the pad side exhibited a brittle nature. After 1000 TCT cycles, more than 60% percentage of fractured area on the pad side showed features of brittleness.

The shear strength for both Sn-3Ag-0.5Cu and Sn-37Pb solder joints did not decrease after 1000 TCT cycles. It is a good indication that the crystal structure of the solder joints are not changed as a result of temperature cycles. Sn-3Ag-0.5Cu solder with Ti/Ni(V)/Cu UBM exhibited slightly higher shear load as compared to the Pb-Sn one. It was due to the strengthening effect from the fine particulate Ag_3Sn phase and bulky $(Cu,Ni)_6Sn_5$ in Sn-3Ag-0.5Cu solders.

Acknowledgements

The joint assemblies preparation from United Microelectronics Corporation is acknowledged. Partial support from National Science Council, Taiwan under Contract No. NSC-96-2922-I-007-205 is also appreciated.

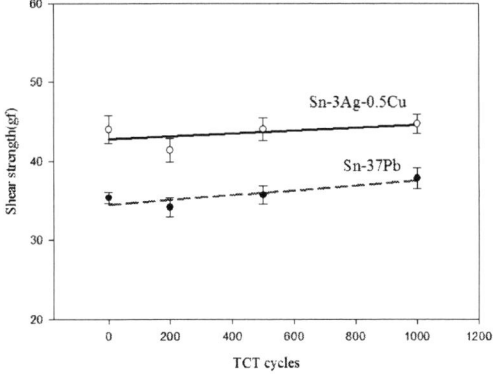

Figure 8 Shear strength of Sn-3Ag-0.5Cu and Sn-37Pb solder joints after TCT cycle test. Each plotted data represented an average value from 30 points in the sample assembly

Reference

1. Nousiainen, J. Putaala, T. Kangasvieri, R. Rautioaho, *Journal of Electronic Materials*, Vol. 35, No. 10 (2006).

2. Z.W. Zhong, *Microelectronics International*, Vol. 25, No.

2 (2008), pp. 9 – 14.

3. E. Martin, C. Larato, A. Clement, M. Saint-Paul, *NDT&E International, Vol.* 41, (2008), pp. 280 – 291.

4. Y.S. Lai, K.M. Chen, C.L. Kao, C.W. Lee, Y.T. Chiu, *Microelectronics Reliability,* Vol. 47, (2007), pp. 1273 – 1279.

5. T.I. Shih, Y.C. Lin, J.G. Duh, T. Hsu, *Journal of Electronic Materials*, Vol. 35, No. 10 (2006), pp. 1773 – 1780.

6. S.W. Yoon, V. Kripesh, S.Y.J. Jeffery, M.K. Iyer, *Journal of Electronic Materials*, Vol. 33, No. 10 (2004), pp. 1144 – 1155.

7. F. Zhang, M. Li, B. Balakrisnan, W. Chen, *Journal of Electronic Materials*, Vol. 31, No. 11 (2002), pp. 1256 – 1263.

8. J.W. Kim, J. Joo, D.J. Quesnel, S.B. Jung, *Materials Science and Technology*, Vol. 21, No.3 (2005), pp. 373.

9. J. Liang, N. Dariavach, P. Callahan, *Soldering & Surface Mount*, Vol. 19, No. 1 (2007), pp. 4 – 14.

10. H.W. Chiang, K. Chang, J.Y. Chen, *Journal of Electronic Materials*, Vol. 35, No. 12 (2006), pp. 2074 – 2080.

Reliability Challenges and Design Considerations for Wafer-Level Packages

Xuejun Fan[1,2], Qiang Han[1]

Department of Engineering Mechanics

[1]South China University of Technology, Guangzhou, China

[2]Department of Mechanical Engineering

Lamar University, PO Box 10028, Beaumont, TX 77710, USA

xuejun.fan@lamar.edu

Abstract

Wafer-Level Packaging (WLP) is essentially a true chip-scale packaging (CSP) technology, since the resulting package is practically of the same size as the die. Furthermore, wafer-level packaging paves the way for true integration of wafer fab, packaging, test, and burn-in at wafer level, for the ultimate streamlining of the manufacturing process undergone by a device from silicon start to customer shipment. There are several WLP technology classifications. Redistribution Layer and Bump technology, the most widely-used WLP technology, extends the conventional wafer fab process with an additional step that deposits a multi-layer thin-film metal rerouting and interconnection system to each device on the wafer. In this paper, an overview of the state of art WLP packaging technologies will be presented. The emphasis will be given to the challenges in reliability and the solutions based on the design.

Introduction

Wafer-Level Packaging (WLP) is essentially a true chip-scale packaging (CSP) technology, since the resulting package is practically of the same size as the die [1-11]. Furthermore, wafer-level packaging paves the way for true integration of wafer fab, packaging, test, and burn-in at wafer level, for the ultimate streamlining of the manufacturing process undergone by a device from silicon start to customer shipment. There are four (4) WLP technology classifications: redistribution layer and bump technology, encapsulated copper post technology, encapsulated wire bond technology, and encapsulated beam lead technology [2]. Redistribution Layer and Bump technology, the most widely-used WLP technology, extends the conventional wafer fab process with an additional step that deposits a multi-layer thin-film metal rerouting and interconnection system to each device on the wafer. This is achieved using the same standard photolithography and thin film deposition techniques employed in the device fabrication itself. This additional level of interconnection redistributes the peripheral bonding pads of each chip to an area array of underbump metal (UBM) pads that are evenly deployed over the chip's surface. The solder balls or bumps used in connecting the device to the application circuit board are subsequently placed over these UBM pads. Aside from providing the WLP's means of external connection, this redistribution technique also improves chip reliability by allowing the use of larger and more robust balls for interconnection and better thermal management of the device's I/O system. Encapsulated Copper Post technology is very similar to redistribution and bump technology in the sense that the chip's bond pads are also rerouted into an area array of interconnection points. In this technology, however, these interconnection points are in the form of electroplated copper posts, instead of pads. These copper posts provide enough stand-off for the active wafer surface to be encapsulated in low-stress epoxy by transfer molding, exposing only the top portions of the posts where the solder balls will be attached.

In this paper, an overview of the state of art WLP packaging technologies will be presented. The emphasis will be given to discuss the challenges in reliability and the solutions based on the design.

Bump on Pad WLP

Bump on pad WLP is a traditional wafer level packaging technology and the process is very similar to a typical flip chip technology. One of those structures is the bump-on-nitride (BON) structure, consisting of solder bump and under bump metal (UBM) seated on the thin inorganic passivation as shown in Fig. 1.

The semiconductor devices utilizing BON WLPs are limited to 5x5 or less area arrays at 0.5mm pitch [5]. There are two reasons why traditional WLPs are predominantly utilized by components with relatively low I/O counts. The first is related to die size. The smaller a given component is, the higher the associated die count per wafer. WLP technologies by their very nature result in a lower cost per die (vs. traditional wirebond) when the die count per wafer is high. The second reason is that the solder joint reliability of WLPs is limited compared to the other CSPs. As the number of I/O per die increases (and thus the Distance to Neutral Point increases), the WLP may not achieve prescribed solder joint reliability requirements. As wafer sizes continue to increase, the effective die size for WLPs to be cost-competitive also increases. To take advantage of this trend the solder joint reliability of WLPs needs to be improved.

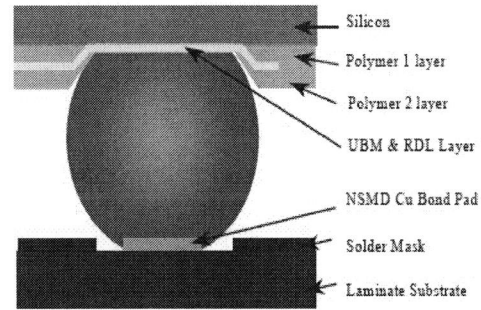

Fig. 1 Schematic diagram of BON structure WLP

Cu Post WLP

Fig. 2 shows a schematic of Cu post WLP structure. The processes involving in making the Cu post is shown in Fig. 3 [6], peripheral pads on the wafer are rearranged in a real array pattern after photolithographic plating. Then metal posts about 100μm high are fabricated on the wafer. After encapsulating the entire surface of the wafer using the new encapsulation method, a package the same size as the chip is fabricated using a wafer-level packaging method. This method involves dividing the wafer into individual semiconductor packages by a standard dicing process. The detailed processes are described in the following,

1. The first step in the formation of posts and the rearrangement of pads is the formation of a polyimide cover film on a device wafer. This layer is a stress buffer against molding pressure. It also improves the adhesive properties between the wafer and the encapsulant. In the next step, the sputtering is applied to fabricate a thin metal film onto the surface of the wafer. The film consists of an adhesion metal layer and a conducting layer (copper in general). These thin metal films are the plating base used in the rerouting process and the metal post-forming process. After forming a patterned resist on the thin metal film surface, a redistribution trace is fabricated by electrolytic plating. After these processes, the resist is reformed and posts are formed by electrolytic plating. Finally, the resist is peeled off, the sputtered film is etched and the post-forming process is completed.

2. The mold die is divided into two pieces, an upper and a lower die. The lower die consists of inner and outer dies. These mold dies are heated to approximately 175 C and a temporary film is held by vacuum to upper die. The wafer on which the posts are formed is set on the inner die of the lower die and an encapsulant tablet is placed on the central part of the wafer with metal posts. The temporary film has three functions. To prevent the encapsulant from making contact with the upper die. 2) To disperse the mold pressure over the entire surface of the wafer. 3) To expose the top of the posts in the next stage. When clamping the mold die, the encapsulant tablet melts by applying heat and pressure. The encapsulant spreads over the entire wafer surface and hardens by held inside the mold die. Even though the encapsulant leaving out mold release agent and has an extremely high adhesive force, the encapsulated wafer can be released easily. This is because only a peripheral part of the mold die comes in contact with the encapsulant.

Table 1 showed the reliability test results of a Cu post WLP package[6]. It clearly indicated that the reliability performance has been improved greatly.

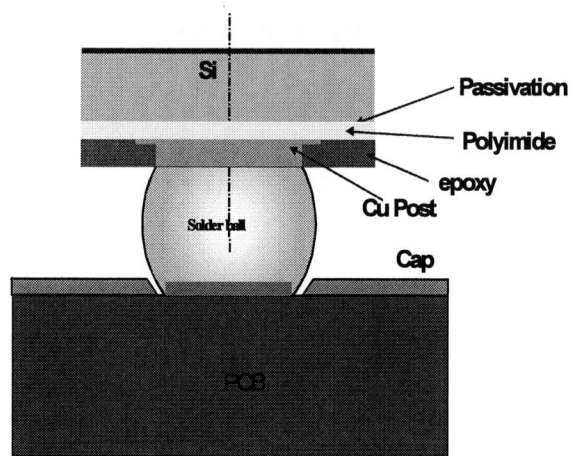

Fig. 2 A schematic of Cu post WLP structure

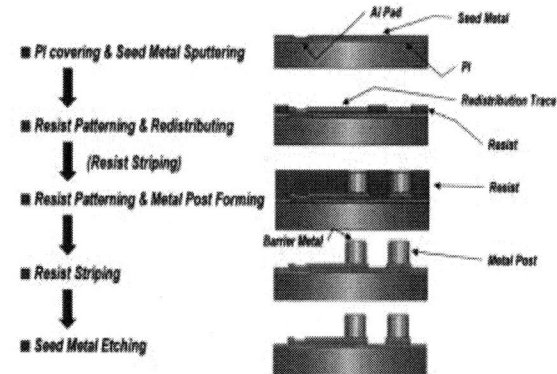

Fig. 3 Schematic diagrams of redistribution trace and metal post forming process.

Table 1 Cu post WLP package reliability

Test	Test Condition		Results
Temperature Cycle	-55/+125°C	N=20	1000cyc. PASS
Pressure Cooker	+121°C/85%RH	N=10	168Hrs PASS
Bending	Bending Span 3mm Bending Speed 5mm/min	N=10	OVER 15mm
Cyclic Bending	Bending Span 3mm	N=4	OVER 10000cyc.
Free fall			OVER 50cyc. OVER 2000cyc.

Bump on Polymer WLP

A typical structure for a bump on polymer (BOP) is shown in Fig. 4. In this structure the solder bump is placed over a layer of polymer dielectric material. There has been a strong performance driver for a WLP having a BOP structure. This technology minimizes the interconnect capacitance for high-speed application. In the case of a traditional BON structure (Fig. 1), the polymer is not subjected to stress originating in the solder bump since the solder bumps and UBM sits on silicon nitride. The typical failure mode has been ductile rupture through the solder at bump shear test and solder fatigue at temperature cycling (TC) tests. In the case of the

932

BOP WLPs, the bump rests on the polymer, and thus any stress applied to the solder bump directly propagates to the underlying polymer. The polymer must be designed to ensure sufficient mechanical toughness and adhesion to adjacent materials to qualify for the BOP structure.

The fabrication steps shown in Fig 5 follow the typical application steps used with photo definable polyimides [3]. The process is straightforward and requires two masks for opening vias in the dielectric layers and two masks for the metallization. The chosen polymer is photosensitive and positive acting, as are most photoresists in use today. Starting from a customer supplied wafer, a layer of polymer is applied over the existing passivation layer. The polymer thickness is targeted for 5 lm nominal after curing. The polymer precursor incorporates its own adhesion promoter, therefore saving an application and drying step. The material can be applied to any passivation material commonly used such as silicon dioxide, silicon nitride, silicon oxinitride, polyimide or similar classes of polymer that are less frequently used for passivation are equally acceptable.

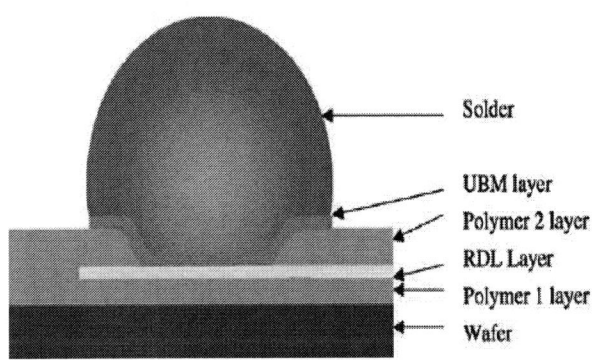

Fig. 4 a schematic of bump on polymer WLP

Because the photosensitizer used is closely related to common positive resists, the required exposure energy is similar to a thick photoresist. No incubation delay or other time constraint before development is necessary; the development can proceed immediately or a day or two later for convenience. The development proceeds in a manner similar to a positive photoresist, using either a puddle on a track system, or the old time honored technique of batch development in a bath, since no surface tension potential problem is expected from the large features prevailing in packaging. Partial curing of the polymer follows. Curing must be tightly controlled because the curing conditions affect the internal stresses and mechanical properties of polymers. A controlled atmosphere and controlled thermal cycle are recommended, even though this polymer is not subject to oxidation during curing and is thermooxidatively stable afterwards. Excellent planarization is obtained due to the high solid content of this material and low shrinkage. Partial curing is used to promote fusing of the subsequent polymer layer that will be applied on top, a technique developed in the 80 г s. The curing must be sufficient to eliminate any possibility of degassing during the following sputtering step and to provide

sufficiently high stiffness to withstand the stresses imposed by the deposited metal layer.

At completion of the first layer curing, the redistribution metal layer can be sputtered over the bonding pads. In contrast to negative acting materials, no plasma etching of bond pads is necessary, since the polymer develops completely and very cleanly without leaving the usual residue in the center of the pad. This is partly due to better photosensitization and partly because the developer is a very slow etchant of aluminum. The developer is able to clean the pad after removing all the photosensitive material, leaving a fresh aluminum surface that later helps obtain a more reliable via contact.

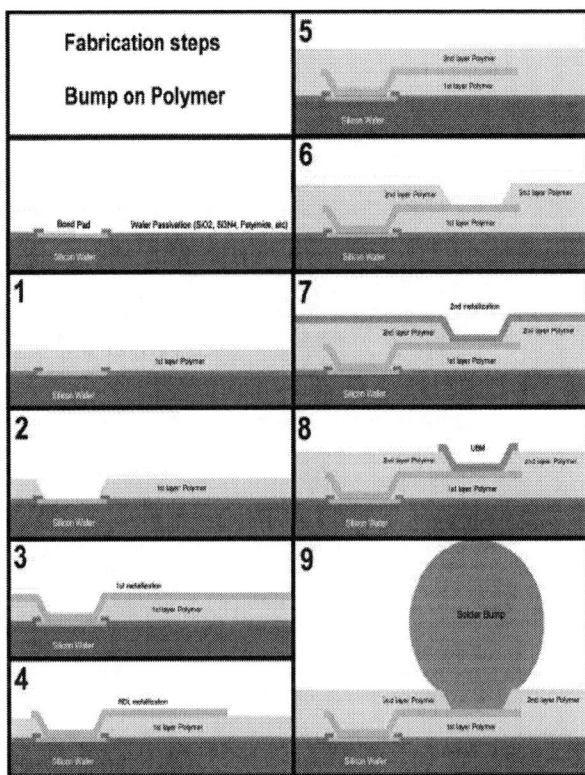

Fig. 5 Schematic of BOP WLP processes [3]

Two sets of TC tests have been carried out. CSP50 is a daisy chain test pattern having 0.5 mm pitch 10x10 array that was used for both test. In this test, the BOP structure consistently showed an approximate average 32% increase in Weibull life over the BON structure, as shown in Fig. 6.

There are a few different bump-on-polymer technologies and terminologies, as shown in Figs. 7-9, such as bump on RDL (Fig. 7), bump on thick repassivation/redistribution (Fig. 8), and bump on silicone bump (fig. 9).

933

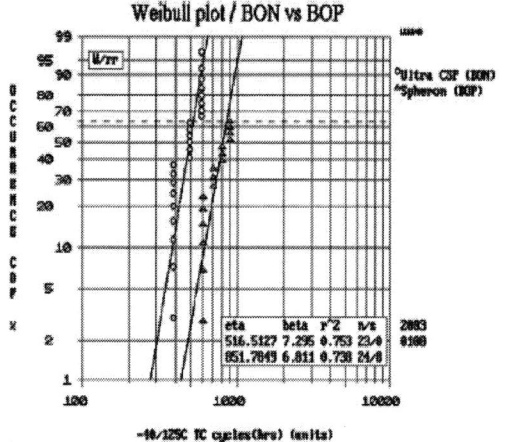

Fig. 6 Thermal cycling test results for BON vs. BOP

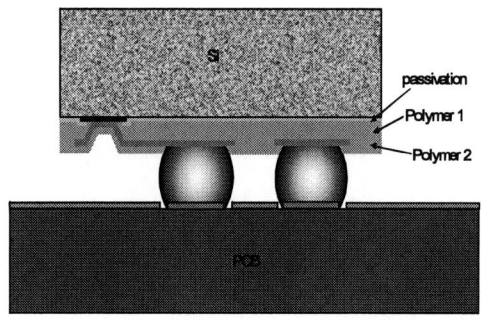

Fig. 7 Schematic of Bump on RDL

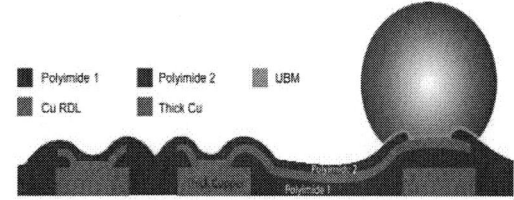

Bump on Thick Repassivation/Redistribution

Fig. 8 Schematic of Bump on Repassivation/Redistribution

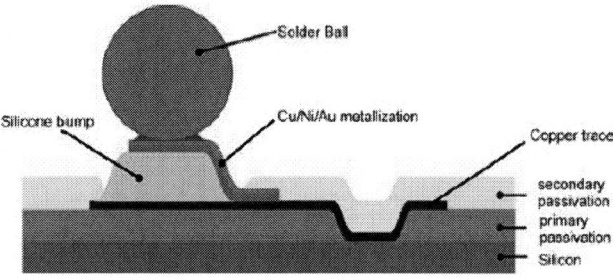

Fig. 9 Schematic of bump on silicone

Fan-Out or Redistributed Chip Packages (RCP) [7,8]

One of the most promising WLP technologies is the fan-out or redistributed chip packages (RCP). A schematic of such packages are shown in Figs. 10, and 11, respectively. Fig. 12 demonstrates the RCP process flow. Singulated die are placed

with active side face-down on a substrate. The die are then encapsulated with a silica-filled epoxy molding compound. After cure, the panel of the die is then ground to the desired panel thickness and then released from the substrate. The die panelization process is done with standard assembly tools such as a pick and place tool. The epoxy panel with die then undergo a redistribution process to route out the signals, power, and ground. The redistribution process is done with standard silicon manufacturing equipment. These processing steps consist of the deposition of copper metallization layers by electroplating techniques. The metal layers are separated by insulating layers comprised of a spin-coated, photoimageable dielectric, and patterned using batch process lithography. The metal layers connect the pads on the die surface to the pads placed on the surface of the package, providing the same function as the metal layers in the substrate of a ball. grid array (BGA), but with a much finer resolution.

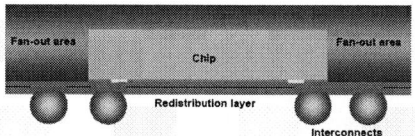

Fig. 10 Schematic of fan out packages

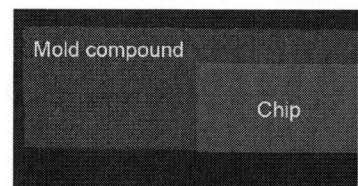

Fig. 11 molded die in a RCP package

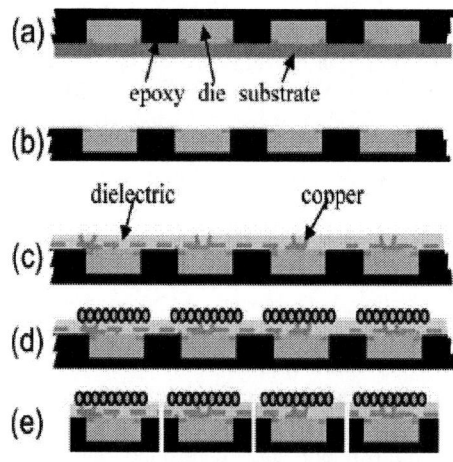

Fig. 12 RCP package process flow

Plastic Cored Solder Balls in WLPs

A plastic core solder ball consists of a large polymer core coated by a Cu layer and covered with eutectic and/or high melting SnPb solder. The main advantages of such a system are higher reliability due to the relaxing of stress by the polymer core and a defined ball height after reflow. These balls could be an alternative method as interconnection technique besides stencil printing and the use of preformed

934

full metallized solder balls. Fig. 13 shows the photos of the plastic core solder balls. Non-melting polymer-core copper and solder coated balls provide larger solder joint heights as well as more compliant 'solder balls' for greater reliability [14].

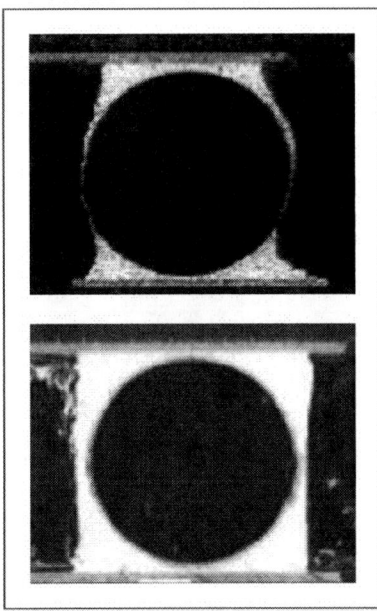

Fig. 13 microphotos of soler coated balls after reflow

Table 2 shows the plastic core material properties

Table 2. Some properties of the polymer cores of the 'SOL' balls.	
Young's Modulus	4.9 GPa (-40 – 60°C)
	4.7 GPa (60 – 200°C)
CTE	40.2 ppm/°C (-40 – 60°C)
	46.2 ppm/°C (60 – 200°C)

Fig. 14 shows the comparison of the thermal cycling results [14]. It clearly indicates that the plastic core solder ball structures improve the fatigue life significantly. The Young's Modulus of plastic core is much smaller than that of metal, so it is superior in relaxing the thermal stress. The result of temperature cycle tests showed that it had excellent reliability and enabled much greater resistance to micro-cracks in the solder layer than that of conventional solder balls. In addition, it is superior in holding the gap flatly between upper and lower substrates, because the particle diameter of polymer core is very uniform.

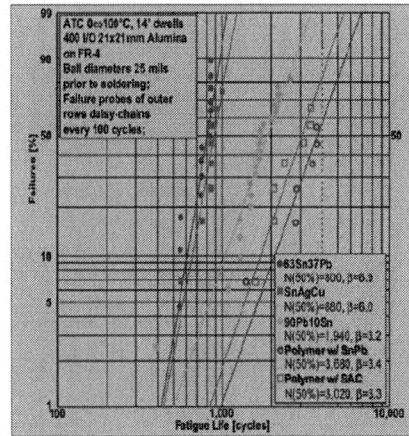

Fig. 14 Fatigue life comparison using plastic cored solder balls

Redistribution Applications in WLPs

Regardless of different classifications of the WLP technologies, redistribution becomes an essential part of each technology. Redistribution requires thin film polymers like secondary passivation and metallization to reroute the typical peripheral pad to an area array configuration. Redistribution technology not only realizes the rerouting, it also provides a mechanism for mechanical protection since such a layer acts as a stress buffer to release the stresses from solder joints. Similar ideas can be applied to the PCB side to enhance the WLP reliability.

References

1. Luu Nguyen, "Wafer-level chip scale packaging", ECTC Professional Development Course Note, 2008
2. Wafer-Level Packaging, www.siliconfareast.com/wl_package.htm
3. John J.H. Reche, Deok-Hoon Kim, "Wafer level packaging having bump-on-polymer structure", *Microelectronics Reliability*, 43 (2003) 879–894
4. Y.P. Wang, L.Y. Hung et al, "High drop performance interconnection: polymer cored solder ball", ECTC 2008
5. Deok-Hoon Kim, Peter Elenius, Michael Johnson, Scott Barrett, "Solder joint reliability of a polymer reinforced wafer level package", ECTC 2002
6. Toshimi Kawahara, "SuperCSPs", IEEE TRANSACTIONS ON ADVANCED PACKAGING, VOL. 23, NO. 2, MAY 2000
7. Georg Meyer-Berg, "Wafer-level ball grid array packaging technology trends", EuroSimE 2008
8. Beth Keser, Craig Amrine, Trung Duong, Scott Hayes, George Leal, William Lytle, Member, Doug Mitchell, and Robert Wenzel, "Advanced packaging: the redistributed chip package", IEEE TRANSACTIONS ON ADVANCED PACKAGING, VOL. 31, NO. 1, FEBRUARY 2008
9. Elenius P, Barrett S, Kim D-H., "Wafer level packaging", APACK 2001, An International Conference on Advances in Packaging, Singapore, 5–7 December 2001.

10. Elenius P. The Ultra CSP wafer level package. In: 4th Pan Pacific Microelectronics Symposium, February 1999.

11. Yang H, Elenius P, Barrett S, Schneider C, Leal J, Moraca, R, et al. Reliability characterization in ultra CSP package development. In: IEEE 50th ECTC, Las Vegas, Nevada, 21–24 May 2000.

12. Reche JJH. High density multichip interconnect and packaging technology. In: IEEE IEMT Workshop on Multichip Interconnection, Orlando, FL, 10–13 October 1988.

13. Kim D-H, Elenius P, Johnson M, Barrett S. Solder joint reliability of a polymer reinforced wafer level package, Microelectron Reliab 2002;42:1837;

14. W. Eaglemier, "Achieving solder joint reliability in a lead-free world",

15. Okinaga, N.; Kuroda, H.; Nagai, Y., "Excellent reliability of solder ball made of a compliant plasticcore", Electronic Components and Technology Conference, 2001. Proceedings., 2001 Page(s):1345 - 1349

Mixed Mode Interface Characterization Considering Thermal Residual Stress

A. Xiao[1], G. Schlottig[12], H. Pape[2], B. Wunderle[3], K M. B. Jansen[1], L. J. Ernst[1]

[1]Delft University of Technology, Mekelweg 2, 2628 CD Delft, the Netherlands
[2]Infineon Technologies AG, 81726 Munich, Germany
[3]Fraunhofer IZM, 13355 Berlin, Germany
Email: A.Xiao@tudelft.nl, Phone: +31 (0)152783726

Abstract

Interfacial delamination has become one of the key reliability issues in the microelectronic industry and therefore is getting more and more attention. The analysis of delamination of a laminate structure with a crack along the interface is central to the characterization of interfacial toughness. Due to the mismatch in mechanical properties of the materials adjacent to the interface and also possible asymmetry of loading and geometry, usually the crack propagates under mixed mode conditions. The present study deals with delamination toughness measurements of an epoxy molding compound - copper lead frame interface as directly obtained from a real production process. As a consequence the specimen dimensions are relatively small and therefore a dedicated small-size test set-up was designed and fabricated. The test setup allows transferring two separated loadings (mode I and mode II) on a single specimen. The setup is flexible and adjustable for measuring specimens with various dimensions. For measurements under various temperatures and moisture conditions, a special climate chamber is designed. The "current crack length" is required for the interpretation of measurement results through FEM-fracture mechanics simulations. Therefore, during testing the "current crack length" is captured using a CCD camera and a micro deformation analysis system (MicroDac). The critical fracture properties are obtained by interpreting the experimental results through dedicated finite element modeling.

1 Introduction

Thin layers of dissimilar materials are used in most microelectronic assemblies in order to achieve specialized functional requirements. Generally, the interface between two different materials is a weak link due to the mismatch in thermo-mechanical properties, such as Young's modules, coefficients of thermal expansion, hygro-swelling, and vapor pressure induced expansion [1]. Furthermore, the residual stresses from the production processes and the changing thermal and moisture conditions often are acting as crack driving factors for interface delamination. Failure of interfaces induces decreased reliability and performance of microelectronic packages. Therefore, the qualified driving mechanisms of delamination related problems are desirable.

Up till now, much research has been contributed to interface characterization for various material interfaces as used in microelectronic devices. However, in the previous researches, a few common problems have not been taken into consideration. Firstly specially fabricated test specimens were generally used for establishing the critical fracture values and therefore usually those specimens do not have the same surface conditions and process history in comparison with real product interfaces. As a consequence there always remained an uncertainty on the applicability of critical fracture properties thus obtained. Secondly, generally, in delamination experiments an "infinitely sharp" initial crack should be present, that due to loading can propagate along the interface. Only then the crack tip singularity can be adequately approximated through FEM-fracture mechanics simulations. The simulation results are then used for the interpretation of the measurement results. In many previous work, researchers tried to initiate an "infinitely sharp" initial crack through adding a small weak layer on the interface (e.g. gold or Teflon), and subsequently applying force or displacement loading. It appears that with this procedure an "infinitely sharp" initial crack is generally not obtained. Testing the sample without a real "infinitely sharp" initial crack then will result into a different type of singularity at the onset of delamination. In those cases FEM-fracture mechanics simulation results can not reliably be used for interpretation of the measurement results. Another deficiency often is the fact that the pre-stress situation due to fabrication of the sample is not properly (or not at all) taken into account. The crack-tip singularities appear to be strongly influenced by the pre-stress due to molding history, cooling down from the molding temperature and further treatments. Therefore it is essential that these effects are properly taken into account.

In the present study, the test specimens are created and separated from real production line packages, such that they really are made with the identical fabrication processes and materials. A new method to initiate an "infinitely sharp" initial crack will be discussed. Since test specimens are obtained from the real production line, the old method, where a small weak layer on the interface is added for crack initiation is not even possible. The newly developed method can well be used for initiation of an "infinitely sharp" initial crack in test specimens obtained from real production line packages.

2 Interface Fracture Evaluation

It is generally agreed that an interfacial crack will propagate when the load at the crack tip exceeds critical strength values. The success of predicting delamination in IC packaging strongly depends on accurate characterization of the critical interface adhesion strength. The crack propagation under pure mode I (opening mode) and pure mode II (sliding mode) have been extensively studied in many literatures. However, more and more attention must be paid to mixed mode loading because it is the most realistic situation. Usually, for composite structures, crack propagates under mixed mode I and mode II combined conditions [2, 3].

978-1-4244-2739-0/08/$25.00 ©2008 IEEE

The contribution of mode III (tearing mode) is often neglected because it is rather small in comparison with mode I and II.

In order to research possible interface delamination generally a fracture mechanics approach is applied. Here, we only recapitulate some basic theory of linear fracture mechanics for interfaces. Some more details are described in [4-6].

Linear elastic fracture mechanics is a theory that describes if and how a crack will grow under given loading conditions when assuming an initial crack with given size and location. It assumes the existence of some detectable cracks and predicts the crack propagation during processing and operational cycles. It applies when the nonlinear deformation of the material is confined to a small region near the crack tip compared to the size of the crack. To predict interface delamination, fracture quantities are needed for comparison to the critical data such as fracture toughness. In general, stress intensity factors (SIF) and energy release rate are used to define the loading state at the crack tip.

A criterion for crack growth can be obtained by regarding the energy balance of the material (1), where U represents the energy per unit of time and volume.

$$U_e = U_i + U_a + U_d + U_k \qquad (1)$$

U_e is the total external mechanical energy that is supplied to the material, U_i is the elastic energy that is stored in the material, U_a is the energy dissipated by crack growth, U_d is the energy dissipation caused by other mechanism, and U_k is the change in kinetic energy. It is assumed that U_d is zero, implying that the crack growth is the only cause of energy dissipation. U_k is zero means that crack growth is slow enough for changing in kinetic energy is negligible. The remaining energy balance is know as the Griffith's energy balance (2), which regards energy per unit of newly created fracture surface, or when the material width is taken to be constant, per unit of crack length a:

$$\frac{dU_e}{da} - \frac{dU_i}{da} = \frac{dU_a}{da} \qquad (2)$$

Dividing the left hand of equation by the material thickness B, it gives the energy release rate (3).

$$G = \frac{1}{B}\left(\frac{dU_e}{da} - \frac{dU_i}{da}\right) \qquad (3)$$

The energy release rate G is known as Griffith's energy balance, which regards energy released per unit of newly created fracture surface when the crack grows a unit of length. The criterion from Griffith states that crack growth occurs when the energy release rate exceeds a critical value $G > G_c$. The critical energy release rate appears to depend on temperature, moisture and mode mixity so that the criterion for fracture is:

$$G(T, C, \psi) > G_c(T, C, \psi) \qquad (4)$$

The mode mixity ψ for a homogeneous material is usually defined as the ratio between mode I to mode II loading and is described by the loading stress state at the crack tip (5).

$$\psi = arctg \frac{K_{II}}{K_I} \qquad (5)$$

Here, K_I and K_{II} represent intensities of mode I and mode II stress states for a crack in a homogeneous material. K_I characterizes the tendency of remote loads to open the crack, while K_{II} characterizes the shear loading.

For an interface crack, due to the elastic mismatch between two materials, the mode mixity can not be simply described by the equation 5. The opening and shearing stresses at the interface ahead of the crack tip, with a distance of r can be calculated from (6).

$$(\sigma_{22} + i\sigma_{12}) = \frac{K}{\sqrt{2\pi r}} r^{i\varepsilon} \qquad (6)$$

Where σ_{12} represents shear stress and σ_{22} represents normal stress. ε is the oscillatory index which is a function of the Young's moduli and the Poisson's ratios. K is the complex stress intensity factor. It is described by:

$$K = K_I + iK_{II} \qquad (7)$$

The mode mixity for an interface crack is described by the relative amount of crack tip stress field shearing to opening and can be expressed as:

$$\psi = \tan^{-1}\left(\frac{\sigma_{12}}{\sigma_{22}}\right)_{r=\hat{L}} \qquad (8)$$

According to the basic solution, stress components along the interface are oscillatory [2, 4] and thus can not well be obtained by numerical solutions. Therefore, often an alternative mode mixity definition is used, where the mode mixity is defined by interface stresses (normal and shear) at a chosen length $\hat{L}$ ahead of the crack tip:

$$\psi = \tan^{-1}\left(\frac{\text{Im}(K\hat{L}^{i\varepsilon})}{\text{Re}(K\hat{L}^{i\varepsilon})}\right) \qquad (9)$$

Here the choice of $\hat{L}$ is somewhat arbitrary, but restricted by the dimensions of test samples and the applications within microelectronics.

Alternatively, Andrew Tay [8] proposed a crack surface displacement extrapolation method (CSDEM) to calculate the mode mixity (10).

$$\psi = \lim_{r \to 0}\left[\tan^{-1}\left(\frac{\delta u_x + 2\varepsilon\delta u_y}{\delta u_y - 2\varepsilon\delta u_x}\right) - \varepsilon\ln\left(\frac{r}{2a}\right)\right] \qquad (10)$$

Where δu_x and δu_y are the components of the crack surface opening displacement, parallel and perpendicular to the crack tip, respectively (see figure 1). ε is the oscillatory index, r is the distance from the crack tip. 2a is the arbitrary length and here it is equal to the total crack length.

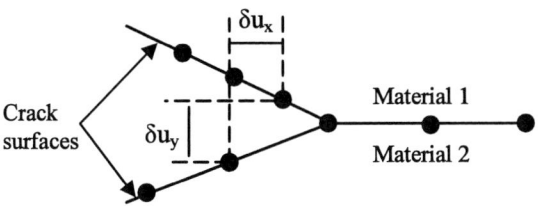

Figure1. Schematic of Crack Tip Opening Displacements

3 Sample Preparation and Fabrication

The present study deals with delamination toughness measurements between epoxy molding compound and copper lead frame interfaces as directly obtained from real production. The sample dimensions are relatively small. Samples are 35mm long, 1.2mm wide and 1mm thick. They are sawn from the runner on the copper lead pad of a matrix lead frame as partly shown in figure 2. The reason that a small production sample is selected because a miniature bi-material sample can be won from real components reproducing the exact process conditions for material characterization. It is noticed that if test samples are created on a larger scale, they often can not be used for interface fracture testing due to the fact that various arbitrary cracks are often present beforehand due to high residual stresses.

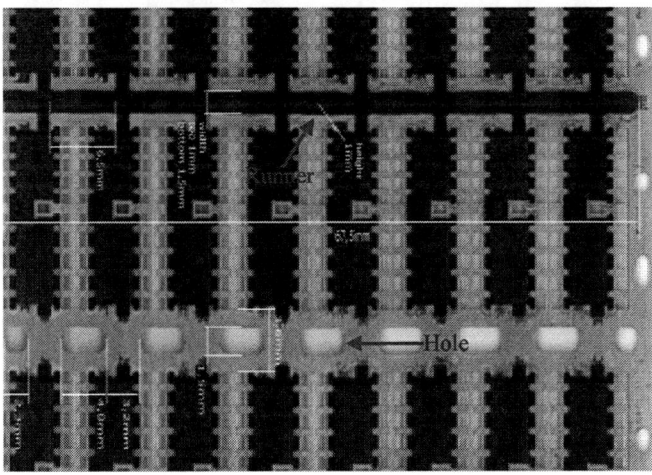

Figure2. Lead-frame Package Set

In the production process of lead frame based packages, there are always some molded runners remaining on matrix lead frames. Usually, these runners are treated as waste. The idea is to take use of these wasted runners or parts of them as interface test samples, because they are available anyway and have the same materials, surfaces and processing as the real product. Small design modifications like closing all or some of the holes shown in figure 2 to get sufficient sample length might be needed and were done.

The lead frame was sawn along both sides of the runners to get bi-material strips for interface testing. Of course this is a critical process which may randomly damage samples of weak interfaces.

Once a good sample is obtained, the next step is to consider a good method for creating a sharp initial crack. The conventional methods for creating an initial crack are razor blade peeling, fatigue loading, adding a weak layer or a high CTE material on the interface. However, a sharp crack is not always obtained using these methods. In this study, a new combined method is developed not only for attaining a sharp crack but also for achieving a more reliable shape for the later loading mode (discussed in section 4).

Based on the literature studies [10-11] and our previous research, the toughness of the copper-epoxy molding compound interface is usually weaker than the toughness of pure epoxy, the method to create a sharp initial crack for such samples is defined by the following processes (as indicated in figure 3). Firstly, a few millimeters of molding compound is polished away, a small copper surface exposes. Secondly, the sample is compressed along its thickness direction. Thirdly, a small hole is drilled on the exposed copper surface for later loading purpose. Finally, the exposed copper is bent downwards while the sample is compressed. The bending angle and bending length are adjustable and reproducible.

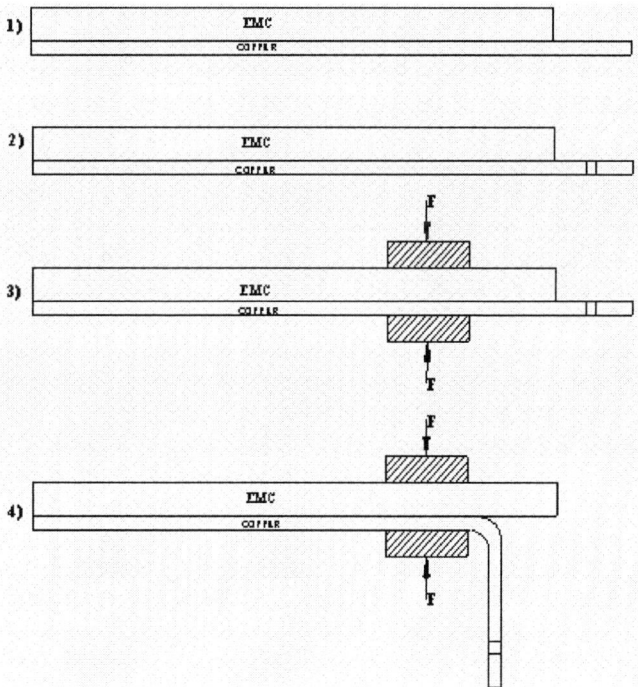

Figure3. Processes of Creating a Sharp Initial Crack

4 Design of the Test Setup

A small-scale experimental setup is designed and fabricated especially for mixed mode bending testing. It allows transferring two separated loads on a single specimen. A schematic drawing is shown in figure 4.

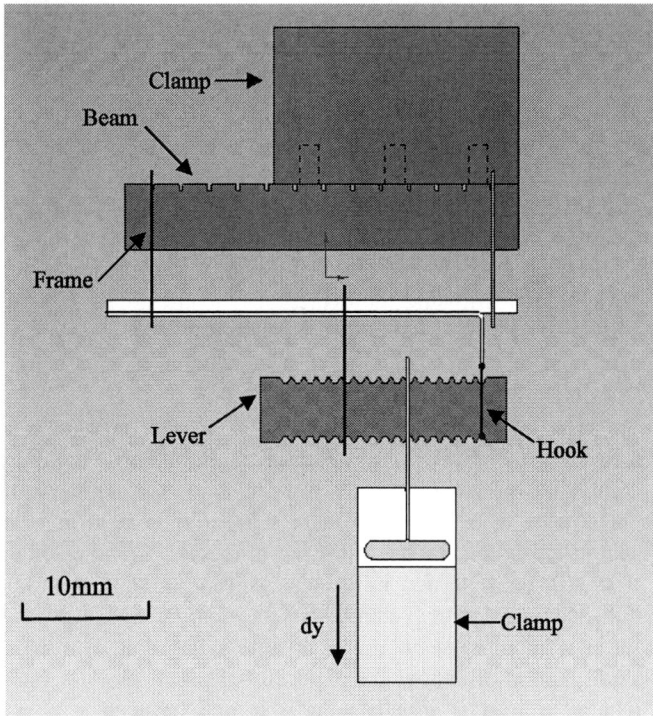

Figure4. Schematic of the Experimental Setup

The setup consists of a loading beam, a lever, four frames, one hook and two clamps. The sample is supported by the two upper frames and the lever is used to transfer the load. The upper frames are first snapped on grooves in the beam. The sample is then inserted in the upper frames, while the mid frame is on the middle of the sample. The hook is crossed through the hole in the copper. Than the hook and the mid frame are subsequently connected with the lever, on which the lower frame is already positioned. The lower frame is finally connected to the lower clamp. By changing the loading position of the lower frame, various (loading) mode mixities can be controlled. The mode II three point bending (TPB) test occurs when not using the lever and directly connecting the middle of the sample with the lower loading frame. The Mode I double cantilever bending (DCB) test occurs when removing the lever, and connecting the hook with the loading frame. The grooves in the beam and lever are used to provide the test abilities for different sample length and also to prevent frame sliding along the horizontal direction during the experiment. When inserting a sample in the setup, it seems that the sample is loaded immediately due to the gravity of the lever. However, the weight of this lever is very small. It is not expected that this mass will propagate the initial crack of the sample. The small load of this lever can simply be taken into account when interpreting the results by FEA analysis (see later).

The advantages of the setup are: 1) It is a rather simple and flexible experimental tool. 2) The loading position and the position of the frames acting on the sample are very well adjusted. The distance between two grooves is accurately defined. For finite element analysis (FEA), it is very important that the modeling situation is exactly the same as in reality. 3) No additional material is used in this setup. Some researches used glues to attach samples on the frames. In this way, it is noted that the non elastic behavior of glues can influence the fracture test results. So it is necessary to include appropriate constitutive behavior of these glues in the FEA. Also these glues often do not survive at high temperatures.

5 Design of the Temperature and Moisture Chamber

A special chamber is designed for testing samples at different temperature and moisture conditions (see also figure 5). The chamber consists of three Teflon parts and one aluminum part. Part number one is used for moisture inlet and part number two is used for moisture outlet. Part number 3 is an isolation layer. It is used for protecting heat loss during temperature testing. Four glass windows are mounted on this layer. The front window is used to observe crack growth inside the chamber and the remaining three windows are used for access with external light resources. These four windows are parallel to the windows bonded on part four. Part number four is a conduction layer. It is used for transferring heat into the chamber during temperature test. Several heating elements and thermal couples are mounted in this part.

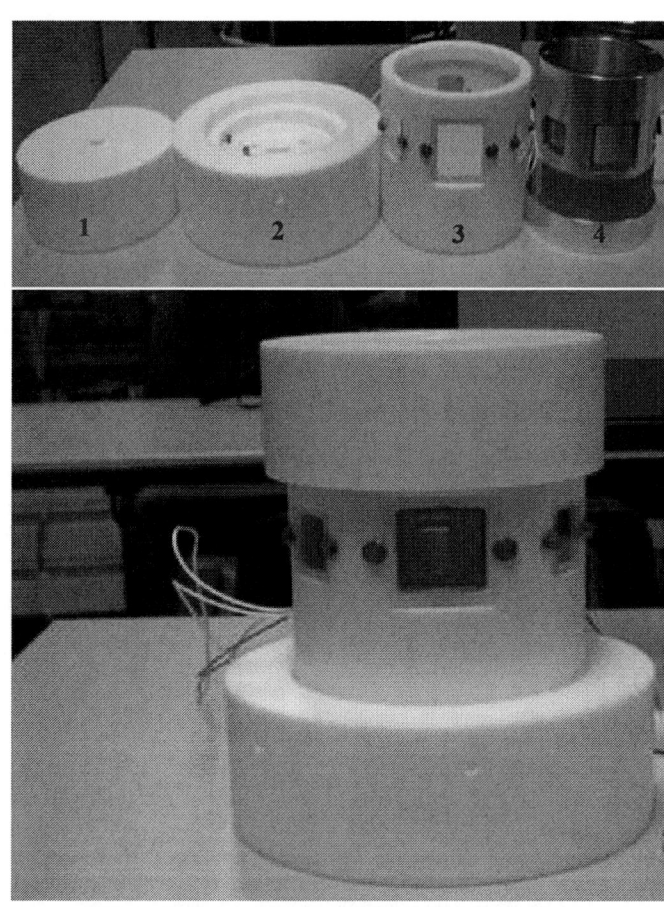

Figure5. Temperature and Moisture Chamber

6 Experiment Procedure and Result

The test specimen is first inserted in the load transfer setup. Then, the setup is clamped in a dynamic mechanical analyzer (DMA) that is used in deformation controlled mode (figure 6). The setup is protected by the chamber and a high resolution CCD camera is placed in front of the chamber (figure 7). The camera is mounted on a XYZ stage.

Figure6. Setup Clamped on Tensile Tester

Figure7. Chamber and Camera

So far several room temperature experiments were done in order to verify the test method. Figures 8 and 9 show the force-displacement results from DCB tests and MMB tests. From the test results, initially, the force -displacement curve represents the opening of the pre-crack. When the pre-crack starts to propagate, the force decreases. It is found that the crack growth initially is not stable.

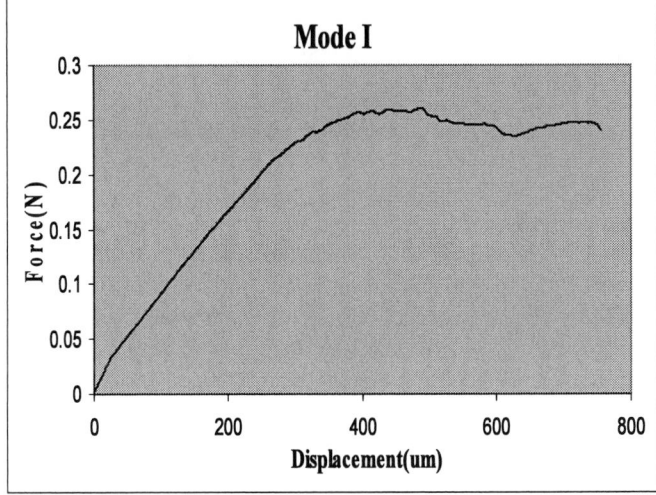

Figure8. DCB Test Result

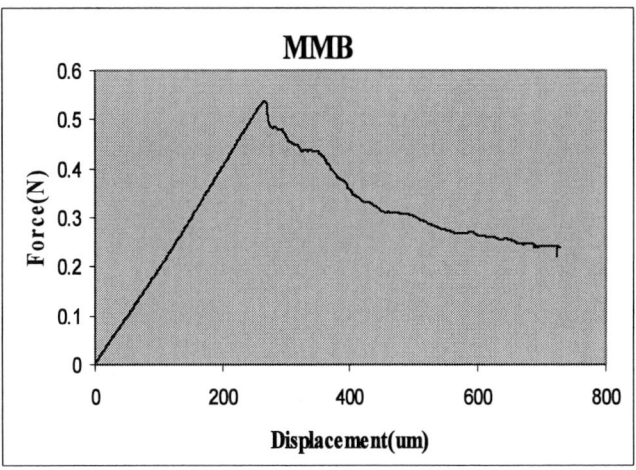

Figure9. MMB Test Result

During the experiment, the CCD camera traces images at fixed intervals (depending on the loading rate). These images are used to analyze the crack tip positions. To capture the "current crack length", a micro deformation analysis system (MicroDac) is used (for a detailed description about the MicroDac system see references 12-13). For image analysis, it appears to be necessary to create a "micro pattern" of sufficient resolution on the sample. In this study a micro powder is applied to the sample sides. In order to get images with sufficient contrast, laser lightning is used to illuminate the sample around the crack tip. A typical image is shown in figure 10.

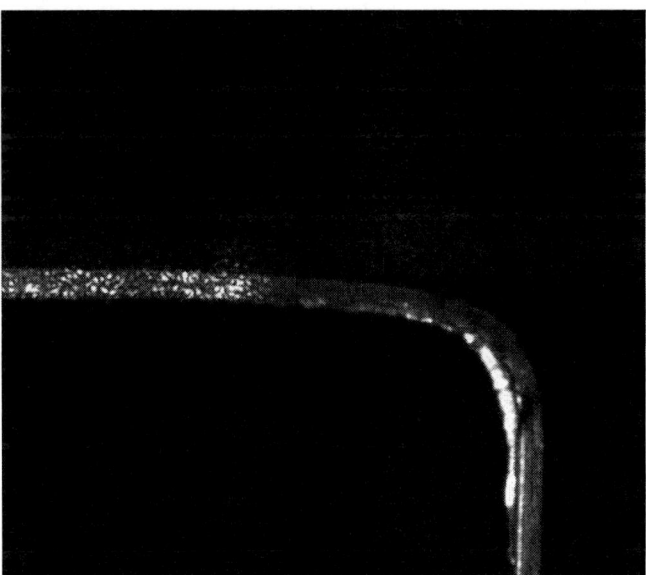

Figure10. Image with Laser Lighting

7 Finite Element Analysis

The (critical) energy release rate and mode mixity is established for various crack lengths through FEM simulations. These FEM simulations are performed on FEM models where the crack length corresponds to experimentally established crack tip positions.

For those crack tip positions the (critical) prescribed displacement is taken from the experimental data as well (such as from figures. 8 and 9).

A 3D FEM model is commonly used to simulate real packages where both lateral dimensions are large compared to the thickness. However, as for the present sample the width is relatively small compared to the height of the sample a 2D model is applied, also because of the limited simulation time.

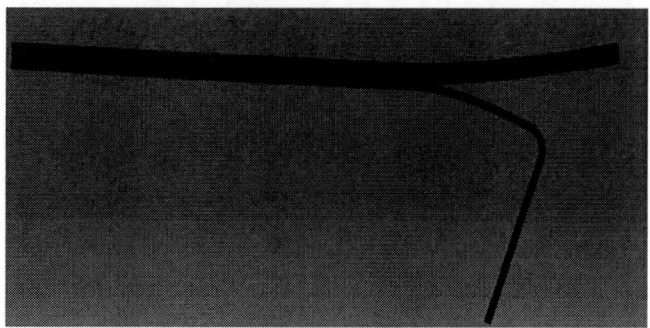

Figure12. Result of Linear model

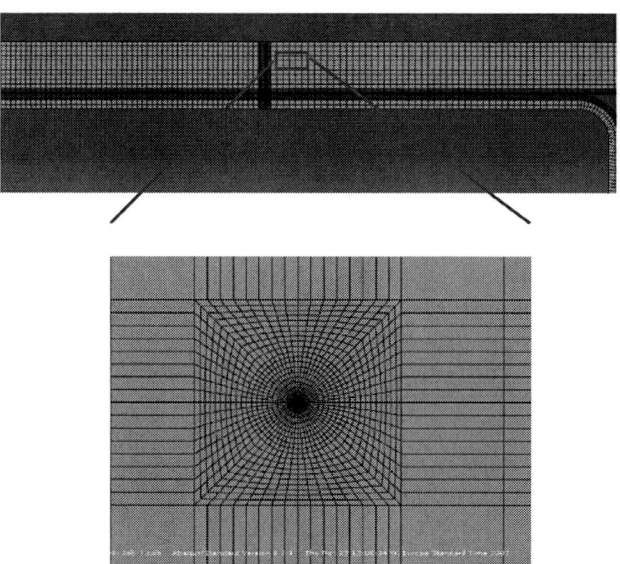

Figure11. Geometry of 2D FEM Model and Crack Tip Mesh

For all FEM simulations an identical crack tip mesh (with different crack tip positions) is used. Quarter-point elements were used around the crack tip to well capture the stress singularity (figure 11). All simulations were performed with the commercial FEM-code ABAQUS. A modified J integral concept [14, 15] is used to establish the (critical) energy release rate.

Since the width of the sample is quite small it is expected that the plane stress assumption is adequate for the 2D-model. A plane strain assumption or a generalized plain strain assumption is inadequate, since these assumptions conflict with the stress-strain distributions in the pre-stress state.

From the measurements it was observed that various parts undergo un-negligible rotations during loading. Therefore, in the FEM simulations all parts (specimen, lever, hook and frames) were included while geometrically non-linear analyses are executed (the model is shown in figure14). Simulation results were also compared to just "linear simulation" results. Figure 12 and 13 show the results from the linear and the nonlinear model, respectively. It is observed that the linear model does not match with the real situation. The nonlinear model is more adequate.

Figure13. Corresponding Result of Nonlinear Model

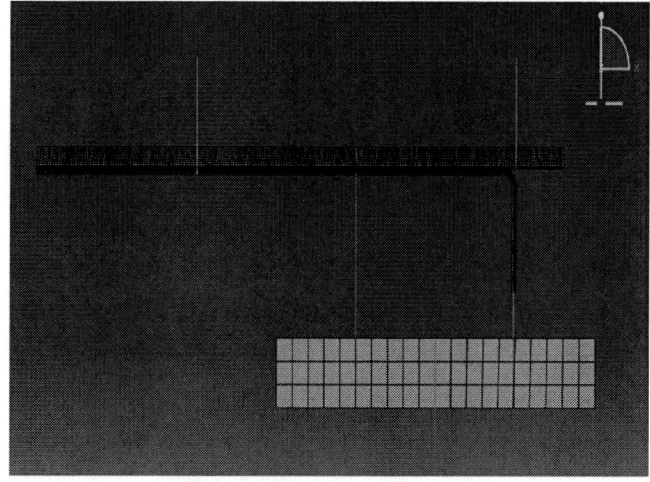

Figure14 Model with the Setup

Based on the experimental results, a few values of the energy release rate of copper-epoxy molding compound during separation was calculated using 2D non linear simulation and mode mixity was calculated using crack surface displacement extrapolation method. As input parameters, the accurate material properties are both experimentally and numerically characterized as functions of temperature of moisture by a parallel project. The simulation result of the cracked sample is shown in figure 15. It is found that when shifting the loading frame close to mode II (to the left corresponding to figure 14), the energy release rate and mode mixity indeed increased.

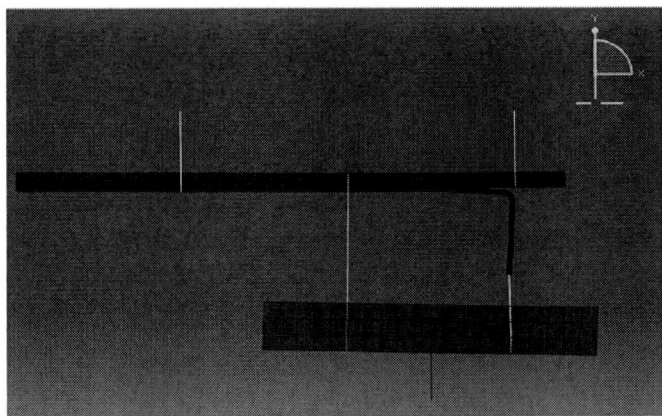

Figure15. Result of the Cracked Sample

To evaluate the influence of initial stress, an industrial cooling down process was described and added in the fracture model. It was found that at certain loading location, with cooling down process, the critical energy release rate increased by 10% and mode mixity increased by 50%. These results prove that the initial stress can not simply be neglected.

In order to check the accuracy of initial stress calculated from the 2D model, both 3D simulation and experiment measurements were applied. It was found that the stress levels from 2D and 3D simulations were comparable. Also the deformation of the sample along the lateral direction from 2D modeling matched quite well with the experiment result.

Conclusions

The paper is focused on delamination of EMC/Copper interface. A small scale loading setup is carefully designed. A special chamber is designed for testing samples at different temperature and moisture conditions. The force-displacement curve is measured using a DMA test facility as tensile tester. The crack length is obtained using a CCD camera and image correlation system. The critical fracture properties are obtained by interpreting the experimental results through dedicated finite element modeling. The finite element model is built to simulate the samples subjected to different loading conditions. The proposed criterion can be easily applied to predict delamination in package design using the mixed mode bending test and finite element method.

At this moment, only a few results were calculated. In the near future, a full range of humidity and temperature dependent interface fracture data, for various EMC-copper interfaces will be established.

References

1. W. D. van Driel et al, *"Driving mechanisms of delamination related reliability problems in exposed pad packages"* 6th. Int. Conf on Thermal, Mechanical and Multiphysics Simulation and Experiments in Micro-Electronics and Micro-Systems, EuroSimE 2005, PP 183-189.
2. L. J. Ernst et al, *"Fracture and Delamination in Microelectronic Devices"* Proceeding Asian Pacific Conference for Fracture and Strength 2006.
3. G. Q. Zhang, W, D. Van Driel, and X. J. Fan *"Mechanics of Microelectronic"* solid mechanics and its applications, volume 141 ISBN-10 1-4020-4934-X (HB).
4. J. W. Hutchinson and Z. Suo, *"Mixed mode cracking in layered materials"* Advances in Applied mechanics, Vol.29, Academic, New York, 1991.
5. A. A. O. Tay, *"Modeling of Interfacial Delamination in Plastic IC Packages Under Hygrothermal Loading"* ASME Journal of Electronic Packaging, vol. 127 (3): 268-275 Sept 2005
6. A. A. O. Tay, Y Ma, T Nakamura, S H Ong, *"A Numerical and Experimental Study of Delamination of Polymer-Metal Interfaces in Plastic Packages at Solder Reflow Temperatures"* In Proceedings of The 9th Intersociety Conference on Thermal and Thermomechanical Phenomena in Electronic Systems, 2004. ITHERM 2004., 1-4 June 2004, pp 245-252
7. H. F. Nied, *"Mechanics of interface fracture with applications in electronic packaging"* IEEE Transactions on Device and Materials Reliability, Vol 3, No.4, Decemeber 2003.
8. Andrew Tay, *"Application of fracture mechanics in microelectronics"* Lecture note of EuroSime 2006 short course.
9. Auersperg. J, B. Michel, *"Towards a Robust Design of Electronics Assemblies under Fracture, Delamination and Fatigue Aspects"* 2007 9th Electronics Packaging Technology Conference, pp. 476-481.
10. H. B. FAN et al, *"A New Method to Predict Delamination in Electronic Packages"* 2005 Electronic Components and Technology Conference, pp. 145-150.
11. Weidong Xie, Suresh K. Sitaraman, *"Investigation of Interfacial Delamination of a Copper-Epoxy Interface Under Monotonic and Cyclic Loading: Experimental Characterization"* IEEE Transactions on Advanced Packaging, Vol. 26, NO. 4, Nov 2003.
12. D. Vogel, R. Dudek, J. Keller, B. Michel, *"Combining DIC Techniques and Finite Element Analysis for Reliability Assessment on Micro and Nano Scale"* 5th Electronics Packaging Technology Conference (EPTC), 2003, Proc. pp.450-455.
13. D. Vogel et al, *"Evaluating microdefect structures by AFM based deformation measurement"* NDE for Health Monitoring and Diagnostics 2003, Proc, of SPIE, Vol.5045, pp. 1-12.
14. Y. T. He, G. Q. Zhang, W. D. van Driel, *"Cracking Prediction of IC's Passivation Layer Using J-Integral"* Proceeding Electronic Components and Technology conference, IEEE 2003.
15. C. C. Lee, C. C. Chiu, K. N. Chiang *"Stability of J Integral Calculation in the Crack Growth of Copper/Low-k Stacked Structures"* IEEE Electronic Components and Technology Conf., pp. 885-891, 2006.
16. TE Tay *"Characterization and analysis of delamination fracture in composites: An overview of developments from 1990 to 2001"* Appl Mech Rev vol 56, no 1, January 2003

The Effect of Strain Rate and Strain Range on Bending Fatigue Test

Minyi Lou, Long Wen, Zhengrong Chen, Jianwei Zhou, Qian Wang, Jaisung Lee
Samsung Semiconductor (China) R&D Co., Ltd
Science Plaza 7F, International Science Park, Suzhou Industrial Park, Suzhou, 215021,China
E-mail: minyi.lou@samsung.com Tel: 86-512-62888288-8813

Abstract

There're many kinds of bending test conditions according to different control modes: displacement control, force control and strain control. In our paper, strain-controlled bending test was chosen, since the lifetime and failure modes were more sensitive relating to different loading strain range and strain rate. ANSYS was used to simulate the bending test to predict board level solder joint reliability. The relationship between loading strain range and lifetime for 60BOC was measured in a similar type to Coffin-Manson equation. 5 failure modes were investigated, component side bulk crack, component side IMC interface crack, board side bulk crack, board side IMC interface crack and pad lift. The simulation results indicated that with the increase of strain range, the fatigue performance of device decreased.

Keywords: Strain-controlled bending fatigue test, strain rate, strain range, lifetime, failure mode

1 Introduction

PWB assemblies experience various mechanical loading conditions during assembly and use. The board cyclic bending during assembly and test operations will cause electrical failures due to circuit board and trace cracks, solder interconnects cracks, and the component cracks. And the actual use conditions such as repeated key-presses in mobile phone also result in a large number of repeated bending cycles, albeit at low magnitude[1].

Since component manufacturers and suppliers cannot evaluate their package performance on actual final products, a board level test method is needed. Failure modes varying with solder composition and pad finish condition have already studied by many researchers, but there's no specific research focusing on the relationship between failure modes and loading strain conditions so far.

In this paper, strain-controlled bending fatigue test was setup. The lifetime and failure modes depending on strain rate and strain range were studied.

2 Experiment

2.1 Sample Preparation

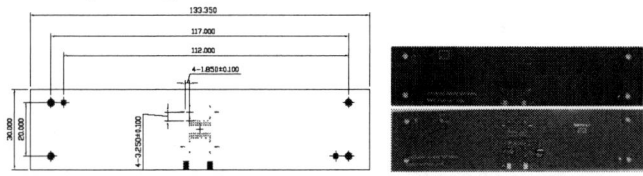

Fig 1. 60BOC Board Level Sample

2.2 Bending Fatigue Test

The bending fatigue system is depicted in Fig 2. MTS 858 system with specific 4-point bending fatigue test fixture was used in this test. This fixture consists of the moving inner roller and the fixed outer roller, which supports the test board and prevents its movement in the test procedure. The test board is placed between these rollers.

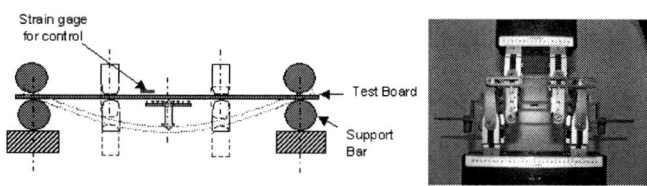

Fig 2. Bending Fatigue Test System

2.3 Test Condition

The test condition was list in table below.

Table 1. Test Condition

Load Span	Support Span	S/B Condition	Failure Criterion	Sampling Rate
55mm	105mm	$\phi 0.45$ (Sn2.5Ag0.5Cu)	1000Ω	1024Hz

1. 8 specimens per loading strain condition were prepared and uniaxial strain gage was attached to every specimen.
2. The failure was all regarded to happen as the monitored resistance of daisy-chain circuit beyond 1000Ω.

2.4 Failure Analysis

2.4.1 Dye and Pry

Failure specimens are immerged into fine particle red ink in a vacuum chamber for 1 hour. Baked at 100℃ for 30 minutes. Then the packages are sheared from the test board by DAGE 5000 and reveal the failure position marked by red particle absorbed.

2.4.2 OM & SEM

Cross-section OM and SEM photos are used to observe the failure modes of the failure specimens.

3 Results Analysis and Discussion

3.1 Lifetime Analysis

Weibull distribution probability plot is drawn as Fig 3. The shape parameter is similar when strain range is±1000με (2000με), ±1250με (2500με) and ±1500με (3000με), which indicates the failure mode is similar during these loading strain range. Then the failure mode changes when strain range increases to 4000με. This is also proved by the later dye-pry analysis and OM/SEM analysis.

978-1-4244-2739-0/08/$25.00 ©2008 IEEE

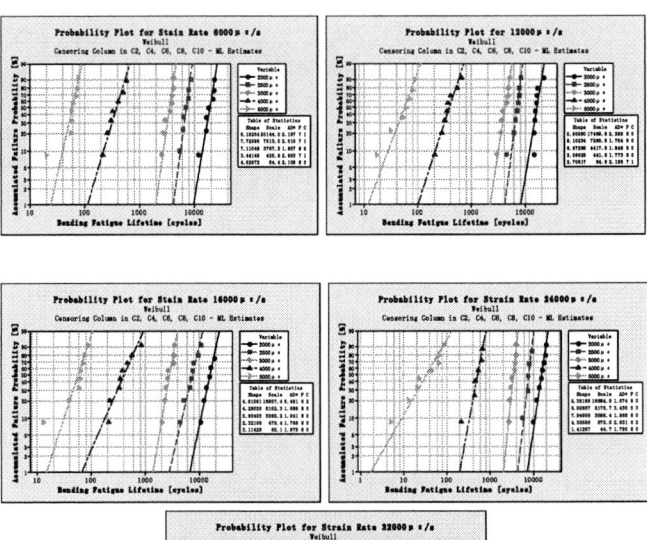

Fig 3. Probability Plot Depending on Strain Rate

3.2 Lifetime Prediction according to Coffin-Manson Equation

Darveaux and Syed[2] have found that for mechanical fatigue, the total strain energy density is a good indicator of solder joint damage and relates well in Coffin-Manson relationship with failure.

$$N_f(63.2\%\,\mathrm{Pr.}) = C(\Delta\varepsilon)^{-\gamma} \qquad (1)$$

Where: N_f is characteristic life of specimen (scale parameter); C and γ are constants, characteristic of the material, and $\Delta\varepsilon$ is the strain range of the bending test.

Using strain-controlled bending test can ensure the constant strain range during the test procedure, so the acquired data can put in the Coffin-Manson equation and further to predict the lifetime.

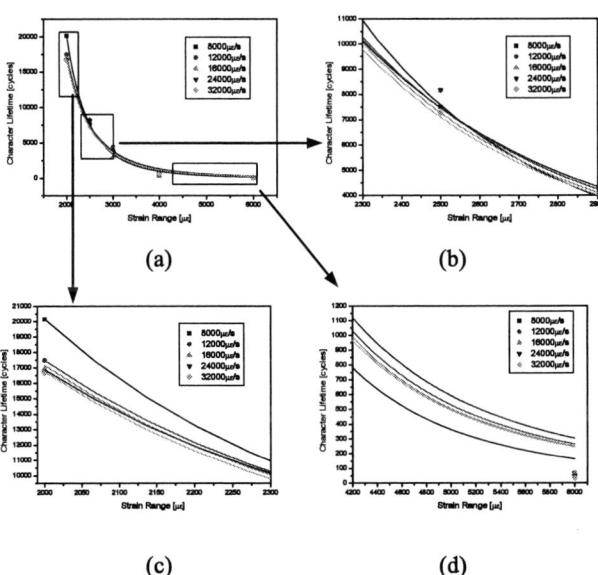

Fig 4. Coffin-Manson Fit under Different Strain Rate

The following phenomena are depicted in Fig 4. The lifetime N_f under small strain range (<2500με) is: $N_f(8000\mu\varepsilon/s) > N_f(12000\mu\varepsilon/s) > N_f(16000\mu\varepsilon/s) > N_f(24000\mu\varepsilon/s) > N_f(32000\ \mu\varepsilon/s)$. Shi et al[3] found that Young's modulus increased quite linear with strain rate (up to 0.3), which caused higher stress at the solder joint and led to solder joint crack easier.

Under middle strain range (2500με ~ 4000με), the lifetime N_f is complex. And under large strain range (>4000με) the lifetime N_f is: $N_f(24000\mu\varepsilon/s) > N_f(12000\mu\varepsilon/s) > N_f(16000\mu\varepsilon/s) \approx N_f(32000\mu\varepsilon/s) > N_f(8000\mu\varepsilon/s)$. Under the middle and large strain range, rise in strain rate increase the modulus of board, which causes higher stress at the solder joint and, at the same time, also increases yield stress of solder joint. It is likely that the two aspects cancel each other to make the lifetime independent of strain rate.

3.3 Failure Mode Analysis

3.3.1 Dye-Pry Analysis

Dye-Pry method was used to observe the failure position. There're 3 main failure positions, component side, board side and pad lift. The typical failure photos are shown in Fig 5.

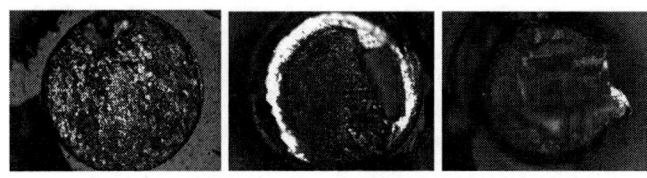

(a) Component Side Crack (b) Board Side Crack (c) Pad Lift

Fig 5. Dye-Pry OM Photo

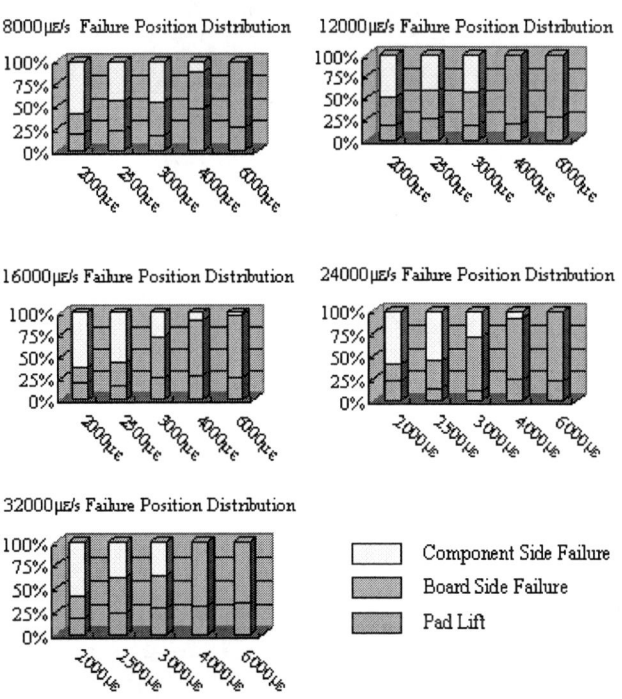

Fig 6. Failure Position Distribution Depend on Strain Rate

The failure position distribution (Fig 6) proves that from small to high strain range, the failure position changes from component side to board side gradually. It's correct under every strain rate. But there's no obvious law depending on different strain rate.

3.3.2 OM & SEM Analysis

SEM photos for the typical failure modes are depicted in Fig 7. 5 failure modes are investigated, component side bulk crack, component side IMC interface crack, board side bulk crack, board side IMC interface crack and pad lift. All the cracks initiate from the corner of the solder ball, but for the crack in solder bulk, it propagates along the copper pad direct in the solder inner. For IMC interface crack, it propagates along the interface between IMC layer and solder bulk. And for the pad lift, it propagates along the board PCB.

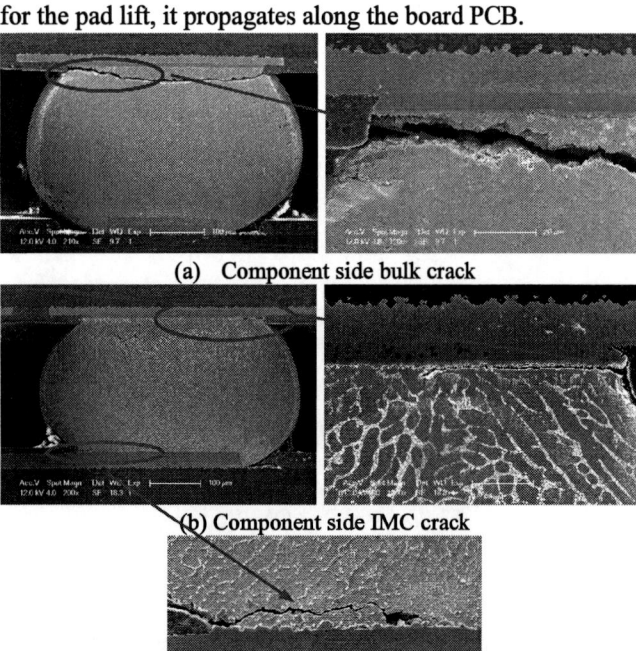

(a) Component side bulk crack

(b) Component side IMC crack

(c) Board side bulk crack

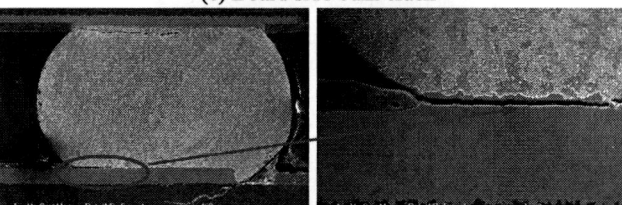

(d) Board side IMC crack

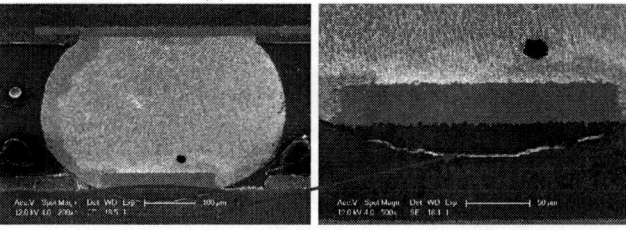

(e) Pad lift
Fig 7. Typical Failure Mode in Bending Fatigue Test

According to the OM photos, it seems that the crack often occurs at component side due to the maximum plastic strain when the solder ball experiences low strain range loading. However, with the increase of strain range, the maximum localized plastic strain occurs at board side because the NSMD pad structure results in strain concentration. This phenomenon is also proved by simulation results.

Compared with strain rate, the large strain rate induces more IMC interface crack rather than bulk crack. The strength of solder joint is rate dependent. The solder bulk alloy increases in strength with the increase of strain rate since it's operating at a high homologous temperature ($T_{test} / T_m > 0.4$). On the other hand, the IMC layer is generally brittle material with higher melting points, so it tends to have lower strength as strain rate increases. These trends are depicted schematically in Fig 8(a). At some strain rate, there's a transition in failure mode from solder bulk failure to IMC interface failure. This occurs when the solder bulk strength has increased beyond the IMC interface strength. The corresponding ductility of the solder joint decreases dramatically when this transition occurs[4]. This is shown schematically in Fig 8(b). With the increase of strain rate, the failure mode changes from solder bulk crack to IMC interface crack.

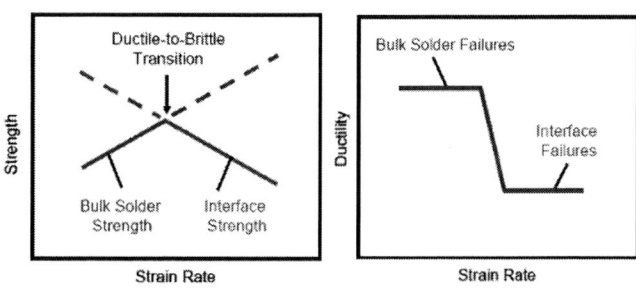

(a) Tensile Strength   (b) Ductility
Fig 8. Solder Joint Ductile-to-brittle Transition with Strain Rate

4 Simulation

4.1 Structural Response Analysis

ANSYS was used to simulate the structural response and internal stress/strain history of package during cyclic loading. To save running time, some simplifications were made and only a quarter model was built. The finite element model is shown in Fig 9.

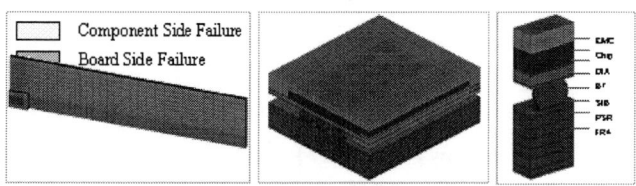

Fig 9. Finite Element Model (1/4)

Especially for the critical part, solder ball, the bilinear plastic material model was used to reflect its mechanical behavior truly, shown in Fig 10. The boundary conditions are shown in Fig 11. The nodes on test board at supporting anvils position were constrained in X&Z direction, while the nodes at loading anvils position were applied a cyclic displacement load which was the displacement history of loading anvils recorded by MTS 858 controller. One loading cycle was simulated using transient analysis.

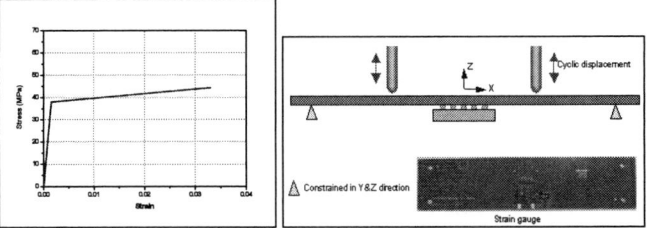

Fig 10. Bilinear Plasticity for S/B   Fig 11. Boundary Conditions

The deflections of test board center as response of input displacement are shown in Fig 12. And the simulation results of board strain in length direction are compared with experimental results. It's observed that the simulation results are coincidental with experimental very well.

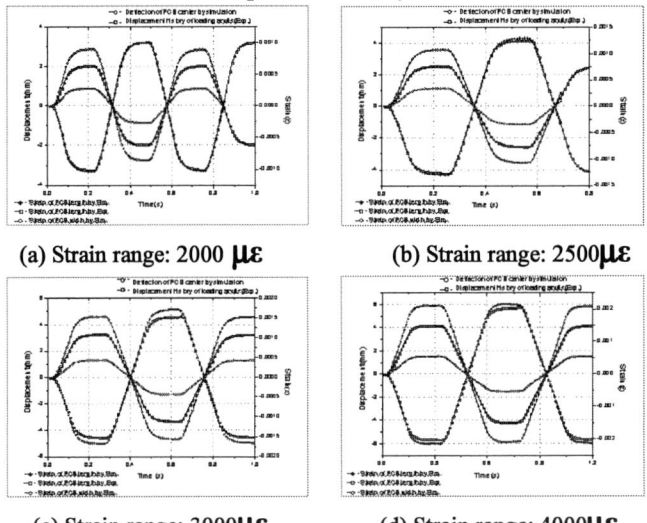

(a) Strain range: 2000 $\mu\varepsilon$ (b) Strain range: 2500$\mu\varepsilon$

(c) Strain range: 3000$\mu\varepsilon$ (d) Strain range: 4000$\mu\varepsilon$

Fig 12. Structural Response under Different Strain Range

During cyclic loading, localized plastic deformation may occur at highest stress site. This plastic deformation induced permanent damage to the component and crack initiated. The stress/strain state of solder ball during cyclic loading was what we concerned mostly. As the solder ball experienced an increasing number of loading cycles, the length of the crack increased. After a certain number of cycles, the crack would lead the solder ball to fail, thus device failed finally. Also, another purpose of investigating the stress/strain state of solder ball was to find out the dominant parameter for fatigue failure. It was the basis of fatigue life prediction methodology selection.

The stress/strain states in solder ball during cyclic bending for different loading conditions, as well as the deflection contour of whole structure, are shown in Table 2. To verify the simulation results, they're also compared with the experimental results.

Table 2. Stress/strain State for Solder Ball (Bending Down)

Strain range	2000με	2500με	3000με	4000με
Deflection	Max. 2.92mm	Max. 3.85mm	Max. 4.67mm	Max. 6.99mm
Stress contour	Max. 43.7MPa	Max. 46.8MPa	Max. 44.1MPa	Max. 47.4MPa
Total strain contour	Max. 0.0235ε	Max. 0.0363ε	Max. 0.0554ε	Max. 0.0816ε
Plastic strain on critical S/B				
Experimental results (failure mode)				

It is investigated that the VonMises stress on solder ball has no significant difference between these four loading conditions. However, the total strain on solder ball increases remarkably with the increase of strain range (Fig 13). Among the total strain, the plastic strain is the primary component, account for more than 90%. As for the elastic strain, which represents the elastic behavior of solder ball under cyclic loading, doesn't change greatly along with strain range due to the deformations of solder ball for all four conditions greatly exceed the range of elasticity. So it seems that the plastic deformation of solder ball is the governing parameter for solder fatigue failure.

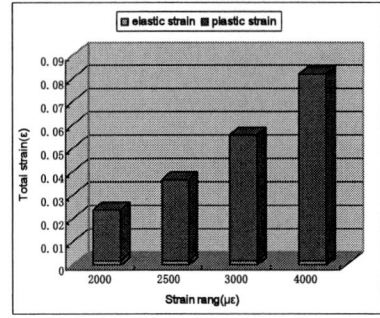

Fig 13. Maximum Strain on S/B depend on Strain Range

It's also observed that the solder balls locating at outmost columns have much more possibility to fail due to occurrence of higher stress and plastic deformation than others.

4.2 Fatigue Life Simulation

Traditional approaches for predicting solder joint fatigue use a fatigue equation with a physics-based damage parameter such as inelastic strain or total plastic work. Coffin-Manson type equation is the most popular one in literature for predicting solder joint fatigue[5]:

$$N_f = C_1(\psi)^{C_2} \qquad (2)$$

Where: N_f is the fatigue life in cycles; C_1 and C_2 are constants found through least squares regression, and ψ is a damage parameter.

In this paper, the ΔW_{avg} for the worst-case solder joint was used to evaluate the fatigue life. A volume-weighted average was used to resolve the inherent problem of singularity in FEM when choosing which nodes/elements to take the damage parameter from. There're two 1mil thick layers of elements at solder joint interface as shown in Fig 14 in which the damage parameter per cycle ψ is volume-averaged over. One layer is on the board side, and another layer is at the substrate side.

The time history plastic work under different strain range during first two bending cycles is shown in Fig 15.

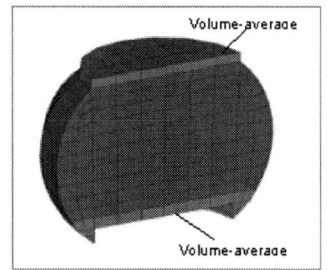

Fig 14. Volume-average
Layers in S/B

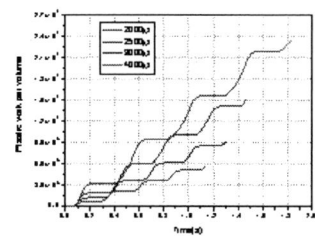

Fig 15. Plastic Work Density
for First Two Cycles

ΔW_{avg} is the element volumetric average of the stabilized change in plastic work between the first cycle and second cycle. The calculated ΔW_{avg} under different strain range is shown in Table 3 the power law regression is conducted to get the Coffin-Manson fatigue model, as shown in Fig 16. It's found that for in our test conditions, the fatigue life N_f was related to ΔW_{avg} in the form such as:

$$N_f = 7342.4(\Delta W_{avg})^{-1.8091}$$

Table 3. ΔW_{avg} for Different Strain Range

Strain Range	ΔW_{avg} (MPa)
2000με (±1000με)	0.6417
2500με (±1250με)	1.0623
3000με (±1500με)	1.6578
4000με (±2000με)	2.3719

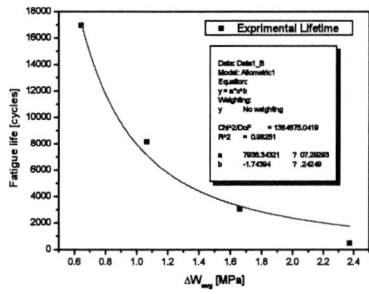

Fig 16. Calibration for Coffin-Manson Fatigue Model

This is a unified finite element modeling methodology for cyclic bending test, and the Coffin-Manson fatigue model can be used in future bending reliability evaluation once the accumulated plastic work density per cycle ΔW_{avg} is calculated using FEM.

Conclusions

1. The relationship between the loading strain range and the bending fatigue lifetime for 60BOC was measured in a similar type to Coffin-Manson equation. Otherwise, the lifetime had scarcely strain rate dependant.

2. From small to high strain range, the failure position changed from component side to board side gradually. At some strain rate, there was a transition in failure mode from solder bulk failure to IMC interface failure.

3. The simulation results indicated that with the increase of strain range, the fatigue performance of device decreased. It seemed that the fatigue crack had more possibility to initiate at component side when experienced small strain range loading. However, with the increase of strain range, the crack easily occurred at board side.

4. The plastic strain of solder ball during cyclic bending was the governing parameter for fatigue failure. The outmost solder ball had more possibility to fail firstly due to much higher plastic strain than others.

5. The Coffin-Manson fatigue model is a unified finite element modeling methodology for cyclic bending test, and can be used in future bending reliability evaluation once the accumulated plastic work density per cycle ΔW_{avg} is calculated using FEM.

References

1. Lei L. Mercado, Betty Phillips, Shubhada Sahasrabudhe, et al, "Use-Condition-Based Cyclic Bend Test Development for Handheld Components", Electronic Components and Technology Conference, (2004), pp. 1279-1287.

2. Darveaux, R., Syed, A., "Reliability of Area Array Solder Joints in Bending", SMTA International, (2000), pp. 313-324.

3. Shi X. Q., Zhou W., Pang H. L. J., et al, "Effect of Temperature and Strain Rate on Mechanical Properties of 63Sn/37Pb Solder Alloy", Journal of Electronic Packaging, Vol.121, (1995), pp. 179-185.

4. Robert Darveaux, Corey Reichman1, "Ductile-to-Brittle Transition Strain Rate", Electronics Packaging Technology Conference, (EPTC '06), pp. 283-289.

5. Yi-Shao Lai, Tong Hong Wang, "A Study of Cyclic Bending Reliability of Bare-die-type chip-scale Packages", 5th. Int. Conf. on Thermal and Mechanical Simulation and Experiments in Micro-electronics and Micro-Systems, (EuroSimE 2004), pp. 313-316.

Dynamic Properties Testing of Solders and Modeling of Electronic Packages Subjected to Drop Impact

Long Wen, Xingming Fu, Jianwei Zhou, Qian Wang, Jaisung Lee
Samsung Semiconductor (China) R&D Co. Ltd
Suzhou Industrial Park, Suzhou, China
E-mail: long.wen@samsung.com

Abstract

The solder joints reliability of electronic packages during board level drop impact is a great concern for semiconductor manufacturers. Many researchers have adopted numerical simulation to investigate the drop performance of electronic package because it is fast and cost-effective. However, the solder balls, which are recognized as vital parts for the integrity of solder joints and the overall function of package, were always assumed to be elastic or simple elastoplastic in previous studies, this is not sufficient to capture the true mechanical behavior of solder balls during drop impact.

To obtain more accurate results in numerical simulation, the strain rate dependent material properties of solders are further studied in this paper and measured using Split Hopkinson Pressure Bar (SHPB) technique. The measured stress-strain data at different strain rates are then incorporated in Johnson-Cook model which can adequately represent the rate-sensitive deformation response during high-rate loading. The accuracy of this material model is evaluated by performing FEM simulation of the SHPB test and comparing the finite element simulation with test results. The drop reliability of a fine pitch BGA package is simulated using validated input acceleration (Input-G) method with Johnson-Cook model for solder balls. It is found that the failure mode and critical solder ball location predicted by this modeling correlate well with test. Comparing with traditional elastic or elastoplastic model for solder balls, this material model has better correlation with experimental measurement of dynamic strain, PCB center deflection. Therefore, the stress/strain distribution in solder ball is more precise. At last, an impact life prediction model based on maximum peeling stress of critical solder joints is proposed for board level drop test to estimate the number of drops to failure. With this model, the drop performance of new packages can be quantified, and further enhanced through modeling.

1. Introduction

With the trend towards miniaturization and multi-functionality in portable electronics devices products like mobile phone, personal digital assistant (PDA) and laptop, miniature IC packages such as fine pitch Ball Grid Array (FBGA) package and Chip Scale Package (CSP) are increasingly being used. However, the inherent vulnerability of these miniature IC packages has brought along new reliability problems. Among them, the drop reliability is a great concern, because it is common for these devices to be subjected to accidental drop impacts resulting in failure of solder interconnection and hence malfunction of product.

Board level drop test is convenient to characterize the solder joints performance, because it is more controllable than product level drop test. A few JEDEC standards [1-3] were released to provide standardized methods to conduct board level drop test. The actual drop test, however, is very expensive and time-consuming, and requires much manpower in measurement and failure analysis. In addition, very limited experimental results can be collected, so it is far from enough to observe the full dynamic responses of whole package during drop test. Especially for solder joint, it is almost impossible to measure its stress or strain response which in turn affects the solder joint reliability.

Numerical modeling is proved a cheap, fast and efficient approach in IC packaging for the purpose of design analysis and optimization. A validated drop impact model can depict the dynamic responses from outside to inside of package and thus enable researchers to understand the physics-of-failure. Early drop test simulation work was reported by Wu [4], who used LS-DYNA to model drop test at both component and product levels. Wong et al [5-6] investigated the mechanics and physics of board level drop test with the intention of providing the fundamental understanding. Submodeling [7] technique was applied to reduce computational time. But limited information about experimental results was provided. Tee et al [8] established a basic drop test model for TFBGA, which had good agreement with actual failure mode, but the model was only useful for qualitative analysis. The Input-G method was developed by Tee and his co-workers [9-10]. This method is more accurate and much faster, and bypasses many technical difficulties in conventional drop model such as contact surface definition. At the same time, Tee [11-12] proposed a life prediction model based on maximum peeling stress in the solders. However, the accuracy of this model is questionable because linearity was assumed for solder ball. The durability model will be revised using more sophisticated material model for solder balls in this paper. Tee et al also applied the similar methodology in the study of board level drop test and simulation of CSP and QFN packages [13-14].

In this paper, the strain rate dependent dynamic properties of solders are considered and measured using Split Hopkinson Pressure Bar technique. Then the drop reliability of a FBGA package is simulated using validated input acceleration (Input-G) method with rate dependent model for solder balls. At last, an impact life prediction model is formulated based on the maximum peeling stress of critical solder joint.

2. Drop Test Procedures and Experiment

JEDEC [1] recommended a standard test procedure and board design for board level drop test of components used in handheld electronic products. Typical setup of board level drop tester is shown in Fig.1. Up to 15 components of same type in a 3 row by 5 column format are mounted on test board which is connected to fixture and drop table with four screws. About 10mm standoff between test board and base plate is added to allow PCB bending. The drop table is dropped from a certain height along the two guiding rods, onto a rigid base covered with one layer of felt. The strike between drop table and rigid base will create an impact pulse which is half-sine acceleration curve with certain G-level and pulse duration. One accelerometer is mounted on top of base plate to measure the impact pulse.

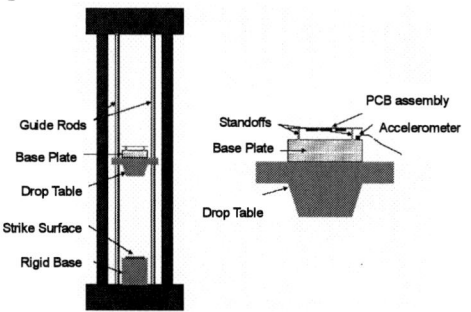

Fig.1 Schematic of a typical drop tester

In this paper, only one package is mounted on the center of test board. A Fine-pitch BGA (FBGA) package with 144 solder balls is selected as a case study. The package size is 14mm×14mm with chip size of 4.78mm×4.78mm×0.2mm and adhesive thickness of 50μm. The schematic vertical profile of package is shown in Fig.2.

Fig.2 Package vertical profile (unit: mm)

The test condition with 1500G peak acceleration and 0.5ms time duration is applied. Good repeatability of both peak acceleration and pulse duration should be ensured during test. The resistance readings of daisy-chain test sample before and after each drop are measured. The threshold resistance of event detector is set to 1000Ω according JEDEC standard.

3. Material Characterization of Solders

In electronic components, solder joints, which act as mechanical support as well as electrical interconnection, therefore, are very important parts for the integrity of electronic devices. For board level drop test, it is normal to produce strain rate of the order of $100s^{-1}$ in solder balls. It has long been known that the strain rate at which a material is loaded greatly affects its mechanical properties. Therefore, to fully understand the processes that take place during drop test and obtain more accurate results in computational simulation, measurements of the strength of solders must be extended to high strain rates.

3.1 Split Hopkinson Pressure Bar

The Split Hopkinson Pressure Bar (SHPB) test is an experimental technique commonly employed in the study of the material properties at high strain rates. The classical SHPB consists of two long bars with high elastic limit, termed incident and transmitted bars respectively, sandwiching a small cylindrical specimen, and a striker bar with similar material as the incident and transmitted bars, shown in Fig.3. The specimen is deformed under the action of compressive stress wave induced by impacting the free end of incident bar with the striker bar launched by a gas gun.

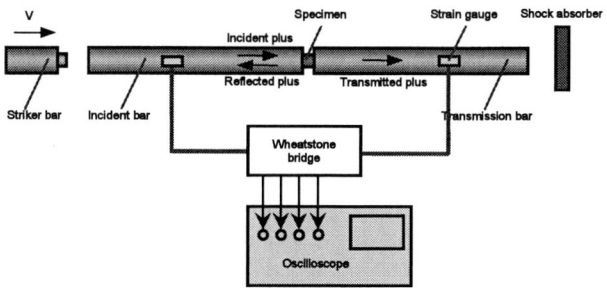

Fig.3 Schematic of SHPB system

When the striker bar moves with the initial velocity and impacts the incident bar, a one dimensional compressive stress wave called incident pulse ε_i, is generated and travels along the incident bar. When the incident pulse reaches the incident/specimen interface, the incident pulse is split into a reflected pulse ε_r which travels back along the incident bar and a transmitted pulse ε_t which propagates through the specimen into the transmitted bar. According to superposition of one-dimension wave and homogeneity of stress and strain in the specimen, the expressions for stress $S(t)$, strain $\varepsilon(t)$ and strain rate $\dot{\varepsilon}(t)$ in the specimen are given as:

$$S(t) = \frac{A}{2A_s} E\left[\varepsilon_i(t) + \varepsilon_r(t) + \varepsilon_t(t)\right] \quad (1)$$

$$\dot{\varepsilon}(t) = \frac{C_0}{L}\left[\varepsilon_i(t) - \varepsilon_r(t) - \varepsilon_t(t)\right] \quad (2)$$

$$\varepsilon(t) = \int_0^t \dot{\varepsilon}(t)\,dt \quad (3)$$

Where, E is the elastic modulus of pressure bar; C_0 is the longitudinal wave speed of the stress wave in pressure bar, A is the cross-section area of pressure bar; A_S is the original cross-section area of specimen; L is the original length of specimen.

During uniform deformation of specimen, the stress at the incident bar/specimen and specimen/transmission bar interface will be equal, thus

$$\varepsilon_i(t) + \varepsilon_r(t) = \varepsilon_t(t) \quad (4)$$

Therefore, the above strain rate and stress equation can be simplified as:

$$S(t) = \frac{A}{A_s} E\varepsilon_t(t) \quad (5)$$

$$\dot{\varepsilon}(t) = -\frac{2C_0}{L}\varepsilon_r(t) \quad (6)$$

A pair of semiconductor strain gauges is attached on each of the incident and transmitted bars at a certain distance from the specimen and is used in combination with a Wheatstone bridge circuit connected with a differential amplifier and a digital oscilloscope to monitor the strain during the test. The outputs from the strain gauges include the three strain waves of interest: the incident wave, reflected wave, and the transmitted wave. The incident wave is used to quantify the pulse duration, while the reflected wave is used to calculate the strain rate which is then integrated to find the strain history of the specimen, and the transmitted wave is used to calculate the stress history in the specimen.

3.2 Experimental Setup and Procedure

The SHPB facility used in this paper comprises of striker, incident and transmitted bars that are made from 16 mm diameter maraging steel having nominal yield strength of 3500 MPa. The striker bar is approximately 200mm long and the incident and transmitted bars are 1.0m long.

The eutectic Sn-2.5Ag-0.5Cu solder alloy in as-cast condition is used in the present study. These specimens are prepared by firstly melting the alloys in vacuum chamber at the temperature above 350℃ and pouring into silica glass tubes to cast into cylinder bar. Then they are naturally cooled in ambient condition. Then rough machining followed with finish machining is performed to prevent from the occurrence of crack on the specimen surface during test.

The cylindrical specimen is 6mm long and 12mm in diameter. All the specimens, before testing, were aged at room temperature for more than two weeks in order to stabilize the microstructure. The amplitude and duration of the loading pulse can be varied in SHPB test by changing the velocity and length of the striker bar. In this study, we change the impact speed of strike bar to obtain the desired strain rate. The firing pressure of gas gun is varied so that the striker bar could impact the incident bar at speed of 7.5, 12.6, 17.5m/s and obtain desired strain rate of 600, 1200, 1800s^{-1} respectively. Over 10 experiments are conducted for each striker impact velocity. In addition, the quasi-static compression tests for solders are also performed on MTS material experiment machine for the purpose of comparison with the results of dynamic tests.

3.3 Experimental Results

The strain-stress curves under quasi-static and dynamic compression mode at different strain rate are shown in Fig.4.

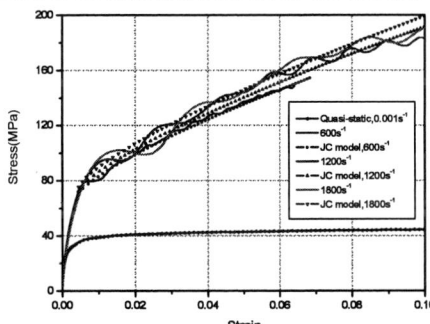

Fig.4 The experimental and simulated stress-strain curves

It is observed that the strain rate has great effect on the flow stress and yield stress. The yield strength at high strain rate (about 70MPa) is almost two times of that in quasi-static (about 38MPa) condition, and the flow stress at high strain rate (120~200MPa) is much higher than that of quasi-static condition (43MPa).

3.4 Constitutive Model

Constitutive laws are very important in the numerical simulation of mechanical behavior of material. The Johnson-Cook material model is used to estimate deformation response and strain rate sensitivity of solders at high strain rate conditions. The formulation for the Johnson-Cook model is empirically based, and represents the flow stress with an equation of the form:

$$\sigma = [A + B(\varepsilon_e^p)^n][1 + C \ln \dot{\varepsilon}^*][1 - T^{*^m}] \qquad (7)$$

Where σ is the effective flow stress; ε_e^p is the effective plastic strain; $\dot{\varepsilon}^*$ is the normalized effective plastic strain, $\dot{\varepsilon}^* = \dot{\varepsilon}_e^p / \dot{\varepsilon}_0$, typically normalized to a strain rate of $\dot{\varepsilon}_0 = 1s^{-1}$; A is the work hardening exponent; C is the strain rate sensitive exponent; B is constant; m is temperature soften exponent, which is not considered in the study, $m=0$.

A regression analysis based on the above experimental results is conducted to obtain the coefficients A, B, C, n which are used as input parameters for finite element code. The curve fitting result is shown in Fig.4.

3.5 Simulation of SHPB Test

To verify the curve fitting results and prove that the Johnson-Cook model can commendably describe the mechanical behavior of solders at high strain rate, the SHPB test was simulated here. 2D axisymmetric model is built. Initial velocity of 12.6m/s for strain rate of 1200s^{-1} is applied to striker bar. Auto surface to surface contact is defined between each interface. Fig.5 shows the Von Mises stress contour on specimen when the stress wave propagates across the specimen.

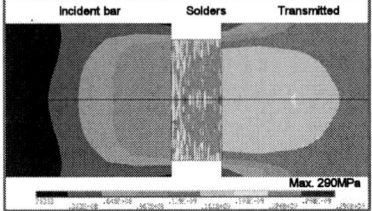

Fig.5 Stress distribution during impact@1200s^{-1}

Fig.6 represents the behavior of the strain wave propagating in the incident bar and transmitted bar. When the striker bar impacts the incident bar, a compressive strain wave which is in a trapezoidal shape is generated in the incident bar. Immediately following impact, this pulse travels along the bar towards the incident-specimen interface at which the pulse is partially reflected into the incident bar and partially transmitted through the specimen and into the transmitted bar.

951

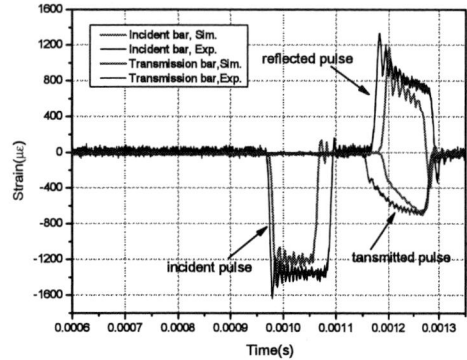

Fig.6 Strain gauge signals from SHPB test

It is concluded from the experimental and simulation results that: 1) during high speed impact, the deformation response of material is not evenly distributed deformation passing through the material like static loading does, but propagates in the form of wave. 2) Johnson-Cook material model can well and truly capture the mechanical behavior of solders during high strain rate conditions.

4. Experimental Analysis of Solder Joint Reliability

(1) Dynamic Responses

The dynamic resistance of daisy-chained solder joints which reflects the solder joint reliability, as well as the dynamic strains on PCB is monitored. Fig. 7 shows the typical dynamic resistance and strains after a certain number of drops, where positive peak of dynamic strain in PCB length direction indicates PCB bending up and negative peak means PCB bending down. The strain in PCB width is offset a certain amount for the clarity of picture. The graph also indicates that the strain in PCB length direction is the dominant one. The strain in width direction is much smaller and has no regular cyclic change as that in length direction.

The dynamic resistance also changes cyclically after certain number of drops. It is obvious that the daisy chain circuit's repeated opening and closing during a drop cycle are corresponding to the vibration of PCB. When PCB bends down, the tensile stress in solder joint results in the crack open, while the compression stress in solder joints as PCB bends up makes the crack close up. the coincidence of cyclic changes of dynamic resistance of solder joints with the cyclic changes of dynamic strains indicates that the dynamic responses of PCB are closely related to solder joint reliability.

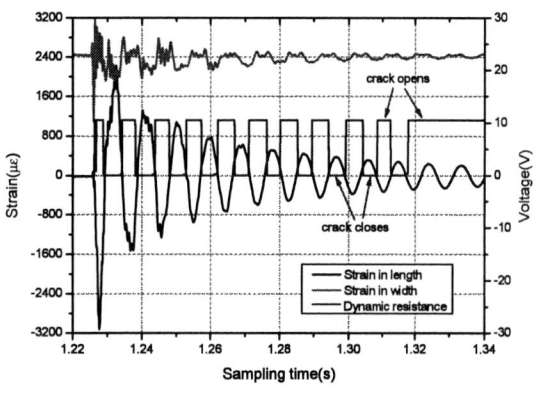

Fig.7 Dynamic responses during drop test

(2) Solder Joint Failure Analysis

Failure analysis has been performed on the samples after reliability test to check the failure location and failure mechanism so that the further design optimization can be conducted. A few techniques are employed to identify the failure modes of failed samples. This includes cross-section polishing and dye& pry test. Two typical failure modes have been observed: a) brittle intermetallic crack at solder/PCB pad interface, and b) PCB pad lift, shown in Fig.8.

It is observed from the dye&pry analysis that the critical solder joint is at the outermost corner, mainly along solder/PCB pad interface. No IMC crack is observed on component side. The possible reason is that the solder joints strength is higher at component side. As for the failure mode of pad life, it is probably because the strength of solder joint at PCB side is higher than that of PCB buildup layer.

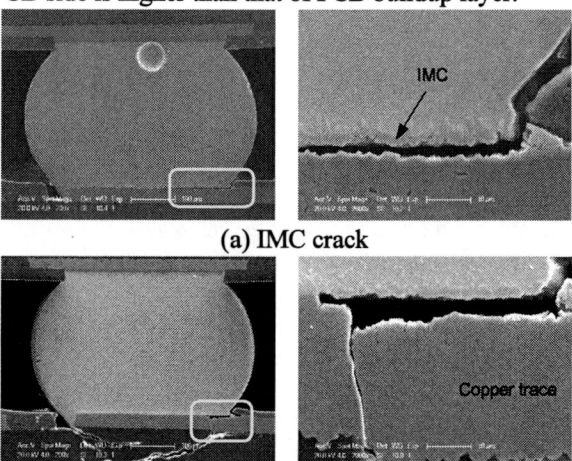

(a) IMC crack

(b) Pad lift (including trace break)

Fig.8 SEM images of typical failure modes

5. Drop Test Simulation

5.1 Finite Element Model

For simplification, one quarter 3D model with one package located at PCB center is used to simulate the board level drop test, see Fig.9. Detailed package geometry, solder balls, and pad design are included in the model. The pad design is SMD on component side and NSMD on PCB side. Fully mapped mesh is established using 154728 hexahedral elements which have 8 nodes for each element. Very fine mesh is defined for solder joints. The drop table, fixture and contact surface are not simulated, but their effects are considered indirectly by using input-G method. For simplification, elastic material model is assumed for all constitutive materials except for solder balls. Rayleigh damping parameters α=320 and β=3E-6 were applied to PCB.

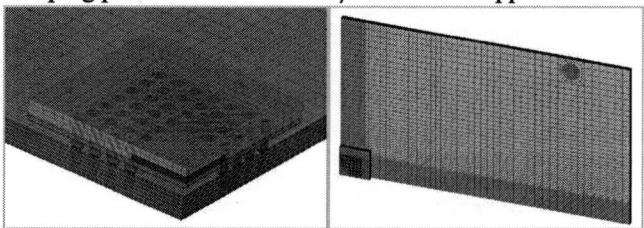

Fig.9 Finite element model

5.2 Input-G Method

In conventional free-fall drop simulation, the drop height or initial velocity before impact should be known. To achieve the required drop test condition of impact pulse, the drop height, strike surface and felt layer should be adjusted in testing. It is very difficult to perform the same adjustment in modeling. Because the velocity just before impact is difficult to calculate due to the friction of guiding rods, and the characteristics of contact felt are hard to describe and the conditions of strike surface may change due to felt worn-out.

Since the actual impact pulse is monitored by an accelerometer during testing (see Fig.10), the disadvantages can be eliminated if the impact pulse is considered as a PCB boundary condition in modeling, because the effects of velocity before impact and contact surface/material are already included in the impact pulse curve. Therefore, the complex variations in friction, contact condition, or other unknown tester parameters can be considered easily and accurately [9]. And the complicated nonlinear behavior in simulation, such as contact definition between drop table and strike surface, can be eliminated, thus reduce calculation time.

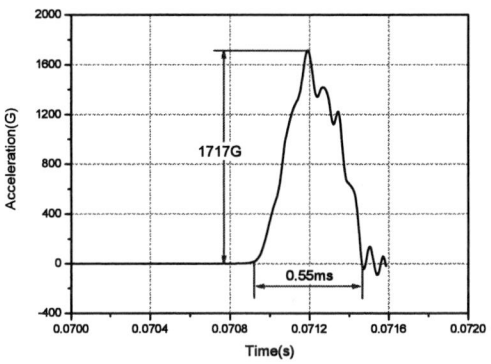

Fig.10 The measured acceleration curve

Fig.11 illustrates the loading and boundary conditions of Input-G method for board level drop test with 4-screw PCB configuration. For the two symmetrical planes, corresponding symmetrical constraint is applied. The measured acceleration curve is applied at the screw region which is constrained in in-plane directions.

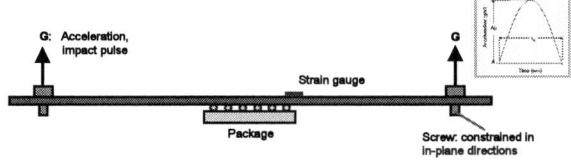

Fig.11 Schematic of Input-G method

5.3 Simulation Results and Discussion

(1) Dynamic Response of PCB

It is observed that the PCB cyclically vibrates, i.e. bends up and down during test due to impact inertial. Fig.12 shows the warpage distribution of PCB during the maximum bending downward. The maximum warpage in length direction is about 3.2mm while the maximum warpage in width direction

is about 0.5mm. It implies that the outmost row of solder balls in the PCB length direction warps more and has higher bending stress level.

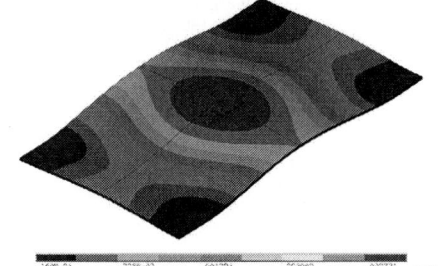

Fig.12 The maximum deflection of PCB

The relative displacement of any part of PCB against time to fixed screw can be obtained from modeling. Fig.13 shows the deflection of PCB center varying with time. The positive value means PCB bends up while the negative value denotes bending down of PCB

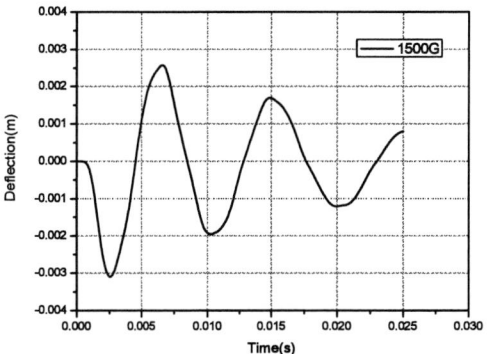

Fig.13 The deflection history of PCB center

(2) Dynamic Strain of PCB

The time-dependent dynamic strain on PCB is a significant index to evaluate the solder joint reliability in actual drop test. To investigate the effect of material model of solder ball, drop simulation using linear elastic and bilinear plastic model for solder ball are appended. The dynamic strain histories in PCB length direction resulted from modeling and test are compared in Fig.14.

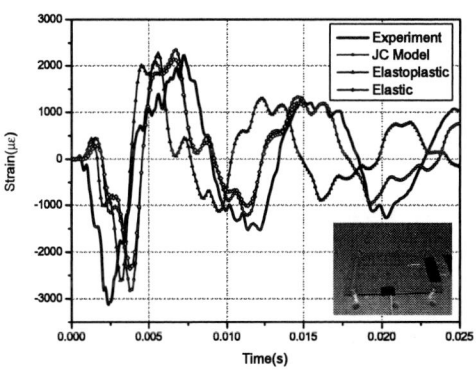

Fig.14 Strain in PCB length direction

953

It is observed from the picture that the simulated dynamic strain resulted from Johnson-Cook model has best correlation with experimental results in amplitude and frequency. The amplitude of PCB vibration decreases gradually with time due the damping of PCB. The period predicted using elastoplastic model is smaller than the experimental measurement.

(3) Stress and Strain Behavior of Critical Solder Joint

Quantitative stress analysis requires sophisticated material model, and thus it is a great challenge to obtain the true stress in solder balls. Fig.15 depicts the variation of stress in the critical solder joint with considering strain rate effect, including vertical normal stress or peeling stress (S_z), second principle stress (S_2), and shear stress (S_{XY}). The cyclic variation of these stresses denotes that the solder balls endure tensile and compressive stress alternately under the vibration of PCB. Among them, the peeling stress has close amplitudes in positive and negative directions and is much higher than other components. It indicates that the peeling stress is the dominant component. Therefore, the solder joint peeling stress is critical during drop impact and can be used as failure criteria for the purpose of drop life prediction.

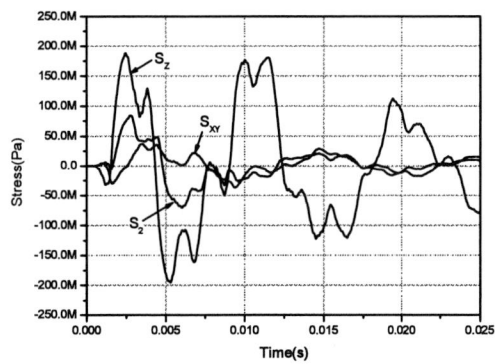

Fig.15 Dynamic stresses of critical solder ball

Fig.16 shows the stress/strain distribution in solder ball when adopting bilinear plastic, strain rate dependent plastic and linear elastic material model.

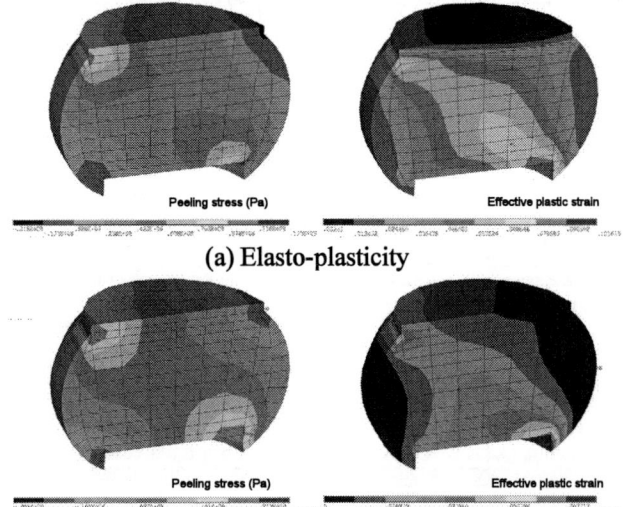

(a) Elasto-plasticity

(b) Strain rate dependent plasticity

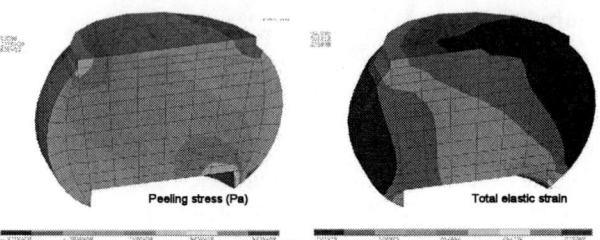

(c) Linear elasticity

Fig.16 The effect of material model on stress/strain

The maximum peeling stress is located at bottom interface for each model, and greatly changes from elastoplastic model to elastic model which are two extreme cases. Like many general metal material, the yield strength and flow stress of solder alloy increases with strain rate. So the strain rate effect becomes very important for high speed drop impact. It is obvious that the stress in solder ball increases from 130MPa to 250MPa when strain rate effect is considered. Therefore, the strain rate effect can not be neglected in numerical simulation, or else the solder alloy is too soft in elastoplastic model which underestimates the stress, or the linear elastic model makes the solders harder and overestimates the stress.

The maximum plastic or elastic strain also concentrated at the solder/PCB pad interface for all three material models. Comparing the strain distribution of the three material models, it implies that the work hardening character or harder solder alloy will induce smaller plastic strain. For elastoplastic and rate dependent material model, both solder ball neck and bottom interface have near same level of plastic strain. The plastic strain of rate dependent model is smaller than that of elastoplastic model due to the consideration of strain rate effect. For elastic model, the elastic strain is much lower.

5.4 Impact Life Prediction Model

When predicting solder joint failure, a metric of the damage occurring, called a damage parameter, must be chosen firstly. The damage parameter is a scalar quantity representing the damage that causes the failure. There are two commonly used damage parameters in failure criteria of drop test, i.e. stress and plastic strain.

Based on previous literatures [12, 14], the failure mode in present study is peel-dominant and the maximum peeling stress can be regarded as a failure criteria for the solder joints under drop impact. Since the maximum peeling stress is the root cause for the brittle crack of solder joint or pad lift. So an impact life prediction model can be concluded base on peeling stress from simulation results and the drop life from actual experiment.

A life prediction model is formulated using power law to relate the maximum peeling stress and mean impact life [12]:

$$N = C_1(\sigma_z)^{C_2} \qquad (8)$$

Where, N is the mean impact life; σ_z is the maximum peeling stress in the critical solder ball; C_1 and C_2 are correction constants.

To obtain the values of parameters in Eq. (8), another two drop conditions with peak acceleration of 2000G and 2900G are tested and simulated. The peeling stresses resulted from strain rate dependent material model are used to fit Eq. (8). Fig.17 shows the experimental impact life and fitting results.

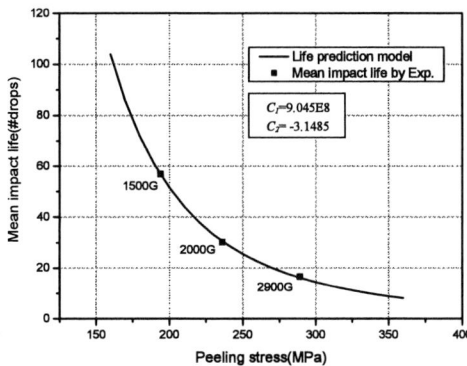

Fig.17 Experimental impact life vs. peeling stress

6. Conclusions

The strain rate dependent dynamic properties of solder alloy is further studied and measured by a Split Hopkinson Pressure Bar facility. It is found that the solder is rate sensitive. The yield strength and flow stress of solders increase with strain rate. Johnson-Cook model can adequately represent the rate-sensitive deformation response of solders during high-rate loading. The drop reliability of a fine pitch BGA package was simulated using validated input acceleration (Input-G) method. It is concluded that the material model of solder ball will affect simulation results. The elastoplastic model with solder material properties tested at quasi-static condition may not suitable for drop impact simulation because it does not include the effect of strain rate and will underestimate the stress of solder ball. The elastic model seems too hard and thus overestimates the stress. Rate dependent material model can result in reasonable results. Based on a strain rate dependent model, the maximum peeling stress on solder ball/ PCB pad interface of critical solder joint is selected as failure criteria to predict the impact life. With this life prediction model, the drop performance of new package can be quantified, and further enhanced through modeling.

References

1. JEDEC Standar JESD22-B111, Board Level Drop Test Method of Components for Handheld Electtronic Products, 2003

2. JEDEC Standard JESD22-B104-B, Mechanical Shock, 2001

3. JEDEC Standard JESD22-B110, Subassembly Mechanical Shock, 2001

4. Wu J, Song G, Yeh C, Drop/impact Simulation and Test Validation of Telecommunication Products, Prodeedings of the InterSociety Conference on Thermal Phenomena, 1998, pp.330-336

5. E.H. Wong, K.M. Lim et al, Drop Impact Test-Mechanics & Physics of Failure, Electronics Packaging Technology Conference, 2002, pp.327-333

6. Jason Wang, Wong E.H. et al, Modelling Solder Joint Reliability of BGA Packages Subject to Drop Impact Loading using Submodelling, ABAQUS Users' Conference, 2002

7. Pekka Marjamaki, Toni Mattila, Finite Element Analysis of Lead-Free Drop Test Boards, Electronic Components and Technology Conference, 2005, pp.462-466

8. Tee TY, Ng HS, Lim CT, Application of Drop Test Simulation in Electronic Packaging, Proceedings of the 4th ASEAN ANSYS Conference, Singapore, 2002

9. Tee TY, Luan JE, Pek E, Lim CT, Novel Numeical and Experimental Analysis of Dynamic Responses under Board Level Drop Test, Proceedings of the EuroSime Conference, Belgium, 2004

10. Tee TY, Luan JE, Pek E, Lim CT, Advanced Experimental and Simulation Techniques for Analysis of Dynamic Responses during Drop Impact, Proceedings of the 54th ECTC Conference, June, 2004

11. Jing-en Luan, TongYan Tee, Effect of Impact Pulse Parameters on Consistency of Board Level Drop Test and Dynamic Responses, Electronic Components and Technology Conference, 2005, pp.665-673

12. Jing-en Luan, TongYan Tee, Drop Impact Life Prediction Model for Lead-free BAG Package and Modulus, Proceedings of the EuroSime Conference, 2005

13. Tee TY, Ng HS, Luan JE, Integrated Modeling and Testing of Fine-pitch CSP under Board Level Drop Test, Bending Test, and Thermal Cycling Test, Proceedings of the ICEP Conference, Janpan, 2004

14. Tee TY, Ng HS, Lim CT, Drop Test and Impact Life Prediction Model for QFN Packages, J Surface Mount Technol 2003, 16(3), pp.31-39

Electromigration in Pb-free Solders

Minhua Lu, Da-Yuan Shih, Paul Lauro
IBM T. J. Watson Research Center
Yorktown Heights, NY 10598, USA
minhua@us.ibm.com

Abstract

Electromigration (EM)-induced damage in lead-free solders strongly depend on the Sn grain orientation in the Pb-free solder joint. Significant damage can develop at a very early stage when the c-axis of a Sn-grain is oriented close to the current direction. Rapid dissolution of both intermetallic compounds (IMC) and under-bump metallurgy (UBM) that led to significant cavitations at interface is caused by fast diffusion of Cu and Ni through the Sn crystal along the c-axis. On the other hand, when the c-axis of a Sn grain is not aligned with the current direction, cavitations at solder-IMC interface are formed mostly due to Sn-self diffusion which is correlated with failures at a much longer stress time. This is a direct proof of the highly anisotropic diffusion behavior of Cu and Ni in Sn, reported by Turnbull and Huntington many years ago. The stable Ag_3Sn network and cyclic twinning in SnAg solder contributed to the better EM performance of Sn1.8Ag compared to that of Sn0.7Cu solder. The ranking of the three surface finishes, from best to worst, is Ni(P)/Cu, Ni(P)/Au, and Cu, when electrons are entering from the tested surfaces. A Ni barrier layer is needed to retard the electromigration damage. However, the addition of Cu at an optimized level to the Sn-Ag solder drastically improved the electromigration performance.

Introduction

As flip chip packaging structures move from Pb-bearing solders to Pb-free solders, as influenced by Europe Union (EU) RoHS regulations, the reliability issues related to Pb-free solder have attracted a lot of attention in the microelectronics industry [1-2]. Electromigration (EM) in C4 interconnections has resurfaced because of the demand for finer pitch interconnections with higher current densities as well as unique Sn alloy properties. Different from the fcc structure of lead, tin (Sn) has a tetragonal crystal structure and tends to form large grains that exhibit highly anisotropic behaviors in mechanical, thermal, electrical, and diffusion properties[3-9]. With the lattice constant of a = b = 5.83Å much larger than that of the c-axis, c = 3.18 Å, the open structure along the c-axis facilitates faster interstitial diffusion of Ni and Cu than along the other two orthogonal directions [3, 4]. Cu diffusivity was measured at about 2×10^{-6} cm^2/sec at 25°C along the c-axis, which is about 500 times faster than that along the a- or b-axis, and is $\sim 10^{12}$ times the rate of Sn self diffusion[3]. For Ni, the anisotropy in diffusivity is even greater. Huntington et. al.[4] reported the diffusivity of Ni along the tetragonal (c-) axis is $\sim 7 \times 10^4$ times that at right angles (a- or b-axis) at 120°C. Therefore, one could anticipate that Sn grain orientation plays an important role in Sn-based Pb-free solder reliability issues such as electromigration.

Experimental methodology plays an important role in EM studies. Although high current density and high temperature stress condition can accelerate the test considerably, overly aggressive stress conditions often lead to failure modes that might not be extendable to field operation conditions. In addition, temperature and current density variations due to current crowding and local Joule heating often complicate and mislead the data interpretation. K. N. Tu et al [10] introduced a wire test structure, which consists of two Cu wires with polished ends joined together by solder. Although the diameter of the wires is much larger than the C4 solder ball dimensions, the straight Cu wire structure offers the advantages of uniform current density and minimal temperature gradients. As a result, pure electromigration effects can be studied with a carefully designed and controlled experiment.

In the study of EM performance of different surface finishes and Pb-free solders, the Cu wire test structure was adopted to eliminate current crowding and temperature non-uniformity caused by local Joule heating. Different from Tu's structure, metal films were deposited on the polished ends of Cu wires to simulate a UBM structure or different pad surface finishes. Since Cu is a very good heat conductor and solder volume is small compared to the wire, the solder temperature is very uniform, almost the same as the Cu wire as demonstrated by FEM modeling [11]. Therefore, the solder temperature can be determined by a temperature coefficient of resistance (TCR)-based temperature measurement. Knowing the temperature of the samples, at least at the beginning of the test, is very important for the interpretation of EM data.

In this paper, the ranking of the EM performance of different Pb-free solders and surface finishes will be reported. EM failure mechanisms, grain orientation dependence, and the effect of Sn microstructure and alloy composition will be discussed as well.

Experiment and Results

The experimental matrix of the electromigration (EM) test consists of two solder alloys, Sn-Ag and Sn-Cu, in combination with three solderable surface finishes, Cu, Ni-Au and Ni-Cu. The opposing pad structure in the solder joints was the same in all experiments and was comprised of a layered structure, simulating Ni-based UBM for controlled collapse chip connection (C4). Since the solder volume of the wire sample is much larger than that of the C4 solder balls, the thicknesses of the coated metal films were increased to ensure: 1) the Ni barrier layer is not consumed during reflow; 2). the Cu film on top of Ni is dissolved completely after reflow; and 3). the intermetallic structure is similar to the IMC in C4 joint. The UBM on the anode side is a three-layer

978-1-4244-2739-0/08/$25.00 ©2008 IEEE

sequentially sputtered film of TiW (1650 Å)/ Ni (6 μm)/Cu (2 μm). The three surface finishes on the cathode side are: a) electroless plated Ni(P) (14 μm) /sputtered Au (500A); b) electroless plated Ni(P) (14 μm) /sputtered Cu (2 μm); c) electrolytically plated Cu (14 μm). Since the focus of this study is on plated Ni and Cu finishes, sputtered Au and Cu is used on Ni(P) to eliminate the complications of the black pad problem[9]. Both Sn0.7Cu and Sn1.8Ag solders are used in the experiment with each pad surface finish. Owing to the contribution of Cu or Au layer deposited on the Ni barrier layers or surface finishes, the final alloy composition of the Sn-Ag solder joint is in fact Sn-Ag-Cu. The final solder compositions of the solders after reflow for different surface finishes are tabulated in Table 1. Ni is neglected since the solubility of Ni in Sn is very low (0.198% at 250°C). The solubility of Cu in Sn is 1.23% at 250°C. All the Cu thin films on top of Ni barrier layers were consumed during reflow. For plated Cu surface finish, since there is an infinite Cu supply, the solder will be saturated with Cu during reflow and that gives a Cu concentration of approximately or greater than 1.2 %, depending on reflow conditions.

Table 1. The Ag and Cu concentration in solder joints for three surface finishes. Ni and Au are neglected in calculation

Solder	Wt%	Cu	Ni(P)/Au	Ni(P)/Cu
Sn1.8Ag	Ag	~ < 1.77	1.79	1.78
	Cu	~ > 1.23	0.55	1.1
Sn0.7Cu	Cu	~ > 2	1.41	1.95

The diameter of the Cu wire is 287 μm and the length of the solder joint is approximately 380 μm. The wires and solder balls are aligned in a silicon V-groove and reflowed at 265°C for 10 minute in forming gas environment to form the solder joints. To prevent the samples from breaking due to the embrittlement of IMC interface after EM stressing, the solder joints are mechanically reinforced by a chip underfill material. All the solder joints were screened by Fein Focus X-ray imaging. Only the void-free samples were used in the electromigration test. The samples with about 2 cm long wires on each side of the solder joint were mounted to a test board. The samples were connected, in series, to a high precision current source, and the voltage drop across each sample is measured with an Agilent 34970A data acquisition system and switching unit controlled by a Labview program. The accuracy of the measurement is about 3 μV, which provides a measurement resolution of approximately 1/1000 of the solder resistance.

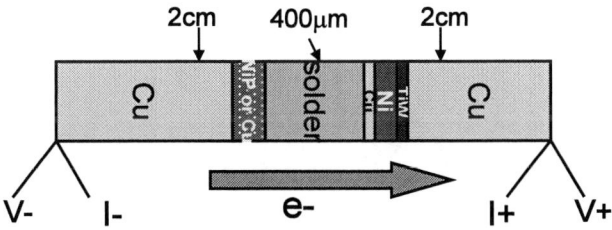

Figure 1 Illustration of sample configuration

The samples were tested in a forming gas environment. TCR calibration was performed on every sample with a 50 mA current at six temperatures from room temperature to 160°C. Since Cu is a very good heat conductor, finite element simulation as expected showed that the temperature of the solder ball is practically the same as the copper wire within a fraction of a degree. Therefore each sample served as its own temperature sensor with the calibrated TCR curves. The temperature of the solder joint with Joule heating is calculated from resistance of the samples at the test stress current level (5A) according to the TCR calibration of each individual sample. This gives a very good measurement of the sample temperature until the EM damage is large enough to cause significant resistance change. Since temperature rise due to Joule heating is extremely sensitive to the heat transfer coefficient at the location of the sample, the final temperatures of the samples with 5A current varied in the range of 136-154°C. Figure 2 is the sample temperature with 5A current at beginning of the EM test. The temperature of the top rack, first 20 samples, are more uniform than that of the lower rack, last 20 samples, mainly due to the oven configuration. The samples were arranged to make sure that every portion of the temperature range had samples from all six types, so that the data can be compared directly.

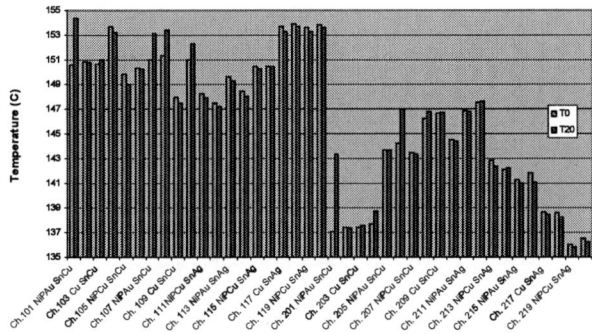

Figure 2 Initial sample temperature with Joule heating.

Figures 3a and 3b show the resistance raw data of Sn-Cu and Sn-Ag solder with Cu surface finish. These data clearly show that Sn-Cu solder joints failed faster than Sn-Ag solder joints. Due to the sample geometry, the solder resistance is about 1/10 of the total sample resistance. Assuming the resistance of the Cu wire does not change during the test, the 1% total resistance change is equivalent to 10% resistance change in solder. In this experiment, a 3% sample resistance change is used as failure criteria, which corresponds to 30% change in solder resistance, including bulk and interface. Figure 4 is a plot of average time to failure for the six types of sample combinations studied. It shows that the performance of the surface finishes from better to worse is Ni(P)/Cu, Ni(P)/Au, and Cu. Although the Ni barrier layer is clearly advantageous in slowing down the electromigation damage process, a thin Cu film, dissolved in the Sn-Ag solder during reflow, is very significant in improving the time to failure performance. But the Sn-Ag or Sn-Ag-Cu solder generally performs better than Sn-Cu solder with all surface finishes. Details will be discussed later.

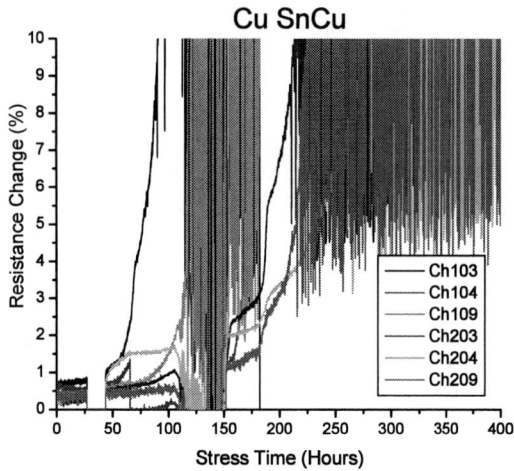

Figure 3a Resistance verse stress time for SnCu solder with plated Cu finish.

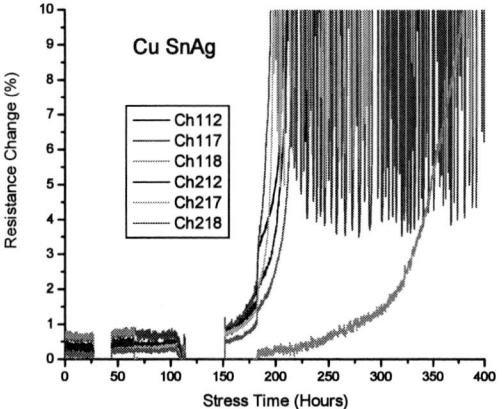

Figure 3b Resistance verse stress time for SnAg solder with plated Cu finish.

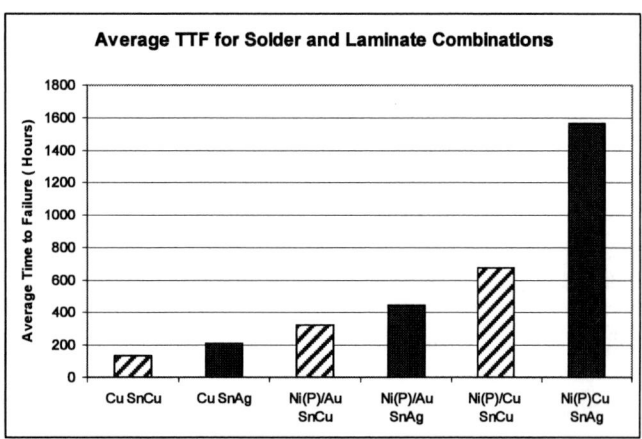

Figure 4 Plot of average time to failure for different solder and laminate finish combinations.

Analysis and Discussion

Although, the resistance data clearly differentiate the electromigration performance of different surface finishes and solders, failure analysis is needed to understand the mechanisms of failure. Samples were taken out after 65, 182, 413, 620, 912 and 1342 hours of EM testing with a 5A current. The samples are cross-sectioned and examined by optical microscopy and SEM. Figure 5 is optical micrographs of the cross-sections, using crossed polarizers, of samples after 65 hours of testing. Electron flow is from top down. The top row is Sn-Cu solder and the bottom row is Sn-Ag solder. The cathode-side surface finish of the columns (left to right) are 14μm plated Cu, 14 μm plated Ni(P) with 500A sputtered gold, and 14 μm plated Ni(P) with 2 μm sputtered Cu. Figure 6 is the cross-section images of the samples after 1342 hours stress. The sample arrangement is the same as in Figure 5. Again, as pointed out before, the composition of the Sn-Ag solder joint is in fact Sn-Ag-Cu or Sn-Ag-Cu-Au due to the contribution of Cu or Au from the Cu or Au films on the Ni barrier layers or surface finishes.

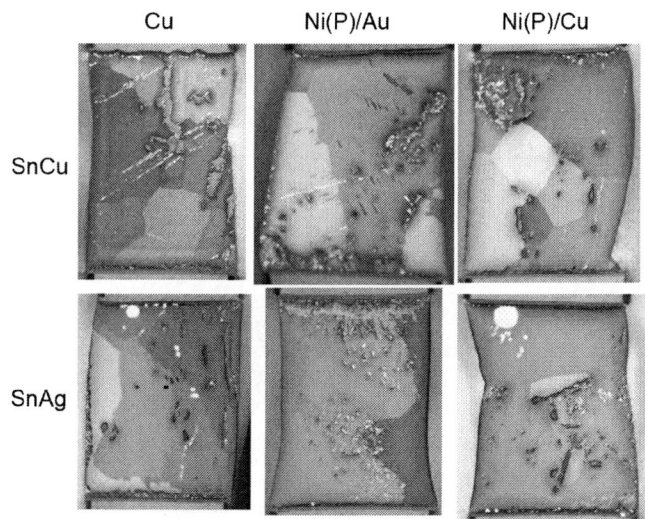

Figure 5 Cross-section images after 65 hours stress.

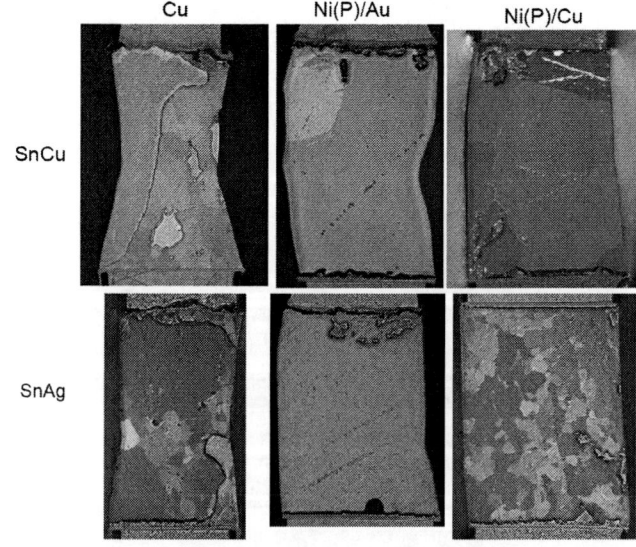

Figure 6 Cross-section images after 1342 hours stress.

The electromigration damage between different surface finishes are compared side to side. For plated Cu, the plated portion of the Cu is completely consumed within 65 hours for Sn-Cu solder. For the Sn-Ag solder, Cu consumption is a little slower than for the Sn-Cu solder; but it is clear that the 14 μm plated Cu is completely consumed before 413 hours. Relating to the resistance change in Figure 3, it seems that Cu wire consumption is associated with drastic resistance increase. The EM current density is sufficient to dramatically enhance the Cu_6Sn_5 intermetallic compound formation at the Cu interface. In turn, the Cu_6Sn_5 intermetallic compound is swept across the solder joint in the direction of the electron current by diffusion processes, enhanced by the EM current. For the surface finishes with plated 14 μm Ni(P), the barrier layer is still intact at 65 hours with some consumption with Sn-Cu solder and almost no consumption with Sn-Ag solder. At 413 hours, Ni(P)/Au started to show severe consumption with Sn-Cu solder and some local attack with Sn-Ag solder. As for Ni(P)/Cu, the barrier layer survived 1342 hours of testing with almost no damage with Sn-Ag solder and some damage with Sn-Cu solder. This means that a Ni barrier layer is capable of slowing down pad consumption and electromigration damage. The Cu thin film deposited on Ni forms $(Cu\text{-}Ni)_6Sn_5$ intermetallic layer at solder reflow provides significant extra protection for the Ni barrier layer. The rank of the surface finishes based on cross-section analysis from best to worst is Ni(P)/Cu, Ni(P)/Au, and Cu. The results are fully consistent with the resistance data.

Furthermore, the interfacial void formation induced by the EM test conditions are examined. Figure 7 is the high magnification optical image of samples after 413 hours of electromigration testing. The top image is the plated Cu with Sn-Cu solder. The plated Cu initially deposited on the Cu wire is consumed, Sn penetrates into Cu wire, Cu_6Sn_5 intermetallic compound (IMC) was swept away from Cu/solder interface and interfacial voids were created. The middle image is the Ni(P)/Au with Sn-Cu solder. In this case, due to the protection from the Ni(P) barrier layer, the Cu wire is little affected. However, the intermetallic compounds have been swept away from the Ni/solder interface, solder has penetrated into the Ni barrier layer, and a thin gap is formed between solder and the remaining barrier layer. In addition voids have formed at the Ni-Cu interface. The cavitations at the solder-Ni interface are similar to the capitation in the top image, but represent a different degree of severity. We classify this as Mode-II electromigration failure mechanism, which is driven by fast diffusion process and associated with drastic IMC movement, surface finish consumption and early failures. This Mode-II failure occurred more frequently in Sn-Cu solder or Cu rich solder system, and it is dependent on the Sn grain orientations, as we will discuss later. The bottom image is Ni(P)/Cu with Sn-Ag solder. Different from the first two cases, the barrier layer and Cu_6Sn_5 intermetallic compound layer are still intact and some voids have formed at the interfacial IMC/solder interface. We called this the Mode-I electromigration failure mechanism, which is driven by Sn diffusion away from the IMC/ solder interface. It is usually occurs later than Mode-II and causes less damage.

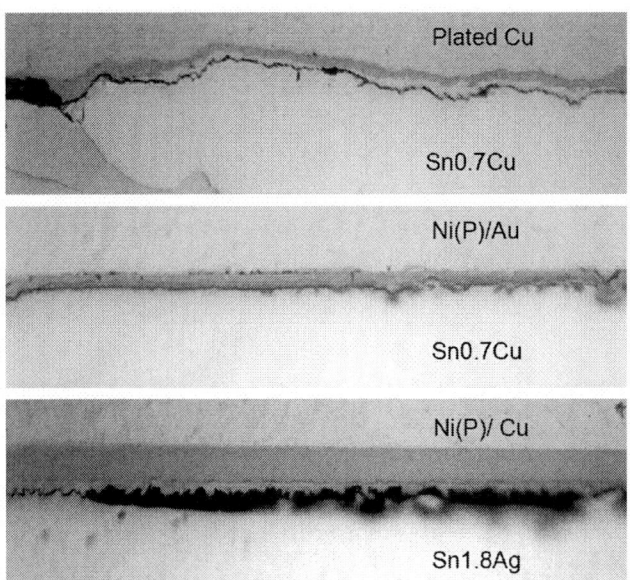

Figure 7 High resolution optical images of the cathode side interfaces after 413 hour stress

To characterize the failure mechanisms in detail, we have also observed a unique example of the electromigration degradation mechanisms, as demonstrated in Figure 8. Figure 8 is a cross-polarized optical microscope image of a Sn-Cu solder with Ni(P) Au surface finish after 555 hours EM testing. The contrast in the image indicated that there are two major grains in the sample, with intermetallic compounds accumulated near the mutual grain boundary. Figure 9 is the SEM image of the cathode interface, at the boxed region in Figure 8. EDS (Energy Dispersive Spectroscopy) was taken at the points indicated by colored arrows. The insert of the spectrum showed the composition of the damaged and undamaged Ni barrier layer. On the right, the Ni barrier is intact. Only Ni and P are detected and Ni concentration is higher than P (blue arrow and spectrum). No Sn is detected. The trace of Cu is from the contamination during sample preparation or Cu diffusion. The adjacent intermetallic is Ni_3Sn_4, which is still intact after EM test. The voids are formed at the IMC/solder interface, which is a typical Mode-I failure caused by Sn self diffusion. On the left, however, the interfacial IMC structure has been swept away and Sn penetrates into the Ni barrier layer and even into the Cu wire as indicated by the yellow arrows. The remaining Ni barrier is P rich indicating Ni depletion, as indicated by the red arrow and spectrum. In addition, Sn is found in Ni. Cu is found in the intermetallic compound that is swept to the grain boundary. This is a typical Mode-II failure, which is brought about by the fast diffusion path for Ni and Cu through the bounding Sn grain, as argued, below.

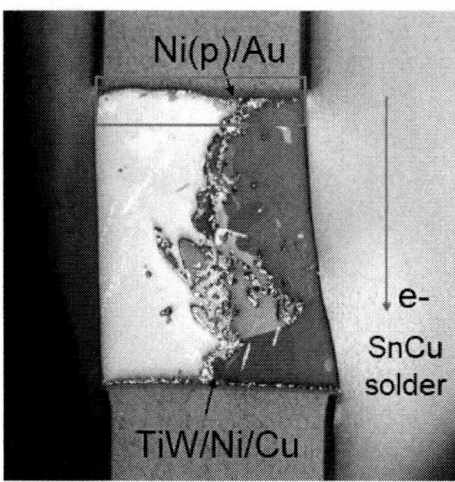

Figure 8 Cross polarized microscope image of a bi-crystal sample showed two failure modes.

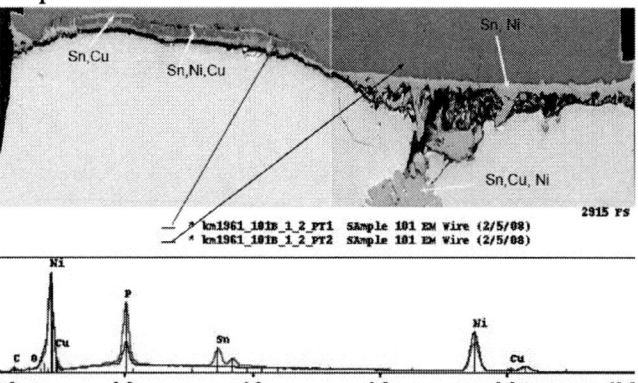

Figure 9 The SEM image of the boxed section in figure 8 and EDS spectrum of the damaged Ni barrier layer on the left and the undamaged Ni layer on the right.

Figure 10 is an EBSD (Electron Backscattering Diffraction) color mapping of the β-Sn grain orientation of the same Sn-Cu solder joint as shown in Figures 8 and 9. The blue colored (right) grain has its c-axis approximately perpendicular to the current direction, while the red colored (left) grain has its c-axis roughly aligned with the current direction. This is a direct proof of the highly anisotropic diffusion behavior in Sn crystal [3-5]. The body centered β-Sn crystal is highly anisotropic in mechanical, thermal, electrical and diffusion properties. The diffusivity along the c-axis is 40 and 70000 times faster than along the a- or b-axis for Cu and Ni at 120°C, respectively. In Mode-I failure, the c-axis of the Sn grain is at a high angle to the current direction, along which the diffusion rate of Cu or Ni in Sn is slow. Failure is characterized by Sn self-diffusion or lattice diffusion resulting in voids formation between IMC and solder. In Mode-II failure, the c-axis is roughly aligned with the current direction, and Cu or Ni diffusivity / transport is very high. As soon as current applied on the sample, the Cu_6Sn_5 IMC dissociates and is diffusively swept to the grain boundary and through the grain to the other side of the solder joint. Once the IMC is removed, the Ni barrier is exposed and Ni will diffuse through the sample and Sn will penetrate in the barrier and then further into the Cu wire under the Ni

barrier layer. Voids are formed at the degraded barrier or Cu interface without Cu_6Sn_5 IMC.

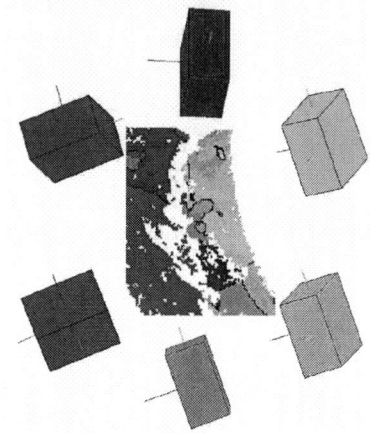

Figure 10 EBSD map of the sample in Figure 8. The colored tetragonal figures indicate the Sn grain orientation of the corresponding colored grains in the map.

Since it is associated with the fast interstitial diffusion process, the Mode-II mechanism is commonly found in early EM failures. Statistically, it is found that Mode-II failure is more common in Sn-Cu solder than in Sn-Ag solder. Figure 11 is a plot of the resistance change over time during EM stress for four samples. Curve A is from the sample with SnCu solder depicted in Figure 8, where about half of the solder joint has a mode–II failure resulting in an early electrical failure. Curve B is from a sample with Sn1.8Ag and Ni(P)/Cu UBM. The final composition of the solder is SnAgCu after reflow. This sample has very little change in resistance after 1342 hours of EM stressing. Failure analysis reveals that the solder has cyclic twinning and hardly any EM damage. Curves A and B are characteristic EM plots for SnCu and SnAgCu solders, respectively. Although Mode-I is common for SnAg(Cu) solder and mode-II is more frequent in SnCu solder, mixed mode failures are observed in both solders. Curves C and D are from the samples with SnCu and SnAg solder, respectively. The longer EM lifetimes of samples C and D than SnCu solders with Mode-II failure are due to the multiple grains (SnCu) and cyclic twinning (SnAg) structure as discussed next.

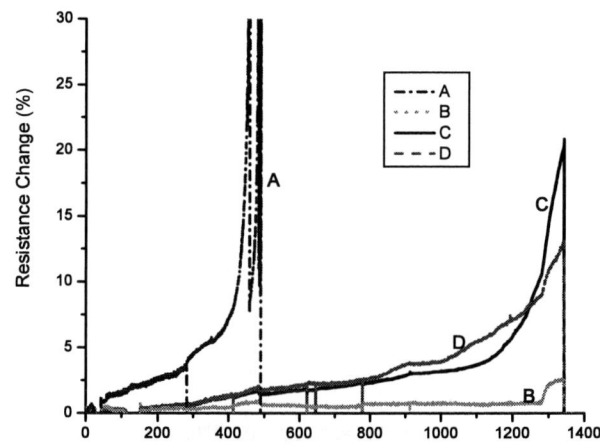

Figure 11 Plot of the resistance change over time during EM stress for four samples.

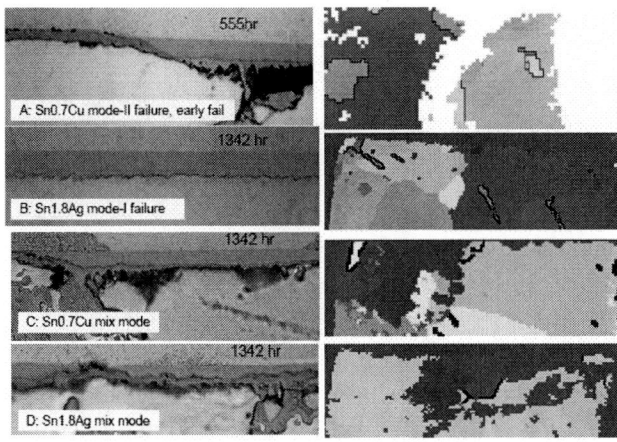

Figure 12 The cross-section images and EBSD maps of the of the four samples shown in figure 11.

Figure 12 is the cross-section images and EBSD maps for the four samples shown in Figure 11. A is the SnCu bi-crystal sample same as in Figure 8 that has one grain with c-axis parallel to current that lead to mode-II early failure. B is a Sn1.8Ag solder with good grain orientations (c-axis is not along current direction) that results minimum EM damage and resistance increase. C is a SnCu solder with multiple grains in the solder joint, where one smaller grain was oriented with its c-axis aligned with the current and the other larger grain was oriented with c-axis away from current direction. As expected, the area where the UBM is adjacent to the larger green grain showed Mode-I type damage and the portion of the UBM adjacent to the smaller red grain showed Mode-II type damage. Even with small twinning, the yellow grain inside the red grain, the UBM consumption appears to be reduced. D is an example of Mode-II damage in SnAg(Cu) solder after 1342 hours of stressing. The small red-colored grain has its c-axis aligned with the current direction. The two larger green-colored grains, where the c-axis is not aligned with current direction, are the cyclic twins of the red grain. This sample exhibits mixed modes of degradation depending on the grain orientation. C and D showed that the presence of multiple grains and twinning structure can mitigate EM damages caused by fast diffusion in Mode-II type failure in SnCu and SnAgCu solder.

Both electrical data and failure analysis indicate that SnAg solder performs better than Sn-Cu solder in electromigration testing. Sn-Ag solders form Ag_3Sn IMC particles during the solidification process. The diffusivity of Ag in Sn is almost 3000 times slower than Cu at 150°C [5], and it has minimal solubility. With slow diffusivity and higher melting temperature, the presence of Ag_3Sn IMC appears to provide a more stable microstructure than the Cu_6Sn_5 IMC against electrical and thermal stress. A Sn-Cu solder joint typically starts with a somewhat smaller grain size (Figure 5), which coarsened significantly into one or two large grains during electromigration (Figure 6) testing. In contrast, Sn-Ag starts with fewer independent grains that have good stability throughout the electromigration test. In the Sn-Ag system many of these grains are crystallographic, cyclic, twin structures. Figure 13 is an EBSD map and SEM

of interlaced twinning in SnAg solder. The red lines indicate a twin boundary that is at about a 60 degree rotation about the a- or b- axis of the Sn crystal. Twinning, especially cyclic, rotational (60 degree) twinning, are more commonly observed in Sn-Ag than in the Sn-Cu system[11]. The required, large angle, rotation in grain orientation resulting from cyclic twinning may be part of the reason for the better EM performance of Sn-Ag solder. Comparing the SEM image and the EBSD map, it appears that the Ag_3Sn IMC particles are decorated at the cyclic twinning boundaries as well small angle grain boundaries, indicating the twins are formed at solidification.

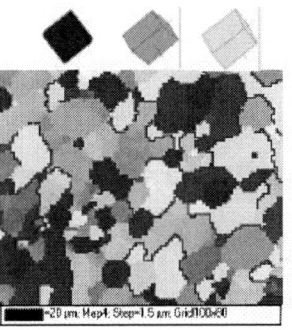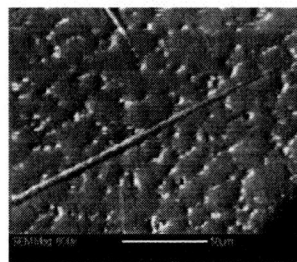

Figure 13 Interlaced twinning in Sn1.8Ag solder alloy, EBSD map is on the left and SEM image is on the right.

As discussed before, Mode-II failure is found more often in SnCu solder than in SnAgCu solder. We believe that it is related to the significant differences in the microstructures between SnCu and SnAgCu solders[7-9]. SnCu solders normally consist of multiple grains with few polysynthetic twins and rarely cyclic twins. Under high temperature and high current stress conditions, significant grain growth and re-orientation were observed, leading to Mode-II type failure where c-axis closely aligned to the current direction is more frequently observed. For SnAg(Cu) solder, the solder joint begins with one or a few large grains which are often associated with beach ball or interlaced cyclic twinning structures[7,8]. The grain structure is much more stable under EM stressing, due mainly to the stable Ag_3Sn IMC network. Cyclic twinning, both interlaced and beach ball twins commonly found in SnAg(Cu) solders, creates randomness in the grain orientation that reduces the propensity of Mode-II failure and thus extends EM lifetime[7-8]. In an extended study of the crystallography of 381µm diameter solder ball with different Ag concentration, it was revealed that a certain amount of Ag, about 1 wt.% or greater, is needed to form more cyclic twins in solder. Figure 14 shows the typical EBSD maps of Sn0.7Cu, Sn0.5Ag, and Sn1.0Ag solders, respectively. All the solder balls have been thermally annealed at 150°C for 500 hours. The black lines mark the twinning boundary with 60±5° rotation about <100> axis [12]. Figure 14(a) is a Sn0.7Cu solder ball, with multiple grains and two sets of laminar twins in orthogonal directions. Figure 14(b) is a Sn0.5Ag solder ball. That consists of multiple grains and laminar twins which is much closer to the SnCu structure than to the Sn1.0Ag structure. Figure 14(c) is a Sn1.0Ag solder ball with beach ball twinning, and Figure 14(d) is another Sn1.0Ag solder ball with two sets of interlaced twins. The interlaced twinning boundaries are decorated with Ag₃Sn IMC networks indicating the twinning is formed during the solidification process. The dense Ag_3Sn IMC network in high Ag content SnAg solders promote a more interfaced twinning structure, the lesser Ag_3Sn IMC network in lower Ag solders results in the SnCu–like multigrain structure. Therefore, the high density Ag_3Sn network or a certain Ag level is needed to ensure cyclic twinning structure which benefits EM stability.

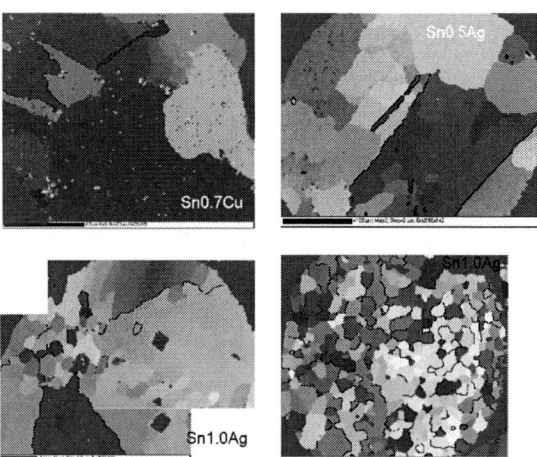

Figure 14 EBSD map of 381µm solder ball after 150°C 1000 hours thermal aging. (a) Sn0.7Cu, (b) Sn0.5Ag, (c) Sn1.0Ag, (d) Sn1.0Ag.

Although the presence of Ag plays an important role in slowing down the electromigration failure process, the role of Cu should not be neglected. The almost three-fold increase of time to failure in Sn-Ag solder with the dissolution of 2 µm Cu film on Ni(P), as shown in Figure 4, indicates the importance of Cu in electromigration test conditions. The formation of Cu_6Sn_5 intermetallic compound layer on the Ni barrier might add additional protection to the Ni barrier layer underneath. In Ni(P)/Au case, the Cu in the system (0.55 wt.%) came from the Cu film on UBM stack on the anode end of solder. When solder first melts at reflow, temporarily, there is no or very low Cu at Ni(P)/Au surface. Au will dissolve rapidly, and we are essentially soldering to the surface of Ni(P) with very low Cu present in the solder. The surface will form Ni_3Sn_4. Since a significant portion of Cu is consumed by formation of Cu-Sn intermetallic on the anode side, Cu content left in solder is probably 0.4 wt% or less, the IMC on the cathode side is predominately Ni_3Sn_4, possibly with some Cu in the later stage of reflow, as shown in Figure 9. For Ni(P)/Cu case, the Cu film on top of the Ni(P) must dissolve before the Ni(P) is exposed to solder. At this point in time, the Cu concentration in the solder is expected to be above 1 wt % and the IMC formation process forms continuous Cu_6Sn_5 layer. Therefore, the IMC formation between the two cases is very different. We believe the Cu_6Sn_5 layer may give extra protection for the Ni barrier layer and thereby improves EM performance. The Cu concentration and IMC formation in this experiment are

consistent with the IMC formation processes described by Kao [13].

An alternative explanation for the improved EM performance could potentially involve a preferred crystallographic texture for the Sn phase that may develop, as a result of the solidification process. As outlined above, the Sn crystallographic orientation and the associated orientation of fast diffusion paths for Ni and Cu critically control the kinetics of the EM failure mechanisms. The possibility of this alternative explanation is presently under evaluation.

Conclusions

In conclusion, the electromigration performance of three surface finishes and two Pb-free solders on Ni UBM has been investigated. The Ni barrier layer is needed to retard the electromigration damage. The ranking of the three surface finishes, from best to worst, is Ni(P)/Cu, Ni(P)/Au, and Cu. Sn-Ag solder has longer EM lifetime than Sn-Cu solder. However, the addition of Cu to the Sn-Ag solder dramatically improved the electromigration performance.

Two failure mechanisms have been identified. Mode-I is probably dominated by Sn self-diffusion resulting in separation between IMC and solder. Mode-II is clearly dominated by a fast diffusion process for Ni and/or Cu in Sn, which depends on grain orientations and probably occurs more often in Sn-Cu than in Sn-Ag solder. A stable Ag_3Sn IMC particulate network and cyclic twinning structure might contribute to the better EM performance in Sn-Ag solder. Additionally, a preferred crystallographic orientation may occur with some solder compositions during the solidification process for the Sn-Ag-Cu solders, such that the c-axis of the Sn grain is not closely aligned with the pad normal direction. Such a crystallographic alignment would reduced EM damage and prolong lifetime.

Acknowledgments

The authors would like to acknowledge Mr. Charles Goldsmith for failure analysis, Dr. Tia Korhonen of Cornell University for EBSD analysis (Figure 10), Ms. S-K. Seo of KAIST for sample preparation (Figure 14), Mr. R. Polastre for data acquisition setup, Drs. C. K. Hu and J. Lloyd, S. Kang, S. Wright, C. Witt, H. Longworth, and T. Wassick for valuable discussions.

References

1. Tu, K. N., et al., "Physics and Material Challenges for Lead-free Solders," J. Appl. Phys., Vol. 93, No. 3, (2003), pp. 1335-1353.
2. Nicholls, L., et al., "Fine Pitch (150µm) Pb-free Filp Chip Bumping and Packaging," *Proc 56th Electronic Components and Technology Conf*, San Diego, CA, May. 2006, pp. 131-138.
3. Dyson, B. F., Anthony, T.R., and Turnbull, D., " Interstitial Diffusion of Copper in Tin," J. Appl. Phys., Vol. 38, No. 8, (1967), p. 3408.

4. Yeh, D.C., and Huntington, H.B., " Extreme Fast-Diffusion System: Nickel in Single-Crystal Tin," Phys. Rev. Lett., Vol. 53, No. 15, (1984) pp.1469-1472.

5. Dyson, B. F., "Diffusion of Gold and Silver in Tin Single Crystals," J. Appl. Phys., Vol 37, No. 6, (1966), pp. 2375-2377.

6. Bieler, T. R., et al., "Influence of Sn Grain Sixe and Orientation on the Thermomechanical Response and Reliability of Pb-free Solder Joints," *Proc 56th Electronic Components and Technology Conf*, San Diego, CA, May. 2006, pp. 1462-1467.

7. Lu, M., et al., "Effect of Sn Grain Orientation on electromigration degradation mechanism in high Sn-based Pb-free Solders," Applied Physics Letters., Vol. 92, No. 21, (2008), p. 211909.

8. Lu, M., et al., "Comparison of Electromigration Performance for Pb-free solders and Surface Finishes with Ni UBM," *Proc 58th Electronic Components and Technology Conf*, Lake Buena Vista, FL, May. 2008, pp. 360-365.

9. Sylvestre, J., and Blander, A., "Large-scale correlation in ther orientaiton of grains for lead- free solder bumps," to be published in J. Electronic Materials.

10. Ren, F., Tu, K.N., "In-situ Study of the Effect of Electromigration on Strain Evolution and Mechanical Porperty Change in Lead-free Solder Joints," *Proc 56th Electronic Components and Technology Conf*, San Diego, CA, May. 2006, pp. 1160-1163.

10. M. Lu, Unpublished data.

11. Chalmers, B., Physical SOC, **47**, (1935), p 733.

12. Ho, C. E., Tsai, R.Y., Lin, Y. L., and Kao, C.R., "Effect of Cu concentration on the Reactions Between Sn-Ag-Cu solders and Ni," J. Electronic Materials, Vol. 31, 2002, p. 548.

Effect of Stand-off Height on the Microstructure and Fracture Mode of Cu/Sn-9Zn/Cu Solder Joint under Tensile Test

Bo Wang[a, b], Fengshun Wu*[a, b], Bin Du[a], Bing An[b], Yiping Wu[a, b]

a Department of Materials Science and Engineering, Huazhong University of Science and Technology, Wuhan, 430074, China

b Wuhan National Laboratory for Optoelectronics, Wuhan, 430074, China

Tel.:+86-27-87558275; E-mail: fengshunwu@mail.hust.edu.cn

Abstract

This study investigates the effect of the stand-off height (SOH) on the microstructure and tensile fracture mode of Cu/Sn9Zn/Cu solder joints. Solder joints with SOH of 100μm, 50μm and 20μm are studied. It is found that as the SOH is reduced, Zn content has a rapid decrease in the solder bulk, while, the intermetallic compound (IMC) layer proportion increases. SOH of the solder joint also has an important effect on the tensile strength and tensile fracture mode. Tensile strength of solder joints decreases with lower SOH, which correlates with the change of microstructure and composition in solder joint. When the SOH is reduced, the fracture path of solder joint transfers from the bulk of the solder joint into the IMC/solder interface, and the fracture mode tends to transform from ductile fracture into brittle fracture.

Keywords: stand-off height (SOH); microstructure; tensile test; fracture mode

1 Introduction

With the development of miniaturization, performance and functionality in electronic products, the solder joint becomes much smaller in size for meeting these requirements. From ball grid array packages (BGA) to wafer level packages (WLP) and even to three-dimension die stacking package (3D), the interconnection pitch and stand-off height (SOH) of solder joints are continuously miniaturized to meet higher integration density. [1] For example, SOH of solder joints in BGA, WLP and 3D packaging types ranges from 100μm to 10μm.

However, the miniaturization of solder joints would lead to some changes in microstructure of solder joints and some new reliability problems. Many studies [2, 3] have reported that the interfacial intermetallic compounds layer proportion, the ratio of interfacial IMC layers volume to the whole solder joint volume, increases with the decreasing stand-off height (SOH), which would have negative effect on the device reliability [4, 5, 6], while few studies are emphasized on the microstructure and composition changes in the solder bulk when the SOH is reduced.

Besides the miniaturization in the packaging industry, Pb-free is another trend of electronic manufacture. Recently Pb-free solder alloy has been widely used for manufacturing electronic products because of the environmental concerns and government legislation. [7] In all of the Pb-free alloy systems, Sn-Zn system alloy has been recognized as a promising solder alloy with the combination of modest melting temperature and some good mechanical properties. [8] The melting temperature of Sn-9wt.%Zn solder is 198°C, relatively close to that of eutectic Sn-Pb solder (183°C), and

much lower than that of the most highly recommended eutectic Sn-Ag-Cu system alloy. This advantage has encouraged electronic packaging industries to use this alloy without the need to replace existing manufacturing lines or electronic components. Additionally, compared with other Pb-free solder alloys, Sn-Zn has the lowest price, which would be another impetus for application.

For research on the reliability of solder joint, shear testing is commonly used as a method to analyze the mechanical properties of solder joints, [9] however, it is not suitable for investigating the failure mechanism. Due to lack of the accurate shear stress conditions, the results are difficult to interpret in terms of fundamental strength properties. And the friction between the shearing surfaces also could affect the failure load, furthermore, the frictional damage impairs fracture surface investigations. However, tensile tests could offer a uniform and equivalent stress across every section layer, such as the solder layer, the intermetallic compounds layer and the substrate layer, which can be reliably used to find out the weak bonding layer in solder joints. In addition tensile fracture surfaces are better preserved and could lend more evidence towards the failure mechanism. [10]

So far quite a lot of work has been focused on the shearing test to study the mechanical properties of Pb-free solder joint, while very a few studies were made to investigate the mechanical characteristics of Cu/Sn-9Zn/Cu solder joint with micro SOH by means of tensile test. Therefore, with the trend of miniaturization and Pb-free solder alloy in electronic packaging, it becomes necessary to study the effect of stand-off height (SOH) on the microstructure and mechanical properties of Cu/Sn-9Zn/Cu solder joint under tensile test. And it is also important to investigate the relationship between microstructure change and mechanical properties of Cu/Sn-9Zn/Cu solder joint.

2 Experimental procedures

In the experiment, the used Cu bars had a length of 30mm, and was cut by spark cutting from a commercial grade Cu (99.99%) wire with a diameter of 0.9mm. Then the cut surfaces of these Cu bars were polished using 1.5μm Al₂O₃ powder pastes, and finally the Cu bars were cleaned in absolute ethyl alcohol by ultrasonic cleaning equipment

Solder joints with 100μm, 50μm and 20μm SOH were got by a clamping apparatus, which could control and regulate the distance of two parallel cut surfaces. After the two Cu bars with parallel surfaces was hold in alignment with the distance of certain SOH, rosin activated flux was coated on the side surfaces. Subsequently, a round sheet of Sn-9Zn solder alloy was put in between the two aligned Cu bars. The round sheet was prepared by rolling the Sn-9Zn solder alloy into ribbon type with the thickness of 0.5mm and then stamping into round with the diameter of 1mm. The clamping apparatus with the specimens was sent into a hot-air convection oven for 200s above the liquidus phase temperature and with the maximum temperature 250°C. After joining, excessive solder flowing out of the solder joint area was ground away by abrasive paper. The advantage of such one-dimensional samples is that tensile test can be applied to study the mechanical properties of solder joints. Furthermore, unlike conventional bulk test samples, these samples have two interfaces with IMC formation, which makes them closer to real solder joints in a device. Figure 1 shows the schematic diagram of stand-off height (SOH) of solder joint.

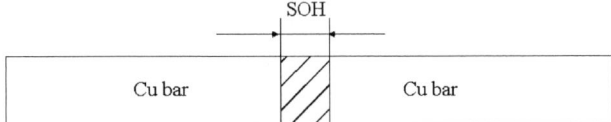

Fig.1 Schematic diagram of stand-off height (SOH) of solder joint

Tensile test was carried out in an INSTRON 4206 tester at room temperature with a constant crosshead speed of 0.1mm/min. Force and displacement were recorded. In order to obtain the average tensile result, at least 15 specimens with the same SOH was prepared as a group for the tensile test.

Both the microstructure and fractograph of solder joint were observed using scanning electron microscope (SEM) in the back scattering mode. Then the compositions of solder joint were analyzed by energy dispersive spectroscopy (EDS).

3 Results and discussion

3.1 effect of SOH on the microstructure and composition of solder joint

Fig.2a-c shows the cross-sectional backscattered electron (BSE) mode SEM images of the Cu/Sn-9Zn/Cu solder joint with 100μm, 50μm and 20μm SOH, respectively.

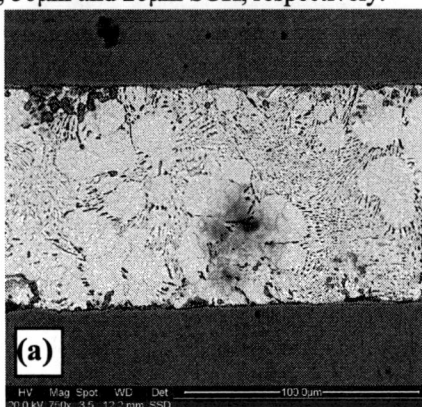

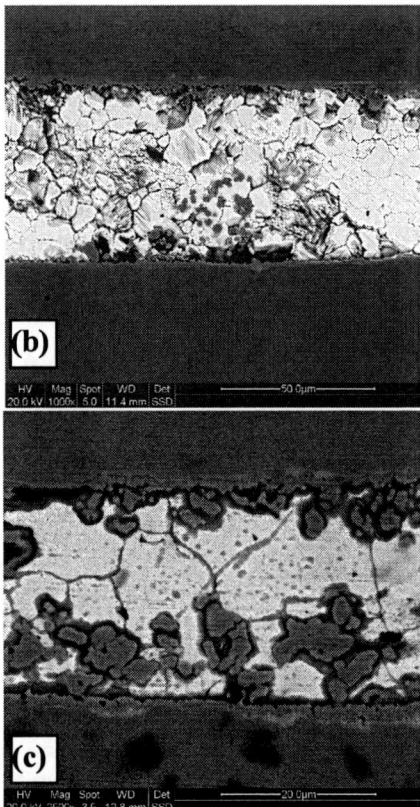

Fig.2a-c. Cross-sectional backscattered electron (BSE) mode SEM images of the Cu/Sn-9Zn/Cu solder joint with (a) 100μm SOH (b) 50μm SOH and (c) 20μm SOH.

From Fig.2a-c, it is found that SOH has a significant effect on the Zn content in the solder joint bulk. Fig.2a shows that eutectic SnZn is observed in solder joint with 100μm SOH, however, fewer eutectic SnZn can be found in solder joint with 50μm SOH, shown in Fig.2b, while, as shown in Fig.2c, no eutectic SnZn can be observed in solder joint with 20μm SOH. Therefore, Zn content in solder bulk has a tendency to decrease with lower SOH, which is also proved by EDS result on the solder bulk. Fig.3 shows the mean Zn content in solder bulk with reduced SOH.

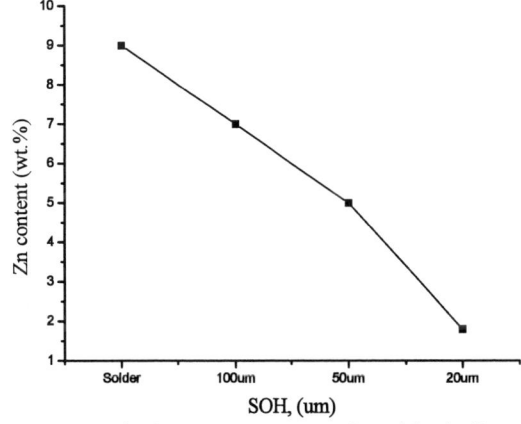

Fig.3 Mean Zn content in solder bulk

It has a rapid decrease in the Zn content with the reduced SOH. We believe that this difference in the Zn content among the three solder joints is due to the effect of SOH. When the SOH is reduced from 100μm to 50μm, and finally to 20μm, the corresponding solder volume in the joint would be reduced too, while, these three kinds of solder joints have the same area of Cu/Sn-9Zn reaction interface. Therefore, The ratio of solder volume to interfacial reaction area in solder joint with 20μm SOH is only about two fifths and two tenths of that in solder joint with 50μm and 100μm SOH, respectively, which means that the corresponding mass of Zn is only two fifths and two tenths too. As in the same liquid reaction time, most Zn in solder bulk could be consumed by the dissolved Cu in solder joint with lower SOH, while, it would be not the case in the solder joint with higher SOH, because it can supply more Zn for the reaction. After reflow process in this study, the composition of solder bulk is transformed from Sn-9wt.%Zn to Sn-2wt.%Zn at the least with the reduced SOH. And the microstructure changes from eutectic phase to hypoeutectic phase. These changes of composition and microstructure would seriously influence the mechanical property of the solder joint.

As we know, solder joint consists of IMC layer and solder bulk layer. From Fig.2, it is found that the thickness of IMC is about 1.8μm at one interface in all of the three solder joints, however, the IMC layer proportion is very different, here, the IMC layer proportion is the ratio of interfacial IMC thickness to SOH. In our study, the IMC layer proportion is about 3.6%, 8%, and 17% in the solder joint with 100μm, 50μm, 20μm SOH, respectively. So, the IMC layer proportion in the solder joint increases with the reduced SOH. In the three solder joints, some ball-type dissociative IMCs closed to the IMC layer at Cu/solder interface can also be observed in the solder bulk. Like in the solder joint with 20μm SOH, those IMCs almost cover half of the solder bulk, and they distribute in the Sn grain boundary unevenly with the size of about 1.9μm. Due to the existence of those IMCs in the solder bulk, the solder joint could become mechanically weak because the dissociative Cu_5Zn_8 IMC could deteriorate the integrity and ductility of Sn matrix phase [11] and easily lead to crack. However, in the solder joint with higher SOH, those dissociative IMC only account for a small percentage of the whole solder bulk. Comparatively speaking, its effect on the mechanical property of solder joint could be lower than that in the 20μm SOH solder joint.

Therefore, due to the reduced SOH, the microstructure of solder joint is changed a lot, and this microstructure change would finally affect the mechanical properties and reliability of solder joint.

3.2 Effect of SOH on the fracture mode and tensile strength of solder joint

Fig.4a-c shows the SEM tensile fracture surface images of solder joint with 100μm, 50μm and 20μm SOH, respectively. As shown in Fig.4, there are many equiaxed dimples at the fracture surface in the solder joint with 100μm and 50μm SOH. While, in the solder joint with 20μm SOH, the fracture morphology is very different, and it could be found that there are much fewer equiaxed dimples, and more than half of the fracture surface is covered by plane part,

which is identified to be Cu-Zn IMC by EDS analysis. So it indicates that when the SOH of solder joint is reduced, a transition from ductile fracture mode to brittle fracture mode occurs.

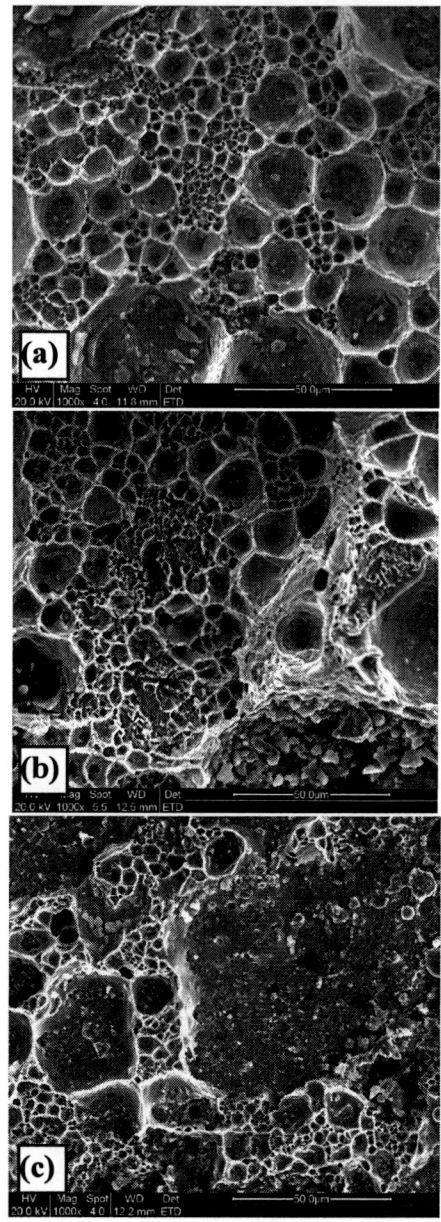

Fig.4a-c. backscattered electron (BSE) mode SEM fracture images of the Cu/Sn-9Zn/Cu solder joint with (a) 100μm SOH (b) 50μm SOH and (c) 20μm SOH.

Fig.5 shows the ultimate tensile strength of solder joint with different SOH, from Fig.5, it is noted that the ultimate tensile strength decreases with the reduced SOH. The composition of solder bulk accounts for the decrease in ultimate tensile strength. Fig.4a-b shows that the fracture occurred in the bulk of solder joint with 100μm and 50μm SOH, and EDS analysis also proves that the composition of fracture surface is close to that of solder bulk. Therefore, the weakest layer in the solder joint is solder bulk layer, accordingly, the composition of solder bulk becomes one of the most important reasons for the ultimate tensile strength of this two solder joints. Fig.5 shows the ultimate tensile strength of solder joint with 100μm is higher than that of 50μm SOH, and they are about 15.3MPa and 13.9MPa, respectively. The ultimate tensile strength of solder joint with 20μm SOH is about 12.3MPa, which is very close to that of pure tin, 12.4MPa. [12] This is because that the solder joint with 100μm SOH has the highest Zn content, which can enhance the ultimate tensile strength of solder joint. As a result, the ultimate tensile strength of solder joint with 100μm SOH shows the highest ultimate tensile strength. However, in the solder joint with 20μm SOH, most Zn is consumed, making the composition of solder bulk close to the pure Sn. Hence, Zn has limited effect on enhancing the ultimate tensile strength. The plane part in the fracture is identified to be Cu-Zn IMC. So, the weakest layer is the interface of solder bulk and IMC layer in the solder joint with 20μm SOH. And the fracture occurs on this interface and shows characteristic of brittle mode.

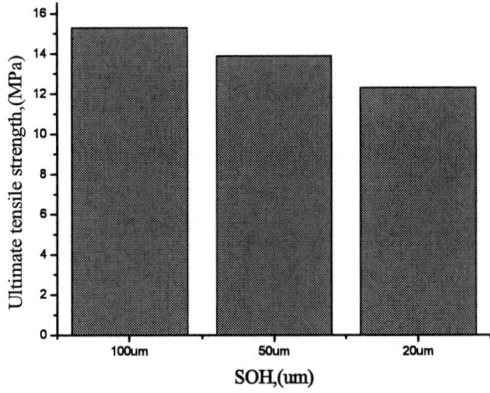

Fig.5. the ultimate tensile strength of solder joint with 100μm, 50μm and 20μm SOH

4 Conclusions

(1) As the SOH is reduced, the composition and microstructure of solder bulk are influenced significantly. For example, Zn content in the solder bulk shows a rapid decrease, from 9wt.% in original solder alloy to 1.8wt.% at the least in the solder joint with 20μm SOH. Correspondingly, the microstructure of solder bulk changes from eutectic microstructure to hypoeutectic microstructure with the reduced SOH.

(2) Due to the change in composition and microstructure of solder bulk with the reduced SOH, the fracture mode and ultimate tensile strength of Cu/Sn9Zn/Cu solder joint are transformed a lot. With the reduced SOH, there is a transition of fracture mode from ductile mode to brittle mode. And the ultimate tensile strength of solder joint is reduced.

Acknowledgments

The authors acknowledge the financial support from National Natural Science Foundation of China (No. 60776033), Natural Science Foundation of Hubei Province (No. 2006ABA091), National High Technology Research and Development Program of China (863 Program) (No. 2006AA04A110).

References

1. R.Plieninger, M.Dittes, K.Pressel, "Modern IC Packaging Trends and their Reliability Implications," *Microelectronics Reliability*, Vol.46, (2006), pp.1868-1873.

2. Fengshun Wu, Mingmin He, Yiping Wu, *et al*, "Effect of Interfacial IMCs Proportion on the Reliability of Miniature Lead-Free Solder Joint," *7th International Conference on Electronics Packaging Technology*, 2006.

3. Ahmed Sharif, Y.C.Chan, Rashed Adnan Islam, "Effect of volume in interfacial reaction between eutectic Sn-Pb solder and Cu metallization in microelectronic packaging," *Materials Science and Engineering. B*, Vol. 106, (2004), pp.120-125.

4. L.Quan, D.R.Frear, D.Grivas, J.W.Morris Jr., *J.Electron. Mater*, Vol. 16, (1987), pp.203.

5. K.H.Prakash, T.Sritharan, "Tensile fracture of tin-lead solder joints in copper," *Materials Science and Engineering. A*, Vol. 379, (2004), pp. 277-285.

6. B.S.Chiou, J.H.Chang, J.G.Duh, *IEEE Trans. CPMT-B* Vol. 18, (1995), pp.537.

7. K.Zeng, K.N.Tu, "Six Cases of Reliability Study of Pb-free Solder Joints in Electronic Packaging Technology," *Materials Science and Engineering. R*, Vol. 38, (2002), pp. 55-105.

8. M.Date, T.Shoji, M.Fujiyoshi, *et al,* "Impact Reliability of Solder Joints," *Electronic Components and Technology Conf*, 2004, pp. 668-674.

9. B.I.Noh, J.M.Koo, J.W.Kim, *et al,* "Effects of number of reflows on the mechanical and electrical properties of BGA package," *Intermetallics*, Vol. 14, (2006), pp. 1375-1378.

10. K.H.Prakash, T.Sritharan, "Tensile fracture of tin-lead solder joints in copper," *Materials Science and Engineering. A*, Vol. 379, (2004), pp. 277-285.

11. Fei-Yi Hung, Truan-Sheng Lui, Li-Hui Chen *et al,* "Vabration fracture behavior of Sn-9Zn-xCu lead-free solders," *J Mater Sci,* Vol.42, (2007), pp. 3865-3873.

12. Solder Date Sheet, Welco Castings, 2 Hillyard Street, Hamilton, Ontario, Canada.

Failure Mode Analysis of Lead-free Solder Joints under Differential Reflow Profiles by High Speed Impact Testing

C. Y. Lin[1], Y. R. Chen[1], and G.S. Shen[1]
D. S. LIU[2], C. Y. KUO[2], and C. L. HSU[2]
1. ChipMOS TECHNOLOGIES LTD, Tainan, Taiwan, China
2. National Chung Cheng University, Department of Mechanical Engineering, Chia-Yi, Taiwan, China

Abstract

The aim of this research is to investigate the mechanical behavior of lead-free solder for high speed impact. A high speed impact test was set up to measure the solder joint reliability. Differential impact speed and room temperature aging effect has been studied with Ni/Au substrate. Furthermore, two different solder alloys (96.5Sn-3Ag-0.5Cu, 98.5Sn-1Ag-0.5C) and three different reflow profiles are considered. This paper focuses on failure mode analysis and investigates the failure characteristics of lead-free solder joints, 96.5Sn-3Ag-0.5Cu and 98.5Sn-1Ag- 0.5C, which are aging at room temperature, respectively, then those solder are impacted at shear rates of 0.3m/s and 1.0m/s.

Four types of failure mode are found in this high speed impact testing result. Mode M1 is the fracture around the interface but not remain the solder on pad. Mode M2 is fracture around the interface and remained the solder on pad. Mode M3 is fracture across the solder ball. Mode M4 is fracture on the substrate with lift the pad. The aging time could increase the interfacial strength, therefore the percentage of M3 and M4 mode failures increases in Ni/Au substrate. According the results, we find that in reflow profile A and reflow profile C, the failure percentage of Mode M2 is increasing; in reflow profile B, the failure percentage of Mode M3 and Mode M4 are large than Mode M1 and Mode M2.The failure mode M2 is the majority in solder alloy 96.5Sn-3Ag- 0.5Cu,and the failure mode M2 and M3 are the majority in solder alloy 98.5Sn-1Ag-0.5C.

Introduction

In recent years, as the electronic packaging towards the rapid development of the lead free process, the mechanical behavior of lead-free solder joints is emphasized urgently. The ball shear test is a widely adopted test method to assess the mechanical strength of solder ball bonds in ball grid array (BGA) packaging. The new standard JESD22-B117 was established for BGA ball shear test in July-2000. [1] Huang et al. studied the solder ball shear test and analysis the computational model under differential shear height and shear speed. They found the shear force-displacement curve trend increase with faster shear height and lower shear height. The results of two dimension finite element model were agreement with testing data. Then they suggest the shear height smaller 25% of the solder ball height and the shear speed slower then 200 μm. [2] Canumalla et al. developed the test method to investigate the package to board interconnection shear strength on vary surface finish. The method is effectively to examine the interconnection quality. [3] Chia et al. applied ball shear test to investigated shear rate effect. They reported the shear strength, total shear work and

shear work up to maximum load are increase with shear rate. However the most frequent failure of movable electronic products is an accidental drop to the ground. Therefore the ability to suffer impact loading on electronic package is current research key point. [4] Data et al. investigated that the effect of aging on impact reliability of SnZn solder by Charpy impact test. They reported the Au/Ni surface finish the failure mode is ductile. [5] Ou et al. presented the impact toughness was increasing trend with the thermal aging by Charpy impact test. [6] Wong et al. studied that the impact characterizations for differential solder, pad finish, solder mask design and thermal aging. They reported the solder mask design is stronger then non-solder mask design. The impact strength and impact energy are important to prevent failure of board level interconnection. And suggest that the high speed shear test can estimate the component level quality. [7] Newman presented the shear and pull strength are increase with test speed. The interfacial fracture rate of pull and shear test are generally increase with test speed. In high speed, the SnAgCu solder balls have higher interfacial fracture rate and failure strength then SnPb/SnPbAg solder balls. [8] Yeh and Lai developed ball impact test and numerical method to correlate the drop test and ball impact test. They also investigate transient structure responses and failure mode of solder joint. [9] Liu et al. discussed the effect of room temperature aging and thermal aging for different failure modes under high speed shear impact testing. In above studies, many studies have been done on speed effect, but only a few studies to divide failure mode on each test. No consider the effect of manufacture process such as reflow profile. Although many test results have been reported, but test condition and sample preparation remains some uncertainties (ex: sample number, manufacture process...etc.). The aim of this research is to investigate the mechanical behavior of lead-free solder for high speed impact under different reflow profiles. A high speed impact test was set up and conducted using a micro-impact system to obtain the force-displacement curves. Differential impact speed, room temperature aging effect has been studied.

Experimental Procedure

Experiments were performed on an Instron micro-impact system. The main structural members of the tester comprise a striking head assembly, velocity control module, a specimen fixture, a xyz stage. The instrument testing velocity capable is 0.2 m/s up to 1.0 m/s. Load sensor is a 45 N load cell that linearity was 0.4% FS and software for continuous data acquisition. Displacement measurements were performed using a linear variable differential transformer (LVDT) at a displacement resolution of 0.1 μm. A vision alignment system was checking the parallel between striking head with specimen by the dual charge coupled device (CCD) cameras with microscope lenses, one providing a top view of specimen and the other a side view. The testing equipment was placed onto an optical table to isolate the equipment from environmental noise. The experimental setup is shown in Figure 1.

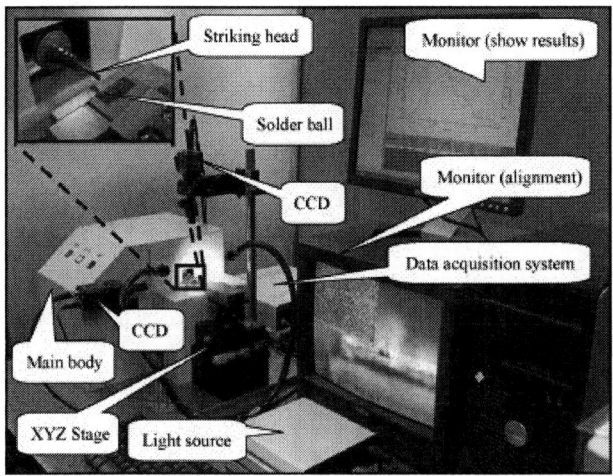

FIGURE 1. SCHEMATIC OF EXPERIMENTAL SETUP FOR MICRO-IMPACT TEST

The solder alloys X and solder alloys Y sample was consists of the bismaleimide triazine (BT) laminate substrate with the electrolytic nickel/gold (NiAu) pad finish. The reflow process of the solder ball on bond pad followed reflow profile. The schematic of three kinds of reflow profiles are shown in Figure 2-4.

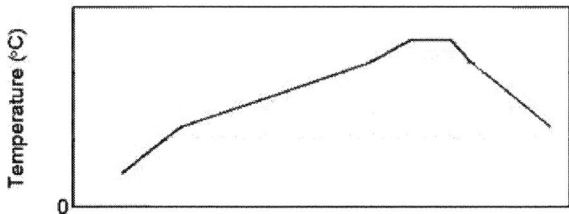

FIGURE 2. SCHEMATIC OF A REFLOW CURVE

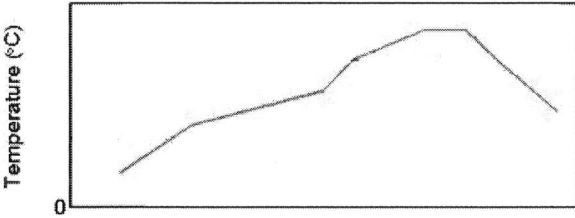

FIGURE 3. SCHEMATIC OF B REFLOW CURVE

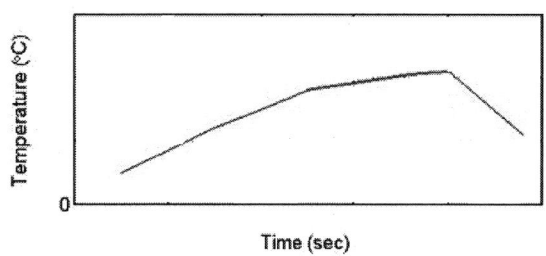

FI GURE 4. SCHEMATIC OF C REFLOW CURVE

The size of sample was 12.5 mm × 5 mm, each sample has 11 balls. Before impact test, the sample has adhered on 18 mm × 10 mm steel pad was shown in Figure 5 then mounted in specimen fixture. The shear height was defined 50 μm by xyz stage and vision alignment system to complete. A series of experiments is conducted to study the effects of impact speed at room temperature with aging range to 168 hours. The impact tests were conducted at testing speed included 0.3 and 1.0 m/s, the designs of experiments are shown in Table 1. Total tested numbers for each testing condition is 66 balls.

FIGURE 5. THE SAMPLE ADHESIVE ON STEEL PAD

TABLE 1. DESIGN OF EXPERIMENT

No.	Solder Alloys	Pad Finish	Reflow Profile	Aging Time (hrs)
1			A	
2	X		B	
3		Ni/Au	C	0,168
4	Y		A	
5			B	

Experimental Results and Discussion

Four types of failure mode are generally separated to examine impact as shown in Figure 6. Mode 1 (M1) was the fracture around the interface but not remain the solder on pad. Mode 2 (M2) was fracture around the interface and remained the solder on pad. Mode 3 (M3) was fracture across the solder ball. Mode 4 (M4) was fracture on the substrate with lift the pad. Above failure modes were captured image by optical microscope (OM) and scanning electron microscope (SEM). The fractographs of specimen and shear direction was shown in Figure 7.

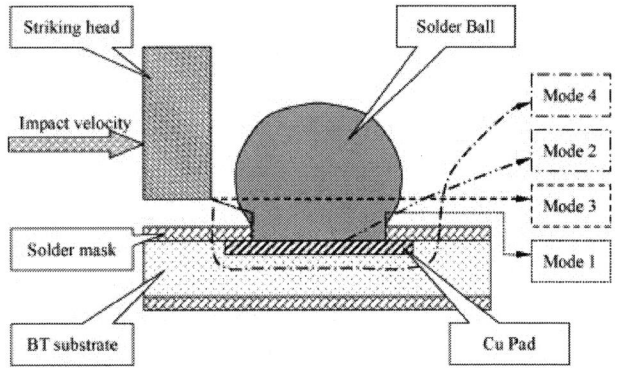

FIGURE 6. SCHEMATIC ILLUSTRATION OF FAILURE MODES

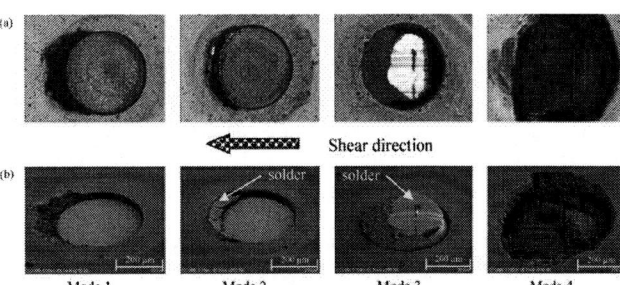

FIGURE 7. (A) OM AND (B) SEM IMAGES FOR DIFFERENTIAL FAILURE MODES

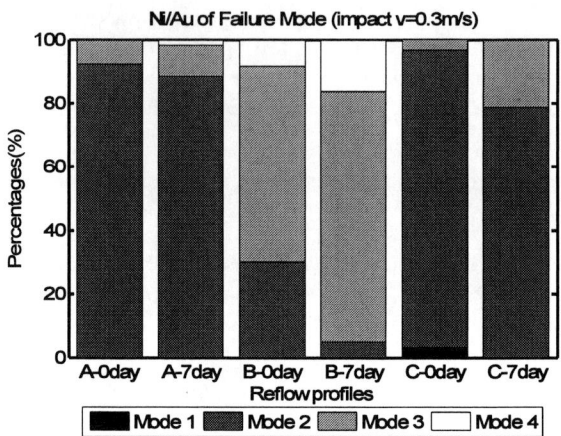

FIGURE 8. THE PERCENTAGE OF THE FOUR KINDS OF FAILURE MODES FOR R.T UNDER IMPACT VELOCITY 0.3 M/S WITH REFLOW PROFILE A, B, C IN SOLDER ALLOYS X

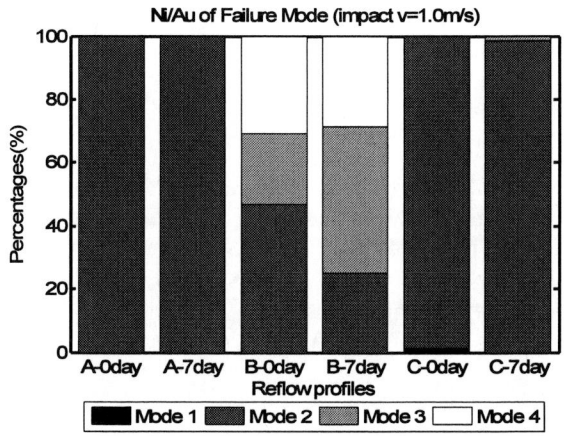

FIGURE 9. THE PERCENTAGE OF THE FOUR KINDS OF FAILURE MODES FOR R.T UNDER IMPACT VELOCITY 1.0 M/S WITH REFLOW PROFILE A, B, C IN Solder Alloys X

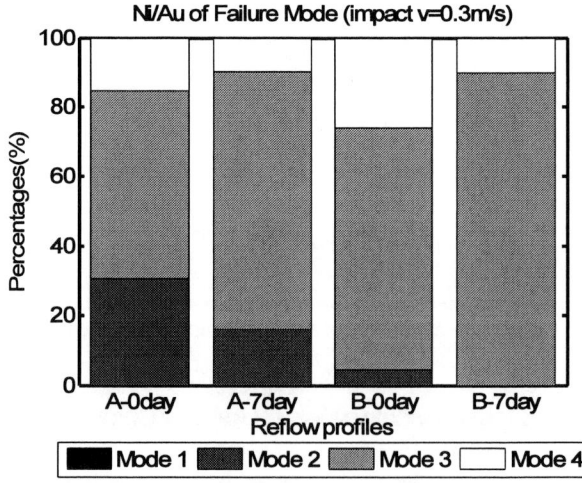

FIGURE 10. THE PERCENTAGE OF THE FOUR KINDS OF FAILURE MODES FOR R.T UNDER IMPACT VELOCITY 0.3 M/S WITH REFLOW PROFILE B, C IN Solder Alloys Y

The percentage of the four kinds of the failure mode at room temperature for differential impact speed from reflowed to 168 hours room temperature aging with reflow profile A,B,C and solder alloys X, was shown in Figure 8 and Figure 9. The percentage of the four kinds of the failure mode at room temperature for differential impact speed from reflowed to 168 hours room temperature aging with reflow profile A, B and solder alloys Y, was shown in Figure 10 and Figure 11. From Figure 8 and Figure 9, the M2 is the majority failure mode in reflow profile A and C with solder alloys X without aging at room temperature. But after aging at room temperature for 168hrs, the M2 is clearly decreasing in reflow profiles C under impact velocity of 0.3 m/s. In reflow profile B, the failure percentage of M3 and M4 are large than M1 and M2 with solder alloys X. After aging at room temperature for 168hrs, the M2 is clearly decreasing in reflow profile B and M3 and M4 are clearly increasing under impact velocity of 0.3 m/s with solder alloys X. For the solder alloys X, the M3 is increasing and M2 is decreasing with room temperature aging under impact velocity 1.0 m/s.

In Figure 10 and Figure 11,under impact velocity 0.3 m/s and 1.0 m/s, in reflow profile B, the failure percentage of M3 and M4 are large than M1 and M2 with solder alloys Y. After aging for 168hrs (7 days) at room temperature, the M2 is clearly decreasing and the M3 is increasing in reflow profiles A, B under impact velocity of 0.3 m/s and the reflow profile A under impact velocity of 1.0 m/s. The effect of room temperature aging is not significantly in reflow profile B under impact velocity 1.0 m/s.

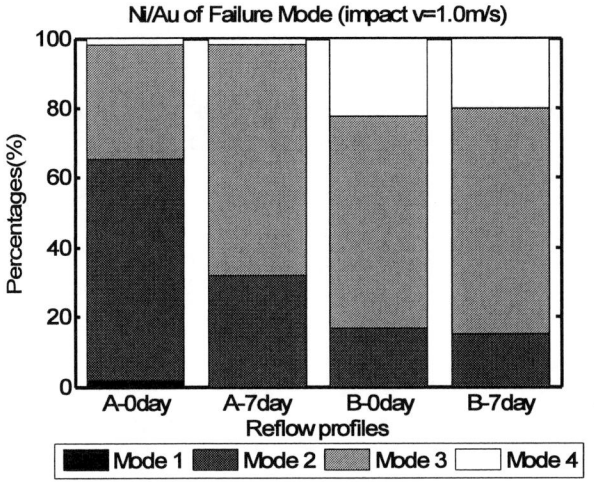

FIGURE 11. THE PERCENTAGE OF THE FOUR KINDS OF FAILURE MODES FOR R.T UNDER IMPACT VELOCITY 1.0 M/S WITH REFLOW PROFILE B, C IN Solder Alloys Y

According the results, we find that in reflow profile A and reflow profile C, the failure percentage of M2 is dominating; in reflow profile B, the failure percentage of M3 and M4 are large than M1 and M2.The failure mode M2 is the majority in solder alloys X, and the failure mode M2 and M3 are the majority in solder alloys Y. The failure modes are summarized in the Table 2.

TABLE 2. FAILURE MODE SUMMARIZE IN THE EXPERIMENT

Solder Alloys	Pad Finish	Reflow Profile	Impact Velocity (m/s)			
			0.3		1.0	
			0 H	168H	0 H	168H
X	Ni/Au	A	M2	M2↘	M2	—
		B	M3	M3↗	M2	M3↗
		C	M2	M2↘	M2	—
Y		A	M3	M3↗	M2	M3↗
		B	M3	M3↗	M3	—

In this investigation, the high impact testing were evaluated the peak load and energy absorption. As a result, the energy absorption capacity of the specimen is defined as the area under the force-displacement curve before the point of peak load. The peak load and energy to peak load of the four kinds of failure modes without room temperature aging results were shown in Figure 12 and Figure 13. The results of room temperature aging 168 hours were shown in Figure 14 and Figure 15.

From failure percentage we can find the failure mode M2 and M3 dominate for solder alloys X. Hence in this section we only compare the peak load and energy absorption results of the failure mode M2 and M3 for solder alloys X. The solder alloys Y were compared the M2, M3 and M4.

For the failure mode M2 with solder alloys X, the peak load and energy to peak both results of the reflow profile B was large then the reflow profile A and C. For mode M3 with solder alloys X, the peak load and energy to peak both results of the reflow profiles A, B was large then reflow profile C.

For the failure mode M2 and M3 with solder alloys Y, the peak load and energy to peak both results is not significantly in different reflow profiles. For the failure mode M4 with solder alloys Y, the peak load and energy to peak both results of the reflow profile A was lightly large then the reflow profile B.

In general the peak load and energy to peak both results of solder alloys Y was large then solder alloys X under the reflow profile A.

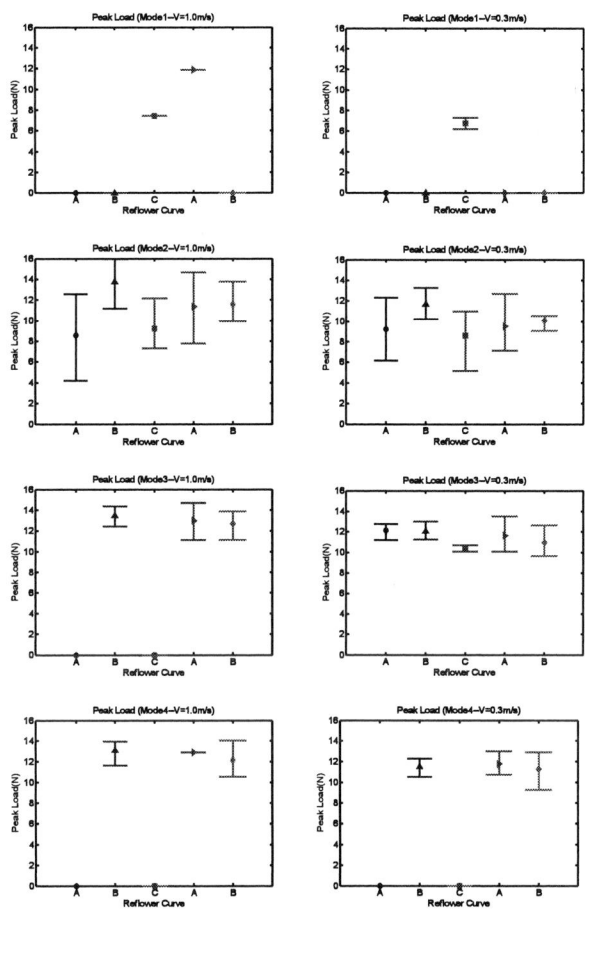

● Solder Alloys X - A ▲ Solder Alloys X - B ■ Solder Alloys X - C ▶ Solder Alloys Y - A ◆ Solder Alloys Y - B

FIGURE 12. THE PEAK LOAD OF THE FOUR KINDS OF FAILURE MODES AS AGING TIME 0 HOURS

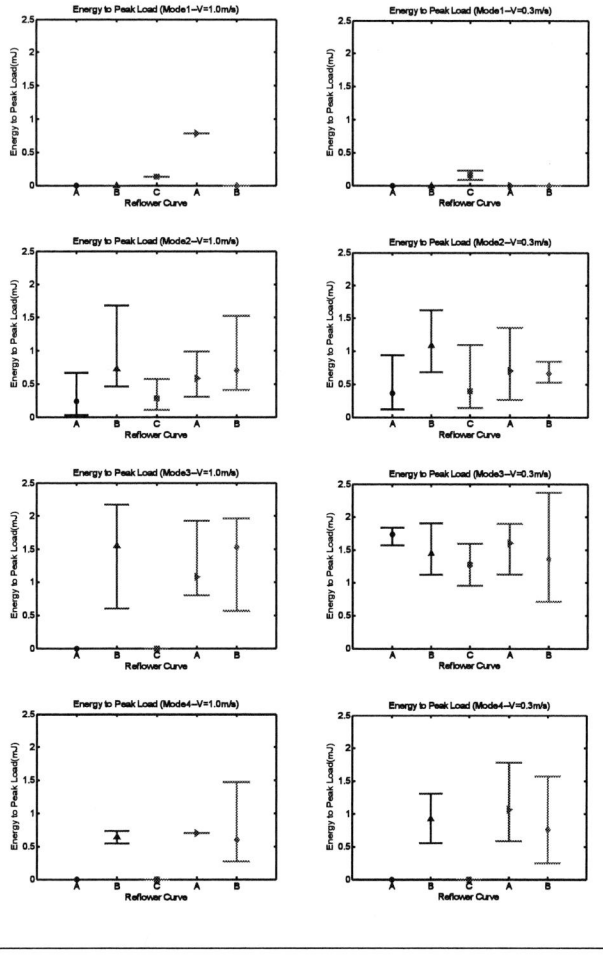

• Solder Alloys X - A   ▲ Solder Alloys X - B   ☷ Solder Alloys X - C   ▷ Solder Alloys Y - A   ◈ Solder Alloys Y - B

FIGURE 13. THE ENERGY TO PEAK LOAD OF THE FOUR KINDS OF FAILURE MODES AS AGING TIME 0 HOURS

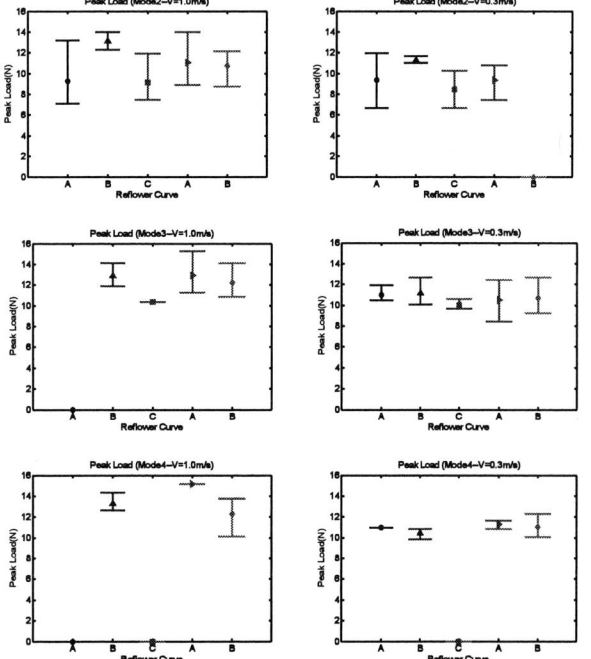

• Solder Alloys X - A   ▲ Solder Alloys X - B   ☷ Solder Alloys X - C   ▷ Solder Alloys Y - A   ◈ Solder Alloys Y - B

FIGURE 14. THE PEAK LOAD OF THE FOUR KINDS OF FAILURE MODES AS AGING TIME 168 HOURS

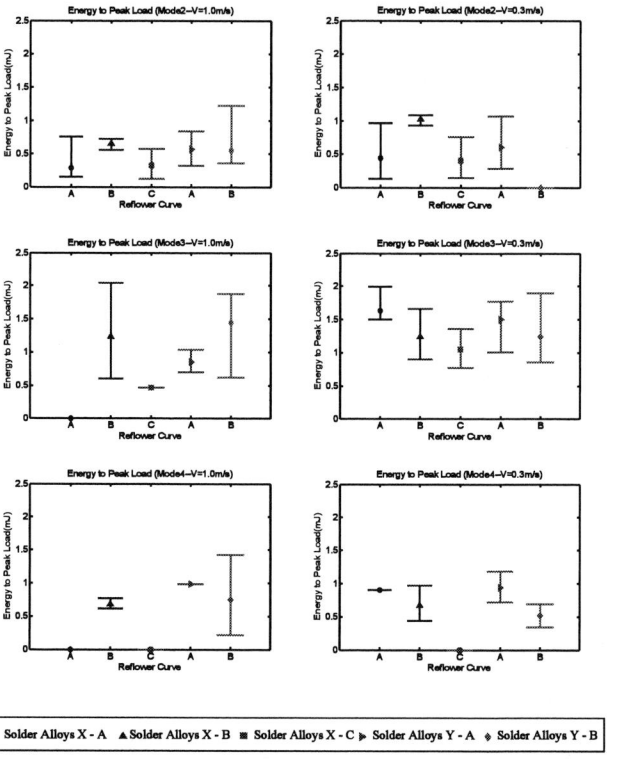

• Solder Alloys X - A   ▲ Solder Alloys X - B   ☷ Solder Alloys X - C   ▷ Solder Alloys Y - A   ◈ Solder Alloys Y - B

FIGURE 15. THE ENERGY TO PEAK LOAD OF THE FOUR KINDS OF FAILURE MODES AS AGING TIME 168 HOURS

Conclusions

Shear strength and failure modes of solder alloys X and solder alloys Y of lead-free solder joints under high speed micro impact test are studied with emphasis on the effects of the reflow profiles. The conclusions base on present study is drawn as follows:

High speed shear could induce different types of failure mode in same impact speed, therefore, enlarge testing sample number to obtain failure mode percentage in each testing condition is necessary.

In reflow profile A and reflow profile C, the failure percentage of Mode M2 is dominating.

In reflow profile B, the failure percentage of Mode M3 and Mode M4 are large than Mode M1 and Mode M2.

The failure mode M2 is the majority in solder alloys X, and the failure mode M2 and M3 are the majority in solder alloys Y.

For the same reflow profile A, the solder alloys Y is better then solder alloys X bonded on Ni/Au pad finish.

In this study, the solder alloys X bonding on Ni/Au pad finish with reflow profile B is better then profile A and profile C.

The influence of reflow profile on the impact characteristics of the solder is significantly.

References

[1] JEDEC, JESD22-B117 (2000), http://www.jedec.org.

[2] X. Huang, S.-W.R. Lee, C.C. Yuan, and S. Hui, Proc. 51st Electronic Components and Technology Conference (Piscataway, NJ: IEEE, 2001), pp. 1065–1071.

[3] S. Canumalla, H.-D. Yang, P. Viswanadham, and T.O. Reinikainen, IEEE Trans. Comp. Packag. Technol. 27, 182 (2004).

[4] J.Y.H. Chia, B. Cotterell, and T.C. Chai, Mater. Sci. Eng. A 417, 259 (2006).

[5] M. Date, T. Shoji, M. Fujiyoshi, K. Sato, and K.N. Tu, Scripta Mater. 51, 641 (2004).

[6] S. Ou, Y. Xu, K.N. Tu, M.O. Alam, and Y.C. Chan, in Ref. 1, pp. 467–471.

[7] E.H. Wong, R. Rajoo, Y.W. Mai, S.K.W. Seah, K.T. Tsai, and L.M. Yap, Proc. 55th Electronic Components and Technology Conference (Piscataway, NJ: IEEE, 2005), pp. 1202–1209.

[8] K. Newman, in Ref. 1, pp. 1194–1201.

[9] C.-L. Yeh and Y.-S. Lai, J. Electron. Mater. 35, 1892, (2006)

Influence of Underfill Methods on the Solder Joint Fatigue of Wafer Level Packaging

Charles REGARD[1,2,3], Christian GAUTIER[1,2], Hélène FREMONT[3], Patrick POIRIER[1,2]

[1] Quality and Analytical Service, NXP Semiconductors, Caen 14000, France
[2] LaMIPS, Université de Caen, Caen 14000, France
[3] IMS, ENSEIRB, UMR 5218 Université Bordeaux 1, Talence 33405 cedex, France

Phone: +33 2 31 45 64 20
Fax: +33 2 31 45 20 50
Charles.regard@nxp.com, etc, as desired

Abstract

To increase miniaturization, CSWLP (chip size wafer level packaging) has been developed. However, the difficulty to get good solder joint reliability leads to manufacture only small CSWLP modules. Different underfill methods are evaluated here, by measurements and simulations: results prove that underfill is necessary, but a bad choice can also decrease the reliability. An original method called "re-enforcement" improves the life time.

Key words: Reliability, Underfill, Thermal shock, Re-enforcement, CSWLP.

Introduction

Nowadays the market trends to dramatically scale down the size of customer application [1]. As we can see in figure 1, to decrease size, cost and time to market Chip Size Wafer Level Package have been developed. The weak point of this kind of package is the solder joint reliability. So, the Coefficient of Thermal Expansion (CTE) mismatch between the die in silicon (CTE around 3 ppm/K) and the PCB (CTE around 20 ppm/K) is very high. This difficulty leads currently to manufacture only small CSWLP modules.

In order to understand the limits of such a technology and overpass them we have done an experimental work with daisy chain test chips. This study takes place in the Amelie European consortium. In a previous article, we have shown that CSWLP meets BGA like failure mechanism at solder joint level even if the stresses are very much higher in case of CSWLP. We have also highlighted the impact of ionic contamination during industrial assembly process [2]. The second step of our study, subject of this paper, is about the use of underfill to improve the life time of solder joint.

Our study is based both on experiments and on simulations. Underfill is a well-known method to improve reliability of BGA package [3]. The question to be solved is "is it the case for WLCSP?" Moreover, the underfilling process step can be included in the BGA manufacturing; it is not the same with CSWLP. Actually, CSWLP are manufactured by the component's provider and assembled by the customer. Thus, exact conditions and parameters to apply underfill should be defined in order to be able to warrant the reliability of CSWLP.

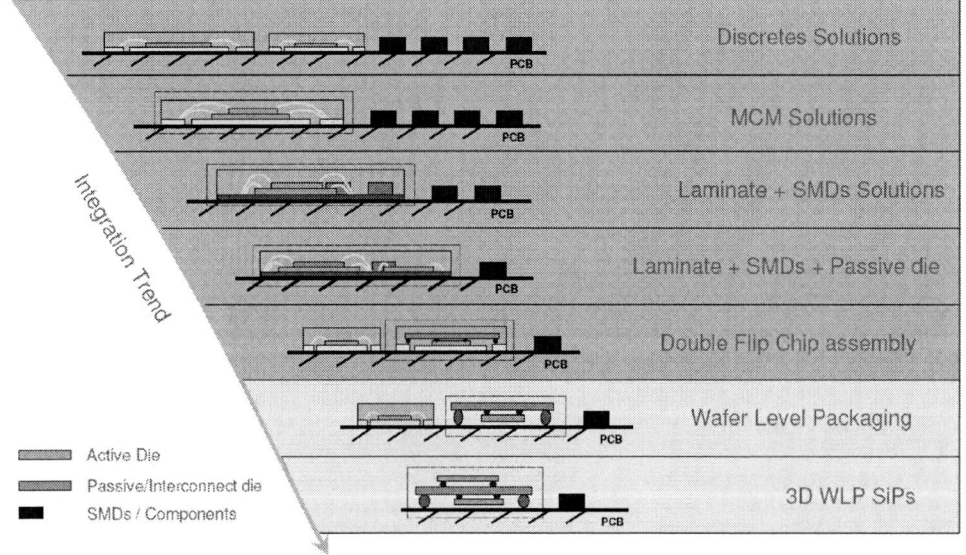

Figure 1: Overview to the integration trend in semiconductor industry

An alternative method of underfill called "re-enforcement" has been proposed in Amelie project. This method is an intermediate method between the non-underfill method and the classical underfill method. It consists in an application of underfill in the last step of CSWLP manufacturing. The underfill is applied around the ball and covers the die side hemisphere This method has the following advantages:

- Integrated to the manufacturing process
- Reworkable
- Fast and cheap.

The objective of this work is to evaluate the efficiency of different methods of underfilling for improvement of solder joint reliability. A design of experiment (DoE), presented in the first part of this document, has been defined. In a second part we present the statistical results, physical analyses and discussions.

Test vehicle and DOE description

Figure 2 shows a schematic cross-section of a WLCSP. Three different underfilling methods are used: no-underfill, underfill and "re-inforcement". Two sizes of the silicon die are considered: 13mm² (3.6mm x 3.6mm) and 31mm² (5.6mm x 5.6mm). All DOE parameters are reported in table 1.

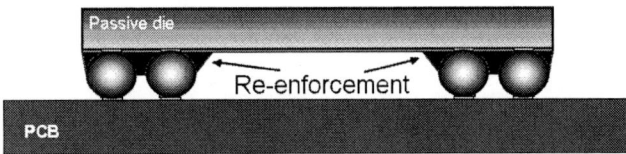

Figure 2: Chip Scale Wafer Level Package cross-section view, here with reinforcement

For each DoE parameter 32 devices were assembled on a standard printed circuit board (PCB), with two rows of 300µm solder balls, according to a 500µm pitch. The PCB is a classical 8-layer 1mm thick FR4 with a high Tg and a medium CTE. The assembly is performed in a real industrial flow. The circuit on the die is daisy chained to follow the electrical continuity.

To avoid ionic contamination before the underfilling step the components were washed after reflow.

Table 1: Table of parameters for Design of Experiment

Balls array	Size	Underfill	Code
7x7	13mm² (3.6mmx3.6mm)	No underfill	7N
		Die side underfill	7R
		Classical underfill	7U
11x11	31mm² (5.6mmx5.6mm)	No underfill	11N
		Die side underfill	11R
		Classical underfill	11U

The stress test performed is a -55°C to +125°C thermal shock, with a ramp of about 40°C/min and a dwell-time of 15 min. We performed up to 2600 cycles and a measurement of resistance was done every 100 cycles for each daisy chain. Only 3 components didn't fail after 2600 cycles to the test. We took out one board at 1000 cycles and one board at 2400 cycles to perform analysis.

Simulations

A three-dimensional, transient FE model for thermo-mechanical analysis of Wafer Level SiP packages with various design parameters has been developed in the University of Greenwich. The influence of the critical parameters and their combinations has been studied and the most influential parameters have been identified. Complete results have been published previously [4], [5], [6] and have been realized before the experimentation in a predictive idea.

These results can be summarized as follows:

1. The highest damage location is found in the solder ball closest the die edge. (figure 3 and 4).

Figure 3: The most sensitive ball [10]

2. The most influential factor among all design parameters of the sbSiP (Silicon based SiP) [7][8] is the size of the package. The smaller size package guarantees higher reliability.

3. The presence of the underfill can improve the reliability of the SiP. This is the second most influential factor.

4. The thickness of the passive die has little effect on reliability of the package.

5. It is not possible to determine if the impact of re-enforcement is positive or negative, but we proved that the stresses are moved from the top to the bottom of the ball (Fig.4).

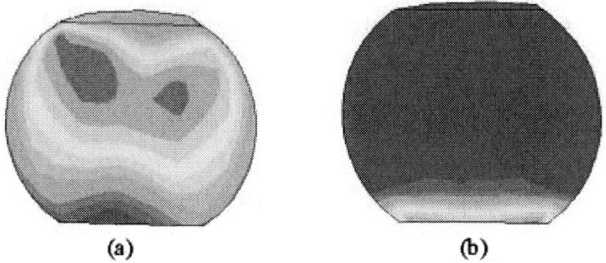

Figure 4: (a) without re-enforcement, (b) with re-enforcement. The stresses are moved [10]

These results showed that we should expect the same failure mechanisms in the WLP as in the BGA. However, the thermo-mechanical simulations are based on known formula on the BGA's solder balls behavior and the point of interest in simulation is the stress locations and intensities. Then it is not surprising to get the same behavior from a CSWLP as from a BGA package. Only change the CTE, Poisson's coefficient, Young's modulus and glass temperature that modify the stresses values but not the effect.

Results and discussions

- ### Electrical results

Influence of size

It can be seen in figure 5 that for the biggest module (31 mm²), 19% of failures occur before 200 cycles and 63% are failed at 400 cycles, whereas for the smallest one (13 mm²), only 3% failed after 400 cycles. We can see here the size limit of CSWLP without underfill.

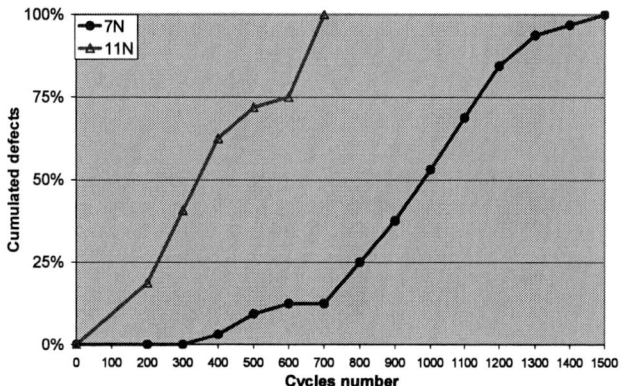

Figure 5: Measurement of size effect on non-underfill samples

Influence of underfill

On the figure 6, we show the impact on the underfill to reduce the size impact on the failure rate. We can see here that the curves are much closer than the previous one (figure 5). After 1000 cycles about 55% of small parts are failed and about 60% of big ones.

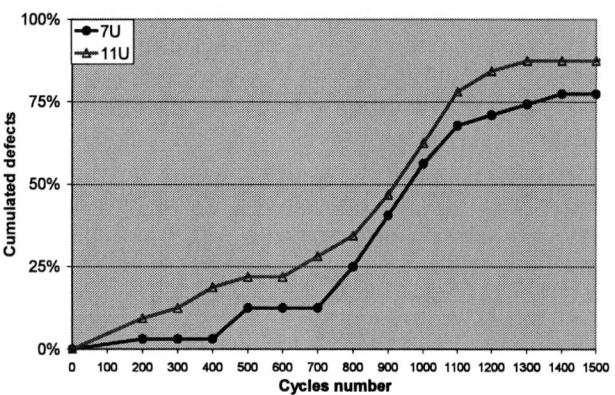

Figure 6: Measurement of underfill effect on failures

Influence of re-enforcement

Another important result in this study is the improvement due to use of "re-enforcement". We can see on figure 7 that the first defect for samples with "re-enforcement" appeared after 900 cycles whereas without underfill, the devices failed before 400 cycles.

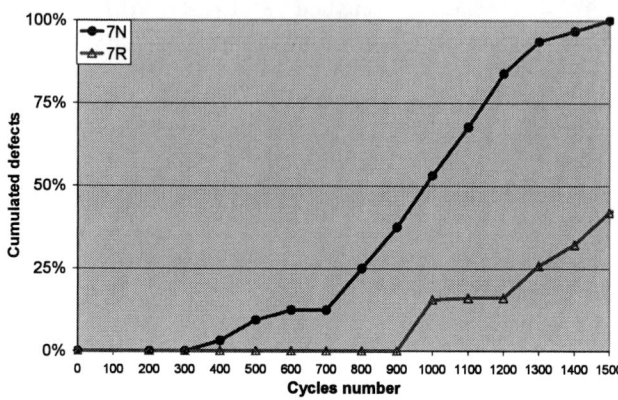

Figure 7: Measurement of "re-enforcement" effect on failures

- ### Statistical analyses

We used a lognormal distribution to characterize the lifetime and the failure mechanism of the components.

Influence of size

As it can be seen on the table 2, whatever the size of the device, the lognormal shape parameters σ are identical : $\sigma=0.34$ for $7*7$ module and $\sigma=0.38$ for $11*11$ module. Consequently the lognormal distribution are parallel (figure 8). We deduce from this result that the failure mechanism on big and small parts is the same.

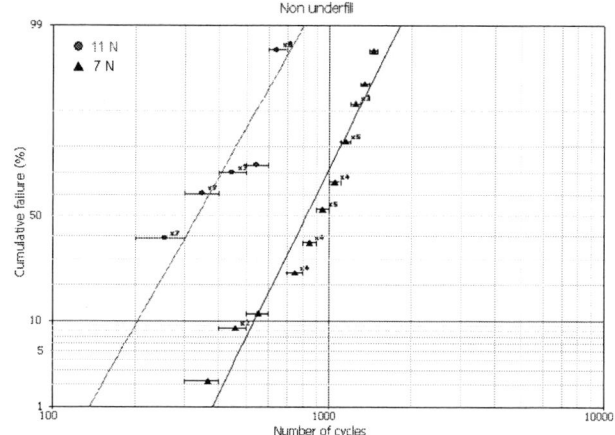

Figure 8: Lognormal plot of 7x7 and 11x11 carrier without underfill

The table 2 shows that by increasing the size of the module from $7*7$ module to $11*11$ module, the median lifetime $N_{50\%}$ (number of cycles for 50% cumulative failure) is divided by a factor 2.5. This result confirms that from the reliability point of view the size is really critical.

Table 2: Comparison of median lifetime ,confidence bounds and lognormal shape parameter σ between 11x11 and 7x7 without underfill

	$N_{50\%}$ (cycles)	$N_{50\%}$ Inf @ 90% CL	$N_{50\%}$ Sup @ 90% CL	σ
7N	839	721	936	0,34
11N	333	290	374	0,38

Influence of underfill

The lognormal distribution of figure 9 and the table 3 show that the median lifetime ratio between 11x11 sample and 7x7 ones is only 1.3. This confirms that by using underfill it is possible to reduce the effect of size and to increase the lifetime of big component.

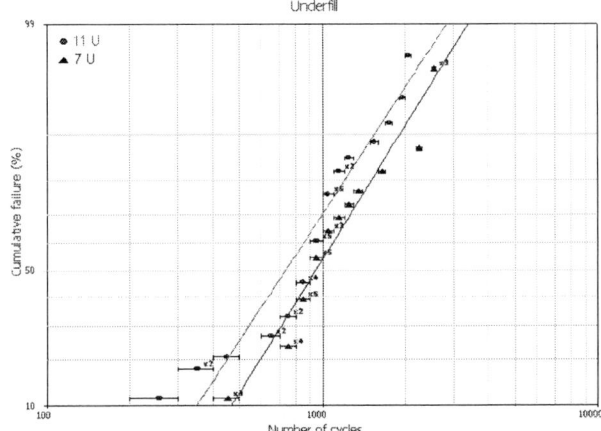

Figure 9: Lognormal plot of 7x7 and 11x11 carrier with underfill

Table 3: Comparison of median lifetime ,confidence bounds and lognormal shape parameter σ between 11x11 and 7x7 with underfill

	N50% (cycles)	N50% Inf @ 90% CL	N50% Sup @ 90% CL	σ
7U	946	789	1113	0,55
11U	736	612	871	0,58

Influence of re-enforcement

Compared to table3, the table 4 shows that whatever the module size, the re-enforcement allows to double the lifetime of WLCSP.

Table 4: Comparison of median lifetime ,confidence bounds and lognormal shape parameter σ between 11x11 and 7x7 with re-enforcement

	N50% (cycles)	N50% Inf @ 90% CL	N50% Sup @ 90% CL	σ
7R	1638	1387	1886	0,47
11R	693	594	775	0,33

The lognormal plot of figure 10 shows that the slope of distribution for parts with re-enforcement (11R) remains the same compared to nude parts (11N) . This confirms that the use of re-enforcement hasn't influence on the failure mechanism. Regarding underfill parts, the slope of distribution is changed. This variation shows that the underfill probably generate another mechanism of failure.

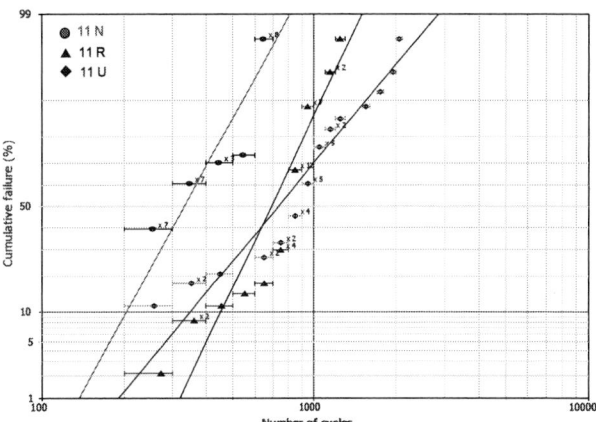

Figure 10: Lognormal plot of 7x7 and 11x11 carrier

- Failure location

As expected after simulation, cracks in solder balls were found along the die edge, in case of nude or underfill parts and at the PCB side in case of re-enforcement parts.

Influence of size:

The importance of the module's size on the reliability has already been shown in different BGA and WLP studies [9, 10]. By simulation it seems also obvious that the size is a critical parameter from the reliability point of view [11].

The impact of the die size is due to CTE mismatch between die and PCB which induces a warpage at the board level. The high stiffness of the silicon die leads to stress at the die interface of solder ball. The warpage is amplified when the die size is increased. As a consequence the resultant stress is higher for a larger die.

As the stress is mainly located at the ball's die interface, we can observe the failure at this interface. The figures 10 and 11 illustrate the cracks in a small component (7x7) and in a big one (11x11).

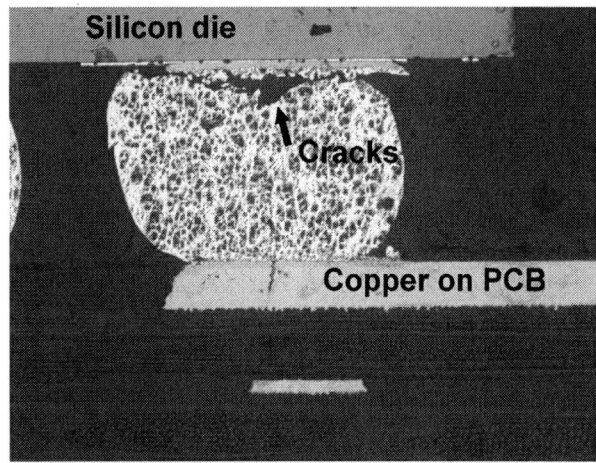

Figure 10: Cracks on corner ball of non-underfill 7x7 carrier after 1300 cycles

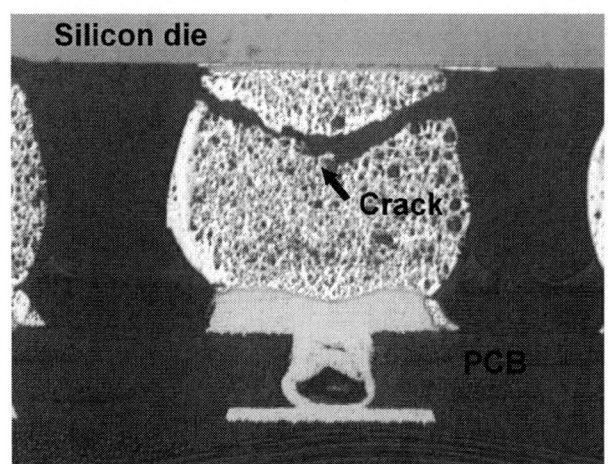

Figure 11: Crack on corner ball of a non-underfill 11x11 carrier after 300 cycles

We can observe that the cracks have similar location and shape. This confirms that the failure mechanism is not changed by increasing the module's size.

Influence of underfill

The underfill acts like an intermediate layer between PCB and die. This layer increases the total stiffness of the system. Then the warpage is attenuated and the stress is more or less diffused in the ball instead of interfaces.

Simulations had shown the influence of the material use for underfilling. Depending on the CTE and the Young's modulus value, the stress in solder joint is more or less important [6]. Strusevitch et al. have clearly shown that a good choice of underfill reduces the impact of size and thickness. On the other hand, a bad choice can reduce the life time.

Even if the underfill used in the DoE is not completely optimized, we observed that the size impact on lifetime is reduced in case of underfilled samples (Fig 9). These experimental results validate the simulation ones. Hence, simulation will help to determine the best underfill to choose, for a given die size and technology.

The cross-sections analysis show that the stress location is not modified by underfill.

The pictures 12 show the cracks on two different components.

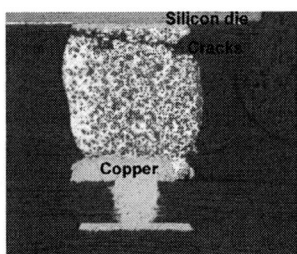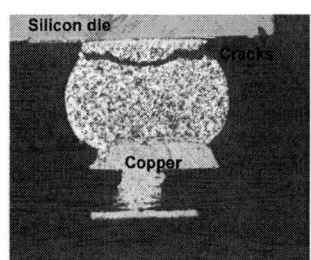

Figure 12: Cracks on underfilled samples in the corner ball after 1100 and 1000 cycles

Influence of "re-enforcement"

The "re-enforcement" decreases the fatigue at the ball/die interface. So this interface becomes stronger and all stresses are removed at the other interface: PCB/ball

interface. Simulations have shown that the total stress is hardly increased [5]. However, from literature it is known that the ball interface at the PCB side is stronger than at the die side. So, even if the stress is higher, it is focused on the more robust interface, and hence the assembly becomes stronger.

Cross-section on the figure 13 confirms that the failure occurred at the ball's bottom side. The top interface is almost not changed. We can see here the effect of the re-enforcement and the interest in future applications to improve lifetime without adding any process step at the customer level.

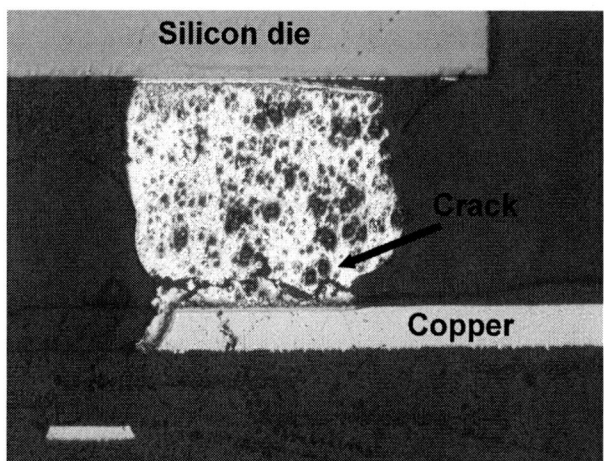

Figure 13: Failure after 1000 cycles at the corner ball of a re-enforced 11x11 carrier

This underfill method doesn't limit the impact of the size because the CTE induced warpage is not decreased in this case. The stresses applied at the die/ball interface depend once again on the re-enforcement's CTE and Young's modulus. Hence, simulation can help to define the most suitable material for re-enforcement.

Conclusions

In order to evaluate the different methods of underfilling, a DoE has been done.

We have shown in this work that the miniaturization via use of Chip Scale Wafer Level Package is very critical from a reliability point of view. So, the degradation level is not proportional to the module's size (increase the size by a factor 2 leads to decrease the lifetime by a factor 2.5). We have seen that the addition of underfill is necessary to increase the life time of solder ball for the large dice. The underfill reduces the influence of the component size but it changes the failure mechanism We have also proved that using "re-enforcement" the solder joint reliability is highly improved (by a factor 2) without influence on failure mechanism. This method is particularly interesting because it allows integrating underfill process in the manufacture of CSWLP. Hence we will continue the experiment to estimate the limit of such a technology.

All these experimental results are in good accordance with simulation ones.

The physical analyses allow us to better understand the failure mechanisms and to explain the exact behavior of the different underfill methods.

Acknowledgments

The authors thank all partners in the Amélie project.

This work was supported by "Conseil Régional de Basse-Normandie"

References

1. T.M. Mak, "Is System in Package the Panacea for Integration?" IEEE Design & Test of Computers, 2006

2. C. Regard, C. Gautier, H. Frémont, A. Val, F. Roullier, P. Schwindenhammer, P. Poirier, « Solder Fatigue of Wafer Level Package Assemblies. Comparison with Flip Chip BGA's", IEEE/EuroSimE, 2008

3. Scott F. Popelar, "A Parametric Study of Flip Chip Reliability Based on Solder Fatigue Modeling," IEEE/CPMT intel Electronics Manufacturing Technology Symposium, 1997.

4. N Strusevitch, S.Stoyanov, D Liu, C Bailey, A Richardson, N Dumas, JM Yannou, V Georgel, "Modeling the Behavior od Solder Joints for Wafer Level SiP", IEEE, EPTC, 2006

5. C Bailey, S.Stoyanov, N Strusevitch, JM Yannou, "Reliability Analysis of SiP Structures", Proceedings of HDP'07

6. S.Stoyanov, JM Yannou, C Bailey, N Strusevitch, "Reliability Based Design Optimisation for System-in-Package", EuroSimE, 2007

7. F. Murray "Silicon Based System-in-Package: a passive integration technology combined with advanced packaging and system based design tools to allow a breakthrough in miniaturization", Proc. BCTM, 2005

8. F. Murray, F. LeCornec, S. Bardy, C. Bunel, J. Verhoeven, E. van den Heuvel, J. Klootwijk, F. Roozeboom, "Silicon-Based System-in-Package: breakthroughs in miniaturization and 'nano'-integration supported by very high quality passives and system level design tools", Mat. Res. Soc. Symp., 2007.

9. SC Chaparala, BD Rogemann, JM Pitarrese, BG Sammakia, "Effects of Geometry and Temperature cycle on the Reliability of WLCSP Solder Joints," IEEE Inter Society Conference on Thermal Phenomena, 2004.

10. N Strusevitch, S.Stoyanov, D Liu, C Bailey, A Richardson, N Dumas, JM Yannou, V Georgel, "Modeling the Behavior od Solder Joints for Wafer Level SiP", IEEE, EPTC2006.

11. N Strusevitch, S.Stoyanov, C Bailey, A Val, JM Yannou, S Bellenger, F Verjus, P Rangheard, C Marie, "Combining the SiP and WLP-CSP trends: state-of-the art and future trends", Interconnex, 2006.

The Role of the Molecular Simulation Approach for IC-backend Developments

C. Yuan[a,c]*, O. van der Sluis[b], W. D. van Driel[a,c], G. Q. Zhang[a,c]

[a] Department of Precision and Microsystem Engineering, Delft University of Technology, The Netherlands
[b] Philips Applied Technology, The Netherlands
[c] NXP Semiconductors, The Netherlands
E-mail: c.a.yuan@tudelft.nl

Abstract

Since recent years, the micro-electronic industry changes the material usage, design and structure, in order to satisfy the customer demands of the higher performance and smaller size. One of the examples is the change of the basic materials from Al/SiO_2 to Cu/low-k in IC interconnect structure. As a consequence, new reliability issues at device/product level have been discovered, and most of the failure modes have the characteristics of multi-scale: the failure of the um or nm induces the malfunction of the device/product. The conventional approach of the failure prediction can be achieved by the well-developed continuum scale theory, e.g., finite element method. Moreover, the nano-meter scaled simulation is demanded in order to link the macro physics to the micro scale. This paper will demonstrate the capability of the molecular simulation of predicting the nano-scaled stiffness and atomic scale failure.

Introduction

As the continuously demand for high performance, multi-functional and small size of micro-electronic products, the complexity of the advanced IC is increasing and feature sizes are decreasing, as indicated in Moore's law. Due to the need for the improving of the electronic performance (e.g., high frequency, large power), new material, structure and process are introduced, but the reliability issues worsen because of the multi-failure modes. Moreover, decreasing the size of the advanced electronic application increases the difficulty of measure/characterize its mechanical behaviours. In order to satisfy time-to-market needs and fulfill the quality/reliability requirements, a reliable/applicable prediction theory/method of the material properties (e.g., stiffness [1,2], interfacial strength) and the further design of material are crucial in the application developments.

Among the materials of advanced IC backend structures, the low-k material (Fig. 1)has a relatively low mechanical stiffness: approximately 5-15 GPa [3–5]. Experiments [5] show that enhancing the Young's modulus of the low-k material increases the interfacial toughness of SiOC:H/Ta/TaN, which is known as the most critical interface in these structures. Among all the enhancement methods, the ultraviolet (UV) treatment is preferred, because it can exhibit an enhancement of mechanical strength without much increase of the dielectric constant. However, the relationship between the chemical composition, porosity, and mechanical properties has so far remained unclear, and a trial-and-error design method is still common practice in the design/fabrication of the low-k material within the industry.

Molecular modelling techniques, which provide a theoretical and numerical framework for many-particle problems, are well developed for the various application fields. In this paper, we demonstrate the capability of the molecular modelling in the prediction of material stiffness and interfacial strength, where a porous/amorphous low-k material is used as a carrier. The simulation results shows good fits with the trend which found in the measurements.

In this paper, the brief overview of the molecular dynamics will be reviewed. Following the mechanical stiffness prediction method and result, and experimental validation. The third part will describe the prediction of the interfacial failure at atomic scale.

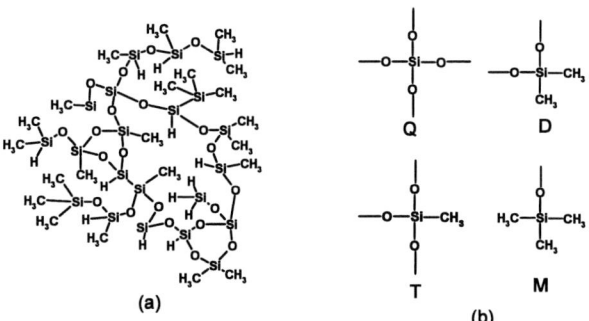

Fig. 1 . Illustration of the chemical structure of SiOC(H). (a) is the illustration of the material. (b) is the illustration of the connection capability of the basic building blocks

Molecular dynamics method

From the quantum mechanics point of view, matters have dual na-tures: particle and wave. However, while the geometry of the system is large enough, the wave nature of individual components becomes un-apparent and the system becomes determined. The molecular dynamics (MD), which is widely used in organic chemistry, is a treatment for the many-particle problems, and a determined response is prescribed. This method assumes the atom(s) as solid particles; their movement is described by coordinate variables. The interactions between the particles are described by the potential functions, also called force fields. When the wave nature of the particle will be ignored or considered implicitly by the potential functions, MD exhibits high efficiency in the simulation of the nano-scaled molecules. The following paragraphs will introduce the basic theory of MD, potential function, time integration scheme, boundary/initial conditions and limitation of MD.

978-1-4244-2739-0/08/$25.00 ©2008 IEEE

Theoretically, MD is based on the Newton's second law of motion,

$$\vec{F}_i = m_i \vec{a}_i \tag{1}$$

for each particle i in a system constituted by N particles. In Eq. (1), m_i is the mass of particle i, $\vec{a}_i = d^2\vec{r}_i/dt^2$ is its acceleration, and $\vec{F}_i$ is the force acting on the particle. Therefore, MD is a deterministic technique: given an initial set of positions and velocities, the subsequent time evolution can be determined.

The interaction force between particles, which is required in Eq. (1), can be defined by the potential functions or force fields:

$$\vec{F}_i = -\frac{\partial}{\partial \vec{r}_i} U(\vec{r}_1, ..., \vec{r}_N) \tag{2}$$

where U is the potential function and $\vec{r}_k, k = 1...N$ is the atomic coordinate.

The reaction forces (i represent the i-th substeps) at the fixed end) can be extracted either by the force of the pseudo-spring of the anchor point (illustrated in Fig. 2) or the energy gradient of the fixed atoms.

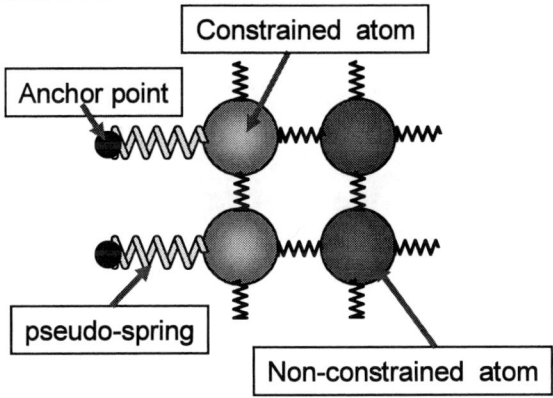

Fig. 2 . Illustration scheme of the constrainted atoms

Molecular model of amorphous/porous SiOC:H

The building blocks (Fig. 1), Q, T, D and M, represent Si atoms having 4, 3, 2, and 1 capabilities to connect other basic blocks, respectively. Instead of building the amorphous structure manually and constructing the model based on amorphous silicon, in this paper the concept of filling the basic blocks into a pre-defined framework [1] is proposed. Additionally, the size of the void is assumed to be the same as the basic blocks. Basic groups are not allowed to connect to pores. In the molecular modeling of the SiOC:H film, we further assume that only a single bond exists between any two atoms. Based on the total nodal number and the concentration of building blocks, the amounts of the void, Q, T, D and M are obtained. These blocks are then sequentially distributed into the each node. The distribution obeys the following rules:

(a) Chemical nature of building blocks: When the pore, Q, T, or D is randomly distributed onto the node, the total link of that node is fixed to 0, 4, 3, and 2, respectively. In the other words, for each node, the exceed links is removed by random choosing from the former linkage status.

(b) Average distribution: The local high concentration of specific type of building blocks will be prevented. Hence, a reliable random number generator is used to obtain the homogeneous distribution.

(c) Minimal numbers of dangling bonds: Because the dangling bonds are not physically favored, reducing them will easily lead to minimal potential energy state.

Mechanical stiffness prediction

In order to implement the molecular simulation, the atomic structure and interaction between atoms should be well defined beforehand. For the low-k material, due to the uncertainty of an amorphous material, it is quite difficult to precisely describe the exact chemical structure. For this purpose, a three-step generation procedure for amorphous/porous low-k material has been developed: (1) define a framework, which is made up of SiO_2 tetrahedral sharing corners; (2) the molecular topology is obtained by distributing the basic building blocks into the pre-defined framework; and (3) the stereochemical structure of low-k is obtained by applying the geometrical optimization procedure on the molecular topology.

In order to predict the Young's modulus of the material, a bar shaped molecular model is established [1]. Along the longitudinal direction, one end of the model is fixed in all degrees of freedom and a displacement with a constant velocity is applied to the opposite end. The Young's modulus can be deduced by $E = \frac{L_0^2}{V_0} \left(\frac{\Delta F}{\Delta d} \right)$, where ΔF, Δd, L, and V are the reaction force, applied displacement, initial length and initial volume of the specimen, respectively. The MD simulation is performed by means of the commercial solver Discover [52] with force field of COMPASS (definition:cff91, version 2.6) [53] and an environmental temperature is 300 K, as illustrated in Fig. 3. Comparing the simulation result to the experiment data, Table I shows good match.

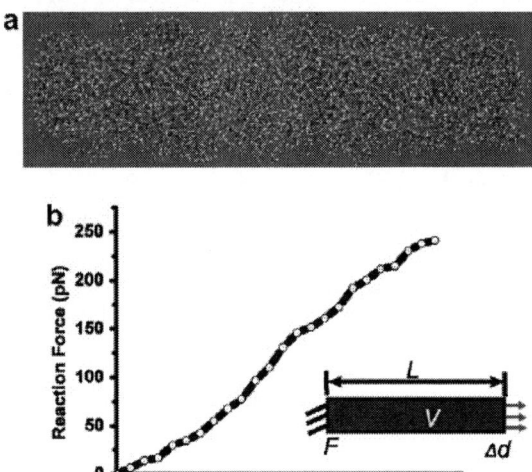

Fig. 3 . Simulation result of SiOC:H. (a) Is the model and (b) shows the reaction force
curve of the model where the boundary condition is illustrated in the inset.

Table I Experimental validation

		Case 1	Case 2
Concentration	Q	15.7%	21.7%
	T	47.7%	49.7%
	D	29.8%	20.5%
	Void	7.0%	8.0%
Young's modulus (GPa)	Simulation result	18.3	21.9
	Experimental result	11.0	16.0

By changing the chemical composition of the low-k material in the model, one can setup a series of simulations. Through the statistic analysis, the relation of the chemical composition and the Young's modulus can be deduced, shown in Fig. 4. The simulation result shows that the increasing the concentration of Q and T will significant increase the mechanical stiffness.

Simulation results on the material/interfacial strength

In order to simulate the nano-scaled interfacial mechanism, the model should also comprise the information of atoms and the atomic interactions between them. As already argued, the physical binding, chemical bond and mechanical interlocking are three main contributors to the interfacial strength. MD is a developing method with the capability of modeling the mechanical response of the nano-scaled system.

Fig. 5 shows the molecular model of TEOS/low-k. A prescribed displacement is applied to the top atom layers with fixed velocity. The bottom atom layers are fixed from which the reaction force can be obtained. Based on the applied displacement and obtained reaction force, depicted in Fig. 5, the interfacial fracture energy can be obtained.

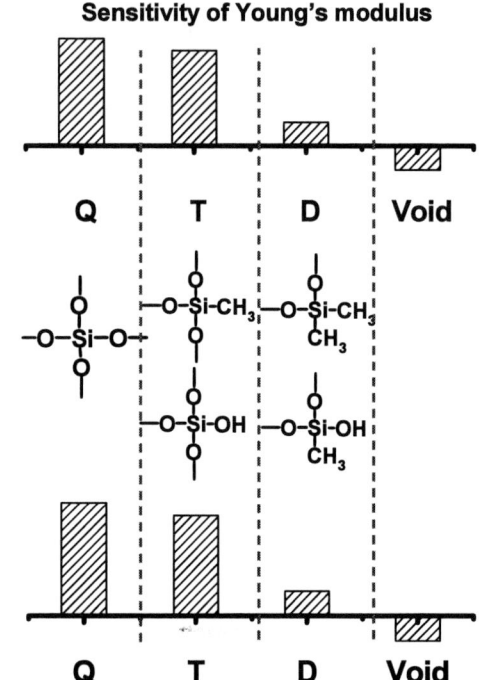

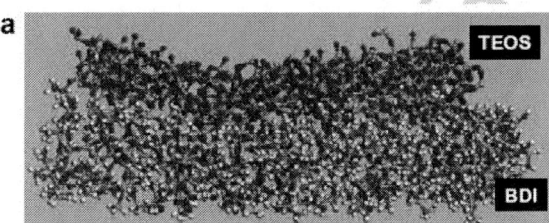

Fig. 4. The sensitivity of Young's modulus and density

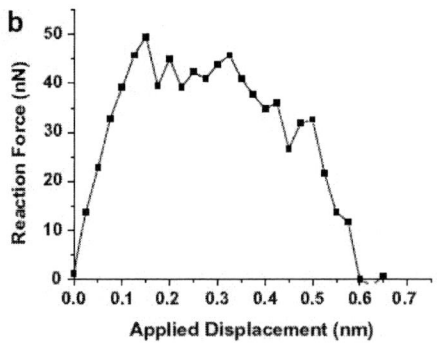

Fig. 5. The sensitivity of Young's modulus and density

Fig. 6 shows the simulation result of the crack initial location and the crack path. The crack initializes near the the edge of the pore due the stretching loading intend to increase the volume of the void. The crack path shows that the failure is not always follow the chemical interface of the system. The crack follows the weakest part of the material.

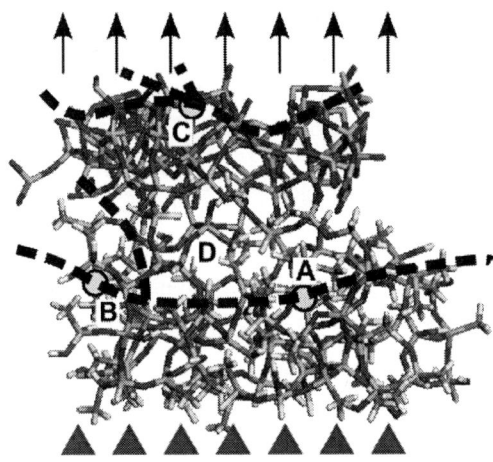

Fig. 6. The simulated crack propagation (blue dashed lines) through the interface (pink dashed line). Points A, B, C are the initial crack locations and point D indicates a pore within the low-*k* material.

Conclusions

The paper demonstrated the molecular dynamics application on the prediction of the low-k material, which can be input to the macro simulation approach. The experimental validation shows that the proposed method can qualitatively represent the trend. Moreover, the simulation results indicate that the slight variation of the chemical configuration can induce significant change of the mechanical stiffness but not the density. However, in order to achieve higher quantitative accuracy, the molecular model should be improved by increasing the geometric size, including defects of realistic size and improving method for the porosity.

Moreover, the nano-scaled fracture of the interfacial system has been depicted. However, there is an enormous challenge to predict the initiation and propagation of an interfacial system in the nano-scaled, due to unknown of defects of the material and interface, lack of sufficient force field, lack of robust multi-scaled modelling technique and lack of exact chemical configuration of interface.

References

1. Cadmus A. Yuan, Olaf van der Sluis, G. Q. Zhang, Leo J. Ernst, Willem D. van Drie, Amy E. Flower, and Richard B. R. van Silfhout, Molecular simulation strategy for mechanical modeling of amorphous/porous low-dielectric constant materials, Applied physics letters, Vol. 92, page 061909, 2008.
2. C.A. Yuan, O. van der Sluis, G.Q. Zhang, L.J. Ernst, W.D. van Driel,R.B.R. van Silfhout, B.J. Thijsse, Chemical–mechanical relationship of amorphous/porouslow-dielectric film materials, Computational Materials Science, Vol. 42, Page 606–613, 2008.
3. A. Grill, D. A. Neumayer, Structure of low dielectric constant to extreme low dielectric constant SiCOH films: Fourier transform infrared spectroscopy characterization, J. Appl. Phys. 94 (10) (2003) 272 6697–6707.
4. K. Maex, M. R. Baklanov, D. Shamiryan, F. Iacopi, S. H. Brongersma, Z. S. Yanovitskaya, Low dielectric constant materials for microelectronics, J. Appl. Phys. 93 (11) (2003) 8793–8841.
5. F. Iacopi, Y. Travaly, B. Eyckens, C. Waldfried, T. Abell, E. P. Guyer, D. M. Gage, R. H. Dauskardt, T. Sajavaara, K. Houthoofd, P. Grobet, P. Jacobs, K. Maex, Short-ranged structural rearrangements and enhancement of mechanical properties of organosilicate glasses induced 277 by ultraviolet radiation, 278 J. Appl. Phys. 99 (11) (2006) 053511.
6. Accelrys, Inc. Materials studio – discover. San Diego, USA: Accelrys, Inc.; 2005.
7. Sun H. An ab initio force–field optimized for condensed phase applicationoverview with details on alkane and benzene compounds. J Phys Chem B 1988;102:7338–64.

Strain-rate and Impact Velocity Effects on Joint Adhesion Strength

Chang-Lin Yeh[1], Yi-Shao Lai[2]

Shanghai Labs, ASE Assembly & Test (Shanghai) Limited[1]

Central Lab, Advanced Semiconductor Engineering, Inc.[2]

No.669, Guoshoujing Rd., Zhangjiang Hi-Tech Park, Pudong New Area, Shanghai 201203, P.R.C.[1]

*Email: chanlin_yeh@aseglobal.com

Abstract

In this paper, numerical studies are carried out on high-speed cold ball pull test by using explicit transient finite element simulations to predict transient response of package-level solder ball subjected to pull loads. The material constitutions of solder alloys are obtained from quasi-static tensile test and Hopkinson's bar test. Erosion technique is adopted for simulations of bulk solder fracturing, and interfacial element for intermetalic compound (IMC) fracturing. Parameter studies on pull velocity effect as well as strain-rate effect are also carried out. Transition points of pull velocity between bulk solder fracturing mode and IMC fracturing mode are identified therefore. From simulation results, transform relationship between pull forces to joint adhesion strengths of solder joints can be set up.

Equipments for joint adhesion measurement

The high-speed cold ball pull (HCBP) test has already received great interests from academic and industrial research institutes and a number of test apparatuses have been implemented [1-4]. In this paper, we employ the DAGE 4000HS test system, shown as Figure 1, to perform HCBP tests. The test system is capable of pulling a solder ball with velocity of 1.0 mm/sec to 1,300 mm/sec.

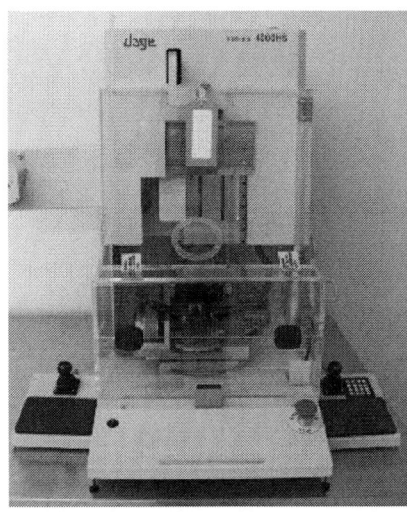

Fig. 1: High-speed ball pull tester

The test vehicle used in this study consists of an FR4 substrate with Au/Ni surface finish. The 96.5Sn-3.0Ag-0.5Cu (SAC305) solder balls of 0.3 mm diameter are mounted on the substrate. In this study tweezers close pressure of 13 psi is applying on solder ball priori to pulling at gripping stage, shown as Figure 2, afterward pull loads based on procedure of JEDEC JESD22-B115 [5] are applying on test vehicles using velocities from 100 mm/sec to 900 mm/sec.

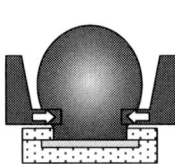 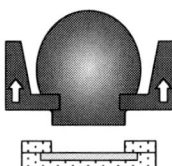

Fig. 2: Gripping stage (left) and pulling stage (right)

Figure 3 and Figure 4 show the two common failure modes that are found after the HCBP test: ductile failure and interfacial break. Failure mode of solder left, shown as Figure 5, in which cover rate of solder residual is between 5% to 50%, is occasionally found at transition pull velocity.

As shown in Figure 3, the majority of the failure modes after higher velocity pull is interfacial break, referring to IMC fracturing around the interface between the solder joint and the substrate pad. In some cases, solder residual that covers less than 5% of the cleavage area is observed, which also corresponds to interfacial break. Failure mode is classified as solder left while solder residual rate is between 5 ~ 50 %, shown as Fig. 5. Occasionally mode of ductile failure is present, shown as Figure 4, in particular when the pull velocity is lower than transition pull velocity. This failure mode implies that the IMC strength is comparably large to resist the pull load.

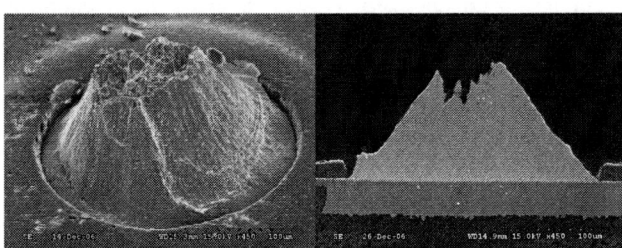

Fig. 3: Ductile failure.

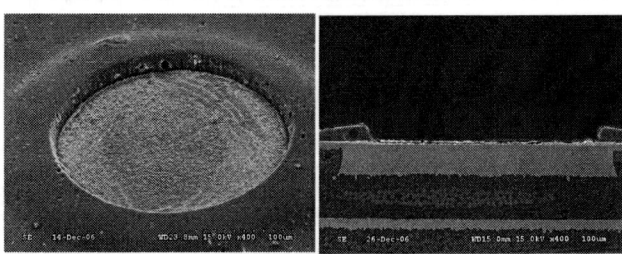

Fig. 4: Interfacial break

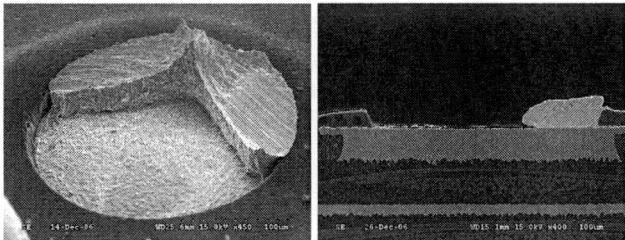

Fig. 5: Solder left.

In this study, pull velocities V_p are set to 100, 300, 500, 700 and 900 mm/sec with 15 samples at each test velocity. Pull force profiles at conditions of 100 and 900 mm/sec are plotted on Figure 6 and Figure 7, respectively. Under low pull velocity condition, $V_p <= 300$ mm/sec, we notice that ductile failure as well as interfacial break occurs, and ductile mode results in a larger pull duration, shown on Figure 6. However only interfacial break is found while pull velocity exceeds transition point.

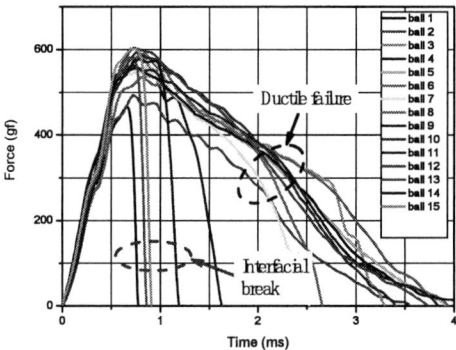

Fig. 6: Pull force profiles, $V_p = 100$ mm/sec.

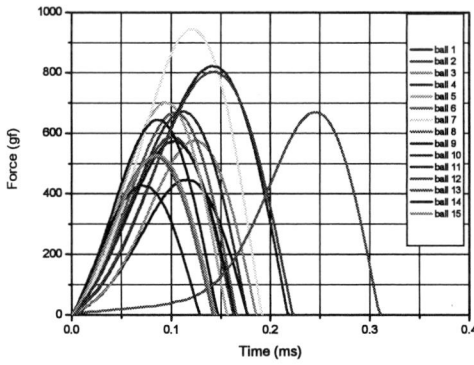

Fig. 7: Pull force profiles, $V_p = 900$ mm/sec.

Numerical simulation method

The physical model of HCBP on a single ball is shown in Figure 8. The tweezers move horizontally to grip the solder ball then pull vertically with a giving constant V_p, where V_p is set from 100 mm/sec to 1,300 mm/sec to examine velocity effect. Figure 9 shows the two-dimensional axisymmetric finite element model for the test vehicle. For components other than the solder joint, linear quadrilateral elements are applied. Since the solder joint is spherical, during the

processes of clamping and pull, force concentrations are expected to occur around the contact region with the tweezers, incurring large deformations over the region. With the reduced integration adopted, to avoid hourglassing, a numerically instable feature particularly for linear quadrilateral solid elements subjected to a concentrated force, linear triangle elements are applied on the regions depicted in the figure. All degrees of freedom at the bottom of the substrate are fixed. The model contains 5,538 elements and 9,030 degrees of freedom. In this paper, the analysis is carried out using LS-DYNA v. 970 for solving and ANSYS v.10 for pre- and post-processing.

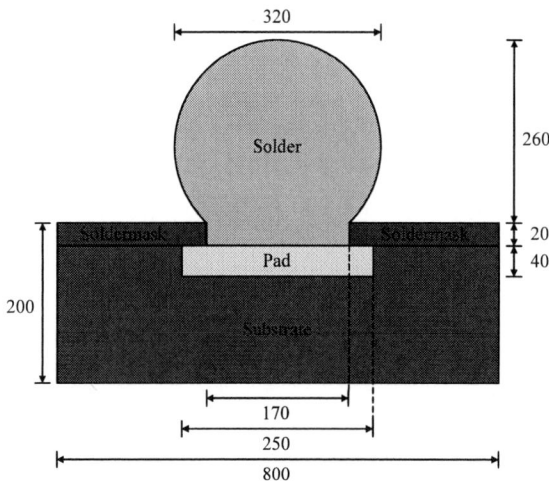

Fig. 8: Schematic drawing of HCBP physical model.

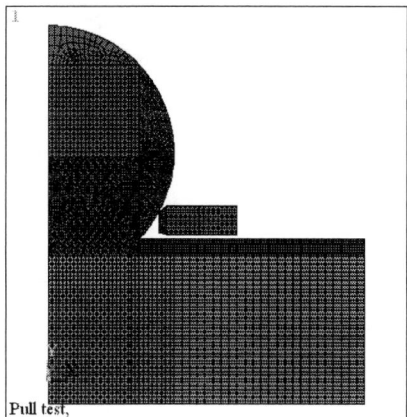

Fig. 8: Schematic drawing of HCBP physical model.

The spot-weld interfacial elements are applied between solder and copper pad and between solder joint and soldermask. Substrate is presumed linearly elastic while the pad and the soldermask are bilinearly elastoplastic.

The spot-weld element links two coincident nodes on adjacent meshes and confine the movements of them until the weld fails, shown as Figure 9.

985

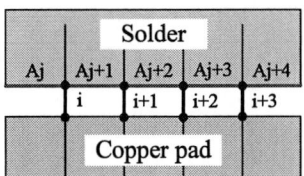

Fig. 9: Schematic of spot-weld elements i and i+n (n=1,2,3..).

The weld failure is characterized by

$$\left(\frac{|f_n|}{S_n}\right)^{C_n} + \left(\frac{|f_s|}{S_s}\right)^{C_s} \geq 1$$

In the equation, the subscripts n and s denote normal and shear, respectively, and f and S the weld force arisen from loadings and the ultimate force when spot-weld fails, respectively. In this paper, we assume $C_n=C_s=2$. The normal and shear ultimate forces of the spot-weld element can be calculated by [6]

$$S_n^i = \frac{1}{2}\sigma_n \sum_{j=1}^{N} A_j^i = arnode(i)*\sigma_n$$

$$S_s^i = \frac{1}{2}\sigma_s \sum_{j=1}^{N} A_j^i = arnode(i)*\sigma_s$$

where A is area of element j subjected to weld element i, which can be calculated by ANSYS ADPL option *arnode*(.).

To identify the material properties of the SAC305 solder alloy, high-strain-rate constitutive models are set up based on Hopkinson's bar test [7]. The corresponding stress-strain curves of SAC305 alloy under different strain-rate conditions are shown on Figure 10.

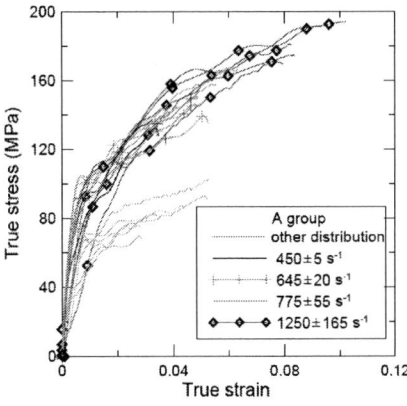

Fig. 10: Strain-stress curves of SAC 305 under different strain-rates conditions.

The eroding technique demolishes elements during the course of the computation based on a strain criterion. At each time step, the equivalent total strain of each element is examined and an element is removed from the finite element model when its equivalent total strain reaches ε_f. A fracture therefore propagates as a series of elements are gradually destructed.

Figure 10 and Figure 11 show plastic strain contours of test vehicles after failure completed. At velocity condition of 100 mm/sec, erosion occurs on contact region of solder ball surface, propagates within solder joint and finally ductile failure accomplishes, shown on Figure 10. When tweezers pull solder ball under higher velocity condition, increase of strain rate of solder leads to solder alloys becoming more rigid and stiff, therefore solder ball elements are hardly eroded due to decrement of plastic strain.

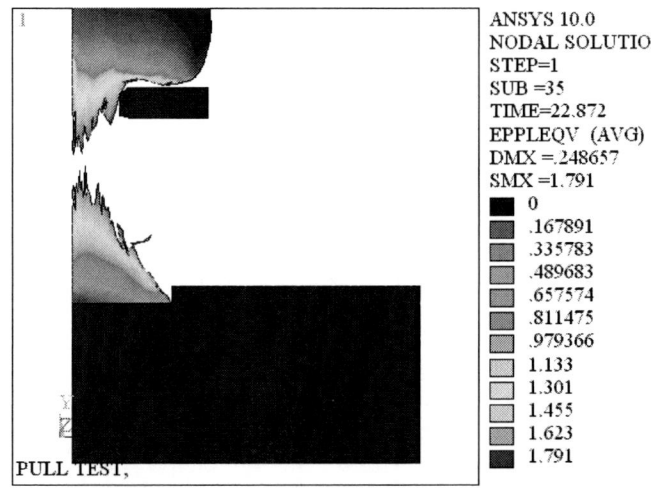

Fig. 10: Plastic contour - ductile failure (V_p = 100 mm/sec).

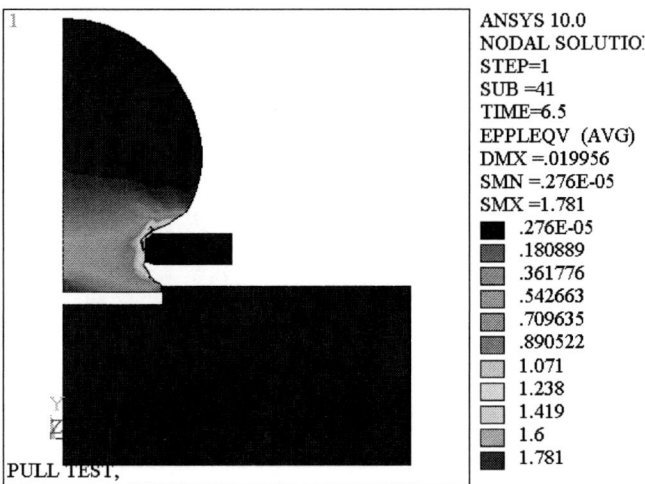

Fig. 11: Plastic contour – interfacial break, (V_p = 900 mm/sec).

Velocity effects

The effect of the pull velocity on the induced maximum adhesion of solder joint is examined by varying the velocity on pull stage from 100 mm/sec to 900 mm/sec with fixed IMC weld strengthes. On this study the normal strength is assumed 500 MPa and shear strength is two times the normal strength.

Figure 12 shows both numerical and experimental results of joint adhesion under different pull velocity. From simulation results we notice that a clear transition pull velocity is found between 400 mm/sec and 500 mm/sec, only mode of ductile failure occurs while pull velocity is less than transition velocity. As pull velocity exceeds transition velocity, the failure mode transfers into interfacial break and joint strength increases significantly.

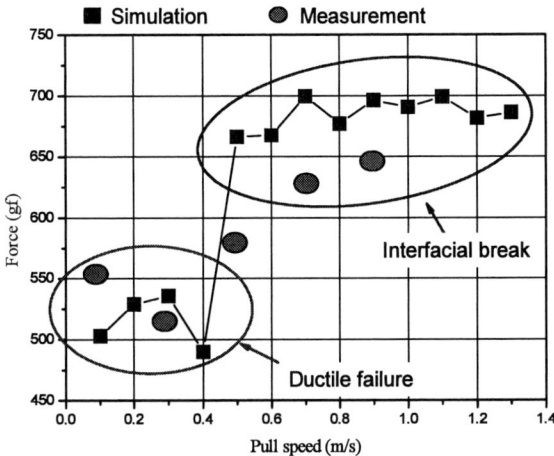

Fig. 12: Pull speed versus solder joint strength.

Conclusions

We investigate experimentally and numerically effects of strain-rate effect of the solder alloy on high-speed cold ball pull test in this paper. A two-dimensional axisymmetric finite element model with utilization of spot-weld elements for interfacial break simulation is proposed. Constitutive model of SAC305 solder alloy is determined based on results of Hopkinson's tests, strain-rate effect on HCBP test can be investigated under different pull velocity conditions.

1. Only modes of ductile failure and interfacial break are found in simulation results. Comparing with experimental results, observations are summarized in the following:

2. Transitional pull velocity V_T is found on both numerical simulations and measurements. Failure mode is dominated by interfacial break while pull speed exceeds V_T. Both ductile failure and interfacial break occur when pull velocity is less than V_T.

3. As failure mode transfers into interfacial break from ductile failure, joint strength of a solder ball increases significantly.

From this study, minor change of joint adhesion occurs while pull velocity is larger than V_T on simulation results. However from experimental results we notice it increases with pull velocity while only interfacial break is found. To obtain more accurate simulation results, a strain-rate dependent interface fracturing criterion is necessary in future study.

Acknowledgments

We would like to appreciate our colleagues, Shu-Hua Lee and Chin-Li Kao of ASE Kaohsiung for conducting high-speed pull tests, Macho Lin of SCHMIDT Scientific Taiwan Ltd. and Bob Sykes of DAGE for technical consulting. We also appreciate Professor S.T. Jenq of NCKU Taiwan and his team for providing Hopkinson's bar test results.

References

1. K. Newman, "BGA Brittle Fracture – Alternative Solder Joint Integrity Test Methods", Proc. 55th *Electronic Component & Technology Conference*, Orlando, FL, June 2005, pp. 1194-1200.

2. F.B. Song, S.W.R. Lee, K. Newman, B. Sykes and S. Clark, "Brittle Failure Mechanism of SnAgCu and SnPb Solder Balls during High Speed Ball Shear and Cold Ball Pull Test," *Proc. 57th Electronic Components & Technology Conference*, Reno, NV, June 2007. pp. 364-372.

3. F.B. Song, S.W.R. Lee, K. Newman, S. Clark and B. Sykes, "Comparison of Joint Strength and Fracture Energy of Lead-free Solder Balls in High Speed Ball Shear/Pull Tests and their Correlation with Board Level Drop Test," *Proc. 9th Electronic Packaging Technology Conference*, Singapore, December 2007, pp. 450-458.

4. K. Newman, "Board-level Solder Joint Reliability of High Performance Computers under Mechanical Loading", *Proc. EuroSime 2008*.

5. JEDEC Standard JESD22-B115, "Solder Ball Pull," May 2007.

6. Yeh, C.-L. and Y.-S. Lai, "Transient fracturing of solder joints subjected to displacement-controlled impact loads," *Microelectr. Reliab.*, in press.

7. S.T. Jenq, et al., "High Strain Rate Mechanical Behavior Analysis for Advanced Solder Materials", NCKU 2007 Report No. 95S060, Taiwan.

Parametric Study on Board-level Electronic Test Device Subjected to JEDEC Vibration Loads

Chang-Lin Yeh[1,*], Yi-Shao Lai[2], Ching-Chun Wang[1]

Shanghai Labs, ASE Assembly & Test (Shanghai) Limited[1]

Central Lab, Advanced Semiconductor Engineering, Inc.[2]

No.669, Guoshoujing Rd., Zhangjiang Hi-Tech Park, Pudong New Area, Shanghai 201203, P.R.C.[1]

*Email: chanlin_yeh@aseglobal.com

Abstract

We derive in this paper equations of motion of board-level IC packages subjected to swept sine vibration loads following the support excitation scheme. Harmonic analysis is performed based on the argument such that at each loading state over the swept sine process, hysteresis responses of solder joints following the isotropic hardening rule vanish fairly quickly so that plasticity is fully developed. Computed and measured acceleration response spectra of a board-level test vehicle are benchmarked. Stress-based failure indices as well as elastoplastic responses and strain rates of solder joints are examined for the test vehicle subjected to swept sine vibration tests of different acceleration levels with vibration frequencies up to 2 kHz.

1. Introduction

The vibration test is intended to evaluate the ability of electronic components to withstand moderate to severe vibrations generated during their transportation or field operations.

Studies concerning vibration tests on assembly-populated boards and corresponding reliability issues have been carried out. Song *et al.* [1] studied resonant vibration fatigue of lead-free and lead-containing solders and hence proposed a fatigue mechanism with respect to microstructures inside the solders. Through experimental measurements, Che *et al.* [2] figured out that the transmissibility, the ratio of the output acceleration level over the input one, is inversely proportional to the input acceleration level for vibration tests of constant acceleration levels. Swept sine vibration tests of different acceleration levels were performed by Perkings and Sitaraman [3]. Analytical solutions obtained from a simplified beam model, and numerical results through modal analysis were compared with the measurements. Kao *et al.* [4] performed harmonic analysis to identify stresses within the solder joints caused by vibration loads of an operating fan with unbalanced mass nearby the package.

In this paper, equations of motion of a board-level test vehicle subjected to swept sine vibration loads are derived following the support excitation scheme [5]. Harmonic analysis is performed based on the argument such that at each loading state over the swept sine process, hysteresis responses of solder joints following the isotropic hardening rule vanish fairly quickly so that plasticity is fully developed [6]. The measured acceleration response spectrum of a board-level test vehicle is used to benchmark numerical solutions. Stress-based failure indices as well as elastoplastic responses and strain rates of solder joints are examined for the test

vehicle subjected to swept sine vibration tests of different acceleration levels with vibration frequencies up to 2 kHz.

2. Kinematics of Swept Sine Vibration Test

We denote time, displacement, velocity, and acceleration as t, u, v, and a, respectively. Fig. 1 shows the schematic diagram of a board-level vibration test, in particular for the one following JEDEC standard JESD22-B103-B [7].

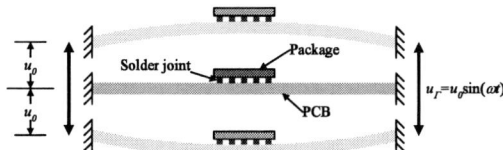

Fig. 1: Schematic of board-level vibration test

The board-level test vehicle is affixed to the shake table at its corners through standoffs. The fixed-ends of the test vehicle are denoted as Γ. A harmonic load with a specific acceleration level, a_0, or a specific displacement amplitude, u_0, and an angular frequency, ω, which is a function of t in a swept sine vibration test, is prescribed to Γ. Therefore, at Γ, we have

$$u_\Gamma(t) = u_0 \sin(\omega t), \qquad (1)$$

or

$$a_\Gamma(t) = a_0 \sin(\omega t). \qquad (2)$$

Eq. (1) or Eq. (2) serves as the boundary condition of the test vehicle. The relationship between a_0 and u_0 is

$$a_0 = -\omega^2 u_0, \qquad (3)$$

where $\omega = 2\pi f$ and f is the frequency of a vibration load. From the free-body analysis, equations of motion of the test vehicle become

$$Ma(t) + Cv(t) + Ku(t) = F_\Gamma(t), \qquad (4)$$

where M, C, and K are mass, damping, and stiffness of the test vehicle, respectively, and $F_\Gamma(t)$ is the vibration force at Γ.

978-1-4244-2739-0/08/$25.00 ©2008 IEEE

The relationship between a_0 and u_0, Eq. (3), suggests that when ω is low, u_0 can be quite large in order to maintain a specific a_0. However, it is difficult for a vibration shaker to generate extra-large displacements. As a resolution, as shown by the vibration spectrum of a swept sine test in Fig. 2, a crossover frequency, ω_c, is defined such that below ω_c, the test is displacement-controlled at a prescribed u_0, Eq. (1), while beyond ω_c, the test is acceleration-controlled at a prescribed a_0, Eq. (2).

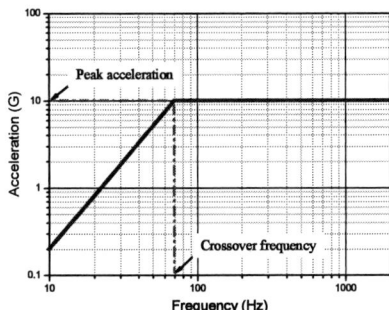

Fig. 2: Vibration spectrum of swept sine test

Table 1 shows the eight swept sine vibration test conditions proposed by JESD22-B103-B [1]. In the table, ω_{min} and ω_{max} denote starting and ending frequencies, respectively.

Table 1: JEDEC swept sine vibration test conditions [7]

Condition	a_0 (G)	$2\,u_0$ (mm)	ω_c (Hz)	$\omega_{min}/\omega_{max}$ (Hz)
1	20	1.5	80	20 / 2000
2	10	1.0	70	10 / 1000
3	3	0.75	45	5 / 500
4	1	0.5	31	5 / 500
5	0.3	0.25	24	5 / 500
6	0.1	0.125	20	5 / 500
7	0.01	0.039	14	5 / 500
8	0.001	0.0127	6.2	5 / 500

According to the support excitation scheme [5], when the local coordinate system is set at Γ, Eq. (4) can be rewritten as

$$M\bar{a}(t) + C\bar{v}(t) + K\bar{u}(t) = -Ma_\Gamma(t), \quad (5)$$

where the hatted symbols denote physical quantities defined according to the local coordinate system. In such a condition, boundary conditions become

$$\bar{u}_\Gamma(t) = \bar{a}_\Gamma(t) \equiv 0. \quad (6)$$

As depicted by the schematic diagram of the test vehicle in Fig. 3, based on the support excitation scheme, external vibration loads have been translated into body forces within the test vehicle, $q(x,t) = -\rho(x)a_\Gamma(t)$, where x is the location and ρ the mass density.

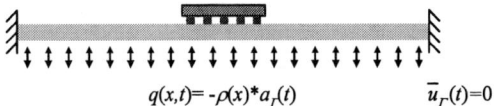

Fig. 3: Schematic of test vehicle subjected to vibration loads following support excitation scheme

Note that if the loading is displacement-controlled, as in the regime where $\omega < \omega_c$ in a swept sine vibration test, $u_\Gamma(t)$ needs to be differentiated twice with respect to t in order to obtain $a_\Gamma(t)$.

3. Vibration Test and Finite Element Analysis

We consider a $6 \times 6 \times 0.67$ mm^3 very-thin-profile fine-pitch ball grid array (VFBGA) chip-scale package interconnected to a $132 \times 77 \times 1$ mm^3 standard JEDEC drop test board [8] through 84 95.5Sn-4Ag-0.5Cu solder joints. The package contains a $4 \times 3.5 \times 0.17$ mm^3 silicon die and a 0.22 mm thick substrate. Diameter, standoff and pitch of a solder joint are 0.34 mm, 0.235 mm and 0.5 mm, respectively. Openings of a solder joint on the package side and the test board side are 0.24 mm and 0.28 mm, respectively. The package is mounted at the center of the test board, Fig. 4, in which the quarter symmetry modeling region for the finite element analysis is also depicted.

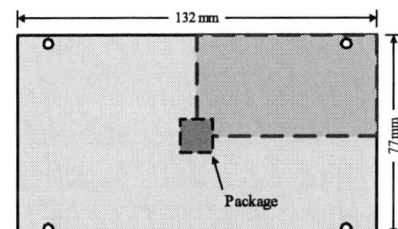

Fig. 4: Board-level test vehicle

An electrical-coil shaker (KD-9363), Fig. 5 (left), capable of producing a maximum force of 1,000 kgf with a frequency range from 1 Hz to 2 kHz, is employed in this study. The board-level test vehicle is affixed to the fixture through four standoffs, as shown in Fig. 5 (right). Vibration loads are applied along the out-of-plane direction of the test vehicle. An accelerometer is mounted on the top surface of the package for verification purposes.

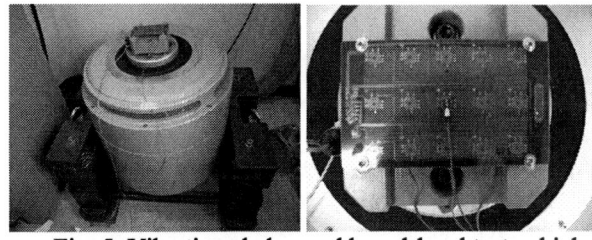

Fig. 5: Vibration shaker and board-level test vehicle

Fig. 6 (a) & (b) shows finite element models for the board-level test vehicle around the package while (c) finite element meshes on the solder joints.

989

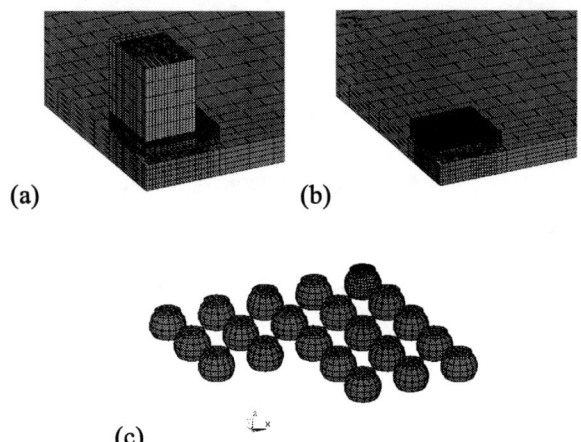

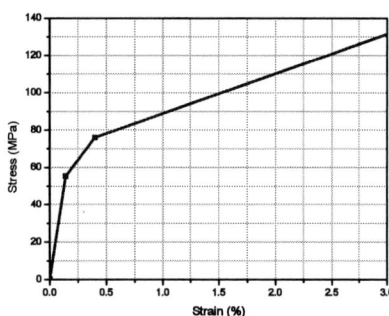

Fig. 6: Finite element models for test vehicles with (a) or without (b) accelerometer, (c) solder joints model.

The accelerometer is simplified as a 4 x 6 x 2.8 mm³ with a mass density of 1.89 g/cm³ in the finite element model. Except for the verification between computed and measured acceleration responses, for reliability evaluations presented in the following sections, no accelerometer is used, however. The finite element model without the accelerometer contains 43,292 linear hexahedral solid elements and 152,388 degrees of freedom. The analysis is carried out using ANSYS v. 8.0.

Trilinear elastoplastic properties are given for the solder alloy, Fig. 7, following Wiese and Rzepka [9]. For the solder alloy, the Poisson's ratio, v, is 0.36 and ρ is 7.44 g/cm³. Isotropic hardening is presumed.

Fig. 7: Trilinear stress-strain relationship for 95.5Sn-4Ag-0.5Cu solder alloy [9]

Mass-weighted damping and stiffness-weighted damping of the test vehicle are assumed 0.018 and 0.0001, respectively.

We assume that the vibration frequency sweeps continuously from 10 Hz to 2 kHz, discretized into 200 intervals. No ω_c is considered. That is, vibrations loads of constant acceleration levels are prescribed to the board-level test vehicle in both displacement-controlled and acceleration-controlled regimes.

Through numerical studies, Yeh et al. [6] has pointed out that when the elastoplastic solder alloy follows isotropic hardening upon yielding, mechanical responses of solder joints eventually turn from elastoplastic to purely elastic after the board-level test vehicle receives several identical pulse loads, prescribed consecutively to the test vehicle. Apply for periodic vibration loads of reasonably high frequencies, raise an argument such that at each loading state over the swept sine process, hysteresis responses of solder joints vanish fairly quickly so that plasticity is fully developed. Following this argument, and according to Eqs. (2) and (5), only the harmonic analysis is required at each loading state in calculating mechanical responses of the board-level test vehicle that contains elastoplastic solder joints.

4. Numerical Results

In this section, the harmonic analysis procedure developed previously is verified by comparing with experimental measurements. Reliability evaluations are also provided.

4.1 Verification

The verification is performed based on a vibration test of a constant $a_0 = 10$ G. Computed and measured acceleration response spectra on the top surface of the package are shown in Fig. 8. These two curves agree with each other.

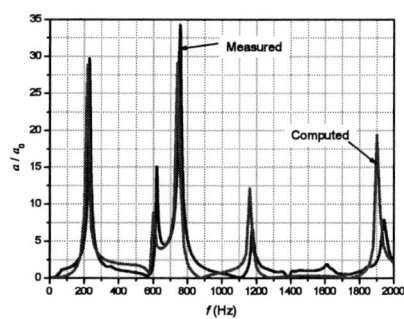

Fig. 8: Acceleration response spectra on top surface of package

For this particular board-level test vehicle, five peaks on the acceleration response spectrum are present below 2 kHz. Peak accelerations at resonance, a_i, and corresponding resonance frequencies, f_i, for which i denotes the i-th natural mode, are summarized in Tables 2 and 3. Interestingly, both experimental and numerical results indicate that the maximum acceleration response occurs at Mode 3, not Mode 1.

Table 2: Resonance frequencies (unit: Hz)

Mode	1	2	3	4	5
Measured	230	621	756	1177	1944
Computed	220	600	740	1160	1920
Discrepancy (%)	4.3	3.4	2.1	1.4	1.2

Table 3: a_i / a_0 on top surface of package

Mode	1	2	3	4	5
Measured	29.68	15.07	34.25	6.66	7.87
Computed	28.91	8.86	29.12	12.11	19.42
Discrepancy (%)	2.6	41.2	15.0	81.9	146.8

The maximum discrepancy between measured and computed resonance frequencies for the first five natural modes is about 4.3%. However, the discrepancy between measured and computed peak accelerations on the top surface of the package is up to 146.8% and it is greater in a higher mode. This indicates that in order to precisely predict mechanical responses of the test vehicle subjected to high-frequency vibration loads, further refinement of the finite element model is necessary.

The accelerometer is for the verification purpose only. In reliability evaluations that follow, no accelerometer is implemented.

4.2 Brittle and ductile fracturing

In general, solder joints are subjected to two types of failures during a board-level reliability test: ductile fracturing within the solder alloy and brittle fracturing around the IMC layer developed between solder alloy and bonding pad. Ductile fracturing usually occurs under a low-strain-rate load such as ball shear at a shearing speed of less than 0.1 mm/s (e.g., [10]) or a typical thermomechanical load (e.g., [11]). On the other hand, brittle fracturing is usually encountered under a high-strain-rate load (e.g., [12-14]).

To distinguish between these two types of failures, different failure indices are considered. For ductile fracturing, the von Mises equivalent stress, σ_{eqv}, is chosen as the failure index while for brittle fracturing, normal components of surface tractions, σ_n, on bonding surfaces of a solder joint are employed. The σ_{eqv} is a broadly used indicator of ductile failures of metals. The adoption of σ_n is based on the consideration that brittle IMC fracturing is a process of interfacial separation between IMC and pad [14]. Owing to limited computational resources, no IMC is actually implemented in the finite element model, however.

Spectra of the maximum σ_{eqv} on package side and test board side of the solder joints under the swept sine vibration test of a constant $a_0 = 10$ G with vibration frequencies up to 2 kHz are shown in Fig. 9. Clearly, the greatest σ_{eqv} occurs at Mode 1. Moreover, the magnitude of σ_{eqv} is greater on the package side than on the test board side.

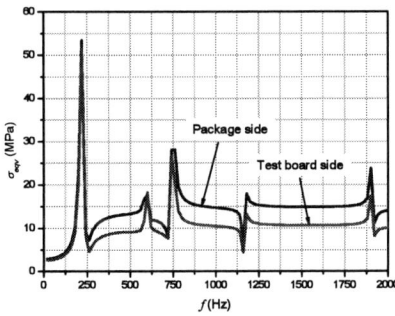

Fig. 9: Spectra of maximum σ_{eqv}

Spectra of the maximum σ_n on package side and test board side of the solder joints are shown in Fig. 10. Similarly, the greatest σ_n occurs at Mode 1 and the magnitude of σ_n is greater on the package side than on the test board side.

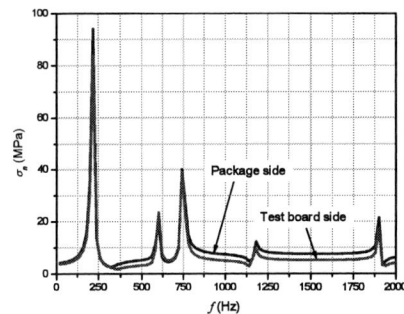

Fig. 10: Spectra of maximum σ_n

From the spectra of σ_{eqv} and σ_n, we note that for this particular board-level test vehicle, these stresses are nearly independent of the vibration frequency in the frequency range from 1.3 kHz to 1.8 kHz.

4.3 Plastic strain and strain rate

In this subsection, elastoplastic responses of solder joints of the board-level test vehicle subjected to swept sine vibration tests of three acceleration levels with vibration frequencies up to 2 kHz are examined.

Fig. 11 shows spectra of the maximum equivalent plastic strain, ε_p, in the solder joints for different a_0. We note that the solder joints do not yield for $a_0 = 3$ G. For $a_0 = 10$ G, yielding of the solder joints occurs at the fundamental resonance frequency, $f_1 = 220$ Hz, and ε_p remains constant after 240 Hz, indicating that the solder joints remain in the elasticity regime. For $a_0 = 20$ G, yielding of the solder joints occurs at the first three resonance frequencies, $f_1 = 220$ Hz, $f_2 = 600$ Hz, and $f_3 = 740$ Hz. Except for small regions around these resonance frequencies, ε_p remains constant over the entire range of vibration frequencies.

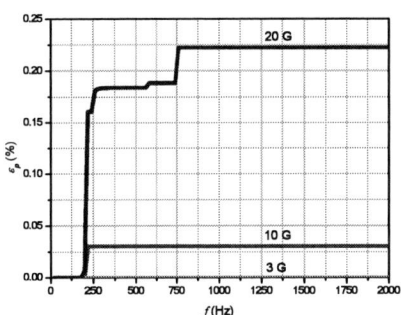

Fig. 11: Spectra of maximum ε_p for different a_0

Fig. 12 shows spectra of the normal component of the maximum total strain, ε_n, for different a_0. Clearly, a larger a_0 leads to a greater ε_n. Moreover, if we disregard ε_n around the resonance frequencies, we note that the magnitude of ε_n, denoted as $|\varepsilon_n|$, is proportional to a_0 but its decay with respect to the increase of the vibration frequency is not apparent. This feature, however, requires further identifications since high vibration frequencies would amplify localized structural responses and, as mentioned previously, our finite element model is not fine enough to capture structural responses particularly well during high vibration frequencies.

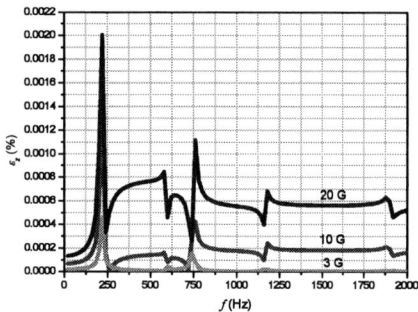

Fig. 12: Spectra of maximum ε_n for different a_0

In the harmonic analysis, ε_n as a function of t can be expressed as

$$\varepsilon_n(t) = |\varepsilon_n| \sin(2\pi f t + \theta), \qquad (7)$$

where θ denotes the phase angle. The strain rate can therefore be obtained as

$$\dot{\varepsilon}_n(t) = 2\pi f |\varepsilon_n| \cos(2\pi f t + \theta). \qquad (8)$$

Fig. 13 shows maximum strain rates in the solder joints of the board-level test vehicle subjected to swept sine vibration tests of different a_0 with vibration frequencies up to 2 kHz.

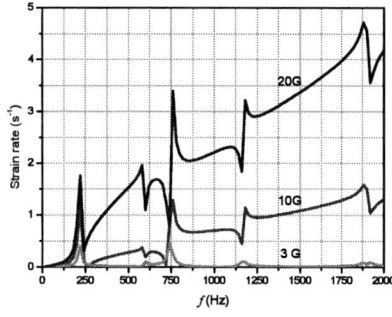

Fig. 13: Maximum strain rates in solder joints for different a_0

It is clear that a larger a_0 leads to greater strain rates. Despite those around resonance frequencies, the strain rate is proportional to a_0 as well as the vibration frequency, as implied by Eq. (8). Apparently, the strain rate is in the range of 10^0 to 10^1 s^{-1} for the specific board-level test vehicle subjected to the swept sine vibration test conditions proposed in this study.

5. Conclusion

In this paper, we propose a harmonic analysis procedure following the support excitation scheme for board-level electronic packages subjected to swept sine vibration tests. Computed and measured acceleration response spectra of a board-level test vehicle are benchmarked, and five resonance frequencies are captured below 2 kHz for the specific test vehicle.

Stress-based failure indices as well as elastoplastic responses and strain rates of solder joints are examined for the test vehicle subjected to swept sine vibration tests of different acceleration levels with vibration frequencies up to 2 kHz. Numerical results are summarized in the following:

1. The outermost solder joint attracts the greatest stresses and strains.
2. Greater σ_{eqv} and σ_n occur on the package side rather than on the test board side of the solder joints.
3. No ε_p is developed for $a_0 = 3$ G. For $a_0 = 10$ G, ε_p evolves only at Mode 1 while for $a_0 = 20$ G, ε_p evolves at the first three resonance modes.
4. Except for small regions around the resonance frequencies, ε_p remains constant over the entire range of vibration frequencies.
5. Despite those around resonance frequencies, the magnitude of ε_n is proportional to the acceleration level but its decay with respect to the increase of the vibration frequency is not apparent. This feature, however, requires further identifications.
6. Despite those around resonance frequencies, the strain rate is proportional to the acceleration level as well as the vibration frequency.
7. The strain rate is in the range of 10^0 to 10^1 s^{-1} for the specific board-level test vehicle subjected to the swept sine vibration test conditions proposed in this study.

Acknowledgement

We are grateful to our colleagues, Ping-Feng Yang and Tsan-Hsien Chen, for measurement supports.

References

1. Song, J. M., T. S. Lui, L. H. Chen and D. Y. Tsai (2003). Resonant vibration behavior of lead-free solders. *J. Electr. Mater.*, 32(12): 1501-1508.
2. Che, F. X., H. L. J. Pang, F. L. Wong, G. H. Lim and T. H. Low (2003). Vibration fatigue test and analysis for flip chip solder joints. *Proc. 5th Electr. Pack. Technol. Conf.*, Singapore, pp. 107-113.
3. Perkins, A. and S. K. Sitaraman (2004). Vibration-induced solder joint fatigue failure of a ceramic column grid array (CCGA) package. *Proc. 54th Electr. Comp. Technol. Conf.*, Las Vegas, NV, pp. 1271-1278.
4. Kao, C.-L., C.-L. Yeh and Y.-S. Lai (2004). Steady-state vibration analysis for printed circuit boards of different package layouts. *Proc. 2004 Taiwan ANSYS Conf.*, Nantou, Taiwan, pp. 67-70.
5. Yeh, C.-L. and Y.-S. Lai (2006). Support excitation scheme for transient analysis of JEDEC board-level drop test. *Microelectr. Reliab.*, 46(2-4): 626-636.

6. Yeh, C.-L., Y.-S. Lai and C.-L. Kao (2005). Prediction of board-level reliability of chip-scale packages under consecutive drops. *Proc. 7th Electr. Pack. Technol. Conf.*, Singapore, pp. 73-80.

7. JEDEC Solid State Technology Association (2002). *JESD22-B103-B: Vibration, Variable Frequency.*

8. JEDEC Solid State Technology Association (2003). *JESD22-B111: Board Level Drop Test Method of Component for Handheld Electronics Products.*

9. Wiese, S. and S. Rzepka (2004). Time-independent elastic-plastic behaviour of solder materials. *Microelectr. Reliab.*, 44(12): 1893-1900.

10. Huang, X., S.-W. R. Lee, C. C. Yuan and S. Hui (2001). Characterization and analysis on the solder ball shear testing conditions. *Proc. 51st Electr. Comp. Technol. Conf.*, Orlando, FL, pp. 1065-1071.

11. Terashima, S., Y. Kariya, T. Hosoi and M. Tanaka (2003). Effect of silver content on thermal fatigue life of Sn-xAg-0.5Cu flip-chip interconnects. *J. Electr. Mater.*, 32(12): 1527-1533.

12. Lai, Y.-S., P.-F. Yang and C.-L. Yeh (2006). Experimental studies of board-level reliability of chip-scale packages subjected to JEDEC drop test condition. *Microelectr. Reliab.*, 46(2-4): 645-650.

13. Yeh, C.-L., Y.-S. Lai, H.-C. Chang and T.-H. Chen (2006). Empirical correlation between package-level ball impact test and board-level drop reliability. *Microelectr. Reliab.*, in press.

14. Lai, Y.-S., H.-C. Chang and C.-L. Yeh (2006). Evaluation of solder joint strength using ball impact test. *Proc. IMAPS Taiwan 2006 Int. Tech. Symp.*, Taipei, Taiwan.

Modeling Techniques for Board Level Drop Test for a Wafer-Level Package

Harpreet S. Dhiman[2], Xuejun Fan[1,2], Tiao Zhou[3]
[1]Department of Engineering Mechanics
South China University of Technology, Guangzhou, China
[2]Department of Mechanical Engineering
Lamar University, PO Box 10028, Beaumont, TX 77710, USA
[3]Maxim Integrated Products, Dallas, TX, USA
xuejun.fan@lamar.edu

Abstract

Reliability performance during drop impact is critical for electronic handheld devices. In this paper, a comprehensive study in efficiency and accuracy of multiple finite element modeling approaches and solution techniques for a wafer-level package (WLP) is presented. JEDEC specified test board is used for the model study. A direct acceleration input method is introduced. Two types of global finite element models for a typical WLP are studied: solder layer and solder bump models. Two different approaches, full implicit dynamics and mode superposition, are applied to solve the JEDEC board dynamic responses. Based on this study, the 8-node solid element with smeared solder layer model, and the full implicit dynamics with either input-G or direct acceleration method are recommend. This combination results in short solution time and produces accurate dynamic solutions for drop test board. It has been found that the fundamental natural frequency of a JEDEC board with WLP typically ranges from 200 to 250 Hz for a large range of array size. There is a large strain gradient close to the component edge for each package on the test board. Due to the rigidity of the silicon chip, the board strain at the center of each component on the opposite side of PCB does not reflect the local bending behaviors of the board. The center of the board between two components might be a stationary point, which does not capture the board bending. With the increase of the chip size, the board strain at edge of each component will increase. The board peak strain at the corner package (U1, U5, U11, and U15) has been found greater than that at the center package (U8), but the bending direction is opposite. The components U6 and U10 have lowest board strains among all components.

1. Introduction

Reliability of handheld electronic devices such as cell phones and PDAs due to drop/impact event is a major concern in electronics industry. During a drop/impact event, the PCB assembly inside the phone casing vibrates causing a flexural/bending motion of the board. The PCB bending results in high stress or strain on solder joints of electronic components due to high level of impact forces. It ultimately leads to the failure in solder joints. The failure can occur at package side or PCB side. Other failure modes such as pad-crater and broken board traces are also observed.

A number of excellent papers have been written on drop characterization and simulations [2-7]. Finite element modeling using commercial software such as ANSYS and ABAQUS have been used extensively for the simulation of drop impact test, and have been successfully applied in product developments [8-9]. Attempts have been made to improve the methodology for simulation and to reduce the computational time [9-10]. The input-G method exerts to the mounting hole location the acceleration impulse measured in experiment. This decouples the board finite element model from the system model. Therefore, the computational efficiency has been significantly improved. An alternative method is the input-D method [11-12]. With this method, the acceleration input is converted to the displacement input through integration over time. There are several approaches in implementing the input-G method. Tee et al. [4] uses explicit dynamics analysis by directly applying acceleration impulse in DYNA-3D. The implicit dynamics method can also be used to solve board dynamic response. For this method however, surface acceleration load is not supported. To get around this difficulty, Syed et al [9] proposed to convert acceleration input into force input by multiplying the acceleration with the large mass. This large mass method with rigid elements effectively applies the acceleration on the support points. The mode superposition method [13] requires shorter computational time than both explicit and implicit methods.

This paper presents a comprehensive study to investigate the efficiency and accuracy of different models and approaches for PCB global dynamic response. A WLP is used for this study. Two kinds of global finite element models for a typical WLP solder ball array are studied: solder layer and solder bump models. Two different solution approaches for JEDEC board dynamic response are compared against each other: full implicit dynamics and mode superposition. For the acceleration input, this paper introduces a direct acceleration input method [14]. The results are compared to the input-G method. Finally, the effect of different parameters such as chip size, acceleration magnitude, and board strain distributions are studied. The effect of boundary conditions on the structure is also taken into account here.

2. JEDEC Board Drop Impact Test

JEDEC drop test standard JESD22-B111 [1] is commonly used for board level drop test for many years. JESD22-B111 recommends 15 components mounted on the PCB in 3 rows of 5 components each. The PCB is mounted on a base plate with 4 mounting screws at the corners. This base plate is then mounted on a drop table. The drop table, guided by guide rods, is allowed to strike on a rigid base from some specified height. A half sine-impulse is produced when the table strikes the rigid base. A layer of felt is used on the strike surface to obtain the desired load conditions. Finite element method is often used to calculate the dynamic response of the drop test board and to correlate the drop impact performance.

It is difficult to include all involved parts in the finite element model. The size of the finite element model will be very large if all details of drop table, PCB, and all components on PCB are included. In this case the numerical difficulties will be encountered to deal with the contact simulation between the felt and drop table block. Since the main interest is focused on the failures of tiny solder balls of each component, the dimension ratio from the table block to the solder ball IMC layer will be too large to handle. In order to reduce the computational time, the Input-G method can be applied to decouple the board dynamic responses from the test system by applying the 'Table-G' directly to the board. This avoids the difficulties in modeling the complex behaviors of contact between the drop table and drop surfaces.

3. Simulation Methodology

To understand the behavior of a structure during drop impact test we will consider mainly two simulation methods used in standard ANSYS (without explicit dynamics solver). The first method is Full Implicit Dynamics which uses the full system matrices to calculate the transient response (no matrix reduction). The second method is called Mode-superposition method. This method sums factored mode shapes (eigenvectors) from a modal analysis to calculate the structure's response. There are pros and cons for both methods. On one hand, the full method allows material and geometric nonlinearities while mode-superposition method considers linear problem only. On the other hand, mode-superposition method in general is faster (theoretically) and less expensive compared to the full method.

To produce the desired G input loading (1500G, 0.5 milliseconds duration, half-sine pulse), we consider two different approaches for each of the above methods here. These two approaches are referred to as Input G with Large Mass method and Direct Input Acceleration method. In input G method, a large mass element is attached to the nodes around the screw holes using rigid elements. The acceleration input is converted into force input by multiplying the acceleration with large mass. This large mass method with rigid elements effectively applies the acceleration on the support points. With direct input acceleration method, on the other hand, half-sine impulse load is directly applied to the model as inertia body force. With this input, the screw hole boundary conditions must be specified.

Some of the main challenges involved in finite element modeling for JEDEC board dynamic responses are:

- Very large finite element models with various mesh densities to transform from millimeter scale to micron scale within one model.
- The level of mesh density also impacts the solution time not only because of model size but also due to time step size requirements during dynamic simulation.

In order to minimize the difficulties involved in modeling and to reduce computational time, two finite element models are built here. The first finite element model is Solder Bump model. In this model the shape of the solder bump is simplified to a rectangular block as shown in figure 1. The complete model consists of a PCB, rectangular solder bumps and chips. A quarter 3D-model is built due to symmetry (figure 1). The size of the chip and number of solder arrays can be varied for the experiments. The second model built is Solder Layer model. The model smears individual solder bumps into a uniform layer with effective material properties.

Such a model can reduce the board model size significantly since the size of finite element model does not depend on the numbers of solder bumps, as shown in Figure 2. 3-D solid elements are used for PCB, silicon chip and solder layer or bumps, which is different from the global model in Syed's paper where the shell element is used for PCB. Since the mesh has been optimized, both models run very effectively with a regular PC by implicit dynamics. For a 6x6 mm chip size the solder bump model has 13161 elements compared to 5569 for the solder layer model. The computational time taken by solder layer model is reduced to 1/3 of that solder bump model. Table 1 lists the material properties used in this paper.

Table1. Material properties

Materials	Modulus (MPa)	Poisson Ratio	Density x 10^{-3} (gm/mm^3)
PCB	22000	0.25	2.1
Solder bumps	51000	0.36	7.2
Chip	130000	0.278	2.5

4. Model Study

In this section, WLP solder ball models, PCB dynamic response solution method, and input acceleration load application options are studied. The test model used for this case and for all the following cases is a 6x6 mm chip size model.

4.1 Comparison between Two Models (Solder Layer vs. Solder Bump)

Two solder ball models and two PCB dynamic response solution methods are examined. Modal analysis is discussed first.

4.1(a) Modal Analysis

Modal analysis is used to determine the natural frequencies and mode shapes of a structure. The natural frequencies and mode shapes are then used to obtain the transient response by superposition.

The nature frequencies for the drop test board are listed in Table 2. The solder balls are modeled two ways, solder bump model (individual cubic blocks), and solder layer model (a homogeneous layer). It is seen from the Table 1 that the lowest five natural frequencies obtained by both solder ball models solder layer are approximately the same. This suggests that solder layer may be used in the drop test simulations without sacrificing the accuracy.

Table 2. Modal analysis of the two models

Mode No.	Frequency (Hz)		Difference
	Model A Solder Bump	Model B Solder Layer	
1	212	210	0.9%
2	551	548	0.3%
3	915	913	0.2%
4	1257	1251	0.4%
5	2193	2181	0.5%

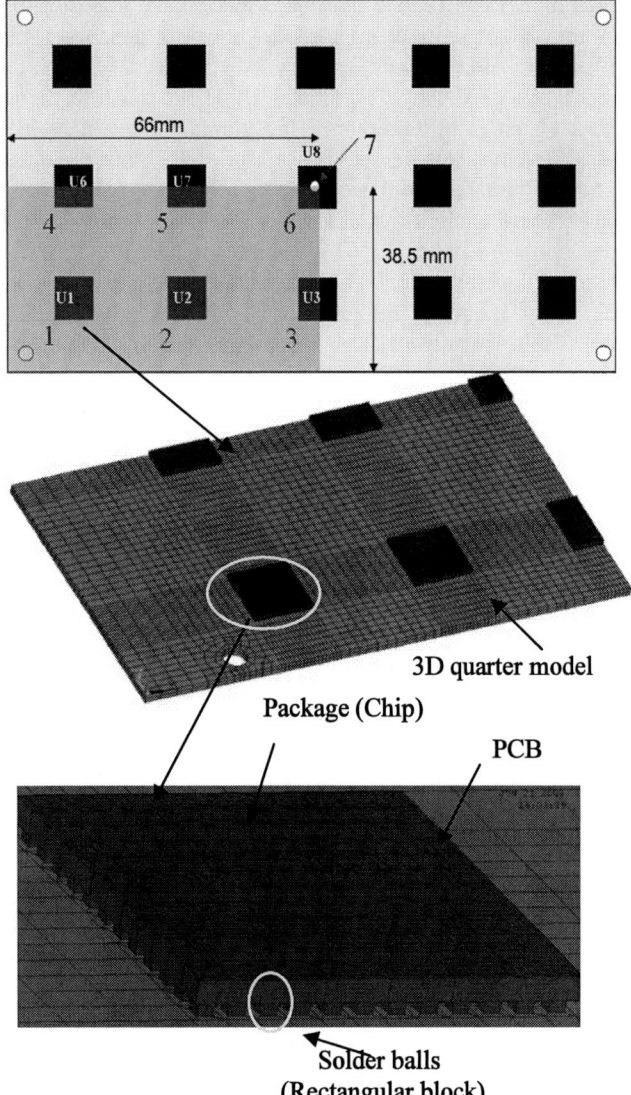

Figure 1. 3D-quarter model of JEDEC board (Solder bump model)

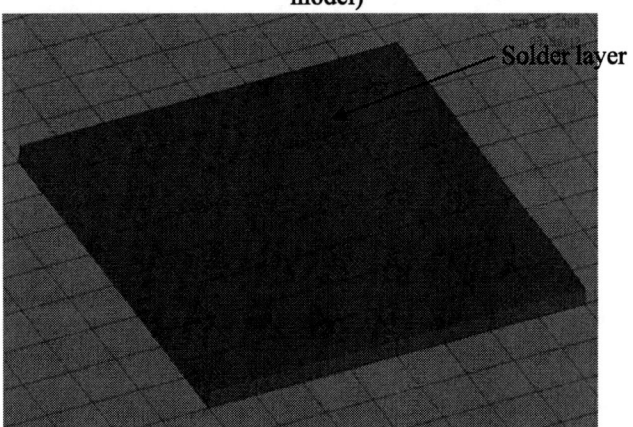

Figure 2 Details at Solder Layer model

4.1(b) Transient Dynamic Analysis

The two solder ball models are then solved with input G method with full implicit dynamics. The acceleration at site 7 for first 7ms during the impact is plotted in Figure 3. Site 7 is at the top of the center component. (Figure 1). It is seen from Figure 3 that the difference between the dynamic response obtained for two models at site 7 is trivial.

The board strain along path 1-3 in Figure 1 is used to examine strain solutions by different approaches. The normal strain in x-direction ε_x at t=1.5 ms is plotted in Figure 4. It is seen that the strain solution for the two solder ball models are almost the same. The difference is mainly at the peaks. This suggests that the solder layer model can be used for board global dynamic response to significantly save computer space and reduce the solution time without sacrificing accuracy.

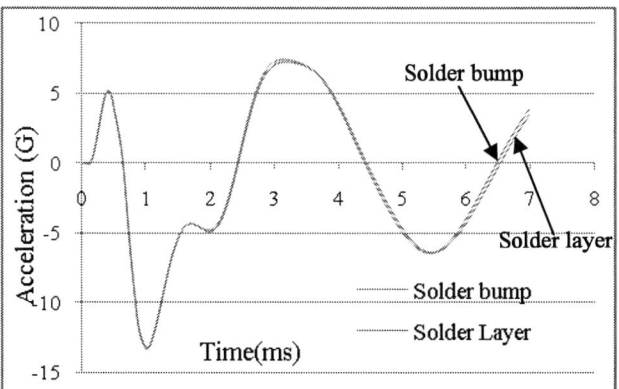

Figure 3. Comparison of acceleration curves for two models (input G method)

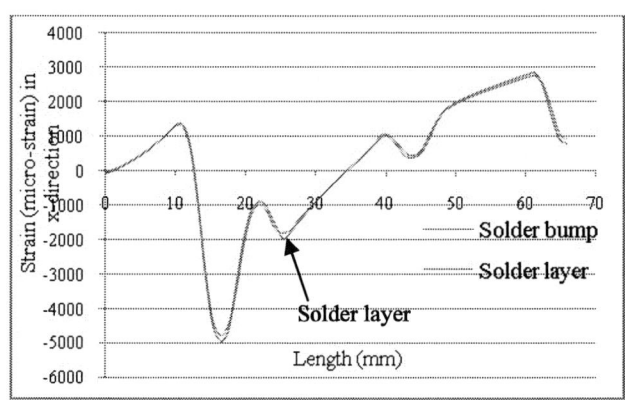

Figure 4. Comparison of board strain for two models

4.2 Comparison of Input G and Direct Input Acceleration Method

Both input G and direct input acceleration are used with a solder bump model. The acceleration history and board strain history results are plotted in Figures 5 to 7. Figure 5 shows the time-dependent acceleration curve at the location 7 (figure 1). It is seen that the acceleration solutions have different values for the initial 0.5 ms acceleration impulse period. However, after this period they overlap. The difference between these two curves is exactly the half-sine acceleration impulse. This suggests that the calculated acceleration from direct acceleration input method includes the initial acceleration applied anywhere in the structure during impact.

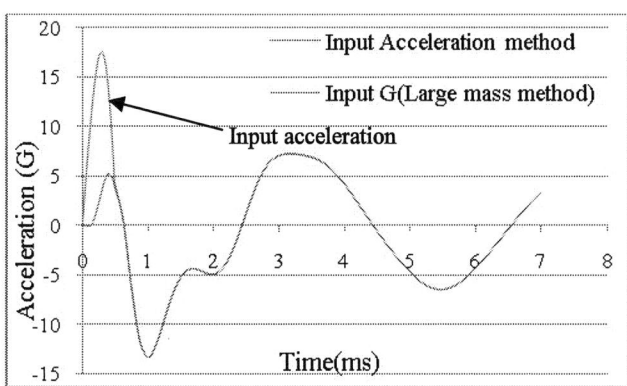

Figure 5. Comparison for input G and input acceleration methods

The board strain along path 1-3 at t=1.5 ms are plotted in Figures 6 and 7. As is seen that the same results are obtained for input G and input acceleration method.

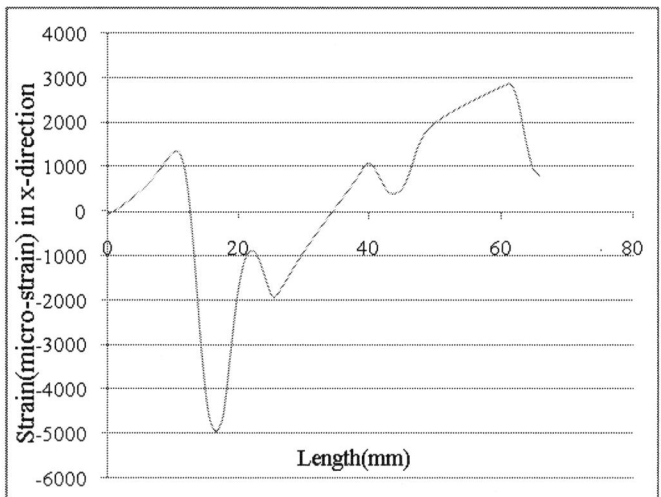

Figure 6. Board strain for input G method (full dynamics)

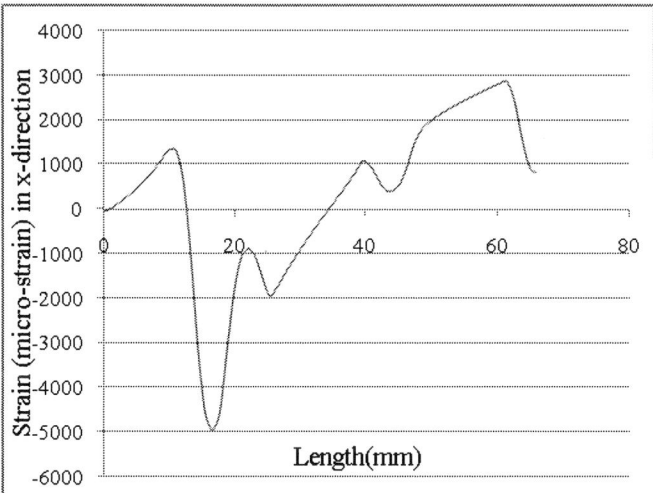

Figure 7. Board strain for input acceleration method (full dynamics)

4.3 Comparison of Full Dynamics and Mode-Superposition Method

The board strains calculated with both full dynamics and mode-superposition methods are is plotted in Figure 8. It is seen that board strains obtained with both solution methods have the same trend. However, peak value of strain is different. It is approximately 5000 micro-strain for full dynamics while it is approximately 4000 micro-strain for mode-superposition method. The board strain and displacement (in z-direction) obtained with the two methods are plotted in figures 9 and 10. It can be seen that solutions with both methods follow the same trends. However, the magnitudes of strain and displacement solutions with these two methods are different. The mode-superposition method always seems to give numerically less value than the full transient analysis.

It is observed from Figures 9 and 10 that the maximum elastic strain occurs near the mounting hole, while the board center has the most deflection. Furthermore, the mounting hole region is bent in the opposite direction compared to the board center.

It should be pointed out here that there are some problems regarding post-processing in ANSYS for the mode-superposition method. The acceleration-time history cannot be plotted. It is sometimes difficult to obtain strain history plots as well. In addition, it takes longer time to post process obtain the full solution by expansion (using EXPASS command). The mode-superposition occupies more memory space than full dynamics. For a case studied, with 6x6mm chip size, mode-superposition method takes more than 4 times the memory space than that of full dynamics. The full dynamics, on the other hand, takes longer time to calculate the full solution. However, it takes overall shorter time since the post processing is fast. Therefore, full dynamics approach is preferred.

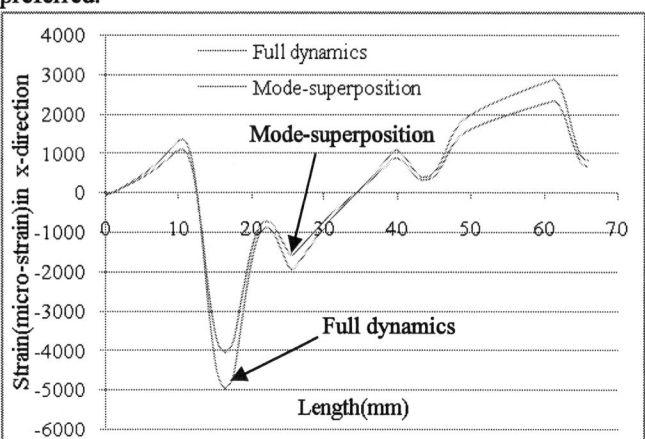

Figure 8. Comparison of board strain full dynamics and mode-superposition method

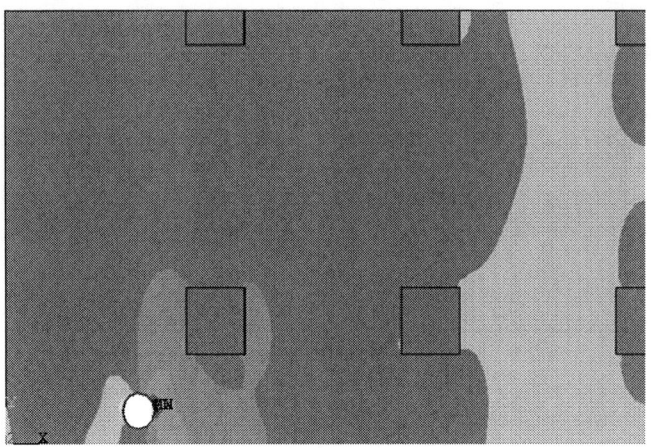

Board strain in x-direction at time 1.5 ms
Full dynamics analysis method

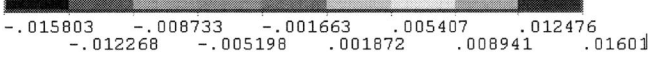

```
-.015803    -.008733    -.001663    .005407    .012476
     -.012268    -.005198    .001872    .008941    .01601
```

Mode superposition
method

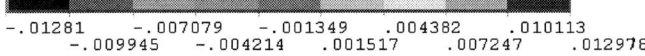

```
-.01281    -.007079    -.001349    .004382    .010113
     -.009945    -.004214    .001517    .007247    .012978
```

Figure 9. Comparison of board strain for input acceleration
and input G

Fig. 11 and 12 plot elastic strain history at location 1 and location 6, respectively for both strains in x- and y- direction. We can see that strain components in x-direction and y-direction at the board corner (1mmx1mm from U1) are much higher than those at board center (1mmx1mm from U8) and bending direction is opposite. Strain in y-direction has higher frequency than the strain in x-direction.

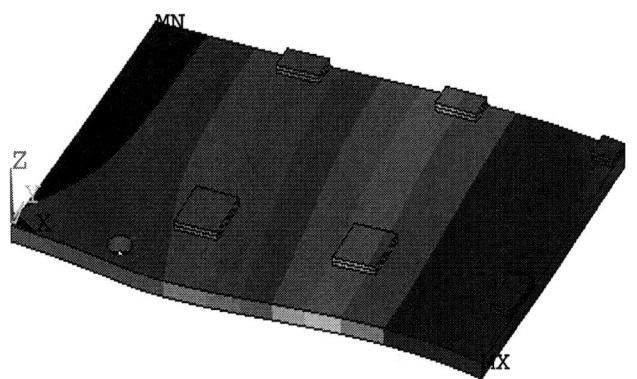

Board displacement in z-direction at time 1.5 ms
Full dynamics analysis method

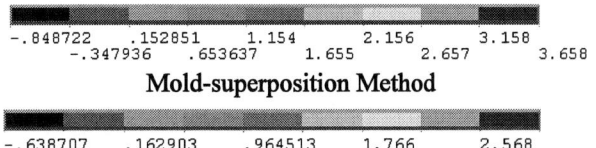

```
-.848722    .152851    1.154    2.156    3.158
     -.347936    .653637    1.655    2.657    3.658
```

Mold-superposition Method

```
-.638707    .162903    .964513    1.766    2.568
     -.237902    .563708    1.365    2.167    2.969
```

Figure10. Comparison of board displacement (in z-direction)
for full dynamics and mode-superposition method

The board displacements in x and y directions for site 1 and 6 (figure 1) are plotted in figures 11 and 12. There are two three observations:

a. the strain in x direction is dominant

b. the strain components in both x and y directions are opposite between locations 1 and 6.

c. Strain component is x direction corresponds to the fundamental natural frequency, and the strain component is y direction corresponds to the 2nd fundamental frequency.

In the following discussion, PDCB strain in x direction ε_x is used to quantify the drop impact performance of WLP. Larger ε_x corresponds to larger PCB bending and worse WLP drop impact performance.

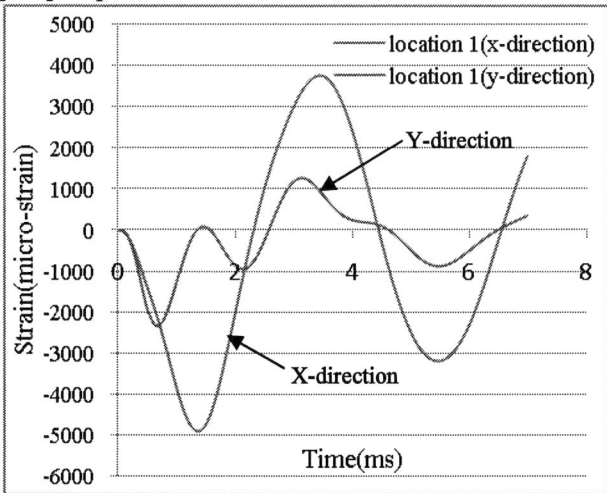

Figure 11. Board strain at location 1

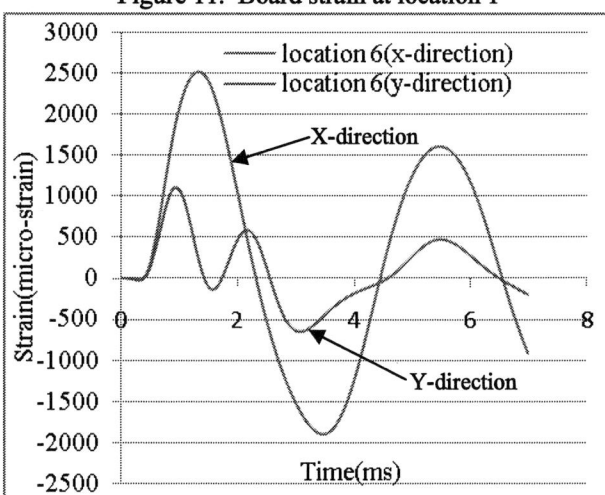

Figure 12. Board strain at location 6

4.4 Comparison of 8 Node vs. 20 Node Finite Element

Two element types, 8 node element (solid 45) and 20 node element (solid 95) may be used for the model. The general expectation is that 20 node element gives more accurate results but the model will take much more computer space and significantly longer time to solve the problem. In order to quantify the effect of element choice comparison is done between the models with these two element types. The results obtained with these two element types are plotted in Figure 13. It is shown that the ε_x difference obtained with 8 node and 20 node elements is trivial (<7%). The results differ mainly at the peaks. Therefore, 8 node element can be used to calculate PCB global dynamics without sacrificing accuracy. The benefit of using 8 node element is significant reduction of computational time. For a typical model, 8 node element reduces the computational type by 2/3, compared to 20- node element. Such results are observed differently when ABAQUS software is used [15].

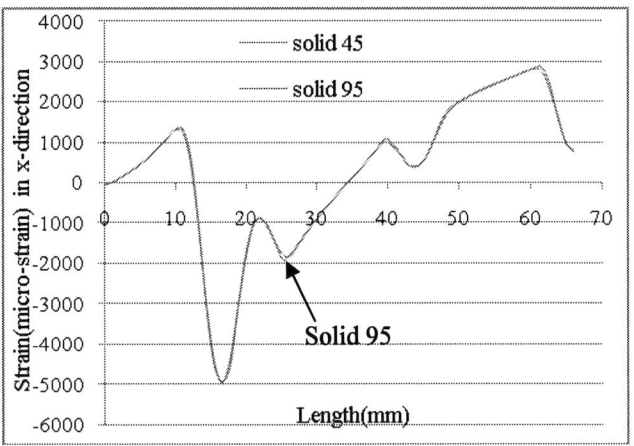

Figure13. Comparison of board strain for 8 node and 20 node elements

5. Results and Discussions

5.1 Effect of the Boundary Condition at Mounting Hole on System Natural Frequency

The boundary conditions at mounting the hole have some effect on the natural frequencies of the system. A 6x6 mm chip size model is studied for this effect. Two types of boundary conditions are considered. The first one is fixing displacement in z-direction only. The second boundary condition is fixing displacement in all directions at mounting hole. The natural frequencies obtained from models with these two different boundary conditions are listed in table 3. It is seen that slightly higher natural frequencies are obtained when displacements of all directions are fixed at mounting hole. This is due to the fact that the board becomes less flexible under this boundary condition. Overall, the fundamental frequency is from 200 to 250 Hz, which is consistent with many test results.

Table 3. Effect of boundary conditions on natural frequency of the model

Mode No.	Natural frequency (Hz)	
	Only Z direction is fixed	All (X,Y,Z) are fixed
1	212	241
2	551	638
3	915	939
4	1257	1338
5	2193	2272

5.2 Effect of Chip Size on System Frequency

To study the effects of chip size on the natural frequency of the system, the chip sizes of 3x3, 4x4, 5x5 and 6x6 mm are considered. The corresponding solder ball arrays are 6x6, 8x8, and 10x10, respectively. For this comparison the displacement in z-direction only is fixed. It is seen in Table 4 that natural frequencies of the system slightly increase with increase in the chip size.

At this point, it is of interest to understand the natural frequency shift. The natural frequency of a system can be expressed by eq. (1). The frequency depends on the stiffness and mass of the system. As the chip size increases the stiffness and mass of the system also increases. In this case however, the board stiffening is dominant leading to increase in the frequency of the system.

$$f := \frac{1}{2\pi}\sqrt{\frac{k}{m}} \qquad \text{eq. (1)}$$

Where k = stiffness of system, N/mm

m = mass of the system, gm

Table 4. Effect of chip size on natural frequency of the model

Mode Number	Natural Frequency (Hz)			
	size (3x3)	size (4x4)	size (5x5)	size (6x6)
1	209.11	209.74	210.59	211.59
2	542.41	544.46	547.24	550.66
3	879.94	887.84	899.59	915.22
4	1234.9	1239.3	1246.4	1256.6
5	2118	2135.8	2161.0	2192.6

5.3 Board Strain Distribution

The board strain in x-direction at 6 different locations is plotted in Figure 14. and following observations are made. The strain location is at points 1, 2, 3, 4, 5 and 6 which are at 1x1 mm away from the their respective package corners at the PCB component side, as defined in Figure 1. Components at center column (U3 and U8), the next column (U2 and U7) are subjected to bending in the same direction. In addition, U3 and U8 are subjected to a larger bending than U2 and U7. However, the outer column components U1 and U6 are subjected to bend in opposite direction. It is important to note that due to the mounting screw effect, the peak board strain at the corner of U1 is greater than that at U8. This observation is from the results showed in Syed's paper. Figure 15 plots the strain in x-direction at 6 different locations in the center of each component. It is found that chip center strain values are much loert compared to the corner strain. It is also noted that the center strain at U8 is greater than that at U1. From the Figure 14, the peak positive strain values at the corner of the component can be ranked in the order of U1>U3>U8>U2>U7>U6.

It is generally expected that drop test failure is due to the peeling stress at solder joint caused by board bending. The board strain is propotional to the degree of board bending. Since there is significantly different board strain values during the drop, the drop impact life of a given WLP is different for different component locations. Based on results of this study, the drop impact lift of a given WLP can be ranked in therms of component locations: U1<U3<U8<U2<U7<U6. This is in agreement with experimental results.

It is important to review the effectivness of the JEDEC board specification JESD22-B111. The JEDEC drop test is mainly used for relative component performance. However, the drop impact performance is not unique for a given test. It dependends on the component locations. In addition, the corner component locations give worst drop impact performance. This has nothing to do with the reliability of the component itself. Rather, it is affected by mounting screw. Including the corner components in JEDEC drop test evaluates the effect of mounting screws very close by, instead of assessing the intrinsic drop impact performance of the components itself. An improvement of JEDEC drop test would be excluding the corner components for drop impact performance relative comparison among packages.

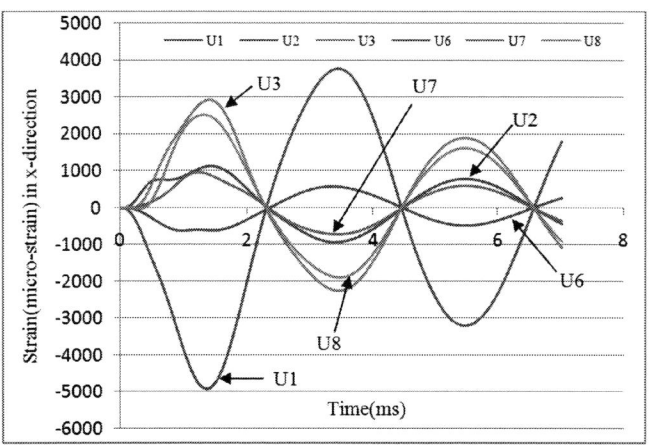

Figure 14. Comparison of board strain at chip corners

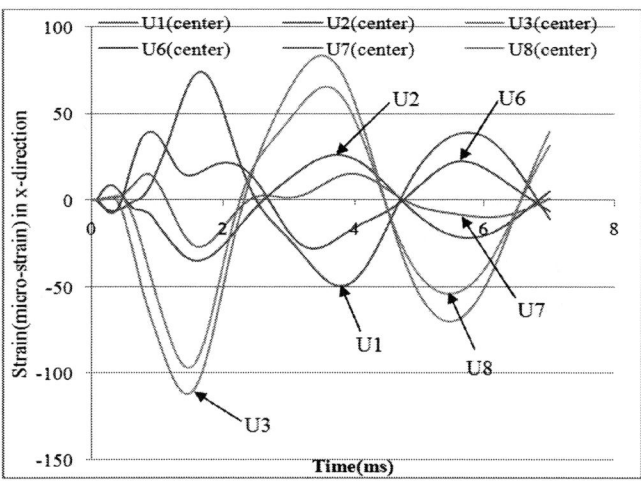

Figure 15. Comparison of board strain at the center of chips

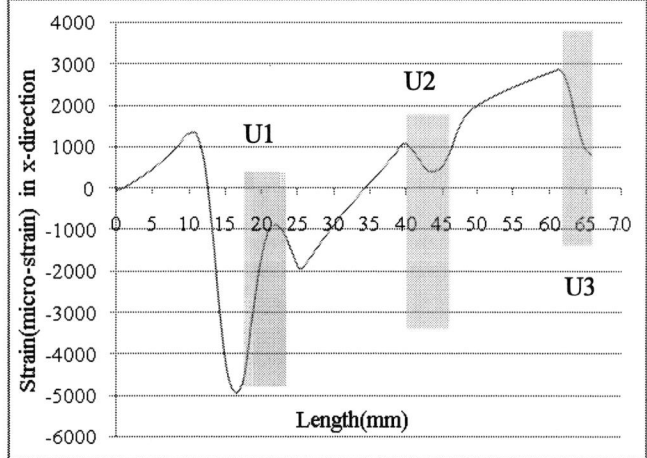

Figure 16. Board strain distribution at specific chip loactions

Figures 16 plots the board strain at t=1.5ms U1-U2-U3 paths, one through center of components and one is 1 mm

away from component edges. It is shown that the board strain amplitued along the path 1 mm away from the component is in general higher the path through the center of the components. This is because the first path captures the maximum strain at component corners. It is also observed that at the center of the chip the strain is minimum due to board stiffening by the component. In figure 17 the shaded region in this figure is where the component is mounted. Again it is seen that the board strain beneath each chip has a 'U' shape pattern, which means that the center strain is the lowest. There is a very large strain gradient close to the package edge. This suggests that the strain gage placed in the center of compoenets on the PCB cannot reflec the board bending behavior, but will be good for the simulation validation. The strain gage at the package corner will be very sensitive to the location due to the large gradient shown in this figure. Between U1 and U2 the vending direction changes. And there is a stationary point (node). When a strain gage is attached in the middle of two component, it might not give useful information since it might be close to the node.

5.4 Effect of Acceleration on Board Strain Disribution

Board strain is studied for the effects of different loading cases of accelerations. Here input acceleration of 1000g, 1500g and 2000g are considered. The board strain history in x-direction at t=1.5ms is plotted in figure 17 for applied loads. It is shown that with increase in the G load the value of strain increases.

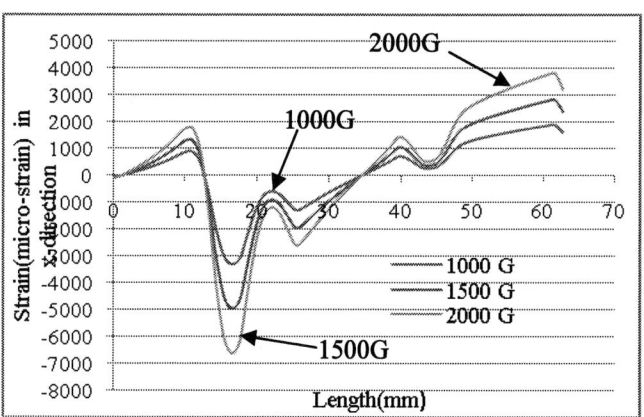

Figure17. Comparison of board strain for different values of acceleration load

5.5 Effect of Chip Size on Board Strain

The effect of chip size on the board strain is studied. The chip sizes 3x3, 4x4, 5x5, and 6x6 are taken into consideration to see how the chip size influences the board strain. The board strain is plotted in figure 18 for different chip sizes. It is seen that as the chip size increases the board strain in the chip region decreases. This is because large components stiffen the board more. The strain amplitude peak near component increases with chip size increase, while strain far away from the component location does not change with the size of the chip. Therefore, it is essential to place the strain gage close to the edge of the components to capture the variations. It can be seen that the board strain amplitude increase with chip size is not significant. The detailed study on the effect of chip size on solder joint stress will be reported in the future.

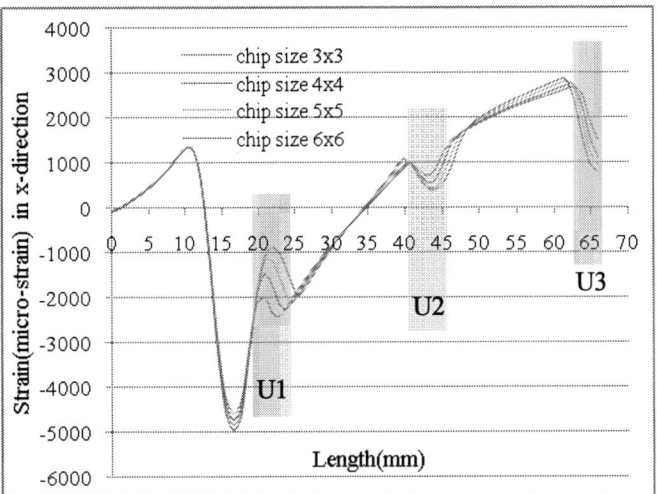

Figure18. Comparison of board strain for different chip sizes

5.6 Effect of Boundary Conditions for Direct Input Acceleration Method

In order to assess the effect of boundary conditions, two boundary conditions at mounting hole are considered. This first case incorporates displacemet components in x, y, and z directions for nodes at the mounting hole being fixed. The second case incorporates displacement in z-direction only is fixed for nodes at the mounting hole. The models with these two sets of bondary conditions are studied with input acceleration method. The board strain is plotted in figure 19. It is seen that slightly different results are obtained for the two different boundary conditions. The theoretical study on the model study will be reported in the future.

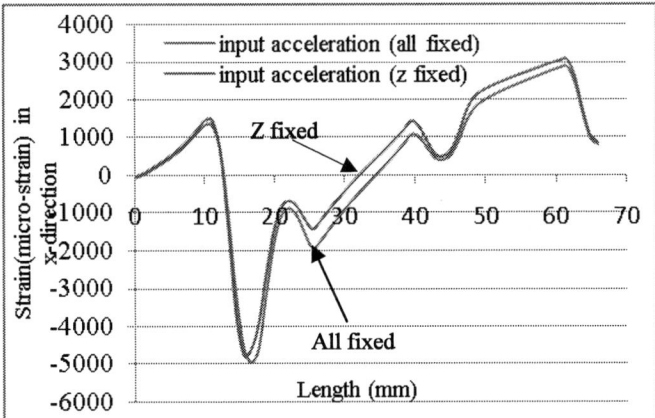

Figure 19. Comparison of board strain for different boundary conditions in input acceleration method.

6. Conclusions

Based on the studies in this paper, it is recommended that, in order to achieve the computational efficieny without the loss of accuracy, a solder layer model for wafer-level packages with direct acceleration input using full transient implicit anaysis provides an accurrate and fast board dynamic response analysis. The mode superposition in ANSYS does not have an advantage in saving computation time since the solution expansion process is needed. The input-G and direct acceleration methods give exactly same results with certain boundary conditions. Neither Input G or Dierect Acceleration Input method truly represents the actual loading conditions during impact. A thereotrcial model study will be reported in the future.

Components near mounting holes may fail first during drop test. This failure rate has nothing to do with the package intrinsic drop impact performance, rather it is due to the effect of the mounting screw. An imporvement to JEDEC drop test procedure may be excluding the corner components when reporting the drop test performance of a package.

This paper focuses on the board global dynamice response analysis only. The solder joint stress analysis and its relationship with the board strain is not included in this paper due to space limitation. The detailed local modeling and solder joint stress analysis will be presented in a separate paper.

References

1. JEDEC Standard JESD22-B111, Board Level Drop Test Method of Components for Handheld Electronic Products.
2. Lim, C.T. and Low, Y.J., "Drop Impact Testing of Portable Electronic Products," 52nd ECTC Conference Proc., 2002, pp. 1270-1274.
3. Lim, C.T. and Low, Y.J., "Drop Impact Survey of Portable Electronic Products," 53rd ECTC Conference Proc., 2003, pp. 113-120
4. Tee, T.Y., Luan, J.E., Pek, E., Lim, C.T., and Zhong, Z.W., "Novel Numerical and Experimental Analysis of Dynamic Responses under Board Level Drop Test," EuroSim Conference Proc., May, 2004.
5. Tee, T.Y., Luan, J.E., Pek, E., Lim, C.T., and Zhong, Z.W., "Advanced Experimental and Simulation Techniques for Analysis of Dynamic Responses During Drop Impact," ECTC 2004, pp. 1089 –1094.
6. Zhu, L., "Modeling Technique for Reliability Assessment of Portable Electronic Product Subjected to Drop Impact Loads," 53rd ECTC Conference Proc., 2003, pp. 100-104.
7. E.H. Wong, K.M. Lim, Norman Lee, Simon Seah, Cindy Hoe, and Jason Wang, "Drop Impact Test - Mechanics and Physics of Failure, 4th EPTC, pp. 327-333, Singapore, 2002.
8. Tee,T.Y., Hu Shen Ng, Lim Chwee Teck, Pek Eric and Zhong Zhaowei, "Board Level Drop Test and Simulation of QFN Packages for Telecommunication Applications," ICEP03 conference, Japan, April 200, pp.221-226.
9. Syed Ahmer, Kim Mo Seung, Lin Wei, Khim Young Jin, Song, Sook Eun, Shin, Hyeon Jae, Panczak Tony, "A Methodology for Droop Performance Prediction and Application for Design Optimization of Chip Scale Packages," 2005 Electronic Components and Technology Conference.
10. Sogo T. and Hara S. "Estimation of Fall Impact strength for BGA Solder Joints," ICEP conference Proceedings, Japan 2001, pp.369-373.
11. Liu,Yumin, Liu,Y., "Board level drop test simulation for an advanced MLP, ICEPT 2007

12. Jing-en Luan and Tong Yan Tee, "Novel board level drop test simulation using Implicit Transient Analysis with Input-G Method", 6th EPTC Conference, Singapore, Dec. 2004, pp. 671-677.

13. Loh Wei Keat; Lee Yung Hsiang; Ajay Munigayah; Tay Tiong We, "Nonlinear dynamic behavior of thin PCB board for solder joint reliability study under shock loading", International Symposium on Electronics Materials and Packaging, 11-14 Dec. 2005 Page(s): 268 - 274

14. Lianxi Shen, "Simulation of drop test board with 15 components using explicit and implicit solvers", 2008 International ANSYS Conference August 26 to 28 in Pittsburgh, Pennsylvania, U.S.A.

15 Xuejun Fan; Min Pei, and Pardeep K. Bhatti, "Effect of finite element modeling techniques on solder joint fatigue life prediction of flip-chip BGA packages", IEEE Electronic Components and Technology Conference (ECTC), 2006, May 30 - June 2, San Diego, CA

Effect of Shear Rate on Lead Free Solder Joint Strength

Zheming Zhang[1,2], Jingshen Wu[1,2], Adam R. Zbrzezny[3], Neil Mclellan[3]

1. Department of Mechanical Engineering, The Hong Kong University of Science and Technology, Clear Water Bay, Kowloon, Hong Kong, China
2. Center for Engineering Materials and Reliability, Fok Ying Tung Graduate School, The Hong Kong University of Science and Technology, Clear Water Bay, Kowloon, Hong Kong, China
3. Package Development Engineering, AMD, Markham, Canada

Abstract

Solder joint strength at high strain rates is a critical reliability requirement for portable electronic devices. Experimental observations showed low shear rate tests of solder joints cannot be used to accurately predict the mode of deformation and failure behaviors under high speed impact condition. Due to strain rate effect, brittle failure of solder joints under dynamic loadings may not take place at low strain rates. Thus, characterization of solder joints under impact becomes critical in package design and manufacturing for high reliability. This is particularly true for lead-free solder in handholds devices.

Present study focused on the deformation and failure behavior of single solder ball joints under impact conditions. With a newly designed single-ball impact tester, impact strength of ball joints of two lead-free solders, i.e. Sn-1Ag-0.5Cu (SAC105) and Sn-1.2Ag-0.5Cu-Ni (LF35), were tested at a speed varying from 0.5 to 3m/s. Peak load and fracture energy obtained at different speeds were compared. The failure modes of the joints were studied by scanning electron microscope (SEM) and correlated to the peak load and total fracture energy which were strain rate dependent and changed with solder ball composition due to different strain hardening effects of the two lead-free solders.

Introduction

Ball grid array (BGA) is considered one of the most popular packaging solutions for high density integrated circuits and has been widely applied in the electronic industry. In the past, BGAs were mostly based on tin-lead alloys. Due to environmental concerns, tin-lead solders have been gradually replaced by lead-free solders in the industry [1]. IHowever, brittle failure of solder joints was frequently observed with BGA units using lead-free solder balls under high speed impact conditions. Hence, characterization of solder joint strength at high speed and understanding the fracture behavior of the joints become an urgent need of the industry.

Solder joint strength is traditionally characterized by the single ball shear test [2] and the drop down test [3]. Due to the limitations of testing devices, the traditional single ball shear test can only be performed at low shear rate [4], generally at 800µm/s or below. At slow strain rates, solder ball joints fracture in most cases in a ductile mode and this does not reflect the true strength of the solder joints under impact loading, in which brittle fracture is very often found. Thus, static single ball shear test cannot be used to evaluate the solder joint strength under a dynamic loading condition [5]. On the other hand, the drop down test is used to test the solder joint strength after BGA units are surface mounted on PCB, i.e. after two reflow processes and under the influences of PCB constraint. Thus, the results of the drop-down test cannot reflect the solder joint strength of a BGA unit before surface mounting. Moreover, drop-down test takes a relatively long time for sample preparation. It is also a complicated and costly test compared to the single ball impact test that can test the joint strength of any solder ball on the BGA unit at any stage of the assembly process at a required strain rate.

In the present study, a single ball impact tester was designed and manufactured. The solder joint strength of a BGA unit using Sn-1Ag-0.5Cu (SAC105) and Sn-1.2Ag-0.5Cu-Ni (LF35) solder balls was studied and compared in terms of impact load and fracture energy at shear rates from 0.5 to 3.0 m/s. The failure mechanisms under these high strain rates were compared with those found in static shear tests.

Experiment Procedures

The newly designed single solder ball impact tester can provide stable speeds from 0.5 to 4.0 m/s. The load data were acquired by a sensor of high sampling rate (> 5MHz) during impact event that lasts about 100µs. The high sampling rate data acquisition system was calibrated using other standard testing systems. Figure 1 schematically shows the working principle of the impact tester.

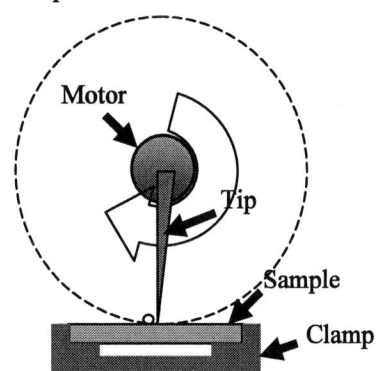

Figure 1 Single solder ball impact tester

Two samples with different alloy compositions, i.e. Sn-1Ag-0.5Cu (SAC105) and Sn-1.2Ag-0.5Cu-Ni (LF35), were used to study the composition effects on the solder joint strength. The diameter of the solder balls is 300µm and solder pad surface was coated by OSP technique.

Load-displacement curve (L-D curve) was obtained during the test. The maximum load point and the fracture energy, which was acquired by integrating the L-D curve, were used to define the solder joint strength in terms of force and fracture energy at different shear rates.

Both static and dynamic shear tests were conducted for the samples. The test conditions were listed in Table 1. Scanning electron microscopy (SEM, JEOL 6390) and EDX (INCA) were applied to investigate the fracture surface after the shear tests.

Table 1 Experimental matrix

Method	Static Shear	High Speed Impact
Equipment	DAGE 4000S	Lab-made impact tester
Shear Rate	0.5mm/s	0.5; 1.0; 3.0m/s
Ram Height	50μm	

Intermetallic component layer

Intermetallic component (IMC) layer forms through Copper diffusion between solder ball and copper pad. Figure 2 shows the geometry of the ball-pad interface schematically.

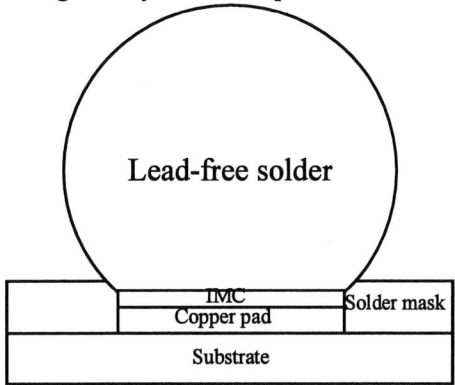

Figure 2 Solder ball geometry

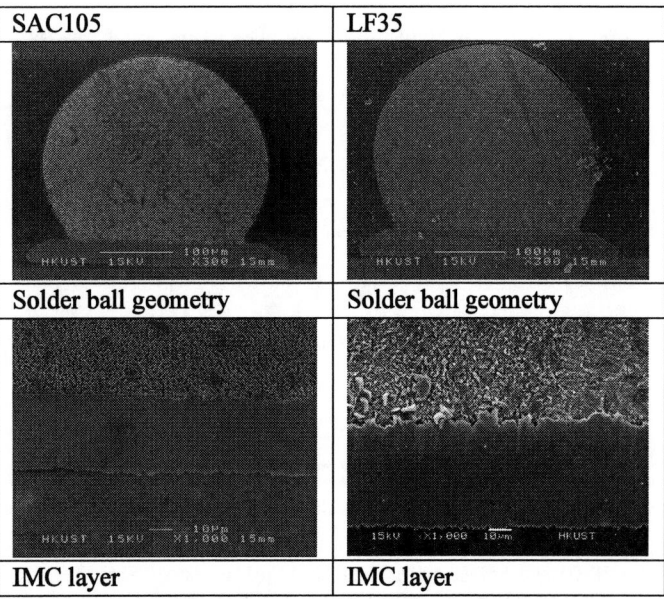

Figure 3 Two solder ball samples

From the cross-section pictures in Figure 3, the average IMC thicknesses of SAC105 and LF35 were about 4μm and 5μm, respectively. EDX observation shows that the IMC composition of both samples was Cu_6Sn_5

Load-displacement curve

The L-D curves recorded by Dage4000S machine and our lab made impact tester are showed in Figure 4-7.

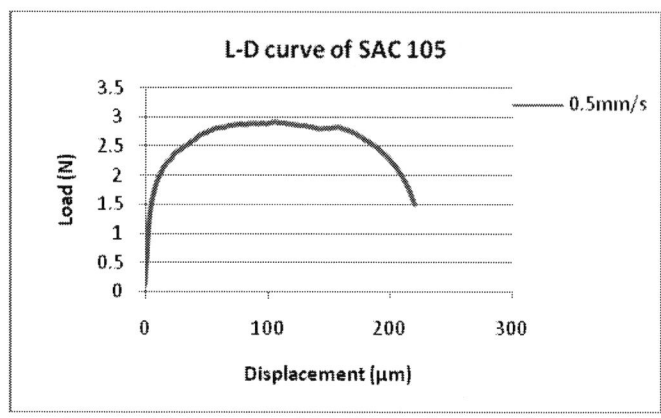

Figure 4 L-D curve of SAC105 at low speed shear

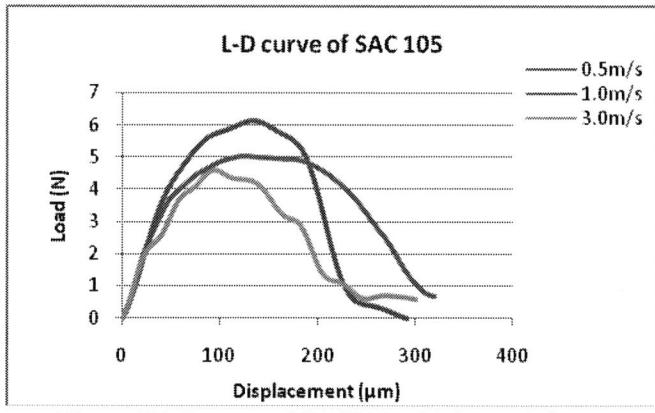

Figure 5 L-D curve of SAC105 at high speed shear

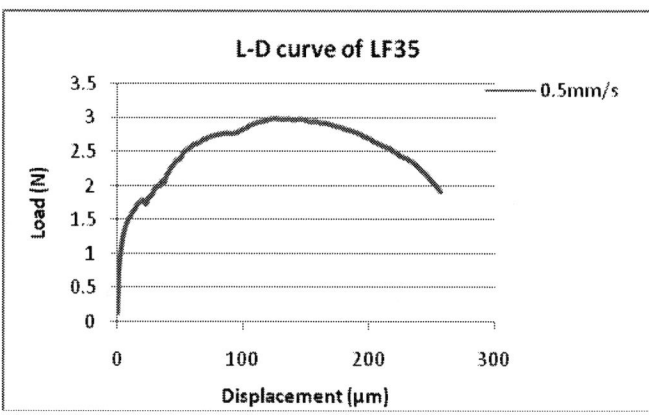

Figure 6 L-D curve of LF35 at low speed shear

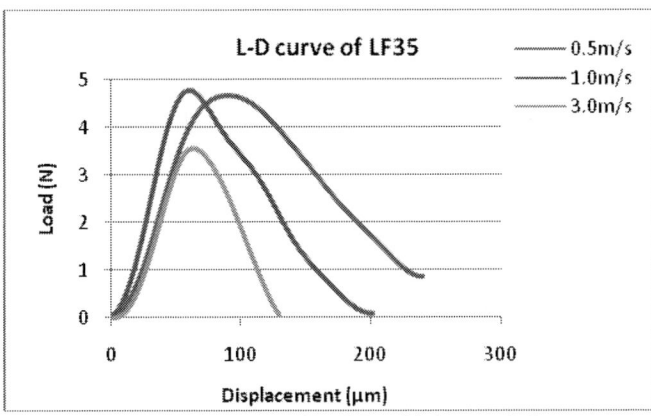

Figure 7 L-D curve of LF35 at high speed shear

Discussion

From the curves, the peak load and fracture energy were extracted and used to benchmark the solder joint strength. The peak load for each experiment was directly read from these graphs. The fracture energy was obtained by integration the L-D curves. Figure 8 and 9 showed the fracture energy and load results obtained at different shear rates.

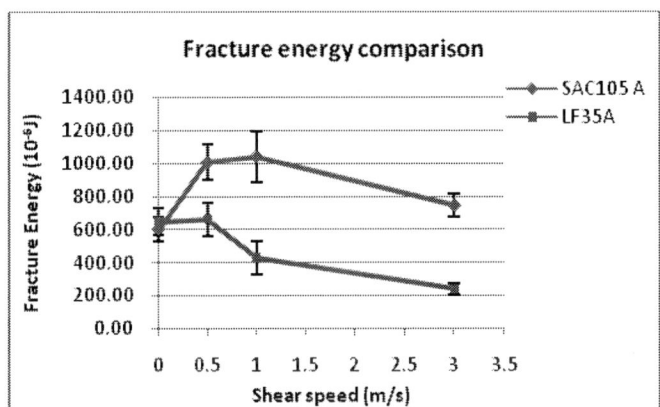

Figure 8 Fracture energy comparisons at difference speed

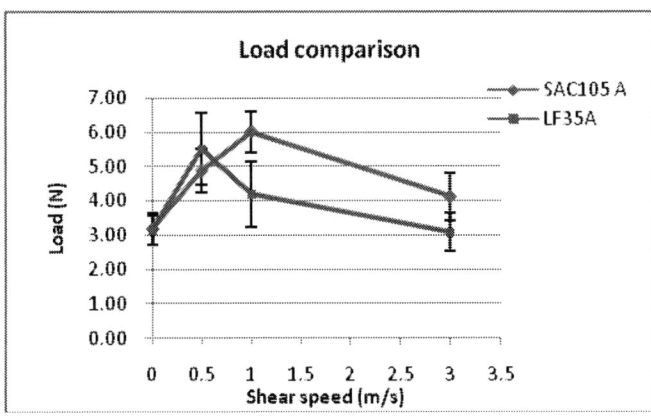

Figure 9 Load comparisons at different speed

The SEM pictures of cross-section were recorded after shear test. Ductile or brittle failure mode can be easily recognized in the fracture surface from Figure 10 and 11. From EDX results, the crack propagated in IMC layer when brittle failure happened.

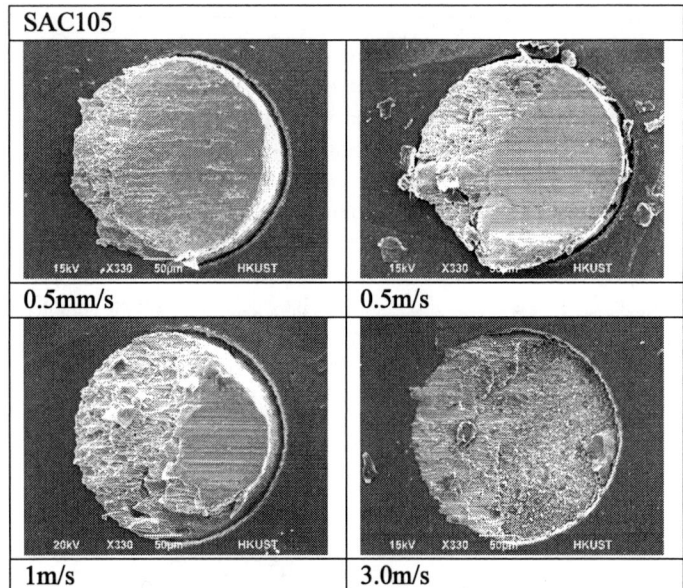

Figure 10 Fracture surface of SAC105

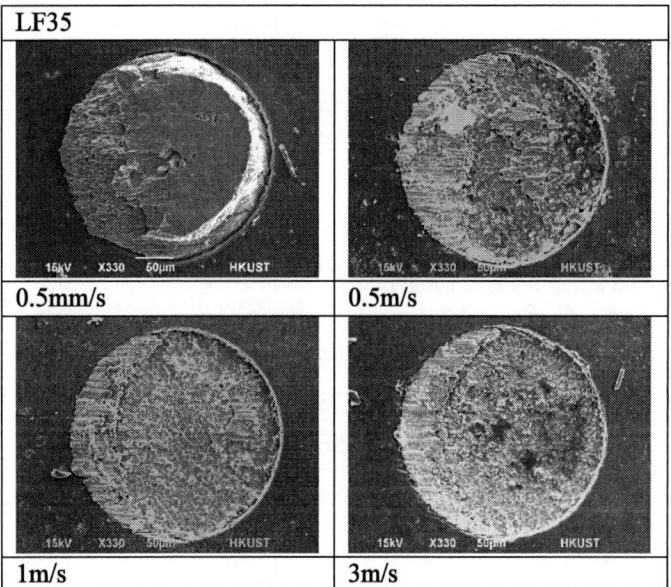

Figure 11 Fracture surface of LF35

The results showed that the shear rate affects the fracture energy and failure mode evidently. Figure 8 and 9 showed that at low speed shear, the two solder joint strengths were virtually the same. With the shear rate increasing, the strength of solder joint increased first, and then decreased.

Figure 10 and 11 showed that brittle failure happened at high shear rate, and the area of brittle fracture surface in both samples increased with shear rate increasing. Compared with fracture energy, it can be found that the maximum fracture energy appeared when the ductile-brittle transition happened.

The effects of alloy composition on the solder joint strength are obvious. The ductile-to-brittle transition of the two solders takes place at different rates. Complete brittle failure happened at 3.0m/s in SAC105 but at 0.5m/s in LF35.

From Figure 8, it is clearly seen that the solder joint strength of SAC105 was a little lower than that of LF35 at shear rate of 0.5 mm/s but much higher at high speed shear tests. Based on a previous work [7], the Vickers hardness of SAC105 was found a little lower than that of LF35 at static situation. It means that the cutting resistance (shear fracture strength) of SAC105 is lower than that of LF35. From the SEM photos shown in Figures 10-11, we noted that the failure of the two solder balls at 0.5 mm/s was due to solder ball cutting-off by the shear-ram. Thus, the higher the cutting resistance is, the higher the shear strength will be.

For the high shear rate test, dynamic mechanical models were used to explain the results. The Split Hopkinson Tensile Bar test results [8] and Cowper-Symonds model was used to get the mechanical property at high strain rate situation. Based on Cowper-Symonds model [9],

$$\frac{\sigma_d}{\sigma_b} = 1 + \left(\frac{\dot{\varepsilon}}{B}\right)^{\frac{1}{q}}$$

Where is the dynamic flow stress and is static flow stress, is the strain rate, B and q are material constants. The material values of different solders were listed in Table 2.

Table 2 Material property of different solders [8]

Material	σ_d	B	q
SAC105	26MPa	106	2.35
LF35	24MPa	34	2.90

Based on our previous finite element analysis [8], the maximum strain rate at 0.5m/s and 2m/s was about 5000 /s and 23000 /s respectively. Therefore, the flow stress of those two materials was listed in Table 3.

Table 3 Flow stress of different solders at different shear rates

Material	0.5m/s	2m/s
SAC105	160MPa	283MPa
LF35	148MPa	261MPa

Flow stress is defined as the instantaneous value of stress required to continue deforming the material [10]. High flow stress indicated the resistance of bulk solder was large. Therefore more energy would be consumed during the impact condition. From the fracture energy in Figure 8, we noted that the fracture energy in SAC105 was higher than that in LF35. However, some other facts may also determine the fracture energy, the flow stress had a big influence to the solder joint strength.

Based on the results above, four different phenomena would be discovered when shear rate increasing:
1. **Ductile failure mode.** At static shear test (in most case, less than 1mm/s), the failure happens inside the bulk solder. The value of break load and fracture energy reflects the shear strength of solder ball. In this case, the interfacial strength between the ball and pad is normally higher than the shear fracture strength of the bulk solder; hence, the test results do not reflect the joint strength at the ball/pad interface.
2. **Hardening mode.** At moderate strain rates (in most cases, less than 1m/s), the failure also happens in bulk

solder. Because of strain rate hardening effect, the solder shear strength increases with the test speed increasing, which leads to a higher impact strength.
3. **Ductile-brittle transition zone.** Further increasing the shear rate, the solder joint would run into ductile-brittle transition zone. The fracture mode starts to change from ductile to brittle. The peak load and fracture energy fluctuate significantly from sample to sample when test was conducted in the ductile-brittle transition zone.
4. **Brittle failure mode.** When increasing the shear rate to a very high level, all solder joint failure takes place in brittle mode. The fracture energy and peak load decreases significantly at very high shear rates.

Conclusions

This work investigated the shear rate effect and solder ball composition on the solder joint strength. The following conclusions can be drawn from this study:
1. High speed ball shear test is a better method to evaluate solder/copper interconnection strength, because at low shear rate, the failure often happens in bulk solder.
2. Ductile and brittle failure modes were identified in high speed impact tests. From the microstructure analysis, it was found that the brittle area of the fracture surface increased with the impact speed. Ductile fracture happened inside bulk solder, while the crack propagated in the IMC layer when brittle failure happened.
3. Before the ductile-brittle transition zone, the fracture energy increased with increasing shear rate. After that zone, the fracture energy would decrease when the shear rate increased.
4. SAC105 solder had a better performance than LF35 under high shear rate condition. However the flow stress of SAC105 was also bigger than that in LF35. It indicated the flow stress had some effects to solder joint strength.
5. Based on the dynamic material property of solder ball and Cowper-Symonds model, solder joint strength caused by strain hardening effect of different solder components can be predicted.
6. Shear rate effect was showed in this study. With increasing shear rate, solder joint failure takes place in four different modes, i.e. ductile mode, hardening mode, ductile-brittle transition zone, and brittle failure mode.

References
1. John H L Pang; Luhua Xu; X Q Shi; W Zhou; S L Ngoh, 2004. Intermetallic growth studies on Sn-Ag-Cu lead-free solder joints. Journal of Electronic Materials.
2. JESD22-B117A, JEDEC Solid State Technology Association, 2006.
3. JESD22-B115, JEDEC Solid State Technology Association, 2007.
4. Xingjia Huang, S.-W. Ricky Lee, Chien Chun Yan, and Sam Hui, 2001. 2001 Electronic Components and Technology Conference.
5. Zheming Zhang, Zhengjian Xu, Jingshen Wu, Tongxi Yu, Adam R. Zbrzezny, Neil Mclellan, Yeong J. Lee,

2007. Impact Strength of Solder Ball Interconnects. Wafer Level Packages IPC/JEDEC International Conference on Lead-Free Electronics 2007.

6. M.Date,T.Shoji, M.Fujiyoshi, K.Sato and K.N.Tu., 2004. Impact reliability of solder Joints. 2004 Electronic Components and Technology Conference.

7. Masamoto Tanaka, Tsutomu Sasaki, Takayuki Kobayashi and Kohei Tatsumi. Improvement in Drop Shock Reliability of Sn-1.2Ag-0.5Cu BGA Interconnects by Ni Addition, 2006. 2006 Electronic Components and Technology Conference.

8. Zhengjian Xu, Zheming Zhang, Jingshen Wu, Tongxi Yu, Adam R. Zbrzezny, Neil Mclellan, Yeong J. Lee, 2007. Mechanical properties of lead-free solder alloys and finite element simulation. Wafer Level Packages IPC/JEDEC International Conference on Lead-Free Electronics 2007.

9. Guoxing Lu and Tongxi Yu Energy Absorption of Structures and Materials Woodhead Publishing Limited, Cambrige, UK.

10. Mikell P. Groover, 2007, "Fundamentals of Modern Manufacturing; Materials, Processes, and Systems," Third Edition, John Wiley & Sons Inc.

Analysis and Comparison of Thermal Stress and Hygrothermal Stress of SiP Device By QFN Packaging

Jiang Haihua, MA Xiao-song, Chen Ning
Guilin University of Electronic Technology, Guangxi, 541004, China
E-mail:hh_jiang@126.com

Abstract:

In this paper, the model is built by using the SIP structure. The SIP structure is consist of two chips which were using QFN flip-structure. The finite element analysis is used to simulate and calculate the distribution of moisture diffusion under thermal and moisture condition, then to simulate the thermal stress and hygrothermal stress. The results showed that thermal stress are mainly distributed in the mismatch of CTE (Coefficient of Thermal Expansion) between the materials. The largest thermal stress lie in the junction of EMC, DA and chips. We can see that it has a larger thermal effect to the SIP devices used QFN packaging inside than general QPN.The distribution of hygrothermal stress, which is similar as thermal stress in trend, is mainly in mismatch of CTE and CME Coefficient of Moisture induced Expansion). However, moisture gradient, which is caused by structure and the uneven diffusing of moisture become larger. It due to moisture difference that hygrothermal stress is much higher than the thermal stress, its stress concentration appears obviously at the junction between the materials. If the interface of the two materials exist defects, they would lead to cracks and cause the failure of the device. In the area of inside device there is less moisture diffused, its hygrothermal stress distribution is similar to general thermal stress, which is due to moisture transfer firstly through the EMC materials then into the device inside. Moreover, the internal bonding material (DA) is thin, only can absorp moisture from EMC, resulting in only a little moisture can diffuse into the DA material, then cause higher stress gradient. From the simulation results, we can see that the greater stress gradient will lead to warpage of the device, it will prone to more damage in the areas stress concentration. It is larger effect on reliability on SiP device in which used QFN packaging than the general QPN device. Therefore, the effect of EMC and DA's moisture on the device's stress can not be ignored. And the reaults show that the maximum hygrothermal stress is 1.56 times as thermal stress. Therefore, the effect of moistrue and thermal in reliability on the device can't be ignored.

1. Introduction:

As microelectronic components packaging density continuously improve, figure size continued to narrow, and expand the scope of application, leading to the working temperatures of electronic devices becomes higher and performance dropped significantly. The issue of thermal has aroused more and more concerns. In addition, in various types of micro-electronics package, there are the vast majority of plastic packaging, accounting for more than 90 percent [1]. As plastic (or epoxy) inherent structure of macromolecules, these have high moisture absorption. The reliability issues of devise become more prominent because of material moisture absorption.With increasing of temperature thermal stress arising even lead to failure. With hygro-mechanical stress

cause by the material moisture absorption coupled with thermo-mechanical stress in the high-temperature welding process, it is possible that lead to cracks and cause the failure of the device such as the delamination and popcorn failures.

In recent years, the QFN (Quad Flat Non-lead Package) package is a new packaging technology.as a packaging form of micro-electronics has been widely used. SiP (System in Package), which is considered the most potential of the 3D package, become hot spots of a package reliability in the electronic research field. So it's very important to study the reliability of the devices (SiP by QFN packaging) caused by thermal and moisture.

In this paper, the model is build by using the SIP structure. The SiP structure is consisted of two chips which were using QFN flip-structure. So the model is not a general sense of the QFN package, and has a certain value of study the reliability of SiP packaging.

In this paper, the finite element analysis is used to simulate and calculate the distribution of moisture diffusion under thermal and moisture condition, then to simulate the thermal stress and hygrothermal stress.

2 Geometry model and Finite element modeling

2.1 Geometry model

Figure1 shows the model used in this paper as analsys model of the SIP device using QFN packaging. There are two chips in this model: Upper die called Passive die and bottom die is active die. Flip Chip structure was used between the two chips. Active die be connected to copper as heatconducting by Die Attach (DA). Heat is transfered by leadframe from Passion Die to outside.

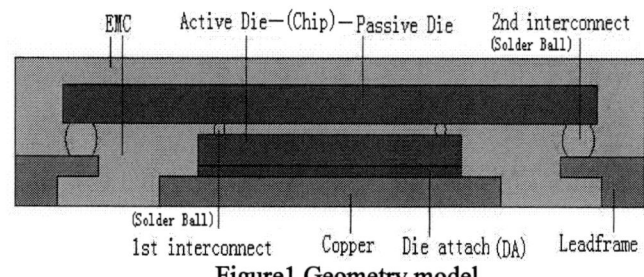

Figure1 Geometry model

2.2 Moisture diffusion modeling

Knowledge of moisture distribution within the package is needed for hygroscopic stress modeling. This can be best done through moisture diffusion modeling. Most commercial FE software is not equipped with moisture diffusion modeling factor. But with an appropriate thermal-moisture analogy, moisture diffusion can be modeled using the thermal diffusion function of the software. The analogous technique for a homogeneous material system has recently been extended to a multi-material system [2] so as to enable

978-1-4244-2739-0/08/$25.00 ©2008 IEEE

modeling of advanced packages. The FEA implementation scheme is presented in Table 1[3], where wetness is defined as W=C/Csat, C is the moisture concentration. and D the moisture diffusion coefficient.

Table 1 FEA thermal-moisture analogy for moisture diffusion modeling of multi-material system

Properties	Thermal	Moisture
Field Variable	Temperature, T	Wetness, W
Density	ρ (kg/cm^3)	1
Conductivity	K(W/m·°C)	$D*C$sat(kg/s·m)
Specific capacity	C(J/kg·°C)	Csat(kg/ m^3)

Csat is the saturated moisture concentration, and D the moisture diffusion coefficient.

For D and Csat, follow Arrhenius formula[4]：

D =D0 exp(QD/RT)　　　　　　　　(1)

Csat=C0 exp(QC/RT)　　　　　　　 (2)

Where D0 and C0 is the diffusion coefficient, Qd and Qc is the activation energy (ev), R is the Boltzmann constant (8.83e-5eV/K), and T is the absolute temperature (K).

Table 2 Moisture and hygroswelling material properties

Material	30°C/RH60%		60°C/RH60%	
	Csat(g/ mm^3)	D (mm^2/s)	Csat(g/ mm^3)	D (mm^2/s)
EMC	7.81×10^{-6}	3.13×10^{-7}	1.34×10^{-5}	1.14×10^{-6}
DA	1.47×10^{-5}	9.39×10^{-8}	2.75×10^{-5}	4.18×10^{-7}

Concidering table2 [5] and Arrhenius formula, we can work out the constant coefficient of Arrhenius formula (Table 3):

Table 3 The constant coefficient of Arrhenius formula

Material	Q_D(eV/K)	$D0$(mm^2/s)	QC(eV/K)	$C0$(mm^2/s)
EMC	-0.375	0.533	-0.157	3.127
DA	-0.474	7.223	-0.182	15.48

Base on the value of table 3 and Arrhenius formula, we can get the value as table 4

Table 4 Moisture and hygroswelling material properties

Material		DA	EMC
85°C RH60%	Csat(g/ mm^3)	4.28×10^{-5}	1.94×10^{-5}
	D (mm^2/s)	1.57×10^{-6}	2.85×10^{-6}
125°C RH60%	Csat(g/ mm^3)	0.0774×10^{-5}	3.24×10^{-5}
	D (mm^2/s)	7.33×10^{-6}	9.66×10^{-6}

2.3 Hygro-mechanical modeling

With the knowledge of moisture distribution in the package, as well as the hygroscopic swelling characteristic of the materials, hygroscopic stress in the package can be readily computed through hygro-mechanical modeling. A simple thermal-hygro analogy has been developed ~Table 5[6]. So that hygro-mechanical modeling can be performed using the thermo-mechanical function of commercial FEA.

Table 5 FEA thermal-hygro analogy for hygro-mechanical modeling

Properties	Thermal	Hygroscopic
Field variable	Temperature, T	Wetness, W
Coef. Of thermal expansion	α	$\beta*C$sat

Where β is the CME, and Csat is the moisture concentration.

2.4 The SiP Finite element modeling

Plane82 elements, which be made of 8 nodes's, is used to establised SiP 2D model. Because of the symmety, half of mode is used to simulate as figure2 (a). As figure2 (b) shows that the element mesh size is my fine, all of them apply quad element. There are 2882 elements and 2985 nodes. It will be benefit to improve result's precision, as well as benefit to get enough data for draw more accurate figure. Symmetry boundary conditions are used on the model.

Figure2（a）The half of model

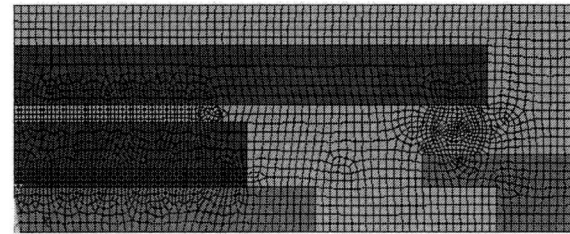

Figure2（b）The SIP FEA model

3 Result Analysis

3.1 Thermal diffusivity analysis

Fig. 3 shows the temperature distribution in the package during 5-min 358K (85°C) heating. Heat is conducted faster in die and copper leadframe than in mold compound (EMC) and die attach (DA). When the external surface is heated to 358K after few min, the internal package reaches this uniform temperature within a few seconds. Therefore, in the subsequent thermo-mechanical and vapor pressure models, temperature distribution during heating can be assumed to be uniform throughout the package.

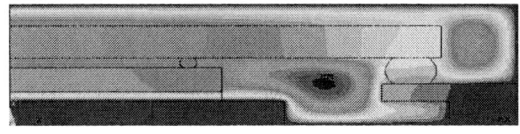

Fig. 3. Package temperature distribution during heating

3.2 Moisture diffusivity analysis

The transient moisture wetness distribution in the device(symmetric half model) is shown in Figure4, The moisture absorption during preconditioning at JEDEC Level 2(85°C/RH60% 168 hours) , and moisture desorption at various hours are characterized.

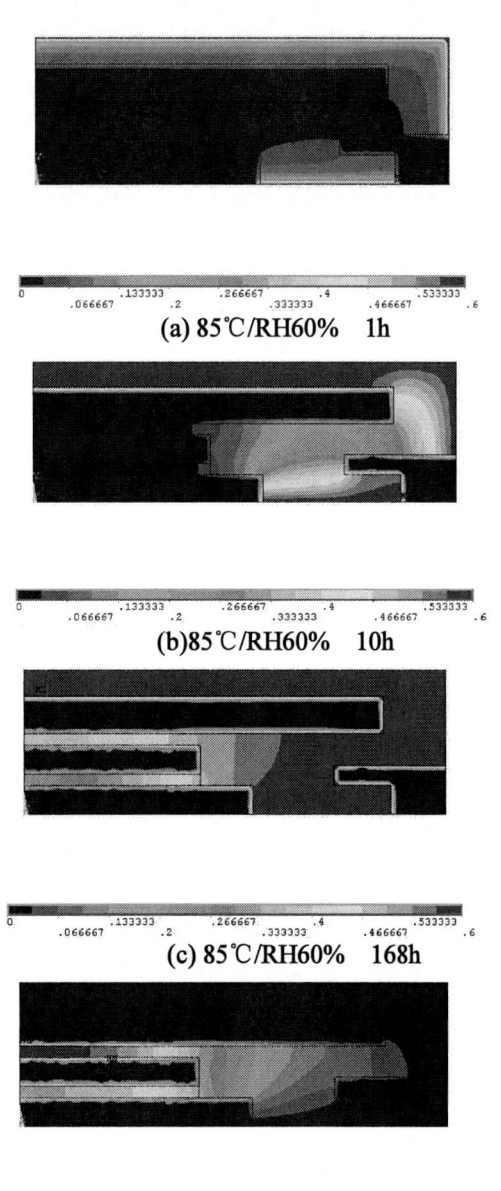

(a) 85℃/RH60%　1h

(b)85℃/RH60%　10h

(c) 85℃/RH60%　168h

(d）85° W=0 20 hours desorption

Figure4 SiP Transient moisture wetness distribution

The solution of moisture model is analogous to thermal diffusion. However, the thermal diffusivity is much faster than the moisture diffusivity.The transient moisture wetness distribution in SiP is shown in Fig. 4(a)-(c), ranging from 1 hour to 168 hours. The moisture diffuses into the package through mold compound (EMC), and gradually spreads into die attach layer(DA). At the end of 168 h of moisture preconditioning under 85 ° C/60%RH, the external EMC material of package is almost fully saturated.However, DA and interior EMC material of the package is still unsaturated. This is caused that DA and interior EMC materials absorb moisture by the narrow interface between these and external EMC material. As well as distribution of moisture exist a great gradient. Fig. 4(d) shows the results of 20 hours 85° C/W=0 desorption. During the proccess of desorption, external package surface loses a significant amount of moisture due to high moisture desorption rate. However, since the lower desorption rate in the interior of package, the

moisture concentration in the interior of the package, including critical die attach and die/mold compound interfaces, still remains relatively high moisture comtent. The local moisture concentration in these critical interfaces determines the strength of interfacial adhesion, which partially lead to moisture induced failures, e.g., delamination and popcorning, in the package.[7]

3.3 Thermo-Mechanical stress and Hygro-Mechanical stress analysis

Considering the forms of water in the device is not certain when the temperature is higher than 100° C, in order to avoid to deal with complex condition, 85 ° C/60%RH heating 5 minutes is applied in this simulation. Assume that interior stress of the package will be zero under initial condition(25° C/W=0).The process of simulationg as followed: Fig. 5 shows the methodology of integrated stress modeling to calculate the package stress induced during the process of heating. The thermal and moisture diffusion models mentioned earlier are related to one another. Result of moisture distribution from the moisture diffusion model is used as input for the hygro-mechanical model. On the other hand, the temperature distribution from the thermal model is applied in thermo-mechanical model. The stress and strain induced thermo-mechanical, and hygro-mechanical models are combined into an equivalent stress model to compute the package stress and strain. Flow chart as following:

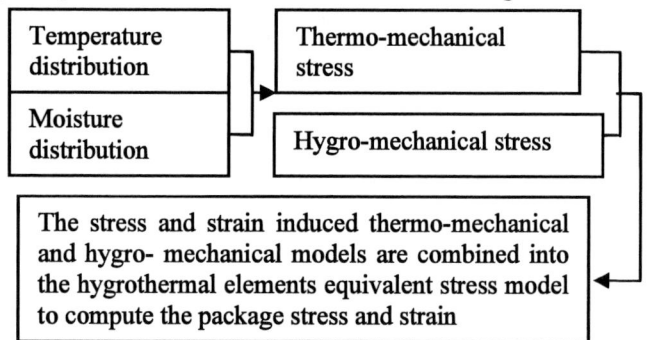

Figure5　Integrated stress flow chart

The simulation results see figure6-figure10. Fig.6、fig.7 show respectively Thermo-mechanical and hygro-mechanical stress warpage distribution. Fig8、9、10 show respectively hygro- mechanical ,Thermo- mechanical and hygrothermal elements equivalent stress distribution, which are combined by FEA software Ansys E_table module. Through these distributions we can clear the stress concentration in the interior of package.

(1) Thermo-mechanical stress analysis:

Fig.6 (a), (b) is reprectively package warpage distributions and Von Mises thermo-mechanical stress distribution. Von Mises thermo-elements equivalent stress show in figure 8. Figure 6 showed the thermal stress is larger around chip, and the largest thermal stress is lie in the junction of EMC, DA and chips. The results shows thermal stress is mainly distributed in the mismatch of CTE (coefficient of thermal expansion) between the materials. Moreover, the thermal stress is obviously larger in great gradient place. Fig.6 (b) shows us there exist stress delamination in the package, which is very badly to material like this plastic material. If the interface of the two materials exist defects, they would lead to cracks and cause the failure of the device.

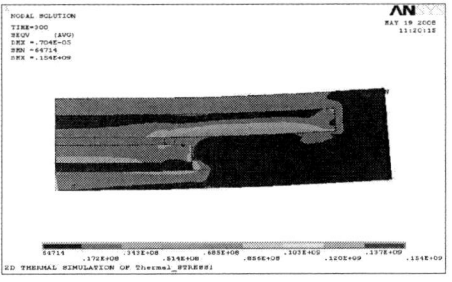

（a）Package warpage distributions

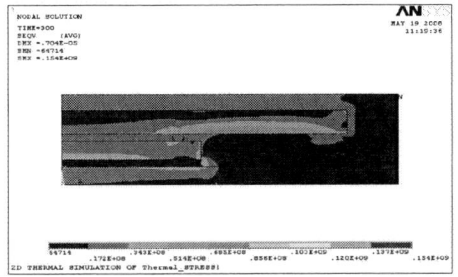

（b）Package Von Mises stress distribution

Figure6 package thermo-warpage and Von Mises stress distribution

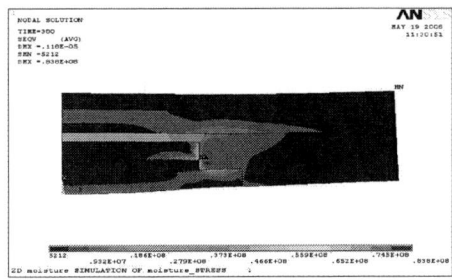

（a）Package warpage distributions

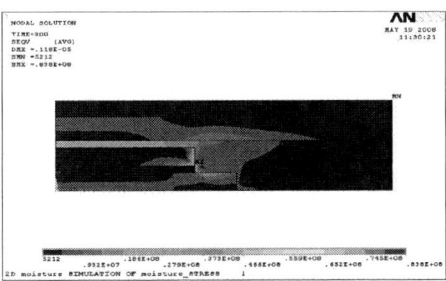

（b）Package Von Mises stress distribution

Figure7 package hygro-warpage and Von Mises stress distribution

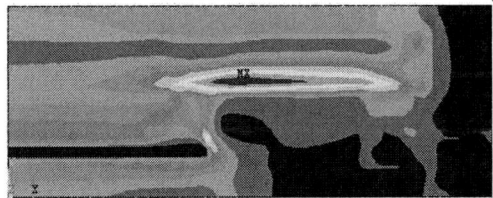

Figure8 Thermo-elements equivalent stress

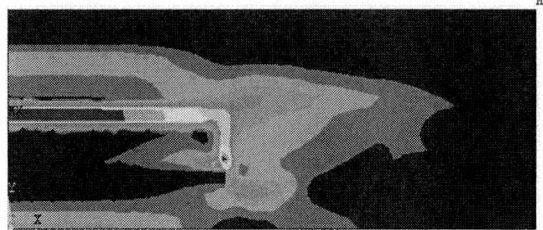

Figure9 Hygro-elements equivalent stress

Figure10 Package hygrothermal elements equivalent stress

(2) Hygro-mechanical stress analysis:

Fig.7 (a), (b) is repectively package warpage distributions and hygro-mechanical Von Mises stress distribution, hypro-elements equivalent stress show in figure9. It is in the region fully more moisture that hygro_mechanical stress is higher. And uneven diffusing of moisture make hygro-mechanical stress gradient larger. Hygro_mechanical stress is mainly distributed in the mismatch of CME (the moisture expansion coefficient of the materials) between the materials. The largest Hygro_mechanical stress is lies in the junction of EMC, DA and chips.

(3) Hygrothermal stress analysis:

Hygrothermal elements equivalent stress show in figure10. Comparing fig.8 and fig 10,The distribution of hygrothermal stress is similar as thermal stress in trend, Hygrothermal stress are mainly distributed in the mismatch of CTE and CME between the materials. However, the hygrothermal stress, which is caused by structure and the uneven diffusing of moisture make hygrothermal stress gradient larger. It is in the region fully more moisture that hygrothermal stress is much higher than the thermal stress, its stress concentration appears obviously at the junction between the materials. From the simulation results, we can see that the greater stress gradient will lead to warping deformation of the entire device, it will prone to more damage in the areas stress concentration. And the maximum hygrothermal stress is 1.56 times as maximum thermal stress. It is larger effect on reliability on SIP device in which used QFN packaging than the general QPN device. The results show the effect of EMC and DA's hyproscopic swelling on the device's stress cannot be ignored.

4 Conclusions

In this paper, the finite element analysis is used to simulate and calculate the distribution of moisture diffusing under thermal and moisture condition, then to simulate the thermal stress and hygrothermal stress. There is a need for comprehensive studies on moisture diffusion, thermal, hygro-mechanical stress and thermo-mechanical stress modeling. The hygrothermal stress model is then established to study the SiP package stress.

(1) Thermal stress is mainly distributed in the mismatch of CTE (coefficient of thermal expansion) between the materials. The largest thermal stress is lies in the junction of EMC, DA and chips. The distribution of hygrothermal stress, which is similar as thermal stress in trend, is mainly in the area mismatch CTE and CME (the thermal and moisture expansion coefficient of the materials).

(2) It is in the region fully more moisture or uneven diffusing of moisture that hygrothermal stress is much higher than the thermal stress, the largest stress concentration appear at the junction of EMC, DA and chips. If the interface of the two materials exist defects, they would lead to cracks and cause the failure of the device.

(3) Comparing the simulation results, we can see that the greater stress gradient will lead to warping deformation of the entire device, it will prone to more damage in the areas stress concentration. It is larger effect on reliability on SIP device in which used QFN packaging than the general QPN device. The effect of EMC and DA's hyproscopic swelling on the device's stress can not be ignored.

5 References:

1. Bi Ke Yun, *Microelectronic Package Technology*, China's electronics production Institute of Technology Branch Series Editorial Board, China University of Science and Technology Publishing House, 2003: 9。

2. Wong, E. H., Teo, Y. C., and Lim, T. B., 1998, ''Moisture Diffusion and Vapor Pressure Modeling of IC Packaging,'' 48th ECTC, pp. 1372–1378.

3. E. H. Wong, R. Rajoo, S. W. Koh, T. B. Lim. The Mechanics and Impact of Hygroscopic Swelling of Polymeric Materials in Electronic Packaging. *Journal of Electronic Packaging*. JUNE 2002, Vol. 124, 122–126.

4. Galloway Jesse E, Miles Barry M. Moisture absorption and desorption predictions for plastic ball array packages [A]. *Intersociety Conference on Thermal Phenomena* [C]. 1996. 180–186.

5. Lam Tim Fai. FEA simulation on moisture absorption in PBGA packages under various moisture pre-conditioning [A]. *Electronic Components and Technology Conference* [C]. 2000.

6. Wu Song, Jiang Ting Biao, Yang Dao Guo, Finite Element Analysis on Moisture Diffusion and Hygrothermal Stresses in PBGA Device [J]. *Electronic Components & Materials*, 2004, 23(6): 42–44.

7. Tong Yan Tee, Zhaowei Zhong, Integrated vapor pressure, hygroswelling, and thermo-mechanical stress modeling of QFN package during reflow with interfacial fracture mechanics analysis, *Microelectronics Reliability*,2004(44):105-114

Crack Growth Analysis of Ball Grid Array Resistor's Solder Joint Subjected to Thermal Cycling and 4 Point Cycling Bending

Xiangzhao, Ye-Yuming, Sun-Fujiang, Tu-Yunhua, Liusang
Huawei Technologies Co., Ltd.
Huawei Industrial Base, Bantian Longgang, Shenzhen 518129 P.R.C.
xiangzhao@huawei.com +86 755 89651218

Abstract

The crack growth rate of two different constructed BGA resistor's solder joint subjected thermal cycling was calculated using a dye penetration method. " The crack percentage " and the number of thermal cycling were recorded until the end of the test that " the crack percentage" greater than 50%. 4-point cycling bending test was also performed to observe failure mode compared with thermal cycling. The experiment indicated that 1) The dyeing appearance of solder was central symmetry and 2)"The crack percentage" and the number of thermal cycling has the exponent relationship and 3)The crack happened on both PCB side and component side, the crack growth direction at component side is from outside to the component central and 4)the contrary crack growth direction was observed in the 4-point bending fatigue test and 5) the failure mode was proved by FEM.

1 Introduction

In the BGA's solder joint reliability evaluation, thermal cycling and 4-point bending are still most widely used stress conditions [1]. There is still no available evidence that those two methods have internal relationship in predicating fatigue life of solder joint. Therefore, understanding crack growth and failure mode under those two stresses are mainly drive force [2] [3]. In the paper, an experiment was designed to research the solder joint crack growth and failure mode. A method that measures the crack growth rate to predict solder joint's fatigue life under thermal cycling loading is proposed.

The Crack growth of two different constructed BGA resistor's solder joint subjected to thermal cycling is present. The crack percentage was determined using a dye penetration method. And the detail information of test was recorded such as crack percentage, fracture appearance, number of thermal cycle. Then the crack growth rate can be calculated which is used to predict the fatigue lift of solder joint. Using finite element analysis (FEA), the crack growth direction and initiation position were predicated. The failure mode under thermal cycle loading is compared with the failure mode under 4-point bending loading.

2 Test Attributes
2.1 Test Vehicles

The two BGA resistor packages consisted of similar construction materials and only varied in substrate thickness and resistance surface process，refer to Figure 1and 2 for a general image of the package structure.

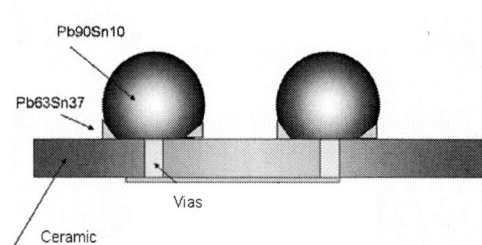

Fig. 1 Structure of Ball Grid Array Resistor of device A

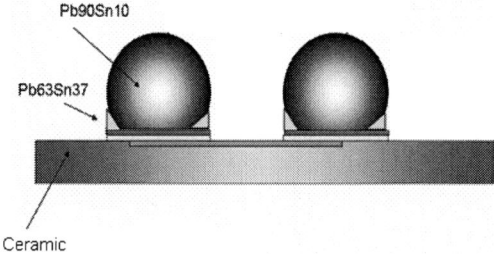

Fig. 2 Structure of Ball Grid Array Resistor of device B

2.2 Test Board Design

The test board for experiment was designed as 332mm x 50mm panels See Figure 3 for a layout of a test board. The board construction was high-Tg FR-4, 2.8mm thick with non-solder mask defined (NSMD) pad with Cu-Osp surface finish. Pad sizes were 0.76mm in diameter.

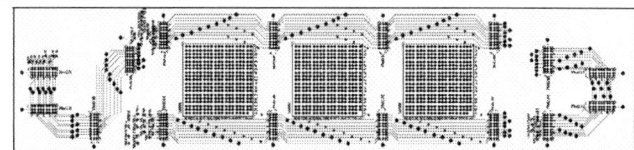

Fig. 3 Test Board Layout

2.3 Assembly Information

A type no-clean eutectic solder paste was used for the assembly work. The paste was deposited using a 0.13 thick stencil with 0.76mm diameter aperture openings. Reflow soldering was performed in a forced convection oven. Peak temperature within the solder joints was approximately 225 ℃.

2.4 Test Setup

A ramp rate of nearly 7℃ /minute with a dwell time of 15 minutes was used in all the thermal Cycling tests. The temperature range was between 0 and 100° C. Resistance readings at specific intervals allowed us to check and verify failures. The crack area was measured for the same set of solder balls in every device at every reading and the area measurement was averaged against temp cycle time.

978-1-4244-2739-0/08/$25.00 ©2008 IEEE   1013

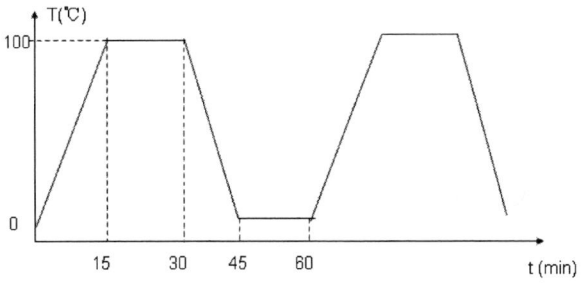

Fig. 4 Accelerate Thermal Cycle Test Profile

For 4-point cycling bending test, the frequency was 1Hz, and the maximum displacement for the loading span was 2mm. The resistance was monitored in-situ. The failure criterion was 20 Ω increment in resistance.

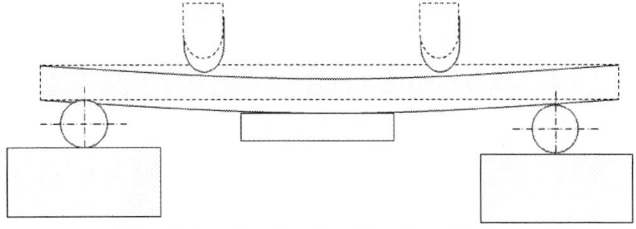

Fig. 5 4-point Cycling Bending Test

2.5 Dye Penetration Inspection

All of the test samples after thermal cycles went through failure analysis at special interval, to verify the failure modes and record the crack length. After the probing of local circuits, dye ink penetrant was applied to some specimens for crack growth inspection. The component was peeled off from the PCB after the ink was dried. The red-inked areas imply crack surfaces in the solder joints before the component was removed. Then the crack length was defined as the maximum red-inked length that goes thought the centre of the circle. And the crack percentage was defined as the ratio of crack length to dimension of pad.

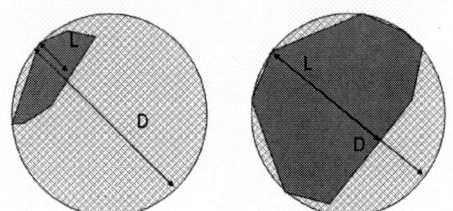

Fig. 6 Crack Growth Measurement

3 Experimental Procedure

According the reliability data form Provider's report [4] [5], at least 1400 cycles could be stand under set thermal cycling condition before first failure. And the test samples were not monitored in-situ during the experiment. Instead, the Dye-Pre test interval was in schedule before test. At least 2 samples were removed form the test board at the special interval for Dye-Pre inspection.

When device was removed from the test board, most solder balls were left at PCB side. But the remained solder ball could be separated from the device body. On the contrary, at the PCB side, the fracture appeared various. So the fracture at component side was used to measure the crack growth.

The crack length was relationship with the distance to the device centre. The crack of solder joint mostly was found at the corner of the device at the early of the experiment. According to the distance between solder joint and package centre, 5 zones were partitioned to research crack growth more careful. See Figure 7 for measurement zone partition.

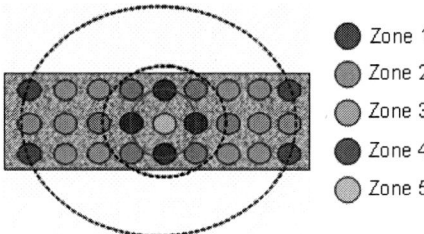

Fig. 7 Measurement zone Partition

4 Data and Analysis

The crack percentage for each component and number of cycle were recorded. And the crack percentage was statistic for each zone defined before. From the data, it appeared that the crack initiated at the corner of the package. And the crack spread from corner to centre.

The crack percentage measurements of zone 1 were plotted versus the number of the cycle. See figure 8. The data was fitted with 2nd order polynomial and the data 0,0 was considered. The figures show excellent fits(0.939,0.976) to the crack percentage.

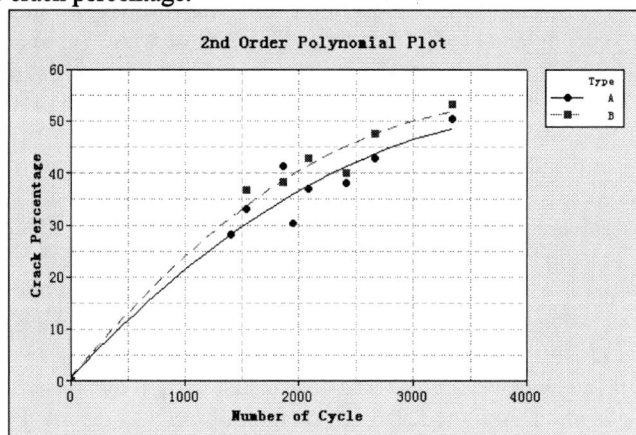

Fig. 8 2nd Order Polynomial Plot for Device A & B

The resulting equations for the polynomial fits are:
A: Crack Percentage $= 0.76 + 0.023 N - 0.000003 N^2$
B: Crack Percentage $= 0.67 + 0.027 N - 0.000003 N^2$
Where N is cycle Number

From the equations and aforementioned figures, some conclusions can be made:1)At the initiation, the crack maybe grow versus the number of the cycle in linear; 2)The fit equation show that the cycle to crack percentage reaching 100% is a significant large number. If the failure criteria are defined as the crack percentage reaching 50%, the predicated fatigue life for device A is 3100Cys and device B is 3400Cys.

The crack percentage measurements of different zones were plotted versus number of the cycle from interval 1400 to 2100.See figure 9. The crack percentage of solder joint (Line Blue) in the centre of component is increasing suddenly from 0 to 37%. It maybe means that initiation stage of solder crack growth is a significant long time. And the mean crack percentage reached at least 20% before inspected from the microscope.

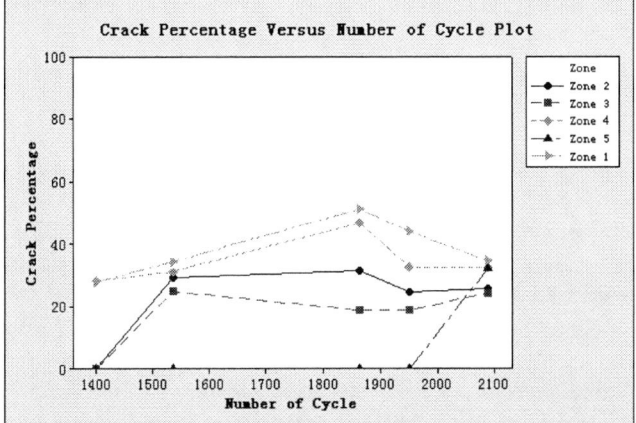

Fig. 9 The crack percentage measurements of different zones were plotted versus number of the cycle from interval 1400 to 2100

5 Failure Analysis

Cross section and Dye-Pry were the main used failure analysis method. The front method was used to observed the crack initiation position. And Dye-Pry was used to observed the crack spread direction and failure mode.

Figure 10 shows that the crack initiated at the corner both PCB side and component side after thermal cycling. And the crack spread inside the solder.

Figure 11 shows the typical fracture under thermal cycling stress after Dye-Pry experiment. Some trends were observed as follows: 1)The dyeing appearance of solder was central symmetry and 2)The crack initiated on both PCB side and component side, the crack growth direction at component side is from outside to the component central.

Fig. 10 Cross-section images of a solder joint

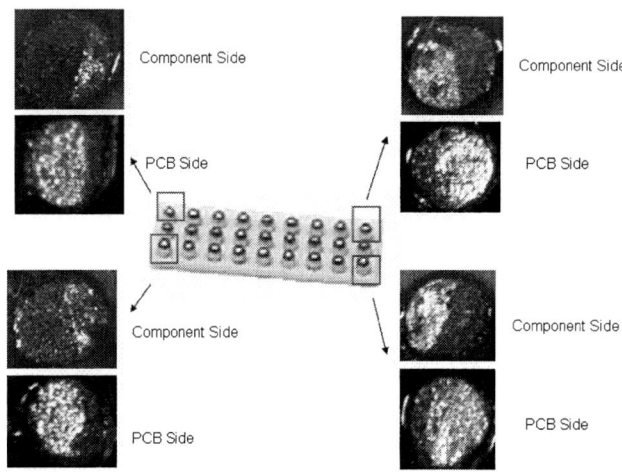

Fig. 11 Images of a solder joint after thermal cycling

Figure 12 shows the typical fracture under 4-point cycling bending test. It indicated that the crack growth direction at component side is from inside to the component outside.

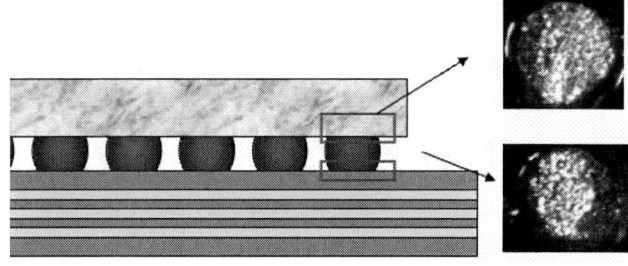

Fig. 12 Images of a solder joint after 4-point cycling bending

A separated solder ball was observed with SEM. Some typical characteristic were observed:1)The crack spread along the grain boundary;2)Wedge-Shaped crack at the surface of the solder ball.

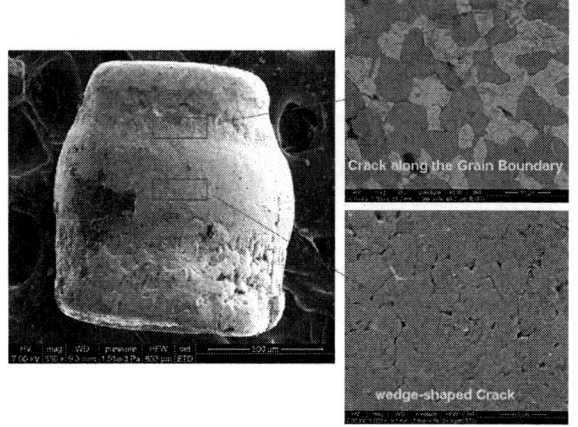

Fig. 13 SEM images of solder ball

6 FEM Analysis

From the failure analysis, under thermal cycling loading crack growth direction from outside to the component central at component side is common failure mode, which is opposite with samples under 4-point bending loading. Figure 14 shows the fracture of a typical solder joint fatigue failure and the FEA prediction of the maximum principal stress. The FEA prediction of the crack initiation location and direction of crack propagation matched exactly with that observed in experimental FA results. Figure 15 indicates the crack growth direction under thermal cycling loading and 4-point cycling bending loading.

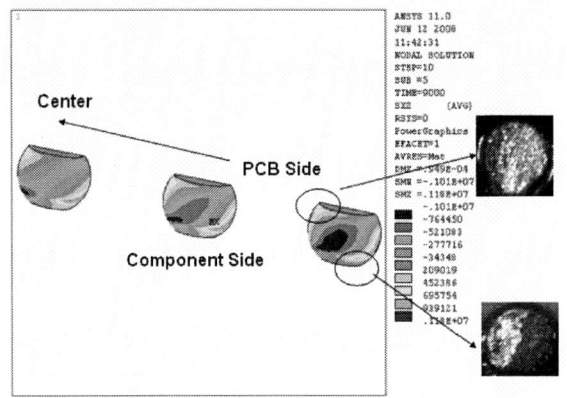

Fig. 14 FEM Prediction of the maximum principal stress

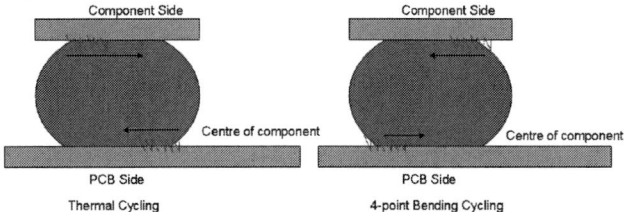

Fig. 15 crack growth direction under different loading

7 Conclusions

The fracture of the samples under thermal cycling loading indicates that 1)Fracture of solder is central symmetry; 2) The crack initiated on both PCB side ;3)At component side, the crack growth direction is from outside to the component central; and 4)At PCB side, the crack growth direction is from inside to the component outside.

The fracture of the samples under 4-point bending cycling loading indicates that 1)At component side, the crack growth direction is from inside to the component outside; and 2)At PCB side, the crack growth direction is from outside to the component inside.

The crack percentage fits versus number of the thermal cycles shows that 1)At the initiation, the crack maybe grow versus the number of the cycle in linear; 2)The fit equation show that the cycle to crack percentage reaching 100% is a significant large number.

The FEA prediction of the crack initiation location and direction of crack spread matched exactly with that observed in experimental FA results.

8 Acknowledgements

The authors would like to acknowledge the help received from Antian, Lisong in failure analyses and methodology development. Special thanks to Shu-Liwen and Huawei management for their support and encouragement.

9 References

1. Yu-Ming Ye, Sang Liu, Li-Min Chen, Jian Zhang, Zhao Xiang, Zhi-Wei Song, Assessment on reliability of 1.0mm pitch BGA package double-side assembly, 8th Electronic Materials and Packaging, 2006, p.1124

2. Dennis Lau, Y. S. Chan, S. W. Ricky Lee, Lifeng Fu, Yuming Ye, Sang Liu, Experimental testing and failure prediction of PBGA package assemblies under 3-Pointbending condition through computational stress analysis, International Conference on Electronics Packaging Technology, 2006, p.687

3. Li-Min Chen, Sang Liu, Yu-Ming Ye, Yun-Hua Tu, Zhi-Wei Song, Zhao Xiang, Zhi Xu, The influence of reflow peak temperature to BGA solder joints' mechanical reliability in backside compatible soldering, 8th Electronic Materials and Packaging, 2006, p.1102

4. CTS ClearONE Terminator Reliability Test Data, CTS CORP, 2001.

5. BGA TEMPERATURE CYCLE TESTING, www.irctt.com.

6. Shaw Fong Wong, Pramod Malatkar, "Vibration Testing and Analysis of Ball Grid Array Package Solder Joints," *2007 Electronic Components and Technology Conference*

7. Jagadeesh Radhakrishnanr, "Effect of Board and Package Attributes on Solder Joint Reliabilityof FCBGA Packages Based on IPC9701 Characterization," *2008 Electronic Components and Technology Conference*

8. IPC 9701- Performance Test Methods and Qualification Requirements for Surface Mount Solder Attachments

9. IPC SM-785 - Guidelines for Accelerated Reliability Testing of Surface Mount Attachments

10. EIAJ ED -4702- Mechanical stress test methods for semiconductor surface mounting devices

11. Yunhua Tu, Sang Liu, Yuming Ye, Limin Chen, Influence of Ni-Sn-Cu ternary intermetallic compound on solder joint reliability, 8th Electronic Materials and Packaging, 2006, p.1058

12. Sang Liu, Yunhua Tu, Limin Chen, Yunxia Xu, Yuan Wang, Yuming Ye, Chaoyun Luo, Study of BGA solder joints interfacial failure mechanism related to wave soldering, 12th Annual Pan Pacific SMTA, 2006, p.143

A Dual-Output Voltage Reference for High-Accuracy Pipelined ADC

Dongfang Cheng*, Xiaohui Li, Jue Zhang , Jiongming Wang
Key Laboratory of Advanced Displays and system Application, Ministry of Education, Shanghai University
Microelectronic Research & Development Centre, Shanghai University
No.149 Yanchang Rd, Shanghai 200072, P.R.China
Email: chengdf@mail.shu.edu.cn* sky_huilx163@hotmail.com

Abstract

A Dual-Output voltage bandgap reference has been designed for high-accuracy pipelined ADC with CSMC0.6 process technology in this work. Based on the temperature and noise analyzing of voltage reference,the traditional circuit has been improved.The stucture with two PN junctions in series,non-operational amplifier stucture and PTAT (Proportional To Absolute Temperature) current supplied by "self-bias" cascode current mirror,were adopted in the circuit designed.A new method to divide voltage was suggested in this study. The simulation results showed that the dual-output voltages were 2.194V and 1.098V respectively when temperature varied from -10℃ to 100℃,and the temperature coefficient is only 7.00ppm/℃;The change of V_{ref} only was 5.326mV, and the PSRR was 49.3dB when source voltage varied from 6V to 9V.The Dual-Output Voltage Reference of the circuit designed can meet the requirement of comparator input voltage in high-accuracy pipelined ADC.

1. Introduction

With the development of Integrated Circuit (IC), an accuracy voltage reference is needed in digital-analogy mixed circuit and analogy circuit. Temperature stability and voice resisting ability are the key factors to influence the inverting accuracy of A/D and D/A, and even influence the system performance in many conditions. Therefore, designing a good voltage reference is very important. There are many kinds of voltage references, such as voltage reference based on negative V_{BE}, voltage reference based on characteristic of reversed breakdown in zener diode, and bandgap voltage reference. Because of the low-temperature coefficient, high power supply rejection ratio, long-term stability and compatibility with standard CMOS technology process, the bandgap voltage reference is popular used.

2. Design theory of bandgap voltage reference

The aim of bandgap voltage reference is to obtain the output of voltage reference with little relationship to the power and technology process parameters, with the certainrelationship to temperature, with little influence to peripheral circuits in whole system, and with high voice resisting ability.

2.1 Conventional circuit structure of bandgap voltage reference

Voltage reference with low-temperature coefficient will be obtained by using the negative-temperature characteristic of V_{BE} (voltage between the base and emitter in bipolar transistor), and the positive–temperature coefficient characteristic of difference between two V_{BE} in two bipolar transistors which worked with different current density[1].

V_{BE} can be expressed as follows.

$$V_{BE} = V_T \ln \frac{I_Q}{I_S} \tag{1}$$

$$I_Q \cong I_S e^{\frac{qV_{BE}}{kT}} \tag{2}$$

Where, I_Q is the bias current under operating point. I_S is reverse saturation current. V_T equals to kT/q, where k is the Boltzmann constant, q is the electrical charge of proton, and T is the absolute temperature.

Under the same bias current (I1=I2)，The difference of V_{BE} between two transistors Q1 and Q2, which have different emitter areas, is proportional to absolute temperature. It is so-called *PTAT* voltage shown as follows.

$$\Delta V_{BE} = V_{BE1} - V_{BE2} = V_T \ln \frac{I_1}{I_{S1}} - V_T \ln \frac{I_2}{I_{S2}} = V_T \ln n \tag{3}$$

Where, *n* equals to I_{S1}/I_{S2}.

By using the negative-temperature characteristic of V_{BE} and the positive-temperature characteristic of $\triangle V_{BE}$, the temperature-independent voltage reference can be obtained through weighting summation between two voltages. The conventional circuit structure of bandgap voltage reference is shown in figure 1.

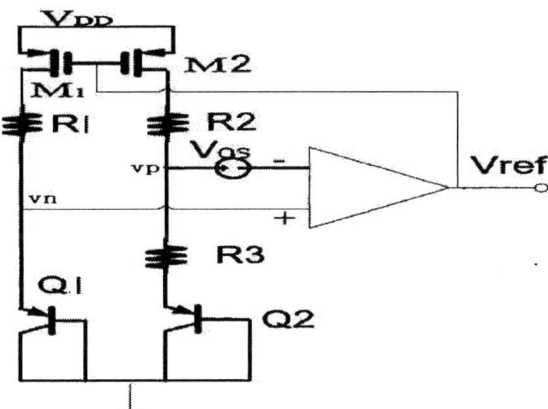

Figure 1 The conventional circuit structure of bandgap voltage reference

It's assumed that the amplifier in figure 1 is ideal, and the offset voltage is zero. If the current flowing through R_1 (R_2) is I_1 (I_2), the relationship between them is shown as follows[2].

978-1-4244-2739-0/08/$25.00 ©2008 IEEE   1017

$$\frac{I_1}{I_2} = \frac{R_2}{R_1} \qquad (4)$$

The voltage clamping effect in ideal amplifier makes $\triangle V_{BE}$ (the voltage in R$_3$) is as follows.

$$\Delta V_{BE} = V_T \ln(\frac{I_1}{I_2} \times \frac{I_{S2}}{I_{S1}}) = V_T \ln(\frac{R_2}{R_1} \times \frac{I_{S2}}{I_{S1}}) \qquad (5)$$

Then, the output voltage reference is shown as V_{ref}.

$$V_{ref} = V_{BE1} + V_{R2} = V_{BE1} + V_T \frac{R_2}{R_3} \ln\left(\frac{R_2 I_{S2}}{R_1 I_{S1}}\right) \qquad (6)$$

The temperature-independent bandgap voltage reference can be obtained through properly selecting the values of R_1, R_2 and R_3.

However, the mismatch threshold voltage and tiny current trans-conductance of MOS transistor, the offset voltage in amplifier is impossible zero. Therefore, the offset voltage in operating amplifier needs to be considered when design a high-accuracy voltage reference. So the equation 4 is rewritten as follows[3].

$$\frac{I_1}{I_2} = \frac{R_2}{R_1}\left(1 - \frac{V_{OS}}{I_2 R_2}\right) \qquad (7)$$

At the same time, the output voltage reference is changed into the following expression.

$$V_{ref} = V_{BE1} + V_{R2} + V_{OS}$$
$$= V_{BE1} + V_T\left(1 + \frac{R_2}{R_3}\right)\left(1 + \frac{V_{OS}}{\Delta V_{BE}}\right)\Delta V_{BE}$$
$$= V_{BE1} - \left(1 + \frac{R_2}{R_3}\right)V_{OS} +$$
$$\frac{R_3}{R_2}V_T \ln\left(\frac{R_2 I_{S1}}{R_1 I_{S2}}(1 - V_{OS})\right) \qquad (8)$$

It can be seen from the equation (8) the offset voltage in the operating amplifier will bring error to bandgap voltage reference. Decreasing the influence to output voltage of bandgap voltage can be implemented by the following two methods.

1). Increasing the coefficient of $\triangle$VBE. For example, adopting the structure with two PN junctions in series will double the $\triangle$VBE, as well as decrease the error caused by offset voltage.

2). Using the structure without the operating amplifier.

2.2 Temperature analyze for the bandgap voltage reference

Essentially, voltage reference is a reference that has a very low sensitivity to temperature. in theory, its temperature coefficient should be zero or very low. Under normal temperature, the temperature coefficient of V_{BE} is approximate to a constant. That is, $\partial V_{BE} / \partial T$ is approximate to -1.5 mV/℃[3].

The differential equation of temperature for equation (3) is as follows.

$$\frac{\partial \Delta V_{BE}}{\partial T} = \frac{k}{q} \ln n$$

$$(9)$$

The differential equation of temperature for equation (6) is as follows.

$$\frac{\partial V_{ref}}{\partial T} = \frac{\partial V_{be1}}{\partial T} + \frac{R_2}{R_3}\frac{k}{q} \ln n \qquad (10)$$

Where, $\partial V_{be1} / \partial T$ and k/q are constants. In order to make $\partial V_{be1} / \partial T$ equal to zero, the value of R1, R2 and n need be selected properly in design.

2.3 Voice analyze for bandgap voltage reference

For the voltage reference needs to be embedded in a chip, crosstalk will generat. Especially, the influence will become obvious when an operating amplifier is used in the circuit of voltage reference. Therefore, it is necessary to consider the voice of voltage reference.

Pipelined ADC converter is implemented based on the comparison of the bandgap voltage reference and the analogy input signal. So, the voice generated from the bandgap voltage reference will significantly influence the converting performance of ADC. Thinking about the input voice voltage (Vop) in amplifier in figure 1, it's known from the KVL that the output voice voltage (Vout) approximately equals to Vop ,because the small-signal leakage current of M1 equals to which of M2. This shows that the voice in amplifier directly emerges in output, which seriously influences the output of voltage reference. The voltage reference supplies voltages for each sub comparison circuit in pipelined ADC. If the voice generated from voltage reference would not be addressed properly, the accuracy and the stability of the whole A/D system will be influenced greatly. Thereby, it's best to use the structure with non-operational amplifier to extremely decrease the influence by voice[4].

3 Analyze for circuit structure and parameters

Based on the temperature and noise analyzing for voltage reference, the traditional circuit has been improved. The structure with two PN junctions in series, non-operational amplifier structure and PTAT current supplied by "self-bias" cascode current mirror[5], were adopted in the circuit designed. The new circuit structure can meet the work requirement of two proportional comparison voltages in pipelined ADC converter. The circuit form is shown in figure 2.

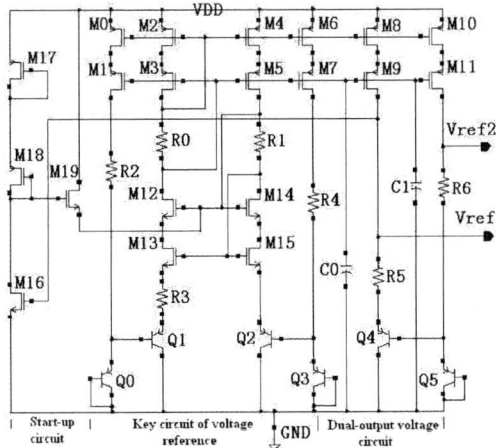

Figure 2 The whole circuit structure of dual-output voltage bandgap reference

3.1 Key circuit of dual-output voltage reference

Compared to the traditional bandgap voltage reference circuit, the proportional voltage output and voltage temperature-shift characteristic were improved in the proposed design, anda structure without amplifier was used. Thus, the proposed circuit can be used as build-in reference source in pipelined ADC converter. It will significantly decrease the output reference dependent of source by using self-biased fold-cascode current mirror structure and long-channel device. R0 and R1 were used to provide the biased voltage for M2~5, M$_{12~15}$, then make those MOS transistors work in saturation region. W/L of M$_{6~11}$ equals to W/L of M$_{2~5}$, and this ensure the I_{PTAT} well copied. The transistors' area of Q0 and Q1, Q2 and Q3 equals to each other, and the former one is n times than the latter. So, the relationship between them will be showed as follows.

$$I_{PTAT} = \frac{2V_T \ln n}{R_3} \qquad (11)$$

$$V_{ref2} = V_{be} + 2\frac{R_6}{R_3}V_T \ln n \qquad (12)$$

3.2 Dual-output voltage circuit

The output voltage in traditional bandgap voltage reference was fixed in 1.025V, which usually can't meet the work requirement. Using the resistor divider will only get the voltage lower than V_{ref}. And the resistor divider will cause the load effect of voltage reference, then influence the accuracy of voltage reference. In this view, a new method is proposed in this circuit. The dual-output voltage can be obtained thought the copy of I_{PTAT} by M8 and M9. The principal of the proposed circuit is as follows.

$$V_{ref} = 2V_{be} + 2\frac{R_5}{R_3}V_T \ln n \qquad (13)$$

Compared the equation (12) and (13), make R5 equals to two times as R6, then Vref equals to two times as V$_{ref2}$. Therefore, two proportional voltages can be obtained. This method not only overcomes the shortcomings of resistor divider, but also ensures that both two output voltages have the same good characteristic of source rejection and temperature shift. using of transistor Q5 can save a bipolar transistor, then reduce the chip area. In addition, the voltage

source rejection ratio and voltage reference stability were increased by using capacitors C0 and C1, because of capacitor's filtering effect.

3.3 Startup circuit

Startup circuit is made up of M$_{16~19}$. When the main circuit is powered on, the Vref on output is zero, M16 turned off. Then M19 turned on because of the high voltage in gate of M19. Finally, M12 turned on, there was current flowing through current mirror, then the circuit is started up. With the increase of V$_{ref}$, M16 turned on and the M19 turned off, and then the startup circuit departed from the main circuit.

Considering the area of the whole chip, the value of n was selected to 6, and the value of R$_3$, R$_5$, and R$_6$ need meet the relationship as follows.

$$\frac{\partial V_{ref}}{\partial T} = 2\frac{\partial V_{be}}{\partial T} + 2\frac{R_6}{R_3}\frac{k}{q}\ln n = 0$$

$$(14)$$

$$\frac{\partial V_{ref2}}{\partial T} = \frac{\partial V_{be}}{\partial T} + 2\frac{R_6}{R_3}\frac{k}{q}\ln n = 0 \qquad (15)$$

So,R$_5$:R$_6$:R$_3$ is selected as 8.08:4.04:1. Theoretical calculation shows that the output of bandgap voltage reference V$_{ref}$ / V$_{ref2}$ is 2.1940V / 1.0960V, which meets the input voltage 2:1 requirement of comparator in Pipelined ADC converter.

4 Simulations

Using the process model of CSMC0.6um BSIM3v3 and the SPECTRE simulation tool, the circuit was simulated.

4.1 Simulation for temperature characteristic

Fixed the voltage source V$_{DD}$(6.0V), the output voltages were scanned in temperature region from -10℃ to 100℃ , results were shown in figure 3.

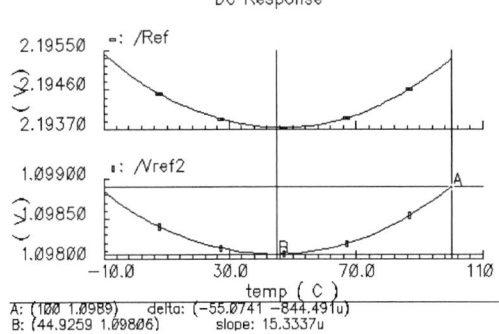

Figure 3 V$_{ref}$ and V$_{ref2}$ temperature characteristic curves

It's shown from the simulation that ($\triangle$V$_{ref}$)$_{max}$=1.692mV and ($\triangle$V$_{ref2}$)$_{max}$=0.844mV under the temperature varying. Calculations show that:

$$t_{Ref} = \frac{1}{V_{ref}}\frac{dV_{ref}}{dT} = \frac{1}{2.194}\frac{1.692\times 10^{-3}}{110}$$

=7.01ppm/℃

The t$_{ref2}$=7.00ppm/℃ can be obtained using the same method.

4.2 PSRR simulation

As the source voltage varying from 0V to 15V, the result was shown in figure 4.

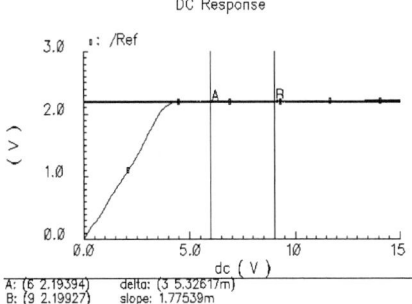

Figure 4 Characteristic curves of voltage source rejection ratio

V_{ref} only changed 5.326mV during source voltage varying from 6V to 9V. Calculation shows that PSRR=46.29dB.

It's known from above that the proposed bandgap voltage reference has simple circuit structure, good characteristics of temperature and high voltage source rejection ratio. So it can properly meet the requirement of proportional input voltage reference in comparator in high-accuracy Pipelined A/D converter.

Conclusions

The bandgap voltage reference with low temperature drift, high source rejection ratio and high-accuracy is proposed in this paper, which is suitable for high-accuracy analogy system. The fold-cascode structure and the long channel device were used to improve the characteristic of source rejection ratio, and the accuracy dual voltage outputs are implemented with the new circuit form. By results of simulation, it's proved that the proposed circuit is suitable as voltage reference unit in high-accuracy system.

References

1. Phillip E.Allen and Douglas R.Holberg, CMOS Analog Circuit Design. Oxford University Press, Inc. 2002
2. GRAY P R, MEYER R G. Analysis and design of analog integrated circuits [M]. New York: John Wiley&Sons, 2001.York: John Wiley &Sons ,2001
3. M.S.J.Steyart and W.M.C.Sansen, Power supply rejection ratio in operational transconductance amplifiers..
4. K-M Tham and K.Nagaraj. A Low Supply Voltage High PSRR Voltage Reference in CMOS process, IEEE J.Solid State Circuits, Vol.30, 1995.
5. Behzard Razavi, Design of Analog CMOS Integrated Circuits. Xi'an: Xi'an Jiaotong University Press, 2004

Study of Interface Reliability in QFN Device under Hygro-Thermal Environment

JIANG Ting-biao[*], NONG Hong-mi, DU Chao
Guilin University of Electronic Technology, Guilin 541004, China
Tel: 0773-5601331, Email: jtb@guet.edu.cn

Abstract

The interface crack caused by moisture absorption is a main reason for the failure of plastic packaging electronic devices. According to the failure of interface cracks in QFN plastic packaging devices caused by hytro-thermal environment, the paper combining with Finite Element Method (FEM), carried on the research by moisture absorption experiment, lead-free reflow soldering experiment, high temperature tidal thermal experiment, and Scanning Electron Microscope (SEM) experiment. The results of study show that: non-moisture absorption devices seldom produce cracks after the lead-free reflow soldering, and moisture absorption devices don't produce any crack during absorbing moisture, but they easily produce cracks after lead-free reflow soldering; The cracks produced by the experiment mainly lay on the interface between Die-Attach material (DA) and the chip, and the cracks at the junction of chip, DA material and Epoxy Molding Compound material (EMC) have the greatest damage; The position and the expansion direction of cracks are closely related to the characteristic and the interface intensity of the two connecting materials. These conclusions have important practical significance to the study and the evaluation criterion of crack.

Keyword: reliability; hygro-thermal; moisture; interface cracks; QFN devices

1 Introduction

Because of low costs and excellent performance, the plastic molding materials especially the thermosetting polymers have been widely used in the field of microelectronics packaging. When exposed to certain circumstances, these microelectronics plastic packaging devices are very easy to absorption moisture[1]. In reflow process, interface cracks which were induced by moisture are very prevalent in plastic packaging devices, especially in lead-free reflow process with higher temperature. Presently, there are many researchers have researched the interface cracks in microelectronics plastic packaging devices: Ferguson[1] had researched the moisture absorption impact of the interface between under-fill epoxy and soldering pad, and pointed out that interface intensity is not only relate to moisture absorption, but also relate to mechanical properties match of materials, load conditions and surface chemical properties of interface; Gektin[2] had researched dry devices and moisture pre-absorption devices under high temperature environment through experiment, the experimental results show that dry devices didn't appear "popcorn" fracture when in reflow process, this prove that the thermal stress in devices isn't the main factor that lead to "popcorn" fracture; Suhling[3] studied the moisture absorption devices in packing process and found that most of those devices appeared "popcorn" fracture, the result show that the main factor of "popcorn" fracture is moisture absorption in devices before high temperature reflow process.

In this study, the interface condition of QFN devices was researched by moisture absorption experiment, lead-free reflow soldering experiment, high temperature tidal thermal experiment and Scanning Electron Microscope (SEM) experiment.

At the same time, the research build up 3D model of QFN device to simulation moisture absorption process and the comprehensive influence in hygro-thermal environment, then combined the results of simulation with the experimental results to analysis the interfacial failure mechanism, the interfacial crack initiation and the crack expansion laws.

2 Experimental Designs

2.1 Sample Preparation

Preparation of specimens for all experiments were QFN devices, which type is RFT6100 and dimension is 6mm×6mm×0.9mm with a total of 40 lead-in pins. The packaging physical structure of QFN device is shown in Fig.1[4]

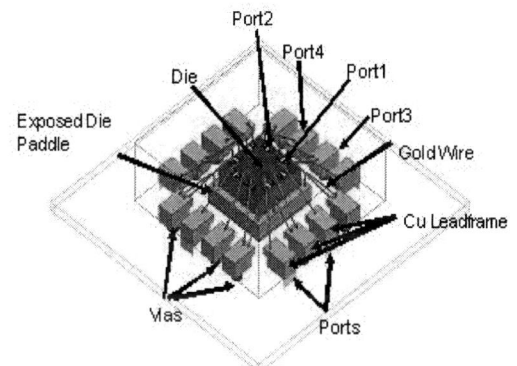

Fig.1 Sketch of QFN devices

2.2 Experimental Methods

(1) Moisture absorption in 85℃/85%RH condition. First put 10 QFN devices in hygro-thermal test chamber, then set the hygro-thermal loading curve. In the first loading stage, the initial condition was 25℃/30%RH, then the hygro-thermal condition up to 85℃/85%RH in 30 minutes. In the second stage, the condition was 85℃/85%RH and maintained 168 hours. The third stage set up 1 hour cooling to room temperature.

(2) Lead-free reflow soldering experiment of 5 moisture absorption devices and 5 non-moisture absorption devices through dynamic infrared heating equipment.

(3) High temperature tidal thermal experiment after lead-free reflow soldering experiment. Put 5 devices in 85℃/85%RH condition of 500 hours and others 1000 hours in the test box of constant temperature and humidity, in order to study the reliability of devices after long time working. According to EIA/JESD22-A110-A level, the condition of

978-1-4244-2739-0/08/$25.00 ©2008 IEEE

85℃/85%RH, 500 hours and 1000 hours can be instead by 130℃/85%RH HAST, 50 hours and 96 hours respective.

(4) SEM experiment. First of all inlaying, cutting, polish-grinding and gold plating film treatment to the pretreatment devices used the metallographic sample inlay machine, the metallographic sample cutting machine, and the polisher respective. Secondly by using the SEM, the interface of the tested devices was scanned, the position and performance of the interface cracking failure was identified by analyzing the microstructure of the interface of the tested devices.

3 Experimental Results and Discussion

3.1 SEM of Non-moisture Absorption Devices

The SEM results are shown in Fig.2. The figure (a) is the structure of EMC in non-moisture absorption devices, and we can see in the figure that the spherulites of EMC distribute evenly and compactly, the size of spherulites is small, the interface between EMC and chip is compact. The figure (b) which was enlarged 5000 times is the interface between EMC and DA. We can see in the figure that the spherulites are adsorbed to the EMC interface, and some small spherulites aggregate to be bigger one. The surface moisture can infiltrate into the devices, not only because the interspaces between the internal elements and the spherulites of EMC and DA, but also because the difference of linear expansion coefficient and thermal stress in devices can easily induce micro-cracks and pores[5]. The figure (c) which was enlarged 6000 times is the interface between DA and chip, the crystal and spherulites of DA distribute evenly in the figure.

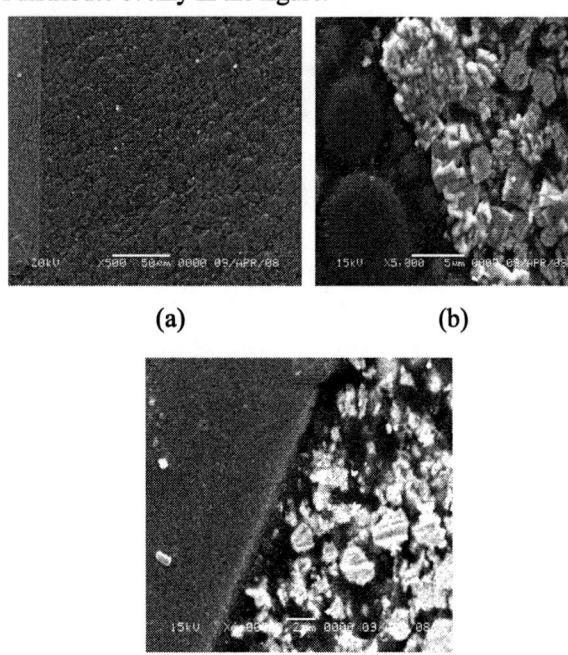

(a)　　　　　　　　(b)

(c)

Fig.2 SEM of non-moisture absorption devices: (a) structure of EMC, (b) interface between EMC and DA, and (c) interface between DA and chip

3.2 Moisture Absorption Experimental Results

We can see in Fig.3 that the distribution of spherulites of EMC become unevenly after moisture absorption, the size of spherulites also become bigger than non-moisture absorption

material. The big spherulites in EMC can induce more interspaces and deficiencies, and reduce the mechanical properties of EMC[6]. In the figures, there is no any crack in the moisture absorption devices, and it proves that the hygro-thermal stress is not enough to cause crack in the devices.

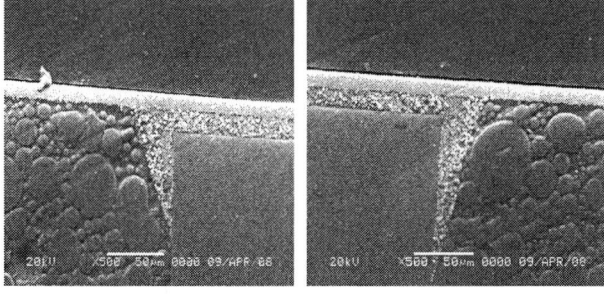

Fig.3 SEM of moisture absorption devices

3.3 Lead-free Reflow Soldering Experimental Results

The internal structure of defective devices after moisture absorption is shown in Fig.4. As can be seen from the figure, there are two defects between DA and chip interface, but there is no any crack in DA or chip. It also shows that the hygro-thermal stress is not enough to cause crack in the moisture absorption devices.

Fig.4 Structure of defective devices after moisture absorption

The cracks after lead-free reflow soldering in moisture absorption devices are shown in Fig.5. As we can see in the first figure, the spherulites in EMC are no aggregation to be bigger one after lead-free reflow soldering, the analytical result prove that the big spherulites in EMC are interaction of moisture and temperature. All moisture absorption devices are crack after lead-free reflow soldering, which confirm that moisture absorption devices are easier to crack than non-moisture absorption devices. As can be seen in Fig.5, the distribution of crack occur mainly on the edge of chip near DA, this indicate that on the edge of chip near DA has larger stress. According to the damage extent of cracks, the maximum stress is at the junction of the chip, DA and EMC. All the cracks have occurred in the base material of chip, and the cracks are not always extend to the internal of chip, but expand with a damage form of wrist and end of turn to conjunct interface.

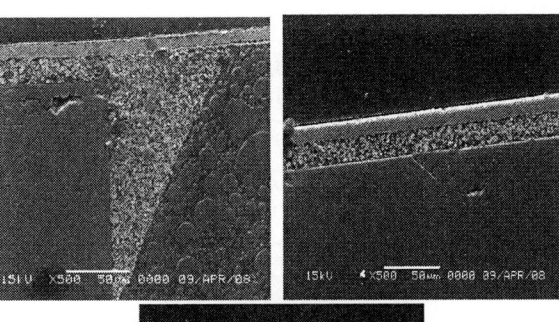

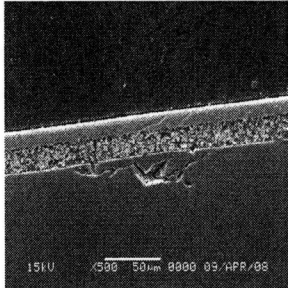

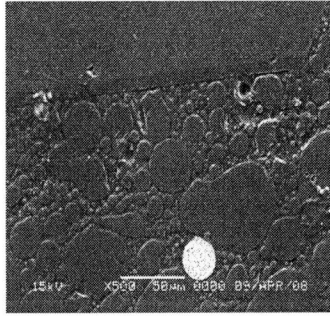

Fig.5 Cracks of moisture absorption devices after lead-free reflow soldering

3.4 High Temperature Tidal Thermal Experimental Results

Fig.6 is the structure of EMC after 50 hours high temperature tidal thermal experiment. There also have some big spherulites as be seen in the figure, and appear aging phenomenon in some part.

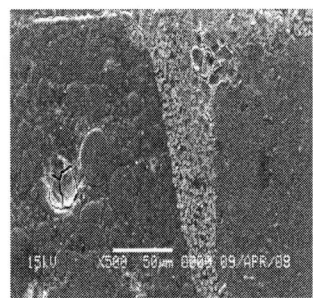

Fig.7 Interface at juncture of DA, EMC and chip after 96 hours high temperature tidal thermal

4 Finite Element Simulations and Analysis

4.1 Modeling Finite Element of QFN Device

The simulation experiments of QFN devices were studied by using the Finite Element Method (FEM). Firstly, the 3D models of QFN devices were established and the moisture diffusion behaviors and hygro-thermal stresses of QFN devices were simulated and analyzed. The models included EMC, Die (chip), DA, pins and soldering boards, and just a quarter of QFN device because of its symmetry.

Secondly, in order to reduce the analytical time and increase the efficiency of analysis, in the simulation we did some assumptions as following: (1) The EMC, chip, DA, pins and soldering boards were considered as isotropic and flexible materials. (2) All of the interfaces were completely connected, and there was not any delamination in the moisture absorption simulation. (3) The residual stress and strain of EMC was ignored in the curing process.

4.2 Analysis of FEM Results

(1) Moisture absorption simulation of QFN device

The simulation of QFN device has experienced moisture absorption in 85℃/85%RH, 168 hours condition and 5 minutes reflow soldering process, the simulation result is shown in Fig.8.

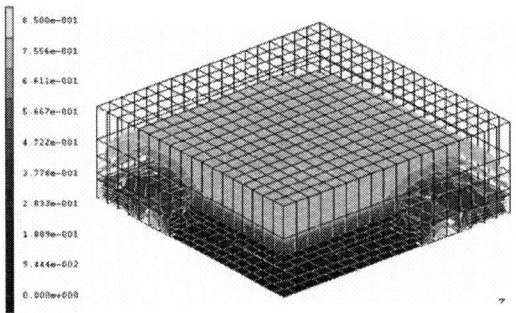

Fig.8 Iso-surface of relative humidity after 168 hours pretreatment

As can be seen in the figure, the distribution of relative humidity is similar to the big spherulites in moisture absorption devices, so the method of FEM is feasible to study the moisture diffusion in devices. After 168 hours' moisture loading in simulation, most cells of EMC are close to or have reached to the saturation, but moisture continued diffuses into device through the interface between EMC and other materials, and gradually diffuses to the chip connection layer. At the same time, DA in which contacted with EMC absorb more moisture, it is because the saturated humidity of DA and EMC

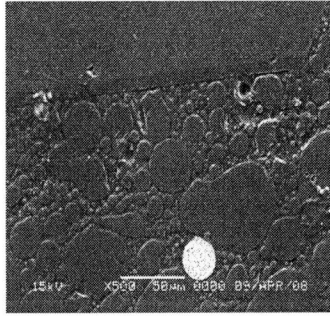

Fig.6 Structure of EMC after 50 hours high temperature tidal thermal

Fig.7 is the interface at juncture of DA, EMC and chip after 96 hours high temperature tidal thermal experiment. The aging phenomenon and crack of DA, EMC and chip can be seen clearly in the figure. Because of the interaction of hygro-moisture stress and vapor pressure, the aging phenomenon is most obviously at the corner between DA and chip. With the aging of DA, material's plasticity gradually reduces and can not bearing the concentrated stress, so it appears some small cracks in DA. But plastic decrease is a gradual process, and hard to create great stress impact[7], so there are no any large cracks.

is similar, so along the DA to the centre of devices, moisture gradually reduces, which similar to the results of moisture absorption experiment.

In order to study the diffusion and distribution of moisture more clearly, the nodes at interface of different materials as shown in Fig.9 are fetched, and observe the curve of relative humidity of different nodes which changed along with time. The different nodes corresponding to the location are listed in Tab.1.

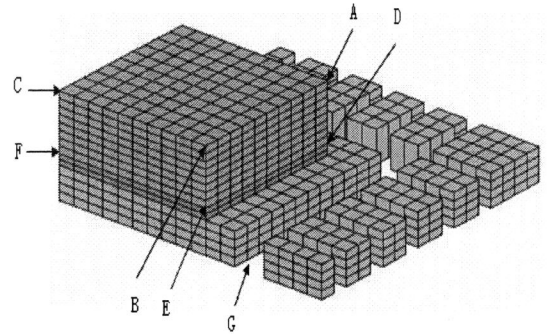

Fig.9 Sketch of different nodes in FEM model

Tab.1 Location of different nodes

Node	Number	Location
A	3211	Interface between EMC and Die (at the outer vertex of Die)
B	4052	Interface between EMC and Die (at the upside edge of Die)
C	2018	Interface between EMC and Die (on the upside surface of Die)
D	3830	Interface between DA and Die (at the outer vertex of Die)
E	4130	Interface between DA and Die (at the downside edge of Die)
F	3988	Interface between DA and Die (in the Die)
G	1020	Arbitrary in EMC

The curves of relative humidity of different nodes which changed along with time are shown in Fig.10. As can be seen in the figure, the moisture diffusion speed has an important relation to the location. Because DA which was in the innermost of device is location between die and soldering board, and because die and soldering board material are non-moisture absorption, the only source of moisture in DA is the edge which contacted with EMC, and the moisture diffusion speed in DA is very slowly. Therefore, the relative humidity in DA is much lower than EMC.

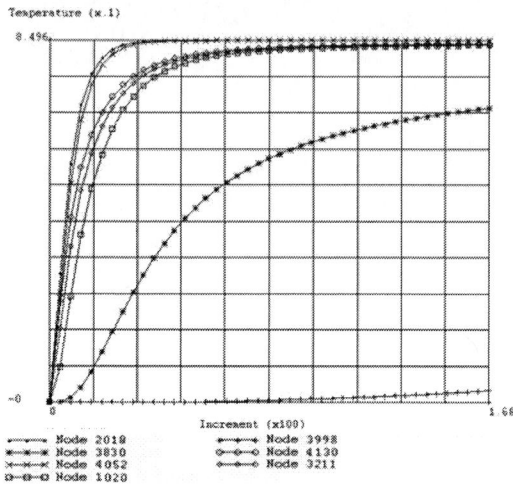

Fig.10 The curves of relative humidity of different nodes after 168 hours in 85℃/85%RH

(2) Comprehensive Influence of Hygro-thermal Stresses in QFN Device

In the process of lead-free reflow soldering, besides the heat-mechanical stress which was induced by the mismatch of Coefficient of Thermal Expansion (CTE), the stresses in QFN devices may also includes wet-mechanical stress which was induced by the mismatch of Coefficient of Moisture Expansion (CME), and may also includes vapor pressure of which was evaporated by high temperature in polymer material micro-voids. Therefore, it is very easy to underestimate the impact of stress on packaging reliability if only consider one kind of stress. In order to observe the impact of stress in the lead-free reflow soldering process, we stimulated the comprehensive influence of hygro-thermal stresses in QFN device, the result is shown in Fig.11.

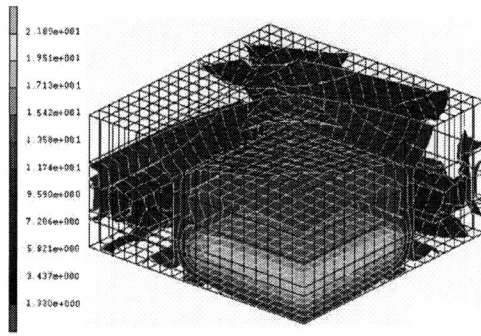

Fig.11 Iso-surface distribution of integrated Von Mises stress (Unit: MPa)

As be shown in Fig.11, the heat stress of the DA edge is large, and the largest stress is at the junction of die, DA and EMC, therefore, if there are initial cracks, packaging flaws or aging phenomenon on these interfaces, the cracks will be induced and extended by integrated stress. This simulation result is agreeable with the experimental results.

5 Conclusions

The research performed for this paper investigated the

interface reliability of QFN devices by experiments and simulations. Among the findings are:

(1) Non-moisture absorption devices seldom produce crack after the lead-free reflow soldering, and moisture absorption devices don't produce crack during absorbing moisture, but they easily produce crack after lead-free reflow soldering. The moisture will reduce the mechanical properties of EMC, but there is no any crack in QFN devices during absorbing moisture, which indicates that the hygro-thermal stress is not enough to cause crack in the devices, and the hygro-thermal stress and vapor pressure which caused by water evaporation leads to the production of the crack during the lead-free reflow soldering.

(2) The cracks produced by the experiment mainly lay on the interface between DA and chip, and the cracks at the junction of chip, DA and EMC have the greatest damage, which is the same as stimulation.

(3) According to the results of the experiment, all of the cracks have taken place in the base material of the chip, and the cracks expand into the inner of the chip with a damage form of wrist and end of turn to the conjunct interface. So the position and the expansion direction of cracks are closely related to the characteristic and the interface intensity of the two connecting materials.

These conclusions have important practical significance to the study and the evaluation criterion of crack, and have reference value for engineering application.

Acknowledgments

The authors would like to acknowledge the National Natural Science Foundation of China (Grant No. 60666002), and also to acknowledge Dr. D.G.Yang for his contribution.

References

1. Pearson R.A., et al., "Adhesion issues in epoxy-based chip attach adhesives," *IEEE transactions on Components, Packaging, and manufacturing Technology, Part A*, Vol.20, No.1 (1997), pp. 31-37.
2. Gektin, V., Bar-Cohen A., "Mechanistic figures of merit for die-attach materials," *Inter-Society Conference on Thermal Phenomena in Electronic systems*, I-THERM V (Cat.No.96CH35940), 1996. OrlandO, FL, USA, 1996, pp. 306-313.
3. M. Kaysar Rahim, Jeffrey C. Suhling, et al. "Fundamental of Delamination Initiation and Growth in Flip Chip Assemblies," *Proceedings of the IEEE Electronic Components and Technology Conference*, Orlando, FL, 2005, pp. 1172 - 1186.
4. Rao R. Tummala, Fundamentals of Micro-system Packaging, MeGraw-Hill (New York, 2001).
5. Tong Yan Tee, Zhong Zhaowei, "Integrated vapor pressure, hygro-swelling, and thermo-mechanical stress modeling of QFN package during reflow with interfacial fracture mechanics analysis," *Microelectronics Reliability*, Vol. 44, No.1 (2004), pp. 105-114.
6. Fan X.J., Zhang G.Q., Ernst L.J., "A Micro-mechanics Approach for Polymeric Material Failures in Microelectronic Packaging," *3rd. Int. Conf. on Benefiting from Thermal and Mechanical Simulation in (Micro-)Electronics*, EuroSIME 2002, pp. 154-164.
7. Guenin B., Packaging: Designing for Thermal Performance, Electronics Cooling, 1997.5.

In-situ Observation on Electrochemical Migration of Lead-free Solder Joints under Water Drop Test

Y.H. Xia[1,2], W. Jillek[2], E. Schmitt[2]

1 College of Mechanical Engineering, Zhejiang University of Technology, Hangzhou, 310032, China
2 Georg-Simon-Ohm University of Applied Sciences, Nuremberg D 90489, Germany
Email: Werner.Jillek@ohm-hochschule.de
xia_yanghua@hotmail.com

Abstract

Electrochemical migration (ECM) of lead-free solder joints were investigated by water drop test method. Nine types solder pastes were employed to compare the ECM susceptible. The effects of applied voltage, electrodes spacing and flux residue on the ECM were in-situ observed. The microstructure and composition of the growing dendrites during ECM were detected. The results revealed that higher voltage and narrower spacing weakened the ECM reliability of solder joints. The flux residua inhibited the occurrence of ECM. The main migration element was Pb in Pb-bearing solder joints. For Sn-Ag-Cu solder joints, it was Sn and Cu to migrate during the ECM process. Zn was the only migration element in the SnZnBi solder joint. Sn-Zn-Bi solder exhibited the best ECM reliability.

1 Introduction

Electrochemical migration (ECM) can be defined as a transport of ions between two metallization stripes under bias through an aqueous electrolyte [1]. The process begins when a thin continuous film of water has been formed and a potential is applied between oppositely charged electrodes. Positive metal ions are formed at the positively biased electrode (the anode), and migrate toward the negatively charged cathode. Over time, these ions accumulate as metallic dendrites with tree or needle-like shape, and eventually create a metal bridge and cause short-circuit failure [2].

Nowadays, as the conductor pitch and package size are reducing continuously, electrochemical migration is more susceptible to being occurred in the electronic assemblies with high density interconnections. This is because that the electric field increases with the narrowing of conductor spacing and the solder joints are more vulnerable to fail under certain environment.

Pb element is banned to use in electrical and electronic products due to the government legislations and environment concerns. As the transition to lead-free solder and process in electronic manufacturing industry, the new solder materials and process emerge more reliability issues. Most of the developed lead-free solder alloys are based on Sn being the primary constituent [3]. The other elements in lead-free solders include Ag, Cu, Zn, etc. Sn-Ag-Cu solders are widely recommended as the replacement of eutectic SnPb solder in reflow soldering [4], however, Silver is known as a mobile element most susceptible to ionic migration [5]. Cu and Zn were also reported to occur migration under certain conditions [1]. Thus, the mobile constituents in lead-free solder alloys bring a new challenge for electrochemical migration in lead-free assemblies. Moreover,

the change in cleaning process for lead-free soldering might enhance the undesired effect of electrochemical migration. No-clean fluxes are widely in use in industry in recent years. The flux residua and contaminant on surface may cause the electrochemical reactions and affect the surface insulation resistance (SIR), which is detrimental to the reliability performance of assemblies.

In this study, several types solder pastes were employed to investigate the electrochemical migration characteristics of lead-free solder joints by a water drop test (WDT) method.

2 Experiment

2.1 Preparation of test samples

Test boards were fabricated by conventional PCB (printed circuit board) process. The layout of test board was shown in Fig.1. The FR4 board had ten pairs of Cu pads with pure tin finish. The thickness of Cu pad and Sn finish were 18μm and ~0.2μm, respectively. Each pair of Cu pads was employed as electrodes in the electrochemical migration tests. The spacing between the electrodes was varied from 0.15mm, 0.35mm, and 0.65mm to 0.95mm.

Tab. 1 Solder paste employed in the experiment

Solder paste No.	Composition (wt%)
1-1*	Sn-4.0Ag-0.5Cu
1-2	Sn-4.0Ag-0.5Cu
2	Sn-37Pb
3	Sn-36Pb-2Ag
4	Sn-3.0Ag-0.5Cu
5	Sn-3.5Ag-
6-1*	Sn-3.5Ag
6-2	Sn-3.5Ag
7	Sn-8Zn-3Bi

(* The solder pastes with the same composition were provided from different suppliers)

Nine types of solder paste from different suppliers were employed (listed in Table I). The compositions of lead-free solder paste included: Sn-4.0Ag-0.5Cu, Sn-3.0Ag-0.5Cu, Sn-

978-1-4244-2739-0/08/$25.00 ©2008 IEEE

3.5Ag-0.75Cu, Sn-3.5Ag, Sn-8Zn-3Bi (wt%). For Sn-4.0Ag-0.5Cu and Sn-3.5Ag solders, there were two types solder paste from different suppliers, respectively. Two Pb-bearing solder paste were used as reference: Sn-37Pb and Sn-36Pb-2Ag (wt%). The solder paste were printed on the test board and then soldered by a soldering gun for 2 min.

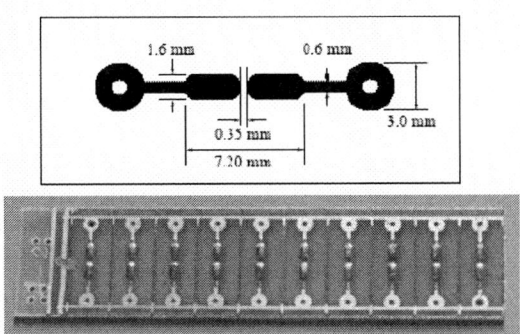

Fig.1 Test vehicle for water drop test

2.2 Water drop test

A DC power source (model KEITHLEY 6487) was used as power source and measure the in-situ resistance change of the test samples. Varying voltage of 3V, 5V and 10V were applied and the change of the resistance as a function of time was recorded. For each voltage, at least 10 samples were tested for each solder alloy. The evolution of microstructure of test samples was real-time monitored and recorded by optical microscope with image process software. We determined the occurrence of the time to short by two criterions: (1) the dendrites connecting the two electrodes exactly, (2) the sudden drop of the resistance.

The microstructure of the residual dendrites was examined by a scanning electron microscopy (SEM) (LEJTZ-AMR 1200) with a voltage of 20 KeV. Energy dispersive x-ray spectroscopy (EDS) (Oxford INCA 200) was also used to determine the composition of dendrites.

3 Results and Discussion

3.1 In-suit observation of dendrite growth during WDT

The growth process of the dendrites in the distilled water was observed under a stereomicroscope. Fig.2 showed the current change with the time during the WDT procedure. The in-situ optical microscopy images at different time interval were also revealed. It took tens of seconds for a dendrite nucleation at the cathode and it reached at the anode quickly. When the two electrodes were connected by a dendrite, the current remarkably increased. Through the whole process, air bubbles were generated continuously at the cathode, especially when the applied voltage was high.

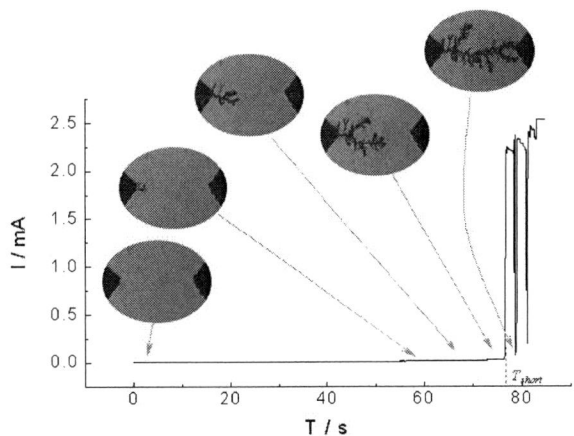

Fig.2 The process of dendrite growth during water drop test (Solder paste: #7, SnZnBi, spacing = 0.95mm, voltage = 5V)

Due to the electrolysis of water, metal ions gradually eluted in the vicinity of the electrodes. Harsanyi et al. [1] summarized the mechanism of the electrochemical process of metal elements. At cathode side, H+ from anode side due to the electric field force would obtain electrons. It is the reason for the formation of the bubbles observed.

3.2 Effect of test parameters
A. Applied voltage

The effect of applied voltage on the time to short and short resistance was investigated. Fig.3 exhibited the time to short under varying applied voltages for different solder alloys. The occurrence of short for all the solders were very quickly (within several minutes). The result showed the higher the voltage was, the shorter the failure time was. Under the same voltage, Sn-37Pb and Sn-36Pb-2Ag solders revealed shorter time to short comparing with other solders, which indicated that the Pb bearing solders had higher ECM susceptibility.

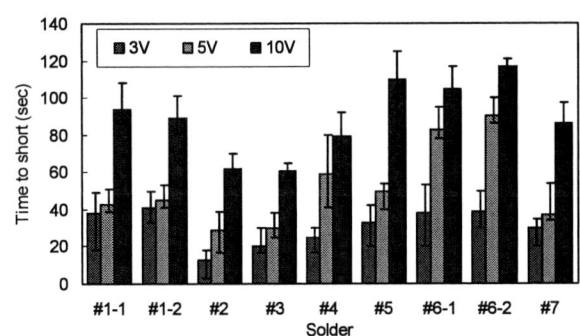

Fig.3 Comparison on time to short under varying applied voltages for different solder pastes (spacing = 0.65mm)

The critical voltage of the occurrence of ECM was detected by gradually increasing the applied voltage from the minimum limit 1.0V. Fig.3 compared the critical voltage of each solders.

The result exhibited that the susceptibility of ECM for all the solder pastes is in the order as following,

Sn-Ag-Cu, Sn-Ag > SnPb > SnPbAg > SnZnBi

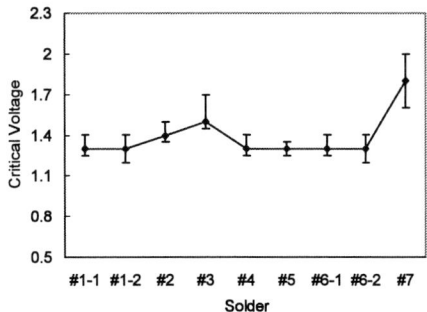

Fig.4 Critical voltage of solders under WDT (spacing = 0.65mm)

B. Spacing

The effect of spacing between the electrodes was also investigated. The ECM behaviours of four types samples with different spacing (0.15mm, 0.35mm, 0.65mm and 0.95mm) were compared. The applied voltage was kept on 2V. Fig.5 showed the typical correlation between spacing and time to short in Sn-3.5Ag-0.75Cu solder joints. It could be found that the time to short significantly increased with the increasing spacing.

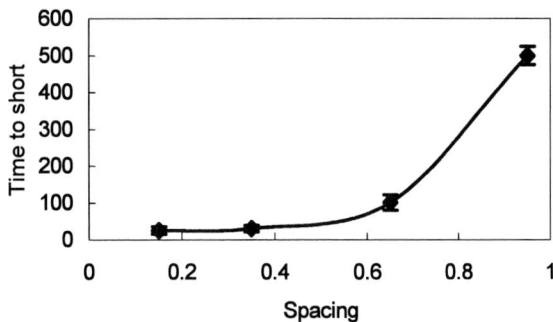

Fig.5 Effect of spacing distance on the time to short (voltage = 2V)

C. Flux residua

The effect of flux residua on the ECM behaviour was investigated. After soldering, flux residua remained around the electrodes in the as-soldered samples. Another set of samples were cleaned by ultrasonic rinse with acetone, ethanol and distilled water in sequence to remove the flux residua. The typical samples with flux residua and flux removal were shown in Fig.6.

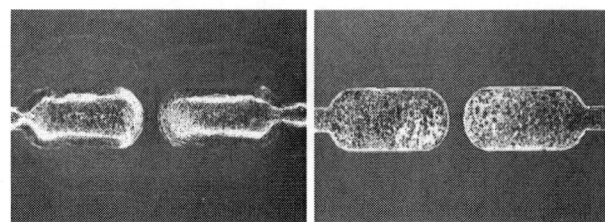

Fig.6 The test samples: a) Flux residue; b) Flux removal.

The ECM behaviors of different solder joints in this experiment were compared in Fig.7. For the solder #1-1, #2 and #6-2, there were no dendrites formation and resistance drop even after 30min applying voltage. Comparing with Fig.6, it was obvious that the time to short in solder joints with flux residue was much longer than those without flux residue. Additionally, it was interestingly found that the lead-free solders from different suppliers exhibited distinct ECM susceptibility. For the Sn-4.0Ag-0.5Cu, no ECM was observed after 30min WDT experiment for solder #1-1 but the dendrites grew to the anode after about 200sec for solder #1-2. For the Sn-3.5Ag solders, comparing with #6-1 and #6-2, the former failed in 15min but the latter kept good after 30min. The difference between the same composition solders was probably attributed to the difference between the flux mixed in the solder paste from suppliers. As we know, the exposed area (contact area with water) was the most serious reason for the electrochemical migration. The thickness of flux residua was not uniform on the surface of solder joints. The flux residua between the cathode and anode was very thick so as to play a role like a waterproof layer. The ECM process was difficult to take place in a short time because the metal was entirely covered by flux residue layer and electrolysis dissolution could not occur.

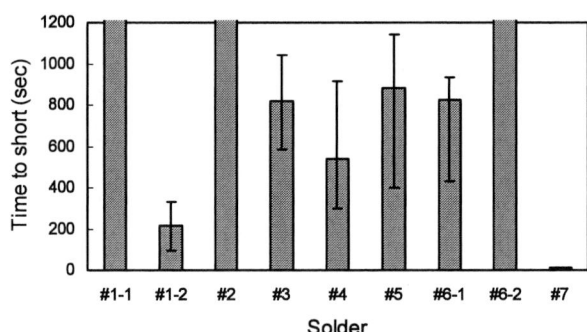

Fig.7 Comparison on time to short for different solder alloys with flux residue (*For the solder #1-1, #2 and #6-2, no ECM occurred even after 30min bias. Spacing = 0.65mm, voltage = 3V)

3.3 Microstructure and composition of dendrites

The microstructure of the dendrites under 3V applied voltage are shown in Fig.8. The microstructure of the dendrites under other applied voltages were similar as these under 3V. In the images, cathode was at the left side and anode was at the right side. The dendrites revealed a tree-like shape for all samples. Table 2 listed the compositions of dendrites in all solder joints samples. The main migration element was Pb in Pb-bearing solder joints. For Sn-3.5Ag-0.5Cu solder joints, it was Sn to migrate during the ECM process. However, the main composition of dendrites was Cu in Sn-3.5Ag and Sn-4Ag-0.5Cu solder joints. For Sn-3.5Ag-0.75Cu solder, although Sn was the dominant element, Cu was also detected in the dendrite. In the SnZnBi sample, Zn was the only migration element.

Tab. 2 The composition of dendrites in solder joints (Applied voltage: 3V; spacing=0.65mm)

Solder	Composition of dendrites (wt%)							
	Cu	Pb	Sn	Zn	Au	C	O	Br
SnPb		79.9			15.6		4.5	
SnPbAg		58.5	7.5		15.6	6.0	12.4	
Sn-3.5Ag	59.4		22.4		13.1		5.1	
Sn-3Ag-0.5Cu			62.7		22.2	10.3		4.83
Sn-3.5Ag-0.75Cu	21.4		61.7		15.6		1.3	
Sn-4Ag-0.5Cu	76.8				14.5		5.4	3.3
Sn-8Zn-3Bi				61.9	25.5		8.1	4.5

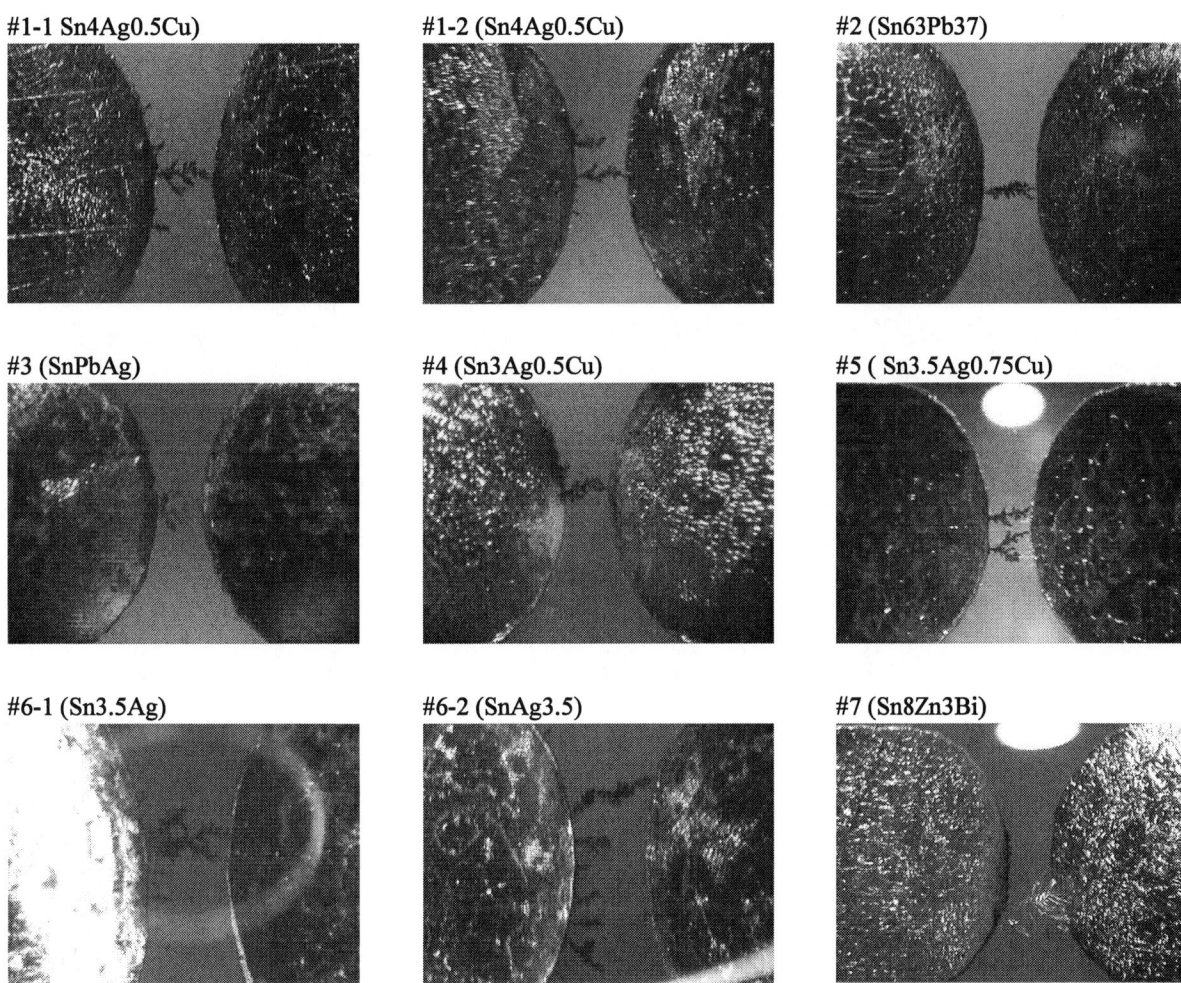

Fig.8 Optical microscopy images of dendrites in solder joints samples (voltage = 3V, spacing =0.35mm)

4 Conclusions

In this paper, the electrochemical migration behaviors of lead-free solder joints were investigated by water drop test method. The effects of electrode spacing, applied voltage, critical voltage and flux residue on ECM were compared. Higher voltage and narrower spacing remarkably weakened the ECM reliability of solder joints. The flux residua would inhibit the occurrence of ECM by preventing the electrolysis of metal ions. The main migration element was Pb in Pb-bearing solder joints. For Sn-Ag-Cu solder joints, it was Sn and Cu to migrate during the ECM process. Zn was the only migration element in the SnZnBi solder joint. The ECM susceptibility of different solders were compared and Sn-Zn-Bi solder revealed the best ECM reliability.

Acknowledgments

The authors are grateful to the work of Mr. Zhang Wei in Shanghai University.

References

1. G. Harsanyi, "Electrochemical processes resulting in migrated short failures in microcircuits," *IEEE Trans CPMT*, Vol. 18, (1995), pp. 602.
2. W.J. Ready, L.J. Turbini, *J. Eletron. Mater*, Vol. 31, (2002), pp. 1208.
3. M. Abtew, G. Selvaduray, *Mater. Sci. Eng. R.* Vol. 27, (2000), pp. 95.
4. IPC website: www.ipc.org/lead-free.htm.
5. M.V. Coleman, A.E. Winster, *Microelectron*, Vol. 4, (1981), pp. 23.

The Reliability Study of Sub 100 Microns SnAg Flip Chip Solder Bump on FR4 Substrate under Thermal Cycling

Xiaoqin Lin, Le Luo

ShangHai Institute of Microsystem and Information Technology, Chinese Academy of Sciences
No.865 Changning Road, Shanghai; 200050, PR China
E-mail: linxq8228@hotmail.com; Tel: 86+21-62511070-5471

Abstract

Due to the advantages of small-footprint, short-lead, high performance, high-packaging-density and thin profile, flip-chip-on-board (FCOB) technology is becoming an attractive choice in today's high density electronic packaging industry. With the trend toward lead-free and miniaturization in consumer electronics, the fatigue reliability of the small size lead-free FC solder joint on low cost PCB substrate are becoming one of the important issues.

In this study, the reliability of sub 100 microns Sn-3.0Ag flip chip solder bump on FR4 substrate was investigated under thermal cycling between -40℃ to 125℃. The influences of the shape of solder joint on the failed plane and the fatigue life were studied. The failed plane of the solder joint was discussed by the stress state analysis of the solder joint. Using the metallography, SEM, and live testing of the resistance, failure character and failure mechanism of the solder joint before and after underfilling were analyzed. The increasement of the fatigue life with the use of underfill was interpreted by plastic mechanics.

1. Introduction

Flip-chip-on-board (FCOB) technology provides the interconnection directly between silicon chips and organic substrates, which attracted great deals of attention in the area of advanced electronic packaging for its short interconnection distance, high density, low-cost and compatibility with SMT. [1,2] With the trend toward lead-free and miniaturization in consumer electronics, small size lead-free solder bump flip chip on low cost PCB substrate and its related interconnection and reliability are becoming one of the important issues. [3-5]

In order to examine the reliability of SnAg lead-free solder joint with the size less than 100 microns on FR4 substrate, the thermal cycling test was studied. Effect of the shape and height of the solder joint on fatigue life and failed plane of the solder joint were discussed. By the analysis of the stressed status of solder joint according to the principle of material mechanics, the failed plane of the solder joint was interpreted. The failure characteristic and failure mechanism of the lead-free solder joint was analyzed before and after underfilling. The influence the underfill on the reliability of the solder joint was analyzed with the principle of plastic mechanics.

2. Experimental Procedures

The test chip was of 4.2mm² with 13×13 area-array Sn3.0Ag solder bumps and with a pitch of 320 microns. The UBM was 80μm in diameter. The FR4 boards with pattern of Non Solder Mask Define (NSMD) and Solder Mask Define (SMD) were designed and manufactured. The Cu traces on PCB were metallized with Ni-Au. The daisy-chained layout on the test chip and FR4 board were prepared to check the integrity of solder bumps after the chip bonding to the PCB, as shown in Fig.1 (a). The process of chip flipping on PCB substrate was preformed by the model 410 Flip Chip Bonder. Then, the assembly was reflowed in a 5-zone reflow oven with a peak temperature of 260℃ holding for 30 seconds in N₂ ambience. A part of assemblies were underfilled using the Hysol FP4549. Thermal cycling test was used to evaluate the reliability of the solder joints between -40 ℃ to 125 ℃ according to JESD22-A104-B.[6] One cycle takes 50 minutes, for the time of increasing and decreasing temperature is 10 minutes, and the holding time at the high and low temperature is 15 minutes. The Keithley digital multimeter was used for real-time and continuous measurement of the electronic resistance of the interlinkage loop of the SnAg solder joint, as shown in Fig.1 (b). A 10% increase in resistance at the same temperature was used as the failure criterion of the flip-chip assembly. The samples were cross-sectioned to observe the shape and the cracks of the solder joints.

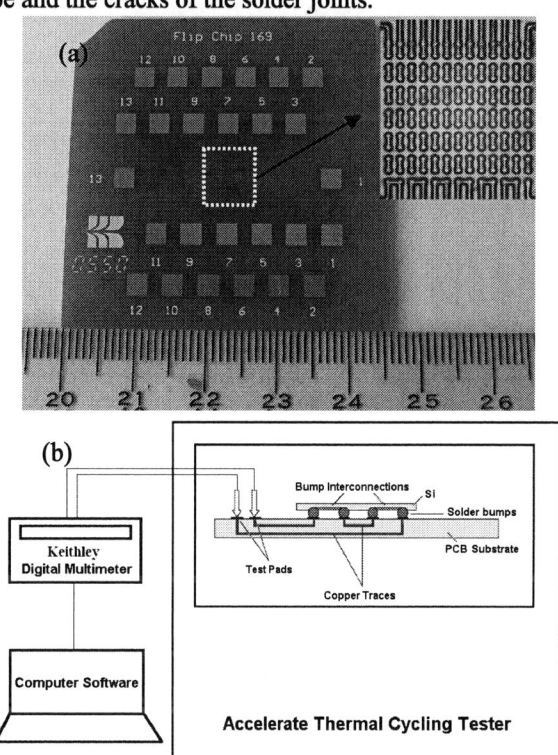

Fig.1 (a) FR4 board with daisy-chained layout, (b) resistance monitoring schematic view

3. Results

Resistance of Interlinkage Loop of the SnAg Solder Joint

Figure 2 compared the resistance of interlinkage loop of the SnAg solder joint approaching failure before and after underfilling. It showed that the resistance would increase to a certain degree before failure. The solder joint before underfilling showed a short time period of resistance increasing keeping for dozens of cycles at the most. While, the solder joint after underfilling showed a long time period of resistance increasing keeping for several hundred cycles.

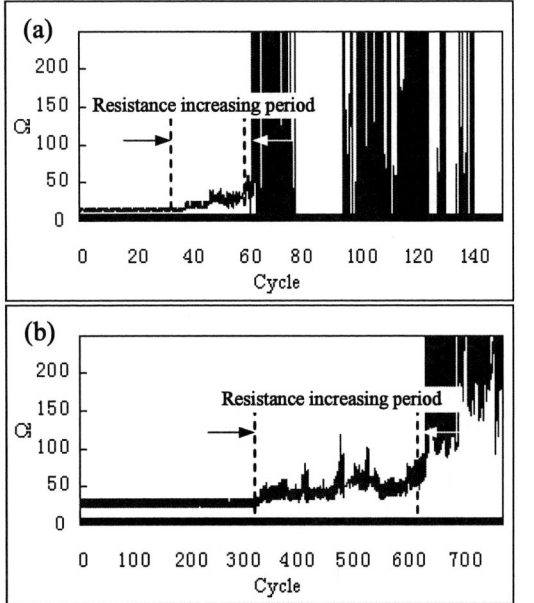

Fig.2 Resistance of interlinkage loop of the SnAg solder joint (a) before underfilling, (b) after underfilling

Weibull Life of SnAg FC Solder Joint

Figure 3 shows the Weibull life distribution of different FC assemblies, in which Eta shows the characteristic life $N_{63.2\%}$, and Beta shows the shape factor β. The plots showed that the fail data fits the Weibull distribution well. Table 1 shows the height of the FC solder joint and its corresponding Weibull life. It can be seen that the Weibull life increased with the height of the solder joint. After underfilling, the Weibull life of sample No.1 increased to 1601 cycles by about 18 times, and the Weibull life of sample No.2 increased to 3287 cycles by about 27 times.

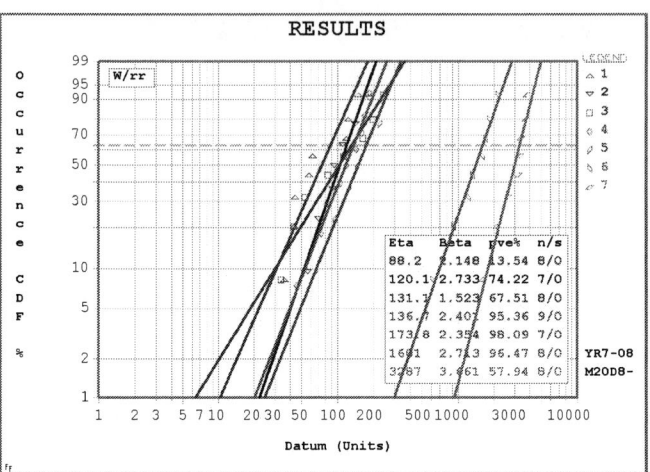

Fig.3 Weibull life of the solder joint

Tab.1 Height and corresponding Weibull life of solder joint

Sample Numbering	No.1	No.2	No.3	No.4	No.5
Type of Patterns of PCB Design	NSMD	SMD	NSMD	SMD	SMD
Height of Solder Joint （μm）	17.5	27.4	20.5	35.6	46.0
Weibull Life of Solder Joint Without Underfill （cycle）	88.2	120.1	131.1	136.7	173.8
Sample Numbering	No.6	No.7	-	-	-
Weibull Life of Solder Joint after Underfilling （cycle）	1601	3287	-	-	-

Fracture Plane of FC Solder Joint

Figure 5 shows the cross-sectional images of the failed solder joint at the position of the chip as shown in Fig.4. It can be seen that the fracture of the solder joint was the primary failure mode of the FC assembly. The solder joint without underfill showed the regular fracture plane, as shown in Fig.5. The position of fracture plane in the solder joint associates not only the shape of the solder joint but also the shear force due to the mismatch of CTE between the chip and PCB substrate. For those bumps at the corner of a chip, the failed plane was with an angle to the cross-sectional plane. However, the solder joint after underfilling showed the irregular fracture plane with multiple cracks, as shown in Fig.6.

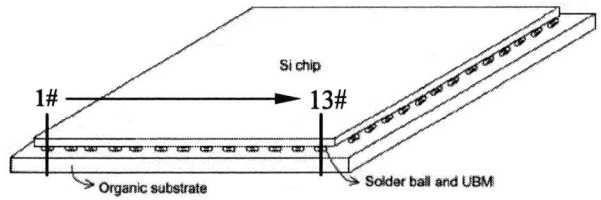

Fig.4 Schematic illustration of the location of the cross-sectional FC solder bump

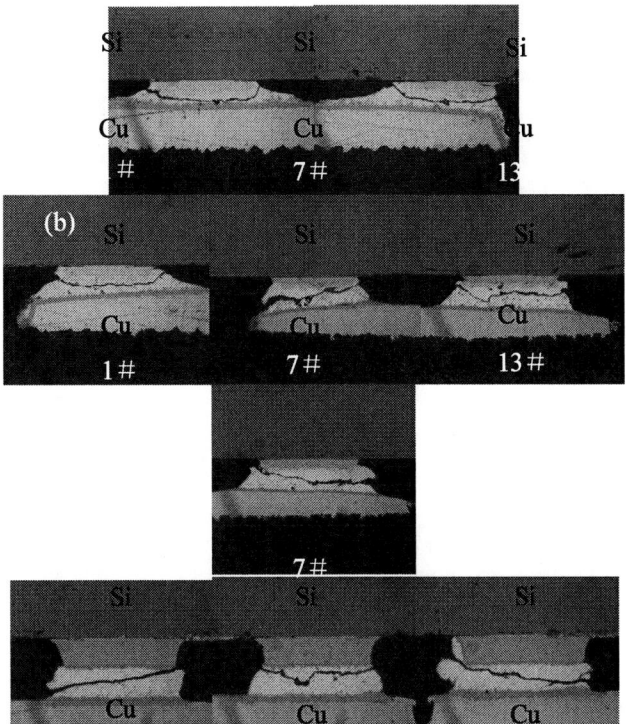

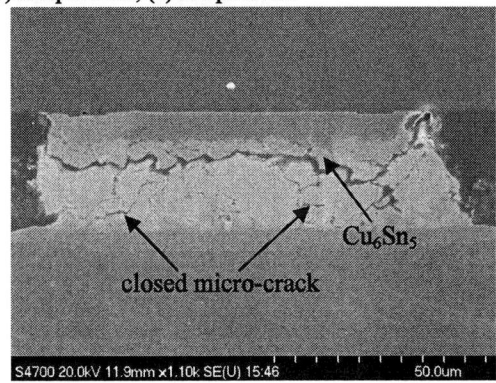

Fig.5 Cross-sectional graphs of failed solder joint (a) sample No.1, (b) sample No.2, (c) sample No.5

Fig.6 Cross-sectional image of failed solder joint after underfilling

4. Discussions

Failure Mechanism of Solder Joint before and after Underfilling

The solder joint before underfilling showed a short time period of resistance increasing keeping for dozens of cycles at the most, as shown in Fig.2 (a). On the other hand, the fracture of the failed solder joint showed the regular plane as shown in Fig.5, according with the transcrystalline fracture, one of the fracture characteristics of the fatigue behavior.[7] So, it can be inferred that the failure of the solder joint without underfill was controlled mainly by fatigue mechanism, and the crack in the solder joint occurred along the stress concentration. While, the solder joint after underfilling showed a long time period of resistance increasing keeping for several hundred cycles. The fracture of the solder joint after underfilling showed the irregular fracture plane along the Cu_6Sn_5/SnAg interface as shown in Fig.6,

which is the fracture characteristic of the creep behavior.[8] So, it can be inferred that the failure of the solder joint with the use of underfill was controlled mainly by creep mechanism, and the crack in the solder joint occurred along the Intermetallic Compound interface.

Failed Plane of the Solder Joint before Underfilling

Figure 5 show that the shape of the solder joint affected the failed plane of the solder joint. This was because that the shape of the solder joint changed the stress concentration position of the solder joint without underfill. The larger wetting area of PCB pad than UBM would result in the trapeziform shape of cross-section of solder joint, in which the fracture tends to occur near the chip side. On the contrary, the larger wetting area of UBM than PCB pad, would result in the inverted-trapeziform shape of cross-section of solder joint, in which the fracture tends to occur near the PCB side. The similar wetting area of PCB with UBM would result in the rectangular shape of cross-section of solder joint, and the fracture tends to occur in the middle of solder joint horizontally or with an angle to the cross-section for those bumps at the corner of the chip (e.g. 1# and 13# position shown in Fig.4).

In order to understand why the fracture plane takes place at different position and direction, it is necessary to study the shear stress state of the diagonal plane of the solder joint. It is known that the stress state of an arbitrary point of the solder joint under shear fore can be shown as Fig.7 (a), in which the shear stresses on opposite and parallel faces of an element are equal in magnitude and opposite in direction, and shear stresses on adjacent faces are equal in magnitude and point towards or away from the common line of intersection of the faces. A part of the hexahedral element was taken as the analyzed object as shown in Fig.7 (b). In which the angle of the oblique section was α, the normal stress and the shear stress suffered to the oblique section was σ_α and τ_α, respectively.

It is specified that clockwise forward from axis to normal line of the section is negative, while counter-clockwise forward from axis to normal line of the section is positive. Balanced equation can be deduced as:

$\Sigma F_n = 0$; That was:

$$\sigma_\alpha dA + (\tau dA \cos\alpha)\sin\alpha + (\tau' dA \sin\alpha)\cos\alpha = 0$$

$\Sigma F_t = 0$; That was:

$$\tau_\alpha dA + (\tau dA \cos\alpha)\cos\alpha + (\tau' dA \sin\alpha)\sin\alpha = 0$$

From the above equation it can be obtained:

$$\sigma_\alpha = -\tau \sin 2\alpha, \quad \tau_\alpha = \tau \cos 2\alpha.$$

When $\alpha = 0°$, $\sigma_{0°} = 0$, $\tau_{0°} = \tau_{max} = \tau$;

when $\alpha = 45°$, $\sigma_{45°} = \sigma_{min} = -\tau$, $\tau_{45°} = 0$;

when $\alpha = -45°$, $\sigma_{-45°} = \sigma_{max} = \tau$, $\tau_{-45°} = 0$;

when $\alpha = 90°$, $\sigma_{90°} = 0$, $\tau_{90°} = -\tau_{max} = -\tau$.

Therefore, the cross-section was suffered the maximum shear stress; the oblique section with the direction of $\alpha = \pm 45°$

1033

was suffered the maximum compressive stress and maximum tensile stress which have the same values with shear stress.

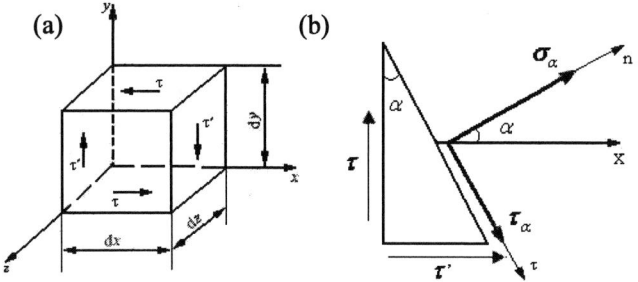

Fig.7 (a) shear stress state of an arbitrary point of the solder joint, (b) stressed state of the oblique section

It can be deduced from the above analysis that the shear stress states of the solder joint on both sides of the neutral plane were shown in Fig.8 (a). For the solder joints on the left side of the neutral plane as shown in Fig.8 (b), the oblique section with the direction of -45° suffered the maximum tensile stress. For the solder joints on the right side of the neutral plane as shown in Fig.8 (c), the oblique section with the direction of -45° suffered the maximum tensile stress. Normally, compression strength is higher than tensile strength for metal and alloy. So, the fracture tends to occur under the tensile stress. In other words, for those solder joints at the corner of the chip, their failed planes were with an angle to the cross-section, because that the oblique section of the solder joint suffered the maximum tension stress. So, it is more importance to protect the corner position of solder joint especially for the peripheral solder joints on the chip.

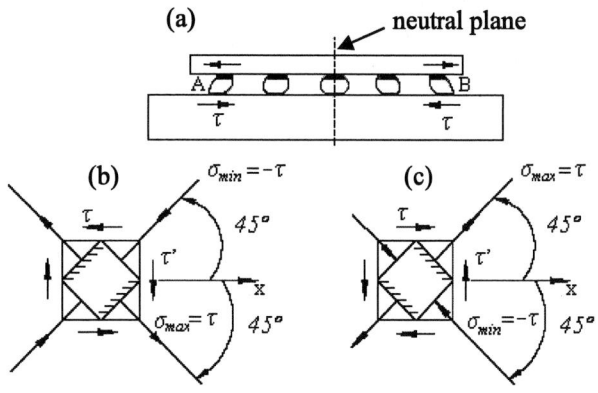

Fig.8 (a) Schematic illustration of shear stress state of FC solder joint, (b) shear stress state of solder joint on the side of A position in (a), (c) the shear stressed state of solder joint on the side of B position in (a)

Impact of Underfill on the Thermal Fatigue Life

Table 1 shows that the thermal fatigue life of the solder joint was improved remarkably with the use of underfill. The SnAg solder joint with the use of underfill failed after more than 3000 thermal cycles, showing the excellent fatigue reliability of the SnAg solder joint. Fig.9 shows the C-SAM image of FCOB of sample No.7 after 1000 cycles. It can be observed the serious underfill delamination between the underfill and the substrate or the chip. However, the Weibull life of the sample No.7 was 3287 cycles. So, it can be inferred that underfill delamination was not the major failure mechanism of

the FC solder joint on board. In order to explain this phenomenon, it needs to understand the mechanism of underfill on the solder joint.

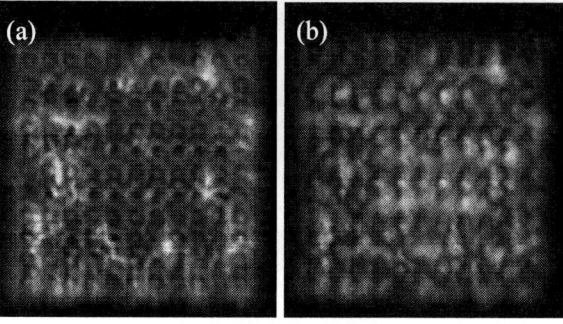

Fig.9 C-SAM image of FCOB of sample No.7 after 1000 cycles (a) at the top interface of underfill, (b) at the bottom interface of underfill

According to the principle of plastic mechanics, the stress state of an arbitrary point of the stressed solder joint can be presented by three primary stress in three-dimensional direction denoted as σ_1、 σ_2 and σ_3 respectively, as shown in Fig.10. It is specified that tensile stress is positive primary stress, while compressive stress is negative primary stress.

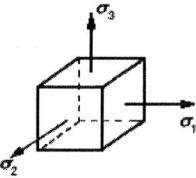

Fig.10 Stress state of one point by three primary stress σ_1、 σ_2 and σ_3

During the thermal cycling process, the solder joint undergoes different stress state due to the CTE mismatch between the chip, underfill, solder, and PCB substrate as shown in Tab.2. Taken the solder joint as analyzed object, due to the solder have a higher CTE than that of Si and PCB substrate, the solder joint without underfill was in a compression stress state during the temperature rising period (Fig.11(a)), and a tensile stress state during the cooling temperature (Fig.11(b)). After underfilling, the solder joint suffered not only the longitudinal stress but also the two-dimensional transverse compression stress state due to the higher CTE of underfill than that of solder. The solder joint after underfilling was in a "three-compression" stress state during the temperature rising period (Fig.11(c)), and a "two-compression and one tensile" stress state during the cooling temperature (Fig.11(d)). According to the knowledge of plastic mechanics, it is known that the plasticity of the alloy is affected by the mean stress $\sigma_m = \frac{1}{3}(\sigma_1 + \sigma_2 + \sigma_3)$ (that is hydrostatic stress). The more directions and larger value of the compression stress, the larger value of the mean stress σ_m, the better plasticity of the alloy can be got. On the contrary, the more directions of the tensile stress state the worse plasticity.

In a word, the hydrostactic-stress σ_m of the solder joint was increased with the use of underfill, so the plasticity of the solder joint was increased. As a result, the fatigue reliability of the solder joint was improved with the use of underfill. The underfill delamination cannot cause the failure of the solder joint directly.

Tab.2 CTE of different materials, ppm/K

Si	Sn-3.0Ag	FR4	FP4549 Underfill	
			CTE（<Tg）	CTE（>Tg）
2.6	30	18		
			45	143

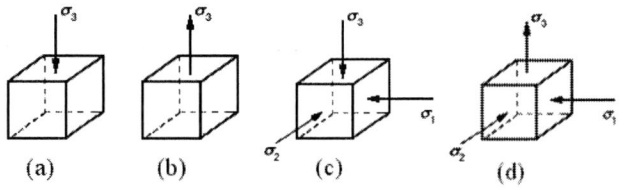

(a)　　(b)　　(c)　　(d)

Fig. 11 Stress state of solder joint during thermal cycling process (a) without underfill, temperature increasing period, (b) without underfill, temperature cooling period, (c) after underfilling, temperature increasing period, (d) after underfilling, temperature cooling period

5. Conclusions

The reliability of the high density FC SnAg solder joint was investigated under thermal cycle between -40℃ to 125℃. The SnAg solder joint with the use of underfill failed after more than 3000 thermal cycles, showing the excellent fatigue reliability of the SnAg solder joint. According to the study of the resistance of the interlinkage loop of the SnAg solder joint and the analysis of the fracture plane, it can be inferred that the failure of the solder joint without underfill was controlled mainly by fatigue mechanism, and the crack in the solder joint occurred along the stress concentration. The failure of the solder joint with the use of underfill was controlled mainly by creep mechanism, and the crack in the solder joint occurred along the Cu_6Sn_5/SnAg interface. By the analysis of the stressed status of solder joint according to the principle of material mechanics, the failed plane of the solder joint was interpreted. For those bumps at the corner of a chip, the failed plane was with an angle to the cross-sectional plane, because that the diagonal plane of the solder joint suffered the largest tension stress. The change of the stress state of the solder joint after underfilling was analyzed with the principle of plastic mechanics. It was demonstrated that the hydrostactic-stress σ_m of the solder joint was increased with the use of underfill, so the plasticity of the solder joint was increased. As a result, the fatigue reliability of the solder joint was improved with the use of underfill.

Acknowledgments

This work was financially supported by National Natural Science Foundation of China under Project No.60676061.

References

1. A.F.J.Baggerman, J.F.J.M.Caers, J.J.Wondergen, et al, "Low-Cost Flip-Chip on Board," IEEE Trans. Comp., Pack. and Manuf. Tech.-Part B, vol.19, No.3 (1996), pp. 736.

2. Z.W.Hou, G.Y.Tian, C.Hatcher, et al, "Lead-Free Solder Flip Chip-on-Laminate Assembly and Reliability," IEEE Trans. Electronic Packaging Manufacturing, Vol.24, No.4 (2001), pp. 282.

3. G.W.Xiao, P.C.H.Chan, A.Teng, et al, "Reliability Study and Failure Analysis of Fine Pitch Solder Bumped Flip Chip on Low-Cost Printed Circuit Board Substrate," Proc. of 51st International Conference on Electronic Components and Technology, 2001, pp. 598.

4. J.Sjoberg, D.A.Geiger, D.Shangguan, et al, "Lead-free solder flip chips on FR-4 substrates with different surface finishes, underfills and fluxes," Proc. of 29th International Conference on Electronics Manufacturing Technology, 2004, pp. 31.

5. P.Wolflick, K.Feldmann, "Lead-Free Low-Cost Flip-Chip Process Chain: Layout Process, Reliability," International Electronics Manufacturing Technology (IEMT) Symposium, 2002, pp. 27.

6. JEDEC standard: JESD22-A104-B, Temperature Cycling, 2000.

7. A. I. Attarwala, J. K. Tien, G. Y. Masada, et al. Confirmation of Creep and Fatigue Damage in Pb/Sn Solder Joints. J Electron Packg, vol.114, No.2 (1992), pp. 109-111.

8. Boon Wong, D. E. Helling. A Mechanistic Method for Solder Joint Failure Prediction under Thermal Cycling. J Electron Packg, vol.112, No.1 (1990), pp. 104-109.

Analysis and Solving of the EMI effect on LC-VCO in mixed-signal ICs

Wenrong Yang*, Jiongming Wang, Jue zhang, Xiaohui Li

Key Laboratory of Advanced Display and System Applications (Shanghai University), Ministry of Education

Microelectronic Research & Development Center, Shanghai University

Shanghai University, Shanghai 200072, P. R. China

*E-mail: yangwr@shu.edu.cn

Abstract

The EMI/EMC problems appear in mixed-signal ICs with the increasing frequency and decreasing process geometries. In this paper, the amplitude modulation (AM) is taken into account the LC-VCO based on the TSMC 0.25um mixed-signal process, and a cascade amplifier circuit, which is not complex, is proposed to optimize the LC-VCO through analyzing the noise coupling transition function of the LC-VCO. Lastly, the efficiency and usefulness of the proposed method is proved in the simulations.

1. Introduction

Voltage-controlled oscillator (VCO) is one of key components in the phase-locked loop (PLL). The electromagnetic interference (EMI) from the digital circuit to the VCO becomes more and more serious as the clock frequency of the digital signal increases. There are three interfered noises. The first one is the power supply line conducted noise, which is caused by the simultaneous switch current of the digital circuit. It will be conducted to the analog circuit through the power network on print circuit board (PCB). The second one is the radiated noise, which is caused by the bone wire of the IC package and will make the inductor of the VCO generate inducted current. The last noise, often called substrate crosstalk noise, is the substrate conducted noise from the digital parts to the analog parts (ex. VCO, amp).

The digital circuit and the analog circuit are integrated on the same substrate (generally P-). Because of the larger and higher frequency current of the power line of the digital parts, the current will be conducted to the analog parts through the shared substrate. The function of analog parts will be impacted, and even be destroyed.

The first noise can be effectively weakened by optimizing the power network on both the chip level and the PCB level or separating the digital power line from the analog parts; the on-chip decoupling capacitor was placed between the VDD and GND of the digital parts to solve the radio emission of the bone wire[1]. Altering the package process can also reduce the radiated emission of the package[2]; many papers proposed the model of the substrate[3]-[4], and used the triple-well or P+ ring to shield the digital circuit from the analog one[5]. Although these methods can decrease the substrate noise, they usually depend on the experiences of the designer.

In this paper the transition function of the substrate noise is obtained firstly, and then the additional cascade amplifier circuit is used to optimize the LC-VCO through the analysis of the transition function. The noise PV (peak voltage) and the phase noise of a traditional LC-VCO and an advanced LC-VCO can be obtained by simulation using Cadence Spectre. The simulations prove the proposed method, the phase noise of the advanced LC-VCO decreases by 6dbc/Hz compared with one of the traditional LC-VCO on a noise frequency of 1.527GHz.

2 the EMI noise coupling analysis

2.1. The ICEM model

It is well-known that the main cause of EMI/EMC in mixed-signal ICs is the coupling from digital ground lines mentioned in the part 1. Ground noise is caused by the supply current and the ground line impedance. The ground noise amplitude is proportional to the di/dt when inductance is dominant in the ground line impedance.

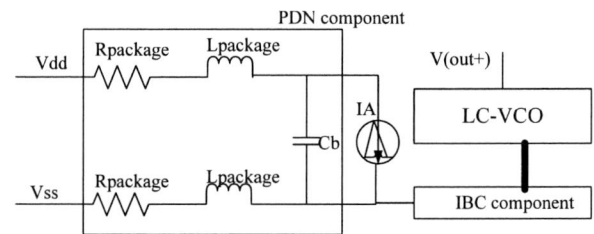

Fig 1 the simple ICEM

An IC EM model called ICEM, proposed by French standardization group (UTE), currently exists in the standardization process. It covers three types of EMC analysis including IC emission analysis, PCB analysis and IC auto-compatibility analysis. Therefore, during the circuit design of mixed-signal ICs, It is more helpful for designers to optimize the victim circuit and improve the auto-compatibility in mixed-signal ICs. As shown in Fig 1, there are three components in the ICEM that is the PDN component (Passive Distribution Network), the IA component (Internal Activity) and the IBC component (Inter-Block Coupling), respectively. The details about the ICEM can be obtained in the IEC62014-3[6].

The relationship between the digital noise and the LC-VCO can be simply expressed as equitation 1, according to the superposition principle.

$$V(out+) = A_0 + A_m \cos(\omega_c t + K_{av} \eth V(t)dt)$$
$$+ V_{noise} H(s)\cos(\omega_c t + K_{av} \eth V(t)dt) \quad (1)$$

The Vnoise can be got according from the Fig 1

$$V_{noise} \approx V_m \exp(-\alpha t)\cos(\beta t)$$

Where Vm , α and β are related to the Lpackage and the frequency of the digital circuit[7],

Then, the V (out+) can be written as:

$$V(out+) = A_0 + A_m \cos(\omega_c t + \int K_{av}(t)dt)$$
$$+ V'_m H(s)\cos((\omega \pm \beta)t + \int K_{av}(t)dt) \quad (2)$$

According to the Hajimiri model of the oscillator:

$$L(\Delta\omega) \approx 10\log_{10}(\frac{2B_m C_m}{4q_{max}\Delta\omega})^2 \quad (3)$$

Where Cm is the amplitude when the angular frequency is $\omega + \triangle \omega$, qmax represents the current amplitude of the VCO in the normal operation. It is obvious that the phase noise of the LC-VCO is proportional to the amplitude of the EMI noise.

2.2 The EMI noise coupling in the LC-VCO

In this section, the substrate coupling transfer function H(s) will be obtained using the high frequency model of the MOS transistor. It can be obtained from the Fig 2:

$$-(\Delta V_{out+} - V_B)(C_{GS1}S + C_{GB1}S + C_{DB2}S + \frac{1}{r_0}) + \quad (4)$$

$$2(\Delta V_{out-} - \Delta V_{out+})C_{GD1}S - G_m(\Delta V_{out-} - V_B) = I_3$$

Because of the symmetry of the two NMOS transistors (M3~M4), the similar equation about the left NMOS transistor (M3) can be obtained:

$$-(\Delta V_{out-} - V_B)(C_{GS2}S + C_{GB2}S + C_{DB1}S + \frac{1}{r_0}) + \quad (5)$$

$$2(\Delta V_{out+} - \Delta V_{out-})C_{GD2}S - G_m(\Delta V_{out+} - V_B) = I_1$$

It is assumed that:

$$A = \frac{1}{r_0} + S(C_{DB1} + C_{GB2} + C_{GS2})$$

$$B = G_m$$

$$C = 2C_{GD1}S = 2C_{GD2}S$$

So the equation (4) and (5) can be simplified:

$$I_1 = (C - B)\Delta V_{out+} - (A + C)\Delta V_{out-} + (A + B)V_B$$
$$I_3 = (C - B)\Delta V_{out-} - (A + C)\Delta V_{out+} + (A + B)V_B \quad (6)$$

And the similar formulations on the two PMOS transistors (M1~M2) can be also obtained.

$$I_1 = (C' + B')\Delta V_{out-} - (C' - A')\Delta V_{out+}$$
$$I_3 = (C' + B')\Delta V_{out+} - (C' - A')\Delta V_{out-} \quad (7)$$

Where:

$$A' = \frac{1}{r'_0} + S(C'_{DB1} + C'_{GB2} + C'_{GS2})$$

$$B' = G'_m$$

$$C' = 2C'_{GD1}S = 2C'_{GD2}S$$

Due to the symmetry of left and right MOS transistors (M1~M4), the Vout+ or Vout- can be got from all above equations.

$$\Delta V_{out+} = \Delta V_{out-} = \frac{A + B}{B' + A + B + A'}V_B \quad or$$

$$H(S) = \frac{\Delta V_{out+}}{V_B} = \frac{\Delta V_{out-}}{V_B} = \frac{A + B}{B' + A + B + A'} \quad (8)$$

And there is a pole point and a zero point of the transfer function H(s):

$$S_{zero} = \frac{1 + r_0 G_m}{r_0(C_{DB1} + C_{GB2} + C_{GS2})} \quad And$$

$$S_{pole} = \frac{r'_0 + r_0 + r_0 r'_0(G_m + G'_m)}{r'_0 r_0(C_{DB1} + C_{GB2} + C_{GS2} + C'_{DB1} + C'_{GB2} + C'_{GS2})}$$

It is well-known that the pole point makes the H(S) amplitude decrease by 20dB/Hz, while the zero point makes it increase by 20dB/Hz. the $\triangle$Vout+ approached to the Vnoise when the noise frequency becomes higher.

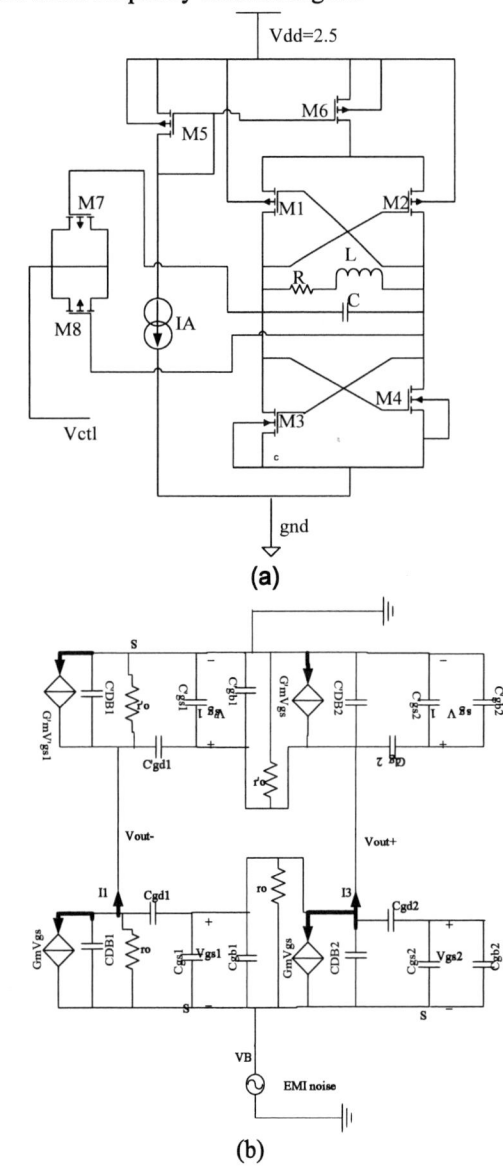

Fig2 (a) The traditional LC-VCO (b) the model of the EMI noise coupling in the LC-VCO

3. The optimization design for the LC-VCO

From the analysis of section 2, it can be seen that EMI noise affects the phase noise of the LC-VCO. The negative feedback is very useful to suppress the noise [8]. However, the negative feedback system usually requires a high performance amplifier, which makes the design very difficult implement in the EM environment.

A pole point introduced near the zero point is effective to suppress the impact of VB on the output voltage according to the equation (8). By doing this, the equation (8) will be

$$\Delta V_{out+} = \Delta V_{out-} = \frac{A_0}{S + S_{new\_pole}} \frac{A+B}{B'+A+B+A'} V_B \quad (9)$$

In order to apply less effect on the V_{out+}, the additional circuit should meet the following requirements:

$$A_0 << 1 \quad (10)$$

$$S_{newpole} \approx \frac{1 + r_0 G_m}{r_0 (C_{DB1} + C_{GB2} + C_{GS2})}$$

In the equation (10), Snewpole is the extra pole point. A similar pole point can be found in the single cascode amplifier. There are three pole points in the amplifier. It is found that the second pole S2 is very close to the zero point. The impacts of the additional more poles are less, because they will decrease the amplitude when the frequency is increased. And the gain of the cascade amplifier can be changed through adjusting the load impedance.

$$S_1 = \frac{1}{R_s[C_{GS1} + (1 + \frac{g_{m1}}{g_{m2} + g_{mb2}})C_{GD1}]}$$

$$S_2 = \frac{g_{m2} + g_{mb2}}{2C_{GD1} + C_{DB1} + C_{SB2} + C_{GS2}} \quad (11)$$

$$S_3 = \frac{1}{R_D(C_{DB2} + C_L + C_{GD2})}$$

Then the schematic of the additional circuit and parameters of the transistors are shown in Fig 3

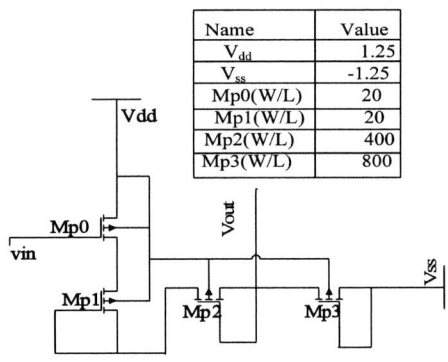

Name	Value
V_{dd}	1.25
V_{ss}	-1.25
Mp0(W/L)	20
Mp1(W/L)	20
Mp2(W/L)	400
Mp3(W/L)	800

Fig 3 the schematic of the additional circuit

The gain of the circuit is similar to:

$$A_0 = \frac{g_{m0}}{g_{m3}} = \frac{M_{P0}(W/L)_0}{M_{P3}(W/L)_3} \quad (12)$$

According to the equation (10), Mp3(W/L)$_3$ is substantially larger than Mp0(W/L)$_0$。

4. Simulations

The transient response and the phase noise of the LC-VCO can be obtained in simulation using Cadence Spectre, the simulation circuit is shown in Fig.1, in which the modified LC-VCO(Fig.4(a)) and conventional LC-VCO is adopted respectively for comparison. The parasitic parameters are shown in table 1 and most of them are estimated value. However, they don't impact much on the simulation result.

Table 1 the parameter of the ICEM

	L(H)	R(Ω)	Cd+Cb(F)
Digital IC package	5n	2	100f
Analog IC package	5n	2	--

The advanced LC-VCO is less sensitive to the EM noise than the traditional one does, as it is indicated in Fig 4(b). And the phase noise of the advanced LC-VCO, shown in the table 2, is a little higher, because the current amplitude of the VCO in normal operation is smaller. But under a noise frequency of 1.527GHz which is obtained from the simulation of Fig 1, the phase noise of the advanced LC-VCO decreases by 6dbc/Hz compared with the traditional LC-VCO.

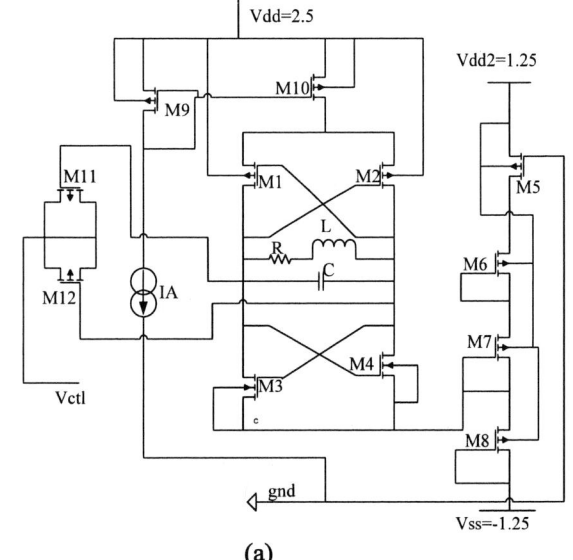

(a)

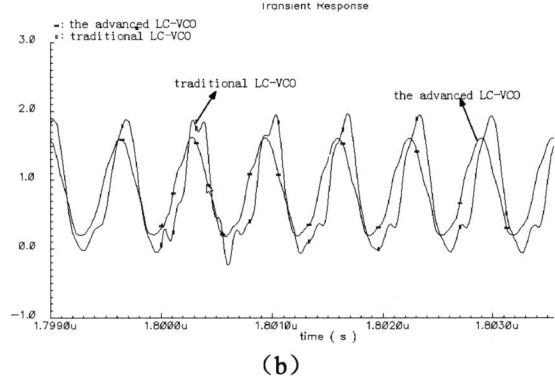

(b)

Fig 4 （a）the advanced LC-VCO （b）the transient response of the advanced LC-VCO and the traditional LC-VCO

Table 2 phase noise of the two types of LC-VCO

The type of the LC-VCO	Phase noise (dbc/Hz)		
	At 1M	At 100M	At 1.527G
The advanced LC-VCO	-143	171.296	173.072
Traditional LC-VCO	142.542	176.399	167.344

5. Conclusions

The transfer functions H(s) of the EMI noise coupling is obtained by using the model of the EMI noise coupling. The zero point of the H(s) is the main cause of the noise coupling, so the zero point's removal is a main method in order to reduce the EMI noise coupling, the addition of a cascade amplifier circuit is proposed to optimize the LC-VCO, and the cascade amplifier circuit is not complicated. The simulation result has proved the efficiency and usefulness of the proposed method.

References

1. J Kim, H Kim, W Ryu, J Kim, YH Yun, SH Kim and SH Ham "Effects of on-chip and off-chip decoupling capacitors on electromagnetic radiated emission" Electronic Components and Technology Conference, 1998. 48th IEEE 25-28 May 1998 Page(s):610 - 614

2. Takahashi, E.Nakayama, T. Saito, Y. "Evaluation of packages by simulating IC emission using a LECCS model" EMC-Zurich 2006. 17th International Zurich Symposium on 27 Feb.-3 March 2006 Page(s):300 – 303

3. H.Lan, Tze W.Chen, Chi On Chui, Parastoo Nikaeen, J.Kim, W.Dutton "Synethesized compact models (SCM)and experimental verifications for substrate noise coupling in mixed signal ICs" Design, Automation and Test in Europe Conference and Exhibition, 2004. Proceedings Publication Date: 16-20 Feb. 2004 Volume: 2, on page(s): 836- 841

4. Samavedam.A.; Mayaram.K and Fiez.T "A Scalable Substrate Noise Coupling Model for Mixed-Signal Ics" Solid-State Circuits, IEEE Journal of Publication Date: Jun 2000 Volume: 35, Issue: 6 On page(s): 895-904

5. Valorge, O.; Andrei, C.; Calmon, F.; Verdier, J.; Gontrand, C.; Dautriche, P.; "a simple way for substrate noise modeling in mixed signal ICs" Circuits and Systems I: Fundamental Theory and Applications, IEEE Transactions on Volume 53, Issue 10, Oct. 2006 Page(s):2167 - 2177

6. IEC62014-3 : Integrated Circuit Electromagnetic Model Cookbook. French committee UTE 47A EMC Task Force May 31, 2002 www.ute-fr.com

7. Steinecke, T. John, W. Koehne, H. Schmidt, M. "EMC Modeling and Simulation on Chip level" Electromagnetic Compatibility, EMC. 2001 IEEE International Symposium on Publication Date: 2001 Volume: 2, On page(s): 1191-1196

8. K.Makie-Fukuda, T.Tsukada "On-chip active guard band filters to suppress substrate-coupling noise in analogy and digital mixed signal ICs" VLSI Circuits, 1999. Digest of Technical Papers. 1999 Symposium on 17-19 June 1999 Page(s):57 - 60

Thermal Behavior Analysis of Lead-free Flip-Chip Ball Grid Array Packages with Different Underfill Material Properties

Hsin-yuan Chen[a], Kuo-yuan Hsu[a], Tsung-shu Lin[b] and Jihperng Leu[a,*]

[a] Department of Materials Science and Engineering, National Chiao-Tung University
1001 University Road. Hsinchu, Taiwan, 30049 China
[b] United Microelectronic Corporation No. 3, Li-Hsin 2nd Road
Hsinchu Science Park, Hsinchu, Taiwan, 30049, China
*E-Mail: jimleu@mail.nctu.edu.tw; Tel: +886-35131420

Abstract

Low-k materials have been introduced in the backend interconnects since 90 nm node for advanced microelectronic products in order to reduce the RC delay. However, the fragile low-k layer is very sensitive to the thermal stress induced by the CTE (coefficient of thermal expansion) mismatch at metal/dielectric level as well as at die/package level. In the die/package interaction, the transition to lead-free solders from conventional SnPb eutectic solder degrades the reliability of flip-chip ball grid array (FC-BGA) packages due to its higher reflow temperature. As a result, the underfill layer becomes more critical in protecting both low-K layer and solder bumps from delaminations and cracks. In this study, regular and high resolution Moiré interferometry were first employed to measure thermal strains of FC-BGA assemblies with CuSn solder and 2 latest underfill materials, which in turn validate our 3D FEA model. Besides, six kinds of FC-BGA assemblies with different solder alloys and underfill materials were also evaluated by thermal cycling test (TCT). A 3D simulation model using ANSYS[TM] was created to predict and analyze the fracture susceptibility of the assemblies. The simulation results showed good agreement with TCT experiments. The underfill material with low CTE, moderate modulus, and high Tg (glass transition temperature) is recommended for low-k/FC-BGA packages.

Introduction

As devices continue scaling down to 45 nm node, copper and ultra low-k dielectrics become the mainstream in the backend interconnects for further reduction in RC delay. Unfortunately, the fragile low-K material possesses weaker mechanical properties than conventional SiO_2 dielectric, and poor adhesion with silicon interface. [1-2] When die is attached to polymer substrate in FC-BGA package, the large CTE mismatch between silicon chip (2.6 ppm/ ℃) and bismaleimide triazine (BT) polymer substrate (14 ppm/℃) results in large die warpage and stress, which may cause low-k delamination in the interconnect layer or die/low-K level during or after thermal cycling test. K. C. Chang et al. illustrated some design guidance for increasing reliability of low-k FC-BGA assemblies. [3] L. Mercado proposed that tiles or slots were added in the interconnect layer to reduce available area for crack growth. [4] M. Rasco et al. evaluated the delamination risks of ultra low-k material for different passivation types. [5]

Due to the rise of environmental awareness, lead content in the electronic components are not allowed to be sold to European Union starting 2006. Thus, the traditional SnPb eutectic solder is gradually replaced by lead-free solder. However, lead-free solder in general requires a higher reflow temperature and new fluxes. Higher reflow temperature may induce larger stress due to larger ΔT, while new fluxes tend to generate by-products which are more difficult to remove and, in turn, degrade the adhesion of solder/underfill/die. [6]

The underfill material has been introduced in the flip chip package to reduce the thermal reliability issues. [7] In general, the underfill material with rigid mechanical properties is good for bump protection, but bad for the low-K delamination issue. [8] It is difficult to find an optimal underfill material to protect both bumps and low-k layer due to its opposite requirements in mechanical properties. Thus, how to balance three major material properties (E, CTE and Tg) of underfill materials to protect both bumps and low-k layer becomes a critical challenge for IC packaging industry. In this study, a 3D finite element analysis (FEA) model using ANSYS[TM] was first established to analyze the warpage and strain distribution in FC-BGA packages with CuSn solder and two latest underfill materials, and validated by experimental results from Moiré interferometry measurement. Such model was then used to analyze the stresses in the bumps and layer-k layer of six FC-BGA package samples with various solder alloys (CuSn and PbSn) and underfill materials (various E, CTE and Tg) and to compare with reliability results from temperature cycling test (TCT). The impact of various underfill material properties on the stress level at the outmost bump and low-k layer was also addressed in this study.

Moiré interferometry

The Moiré interferometry is a powerful tool for measuring the thermal deformation of FC-BGA packages. It can provide *in-situ*, whole field, and in-plane displacement. Two incident coherent laser beams interact at deformed grating surface and produce an interference pattern which stands for the displacement contour. The relationship between fringe order and displacement can be expressed as:

$$U_{(x,y)} = \frac{1}{f} N_x(x,y) \tag{1}$$

$$V_{(x,y)} = \frac{1}{f} N_y(x,y) \tag{2}$$

where f = 2400 lines/mm is the frequency of virtual reference grating, N_x and N_y are the fringe orders of U field and V field, respectively.

The strains can then be determined as:

$$\varepsilon_x = \frac{\partial U}{\partial x} = \frac{1}{f}\left[\frac{\partial N_x}{\partial x}\right]$$

$$\varepsilon_y = \frac{\partial V}{\partial y} = \frac{1}{f}\left[\frac{\partial N_y}{\partial y}\right]$$

$$\gamma_{x,y} = \frac{\partial U}{\partial y} + \frac{\partial V}{\partial x} = \frac{1}{f}\left[\frac{\partial N_x}{\partial y} + \frac{\partial N_y}{\partial x}\right] \quad (3)$$

The resolution of regular Moiré interferometry, which depends on the grating frequency, is 0.417 μm in this study. However, such sensitivity is not enough to measure the deformation of solder bumps. As a result, a phase shifting Moiré interferometry was employed to enhance the resolution to 26 nm. Each fringe spacing is equal to a 2π phase angle difference and 0.417 μm displacement. The unknown phase angle can be extracted from four precisely phase-shifted interference patterns by phase shifting Moiré interferometry. The intensity of the four patterns can be expressed as: [11]

$$I_1(x,y) = I_0(x,y) + I'(x,y)\cos[\phi(x,y)]$$

$$I_2(x,y) = I_0(x,y) + I'(x,y)\cos[\phi(x,y) + \pi/2]$$

$$I_3(x,y) = I_0(x,y) + I'(x,y)\cos[\phi(x,y) + \pi]$$

$$I_4(x,y) = I_0(x,y) + I'(x,y)\cos[\phi(x,y) + 3\pi/2]$$

(4)

where $I_0(x,y)$ and $I'(x,y)$ are the background and periodically varying intensities in the interference pattern. $\phi(x,y)$ is the unknown phase angle of the interference pattern at each pixel location. Each subsequent pattern is obtained by shifting a phase angle of exactly $\pi/2$ of the fringe period. The different intensities of the four patterns can be used to determine the unknown phase angle by the following equation:

$$\phi = \arctan\frac{I_4 - I_2}{I_1 - I_3}$$

(5)

After the phase angle is solved, the continuous displacement of U field can be determined as

$$u = \frac{\phi}{4\pi f}$$

(6)

The displacement of V field is similar to Eq. (6).

(a)

(b)

Figure 1. The experimental FC-BGA assembly (a) before and (b) after cutting

Table 1. The material properties and dimensions of key components in FC-BGA

Material	Properties				
	Thickness (mm)	Young's Modulus (kg/mm2)	CTE(ppm/℃)	ν	Tg(℃)
Die (16.35*16.35 mm)	0.75	16000	2.8	0.3	-
Underfill 1		E1=826.5, E2=30.6	CTE1=26 CTE2=91	0.35	125
Underfill 2		E1=969, E2=11	CTE1=27 CTE2=92	0.35	100
Underfill 3		E1=800, E2=4.7	CTE1=32 CTE2=102	0.35	80
Underfill 4	0.1	E1=700, E2=4.7	CTE1=32 CTE2=110	0.35	70
Sn0.7Cu		2600	22	0.35	-
Sn37Pb		2730	23.5	0.35	-
Sn95Pb		2388	29.1	0.35	-
BT Core	0.8	2451	X=Y=14, Z=58	0.28	-
Heat Sink, Cu	Width=4, thick=0.5	12100	16.3	0.3	-
Glue	0.1	700	CTE1=48 CTE2=99	0.35	75

Experimental procedures

The exterior of FC-BGA packaging sample with 40 x 40 mm dimensions was shown in Fig. 1(a). In this study, the thermal deformation of FC-BGA assemblies with UF-1 and UF-2 were measured and compared by Moiré interferometry. The solder bumps in these samples were SnCu alloy with a pitch of 200 μm. The samples were cut by a low speed diamond saw and polished with 1200 grid abrasive paper to the cross-section of the first bump row as shown in Fig. 1(b). The specimen grating was replicated on the cross-section of each assembly at 85 ℃ as the zero-displacement reference state. The in-plane thermal deformation of assemblies was measured by Moiré interferometry at room temperature (25 ℃). This represented a -60℃ thermal loading applied in the assemblies. The fringe patterns were captured by a CCD camera. A program developed by UT-Austin group was used to analyze the four continuous fringe patterns obtained by high resolution Moiré. [11]

A simulation model using ANSYS^{TM} program was employed to predict the die warpage and stress distribution. The dimensions and material properties of key components in this study are listed in Table 1. Since the packaging samples were symmetric assemblies, a 3D 1/4 model with boundary conditions was created for full assembly as shown in Fig. 2(a), while a 1/2 model was used for the cut assembly shown by Fig. 2(b). To reduce elements and calculation time, only 10 rows of bumps near the sectioned plane were established. The accuracy of 1/2 symmetric model was validated by the experimental data from Moiré. Then, a 1/4 model was used to predict the stress distribution after one TCT cycle from 125 to -55℃.

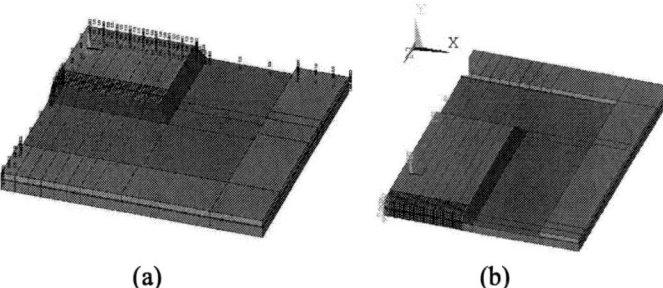

(a)   (b)

Figure 2. 3D FEA model and boundary conditions with a (a) 1/4 and (b) 1/2 symmetry

(a)
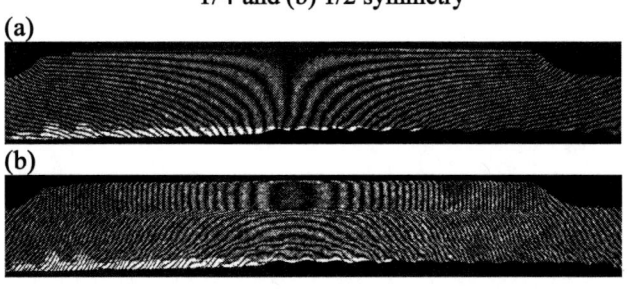

(b)

(c)
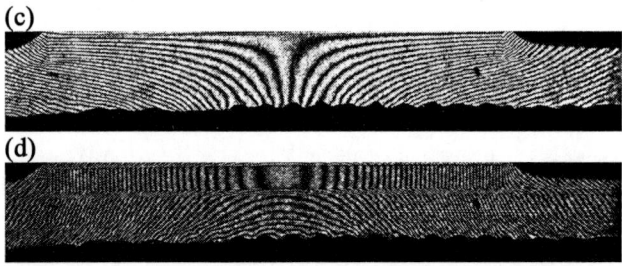

(d)

Figure 3. Regular Moiré patterns of (a) U field for UF-1 (b) V field for UF-1 (c) U field for UF-2 (d) V field for UF-2

Results and Discussion

Figure 3 showed the interference images of the packages with UF-1 and UF-2 by regular Moiré interferometry. Each fringe interval was equal to 0.417 μm. The relative displacement could be readily calculated by counting the number of fringes from the neutral line. The fringe patterns of U field showed larger compressive strains at the bottom of package and almost zero strain at the top edge of silicon chip. The fringe patterns of V field also showed much higher strain gradient at print circuit board than at silicon die. This can be attributed to greater CTE and smaller elastic modulus of substrate than those of silicon die. The patterns showed less fringe orders in U field than V field. It revealed that the package was under bending after cooling from 85 ℃ to 25 ℃. The V field displacement, namely the die warpage was 16.06 and 16.26 for UF-1 and UF-2, respectively. According to Eq. 3, the term $\left[\frac{\partial N_x}{\partial y} + \frac{\partial N_y}{\partial x}\right]$ implied that the shear strain value was proportional to the fringe density. Thus, the shear strain increased gradually from the center to the die edge at which delamination risk was highest.

Table 2 summarized the difference between simulation data and experimental result from Moiré interferometry. Excellent agreement was found between simulation and data

from Moiré. The errors of displacement in the V field for the assemblies with UF-1 and UF-2 were 4.97% and 4.91%, respectively. For U field, the error was less than 3% for the both assemblies. Moreover, the assembly with UF-1 had smaller displacement in the V field than the assembly with UF-2, which could be attributed to the lower elastic modulus of UF-1. The compliant underfill material absorbed the contraction force induced by the CTE mismatch between silicon chip and plastic substrate. Therefore, the displacement results indicated that the underfill material with higher elastic modulus induced larger die warpage.

Table 2 The displacement at die edge as obtained by regular Moiré interferometry and simulation for UF-1 and UF-2, and their comparison

Underfill	Fringe counts		Displacement (μm)		Error rate
			Moiré	Simulation	
UF-1	U field	6	2.50	2.43	2.88%
	V field	38.5	16.06	16.90	4.97%
UF-2	U field	6	2.50	2.43	2.88%
	V field	39	16.26	17.07	4.75%

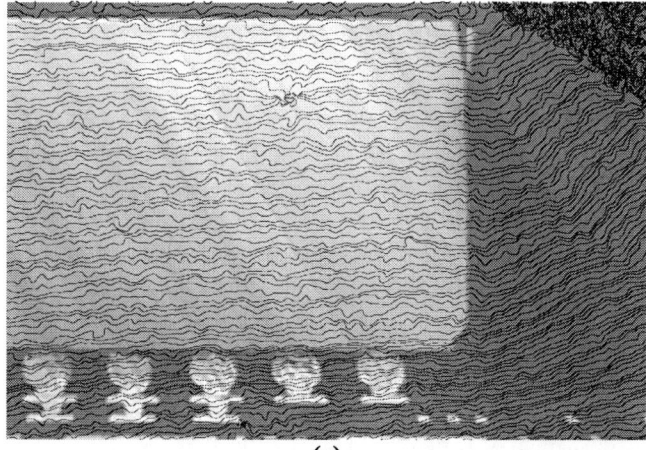

(a)

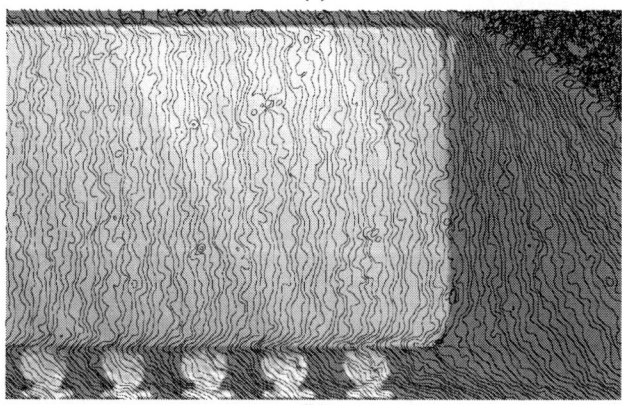

(b)

Figure 4. The contour map of UF-1 assembly (a) U field (b) V field

The resolution of regular Moiré interferometry is not enough to measure the thermal deformation of solder bumps with 110 μm diameter. Therefore, a high resolution Moiré interferometry was employed to analyze the thermo-mechanical behaviors of underfill layer. The contour maps of UF-1 assembly were shown in Fig. 4(a) for U field and Fig. 4(b) for V filed using Moiré analysis software. The contour interval corresponds to 52 nm. The cross-sectional image of the assembly was superimposed onto the phase contour maps with the help of the obvious turns at die/underfill interface of V field contour maps for relating with their relative positions. The undulate contour lines were observed in underfill layer from the U field phase contour map due to different mechanical properties between bumps and underfill material.

Fig. 5 showed the shear strain data along the plane of bump center near die edge for two different underfill materials based on high-resolution Moiré measurement. The shear strain distributions showed that the highest shear strain located near the bottom of die edge. The difference between solder bumps and underfill was readily observed. The shear strain increased in solder bump region and decreased in underfill region. This may be attributed to the compliant material properties of underfill material. Moreover, the outmost solder bump had the highest shear strain compared to other ones. It implied that the outmost solder bump has highest risk of crack or delamination. In addition, the UF-1 has higher average shear strain than UF-2 due to its lower modulus.

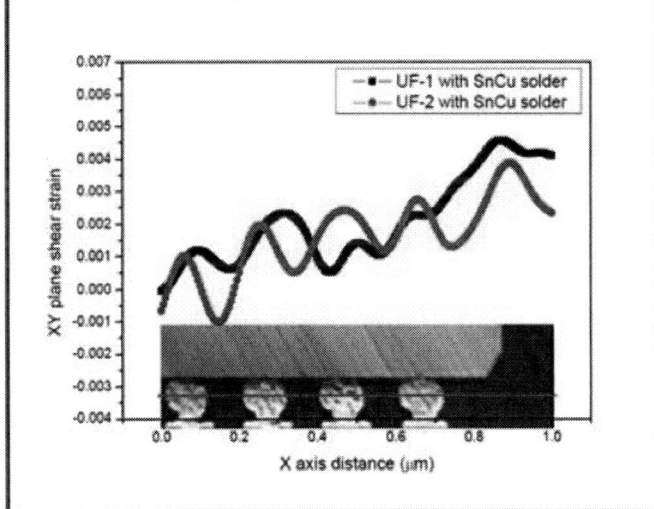

Figure 5. The shear strain near the die edge as measured by high resolution Moiré interferometry

Thermal reliability test

Temperature cycling test (TCT) under 125 to -55 ℃ for 1000 cycles was then carried out for the assemblies with UF-1, UF-2 and SnCu solder. For comparative study, two more underfill materials with higher CTE but lower modulus (UF-3 and UF-4) were introduced along with eutectic SnPb, high-lead (Sn95Pb) and SnCu solders. In total, 6 FC-PGA package assemblies underwent the same TCT. Table 3 summarized the temperature cycling test results and possible low-k layer delaimation for these 6 different packages. TCT results showed that no low-k delamination was found in all 6 samples. Furthermore, underfill materials with higher CTE and lower

modulus such as UF-3 and UF-4 were more vulnerable in conjunction with SnCu solder and high-lead Sn95Pb, respectively. In contrast, UF-3 worked well with SnPb solders, but failed for SnCu with higher reflow temperature.

To further analyze the impact of underfill materials properties on the FC-BGA packages based on SnPb or SnCu solders and its correlation with TCT results, a 3D FEA model previously validated by Moiré measurement was attempted for TCT under 125 to -55 ℃ cycling. However, previous simulation model was for package parts sectioned through the first row of bumps, instead of a complete FC-BGA package. Therefore, a quarter symmetric 3D FEA model illustrated in Fig. 2(a) was created to study the thermal induced stresses for different packaging samples for single cooling cycle from 125 to -55 ℃.

First, failure analysis of sample C after TCT showed bump crack near the bottom as illustrated by SEM in Fig. 6(b). Based on a 3D FEA model of sample C after TCT, the von Mises stress (σ_e) distribution of the outmost solder bump was illustrated in Figure 6(a). The maximum stress located at the right bottom corner of the outmost solder bump, which correlated well with failure analysis shown in Fig. 6(b).

Table 3. TCT1000 results for different packaging samples

Item	underfill	bump alloy	bump failure	low-k delamination
A	UF-1	Sn0.7Cu	Passed	Passed
B	UF-2	Sn0.7Cu	Passed	
C	UF-3	Sn0.7Cu	Failed	
D	UF-3	Sn37Pb	Passed	
E	UF-3	Sn95Pb	Passed	
F	UF-4	Sn95Pb	Failed	

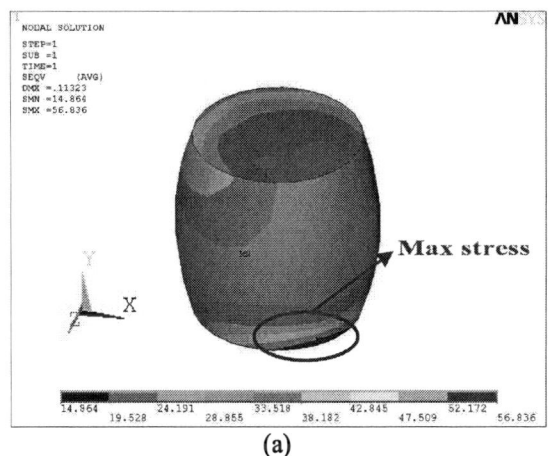

(a)

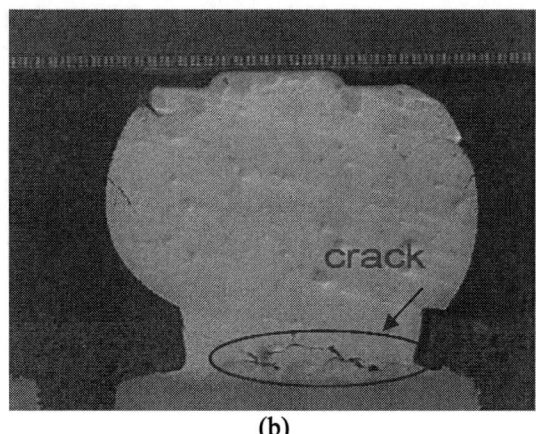

(b)

Figure 6. (a) The max von Mises stress position of sample C from simulation (b) The SEM graph of sample C after TCT 1000 cycles

FEA analysis was then carried out for all 6 different packages. The maximum σ_e of the outmost solder bump for the six packaging samples were shown in Fig. 7. For samples with SnCu solders (A through C), maximum von Mises stress in UF-3 is much higher (56.8 kg/mm^2) than A (UF-1) or B (UF-2) sample. This agreed well with the crack formation found only in UF-3. Referring to materials database listed in Table 1, UF-3 material had the lower Tg, elastic modulus, but higher CTE than UF-1 and UF-2. It can be inferred that UF-3 materials was too compliant to protect the solder bump from cracking.

When UF-3 material was applied in the eutectic packaging sample (D), sample D showed a slightly higher von Mises stress than C, but no bump failure. This implied that weaker solder joints strength existed in the lead-free CuSn solder bumps. Since lead-free solder introduced new fluxes in the process and high reflow temperature, it was suspected that the residues and byproducts hard may degrade the adhesion between die and underfill. [6] Furthermore, Sn0.7Cu alloy may form a brittle Cu-Sn intermetallic compound (IMC) layer at the solder/Cu pads interface. The sustained growth of Cu-Sn layer during reflow or reliability temperature cycling test may decrease the interface strength of solder joints. If there was a crack occurring in this region, it would propagate easily along the IMC layer. [12] Therefore, more attention in the barrier metallization and underfill was required to eliminate the bump cracking issue.

Excellent correlation was also found in sample F (UF-4) based on high-lead Sn95Pb solder with bump cracking. Since high-lead alloy possessed lower mechanical strength than conventional Sn37Pb eutectic alloy, high-lead packaging assembly require more rigid underfill material to provide good protection for solder bumps. The UF-4 material failed to deliver the protection because its Tg and elastic modulus were the lowest among 4 underfill materials used in this study.

Although there was no low-K delamination observed after TCT 1000 cycles, low-K delamination risks in these 6 samples was still assessed by FEA. Since the shear stress (σ_{xy}) and σ_e did not play an important role for the low-K delamination issue, first principal stress, σ_1 was used as the delamination potential index. [13] Figure 8 showed the maximum σ_1 in low-K layer for samples A through F. Clearly,

sample A (UF-1 with SnCu) had the lowest stress than other samples in low-K layer. The σ_1 value appeared to increase with decreasing Tg or increasing CTE among samples A, B and C. The effect of underfill materials' properties such as Tg, CTE and E on the die/package reliability warranted our attention in the following section.

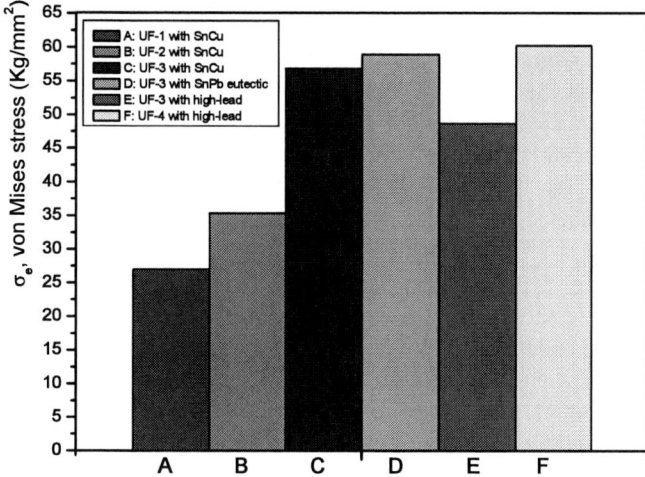

Figure 7. The max von Mises in the outmost bump

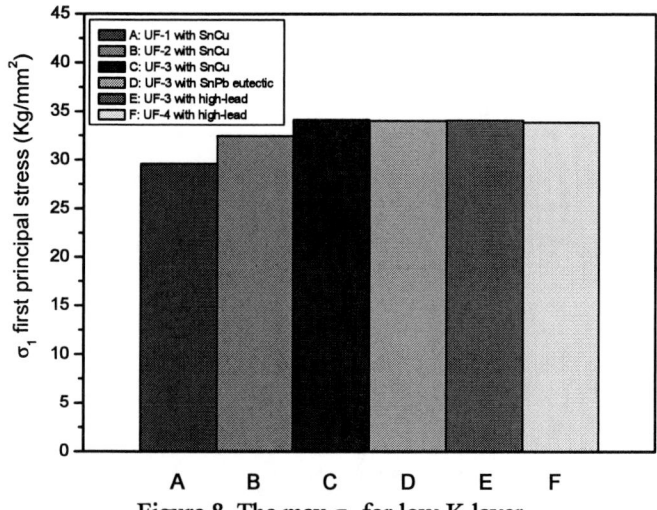

Figure 8. The max σ_1 for low-K layer

The effects of different underfill mechanical properties

In the aforementioned study, we predicted the stresses of bump and low-k layer by FEA. The thermal reliability test results correlated well with simulation results. Also, the lead-free and high lead solder had higher bump crack risk. The question was how to modify the underfill mechanical properties to protect both bump and fragile low-k.

The Tg, CTE, and E were the major properties of underfill material which dictates the mechanical integrity of die/package. Most of the underfill materials were composed by epoxy resin and silica filler. The CTE and modulus in underfill materials were inversely related because a higher percentage (~60-70%) silica was typically added. Recently, new underfill materials, UF-1 and UF-2 with independent E and CTE have been developed, although their Tg's were not the same. We chose sample B as a reference to understand how CTE and E may affect the stress distribution in FC-BGA packages using finite-element analysis. Fig. 9 and Fig. 10 showed the impact of different modulus and CTE of underfill materials on the solder bump (σ_e) and low-k layer stresses (σ_1), respectively. The modulus of underfill material showed opposite trend for the protection of solder bump and low-k layer protection as illustrated in Fig. 9. The underfill material with high modulus reduced bump stress but increased stress for fragile low-k layer. As shown by Fig. 10, the CTE of underfill material should be lower to reduce the thermal stress in both bump and low-K layer. However, the lower CTE limit of underfill material is about 20 ppm/℃ because silica filler content greater than 70% will cause capillary flow issue. [14]

The Tg temperature of underfill material was also a critical parameter for TCT reliability because CTE and modulus changed acutely for temperature above Tg. An optimal underfill material should provide enough protection to solder bumps at higher temperature region and mitigate the stress to fragile low-K layer at lower temperature during thermal cycling test. Based on FEA results, UF-1 with high Tg, moderate E and lower CTE, can simultaneously provide the lowest stress in both bump and low-K layer.

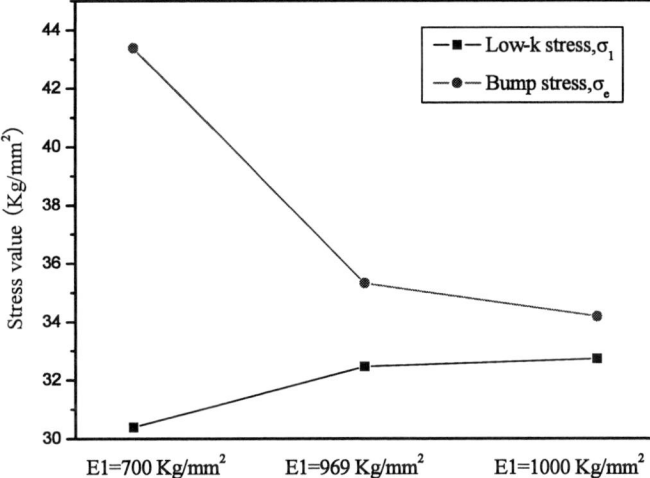

Figure 9. The max stress comparison of different E1 of underfill

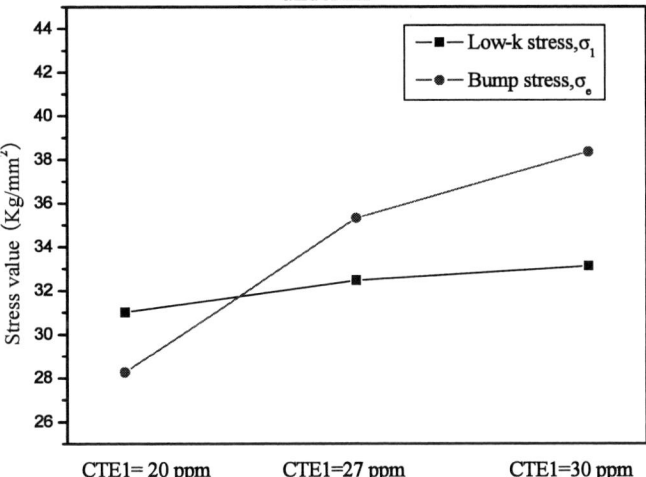

Figure 10. The max stress comparison of different CTE1 of underfill

Conclusions

In this study, the high resolution Moiré interferometry was used to measure the thermal deformation of solder bumps. Based on the Moiré measurement, the underfill material with lower modulus was found to induce larger shear strain in bump/underfill layer. The TCT results illustrated that lead-free and high-lead solder bump have higher crack potential due to the higher reflow temperature and weaker mechanical strength of Sn95Pb alloy, respectively. A simplified 3D FEA model with high accuracy has been established in this study. The displacement difference between FEA prediction and Moiré interferometry measurement were less than 5%. Furthermore, the actual bump crack position after TCT 1000 cycles correlated well with FEA simulation result.

There is a need to redesign underfill material by adjusting its mechanical and thermal properties to protect both solder bumps and fragile low-K layer during thermal fatigue tests. The E and CTE of underfill material should be seen as two independent mechanical properties. The lower CTE implies a lower contraction force during a cooling process. The underfill material with higher E can provide good protection but raise the low-K layer delamination risk. Higher Tg can make underfill material retain its low CTE at elevated temperature. Therefore, the underfill material with high Tg, moderate E and low CTE is recommended for low-K FC-BGA packages.

Acknowledgments

The authors at NCTU appreciate Dr. K. M. Chen of UMC in providing FC-BGA package samples and the financial support by Semiconductor Research Corporation through UMC Customization Program (SRC Contract No: SRC 2005-KJ1303). The support from National Center for High-Performance Computing is also acknowledged.

References

1. Chen, K. M. *et al*, "Effects of Underfill Materials on the reliability of low-k flip-chip packaging," *IEEE Mircoelectronics Reliability*, Vol. 46, (2006), pp. 155-163.
2. Shen, L. *et al*, "Comparison of Mechanical Properties of Porous and Non-porous Low-K Dielectric Films," *Microelectronic Engineering*, Vol. 71, Issue. 3 (2004), pp. 221-228.
3. Chang, K. C. *et al*, "Design Guidance for The Mechanical Reliability of Low-K Flip Chip BGA Package," *International Microelectronics And Packaging Society Conf*, Austin, Texas, June. 2004.
4. Mercado, L. *et al*, "A Simulation Method for Predicting Packaging Mechanical Reliability with Low-k Dielectrics," in *Proc. IEEE 2002 Int. Interconnect Technology Conf*, pp. 119-121.
5. Rasco, M. *et al*, "Packaging Assessment of Porous Ultra Low-k Materials," in *Proc. IEEE 2002 Int. Interconnect Technology Conf.*, pp. 113-115.
6. Lee, K. W. *et al*, "Adhesion Failure at the Chip-Underfill Interface of Electronic Packages," *29th Annual Adhesion Society Meeting*, Jacksonville, FL, Feb. 2006.
7. Cheng, H. C., *et al*, "A Study of Factors Affecting Solder Joint Fatigue Life of Thermally Enhanced Ball Grid Array

Assemblies", *Journal of Chinese Institute of Engineers*, Vol. 24, (2001), pp. 439-451.

8. Pang, J. H. L and D. Y. R. Chong, "Flip Chip on Board Solder Joint Reliability Analysis Using 2-D and 3-D FEA Models," *IEEE Transactions on Advanced Packaging*, Vol. 24, No.4 (2001), pp. 499-506.

9. Suryanarayana, D. *et al*, "Flip-chip Solder Bump Fatigue Enhancement by Polymer Encapsulation," *Proc 40th Electronic Components and Technology Conf*, 1990, pp.338-344.

10. Lee, W. H. *et al*, "Underfill Selection Strategy for Low-K, High Lead/Lead-free Flip Chip Application," *Microsystems, Packaging, Assembly and Circuits Technology*, (2007), pp. 338-341.

11. Wang, G. *et al*, "Thermal Deformation Analysis on Flip-Chip Packages Using High Resolution Moiré Interferometry," in *Proc. IEEE 2002 Int. society conference on thermal phenomena*, pp. 869-875.

12. Tu, P. L. et al, "Effect of Intermetallic Compounds on the Thermal Fatigue of Surface Mount Solder Joints," *IEEE Trans. on Components, Packaging, and Manufacturing Technology - Part B*, Vol. 20, No.1 (1997), pp. 87-93.

13. Wang, T. H. et al, " Underfill Selection for Reducing Cu/low-K Delamination Risk of Flip-chip Assembly," in *Proc. IEEE 2006Electronic Packaging Techology Conf.*, pp. 233-236.

14 Carson, G. and Todd, M., "Toward Determining Optimal Mechanical Properties in Adhesives for Flip Chip and Wirebonded Low-K Die," in *Proc. IEEE 2004 Int. Electronics Manufacturing Technology Symposium*, pp. 10-12.

Enrichment and Removal of Heavy Metals Contained in PCB Boards by Multiwalled Carbon Nanotubes for WEEE Directive

L. Hua[1], H. N. Hou[1]

1. Department of Chemistry and Life Science, Hubei University of Education, China

txuehua@163.com, Tel: 086-027-87943696

Abstract

PCB board is an important part of electronic and electrical equipment. In place of piling e-wastes, much of heavy metals in PCB boards percolate into soil, air, river etc. which is a great threaten to environment. In order to removal the hazardous materials for WEEE directive, in this paper, a study on enrichment of lead, cadmium ions by multiwalled carbon nanotubes (MWCNTs) as a solid-phase extraction adsorbent was employed. ICP-OES was used to determine the adsorbed concentrations. Some valuable guidelines can be drawn from the following discussions. MWCNTs has proven to be a promising materials for the removal of contaminants owing to its amazing effects of enrichment, the objective content was concentrated about 50-100 fold, and limit of detection (LOD) was $0.5 \ \mu g \cdot kg^{-1}$ for Pb, $0.2 \ \mu g \cdot kg^{-1}$ for Cd.

The ion exchange or hydrogen binding mechanism can very well explain the heavy metals such as Pb, Cd adsorption onto CNTs. Sorption can be modeled by Freundlich isotherms from which thermodynamic parameters such as free energy change (ΔG), enthalpy change (ΔH), and entropy change (ΔS) can be calculated. $\Delta G < 0$, $\Delta S > 0$ indicated the process to be feasible and spontaneous nature. $\Delta H > 0$ suggested that the process to be an endothermic nature.

Enrichment can be influenced by factors as contact time, temperature, pH and initial concentration of adsorbate, etc. Sorption increased with increasing contact time, and temperature, initial concentration of adsorbate. For each of analyte, there is a neutral pH beyond which MWCNTs will be either positively or negatively charged. Desorption studies have shown the applicability to regenerate the CNTs used. The process is economically feasible and easy to carry out. All those add more credits to MWCNTs for removing pollutants from e-wastes, which is meaningful for complying with WEEE directive.

Keywords: Enrichment and removal, Heavy metals, PCB boards, Multiwalled carbon nanotubes, WEEE directive

Introduction

The pollution due to the release of heavy metals contained in PCB boards of the old telephones, mobiles, personal computers, other electronic and electrical products has been an increasing worldwide concern for the last few decades, the trend of pollution is even increasing in certain areas of the world, for example, in Guangdong province of China. The effects of high-dose exposure to heavy metals can cause to anemia, abdominal colic, peripheral neuropathy, central neuropathy with toxic encephalopathy, nephropathy, sterility, and other diseases. Therefore, in order to removal hazardous substances and reuse, recycle the useful substances in the e-wastes, the WEEE directive was published in 2003 in official Journal of the European Union, which is meaningful for energy and resource conservation [1]. In today's global economy, this type of regulation affects all of us, measurement of possible contaminants and removal them at low concentration is increasingly important on a worldwide basis. Unlike organic compound, divalent metal cation ions like Pb^{2+}, Cd^{2+}, Hg^{2+} are non-biodegradable and, therefore, must be removed from e-wastes before discharging them into water, air, soil etc. Many methods are used to preconcentrate and removal heavy metals ions, solid phase extraction (SPE) method, Especially the adsorption on oxided or actived carbon nanotubes (CNTs) has attracted a particular attention [2,3]. Carbon nanotubes (CNTs), with nano-sized diameter and tubular microstructure, have been the worldwide hotspot of study since their discovery due to their unique morphologies and various potential applications [4–7]. Because of their relatively large specific surface areas and easily modified surfaces, CNTs have attracted researchers' interests as a new type of sorbent [8].

The aim of the presented work is to establish a speciation procedure based on adsorption of metal ions contained in PCB boards on oxidized multiwalled carbon nanotubes (MWCNTs), the adsorption thermodynamic and kinetic were studied systematically. At this point, MWCNTs have been proposed as a novel solid phase extractor for various inorganic and organic materials at trace levels [9-11]. It is expected to provide a technologic support for electronic and electrical industries in South China when they comply with WEEE directive.

1 Experimental

1.1 Instrumental

Optima 5300™ DV Inductively coupled plasma-optical emission spectrometry (ICP-OES, PerkinElmer Corp., USA) was employed to determine concentrations of heavy metals. The operational condition was adjusted to give the optimum response for sample emission. Anton Paar Multiwave 3000 SOLV digestion system (Anton Paar GmbH, Austria-Europe) was used to prepare samples. An Orion Star 3 pH meter (Thermo Electron Corp.) was employed for monitoring the pH value in the aqueous solutions.

1.2 Materials and reagents

All chemicals used in this work, were of analytical reagent grade and were used without further purification. Distilled water (DI water) was used for all dilutions and washing. All the plastic and glasswares were cleaned by soaking in dilute HNO_3 (1+9) and were rinsed with distilled water prior to use. Multiwalled carbon nanotube with $5 \sim 10nm$ in outer diameter, $10 \sim 50\mu m$ in length was purchased from Carbon Nanometer Harbor Corp., Shenzhen in China. Several grams of MWCNTs were immersed in concentrated HNO_3 for 24h at room temperature to removal the surface remnants, rinsed with distilled water until the neutral pH, filtered through 0.45 μm membrane filter. Then they were oxidized by refluxing the as-prepared CNTs with 10% (v/v) KMnO₄ solution for 12h, the resulting CNTs were washed with distilled water and dried for further use. All initial pH was adjusted by NaOH and HNO₃ (3%) solutions. The element standard solutions used for calibration were produced by diluting a stock solution of 1000 mg L⁻¹ of the given element. The solution was further diluted to the required concentration before used. Stock solutions of diverse elements were prepared from high purity compounds. Adsorption process was done in glass beakers by bathing.

1.3 Adsorption experiments

PCB board samples were digested by microwave system and diluting with DI water, white closed microwave vessels can stand the maximum pressure 1200 psi, maximum temperature 280°C. In adsorption experiments, 400mg of CNTs were introduced into flasks, to which 100mL of metal ion solutions containing varying concentrations at various pH were added. The initial metal concentration was increased from 1~500 μg/L, with pH of all solutions were kept at the appropriate optimal value. The solutions were stirred for 3h at a fixed temperature (25°C), the two phases were separated by filtration through a 0.45 μm membrane filter, the final ion concentration was determined by ICP-OES spectrometry with reference to US EPA Method 6010B. The solid-phase concentrations of metal ions on CNTs were determined after rinsing CNTs with concentrated HNO₃ or HCl. The kinetic sorption data (discussed in part 2.3) suggested that 100mins was enough to achieve the sorption equilibrium. The adsorption capacity q_t (mg/g CNTs) finally, q_e at equilibrium, adsorption rate S(%) are calculated as follows:

$$S(\%) = \frac{C_0 - C_e}{C_e} \times 100 \quad (1)$$

$$q_e = \frac{C_0 - C_e}{M} \times V \quad (2)$$

$$q_t = \frac{C_0 - C_t}{M} \times V \quad (3)$$

Where C_0, C_t, C_e are the initial, final, equilibrium concentrations (μg/L) of metal ions in the aqueous solution, respectively, C_{Ae} is the concentrations (μg/L) of metal ions on solid phase on CNTs, V is the volume of metal ion solution and M is the weight of CNTs.

To study the effects of pH, contact time, experiment temperature on metal ions adsorption, various amounts of oxidized MWCNTs were dispersed into solutions containing different amounts of each of heavy metal ions. The initial pH values were adjusted in a loose range. After the suspensions were shaken for 3h at room temperature, they were filtered. The amounts of metal ions adsorbed on CNTs were calculated as the difference between initial and final concentrations at t, equilibrium minute.

2. Result and discussion

2.1 Effect of pH

The pH value plays an important role in adsorption of particular ions, in order to evaluate the effect of pH on the adsorption on CNTs, prepared in different ways, a series of sample solutions containing single component at concentration of 100 μg/L were adjusted to a pH range of 0.5~14. The result showed in the Fig.1 that the adsorption of heavy metals on the oxizided MWCNTs is strongly dependent on pH values. The interaction between the oxygen functional groups such as hydroxyl (-OH), carboxyl (-COOH), carbonyl (>C=O) treated by KMnO₄ on the surface of CNTs and metal ions was greatly dependent on pH of enrichment process. These oxygen-containing functional groups present abundantly on the external and internal surfaces of oxidized CNTs pores, which can provide numerous chemical sorption sites and thereby increase the sorption capacity of oxidized CNTs [12]. The final pH decreased meant that adsorption is a process of H⁺ release, which is formed by ionization of hydroxyl, carboxyl groups. From Fig.1, as can be seen that for Pb(II) the optimal adsorption was in the pH range of 5~10, the optimal adsorption was in the range of 9~11 for Cd, when pH exceeds 11, the removal drop suddenly. According to Dubinin [13], the water molecules adsorbed to the oxygen groups on carbon surface become secondary adsorption centers, which retain other water molecules by means of hydrogen bonds. As a result, complexes of associated water form. At higher pH values, these oxygen groups can easily be ionized, formation of water cluster hinders it access to the surface of CNTs, which cause to decrease of sorption capacity. While in all pH range, adsorption of Hg is little and fluctuates very little. The removal rate of Pb(II) and Cd(II) remaining almost constant in such a wide pH range suggests that MWCNTs are excellent adsorbents.

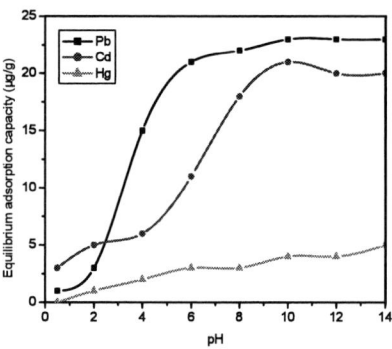

Fig.1 The equilibrium sorption capacity as a function of pH in the solutions

2.2 Sorption equilibrium time

The removal of Pb²⁺, Cd²⁺ from PCB board samples by MWCNTs at various pHs as a function of contact time is shown in Fig.2. As can be seen, the adsorption of Pb²⁺, Cd²⁺ increases quickly with time and then reaches equilibrium soon. The larger adsorption capacity of oxidized MWCNTs for metal ions and shorter equilibrium time are mainly due to the oxygenous functional groups on the surface of oxidized MWCNTs which can react with Pb(II), Cd(II) to form salt or complex deposited on the surface of MWCNTs. For Pb, the contact time to reach equilibrium is equal to 22mins for C_0=200μg/L, equal to 50min for C_0=300μg/L, and 60min for C_0=500μg/L. For Cd, the equilibrium time is longer, the equilibrium time is 35min for C_0= 200μg/L, 60min for C_0=300μg/L, 70min for C_0=500μg/L. The final sorption capacity of Pb(II), Cd(II) reach 12, 14.8 μg/g for C_0=200μg/L, achieve 14.5, 17.9μg/g for C_0=300μg/L and 17.8, 22μg/g for C_0=500μg/L. It is obvious that the sorption capacity of Pb(II) is larger than that of Cd(II). The longer contact time to reach equilibrium for lower initial ion concentration may be explained by the fact that diffusivity mechanisms control the sorption of metal ions onto MWCNTs. Reid et al. indicated that the mass diffusivity decreases with decreasing concentration under very dilute solution and causes the decrease in diffusion flux of adsorbate onto the surface of the adsorbent [14].

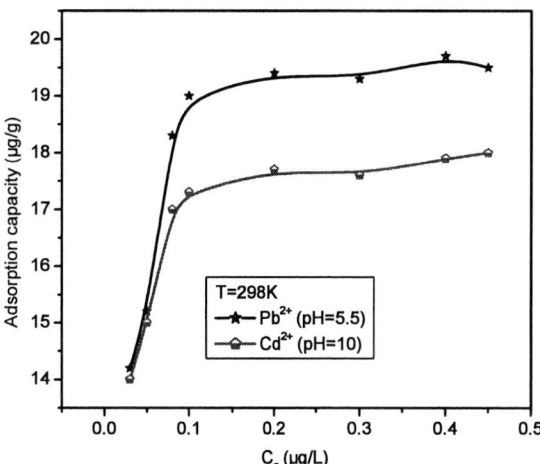

Fig.3 The curve of adsorption capacity as a function of initial concentrations

2.4 Sorption isotherms

Fig.4 is the sorption isotherms, the normal models such as Freundlich and Langmuir were employed for adsorption, however, the experiment data showed the sorption of heavy metal ions on MWCNTs is better approximated by Freundlich model, wherein formula (4) represents the relationship between the amount of heavy metals adsorbed by per unit mass of adsorbent (q_e) and the concentration of metal ions at equilibrium (C_e). The logarithmic equation is as formula (5).

$$q_e = KC_e^{1/n} \qquad (4)$$

$$Lnq_e = Lnk + \frac{1}{n}LnC_{Ae} \qquad (5)$$

Where k and n are Freundlich constants representing the adsorption capacity and intensity of the adsorption, respectively. Values of k and n can be obtained by plotting Lnq_e versus LnC_{Ae}.

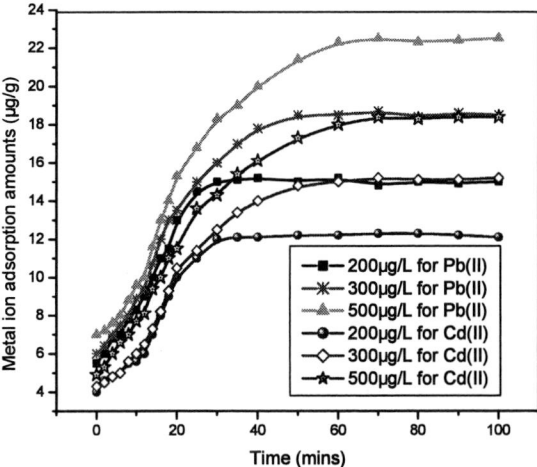

Fig.2 The sorption amounts of Pb(II), Cd(II) as a function of time (pH=5.5 for Pb^{2+}, pH=10 for Cd^{2+})

2.3 Effect of initial concentration

In order to demonstrate the influence of initial concentration of analytes in enrichment efficiency, a series of solutions containing various ions with different concentrations were experimented. The result was shown in Fig.3, from here, the sorption capacity increases with increasing of initial concentrations in the beginning, after concentration is equal to equilibrium concentration, the sorption increases with initial concentrations no longer. It can be explained that the impetus of initial high concentration is larger than that of low concentration, sorption onto or in the inner of CNTs was a diffusivity process. When at equilibrium, the inner and outer pressure around ions is same as each other, so sorption is kept constant.

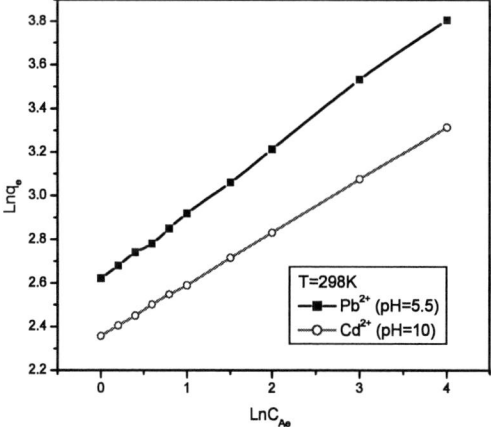

Fig.4 The curve of Lnq_e versus LnC_{Ae}

The linear relative between Lnq_e and LnC_{Ae} showed that Freundlich isotherm model match the experiment data very well, with the correlation coefficient values of 0.993, 0.989, respectively. From here, constants k and n was obtained from the intercept and slope of two sorption isotherm lines, they are listed in Table 1.

1049

Table 1 Parameters in Freundlich sorption isotherms of Pb^{2+}, Cd^{2+} on MWCNTs

Metal ions	k	n	pH	Temperature (K)
Pb^{2+}	13.764	3.3802	5.5	298
Cd^{2+}	10.561	4.1764	10	298

2.5 Effect of temperature and thermodynamic parameters

Fig.5 showed the sorption isotherms of Pb^{2+}, Cd^{2+} on oxidized MWCNTs at 298K, 313K, 323K, respectively. It can be seen that sorption of two ions onto CNTs increases with the increase in temperature. This indicates that adsorption to CNTs is an endothermic reaction, similar to chemical sorption. The thermodynamic parameters were calculated from the variation of the thermodynamic equilibrium constant K_s with the change in temperature. The calculation method was discussed in literature [15], [16]. K_s is defined as follows (6):

$$K_s = \frac{a_s}{a_e} = \frac{v_s}{v_e}\frac{C_{Ae}}{C_e} \qquad (6)$$

where a_s is the activity of adsorbed solute, a_e is the activity of the solute in solution at equilibrium, C_{Ae} is the surface concentration of heavy metal ions on the per gram of CNTs (μg/L), C_e is the concentration of metal ions at equilibrium (μg/L), v_s is the activity coefficient of the adsorbed solute and v_e is the activity coefficient of the solute in solution. As the concentration of the solute in the solution approaches zero[17], the activity coefficient approaches unity, reducing Eq. (6) to the following form (7):

$$K_s = \frac{a_s}{a_e} = \frac{C_{Ae}}{C_e} \qquad (7)$$

Values of K_s are obtained by plotting $Ln(C_{Ae}/C_e)$ versus C_{Ae} and extrapolating C_{Ae} to zero. The straight line obtained is fitted to the points based on a least-squares analysis. Its intercept with the vertical axis gives the values of K_s. Free energy changes (ΔG) for interac tions are calculated from the relationship (8), (9):

$$\Delta G = -RTLnK_s \quad (8) \Rightarrow LnK_s = \frac{-\Delta G}{RT} \qquad (9)$$

Wherein, R is the universal gas constant and T is the temperature in Kelvin. The relationship among Free energy changes (ΔG), enthalpy change (ΔH), and entropy change (ΔS) is shown in Eq. (10). The relationship among K_s, ΔH, and ΔS is as follows (11):

$$\Delta G = \Delta H - T\Delta S \qquad (10)$$

$$LogK_s = -\frac{\Delta H}{2.303R}\left(\frac{1}{T}\right) + \frac{\Delta S}{2.303R} \qquad (11)$$

By plotting $LogK_s$ versus $\frac{1}{T}$, as shown in Fig.6, ΔH is obtained from slope, ΔS is obtained from intercept, the data is shown in Table 2. A positive enthalpy change suggests that

the interaction of Pb^{2+}, Cd^{2+} adsorbed by CNTs is endothermic, which supported by the increasing adsorption of Pb^{2+}, Cd^{2+} with the increase in temperature; a negative free energy change and a positive entropy change indicate that the adsorption reaction is a spontaneous process [18]. The larger absolute value of free energy change shows that adsorption on MWCNTs is quick and also indicates that CNTs is a potential sorbent. The positive entropy change may be due to the release of water molecule produced by ion exchange reaction between the adsorbate and the functional oxygen groups on the surfaces of CNTs [19].

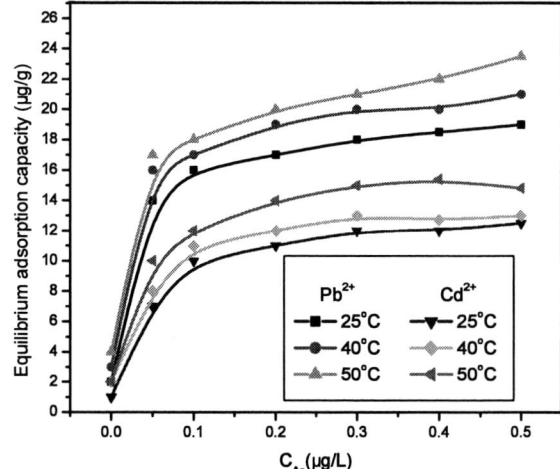

Fig.5 The equilibrium capacity as a function of temperature at various initial concentrations of Pb^{2+}, Cd^{2+} on CNTs

Table 2 The thermodynamic parameters of Pb^{2+}, Cd^{2+} adsorptions on MWCNTs

Ions	Temperature (°C)	Constants $LogK_s$	Free energy changes ΔG^o (kJ·mol⁻¹)	enthalpy change ΔH^o (kJ·mol⁻¹)	entropy change ΔS^o (J·mol⁻¹·K⁻¹)
Pb^{2+}	25	2.793	-15.937	0.138	347.947
	40	3.514	-21.061	0.138	347.893
	60	4.387	-27.973	0.138	347.781
Cd^{2+}	25	2.176	-12.416	0.146	213.963
	40	2.561	-15.353	0.146	213.960
	60	3.074	-19.6	0.146	213.959

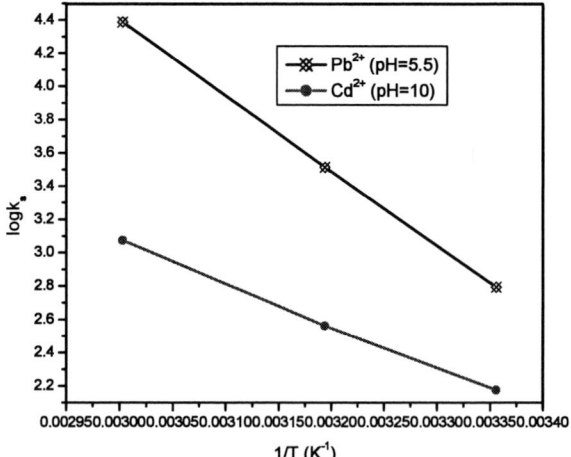

Fig.6 The plot of $LogK_s$ versus $\frac{1}{T}$

2.6 Desorption experiment

The longevity of MWCNTs was checked by subjecting the sorbent to several loading and elution experiments. The capacity of sorbent was found to be practically constant (variation of 1-3%) after repeated use more than five times, thus use of sorbent many times is feasible. MWCNTs is also nontoxic, the disposal CNTs is safe to environment.

2.7 Limit of detection

The detection limit (LOD) of the present work was calculated under optimal experimental conditions after application of the enrichment procedure to blank solutions. The detection limits of Pb and Cd based on three times the standard deviations of the blank (3σ) were 0.5 $\mu g \cdot kg^{-1}$ and 0.2 $\mu g \cdot kg^{-1}$.

3. Conclusions

The enrichment process of two metal ions Pb^{2+}, Cd^{2+} contained in PCBs of electronic and electrical equipment by MWCNTs was discussed in this paper, the result showed that sorption is greatly dependent on pH, temperature of experiment, contact time, initial concentrations of analytes etc, Freundlich model fits well to the sorption of Pb^{2+}, Cd^{2+} onto oxidized MWCNTs. In a board pH range, CNTs express high enrichment efficiency. Their active sites on the surface of oxidized CNTs such as hydroxyl (-OH), carboxyl (-COOH), carbonyl (>C=O) caused by $KMnO_4$ function contribute to the high enrichment capacity. The thermodynamic parameters such as free energy change (ΔG), enthalpy change (ΔH), and entropy change (ΔS) were discussed, a negative free energy change and a positive entropy change indicate that the adsorption reaction is a spontaneous process. The larger absolute value of free energy change shows that adsorption on MWCNTs is quick and also indicates that CNTs is a potential sorbent. Desorption studies have shown the applicability to regenerate the MWCNTs used. The process is economically feasible and easy to carry out. All those add more credits to MWCNTs for removing pollutants from e-wastes, which is meaningful for complying with WEEE directive and protecting environment.

References

1. European Parliament/Council of the European Union, "Directive 2002/96/EC of the European Parliament and of the Council on waste electrical and electronic equipment", Official Journal of the European Union, 1, 27, (2003), pp. 24-37.
2. Benjamin, M. M., Leckie, J. O., J. Colloid Interface Sci. 79, (1981), pp. 209.
3. Chen, J. P., Lin, M., Water Res. 35, (2001), pp. 2385.
4. S. IiJima, Helical microtubules of graphitic carbon, Nature 354, (1991), pp. 56–58.
5. Wagner, H. D., Lourie, O., Feldman, Y., et al, "Stress-induced fragmentation of multiwall carbon nanotubes in a polymer matrix", Appl. Phys. Lett. 72, (1998), pp. 188–190.
6. Wang, Q. H., Setlur, A. A., Lauerhaas, J. M., et al, "A nanotube-based field-emission flat panel display", Appl. Phys. Lett. 72, (1998), pp. 2912–2913.

7. Dillon, A. C., Jones, K. M., Bekkedahl, T. A., et al, "Storage of hydrogen in single-walled carbon nanotubes", Nature 386, (1997), pp. 377–379.
8. Long, R. Q., Yang, R. T., "Carbon nanotubes as superior sorbent for dioxin removal", J. Am. Chem. Soc. 123, (2001), pp. 2058–2059.
9. Zheng, F., Baldwin, D. L., Fifield, L. S., et al, "Single-walled carbon nanotube paper as a sorbent for organic vapor preconcentration", Anal. Chem. 78, (2006), pp. 2442–2446.
10. Zhou, Q. X., Xiao, J. P., Wang, W.D., et al, "Determination of atrazine and simazine in environmental water samples using multiwalled carbon nanotubes as the adsorbents for preconcentration prior to high performance liquid chromatography with diode array detector", Talanta 68, (2006), pp. 1309–1315.
11. Liang, P., Ding, Q., Song, F., "Application of multiwalled carbon nanotubes as solid phase extraction sorbent for preconcentration of trace copper in water samples", J. Sep. Sci. 28, (2005), pp. 2339–2343.
12. Lu, C., Chung, Y. L., Chang, K. F., "Adsorption thermodynamic and kinetic studies of trihalomethanes on multiwalled carbon namotubes". J. Hazard. Mater. 138, (2006), pp. 304-310.
13. Dubinin, M.M., in: P.L. Walker Jr. (Ed.), Chemistry and Physics of Carbon, Marcel Dekker, New York, 1966.
14. Reid, R.C., Prausnitz, J.M., Poling, B.E., 1988. The Properties of Gases & Liquids, fourth ed. McGraw-Hill, New York.
15. Yan, H. L., Zechao, D., et a, "Adsorption thermodynamic, kinetic and desorption studies of Pb^{2+} on carbon nanotubes", Water Research 39, (2005), pp. 605-609.
16. Benguella, B., Benaissa, H., "Cadmium removal from aqueous solutions by chitin: kinetic and equilibrium studies". Water Res. 36, (2002), pp. 2463–2474.
17. Niwas, R., Gupta, U., Khan, A. A., et al, "The adsorption of phosphamidon on the surface of styrene supported zirconium tungstophophate: a thermodynamic study", Colloid. Surf. A: Physicochem. Eng. Aspects 164, (2000), pp. 115–119.
18. Liang, R., Chen, B., "Study of the effects of ionic strength and temperature on the adsorption of anionic dye on activated carbon with flow-injection spectrophotometry", Chem. Bull. 67, (2004), pp. 1–8.
19. Li, Y. H., Wang, S. G., Cao, A. Y., et al, "Adsorption of fluoride from water by amorphous alumina supported on carbon nanotubes", Chem. Phys. Lett. 350, (2001), pp. 412–416.

A Study on Application of N&K Analyzer in OLED Failure Analysis

Qiang Fang, Yafang Peng, HK Yu
Department of Material Science, Fudan university
No. 220 Handan Road, Shanghai, 200443, China
Feyme.fq@gmail.com, +86-21-65642110

Abstract

The thickness of each layer of the OLED (Organic Light Emitting Diode) with the multi-layer structure of ITO/NPB/Alq3/LiF/Al was studied by N&K Analyzer before and after aging experiments. Through comparing the film thickness when kept in ambient atmosphere with different exposure duration and failed samples, main thickness changes occur in the Alq3 and LiF layers while the ITO and NPB layer had no dramatic changes in thickness. The trend is Alq3 layer decrease and LiF layer increase in the process of failure. Before electrical failure, the devices exhibit good uniformity in film thickness; but when electrical failure occurs, this uniformity has obviously been destroyed. So it is proved that N&K Analyzer is an effective new method for OLED failure analysis especially in the condition that it is very difficult to characterize the OLED's multi-ultra thin-layer structure by traditional means.

I. Introduction

A OLED (Organic Light Emitting Diode) is a flat display technology, made by placing a series of organic thin films between two conductors. So it is a kind of device with multi-layer organic thin films whose thickness ranges from tens nanometers to hundreds nanometers. That is so thin that the structure of the OLED is difficult to characterize and to do failure analysis by traditional means. N&K Analyzer is a new method to characterize thin films by optical means which was introduced to do failure analysis on OLED.

Actually the N&K Analyzer cannot measure the thickness of the thin film directly. In order to characterize any thin films with optical means, certain quantities such as reflectance, or transmittance or phase shift are measured. However, these quantities along cannot be directly used to properly characterize materials because their values depend on the values of several primary quantities. These primary quantities which relate to material characteristics are film thickness, the spectra of n and k. (n: refractive Index; k: extinction coefficient) as well as interface roughness. The Forouhi-Bloomer (F-B) dispersion equations for n and k, given below, can be incorporated into the Fresnel cuefficients for reflection and transmission, to obtain equations for reflectance from a multilayer thin film structure. By comparing such equations with actual measurements for reflectance, film thickness, n and k spectra, as well as interface roughness can be determined[1]. So N&K Analyzer just bases on the Forouhi-Bloomer Model. N&K Analyzer receives the reflectance spectra at different wavelength on the material surface and then calculates and fits the data to give the film thickness.

The Forouhi-Bloomer equations[2,3] are as follows:

$$k(E) = \sum_{i=1}^{q} \frac{A_i(E - E_g)^2}{E^2 - B_iE + C_i}$$

$$n(E) = n(\infty) + \sum_{i}^{q} \frac{B_{0i}E + C_{0i}}{E^2 - B_iE + C_i}$$

Bringing the k(E) and n(E) to Fresnel Formula, it will get the film thickness.

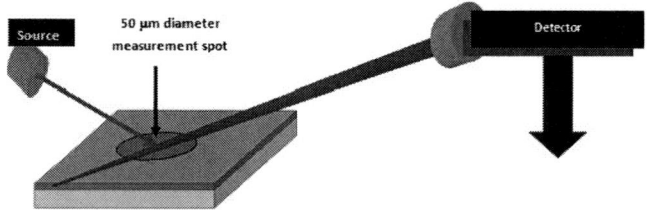

Fig. 1 N&K thin film measurement system

II. Experiments

Tris-(8-hydroxyquinoline) aluminum (Alq3) based devices were prepare for failure analysis. Alq3 was chosen for Electron-Transport material and N,N-bis(naphthalen-1-y)-N,N' -bis(phenyl) benzidine (NPB) as Hole-Transport Material, Al as cathode. There was a LiF thin layer about 50 Å introduced between the Al cathode and Alq3 to enhance the light emitting performance. So our experimental OLED samples were with the structure of indium tin oxide (ITO)/ N,N-bis(naphthalen-1-y)-N,N'-bis(phenyl) benzidine (NPB, 800Å)/tris-(8-hydroxyquinoline) aluminum (Alq3, 700Å)/LiF(50 Å)/Al. Film was grown by vacuum evaporation while the film thickness was controlled by Sigma Instrument SQM-160 Quartz crystal oscillation deposition thickness monitor . For the convenience of measurement, the samples were not packaged.

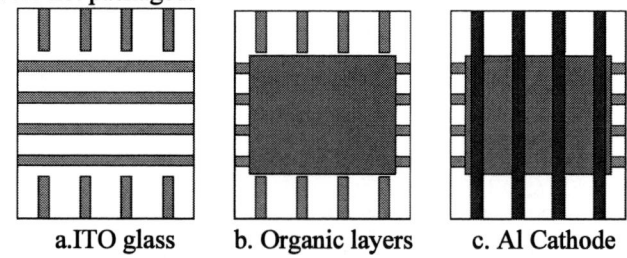

a.ITO glass   b. Organic layers   c. Al Cathode
Fig. 2 Layout of the OLED samples

At the same time when the sample preparation finished, select several dots to measure the thickness of the thin film layers by N&K Littlefoot 8000 Analyzer, that were the thickness of NPB, Alq3 and LiF while the thickness of ITO was fixed. Then expose the devices into the ambient atmosphere lasting for 7 days, 10 day and 40 days, check the thick the thickness of organic layers again to find out whether there were changes in the film thickness after atmospheric

978-1-4244-2739-0/08/$25.00 ©2008 IEEE   1052

corrosion. To accelerate the failure of the aged device, the devices exposed for 40 days were illuminated in atmosphere through to gone out (lasting about 2 hours), Then check the film thickness of light-emitting areas.

Discussion 3 Section Header

3.1 Reflectivity changes with atmospheric exposure time duration

The reflectance spectra of the same dot on one specific sample with the exposure time 7days, 10days, 40days without electrifying and exposure time 40days with electrifying to failure are shown in Figure 3. Red curve presents the spectra with 7days exposure while the green one 10days, ultramarine one 40days and Cambridge blue 40days with electrical failure. Obviously, there were obvious spectrum offset in reflectivity with different exposure time. That means the device structure or the property of the material inside had changed with exposure time. In order to analyze the root cause of this phenomenon, the film thickness of the device exposed by 10days and 40days were checked.

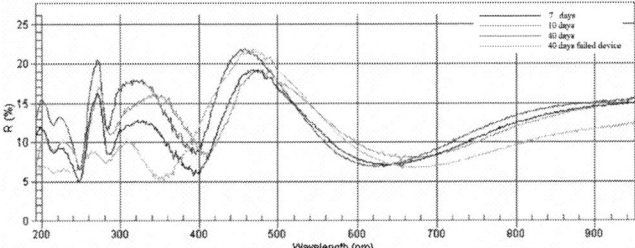

Fig. 3 Reflectance spectra with different exposure durations

3.2 Film thickness after 10days atmospheric exposure

Because the ITO was deposited in advance by electroplating method, the thickness of this layer was fix to 606.8 Å which was the average value of several measurements by N&K Analyzer and close to actual situation. Then in film thickness measurements we ignored the ITO layer and focused on the changes of the organic layers and LiF layer.

Film thickness measured by N&K Analyzer was shown in Table 1. From the data presented we can see high film thickness uniformity especially in the organic layers. There were thickness decrease in NPB and Alq3 layers and increase in LiF layer.

Table 1. Film thickness after 10days atmospheric exposure

	1st layer ITO	2nd layer NPB	3rd layer Alq3	4th layer LiF
1	608	701.4	524.4	69.9
2	608	701.4	523.5	59.8
3	608	705.8	521.7	56.9
4	608	693.7	524	73.6
average	608	703.1	523.7	62.3
original		800	700	50

3.3 Film thickness after 40days atmospheric exposure and electrified

The thickness of Alq3 and LiF layer after 40days atmospheric exposure was shown in table 2. Here we found the average thickness of NPB layer was 709nm as well as the

ITO 599 Å which almost had no difference with the device exposed for 10days. So for convenience, we just check the changes in Alq3 and LiF layer. Data gained also showed good uniformity. In contrast to the device with shorter exposure time, the Alq3 layer thickness decrease urther (473 Å average) and LiF layer increase further(84 Å average).

Table 2. Film thickness after 40days atmospheric exposure

	3rd layer (Alq3)	4th layer (LiF)
5	472.7	86.7
6	471.2	85.5
7	468.5	83.3
8	496.6	84.2
9	472.5	82.1
10	474.3	84.9
11	472.2	81.7
12	473.2	81.6
13	477.1	81.4
14	476.6	83.3
15	477.6	84.1
16	473.0	83.8
average	473.2	83.6

Select one device exposed for 40days to apply 10V voltage and make it light up, after going out to fail, measure the film thickness again. The results was shown in table 3. Then the thickness data showed great fluctuation with some dot ultra thin or ultra thick and the film uniformity had been broken. In general, the thickness of Alq3 layer decrease and LiF layer increase further.

Table 4. Film thickness of failed samples after 40days atmospheric exposure and electrified

	3rd layer (Alq3)	4th layer (LiF)
1	521.6	121.7
2	306.7	153.3
3	453.3	100.7
4	401.2	96.2
average	420.7	118.0

To summarize, in the failure process of OLED with the multi-layer structure ITO/NPB/Alq3/LiF/Al, main thickness changes happen at the Alq3 and LiF layers, in general, Alq3 layer decreases and LiF layer increases while the ITO and NPB layers had no dramatic changes.

Table 5. Film thickness changes with exposure duration

	1st layer (ITO)	2nd layer (NPB)	3rd layer (Alq3)	4th layer (LiF)
10days	608	703.1	523.7	62.3
40days	599.8	709.3	473.2	83.6
40days fail	599.8	709.3	420.7	118

Conclusions

N&K Analyzer is proved to be an effective new method for failure analysis of OLED. For the multi-layer structure OLED, The thickness of each layer and film uniformity can be conveniently obtained using N&K Analyzer. Through comparing the structure changes of different samples with different atmospheric exposure duration, we found out that for devices with multi-layer structure of ITO/NPB/Alq3/LiF/Al, the main causes of failure are rooted in Alq3 and LiF layer with the trend of obvious decrease in Alq3 and increase in LiF.

References

1. G. G. Li, A. R. Forouhi, A. Auberton-Herve, and A. Wittkower. "Optical Charzcterization of Silicon-on-Insulator," *Proceedings 1995 IEEE International SOI Conference*, Oct. 1995

2. Forouhi A. R. and Bloomer I, "Optical Dispersion Relations for Amorphous Semiconductors and Amorphous Dielectrics," Phy. Rev. B, 1986,34:7018

3. Forouhi A. R. and Bloomer I, "Optical Properties of Crystalline Semiconductors and Dielectrics," Phy. Rev. B, 1988,38: 1865

Failure Analysis of the First Wire's Bond

Ren Chunling, Huang Qiang, Ding Rongzheng, Jiang Changshun
China Electronic Technology Group Corporation No.58 Institute
Wuxi China 214035
rcl1981@sohu.com, 0510-85815818

Abstract

Wire bonding was an important process in electronics packaging technology. The bonding quality had a great influence on the performance of integrated circuit. In this paper, the first golden wire's bond failure was analyzed in detail. It was found that the results of the destructive bond pull test were in accordance with the failure criteria specified in the MIL-STD-883G Method 2011.7 after 150 ℃ high temperature stability baking for 24 hours. But the first wire's bond peeled between the bond and the pad interfaces when the circuits were in use. Material and structure of the pad were analyzed and the bonding parameters' window was tested. The results showed that the thickness of the pad was three times as thick as conventional one, so the power used conventionally became the power limit for the circuit. This could not even be detected in the on-line bond pull testing. Finally, the defect was eliminated by adding golden ball peeling test and improving the bonding parameters. The bonding reliability was improved remarkably.

1 Introduction

As an important packaging process, thermosic bonding was broadly used in the I/O port connection of microelectronic packaging. Wire bonding failure was a potential danger in the IC. One failure bonding will directly result in IC failure. So how to ensure the bonding quality is considerably important.

In this paper, the reason was discussed that a group of IC wasn't abnormal in the packaging and after the packaging, but some were failed in use. The result showed that the first wire was peeled off from the bonding site. The wires' tensile strength was tested and it was found that the first wire's tensile strength value was dissatisfied with the method 2011.7 of MIL-883G [1]. The first wire's tensile strength was lower than the criterion, and the Au ball was peeled off from the pad. The proportion was about 40%. The bonding power used was lowered about 5% because of Al layer's drape under the microscope's observed. But destroy pull test satisfied with the process's demands after wire bonding. The failure was tested and analyzed in this paper.

2 Experiments

The technology for the IC was Au wire ball bonding. Wire tensile strength was tested, then the first wire's order was changed and the wire's tensile strength was tested. Wire tensile strength tests' results under different powers were compared. The structure and material of the IC pad were analyzed; cross section was analyzed using SEM. High temperature storage test was in the condition of 300 ℃ and 1 hour.

3 Results and Discussions

3.1 Bond pull destructive test

Failure circuit was tested by bond wire pull destructive test. The results were that the first wire's bond accorded with the criteria GJB548B-2005 and the failure rate reached 40%. Failure samples were observed under microscopes and it was found that failure mode was that the first order bond was peeled off from the Al bond areas. As shown in Fig.1, bonding areas had the ring gold remnants leave, Bonding zones around the Al layer were clearly extruded from the bond district and the middle aluminum district had marked indentation but no gold residue, while other bonds were normal.

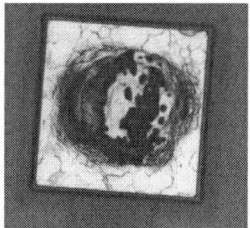

Fig. 1 Failure surface between the first wire's bond and the Al pad

Bond power reduced to 95% of the conventional power and other parameters unchanged. Bond parameters were also within the process window. Bond ball size, bond uniformity and bond ball height were in technology demands.

3.2 Material structural

Failure appeared on the first wire's bond, there may be the existence of structural differences. The first order of the chip bonding district's aluminum material was compared with other bond areas' material by spectrum analysis, the results showed that material composition of all the pads were same, no silicon were found, which was different with Al-Si-Cu.

The chip's first order bonding area's cross section were analyzed, as showed in the Fig.3. The Al-layer bonding area's structure was three layers and the layers were three times thicker than the conventional chip's aluminum layers. Bond areas had no grid hole to strengthen the structure.

The chips' bond pads had no test probe points. From the Fig.3 and Fig.4, the aluminum layer of the first order's pad was the same with others.

978-1-4244-2739-0/08/$25.00 ©2008 IEEE

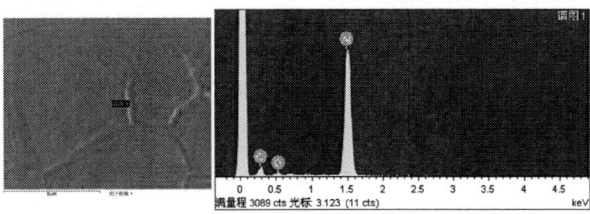

Elements	Weight percent(%)	Atom percent(%)
C	9.32	18.63
O	1.11	1.66
Al	89.57	79.71
total	100.00	100.00

The first order bond district (naked chip)

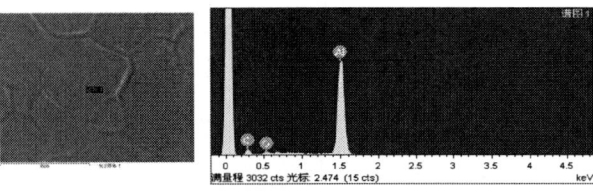

Elements	Weight percent(%)	Atom percent(%)
C	7.85	15.93
O	1.23	1.88
Al	90.92	82.19
total	100.00	100.00

The second order bond district (nakedness chip)

Fig.2 spectrum of the first order bond area's aluminum layer and second order bond area's aluminum layer

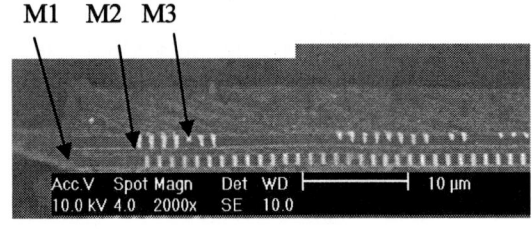

M3	3μm around
M2	700nm around
M1	700nm around

Fig.3 Cross section of bonding pad

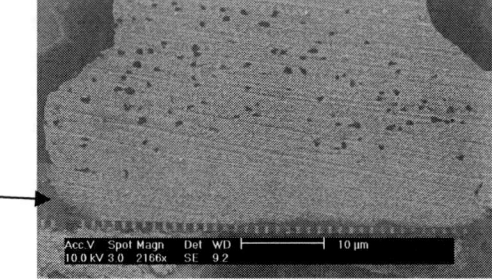

Extruded aluminium

a

Extruded aluminium

Gold

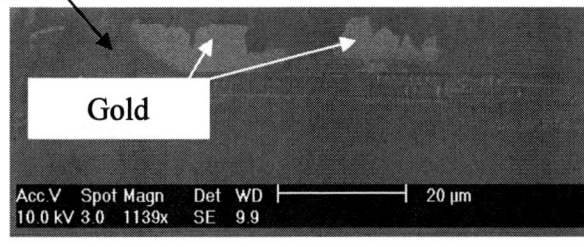

b

Fig.4 Cross section of bonding pad: a. normal bond, b. pulled bond

The chip had the same structure, so the chip structure was not the reason for the failure.

3.3 Different first wire's bond test

In order to analyze whether the failure resulted from the first order wire's bond. Selecting the chips' different pads as the first wire's bond bonded wire with the same parameters. It was found that other pads as the first order bond also resulted in the same failure. And the failure didn't appear on the other order bond site.

In order to analyze and confirm the first order bond's failure, every five minutes, a wire was bonded in the same chip on a different pad. After bonding, the simple's tensile strength was tested, and it was found that the failure was random, as showed in the Fig.5, and the bond failure rate arrived probably to 15 percent, but the failure position was uncertain. This showed that the main factors of the failure were capillary and the first gold ball with low-temperature.

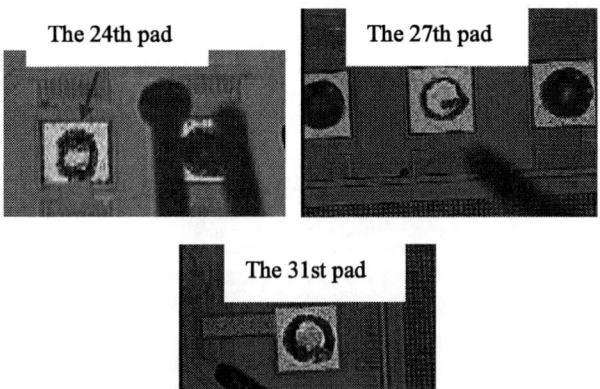

Fig.5 Pulled bond of different pads as the first wire's bond

3.4 Tests under different parameters and high temperature stored

Bond power was one major factor to cause failure. Therefore the tests were done under different bond parameters. Other conditions were the same as the failure. The gold ball's diameter, ball's shape and thickness were all satisfied with the criterion. The bond parameters were shown as in the table 1.

Table 1 bonding test under different bond parameters

Serial	Test items	Test results
1	Bonding process parameters of tolerance in the lower bound of 5%	Seven samples with the failure mode, three bond site was the first order bond.
2	repeat No. 1 test, storage: 300 ℃ 1 h	Two bond failure in six sample, which was the first wire's bond.
3	Bonding process parameters on the side of the middle 5%	Storage, no storage of six samples had no peeled failure model.
4	Between No. 1 and No.3 technical parameters	Storage and no storage of six samples had no peeled failure mode, after peeling off the golden ball, no sufficient bond was observed.
5	Bonding process parameters as the serial NO.1 and bond one wire every five minute	Bond site 32, and failure mode was five pads; the failure rate was 15.6%.

Test 1 and test 2 had no peeled failure mode. When storing, the bond was not sufficient, as shown in Fig.6. As the first order of bonding, capillary and the golden ball in the cold condition, bond parameters in the lower bound of the windows, the rate of peeled failure mode would be possible.

No sufficient adhesion phenomenon under golden ball

Fig.6 Peeled off the gold ball from the pad by manual

It was different directly to compare bond wire's tensile strength with storage of 300 ℃ after one hour; the failure mode after storage was decreased and concentrated in the first wire's bond. This was mainly due to the defects that would eliminate little in high-temperature, but this flaw was generally not completely eliminated, the defects always existed. When the temperature and time changed, the defects would gradually appear.

4 Failure mechanism analyses

Conventional chip's bonding pads' material with Al-1.0% Si-0.5% Cu had high hardness, and its Au ball's bond strength was stronger than pure Al pads'. As shown in Fig.8, if the bond areas' hardness was different, the gold wire bond's strength was also different under the same bonding parameters. The greater hardness, the more ability to cut off and remove the oxide film, so the effective area of the bond was increased and the joint was enhanced.

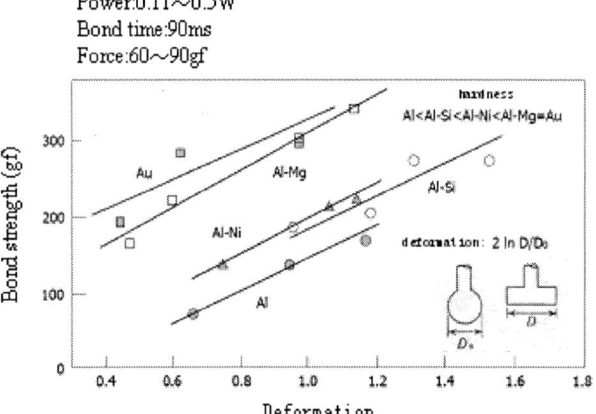

Fig.7 Alloy elements influence on the ultrasonic pressing properties

Failure sample's bonding area material was pure aluminum; its hardness was less than Al-Si alloy's. The pure Al layer was three microns thickness and three times thicker than the conventional pad. Because the thick layer was pure aluminum, the bond parameters were also conventional parameters in the processing, so the soft aluminum layer was extruded out easily. As shown in the Fig.8. The main reason was that the conventional parameters were the lower bound for the pure Al layer.

In the bonding process, there was an interval between the circuit's wire bonding. So the capillary and gold ball of the first order's bonding would be cold. When the gold ball was in the cold state, the valid power was lost on the gold ball's extrusion and the aluminum extrusive deformation. The valid power was not adequate for the first order bond, as shown in Fig.8. When the ultrasonic bonding power, time, force were in the parameters' lower bound of the window, it would easily lead to bonding defects. And the subsequent bond due to capillary and the gold ball on the accumulation of heat and just fired in 0.05 s with bond completed, the bond-effective energy to be assured that it will not lead to major bond defects, the defects rate was very low. Items 5 of table 1 test results showed that when the capillary and wire was cooled, the failure rate of any pad as the first order's bonding point was the same. Test 1 and test 2 showed that the lower bonding power increased the risk of failure; Test 3 and test 4 showed that the increasing power reduced the failure.

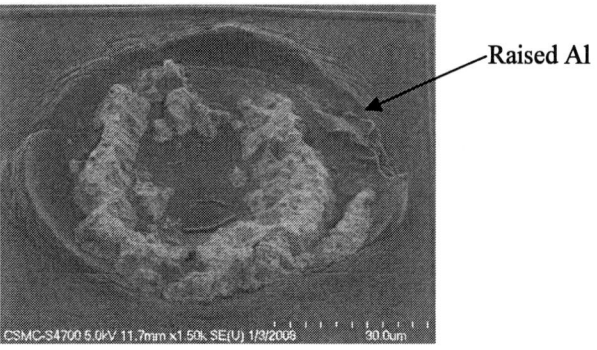

Raised Al

Fig.8 the photo of the sample's failure bond

In the lower bound of the process windows, the bond's defects were not found by using a normal bond tensile strength test, even after 150 ℃ high-temperature storage for 24 hours. If only the temperature changed in a long time, its bonding defects would be gradually expand and wire bonding strength would be weakened, even the wire's bond peeled off from the interface.

5 Conclusions

As the chip manufacturing process improving, in order to reach different needs, materials and structure of the chips' pad were a larger difference, so the bond parameters must be different. In order to improve the reliability of bonding, the bond parameters should be optimized. At the same time, improving the quality of bonding was effective by different means. If the chip designer considered what kinds of bonding process and the reliability of the circuit design and manufacture, the reliability of the circuit would be assured, which can greatly reduce the unusual situation.

References

1. MIL-STD-883G 28 FEBRUARY 2006.
2. Industrial production of electronic technical manuals, semiconductors and integrated circuits Volume 7, 1991, Defense Industry Press.

Investigation of Electromigration in Copper Interconnection of ULSI

Dechun Lǔ, Shengxiang Bao, Lili Ma, Zhibo Du
State Key Laboratory of Electronic Thin Films and Integrated Devices
University of Electronic Science and Technology of China, Chengdu 610054, China
E-mail: lvdch2008@yahoo.cn

Abstract

With the development of higher integration and the improvement of integrated density of devices，copper interconnect technology become the current important connection technology. Its excellent mechanical and electrical characteristics attract the high-speed, power management devices and fine pitch applications. Copper interconnection has gained considerable attention because of its economic advantage, strong resistance to sweeping and superior electrical performance.

The design and application of novel test interconnection to study electromigration(EM). The results show that the size, shape and microstructure of interconnection metallic line how to play an important role in the process of EM. Also, the temperature, current density and alloy elements have strongly effects on Mean Time of Failure (MTF) of EM. Through the EM experiment，the EM resistance of copper interconnection with different width was compared; The failure mechanism was explored. The failure distribution is concentrated at or above line-lengths longer than 150um with a very distinct change. The evidences show that long length line of EM damage not only exists, but also present is damage that occurs in shorter-length interconnects.

1 Introduction

EM damage is one of the major causes for the failure of interconnection. As it is known, EM is the electric current that induces transportation of atoms and can cause failure in micro interconnect lines of integrated circuits through the formation of voids and hillocks. Now with the development of small scale packaging technique, the package's reliability due to EM in interconnections under high direct current density becomes a serious concern, especially for the high density microelectronic packaging and power electronic packaging.

In recent years, the line width of thin-film interconnects has been shrunk into the submicron regime. It gives rise to serious concerns about EM-induced failures. EM-induced void can grow and lead to resistance increase or even catastrophic open of interconnect. EM-induced hillocks can cause both intra-level and inter-level metal shorting. The time and the location of void-open or extrusion-short are statistical nature, which depending on the spatial distribution of current density and temperature. In high-speed circuit a multi-layer layout of interconnects is used to decrease the chip area, shorten the transmission distance of signals and reduce the RC delay. The characteristics of via are high aspect ratio (the height divided by the diameter), small sectional area, high current density, obvious thermal effect and concentrate stress. The high aspect ratio results in bad covering of deposited film on the step and thinner metal film deposited on the bottom and the side wall of the via, which both become the feeble tache of ULSI reliability and the sensitive position of failure. And they hold the most part of EM failures in copper interconnects therefore become focus of the study on ULSI reliability.

The paper is organized as follows: first, we introduce EM-induced failure model in section 3. Section 4 shows some factors influence EM performance and the temperature, current density and alloy elements have effects on MTF of EM; we discuss the effects of via contact test structure on EM life time. The conclusions are drawn in section 5.

2 Experimental Details

EM testing was performed using via contact test structures with various lengths and widths between vias under different test conditions. A schematic diagram of a cross section of via contact test structure is shown in Fig.1. Two-level metal structures fabricated using via first dual damascene scheme to be used for the present investigation. A fluorinated silicon oxide layer was used to isolate the interconnect structures. A dielectric layer of Si_3N_4 was used to stop dielectric diffusion barrier. First metal (M1) and second metal (M2) consist of a Ti/TiN/Cu//TiN/Ti stack. The structure had line widths in the range of 1~10um and line lengths between vias vary from 10 to 400um and the Cu thickness is 0.55um. The test structures were patterned using conventional photoresist and reactive ionic etch processing. The stress condition was a current range from 6 to 10mA at an oven temperature of 200°C. The current was flown upstream alone the M1-M2-M1 direction as seen from the schematic. A constant current was applied through the structure at a fixed temperature. Activation energy (Ea) was done using temperature and current densities in the range of 200-350^{0}C and 1E5- 3E6A/cm^2, respectively.

Resulting structures were filled with electrochemically plated copper. In some cases the structures were filled with a dilute alloy of Cu. Annealing at temperatures in the range of 200-350°C for periods of 30sec-30min, stabilized the microstructure of the electroplated copper prior to planar using chemical mechanical polishing. In all other cases, a Si_3N_4 layer passivated the polished surface.

The leads between the samples and measurement system were covered with aluminum foil to avoid external electromagnetic interference. The voltage was acquired once per minute or per 10 minutes automatically. The resistance was obtained by dividing the voltage by the current.

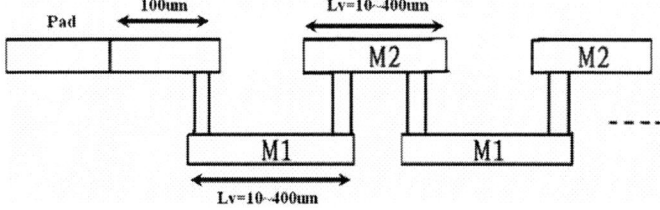

Fig 1 Schematic cross section of structures used for EM

characterization M1 and M2 line length between vias varied from 10 to 400um.

3 EM-induced failure models

EM-induced MTF under current stress has been well established by Black's equation:

$$MTF = A \square J^{-n} \square \exp(Ea/kT)$$

(1)

Ea, and k are activation energy and Boltzmann's constant, respectively. J is the current density. n is the current density exponent. A is a proportionality constant dependent on the physical dimension of interconnect. In the digital circuit environment, the interconnects will experience unidirectional pulsed or bidirectional current stress. Many models have been proposed to estimate EM-induced failure lifetime under the bidirectional current stress. There we use the average current recovery model:

$$MTF_{ac} = A \square J_{eff}^{-n} \square \exp(Ea/kT)$$

(2)

The effective current density J_{eff} is defined as

$$J_{eff} = \left| \overline{J^+} \right| - \gamma \left| \overline{J^-} \right|$$

(3)

Where, $\overline{J^+}$ and $\overline{J^-}$ are the time-averaged current density only including positive current or negative current, respectively, and $\left| \overline{J^+} \right| \geq \gamma \left| \overline{J^-} \right|$ is assumed. γ, which is in the range [0, 1], represents the degree of damage recovery due to opposite polarity current. The experimental results show that the value of γ is around 0.9 in most of interconnect materials. γ, A, E_a and n in Eq. (2) can be obtained from measurement. The unknown parameters for the interconnects in a digital circuit are effective current density J_{eff} and the temperature T.

According to the stress conditions, we can divide all interconnects into two categories. The first one is the signal lines between gates, which is under symmetrical AC current stress (i.e., $\left| \overline{J^+} \right| \approx \gamma \left| \overline{J^-} \right|$). Due to healing effect, they have very large MTFs. The others are power and ground bases. Basically they experience unsymmetrical AC or unidirectional current stress, which may lead to serious EM-reliability problem.

4 Results and Discussion

From the Eq.(1), we can draw a conclusion that MTF has the relations with current density(J) and temperature(T). n is not a constant but affected by current density(J). The reason is that with the increase of current density, the digital circuit creates more Joule heating, and then the temperature gradient increases. It's proved in the experiment that when $J < 2 \times 10^5 A/cm^2$, the circuit has a relatively higher thermal losing rate, so the temperature of the circuit increases too slowly to affect parameter n; when $2 \times 10^5 A/cm^2 < J < 10^6 A/cm^2$, the peak value of the increased temperature of Cu is proportional to J^2;

when $J > 10^6 A/cm^2$, values of n greater than 2 can probably be attributed to Joule heating effects which result in a temperature gradient induced flux divergence. The thermal conductivity of Ti (22W/m-K) is much smaller than that of Cu (395W/m-K). Besides, the resistivity of Ti film is much larger than that of Cu. One would expect that the introduction of the Ti barrier layer degrades the power dissipation ability of Cu films and results in a larger n. However, there are many factors, such as: electric field, temperature gradient, residual stress, etc., which would influence the migration of Cu.

Fig.2 shows length-dependent failure analysis on samples tested at T=225^{0}C and j=1$\times$10^6 A/cm^2. The evidence of EM-related damage, such as final failure site and extrusion damage was tabulated at each line length. A total of test-structures with interconnects for a given length were examined. Within the sampling of 188 interconnects, 24 damage sites were readily identified. The results clearly show that the failure distribution is concentrated at or above line-lengths longer than 150um with a very distinct change. Thus, evidence of length-dependent damage formation exists, but also present is damage that occurs in shorter-length interconnects. These short-length failures appear to be derived from early failures that occur prior to onset of steady state back stress or large amounts of material leakage from the interconnect below the via. Hence, interconnect immortality below the threshold product might not be absolute.

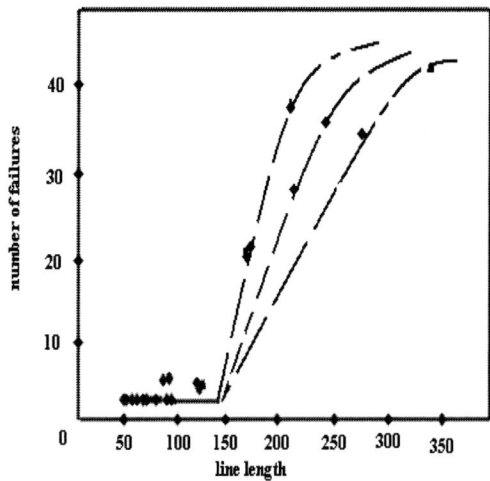

Fig.2 EM damage distribution of individual interconnections as a function of line-length

Fig.3 illustrate the dependence of the lognomal failure distribution using the via contact test structure at different stress current. Since the current density exponent n is about 2, different EM lifetimes were obtained from the time to failure data at each stress current condition, particularly at low failure rate. As the temperature increases, the number of failure samples goes up.

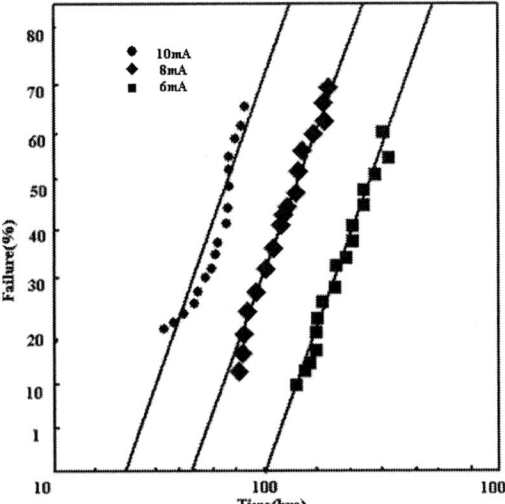

Fig.3 Failure distribution as a function of the stress current

Ongoing tests show that at the beginning the life time of EM decreases and then the MTF will increase when it reaches a minimum value with the increase of the length of line. But when the length reaches a certain range, the decrease of the life time becomes inconspicuous, and finally reaches a constant. The phenomenon can be explained as fellows. The EM invalidation of the line always causes by a certain fatal defect. With the increase of the line length, the number of the defects increases, and the EM failure ratio increases.

If the current direction is along the width of interconnection, as the line-width turns wider or bigger, the life time of EM becomes shorter. From the Fig.5 we can come to get a conclusion that when the range of line-width vary increasingly, the life time was shorten and it reached saturation when reached certain value. EM failure will not only appear the narrow line region and the wide line region, but also will appear the variation regions. So the life time is shorter than the average line width of interconnection.

In order to investigate the impact of fill metallurgy on the EM performance, a dilute alloy of CuTi (0.1-0.2Wt.%Ti) was co-deposited using electrochemical deposition. Alloying resulted in about 1-3% increased resistivity compared to pure copper. EM data show that the MTF for the alloy fill samples were higher than that of samples with pure copper-fill. The activation energy for the alloy-fill specimens was determined to be 0.81eV. The CuTi films exhibited smaller grain sizes with slower self- annealing. Compared to pure copper films the grain size of CuTi film was smaller, even after annealing at higher temperatures. These observations indicate that the addition of Ti retards the transport kinetics along lattice and grain boundaries. Despite this, the activation energy of CuTi samples was similar to copper samples. This indicates that the copper/barrier (dielectric& metal) interfaces play more dominant roles in the failure of copper interconnects rather than grain boundary diffusion.

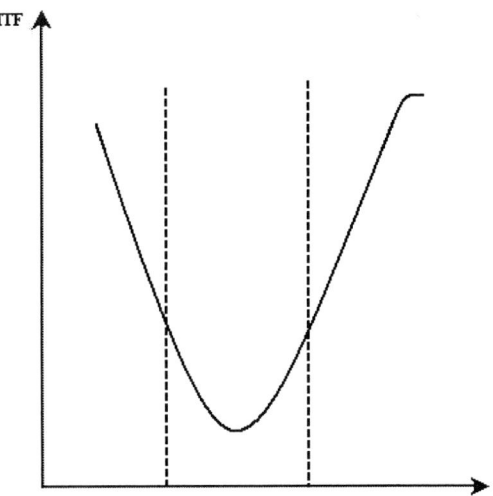

Fig. 4 The relationship of MTF and line-width of Cu interconnection

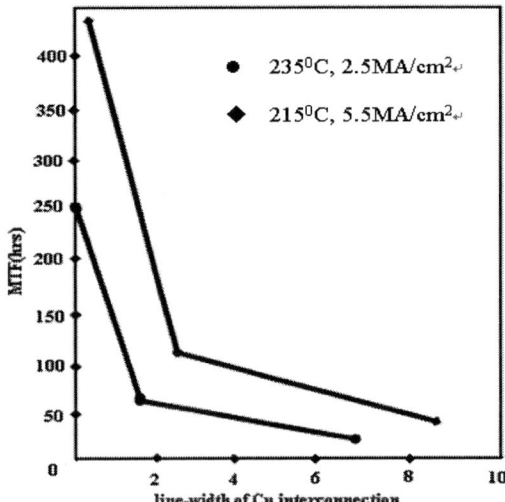

Fig. 5 the relation between MTF and line-length of Cu interconnection

The geometry of the metallization also affects the EM. Cu metallization with zigzag patterns were fabricated and the EM tested. The activation energy of a zigzag Cu film (0.5eV) is smaller than that of a straight one (0.77eV). EM occurs at high currents, a bend in metallization results in current crowding and, consequently, reduces the activation energy of the zigzag film. The current exponent for the zigzag film is 3.35, which is comparable to that of straight Cu film. From this standpoint, it is necessary to reduce the thickness of barrier layer along the sidewalls.

5 Conclusions

1) There are many factors, such as: the size, shape and microstructure of interconnection metallic etc., which would influence the EM of Cu.

2) MTF has the relations with current density (J), temperature (T) and alloy elements. As the line-width turns wider or bigger, the life time of EM turns shorter. EM failure will not only appear the narrow line region and the wide line region, but will appear the variation regions of length and width.

3) The EM damage is not only to exist long length line, but also occurs in shorter-length interconnects.

References

1. Oliver Aubel, "Highly Accelerated Electromigration Lifetime Test (HALT) of Copper," IEEE Transactions on Device and Materials Reliability, Vol. 3, No. 4 (2003), pp.213-217.

2. Pawan Kapur, James P. McVittie, "Technology and Reliability Constrained Future Copper Interconnects," IEEE Transactions on Electron Devices, Vol. 49, No. 4(2002), pp. 590-597.

3. Shingo Kaimori, Tsuyoshi Nonaka, Akira Mizoguchi, "the Development of Cu Bonding Wire with Oxidation-Resistant Metal Coating," IEEE Transaction on Advanced Packing, Vol. 29, No. 2(2006), pp.227-232.

4. Surasit Chungpaiboonpatana, Frank G. Shi, "Packaging of Copper/Low-k IC Devices: A Novel Direct Fine Pitch Gold Wire-bond Ball Interconnects onto Copper/Low-k Terminal Pads," IEEE Transaction on Advanced Packing, Vol. 27, No. 3(2004), pp, 476-489.

5. Pawan Kapur, Gaurav Chandra, James P. McVittie, Krishna C. Saraswat, "Technology and Reliability Constrained Future Copper Interconnects," IEEE Transactions on Electron Devices, Vol. 49, No. 4 (2002), pp. 598-604.

6. Baozhen Li, Timothy D., Sullivan,Tom C., Lee,Dinesh Badami, "Reliability Challenges for Copper Interconnect," Microelectronics Reliability, 44(2004), pp. 365–380.

7. M. Hauschildt, M. Gall, S. Thrasher, P. Justison, L. Michaelson,R. Hernandez, H. Kawasaki. P. S. Ho., "Analysis of Electromigration Statistics for Cu Interconnects," Applied Physics Letters 88, 211907 (2006).

8. Song Dengyuan, Zong Xiaoping, Sun Rongxia, "Copper Connections for IC and Studies on Related Problems," Semiconductor Technology, 2001, 26(2): 292-232.

9. Guotao Wang, Caroline Merrill, Jie-Hua Zhao, Steven K. Groothuis , "Packaging Effects on Reliability of Cu/Low-k Interconnects," IEEE Transactions on Device and Materials Reliability, Vol. 3, No.4, 12 (2003), pp.119-128.

10. Huang Yong, "En Yunfei, Reliability Analysis of Cu Interconnect," Electronics Quality, Vol. 9 (2006), pp.29-34.

11. Dion M.J., "Electromigration lifetime enhancement of lines with multiple branches," In: Proceedings of the 38th International Reliability Physics Symposium Proceedings, San Jose, CA, 2000, pp. 324~332.

12. L.M. Ting and C.D. Graas, "Impact of test structure design on electromigration lifetime measurements," Proc 33th Annual International Reliability Physics Symposium. IEEE, 1995, pp326-332.

13. A. S. Oates, "Electromigration failure distribution of contacts and via contacts as a function of stress conditions in submicron IC metallizations," Proc 29th Annual International Reliability Physics Symposium. IEEE, 1996, pp164-171.

Study on MCM Interconnect Test Generation using Ant Algorithm and Particle Swarm Optimization Algorithm

Chen Lei
Guilin University of Electronic Technology
Guilin, Guangxi, P.R. China 541004
chenlei@guet.edu.cn

Abstract

A new approach based on ant algorithm (AA) and particle swarm optimization (PSO) algorithm is proposed for Multi-chip Module (MCM) interconnect test generation in this paper. Using the pheromone-updating rule and state transition rule, AA generates the initial candidate test vectors. PSO is employed to evolve the candidates generated by AA. The optimized search is guided by the swarm intelligent generated from cooperation and competition among particles of swarm, in order to get the best test vector with the high fault coverage. The international standard MCM benchmark circuit provided by the MCNC group was used to verify the approach. Comparing with the evolutionary algorithms and the deterministic algorithms, experimental results demonstrate that the approach can achieve high fault coverage and short execution time.

Keywords: MCM (Multi-chip Module), interconnect test, Ant algorithm, Particle swarm optimization, test generation.

1. Introduction

The development of digital integrated circuit has put forward urgent demands for test technology. With rapid development of very large scale integration (VLSI), Multi-chip Module (MCM) and Multi-layer Printed Circuit Boards (MPCB), interconnect test technology has become a bottleneck in the application of these circuits. The high reliability of MCM is due to that bare integrated circuit chips are welded and interconnected under high density and small dimension condition [1]. But it is also hard to resolve the problem of MCM interconnect test. Interconnect test technology has become a bottleneck in the application of MCM. Test generation is one key technologies of MCM interconnect test, so study on novel method of test generation to acquire better test set is significant.

Various deterministic interconnecting algorithms have been studied during recent years. We describe the performance of some representative algorithms as follows. For Counting Sequence Algorithm (CSA) [2] , log2N vectors are optimal for detecting all shorts in a circuit of N nets, while Modified Counting Sequence Algorithm (MCSA) needs $\lceil \log_2 (N + 2) \rceil$ [3] vectors for testing all faults. In order to make the fault coverage rate equal to 100%, True/Compliment Algorithm (T/CA) generate 2log2(N+2) [4] test vectors; Walking One's Algorithm (WOA) is a very common test approach for interconnect testing, whose test set length is N[4] .

We consider the following classes of faults in MCM interconnect test [4]:

- Two-Net OR-type Short. If the drivers are such that a '1' dominates, then the resultant logic value is an OR of the logic values on the individual nets.
- Two-Net AND-type Short. If the drivers are such that a '0' dominates, then the resultant logic value is an AND of the logic values on the individual nets.
- Single-Net Faults. These are stuck-at-one, stuck-at-zero, and open faults on single nets.

The fault model allows for single or multiple Occurrences of either two-net faults and for single-net faults with deterministic behavior. The logic value on the net can also be non-deterministic or undefined. This behavior is not included in this fault model and is not considered in the remainder of this paper.

Particle swarm optimization is an evolutionary computation technique developed by Dr. Eberhart and Dr. Kennedy in 1995 [5, 6], inspired by social behavior of bird flocking or fish schooling. Particle swarm optimization is a population-based, self-adaptive search optimization technique. It is attached importance because it has general convergence similar to Genetic method, faster convergence velocity and small computational cost. As a kind of intelligent algorithm, it can be used to solve various optimization problems and shows great potential in practice. Now, it has been widely applied in many other areas, such as artificial neural network and fuzzy system control.

In this paper, a hybrid optimization scheme of PSO and AA [7, 8, 9] is presented for the MCM interconnect test generation problem. Employing the pheromone-updating rule and state transition rule, AA generates the initial candidate test vectors. Then PSO evolves these initial candidates. By combing the characteristics of interconnect test, this paper made the velocity-position model of discrete particles swarm optimization [10] for automatic test generation. The optimized search is guided by the swarm intelligent generated from cooperation and competition among particles of swarm. A fault simulator is employed to compute the fitness of each candidate vector. The results of experiments on the international standard MCM circuit prove that the scheme is able to achieve very good performances, comparing with other algorithms.

The article is organized as follows. Section 2 is dedicated to the study of a hybrid optimization scheme for MCM test generation, which is based on the AA and PSO. The velocity updating equation and position updating equation of PSO are then described in this section, where an AA framework is also given in details. Section 3 provides an overview of results on a set of standard test problems and comparisons of those using well-known interconnect test generation

978-1-4244-2739-0/08/$25.00 ©2008 IEEE

algorithms. Section 4 briefly summarizes the main results and indicates directions for further research.

2. AA-PSO-based Hybrid Optimization Approach

Informally, the new optimization scheme for the MCM interconnects test generation works as follows: AA is utilized to generate an initial population of individual test vectors for PSO, by employing the pheromone-updating rule and state

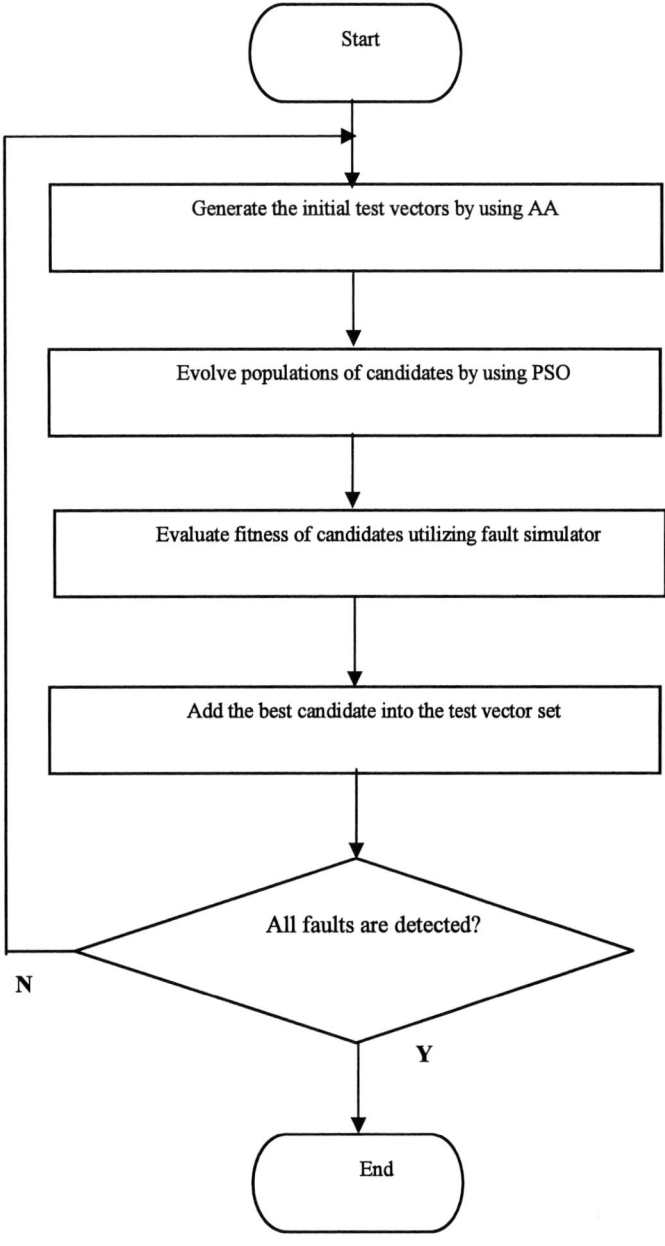

Fig.1 MCM test generation process based on AA-PSO

transition rule. After the AA-based optimization is completed, the best ants are selected and evolved by PSO. Each particle is evolved by employing the velocity updating equation and position updating equation specially presented for MCM interconnect test. The swarm searches for optima solution by repeating the above operation during a sufficient number of iterations. After the evolutionary operation is

completed, the best individual is selected and added to the test set. Then an interconnect fault simulator is used to update the fault list of the circuit. The process is iterated until all faults are detected. The whole process of the generation scheme is illustrated in Fig.1.

Ant Algorithm

Informally, AA used in MCM interconnecting test generation works as follows: Each ant generates a complete test vector by choosing the input value of each net, i.e. 0 or 1, according to a probabilistic state transition rule. Then an ant represents a test vector. Once all ants have completed their tours, the pheromone-updating rule is applied.

In the following we discuss the state transition rule and pheromone-updating rule of AA.

The state transition rule for interconnecting test works as follows: Once all ants have built their tours, the value of the net is updated on all edges according to Eq. 1, i.e. the value of test vector will be changed form 0 to 1 or form 1 to 0:

$$P_k(i) = \begin{cases} 0 & \tau_k(i) > threshold \\ 1 & otherwise \end{cases} \qquad (1)$$

Where $threshold = \dfrac{vector\_faul(k)}{all\_faul}$, vector_faul(k) is the number of faults from net k ($k \in [1, 2 \cdots n]$), n is the total number of ants; all_faul means the total number of faults in the tested circuit. 0 means that if $\tau_k(i)$ is bigger than threshold, the value of No. i net will be same; 1 means that if $\tau_k(i)$ is less than threshold, the value of No. i net will be reversed, i.e. the input value of the net will be changed form 0 to 1 or form 1 to 0.

The pheromone-updating rule is intended to allocate a greater amount of pheromone to candidates that can detect more faults. After building a test vector of the test generation set, pheromone is updated on all nets by applying the pheromone-updating rule of Eq.2.

$$\tau_k(i) = (1-\rho)\cdot\tau_k(i-1) + \Delta\tau_k \qquad (2)$$

Where $\Delta\tau_k(i) = \dfrac{tested\_fault(k)}{all\_fault(k)}$, ρ is the pheromone decay parameter, $i \in [1, 2 \cdots n]$) is the number of nets, tested_fault(k) is the number of faults which can be detected by ant k ($k \in [1, 2 \cdots m]$), all_fault(k) means the total number of faults in the tested circuit when ant k is generating a test vector.

Particle Swarm Optimization algorithm

In the following we discuss the velocity updating equation and position updating equation of PSO.

The velocity updating equation of PSO for interconnecting test works as follows: Once all particles have moved to their new positions, the velocities of the particles are updated according to Eq. 3.

$$v_{k+1} = w \cdot v_k + c_1 \cdot r_1 \cdot pbest_k + c_2 \cdot r_2 \cdot gbest \quad (3)$$

where v_{k+1} is the velocity of current particle, w is inertia weight which balances the global exploitation and local exploration abilities of the particles, c_1 and c_2 are acceleration constants, r_1 and r_2 are random values between 0 and 1, pbest is the best position found by the current particle itself, gbest represents the best position found so far by the whole swarm.

In order to avoid premature stagnation, the velocities of the particles are limited in [V_{min}, V_{max}]. If v is smaller than V_{min}, an element of the velocity is set equal to V_{min}; if v is greater than V_{max}, and then set equal to V_{max}.

In PSO, positions of the particles are candidate solutions to the problem, and the moves of the particles are regarded as the search process of better solutions. The position updating equation of PSO is given in Eq.4.

$$x_{k+1}(i) = \begin{cases} x_k(i) & if \quad v_{k+1} > \beta \\ \overline{x_k(i)} & if \quad v_{k+1} \leq \beta \end{cases} \quad (4)$$

Where $\beta = \dfrac{vector\_faul(k)}{all\_faul}$ is the threshold which determines whether or nor to change the positions of the particles, vector_faul(k) is the number of faults from net i ($i \in [1, 2 \cdots n]$), n is the total number of nets for the tested circuit; all_faul means the total number of faults in the tested circuit. If $v_{k+1}(i)$ is bigger than β, the value of No. i net will be same; if is less than threshold β, the value of No. i net will be reversed, i.e. the input value of the net will be changed form 0 to 1 or form 1 to 0.

An accurate fitness function is essential to achieve a high quality test set. The fitness of a candidate particle is a measure of the number of faults tested, which can be calculated by applying the rule given by Eq. 5.

$$fitness = \dfrac{tested\_fault(k)}{all\_fault(k)} \quad (5)$$

Where tested_fault(k) is the number of faults which can be detected by candidate particle k, all_fault(k) means the total number of faults in the tested circuit when particle k is generating a test vector. If the fitness value is better than the best fitness value pbest or gbest, the current value will be set as the new pbest or gbest.

3. Simulation Results

The AA-PSO-based test generation scheme was implemented by using the interconnecting circuit fault simulator, which was written in C++ language. Using the proposed generator on a PIV1.6 computer with 128 MB memory, test vectors are generated for the mcc1-75 MCM interconnecting circuit provided by the MCNC group, which contains 799 nets and 320399 faults.

Given that the fault coverage rate of all algorithms is equal to 100%, results in the following tables are averaged over ten runs.

In all experiments of the following sections, the parameters of PSO are set to the following values: w=0.5, r_1 = r_2 =0.2, c_1 =0.3, c_2 =0.7, the iteration number 10, the number of particles 8. The parameters of AA are set to the following values: ρ =0.8, m=8, $\tau_0 = (n)^{-1}$, where n is the number of interconnecting nets, m is the number of ants. These values were obtained by a preliminary optimization phase, where the experimental optimal values of the parameters were largely independent of the problem.

Test results compared with other algorithms are shown in Table 1. In the Table 1, the parameters of PSO are set the same as the above. The parameters of GA are set as follows: the population size equal to 30, the number of generation 30. We use a crossover rate of 0.6, a mutation of 0.01.

Table 1. MCM interconnect test results comparison

Algorithms	WOA	T/CA	MCSA	GA	PSO	AA-PSO
CPU time[s]	8	5	9	24	11	14
Test set length	799	20	10	39	14	11

Results in Table 1 demonstrate that test set length of AA-PSO is only 1.4% that of WOA, 28% that of GA, 55% that of T/CA, 78% that of PSO. Test set length is shorter than other algorithms except for MCSA. And the execution time of AA-PSO and MCSA is 14.0s and 9.0s respectively. The results indicate that the performance of the scheme in execution time, test set length and fault coverage is comparable to other interconnect generation algorithms.

Furthermore, tests are generated for the circuits with different net number by using different algorithms. Results in Table 2, where the fault coverage rate of all algorithms is equal to 100%, and results of all algorithms are averaged over ten runs, show that the optimized scheme can also achieve good performances.

Table 2. Test results of different net number

NET	100	200	400	500	600	799	1000	2000
PSE	8	10	10	11	13	14	15	16
AAPSO	7	9	10	10	11	11	12	13
WOA	100	200	400	500	600	799	1000	2000
T/CA	14	16	18	18	20	20	20	22
MCSA	7	8	9	9	10	10	10	11

Therefore, comparing with other algorithms, the optimized approach based on PSO and AA can achieve very good performances in execution time, fault coverage rate and test set length.

4. Conclusions

In this paper, a novel optimization approach based on PSO and AA is developed for the MCM interconnect test generation applications. Employing the pheromone-updating rule and state transition rule, AA generates the initial candidate test vectors. Employing the velocity updating equation specially presented for MCM interconnect test, PSO evolves the candidates generated by AA. A fault simulator is employed to compute the fitness of each candidate vector. After the AA-PSO-based optimization is complete, the best individual is selected and added to the test set. Then the simulator is used to update the fault list of the tested circuit. The process is iterated until all faults are detected.

The international standard MCM benchmark circuit was used to verify the approach. Comparing with the other algorithms, experimental results demonstrate that the approach is able to achieve very good performances in execution time and fault coverage.

References

1. Doane D A, Franzon P D., Multichip Module Technologies and Alternatives:the Basics, Brooks Publishing Company (New York, 1995), pp. 1-85.
2. W.K.Kautz, "Testing of faults in wiring interconnects," IEEE Transaction on Computer, Vol C-23, No.4, (1974), pp.358-363.
3. P. Goel and M. T. McMahon, "Electronic chip-in-place test," Proceeding of International Test Conference, (1982), pp.83-90.
4. Najmi Jarwalw and Chi W. Yau, "A New framework for analyzing test generation and diagnosis algorithms for wiring interconnects," Proceedings of International Test Conference, (1989), pp.63-70.
5. J.Kennedy, R.C.Eberhart., "Particle swarm optimization," Proceedings of IEEE International Conference on Neural Networks, Piscataway, NJ (1995), pp. 1942-1948.
6. Eberhart, R.C., Kennedy , "A New Optimizer Using Particles Swarm Theory," Proceedings of the Sixth Intemational Symposium on Micro Machine and Human Science, (Nagoya, Japan), IEEE Service Center, Piscataway, NJ (1995) , pp.39-43.
7. Schoonderwoerd R., Holland O., Bruten J., "Ant-based Load Balancing in Telecommunications Networks," Adaptive Behavior, vol. 5, issue 2,(1997), pp.169~207.
8. Dorigo M., Gambarddella L. M., "Ant Colony System: A Cooperative Learning Approach to the Traveling Salesman Problem," IEEE Trans. Evol. Comp, vol. 1,(1997), pp.53~56.
9. Stutzle T, Hoos HH: MAX-MIN ant system and local search for the traveling salesman problem. IEEE Int'l Conf. on Evolutionary Computation. Indianapolis, (1997), p. 309~314..
10. Y. Shi, R.C. Eberhart., "A modified particle swarm optimizer," Proceedings of the IEEE Congress on Evolutionary Computation, Piscataway, NJ(1998), pp. 69-73.

MCM Interconnect Test Scheme based on Adaptive Genetic Algorithm

Chen Lei
Guilin University of Electronic Technology
Guilin, Guangxi, P.R. China 541004
chenlei@guet.edu.cn

Abstract

Interconnect test technology has become a bottleneck in the application of Multi-chip Module (MCM), so study on new methods of test generation to acquire better test set is significant. This paper presents a novel optimization approach of adaptive genetic algorithm (AGA) for the MCM interconnect test generation problem. By combing the characteristics of MCM interconnect test, an accurate fitness function is designed to compute the fitness of each candidate vector. AGA is composed of populations of chromosomes and three evolutionary operators: selection, crossover and mutation.

The international standard MCM benchmark circuit was used to verify the approach. Comparing with not only the evolutionary algorithms, but also the deterministic algorithms, experimental results demonstrate that the hybrid approach can achieve high fault coverage, short CPU time and compact test set, which shows that it is a novel optimized method deserving research.

Keywords: MCM (Multi-chip Module), interconnect test, adaptive genetic algorithm, test generation.

1. Introduction

The development of digital integrated circuit has put forward urgent demands for test technology. The high reliability of MCM is due to that bare integrated circuit chips are welded and interconnected under high density and small dimension condition [1]. But it is also hard to resolve the problem of MCM interconnect test. Interconnect test technology has become a bottleneck in the application of MCM. Test generation is one key technologies of MCM interconnect test, so study on novel method of test generation to acquire better test set is significant.

Various deterministic interconnecting algorithms have been studied during recent years. We describe the performance of some representative algorithms as follows. For Counting Sequence Algorithm (CSA) [2] , log2N vectors are optimal for detecting all shorts in a circuit of N nets, while Modified Counting Sequence Algorithm (MCSA) needs $\lceil \log_2(N+2) \rceil$ [3] vectors for testing all faults. In order to make the fault coverage rate equal to 100%,

True/Compliment Algorithm (T/CA) generate 2log2(N+2) [4] test vectors; Walking One's Algorithm (WOA) is a very common test approach for interconnect testing, whose test set length is N[4] .

We consider the following classes of faults in MCM interconnect test [4]:

- Two-Net OR-type Short. If the drivers are such that a '1' dominates, then the resultant logic value is an OR of the logic values on the individual nets.
- Two-Net AND-type Short. If the drivers are such that a '0' dominates, then the resultant logic value is an AND of the logic values on the individual nets.
- Single-Net Faults. These are stuck-at-one, stuck-at-zero, and open faults on single nets.

The fault model allows for single or multiple Occurrences of either two-net faults and for single-net faults with deterministic behavior. The logic value on the net can also be non-deterministic or undefined. This behavior is not included in this fault model and is not considered in the remainder of this paper.

In this paper, an optimization scheme of AGA [5, 6] is presented for the MCM interconnect test generation problem. The AGA evolves these initial candidates by utilizing genetic operator. The implementation methods of the objective function, selection operator, crossover operator, and mutation operator of AGA are discussed in details. In order to apply AGA into the interconnect test, by combing the characteristics of MCM interconnect test generation, an accurate fitness function is designed to compute the fitness of each candidate vector. The fault simulator is used to update the fault list of the tested circuit. The process is iterated until all faults are detected. The results of experiments on the international standard MCM circuit prove that the scheme is able to achieve very good performances, comparing with other algorithms.

The article is organized as follows. Section 2 is dedicated to the study of the optimization scheme for MCM test generation, which is based on the AGA. An AGA framework is given in details. Section 3 provides an overview of results on a set of standard test problems and comparisons of those using well-known deterministic algorithms. Section 4 briefly summarizes the main results and indicates directions for further research.

978-1-4244-2739-0/08/$25.00 ©2008 IEEE

2. AGA-based Optimization Approach

Informally, the new optimization scheme for the MCM interconnects test generation works as follows: During generation of individuals, each character of a chromosome in the population is mapped to an input of a net of a circuit. So a binary code is utilized and the chromosome represents a test vector. The initial population is generated randomly in this paper. The selection scheme in the paper is binary tournament selection without replacement, where two individuals are selected by the roulette wheel approach, and the better individual is selected from the two. After two chromosomes are selected, the crossover operator is employed to generate two offspring. We use the uniform crossover scheme, where each chromosome position is crossed with an adaptive probability. In the binary code, mutation is done by flipping a bit. As the new individuals are generated, each character is mutated with an adaptive rate. After the genetic operation is completed, the best individual is selected and added to the test set. Then an interconnect fault simulator is used to update the fault list of the circuit. The process is iterated until all faults are detected. The whole process of the generation scheme is illustrated in Fig.1.

An accurate fitness function is essential to achieve a high quality test set. The fitness of a candidate vector is a measure of the number of faults tested, which can be calculated by applying the rule given by Eq. 1.

$$fitness = \frac{tested\_fault(k)}{all\_fault} \quad (1)$$

Where tested_fault(k) is the number of faults, which can be detected by candidate k; all_fault means the total number of faults in the tested circuit during the AGA generation process.

3. Simulation Results

The AGA-based test generation scheme was implemented by using the interconnecting circuit fault simulator, which was written in C++ language. Using the MMASAGA-based generator on a PIV1.6 computer with 128 MB memory, test vectors are generated for the mcc1-75 MCM interconnecting circuit provided by the MCNC group, which contains 799 nets and 320399 faults.

Given that the fault coverage rate of all algorithms is equal to 100%, results in the following tables are averaged over ten runs.

In all experiments of the following sections, here we set the parameters of AGA: the population size equal to 10, the number of generation 8.

Test results compared with other deterministic algorithms are shown in Table1.The parameters of GA are set as follows. Here we set the population size equal to 30, the number of generation 30. The other GA parameters of interest are the crossover and mutation rates. We use a crossover rate of 0.6, a mutation of 0.01.

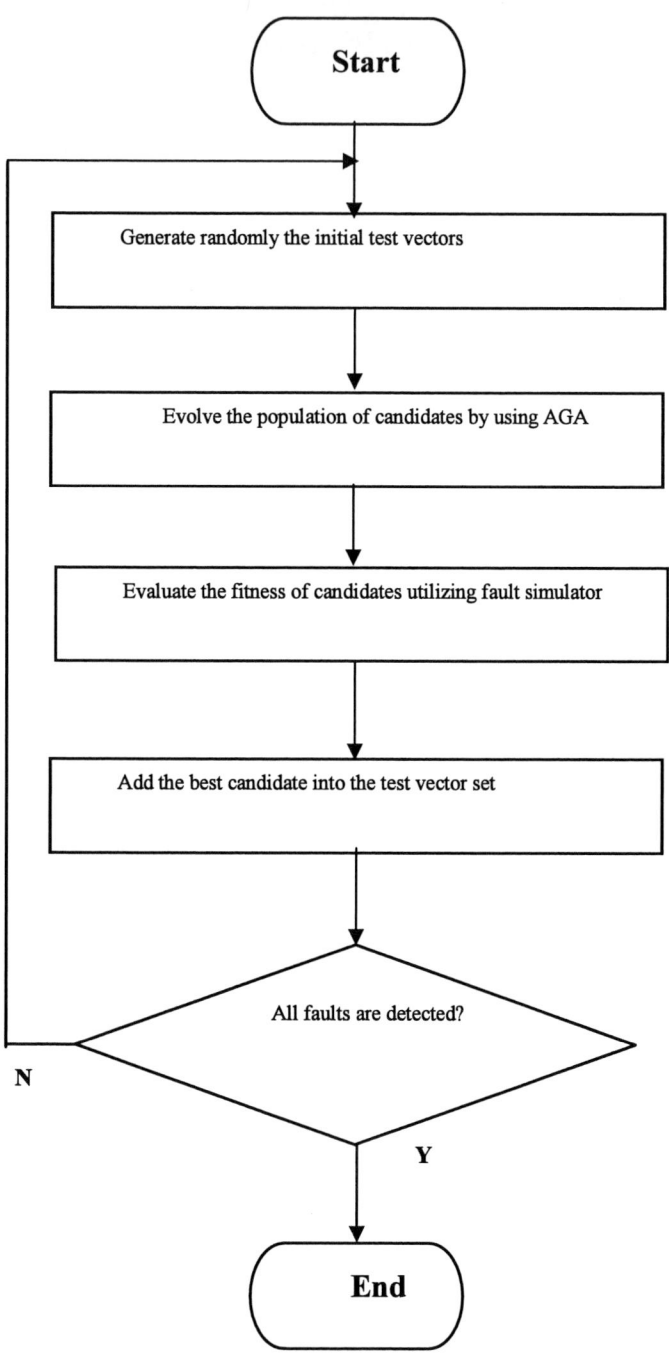

Fig. 1　MCM test generation process based on AGA

Table 1 MCM interconnect test results comparison

Algorithms	WOA	T/CA	MCSA	AA	GA	AGA
execution time[s]	8	5	9	8	24	11
Test set length	799	20	10	20	40	11

For AGA, Test set length is shorter than other algorithms except for MCSA. Given that the test set length is equal to 10, fault coverage rate of AGA and MCSA is 99.96% and 100% respectively. And execution time of AGA and MCSA is 13.2s and 9.0s respectively. The results indicate that the performance of the scheme in execution time and fault coverage is comparable to other deterministic algorithms.

Furthermore, tests are generated for the circuits with different net number by using different algorithms. Results in Table 2 show that the proposed scheme can also achieve good performances.

Table 2 Test results of different net number

NET	100	200	500	799	1000	2000
AGA	7	8	10	11	11	13
WOA	100	200	500	799	1000	2000
T/CA	14	16	18	20	20	22
MCSA	7	8	9	10	10	11
AA	9	10	16	20	23	27
GA	12	16	34	40	43	48

Therefore, comparing with other algorithms, the proposed scheme can achieve very good performances in execution time, fault coverage rate and test set length.

4. Conclusions

In this paper, a new optimization approach based on AGA is developed for the MCM interconnect test generation problem. The implementation methods of the objective function, selection operator, crossover operator, and mutation operator of AGA are discussed in details. In order to apply AGA into the interconnect test, an accurate fitness function is designed to compute the fitness of each candidate vector. The best individual is selected and added to the test set. Then the fault simulator is used to update the fault list of the tested circuit. The process is iterated until all faults are detected.

The international standard MCM benchmark circuit provided by the MCNC group was used to verify the approach. Comparing with not only the evolutionary algorithms, but also the deterministic algorithms, experimental results demonstrate that the approach can achieve high fault coverage and short CPU time, which shows that it is a new optimized method deserving research.

References

1. Doane D A, Franzon P D., Multichip Module Technologies and Alternatives:the Basics, Brooks Publishing Company (New York, 1995), pp. 1-85.
2. W.K.Kautz, "Testing of faults in wiring interconnects," IEEE Transaction on Computer, Vol C-23, No.4, (1974), pp.358-363.
3. P. Goel and M. T. McMahon, "Electronic chip-in-place test," Proceeding of International Test Conference, (1982), pp.83-90.
4. Najmi Jarwalw and Chi W. Yau, "A New framework for analyzing test generation and diagnosis algorithms for wiring interconnects," Proceedings of International Test Conference, (1989), pp.63-70.
5. Elizabeth M.Rudnick et al, "A Genetic Algorithm Framework for Test Generation," IEEE Translations on Computer-aided Design of Integrated Circuits and Systems,VOL 16, (1997), pp.698-704 .
6. Srinivas M, Patnaik L.M: Adaptive probabilities of crossover and mutations in Gas. IEEE Trans. on SMC, vol. 24, No.4, (1997), p.656-667.

Investigation on Fatigue-Creep Interaction Damage Model for Solder

Na Liu, Xiaoyan Li, Yongchang Yan
School of Materials Science and Engineering, Beijing University of Technology, Beijing, 100022, China
Email: liunayang@emails.bjut.edu.cn

Abstract

It is well known, reliability and workability are the more important issues in the field of chip size package (CSP). Creep and fatigue behaviors are the main loads of the solder joints, the reliability of which should take account of those two main loads. Based on the theory of continuum damage mechanics (CDM), this paper focuses on damage evolution of interaction between the fatigue and creep. A new damage model of fatigue-creep interaction has been developed. And the new fatigue-creep interaction damage model in this paper does not require the simple fatigue model and simple creep model.

1. Introduction

Solder joint interconnections play an important role in electronic packaging, since they provide electrical connections and also are the sole means of mechanically attaching the electronic components to the printed circuit board (PCB). In addition, the failure of solder joints is recognized as the major reason leading to electronic system damage. It is well known that creep and fatigue are the main loads on solder joints, so the solder joints reliability should be based on analysis of these two main loads [1-2]. When creep and fatigue operate simultaneously, it is therefore necessary to consider not only their individual effects, but also effects of their interaction, so as to obtain a more accurate prediction of component life [3-4].

In past decades, a stronger theoretical foundation in continuum damage mechanics（CDM）has been developed. And CDM has been used successfully in many engineering fields to analyses creep, fatigue, ductile fracture and composite failure [5]. Creep failure, as a function of time, usually occurs at high temperature. There are two kinds of creep failures, ductile transgranular creep failure under action of stress, and brittle intergranular creep fractures that are an accumulation and growth of microcracks on intergranular. However, fatigue failure under the action of cyclic loading causes a gradual deterioration of the structure of a material [6]. Additionally, as the temperature increases, the interaction between the processes of creep and fatigue can lead to significant reductions in product life. As discussed in many references, it may not be possible to obtain a reliable prediction of the processes by means of individual deformation behavior analysis. Unified constitutive models representing the damage induced by creep and fatigue, and the damage caused by their interaction, have thus been proposed [3].

Based on the theory of CDM, this paper carried out the researches on the damage evolution caused by fatigue, creep and fatigue-creep interaction, individually. A new damage model has been developed for fatigue-creep interaction.

2. CDM of fatigue-creep interaction

2.1 Damage theory of fatigue-creep interaction

The CDM model relied on the assumption that damage decreased the effective cross-sectional area and increased the effective stress (Kachanov, described in Ker and Zioupos 1997) .The damage variable D is an internal variable of an irreversible damage process, which describe the decrease of effective cross-sectional area. And takes into account the weakness of the material due to the presence of voids or micro-cracks, the damage variable D is defined as

$$D = 1 - \frac{\widetilde{A}}{A_0} \tag{1}$$

Where A_0 is the initial area of the undamaged section, $\widetilde{A}$ is the effective cross-sectional area.

The decrease of the effective area brings on an increase of effective stress. The effective stress can be defined as

$$\widetilde{\sigma} = \sigma / (1 - D) \tag{2}$$

where $\widetilde{\sigma}$ is effective stress, σ is nominal stress [5-6].

Using D_f to represent fatigue damage and D_c to represent creep damage, the incremental form of these two kinds of damage can be written as follows

$$dD_f = F_f(\Delta P, D_f, D_c)dN \tag{3}$$

$$dD_c = F_c(\sigma_{eq}, D_c, D_f)dt \tag{4}$$

where ΔP is the strain range, σ_{eq} is the effective stress.

Although different defects in a material cannot be added directly, according to the definition of effective stress in Damage Mechanics, the decrement of effective area made by different defects can be added [4-11]. If it is assumed that

$$D = D_f + D_c \tag{5}$$

equations (3) and (4) can be written as

$$dD_f = F_f(\Delta P, D_f + D_c)dN = F_f(\Delta p, D)dN \tag{6}$$

$$dD_c = F_c(\sigma_{eq}, D_c + D_f)dt = F_c(\sigma_{eq}, D)dt \tag{7}$$

2.2 Damage theory of fatigue

Based on the theories of CDM, constitutive equation for fatigue damage evolution can be described by an appropriate dissipation potential. A damage model can be obtained by differentiating the constitutive equation for creep-fatigue damage evolution. According to test, the coefficients of the damage model can be determined and the damage model can be verified.

Postulating the dissipation potential consists of three parts,

$$\phi = \phi_p + \phi_D(Y, \dot{p}, \dot{k}; \varepsilon^e, D) + \phi_k \qquad (8)$$

where ϕ_p corresponds to the plastic part, ϕ_D corresponds to the damage dissipation part, and ϕ_k corresponds to the micro-plastic dissipation part[5].

On the assumption that damage is isotropic, the damage dissipation potential, which is a scalar convex function of the state variables in case of isotropic plasticity, and isotropic damage can be structured as follow[12]:

$$\phi = \frac{Y^2}{2S_0} \frac{\Delta \dot{r}}{(1-D)^{\alpha_0}} \qquad (9)$$

where Y is the damage energy release rate, S_0 and α_0 are material and temperature dependent, $\Delta \dot{r}$ is the strain rate.

Fatigue damage can be defined as follow [6]

$$D = \frac{N}{N_f} \qquad (10)$$

Simultaneously equations (9) and (10), the dissipation potential can be expressed as

$$\phi = \frac{Y^2}{2S_0} \frac{\Delta \dot{r}}{(1 - N/N_f)^{\alpha_0}} \qquad (11)$$

Differentiating the dissipation potential (11) to obtain the damage growth rate [11-12]

$$\dot{D} = -\frac{\partial \phi}{\partial Y} \qquad (12)$$

Y corresponds to the variation of internal energy density due to damage growth at constant stress and is given by

$$Y = -\frac{\sigma_{eq}^2 R_v}{[2E(1-D)^2]} \qquad (13)$$

where R_v is the triaxial coefficient, E is the Young's modulus. When considering the specimen as a single axis the numerical value of R_v equal to 1.

Combining equations（11）,（12）and（13）, the constitutive equation for fatigue damage evolution is obtained as

$$\dot{D} = \frac{K^2 R_v}{2ES_0} \frac{\Delta \dot{r}}{(1 - N/N_f)^{1-\beta(\sigma,T)}} \qquad (14)$$

where K and $\beta(\sigma,T)$ are material and temperature dependent.

2.4 Damage theory of creep

At high temperatures and under action of stress, creep damage comes into being either by ductile transgranular or brittle intergranular fractures [6]. The uniaxial form of damage rate can be described as

$$\dot{D} = -\frac{Y}{S_0} \frac{\dot{p}}{(1-D)^{\alpha_0}} \qquad (15)$$

As we known, creep consists of three stages, and creep accelerates in the third stage, so creep behavior mainly occurs in this stage. However it is difficult to calculate $\dot{p}$ in this stage. Based on Odgvist's law of, $\dot{p}$ can be expressed by

$$\dot{p} = \frac{1}{1-D} [\frac{\sigma_{eq}}{K(1-d)}]^{N^*} \qquad (16)$$

Simultaneous equations (15) and (16), the constitutive equation for creep damage evolution can be obtained as

$$\dot{D} = [\frac{\sigma}{A}]^r \frac{1}{(1-D)^{\alpha}} \qquad (17)$$

where r, A and α are temperature-dependent material constants.

2.5 Damage model of fatigue-creep interaction

Integrating equation (14) under the boundary conditions: $D = D_{f0}$ at $N = 0$, $D = D_{ff}$ at $N = N_f$, where D_0 is the initial fatigue damage, D_f is the failure fatigue damage and N_f is the fatigue damage cycles. The equation for fatigue damage evolution can be obtained as

$$D_f = D_{ff} - (D_{ff} - D_{f0})(1 - N/N_f)^{\beta(\sigma,T)} \qquad (18)$$

Integrating equation (17) under the boundary conditions: $D = D_{c0}$ at $t = 0$, $D = D_{cc}$ at $t = t_c$, where D_{c0} is the initial creep damage, D_{cc} is the failure creep damage, and the t_c is the time of creep failure, the equation for creep damage evolution can be developed as

$$D_c = D_{cc} - (D_{cc} - D_{c0})(1 - t/t_c)^{1/[k(\sigma,T)+1]} \qquad (19)$$

The relationship of t and N can be can be written as

$$t = T_{F-C} N \qquad (20)$$

where T_{F-C} is the time of fatigue-creep interaction.

Substituting equation (20) into equation (19), we can obtain

$$D_c = D_{cc} - (D_{cc} - D_{c0})(1 - N/N_c)^{\alpha(\sigma,T)} \qquad (21)$$

where $\alpha(\sigma,T)$ is material and temperature dependent.

Substituting equation (18) into equation (5), we can obtain

$$D_f = (D_{ff} + D_{cc}) - [(D_{ff} - D_{f0})(1 - N/N_f)^{\beta(\sigma,T)} + (D_{cc} - D_{c0})(1 - N/N_D)^{\alpha(\sigma,T)}]$$

(22)

where $D_{ff} + D_{cc} = D_{fc}$ is failure fatigue-creep interaction damage, $D_{f0} + D_{c0} = D_0$ is the initial damage and N_D is the cycle for damage. Postulating that $\beta(\sigma,T)$ is equal to $\alpha(\sigma,T)$ [8-11], the damage evolution equation for fatigue-creep interaction can be written as

$$D = D_{fc} - (D_{fc} - D_0)(1 - N/N_D)^{\beta(\sigma,T)}$$

(23)

where N_D is the numerical value of fatigue-creep interaction damage.

3.Conclusions

1. In previous years, many fatigue-creep interaction models have been developed to predict the fatigue life of solder joints under thermal cycle conditions. Most of them are based on a series of simple fatigue data and simple creep data, which were coupled to develop the damage models. The new fatigue-creep interaction damage model of in this paper does not require the simple fatigue model and simple creep model.

2. As we all know, the equation (23) was developed for predict the fatigue life. Those unknown quantities, such as D_{fc}, D_0, N_D and $\beta(\sigma,T)$ should be obtained by fatigue-creep interaction test for solder. Meanwhile, in order to simulate the fatigue-creep interaction better, an experiment has been designed to determine the values of the variables in damage equation. The laboratory equipments made up of two kinds of alloy, INVAR alloy and aluminum alloy. The solder joint samples were fixed between the two alloys, along with the module, were put into the temperature loading stove. The thermal expansion coefficient of aluminum ally is $23 \times 10^{-6} m/℃$ and that of INVAR ally nearly close to zero. Due to the thermal expansion coefficient mismatch of aluminum ally and INVAR ally and the change of the temperature, the shearing force is caused between the two alloys. Therefore, both the variable temperature and shearing force result in fatigue-creep interaction.

Acknowledgments

This study was partially supported by National Natural Science Foundation of China （No. 50475043), the Natural Science Foundation of Beijing (No.2052006，No.2082003), which was acknowledged.

References

1. Gregor Massiot, "A Review of Creep Fatigue Failure Models in Solder Material-Simplified use of a Continuous Damage Mechanical Approach," *5th.Int.Conf.on Thermal and Mechanical Simulation and Experiments in Micro-electronics and Micro-Systems*, EuroSimE, 2004, pp.465 - 472.

2. X. P. Zhang, C. S. H. Lim, et al, "Thermal Fatigue and Creep Fracture Behaviors of a Nanocomposite Solder in Microelectronic/ optoelectronic Packaging," *Key Engineering Materials*, Vol. 312, (2006), pp. 237 - 242.

3. Tae-Won Kim, Dong-Hwan Kang, Jong-Taek Yeomb, et al, "Continuum damage mechanics-based creep-fatigue interacted life prediction of nickel-based superalloy at high temperature," *Scripta Materialia*, Vol. 57, No. 12 (2007), pp. 1149 - 1152.

4. Jing JianPing, Meng Guang, Sun Yi, et al, "An effective continuum damage mechanics model for creep–fatigue life assessment of a steam turbine rotor," *International Journal of Pressure Vessels and Piping, International Journal of Pressure Vessels and Piping,* Vol. 80, No. 6 (2003), pp. 389 - 396.

5. Zhaoxia.Li, Damage Machanics and Applications, Science publisher (Beijing, 2002), pp.16 - 31.

6. L.M.Kachanov, Introduction to Continuum Damage Mechanics, Martinus Nijhoff (Dordrecht, 1986), pp.1 - 57.

7. A. El Gharad, H. Zedira, Z. Azari, G. Pluvinage, "A synergistic creep fatigue failure model damage," *Engineering Fracture Mechanics*, Vol. 73, No. 6 (2006), pp. 750 - 770.

8. Timothy D. Schwaba, Clifton R. Johnstonb, Thomas R. Oxlanda, et al, "Continuum damage mechanics (CDM) modelling demonstrates that ligament fatigue damage accumulates by different mechanisms than creep damage," *Journal of Biomechanics*, Vol. 40, No. 14 (2007), pp. 3279 - 3284.

9. Chen Zhiping, Jiang Jialing, Chen Ling, "Research on fatigue-creep interaction damage of steel 1.25Cr0.5Mo," *Acta metallurgica sinica*, Vol. 43, No. 6 (2007), pp.637-642.

10. Soo Woo Nam, "Assessment of damage and life prediction of austenitic stainless steel under high temperature creep–fatigue interaction condition," *Materials Science and Engineering*, Vol. 322, No. 1 (2002), pp. 64 - 72.

11. Baidurya Bhattacharya, Bruce Ellingwood, "A new CDM-based approach to structural deterioration," *International Journal of Solids and Structures,* Vol. 36, No. 12 (1999), pp. 1757 - 1779.

12. C. Sommitsch, P. Polt, G. Ruf, et al, "On the modelling of the interaction of materials softening and ductile damage during hot working of Alloy 80A," *Journal of Materials Processing Technology*, Vol. 177, No. 1 (2006), pp. 282 - 286

Study of Plasticity Damage Mechanics Constitutive Model for SnAgCu Solder Joint

Xiao-yan Li, Yong-chang Yan，Na Liu

School of Materials Science and Engineering, Beijing University of Technology, Beijing, 100022, China

Email: yongchangyan@vip.sina.com

Abstract

A thermodynamics-based damage mechanics rate dependent constitutive model is used to simulate experiments conducted on thin layer eutectic SnAgCu(SAC) solder joints. The non-damage constitutive is measured by bulk tensile test. The relationship between true stress and strain is $\sigma=85.26\varepsilon0.3536$. Damage evolution equation is proposed based Lemaitre ductile damage theory and the constant in the equation is measured by unloading elastic modulus method. The damage evolution equation is $D=1.0689\varepsilon P-0.0008$. Simulation (using software Ansys 9.0) of shear test of solder joint between Cu sticks employing damage mechanics rate independent constitutive is uniform to practicable test.

1. Introduction

Due to environment concerns, legislation to ban lead in electronic products required lead-free solder application. The impending ban on lead initiates much research work on lead-free solder [1,2]. Among the diversity of alternative lead-free solders the SnAgCu (SAC) solder seems to make the race [3]. For a long time, constitutive modeling of SAC has focused onto the elastic and creep properties. Indeed, under the thermal cycling condition, the creep model is proper. However, in applications, such as hand-held electronic devices or automotive products, the pure mechanical impact, like shock, bending and twisting may even matter more than sole thermo-mechanical fatigue [4]. So the elastic-plasticity property of solder must be taken into consideration.

2. Theory of damage

The concept of continuity was proposed first by Kachonov during research of metal creep rapture in 1958. Then the concept of damage efficient was proposed by Rabtonov[5] in 1963. In 1970s, Lemaitre [6] studied the influence of damage to elastic and plastic of metal from the aspect of thermodynamic of irreversible process; Hult and Leckie studied the coupling of damage and creep. The term of Contimuum Damage Mechanics was first introduced by J. Hult in 1972[7]. From then on, the Damage Mechanics is developed and widely used in engineering field. In recent years, several investigations are carried out to examine the effects of damage on behavior of solder alloy (Basaran and Yan, 1998; Qian et al., 1999; Wei et al., 1999).

If we assume the distribution of damage is no difference to the all direction of material performance, damage variable is a scalar:

$$D = \frac{\delta S_D}{\delta S} \tag{1}$$

where, D is damage variable, δS represents the sectional area of a mass, and δS_D is damaged area in the sectional area.

Damage is a dissipation process which is always associated with strain and also involved in dissipation. Here, potential dissipation is assumed to be related to accumulating plasticity strain P :

$$\overset{*}{\varphi}_D = \frac{1}{2}\frac{Y^2}{S_0}\frac{\overset{\bullet}{p}}{(1-D)^{\alpha_0}p^{2n}} \tag{2}$$

where, Y is strain energy rate, S_0 is material constant, α0 is exponent of material damage, n is strain hardening exponent. Damage velocity is the differential of potential dissipation to strain energy rate:

$$\overset{\bullet}{D} = -\frac{\partial \overset{*}{\varphi}_D}{\partial \overline{Y}} = \frac{\overline{Y}}{S_0}\frac{\overset{\bullet}{p}}{(1-D)^{\alpha_0}p^{2n}} \tag{3}$$

where, $\overline{Y}=-Y$. According to the principle of strain equivalence and Ramberg-Osgood hardening rule, the constitutive of material coupling damage is:

$$\frac{\sigma_{eq}}{1-D} = Kp^n \tag{4}$$

where, K is coefficient of hardening, σ_{eq} is Von-Mises equivalent stress. According to Lemaitre damage theory, strain energy rate can be expressed as follows:

$$\overline{Y} = \frac{\sigma_{eq}{}^2}{2E(1-D)^2} f(\frac{\sigma_m}{\sigma_{eq}}) \tag{5}$$

$$f(\frac{\sigma_m}{\sigma_{eq}}) = \frac{2}{3}(1+v)+3(1-2v)(\frac{\sigma_m}{\sigma_{eq}})^2 \tag{6}$$

Where σ_m is hydrostatic stress; E is elastic modulus of material; v is Poisson ratio. Put Eq.(4) and Eq.(5) into Eq.(3), damage velocity can be derived:

$$\overset{\bullet}{D}(1-D)^{\alpha_0} = cf\overset{\bullet}{p} \tag{7}$$

Where $c = \frac{K^2}{2ES_0}$.

3. Experiment and results

The specimens (Fig.1) in tensile test and unloading elastic modulus method are made of bulk material. The specimens are

978-1-4244-2739-0/08/$25.00 ©2008 IEEE

casted in a cuboid mould, and then are processed into the shape in Fig.1 by line cutting.

Solder joint specimens (Fig.2) are composed of two Cu sticks (φ3mm) which were connected by SAC solder joint. Fixed Cu sticks in a mould (Fig.3), the gap length between two sticks is 0.8mm. Before welding, put solder beside the gap. When the temperature is increased to 221℃, the solder will be melted and dispersed to gap. After cooling, the specimen is completed.

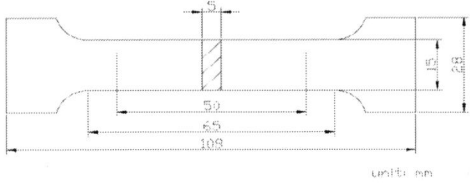

Fig.1 the specimen of tensile test

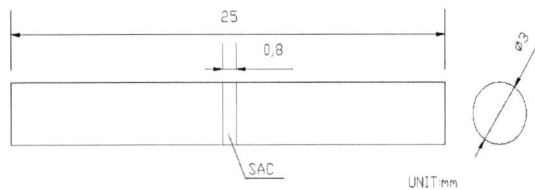

Fig. 2 SAC solder joint

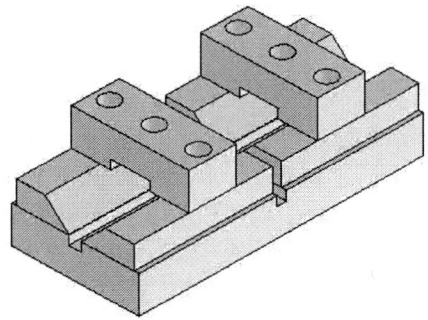

Fig.3 mould of assembling solder joint

Test program. The aim of tensile test is to earn the constant value of K, n in Ramberg-Osgood relationship. The tensiometer with 50mm length is fixed in the middle of the specimen. The velocity of load is 2mm/min. We can earn true stress σ and true strain ε:

$$\sigma = F(\frac{L}{L_e S_0}) \qquad (8)$$

$$\varepsilon = \ln(\frac{L}{L_e}) \qquad (9)$$

where F is load, Le is original length of tensiometer, So is sectional area of specimen before deforming and L is the length of tensiometer after deforming. The relationship between true stress and true strain can be expressed as follows:

$$\ln \sigma = n \ln \varepsilon + \ln K \qquad (10)$$

The tensile date can be fitted by method of least squares. The coefficient of hardening K=82.26, and the strain hardening exponent n=0.3536, see Fig.4.

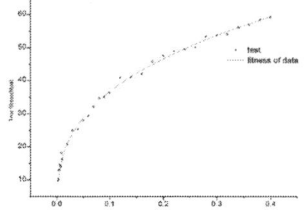

Fig.4 Fitness of material constitutive relationship, K=82.26, n=0.3536

The unload elastic modulus method is used to determine the constant in Eq.(7). The course of test is shown in Fig.5. The variation of elastic modulus vs. plastic strain is shown in Fig.6.

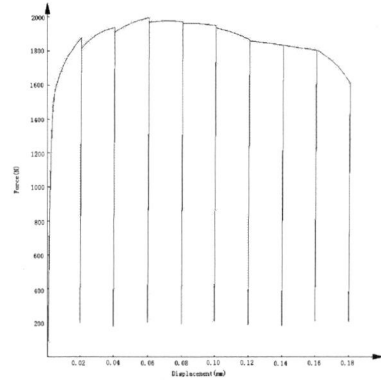

Fig.5 Curse of unload elastic modulus method

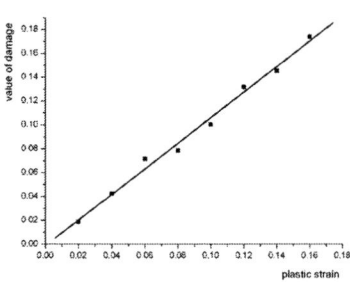

Fig.6 value of damage vs. plastic strain during unloading elastic modulus method

The relationship between damage and plastic strain is linear. So α0 in Eq.(7) equal to zero. By integration, with D=0 for ε＜ε0 (ε0 being the one dimensional strain threshold):

$$D = cf(\varepsilon - \varepsilon_0) \qquad (11)$$

Fitting the point, the value of cf is equal to 1.0689, and $\varepsilon_0 = 0.000748$.

Solder joint specimen is used in shear test. The distance between two grip holders is 4.8mm, and load velocity is 0.001mm/s.

4. Simulation

The aim of simulation is to verify the damage model through FEM. Among shear test, displacement-load curse will be achieved. Through FEM, the variation of load can be simulated as the displacement of solder joint increase.

The software of Ansys 9.0 is used to simulate the behavior of SAC solder joint. The material constitutive of Cu is elastic and its elastic modulus is 129.8GPa. The constitutive of SAC is elastic-plastic coupling with damage evolution, as shown in Eq.(4). Fig 7 shows the relationship between stress and strain. The elastic modulus of SAC is 37.798GPa.

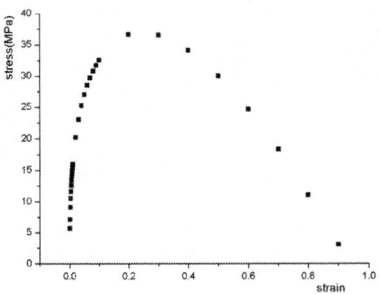

Fig.7 material model of SAC in FEM analysis

The boundary condition of FEM is identical to shear test. As the relationship between stress-strain is decreased when the value of strain is large, it is important to select suitable solution control. Arc-length options can be selected because it can guarantee convergence.

The relationship between displacement and load refers to Fig.8. In general, FEM can describe the behavior of solder joint during shear test when the damage constitutive of SAC is selected. The maximum value of load (149.0778N) for simulation is similar to the maximum value of shear test (147.68N). During small displacement, the load increases as the displacement increases. In shear test, there is a platform when displacement is between 0.35mm and 0.45mm. When the displacement is larger than 0.45mm, the load begins to decrease as the displacement increase. The result of formulation describes the similar tendency during loading process, even they have similar platform range.

It can be observed that in smaller displacement the difference between formulation and test is much larger than in larger displacement. That means, in condition of small stress and strain, the plastic damage constitutive model can not describe the behavior of SAC very well because of the existing of creeping. The relationship between the ratio of plastic deform and force is described in Fig.9. It can be seen that when the force increases, the ratio of plastic deforming also increases, the difference between formulation and test becomes smaller.

In large displacement (between 0.35mm and 0.85mm), the curve of formulation and test is primarily uniform. That means in this stage the value of creeping can be ignored, and the damage plastic constitutive can describe the behavior of material very well.

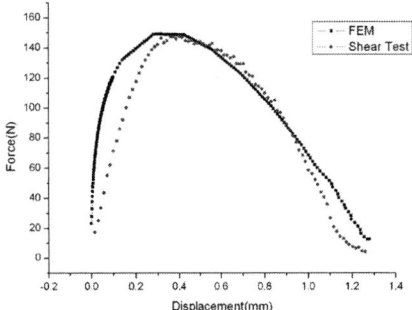

Fig.8 relationship between displacement and force (shear test vs. FEM)

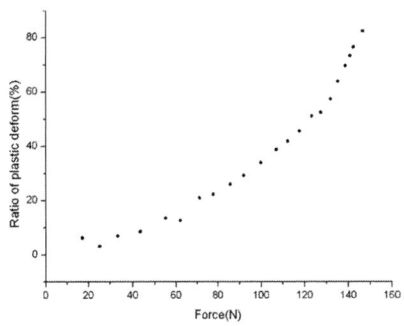

Fig.9 relationship between the ratio of plastic deform and force

When the displacement is larger than 0.85mm, the force decreasing velocity in test is faster than that in formulation. That can be explained that the cavity have been created in this stage, but material is also continuous in condition of formulation.

Through the analysis above, the conclusion can be achieved:

(1) The thermodynamics-based damage mechanics rate dependent constitutive model can describe the behavior of SAC joint, especially in condition of large strain and stress;

(2) In the condition of fatigue (generally small strain or small stress), the behavior of SAC joint described by plastic damage constitutive is not enough. The creeping deforming must be taken into consideration;

(3) The plastic damage constitutive can be used in the condition of tensile and shear. For example, the fracture strengthen can be determined by the use of plastic damage constitutive in formulation.

Acknowledgments

This study was partially supported by National Natural Science Foundation of China （No. 50475043), the Natural Science Foundation of Beijing (No.2052006，No.2082003), which was acknowledged.

Reference

1. Richards B P., Lead-free legislation, National PhysicalLaboratory (UK, 2002).
2. Pang, John H.L., Yeo Alfred, Low, T.H. et al, "Lead-free 96.Sn-3.5Ag Flip Chip Solder Joint Reliability Analysis," *Ninth Intersociety Conference on Thermal and Thermomechanical Phenomena in Electronic Systems*, Las Vegas, United States, 2004, pp. 160 - 164.
3. Wiese S., Meusel E., "Characterization of Lead-free Solders in Flip Chip Joints," *Journal of Electronic Packaging*, Vol. 125, No.4 (2003), pp. 531 - 538.
4. Wiese S., Rezpka S., Meusel E., "Time-independent Plastic Behavior of Solders and its Effect on FEM Simulations for Electronics Packages," *8th International Advanced Packaging Materials Symposium*, Stone Mountain, GA, USA, 2002, pp. 104 - 111.
5. Robotnov Y.N., "On the equations of state for creep," *Progress in Applied mechanics*, (1963), pp. 307 - 315.
6. Lemaitre J., "A continuum damage mechanics model for ductile fracture," *Journal of Engineer Materials and Technology*, Vol. 107, (1985), pp. 83 - 89.
7. Lemaitre J., Desmorat R., *Engineering Damage Mechanics*, (2004), pp. 1.

A Design for Increasing the Immunity to RFI of Protection IC of Lithium-ion Battery

Dongfang Cheng Jue Zhang Xiaohui Li Jiongming Wang

Key Laboratory of Advanced Displays and system Application, Ministry of Education, Shanghai University

Microelectronic Research & Development Centre, Shanghai University

No.149 Yanchang Rd, Shanghai 200072, P.R.China

E-mail: chengdf@mail.shu.edu.cn   Tel: 086-021-56331206-112   Fax: 086-021-56331272

Abstract

Illustrated by the case of a lithium-ion battery protection IC, the paper focuses on the design of internal immunity to RFI. With analysis of the chip's three major elements of electromagnetic compatibility (EMC), the qualitative and quasi-quantitative analysis results of RFI influence to the chip are given out. By using a simple filter circuit and available material physical construction which can isolate, absorb and consume the RFI energy, the protection IC's electromagnetic susceptibility has been reduced effectively. The simulation of the devised structure in the time domain is gained by Winspice, and tool IC_EMC makes it possible of the conversion from time domain to frequency domain in which the spectrum analysis is completed. The design has passed the simulation verification and the layout implement of the devised construction designed is also available.

1 Introduction

With the rapid development of compact portable communication and computing systems, lithium-ion batteries have been the most popular power supply for portable systems. Operating near the radio frequency source, the protection IC in lithium-ion battery package becomes the victim of radio frequency interference (RFI) whose immunity to RFI should be improved. In fact, with the EMI radiation environment and the increasing of devices' integration density and analog/digital mixed integrate circuits' operating speed, any close loop in the circuit is susceptible to interference signals which will result in the failure of the IC[1].

Furthermore, the lack of EMI immunity forces the IC designers to reduce circuit susceptibility by means of a posterior layout adjustments, filters, changing in the operating frequency, shielding, etc, that are sometimes limited, seldom viable and often complex and expensive[2]. Thus, in recent years, EMI were carefully investigated [3] both theoretically and experimentally to find possible prevention methodologies for the specifically cases.

The paper focuses on the design of internal immunity to RFI which is illustrated by the case of a lithium-ion battery protection IC. Through the analysis of the electromagnetic interference (EMI) source, propagation path and the susceptive object which are the three major elements of electromagnetic compatibility (EMC), the qualitative and quasi-quantitative analysis results of RFI influencing to the chip are given out. By using a simple filter circuit and available material physical construction which can isolate, absorb and consume the RFI energy, the protection IC's electromagnetic susceptibility has been reduced effectively.

2 Qualitative Analysis of Three Major Elements

2.1 Source

EMC problems associated with integrated circuits can generally be classified as intra-chip or externally-coupled.

As shown in Fig. 1, two MOSFET are used as the current controller devices in the protection circuit. The switch characteristic of the MOSFET is the high dv/dt during the turn-off delay and the large change of the current in rise/fall during the open-close reversion. They are easy to generate the RF energy. The high di/dt is the immediacy cause of EMI, so it is facility to be the interface source. Changing the external series Gate resistor, the di/dt can be under control.

However, due to the slow operating speed of the MOSFET, it was concluded that the protection system itself would emit very little interference. The major source of interference is the external RFI (radio frequency interference) generated by the nearby high frequency receiving and emission switch circuit in the portable communication system.

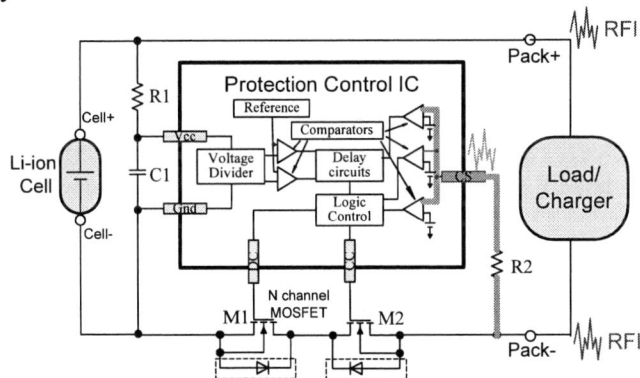

Figure 1   conduction path of RFI influence on protection IC

2.2 Coupling Paths

The RFI coupling path can be classified as conduction coupling and radiation coupling.

In the range of the environment frequencies (GSM900/ GSM1800/3G/4G) and the chip size, conduction seems to be the most relevant way of propagation.[4] And the way to couple noise into the protection IC is mainly via the package leads pins. The protection IC is found suffering from RF-disturbance voltages induced by EM-energy penetrating these analog areas via the IC-package leads that are connected via PCB-tracks to attached cables acting as effective antennas. RF-disturbance contains close loop's induction to the radiation field and the voltage drops on MOSFETs generated by high frequency operating current of the receiving and

978-1-4244-2739-0/08/$25.00 ©2008 IEEE

emission machine what is the load of the battery (Fig.1). The performance may be affected by this RF-disturbance which will bring "black screen" (turn-off) especially under a worse signal condition. Sufficient experiments have proved this. Otherwise, radiation interference which influences the circuit correspondingly slightly can be effectively shielded by a top-metal layer connecting to the ground in the chip.

2.3 Susceptibility Equipment

The protection IC operating near the radio frequency source is the susceptibility equipment. Where IC's input pin is the most susceptive which may affect the operation of protection IC up to cause serious failures in the system. According to Fig. 1, DC pin and CC pin are unsusceptible output pins which connect to the gate of the MOSFET with the typical C_{gs} value $0.76nF$. There is a on-chip capacitor between Vcc and Gnd whose typical value is $0.1\mu F$. This decoupling capacitor can filter the transient voltage change from the battery. Whereas the CS pin dose not have an effective filtering mechanism if it is not connected with a external capacitor. Because of the required high sensitivity of the input checking comparators, their susceptibility would be a problem. Thus CS pin is the major susceptibility equipment.

The objective of common restraining EMI methods is to cut the coupling path between the EMI source and susceptibility equipment. So, for the design of increasing the immunity of protection IC of lithium-ion battery, the key is to design a high frequency grounding path at the input pin Cs which can effectively restrain RFI signal and have no affect on the direct current input signal. With the assistant of grounding, shielding and other technology methods, the immunity will be better.

According to the characteristic of high frequency electromagnetic wave and the common design rule of the RF circuit/IC, a special structure has been adopted. The devised structure implements filter through the effectively attenuation of the interference electromagnetic wave coupled to the pin of chip. This method has the advantage of low cost with small area and technology compatibility. As the widely used of microstrip transmission line in RFIC design and the following quasi-quantitative analysis, the design can be considered to have a good anticipation.

3 Quasi-quantitative Analysis

3.1. Electrical Model and Skin Effect

The electrical dimension k of any chip depends on its physical dimension L, inspire source's frequency f and the transmit speed v of wave in the material [5].

$$k = \frac{L}{\lambda} = \frac{Lf}{v} \qquad (1.1)$$

If $k \square 1$ or $L < \frac{1}{10}\lambda$, it is regard as *electrically small* model in which lumped parameter circuit model and Kirchhoff voltage and current law is available. For *electrically big* model, only Maxwell's equations are available.

Illustrated by the case of frequency1GHz, the electrical dimension k of the protection IC is $k = 3.33 \times 10^{-3} \square 1$ which belongs to electrically small model.

On the other hand, electromagnetic wave just exists in the surface of the metal because of its fast attenuation in good conductor which is called *Skin Effect*. Skin depth is usually used to characterize the skin degree, represented by δ:

$$\delta = \frac{1}{\alpha} = \sqrt{\frac{2}{\omega\mu\gamma}} = \frac{1}{\sqrt{\pi f \mu\gamma}} \qquad (1.2)$$

In which f is the frequency of electromagnetic wave, μ is the magnetic permeability of the material and γ is the bulk conductivity of the material.

Still illustrated by the case of frequency 1GHz, the skin depth of aluminum is $\delta = 2.62\mu m$. As the thickness of aluminum interconnect-line under the current CMOS technology is far less than this value, currents can be approximately considered uniformly distributing in the lead and the skin effect of lead is neglected.

3.2. Transmission Line Model

Under high frequency operation condition, the interconnect-line in the integrated circuit is similar with the microstrip transmission line whose circuit parameters are uniformly distributing along the line.

A transmission line can be described by the following four parameters: series resistance per unit length $R_S(\Omega/m)$ (due to conductors), series inductance per unit length $L_S(H/m)$ (due to the mutual inductance between the conductors) and shunt conductance per unit length $G_P(S/m)$ (due to non-ideal isolation), shunt capacitance per unit length $C_P(F/m)$ (capacitance between the conductors). All these distributed parameters are complex functions of dimension, shape, thickness and environment material of the microstrip.

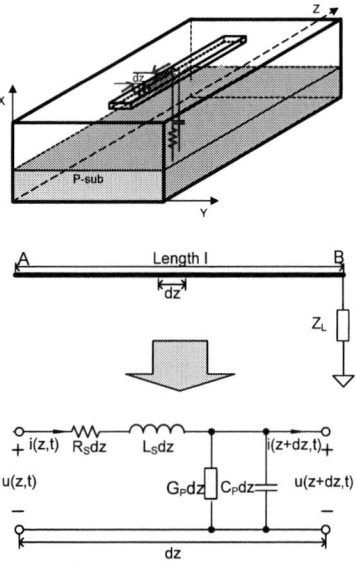

Figure 2　the equivalent microstrip model of interconnect line on substrate

A small section of a dz-long transmission line has thus the equivalent circuit as Fig. 2.

The characteristic impedance Z_0 of transmission line is:

$$Z_0 = \sqrt{\frac{R_S + j\omega L_S}{G_P + j\omega C_P}} \qquad (1.3)$$

In which ω is the operating frequency.

Seen from net A to B, the input impedance Z_A is the function of characteristic impedance Z_0, load impedance Z_L, propagation constant γ and the length of transmission line l:

$$Z_A = Z_0 \frac{Z_L \cosh \gamma l + Z_0 \sinh \gamma l}{Z_0 \cosh \gamma l + Z_L \sinh \gamma l} \qquad (1.4)$$

$$\gamma = \sqrt{(R_S + j\omega L_S)(G_P + j\omega C_P)} = \alpha + j\beta \quad (1.5)$$

In which, the real part α (Np/m) is attenuation constant which represents the reduced degree of voltage or current via per unit length transmission line. The imaginary part β (rad/m or $°/m$) is phase constant which represents the phase change of voltage or current via per unit length transmission line.

For low loss microstrip:

$$\begin{cases} R_S \ll L_S \omega \\ G_P \ll C_P \omega \end{cases} \qquad (1.6)$$

Then

$$\gamma = j\omega\sqrt{L_S C_P} = j\beta \qquad (1.7)$$

$$Z_A = Z_0 \frac{Z_L \cos \beta l + j Z_0 \sin \beta l}{Z_0 \cos \beta l + j Z_L \sin \beta l} \qquad (1.8)$$

βl (l is the transmission line length) is called the electrical length of the transmission line.

When $\beta l = 2\pi$, the characteristic wavelength is:

$$\lambda = \frac{2\pi}{\beta} = \frac{2\pi}{\omega\sqrt{L_S C_P}} \qquad (1.9)$$

3.3. RF Grounding

3.3.1. In Fig.2, if B is a RF grounding net, $Z_L = 0$,then

$$Z_A = Z_0 j \tan \beta l \qquad (1.10)$$

This means the change of parameters can turn the microstrip to lossless short line. So, net A can be grounded via net B under the condition:

(a) if $l \ll \dfrac{\lambda}{4}$

then $Z_A \to 0$

(b) if $l = \dfrac{n\pi}{\omega\sqrt{L_S C_P}} = n\dfrac{\lambda}{2}, n = 0, 1, 2, 3, 4, \cdots$

then, $Z_A = 0$

It can be concluded that in order to grounding net A the length of microstrip should be far less than quarter wavelength (as (a) shown) or integral multiple of half wavelength (as (b) shown). Strictly speaking, the grounding condition (b) is accurate for one fixed frequency. However, the approximation in pass band can also satisfy the condition well as long as the bandwidth is not very wide. So, the shortest length of lossless line is $n = 1, l = \dfrac{\lambda}{2}$.

3.3.2.   In Fig.2, if B is connected with a opening load for RF, so called microstrip terminal open $Z_L \to \infty$,

When $l = (2n+1)\dfrac{\lambda}{4}, n = 0, 1, 2, 3, 4, \cdots$ (1.11)

$$Z_A = \frac{Z_0^2}{Z_L} \to 0 \qquad (1.12)$$

This means net A can also be grounded effectively.

It can be concluded that at a opening net B if the length of microstrip between A and B is integral multiple of quarter wavelength, A is a grounding net. The shortest length of lossless line is $n = 0, l = \dfrac{\lambda}{4}$.

Obviously, for the requirement of RF grounding in the integrate circuit design, quarter wavelength grounding has more significant advantage than half wavelength grounding. However, it should be observed that the infinite large impedance of opening net B is a theory approximation. In fact, especially in the band of RF, there are always some disturbed capacitances.

4   Devised Methods

A path for RF grounding can be obtained by utilizing the parasitic inductance on lead and the parasitic capacitance between lead layer and substrate (ground) which performs as a high frequency filter at input stage [6]. This structure can effectively restrain RFI signal and have no affect on the direct current input of the lithium-ion battery protection IC then increases the immunity of the protection IC.

The frequency ranges from 100MHz up to 4 GHz, to account for the spectrum of most of the current possible interfering signals, mainly including the cellular phone bands. In the normal condition, the lithium-ion battery protection IC operates in direct current. So, a larger capacitance (approximately several tens pF) is need to implement low-pass filter while several hundreds ohm of resistance is need to increase the high frequency consumption on the lead.

The simulation of devised structure in the time domain is implemented by Winspice, and tool IC_EMC makes it possible of the conversion from time domain to frequency domain under which the spectrum analysis is completed.

4.1 Parameters Extraction

The distributed parameters of transmission line can be gained by Interconnect Parameters tool of IC_EMC combined with the selected manufacture technology (Fig.3).

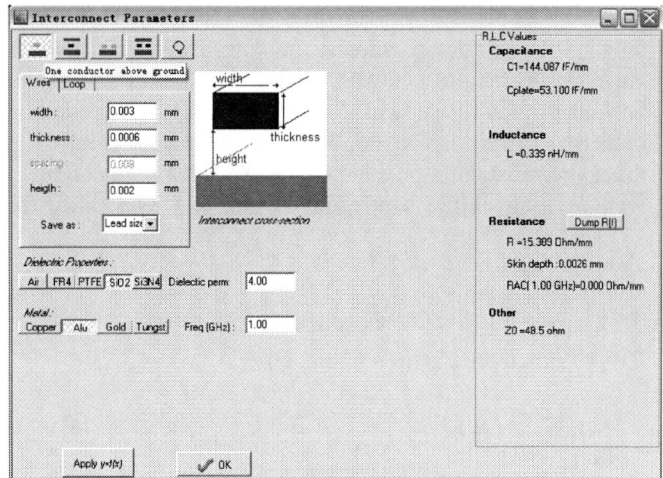

Figure 3 extract distributed parameters of transmission line

Table 1   the selected parameters and the extracted parameters

Input			Output		
Width	3um		Capacitance	C1	144.087fF/mm
Thickness	0.6um			Cplate	53.100fF/mm
Height	2um		Inductance		0.338nH/mm
Dielectric properties	SiO2		Resistance	Skin depth	0.0026mm
Metal	Al			R	15.389Ohm/mm
Frequency	1GHz		Z0		48.5Ohm

From the parameters listed in table 2/3, it can be found that a traditional several tens pF capacitance will occupy a great piece of chip area. In order to control the area and decrease the cost, we adopt a special structure by utilizing parasitical n+/p-substrate (cj=4.35e-04+cjn; cjsw=3.65e-10+cjswn) junction capacitance to get the larger capacitance we need.

Table 2 0.6um DPDM CMOS PARASITIC CAP

No.	PARASITIC CAP	Attribute (pF/sq micron)
A	Metal1 to poly1	6.28E-17
C	Metal1 to n+	5.75E-17
D	Metal1 to p+	5.75E-17
E	Metal1 to p-substrate	3.45E-17
F	Metal1 to n-well	3.45E-17
G	Poly1 to p-substrate	8.63E-17
H	Poly1 to n-well	8.63E-17
I	Metal2 to poly1	2.88E-17
K	Metal2 to n+	2.76E-17
L	Metal2 to p+	2.76E-17
M	Metal2 to p-substrate	2.47E-17
N	Metal2 to n-well	2.47E-17
T	Metal1 to Metal2	2.47E-17

Table 3 0.6um DPDM CMOS Available Cap

No.	Device CAP	attribute(pF)
C1	poly1/n+ (nncap)	area * 0.0026 * 1e-12
C2	poly1/poly2 (ppcap)	area * 0.0007 * 1e-12

4.2 Simulation

As shown in Fig.4, a comparatively more accurate two-π model [7] is used to simulate the designed transmission line.

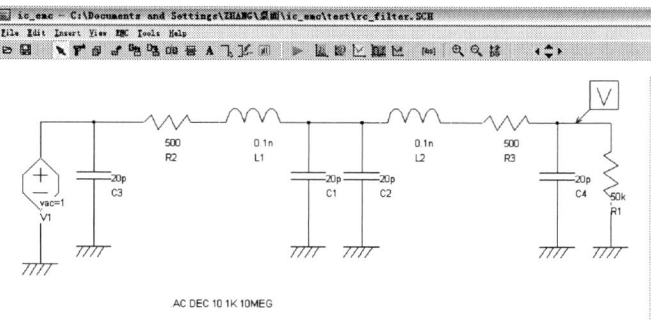

Figure 4   the two-π model simulation circuit

In which, C=20pF, L=0.1nH, R=500Ohm, RL=50KOhm

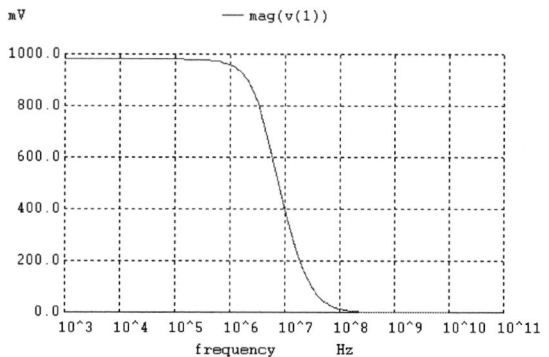

(a) Simulation result of WinSpice v1.05.07a

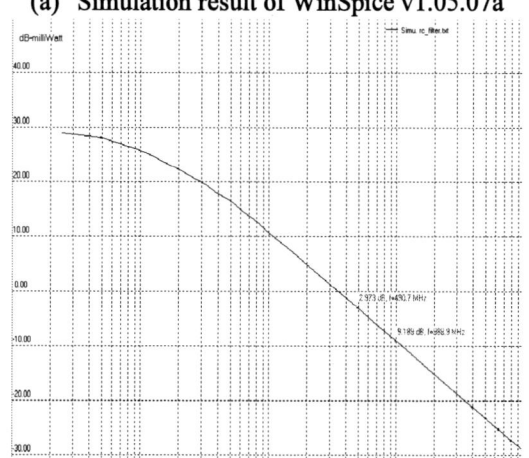

(b) Simulation result of ic_emc's susceptibility vs. frequency

Figure 5   Simulation result of circuit

From the simulation result in Fig.5 (a) & (b), we can see the designed structure can effectively filter RFI relying on the disturbed parameters and parasitic capacitance of the transmission line. Fig.5 (b) shows that the -3dB susceptibility appears at the frequency of 485MHz, and for 1GHz, the susceptibility is reduced to -9.393dB. Therefore, the designed structure has improved the immunity to RFI of the lithium-ion battery protection IC, especially under the frequency above 485MHz. Limited by the value of capacitance, otherwise the cut-off frequency can be further decreased.

4.3 Layout Design

The layout should be designed specially to meet the needs of the parameters. This can help to decrease the increase of the chip's area and cost cause by adding the special filter component.

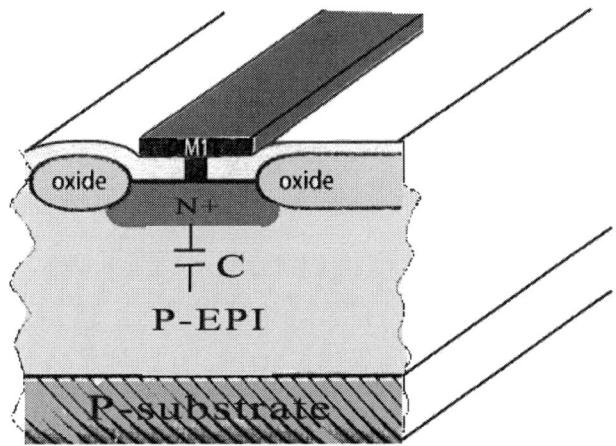

Figure 6   sketch map of parasitic capacitance

Figure 7   layout implement

The part inside the highlight double-line frame in Fig.7 is the designed RFI filtering structure which contains the transmission line (in the middle of the figure) with a larger parasitic capacitance shown in Fig.6 and poly1 resistance (in the top of the figure) with the larger high frequency consumption. The structure improves the immunity of the internal chip and ensures the susceptible devices and modules such as comparators can operate normally.

Conclusions

Different from the traditional design rule such as posteriori layout adjustments, filters, changing in the operating frequency and shielding, etc, the paper focuses on the design of internal immunity to RFI. With analysis of the chip's three major elements of electromagnetic compatibility (EMC), the qualitative and quasi-quantitative analysis results of RFI influence to the chip are given out. By using a simple filter circuit and available material physical construction which can isolate, absorb and consume the RFI energy, the protection IC's electromagnetic susceptibility has been reduced effectively. The simulation of devised structure in the time domain is gained by Winspice, and tool IC_EMC makes it possible of the conversion from time domain to frequency domain under which the spectrum analysis is completed. The design has passed the simulation verification and the layout implement of the devised construction designed is also available.

References

1. Edited by Sonia Ben Dhia, Mohamed Ramdani and Etienne Sicard, Electromagnetic Compatibility of Integrated Circuits: Techniques for Low Emission and Susceptibility, Springer Science Business Media, Inc. (2006), pp. 65-67.
2. Anna Richelli, Luigi Colalongo, and Zsolt M. Kovács-Vajna, Senior Member, IEEE ,"Increasing the Immunity to Electromagnetic Interferences of CMOS OpAmps," *IEEE transactions on reliability*, vol. 52, NO. 3 (2003).
3. G. Masetti, S. Graffi, D. Golzio, and Z. M. Kovács-Vajna, "Failures induced on analog integrated circuits from conveyed electromagnetic interferences: A review," *Microelectronics and Reliability*, vol. 36, no. 7/8 (1996), pp. 995-972.
4. Boris Traa, "RF-Susceptibility Analysis of Complex Integrated Analog Circuits", *IEEE International Symposium* on Volume 2, 19-23 (2002), pp. 987-992
5. Richard Li, translated by Wang Zhi Gong, Key Issues in RF/RFIC Circuit Design, Higher Education Press, (2007), pp. 71-76
6. Edited by Zhao Yang, See Kye Yak, Fundamental of Electromagnetic Compatibility and Application, China Machine Press, (2006), pp. 33-41
7. Clayton R.Paul, Introduction to Electromagnetic Compatibility, China Machine Press, (2006), pp.106-108.

The Effect of the Different Teflon Films on Anisotropic Conductive Adhesive Film (ACF) Bonding

Jun Zhang[1], Y.C. LIN[2], Liugang Huang[1]

[1]School of Chemical Engineering & Technology, Zhengzhou University, Zhengzhou, 450001, P.R.China

[2] School of Mechanical & Electrical Engineering, Central South University, Changsha 410083, China

Corresponding author: yclin@mail.csu.edu.cn phone: 0731-8877915

Abstract

New interconnect materials are always necessary as a result of evolving packaging technologies and increasing performance and environmental demands on electronic systems. Polymer-based conductive-adhesive materials have become widely used in many electronic packaging interconnect applications. Among all the conductive-adhesive materials, the anisotropic conductive adhesives (ACA) (or anisotropic conductive adhesive films, ACF) have gained popularity as a potential replacement for solder interconnects. For ACF interconnection, thermo-compression (T/C) bonding is the most common method. In this study, the effects of the some important processing parameters, including the increasing rate of bonding temperature and different Teflon films, on the reliability of the ACF joints were investigated. Results show that the performances of the ACF joints were affected by the distribution of conductive particles and the curing degree of the ACF, which was determined by the bonding temperature ramp rates. The bonding strengths of ACF joints are different for the different Teflon film's thickness and kinds.

Introduction

Electronic packages nowadays are becoming smaller, lighter with higher in/out (I/O) count and better performance that are more cost competitive. The trend in electronic packaging runs from bulky plastic ball grid array (PBGA) with solder joints to miniature flip chip with polymer-based conductive adhesive interconnects. Meanwhile, the environmentally friendly manufacturing is another most important goal for the electronic packaging industry. In particular, the use of electrically conductive adhesives (ECA) instead of soldering and underfill encapsulation helps to achieve such goals [1-4]. The ECA mainly consist of an organic/polymeric binder matrices and metal filler. The conductive fillers provide the electrical properties and the polymeric matrices provide the physical and mechanical properties. Therefore, electrical and mechanical properties of ECA are provided by different components, which is different from the case for metallic solders that provide both electrical and mechanical properties. Among all the electrically conductive adhesives materials, the anisotropic conductive adhesives (ACA) (or anisotropic conductive adhesive films, ACF) have gained popularity as a potential replacement for solder interconnects in surface mount technology processes. The interest in using ACA instead of solder, comes partly from the fact that the use of ACA for the direct interconnection of flipped silicon chips to printed circuits (flip chip packaging), offers numerous advantages such as reduced thickness, improved environmental compatibility, lowered assembly process temperature, increased metallization options, reduced cost, and decreased

equipment needs. ACAs have been widely used for packaging technologies in flat panel displays (FPDs) such as liquid crystal displays (LCDs) for last two decades. So far, various packaging technologies such as tape carrier package (TCP) on LCD panel or PWB, chip on flex (COF) and chip on glass (COG) using ACAs have been realized to meet the requirement of fine pitch capability and make the flat panel displays smaller, lighter and thinner for high performance consumer products [6-12].

The principle of ACA joints are that the electrical connections are established through conductive particles and the mechanical interconnections are maintained by the cured adhesive. For example, Fig. 1 shows a typical chip-on-glass (COG) process based on ACF's thermo-compress interconnection techniques. First, an ACF is laminated to a glass substrate with ITO (indium tin oxide) tracks. Pressure and temperature are applied during the lamination process to ensure positioning accuracy, uniformity, etc. Then, the bumps on integrated circuits (ICs) are aligned with the tracks on the glass. Finally, the IC chip is pressed onto the glass at a specified high temperature and pressure. The conductive particles are trapped between the bumps and tracks, while the adhesive resin is squeezed out. The interconnections are established by the compressive force between the electrodes due to the shrinkage of the adhesive after curing. Consequently, the electrical conduction is restricted to the z-direction and the electrical isolation is maintained in the x–y plane.

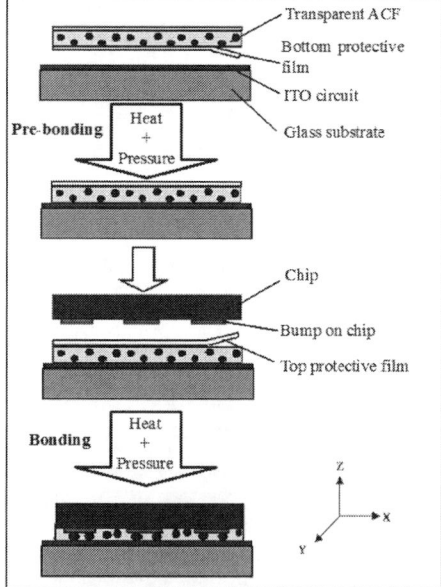

Fig. 1 Interconnection process for a COG assembly

In this study, the PI film, Suneast pad and ShinEtsu silicone were tested with three different thicknesses. The effects of Teflon film thickness changes on the rate of heat-

978-1-4244-2739-0/08/$25.00 ©2008 IEEE

up during the bonding progress were investigated by experiments. The bonding strengths of the ACF joints with different processing parameters and the damage of conductive particles were analyzed.

Specimen preparation and experiment

Specimen preparation

A commercial Hitachi anisotropic conductive film, ANISOLM AC-7106-25, is used to make the electrical interconnection between electrolysis Tin coated copper pads on a polyamide flexible circuit and ITO glass plate substrate. The ACF used in this study is 25 lm in thickness and 1.5 mm in width. The ACF is made of insulating epoxy resin in which nickel and gold plated polymer particles are dispersed. The particles are 5 μm in diameter. The specifications of ACF are summarized in Table 1.

Table 1 Specifications of the ACF

Description	Specification
Film thickness (μm)	25
Film width (mm)	1.5
Conductive particle	Au/Ni coated polymer
Particle size (μm)	5
Pre-bonding temperature (°C)	80 ± 10
Pre-bonding time (s)	5
Pre-bonding pressure (MPa)	0.1
Bonding temperature (°C)	180
Bonding time (s)	18
Bonding pressure (MPa)	0.15
T_g (°C)	145

The ACF interconnection is a blank ITO (In$_2$O$_3$: 90%; SnO$_2$: 10%) deposited glass and flexible printed circuit (FPC) with patterned metallization I/O pads. The metal pad on flexible film consists of approximately 35μm thick copper and 2 μm electrolytically plated Tin. The bonding width of the specimen is 2.1 mm and the length is 7.4 mm. The structure and the dimensions of the specimens are shown in Fig. 2.

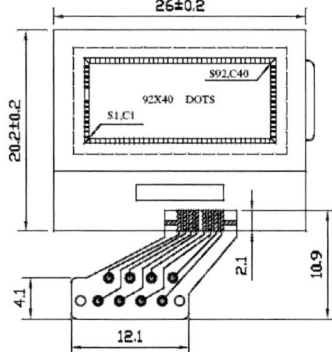

Fig. 2 The geometry and the dimensions of the specimens

Specimen fabrication

The equipment used in the fabrication of the specimens is TCW-125 bonder. The bonding temperature, bonding pressure and curing time of the specimens can be adjusted directly by the bonding machine as shown in Fig. 3.

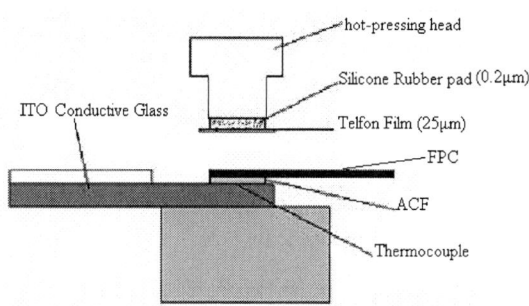

Fig. 3 Specimen production

In order to level off the ACF and heat head, the equalizing Teflon film is used between ACF and heat head when bonding is processing. The Teflon film used in this study are PI film, Suneast pad and ShinEtsu silicone, and their thicknesses are 0.05mm, 0.1mm and 0.2mm, respectively. Before the specimens are fabricated, the thermocouple is used to probe the bonding temperature amp. The different thickness of the Teflon film results in various bonding temperature ramp. The heat-up time is 6s, 9s, and 12s, respectively.

First, the pre-bonding of ACF joints was carried out on the pre-bonding machine, OCF-4000A. The pre-bonding tempera-ture was 80 ℃, and the pre-bonding pressure was 0.1MPa. Then, anisotropic conductive film was cut to the size corresponding to FPC automatically, while the upside and downside protective film of anisotropic conductive film was removed. After pre-bonding, specimen was sent to the bonding machine, TCW-125. The thickness of silicone rubber pad is 0.2 μ m. The bonding pressure and temperatures were provided at Table 2. While bonding is processing, FPC circuit with the circuit of conductive glass should be firstly arranged in the right place, and ensure that no dislocation lines existed by 50 times microscope. The adjustments of the bonding temperatures must be achieved by thermocouple, and make sure temperature increasing rate and bonding curing time unchanged. The thermocouple can measure the bonding actual temperature, shown in Fig. 3.

While producing the specimens, bonding machine setting was formulated according to the experimental adhesive parameters, and then the adhesive parameters requirements of the specimens was achieved, and individually marked and recorded. The specimen production process parameters are shown as Table 2.

Observation of specimen conductive particles and measurement of adhesive strength

Once the specimens were prepared, the bonding strength was measured by the 90°-peel tests. The 90°-peel tests were carried out by TESTOMETRIC tensile test machine, as shown in Fig. 4. Tensile speed was 1 mm/min. To reduce the effects of the angle changes on peel strength in the tensile progress, it must be ensured that the bonding surface is vertical with tensile soft band. When peeling, the tensile angel changes must be in 1.4°. The photographs of conductive particles after bonded were taken by optical microscopy. Then, the damage

situation of external insulating layer of the conductive particles can be obtained.

Table 2 Bonding process parameters

Teflon film	Temperature (°C)	Time(s)	Pressure (MPa)	Teflon thickness(mm)	Specimen number
PI film	180	22.5	0.15	0.05	12
Suneast film	335	22.5	0.15	0.1	12
ShinEtsu film	295	22.5	0.15	0.2	12

Fig.4 Configuration of the special clamp installation

Experiment results and analysis

Because the existence of Teflon film, the rising speed of temperature and conductive adhesive bonding time will be influenced. These two parameters may also affect the damage of particles' external insulation and the adhesive strength. In order to investigate the effects of the different Teflon film on the reliability of ACF joints, three groups of specimen 90°peel strength tests have been carried out, and the test results shown in Fig.5. It is obvious that the bonding strength used PI film is the highest, and the average bonding strength is 1.1MPa. The adhesive strength of ACF joints used Suneast and ShinEtsu films were very close, and their average bonding strength were 0.83MPa and 0.81Mpa respectively.

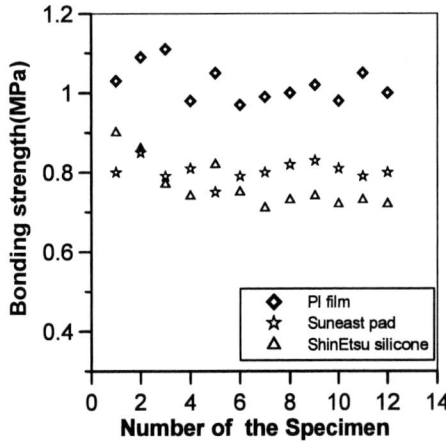

Fig. 5 Results of 90°peel strength test

The different thickness of the Teflon film results in various bonding temperature ramp. The heat-up times were measured as 6s, 9s, and 12s for the PI film, Suneast pad and ShinEtsu silicone, respectively, and the temperature curve is shown in Fig.6. This is because the different thicknesses of Teflon films lead to not only the change of temperature, also change in the flow velocity of ACF Colloidal and its curing degree. As a result, the bonding strength varies. Furthermore,

as the bonding time was constant, the increasing of heating rate causes curing time decreases. As shown in Fig.7, the relationship of the curing time and temperature ramp, the faster heating up, in other words, the steeper Plot of ACF bonding heat-up curves, the better curing situation cause the specimen bond strength higher, so the bonding strength of PI film is the highest. The heating-up rate of Suneast film is close to ShinEtsu film, so their bonding strengths are near.

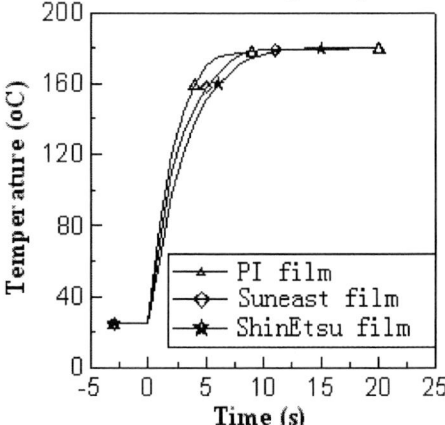

Fig. 6 Plot of ACF bonding heat-up

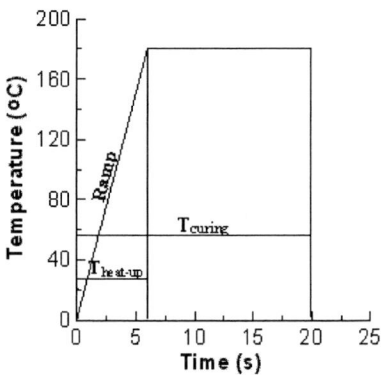

Fig.7 The relationship of the curing time, heat up time, and temperature ramp

Although the variation of Teflon film thickness can level off the bonding interface, it has a great influence on the rate of heat-up. At the same time, different harnesses of the Teflon films can also impact deformation degree of conductive particles. As shown in Fig. 8, the varieties of conductive particles were observed by optical microscopy. In Fig. 8 (a), for the hardness of PI film is moderate, the damage extent of external insulation layer of conductive particles was better. The hardness of SunEast film material is lower, so the external insulation layer of conductive particles wasn't been broken yet, shown in Fig. 8 (b). Because the hardness of ShinEtsu film is higher, the damage of external insulation layer of conductive particles was serious, shown in Fig. 8 (c). Through the experimental results, the best thickness of Teflon film is 0.05mm, and its hardness should be moderate. Therefore, PI film is the best of Teflon film material in the three experimental materials.

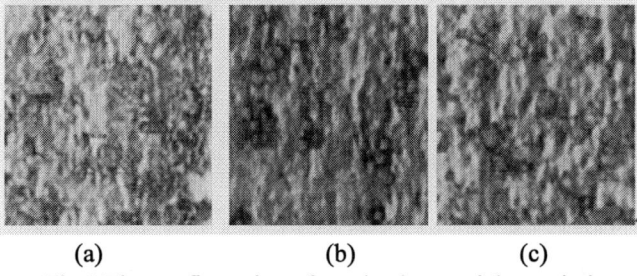

| (a) | (b) | (c) |

Fig 8 The configuration of conductive particles varieties for different the Teflon films. (a) PI film (b) SunEast pad (c) Shin silicone

Conclusions

Results of the peeling tests show that the bonding strength of PI film is the highest, while Suneast film and ShinEtsu film is lower and very close. Different thicknesses of Teflon films can influence the heat-up rate and the bonding curing time while the anisotropic conductive film is used. The variety of temperature increase ramp will affect the curing time of the ACF, and then affect its bonding strength. When bonding, temperature increase ramp of PI film specimen was six seconds, so it could meet the bonding requirements. The thickness of Teflon film needs to be no more than 0.05mm. The different hardness of Teflon films will influence the deformation degree of conductive particles while bonding, the hardness of PI film is moderate in the three kinds of experimental materials. It is the best Teflon film material.

Acknowledgments

This work was supported by China Postdoctoral Science Foundation (Grant No.20070410302), the Postdoctoral Science Foundation of Central South University, and 973 Program (Grant No. 2003CB716202).

References

1. Li, Y., Wong, C.P., "Recent advances of conductive adhesives as a lead-free alternative in electronic packaging: Materials, processing, reliability and applications," *Materials Science and Engineering R*, Vol. 51 (2006), pp.1–35.

2. Lin, Y.C., Zhong, J., "A review of the influencing factors on anisotropic conductive adhesives joining technology in electrical applications," *Journal of Materials Science*, Vol. 43, No. 9, (2008), pp. 3072-3093.

3. Lau, J., Wong, C.P., Lee, N.C., Lee, S.W.R., <u>Electronics Manufacturing: with Lead-free, Halogen-free, and Conductive adhesive Materials,</u> McGraw Hill (New York, 2002).

4. Lu, D., Wong, C.P., "Conductive adhesives for solder replacement in electronic packaging," *Proc the 2000 International Symposium on Advanced Packaging Materials*, 2000, pp. 24–31

5. Chan, Y.C., Luk, D.Y., "Effects of bonding parameters on the reliability performance of anisotropic conductive adhesive interconnects for flip-chip-on-flex packages assembly. I. Different bonding temperature", *Microelectr. Reliab.*, Vol. 42, No. 8, (2002), pp.1185-94.

6. Lai, Z.H., Liu, J., "Anisotropically conductive adhesive flip-chip bonding on rigid and flexible printed circuit substrates", *IEEE Transactions on Components, Packaging and Manufacturing Technology, Part B: Advanced Packaging*, Vol. 19, No. 3, (1996), pp. 644-60.

7. Lin, Y.C., Chen, X., Wang Z.P., "Effects of hygrothermal aging on anisotropic conductive joints: experiments and theoretical analyses," *J. Adhesion Sci. Technol.*Vol. 20, (2006), pp. 1383-1399.

8. Lin, Y.C., Chen, X., Zhang, H.J., Wang Z.P., "Effects of hygrothermal aging on epoxy-based anisotropic conductive film," *Mater. Lett.* Vol.60, (2006), pp. 2958-2963.

9. Wu, C.M.L., Liu, J., Yeung, N.H., "The effects of bump height on the reliability of ACF in flip chip", *Soldering & Surface Mount Technology*, Vol. 13, No. 1, (2001), pp. 25-30.

10. Cao, L.Q., Li,S.m., Lai, Z.h., Liu, J., "Formulation and Characterization of Anisotropic Conductive Adhesive Paste for Microelectronics Packaging Applications," *Journal of Electronic Materials*, Vol. 34, No. 11, (2005), pp. 1420-1427.

Reliability Study of Flexible Display Module by Experiments

Quayle Chen, Leon Xu, Antti Salo
Nokia Research Center / Beijing
7th floor, Nokia Building 2, NO.5 Dong Huan
Zhong Lu, BDA, Beijing, 100176, PR.China
Quayle.chen@nokia.com,
Tel: +86 10 87112566 Fax: +86 10 87114754
Gustavo Neto, Germano Freitas
Instituto Nokia de Technologia (INdT/Manaus)
Avenida Torquato Tapajós, 7200, km 12
Bairro Colônia Terra Nova, Manaus – AM
CEP 69093-415, Brasil

Abstract

Flexible display module reliability were investigated herein with experiments, such as bending, twisting and ball drop. The pretests of all the three experiments were carried out firstly to primarily understand the flexibility and mechanical behavior of the display. Based on the pretest results, the corresponding fatigue test setup method and process were put forward. Then, the fatigue tests were executed. At last, through the failure analysis, the flexibility and reliability of the flexible display in different use cases were evaluated. Suggestions about how to use the display and improve the reliability through change the design were given also.

Introduction

Flexible Display, with the tremendous requirements in portable device industry and their advantageous features in comparison with the conventional glass based flat panel displays, such as weight, thickness, ruggedness and so forth, is now attracting an enormous interest research both in university and industry. [7] Currently, many different applications can be found in the industry. Especially in e-paper application, many large companies, such as Fujitsu, Panasonic, Philip etc, had already developed the corresponding products. While with the display thinner and flexible, the reliability issue is uncovered. To solve this issue, some experiment test methods were studied, especially the bend-testing system and analysis methodology. [4,9,10] Most of the flexible display reliability studies are focused on the functional impaction with the substrate bending. That's with the ITO layer or transparent conducting substrates bended, the changes of electronic current and voltage result in different display effect. [1,2,3,6,8] While the flexible display mechanical performance varies significant with the different realization methods, fabrication process, material and applications. [5] It is also significant different with the mechanical performance of single ITO or substrate film layer. This paper is aim to develop a methodology to study the reliability of flexible display module, as well as evaluate one kind of new STN display that will be applied on our products.

For the practical application, one kind of STN display is considered. In order to understand its mechanical reliability, as well as considering the practical usability, new fatigue test processes were developed based on the pretest of the display. Normally, bending, twisting and ball drop test are the basic methods to study the mechanical performance of the glass based display. In this paper, the three testing methods are also taken into the study. In order to get the basic behavior of the STN display, a slowly static loading to failure process was put forward, as well as define the test setup based on the analysis of the practical display architecture. Based on the test results, the process was revised. For bending test, three cases that with the different bending center line location along screen longer side, shorter side and diagonal were considered. For twisting test, both the clockwise twisting and anti-clockwise twisting are involved in one cycle test. In order to get the flexible display point impact performance, ball drop tests are taken in two cases with different sample support methods, partial support and full solid support. For each case, one sample is prepared.

Pretest and Process

As Figure 1 shown, the flexible STN display module includes the screen, IC driver and flexible printed circuit. There are lot of copper traces on the edge of the screen that provide the electric signal between display and driver. Three points bending and four points bending are the normal experiments used much more in the industry. Considering the practical application of the display, three points bending are more close to the practical application case than four points bending. Therefore, three points bending was taken in the test. According to the display architecture, the display will be supported by two rollers with the screen face downwards. The load roller is located between the two support rollers. Figure 2 shows the test setup. In order to control the sample movement in the horizontal plane, two stoppers are applied to stop the display movement during the test. With the load roller move down, the display is bended. To deeply and roundly understand the bend-ability of the display, different bending center lines are considered. That's bending along with the screen longer side, shorter side and diagonal.

Figure 1 STN Display Architecture

Figure 2 Three points bending test setup

The load condition in the pretest is that static slowly bend to failure step by step. In every increment step, the load roller move down one mini-meter. After completed one increment step and the display will be powered on to check its function. If failure detected, then stop the test and record the displacement of the load roller. Three samples are separately used in the three test cases.

In case of bending along screen longer side, the functional failure occurred when the load roller moving down to 10mm. So in the practical cyclic test, we set the bending distance to 6mm. It is also the limitation of display bending in the practical application cases. For bending along screen shorter side, the failure occurred at the bending distance equal to 2mm. It's the IC driver crack in the component center where is the cross between the component and the bending center line. The IC driver is narrow and layout along the screen shorter side and it is brittle. It shows that to bend the display along screen shorter side doesn't make sense. So we decided to give up the fatigue cyclic test of this bending. It is better to provide protection of this component during the application, fabrication or design process. For bending along diagonal, there is no any functional failure found after bend to 13mm.

In fact, the display mostly will be bended along the screen longer side, and there is no much more opportunity for user bending the display along the shorter side and diagonal. While for the display component reliability study, we also had done the bending test with these two cases. So the reliability study mainly focused on the test of bending along screen longer side. In the practical application, the most seriously bending along the screen longer side is not in the center of the display, but the interface between screen and the FPC. For this reason, we use all the samples that plan to be

bended along screen shorter side in the bending along screen longer side with the bending center line located in the interface between the screen and FPC.

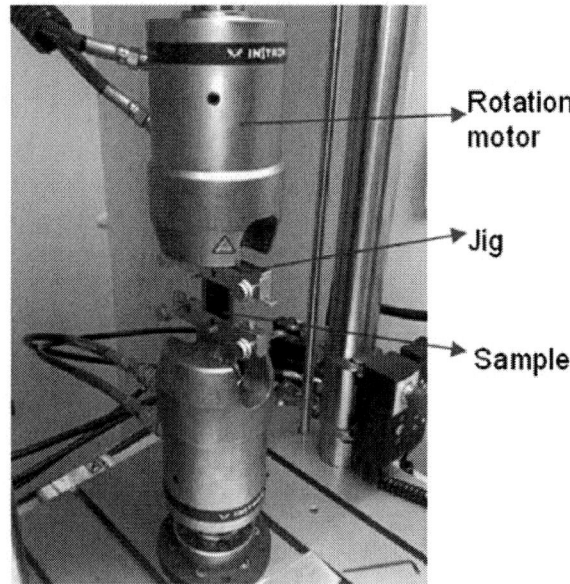

Figure 3 Twisting test setup

As Figure 3 show the twisting test setup. Two jigs are used to hold the display. The top jig is fixed with a rotational driver and the bottom jig is fixed on the ground. Only the top jig can rotate with the Z axis. With the jig rotation, the display will be twisted in clockwise and anti-clockwise.

In the pretest, one loading step includes a completely cycle with twisting angle change from 0^0, 2^0, 0^0, -2^0, 0^0. And in every step to increase 2^0, so that in the next step the twisting angle will change with 0^0, 4^0, 0^0, -4^0, 0^0. Figure 4 shows the load profile. The cycle time is 0.5 seconds and the frequency equal to 2Hz. During the test, when the twisting angle increased to 30^0, the failure was detected. Actually, the display will not be twisted with more than 10^0. Consequently, the twisting limitation angle is set to 20^0 in the cyclic tests. And the load profile is sinusoid with color of cyan indicated in Figure 4.

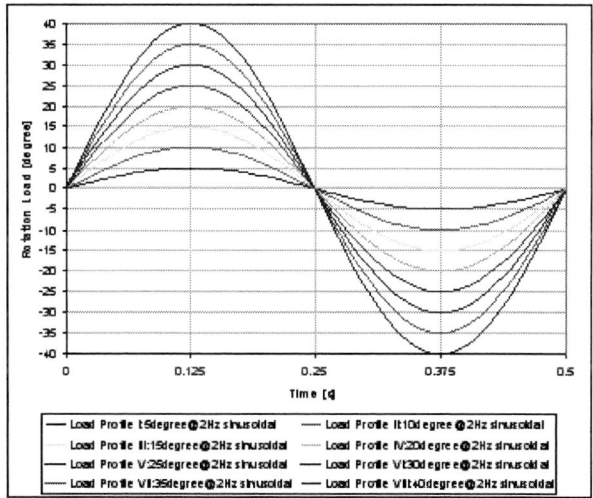

Figure 4 Load profile of twisting

Ball drop test is a mainly method to study the reliability with the pressure stress concentration. It is mostly used on the glass based display to evaluate the lens anti-drop capability and performance in the industry. For the flexible display, this method was taken here also to preliminary study the corresponding reliability. For this study, two extreme sample support conditions were considered. One is support sample with solid steel block called full support. The other is that to support the sample with two rollers called partial support. A steel ball with the diameter of 12mm and weight of 7.94g was taken for the drop. Figure 5 shows the experiment setup.

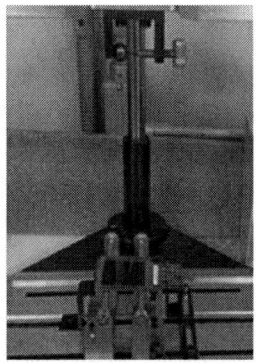

(a) Solid steel block support   (b) Two rollers support

Figure 5 Ball drop test setup

In case of drop test with full support, ball drop from 40mm-70mm height will damage the flexible display. While in partial support, the drop height can be increased to 670mm while not destroy the display during the impaction. It's very clear that the display will be seriously damaged by the ball in case of full support. While, in case of partial support, the dropped ball will be rebounded by the display because of the flexibility of the display. The system of the display and the support rollers acts as a spring. During the test, we found that the impact point of ball drop can't be centralized in one point at every time. The error is limited within 5mm. For improving the accuracy of impact point, a magnetic control system has been put forward to avoid the impaction of the experiment operator.

Ball drop test is different with the fatigue test. It results in the stress concentration instantaneously. However, the fatigue test is another method that results in the stress concentration by accumulating. In this study, a united test method that integrated the two ways is set forward. The drop impact point is set on the center of the screen. The initial drop height is 30mm. and in every drop height, we will drop ball three times, then increase the drop height with 10mm. The maximum drop height is set to 160mm. If no functional failure found, then stop the test. Because the ball drop test with rollers support is not the really pressure stress concentration, it is meaningless with this kind of tests. So we give up it.

Bending Fatigue Test

Figure 6 and 7 shows load profile and the sample location method of the bending fatigue test respectively. In order to reduce the cycle time of whole test without any great impaction to the test, the unloading speed is set to double of the loading speed. The support span is 30mm. According to the practical application, the maximum cycle is set to 20000 times. And the sample checking starts from 10000 cycles to reduce the whole cycle time. The frequency of sample checking is once per 500 cycles. Both include the appearance checking and functional checking should be done. Because the flexible display is also one kind of liquid crystal display, it is very difficult to define the failure in most of the time. The liquid crystal will move in the display inner frame with the deformation of the display. So there are some feints of failure because of this movement when powered the display on. After the load is removed or the change of the environmental, the failure can disappear. For this reason, the failure check rules are set as follow: If only the appearance failure detected, don't stop and continue the test as well as record the appearance failure mode. Once the functional failure detected, stop the test and record the phenomenon and state. This rule is also used in the fatigue test of twisting and ball drop.

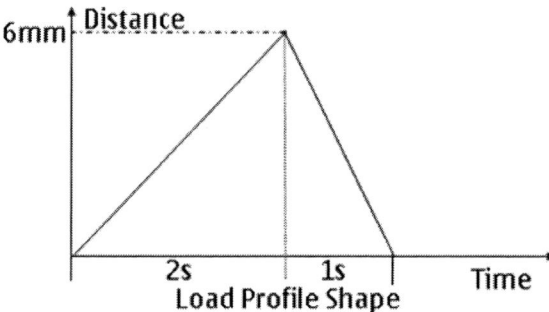

Figure 6 Load profile of bending cyclic test

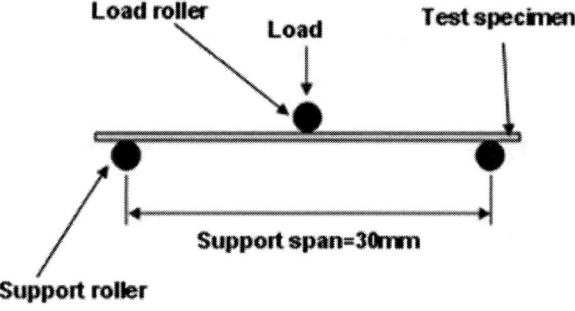

Figure 7 Sample location of bending cyclic test

Seven samples had been used to the bending fatigue test along the screen longer side. All the samples had failed during the 20000 cyclic tests. As Table 1 shown, almost all the samples are functional failed after 11000 bending cycles. And the failure modes of all the samples are the same, as Figure 8 demonstrated the typical failure mode. There is a black line on the display when powered it on. It indicates that the root cause of all the failures may be the same.

Table 1. Survive cycles of the sample during the test

Sample	Number of Cycles
Sample -273	$10500-11000$
Sample -274	$10000-10500$
Sample -276	$10000-10500$
Sample -275	$0-10000$
Sample -269	$10000-10500$
Sample -277	$10000-10500$
Sample -269	$11000-11500$

Figure 8 Functional failure modes

The second bending fatigue test we had done is to bend the display along longer side. The bending line is located nearby screen and FPC connection place. Figure 9 shows the bending center line and setup. It is different with the bending along longer side in the screen center. In this setup, there is only one support roller. And the display will be fixed at the FPC site. With the load roller moving down, the display will be bended.

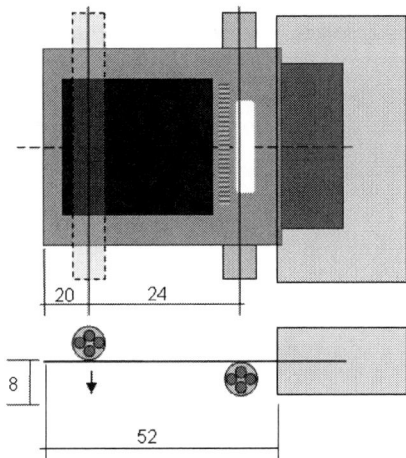

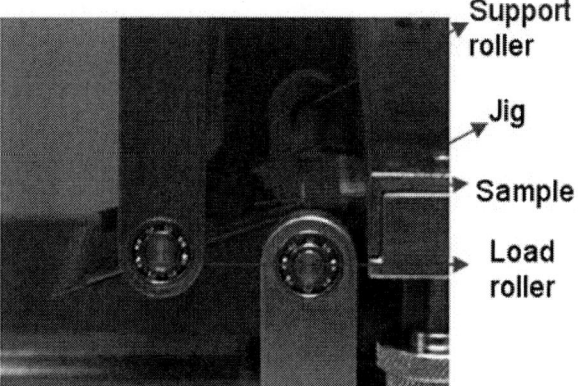

Figure 9 Bending at the connection area setup

The distance between the two rollers is 24mm. The maximum displacement of load roller is 8mm. The test frequency and other parameters are the same as the above bending fatigue test along screen longer side.

`

Five samples had been done this kind of fatigue test. As Table 2 shown, all the samples haven't any damage after the bending of 20000 cycles.

Table 2. Survive cycles of sample during the test

Sample	Number of Cycles
Sample 279	$S-20000$
Sample 290	$S-20000$
Sample 285	$S-20000$
Sample 257	$S-20000$
Sample 267	$S-20000$

The third bending fatigue test is to bend sample along the screen diagonal, as Figure 10 shows the setup. It is the same as the bending along screen longer side, just rotate sample with 45 degree in the horizontal plane. The bending process is also the same as bending along screen longer side.

Figure 10 Bending along Diagonal setup

Five samples had been used for bending along screen diagonal. The test shows that all the samples can survive the cyclic bending about 20000 cycles as Table 3 shown.

1089

Table 3. Survive cycles of bending along diagonal

Sample	Number of Cycles
Sample 291	S – 20000
Sample 266	S – 20000
Sample 289	S – 20000
Sample 287	S – 20000
Sample 270	S – 20000

Twisting fatigue test

The twisting fatigue test setup hasn't change with the pretest both include the sample location and load profile. The sample checking process followed the bending test along screen longer side. Nine samples had been used for the twisting fatigue tests. The survived cycles of every sample shows in Table 4.

Table 4. Survive cycles in twisting fatigue test

Sample	Number of Cycles
Sample 267	15000 – 15500
Sample 254	S – 20000
Sample 253	S – 20000
Sample 260	19000 – 20000
Sample 263	S – 20000
Sample 262	S – 20000
Sample 281	12000 – 13000
Sample 264	13000 – 14000
Sample 250	S – 20000

It indicates that 5 samples from 9 (55.56%) can survive during 20000 cycles fatigue test. Two samples failed before 15000 test cycles, two samples failed after 15000 test cycles. The failure modes are also the black line when power the display on. While the difference of failure mode between bending test and twisting test are the line position and direction. There is no consistency about the failures from the twisting tests. The black line direction may be horizontal or vertical, and the position may be in the screen center or edge.

Ball drop test

Only the ball drop test with full support had been done. Eight samples had been taken for this test. The Experiments tell that only 2 sample can survive in the test. As Table 5 shown, five samples from eight had failed during the drop height increased to 40mm, one sample failed during the drop height increase to 150mm. The test indicates that the performance of samples varied from sample to sample. As Figure 11 shown, the failure mode are the broken lines on the impact point that appear when power the display on. For the 2 survived samples, there is no functional failure, while there are serious appearance failures. These are not only the movement of liquid crystal, but material damage of ITO.

Table 5. State of failure detected in ball drop test

Sample	Drop height & cycle times
Sample 286	150mm, 2
Sample 255	40mm, 2
Sample 256	40mm, 3
Sample 252	40mm, 2
Sample 271	160mm, not fail
Sample 268	40mm, 1
Sample 294	160mm, not fail
Sample 259	40mm, 2

Figure 11 Failure mode of ball drop test

Failure analysis

As mentioned above, most of the failure is the black line, especially in the bending tests. According to the architecture of the display, three of four sides of screen layout many copper traces to contact with the display and driver. So the first suspicion of the potential root cause maybe the connection failed during the tests. That may result in the disconnection of electronic signal between display and driver. To verify this, finger push test had been done. That is to use the finger to push the corresponding copper trace place where may result in the black line failure to see if the failure can be remove or repeat. As Figure 12 shown, when we press the copper trace at the side of the display, the failure can appear or disappear. So in other words, the displays haven't physical failure, it is only the connection broken. For failures from ball drop test, the failure mode can't be changed whatever we press the copper trace. The physical material damage occurred on the screen because of stress concentration. The electrical signal is broken at the impact area. So broken line can be detected when power the display on.

Figure 12 Finger push test

Conclusions and Discussion

Through the pretest, the corresponding fatigue test setup and process had been developed. Through series fatigue tests, the flexible display reliability had been evaluated. This STN display is suitable for applied in bending along screen longer side application cases. The connection of copper trace with screen should be improved during the display fabrication, design or application. It is suggested to use high performance glue or add component to protect the FPC copper trace peel off from the display. It is better to take some action to protect the IC driver during the application or display fabrication. The display can not support instantaneously shock on the screen. All the test results are based on the specified flexible display and finite samples. More samples are necessary for deeply reliability study. And the experiment process or methodology can be used for the corresponding study with the similar cases.

Acknowledgments

The experiments had been done in Institute of Nokia Technology in the end of last year and beginning of this year. The authors would like to thank Joaci, Luciano, Cleber etc engineers in Institute of Nokia Technology for providing experiment convenience and helpful discussion. At the same time to appreciate Tommi Reinikainen for great support.

References

1. S.P. Gorkhali, D.R. Cairns, G.P. Crawford, "Reliability of transparent conducting substrates for rollable displays: A cyclic loading investigation ", *Journal of the Society for Information Display* Vol. 12, No. 1 (2004), pp. 45-49. [A reference to a journal article ...]

2. Jack Hou, Yajuan Chen etc, "Reliability and Performance of Flexible Electrophoretic Displays by Roll-to-Roll Manufacturing Processes", *Society of Information Display, 2004 DIGEST,* pp. 1066-1069. [A reference to a journal article ...]

3. G.M. Danner, J. Atkinson etc, "Reliability Performance for Microcapsulated Electrophoretic Displays with Simulated Active Matrix Drive", *Society of Information Display, 2003 DIGEST,* pp. 573~575. [A reference to a journal article ...]

4. Sonia Grego, Jay Lewis etc, "Development and evaluation of bend-testing techniques for flexible-display", *Journal of the Society of Information Display,* Vol. 13, No. 7 (2005), pp. 575-581. [A reference to a journal article ...]

5. Peter J.Slikkerveer, "Bending the display rules: Options and Challenges for Flexible Displays", *EURODISPLAY* (2002). pp. 273-276. [A reference to a presentation at a Conference...]

6. Darran R. Cairns, Victoria L. Shier etc, "Mechanical Reliability of Indium Tin Oxide Electrodes on Polymer Substrates for Lightweight Flexible Display", *www.sidmembers.org - /proc/ASID2000/PA-24.pdf.* [A reference to a website article ...]

7. T. Sakai, A.J.J. van der Horst etc, "A Study on Roll-to-Roll Method for Flexible TFT Backplane Manufacturing", 137-140, *International Display Workshops (2007).* [A reference to a conference article ...]

8. G.V.Brodovoy, Yu.V Kolomzarov etc, "Improvement of the performance reliability of cockpit TN LCDs", 24~29, *Society of Information Display.* [A reference to a article at a technical society...]

9. Alexander Ptchelintsev, "Fatigue Analysis and Optimization of Flexible Printed Circuits", *2006 ABAQUS User's Conference*, pp. 405-416. [A reference to a conference article ...]

10. Yves Leterrier, Piet Bouten, Xin Jiang, "Layer mechanics Experiment methods and models" *FLEXled-epfl-0209-002/Leterrier*, September 2002. [A reference to a research report...]

Thermal Fatigue Life Analysis and Forecast of PBGA Solder Joints On the Flexible PCB Based on Finite Element Analysis

Huang Chunyue
School of Mechanical & Electrical Engineering
Guilin University of Electronic Technology, Guilin 541004, China

Abstract

Thermal fatigue life of PBGA (Plastic Ball Grid Array) solder joint on the FPC (Flexible Printed Circuit) was analysed based on finite element analysis. According to symmetry theory, a quarter finite element model of 144-PIN PBGA was established. Both the stress and strain of lead and lead-free PBGA solder joints on the flexible PCB with the basic material of polyimide were studied by non-linear finite element analysis (FEA) respectively under -55~125°C thermal cycling. Based on the plastic strain calculated by FEA and Coffin-Manson formula, the thermal fatigue life of key lead and lead-free solder joints on the flexible PCB was calculated respectively. The results show that: (1) In the process of thermal cycling loading, the outside solder joint which is the farthest away from the inter-connect center of PBGA undergoes the largest alternating stress and strain; (2) Temperature has great impact on stress and strain in solder joint, and in the process of temperature remaining, the stress and strain has little change, however, it arise rapidly in the process of temperature decreasing while drops fast in heating-up period. (3) Plastic strain of dangerous point is accumulated in the process of temperature cycling, which will ultimately cause solder joint failure. (4) Compared with the lead-free solder joints, the lead ones of PBGA obviously have the longer fatigue life.

1. Introduction

The Flexible Printed Circuit Board (FPC) is a kind of copper-clad laminates taking flexible insulation film as the base material and PCB production produced in a way similar with the one of producing the Printed board. It's light, thin, short, small, and flexible in structure. It can be static or dynamic flexural, as well as curly or folding, which means that it can be in accordance with any requirements of spatial arrangement, and move or telescopic in the three-dimensional space with no restrict to integrated component assembly and conduct joining. FPC is widely used in the filed of computers and the supporting systems, medical instruments, military affairs and space technology, consumer civilian industry products and cars, and so on. [1][2][3] The solder joint connecting the PBGA element packaged by SMT (Surface Mount Technology) and the flexible PCB(Printed Circuit Board) takes in charge of both electrical and mechanical connection. So its quality has a direct influence on performance and reliability of the products. Considering the possible fatigue failure of the solder joint caused by the different thermal expansion coefficient of the plastic packaging material, the joint, and the FPC, reliability of the PBGA solder joint under thermal cycling loading is essential.[4][5] In this paper, the 144—PIN PBGA solder joint was taken as the research subject, and a quarter finite element model of 144-PIN PBGA was established. According to the result of the FEA (Finite Element Analysis), we gained the position of key solder joints of PBGA on the flexible PCB under the thermal cycling loading, and the changing regularity of stress and strain. Then based on the Coffin-Manson formula, the thermal fatigue life of key lead solder joints and lead-free ones on the flexible PCB was calculated respectively. The results show that, compared with the lead-free PBGA solder joints, the lead ones obviously have longer fatigue life.

2. Finite element analysis of Stress and Strain in PBGA Solder Joint Under the Thermal Cycle

2.1 Finite Element Analysis Model of the PBGA Joint

2.1.1 Establishment of Geometric Model and Choice of Parameters

Structure dimensions of 144-PIN PLASTIC BGA (13x13) are showed in Fig 1.

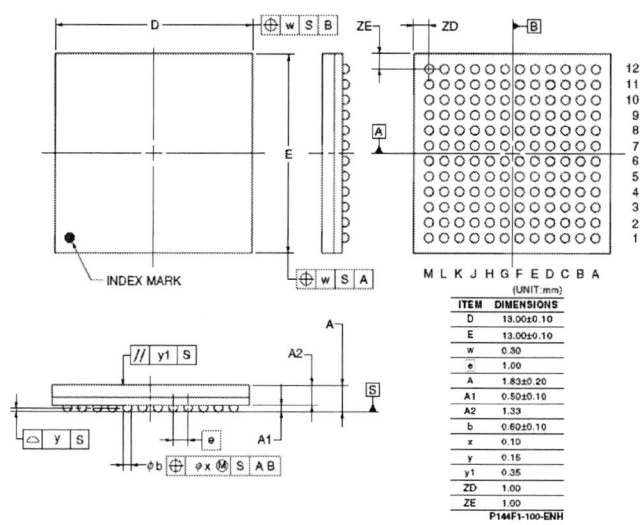

Fig. 1 144-pin plastic BGA (13x13) structure size

Taking symmetry theory into account, a quarter finite element model cut along the diagonal of 144-PIN PBGA was taken as research subject (Fig. 2).

A PBGA element consists of EMC, BT and solder joints. To simplify the model, influence of copper plated conduct, pad and inner Die was ignored. In the model, FPC, EMC, BT and solder are assumed as the material with isotropic properties. And only elastic deformation of the first three and viscoplasticity behavior of the last were taken into consideration. Table 1 and table 2 show their parameters.

978-1-4244-2739-0/08/$25.00 ©2008 IEEE

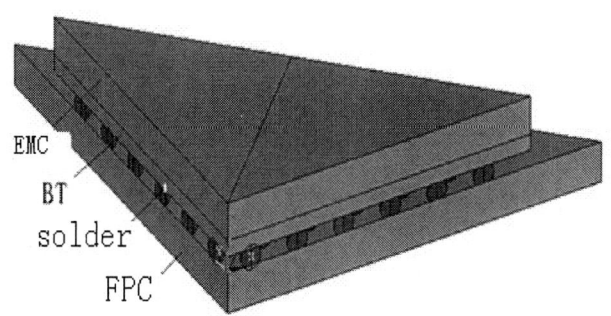

Fig. 2 PBGA simplified model

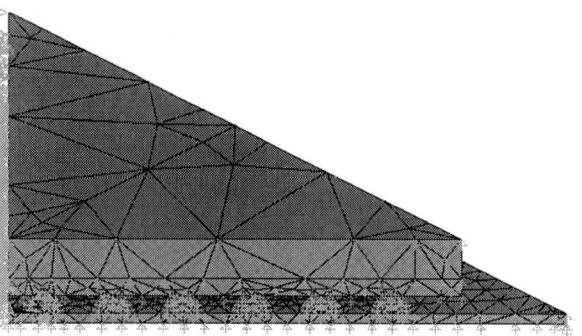

Fig. 3 Element model of PBGA

Table 1 Material parameters of PBGA components

material	young's modulu（Gpa）	Poisson ratio	Factor of expansion （ppm/℃）
EMC	15.435	0.250	15.0
BT	17.8	0.390	15.0
Polyimide	3.724	0.335	15.0
63Sn37Pb	34474-152t	0.35	21.0
96.5Sn3.5Ag	52708-67.14t-0.0587t2	0.4	21.85+0.02039t

Table 2 Material parameters of solder Anand sticky plastic model

ANSYS	material parameters	solder	
		63Sn37Pb	96.5Sn3.5Ag
C1	s_0 (Mpa)	12.41	39.09
C2	Q/R (°K)	9400	8900
C3	A (1/s)	4.0E6	2.23E4
C4	ξ	1.5	6
C5	m	0.303	0.182
C6	h_0 (Mpa)	1378.95	3321.15
C7	$\hat{s}$ (Mpa)	13.79	73.81
C8	n	0.07	0.018
C9	a	1.30	1.82

2.1.2 Establishment, constraint, load and solution of finite element model

VISCO107 Hexahedral elements proper for viscoplasticity matters are used to plot solder joints, and SOLID45 elements to the ship, base board, BT and MOLD. Fig 3 shows the finite element mesh model generated by mesh auto-generation, which was divided into 253652 elements including 3211 elements in flexible PCB, 316 in EMC, 4647 in BT, and 17186 in the solder joint. According to the symmetry theory and the loading, the constraint was applied as fig 3: symmetry constraint was applied on symmetric plane, and homonymic constraint on endpoints of the hypotenuse.

According to United States Military Standard MIL-STD-883, -55~125°C thermal cycling loading with rate of temperature rising or reducing at 36℃/min was applied on all nodes. It consists of three thermal cycling at 30min/cycle, and the temperature remaining at high or low point takes 10 minutes. Supposing there was no stress in the model at 125 ℃, fig 4 shows the curve of thermal cycling loading versus time.

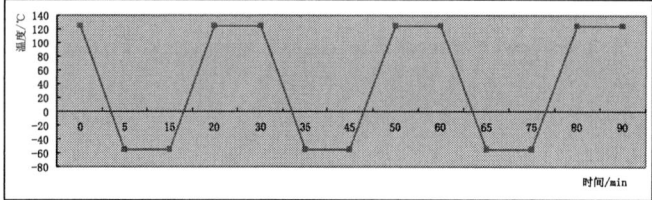

Fig. 4 Temperature cycle curve

2.2 Finite element analysis and results

2.2.1 Analysis of the position of the key solder joint

Fig 5 (a), (b) show respectively Von Mises plastic strain image of the model with lead solder joints, and the one with lead-free solder joints.

Fig 6 (a), (b) show respectively Von Mises stress image of the model with lead solder joints, and the one with lead-free solder joints.

Fig. 7 (a), (b) show respectively Von Mises plastic strain image of the lead solder joints and the one of the lead-free ones.

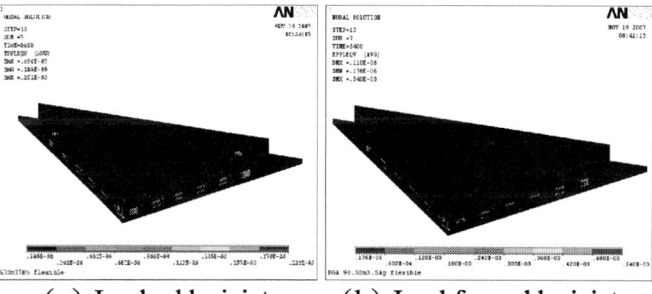

（a）Lead solder joint　　（b）Lead-free solder joint

Fig. 5 Plastic strain image of the PBGA element with lead solder joint and lead-free ones

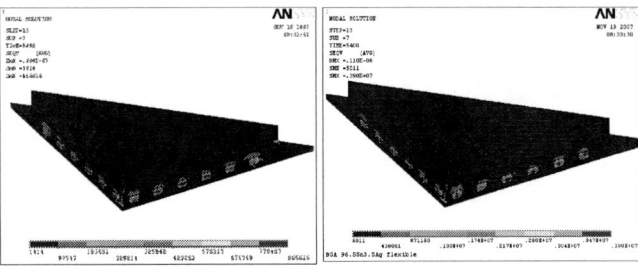

（a）lead solder joint   （b）lead-free solder joint

Fig. 6 Stress image of the PBGA element with lead
solder joint and lead-free ones

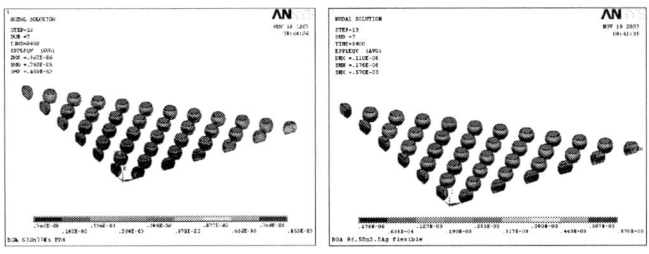

（a）Lead-free solder joint （b）Lead solder joint

Fig. 7 Plastic strain image of the lead solder joint and
the lead-free ones

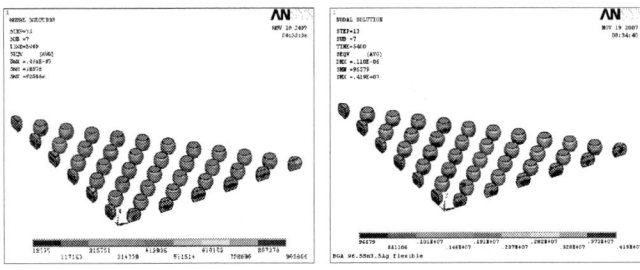

（a）Lead-free solder joint （b）Lead solder joint

Fig. 8 Stress image of the lead solder joints
and the lead-free ones

Fig. 8 (a), (b) show respectively Von Mises stress image of the lead solder joints and the one of the lead-free ones.

From the equivalent stress and plastic strain distribution, we got a conclusion: Because of the different thermal expansion coefficient of upper and lower substrates, in the process of thermal cycling loading, micro warp deformation in assembly FPC is caused in some degree and deformation in the edge of FPC is larger than in the center, which causes the largest stress and strain in the outermost solder joint. So the farther the solder joint from the center, the larger the stress and strain within it, and the outermost solder joint is the most dangerous.

In view of the conclusion above, the outermost solder joint will be chosen to be the search subject in the following study.

The maximum stress/stain of the lead solder joints and the lead-free ones after three cycles are shown in table 3, from which we know that, on FPC with polyimide as base substrate, compared with the lead PBGA solder joints, the lead-free ones obviously have the larger maximum stress and plastic strain.

Table 3 maximum stress and plastic strain of lead solder
joints and the lead-free ones

	solder	stress $(10^7 Pa)$	plastic strain (10^{-3})
Model one	63Sn37Pb	4.4631	1.0991
Model two	96.5Sn3.5Ag	2.9886	1.6139

2.2.2 Process analysis of variation of stress and strain

After identification of key solder joint and it's maximum plastic strain point, curves of stress and plastic strain versus loading time at this point was drawn by the time course of post-processor. Curves of stress and plastic strain versus loading time at the maximum plastic strain point of the lead solder joint are shown in fig 9 (a)、(b) and curves of the lead-free ones in fig 10 (a)、(b). [6]

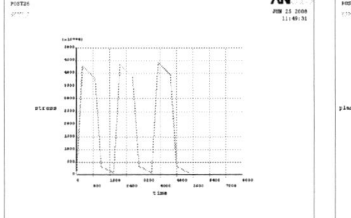 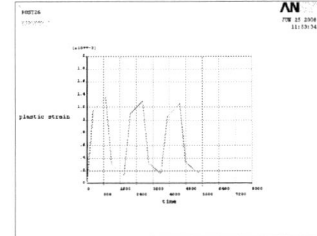

（a）Curve of von mises   （b）Cure of von mises
stress versus time   strain versus time

Fig. 9 Cures of stress and plastic strain versus time at the
maximum plastic strain point of the lead solder joint

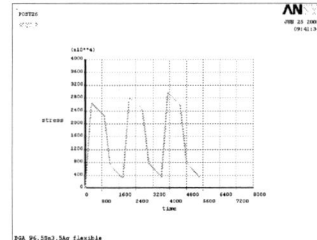 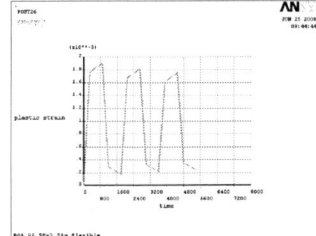

（a）Curve of von mises   （b）Cure of von mises
stress versus time   strain versus time

Fig. 10 Cures of stress and plastic strain versus time at
the maximum plastic strain point of the lead solder joint

From curves of von Mises stress versus time in fig 9 (a) and fig 10 (a), we got the conclusion: stress relaxation occurs in temperature remaining process. Stress response cures are extremely similar under thermal cycling loading, which means, in the three cycles, stress response of the dangerous point is relatively stable. At the end of the first cycle, residual stress of the lead-free solder joint is obviously bigger than the lead ones'.

From curves of von Mises plastic strain versus time in fig 9 (b) and fig 10 (b), we got the conclusion: under thermal cycling loading, plastic strain of the dangerous point is a cumulative process, which will ultimately cause failure of the solder joint. This is also called temperature ratcheting. At the end of the first cycle, plastic strain of the lead-free solder joint is obviously bigger than the lead ones', but, at the last two cycles plastic strain accumulation of the lead solder joint is bigger than the lead-free ones'.

2.3 Thermal fatigue life prediction of the PBGA solder joints

Based on the identified scope of the largest plastic strain and Coffin-Manson formula (equation 1), thermal fatigue life of the joint can be calculated. [7]

$$N_f = \frac{1}{2}(\frac{\Delta\gamma}{2\varepsilon_f'})^{(1/C)} \qquad (1)$$

Where N_f is the number of cycles to failure, $\Delta\gamma$ is shear strain range, which can be calculated by equation (2), ε_f' is fatigue ductility coefficient, C is fatigue ductility exponent which can be calculated by equation (3).

$$\Delta\gamma = \sqrt{3}\Delta\varepsilon \qquad (2)$$

Where $\Delta\varepsilon$ is strain range that can be obtained from hysteresis curves shown in fig 11(a)、(b). As shown in fig 11, hysteresis curves of the stress and strain tend towards stability at the first few cycles, and there is difference between the stable hysteresis curves and the initial ones, but not very obvious. Von mises plastic strain range at the third cycle is taken as $\Delta\varepsilon$.

$$C = -0.442 - 6\times10^{-4}Tm + 1.74\times10^{-2}\ln(1+f) \qquad (3)$$

$$Tm = \frac{1}{2}(T_{max} + T_{min}) = \frac{1}{2}(125 - 55) = 35$$

Where T_m is the average of the thermal cycling，f is the thermal cycling frequency. $f = 48(cycle/day)$ ， so $C = -0.395$

$\Delta\varepsilon$, as well as the fatigue life of the lead and lead-free solder joint, are shown in Table 4.

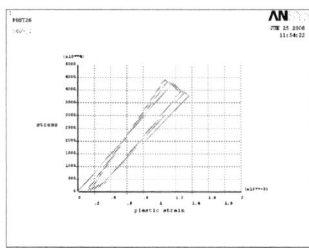 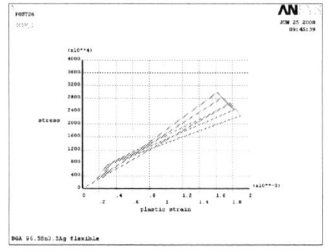

（a）Lead joints　　　（b）Lead-free joints

Fig. 11 Hysteresis curves at the maximum plastic point of lead joints and lead-free ones

Table 4 Thermal fatigue life values of solder joints

	solder	$\Delta\varepsilon$ (10^{-3})	Life (cycle)
Model one	63Sn37Pb	0.911338	2.0591e+6
Model two	96.5Sn3.5Ag	1.534745	5.5048e+5

3. Conclusion

(1) In the process of thermal cycling loading, the outermost solder joint of PBGA chip which undergoes the largest alternating stress and strain, is the most dangerous point and the focus of reliability analysis.

(2) Temperature has great impact on stress and strain within solder joints, and in the process of temperature remaining, the stress and strain has little change, however, it arises rapidly in the temperature decreasing period while drops fast in the heating-up period. Therefore in the temperature rising and dropping process, the stress and strain under the thermal cycling loading change sharply which causes the solder joints failure finally.

(3) Compared with the lead-free solder joints, the lead ones of PBGA on the FCB with base material of polyimide obviously have longer fatigue life.

Acknowledgments

This research is sponsored by the Science Foundation of Guangxi Zhuang Autonomous Region Government (Grant No. 0832083) and Director Project Found of GuangXi Key Laboratory of Manufacturing System & Advanced Manufacturing Technology (Grant No.07109008_011_Z_).

References

1. Cai Jiqing, "New Application and Material Technology of FPC", Printed Circuit Information, No. 3 (2004), pp. 41-45.
2. Chen Bing, Chai Zhiqiang, Flexible printed circuit technology, Science Press(Bei jing, 2005), pp.1-76.
3. Joseph Fjelstad, Flexible Circuit Technology. 2006. http://www.flexiblecircuittechnology.com/.
4. Liu Changkang, Zhou Dejian, Pan Kailin, "Thermal Fatigue Life Analysis on the PBGA Solder Joint", Journal of mechanical strength, Vol. 21, No. 3 (1999), pp. 212-214.
5. Huang Chun yue, Zhou Dejian, Li Chunquan, "The FEM Analysis of Stress and Strain in CCGA Solder Joint Under the Thermal Cycle", Journal of Guilin Institute of Electronic Technology, Vol. 21, No. 3 (2001), pp. 22-28.
6. Tan Changjian, Zhang Juan, Senior Engineering Application Examples of Ansys and Secondary Development, Publishing House of Electronics Industry(Bei Jing, 2006),pp329-344.
7. Engelmaier W. , "Fatigue life of leadless chip carrier solder joint during power cycling",IEEE CHMT,Vol. 6, No. 3 (1983), pp. 232-237.

This page intentionally left blank.

Author Index

A

An, Bing 129, 914, 964, 534
An, Rong 314
Andersson, Cristina 600
Aschenbrenner, Rolf 84
Azarian, Michael H. 1

B

Bai, Shuo 563
Bailey, C. 509
Bailey, Chris 250
Bao, Shengxiang 1059
Bei, Liu 826
Bian, Lifei 703
Bin, Liu 183
Biswas, Kalyan 299
Bo, Tao 885
Buchwalter, S. 577
Busby, J. 577

C

Cai, Jian 121
Cai, Miao 856
Cai, Xiong-Hui 129
Chai, Tc 299
Changshun, Jiang 1055
Chen, Fei 456
Chen, Guoqin 748, 751
Chen, Haibin 319
Chen, Hsin-Yuan 1040
Chen, Jin 394
Chen, Jing 660
Chen, Lih-Shan 540
Chen, Mingxiang 152
Chen, Quayle 1086
Chen, Su 748
Chen, Wei 804
Chen, Wenqing 461
Chen, Xi 740, 744
Chen, Xin 788
Chen, Xuefan 889
Chen, Y. R. 968
Chen, Yu-Ren 290
Chen, Zhengrong 944
Chen, Zhi 558
Chen, Zhuo 117
Chena, Qian 125
Cheng, Dongfang 1017, 1077
Cheng, Hung-Hsiang 157, 204, 295
Cheng, Ting 464
Cheng, Ting 493
Cheng, Xingming 679
Cheng, Yuan-Yuan 337
Cheng, Zhaonian 192, 523, 600, 725

Cheng, Zhong 774
Chiang, Kuo-Ning 23, 279
Chiang, Shih-Ying 279
Chiu, Chi-Tsung 157
Chou, Chan-Yen 279
Chou, J. H. 268
Chu, Huabin 626
Chuan, Tang 720
Chuan-Pei, Xu 794
Chung, T.F. 134
Chung, Tom 28
Chunling, Ren 1055
Chunyue, Huang 1092
Chunyue, ... 431
Conway, Paul P. 57
Cui, C.Q. 134
Cui, Shan-Ling 804

D

Dan, Huang 713
Dang, B. 577
Deng, Guiling 438
Deng, Zongquan 751
Desmulliez, M.P.Y. 509
Dhiman, Harpreet S. 994
Ding, Dongyan 558, 563
Ding, Guifu 304, 699
Dong, Zhang 410
Dou, Xinyu 121
Du, Bin 964
Du, Chao 1021
Du, Ruxu 605
Du, Xinyu 679
Du, Zhibo 1059
Dudek, Rainer 145
Duh, Jenq Gong 873, 927

E

Ernst, L. J. 937, 271, 636

F

Fan, H. B. 285
Fan, Jing-Yu 192
Fan, Pingyue 905
Fan, Xuejun 496, 931, 994
Fan, Yanhua 609
Fan, Yi 523
Fang, Huajing 461
Fang, Qiang 1052
Fei, Qin 171, 178, 183, 613
Feng, Lin 415, 419
Feng, Ran 331, 344, 398, 483
Feng, Tao 121
Frémont, Hélène 636, 974

A-1

Author Index

Fu, Ran 707
Fu, Shen-Li 540
Fu, Xingming 949
Fu, Yifeng 523, 641
Fu, Yonggao 765

G

Gan, Zhiyin 464, 493, 667
Gao, Guohua 187
Gao, Jian 788
Gao, Jie 735
Gao, Wei 50, 211, 223
Gao, Xiang 558
Gao, Yulai 600
Gao, Ziyang 28
Garant, E. Perfecto J. 577
Gautier, Christian 636, 974
Geng, Fei 438, 111, 165
Gruber, P. 577
Gu, Xiaolong 758
Guan, Rongfeng 324
Gui-Ling, Deng 807
Guiling, Deng 813
Guo, J. J. 848
Guo, J.D. 863
Guo, Jianjun 758
Guo, X.M. 544
Guo, Yunxia 256
Guoliang, Wang 72
Guoyuan, Li 720

H

Haihua, Jiang 1008, 729
Han, Ding 885
Han, Lei 106, 778
Han, Lei 798
Han, Qiang 931
Hao, Hu 807
Hao, Wang 394, 774
Hao, Yilong 534
He, Jie 717
He, Yandong 534
He, Yanping 558
Hochstenbach, P 651
Hong, Zhao 468
Hongjun, Liu 72, 75
Hou, H. N. 1047
Hou, Zhezhe 589
Hsieh, Ming-Liang 540
Hsu, C. L. 968
Hsu, Hsiang-Chen 540
Hsu, Kuo-Yuan 1040, 528
Hu, Anmin 735, 740
Hu, Anming 117, 744
Hu, Dyi-Chung 23

Hu, Jun 626
Hu, L. 1047
Hu, W. G. 632
Hu, Zhili 544
Huang, Dejian 431
Huang, Liugang 1082
Huang, Louie 39, 572
Huang, Mian 379, 605, 609
Huang, Qing-An 80, 99
Huang, Qiuping 111
Huang, Suyi 464, 493
Hughlett, E. 577
Hui, Dong 410
Hui, Shi 398, 424
Hung, Chih-Pin 157
Hung, Mike 39, 572

I

Inoue, Masahiro 655

J

Jansen, K M. B. 937, 636
Ji, Hongjun 592
Ji, Li-Na 867
Jia, Wang 415, 419
Jian, Zhang 468
Jian, Zhou 713
Jiang, Sijia 655
Jiang, Ting-Biao 1021
Jianqiangwang 589
Jiao, Li 344, 424
Jiao, Qinghua 754
Jide-Zhao, ... 449
Jie, Yingliang 626
Jillek, W. 1026
Jin, Li 774
Jin, Ling 626
Jin, Xing 836
Jin, Yufeng 256, 515, 534
Jinyi, Zhang 398, 410, 415, 419, 424
Johan, Liu 75
Johnson, Mark 250
Jun, Cai 483
Jun, Hu 103
Junquan, Wang 813

K

Kai, Qiao 140
Kai, Zhou Yi 419
Kai-Lin, Pan 308, 832
Kang, S. 577
Kaulfersch, Eberhard 145
Ke, Li 161
Ke, Yan 371

Author Index

Kelly, Robert 788
Kettner, Paul 43
Kim, Bioh 43
Kim, Daewon 592
Kim, Hongbae 584
Kim, Jong-Hoon 841
Kim, Jongmyung 584, 592
Kim, Keun-Soo 841
Knauf, Benedikt J. 57
Knickerbocker, J. 577
Knickerbocker, S. 577
Kong, Lingwen 763
Kui, Dai 794
Kuo, C. Y. 968
Kwan, Kenneth 319
Kwon, Henri Hk. 121

L

Lai, Chi-Chang 295
Lai, Xiao-Wei 129
Lai, Yi-Shao 984, 988
Lam, Angus 319
Lauro, Paul 956
Le, Kriangsak Sae 260
Lee, Aching 39
Lee, Jaisung 567, 944, 949
Lee, P. H. 268
Lee, S. W. Ricky 503
Lee, Yong-Won 841
Lei, Chen 1063, 1067
Leng, Yi 551
Leu, Jihperng 1040, 528
Leung, Vincent Chi-Kuen 93
Li, Baoxia 50
Li, C Y 519
Li, Dou-Xing 617
Li, Fenghui 769
Li, Gongke 909
Li, Han 103
Li, Jun 211, 223
Li, Junhui 106
Li, Kejia 515
Li, Lei 227, 379, 605, 609
Li, Lin-Kai 717
Li, Ming 117, 375, 558, 563, 735, 740, 744, 836, 852
Li, Mingyu 584, 592
Li, Peng 896
Li, Qingxia 551
Li, Quan-Yong 856
Li, Ren 720
Li, Song 208
Li, Tiezhu 360, 364
Li, Xiaohui 1017, 1036, 1077
Li, Xiaoyan 1070, 1073, 765, 769
Li, Xueping 699

Li, Yan.. 563
Li, Yingliang... 449
Li, Ying-Liang 452
Li, Yuanxun 703
Li, Zhihua 407
Li, Zhihua 50
Liang, Le.. 597, 684
Liang, Lihua 197, 889
Liao, Cheng 223
Liao, Hongguang 256, 515
Liao, Xiao-Ping 327
Lijun, Yang.. 826
Lin, Bryan 572
Lin, C. Y... 968
Lin, Changyong 152
Lin, Feng... 479
Lin, Han.. 379
Lin, Penrong 684
Lin, Tsung-Shu 1040
Lin, Tzu-Chih 157
Lin, Xiaoqin 1031
Lin, Y.C. 1082
Lina, Pengrong 597
Ling, Jin 103, 183
Ling, Li 826
Liu, Bin... 852
Liu, Changqing 57
Liu, D. S. 968
Liu, Deming 707
Liu, Jianhui 763
Liu, Jing.. 235
Liu, Johan 192, 523, 544, 600, 641, 655, 725, 754
Liu, Junwen 99
Liu, Li 754
Liu, Lilin 707
Liu, Na 1070, 1073
Liu, Rui.. 304, 699
Liu, Sheng152, 216, 243, 354, 386, 456, 464, 493, 551, 667
Liu, Shiguo 299
Liu, Wenming 152
Liu, Y. 632
Liu, Yang 783
Liu, Yingli 703
Liu, Zhen 717
Liu, Zhi-Quan 617, 629, 664
Liu, Zhongyuan 456
Liu, Zongyuan 243, 354, 386
Liusang, ... 1013
Loh, Wei Sun 250
Lou, Minyi 914, 944
Lu, Dafu 703
Lu, Dechun 1059
Lu, Hao 368
Lu, Hua 250
Lu, M. 577
Lu, Minfei 763

Author Index

Lu, Minhua 956
Lu, Sheng...660
Lu, Xiu Zhen...725
Lui, Tung-Chin..28
Luo, Fei...660
Luo, Le.. 1031, 111, 165
Luo, Xiaobing 216, 243, 354, 386, 456, 464, 493, 667
Luo, Yi..34
Lv, Lei...778
Lv, Yao...46, 50

M

Ma, Lili...1059
Ma, Vivian Wei 93
Ma, Xiaosong...636
Ma, Xiao-Song...729
Mao, Dali 117, 558, 563, 735, 740, 744, 852
Mao, Jiwang..740, 852
Mao, Shengping....................................304, 699
Maslyk, Dan...63
Mcelroy, J. B. 648
Mclellan, Neil 1003
Mei, Hujie..679
Meihua, Xu 161, 483
Meng, Songhe..748
Miao, Min..256
Michel, Bernd...145
Ming-Shan, Yang..717
Minliang, Zhang 72, 75
Muc, Xiu..667

N

Na, Chen 171, 613
Ni, Li...308, 832
Ning, Chen 1008, 729
Nong, Hong-Mi...1021

O

Ou, Zhaohui..479

P

Pan, Kailin 140
Pan, Kai-Lin...896
Pan, Yingfeng...551
Pang, Xue-Yong 629, 664
Pape, H..937
Pargfrieder, Stefan 43
Parrott, A.K..509
Pearl, Agyakwa..250
Pecht, Michael 1
Pei, Li Pei..713
Peng, Chung-Nan..927
Peng, Jin..882

Peng, Li 832
Peng, Yafang..1052
Pfahl, R. C..648
Poh, Edith S. W..671
Poirier, Patrick 974
Pun, Kelvin...134

Q

Qi, Fangjuan 589
Qi, Quan..66
Qian, Bin-Feng......................................231, 239
Qiang, Huang...1055
Qiao, Jiaping 819
Qin, Cha..589
Qin, Liancheng..909
Qin, Ming..80
Qin, Suqiong...679
Qiu, Shen-Nan 340

R

Ran, Feng..............................235, 360, 364, 403
Regard, Charles 636, 974
Ridout, Steve 250
Rizvi, J...509
Röllig, Mike...145
Rongen, R.T.H...271
Rongzheng, Ding...1055
Ruhmer, R. Weisman K.....................................577
Rzepka, Sven 145

S

Salo, Antti..1086
Schlottig, G..937
Schmitt, E..1026
Semkow, K...577
Sham, Man-Lung...28
Shang, J. K......................................848, 863, 878
Shang, Jian-Ku.......................................617, 629, 664
Shang, Jintang..99
Shang, Pan-Ju..617, 664
Shen, G. S...290, 968
Shen, Guang-Ping...80
Shen, Xuwei...783
Shen, Yidong..744
Shi, Daniel Xun-Qing......................................93
Shi, Daniel..86
Shi, Ying...754
Shih, D.-Y...577
Shih, Da-Yuan...956
Sinclair, K.I...509
Song, Jing 327, 900, 99
Suganuma, Katsuaki......................................841
Sun, F. L..632, 651
Sun, Fenglian...783

Author Index

Sun, Li 589
Sun, Peng 86, 93
Sun, Rong 605, 609
Sun, Wei 260, 671
Sun, Yiqin 379
Sun, Yongqiao 319
Sun, Young 515
Sundlof, B. 577
Sun-Fujiang, ... 1013

T

Tai, Yu-Che 295
Tan, H.B. 260, 671
Tan, Wei 679
Tang, Jia-Jie 165
Tang, Jieying 900, 99
Tang, Jun 304, 699
Tang, Zhi-Jie 331
Tao, Wenkai 600
Tao, Yuan 125
Te Gan, Richard 671
Teng, Hui 688
Tian, Dalei 324
Tian, Dewen 472
Tian, Xiao-Wei 621
Tian, Yanhong 314, 472, 684, 597
Tilford, T. 509
Tong, An 171, 178, 183, 613
Tsai, Mars 23
Tseng, Andy 572
Tseng, H. C. 268
Tsui, Anthony C. 688
Tu-Yunhua, ... 1013

V

Van Der Sluis, O. 636, 980
Van Driel, W. D. 651, 980, 636
Van Soestbergen, M. 271

W

Wan, Lixi 211, 223, 227, 379, 407, 46, 50, 54, 605, 609
Wan, Xin 256
Wang, Baohua 660
Wang, Bo 964
Wang, Chen-Chao 157, 295
Wang, Ching-Chun 988
Wang, Chunqing 314, 472, 592, 684
Wang, Fuliang 819
Wang, Hong 304, 699
Wang, Honghui 187
Wang, Jiaji 905
Wang, Jiongming 1017, 1036, 1077, 337
Wang, Junquan 438
Wang, Kai Jheng 873

Wang, Kai 243, 354, 386, 456, 667
Wang, Lei 567
Wang, Li-Ding 34
Wang, Lifeng 783
Wang, Ling 765
Wang, Qian 121, 567, 944, 949
Wang, Shao-Qing 629
Wang, Te-Chun 428
Wang, Teng 523, 641
Wang, Vicky 63
Wang, Weiqiang 1
Wang, X.J. 878
Wang, Xiao-Dong 34
Wang, Xiaojing 523
Wang, Xi-Chuan 231
Wang, Xing 324
Wang, Yu 852
Wang, Yunfeng 227
Wang, Z. G. 848, 878
Wanga, Chunqing 597
Wangyao-Ming, ... 382
Webb, D. Patrick 57
Wei, B. 519
Wei, De-Fang 452
Wei, Luo Xiao 424
Wen, Long 944, 949
Wen, Tao 375
Wen, Zhang 75
Weng, Xue-Tao 208
Wong, Cell K Y 692, 285
Wu, Dejian 43
Wu, Feng-Shun 125, 129, 914, 964
Wu, Gaohui 748, 751
Wu, Jian 360, 364
Wu, Jian 80
Wu, Jingshen 1003, 319
Wu, Li-Ming 804
Wu, Sung-Mao 295
Wu, Yanan 754
Wu, Yanhong 111
Wu, Yi-Ping 125, 129
Wu, Yiping 914, 964
Wunderle, B. 937

X

Xi, Hongjia 914
Xi, Yanyan 152
Xia, Y.H. 1026
Xia, Yangjian 197
Xiangguang, Xuan 161
Xiangzhao, ... 1013
Xian-Wei, Rong 468
Xiao, A. 937
Xiaodong, Yang 410
Xiaojing, Wang 72, 75

Author Index

Xiao-Song, Ma 1008
Xie, Bin 86, 93
Xie, Guangchao 679
Xie, Jinghua 438
Xie, Xiaoqiang 567
Xin, Yibo 461
Xiong, Wei 216, 493
Xiu, Ziyang 748, 751
Xu, Gaowei 111
Xu, Hongbo 584
Xu, Leon 1086
Xu, Li 725
Xu, Liyuan 900
Xu, Meihua 235, 360, 364, 403
Xu, Mei-Hua 331
Xu, Po 117
Xuan, Xiangguang 403
Xuanxiang-Guang, ... 350, 382
Xue, Ke 319
Xuefeng, Shu 920, 923

Y

Yan, Guoguang 334
Yan, Li-Min 239, 340
Yan, Sun 415
Yan, T. L. 632
Yan, Yongchang 1070
Yan, Yong-Chang 1073
Yang, Chengyue 50
Yang, D. G. 909
Yang, Daoguo 140
Yang, Dao-Guo 856
Yang, Dian 344
Yang, Guoji 187
Yang, Hongbo 688
Yang, Le 327
Yang, Lianfa 140
Yang, Shihua 684
Yang, Shin-Yueh 279
Yang, Wang 826
Yang, Wen-Kung 23
Yang, Wenrong 1036
Yang, Wen-Rong 337
Yang, Xiaohua 769
Yang, Zhen-Guo 621, 867
Yang, Zhijun 788
Yang, Zhiqin 763
Yang, Zongbao 703
Yanga, Shihua 597
Yanhui, Jiang 415, 419
Yeh, Chang-Lin 204, 984, 988
Yew, Ming-Chih 23, 279
Ye-Xiang, Ning 308, 896
Ye-Yuming, ... 1013
Yi, Fuxi 909

Yi, Yang 410
Yiming, Jiang 774
Yin, C. 509
Yin, Guangfu 605
Yin, Yansheng 609
Yiping, Wu 885
Yngve, Wang 183
Yu, Chun 368
Yu, Chun-Fai 23
Yu, Hk 1052
Yu, Roy 13
Yu, Shuhui 605, 609
Yu, Zhang 923
Yu, Zhiyuan 379
Yuan, C. 980
Yue, Cong 192, 725
Yue-Li, Hu 371, 350, 382
Yuen, John 319
Yuen, Matthew M F 692, 285
Yule, Fan 161
Yunxin, Wu 823
Yuxi, Jiang 344

Z

Zbrzezny, Adam R. 1003
Zeng, Lu 765
Zeng, Qiulian 758
Zhai, Qijie 600
Zhang, G. Q. 651, 980, 271, 636
Zhang, Hua 80
Zhang, Huaiwu 703
Zhang, Huiliang 688
Zhang, J H 519
Zhang, Jue 1017, 1036, 1077
Zhang, Jun 1082
Zhang, Kouchi 140
Zhang, Lili 600
Zhang, Li-Na 798
Zhang, Liqing 584
Zhang, Qin 667
Zhang, Rong 503
Zhang, Shu-Qiang 157, 204
Zhang, Tong-Yi 707
Zhang, X.F. 863
Zhang, Xiaowu 299
Zhang, Xu 50
Zhang, Yan 192, 523, 600, 725
Zhang, Yuanxiang 197
Zhang, Z. F. 848
Zhang, Zheming 1003
Zhang, Zhen-Qiang 34
Zhang, Zhikun 655, 754
Zhang, Zong-Bo 34
Zhangjin, Chen 394
Zhangyi-Chi, ... 350

Author Index

Zhanhui, Li 823
Zhao, Ji-De 452
Zhao, Liwei 256
Zhao, Wu 431
Zhao, Xiaolin 304
Zhao, Xinbing 758
Zhao, Zhenqing 567
Zhaohua, Zhou 443
Zhe, Wang 826
Zhendong, Zhao 923
Zheng, Xu 605
Zheng, Yin-Guang 204, 428
Zhigang, Li 920, 923
Zhili, Long 823
Zhong, Jue 106
Zhong, Li-Jun 856
Zhou, Hua 431
Zhou, Jian 111
Zhou, Jiang 487
Zhou, Jianwei 944, 949
Zhou, Ming 688
Zhou, Tiao 994
Zhouping, Yin 885
Zhu, Haiqing 187
Zhu, Q. S. 848
Zhu, Swen 43
Zhu, W.H. 260, 671
Zhu, Zhi-Jun 125
Zou, Changhui 819

A-8

9781424427390